- Main Group metals
- Transition metals, lanthanide series, actinide series
- Metalloids
- Nonmetals, noble gases

	1B	2B	3A	4A	5A	6A	7A		
								2 **He** Helium 4.0026 G	1
			5 **B** Boron 10.811 S	6 **C** Carbon 12.011 S	7 **N** Nitrogen 14.0067 G	8 **O** Oxygen 15.9994 G	9 **F** Fluorine 18.9984 G	10 **Ne** Neon 20.1797 G	2
			13 **Al** Aluminum 26.9815 S	14 **Si** Silicon 28.0855 S	15 **P** Phosphorus 30.9738 S	16 **S** Sulfur 32.066 S	17 **Cl** Chlorine 35.4527 G	18 **Ar** Argon 39.948 G	3
28 **Ni** Nickel 58.693 S	29 **Cu** Copper 63.546 S	30 **Zn** Zinc 65.39 S	31 **Ga** Gallium 69.723 S	32 **Ge** Germanium 72.61 S	33 **As** Arsenic 74.9216 S	34 **Se** Selenium 78.96 S	35 **Br** Bromine 79.904 L	36 **Kr** Krypton 83.80 G	4
46 **Pd** Palladium 106.42 S	47 **Ag** Silver 107.8682 S	48 **Cd** Cadmium 112.411 S	49 **In** Indium 114.82 S	50 **Sn** Tin 118.710 S	51 **Sb** Antimony 121.757 S	52 **Te** Tellurium 127.60 S	53 **I** Iodine 126.9045 S	54 **Xe** Xenon 131.29 G	5
78 **Pt** Platinum 195.08 S	79 **Au** Gold 196.9665 S	80 **Hg** Mercury 200.59 L	81 **Tl** Thallium 204.3833 S	82 **Pb** Lead 207.2 S	83 **Bi** Bismuth 208.9804 S	84 **Po** Polonium (209) S	85 **At** Astatine (210) S	86 **Rn** Radon (222) G	6
110 — — (269) X	111 — — (272) X	112 — — (277) X							7

63 **Eu** Europium 151.965 S	64 **Gd** Gadolinium 157.25 S	65 **Tb** Terbium 158.9253 S	66 **Dy** Dysprosium 162.50 S	67 **Ho** Holmium 164.9303 S	68 **Er** Erbium 167.26 S	69 **Tm** Thulium 168.9342 S	70 **Yb** Ytterbium 173.04 S	71 **Lu** Lutetium 174.967 S
95 **Am** Americium (243) X	96 **Cm** Curium (247) X	97 **Bk** Berkelium (247) X	98 **Cf** Californium (251) X	99 **Es** Einsteinium (252) X	100 **Fm** Fermium (257) X	101 **Md** Mendelevium (258) X	102 **No** Nobelium (259) X	103 **Lr** Lawrencium (260) X

THE CHEMICAL WORLD

Concepts and Applications

Second Edition

Here's what your colleagues are saying about the second edition of *The Chemical World: Concepts and Applications*

"The best feature of the text is the very successful integration of organic chemistry, biochemistry, and environmental topics into the presentation of chemical principles. . . . The authors have managed to do this **without sacrificing the rigor** that has become the trademark of chemistry courses in general."

—*David Miller*
California State University, Northridge

"I was impressed with the implementation of both the emphasis on conceptual understanding and especially the sections that included biochemistry and organic examples. I was particularly impressed with the discussions of free energy and biology and of the production of ATP to store energy."

—*Jimmy Reeves*
University of North Carolina, Wilmington

"The integration of 'real chemistry' and the decrease of algorithmic/mathematical discussions are particularly attractive."

—*Richard Stolzberg*
University of Alaska, Fairbanks

"The authors have done a beautiful job of being consistent in the level and style of writing. The coverage is very thorough and **the level of rigor has not been sacrificed** in order to make the material more relevant. . . . The integration of organic, biochemistry, and environmental chemistry goes far beyond what any other text for chemistry majors has done. This is an excellent step toward the integration of relevant topics and chemical principles."

—*Miles Koppang, University of South Dakota*

"I consider the conceptual exercises to be a major and welcome addition to the new text. These **questions force students to apply the concepts** they have just read about and, consequently, enhance their basic understanding of the material. These are the kinds of questions that I typically ask on homework assignments and on exams."

—*Patrick Holt, Bellarmine College*

"Integrating organic chemistry, clinical and biochemistry, and environmental chemistry is a welcome change and should be promoted. Clearly this is a bold, much-needed addition to freshman chemistry."

—*Joseph Sneddon, McNeese State University*

"The conceptual problems in *The Chemical World* will make students stop and think instead of merely punching numbers on their calculators to satisfy a formula."

—*Robert Profilet, Texas Tech University*

"The development of Principles of Reactivity is excellent, with a particularly good approach to making the origin of the Gibbs free energy more than a just-so story. . . . The chapter on chemical kinetics is excellent—**it is rigorous without being mathematically overwhelming,** and the examples chosen are both clear and pertinent to actual chemical research."

—*Ruben Puentedura, Bennington College*

"I am very happy to see so many end-of-chapter questions. They are good ones and are well-linked to the chapter material."

—*Donald Williams, Hope College*

"The biggest strength of the text is the readability, the historical information, the blending of worked-out examples with the text, and the large number of conceptual problems found within each chapter."

—*Sidney Young, University of South Alabama*

THE CHEMICAL WORLD

Concepts and Applications

Second Edition

JOHN W. MOORE
Professor of Chemistry
Director, Institute for Chemical Education
University of Wisconsin-Madison
Madison, Wisconsin

CONRAD L. STANITSKI
Professor of Chemistry
University of Central Arkansas
Conway, Arkansas

JAMES L. WOOD
Infocode, Inc., and
Adjunct Professor
David Lipscomb University
Nashville, Tennessee

JOHN C. KOTZ
SUNY Distinguished Teaching Professor
State University of New York
College at Oneonta
Oneonta, New York

MELVIN D. JOESTEN
Professor of Chemistry
Vanderbilt University
Nashville, Tennessee

SAUNDERS GOLDEN SUNBURST SERIES
Saunders College Publishing
Harcourt Brace College Publishers

FORT WORTH PHILADELPHIA SAN DIEGO NEW YORK ORLANDO AUSTIN
SAN ANTONIO TORONTO MONTREAL LONDON SYDNEY TOKYO

Vice President/Publisher: Emily Barrosse
Vice President/Publisher: John Vondeling
Contributing Editor: Mary E. Castellion
Associate Editor: Marc Sherman
Project Editors: Robin C. Bonner, Bonnie Boehme
Production Manager: Charlene Catlett Squibb
Art Director: Caroline McGowan
Text and Cover Designer: Ruth Hoover
Product Manager: Angus McDonald

Cover credit: Photograph of Weddell Sea, Minden Pictures, ©Frans Lanting; molecular ice structure rendered by Rolin Graphics, from a computer-generated diagram by John W. Moore.

Printed in the United States of America

THE CHEMICAL WORLD: CONCEPTS AND APPLICATIONS
Second Edition

0-03-019094-0

Library of Congress Catalog Number: 96-071994
8901234567 032 10 9 8 7 6 5 4 3 2 1

PREFACE

Why write a general chemistry textbook? Aren't there so many available as to shift the equilibrium toward leaving well enough alone? And isn't it almost impossible to do something new and better? We think not, and this book's first edition was constructed to express clearly the directions we thought general chemistry courses ought to be taking. This second edition reflects the continuing development of the philosophy embodied there.

Our title, *The Chemical World: Concepts and Applications,* indicates that this book is about the facts, theories, and models of chemistry and how they can be applied to understanding the world around us. We have described these concepts and facts in a way that is interesting and accessible to a broad range of students who plan careers in chemistry, other natural sciences, engineering, and related fields. We believe that by integrating applications, facts, and concepts we can motivate students to become actively engaged and involved with the material, thereby enhancing their long-term understanding.

Because the first edition of *The Chemical World* was not a clone of the typical textbook, those who reviewed it and used it paid special attention to the new approaches we took. We appreciate the very useful feedback and suggestions we have received from a variety of sources. This second edition reflects what we have learned from faculty at diverse institutions who are teaching students with a broad range of abilities and motivations. Based on this input and our own observations as we used the book, we have developed further the philosophy that led to and infused the first edition.

Goals of This Book

Our principal goals are to help students develop

- a broad overview of chemistry and chemical reactions,
- an understanding of the most important concepts and models that chemists, and those in chemistry-related fields, use,
- the ability to apply the facts, concepts, and models of chemistry appropriately to new situations in chemistry, other sciences and engineering, and other disciplines,
- knowledge of the many practical applications of chemistry in our society and our environment,

- appreciation of the many ways that chemistry impacts the daily lives of everyone, students included, and
- motivation for studying in ways that help them achieve long-term retention of facts and concepts.

Audience

The Chemical World: Concepts and Applications is intended for general chemistry courses for students who expect to pursue further study in science or science-related disciplines. Those planning to major in chemistry, biochemistry, biological sciences, engineering, geological sciences, materials science, physics, and many related areas should benefit from this book and its approach. We assume that the students who will use this book have had a basic foundation in mathematics (algebra and geometry) and in general science. Most will also have had a chemistry course before coming to college.

Philosophy, Approach, and Special Features

Our intent is that this book can be used in its entirety in a two-semester course. We provide thorough, concept-rich treatments of those subjects that we (and users of the first edition) have identified as the most important for chemistry students to learn and understand. We have attempted to distill the essence from the encyclopedic liquor that is called general chemistry. And we have used some of the space made available to incorporate new topics and ideas from modern chemistry and related fields. The criterion for including a principle or concept continues to be that it is used later in the book or will be used by a large fraction of students in subsequent studies. By concentrating on the most important ideas and integrating them with applications as well as other concepts, we have aimed for a whole that provides students with much better understanding and longer-term retention of facts and principles than could be achieved by disconnected, individual parts.

Whenever possible we include practical applications, especially those applications that students will revisit when they study other natural science and engineering disciplines. Applications have been integrated where they are relevant, not relegated to separate chapters and separated from the principles and facts on which they are based. We have included **numerous cross references (indicated by ⇐)** that will help students link a concept being developed in the chapter they are reading with an earlier, related principle or fact.

Of particular importance is the **integration of organic chemistry and biochemistry** throughout the book. In many areas, such as stoichiometry and molecular formulas, organic compounds provide excellent examples. To take advantage of this we have incorporated basic organic topics into the text beginning with Chapter 3 and used them wherever they are appropriate. In the discussion of molecules and the properties of molecular compounds, the concepts of structural formulas, functional groups, and isomers are developed naturally and effectively.

Many of the principles that students encounter in general chemistry are directly applicable to biochemistry, and a large fraction of the students in most general chemistry courses are planning careers in biological or medical areas that make constant use of biochemistry. Therefore we have chosen to deal with biochemical topics in juxtaposition with the general chemistry principles that underlie them. For example, elements essential to life and two classes of biomolecules are introduced in Chapter 3 and used in examples involving moles and formulas of molecular compounds. Metabolism and Gibbs free energy are closely coupled in Chapter 7, DNA structure is discussed in Chapter 10 in conjunction with hydrogen bonding and other noncovalent interactions, proteins are covered in Chapter 11, and enzymes are dealt with in Chapter 12 on kinetics.

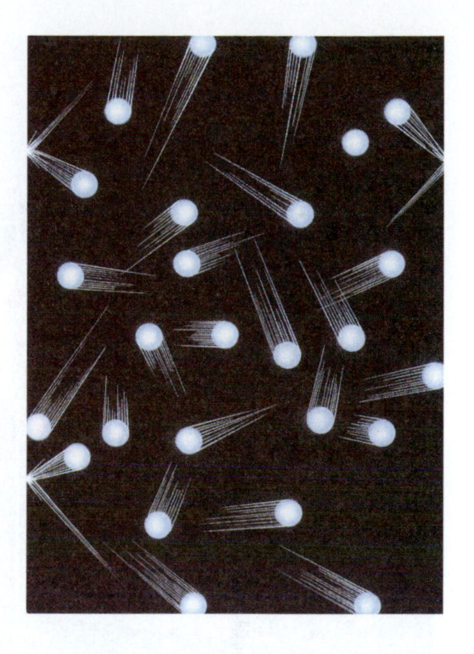

Environmental topics are also integrated. Stratospheric ozone depletion provides an excellent reinforcement of the concept of catalysis in Chapter 12 (kinetics), atmospheric chemistry and air pollution are discussed in Chapter 14 (gases), and the chemistry of the aqueous environment is developed naturally in Chapter 16 (solutions). Units are introduced on a need-to-know basis. Celsius temperature is defined in Chapter 1 with the introduction to the properties of matter. Length, mass, and volume units are introduced in Chapter 2 in the context of the sizes of atoms. Energy units are defined in conjunction with energy, work, and heat in Chapter 6 (thermochemistry), and the thermodynamic temperature scale is introduced in Chapter 7 (thermodynamics), where the need for it can be made clear.

We continue to believe that a sound conceptual foundation is the best means by which students can approach and solve a wide variety of real-world problems. Three features have been designed to help students test their conceptual understanding: a large number of **Conceptual Exercises (indicated by**) are included within each chapter; a separate section of the end-of-chapter questions, headed **Applying Concepts,** has been developed by Patricia Metz, Texas Tech University; and each chapter has a set of **Conceptual Challenge Problems,** most of them written by H. Graden Kirksey, University of Memphis, that will make individual students think in new ways about the concepts they have learned. The Conceptual Challenge Problems are also suitable for students to work on in cooperative groups.

To support our intent that students develop long-term understanding rather than rote memorization, we have included the many numerical **Exercises** (with answers only available in the back of the book) that characterized the first edition, expanding them with the Conceptual Exercises described previously. We have also developed a new form of example that we call a **Problem-Solving Example.** It consists of four parts: a Question (problem); an Answer, stated briefly; an Explanation that provides significant help for students whose answer did not agree with ours; and additional Problem-Solving Practice that provides similar questions or problems. We encourage students first to work out an answer for the question or problem without looking at either the Answer or the Explanation, then to check their answer, next to repeat their work if their answer did not agree with ours, and only after that to look at the Explanation.

We believe that our approach to problem solving will enable students to learn strategies and techniques that will help them approach and solve real problems they will encounter later in their careers. To help them evaluate how well they have developed such abilities, the **end-of-chapter questions have been revised and expanded** by Patricia Metz, Texas Tech University; Melinda Oliver, Louisiana State University; and Allan Smith, Drexel University. In addition to the Conceptual Challenge Problems and Applying Concepts questions, we have included questions keyed to sections within each chapter. There is also an extensive set of general questions that are not identified with chapter sections and that often draw on several related concepts for their solution. To further emphasize that problem solving is a crucial aspect of learning science, we have keyed a question to the photograph with which each chapter opens. Students answer that question as part of a **Summary Problem** that ends the chapter and ties together many or all of the major concepts discussed in the chapter.

There is considerable evidence in the literature that students learn better when they can see the relationships among macroscopic phenomena, submicroscopic models involving atoms, molecules and/or ions (we call this the nanoscale), and symbolic

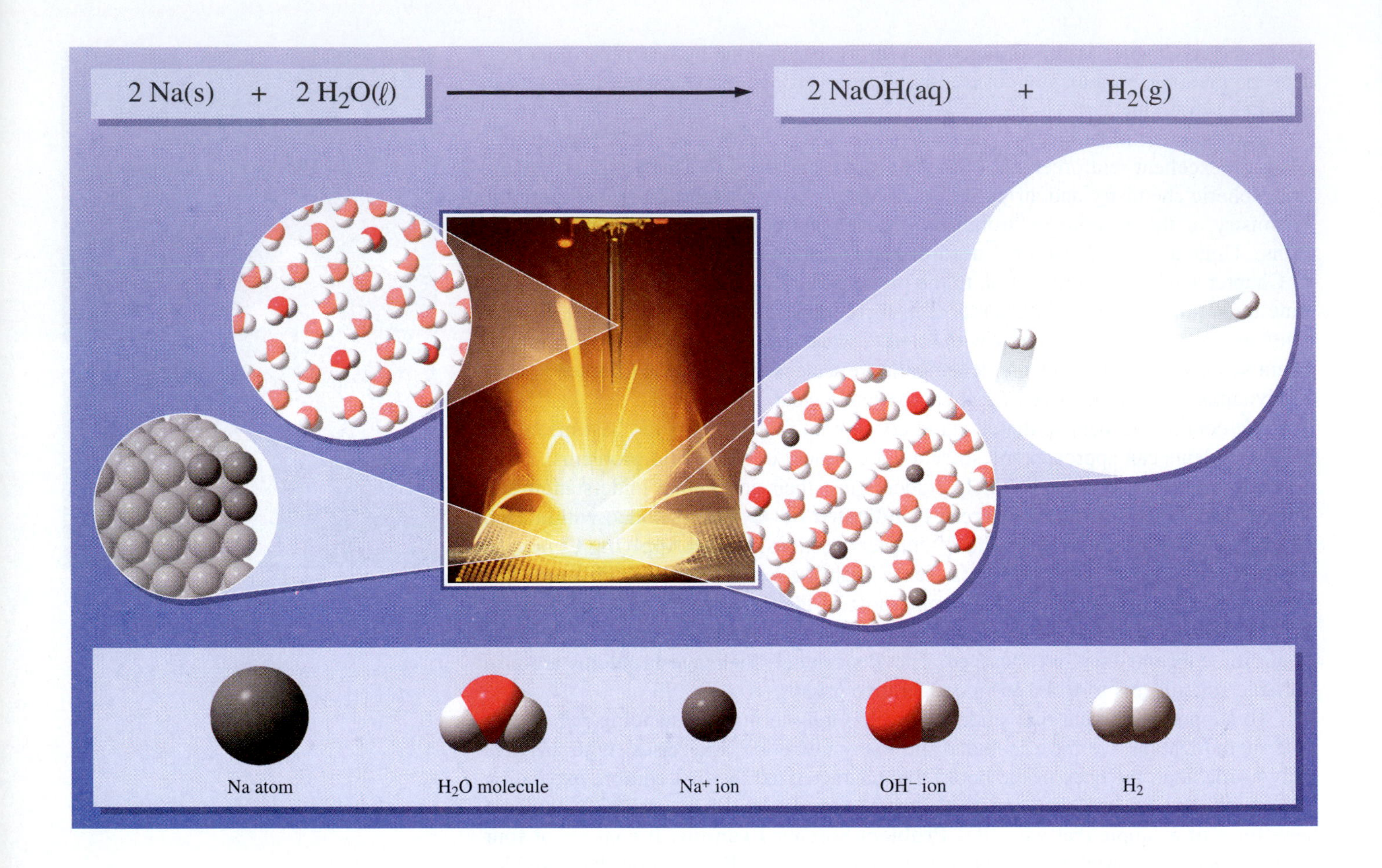

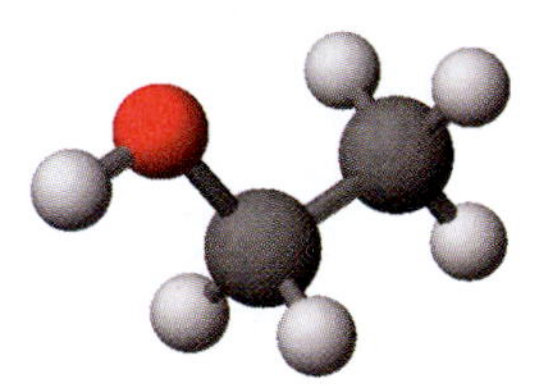

representations such as formulas and equations. Therefore we have embarked upon an **innovative illustration program** that juxtaposes all three domains. Excellent color photographs of substances and reactions, many by Charles D. Winters, are presented together with greatly magnified illustrations of the atoms, molecules, and/or ions involved. Often these are also accompanied by the symbolic formula for a substance or the equation for a reaction. The **nanoscale views of atoms, molecules, and ions** have been generated using CAChe Scientific molecular modeling software and then combined by a skilled artist with the photographs and formulas or equations. These computer-generated illustrations also appear in exercises, examples, and end-of-chapter problems to make certain that students are tested on the ideas they represent. The result, we think, provides an exceptionally effective way for students to learn how chemists think about the nanoscale world of atoms, molecules, and ions.

Another important aspect of our approach is that students should be involved in doing chemistry, and they ought to learn that common household materials are also chemicals. **Chemistry You Can Do,** which was a popular feature of the first edition, has been retained, and there is at least one take-home/dorm-room experiment in each chapter. These illustrate a topic included in the chapter and can be done with simple equipment and familiar chemicals available at home or on a college campus.

A Capstone Experience via the World Wide Web

We believe that it is highly effective for students to end a one-year general chemistry course with a capstone experience that integrates and coordinates the chemistry concepts learned earlier with a variety of other areas and gives students a realistic picture of the broad applicability and importance of our discipline. To this end we provide a modular Chapter 20, "The Chemical World." It consists of three sections, any of which can be used as a case study to wrap up a first-year course. Each section is independent of the other two and deals with an application of chemistry to an area of considerable practical importance that connects chemistry with a variety of related subjects. The three modules are titled

- Metals in Modern Society
- The Atmospheric Environment
- Biochemical Structure and Function

We expect that few if any faculty would use all three of these modules in the same year, but many will use different modules in different years. In addition, it is important that each module be up to date, because both the science and its societal consequences are developing continually. For these reasons we have decided on a highly innovative approach. Chapter 20 does not appear in this textbook, but rather is available to adopters of the book via the Saunders College Publishing World Wide Web site (www.saunderscollege.com). (Printed copies will also be made available if there is a need for them.) In addition to text, color graphics, and the other features of a printed book, the Web version provides animations, videos, and links to other relevant Web sites. It can be used directly on the Web or printed for more extended examination and study. We strongly encourage adopters of this book to utilize this Web site and to provide us with collective knowledge that will enhance both its content and presentation.

Organization

The most important change in organization from the first edition to this one is integration of biochemistry, organic chemistry, environmental chemistry, and applications of chemistry with the principles and concepts that are the essence of a general chemistry course.

Chapter 1, "The Nature of Chemistry," is an overview of science, its methods, and its practitioners that uses as an example the chemistry behind the hole in the ozone layer and the recent winners of the Nobel Prize who discovered much of that science. The chapter also introduces qualitatively the kinetic-molecular theory and describes mixtures, compounds, elements, and separation and purification processes. Chapter 2, "Elements and Atoms," deals with the atomic theory, atomic structure and atomic weight, the mole, and a brief introduction to the periodic table. "Chemical Compounds" (Chapter 3) describes properties and nomenclature of molecular compounds and ionic compounds; it introduces carbohydrates, fats, and oils; and it revisits the mole. "Chemical Reactions" (Chapter 4) introduces chemical equations and categories of chemical reactions, providing a firm base of descriptive chemistry on which subsequent chapters can build. In Chapter 5, "Relating Quantities of Reactants and Products," stoichiometry is applied to reactions among pure substances and in solution.

The next two chapters begin to develop the principles on which chemical reactivity is based. We have split thermodynamics from thermochemistry, treating energy,

enthalpy, calorimetry, and fuels in Chapter 6, "Energy Transfer and Chemical Reactions." Entropy and Gibbs free energy are discussed in Chapter 7, "Directionality of Chemical Reactions," which could easily be taught in the second semester, long after Chapter 6. Chapter 7 also applies thermodynamic ideas to (free) energy resources, conservation of (free) energy, coupling of reactions, and metabolism and biochemical energetics.

The next three chapters, "Electron Configurations, Periodicity, and Properties of Elements" (Chapter 8), "Covalent Bonding" (Chapter 9), and "Molecular Structures" (Chapter 10), develop the electronic structure of atoms, chemical bonding, molecular structure including hybrid orbitals, isomerism, and the consequences of noncovalent interactions between molecules. The ideas developed here are applied to spectroscopy of atoms and molecules (and MRI as well), descriptive chemistry of elements in the second row of the periodic table, chiral drugs, and the structure of DNA. Chapter 11, "Energy, Organic Chemicals, and Polymers" continues to apply the ideas of bonding and structure to fossil fuels, alcohols, carboxylic acids, organic polymers, plastics and recycling, and proteins.

Chapter 12, "Chemical Kinetics," and Chapter 13, "Chemical Equilibrium," develop further the principles that underlie chemical reactivity and apply them to biological and industrial catalysts, environmental systems such as stratospheric ozone where catalysis is important, and the control of chemical reactions to obtain the substances we want. "Gases and the Atmosphere" (Chapter 14) and "The Liquid State, the Solid State, and Modern Materials" (Chapter 15) deal with the three major states of matter and changes among them. The chemistry developed in Chapter 14 is applied to obtaining chemicals from the atmosphere, gas-phase reactions in the environment, and air pollution. In Chapter 15 semiconductors, superconductors, and ceramics exemplify the importance of materials science and its relation to chemistry.

The next three chapters are devoted largely to chemistry in aqueous solutions. Chapter 16, "Water and Solution Chemistry," develops the fundamentals of solubility and applies them to hard water, water pollution, and water purification. Chapter 17, "Acids and Bases," describes aqueous equilibria involving acids and bases, and applies these ideas to household chemicals and acid rain. Chapter 18, "Electrochemistry and Its Applications," does what its title implies. The many practical applications of both voltaic cells and electrolysis are emphasized.

"Nuclear Chemistry" is the subject of Chapter 19—one that is constantly in the news. Chapter 20, "The Chemical World," is our innovative Web-based set of three modules: Metals in Modern Society; The Atmospheric Environment; and Biochemical Structure and Function. Any one of these provides a capstone experience for students in a first-year college course, bringing to bear many of the principles, concepts, models, and facts learned in previous chapters.

New to This Edition

Many of the book's features have already been mentioned to illustrate how we have implemented our philosophy. New features in this edition are

- Chapter-opening photograph and question keyed to the Summary Problem at the chapter's end
- Computer-generated art that relates the nanoscale of atoms, molecules, and ions, the macroscale of the laboratory, and the symbolic formulas and equations of chemistry
- Conceptual Exercises, which are indicated by the symbol

- Connections among related concepts and applications, which are indicated by the ⇐ symbol
- Summary Problem at the end of each chapter to consolidate and reinforce concepts
- Chapter-end Conceptual Challenge Problems that require considerable thought and are suitable for cooperative group work
- Expanded and revised chapter-end problems, including a new section, Applying Concepts
- Key terms list at the end of each chapter
- Complete glossary of key terms at the end of the book
- Large number of Problem-Solving Examples with answers, explanations, and Problem-Solving Practice questions—structured to avoid rote problem solving
- An applications-based Web site (*see p. xiii*)

Integrated Material

Integrated organic chemistry, biochemistry, environmental chemistry, and applications include

- Introduction to organic chemistry, fats, carbohydrates, and dietary essential elements (Chapter 3)
- Patterns of chemical reactions related to environmental applications and industrial products (Chapter 4)
- Stoichiometry applied to industrial chemical processes (Chapter 5)
- Bond energies linked to fossil fuels and hydrogen use (Chapter 6)
- Gibbs free energy in relation to fuel resources and to energy transfer in biological systems—ADP/ATP, photosynthesis (Chapter 7)
- Application of nuclear magnetic resonance to medical magnetic resonance imaging (MRI) (Chapter 8)
- Isomerism in hydrocarbons; coordination compounds and life (Chapter 9)
- Organic molecules and biomolecules as examples of molecular structure, and their structural determination using uv-visible and infrared spectroscopy (Chapters 9 and 10)
- Overview of DNA structure and function with molecular structure and noncovalent forces (Chapter 10)
- Petroleum refining, additional organic chemistry, and polymers (Chapter 11)
- Enzymes and stratospheric ozone depletion with kinetics (Chapter 12)
- Applications of equilibrium principles to geological and atmospheric processes (Chapter 13)
- Atmospheric chemistry and air pollution with gases (Chapter 14)
- Ceramics and composite materials with solid state (Chapter 15)
- Chemistry of the aqueous environment with solutions (Chapter 16)
- Household chemicals with acids and bases (Chapter 17)
- Newly developed batteries (Chapter 18)

Features Retained from the First Edition

- Introductory paragraph giving an overview of each chapter and providing practical reasons to study the material
- Numerous Exercises that encourage students to study actively, not just read; answers are provided in an appendix
- A Chemistry You Can Do take-home or dorm-room experiment in every chapter

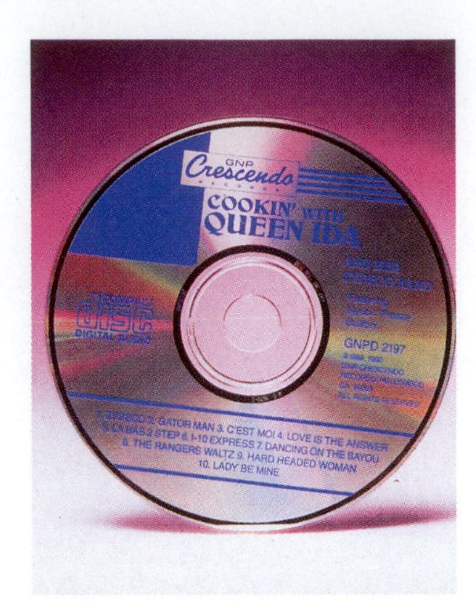

- A Portrait of a Scientist box in every chapter to show the human side of science and scientists
- A Chemistry in the News box in every chapter to relate the chemistry being studied to recently breaking news and to emphasize the applicability of chemistry to everyday life
- Plentiful margin notes that highlight important points and steer students around potential pitfalls
- In Closing section at the end of each chapter to list important points students should have mastered and to link them to sections in the chapter
- Numerous Questions for Review and Thought (some illustrated with photographs or nanoscale art) at the end of each chapter, with answers to boldface-numbered questions in an appendix
- Appendices on problem solving, mathematical operations, units and conversion factors, physical constants, nomenclature, ionization constants for acids and bases, solubility product constants, reduction potentials, thermodynamic data, answers to problem-solving practices and exercises, and answers to selected chapter-end questions

Supporting Materials

Written Materials for the Student and Instructor

The **Student Study Guide** by Dean Nelson of the University of Wisconsin-Eau Claire has been designed around the key objectives of the book. Each chapter of the study guide includes study hints, true/false questions, a list of symbols presented in the chapter, a list of terminology, concept maps of the chapter topics, a concept test, a discussion of how the chapter material relates to other chapters, crossword puzzles, and supplementary reading of current interest.

The **Student Solutions Manual** by Paul Kelter and James Carr of the University of Nebraska-Lincoln gives detailed solutions to designated end-of-chapter questions. The manual also contains strategies for problem solving.

The **Instructor's Resource Manual** by Patricia A. Metz of Texas Tech University focuses on four main areas: conceptual understanding and how to help students develop it; pedagogical approaches and learning strategies; cooperative learning and suggested group activities; and technology-based enhancement of learning. In addition, there is background information regarding the Chemistry You Can Do experiments, suggestions for chemical demonstrations that illustrate important points, and answers for the Summary Problems and Conceptual Challenge Problems. Solutions in the Instructor's Manual are contributed by Paul Kelter and James Carr of the University of Nebraska, Gary Michels of Creighton University, and by Therese Michels of Dana College.

A **Test Bank** by Ronald Clark of Florida State University provides hundreds of conceptual and numerical problems for use by the instructor. It is available in written, Macintosh, IBM, and Windows versions.

Overhead Transparencies of 150 figures from the book are available to adopters of the book. In addition, these transparencies as well as those from all other Saunders College Publishing chemistry textbooks are available on the Shakhashiri Chemical Demonstrations videodisc mentioned on p. xiv. Both the transparencies and the videodisc are available at no charge to the adopters of this book.

Laboratory Manuals

Saunders offers many excellent general chemistry laboratory manuals, all of which can be used in conjunction with *The Chemical World.* An instructor's manual is available for each title. In the event that none of these laboratory manuals meets your needs individually, selected experiments can be custom published as a separate laboratory manual, in accordance with Saunders' custom publishing policy. Please contact your local sales representative for more information on custom publishing.

Laboratory Experiments for General Chemistry (1998) by Joe March of the University of Wisconsin–Madison and David Shaw of Madison Area Technical College, evolved in conjunction with the NSF-supported New Traditions Project. This guided-inquiry approach has more than 25 experiments.

Chemical Principles in the Laboratory, sixth edition (1996), by Emil J. Slowinski and Wayne Wolsey of Macalaster College and William J. Masterton of the University of Connecticut provides detailed directions and advance study assignments. These thoroughly class-tested experiments were chosen with regard to cost and safety. The manual includes 43 experiments. An Alternate Version containing Qualitative Analysis (1996) is also available.

Standard and Microscale Experiments in General Chemistry, third edition, by Carl B. Bishop and Muriel B. Bishop, both of Clemson University, contains descriptive, quantitative, and instrumental experiments. This manual provides detailed instructions and incorporates helpful notes into the experiments to help students gain confidence. Alternate procedures have been added for several experiments, providing the instructor with a choice between macro and **microscale** uses. Approximately 30% of the experiments are available in microscale.

Laboratory Experiments for General Chemistry, third edition (1998, available July 1997), by Harold R. Hunt and Toby F. Block of Georgia Institute of Technology contains experiments that are designed to minimize waste of materials and stress safety. Pre-lab exercises and post-lab questions are included. The manual includes 42 experiments.

Experiments in General Chemistry, (1991) by Frank Milio and Nordulf Debye of Towson State University and Clyde Metz of the College of Charleston, is the result of many years of collaboration on the development of laboratory-tested experiments in which attention has been given to cost and safety. This manual contains 44 experiments.

Saunders Chemistry Web Resource Center

Based on feedback from hundreds of chemistry educators and students, Saunders has created a unique Web site that encourages students to probe and question. An **algorithmic problem-solving program** allows students to improve their skills and prepare for exams with new sets of problems every time they log on. The variety of problems, including multiple choice and numeric input, is unmatched by other Web sites. (See www.saunderscollege.com.)

Multimedia Materials

Saunders Interactive General Chemistry CD-ROM is a revolutionary interactive tool. This multimedia presentation serves as a companion to the text. Divided into chapters, the CD-ROM presents ideas and concepts with which the user can interact in several different ways, for example, by watching a reaction in progress, changing a variable in an experiment and observing the results, and listening to tips and suggestions for understanding concepts or solving problems. Students navigate through

the CD-ROM using original animation and graphics, interactive tools, pop-up definitions, over 100 video clips of chemical experiments, which are enhanced by sound effects and narration, and over 100 molecular models and animations.

Chemistry 1998 MediaActive contains imagery from various Saunders texts in general and organic chemistry in a presentation CD-ROM that can be used in conjunction with commercial presentation packages, such as Power-Point™, Persuasion™, and Podium™. Available for both Windows and Macintosh platforms.

Shakhashiri Chemical Demonstration Videotapes contain a unique set of 50 three- to five-minute chemical experiments performed by Bassam Shakhashiri. These videos bring the drama of chemistry into the classroom. An instructor's manual describes each experiment and includes questions for discussion. The videotapes are available free to adopters of the text. **Saunders General Chemistry Videodisc** contains most of the Shakhashiri demonstrations. In addition to this live-action footage, almost 3000 still images taken from eight of Saunders' general chemistry textbooks, including *The Chemical World,* are included on the two-disc set. This videodisc is free upon adoption of any Saunders general chemistry text.

The **World of Chemistry Videotapes** are based on the television series and telecourse, "The World of Chemistry," with Roald Hoffmann as host. These 26 tapes are each about 30 minutes long and provide introductory material on the principles and applications of chemistry. The tapes may be ordered through the Annenberg Foundation at 1-800-LEARNER ($350).

The following items may be purchased from the *Journal of Chemical Education: Software* at JCE: Software, Department of Chemistry, University of Wisconsin-Madison, 1101 University Ave., Madison, WI 53706, (800) 991-5534.

The Periodic Table Live! CD provides on one CD-ROM for Macintosh and Windows PCs a broad range of textual, numeric, and visual information about the chemical elements. Essentially all of the video material from the Periodic Table Videodisc is on the CD and can be accessed easily via a periodic-table-based interface. In addition there is information about macroscale and nanoscale physical properties, solid-state structures, and discovery and applications of each element.

The JCE: Software General Chemistry Collection is a CD-ROM for Macintosh and Windows. Intended for use by students, it includes nearly 40 software programs that correlate with subjects described in this book.

Inorganic Molecules: A Visual Database by Charles Ophardt is a Macintosh-only CD-ROM that contains images of a broad range of molecules specifically designed for classroom presentation by a teacher. Ball-and-stick and space-filling models, Lewis diagrams, VSEPR geometries, orbital hybridization, molecular orbitals, and dipole moments are illustrated.

The Periodic Table Videodisc: Reactions of the Elements (second edition) by Alton J. Banks is a *visual* database; it shows still and motion images of the elements, their uses, and their reactions with air, water, acids, and bases. It is a particularly useful way to demonstrate chemical reactions in a large lecture room. The videodisc can be operated from a videodisc player by a hand-controlled keypad, a barcode reader, or an interfaced computer.

ChemDemos I and II by John W. Moore *et al.* are two videodiscs that contain a broad range of chemical demonstrations roughly correlated with the order of presentation of chemistry content in this and other texts. They are particularly useful for demonstration of chemical reactions in a large lecture room. The videodiscs can be operated from a videodisc player by a hand-controlled keypad, a barcode reader, or an interfaced computer.

The World of Chemistry: Selected Demonstrations and Animations I and II are two videodiscs that show demonstrations together with animated sequences that give a nanoscale interpretation of the demonstrations, both selected from the "World of Chemistry" television series and telecourse.

HIV-1 Protease: An Enzyme at Work by Erica Bode Jacobsen *et al.* is a VHS videotape in NTSC format that uses a very timely example (protease inhibitor treatment of AIDS) and state-of-the-art molecular modeling to show enzyme action. It includes class-tested teaching materials designed to help students learn about enzyme catalysis and inhibition.

Saunders College Publishing may provide complimentary instructional aids and supplements or supplement packages to those adopters qualified under our adoption policy. Please contact your sales representative for more information. If as an adopter or potential user you receive supplements you do not need, please return them to your sales representative or send them to:

Attn: Returns Department
Troy Warehouse
465 South Lincoln Drive
Troy, MO 63379

The New Traditions Project

The philosophy and approach taken in this book and many of the specific implementations of that philosophy are congruent with those of the systemic curriculum project Establishing New Traditions: Revitalizing the Curriculum, and this book will be an excellent resource for those who adopt and adapt the New Traditions approach. The New Traditions project is funded by the National Science Foundation, Directorate for Education and Human Resources, Division of Undergraduate Education, grant DUE-9455928.

REVIEWERS

Reviewers play a vitally important role in the preparation of a textbook. Our reviewers have risen to the challenge of evaluating a text that aims to find new ways to generate interest and excitement about chemistry for our students while maintaining rigor and appropriate content coverage. We deeply appreciate the effort they put into their tasks, and we have paid close attention to every comment.

The individuals listed below have played a significant role in shaping this book, and we thank them for their efforts. John DeKorte, Glendale Community College, Arizona, gets our special thanks for taking great care to check every word of the final draft for accuracy, and doing so quickly and cheerfully. Rob Profilet, Texas Tech University and Texaco, checked selected end-of-chapter questions promptly and enthusiastically.

Thomas Ballantine, *University of North Dakota*
Patrick Holt, *Bellarmine College*
Albert Jache, *Marquette University*
Paul Kelter, *University of Nebraska*
Miles Koppang, *University of South Dakota*
David Miller, *California State University-Northridge*
Dean Nelson, *University of Wisconsin-Eau Claire*
Jose Ivan Padovani, *University of Puerto Rico*
Richard Peterson, *Montana State University*
Bernard Powell, *University of Texas-San Antonio*
Robert Profilet, *Texas Tech University*
Ruben Puentedura, *Bennington College*
Jimmy Reeves, *University of North Carolina-Wilmington*
Thomas Ridgway, *University of Cincinnati*
Joseph Sneddon, *McNeese State University*
Richard Stolzberg, *University of Alaska-Fairbanks*
Donald Williams, *Hope College*
Sidney Young, *University of South Alabama*

The following reviewers based their comments on the first edition of *The Chemical World:*

Loren Carter, *Boise State University*
Wally Cordes, *University of Arkansas*
Frank Darrow, *Ithaca College*
James Marshall, *University of North Texas*
Ronald Martin, *The University of Western Ontario*
Mary O'Brien, *Edmonds Community College*
Melinda Oliver, *Louisiana State University*
Richard Peterson, *Montana State University*
Amy Phelps, *University of Northern Iowa*
Jimmy Reeves, *University of North Carolina-Wilmington*
Philip Watson, *Oregon State University*
John Woolcock, *Indiana University of Pennsylvania*
Sidney Young, *University of South Alabama*

The following professors added helpful comments in a survey about the first edition:

Robert Allendoerfer, *State University*

of New York at Buffalo
Robert Browne, *Douglas College*
Ken Coskran, *State University of New York at Potsdam*
James Fennessey, *Northland College*
Lorie Juhl, *Eastern Idaho Technical College*
Roger Kugel, *Saint Mary's College*
Alan Millen, *Brunswick Community College*
Andrew Price, *Ursinus College*
James Walsh, *John Carroll University*
Carla Wifler, *University of Wisconsin-Fond du Lac*
John Woolcock, *Indiana University of Pennsylvania*

Finally, the original reviewers of the first edition of *The Chemical World* deserve our thanks for helping us develop a new approach to teaching general chemistry:

Jon M. Bellama, *University of Maryland*
Rathindra N. Bose, *Kent State University*
Mary K. Campbell, *Mount Holyoke College*
Dewey Carpenter, *Louisiana State University*
Coran L. Cluff, *Brigham Young University*
John DeKorte, *Northern Arizona University*
Gary Edvenson, *Moorehead State University*
John Fortman, *Wright State University*
Ron Furstenau, *U.S. Air Force Academy*
Roy Garvey, *North Dakota State University*
Tom Greenbowe, *Iowa State University of Science & Technology*
Henry Heikkinen, *University of Northern Colorado*
Paul Hladky, *University of Wisconsin, Stevens Point*
John Hostettler, *San Jose State University*
James House, *Illinois State University*
Dave Johnson, *University of Oregon*
Stanley Johnson, *Orange Coast College*
Paul Karol, *Carnegie Mellon University*
Doris Kimbrough, *University of Colorado, Denver*
Donald Kleinfelter, *University of Tennessee, Knoxville*
Anna McKenna, *College of St. Benedict*
Clyde Metz, *College of Charleston*
Patricia Metz, *Texas Tech University*
Dean Nelson, *University of Wisconsin-Eau Claire*
Ronald O. Ragdale, *University of Utah*
Dave Robson, *Chem Matters Magazine*
Tom Rowland, *University of Puget Sound*
Arlene Russell, *University of California, Los Angeles*
Barbara Sawrey, *University of California, San Diego*
Conrad Stanitski, *University of Central Arkansas*
George Stanley, *Louisiana State University*
Gail Steehler, *Roanoke College*
Tamar Y. Susskind, *Oakland Community College*
Donald Titus, *Temple University*
Raymond Trautman, *San Francisco State University*
Susan Weiner, *West Valley College*
Deborah Wiegand, *University of Washington*
Donald Williams, *Hope College*
Stanley Williamson, *University of California, Santa Cruz*
R. Terrell Wilson, *Virginia Military Institute*

ACKNOWLEDGMENTS

We have had outstanding assistance in meeting the goals we set for ourselves in revising this book, and we thank everyone who played a role. First and foremost is our Publisher, John Vondeling, who made the whole thing possible, starting with the first edition and his belief that the time was ripe for a new approach to general chemistry. John has been instrumental in creating the new alignment of authors for this edition and the talented, dedicated team at Saunders College Publishing who supported us every step of the way. With his usual patience and good cheer, John has provided guidance when and where it was needed most.

How we test students' understanding of the concepts of chemistry is extremely important. We have collaborated with a team of knowledgeable faculty who helped update and expand the chapter-end questions and problems in this text. For this endeavor, our thanks to: H. Graden Kirksey, University of Memphis, for Conceptual Challenge Problems; Patricia Metz, Texas Tech University, for Applying Concepts problems, reviewing all problems in the book, and writing many new ones; and Melinda Oliver, Louisiana State University, and Allan Smith, Drexel University, for also writing many new problems for this edition.

Our Contributing Editor, Mary E. Castellion, has played a key role throughout this project, from helping us to establish our goals and reorganize content, through getting the words and graphics onto the pages, to keeping tabs on the contributions of our large team. She brought to these tasks her unique combination of experience as both an author and an editor, and we thank her for her hard work and ability to help each of us produce the very best result.

Also helpful in the text development was Jennifer Bortel, who was with us at the start, but soon moved on to a higher position at Saunders. Senior Associate Editor Marc Sherman, her replacement, commissioned reviews and helped coordinate the editorial-to-production stage of the project.

Other valuable contributors at Saunders include Bonnie Boehme and Robin Bonner, Project Editors, who took painstaking care to ensure quality; Charlene Squibb, Senior Production Manager, who cheerfully supervised the entire production process; Caroline McGowan, Art Director, who spent many hours crafting the integration of text and artwork; Sue Kinney, Illustration Supervisor, who managed the artwork; Ruth Hoover, who designed the text and cover; George Semple, Manager of Photos and Permissions, who obtained excellent photographs under tight deadlines; and Rebecca Rhynhart, Editorial Assistant, who kept the manuscript flow organized. Our copy editor, Maureen Iannuzzi, did an excellent job of keeping the text consistent and prompting us for things we had forgotten, all while keeping to a very tight schedule.

Many of the color photographs for this edition are the product of the creative eye and mind of Charles D. Winters of Oneonta, New York. He spends countless hours in his studio to get a photograph just right.

The beautiful and meaningful artwork in the book was rendered by Rolin Graphics, as well as George Kelvin, perhaps the finest scientific illustrator in the United States. His drawings not only illustrate the principles of chemistry, but are truly works of art.

Other excellent illustrations in this book have been generated by computer, using the CAChe Scientific software package from Oxford Molecular, Inc. Some were printed directly from computer files. The computer work, which involved constructing molecules and crystal lattices, sizing files appropriately, and converting files to a format that the compositor could use, was done by Randy Wildman at the University of Wisconsin-Madison. Randy has worked very hard on this project and has contributed many creative ideas. We are also grateful to Charlene Squibb and Caroline McGowan, whose legwork showed that our computer files could be printed successfully. Caroline oversaw all aspects of converting computer-generated art into final form. Mary E. Castellion helped select figures appropriate for nanoscale representation using CAChe, and the whole process was overseen by one of the authors (JWM), to whom the other authors are very grateful.

While authors are concentrating on putting a book together, others are focused on spreading the word. We are pleased to work with Angus McDonald, Executive Product Manager at Saunders, and thank him for conducting market research, helping us refine and test ideas, and getting the book into the right hands. Angus was ably assisted by Caren Hefner, Marketing Coordinator.

Nicholas Speckman, Acquisitions Editor for Technology, and Christine Livecchi, Developmental Editor for Technology, have been responsible for refining the Saunders Web site and the media ancillaries that accompany the text.

Many of the Chemistry You Can Do take-home experiments in this book were adapted from activities published by the Institute for Chemical Education as "Fun with Chemistry," Volumes I and II. These were originally collected for ICE by Mickey and Jerry Sarquis of Miami University, Ohio.

Finally, thanks to our colleagues, students, families, and the many unnamed others from whom we received information, advice, and support.

We have pledged that *The Chemical World* is a book that will evolve with the changing times in chemical education. To help us meet this goal, we urge you to communicate with us or our editors at Saunders College Publishing about anything you believe we should know: something you particularly like or dislike, a better way to explain a concept, a more relevant application or demonstration, a new and creative way to use molecular structure art, or correction of an error that has crept in despite our best efforts.

We hope that using this book results in a lively and productive experience for both faculty and students.

John W. Moore — *Madison, Wisconsin*
Conrad L. Stanitski — *Conway, Arkansas*
James L. Wood — *Nashville, Tennessee*

CONTENTS OVERVIEW

Available on the World Wide Web at http://www.saunderscollege.com. This chapter serves as a capstone experience that integrates and coordinates chemistry concepts learned earlier with a variety of other areas to give students a realistic picture of the broad applicability and importance of the chemistry discipline.

For more detailed information see p. ix of the Preface.

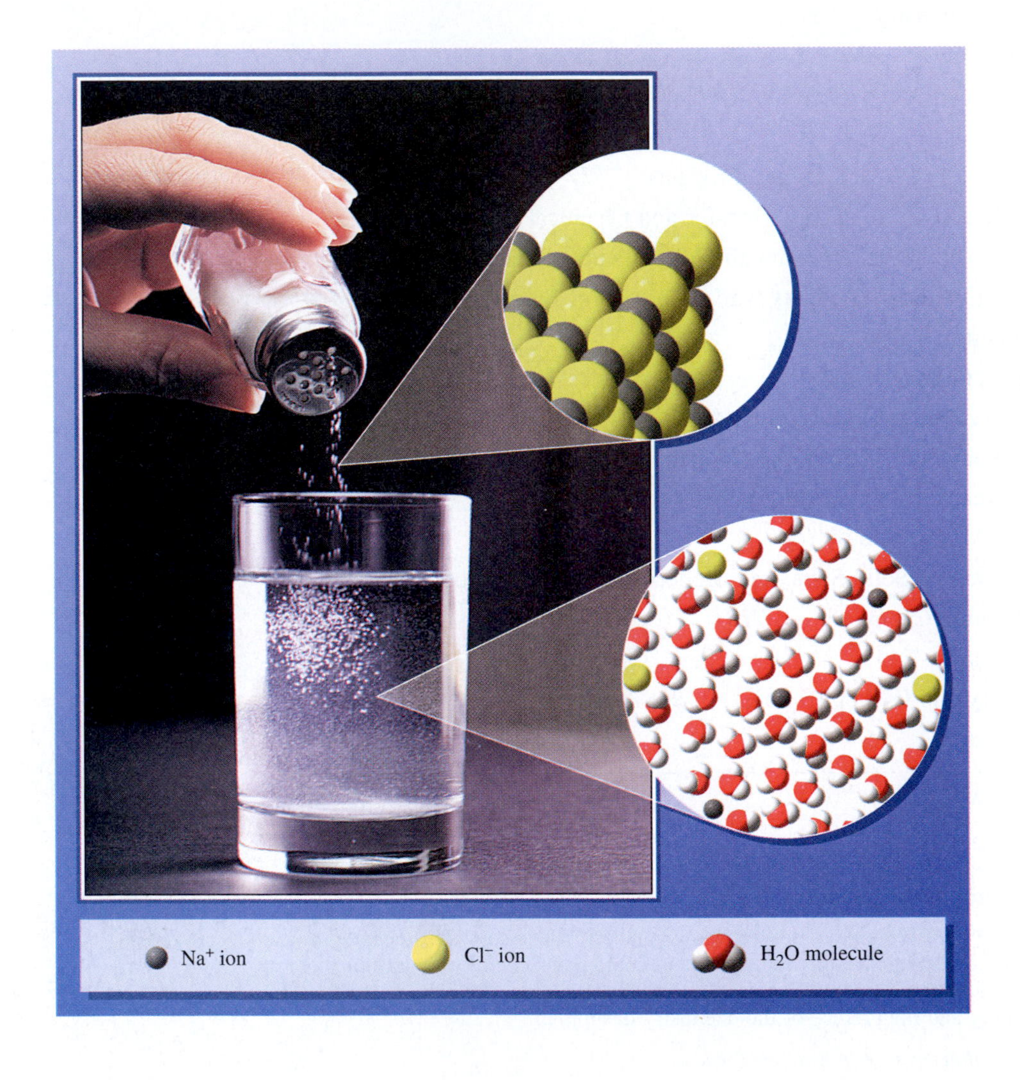

Na^+ ion
Cl^- ion
H_2O molecule

CONTENTS

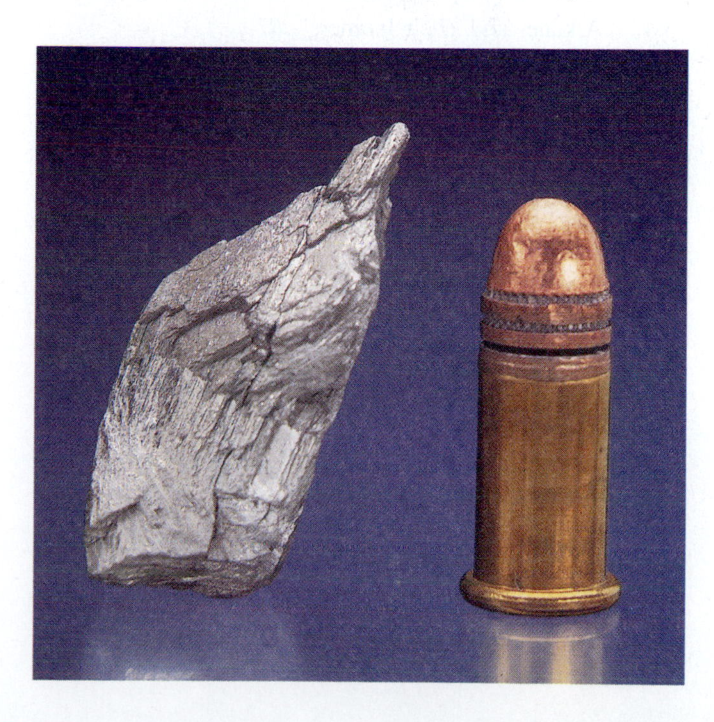

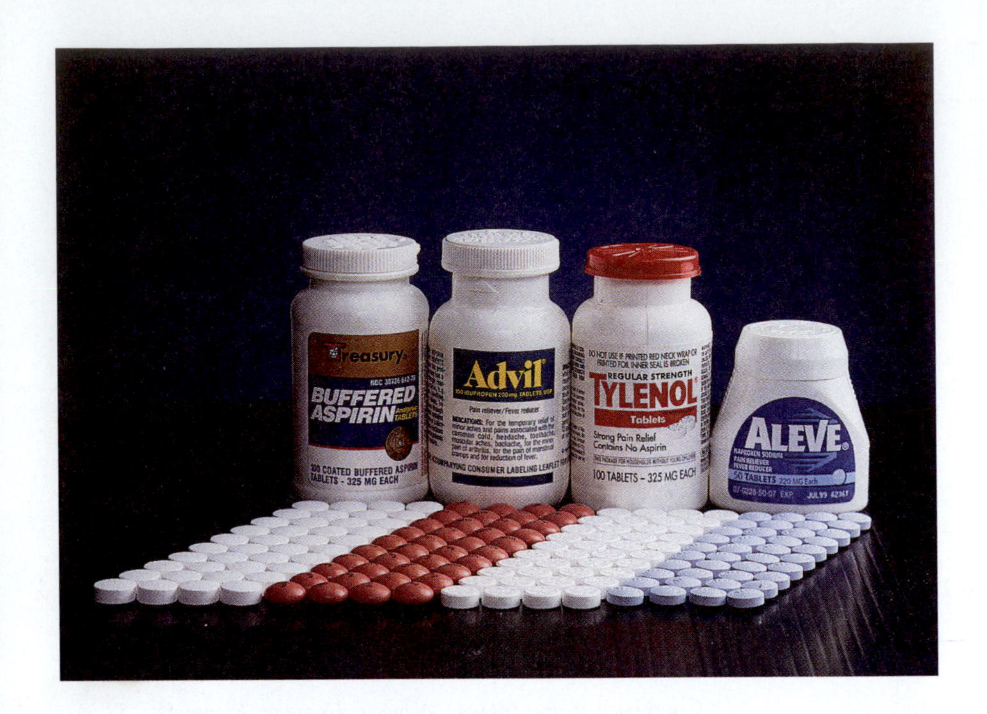

3 CHEMICAL COMPOUNDS *78*

4 CHEMICAL REACTIONS *130*

5 RELATING QUANTITIES OF REACTANTS AND PRODUCTS *179*

6 PRINCIPLES OF REACTIVITY I: ENERGY TRANSFER AND CHEMICAL REACTIONS *219*

Lighter
Butane

9 COVALENT BONDING *353*

10 MOLECULAR STRUCTURES *397*

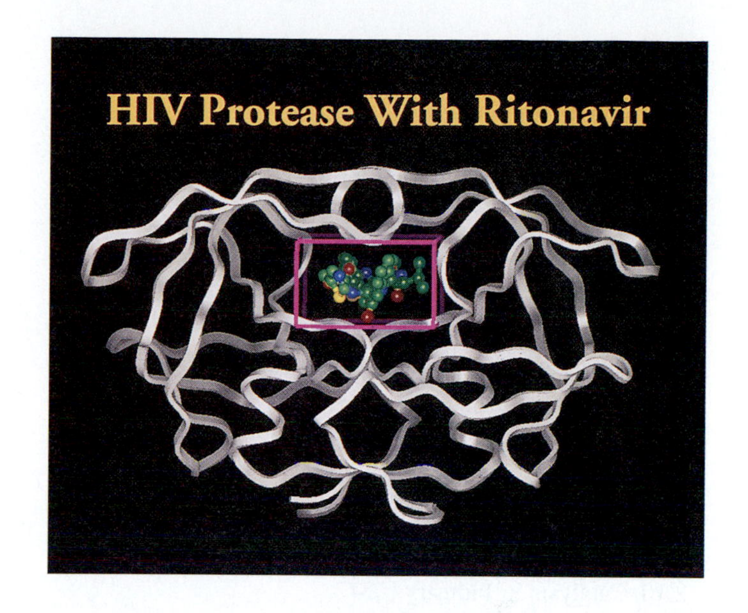

11 ENERGY, ORGANIC CHEMICALS, AND POLYMERS *449*

12 PRINCIPLES OF REACTIVITY III: CHEMICAL KINETICS *511*

13 PRINCIPLES OF REACTIVITY IV: CHEMICAL EQUILIBRIUM *566*

14 GASES AND THE ATMOSPHERE *609*

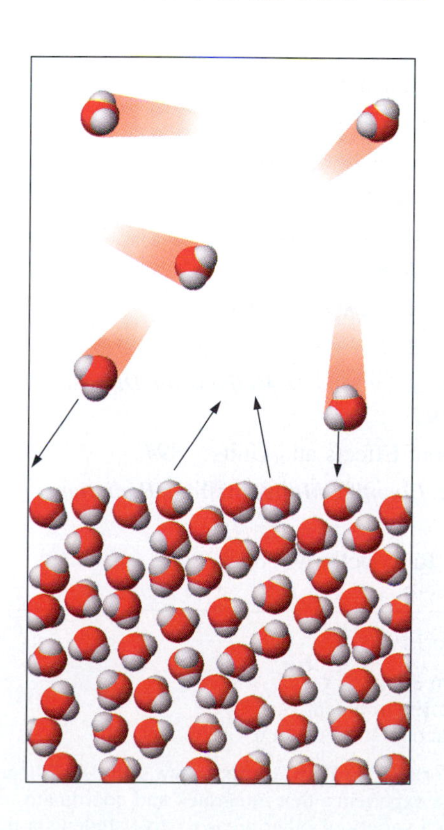

INDEX TO SPECIAL FEATURES

Chemistry in the News

Chemistry You Can Do

Portrait of a Scientist

CHAPTER

1

The Nature of Chemistry

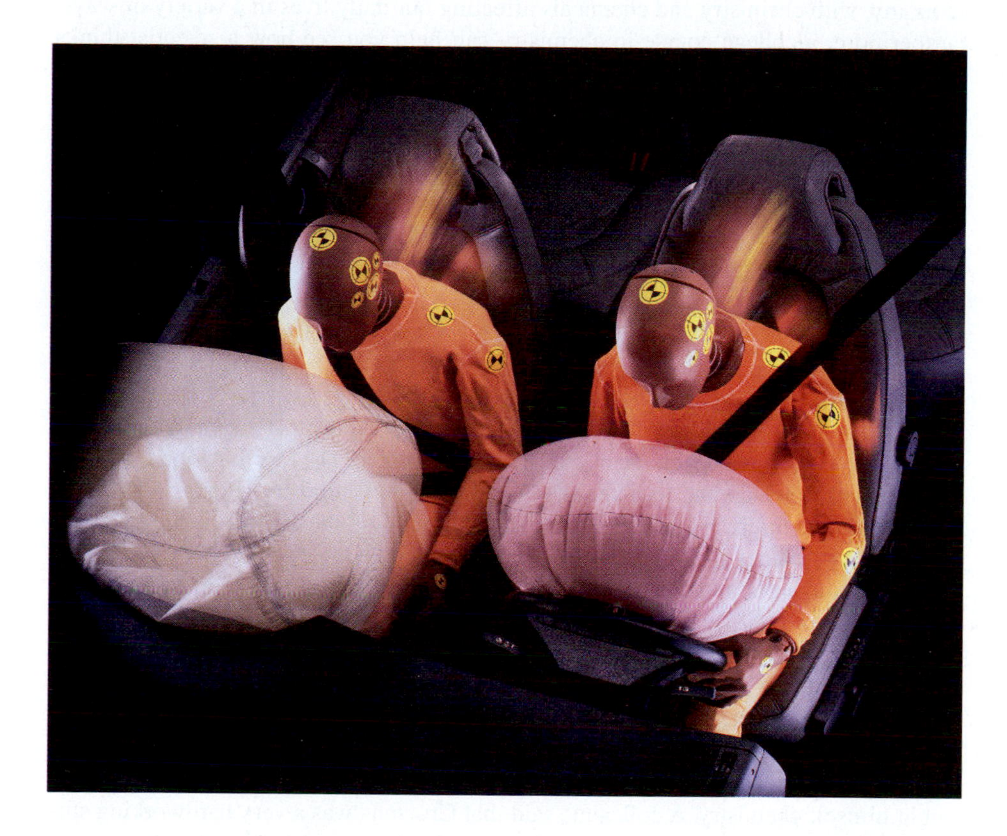

These automobile air bags illustrate not only uses of chemistry but also the balancing of risks and benefits. In an accident, air bags are inflated by nitrogen gas that comes from a chemical change, and there is abundant evidence that a driver's-side air bag saves lives. There is, however, some question about the benefits of passenger's-side air bags, because injury or even death of some infants has been attributed to them. *(Saab USA, Inc.)*

Why study chemistry? Each person probably has a different answer. Some find the subject itself fascinating, but many are in a college chemistry course because someone else has decided it is useful as part of a background for a particular career. But why should it be so useful? Chemistry is central to our understanding of biology, geology, materials science, medicine, many branches of engineering, and other sciences. In addition, chemistry plays a major role in our economy, with chemistry and chemicals affecting our daily lives in a variety of ways. Furthermore, a college course in chemistry can help you see how a scientist thinks about the world and how to solve problems. The knowledge and skills developed in such a course will benefit you in many career paths and will help you become a better informed citizen in a world that is becoming technologically more and more complex—and interesting. Therefore, to begin your study of chemistry, this chapter discusses some of the most fundamental ideas used by practicing chemists. It also introduces you to a problem confronting chemists, and indeed all citizens, almost every day—the weighing of the benefits of a discovery or a practice against its risks to individuals and society.

1.1 SCIENCE AND THE OZONE LAYER

In October 1995 it was announced that Dr. Paul Crutzen, Dr. Mario Molina, and Dr. F. Sherwood Rowland had received the Nobel Prize in Chemistry (Figure 1.1). Their story, and their discovery of the effects of pollutants on the earth's ozone layer, illustrates how science is done.

Paul Crutzen is now at the Max Planck Institute for Chemistry in Mainz, Germany, but he is a native of Amsterdam, the Netherlands. His early schooling was interrupted by World War II, but he eventually qualified as a construction engineer and moved to Sweden. There he applied for a job as a computer programmer in the meteorology department at Stockholm University—even though he had no experience in either field—and he was soon using computers to develop weather models. He began learning meteorology and, because the atmosphere is a soup of chemicals, he taught himself chemistry. A colleague said that Crutzen "was a very hardworking student and very well focused." Crutzen became fascinated with the role of ozone, a form of oxygen, in the stratosphere. Although he assumed the chemistry of this gas

Modeling of the behavior of natural systems is important in the physical, biological, and social sciences. Crutzen modeled weather systems. Chemists develop models of molecules and of solids, liquids, and gases. Such models are an important theme of this book.

(a)

(b)

Figure 1.1 Recipients of the 1995 Nobel Prize in Chemistry. Paul Crutzen (a), Mario Molina (b; left), and F. Sherwood Rowland (b; right) shared the Nobel Prize in Chemistry for their studies of the chemistry of earth's ozone layer and the effects of pollutants such as chlorofluorocarbons on the ozone. *(Molina and Rowland: Bettmann Archive)*

was well known, it was not, and his first study concerned the interaction between ozone and nitrogen oxides. When he published the study, other atmospheric scientists saw immediately that it was important work, and he has spent much of the rest of his career working in the field of atmospheric chemistry.

As often happens in science, one person's work meshes in unexpected ways with that of other scientists. Sherwood Rowland at the University of California–Irvine heard about some studies by the British scientist James Lovelock, who had recently found measurable quantities of the *chlorofluorocarbons* (CFC-11 and CFC-12) in the atmosphere. Rowland was curious—what was the fate of these compounds? One reason they were attractive items of commerce was that they were thought to be inert; that is, they did not interact with other chemicals to give still other new chemicals. If they were indeed inert, would they build up in the atmosphere as more and more CFCs were manufactured? Or, did they produce something new by interacting with chemicals in the atmosphere? Finally, because the CFCs were found widely in the atmosphere, if there was a problem, was the problem a global one?

Ozone, nitrogen oxides (which are formed by the combination of nitrogen and oxygen), and CFCs (called chlorofluorocarbons because they contain chlorine, fluorine, and carbon) are chemicals that will be described further in later sections of this book.

What happened next is typical of the way science works: we perform experiments to probe our ideas. Mario Molina came to Rowland's laboratory as a research assistant after finishing his Ph.D. research, and Rowland gave him the option of studying the atmospheric fate of CFCs. They quickly learned that there were no processes in the lower atmosphere that would affect the CFCs. So, if they were to be destroyed and disappear from earth's atmosphere, the only mechanism was destruction by high-energy ultraviolet radiation in the middle levels of the stratosphere, that is, about 25 to 30 kilometers above the earth's surface. Rowland and Molina knew the earth has a stratospheric ozone shield. From Crutzen's work they also knew a great deal about the atmospheric chemistry of ozone. High-energy ultraviolet radiation from the sun strikes the ozone molecules in the stratosphere, and they are ripped apart to form oxygen molecules and oxygen atoms, which eventually come back together to form ozone molecules. As a result, the amount of ozone in the stratosphere stays about the same.

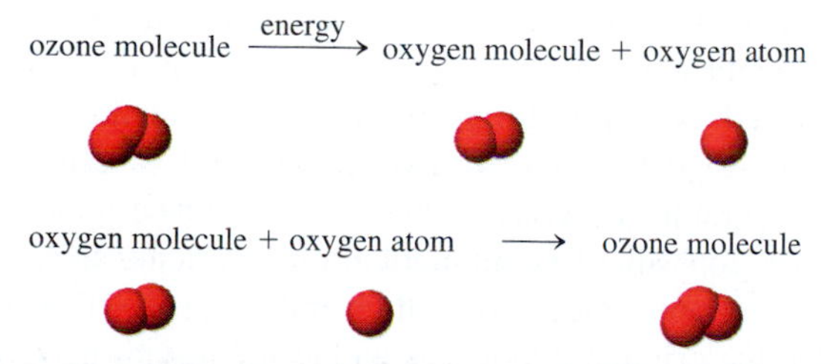

Atoms and molecules are so fundamental to chemistry that it's almost impossible to write about chemistry without referring to them. In the picture at the left, each sphere represents an atom, and touching spheres are connected in a molecule. Explanations of the nature of these tiny particles of matter are given in Chapters 2 and 3.

This important process is the reason that very high-energy ultraviolet radiation does not reach the earth's surface. Energy that could produce severe damage to living plants and animals is diverted because it splits ozone molecules. Rowland and Molina reasoned that CFCs could be destroyed only after they had risen above the ozone shield and that, as a consequence, CFCs could exist for a very long time in the lower atmosphere. In the 1970s around 700,000 metric tons of CFCs were being produced each year, so the level of CFCs in the atmosphere could become very large indeed.

But what if the CFCs rose, by normal atmospheric mixing processes, above the ozone layer to an altitude where they could be bombarded with high-energy radiation? What Rowland and Molina learned next set off a chain of events of worldwide importance. Their experiments quickly showed that the chlorine atoms released by the destruction of CFCs in the stratosphere could interact with the ozone and disturb the delicate balance between ozone formation and destruction so that destruction would be favored. The result would be substantial depletion of the stratospheric ozone shield.

Their continued experiments, and experiments by many other scientists, led many to conclude that there was a potential problem. There was enough evidence of a problem that manufacturers, beginning in about 1977, started removing CFCs as propellants in spray cans of paint and underarm deodorant. On the other hand, no observations had yet been made of actual conditions in the atmosphere. These came in 1985. First Joseph Farman of the British Antarctic Survey, and then Susan Solomon, leader of the National Ozone Expedition to the Antarctic, found evidence that the ozone layer over the Antarctic was in fact depleted.

The problem of CFCs remains because thousands of tons of CFCs are still in use in refrigerators and air conditioners around the world. How can these be destroyed safely as they are replaced with environmentally safe alternatives? A very simple chemical process was recently announced by Professor Robert Crabtree and graduate student Juan Burdeniuc of Yale University. They use a chemical found in natural substances in a process described further in Chemistry in the News: CFC Disposal Using Redox Reactions, in Chapter 4.

In 1987, 43 nations of the world, acting under the auspices of the United Nations Environmental Program, agreed in Montreal, Canada, to cut back CFC production by 50% by the end of the 20th century. In 1990, with evidence accumulating of rapid ozone loss, 93 nations agreed to halt all CFC production by 2000. By 1994, 134 nations had agreed to this plan, which is now referred to as the Montreal Protocol.

As this book is being written, according to Rowland and Molina, "the world is well on its way in transition to a CFC-free economy, although not yet to a CFC-free atmosphere." There is experimental evidence, however, that chlorine levels in the atmosphere are decreasing. It has been calculated that the ozone layer will begin recovering by 2000 and that it should return to 1979 levels by around 2050.

1.2 HOW SCIENCE IS DONE

The story of the atmospheric fate of CFCs illustrates the way science is done. After seeing a problem—one that might well be of crucial significance to human society—Rowland and Molina collected experimental information and surveyed the work of other scientists such as Crutzen. Often even seemingly unrelated studies from other laboratories shed some light on the problem being studied. From these initial studies, Rowland and Molina then developed a **hypothesis**—a tentative explanation or prediction of experimental observations. Next, they performed experiments designed to give results that might confirm their hypothesis or eliminate erroneous explanations; that is, they tested the claim. In doing so, they collected both quantitative and qualitative information or data. **Quantitative** information usually means numerical data, such as the temperature at which a chemical substance melts. **Qualitative** information, in contrast, consists of nonnumerical observations, such as the color of a substance or its physical appearance. With their initial results, Rowland and Molina revised and extended the original hypothesis and tested it with new experiments. After they had done a number of experiments and continually checked to make sure the results were truly reproducible, a pattern began to emerge. At this point, they prepared the results for publication in the chemical literature.

Conceptual Challenge Problem CP-1.A at the end of the chapter relates to topics covered in this section.

Conceptual Challenge Problem CP-1.B at the end of the chapter relates to topics covered in this section.

Often a scientist is eventually able to summarize the experimental observations in the form of a **law** or general set of rules, concise verbal or mathematical statements of a relation that always occurs under the same conditions. You will see many laws in chemistry, and we base much of what we do on them because they help us predict what may occur under a new set of circumstances.

Keep in mind, though, that a result might be different from what is expected based on a law or general rule. Then a scientist gets excited because experiments that do not follow the laws of chemistry are the most interesting. Most importantly, we know that understanding the exceptions almost invariably gives new insights.

Once scientists have done enough reproducible experiments to lead to a law, they may be able to formulate a theory to explain their observations. A **theory** is a unifying principle that explains a body of facts and the laws based on them. It is capable of suggesting new hypotheses. Theories abound to account for the way our economy or society works, for example. In chemistry, excellent examples of theories are those developed to account for chemical bonding. It is a fact that atoms are held together or bonded to one another, as in a molecule of ozone. But how and why? Several theories are currently used in chemistry to answer these questions, but are they correct? What are their limits? Can the theories be improved, or are completely new theories necessary? Laws summarize the facts of nature and rarely change. Theories are inventions of the human mind. Theories can and do change as new facts are uncovered.

Exercise 1.1 Quantitative and Qualitative

Identify each of the following descriptions as qualitative or quantitative information.
(a) A sample of lead has a mass of 0.123 g and melts at 327 °C.
(b) An iron-containing compound is green.
(c) Your textbook has a yellow cover and a mass of 1.8 kilograms.

People outside of science usually have the idea that science is an intensely logical field. They picture a white-coated chemist moving logically from hypothesis to experiment and then to laws and theories without human emotion or foibles. Nothing could be further from the truth! Often, scientific results and understanding arise quite by accident. Creativity and insight are needed to transform a fortunate accident into useful and exciting results.

Serendipity is the accidental discovery of things not sought for. For an excellent account of such events in chemistry see *Serendipity* by R. M. Roberts, John Wiley, New York, 1989.

1.3 SIMPLE BUT ELEGANT EXPERIMENTS: EXPLAINING THE *CHALLENGER* EXPLOSION

Another person who understood very well how science works was Richard Feynman, the recipient of the Nobel Prize in Physics in 1965 and one of the most original thinkers of this century. Feynman said that "Scientific knowledge is a body of statements of varying degrees of certainty—some most unsure, some nearly unsure, but none *absolutely* certain. . . . Now, we scientists are used to this, and we take it for granted that it is perfectly consistent to be unsure, that it is possible to live and *not* know." It is well to remember Feynman's statement as you read statements about our environment, about health care and medicine, and about nutrition.

The remarks by Richard Feynman (1919–1988) and the story of his role in the *Challenger* disaster are found in his book *What Do You Care What Other People Think?* (W.W. Norton and Company, New York, 1988). See also his autobiography *Surely You're Joking Mr. Feynman* (W.W. Norton and Company, New York, 1985), in which he tells about his life in physics and his passion for the bongo drums.

Richard Feynman gave the United States a lesson in how science is done when he used a simple experiment to uncover the reason for the disastrous explosion of the space shuttle *Challenger*. The *Challenger* was launched on Tuesday, January 28, 1986. The day was unusually cold for Florida; the temperature at the time of launch was 29 °F. The world watched in horror when, after a minute or so of flight, the shuttle and its rockets exploded in a fire ball, killing the astronauts on board.

To understand the reason for the explosion, and the importance of Feynman's experiment, recall something of the design of the shuttle. The main engine is fueled by liquid hydrogen and oxygen, which are contained in the large tank strapped onto the belly of the shuttle (Figure 1.2a). The hydrogen and oxygen combine to give water, and the energy released provides the thrust to boost the shuttle into orbit.

hydrogen molecules + oxygen molecule ⟶ water molecules (energy released)

(a)

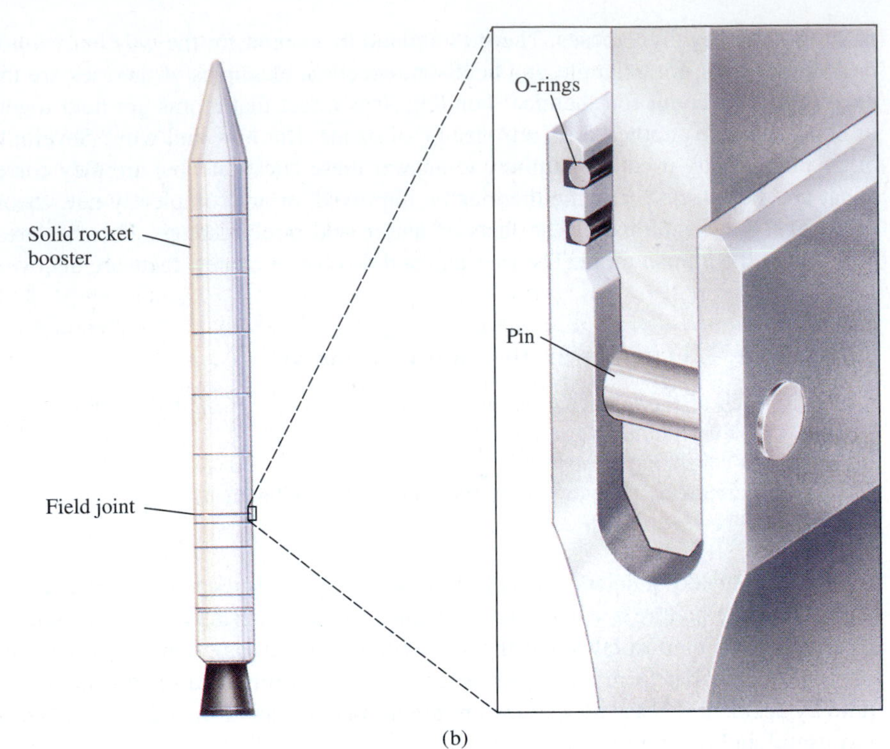

(b)

Figure 1.2 The space shuttle *Challenger* accident. (a) The space shuttle *Atlantis*, lifting off on August 2, 1991. The large orange tank on the belly of the orbiter contains hydrogen and oxygen, which are burned in the orbiter's main engines. The long tubes on either side of the hydrogen-oxygen tank (only one is seen here) are the solid fuels boosters. (b) Each solid-fuel booster rocket is made of a number of sections bolted together. The joints between sections are complicated and involve, among other things, O-rings made of a special rubber. The function of the O-rings is to seal the joints and prevent hot gases from leaking out from the sides of the rocket. *(a, NASA)*

However, to provide additional thrust at the time of launch, there are solid fuel booster rockets strapped on each side of the shuttle. The solid fuel rockets are made in sections, and the sections are shipped from the factory to the Kennedy Space Center, where they are joined to make the completed booster rocket. The joint between the sections was designed so that hot gases from the burning solid fuel would not leak through the walls of the rocket. Part of the design to close the joint, and yet make it somewhat flexible, was the use of a thin O-ring made of a special rubber (Figure 1.2b). From the beginning, Feynman and others thought that a possible cause of the *Challenger* explosion was that the solid fuel had burned through the wall of the booster rocket and then burned into the tank holding the liquid hydrogen, thus exploding the hydrogen. But how did this happen?

When the shuttle is launched, and the solid fuel begins to burn, the walls of the rocket casing move slightly outward. If this movement caused the joints between sections to open up, then the fuel would burn through the joint. But the O-rings were supposed to prevent this. Based on information from engineers involved in the design of the solid rocket boosters, Feynman's hypothesis was that, due to the unusually cold weather, the rubber O-rings had not expanded properly, and flame burned through the joint. To prove his point, Feynman did a dramatic, but very simple, experiment.

During a public hearing Feynman took a sample of the rubber O-ring, held tightly in a C-clamp, and then put it into a glass of ice water (Figure 1.3). Everyone could make the qualitative observation that the rubber did not spring back to its original shape! The poor resilience of the rubber at low temperatures doomed the *Challenger.* Feynman's hypothesis was supported by his elegantly simple experiment.

Not all experiments are as simple as Feynman's, or so dramatically illustrate the point. Some take days or even months to complete and involve complex and expensive instruments. Nonetheless, it is often true that the very best experiments, the ones that produce the most useful and persuasive results, are the simplest.

The O-rings in the *Challenger*'s rockets were flexible at ordinary temperatures but were less elastic at lower temperatures. This property of the rings, and the chemical property that they could burn, determined the *Challenger*'s fate.

Figure 1.3 A simple experiment. Richard Feynman shows experimentally that a cold O-ring, of the type used in the space shuttle *Challenger*, does not snap back quickly to its original shape. *(Marilynn K. Yee/NYT Pictures)*

1.4 PHYSICAL PROPERTIES OF MATTER

You put about half a pound of sugar (chemical name, sucrose) along with half a cup of water and half a cup of corn syrup (a glucose solution) in a pan and heat the mixture while stirring steadily. You continue to boil the syrup and watch as it slowly turns brown; a cloud of vapor rises above the bubbling mixture. When the temperature reaches 140 °C, you toss in a handful of peanuts and a pinch of sodium bicarbonate and pour the liquid mixture quickly onto a piece of aluminum foil (Figure 1.4). When it cools, you smash the solid into small pieces and pop some into your mouth! You have just made peanut brittle—and have carried out a wonderful kitchen experiment, in the course of which you have done many of the things that chemists pay attention to every day.

Some Physical Properties
- Temperature at which a substance melts or boils
- Density
- Color
- Vapor pressure of a solid or liquid
- Thermal conductivity
- Electrical conductivity

Chemistry is the study of matter and the changes it can undergo. In making peanut brittle, you have observed physical and chemical changes and different states of matter, have made qualitative and quantitative observations, and have made measurements. That is what chemistry is all about: making sense of the material world through qualitative and quantitative observations, and using the knowledge so gained to adapt the material world to our own interests.

Your friends can recognize you by your physical appearance: your height and weight and the color of your eyes and hair. Chemical substances can also be recognized by their physical appearance. While making peanut brittle you could distinguish sugar from water because you know that sugar consists of small white particles of solid, while water is a colorless liquid. Corn syrup is also a liquid, but it comes in light and dark colors and is much thicker (pours more slowly) than water. Properties such as these, which can be observed and measured without changing the composition of a substance, are called **physical properties.** Physical properties allow us to classify and identify substances of the material world.

Figure 1.4 Making peanut brittle. For more information on making peanut brittle and the chemistry of the process, see "Peanut Brittle" by E. Catelli in *Chem Matters*, December 1991, page 4. *(C. D. Winters)*

Temperature

Two very useful physical properties of pure chemical elements and compounds are the temperatures at which a solid melts (its **melting point**) and a liquid boils (its **boiling point**). **Temperature** is the property of matter that determines whether there can

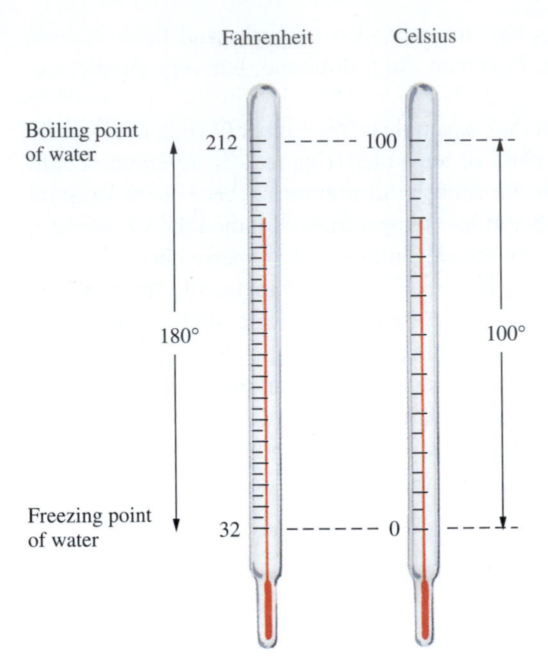

Figure 1.5 **A comparison of the Fahrenheit and Celsius temperature scales.**

be energy transfer as heat from one body to another and the direction of that transfer: energy transfers spontaneously only from a hotter object to a cooler one. The number that represents an object's temperature depends on the unit chosen for the measurement. (This is just like the fact that the number representing the distance between two points depends on its unit of measurement. The length of a football field could be expressed as 100 yards or 91.4 meters, for example.)

In the United States, everyday temperatures are reported using the Fahrenheit scale, but the **Celsius** scale is used in most other countries and in science. Both are based on the properties of water. The size of the Celsius degree is defined by assigning zero as the freezing point of pure water (0 °C) and 100 as its boiling point (100 °C), as illustrated in Figure 1.5. The number of units between the freezing and boiling points of water is 180 Fahrenheit degrees and 100 Celsius degrees. Comparing the two units, the Celsius degree is almost twice as large as the Fahrenheit degree; it takes only 5 Celsius degrees to cover the same temperature range as 9 Fahrenheit degrees, and it is this relationship that can be used to convert a temperature on one scale to a temperature on the other (see Appendix C). However, because all temperatures in scientific studies are measured in Celsius units, there is really little need to make conversions to and from the Fahrenheit scale when studying chemistry.

To be entirely correct, we must specify that water boils at 100 °C and freezes at 0 °C only when the pressure of the surrounding atmosphere is 1 standard atmosphere. We shall discuss pressure and its effect on boiling point in Section 15.2.

To help you think in terms of the Celsius scale, it is useful to know that water freezes at 0 °C and boils at 100 °C, a comfortable room temperature is about 22 °C, your body temperature is 37 °C, and the hottest water you could put your hand into without serious burns is about 60 °C.

Exercise 1.2 Temperature

(a) Which is the higher temperature, 37 °C or 85 °F?
(b) Which is the lower temperature, 20 °F or 0 °C?

Physical Change

Changes in the physical properties of a substance, **physical changes,** are of concern in chemistry. The same substance is present before and after a physical change even though the substance's physical state or the gross size and shape of its pieces may have changed. An example is the melting of a solid (Figure 1.6), and the temperature at which this occurs is often so characteristic that it can be used to identify the substance. For example, tin and lead resemble one another in outward appearance, but tin melts at 231.8 °C while lead melts at a temperature nearly 100 Celsius degrees higher (327.5 °C).

Figure 1.6 Physical change. When ice melts, it changes—physically—from a solid to a liquid. The substance is still water. *(C. D. Winters)*

Exercise 1.3 Physical Properties and Changes

(a) Identify the physical property or change(s) in each of the following statements:
   (i) The blue chemical compound azulene melts at 99 °C.
   (ii) The colorless crystals of salt are cubic.
(b) The boiling point of a substance is 15 °C. If you hold a sample of the substance in your hand, will it boil?

1.5 STATES OF MATTER AND A MODEL TO EXPLAIN THEM

An easily observed and very useful property of matter is its physical state. Is it a solid, liquid, or gas? A **solid** can be recognized because it has a rigid shape and a fixed volume that changes very little as temperature and pressure change (Figure 1.7). Like a solid, a **liquid** has a fixed volume, but a liquid is fluid—it takes on the shape of its container and has no definite form of its own. **Gases** are also fluid, but gases expand to fill whatever containers they occupy, and their volumes vary considerably with temperature and pressure. For most substances, when compared at the same conditions, the volume of the solid is slightly less than the volume of the same mass of liquid, but the volume of the same mass of gas is much, much larger. As the temperature is raised, most solids melt to form liquids; eventually, if the temperature is raised enough, most liquids boil to form gases.

Figure 1.7 A quartz crystal, an example of a substance in the solid state. *(C. D. Winters)*

All the physical properties just described can be observed by the unaided human senses and refer to samples of matter large enough to be seen, measured, and handled. Such samples are macroscopic; their size places them on the **macroscale.** By contrast, samples of matter so small that they have to be viewed with a microscope are on the **microscale** of matter. Viruses and bacteria, for example, are matter at the microscale. The matter that really interests chemists, however, is at the **nanoscale.** The term comes from the prefix "nano," which indicates something one billion times smaller than something else. For example, a line one billion (10^9) times shorter than 1 meter is 1 *nano*meter (1×10^{-9} m) long. The sizes of atoms and molecules place them on the nanoscale. It is a fundamental idea of chemistry that matter is the way it is because of the nature of its parts, and those parts are very, very tiny. Therefore, we need to use imagination to discover useful theories that connect the behavior of those tiny, nanoscale parts to the observed behavior of chemical substances at the macroscale. Chemistry enables you to "see" in the things all around you nanoscale structure that cannot be seen with your eyes.

The use of 10^9 to represent 1,000,000,000 or 1 billion is called scientific notation and is explained in Appendix B.2.

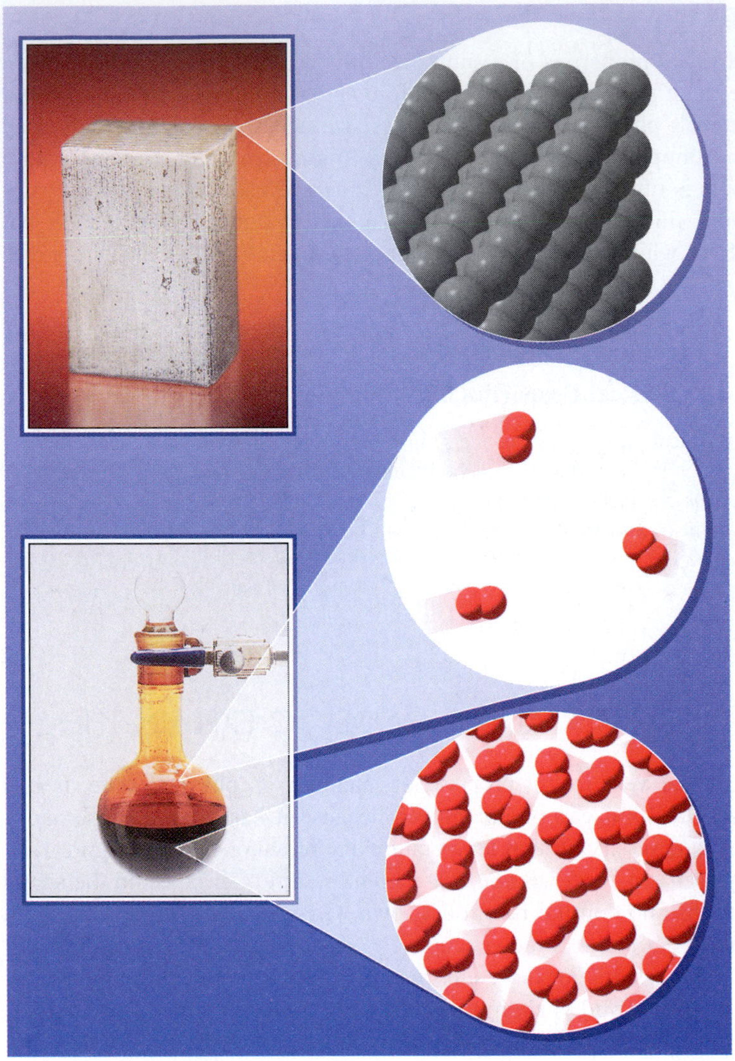

Figure 1.8 A nanoscale representation of the three states of matter. In the solid block of iron (*top*), the particles are close together and almost totally restricted to specific locations. In the liquid bromine in the flask (*bottom*), the particles are still reasonably close together and interact with one another. There is more motion of the particles than in the solid phase, although the particles move only very small distances before colliding with others. In the gaseous bromine above the liquid, bromine molecules move rapidly, traveling over distances much larger than the sizes of the particles themselves between collisions. There is little interaction between gas-phase particles. *(C. D. Winters)*

Kinetic-Molecular Theory

An example of a theory that explains matter at the nanoscale is the **kinetic-molecular theory** of matter. According to this theory, all matter consists of extremely tiny particles that are in constant motion. In a solid these particles are packed closely together in a regular array as shown in Figure 1.8a. These particles vibrate back and forth about their average positions, but seldom does a particle in a solid squeeze past its immediate neighbors to come into contact with a new set of particles. Because the particles are packed so tightly and in such a regular arrangement, a solid is rigid, its volume is fixed, and the volume of a given mass is small. The external shape of a solid often reflects the internal arrangement of its particles. This relation between the observable structure of the solid and the arrangement of the particles from which it is made is one reason scientists have long been fascinated by the shapes of crystals and minerals.

The kinetic-molecular theory of matter can also interpret the properties of liquids and gases, as shown in Figure 1.8b. Liquids and gases are fluid because the atoms or molecules are arranged more randomly than in solids. They are not confined to specific locations but rather can move past one another. Because the particles are a little farther apart in a liquid than in the corresponding solid, the volume is a little bigger. No particle goes very far without bumping into another—the particles in a liquid interact with each other constantly. In a gas the particles are very far from one another and moving quite rapidly. (In air at room temperature, for example, the average molecule is going faster than 1000 miles per hour.) A particle hits another particle every so often, but most of the time each is quite independent of the others. The particles fly about to fill any container they are in; hence, a gas has no fixed shape or volume.

In addition, the kinetic-molecular theory includes the concept that the higher the temperature is the more active the particles are. A solid melts when its temperature is raised to the point where the particles vibrate fast enough and far enough to push each other out of the way and move out of their regularly spaced positions. The substance becomes a liquid because the particles are now behaving as they do in a liquid. As temperature goes even higher, the particles move even faster until finally they can escape the clutches of their comrades and become independent; the substance becomes a gas. Increasing temperature corresponds to faster and faster motions of atoms and molecules, a general rule that you will find useful in many future discussions of chemistry.

Conceptual Challenge Problem CP-1.C at the end of the chapter relates to topics covered in this section.

Exercise 1.4 **Kinetic-Molecular Theory**

An ice cube sitting in the sun slowly melts, and the liquid water eventually evaporates. Are these chemical or physical changes? Describe them in terms of the kinetic-molecular theory.

1.6 SUBSTANCES, MIXTURES, AND SEPARATIONS

A **substance** is matter of a particular kind. Each substance has a well-defined set of characteristic properties that are different from the properties of any other substance. It also has the same composition—it has the same stuff in the same proportions—as every other sample of that substance (Figure 1.9).

Most natural samples of matter consist of two or more substances; that is, they are mixtures. Peanut brittle is obviously a mixture, because one can see that the peanuts are different from the surrounding material. A mixture in which the uneven texture of the material is visible is called a **heterogeneous** mixture. Some heterogeneous mixtures may first appear completely uniform but on closer examination are not. For example, blood appears smooth in texture to the unaided eye, but magnification reveals particles within the liquid (Figure 1.10a). In a heterogeneous mixture the properties in one region are different from the properties in another region.

A **homogeneous** mixture, or **solution,** is completely uniform at the macroscopic level and consists of two or more substances in the same state (Figure 1.10b). No amount of optical magnification will reveal a solution to have properties in one region different from those in another, because heterogeneity exists in a solution only at the nanoscale, where the individual atoms or molecules are too small to be seen

(Text cont. on page 13)

Figure 1.9 **Some examples of pure substances:** bromine, copper, and silicon on a sheet of aluminum. *(C. D. Winters)*

CHEMISTRY YOU CAN DO

Crystalline Solids

This experiment will allow you to observe properties of crystals of ordinary table salt. Other solids may be included as well. You will need

- a small quantity of salt—one or two servings from a fast-food restaurant should be enough
- a small quantity of water
- a small container, such as a film canister, medicine cup, or plastic glass
- a mirror, or something with a smooth, flat surface that can be washed clean
- a towel or tissue and some soap or detergent
- a magnifying glass

Place about two thirds of the salt into the bottom of the container and just cover with water. Swirl the water around for a minute or so. Allow the container to stand for 5 minutes. Repeat the swirling or stirring every 5 minutes or so for half an hour. If all of the salt dissolves in the water, add more salt. Allow the mixture to stand for at least 5 minutes so that all the solid salt falls to the bottom of the container.

While the salt is dissolving, wash the mirror with soap, dry it with a towel, and allow it to stand and dry completely. Place the mirror on a horizontal surface in a place where it can remain undisturbed for several hours. Carefully pour a little salt water onto the mirror. Allow the liquid to stand, and observe it every 15 minutes for several hours; then let it stand overnight. If crystals form, observe them carefully. What shape are they? Do crystals of different shapes form from the same solution? Compare your crystals with those other students have grown.

You can repeat the experiment with several other solids. (Sugar, Epsom salts, sodium bicarbonate, instant coffee, and others will dissolve in water and may form crystals as the water evaporates.) In each case it is important that the quantity of water added is not enough to dissolve all of the solid. Observe the results as the water slowly evaporates from the solution on the mirror. Do all of the solids form crystals as they come out of solution? What shapes are the crystals? It is handy to keep some notes on your observations.

Equipment needed for Chemistry You Can Do. *(C.D. Winters)*

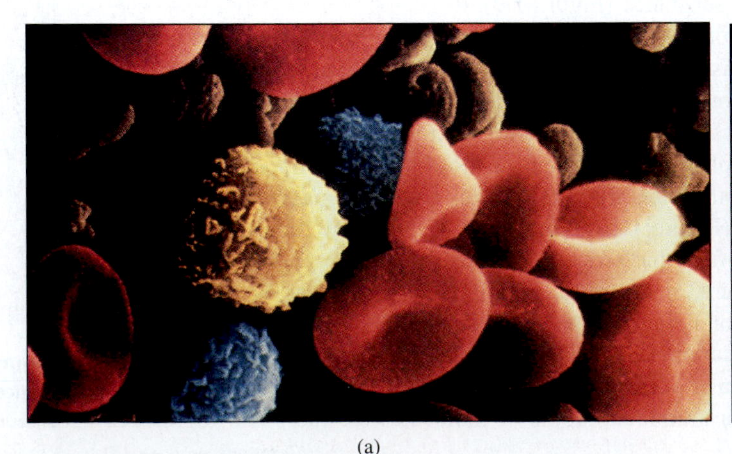

(a)

(b)

Figure 1.10 Heterogeneous and homogeneous mixtures: (a) Blood appears to be a homogeneous fluid, but closer inspection shows it contains tiny particles. (b) When the red compound cobalt(II) chloride is stirred into water, it dissolves to form a homogeneous solution. *(a, Ken Edward/Science Source/Photo Researchers b, C. D. Winters)*

with ordinary light. Examples of solutions are clear air (mostly a mixture of nitrogen and oxygen gases), sugar water, and some brass alloys (which are homogeneous mixtures of copper and zinc). The properties of a homogeneous mixture are the same everywhere in any particular sample, but they can vary from one sample to another depending on how much of one component is present relative to another component.

Figure 1.11 **Separating iron and sulfur.** The iron chips in the iron–sulfur mixture may be removed by stirring the heterogeneous mixture with a magnet. *(C. D. Winters)*

Separation, Purification, and Pure Substances

When a mixture is separated into its components, the components are said to have been purified. However, most efforts at separation are not complete in a single step, and repetition is almost always necessary to give an increasingly purer substance. For example, the iron can be separated from a heterogeneous mixture of iron and sulfur by repeatedly stirring the mixture with a magnet. When the mixture is stirred the first time and the magnet is removed, much of the iron is removed with it, leaving the sulfur in a higher state of purity (Figure 1.11). After just one stirring, however, the sulfur may still have a dirty appearance due to a small amount of iron that remains. Repeated stirrings with the magnet, or perhaps the use of a very strong magnet, will finally leave a bright yellow sample of sulfur that apparently cannot be purified further, at least by this technique. In this purification process a property of the mixture, its color, is a measure of the extent of purification. (The color depends on the relative quantities of iron and sulfur in the mixture.) After the bright yellow color is obtained, it is assumed that the sulfur has been purified by removing all the iron.

Drawing a conclusion based on one property of the mixture may be misleading, because other methods of purification might change some other properties of the sample. It is safe to call sulfur pure only when a variety of methods of purification fail to change its properties. Each substance has a set of physical properties by which it can be recognized, just as you can be recognized by a set of characteristics such as your hair or eye color or your height. Thus, a **pure substance** is a sample of matter with properties that cannot be changed by further purification.

Only a few substances occur in nature in pure form. Gold, diamonds, and sulfur are examples, but these substances are special cases. We live in a world of mixtures; all living things, the air and food we depend on, and many products of technology, are mixtures.

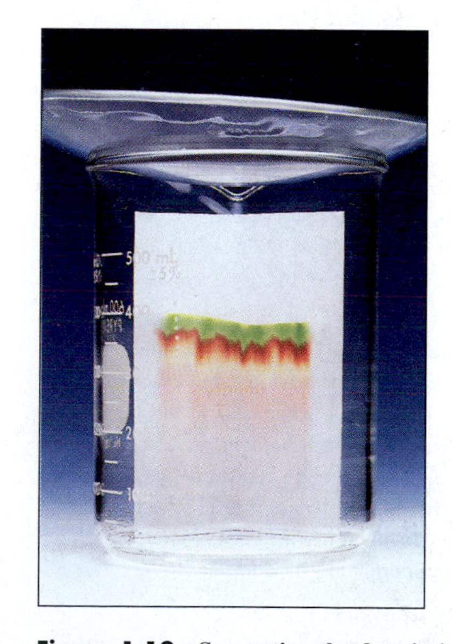

Figure 1.12 **Separating the dyes in ink.** Here brown ink from a felt-tip pen is separated into the different colored dyes that are combined to make brown. Separation occurs because the different dyes, when dissolved in water, interact differently with the water and adhere differently to the surface of paper. *(C. D. Winters)*

Although substances are seldom found pure in the natural world, it is possible, using modern purification techniques, to separate many of them from naturally occurring mixtures. Familiar substances obtained in this manner include refined sugar, table salt (sodium chloride), copper, nitrogen gas, and carbon dioxide, to mention just a few. However, there are other, less familiar but equally important ones such as taxol, a promising drug derived from the bark of the yew tree and used for the treatment of advanced cases of ovarian, breast, and other cancers. In all, over 12 million pure substances have been obtained from natural mixtures or created by chemists in the laboratory.

Many methods have been developed to separate mixtures into pure substances. Different physical and chemical properties of the substances—such as solubility in water, density, the temperature at which a substance melts or boils, or its ability to react with other chemicals—are exploited to make the separation (Figure 1.12). For example, pure nitrogen can be separated from other components of air because each component of air has a different temperature at which it boils at a given pressure. The element chlorine can be separated from sodium chloride (salt) by a chemical change brought about by passing electricity through a solution of salt in water.

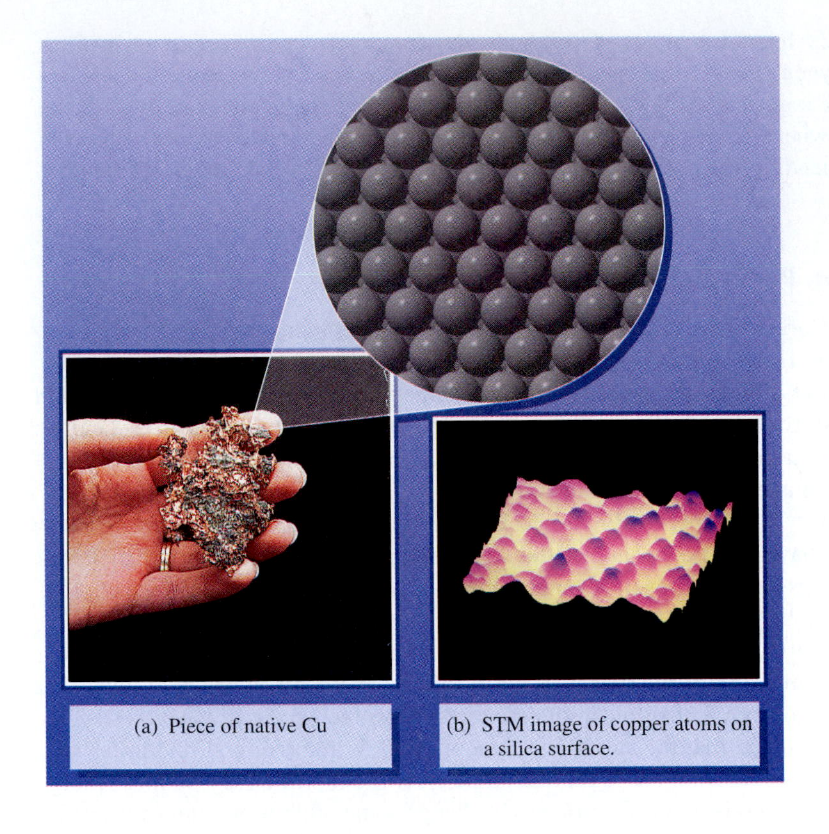

(a) Piece of native Cu
(b) STM image of copper atoms on a silica surface.

Figure 1.13 Elements, atoms, and the nanoscale world of chemistry. (a) A macroscopic piece of copper metal. In the nanoscale, magnified representation of a tiny portion of the surface of the copper, it is seen that all the atoms are the same. A sample of copper is composed of one kind of atom. (b) A scanning tunneling microscopy (STM) image, enhanced by a computer, of a layer of copper atoms on the surface of silica (a compound of silicon and oxygen, SiO_2). The image is 1.7 nanometers square, and the rows of atoms are separated by about 0.44 nanometer. [*(a) C. D. Winters, (b) X. Xu, S. M. Vesecky, and D. W. Goodman,* Science, *1992, 258, 788.*]

1.7 CHEMICAL ELEMENTS

Pure table sugar will decompose when heated in a complex series of changes (caramelization) that produces the brown color and flavor of peanut brittle or caramel candy. If heated for a longer time at a high enough temperature, sucrose can be converted completely to two other substances, carbon and water. Furthermore, if the water is collected, it can be decomposed still further to pure hydrogen and oxygen by passing an electric current through it. Substances like carbon, hydrogen, and oxygen, however, cannot be decomposed further into two or more new substances. Such substances are called **elements.**

There are now 112 known chemical elements, the latest having been discovered in late 1995. (See Chemistry in the News: *Newest, Heaviest Elements* in Chapter 2.) Of all these, though, only 25 are known to be essential to human life (Section 3.10). There is much more about elements and atoms beginning in Chapter 2.

The existence of elements can be explained by a nanoscale model, just as the existence of solids, liquids, and gases can (Figure 1.13). According to the model, an element cannot be decomposed further because at the nanoscale it consists of one and only one kind of atom. There are a *very* great many atoms in any visible sample of an element, but all of them are the same kind of atom. For example, the sample of copper in Figure 1.13 is made up entirely of copper atoms.

1.8 CHEMICAL COMPOUNDS

Pure substances like sucrose and water that can be decomposed into two or more different pure substances are referred to as **chemical compounds.** Each compound is composed of atoms of two or more elements combined in a manner distinctive to that compound. In any sample of the same compound, the elements are combined in the same proportion by mass, and so the compound has the same composition.

When elements are a part of a compound, their original characteristic properties, such as color, hardness, and melting point are replaced by the characteristic properties of the compound. Consider ordinary table sugar, sucrose, which is composed of the following three elements:

- carbon, which is usually a black powder, but is also found in the form of diamonds;
- hydrogen, the lightest gas known; and
- oxygen, a gas necessary for human and animal respiration.

Sucrose, a white, crystalline powder, has properties completely unlike any of these three elements.

If a compound consists of two or more different elements, how is it different from a mixture? There are two ways: (1) a compound has specific properties, and (2) a compound has a specific composition. Both the properties and the composition of a mixture can vary. A solution of sugar in water can be very sweet or only a little sweet, depending on how much sugar has been dissolved. Other properties, such as its viscosity (resistance to flow) can also vary. There is no particular composition of a sugar solution that is favored over any other; and as the composition changes, properties change too.

On the other hand, 100.0 grams (g) of pure water always contains 11.2 grams of hydrogen and 88.8 grams of oxygen. Pure water always melts at 0.0 °C and boils at 100 °C (at one atmosphere pressure), and it is always a colorless liquid at room temperature.

The nanoscale model of how atoms combine to form chemical compounds is discussed in Chapter 3 and later chapters.

1.9 CHEMICAL CHANGES AND CHEMICAL PROPERTIES

Heating sucrose to make peanut brittle eventually produces a brown color as the compound decomposes to give water and other new compounds. Treating the sucrose with another chemical compound, sulfuric acid (Figure 1.14), produces an element (car-

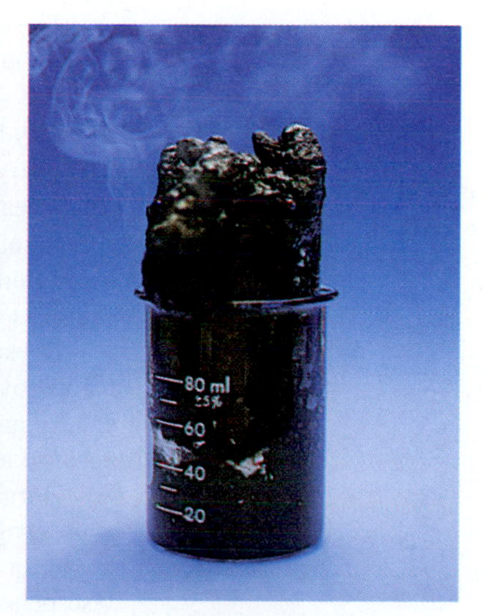

Figure 1.14 Chemical change. Sucrose (in the beaker) can be decomposed using sulfuric acid (in the graduated cylinder) to form the chemical element carbon (the black solid) and water (seen as steam emerging from the beaker). *(C. D. Winters)*

Figure 1.15 Chemical change. (Left) Metallic sodium reacts readily with water to produce hydrogen gas and lye (whose chemical name is sodium hydroxide). (Center) In this chemical change an acid has been added to the green solid compound nickel(II) carbonate. The colorless gas carbon dioxide, seen here as bubbles rising from the solid, is one result. (Right) The chemical reaction in a light stick evolves energy in the form of light. *(C. D. Winters)*

bon) and another compound (water). These are examples of a **chemical change** or **chemical reaction** because one or more substances (the **reactants**) are transformed into one or more different substances (the **products**).

$$\underset{\text{reactant}}{\text{Sucrose}} \rightarrow \underbrace{\text{carbon} + \text{water}}_{\text{products}}$$

The elements and the compounds they form have **chemical properties,** which describe the kinds of chemical reactions they can undergo. It is a chemical property of metallic sodium, for example, that it reacts extremely rapidly with water to produce a gas (hydrogen) and a solution of a compound commonly called lye (Figure 1.15, left). It is also a chemical property of metal carbonates that they produce carbon dioxide when treated with an acid (Figure 1.15, center). Understanding chemical properties—how an element or compound will behave when heated, cooled, irradiated with light, or treated with another element or compound—is one of the main objectives of chemists and biochemists, and even of geologists and chemical engineers.

Physical changes and, to a greater extent, chemical reactions are often accompanied by transfers of energy. The chemical reaction in a "light stick" releases light energy and a little heat (Figure 1.15, right); and a battery makes a calculator work because a chemical reaction forces electric current to flow through the circuits.

Burning coal or oil, or the combustion of gasoline in your car's engine, is a chemical change. Such a reaction can transfer energy that can generate electricity, heat a home, or run an automobile. Chemists and engineers are working every day to devise better batteries to power toys, tools, stereos, laptop computers, and even automobiles. The use of chemical changes to provide energy is an important theme of this book.

Exercise 1.5 Chemical and Physical Changes

Describe the chemical and physical changes implicit in the following statement: Propane gas burns, and the heat of the combustion reaction is used to cook an egg and boil water.

1.10 CLASSIFYING MATTER

The discussion of separating mixtures to obtain elements or compounds and decomposing compounds to obtain elements leads to a useful way to classify matter (Figure 1.16). Heterogeneous mixtures such as sand in water can be separated using simple manipulation, for example, by allowing the sand to settle and by pouring off the water. Homogeneous mixtures are somewhat more difficult to separate, but physical processes will serve. For example, salt water can be purified by heating to evaporate the water, collecting the salt crystals, and eventually condensing the water vapor back to liquid. Most difficult of all is separation of the elements that form a compound. This requires a chemical reaction and may involve inputs of energy or of other substances.

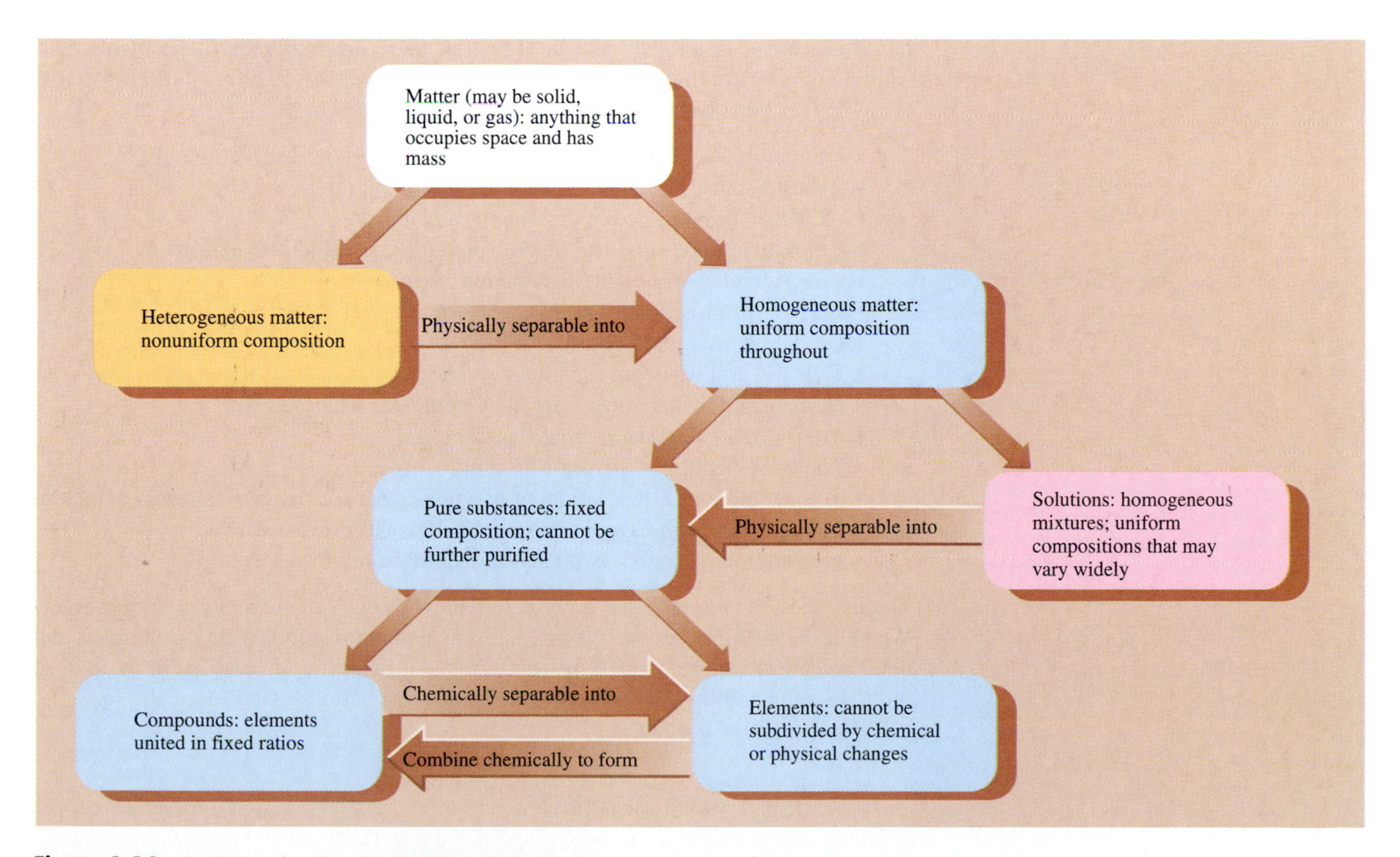

Figure 1.16 A scheme for the classification of matter.

Exercise 1.6 Classifying Matter

Classify each of the following with regard to the type of matter represented.
(a) Sugar dissolved in water
(b) Concrete (such as that used to make a road)
(c) A sample of water from a muddy river
(d) The diamond in a ring
(e) A penny
(f) Table salt

1.11 WHY CARE ABOUT SCIENCE?

The *Challenger* disaster and Feynman's experiment; the debate over CFCs in the stratosphere; the search for cures for AIDS or cancer; gene therapy; and genetic counseling all bring us to the reasons you should care about science and the way it is done. Many in the United States were saddened by the *Challenger* explosion, not only because of the deaths of the astronauts but also because, unless solved, it could clearly affect the future of our space program and its role in our national goals. Further, such engineering failures can lead uninformed citizens to lose faith in all technology, to the detriment of our economy and society. Professor Feynman and his common-sense approach uncovered the reasons for the *Challenger* explosion, and the space program has since moved cautiously ahead. However, there are hundreds of problems confronting our society that depend in some way on science for their resolutions. Almost every day we read or hear statements such as the following:

- We could develop an AIDS vaccine if we just moved more quickly.
- Asbestos must be removed from all public buildings.
- All pesticides and herbicides should be banned.
- Chlorine and the products made from chlorine should be banned.
- The Antarctic ozone hole is becoming larger.
- Burning tropical forests is an ecological disaster.
- Global warming is a reality.
- Global warming is not a reality.
- Homeopathic "medicines" are a sure cure for whatever ails us.
- You should substitute margarine for butter.
- There are health risks from eating margarine.
- You should check the basement of your house for dangerous levels of radon.
- Eating foods with high levels of selenium will prevent cancer.
- Incineration of wastes is highly damaging to the local ecology.

The list goes on and on. What are you to believe? How do you analyze the problem? We believe that some knowledge of chemistry and the methods used by chemists and other scientists can be of great benefit. This can help you at least begin to analyze the risks and benefits from your own decisions, the actions of business and industry, and of governmental policies.

1.12 RISKS AND BENEFITS

The problems listed above, all related directly to chemistry and chemicals, will be with us for many years, and others will surely be added. No matter where you live

TABLE 1.1 Estimates of Risk: Activities That Produce One Additional Death per One Million People Exposed to the Risk

Activity	Cause of Death
Smoking 1.4 cigarettes	Cancer, lung disease
Living 2 months with a cigarette smoker	Cancer, lung disease
Eating 40 tablespoons of peanut butter	Liver cancer caused by the natural carcinogen aflatoxin B
Drinking 40 cans of saccharin-sweetened soda	Cancer
Eating 100 charcoal-broiled steaks	Cancer
Traveling 6 minutes by canoe	Accident
Traveling 10 minutes by bicycle	Accident
Traveling 300 miles by car	Accident
Traveling 1000 miles by jet aircraft	Accident
Drinking Miami drinking water for 1 year	Cancer from chloroform
Living 2 months in Denver	Cancer caused by cosmic radiation
1 Chest x ray in a good hospital	Cancer
Living 5 years at the boundary of a typical nuclear power plant	Cancer

From L. Gough and M. Gough: "Risky business," *Chem Matters,* pp. 10–12, December, 1993.

or what your occupation, you are exposed every day to chemicals, some of which may be hazardous. Some you have decided you can tolerate because the risk they pose seems to be outweighed by the benefits they offer. Examples might include having an occasional beer, which contains toxic alcohol, or using a herbicide on your garden or lawn. In other cases, however, it is difficult for us to weigh the risks and benefits. This means that we will hear more about the assessment and management of risks.

"Risk assessment" is a process that brings together the scientific disciplines of chemistry, biology, toxicology, epidemiology, and statistics to attempt to determine the severity of a health risk from exposure to a particular chemical. The first step is to try to identify the hazard by doing animal tests or by examining data for human exposure and to find the relation between the amount of exposure and the chances of experiencing an adverse health effect. Also involved is an attempt to determine how people are exposed and to how much. Once these things are known, one can then try to make an estimate of the overall risk.

In recent years a new discipline called "risk communication" has grown up to facilitate the accurate transfer of information between the scientists who assess risk, the government agencies that manage risk, and the public. Those involved in risk communication have found that the response of people to risks depends on a number of interesting factors. For example, people accept voluntary risks, such as smoking or playing in a high school football game, much more readily than being in a building that contains asbestos. What is interesting about this is that, while the annual death rate per million people is 10 for high school football or 40 for a long-term smoker, the death rate from asbestos in schools is estimated to be in the range 0.005 to 0.093 per million people.

It is also interesting that people often conclude that anything manmade is "bad," while anything natural is "good." However, this is not always confirmed by risk assessment, as you can see in Table 1.1.

"Risk management" involves ethics, equity, economics, and other matters that are part of government and political and social interactions. The removal of asbestos from public buildings is an example of the management of a risk that many scientists believe to be quite low for the general public. The public has not been willing to tolerate the risk and so the political system has responded.

Another example of risk management is the ban on chlorofluorocarbons (CFCs) in many nations because of the damage CFCs are causing to the earth's ozone layer. Here ethics and economics clearly clash. After so many years of development, refrigeration equipment that uses CFCs is efficient, relatively inexpensive, and widespread. Its use is spreading to Third World countries such as the nations of Africa and many in South America, and its availability has a profound effect on their economies. Now, just as these countries are beginning to develop, the countries of the so-called developed world tell them that CFC-based refrigeration equipment should no longer be used. What is better for the greater good: a ban on equipment and processes using CFCs, which will have a long-term beneficial effect on the ozone layer, or some limited, reasonable use of CFCs in the refrigeration equipment that is vital to the development of Third World countries? Are the alternatives to CFCs economically feasible? Do they themselves have damaging environmental effects? Can the technology be transferred readily and efficiently to Third World countries?

For a discussion of epidemiology and the connection between diet, lifestyle, and environmental factors, see "Epidemiology Faces Its Limits," G. Taubes, *Science,* Vol. 269, July 14, 1995, pp. 164–169.

When you size up the risks in your own life, keep in mind that epidemiologists have found very few environmental agents that are strongly linked to cancer, for example. These are cigarette smoke, alcohol, ionizing radiation, a few drugs, a handful of occupational carcinogens, and perhaps three viruses (hepatitis B, human T cell leukemia, and human papillomavirus). Every year, though, dozens of papers are published in the scientific literature that report potential environmental causes of cancer. Experts say, however, that you should take these new reports seriously only if there is *at least* a twofold increased risk. For example, if you use high-alcohol mouthwash, it has been reported that there is a 1.5 increased risk of mouth cancer. Another study states that if you consume olive oil once a day or less, there is a relative risk for breast cancer of 1.25. The problem is that this report was contradicated by a study that showed that the breast cancer risk was reduced by 25%! Risk analysis and management is difficult; it is not an exact science.

Conceptual Challenge Problem CP-1.D at the end of the chapter relates to topics covered in this section.

Analysis and Detection

A part per million (ppm) means we can find 1 gram of a substance in one million grams of material. That's about a tenth of a drop in a bucket.

Finally, as you study chemistry you will learn methods of detecting and analyzing substances, and it will become apparent that special methods are needed to detect very low levels of a substance. Chemists, however, are inventing more and more sensitive methods of measurement. For example, in 1960 mercury could be detected at a concentration of 1 part per million, in 1970 the detection limit was 1 part per billion, and by 1980 the limit had dropped to 1 part per trillion.

The ability to detect lower and lower levels of environmental pollutants has an important effect. A few decades ago, when food, air, or water was tested, toxic substances were often not found. Today, most materials contain detectable amounts of numerous toxic substances. The news media report these findings, and the public becomes alarmed. Demands are made for regulation or corrective action. This often means a demand for reduction of the pollutants below the detection limit. But there are two problems with this. First, although we expect that chemistry will push detection limits lower and lower, measuring zero in a chemical analysis will always be impossible. Second, it is not clear that miniscule amounts of toxic substances are a genuine health threat.

Risks, Benefits, and Chemistry

Why study chemistry? The reasons are clear. You will be called upon to make many decisions in your life for your own good, or for the good of those in your community—whether that is your local community or the global community. An understanding of the nature of risk, of science in general, and of chemistry in particular, can only serve to help in these decisions.

Exercise 1.7 Risks

Name five risks you have taken today and rank them in relative order of their danger. Compare your list with the list of a friend. Are any of the items on your list the same as your friend's?

SOME READINGS FOR THIS CHAPTER

If you wish to learn more about some of the subjects introduced here, about the importance of chemistry in our world and the thoughts of some scientists who have contemplated the role of science in our society, consider some of the following books and articles.

- Rachel Carson: *Silent Spring,* New York, Houghton Mifflin, 1962.
- Richard Feynman: *What Do You Care What Other People Think?* New York, W.W. Norton and Company, 1988; and *Surely You're Joking Mr. Feynman,* New York, W.W. Norton and Company, 1985.
- Thomas S. Kuhn: *The Structure of Scientific Revolutions,* Chicago, The University of Chicago Press, 1970.
- Primo Levi: *The Periodic Table,* New York, Schocken Books, 1984.
- Lewis Thomas: *The Lives of a Cell,* New York, Penguin Books, 1978.
- Sharon D. McGrayne: *Nobel Prize Women in Science,* New York, Birch Lane Press, 1993.

IN CLOSING

Having studied this chapter, you should be able to . . .

- describe the approach used by scientists in solving problems (Sections 1.1–1.3).
- understand the differences among a hypothesis, a theory, and a law (Section 1.2).
- define quantitative and qualitative observations (Section 1.2).
- have a qualitative sense of temperatures on the Celsius scale (Section 1.2).
- identify the physical properties of matter or the physical change occurring in a sample of matter (Section 1.4).
- characterize the three states of matter—gases, liquids, and solids—and appreciate the differences among them (Section 1.5).
- understand the difference between matter on the macroscale and the nanoscale (Section 1.5).
- describe the kinetic-molecular theory at the nanoscale (Section 1.5).
- explain the difference between homogeneous and heterogeneous mixtures (Section 1.6).

- understand the difference between a chemical element and a chemical compound (Sections 1.7 and 1.8).
- identify the chemical properties of matter and chemical changes occurring in samples of matter (Section 1.9).
- classify matter (Figure 1.10).
- appreciate the balance of benefits and risks in our environment (Section 1.12).

Key Terms

The following terms were defined and given in boldface type in this chapter. You should be sure to understand each of these terms and the concepts with which they are associated. (The number of the section where each term is introduced is given in parentheses.)

boiling point *(1.4)*
Celsius *(1.4)*
chemical change *(1.9)*
chemical compound *(1.8)*
chemical property *(1.9)*
chemical reaction *(1.9)*
chemistry *(1.4)*
element *(1.7)*
gas *(1.5)*
heterogeneous *(1.6)*
homogeneous *(1.6)*
hypothesis *(1.2)*
kinetic-molecular theory *(1.5)*
law *(1.2)*
liquid *(1.5)*
macroscale *(1.5)*
melting point *(1.4)*
microscale *(1.5)*
nanoscale *(1.5)*
physical changes *(1.4)*
physical properties *(1.4)*
product *(1.9)*
pure substance *(1.6)*
qualitative *(1.2)*
quantitative *(1.2)*
reactant *(1.9)*
solid *(1.5)*
solution *(1.6)*
substance *(1.6)*
temperature *(1.4)*
theory *(1.2)*

Questions for Review and Thought

Conceptual Challenge Problems

CP-1.A Some people use expressions such as "a rolling stone gathers no moss" and "where there is no light there is no life." Why do you believe these are "laws of nature"?

CP-1.B Parents teach their children to wash their hands before eating. (a) Do all parents accept the germ theory of disease? (b) Are all diseases caused by germs?

CP-1.C In Section 1.5 you read that, on an atomic scale, all matter is in constant motion. (For example, the average speed of a molecule of nitrogen or oxygen in the air is greater than 1000 miles per hour at room temperature.) (a) What evidence can you put forward that supports the kinetic-molecular theory? (b) Suppose you accept the notion that molecules of air are moving at speeds near 1000 miles per hour. What can you then reason about the paths that these molecules take when moving at this speed?

CP-1.D The life expectancy of United States citizens in 1992 was 76 years. In 1916 the life expectancy was only 52 years. This is an increase of 46% in a lifetime. (a) Could this astonishing increase occur again? (b) To what single source would you attribute this noteworthy increase in life expectancy? Why did you identify this one source as being most influential?

Review Questions

1. In the accompanying photo, you see crystals of the mineral halite, a form of ordinary salt. Are these crystals in the macroscale or nanoscale world? How would you describe the shape of these crystals? What might this tell you about the arrangement of the atoms deep inside the crystal?

A sample of a halite (sodium chloride) crystal. *(C. D. Winters)*

2. Galena, pictured below, is a black mineral that contains lead and sulfur and that shares its name with a number of towns in the United States; they are located in Alaska, Illinois, Kansas, Maryland, Missouri, and Ohio. How would you describe the shape of the galena crystals? What might this tell you about the arrangement of the atoms deep inside the crystal?

A sample of a galena or lead sulfide crystal. *(C. D. Winters)*

3. What are the states of matter and how do they differ from one another?
4. In the accompanying photo, you see a crystal of the mineral calcite surrounded by piles of calcium and carbon, two of the elements that combine to make the mineral. (The other element combined in calcite is oxygen.) Based on the photo, describe some of the physical properties of the elements and the mineral. Are any the same? Are any properties different?

A sample of calcite (the clear crystal) and two of the elements of which it is composed, calcium metal and carbon. Calcium is normally a shiny metal, but these chips are covered with a thin film of white calcium oxide. *(C. D. Winters)*

5. Small chips of iron are mixed with sand (see photo). Is this a homogeneous or heterogeneous mixture? Suggest a way to separate the iron and sand from each other.

A mixture of iron and sand. *(C. D. Winters)*

6. In each case, describe the underlined property as a physical or chemical property.
   (a) The normal color of the element bromine is red-orange.
   (b) Iron is transformed into rust in the presence of air and water.
   (c) Dynamite can explode.
   (d) Aluminum metal, the shiny "foil" you use in the kitchen, melts at 387 °C.
7. In each case, describe the change as a chemical or physical change. Give a reason for your choice.
   (a) A cup of household bleach changes the color of your favorite T-shirt from purple to pink.
   (b) The fuels in the space shuttle (hydrogen and oxygen) combine to give water and provide the energy to lift the shuttle into space.
   (c) An ice cube in your glass of lemonade melts.
8. While camping in the mountains you build a small fire out of tree limbs you find on the ground near your campsite. The dry wood crackles and burns brightly and warms you. Before slipping into your sleeping bag for the night, you put the fire out by dousing it with cold water from a nearby stream. Steam rises when the water hits the hot coals. Describe the physical and chemical changes in this scene.
9. When Feynman said the O-rings on the *Challenger* did not expand properly because of the cold weather, was he stating a theory or a hypothesis?

General Questions

10. List three specific examples of how chemistry is used in your major study area in college.
11. Find four newspaper or magazine articles, advertisements, or cartoons that are directly related to chemistry.
12. What was Rowland and Molina's hypothesis about the ozone layer?

13. Solid gallium has a melting point of 29.8 °C. If you hold this metal in your hand, what will be its physical state? That is, will it be a solid or a liquid? Explain briefly.

14. You open a can of soft drink, and the carbon dioxide gas expands rapidly as it rushes from the can. Describe these changes in terms of the kinetic-molecular theory.

15. After you wash your clothes, you hang them on a line in the sun to dry. Describe the change or changes that occur in terms of the kinetic-molecular theory. Are the changes that occur physical or chemical changes?

16. The *Challenger* accident was caused by leaking O-ring seals. The rubber O-rings were less flexible below 0 °C than around ordinary temperatures such as 20 °C. Use the kinetic-molecular theory to speculate on the loss of flexibility at lower temperatures.

17. Consider each of the following statements and decide (i) whether the benefits outweigh the risks or the risks outweigh the benefits and (ii) who benefits and who is at risk. Give several reasons for your choice.
(a) Several new anti-AIDS drugs, which could prolong the lives of many individuals who are HIV-positive, have been discovered. The FDA, however, will not approve their use until they have been tested thoroughly.
(b) Potato chips cooked in Olestra, a fake fat, have fewer calories and zero grams of fat compared with regular potato chips. They are also known to cause gastrointestinal side effects in many people.
(c) We live in an age of plastics, but, as we all know, plastics are causing a major landfill problem in the United States. We should do away with plastics and return to the “good old days” of natural materials.
(d) Much of the country gets its electrical energy from coal-fired power plants. Coal-fired plants are a major contributor to acid rain and the greenhouse effect.
(e) As the world population grows, so does the demand for food. The increased use of pesticides has helped farmers get more crops per acre. This also increases the chance of pesticides leaching out of the soil and ending up in our rivers, lakes, and aquifers.

Applying Concepts

18. Identify the information in each sentence as qualitative or quantitative.
(a) The element gallium melts at 29.8 °C.
(b) A chemical compound containing cobalt and chlorine is blue.
(c) Aluminum metal is a conductor of electricity.

19. Identify the information in each sentence as qualitative or quantitative.
(a) The chemical compound ethanol boils at 79 °C.
(b) A chemical compound containing lead and sulfur forms shiny plate-like yellow crystals.

20. Classify the information in each of the following statements as quantitative or qualitative, and as relating to a physical or chemical property.
(a) A white chemical compound has a mass of 1.456 grams, and when placed in water containing a dye, causes the red color of the dye to fade to colorless.
(b) A sample of lithium metal, with a mass of 0.6 gram, was placed in water. The metal reacted with the water to produce the compound lithium hydroxide and the element hydrogen.

21. Classify the information in each of the following statements as quantitative or qualitative, and as relating to a physical or chemical property.
(a) A liter of water, colored with a purple dye, was passed through a charcoal filter. The charcoal adsorbed the dye, and colorless water came through.
(b) When a white powder dissolved in a test tube of water, the test tube felt cold. Hydrochloric acid was then added, and a white powder formed.

22. Which has the higher temperature, a sample of water at 65 °C or a sample of iron at 65 °F?

23. The following temperatures are measured at various locations during the winter in North America: −10 °C at Montreal, 28 °F at Chicago, 20 °C at Charlotte, and 40 °F at Philadelphia. What city is the warmest? What city is the coldest?

24. Classify each of the following with regard to the type of matter represented.
(a) A piece of newspaper
(b) Solid, granulated sugar
(c) Freshly squeezed orange juice
(d) Gold jewelry

25. Classify each of the following with regard to the type of matter represented.
(a) A cup of coffee
(b) A soft drink such as a Coke or Pepsi
(c) A piece of Dry Ice (a solid form of carbon dioxide)

26. For each of the changes described, decide if an element formed a compound or if a compound formed elements (or other compounds).
(a) Upon heating, a blue powder turned white and lost mass.
(b) A white solid forms three different gases when heated. The total mass of the gases is the same as that of the solid.
(c) After a reddish colored metal is placed in a flame, it turns black and has a higher mass.
(d) A white solid is heated in oxygen and forms two gases. The mass of gases is the same as the masses of the solid and the oxygen.

27. This is a nanoscale view of a thermometer registering 10 °C.

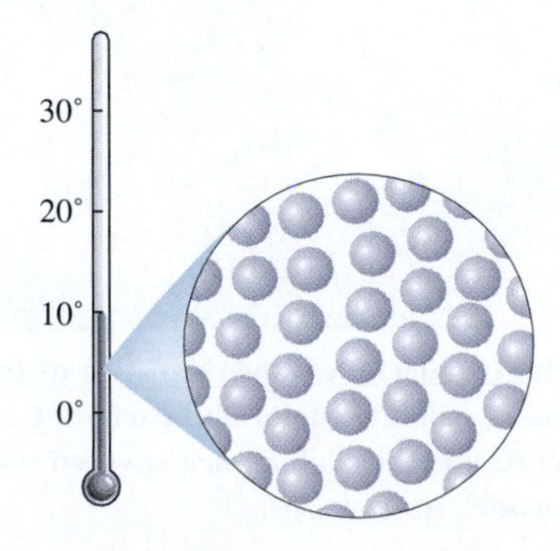

Which nanoscale drawing best represents the liquid in this same thermometer at 20 °C? (Assume that the same volume of liquid is shown in each nanoscale drawing.)

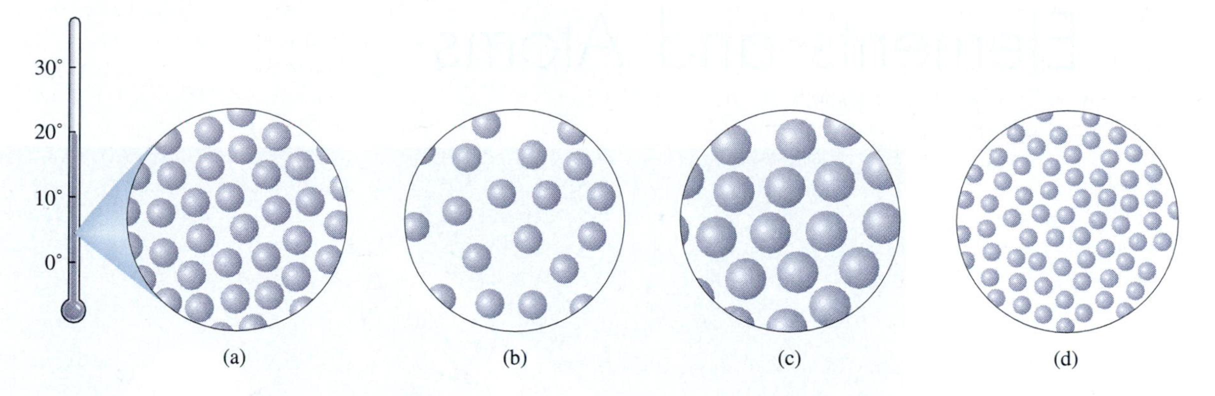

CHAPTER 2

Elements and Atoms

The pure, lustrous element shown on the left of the photo above is mixed with lead to form a bullet, such as pictured on the right side of the photo. • What is this shiny element, which is used in ammunition, including the bullet that killed President John Kennedy? The answer is part of the Summary Problem.* *(C.D. Winters)*

*This and each subsequent chapter begins with a photo, comment, or question that is linked to the Summary Problem at the end of the chapter. The Summary Problem reviews many of the key principles in the chapter.

A sample of matter can be classified as either a mixture or a pure substance by its properties and composition. Among the vast array of more than 12 million known pure substances, only 112 are elements—pure substances that are the building blocks of all materials. At the *nanoscale* level, each element consists of atoms of a single kind. Therefore, to understand the properties of elements we need to understand the properties of atoms. Atoms are made up of even smaller particles (protons, electrons, neutrons), and the masses of atoms differ from one element to another. When elements are arranged in the order of increasing number of protons in their atoms, there is a repeating pattern of similar properties. This is the basis for the periodic table ***(⇐ inside front cover),*** which summarizes a great deal of information about the properties of the elements. By their positions in the table, we can see which elements are similar to each other and which are different.

We turn now to the atom and its structure as the basis for understanding the properties of the elements. The story covers about 2500 years, but the major discoveries have occurred much more recently, within the past 200 years.

2.1 ORIGINS OF ATOMIC THEORY

The early view of matter as tiny particles in constant motion ***(⇐ p. 10)*** is an interpretation sufficient to explain many macroscale properties, such as differences among solids, liquids, and gases. For a long time, though, the identity of the particles was a mystery. With discoveries during the 1800s, scientists became convinced that these tiny particles are either atoms, or atoms as they are combined in chemical compounds. The first description of atoms goes back to the Greek philosopher Democritus (460–370 BC). Democritus reasoned that if a piece of matter such as gold were divided into smaller and smaller pieces, one would ultimately arrive at a tiny particle of gold that could not be divided further, but would still retain the properties of gold. He used the word "atom" (Greek, "uncuttable") to describe this indivisible, ultimate particle of matter. Democritus taught that atoms are hard, move spontaneously, and link to one another by some kind of hook-and-eye connection.

Democritus also used his concept of atoms to explain the properties of substances. For example, the denseness and softness of lead could be interpreted as lead atoms packed very closely together, like marbles that moved easily past one another. Iron, less dense and harder than lead, would have atoms shaped like corkscrews that would entangle in a rigid, but lightweight structure. Democritus was able to use atoms to explain other well-known phenomena, such as clothes drying, moisture appearing on the outside of a vessel of cold water, and crystals growing from a solution. He imag-

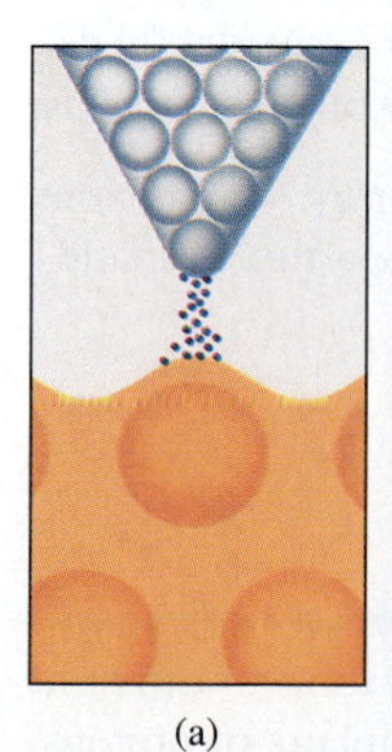

(a)

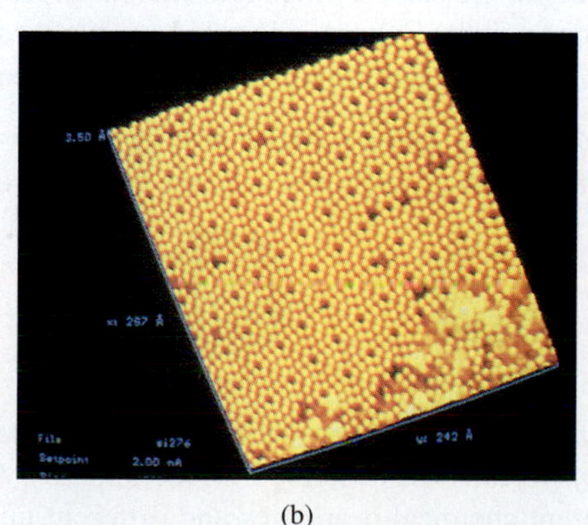

(b)

(c)

Nanoscale to microscale views of silicon. (a) Illustration of the tip of a scanning tunneling microscope (STM) moving along a surface of silicon atoms; (b) STM image of the atoms in the silicon surface; (c) highly purified crystals of silicon—Essentially all the atoms in this sample are silicon atoms. *(b, John Ozcomert/Michael Trenary)*

Jacob Bronowski, in a television series and book titled *The Ascent of Man,* had this to say about the importance of imagination: "There are many gifts that are unique in man; but at the center of them all, the root from which all knowledge grows, lies the ability to draw conclusions from what we see to what we do not see."

ined that such events resulted from the scattering or collecting of atoms. All atomic theory has been built on the assumption of Democritus: *the properties of matter that we can see are explained by the properties and behavior of atoms that we cannot see.*

Exercise 2.1 **Atomic Theory**

Use the idea that matter consists of tiny particles in motion to interpret each observation.

(a) Wet clothes hung on a line eventually dry.
(b) Moisture appears on the outside of a glass of ice water.
(c) Evaporation of a salt solution forms crystals.

2.2 THE MODERN ATOMIC THEORY

In 1803 John Dalton linked the existence of *elements,* substances that cannot be decomposed by chemical reactions into simpler substances (← ***p. 14***), to the indivisibility of atoms. An **atom** is the smallest particle of an element that retains the chemical properties of that element. An element is matter consisting of only one kind of atom. In contrast, compounds can be broken down by chemical reactions into two or more different substances, which may be elements or different compounds. Therefore, compounds must contain two or more different kinds of atoms (for example, the hydrogen and oxygen atoms combined in water). Dalton also said that each kind of atom must have its own properties—in particular, a characteristic mass. This idea allowed his theory to account for the masses of different elements that combine in chemical reactions to form compounds. Unlike earlier concepts about atoms, Dalton's ideas could be used to quantitatively interpret known chemical facts.

Robert Boyle (1661) was the first to propose that elements could not be decomposed to simpler substances by chemical means.

Two laws known at Dalton's time summarized well-verified previously unexplained observations. One was based on experiments in which the starting materials were carefully weighed before a chemical reaction, and the reaction products were carefully weighed afterwards. The results led to the **law of conservation of matter:** *there is no detectable change in mass in an ordinary chemical reaction.* The atomic theory accounts for this law—mass is conserved because atoms are conserved.

The second law was based on the observation that when a chemical compound is broken down to its elements, the proportions of the elements by mass are always the same. Water, for example, always contains 1 gram of hydrogen for every 8 grams of oxygen. The **law of constant composition** summarizes such observations: *A chemical compound always contains the same elements in the same proportions by mass.*

The *modern* atomic theory includes Dalton's original ideas, modified to account for discoveries since his time. The modern atomic theory is based on the following assumptions:

- ***Elements are composed of atoms, which are extremely tiny.*** Interactions among these ultimate chemical particles account for the properties of matter.

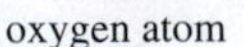

- ***All atoms of a given element have the same chemical properties and contain the same number of protons*** (as described in Section 2.5). Atoms of different elements have different chemical properties and different numbers of protons.

Portrait of a Scientist • *John Dalton (1766–1844)*

John Dalton was born about the fifth of September, 1766, in the village of Eaglesfield in Cumberland, England. His family was quite poor, and his formal schooling ended at age 11. However, he was clearly a bright young man, and with the help of influential patrons he began a teaching career at the age of 12. Shortly thereafter he made his first attempts at scientific investigation: weather observations, an interest that was to last his lifetime.

In 1793 Dalton moved to Manchester, England, where he was a tutor at the New College, but he left there in 1799 to pursue scientific inquiry on a full-time basis. It was not long afterward, on October 21, 1803, that he presented a paper introducing his "Chemical Atomic Theory" to the Literary and Philosophical Society of Manchester. He then lectured in London and in other cities in England and Scotland, and his reputation as a scientist rose rapidly. In 1810 he was nominated for membership in the Royal Society, the top scientific society in Britain, and many other honors followed.

John Dalton. (*Oesper Collection in the History of Chemistry, University of Cincinnati*)

- ***Compounds are formed by the chemical combination of two or more different kinds of atoms.*** Atoms combine in the ratio of small whole numbers, for example, one atom of A with one atom of B, or two atoms of A with one atom of B, such as in carbon monoxide (CO) and carbon dioxide (CO_2).

carbon monoxide

carbon dioxide

- ***Atoms are the units of chemical change.*** Chemical reactions result only in the combination, separation, or rearrangement of atoms. Atoms are not created, destroyed, or converted into other kinds of atoms during a chemical reaction.

Although it was not until the 1860s that a consistent set of relative masses of atoms was agreed upon, Dalton's idea that the different masses of atoms of different elements are crucial to quantitative chemistry was accepted from the early 1800s on.

2.3 THE CHEMICAL ELEMENTS

Every element has been given a unique *name* and a *symbol* derived from the name. These names and symbols are listed in the tables in the front of the book. The first letter of each symbol is capitalized; the second letter, if there is one, is lower case, as in He, the symbol for helium. Elements discovered a long time ago have names and symbols with Latin or other origins, such as Au for gold (from *aurum* meaning "bright dawn"). The names of more recently discovered elements are derived from their place of discovery or for a person or place of significance (Table 2.1).

Ancient people knew of nine elements—gold (Au), silver (Ag), copper (Cu), tin (Sn), lead (Pb), mercury (Hg), iron (Fe), sulfur (S), and carbon (C). Most of the other elements were discovered during the 1800s, as one by one they were identified in the earth's atmosphere or in minerals in the earth's crust. Currently, 112 elements are known, 90 of which occur in nature. The remaining elements such as technetium (Tc),

TABLE 2.1 The Names of Some Chemical Elements

Element	Symbol	Date of Discovery	Discoverer	Derivation of Name or Symbol
Aluminum	Al	1827	F. Wohler	L, *alumen* (alum)
Berkelium	Bk	1950	G. T. Seaborg, S. G. Thompson, and A. Ghiorso	Berkeley, CA, was site of Seaborg's laboratory
Carbon	C		Ancient	L, *carbo* (charcoal)
Calcium	Ca	1808	H. Davy	L, *calx* (lime)
Copper	Cu		Ancient	L, *cuprum* (from island of Cyprus)
Hydrogen	H	1766	H. Cavendish	Gk, *hydro* (water) and *genes* (former)
Helium	He	1868	P. Janssen, et al. W. Ramsay (1895) from uranium mineral clevite	Gk, *helios* (sun)
Iodine	I	1811	B. Courtois	Gk, *violet*
Iron	Fe		Ancient	L, *ferrum*
Gold	Au		Ancient	L, *aurum* (bright)
Mercury	Hg		Ancient	L, *hydrargyrum*
Mendeleevium	Md	1955	A. Ghiorso, B. Harvey, G. Choppin, S. Thompson, and G. Seaborg	Honoring Dmitri Mendeleev, periodic table developer
Sodium	Na	1807	H. Davy	L, *sodanum*
Phosphorus	P	1669	H. Brand	Gk, *phosphorus*
Polonium	Po	1898	M. Curie	Poland, homeland of discoverer
Sulfur	S		Ancient	L, *sulphurium*
Tin	Sn		Ancient	L, *stannum*

In Section 2.12 we will describe the periodic table, a powerful tool in comparing and contrasting the properties of all the elements.

neptunium (Np), and mendeleevium (Md) have been made synthetically since the 1930s, by a variety of elegant (and expensive) methods. Three new elements have been created since 1994 (Section 2.8), bringing the total of known elements to 112.

Types of Elements

Elements can be divided roughly into two categories—metals and nonmetals. The vast majority of the elements are metals; only 18 are nonmetals. Because you use metals, you are probably familiar with many of their properties. **Metals** are solids (except for mercury, which is a liquid), conduct electricity, are ductile (can be drawn into wires), are malleable (can be rolled into sheets), and can form alloys (solutions of one or more metals in another metal) (Figure 2.1). Iron (Fe) and aluminum (Al) are used in automobile parts because of their ductility, malleability, and relatively low cost. Copper (Cu) is used in electrical wiring because it conducts electricity better than most metals. Gold (Au) is used for the vital electrical contacts in automobile air bags and in some computers because it does not corrode and is an excellent electrical conductor.

In contrast, except for graphite, a form of carbon, none of the **nonmetals** conduct electricity, the main property that distinguishes them from metals. Nonmetals are also more diverse in their physical properties than are metals (Figure 2.2). At room temperature some nonmetals are solids (phosphorus, sulfur, iodine); bromine is a liquid; and others, like nitrogen and oxygen, are gases.

Figure 2.1 Some metallic elements—iron, aluminum, copper, and gold. The steel ball is principally iron. The rod is made of aluminum. Gold makes up the inner coil nested in a ring of copper wire. Metals are malleable, ductile, and conduct electricity. *(C.D. Winters)*

Figure 2.2 Some nonmetallic elements—(a) white phosphorus, (b) sulfur, (c) bromine, (d) bromine, and (e) carbon. Nonmetals do not conduct electricity. Bromine is the only nonmetallic element that is a liquid at room temperature. *(c, d, e, Larry Cameron)*

(a)

(b)

Figure 2.3 **Metalloids**—Four of the six metalloids. (a) boron (*top*) and silicon (*bottom*); (b) arsenic (*left*) and antimony (*right*). Metalloids have metallic and nonmetallic properties. *(a, Chip Clark; b, C.D. Winters)*

A few elements—boron, silicon, germanium, arsenic, antimony, and tellurium—are classified as **metalloids,** elements that have some typically metallic properties and other properties that are characteristic of nonmetals (Figure 2.3). For example, metalloids do not conduct electricity as well as metals, but are not nonconductors like the nonmetals. Many metalloids are semiconductors and are essential for the electronics industry.

The elemental gases helium (He), neon (Ne), argon (Ar), krypton (Kr), xenon (Xe), and radon (Rn) exist as individual atoms. Other nonmetals exist as molecules. A **molecule** is a unit of matter consisting of two or more atoms joined together by a chemical bond (Section 8.1). Chlorine molecules contain two chlorine atoms per molecule (represented by Cl_2), meaning that Cl_2 molecules are the simplest unit in chlorine gas. With the aid of computers, molecules can be drawn in a variety of ways. The representation of a Cl_2 molecule on the left below is a *space-filling model* in which the atoms nestle together. This kind of model is closest to showing molecules as they actually exist. Sometimes, to show more clearly the individual atoms and how they are connected, we use *ball-and-stick models* like the one on the right.

Space-filling model of Cl_2

Ball-and-stick model of Cl_2

Molecules, like Cl_2, that contain two atoms per molecule are called **diatomic** molecules. Oxygen and nitrogen also exist as diatomic molecules, as do hydrogen (H_2), fluorine (F_2), bromine (Br_2), and iodine (I_2).

Diatomic molecules: H_2, O_2, N_2, F_2, Cl_2, Br_2, and I_2. You need to remember these diatomic molecules because they will be encountered frequently.

In metals in the solid state, all atoms are closely packed. Because of this, metals are more dense than nonmetals.

CHEMISTRY YOU CAN DO

Preparing a Pure Sample of an Element

You will need the following items to do this experiment:

- two glasses or plastic cups that will each hold about 250 mL of liquid
- approximately 100 mL (about 3.5 oz) of vinegar
- soap
- an iron nail, paper clip, or other similar-sized piece of iron
- something abrasive, such as a piece of steel wool, Brillo, sandpaper, or nail file
- about 40 to 50 cm of thin string or thread
- some table salt
- a magnifying glass (optional)
- 15 to 20 dull pennies (Shiny pennies will not work.)

Wash the piece of iron with soap, dry it, and clean the surface further with steel wool or a nail file until shiny. Tie one end of the string around one end of the piece of iron.

Place the pennies in one cup (A) and pour in enough vinegar to cover them. Sprinkle on a little salt, swirl the liquid around so it contacts all the pennies, and observe what happens. When nothing more seems to be happening, pour the liquid into the second cup (B), leaving the pennies in cup A (that is, decant off the liquid). Suspend the piece of iron from the thread so that it is half submerged in the liquid in cup B.

Observe the piece of iron over a period of 10 minutes or so, and then use the thread to pull it out of the liquid. Observe it carefully, using a magnifying glass if you have one. Compare the part that was submerged with the part that remained above the surface of the liquid.

1. What did you observe happening to the pennies?
2. How could you account for what happened to the pennies in terms of a microscopic model? Cite observations that support your conclusion.
3. What did you observe happening to the piece of iron?
4. Interpret the experiment in terms of a microscopic model, citing observations that support your conclusions.
5. Would this method be of use in purifying copper? If so, can you suggest ways that it could be used effectively to obtain copper from ores?

Exercise 2.2 Elements

Use the table in the inside front cover to answer the following.
(a) Three elements are named for planets in our solar system. Give their names and symbols.
(b) One element is named for a state in the United States. Name the element and give its symbol.
(c) The symbols for two elements are the first letters of the surnames of two of the authors of this book. A third element's symbol is that of the first two letters of another author of this book. Identify these three elements.

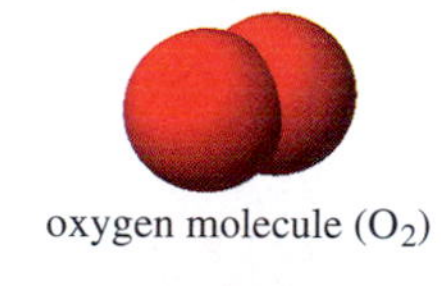

oxygen molecule (O_2)

ozone molecule (O_3)

Space-filling models of O_2 and O_3.

You can smell the odor of ozone in an intense thunderstorm. Lightning flashes provide the energy to convert O_2 in the atmosphere to ozone, O_3.

The excessive depletion of stratospheric ozone creating an "ozone hole" is discussed in Section 12.11.

Oxygen, phosphorus, sulfur, and carbon are interesting because all of them exist as **allotropes,** different forms of the *same* element in the *same* physical state at the same temperature and pressure. The allotropes of oxygen are O_2, sometimes called dioxygen, and O_3, ozone. Dioxygen is by far the more common allotropic form, a major component of the atmosphere. Ozone is a highly reactive blue gas first detected by its characteristic pungent odor. Its name comes from *ozein,* a Greek word meaning "to smell."

The most common elemental phosphorus allotrope, white phosphorus, consists of four-atom (tetratomic) molecules (P_4) in which P atoms are arranged at the cor-

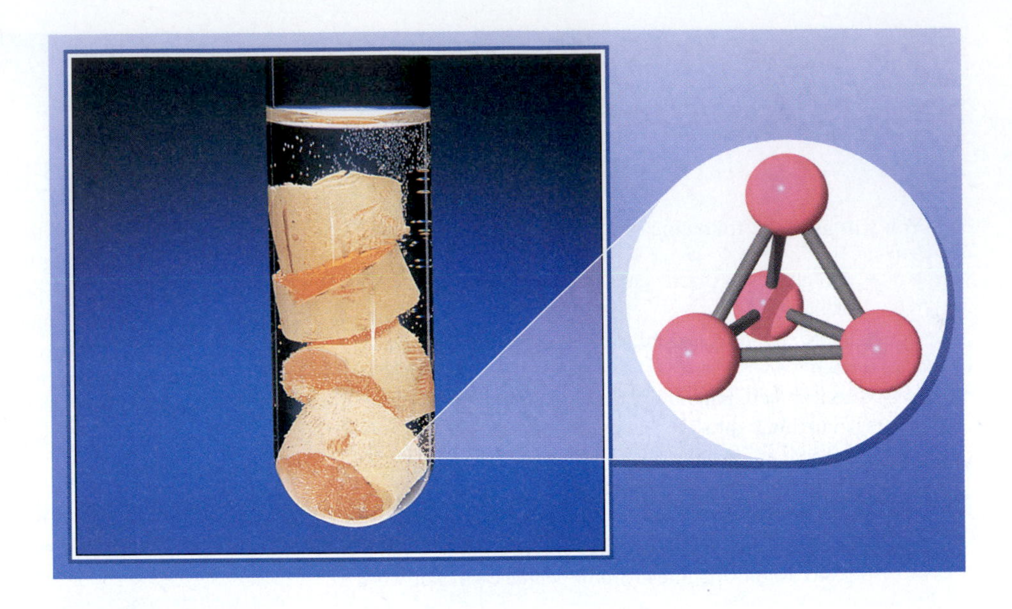

Figure 2.4 White phosphorus. This allotropic form of phosphorus contains P_4 molecules; another allotrope consists of P_4 units joined to each other in chains. Black phosphorous has a layer structure in which each P atom is bonded to three others. *(C.D. Winters)*

ners of a tetrahedron. Two other allotropes, red phosphorus and the less common black form, consist of more complex networks of phosphorus atoms (Figure 2.4).

The common lemon-yellow allotrope of sulfur (Figure 2.5) consists of crown-shaped S_8 molecules. When heated above 150 °C, the S_8 rings break into S_8 chains, which combine with each other to give long, tangled strands of S atoms, forming a viscous, syrupy liquid. At even higher temperatures, the strands break to give a liquid that, poured slowly into water, forms a flexible orange-red thread, an allotropic form sometimes called "plastic sulfur."

Diamond and graphite, known for centuries, are two allotropes of carbon containing extended networks of carbon atoms. Diamond and graphite had long been considered the only allotropes of carbon with well-defined structures. Therefore, it was a surprise in the 1980s when another carbon allotrope was discovered in soot (the black stuff in candle flames and on the bottom of campfire pots) and produced when carbon-containing materials are burned with very little oxygen. The new allotrope consists of 60-carbon atom cages and represents a new class of molecules.

The C_{60} molecule resembles a hollow soccer ball made up of five-membered rings linked to six-membered rings (Figure 2.6). This molecular structure of carbon pentagons and hexagons reminded its discoverers of a geodesic dome, a structure popularized years ago by the innovative American philosopher and engineer R. Buckminster Fuller. Therefore, the official name of the C_{60} allotrope is buckminsterfullerene, but chemists often call it simply a "buckyball." (See the accompanying Portrait of a Scientist.) We know now that C_{60} buckyballs are only one member of a larger molecular family of even-numbered carbon cages, generally called "fullerenes." Some fullerenes have fewer than 60 C atoms (called "buckybabies"), while others have an even larger number—C_{70} and giant fullerenes such as C_{240}, C_{540}, and C_{960}.

C_{60} is thought to be the only molecule made of a single kind of atom to form a spherical cage.

Another "twist" in the buckyball story, so to speak, is that carbon atoms can also form concentric tubes that resemble rolled-up chicken wire. These single- and multi-walled *nanotubes* of only carbon atoms are excellent electrical conductors and extremely strong. Imagine the exciting applications for such properties, including making buckyfibers that could substitute for the metal wires now used to transmit electrical power.

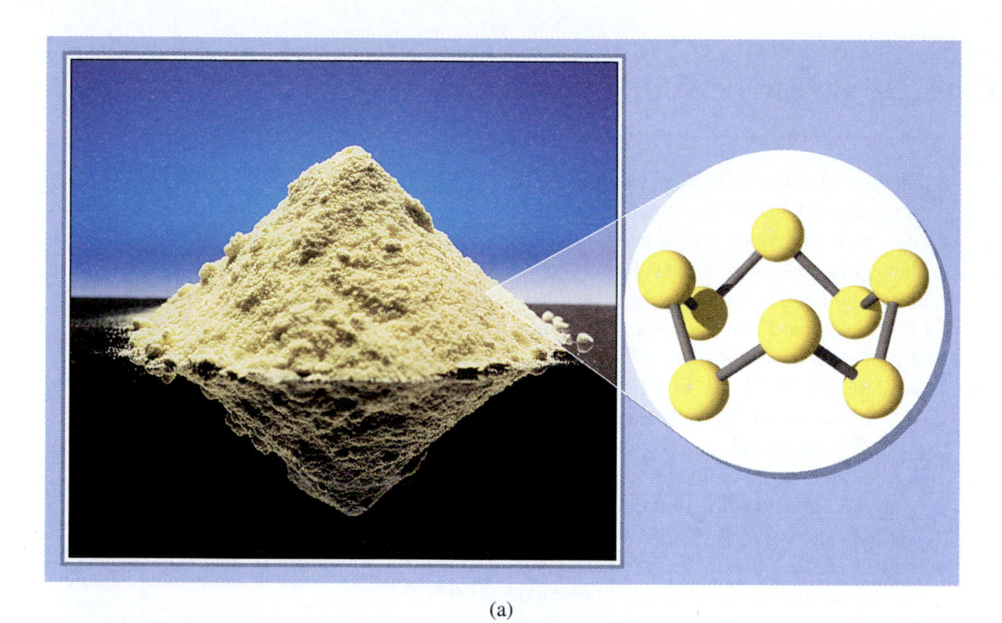

(a)

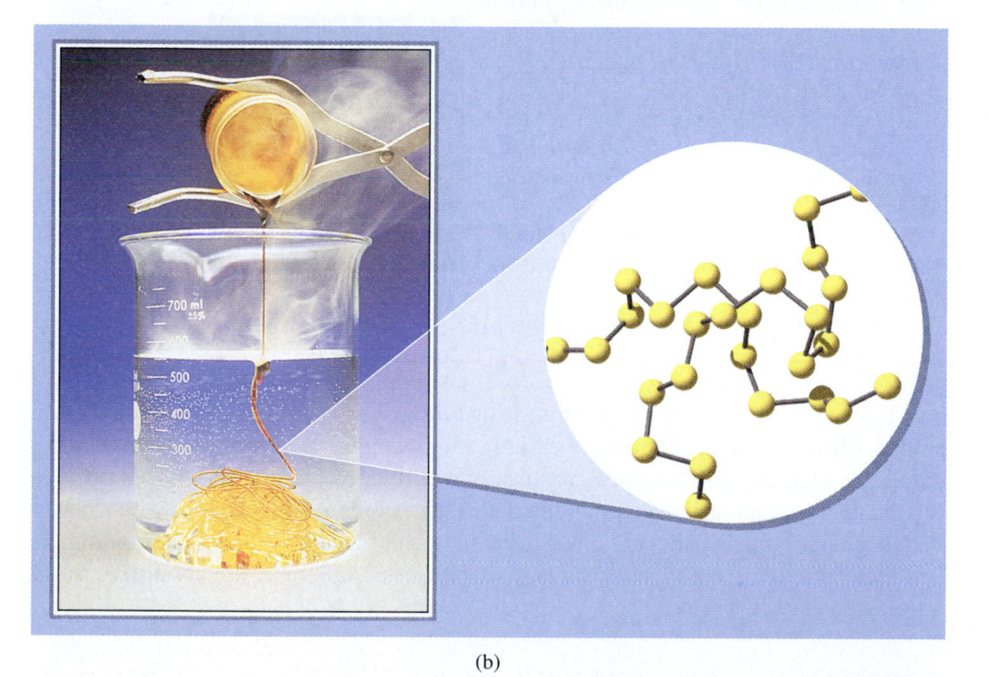

(b)

Figure 2.5 Sulfur allotropes. (a) At room temperature, sulfur is a bright yellow solid. At the nanoscale level, it consists of eight-membered puckered rings of sulfur atoms. (b) When melted, the sulfur rings break open to form long chains of sulfur atoms called "plastic sulfur." *(C.D. Winters)*

(a)

(b)

(c)

Figure 2.6 Models for fullerenes (a) Geodesic domes at Elmira College, Elmira, N.Y. Geodesic domes, such as those designed originally by R. Buckminster Fuller, contain linked hexagons and pentagons. *(Grant Heilman)* (b) A soccer ball is a model of the C_{60} structure. (c) The C_{60} fullerene molecule, which is made up of five membered rings (black rings on the soccer ball) and six-membered rings (white rings on the ball). *(C.D. Winters)*

Portrait of a Scientist • *Richard E. Smalley (1943–)*

For a number of years, the research of Professor Richard Smalley of Rice University and his co-workers focused on atom clusters, small clumps of ten or more atoms, and their possible application to semiconductors (Section 15. 11). This work led them to the study of carbon atom clusters and, ultimately, to buckyballs in 1985. The researchers were trying to make a carbon atom cluster with no dangling, unused bonds. C_{60} is exactly that kind of cluster—a molecule with no "loose ends." Since 1985 this immense new field of research has resulted in publication of hundreds of technical papers in scientific journals.

When experimental evidence verified the existence of the C_{60} molecules, Smalley, that night in his kitchen, assembled paper hexagons and pentagons to make a model of how the 60 carbon atoms could be arranged intact. The model looked like a soccer ball (see photo) and turned out to be the correct structure, but what name to give it?

Smalley recalls how the naming occurred: " . . . I asked Harry (Kroto) if he remembered who was the architect who worked with big domes. Didn't the structure of these domes look something like a curved lattice of hexagons? He said it was Buckminster Fuller. Within a few moments we drew a ball on the blackboard and shouted, with rather Monty Pythonesque humor, 'IT'S BUCKMINSTER FULLER . . . ENE!' "* Smalley adds, "So we gathered our courage and sent the paper in, after adding a paragraph at the end somewhat apologizing for the title and leaving the ultimate name of the molecule to be settled by consensus." The playful name was used in the title of the research paper submitted to the prestigious journal *Nature*. The journal quickly accepted the exciting news of the surprising new form of carbon, frivolous title and all (including a photo of a soccer ball)!

Richard Smalley, Harry Kroto, and Robert Curl, Jr., received the 1996 Nobel Prize in Chemistry for their discovery of fullerenes.

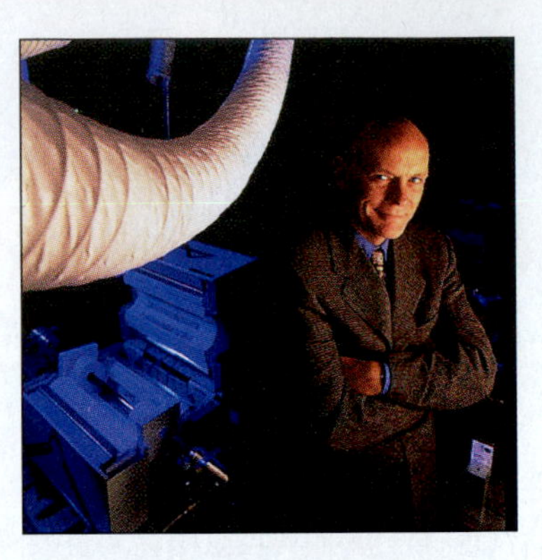

Richard Smalley *(Paul S. Howell/Rice University)*

The paper model of a buckyball created by Richard Smalley. *(Rice University)*

*Source: *Buckminsterfullerenes*, p. vi. W. E. Billups and M. A. Ciufolini, eds. New York, VCH Publishers, 1993.*

Dozens of uses have been proposed for fullerenes, buckytubes, and buckyfibers, among them: microscopic ball bearings, lightweight batteries, new lubricants, nanoscale electric switches, new plastics, and antitumor therapy for cancer patients (by enclosing a radioactive atom within the cage). All these applications await a cheap way of making buckyballs and other fullerenes, the biggest hurdle to their commercial application. Currently, buckyballs, the "cheapest" fullerene, are about five times more expensive than gold.

Exercise 2.3 Allotropes

A student says that tin and lead are allotropes because they are both dull gray metals. Why is the statement wrong?

2.4 ATOMIC STRUCTURE

Dalton's atomic theory said nothing about Democritus' idea that atoms have structure. Knowledge of atomic structure is important because it gives us insights into how and why atoms bond together to form molecules. For example, finding buckyballs was a surprise. But once they were found, their properties could be interpreted by

(a)

(b)

Figure 2.7 An electroscope. (a) With no electrical influence, the foil leaf hangs straight down. (b) A rubber rod that has been rubbed with fur is brought close to the bulb of the electroscope. The electric charge that has built up on the rod induces an opposite charge on the bulb of the electroscope, and the movable leaf diverges from the stationary leaf. The reason for this observation is that the same charge has flowed onto both leaves. Since the leaves repel one another, we conclude that like charges repel. *(C.D. Winters)*

applying knowledge of how carbon atoms can connect to each other. Thus, Democritus' fundamental notion of being able to better understand how substances behave by knowing about the structures of atoms turned out to be correct. The next few sections describe the experiments that support the idea that atoms are made of even smaller (subatomic) particles. By **atomic structure** we refer to the identity and arrangement of these particles.

Divisible Atoms

Electricity played an important role in many of the experiments from which the theory of atomic structure was derived. Electric charge was first observed and recorded by the ancient Egyptians, who noted that amber, when rubbed with wool or silk, attracted small objects. Your hair is attracted to a comb on a dry day because of electric charge. Two types of electric charge had been discovered by the time of Benjamin Franklin (1706–1790), who named them positive (+) and negative (−) because they appear as opposites and can neutralize each other. Experiments with an electroscope (Figure 2.7) show that *like charges repel one another and unlike charges attract one another.* Franklin also concluded that charge is conserved: if a negative charge appears somewhere, a positive charge of the same size must appear somewhere else. Rubbing one substance over another builds up a charge, implying that the rubbing separates positive and negative charges. Thus, it is reasonable to assume that atoms must contain positive and negative charges.

Exercise 2.4 Electric Charge

Explain why the buildup of electric charge that occurs when substances are rubbed together indicates that atoms contain positive and negative charges.

Radioactivity

In 1896 Henri Becquerel discovered that a uranium ore emitted rays that exposed a photographic plate, even though the plate was covered by a protective black paper.

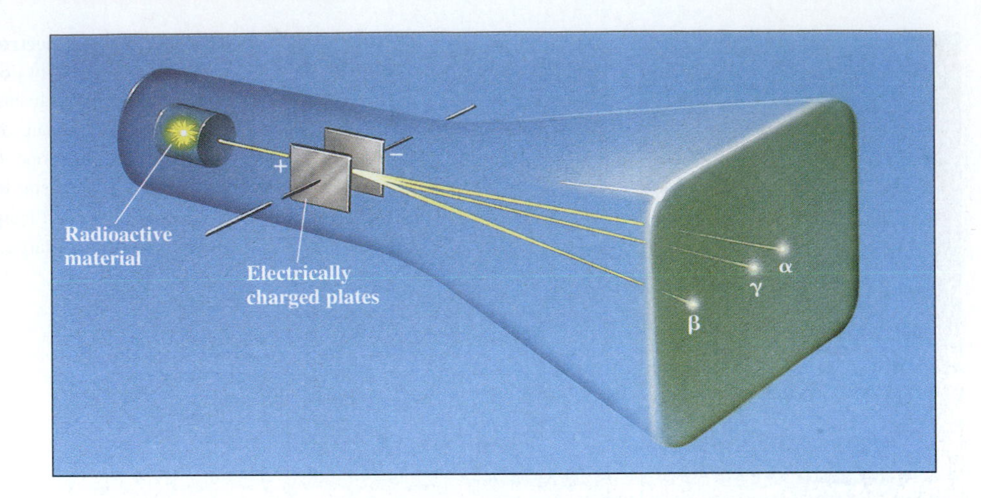

Figure 2.8 Separation of α, β, and γ rays from a radioactive element by an electrical field. Positively charged α particles are attracted toward the negative plate, while the negatively charged β particles are attracted toward the positive plate. (Note that the heavier α particles are deflected less than the lighter β particles.) Gamma rays have no electrical charge and pass undeflected between the charged plates.

In 1898 Marie Curie and her husband, Pierre, isolated the new elements polonium and radium, which also emitted the same kind of rays. The following year Marie suggested that atoms of such elements spontaneously emit these rays and named this phenomenon **radioactivity.**

Radioactive elements can spontaneously emit three different kinds of radiation: alpha (α), beta (β), and gamma (γ) rays. These radiations behave differently when passed between electrically charged plates, as shown in Figure 2.8. Alpha and beta rays are deflected, while gamma rays pass straight through. Deflection by the charged plates can be explained if it is assumed that alpha rays are positively charged and beta rays are negatively charged. Alpha particles are deflected less and so must be heavier than beta particles. Gamma rays have no detectable charge or mass—they behave like light rays. If radioactive atoms can break apart to produce subatomic particles, there must be something smaller inside the atom.

2.5 SUBATOMIC PARTICLES

Electrons

Further evidence that atoms are composed of subatomic particles came from experiments with specially constructed glass tubes called cathode-ray tubes. Most of the air has been removed from these tubes and a metal electrode sealed into each end (Figure 2.9). When sufficiently high voltage is applied to the electrodes, a beam of rays flows from the negatively charged electrode to the positive electrode. These rays, known as cathode rays, travel in straight lines, are attracted toward positively charged plates, can be deflected by a magnetic field, cast sharp shadows, can heat metal objects red hot, and cause gases and fluorescent materials to glow. When cathode rays strike a fluorescent screen, the energy transferred causes light to be given off as tiny flashes. Thus, the properties of a cathode ray are those of a beam of negatively charged particles, each of which produces a light flash when it hits a fluorescent screen. Sir Joseph John Thomson suggested that cathode rays consist of the same particles that had earlier been named **electrons** and been suggested to be the carriers of electricity.

In 1897 Thomson used a specially designed cathode-ray tube (Figure 2.10) to simultaneously apply electric and magnetic fields to a beam of cathode rays. By balancing the electric field against the magnetic field and using basic laws of electricity

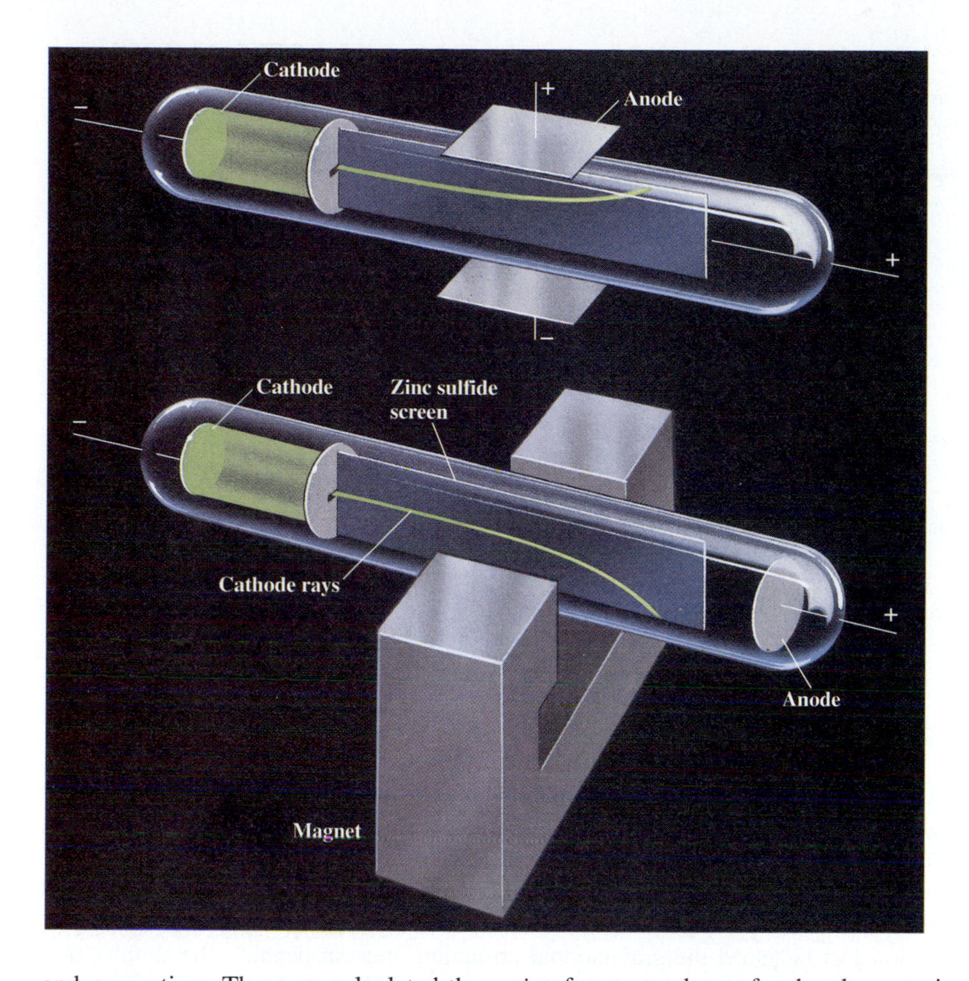

Figure 2.9 Deflection of a cathode ray by an electric field *(top)* and by a magnetic field *(bottom)*. When an external electric field is applied, the cathode ray is deflected toward the positive pole. When a magnetic field is applied, the cathode ray is deflected from its normal straight path into a curved path. In both cases, the curvature is related to the mass of the particles of the cathode rays and the magnitude of the field.

and magnetism, Thomson calculated the *ratio* of mass to charge for the electrons in the cathode-ray beam: 5.60×10^{-9} grams per coulomb (g/C). The coulomb, C, is a fundamental unit of electrical charge.

Thomson obtained the same mass-to-charge ratio in separate experiments with 20 different metals and several different gaseous elements in the cathode-ray tube. These results suggested that electrons are present in all kinds of matter and that, presumably, they exist in atoms of all elements. However, Thomson was not able to independently determine either the electron's mass or charge.

The deflection of cathode rays by charged plates is used to create the picture on a television picture tube or a computer's CRT (cathode-ray tube) screen.

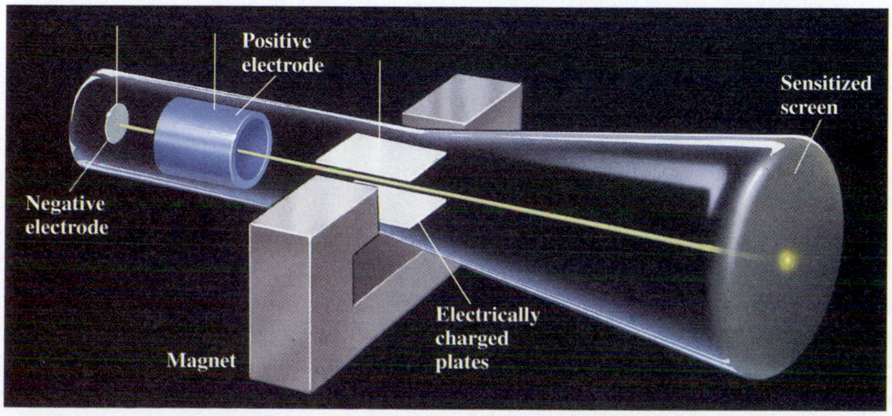

Figure 2.10 Thomson's experiment to measure the charge/mass ratio of the electron. A beam of electrons (cathode rays) passes through an electric field and a magnetic field. The experiment is arranged so that the electric field causes the beam to be deflected in one direction, while the magnetic field deflects the beam in the opposite direction. By balancing the effects of these fields, the mass-to-charge ratio of the electron can be determined.

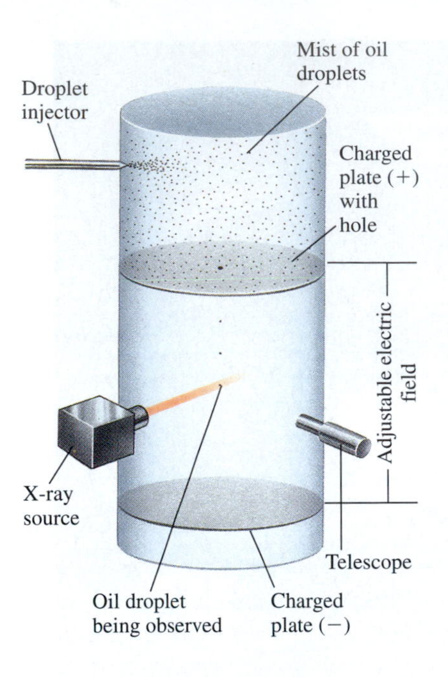

Figure 2.11 Millikan oil-drop experiment. Tiny oil droplets fall through the hole in the upper plate and settle slowly through the air. X rays cause air molecules to give up electrons to the oil droplets, which become negatively charged. From the known mass of the droplets and the applied voltage at which the charged droplets were held stationary, Millikan could calculate the charges on the droplets.

Fourteen years later, Robert Andrews Millikan cleverly measured the charge of an electron with the apparatus shown in Figure 2.11. Tiny oil droplets were sprayed into a chamber. As they settled slowly through the air, the droplets were exposed to x rays, which caused electrons to be transferred from gas molecules in the air to the droplets. Using a small telescope to observe an individual droplet, Millikan adjusted the electric charge on plates above and below the droplet so that the electrostatic attraction just balanced the gravitational attraction, thus suspending the droplet motionless. From equations describing these forces, Millikan calculated the charge on the droplet. Different droplets had different charges, but Millikan found that each was a whole-number multiple of the smallest charge. That smallest charge was 1.60×10^{-19} C. Millikan assumed this to be the fundamental unit of charge, the charge on an electron. From this value and the mass-to-charge ratio determined by Thomson, the mass of an electron could be calculated: $(1.60 \times 10^{-19}\ \text{C}) \times (5.60 \times 10^{-9}\ \text{g/C}) = 8.96 \times 10^{-28}$ g. The currently accepted most accurate mass value is 9.109389×10^{-28} g, and the currently accepted value of the electron's charge is $-1.60217733 \times 10^{-19}$ C. This much charge is represented as -1 when used with electrons, atoms, and other nanoscale particles.

Conceptual Challenge Problem CP-2.A at the end of the chapter relates to topics covered in this section.

See Appendix B.2 for a review of scientific notation, which is used to represent very small or very large numbers as powers of 10. For example, 0.000001 is 1×10^{-6}, and 2,000,000 is 2×10^{6}.

Other experiments provided further evidence that the electron is a fundamental particle of matter by showing that beta particles emitted by radioactive elements have the same properties as cathode rays, which are streams of electrons.

Protons

Ions are atoms or groups of atoms that have positive or negative charges. Positive ions are formed by loss of one or more electrons from neutral atoms; negative ions are formed when extra electrons are added to neutral atoms (Section 3.4).

The experimental evidence for a fundamental positive particle first came from studies of so-called canal rays, which were observed in a special cathode-ray tube (Figure 2.12). At high voltage, cathode rays are formed as in any cathode-ray tube. The cathode-ray electrons collide with gaseous atoms in the tube, knocking out electrons from neutral atoms, thereby converting the atoms to positively charged particles called positive **ions.** The ions are attracted toward a negatively charged plate, confirming

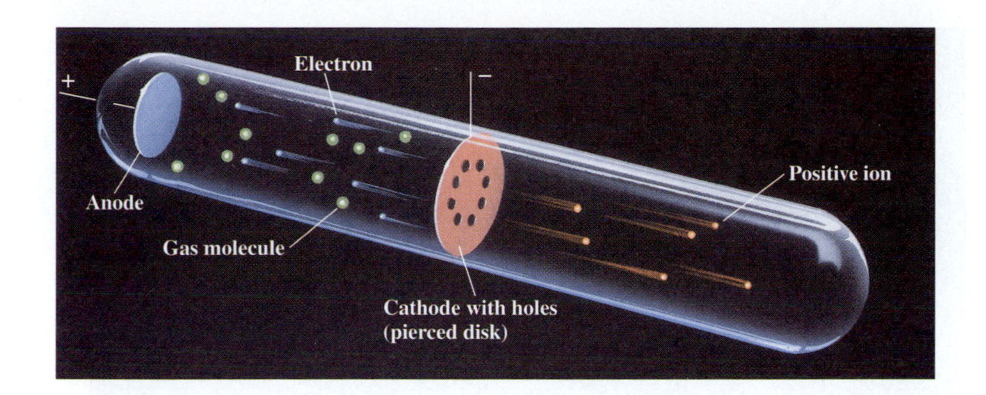

Figure 2.12 Cathode-ray tube with perforated cathode. Electrons collide with gas molecules and produce positive ions, which are attracted to the negative cathode. Some of the positive ions pass through the holes and form a positive ray. Like cathode rays, positive rays (or "canal rays") are deflected by electric and magnetic fields, but much less so for a given value of the field because positive ions are much heavier than electrons.

that they must be composed of positively charged particles. But what causes the positive charge?

The answer came from recognizing that each gas used in the tube gives a different mass-to-charge ratio for its positively charged particles. Denser gases have larger mass-to-charge ratios than lighter ones (unlike the cathode rays, which have the same mass-to-charge ratio no matter what the gas is). The smallest mass-to-charge ratio is obtained with hydrogen, indicating that hydrogen provides positive particles with the smallest mass.

As mass increases, mass-to-charge ratio increases for a given amount of charge.

The hydrogen particles were considered to be the fundamental positively charged particles of atomic structure and were called **protons** (from a Greek word meaning "the primary one"). The mass of a proton is known from experiment to be 1.672623×10^{-24} g, about 1800 times the mass of an electron. The charge on a proton ($+1.60217733 \times 10^{-19}$ C) is equal in size, but opposite in sign, to the charge on an electron. The proton's charge is designated as $+1$ when referring to atoms, ions, or molecules.

Neutrons

Because atoms are electrically neutral, they must contain equal numbers of protons and electrons. However, most neutral atoms have masses greater than the sum of the masses of their protons and electrons. This additional mass indicates that subatomic particles with mass but no charge must be present. Because they have no charge, detecting these particles was difficult. It was not until 1932, many years after the discovery of the proton, that James Chadwick devised a clever experiment that detected the neutral particles by having them knock protons out of wax; he then detected the protons (Figure 2.13). The neutral subatomic particles are called **neutrons.** They have no electric charge and a mass of $1.6749286 \times 10^{-24}$ g, nearly the same as the mass of a proton.

In 1920, Ernest Rutherford proposed that the nucleus might contain an uncharged particle whose mass approximated that of a proton.

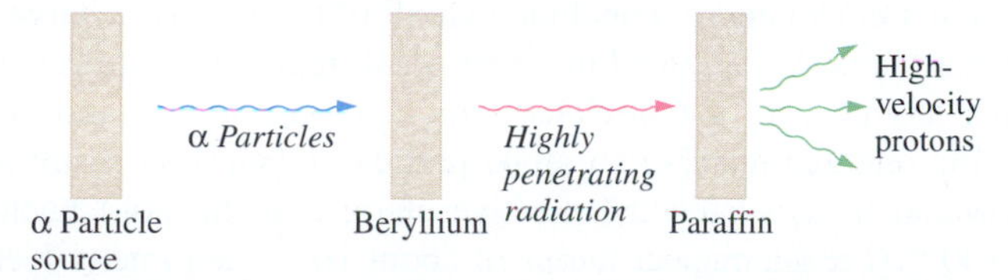

Figure 2.13 Experimental discovery of the neutron. Chadwick recognized that only a neutral particle with a mass close to that of a proton could knock protons of the measured energy out of the paraffin. Neutrons are the highly penetrating radiation ejected from beryllium.

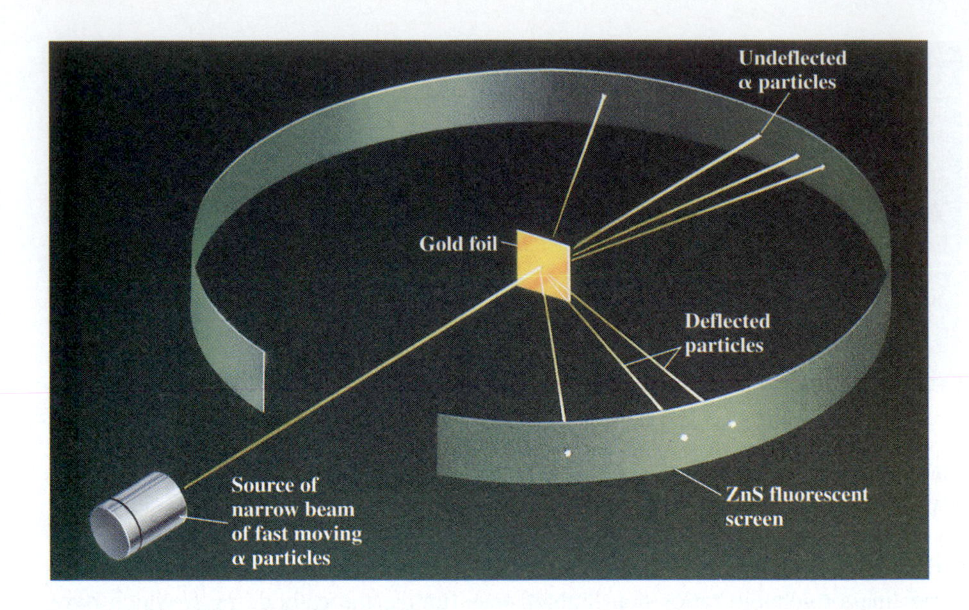

Figure 2.14 The Rutherford experiment. A beam of positively charged α particles was directed at a very thin piece of gold foil. A luminescent screen coated with zinc sulfide (ZnS) was used to detect particles passing through or deflected by the foil. Most particles passed straight through. Some were deflected to some extent, and a few were even deflected backward. (Note that a circular luminescent screen is shown for simplicity; actually a smaller, movable screen was used.)

2.6 THE NUCLEAR ATOM

Once it was known that there were subatomic particles, the next question scientists wanted to answer was, how are these particles arranged in an atom? By the early 1900s J. J. Thomson had proposed that an atom was a positively charged sphere within which thousands of electrons were imbedded. In about 1910 Ernest Rutherford (1871–1937) decided to test Thomson's atomic model. Earlier, Rutherford had discovered that alpha rays are positively charged particles with the same mass as helium atoms and a +2 charge. He reasoned that, if Thomson's atomic model were correct, a beam of the relatively massive alpha particles would be deflected very little as they passed through the atoms in a very thin sheet of gold foil. Rutherford's associate, Hans Geiger, and a young student, Ernest Marsden, set up the apparatus diagrammed in Figure 2.14, and observed what happened when alpha particles hit the gold foil. Almost all the particles passed through undeflected. But Geiger and Marsden were amazed to find that a very few alpha particles were deflected through large angles; some came almost straight back! Rutherford described this unexpected result by saying, "It was about as credible as if you had fired a 15-inch [artillery] shell at a piece of paper and it came back and hit you."

Alpha particles are four times heavier than the lightest atoms, which are hydrogen atoms.

Geiger went on to develop the Geiger counter to detect radioactive emissions.

Earlier experiments, such as Millikan's determination of the electron's charge, were carefully designed to give a desired result. Rutherford's experiment was one of those exciting surprises ***(⇐ p. 4).*** The only way to account for this behavior was to discard Thomson's model and to conclude that all of the positive charge and most of the mass of the atom is concentrated in a very small region (Figure 2.15). Rutherford called this tiny atomic core the **nucleus.** Only such a region could be sufficiently dense and highly charged to repel an alpha particle. From their results, Rutherford, Geiger, and Marsden calculated the positive charge on the gold nucleus to be in the range of 100 ± 20 and a nuclear radius of about 10^{-12} centimeters (cm). The currently accepted values are a charge of +79 and a radius of about 10^{-13} cm.

(Text continues on page 44.)

Portrait of a Scientist • *Ernest Rutherford (1871–1937)*

Lord Rutherford, one of the most interesting people in the history of science, was born in New Zealand in 1871, but went to Cambridge University in England to pursue his Ph.D. in physics in 1895. His original aim was to study radio waves and make his fortune in that field so as to marry his fiancée back in New Zealand. However, his Cambridge professor, J. J. Thomson, convinced him to work on the newly discovered phenomenon of radioactivity. Rutherford discovered alpha and beta radiation while at Cambridge. In 1899 he moved to McGill University in Canada, where he proved that alpha radiation is composed of helium nuclei and that beta radiation consists of electrons. For this work he received the Nobel Prize in Chemistry in 1908.

In 1903 Rutherford and his young wife visited Pierre and Marie Curie in Paris, on the very day that Madame Curie received her doctorate in physics (see Chapter 19). That evening during a party, Pierre Curie brought out a tube coated with a phosphor and containing a large quantity of a radioactive radium compound in solution. The phosphor glowed brilliantly from the radiation given off by the radium. Rutherford later said the light was so bright that he could clearly see Pierre Curie's hands were "in a very inflamed and painful state due to exposure to radium rays."

In 1919 Rutherford moved back to Cambridge and assumed the position formerly held by J. J. Thomson. Not only was Rutherford responsible for very important work in physics and chemistry, but he also guided the work of no fewer than ten future Nobel Prize recipients.

Ernest Rutherford. (*Oesper Collection in the History of Chemistry, University of Cincinnati*)

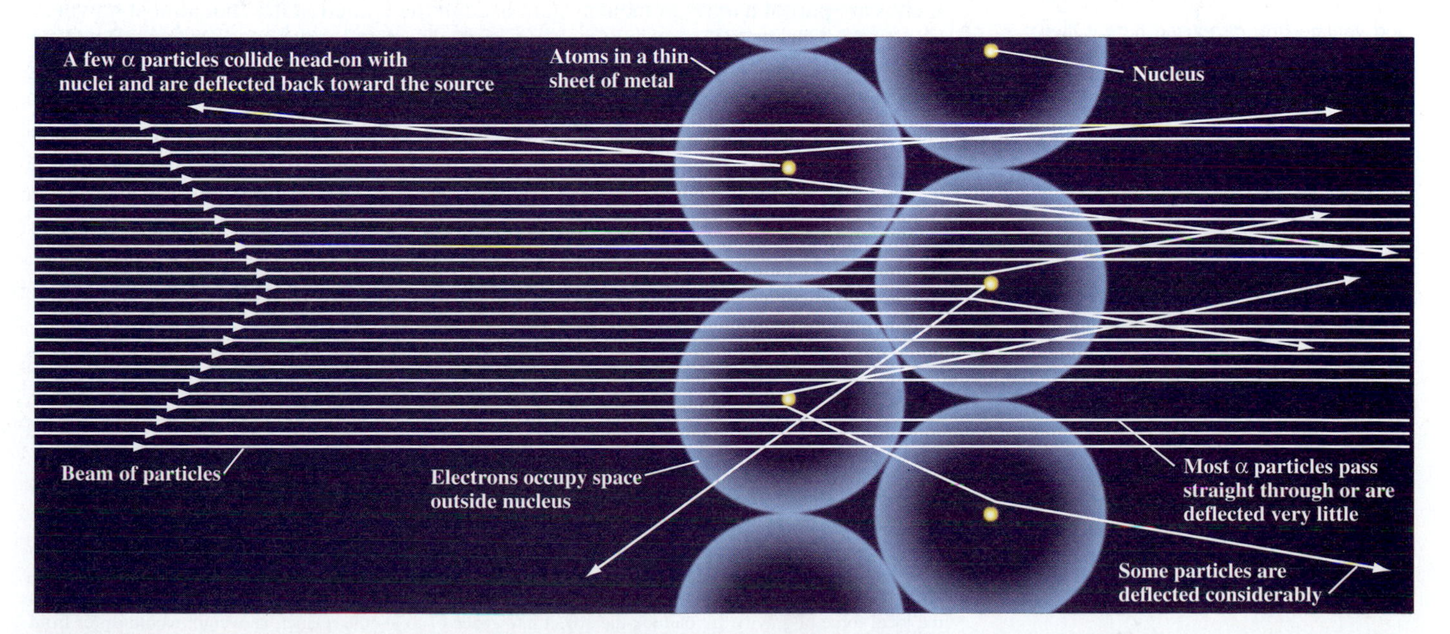

Figure 2.15 **Rutherford's interpretation of the results of the experiment done by Geiger and Marsden.**

This makes the nucleus about 100,000 times *smaller* than the atom. Most of the volume of the atom is occupied by electrons. Somehow, the negative electrons occupy the space outside the nucleus, but their arrangement was completely unknown to Rutherford and other scientists of his time. This arrangement is now well understood and is the subject of Chapter 8.

In summary, there are three primary subatomic constituents: protons, electrons, and neutrons. Protons and neutrons make up the nucleus, providing most of an atom's mass and all of its positive charge. The nuclear radius is about 100,000 times smaller than that of the atom itself. Negatively charged electrons outside the nucleus occupy most of the volume of an atom, but contribute very little mass. A neutral atom has no net electric charge, because the number of electrons outside the nucleus equals the number of protons inside it. To chemists, the electrons are the important subatomic particles because they are the first part of an atom to contact another atom. The electrons largely dictate how atoms combine to form chemical compounds.

Exercise 2.5 Describing Atoms

If an atom had a radius of 100 meters, it would approximately fill a football stadium. What size would the radius of the nucleus of such an atom be? What object is about that size?

2.7 THE SIZES OF ATOMS AND UNITS TO REPRESENT THEM

The International System of units (or SI units) is the officially recommended measurement system in science. It is derived from the metric system and described in Appendix C. The units for mass, length, and volume are introduced here. Other units are introduced as they are needed in later chapters.

Atoms are extremely small. One teaspoon of water contains about three times as many atoms as the Atlantic Ocean contains teaspoons of water. To state the size of an object on the macroscale (for example, yourself), in the United States we would give your weight in pounds and your height in feet and inches.* The pounds, feet, and inches are part of a measurement system used in the United States, but almost nowhere else in the world. Most other countries use the **metric system** of units for recording and reporting measurements. The metric system is a decimal system that adjusts the size of its basic units by multiplying or dividing them by multiples of 10.

In the metric system, your weight, or mass, would be given in kilograms. The **mass** of an object is a fundamental measure of the quantity of matter in that object. The metric units for mass are *grams* or multiples or fractions of a gram. The prefixes listed in Table 2.2 are used with all metric units. A *kilo*gram, for example, is equal to 1000 grams and is a convenient size for measuring the mass of people. For example, the authors of this book weigh 80 kg, 79 kg, and 77 kg.

For objects very much smaller than people, prefixes that represent negative powers of 10 are used. For example, 1 *milli*gram equals 1×10^{-3} g.

$$1\ \textit{milli}\text{gram (mg)} = \frac{1}{1000} \times 1\text{ g} = 0.001\text{ g} \qquad \text{or} \qquad 1 \times 10^{-3}\text{ g}$$

Although individual atoms are too small to be weighed, chemists know from experiments that an atom's mass is on the order of 1×10^{-22} g. For example, a sample of

*Strictly speaking, the pound is not a unit of mass, but rather of *weight.* The weight of an object depends on the local force of gravity. If you took the object into space or to another planet, its weight would differ from its terrestrial value, although its mass would remain the same. For measurements made on the earth's surface, however, this distinction between mass and weight is not generally useful.

TABLE 2.2 Selected Prefixes in the Metric System

Prefix	Abbreviation	Meaning	Example
mega	M	10^6	1 megaton = 1×10^6 tons
kilo	k	10^3	1 kilogram (kg) = 1×10^3 g
deci	d	10^{-1}	1 decimeter (dm) = 0.1 m
centi	c	10^{-2}	1 centimeter (cm) = 0.01 m
milli	m	10^{-3}	1 millimeter (mm) = 0.001 m
micro	μ	10^{-6}	1 micrometer (μ m) = 1×10^{-6} m
nano	n	10^{-9}	1 nanometer (nm) = 1×10^{-9} m
pico	p	10^{-12}	1 picometer (pm) = 1×10^{-12} m

copper that weighs one *nano*gram (1 ng = 1×10^{-9} g) contains about 9×10^{12} copper atoms. The most sensitive laboratory balances can weigh samples of about 0.0000001 g (1×10^{-7} g, or 0.1 μg).

Your height in metric units would be given in *meters,* the metric unit for length. The authors of this textbook are 1.93 m, 1.83 m, and 1.83 m tall. You might be taller or shorter than this, but atoms or molecules aren't nearly that big. Their sizes are often reported in *pico*meters (1 pm = 1×10^{-12} m).

The radius of a typical atom is only between 30 and 300 pm. For example, the radius of a copper atom is 128 pm. To get a feeling for these dimensions, consider how many copper atoms it would take to form a single file of copper atoms just touching each other across the diameter of a U.S. penny, a distance of 1.90×10^{-2} m. This distance can be calculated in picometers by using a conversion factor based on 1 pm = 1×10^{-12} m. A **conversion factor** is a relationship between two measurement units that is derived from the equality between the units, and is used in the following way.

$$\text{Data given} \times \underbrace{\frac{\text{desired data units}}{\text{given data units}}}_{\uparrow\ \text{conversion factor}} = \text{answer in desired units}$$

$$1.90 \times 10^{-2}\,\cancel{\text{m}} \times \frac{1\ \text{pm}}{1 \times 10^{-12}\,\cancel{\text{m}}} = 1.90 \times 10^{10}\ \text{pm}$$

Note that the units (meters) cancel, leaving those of the correct answer (pm).

Every conversion factor can be used in two ways. We just converted meters to picometers by using

$$\frac{1\ \text{pm}}{1 \times 10^{-12}\ \text{m}}$$

Picometers can be converted to meters by inverting the conversion factor, such as in the conversion

$$8.70 \times 10^{10}\,\cancel{\text{pm}} \times \frac{1 \times 10^{-12}\ \text{m}}{1\,\cancel{\text{pm}}} = 8.70 \times 10^{1}\ \text{m}$$

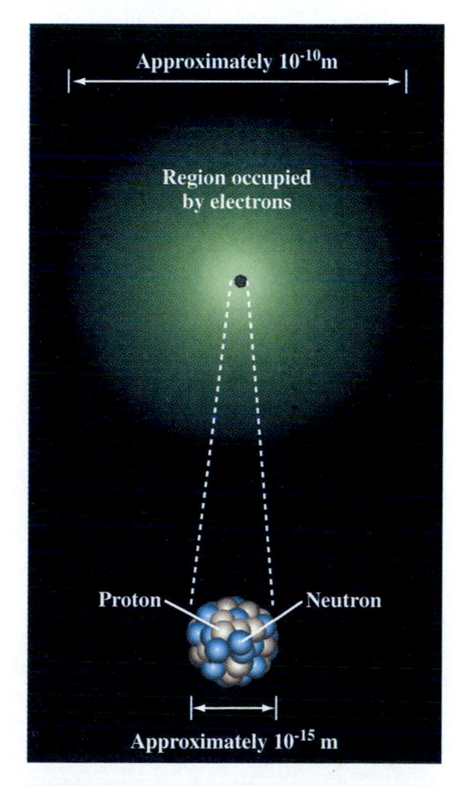

Relative sizes of the atomic nucleus and an atom.

TABLE 2.3 Some Common Unit Equalities

Length: SI Unit, Meter (m)	
1 kilometer	= 1000 meters = 0.062137 mile
1 meter	= 100 centimeters
1 centimeter	= 10 millimeters
1 nanometer	$= 1 \times 10^{-9}$ meter
1 picometer	$= 1 \times 10^{-12}$ meter
1 inch	= 2.54 centimeters (exactly)
1 angstrom	$= 1 \times 10^{-10}$ meter
Volume: SI Unit, Cubic Meter (m^3)	
1 liter (L)	$= 1 \times 10^{-10}$ m^3 = 1000 mL
	= 1000 cm^3 = 1.056710 quarts
1 gallon	= 4.00 quarts
Mass: SI Unit, Kilogram (kg)	
1 kilogram	= 1000 grams
1 gram	= 1000 milligrams
1 pound	= 453.59237 grams = 16 ounces
1 ton (metric)	= 1000 pounds
1 ton (American)	= 2200 pounds

The number of copper atoms needed to stretch across a penny, 1.90×10^{-2} m, can be calculated using a conversion factor linking a distance in picometers with the diameter of a copper atom. The diameter of a Cu atom is twice the radius, 2×128 pm = 256 pm. Therefore,

$$1.90 \times 10^{10}\ \text{pm} \times \frac{1\ \text{Cu atom}}{256\ \text{pm}} = 7.42 \times 10^{7}\ \text{Cu atoms}$$

It takes 74 million copper atoms to reach the required distance!

To put this number into perspective, it would take 1000 people, each counting 1 Cu atom per second nonstop, nearly *21 hours* to count this number of atoms (if they could). Atoms are indeed small!

The use of units in this way to solve chemical calculations is sometimes called dimensional analysis. Conversion factors are the basis for dimensional analysis, a commonly used problem-solving technique. It is covered in detail in Appendix A.3

In chemistry, the most commonly used mass units are the kilogram, the gram, and the milligram. The most commonly used length units are the centimeter, the millimeter, and the nanometer. The relationship of these units to each other and some other units is given in Table 2.3.

Example 2.1 illustrates the use of dimensional analysis in a unit-conversion problem. Notice that in this example, and throughout the book, the answer is given before the explanation of how the answer is found. We **urge** you to first try to answer the problem on your own. Then, check to see if your answer matches the correct one. If it doesn't match, try again. *Then,* read the Explanation to find out why your reasoning differs from the authors'! If your answer is correct, but your reasoning differs from the explanation, you might have discovered another way to solve the problem.

PROBLEM-SOLVING EXAMPLE 2.1 Nanoscale Distances

"Buckytubes" are single-walled fullerenes that are reported to be 11 Angstroms (Å) in diameter and 100 μm (micrometers) long. What are these distances in nanometers? In picometers? 1.0 Å = 1.0×10^{-10} m; 1.0 μm = 1.0×10^{-6} m.

Answer Diameters: 1.1 nm and 1.1×10^{3} pm; lengths: 1.1×10^{5} nm and 1.0×10^{8} pm.

Explanation We use the conversion factors given in the problem and those in Table 2.3 to calculate the answer. In the case of Angstroms, we work through meters to get to nanometers (1.0×10^{-9} m), and then picometers (1.0 pm = 1.0×10^{-12} m). We set up the conversion factor so that the units of Angstroms cancel, leaving meters as the unit.

$$11\,\cancel{\text{Å}} \times \frac{1.0 \times 10^{-10}\ \text{m}}{1.0\,\cancel{\text{Å}}} = 1.1 \times 10^{-9}\ \text{m}$$

The conversion of meters to nanometers (1.0 nm = 1.0×10^{-9} m) is next, followed by the change to picometers. The correct conversion factors are 1 nm per 1.0×10^{-9} m and 1.0 pm per 1.0×10^{-12} m

$$1.1 \times 10^{-9}\,\cancel{\text{m}} \times \frac{1.0\ \text{nm}}{1.0 \times 10^{-9}\,\cancel{\text{m}}} = 1.1\ \text{nm in diameter}$$

$$1.1 \times 10^{-9}\,\cancel{\text{m}} \times \frac{1.0\ \text{pm}}{1.0 \times 10^{-12}\,\cancel{\text{m}}} = 1.1 \times 10^{3}\ \text{pm diameter}$$

Below we have combined the separate steps into a single math setup. With this method it is easier to check that the answer contains the correct units, and the calculation is easier to enter into a calculator. But note that the individual conversions are the same as before.

$$11\,\cancel{\text{Å}} \times \underbrace{\frac{1.0 \times 10^{-10}\,\cancel{\text{m}}}{1.0\,\cancel{\text{Å}}}}_{\text{Å to m}} \times \underbrace{\frac{1.0\ \text{nm}}{1.0 \times 10^{-9}\,\cancel{\text{m}}}}_{\text{m to nm}} = 1.1\ \text{nm}$$

We can make conversions from micrometers to nanometers and picometers in the same way, through meters.

$$100\ \cancel{\text{micrometers}} \times \frac{1.0 \times 10^{-6}\,\cancel{\text{m}}}{1.0\ \cancel{\text{micrometer}}} \times \frac{1\ \text{nm}}{1.0 \times 10^{-9}\,\cancel{\text{m}}} = 1.0 \times 10^{5}\ \text{nm}$$

$$100\ \cancel{\text{micrometers}} \times \frac{1.0 \times 10^{-6}\,\cancel{\text{m}}}{1.0\ \cancel{\text{micrometer}}} \times \frac{1.0\ \text{pm}}{1.0 \times 10^{-12}\,\cancel{\text{m}}} = 1.0 \times 10^{8}\ \text{pm}$$

Problem-Solving Practice 2.1

Do the following conversions using factors based on equalities in Table 2.3:
(a) 650 mg to kilograms
(b) 4.8×10^{-8} m^{3} to millimeters

When working the Problem-Solving Examples in this book, we suggest that you first cover the Answer and the Explanation with a piece of paper and solve the problem on your own.

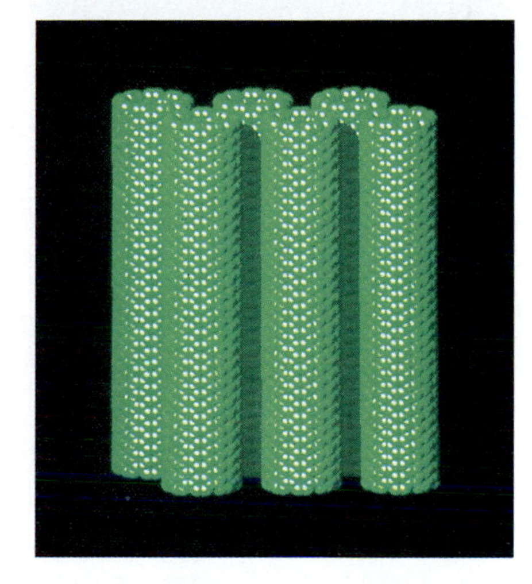

Buckytubes. *(Biosym Technologies/Science Photo Library)*

Table 2.3 also lists the relationships of the liter (L) and milliliter (mL)—the most common volume units of chemistry—to other volume units. There are 1000 mL in 1 L. One liter is a bit larger than a quart, and the teaspoon of water mentioned at the beginning of this section has a volume of about 5 mL. *Chemists often use the terms milliliter and cubic centimeter (or "cc") interchangeably because they are equivalent* (1 mL = 1 cm^{3}).

As illustrated in Problem-Solving Example 2.2, two or more steps of a calculation using dimensional analysis are best written in a single setup and entered into a calculator as a single calculation.

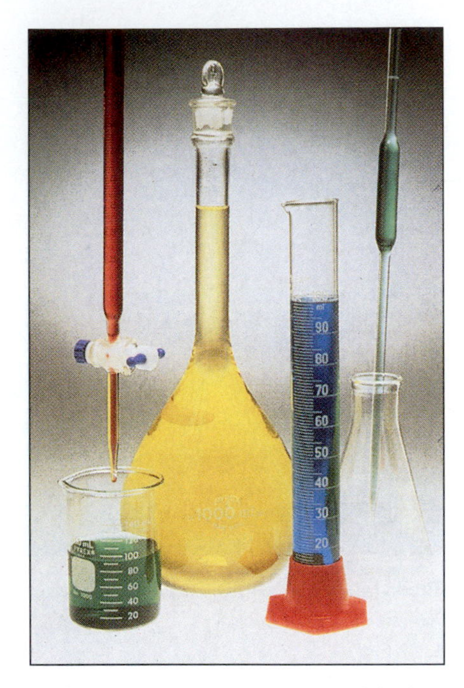

Devices that measure the volume of liquids.
(C.D. Winters)

PROBLEM-SOLVING EXAMPLE 2.2 Volume Units

In a chemical analysis, a chemist uses 30 μL (microliter) of sample for an analysis ($1\ \mu L = 1 \times 10^{-6}$ L). What is this volume in cm^3? In mL?

Answer 3.0×10^{-2} cm^3; 3.0×10^{-2} mL

Explanation We need to use conversion factors based on two equalities to convert microliters to milliliters: $1\ \mu L = 1 \times 10^{-6}$ L, and 1 L = 1000 mL. We multiply the conversion factors to cancel μL and L, leaving only cm^3.

$$30\ \cancel{\mu L} \times \frac{1 \times 10^{-6}\ \cancel{L}}{1\ \cancel{\mu L}} \times \frac{1000\ \text{mL}}{1\ \cancel{L}} = 3.0 \times 10^{-2}\ \text{mL}$$

Because 1 mL and 1 cm^3 are equivalent, the sample size can also be expressed as 3.0×10^{-2} cm^3.

Problem-Solving Practice 2.2

A patient's blood cholesterol level measures 165 mg/dL. Express this value in g/L (1 deciliter (dL) = 1×10^{-1} L).

2.8 ISOTOPES

Experiments done in the early part of the 20th century found that *atoms of the same element have the same numbers of protons in the nucleus.* This number is called the **atomic number** and is given the symbol *Z*. In the periodic table at the front of this book, the atomic number for each element is written above the element's symbol. For example, copper has a nucleus containing 29 protons, so its atomic number is 29 ($Z = 29$); lead (Pb) has 82 nuclear protons, so the atomic number for lead is 82.

Just as altitude is relative, based on sea level as its reference standard (zero altitude), a scale of atomic masses is established relative to a standard. This standard is the mass of a carbon atom that has six protons and six neutrons in its nucleus. Such an atom is defined to have a mass of exactly 12 **atomic mass units** (12 amu). *The mass of every other element is established relative to this mass.* Thus, for example, experiment shows that a gold atom, on the average, is 16.4 times heavier than a carbon atom, so a gold atom has an average mass of 16.4×12.0 amu, or 197 amu.

1 atomic mass unit (amu) = 1/12 the mass of a carbon atom having 6 protons and 6 neutrons in the nucleus.

The masses of the fundamental subatomic particles in amu have been determined experimentally (Table 2.4). Notice that the proton and neutron have masses very close to 1 amu, while the electron mass is about 1800 times lighter.

TABLE 2.4 Properties of Subatomic Particles

Particle	Mass		Charge
	Grams	Atomic Mass Units	
Electron	9.109389×10^{-28}	0.0005485799	−1
Proton	1.672623×10^{-24}	1.007276	+1
Neutron	1.674929×10^{-24}	1.008665	0

Once a relative scale of atomic masses has been established, we can estimate the mass of any atom whose nuclear composition is known. The proton and neutron have masses so close to 1 amu that the difference can essentially be ignored. Electrons are so light that even a large number of them will not greatly affect the mass of the atom. Therefore, to estimate an atom's mass, we add up its number of protons and neutrons. This sum, called the **mass number** of that *particular* atom, is given the symbol A. For example, a copper atom that has 29 protons and 34 neutrons in its nucleus has a mass number, A, of 63. A lead atom that has 82 protons and 126 neutrons has $A = 208$. With this information, an atom of known composition, such as this lead atom, can be represented by the notation

$$^{A}_{Z}X \quad \begin{array}{l} \leftarrow \text{Mass number} \searrow \\ \leftarrow \text{Element symbol} \rightarrow {}^{208}_{82}\text{Pb} \\ \leftarrow \text{Atomic number} \nearrow \end{array}$$

Because each element is identified by its own unique symbol and atomic number, the symbol also tells you what the atomic number must be, so the subscript Z is optional. For example, the copper atom described above would have the symbol $^{63}_{29}\text{Cu}$ or just ^{63}Cu or copper-63. We would simply say "copper-63."

The number of neutrons is calculated by subtracting Z from A.

Each element has a unique atomic number.

PROBLEM-SOLVING EXAMPLE 2.3 Atomic Nuclei

Iodine-127 is important in diagnostic medicine because it concentrates in the thyroid gland, where its quantity can be detected and used to measure thyroid gland activity. How many neutrons are there in an iodine atom of mass number 127?

Answer There are 74 neutrons.

Explanation The table of elements (inside front cover) shows that the atomic number of iodine (I) is 53. Therefore, the atom has 53 protons in the nucleus. Because the mass number of the atom is the sum of the numbers of protons and neutrons in the nucleus,

$$\text{Mass number} = \text{number of protons} + \text{number of neutrons}$$
$$127 = 53 + \text{number of neutrons}$$
$$\text{Number of neutrons} = 127 - 53 = 74$$

Problem-Solving Practice 2.3

(a) What is the mass number of a phosphorus atom with 16 neutrons?
(b) How many protons, neutrons, and electrons are there in a neutral neon-22 atom?
(c) Write the symbol for the isotope with 82 protons and 125 neutrons.

Atomic masses are now determined experimentally using mass spectrometers. Figure 2.16 illustrates the use of a mass spectrometer to determine the atomic masses of neon atoms. The heavier neon-22 ions are deflected less than the lighter neon-21 and neon-20 ions, therefore allowing the ions of different mass to be separated. Their masses are calculated from their charges, the magnetic field strength, and the accelerating voltage.

Although an atom's mass approximately equals its mass number, the actual mass is not an integral number. For example, the actual mass of a gold-196 atom is 195.9231 amu, slightly less than the mass number of 196.

The actual mass of an atom is slightly less than the sum of the masses of its protons, neutrons, and electrons. The difference, known as the mass defect, is related to the energy binding nuclear particles together, a topic discussed in Chapter 19.

Mass spectrometric analysis of most naturally occurring elements reveals that not all atoms of an element have the same mass number. For example, the silicon atoms

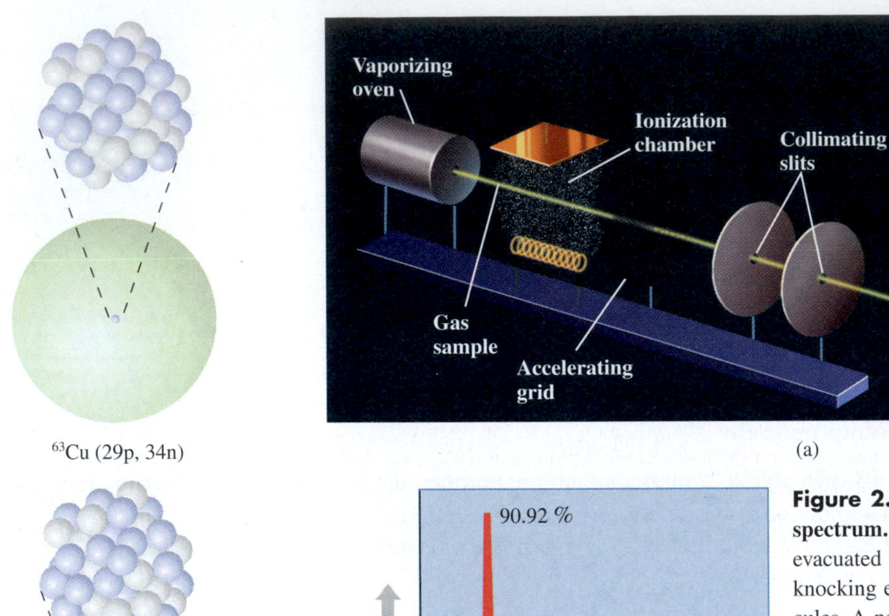

(a)

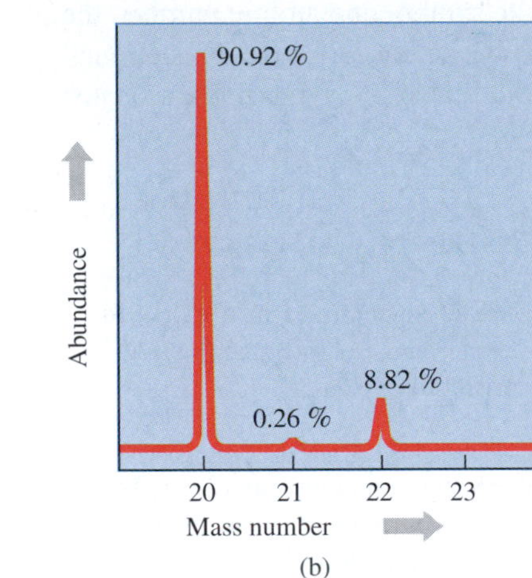

(b)

Figure 2.16 Mass spectrometer and a mass spectrum. (a) A gas sample is vaporized into an evacuated tube. An electron beam ionizes the gas by knocking electrons from the neutral atoms or molecules. A negatively charged grid accelerates positive ions into the rest of the apparatus. The slits block all but a narrow beam of ions, which move into a magnetic field perpendicular to their path. Each ion follows a curved path determined by its mass-to-charge ratio. A detector determines the angle by which each kind of ion has been deflected. (b) Result of separating the ions formed by the different isotopes of neon in a mass spectrometer. The principal peak corresponds to the most abundant isotope, neon-20. The peak height indicates the percent relative abundance of the isotopes as shown.

^{63}Cu (29p, 34n)

^{65}Cu (29p, 36n)

Copper isotopes. Copper-63 and copper-65 atoms each contain 29 protons; copper-63 atoms have 34 neutrons and copper-65 atoms contain 36 neutrons.

Two different elements can't have the same atomic number. If two atoms differ in their number of protons, they are atoms of different elements. If only their number of neutrons differs, they are isotopes.

Frederick Soddy, Ernest Rutherford's assistant, coined the word "isotope" to describe the different forms of the same element.

in computer chips all have 14 protons, but some silicon nuclei have 14 neutrons, others have 15, and still others have 16. Thus, naturally occurring silicon (atomic number 14) is always a mixture of silicon-28, silicon-29, and silicon-30 atoms. Such different atoms of the *same* element are called isotopes. **Isotopes** are atoms with the *same* atomic number (Z) but different mass numbers (A). All the silicon isotopes have 14 protons; that's what makes them silicon atoms. But these isotopes differ in their mass numbers because they have different numbers of neutrons.

PROBLEM-SOLVING EXAMPLE 2.4 Isotopes

Copper has two isotopes, one with 34 neutrons and the other with 36 neutrons. What are the mass numbers and notations of these isotopes?

Answer The mass numbers are 63 and 65. The symbols are $^{63}_{29}\text{Cu}$ and $^{65}_{29}\text{Cu}$.

Explanation Copper has an atomic number of 29, so it has 29 nuclear protons. Therefore, mass numbers of the two isotopes are

Isotope 1: Cu = 29 protons + 34 neutrons = 63 (copper-63)
Isotope 2: Cu = 29 protons + 36 neutrons = 65 (copper-65)

CHEMISTRY IN THE NEWS*

Newest, Heaviest Elements

How many elements are known? The number depends on what year you ask the question. Three new ones—elements 110, 111, and 112—have been synthesized since 1994, bringing to 112 the total of known elements. The creation of these newest elements is a tale of cutting-edge science done by a team of international researchers working in Darmstadt, Germany, at the Society for Heavy-Ion Research (GSI).

GSI researchers have synthesized elements 107 to 112 by bombarding lead-208 or bismuth-209 nuclei for two to three weeks with a beam of nuclei with atomic numbers from 24 (Cr) to 30 (Zn) (see table). The bombarding nuclei that have just the proper energies combine with the lead-208 or bismuth-209 target nuclei to form atoms of the new element, plus neutrons. The lighter, positively charged nuclei must be accelerated to velocities that give them the correct energy to overcome the repulsive positive charge of the Pb or Bi target nuclei.

The energy of the bombarding particles must be just right—if too energetic, the combination splits into two lighter nuclei; not enough energy and the combination doesn't occur. When atoms of a new element are formed, they are separated from the beam by electric and magnetic fields along an 11-m heavy-ion separator. Very sophisticated, ultrasensitive detectors are needed because the new atoms are not mass produced. A 2-week "run," for example, produced just four atoms of element 110, and it took 18 days to create three atoms of element 111.

In spite of having so few atoms to work with, researchers can identify the new elements by the energies of the alpha particles they emit while undergoing radioactive decay to nuclei of lower atomic number.

The GSI team hopes to extend even further the number of known elements by bombarding lead-208 with germanium-76 to get element 114, and bismuth-209 with zinc-70 or germanium-76 to get 113 and 115, respectively.

Source: M. Freemantle: "Heavy-Ion Research Institute Explores Limits of Periodic Table," Chemical & Engineering News, *March 13, p. 37, 1995; and February 26, p. 6, 1996.*
M. Rouhi: "Element 112: New Element Made from Zinc and Lead," Chemical & Engineering News, *February 26, p. 6, 1996.*

Elements Created at GSI

Bombarding Particle (Beam)	Target Nucleus	Atomic Number of New Element	Mass Number of New Element	New Element Created
Cr-54	Bi-209	107	262	Feb. 1981
Fe-58	Pb-208	108	265	March 1984
Fe-58	Bi-209	109	266	Sept. 1982
Ni-62	Pb-208	110	269	Nov. 1994
Ni-64	Bi-209	111	272	Dec. 1994
Zn-70	Pb-208	112	277	Feb. 1996

Placing the atomic number at the bottom left and the mass number at the top left gives $^{63}_{29}Cu$ and $^{65}_{29}Cu$.

Problem-Solving Practice 2.4

Naturally occurring magnesium has three isotopes with 12, 13, and 14 neutrons, respectively. What are the mass numbers and notations of these three isotopes?

We generally refer to a particular isotope by giving its mass number (for example, $^{238}_{92}U$, U-238). But some isotopes have distinctive names and symbols because of their importance, such as the isotopes of hydrogen, all with just one proton. When

Hydrogen isotopes. Hydrogen, deuterium, and tritium each contain one proton. Hydrogen has no neutrons; deuterium and tritium have one and two neutrons, respectively.

$^{238}_{92}U$, $^{2}_{1}H$, and $^{3}_{1}H$ are also represented by uranium-238 (U-238), hydrogen-2 (H-2), and hydrogen-3 (H-3), respectively.

that is the only nuclear particle, the element is called simply "hydrogen." With one neutron and one proton present, the isotope $^{2}_{1}H$ is called deuterium or "heavy" hydrogen (symbol, D). When two neutrons are present, it is $^{3}_{1}H$, or tritium (symbol, T).

Exercise 2.6 Allotropes and Isotopes

A student in your chemistry class tells you that nitrogen-14 and nitrogen-15 are allotropes because they are forms of the same element. How would you refute this statement?

2.9 ISOTOPES AND ATOMIC WEIGHT

As described in Example 2.4, copper has two naturally occurring isotopes, copper-63 and copper-65. These two isotopes have atomic masses of 62.9296 amu and 64.9278 amu, respectively. In a macroscopic collection of naturally occurring copper atoms, the average mass of the atoms is not 63 amu (all copper-63) or 65 amu (all copper-65). Rather, the average atomic mass is in between, its value depending on the proportion of each isotope in the mixture, which can be found with a mass spectrometer. The average atomic mass is analogous to the average mass of people in a room; it depends on the proportion of heavier and lighter people in the room.

The proportion of atoms of each isotope in a natural sample of an element is called the **percent abundance,** the percentage of atoms of a *particular* isotope.

The concept of percent is widely used in chemistry and it is worth briefly reviewing here. For example, our atmosphere contains 78% nitrogen and 22% oxygen. And although we think of a penny as being copper, you might be surprised to find that all U.S. pennies minted after 1982 are actually only 2.4% copper; the rest is zinc.

Sales taxes, tips to servers, and bargain sales in stores all use percent. For example, suppose you buy $130 of merchandise at a store where the local sales tax is 5.0%. This means that you pay $5 for every $100 of the merchandise purchased. The word "percent" literally means per 100, or parts per 100 parts. A 5.0% sales tax means $5 tax/$100 spent. You find the sales tax you owe on your purchase by multiplying its price by the value of the percent divided by 100. The sales tax is $6.50:

Multiplying by 0.05 or 0.95 is the same as multiplying by 5/100 or 95/100, respectively.

$$\$130 \text{ spent} \times \frac{\$5 \text{ tax}}{\$100 \text{ spent}} = \$6.50 \text{ tax}$$

Notice that the "$ spent" cancels out, leaving "$ tax," the unit we want. The conversion factor in a percent calculation is always the value of the percent divided by 100.

U.S. Pennies. Part of the surface of the pre-1982 penny on the left has been removed to show that the coin is solid copper. The post-1982 penny on the right shows the inner zinc filling, covered by a thin coating of copper *(C.D. Winters)*

PROBLEM-SOLVING EXAMPLE 2.5 Applying Percent

Currently, U.S. pennies are 2.40% copper and 97.60% zinc. What mass of each element is in a penny weighing 2.485 g?

Answer 0.0596 g Cu; 2.425 g Zn

Explanation Let's start by calculating the mass of copper. Its percentage, 2.40% copper, means every 100 g of penny contains 2.40 g of copper.

$$2.485 \text{ g penny} \times \frac{2.40 \text{ g copper}}{100 \text{ g penny}} = 0.0596 \text{ g copper}$$

The zinc percentage is found the same way, using the conversion factor of 97.60 g of zinc per 100 g of penny.

$$2.485 \text{ g } \cancel{\text{penny}} \times \frac{97.60 \text{ g zinc}}{100 \text{ g } \cancel{\text{penny}}} = 2.425 \text{ g zinc}$$

We also could have directly obtained this value by recognizing that the masses of zinc and copper must add up to the mass of the penny. Therefore,

$$2.485 \text{ g penny} = 0.0596 \text{ g copper} + x \text{ g zinc}$$

Solving for x, the mass of zinc, gives 2.485 g − 0.0596 g = 2.425 g of zinc.

The sum of the percents in a sample must be 100.

Problem-Solving Practice 2.5

Many heating devices such as hair dryers contain Nichrome wire, an alloy containing 80% nickel and 20% chromium, which gets hot when an electric current passes through it. If a heating device contains 75 g of Nichrome wire, how many grams of nickel and how many grams of chromium does it contain?

We return now to the percent abundance of isotopes, which for each isotope present in a sample is given as follows:

$$\text{Percent abundance} = \frac{\text{Number of atoms of a given isotope}}{\text{Total number of atoms of all isotopes of that element}} \times 100\%$$

Table 2.5 gives information about percent abundance for naturally occurring isotopes of hydrogen, boron, and bromine. The percent abundance and isotopic mass of each isotope can be used to find the average mass of atoms of that element. The **atomic weight** of an element is the average mass of a representative sample of atoms of the element, expressed in atomic mass units.

Boron, for example, is a relatively rare element present in compounds used in laundry detergents, mild antiseptics, and Pyrex cookware. It has two naturally occurring isotopes: B-10, with a mass of 10.0129 amu and 19.91% abundance, and B-11, with a mass of 11.0093 amu and 80.09% abundance. Stating their percent abundances another way, if you could count 10,000 boron atoms from an "average" natural sample, 1991 of them would be B-10, with a mass of 10.0129 amu, and 8009 of

TABLE 2.5 Masses of the Stable Isotopes of Some Elements

Element	Symbol	Atomic Weight	Mass Number	Isotopic Mass	Percent Abundance
Hydrogen	H	10.00794	1	1.007825	99.9855
	D		2	2.0141022	0.0145
Boron	B	10.811	10	10.012939	19.91
			11	11.009305	80.09
Bromine	Br	79.904	79	78.918336	50.69
			81	80.916289	49.31

1 amu = 1/12 mass of a carbon-12 atom.

You might wonder why we chose to interpret 19.91% as 1991 atoms out of 10,000 atoms rather than 19.91 atoms of B-10 out of 100 boron atoms. This is because we wanted to avoid partial atoms, so we "scaled up" the ratio to base it on 10,000 atoms to obtain whole numbers of atoms.

"Atomic weight" is so commonly used that it has become accepted, even though it is more properly a mass than a weight.

Conceptual Challenge Problem CP-2.B at the end of the chapter relates to topics covered in this section.

them would be B-11, with a mass of 11.0093 amu. The atomic weight of an element is found from the percent abundance data as shown by the following calculation for boron. The mass of each isotope is multiplied by its percent abundance expressed as a decimal fraction.

$$\begin{aligned}\text{Atomic weight} &= (\text{percent abundance B-10})\ (\text{isotopic mass B-10}) \\ &\quad + (\text{percent abundance B-11})\ (\text{isotopic mass B-11}) \\ &= (19.91\%)\ (10.0129\text{ amu}) + (80.09\%)\ (11.093\text{ amu}) \\ &= (0.1991)\ (10.0129\text{ amu}) + (0.8009)\ (11.0093\text{ amu}) \\ &= 10.81\text{ amu}\end{aligned}$$

Thus, atomic weight is a special kind of average that accounts for the proportion of each component, not just the usual arithmetic average where the values are simply summed and divided by the number of values.

The atomic weight of each stable element has been determined and these are the masses that appear in the tables in the front of the book. In the periodic table, each element's box contains the atomic number (number of protons), the symbol, and the atomic weight.

For example, the periodic table entry for copper:

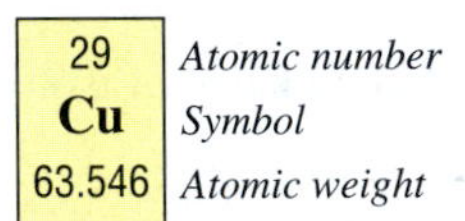

Exercise 2.7 Atomic Weight

Verify that the atomic weight of lithium is 6.941 amu, given the following information:

$^{6}_{3}Li$, mass = 6.015121 amu, percent abundance = 7.500%
$^{7}_{3}Li$, mass = 7.016003 amu, percent abundance = 92.50%

Exercise 2.8 Isotopic Abundance

Naturally occurring magnesium contains three isotopes: magnesium-24 (78.70%), magnesium-25 (10.13%), and magnesium-26 (11.17%). Make a simple *estimate* of the atomic weight of magnesium and compare it to the weight of the *arithmetic* average of the atomic masses. Which is larger? Why?

Exercise 2.9 Percent Abundance

Consider the accurate atomic weight of boron, 10.811 amu. If you knew only this value and not the percent abundances, make the case that the percent abundance of each of the two boron isotopes cannot be 50%.

2.10 AMOUNTS OF SUBSTANCES—THE MOLE

Earlier we discussed the extremely small size of atoms—they are too small to be seen directly or to be weighed individually on the most sensitive balance. However, in working with chemicals, it is useful to know how many atoms, molecules, or other nanoscale units of an element or compound you have. Ten grams of lithium, the least dense metal, is quite a different amount of an element than 10 grams of lead—there are many more lithium atoms than lead atoms in these samples. Because atoms or molecules are very small (the nanoscale level), it is easier to deal with a large number of them (the macroscale level) than to work with one or two at a time. Therefore, chemists have defined a convenient unit of matter that contains a known number of particles. This chemical-counting unit is called the **mole.**

The word "mole" was apparently introduced around 1896 by Wilhelm Ostwald, who derived the term from the Latin word *moles,* meaning a "heap" or "pile." The mole, whose symbol is **mol,** is defined as the amount of substance that contains as many atoms, molecules, ions, or other nanoscale units as there are atoms in *exactly* 12 g of the carbon-12 isotope. Thus, a mole is a heap or pile of stuff that contains a particular number of nanoscale particles.

One mole of carbon weighs 12.01 g, not exactly 12 g. The difference is due to the fact that naturally occurring carbon contains carbon-12 (98.89%) and carbon-13 (1.11%). One mole of carbon-12 weighs exactly 12 g.

The essential point to understanding the mole is that *one mole always contains the same number of particles,* no matter what the substance is. But how many particles are in a mole? Many ingenious experiments over the years have established that number as

$$1 \text{ mol} = 6.0221367 \times 10^{23} \text{ particles}$$

The mole is the connection between the macroscale and nanoscale worlds, the visible and the not directly visible.

The number of particles in a mole is commonly known as **Avogadro's number** in honor of Amedeo Avogadro (1776–1856), an Italian physicist (and lawyer) who conceived the basic idea but never determined the number; the experimental determination came later. There is nothing particularly special about the value of Avogadro's number. It is determined on a relative basis by defining the mole as the number of particles in exactly 12 g of carbon-12. If, instead, one mole were defined in terms of 10 g of carbon, then Avogadro's number would have a different value. It is interesting that the number was revised in the 1980s to the new value (from 6.022045×10^{23}) as a result of new instrumentation and better measurements. Even the most fundamental numerical values and ideas of science are under constant scrutiny.

Avogadro's complete name was Lorenzo Romano Amedeo Carlo Avogadro di Quarequa e di Cerreto, quite a mouthful.

A difficulty in comprehending Avogadro's number is its size. It may help to write it out in full as

$$6.022 \times 10^{23} = 602{,}200{,}000{,}000{,}000{,}000{,}000{,}000$$

or as $602{,}200 \times 1$ million $\times 1$ million $\times 1$ million. It also may help you to think of it this way: if you poured Avogadro's number of marshmallows over the continental United States, the marshmallows would cover the country to a depth of 650 miles! Or, if a mole of pennies could be divided equally among every man, woman, and child in the United States, your share alone could pay off the national debt (about $3 trillion or $\$3 \times 10^{12}$) and you would still have about 20 trillion *dollars* left to buy some pizza for you and your friends.

And so chemists heap up particles by the mole: 1 mol contains 6.022×10^{23} particles. You can think of the mole simply as a counting unit, analogous to other units we use to count collections of ordinary items such as donuts or bagels by the dozen, socks and shoes by the pair, and sheets of paper by the ream (500 sheets). Atoms, molecules, and other particles in chemistry are counted by the mole.

Although Avogadro's number is known to eight significant figures, we will generally use it rounded off to 6.022×10^{23}.

If the mole connects the nanoscale amounts of a substance (number of particles) with the macroscale amount (a mole), how is the connection actually made? The answer is, by weighing. The experimental determination of masses of atoms of all elements makes this possible ***(← p. 50).*** For example, experiments show that an alu-

Figure 2.17 **One-mole quantities of some common elements:** Iron (rod) (55.847 g), mercury (in cylinder) (200.59 g), copper (wire) (63.546 g), sodium (under oil to protect it from air) (22.9898 g), aluminum (powder) (26.9815 g), and nitrogen (gas in balloons) (28.0134 g). *(Larry Cameron)*

minum-27 atom is 2.25 times heavier than a carbon-12 atom, or that a mercury-200 atom is 16.66 times heavier than a carbon-12 atom. A mole of carbon-12 atoms has a mass of 12.00 g. This means that a mole of aluminum-27 atoms has a mass of 27.0 g, 2.25 times greater than the mass of a mole of carbon-12 atoms (12.00 g). Likewise, the mass of a mole of mercury-200 atoms is 200.0 g = 12.00 g × 16.66.

The different masses of the elements shown in Figure 2.17 each contain one mole of atoms. For each element in Figure 2.17, the pictured mass is the numerical equivalent in grams (the macroscale) of the atomic weight in atomic mass units (the nanoscale). The **molar mass** of any substance is the mass, in grams, of one mole of that substance. Molar mass has the units of grams per mole (g/mol).

Conceptual Challenge Problem CP-2.C at the end of the chapter relates to topics covered in this section.

Moles of Atoms

For all elements as they occur naturally, except those that are diatomic or polyatomic (e.g., Cl_2 or P_4) *(← p. 32)*, the molar mass is the mass in grams numerically equal to the atomic weight of the element in atomic mass units. We say that the molar mass of, for example, copper, is 63.55 g/mol and that of aluminum is 26.98 g/mol.

$$\begin{aligned}\text{1 molar mass of copper (Cu)} &= \text{mass per 1 mol of Cu atoms}\\ &= \text{mass of } 6.022 \times 10^{23} \text{ Cu atoms}\\ &= 63.55 \text{ g/mol}\end{aligned}$$

$$\begin{aligned}\text{1 molar mass of aluminum (Al)} &= \text{mass per 1 mol of Al atoms}\\ &= \text{mass of } 6.022 \times 10^{23} \text{ Al atoms}\\ &= 26.98 \text{ g/mol}\end{aligned}$$

The atomic weights are given in the alphabetical table of elements in the front of this book.

Each molar mass contains Avogadro's number of atoms; the molar masses of different elements differ, however, because the atoms of different elements have different masses. Think of a mole as analogous to a dozen. We could have one dozen golf balls, a dozen baseballs, or a dozen bowling balls—in each case 12 items. But those three different dozens do not weigh the same because their individual units have different masses: 1 golf ball = 45 g; 1 baseball = 139 g; and 1 bowling ball = 7200 g. This is analogous to the different masses of a mole of different elements. Because individual atoms of different elements have different masses, Avogadro's number of atoms of one element will have a different mass from Avogadro's number of atoms of another element.

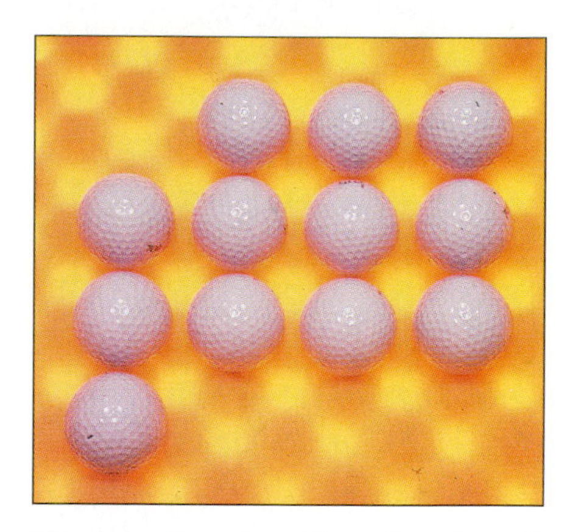

The mass of one dozen items depends on the mass of the individual items. The mass of one dozen golf balls is 540 g (12 baseballs = 1.668 g; 12 bowling balls = 86,400 g). The mass of one mole of an element depends on the atomic weight of its atoms. *(C.D. Winters)*

Exercise 2.10 Grams, Moles, and Avogadro's Number

You have a 10.00-g sample of lithium and a 10.00-g sample of iridium. How many more atoms are in the lithium sample than in the iridium sample?

2.11 MOLAR MASS AND PROBLEM SOLVING

Understanding the idea of a mole and applying it is *essential* to doing quantitative chemistry. In particular, it is *absolutely necessary* to be able to make two basic conversions: *"moles → mass" and "mass → moles" for elements and compounds.* In making these and many other calculations in chemistry, it is most helpful to use dimensional analysis in the same way it is used in unit conversions (⇐ *p. 45*). Write the units with all quantities in a calculation and cancel the units except for the desired one along with calculating the final answer. If the problem is set up properly, the answer will have the desired units.

If you have learned dimensional analysis (also called factor-label method) in a previous science course but are a bit out of practice with it, Appendix A provides a good review of the topic. If you are not familiar with dimensional analysis, read Appendix A now.

Let's now apply dimensional analysis to converting mass to moles, or moles to mass. *In either case, the conversion factor is provided by the molar mass of the substance, the number of grams in one mole, that is, grams per mole (g/mol).*

Mass ↔ moles conversions for substance A

Mass A → moles

$$\text{Grams A} \times \frac{1 \text{ mol A}}{\text{grams A}} = \text{moles A}$$

↑ molar mass

Moles A → mass A

$$\text{Moles A} \times \frac{\text{grams A}}{1 \text{ mol A}} = \text{grams A}$$

↑ 1/molar mass

Suppose that you need 0.250 mol of copper for an experiment. How many grams of copper should you use? The key relationship is the molar mass of copper, a value that relates grams and moles. From the tables in the front of the book, the atomic weight of copper is 63.546 amu, so the molar mass of copper is 63.546 g. To calculate the mass of 0.250 mol Cu, the conversion factor is 63.546 g Cu/1 mol Cu.

$$0.250 \cancel{\text{mol Cu}} \times \frac{63.55 \text{ g Cu}}{1 \cancel{\text{mol Cu}}} = 15.9 \text{ g Cu}$$

In this book we will, when possible, *use one more significant figure in the molar mass than in any of the other data in the problem* (see Appendix A.3). In the problem just completed, note that four significant figures were used in the molar mass of copper, when three were in the sample mass. Using one more significant figure of the molar mass guarantees that its precision is greater than that of the other numbers and does not limit the precision of the result.

Appendix A.3 reviews the use of significant figures. If you are unfamiliar with how to use significant figures, read Appendix 3 now.

Frequently a problem requires converting the mass of a substance to its equivalent number of moles, such as calculating the number of moles of bromine in 10.00 g of bromine. Because bromine is a diatomic element (← *p. 32*), it consists of Br_2 molecules. Therefore, there are two moles of atoms in one mole of Br_2 molecules; and the molar mass of bromine is twice its atomic mass = $2 \times$ (79.904 g/mol) = 159.81 g/mol. To calculate the moles of bromine in 10.00 g Br_2, the conversion factor 1 mol Br_2/159.81 g Br_2 is used.

$$10.00\ \cancel{\text{g Br}_2} \times \frac{1\ \text{mol Br}_2}{159.81\ \cancel{\text{g Br}_2}} = 6.26 \times 10^{-2}\ \text{mol Br}_2$$

Problem-Solving Example 2.6 Mass and Moles

(a) Titanium (Ti) is a metal used to build airplanes. How many moles of titanium are in a 100 g sample of the pure metal?
(b) Aluminum is also used in airplane manufacturing. A piece of aluminum contains 2.16 mol Al. Is the mass of aluminum greater or less than the mass of titanium in Part a?

Answer (a) 2.09 mol Ti; (b) 58.3 g Al, which is less than the mass of titanium.

Explanation
(a) This is a mass-to-moles conversion that is solved using the conversion factor 1 mol Ti/47.88 g Ti.

$$100\ \cancel{\text{g Ti}} \times \frac{1\ \text{mol Ti}}{47.88\ \cancel{\text{g Ti}}} = 2.09\ \text{mol Ti}$$

(b) Calculate the mass of aluminum by converting moles to grams of aluminum. Then compare this mass to the mass of titanium.

$$2.16\ \cancel{\text{mol Al}} \times \frac{26.98\ \text{g Al}}{1\ \cancel{\text{mol Al}}} = 58.3\ \text{g Al}$$

This mass is less than the mass of titanium.

Problem-Solving Practice 2.6

Calculate
(a) The number of moles in 1.00 mg of molybdenum (Mo).
(b) The number of grams in 5.00×10^{-3} mol of gold (Au).

Conceptual Challenge Problems CP-2.D and CP-2.E at the end of the chapter relate to topics covered in this section.

A fundamental physical property of matter is **density,** the ratio of an object's mass to its volume. Density is a distinctive physical property (← *p. 7*) that can be used to help identify a substance:

$$\text{Density} = \frac{\text{mass}}{\text{volume}}$$

Density is commonly expressed in grams per cubic centimeter (g/cm^3), which is the same as grams per milliliter (g/mL). Lithium metal, for example, has a density of 0.534 g/cm^3, making it the least dense ("lightest") metal. In contrast, the density of the metallic element osmium is 22.5 g/cm^3, making it one of the most dense ("heaviest") metals.

You can calculate the density from the mass and volume of a sample. Suppose that you need to determine the approximate density of mercury (Hg), the only metallic element that is a liquid at room temperature. You would first weigh a clean, dry, graduated cylinder and then add some mercury to it. Reading from the volume markings on the cylinder, you would obtain the volume of mercury in the cylinder; suppose it is 8.3 mL. You would then weigh the cylinder with the liquid mercury and subtract the weight of the cylinder from the total weight to find the weight of the mercury; suppose it is 113 g. You can now calculate the approximate density of mercury.

$$\text{Density of Hg} = \frac{113 \text{ g}}{8.3 \text{ mL}} = 14 \text{ g/mL}$$

The density of a substance is a conversion factor for finding mass from volume or volume from mass.

PROBLEM-SOLVING EXAMPLE 2.7 Density, Grams, and Moles

A student measured 25.0 mL of liquid bromine (Br_2) at 25 °C. The density of bromine at that temperature is 3.103 g/mL. How many moles of Br_2 are in the sample?

Answer 0.485 mol Br_2

Explanation In this case, the quantity of bromine is expressed as a volume (mL). In order to calculate moles, you first have to convert the volume to a mass. Density is the conversion factor for converting volume to mass.

$$\text{Mass of Br}_2 = 25.0 \text{ mL Br}_2 \times \frac{3.103 \text{ g Br}_2}{1 \text{ mL Br}_2} = 77.58 \text{ g Br}_2$$

The next step is to convert grams to moles of bromine using the molar mass of bromine, which is twice its atomic molar mass because bromine is diatomic: 2 × (79.904 g/mol) = 159.81 g/mol Br_2.

$$77.58 \text{ g Br}_2 \times \frac{1 \text{ mol Br}_2}{159.8 \text{ g Br}_2} = 0.485 \text{ mol Br}_2$$

An important note: In calculations with multiple steps, use one more significant figure in intermediate steps than the lowest number of significant figures given in the problem. This reduces premature rounding off that can introduce errors (Appendix A.3). Therefore, we kept four significant figures in the answer to the first step and used it in the second step (77.58 g Br_2, rather than 77.6 g Br_2; if we had used 77.6 g Br_2, the final answer would have been 0.486 mol Br_2). But, remember that you need to decrease the significant figures by one in the *final* answer (0.485 mol Br_2 in this example) to agree with the least number of significant figures in the problem (25.0 mL).

In the stepwise solution of this problem, we retained one more than the appropriate number of significant figures in the result of the first step (77.58 g Br_2). As described in Appendix A.3, rounding should be done at the end of a calculation. Had we rounded to 77.6 g Br_2 in the first step, an incorrect answer, 0.486 mol Br_2, would have been calculated.

Problem-Solving Practice 2.7

Calculate

(a) The mass, in grams, of 0.550 mol of magnesium.

(b) The number of moles of lead in a 200.0-cm^3 sample (Pb; density = 11.34 g/cm^3).

Exercise 2.11 Density and Moles

The metals osmium and iridium have the same density, 22.5 g/cm^3. Based on this information, a student concludes that equal masses of these two elements contain the same number of moles. Explain the error in his thinking.

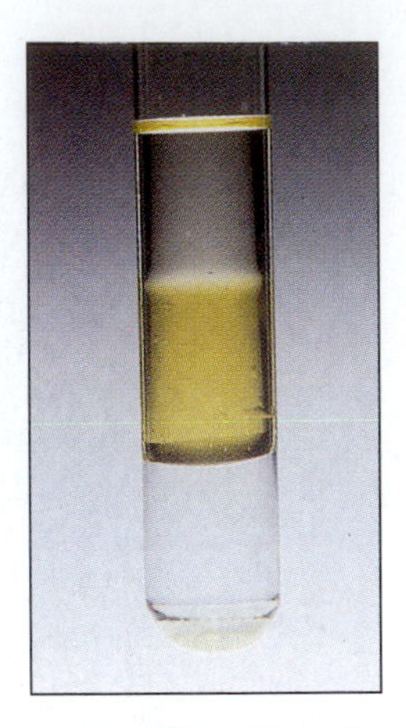

Liquid densities differ. Vegetable oil (middle layer), water (bottom layer), and kerosene (top layer) have different densities.
(C.D. Winters)

Exercise 2.12 Density of Liquids

When vegetable oil, water, and kerosene are put into a test tube, they form three layers, as given in the photo.
(a) Which is the most dense liquid? The least dense liquid?
(b) An additional 5.0 mL of vegetable oil is added to the container. Explain whether this addition causes a permanent change of which liquids are in the top, middle, or bottom layers, and if so, why the change occurs.

2.12 THE PERIODIC TABLE

We have already used the periodic table of the elements (Figure 2.18 and in the front of the book) to obtain the atomic numbers and atomic weights of elements. But it is much more valuable than this. Elements have physical properties such as mass, volume, and density, and chemical properties that describe how elements react when they combine to form compounds. The periodic table is an exceptionally useful tool in chemistry whose principal use is to organize and interrelate the chemical and physical properties of the elements. We can, for example, use the periodic table to classify elements as metals, nonmetals, or metalloids by their positions in the periodic table (Figure 2.18). You should become familiar with its main features and terminology. It also has an interesting history.*

Mendeleev and the Periodic Table

On the evening of February 17, 1869, at the University of St. Petersburg in Russia, a 35-year-old professor of general chemistry, Dmitri Ivanovitch Mendeleev, was writing a chapter of his soon-to-be-famous chemistry textbook, *Principles of Chemistry.* Puzzling over a way by which students could better understand and relate elemental properties to each other, he had written the properties of each known element on a separate card. While shuffling the cards trying to gather his thoughts before writing his manuscript, he realized that if the elements were arranged in order of increasing atomic weight, there were properties that repeated several times! That is, Mendeleev saw that there was a periodicity to the properties of elements (Figure 2.19). He summarized this in a table, the most complete version of which was published in 1871 (Figure 2.20).

He lined up the elements in horizontal rows in order of increasing atomic weight, starting a new row when he came to an element with properties similar to one already in the row. He aligned that element under the element with similar properties in the previous row. The resulting columns contained elements with similar properties.

The great impact of Mendeleev's periodic table went far beyond his textbook because he did something daringly creative. He realized that elements beyond calcium (Ca) would align properly only if he left empty spaces in certain places. He did so and boldly predicted that new elements would be discovered to occupy those spaces. For example, because arsenic forms compounds similar to those of phosphorus (P) and antimony (Sb), he put arsenic under phosphorus, not under aluminum, and next to zinc. In doing so, he left spaces between zinc and arsenic for two undiscovered elements.

Using the properties of the known elements above, below, before, and after an empty space in his table, Mendeleev predicted chemical and physical properties for the not-yet-discovered elements. The missing elements were soon discovered: gallium (Ga) in 1875 and germanium (Ge) in 1886.

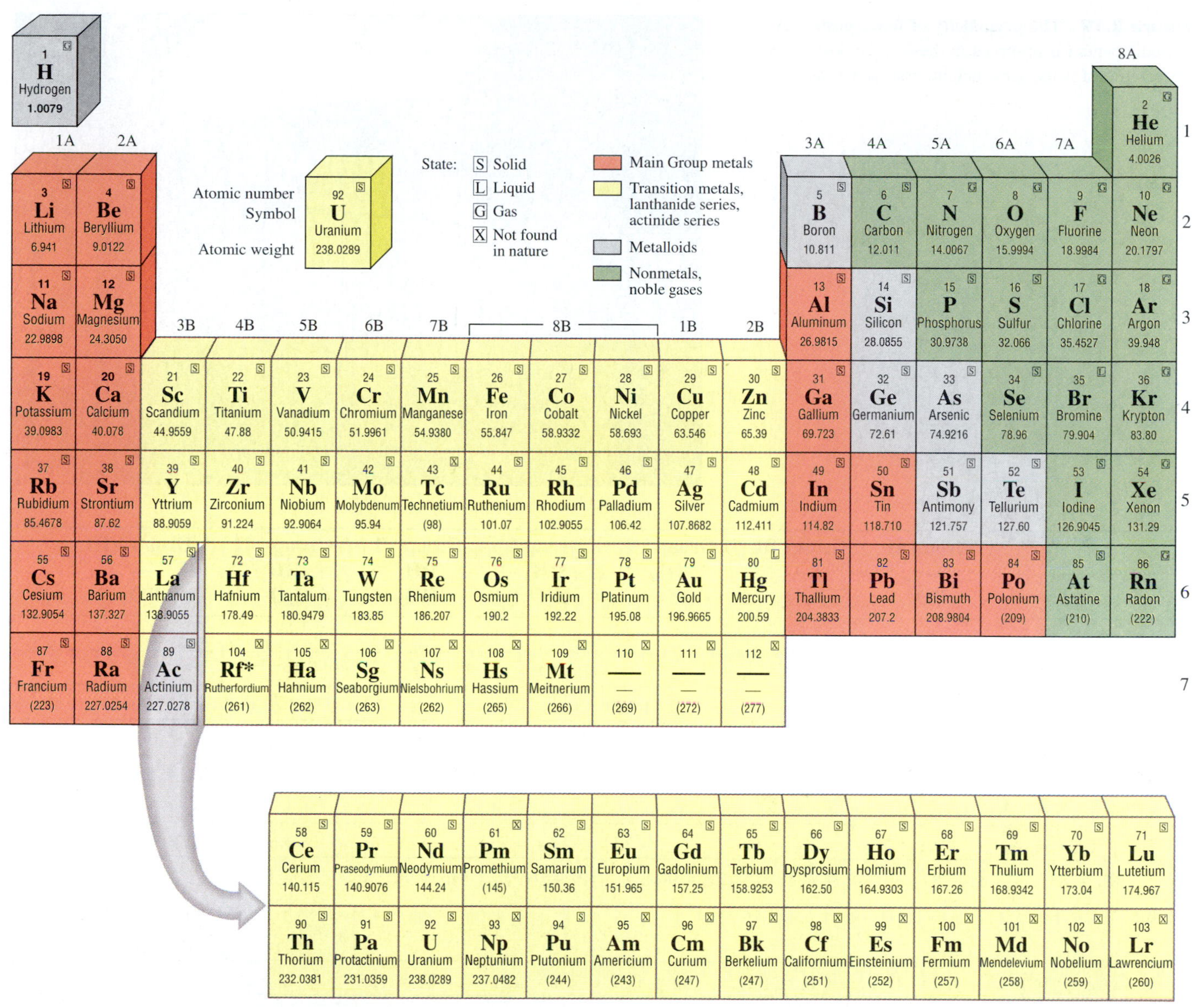

*The names of elements 104–109 are controversial. We will use the names recommended by their discoverers. Names of elements 110–112 have not been proposed, pending resolution of the controversy naming elements 104–109.

Figure 2.18 Modern periodic table of the elements. Elements are listed in ascending order of atomic number. The following points are important: (1) Metals are shown in red and yellow, the metalloids in gray, and the nonmetals in green. (2) Periods are horizontal rows of elements, and groups are vertical columns. (3) Groups are labeled by a number from 1 to 8 with a label of A (main group elements in red) or B (transition elements in yellow), the system used most commonly at present in the United States. The new international system is to number the groups from 1 to 18. (4) Some groups have common names: Group 1A = alkali metals; Group 2A = alkaline earth metals; Group 7A = halogens; Group 8A = noble gases.

(Text continues on page 63.)

Figure 2.19 The periodicity of fence posts. A periodic pattern is observed in these fence posts; every sixteenth post is the same height. *(George Semple)*

Row	Group I — R_2O	Group II — RO	Group III — R_2O_3	Group IV RH_4 RO_2	Group V RH_3 R_2O_5	Group VI RH_2 RO_3	Group VII RH R_2O_7	Group VIII — RO_4
1	H = 1							
2	Li = 7	Be = 9.4	B = 11	C = 12	N = 14	O = 16	F = 19	
3	Na = 23	Mg = 24	Al = 27.3	Si = 28	P = 31	S = 32	Cl = 35.5	
4	K = 39	Ca = 40	____ = 44	Ti = 48	V = 51	Cr = 52	Mn = 55	Fe = 56, Co = 59, Ni = 59, Cu = 63
5	(Cu = 63)	Zn = 65	____ = 68	____ = 72	As = 75	Se = 78	Br = 80	
6	Rb = 83	Sr = 87	?Yt = 88	Zr = 90	Nb = 94	Mo = 96	____ = 100	Ru = 104, Rh = 104, Pd = 106, Ag = 108
7	(Ag = 108)	Cd = 112	In = 113	Sn = 118	Sb = 122	Te = 125	I = 127	
8	Cs = 133	Ba = 137	?Di = 138	?Ce = 140				
9								
10			?Er = 178	?La = 180	Ta = 182	W = 184		Os = 195, Ir = 197, Pt = 198, Au = 199
11	(Au = 199)	Hg = 200	Tl = 204	Pb = 207	Bi = 208			
12				Th = 231		U = 240		

Figure 2.20 Mendeleev's periodic table. This English translation of Mendeleev's periodic table was published in 1871. The formulas for simple oxides, chlorides, and hydrides are shown under each group heading. R represents the element in each group. Mendeleev predicted several elements that were unknown by leaving empty spaces where the elements would fall, based on their predicted properties.

Portrait of a Scientist • *Dmitri Ivanovitch Mendeleev (1834–1907)*

Born in Tobolsk, Siberia, Mendeleev was educated in St. Petersburg where he lived virtually all his adult life. He taught at St. Petersburg University and while there wrote books and published his concept of chemical periodicity.

It is interesting that Mendeleev did little else with chemical periodicity after his initial articles. He went on to other interests, among them studying the natural resources of Russia and their commercial applications. In 1876 he visited the United States to study the fledgling oil industry and was much impressed with the industry, but not with the country. He found Americans uninterested in science, and he felt the country carried on the worst features of European civilization.

By the end of the 19th century, political unrest was growing in Russia, and Mendeleev lost his position at the university. He was appointed Chief of the Chamber of Weights and Measures for Russia, however, and established an inspection system for guaranteeing the honesty of weights and measures used in Russian commerce.

All pictures of Mendeleev show him with long hair. He made it a rule to cut his hair only once a year, in the spring, regardless of whether he had to appear at an important function.

Dmitri Ivanovitch Mendeleev. *(Oesper Collection in the History of Chemistry, University of Cincinnati)*

Do we have Mendeleev to thank for computers? This may seem like a strange question, but to some degree we do. Mendeleev's remarkably accurate predictions led to the discovery of germanium, a mainstay in the computer and electronics industry. He even suggested the geological regions in which minerals containing the elements could be found. Mendeleev's prediction of the properties of ekasilicon is shown in Table 2.6. The term *eka* comes from Sanskrit, and means "one"; thus, *ekasilicon* means "one place away from silicon," the position eventually filled by germanium. He also

TABLE 2.6 Some of Mendeleev's Predicted Properties of Ekasilicon and the Corresponding Observed Properties of Germanium

Property	Predicted Properties of Ekasilicon	Observed Properties of Germanium (Ge)
Atomic weight	72	72.6
Color of element	Gray	Gray
Density of element (g/mL)	5.5	5.36
Formula of oxide	EsO_2	GeO_2
Density of oxide (g/mL)	4.7	4.228
Formula of chloride	$EsCl_4$	$GeCl_4$
Density of chloride	1.9	1.884
Boiling point of chloride (°C)	< 100	84

predicted the properties of ekaboron (scandium) and ekaaluminum (gallium). His successful predictions of the missing elements' properties and the discovery of the elements stimulated a flurry of prospecting for other new elements in the 1870s and 1880s. In that time ten new elements, more than the number of elements known to the ancients, were discovered.

It gives a useful perspective to realize that Mendeleev did this long before electrons, protons, and nuclei were known.

Periodicity and Atomic Number

The "periodicity" of the elements is a recurrence of similar properties at regular intervals when the elements are arranged in the correct order.

Mendeleev used atomic weight as the basis for elements' periodicity in spite of the fact that, based on their properties, he placed some element pairs such as cobalt (Co)–nickel (Ni) and tellurium (Te)–iodine (I) in reverse order in terms of increasing atomic weight. This seeming inconsistency could not be adequately explained by Mendeleev, but provided a clue to something even more basic. In spite of his successes with the table, periodicity of the elements is not based on atomic weight. It depends on something more fundamental, a property unique to each element. This was discovered in 1913 by H.G.J. Moseley, a young physicist working with Ernest Rutherford.

The wavelength of an x ray, or of any other type of radiation, is the distance between two crests or two troughs of the wave. See Section 8.1.

Moseley bombarded many different metals with cathode-ray electrons and found that the wavelengths of x rays emitted from a specific element are related precisely to the atomic number of that element, its number of protons. He quickly realized that other atomic properties may be similarly related to atomic number and not, as Mendeleev had believed, to atomic weight, thereby removing the inconsistencies in the Mendeleev table. Arranging the elements in order of increasing atomic number gives the **law of chemical periodicity:** *The properties of the elements are periodic functions of atomic number.*

Periodic Table Features

Elements in the **periodic table** are arranged according to atomic number so that elements with *similar chemical properties occur in vertical columns* called **groups.** The table commonly used in the United States has groups numbered 1 through 8, with each number followed by a letter A or B. The A groups (Groups 1A and 2A at the left of the table and Groups 3A-8 at the right) are collectively known as **main group elements;** the B groups (in the middle between the A groups) are called **transition elements.** There is a movement to adopt a new set of group designations as an international standard; in this designation the groups are simply numbered 1 through 18 from left to right. We will use the "A/B table" in this textbook.

The horizontal rows of the table are called **periods,** and they are numbered beginning with 1 for the period containing only H and He. For example, sodium (Na), is in Group 1A and is the first element in the third period. Silver (Ag) is in Group 2B and in the fifth period.

The table helps us recognize that most elements are metals (red and yellow in Figure 2.18 and in the front of the book), far fewer are nonmetals (green), and even fewer are metalloids (gray) (⇐ ***p. 32***). Elements gradually become less metallic from left to right across a period; eventually one or more nonmetals are reached. The six metalloids (B, Si, Ge, As, Sb, and Te) fall along a zigzag line passing between aluminum and silicon, germanium and arsenic, antimony and tellurium. Photographs of a number of elements are given in Figures 2.1, 2.2, and 2.3. Certain regions of the periodic table are identified by distinctive names. Knowing these names and the general properties of the elements in these regions is very helpful.

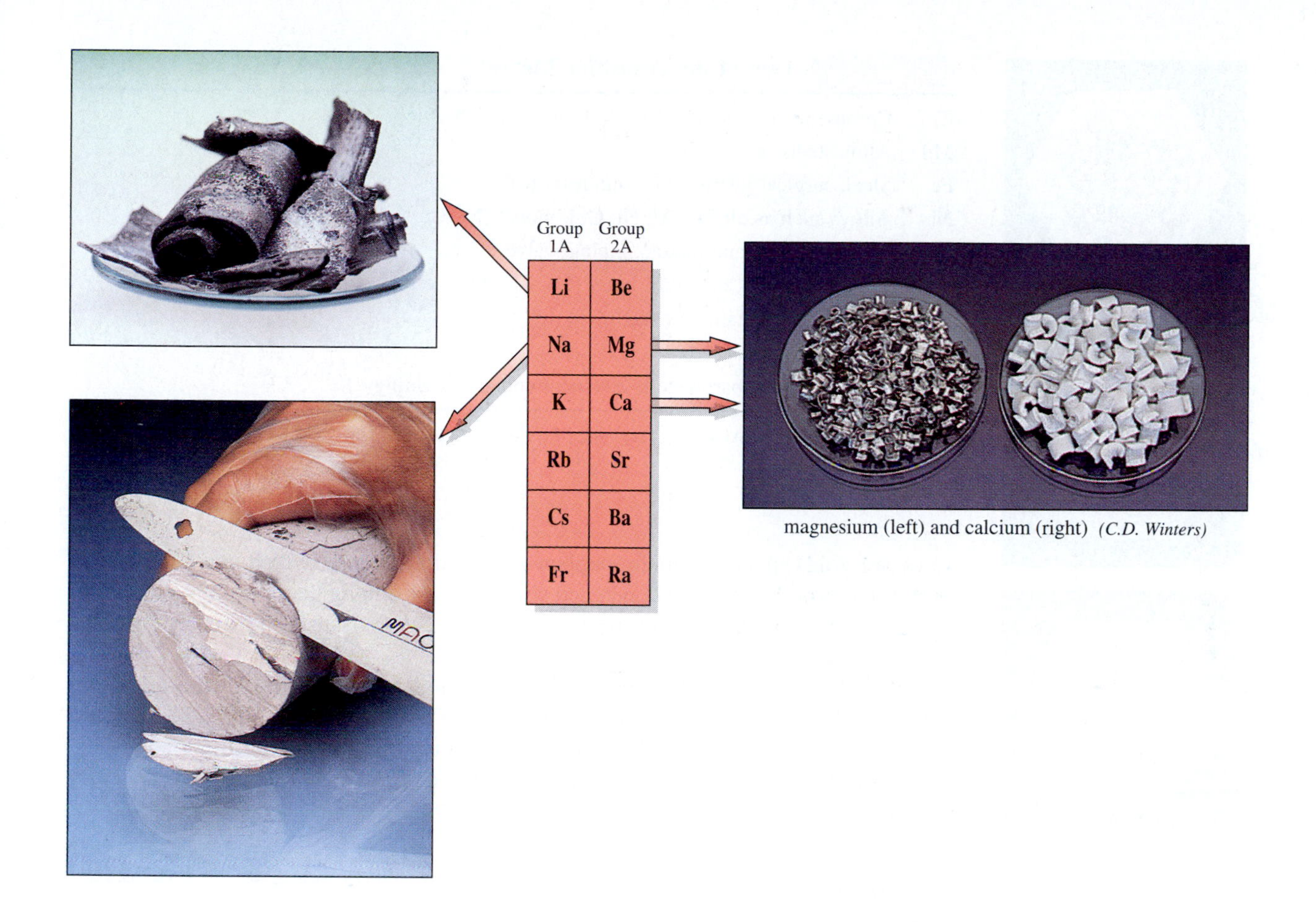

magnesium (left) and calcium (right) *(C.D. Winters)*

The Alkali Metals (Group 1A) and Alkaline Earth Metals (Group 2A)

Elements in the leftmost column (Group 1A) are called the **alkali metals** (except hydrogen) because their water solutions are alkaline (basic). Elements in Group 2A, known as **alkaline earth metals,** are extracted from minerals (earths) and also produce alkaline aqueous solutions (except beryllium).

"Alkali" comes from the Arabic language. Ancient Arabian chemists discovered that ashes of certain plants, which they called al-qali, gave water solutions that felt slippery and burned the skin.

Alkali metals and alkaline earth metals are very reactive and, consequently, are found in nature only combined in compounds, never as the metallic element. Their compounds are plentiful and many are significant to human and plant life. Sodium (Na) in sodium chloride (table salt) is a fundamental part of human and animal diets, and civilizations quested for salt as a dietary necessity and commercial commodity. Today, sodium chloride is commercially important as a source of two of the top ten industrial chemicals—sodium hydroxide and chlorine (see table of top 25 chemicals, inside back cover). Magnesium (Mg) and calcium (Ca), the sixth and fifth most abundant elements in the earth's crust, respectively, are present in a vast array of compounds, never as the uncombined metal, because of their reactivity.

Elements found in nature uncombined are sometimes called "free" elements.

Gold and silver as free metals in nature triggered the great gold and silver "rushes" of the 1800s in the United States. The original "49ers" weren't football players, but people who rushed to prospect for gold after the 1849 discovery of gold at Sutter's Mill in California, near San Francisco.

Transition Elements, Lanthanides, and Actinides

The transition elements fill the middle of the periodic table in periods 4, 5, 6, and 7. All are metals, and most are found only in compounds in nature. The notable excep-

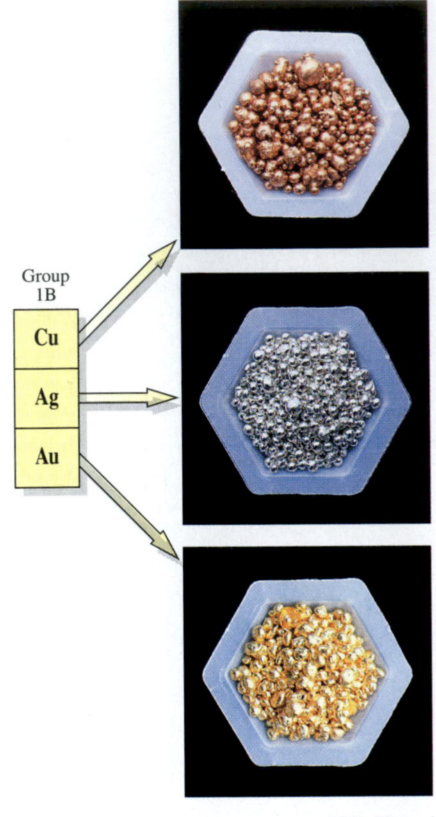

(C.D. Winters)

TABLE 2.7 Uses of the Transition Metals

Cr	Chrome plate, stainless steel (Cr, Ni, Fe), nichrome (Ni, Cr)
Mn	Alloy steels (rails, safes)
Fe	Steel, surgical instruments, stainless steel
Ni	Alloys such as alnico (Al, Ni, Co), monel (Ni, Cu, Fe), coinage (Ni, Cu)
Cu	Electric wiring and motors, piping, valves
Zn	Galvanized iron, dry cells, brass (Cu, Zn), bronze (Cu, Zn, Pb)
Ag	Jewelry, photography as AgBr
Au	Jewelry, gold leaf
Hg	Thermometers, barometers, electrical contacts, lighting

W.L. Masterton, E.J. Slowinski, and C.L. Stanitski: *Chemical Principles,* 6th ed., p. 750. Philadelphia, Saunders College Publishing, 1985.

tions are gold, silver, platinum, copper, and liquid mercury, which can be found in elemental form. Iron, zinc, copper, and chromium are among the most important commercial metals (Table 2.7). Because of their vivid colors, transition metal compounds are used for pigments.

Listed in two rows at the bottom of the periodic table are the **lanthanides** (beginning with the element lanthanum [La]) and the **actinides** (beginning with actinium [Ac]). These elements are relatively rare and not as commercially important as the earlier transition elements, although some lanthanides are used in color television picture tubes.

Group 3A to 6A

These four groups have no special name, but contain the most abundant elements in the earth's crust and atmosphere (Table 2.8).

Groups 4A to 6A begin with one or more nonmetal elements, followed by one or two metalloids, and end with a metallic element. Group 4A, for example, contains carbon, a nonmetal, then two metalloids—silicon (Si) and germanium (Ge)—and finishes with two metals, tin (Sn) and lead (Pb). Group 3A starts with boron (B), a metalloid, rather than a metal or nonmetal, indicating Group 3A as the shift from metallic to nonmetallic character.

TABLE 2.8 Selected Group 3A–6A Elements

Group 3A	Group 4A	Group 5A	Group 6A
Aluminum: Most abundant *metal* in earth's crust (7.4%). Always found combined naturally, especially with silicon and oxygen (clay minerals).	*Silicon:* Second most abundant element in earth's crust (25.7%). Always found combined naturally, generally with oxygen in quartz and silicate minerals.	*Nitrogen:* Most abundant element in earth's *atmosphere* (78.1%), but not abundant in earth's crust because of relatively low chemical reactivity.	*Oxygen:* Most abundant element in earth's crust (49.5%) because of high chemical reactivity; second most abundant element in earth's atmosphere (21.0%).

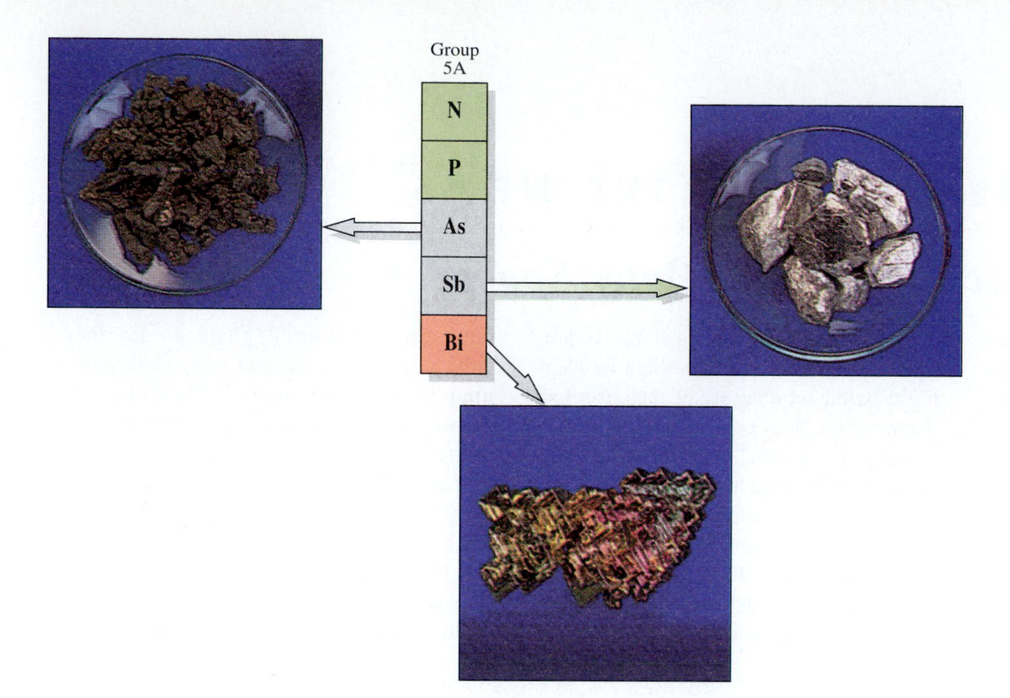

(C.D. Winters)

The Halogens, Group 7A

The elements in this group consist of diatomic molecules and are highly reactive. The group name, **halogens,** comes from the Greek words *hals,* meaning "salt," and *genes,* meaning "forming." The halogens all form salts—compounds similar to sodium chloride (NaCl)—by reacting vigorously with alkali metals, and with other metals as well. They also react with most nonmetals. Carbon compounds containing chlorine and fluorine are very unreactive. As you have seen (← *p. 3*), the chlorofluorocarbons (CFCs) are responsible for the loss of stratospheric ozone and the seasonal ozone "hole" in the southern hemisphere.

The Noble Gases, Group 8A

The **noble gases** at the far right of the periodic table are the least reactive elements, hence the term "noble." Their lack of chemical activity, as well as their rarity, prevented them from being discovered until just about a century ago. Thus, they were not known when Mendeleev developed his periodic table. Until 1962 they were called the *inert* gases, because they were thought not to combine with any element, that is, not to form compounds. In 1962, this basic canon of chemistry was overturned when compounds of xenon with fluorine and with oxygen were synthesized. Since then other xenon compounds have been made, as well as compounds of fluorine with krypton and with radon.

Once the arrangement of electrons in atoms was understood (Section 8.5), the place where the noble gases fit into the periodic table was obvious.

Exercise 2.13 The Periodic Table

1. How many (a) metals; (b) nonmetals; and (c) metalloids are in the fourth period of the periodic table? Give the name and symbol in each case.
2. Which groups of the periodic table contain (a) only metals; (b) only nonmetals; (c) only metalloids?
3. Which period contains the most metals?

CHEMISTRY IN THE NEWS

Buckyballs from Outer Space?

Nearly two billion years ago a meteorite crashed into earth near Sudbury, Ontario. Recently, scientists from the Scripps Institute of Oceanography in La Jolla, California, and the University of Rochester studying the meteor's crater found helium trapped inside C_{60} and C_{70} buckyballs there. This discovery raises two interesting questions. Where did the buckyballs originate? Do the helium isotopes shed light on the origin of the buckyballs?

Two possible scenarios for helium-filled buckyball formation could be identified based on analysis of their isotopic composition. They could have formed during impact, in which case the buckyballs would have trapped helium having the relative amounts of helium-3 and helium-4 in the earth's atmosphere at that time. A second scenario is that the buckyballs, already combined with helium from outer space, were in the meteor as it traveled through outer space before it plunged into the earth's atmosphere, and they survived upon impact. If so, the He-3/He-4 ratio of the trapped helium would resemble the ratio found in other meteorites and interplanetary dust particles, a much different He-3/He-4 ratio than that found in the earth's atmosphere. Dr. Robert Podera of the University of Rochester says, "They [the buckyballs] would then have an isotopic signature of helium and neon and ar-

(C.D. Winters)

gon that would be consistent with formation in the cosmos." The He-3/He-4 ratio in the Sudbury fullerenes isconsistent with helium isotope composition of *extraterrestrial* origins, implying that the helium-packed buckyballs survived the turbulent meteor ride and impact.

Although fullerenes are temperature-resistant, they are sensitive to reaction with oxygen, and some scientists doubt that the C_{60} and C_{70} molecules would have survived entry into the earth's current atmosphere. But the earth's atmosphere had far less oxygen two billion years ago, a situation that would have favored the helium-laden buckyballs surviving the crash. To obtain possible further evidence of extraterrestrial buckyball formation, further research is needed to study isotopic ratios among other noble gases trapped in the buckyballs—neon and argon—to compare them with those of terrestrial origin.

Source: Chemical & Engineering News, *April 15, p. 6, 1996;* Science, *Vol. 272, p. 249, 1996.*

Summary Problem

An atom of the element pictured in the opening photo contains 51 protons and 70 neutrons.

(a) Identify the element and give its symbol.
(b) What is this atom's (i) atomic number? (ii) mass number?
(c) This element has two naturally occurring isotopes.

Isotope	Mass Number	Percent Abundance	Isotopic Mass (amu)
1	121	57.25	120.9038
2	123	42.75	122.9041

Calculate the atomic weight of the element.

(d) This element has a density of 6.69 g/cm^3. Calculate the volume of a 43.5-g sample of the element.

(e) This element is in which group of the periodic table? Is this element a metal, a metalloid, or a nonmetal? Explain your choice.

(f) A 357-caliber bullet contains 1.07×10^{-6} g of the element. (i) How many moles of the element are in this mass? (ii) How many atoms of the element are in the sample? (iii) Atoms of this element have an atomic diameter of 282 pm. If all the atoms of this element in the sample were laid side by side, how many meters would the chain of atoms be?

In Closing

Having studied this chapter, you should be able to . . .

- explain the historical development of the atomic theory and identify some of the scientists who made important contributions (Sections 2.1–2.2).

- describe in general the discovery and naming of elements, define allotropes, and locate metals, nonmetals, and metalloids on the periodic table (Section 2.3 and 2.12).
- describe radioactivity, electrons, protons, and neutrons, and the general structure of the atom (Sections 2.4–2.6).
- use conversion factors for the units for mass, volume, and length common in chemistry (Sections 2.7).
- define isotope and give the mass number and number of neutrons for a specific isotope (Section 2.8).
- calculate the atomic weight of an element from isotopic abundances (Section 2.9).
- explain the difference between the atomic number and atomic weight of an element and find this information for any element (Sections 2.8–2.9).
- relate amounts of substances to the mole, Avogadro's number, and molar mass (Section 2.10).
- do mass–mole conversions for elements (Section 2.11).
- identify the periodic table location of groups, periods, alkali metals, alkaline earth metals, halogens, noble gases, transition elements, lanthanides, and actinides (Section 2.12).

Key Terms

The following terms were defined and given in boldface type in this chapter. You should be sure to understand each of these terms and the concepts with which they are associated.

actinides *(2.12)*
alkali metals *(2.12)*
alkaline earth metals *(2.12)*
allotropes *(2.3)*
atom *(2.2)*
atomic mass units (amu) *(2.8)*
atomic number *(2.8)*
atomic structure *(2.4)*
atomic weight *(2.9)*
Avogadro's number *(2.10)*
conversion factor *(2.7)*
density *(2.11)*
diatomic *(2.3)*
electron *(2.5)*
group *(2.12)*
halogens *(2.12)*
ion *(2.5)*
isotopes *(2.8)*
lanthanides *(2.12)*
law of chemical periodicity *(2.12)*
law of conservation of matter *(2.2)*
law of constant composition *(2.2)*
main group elements *(2.12)*
mass *(2.7)*
mass number *(2.8)*
metal *(2.3)*
metalloid *(2.3)*
metric system *(2.7)*
molar mass *(2.10)*
mole (mol) *(2.10)*
molecule *(2.3)*
neutron *(2.5)*
noble gases *(2.12)*
nonmetal *(2.3)*
nucleus *(2.6)*
percent abundance *(2.9)*
period *(2.12)*
periodic table *(2.12)*
proton *(2.5)*
radioactivity *(2.4)*
transition elements *(2.12)*

Questions for Review and Thought

Conceptual Challenge Problems

CP-2.A. Suppose that you are faced with a problem similar to the one faced by Robert Millikan when he analyzed data from his oil-drop experiment. Below are the masses of three stacks of dimes. What do you conclude to be the mass of a dime and what is your argument?

Stack 1 = 9.12 g   stack 2 = 15.96 g   stack 3 = 27.36 g

CP-2.B. In Section 2.1 Jacob Bronowski was quoted in observing how gifted we are because we "can draw conclusions from what we see to what we do not see." Suppose you knew that the mass of osmium atoms is 27 times greater than the mass of lithium atoms, but the volume of lithium atoms is 1.4 times larger than that for osmium atoms. What gifted conclusions can you draw and defend concerning something "we do not see": how atoms are constructed?

CP-2.C. The age of the universe is unknown, but some conclude from measuring Hubble's constant that it is about 18 billion years old, which is about four times the age of the Earth. If so, what is the age of the universe in seconds? If you had a sample of carbon with the same number of carbon atoms as there have been seconds since the universe began, could you measure this sample on a laboratory balance that can detect masses as small as 0.1 mg?

CP-2.D. In Section 2.11 you read that osmium has a density of 22.5 g/cm^3. This is 42 times greater than the density of lithium, the least dense metal, which has a density of 0.534 g/cm^3. By using these density data and the molar masses of these two metals, how would you present a valid argument to show which of these two metals has the larger atoms? How much larger are they than those of the other element?

CP-2.E. In Section 2.7 you were told that atoms have radii between 30 and 300 pm. Based upon the argument that you presented to Question CP-2.D, what would you predict to be the radii of osmium and lithium atoms?

Review Questions

1. Given each of John Dalton's original postulates below, decide whether each postulate is still true based on our present views of atomic theory. For the postulates that are not true, revise each so that the postulate adheres to present views about atomic theory.

   - All matter is made of atoms. These indivisible and indestructible objects are the ultimate chemical particles.
   - All atoms of a given element are identical, both in mass and in properties. Atoms of different elements have different masses and different properties.
   - Compounds are formed by combination of two or more different kinds of atoms. Atoms combine in the ratio of small whole numbers, for example, one atom of A with one atom of B, or two atoms of A with one atom of B.
   - Atoms are the units of chemical change. A chemical reaction involves only combination, separation, or rearrangement of atoms; but atoms are not created, destroyed, divided into parts, or converted into other kinds of atoms during a chemical reaction.

2. Two phenomena played an important role in the discovery of atomic structure.
   (a) The first, ____________, was investigated by Benjamin Franklin using such experiments as the infamous key on a kitestring in a thunderstorm. This particular phenomenon consists of two parts which are *conserved,* meaning you cannot find one part without finding one of its counterpart. These two parts are a ____________ charge and a ____________ charge. How is this phenomenon similar to the structure of an atom?
   (b) The second, ____________, was investigated by many scientists including Marie Curie, who named it; but Henri Becquerel discovered it. Three types of this phenomenon can be observed coming from some elements. The types are ____________, ____________, and ____________. What are the respective charges on each type of radiation? How does each type behave when passing through electrically charged plates?
3. (a) List the properties exhibited by cathode rays.
   (b) What is the charge on the particles making up cathode rays?
   (c) Which subatomic particles are cathode rays most like?
4. Thomson was able to manipulate cathode rays using both electric and magnetic fields to determine the ____________-to-____________ ratio for an electron. What is the fundamental unit for charge?
5. Millikan was able to determine the charge on an electron using his famous oil-drop experiment. Describe the experiment and how Millikan was able to calculate the charge on an electron using his results and the ratio discovered earlier by Thomson.
6. Cathode rays and beta particles consist of which subatomic particle?
7. Canal rays were used to discover the evidence needed to prove the existence of positive ions. Explain why hydrogen has a smaller mass-to-charge ratio than helium or neon. This result is exactly opposite from that obtained when using the cathode-ray tubes for examining electrons; all the electrons had the same mass-to-charge ratio. What do these two pieces of evidence tell us about the relative sizes of the negatively charged particles and the positively charged particles in atoms?
8. The positively charged particle in an atom is called the proton.
   (a) How much heavier is a proton than an electron?
   (b) What is the difference in the charge on a proton and an electron?
9. Most neutral atoms have masses greater than the combined masses of the protons and electrons in the atom. The third subatomic particle eluded scientists because it is uncharged. Explain how Chadwick accounted for the existence of neutrons as well as for their mass.
10. By the early 1900s, J.J. Thomson had described an atom as a bread pudding with raisins as the electrons. Ernest Rutherford's famous gold-foil experiment refuted this model.
    (a) Explain what the results from the gold-foil experiment should have been if Thomson were correct about his bread pudding analogy of an atom.
    (b) The results of the gold-foil experiment enabled Rutherford to calculate that the nucleus was much smaller than the atom. How much smaller?
11. In any given *neutral* atom, how many protons are there compared with the number of electrons?

Units and Unit Conversions

12. If the nucleus of an atom were the size of a medium-sized orange (let us say with a diameter of about 6 cm), what would be the diameter of the atom?

13. The average lead pencil, new and unused, is 19 cm long. What is its length in millimeters? In meters? In inches?

14. An excellent height in the pole vault is 18 feet, 11.5 inches. What is this height in meters?

15. The maximum speed limit in many states is 65 miles per hour. What is this speed in kilometers per hour?

16. A sailboat has a length of 36 feet 7 inches and is 12 feet at its widest point. What are these distances in meters? In centimeters?

17. A Volkswagen engine has a displacement of 120 cubic inches. What is this volume in cubic centimeters? In liters?

18. An automobile engine has a displacement of 250. in^3. What is this volume in cubic centimeters? In liters?

19. Suppose your bedroom is 18 ft long, 15 ft wide, and the distance from floor to ceiling is 8 ft, 6 in. You need to know the volume of the room in metric units for some scientific calculations. What is the room's volume in cubic meters? In liters?

20. Ethylene glycol is a liquid (density = 1.1135 g/cm^3 at 20 °C) that is the base of the antifreeze you use in the radiator of your car. If you need 500 mL of this liquid, what mass of it, in grams, is required?

21. A piece of silver metal has a mass of 2.365 g. If the density of silver is 10.5 g/cm^3, what is the volume of the silver?

22. The "cup" is a volume widely used by cooks in the United States. One cup is equivalent to 225 mL. If 1 c of olive oil has a mass of 205 g, what is the density of the oil?

23. Peanut oil has a density of 0.92 g/cm^3. If a recipe calls for 1c of peanut oil (1c = 225 mL), what mass of peanut oil (in grams) are you using?

24. A 37.5-g sample of unknown metal placed in a graduated cylinder containing water caused the water level to rise 13.9 mL. Which metal listed below is most likely the sample? (*d* is density.)
(a) Mg, $d = 1.74$ g/cm^3
(b) Fe, $d = 7.87$ g/cm^3
(c) Ag, $d = 10.5$ g/cm^3
(d) Al, $d = 2.70$ g/cm^3
(e) Cu, $d = 8.96$ g/cm^3
(f) Pb, $d = 11.3$ g/cm^3

25. A crystal of fluorite (a mineral that contains calcium and fluorine) has a mass of 2.83 g. What is this mass in kilograms? In pounds? Give the symbols for the elements in this crystal.

Percent

26. Silver jewelry is actually a mixture of silver and copper. If a bracelet with a mass of 17.6 g contains 14.1 g of silver, what is the percentage of silver? Of copper?

27. The solder once used by plumbers to fasten copper pipes together consists of 67% lead and 33% tin. What is the mass of lead (in grams) in a 1.00-lb block of solder? What is the mass of tin?

28. Automobile batteries are filled with sulfuric acid. What is the mass of the acid (in grams) in 500. mL of the battery acid solution if the density of the solution is 1.285 g/cm^3 and the solution is 38.08% sulfuric acid by mass?

Isotopes

29. What is wrong with the following statement?
Atoms of the same element always have the same mass number but atoms of the same element can have different atomic numbers due to the presence of *isotopes.*

30. Americium-241 is used in household smoke detectors and in bone-mineral analysis. Give the number of electrons, protons, and neutrons in an atom of americium-241.

31. The artificial radioactive element technetium is used in many medical studies. Give the number of electrons, protons, and neutrons in an atom of technetium-99.

32. Atoms of the same element have the same number of protons in the nucleus and, therefore, all the atoms for any given element would have the same atomic ________.

33. What is the definition of the atomic mass unit?

34. To estimate an atom's mass, one only has to add up the number of protons and neutrons in the nucleus. This estimate, a whole number, is referred to as the ____________ number for a given atom. Why can we essentially ignore the mass of the electrons?

35. What is the difference between the mass number and the atomic number of an atom?

36. When you subtract the atomic number from the mass number for an atom, what do you obtain?

37. Give the mass number of each of the following atoms: (a) beryllium with 5 neutrons, (b) titanium with 26 neutrons, and (c) gallium with 39 neutrons.

38. Give the mass number of (a) an iron atom with 30 neutrons, (b) an americium atom with 148 neutrons, and (c) a tungsten atom with 110 neutrons.

39. Give the complete symbol ${}^{A}_{Z}X$ for each of the following atoms: (a) sodium with 12 neutrons; (b) argon with 21 neutrons; and (c) gallium with 38 neutrons.

40. Give the complete symbol ${}^{A}_{Z}X$ for each of the following atoms: (a) nitrogen with 8 neutrons; (b) zinc with 34 neutrons; and (c) xenon with 75 neutrons.

41. How many electrons, protons, and neutrons are there in an atom of (a) calcium-40, ${}^{40}_{20}Ca$; (b) tin-119, ${}^{119}_{50}Sn$; and (c) plutonium-244, ${}^{244}_{94}Pu$?

42. How many electrons, protons, and neutrons are there in an atom of (a) carbon-13, ${}^{13}_{6}C$; (b) chromium-50, ${}^{50}_{24}Cr$; and (c) bismuth-205, ${}^{205}_{83}Bi$?

43. Fill in the following table:

Z	A	Number of Neutrons	Element
35	81	________	________
________	________	62	Pd
77	________	115	________
________	151	________	Eu

44. Fill in the following table:

Z	A	Number of Neutrons	Element
60	144	______	______
______	______	12	Mg
64	______	94	______
______	37	______	Cl

45. Which of the following are isotopes of element *X*, whose atomic number is 9: ${}^{18}_{9}X$, ${}^{20}_{9}X$, ${}^{9}_{4}X$, ${}^{15}_{9}X$?

46. Cobalt has three radioactive isotopes used in medical studies. Atoms of these isotopes have 30, 31, and 33 neutrons, respectively. Give the symbol for each of these isotopes.

Atomic Weight

47. Verify that the atomic weight of lithium is 6.941 amu, given the following information:

^{6}Li, exact mass = 6.015121 amu
percent abundance = 7.500%
^{7}Li, exact mass = 7.016003 amu
percent abundance = 92.50%

48. Verify that the atomic weight of magnesium is 24.3050 amu, given the following information:

^{24}Mg, exact mass = 23.985042 amu
percent abundance = 78.99%
^{25}Mg, exact mass = 24.98537 amu
percent abundance = 10.00%
^{26}Mg, exact mass = 25.982593 amu
percent abundance = 11.01%

49. Gallium has two naturally occurring isotopes, ^{69}Ga and ^{71}Ga, with masses of 68.9257 amu and 70.9249 amu, respectively. Calculate the abundances of these isotopes of gallium.

50. Copper has two stable isotopes, ^{63}Cu and ^{65}Cu, with masses of 62.939598 amu and 64.927793 amu, respectively. Calculate the abundances of these isotopes of copper.

51. Lithium has two stable isotopes, ^{6}Li and ^{7}Li. Knowing that the atomic weight of lithium is 6.941, which is the more abundant isotope?

52. Argon has three naturally occurring isotopes: 0.337% ^{36}Ar, 0.063% ^{38}Ar, and 99.60% ^{40}Ar. What would you estimate the atomic weight of argon to be? If the masses of the isotopes are 35.968, 37.963, and 39.962, respectively, what is the atomic weight of natural argon?

The Mole

53. The "mole" is simply a convenient unit for counting molecules and atoms. Name four "counting units" (such as a dozen for eggs and cookies) that you commonly encounter.

54. If you divide Avogadro's number of pennies among the 250 million men, women, and children in the United States, and if each person could count one penny each second every day of the year for 8 hours a day, how long would it take to count all of the pennies? How long would it take you to get bored counting pennies?

55. Why do you think it is more convenient to use some chemical counting unit when doing calculations (chemists have adopted the unit of the mole, but it could have been something different) rather than using individual molecules?

56. Calculate the number of grams in
(a) 2.5 mol of boron
(b) 0.015 mol of oxygen
(c) 1.25×10^{-3} mol of iron
(d) 653 mol of helium

57. Calculate the number of grams in
(a) 6.03 mol of gold
(b) 0.045 mol of uranium
(c) 15.6 mol of Ne
(d) 3.63×10^{-4} mol of plutonium

58. Calculate the number of moles represented by each of the following:
(a) 127.08 g of Cu
(b) 20.0 g of calcium
(c) 16.75 g of Al
(d) 0.012 g of potassium
(e) 5.0 mg of americium

59. Calculate the number of moles represented by each of the following:
(a) 16.0 g of Na
(b) 0.0034 g of platinum
(c) 1.54 g of P
(d) 0.876 g of arsenic
(e) 0.983 g of Xe

60. A piece of sodium metal, Na, if put into a bucket of water, produces a dangerously violent explosion from the reaction of sodium with water. If the piece contains 50.4 g of sodium, how many moles of sodium does that represent?

61. Krypton really does not give Superman his strength. If you have 0.00789 g of the gaseous element, how many moles does this represent?

62. A certain brand of gasoline has a density of 0.67 g/cm^3. If you have 12 gallons of this gasoline in your car's gas tank, how many pounds of gasoline are in the tank?

63. In an experiment, you need 0.125 mol of sodium metal. Sodium can be cut easily with a knife, so if you cut out a block of sodium, what should be the volume of the block in cubic centimeters? If you cut a perfect cube, what will be the length of the edge of the cube? (The density of sodium is 0.968 g/cm^3). (*Caution:* Sodium is *very* reactive with water. The metal should be handled only by a knowledgeable chemist.)

64. If you have a 35.67-g piece of chromium metal on your car, how many atoms of chromium do you have?

65. If you have a ring that contains 1.94 g of gold, how many atoms of gold are in the ring?

66. What is the average mass of one copper atom?

67. What is the average mass of one atom of titanium?

The Periodic Table

68. What was incorrect about Mendeleev's original concept of the periodic table? What is the modern "law of chemical periodicity" and how is this related to Mendeleev's ideas?
69. What is the difference between a group and a period in the periodic table?
70. Name and give symbols for (a) three elements that are metals; (b) four elements that are nonmetals; and (c) two elements that are metalloids. In each case, also locate the element in the periodic table by giving the group and period in which the element is found.
71. Name and give symbols for three transition metals in the fourth period. Look up each of your choices in a dictionary and make a list of the properties that the dictionary entry gives. Also list the uses of the element as given by the dictionary.
72. Name two halogens. Look up each of your choices in a dictionary (or a book such as *The Handbook of Chemistry and Physics*) and make a list of the properties that the dictionary entry gives. Also list any uses of the element that are given by the dictionary.
73. Name three transition elements, two halogens, and one alkali metal.
74. Name an alkali metal, an alkaline earth metal, and a halogen.
75. Name an element discovered by Madame Curie. Give its name, symbol, and atomic number. Use a dictionary to find the origin of the name of this element.
76. How many elements are there in Group 4A of the periodic table? Give the name and symbol of each of these elements. Tell whether each is a metal, nonmetal, or metalloid.
77. How many elements are there in the fourth period of the periodic table? Give the name and symbol of each of these elements. Tell whether each is a metal, metalloid, or nonmetal.
78. Which single period in the periodic table contains the most metals, metalloids, *and* nonmetals?
79. How many periods of the periodic table have 8 elements, how many have 18 elements, and how many have 32 elements?
80. The chart at the right is a plot of the logarithm of the relative abundances of elements 1 through 36 in the solar system. The abundances are given on a scale that gives silicon a relative abundance of 1.00×10^6 the logarithm of which is 6.
    (a) What is the most abundant metal?
    (b) What is the most abundant nonmetal?
    (c) What is the most abundant metalloid?
    (d) Which of the transition elements is most abundant?
    (e) How many halogens are considered on this plot and which is the most abundant?

81. Again consider the plot of relative abundance versus atomic number. Can you uncover any relation between abundance and atomic number? Is there any difference between elements of even atomic number and those of odd atomic number?

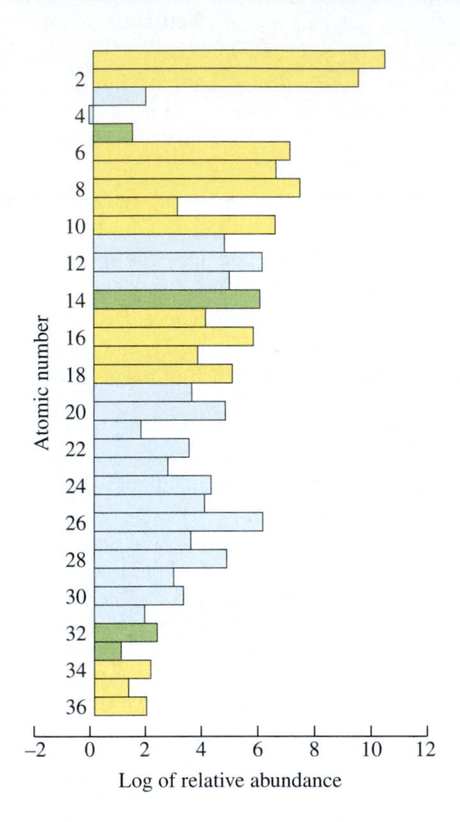

General Questions

82. In the beautifully written autobiography by Primo Levi, *The Periodic Table,* Levi says of zinc that "it is not an element which says much to the imagination, it is gray and its salts are colorless, it is not toxic, nor does it produce striking chromatic reactions; in short, it is a boring metal. It has been known to humanity for two or three centuries, so it is not a veteran covered with glory like copper, nor even one of these newly minted elements which are still surrounded with the glamour of their discovery." From this description, and from reading this chapter, make a list of the properties of zinc. For example, include in your list the position of the element in the periodic table, and tell how many electrons and protons an atom of zinc has. What are its atomic number and atomic weight? Zinc is important in our economy. Check your dictionary (or a book such as *The Handbook of Chemistry and Physics*) and make a list of the uses of the element.

83. The density of a solution of sulfuric acid is 1.285 g/cm^3, and it is 38.08% acid by mass. What volume of the acid solution (in mL) do you need to supply 125 g of sulfuric acid?

84. Molecular distances are usually given in nanometers (1 nm = 1×10^{-9} m) or in picometers (1 pm = 1×10^{-12} m). However, a commonly used unit has been the angstrom, where 1Å = 1×10^{-10} m. (The angstrom unit is not an SI unit.) If the distance between the Pt atom and the N atom in the cancer chemotherapy drug cisplatin is 1.97 Å, what is the distance in nm? In pm?

85. The separation between carbon atoms in diamond is 0.154 nm. (a) What is their separation in meters? (b) What is the carbon atom separation in angstrom units (where 1 Å = 10^{-10} m)?

86. The smallest repeating unit of a crystal of common salt is a cube with an edge length of 0.563 nm. What is the volume of this cube in nm^3? In cm^3?

87. The cancer drug cisplatin contains 65.0% platinum. If you have 1.53 g of the compound, how many grams of platinum does this sample contain?

88. At 25 °C, the density of water is 0.997 g/cm^3, whereas the density of ice at −10 °C is 0.917 g/cm^3. (a) If a plastic soft-drink bottle (volume = 250 mL) is filled with pure water, capped, and then frozen at −10 °C, what volume will the solid occupy? (b) Could the ice be contained within the bottle?

89. A common fertilizer used on lawns is designated as "16-4-8." These numbers mean that the fertilizer contains 16% nitrogen-containing compounds, 4.0% phosphorus-containing compounds, and 8.0% potassium-containing compounds. You buy a 40.0-lb bag of this fertilizer and use all of it on your lawn. How many grams of the phosphorus-containing compound are you putting on your lawn? If the phosphorus-containing compound consists of 43.64% phosphorus (the rest is oxygen), how many grams of phosphorus are there in 40.0 lb of fertilizer?

90. The fluoridation of city water supplies has been practiced in the United States for several decades, because it is believed that fluoride prevents tooth decay, especially in young children. This is done by continuously adding sodium fluoride to water as it comes from a reservoir. Assume you live in a medium-sized city of 150,000 people and that each person uses 175 gal of water per day. How many tons of sodium fluoride would you have to add to the water supply each year (365 days) in order to have the required fluoride concentration of 1 part per million (that is, 1 ton of fluoride per million tons of water)? (Sodium fluoride is 45.0% fluoride, and one U.S. gallon of water has a mass of 8.34 lb.)

91. Which occupies a larger volume, 600 g of water (with a density of 0.995 g/cm^3) or 600 g of lead (with a density of 11.34 g/cm^3)?

92. You can identify a metal by carefully determining its density. An unknown piece of metal, with a mass of 29.454 g, is 2.35 cm long, 1.34 cm wide, and 1.05 cm thick. Which of the following is the element?
(a) nickel, 8.91 g/cm^3 (c) zinc, 7.14 g/cm^3
(b) titanium, 4.50 g/cm^3 (d) tin, 7.23 g/cm^3

93. Name three elements that you have encountered today. (Name only those that you have seen as elements, not those combined into compounds.) Give the location of each of these elements in the periodic table by specifying the group and period in which it is found.

94. Potassium has three stable isotopes, ^{39}K, ^{40}K, and ^{41}K; but ^{40}K has a very low natural abundance. Which of the other two is the more abundant?

95. Figure 2.16 on p. 50 shows the mass spectrum of the isotopes of neon. What are the symbols of the isotopes? Which is the more abundant isotope? How many protons, neutrons, and electrons does this isotope have? Without looking at a periodic table, give the approximate atomic weight of neon.

96. When an athlete tears ligaments and tendons, they can be surgically attached to bone to keep them in place until they reattach themselves. A problem with current techniques, though, is that the screws and washers used are often too big to be positioned accurately or properly. Therefore, a new titanium-containing device has been invented and is coming into use. (See figure below.)
(a) What are the symbol, atomic number, and atomic weight of titanium?
(b) In what group and period is it found? Name the other elements of its group.
(c) What chemical properties do you suppose make titanium an excellent choice for this and other surgical applications?
(d) Use a dictionary (or a book such as *The Handbook of Chemistry and Physics*) to make a list of the properties of the element and its uses.

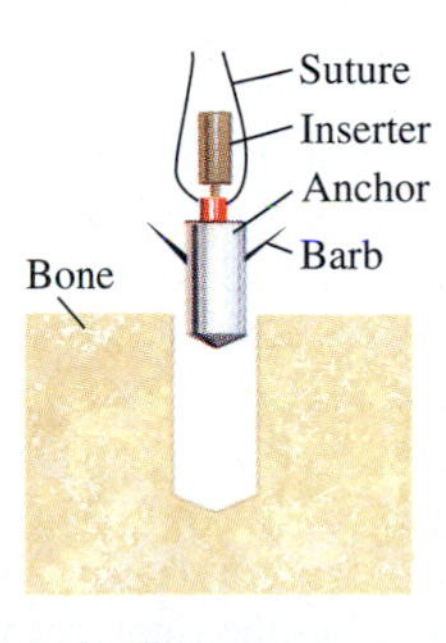

1. Sutures are attached to the anchor, which is then placed on the inserter.

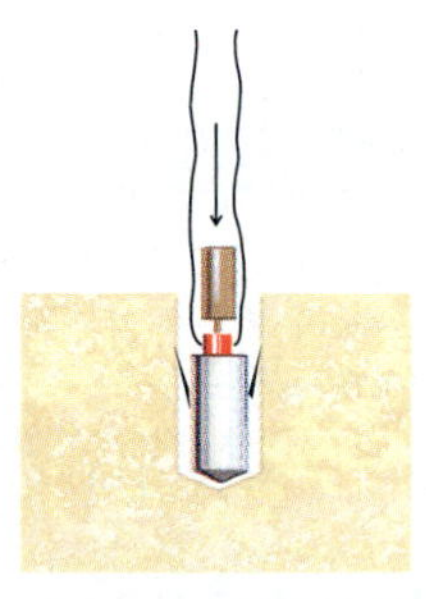

2. The anchor is pushed into the hole, causing the wire barbs to compress.

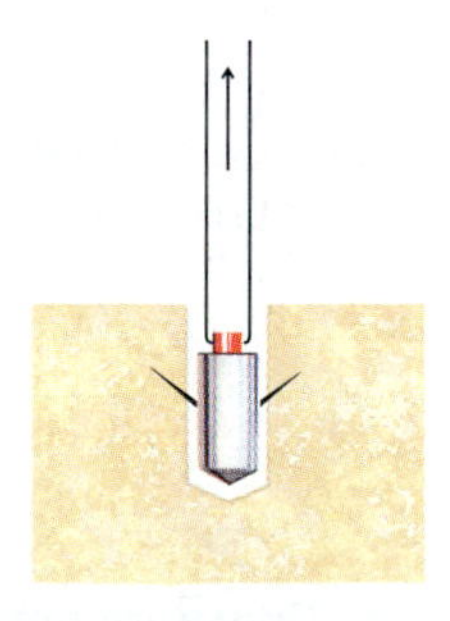

3. Tension is applied, causing the barbs to spread out, securing the anchor.

97. The following plot shows the variation in density with atomic number for the first 36 elements. Use this plot to answer the following questions:
(a) What three elements in this series have the highest values of density? What is their approximate density? Are these elements metals or nonmetals?
(b) Which element in the second period has the greatest density? Which element in the third period has the largest density? What do these two elements have in common?
(c) Some elements have densities so low that they do not show up in the plot. What elements are these? What property do they have in common?

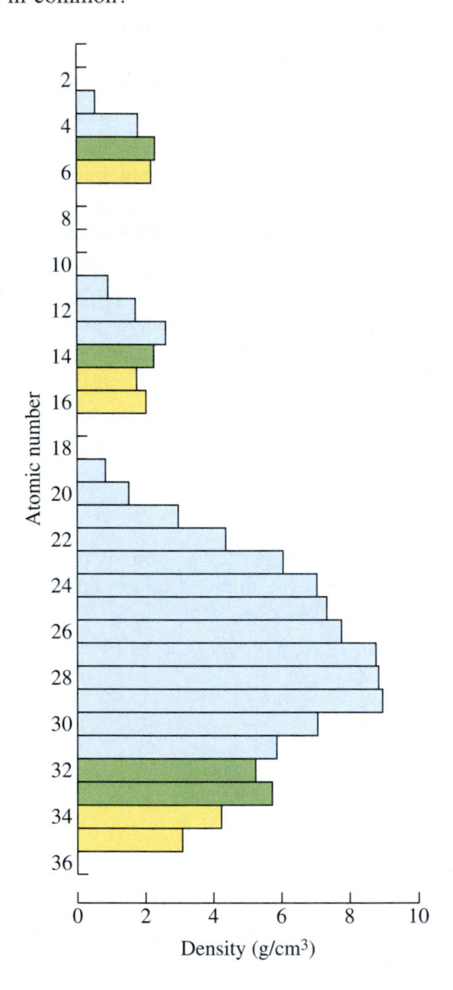

98. Crossword puzzle: In the 2 × 2 crossword to the right, each letter must be correct four ways: horizontally, vertically, diagonally, and by itself. Instead of words, use symbols of elements. When the puzzle is complete, the four spaces will contain the overlapping symbols of ten elements. There is only one correct solution.*

*This puzzle appeared in *Chemical and Engineering News,* December 14, 1987, p. 86 (submitted by S.J. Cyvin), and in *Chem Matters,* October, 1988.

Horizontal
1–2: Two-letter symbol for a metal used in ancient times
3–4: Two-letter symbol for a metal that burns in air and is found in Group 5A

Vertical
1–3: Two-letter symbol for a metalloid
2–4: Two-letter symbol for a metal used in U.S. coins

Single squares (all one-letter symbols)
1: A colorful nonmetal
2: A colorless gaseous nonmetal
3: An element that makes fireworks green
4: An element that has medicinal uses

Diagonal
1–4: Two-letter symbol for an element used in electronics
2–3: Two-letter symbol for a metal used with Zr to make wires for superconducting magnets

1	2
3	4

99. Draw a picture showing the approximate positions of all protons, electrons, and neutrons in an atom of helium-4. Make certain that your diagram indicates both the number and position of each type of particle.

100. Gems and precious stones are measured in carats, a weight unit equivalent to 200 mg. If you have a 2.3-carat diamond in a ring, how many moles of carbon do you have?

101. The international markets in precious metals operate in the weight unit "troy ounce" (where 1 troy ounce is equivalent to 31.1 g). Platinum sells for $325 per troy ounce. (a) How many moles are there in 1 troy ounce? (b) If you have $5000 to spend, how many grams and how many moles of platinum can be purchased?

102. Gold prices fluctuate, depending on the international situation. If gold currently sells for $338.70 per troy ounce, how much must you spend to purchase 1.00 mol of gold (1 troy ounce is equivalent to 31.1 g)?

103. The Statue of Liberty in New York harbor is made of 2.00×10^5 lb of copper sheets bolted to an iron framework. How many grams and how many moles of copper does this represent (1 lb = 454 g)?

104. A piece of copper wire is 25 ft long and has a diameter of 2.0 mm. Copper has a density of 8.92 g/cm^3. How many moles of copper and how many atoms of copper are there in the piece of wire?

Applying Concepts

105. Which sets of values are possible? Why are the others not possible?

	Mass Number	Atomic Number	Number of Protons	Number of Neutrons
(a)	19	42	19	23
(b)	235	92	92	143
(c)	53	131	131	79
(d)	32	15	15	15
(e)	14	7	7	7
(f)	40	18	18	40

106. Which pair has the greater number of particles? Explain why.
(a) 1 mol Cl or 1 mol Cl_2
(b) 1 molecule O_2 or 1 mol O_2
(c) 1 nitrogen atom or 1 nitrogen molecule
(d) 6.022×10^{23} fluorine molecules or 1 mol of fluorine molecules
(e) 20.2 g Ne or 1 mol Ne
(f) 1 molecule Br_2 or 159.8 g Br_2
(g) 107.9 g Ag or 6.9 g Li
(h) 58.9 g Co or 58.9 g Cu
(i) 1 g calcium or 6.022×10^{23} calcium atoms
(j) 1 g of chlorine atoms or 1 g of chlorine molecules

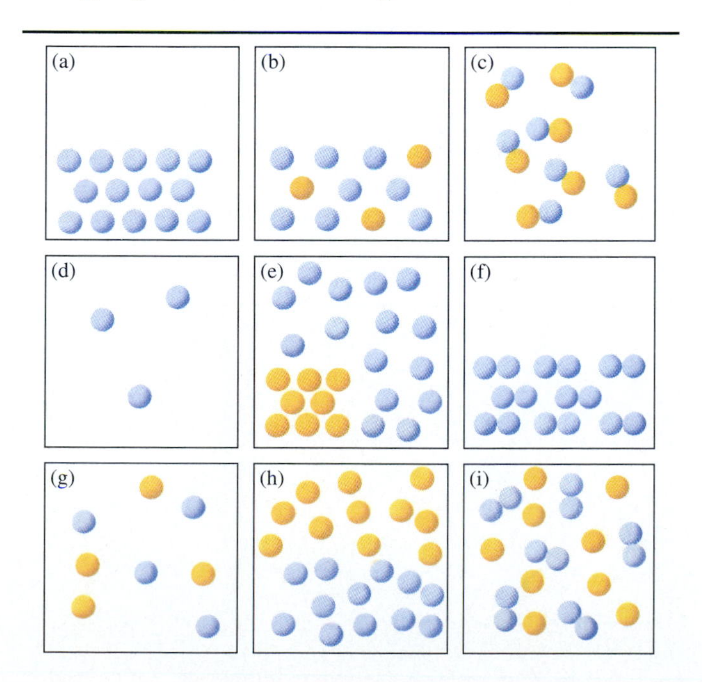

107. Which has the greater mass? Explain why.
(a) 1 mol iron or 1 mol aluminum
(b) 6.022×10^{23} lead atoms or 1 mol lead
(c) 1 copper atom or 1 mol copper
(d) 1 mol Cl or 1 mol Cl_2
(e) 1 g of oxygen atoms or 1 g of oxygen molecules
(f) 24.3 g Mg or 1 mol Mg
(g) 1 mol Na or 1 g Na
(h) 4.0 g He or 6.022×10^{23} He atoms
(i) 1 molecule I_2 or 1 mol I_2
(j) 1 oxygen molecule or 1 oxygen atom

108. A group of astronauts in a spaceship accidentally encounters a spacewarp that traps them in an alternative universe where the chemical elements are quite different from the ones they are used to. The astronauts find these properties for the elements that they have discovered:

Atomic Symbol	Atomic Weight	State	Color	Electrical Conductivity	Electrical Reactivity
A	3.2	Solid	Silvery	High	Medium
D	13.5	Gas	Colorless	Very low	Very high
E	5.31	Solid	Golden	Very high	Medium
G	15.43	Solid	Silvery	High	Medium
J	27.89	Solid	Silvery	High	Medium
L	21.57	Liquid	Colorless	Very low	Medium
M	11.23	Gas	Colorless	Very low	Very low
Q	8.97	Liquid	Colorless	Very low	Medium
R	1.02	Gas	Colorless	Very low	Very high
T	33.85	Solid	Colorless	Very low	Medium
X	23.68	Gas	Colorless	Very low	Very low
Z	36.2	Gas	Colorless	Very low	Medium
Ab	29.85	Solid	Golden	Very high	Medium

(a) Arrange these elements into a periodic table.
(b) If a new element, X, with atomic weight 25.84 is discovered, what would its properties be? Where would it fit in the periodic table you constructed?
(c) Are there any elements that have not yet been discovered? If so, what would their properties be?

109. Answer the following questions using figures (a) to (i), to the left. (Each question may have more than one answer.)
(a) Which represents nanoscale particles in a sample of solid?
(b) Which represents nanoscale particles in a sample of liquid?
(c) Which represents nanoscale particles in a sample of gas?
(d) Which represents nanoscale particles in a sample of an element?
(e) Which represents nanoscale particles in a sample of a compound?
(f) Which represents nanoscale particles in a sample of a pure substance?
(g) Which represents nanoscale particles in a sample of a mixture?

CHAPTER

3 Chemical Compounds

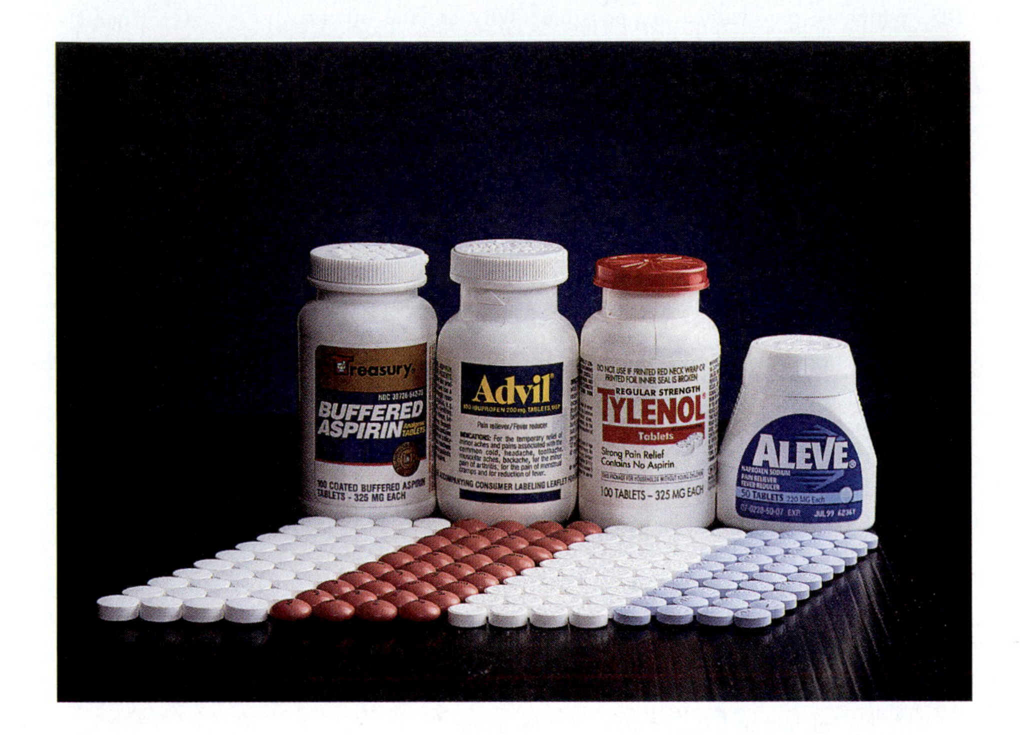

Since aspirin was first synthesized nearly a century ago, other pain relievers have been developed, especially in the last 20 years, including ibuprofen (Advil), acetaminophen (Tylenol), and naproxen (Aleve). These drugs, like all compounds, contain atoms of at least two elements. Aspirin, ibuprofen, and naproxen contain carbon, hydrogen, and oxygen; acetaminophen also contains nitrogen. • What is the formula of ibuprofen? The answer to this question is part of the Summary Problem at the end of this chapter. *(George Semple)*

Civilization has fought a long and arduous battle against sickness, pain, and disease. More than 2300 years ago, the famous Greek physician Hippocrates described the painkilling effects of a tea made from willow bark. Nearly a century ago the substance responsible for willow bark's curative powers was isolated and identified as salicylic acid. However, salicylic acid had some troubling side effects, including stomach distress. In an attempt to decrease the side effects, Felix Hofman, a German chemist, experimented with molecules closely related to those of salicylic acid. In 1898 his "molecular tinkering" led to a new compound that has become one of the world's most widely used drugs, aspirin.

The synthesis of new compounds for medications has achieved remarkable success in the past 50 years and especially within the past three decades. Chemical synthesis also produces new compounds used in fields as diverse as cosmetics, electronics technology, and building materials.

Arguably, the most defining thing chemists do is make new chemical compounds, substances that on the nanoscale consist of new, unique combinations of atoms. These compounds may have properties similar to existing compounds, or they may be very different. Often chemists can custom design a new compound to have desirable properties. All compounds contain atoms of at least two elements and some compounds contain many more. This chapter deals with two major, general types of chemical compounds—those consisting of discrete molecules and those made of positively and negatively charged atoms called ions. We will also examine how compounds are represented by symbols, formulas, and names.

3.1 MOLECULAR COMPOUNDS

In a **molecular compound,** at the nanoscale level, atoms of two or more different elements are combined into the independent units known as molecules (← *p. 32*). Daily we inhale, exhale, metabolize, and in other ways use thousands of molecular compounds. Water, carbon dioxide, sucrose (table sugar), ethyl alcohol, and caffeine are among the many common molecular compounds.

Molecular Formulas

The composition of a molecular compound is represented in writing by its **molecular formula** in which the number and kinds of atoms combined in one molecule of the compound are indicated by the subscripts and the element symbols. For example, the formula for water, H_2O, shows that there are three atoms per molecule, two hydrogen atoms and one oxygen atom. The subscript to the right of the element's symbol indicates the number of atoms of that element in the molecule. If the subscript is omitted, it is understood to be one, as for O in H_2O. These same principles apply to the formulas of all molecules.

Some elements are also composed of molecules (← *p. 32*.) In oxygen, for example, two oxygen atoms are joined in an O_2 molecule.

Ethanol and octane are examples of **organic compounds.** Organic compounds invariably contain carbon, usually contain hydrogen, and may also contain oxygen, nitrogen, sulfur, or phosphorus. Such compounds are of great interest because they are the basis of the clothes we wear, the food we eat, and the living organisms in our environment.

Atom colors in molecular models:

H	Light Gray
C	Dark Gray
N	Blue
O	Red
S	Yellow

Name	Molecular Formula	Number and Kind of Atoms	Molecular Model
Carbon dioxide	CO_2	3 total: 1 carbon, 2 oxygen	
Ammonia	NH_3	4 total: 1 nitrogen, 3 hydrogen	
Sulfuric acid	H_2SO_4	7 total: 2 hydrogen, 1 sulfur, 4 oxygen	
Ethanol	C_2H_6O	9 total: 2 carbon, 6 hydrogen, 1 oxygen	
Octane	C_8H_{18}	26 total: 8 carbon, 18 hydrogen	

Ethanol (commonly known as ethyl alcohol) is made by the fermentation of sugar, starch, or cellulose. Alternatively, it can be synthesized from simpler molecules, a process that yields about 550 million pounds per year in the United States. This alcohol is found in many popular beverages. It is also commonly used to dissolve other chemicals, so you find it in toilet articles and pharmaceuticals. In addition, it is used as a germicide, and it may soon serve more widely in the United States as an automotive fuel.

There may be several ways to write a molecular formula of a compound. The formula given previously for ethanol, C_2H_6O, is an example of a simple molecular formula. For an organic compound, the symbols of the elements other than carbon are frequently written in alphabetical order, each with a subscript indicating the total number of atoms of that type in the molecule. The formula can also be written in a modified form to show how the atoms are grouped together in the molecule. Such formulas, called **condensed formulas,** emphasize chemically important groups of atoms in the molecule. For ethanol the condensed formula is CH_3CH_2OH. This indicates that the end carbon atom has three hydrogen atoms attached to it. This carbon atom is also attached to a second carbon atom that is joined to two hydrogen atoms and an –OH group. The OH attached to the C atom is important because it is a group present in all alcohols. The –OH group is an example of a **functional group,** which is a distinctive group of atoms that, as part of an organic molecule, imparts specific properties to the molecule. A functional group is often the part of the molecule that changes when it reacts with another atom or molecule.

Functional groups are responsible for the characteristic chemical behavior of organic compounds. For example, the chemistry of alcohols depends on the presence of an –OH group attached to a C atom. Thus, ethanol (CH_3CH_2OH) is similar to propanol ($CH_3CH_2CH_2OH$) and other alcohols.

You will also see the formulas of molecules written to show how the atoms of the molecule are connected to one another. These are called **structural formulas.** The three different ways of writing molecular formulas are illustrated below for ethanol:

Organic compounds are divided into classes according to their functional groups. For example, compounds with –OH groups are classified as alcohols. A complete list of classes of organic compounds is given in Table E.2 in Appendix E.2, Functional Groups.

structural formula

```
    H  H
    |  |
 H—C—C—O—H
    |  |
    H  H
```

As illustrated earlier for molecular elements (← *p. 32*), molecular compounds are also represented by space-filling or ball-and-stick models:

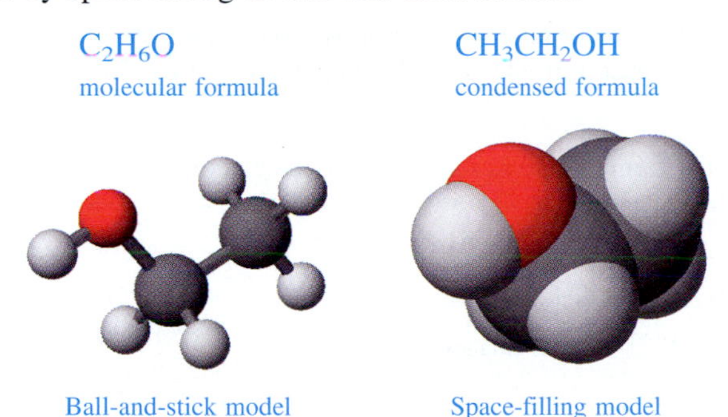

Ball-and-stick model

Space-filling model

Some additional examples of structural formulas are given below. You can determine the molecular formula of the compound from the structural formula or condensed formula simply by counting up the atoms.

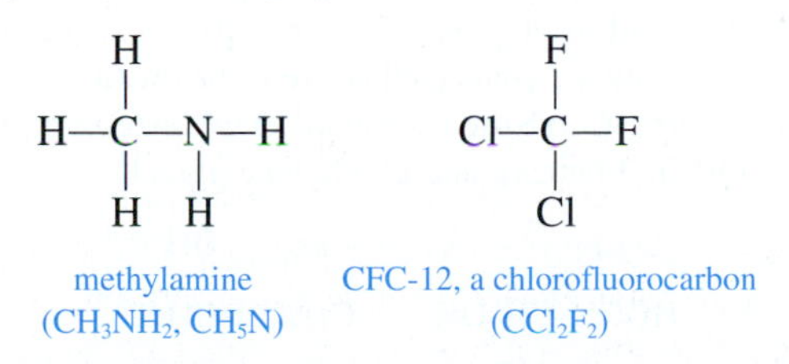

methylamine (CH_3NH_2, CH_5N)

CFC-12, a chlorofluorocarbon (CCl_2F_2)

Methylamine contains the amine functional group, $–NH_2$, and is the simplest member of the class of organic compounds known as amines.

CFC-12 is an industrial code used by manufacturers to designate the compound whose structural formula is shown.

PROBLEM-SOLVING EXAMPLE 3.1 Condensed and Molecular Formulas

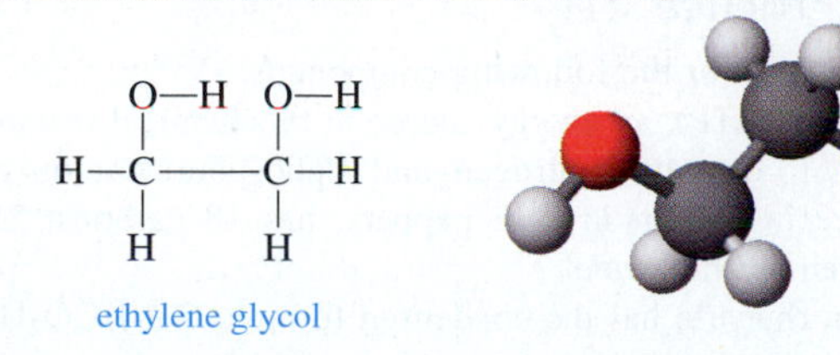

ethylene glycol

```
H  H  O—H  H
|  |  |    |
H—C—C—C————C—H
|  |  |    |
H  H  H    H
```

2-butanol

```
   H  F
   |  |
H—C—C—F
   |  |
   F  F
```

HCFC-134a

The structural formulas and molecular models are given above for ethylene glycol, 2-butanol, and HCFC-134a (this is an industrial code notation for this compound), respectively.

(a) Write the molecular formula for each of the compounds above.

(b) Write the condensed formulas for ethylene glycol and for 2-butanol.

Answer

(a) The molecular formulas are as follows: ethylene glycol, $C_2H_6O_2$; 2-butanol, $C_4H_{10}O$; and HCFC-134a, $C_2H_2F_4$.

(b)

$HOCH_2CH_2OH$ (ethylene glycol) $CH_3CH_2CH(OH)CH_3$ (2-butanol; OH drawn above the CH carbon)

Explanation

(a) Simply count up each of the atoms in each compound to obtain the molecular formulas: Ethylene glycol is $C_2H_6O_2$; 2-butanol is $C_4H_{10}O$; and HCFC-134a is $C_2H_2F_4$.

(b) In condensed formulas, each carbon atom and its hydrogen atoms are written without connecting lines (CH_3, CH_2, or CH). Other groups are usually written on the same line with the carbon and hydrogen atoms if the groups are at the beginning or end of the molecule. Otherwise, they are connected above or below the line by straight lines to the respective carbon atoms. Condensed formulas emphasize important groups in molecules, such as the –OH in 2-butanol and in ethylene glycol.

$HOCH_2CH_2OH$ (ethylene glycol) $CH_3CH_2CH(OH)CH_3$ (2-butanol; OH drawn above the CH carbon)

Problem-Solving Practice 3.1

Write the molecular formulas for the following compounds.

(a) Adenosine triphosphate (ATP), an energy source in biochemical reactions, contains 10 carbon, 11 hydrogen, 13 oxygen, 5 nitrogen, and 3 phosphorus atoms per molecule.

(b) Capsaicin, the active ingredient in chili peppers, has 18 carbons, 27 hydrogens, 3 oxygens, and 1 nitrogen per molecule.

(c) Oxalic acid, found in rhubarb, has the condensed formula CO_2HCO_2H.

Exercise 3.1 Structural, Condensed, and Molecular Formulas

A model is given below of propylene glycol, used in some "environmentally friendly" antifreezes:

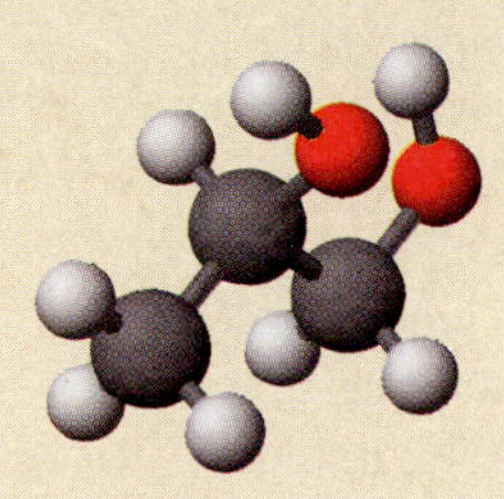

Write the structural formula, the condensed formula, and the molecular formula of propylene glycol.

3.2 NAMING BINARY MOLECULAR COMPOUNDS

Each element has a unique name and so does each chemical compound, regardless of the type of compound it is. This is useful because if several compounds shared the same name, there would be no way to know which specific compound to use in an experiment. To avoid such confusion, each compound is named in a systematic way. We will begin by applying the rules to name simple binary compounds systematically, and then introduce other naming rules as we need them. **Binary molecular compounds** consist of molecules that contain atoms of only two elements.

There is a binary compound of hydrogen with every nonmetal (except the noble gases). For compounds with oxygen, sulfur, and the halogens, the hydrogen is written first in the formula and named first. The other nonmetal is then named, with the element's name changed to end in *-ide*. For example, HCl is named hydrogen chloride.

Formula	Name
HBr	Hydrogen bromide
HI	Hydrogen iodide
H_2Se	Hydrogen selenide

(Hydrogen compounds with carbon are discussed in Section 3.3.)

Many other binary compounds contain nonmetallic elements from Groups 4A, 5A, 6A, and 7A. In these compounds the elements are listed and named in the order of the group number. In such compounds, prefixes are used to designate the number of a particular kind of atom. The prefixes used (listed in Table 3.1) include *mono-* (1), *di-* (2), *tri-* (3), *tetra-* (4), *penta-* (5), *hexa-* (6), and so on. Table 3.2 illustrates how these prefixes are applied.

TABLE 3.1 Prefixes Used in Naming Binary Compounds of Nonmetals

Prefix	Number
Mono-	1
Di-	2
Tri-	3
Tetra-	4
Penta-	5
Hexa-	6
Hepta-	7
Octa-	8
Nona-	9
Deca-	10

TABLE 3.2 Examples of Naming Binary Compounds

Compound	Name	Use
CO	Carbon monoxide	Steel manufacturing
NO_2	Nitrogen dioxide	Preparation of nitric acid
N_2O	Dinitrogen oxide	Anesthetic, spray can propellant
N_2O_5	Dinitrogen pentaoxide	Forms nitric acid
PBr_3	Phosphorus tribromide	Forms phosphorous acid
PBr_5	Phosphorus pentabromide	Forms phosphoric acid
SF_6	Sulfur hexafluoride	Transformer insulator
P_4O_{10}	Tetraphosphorus decaoxide	Drying agent

A number of binary nonmetal compounds were discovered and named years ago, before naming rules were developed. These common names are still used today. Such nonsystematic names must simply be learned.

By tradition, the formula H_2O is written with H before O; and the hydrogen compounds of Groups 6A and 7A are H_2S, H_2Se, and HF, HCl, HBr, and HI. Other H-containing compounds are generally written with the H atom after the other atom.

Compound	Common Name
H_2O	Water
NH_3	Ammonia
N_2H_4	Hydrazine
PH_3	Phosphine
NO	Nitric oxide
N_2O	Nitrous oxide ("laughing gas")

Exercise 3.2 Names and Formulas of Compounds

Give the formula for each of the following binary nonmetal compounds:

(a) Carbon disulfide
(b) Phosphorus trichloride
(c) Sulfur dibromide
(d) Selenium dioxide
(e) Oxygen difluoride
(f) Xenon trioxide

3.3 ORGANIC COMPOUNDS: HYDROCARBONS

Hydrocarbons are organic compounds composed of only carbon and hydrogen. They are the simplest of the incredible variety of organic compounds that result from the unique ability of carbon atoms to bond (connect) to each other. Organic compounds come in many sizes and shapes because each carbon atom can connect with up to four other carbon atoms.

The molecular formulas of hydrocarbons are traditionally written with the C first, then the H. An example is methane, CH_4, the simplest member of the major class of hydrocarbons called **alkanes**, which are useful as fuels and lubricants. An alkane is a hydrocarbon that has the general formula C_nH_{2n+2}, where n is a whole number. The names of all alkanes end in *-ane* (Table 3.3). The first four alkanes ($n = 1$ to 4) have common names that must just be remembered. When n is five or greater, the prefix (← ***p. 83, Table 3.1***) indicates the number of carbon atoms in the molecule. For example, the six-carbon alkane is *hexane*.

TABLE 3.3 Examples of Alkane Hydrocarbons, C_nH_{2n+2}*

Molecular Formula	Name	State at Room Temperature
CH_4	Methane	**Gas**
C_2H_6	Ethane	
C_3H_8	Propane	
C_4H_{10}	Butane	
C_5H_{12}	Pentane	**Liquid**
C_6H_{14}	Hexane	
C_7H_{16}	Heptane (hept = 7)	
C_8H_{18}	Octane (oct = 8)	
C_9H_{20}	Nonane (non = 9)	
$C_{10}H_{22}$	Decane (dec = 10)	
$C_{17}H_{36}$	Heptadecane (heptadec = 17)	**Solid**
$C_{18}H_{38}$	Octadecane (octadec = 18)	
$C_{19}H_{40}$	Nonadecane (nonadec = 19)	
$C_{20}H_{42}$	Eicosane (eicos = 20)	

*This table lists only selected alkanes. Liquid compounds with 11 to 16 carbon atoms are well known.

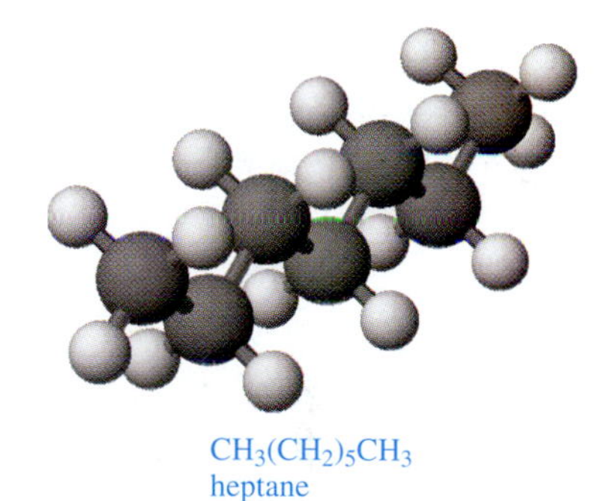

$CH_3(CH_2)_5CH_3$
heptane

The first four alkanes are commercially important. Methane, the simplest alkane, comprises about 85% of natural gas. It is also thought to contribute to global warming through the "greenhouse effect" (Section 11.4). Ethane, propane, and butane are burned to provide heat for homes and in industry. In these simple alkanes, the carbon atoms are connected in unbranched straight chains and each carbon atom is connected to two or three hydrogen atoms, as shown in their structural and condensed formulas.

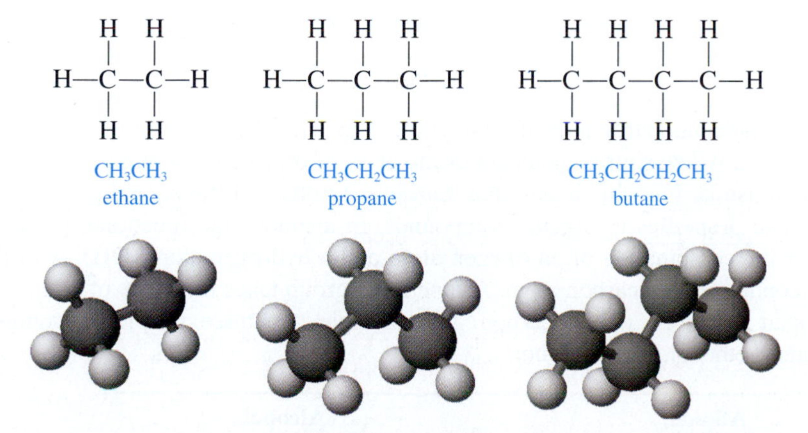

CH_3CH_3 ethane | $CH_3CH_2CH_3$ propane | $CH_3CH_2CH_2CH_3$ butane

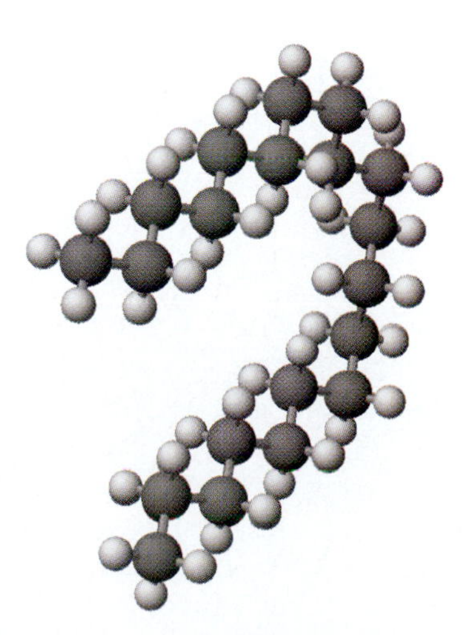

$C_{20}H_{42}$
eicosane

Larger alkanes are characterized by increasingly longer chains of carbon atoms, with their affiliated hydrogens. Two such alkanes are heptane (C_7H_{16}) and eicosane ($C_{20}H_{42}$), found in petroleum and paraffin wax, respectively.

Butane ($CH_3CH_2CH_2CH_3$) is the fuel in this lighter. Butane molecules are present in the liquid and gaseous states in the lighter. *(C.D. Winters)*

Exercise 3.3 Alkane Molecular Formulas

(a) Using the general formula for alkanes, write the molecular formulas for alkanes with 16 carbon atoms and with 28 carbon atoms.
(b) How many carbon and hydrogen atoms are in tetradecane?

The molecular structures of hydrocarbons provide the framework for all organic compounds. A few simple members of the class of organic compounds known as **alcohols** illustrate how this is possible. Earlier we noted that functional groups impart distinctive properties to organic compounds. In alcohols, the functional group consists of the combination of an oxygen atom and a hydrogen atom (OH), with the O atom connected to a carbon atom. A functional group takes the place of one or more hydrogen atoms in a hydrocarbon. For example, the three simplest alcohols are methanol, ethanol, and propanol.

Alkane		Alcohol	
Methane	CH_4	Methanol	CH_3OH
Ethane	CH_3CH_3	Ethanol	CH_3CH_2OH
Propane	$CH_3CH_2CH_3$	Propanol	$CH_3CH_2CH_2OH$

Exercise 3.4 Structural Formulas

Write the structural formula for propanol.

Exercise 3.5 Structural Formulas

Another alcohol has the same molecular formula as propanol, but the –OH group is connected to a different carbon atom. Write the structural formula for this molecular relative of propanol.

3.4 ALKANES AND THEIR ISOMERS

Two or more compounds with the same molecular formula but different arrangements of atoms are called **isomers.** Isomers differ in one or more physical or chemical properties such as boiling point, color, solubility, reactivity, and density. Several different kinds of isomerism are possible, particularly in organic compounds. **Constitutional isomers** (also called structural isomers) are compounds with the same formula that differ in the order in which their atoms are bonded together.

In this context, "straight chain" means a chain of carbon atoms with no branches to other carbon atoms; the carbon atoms are in an unbranched sequence. As you can see from the molecular structures, the chain is not actually straight, but rather a zig-zag.

Straight- and Branched-Chain Isomers of Alkanes

The first three alkanes—methane (CH_4), ethane (CH_3CH_3), and propane ($CH_3CH_2CH_3$)—have only one possible structural arrangement. Constitutional isomers are possible for butane ($CH_3CH_2CH_2CH_3$) and higher members of the series because of the possibility of chain branching. Two structural arrangements are possible for butane.

Molecular Formula	Condensed Formula	Structural Formula	Molecular Model
C_4H_{10}	Butane $CH_3(CH_2)_2CH_3$ Melting point −138 °C Boiling point −0.5 °C Density 0.58 g/L	H H H H H—C—C—C—C—H H H H H	
C_4H_{10}	Methylpropane $CH_3CH(CH_3)CH_3$ Melting point −159 °C Boiling point −11.6 °C Density 0.60 g/L	H H H—C—H H H—C——C——C—H H H H	

Historically, straight-chain hydrocarbons were referred to as *normal* hydrocarbons, and *n*- was used as a prefix in their names. The current practice is not to use *n*-. If a hydrocarbon's name is given without indication that the compound is branched-chain, assume it is a straight-chain hydrocarbon.

These are constitutional isomers because they have the same molecular formula, but they are different compounds with different properties. Constitutional isomerism is like the results of a child building many different structures with the same collection of building blocks and using all of the blocks in each structure.

Methylpropane, the branched-chain isomer of butane, has a "methyl" group ($—CH_3$) attached to the central carbon atom. This is the simplest example of the fragments of alkanes known as **alkyl groups.** In this case, removal of an H atom from methane gives the **methyl group.**

$$\mathrm{H{-}C(H)_2{-}H} \xrightarrow{-\mathrm{H}} \mathrm{H{-}C(H)_2{-}} \quad \text{or} \quad —\mathrm{CH_3} \left(\begin{array}{l}\text{also written}\\ \text{as } \mathrm{CH_3}—\end{array}\right)$$

Removal of an H atom from ethane gives an ethyl group:

$$\mathrm{H{-}C(H)_2{-}C(H)_2{-}H} \xrightarrow{-\mathrm{H}} \mathrm{H{-}C(H)_2{-}C(H)_2{-}} \quad \text{or} \quad —\mathrm{C_2H_5} \left(\begin{array}{l}\text{also written}\\ \text{as } \mathrm{CH_3CH_2}—\end{array}\right)$$

Notice that more than one alkyl group is possible when an H atom is removed from C_3H_8.

$$\mathrm{H{-}C(H)_2{-}C(H)_2{-}C(H)_2{-}H}$$

— H from either end carbon

$$\mathrm{H{-}C(H)_2{-}C(H)_2{-}C(H)_2{-}} \quad \text{or} \quad \mathrm{CH_3CH_2CH_2}—$$

propyl group

— H from middle carbon

$$\mathrm{H{-}C(H)_2{-}C(H){-}C(H)_2{-}H} \quad \text{or} \quad \mathrm{(CH_3)_2CH}—$$

isopropyl group

Alkyl groups are named by dropping *-ane* from the parent alkane name and adding *-yl*. Theoretically, any alkane can be converted to an alkyl group. Some of the more common examples of alkyl groups are given in Table 3.4.

Exercise 3.6 Straight-Chain and Branched-Chain Isomers

Three constitutional isomers are possible for the isomeric pentanes (five-carbon alkanes). Write structural and condensed formulas for these isomers.

Naming Branched-Chain Alkanes

Many alkanes and other organic compounds have both common names and systematic names. Usually the common name was assigned first and is widely known. Many consumer products are labeled with the common names; when only a few isomers are possible, the common name adequately identifies the product for the consumer.

TABLE 3.4 Some Common Alkyl Groups

Name	Condensed Structural Representation
Methyl	$CH_3—$
Ethyl	$CH_3CH_2—$
Propyl	$CH_3CH_2CH_2—$
Isopropyl	$CH_3CH(CH_3)—$ or $(CH_3)_2CH—$
Butyl	$CH_3CH_2CH_2CH_2—$
sec-Butyl*	$—CH(CH_3)CH_2CH_3$
t-Butyl†	$CH_3C(CH_3)_2—$ or $(CH_3)_3C—$

**sec* stands for *secondary,* which means that the central C atom is bonded to two other C atoms and an H atom.

†*t* stands for *tertiary,* sometimes abbreviated *tert,* which means that the central C atom is bonded to three other C atoms.

However, as the examples in this section will illustrate, a system of common names quickly fails when several constitutional isomers are possible. An elaborate formal system has been developed for naming organic compounds.

The rules for systematic names were formulated by the International Union of Pure and Applied Chemistry and are described in Appendix E.

For example, 2,2,4-trimethylpentane is the systematic name of the following branched-chain isomer of octane, $CH_3CH_2CH_2CH_2CH_2CH_2CH_2CH_3$.

$$\underset{1}{CH_3}—\underset{2}{C}(CH_3)_2—\underset{3}{CH_2}—\underset{4}{CH}(CH_3)—\underset{5}{CH_3}$$

2,2,4-trimethylpentane

The "pentane" part, which means a straight (unbranched) five-carbon chain, identifies the longest chain in the molecule. The numbers "2,2,4-" indicate the locations of the three (*tri-*) methyl groups, according to the numbering of the pentane chain from one end to the other. This particular isomer of octane is used as a standard in assigning the "octane ratings" of various types of gasoline.

Exercise 3.7 Branched-Chain Alkanes

Draw the condensed structural formula for each of the following compounds: (a) 2-methylpentane; (b) 3-methylpentane; (c) 2,2-dimethylbutane; and (d) 2,3-dimethylbutane.

In this textbook we will not emphasize nor expect you to learn the extensive systematic methods for naming organic compounds. Such naming will be covered thoroughly if you take an organic chemistry course.

3.5 IONS AND IONIC COMPOUNDS

An atom or group of atoms that has a positive or negative electrical charge is called an **ion.** A compound whose nanoscale composition consists of ions is classified as an

Ions were first mentioned in Section 2.5 (← *p. 40*).

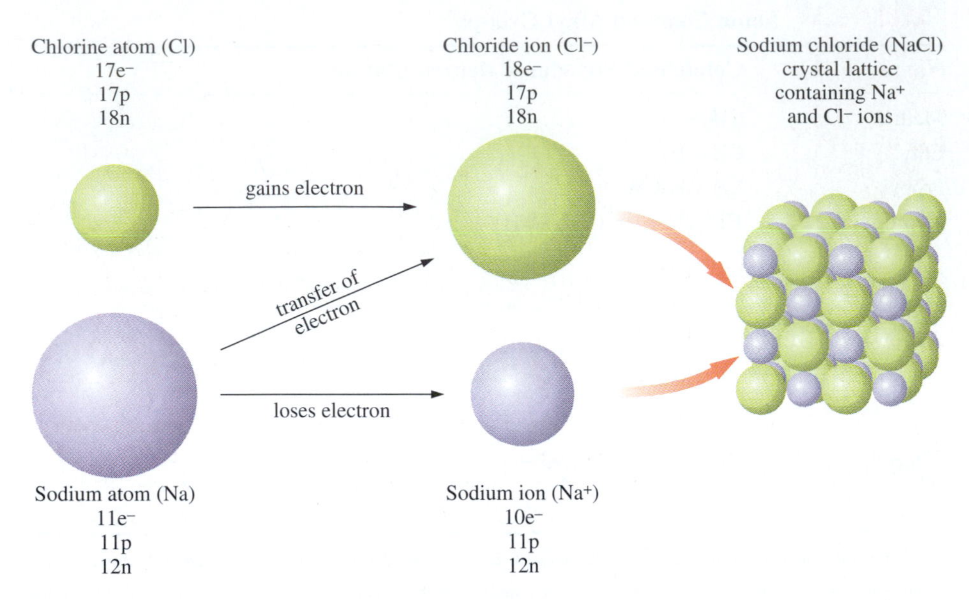

Figure 3.1 **Ion formation.** A neutral sodium atom loses one electron to form a sodium ion (Na^+); and a neutral chlorine atom gains one electron to become a Cl^- ion. Na^+ and Cl^- ions attract each other to form an NaCl crystal.

ionic compound. Many common substances such as table salt (NaCl), lime (CaO), lye (NaOH), and baking soda ($NaHCO_3$) are ionic compounds.

When metals react with nonmetals, the metal atoms typically lose electrons to form positive ions. Any positive ion is referred to as a **cation** (pronounced cat-ion). Cations always have fewer electrons than protons. For example, Figure 3.1a shows how a neutral sodium atom (11 protons [11+] and 11 electrons [11−]) loses one electron to become a sodium ion. The sodium ion, with 11 protons but only 10 electrons, has a net 1+ charge, symbolized Na^+. *The quantity of positive charge on a metal ion equals the number of electrons lost.* For example, when a neutral magnesium atom loses two electrons, it forms a 2+ magnesium ion, Mg^{2+}.

Conversely, *atoms of nonmetals commonly gain electrons to form negatively charged ions called* **anions** (pronounced ann-ions). Figure 3.1b shows how a neutral chlorine atom (17+, 17−) can gain an electron to form a chlor*ide* ion, Cl^-. With 17 protons and 18 electrons, the chloride ion has a *net* 1− charge. Chloride ions can join with sodium ions to form sodium chloride (table salt) (Figure 3.1c). *When electrons are added to a nonmetal atom, the charge on the ion formed equals the number of electrons gained.* For example, a sulfur atom that gains two electrons forms a sulf*ide* ion, S^{2-}.

The terms "cation" and "anion" are derived from the Greek words *ion* (traveling), *cat* (down), and *an* (up).

Monatomic Ions

It is extremely important that you know the ions commonly formed by the elements shown in Figure 3.2 so that you can recognize ionic compounds and their formulas and write their formulas for reaction products (Section 4.4).

A **monatomic ion** is a single atom that has lost or gained electrons. Electron loss creates a positively charged monatomic cation; electron gain creates a negatively charged monatomic anion. Typical charges of monatomic ions are given in Figure 3.2. If you look carefully at Figure 3.2, you will notice that *metals of Groups 1A, 2A, and 3A*

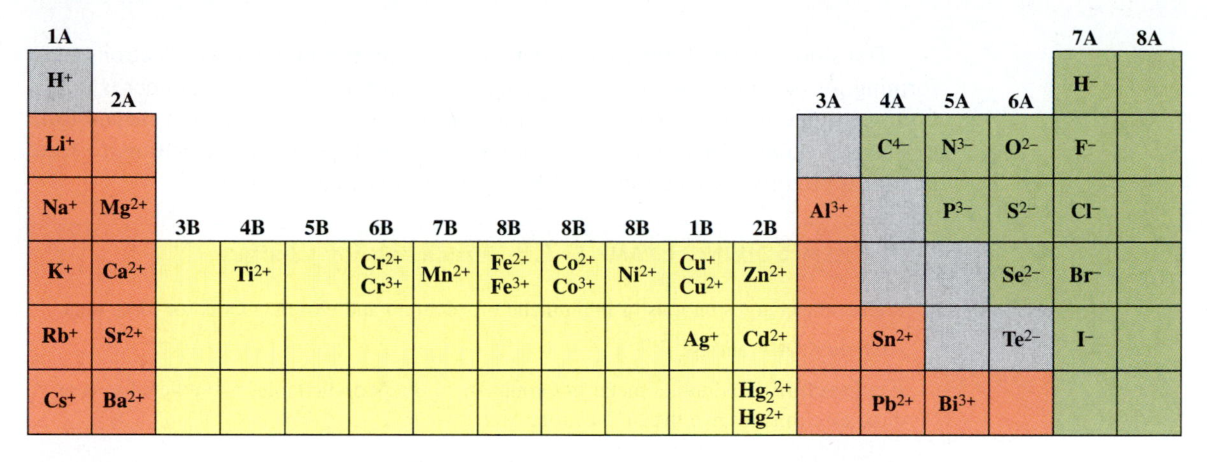

Figure 3.2 Charges on some common monatomic cations and anions. Note that metals usually form cations. The cation charge is given by the group number in the case of the main group elements *(red)*. For transition metals *(yellow)*, the positive charge is variable, and other ions in addition to those illustrated are possible. Nonmetals (green) generally form anions with a charge equal to 8 minus the group number.

form monatomic ions with charges equal to the group number, as shown in the following examples:

Group	Neutral Metal Atom	Electrons Lost	Metal Ion
1A	K (19 protons, 19 electrons)	1	K^+ (19 protons, 18 electrons)
2A	Mg (12 protons, 12 electrons)	2	Mg^{2+} (12 protons, 10 electrons)
3A	Al (13 protons, 13 electrons)	3	Al^{3+} (13 protons, 10 electrons)

When an atom of a nonmetal forms an anion by gaining electrons, *the negative charge usually equals 8 minus the group number of the nonmetal.* For example, oxygen in Group 6A can gain $8 - 6 = 2$ electrons to form a 2− oxide ion, O^{2-}.

In Section 8.5 we will explain the basis for the (8 − group number) relationship for nonmetals.

Group	Neutral Nonmetal Atom	Electrons Gained = 8 − Group No.	Nonmetal Ion
5A	N (7 protons, 7 electrons)	$3 = (8 - 5)$	N^{3-} (7 protons, 10 electrons)
6A	S (16 protons, 16 electrons)	$2 = (8 - 6)$	S^{2-} (16 protons, 18 electrons)
7A	F (9 protons, 9 electrons)	$1 = (8 - 7)$	F^- (9 protons, 10 electrons)

You might have noticed in Figure 3.2 that hydrogen appears at two locations in the table. Because a neutral hydrogen atom can either gain or lose an electron, it can form either a cation or an anion. When it loses an electron, it forms a hydrogen ion, H^+ (1 proton, 0 electrons). By gaining an electron, a neutral hydrogen atom forms a hydride ion, H^- (1 proton, 2 electrons).

Noble gas atoms do not easily lose or gain electrons and have no common ions that can be listed in Figure 3.2.

Transition metals form cations but can lose varying numbers of electrons, thus forming ions of different charges (Figure 3.2). Therefore, the group number is not an accurate guide to the charges on transition metal ions. Many transition metals form 2+ and 3+ ions. For example, iron atoms can lose two or three electrons to form Fe^{2+} (26 protons, 24 electrons) and Fe^{3+} (26 protons, 23 electrons).

PROBLEM-SOLVING EXAMPLE 3.2 Predicting Ion Charges

Predict the charges on ions of indium and of selenium and write symbols for these ions.

Answer In^{3+} and Se^{2-}

Explanation Indium, a metal in Group 3A of the periodic table, is predicted to lose three electrons to give the In^{3+} cation.

$$In \rightarrow In^{3+} + 3e^-$$

Selenium, a Group 6A nonmetal, is predicted to gain electrons to give an anion. The number of electrons gained is $8 - 6 = 2$. Therefore,

$$Se + 2e^- \rightarrow Se^{2-}$$

Problem-Solving Practice 3.2

Indicate whether each of the following ion charges is likely in an ionic compound, and explain your answer.

(a) Ca^{4+} (b) Cr^{2+} (c) Sr^-

Polyatomic Ions

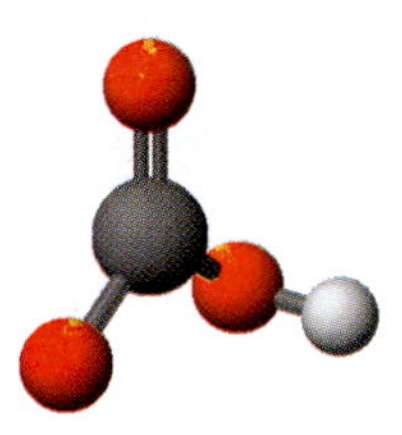

bicarbonate ion (HCO_3^-)

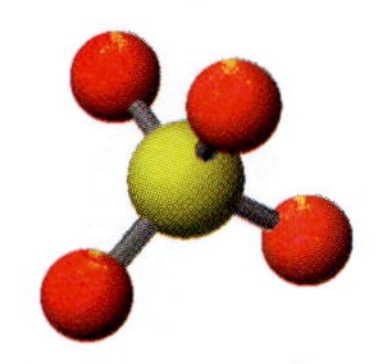

sulfate ion (SO_4^{2-})

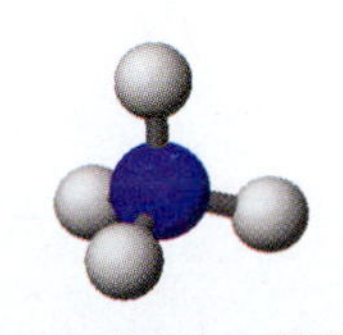

ammonium ion (NH_4^+)

The term **polyatomic ion** refers to a *unit* of two or more atoms that bears a charge. Polyatomic ions abound in many places—oceans, minerals, living cells, and foods. For example, hydrogen carbonate ion, HCO_3^-, is a common polyatomic anion that is present in rain water, sea water, blood, and baking soda. It consists collectively of one carbon atom, three oxygen atoms, and a hydrogen atom, with one unit of negative charge spread over the group of five atoms. Sulfate ion, SO_4^{2-}, is a polyatomic ion with one sulfur and four oxygen atoms, collectively having a 2− charge. One of the most common polyatomic cations is NH_4^+, the ammonium ion. In this case, four hydrogen atoms are connected to a nitrogen atom, and the group bears a 1+ charge. In many chemical reactions the polyatomic ion unit remains intact. *It is important to know the names, formulas, and charges of the common polyatomic ions listed in Table 3.5.*

Ionic Compounds

Many naturally occurring substances, including commercially important minerals, are ionic compounds. In these compounds, cations (positively charged) and anions (negatively charged) are attracted to each other by electrostatic forces—the forces of attraction between positive and negative charges. The strength of the electrostatic attraction dictates many of the properties of ionic compounds.

The force of attraction between oppositely charged ions increases with charge and decreases with the distance between the ions. Therefore, the attractive force between 2+ and 2− ions—for example, Mg^{2+} and O^{2-}—is greater than that between

TABLE 3.5 **Some Common Polyatomic Ions**

Cation (1+)			
NH_4^+	Ammonium ion		
Anions (1−)			
OH^-	Hydroxide ion	NO_2^-	Nitrite ion
CN^-	Cyanide ion	NO_3^-	Nitrate ion
$CH_3CO_2^-$	Acetate ion	HCO_3^-	Hydrogen carbonate ion (or bicarbonate ion)
HSO_4^-	Hydrogen sulfate ion	$H_2PO_4^-$	Dihydrogen phosphate ion
ClO_3^-	Chlorate ion	MnO_4^-	Permanganate ion
ClO_4^-	Perchlorate ion		
Anions (2−)			
CO_3^{2-}	Carbonate ion	SO_3^{2-}	Sulfite ion
HPO_4^{2-}	Hydrogen phosphate	SO_4^{2-}	Sulfate ion
Anions (3−)			
PO_4^{3-}	Phosphate ion		

Potassium dichromate $K_2Cr_2O_7$. This beautiful orange-red compound contains potassium ions (K^+) and dichromate ions ($Cr_2O_7^{2-}$). *(Marna G. Clarke)*

1+ and 1− ions (Na^+ and Cl^-). The attractive force also increases as the distance between the ions decreases. Thus, smaller, oppositely charged ions will attract each other more strongly than do larger ones.

We have spent a good deal of time distinguishing between molecular and ionic compounds. Being able to classify compounds as ionic or molecular is very useful because, as you will see, these two types of compounds have quite different properties and thus different uses. These generalizations will help you to predict whether a compound is ionic.

1. Metals (the elements in the red and yellow areas in Figure 3.2) often form ionic compounds; metals lose electrons to form positive ions.
2. Nonmetals (the elements in the green area of Figure 3.2) can form both ionic and molecular compounds. They form monatomic negative ions *only* when combined with metals. Nonmetals form molecular compounds when combined with nonmetals or metalloids.
3. It is difficult to predict when the metalloids (the elements in the gray area of Figure 3.2) will form ions. These elements form both ionic and molecular compounds.

For example, the compounds formed when calcium combines with chlorine, sodium with sulfur, and strontium with bromine have the formulas $CaCl_2$, Na_2S, and $SrBr_2$, respectively. The metals form cations, and the nonmetals form anions. These ionic compounds consist of Ca^{2+} and Cl^-; Na^+ and S^{2-}; and Sr^{2+} and Br^-. On the other hand, although the nonmetal chlorine can combine with the metalloid boron to form a compound (BCl_3), and with carbon (a nonmetal) to form CCl_4, these two compounds are molecular, not ionic.

Exercise 3.8 Ionic and Molecular Compounds

Predict whether each of these compounds is likely to be ionic or molecular: (a) FeS; (b) $CoCl_2$; (c) $COCl_2$; (d) $(NH_2)_2CO$; (e) $(NH_4)_2CO_3$; (f) C_8H_{18}.

As with the formulas for molecular compounds, a subscript of 1 in formulas of ionic compounds is *understood* to be there, and not written.

Writing Formulas for Ionic Compounds

All compounds, including ionic compounds, are electrically neutral. Therefore, when cations and anions combine to form an ionic compound, there must be zero net charge. The total positive charge of all the cations must equal the total negative charge of all the anions. For example, consider the ionic compound formed when potassium reacts with sulfur. Potassium is in Group 1A, so a potassium metal atom loses one electron to become a K^+ ion. Sulfur is a Group 6A nonmetal, so a sulfur atom gains two electrons to become an S^{2-} ion. To make the compound electrically neutral, two K^+ ions (total charge is 2+) are needed for each S^{2-} ion. Consequently, the compound has the formula K_2S. Subscripts are used in ionic compounds the same way they are with molecular compounds. In this case, the subscript 2 indicates two K^+ ions for every S^{2-} ion.

Similarly, aluminum oxide, a combination of Al^{3+} and O^{2-} ions, has the formula Al_2O_3: 2 Al^{3+} = 6+; 3 × O^{2-} = 6−; total charge = 0.

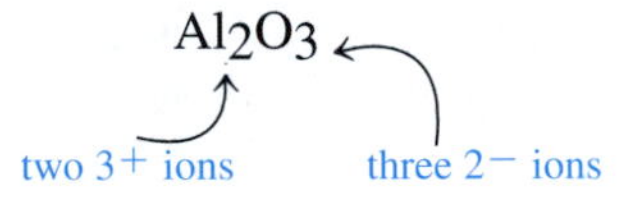

Notice that in writing the formulas for ionic compounds, *the cation symbol is written first, followed by the anion symbol.* The charges of the ions are *not* included in the formulas of ionic compounds.

Let's now consider several ionic compounds of magnesium, a Group 2A metal that forms Mg^{2+} ions.

Combining Ions	Overall Charge	Formula
Mg^{2+} and Br^-	$(2+) + 2 \times (1-) = 0$	$MgBr_2$
Mg^{2+} and SO_4^{2-}	$(2+) + (2-) = 0$	$MgSO_4$
Mg^{2+} and OH^-	$(2+) + 2 \times (1-) = 0$	$Mg(OH)_2$
Mg^{2+} and PO_4^{3-}	$3 \times (2+) + 2 \times (3-) = 0$	$Mg_3(PO_4)_2$

Notice in the latter two cases that when a polyatomic ion occurs more than once in the formula, the polyatomic ion's formula is put in parentheses followed by the necessary subscript.

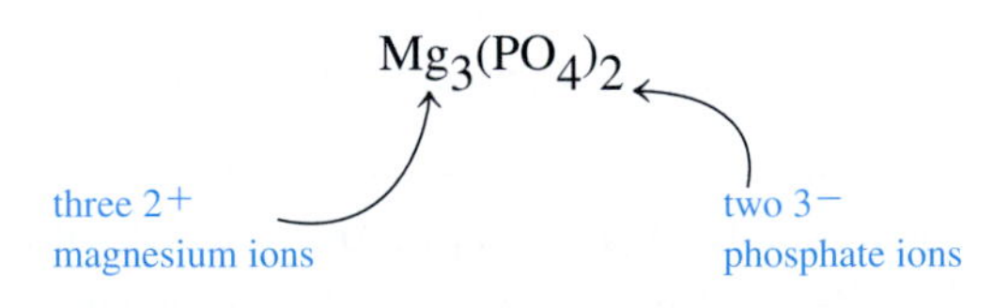

Total: 6+ Total: 6−

PROBLEM-SOLVING EXAMPLE 3.3 Ions in Ionic Compounds

For each compound, give the symbol or formula of each ion present and indicate how many of each ion are present: (a) Li_2S; (b) Na_2SO_3; (c) $Ca(CH_3CO_2)_2$; (d) $Al_2(SO_4)_3$.

Answer (a) Two Li^+, one S^{2-}; (b) Two Na^+, one SO_3^{2-}; (c) One Ca^{2+}, two $CH_3CO_2^-$; (d) Two Al^{3+}, three SO_4^{2-}

Explanation
(a) Li_2S contains two Li^+ ions and one S^{2-} ion. Lithium is a Group 1A element and *always* forms 1+ ions. The S^{2-} ion is formed from sulfur, a Group 6A element, by gaining two electrons ($8 - 6 = 2$).
(b) Na_2SO_3 is composed of two Na^+ ions that offset the 2− charge of the single sulfite polyatomic ion, SO_3^{2-} (see Table 3.5).
(c) $Ca(CH_3CO_2)_2$ contains one Ca^{2+} ion counterbalancing two 1− acetate ions, $CH_3CO_2^-$. By recognizing that calcium is a Group 2A element that always forms 2+ ions, you can figure out that the charge on the acetate ion must be 1− for the net charge of the compound to be 0 (see Table 3.5).
(d) This is similar to the $Mg_3(PO_4)_2$ example, where a 3:2 combination of ions achieves a net charge of 0. Here, it is a 2:3 combination of two Al^{3+} ions ($2 \times 3+ = 6+$), joined with three 2− sulfate ions, SO_4^{2-}, ($3 \times 2- = 6-$).

Problem-Solving Practice 3.3

Determine how many ions and how many atoms there are in (a) $In_2(SO_4)_3$; (b) $(NH_4)_3PO_4$.

PROBLEM-SOLVING EXAMPLE 3.4 Formulas of Ionic Compounds

Write formulas for ionic compounds composed of (a) barium and iodide ions; (b) lithium and carbonate ions; (c) Fe^{3+} and nitrate ions.

Answer (a) BaI_2; (b) Li_2CO_3; (c) $Fe(NO_3)_3$

Explanation In each case, use Figure 3.2 and Table 3.5 for information about the symbols and charges of these ions.
(a) Barium is a Group 2A metal, so it forms 2+ ions. Iodine is a Group 7A nonmetal that forms 1− ions. Therefore, we need two iodide 1− ions for every Ba^{2+} ion. The formula is BaI_2.
(b) Carbonate is a 2− polyatomic ion that combines with two Li^+ ions to form Li_2CO_3.
(c) Because iron in this case has a 3+ charge, it will require three 1− nitrate ions to balance the Fe^{3+}, giving a formula of $Fe(NO_3)_3$. Notice that the polyatomic ion is enclosed in parentheses followed by the subscript. It would be confusing to write $FeNO_{33}$.

It is important for you to recognize that the subscript 2 in BaI_2 comes from the balancing of charges in this ionic compound. It is *not because* iodine is a diatomic molecule; the I_2 molecule is not involved here in any way.

Problem-Solving Practice 3.4

(a) Identify each constituent ion and tell how many there are in these ionic compounds:

(a) CaF_2; (b) $CoCl_2$; (c) $K_2(HPO_4)$.

(b) Copper is a transition metal that can form two compounds with bromine that contain either Cu(I) or Cu(II). Write the formulas of these compounds.
(c) Write the formulas for the ionic compounds formed from the possible combinations of K^+ and Sr^{2+} with O^{2-} and SO_4^{2-}.

3.6 NAMING IONIC COMPOUNDS

A systematic way of naming simple molecular compounds has been explained and applied (← *p. 83*). Ionic compounds can also be named unambiguously by following the rules given in the following paragraphs, rules that you should learn thoroughly.

Copper(I) oxide (*left*) and copper(II) oxide. The different copper ion charges result in different colors. *(C.D. Winters)*

The Stock system came from Alfred Stock (1876–1946), a German chemist famous for his work on the hydrogen compounds of boron and silicon. Because he worked closely with mercury metal for many years, he contracted mercury poisoning, which leads to memory loss and severe physical disabilities.

Another cation that you will see on occasion is Hg_2^{2+}, named mercury(I) ion. The reason for the Roman numeral (I) is that the ion is composed of two Hg^+ ions bonded together, having a collective 2+ charge.

Naming Positive Ions

Virtually all of the cations used in this book are metal ions that can be named by the rules given below. The ammonium ion (NH_4^+) is the major exception—a polyatomic ion composed of nonmetal atoms.

1a. ***For a metal cation, the name is simply the name of the metal plus the word "ion."*** For example, Mg^{2+} is the magnesium ion.

1b. ***For metals that can form more than one kind of positive ion, especially transition metals, the name of each ion must indicate its charge.*** To do so, a Roman numeral in parentheses is given immediately following the ion's name (the Stock system). For example, Cu^{2+} is the copper(II) ion and Cu^+ is the copper(I) ion.

An older system differentiated such ions by using suffixes; *-ic* for the more highly charged species and *-ous* for the ion of lower charge. These suffixes are added to the Latin name of the metal, for example, cupr*ic* for Cu^{2+} and cupr*ous* for Cu^+.

Naming Negative Ions

The following rules apply to naming monatomic and polyatomic anions.

2a. ***A* monatomic anion *is named by adding* -ide *to the stem of the name of the nonmetal element from which the ion is derived.*** For example, a *phosph*orus atom gives a *phosph*ide ion, and a *chlor*ine atom forms a *chlor*ide ion. Anions of Group 7A elements, the halogens, are called collectively the **halide ions.**

2b. ***The names of the most common polyatomic ions are given in Table 3.5.*** Most must simply be memorized. There are, however, some guidelines that can help, especially for related series of **oxoanions,** which are polyatomic ions that contain oxygen.

For oxoanions with a nonmetal in common in addition to oxygen, the oxoanion with the greater number of oxygen atoms is given the suffix *-ate*. The oxoanion with the smaller number of oxygen atoms is given the suffix *-ite*.

NO_3^- is the nitr*ate* ion, whereas NO_2^- is the nitr*ite* ion.

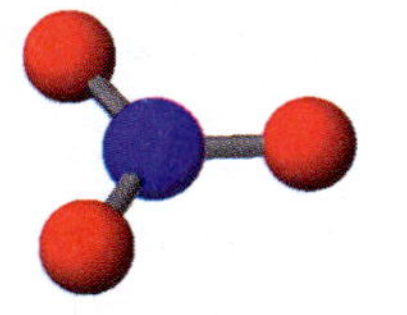

nitrate ion

nitrite ion

SO_4^{2-} is the sulf*ate* ion, whereas SO_3^{2-} is the sulf*ite* ion.

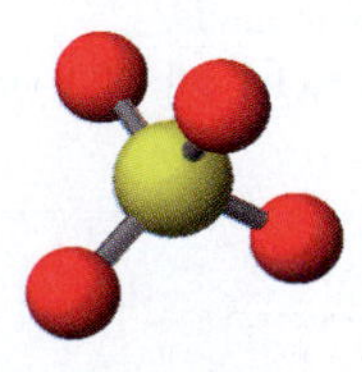

sulfate ion

sulfite ion

Note that *-ate* and *-ite* suffixes do not relate to the ion's charge, but to the relative number of oxygen atoms.

When more than two different oxoanions of a given nonmetal exist, a more extended naming scheme must be used. Here, the ion with the largest number of oxygen atoms has the prefix *per-* and the suffix *-ate*. The ion with the smallest number of oxygen atoms has the prefix *hypo-* and the suffix *-ite*. The oxoanions of chlorine are good examples:

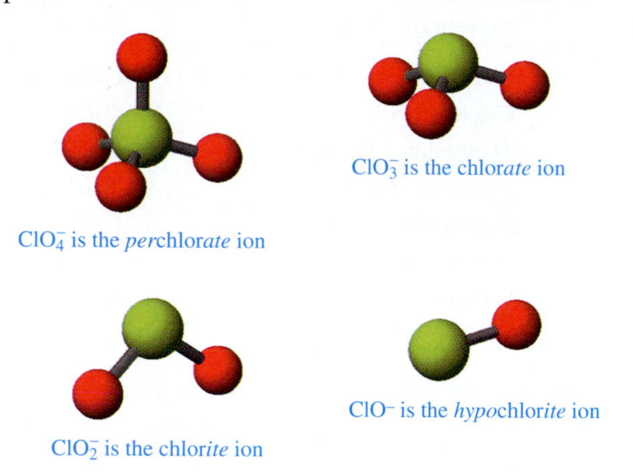

ClO_4^- is the *per*chlor*ate* ion

ClO_3^- is the chlor*ate* ion

ClO_2^- is the chlor*ite* ion

ClO^- is the *hypo*chlor*ite* ion

Oxoanions containing hydrogen are named simply by adding the word "hydrogen" before the name of the oxoanion, for example, hydrogen sulfate ion, HSO_4^-. If there are two hydrogens, we say *di*hydrogen, for example, *di*hydrogen phosphate for $H_2PO_4^-$ (and *mono*hydrogen phosphate for HPO_4^{2-}). Because many hydrogen-containing oxoanions have common names that are used often, you should know them. For example, the hydrogen carbonate ion, HCO_3^-, is often called bicarbonate ion.

Ion	Systematic Name	Common Name
HCO_3^-	Hydrogen carbonate ion	Bicarbonate ion
HSO_4^-	Hydrogen sulfate ion	Bisulfate ion
HSO_3^-	Hydrogen sulfite ion	Bisulfite ion

Naming Ionic Compounds

Table 3.6 lists a number of common ionic compounds. We will use these compounds to demonstrate the rules for systematically naming ionic compounds. *One basic naming rule is by now probably apparent—the name of the cation comes first, then the name of the anion.*

Notice the following from Table 3.6:

- Calcium oxide, CaO, is named from calcium for Ca^{2+} (Rule 1a) and oxide (Rule 2a). Likewise, sodium chloride is derived from sodium (Na^+, Rule 1a) and chloride (Cl^-, Rule 2a).
- The name ammonium carbonate for $(NH_4)_2CO_3$ comes from Rule 2b: ammonium for NH_4^+ and carbonate for CO_3^{2-}.
- Copper(II) sulfate indicates that Cu^{2+} is present. To derive the name from the formula, recall that the compound has no net charge. In the formula there is

TABLE 3.6 Common Ionic Compounds

Common Name	Systematic Name	Formula
Baking soda	Sodium hydrogen carbonate	$NaHCO_3$
Lime	Calcium oxide	CaO
Milk of magnesia	Magnesium hydroxide	$Mg(OH)_2$
Table salt	Sodium chloride	$NaCl$
Blue stone	Copper(II) sulfate	$CuSO_4$
Smelling salts	Ammonium carbonate	$(NH_4)_2CO_3$
Lye	Sodium hydroxide	$NaOH$

one copper ion and one sulfate ion. Because sulfate has a charge of 2−, the copper ion must have a compensating 2+ charge. Thus, the copper ion in $CuSO_4$ is Cu^{2+}, not Cu^+, the other possibility.

Exercise 3.9 Names of Ionic Compounds

Name the following ionic compounds:

(a) KNO_2 (d) $Mn_2(SO_4)_3$
(b) $NaHSO_3$ (e) Ba_3N_2
(c) $Mn(OH)_2$ (f) LiH

PROBLEM-SOLVING EXAMPLE 3.5 Names and Formulas of Ionic Compounds

Write the formula for ammonium monohydrogen phosphate and the name of $Fe(ClO_4)_2$.

Answer $(NH_4)_2HPO_4$; iron(II) perchlorate

Explanation Ammonium monohydrogen phosphate contains ammonium ions, NH_4^+, and monohydrogen phosphate ions, HPO_4^{2-}. Thus, two ammonium ions are needed for each monohydrogen phosphate ion, and the formula is $(NH_4)_2HPO_4$. Recall that where a polyatomic ion occurs more than once (ammonium ion in this compound), it is put into parentheses followed by the necessary subscript.

Recognize that Fe(II) indicates Fe^{2+} (Rule 1b), and that ClO_4^- is perchlorate ion. Therefore, this compound is called iron(II) perchlorate.

Problem-Solving Practice 3.5

Write the formula for each of the following ionic compounds.

(a) Potassium dihydrogen phosphate (d) Ammonium perchlorate
(b) Copper(I) hydroxide (e) Titanium(IV) chloride
(c) Sodium hypochlorite (f) Iron(II) sulfite

3.7 PROPERTIES OF IONIC COMPOUNDS

Like those of any compound, the properties of an ionic compound differ significantly from those of the component elements. Consider that most ordinary of ionic compounds, table salt (sodium chloride, NaCl), composed of Na^+ and Cl^- ions. Sodium

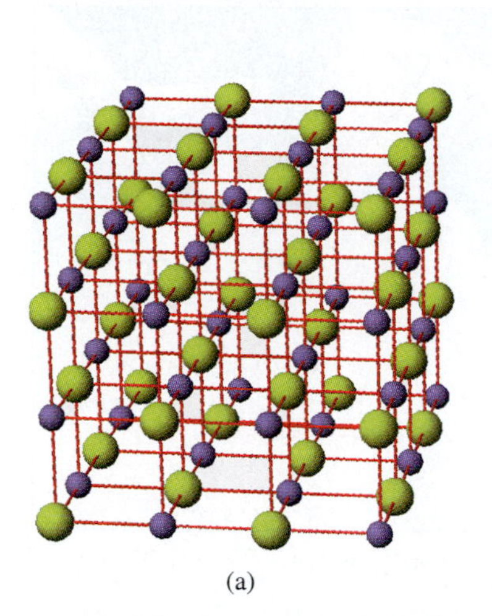
(a)

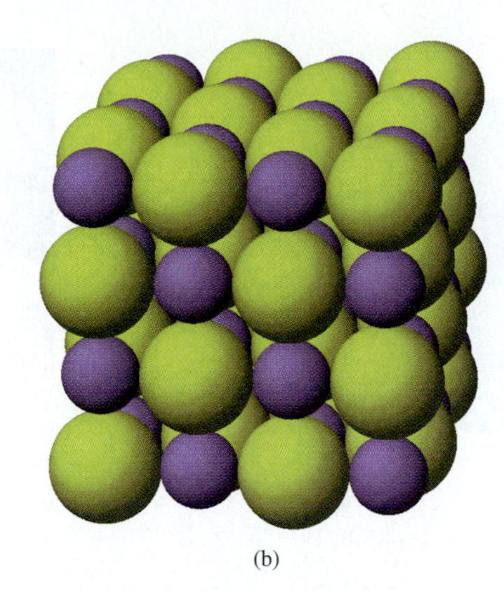
(b)

Figure 3.3 Two models of a sodium chloride crystal. (a) A ball-and-stick model; (b) a space-filling model. The lines between the ions in the ball-and-stick model are not chemical bonds but are simply reference lines to show the relationship of Na^+ (blue-gray) and Cl^- (yellow-green). This model illustrates clearly how the ions are arranged, although it shows the ions too far apart. The space-filling model more correctly shows how the ions are packed together, but it is difficult in this model to see the locations of ions other than those on the faces of the crystal.

chloride is a white, crystalline, water-soluble solid, very different from its component elements, metallic sodium and gaseous chlorine. Metallic sodium reacts violently with water. Chlorine is a diatomic, toxic gas that reacts with water. Sodium ions and chloride ions do not undergo such reactions; NaCl dissolves uneventfully in water. You eat sodium ions and chloride ions daily in food, but you would not want to eat elemental sodium or inhale chlorine gas. When the two *elements* join chemically, each sodium atom loses an electron to a chlorine atom, forming sodium ions, Na^+, and chloride ions, Cl^-. The ions in NaCl, a typical ionic compound, have considerably different properties from those of the neutral elemental sodium and chlorine atoms from which they are formed.

In ionic solids, cations and anions are held in an orderly array called a **crystal lattice,** such as that of NaCl (Figure 3.3). In a crystal lattice, each cation is surrounded by anions, and each anion is surrounded by cations. This arrangement maximizes the attraction between cations and anions and minimizes the repulsion between ions of like charge. In NaCl, as shown in Figure 3.3, six chloride ions surround each sodium ion, and six sodium ions surround each chloride ion. Therefore, the formula must show equal numbers of sodium ions and chloride ions. It could be written Na_6Cl_6, but the formula is always simplified to NaCl.

There are no discrete NaCl molecules in sodium chloride, nor are there molecules in other ionic compounds. The formula of an ionic compound indicates only the smallest whole number ratio of the number of cations to the number of anions in the compound. In NaCl that ratio is 1:1, and the formula indicates that there is one Na^+ for each Cl^-. An Na^+Cl^- pair is referred to as a **formula unit** of sodium chloride.

The regular array of ions in a crystal lattice gives ionic compounds two characteristic properties—high melting points and distinctive crystalline shapes. The melting points are related to the charges on the ions. For ions of similar size, such as O^{2-} and F^-, *the higher the charges, the higher the melting point,* because of the greater

Figure 3.4 Cleavage of an ionic crystal. (a) When an external force (*large gray arrow at top*) causes one layer of ions to shift slightly with respect to another, positive ions are brought close to other positive ions, and negative ions become nearest neighbors to other negative ions. The strong repulsive forces produced by this arrangement of ions (*double-headed black arrows*) cause the two layers to split apart. (b) A sharp blow on a knife edge lying along a plane of a salt crystal causes the crystal to split. *(b, C.D. Winters)*

attraction between ions of higher charge. For example, CaO (Ca^{2+} and O^{2-} ions) melts at 2572 °C, while NaF (Na^+ and F^-) melts at 993 °C.

Crystalline ionic solids have characteristic shapes because the ions are held rather rigidly in position by strong attractive forces. Such alignment of ions creates planes of ions within the crystal. Ionic crystals can be cleaved by an outside force that causes the planes of ions to shift slightly, bringing ions of like charge closer together (Figure 3.4). This repulsion causes the layers on opposite sides of the cleavage plane to separate, and the crystal splits apart.

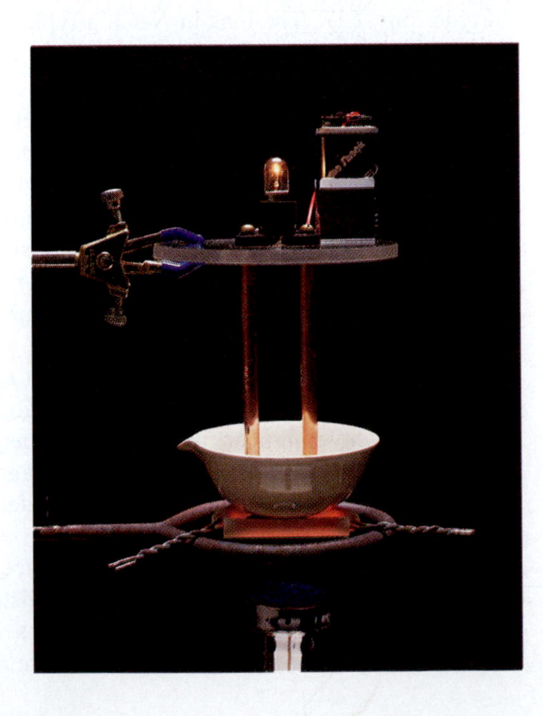

Figure 3.5 The conductivity of a molten salt. When an ionic compound is a solid, it will not conduct electricity. However, when the salt is melted, ions are free to move and migrate to the electrodes dipping into the melt (a). The lighted bulb shows that the electric circuit is complete (b). *(b, C.D. Winters)*

TABLE 3.7 Properties of Molecular and Ionic Compounds

Molecular Compounds	Ionic Compounds
Many are formed by combination of nonmetals with other nonmetals or with less reactive metals	Many are formed by combination of reactive metals with reactive nonmetals
Gases, liquids, and solids	Crystalline solids
Solids are brittle and weak, or soft and waxy	Hard and brittle
Low melting points	High melting points
Low boiling points (−250 °C to 600 °C)	High boiling points (700 °C to 3500 °C)
Poor conductors of heat and electricity	Good conductors of electricity when molten; poor conductors of heat and electricity when solid
Many are insoluble in water but soluble in organic solvents	Many are soluble in water
Examples: Hydrocarbons	Examples: NaCl, CaF_2

Because the ions in a crystal are only vibrating within fixed positions, ionic solids do not conduct electricity. However, when the solid melts, ions can move. Thus, molten ionic compounds conduct an electric current. (An electric current is the movement of charged particles from one place to another.) In a metal wire, the electric current is due to the movement of electrons through the wire. A molten ionic compound conducts electricity because cations and anions move in opposite directions toward positively and negatively charged **electrodes,** as shown in Figure 3.5. Ions moving through the liquid complete the electrical circuit and light the bulb.

The general properties of molecular and ionic compounds are summarized in Table 3.7. In particular, note the differences in physical state, electrical conductivity, melting point, and water solubility that result from the presence of ions or molecules.

3.8 IONIC COMPOUNDS IN AQUEOUS SOLUTIONS: ELECTROLYTES

Many ionic compounds are soluble in water. As a result, the oceans, rivers, lakes, and even the tap water in our residences contain many kinds of ions in solution. This makes the solubilities of ionic compounds and the properties of ions in solution of great practical interest. Of course, the oceans contain much higher concentrations of dissolved ions than do fresh water or municipal water supplies.

When an ionic compound dissolves in water, it *dissociates*—the oppositely charged ions separate from each other. For example, when solid NaCl dissolves in water, it dissociates into Na^+ and Cl^- ions that become uniformly mixed with water molecules and dispersed throughout the solution (Figure 3.6).

Aqueous solutions of ionic compounds, like molten ionic compounds, conduct electricity because the cations and anions are free to move about in the solution carrying electrical charges with them. All substances that conduct electricity when dissolved in water are called **electrolytes.** The electrical conductivity of the solution can be demonstrated by a simple experiment in which two electrodes, connected to a battery that supplies electricity, are immersed in an ionic solution. The positively charged

(Text continues on page 103.)

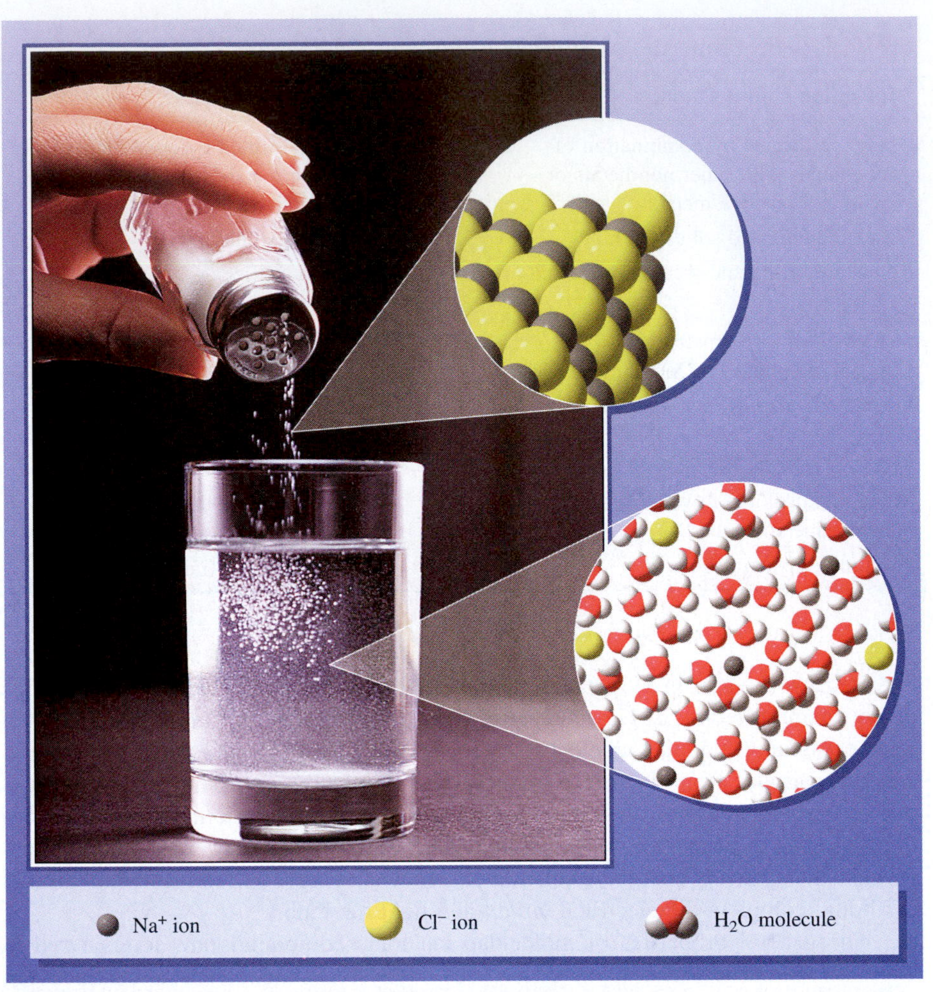

Figure 3.6 A model for the process of dissolving NaCl in water. The crystal dissociates to give Na^+ and Cl^- in aqueous (water) solution. These ions, which are surrouded by water molecules, are free to move about. Such solutions conduct electricity, so the substance dissolved is called an electrolyte. *(C.D. Winters)*

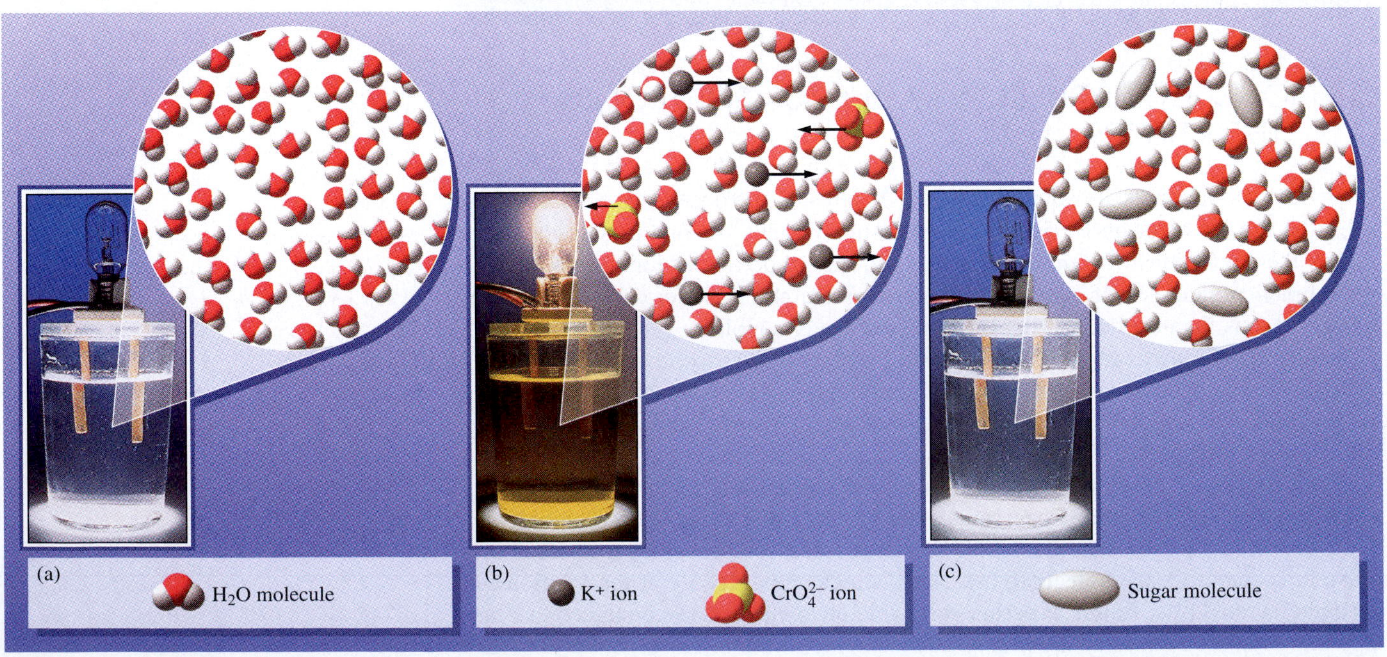

Figure 3.7 Conductivity of water and solutions. When an electrolyte is dissolved in water and provides ions that move about, the electrical circuit is completed between the electrodes, and the light bulb in the circuit glows. (a) Pure water is a nonelectrolyte and the bulb does not glow; (b) dilute K_2CrO_4, potassium chromate, is a strong electrolyte and the bulb glows brightly, indicating that the ions of every K_2CrO_4 unit have dissociated: $2K^+$ and CrO_4^{2-}. The CrO_4^{2-} ions move toward the positive electrode on the left, and the K^+ ions move toward the negative electrode on the right, transferring electric charge through the solution. (c) A sucrose solution does not conduct an electric current, indicating that sucrose is a nonelectrolyte; sucrose molecules do not ionize in aqueous solution. *(C.D. Winters)*

Portrait of a Scientist • *Michael Faraday (1791–1867)*

The terms "anion," "cation," "electrode," and "electrolyte" originated with Michael Faraday, one of the most influential persons in the history of chemistry. Faraday was apprenticed to a bookbinder in London, England, when he was only 13. This suited him, however, as he enjoyed reading the books sent to the shop for binding. One of these chanced to be a small book on chemistry, and his appetite for science was whetted. He soon began performing experiments on electricity, and in 1812 a patron of the shop invited Faraday to accompany him to a lecture at the Royal Institution by one of the most famous chemists of his day, Sir Humphry Davy. Faraday was so intrigued by Davy's lecture that he wrote to ask Davy for a position as an assistant. Faraday was accepted and began his work in 1813. His work was very fruitful, and Faraday was so talented that he was made the Director of the Laboratory of the Royal Institution approximately 12 years later.

It has been said that Faraday's contributions are so enormous that had there been Nobel Prizes when he was alive, he would have received at least six. These could have been for contributions such as

- the explanation of electromagnetic induction, which led to the first transformer and electric motor
- the proposal of the laws of electrolysis (the effect of electric current on chemicals)
- the discovery of the magnetic properties of matter
- the discovery of benzene and other organic chemicals (which led to important chemical industries)
- the discovery of the "faraday effect" (the rotation of the plane of polarized light by a magnetic field)
- the introduction of the notion of electric and magnetic fields

In addition to making discoveries that had profound effects on science, Faraday was an educator. He wrote and spoke about his work in memorable ways, especially in lectures to the general public that helped to popularize science.

Michael Faraday (*Oesper Collection in the History of Chemistry/University of Cincinnati*)

cations migrate to the negative electrode, and the negatively charged anions move to the positive electrode. So, a light bulb inserted into the circuit lights when the electrodes are immersed in the ionic solution (Figure 3.7b), indicating a complete circuit. The movement of the ions constitutes an electric current. The bulb will not light when the electrodes are not in the solution; the circuit is not complete.

Water-soluble ionic compounds are known as **strong electrolytes** because they are completely (100%) dissociated in solution. The high concentration of ions in solution makes the solution a good conductor of electricity (Figure 3.7b).

Substances whose solutions do not conduct electricity are called **nonelectrolytes.** These are generally molecular compounds that do not ionize in solution, such as sugar, ethanol, and ethylene (Figure 3.7c).

We use the term "dissociate" for ionic compounds that separate into their constituent ions in water. The term "ionize" is used for nonionic compounds, like acetic acid, whose molecules react with water to form ions.

Some substances, which are described in Section 4.5, are only partly converted into ions in solution and are known as weak electrolytes.

3.9 IONIC COMPOUNDS IN AQUEOUS SOLUTIONS: SOLUBILITY

Many of the ionic compounds that you frequently encounter, such as table salt, baking soda, and household-plant fertilizers, are soluble in water. It is therefore tempting to conclude that all ionic compounds are soluble in water, but such is not the case. Although many ionic compounds are water-soluble, some are only slightly soluble, and others dissolve hardly at all.

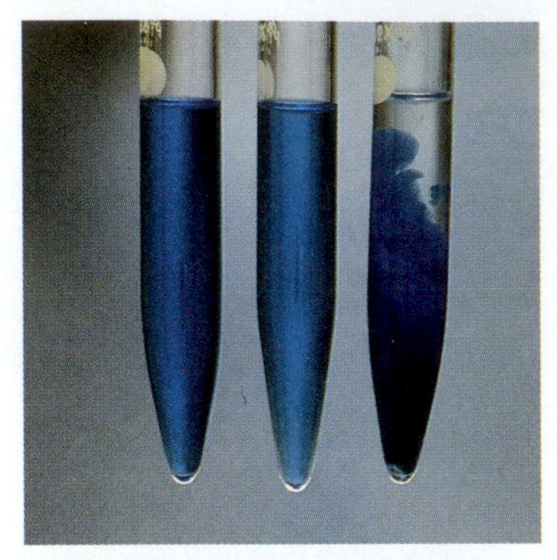

$Cu(NO_3)_2/CuSO_4/Cu(OH)_2$

$AgNO_3/AgCl/AgOH$

$HgS/CdS/Sb_2S_3/PbS$

Figure 3.8 Illustration of some solubility guidelines of Table 3.8. $AgNO_3$ and $Cu(NO_3)_2$, like all nitrates, are soluble. $Cu(OH)_2$ and AgOH, like most hydroxides, are insoluble. CdS, Sb_2Sb_3, and PbS, like nearly all sulfides, are insoluble, but $(NH_4)_2S$ is an exception (it is soluble). *(C.D. Winters)*

TABLE 3.8 Solubility Rules for Ionic Compounds

Ammonium, Sodium, Potassium; NH_4^+, Na^+, K^+	All ammonium, sodium, and potassium salts are soluble.
Nitrates, NO_3^-	All nitrates are soluble.
Chlorides, Cl^-	All chlorides are soluble except AgCl, Hg_2Cl_2, and $PbCl_2$.
Sulfates, SO_4^{2-}	Most sulfates are soluble; exceptions include $SrSO_4$, $BaSO_4$, and $PbSO_4$.
Chlorates, ClO_3^-	All chlorates are soluble.
Perchlorates, ClO_4^-	All perchlorates are soluble.
Acetates, $CH_3CO_2^-$	All acetates are soluble.
Phosphates, PO_4^{3-}	All phosphates are insoluble except those of NH_4^+ and Group 1A elements (alkali metal cations).
Carbonates, CO_3^{2-}	All carbonates are insoluble except those of NH_4^+ and Group 1A elements (alkali metal cations).
Hydroxides, OH^-	All hydroxides are insoluble except those of NH_4^+, Group 1A (alkali metal cations), $Sr(OH)_2$, and $Ba(OH)_2$; $Ca(OH)_2$ is slightly soluble.
Oxides, O^{2-}	All oxides are insoluble except those of Group 1A (alkali metal cations).
Oxalates, $C_2O_4^{2-}$	All oxalates are insoluble except those of NH_4^+ and Group 1A (alkali metal cations).
Sulfides, S^{2-}	All sulfides are insoluble except those of NH_4^+, Group 1A (alkali metal cations), and Group 2A (MgS, CaS, and BaS are sparingly soluble).

The solubility rules given in Table 3.8 are general guidelines to predicting the water solubilities of ionic compounds based on the ions they contain. *If a compound contains at least one of the ions indicated for soluble compounds in Table 3.8, then the compound is at least moderately soluble.*

Figure 3.8 shows examples illustrating the solubility rules, especially comparing nitrates, chlorides, and hydroxides of various metal ions.

Applying the solubility rules, for example, suppose you want to know whether $NiSO_4$ is soluble. $NiSO_4$ contains Ni^{2+} and SO_4^{2-} ions. Although Ni^{2+} is not mentioned in Table 3.8, substances containing SO_4^{2-} are described as soluble (except for $SrSO_4$, $BaSO_4$, and $PbSO_4$). Because $NiSO_4$ contains an ion (SO_4^{2-}) that indicates solubility, it is predicted to be soluble.

PROBLEM-SOLVING EXAMPLE 3.6 Using Solubility Rules

Indicate which ions are present in each of the following compounds. Then predict whether each is likely to be water-soluble.

(a) NH_4Cl (b) $Mg_3(PO_4)_2$ (c) $Ba(OH)_2$ (d) $Ca(NO_3)_2$

Answer (a) NH_4^+, Cl^-; (b) Mg^{2+}, PO_4^{3-}; (c) Ba^{2+}, OH^-; (d) NH_4Cl, $Ba(OH)_2$, and $Ca(NO_3)_2$ are soluble; $Mg_3(PO_4)_2$ is insoluble.

Explanation The use of solubility rules requires identifying the ions present and checking their effects on water solubility (Table 3.8).
(a) NH_4Cl contains NH_4^+ and Cl^- ions. The presence of NH_4^+ means that the compound is soluble.
(b) $Mg_3(PO_4)_2$ is composed of Mg^{2+} and PO_4^{3-} ions. Most phosphates are insoluble, with the exception of those containing ammonium ion or alkali metal cations. Because Mg^{2+} is not an alkali metal ion, $Mg_3(PO_4)_2$ is insoluble.
(c) Most hydroxides are insoluble, but not $Ba(OH)_2$. It is soluble.
(d) $Ca(NO_3)_2$ contains Ca^{2+} and NO_3^- (nitrate) ions. Table 3.8 indicates that all nitrates are soluble, and so $Ca(NO_3)_2$ is soluble.

Problem-Solving Practice 3.6

Predict whether each of the following compounds is likely to be water-soluble.
(a) NaF (b) $Ca(CH_3CO_2)_2$ (c) $SrCl_2$ (d) MgO (e) $PbCl_2$ (f) HgS

3.10 THE BIOLOGICAL PERIODIC TABLE

Most of the more than 100 known elements are not directly involved with our personal health and well-being. However, about 26 of them are absolutely essential to us. Among these elements, indicated in Figure 3.9, are 13 metals, 10 nonmetals, and possibly 3 metalloids (B, Si, As), each with its role to play at the cellular level. All are part of a well-balanced diet.

Table 3.9 lists the essential elements in the order of their relative abundances per million atoms in the body. For example, among each million atoms there are nearly 95,000 carbon atoms but only about 400 sodium ions. Thus, the essential elements

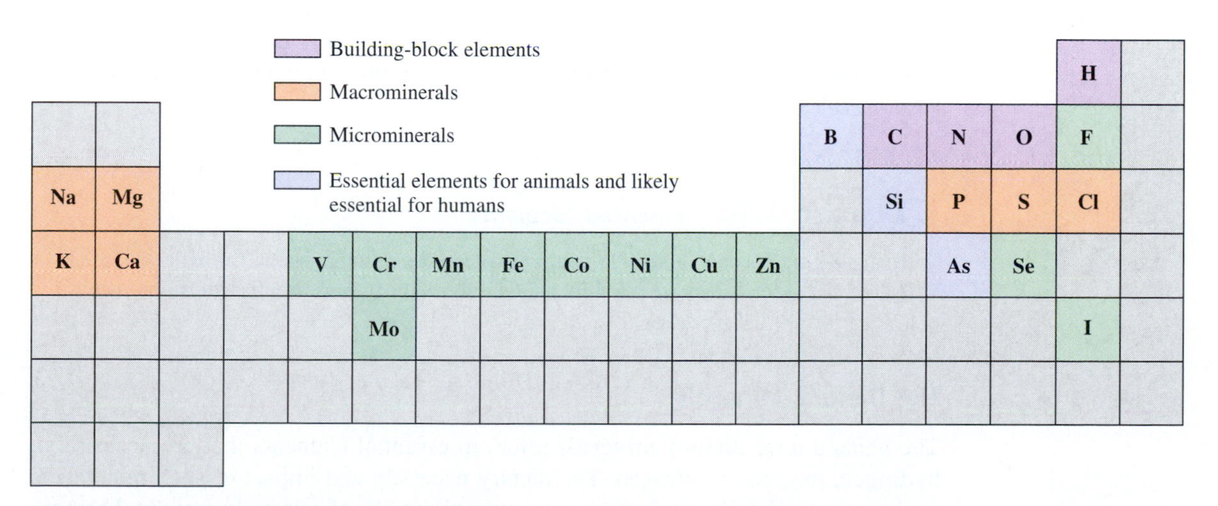

Figure 3.9 Elements essential to human health. Four elements, C, H, N, and O (violet), form the many organic compounds that make up living organisms. The macrominerals (orange) are required in relatively large amounts; microminerals (green) are required in lesser amounts. Some elements, B, Si, and As (blue), are essential in animals and are likely to be essential for humans, but have not yet been demonstrated to be so.

TABLE 3.9 Major Elements of the Human Body

Element	Symbol	Relative Abundance in Atoms/Million Atoms in the Body
Oxygen	O	255,000
Carbon	C	94,500
Hydrogen	H	630,000
Nitrogen	N	13,500
Calcium	Ca	3100
Phosphorus	P	2200
Chlorine	Cl	570
Potassium	K	260
Sulfur	S	490
Sodium	Na	410
Magnesium	Mg	130

are not needed in equal amounts; nevertheless, each is necessary. Table 3.9 shows the preeminence of four of the nonmetals—oxygen, carbon, hydrogen, and nitrogen. They contribute most of the atoms in the biologically significant chemicals—the biochemicals—composing all plants and animals. With few exceptions, biochemicals are organic compounds characterized by frameworks of carbon atoms with associated hydrogen atoms and functional groups containing oxygen, nitrogen, phosphorus, and sometimes sulfur. Clearly, carbon, hydrogen, oxygen, and nitrogen atoms are the major building blocks for compounds in the body. In the next section we will introduce you to two major types of biochemicals: fats and sugars.

If you weigh 150 lb, about 90 lb (60%) is water, 30 lb is fat, and the remaining 30 lb is a combination of proteins, carbohydrates, and calcium, phosphorus, and other dietary minerals.

Nonmetals are also present as anions in body fluids, including chloride ion (Cl^-), phosphorus in three ionic forms (PO_4^{3-}, HPO_4^{2-}, and $H_2PO_4^-$), and carbon as bicarbonate ion (HCO_3^-) and carbonate ion (CO_3^{2-}). Metals are present in the body as cations in solution (for example, Na^+, K^+), in solids (Ca^{2+} in bones and teeth), and incorporated into large biomolecules (for example, Fe^{2+} in hemoglobin and Co^{3+} in vitamin B-12).

Exercise 3.10 Essential Elements

Using Figure 3.9 and Table 3.9, identify (a) the essential nonmetals; (b) the essential alkaline earth metals; (c) the essential halide ions; and (d) four essential transition metals.

The Dietary Minerals

The general term **dietary minerals** refers to essential elements that are not carbon, hydrogen, oxygen, or nitrogen. The dietary necessity and impact of such minerals is far beyond their collective presence as only about 4% of our body weight. They exemplify the old saying that "good things come in small packages." Because the body uses them efficiently, recycling them through many reactions, dietary minerals are required in only small amounts. But their absence from your diet can cause significant health problems.

CHEMISTRY YOU CAN DO

Pumping Iron: How Strong Is Your Breakfast Cereal?

Iron is an essential dietary mineral that enters our diets in many ways. To prevent dietary iron deficiency through poor diets, the U.S. Food and Drug Administration (FDA) allows food manufacturers to "fortify" (add iron to) their products. One way of fortifying iron in a cereal is to add an iron compound, such as iron(III) phosphate, that dissolves in the stomach acid (HCl). The iron in Total cereal contains particles of pure iron metal baked into the flakes. Can you detect metallic iron in a cereal? You can find out with the following experiment for which you will need

- a reasonably strong magnet. You might find a good one holding things to your refrigerator door or in a toy store or hobby shop. Agricultural supply stores will have "cow magnets," which will work well.
- a plastic Ziplock freezer bag (quart size) and a rolling pin (or something else to crush the cereal).
- one serving of Total (30–35 g).

Put the serving of Total into the plastic bag and crush the cereal into small particles. Place the magnet into the bag and mix it well with the cereal for several minutes. Carefully remove the magnet from the bag and examine the magnet closely. Is anything clinging to the magnet that was not there before?

Think about the following questions:

- Based on this experiment, is there metallic iron in the cereal?
- What happens to the metallic iron after you swallow it?
- Does this iron contribute to your daily iron requirement?

The 22 dietary minerals indicated in Figure 3.9 are classified into the relatively more abundant **macrominerals** and the less plentiful **microminerals.** Macrominerals are those present in quantities greater than 0.08 g (80 mg) per kilogram of body weight, which is more than 5 g for a 60-kg (132-lb) individual for example. Microminerals are present in smaller, sometimes far smaller, amounts. For example, the daily total intake of iodine is only $150 \mu g$.

The cationic macrominerals are Na^+, K^+, Mg^{2+}, and Ca^{2+}. Chloride (Cl^-) is an anionic macromineral, the most abundant of the halide ions. You might be surprised that some of the essential elements, such as arsenic and selenium, are normally thought of as toxic. They are toxic, but only at higher concentrations. Selenium is well established as an essential micromineral in trace amounts, 55–70 μg per day. Arsenic has been established as essential for animals and is probably essential for humans.

There are also a number of essential transition metals—vanadium, chromium, manganese, cobalt, nickel, copper, zinc, molybdenum, and cadmium—in addition to the more familiar iron. Each transition metal plays a specific metabolic role. Many of these elements have become recognized as micronutrients only relatively recently as a result of nutritional research. Improved methods of chemical analyses have made possible the detection of smaller quantities of essential dietary substances. As a result, the list of essential elements continues to expand. The initial studies are usually done with animals by eliminating a suspected trace mineral from their diets and observing any changes in the test animals' health.

3.11 BIOMOLECULES: CARBOHYDRATES AND FATS

One benefit of understanding organic molecules is the application of this understanding to the molecules of life—organic molecules that we eat and that make up our bodies. You have already learned enough chemistry to consider a few simple biomolecules. In this section, we're going to examine the structural formulas of two simple carbohydrates, glucose and sucrose, and tristearin, a typical solid fat.

Carbohydrates

Carbohydrates make up about half of the average human diet. The word "carbohydrate" literally means "hydrate of carbon." **Carbohydrates** have the general formula $C_x(H_2O)_y$ in which x and y are whole numbers. The molecular formulas for glucose, $C_6H_{12}O_6$ ($x = 6$, $y = 6$), and sucrose (table sugar), $C_{12}H_{22}O_{11}$ ($x = 12$, $y = 11$), illustrate the general formula. Carbohydrates originally were thought to be simple combinations of carbon and water, but this is not the case. We now know that the carbon, hydrogen, and oxygen atoms in carbohydrates are arranged primarily in alcohol, aldehyde, and ketone functional groups (in color):

Some atoms in molecules are connected by a "double" bond (represented by ═, rather than a single bond represented by —). The nature of these bonds is described in Chapter 9.

Like alcohols, aldehydes and ketones are major classes of organic compounds. The simplest aldehyde and ketone are formaldehyde, H—C(═O)—H, and acetone, $CH_3C(=O)CH_3$, respectively.

—C—O—H (alcohol group)

—C—C(═O)—H (aldehyde group)

—C—C(═O)—C— (ketone group)

Carbohydrates are divided into three classes. **Monosaccharides** (from the Latin *saccarum,* "sugar") are the simplest carbohydrates, and glucose is the most important one. Its structural formula can be written with the six carbon atoms in a straight chain; or the first carbon can join with the oxygen on the fifth carbon to form a ring. Note that each carbon atom in glucose is connected to an oxygen atom and that all but one of the oxygen atoms is in an –OH group. The other oxygen atom is part of an aldehyde group in the chain structure.

```
                    H
                    |
       H   H   H    O   H   O
       |   |   |    |   |   ||
H—O—C—C—C—C—C—C—H
       |   |   |    |   |
       H   O   O    H   O
           |   |        |
           H   H        H
```

Chain structural formula and ball-and-stick molecular model for glucose

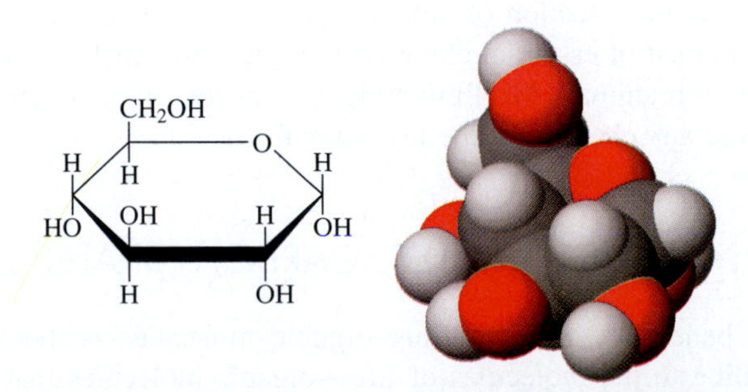

Ring structural formula and space-filling molecular model for glucose

For simplicity, the structural formulas of glucose and other carbohydrates are drawn without the carbon atoms in the ring. The carbon atoms are represented by the junction of the lines in the plane of the ring, as shown above.

Plants make glucose by photosynthesis (Section 7.xx). It is stored as starch or used to make cellulose for structural support. Animals rely on glucose as their major energy source—we metabolize it to produce the energy needed to stay alive (Section 7.xx), just as furnaces burn natural gas and fuel oil to heat where we live and work. Although glucose is present in fruits and honey, we don't have to eat our daily supply of glucose because chemical reactions in our bodies make glucose from more plentiful molecules in our diets. Serious health problems arise when the glucose concentration in the blood is either too high or too low. For example, diabetes results from the body's failure to regulate glucose concentration at the proper level.

Sucrose is representative of the second class of carbohydrates known as **disaccharides.** As you might guess from the name, disaccharides are molecules in which two monosaccharides are connected to each other. A sucrose molecule is made of two monosaccharide molecules in their ring forms—glucose and fructose.

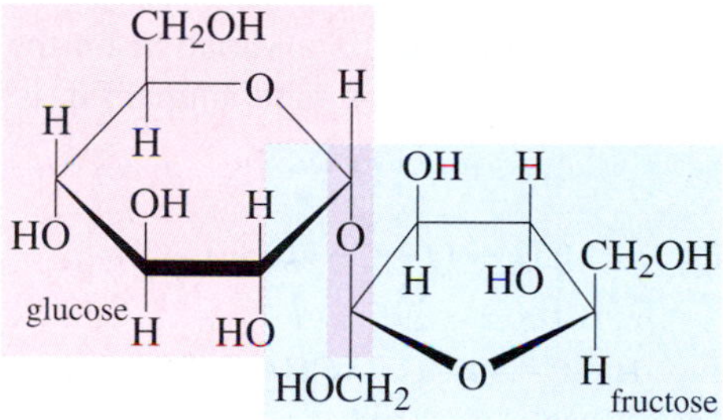

Note that, like sucr*ose,* gluc*ose,* and fruct*ose,* the names of all mono- and disaccharides end in *-ose.*

Monosaccharides and disaccharides together are referred to as **simple sugars.** About half of the simple sugars in our diet come from milk and fresh fruit; the remainder come from sweeteners added to prepared foods. High-purity sucrose is produced on an enormous scale from sugar beets and sugar cane. Table 3.10 compares the sweetness of common sugars and artificial sweeteners with that of sucrose.

The structure 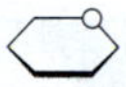 represents the five carbon atoms and the oxygen atom in the six-membered glucose ring.

Glucose is also known as dextrose or blood sugar.

TABLE 3.10 **Sweetness of Common Sugars and Artificial Sweeteners Relative to Sucrose**

Substance	Sweetness Relative to Sucrose at 1.00
Lactose (milk sugar, a disaccharide)	0.16
Galactose (a monosaccharide in milk sugar)	0.32
Maltose (a disaccharide used in beer making)	0.33
Glucose (dextrose, a common monosaccharide)	0.74
Sucrose (table sugar, a disaccharide)	1.00
Fructose (fruit sugar, a monosaccharide)	1.74
Aspartame (artificial sweetener, Nutrasweet)	180
Saccharin (artificial sweetener)	300

The open-chain form of fructose contains a ketone group; the ring structure does not.

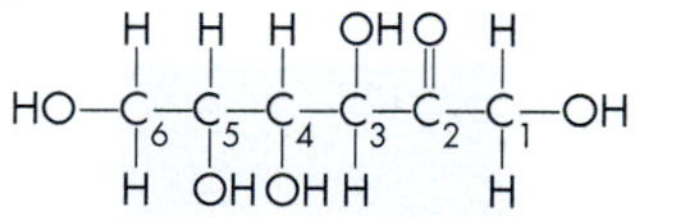

From Table 3.10 you can see that aspartame is nearly 200 times sweeter than sucrose. This means you need only a "pinch" of aspartame to be equivalent to a teaspoon of table sugar.

Dietary fiber is composed of polysaccharides. *(C.D. Winters)*

Polysaccharides are polymers of monosaccharides. The chemistry of natural and synthetic polymers is discussed in Section 11.1.

The carboxylic acid group is present in all organic acids. Acids of all kinds are an important class of compounds, as is discussed in Chapter 17.

Cooking oils are liquids at room temperature; butter is a solid. The difference between liquid oils and solid fats relates to how the carbon atoms in the fatty acids are connected to each other. *(C.D. Winters)*

The third class of carbohydrates comprises the **polysaccharides,** which consist of many monosaccharide molecules joined together, up to several thousand of them. Cellulose is a polysaccharide built from glucose units. Together with other carbohydrates, cellulose forms wood, cotton, and paper. Cellulose also acts as a dietary fiber in our foods because we cannot digest it. We can, however, digest starch, another glucose-based polysaccharide. The glucose units in cellulose are joined in a slightly different arrangement from that in starch. This slight difference in molecular shapes plays a significant role in how the body handles these two biomolecules.

Exercise 3.11 Carbohydrates

Which of the following four compounds might be a carbohydrate?
(a) $C_9H_8O_4$ (b) $C_5H_{10}O_5$ (c) $C_5H_7NO_4$ (d) $C_6H_{10}O_5$

Fats and Oils

The structure of all fat and oil molecules is a variation on the same structure. The condensed formula for tristearin, a very common animal fat, is shown below.

$$\begin{array}{l} CH_2{-}O{-}\overset{\displaystyle O}{\overset{\|}{C}}{-}(CH_2)_{16}CH_3 \\ \quad | \\ H{-}C{-}O{-}\overset{\displaystyle O}{\overset{\|}{C}}{-}(CH_2)_{16}CH_3 \\ \quad | \\ CH_2{-}O{-}\overset{\displaystyle O}{\overset{\|}{C}}{-}(CH_2)_{16}CH_3 \end{array}$$

tristearin, a natural fat

To illustrate how all fats and oils are similar, we can separate the tristearin molecule into its component parts: glycerol and stearic acid.

$$\begin{array}{l} CH_2OH \\ | \\ CHOH \\ | \\ CH_2OH \end{array} \qquad CH_3(CH_2)_{16}\overset{\displaystyle O}{\overset{\|}{C}}{-}OH$$

glycerol — stearic acid, a fatty acid

Glycerol has three alcohol functional groups, the –OH groups. Stearic acid is representative of fatty acids, all of which are molecules with long chains of carbon atoms with attached hydrogen atoms. At the end of the carbon chain is a **carboxylic acid** functional group, $-CO_2H$. Glycerol is a syrupy liquid used in hydraulic fluid, in cosmetics, and as a sweetener. Stearic acid, a colorless, waxy solid obtained from animal fat, is used in soaps, lubricants, shoe polish, and other products.

Earlier we noted that the functional groups are where the chemical action is in organic molecules. In tristearin formation, the carboxylic acid groups of three molecules of stearic acid react with the three –OH groups of a glycerol molecule. A water molecule is formed and released where each stearic acid molecule is joined to the glycerol molecule. All fats and oils have the same basic molecular structure—three long hydrocarbon chains from three fatty acids linked to glycerol. The fatty acids usually have an even number of carbon atoms, ranging from about 4 to 20.

(Text continues on page 112.)

CHEMISTRY IN THE NEWS

Olestra: Sugar + Fat = Low Calories

Imagine being able to eat potato chips that taste good, contain no digestible fat, and have only half the calories of regular potato chips. That is now possible with Olestra, a fat substitute developed by Procter and Gamble over two decades at a cost of nearly $300 million. On January 24, 1996, the U.S. Food and Drug Administration (FDA) approved the use of Olestra in savory snack foods, the first synthetic nutrient to gain FDA approval in 20 years since aspartame, now a widely used artificial sweetener (NutraSweet).

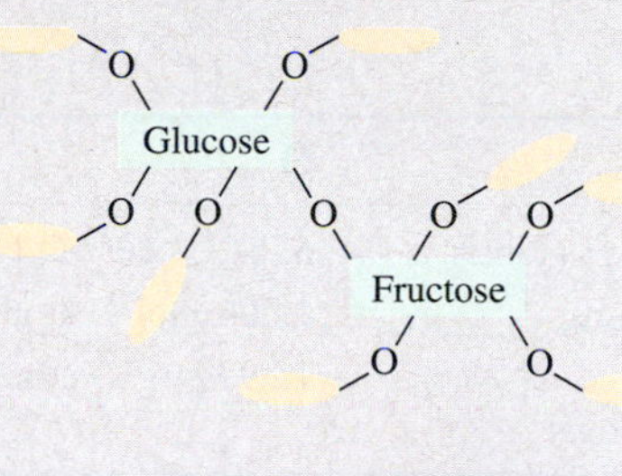

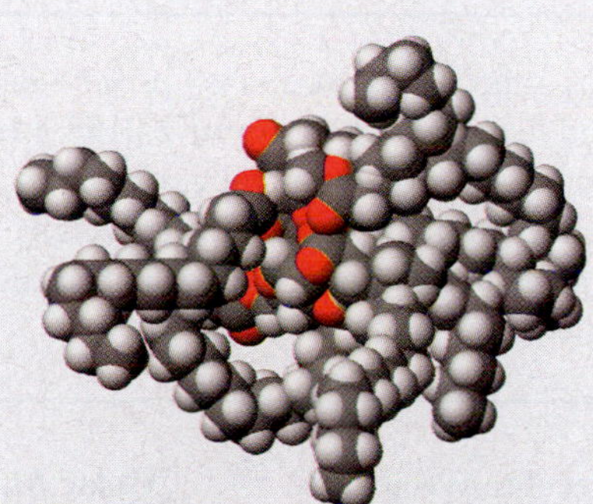

A stylized structural formula of Olestra. The $CH_3(CH_2)_{16}CO$ groups (represented by the colored ellipses) join to oxygen atoms in the $-OH$ groups of sucrose. The molecular model below shows how these groups surround the sucrose molecules.

Olestra consists of a sucrose (table sugar) molecule bonded to six, seven, or eight naturally occurring fatty acid molecules, which have the general formula $CH_3(CH_2)_nCO_2H$, where n is typically 16. Alcohol groups ($-OH$) on the sucrose molecule combine with the acid group ($-CO_2H$) of each fatty acid molecule. The result is elimination of water and addition of the $CH_3(CH_2)_{16}\overset{O}{\overset{\|}{C}}$ fragment of the acid to the oxygen of the $-OH$ groups in the sucrose molecule, yielding molecules like the one shown below, which has eight fatty acid groups connected to sucrose. (See the structural formula for sucrose on p. 109.)

Snack foods prepared with Olestra taste fatty. Because digestive enzymes cannot break down Olestra, however, it passes through the body undigested, thereby contributing no dietary calories (but other substances in the chips do). A 1-ounce serving of potato chips with Olestra contains no digestible fat and 75 Calories, compared with a similar serving of regular potato chips that weighs in at 10 g of digestible fat and 150 Calories. Currently, Olestra is limited to use in salty snack foods; and before being given the green light to be used in other typically fat-laden foods like pastries and doughnuts, Olestra must undergo additional FDA review.

In announcing the approval of Olestra, FDA Commissioner Dr. David Kessler called the new fat substitute "probably one of the most extensively studied food substances to date," saying that the "research data on Olestra demonstrate reasonable certainty of no harm for use in certain snack foods" (like potato chips, crackers, and tortilla chips). But Olestra is controversial because it slightly inhibits the absorption of the fat-soluble vitamins—A, D, E, and K. This is countered by adding these vitamins to Olestra in foods, but critics are concerned about possible long-term effects of inhibiting vitamin absorption. Another possible drawback is that some people who eat Olestra, especially in large amounts, have gastrointestinal discomfort and loose stools because Olestra is oily and not digested. Packages carry a label warning consumers of such possible side effects and provide a toll-free number for complaints to the manufacturer, which will be forwarded to the FDA. The possible side effects seemed not to affect sales of Olestra-containing snack chips in three test markets where 70,000 bags were sold in the first month (April 1996).

Source: Adapted from Journal of Chemical Education, *Vol. 68, pp. 476–479, June 1991; and the* New York Times, *May 20, 1996.*

Within our bodies, fat is a form of stored energy. Food that does not release its stored energy as it is digested ends up being converted to fat that is stored. Fats and oils are the most concentrated energy source in our diets. One gram of fat releases more than twice the amount of energy available from 1 g of carbohydrate. The average American diet currently derives about 37% of its calories from fats and oils. Nutritionists urge that this be reduced to 30% because of mounting evidence that the risk of heart disease increases with the quantity of fats and oils consumed.

3.12 MOLES OF COMPOUNDS

Molar Mass of Molecular Compounds

One mole of a compound does *not* mean one molecule. It means 6.022×10^{23} molecules.

The most recognizable formula, H_2O, shows us that there are two H atoms for every O atom in a water molecule. In two water molecules there are a total of four H atoms and two O atoms; in a dozen water molecules there are 2 dozen H atoms and 1 dozen O atoms. We can extend this until we have one mole of water molecules (Avogadro's number, 6.022×10^{23}), each containing two hydrogen atoms and one oxygen atom (← *p. 56*). We can also say that in 1.000 mol of water there are 2.000 mol of H atoms and 1.000 mol of O atoms:

H_2O	H	O
6.022×10^{23} water molecules	2 (6.022×10^{23} H atoms)	6.022×10^{23} O atoms
1.000 mol H_2O molecules	2.000 mol H atoms	1.000 mol O atoms
18.0152 g H_2O	2 (1.0079 g H) = 2.0158 g H atoms	15.9994 g O atoms

The mass of one mole of water molecules—the *molar mass* (← *p. 56*)—is the sum of the masses of two moles of H atoms and one mole of O atoms: 2.0158 g H + 15.9994 g O = 18.0152 g of water in a mole of water. For chemical compounds, the molar mass, in grams per mole, is numerically the same as the molecular weight, the sum of the atomic weights of *all* the atoms in the compound's formula. The molar masses of several molecular compounds are given below.

Compound	Structural Formula	Molecular Weight	Molar Mass
NH_3	H—N(—H)—H	14.0067 amu, N + 3(1.0079 amu, H) = 17.0304 amu	17.0304 g/mol
CHF_3	F—C(H)(F)—F	12.011 amu, C + 1.0079 amu, H + 3(18.9984 amu, F) = 70.014 amu	70.014 g/mol
SO_2	O—S=O	32.066 amu, S + 2(15.9994 amu, O) = 64.065 amu	64.065 g/mol
$C_3H_8O_3$ (glycerol)	CH_2OH—CHOH—CH_2OH	3(12.011 amu, C) + 8(1.0079 amu, H) + 3(15.9994 amu, O) = 92.094 amu	92.094 g/mol

Figure 3.10 One-mole quantities of some ionic compounds. Clockwise from front right, they are NaCl, white (58.44 g/mol); $CuSO_4 \cdot 5\ H_2O$, blue (249.7 g/mol); $NiCl_2 \cdot 6\ H_2O$, green (237.70 g/mol); $K_2Cr_2O_7$, orange (294.2 g/mol); and $CoCl_2 \cdot 6\ H_2O$, red (237.9 g/mol).
(C.D. Winters)

Molar Mass of Ionic Compounds

Because ionic compounds do not contain individual molecules, the formulas for ionic compounds indicate formula units. Therefore, the term "formula weight" is sometimes used for ionic compounds instead of molecular weight. Like the molecular weight, the formula weight is the sum of the atomic weights of *all* the atoms in the compound's formula. The molar mass of an ionic compound, expressed in grams per mole (g/mol), is numerically equivalent to its formula weight. The term "molar mass" is used for both molecular and ionic compounds. Figure 3.10 pictures one mole of several ionic compounds.

Compound	Formula Weight	Molar Mass
NaCl	22.9898 amu, Na + 35.4527 amu, Cl = 58.4425 amu	58.4425 g/mol
MgO	24.3050 amu, Mg + 15.9994 amu, O = 40.3044 amu	40.3044 g/mol
K_2S	2(39.0983 amu, K) + 32.066 amu, S = 110.263 amu	110.263 g/mol
$Ca(NO_3)_2$	40.078 amu Ca + 2(14.0067 amu, N) + 6(15.9994 amu, O) = 164.088 amu	164.088 g/mol
$Mg_3(PO_4)_2$	3(24.3050 amu, Mg) + 2(30.9738 amu, P) + 8(15.9994 amu, O) = 262.8578 amu	262.8578 g/mol

Notice that $Mg_3(PO_4)_2$ has 2 P atoms and 2×4 O atoms because there are two PO_4^{3-} ions in the formula.

Exercise 3.12 Molar Masses

Calculate the molar mass of each of the following compounds:
(a) K_2HPO_4; (b) $C_{27}H_{46}O$ (cholesterol); (c) $Mn_2(SO_4)_3$; (d) $C_8H_{10}N_4O_2$ (caffeine).

Mass-Mole Conversions

As you might expect, it is essential to be able to do "mass-to-mole" and "mole-to-mass" conversions for compounds, just as we did for elements in Section 2.11. The key to such conversions is using the molar mass of the compound as a conversion factor. This is analogous to mass-mole conversions for elements using the molar mass of the element.

PROBLEM-SOLVING EXAMPLE 3.7 Grams and Moles

Aspartame is a widely used artificial sweetener (NutraSweet) that is 200 times sweeter than sucrose (← *p. 109, Table 3.10*). One sample of aspartame, $C_{14}H_{18}N_2O_5$, weighs 1.80 g; another contains 0.220 mol of aspartame. To see which sample is larger, answer these questions:
(a) Calculate the molar mass of aspartame.
(b) How many moles of aspartame are in the 1.80-g sample?
(c) How many grams of aspartame are in the other sample?

Answers (a) 294.3 g/mol; (b) 6.11×10^{-3} mol; (c) 64.8 g. The 0.220-mol sample is larger.

Explanation Determining which sample has the larger mass requires using steps a, b, and c to find the mass of the 0.220-mol sample.
(a) The molar mass is numerically the same as the molecular weight, the sum of the atomic weights of all the atoms in the formula:

14 C atoms (12.011 amu/C atom) + 18 H atoms (1.008 amu/H atom) + 2 N atoms (14.007 amu/N atom) + 5 O atoms(15.999 amu/O atom) = 294.3 amu.

The molar mass of aspartame is 294.3 g/mol.
(b) Use the molar mass to calculate moles from grams.

$$1.80 \text{ g aspartame} \times \frac{1 \text{ mol aspartame}}{294.3 \text{ g aspartame}} = 6.11 \times 10^{-3} \text{ mol aspartame}$$

(c) Converting moles to grams is the opposite of item (b):

$$0.220 \text{ mol aspartame} \times \frac{294.3 \text{ g aspartame}}{1 \text{ mol aspartame}} = 64.8 \text{ g aspartame}$$

This sample is the larger one.

Problem-Solving Practice 3.7

(a) Calculate the number of moles in 10.0 g of each of the following compounds: $C_{27}H_{46}O$ (cholesterol); $Mn_2(SO_4)_3$.
(b) Calculate the number of grams in 0.25 mol of each of the following compounds: K_2HPO_4; $C_8H_{10}O_2N_4$ (caffeine).

TABLE 3.11 Some Common Hydrated Ionic Compounds

Compound	Systematic Name	Common Name	Uses
$Na_2CO_3 \cdot 10\ H_2O$	Sodium carbonate decahydrate	Washing soda	Water softener
$Na_2S_2O_3 \cdot 5\ H_2O$	Sodium thiosulfate pentahydrate	Hypo	Photography
$MgSO_4 \cdot 7\ H_2O$	Magnesium sulfate heptahydrate	Epsom salt	Cathartic dyeing and tanning
$CaSO_4 \cdot 2\ H_2O$	Calcium sulfate dihydrate	Gypsum	Wallboard
$CaSO_4 \cdot \frac{1}{2}\ H_2O$	Calcium sulfate hemihydrate	Plaster of Paris	Casts, molds
$CuSO_4 \cdot 5\ H_2O$	Copper(II) sulfate pentahydrate	Blue vitriol	Insecticide

Exercise 3.13 Moles and Formulas

Is the following statement true? "Two different compounds have the same formula. Therefore, 100 g of each compound contains the same number of moles." Justify your answer.

Moles of Ionic Hydrates

When precipitated from aqueous solution, many ionic compounds incorporate water molecules in the solid ionic crystal. Such compounds, called **ionic hydrates** or hydrated compounds, have water molecules trapped within the crystal lattice. The associated water is called the **water of hydration.** The formula for a beautiful deep blue compound is $CuSO_4 \cdot 5\ H_2O$; it is named copper(II) sulfate *penta*hydrate. The dot and the term *penta*hydrate indicate five moles of water associated with every mole of copper(II) sulfate. The molar mass of a hydrate includes the mass of the water of hydration. Thus, the molar mass of $CuSO_4 \cdot 5\ H_2O$ is 249.6 g/mol: 159.5 g $CuSO_4$ + 90.1 g (for 5 mol H_2O) = 249.6 g. There are many ionic hydrates, including a number of common substances (Table 3.11).

One commonly used hydrate may well be in the walls of your room. Plasterboard (sometimes called wallboard or gypsum board) contains hydrated calcium sulfate, or gypsum, $CaSO_4 \cdot 2\ H_2O$, as well as unhydrated $CaSO_4$, sandwiched between paper. Gypsum is a natural mineral that can be mined, but it is also obtained as a byproduct of removing sulfur dioxide from exhaust gases of electric power plants.

Heating gypsum to 120 to 180 °C drives off some of the water of hydration to form calcium sulfate hemihydrate, $CaSO_4 \cdot \frac{1}{2}\ H_2O$, commonly called "plaster of Paris" (Figure 3.11). This compound is widely used in medical casts for broken limbs. When added to water, it forms a thick slurry that can be poured into a mold or spread out over a part of a body. As it hardens, it takes on additional water of hydration and its volume increases, forming a rigid protective cast.

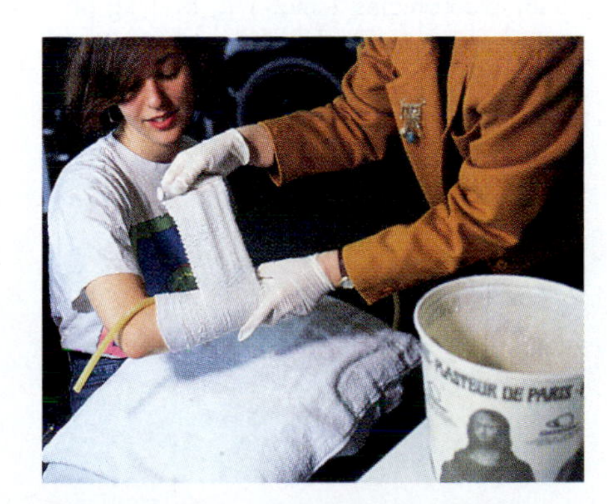

Figure 3.11 Applying a cast of Plaster of Paris, $CaSO_4 \cdot \frac{1}{2}\ H_2O$. *(C.D. Winters)*

Exercise 3.14 Moles of an Ionic Hydrate

A home remedy calls for using 2 teaspoons (20 g) of Epsom salt (see Table 3.11). Calculate the number of moles of the hydrate this mass represents.

3.13 PERCENT COMPOSITION

We saw in the previous section that the composition of water (or any other compound) can be expressed as either *(1) the number of atoms of each type per molecule or formula unit* (2 hydrogen and 1 oxygen per molecule of water) or *(2) the mass of each element in a mole of the compound* (for water: 2.0158 g H atoms and 15.9994 g O atoms in 1 mol of water, 18.0152 g H_2O). We can apply the latter relationship to state the composition of a compound from another perspective, the mass of each element in the compound compared with the total mass of the compound. This relationship expresses the **mass percent** of each element, also called the **percent composition by mass.** For carbon dioxide (CO_2) the mass percents are 27.29% C and 72.71% O.

$$\% \text{ C in } CO_2 = \frac{\text{mass of C in 1 mol } CO_2}{\text{mass of } CO_2 \text{ in 1 mol } CO_2} \times 100\%$$

$$= \frac{12.011 \text{ g C}}{44.011 \text{ g } CO_2} \times 100\%$$

$$= 27.29\% \text{ (also expressed as 27.29 g C per 100.0 g } CO_2)$$

$$\% \text{ O in } CO_2 = \frac{\text{mass of O in 1 mol } CO_2}{\text{mass of } CO_2 \text{ in 1 mol } CO_2} \times 100\%$$

$$= \frac{32.00 \text{ g O}}{44.011 \text{ g } CO_2} \times 100\%$$

$$= 72.71\% \text{ O (or 72.71 g O per 100.0 g } CO_2)$$

It is important to recognize that the percent composition of a compound is independent of the quantity of the compound. The percent composition remains the same whether a sample contains 1 mg, 1 g, or 1 kg of the compound.

Note that the total of the percentages adds up to 100%. Therefore, once we calculated the percent of carbon in carbon dioxide, we also could have determined the percent oxygen simply by subtracting the percent carbon from 100%: 100% − 27.29% C = 72.71% O. Calculating all percentages and adding them to check that they give 100% is a good way to check for errors.

Exercise 3.15 Percent Composition

For each of the following compounds, express the composition as the mass of each element in 1.000 mol of the compound, and the mass percent of each element: (a) SF_6; (b) $C_{12}H_{22}O_{11}$; (c) $Al_2(SO_4)_3$; (d) $U(OTeF_5)_6$.

3.14 DETERMINING EMPIRICAL AND MOLECULAR FORMULAS

A formula can be used to derive the percent composition of a compound, and so the reverse process also works—we can determine the formula of a compound from mass percent data. In doing so, keep in mind that the subscripts in a formula indicate the relative numbers of moles of each element in that compound.

We can apply this method to finding the formula of diborane. Finding the formula of a new compound first requires experiments to determine the composition. Experiments show that diborane has the following percent composition: 78.13% B and 21.87% H. Based on these percentages, a 100.0-g diborane sample contains 78.14 g of B and 21.86 g of H. From this information we can calculate the number of moles of each element in the sample:

$$78.13 \text{ g B} \times \frac{1 \text{ mol B}}{10.811 \text{ g B}} = 7.227 \text{ mol B}$$

$$21.87 \text{ g H} \times \frac{1 \text{ mol H}}{1.0079 \text{ g H}} = 21.70 \text{ mol H}$$

To determine the formula from these data, we next need to find the number of moles of each element *relative to the other,* in this case the ratio of moles of hydrogen to moles of boron. Looking at the numbers reveals that there are about three times as many moles of H atoms as there are moles of B atoms. To calculate the ratio exactly, divide the smaller number of moles into the larger number of moles. For diborane that ratio is

$$\frac{21.70 \text{ mol H}}{7.227 \text{ mol B}} = \frac{3.003 \text{ mol H}}{1.000 \text{ mol B}}$$

This confirms that there are three moles of H atoms for every one mole of B atoms, and that there are three hydrogen atoms for each boron atom, represented by the molecular formula BH_3.

For a molecular compound like diborane, the molecular formula must also accurately reflect the *total number of atoms in a molecule of the compound.* The calculation we have done gives the *simplest possible ratio of atoms in a molecule* and BH_3 is the simplest formula for diborane, which can't have fewer than one boron atom per molecule. A formula that reports the simplest possible ratio is called the **empirical formula.** However, multiples of the simplest formula are possible, such as B_2H_6, and B_3H_9, and so on.

To determine the actual molecular formula from the empirical formula requires that we *experimentally determine* the molar mass of the compound. The experimental molar mass is then compared with the molar mass predicted by the empirical formula. If the two molar masses are the same, the empirical and molecular formulas are the same. However, if the experimental molar mass is some multiple of the empirical formula's molar mass, then the molecular formula is that multiple of the empirical formula. In the case of diborane, experiments indicate that the molar mass of diborane is 27.67 g/mol. This compares with the molar mass of 13.84 g/mol for BH_3. Because these two molar masses differ, the molecular formula is a multiple of the empirical formula. That multiple is 27.67/13.84 = 2.00. Thus, the molecular formula of diborane is B_2H_6, two times BH_3.

PROBLEM-SOLVING EXAMPLE 3.8 Molecular Formula from Percent Composition Data

When oxygen reacts with phosphorus, two possible oxides can form. One oxide contains 56.34% P and 43.66% O, and its molar mass is 219.90 g/mol. Determine its molecular formula.

Answer The oxide is P_4O_6.

Explanation Calculate the relative number of moles of each element, and then determine the empirical formula. A 100.0-g sample of this phosphorus oxide contains 56.34 g of P and 43.66 g of O.

$$56.34 \text{ g P} \times \frac{1 \text{ mol P}}{30.9738 \text{ g P}} = 1.819 \text{ mol P}$$

$$43.66 \text{ g O} \times \frac{1 \text{ mol O}}{15.9994 \text{ g O}} = 2.729 \text{ mol O}$$

Thus, the mole ratio (and atom ratio) is

$$\frac{2.729 \text{ mol O}}{1.819 \text{ mol P}} = \frac{1.500 \text{ mol O}}{1.000 \text{ mol P}}$$

There are 1.5 oxygen atoms for every 1 phosphorus atom in the molecule. Because we can't have half atoms, we double the numbers to convert them to whole numbers, 3 oxygen atoms for every 2 phosphorus atoms. This gives an empirical formula of P_2O_3. The molar mass corresponding to this empirical formula is

$$\left(2 \text{ mol P} \times \frac{30.9738 \text{ g P}}{1 \text{ mol P}}\right) + \left(3 \text{ mol O} \times \frac{15.9994 \text{ g O}}{1 \text{ mol O}}\right)$$
$$= 109.95 \text{ g } P_2O_3/\text{per mole of } P_2O_3$$

compared with a known molar mass of 219.90 g/mol. We can see that the experimental molar mass is

$$\frac{219.90}{109.95} = 2 \text{ times the empirical formula mass,}$$

and so the molecular formula of the oxide is P_4O_6.

Problem-Solving Practice 3.8

The other phosphorus oxide contains 43.64% P and 56.36% O. Determine its empirical formula. The molar mass of the compound is 283.89 g/mol. Calculate its molecular formula from its empirical formula.

Conceptual Challenge Problem CP-3.A at the end of the chapter relates to topics covered in this section.

PROBLEM-SOLVING EXAMPLE 3.9 Determining a Molecular Formula

Rubbing alcohol, also commonly called isopropyl alcohol, has a molar mass of 60.096 g/mol. It contains 59.96% C, 13.42% H, and the rest is oxygen. What are the empirical and molecular formulas of isopropyl alcohol?

Answer Both formulas are C_3H_8O.

Explanation First find the number of moles of each element in 100.0 g of the compound.

$$59.96 \text{ g C} \times \frac{1 \text{ mol C}}{12.011 \text{ g C}} = 4.992 \text{ mol C}$$

$$13.42 \text{ g H} \times \frac{1 \text{ mol H}}{1.0079 \text{ g H}} = 13.32 \text{ mol H}$$

That leaves moles of oxygen to be calculated. The mass of oxygen in a 100.0-g isopropyl alcohol sample must be

$$\text{Mass of oxygen} = 100.0 \text{ g} - 59.96 \text{ g C} - 13.42 \text{ g H}$$
$$= 26.62 \text{ g}$$

Converting this to moles of oxygen:

$$26.62 \text{ g O} \times \frac{1 \text{ mol O}}{15.9994 \text{ g O}} = 1.664 \text{ mol O}$$

Base the mole ratio on the smallest number of moles present, in this case, moles of oxygen.

$$\frac{4.992 \text{ mol C}}{1.664 \text{ mol O}} = \frac{3.000 \text{ mol C}}{1.000 \text{ mol O}}$$

$$\frac{13.32 \text{ mol H}}{1.664 \text{ mol O}} = \frac{8.000 \text{ mol H}}{1.000 \text{ mol O}}$$

Therefore, the empirical formula has an atom ratio of 3 carbons to 8 hydrogens to 1 oxygen, C_3H_8O. The molar mass for this empirical formula is 60.096 g/mol, the same as the experimental molar mass of the compound, indicating that the molecular formula is the same as the empirical formula.

Problem-Solving Practice 3.9

Vitamin C (ascorbic acid) contains 40.9% C, 4.58% H, and 54.5% O and has a molar mass of 176.13 g/mol. From these data, determine its empirical and molecular formulas.

Determining Formulas from Experimental Data

There are well over one million molecular and ionic compounds that are white, crystalline solids. Their properties (other than color) help to differentiate them. The properties of a newly synthesized compound may be similar to or very different from *known* compounds. A major step in determining the identity of a *new* compound is to determine its molecular formula, which can be derived from percent composition and molar mass data as just illustrated. The molar mass can be determined using a variety of methods, some of which are discussed in Section 15.3.

Although we have just discussed using percent composition data to obtain empirical formulas, we haven't indicated how such data are obtained. Several methods are used, all involving *quantitative chemical analysis,* experimental techniques to measure how much of each element is in a compound.

One technique to determine the formula of a binary compound formed by direct combination of its two elements is to measure the mass of reactants that are converted to the product compound. The formulas for binary compounds such as those formed by combining an element directly with oxygen, sulfur, or a halogen can be determined this way. Such a case is the reaction of solid red phosphorus with liquid bromine to produce a phosphorus bromide.

$$P_4 + Br_2 \xrightarrow{\text{produces}} P_xBr_y$$

By knowing the masses of phosphorus and bromine that react, the values of x and y can be calculated, and the empirical formula determined for the phosphorus bromide.

The reaction is carried out and it is found that 0.347 g of P_4 reacts with 0.860 mL of Br_2. The density of bromine is 3.12 g/mL. We use these values to calculate the moles of P_4 and Br_2 that combined:

$$0.347 \text{ g } P_4 \times \frac{1 \text{ mol } P_4}{123.92 \text{ g } P_4} = 2.800 \times 10^{-3} \text{ mol } P_4$$

To determine the moles of bromine, we first use its density to convert milliliters of bromine to grams, then to moles:

$$0.860 \text{ mL } Br_2 \times \frac{3.12 \text{ g } Br_2}{1 \text{ mL } Br_2} \times \frac{1 \text{ mol } Br_2}{159.8 \text{ g } Br_2} = 1.679 \times 10^{-2} \text{ mol } Br_2$$

Remember that if a calculation is done stepwise, always carry one more significant figure in each intermediate result and round the final result.

Notice that we connected the conversion factors rather than doing individual calculations of milliliters to grams, and then grams to moles. Connecting several conversion factors saves time in a calculation and should be used, where possible.

The mole ratio of bromine atoms to phosphorus atoms in a molecule of the compound can be calculated from the moles of atoms of each element. Recall that bromine is diatomic with two moles of Br atoms in one mole of Br_2. There are four moles of phosphorus atoms in one mole of P_4 molecules.

$$1.679 \times 10^{-2} \text{ mol } Br_2 \text{ molecules} \times \frac{2 \text{ mol Br atoms}}{1 \text{ mol } Br_2 \text{ molecule}} = 3.36 \times 10^{-2} \text{ mol Br atoms}$$

$$2.800 \times 10^{-3} \text{ mol } P_4 \text{ molecules} \times \frac{4 \text{ mol P atoms}}{1 \text{ mol } P_4 \text{ molecules}} = 1.12 \times 10^{-2} \text{ mol P atoms}$$

$$\frac{3.36 \times 10^{-2} \text{ mol Br atoms}}{1.12 \times 10^{-2} \text{ mol P atoms}} = \frac{3.00 \text{ mol Br atoms}}{1.00 \text{ mol P atoms}}$$

Conceptual Challenge Problems CP-3.B and CP-3.C at the end of the chapter relate to topics covered in this section.

The mole ratio in the compound is 3.00 mol of bromine atoms for 1.00 mol of phosphorus atoms. Consequently, in a molecule of this compound there are three bromine atoms for every phosphorus atom. Therefore, the empirical formula is PBr_3. By other experimental methods, the known molar mass of this compound is found to be the same as the empirical formula molar mass. Thus, the molecular formula is also PBr_3.

Exercise 3.16 Empirical Formula of a Binary Compound

The reaction of 0.569 g of tin with 2.434 g of iodine formed Sn_xI_y. What is the empirical formula of this tin iodide?

Summary Problem

Part I

During each launch of the Space Shuttle, the booster rocket uses about 1.5×10^6 pounds of ammonium perchlorate as fuel.

1. Write the chemical formula for (a) ammonium perchlorate; (b) ammonium chlorate; (c) ammonium chlorite.
2. (a) Write the equation for the dissociation of ammonium perchlorate in water. (b) Would this aqueous solution conduct an electric current? Explain your answer.
3. Use Table 3.8 on page 104 to determine what substance, if any, could be added to ammonium chlorate to form a precipitate containing (a) ammonium ions; (b) chlorate ions.
4. How many moles of ammonium perchlorate are used in Space Shuttle booster rockets during a launch?

Part II

Chemical analysis of ibuprofen (Advil) indicates that it contains 75.69% carbon, 8.80% hydrogen, and the remainder, oxygen. The empirical formula is also the molecular formula.

1. Determine the molecular formula of ibuprofen.
2. Two 200-mg ibuprofen tablets were taken by a patient to relieve pain. Calculate the number of moles of ibuprofen contained in the two tablets.

In Closing

Having studied this chapter, you should be able to . . .

- interpret the meaning of molecular formulas, condensed formulas, and structural formulas (Section 3.1).
- name binary molecular compounds, including straight-chain alkanes (Sections 3.2 and 3.3).
- write structural formulas for and identify straight- and branched-chain alkane constitutional isomers (Section 3.4).
- predict the charges on monatomic ions of metals and nonmetals (Section 3.5).
- give the names or formulas of polyatomic ions, knowing their formulas or names, respectively (Section 3.5).
- describe the properties of ionic compounds (Sections 3.5 and 3.7).
- write the formulas of ionic compounds (Section 3.5).
- name ionic compounds (Section 3.6).
- describe electrolytes in aqueous solution and summarize the differences between electrolytes and nonelectrolytes (Section 3.8).
- predict the solubility of ionic compounds in water (Section 3.9).
- identify biologically important elements (Section 3.10).
- identify the important functional groups in carbohydrates and fats (Section 3.11).
- thoroughly explain the use of the mole concept for chemical compounds (Section 3.12).
- calculate the molar mass of a compound (Section 3.12).
- calculate the number of moles of a compound given the mass, and vice versa (Section 3.12).
- explain the formula of a hydrated ionic compound and calculate its molar mass (Section 3.12).
- express molecular composition in terms of percent composition (Section 3.13).
- use percent composition and molar mass to determine the empirical and molecular formulas of a compound (Section 3.13).

Key Terms

The following terms were defined and given in boldface type in this chapter. You should be sure to understand each of these terms and the concepts with which they are associated.

alcohol *(3.3)*
alkanes *(3.3)*
alkyl groups *(3.4)*
anions *(3.5)*
binary molecular compound *(3.2)*
carbohydrates *(3.11)*
carboxylic acid *(3.11)*
cations *(3.5)*
condensed formula *(3.1)*
constitutional isomers *(3.4)*
crystal lattice *(3.7)*
dietary minerals *(3.10)*
disaccharides *(3.11)*
electrodes *(3.7)*
electrolytes *(3.8)*
empirical formula *(3.13)*
formula unit *(3.7)*
functional group *(3.1)*
halide ions *(3.6)*
hydrocarbons *(3.3)*
ionic compound *(3.5)*
ionic hydrate *(3.12)*
ions *(3.5)*
isomers *(3.4)*
macrominerals *(3.10)*
mass percent *(3.13)*
methyl group *(3.4)*
microminerals *(3.10)*
molar mass *(3.12)*
molecular compound *(3.1)*
molecular formula *(3.1)*
monatomic ion *(3.5)*
monosaccharides *(3.11)*
nonelectrolyte *(3.8)*
organic compounds *(3.1)*
oxoanions *(3.6)*
percent composition by mass *(3.13)*
polyatomic ion *(3.5)*
polysaccharides *(3.11)*
simple sugars *(3.11)*
strong electrolyte *(3.8)*
structural formula *(3.1)*
water of hydration *(3.12)*

Questions for Review and Thought

Conceptual Challenge Problems

CP-3.A. A chemist analyzes three compounds and reports the following data for the percent by mass of the elements Ex, Ey, and Ez in each compound.

Compound	% Ex	% Ey	%Ez
A	37.485	12.583	49.931
B	40.002	6.7142	53.284
C	40.685	5.1216	54.193

Assume that you accept the notion that the numbers of atoms of the elements in compounds are in small whole number ratios, and that the number of atoms in a sample of any element is directly proportional to that sample's mass. What is possible for you to know about the empirical formulas for these three compounds?

CP-3.B. The following table displays on each horizontal row an empirical formula for one of the three compounds noted in CP-3.A.

Compound A	Compound B	Compound C
______	$ExEy_2Ez$	______
$Ex_6Ey_8Ez_3$	______	______
______	______	Ex_3Ey_2Ez
______	$Ex_9Ey_2Ez_6$	______
______	______	$ExEy_2Ez_3$
$Ex_3Ey_8Ez_3$	______	______

Based only on what was learned in that problem, what is the empirical formula for the other two compounds in that row?

CP-3.C. (a) Suppose that a chemist now determines that the ratio of the masses of equal numbers of atoms of Ez and Ex atoms is 1.3320 g Ez/1 g Ex. With this added information, what can now be known about the formulas for compounds A, B, and C in CP-3.A?

(b) Suppose that this chemist further determines that the ratio of the masses of equal numbers of atoms of Ex and Ey is 11.916 g Ex/1 g Ey. What is the ratio of the masses of equal numbers of Ez and Ey atoms?

(c) If the mass ratios of equal numbers of atoms of Ex, Ey, and Ez are known, what can be known about the formulas of the three compounds A, B, and C?

Review Questions

1. A dictionary defines the word "compound" as a "combination of two or more parts." What are the "parts" of a chemical compound? Identify three pure (or nearly pure) compounds you have encountered today. What is the difference between a compound and a mixture?
2. Nobel Prize–winning chemist Roald Hoffmann has said that "Today chemistry is the science of molecules and their transformations." What does that statement mean in the context of your own experience?
3. For each of the following structural formulas, write the molecular formulas and condensed formulas.

```
       H  H           H              H  H  O
       |  |          /               |  |  ‖
(a) H—C—C—C=C          (b) H—C—C—C—OH
       |  |  |       \               |  |
       H  H  H        H              H  H

       H  H  H  O
       |  |  |  ‖
(c) H—N—C—C—C—OH
          |  |
          H  H
```

4. Given the following condensed formulas, write the structural and molecular formulas.
   (a) CH_3OH (b) $CH_3CH_2NH_2$ (c) $CH_3CH_2SCH_2CH_3$
5. Give a molecular formula for each of the following organic acids.

```
   H  O  O              O  H  H     H     O
   |  ‖  ‖              ‖  |  |     |     ‖
H—C—C—C—OH         HO—C—C—C——C——C—OH
   |                       |  |     |
   H                       H  C=O  OH
                              |
                              OH
(a) pyruvic acid        (b) isocitric acid
```

6. Give a molecular formula for each of the following molecules.

```
   H  CH3                     OH     CH3
   |  |                       |      |
H—C—C—H                 CH3CHCH2CHCH2CH3
   |  |                 (b) 4-methyl-2-hexanol
   H  |  O
H     |  ‖
 \N—C—C—OH
 /    |
H     H
  (a) valine
```

7. Give the name for each of the following binary nonmetal compounds:
   (a) NF_3 (c) BBr_3
   (b) HI (d) C_6H_{14}
8. Give the name for each of the following binary nonmetal compounds:
   (a) C_8H_{18} (c) OF_2
   (b) P_2S_3 (d) XeF_4
9. Give the formula for each of the following nonmetal compounds:
   (a) Sulfur trioxide
   (b) Dinitrogen pentaoxide
   (c) Phosphorous pentachloride
   (d) Silicon tetrachloride
   (e) Diboron trioxide (commonly called boric oxide)

Molecular and Structural Formulas

10. Give the formula for each of the following nonmetal compounds:
   (a) Bromine trifluoride (d) Pentadecane
   (b) Xenon difluoride (e) Hydrazine
   (c) Diphosphorus tetrafluoride

11. Write structural formulas for the following alkanes.
   (a) Butane (d) Octane
   (b) Nonane (e) Octadecane
   (c) Hexane

12. Write the molecular, condensed, and structural formulas for the simplest alcohols derived from butane and pentane.

13. Octane is an alkane (Table 3.3). For the sake of this problem, we will assume that gasoline, a complex mixture of hydrocarbons, is represented by octane. If you fill the tank of your car with 18 gal of gasoline, how many grams and how many pounds of gasoline have you put into the car? Information you may need is (a) the density of octane, 0.692 g/cm^3; (b) the volume of 1 gal in milliliters (3790 mL).

14. Which of the following molecules contains more O atoms, and which contains more atoms of all kinds: (a) sucrose, $C_{12}H_{22}O_{11}$, or (b) glutathione, $C_{10}H_{17}N_3O_6S$ (the major low-molecular-weight sulfur-containing compound in plant or animal cells)?

15. Write the molecular formula of each of the following compounds: (a) benzene, a liquid hydrocarbon, has six carbon atoms and six hydrogen atoms per molecule; (b) vitamin C has six carbon atoms, eight hydrogen atoms, and six oxygen atoms per molecule.

16. Write the formula for
   (a) A molecule of the organic compound, heptane, which has 7 carbon atoms and 16 hydrogen atoms.
   (b) A molecule of acrylonitrile, the basis of Orlon and Acrilan fibers, which has three carbon atoms, three hydrogen atoms, and one nitrogen atom.
   (c) A molecule of Fenclorac, an anti-inflammatory drug, which has 14 carbon atoms, 16 hydrogen atoms, 2 chlorine atoms, and 2 oxygen atoms.

17. Give the total number of atoms of each element in one formula unit of each of the following compounds:
   (a) CaC_2O_4 (d) $Pt(NH_3)_2Cl_2$
   (b) $C_6H_5CHCH_2$ (e) $K_4Fe(CN)_6$
   (c) $(NH_4)_2SO_4$

18. Give the total number of atoms of each element in each of the molecules below.
   (a) $CO_2(CO)_8$ (d) $C_{10}H_9NH_2Fe$
   (b) $HOOCCH_2CH_2COOH$ (e) $C_6H_2CH_3(NO_2)_3$
   (c) $CH_3NH_2CH_2COOH$

Constitutional Isomers

19. Draw condensed structures for the five constitutional isomers of C_6H_{14}.

20. Draw the structural formula and condensed formula of 2,2-dimethylpropane. Explain why there is no compound named 2,3-dimethylpropane.

Predicting Ion Charges

21. For each of the following metals, write the chemical symbol for the corresponding ion (with charge).
(a) Lithium
(b) Strontium
(c) Aluminum
(d) Calcium
(e) Zinc

22. For each of the following nonmetals, write the chemical symbol for the corresponding ion (with charge).
(a) Nitrogen
(b) Sulfur
(c) Chlorine
(d) Iodine
(e) Phosphorus

23. Predict the charges of the ions in an ionic compound composed of barium and bromine.

24. Predict the charges for ions of the following elements:
(a) Magnesium (c) Iron
(b) Zinc (d) Gallium

25. Predict the charges for ions of the following elements:
(a) Selenium (c) Nickel
(b) Fluorine (d) Nitrogen

26. Cobalt is a transition metal and so can form ions with at least two different charges. Write the formulas for the compounds formed between cobalt ions and oxide ion.

27. Although not a transition element, lead can also form two cations: Pb^{2+} and Pb^{4+}. Write the formulas for the compounds of these ions with the chloride ion.

28. Which of the following are the correct formulas of compounds? For those that are not, give the correct formula.
(a) $AlCl$ (c) Ga_2O_3
(b) NaF_2 (d) MgS

29. Which of the following are the correct formulas of compounds? For those that are not, give the correct formula.
(a) Ca_2O (c) Fe_2O_5
(b) $SrCl_2$ (d) K_2O

Polyatomic Ions

30. For each of the following compounds, tell what ions are present and how many there are per formula unit:
(a) $Pb(NO_3)_2$ (b) $NiCO_3$ (c) $(NH_4)_3PO_4$
(d) K_2SO_4

31. For each of the following compounds, tell what ions are present and how many there are per formula unit:
(a) $Ca(CH_3CO_2)_2$ (b) $Co_2(SO_4)_3$ (c) $Al(OH)_3$
(d) $(NH_4)_2CO_3$

32. Determine the chemical formulas for barium sulfate, magnesium nitrate, and sodium acetate. Each compound contains a monatomic cation and a polyatomic anion. What are the names and electrical charges of these ions?

33. Write the chemical formulas for the following compounds:
(a) Nickel(II) nitrate (b) Sodium bicarbonate
(c) Lithium hypochlorite (d) Magnesium chlorate
(e) Calcium sulfite

Ionic Compounds

34. Determine which of the following substances are ionic:
(a) CF_4 (d) SiO_2
(b) $SrBr_2$ (e) KCN
(c) $Co(NO_3)_3$ (f) SCl_2

35. Which of the following are ionic? Write the formula for each.
(a) Methane (d) Hydrogen selenide
(b) Dinitrogen pentaoxide (e) Sodium perchlorate
(c) Ammonium sulfide

36. Give the formula for each of the following ionic compounds:
(a) Ammonium carbonate (c) Copper(II) bromide
(b) Calcium iodide (d) Aluminum phosphate

37. Give the formula for each of the following ionic compounds:
(a) Calcium hydrogen carbonate
(b) Potassium permanganate
(c) Magnesium perchlorate
(d) Ammonium hydrogen phosphate

38. Name each of the following ionic compounds:
(a) K_2S (b) $NiSO_4$ (c) $(NH_4)_3PO_4$

39. Name each of the following ionic compounds:
(a) $Ca(CH_3CO_2)_2$ (b) $Co_2(SO_4)_3$ (c) $Al(OH)_3$

40. Name each of the following ionic compounds:
(a) KH_2PO_4 (b) $CuSO_4$ (c) $CrCl_6$

41. Solid magnesium oxide melts at 2800 °C. This property, combined with the fact that magnesium oxide is not an electrical conductor, makes it an ideal heat insulator for electric wires in cooking ovens and toasters (see photo). In contrast, solid NaCl melts at the relatively low temperature of 801 °C. What is the formula of magnesium oxide? Suggest a reason that it has a melting temperature so much higher than that of NaCl.

42. Assume you have an unlabeled bottle containing a white crystalline powder. The powder melts at 310 °C. You are told that it could be NH_3, NO_2, or $NaNO_3$. What do you think it is and why?

Electrolytes

43. What is an electrolyte? How can we differentiate between a weak and a strong electrolyte? Give an example of each.

44. Tell how to predict that $Ni(NO_3)_2$ is soluble in water while $NiCO_3$ is not soluble in water.

45. Predict whether each of the following compounds is likely to be water-soluble. Indicate which ions are present for the water-soluble compounds.
(a) $Fe(ClO_4)_2$ (b) Na_2SO_4 (c) KBr (d) Na_2CO_3

46. Predict whether each of the following compounds is likely to be water-soluble. Indicate which ions are present for the water-soluble compounds.
(a) Potassium hydrogen phosphate (b) Sodium hypochlorite
(c) Magnesium chloride (d) Calcium hydroxide
(e) Aluminum bromide

Biological Periodic Table

47. Make a list of the top ten most abundant essential elements needed by the human body.

48. Which types of compounds contain the majority of the oxygen found in the human body?

49. How are metals found in the body: as atoms or as ions? What are two uses for metals in the human body?

50. Distinguish between macrominerals and microminerals.

51. Which minerals are essential at smaller concentrations but are toxic at higher concentrations?

Carbohydrates and Fats

52. Name the monosaccharide commonly found in fresh fruit. What is the structural formula for this sugar?

53. Why are fats and oils known as triglycerides?

Moles of Compounds

54. Fill in the table for 1 mol of methanol, CH_3OH:

	CH_3OH	Carbon	Hydrogen	Oxygen
No. of moles	______	______	______	______
No. of molecules or atoms	______	______	______	______
Molar mass	______	______	______	______

55. Fill in the following table for 1 mol of glucose, $C_6H_{12}O_6$:

	$C_6H_{12}O_6$	Carbon	Hydrogen	Oxygen
No. of moles	______	______	______	______
No. of molecules or atoms	______	______	______	______
Molar mass	______	______	______	______

56. Calculate the molar mass of each of the following compounds:
(a) Fe_2O_3, iron(III) oxide
(b) BF_3, boron trifluoride
(c) N_2O, dinitrogen oxide (laughing gas)
(d) $MnCl_2 \cdot 4\ H_2O$, manganese(II) chloride tetrahydrate
(e) $C_6H_8O_6$, ascorbic acid

57. Calculate the molar mass of each of the following compounds:
(a) $B_{10}H_{14}$, a boron hydride once considered as a rocket fuel
(b) $C_6H_2(CH_3)(NO_2)_3$, TNT, an explosive
(c) $PtCl_2(NH_3)_2$, a cancer chemotherapy agent called cisplatin
(d) $CH_3(CH_2)_3SH$, a compound that has a skunk-like odor
(e) $C_{20}H_{24}N_2O_2$, quinine, used as an antimalarial drug

58. How many moles are represented by 1.00 g of each of the following compounds?
(a) CH_3OH, methanol
(b) Cl_2CO, phosgene, a poisonous gas
(c) NH_4NO_3, ammonium nitrate
(d) $MgSO_4 \cdot 7\ H_2O$, magnesium sulfate heptahydrate (Epsom salt)
(e) $AgCH_3CO_2$, silver acetate

59. Assume you have 0.250 g of each of the following compounds. How many moles of each are represented?
(a) $C_7H_5NO_3S$, saccharin, an artificial sweetener
(b) $C_{13}H_{20}N_2O_2$, procaine, a pain killer used by dentists
(c) $C_{20}H_{14}O_4$, phenolphthalein, a dye

60. An Alka-Seltzer tablet contains 324 mg of aspirin ($C_9H_8O_4$), 1904 mg of $NaHCO_3$, and 1000 mg of citric acid ($C_6H_8O_7$). (The last two compounds react with each other to provide the "fizz," bubbles of CO_2, when the tablet is put into water.)
(a) Calculate the number of moles of each substance in the tablet.
(b) If you take one tablet, how many molecules of aspirin are you consuming?

61. The use of CFCs (chlorofluorocarbons) is being curtailed, because there is strong evidence that they cause environmental damage. If a spray can contains 250 g of one of these compounds, CCl_2F_2, how many molecules of this CFC are you releasing to the air when you empty the can?

62. Sulfur trioxide, SO_3, is made in enormous quantities by combining oxygen and sulfur dioxide, SO_2. The trioxide is not usually isolated but is converted to sulfuric acid. If you have 1.00 lb (454 g) of sulfur trioxide, how many moles does this represent? How many molecules? How many sulfur atoms? How many oxygen atoms?

63. CFCs (chlorofluorocarbons) are implicated in decreasing ozone in the stratosphere. A CFC substitute is CF_3CH_2F. If you have 25.5 g of this compound, how many moles does this represent? How many atoms of fluorine are contained in 25.5 g of the compound?

Percent Composition

64. Calculate the molar mass of each of these compounds and the weight percent of each element.
(a) PbS, lead(II) sulfide, galena
(b) C_2H_6, ethane, a hydrocarbon fuel
(c) CH_3CO_2H, acetic acid, an important ingredient in vinegar
(d) NH_4NO_3, ammonium nitrate, a fertilizer

65. Calculate the molar mass of each of these compounds and the weight percent of each element.
(a) $MgCO_3$, magnesium carbonate
(b) C_6H_5OH, phenol, an organic compound used in some cleaners
(c) $C_2H_3O_5N$, peroxyacetyl nitrate, an objectionable compound in photochemical smog
(d) $C_4H_{10}O_3NPS$, acephate, an insecticide

66. A certain metal, M, forms two oxides, M_2O and MO. If the percent by weight of M in M_2O is 73.4%, what is its percent by weight in MO?

67. The copper-containing compound $Cu(NH_3)_4SO_4 \cdot H_2O$ is a beautiful blue solid. Calculate the molar mass of the compound and the mass percent of each element.

68. Chemists often express the composition of compounds in terms of the percentage of a particular element that is present. Look for some food product that gives the composition in terms of percentages. What data are given? Is percent by weight of an element or of a compound listed?

69. If a food label gives the percent composition of a particular compound in a food as 14.5%, what does that indicate about the compound in terms of the food?

Empirical and Molecular Formulas

70. What is the difference between an empirical formula and a molecular formula? Use the compound ethane, C_2H_6, to illustrate your answer.

71. The empirical formula of maleic acid is CHO. Its molar mass is 116.1 g/mol. What is its molecular formula?

72. A well-known reagent in analytical chemistry, dimethylglyoxime, has the empirical formula C_2H_4NO. If its molar mass is 116.1 g/mol, what is the molecular formula of the compound?

73. Acetylene is a colorless gas that is used as a fuel in welding torches, among other things. It is 92.26% C and 7.74% H. Its molar mass is 26.02 g/mol. Calculate the empirical and molecular formulas.

74. There is a large family of boron-hydrogen compounds called boron hydrides. All have the formula B_xH_y and almost all react with air and burn or explode. One member of this family contains 88.5% B; the remainder is hydrogen. Which of the following is its empirical formula: BH_3, B_2H_5, B_5H_7, B_5H_{11}, or BH_2?

75. Nitrogen and oxygen form an extensive series of at least seven oxides of general formula N_xO_y. One of them is a blue solid that comes apart or "dissociates," reversibly, in the gas phase. It contains 36.84% N. What is the empirical formula of this oxide?

76. Cumene is a hydrocarbon, a compound composed only of C and H. It is 89.94% carbon, and the molar mass is 120.2 g/mol. What are the empirical and molecular formulas of cumene?

77. Acetic acid is the important ingredient in vinegar. It is composed of carbon (40.0%), hydrogen (6.71%), and oxygen (53.29%). Its molar mass is 60.0 g/mol. Determine the empirical and molecular formulas of the acid.

78. An analysis of nicotine, a poisonous compound found in tobacco leaves, shows that it is 74.0% C, 8.65% H, and 17.35% N. Its molar mass is 162 g/mol. What are the empirical and molecular formulas of nicotine?

79. Cacodyl, a compound containing arsenic, was reported in 1842 by the German chemist Bunsen. It has an almost intolerable garlic-like odor. Its molar mass is 210 g/mol, and it is 22.88% C, 5.76% H, and 71.36% As. Determine its empirical and molecular formulas.

80. The action of bacteria on meat and fish produces a poisonous compound called cadaverine. As its name and origin imply, it stinks! It is 58.77% C, 13.81% H, and 27.42% N. Its molar mass is 102.2 g/mol. Determine the molecular formula of cadaverine.

81. If Epsom salt, $MgSO_4 \cdot x\ H_2O$, is heated to 250 °C, all the water of hydration is lost. After heating a 1.687-g sample of the hydrate, 0.824 g of $MgSO_4$ remains. How many molecules of water are there per formula unit of $MgSO_4$?

82. The alum used in cooking is potassium aluminum sulfate hydrate, $KAl(SO_4)_2 \cdot x\ H_2O$. To find the value of *x*, you can heat a sample of the compound to drive off all of the water and leave only $KAl(SO_4)_2$. Assume that you heat 4.74 g of the hydrated compound and that it loses 2.16 g of water. What is the value of *x*?

General Questions

83. Draw a diagram showing the crystal lattice of sodium chloride (NaCl). Show clearly why such a crystal can be cleaved easily by tapping on a knife blade properly aligned along the crystal. Describe in words why the cleavage occurs as it does.

84. Give the molecular formula for each of the following molecules.

CH_3
C
NO_2—C C—NO_2
C C
C
NO_2

trinitrotoluene, TNT

```
     H  NH2
     |  |
HO—C—C—H
     |  |
     H  C=O
        |
        OH
```

serine, an essential amino acid

85. Calculate the mass of one molecule of nitrogen. Now assume that someone decided to make Avogadro's number have a simpler value, say 1.000×10^{20}. Now what is the mass of a molecule of nitrogen?

86. Which of the following pairs of elements are likely to form ionic compounds? Write appropriate formulas for the compounds you expect to form, and give the name of each.
(a) Chlorine and bromine
(b) Lithium and tellurium
(c) Sodium and argon
(d) Magnesium and fluorine
(e) Nitrogen and bromine
(f) Indium and sulfur
(g) Selenium and bromine

87. Name each of the following compounds and tell which ones are best described as ionic:
(a) $ClBr_3$
(b) NCl_3
(c) $CaSO_4$
(d) C_7H_{16}
(e) XeF_4
(f) OF_2
(g) NaI
(h) Al_2S_3
(i) PCl_5
(j) K_3PO_4

88. Write the formula for each of the following compounds and tell which ones are best described as ionic:
(a) Sodium hypochlorite
(b) Aluminum perchlorate
(c) Potassium permanganate
(d) Potassium dihydrogen phosphate
(e) Chlorine trifluoride
(f) Boron tribromide
(g) Calcium acetate
(h) Sodium sulfite
(i) Disulfur tetrachloride
(j) Phosphorus trifluoride

89. Precious metals such as gold and platinum are sold in units of "troy ounces," where 1 troy ounce is equivalent to 31.1 g. If you have a block of platinum with a mass of 15.0 troy ounces, how many moles of the metal do you have? What is the size of the block in cubic centimeters? (The density of platinum is 21.45 g/cm^3 at 20 °C.)

90. "Dilithium" is the fuel for the *Starship Enterprise.* However, because its density is quite low, you will need a large space to store a large mass. As an estimate for the volume required, we shall use the element lithium. If you want to have 256 mol for an interplanetary trip, what must the volume of a piece of lithium be? If the piece of lithium is a cube, what is the dimension of an edge of the cube? (The density of lithium is 0.534 g/cm^3 at 20 °C.)

91. Fluorocarbonyl hypofluorite was recently isolated, and analysis showed it to be 14.6% C, 39.0% O, and 46.3% F. If the molar mass of the compound is 82 g/mol, determine the empirical and molecular formulas of the compound.

92. Azulene, a beautiful blue hydrocarbon, is 93.71% C and has a molar mass of 128.16 g/mol. What are the empirical and molecular formulas of azulene?

93. A major oil company has used a gasoline additive called MMT to boost the octane rating of its gasoline. What is the empirical formula of MMT if it is 49.5% C, 3.2% H, 22.0% O, and 25.2% Mn?

94. Direct reaction of iodine (I_2) and chlorine (Cl_2) produces an iodine chloride, I_xCl_y, a bright yellow solid. If you completely use up 0.678 g of iodine, and produce 1.246 g of I_xCl_y, what is the empirical formula of the compound? A later experiment shows that the molar mass of I_xCl_y is 467 g/mol. What is the molecular formula of the compound?

95. Pepto-Bismol, which helps provide relief for an upset stomach, contains 300 mg of bismuth subsalicylate, $C_7H_5BiO_4$, per tablet. If you take two tablets for your stomach distress, how many moles of the "active ingredient" are you taking? How many grams of Bi are you consuming in two tablets?

96. Iron pyrite, often called "fool's gold," has the formula FeS_2. If you could convert 15.8 kg of iron pyrite to iron metal and remove the sulfur, how many kilograms of the metal could you obtain?

Iron pyrite, or fool's gold. *(C.D. Winters)*

97. Ilmenite is a mineral that is an oxide of iron and titanium, $FeTiO_3$. If an ore that contains ilmenite is 6.75% titanium, what is the mass (in grams) of ilmenite in 1.00 metric ton (exactly 1000 kg) of the ore?

98. Stibnite, Sb_2S_3, is a dark gray mineral from which antimony metal is obtained. If you have one pound of an ore that contains 10.6% antimony, what mass of Sb_2S_3 (in grams) is there in the ore?

99. Draw diagrams in the following boxes to indicate the arrangement of nanoscale particles of each substance. Consider each box to hold a very tiny portion of each substance. Each drawing should contain at least 16 particles, and it need not be three-dimensional.

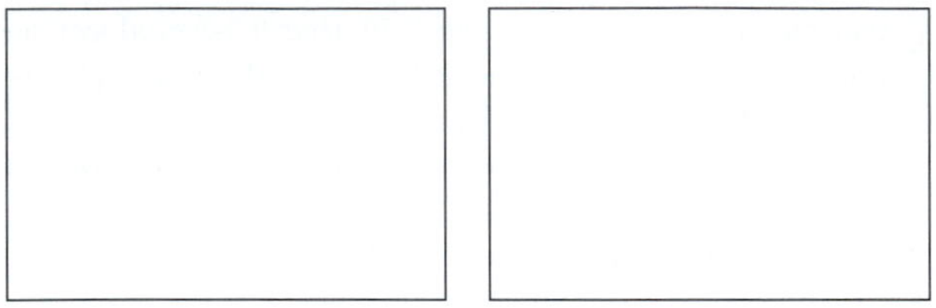

$Br_2(\ell)$ LiF(s)

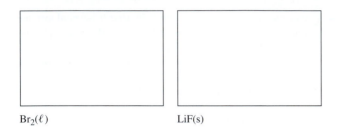

100. A piece of nickel foil, 0.550 mm thick and 1.25 cm square, was allowed to react with fluorine, F_2, to give a nickel fluoride. (a) How many moles of nickel foil were used? (b) If you isolate 1.261 g of the nickel fluoride, what is its formula? (c) What is its name? (The density of nickel is 8.908 g/cm^3.)

101. Uranium is used as a fuel, primarily in the form of uranium(IV) oxide, in nuclear power plants. This question considers some uranium chemistry.
(a) A small sample of uranium metal (0.169 g) is heated to 800 to 900 °C in air to give 0.199 g of a dark-green oxide, U_xO_y. How many moles of uranium metal were used? What is the empirical formula of the oxide U_xO_y? What is the name of the oxide? How many moles of U_xO_y must have been obtained?
(b) The oxide U_xO_y is obtained if $UO_2(NO_3)\cdot n\ H_2O$ is heated to temperatures greater than 800 °C in the air. However, if you heat it gently, only the water of hydration is lost. If you have 0.865 g of $UO_2(NO_3)\cdot n\ H_2O$, and obtain 0.679 g of $UO_2(NO_3)$ on heating, how many molecules of water of hydration were there in each formula unit of the original compound?

102. Draw diagrams of each nanoscale situation listed below. Represent atoms or monatomic ions as circles; represent molecules or polyatomic ions by overlapping circles for the atoms that make up the molecule or ion; and distinguish among different kinds of atoms by labeling or shading the circles. In each case draw representations of at least five nanoscale particles. Your diagrams can be two-dimensional.
(a) A crystal of solid sodium chloride
(b) The sodium chloride from part a after it has been melted
(c) A sample of molten aluminum oxide, Al_2O_3

103. Draw diagrams of each nanoscale situation listed below. Represent atoms or monatomic ions as circles; represent molecules or polyatomic ions by overlapping circles for the atoms that make up the molecule or ion; and distinguish among different kinds of atoms by labeling or shading the circles. In each case draw representations of at least five nanoscale particles. Your diagrams can be two-dimensional.
(a) A sample of solid lithium nitrate, $LiNO_3$
(b) A sample of molten lithium nitrate
(c) The same sample of lithium nitrate after electrodes have been placed into it and a direct current applied to the electrodes
(d) A sample of solid lithium nitrate in contact with a solution of lithium nitrate in water

104. Assume that each of the ions listed below is in aqueous (water) solution. Draw a submicroscopic picture of the ion and the water molecules in its immediate vicinity.
(a) Cl^- (c) Mg^{2+}
(b) K^+

Applying Concepts

105. When asked to draw all the possible constitutional isomers for C_3H_8O, a student drew the structures below. The student's instructor said some of the structures were identical. How many actual isomers are there? Which structures are identical?

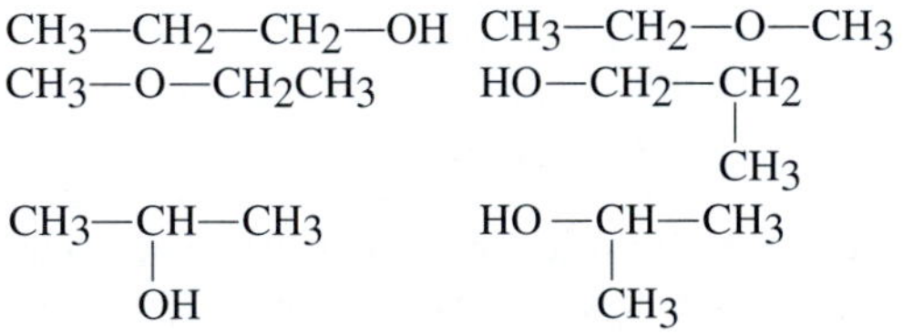

106. A student incorrectly named the structure below as 2,4,4-trimethylbutane. What is the correct name?

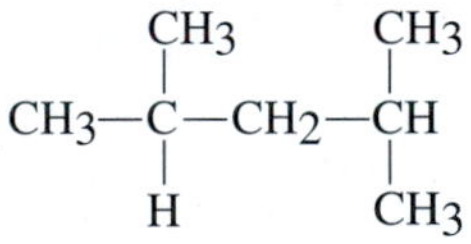

107. The statement that "metals form positive ions by losing electrons" is difficult to grasp for some students because positive signs indicate a gain and negative signs a loss. How would you explain this contradiction to a classmate?

108. The formula for thallium nitrate is $TlNO_3$. Based on this information, what would be the formulas for thallium carbonate and thallium sulfate?

109. The name given with each formula below is incorrect. What are the correct names?
(a) CaF_2, calcium difluoride
(b) CuO, copper oxide
(c) $NaNO_3$, sodium nitroxide
(d) NI_3, nitrogen iodide
(e) $FeCl_3$, iron(I) chloride
(f) Li_2SO_4, dilithium sulfate

110. Based on the guidelines for naming oxyanions in a series, how would you name the following?
(a) BrO_4^-, BrO_3^- BrO_2^-, BrO^-
(b) SeO_4^{2-}, SeO_3^{2-}

111. Which illustration best represents $CaCl_2$ dissolved in water?

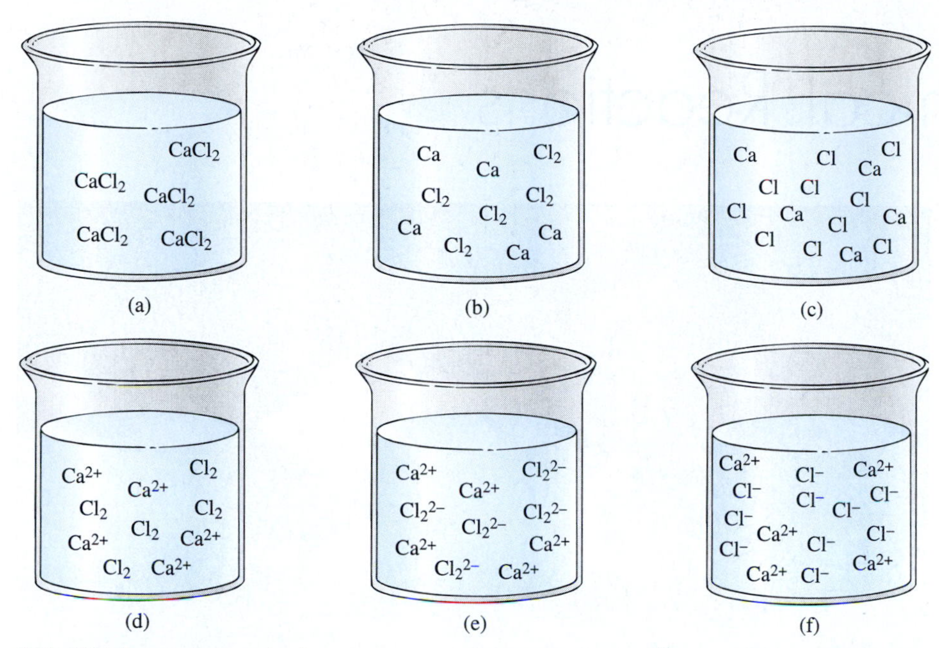

112. Which sample has the largest amount of NH_3: 6.022×10^{24} molecules of NH_3, 0.1 mol of NH_3, or 17.03 g of NH_3?

113. One molecule of an unknown compound has a mass of 7.308×10^{-23} g, and 27.3% of that mass is due to carbon and the rest oxygen. What is the compound?

CHAPTER

4 Chemical Reactions

The sand of this beach contains the two most abundant elements of the earth's crust—silicon and oxygen. • How is pure silicon obtained from sand? The Summary Problem at the end of this chapter identifies chemical reactions used to extract and purify silicon. *(photo by Bob Krist for the Puerto Rico Tourism Co.)*

Silicon is the second most abundant element in the earth's crust, always found there combined in a wide variety of compounds. The elemental silicon used in computer chips must first be extracted from its principal natural sources—quartz and sand (SiO_2)—and then purified to a high degree.

A major emphasis of chemistry is understanding how compounds such as SiO_2 react and what products are formed. A chemical compound may consist of molecules or oppositely charged ions, and the compound's properties can be interpreted in terms of the behavior of those molecules or ions. The *chemical* properties of a compound are the transformations that the molecules or ions can undergo when the substance reacts. A central focus of chemistry is knowing answers to the following questions: When two compounds are mixed together, will a chemical reaction occur? If a chemical reaction is possible, what will the products be?

To understand chemical reactions, the first step is to be able to write balanced chemical equations—equations that are consistent with the law of conservation of matter and the existence of atoms. This chapter begins with a discussion of how to write balanced chemical equations.

Many of the very large number of known chemical reactions can be assigned to a few general categories: combination, decomposition, displacement, exchange (which includes precipitation, acid-base, and gas formation) and oxidation-reduction reactions. You will be introduced to these kinds of reactions in this chapter. The ability to recognize when a given type of reaction may occur will enable you to predict the kinds of compounds that can be produced. Chapters 5, 6, and 7 extend our consideration of chemical reactions to the quantities of reactants and products, and the energy associated with the reactions.

4.1 CHEMICAL EQUATIONS

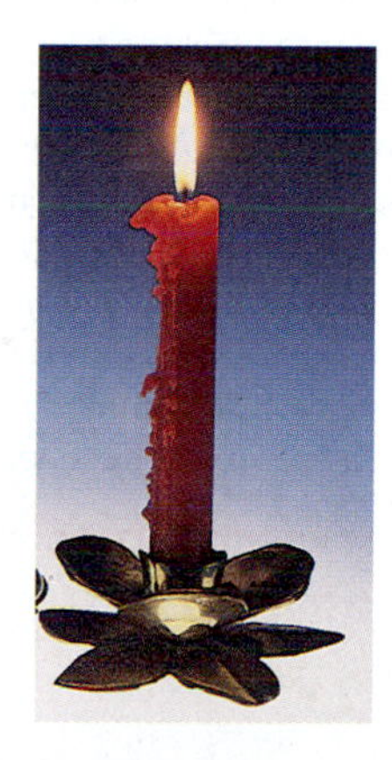

Burning hydrocarbons from candle wax create a candle flame. *(C.D. Winters)*

A candle flame can create a mood as well as provide light. It also is the result of a chemical reaction, a process in which reactants are converted into products (⇐ p. 16). The reactants and products can be elements, compounds, or both. In equation form we write

$$\text{Reactant(s)} \longrightarrow \text{Product(s)}$$

where the arrow means "forms" or "yields."

In a burning candle, the reactants are the hydrocarbons from the candle wax and oxygen from the air. Such reactions in which an element or compound burns in air or oxygen are called **combustion reactions** (Section 4.2). The products of the *complete* combustion of hydrocarbons (⇐ p. 84) are always carbon dioxide and water.

$$C_{25}H_{52}(s) + 38\ O_2(g) \longrightarrow 25\ CO_2(g) + 26\ H_2O\ (\ell)$$

a hydrocarbon in candle wax; diatomic oxygen; carbon dioxide; water

This **balanced chemical equation** indicates the relative amounts of reactants and products so that the number of atoms of each element in the reactants equals the number of atoms of each element in the products. In the next section we'll discuss how to write balanced equations.

Usually the physical states of the reactants and products are also indicated in a chemical equation by one of the following symbols after each reactant and product: (s) for solid, (g) for gas, and (ℓ) for liquid. The symbol (aq) is used to represent a

Portrait of a Scientist • *Antoine Laurent Lavoisier (1743–1794)*

On Monday, August 7, 1774, the Englishman Joseph Priestley (1733–1804) became the first person to isolate oxygen. He heated solid mercury(II) oxide, HgO, causing the oxide to decompose to mercury and oxygen.

$$2\ HgO(s) \xrightarrow{\text{heat}} 2\ Hg(\ell) + O_2(g)$$

Priestley did not immediately understand the significance of the discovery, but he mentioned it to the French chemist Lavoisier in October, 1774. One of Lavoisier's contributions to science was his recognition of the importance of exact scientific measurements and of carefully planned experiments, and he applied these methods to the study of oxygen. From this work he came to believe Priestley's gas was present in all acids and so he named it "oxygen," from two Greek words that mean "to form an acid." In addition, Lavoisier observed that the heat produced by a guinea pig in exhaling a given amount of carbon dioxide is similar to the quantity of heat produced by burning carbon to give the same amount of carbon dioxide. From this and other experiments he concluded that "Respiration is a combustion, slow it is true, but otherwise perfectly similar to that of charcoal." Although he did not understand the details of the process, this was an important step in the development of biochemistry.

Lavoisier also introduced principles for naming chemical substances that are still in use today. Further, he wrote a textbook in which he applied for the first time the principle of the conservation of matter to chemistry and used the idea to write early versions of chemical equations.

As Lavoisier was an aristocrat, he came under suspicion during the Reign of Terror of the French Revolution. He was an investor in the Ferme Générale, the infamous tax-collecting organization in 18th-century France. Tobacco was a monopoly product of the Ferme Générale, and it was common to cheat the purchaser by adding water to the tobacco, a practice that Lavoisier opposed. Nonetheless, because of his involvement with the Ferme, his career was cut short by the guillotine on May 8, 1794, on the charge of "adding water to the people's tobacco."

For a fascinating account of Lavoisier's life and his friendship with Benjamin Franklin, see "The Passion of Antoine Lavoisier" in *Bully for Brontosaurus* by Stephen Jay Gould (Norton, 1991).

Antoine Lavoisier and his wife, painted in 1788 by Jacques-Louise David.

(The Bettmann Archive)

substance dissolved in water, an **aqueous solution,** as illustrated by the equation for the reaction of metallic zinc with hydrochloric acid, which is an aqueous solution of hydrogen chloride. When zinc and hydrochloric acid are mixed, they react to produce hydrogen gas and an aqueous solution of zinc chloride, a soluble ionic compound ***(⇐ p. 104).***

$$\underbrace{Zn(s) + 2\ HCl\ (aq)}_{\text{reactants}} \longrightarrow \underbrace{H_2(g) + ZnCl_2(aq)}_{\text{products}}$$

In the 18th century, the great French scientist Antoine Lavoisier introduced the law of conservation of matter, which later became part of Dalton's atomic theory ***(⇐ p. 28).*** Lavoisier showed that *matter is not created or destroyed in chemical reactions.* Therefore, *if you use 5 g of reactants they will form 5 g of products if the reaction is complete; and if you use 500 mg of reactants, they will form 500 mg of products; and so on.* Combined with Dalton's atomic theory, this also means that if there are 1000 atoms of a particular element in the reactants, then those 1000 atoms must appear in the products.

There must be an accounting for *all* atoms in a chemical reaction.

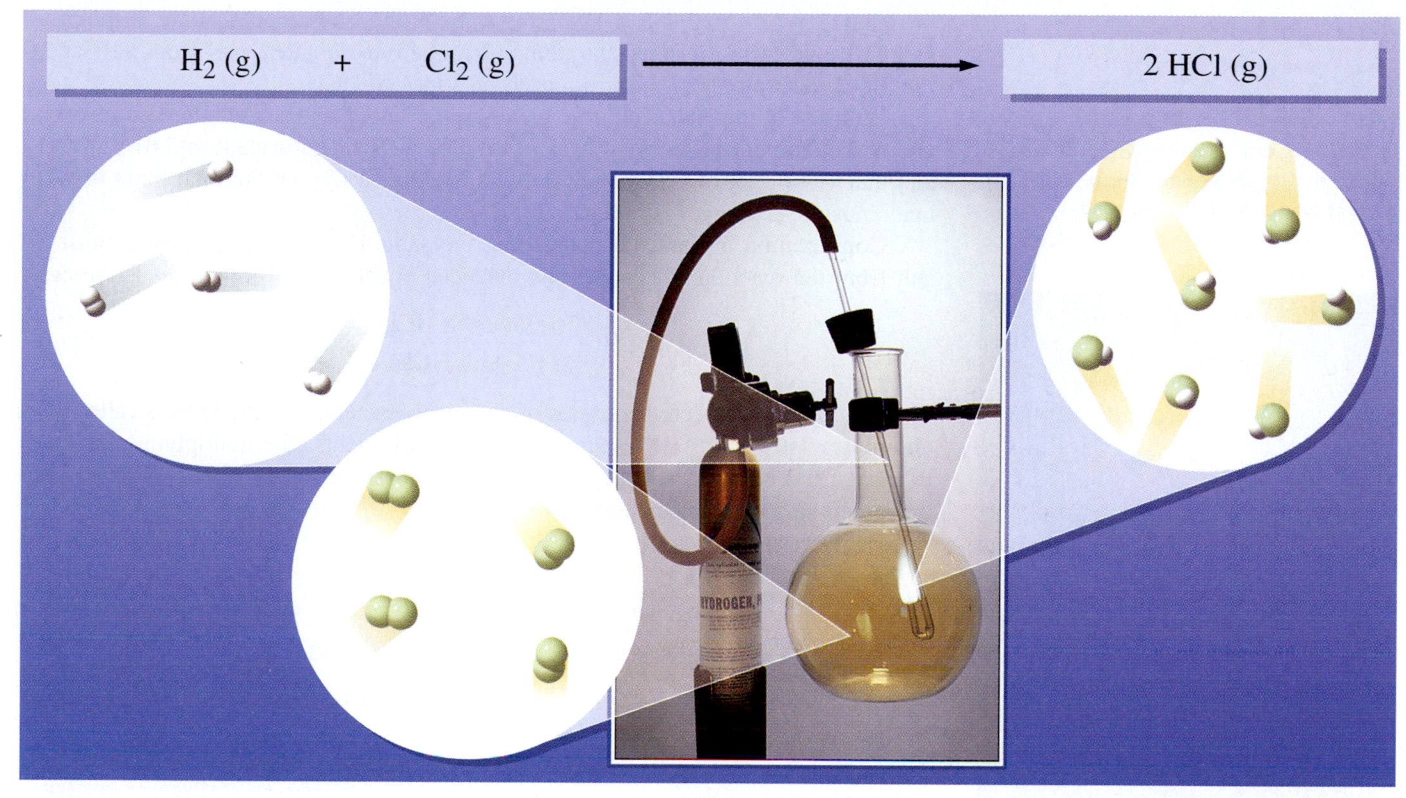

Figure 4.1 Hydrogen (H_2) and chlorine (Cl_2) react to form hydrogen chloride (HCl). Two molecules of HCl are formed when one H_2 molecule reacts with a Cl_2 molecule. This ratio is maintained when the reaction is carried out on a larger scale. *(C.D. Winters)*

Consider, for example, the reaction between gaseous hydrogen and chlorine to produce hydrogen chloride gas:

$$H_2(g) + Cl_2(g) \longrightarrow 2\ HCl(g)$$

When applied to this reaction, the conservation of matter means that one diatomic molecule of H_2 (two atoms of hydrogen) and one diatomic molecule of Cl_2 (two atoms of Cl) must produce *two* molecules of HCl. The numbers in front of the formulas—the **coefficients**—in the balanced equation are put there to show how matter is conserved. The 2 HCl indicates that two HCl molecules are formed, each containing one hydrogen atom and one chlorine atom. Note how *in an equation the symbol of an element or the formula of a compound is multiplied through by the coefficient that precedes it.* The equality of the number of atoms of each kind in the reactants and in the products is what makes the equation "balanced."

The balanced equation for the reaction of hydrogen with chlorine also implies that 6 molecules of H_2, for example, will react with 6 Cl_2 molecules to give 12 molecules of HCl (Figure 4.1). If we continue to scale up the reaction, we reach the point where 6.022×10^{23} H_2 molecules (1 mol of H_2; $2 \times 6.022 \times 10^{23}$ H atoms) reacting with 6.022×10^{23} Cl_2 molecules (1 mol of Cl_2; $2 \times 6.022 \times 10^{23}$ Cl atoms) will produce $2 \times 6.022 \times 10^{23}$ molecules of HCl (2 mol of HCl). As demanded by the conservation of matter, the total number of atoms in the reactants is 2.408×10^{24}

Recall that there are 6.022×10^{23} atoms in a mole of any element (← p. 55) or 6.022×10^{23} molecules (← p. 112) in a mole of any molecular compound.

($4 \times 6.022 \times 10^{23} = 2.408 \times 10^{24}$), the same number of atoms that are in the product. There are 4 mol of reactant atoms (2 mol from H_2 and 2 mol from Cl_2) and 4 mol of product atoms (2 mol from each of 2 mol of HCl).

With the atoms balanced, the masses represented by the equation are also balanced. The molar masses show that 1.000 mol of H_2 is equivalent to 2.016 g of H_2, and that 1.000 mol of Cl_2 is equivalent to 70.90 g of Cl_2, so the total mass of reactants must be 2.016 g + 70.90 g = 72.92 g.

Conservation of matter demands that the same mass, 72.92 g of HCl, must result from the reaction. Of course, the balanced equation shows that this is the case.

$$2.000 \text{ mol HCl} \times \frac{36.458 \text{ g HCl}}{1 \text{ mol HCl}} = 72.92 \text{ g HCl}$$

The relationship between the masses of chemical reactants and products is called **stoichiometry** (stoy-key-AHM-uh-tree), and the coefficients (the multiplying numbers) in a balanced equation are the **stoichiometric coefficients.** Chapter 5 is devoted to the stoichiometry of chemical reactions.

Figure 4.2 Powdered iron burns in air to form the iron oxides FeO and Fe_2O_3. The flame and the energy released during the reaction heat the particles to incandescence. *(C.D. Winters)*

Exercise 4.1 Stoichiometric Coefficients

Heating iron metal in oxygen forms iron(III) oxide, Fe_2O_3 (Figure 4.2).

$$4\,Fe(s) + 3\,O_2(g) \longrightarrow 2\,Fe_2O_3(s)$$

(a) If 2.50 g of Fe_2O_3 are formed by this reaction, what is the maximum *total* mass of iron metal and oxygen that reacted?
(b) Identify the stoichiometric coefficients in this equation.
(c) If 10,000 O atoms reacted, how many Fe atoms were needed to react with this amount of oxygen?

Combination reactions, in which two or more reactants combine to form a single product, are one of the four major types of chemical reactions, as discussed in Section 4.3.

4.2 BALANCING CHEMICAL EQUATIONS

To balance a chemical equation means using coefficients so that the same number of atoms of each element appears on each side of the equation. We will begin with one of the general classes of reactions, the combination of reactants to produce a single product, to illustrate how to balance chemical equations by a largely trial-and-error process.

Many metals and nonmetals react directly with oxygen to form **oxides,** compounds in which oxygen has combined with another element. The rusting of iron, the formation of the air pollutants nitrogen dioxide and carbon monoxide, and the formation of a protective oxide coating on aluminum exposed to air are examples of oxide formation.

The temperature in the combustion chambers of an automobile engine is high enough that nitrogen and oxygen can combine directly, a reaction that does not occur in the air at lower temperatures.

$$N_2(g) + O_2(g) \longrightarrow 2\,NO(g)$$

nitrogen monoxide

This equation is balanced; the numbers of N and O atoms are the same on each side of the equation.

Now consider the case where nitrogen monoxide, NO, as it comes from the automobile engine, reacts with oxygen to form nitrogen dioxide, NO_2, a major air pollutant (Figure 4.3).

(unbalanced equation) $NO(g) + O_2(g) \longrightarrow NO_2(g)$
nitrogen monoxide — nitrogen dioxide

In this unbalanced equation, the N atoms are balanced—one in NO on the left and one in NO_2 on the right. But the oxygens are not balanced; there are three on the left (one in NO and two in O_2) and two on the right (in NO_2). A coefficient of 2 in front of NO_2 gives four O atoms on the right, but the O atoms are still not balanced, and now the N atoms are also not balanced.

(unbalanced equation) $NO(g) + O_2(g) \longrightarrow 2\ NO_2(g)$

atom count:	1 N		2 N
	1 O	2 O	4 O

To get four O atoms on the left and also two N atoms on the right, we can put a 2 in front of NO. This produces a balanced equation.

(balanced equation) $2\ NO(g) + O_2(g) \longrightarrow 2\ NO_2(g)$

atom count:	2 N		=	2 N
	2 O	+ 2 O	=	4 O

$2\ NO(g) + O_2(g) \longrightarrow 2\ NO_2(g)$

2 N		=	2 N
2 O	+ 2 O	=	4 O

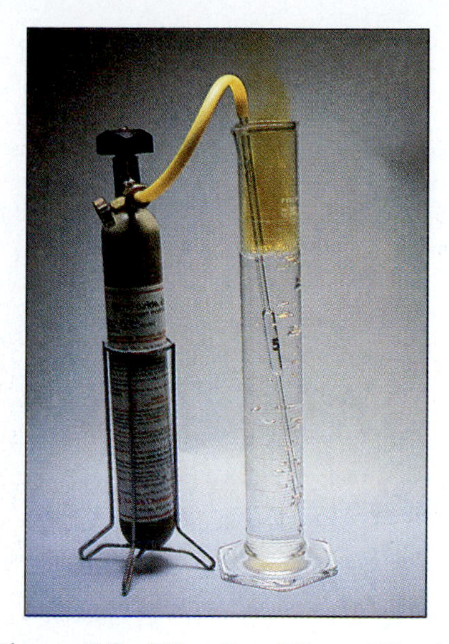

Figure 4.3 $NO + O_2$. The gas NO, nitrogen monoxide, is stored in a tank. It is bubbled through water, where it is evident that the gas is colorless. However, as soon as the bubbles of NO enter the atmosphere, the NO is oxidized to brown NO_2, nitrogen dioxide. *(C.D. Winters)*

Remember that a coefficient multiplies the number of atoms of each element in the symbol or formula that immediately follows the coefficient.

Below is a series of steps for organizing a trial-and-error approach so that there are fewer trials and errors.

Step 1: Write an initial unbalanced equation containing the correct formulas for the reactants on the left and the products on the right of the arrow. We began in this way with the reaction of NO and O_2. It is important to remember that *after the **correct** formulas for the reactants and products are written, the subscripts cannot be changed in order to balance equations.* Balancing is done by adjusting coefficients, *not* subscripts. Changing subscripts changes the original compound to something else, for example, H_2O and H_2O_2 (hydrogen peroxide), a compound *very* different than water.

Step 2: Check the equation to see which atoms are not in balance. Then start by balancing atoms of one of the elements. If possible, first balance atoms other than those of hydrogen or oxygen. It is best to begin with atoms that appear in only one formula on each side of the equation. Notice that we did this by beginning with N atoms in the reaction of NO with O_2.

Step 3: Balance atoms of the remaining elements. This step may require adjusting coefficients used in Step 2.

Step 4: Verify that the number of atoms of each element is balanced.

To illustrate using these steps, we will balance the equation for the formation of ammonia from nitrogen and hydrogen. Millions of tons of ammonia (NH_3) are manufactured worldwide annually by this reaction using nitrogen extracted from air and hydrogen obtained from natural gas. We begin by writing the unbalanced equation (Step 1).

$$N_2(g) + H_2(g) \longrightarrow NH_3(g)$$

Clearly, nitrogen and hydrogen are unbalanced. There are two nitrogen atoms on the left and only one on the right; two hydrogen atoms on the left and three on the right.

We start by using a coefficient of 2 on the right to balance the nitrogen atoms (Step 2); 2 NH_3 indicates two ammonia molecules, each containing a nitrogen atom and three hydrogen atoms. On the right we now have two nitrogen atoms and six hydrogen atoms.

(unbalanced equation) $N_2(g) + H_2(g) \longrightarrow 2\ NH_3(g)$

To balance the six hydrogen atoms on the right, we use a coefficient of 3 for the H_2 on the left to furnish six hydrogen atoms (Step 3).

We can verify that the equation is balanced (Step 4) by doing an atom count to check that the numbers of nitrogen and hydrogen atoms are the same on each side of the equation.

(balanced equation) $N_2 + 3\ H_2 \longrightarrow 2\ NH_3$

atom count: $2\ N + 6\ H = 2 \times N + 2 \times 3\ H$

$2\ N + 6\ H = 2\ N + 6\ H$

(balanced equation) $N_2 + 3\ H_2 \longrightarrow 2\ NH_3$

atom count: $2\ N + 6\ H = 2 \times N + 2 \times 3\ H$

$2\ N + 6\ H = 2\ N + 6\ H$

The physical states of the reactants and products are generally also included in the balanced equation. Thus, the final equation for ammonia formation is

$$N_2(g) + 3\ H_2(g) \longrightarrow 2\ NH_3(g)$$

PROBLEM-SOLVING EXAMPLE 4.1 Balancing A Chemical Equation

Chromium metal reacts with gaseous chlorine to form solid chromium(III) chloride, $CrCl_3$. Write the balanced equation for the formation of $CrCl_3$.

Answer $2\ Cr(s) + 3\ Cl_2(g) \rightarrow 2\ CrCl_3(s)$

Explanation

Step 1: *Write an initial unbalanced equation containing the correct formulas of all reactants and products.* The unbalanced equation is

unbalanced equation: $Cr + Cl_2 \longrightarrow CrCl_3$

Step 2: *Start by balancing atoms of one of the elements.* Chlorine is unbalanced; two chlorine atoms are on the left and three on the right. In such cases we can use coefficients of 3 on the left and 2 on the right to have six chlorine atoms on each side.

unbalanced equation: $Cr + 3\ Cl_2 \longrightarrow 2\ CrCl_3$

Note that the chromium atoms are now unbalanced.

Step 3: *Balance atoms of the remaining elements.* Chromium atoms are balanced by using a coefficient of 2 on the left that balances the equation as well.

$$2\ Cr(s) + 3\ Cl_2(g) \longrightarrow 2\ CrCl_3(s)$$

Step 4: *Verify that the number of atoms of each element is balanced.* As verification, there are two chromium atoms and six chlorine atoms on each side of the equation.

Problem-Solving Practice 4.1

Balance these equations.

(a) $Xe(g) + F_2(g) \rightarrow XeF_4(g)$

(b) $As_2O_3(s) + H_2(g) \rightarrow As(s) + H_2O(\ell)$

We now turn to the combustion of ethanol (C_2H_6O) and octane (C_8H_{18}) to illustrate balancing somewhat more complex chemical equations for combustion reactions. We will again assume that complete combustion occurs, meaning that the only products are carbon dioxide and water.

PROBLEM-SOLVING EXAMPLE 4.2 Balancing the Equation for a Combustion Reaction

Write the balanced equation for the combustion of ethanol, C_2H_6O.

Answer $C_2H_6O(\ell) + 3\ O_2(g) \rightarrow 2\ CO_2(g) + 3\ H_2O(\ell)$

Explanation

Step 1: *Write an initial unbalanced equation.*

(unbalanced equation) $C_2H_6O(\ell) + O_2(g) \longrightarrow CO_2(g) + H_2O(\ell)$

Step 2: *Start by balancing atoms of one of the elements.* None of the elements are balanced, so we could start with C, H, or O. We start with carbon because it appears in only one formula on each side of the equation. We have two C atoms in each C_2H_6O reactant molecule and only one in the carbon-containing product, CO_2. Therefore, a coefficient of 2 is used with CO_2 to balance the carbon atoms.

(unbalanced equation) $C_2H_6O + O_2 \longrightarrow 2\ CO_2 + H_2O$

Step 3: *Balance atoms of the remaining elements.* There are six H atoms as reactants. Because each water molecule contains two H atoms, the six H atoms in the reactants will combine with oxygen to form three water molecules.

(unbalanced equation) $C_2H_6O + O_2 \longrightarrow 2\ CO_2 + 3\ H_2O$

The products now contain seven oxygen atoms ($2 \times 2 = 4$ in 2 CO_2 plus $3 \times 1 = 3$ in 3 H_2O). The seven oxygens can be balanced by one atom in one ethanol molecule and six oxygen atoms in three diatomic O_2 molecules.

(balanced equation) $C_2H_6O(\ell) + 3\ O_2(g) \longrightarrow 2\ CO_2(g) + 3\ H_2O(\ell)$

Step 4: *Verify that the number of atoms of each element is balanced.* An atom count for each molecule verifies this number.

$$C_2H_6O(\ell) + 3\ O_2(g) \longrightarrow 2\ CO_2(g) + 3\ H_2O(\ell)$$

atom count:				
2 C		= 2 C		
6 H		=		6 H
1 O	+ 6 O	= 4 O		+ 3 O

In balancing the equation for a combustion reaction of compounds containing oxygen, we generally leave the oxygen atoms until last because oxygen is found in all the reactants and the products.

Problem-Solving Practice 4.2

(a) Write a balanced chemical equation to represent the combustion of propane, C_3H_8.
(b) When propane burns in an insufficient amount of oxygen, water forms but carbon monoxide rather than carbon dioxide is produced. Write a balanced chemical equation to represent the incomplete combustion of propane. Use this equation and the one from part (a) to interpret what is meant by the term "insufficient amount of oxygen."

PROBLEM-SOLVING EXAMPLE 4.3 Balancing a Combustion Reaction Equation

Write a balanced equation for the complete combustion of C_8H_{18}, a gasoline component.

Answer $2\ C_8H_{18}(\ell) + 25\ O_2(g) \rightarrow 16\ CO_2(g) + 18\ H_2O(\ell)$

Explanation

Step 1: *Write an initial unbalanced equation.* Because this is a complete combustion reaction, CO_2 and H_2O are the only products. Thus, the initial equation is

$$\text{(unbalanced equation)}\quad C_8H_{18}(\ell) + O_2(g) \longrightarrow CO_2(g) + H_2O(\ell)$$

Step 2: *Start by balancing the atoms of one of the elements.* C, H, and O are all unbalanced in the above equation, and we can start by balancing carbon atoms. The eight C atoms in C_8H_{18} will produce eight CO_2 molecules.

$$\text{(unbalanced equation)}\quad C_8H_{18} + O_2 \longrightarrow 8\ CO_2 + H_2O$$

Step 3: *Balance atoms of the remaining elements.* We next balance the hydrogen atoms. The 18 H atoms in the reactants will combine with oxygen to form 9 water molecules, each containing 2 H atoms.

$$C_8H_{18} + O_2 \longrightarrow 8\ CO_2 + 9\ H_2O$$

Oxygen is the remaining element to balance. At this point, there are 25 oxygen atoms in the products (8×2 in 8 CO_2, plus 9×1 in 9 H_2O), but only 2 in the reactants. In this case, there is an even number of O atoms in the reactants and an odd number of them in the products. A stoichiometric coefficient of 25/2 for O_2 on the left can give us 25 oxygen atoms because

$$\frac{25}{2}\ O_2 \times \frac{2\ \text{oxygen atoms}}{1\ O_2\ \text{molecule}} = 25\ \text{oxygen atoms}$$

Therefore the equation could be balanced as follows:

$$C_8H_{18}(\ell) + \frac{25}{2}\ O_2(g) \longrightarrow 8\ CO_2(g) + 9\ H_2O(\ell)$$

Except in special circumstances, however, it is customary to use only whole-number coefficients. Thus, multiplying every coefficient in the equation by 2 gives the balanced equation with whole number coefficients.

$$2\ C_8H_{18}(\ell) + 25\ O_2(g) \longrightarrow 16\ CO_2(g) + 18\ H_2O(\ell)$$

Fractional coefficients are sometimes needed for one or more reactants or products when it is necessary to write an equation for a specific whole number of moles of another reactant or product.

Step 4: *Verify that the number of atoms of each element is balanced.*

$$2\ C_8H_{18}(\ell) + 25\ O_2(g) \longrightarrow 16\ CO_2(g) + 18\ H_2O(\ell)$$

atom count:				
16 C		=	16 C	
36 H		=		36 H
	50 O	=	32 O	+ 18 O

Problem-Solving Practice Exercise 4.3

Methyl-*tert*-butyl ether, $C_5H_{12}O$, is a common gasoline additive used to enhance octane rating. Write the balanced equation for the combustion of methyl-*tert*-butyl ether
(a) Completely to carbon dioxide and water.
(b) Incompletely to carbon monoxide and water.

4.3 PATTERNS OF CHEMICAL REACTIONS

The classification of reaction patterns given in Figure 4.4 applies mainly to elements and inorganic compounds.

Many simple chemical reactions fall into one of the reaction patterns illustrated in Figure 4.4. It is important that you understand why learning to recognize these reaction patterns is useful. They serve as a guide to predict what might happen when chemicals are mixed together or have energy transferred to them. Also, they can help you recognize what there is to be learned by reading a chemical equation.

Take a look at the equation below. What does it mean to you at this stage in your study of chemistry?

$$Cl_2(g) + 2\ KBr(aq) \longrightarrow 2\ KCl(aq) + Br_2(\ell)$$

It's easy to let your eye slide by an equation on the printed page, but *don't* do that; read the equation. In this case it shows you that gaseous diatomic chlorine mixed with an aqueous solution of potassium bromide reacts to produce an aqueous solution of potassium chloride plus diatomic bromine. After you learn to recognize reaction patterns, you will see that this is a *displacement* reaction—chlorine has displaced bromine so that the resulting compound in solution is KCl instead of the KBr originally present. Because chlorine displaces bromine, chlorine is what chemists describe as "more active" than bromine. The occurrence of this reaction implies that when chlorine is mixed with a solution of a different bromine compound, displacement might also take place.

Throughout the rest of this chapter and the rest of this book, you should read and interpret chemical equations as we have just illustrated. Note the physical states of reactants and products. Mentally classify the reactions as described in the following sections and look for what can be learned from each example. Most importantly, *don't* think that the equations are there to be memorized. There are far too many chemical reactions for that. Look instead for patterns, classes of reactions and the kinds of substances that undergo them, and information that can be applied in other situations.

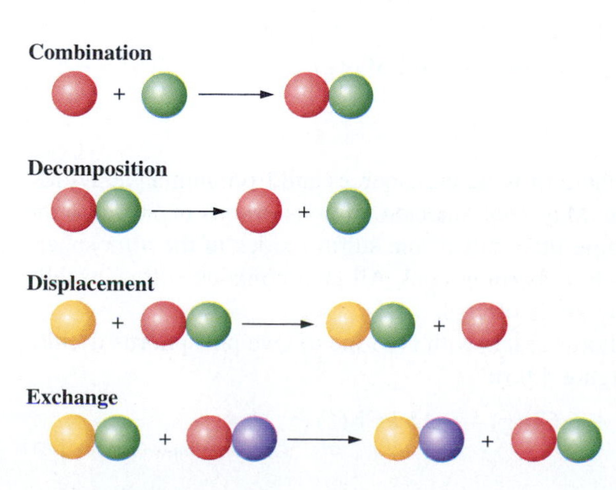

Figure 4.4 Four general types of chemical reactions. The spheres may represent atoms or groups of atoms.

Doing this will give you insight into how chemistry is used every day in a wide variety of applications.

Combination Reactions

In a **combination reaction,** two or more substances react to form a single product.

$$X + Z \longrightarrow XZ$$

Recall that the halogens are F_2, Cl_2, Br_2, and I_2.

Oxygen and the halogens (Group 7A) are such reactive elements that they undergo combination reactions with most other elements. Thus, if one of two possible reactants is oxygen or a halogen and the other is another element, it is reasonable to expect that a combination reaction will occur.

The reaction of a metal and oxygen produces an *ionic compound,* a metal oxide. As for any ionic compound, the formula of the metal oxide can be predicted from the fact that it must be electrically neutral. Because you can predict a reasonable positive charge for the metal ion by knowing the position of the metal in the periodic table and by using the guidelines in Section 3.4, you can determine the formula of the metal oxide. (Recall that oxygen forms the O^{2-} ion.) For example, when aluminum (Al, Group 3A; forms 3+ ions) reacts with O_2, the product must be aluminum oxide Al_2O_3 (2 Al^{3+} and 3 O^{2-} ions), a compound also known as *alumina,* or *corundum.*

$$4\ Al(s) + 3\ O_2(g) \longrightarrow 2\ Al_2O_3(s)$$

aluminum oxide

The halogens also combine with metals to form ionic compounds with formulas that are predictable based on the charges of the ions formed. The halogens all form 1− ions in ionic compounds. For example, sodium combines with chlorine and zinc combines with iodine to form sodium chloride and zinc iodide, respectively (Figure 4.5).

$$2\ Na(s) + Cl_2(g) \longrightarrow 2\ NaCl(s)$$
$$Zn(s) + I_2(s) \longrightarrow ZnI_2(s)$$

When nonmetals or metalloids combine with oxygen or chlorine, the compounds formed are not ionic but are composed of *molecules.* For example, sulfur, the Group 6A neighbor of oxygen, combines with oxygen to form two oxides, SO_2 and SO_3, in reactions of great environmental and industrial significance (Figure 4.6a).

$$S_8(s) + 8\ O_2(g) \longrightarrow 8\ SO_2(g)$$

sulfur dioxide
(colorless; choking odor)

$$2\ SO_2(g) + O_2(g) \longrightarrow 2\ SO_3(g)$$

sulfur trioxide
(colorless; and even more choking)

Sulfur dioxide enters the atmosphere from natural sources and from human activities. The eruption of Mt. St. Helens in May 1980 injected millions of tons of SO_2 into the atmosphere, for example; but more than half of the sulfur oxides in the atmosphere come from human activities such as burning coal. All coal contains sulfur, usually from 1 to 4% by weight in many kinds of coal.

As another example, phosphorus reacts with chlorine to give phosphorus trichloride, a molecular compound (Figure 4.6b):

$$P_4(s) + 6\ Cl_2(g) \longrightarrow 4\ PCl_3(\ell)$$

(Text continues on page 142.)

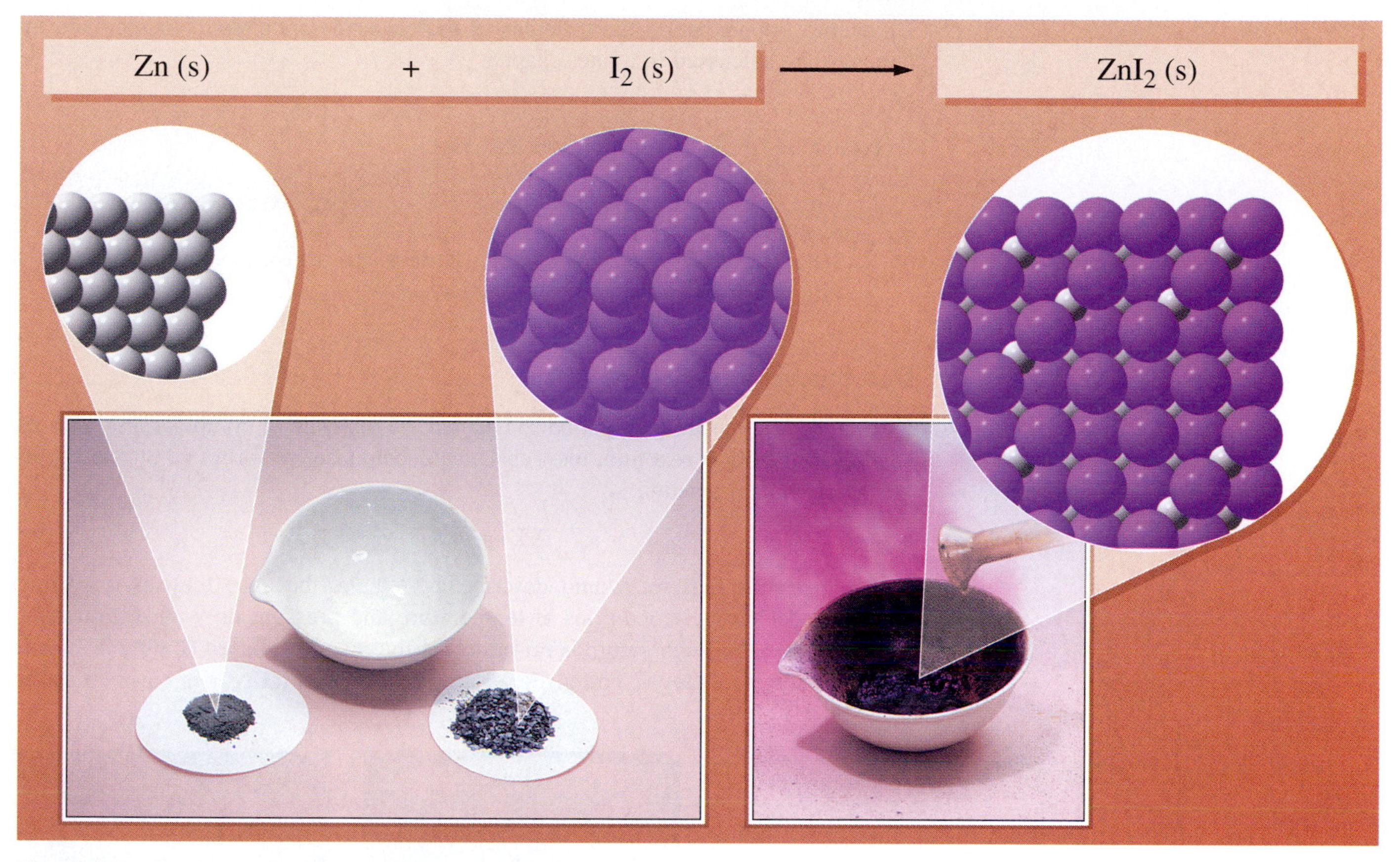

Figure 4.5 Combination of zinc and iodine. Gray, powdered zinc metal (far left) reacts with grayish-purple iodine in a vigorous combination reaction. Zinc atoms react with diatomic iodine molecules to form zinc iodide, an ionic compound, and the heat of the reaction is great enough that excess iodine sublimes as a purple vapor. *(C.D. Winters)*

Figure 4.6 Combination reactions. (a) Sulfur burns in pure oxygen with a bright blue flame to give sulfur dioxide, SO_2. *(C.D. Winters)* (b) White phosphorus burns in chlorine gas to produce phosphorous trichloride, PCl_3. *(C.D. Winters)*

Guidelines for predicting the formulas of molecular compounds of oxygen and the halogens are discussed in a later chapter.

Exercise 4.2 Combination Reactions

1. Write a balanced equation for the reaction of copper with oxygen in air to form copper(II) oxide.
2. Sulfur reacts with fluorine to give SF_6. Write a balanced equation for this reaction.
3. Write a balanced equation for the reaction of barium with bromine.

Decomposition Reactions

Decomposition reactions might be considered the opposite of combination reactions. In a **decomposition reaction,** one substance decomposes to form two or more products. The general reaction is

$$XZ \longrightarrow X + Z$$

Compounds that we would describe as "stable" because they exist without change under normal conditions of temperature and pressure only undergo decomposition when the temperature is raised to get the reaction started (Figure 4.7) or to transfer energy to keep it going. For example, a few metal oxides can be decom-

Figure 4.7 The dichromate volcano. Ammonium dichromate decomposes according to the equation $(NH_4)_2Cr_2O_7(s) \longrightarrow N_2(g) + Cr_2O_3(s) + 4\ H_2O(\ell)$. In the first photo, the dichromate compound is placed in a beaker with an alcohol-soaked paper wick. After ignition, the heat evolved by the reaction allows it to continue. On completion, the beaker contains a pile of green chromium(III) oxide, and water droplets cling to the side of the beaker (*Note:* Compounds containing the Cr(VI) ion, such as dichromates, are carcinogenic. These compounds should be handled only under properly supervised conditions.) *(C.D. Winters)*

posed by heating to give the metal and oxygen, the reverse of combination reactions. One of the best-known decomposition reactions is the reaction by which Joseph Priestley discovered oxygen in 1774 (← ***p. 132, Portrait of a Scientist***).

$$2\ HgO(s) \xrightarrow{\text{heat}} 2\ Hg(\ell) + O_2(g)$$

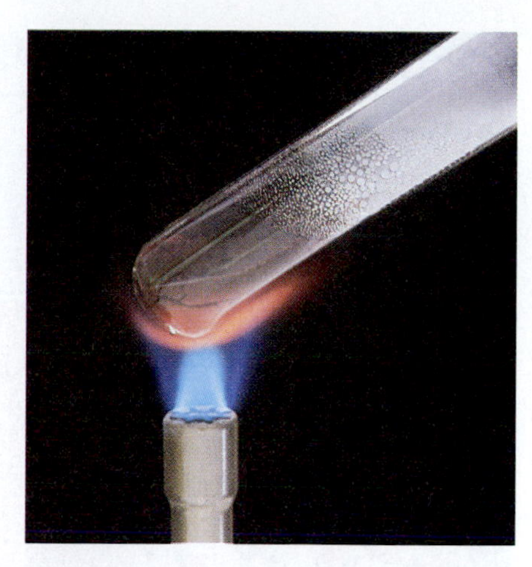

Decomposition of HgO. When heated, red mercury(II) oxide decomposes into liquid mercury and oxygen gas. *(C.D. Winters)*

A very common, and important, type of decomposition reaction is illustrated by the chemistry of *metal carbonates,* and calcium carbonate in particular. Carbonates decompose to give oxides plus carbon dioxide:

$$\underset{\text{limestone}}{CaCO_3(s)} \xrightarrow{800\text{–}1000\ ^\circ C} \underset{\text{lime}}{CaO(s)} + CO_2(g)$$

Calcium is the fifth most abundant element in the earth's crust and is the third most abundant metal (after Al and Fe). Most naturally occurring calcium is in the form of calcium carbonate ($CaCO_3$), from the fossilized remains of early marine life (Figure 4.8). Limestone, a form of calcium carbonate, is one of the basic raw materials of industry. Lime (calcium oxide), made by the decomposition reaction shown above, is a raw material needed for the manufacture of chemicals, in water treatment, and in the paper industry (see Table of Top 25 Chemicals, inside back cover).

Some compounds are sufficiently unstable and decompose with such force that their decomposition reactions are explosions. Sodium azide, NaN_3, is used to inflate automobile air bags (Figure 4.9). In emergency circumstances where inflation is required, an electronic detonator rapidly decomposes the sodium azide, forming nitrogen gas that inflates the bag in a few milliseconds.

$$2\ NaN_3(s) \longrightarrow 2\ Na(s) + 3\ N_2(g)$$

Nitroglycerine is an explosive so sensitive that it decomposes violently at the slightest shock.

$$4\ C_3H_5(NO_3)_3(\ell) \longrightarrow 12\ CO_2(g) + 10\ H_2O(\ell) + 6\ N_2(g) + O_2(g)$$

Alfred Nobel, the originator of the Nobel Prizes, made his fortune by developing a safer way for handling nitroglycerine by mixing it with a clay-like material to form dynamite.

Figure 4.8 A bed of limestone, primarily composed of calcium carbonate ($CaCO_3$), along the Verde River in Arizona. *(Image Enterprises/James Cowlin)*

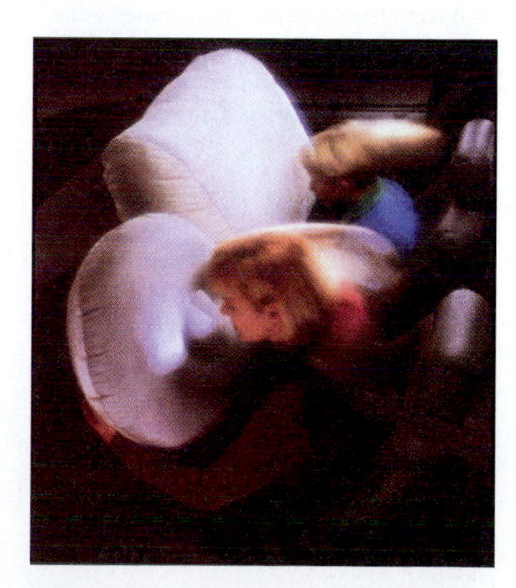

Figure 4.9 Chemical decomposition in action. Automobile air bags are rapidly inflated when sodium azide decomposes to form nitrogen gas: $2\ NaN_3(s) \rightarrow 2\ Na(s) + 3\ N_2(g)$. *(American Plastics Council)*

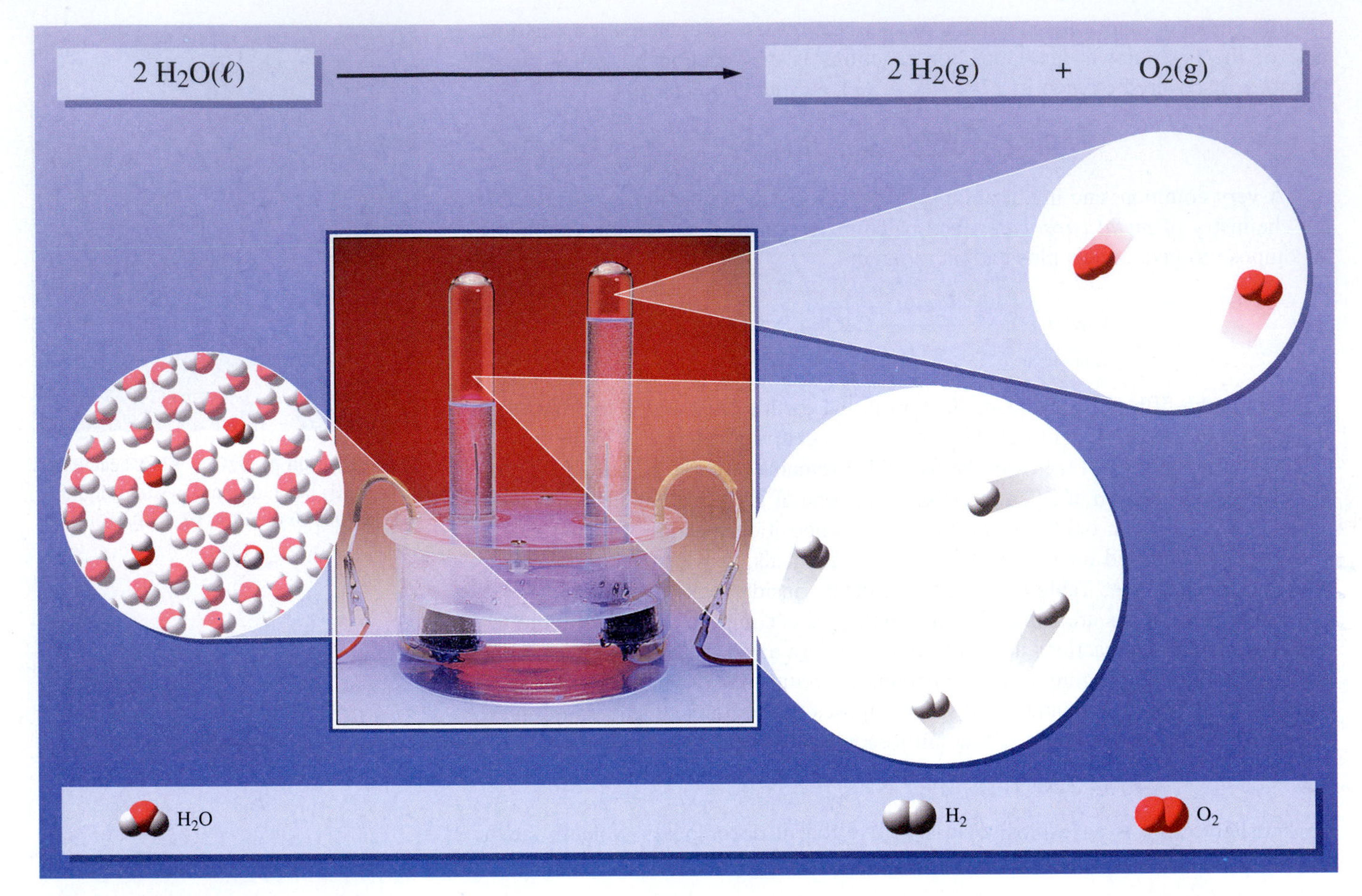

Figure 4.10 Decomposition of water. A direct electric current decomposes water into gaseous hydrogen (H_2) and oxygen (O_2). (In the nanoscale picture, only those water molecules that react are shown in full color.) *(C.D. Winters)*

Water, by contrast, is such a stable compound that it can be decomposed to hydrogen and oxygen only at a very high temperature or by using an electric current, a process called **electrolysis** (Figure 4.10).

$$2\ H_2O(\ell) \xrightarrow{\text{direct current}} 2\ H_2(g) + O_2(g)$$

Exercise 4.3 Combination and Decomposition Reactions

Predict the products formed by these reactions:
(a) Strontium with phosphorus
(b) Calcium with fluorine
(c) The thermal decomposition of magnesium carbonate

Displacement Reactions

Displacement reactions are those in which one element reacts with a compound to form a new compound and releases a different element. The element released is said to have been displaced. The general case for a **displacement reaction** is

$$A + XZ \longrightarrow AZ + X$$

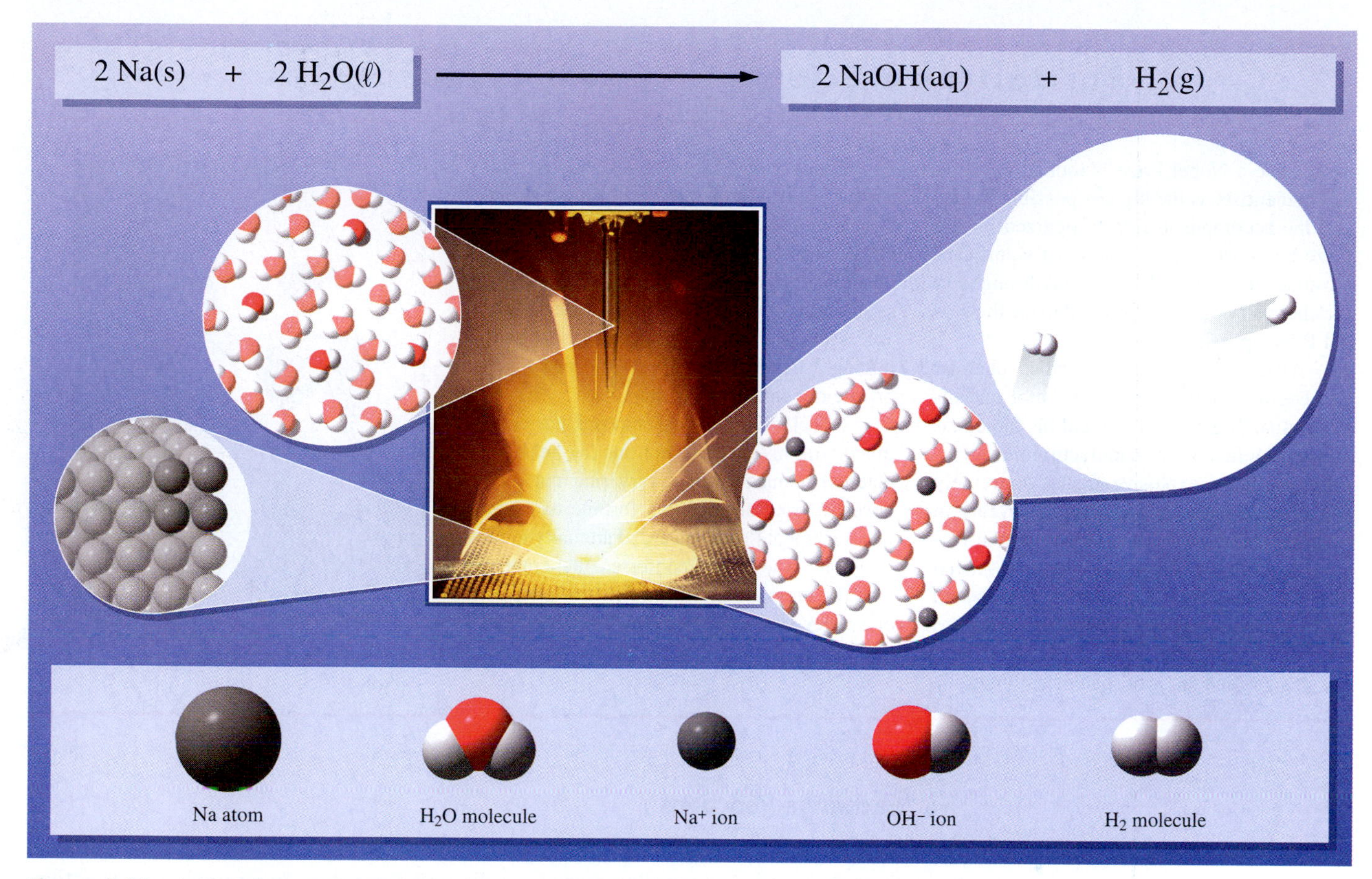

Figure 4.11 A displacement reaction. When liquid water drips from a buret onto a sample of solid sodium metal, the sodium displaces hydrogen gas from the water, and an aqueous solution of sodium hydroxide is formed. The hydrogen gas burns, producing the flame shown in the photograph. (In the nanoscale pictures, only those molecules that undergo reaction are shown in full color. Other water molecules are paler in color.) *(C.D. Winters)*

The reaction of metallic sodium with water is such a reaction.

$$2\ Na(s) + 2\ H_2O(\ell) \longrightarrow 2\ NaOH(aq) + H_2(g)$$

Here sodium displaces hydrogen from water. All of the alkali metals, which are very reactive elements, react in this way when exposed to water (Figure 4.11). The alkaline earth metals calcium, strontium, and barium also displace hydrogen from water to form hydroxides; for example,

$$Ca(s) + 2\ H_2O(\ell) \longrightarrow Ca(OH)_2(aq) + H_2(g)$$

Magnesium undergoes the same reaction, but because it is a less reactive metal, it requires a higher temperature to react (Section 4.9).

Iron metal can be displaced from Fe_3O_4 by hydrogen:

$$4\ H_2(g) + Fe_3O_4(s) \longrightarrow 3\ Fe(s) + 4\ H_2O(g)$$

In terms of the general displacement reaction, A = H, X = Fe, and Z = O. Hydrogen can release many metals from their metal oxides this way. Section 4.9 will indicate how to predict when displacement reactions will occur according to the relative reactivities of the metals.

Portrait of a Scientist • *Alfred Bernhard Nobel (1833–1896)*

To receive a Nobel Prize is looked upon by most scientists as the highest possible honor, and the accomplishments recognized by the Prize are always among the most significant advances in science. It was breakthroughs in the chemistry of explosives that made the Nobel Prizes possible.

Alfred Nobel was a Swedish chemist and engineer, the son of an inventor and industrialist. Together Alfred and his father established a factory to manufacture nitroglycerine, a liquid explosive that is extremely sensitive to light and heat. After 2 years in operation, there was an explosion at the factory that killed five people, including Alfred Nobel's younger brother. Perhaps motivated by this event, Alfred searched for a way to handle nitroglycerin more safely. He found the answer and in 1867 patented dynamite. By mixing nitroglycerin with a soft, absorbent, nonflammable natural material known as kieselguhr, or diatomaceous earth, Nobel had stabilized the explosive power of nitroglycerin. His dynamite could be shipped safely and would explode only when desirable. He also invented the blasting caps (containing a mercury compound) needed to set off the dynamite explosion.

Nobel's talent as an entrepreneur combined with his many inventions (he held 355 patents) to make him a very rich man. He never married and left his fortune to establish the Nobel Prizes, awarded annually to individuals who "have conferred the greatest benefits on mankind in the fields of physics, chemistry, physiology or medicine, literature and peace."

Alfred Bernhard Nobel. *(The Granger Collection)*

Exchange Reactions

Exchange reactions include several kinds of reactions that take place between reactants that are ionic compounds and are dissolved in water. In an **exchange reaction,** there is an interchange of partners between two compounds. In general:

$$AD + XZ \longrightarrow AZ + XD$$

These reactions are also called metathesis or double displacement reactions.

Recall that ionic compounds dissociate in water to release their component ions into solution *(← p. 104)*. What might happen if two aqueous solutions are mixed together, one containing dissolved calcium nitrate, $Ca(NO_3)_2$, and the other containing dissolved sodium chloride, NaCl? These are both soluble ionic compounds, so the new solution is going to contain Ca^{2+}, NO_3^-, Na^+, and Cl^-. To decide whether a reaction will occur requires deciding whether any of these ions can react with each other to form a new compound. For an exchange reaction to occur, the calcium ion and the chloride ion would have to form calcium chloride ($CaCl_2$), and the sodium ion and the nitrate ion would have to form sodium nitrate ($NaNO_3$). Is this a possible chemical reaction? The answer is yes *if* either of these compounds is insoluble. Checking the solubility rules *(← p. 104)* shows that both of these compounds are water-soluble. No reaction to remove the ions from solution is possible; therefore, the answer is no.

A mixture of solutions of lead(II) nitrate and potassium chromate provides an example of a case in which an insoluble product—a **precipitate**—can form. In the resulting reaction, aqueous Pb^{2+} ions and K^+ ions exchange partners to form insoluble lead chromate and water-soluble potassium nitrate (Figure 4.12).

$$\underset{\text{lead nitrate}}{Pb(NO_3)_2(aq)} + \underset{\text{potassium chromate}}{K_2CrO_4(aq)} \longrightarrow \underset{\text{lead chromate}}{PbCrO_4(s)} + \underset{\text{potassium nitrate}}{2\ KNO_3(aq)}$$

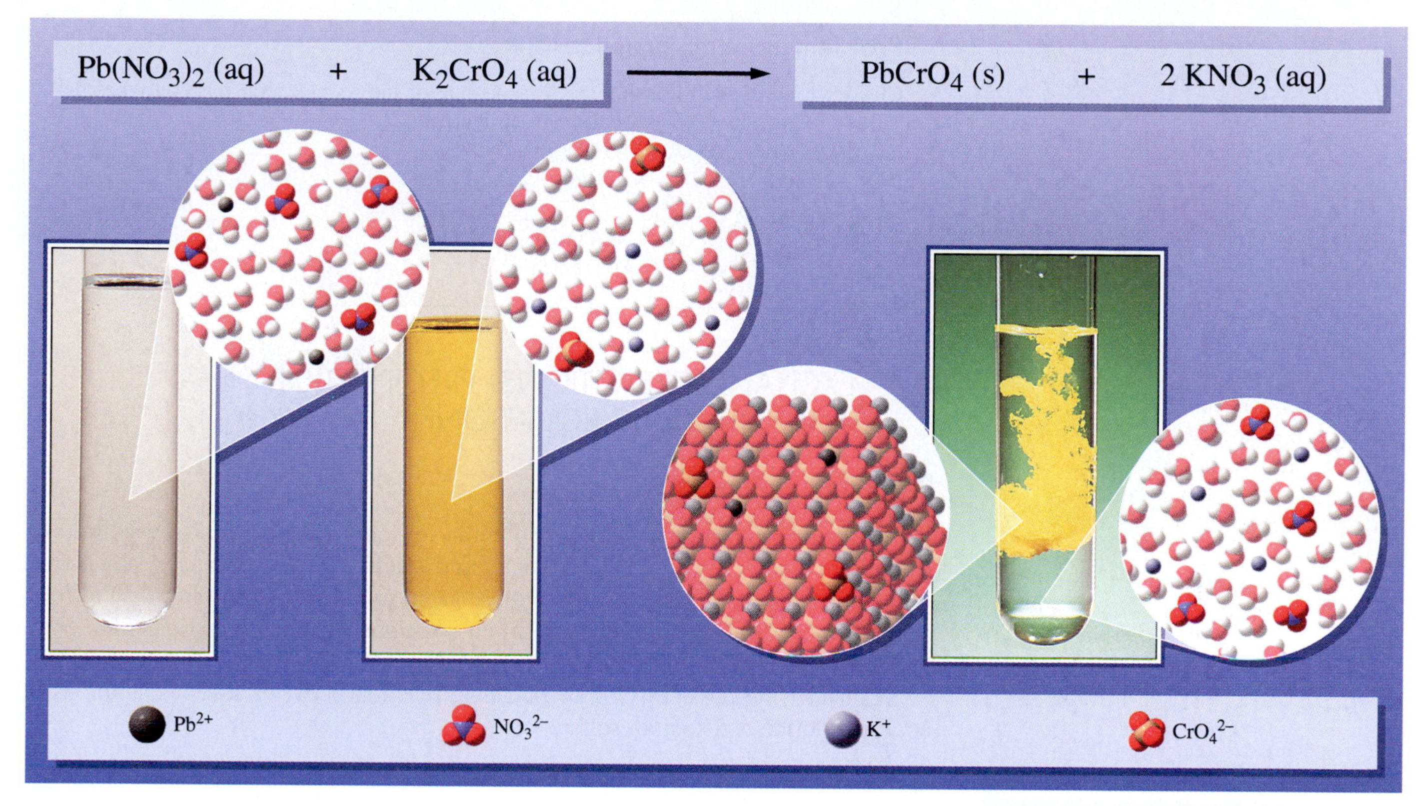

Figure 4.12 Precipitation of lead(II) chromate. Adding aqueous potassium chromate (K_2CrO_4) to a solution of lead(II) nitrate ($Pb(NO_3)_2$) results in an exchange reaction that forms a yellow precipitate of lead(II) chromate (Pb_2CrO_4) and leaves water-soluble potassium nitrate (KNO_3) in solution. *(C.D. Winters)*

Exchange reactions, which are further examined in the next three sections, occur when reactant ions are removed from solution by the formation of one of three types of product: (1) a precipitate (Section 4.4); (2) a molecular (non-ionic) compound (Section 4.5); or (3) a gas (Section 4.6).

PROBLEM-SOLVING EXAMPLE 4.4 Classifying Reactions by Type

Classify the following reactions as one of the four general reaction types described in this section.

(a) $Pt(s) + 2\ F_2(g) \rightarrow PtF_4(\ell)$

(b) $3\ Fe(s) + 4\ H_2O(g) \rightarrow Fe_3O_4(s) + 4\ H_2(g)$

(c) $2\ H_3BO_3(s) \rightarrow B_2O_3(s) + 3\ H_2O(\ell)$

(d) $B_2O_3(s) + 6\ HF(\ell) \rightarrow 2\ BF_3(g) + 3\ H_2O(\ell)$

Answer (a) Combination; (b) displacement; (c) decomposition; (d) exchange

Explanation

(a) This reaction is the direct combination of platinum and fluorine to form platinum tetrafluoride.

(b) The general equation for a displacement reaction, $A + XZ \rightarrow AZ + X$, matches what occurs in the given reaction: iron (A) displaces hydrogen (X) from water (XZ) to form Fe_3O_4 (AZ) plus H_2 (X).

(c) In this reaction, a single substance, H_3BO_3 (boric acid), decomposes to form two products, B_2O_3 and H_2O.

(d) The reactants interchange partners in this exchange reaction. Applying the general equation for an exchange reaction ($AD + XZ \rightarrow AZ + XD$) to this case, we find A = boron, D = oxygen, X = hydrogen, and Z = fluorine.

Problem-Solving Practice 4.4

Classify each of the following reactions as one of the four general reaction types described in Section 4.3. Also, balance each equation.

(a) $Al(OH)_3(s) \rightarrow Al_2O_3(s) + H_2O(g)$

(b) $Na_2O(s) + H_2O(\ell) \rightarrow NaOH(aq)$

(c) $S_8(s) + F_2(g) \rightarrow SF_6(g)$

(d) $NaOH(aq) + H_3PO_4 \rightarrow Na_3PO_4(aq) + H_2O(\ell)$

(e) $C(s) + Fe_2O_3(s) \rightarrow CO(g) + Fe(\ell)$

4.4 EXCHANGE REACTIONS: PRECIPITATION AND NET IONIC EQUATIONS

Precipitation Reactions

In Section 3.8 you studied the water solubility of ionic compounds and used solubility rules to predict whether an ionic compound would dissolve in water. As illustrated in the preceding section, you can use those same solubility rules (Table 4.1) to forecast whether a precipitate will form when solutions of ionic compounds are combined.

Consider the possibility of an exchange reaction between aqueous solutions of barium chloride and sodium sulfate.

$$BaCl_2(aq) + Na_2SO_4(aq) \longrightarrow ? + ?$$

If the barium ions and sodium ions exchange partners to form $BaSO_4$ and NaCl, the equation will be

$$\underset{\text{barium chloride}}{BaCl_2(aq)} + \underset{\text{sodium sulfate}}{Na_2SO_4(aq)} \longrightarrow \underset{\text{barium sulfate}}{BaSO_4} + \underset{\text{sodium chloride}}{2\ NaCl}$$

Will a precipitate be formed? Checking Table 4.1, we find that although NaCl is soluble, $BaSO_4$ is not soluble (sulfates of Sr^{2+}, Ba^{2+}, and Pb^{2+} are insoluble). Therefore, an exchange reaction will occur, and barium sulfate will precipitate from the solution. The precipitate formation is indicated by an (s) in the overall equation (Figure 4.13). Because it is soluble, NaCl remains dissolved in solution.

$$BaCl_2(aq) + Na_2SO_4(aq) \longrightarrow BaSO_4(s) + 2\ NaCl(aq)$$

Figure 4.13 Precipitation of barium sulfate. The mixing of aqueous solutions of barium chloride ($BaCl_2$) and sodium sulfate (Na_2SO_4) form a precipitate of barium sulfate ($BaSO_4$). Sodium chloride (NaCl), the other product of this exchange reaction, is soluble and Na^+ and Cl^- ions remain in solution. *(C.D. Winters)*

PROBLEM-SOLVING EXAMPLE 4.5 Exchange Reactions

For each of the following pairs of ionic compounds, decide whether an exchange reaction will occur and write a balanced equation for those reactions that will occur.

(a) $AgNO_3$ and NaBr (b) $CaCl_2$ and $(NH_4)_3PO_4$ (c) KI and $MgCl_2$

Answer Precipitates will form in (a), AgBr, and in (b), $Ca_3(PO_4)_2$.

(a) $AgNO_3(aq) + NaBr(aq) \rightarrow AgBr(s) + NaNO_3(aq)$

(b) $3\ CaCl_2(aq) + 2\ (NH_4)_3PO_4(aq) \rightarrow 6\ NH_4Cl(aq) + Ca_3(PO_4)_2(s)$

(c) No reaction occurs; both products are soluble.

Explanation In each case, consider which cation-anion combinations can be formed and whether the possible compounds will precipitate.

(a) An exchange reaction between $AgNO_3$ and NaBr could form AgBr and $NaNO_3$. Table

4.1 indicates that $NaNO_3$ is soluble (all nitrates are soluble) and remains dissolved in solution, but AgBr is not soluble and therefore precipitates.

$$AgNO_3(aq) + NaBr(aq) \longrightarrow AgBr(s) + NaNO_3(aq)$$

(b) The exchange of Ca^{2+} for NH_4^+ forms insoluble $Ca_3(PO_4)_2$, leaving soluble NH_4Cl in solution.

$$3\ CaCl_2(aq) + 2\ (NH_4)_3PO_4(aq) \longrightarrow 6\ NH_4Cl(aq) + Ca_3(PO_4)_2(s)$$

(c) No precipitate forms because each possible product, KCl and MgI_2, is soluble. Thus, all four of the ions—K^+, I^-, and Mg^{2+}, and Cl^-—remain in the solution. No exchange reaction is possible because no product is formed that removes ions from the solution.

Problem-Solving Practice 4.5

Predict the products and write a balanced chemical equation for the exchange reaction in aqueous solution between the following ionic compounds. Use Table 4.1 to determine solubilities and indicate in the equation whether a precipitate forms.
(a) $NiCl_2$ and NaOH (b) K_2CO_3 and $CaBr_2$

TABLE 4.1 Solubility Rules for Ionic Compounds

Ammonium, Sodium, Potassium; NH_4^+, Na^+, K^+	All ammonium, sodium, and potassium salts are soluble.
Nitrates, NO_3^-	All nitrates are soluble.
Chlorides, Cl^-	All chlorides are soluble except AgCl, Hg_2Cl_2, and $PbCl_2$.
Sulfates, SO_4^{2-}	Most sulfates are soluble; exceptions include $SrSO_4$, $BaSO_4$, and $PbSO_4$.
Chlorates, ClO_3^-	All chlorates are soluble.
Perchlorates, ClO_4^-	All perchlorates are soluble.
Acetates, $CH_3CO_2^-$	All acetates are soluble.
Phosphates, PO_4^{3-}	All phosphates are insoluble except those of NH_4^+ and Group 1A elements (alkali metal cations).
Carbonates, CO_3^{2-}	All carbonates are insoluble except those of NH_4^+ and Group 1A elements (alkali metal cations).
Hydroxides, OH^-	All hydroxides are insoluble except those of NH_4^+, Group 1A (alkali metal cations), $Sr(OH)_2$, and $Ba(OH)_2$; $Ca(OH)_2$ is slightly soluble.
Oxides, O^{2-}	All oxides are insoluble except those of Group 1A (alkali metal cations).
Oxalates, $C_2O_4^{2-}$	All oxalates are insoluble except those of NH_4^+ and Group 1A (alkali metal cations).
Sulfides, S^{2-}	All sulfides are insoluble except those of NH_4^+, Group 1A (alkali metal cations), and Group 2A (MgS, CaS, and BaS are sparingly soluble).

Net Ionic Equations

In writing equations for exchange reactions in the preceding section, we have used overall equations. But there is another way to represent what actually happens. In each case in which a precipitate formed, the product that did not precipitate remained in solution. Therefore, its ions were in solution as reactants and remained there after the reaction (they did not form a precipitate). Such ions are commonly called **spectator ions** because, like the spectators at a play or game, they are present but they are not involved in the real action. Consequently, the spectator ions can be left out of the equation that represents the chemical change that occurs. An equation that includes only the symbols or formulas of ions in solution or compounds that undergo change is called a **net ionic equation.** We will use the reaction of aqueous NaCl with $AgNO_3$ to form AgCl and $NaNO_3$ to illustrate the general steps for writing a net ionic equation.

Step 1: Write the overall balanced equation using the correct formulas for the reactants and products.

$$\underset{\text{silver nitrate}}{AgNO_3} + \underset{\text{sodium chloride}}{NaCl} \longrightarrow \underset{\text{silver chloride}}{AgCl} + \underset{\text{sodium nitrate}}{NaNO_3}$$

Step 2: Use the general guidelines in Table 4.1 to determine the solubilities of reactants and products. In this case, the guidelines indicate that nitrates are soluble, so $AgNO_3$ and $NaNO_3$ are soluble. NaCl is water soluble because almost all chlorides are soluble. However, AgCl is one of the three insoluble chlorides (AgCl, Hg_2Cl_2, and $PbCl_2$). Using this information we can write

$$AgNO_3(aq) + NaCl(aq) \longrightarrow AgCl(s) + NaNO_3(aq)$$

Step 3: Recognize that all soluble ionic compounds dissociate into their component ions in aqueous solution. Therefore we have

$$AgNO_3(aq) \longrightarrow Ag^+(aq) + NO_3^-(aq)$$
$$NaCl(aq) \longrightarrow Na^+(aq) + Cl^-(aq)$$
$$NaNO_3(aq) \longrightarrow Na^+(aq) + NO_3^-(aq)$$

Step 4: Write a complete ionic equation with the ions in solution from each soluble compound shown separately.

$$Ag^+(aq) + NO_3^-(aq) + Na^+(aq) + Cl^-(aq) \longrightarrow AgCl(s) + Na^+(aq) + NO_3^-(aq)$$

Note that the precipitate is represented by its complete formula.

Step 5: Cancel out the spectator ions from each side of the complete ionic equation to create the net ionic equation. Sodium ions and nitrate ions are the spectator ions in this example and we cancel them from the complete ionic equation to give the net ionic equation (Figure 4.14).

Complete ionic equation:

$$Ag^+(aq) + \cancel{NO_3^-(aq)} + \cancel{Na^+(aq)} + Cl^-(aq) \longrightarrow AgCl(s) + \cancel{Na^+(aq)} + \cancel{NO_3^-(aq)}$$

Net ionic equation:

$$Ag^+(aq) + Cl^-(aq) \longrightarrow AgCl(s)$$

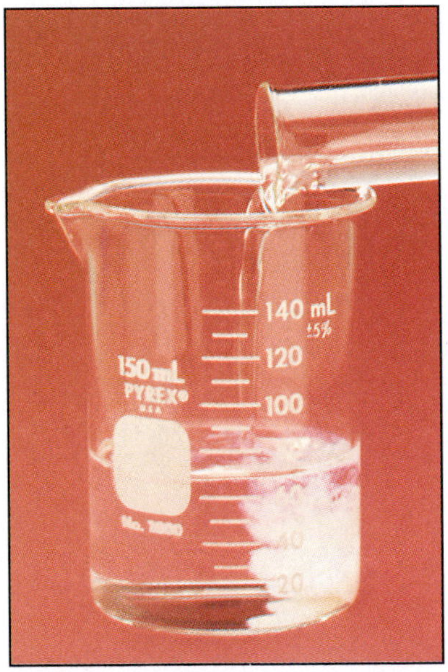

Figure 4.14 Precipitation of silver chloride. Adding aqueous sodium chloride, NaCl, to aqueous silver nitrate, $AgNO_3$, leads to a precipitate of white silver chloride, AgCl, and an aqueous solution of sodium nitrate, $NaNO_3$. *(Larry Cameron)*

Step 6: Check to see that the sum of the charges is the same on each side of the net ionic equation. The charge on each side of a net ionic equation must be the same for the equation to be correctly balanced. For the equation in Step 5 the sum of charges is zero on each side $(1+ + 1-) = 0$ on the left; AgCl is an ionic compound with no net charge).

If the charge is not the same on both sides of a balanced equation, then electrons are being created or destroyed, which is impossible according to the Law of Conservation of Matter.

Problem-Solving Example 4.6 Net Ionic Equations

Write the net ionic equation for the reaction that occurs when aqueous solutions of magnesium sulfate ($MgSO_4$) and sodium oxalate ($Na_2C_2O_4$) are mixed.

Answer $Mg^{2+}(aq) + C_2O_4^{2-}(aq) \rightarrow MgC_2O_4(s)$

Explanation

Step 1: *Write the overall balanced equation.*

$$MgSO_4 + Na_2C_2O_4 \longrightarrow MgC_2O_4 + Na_2SO_4$$

Step 2: *Determine the solubilities of reactants and products.* Using the solubility rules in Table 4.1 shows that all the reactants and products, except MgC_2O_4, are soluble.

$$MgSO_4(aq) + Na_2C_2O_4(aq) \longrightarrow MgC_2O_4(s) + Na_2SO_4(aq)$$

Step 3: *Write equations for the soluble compounds that dissociate in solution.*

$$MgSO_4(aq) \longrightarrow Mg^{2+}(aq) + SO_4^{2-}(aq)$$

$$Na_2C_2O_4(aq) \longrightarrow 2\ Na^+(aq) + C_2O_4^{2-}(aq)$$

$$Na_2SO_4(aq) \longrightarrow 2\ Na^+(aq) + SO_4^{2-}(aq)$$

Step 4: *Write the complete ionic equation.*

$$Mg^{2+}(aq) + SO_4^{2-}(aq) + 2\ Na^+(aq) + C_2O_4^{2-}(aq) \longrightarrow MgC_2O_4(s) + 2\ Na^+(aq) + SO_4^{2-}(aq)$$

Steps 5 and 6: *Cancel spectator ions; check to see that charge is balanced.* To create the net ionic equation, we cancel out all spectator ions from each side of the equation and check that the sum of charges is equal on each side of the net ionic equation.

Net ionic equation: $Mg^{2+}(aq) + C_2O_4^{2-}(aq) \longrightarrow MgC_2O_4(s)$

$$\text{Net charge} = (2+) + (2-) = 0$$

Problem-Solving Practice 4.6

Write a balanced equation for the reaction (if any) for each of the following ionic compound pairs in aqueous solution. Then write their balanced net ionic equations.

(a) NaF and $Ca(CH_3COO)_2$ (b) $(NH_4)_2S$ and $FeCl_2$

Exercise 4.4 Net Ionic Equations

It is possible for an exchange reaction to occur in aqueous solution where both products precipitate. Using Table 4.1, identify the reactants and products of such a reaction.

4.5 ACIDS, BASES, AND EXCHANGE REACTIONS

Litmus is a dye derived from certain lichens. Phenolphthalein is a synthetic dye.

Acids and bases are two important classes of compounds, so important that there is a whole chapter about them later (Chapter 17). Here we focus on a few of their general properties and how acids and bases react with each other. Acids have a number of properties in common, and so do bases. Some properties of acids are related to properties of bases. Solutions of acids change the colors of pigments in specific ways; for example, all acids change the color of litmus from blue to red and cause the dye phenolphthalein to be colorless, but bases turn red litmus to blue and make phenolphthalein pink. If an acid has made litmus red, adding a base will reverse the effect, making the litmus blue again. Thus, acids and bases seem to be opposites. A base can *neutralize* the effect of an acid, and an acid can neutralize the effect of a base.

Acids have other characteristic properties. They taste sour, they produce bubbles of gas when reacting with limestone, and they dissolve many metals while producing a flammable gas. Although tasting substances is never done in a chemistry laboratory, you have probably experienced the sour taste of at least one acid—vinegar, which is a solution of acetic acid in water. Bases, in contrast, have a bitter taste. Rather than dissolving metals, bases often cause metal ions to form insoluble compounds that precipitate from solution. Such precipitates can be made to dissolve by adding an acid, another case in which an acid counteracts a property of a base.

The properties of acids can be explained by a common feature of acid molecules, and a different common feature can explain the properties of bases. An **acid** is any substance that, dissolved in water, increases the concentration of hydrogen ions, H^+. A "naked" H^+ ion, however, can't exist in water. Because it is just a proton, H^+ is the smallest possible ion and is strongly attracted to any negative charge in the vicinity. The oxygen end of a water molecule is negatively charged, so H^+ and H_2O combine to form H_3O^+, known as the **hydronium ion.** In Chapter 17 the importance of the hydronium ion to acid-base chemistry will be explored. For now, we represent hydronium ion as $H^+(aq)$. The properties that acids have in common are those of hydrogen ions dissolved in water.

One of the most common acids is hydrochloric acid, which ionizes in aqueous solution to form hydrogen ions and chloride ions. (See Table 4.2 for other common acids.) Because it is completely converted to ions in aqueous solution, HCl is a **strong electrolyte** and is classified as a **strong acid** (Figure 4.15a).

$$HCl(aq) \longrightarrow H^+(aq) + Cl^-(aq)$$

(a)

(b)

(c)

Acids and bases. (a) Many common foods and household products are acidic or basic. Citrus fruits contain citric acid, and household ammonia and oven cleaner are basic. (b) The acid in lemon juice turns blue litmus paper red, while (c) household ammonia turns red litmus paper blue. *(C.D. Winters)*

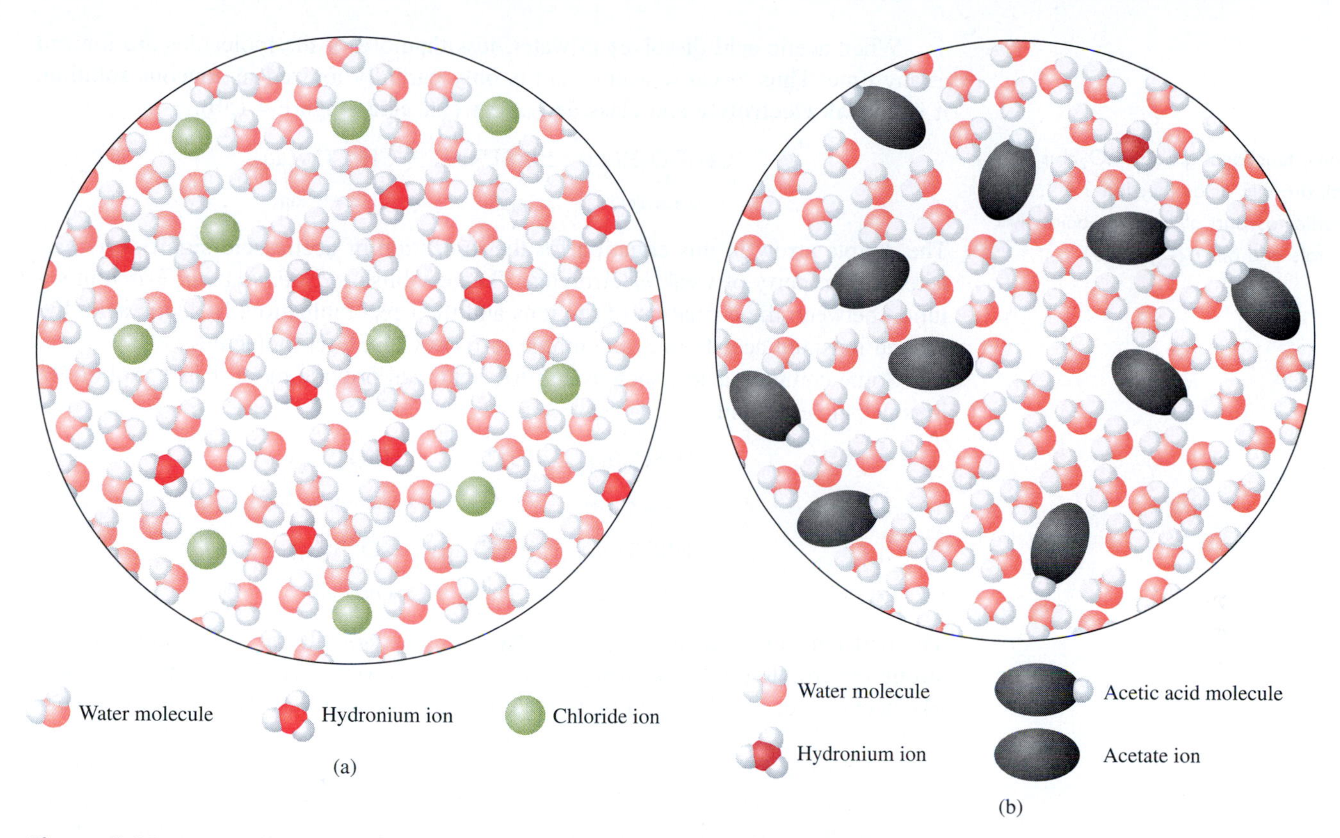

Figure 4.15 The ionization of acids in water. (a) A strong acid like hydrochloric acid (HCl) is completely ionized in water; all the HCl molecules ionize to form $H_3O^+(aq)$ and $Cl^-(aq)$ ions. (b) Weak acids such as acetic acid (CH_3CO_2H) are only slightly ionized in water. Non-ionized acetic acid molecules far outnumber aqueous H_3O^+ and $CH_3CO_2^-$ ions formed by the ionization of acetic acid molecules.

TABLE 4.2 Common Acids and Bases

Strong Acids (Strong Electrolytes)		Strong Bases (Strong Electrolytes)	
HCl	hydrochloric acid	LiOH	lithium hydroxide
HNO_3	nitric acid	NaOH	sodium hydroxide
H_2SO_4	sulfuric acid	KOH	potassium hydroxide
$HClO_4$	perchloric acid	$Ca(OH)_2$	calcium hydroxide
HBr	hydrobromic acid	$Ba(OH)_2$	barium hydroxide
HI	hydroiodic acid	$Sr(OH)_2$	strontium hydroxide
Weak Acids (Weak Electrolytes)		**Weak Bases (Weak Electrolytes)**	
H_3PO_4	phosphoric acid	NH_3	ammonia
CH_3CO_2H	acetic acid		
H_2CO_3	carbonic acid		

The organic functional group $-CO_2H$ is present in all organic acids (Section 17.). You saw it earlier as part of the fatty acids that make up fats and oils (← *p. 110*).

When acetic acid dissolves in water, less than 5% of the molecules are ionized at any time. Thus, because acetic acid is only partially ionized in aqueous solution, it is a **weak electrolyte** and classified as a **weak acid** (Figure 4.15b).

$$CH_3CO_2H(aq) \rightleftharpoons H^+(aq) + CH_3CO_2^-(aq)$$

acetic acid — acetate ion

The double arrow in this equation for the ionization of acetic acid signifies a characteristic property of weak electrolytes. They establish a *dynamic equilibrium* in solution between the formation of the ions and their recombination. In acetic acid, hydrogen ions and acetate ions recombine to form CH_3CO_2H molecules.

Some common acids, such as sulfuric acid, can provide more than 1 mol of H^+ ions per mole of acid:

$$H_2SO_4(aq) \longrightarrow H^+(aq) + HSO_4^-(aq)$$

sulfuric acid — hydrogen sulfate ion

$$HSO_4^-(aq) \rightleftharpoons H^+(aq) + SO_4^{2-}(aq)$$

hydrogen sulfate ion — sulfate ion

The first ionization reaction is essentially complete, so sulfuric acid is considered a strong electrolyte (and so a strong acid as well). However, the hydrogen sulfate ion, like acetic acid, is only partially ionized, so it is a weak electrolyte and also a weak acid.

A **base** is a substance that increases the concentration of the **hydroxide ion,** OH^-, when dissolved in pure water. The properties that bases have in common are the properties of $OH^-(aq)$. Compounds that contain hydroxide ions, such as sodium hydroxide or potassium hydroxide, are obvious bases. As ionic compounds they are strong electrolytes and **strong bases.**

$$NaOH(s) \xrightarrow{H_2O} Na^+(aq) + OH^-(aq)$$

strong electrolyte = 100% dissociated — hydroxide ion

Ammonia, NH_3, is another very common base. Although the compound does not have an OH^- ion as part of its formula, it produces the ion on reaction with water.

$$NH_3(aq) + H_2O(\ell) \rightleftharpoons NH_4^+ + OH^-$$

ammonia, weak base weak electrolyte — ammonium ion — hydroxide ion

In the equilibrium between NH_3 and the NH_4^+ and OH^- ions, only a small concentration of the ions is present, so ammonia is a weak electrolyte ($< 5\%$ ionized) and a **weak base.**

Exercise 4.5 Acids and Bases

(a) What ions are produced when perchloric acid dissolves in water? (b) Calcium hydroxide is not very soluble in water. What little does dissolve, however, is dissociated. What ions are produced? Write an equation for the dissociation of calcium hydroxide.

Neutralization Reactions

When solutions of a strong acid and a strong metal hydroxide base are mixed (for example, hydrochloric acid and sodium hydroxide), the ions in solution are the hydrogen ion and the anion from the acid, the metal cation and the hydroxide ion from the base:

From hydrochloric acid: H^+, Cl^-
From sodium hydroxide: Na^+, OH^-

As in precipitation reactions, an exchange reaction will occur if two of these ions can react with each other to form a compound that removes ions from solution. In an acid-base reaction, that compound is water, H_2O.

As has been pointed out, when an acid and base react, they neutralize each other. This happens because the hydrogen ions from the acid react with hydroxide ions from the base to form water. The product water is a molecular compound that is neither acidic nor basic; it is neutral. If the water is evaporated from this acid-base reaction, a solid remains consisting of the cation from the base and the anion of the acid. The general term for this kind of ionic compound is a **salt.** In the case of hydrochloric acid plus sodium hydroxide, the salt is sodium chloride, and evaporation of the product water would leave the solid salt behind.

$$\mathrm{HCl(aq) + NaOH(aq) \longrightarrow \underset{(H_2O)}{H\,OH(\ell)} + Na\,Cl(aq)}$$

Thus, the overall general equation for the reaction of an acid and a base is

$$\mathrm{acid(aq) + base(aq) \longrightarrow \underset{(H_2O)}{HOH(\ell)} + salt(aq)}$$

If we represent acids in general by H*X,* metallic hydroxide bases as (*M*OH), and the salt as *MX,* the equation becomes

$$\underset{\text{acid}}{\mathrm{H}X\mathrm{(aq)}} + \underset{\text{base}}{M\mathrm{OH(aq)}} \longrightarrow \underset{(\mathrm{H_2O})}{\mathrm{HOH}(\ell)} + \underset{\text{salt}}{MX\mathrm{(aq)}}$$

You should recognize this as an exchange reaction in which H^+ ions from the aqueous acid and M^+ ions from the metal hydroxide exchange partners.

The particular salt that forms depends on the acid and base used. From the general equation you can recognize that the *salt is formed from the anion of the acid* (X^{m-}) *and the cation of the base* (M^{n+}). For example, magnesium chloride is formed when a commercial antacid containing magnesium hydroxide is taken to neutralize excess hydrochloric acid in the stomach.

$$\underset{\text{hydrochloric acid}}{\mathrm{2\,HCl(aq)}} + \underset{\text{magnesium hydroxide}}{\mathrm{Mg(OH)_2(aq)}} \longrightarrow \mathrm{2\,H_2O(\ell)} + \underset{\text{magnesium chloride}}{\mathrm{MgCl_2(aq)}}$$

Organic acids, such as acetic acid and propionic acid, which contain the acid functional group $-CO_2H$, also neutralize bases to form salts. The reaction of propionic acid, $CH_3CH_2CO_2H$, and sodium hydroxide produces the salt sodium propionate, $NaCH_3CH_2CO_2$, containing sodium ions (Na^+) and propionate ions $CH_3CH_2CO_2^-$.

$$\underset{\text{propionic acid}}{\mathrm{CH_3CH_2CO_2H(aq)}} + \mathrm{NaOH(aq)} \longrightarrow \mathrm{H_2O(\ell)} + \underset{\text{sodium propionate}}{\mathrm{NaCH_3CH_2CO_2(aq)}}$$

Sodium propionate is commonly used as a food preservative.

PROBLEM-SOLVING EXAMPLE 4.7 Acids, Bases, and Salts

Identify the base and acid that could be used to form each of the following salts.
(a) $NaNO_3$ (b) Cs_3PO_4

Answer

(a) Sodium hydroxide (NaOH) and nitric acid (HNO_3); (b) cesium hydroxide (CsOH) and phosphoric acid (H_3PO_4).

Explanation

(a) A salt is formed from the cation of a base and the anion of an acid. $NaNO_3$ contains sodium and nitrate ions; Na^+ is derived from the base NaOH, and NO_3^- comes from nitric acid, HNO_3. The reaction of NaOH and HNO_3 produces sodium nitrate (and water).

$$NaOH(aq) + HNO_3(aq) \longrightarrow NaNO_3(aq) + H_2O(\ell)$$

(b) Cesium phosphate is formed from cesium ions (Cs^+) and phosphate ions (PO_4^{3-}) that come from cesium hydroxide (CsOH) and phosphoric acid (H_3PO_4), respectively.

$$3\ CsOH(aq) + H_3PO_4(aq) \longrightarrow Cs_3PO_4(aq) + 3\ H_2O(\ell)$$

Problem-Solving Practice 4.7

Identify the acid and base that form
(a) $MgSO_4$; (b) $SrCO_3$.

Net Ionic Equations for Acid-Base Reactions

As with precipitation reactions, net ionic equations can be written for acid-base reactions. This should not be surprising because precipitation and acid-base neutralization reactions are both exchange reactions.

Consider the reaction given earlier of magnesium hydroxide with hydrochloric acid to relieve excess stomach acid (HCl).

First we write the correct overall balanced equation, using the correct formulas for the reactants and products.

$$2\ HCl(aq) + Mg(OH)_2(aq) \longrightarrow 2\ H_2O(\ell) + MgCl_2(aq)$$

Next, we recognize that the acid and base furnish hydrogen ions and hydroxide ions, respectively. In this case the reactions are

$$2\ HCl(aq) \longrightarrow 2\ H^+(aq) + 2\ Cl^-(aq)$$

and

$$Mg(OH)_2(s) \longrightarrow Mg^{2+}(aq) + 2\ OH^-(aq)$$

Although magnesium hydroxide is not very soluble, what little dissolves is completely dissociated.

Note that we retain the coefficients from the balanced overall equation (first step).

We now use this information to write a complete ionic equation. To do so, recall that the cation of the base and the anion of the acid form the salt, in this case $MgCl_2$. We also use Table 4.1 to check the solubility of the salt. In this case, we find that magnesium chloride is soluble, so the Mg^{2+} and Cl^- ions remain in solution. The complete ionic equation is

$$\cancel{Mg^{2+}}(aq) + 2\ OH^-(aq) + 2\ H^+(aq) + 2\ \cancel{Cl^-}(aq) \longrightarrow \cancel{Mg^{2+}}(aq) + 2\ \cancel{Cl^-}(aq) + 2\ H_2O(\ell)$$

Cancelling spectator ions from each side of the complete ionic equation yields the net ionic equation. In this case magnesium ions and chloride ions are the spectator ions. Removing them leaves us with the net ionic equation:

$$2\ H^+(aq) + 2\ OH^-(aq) \longrightarrow 2\ H_2O(\ell)$$

or simply

$$H^+(aq) + OH^-(aq) \longrightarrow H_2O(\ell)$$

This is the net ionic equation for the neutralization reaction between a strong acid and a strong base that yields a soluble salt. Note that, as always, there is *conservation of charge* in the net ionic equation. On the left, $(1+) + (1-) = 0$; water has no net charge.

Exercise 4.6 Neutralization and Net Ionic Equations

Write balanced complete ionic equations and net ionic equations for the neutralization reactions of the following acids and bases:
(a) HCl and KOH; (b) $Ba(OH)_2$ and H_2SO_4 (Remember that sulfuric acid dissociates to provide 2 mol of H^+ per mole of sulfuric acid.); (c) CH_3CO_2H and $Ca(OH)_2$.

Exercise 4.7 Net Ionic Equations and Antacids

The commercial antacids Maalox, Di-Gel tablets, and Mylanta contain aluminum hydroxide and magnesium hydroxide that react with excess hydrochloric acid in the stomach. Write balanced complete ionic equations and net ionic equations for the soothing neutralization reaction of aluminum hydroxide with HCl.

Figure 4.16 Reaction of calcium carbonate with an acid. A piece of coral that is largely calcium carbonate, $CaCO_3$, reacts readily with hydrochloric acid to give CO_2 gas and aqueous calcium chloride. *(C.D. Winters)*

4.6 GAS-FORMING REACTIONS

The formation of a gas is the third factor causing exchange reactions to occur. Escape of the gas drives the reaction to form products from the reactants. Acids are involved in many gas-forming exchange reactions.

Precipitate formation and removal of ions by formation of a molecular substance such as water are the other two factors driving exchange reactions.

The reaction of metal carbonates with acids is an excellent example of a gas-forming exchange reaction (Figure 4.16).

$$CaCO_3(s) + 2\ HCl(aq) \longrightarrow CaCl_2(aq) + H_2CO_3(aq)$$
$$H_2CO_3(aq) \longrightarrow H_2O(\ell) + CO_2(g)$$

A salt and H_2CO_3 (carbonic acid) are *always* the products from an acid reacting with a metal carbonate, and their formation illustrates the exchange reaction pattern. Carbonic acid is unstable, however, and much of it is rapidly converted to water and CO_2 gas. If the reaction is done in an open container, most of the gas bubbles out of the solution.

Carbon dioxide is always released when acids react with a metal carbonate or a metal hydrogen carbonate. For example, excess hydrochloric acid in the stomach is neutralized by ingesting commercial antacids such as Alka-Seltzer ($NaHCO_3$), Tums ($CaCO_3$), or Di-Gel liquid ($MgCO_3$). Taking an Alka-Seltzer or a Tums to relieve excess stomach acid produces the following helpful reactions.

Alka-Seltzer: $NaHCO_3(aq) + HCl(aq) \longrightarrow NaCl(aq) + H_2O(\ell) + CO_2(g)$

Tums: $CaCO_3(aq) + 2\ HCl(aq) \longrightarrow CaCl_2(aq) + H_2O(\ell) + CO_2(g)$

The net ionic equations for these two reactions are:

$$HCO_3^-(aq) + H^+(aq) \longrightarrow H_2O(\ell) + CO_2(g)$$

$$CO_3^{2-}(aq) + 2\ H^+(aq) \longrightarrow H_2O(\ell) + CO_2(g)$$

The released carbon dioxide sometimes produces a burp.

Acids also react by exchange reactions with metal sulfites or sulfides to produce foul-smelling gaseous SO_2 or H_2S, respectively. With sulfites, the initial product is sulfurous acid which, like carbonic acid, quickly decomposes.

$$CaSO_3(aq) + 2\ HCl(aq) \longrightarrow CaCl_2(aq) + H_2SO_3(aq)$$

$$H_2SO_3(aq) \longrightarrow H_2O(\ell) + SO_2(g)$$

$$Na_2S(aq) + 2\ HCl(aq) \longrightarrow 2\ NaCl(aq) + H_2S(g)$$

Exercise 4.8 Gas-Forming Reactions

Predict the products and write the balanced overall equation and the net ionic equation for each of the following gas-generating reactions.

(a) $Na_2CO_3(aq) + H_2SO_4(aq) \rightarrow$

(b) $FeS(s) + HCl \rightarrow$

(c) $K_2SO_3(aq) + HCl$

Exercise 4.9 Exchange Reaction Classification

Identify each of the following exchange reactions as a precipitation reaction, an acid-base reaction, or a gas-forming reaction. Predict the products of each reaction and write an overall balanced equation for the reaction.

(a) $NiCO_3(s) + H_2SO_4(aq) \rightarrow$

(b) $Sr(OH)_2(s) + HNO_3(aq) \rightarrow$

(c) $BaCl_2(aq) + Na_2C_2O_4(aq) \rightarrow$

(d) $PbCO_3(s) + H_2SO_4(aq) \rightarrow$

4.7 OXIDATION-REDUCTION REACTIONS

Now, we turn to oxidation-reduction reactions, which are classified not by any specific reaction pattern, but by what happens with electrons at the nanoscale level as a result of the reaction. Virtually every reaction can be identified as either an oxidation-reduction reaction or not an oxidation-reduction reaction.

The terms "oxidation" and "reduction" come from reactions that have been known for centuries. Ancient civilizations learned how to change metal oxides and sulfides to the metal, that is, how to *reduce* ore to the metal. For example, cassiterite or tin(IV) oxide, SnO_2, is a tin ore discovered in Britain centuries ago, and it is very easily reduced to tin by heating with carbon.

SnO_2 loses oxygen and is reduced

$$SnO_2(s) + 2\ C(s) \longrightarrow Sn(s) + 2\ CO(g)$$

C is the agent of reduction

In this reaction carbon is the agent that brings about the reduction of tin ore to tin metal, so carbon is called the **reducing agent.**

When SnO_2 is reduced by carbon, oxygen is removed from the tin and added to the carbon, which is "oxidized" by the addition of oxygen. In fact, *any process in which oxygen is added to another substance is an oxidation.* This too is a process known for centuries. In the reaction with magnesium, oxygen is the **oxidizing agent** because it is the agent that is responsible for the oxidation.

Mg combines with oxygen and is oxidized

$$2\,Mg(s) + O_2(g) \longrightarrow 2\,MgO(s)$$

O_2 is the agent of oxidation

The experimental observations we have just outlined point to several fundamental conclusions: (a) If one substance is oxidized, another substance in the same reaction must be reduced. For this reason, we refer to such reactions as **oxidation-reduction reactions,** or **redox reactions** for short. (b) The reducing agent is itself oxidized, and the oxidizing agent is reduced. (c) Oxidation is the reverse of reduction. For example, the reactions we have described show that addition of oxygen is oxidation and removal of oxygen is reduction.

Redox Reactions and Electron Transfer

Many oxidation and reduction reactions involve direct transfer of electrons from one reactant to another. When a substance *accepts electrons,* it is said to be **reduced.** The language is descriptive because in a reduction there is a decrease in the real or apparent electric charge on an atom of the substance when it takes on electrons. In the following net ionic equation, Ag^+ is reduced to uncharged Ag(s) by accepting electrons from copper metal. Since copper metal supplies the electrons and causes the Ag^+ ion to be reduced, Cu is the reducing agent (Figure 4.17).

Figure 4.17 Oxidation of copper metal by silver ion. Copper wire was formed into a "tree" by wrapping short lengths around a central "trunk." The tree was immersed in a solution of silver nitrate, $AgNO_3$. With time, the copper reduces Ag^+ ions to silver metal crystals, and the copper metal is oxidized to Cu^{2+} ions. The blue color of the solution is due to the presence of aqueous copper(II) ion. *(C.D. Winters)*

Ag^+ accepts electrons and is reduced to Ag; Ag^+ is the oxidizing agent

$+2e^-$ from Cu

$$2\,Ag^+(aq) + Cu(s) \longrightarrow 2\,Ag(s) + Cu^{2+}(aq)$$

$-2e^-$ to Ag

Cu donates electrons and is oxidized to Cu^{2+}; Cu is the reducing agent

When a substance *releases electrons,* it is said to be **oxidized.** In **oxidation,** the real or apparent electric charge on an atom of the substance increases when it gives up electrons. In our example, copper metal releases electrons on going to Cu^{2+}; its electric charge has increased, and it is said to have been oxidized. In order for this to happen, there must be something available to take the electrons offered by the copper. In this case, Ag^+ is the electron acceptor and its charge is reduced (to zero in the element). Therefore, Ag^+ is the "agent" that causes Cu metal to be oxidized, so Ag^+ is the oxidizing agent. In every oxidation-reduction reaction, something is re-

Figure 4.18 $Mg(s) + O_2(g)$. A piece of magnesium ribbon burns in air, oxidizing the metal to the white solid magnesium oxide, MgO. *(C.D. Winters)*

duced (and therefore is the oxidizing agent) and something is oxidized (and therefore is the reducing agent).

In the reaction of magnesium and oxygen (Figure 4.18), oxygen gains electrons on going to the oxide ion. The charge of the O atoms changes from 0 to -2.

Mg releases $2e^-$ per atom;
Mg is oxidized and is the reducing agent.

$$2\,Mg(s) + O_2(g) \longrightarrow 2\,[Mg^{2+}, O^{2-}]$$

O_2 gains $4e^-$ per molecule;
O_2 is reduced and is the oxidizing agent.

In the same reaction, magnesium is the reducing agent because it loses two electrons per atom on forming the Mg^{2+} ion. All redox reactions can be analyzed in a similar manner. You will learn a formal bookkeeping system for this kind of analysis in Section 4.8.

Common Oxidizing and Reducing Agents

Oxidation is the loss of electrons.
$X \rightarrow X^+ + e^-$
X loses one or more electrons, is oxidized, and is the reducing agent (reduces something else).

Reduction is the gain of electrons.
$Y + e^- \longrightarrow Y^-$
Y gains one or more electrons, is reduced, and is the oxidizing agent (oxidizes something else).

Like oxygen, the halogens (F_2, Cl_2, Br_2, and I_2) are always oxidizing agents in their reactions with metals and most nonmetals (for example, the reaction of sodium metal with chlorine).

Na releases $1e^-$ per atom;
Na is oxidized and is the reducing agent.

$$2\,Na(s) + Cl_2(g) \longrightarrow 2\,[Na^+, Cl^-]$$

Cl_2 gains $2e^-$ per molecule;
Cl_2 is reduced and is the oxidizing agent.

Here sodium begins as the element, but it ends up as the Na^+ ion after combining with chlorine. Thus, sodium is oxidized and is the reducing agent. Chlorine ends up as Cl^-; having acquired two electrons per Cl_2 molecule, it has been reduced and therefore is the oxidizing agent.

When the halogens are combined with metals they often end up as halide ions, X^- (that is, as F^-, Cl^-, Br^-, or I^-) in the product. Therefore, like oxygen, the halogens are oxidizing agents in their reactions with metals.

Reduction reaction

$$X_2 + 2e^- \longrightarrow 2\,X^-$$

oxidizing agent

That is, a halogen will always oxidize a metal to give a metal halide (or, conversely, a metal will always reduce a halogen to a halide), and the formula of the product can be predicted from the charge on the metal ion and the charge of the halide. The halogens in decreasing order of oxidizing ability are as follows:

Oxidizing Agent	Usual Reduction Product
F_2	F^-
Cl_2	Cl^-
Br_2	Br^-
I_2	I^-

Chlorine is widely used as an oxidizing agent in water and sewage treatment. A common contaminant of water is hydrogen sulfide, H_2S, which may come from the decay of organic matter or from underground mineral deposits. Hydrogen sulfide gives a thoroughly unpleasant rotten-egg odor to the water, but chlorine oxidizes H_2S to insoluble, elemental sulfur, which is easily removed.

$$8\,Cl_2(g) + 8\,H_2S(aq) \longrightarrow S_8(s) + 16\,HCl(aq)$$

Oxidation and reduction occur readily when a strong oxidizing agent comes into contact with a strong reducing agent. Knowing the easily recognized oxidizing and reducing agents enables you to predict that a reaction will take place when they are combined, and in some cases to predict what the products will be. Table 4.3 and the following points provide some guidelines.

- If something has combined with oxygen, it has been oxidized. In the process the oxygen, O_2, is changed to the oxide ion, O^{2-} (as in a metal oxide), by adding electrons or is combined in a molecule such as CO_2 or H_2O (as occurs in the combustion reaction of a hydrocarbon). Therefore, the oxygen has been reduced. Since it must have taken on electrons, oxygen is the oxidizing agent, and it is a fairly strong one.
- If something has combined with a halogen, it has been oxidized. In the process the halogen, X_2, is changed to halide ions, X^-, by adding electrons, or it is combined in a molecule such as HCl. Therefore, the halogen has been reduced, and it is the oxidizing agent. Among the halogens, fluorine and chlorine are particularly strong oxidizing agents.

TABLE 4.3 Common Oxidizing and Reducing Agents

Oxidizing Agent	Reaction Product	Reducing Agent	Reaction Product
O_2(oxygen)	O^{2-}(oxide ion)	H_2(hydrogen)	H^+ (hydrogen ion) or H combined in H_2O
F_2, Cl_2, Br_2, or I_2 (halogens)	F^-, Cl^-, Br^-, or I^- (halide ion)	*M*, metals such as Na, K, Fe, or Al	M^{n+}, metal ions such as Na^+, K^+ or Al^{3+}
HNO_3 (nitric acid)	Nitrogen oxides such as NO and NO_2	C (carbon), used to reduce metal oxides	CO and CO_2
$Cr_2O_7^{2-}$ (dichromate ion)	Cr^{3+}(chromium(III) ion), in acid solution		
MnO_4^- (permanganate ion)	Mn^{2+} (manganese(II) ion), in acid solution		

Figure 4.19 **Cu(s) + HNO_3(aq).** Copper reacts vigorously with concentrated nitric acid to give brown NO_2 gas. *(C.D. Winters)*

There are exceptions to the guideline that metals are always positively charged in compounds. However, you probably will not encounter these exceptions in introductory chemistry.

- If a metal combines with something, it has been oxidized. In the process, the metal has lost electrons, usually to form a positive ion (as in metal oxides or halides, for example).

Oxidation reaction

$$\underset{\text{reducing agent}}{\mathbf{M}} \longrightarrow \mathbf{M}^{n+} + ne^-$$

Therefore, the metal (an electron donor) has been oxidized and has functioned as a reducing agent. Most metals are reasonably good reducing agents, and metals such as sodium, magnesium, and aluminum from Groups 1A, 2A, and 3A are particularly good ones.

- Other common oxidizing and reducing agents are listed in Table 4.3, and some are described below. When one of these agents takes part in a reaction, it is reasonably certain that it is a redox reaction. (Nitric acid can be an exception. In addition to being a good oxidizing agent, it is also an acid and functions only in this role in reactions such as the decomposition of a metal carbonate.)

Figure 4.19 illustrates one of the best oxidizing agents, concentrated nitric acid, HNO_3. Here the acid oxidizes copper metal to give copper(II) nitrate, and the acid is reduced to the brown gas NO_2. The net ionic equation for the reaction is

$$\underset{\text{reducing agent}}{Cu(s)} + \underset{\text{oxidizing agent}}{2\ NO_3^-(aq)} + 4\ H^+(aq) \longrightarrow Cu^{2+}(aq) + 2\ NO_2(g) + 2\ H_2O(\ell)$$

The metal is clearly the reducing agent, since it is the substance oxidized. The most common reducing agents in the laboratory, in fact, are metals. Some metal ions such as Fe^{2+} can also be reducing agents because they can be oxidized to ions of higher charge. Aqueous Fe^{2+} ion reacts readily with the strong oxidizing agent MnO_4^-, the permanganate ion. The Fe^{2+} ion is oxidized to Fe^{3+}, and the MnO_4^- ion is reduced to the Mn^{2+} ion.

$$5\ Fe^{2+}(aq) + MnO_4^-(aq) + 8\ H^+(aq) \longrightarrow 5\ Fe^{3+}(aq) + Mn^{2+}(aq) + 4\ H_2O(\ell)$$

Carbon can reduce many metal oxides to metals, and it is widely used in the metals industry as a reducing agent. For example, titanium is produced by treating a mineral containing titanium(IV) oxide with carbon and chlorine.

$$TiO_2(s) + C(s) + 2\ Cl_2(g) \longrightarrow TiCl_4(\ell) + CO_2(g)$$

In effect, the carbon reduces the oxide to titanium metal, and the chlorine then oxidizes it to titanium(IV) chloride. Because $TiCl_4$ is easily converted to a gas, it can be removed from the reaction mixture and recovered. The $TiCl_4$ is then reduced with another metal, such as magnesium, to give titanium metal.

$$TiCl_4(\ell) + 2\ Mg(s) \longrightarrow Ti(s) + 2\ MgCl_2(s)$$

Finally, H_2 gas is a common reducing agent, widely used in the laboratory and in industry. For example, it readily reduces copper(II) oxide to copper metal (Figure 4.20).

$$\underset{\text{reducing agent}}{H_2(g)} + \underset{\text{oxidizing agent}}{CuO(s)} \longrightarrow Cu(s) + H_2O(g)$$

Balancing equations for the reactions of metals with oxygen and halogens, or those between metal oxides and hydrogen, is usually straightforward; and you should be able to predict the outcome of such reactions and balance the equations. However, it is somewhat more difficult to balance reactions involving species such as HNO_3, $Cr_2O_7^{2-}$, and MnO_4^-, and you will not be asked to do so at this time.

It is *important* to be aware that it can be dangerous to mix a strong oxidizing agent with a strong reducing agent; a violent reaction, even an explosion, may take place. Chemicals are no longer stored on laboratory shelves in alphabetical order, because such an ordering may place a strong oxidizing agent next to a strong reducing agent. Swimming pool chemicals that contain chlorine and are strong oxidizing agents should not be stored in the hardware store or the garage next to easily oxidized materials.

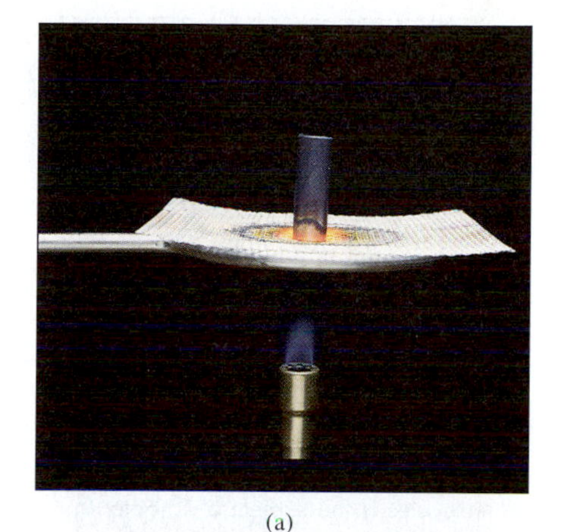

(a)

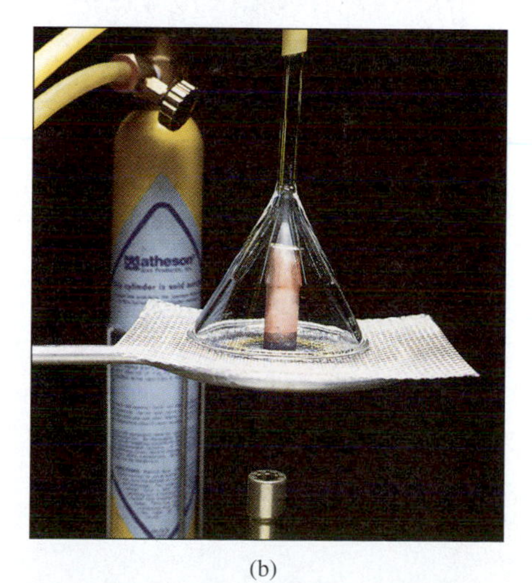

(b)

Figure 4.20 Reduction of copper oxide with hydrogen. (a) A piece of copper has been heated in air to form a film of black copper(II) oxide on the surface. (b) When the hot copper metal, with its film of CuO, is placed in a stream of hydrogen gas (from the yellow tank at the rear), the oxide is reduced to copper metal, and water forms as the byproduct. *(C.D. Winters)*

Exercise 4.10 Oxidation-Reduction Reactions

Decide which of the following reactions are oxidation-reduction reactions. In each case explain your choice and identify the oxidizing and reducing agents.

(a) $NaOH(aq) + HNO_3(aq) \rightarrow NaNO_3(aq) + H_2O(\ell)$
(b) $4\ Cr(s) + 3\ O_2(g) \rightarrow 2\ Cr_2O_3(s)$
(c) $NiCO_3(s) + 2\ HCl(aq) \rightarrow NiCl_2(aq) + H_2O(\ell) + CO_2(g)$
(d) $Cu(s) + Cl_2(g) \rightarrow CuCl_2(s)$

4.8 OXIDATION NUMBERS AND REDOX REACTIONS

An arbitrary bookkeeping system has been devised for keeping track of electrons in redox reactions. It extends the obvious oxidation and reduction case when neutral atoms become ions to reactions in which the changes are less obvious. The system is

TABLE 4.4 Recognizing Oxidation-Reduction Reactions

	Oxidation	Reduction
In Terms of Oxygen	Gain of oxygen	Loss of Oxygen
In Terms of Halogen	Gain of halogen	Loss of halogen
In Terms of Electrons	Loss of electrons	Gain of electrons
In Terms of Oxidation Numbers	Increase of oxidation number	Decrease of oxidation number

set up so that *oxidation numbers always change in redox reactions.* As a result, oxidation and reduction can be determined in the ways shown in Table 4.4.

An **oxidation number** compares the charge of an uncombined atom with its actual charge or its relative charge in a compound. All neutral atoms have an equal number of protons and electrons and, thus, have no net charge. When sodium metal atoms (0 net charge) combine with chlorine atoms (0 net charge) to form sodium chloride, each sodium atom loses an electron to form a sodium ion, Na^+, and each chlorine atom gains an electron forming a chloride ion, Cl^-. Therefore, Na^+ has an oxidation number of +1 because it has one *less* electron than a sodium atom and Cl^- has an oxidation number of −1 because it has one *more* electron than a chlorine atom. Oxidation numbers of atoms in molecular compounds are assigned *as though* electrons were completely transferred to form ions. In phosphorus trichloride (PCl_3), for example, chlorine is assigned an oxidation number of −1 even though it is not a Cl^- ion; the chlorine is directly bonded to the phosphorus. The chlorine atoms in PCl_3 are thought of as "owning" more electrons than they have in Cl_2.

How electrons participate in bonding atoms in molecules is the subject of Chapter 9.

Oxidation numbers are also called oxidation states.

You can use the following set of rules to determine oxidation numbers.

Rule 1: The oxidation number of an atom of a pure element is 0. When the atoms are not combined with those of any other element (for example, oxygen in O_2, sulfur in S_8, iron in metallic Fe, or chlorine in Cl_2), the oxidation number is 0.

Rule 2: The oxidation number of a monatomic ion equals its charge. Thus, the oxidation number of Cu^{2+} is +2; that of S^{2-} is −2.

Rule 3: Some elements have the same oxidation number in almost all their compounds and can be used as references for oxidation numbers of other elements in compounds.

(a) Hydrogen has an oxidation number of +1, unless it is combined with a metal, in which case its oxidation number is −1.

(b) Fluorine has an oxidation number of −1 in all its compounds.

(c) In most of its compounds, oxygen has an oxidation number of −2. Peroxides, such as hydrogen peroxide, H_2O_2, are exceptions in which −1 is the oxidation number for oxygen (and hydrogen is +1).

(d) In binary compounds (compounds of two elements) atoms of Group VIA (Group 16) elements (O, S, Se, Te) have an oxidation number of −2, except when combined with oxygen or halogens.

Rule 4: The sum of the oxidation numbers in a neutral compound is 0; the sum of the oxidation numbers in a polyatomic ion equals the charge on the ion. For example, in SO_2, the oxidation number of oxygen is −2; and with two O atoms, the total for oxygen is −4. Because the sum of the oxidation numbers must equal zero, the oxidation number of sulfur must be +4: $(+4) + 2(-2) = 0$. In SO_3^{2-}, the net charge is 2−, the charge on the polyatomic sulfite ion. Because each oxygen is −2, the oxidation number of sulfur in sulfite must be +4: (+4) + 3(−2) = −2.

$$\overset{+4\ -6}{(SO_3^{2-})^{2-}}$$
$$(3 \times -2)$$

Now, let's apply these rules to the equations for simple combination and displacement reactions involving sulfur and oxygen.

$$\textit{Combination: } \overset{0}{S_8(s)} + 8\ \overset{0}{O_2(g)} \longrightarrow 8\ \overset{+4\ -4}{SO_2(g)} \quad (2 \times -2)$$

$$\textit{Combination: } \overset{+2\,-2}{ZnS(s)} + 2\ \overset{0}{O_2(aq)} \longrightarrow \overset{+2\ +6\ -8}{ZnSO_4(aq)} \quad (4 \times -2)$$

$$\textit{Displacement: } \overset{+2\ \ 2-}{Cu_2S(s)} + \overset{0}{O_2(g)} \longrightarrow 2\ \overset{0}{Cu(s)} + \overset{+4\ -4}{SO_2(g)} \quad (2 \times +1) \quad (2 \times -2)$$

These are all oxidation-reduction reactions as shown by the fact that *there has been a change in the oxidation numbers of atoms from reactants to products.* That is one way to tell if a reaction is an oxidation-reduction reaction.

It should be apparent from the examples just given that *every reaction in which an element becomes combined in a compound is a redox reaction.* The oxidation number of the element must increase or decrease from its original value of zero. Combination reactions and displacement reactions in which one element displaces another are all redox reactions.

Conceptual Challenge Problem CP-4.A at the end of the chapter relates to topics covered in this section.

Those decomposition reactions in which elemental gases are produced are also redox reactions. Millions of tons of ammonium nitrate, NH_4NO_3, are used as fertilizer to supply nitrogen to crops. Ammonium nitrate is also used as an explosive that is decomposed by heat. In 1947, a load of NH_4NO_3 en route to be used as a fertilizer was accidently detonated on a ship in the Texas City, Texas, harbor blowing up the ship and adjacent buildings and killing 570 people.

$$2\ NH_4NO_3(s) \longrightarrow 2\ N_2(g) + 4\ H_2O(\ell) + O_2(g)$$

Like a number of other explosives, ammonium nitrate contains an element with two different oxidation numbers, in effect having an oxidizing and reducing agent in the same compound. Nitrogen in ammonium nitrate has two different oxidation numbers. Recalling that ammonium nitrate contains ammonium (NH_4^+) and nitrate (NO_3^-) ions, we can determine these numbers:

Ammonium nitrate was used in the 1995 bombing of the Federal Building in Oklahoma City, Oklahoma.

$$\overset{-3\ +4}{[NH_4]^+} \qquad \overset{+5\ -6}{[NO_3]^-}$$
$$(4 \times +1) \qquad (3 \times -2)$$

Note that nitrogen is −3 in ammonium ion and +5 in nitrate ion. Therefore, in the decomposition of ammonium nitrate to N_2, the N in the ammonium ion is oxidized from −3 to zero, and the ammonium ion is the reducing agent. The N in the nitrate ion is reduced from +5 to zero, and the nitrate ion is the oxidizing agent.

PROBLEM-SOLVING EXAMPLE 4.8 Applying Oxidation Numbers

Henry Cavendish discovered hydrogen in 1671 by dripping sulfuric acid down an iron gun barrel.

$$Fe(s) + H_2SO_4(aq) \longrightarrow FeSO_4(aq) + H_2(g)$$

Assign oxidation numbers for each atom in this equation. Identify what has been oxidized and what has been reduced.

Answer

$$\overset{0}{Fe}(s) + \overset{+2\ +6\ -8}{H_2SO_4}(aq) \longrightarrow \overset{+2\ +6\ -8}{FeSO_4}(aq) + \overset{0}{H_2}(g)$$

Fe is oxidized; H^+ is reduced.

Explanation Because iron is in its elemental state as a reactant, its oxidation number is 0 (Rule 1). The same rule applies for the product H_2; H has a 0 oxidation number.

For sulfuric acid, we start by recognizing that it is a compound and has no net charge (Rule 4). Therefore, because the oxidation state of each oxygen is −2 (Rule 3c) for a total of −8, and that of each hydrogen is +1 (Rule 3a), for a total of +2, the oxidation state of sulfur must be +6: $0 = 2(+1) + (+6) + 4(-2)$.

In the case of $FeSO_4$, we can assign an oxidation number to Fe by recognizing that it must be +2 to balance the charge of SO_4^{2-}. In the sulfate ion, given that oxygen is −2, the oxidation number of sulfur must be +6 to give the ion a net −2 charge. The iron has been oxidized as shown by its oxidation number change from 0 to +2. The H^+ has been reduced as indicated by the change in its oxidation number from +1 to 0.

Problem-Solving Practice 4.8

Determine the oxidation number for each atom in this equation:

$$Sb_2S_3(s) + Fe(s) \longrightarrow 3\ FeS(s) + 2\ Sb(s).$$

Cite the oxidation number rule(s) for why you chose your answers.

Conceptual Challenge Problem CP-4.B at the end of the chapter relates to topics covered in this section.

Exchange reactions of ionic compounds in aqueous solution are not redox reactions because no change of oxidation numbers occurs. Consider, for example, the precipitation of barium sulfate when aqueous solutions of barium chloride and sulfuric acid are mixed.

$$Ba^{2+}(aq) + 2\ Cl^-(aq) + 2\ H^+(aq) + SO_4^{2-}(aq) \longrightarrow BaSO_4(s) + 2\ HCl(aq)$$

Net ionic equation: $Ba^{2+}(aq) + SO_4^{2-}(aq) \longrightarrow BaSO_4(s)$

The oxidation numbers of the reactant and product atoms remain unchanged.

CHEMISTRY IN THE NEWS

CFC Disposal Using Redox Reactions

Chlorofluorocarbons (CFCs) are molecular compounds of carbon, chlorine, and fluorine used as cooling fluids in air conditioners and refrigerators. CFCs have been found to decrease the protective stratospheric ozone layer *(← p. 3)*. Vented from leaky air conditioners and refrigerators, CFCs move through the lower atmosphere virtually unchanged into the stratosphere due to their chemical inertness. A ban on CFC production in the United States began in 1996 as part of a global effort to decrease the amount of CFCs available for stratospheric ozone depletion.

A convenient process has been developed recently that could be used to dispose of stockpiled CFCs *(← p. 3)*. The process, developed by chemists Robert Crabtree and Juan Berdeniuc at Yale University, uses sodium oxalate, $Na_2C_2O_4$, a cheap, readily available substance that is a good reducing agent. The CFCs are vaporized and passed through a layer of powdered sodium oxalate at 270 °C. When CF_2Cl_2, known as CFC-12, is used, the reaction that occurs is

$$CF_2Cl_2(g) + 2\ Na_2C_2O_4(s) \longrightarrow 2\ NaF(s) + 2\ NaCl(s) + C(s) + 4\ O_2(g)$$

The products can be separated and disposed of or recycled easily.

Yale University chemists Juan Burdeniuc and Robert H. Crabtree. *(Michael Marshland/Yale University Office of Public Affaira)*

Source: Science, *Vol. 271, 19 January, pp. 340–341, 1996.*

Exercise 4.11 Redox in CFC Disposal

The Chemistry in the News article described a redox reaction for the disposal of chlorofluorocarbons, CFCs, by their reaction with sodium oxalate, $Na_2C_2O_4$:

$$CF_2Cl_2(g) + 2\ Na_2C_2O_4(s) \longrightarrow 2\ NaF(s) + 2\ NaCl(s) + C(s) + 4\ CO_2(g)$$

(a) What species are oxidized in this reaction?
(b) What species are reduced?

4.9 DISPLACEMENT REACTIONS, REDOX, AND THE ACTIVITY SERIES

Displacement reactions, like combination reactions, are oxidation-reduction reactions. For example, in the reaction of hydrochloric acid with iron (Figure 4.21),

$$Fe(s) + 2\ HCl(aq) \longrightarrow FeCl_2(aq) + H_2(g)$$

metallic iron is the reducing agent; it is oxidized in going from an oxidation number of 0 in Fe(s) to +2 in $FeCl_2$. Hydrogen ions, H^+, in hydrochloric acid are reduced to hydrogen gas (H_2) in which hydrogen has an oxidation number of 0.

Extensive studies with many metals have led to a **metal activity series,** a ranking of relative reactivity of metals in displacement and other kinds of reactions (Table 4.5). The most reactive metals are at the top of the series, and activity decreases down

TABLE 4.5 Activity Series of Metals

Displace H_2 from $H_2O(\ell)$, steam, or acid	Li
	K
	Ba
	Sr
	Ca
	Na
Displace H_2 from steam or acid	Mg
	Al
	Zn
	Cr
Displace H_2 from acid	Fe
	Ni
	Sn
	Pb
	H_2
Do not displace H_2 from $H_2O(\ell)$, steam, or acid	Sb
	Cu
	Hg
	Ag
	Pd
	Pt
	Au

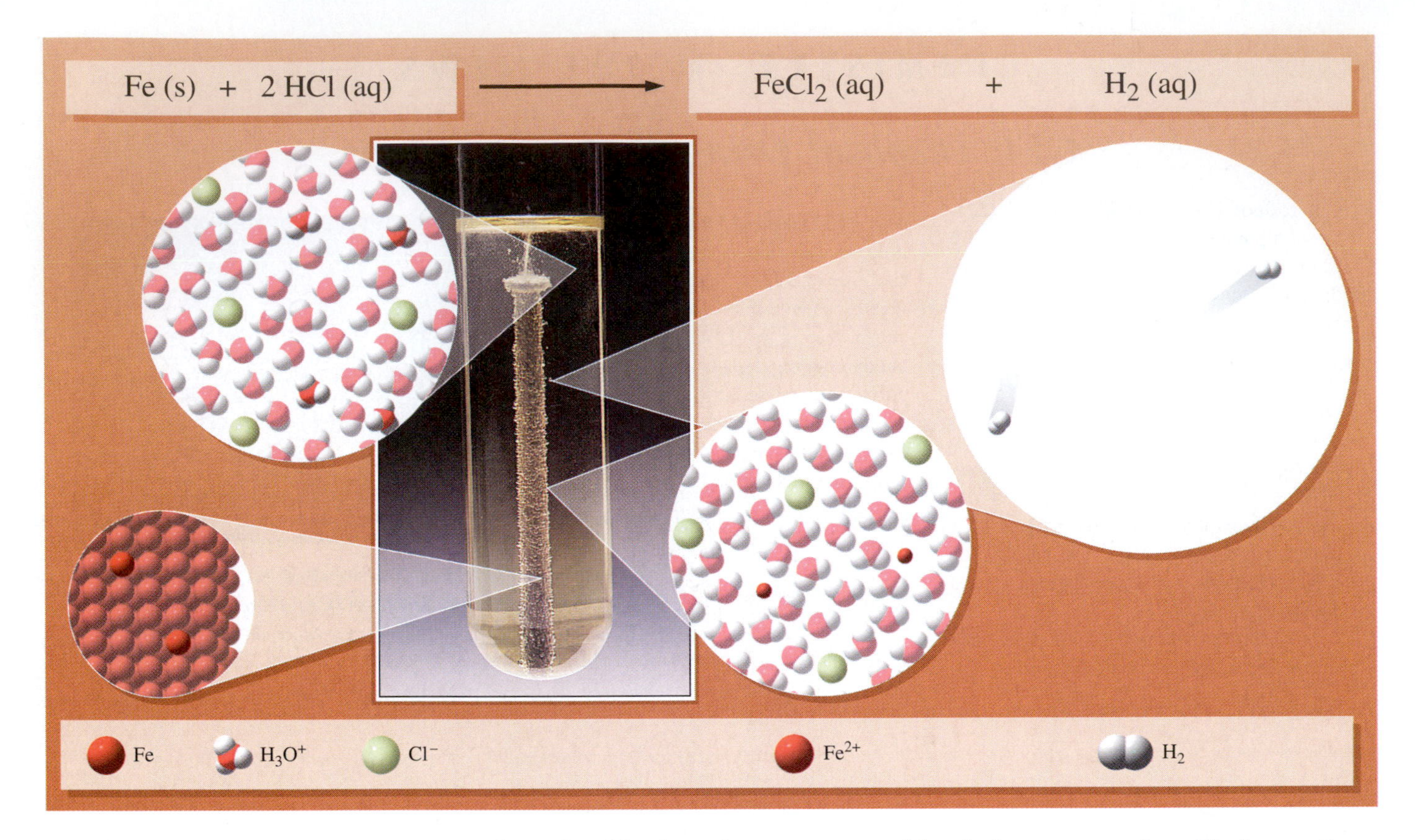

Figure 4.21 Metal + acid displacement reaction. An iron nail reacts with hydrochloric acid to form a solution of iron(II) chloride, $FeCl_2(aq)$, and hydrogen gas. *(C.D. Winters)*

the series. Metals at the top are powerful reducing agents and readily form cations. The metals at the lower end of the series are poor reducing agents. However, their cations (Au^+, Ag^+) are powerful oxidizing agents that readily react to form the free metal.

A metal higher in the activity series will displace an element below it in the series from its compounds. For example, zinc displaces copper from copper(II) sulfate solution, and copper metal displaces silver from silver nitrate solution (Figure 4.22).

$$Zn(s) + CuSO_4(aq) \longrightarrow ZnSO_4(aq) + Cu(s)$$

$$Cu(s) + 2\ AgNO_3(aq) \longrightarrow Cu(NO_3)_2(aq) + 2\ Ag(s)$$

In each case, the elemental metal (Zn, Cu) is the reducing agent and is oxidized; Cu^{2+} ions and Ag^+ ions are oxidizing agents and are reduced to Cu(s) and Ag(s).

Metals above hydrogen in the series react with acids whose anions are not oxidizing agents, such as hydrochloric acid, to form hydrogen (H_2) and the metal salt containing the cation of the metal and the anion of the acid, for example $FeCl_2$ from iron and hydrochloric acid, and $ZnBr_2$ from zinc and hydrobromic acid.

When an acid like HNO_3, whose nitrate ion is a strong oxidizing agent, reacts with a metal, the anion is also reduced, and different products are formed.

$$Fe(s) + 2\ HCl(aq) \longrightarrow FeCl_2(aq) + H_2(g)$$

$$Zn(s) + 2\ HBr(aq) \longrightarrow ZnBr_2(aq) + H_2(g)$$

Metals below hydrogen do not displace hydrogen from acids in this way.

Very reactive metals, those at the top of the activity series, from lithium (Li) to sodium (Na), can displace hydrogen from water. Some do so violently (Figure 4.23). Metals of intermediate activity (Mg–Fe) displace hydrogen from steam, but not from liquid water at room temperature.

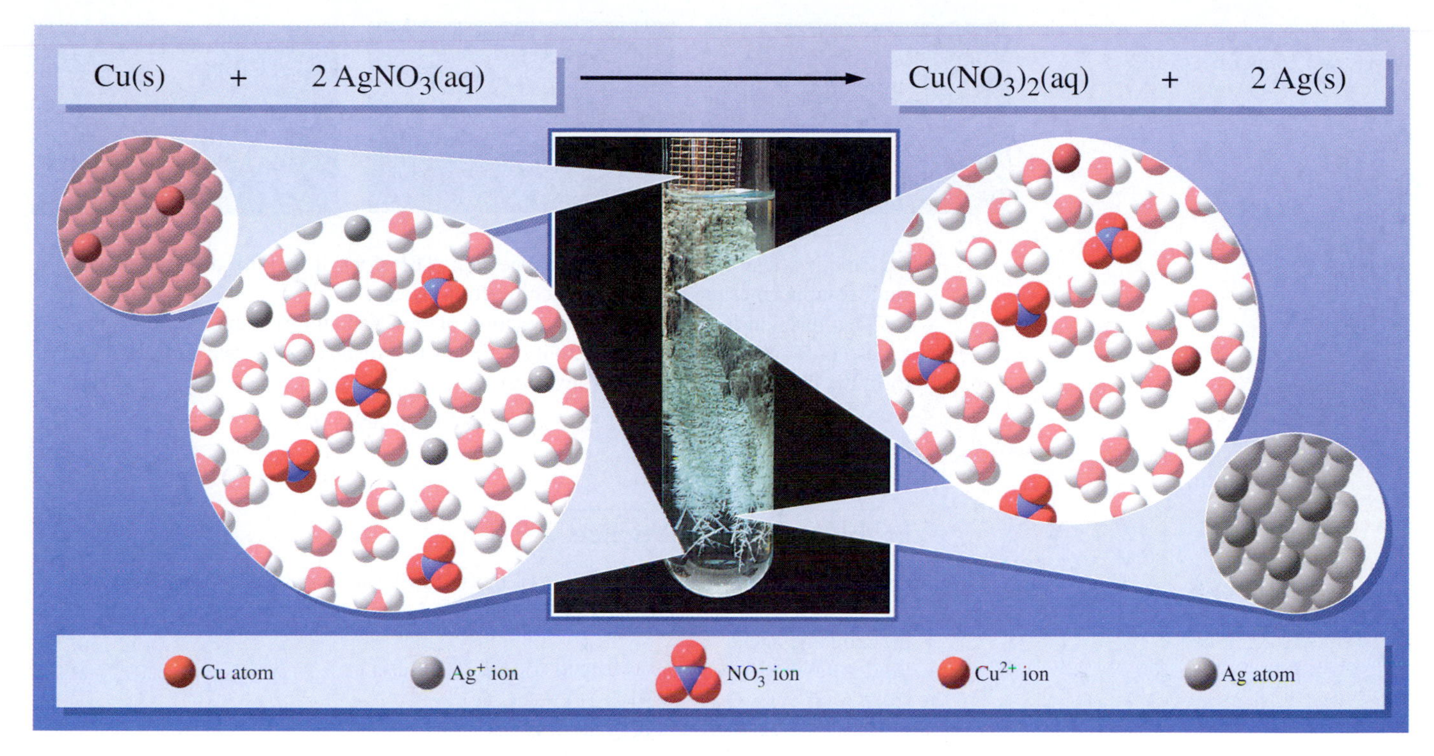

Figure 4.22 Metal + aqueous metal salt displacement reaction. The oxidation of copper metal by silver ion. A clean piece of copper screen is placed in a solution of silver nitrate, $AgNO_3$. With time, the copper reduces Ag^+ ions to silver metal crystals, and the copper metal is oxidized to Cu^{2+} ions. The blue color of the solution is due to the presence of aqueous copper(II) ion. (Atoms or ions that take part in the reaction have been highlighted in the nanoscale pictures.) *(C.D. Winters)*

Elements very low in the activity series are unreactive. They are sometimes called noble metals (Au, Ag, Pt), and they are prized for their nonreactivity. It is no accident that gold and silver have been extensively used for coinage since antiquity. These metals do not react with air, water, or even common acids, thus maintaining their luster (and value) for many years. Their low reactivity is also why they occur naturally as free metals and have been known as elements since antiquity.

Conceptual Challenge Problem CP-4.C at the end of the chapter relates to topics covered in this section.

Figure 4.23 Potassium, an active metal. When a drop of water falls into a sample of potassium metal, it reacts vigorously with water to give hydrogen gas and a solution of potassium hydroxide. *(C.D. Winters)*

PROBLEM-SOLVING EXAMPLE 4.9 Activity Series of Metals

Use the activity series (Table 4.5) to predict whether any of the following reactions will occur. Complete and balance the equations for those reactions that will occur.

(a) $Fe(s) + CuCl_2(aq) \rightarrow$

(b) $Zn(s) + MgCl_2(aq) \rightarrow$

(c) Manganese + hydrochloric acid $\rightarrow$

Answer

(a) $Fe(s) + CuCl_2(aq) \rightarrow FeCl_2(aq) + Cu(s)$

(b) No reaction

(c) $Mn(s) + 2\ HCl(aq) \rightarrow MnCl_2(aq) + H_2(g)$

Explanation

(a) Iron is above copper in the activity series. Therefore it will displace copper ions from a solution of copper(II)chloride to form metallic copper and Fe^{2+} ions.

$$Fe(s) + CuCl_2(aq) \longrightarrow FeCl_2(aq) + Cu(s)$$

Natural deposits of gold, silver, and copper. *(C.D. Winters)*

(b) Zinc is a less active metal than magnesium. Therefore, it will not displace magnesium from magnesium chloride, and no reaction occurs.
(c) This reaction is analogous to the reaction of iron and sulfuric acid. Because manganese is above hydrogen in the activity series, it will displace hydrogen from HCl and form the metal salt ($MnCl_2$) plus hydrogen gas.

$$Mn(s) + 2\ HCl(aq) \longrightarrow MnCl_2(aq) + H_2(g)$$

Problem-Solving Practice 4.9

Use Table 4.5 to predict whether each of the following reactions will occur. If reaction occurs, identify what has been oxidized or reduced, and what is the oxidizing agent and the reducing agent.
(a) $2\ Al(s) + 3\ CuSO_4(aq) \rightarrow Al_2(SO_4)_3(aq) + 3\ Cu(s)$
(b) $2\ Al(s) + Cr_2O_3(s) \rightarrow Al_2O_3(\ell) + 2\ Cr(\ell)$
(c) $Pt(s) + 2\ HCl(aq) \rightarrow PtCl_2(aq) + H_2(g)$
(d) $Au(s) + 3\ AgNO_3(ag) \rightarrow Au(NO_3)_3 + 3\ Ag(s)$

Summary Problem

Silicon is the second most abundant element in the earth's crust (27.5%), exceeded only by oxygen (49.5%). Although silicon is not found free in nature, silicon compounds make up much of earth's structural materials such as sand and quartz (SiO_2) and clays.

Elemental silicon was a laboratory curiosity until it began to be used in commercial electronic devices in the 1950s. It was highly purified silicon that made possible the computer chip, thereby paving the way for the personal-computer revolution. The technology was developed in the region now known as "Silicon Valley" near San Francisco, California. Ultra–high purity silicon is required for making computer chips—silicon containing no more than 1 atom of an impurity like boron, aluminum, or arsenic per billion silicon atoms. To first obtain silicon from its principal natural sources, quartz or sand, and then purify it to such a high degree requires several steps.

The first step is production of the element from SiO_2 by a high-temperature reaction (about 3000 °C) using high-purity coke (carbon). Carbon monoxide is also formed. The silicon produced by this step is only 96 to 99% pure, not pure enough to use in computer chips. This silicon is then reacted with pure chlorine to produce liquid silicon tetrachloride, which is purified. Very–high purity silicon is obtained by treating the purified silicon tetrachloride with very pure magnesium. The very pure silicon is further purified to ultra-pure silicon by a special heating process called zone refining.

(a) Write balanced chemical equations for each of the three steps to convert SiO_2 to very–high purity silicon.

CHEMISTRY YOU CAN DO

Pennies, Redox, and the Activity Series of Metals

In this experiment, you will use pennies to test the reactivity of copper and zinc with acid. Post-1982 pennies are a copper and zinc "sandwich" with zinc in the middle covered by a layer of copper. Pre-1982 pennies do not have this composition.

To do this experiment you will need

- two glasses or plastic cups that will each hold 50 mL (about 1.5 oz) of liquid
- about 100 mL of "pickling" vinegar, such as Heinz Ultra-strength brand (Regular vinegar is only about 4–5% acetic acid.)
- an abrasive such as a piece of sandpaper, steel wool, or a Brillo pad
- a small file
- four pennies—two pre-1982 and two post-1982

Clean the pennies with the abrasive until all the surfaces (including the edges) are shiny. Use the file to make two cuts into the edge of each penny, one across from the other. If you look carefully, you might observe a shiny metal where you cut into the post-1982 pennies.

Caution: Keep the vinegar from your skin and clothes and especially your eyes. If vinegar spills on you, rinse it off with flowing water.

Place the two pre-1982 pennies into one of the cups and the post-1982 pennies into the other cup. Add the same volume of vinegar to each cup, making sure that the pennies are covered by the liquid. Let the pennies remain in the liquid for several hours (even overnight), and periodically observe any changes in the pennies. After several hours, pour off the vinegar and remove the pennies. Dry them carefully and observe any changes that have occurred.

- What difference did you observe between the pre-1982 pennies and the post-1982 ones?
- Which is the more reactive element—copper or zinc?
- What happened to the zinc in the post-1982 pennies? Interpret the change in redox terms, and write a chemical equation to represent the reaction.
- How could this experiment be modified to determine the percent zinc and percent copper in post-1982 pennies?

(b) Identify which step involves a direct combination reaction and which steps are displacement reactions.
(c) Identify which of the reactions involve oxidation-reduction. Use oxidation numbers to determine which substances have been oxidized or reduced.
(d) Determine what are the oxidizing and reducing agents in the redox reactions.
(e) Decorative etching of glass (SiO_2) and "frosting" light bulbs are done using an exchange reaction of glass with hydrofluoric acid, HF. Write a balanced equation for this artful reaction.

In Closing

Having studied this chapter, you should be able to . . .

- interpret the information conveyed by a balanced chemical equation (Section 4.1).
- balance simple chemical equations (Section 4.2).
- recognize the general reaction types: combination, decomposition, displacement, and exchange (Section 4.3).
- predict products of common types of chemical reactions: combination, decomposition, precipitation, acid-base, and gas-forming (Sections 4.4–4.6).
- write a net ionic equation and know how to arrive at such an equation for a given reaction in aqueous solution (Section 4.4).
- recognize common acids and bases and understand neutralization reactions (Section 4.5).

- recognize oxidation-reduction reactions and common oxidizing and reducing agents (Section 4.7).
- assign oxidation numbers to reactants and products (Section 4.8), and identify what has been oxidized or reduced in a redox reaction (Section 4.8).
- use the activity series to predict products of displacement redox reactions (Section 4.9).

Key Terms

The following terms were defined and given in boldface type in this chapter. You should be sure to understand each of these terms and the concepts with which it is associated.

acid *(4.5)*
aqueous solution *(4.1)*
balanced chemical equation *(4.1)*
base *(4.5)*
coefficients *(4.1)*
combination reactions *(4.3)*
combustion reactions *(4.1)*
decomposition reaction *(4.3)*
displacement reaction *(4.3)*
electrolysis *(4.3)*
exchange reaction *(4.3)*
hydronium ion *(4.5)*
hydroxide ion *(4.5)*
metal activity series *(4.9)*
net ionic equation *(4.4)*
oxidation *(4.7)*
oxidation number *(4.8)*
oxidation-reduction reaction *(4.7)*
oxides *(4.2)*
oxidized *(4.7)*
oxidizing agent *(4.7)*
precipitate *(4.3)*
redox reactions *(4.7)*
reduced *(4.7)*
reducing agent *(4.7)*
reduction *(4.7)*
salt *(4.5)*
spectator ions *(4.4)*
stoichiometric coefficients *(4.1)*
stoichiometry *(4.1)*
strong acid *(4.5)*
strong base *(4.5)*
strong electrolyte *(4.5)*
weak acid *(4.5)*
weak base *(4.5)*
weak electrolyte *(4.5)*

Questions for Review and Thought

Conceptual Challenge Problems

CP-4.A. There is a conservation of the number of electrons exchanged during redox reactions, which is tantamount to stating that electric charge is conserved during chemical reactions. The assignment of oxidation numbers is an arbitrary, yet clever, way to do the bookkeeping for these electrons. What makes it possible to assign the same oxidation number to all elements that are not bound to other elements in chemical compounds?

CP-4.B. Consider the following two redox reactions:

1. $HIO_3 + FeI_2 + HCl \rightarrow FeCl_3 + ICl + H_2O$
2. $CuSCN + KIO_3 + HCl \rightarrow CuSO_4 + KCl + HCN + ICl + H_2O$

(a) Identify the species that have been oxidized or reduced in each of the reactions.

(b) After you have correctly identified the species that have been oxidized or reduced in each equation, you might like to try using oxidation numbers to balance each equation. This will be a challenge because, as you have discovered, there is more than one kind of atom that is oxidized or reduced, although in all cases the product of the oxidation and reduction is unambiguous. Record the initial and final oxidation state of each kind of atom that is oxidized or reduced in each equation. Then, decide on the coefficients that will equalize the oxidation number changes and satisfy any other atom balancing needed. Finally, balance the equation by adding the correct coefficients to it.

CP-4.C. A student was given four metals (A, B, C, and D) and solutions of their corresponding salts (AZ, BZ, CZ, and DZ). The student was asked to determine the relative reactivity of the four metals by reacting the metals with the solutions. The student's laboratory observations are as follows:

Metal	AZ(aq)	BZ(aq)	CZ(aq)	DZ(aq)
A	No reaction	No reaction	No reaction	No reaction
B	Reaction	No reaction	Reaction	No reaction
C	Reaction	No reaction	No reaction	No reaction
D	Reaction	Reaction	Reaction	No reaction

Arrange the four metals in order of decreasing activity.

Review Questions

1. What information is provided by a balanced chemical equation?
2. Find in this chapter one example of each of the following reaction types and write the balanced equation for the reaction: (a) combustion; (b) combination; (c) exchange; (d) a decomposition reaction; and (e) an oxidation-reduction reaction. Name the products of each reaction.
3. Classify each of the following reactions as a combination, decomposition, exchange, acid-base, or oxidation-reduction reaction:
   (a) $MgO(s) + 2\ HCl(aq) \rightarrow MgCl_2(aq) + H_2O(\ell)$
   (b) $2\ NaHCO_3(s) \xrightarrow{heat} Na_2CO_3(s) + CO_2(g) + H_2O(g)$
   (c) $CaO(s) + SO_2(g) \rightarrow CaSO_3(s)$
   (d) $3\ Cu(s) + 8\ HNO_3(aq) \rightarrow 3\ Cu(NO_3)_2(aq) + 2\ NO(g) + 4\ H_2O(\ell)$
   (e) $2\ NO(g) + O_2(g) \rightarrow 2\ NO_2(g)$
4. Find two examples in this chapter of the reaction of a metal with a halogen, write a balanced equation for each example, and name the product.
5. Find two examples of acid-base reactions in this chapter. Write balanced equations for these reactions and name the reactants and products.
6. Find two examples of precipitation reactions in this chapter. Write balanced equations for these reactions and name the reactants and products.
7. Find an example of a gas-forming reaction in this chapter. Write a balanced equation for the reaction and name the reactants and products.
8. Explain the difference between oxidation and reduction. Give an example of each.
9. For each of the following, does the oxidation number increase or decrease in the course of a redox reaction?
   (a) An oxidizing agent (c) A substance undergoing oxidation
   (b) A reducing agent (d) A substance undergoing reduction
10. Explain the difference between an oxidizing agent and a reducing agent. Give an example of each.

Stoichiometry

11. For the following reaction, fill in the table below:

$$2\ C_2H_6(g) + 7\ O_2(g) \longrightarrow 4\ CO_2(g) + 6\ H_2O(g)$$

	$C_2H_6(g)$	$O_2(g)$	$CO_2(g)$	$H_2O(g)$
No. of Molecules				
No. of Atoms				
No. of moles of Molecules				
Mass				
TOTAL Mass of Reactants =				
TOTAL Mass of Products =				

12. Magnesium metal burns brightly in the presence of oxygen to produce a white powdery substance, MgO. The equation below represents this reaction:

$$Mg(s) + O_2(g) \longrightarrow MgO(s)$$

(a) If 1.00 g of MgO(s) is formed by this reaction, what is the maximum *total* mass of magnesium metal and oxygen that reacted?
(b) Identify the stoichiometric coefficients in this equation.
(c) If 50 *atoms* of oxygen reacted, how many magnesium atoms were needed to react with this much oxygen?

13. Balance the following *combination* reaction by adding coefficients as needed:

$$Fe(s) + O_2(g) \longrightarrow Fe_2O_3(s)$$

14. The following diagram shows A (blue spheres) reacting with B (red spheres). Which equation best describes the stoichiometry of the reaction depicted in this diagram?

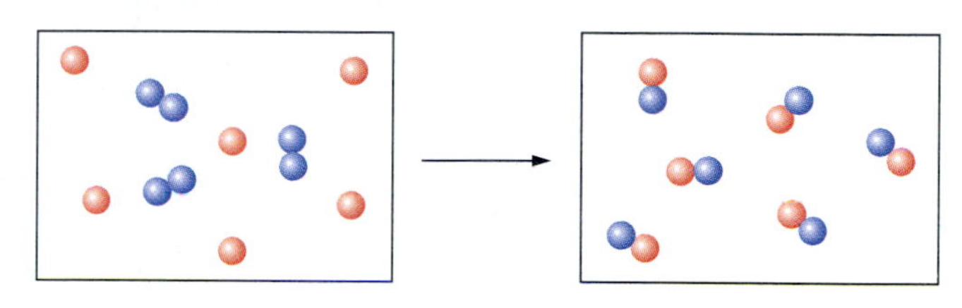

(a) $3\ A_2 + 6\ B \rightarrow 6\ AB$ (b) $A_2 + 2\ B \rightarrow 2\ AB$
(c) $2\ A + B \rightarrow AB$ (d) $3\ A + 6\ B \rightarrow 6\ AB$

15. Given the following equation,

$$4\ A_2 + 3\ B \longrightarrow B_3A_8$$

use a diagram to illustrate the amount of reactant A and product (B_3A_8) that would be needed/produced from the reaction of six atoms of B.

16. Balance the following equation and determine which box represents reactants and which box represents products?

$$Sb + Cl_2 \longrightarrow SbCl_3$$

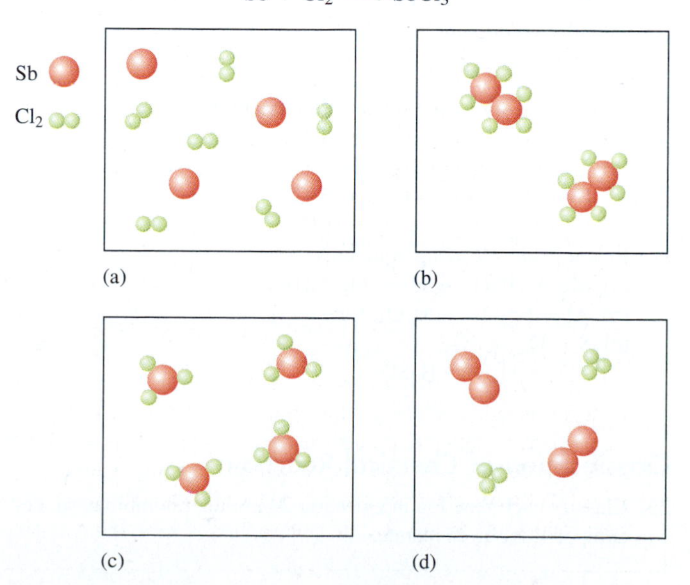

Balancing Equations

17. Balance the following equations.
    (a) $Al(s) + O_2(g) \rightarrow Al_2O_3(s)$
    (b) $N_2(g) + H_2(g) \rightarrow NH_3(g)$
    (c) $C_6H_6(\ell) + O_2(g) \rightarrow H_2O(\ell) + CO_2(g)$
18. Balance the following equations.
    (a) $Fe(s) + Cl_2(g) \rightarrow FeCl_3(s)$
    (b) $SiO_2(s) + C(s) \rightarrow Si(s) + CO(g)$
    (c) $Fe(s) + H_2O(g) \rightarrow Fe_3O_4(s) + H_2(g)$
19. Balance the following equations.
    (a) $UO_2(s) + HF(\ell) \rightarrow UF_4(s) + H_2O(\ell)$
    (b) $B_2O_3(s) + HF(\ell) \rightarrow BF_3(g) + H_2O(\ell)$
    (c) $BF_3(g) + H_2O(\ell) \rightarrow HF(\ell) + H_3BO_3(s)$
20. Balance the following equations.
    (a) $MgO(s) + Fe(s) \rightarrow Fe_2O_3(s) + Mg(s)$
    (b) $H_3BO_3(s) \rightarrow B_2O_3(s) + H_2O(\ell)$
    (c) $NaNO_3(s) + H_2SO_4(aq) \rightarrow Na_2SO_4(aq) + HNO_3(g)$
21. Balance the following equations.
    (a) Reaction to produce hydrazine, N_2H_4, a good industrial reducing agent:

$$H_2NCl(aq) + NH_3(g) \longrightarrow NH_4Cl(aq) + N_2H_4(aq)$$

    (b) Reaction of the fuel (dimethylhydrazine and dinitrogen tetraoxide) used in the moon lander and space shuttle:

$$(CH_3)_2N_2H_2(\ell) + N_2O_4(g) \longrightarrow N_2(g) + H_2O(g) + CO_2(g)$$

    (c) Reaction of calcium carbide to produce acetylene, C_2H_2:

$$CaC_2(s) + H_2O(\ell) \longrightarrow Ca(OH)_2(s) + C_2H_2(g)$$

22. Balance the following equations.
    (a) Reaction of calcium cyanamide to produce ammonia:

$$CaNCN(s) + H_2O(\ell) \longrightarrow CaCO_3(s) + NH_3(g)$$

    (b) Reaction to produce diborane, B_2H_6:

$$NaBH_4(s) + H_2SO_4(aq) \longrightarrow B_2H_6(g) + H_2(g) + Na_2SO_4(aq)$$

    (c) Reaction to rid water of hydrogen sulfide, H_2S, a foul-smelling compound:

$$H_2S(aq) + Cl_2(aq) \longrightarrow S_8(s) + HCl(aq)$$

23. Balance the following combustion reactions:
    (a) $C_6H_{12}O_6 + O_2 \rightarrow CO_2 + H_2O$
    (b) $C_5H_{12} + O_2 \rightarrow CO_2 + H_2O$
    (c) $C_7H_{14}O_2 + O_2 \rightarrow CO_2 + H_2O$
    (d) $C_2H_4O_2 + O_2 \rightarrow CO_2 + H_2O$
24. Balance the following equations:
    (a) $Mg + HNO_3 \rightarrow H_2 + Mg(NO_3)_2$
    (b) $Al + Fe_2O_3 \rightarrow Al_2O_3 + Fe$
    (c) $S + O_2 \rightarrow SO_3$
    (d) $SO_3 + H_2O \rightarrow H_2SO_4$

Classification of Chemical Reactions

25. Classify each reaction in Question 24 as either combination, decomposition, or exchange.
26. Write a balanced equation for the formation of each of the following compounds from the elements.
    (a) Carbon monoxide
    (b) Nickel(II) oxide
    (c) Chromium(III) oxide
27. Write a balanced equation for the formation of each of the following compounds from the elements.
    (a) Copper(I) oxide
    (b) Arsenic(III) oxide
    (c) Zinc oxide
28. Write a balanced equation for each of the following decomposition reactions. Name each product.
    (a) $BeCO_3(s) \xrightarrow{\text{heat}}$
    (b) $NiCO_3(s) \xrightarrow{\text{heat}}$
    (c) $Al_2(CO_3)_3(s) \xrightarrow{\text{heat}}$
29. Write a balanced equation for each of the following decomposition reactions. Name each product.
    (a) $ZnCO_3(s) \xrightarrow{\text{heat}}$
    (b) $MnCO_3(s) \xrightarrow{\text{heat}}$
    (c) $PbCO_3(s) \xrightarrow{\text{heat}}$
30. Write a balanced equation for the following combustion reactions.
    (a) $C_4H_{10}(g) + O_2(g) \rightarrow$
    (b) $C_6H_{12}O_6(s) + O_2(g) \rightarrow$
    (c) $C_4H_8O(\ell) + O_2(g) \rightarrow$
31. Complete and balance the following equations involving oxygen reacting with an element. Name the product in each case.
    (a) $Mg(s) + O_2(g) \rightarrow$
    (b) $Ca(s) + O_2(g) \rightarrow$
    (c) $In(s) + O_2(g) \rightarrow$
32. Complete and balance the following equations involving oxygen reacting with an element.
    (a) $Ti(s) + O_2(g) \rightarrow$ titanium(IV) oxide
    (b) $S_8(s) + O_2(g) \rightarrow$ sulfur dioxide
    (c) $Se(s) + O_2(g) \rightarrow$ selenium dioxide
33. Complete and balance the following equations involving the reaction of a halogen with a metal. Name the product in each case.
    (a) $K(s) + Cl_2(g) \rightarrow$
    (b) $Mg(s) + Br_2(\ell) \rightarrow$
    (c) $Al(s) + F_2(g) \rightarrow$
34. Complete and balance the following equations involving the reaction of a halogen with a metal.
    (a) $Cr(s) + Cl_2(g) \rightarrow$ chromium(III) chloride
    (b) $Cu(s) + Br_2(\ell) \rightarrow$ copper(II) bromide
    (c) $Pt(s) + F_2(g) \rightarrow$ platinum(IV) fluoride

Exchange Reactions

35. Write a balanced equation for the reaction of nitric acid with calcium hydroxide.
36. For each of the following pairs of ionic compounds, write a balanced equation reflecting whether precipitation will occur in

aqueous solution. For those combinations that do not produce a precipitate, write "no reaction" on the products side.
(a) $MnCl_2 + Na_2S$ (b) $HNO_3 + CuSO_4$
(c) $NaOH + HClO_4$ (d) $Hg(NO_3)_2 + Na_2S$
(e) $Pb(NO_3)_2 + HCl$ (f) $BaCl_2 + H_2SO_4$

37. Name the spectator ions in the reaction of nitric acid and magnesium hydroxide, and write the net ionic equation from the following complete ionic equation:

$$2\ H^+(aq) + 2\ NO_3^-(aq) + Mg(OH)_2(s) \longrightarrow 2\ H_2O(\ell) + Mg^{2+}(aq) + 2\ NO_3^-(aq)$$

What type of reaction is this?

38. Name the water-insoluble product in each reaction.
(a) $CuCl_2(aq) + H_2S(aq) \rightarrow CuS + 2\ HCl$
(b) $CaCl_2(aq) + K_2CO_3(aq) \rightarrow 2\ KCl + CaCO_3$
(c) $AgNO_3(aq) + NaI(aq) \rightarrow AgI + NaNO_3$

39. Name the spectator ions in the reactions from Question 38 and write the net ionic equations for those reactions.

40. Balance each of the following equations, and then write the complete and net ionic equations.
(a) $Zn(s) + HCl(aq) \rightarrow H_2(g) + ZnCl_2(aq)$
(b) $Mg(OH)_2(s) + HCl(aq) \rightarrow MgCl_2(aq) + H_2O(\ell)$
(c) $HNO_3(aq) + CaCO_3(s) \rightarrow Ca(NO_3)_2(aq) + H_2O(\ell) + CO_2(g)$
(d) $HCl(aq) + MnO_2(s) \rightarrow MnCl_2(aq) + Cl_2(g) + H_2O(\ell)$

41. Balance each of the following equations, and then write the complete and net ionic equations.
(a) $(NH_4)_2CO_3(aq) + Cu(NO_3)_2(aq) \rightarrow CuCO_3(s) + NH_4NO_3(aq)$
(b) $Pb(NO_3)_2(aq) + HCl(aq) \rightarrow PbCl_2(s) + HNO_3(aq)$
(c) $BaCO_3(s) + HCl(aq) \rightarrow BaCl_2(aq) + H_2O(\ell) + CO_2(g)$

42. Balance each of the following equations, and then write the complete and net ionic equations. Refer to Tables 4.1 and 4.2 for information on solubility and on acids and bases. Show phases for all reactants and products.
(a) $Ca(OH)_2 + HNO_3 \rightarrow Ca(NO_3)_2 + H_2O$
(b) $BaCl_2 + Na_2CO_3 \rightarrow BaCO_3 + NaCl$
(c) $Na_3PO_4 + Ni(NO_3)_2 \rightarrow Ni_3(PO_4)_2 + NaNO_3$

43. Balance each of the following equations, and then write the complete and net ionic equations. Refer to Tables 4.1 and 4.2 for information on solubility and on acids and bases. Show phases for all reactants and products.
(a) $ZnCl_2 + KOH \rightarrow KCl + Zn(OH)_2$
(b) $AgNO_3 + KI \rightarrow AgI + KNO_3$
(c) $NaOH + FeCl_2 \rightarrow Fe(OH)_2 + NaCl$

44. Barium hydroxide is used in lubricating oils and greases. Write a balanced equation for the reaction of this hydroxide with nitric acid to give barium nitrate, a compound used in pyrotechnics such as green flares.

45. Aluminum is obtained from bauxite, which is not a specific mineral but a name applied to a mixture of minerals. One of those minerals, which can dissolve in acids, is gibbsite, $Al(OH)_3$. Write a balanced equation for the reaction of gibbsite with sulfuric acid.

46. Balance the equation for the following precipitation reaction and then write the complete and net ionic equations.

$$CdCl_2 + NaOH \longrightarrow Cd(OH)_2 + NaCl$$

47. Balance the equation for the following precipitation reaction and then write the complete and net ionic equations.

$$Ni(NO_3)_2 + Na_2CO_3 \longrightarrow NiCO_3 + NaNO_3$$

48. Write an overall, balanced equation for the precipitation reaction that occurs when aqueous lead(II) nitrate is mixed with an aqueous solution of potassium chloride. Name each reactant and product. Indicate the state of each substance (s, ℓ, g, or aq).

49. Write an overall, balanced equation for the precipitation reaction that occurs when aqueous copper(II) nitrate is mixed with an aqueous solution of sodium carbonate. Name each reactant and product. Indicate the state of each substance (s, ℓ, g, or aq).

50. The beautiful mineral rhodochrosite is manganese(II) carbonate. Write an overall, balanced equation for the reaction of the mineral with hydrochloric acid. Name each reactant and product.

Acids, Bases and Salts

51. Classify each of the following as an acid or a base. What ions are produced when each is dissolved in water?
(a) KOH (d) HBr
(b) $Mg(OH)_2$ (e) LiOH
(c) HClO (f) H_2SO_3

52. For each acid and base in Question 51, which are strong and which are weak?

53. Identify the acid and base used to form the following salts.
(a) $NaNO_2$ (d) $Mg_3(PO_4)_2$
(b) $CaSO_4$ (e) $NaCH_3CO_2$
(c) NaI

54. For each salt in Question 53, write the overall neutralization reaction that formed each salt. Write the complete and net ionic equations for each neutralization reaction.

55. Classify each of the following exchange reactions; that is, tell whether the reaction is an acid-base reaction, a precipitation, or a gas-forming reaction. Predict the products of the reaction, and then balance the completed equation.
(a) $MnCl_2(aq) + Na_2S(aq) \rightarrow$
(b) $Na_2CO_3(aq) + ZnCl_2(aq) \rightarrow$
(c) $K_2CO_3(aq) + HClO_4(aq) \rightarrow$

56. Classify each of the following exchange reactions; that is, tell whether the reaction is an acid-base reaction, a precipitation, or a gas-forming reaction. Predict the products of the reaction, and then balance the completed equation.
(a) $Fe(OH)_3(s) + HNO_3(aq) \rightarrow$
(b) $FeCO_3(s) + H_2SO_4(aq) \rightarrow$
(c) $FeCl_2(aq) + (NH_4)_2S(aq) \rightarrow$
(d) $Fe(NO_3)_2(aq) + Na_2CO_3(aq) \rightarrow$

Oxidation-Reduction Reactions

57. Assign oxidation numbers to each element in the following compounds:
(a) SO_3 (b) HNO_3 (c) $KMnO_4$
(d) H_2O (e) LiOH (f) CH_2Cl_2
58. Assign oxidation numbers to each element in the following compounds:
(a) $Fe(OH)_3$ (b) $HClO_3$ (c) $CuCl_2$
(d) K_2CrO_4 (e) $Ni(OH)_2$ (f) N_2H_4
59. Assign oxidation numbers to each element in the following ions:
(a) SO_4^{2-} (b) NO_3^- (c) MnO_4^-
(d) $Cr(OH)_4^-$ (e) $H_2PO_4^-$ (f) $S_2O_3^{2-}$
60. Which of the following reactions are oxidation-reduction reactions? Explain your answer briefly. Classify the remaining reactions.
(a) $CdCl_2(aq) + Na_2S(aq) \rightarrow CdS(s) + 2\ NaCl(aq)$
(b) $2\ Ca(s) + O_2(g) \rightarrow 2\ CaO(s)$
(c) $Ca(OH)_2(s) + 2\ HCl(aq) \rightarrow CaCl_2(aq) + 2\ H_2O(\ell)$
61. Which of the following reactions are oxidation-reduction reactions? Explain your answer in each case. Classify the remaining reactions.
(a) $Zn(s) + 2\ NO_3^-(aq) + 4\ H_3O^+(aq) \rightarrow Zn^{2+}(aq) + 2\ NO_2(g) + 6\ H_2O(\ell)$
(b) $Zn(OH)_2(s) + H_2SO_4(aq) \rightarrow ZnSO_4(aq) + 2\ H_2O(\ell)$
(c) $Ca(s) + 2\ H_2O(\ell) \rightarrow Ca(OH)_2(s) + H_2(g)$
62. Which region of the periodic table has the best reducing agents? The best oxidizing agents?
63. Which of the following substances are common oxidizing agents?
(a) Zn (d) MnO_4^-
(b) O_2 (e) H_2
(c) HNO_3 (f) H^+
64. Which of the following substances are reducing agents?
(a) Ca (d) Al
(b) Ca^{2+} (e) Br_2
(c) $Cr_2O_7^{2-}$ (f) H_2
65. Identify the products of the following redox combination reactions:
(a) $C(s) + O_2(g) \rightarrow$ (d) $Mg(s) + N_2(g) \rightarrow$
(b) $P_4(s) + Cl_2(g) \rightarrow$ (e) $FeO(s) + O_2(g) \rightarrow$
(c) $Ti(s) + Cl_2(g) \rightarrow$ (f) $NO(g) + O_2(g) \rightarrow$
66. Complete and balance the following equations for redox displacement reactions:
(a) $K(s) + H_2O(\ell) \rightarrow$
(b) $Mg(s) + HBr(aq) \rightarrow$
(c) $NaBr(aq) + Cl_2(aq) \rightarrow$
(d) $WO_3(s) + H_2(g) \rightarrow$
(e) $H_2S(aq) + Cl_2(aq) \rightarrow$

Activity Series

67. Give an example of a displacement reaction that is also a redox reaction and identify which species is (a) oxidized; (b) reduced; (c) the reducing agent; and (d) the oxidizing agent.
68. (a) In what groups of the periodic table are the most reactive metals found? Where do we find the least reactive metals?
(b) Silver (Ag) does not react with 1 M HCl solution. Will Ag react with a solution of aluminum nitrate, $Al(NO_3)_3$? If so, write a chemical equation for the reaction.
(c) Lead (Pb) will react very slowly with 1 M HCl solution. Aluminum will react with lead(II) sulfate solution ($PbSO_4$). Will Pb react with an $AgNO_3$ solution? If so, write a chemical equation for the reaction.
(d) On the basis of the information obtained in answering parts (a), (b), and (c), arrange Ag, Al, and Pb in decreasing order of reactivity.
69. Use the activity series of metals (Table 4.5) to predict the outcome of each of the following reactions. If no reaction occurs, write N.R.
(a) $Na^+(aq) + Zn(s) \rightarrow$
(b) $HCl(aq) + Pt(s) \rightarrow$
(c) $Ag^+(aq) + Au(s) \rightarrow$
(d) $Au^{3+}(aq) + Ag(s) \rightarrow$
70. Using the activity series of metals (Table 4.5), predict whether the following reactions will occur in aqueous solution:
(a) $Mg(s) + Ca(s) \rightarrow Mg^{2+} + Ca^{2+}$
(b) $2\ Al^{3+} + 3\ Pb^{2+} \rightarrow 2\ Al(s) + 3\ Pb(s)$
(c) $H_2(g) + Zn^{2+} \rightarrow 2\ H^+ + Zn(s)$
(d) $Mg(s) + Cu^{2+} \rightarrow Mg^{2+} + Cu(s)$
(e) $Pb(s) + 2\ H^+ \rightarrow H_2(g) + Pb^{2+}$
(f) $2\ Ag^+ + Cu(s) \rightarrow 2\ Ag(s) + Cu^{2+}$
(g) $2\ Al^{3+} + 3\ Zn(s) \rightarrow 3\ Zn^{2+} + 2\ Al(s)$

Halogens in Redox Reactions

71. Which halogen is the strongest oxidizing agent? Which is the strongest reducing agent?
72. Predict the products of the following halogen displacement reactions. If no reaction occurs, write N.R.
(a) $I_2(s) + NaBr(aq) \rightarrow$
(b) $Br_2(\ell) + NaI(aq) \rightarrow$
(c) $F_2(g) + NaCl(aq) \rightarrow$
(d) $Cl_2(g) + NaBr(aq) \rightarrow$
(e) $Br_2(\ell) + NaCl(aq) \rightarrow$
(f) $Cl_2(g) + NaF(aq) \rightarrow$
73. For the reactions in Question 72 that occurred, identify the species oxidized or reduced as well as the oxidizing and reducing agents.
74. For the reactions in Question 72 that do not occur, rewrite the equation so that a reaction does occur (consider the halogen activity series).

General Questions

75. Name the spectator ions in the reaction of calcium carbonate and hydrochloric acid and write the net ionic equation.

$$CaCO_3(s) + 2\ H^+(aq) + 2\ Cl^-(aq) \longrightarrow CO_2(g) + Ca^{2+}(aq) + 2\ Cl^-(aq) + H_2O(\ell)$$

What type of reaction is this?

76. Magnesium metal reacts readily with HNO_3, as shown in the following equation:

$$Mg(s) + HNO_3(aq) \longrightarrow Mg(NO_3)_2(aq) + NO_2(g) + H_2O(\ell)$$

(a) Balance the equation for the reaction.
(b) Name each reactant and product.
(c) Write the net ionic equation for the reaction.
(d) What type of reaction does this appear to be?

77. Aqueous solutions of $(NH_4)_2S$ and $Hg(NO_3)_2$ react to give HgS and NH_4NO_3.
(a) Write the overall balanced equation for the reaction. Indicate the state (s or aq) for each compound.
(b) Name each compound.
(c) Write the net ionic equation for the reaction.
(d) What type of reaction does this appear to be?

78. Classify the following reactions and predict the products formed:
(a) $Li(s) + H_2O(\ell) \rightarrow$
(b) $Ag_2O(s) \xrightarrow{heat}$
(c) $Li_2O(s) + H_2O(\ell) \rightarrow$
(d) $I_2(s) + Cl^-(aq) \rightarrow$
(e) $Cu(s) + HCl(aq) \rightarrow$
(f) $BaCO_3(s) \xrightarrow{heat}$

79. Classify the following reactions and predict the products formed:
(a) $SO_3(g) + H_2O(\ell) \rightarrow$
(b) $Sr(s) + H_2(g) \rightarrow$
(c) $Mg(s) + H_2SO_4(aq, dilute) \rightarrow$
(d) $Na_3PO_4(aq) + AgNO_3(aq) \rightarrow$
(e) $Ca(HCO_3)_2(s) \xrightarrow{heat}$
(f) $Fe^{3+}(aq) + Sn^{2+}(aq) \rightarrow$

80. Azurite is a copper-containing mineral that often forms beautiful crystals. Its formula is $Cu_3(CO_3)_2(OH)_2$. Write a balanced equation for the reaction of this mineral with hydrochloric acid.

81. What species (atoms, molecules, ions) are present in an aqueous solution of each of the following compounds?
(a) NH_3 (c) NaOH
(b) CH_3CO_2H (d) HBr

82. Use the activity series to predict whether the following reactions will occur:
(a) $Fe(s) + Mg^{2+} \rightarrow Mg(s) + Fe^{2+}$
(b) $Ni(s) + Cu^{2+} \rightarrow Ni^{2+} + Cu(s)$
(c) $Cu(s) + 2\ H^+ \rightarrow Cu^{2+} + H_2(g)$
(d) $Mg(s) + H_2O(g) \rightarrow MgO(s) + H_2(g)$

83. Determine which of the following are redox reactions. Identify the oxidizing and reducing agent in each of the redox reactions.
(a) $NaOH(aq) + H_3PO_4(aq) \rightarrow NaH_2PO_4(aq) + H_2O(\ell)$
(b) $NH_3(g) + CO_2(g) + H_2O(\ell) \rightarrow NH_4HCO_3(aq)$
(c) $TiCl_4(g) + 2\ Mg(\ell) \xrightarrow{heat} Ti(s) + 2\ MgCl_2(\ell)$
(d) $NaCl(s) + NaHSO_4(aq) \xrightarrow{heat} HCl(g) + Na_2SO_4(s)$

84. Identify the substance oxidized, reduced, the reducing agent, and the oxidizing agent in the equations in Problem 83. For each oxidized or reduced substance, identify the change in its oxidation number.

85. Much has been written about chlorofluorocarbons and their impact on our environment. Their manufacture begins with the preparation of HF from the mineral fluorspar, CaF_2, according to the following *unbalanced* equation.

$$CaF_2(s) + H_2SO_4(aq) \longrightarrow HF(g) + CaSO_4(s)$$

The HF is combined with, for example, CCl_4 in the presence of $SbCl_5$ to make CCl_2F_2, called dichlorodifluoromethane or CFC-12, and other chlorofluorocarbons.

$$2\ HF(g) + CCl_4(\ell) \longrightarrow CCl_2F_2(g) + 2\ HCl(g)$$

(a) Balance the first equation above and name each substance.
(b) Is the first reaction best classified as an acid-base reaction, an oxidation-reduction, or a precipitation reaction?
(c) Give the names of the compounds CCl_4, $SbCl_5$, and HCl.
(d) Another chlorofluorocarbon produced in the reaction is composed of 8.74% C, 77.43% Cl, and 13.83% F. What is the empirical formula of the compound?

Applying Concepts

86. Chemical equations can be interpreted on either a particulate level (atoms, molecules, ions) or a mole level (moles of reactants and products). Write word statements to describe the combustion of butane on a particulate level and a mole level.

$$2\ C_4H_{10}(g) + 13\ O_2(g) \longrightarrow 8\ CO_2(g) + 10\ H_2O(\ell)$$

87. Write word statements to describe the following reaction on a particulate level and a mole level.

$$P_4(s) + 6\ Cl_2(g) \longrightarrow 4\ PCl_3(\ell)$$

88. What is the single product of this reaction?

$$4\ A_2 + AB_3 \longrightarrow 3\_\_\_\_\_\_$$

89. What is the single product of this hypothetical reaction?

$$3\ A_2B_3 + B_3 \longrightarrow 6\_\_\_\_\_\_$$

90. When the following pairs of reactants are combined in a beaker: (a) describe in words what the contents of the beaker would look like before and after any reaction occurs; (b) use different circles for atoms, molecules, and ions to draw a nanoscale (particulate-level) diagram of what the contents would look like; and (c) write a chemical equation to represent symbolically what the contents would look like.

$LiCl(aq)$ and $AgNO_3(aq)$

$NaOH(aq)$ and $HCl(aq)$

91. When the following pairs of reactants are combined in a beaker: (a) describe in words what the contents of the beaker would look like before and after any reaction occurs; (b) use different circles for atoms, molecules, and ions to draw a particulate-level diagram of what the contents would look like; and (c) write a chemical equation to represent symbolically what the contents would look like.

$CaCO_3(s)$ and $HCl(aq)$

$NH_4NO_3(aq)$ and $KOH(aq)$

92. Explain how you could prepare barium sulfate by (a) an acid-base reaction; (b) a precipitation reaction; and (c) a gas-forming reaction. The materials you have to start with are $BaCO_3$, $Ba(OH)_2$, Na_2SO_4, and H_2SO_4.

93. Students were asked to prepare nickel sulfate by reacting a nickel compound with a sulfate compound in water and then evaporating the water. Three students chose these pairs of reactants:

Student 1	$Ni(OH)_2$ and H_2SO_4
Student 2	$Ni(NO_3)_2$ and Na_2SO_4
Student 3	$NiCO_3$ and H_2SO_4

Comment on each student's choice of reactants and how successful you think each student will be at preparing nickel sulfate by the procedure indicated.

94. An unknown solution contains either lead ions or barium ions, but not both. Which one of the following aqueous solutions could you use to tell whether the ions present are Pb^{2+} or Ba^{2+}? Explain the reasoning behind the choice.

$HCl(aq)$, $H_2SO_4(aq)$, $H_3PO_4(aq)$

95. An unknown solution contains either calcium ions or strontium ions, but not both. Which one of the following solutions could you use to tell whether the ions present are Ca^{2+} or Sr^{2+}? Explain the reasoning behind the choice.

$NaOH(aq)$, $H_2SO_4(aq)$, $H_2S(aq)$

96. When given an oxidation-reduction reaction and asked what is oxidized or what is reduced, why should you never choose one of the products for your answer?

97. When given an oxidation-reduction reaction and asked what is the oxidizing agent or what is the reducing agent, why should you never choose one of the products for your answer?

CHAPTER 5

Relating Quantities of Reactants and Products

Phosphoric acid (H_3PO_4) is a major industrial chemical used to manufacture vast quantities of detergents, fertilizers, and compounds used in water treatment. It is also added to soft drinks to give them a tart flavor. In 1995 more than ten billion kilograms of phosphoric acid were produced in the United States. • How is phosphoric acid produced? Methods for its production are described in the Summary Problem at the end of this chapter. *(George Semple).*

One reason for studying chemical reactions is the hope of improving a known chemical process such as phosphoric acid manufacturing, or finding some new material that can be helpful to society. If a product has the potential for curing a disease, improving a polymer used for clothing or sports equipment, or treating polluted water or air, then the efficiency of the reaction is of interest. Can the product be made in sufficient quantity and purity that it can be sold at a reasonable price? To answer questions such as these, *reactions must be studied quantitatively.*

In phosphoric acid production, the quantity of product formed from a certain amount of reactants is chemically, and certainly economically, important. The quantity of product formed can make the difference between financial success or ruin for a chemical manufacturer. This chapter applies the stoichiometric principles introduced in Section 4.2 (← *p. 135*) to working quantitatively with amounts of reactants used and products formed.

5.1 THE MOLE AND CHEMICAL REACTIONS: THE MACRO-NANO CONNECTION

In a balanced chemical equation, each stoichiometric coefficient indicates the number of atoms, ions, or molecules, or a number of moles of reactants or products. These coefficients, together with the relationship between moles and grams of substances (← ***pp. 57 and 114***), allow us to calculate the masses of reactants and products involved in a reaction.

As a simple example, consider the formation of aluminum bromide, Al_2Br_6, by the reaction of aluminum with bromine (Figure 5.1). The balanced equation provides information about the numbers of atoms of aluminum and molecules of bromine and aluminum bromide. It also relates moles of aluminum atoms, bromine molecules, and aluminum bromide molecules. From the mole relationships we can calculate the mass of each reactant and the product.

The reaction of aluminum with bromine forms Al_2Br_6, a molecular compound, rather than $AlBr_3$, the ionic compound that would be predicted from the reaction of a metal and a nonmetal.

$$2\,Al(s) \quad + \quad 3\,Br_2(\ell) \quad \longrightarrow \quad Al_2Br_6(s)$$

2 Al(s)	3 $Br_2(\ell)$	$Al_2Br_6(s)$
2 Al atoms	3 Br_2 molecules	1 Al_2Br_6 molecule
2 mol Al atoms	3 mol Br_2 molecules	1 mol Al_2Br_6 molecules
54.0 g Al	479 g Br_2	533 g Al_2Br_6

The stoichiometric factor is also called a *mole ratio.*

The stoichiometric coefficients in a balanced chemical equation provide a **stoichiometric factor,** which is a mole-to-mole ratio relating moles of a reactant or product to moles of another reactant or product. Stoichiometric factors are extremely useful. We can, for example, calculate the number of moles of bromine required to react with 3.71×10^{-2} mol Al by using the proper stoichiometric factor from the balanced equation above. In such problems, a reliable method is to recognize and use what is given (known) to calculate the answer (what is to be found). In this case, the number of moles of aluminum is given and the moles of Br_2 are to be determined. The stoichiometric factor $\frac{3 \text{ mol } Br_2}{2 \text{ mol Al}}$ from the balanced equation relates these two quantities, and we can use it to directly calculate the required number of moles of Br_2.

$$3.71 \times 10^{-2} \text{ mol Al} \times \frac{3 \text{ mol } Br_2}{2 \text{ mol Al}} = 5.57 \times 10^{-2} \text{ mol } Br_2 \text{ required}$$

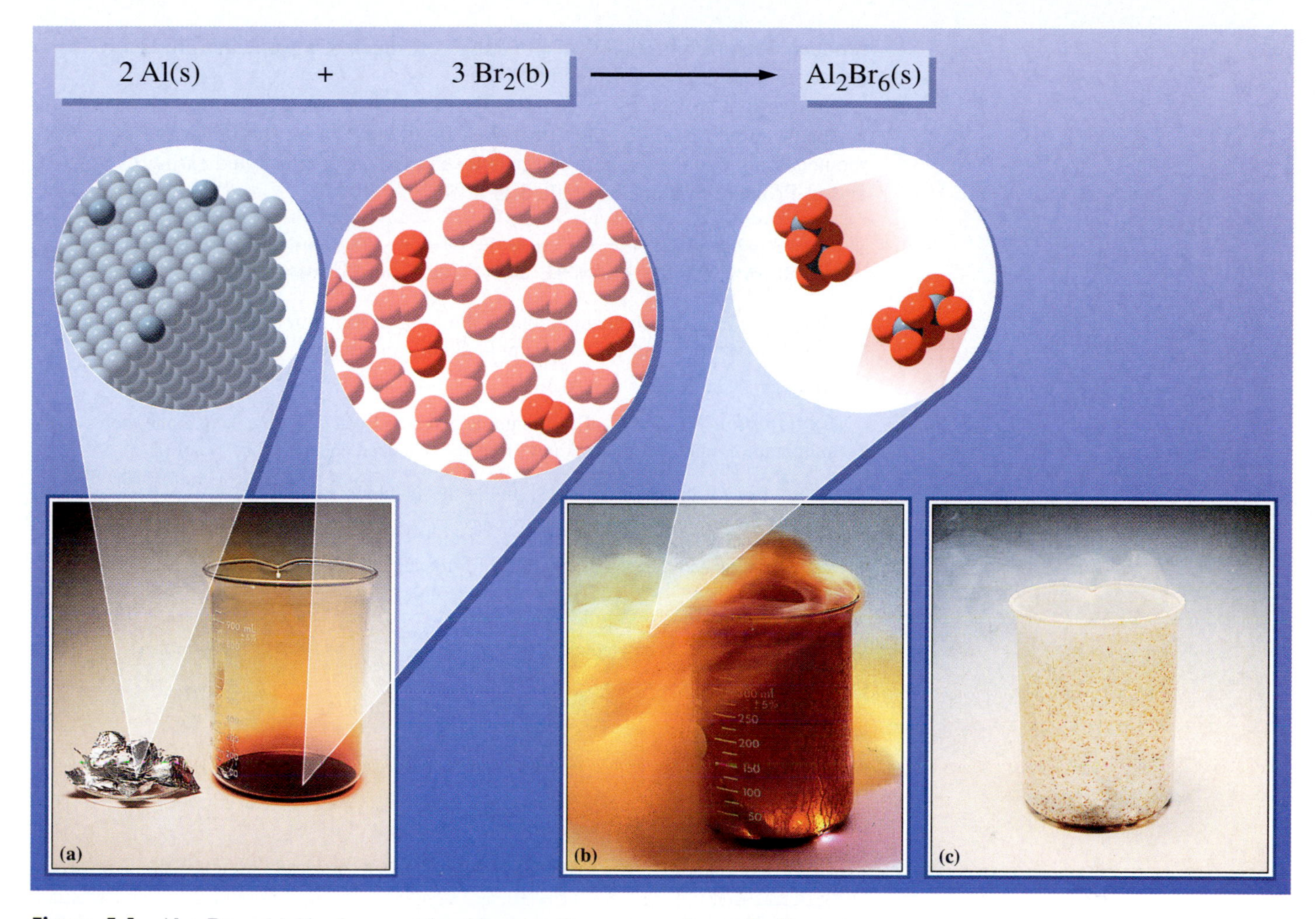

Figure 5.1 **Al + Br_2.** (a) Aluminum metal and bromine, Br_2, an orange-brown liquid, react so vigorously (b) that the aluminum becomes molten and glows white hot. The vapor in (b) consists of vaporized Br_2 and some of the product, white Al_2Br_6. At the end of the reaction (c), the beaker is coated with condensed aluminum bromide and the products of its reaction with atmospheric moisture. (*Note:* This reaction is dangerous! Under no circumstances should it be done except under properly supervised conditions.) *(C.D. Winters)*

Exercise 5.1 Stoichiometric Factors

Write all the possible stoichiometric factors that can be obtained from the balanced equation of the reaction of aluminum with bromine.

Calculations such as that above are based on the complete reaction of all the available reactants. This, however, does not happen very often; usually one reactant is provided in excess, and that excess remains unreacted. In some cases, competing reactions occur and other products ("side products") form in addition to the desired product. Such factors and others affecting how much of a particular product forms in a reaction are discussed in Section 5.3.

Often in a stoichiometry problem, the known or unknown information, or both, is a mass rather than a number of moles. In all such cases, the molar mass provides the necessary connection between masses and moles, as illustrated earlier (← *p. 180*) and in Problem-Solving Example 5.1.

PROBLEM-SOLVING EXAMPLE 5.1 Moles and Grams in Chemical Reactions

How many moles of aluminum bromide are formed from 5.02 g of aluminum? Assume complete conversion of the aluminum to aluminum bromide.

Answer 9.30×10^{-2} mol Al_2Br_6

Explanation The mass of aluminum is given and you want to calculate the moles of aluminum bromide. You also know from the balanced equation that 1 mol of Al_2Br_6 forms from every 2 mol Al giving the stoichiometric factor $\frac{1 \text{ mol } Al_2Br_6}{2 \text{ mol Al}}$.

Therefore, if you calculate the number of moles of aluminum in 5.02 g of Al, you can use that value and the stoichiometric factor to calculate the moles of aluminum bromide formed.

$$\text{Moles of Al reacted: } 5.02 \text{ g Al} \times \frac{1 \text{ mol Al}}{26.98 \text{ g Al}} = 1.861 \times 10^{-1} \text{ mol Al}$$

$$\text{Moles of } Al_2Br_6 \text{ formed: } 1.861 \times 10^{-1} \text{ mol Al} \times \frac{1 \text{ mol } Al_2Br_6}{2 \text{ mol Al}}$$

$$= 9.30 \times 10^{-2} \text{ mol } Al_2Br_6.$$

Problem-Solving Practice 5.1

How many grams of bromine (Br_2) are required to react completely with 0.0500 mol of Al?

A series of steps can be used to calculate the stoichiometric amount of a reactant or product in any chemical reaction regardless of whether the known or unknown is in grams or moles. Consider the following question: How many grams of bromine are needed to react completely with 1.00 g of aluminum? Figure 5.2 illustrates the steps given below.

Step 1: Write the correct formulas for reactants and products and balance the chemical equation. The balanced equation has been given:

$$2\,Al(s) + 3\,Br_2(\ell) \longrightarrow Al_2Br_6(s)$$

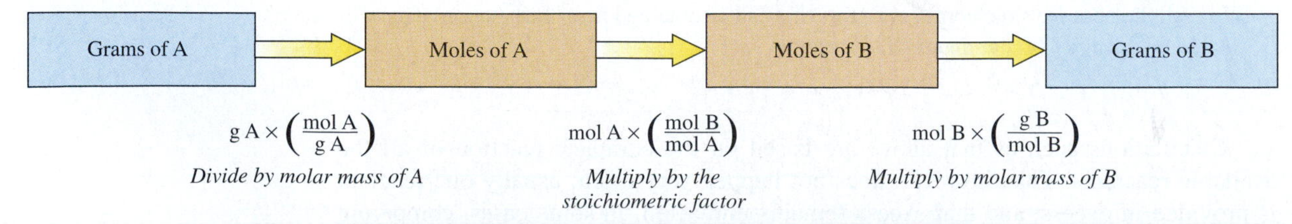

Figure 5.2 Stoichiometric relationships in a chemical reaction. The mass or amount in moles of one reactant or product (A) are related to the mass or amount in moles of another reactant or product (B) by the calculations shown.

Step 2: Decide what information about the problem is known and what is unknown. Map out a strategy for answering the question. You know the mass of aluminum and you want to calculate the mass of bromine. You also know that you can use molar mass to calculate moles of aluminum from grams of aluminum. Then you can use the stoichiometric factor

$$\frac{3 \text{ mol Br}_2}{2 \text{ mol Al}}$$

from the balanced equation to determine the moles of bromine needed. Once you have this result, you can use the molar mass of bromine to calculate the grams of bromine required.

Step 3: Calculate moles from grams (if necessary). The known mass (1.00 g Al) must be converted to moles because the balanced equation gives mole relationships directly, not mass relationships (Figure 5.2).

Grams of reactant A → Moles of reactant A

$$\text{g A} \times \left(\frac{\text{mol A}}{\text{g A}}\right)$$

Calculating moles of aluminum:

$$1.00 \text{ g Al} \times \frac{1 \text{ mol Al}}{26.98 \text{ g Al}} = 3.71 \times 10^{-2} \text{ mol Al}$$

Step 4: Use the stoichiometric factor to calculate the unknown number of moles and, if necessary, convert the result to grams.

In multistep calculations remember to carry one additional significant figure in intermediate steps before rounding to the final value.

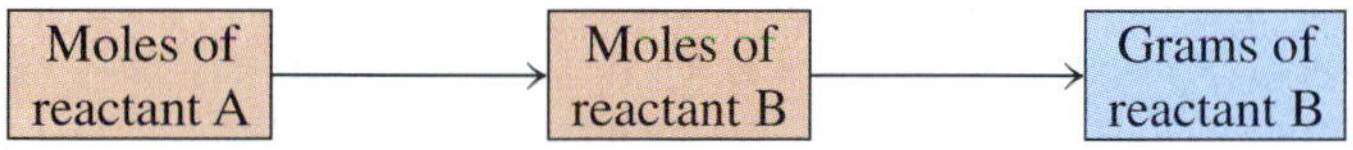

Moles of reactant A → Moles of reactant B → Grams of reactant B

$$\text{mol A} \times \left(\frac{\text{mol B}}{\text{mol A}}\right) \qquad \text{mol B} \times \left(\frac{\text{g B}}{\text{mol B}}\right)$$

To calculate grams of bromine:

$$3.706 \times 10^{-2} \text{ mol Al} \times \frac{3 \text{ mol Br}_2}{2 \text{ mol Al}} = 5.560 \times 10^{-2} \text{ mol Br}_2$$

$$5.560 \times 10^{-2} \text{ mol Br}_2 \times \frac{159.8 \text{ g Br}_2}{1 \text{ mol Br}_2} = 8.88 \text{ g Br}_2$$

Step 5: Check the answer to see if it is reasonable. Because 1.00 g Al is much less than 1 mol Al (26.98 g Al), the mass of Br_2 should be less than 1 mol Br_2 (159.8 g). Therefore, 0.0556 mol Br_2 and 8.88 g Br_2 are reasonable answers.

Exercise 5.2 Moles and Grams in Chemical Reactions

Verify that 9.89 g Al_2Br_6 is produced by the reaction of sufficient aluminum with 5.57×10^{-2} mol of Br_2.

Problem-Solving Examples 5.2 and 5.3 illustrate the further application of the procedure summarized in Figure 5.2.

Magnesium oxide. Magnesium oxide is produced when magnesium metal burns as in this display of fireworks. *(Peter Skinner/Photo Researchers, Inc.)*

PROBLEM-SOLVING EXAMPLE 5.2 Moles and Grams

Magnesium metal in fireworks reacts with oxygen in air to produce a brilliant white flash. The product of this combination reaction is magnesium oxide, MgO. How many grams of magnesium oxide will be formed by the reaction of 0.500 mol of magnesium with 0.250 mol oxygen gas?

Answer 20.2 g MgO

Explanation We can apply the steps given before to calculate the mass of MgO formed.

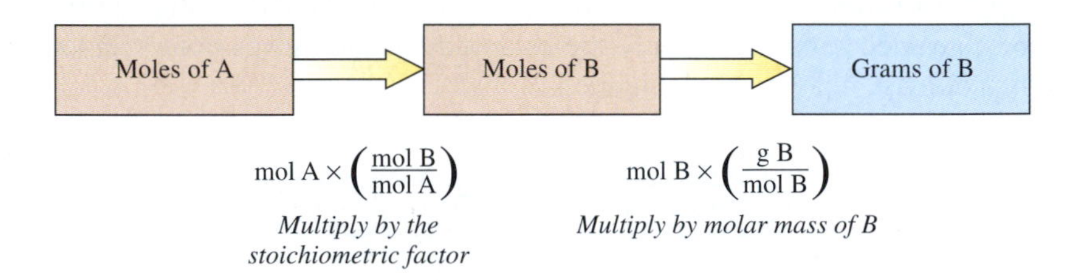

$\text{mol A} \times \left(\frac{\text{mol B}}{\text{mol A}}\right)$ — *Multiply by the stoichiometric factor*

$\text{mol B} \times \left(\frac{\text{g B}}{\text{mol B}}\right)$ — *Multiply by molar mass of B*

Step 1: Magnesium metal and diatomic oxygen gas are the reactants; solid magnesium oxide is the product.

$$2\ \text{Mg(s)} + \text{O}_2\text{(g)} \longrightarrow 2\ \text{MgO(s)}$$

Step 2: We already know how many moles of Mg and O_2 are present; they were given in the problem. Thus, all that is needed is the stoichiometric factor to determine the number of moles of MgO. We can use the molar mass of MgO to determine how many grams of it will be formed.

Step 3: The moles of reactants are known.

Step 4: Because the moles of each reactant are known in this case, we can use either one of two stoichiometric factors to calculate the moles of MgO formed; both give the same answer:

$$0.500\ \text{mol Mg} \times \frac{2\ \text{mol MgO}}{2\ \text{mol Mg}} = 0.500\ \text{mol MgO}$$

or

$$0.250\ \text{mol O}_2 \times \frac{2\ \text{mol MgO}}{1\ \text{mol O}_2} = 0.500\ \text{mol MgO}$$

The grams of MgO formed by the reaction are

$$0.500\ \text{mol MgO} \times \frac{40.31\ \text{g MgO}}{1\ \text{mol MgO}} = 20.2\ \text{g MgO}$$

Step 5: The answer of 20.2 g MgO is reasonable because 1 mol of oxygen would produce 2 mol of MgO (80.6 g); therefore, one fourth of a mole of oxygen will generate one half of a mole of magnesium oxide (20.2 g).

Problem-Solving Practice 5.2

Tin is extracted from its ore cassiterite, SnO_2, by reduction with carbon from coal.

$$\text{SnO}_2\text{(s)} + 2\ \text{C(s)} \longrightarrow \text{Sn}(\ell) + 2\ \text{CO(g)}$$

(a) What mass of tin can be produced from 0.300 mol of cassiterite?
(b) How many moles of carbon are required to produce this much tin?

Conceptual Challenge Problem CP-5.A at the end of the chapter relates to topics covered in this section.

Problem-Solving Example 5.3 Grams, Moles, and Grams

A popular candy bar contains 21.1 g of glucose, $C_6H_{12}O_6$. When the candy bar is eaten, the glucose is metabolized according to the overall reaction

$$C_6H_{12}O_6(s) + O_2(g) \longrightarrow CO_2(g) + H_2O(\ell)$$

(unbalanced equation)

(a) Balance the chemical equation.
(b) What mass of oxygen is consumed by the reaction?
(c) How many grams of carbon dioxide and of water are produced by the metabolism of the 21.1 g of glucose in the candy bar?

Answer (a) $C_6H_{12}O_6(s) + 6\ O_2(g) \rightarrow 6\ CO_2(g) + 6\ H_2O\ (\ell)$; (b) 22.5 g O_2; (c) 30.9 g CO_2 and 12.7 g H_2O

Explanation First, balance the equation.

$$C_6H_{12}O_6(s) + 6\ O_2(g) \longrightarrow 6\ CO_2(g) + 6\ H_2O(\ell)$$

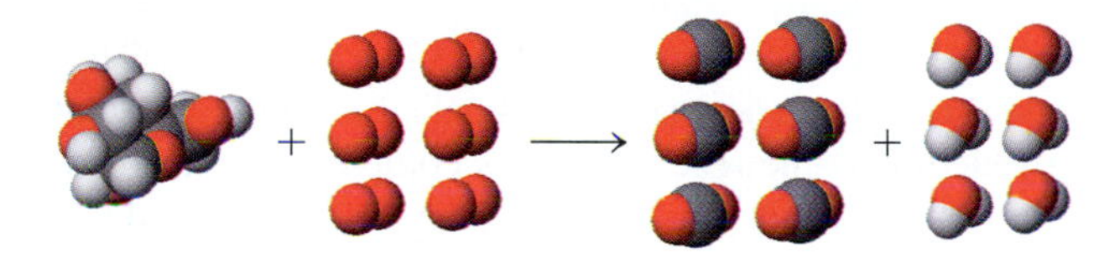

Then use the molar mass to calculate moles of glucose from grams of glucose:

$$21.1 \text{ g glucose} \times \frac{1 \text{ mol glucose}}{180.2 \text{ g glucose}} = 0.1170 \text{ mol glucose}$$

Next, use the correct stoichiometric factors and molar masses to calculate the number of grams of oxygen, carbon dioxide, and water.

$$0.1170 \text{ mol glucose} \times \frac{6 \text{ mol } O_2}{1 \text{ mol glucose}} \times \frac{31.99 \text{ g } O_2}{1 \text{ mol } O_2} = 22.5 \text{ g } O_2$$

$$0.1170 \text{ mol glucose} \times \frac{6 \text{ mol } CO_2}{1 \text{ mol glucose}} \times \frac{44.01 \text{ g } CO_2}{1 \text{ mol } CO_2} = 30.9 \text{ g } CO_2$$

and

$$0.1170 \text{ mol glucose} \times \frac{6 \text{ mol } H_2O}{1 \text{ mol glucose}} \times \frac{18.02 \text{ g } H_2O}{1 \text{ mol } H_2O} = 12.7 \text{ g } H_2O$$

The mass of water can also be found by using the conservation of mass:

Total mass of reactants = total mass of products
Total mass of reactants = 21.1 g glucose + 22.5 g O_2 = 43.6 g
Total mass of products = CO_2 + ? g of H_2O = 43.6 g
Mass of H_2O = 43.6 g − 30.9 g = 12.7 g

These are reasonable answers because approximately 0.1 mol of glucose should require approximately 0.6 mol of O_2, and produce approximately 0.6 mol of CO_2 and 0.6 mol of H_2O. Therefore, none of the calculated masses should be larger than the molar masses, and none are.

Problem-Solving Practice 5.3

A lump of coke (carbon) weighs 57 g.
(a) What mass of oxygen is required to burn it to carbon monoxide?
$2\ C(s) + O_2(g) \rightarrow 2\ CO\ (g)$
(b) How many grams of CO are produced?

5.2 REACTIONS WITH ONE REACTANT IN LIMITED SUPPLY

In the previous section, we assumed that exactly stoichiometric amounts of reactants were present; none of the reactants was left when the reaction was over. However, this is rarely the case when chemists carry out an actual synthesis, whether for small quantities in a laboratory or on a large scale in an industrial process. Generally one reactant is more expensive or less readily available than others. The cheaper or more available reactants are used in excess to ensure that the more expensive material is completely converted to product.

The industrial production of methanol, CH_3OH, is such a case. Methanol, one of the top 25 chemicals produced in the United States, is manufactured by the reaction of hydrogen and carbon monoxide.

$$CO(g) + 2\ H_2(g) \longrightarrow CH_3OH(\ell)$$

Carbon monoxide is manufactured cheaply by burning coke (carbon) in a limited supply of air so that there is insufficient oxygen to form carbon dioxide. Hydrogen is more expensive to manufacture. Therefore, methanol synthesis uses an excess of carbon monoxide, and the amount of methanol produced is dictated by the amount of hydrogen available. Hydrogen acts as what is called a limiting reactant.

A **limiting reactant** is the reactant that is completely consumed during a reaction. Once the limiting reactant has been used up, no more product can form. *The limiting reactant must be used as the basis for calculating the maximum possible amount of product(s)* because the limiting reactant limits the amount of product(s) that can be formed.

We can make an analogy to a chemistry "limiting reactant" in the assembling of a booklet containing colored sheets of paper. Each booklet must have 5 yellow sheets, 3 blue sheets, and 2 pink sheets. You have in stock 400 yellow sheets, 300 blue sheets, and 200 pink sheets. How many complete pamphlets can be assembled?

Each pamphlet must have the pages in the ratio of 5 yellow: 3 blue: 2 pink (analogous to coefficients in a balanced chemical equation). Using the stock on hand and the 5:3:2 requirement, only 80 complete pamphlets can be assembled.

Sheets in Stock	Required for 1 Pamphlet	Assembling
Yellow: 400	5	$400 \text{ yellow sheets} \times \frac{1 \text{ pamphlet}}{5 \text{ yellow sheets}} = 80 \text{ pamphlets};$ 0 yellow sheets left
Blue: 300	3	$80 \text{ pamphlets} \times \frac{3 \text{ blue sheets}}{1 \text{ pamphlet}} = 240 \text{ blue sheets}$ used; 60 blue sheets left
Pink: 400	2	$80 \text{ pamphlets} \times \frac{2 \text{ pink sheets}}{1 \text{ pamphlet}} = 160 \text{ pink sheets}$ used; 240 pink sheets left

The 80 pamphlets use all the yellow sheets, making them the "limiting reactant" (Figure 5.3). Overall, the 80 pamphlets contain a total of 400 yellow, 240 blue, and 160 pink sheets; 60 blue sheets and 240 pink sheets are left over.

In determining how many pamphlets could be assembled, the "limiting reactant" was the yellow sheets. Similarly, the limiting reactant must be identified in a chemical reaction in order to determine how much product(s) will be produced if all the reactants are converted to the desired product(s). Problem-Solving Example 5.4 il-

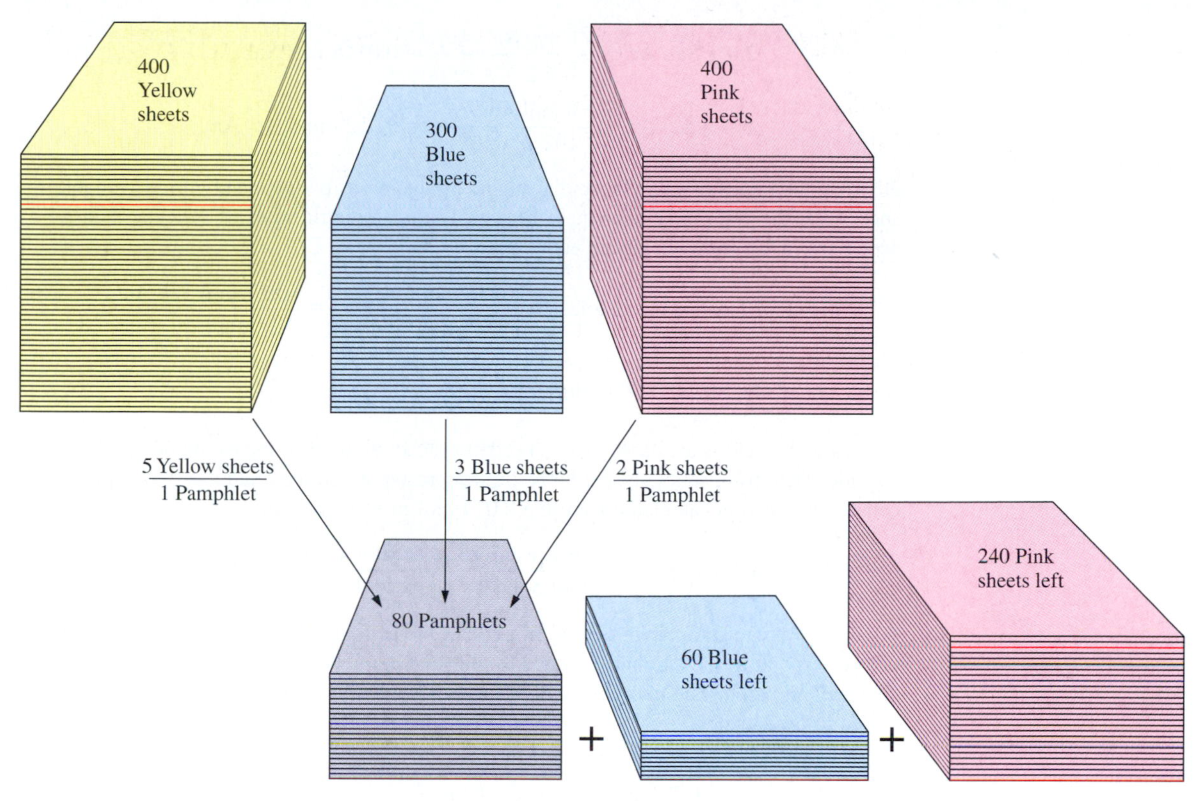

Figure 5.3 Limiting reactant in pamphlet assembly. Assembling a ten-page pamphlet that requires 5 yellow sheets of paper, 3 blue sheets, and 2 pink sheets per pamphlet. The yellow sheets are the "limiting reactant" because all 400 yellow sheets are used when 80 pamphlets are produced, leaving 60 blue sheets and 240 pink sheets unused.

lustrates two methods for identifying a limiting reactant—the ratio method and the mass method.

PROBLEM-SOLVING EXAMPLE 5.4 Limiting Reactant

In a general chemistry laboratory experiment, a student prepares aspirin (acetylsalicylic acid, $C_9H_8O_4$) by reacting salicylic acid, $C_7H_6O_3$ (abbreviated Sal. Ac.), with acetic anhydride, $C_4H_6O_3$ (Ac. An.):

$$2\ C_7H_6O_3(s) + C_4H_6O_3(\ell) \longrightarrow 2\ C_9H_8O_4(s) + H_2O(\ell)$$

salicylic acid (Sal. Ac.)	acetic anhydride (Ac. An.)	aspirin
138 g/mol	102 g/mol	180 g/mol

Her laboratory book directs her to use 2.0 g of salicyclic acid and 5.4 g of acetic anhydride. (a) What is the limiting reactant? (b) What is the maximum mass of aspirin she can prepare?

Answer (a) Salicyclic acid is the limiting reactant; (b) 2.5 g aspirin.

Explanation *Method 1 (ratio method):* The limiting reactant can be identified by finding the number of moles of each reactant available, and then comparing their mole ratio with the mole ratio from the stoichiometric coefficients. The moles of each reactant are

$$2.0 \text{ g Sal. Ac.} \times \frac{1 \text{ mol Sal. Ac.}}{138 \text{ g Sal. Ac.}} = 0.0145 \text{ mol Sal. Ac.}$$

$$5.4 \text{ g Ac. An.} \times \frac{1 \text{ mol Ac. An.}}{102 \text{ g Ac. An.}} = 0.0529 \text{ mol Ac. An.}$$

Dividing the moles of salicylic acid available by the moles of acetic anhydride available shows that the ratio is smaller than the 2-to-1 stoichiometric ratio of salicylic acid to acetic anhydride in the balanced equation.

$$\text{Moles of reactants available} = \frac{0.0145 \text{ mol Sal. Ac.}}{0.0529 \text{ mol Ac. An}} = \frac{0.27 \text{ mol Sal. Ac.}}{1.0 \text{ mol Ac. An.}}$$

$$\text{Stoichiometric ratio} = \frac{2 \text{ mol Sal Ac.}}{1 \text{ mol Ac. An.}}$$

Less salicylic acid is available than is needed to react with all of the available acetic anhydride. Therefore, salicylic acid is the limiting reactant and the quantity of aspirin that can be produced is calculated from the 0.014 mol of salicylic acid.

$$0.014 \text{ mol Sal. Ac.} \times \frac{2 \text{ mol aspirin}}{2 \text{ mol Sal. Ac.}} \times \frac{180 \text{ g aspirin}}{1 \text{ mol aspirin}} = 2.5 \text{ g aspirin}$$

We can summarize the information of the example in a table. The change values correspond to the amounts dictated by the stoichiometric factors from the balanced chemical equation. The negative sign indicates a depletion of that reactant as it is converted to products.

	$2\ C_7H_6O_3$ Salicylic Acid	$+\ C_4H_6O_3$ Acetic Anhydride	$\longrightarrow 2\ C_9H_8O_4\ +$ Aspirin	H_2O Water
Initial amt.	0.014 mol	0.053 mol	0 mol	0 mol
Change as reaction occurs	−0.014 mol	−0.0073 mol	0.014 mol	0.0073 mol
Amt. at end of reaction	0 mol	0.046 mol	0.014 mol	0.0073 mol

Method 2 (mass method): Another way to identify salicyclic acid as the limiting reactant is to calculate the quantity of aspirin that would be produced from 0.014 mol of salicylic acid and unlimited acetic anhydride, or from 0.053 mol of acetic anhydride and unlimited salicylic acid. The limiting reactant produces the lesser quantity of aspirin.

Quantity of aspirin produced from 0.014 mol of salicyclic acid and unlimited acetic anhydride

$$0.014 \text{ mol Sal. Ac.} \times \frac{2 \text{ mol aspirin}}{2 \text{ mol Sal. Ac.}} \times \frac{180 \text{ g aspirin}}{1 \text{ mol aspirin}} = 2.5 \text{ g aspirin}$$

Quantity of aspirin produced from 0.053 mol of acetic anhydride and unlimited salicylic acid

$$0.053 \text{ mol Ac. An.} \times \frac{2 \text{ mol aspirin}}{1 \text{ mol acetic anhydride}} \times \frac{180 \text{ g aspirin}}{1 \text{ mol aspirin}} = 19 \text{ g aspirin}$$

This comparison shows that the amount of available salicylic acid would produce less aspirin than the amount of acetic acid available, providing further proof that salicylic acid is the limiting reactant. Using excess acetic anhydride makes economic sense as well as chemical sense because acetic anhydride costs only about half as much as salicylic acid.

Problem-Solving Practice 5.4

Carbon disulfide reacts with oxygen to form carbon dioxide and sulfur dioxide:

$$CS_2(\ell) + O_2(g) \longrightarrow CO_2(g) + SO_2(g)$$

A mixture of 3.5 g of CS_2 and 1.75 g O_2 is reacted.
(a) Balance the equation.
(b) What is the limiting reactant?
(c) What is the maximum number of grams of sulfur dioxide that can be formed?

PROBLEM-SOLVING EXAMPLE 5.5 Limiting Reactant

Silicon carbide, SiC, also known as carborundum (Figure 5.4), is an important industrial abrasive made by the high-temperature reaction of SiO_2 with carbon.

$$SiO_2(s) + 3\ C(s) \longrightarrow SiC(s) + 2\ CO(g)$$

(a) Determine the limiting reactant when a mixture of 5.00×10^3 g of SiO_2 and 5.00×10^3 g of carbon react. (b) What is the maximum number of grams of carborundum that can form? (c) How many grams of the excess reactant remain after the reaction is complete?

Answer (a) SiO_2 is the limiting reactant; (b) 3.34×10^3 g SiC; (c) 2.01×10^3 g C remain.

Explanation We begin by determining how many moles of each reactant are available.

$$5.00 \times 10^3 \text{ g } SiO_2 \times \frac{1 \text{ mol } SiO_2}{60.1 \text{ g } SiO_2} = 83.19 \text{ mol } SiO_2$$

$$5.00 \times 10^3 \text{ g C} \times \frac{1 \text{ mol C}}{12.01 \text{ g C}} = 416.3 \text{ mol C}$$

The masses of carborundum produced, based on the masses available of each reactant, are

$$83.19 \text{ mol } SiO_2 \times \frac{1 \text{ mol SiC}}{1 \text{ mol } SiO_2} \times \frac{40.10 \text{ g SiC}}{1 \text{ mol SiC}} = 3.34 \times 10^3 \text{ g SiC}$$

$$416.3 \text{ mol C} \times \frac{1 \text{ mol SiC}}{3 \text{ mol C}} \times \frac{40.10 \text{ g SiC}}{1 \text{ mol SiC}} = 5.56 \times 10^3 \text{ g SiC}$$

Clearly, the quantity of carborundum that can be formed using the given quantities of carbon and silicon dioxide is controlled by the quantity of silicon dioxide; it is the limiting reactant. The reaction used only 250. mol of carbon of the 416.3 mol available

$$83.19 \text{ mol } SiO_2 \times \frac{3 \text{ mol C}}{1 \text{ mol } SiO_2} = 250. \text{ mol C}$$

This leaves 166. mol of carbon (2.00×10^3 g C) unreacted.

$$166. \text{ mol C unreacted} \times \frac{12.01 \text{ g C}}{1.00 \text{ mol C}} = 2.00 \times 10^3 \text{ g C unreacted}$$

Problem-Solving Practice 5.5

Construct a table similar to that in Problem-Solving Example 5.4 to summarize the carborundum synthesis described in Problem-Solving Example 5.5.

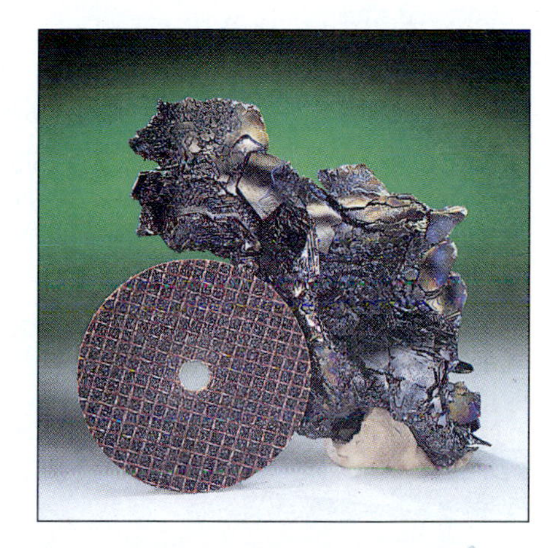

Figure 5.4 Silicon carbide, SiC. The grinding wheel (left) is coated with SiC. Naturally occurring silicon carbide (right), is also known as carborundum, and is one of the hardest substances known, making it valuable as an abrasive. *(C.D. Winters)*

It is useful, although not necessary, to calculate the amount of excess reactant remaining to verify that the reactant in excess is not the limiting reactant.

CHEMISTRY YOU CAN DO

Vinegar and Baking Soda: A Stoichiometry Experiment

This experiment focuses on the reactions of metal carbonates with acid. For example, limestone reacts with hydrochloric acid to give calcium chloride, carbon dioxide, and water:

$$CaCO_3(s) + 2\ HCl(aq) \longrightarrow CaCl_2(aq) + CO_2(g) + H_2O(\ell)$$

In a similar way, baking soda (sodium bicarbonate) and vinegar (aqueous acetic acid) react to give sodium acetate, carbon dioxide, and water:

$$NaHCO_3(s) + CH_3CO_2H(aq) \longrightarrow NaCH_3CO_2(aq) + CO_2(g) + H_2O(\ell)$$

In this experiment we want to explore the relation between the quantity of acid or carbonate used and the quantity of carbon dioxide evolved. To do the experiment you need some baking soda, vinegar, a small balloon, and a bottle with a narrow neck and a volume of about 100 mL. The balloon should fit tightly but easily over the top of the bottle. (It may slip on more easily if the bottle and balloon are wet.) Inflate the balloon ahead of time to check that there are no leaks.

Place 1 level teaspoon of baking soda in the balloon. (You can make a funnel out of a rolled-up piece of paper to help get the baking soda into the balloon.) Add 3 teaspoons of vinegar to the bottle, and then slip the balloon over the neck of the bottle. Turn the balloon over so that the baking soda runs into the bottle, and then shake the bottle to make sure that the vinegar and baking soda are well mixed. What do you see? Does the balloon inflate? If so, why?

Now repeat the experiment several times using the following amounts of vinegar and baking soda:

Baking Soda	Vinegar
1 tsp	1 tsp
1 tsp	4 tsp
1 tsp	7 tsp
1 tsp	10 tsp

Be sure to use a new balloon each time and rinse out the bottle between tests. In each test, record how much the balloon inflates.

Is there a relationship between the quantity of vinegar and baking soda used and the extent to which the balloon inflates? If so, how can you explain this connection?

At which point does an increase in the quantity of vinegar not increase the volume of the balloon? Based on what we know about chemical reactions, how could increasing the amount of one reactant not have an effect on the balloon's size?

The set-up for the study of the reaction of baking soda with acetic acid. *(C.D. Winters)*

Exercise 5.3 Limiting Reactant

Urea is used as a fertilizer because it can react with water to release ammonia, which provides nitrogen to plants.

$$(NH_2)_2CO(s) + H_2O(\ell) \rightarrow 2\ NH_3(aq) + CO_2(g)$$

(a) Determine the limiting reactant when 300 g of urea and 100 g of water are combined.

Figure 5.5 Popcorn yield. We began with 20 pieces of popcorn and found that only 16 of them popped. The percent yield of popcorn from our "reaction" was (16/20) × 100%, or 80%. *(C.D. Winters)*

(b) How many grams of ammonia and how many grams of carbon dioxide form?
(c)What mass of the excess reactant remains after reaction?

5.3 EVALUATING THE SUCCESS OF A SYNTHESIS: PERCENT YIELD

The theoretical yield must be based on the amount of the limiting reactant.

A reaction that forms the quantity of the desired product calculated from a given quantity of limiting reactant is said to have a 100 percent yield. This maximum possible quantity of product formed when all of the limiting reactant is converted into the desired product(s) is called the **theoretical yield.** Many times the **actual yield,** the quantity of desired product actually obtained from a synthesis in a laboratory or industrial chemical plant, is less than the theoretical yield.

The efficiency of a particular synthesis method is evaluated by calculating the **percent yield** (Figure 5.5), which is defined as

$$\text{Percent yield} = \frac{\text{actual yield}}{\text{theoretical yield}} \times 100\%$$

Percent yield can be applied, for example, to the synthesis of aspirin as described in Problem-Solving Example 5.4 (Figure 5.6). Suppose a student carried out the synthesis and obtained 2.2 g of aspirin rather than the theoretical yield of 2.7 g. What is the percent yield of this reaction?

$$\text{Percent yield} = \frac{\text{actual yield}}{\text{theoretical yield}} \times 100\% = \frac{2.2\ \text{g}}{2.7\ \text{g}} \times 100\% = 82\%$$

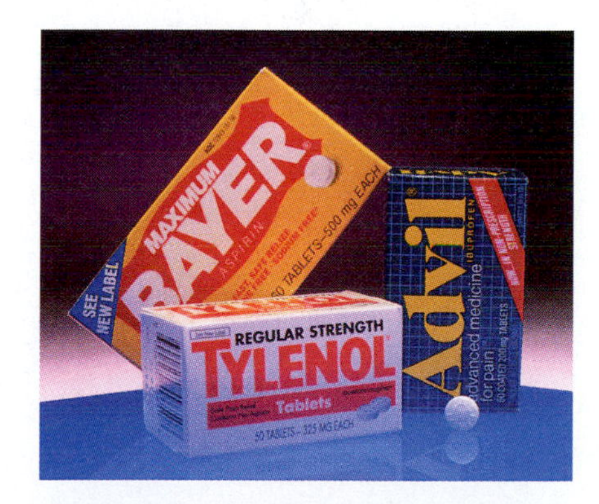

Figure 5.6 Nonprescription pain remedies. Aspirin, a white solid ($C_9H_8O_4$), is made by the reaction of salicylic acid ($C_7H_6O_3$) and acetic anhydride ($C_4H_6O_3$). Other pain-relief medications contain other active compounds such as acetaminophen ($C_8H_9NO_2$) and ibuprofen ($C_{13}H_{18}O_2$), which are quite different from aspirin. *(C.D. Winters)*

Although we hope to obtain as close to the theoretical yield as possible, few reactions or experimental manipulations are so efficient, in spite of controlled experimental conditions and careful laboratory techniques. Side reactions may occur that form products other than the desired one, and during the isolation and purification of the desired product, some of it may be lost. When chemists report the synthesis of a new compound, or the development of a new synthesis, they state the percent yield of the reaction or the overall series of reactions. Other chemists who wish to repeat the synthesis then have an idea of how much product can be expected from a certain amount of reactants, the starting materials.

PROBLEM-SOLVING EXAMPLE 5.6 Percent Yield

Acetic acid (CH_3CO_2H) is produced industrially by the direct combination of methanol (CH_3OH) with carbon monoxide (CO):

$$CH_3OH(\ell) + CO(g) \longrightarrow CH_3CO_2H(\ell)$$

How many grams of methanol are needed to react with excess carbon monoxide to prepare 5.0×10^3 g of acetic acid if the expected yield is 88%?

Answer 3.0×10^3 g of methanol

Explanation The first step is to calculate the theoretical yield. Expressing the 88% yield as a decimal fraction gives

$$\frac{\text{Actual yield}}{\text{theoretical yield}} = 88\% = 0.88$$

$$\text{Theoretical yield} = \frac{5.0 \times 10^3 \text{ g acetic acid}}{0.88} = 5.7 \times 10^3 \text{ g acetic acid}$$

Carbon monoxide is in excess and so methanol, the limiting reactant, determines the maximum amount of acetic acid that can form. The mass of methanol needed can be calculated from the theoretical yield of acetic acid and the 1:1 stoichiometric factor for methanol and acetic acid:

$$5.7 \times 10^3 \text{ g acetic acid} \times \frac{1 \text{ mol acetic acid}}{60.0 \text{ g acetic acid}} \times \underbrace{\frac{1 \text{ mol methanol}}{1 \text{ mol acetic acid}}}_{\text{stoichiometric factor}} \times \frac{32.0 \text{ g methanol}}{1 \text{ mol methanol}}$$

$$= 3.0 \times 10^3 \text{ g methanol}$$

Problem-Solving Practice 5.6

You heat 2.50 g of copper with an excess of sulfur and synthesize 2.53 g of copper(I) sulfide, Cu_2S:

$$16\ Cu(s) + S_8(s) \rightarrow 8\ Cu_2S(s)$$

Your laboratory instructor expects students to have at least a 60% yield for this reaction. Did your synthesis meet this standard?

Exercise 5.4 Percent Yield

Percent yield can be reduced by side reactions that produce undesired product(s) and by poor laboratory technique in isolating and purifying the desired product. Identify two other factors that could lead to a low percent yield.

5.4 CHEMICAL EQUATIONS AND CHEMICAL ANALYSIS

With an increased awareness of environmental problems in recent years, we need to know what chemicals are in our environment and in what quantities. This need relies on analytical chemistry, a field in which chemists creatively identify pure substances and measure the quantities of components of mixtures. Although analytical chemistry is often done now by instrumental methods, classical chemical reactions and stoichiometry still play a central role.

The analysis of mixtures is often challenging. It can take a great deal of imagination to figure out how to use chemistry to determine what, and how much, is there. We can illustrate how analytical chemistry problems are solved, though, with a reasonably straightforward example. We'll examine the problem of determining the purity of a sample of potassium chloride (KCl), an ionic compound used as salt substitute by individuals who must limit their intake of sodium chloride (NaCl).

The purity of a KCl sample can be determined by dissolving the sample in water and adding an excess of silver nitrate ($AgNO_3$) solution to precipitate AgCl (Figure 5.7). The precipitated AgCl can be dried and weighed. The purity of the sample is defined as the fraction of the sample (expressed as a percentage) that is KCl.

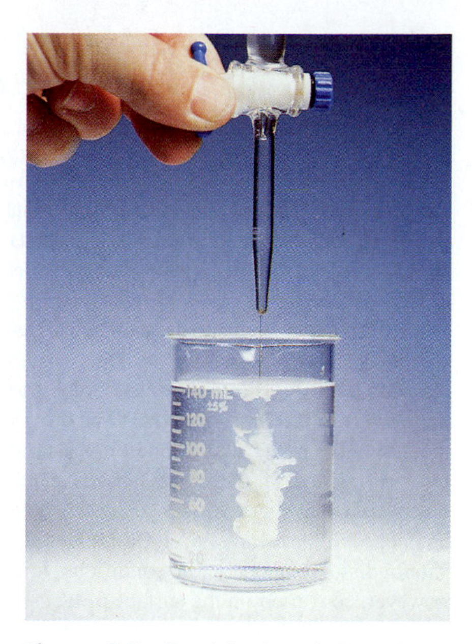

Figure 5.7 Precipitation of silver chloride. Adding aqueous potassium chloride, KCl, to aqueous silver nitrate, $AgNO_3$, leads to a precipitate of white silver chloride, AgCl, and an aqueous solution of potassium nitrate, KNO_3. *(C.D. Winters)*

$$\mathrm{KCl(aq) + AgNO_3(aq) \longrightarrow AgCl(s) + KNO_3(aq)}$$

In a particular case, 147 mg of impure KCl was dissolved in water and an excess of silver nitrate solution was added until no more silver chloride precipitated. The mass of dried AgCl was 240 mg. What is the purity of the KCl?

We can link the mass of AgCl to the mass of KCl through moles in the balanced equation: 1 mol KCl to 1 mol AgCl. We first find the number of moles of AgCl that was formed.

$$240\text{ mg AgCl} \times \frac{1\text{ g AgCl}}{10^3\text{ mg AgCl}} \times \frac{1\text{ mol AgCl}}{143.4\text{ g AgCl}} = 1.674 \times 10^{-3}\text{ mol AgCl}$$

Next, using the stoichiometric factor and molar mass of KCl, we find the mass of KCl that was present in the sample:

$$1.674 \times 10^{-3}\text{ mol AgCl} \times \frac{1\text{ mol KCl}}{1\text{ mol AgCl}} \times \frac{74.55\text{ g KCl}}{1\text{ mol KCl}} \times \frac{10^3\text{ mg KCl}}{1\text{ g KCl}} = 124.8\text{ mg KCl}$$

This mass is less than the KCl sample mass, indicating that the sample was not pure KCl. Its percent purity is 85.0%:

$$\frac{\text{Mass of KCl}}{\text{Mass of sample}} \times 100\% = \frac{124.8\text{ mg KCl}}{147\text{ mg sample}} \times 100\% = 85.0\%$$

Exercise 5.5 Analysis by Precipitation

Suggest two or more reasons why an analysis for KCl by precipitating AgCl as described could give incorrect results.

PROBLEM-SOLVING EXAMPLE 5.7 Evaluating an Ore

The weight percent of titanium dioxide (TiO_2) in an ore can be evaluated by reacting the ore with bromine trifluoride and measuring the mass of oxygen gas evolved.

$$\mathrm{3\ TiO_2(s) + 4\ BrF_3(\ell) \longrightarrow 3\ TiF_4(s) + 2\ Br_2(\ell) + 3\ O_2(g)}$$

Portrait of a Scientist • *Marion David Francis (1923–)*

There may seem to be little, if any, relationship between toothpaste and chemistry, but the research career of Marion David Francis at Procter & Gamble (1956–1993) was closely tied to them.

Although a great deal of chemical research begins with seeking answers to the question "Why?", Francis instead sought to answer the question "Why not?": Why don't teeth dissolve in the acidic conditions caused by food residues in the mouth? His research discovered that the main substance in tooth enamel, hydroxyapatite, $Ca_5(PO_4)_3OH$, should dissolve under acidic conditions. It is, however, converted on the surface of teeth to a very thin film of acid-insoluble calcium monohydrogen phosphate dihydrate, $CaHPO_4 \cdot 2\,H_2O$.

His further research on the chemistry of tooth enamel and decay led Procter & Gamble to first add stannous fluoride, SnF_2, then later sodium fluoride, NaF, to its Crest toothpaste. In 1960 this innovation earned the rare endorsement of the American Dental Association: "Crest has been shown to be an effective decay-preventative dentifrice that can be of significant value when used in a conscientiously applied program of oral hygiene and professional care." That's a rather long-winded way of saying that fluoridating toothpaste works, as evidenced by its use in Crest and many other commercial toothpaste brands. Fluoridation is effective because fluoride ions replace hydroxide ions in hydroxyapatite to form fluoroapatite, $[Ca_3(PO_4)_2]_3 \cdot CaF_2$, which is more acid-resistant.

Tartar on teeth, an undesirable buildup of an insoluble calcium-based substance, also attracted Francis' attention. This time instead of trying to find something that would prevent dissolving, he sought to discover a substance that would dissolve tartar, preventing it from building up. That substance turned out to be sodium pyrophosphate, $Na_4P_4O_7$, also used in laundry detergents. The pyrophosphate ion, $P_4O_7^{4-}$, blocks calcium tartar deposits by bonding to calcium ions. Sodium pyrophosphate is an active ingredient in some mouthwash products.

In 1996 Marion Francis was awarded the Perkin Medal, a prestigious honor given annually by the American Section of the Society of Chemical Industry. Francis was honored for his outstanding application of chemistry to developing exceptional tartar- and cavity-preventing toothpaste additives. His broad training allowed him to employ fundamental principles from analytical chemistry, physics, and biochemistry to develop these substances that reduce the public health problem of tooth decay.

Marion David Francis *(Courtesy Procter & Gamble)*

If 2.3676 g of a TiO_2-containing ore evolves 0.143 g of O_2, what is the mass percent of TiO_2 in the ore sample?

Answer 15.0% TiO_2

Explanation The mass percent of TiO_2 is

$$\text{Mass percent TiO}_2 = \frac{\text{mass TiO}_2}{\text{mass sample}} \times 100\%$$

The mass of the sample is given, and the mass of TiO_2 can be determined from the known mass of oxygen and the balanced equation.

$$0.143\text{ g O}_2 \times \frac{1\text{ mol O}_2}{32.00\text{ g O}_2} \times \frac{3\text{ mol TiO}_2}{3\text{ mol O}_2} \times \frac{79.88\text{ g TiO}_2}{1\text{ mol TiO}_2} = 0.3569\text{ g TiO}_2$$

The mass percent of TiO_2 is

$$\frac{0.3569\text{ g TiO}_2}{2.376\text{ g sample}} \times 100\% = 15.0\%$$

Titanium dioxide is a valuable commercial product so widely used in paints and pigments that an ore, even with only 15% TiO_2, can be mined profitably.

Problem-Solving Practice 5.7

The purity of magnesium metal can be determined by reacting the metal with excess hydrochloric acid to form $MgCl_2$, evaporating the water from the resulting solution, and weighing the solid $MgCl_2$ formed.

$$Mg(s) + 2\ HCl(aq) \longrightarrow MgCl_2(aq) + H_2(g)$$

Calculate the percentage of magnesium in a 1.72-g sample that produced 6.46 g $MgCl_2$ when reacted with excess HCl.

5.5 PERCENT COMPOSITION AND EMPIRICAL FORMULAS

In Section 3.14, percent composition data were used to derive empirical and molecular formulas, but nothing was mentioned about how such data are obtained. **Combustion analysis** is one way of obtaining percent composition data for those compounds that burn in oxygen. This analysis method is often used for organic compounds, many of which contain carbon and hydrogen. Carbon is converted to carbon dioxide, and hydrogen to water. These combustion products are collected, weighed, and used to calculate the amounts of carbon and hydrogen in the original substance (Figure 5.8). Many carbon and hydrogen compounds also contain oxygen. In such cases, the mass of oxygen in the sample can be determined simply by difference.

$$\text{Mass of oxygen} = \text{mass sample} - (\text{mass C} + \text{mass H})$$

An analytical chemist used combustion analysis to determine the empirical formula of MTBE, a carbon, hydrogen, and oxygen compound used to enhance the octane ratings of another company's gasolines. Combustion of 0.2250 g of pure MTBE produced 0.5616 g of CO_2 and 0.2749 g of H_2O. A different experiment determined that the molar mass of MTBE is 88.15 g/mol.

MTBE, a gasoline additive, is methyl tertiary-butyl ether.

The task here is to determine *a, b,* and *c* in the empirical formula of MTBE, $C_aH_bO_c$. Recall from Chapter 3 that the subscripts in a chemical formula tell how many moles of atoms of each element are in 1 mol of the compound. We begin by determining how many grams of carbon, hydrogen, and oxygen were in the original

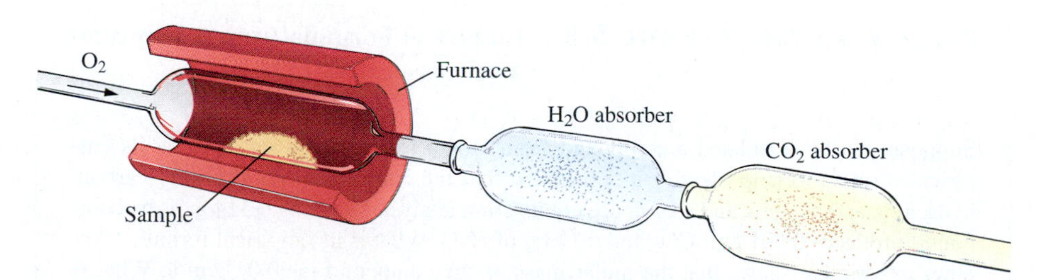

Figure 5.8 Combustion analysis. If a compound containing C and H is burned in oxygen, CO_2 and H_2O are formed, and the mass of each can be determined. The H_2O is absorbed by magnesium perchlorate, and the CO_2 is absorbed by finely divided NaOH supported on asbestos. The mass of each absorbent before and after combustion will give the masses of CO_2 and H_2O. Only a few milligrams of a combustible compound are needed for analysis.

sample. *All* of the carbon in the CO_2 and all of the hydrogen in the H_2O came from the MTBE sample that was burned, and so we can work backwards to assess the composition of MTBE.

First, we calculate the masses of carbon and hydrogen in the original sample.

$$0.5616 \text{ g } CO_2 \times \frac{1 \text{ mol } CO_2}{44.009 \text{ g } CO_2} \times \frac{1 \text{ mol C}}{1 \text{ mol } CO_2} \times \frac{12.011 \text{ g C}}{1 \text{ mol C}}$$

$$= 0.15327 \text{ g C in the original sample burned to } CO_2$$

$$0.2749 \text{ g } H_2O \times \frac{1 \text{ mol } H_2O}{18.015 \text{ g } H_2O} \times \frac{2 \text{ mol H}}{1 \text{ mol } H_2O} \times \frac{1.0079 \text{ g H}}{1 \text{ mol H}}$$

$$= 0.030760 \text{ g H in the original sample burned to } H_2O$$

The amount of oxygen in the original sample can be calculated by difference:

Mass of O in sample =
0.2250 g sample − (0.15327 g C in sample + 0.030760 g H in sample)
= 0.04097 g O in original sample

From the mass data, we can now calculate how many moles of each element were in the sample.

$$0.15327 \text{ g C} \times \frac{1 \text{ mol C}}{12.011 \text{ g C}} = 0.012760 \text{ mol C}$$

$$0.030760 \text{ g H} \times \frac{1 \text{ mol H}}{1.0079 \text{ g H}} = 0.030519 \text{ mol H}$$

$$0.04097 \text{ g O} \times \frac{1 \text{ mol O}}{15.999 \text{ g O}} = 0.0025607 \text{ mol O}$$

Next, we find the mole ratios of the elements in the compound.

$$\frac{0.012760 \text{ mol C}}{0.0025607 \text{ mol O}} = \frac{4.983 \text{ mol C}}{1.000 \text{ mol O}} \quad \text{and} \quad \frac{0.030519 \text{ mol H}}{0.0025607 \text{ mol O}} = \frac{11.92 \text{ mol H}}{1.000 \text{ mol O}}$$

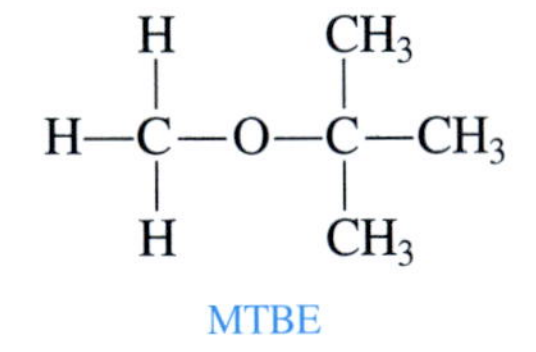

MTBE

The ratios are very close to 5.00 mol C : 12.0 mol H : 1.00 mol O, which gives the empirical formula of MTBE as $C_5H_{12}O$. From this we can calculate an empirical molar mass of 88.15 g. Because the calculated and the experimental molar masses are the same, the molecular formula of MTBE is also $C_5H_{12}O$.

PROBLEM-SOLVING EXAMPLE 5.8 Empirical Formula from Combustion Analysis

Suppose you have isolated a compound from clover leaves and want to know its empirical formula to help identify it. You know that the compound contains only carbon, hydrogen, and oxygen, and so you use combustion analysis. Burning 0.514 g of the compound produces 0.501 g of CO_2 and 0.103 g of H_2O. What is its empirical formula? Another experiment shows that the molar mass of the compound is 90.04 g/mol. What is its molecular formula?

Answer The empirical formula is CHO_2; the molecular formula is $C_2H_2O_4$.

Explanation All of the carbon and hydrogen in the compound are burned to carbon dioxide and water, respectively. Therefore, use the masses of CO_2 and H_2O to find how many moles of C and H, respectively, were in the unknown compound.

$$0.501 \text{ g } CO_2 \times \frac{1 \text{ mol } CO_2}{44.01 \text{ g } CO_2} = 0.01138 \text{ mol } CO_2$$

$$0.103 \text{ g } H_2O \times \frac{1 \text{ mol } H_2O}{18.02 \text{ g } H_2O} = 0.005716 \text{ mol } H_2O$$

The moles of CO_2 and H_2O generated from the unknown compound can now be converted to the masses of C and H that were originally in the sample of the unknown compound that was burned.

$$0.01138 \text{ mol } CO_2 \times \frac{1 \text{ mol C}}{1 \text{ mol } CO_2} \times \frac{12.01 \text{ g C}}{1 \text{ mol C}} = 0.1366 \text{ g C formerly in the sample burned}$$

$$0.005716 \text{ mol } H_2O \times \frac{2 \text{ mol H}}{1 \text{ mol } H_2O} \times \frac{1.008 \text{ g H}}{1 \text{ mol H}} = 0.01152 \text{ g H formerly in the sample burned}$$

These calculations reveal that the 0.514-g sample of the unknown compound contains 0.1366 g C and 0.01152 g H; the remaining mass, 0.3659 g, must be oxygen.

$$0.1366 \text{ g C} + 0.01152 \text{ g H} + 0.3659 \text{ g O} = 0.514 \text{ g sample}$$

Finding the number of moles of each element in the unknown compound reveals its empirical formula (Section 3.14).

$$0.1366 \text{ g C} \times \frac{1 \text{ mol C}}{12.01 \text{ g C}} = 0.01137 \text{ mol C}$$

$$0.01152 \text{ g H} \times \frac{1 \text{ mol H}}{1.008 \text{ g H}} = 0.01142 \text{ mol H}$$

$$0.3659 \text{ g O} \times \frac{1 \text{ mol O}}{16.00 \text{ g O}} = 0.02286 \text{ mol O}$$

To find the mole ratios of the elements, we divide the moles of each element by the *smallest* number of moles.

$$\frac{0.01142 \text{ mol H}}{0.01137 \text{ mol C}} = \frac{1.004 \text{ mol H}}{1.000 \text{ mol C}} \qquad \frac{0.12286 \text{ mol O}}{0.01137 \text{ mol C}} = \frac{2.010 \text{ mol O}}{1.000 \text{ mol C}}$$

The mole ratios indicate that for every C atom in the molecule, there are one H atom and two O atoms. Therefore the *empirical formula* of the unknown compound is CHO_2 with an empirical formula molar mass of 45.02 g/mol.

Finally, to determine the molecular formula, compare the empirical formula molar mass with the experimental molar mass.

$$\frac{90.04 \text{ g/mol unknown compound}}{45.02 \text{ g/mol of } CHO_2} = \frac{2.000}{1.000 \text{ mol unknown compound}}$$

Therefore, the molecular formula of the unknown compound is $C_2H_2O_4$, twice the empirical formula.

If after dividing by the smallest number of moles, the ratios are not whole numbers, multiply each coefficient by a number that converts the fractions to whole numbers. For example, multiplying $NO_{2.5}$ by 2 changes it to N_2O_5.

Problem-Solving Practice 5.8

Phenol is a compound of carbon, hydrogen, and oxygen that is used commonly as a disinfectant. Combustion analysis of a 175-mg sample of phenol yielded 491 mg CO_2 and 46.5 mg H_2O.

(a) Calculate the empirical formula of phenol.
(b) What other information is necessary to determine whether the empirical formula is the actual molecular formula?

(a)

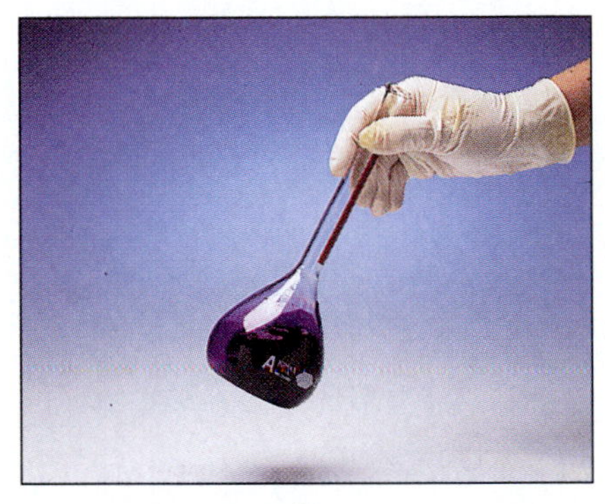

(b)

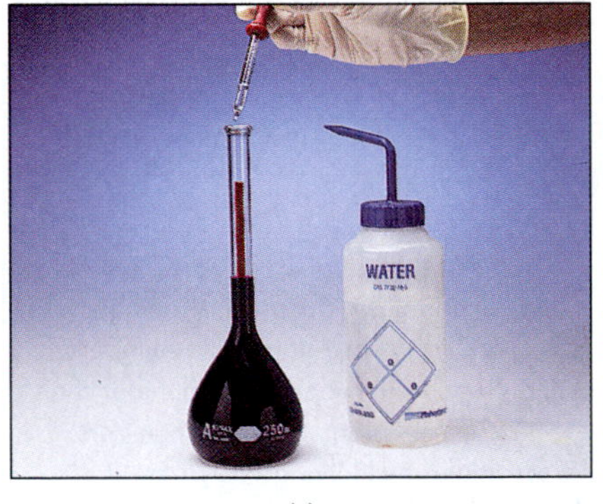

(c)

Figure 5.9 Solution preparation from a solid solute. Making a 0.0100 M aqueous solution of $KMnO_4$. (a) Initially, 1.58 g (0.0100 mol) of solid $KMnO_4$ is placed in a 1.00-L volumetric flask and approximately 950 ml of distilled water is added. (b) The flask is shaken to hasten dissolving of the $KMnO_4$. (c) After the solid dissolves, sufficient water is added to fill the flask to the mark etched in the neck of the volumetric flask indicating a volume of 1.00 L. The flask is shaken again to thoroughly mix its contents. The flask now contains 1.00 L of 0.0100 M $KMnO_4$ solution. *(C.D. Winters)*

Exercise 5.6 Formula From Combustion Analysis

Nicotine, a compound found in cigarettes, contains C, H, and N. Outline a method by which you could use combustion analysis to determine the empirical formula of nicotine.

5.6 A SOLUTION FOR SOLUTIONS

Many of the chemicals in your body or in a plant are dissolved in water; that is, they are in an aqueous solution. Just as a living system uses chemistry in solution, so do chemists, and we need to do our work quantitatively. To accomplish this, we continue to use balanced equations and moles, but we measure volumes of solution rather than masses of solids, liquids, and gases. A solution is a homogeneous mixture of a **solute,** the substance that has been dissolved, and the **solvent,** the substance in which the solute has been dissolved. To know the quantity of solute in a given volume of a liquid solution requires knowing the **concentration** of the solution—the relative amounts of solute and solvent. Molarity, which relates the amount of solute in moles to the solution volume in liters, is the most useful of the many ways of expressing solution concentration for studying chemical reactions in solution.

Molarity

The **molarity** of a solution is defined as moles of solute per liter of solution.

$$\text{Molarity} = \frac{\text{moles of solute}}{\text{liters of solution}}$$

Note that the volume term is liters of *solution, not* liters of solvent (such as water).

If, for example, 40.0 g (1.00 mol) of NaOH is dissolved in sufficient water to produce a solution whose total volume is 1.00 L, the solution has a concentration of 1.00 mol NaOH/1.00 L of solution, which is a 1.00 *molar solution.* The molarity of this solution is reported as 1.00 M, where the capital M stands for moles/liter. Molarity is also represented by square brackets around the formula of a compound or ion, such as [NaOH] or $[OH^-]$. The brackets indicate moles of the compound or ion per liter of solution.

A solution of known molarity can be made by adding the required amount of solute to a volumetric flask, adding some solvent to dissolve all the solute, and then adding sufficient solvent with continual mixing, to fill the flask "to the mark." As shown in Figure 5.9, the etched marking indicates the liquid level equal to the specified volume of the flask.

PROBLEM-SOLVING EXAMPLE 5.9 Molarity

Suppose 0.275 g of K_2CrO_4 is placed into a 500-mL volumetric flask and water is added until the *solution* volume is exactly 500 mL. What is the molarity of K_2CrO_4 in this solution?

Answer 0.00283 M

Explanation To calculate the molarity, we need the solution volume in liters and the moles of solute. In this case the volume is 500 mL × (1 L/1000 mL) = 0.500 L. Use the molar mass of K_2CrO_4 (194.2 g/mol) to obtain moles from grams.

$$0.275 \text{ g } K_2CrO_4 \times \frac{1 \text{ mol } K_2CrO_4}{194.2 \text{ g } K_2CrO_4} = 1.416 \times 10^{-3} \text{ mol } K_2CrO_4$$

We can now calculate the molarity.

$$\text{Molarity of } K_2CrO_4 = \frac{1.416 \times 10^{-3} \text{ mol } K_2CrO_4}{0.500 \text{ L solution}} = 2.83 \times 10^{-3} \text{ mol/L}$$

We could say that the molarity is 0.00283 M, or that $[K_2CrO_4] = 0.00283$ M.

Problem-Solving Practice 5.9

Calculate the molarity of sodium sulfate in a solution that contains 36.0 g of Na_2SO_4 in 750. mL of solution.

Exercise 5.7 Cholesterol Molarity

A blood serum cholesterol level greater than 240 mg of cholesterol per deciliter (0.100 L) of blood generally indicates the need for medical intervention. Calculate this serum cholesterol level in terms of molarity. Cholesterol's formula is $C_{27}H_{46}O$.

Sometimes the molarity of a particular ion in a solution is required, a value that depends on the formula of the solute. For example, potassium chromate is a soluble ionic compound and a strong electrolyte that is completely dissociated in solution to form 2 mol of K^+ ions and 1 mol of CrO_4^{2-} ions for each mole of K_2CrO_4 that dissolves:

$$K_2CrO_4(aq) \longrightarrow 2\ K^+(aq) + CrO_4^{2-}(aq)$$

1 mol   2 mol   1 mol

100% dissociation

The K^+ concentration is twice the K_2CrO_4 concentration because each mole of K_2CrO_4 contains 2 mol of K^+. Therefore, the 0.00284 M K_2CrO_4 has a K^+ concentration of 2×0.00283 M = 0.00566 M, and a CrO_4^{2-} concentration of 0.00283 M.

Exercise 5.8 Molarity

A student dissolves 6.37 g of aluminum nitrate in sufficient water to make 250 mL of solution.

(a) Calculate the molarity of aluminum nitrate in this solution.
(b) Calculate the molar concentration of aluminum ions and of nitrate ions in this solution.

Exercise 5.9 Molarity

A student is in a hurry to prepare a 0.100 M solution and adds 1.0 L of water to 0.100 mol of solute. He is annoyed to learn that the solute concentration is not 0.100 M. What was his error?

Preparing a Solution of Known Molarity by Diluting a More Concentrated One

Consider two cases: A teaspoon of sugar ($C_{12}H_{22}O_{11}$) is dissolved in a glass of water and then the glass of sugar solution is poured into a swimming pool full of water. The swimming pool and the glass contain the same number of moles of sugar, but the concentration of sugar in the swimming pool is far less because the volume of water in the pool is much greater than that in the glass.

Frequently a solution needs to be available in a wide variety of molarities such as hydrochloric acid, for example, that is 6 M, 1.0 M, and 0.050 M. To make these solutions, chemists use a concentrated solution of known molarity and dilute samples of it with water to make solutions of lesser molarity. **Diluting a solution does *not* change the number of moles of solute in the sample; it does increase the volume, which lowers the molarity.** ***Therefore, the moles of solute in the dilute solution must be the same as the number of moles of solute in the sample of the more concentrated solution.***

For example, there is 0.15 mol of HCl in 250 mL (0.250 L) of a 0.600 M HCl solution.

$$0.250 \text{ L} \times \frac{0.600 \text{ mol HCl}}{\text{L}} = 0.150 \text{ mol HCl}$$

If the 0.600 M solution were diluted to 400 mL, the HCl concentration would decrease to 0.375 M, but the number of moles of HCl remains unchanged—0.150 mol HCl.

$$0.400 \text{ L} \times \frac{0.375 \text{ mol HCl}}{\text{L}} = 0.150 \text{ mol HCl}$$

A quick and useful check on a dilution calculation is to make certain that the molarity of the diluted solution has not become larger than for the original solution.

Consequently, the moles in each case are equal and a simple relationship applies:

$$\text{Molarity (conc)} \times \text{V (conc)} = \text{Molarity (dil)} \times \text{V (dil)}$$

where Molarity (conc) and V (conc) represent the molarity and the volume (in liters) of the concentrated solution; Molarity (dil) and V (dil) represent the molarity and volume of the dilute solution. *Multiplying a volume in liters by a solute's molarity (moles/liter) yields moles of solute.*

We can calculate, for example, the concentration of a hydrochloric acid solution made by diluting 25.0 mL of 6.0 M HCl to 500 mL. In this case, we want to determine Molarity (dil); Molarity (conc) = 6.0 M, V (conc) = 0.0250 L, and V (dil) = 0.500 L. We algebraically rearrange the relationship to get the concentration of the diluted HCl.

$$\text{Molarity (dil)} = \frac{\text{Molarity (conc)} \times \text{V (conc)}}{\text{V (dil)}} = \frac{\frac{6.0 \text{ mol}}{\text{L}} \times 0.0250 \text{ L}}{0.500 \text{ L}} = 0.30 \text{ mol/L}$$

A diluted solution will always be less concentrated (lower molarity) than the more concentrated stock solution (Figure 5.10).

PROBLEM-SOLVING EXAMPLE 5.10 Solution Concentration and Dilution

How can 400 mL of 2.00 M HNO_3 be prepared from concentrated nitric acid, 16.0 M HNO_3?

Answer Add 50.0 mL of the concentrated acid to enough water to make up a total volume of 400 mL of solution.

Explanation This is a dilution problem in which the concentrations of the concentrated (16.0 M) and less concentrated (2.00 M) solutions are given, as well as the volume of the diluted solution (0.400 L). The volume of the concentrated nitric acid

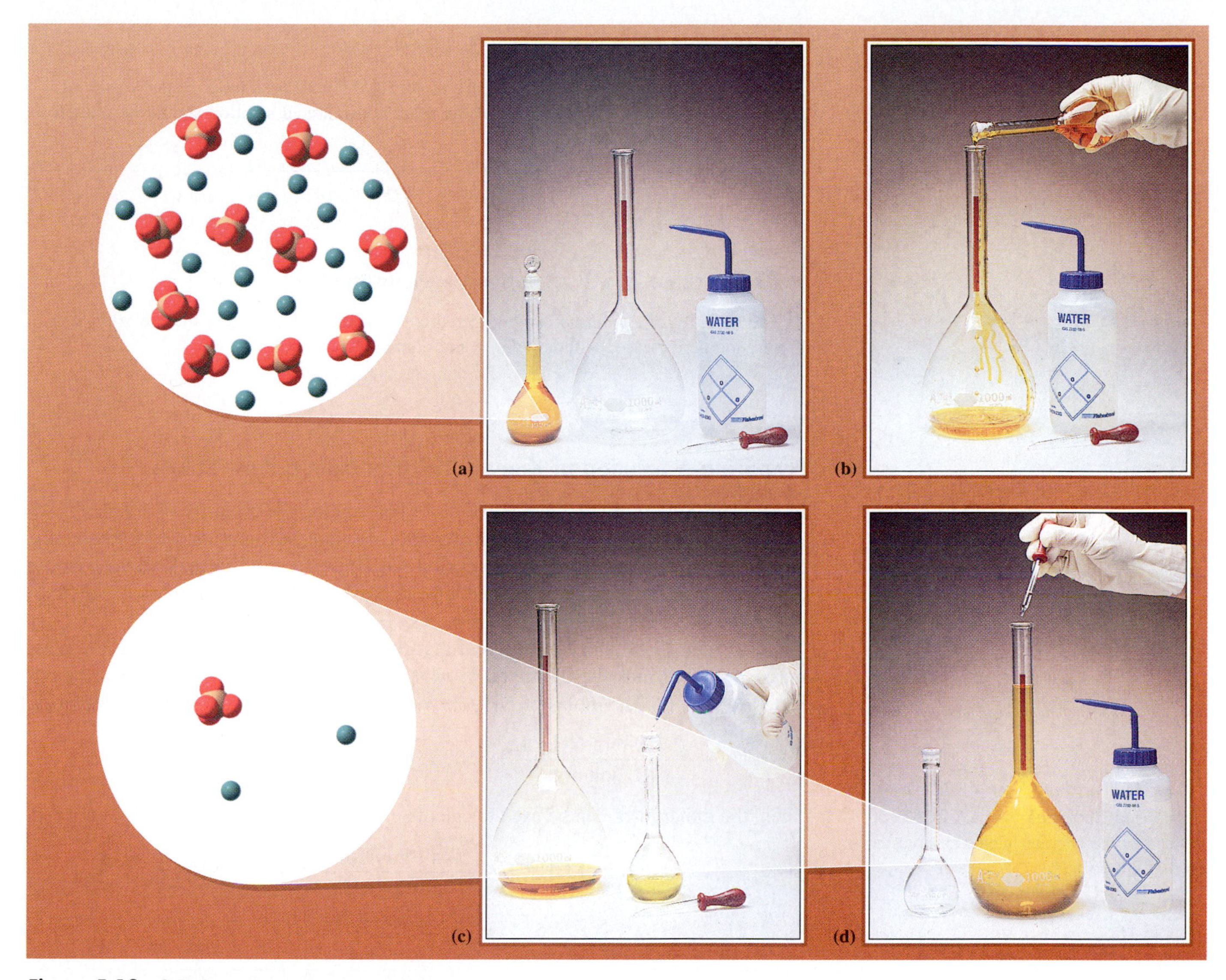

Figure 5.10 Solution preparation by dilution. (a) A 100.0-mL volumetric flask has been filled to the mark to make a 0.100 M $K_2Cr_2O_7$ solution. (b) This is transferred to a 1.000-L volumetric flask. (c) All of the initial solution is rinsed out of the 100.0-mL flask. (d) The 1.000-L flask is then filled with distilled water to the mark on the neck, and shaken thoroughly. The concentration of the now-diluted solution is 0.0100 M. *(C.D. Winters)*

to be diluted, V (conc), is needed, and can be calculated from the relationship Molarity (conc) × V (conc) = Molarity (dil) × V (dil).

$$\text{V (conc)} = \frac{\text{Molarity (dil)} \times \text{V (dil)}}{\text{Molarity (conc)}} = \frac{\frac{2.00 \text{ mol}}{\text{L}} \times 0.400 \text{ L}}{16.0 \text{ mol/L}} = 0.0500 \text{ L} = 50.0 \text{ mL}$$

Therefore, 50.0 mL of concentrated nitric acid are added slowly, with stirring, to about 300 mL of pure water. When the solution has cooled to room temperature, sufficient water is added to bring the final volume to 400 mL, resulting in a 2.00 M nitric acid solution.

Caution must be used when diluting a concentrated acid. The more concentrated acid should be added slowly to the solvent (water) so that the heat generated during the dilution is slowly dissipated. If water is added to the acid, the heat generated by the dissolving could be sufficient to vaporize the solution, spraying the acid over you and those in the area. *(C.D. Winters)*

Problem-Solving Practice 5.10

A laboratory procedure calls for 50 mL of 0.150 M NaOH. You have available 100 mL of 0.500 M NaOH. What volume of the more concentrated solution should be diluted to make the desired solution?

Exercise 5.10 Solution Concentration

The molarity of a solution can be decreased by dilution. How could the molarity of a solution be increased without adding additional solute?

Preparing a Solution of Known Molarity from a Pure Solute

In Problem-Solving Example 5.9, we described finding the molarity of a K_2CrO_4 solution that was prepared from known amounts of solute and solution. More frequently, a solid or liquid solute (sometimes even a gas) must be used to make up a solution of known molarity. The problem becomes one of calculating what mass of solute to use to provide the proper number of moles.

Consider a laboratory experiment that requires 2.00 L of 0.750 M NH_4Cl. What mass of NH_4Cl must be dissolved in water so that the solution volume can be adjusted to 2.00 L? The moles of NH_4Cl required can be calculated from the molarity:

$$\frac{0.750 \text{ mol } NH_4Cl}{1 \text{ L solution}} \times 2.00 \text{ L solution} = 1.500 \text{ mol } NH_4Cl$$

Conceptual Challenge Problem CP-5.B at the end of the chapter relates to topics covered in this section.

Then, the molar mass can be used to calculate the number of grams of NH_4Cl needed.

$$1.500 \text{ mol } NH_4Cl \times \frac{53.49 \text{ g } NH_4Cl}{1 \text{ mol } NH_4Cl} = 80.2 \text{ g } NH_4Cl$$

PROBLEM-SOLVING EXAMPLE 5.11 Solute Mass and Molarity

How is 250 mL of 0.0150 M $Ce(SO_4)_2$ solution prepared from solid $Ce(SO_4)_2$?

Answer Dissolve 1.25 g of solid $Ce(SO_4)_2$ in water and add enough water to make 250 mL of solution.

Explanation First find the number of moles of the solute, $Ce(SO_4)_2$, in 250 mL (0.250 L) of 0.0150 M $Ce(SO_4)_2$ solution. From this calculate the number of grams.

$$0.250 \text{ L solution} \times \frac{0.0150 \text{ mol } Ce(SO_4)_2}{1 \text{ L solution}} = 3.750 \times 10^{-3} \text{ mol } Ce(SO_4)_2$$

$$3.750 \times 10^{-3} \text{ mol } Ce(SO_4)_2 \times \frac{332.2 \text{ g } Ce(SO_4)_2}{1 \text{ mol } Ce(SO_4)_2} = 1.25 \text{ g } Ce(SO_4)_2$$

The solution is prepared by putting 1.25 g of $Ce(SO_4)_2$ into a container and adding distilled water until the solution volume is 250 mL, resulting in a 0.0150 M $Ce(SO_4)_2$ solution.

Problem-Solving Practice 5.11

Describe how you would prepare

(a) 1.0 L of 0.125 M Na_2CO_3 from solid Na_2CO_3.
(b) 100 mL of 0.0500 M Na_2CO_3 from a 0.125 M Na_2CO_3 solution.
(c) 500 mL of 0.0215 M $KMnO_4$ from solid $KMnO_4$.
(d) 250 mL of 0.00450 M $KMnO_4$ from 0.0215 M $KMnO_4$.

5.7 MOLARITY AND REACTIONS IN AQUEOUS SOLUTIONS

Many kinds of reactions—acid-base *(← p. 152)*, precipitation *(← p. 148)*, redox *(← p. 158)*—occur in aqueous solutions. In such reactions, molarity is the concentration unit of choice because it quantitatively relates a volume of one reactant and the moles of reactant contained in solution to the volume and corresponding moles of another reactant or a product in solution. This allows us to make conversions between volumes of solutions and moles of reactants and products as given by the stoichiometric coefficients. Molarity is used to link mass, moles, and volume of solution (Figure 5.11).

Conceptual Challenge Problem CP-5.C at the end of the chapter relates to topics covered in this section.

PROBLEM-SOLVING EXAMPLE 5.12 Solution Reaction Stoichiometry

A major industrial use of hydrochloric acid is for "pickling," the removal of rust from metals, especially steel (largely iron) before it is used to fabricate steel products. The rust is removed by dipping the steel into very large baths of hydrochloric acid. The acid reacts in an exchange reaction with rust, which is essentially Fe_2O_3, leaving behind a clean steel surface.

$$Fe_2O_3(s) + 6\ HCl(aq) \longrightarrow 2\ FeCl_3(aq) + 3\ H_2O(\ell)$$

Once the rust is taken off, the steel is removed from the acid bath and rinsed before the acid reacts significantly with the iron in the steel.

How many pounds of rust can be removed when rust-covered steel reacts with 800 L of 12.0 M HCl? Assume that only the rust reacts with the HCl (1.000 lb = 454.6 g).

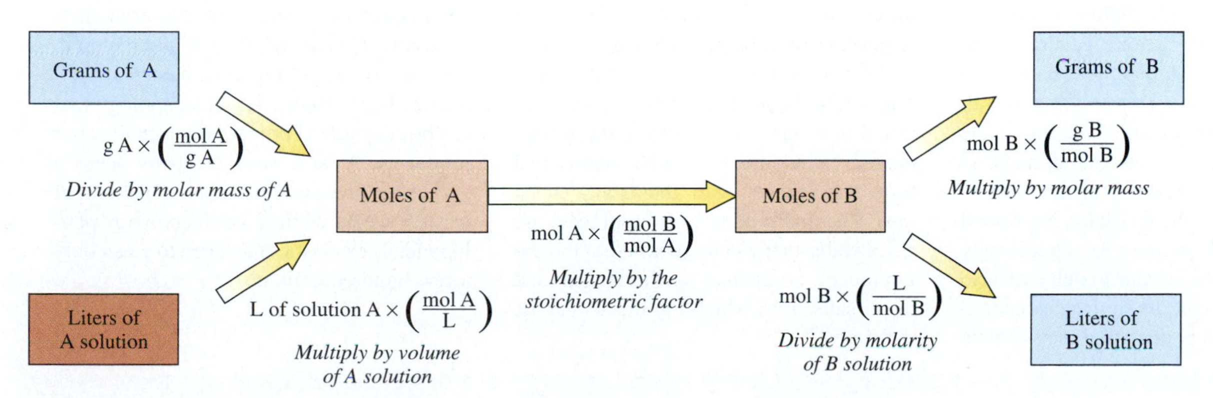

Figure 5.11 Stoichiometric relationships for a chemical reaction in aqueous solution. Molarity and volume provide the connection between moles of a reactant or a product and moles of another reactant or product.

(Text continues on page 205.)

CHEMISTRY IN THE NEWS

Solving a Problem in Industrial Chemistry

Many of the reactions you have seen in Chapter 4 and in this chapter involve acids. One reason for the emphasis on acid chemistry is that acids are some of the most important chemicals in our economy. Five acids are listed among the top 50 chemicals produced in the United States in 1995.

Acid	Billions of Pounds	Rank
Sulfuric acid	95.4	1
Phosphoric acid	26.2	7
Nitric acid	17.2	14
Hydrochloric acid	7.3	27
Acetic acid	4.7	33

A transportable acid recovery pilot-plant. Used acids that contain dissolved metal ions are transformed in this equipment into reusable acid, reclaimable metal salts, and clean water. *(The WADR pilot-plant was developed for the Department of Energy by Pacific Northwest Laboratory. Courtesy of Viatec/Recovery Systems, Inc., Richland, WA.)*

Over 15,000 companies use these acids to make other chemicals, to clean and refinish metals, to plate metals onto other metals or onto plastics, and in many other applications. A problem faced by all of these industries is what to do with acid-containing waste. For example, when acids are used to wash a metal surface, the washings contain unused acid along with ions of metals such as copper(II), vanadium(II), silver(I), nickel(II), and lead(II). It is estimated that over eight billion pounds of acid-containing wastes are generated annually, and they cannot simply be flushed into the nearest lake or river. Not only will the acid damage aquatic life, but many metals are toxic to plants and animals.

A process developed at the Department of Energy's (DOE) Pacific Northwest Laboratory holds promise for significantly reducing the volume and toxicity of acid waste. Typically the waste from a chemical or metallurgical operation may contain sulfuric, phosphoric, and nitric acids, along with metal ions dissolved in the aqueous acid. In the DOE process, the mixture is heated, the acids vaporize, and the metal ions remain in the liquid phase. The vapor is purified to yield a clean acid, in some cases cleaner than industrial-grade acids. The solution containing metal ions is collected in a tank, and the metal ions are removed by adding salts with anions that form precipitates with the heavy metal ions. The end products of the process are clean water that can be returned to the environment, acids that can be reused, and metal salts from which the metals can be reclaimed.

The process has received several awards for outstanding new technology. It has moved out of the laboratory and into the next stage of development in which it is being adapted for a variety of industrial environments. Savings of $1 to $5 per gallon in the cost of processing acid wastes are expected. Furthermore, it is a technology that can be adapted to both large and small waste producers. A large steel company needs a system to process thousands of gallons of waste a week, while a small company plating metals or plastic may need to clean only a few hundred gallons.

Answer 562 lb of Fe_2O_3

Explanation First, calculate the number of moles of HCl available in 800 L:

$$800 \text{ L HCl} \times \frac{12.0 \text{ mol HCl}}{1 \text{ L solution}} = 9.600 \times 10^3 \text{ mol HCl}$$

The mass of Fe_2O_3 can be determined using the remaining steps.

$$9.600 \times 10^3 \text{ mol HCl} \times \frac{1 \text{ mol } Fe_2O_3}{6 \text{ mol HCl}} \times \frac{159.7 \text{ g } Fe_2O_3}{1 \text{ mol } Fe_2O_3} \times \frac{1 \text{ lb } Fe_2O_3}{454.6 \text{ g } Fe_2O_3} = 562 \text{ lb } Fe_2O_3$$

The solution remaining in the acid bath presents a real disposal challenge because of the metal ions it contains. This is discussed in Chemistry in the News: Solving a Problem in Industrial Chemistry.

Problem-Solving Practice 5.12

In 1995, 1.2×10^{10} kg of sodium hydroxide (NaOH) was produced in the United States by passing an electric current through brine, an aqueous solution of sodium chloride.

$$2 \text{ NaCl(aq)} + 2 \text{ } H_2O(\ell) \longrightarrow 2 \text{ NaOH(aq)} + Cl_2(g) + H_2(g)$$

What volume of brine is needed to produce this mass of NaOH? (The density of brine is 1.2 g/mL; 1.0 L of brine contains 360 g of dissolved NaCl.)

Another application of solution stoichiometry occurs in photography. When silver bromide in black-and-white photographic film is exposed to light, silver ions in the silver bromide are reduced to metallic silver. When the photograph is developed, this creates black regions on the negative. If left on the film, the unreacted silver bromide would ruin the picture because it would darken when exposed to light. AgBr(s) is dissolved away from the film with a solution of the "fixer," sodium thiosulfate ($Na_2S_2O_3$). Aqueous thiosulfate ions ($S_2O_3^{2-}$) combine with silver ions from AgBr to form a soluble product (Figure 5.12). The net ionic equation for this displacement reaction is

$$\text{AgBr(s)} + 2 \text{ } S_2O_3^{2-} \longrightarrow Ag(S_2O_3)_2^{3-}(aq) + Br^-(aq)$$

Sodium ions are spectator ions in this reaction.

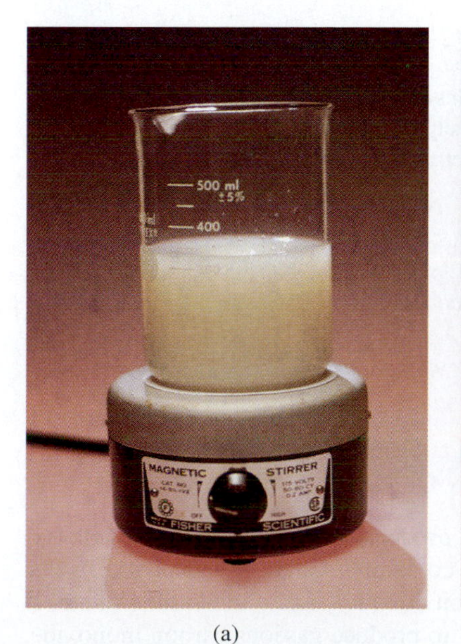

(a)

(b)

Figure 5.12 Dissolving silver bromide with thiosulfate. (a) A precipitate of AgBr formed by adding $AgNO_3(aq)$ to KBr(aq). (b) On adding sodium thiosulfate, $Na_2S_2O_3(aq)$, the solid AgBr dissolves. *(C.D. Winters)*

If you need to dissolve 50.0 mg (0.0500 g) of AgBr, how many milliliters of 0.0150 M $Na_2S_2O_3$ should you use? Use the diagram below and follow the appropriate path (shown in color). The first step is to find the moles of $Na_2S_2O_3$ needed, considering that 2 mol $S_2O_3^{2-}$ is required for 1 mol AgBr and that 1 mol of $Na_2S_2O_3$ contains 1 mol of $S_2O_3^{2-}$ ions.

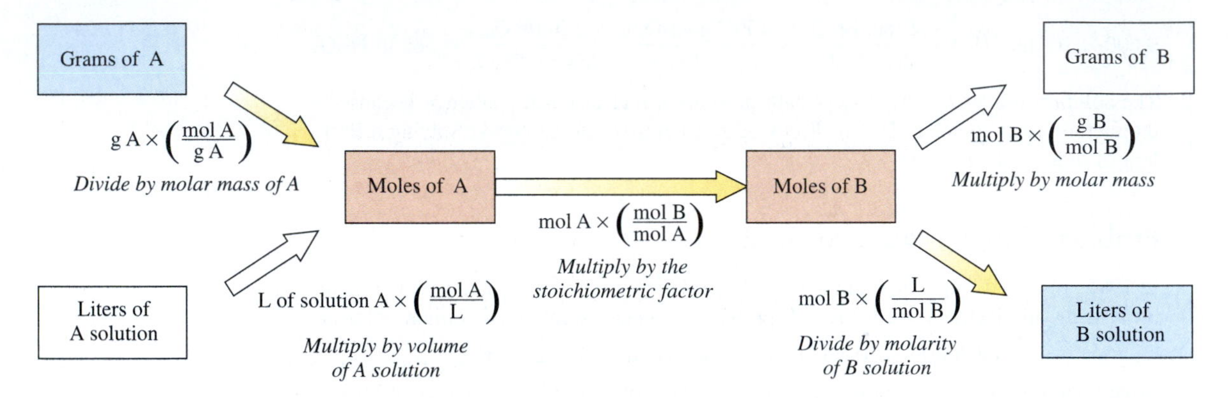

Moles of $Na_2S_2O_3$ needed:

$$0.0500 \text{ g AgBr} \times \frac{1 \text{ mol AgBr}}{187.8 \text{ g AgBr}} \times \frac{2 \text{ mol } S_2O_3^{2-}}{1 \text{ mol AgBr}} \times \frac{1 \text{ mol } Na_2S_2O_3}{1 \text{ mol } S_2O_3^{2-}}$$

$$= 5.324 \times 10^{-4} \text{ mol } Na_2S_2O_3$$

A mole of $Na_2S_2O_3$ dissociates into 2 mol of sodium ions and 1 mol of thiosulfate ions. Thus, the 0.0150 mol of $Na_2S_2O_3$ in 1 L of 0.150 M $Na_2S_2O_3$ solution dissociates into 0.0300 mol Na^+ and 0.0150 mol $S_2O_3^{2-}$ ions.

Conceptual Challenge Problem CP-5.D at the end of the chapter relates to topics covered in this section.

The volume of 0.0150 M $Na_2S_2O_3$ required is obtained using the molarity of the solution.

$$5.324 \times 10^{-4} \text{ mol } Na_2S_2O_3 \times \frac{1 \text{ L solution}}{0.0150 \text{ mol } Na_2S_2O_3}$$

$$= 0.0355 \text{ L of } 0.0150 \text{ M } Na_2S_2O_3 \text{ solution or } 35.5 \text{ mL}$$

Exercise 5.11 Molarity

Sodium chloride is used in intravenous solutions for medical applications. The NaCl concentration in such solutions must be accurately known, and can be assessed by reacting the solution with an experimentally determined amount of $AgNO_3$ solution of known concentration. The net ionic equation is

$$Ag^+(aq) + Cl^-(aq) \longrightarrow AgCl(s)$$

Suppose that a chemical technician uses 19.3 mL of 0.200 M $AgNO_3$ to convert to AgCl all the NaCl in a 25.0-mL sample of an intravenous solution. Calculate the molarity of NaCl in the solution.

Summary Problem

One of the ways phosphoric acid (H_3PO_4) is produced industrially begins with extracting elemental phosphorus from minerals containing calcium phosphate. In this process, a mixture of calcium phosphate, silicon dioxide (sand), and carbon (coke) is heated in an electric furnace to 1400–1500 °C to produce gaseous carbon monoxide, phosphorus vapor (P_4), and a compound of calcium, silicon, and oxygen. This compound has a percent composition by mass of 34.50% calcium, 24.18% silicon, and 41.32% oxygen.

The phosphorus is condensed, purified, and reacted with oxygen in air to produce tetraphosphorus decaoxide, which reacts with water to produce H_3PO_4.

Phosphoric acid is also produced by the so-called "wet process" in which calcium phosphate reacts with sulfuric acid and water to form calcium sulfate dihydrate and phosphoric acid.

(a) Use the mass percent composition data to determine the formula of the calcium, silicon, and oxygen compound formed in the extraction of phosphorus from calcium phosphate.
(b) Write balanced equations for the formation of
  (i) Elemental phosphorus from coke, sand, and calcium phosphate.
  (ii) Tetraphosphorus decaoxide from phosphorus and oxygen.
  (iii) Phosphoric acid from tetraphosphorus decaoxide and water.
  (iv) Phosphoric acid by the wet process.
(c) Classify reactions ii through iv in part (b) according to the reaction types described in Chapter 4.
(d) If the three steps in the formation of H_3PO_4 starting from coke, sand, and calcium phosphate have percent yields of 70.%, 85.%, and 95.%, respectively, what mass of H_3PO_4 can be formed from 5.00×10^5 g of calcium phosphate? Assume an excess of the other reactants.
(e) How many liters of 14.7 M phosphoric acid could be formed from this mass of phosphoric acid?
(f) Suppose that 5×10^4 g of calcium phosphate and 2.0×10^4 g of H_2SO_4 were used to make phosphoric acid by the wet process.
  (i) Identify the limiting reactant.
  (ii) Calculate the mass of phosphoric acid that could be produced from this reaction mixture, assuming a 90% yield.
(g) Predict the products of the following reactions:
  (i) $H_3PO_4(aq) + 3\ NaOH(aq) \rightarrow$
  (ii) $P_4(s) + 3\ O_2(g) \rightarrow$
  (iii) The exchange reaction of hydrochloric acid with calcium phosphide
(h) The first baking powder was a mixture of calcium dihydrogen phosphate and sodium hydrogen carbonate, which react during baking to form calcium monohydrogen phosphate, sodium dihydrogen phosphate, and two other products.
  (i) Identify the two other products.
  (ii) Write the balanced chemical equation for the reaction.

In Closing

Having studied this chapter, you should be able to . . .

- Use stoichiometric factors to calculate the number of moles or number of grams of one reactant or product from the number of moles or number of grams of another reactant or product by using the balanced chemical equation (Section 5.1).
- determine which of two reactants is the limiting reactant (Section 5.2).
- explain the differences among actual yield, theoretical yield, and percent yield; and calculate theoretical and percent yields (Section 5.3).

- use stoichiometry principles in the chemical analysis of a mixture (Section 5.4) or to find the empirical formula of an unknown compound using combustion analysis (Section 5.5).
- define molarity and calculate concentrations (Section 5.6).
- describe how to prepare a solution of a given molarity from the solute and water or by dilution of a more concentrated solution (Section 5.6).
- solve stoichiometry problems using solution molarities (Section 5.7).

KEY TERMS

The following terms were defined and given in boldface type in this chapter. You should be sure to understand each of these terms and the concepts with which they are associated.

actual yield *(5.3)*
combustion analysis *(5.5)*
concentration *(5.6)*
limiting reactant *(5.2)*
molarity *(5.6)*
mole ratio *(5.2)*
percent yield *(5.3)*
solute *(5.6)*
solvent *(5.6)*
stoichiometric factor *(5.1)*
theoretical yield *(5.3)*

QUESTIONS FOR REVIEW AND THOUGHT

Conceptual Challenge Problems

CP-5.A. In Example 5.3 it was not possible to find the mass of O_2 directly from a knowledge of the mass of glucose. Are there chemical reactions in which the mass of a product or another reactant can be known directly from a knowledge of the mass of a reactant? Can you cite a couple of these reactions?

CP-5.B. How would you prepare 1 L of 1.00×10^{-6} M NaCl (molar mass = 58.44 g/mol) solution by using a balance that can measure mass only to 0.01 g?

CP-5.C. How could you show that when baking soda reacts with the acetic acid, CH_3COOH, in vinegar, the carbon and oxygen atoms in the carbon dioxide produced all come from the baking soda alone and none from the acetic acid in vinegar?

CP-5.D. A 0.250-g sample of potassium chloride, KCl, was analyzed for its purity by dissolving the sample in 50 mL of water and adding 100 mL of 0.10 M $AgNO_3$ to it. The mass of AgCl precipitated from this mixture was measured to be 0.492 g. What explanation can you propose to reconcile these laboratory data?

Review Questions

1. This chapter introduced the quantitative aspects of chemical reactions. Which of these questions address quantitative issues?
   (a) Is this chemical process beneficial to society?
   (b) Does the yield justify the production costs?
   (c) How pure is the product?
   (d) Does the chemical reaction pollute the environment?
   (e) Will the reaction occur?
   (f) How much of the reactants are wasted in the synthesis?
2. Complete the table for the reaction

$$3\ H_2(g) + N_2(g) \longrightarrow 2\ NH_3(g)$$

H_2	N_2	NH_3
?? mol	1 mol	?? mol
3 molecules	?? molecules	?? molecules
?? g	?? g	34.08 g

3. What is meant by the statement, "the reactants were present in stoichiometric amounts"?
4. Write all the possible stoichiometric factors for the reaction

$$3\ MgO(s) + 2\ Fe(s) \longrightarrow Fe_2O_3(s) + 3\ Mg(s)$$

5. If a 10.0-g mass of carbon is combined with an exact stoichiometric amount of oxygen (26.6 g) to make carbon dioxide, what mass in grams of CO_2 can be isolated?
6. Given the reaction

$$2\ Fe(s) + 3\ Cl_2(g) \longrightarrow 2\ FeCl_3(s)$$

fill in the missing conversion factors for the scheme

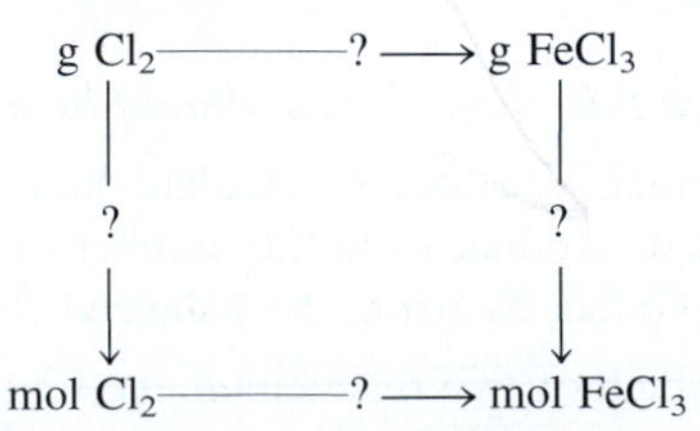

7. If you are making cheeseburgers with 2 slices of cheese, 1 hamburger patty, and 1 bun, how many cheeseburgers can you make

from 8 buns, $\frac{1}{2}$ dozen patties, and 20 slices of cheese? What is the limiting "reactant"? What "reactants" are in excess?

8. When an exam question asks, "What is the limiting reactant?", students often guess the reactant with the smallest mass. Why is this not a good strategy?
9. Why can't the product of a reaction ever be the limiting reactant?
10. A cookie recipe states that it makes a batch of 5 doz cookies; however, you only get 4 doz per batch. What are your actual and theoretical yields? What is your percent yield?
11. Does the limiting reactant determine the theoretical yield, actual yield, or both? Why?
12. A laboratory procedure for making a 0.5 M NaCl solution involves dissolving 29.22 g of NaCl in enough water to make 1 L of solution. What are the solute, solvent, concentration, and molarity in this statement?

The Mole and Chemical Reactions

13. Chlorine can be produced in the laboratory by the reaction of hydrochloric acid with manganese(IV) oxide.

$$4\ HCl(aq) + MnO_2(s) \longrightarrow Cl_2(g) + 2\ H_2O(\ell) + MnCl_2(aq)$$

How many moles of HCl are needed to form 12.5 mol of Cl_2?

14. Methane, CH_4, is the major component of natural gas. How many moles of oxygen are needed to burn 16.5 mol of CH_4?

$$CH_4(g) + 2\ O_2(g) \longrightarrow CO_2(g) + 2\ H_2O(\ell)$$

15. Nitrogen monoxide is oxidized in air to give brown nitrogen dioxide.

$$2\ NO(g) + O_2(g) \longrightarrow 2\ NO_2(g)$$

Starting with 2.2 mol of NO, how many moles and how many grams of O_2 are required for complete reaction? What mass of NO_2, in grams, is produced?

16. Aluminum reacts with oxygen to give aluminum oxide.

$$4\ Al(s) + 3\ O_2(g) \longrightarrow 2\ Al_2O_3(s)$$

If you have 6.0 mol of Al, how many moles and how many grams of O_2 are needed for complete reaction? What mass of Al_2O_3, in grams, is produced?

17. Many metals react with halogens to give metal halides. For example, iron gives iron(II) chloride, $FeCl_2$.

$$Fe(s) + Cl_2(g) \longrightarrow FeCl_2(s)$$

Beginning with 10.0 g of iron, what mass of Cl_2, in grams, is required for complete reaction? What quantity of $FeCl_2$, in moles and in grams, is expected?

18. Like many metals, manganese reacts with a halogen to give a metal halide.

$$2\ Mn(s) + 3\ F_2(g) \longrightarrow 2\ MnF_3(s)$$

(a) If you begin with 5.12 g of Mn, what mass in grams of F_2 is required for complete reaction? (b) What amount in moles and in grams of the red solid MnF_3 is expected?

19. The final step in the manufacture of platinum metal (for use in automotive catalytic converters and other products) is the reaction

$$3\ (NH_4)_2PtCl_6(s) \longrightarrow 3\ Pt(s) + 2\ NH_4Cl(s) + 2\ N_2(g) + 16\ HCl(g)$$

Complete this table of reaction quantities for the reaction of 12.35 g of $(NH_4)_2PtCl_6$.

$(NH_4)_2PtCl_6$	Pt	HCl
12.35 g	?? g	?? g
?? mol	?? mol	?? mol

20. Disulfur dichloride, S_2Cl_2, is used to vulcanize rubber. It can be made by treating molten sulfur with gaseous chlorine.

$$S_8(\ell) + 4\ Cl_2(g) \longrightarrow 4\ S_2Cl_2(g)$$

Complete this table of reaction quantities for the production of 103.5 g of S_2Cl_2.

S_8	Cl_2	S_2Cl_2
?? g	?? g	103.5 g
?? mol	?? mol	?? mol

21. Many metal halides react with water to produce the metal oxide (or hydroxide) and the appropriate hydrogen halide. For example,

$$TiCl_4(\ell) + 2\ H_2O(g) \longrightarrow TiO_2(s) + 4\ HCl(g)$$

(a) If you begin with 14.0 g of $TiCl_4$, how many moles of water are required for complete reaction? (b) How many grams of each product are expected?

Titanium tetrachloride ($TiCl_4$), a liquid. When exposed to air it forms a dense fog of titanium(IV) oxide, TiO_2. *(C.D. Winters)*

22. Gaseous sulfur dioxide, SO_2, can be removed from smokestacks by treatment with limestone and oxygen.

$$2\ SO_2(g) + 2\ CaCO_3(s) + O_2(g) \longrightarrow 2\ CaSO_4(s) + 2\ CO_2(g)$$

(a) How many moles each of $CaCO_3$ and O_2 are required to remove 150. g of SO_2? (b) What mass of $CaSO_4$ is formed when 150 g of SO_2 is consumed completely?

23. If you want to synthesize 1.45 g of the semiconducting material GaAs, what masses of Ga and of As, in grams, are required?

24. Ammonium nitrate, NH_4NO_3, is a common fertilizer and explosive. It was used in the Oklahoma City bombing. When heated, it decomposes into gaseous products

$$2\ NH_4NO_3(s) \longrightarrow 2\ N_2(g) + 4\ H_2O(g) + O_2(g)$$

How many grams of each product are formed from 1.0 kg of NH_4NO_3?

25. Iron reacts with oxygen to give iron(III) oxide, Fe_2O_3. (a) Write a balanced equation for this reaction. (b) If an ordinary iron nail (assumed to be pure iron) has a mass of 5.58 g, what mass in grams of Fe_2O_3 would be produced if the nail is converted completely to this oxide? (c) What mass of O_2 (in grams) is required for the reaction?

26. Freons, such as CCl_2F_2, have been banned from use in automobile air conditioners because the Freons are destroying the ozone layer. Researchers at MIT have found an environmentally safe way to decompose Freons by treating them with sodium oxalate, $Na_2C_2O_4$. The products of the reaction are carbon, carbon dioxide, sodium chloride, sodium fluoride. (a) Write a balanced equation for this reaction. (b) What mass of $Na_2C_2O_4$ is needed to remove 76.8 g of CCl_2F_2? (c) What mass of CO_2 is produced?

27. Careful decomposition of ammonium nitrate, NH_4NO_3, gives laughing gas (dinitrogen monoxide, N_2O) and water. (a) Write a balanced equation for this reaction. (b) Beginning with 10.0 g of NH_4NO_3, what masses of N_2O and water are expected?

28. Cisplatin, $Pt(NH_3)_2Cl_2$, can be made by the reaction of K_2PtCl_4 with ammonia, NH_3. Besides cisplatin, the other product is KCl. (a) Write a balanced equation for this reaction. (b) In order to obtain 2.50 g of cisplatin, what masses in grams of K_2PtCl_4 and ammonia do you need?

Limiting Reactant

29. Aluminum chloride, $AlCl_3$, is an inexpensive reagent used in many industrial processes. It is made by treating scrap aluminum with chlorine according to the balanced equation

$$2\ Al(s) + 3\ Cl_2(g) \longrightarrow 2\ AlCl_3(s)$$

(a) Which reactant is limiting if 2.70 g of Al and 4.05 g of Cl_2 are mixed? (b) What mass of $AlCl_3$ can be produced? (c) What mass of the excess reactant will remain when the reaction is completed?

30. Methanol, CH_3OH, is a clean-burning, easily handled fuel. It can be made by the direct reaction of CO and H_2 (obtained from heating coal with steam).

$$CO(g) + 2\ H_2(g) \longrightarrow CH_3OH(\ell)$$

(a) Starting with a mixture of 12.0 g of H_2 and 74.5 g of CO, which is the limiting reactant? (b) What mass of the excess reactant, in grams, is left after reaction is complete? (c) What mass of methanol can be obtained in theory?

31. The reaction of methane and water is one way to prepare hydrogen.

$$CH_4(g) + 2\ H_2O(g) \longrightarrow CO_2(g) + 4\ H_2(g)$$

Construct a table, like the one on page 188, for a reaction between 995 g of CH_4 and 2510 g of water.

32. Ammonia gas can be prepared by the following reaction:

$$CaO(s) + 2\ NH_4Cl(s) \longrightarrow 2\ NH_3(g) + H_2O(g) + CaCl_2(s)$$

Construct a table, like the one on page 188, for an ammonia synthesis that begins with 112 g of CaO and 224 g of NH_4Cl.

33. The equation for one of the reactions in the process of turning iron ore into the metal is

$$Fe_2O_3(s) + 3\ CO(g) \longrightarrow 2\ Fe(s) + 3\ CO_2(g)$$

If you start with 2.00 kg of each reactant, what is the maximum amount of iron you can produce?

34. Aspirin is produced by the reaction of salicylic acid and acetic anhydride.

$$\underset{\text{salicylic acid}}{2\ C_7H_6O_3(s)} + \underset{\text{acetic anhydride}}{C_4H_6O_3(\ell)} \longrightarrow \underset{\text{aspirin}}{2\ C_9H_8O_4(s)} + H_2O(\ell)$$

If you mix 100 g of each of the reactants, what is the maximum mass of aspirin that can be obtained?

Percent Yield

35. Ammonia gas can be prepared by the reaction of calcium oxide with ammonium chloride.

$$CaO(s) + 2\ NH_4Cl(s) \longrightarrow 2\ NH_3(g) + H_2O(g) + CaCl_2(s)$$

If exactly 100 g of ammonia is isolated, but the theoretical yield is 136 g, what is the percent yield of this gas?

36. Quicklime, CaO, is formed when calcium hydroxide is heated:

$$Ca(OH)_2(s) \longrightarrow CaO(s) + H_2O(\ell)$$

If the theoretical yield is 65.5 g but only 36.7 g of quicklime is produced, what is the percent yield?

37. Diborane, B_2H_6, is valuable for the synthesis of new organic compounds. The boron compound can be made by the reaction

$$2\ NaBH_4(s) + I_2(s) \longrightarrow B_2H_6(g) + 2\ NaI(s) + H_2(g)$$

Suppose you use 1.203 g of $NaBH_4$ and excess iodine, and you isolate 0.295 g of B_2H_6. What is the percent yield of B_2H_6?

38. Methanol, CH_3OH, is used in racing cars because it is a clean-burning fuel. It can be made by the reaction

$$CO(g) + 2\ H_2(g) \longrightarrow CH_3OH(\ell)$$

What is the percent yield if 5.0×10^3 g of H_2 reacts with excess CO to form 3.5×10^3 g of CH_3OH?

39. Disulfur dichloride, which has a revolting smell, can be prepared by directly combining S_8 and Cl_2, but it can also be made by the reaction

$$3\ SCl_2(\ell) + 4\ NaF(s) \longrightarrow SF_4(g) + S_2Cl_2(\ell) + 4\ NaCl(s)$$

What mass of SCl_2 is needed to react with excess NaF to prepare 1.19 g of S_2Cl_2 if the expected yield is 51%?

40. The ceramic silicon nitride, Si_3N_4, is made by heating silicon and nitrogen at an elevated temperature:

$$3\ Si(s) + 2\ N_2(g) \longrightarrow Si_3N_4(s)$$

How many grams of silicon must combine with excess N_2 to produce 1.0 kg of Si_3N_4 if this process is 92% efficient?

Chemical Analysis

41. A mixture of $CuSO_4$ and $CuSO_4{\cdot}5\ H_2O$ has a mass of 1.245 g; but after heating to drive off all the water, the mass is only 0.832 g. What is the weight percent of $CuSO_4{\cdot}5\ H_2O$ in the mixture?

Dehydrating hydrated copper(II) sulfate. *(C.D. Winters)*

42. A sample of limestone and other soil materials is heated, and the limestone decomposes to give calcium oxide and carbon dioxide. A 1.506-g sample of limestone-containing material gives 0.711 g of CaO, in addition to gaseous CO_2, after being strongly heated. What was the weight percent of $CaCO_3$ in the original sample?

43. A 1.25-g sample contains some of the very reactive compound $Al(C_6H_5)_3$. On treating the compound with HCl, 0.951 g of C_6H_6 is isolated.

$$Al(C_6H_5)_3(s) + 3\ HCl(aq) \longrightarrow AlCl_3(aq) + 3\ C_6H_6(\ell)$$

What was the weight percent of $Al(C_6H_5)_3$ in the original 1.25-g sample?

44. Bromine trifluoride reacts with metal oxides to evolve oxygen quantitatively. For example,

$$3\ TiO_2(s) + 4\ BrF_3(\ell) \longrightarrow 3\ TiF_4(s) + 2\ Br_2(\ell) + 3\ O_2(g)$$

Suppose you wish to use this reaction to determine the weight percent of TiO_2 in a sample of ore. To do this, the O_2 gas from the reaction is collected. If 2.367 g of the TiO_2-containing ore evolves 0.143 g of O_2, what is the weight percent of TiO_2 in the sample?

45. The active ingredient in calcium tablets is $CaCO_3$. When a 1.00-g tablet is treated with excess hydrochloric acid, 0.25 g of CO_2 is produced.

$$CaCO_3(s) + 2\ HCl(\ell) \longrightarrow CaCl_2(aq) + H_2O(\ell) + CO_2(g)$$

What is the weight percent of $CaCO_3$ in the tablet? What is the weight percent of Ca in the tablet?

46. Phosphate fertilizers can be made by treating phosphate rock, $Ca_3(PO_4)_2$, with phosphoric acid, H_3PO_4.

$$Ca_3(PO_4)_2(s) + 4\ H_3PO_4(\ell) \longrightarrow 3\ Ca(H_2PO_4)_2(s)$$

If a 556-g sample of phosphate rock is 95% $Ca_3(PO_4)_2$, how many grams of $Ca(H_2PO_4)_2$ can be produced?

Empirical Formulas

47. Styrene, the building block of polystyrene, is a hydrocarbon (a compound consisting only of C and H). If 0.438 g of the compound is burned and produces 1.481 g of CO_2 and 0.303 g of H_2O, what is the empirical formula of the compound?

48. Mesitylene is a liquid hydrocarbon. If 0.115 g of the compound is burned in pure O_2 to give 0.379 g of CO_2 and 0.1035 g of H_2O, what is the empirical formula of the compound?

49. Propionic acid, an organic acid, contains only C, H, and O. If 0.236 g of the acid burns completely in O_2 and gives 0.421 g of CO_2 and 0.172 g of H_2O, what is the empirical formula of the acid?

50. Quinone, which is used in the dye industry and in photography, is an organic compound containing only C, H, and O. What is the empirical formula of the compound if 0.105 g of the compound gives 0.257 g of CO_2 and 0.0350 g of H_2O when burned completely?

Solution Concentrations

51. You have a 0.12 M solution of $BaCl_2$. What ions exist in solution and what are their concentrations?

52. A flask contains 0.25 M $(NH_4)_2SO_4$. What ions exist in the solution and what are their concentrations?

53. Assume 6.73 g of Na_2CO_3 is dissolved in enough water to make 250. mL of solution. (a) What is the molarity of the sodium carbonate? (b) What are the concentrations of the Na^+ and CO_3^{2-} ions?

54. Some $K_2Cr_2O_7$, with a mass of 2.335 g, is dissolved in enough water to make 500. mL of solution. (a) What is the molarity of the potassium dichromate? (b) What are the concentrations of the K^+ and $Cr_2O_7^{2-}$ ions?

55. What is the mass, in grams, of solute in 250. mL of a 0.0125 M solution of $KMnO_4$?

56. What is the mass, in grams, of solute in 100. mL of a 1.023×10^{-3} M solution of Na_3PO_4?

57. What volume of 0.123 M NaOH, in milliliters, contains 25.0 g of NaOH?

58. What volume of 2.06 M $KMnO_4$, in liters, contains 322 g of solute?

59. If 6.00 mL of 0.0250 M $CuSO_4$ is diluted to 10.0 mL with pure water, what is the concentration of copper(II) sulfate in the diluted solution?

60. If you dilute 25.0 mL of 1.50 M HCl to 500. mL, what is the molar concentration of the diluted HCl?

61. If you need 1.00 L of 0.125 M H_2SO_4, which method would you use to prepare this solution?
(a) Dilute 36.0 mL of 1.25 M H_2SO_4 to a volume of 1.00 L.
(b) Dilute 20.8 mL of 6.00 M H_2SO_4 to a volume of 1.00 L.
(c) Add 950. mL of water to 50.0 mL of 3.00 M H_2SO_4.
(d) Add 500. mL of water to 500. mL of 0.500 M H_2SO_4.

62. If you need 300. mL of 0.500 M K_2CrO_7, which method would you use to prepare this solution?
(a) Dilute 250. mL of 0.600 M K_2CrO_7 to 300. mL.
(b) Add 50.0 mL of water to 250. mL of 0.250 M K_2CrO_7.
(c) Dilute 125 mL of 1.00 M K_2CrO_7 to 300. mL.
(d) Add 30.0 mL of 1.50 M K_2CrO_7 to 270. mL of water.

Calculations for Reactions in Solution

63. What mass in grams of Na_2CO_3 is required for complete reaction with 25.0 mL of 0.155 M HNO_3?

$$Na_2CO_3(aq) + 2\ HNO_3(aq) \longrightarrow 2\ NaNO_3(aq) + CO_2(g) + H_2O(\ell)$$

64. Hydrazine, N_2H_4, a base like ammonia, can react with an acid such as sulfuric acid.

$$2\ N_2H_4(aq) + H_2SO_4(aq) \longrightarrow 2\ N_2H_5^+(aq) + SO_4^{2-}(aq)$$

What mass of hydrazine can be consumed by 250. mL of 0.225 M H_2SO_4?

65. What volume in milliliters of 0.125 M HNO_3 is required to react completely with 1.30 g of $Ba(OH)_2$?

$$2\ HNO_3(aq) + Ba(OH)_2(s) \longrightarrow Ba(NO_3)_2(aq) + 2\ H_2O(\ell)$$

66. Diborane, B_2H_6, can be produced by the following reaction.

$$2\ NaBH_4(aq) + H_2SO_4(aq) \longrightarrow 2\ H_2(g) + Na_2SO_4(aq) + B_2H_6(g)$$

What volume in milliliters of 0.0875 M H_2SO_4 should be used to completely consume 1.35 g of $NaBH_4$?

67. What volume in milliliters of 0.512 M NaOH is required to react completely with 25.0 mL of 0.234 M H_2SO_4?

68. What volume in milliliters of 0.812 M HCl would be required to completely neutralize 15.0 mL of 0.635 M NaOH?

69. What is the maximum mass, in grams, of AgCl that can be precipitated by mixing 50.0 mL of 0.025 M $AgNO_3$ solution with 100.0 mL of 0.025 M NaCl solution? Which reactant is in excess? What is the concentration of the excess reactant remaining in solution after the maximum mass of AgCl has been precipitated?

70. Suppose you mix 25.0 mL of 0.234 M $FeCl_3$ solution with 42.5 mL of 0.453 M NaOH. (a) What is the maximum mass, in grams, of $Fe(OH)_3$ that will precipitate? (b) Which reactant is in excess? (c) What is the concentration of the excess reactant remaining in solution after the maximum mass of $Fe(OH)_3$ has precipitated?

71. A soft drink contains an unknown amount of citric acid, $C_3H_5O(COOH)_3$. A volume of 10.0 mL of the soft drink requires 6.42 mL of 9.580×10^{-2} M NaOH to neutralize the citric acid completely.

$$C_3H_5O(COOH)_3(aq) + 3\ NaOH(aq) \longrightarrow Na_3C_3H_5O(COO)_3(aq) + 3\ H_2O(\ell)$$

(a) Which step in the following calculations for mass of citric acid in 1 mL of soft drink is not correct? (b) What is the correct answer?
(i) Moles NaOH =
(6.42 mL) (1 L/1000 mL) (9.580×10^{-2} mol/L)
(ii) Moles citric acid =
(6.15×10^{-4} mol NaOH) (3 mol citric acid/1 mol NaOH)
(iii) Mass citric acid in sample =
(1.85×10^{-3} mol citric acid) (192.12 g/mol citric acid)
(iv) Mass citric acid in 1 mL soft drink =
(0.354 g citric acid)/(10 mL soft drink)

72. Vitamin C is the compound $C_6H_8O_6$. Besides being an acid, it is also a reducing agent that reacts readily with bromine, Br_2, a good oxidizing agent.

$$C_6H_8O_6(aq) + Br_2(aq) \longrightarrow 2\ HBr(aq) + C_6H_6O_6(aq)$$

Suppose a 1.00-g chewable vitamin C tablet requires 27.85 mL of 0.102 M Br_2 to react completely. (a) Which step in the following calculations for mass, in grams, of vitamin C in the tablet is incorrect? (b) What is the correct answer?
(i) Mole Br_2 = (27.85 mL) (0.102 mol/L)
(ii) Moles $C_6H_8O_6$ =
(2.84 mol Br_2) (1 mol $C_6H_8O_6$/1 mol Br_2)
(iii) Mass $C_6H_8O_6$ = (2.84 mol $C_6H_8O_6$) (176 g/mol $C_6H_8O_6$)
(iv) Mass $C_6H_8O_6$ = (500 g $C_6H_8O_6$)/(1 g tablet)

73. If a volume of 32.45 mL of HCl is used to completely neutralize 2.050 g of Na_2CO_3 according to the following equation, what is the molarity of the HCl?

$$Na_2CO_3(aq) + 2\ HCl(aq) \longrightarrow 2\ NaCl(aq) + CO_2(g) + H_2O(\ell)$$

74. Potassium acid phthalate, $KHC_8H_4O_4$, is used to standardize solutions of bases. The acidic anion reacts with bases according to the following net ionic equation:

$$HC_8H_4O_4^-(aq) + OH^-(aq) \longrightarrow H_2O(\ell) + C_8H_4O_4^{2-}(aq)$$

If a 0.902-g sample of potassium acid phthalate requires 26.45 mL of NaOH to react, what is the molarity of the NaOH?

75. Sodium thiosulfate, $Na_2S_2O_3$, is used as a "fixer" in black-and-white photography. Assume you have a bottle of sodium thiosulfate and want to determine its purity. The thiosulfate ion can be oxidized with I_2 according to the equation

$$I_2(aq) + 2\ S_2O_3^{2-}(aq) \longrightarrow 2\ I^-(aq) + S_4O_6^{2-}(aq)$$

If you use 40.21 mL of 0.246 M I_2 to completely react a 3.232-g sample of impure $Na_2S_2O_3$, what is the percent purity of the $Na_2S_2O_3$?

76. A sample of a mixture of oxalic acid, $H_2C_2O_4$, and sodium chloride has a mass of 4.554 g. If a volume of 29.58 mL of 0.550 M NaOH is required to neutralize all the $H_2C_2O_4$, what is the weight percent of oxalic acid in the mixture? Oxalic acid and NaOH react according to the equation

$$H_2C_2O_4(aq) + 2\ NaOH(aq) \longrightarrow Na_2C_2O_4(aq) + 2\ H_2O(\ell)$$

77. You are given 0.954 g of an unknown acid, H_2A, which reacts with NaOH according to the balanced equation

$$H_2A(aq) + 2\ NaOH(aq) \longrightarrow Na_2A(aq) + 2\ H_2O(\ell)$$

If a volume of 36.04 mL of 0.509 M NaOH is required to react all of the acid, what is the molar mass of the acid?

78. You are given an acid and told only that it could be citric acid (molar mass = 192.1 g/mol) or tartaric acid (molar mass = 150.1 g/mol). To determine which acid you have, you react it with NaOH. The appropriate reactions are

Citric acid: $C_6H_8O_7 + 3\ NaOH \longrightarrow Na_3C_6H_5O_7 + 3\ H_2O$

Tartaric acid: $C_4H_6O_6 + 2\ NaOH \longrightarrow Na_2C_4H_4O_6 + 2\ H_2O$

You find that a 0.956-g sample requires 29.1 mL of 0.513 M NaOH for a stoichiometric amount. What is the unknown acid?

General Questions

79. Nitrogen gas can be prepared in the laboratory by the reaction of ammonia with copper(II) oxide according to the following unbalanced equation.

$$NH_3(g) + CuO(s) \longrightarrow N_2(g) + Cu(s) + H_2O(g)$$

If 26.3 g of gaseous NH_3 is passed over a bed of solid CuO (in stoichiometric excess), what mass in grams of N_2 can be isolated?

80. In an experiment, 1.056 g of a metal carbonate containing an unknown metal M was heated to give the metal oxide and 0.376 g CO_2.

$$MCO_3(s) \xrightarrow{\text{heat}} MO(s) + CO_2(g)$$

What is the identity of the metal M?

(a) M = Ni (c) M = Zn
(b) M = Cu (d) M = Ba

81. Aluminum bromide is a valuable laboratory chemical. What is the maximum theoretical yield in grams of Al_2Br_6 if 25.0 mL of liquid bromine (density = 3.1023 g/mL) and excess aluminum metal are used?

$$2\ Al(s) + 3\ Br_2(\ell) \longrightarrow Al_2Br_6(s)$$

82. Uranium(VI) oxide reacts with bromine trifluoride to give uranium(IV) fluoride, an important step in the purification of uranium ore.

$$6\ UO_3(s) + 8\ BrF_3(\ell) \longrightarrow 6\ UF_4(s) + 4\ Br_2(\ell) + 9\ O_2(g)$$

If you begin with 365 g each of UO_3 and BrF_3, what is the maximum yield, in grams, for UF_4?

83. The cancer chemotherapy agent cisplatin is made by the following reaction:

$$(NH_4)_2PtCl_4(s) + 2\ NH_3(aq) \longrightarrow 2\ NH_4Cl(aq) + Pt(NH_3)_2Cl_2(s)$$

Assume that 15.5 g of $(NH_4)_2PtCl_4$ is combined with 120. mL of 1.25 M aqueous NH_3 to make cisplatin. What is the theoretical mass in grams of cisplatin that can be formed?

84. Diborane, B_2H_6, can be produced by the following reaction:

$$2\ NaBH_4(aq) + H_2SO_4(aq) \longrightarrow 2\ H_2(g) + Na_2SO_4(aq) + B_2H_6(g)$$

What is the maximum yield, in grams, of B_2H_6 that can be prepared starting with 250. mL of 0.0875 M H_2SO_4 and 1.55 g of $NaBH_4$?

85. Silicon and hydrogen form a series of interesting compounds, Si_xH_y. To find the formula of one of them, a 6.22-g sample of the compound is burned in oxygen. On doing so, all of the Si is converted to 11.64 g of SiO_2 and all of the H to 6.980 g of H_2O. What is the empirical formula of the silicon compound?

86. Boron forms an extensive series of compounds with hydrogen, all with the general formula B_xH_y. To analyze one of these compounds, you burn it in air and isolate the boron in the form of B_2O_3 and the hydrogen in the form of water. If 0.148 g of B_xH_y gives 0.422 g of B_2O_3 when burned in excess O_2, what is the empirical formula of B_xH_y?

87. What is the limiting reactant for the reaction

$$4\ KOH + 2\ MnO_2 + O_2 + Cl_2 \longrightarrow 2\ KMnO_4 + 2\ KCl + 2\ H_2O$$

if 5 mol of each reactant are present? What is the limiting reactant when 5 g of each reactant are present?

88. The Hargraves process is an industrial method for making sodium sulfate for use in papermaking.

$$4\ NaCl + 2\ SO_2 + 2\ H_2O + O_2 \longrightarrow 2\ Na_2SO_4 + 4\ HCl$$

(a) If you start with 10 mol of each reactant, which one will determine the amount of Na_2SO_4 produced? (b) What if you start with 100 g of each reactant?

89. What are the concentrations of ions in a solution made by diluting 10.0 mL of 2.56 M HCl to 250 mL?

90. What is the molarity of each ion after 5.00 mL of 3.45 M Na_2CO_3 is diluted to 100 mL?

91. Half a liter of 2.50 M HCl is mixed with 250 mL of 3.75 M HCl. What is the concentration of hydrochloric acid in the resulting solution?

92. If 2.75 L of 0.0193 M NaOH is combined with 275 mL of 1.93 M NaOH, what is the final molarity?

93. The lead content of a sample can be estimated by converting the lead to PbO_2

$$\text{Pb in sample} + \text{oxidizing agent} \longrightarrow PbO_2(s)$$

and then dissolving the PbO_2 in an acid solution of KI. This liberates I_2 according to the equation

$$PbO_2(s) + 4\ H^+(aq) + 2\ I^-(aq) \longrightarrow Pb^{2+}(aq) + I_2(aq) + 2\ H_2O(\ell)$$

The liberated I_2 is then completely reacted with $Na_2S_2O_3$.

$$I_2(aq) + 2\ S_2O_3^{2-}(aq) \longrightarrow 2\ I^-(aq) + S_4O_6^{2-}(aq)$$

The amount of I_2 is related to the amount of lead in the sample. If 0.576 g of lead-containing mineral requires 35.23 mL of 0.0500 M $Na_2S_2O_3$ for the liberation of I_2, calculate the weight percent of lead in the mineral.

94. You wish to determine the weight percent of copper in a copper-containing alloy. After dissolving a sample in acid, you add an excess of KI, and the Cu^{2+} and I^- undergo the reaction

$$2\ Cu^{2+}(aq) + 5\ I^-(aq) \longrightarrow 2\ CuI(s) + I_3^-(aq)$$

The liberated I_3^- is combined with sodium thiosulfate according to the equation

$$I_3^-(aq) + 2\ S_2O_3^{2-}(aq) \longrightarrow S_4O_6^{2-}(aq) + 3\ I^-(aq)$$

If a volume of 26.32 mL of 0.101 M $Na_2S_2O_3$ is required for a stoichiometric ratio, what is the weight percent of Cu in 0.251 g of the alloy?

95. Cobalt(III) ion forms many compounds with ammonia. To find the formula of one of these compounds, you completely react the NH_3 with hydrochloric acid.

$$Co(NH_3)_xCl_3(aq) + x\ HCl\ (aq) \longrightarrow x\ NH_4^+(aq) + Co^{3+}(aq) + (x + 3)Cl^-(aq)$$

Assume that 23.63 mL of 1.500 M HCl is used to react 1.580 g of $Co(NH_3)_xCl_3$. What is the value of x? (a) 2 (b) 3 (c) 4 (d) 6

96. To find the formula of a compound composed of iron and carbon monoxide, $Fe_x(CO)_y$, the compound is burned in pure oxygen, a reaction that proceeds according to the following unbalanced equation.

$$Fe_x(CO)_y(s) + O_2(g) \longrightarrow Fe_2O_3(s) + CO_2(g)$$

If you burn 1.959 g of $Fe_x(CO)_y$ and find 0.799 g of Fe_2O_3 and 2.200 g of CO_2, what is the empirical formula of $Fe_x(CO)_y$?
(a) $Fe(CO)_4$ (c) $Fe(CO)_5$
(b) $Fe_2(CO)_9$ (d) $Fe(CO)_6$

Applying Concepts

97. In a reaction, 1.2 g of element *A* reacts with exactly 3.2 grams of oxygen to form an oxide, AO_x; 2.4 g of element *A* reacts with exactly 3.2 g of oxygen to form a second oxide, AO_y.
(a) What is the ratio x/y?
(b) If $x = 2$, what is the identity of element *A*?

98. If 1.5 mol Cu reacts with a solution containing 4.0 mol $AgNO_3$, what ions will be present in the solution at the end of the reaction?

$$Cu(s) + 2\ AgNO_3(aq) \longrightarrow Cu(NO_3)_2(aq) + 2\ Ag(s)$$

99. Ammonia is formed in a direct reaction of nitrogen and hydrogen.

$$N_2(g) + 3\ H_2(g) \longrightarrow 2\ NH_3(g)$$

A tiny portion of the starting mixture is represented by the following diagram, where the blue circles represent N and the white circles represent H. Which of the following represents the product mixture?

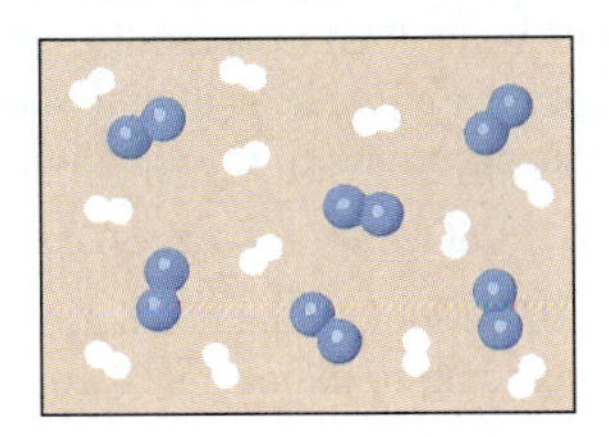

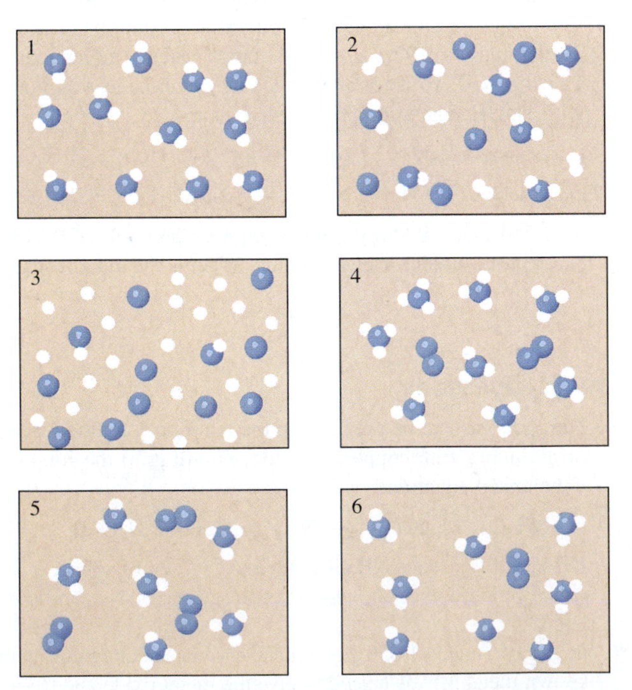

For the reaction of the given sample, which of the following is true?

(a) N_2 is the limiting reactant.
(b) H_2 is the limiting reactant.
(c) NH_3 is the limiting reactant.
(d) No reactant is limiting; they are present in the correct stoichiometric ratio.

100. Carbon monoxide burns readily in oxygen to form carbon dioxide:

$$2\,CO(g) + O_2(g) \longrightarrow 2\,CO_2(g)$$

The box on the left represents a tiny portion of a mixture of CO and O_2. If these molecules react to form CO_2, what should the contents of the box on the right look like?

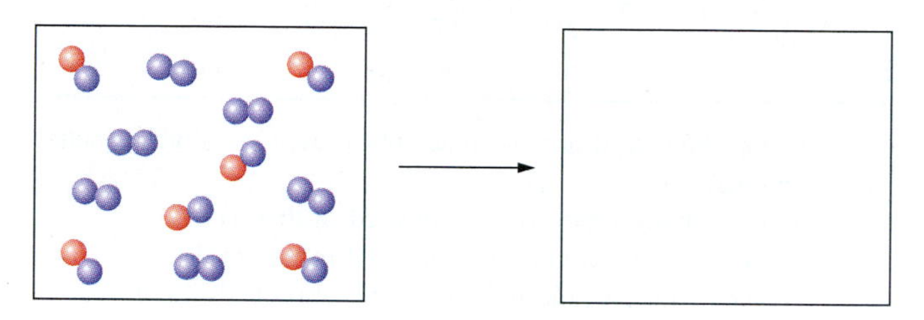

101. Which chemical equation best represents the reaction taking place in this illustration?
(a) $X_2 + Y_2 \longrightarrow XY_3$
(b) $X_2 + 3\,Y_2 \longrightarrow 2\,XY_3$
(c) $6\,X_2 + 6\,Y_2 \longrightarrow 4\,XY_3 + 4\,X_2$
(d) $6\,X_2 + 6\,Y_2 \longrightarrow 4\,X_3Y + 4\,Y_2$

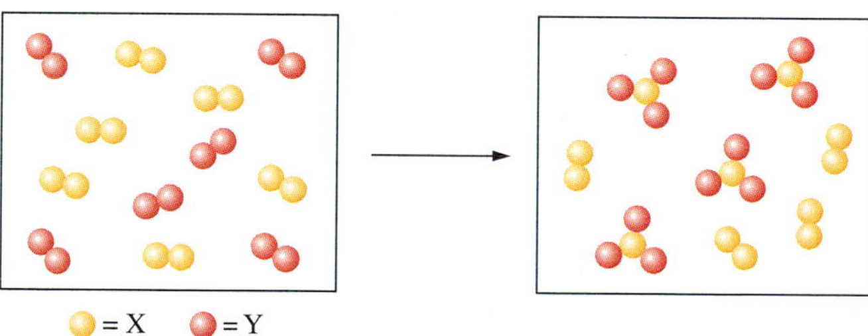

102. A student set up an experiment, like the one described in Chemistry You Can Do on page 190 for six different trials between acetic acid, CH_3CO_2H, and sodium bicarbonate, $NaHCO_3$. (see figure, bottom of page).

$$CH_3CO_2H(aq) + NaHCO_3(s) \longrightarrow NaCH_3CO_2(aq) + CO_2(g) + H_2O(\ell)$$

The volume of acetic acid is kept constant, but the mass of sodium bicarbonate increased with each trial. The results of the tests are shown above. In which trial(s) is the acetic acid the limiting reactant? In which trial(s) is sodium bicarbonate the limiting reactant?

103. A weighed sample of a metal is added to liquid bromine and allowed to react completely. The product substance is then separated from any leftover reactants and weighed. This experiment is repeated with several masses of the metal but with the same volume of bromine. The following graph indicates the results. Explain why the graph has the shape that it does.

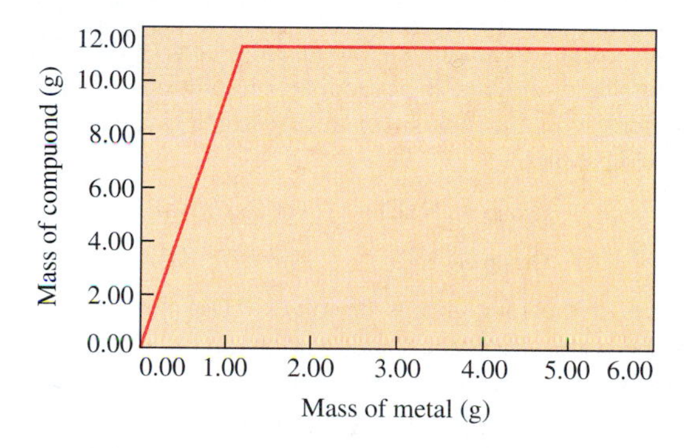

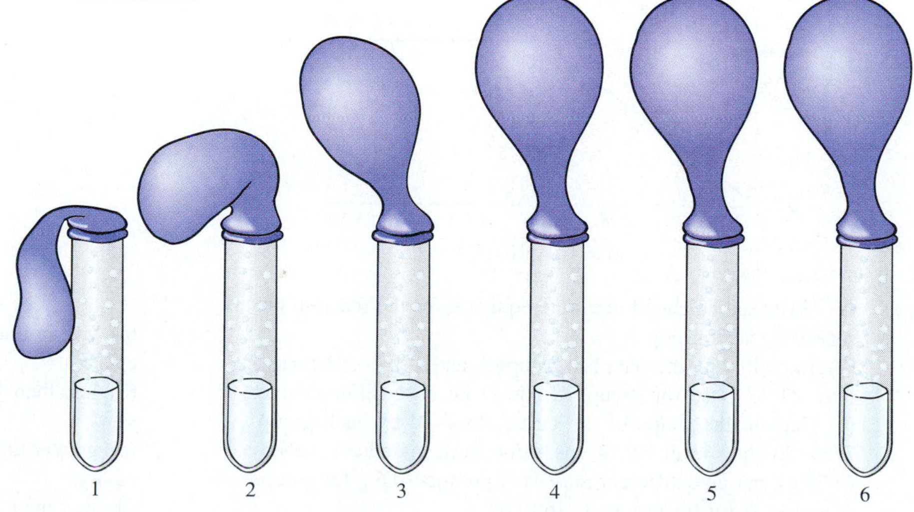

Reaction of acetic acid with sodium bicarbonate.

104. A series of experimental measurements like the ones described in Question 103 is carried out for iron reacting with bromine. The following graph is obtained. What is the empirical formula of the compound formed between iron and bromine? Write a balanced equation for the reaction between iron and bromine. Name the product.

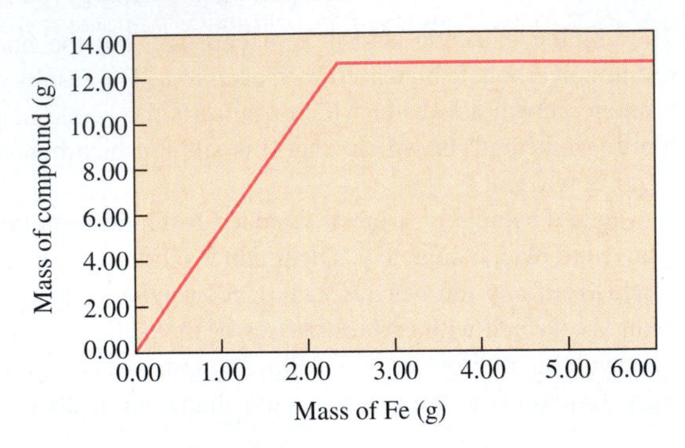

105. Four groups of students from an introductory chemistry laboratory are studying the reactions of solutions of alkali metal halides with aqueous silver nitrate, $AgNO_3$. They use the following salts.

Group A: NaCl Group C: NaBr

Group B: KCl Group D: KBr

Each of the four groups dissolves 0.004 mol of their salt in some water. Each then adds various masses of silver nitrate, $AgNO_3$, to their solutions. After each group collects the precipitated silver halide, the mass of this product is plotted versus the mass of $AgNO_3$ added. The results are given on the following graph.

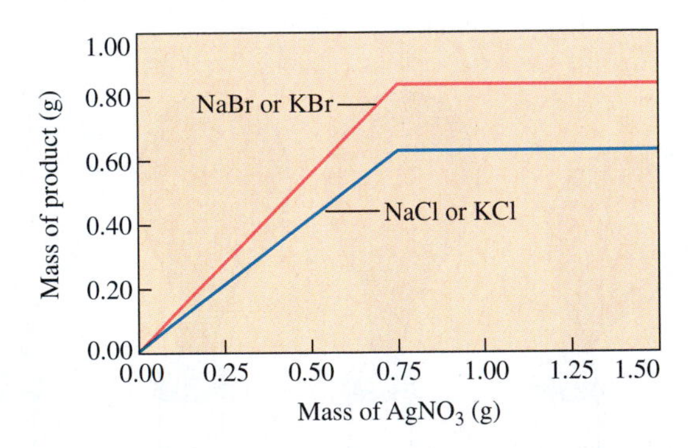

(a) Write the balanced net ionic equation for the reaction observed by each group.
(b) Explain why the data for groups A and B lie on the same line, while those for groups C and D lie on a different line.
(c) Explain the shape of the curve observed by each group. Why do they level off at the same mass of added $AgNO_3$ (0.75 g), but give different masses of product (0.6 g for groups A and B; 0.8 g for groups C and D)?

106. One way to determine the stoichiometric relationships among reactants is continuous variations. In this process, a series of reactions is carried out in which the reactants are varied systematically, while keeping the total volume of each reaction mixture constant. When the reactants combine stoichiometrically, they react completely; none is in excess. The following data were collected to determine the stoichiometric relationship for the reaction.

$$m\,X^{n+} + n\,Y^{m-} \longrightarrow X_mY_n$$

Trial	**A**	**B**	**C**	**D**	**E**
0.10 M X^{n+}	7 mL	6 mL	5 mL	4 mL	3 mL
0.20 M Y^{m-}	3 mL	4 mL	5 mL	6 mL	7 mL
Excess X^{n+} present?	yes	yes	yes	no	no
Excess Y^{m-} present?	no	no	no	no	yes

(a) In which trial are the reactants present in stoichiometric amounts?
(b) How many moles of X^{n+} reacted in that trial?
(c) How many moles of Y^{m-} reacted in that trial?
(d) What is the whole-number mole ratio of X^{n+} to Y^{m-}?
(e) What is the chemical formula for the product X_mY_n?

107. You prepared a NaCl solution by adding 58.44 g of NaCl to a 1-L volumetric flask and then adding water to dissolve it. When finished, the final volume in your flask looked like this:
The solution you prepared is

(a) greater than 1 M because you added more solvent than necessary.
(b) less than 1 M because you added less solvent than necessary.
(c) greater than 1 M because you added less solvent than necessary.
(d) less than 1 M because you added more solvent than necessary.
(e) is 1 M because the amount of solute, not solvent, determines the concentration.

108. The drawings below represent beakers of aqueous solutions. Each ● represents a dissolved solute particle.
(a) Which solution is most concentrated?
(b) Which solution is least concentrated?
(c) Which two solutions have the same concentration?
(d) When solutions E and F are combined, the resulting solution has the same concentration as solution ________.

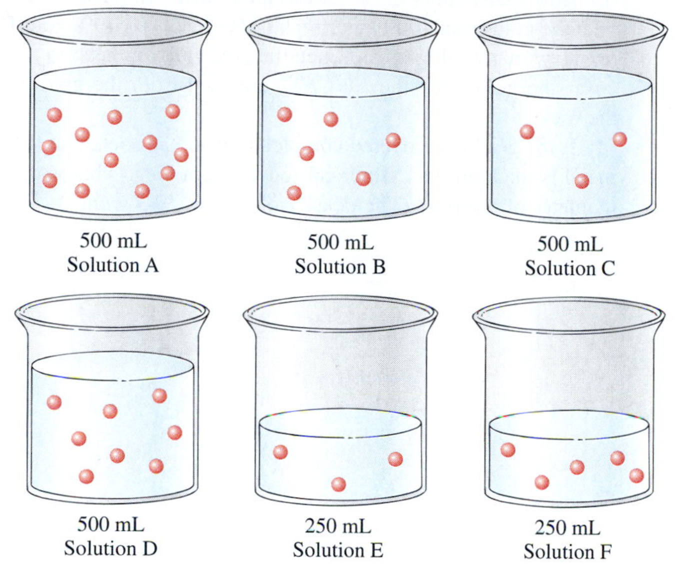

(e) When solutions C and E are combined, the resulting solution has the same concentration as solution ________.
(f) If you evaporate half of the water off solution B, the resulting solution will have the same concentration as solution ________.
(g) If you place half of solution A in another beaker and then add 250 mL of water, the resulting solution will have the same concentration as solution ________.

109. Ten milliliters of a solution of an acid is mixed with 10 mL of a solution of a base. When the mixture was tested with litmus paper, the blue litmus turned red, and the red litmus remained red. Which of the following interpretations is (are) correct?
(a) The mixture contains more hydrogen ions than hydroxide ions.
(b) The mixture contains more hydroxide ions than hydrogen ions.
(c) When an acid and a base react, water is formed, and so the mixture cannot be acidic or basic.
(d) If the acid was HCl and the base was NaOH, the concentration of HCl in the initial acidic solution must have been greater than the concentration of NaOH in the initial basic solution.
(e) If the acid was H_2SO_4 and the base was NaOH, the concentration of H_2SO_4 in the initial acidic solution must have been greater than the concentration of NaOH in the initial basic solution.

110. A chemical company was interested in characterizing a competitor's organic acid (it consists of C, H, and O). After determining that it was a diacid, H_2X, a 0.1235-g sample was neutralized with 15.55 mL of 0.1087 M NaOH. Next, a 0.3469-g sample was burned completely in pure oxygen producing 0.6268 g of CO_2 and 0.2138 g of H_2O.
(a) What is the molar mass of H_2X?
(b) What is the empirical formula for the diacid?
(c) What is the molecular formula for the diacid?

111. Various masses of the three Group 2A elements magnesium, calcium, and strontium were allowed to react with liquid bromine, Br_2. After the reaction was complete, the reaction product was freed of excess reactant(s) and weighed. In each case the mass of product was plotted against the mass of metal used in the reaction shown below.
(a) Based on your knowledge of the reactions of metals with halogens, what product is predicted for each reaction? What

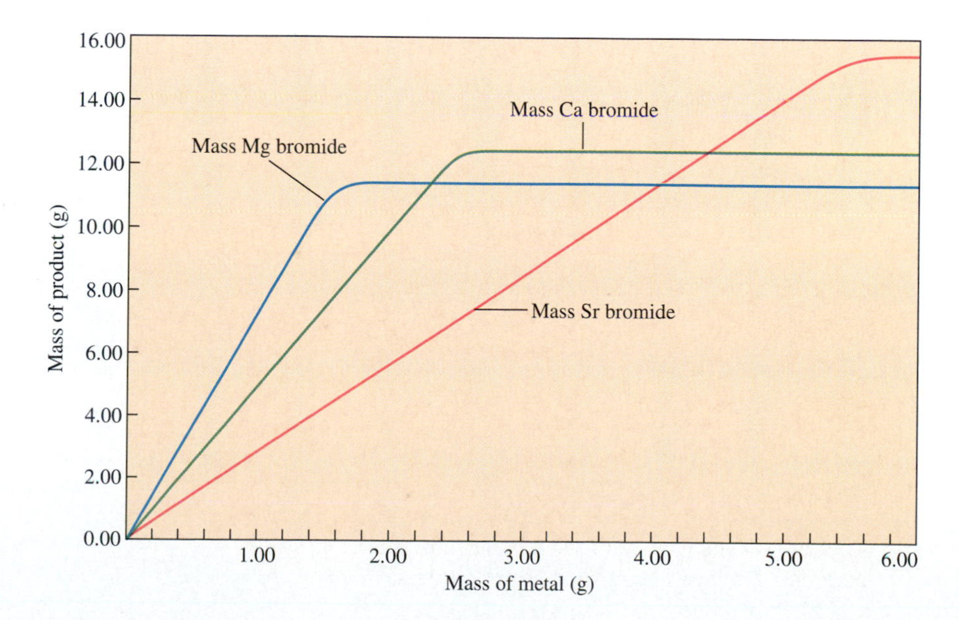

are the name and formula for the reaction product in each case?

(b) Write a balanced equation for the reaction occurring in each case.

(c) What kind of reaction occurs between the metals and bromine; that is, is the reaction a gas-forming reaction, a precipitation reaction, or an oxidation-reduction?

(d) Each plot shows that the mass of product increases with increasing mass of metal used, but the plot levels out at some point. Use these plots to verify your prediction of the formula of each product and explain why the plots become level at different masses of metal and different masses of product.

112. Gold can be dissolved from gold-bearing rock by treating the rock with sodium cyanide in the presence of the oxygen in air.

$$4\ Au(s) + 8\ NaCN(aq) + O_2(g) + 2\ H_2O(\ell) \longrightarrow 4\ NaAu(CN)_2(aq) + 4\ NaOH(aq)$$

Once the gold is in solution in the form of the $Au(CN)_2^-$ ion, it can be precipitated as the metal according to the following unbalanced equation:

$$Au(CN)_2^-(aq) + Zn(s) \longrightarrow Zn^{2+}(aq) + Au(s) + CN^-(aq)$$

(a) Are the two reactions above acid-base or oxidation-reduction reactions? Briefly describe your reasoning.

(b) How many liters of 0.075 M NaCN will you need to extract the gold from 1000 kg of rock if the rock is 0.019% gold?

(c) How many kilograms of metallic zinc will you need to recover the gold from the $Au(CN)_2^-$ obtained from the gold in the rock?

(d) If the gold is recovered completely from the rock, and the metal is made into a cylindrical rod 15.0 cm long, what is the diameter of the rod? (The density of gold is 19.3 g/cm^3.)

CHAPTER 6

Principles of Reactivity I: Energy Transfer and Chemical Reactions

Coal-fired electric power generating plants like this one transform energy stored in fuels into electricity, a form of energy that is more convenient for society. They also emit sulfur oxides and other pollutants into the atmosphere. • How can sulfur oxides be trapped before they are emitted? And how much energy transfer is associated with a trapping reaction? These questions are answered in the Summary Problem at the end of the chapter. *(Martin Bond/Science Photo Library Photo Researchers, Inc.)*

In our industrialized, high-technology, appliance-oriented society, the average use of energy per person is at nearly its highest point in history. The United States, with only 5% of the world's population, consumes 30% of the world's energy resources; and in every year since 1958 we have consumed more energy resources than have been produced within our borders. Most of this energy comes from fossil fuels: coal, petroleum, and natural gas. A little comes from hydroelectric power plants, nuclear power plants, solar and wind collectors, and burning wood and other plant material. World energy use is growing rapidly, despite occasional shortages like the one that occurred in 1973 as a result of an embargo on exports by Middle Eastern oil-producing countries.

Joseph Romm and Charles Curtis clearly describe the importance of energy in our daily lives, to our political system, and to our collective future; see *The Atlantic Monthly,* April 1996, pp. 59–74.

For the past hundred years or so the majority of the energy we use has come from combustion of fossil fuels, and this will continue to be true well into the future. Consequently it is very important to understand how energy and chemical reactions are related, and how chemistry can be used to alter our dependence on fossil fuels. The fastest-growing new industries in the next quarter century may well be those that capitalize on such new chemistry.

It may surprise you to learn that a significant portion of our energy resources are used to cause desirable chemical reactions to occur—reactions that transform inexpensive, readily available substances into new substances with more useful properties. Important examples are the manufacture of rubber from petroleum and the manufacture of medicines from simple chemicals. The relation between chemistry and energy involves more than how much heat we can get by burning a fuel, although that is important to know. Knowledge of chemical energetics also enables us to predict what will happen when potential reactants are mixed. Will most of the reactants be converted to products? Will some be converted? Or virtually none?

"A theory is the more impressive the greater the simplicity of its premises is, the more different kinds of things it relates, and the more extended is its area of applicability. Therefore, the deep impression which classical thermodynamics made upon me. It is the only physical theory of universal content concerning which I am convinced that, within the framework of the applicability of its basic concepts, it will never be overthrown." *Albert Einstein. From "Albert Einstein: Philosopher-Scientist," P.A. Schlipp, ed. In The Library of Living Philosophers,* Vol. VII, Autobiographical notes, p. 33. LaSalle, IL, Open Court Publishing Co., 3rd ed., 1969.

If the surroundings are heated as a result of a reaction, and if the products of the reaction are more dispersed or mixed up than the reactants, then most of the reactants will be transformed into products. If the surroundings must transfer energy into reactants and products as a reaction occurs and the products are more ordered than the reactants, then almost all of the reactants will remain unchanged. But how do we decide whether a reaction will heat its surroundings and whether its products are more disordered? Answering these questions requires knowledge of **thermodynamics,** the science of heat, work, and transformations of one to the other. This is the story to be told in this and the next chapter.

6.1 CONSERVATION OF ENERGY

Just what is energy and where does the energy we use come from? And how can chemical reactions result in the transfer of energy to or from their surroundings? **Energy** is defined as the capacity to do work, and working is something you experience all the time. If you climb a mountain or a staircase, you work against the force of gravity as you move upward. You can do this work because you have the energy or capacity to do so. The energy is provided by food you have eaten; it is stored in chemical compounds and made available when the compounds undergo chemical reactions (metabolism) within your body. The energy contributes to working against the force of gravity as you climb and to heating your body (climbing makes you hotter as well as higher).

Both working and heating are *processes* by which energy is transferred from one form or one place to another. However, we often talk about heat and work as if they were forms of energy. More important is the quantity of energy transferred from one object or sample to another by working or heating processes. To emphasize this, we often will use the words "working" and "heating" where many people would use "work" and "heat."

Working and energy. This climber works against the force of gravity to increase her potential energy. *(P.H. Royer/Photo Researchers, Inc.)*

Energy can be assigned to one of two classes: kinetic or potential. **Kinetic energy** is energy that something has because it is moving. Examples are

- energy of motion of a macroscale object, such as a moving baseball or automobile; this is often called mechanical energy.
- energy of motion of nanoscale objects such as atoms, molecules, or ions; this is often called thermal energy.
- energy of motion of electrons moving through an electrical conductor; this is often called electrical energy.
- energy of periodic motion of nanoscale particles when macroscale sample is alternately compressed and expanded (as when a sound wave passes through air).

All matter has thermal energy because, according to the kinetic-molecular theory *(← p. 10)*, all matter consists of nanoscale particles that are in constant motion (Figure 6.1). Thermal energy increases with temperature for any sample of matter.

Thermal energy is often referred to as heat. However, it is helpful in understanding thermodynamics to distinguish the *process* of heating (or heat transfer), in which energy is transferred from a higher-temperature sample to a lower-temperature sample, from the energy itself.

The kinetic-molecular theory was described qualitatively in Chapter 1 ***(← p. 10)***. A corollary to this theory is that molecules move faster on average the higher the temperature is.

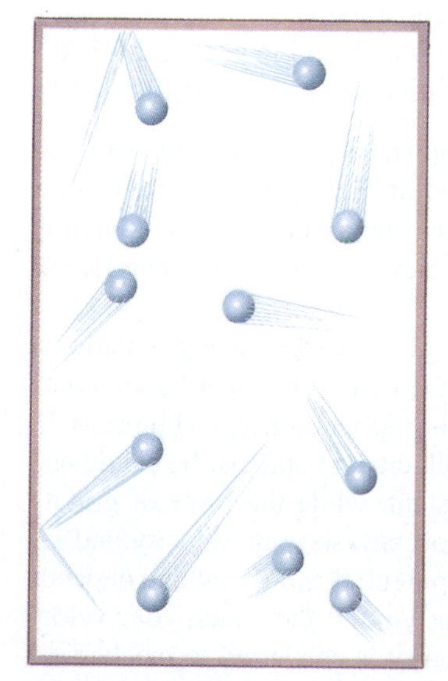

Figure 6.1 Thermal energy. According to the kinetic-molecular theory, atoms, molecules, and ions in chemical elements and chemical compounds are in constant motion. Here atoms of gaseous helium are shown on the nanoscale. Each atom has kinetic energy that depends on how fast it is moving. The thermal energy of the sample is the sum of the kinetic energies of all the helium atoms. The higher the temperature of the helium the higher its thermal energy.

Figure 6.2 Gravitational potential energy. Water on the brink of a waterfall has potential energy (stored energy that could be used to do work) because of its position relative to the earth; that energy could be used to generate electricity, for example, as it is at Niagara Falls (pictured). *(N.Y. Department of Economic Development, Albany, N.Y.)*

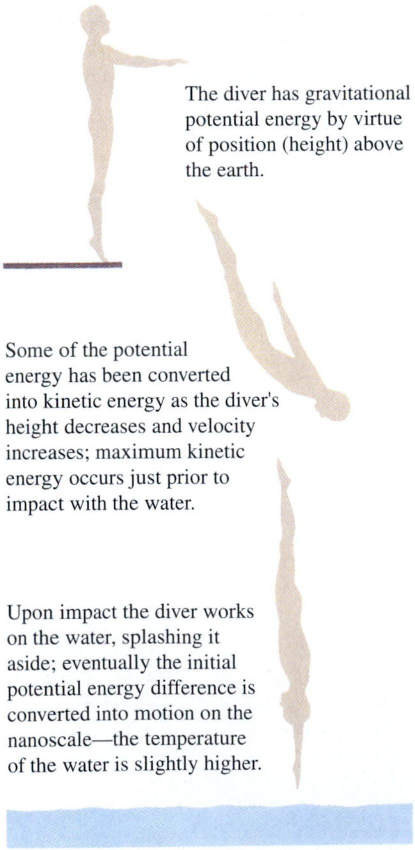

Potential energy is energy that something has a result of its position. Examples are

- energy that a ball has when held well above the floor or that water has at the top of a waterfall (Figure 6.2); this is often called gravitational energy.
- energy that charged particles (such as positive and negative ions) have because of their positions; this is often called electrostatic energy.
- energy resulting from attractions among electrons and atomic nuclei; this is often called chemical potential energy and is the kind of energy stored in foods.

Potential energy can be converted to kinetic energy and *vice versa.* As droplets of water fall over a waterfall, the potential energy they had at the top is converted to kinetic energy—they move faster and faster. Conversely, the kinetic energy of falling water could drive a water wheel to pump water to an elevated reservoir where its potential energy would be higher.

When you stand above a pool of water, poised to dive in, you have gravitational potential energy because of your position relative to the earth. Once you jump, some of that potential energy is converted progressively into kinetic energy (Figure 6.3). Kinetic energy (E_k) depends on your mass (m) and velocity (v) and can be calculated as $E_k = (1/2)mv^2$. During the dive your mass is constant, while the force of gravity accelerates your body to move faster and faster as you fall, so your velocity and kinetic energy increase. This happens at the expense of potential energy (which depends on how high above the earth you are). At the moment you hit the water, your velocity is abruptly reduced, and much of your kinetic energy is converted to mechanical energy of the water, which splashes as your body does work on it to move it aside. Eventually you float on the surface and the water becomes still again. However, the water molecules are moving a little faster in the vicinity of your point of impact; the temperature of the water is higher.

The diver has gravitational potential energy by virtue of position (height) above the earth.

Some of the potential energy has been converted into kinetic energy as the diver's height decreases and velocity increases; maximum kinetic energy occurs just prior to impact with the water.

Upon impact the diver works on the water, splashing it aside; eventually the initial potential energy difference is converted into motion on the nanoscale—the temperature of the water is slightly higher.

Figure 6.3 Energy transformations. Potential and kinetic energy are interconverted when someone dives into water. These interconversions are governed by the law of conservation of energy.

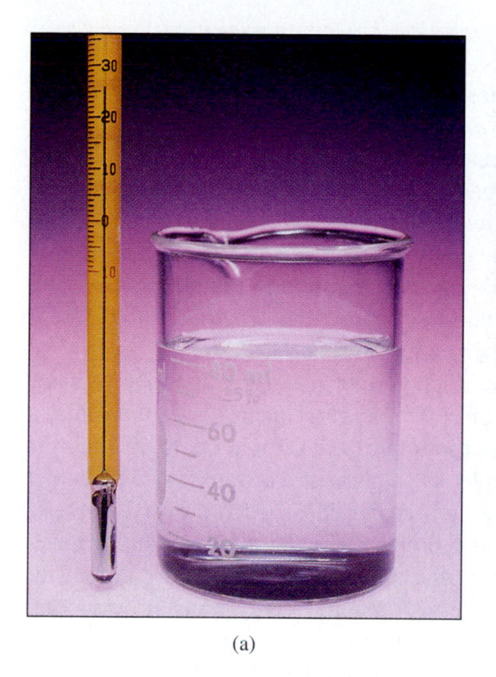

(a)

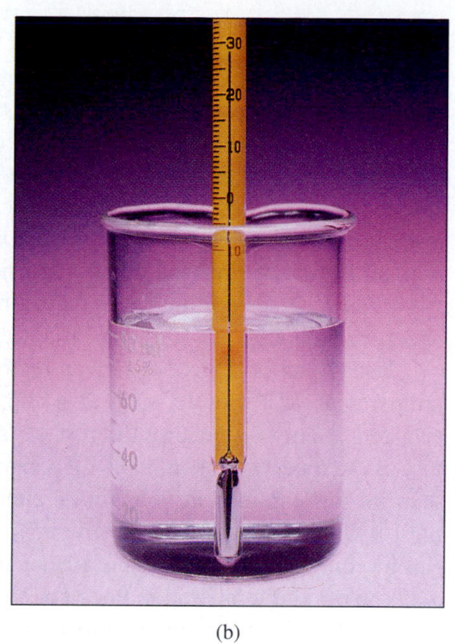

(b)

Figure 6.4 Measuring temperature. The mercury in a thermometer (a) expands (b) because the mercury atoms are moving faster (have more energy) after the warm water transfers energy to (heats) the mercury; the temperature and the volume of the mercury have increased. *(C.D. Winters)*

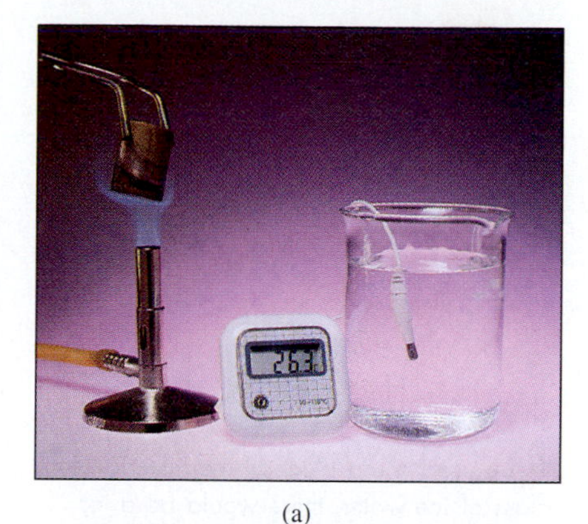

(a)

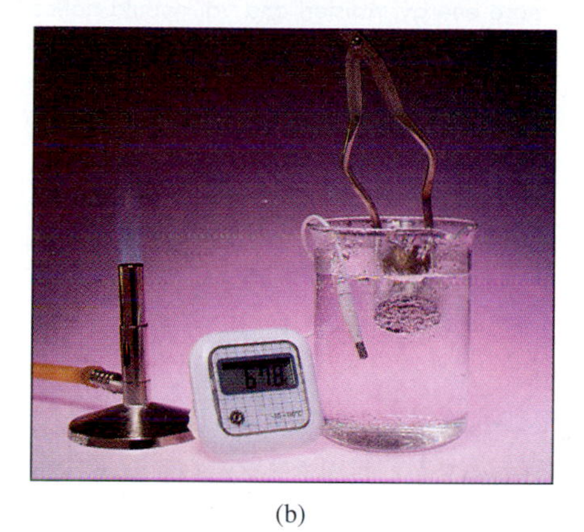

(b)

Figure 6.5 Energy transfer by heating. Water in a beaker (a) is heated when a hotter sample (a brass bar) is plunged into the water (b). There is a transfer of energy from the hotter metal bar to the cooler water. Eventually enough energy is transferred so that the bar and the water reach the same temperature. *(C.D. Winters)*

This series of energy conversions caused by the diver, from potential to kinetic and from macroscale kinetic to nanoscale kinetic (or thermal), is governed by the **law of conservation of energy,** which states that *energy can neither be created nor destroyed—the total energy of the universe is constant.* This is also called the **first law of thermodynamics.** In a great many experiments in which heating and working have been measured, the total energy has always been found to be the same before and after an event. These experiments are summarized by the law of conservation of energy.

Temperature, Energy, and Heating

Transferring energy to a sample of matter usually increases the sample's temperature. The temperature increase can be measured with a thermometer. Figure 6.4 shows a thermometer containing mercury. When the thermometer is placed into hot water, energy is transferred from the water to the thermometer (the thermometer is heated). The increased energy of the mercury atoms means that they are moving about more rapidly, which slightly increases the volume of the spaces between the atoms. Consequently, the mercury expands (as most substances do upon heating), and the column of mercury rises higher in the thermometer tube.

Conceptual Challenge Problems CP-6.A and CP-6.B at the end of the chapter relate to topics discussed in this section.

Energy transfer by heating happens whenever two samples of matter at different temperatures are brought into contact. *Energy is always transferred from the hotter to the cooler sample.* For example, a piece of metal being heated in a Bunsen burner flame and a beaker of cold water (Figure 6.5a) are two samples of matter with different temperatures or hotness. When the hot metal is plunged into the cold water (Figure 6.5b), energy is transferred from the metal to the water until the two samples reach the same temperature. The quantity of energy transferred may be sufficient to heat the water right next to the metal at a temperature above 100 °C. This causes some of the water to boil and you hear a sizzle.

Transferring energy by heating is a process, but it is common to talk about that process as if heat were a form of energy. It is often said that one sample transfers heat to another when what is meant is that one sample transfers energy by heating the other.

Internal Energy

The **internal energy** of a sample of matter is the *sum of the individual energies of all nanoscale particles (atoms, molecules, or ions)* in that sample. Increasing the temperature increases the internal energy because it increases the average speed of motion of nanoscale particles. The faster a particle moves the more energy it has, because $E_k = (1/2)mv^2$. *The total internal energy of a sample of matter depends on temperature, the type of particles, and how many of them there are in the sample.* For a given substance, internal energy depends on temperature and the size of the sample. Thus, a cup of boiling water may contain less energy than a bathtub full of warm water, even though the former is at a higher temperature.

If you used a big block of ice instead of a glass of ice water, there would be a lot more energy transfer, and you would notice a change in the temperature of the room.

Usually most objects in a given region, such as your room, are at about the same temperature. If a sample, such as a cup of coffee, is very much hotter than this, it transfers energy by heating the rest of the room until the sample cools off (and the rest of the room warms up a bit). If a sample, say a glass of ice water, is much cooler than its surroundings in your room, energy is transferred to it from everything else until it warms up (and your room cools off a little). Since the total amount of material in your room is very much greater than that in a cup of coffee or a glass of ice water, the room temperature does not change very much. Energy transfer occurs until everything is at the same temperature.

It is interesting (and useful) to think about *why* this happens. According to the law of energy conservation, it could just as well happen that energy would transfer from the rest of your room to a hot cup of coffee. The coffee would then get hotter and hotter, eventually boiling. But you know from experience that this never happens. It also never happens that a glass of ice water freezes, warming the rest of the room around it. There is *directionality* in heating: energy always transfers from hotter to colder, never the reverse. This directionality corresponds to the *spreading out of energy* over the greatest possible number of atoms, molecules, or ions. When a relatively small number of particles in a hot cup of coffee transfer energy to a large number of particles surrounding the cup, the total thermal energy is spread over the maximum number of particles. The same is true when the large number of particles in the surroundings heats a glass of ice water by transferring energy to the relatively few particles within the glass. Concentrating energy in only a few particles at the expense of many, or even concentrating energy over a large number of particles at the expense of a few, has never been observed to happen of its own accord. Energy transfers always occur in the way that spreads energy over as many particles as possible. This same idea will be used in Chapter 7 to help us predict directionality in the conversion of chemical reactants to products.

Exercise 6.1 Energy Transfers

You toss a rubber ball up into the air. It falls to the floor, bounces for a while, and eventually comes to rest. As in jumping into a pool of water, several energy transfers are involved. Describe them and the changes they cause. Describe also how this example corresponds to the spreading of energy over a larger number of nanoscale particles.

6.2 ENERGY UNITS

Since heating involves transfer of energy, some energy units have been designed to measure heating. A **calorie** (abbreviated "cal") was originally defined as the quantity of energy required to raise the temperature of 1.00 g of pure liquid water by 1.00 degree Celsius, from 14.50 °C to 15.50 °C. This is a very small quantity of energy, and so the **kilocalorie** (1 kcal = 1000 cal) is often used. Most of us are used to calories because we hear about counting them for dieting or read about them on food containers. However, the "calorie" used to describe food is a Calorie with a capital C, a unit equal to the kilocalorie. Thus, a breakfast cereal that gives you 100 Calories of nutritional energy really provides 100 kcal or 100×10^3 calories (with a small c).

1 kilocalorie = 1 kcal = 1000 cal = 1 Calorie

Another common energy unit is the **joule (J).** The joule is a derived unit, which means that it can be expressed as a combination of other units. $1 \text{ J} = 1 \text{ kg·m}^2/\text{s}^2$. One calorie is now defined as exactly 4.184 J. The joule is the preferred energy unit in science, because it is derived directly from the units used in the calculation of potential and kinetic energy. If a 2.0-kg object (about 4 lb) is moving with a velocity of 1.0 meter per second (roughly 2 miles per hour), its kinetic energy is

The joule is the unit of energy used in the International System of units (SI units). SI units are described in Appendix C.

1 calorie = 4.184 J exactly

$$E_k = \frac{1}{2} mv^2 = (\tfrac{1}{2}) \times (2.0 \text{ kg}) \times (1.0 \text{ m/s})^2 = 1.0 \text{ kg·m}^2/\text{s}^2 = 1.0 \text{ J}$$

As another example, if you drop a six-pack of soft drinks on your foot, the kinetic energy at the moment of impact is between 4 and 10 J (about a calorie or two). In many countries food energy is reported in joules rather than Calories. For example, the label on the packet of non-sugar sweetener shown in Figure 6.6 indicates that it provides 16 kJ of nutritional energy.

The joule is named for James P. Joule (1818–1889). Joule was the son of a brewer who lived near Manchester, England, and was a student of John Dalton (← p. 29).

Like the calorie, the joule is relatively small, and so we often use the kilojoule (1 kJ = 1000 J). Consider, for example, the rate at which solar energy enters the earth's atmosphere: about 8×10^{15} kJ/min for the entire globe or about 8 J/min for every square centimeter. This is an enormous quantity of energy, even though it is only 3 ten-millionths (0.0000003) of the total energy emitted by the sun. On average, about half of this energy, 4 J/(min·cm^2), reaches the earth's surface; the rest is reradiated from the upper atmosphere into outer space or is absorbed and scattered by the lower portion of the atmosphere. The exact quantity that reaches the surface de-

1 kilojoule = 1 kJ = 1000 joules

Figure 6.6 Food energy. A packet of artificial sweetener from Australia. As its label shows, the sweetener in the packet supplies 16 kJ of nutritional energy. It is equivalent in sweetness to 2 level teaspoons of sugar, which would supply 140 kJ of nutritional energy. *(C.D. Winters)*

pends on location, the season, and weather conditions. However, the average of 4 J/(min · cm^2) provides the roof of an average American house with about 100,000 kJ/day. This is equivalent to the heating provided by burning about 32 lb of coal, or to 120 kW-h (kilowatt-hours) of electrical energy. This is more than enough to heat the house on a cold winter day—provided the sun is shining.

PROBLEM-SOLVING EXAMPLE 6.1 Energy Units

A single Cheetos snack has a food energy of 7.0×10^3 cal. What is this energy in units of joules?

Answer 29×10^3 J

Explanation To find the energy in joules, we use the definition 1 cal = 4.184 J.

$$7.0 \times 10^3 \text{cal} \times \frac{4.184 \text{ J}}{1.000 \text{ cal}} = 29 \times 10^3 \text{ J} = 29 \text{ kJ}$$

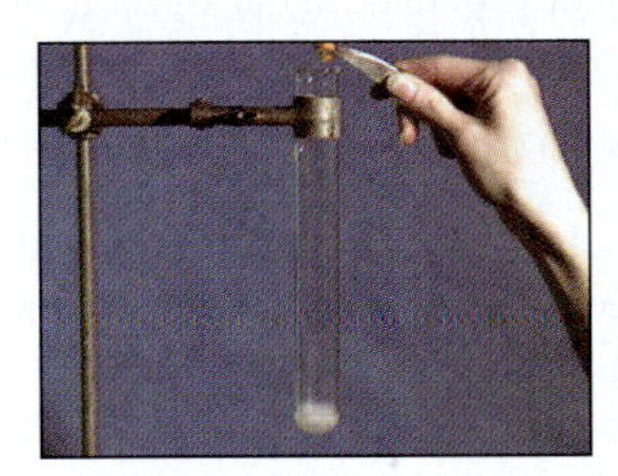

Food energy. A single Cheetos burns in oxygen generated by thermal decomposition of potassium chlorate. *(Used with permission from Chem Demos I Videodisc,* JCE: Software, *SP-8,* ***1994.****)*

Problem-Solving Practice 6.1

(a) If you eat a hot dog, it will provide 160 Calories of energy. What is this energy in joules?

(b) A watt is a unit of power that corresponds to transfer of 1 J of energy in 1 s. The energy used by an x-watt light bulb operating for y s is $x \times y$ joules. If you turn on a 75-watt bulb for 3.0 h, how many joules of electrical energy will be transformed into light and heat?

(c) The packet of non-sugar sweetener in Figure 6.6 provides 16 kJ of nutritional energy. What is this energy in kilocalories?

6.3 HEAT CAPACITY AND SPECIFIC HEAT CAPACITY

Joseph Black (1728–1799) was a professor in Glasgow and Edinborough, Scotland. Among his important studies was that of heat, which later led his student James Watt to the invention of the steam engine.

In the 1770s Joseph Black of the University of Glasgow (Scotland) was the first to distinguish clearly between temperature and heat capacity. The **heat capacity** of a sample of matter is *the quantity of energy required to increase the temperature of that sample by one degree.* Heat capacity depends on the size of the sample and the substance of which it is made (or substances, if it is not pure).

Specific Heat Capacity

To make useful comparisons among samples of different sizes, the **specific heat capacity** (which is often just called *specific heat*) is defined as the quantity of energy needed to increase the temperature of one gram of a substance by one degree. For water at 15 °C, the specific heat capacity is 1.00 cal/g·°C or 4.184 J/g·°C; for common window glass it is only about 0.8 J/g·°C. That is, it takes about five times as much heat to raise the temperature of a gram of water by 1 °C as it does for a gram of glass. Thus, like density, specific heat capacity is a property that can be used to distinguish one substance from another. It can also be used to distinguish a pure substance from a solution or mixture, because the specific heat capacity of a mixture will vary with the proportions of the mixture's components.

Conceptual Challenge Problem CP-6.C at the end of the chapter relates to topics discussed in this section.

The specific heat capacity, c, of a substance can be determined experimentally by measuring the quantity of energy transferred by a known mass of the substance as its temperature rises or falls.

$$\text{Specific heat capacity} = \frac{\text{quantity of energy transferred by heating}}{\text{(sample of mass} \times \text{(temperature change)}} \quad \text{or } c = \frac{q}{m \times \Delta T}$$

The symbol q designates the quantity of energy transferred between samples by heating. The symbol Δ (the Greek letter delta) means "change in" and is calculated as (final value) − (initial value).

The Greek letter Δ (capital delta) represents a change. A change is calculated by subtracting the initial value of a quantity from the final value. In this case we are calculating a temperature change because the Δ precedes the letter T. Therefore the initial temperature is subtracted from the final temperature. Suppose that for a 25.0-g sample of ethylene glycol (a compound used as antifreeze in automobile engines) it takes 60.5 J to change the temperature from 24.2 °C to 25.2 °C. Thus

$$\Delta T = (25.2\ °\text{C} - 24.2\ °\text{C}) = 1.0\ °\text{C}$$

and the specific heat capacity of ethylene glycol is

$$\text{Specific heat capacity of ethylene glycol} = c_{\text{ethylene glycol}}$$

$$= \frac{60.5\ \text{J}}{(25.0\ \text{g}) \times (1.00\ °\text{C})} = 2.42\ \text{J/g·°C}$$

The specific heat capacities of many substances have been determined, and a few values are listed in Table 6.1. Notice that water has one of the highest values. This is important because a high specific heat capacity means that a great deal of energy must be transferred to a large body of water to raise its temperature by just a degree. Conversely, a lot of energy must be transferred away from the water before its temperature falls by one degree. Thus, a lake or ocean can store an enormous quantity of energy and thereby moderate local temperatures. This has a profound influence on weather near lakes or oceans.

The high specific heat capacity of water helps to keep your body temperature relatively constant.

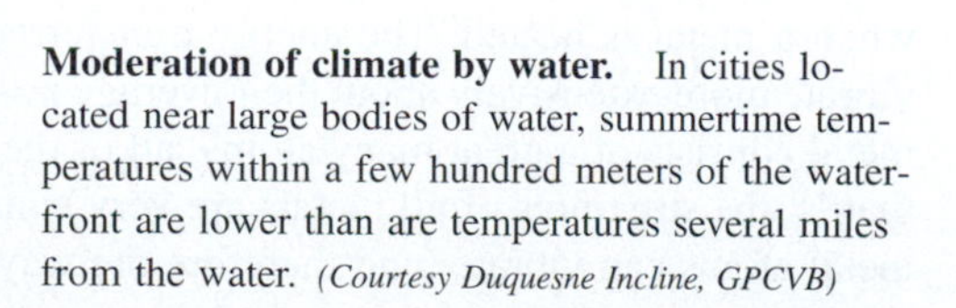

Moderation of climate by water. In cities located near large bodies of water, summertime temperatures within a few hundred meters of the waterfront are lower than are temperatures several miles from the water. *(Courtesy Duquesne Incline, GPCVB)*

When the specific heat capacity of a substance is known, you can calculate the energy transferred to or from that substance by heating, provided that you also know its mass and how much its temperature changed. Conversely, you can calculate the temperature change that should occur when a given quantity of energy is transferred to or from a sample of known mass. The equation that defines specific heat capacity allows you to do this, but it is more convenient to rewrite that equation as

$$q = c \times m \times \Delta T \qquad \text{or} \qquad \Delta T = q/(c \times m)$$

Exercise 6.2 Using Specific Heat Capacity

If 24.1 kJ is supplied to warm a piece of aluminum with a mass of 250. g from an initial temperature of 5.0 °C, what would be the final temperature of the aluminum? The specific heat capacity of aluminum is 0.902 J/g · °C.

Exercise 6.3 Specific Heat Capacity and Temperature Change

Suppose you heat a small piece of rock (granite) and a glass of water in a Bunsen burner flame for 3 min. If the rock and the water have the same mass, and are both heated at the same rate, which has the higher temperature when heating is stopped?

Molar Heat Capacity

It is often useful to know the heat capacity of a sample in terms of the same number of particles instead of the same mass. For this purpose we use the **molar heat capacity,** which is *the quantity of energy that must be transferred to increase the temperature of one mole of a substance by 1 °C.* The molar heat capacity is easily calculated from the specific heat capacity by using the molar mass of the substance. For example, the specific heat capacity of liquid ethanol is given in Table 6.1 as 2.46 J/g·°C. The formula of ethanol is CH_3CH_2OH, and so its molar mass is 46.07 g/mol. The molar heat capacity can be calculated as

$$c_{\mathrm{m}} = \frac{2.46\ \mathrm{J}}{\mathrm{g}\cdot{}^\circ\mathrm{C}} \times \frac{46.07\ \mathrm{g}}{\mathrm{mol}} = 113\ \mathrm{J/mol}\cdot{}^\circ\mathrm{C}$$

Exercise 6.4 Molar Heat Capacity

Calculate the molar heat capacities of all of the metals listed in Table 6.1. Compare these with the value just calculated for ethanol. Based on your results, suggest a way to predict the molar heat capacity of a metal. Can this same rule be applied to other kinds of substances?

As you should have found in Exercise 6.4, molar heat capacities of metals are very similar. This can be explained if we consider what happens on the nanoscale when a metal is heated. The energy transferred by heating a solid makes the atoms vibrate more extensively about their average positions in the solid crystal lattice. Every metal consists of a great many atoms, all of the same kind and packed close together; that is, the structures of all metals are very similar. Because of this, the ways that the metal atoms can vibrate (and therefore the ways that their energies can be increased)

TABLE 6.1 Specific Heat Capacities for Some Elements, Compounds, and Common Solids

Symbol or Formula	Name	Specific Heat Capacity (J/g · °C)
Elements		
Al	Aluminum	0.902
C	Carbon (graphite)	0.720
Fe	Iron	0.451
Cu	Copper	0.385
Au	Gold	0.128
Compounds		
$NH_3(\ell)$	Ammonia	4.70
$H_2O(\ell)$	Water (liquid)	4.184
$C_2H_5OH(\ell)$	Ethanol	2.46
$(CH_2OH)_2(\ell)$	Ethylene glycol (antifreeze)	2.42
$H_2O(s)$	Water (ice)	2.06
$CCl_4(\ell)$	Carbon tetrachloride	0.861
$CCl_2F_2(\ell)$	A chlorofluorocarbon (CFC)	0.598
Common Solids		
	Wood	1.76
	Concrete	0.88
	Glass	0.84
	Granite	0.79

are very similar. Thus, no matter what the metal, nearly the same quantity of energy must be transferred per metal atom to increase the temperature by one degree. The quantity of energy per mole is therefore very similar for all metals.

Sign Conventions in Energy Calculations

When ΔT is calculated as (final temperature − initial temperature), it will have an algebraic sign: positive (+) for an increase in temperature and negative (−) for a decrease in temperature. This is true whenever a change in a quantity is indicated by a Δ: a positive value indicates an increase in the quantity and a negative value a decrease. From the equation $q = c \times m \times \Delta T$, you can see that if ΔT has a positive sign, then so does q. This means that if the temperature of the substance increased, energy was transferred to it from something else. The opposite case, a decrease in the temperature of the substance, means that ΔT has a negative sign and so does q; energy was transferred from the substance to something else. Thus the *magnitude* of q indicates the *quantity* of energy transferred, and the *sign* of q indicates the *direction* in which it is transferred.

A good analogy is your bank account. Assume you have a balance of \$26 in your account (B_{initial}), and after a withdrawal you have \$20 (B_{final}). The cash flow is thus

$$\text{Cash flow} = B_{\text{final}} - B_{\text{initial}} = \$20 - \$26 = -\$6$$

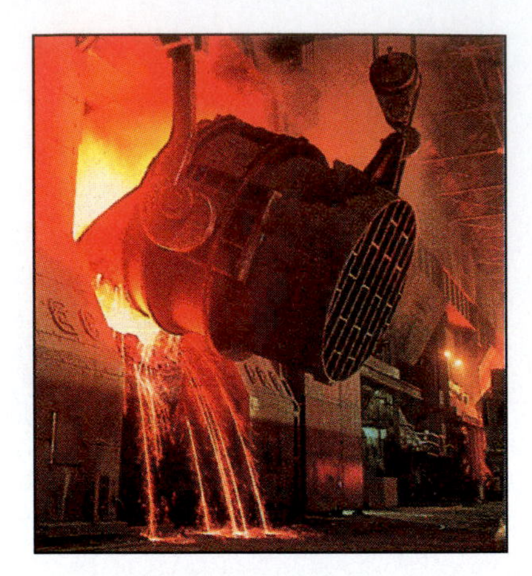

Energy transfer by heating. The red-hot iron shown here is at a much higher temperature than its surroundings and is transferring energy rapidly by heating its surroundings. If you were nearby, your skin would feel warm as a result of some of this energy transfer. *(Bethlehem Steel)*

The negative sign on the $6 indicates that a withdrawal has been made; the cash itself is not negative, but your balance went down. Thus, when we talk about the magnitude of a quantity of energy we use an unsigned number. However, when we want to indicate the direction of transfer in a process, we attach a negative sign (energy transferred from the substance) or a positive sign (energy transferred to the substance) to the value of q. The table below summarizes this convention.

Change in T of Sample	Sign of ΔT	Sign of q	Direction of Energy Transfer
Increase	+	+	To substance
Decrease	−	−	From substance

As an example of the use of specific heat capacity, you can find the quantity of energy transferred from a cup of coffee to your body and the surrounding air when the temperature of the coffee drops from 60.0 °C to 37.0 °C (normal body temperature). Assume the cup holds 250. mL of coffee with a density of 1.00 g/mL, so the coffee has a mass of 250. g; also assume the specific heat capacity of coffee is the same as that of water. The quantity of energy transferred can be obtained from $q = c \times m \times \Delta T$.

It is reasonable to assume that the heat capacity of coffee is the same as for water because coffee is mainly water.

$$\text{Thermal energy transferred} = q = (4.184\ \text{J/g}\cdot{}^\circ\text{C})(250.\ \text{g})(\underset{\substack{\uparrow \\ \text{final temp.}}}{37.0\ {}^\circ\text{C}} - \underset{\substack{\uparrow \\ \text{initial temp.}}}{60.0\ {}^\circ\text{C}})$$

$$= -24.1 \times 10^3\ \text{J} = -24.1\ \text{kJ}$$

Notice that the final answer has a negative value. The negative sign indicates that 24.1 kJ is transferred from the coffee to the surroundings, and the temperature of the coffee decreases.

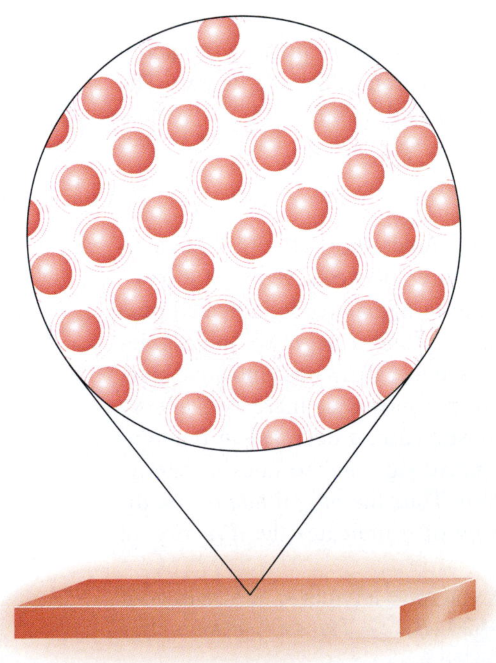

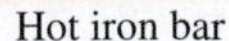

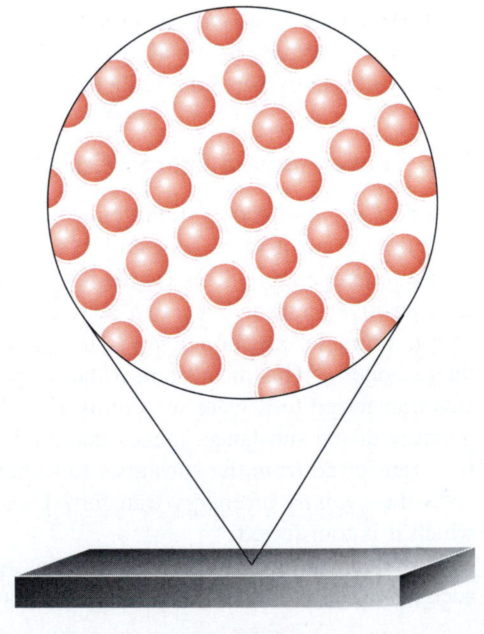

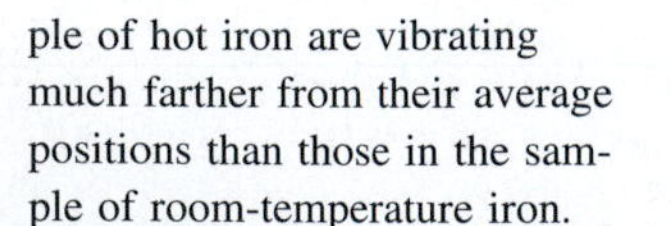

Hot and cold iron. On the nanoscale the atoms in the sample of hot iron are vibrating much farther from their average positions than those in the sample of room-temperature iron.

PROBLEM-SOLVING EXAMPLE 6.2 Transfer of Energy Between Samples by Heating

Suppose a 55.0-g piece of iron is heated in the flame of a Bunsen burner to 425 °C and then plunged into a beaker of water. The beaker holds 600 mL of water (density = 1.00 g/mL), and its temperature before the iron is dropped in is 25 °C. What is the final temperature of both the water and the piece of iron? (Assume that there is no energy transfer to the glass beaker or to the air or to anything else but the water.)

Answer T_{final} = 29 °C

Explanation The most important aspects of this problem are that (1) the water and the iron bar will end up at the same temperature (T_{final} is the same for both), and (2) the law of energy conservation requires that the total quantity of energy be the same before and after the experiment. This means that the net quantity of energy transferred must be zero. Since the energy transfer involves only the iron and the water, summing the transfers must give zero:

$$q_{iron} + q_{water} = 0 \qquad \text{or, by algebraic rearrangement} \qquad q_{water} = -q_{iron}$$

The quantity of energy transferred to the water and the quantity transferred from the iron are equal in size but opposite in algebraic sign. The minus sign in front of q_{iron} indicates that energy was transferred from the iron as its temperature dropped. Conversely, q_{water} has a plus sign because energy was transferred into the water to raise its temperature. Substituting the specific heat capacities of iron and water from Table 6.1 and the masses of iron and water given in the problem statement into the equation $q = c \times m \times \Delta T$, where $T_{initial}$ for the iron is 425 °C and $T_{initial}$ for the water is 25 °C, we have

The quantities of energy transferred have opposite signs because they take place from the iron (negative), to the water (positive).

$$q_{water} = -q_{iron}$$

$$(4.184\ \text{J/g·°C})(600.\ \text{g})(T_{final} - 25\ \text{°C}) = -(0.451\ \text{J/g·°C})(55.0\ \text{g})(T_{final} - 425\ \text{°C})$$

$$(2510.\ \text{J/°C})\ T_{final} - 6.276 \times 10^4\ \text{J} = -(24.80\ \text{J/°C})\ T_{final} + 1.054 \times 10^4\ \text{J}$$

$$(2535\ \text{J/°C})\ T_{final} = 7.330 \times 10^4\ \text{J}$$

Solving, we find T_{final} = 29 °C. The iron has indeed cooled down (ΔT_{iron} = −396 °C) and the water has warmed up (ΔT_{water} = 4 °C) to the same final temperature (T_{final}), which lies between the two initial values.

ΔT_{iron} = 29 °C − 425 °C = −396 °C
ΔT_{water} = 29 °C − 25 °C = 4 °C

A final comment—don't be confused by the fact that transferring the same quantity of energy has resulted in two different values of ΔT; this is because the specific heat capacities and masses of iron and water are so different.

Problem-Solving Practice 6.2

A 400.-g piece of iron is heated in a flame and then immersed in 1000. g of water in a beaker. The initial temperature of the water was 20.0 °C, and both the iron and the water are at 32.8 °C at the end of the experiment. What was the original temperature of the hot iron bar? (Assume that no energy transfer occurs other than between the water and the iron.)

6.4 ENERGY TRANSFERS AND CHANGES OF STATE

So far we have described transfers of energy between samples of matter as a result of temperature differences. But energy transfers also accompany physical or chemical changes, *even though there may be no change in temperature.*

Conceptual Challenge Problem CP-6.D at the end of the chapter relates to topics discussed in this section.

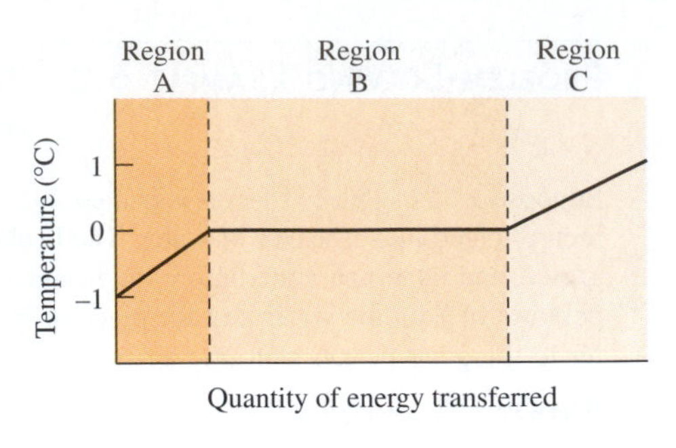

Figure 6.7 Heating graph. When a sample of ice is heated at a constant rate, the temperature does not always increase at a constant rate. In region A of the graph, ice is warming from −1 °C to 0 °C; in region B, ice is melting and the temperature remains constant at 0 °C until all the ice changes to liquid; in region C, liquid water is warming from 0 °C to 1 °C.

Changes of state (between solid and liquid, liquid and gas, or solid and gas) take place without a change in temperature. This is why melting points and boiling points can be measured relatively easily: the temperature remains constant until all of one phase disappears.

Fusion means melting. If too much current flows, an electrical fuse melts (fuses) and thereby cuts off electric current that might otherwise start a fire.

As an example, consider what happens when ice is heated at a slow, constant rate from −1 °C to +1 °C. A graph of temperature as a function of quantity of energy transferred is shown in Figure 6.7. When the temperature reaches 0 °C, it remains constant as long as solid ice is changing into liquid water, but the sample is still being heated. As long as the ice is melting, thermal energy must be continually supplied because the potential energy of the water molecules is higher in the liquid water than in the solid ice. Melting a solid is an example of a **change of state** or **phase change,** a physical process in which one state of matter is transformed into another. During a change of state, the temperature remains constant, but energy must be continually transferred because the particles that make up the sample have different quantities of potential energy before and after the change.

The quantity of thermal energy that must be transferred to melt a solid is called the **heat of fusion.** For ice it is 333 J/g at 0 °C, which is the same quantity of energy that would be required to raise the temperature of a 1.00-g block of iron from 0 °C to 738 °C. As shown in Figure 6.8, 333 J only converts the 1.00 g of ice to liquid water at 0 °C, but it can heat 1.00 g of iron red hot.

The quantity of energy that must be transferred to convert a liquid to vapor is called the **heat of vaporization.** For water it is 2260 J/g at 100 °C. This is considerably larger than the heat of fusion, and so the same 333 J that melted 1.00 g of ice will boil only 0.147 g of water at 100 °C.

$$333 \text{ J} \times \frac{1.00 \text{ g water vaporized}}{2260 \text{ J}} = 0.147 \text{ g liquid water vaporized}$$

Exercise 6.5 Heating Graph

Assume that a 1.0-g sample of ice at −5 °C is heated at a uniform rate until the temperature is 105 °C. Draw a graph like the one in Figure 6.7 to show how temperature varies with energy transferred. Your graph should be approximately to the correct scale.

Exercise 6.6 Changes of State

Assume you have 1 cup of ice (237 g) at 0.0 °C. How much heating is required to melt the ice, warm the resulting water to 100.0 °C, and then boil the water to vapor at 100.0 °C?

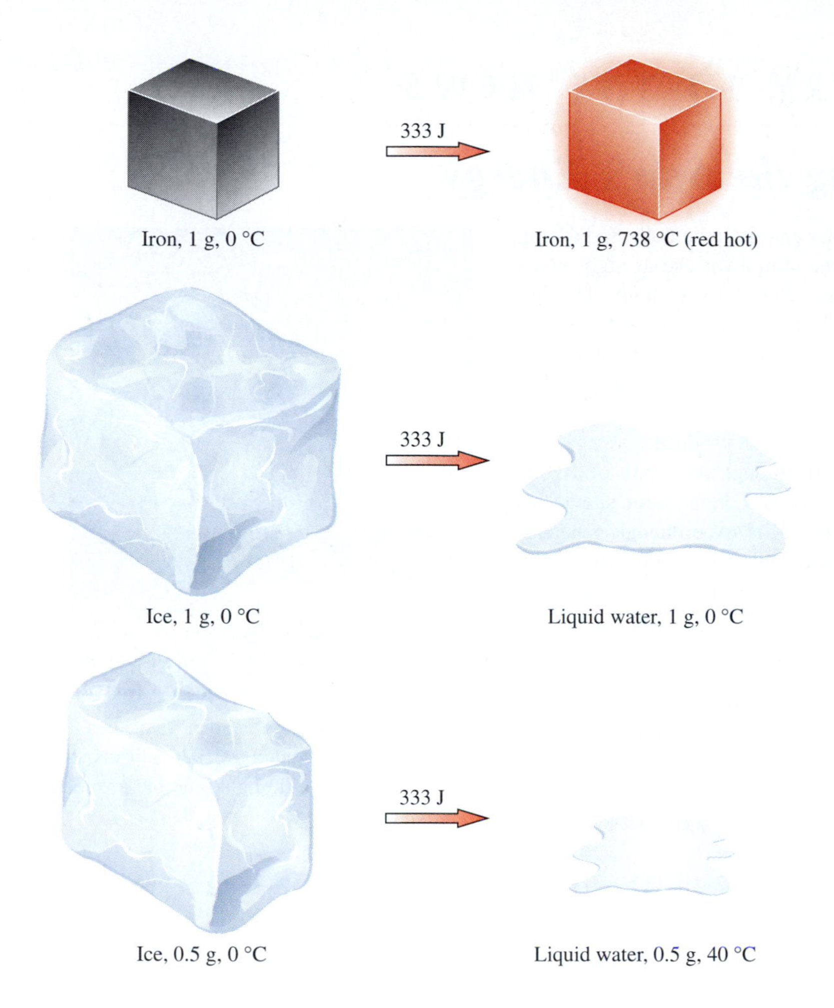

Figure 6.8 Heating, temperature, and phase changes. Heating a substance can cause a temperature change, a phase change, or both. Here 333 J of energy has been transferred to each of three samples: a 1-g block of iron at 0 °C; a 1-g block of ice at 0 °C; and a 0.5-g block of ice at 0 °C. The iron block becomes red hot; its temperature increases to 738 °C. The 1-g block of ice melts, resulting in 1 g of liquid water at 0 °C. The 0.5-g block of ice melts, and there is enough energy to heat the liquid water to 40 °C.

Thermodynamics and Changes of State

To make our discussion of phase changes and other processes more precise in thermodynamic terms, we will define two regions, the system and the surroundings. *The region of primary concern* is designated as the **system.** The system might be the atoms, molecules, or ions of a substance involved in a change of state, or of one or more substances undergoing a reaction. *Everything that can exchange energy with the system* is called the **surroundings.** A system may be delineated by an actual physical boundary, such as the surface of a flask or the membrane of a cell in your body. Alternatively, the boundary may be purely imaginary. For example, you could study the solar system within its surroundings, the rest of the galaxy.

As an example of system and surroundings, consider Dry Ice (carbon dioxide), which undergoes a change of state from solid to gas, a process called **sublimation,** at −78 °C and normal atmospheric pressure.

$$CO_2(\text{solid}, -78°\text{C}) \longrightarrow CO_2(\text{gas}, -78\ °\text{C})$$

CHEMISTRY IN THE NEWS

Capturing the Sun's Energy

In the Mojave Desert near Daggett, California, a new $39 million solar power plant was dedicated on June 5, 1996. Called Solar Two, it demonstrates the latest technology for converting sunlight into electricity. It consists of 1926 mirrors surrounding a tower nearly 100 m high. Each mirror can be adjusted by two computer-controlled electric motors to reflect sunlight to the top of the tower. There an array of pipes coated with special energy-absorbing paint collects the solar energy. The top of the tower glows because of the concentrated sunlight, which heats it to nearly 600 °C.

A mixture of molten sodium nitrate and potassium nitrate circulates through the pipes and is heated to 565 °C. More than 1.3 million kilograms of molten salts is used. Once heated, the mixture of salts is stored in specially insulated tanks. This allows solar energy to be collected during the day and stored for use at night when demand for electricity is high. The hot molten salts are used to boil water, which then passes through steam generators similar to those used in coal-burning electric power plants.

Solar Two can produce 10 megawatts of electricity, enough for a city of 10,000 homes. It was built on the same site as Solar One, an earlier experimental generator that used the sun's energy directly to boil water, generate steam, and produce electricity. By using the large heat capacity of the 1.3×10^6 kg of molten salt mixture, Solar Two is able to store solar energy for use when there is high demand; therefore, it is considerably more attractive as a means of converting solar energy for human use.

Solar energy collection. Solar Two's central tower is surrounded by 1926 adjustable mirrors that reflect sunlight onto the tower, nearly 100 m above the ground. *(Randy Montoya, Sandia National Laboratories)*

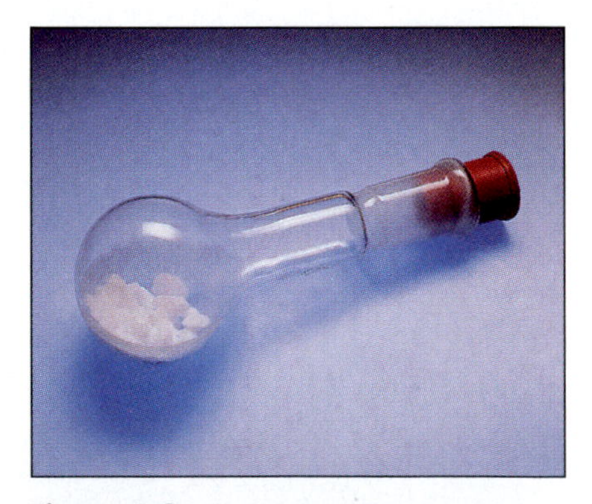

Figure 6.9 Phase change in a constant-volume system. There is a transfer of thermal energy into the solid CO_2 in the sealed flask at −78 °C, but there is no mechanical connection between the system and its surroundings, and so there is no energy transferred by doing work. Therefore, q_v = thermal energy transferred at constant volume = ΔE. The pressure in the container increases as the solid changes to gas; unless the container is strong, it could shatter. *(C.D. Winters)*

In Figures 6.9 and 6.10, a CO_2 sample (both solid and vapor) is the system, and it interacts with its surroundings—the flask or plastic bag, the table on which they rest, the surrounding air, and, in one case, a book.

Suppose that our system and its immediate surroundings are at −78 °C, the sublimation temperature of solid CO_2. The CO_2 molecules are much more widely separated in the gas phase than they are in the solid. Just as the potential energy of a ball is greater when it is farther above the earth, which attracts it, the potential energy of a CO_2 molecule is greater when it is farther from another CO_2 molecule, which attracts it. Because work has to be done to overcome the attractive forces between the CO_2 molecules, the internal energy of a mole of CO_2 gas at −78 °C is greater than the internal energy of a mole of CO_2 solid. As the CO_2 sublimes, energy must be supplied from the surroundings to increase the energy of the system. How the extra energy is supplied depends on whether the CO_2 is in a rigid container or a flexible one.

Suppose that the CO_2(s) and CO_2(g) are inside a *rigid container* so that the total volume cannot change (Figure 6.9). The energy needed for one molecule to escape from solid to gas can be supplied by neighboring molecules, which consequently vibrate a little less about their positions in the crystal lattice. As more and more molecules escape from the lattice, the motion of the remaining solid-state molecules is more and more diminished, which corresponds to a lowering of the temperature. How-

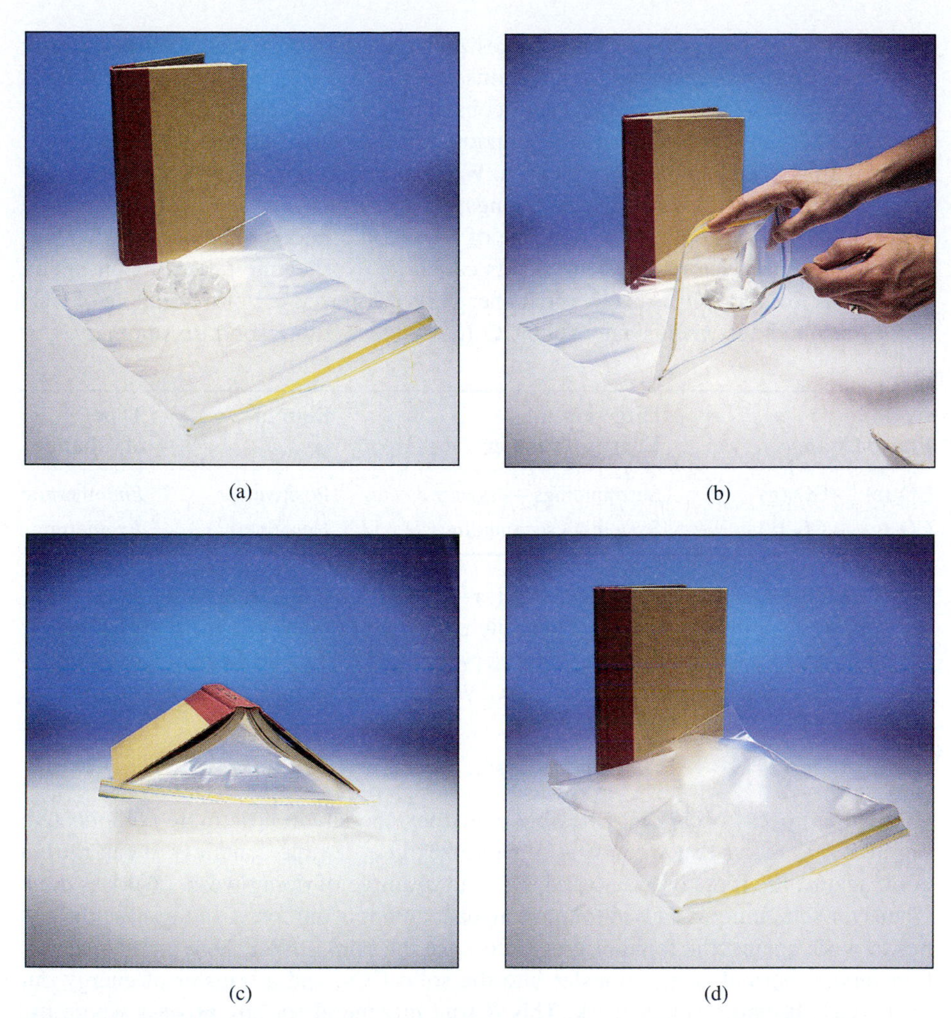

Figure 6.10 Phase change in a constant-pressure system. (a) Pieces of Dry Ice and an empty plastic bag. (b) Crushed Dry Ice is being placed inside the plastic bag. (c) The bag has been sealed. The Dry Ice has sublimed, converting $CO_2(s)$ at −78 °C to $CO_2(g)$ at −78 °C. The gas has done work by raising a book that has been placed on the bag. (d) A similar experiment but without the book. The expanding CO_2 in this case has pushed aside the air that formerly occupied the space taken up by the partially inflated bag. Pushing aside the atmosphere requires work just as pushing the book does. Work must always be done by any constant-pressure reaction whose products occupy greater volume than the reactants. The transfer of energy by heating the system under constant-pressure conditions, q_p, is identified as the enthalpy change, ΔH. *(C.D. Winters)*

ever, atoms and molecules in the surroundings (the rigid container) are in contact with CO_2 molecules in the solid, and energy transfer can occur. As soon as the temperature of solid CO_2 drops slightly below the temperature of the surroundings, there is a net transfer of energy into the CO_2, and the temperature of $CO_2(s)$ is restored to −78 °C. In order for the temperature not to drop below −78 °C, the energy transfer in from the surroundings must equal exactly the energy required to overcome attractive forces between CO_2 molecules as solid CO_2 sublimes.

What has just been described is transfer of energy from surroundings to system while the process $CO_2(s) \rightarrow CO_2(g)$ takes place. In such a process, thermal energy transferred from the surroundings maintains constant temperature. When energy must be transferred *into* a system to maintain constant temperature as a process occurs, the process is said to be **endothermic.** An endothermic process that you are certainly familiar with is evaporation of perspiration. When water on your skin vaporizes, energy has to be transferred from the surroundings, that is, from your skin and flesh, to the water; thus you are cooled. The opposite of an endothermic process occurs when energy is transferred *out of* a system; this is called an **exothermic** process. Conversion of CO_2 gas to solid CO_2 is exothermic; energy is transferred *out* of the sample of CO_2 to the surroundings when some of the $CO_2(g)$ condenses to a solid. In summary,

Usually the surroundings contain a great deal more matter than the system and hence have a much greater heat capacity. Consequently the change in temperature of the surroundings is often so small that it cannot be measured.

Phase Change	Direction of Energy Transfer	Sign of q_{system}	Type of Change
$CO_2(s) \rightarrow CO_2(g)$	Surroundings $\rightarrow$ system	Positive	*Endo*thermic
$CO_2(g) \rightarrow CO_2(s)$	System $\rightarrow$ surroundings	Negative	*Exo*thermic

When the endothermic process $CO_2(s) \rightarrow CO_2(g)$ occurs inside a rigid container, the volume remains constant, and the only exchange of energy is the thermal energy transferred into the system. Thus, by conservation of energy, the *thermal energy transferred at constant volume, q_v, must equal the change in internal energy of the system, ΔE.*

$$\Delta E = q_v$$

If the process occurs in a flexible container, the situation is more complicated. Figure 6.10c shows that when $CO_2(s)$ vaporizes in a flexible container, it can raise a book against the force of gravity, which is an example of doing work. (Work is done whenever something is caused to move against an opposing force. In this case the gas has to work against the force of gravity to raise the book.) Now there are two energy transfers: a thermal energy transfer into the solid CO_2, and a transfer of energy out of the CO_2 because it does work. This is true in general for any process where the volume is not constant: there are both heat and work transfers of energy. This is represented schematically in Figure 6.11. In the figure, transfer of thermal energy into the system is represented by q and heat transfer out by $-q$; work transfer of energy into the system is represented by w and work transfer out by $-w$. In the case of sub-

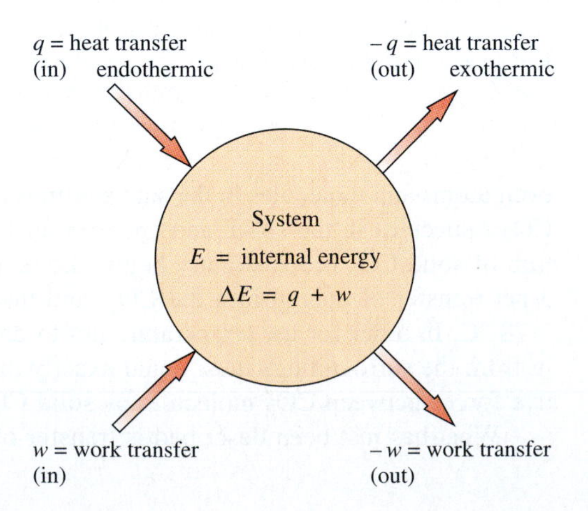

Figure 6.11 System and surroundings. Schematic diagram showing energy transfers between a thermodynamic system and its surroundings.

Portrait of a Scientist • *James Prescott Joule (1818–1889)*

James P. Joule. *(Oesper Collection in the History of Chemistry, University of Cincinnati)*

The person who established that heating and working were both processes by which energy is transferred from one sample of matter to another was James Joule. Born into a wealthy brewing family in Salford, England, in 1818, Joule received some formal training in mathematics from John Dalton, but most of his education was obtained through his own studies, without a teacher.

Between 1837 and 1847, Joule experimented with heating, chemical reactions, mechanical devices, and other means of energy transfer. His experiments were important in establishing the law of conservation of energy. In addition they disproved the "caloric theory." This was the idea that heat is a kind of fluid that flows from one place to another in the same way that water flows from a bottle when poured. The scientists of Joule's day referred to heating a beaker with a Bunsen burner as a flow of "caloric fluid" from the flame to the water. However, in cases where both heating and working were involved (for example, raising a book by expansion of carbon dioxide gas) the quantity of caloric fluid apparently changed. This was difficult for the caloric theory to explain, because the quantity of water moving from one place to another did not change, and "caloric fluid" was supposed to behave like any other fluid.

Joule set up an apparatus in which he could carefully measure the quantity of mechanical energy input to a thermally insulated sample of water. Weights were allowed to fall under the influence of gravity, and their fall turned a paddle in the water. The friction of paddle on water raised the temperature of the water. Joule was able to show that the quantity of heating of the water by the falling weights was directly proportional to the reduction in the potential energy of the weights when they moved from a higher to a lower position. The modern-day equivalent of Joule's experimental result is that 4.184 J of mechanical energy can raise the temperature of 1 g of water by 1 °C. The same quantity of heating can also raise the temperature of 1 g of water by 1 °C. This is known as the mechanical equivalent of heat and was an important advance in our understanding of thermodynamics.

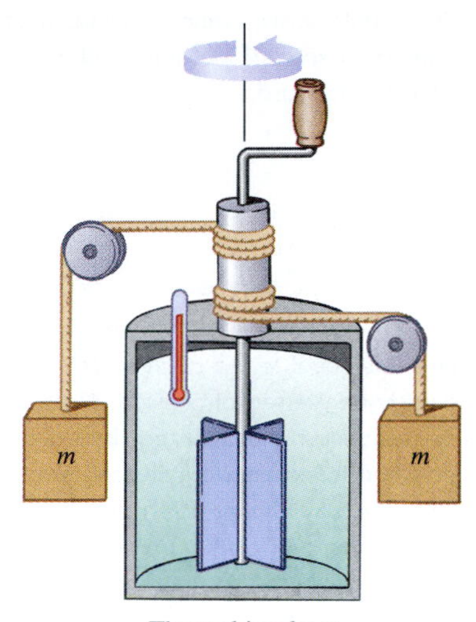

Joule's apparatus for mechanical equivalent of heat.

limation of CO_2 (Figure 6.10), w is negative, because energy is transferred out of the system to raise the book. The change in energy of such a system is

$$\Delta E = q_p + w$$

where q_p is the thermal energy transfer at constant pressure. The energy of the system is changed by the quantity of energy transferred by heating and by the quantity of energy transferred by working. This is another statement of the first law of thermodynamics. Energy may be transferred by working or heating, but no energy can be lost, nor can heating or working be obtained from nothing. *All the energy transferred between a system and its surroundings must be accounted for as heating and working, and the total quantity of energy does not change.*

Even if the book had not been on top of the plastic bag in Figure 6.10c, work would have been done by the expanding gas. Whenever a gas expands into the atmosphere it has to push back the atmosphere itself. Instead of raising a book, the expanding gas raises a part of the atmosphere. For sublimation of CO_2 at constant pressure, the transfer of thermal energy from the surroundings has two effects: (1) it overcomes the forces holding the molecules together in the solid state at −78 °C, and (2) it does work on the atmosphere as the gas expands. The energy that allows the system to perform work on its surroundings has to come from somewhere, and that somewhere was transfer of thermal energy, *q*, to the system from its surroundings. The system simply converts part of that thermal energy into work.

Systems that convert heat into work are called heat engines: an example is the engine in an automobile, which converts heat from the combustion of fuel into work to move the car forward.

In plants and animals, as well as in laboratories, reactions usually occur in contact with the atmosphere, that is, at constant atmospheric pressure. They do not occur in closed containers (at constant volume). Therefore we usually deal with processes at constant pressure, and it is convenient to define *the quantity of thermal energy transferred into a system at constant pressure,* q_p, as the **enthalpy change** of the system. Enthalpy change is symbolized by ΔH, which as usual represents $H_{\text{final}} - H_{\text{initial}}$. Thus,

$$\Delta H = q_p = \text{quantity of thermal energy transferred into a system at constant pressure}$$

Since $\Delta E = q_p + w$ at constant pressure, and $q_p = \Delta H$, then for a constant-pressure process $\Delta E = \Delta H + w$. ΔH accounts for all the energy transferred except the quantity that does the work of pushing back the atmosphere, which is usually relatively small. That is, ΔH is closely related to the change in the internal energy of the system but is slightly different in magnitude.

Because it is equal to the quantity of thermal energy transferred at constant pressure, and because most chemical reactions are carried out at atmospheric (constant) pressure, the enthalpy change for a process is often called the *heat of that process.* For example, the enthalpy change for sublimation is also called the heat of sublimation.

Thermochemical Equations

We mentioned earlier that evaporation of water cools you when perspiration evaporates from your skin. A small extension of the typical chemical equations written in earlier chapters allows us to describe this endothermic process concisely. We write what is called a **thermochemical equation.** This is *a balanced chemical equation together with the corresponding value of the enthalpy change.* For evaporation of water the thermochemical equation can be written.

$$H_2O(\ell) \longrightarrow H_2O(g) \qquad \Delta H^\circ = +44.0 \text{ kJ}$$

Because the value of the enthalpy change depends on the pressure at which the process is carried out, all enthalpy changes are reported at the same standard pressure. The value of the enthalpy change also varies slightly with temperature. For thermochemical equations in this book, the temperature can be assumed to be 25 °C, unless some other temperature is specified. The symbol ΔH° (pronounced delta-aitch-standard) represents the **standard enthalpy change,** which is defined as the *enthalpy change at the standard pressure of 1 bar and a specified temperature.* The bar is a unit of pressure that is very close to the pressure of the earth's atmosphere at sea level; you may have heard the bar used in a weather report.

In 1982 the International Union of Pure and Applied Chemistry chose a pressure of 1 bar as the standard for tabulating information for thermochemical equations. This pressure is very close to the standard atmosphere: 1 bar = 0.98692 atm = 1 kg/m·s^2. (Pressure units are discussed further in Section 14.1.)

The thermochemical equation given above indicates that when *one mole* of liquid water (at 25 °C and 1 bar) evaporates to form *one mole* of water vapor (at 25 °C and 1 bar), 44.0 kJ of energy must be transferred from the surroundings to the system to maintain the temperature at 25 °C. The size of the enthalpy change depends on how much process (in this case evaporation) takes place. If more sweat evaporates, you will be cooled more. If 2 mol $H_2O(\ell)$ is converted to 2 mol $H_2O(g)$, 88 kJ of energy is transferred, and if 0.5 mol $H_2O(\ell)$ is converted to 0.5 mol $H_2O(g)$, only 22 kJ is required. The numerical value of ΔH° corresponds to the reaction as written, with the coefficients indicating moles of each reactant and moles of each product. If this equation had been written

$$2\ H_2O(\ell) \longrightarrow 2\ H_2O(g) \qquad \Delta H^\circ = +88.0 \text{ kJ}$$

then the process would correspond to evaporating 2 mol $H_2O(\ell)$ to form 2 mol $H_2O(g)$, both at 25 °C and 1 bar. Then the enthalpy change is twice as great as for the case where there is a coefficient of 1 on each side of the equation.

Exercise 6.7 Interpreting Thermochemical Equations

What part of the thermochemical equation for vaporization of water indicates that energy is transferred from the surroundings to the system when the evaporation process occurs?

Now consider water vapor condensing to form liquid. If 44.0 kJ of energy is required to do the work of separating the water molecules in 1 mol of the liquid as it vaporizes, the same quantity of energy will be released when the molecules move closer together as the vapor condenses to form liquid.

$$H_2O(g) \longrightarrow H_2O(\ell) \qquad \Delta H° = -44.0 \text{ kJ}$$

This thermochemical equation indicates that 44.0 kJ of energy is transferred *to* the surroundings *from* the system when 1 mol of water vapor condenses to liquid at 25 °C and 1 bar.

The idea here is similar to the example given earlier of water falling over a waterfall. The decrease in potential energy of the water when it falls from top to bottom of the waterfall is exactly equal to the increase in potential energy that would be required to take the same quantity of water from the bottom of the fall to the top. The signs are opposite because in one case potential energy is transferred *from* the water and in the other it is transferred *to* the water.

Exercise 6.8 Thermochemical Equations

Why is it essential to specify the state of each reactant and each product in a thermochemical equation?

Several very useful ideas that apply to all of thermodynamics are noted in Table 6.2.

TABLE 6.2 Characteristics of Thermochemical Equations

Sign of $\Delta H°$	When transfer of energy occurs at constant pressure from a system to its surroundings, $\Delta H°$ has a negative value (exothermic process). Conversely, when energy is transferred to the system from the surroundings, $\Delta H°$ has a positive value (endothermic process).
Opposite processes	For changes that are the reverse of each other, the $\Delta H°$ values are numerically the same, but their signs are opposite. Thus, for evaporation of 1 mol of water $\Delta H° = +44.0$ kJ, while for the condensation of 1 mol of water $\Delta H° = -44.0$ kJ.
Quantity of material	The change in energy or enthalpy is directly proportional to the quantity of material undergoing a change. If 2 mol of water were evaporated, twice as much energy transfer, or 88.0 kJ, would be required.
Balanced equation represents moles	The value of $\Delta H°$ is always associated with a balanced equation for which the coefficients are read as moles, so that the equation shows the macroscale quantity of material to which the value of $\Delta H°$ applies. This makes it possible to calculate transfers of thermal energy for any quantity of substance.

PROBLEM-SOLVING EXAMPLE 6.3 Changes of State and $\Delta H°$

Calculate the energy transferred to the surroundings when water vapor in the air condenses to give rain in a thunderstorm. Suppose that one inch of rain falls over one square mile of ground, so that $6.6 \times 10^{10} cm^3$ 1 or 6.6×10^{10} g of water (assuming a density of
1.0 g/cm^3) has fallen.

Answer 1.6×10^{11} kJ

Explanation The standard enthalpy of condensation for 1 mol of water at 25 °C is −44.0 kJ. Therefore, we first calculate how many moles of water condensed.

$$\text{Amount of water condensed} = 6.6 \times 10^{10} \text{ g water} \times \frac{1 \text{ mol}}{18.0 \text{ g}} = 3.66 \times 10^{9} \text{ mol water}$$

Next, calculate the quantity of energy transferred from the fact that 44.0 kJ is transferred per mole of water.

$$\text{Quantity of energy transferred} = 3.66 \times 10^{9} \text{ mol water} \times \frac{-44.0 \text{ kJ}}{1 \text{ mol}}$$
$$= -1.6 \times 10^{11} \text{ kJ}$$

The negative sign indicates that energy is transferred from the water to the surrounding air.

Since the explosion of 1000 tons of dynamite is equivalent to 4.2×10^{9} kJ, the energy transferred by our hypothetical thunderstorm is about the same as that released when 38,000 tons of dynamite explodes! A great deal of energy can be stored in water vapor, which is one reason why storms can cause so much damage.

Like all examples in this chapter, this one assumes that the temperature of the system remains constant, so that all the energy associated with the phase change transfers to or from the surroundings.

Problem-Solving Practice 6.3

The enthalpy change for sublimation of 1 mol of solid iodine at 25 °C and 1 bar is 62.4 kJ.

$$I_2(s) \longrightarrow I_2(g) \qquad \Delta H° = +62.4 \text{ kJ}$$

(a) What quantity of energy must be transferred to vaporize 10.0 g of solid iodine?
(b) If 3.45 g of iodine vapor condenses to solid iodine, what quantity of energy is transferred? Is the process exothermic or endothermic?

Figure 6.12 Portable hand warmer. Opening the tightly sealed outer package exposes iron to air and the oxidation reaction warms its surroundings. The increase in temperature is sufficient to keep your fingers or toes toasty warm in cold weather. *(C.D. Winters)*

6.5 ENTHALPY CHANGES FOR CHEMICAL REACTIONS

Like phase changes, chemical reactions can be exothermic or endothermic, but reactions usually involve much larger energy transfers than do phase changes. Indeed, a temperature change is one piece of evidence that a chemical reaction has taken place. The following Chemistry You Can Do involves an exothermic reaction of iron with oxygen from the air, and the same reaction is used in portable hand-warming packets like the one shown in Figure 6.12. *The large energy transfers that occur during chemical reactions are the result of breaking and making chemical bonds as reactants are converted into products.* These energy transfers have important applications in living systems, in industrial processes, in heating or cooling your home, and in many other situations.

During a phase change, energy transfer occurs because atoms, ions, or molecules that are attracted to one another become more widely separated or move closer to-

CHEMISTRY YOU CAN DO

Rusting and Heating

Chemical reactions can heat their surroundings, and a simple experiment demonstrates this very well. To do it you will need a steel wool pad (without soap), $\frac{1}{4}$ cup of vinegar, a cooking or outdoor thermometer, and a large jar with a lid. (The thermometer must fit inside the jar.)

Soak the steel wool pad in vinegar for several minutes. While doing so, place the thermometer in the jar, close the lid, and let it stand for several minutes. Read the temperature.

Squeeze the excess vinegar out of the steel wool pad, wrap the pad around the bulb of the thermometer, and place both in the jar. Close the lid. After about 5 min, read the temperature again. What has happened?

Repeat the experiment with another steel pad, but wash it with water instead of vinegar. Try a third pad that is not washed at all. Allow each pad to stand in air for a few hours or for a day and observe the pad carefully. Do you see any change in the metal? Suggest an explanation for your observations of temperature changes and appearance of the steel wool.

(C.D. Winters)

gether. Separating such nanoscale particles usually increases their potential energy, and so energy usually must be transferred in from the surroundings. Moving nanoscale particles closer together usually has the opposite effect: their potential energy goes down and that energy difference is transferred from the system of particles to its surroundings.

During a chemical reaction, the situation is related, but different, because during a chemical reaction chemical compounds are created or broken down. It is atoms that move farther apart or closer together. Atoms are attracted to each other by chemical bonds, which involve large attractive forces. When existing chemical bonds are broken and new chemical bonds are formed, the energy differences are usually much greater than for phase changes. Because the atoms are the same before and after a chemical reaction, but their connections with each other have changed, *it is useful and convenient to define the system for a chemical reaction as all of the atoms in the reactants (and of course the products) of the reaction.*

As an example, consider the reaction of hydrogen gas with chlorine gas to form hydrogen chloride gas.

$$H_2(g) + Cl_2(g) \longrightarrow 2\ HCl(g)$$

In order for this reaction to occur it is necessary to separate the two hydrogen atoms in a H_2 molecule and to separate the two chlorine atoms in a Cl_2 molecule, so that the atoms can combine in a different way—as two HCl molecules. The simplest way to imagine these separations and combinations of atoms is to separate all the atoms in the reactant molecules and then put them back together to form product molecules.

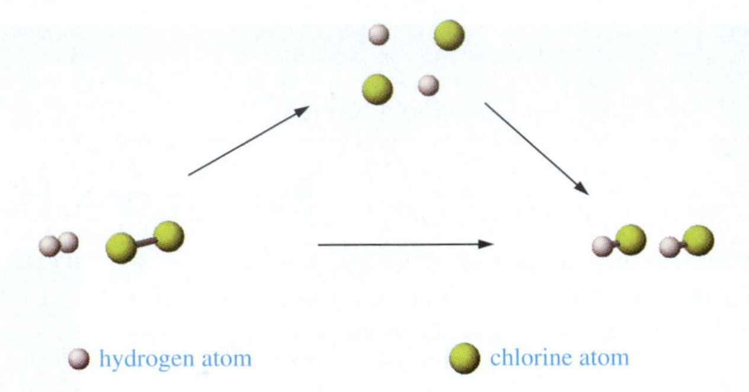

Figure 6.13 Rearranging atoms in a chemical reaction. Carrying out a reaction by separating all the atoms of the reactant molecules and then putting them together to form product molecules. In the upper pathway a H_2 molecule and a Cl_2 molecule are broken into two H atoms and two Cl atoms. These atoms are rearranged into two HCl molecules. In the lower pathway the H_2 molecule and the Cl_2 molecule are transformed directly into two HCl molecules.

This is diagrammed for the reaction of hydrogen with chlorine in Figure 6.13. In the upper pathway the first step is to break all bonds in the reactant H_2 and Cl_2 molecules. The second step is to form the bonds in the two product HCl molecules.

The reaction of hydrogen with chlorine actually occurs by a complicated series of steps. However, the enthalpy change for a reaction is like the change in altitude when you climb a mountain. No matter which route you take to the summit (which atoms you separate or combine first), the difference in altitude between the summit and where you started to climb (the enthalpy difference between products and reactants) is the same. This means that we can concentrate on products and reactants and not worry about exactly what happens in between.

Exercise 6.9 Reaction Pathways and Enthalpy Change

Suppose that the enthalpy change were different depending on the pathway a reaction took from reactants to products. For example, suppose that 190 kJ were released when a mole of hydrogen gas and a mole of chlorine gas combine to form two moles of hydrogen chloride (the lower pathway in Figure 6.13), but suppose that only 185 kJ were released when the same reactant molecules are broken into atoms and the atoms are recombined to form hydrogen chloride (the upper pathway in Figure 6.13). Would this violate the first law of thermodynamics? Explain why or why not.

The bond energies in Table 6.3 are for *gas-phase reactions.* If liquids or solids are involved there are additional energy transfers for the phase changes needed to convert the liquids or solids to the gas phase. We shall restrict our use of bond energies to gas-phase reactions for that reason.

Bond Energies

Separating atoms that are bonded together requires a transfer of energy into the system, because work must be done against the force holding the atoms together. *The enthalpy change that occurs when two bonded atoms in a gas-phase molecule are separated completely* is called the **bond energy** or, more correctly, the **bond enthalpy.** (The two terms are often used interchangeably.) The bond energy is usually expressed per mole of bonds. For example, the bond energy for a Cl_2 molecule is 243 kJ/mol, and so we can write

$$Cl_2(g) \longrightarrow 2\ Cl(g) \qquad \Delta H^\circ = 243\ \text{kJ}$$

TABLE 6.3 Average Bond Energies (in kJ/mol)

Single Bonds

	I	Br	Cl	S	P	Si	F	O	N	C	H
H	297	368	431	339	318	293	569	463	389	414	436
C	238	276	330	259	264	289	439	351	293	347	
N	—	243?	201	—	209	—	272	201	159		
O	201	—	205	—	351	368	184	138			
F	—	197	255	285	490	540	159				
Si	213	289	360	226	213	176					
P	213	272	331	230	213						
S	—	213	251	213							
Cl	209	218	243								
Br	180	192									
I	151										

Multiple Bonds

N$=$N	418	C$=$C	611
N$\equiv$N	946	C$\equiv$C	837
C$=$N	615	C$=$O (in CO_2, O$=$C$=$O)	803
C$\equiv$N	891	C$=$O (as in $H_2C{=}O$)	745
O$=$O (in O_2)	498	C$\equiv$O	1075

Bond energies are always positive, because there is always a transfer of energy into the system (in this case the mole of Cl_2 molecules) in order to increase the potential energy of the atoms that are being separated. *Bond breaking* is always *endothermic.* Conversely, when atoms come together to form a bond, energy will invariably be transferred to the surroundings, because the potential energy of the atoms is lower when they are bonded together than when they are separated. The *formation of bonds* from separated atoms is always *exothermic.*

Bond energies provide a way to see what makes a process exothermic or endothermic. If the total energy transferred out of the system when new bonds are formed is greater than the total energy required to break all of the bonds in the reactants, then the reaction is exothermic. If the opposite is true, the reaction is endothermic. If weak bonds are broken and strong bonds are formed, then a reaction will be exothermic. If the bonds are of about the same strength, but more bonds are formed than broken, then the reaction will be exothermic. Data on the strengths of bonds between atoms in gas-phase molecules are summarized in Table 6.3, which can be used to quickly estimate whether a given reaction is exothermic or endothermic.

Conceptual Challenge Problems CP-6.E and CP-6.F at the end of the chapter relate to topics discussed in this section.

We can use data from Table 6.3 to estimate $\Delta H°$ for the reaction of hydrogen gas with chlorine gas. As shown in Figure 6.14, the reaction could be carried out by breaking the $H_2(g)$ and $Cl_2(g)$ molecules into atoms and then forming the product HCl(g) molecules. Breaking the bond of H_2 requires an *input* to the system of 436 kJ/mol; breaking the bond in Cl_2 requires 243 kJ/mol. In both of these cases the sign of the enthalpy change will be positive. Forming a bond in HCl transfers 431 kJ/mol out of the system. Since 2 mol of HCl bonds is formed, there will be 2 mol $\times$ 431 kJ/mol $=$ 862 kJ transferred out. In this case the sign of the enthalpy change will be negative.

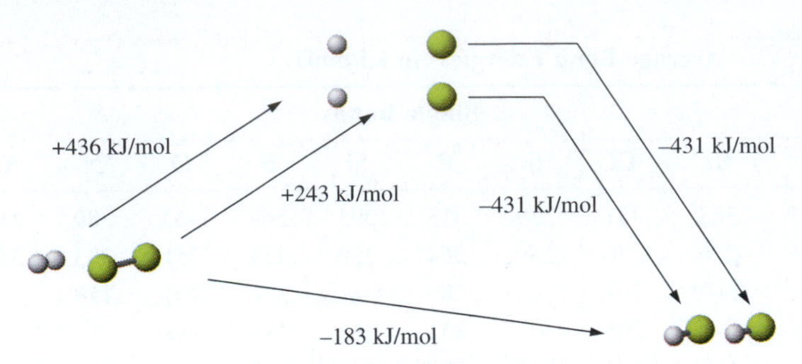

Figure 6.14 Stepwise energy changes in a reaction. Breaking a mole of H_2 molecules into H atoms requires 436 kJ. Breaking a mole of Cl_2 molecules into Cl atoms requires 243 kJ. Putting 2 mol of H atoms together with 2 mol of Cl atoms to form 2 mol of HCl provides 2×431 kJ = 862 kJ, and so the reaction is exothermic by 183 kJ.

If we represent a bond energy by the letter *D*, with a subscript to shown which bond it refers to, the net transfer of energy is

$$\Delta H^\circ = \{(1 \text{ mol H—H}) \times D_{\text{H—H}} + (1 \text{ mol Cl—Cl}) \times D_{\text{Cl—Cl}}\} - (2 \text{ mol H—Cl}) \times D_{\text{H—Cl}}$$

$$= \{(1 \text{ mol H—H}) \times (436 \text{ kJ/mol}) + (1 \text{ mol Cl—Cl}) \times (243 \text{ kJ/mol})\} - (2 \text{ mol H—Cl}) \times (431 \text{ kJ/mol})$$

$$= -183 \text{ kJ/mol}$$

This differs only very slightly from the experimentally determined value of −184.614 kJ.

Notice how the calculation was done. We thought of the process in terms of breaking all the bonds in each reactant molecule and then forming all the bonds in the product molecules, as diagrammed in Figure 6.14. Each bond energy was multiplied by the number of moles of bonds that were broken or formed. For bonds in reactant molecules, we added the bond energies because breaking bonds is endothermic. For products, we subtracted the bond energies because bond formation is an exothermic process. The way this calculation was done can be summarized in the following equation:

$$\Delta H^\circ = \Sigma\{(\text{moles of bonds}) \times D(\text{bonds broken})\} - \Sigma\{(\text{moles of bonds}) \times D(\text{bonds formed})\}$$

The Σ (Greek capital letter sigma) represents summation. We add the bond energies for all bonds broken, and we subtract the bond energies for all bonds formed. This equation and the values in Table 6.3 allow us to estimate enthalpy changes for a wide variety of gas-phase reactions.

There are several important points about the bond energies in Table 6.3:

- The energies listed are often *average* bond energies and may vary from molecule to molecule. For example, the energy of a C—H bond is given as 414 kJ/mol, but C—H bond strengths are affected by other atoms and bonds in the same molecule. Depending on the molecule, the energy required to break a mole of C—H bonds may vary by 30 to 40 kJ/mol, and so the values in Table 6.3 can only be used to *estimate* an enthalpy change, not to calculate it exactly.
- The energies in Table 6.3 are for breaking bonds in molecules in the gaseous state. If a reactant or product is in the liquid or solid state, the energy required

to convert it to or from the gas phase will also contribute to the enthalpy change and must be accounted for.

- Multiple bonds, shown as double and triple lines between atoms, are listed at the bottom of Table 6.3. In some cases different energies are given for multiple bonds in specific molecules such as $O_2(g)$ or $CO_2(g)$. Multiple bonds, and bonds in general, will be discussed more fully in Chapter 9.

PROBLEM-SOLVING EXAMPLE 6.4 Estimating $\Delta H°$ from Bond Energies

Estimate $\Delta H°$ for burning a mole of methane according to the following equation:

$$CH_4(g) + 2\ O_2(g) \longrightarrow CO_2(g) + 2\ H_2O(g)$$

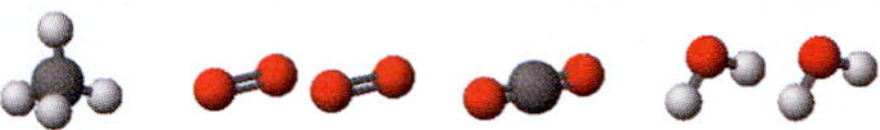

Answer $\Delta H° = -806$ kJ

Explanation From the pictures of the molecules and their bonds we can see that there are four C—H bonds in a CH_4 molecule. The O_2 molecule and the CO_2 molecule are special cases; their bond energies are listed at the bottom of Table 6.3. In each H_2O molecule there are two O—H bonds.

$$\Delta H° = \Sigma\{(\text{moles of bonds}) \times D(\text{bonds broken})\} - \Sigma\{(\text{moles of bonds}) \times D(\text{bonds formed})\}$$

$$= \{(4\text{ mol C—H}) \times D_{C—H} + (2\text{ mol O═O}) \times D_{O═O}\} - \{(2\text{ mol C═O}) \times D_{C═O} + (4\text{ mol O—H}) \times D_{O—H}\}$$

$$= \{(4\text{ mol C—H}) \times (414\text{ kJ/mol}) + (2\text{ mol O═O}) \times (498\text{ kJ/mol})\} - \{(2\text{ mol C═O}) \times (803\text{ kJ/mol}) + (4\text{ mol O—H}) \times (463\text{ kJ/mol})\}$$

$$= -806\text{ kJ}$$

The experimental value for the enthalpy of combustion of methane is −802 kJ, and so the average bond energies give a good estimate.

Problem-Solving Practice 6.4

Use values from Table 6.3 to estimate the enthalpy change when hydrogen and oxygen combine according to the following equation:

$$2\ H_2(g) + O_2(g) \longrightarrow 2\ H_2O(g)$$

Exercise 6.10 Enthalpy Change and Bond Energies

Consider the exothermic reactions in Problem-Solving Example 6.4 and Problem-Solving Practice 6.4. In which case is formation of stronger bonds the most important factor in making the reaction exothermic? In which case is formation of a larger number of bonds most important?

6.6 USING THERMOCHEMICAL EQUATIONS FOR CHEMICAL REACTIONS

To learn more about enthalpy changes and thermochemical equations, consider the formation of water vapor from its elements.

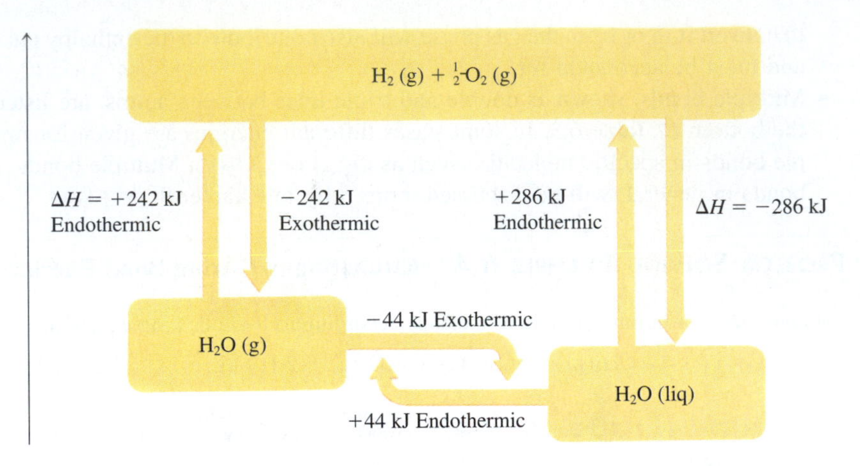

Figure 6.15 Enthalpy diagram. Water vapor ($H_2O(g)$), liquid water ($H_2O(\ell)$), and the elements $H_2(g)$ and $\frac{1}{2}$ $O_2(g)$ all have different enthalpy values. The diagram shows how these are related.

$$H_2(g) + \tfrac{1}{2}\,O_2(g) \longrightarrow H_2O(g) \qquad \Delta H^\circ = -241.8\text{ kJ}$$

Change is exothermic; ΔH° is negative. Formation of 1 mol of water vapor transfers 241.8 kJ of energy to the surroundings.

Note that in order to write an equation for the formation of 1 mol of H_2O it is necessary to use a fractional coefficient for O_2. This is acceptable in a thermochemical equation because coefficients are taken to mean moles and not molecules. The left side of Figure 6.14 shows that the enthalpy of the reactants of this reaction is greater than that of the product. Because the product has less enthalpy, 241.8 kJ must be transferred to the surroundings (law of conservation of energy). That is, the enthalpy change for the *exo*thermic formation of 1 mol of water vapor is $\Delta H^\circ = +241.8$ kJ. The same important characteristics of thermochemical equations that were given in Table 6.2 for phase changes apply to thermochemical equations for chemical reactions; take a moment to go back and review them now.

The relationship of reaction heat transfer and enthalpy change:
- Reactant → product with transfer of thermal energy to surroundings
- ΔH is negative; reaction is *exo*thermic.
- Reactant → product, with transfer of thermal energy from surroundings
- ΔH is positive; reaction is *endo*thermic.

Since energy is transferred when a substance undergoes a change of state, the energy transfer associated with a chemical reaction depends on the physical states (solid, liquid, or gas) of the reactants and products. For example, the formation of 1 mol of *liquid* water from H_2 and O_2

$$H_2(g) + \tfrac{1}{2}\,O_2(g) \longrightarrow H_2O(\ell) \qquad \Delta H^\circ = -285.8\text{ kJ}$$

is even more exothermic than formation of water vapor, as shown in Figure 6.15. This is because the enthalpy of liquid water is lower than for water vapor by the heat of vaporization, 44 kJ. Thus, *it is always necessary to specify the state of a reactant and each product in the thermochemical equation for a chemical reaction.*

Conceptual Challenge Problem CP-6.G at the end of the chapter relates to topics discussed in this section

Enthalpy changes for reactions have many practical applications. For instance, when enthalpies of combustion are known, the quantity of energy transferred by the combustion of a given mass of fuel can be calculated. Suppose you are designing a heating system, and you want to know how much heating can be provided per pound (454 g) of propane, C_3H_8, burned in a furnace. The reaction that occurs is *exothermic:*

Propane is a major component of bottled or LP gas, which is used to heat some houses.

$$C_3H_8(g) + 5\,O_2(g) \longrightarrow 3\,CO_2(g) + 4\,H_2O(\ell) \qquad \Delta H^\circ = -2220\text{ kJ}$$

that is, *2220 kJ of energy transfer to the surroundings occurs per mole of propane burned.* The first step in solving this problem is to find how many moles of propane are present in a 1-lb sample.

$$\text{Amount of propane} = 454 \text{ g} \times \frac{1 \text{ mol propane}}{44.10 \text{ g}} = 10.29 \text{ mol propane}$$

Then you can multiply by the energy transferred per mole of propane to find the total energy transferred.

$$\text{Energy transferred} = 10.29 \text{ mol propane} \times \frac{2220 \text{ kJ evolved}}{1.00 \text{ mol propane}} = 22{,}900 \text{ kJ}$$

This is a substantial quantity of energy. When your body completely "burns" 454 g of milk, only about 1400 kJ is evolved; milk is largely water, which is not metabolized, and so only a small part of the milk contributes energy to your body.

PROBLEM-SOLVING EXAMPLE 6.5 Calculating Energy Transferred

The reaction of iron with oxygen from the air provides the energy transferred by the hot pack shown in Figure 6.12. Assuming that the iron is converted to iron(III) oxide, how much heating can be provided by a hot pack that contains 45.4 g of iron? The thermochemical equation is

$$2\,Fe(s) + \tfrac{3}{2}\,O_2(g) \longrightarrow Fe_2O_3(s) \qquad \Delta H° = -824.2 \text{ kJ}$$

Answer 670. kJ

Explanation Begin by calculating how many moles of iron are present.

$$\text{Amount of iron} = 45.4 \text{ g} \times \frac{1 \text{ mol Fe}}{55.84 \text{ g}} = 0.8130 \text{ mol Fe}$$

Then use the enthalpy change to calculate the energy transferred. Because Fe has a coefficient of 2 in the thermochemical equation, the energy transfer is per 2 mol Fe, and this must be taken into account.

$$\text{Energy transferred} = 0.8130 \text{ mol Fe} \times \frac{-824.2 \text{ kJ}}{2 \text{ mol Fe}} = -670. \text{ kJ}$$

Thus 670. kJ is transferred by the reaction to heat your fingers. This is equal to 160. kcal. It corresponds to the heat your body could obtain from about half a pound of milk.

Problem-Solving Practice 6.5

How much thermal energy transfer would be required to maintain constant temperature during decomposition of 12.6 g of liquid water to the elements hydrogen and oxygen?

$$H_2O(\ell) \longrightarrow H_2(g) + \tfrac{1}{2}\,O_2(g) \qquad \Delta H° = +285.8 \text{ kJ}$$

Liquefied petroleum (LP) gas is used for heating. *(Tom Tracy/Tony Stone Images)*

6.7 MEASURING ENTHALPY CHANGES FOR REACTIONS: CALORIMETRY

A thermochemical equation tells us how much energy is transferred as a chemical process occurs. Knowing this enables us to calculate the heat obtainable when a fuel is burned, as was done in the preceding section. Also, when reactions are carried out on a larger scale, say in a chemical plant that manufactures sulfuric acid, the surroundings must have enough cooling capacity to prevent an exothermic reaction from overheating, speeding up, and possibly damaging the plant. For these and many other reasons it is useful to know as many $\Delta H°$ values as possible.

For many reactions, direct experimental measurements can be made by using a **calorimeter,** which is a device that measures heat transfers. Calorimetric measurements can be made at constant volume or at constant pressure. Often, in finding heats of combustion or the caloric value of foods, where at least one of the reactants is a gas, the measurement is done at a constant volume in a *bomb calorimeter* (Figure 6.16). The "bomb" is a cylinder about the size of a large fruit-juice can with heavy steel walls so that it can contain high pressures. A weighed sample of a combustible solid or liquid is placed in a dish inside the bomb, which is filled with pure O_2 and then placed in a water-filled container with well-insulated walls. The sample is ignited, usually by an electrical spark. The heat generated when the sample burns warms the bomb and the water around it.

In this configuration, the oxygen and the compound represent the *system* and the bomb and the water around it are the *surroundings.* From the law of the conservation of energy, we can say that energy transferred from the system = energy transferred to the surroundings,

$$-q_{\text{reaction}} = (q_{\text{bomb}} + q_{\text{water}})$$

where q_{reaction} has a negative value because the combustion reaction is exothermic. The temperature change of the water, which is the same as for the bomb, is measured. Then the total energy transfer, $q_{\text{bomb}} + q_{\text{water}}$, can be calculated from the heat capacities of the bomb and the water. According to the equation above, this total gives the energy evolved by combustion of the compound. Since the bomb is rigid, the heat transfer is measured at *constant volume* and is therefore equivalent to ΔE, the change

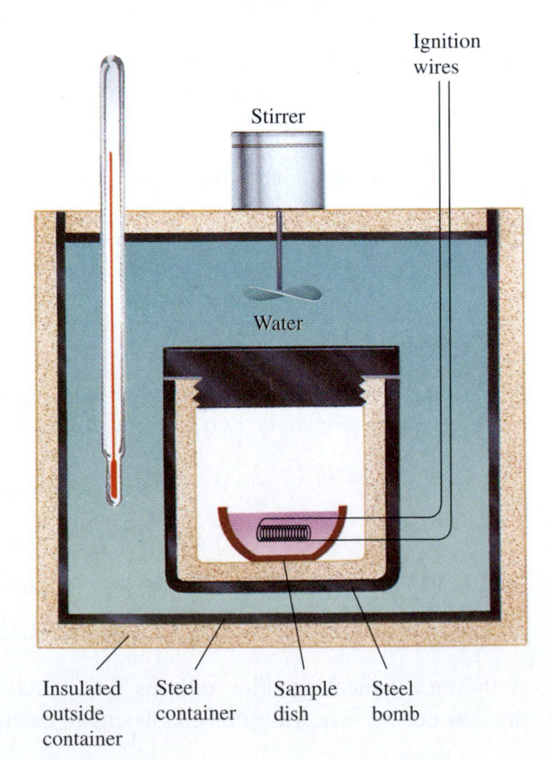

Figure 6.16 Combustion calorimeter. A combustible sample is burned in pure oxygen in a steel "bomb." Energy is transferred into the bomb and the water surrounding it as a result of the reaction. The heat transfer warms both to the same temperature. From the measured temperature increase, the quantity of energy transferred by the reaction can be calculated.

in internal energy (← *p. 224*). Using $\Delta E = \Delta H + w$ and calculating the work that would be done under conditions of constant pressure allows us to calculate ΔH (and ΔH°) from ΔE.

When reactions take place in solution, it is much easier to use a calorimeter that is open to the atmosphere. An example, often encountered in introductory chemistry courses, is the *coffee-cup calorimeter* shown in Figure 6.17. The nested coffee cups provide good thermal insulation; reactions can occur when solutions are poured together in the inner cup. Because a coffee-cup calorimeter is a constant-pressure device, the measured heat transfer is q_p, which equals ΔH°.

In a coffee-cup calorimeter the masses of substances other than the solvent water are often so small that their heat capacities can be ignored; all of the energy of a reaction can be assumed to be transferred to the water.

PROBLEM-SOLVING EXAMPLE 6.6 Measuring Enthalpy Change with a Coffee-Cup Calorimeter

When 0.800 g of magnesium is added to 250. mL of 0.40 M HCl in a coffee-cup calorimeter at 1 bar, the temperature of the solution increases from 23.4 °C to 37.9 °C. Determine the enthalpy change for the reaction below and complete the thermochemical equation. Assume that the heat capacities of the coffee cups, the temperature probe, and the stirrer are negligible and that the solution has the same density and the same specific heat capacity as water.

$$Mg(s) + 2\ HCl(aq) \longrightarrow H_2(g) + MgCl_2(aq) \qquad \Delta H^\circ = ?$$

Answer $\Delta H^\circ = 462$ kJ

Explanation Use the definition of specific heat capacity to calculate q_p, the heat transfer. The mass of solution is the mass of the HCl solution plus the mass of the Mg, 250.8 g.

$$q_p = c \times m \times \Delta T = (4.184\ \text{J/g}\cdot{}^\circ\text{C}) \times (250.8\ \text{g}) \times (37.9 - 23.4)\ {}^\circ\text{C}$$
$$= 1.521 \times 10^4\ \text{J} = 15.21\ \text{kJ}$$

This is the calculated heat transfer for 0.800 g Mg or for 2.50 mL of 0.40 M HCl, whichever is the limiting reactant. Determine which was the limiting reactant by calculating the number of moles of Mg, the number of moles of HCl, and applying the stoichiometric factor:

$$\text{Amount of Mg} = 0.800\ \text{g} \times \frac{1\ \text{mol}}{24.31\ \text{g}} = 3.291 \times 10^{-2}\ \text{mol Mg}$$

$$\text{Amount of HCl} = 250.\ \text{mL} \times \frac{0.40\ \text{mol}}{1000\ \text{mL}} = 1.00 \times 10^{-1}\ \text{mol HCl}$$

Since 2 mol HCl is required for each 1 mol Mg, and since there are more than twice as many moles of HCl as of Mg, Mg is the limiting reactant and should be used to calculate the enthalpy change. Because the thermochemical equation involves 1 mol Mg, the heat transfer must be scaled to this quantity of Mg.

$$\Delta H^\circ = 15.21\ \text{kJ} \times \frac{1\ \text{mol Mg}}{3.291 \times 10^{-2}\ \text{mol Mg}} = 462\ \text{kJ}$$

Problem-Solving Practice 6.6

Suppose that 1.00 mL of 1.0 M HCl and 100. mL of 1.0 M NaOH, both at 20.4 °C, are mixed in a coffee-cup calorimeter. If the thermochemical equation for the neutralization reaction is

$$H^+(aq) + OH^-(aq) \longrightarrow H_2O(\ell) \qquad \Delta H^\circ = -58.7\ \text{kJ}$$

to what final temperature will the solution in the calorimeter be heated? Assume that the specific heat capacity and density of the solution are the same as for water, and that the heat capacities of all other substances in contact with the reaction are negligible.

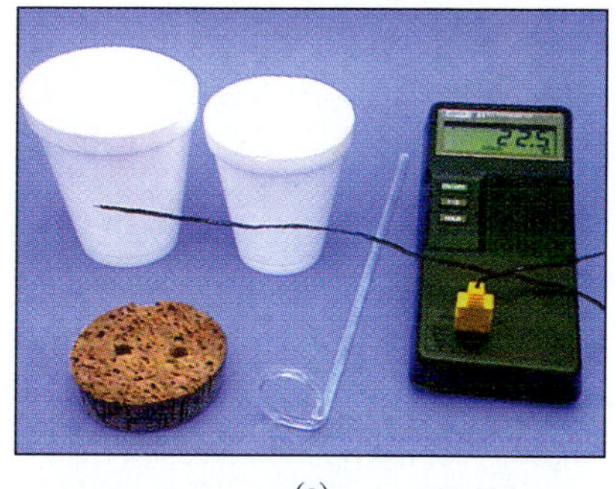

(a)

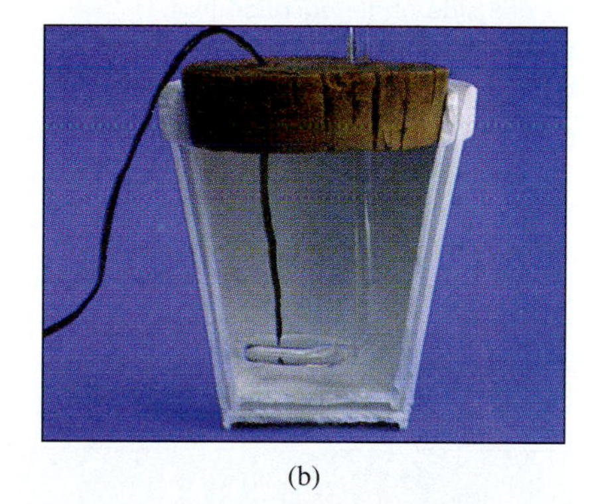

(b)

Figure 6.17 Coffee-cup calorimeter. (a) A constant-pressure calorimeter can be made from two coffee cups that are good thermal insulators, a cork or other insulating lid, a temperature probe, and a stirrer. (b) Close-up of the nested cups that make up the calorimeter. A reaction carried out in an aqueous solution within the calorimeter will change the temperature of the solution, because the thermal insulation is extremely good and essentially no energy transfer can occur to or from anything outside the calorimeter. *(Jerrold J. Jacobsen)*

6.8 HESS'S LAW

Calorimetry works well for some reactions, but for many others it is not a simple task. Besides, it would be very time-consuming to measure values for every conceivable reaction, and it would take a great deal of space to tabulate so many values. Fortunately there is a better way. It is based on **Hess's law,** which states that, *if the equation for a reaction is the sum of the equations for two or more other reactions, then* ΔH° *for the first reaction must be the sum of* ΔH° *values of the other reactions.* Hess's law is another way of stating the law of conservation of energy. It works even if the overall reaction does not actually occur by way of the separate equations that are summed.

For example, in Figure 6.15 (← *p. 246*) we noted that the formation of liquid water from its elements $H_2(g)$ and $O_2(g)$ could be thought of as two successive changes: (1) formation of water vapor from the elements and (2) condensation of water vapor to liquid water. As shown below, the equation for decomposition of liquid water can be obtained by adding algebraically the chemical equations for these two steps. Therefore, according to Hess's law, the ΔH° value can be found by adding the ΔH° values for the two steps.

Note that it takes 1 mol $H_2O(g)$ to cancel 1 mol $H_2O(g)$. If the coefficient of $H_2O(g)$ had been different on one side of the equation from the coefficient on the other side, $H_2O(g)$ could not have been completely canceled.

(1) $H_2(g) + \frac{1}{2}O_2(g) \longrightarrow \cancel{H_2O(g)}$ $\qquad \Delta H_1^\circ = -241.8 \text{ kJ}$

(2) $\cancel{H_2O(g)} \longrightarrow H_2O(\ell)$ $\qquad \Delta H_2^\circ = -44.0 \text{ kJ}$

(1) + (2) $H_2(g) + \frac{1}{2}O_2(g) \rightarrow H_2O(\ell)$ $\qquad \Delta H^\circ = \Delta H_1^\circ + \Delta H_2^\circ = -285.8 \text{ kJ}$

Here, 1 mol $H_2O(g)$ is a product of the first reaction and a reactant in the second. Thus $H_2O(g)$ can be canceled out. This is similar to adding two algebraic equations: if the same quantity or term appears on both sides of the equation, it cancels. The net result is an equation for the overall reaction and its associated enthalpy change. This overall enthalpy change applies even if the liquid water is formed directly from hydrogen and oxygen.

Some common products contained in polyethylene bottles. *(C.D. Winters)*

PROBLEM-SOLVING EXAMPLE 6.7 Using Hess's Law

In designing a chemical plant for manufacturing the plastic polyethylene, you need to know the enthalpy change for the removal of H_2 from C_2H_6 (ethane) to give C_2H_4 (ethylene), a key step in the process.

$$\underset{\text{ethane}}{C_2H_6(g)} \longrightarrow \underset{\text{ethylene}}{C_2H_4(g)} + H_2(g) \qquad \Delta H^\circ = ?$$

From experiments you know the following thermochemical equations:

(a) $2\ C_2H_6(g) + 7\ O_2(g) \longrightarrow 4\ CO_2(g) + 6\ H_2O(\ell)$ $\qquad \Delta H_a^\circ = -3119.4 \text{ kJ}$

(b) $C_2H_4(g) + 3\ O_2(g) \longrightarrow 2\ CO_2(g) + 2\ H_2O(\ell)$ $\qquad \Delta H_b^\circ = -1410.9 \text{ kJ}$

(c) $H_2(g) + \frac{1}{2}O_2(g) \longrightarrow H_2O(\ell)$ $\qquad \Delta H_c^\circ = -285.8 \text{ kJ}$

Use this information to find the value of ΔH° for the formation of ethylene from ethane.

Answer $\Delta H^\circ = +137.0$ kJ

Explanation Reactions (a), (b), and (c) as written cannot be added to give the desired equation. Equation (a) involves 2 mol of ethane on the reactant side, but only 1 mol is required in the desired equation. Equation (b) has C_2H_4 as a reactant, but it is a product in the desired equation. Equation (c) has H_2 as a reactant, but it is a product in the desired equation. Therefore, we must change each equation in some way so that summing all three gives the desired equation.

First, since the desired equation has only 1 mol of ethane on the reactant side, we multiply equation (a) by $\frac{1}{2}$ to give an equation (a′) that also has 1 mol of ethane on the reactant side. Halving the equation also halves the enthalpy change.

$$(a') = \tfrac{1}{2}(a) \qquad C_2H_6(g) + \tfrac{7}{2}\,O_2(g) \longrightarrow 2\,CO_2(g) + 3\,H_2O(\ell) \qquad \Delta H^\circ_{a'} = -1559.7\text{ kJ}$$

Next, we reverse equation (b) so that C_2H_4 is on the product side, giving equation (b′). This also reverses the sign of the enthalpy change. (If combustion of C_2H_4 is exothermic, the reverse reaction must be endothermic.)

$$(b') = -(b) \qquad 2\,CO_2(g) + 2\,H_2O(\ell) \longrightarrow C_2H_4(g) + 3\,O_2(g) \qquad \Delta H^\circ_{b'} = -\Delta H^\circ_b = +1410.9\text{ kJ}$$

To get $H_2(g)$ on the product side, we also reverse equation (c) and the sign of its enthalpy change.

$$(c') = -(c) \qquad H_2O(\ell) \longrightarrow H_2(g) + \tfrac{1}{2}\,O_2(g) \qquad \Delta H^\circ_{c'} = -\Delta H^\circ_c = +285.8\text{ kJ}$$

Now it is possible to add equations (a′), (b′), and (c′) to give the desired equation.

$$(a') \qquad C_2H_6(g) + \cancel{\tfrac{7}{2}\,O_2(g)} \longrightarrow \cancel{2\,CO_2(g)} + \cancel{3\,H_2O(\ell)} \qquad \Delta H^\circ_{a'} = -1559.7\text{ kJ}$$

$$(b') \qquad \cancel{2\,CO_2(g)} + \cancel{2H_2O(\ell)} \longrightarrow C_2H_4(g) + \cancel{3\,O_2(g)} \qquad \Delta H^\circ_{b'} = +1410.9\text{ kJ}$$

$$(c') \qquad \cancel{H_2O(\ell)} \longrightarrow H_2(g) + \cancel{\tfrac{1}{2}\,O_2(g)} \qquad \Delta H^\circ_{c'} = +285.8\text{ kJ}$$

$$\text{Net} \qquad C_2H_6(g) \rightarrow C_2H_4(g) + H_2(g) \qquad \Delta H^\circ_{net} = 137.0\text{ kJ}$$

When the equations are added you see that there are $\frac{7}{2}$ mol $O_2(g)$ on the reactant side and $3 + \frac{1}{2} = \frac{7}{2}$ mol $O_2(g)$ on the product side, so these cancel out. There are 3 mol H_2O on each side, and there are 2 mol CO_2 on each side, and so these also cancel. We are left with the equation for the conversion of ethane to ethylene and hydrogen. Notice that the net result is an endothermic reaction, as we might expect when a molecule is broken down into simpler molecules.

Problem-Solving Practice 6.7

Given these two thermochemical equations,

$$C(s) + O_2(g) \longrightarrow CO_2(g) \quad \Delta H^\circ = -393.5\text{ kJ}$$

$$CO(g) + \tfrac{1}{2}\,O_2(g) \longrightarrow CO_2(g) \quad \Delta H^\circ = -283.0\text{ kJ}$$

calculate ΔH° for the reaction

$$C(s) + \tfrac{1}{2}\,O_2(g) \longrightarrow CO(g)$$

6.9 STANDARD MOLAR ENTHALPIES OF FORMATION

Hess's law makes it possible to tabulate ΔH° values for a relatively few reactions and, by suitable combinations of these few reactions, calculate ΔH° values for a great many other reactions. To make such a tabulation we use standard states. The **standard state** of an element or compound is the most stable form of the substance in the physical state in which it exists at 1 bar and the specified temperature. Thus, at 25 °C the standard state for hydrogen is $H_2(g)$ and for sodium chloride is NaCl(s). For an element that can exist in several different allotropic forms (p. 33) at 1 bar and 25 °C, the most stable form is usually selected as the standard. For example, graphite, not diamond or buckminsterfullerene, is the standard for carbon; $O_2(g)$, not $O_3(g)$, is the standard for oxygen.

The word "molar" means "per mole." Thus the standard molar enthalpy of formation is the standard enthalpy of formation per mole of compound formed.

It is common to use the term "heat of formation" interchangeably with "enthalpy of formation." Understand that it is only the heat of reaction at constant pressure that is equivalent to the enthalpy change. If heat of reaction is measured under other conditions it may not equal the enthalpy change. For example, when measured at constant volume in a bomb calorimeter, heat of reaction corresponds to the energy change.

The *standard enthalpy change for formation of one mole of a compound from its elements* is called the **standard molar enthalpy of formation,** ΔH_f°. The subscript *f* indicates *f*ormation of the compound from the most stable form of its elements. The superscript ° indicates that all substances are in their standard states. Some examples are

$$H_2(g) + \tfrac{1}{2}\,O_2(g) \longrightarrow H_2O(\ell) \qquad \Delta H_f^\circ = -285.8 \text{ kJ/mol}$$
$$2\ C(\text{graphite}) + 2\ H_2(g) \longrightarrow C_2H_4(g) \qquad \Delta H_f^\circ = 52.26 \text{ kJ/mol}$$

As another example, the following equation shows that 277.7 kJ would be evolved if 2 mol of graphite, the standard-state form for carbon, were to combine with 3 mol of gaseous hydrogen and $\frac{1}{2}$ mol of gaseous oxygen to form 1 mol of liquid ethanol at 25 °C.

$$2\ C(\text{graphite}) + 3\ H_2(g) + \tfrac{1}{2}\,O_2(g) \longrightarrow C_2H_5OH(\ell) \qquad \Delta H_f^\circ = -277.7 \text{ kJ/mol}$$

Finally, it is important to understand that a standard molar enthalpy of formation, ΔH_f°, is just a special case of the standard enthalpy of reaction, ΔH°. It is the case where *1 mol of a compound in its standard state is formed directly from its elements in their standard states.* An example that is *not* a standard molar enthalpy of formation is the enthalpy change for the exothermic reaction

$$MgO(s) + SO_3(g) \longrightarrow MgSO_4(s) \qquad \Delta H^\circ = -287.48 \text{ kJ}$$

In this case magnesium sulfate has been formed from two other *compounds,* not from its elements. And neither is ΔH° for the following reaction a standard molar enthalpy of formation.

$$P_4(s) + 6\ Cl_2(g) \longrightarrow 4\ PCl_3(\ell) \qquad \Delta H^\circ = -1278.8 \text{ kJ}$$

In this equation, 4 mol of product is formed, not 1 mol, even though it has formed from the elements. Therefore $\Delta H^\circ = 4 \text{ mol} \times \Delta H_f^\circ[PCl_3(\ell)]$.

Exercise 6.11 Writing Equations to Define Enthalpies of Formation

(a) The standard molar enthalpy of formation of gaseous ammonia at 25 °C is −46.11 kJ/mol. Write the balanced equation for the formation reaction. (You will need to use fractional coefficients.)
(b) Write the equation that corresponds to the standard molar enthalpy of formation of nitrogen gas. What process, if any, takes place in this equation? What does this imply about the enthalpy change?

Table 6.4 and Appendix J list values of ΔH_f°, obtained from the National Institute for Standards and Technology (NIST), for many compounds. Notice that there are no values listed in these tables for elements such as C(graphite) or $O_2(g)$. *Stan-*

TABLE 6.4 **Selected Standard Molar Enthalpies of Formation at 298.15 K**

Substance	Name	Standard Molar Enthalpy of Formation (kJ/mol)	Substance	Name	Standard Molar Enthalpy of Formation (kJ/mol)
$Al_2O_3(s)$	Aluminum oxide	−1675.7	$HI(g)$	Hydrogen iodide	+26.48
$BaCO_3(s)$	Barium carbonate	−1216.3	$KF(s)$	Potassium fluoride	−567.27
$CaCO_3(s)$	Calcium carbonate	−1206.92	$KCl(s)$	Potassium chloride	−436.747
$CaO(s)$	Calcium oxide	−635.09	$KBr(s)$	Potassium bromide	−393.8
$CCl_4(\ell)$	Carbon tetrachloride	−135.44	$MgO(s)$	Magnesium oxide	−601.70
$CH_4(g)$	Methane	−74.81	$MgSO_4(s)$	Magnesium sulfate	−1284.9
$C_2H_5OH(\ell)$	Ethyl alcohol	−277.69	$Mg(OH)_2(s)$	Magnesium hydroxide	−924.54
$CO(g)$	Carbon monoxide	−110.525	$NaF(s)$	Sodium fluoride	−573.647
$CO_2(g)$	Carbon dioxide	−393.509	$NaCl(s)$	Sodium chloride	−411.153
$C_2H_2(g)$	Acetylene	+226.73	$NaBr(s)$	Sodium bromide	−361.062
$C_2H_4(g)$	Ethylene	+52.26	$NaI(s)$	Sodium iodide	−287.78
$C_2H_6(g)$	Ethane	−84.68	$NH_3(g)$	Ammonia	−46.11
$C_3H_8(g)$	Propane	−103.8	$NO(g)$	Nitrogen monoxide	+90.25
$C_4H_{10}(g)$	Butane	−888.0	$NO_2(g)$	Nitrogen dioxide	+33.18
$C_6H_{12}O_6(s)$	α-D-glucose	−1274.4	$PCl_3(\ell)$	Phosphorus trichloride	−319.7
$CuSO_4(s)$	Copper(II) sulfate	−771.36	$PCl_5(s)$	Phosphorus pentachloride	−443.5
$H_2O(g)$	Water vapor	−241.818	$SiO_2(s)$	Silicon dioxide (quartz)	−910.94
$H_2O(\ell)$	Liquid water	−285.830	$SnCl_2(s)$	Tin(II) chloride	−325.1
$HF(g)$	Hydrogen fluoride	−271.1	$SnCl_4(\ell)$	Tin(IV) chloride	−511.3
$HCl(g)$	Hydrogen chloride	−92.307	$SO_2(g)$	Sulfur dioxide	−296.830
$HBr(g)$	Hydrogen bromide	−36.40	$SO_3(g)$	Sulfur trioxide	−395.72

From D.D. Wagman, W.H. Evans, V.B. Parker, R.H. Schumm, I. Halow, S.M. Bailey, K.L. Churney, and R. Nuttall: "The NBS Tables of Chemical Thermodynamic Properties," *J. Phys. Chem. Ref. Data,* **1982,** *11,* Suppl. 2.

dard enthalpies of formation for the elements in their standard states are zero, because forming an element in its standard state from the same element in its standard state involves no chemical or physical change. This equation

$$O_2(g) \longrightarrow O_2(g)$$

for example, has exactly the same mole of oxygen as product that is there as reactant. Since there is no change, no energy can be transferred. $\Delta H°$ for this reaction must be zero, and hence $\Delta H_f°$ for oxygen gas also is zero.

You can use Hess's law to find the standard enthalpy change for any reaction if there is a set of reactions whose enthalpies are known and whose equations, when added together, will give the equation for the desired reaction. For example, suppose you are a chemical engineer and want to know how much heating is required to decompose limestone (calcium carbonate) to lime (calcium oxide) and carbon dioxide.

$$CaCO_3(s) \longrightarrow CaO(s) + CO_2(g) \qquad \Delta H° = ?$$

As a first approximation you can assume that all substances are in their standard states at 25 °C, and look up standard molar enthalpies of formation in a table such as Table 6.4 or Appendix J. You find the following values:

(a) $\Delta H_f^\circ[CaCO_3(s)] = -1206.9$ kJ/mol
(b) $\Delta H_f^\circ[CaO(s)] = -635.1$ kJ/mol
(c) $\Delta H_f^\circ[CO_2(g)] = -393.5$ kJ/mol

These standard molar enthalpies of formation correspond to the following equations:

(a) $Ca(s) + C(graphite) + \frac{3}{2}\,O_2(g) \longrightarrow CaCO_3(s) \qquad \Delta H_a^\circ = -1206.9$ kJ
(b) $Ca(s) + \frac{1}{2}\,O_2(g) \longrightarrow CaO(s) \qquad \Delta H_b^\circ = -635.1$ kJ
(c) $Ca(graphite) + O_2(g) \longrightarrow CO_2(g) \qquad \Delta H_c^\circ = -393.5$ kJ

Now add the three equations in such a way that the resulting equation is the one given above for the decomposition of limestone. In equation (a), $CaCO_3(s)$ is a product, but it must appear in the desired equation as a reactant. Therefore, equation (a) must be reversed *and* the sign of ΔH_a° must also be reversed. On the other hand, CaO(s) and $CO_2(g)$ must appear as products in the desired equation, so equations (b) and (c) can be added with the same direction and sign of ΔH° as they have in the ΔH_f° equations:

(a′) = $-$(a) $CaCO_3(s) \longrightarrow Ca(s) + C(graphite) + \frac{3}{2}\,O_2(g) \qquad \Delta H_{a'}^\circ = +1206.9$ kJ
(b) $Ca(s) + \frac{1}{2}O_2(g) \longrightarrow CaO(s) \qquad \Delta H_{b'}^\circ = -635.1$ kJ
(c) $C(graphite) + O_2(g) \longrightarrow CO_2(g) \qquad \Delta H_{c'}^\circ = -393.5$ kJ

$CaCO_3(s) \longrightarrow CaO(s) + CO_2(g) \qquad \Delta H^\circ = +178.3$ kJ

Lime production. This plant makes lime for use in cement by grinding and roasting limestone. *(Derek Ditchburn/Visuals Unlimited)*

When the equations are added in this fashion, 1 mol of each of C(graphite) and Ca(s) and $\frac{3}{2}$ mol of $O_2(g)$ appear on opposite sides and so are canceled out. Thus, the sum of these equations is the desired one for the decomposition of calcium carbonate, and the sum of the enthalpy changes of the three equations gives that for the desired equation.

Another very useful conclusion can be drawn from this example. The mathematics of the problem can be summarized by the expression

$$\Delta H^\circ = \Sigma\{(\text{moles of product}) \times \Delta H_f^\circ\,(\text{product})\} - \Sigma\{(\text{moles of reactant}) \times \Delta H_f^\circ(\text{reactant})\}$$

As in the case of bond energies, Σ means "to sum." This equation says to multiply the standard molar enthalpy of formation of each product by the number of moles of that product and then sum over all products. Then multiply the standard molar enthalpy of formation of each reactant by the number of moles of that reactant and sum over all reactants. Finally, subtract the sum for the reactants from the sum for the products to get the standard enthalpy change of the reaction. This is a useful shortcut to writing the equations for all appropriate formation reactions and applying Hess's law, as we did above. Returning to the equation for decomposition of limestone, we calculate the standard enthalpy change as follows:

$$\Delta H^\circ = \Delta H_f^\circ[CaO(s)] + \Delta H_f^\circ[CO_2(g)] - \Delta H_f^\circ[CaCO_3(s)]$$
$$= -635.1 \text{ kJ} - 393.5 \text{ kJ} - (-1206.9 \text{ kJ}) = 178.3 \text{ kJ}$$

PROBLEM-SOLVING EXAMPLE 6.8 Using Standard Molar Enthalpies of Formation

Benzene, C_6H_6, is an important hydrocarbon. Calculate its enthalpy of combustion; that is, find the value of $\Delta H°$ for the reaction

$$C_6H_6(\ell) + \frac{15}{2}\, O_2(g) \longrightarrow 6\, CO_2(g) + 3\, H_2O(\ell)$$

For benzene, $\Delta H_f^\circ[C_6H_6(\ell)] = 49.0$ kJ/mol. Use Table 6.4 for any other values you may need.

Answer $\Delta H° = -3169.5$ kJ

Explanation To calculate $\Delta H°$ you need standard molar enthalpies of formation for all compounds (and possibly elements, if they are not in their standard states) involved in the reaction. Since $O_2(g)$ is in its standard state, it is not included. From Table 6.4,

$$C(\text{graphite}) + O_2(g) \longrightarrow CO_2(g) \qquad \Delta H_f^\circ = -393.509 \text{ kJ/mol}$$

$$H_2(g) + \frac{1}{2}\, O_2(g) \longrightarrow H_2O(\ell) \qquad \Delta H_f^\circ = -285.830 \text{ kJ/mol}$$

Using the equation given above,

$$\begin{aligned}\Delta H° &= 6 \text{ mol} \times \Delta H_f^\circ[CO_2(g)] + 3 \text{ mol} \times \Delta H_f^\circ[H_2O(\ell)] - \Delta H_f^\circ[C_6H_6(\ell)] \\ &= 6 \text{ mol} \times (-393.509 \text{ kJ/mol}) + 3 \text{ mol} \times (-285.830 \text{ kJ/mol}) \\ &\qquad - 1 \text{ mol} \times (49.0 \text{ kJ/mol}) \\ &= -3169.5 \text{ kJ}\end{aligned}$$

Problem-Solving Practice 6.8

Nitroglycerin is a powerful explosive because it decomposes exothermically and four different gases are formed.

$$2\, C_3H_5(NO_3)_3(\ell) \longrightarrow 3\, N_2(g) + \frac{1}{2}\, O_2(g) + 6\, CO_2(g) + 5\, H_2O(g)$$

For nitroglycerin, $\Delta H_f^\circ[C_3H_5(NO_3)_3(\ell)] = -364$ kJ/mol. Using data from Table 6.4, calculate the transfer of thermal energy when 10.0 g of nitroglycerin explodes.

6.10 CHEMICAL FUELS

A **chemical fuel** is any substance that will react exothermically with atmospheric oxygen and is available at reasonable cost and in reasonable quantity. It is desirable that when a fuel burns, the products create as little environmental damage as possible. As indicated in Figure 6.18, most of the fuels that supplied us with thermal energy in 1995 were fossil fuels: coal, petroleum, and natural gas. Biomass fuels consisting of wood, peat, and other plant matter were a distant second among chemical fuels, and some energy came from nuclear reactors and hydroelectric power plants. For specific applications, other fuels are sometimes chosen because of their special properties. For example, hydrogen is used as the fuel for the space shuttle, and hydrazine (N_2H_4) is used as a rocket fuel in some applications.

What nanoscale characteristics make a good fuel? If some or all of the bonds in the molecules of a fuel are weak, or if the bonds in the products of its combustion

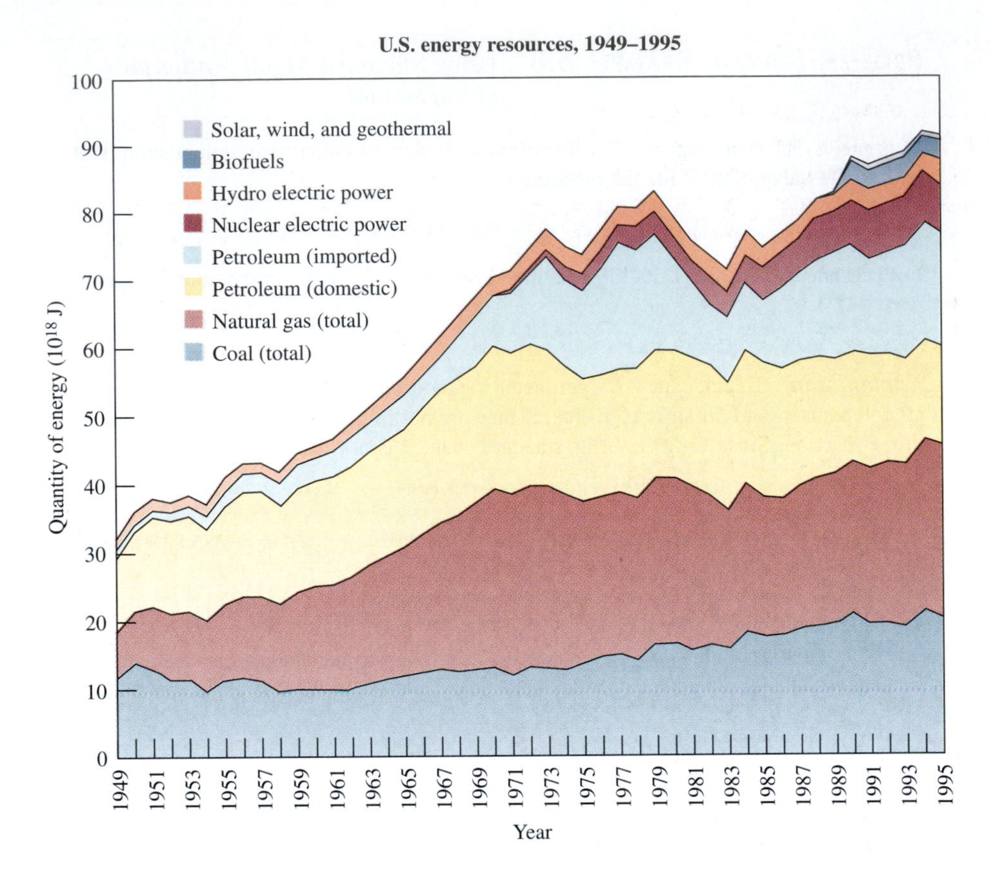

Figure 6.18 Energy resources of the United States. Use of energy resources in the United States is plotted from 1949 to 1995. (An energy resource is a naturally occurring fuel, such as petroleum, or a continuous supply, such as sunlight.) In 1949 coal and petroleum were almost equally important, with natural gas third. Today petroleum and natural gas are used in greater quantity than coal, and more than half of the petroleum is imported. Nuclear electric power did not exist in 1949 but contributes significantly today, while hydroelectric electricity generation has grown slightly. *(Data from Annual Energy Review, 1995, U.S. Department of Energy, Energy Information Administration. http://www.eia.deo.gov/emeu/aer/aergs/aer2.html.)*

are strong, then the combustion reaction will be exothermic. An example of a molecule with a weak bond is hydrazine.

$$N_2H_4(g) + O_2(g) \longrightarrow N_2(g) + 2\,H_2O(g)$$

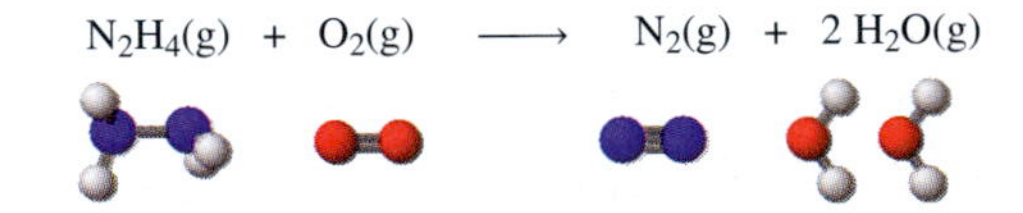

Its N–N bond energy is only 159 kJ/mol, although its four N–H bonds are reasonably strong at 389 kJ/mol. When hydrazine burns, it forms N_2, which has a very strong bond at 946 kJ/mol, and H_2O, which also has strong O–H bonds at 463 kJ/mol. In this case there are actually fewer bonds after the reaction than before, but the bonds are much stronger, and so the reaction is exothermic.

Exercise 6.12 Using Bond Energies to Evaluate a Fuel

Based on the molecular structures given above, use bond energies to estimate the thermal energy transferred to the surroundings when 500 lb of hydrazine burns completely in air.

Coal, petroleum, and natural gas consist mainly of hydrocarbon molecules. When these fuels burn completely in air they produce water and carbon dioxide. A carbon dioxide molecule contains two very strong carbon-oxygen bonds (803 kJ/mol each), and a water molecule contains two very strong O–H bonds. As shown by the molecular structures in Problem-Solving Example 6.6 (← *p. 249*), when methane (CH_4) burns, the number of bonds in reactant molecules is the same as the number of bonds in product molecules. Because the bonds formed are stronger than the bonds broken, the reaction is exothermic. You can verify this by using bond energies from Table 6.3.

Another very good fuel is hydrogen. It burns in air to produce only water. This is a big advantage from an environmental point of view, because it avoids potential problems that may be caused by increasing CO_2 levels in the atmosphere resulting from combustion of hydrocarbon fuels. The thermochemical equation for combustion of hydrogen corresponds to the formation of water from its elements, and so the standard enthalpy change is just ΔH_f°:

$$H_2(g) + \tfrac{1}{2}O_2(g) \longrightarrow H_2O(g) \qquad \Delta H^\circ = -241.818 \text{ kJ}$$

Because hydrogen does not occur naturally as the element on earth but is always combined in compounds, hydrogen does not meet the specification of availability in reasonable quantity mentioned at the beginning of this section. It is manufactured as a byproduct of petroleum refining at present, which makes it too expensive for most fuel applications. However, considerable research is aimed at finding ways to produce hydrogen either by electrolysis of water or chemically. For example, electricity supplied by solar cells could be used to electrolyze water and produce hydrogen, which could then be used as fuel. Or solar energy might be used directly to cause a series of chemical reactions in which hydrogen was produced from water. At present none of these ways of producing hydrogen is cheap enough to be competitive with fossil fuels, but as supplies of fossil fuels become exhausted, hydrogen may become much more important.

The fuel that was displaced by fossil fuels during the industrial revolution was wood. It is now referred to more generically as biomass, because plant matter other than wood can also be burned. Biomass is very important as a fuel in many less-developed countries, and it is a renewable resource. Coal, petroleum, and natural gas will eventually be used up, but it is possible to continue growing plants to create biomass. Although biomass is a mixture of materials, it is primarily carbohydrate (cellulose in wood, for example) and can be represented by its empirical formula CH_2O. Combustion of biomass is highly exothermic:

$$CH_2O(s) + O_2(g) \longrightarrow CO_2(g) + H_2O(g) \qquad \Delta H^\circ = -425 \text{ kJ}$$

Exercise 6.13 Comparing Fuels

An important criterion for a fuel is the energy transfer that can be provided by a given mass of fuel. Evaluate each of the following fuels on this basis. Use data from Table 6.4 or Appendix J. Which fuel provides the most thermal energy per gram?

(a) Methane, $CH_4(g)$
(b) Octane, $C_8H_{18}(\ell)$
(c) Hydrazine, $N_2H_4(\ell)$
(d) Hydrogen, $H_2(g)$
(e) Biomass; assume that the fuel is entirely carbohydrate.

Summary Problem

Sulfur dioxide, SO_2, is a major pollutant emitted by power plants such as the one pictured at the beginning of this chapter. It is also often found in wine.

(a) In winemaking, SO_2 is commonly added to kill microorganisms in the grape juice when it is put into vats before fermentation. Further, it is used to neutralize byproducts of the fermentation process, enhance wine flavor, and prevent oxidation. Wine usually contains 80 to 150 ppm SO_2 (1 ppm = 1 part per million = 1 g of SO_2 per 10^6 g of wine). The United States produced 440 million gallons of wine in 1987. Assuming the density of wine is 1.00 g/cm^3, and that the average bottle of wine contains 100. ppm of SO_2, how many grams and how many moles of SO_2 are contained in this wine?

(b) When SO_2 is given off by an oil- or coal-burning power plant, it can be trapped by reaction with MgO in air to form $MgSO_4$.

$$MgO(s) + SO_2(g) + \frac{1}{2}O_2(g) \longrightarrow MgSO_4(s)$$

If 20. million tons of SO_2 is given off by coal-burning power plants each year, how much MgO would you have to supply to remove all of this SO_2? How much $MgSO_4$ would be produced?

(c) If ΔH_f° for $MgSO_4$(s) is −1284.9 kJ/mol, how much heat (at constant pressure) is evolved or absorbed per mole of $MgSO_4$ by the reaction in part (b)?

(d) Sulfuric acid comes from the oxidation of sulfur, first to SO_2 and then to SO_3. The SO_3 is then absorbed by water to make H_2SO_4.

$$S(s) + O_2(g) \longrightarrow SO_2(g) \qquad \Delta H^\circ = -296.8 \text{ kJ}$$

$$SO_2(g) + \frac{1}{2}O_2(g) \longrightarrow SO_3(g) \qquad \Delta H^\circ = -98.9 \text{ kJ}$$

$$SO_3(g) + H_2O(\text{in } 98\%\ H_2SO_4) \longrightarrow H_2SO_4(\ell) \qquad \Delta H^\circ = -132.5 \text{ kJ}$$

The typical plant produces 750. tons of H_2SO_4 per day (1 ton = 9.08×10^5 g). Calculate the heating produced by these reactions per day.

In Closing

Having studied this chapter, you should be able to . . .

- describe the various forms of energy and the nature of heating and energy transfer (Section 6.1).
- know typical energy units and be able to convert from one unit to another (Section 6.2).
- use specific heat capacity and the sign conventions for transfer of thermal energy (Section 6.3).
- recognize and use the language of thermodynamics: the system and its surroundings, exothermic and endothermic reactions, and the first law of thermodynamics (the law of energy conservation) (Section 6.4).
- use bond energies to estimate the standard enthalpy change for a reaction (Section 6.5).
- use the fact that the standard enthalpy change for a reaction, ΔH°, is proportional to the quantity of reactants consumed or products produced when the reaction occurs (Section 6.6).

- describe how one can measure the quantity of thermal energy transferred in a reaction by using calorimetry (Section 6.7).
- apply Hess's law to find the enthalpy change for a reaction (Sections 6.8 and 6.9).
- use standard molar enthalpies of reaction to calculate the thermal energy transfer when a reaction takes place (Section 6.9).
- be able to define and give examples of some chemical fuels and to evaluate their abilities to provide heating (Section 6.10).

Key Terms

The following terms were defined and given in boldface type in this chapter. You should be sure to understand each of these terms and the concepts with which they are associated.

bond energy *(6.5)*
bond enthalpy *(6.5)*
calorie *(6.2)*
calorimeter *(6.7)*
change of state *(6.4)*
chemical fuel *(6.10)*
endothermic *(6.4)*
energy *(6.1)*
enthalpy change *(6.4)*
exothermic *(6.4)*
first law of thermodynamics *(6.1)*
heat capacity *(6.3)*
heat of fusion *(6.4)*
heat of vaporization *(6.4)*
Hess's law *(6.8)*
internal energy *(6.1)*
joule (J) *(6.2)*
kilocalorie *(6.2)*
kinetic energy *(6.1)*
law of conservation of energy *(6.1)*
molar heat capacity *(6.3)*
phase change *(6.4)*
potential energy *(6.1)*
specific heat capacity *(6.3)*
standard enthalpy change *(6.4)*
standard molar enthalpy of formation *(6.9)*
standard state *(6.9)*
sublimation *(6.4)*
surroundings *(6.4)*
system *(6.4)*
thermochemical equation *(6.4)*
thermodynamics *(Introduction)*

Questions for Review and Thought

Conceptual Challenge Problems

CP-6.A. Suppose a scientist discovered that energy was not conserved, but $1 \times 10^{-7}\%$ of the energy transferred from one system vanishes before it enters another system. How would this affect electric utilities, thermochemical experiments in scientific laboratories, and scientific thinking?

CP-6.B. Suppose that someone were to tell your teacher during class that energy is not always conserved. This person states that he or she had previously learned that in the case of nuclear reactions mass is converted into energy according to Einstein's equation $E = mc^2$. Hence, energy is continuously produced as mass is changed into energy. Your teacher quickly responds by making the following assignment to the class, "Please write a paragraph or two to refute or clarify this student's thesis." What would you say?

CP-6.C. The specific heat capacities at 25 °C for three metals with widely differing molar masses are 3.6 J/g·°C for Li, 0.25 J/g·°C for Ag, and 0.11 J/g·°C for Th. Suppose that you have three samples, one of each metal and each containing the same number of atoms.

(a) Is the energy transfer required to increase the temperature of each sample by 1 °C significantly different from one sample to the next?

(b) What interpretation can you make about temperature based on the result you found in part (a)?

CP-6.D. During one of your chemistry classes a student asked the professor, "Why does hot water freeze more quickly than cold water?"

(a) What do you expect the professor to say in answer to the student's question?

(b) In one experiment, two 100.-g samples of water were placed in identical containers on the same surface 1 decimeter apart in a room at −25 °C. One sample had an initial temperature of 78 °C, while the second was at 24 °C. The second sample took 151 min to freeze, and the first took 166 min (only 10% longer) to freeze. Clearly the cooler sample froze more quickly, but not nearly as quickly as one might have expected. How can this be so?

CP-6.E. The mere fact that it is possible to tabulate bond energies as was done in Table 6.3 has implications about the nature of the bonds between atoms in molecules. Think about each of the following questions and then write a brief paragraph about the conclusions you can draw about bonds.

(a) What is the factor most important in determining how strong a bond is?

(b) Suppose that two atoms are bonded to each other. How does the presence of other atoms connected to one of the two bonded atoms affect the bond strength?

CP-6.F. Use the bond energies in Table 6.3 to calculate the enthalpy of combustion of ethane, CH_3CH_3. Use the standard molar enthalpies of formation in Table 6.4 to calculate the enthalpy of combustion of ethane. Which value is correct? Why are the values not the same?

CP-6.G. On March 4, 1996, five railroad tank cars carrying liquid propane derailed in Weyauwega, Wisconsin, forcing evacuation of the town for over a week. Residents who lived within the square mile centered on the accident were unable to return to their homes for over two weeks. Evaluate whether this evacuation was reasonable and necessary by considering the following questions.

(a) Estimate the volume of a railroad tank car. Obtain the density of liquid propane (C_3H_8) at or near its boiling point and calculate the mass of propane in the five tank cars. Obtain the data you need in order to calculate the energy transfer if all of that propane burned at once. (Assume that the reaction takes place at room temperature.)

(b) The enthalpy of decomposition of TNT ($C_7H_5H_3O_6$) to water, nitrogen, carbon monoxide, and carbon is -1066.1 kJ/mol. How many metric kilotons (1 metric ton = 1 Mg = 1×10^6g) of TNT would provide energy transfer equivalent to that produced by combustion of propane in the five tank cars?

(c) Find the energy transfer (in kilotons of TNT) resulting from the nuclear fission bombs dropped on Hiroshima and Nagasaki, Japan, in 1945, and the energy transfer for modern fission weapons. What is the largest nuclear weapon thought to have been detonated to date? Compare the energy of the Hiroshima and Nagasaki bombs with the Weyauwega propane spill. What can you conclude about the wisdom of evacuating the town?

(d) Compare the energy that would have been released by burning the propane with the energy of a hurricane.

Review Questions

1. Name two laws stated in this chapter and explain each in your own words.
2. For each of the following, define a system and its surroundings and give the direction of heat transfer:
   (a) Propane is burning in a Bunsen burner in the laboratory.
   (b) Water drops, sitting on your skin after a dip in the ocean, evaporate.
   (c) Water, originally at 25 °C, is placed in a freezing compartment of a refrigerator.
   (d) Two chemicals are mixed in a flask sitting on a laboratory bench. A reaction occurs and heat is evolved.
3. What is the value of the standard enthalpy of formation for any element under standard conditions?
4. Criticize the following statements:
   (a) An enthalpy of formation refers to a reaction in which 1 mol of one or more reactants produces some quantity of product.
   (b) The standard enthalpy of formation of O_2 as a gas at 25 °C and a pressure of 1 atm is 15.0 kJ/mol.
5. Explain how a coffee-cup calorimeter may be used to measure the enthalpy change of (a) a change in state and (b) a chemical reaction.
6. Describe how energy is interconverted from one form to another in the following processes:
   (a) At a July 4 celebration, a match is lit, which in turn ignites the fuse of a rocket flare which fires off and explodes at an altitude of 1000 ft.
   (b) A gallon of gasoline is pumped from an underground storage tank into the fuel tank of your car, and you use it up by driving 25 mi.

Energy Units

7. (a) A 2-inch piece of two-layer chocolate cake with frosting provides 1670 kJ of energy. What is this in Cal?
   (b) If you were on a diet that calls for eating no more than 1200 Cal per day, how many joules would this be?

8. Melting lead requires 5.91 cal per gram. How many joules are required to melt 1.00 lb (454 g) of lead?

9. Sulfur dioxide, SO_2, is found in wines and in polluted air. If a 32.1-g sample of sulfur is burned in the air to get 64.1 g of SO_2, 297 kJ of energy is released. Express this energy in (a) joules, (b) calories, and (c) kilocalories.

10. When an electrical appliance whose power usage is X watts is run for Y seconds, it uses $X \times Y$ J of energy. The energy unit used by electrical utilities in their monthly bills is the *kilowatt hour* (1 kilowatt used for one hour). How many joules are there in a kilowatt hour? If electricity costs $.09 per kilowatt hour, how much does it cost per megajoule?

11. A 100-watt light bulb is left on 14 h. How many joules of energy are used? With electricity at $.09 per kW-h, how much does it cost to leave the light on for 14 h?

12. On a sunny day, solar energy reaches the earth at a rate of 4.0 J/(min · cm^2). Suppose a house has a square flat roof of dimensions 12 m by 12 m. How much solar energy reaches this roof in 1 h? (*Note:* This is why roofs painted with light-reflecting paint are better than black, unpainted roofs in warm climates: they reflect most of this energy rather than absorb it.)

Heat Capacity and Specific Heat Capacity

13. Which requires more energy: (a) warming 15.0 g of water from 25 °C to 37 °C or (b) warming 60.0 g of aluminum from 25 °C to 37 °C?
14. You hold a gram of copper in one hand and a gram of aluminum in the other. Each metal was originally at 0 °C. (Both metals are in the shape of a little ball that fits into your hand.) If they both take up heat at the same rate, which will warm to your body temperature first?
15. How much thermal energy is required to heat all the aluminum in a roll of aluminum foil (500. g) from room temperature (25 °C) to the temperature of a hot oven (250 °C)? Report your result in kilojoules.
16. How much thermal energy is required to heat all of the water in a swimming pool by 1 °C if the dimensions are 4 ft deep by 20 ft wide by 75 ft long? Report your result in megajoules.
17. Ethylene glycol, $(CH_2OH)_2$, is often used as an antifreeze in cars.
(a) Which requires more thermal energy to warm from 25.0 °C to 100.0 °C, pure water or an equal mass of pure ethylene glycol?
(b) If the cooling system in an automobile has a capacity of 5.00 quarts of liquid, compare the quantity of thermal energy the liquid in the system absorbs when its temperature is raised from 25.0 °C to 100.0 °C for water and ethylene glycol. (The densities of water and ethylene glycol are 1.00 g/cm^3 and 1.113 g/cm^3. 1 quart = 0.946 L. Report your results in joules.)
18. One way to cool a cup of coffee is to plunge an ice-cold piece of aluminum into it. Suppose a 20.0-g piece of aluminum is stored in the refrigerator at 32 °F (0.0 °C) and then dropped into a cup of coffee. The coffee's temperature drops from 90.0 °C to 75.0 °C. How much energy (in kilojoules) did the piece of aluminum absorb?
19. A piece of iron (400. g) is heated in a flame and then plunged into a beaker containing 1.00 kg of water. The original temperature of the water was 20.0 °C, but it is 32.8 °C after the iron bar is dropped in. What was the original temperature of the hot iron bar?
20. A 192-g piece of copper was heated to 100.0 °C in a boiling water bath and then was dropped into a beaker containing 750. mL of water (density = 1.00 g/cm^3) at 4.0 °C. What is the final temperature of the copper and water after they come to thermal equilibrium?

Energy Transfers and Changes of State

21. The thermal energy required to melt 1.00 g of ice at 0 °C is 333 J. If one ice cube has a mass of 62.0 g, and a tray contains 20 ice cubes, how much energy is required to melt a tray of ice cubes at 0 °C?
22. How much energy (in joules) would be required to raise the temperature of 1.00 lb of lead (1.00 lb = 454 g) from room temperature (25 °C) to its melting point, 327 °C, and then melt the lead at 327 °C? The specific heat capacity of lead is 0.159 J/g·°C, and its enthalpy of fusion is 24.7 J/g.
23. The hydrocarbon benzene, C_6H_6, boils at 80.1 °C. How much energy is required to heat 1.00 kg of this liquid from 20.0 °C to the boiling point and then change the liquid completely to a vapor at that temperature? (The specific heat capacity of liquid C_6H_6 is 1.74 J/g·°C, and the enthalpy of vaporization is 395 J/g. Report your result in joules.)
24. Calculate the quantity of heating required to convert the water in four ice cubes (60.1 g each) from $H_2O(s)$ at 0 °C to $H_2O(g)$ at 100 °C. The enthalpy of fusion of ice at 0 °C is 333 J/g and the enthalpy of vaporization of liquid water at 100 °C is 2260 J/g.
25. Mercury, with a freezing point of −39 °C, is the only metal that is liquid at room temperature. How much thermal energy must be transferred to its surroundings if 1.00 mL of the mercury is cooled from room temperature (23.0 °C) to −39 °C and then frozen to a solid? (The density of mercury is 13.6 g/cm^3. Its specific heat capacity is 0.138 J/g·°C, and its enthalpy of fusion is 11 J/g.)
26. On a cold day in winter, ice can sublime (go directly from solid to gas without melting). The heat of sublimation is approximately equal to the sum of the heat of fusion and the heat of vaporization (see Question 24). How much thermal energy in joules does it take to sublime 0.1 g of frost on a windowpane?
27. Draw a cooling graph for steam-to-water-to-ice.
28. Draw a heating graph for converting Dry Ice to carbon dioxide gas.

Enthalpy Changes for Chemical Reactions

29. Energy is stored in the body in adenosine triphosphate, ATP, which is formed by the reaction between adenosine diphosphate, ADP, and phosphoric acid.

$$ADP(aq) + H_3PO_4(aq) \longrightarrow ATP(aq) + H_2O(\ell) \qquad \Delta H° = +38 \text{ kJ}$$

Is the reaction endothermic or exothermic?
30. Calcium carbide, CaC_2, is manufactured by reducing lime with carbon at high temperature. (The carbide is used in turn to make acetylene, an industrially important organic chemical.)

$$CaO(s) + 3\ C(s) \longrightarrow CaC_2(s) + CO(g) \qquad \Delta H° = +464.8\text{kJ}$$

Is the reaction endothermic or exothermic?
31. A diamond can be considered a giant all-carbon supermolecule in which almost every carbon atom is bonded to four other carbons (← *p. 34*). When a diamond cutter cleaves a diamond, carbon-carbon bonds must be broken. Is the cleavage of a diamond endothermic or exothermic? Explain.
32. When table salt is dissolved in water, the temperature drops slightly. Write a chemical equation for this process, and indicate if it is exothermic or endothermic.
33. Write a word statement for the thermochemical equation

$$H_2O(s) \longrightarrow H_2O(\ell) \qquad \Delta H° = +6.0 \text{ kJ}$$

34. Write a word statement for the thermochemical equation

$$HI(\ell) \longrightarrow HI(s) \qquad \Delta H° = -2.87 \text{ kJ}$$

35. Estimate $\Delta H°$ for forming 2 mol of ammonia from molecular nitrogen and molecular hydrogen. Is this reaction exothermic or endothermic? (N_2 has a triple bond.)

36. Estimate $\Delta H°$ for the conversion of 1 mol of carbon monoxide to carbon dioxide by combination with molecular oxygen. Is this reaction exothermic or endothermic? (CO has a triple bond.)

37. Which of the four molecules HF, HCl, HBr, and HI has the strongest chemical bond?

38. Using bond energies, estimate $\Delta H°$ for the reaction of molecular hydrogen with each of the gaseous molecular halogens: fluorine, chlorine, bromine, and iodine. Which is the most exothermic reaction?

Using Thermochemical Equations for Chemical Reactions

39. "Gasohol," a mixture of gasoline and ethanol, C_2H_5OH, is used as automobile fuel. The alcohol releases energy in a combustion reaction with O_2.

$$C_2H_5OH(\ell) + 3\ O_2(g) \longrightarrow 2\ CO_2(g) + 3\ H_2O(\ell)$$

If 0.115 g of alcohol evolves 3.62 kJ when burned at constant pressure, what is the molar enthalpy of combustion for ethanol?

40. White phosphorus, P_4, ignites in air to produce heat, light, and P_4O_{10}

$$P_4(s) + 5\ O_2(g) \longrightarrow P_4O_{10}(s)$$

If 3.56 g of P_4 is burned, 85.8 kJ of thermal energy is evolved at constant pressure. What is the molar enthalpy of combustion of P_4?

41. A laboratory "volcano" can be made from ammonium dichromate (Figure 4.7). When ignited, the compound decomposes in a fiery display.

$$(NH_4)_2Cr_2O_7(s) \rightarrow N_2(g) + 4\ H_2O(g) + Cr_2O_3(s)$$

If the decomposition produces 315 kJ per mole of ammonium dichromate at constant pressure, how much thermal energy would be produced by 28.4 g (1.00 ounce) of the solid?

42. The thermite reaction, the reaction between aluminum and iron(III) oxide,

$$2\ Al(s) + Fe_2O_3(s) \longrightarrow Al_2O_3(s) + 2\ Fe(s) \qquad \Delta H° = -851.5\ kJ$$

produces a tremendous amount of heat. If you begin with 10.0 g of Al and excess Fe_2O_3, how much energy is evolved at constant pressure?

43. When wood is burned we may assume that the reaction is the combustion of cellulose (empirical formula, CH_2O).

$$CH_2O(s) + O_2(g) \longrightarrow CO_2(g) + H_2O(g) \qquad \Delta H° = -425\ kJ$$

How much energy is released when a 10-lb wood log burns completely? (Assume 100% efficient burning via the reaction above.)

44. A plant takes CO_2 and H_2O from its surroundings and makes cellulose by the reverse of the reaction in the previous problem. The energy provided for this process comes from the sun via photosynthesis. How much energy does it take for a plant to make 100 g of cellulose?

Measuring Enthalpy Changes for Reactions: Calorimetry

45. Suppose you add a small ice cube to room-temperature water in a coffee-cup calorimeter. What is the final temperature when all of the ice is melted? Assume that you have 200 mL of water at 25 °C and that the ice cube weighs 15.0 g.

46. A coffee-cup calorimeter can be used to investigate the "cold-pack reaction," the process that occurs when solid ammonium nitrate dissolves in water:

$$NH_4NO_3(s) \longrightarrow NH_4^+(aq) + NO_3^-(aq)$$

25.0 g of solid NH_4NO_3 at 23.0 °C is added to 250 mL of H_2O at the same temperature, and after the solid is all dissolved the temperature is measured to be 15.6 °C. Calculate the enthalpy change for the cold pack reaction. (*Hint:* Solve for kJ/mol of NH_4NO_3.) Is it endothermic or exothermic?

47. How much thermal energy is evolved by a reaction in a bomb calorimeter (Figure 6.16) in which the temperature of the bomb and water increases from 19.50 °C to 22.83 °C? The bomb has a heat capacity of 650 J/°C; the calorimeter contains 320. g of water. Report your result in kJ.

48. Sulfur (2.56 g) was burned in a bomb calorimeter with excess $O_2(g)$. The temperature increased from 21.25 °C to 26.72 °C. The bomb had a heat capacity of 923 J/°C and the calorimeter contained 815 g of water. Calculate the heat evolved, per mole of SO_2 formed, in the course of the reaction

$$S(s) + O_2(g) \longrightarrow SO_2(g).$$

49. You can find the quantity of thermal energy evolved during combustion of carbon by carrying out the reaction in a combustion calorimeter. Suppose you burn 0.300 g of C(graphite) in an excess of $O_2(g)$ to give $CO_2(g)$.

$$C(graphite) + O_2(g) \longrightarrow CO_2(g)$$

The temperature of the calorimeter, which contains 775 g of water, increases from 25.00 °C to 27.38 °C. The heat capacity of the bomb is 893 J/°C. What quantity of thermal energy is evolved per mole of C?

50. Benzoic acid, $C_7H_6O_2$, occurs naturally in many berries. Suppose you burn 1.500 g of the compound in a combustion calorimeter and find that the temperature of the calorimeter increases from 22.50 °C to 31.69 °C. The calorimeter contains 775 g of water, and the bomb has a heat capacity of 893 J/°C. How much heat is evolved per mole of benzoic acid?

Hess's Law

51. Calculate the standard enthalpy change, $\Delta H°$, for the formation of 1 mol of strontium carbonate (the material that gives the red color in fireworks) from its elements.

$$Sr(s) + C(graphite) + \tfrac{3}{2}\ O_2(g) \longrightarrow SrCO_3(s)$$

The information available is

$$Sr(s) + \frac{1}{2} O_2(g) \longrightarrow SrO(s) \quad \Delta H° = -592 \text{ kJ}$$

$$SrO(s) + CO_2(g) \longrightarrow SrCO_3(s) \quad \Delta H° = -234 \text{ kJ}$$

$$C(\text{graphite}) + O_2(g) \longrightarrow CO_2(g) \quad \Delta H° = -394 \text{ kJ}$$

52. What is the standard enthalpy change for the reaction of lead(II) chloride with chlorine to give lead(IV) chloride?

$$PbCl_2(s) + Cl_2(g) \longrightarrow PbCl_4(\ell)$$

It is known that $PbCl_2(s)$ can be formed from the metal and $Cl_2(g)$,

$$Pb(s) + Cl_2(g) \longrightarrow PbCl_2(s) \quad \Delta H° = -359.4 \text{ kJ}$$

and that $PbCl_4(\ell)$ can be formed directly from the elements.

$$Pb(s) + 2\, Cl_2(g) \longrightarrow PbCl_4(\ell) \quad \Delta H° = -329.3 \text{ kJ}$$

53. Using the following reactions, find the standard enthalpy change for the formation of 1 mol of PbO(s) from lead metal and oxygen gas.

$$PbO(s) + C(\text{graphite}) \longrightarrow Pb(s) + CO(g) \quad \Delta H° = +106.8 \text{ kJ}$$

$$2\, C(\text{graphite}) + O_2(g) \longrightarrow 2\, CO(g) \quad \Delta H° = -221.0 \text{ kJ}$$

If 250 g of lead reacts with oxygen to form lead(II) oxide, what quantity of thermal energy (in kJ) is absorbed or evolved?

54. Three reactions very important to the semiconductor industry are

(a) the reduction of silicon dioxide to crude silicon,

$$SiO_2(s) + 2\, C(s) \longrightarrow Si(s) + 2\, CO(g) \quad \Delta H° = +689.9 \text{ kJ}$$

(b) the formation of silicon tetrachloride from crude silicon,

$$Si(s) + 2\, Cl_2(g) \longrightarrow SiCl_4(g) \quad \Delta H° = -657.0 \text{ kJ}$$

(c) and the reduction of silicon tetrachloride to pure silicon with magnesium,

$$SiCl_4(g) + 2\, Mg(s) \longrightarrow 2\, MgCl_2(s) + Si(s) \quad \Delta H° = -625.6 \text{ kJ}$$

What is the overall enthalpy change for changing 1.00 mol of sand (SiO_2) into very pure silicon?

Standard Molar Enthalpies of Formation

55. The standard molar enthalpy of formation of AgCl(s) is −127.1 kJ/mol. Write a balanced equation for which the enthalpy of reaction is −127.1 kJ.

56. The standard molar enthalpy of formation of methanol, $CH_3OH(\ell)$, is −238.7 kJ/mol. Write a balanced equation for which the enthalpy of reaction is −238.7 kJ.

57. For each compound below, write a balanced equation depicting the formation of 1 mol of the compound. Standard molar enthalpies of formation are found in Appendix J.
(a) $Al_2O_3(s)$ (b) $TiCl_4(\ell)$ (c) $NH_4NO_3(s)$

58. The standard molar enthalpy of formation of glucose, $C_6H_{12}O_6(s)$, is −1260 kJ/mol. (a) Is the formation of glucose from its elements exothermic or endothermic? (b) Write a balanced equation depicting the formation of glucose from its elements and for which the enthalpy of reaction is −1260 kJ.

59. An important reaction in the production of sulfuric acid is

$$SO_2(g) + \frac{1}{2} O_2(g) \longrightarrow SO_3(g)$$

It is also a key reaction in the formation of acid rain, beginning with the air pollutant SO_2. Using the data in Table 6.4, calculate the enthalpy change for the reaction.

60. In photosynthesis, the sun's energy brings about the combination of CO_2 and H_2O to form O_2 and a carbon-containing compound such as a sugar. In its simplest form, the reaction could be written

$$6\, CO_2(g) + 6\, H_2O(\ell) \longrightarrow 6\, O_2(g) + C_6H_{12}O_6(s)$$

Using the enthalpies of formation in Table 6.4, (a) calculate the enthalpy of reaction and (b) decide whether the reaction is exothermic or endothermic.

61. The first step in the production of nitric acid from ammonia involves the oxidation of NH_3.

$$4\, NH_3(g) + 5\, O_2(g) \longrightarrow 4\, NO(g) + 6\, H_2O(g)$$

Use the information in Table 6.4 or Appendix J to find the enthalpy change for this reaction. Is the reaction exothermic or endothermic?

62. The Romans used CaO as mortar in stone structures. The CaO was mixed with water to give $Ca(OH)_2$, and this slowly reacted with CO_2 in the air to give limestone.

$$Ca(OH)_2(s) + CO_2(g) \longrightarrow CaCO_3(s) + H_2O(g)$$

Calculate the enthalpy change for this reaction.

63. A key reaction in the processing of uranium for use as fuel in nuclear power plants is the following:

$$UO_2(s) + 4\, HF(g) \longrightarrow UF_4(s) + 2\, H_2O(g)$$

Calculate the enthalpy change, $\Delta H°$, for the reaction using the data in Table 6.4, Appendix J, and the following: ΔH_f° for $UO_2(s) = -1085$ kJ/mol; ΔH_f° for $UF_4(s) = -1914$ kJ/mol.

64. Oxygen difluoride, OF_2, is a colorless, very poisonous gas that reacts rapidly and exothermically with water vapor to produce O_2 and HF.

$$OF_2(g) + H_2O(g) \longrightarrow 2\, HF(g) + O_2(g) \quad \Delta H° = -318 \text{ kJ}$$

Using this information and Table 6.4 or Appendix J, calculate the molar enthalpy of formation of $OF_2(g)$.

65. Iron can react with oxygen to give iron(III) oxide. If 5.58 g of Fe is heated in pure O_2 to give $Fe_2O_3(s)$, how much thermal energy is transferred out of this system (at constant pressure)?

66. The formation of aluminum oxide from its elements is highly exothermic. If 2.70 g Al metal is burned in pure O_2 to give Al_2O_3, how much thermal energy is evolved in the process (at constant pressure)?

Chemical Fuels

67. If you want to convert 56.0 g of ice (at 0 °C) to water at 75.0 °C, how many grams of propane (C_3H_8) would you have to burn in order to supply the energy to melt the ice and then warm it to the final temperature (at constant pressure)?

68. Suppose you want to heat your house with natural gas (CH_4). Assume your house has 1800 ft^2 of floor area and that the ceilings are 8.0 ft from the floors. The air in the house has a molar heat capacity of 29.1 J/mol · °C. (The number of moles of air in the house can be found by assuming that the average molar mass of air is 28.9 g/mol and that the density of air at these temperatures is about 1.22 g/L.) How many grams of methane do you have to burn to heat the air from 15.0 °C to 22.0 °C?

69. Companies around the world are constantly searching for compounds that can be used as substitutes for gasoline in automobiles. Perhaps the most promising of these is methanol, CH_3OH, a compound that can be made relatively inexpensively from coal. The alcohol has a smaller energy content than gasoline, but, with its higher octane rating, it burns more efficiently than gasoline in internal combustion engines. (It also contributes smaller quantities of some air pollutants.) Compare the quantity of thermal energy produced per gram of CH_3OH and C_8H_{18} (octane), the latter being representative of the compounds in gasoline. The ΔH_f° for octane is −250.1 kJ/mol.

70. Hydrazine and 1,1-dimethylhydrazine both react spontaneously with O_2 and can be used as rocket fuels.

$$N_2H_4(\ell) + O_2(g) \longrightarrow N_2(g) + 2\ H_2O(g)$$
hydrazine

$$N_2H_2(CH_3)_2(\ell) + 4\ O_2(g) \longrightarrow 2\ CO_2(g) + 4\ H_2O(g) + N_2(g)$$
1,1-dimethylhydrazine

The molar enthalpy of formation of liquid hydrazine is +50.6 kJ/mol, and that of liquid dimethylhydrazine is +49.2 kJ/mol. By doing appropriate calculations, decide whether the reaction of hydrazine or dimethylhydrazine with oxygen gives more heat per gram (at constant pressure). (Other enthalpy of formation data can be obtained from Table 6.4.)

71. The four hydrocarbons of lowest molar mass are methane, ethane, propane, and butane. All are used extensively as fuels in our economy. Calculate the thermal energy transferred to the surroundings per gram of these four fuels and rank them by this quantity.

General Questions

72. The specific heat of copper metal is 0.385 J/g·°C, while it is 0.128 J/g·°C for gold. Assume you place 100. g of each metal, originally at 25 °C, in a boiling water bath at 100 °C. If each metal is heated at the same rate, which piece of metal reaches 100 °C first?

73. Calculate the molar heat capacity, in J/mol·°C, for the four metals in Table 6.1. What observation can you make about these values? Are they widely different or very similar? Using this information, can you calculate the heat capacity in the units J/g·°C for silver? (The correct value for silver is 0.23 J/g·°C.)

74. Suppose you add 100.0 g of water at 60.0 °C to 100.0 g of ice at 0.00 °C. Some of the ice melts and cools the warm water to 0.00 °C. When the ice/water mixture has come to a uniform temperature of 0.00 °C, how much ice has melted?

75. The combustion of diborane, B_2H_6, proceeds according to the equation

$$B_2H_6(g) + 3\ O_2(g) \longrightarrow B_2O_3(s) + 3\ H_2O(\ell)$$

and 2166 kJ is liberated per mole of $B_2H_6(g)$ (at constant pressure). Calculate the molar enthalpy of formation of $B_2H_6(g)$ using this information, the data in Table 6.4, and the fact that ΔH_f° for $B_2O_3(s)$ is −1273 kJ/mol.

76. When we estimate ΔH° from bond energies we assume that all bonds between the same two atoms have the same energy, regardless of the molecule in which they occur. The purpose of this problem is to show you that this is only an approximation. You will need the following standard enthalpies of formation:

C(g)	ΔH° = 716.7 kJ/mol
CH(g)	ΔH° = 596.3 kJ/mol
CH_2(g)	ΔH° = 392.5 kJ/mol
CH_3(g)	ΔH° = 146.0 kJ/mol
H(g)	ΔH° = 218.0 kJ/mol

(a) What is the average C–H bond energy in methane, CH_4?
(b) Using bond energies, estimate ΔH° for the reaction

$$CH_4(g) \longrightarrow C(g) + 2\ H_2(g)$$

(c) By heating CH_4 in a flame it is possible to produce the reactive gaseous species CH_3, CH_2, CH, and even carbon atoms, C. Experiments give the following values of ΔH° for the reactions shown:

$$CH_3(g) \longrightarrow C(g) + H_2(g) + H(g) \qquad \Delta H^\circ = 788.7\ kJ$$
$$CH_2(g) \longrightarrow C(g) + H_2(g) \qquad \Delta H^\circ = 324.2\ kJ$$
$$CH(g) \longrightarrow C(g) + H(g) \qquad \Delta H^\circ = 338.3\ kJ$$

For each of these reactions, draw a diagram similar to Figure 6.14. Then calculate the *average* C—H bond energy in CH_3, CH_2, and CH. Comment on any trends you see.

77. In principle, copper could be used to generate valuable hydrogen gas from water.

$$Cu(s) + H_2O(g) \longrightarrow CuO(s) + H_2(g)$$

(a) Is the reaction exothermic or endothermic?
(b) If 2.00 g of copper metal reacts with excess water vapor at constant pressure, how much thermal energy transfer is involved (either into or out of the system) in the reaction?

78. P_4 ignites in air to give P_4O_{10} and a large quantity of thermal energy transfer to the surroundings. This is an important reaction, since the phosphorus oxide can then be treated with water to give phosphoric acid for use in making detergents, toothpaste, soft drinks, and other consumer products. About 500,000 tons of elemental phosphorus is made annually in the United States. If you oxidize just one ton of P_4 (9.08×10^5 g) to the oxide, how much thermal energy (in kJ) is evolved at constant pressure?

79. The enthalpy changes of the following two reactions are available.

$$2\ C(graphite) + 2\ H_2(g) \longrightarrow C_2H_4(g) \qquad \Delta H° = +52.3\ kJ$$

$$C_2H_4Cl_2(\ell) \longrightarrow Cl_2(g) + C_2H_4(g) \qquad \Delta H° = +217.5\ kJ$$

Calculate the molar enthalpy of formation of $C_2H_4Cl_2(\ell)$.

80. Given the following information and the data in Table 6.4, calculate the molar enthalpy of formation for liquid hydrazine, N_2H_4.

$$N_2H_4(\ell) + O_2(g) \longrightarrow N_2(g) + 2\ H_2O(g) \qquad \Delta H° = -534\ kJ$$

81. The combination of coke and steam produces a mixture called coal gas, which can be used as a fuel or as a starting material for other reactions. If we assume coke can be represented by graphite, the equation for the production of coal gas is

$$2\ C(s) + 2\ H_2O(g) \longrightarrow CH_4(g) + CO_2(g)$$

Determine the standard enthalpy change for this reaction from the following standard enthalpies of reaction:

$$C(s) + H_2O(s) \longrightarrow CO(g) + H_2(g) \qquad \Delta H° = +131.3\ kJ$$

$$CO(g) + H_2O(g) \longrightarrow CO_2(g) + H_2(g) \qquad \Delta H° = -41.2\ kJ$$

$$CH_4(g) + H_2O(g) \longrightarrow 3\ H_2(g) + CO(g) \qquad \Delta H° = +206.1\ kJ$$

82. Some years ago Texas City, Texas, was devastated by the explosion of a shipload of ammonium nitrate, a compound intended to be used as a fertilizer. When heated, however, ammonium nitrate can decompose exothermically to N_2O and water.

$$NH_4NO_3(s) \longrightarrow N_2O(g) + 2\ H_2O(g)$$

If the heat from this exothermic reaction is contained, higher temperatures are generated, at which point ammonium nitrate can decompose explosively to N_2, H_2O, and O_2.

$$2\ NH_4NO_3(s) \longrightarrow 2\ N_2(g) + 4\ H_2O(g) + O_2(g)$$

If oxidizable materials are present, fires can break out, as was the case at Texas City. Using the information in Appendix J, answer the following questions.
(a) How much thermal energy is evolved (at constant pressure and under standard conditions) by the first reaction?
(b) If 8.00 kg of ammonium nitrate explodes (the second reaction), how much thermal energy is evolved (at constant pressure and under standard conditions)?

83. Uranium-235 is used as a fuel in nuclear power plants. Since natural uranium contains only a small amount of this isotope, the uranium must be enriched in uranium-235 before it can be used. To do this, uranium(IV) oxide is first converted to a gaseous compound, UF_6, and the isotopes are separated by a gaseous diffusion technique. Some key reactions are

$$UO_2(s) + 4\ HF(g) \longrightarrow UF_4(s) + 2\ H_2O(g)$$

$$UF_4(s) + F_2(g) \longrightarrow UF_6(g)$$

How much thermal energy (at constant pressure) would be involved in producing 225 tons of $UF_6(g)$ from UO_2? (1 ton = 9.08×10^5 g) Some necessary standard enthalpies of formation are:

$$\Delta H_f°[UO_2(s)] = -1085\ kJ/mol$$
$$\Delta H_f°[UF_4(s)] = -1914\ kJ/mol$$
$$\Delta H_f°[UF_6(g)] = -2147\ kJ/mol$$

84. One method of producing H_2 on a large scale is the following chemical cycle.

Step 1: $SO_2(g) + 2\ H_2O(g) + Br_2(g) \longrightarrow H_2SO_4(\ell) + 2\ HBr(g)$

Step 2: $H_2SO_4(\ell) \longrightarrow H_2O(g) + SO_2(g) + \frac{1}{2}\ O_2(g)$

Step 3: $2\ HBr(g) \longrightarrow H_2(g) + Br_2(g)$

Using the table of standard enthalpies of formation in Appendix J, calculate $\Delta H°$ for each step. What is the equation for the overall process and what is its enthalpy change? Is the overall process exothermic or endothermic?

85. One reaction involved in the conversion of iron ore to the metal is

$$FeO(s) + CO(g) \longrightarrow Fe(s) + CO_2(g)$$

Calculate the standard enthalpy change for this reaction from the following reactions of iron oxides with CO:

$$3\ Fe_2O_3(s) + CO(g) \longrightarrow 2\ Fe_3O_4(s) + CO_2(g) \qquad \Delta H° = -47\ kJ$$

$$Fe_2O_3(s) + 3\ CO(g) \longrightarrow 2\ Fe(s) + 3\ CO_2(g) \qquad \Delta H° = -25\ kJ$$

$$Fe_3O_4(s) + CO(g) \longrightarrow 3\ FeO(s) + CO_2(g) \qquad \Delta H° = +19\ kJ$$

Applying Concepts

86. Based on your experience, when ice melts to liquid water is the process exothermic or endothermic (with respect to the ice)? When liquid water freezes to ice at 0 °C, is this exothermic or endothermic (with respect to the liquid water)?

87. You pick up a six-pack of soft drinks from the floor, but it slips from your hand and smashes onto your foot. Comment on the work and energy involved in this sequence. What forms of energy are involved and at what stages of the process?

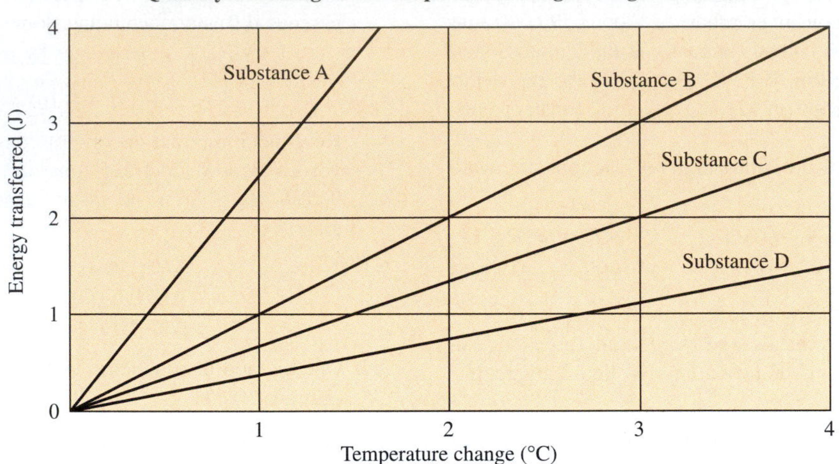

88. Consider the graph above. Which substance has the highest specific heat capacity?

89. Based on the graph above, how much heat would need to be transferred to 10 g of substance B to raise its temperature from 35 °C to 38 °C?

90. The sketch below shows two identical beakers with different volumes of water at the same temperature.

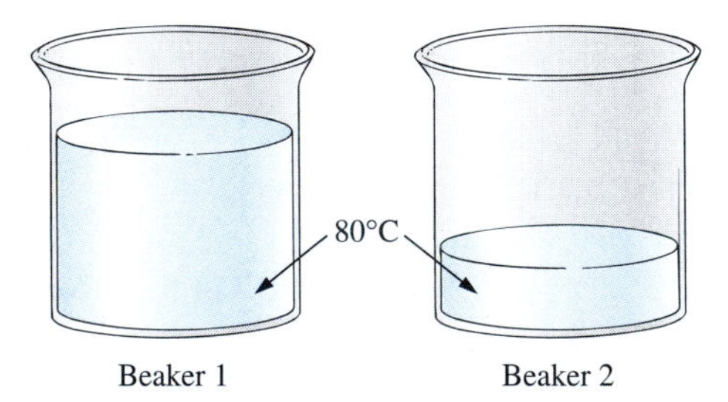

Is the thermal energy content of beaker 1 greater than, less than, or equal to that of beaker 2? Explain your reasoning.

91. If the same quantity of thermal energy were transferred to each beaker in question 90, would the temperature of beaker 1 be greater than, less than, or equal to that of beaker 2? Explain your reasoning.

92. Thermochemistry is often confusing because enthalpy change is referred to in different ways, for example: enthalpy of combustion, enthalpy of formation, enthalpy of reaction, and enthalpy of decomposition. What is similar about each of these situations? What is different?

93. In this chapter, the symbols ΔH_f° and ΔH° were used to denote a change in enthalpy. What is similar and what is different about the enthalpy changes they represent?

94. Consider the following equation

$$S(s) + 3\ O_2(g) \longrightarrow 2\ SO_3(g) \qquad \Delta H^\circ = -791\text{ kJ}$$

and the enthalpy value for $SO_3(g)$ listed in Table 6.4. Why are the enthalpy values different?

95. A student had five beakers each containing 100. mL of 0.500 M NaOH (aq), all at room temperature (20.0 °C). The student planned to add a carefully weighed quantity of solid ascorbic acid, $C_6H_8O_6$, to each beaker, stir until it dissolved, and measure the increase in temperature. After the fourth experiment, the student was interrupted and called away. The data table looked like this:

Experiment	Mass of Ascorbic Acid (g)	Final Temperature (°C)
1	2.20	21.7
2	4.40	23.3
3	8.81	26.7
4	13.22	26.6
5	17.62	—

(a) Predict the temperature the student would have observed in experiment 5. Explain why you predicted this temperature.
(b) For each experiment indicate which is the limiting reactant, sodium hydroxide or ascorbic acid.
(c) When ascorbic acid reacts with NaOH, how many hydrogen ions are involved? One, as in the case of HCl? Or two, as in the case of H_2SO_4? Or three, as in the case of phosphoric acid, H_3PO_4? Explain clearly how you can tell, based on the student's calorimeter data.

96. In their home laboratory two students do an experiment (a rather dangerous one—don't try it!) with drain cleaner (Drano, a solid) and toilet bowl cleaner (The Works, a liquid solution). The students measure 1 teaspoon (tsp) of Drano into each of four styrofoam coffee cups and dissolve the solid in half a cup of water. Then they go have lunch. When they return, they measure the temperature of the solution in each of the four cups and find it to be 22.3 °C. Next they measure into separate small empty cups 1, 2, 3, and 4 tablespoons (Tbsp) of The Works. In each

cup they add enough water to make the total volume 4 Tbsp. After a few minutes they measure the temperature of each cup and find it to be 22.3 °C. Finally the two students take each cup of The Works, pour it into a cup of Drano solution, and measure the temperature over a period of a few minutes. Their results are reported in the table below.

Expt. Number	Volume of The Works (Tbsp)	Highest Temperature (°C)
1	1	28.0
2	2	33.6
3	3	39.3
4	4	39.4

Discuss these results and interpret them in terms of the thermochemistry and stoichiometry of the reaction. Is the reaction exothermic or endothermic? Why is more energy transferred in some cases than others? For each experiment, which reactant, Drano or The Works, is limiting? Why are the final temperatures nearly the same in experiments 3 and 4? What can you conclude about the stoichiometric ratio between the two reactants?

CHAPTER 7

Principles of Reactivity II: Directionality of Chemical Reactions

In a blast furnace such as the one that produced the molten iron pictured here, oxidation of carbon to carbon monoxide and carbon dioxide is coupled to reduction of iron oxides to iron. Reduction of iron ore to iron would not occur of its own accord, but when coupled to oxidation of carbon, which occurs quite readily, the reduction can be carried out. • What reactions take place in a blast furnace? At what temperatures will these reactions produce iron? These questions are answered in the Summary Problem at the end of the chapter. *(Courtesy Bethlehem Steel Corporation)*

Some chemical reactions begin as soon as the reactants come into contact and continue until at least one reactant (the limiting reactant) is completely consumed. Drop a clean piece of potassium into water and it reacts violently. Sparks fly, and you see a purple flame characteristic of very hot potassium salts. Other reactions happen much more slowly, but reactants are still converted completely to products. An example is the reaction of iron with oxygen at room temperature. After many years and enough flaking of iron(III) oxide from its surface, a piece of iron exposed to air will rust away. Still other reactions are even slower at room temperature, but nevertheless reactants are slowly converted to products. For example, gasoline reacts so slowly with air at room temperature that it can be stored for long periods, but if the temperature is raised by a spark or flame, gasoline vapor burns rapidly and is essentially all converted to CO_2 and H_2O.

If, when a reaction appears to be over, *products predominate over reactants,* we designate the reaction a **product-favored system.** Examples are the reaction of potassium with water, rusting of iron, and combustion of gasoline. If a system is product-favored, most of the reactants will eventually be converted to products without continuous outside intervention, although "eventually" may mean a very, very long time. (How fast reactants are converted to products is the subject of Chapter 12, Chemical Kinetics.)

The term "product-favored" refers to reactions that are often said to be "spontaneous"; many people will use the two terms interchangeably. We prefer product-favored because some reactions do begin spontaneously, but produce only tiny quantities of products. Product-favored describes clearly a situation where products predominate over reactants.

There are other reactions that have virtually no tendency to occur by themselves. (Some examples are the reactions for which we wrote "N.R." for "no reaction" in Chapter 4.) For example, nitrogen and oxygen have coexisted in the earth's atmosphere for at least a billion years without appreciable concentrations of nitrogen oxides such as N_2O, NO, or NO_2 building up. (The concentration of N_2 in the atmosphere is more than two million times the concentration of N_2O, the most abundant oxide of nitrogen.) Similarly, deposits of salt (NaCl) have existed on earth for millions of years without forming the elements Na and Cl_2. We label *chemical systems in which little or none of the reactants are converted to products* as **reactant-favored systems.** A reactant-favored system always involves a transformation that is exactly the opposite of the transformation in a product-favored system. For example, Na(s) + Cl_2(g) react to form NaCl(s), but NaCl(s) does not react in the opposite direction to form Na(s) + Cl_2(g). Reactants in a reactant-favored system are not transformed into appreciable quantities of products, unless there is some continuous outside intervention.

What do we mean by continuous outside intervention? Usually it is some flow of energy. For example, if enough energy is provided to N_2 and O_2 to keep them at a very high temperature, small but significant quantities of NO can be formed from air. Such high temperatures are found in lightning bolts, power plants, and automobile engines. The large number of automobiles and power plants can produce enough NO and other nitrogen oxides to cause significant air pollution problems. Salt can be decomposed to its elements by continuously heating it to keep it molten and passing electricity through it to separate the ions and form the elements:

The air in the immediate vicinity of a lightning bolt can be heated enough to cause a small fraction of the nitrogen and oxygen to combine to form NO, but this takes place only while the lightning is present. A similar reaction can occur in the engine of an automobile, but again only a small fraction of the air is converted to nitrogen oxides, and only while the temperature is high.

$$2\,\mathrm{NaCl}(\ell) \xrightarrow{\text{electricity}} 2\,\mathrm{Na}(\ell) + \mathrm{Cl_2(g)}$$

In each case, a reactant-favored system can be forced to produce products if energy is continuously supplied. This is in contrast to the situation for a product-favored system such as combustion of gasoline, which requires only a brief spark to initiate the reaction. Once started, gasoline combustion continues of its own accord without a supply of energy from outside.

7.1 PROBABILITY AND CHEMICAL REACTIONS

Most exothermic reactions are product-favored at room temperature, and so we can use them to produce substances that we would like to have. The reason that exothermic reactions are usually product-favored is very similar to the reason that there is a one-way transfer of energy from a hotter to a colder sample (← *p. 223*). When an exothermic reaction takes place, energy is transferred to the surroundings, as seen in Figure 7.1 for the reaction of potassium with water. Chemical potential energy that has been stored in bonds between relatively few atoms and molecules (the reactants) spreads over many more atoms and molecules as the surroundings (and the products) are heated. Because there are a great many more atoms, molecules, and ions in the surroundings than in the reactants and products, it is always true that after an exothermic reaction, energy will be distributed more randomly—dispersed over a much larger number of particles—than it was before.

But why is dispersal of energy favored? The answer lies in probability. Energy is much more likely to be dispersed than concentrated. To better understand energy dispersal and probability, consider the hypothetical case of a very small sample of matter consisting of two atoms, A and B; and suppose that this sample contains two units of energy, each designated by *. There are three ways the energy can be distributed over the two atoms: atom A could have both units of energy; atom A and atom B could each have one; or atom B could have both. Designate these three situations as A^{**}, A^*B^*, and B^{**}.

Now suppose that atoms A and B come into contact with two other atoms, C and D, that have no energy. Consider the possibilities for distributing the two units of energy over all four atoms. There are ten: A^{**}, A^*B^*, A^*C^*, A^*D^*, B^{**}, B^*C^*, B^*D^*, C^{**}, C^*D^*, and D^{**}. Only three of these (A^{**}, A^*B^*, and B^{**}) have all the energy in atoms A and B, which was the initial situation. When all four atoms are in contact, there are seven chances out of ten that some energy will have transferred from A and

Figure 7.1 Product-favored reaction. Potassium reacts exothermically with water to give hydrogen gas and a solution of potassium hydroxide. Like this one, many reactions begin as soon as the reactants come into contact, and continue until one or more reactants have been consumed. We designate reactions such as these as "product-favored." *(C.D. Winters)*

B to C and D. Thus, there is a 70% probability that the energy will become spread out over more than just the two atoms A and B. When this dispersal of energy occurs, we say that the atoms A, B, C, and D have gone from a situation where the energy was ordered to one where it has become more disordered.

Conceptual Challenge Problem CP-7.A at the end of the chapter relates to topics discussed in this section.

Exercise 7.1 Probability of Energy Dispersal

Suppose that you have three units of energy to distribute over two atoms, A and B. Designate each possible arrangement. Now suppose that atoms A and B come into contact with three more atoms, C, D, and E. From the possible arrangements of energy over the five atoms, calculate the probability that all the energy will remain confined to atoms A and B.

The probability that energy will become disordered becomes overwhelming when large numbers of atoms or molecules are involved. For example, suppose that atoms A and B had been brought into contact with a mole of other atoms. There would still be only three arrangements in which all the energy was associated with atoms A and B, but there would be many, many more arrangements (more than 10^{47}) in which all the energy has been transferred to other atoms. In such a case it is essentially certain that energy will be transferred. *If energy can be dispersed* over a very much larger number of particles, *it will be.*

Recall that for the same amount of substance or number of particles, higher thermal energy corresponds to higher temperatures. Therefore, a substance at a higher temperature has greater energy per particle on average. Dispersal of energy over a larger number of particles corresponds to transfer of energy from a substance at a higher temperature to another at a lower temperature.

Just as there is a tendency for concentrated energy to disperse, concentrated matter also tends to disperse, unless there is something to prevent it. For example, in a gas at room temperature there are only weak forces attracting the molecules together. Suppose a sample of argon gas is confined within one flask that is connected through a stopcock to a second flask of equal size from which all gas molecules have been removed (Figure 7.2). What will happen if the stopcock is opened? Whoosh! The confined argon will expand to fill the entire volume available to it.

Dispersal of the gas can be analyzed in the same way as dispersal of energy. Suppose that a single argon gas atom occupies the two-flask system with the stopcock open. The atom will move around within one flask until it hits the opening to the other flask; then it will occupy the second flask for a while before returning to the first. Since the volumes are equal and the argon atom's motion is random, it will spend half its time in each flask, on average. The probability is 1/2 that it can be found in flask A and 1/2 that it is in flask B. Now consider two atoms within the same two-

The conclusions drawn about dispersal of argon atoms in this paragraph apply in general to particles of a gas, whether they are atoms or molecules.

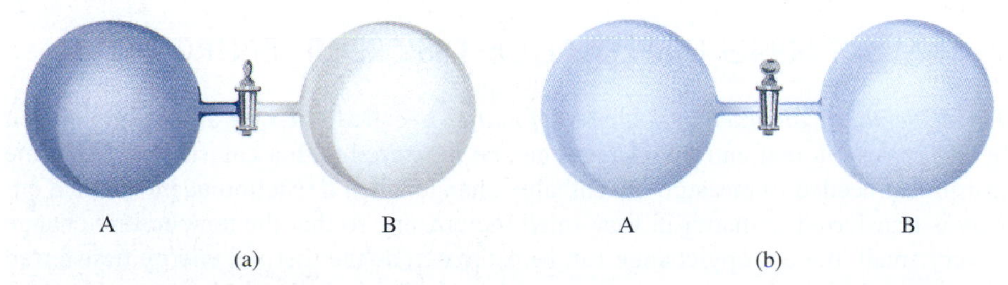

Figure 7.2 Expansion of a gas. This is a highly probable process. (a) A gas is confined in one flask, A. There are no atoms or molecules of gas in flask B. (b) When the valve between the flasks is opened, the gas particles rush into flask B and eventually become evenly distributed throughout both flasks.

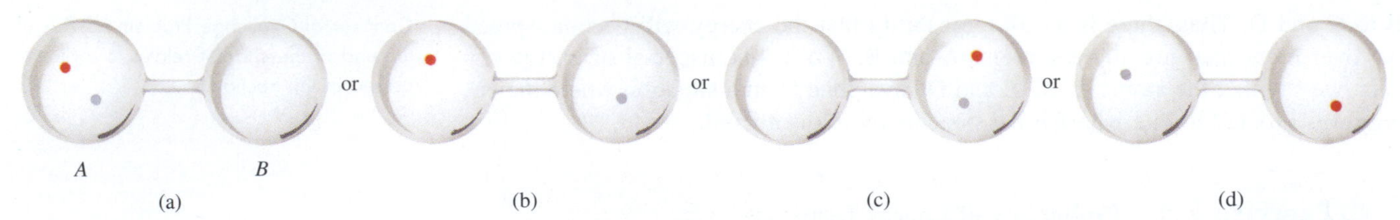

Figure 7.3 Gas expansion and probability. When two molecules occupy two flasks of equal volume, four different, equally probable arrangements are possible.

The low probability that a lot of energy will be associated with only a few particles makes a substance with a lot of chemical potential energy valuable. We call substances like coal, oil, and natural gas "energy resources" and sometimes fight wars over them because of their concentrated energy.

Atoms or molecules of a material that is a solid at room temperature (like the glass flask) do not disperse, because there are strong attractive forces between them. Their tendency to disperse becomes more obvious if the temperature is raised so that they can vaporize.

Conceptual Challenge Problem CP-7.B at the end of the chapter relates to topics discussed in this section.

flask system. There are four equally probable arrangements, as shown in Figure 7.3, but only one of them has both atoms in flask A. The probability that both atoms are in flask A is thus 1/4 or $(1/2)^2$. By making a similar diagram you can verify that for three atoms there are eight arrangements. Only one of these eight corresponds to all three atoms in flask A, giving a probability of 1/8 or $(1/2)^3$. In general (with the stopcock open) the probability that all argon atoms will be in the flask is $(1/2)^n$, where n is the number of atoms.

As the number of particles increases, the probability of their all being in the same flask (a more highly ordered state) gets very, very small. If a mole of gas were present, the probability that all particles were in flask A would be so incredibly small that it is absolutely certain that many particles would move from flask A to flask B. This two-flask system is a very simple example of a product-favored system—the final arrangement with gas in both flasks is much more probable than the initial one, and so the process occurs of its own accord. On the other hand, if we wanted to reverse the process by concentrating all the particles into flask A, a continuous outside influence such as a pump would be required—the pump could do work on the gas to force it into a highly improbable arrangement. The work done by the pump would be stored in the gas and could later be used for some other purpose.

To summarize, there are two ways that the final state of a system can be more probable than the initial one: having energy dispersed over a greater number of atoms, molecules, or ions, and having the atoms and molecules themselves more disordered. If both of these happen, then a reaction will definitely be product-favored, since the products and the distribution of energy will both be more probable. If only one of them happens, then quantitative information is needed to decide which effect is greater. The remainder of this chapter is devoted to developing that quantitative information. If a process would spread out neither matter nor energy, then that process will be reactant-favored—the initial substances will remain no matter how long we wait.

7.2 MEASURING DISPERSAL OF DISORDER: ENTROPY

The equation $\Delta S = q/T$ applies only to processes taking place at constant temperature and pressure. An example of such a process would be melting of ice at 0 °C and normal atmospheric pressure. However, if the temperature change is very small, the equation still works quite well. If a substance is heated in small increments, the equation is accurate.

The dispersal or disorder in a sample of matter is called **entropy** and is symbolized by S. It turns out that entropy changes can be measured with a calorimeter, the same instrument needed to measure the enthalpy change when a reaction occurs. When energy is transferred to matter in very small increments, so that the temperature change is very small, the entropy change can be estimated as the thermal energy transferred divided by the temperature,

$$\Delta S = q/T$$

In the equation $\Delta S = q/T$, it is necessary to use what is called the **absolute temperature scale**—the *scale on which the zero of temperature is the lowest possible*

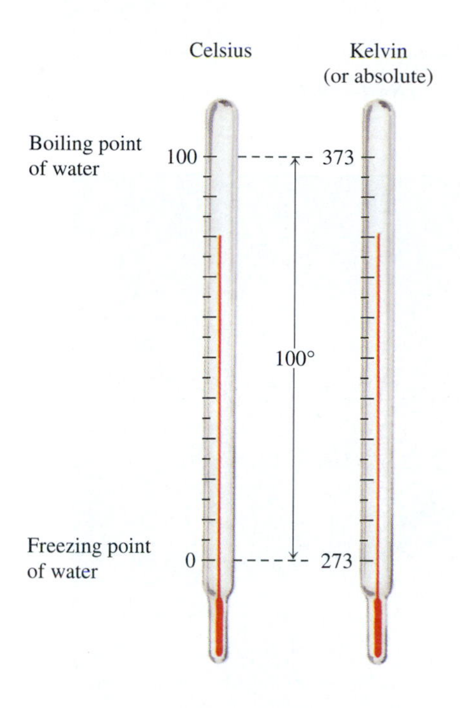

Figure 7.4 Temperature scales. The zero on the Celsius temperature scale is different from the zero on the absolute (Kelvin) scale. Zero on the absolute scale is the minimum possible temperature (0 K or −273.15 °C). The symbol for the absolute temperature unit is just K—there is no degree sign. Also, the sizes of the kelvin and the degree Celsius are the same, so that calculated changes in temperature are the same on both scales.

temperature. The unit of temperature on the absolute scale is called the kelvin and its symbol is K. The relationship between the absolute scale and the Celsius scale is shown in Figure 7.4. The kelvin is the same size as a degree Celsius *(← p. 8)* and so when ΔT is calculated by subtracting one temperature from another, the result will be the same on both scales, even though the numbers involved are different. It is generally agreed that the lowest possible temperature is −273.15 °C. This means that if t °C represents a temperature on the Celsius scale and T K a temperature on the absolute scale,

$$T = t + 273.15$$

Thus, 25.00 °C (a typical room temperature) is the same as 298.15 K, because 25.00 + 273.15 = 298.15.

The absolute temperature scale is also called the **Kelvin temperature scale** or the **thermodynamic temperature scale.** Its origin is described in Section 14.2. The kelvin is the internationally agreed unit of temperature (Appendix C). Except when calculating temperature differences, the kelvin unit should be used in all calculations involving temperature.

Exercise 7.2 The Importance of Absolute Temperature

Consider what would happen if the Celsius temperature scale were used when calculating entropy change by means of $\Delta S = q/T$. Suppose, for example, that energy were transferred to $H_2O(s)$ at a temperature 10° below its melting point and we wanted to calculate the entropy change. Would the calculated value agree with the fact that transfer of thermal energy *to* a sample increases its entropy?

Exercise 7.3 Calculating Temperature Changes

Use algebra to show that the same value must always be obtained when ΔT is calculated using either the Celsius or the absolute temperature scale. Use T K to represent an absolute temperature, and t °C to represent a Celsius temperature. Given the equation $T = t + 273.15$, show that $T_2 - T_1 = t_2 - t_1$.

To see that calculated temperature differences are the same for Celsius and absolute temperatures, calculate ΔT for a temperature increase from 0.00 °C to 25.00 °C. In one case you calculate (25.00 − 0.00) = 25.00. In the other case you calculate (298.15 − 273.15) = 25.00. Since calculated values of ΔT are the same on both temperature scales, either scale may be used in equations like $q = c \times m \times \Delta T$ *(p. 228).*

The equation $\Delta S = q/T$ can be used to determine the actual quantity of entropy in any sample of matter at any temperature, if we assume that in a perfect crystal at

Portrait of a Scientist • *Ludwig Boltzmann (1844–1906)*

Ludwig Boltzmann was an Austrian mathematician who gave us the useful interpretation of entropy as probability (and who also did much of the work on the kinetic theory of gases). Engraved on his tombstone in Vienna is his equation relating entropy and "chaos,"

$$S = k \log W$$

(where k is a fundamental constant of nature, now called Boltzmann's constant, and log means take a natural logarithm, the sort we would represent by *ln*). Boltzmann used the symbol W to represent what he called thermodynamic probability—the number of different ways that atoms or molecules can be arranged on the nanoscale, all corresponding to the same macroscale state. Therefore, his equation tells us that if there are only a few ways to arrange the atoms of a substance—that is, if there are only a few places in which we can put our atoms or molecules—then the entropy is low. On the other hand, the entropy is high if there are many possible arrangements ($W >> 1$), that is, if the level of chaos or disorder is high.

Boltzmann's tombstone. *(Oesper Collection in the History of Chemistry, University of Cincinnati)*

Though it is impossible to cool anything all the way to absolute zero, it is possible to get very close, and there are ways of estimating the disorder that is already in a substance near 0 K; thus, accurate entropy values can be obtained for many substances.

The process of introducing small quantities of energy, calculating an entropy increase, and then summing these small entropy increases is actually done by measuring the heat capacity of a substance as a function of temperature and then using integral calculus to calculate the integral of the function q/T between the limits of 0 K and the desired temperature.

the absolute zero of temperature (0 K or −273.15 °C) there is minimum disorder. A perfect crystal is one in which every nanoscale particle is in exactly the right position in the crystal lattice. There are no empty spaces or discontinuities. Because decreasing temperature corresponds to decreased molecular motion, the minimum possible temperature can reasonably be expected to correspond to minimum motion and thus minimum disorder. Thus it is logical to set the zero of the entropy scale for each substance as a perfect crystal of that substance at 0 K. By starting as close as possible to absolute zero and repeatedly introducing small quantities of energy, an entropy change can be determined for each small increase in temperature. These entropy changes can then be added to give the total (or absolute) entropy of a substance at any desired temperature.

The results of such measurements for several substances at 298.15 K are given in Table 7.1. These are standard molar entropy values, and so they apply to 1 mol of each substance at the standard pressure of 1 bar and the specified temperature of 25 °C. The units are joules per kelvin per mole (J/K·mol). Because there is an actual zero on the entropy scale, the values in Table 7.1 are not measured relative to elements in their most stable form under standard-state conditions. Therefore absolute entropies can be determined for elements, as well as compounds. This is different from the situation for standard molar enthalpies of formation ***(← Table 6.4, p. 253)***, where there are no values for elements.

Qualitative Rules for Entropy

Some useful generalizations can be drawn from the examples given in Table 7.1.

- ***Entropies of gases are generally much larger than those of liquids, which in turn are generally larger than those of solids.*** In a solid the particles can vi-

TABLE 7.1 **Some Standard Molar Entropy Values at 298.15 K***

Compound or Element	Entropy, $S°$(J/K·*mol*)	Compound or Element	Entropy, $S°$(J/K·*mol*)	Compound or Element	Entropy, $S°$(J/K·*mol*)
C(graphite)	5.740	Ar(g)	154.7	Cl_2(g)	223.036
C(g)	158.096	H_2(g)	130.684	Br_2(ℓ)	152.2
CH_4(g)	186.64	O_2(g)	205.138	I_2(s)	116.135
C_2H_6(g)	229.60	N_2(g)	191.61	NaCl(s)	72.8
C_3H_8(g)	269.9	H_2O(g)	188.825	NaF(s)	51.5
CH_3OH(ℓ)	126.8	H_2O(ℓ)	69.91	MgO(s)	26.94
CO(g)	197.674	NH_3(g)	192.45	$MgCO_3$(s)	65.854
CO_2(g)	213.74	HCl(g)	186.908		
Ca(s)	41.42	F_2(g)	202.78		

*Taken from "The NBS Tables of Chemical Thermodynamic Properties," 1982.

brate only around their lattice positions. When a solid melts, its particles can move around more freely, and there is an increase in molar entropy. When a liquid vaporizes, the position restrictions due to forces between the particles nearly disappear, and there is another large entropy increase (Figure 7.5). For example, the entropies (in J/K·mol) of the halogens I_2(s), Br_2(ℓ), and Cl_2(g) are 116.1, 152.2, and 223.0. Similarly, the entropies of C(s, graphite) and C(g) are 5.7 and 158.1.

Conceptual Challenge Problem CP-7.C at the end of the chapter relates to topics discussed in this section.

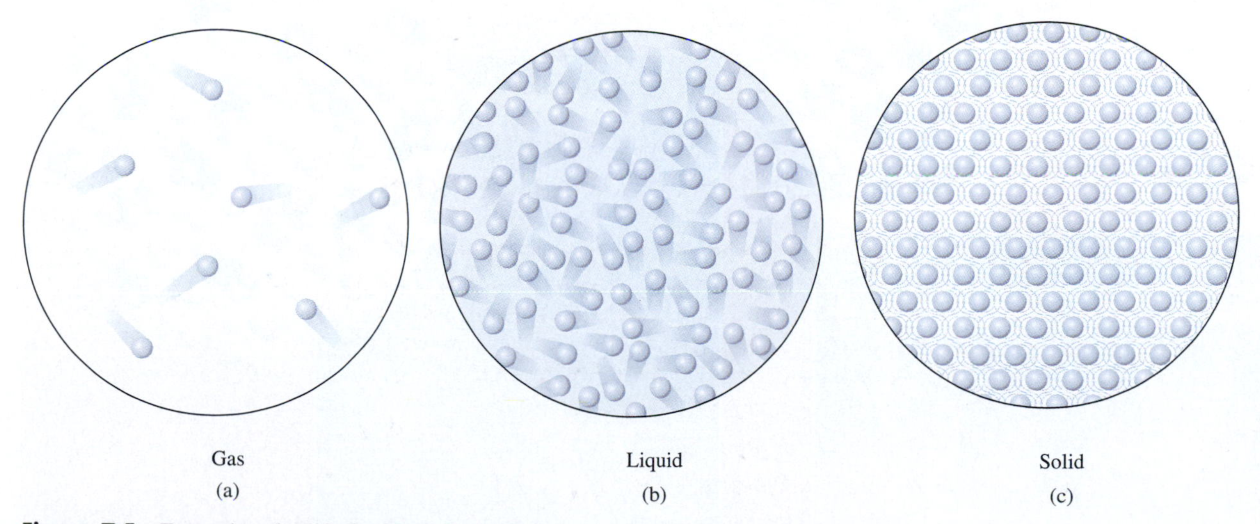

Figure 7.5 **Entropies of solid, liquid, and gas phases.** (a) The particles in a sample of gas move in random directions at random speeds. (b) The particles in the liquid are more concentrated than in a gas, but move randomly and have no fixed positions. (c) The particles in a solid are constrained to vibrate about specific positions.

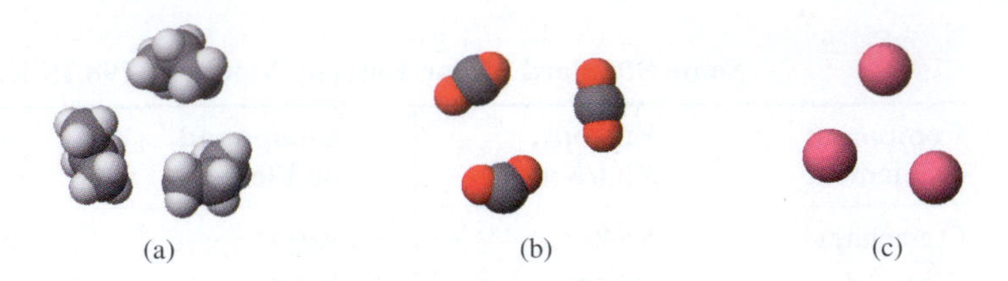

Figure 7.6 Entropy and molecular structure. Three particles of similar molar mass are shown. There are many more ways to arrange the atoms in $CH_3CH_2CH_3$ (a) than there are in CO_2 (b). There are more ways to arrange three atoms in CO_2 than for the single Ar atom (c).

- ***Entropies of more complex molecules are larger than those of simpler molecules, especially in a series of closely related compounds.*** In a more complicated molecule there are more ways for the atoms to move about in three-dimensional space and hence there is greater entropy. For example, the entropies (in J/K · mol) of methane (CH_4), ethane (CH_3CH_3), and propane ($CH_3CH_2CH_3$) are 186.26, 229.6, and 269.9. For atoms or molecules of similar molar mass, we have Ar, CO_2, and $CH_3CH_2CH_3$ with entropies of 154.7, 213.74, and 269.9 (Figure 7.6).
- ***Entropies of ionic solids are larger the weaker the attractions among the ions.*** The weaker such forces, the easier it is for ions to vibrate about their lattice positions. Examples are NaF(s) and MgO(s) with entropies of 51.5 and 26.8 J/K · mol; the 2+ and 2− charges on the magnesium and oxide ions result in greater attractive forces (← *p. 92–93*) and hence lower entropy.
- ***Entropy usually increases when a pure liquid or solid dissolves in a solvent.*** Since Table 7.1 refers only to pure substances, no example values are avail-

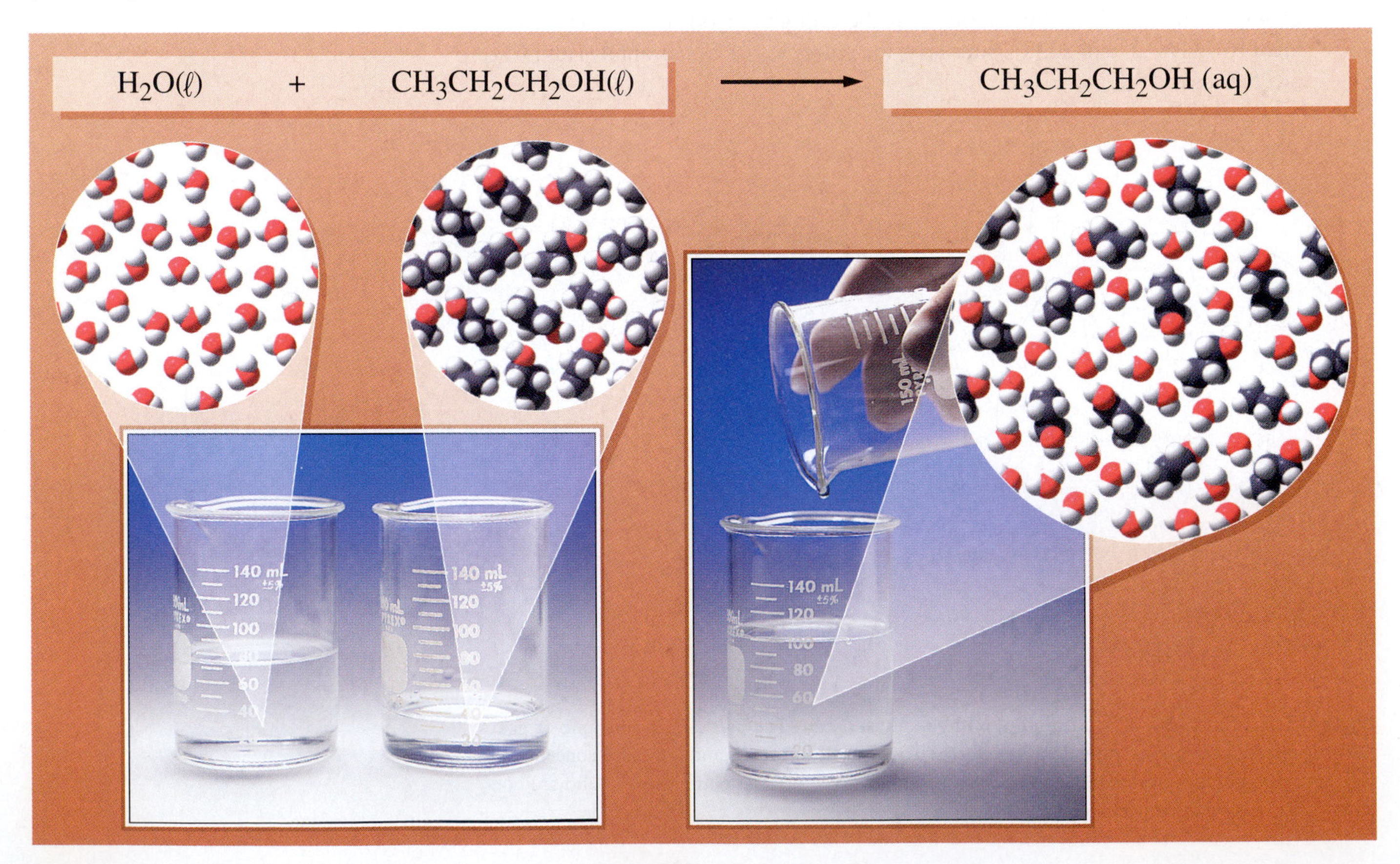

Figure 7.7 Entropy and dissolving. There is usually an increase in entropy when a solid or liquid dissolves in a liquid solvent, because the solute particles become dispersed among the solvent particles. (a) Nanoscale views of water and propanol before mixing. (b) Nanoscale view of a solution of propanol in water, where the entropy is higher because the molecules are mixed. *(C.D. Winters)*

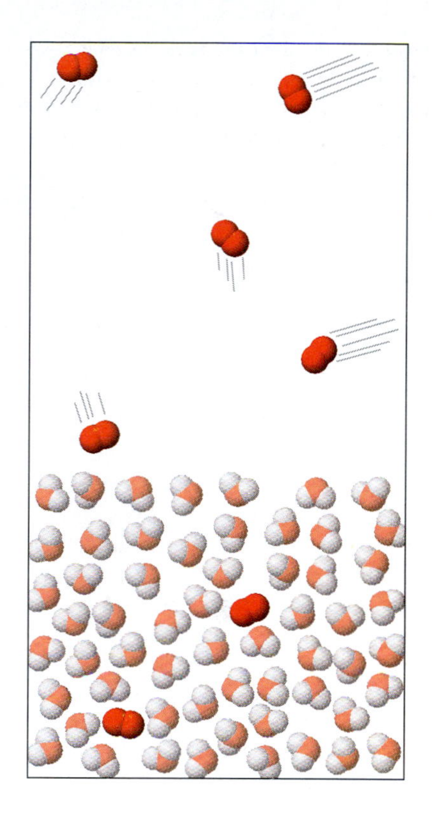

Figure 7.8 Entropy of solution of a gas. The very large entropy of the gas exceeds that of the solution. Even though particles are dispersed among each other in the liquid solution, the gas particles are much more widely spread out and have much higher entropy.

able, but matter usually becomes more dispersed or disordered when a substance dissolves and different kinds of molecules mix together (Figure 7.7).

- ***Entropy decreases when a gas dissolves in a liquid.*** Although gas molecules are dispersed among solvent molecules in solution, the very large entropy of the gas phase results in a decrease in entropy when the widely separated gas particles become crowded together with solvent particles in the liquid solution (Figure 7.8).

These generalizations can be used to estimate whether there will be an increase in disorder of the substances involved when reactants are converted to products. Such predictions are easier to make for entropy changes than for enthalpy changes, where a knowledge of bond energies is needed. For both of the processes $H_2O(s) \rightarrow H_2O(\ell)$ and $H_2O(\ell) \rightarrow H_2O(g)$, we expect an entropy increase, because water molecules in the solid are more ordered than in the liquid and much more ordered than in the gas. This is confirmed by entropy measurements. At 275.13 K (0 °C) and 1 bar, for example, 1 mol of ice can be converted to liquid water with no temperature change. The thermal energy transferred is 6.02 kJ, and the transfer is into the water, so $q = 6020$ J. This gives

$$\Delta S° \;[H_2O(s) \longrightarrow H_2O(\ell)] = \frac{q}{T} = \frac{6020 \text{ J}}{273.15 \text{ K}} = 22.0 \text{ J/K}$$

Similarly, in order to boil a mole of water at 100 °C we must transfer 40.7 kJ into the water, q must be positive, and so must ΔS; the value in this case is 109 J/K.

$$\Delta S° \;[H_2O(\ell) \longrightarrow H_2O(g)] = \frac{q}{T} = \frac{40{,}700 \text{ J}}{373.15 \text{ K}} = 109 \text{ J/K}$$

Exercise 7.4 Calculating $\Delta S°$ Values for Phase Changes

The enthalpy of vaporization of benzene (C_6H_6) is 30.8 kJ/mol at the boiling point of 80.1 °C. Calculate the entropy change for benzene going from (a) liquid to vapor and (b) vapor to liquid at 80.1 °C.

For the decomposition of iron(III) oxide to its elements,

$$2\,Fe_2O_3(s) \longrightarrow 4\,Fe(s) + 3\,O_2(g) \qquad \Delta S° = +551.7 \text{ J/K}$$

we would also predict an increase in entropy, because 3 mol of gaseous oxygen is present in the products, and the reactant is a solid. Since gases have much higher entropy than solids or liquids, gaseous substances are most important in determining entropy changes. An example where a decrease in entropy would be predicted is

$$2\,CO(g) + O_2(g) \longrightarrow 2\,CO_2(g) \qquad \Delta S° = -173.0 \text{ J/K}$$

Here there is 3 mol of gaseous substances (two of CO and one of O_2) at the beginning but only 2 mol of gaseous substance at the end of the reaction. Two moles of gas almost always contains less entropy than 3 mol of gas, and so $\Delta S°$ is negative. Another example where entropy decreases is the process

$$Ag^+(aq) + Cl^-(aq) \longrightarrow AgCl(s)$$

Notice that entropy changes are usually reported in units of joules per kelvin (J/K), whereas enthalpy changes are usually given in kJ. This means that you need to be careful about the units to avoid being off by a factor of 1000.

Here the reactant ions are free to move about among water molecules in aqueous solution, but those same ions are held in a crystal lattice in the solid, a situation with much greater constraint.

Exercise 7.5 Predicting Entropy Changes

For each of the following processes, tell whether you would expect entropy to be greater for the products than for the reactants and explain how you arrived at your prediction.

(a) $2\ CO_2(g) \longrightarrow 2\ CO(g) + O_2(g)$

(b) $NaCl(s) \longrightarrow NaCl(aq)$

(c) $MgCO_3(s) \xrightarrow{\text{heat}} MgO(s) + CO_2(g)$

7.3 CALCULATING ENTROPY CHANGES

The entropy values given in Table 7.1 can be used to calculate the entropy changes for physical and chemical processes. To do so, we assume that each reactant and each product is present in the amount required by its stoichiometric coefficient in the equation for the process. All substances are assumed to be at the standard pressure of 1 bar and at a specified temperature (in Table 7.1 it is 298.15 K). Then, to see whether there is an increase or decrease in entropy, we can just multiply the entropy of each product by the number of moles of that product and sum over all products and then subtract the sum of the entropies of reactants, each multiplied by the appropriate number of moles.

Notice that the equation for calculating ΔS has the same form as that for calculating ΔH for a reaction (← p. 254).

$$\Delta S^\circ = \Sigma\{(\text{moles of products}) \times S^\circ(\text{product})\} - \Sigma\{(\text{moles of reactants}) \times S^\circ(\text{reactant})\}$$

PROBLEM-SOLVING EXAMPLE 7.1 Calculating an Entropy Change from Tabulated Values

Calculate the entropy change for the reaction

$$CO(g) + 2\ H_2(g) \longrightarrow \underset{\text{methanol}}{CH_3OH(\ell)}$$

which is being evaluated as a possible way to manufacture liquid methanol for use in motor fuel.

Answer -332.3 J/K · mol

Explanation We use information in Table 7.1, subtracting the entropies of reactants from the entropy of the product.

$$\begin{aligned}\Delta S^\circ &= \Sigma\{(\text{moles of product}) \times S^\circ(\text{product})\} - \Sigma\{(\text{moles of reactant}) \times S^\circ(\text{reactant})\}\\ &= (1\ \text{mol}) \times S^\circ[CH_3OH(\ell)] - \{(1\ \text{mol}) \times S^\circ[CO(g)] + (2\ \text{mol}) \times S^\circ[H_2(g)]\}\\ &= (1\ \text{mol}) \times (126.8\ \text{J/K} \cdot \text{mol}) - \{(1\ \text{mol}) \times (197.7\ \text{J/K} \cdot \text{mol})\\ &\qquad + (2\ \text{mol}) \times (130.7\ \text{J/K} \cdot \text{mol})\}\\ &= -\ 332.3\ \text{J/K}\end{aligned}$$

Notice that this calculation gives the entropy change for the *system,* since it involves dispersal of the atoms and molecules that make up the system. In this case there is a large decrease in entropy because 3 mol of gaseous material is converted to 1 mol of a more complicated, but liquid-phase, product.

Problem-Solving Practice 7.1

Use absolute entropies from Table 7.1 to calculate the entropy change for each of the following processes, hence verifying the predictions made in Exercise 7.5.

(a) $2\ CO_2(g) \longrightarrow 2\ CO(g) + O_2(g)$

(b) $NaCl(s) \longrightarrow NaCl(aq)$

(c) $MgCO_3(s) \xrightarrow{\text{heat}} MgO(s) + CO_2(g)$

7.4 ENTROPY AND THE SECOND LAW OF THERMODYNAMICS

A great deal of experience with many chemical reactions and other processes in which energy is transferred has led to the **second law of thermodynamics,** which states that *the total entropy of the universe is continually increasing.* Whenever a product-favored chemical or physical process occurs, matter, energy, or both become more dispersed or disordered. Evaluating whether this will happen during a proposed chemical reaction allows us to predict whether reactants will form appreciable quantities of products.

Predicting whether a reaction is product-favored can be done in three steps: (1) calculate how much entropy is created or destroyed by dispersal or concentration of energy, (2) calculate how much entropy is created or destroyed by dispersal or concentration of matter, and (3) add the two results. Calculations (1) and (2) can be carried out by assuming that reactants under standard conditions are converted into products under standard conditions. Let us apply these steps to the manufacture of liquid methanol by the reaction

$$CO(g) + 2\ H_2(g) \longrightarrow \underset{\text{methanol}}{CH_3OH(\ell)}$$

for which we calculated the entropy change in Problem-Solving Example 7.1. If the reaction is product-favored, it would be a good way to produce methanol for use as automotive fuel. The reactants can be obtained from plentiful resources: coal and water. We could base our prediction upon having as reactants 1 mol CO(g) and 2 mol $H_2(g)$, and as product 1 mol of liquid methanol, with all substances at 1 bar and 25 °C. Then data from Tables 6.4 (← *p. 253*) and 7.1 will apply. If the entropy of the universe is predicted to be higher after the product has been produced, then the reaction is product-favored under these conditions and might be useful. If not, perhaps some other conditions could be used, or perhaps we should consider some other reaction altogether.

We shall consider nonstandard conditions and the effect of temperature in Chapter 13; rules will be developed to suggest how to change the conditions under which a reaction is carried out so as to produce more products than could be obtained under standard conditions.

Dispersal or concentration of energy by a chemical reaction can be evaluated by calculating $\Delta H°$ and assuming that this quantity of thermal energy is transferred to or from the surroundings. If the energy transfer is slow and occurs at a constant temperature, the entropy change for the surroundings can be calculated as:

$$\Delta S_{\text{surroundings}} = \frac{q_{\text{surroundings}}}{T} = \frac{-\Delta H_{\text{system}}}{T} = \frac{-\Delta H°}{T}$$

The minus sign in this equation comes from the fact that $-\Delta H_{\text{system}}$ is the energy transferred out of the system, and hence equals the energy transferred into the surroundings, $q_{\text{surroundings}}$.

The equation says that for an exothermic reaction (negative $\Delta H°$) there will be an increase in entropy of the surroundings, a fact that we have already mentioned. For the proposed methanol-producing reaction, $\Delta H°$ (calculated from Table 6.4) is -128.1 kJ, and so the entropy change is

$$\Delta S_{\text{surroundings}} = \frac{-(-128.1\ \text{kJ})}{298\ \text{K}} \times \frac{1000\ \text{J}}{\text{kJ}} = 430.\ \text{J/K}$$

CHEMISTRY YOU CAN DO

Rubber Bands and Thermodynamics

Here are two experiments that you can analyze with the laws of thermodynamics. For the first experiment you need a wide rubber band and a stiff upper lip. (Your lip is a good detector of small temperature changes.) For the second experiment you will need the same rubber band, something heavy enough to stretch the rubber band, like a hammer or a coffee mug, and a hair dryer.

In this experiment the rubber band is the thermodynamic system; you and your lip are part of the surroundings. Hold the rubber band against your lip and quickly stretch it to its limit. What did you feel? Was the process of stretching the rubber band exothermic or endothermic? What was the sign of q for this process? When you stretched the rubber band you had to work to do it. For the rubber band, what is the sign of w for the stretching process? Can you determine the sign of ΔE for the rubber band? If so, what is it? If not, why not? What kinds of experiments might you devise to make possible a determination of both the sign and the value of ΔE? Since you carried out this experiment at constant (atmospheric) pressure, what was the sign of ΔH?

Now for the second experiment. Use the rubber band to hang the hammer or coffee mug from a door knob. (You could do this on a larger scale with a bungee cord!) Once the suspended object has stopped bouncing up and down, turn on the hair dryer and

The entropy change for dispersal of matter is just the entropy change for the system itself (the atoms involved in the methanol-producing reaction). This entropy change can be evaluated from the absolute entropies of the products and reactants as described in the previous section and has already been calculated in Problem-Solving Example 7.1 to be $\Delta S_{\text{system}} = \Delta S° = -332.3$ J/K.

The *total* entropy change includes the changes for both system and surroundings. It is referred to as $\Delta S_{\text{universe}}$, where the universe includes everything. It is the sum of the entropy change for the system and the entropy change for the surroundings. (We assume that nothing else but our reaction happens, and so there are no other entropy changes.) This total entropy change is

$$\Delta S_{\text{universe}} = \Delta S_{\text{system}} + \Delta S_{\text{surroundings}} = (-332.3 + 430.)\ \text{J/K} = 98\ \text{J/K}$$

The process we are evaluating is accompanied by an increase in entropy of the universe. This means that if we had CO(g), H_2(g), and $CH_3OH(\ell)$, each at 1 bar pressure and all in contact with each other, some of the CO(g) and some of the H_2(g) would react to form $CH_3OH(\ell)$. The process is product-favored and might be useful for manufacturing methanol.

A product-favored reaction. Decomposition of hydrogen peroxide is catalzyed by manganese(II) oxide. *(C.D. Winters)*

Exercise 7.6 Effect of Temperature on Entropy Change

The reaction of carbon monoxide with hydrogen to form methanol is quite slow at room temperature. As a general rule, reactions go faster at higher temperatures. Suppose that you tried to speed up this reaction by increasing the temperature. Assuming that $\Delta H°$ does not change very much as the temperature changes, what effect would increasing the temperature have on $\Delta S_{\text{surroundings}}$? Assuming that $\Delta S°$ for a reaction system does not change much as the temperature changes, what effect would increasing the temperature have on $\Delta S_{\text{universe}}$?

Predictions of the sort we have just made by calculating $\Delta S_{\text{universe}}$ can also be made qualitatively, without calculating, if we know whether a reaction is exothermic and if we can predict whether there is a dispersal of matter when the reaction takes place. *A reaction is sure to be product-favored if it is exothermic and the atoms of the reaction system are more disordered afterwards than before.* Also, *a reaction is certainly **not** product-favored if it is endothermic and there is a decrease in entropy*

heat the rubber band. What happens? Is this what you expected? Do most objects do this when heated?

Now think about what happened when you released the stretched rubber band in the first experiment. What is the sign of the enthalpy change for unstretching the rubber band? Was what happened when you released the rubber band a product-favored process? Is that process favored by the enthalpy change? What must be the sign of the entropy change for unstretching a rubber band? Use your conclusions about the entropy and enthalpy changes for unstretching the rubber band to account for the observation you made in the second experiment.

(C.D. Winters)

for the system. There are two other possible cases, as indicated in Table 7.2, but they are more difficult to predict without quantitative information.

As examples, consider the reactions of carbonates with acids. These reactions are product-favored because they are exothermic and produce highly disordered gases and solutions. Reaction of limestone with hydrochloric acid is typical:

$$CaCO_3(s) + 2\ HCl(aq) \longrightarrow CaCl_2(aq) + H_2O(\ell) + CO_2(g) \qquad \text{(exothermic)}$$

Similarly, combustion reactions of hydrocarbons are product-favored because they are exothermic and produce a larger number of gas phase product molecules than there were gas phase reactant molecules.

$$\underset{\text{butane}}{2\ C_4H_{10}(g)} + 13\ O_2(g) \longrightarrow 8\ CO_2(g) + 10\ H_2O(g) \qquad \text{(exothermic)}$$

But what about a reaction such as the production of ethylene, C_2H_4, from ethane, C_2H_6? Although entropy is predicted to increase (one gas-phase molecule forms two), the reaction is very endothermic,

$$C_2H_6(g) \longrightarrow H_2(g) + C_2H_4(g) \qquad \Delta H^\circ = +137\ \text{kJ} \qquad \Delta S^\circ = +121\ \text{J/K}$$

Enthalpy change predicts that this process is reactant-favored, while entropy change predicts the opposite. Which is more important? It depends on the temperature.

Calculating $\Delta S_{surroundings}$ $(= -\Delta H^\circ/T)$ involves dividing the enthalpy change by the temperature. Since ΔH° stays pretty much the same at different temperatures, the bigger the temperature the smaller the absolute value of $\Delta S_{surroundings}$ $(= -\Delta H^\circ/T)$. At room temperature $\Delta S_{surroundings}$ is usually bigger in absolute value than ΔS_{system}, and so exothermic reactions are expected to be product-favored and endothermic re-

The absolute value of an algebraic number or variable is the value without the sign.

TABLE 7.2 **Predicting Whether a Reaction is Product-Favored**

Sign of ΔH_{system}	Sign of ΔS_{system}	Product-Favored?
− (Exothermic)	+	Yes
− (Exothermic)	−	Yes at low *T*; no at high *T*
+ (Endothermic)	+	No at low *T*; yes at high *T*
+ (Endothermic)	−	No

actions (like this one) are expected to be reactant-favored. The ethylene-producing reaction is indeed reactant-favored at 25 °C, because $\Delta S_{surroundings} = -137$ kJ/298 K = -460 J/K, $\Delta S_{system} = 121$ J/K and $\Delta S_{universe} = -339$ J/K. To make a successful industrial process, chemical engineers have designed plants that carry it out at about 1000 °C. At this higher temperature, $\Delta S_{surroundings}$ (= -137 kJ/1273 K = -108 J/K) is smaller in magnitude than $\Delta S°$ (= 121 J/K). The entropy term is more important than the enthalpy term, $\Delta S_{universe} = 13$ J/K, and products are predicted to predominate over reactants.

Exercise 7.7 Predicting the Direction of a Reaction

Classify each of the following reactions into one of the four types summarized in Table 7.2. Hence predict whether each reaction is product-favored or reactant-favored at room temperature.

Reaction	$\Delta H°$ (298 K) kJ	$\Delta S°$ (298 K) J/K
(a) $CH_4(g) + 2\ O_2(g) \longrightarrow 2\ H_2O(\ell) + CO_2(g)$	−890.3	−243.0
(b) $2\ Fe_2O_3(s) + 3\ C(graphite) \longrightarrow 4\ Fe(s) + 3\ CO_2(g)$	+467.9	+560.3
(c) $C(graphite) + O_2(g) \longrightarrow CO_2(g)$	−393.5	2.9
(d) $2\ Ag(s) + 3\ N_2(g) \longrightarrow 2\ AgN_3(s)$	+617.6	−451.5

Exercise 7.8 Product- or Reactant-Favored?

Is the combination reaction of hydrogen and chlorine to give hydrogen chloride gas predicted to be product-favored or reactant-favored at 298 K?

$$H_2(g) + Cl_2(g) \longrightarrow 2\ HCl(g)$$

Answer the question by calculating the value for $\Delta S_{universe}$.

7.5 GIBBS FREE ENERGY

Calculations of the sort done in the previous section would be simpler if we did not have to separately evaluate the entropy change of the surroundings from a table of $\Delta H_f°$ values and the entropy change of the system from a table of $S°$ values. To simplify such calculations, a new thermodynamic function was defined by J. Willard Gibbs (1838–1903). It is now called the **Gibbs free energy** and given the symbol G. Gibbs defined his free energy so that $\Delta G_{system} = -T\Delta S_{universe}$. Notice that because of the minus sign, if the entropy of the universe increases, the free energy of the system must decrease. That is, *a decrease in Gibbs free energy of a system is characteristic of a product-favored process.*

In the previous section we showed that the total entropy change accompanying a chemical reaction carried out at constant temperature and pressure is

$$\Delta S_{universe} = \Delta S_{surroundings} + \Delta S_{system} = \frac{-\Delta H_{system}}{T} + \Delta S_{system}$$

Combining this algebraically with Gibbs's definition of free energy, we have

$$\Delta G_{system} = -T\Delta S_{universe} = -T\left[\frac{-\Delta H_{system}}{T} + \Delta S_{system}\right] = \Delta H_{system} - T\Delta S_{system}$$

TABLE 7.3 **Using $\Delta G^\circ = \Delta H^\circ - T\Delta S^\circ$ to Predict Whether a Process is Product-Favored**

Sign of ΔH°	Sign of ΔS°	Sign of ΔG°	Product-Favored?
− (Exothermic)	+	−	Yes
− (Exothermic)	−	Depends on T	Yes at low T; no at high T
+ (Endothermic)	+	Depends on T	No at low T; yes at high T
+ (Endothermic)	−	+	No

or, under standard-state conditions,

$$\Delta G^\circ_{\text{system}} = \Delta H^\circ_{\text{system}} - T\,\Delta S^\circ_{\text{system}}$$

The Gibbs free energy change provides a way of predicting whether a reaction will be product-favored that depends only on the system—the chemical substances undergoing reaction. Therefore, we can tabulate values of ΔG_f° for a variety of substances, and from them calculate

$$\Delta G^\circ = \Sigma\{(\text{moles of product}) \times \Delta G_f^\circ(\text{product})\} - \Sigma\{(\text{moles of reactant}) \times \Delta G_f^\circ(\text{reactant})\}$$

for a great many reactions. The calculation is similar to using ΔH_f° values from Table 6.4 to calculate ΔH° for a reaction. As was the case for ΔH_f° values, there are no ΔG_f° values for elements, because forming an element from itself constitutes no change at all. Appendix J contains a table that includes ΔG_f° values for many compounds.

The equation $\Delta G^\circ = \Delta H^\circ - T\,\Delta S^\circ$ confirms the conclusion we drew earlier and summarized in Table 7.2 for reactions under standard conditions. Table 7.3 interprets those conclusions in terms of the Gibbs free energy for a reaction system.

PROBLEM-SOLVING EXAMPLE 7.2 Using Standard Free Energies of Formation

Calculate the standard Gibbs free energy change for combustion of methane using values of ΔG_f° from Appendix J.

Answer −800.9 kJ

Explanation Write a balanced equation for the combustion reaction and look up ΔG_f° values in Appendix J.

$$CH_4(g) + 2\,O_2(g) \longrightarrow 2\,H_2O(g) + CO_2(g)$$

ΔG_f° (kJ/mol): −50.7 0 −228.6 −394.4

(Notice that elements in their standard states have $\Delta G_f^\circ = 0$, just as they have $\Delta H_f^\circ = 0$.) Now calculate

$$\begin{aligned}\Delta G^\circ &= \Sigma\{(\text{moles of product}) \times \Delta G_f^\circ(\text{product})\} - \Sigma\{(\text{moles of reactant}) \times \Delta G_f^\circ(\text{reactant})\}\\ &= \{2\text{ mol} \times \Delta G_f^\circ[H_2O(g)]\} + \{1\text{ mol} \times \Delta G_f^\circ[CO_2(g)]\} - \{1\text{ mol} \times \Delta G_f^\circ[CH_4(g)]\} - \{2\text{ mol} \times \Delta G_f^\circ[O_2(g)]\}\\ &= \{2\text{ mol} \times (-228.6\text{ kJ/mol})\} + \{1\text{ mol} \times (-394.4\text{ kJ/mol})\} - \{1\text{ mol} \times (-50.7\text{ kJ/mol})\} - \{2\text{ mol} \times (0\text{ kJ/mol})\}\\ &= -800.9\text{ kJ}\end{aligned}$$

Combustion of natural gas. *(Rick Poley/Visuals Unlimited)*

The free energy change for this combustion reaction, $\Delta G°$, is a large negative number, clearly indicating the reaction is product-favored under standard conditions.

Problem-Solving Practice 7.2

In the text we concluded that the reaction to produce methanol from CO to H_2 is product-favored.

$$CO(g) + 2\ H_2(g) \longrightarrow CH_3OH(\ell)$$

(a) Verify this result by calculating $\Delta G°$ from $\Delta H°$ and $\Delta S°$ for the system. Use values of $\Delta H_f°$ and $S°$ in Appendix J.
(b) Compare your result in part (a) with the calculated value of $\Delta G°$ obtained from $\Delta G_f°$ values from Appendix J.
(c) Is the sign of $\Delta G°$ positive or negative? Is the reaction product-favored? At all temperatures?

7.6 GIBBS FREE ENERGY AND ENERGY RESOURCES

A major goal of chemists is to assemble small molecules into larger ones that can be sold for much more than the cost of the reactants. An example is synthesis of methane (the main component of natural gas) from coal (mainly carbon) and water (steam).

$$2\ C(s) + 2\ H_2O(g) \longrightarrow CH_4(g) + CO_2(g)$$
$$\Delta H° = 15.441\text{ kJ} \quad \Delta S° = 16.614\text{ J/K} \quad \Delta G° = 12.065\text{ kJ}$$

This is a good way to make a conveniently used gaseous fuel from coal, which, although relatively plentiful, is hard to use because it is a solid. It is an example of a coal-gasification process—the subject of considerable research and development effort. This process is hard to carry out, though. Its Gibbs free energy change is positive, it is reactant-favored, and the entropy of the universe would decrease if it occurred. Does this mean that we can never obtain methane from coal this way? No. It means that this reaction cannot occur *unless there is some continuous outside intervention,* which we identified earlier as a transfer of energy.

Roald Hoffmann, who shared the Nobel Prize in Chemistry in 1981, has said that "One amusing way to describe synthetic chemistry, the making of molecules that is at the intellectual and economic center of chemistry, is that it is the local defeat of entropy." (*American Scientist,* Nov–Dec. 1987, pp. 619–621.)

For example, a dead car battery will not charge itself. It is a reactant-favored system. But a battery can be charged if it is connected to a charger that is in turn powered by electricity generated in a power plant. Probably the power plant generates electricity by burning coal, which is mainly carbon and burns in air according to the equation

$$C(s) + O_2(g) \longrightarrow CO_2(g) \qquad \Delta G° = -394.4\text{ kJ}$$

The large negative Gibbs free energy change for combustion of coal more than offsets the positive Gibbs free energy change of the battery-charging reactions, and so there is an overall decrease in Gibbs free energy, even though the battery-charging part we are interested in has an increase.

This is an example of the coupling of a product-favored reaction with a reactant-favored process to cause the latter to take place. There are a great many situations similar to charging a battery or synthesizing natural gas. Other examples are obtaining aluminum or iron from their ores; synthesizing large, complicated molecules from simple reactants to make medicines, plastics, and other useful materials; and maintaining a comfortable temperature in a house on a day when the outside te mperature is below zero. All of these processes involve decreasing entropy (increasing order) in the area of our interest, but all can be made to occur provided there is a larger increase in entropy at a power plant or somewhere else.

CHEMISTRY IN THE NEWS

Generating Oxygen—and Heat

Oxygen canister *(AP/Wide World Photos)*

On Saturday, May 11, 1996, ValuJet Airlines Flight 592 crashed in Florida's Everglades with the loss of 110 lives. The difficulties of recovering the aircraft and flight recorders from the swamp may prevent us from ever knowing exactly what caused the crash, but oxygen-generating canisters have been strongly implicated. It has been theorized that at least one of 50 to 60 spent oxygen generators stored in a cargo hold may have been activated accidentally, and the resulting release of oxygen and increase in temperature started a fire that ultimately destroyed the plane.

Oxygen-generators are designed to provide pure oxygen if there is an emergency loss of cabin pressure. Each canister generates enough oxygen to last three or four passengers 15 min or so—enough time for the plane to move to a lower altitude where there is sufficient oxygen to breathe. Just how do such canisters work, and why do they generate so much heat that they could start a deadly fire?

In order to generate oxygen to flow to each mask, a product-favored chemical reaction is used. The reaction chosen is decomposition of sodium chlorate, which is exothermic and has a large entropy increase because oxygen gas is formed.

$$2\ NaClO_3(s) \rightarrow 2\ NaCl(s) + 3\ O_2(g)$$
$$\Delta H° = -90.8\ \text{kJ} \qquad \Delta S° = 512.9\ \text{J/K}$$

This reaction is very slow at room temperature, and sodium chlorate can be safely stored in a bottle in the lab or in an oxygen canister in an airplane. This is an advantage, because the canister can be installed in a plane and left for some time without having to check whether it will work. (A tank of oxygen under pressure would need to be checked more often to make certain it had not leaked, and it also would have to be larger and heavier to store the same quantity of oxygen.) The canisters in the cargo hold of ValuJet Flight 592 had exceeded their shelf life and were being returned to the airline's headquarters for regeneration.

If cabin pressure suddenly drops, passengers are told to pull down their oxygen masks, place them over nose and mouth, and breathe normally. Pulling the mask causes a small explosion that drives powdered iron into sodium chlorate. Mixing a good reducing agent (iron) with a strong oxidizing agent (chlorate ion) results in an exothermic, product-favored reaction that produces iron(III) oxide, sodium chloride, and oxygen. The heat of this reaction warms sodium chlorate to the point where it decomposes rapidly according to the equation above. Since this reaction is also exothermic and product-favored, the temperature rises, reaching about 200 to 250 °C. This is enough to ignite other materials nearby, and oxygen released from the canister makes combustion very vigorous. (The vigorous reaction of a single Cheetos with oxygen generated from potassium chlorate was shown in Section 6.2). Apparently, used tires were also stored in the cargo hold of the ill-fated plane, and they may have burned rapidly.

Oxygen generators are safe when installed normally in an airplane, because they are surrounded by non-combustible insulating material, and the oxygen they generate flows through tubing to a different location. As cargo, however, spent oxygen generators are considered hazardous waste. They have been blamed for at least two other accidents, both non-fatal. Oxygen canisters in a disposal barrel at Seattle-Tacoma airport exploded in July 1994, and in August 1986 a canister in an empty airliner at Chicago's O'Hare airport started a cargo-hold fire that destroyed the plane. Highly exothermic product-favored reactions can benefit humankind greatly because of the Gibbs free energy they can release, but they also have the potential for disaster. Like fire, the first of this class of reactions to be harnessed, they require careful, intelligent handling.

This news feature includes information from The New York Times, *May 16, 1996, the* Wisconsin State Journal, *May 12, 1996, the* Capital Times, *May 13, 14, 15, 16, 27, and June 3, 1996, and the Internet, ChemEd-L list server.*

The Gibbs free energy change indicates a chemical reaction's capacity to drive a reactant-favored system so that it produces products. The word "free" in the name indicates not "zero cost," but rather "available." Gibbs free energy is available to do useful tasks that would not happen on their own. Another way of saying this is that Gibbs free energy is a measure of the *quality* of the energy contained in a chemical

All plants and animals, and indeed earth's ecosystem as a whole, depend on coupling of chemical reactions for their very existence. We will consider this topic in more detail in the next section.

system. If it contains a lot of Gibbs free energy, a chemical system can do a lot of useful work for us; the energy is of high quality—potentially useful to humankind. When the system's reactants are transformed into products, that available free energy can do useful work, but only if the system is coupled to some other, reactant-favored process we want to carry out. If systems are not coupled, then the free energy will be wasted.

The law of conservation of energy is the reason we have been careful not to say that when a lump of coal is burned its energy has been used up. What has been used up is an energy resource: the coal's capacity to transfer energy by heating its surroundings when it burns. If the coal is burned in a power plant, its chemical energy is changed to an equal quantity of energy in other forms. These are mainly electricity, which can be very useful, and thermal energy in the gases going up the smokestack and in the immediate surroundings of the plant. It is not the coal's energy, but rather its ability to store energy and release it in the form that people can use that has been used up. That is, it is the coal's Gibbs free energy that has been used up (decreased).

Energy conservation does not really consist of conserving energy—nature takes care of that automatically, as recognized in the law of conservation of energy. Nature does not, however, automatically conserve Gibbs free energy. Systems with high Gibbs free energies are energy resources, and it is their *free* energy that we must take pains to conserve. Once a product-favored reaction has taken place, its free energy cannot be restored, except at the expense of another product-favored reaction coupled to it. Analysis of chemical systems in terms of Gibbs free energy can lead to important insights into how energy resources can be conserved effectively.

Exercise 7.9 Using Gibbs Free Energy

One way to produce iron metal is to reduce iron(III) oxide with aluminum. You can think about the reaction as occurring in two steps. The first is the loss of oxygen from iron(III) oxide,

$$\text{(i) } Fe_2O_3(s) \longrightarrow 2\ Fe(s) + \tfrac{3}{2}\ O_2(g)$$

and the second is the combination of aluminum with the oxygen.

$$\text{(ii) } 2\ Al(s) + \tfrac{3}{2}\ O_2(g) \longrightarrow Al_2O_3(s)$$

(a) Calculate the enthalpy, entropy, and free energy changes for each step. Decide whether each step is product- or reactant-favored. Comment on the signs of ΔH°, ΔS°, and ΔG° for each step.
(b) What is the net reaction that occurs when aluminum is combined with iron(III) oxide? What are the enthalpy, entropy, and free energy changes for the net reaction? Is it product- or reactant-favored? Comment on the signs of ΔH°, ΔS°, and ΔG° for the net reaction.
(c) Discuss briefly the effect of coupling reaction (i) with reaction (ii), that is, adding equations (i) and (ii), on our ability to obtain iron metal from iron(III) oxide.
(d) Suggest a reaction other than oxidation of aluminum that might be used to reduce iron(III) oxide to iron. Test your selection by calculating the Gibbs free energy change for the coupled system.

Thermite reaction. The reaction of aluminum with iron(III) oxide is very product-favored, releasing a large quantity of Gibbs free energy. *(C.D. Winters)*

7.7 FREE ENERGY AND BIOLOGICAL SYSTEMS

Have you ever thought about how unlikely it is that a human being can exist? Your body contains about 100 trillion cells, all working together to make you what you are. Each of those cells contains trillions of molecules, and many of the molecules contain hundreds of thousands of atoms. Those molecules and cells are arranged in structures such as organs, bones, and skin that provide for all the functions of your body and that determine its overall shape and size. When you need to, you can synthesize molecules on very short notice. For example, it does not take long to generate the shot of adrenalin your body makes when you are scared. Your body is a very highly organized system, which means that its entropy must be very low. This in turn means that, thermodynamically speaking, you are very, very improbable!

How can it be, then, that you exist? How can all the molecules from which you are made be synthesized and organized into the organs and other tissues of your body? The answer lies in the coupling of reactions described in the preceding section. Since you are a very low-entropy system, you must be very high in Gibbs free energy. Your body extracts that free energy from the food you eat. In the processes of metabolism, foods rich in free energy are oxidized and their oxidation is coupled to other reactions that store free energy in specific molecules within your body. Later those molecules can release the free energy to cause muscles to contract, nerve signals to be sent, important molecules to be synthesized, and other processes to occur. **Metabolism** refers to all of the chemical changes that occur as food nutrients are converted by an organism into Gibbs free energy and the complex chemical constituents of living cells. **Nutrients** are the chemical raw materials needed for survival of an organism.

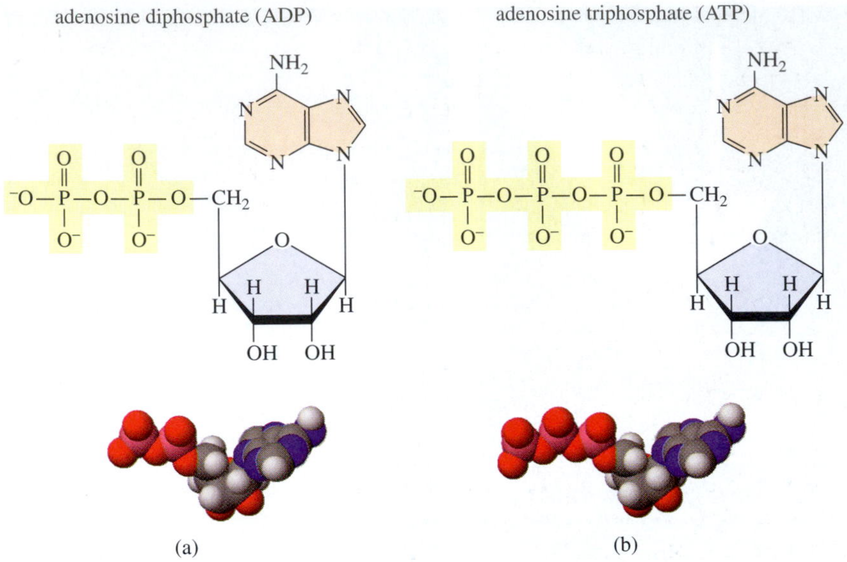

Figure 7.9 Biochemical free energy storage. Structures of (a) adenosine diphosphate (ADP) and (b) adenosine triphosphate (ATP). Notice that ATP has one more phosphate (PO_4) group at the left end of the molecule, but otherwise the structures are identical. Notice also that ADP has three negatively charged oxygen atoms and ATP has four.

As an example of metabolism, consider the single nutrient glucose, also known as dextrose or blood sugar *(**← p. 108**).* Glucose can be oxidized to carbon dioxide and water according to the equation

$$C_6H_{12}O_6(aq) + 6\ O_2(g) \longrightarrow 6\ CO_2(g) + 6\ H_2O(\ell) \qquad \Delta G^{\circ\prime} = -2870\ \text{kJ}$$

The prime symbol (′) on $\Delta G^{\circ\prime}$ indicates that the value of the Gibbs free energy change is for the same concentration of H^+ ions as in a typical cell, namely a neutral solution. When aqueous solutions are involved, ΔG° values in tables such as Appendix J refer to concentrations of 1 mol/L, which for an acid would destroy a typical cell. Consequently, biochemists have calculated a set of $\Delta G^{\circ\prime}$ values that apply to neutral solutions and are indicated by the prime.

Thus a large quantity of Gibbs free energy can be released when glucose is oxidized. This reaction is strongly product-favored. This is an example of an **exergonic** reaction—*one that releases free energy.* The same quantity of free energy is available whether glucose is burned in air or reacts in your body. However, burning glucose would release all of the free energy as thermal energy. This would not be appropriate in your body because it would raise the temperature rapidly, which in turn would kill cells. Instead, your body makes use of a large number of reactions that allow the free energy to be released in small steps and stored in small quantities that can be used later.

By far the most important way that Gibbs free energy is stored in your body is through formation of adenosine triphosphate (ATP) from adenosine diphosphate (ADP).

Notice from Figure 7.9 that ADP carries three units of negative charge and ATP carries four.

$$ADP^{3-}(aq) + H_2PO_4^-(aq) \longrightarrow ATP^{4-}(aq) + H_2O(\ell) \qquad \Delta G^{\circ\prime} = 30.5\ \text{kJ}$$

The words exergonic and endergonic have nearly the same prefixes as exothermic and endothaermic. In both cases "ex" means *out* and "end" means *into.* "Thermic" indicates that it is thermal energy that is released or taken up. "Ergonic" indicates that it is free energy that is released or used up.

The structures of ADP and ATP are shown in Figure 7.9; note that they are closely related. This is an example of an **endergonic** reaction—*one that uses up free energy* and is therefore reactant-favored. In a typical human cell, this reaction takes place 32 times for each molecule of glucose that is oxidized. In bacterial cells, it takes place 38 times for each molecule of glucose oxidized. That is, in a human cell the Gibbs free energy released by the exergonic glucose oxidation is used to force the endergonic, reactant-favored process of forming ATP from ADP to occur 32 times. The overall process is

$$C_6H_{12}O_6(aq) + 6\ O_2(g) + 32\ ADP^{3-}(aq) + 32\ H_2PO_4^-(aq) \longrightarrow$$
$$6\ CO_2(g) + 32\ ATP^{4-}(aq) + 38\ H_2O(\ell) \qquad \Delta G^{\circ\prime} = -1894\ \text{kJ}$$

Since it is exergonic, it must be product-favored, and therefore appreciable quantities of products can be obtained.

Exercise 7.10 Coupled Metabolic Reactions

Add the Gibbs free energy change for oxidation of glucose to the appropriate free energy change for 32 conversions of ADP to ATP. Hence verify that the Gibbs free energy change given above for the overall reaction is correct. What happens to the 1894 kJ of free energy released by the overall reaction?

The metabolic process by which the Gibbs free energy contained in nutrients is stored in ATP is far more complicated than the overall equation given above makes it seem. It can be divided into three stages that were first clearly identified by Hans Krebs. The first stage is digestion, which breaks apart large molecules, such as carbohydrates (polysaccharides), fats, or proteins, into smaller molecules, such as glucose, glycerol and fatty acids, or amino acids. These smaller molecules are more easily transferred into the blood by the digestive system. In the second stage, the smaller molecules are changed into a few simple units that play a central role in metabolism. The most important of these is the acetyl group in acetyl coenzyme A (acetyl CoA). The structure of acetyl CoA is shown in Figure 7.10. The third stage consists of oxidation of the acetyl group from acetyl CoA to form carbon dioxide and water. This takes place in an eight-step cycle of reactions called the citric-acid cycle, which also

Because ATP is high in free energy, it is said to be a high-energy molecule (or ion). Sometimes the bonds in ATP are called high-energy bonds, but this is a misnomer: actually, the bonds have low bond energies and can break easily to form ADP and release free energy.

The citric-acid cycle is also known as the Krebs cycle or the tricarboxylic acid (TCA) cycle.

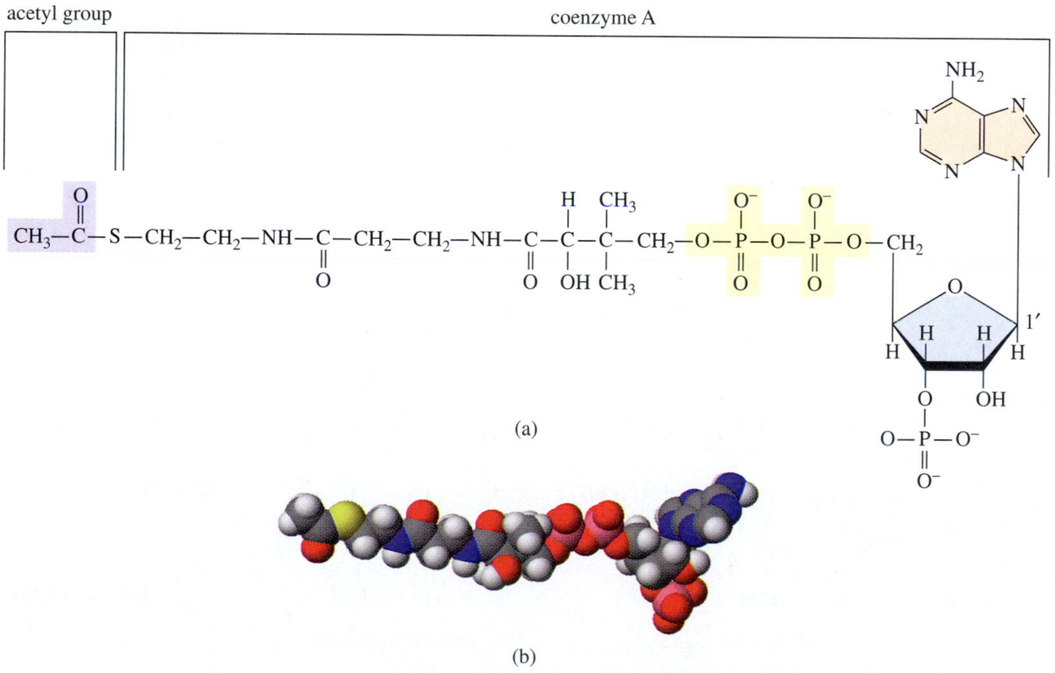

acetyl group

coenzyme A

(a)

(b)

Figure 7.10 Structure of acetyl coenzyme A. The acetyl group ($CH_3\overset{\overset{O}{\|}}{C}—$) at the far left end of the molecule could have been formed from glucose that came originally from starch.

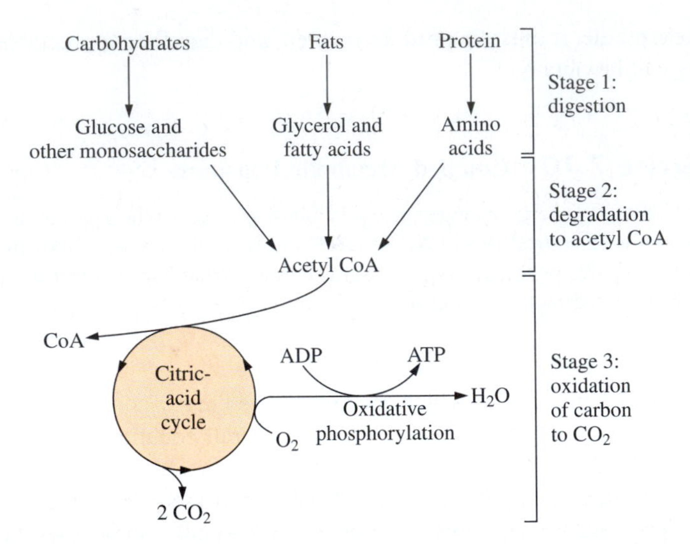

Figure 7.11 **Free energy and nutrients.** Extraction of Gibbs free energy from nutrients is a three-stage process. In stage 1, digestion, large molecules are broken down into smaller ones. In stage 2, smaller molecules are converted to acetyl groups attached to coenzyme A. In stage 3, the citric-acid cycle, the acetyl groups are oxidized to carbon dioxide and water.

produces substances that cause ADP to be transformed into ATP, a process called oxidative phosphorylation. The overall three-stage process is diagrammed in Figure 7.11.

Because conversion of ADP to ATP is endergonic, ATP contains stored Gibbs free energy. In your body, ATP generated from glucose or other nutrients is a convenient and readily available free energy resource, just as electricity generated from coal or natural gas is a convenient and readily available free energy resource in modern society. ATP can release free energy in packets of 30.5 kJ for each ATP converted to ADP. This size is convenient for driving many biochemical processes in your body. For example, as part of the metabolism of glucose, it is necessary to attach a phosphate group to the glucose molecule:

$$\text{glucose} + H_2PO_4^- \longrightarrow \text{glucose 6-phosphate} + H_2O + H^+ \qquad \Delta G^{0\prime} = 13.8 \text{ kJ}$$

glucose

glucose 6-phosphate

This reaction is endergonic by 13.8 kJ and therefore is not product-favored. It will not occur unless forced to do so.

The endergonic reaction can be forced to occur by coupling it to the transformation of ATP to ADP.

$$ATP(aq)^{4-} + H_2O(\ell) \longrightarrow ADP^{3-}(aq) + H_2PO_4^-(aq) \qquad \Delta G^\circ = -30.5 \text{ kJ}$$

Notice that $H_2PO_4^-(aq)$ produced by this reaction can be used to react with glucose in the first reaction, coupling the two directly. Also, water produced in the first reaction is used up in the second one. The overall process is

$$\text{Glucose} + ATP^{4-} \longrightarrow \text{glucose 6-phosphate}^{2-} + ADP^{3-} + H^+$$
$$\Delta G^\circ = (-30.5 + 13.8) \text{ kJ} = -16.7 \text{ kJ}$$

The "6" in glucose 6-phosphate indicates that the phosphate group has been added to carbon number 6 of glucose.

The negative value of ΔG° indicates that the overall process is exergonic and product-favored. Thus the ATP → ADP transformation can force glucose to combine with dihydrogen phosphate. The 16.7 kJ of free energy released appears as thermal energy transferred from the system to its surroundings.

In biochemistry it is conventional to write these equations in a shorthand notation that indicates that they are coupled. The process just described would be represented as follows:

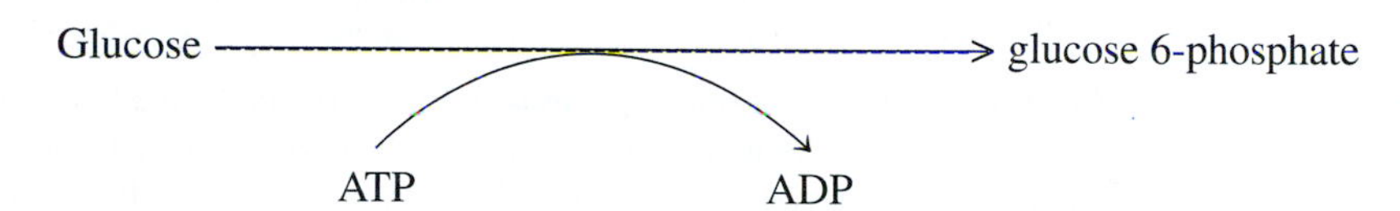

The curved line indicates that the transformation of ATP to ADP occurs simultaneously with the glucose reaction, and that the two are coupled.

Photosynthesis and Gibbs Free Energy

You may be wondering where the nutrients you take into your body get the Gibbs free energy they so obviously have. The answer is from solar energy via photosynthesis. **Photosynthesis** is a series of reactions in a green plant that combines carbon dioxide with water to form carbohydrate and oxygen. The carbohydrate and other constituents you consume in vegetables are derived from photosynthesis. If you eat meat, the animal from which it came probably ate vegetables, and therefore derived its nutrients from plant photosynthesis. The overall reaction in photosynthesis is just the opposite of oxidation of glucose:

$$6\ CO_2 + 6\ H_2O \longrightarrow C_6H_{12}O_6 + O_2 \qquad \Delta G^{\circ\prime} = 2870 \text{ kJ}$$

It is endergonic and can only occur because of an influx of free energy in the form of sunlight. That is, the energy in the sunlight causes this reactant-favored process to form appreciable quantities of products, and the sunlight's free energy is stored in the glucose and oxygen that are formed. This is diagrammed in Figure 7.12.

Organisms that can carry out photosynthesis are called **phototrophs** (literally, "light-feeders") because they can use sunlight to supply needed energy. Phototrophs include all green plants, all algae, and some groups of bacteria. The phototrophs capture light by means of photosynthetic pigment systems and store the light energy in chemical bonds in molecules such as glucose. All other organisms belong to the class of **chemotrophs** (literally, "chemical-feeders"), which must depend on the chemical bonds created by the phototrophs for their energy. All animals, fungi, and most bac-

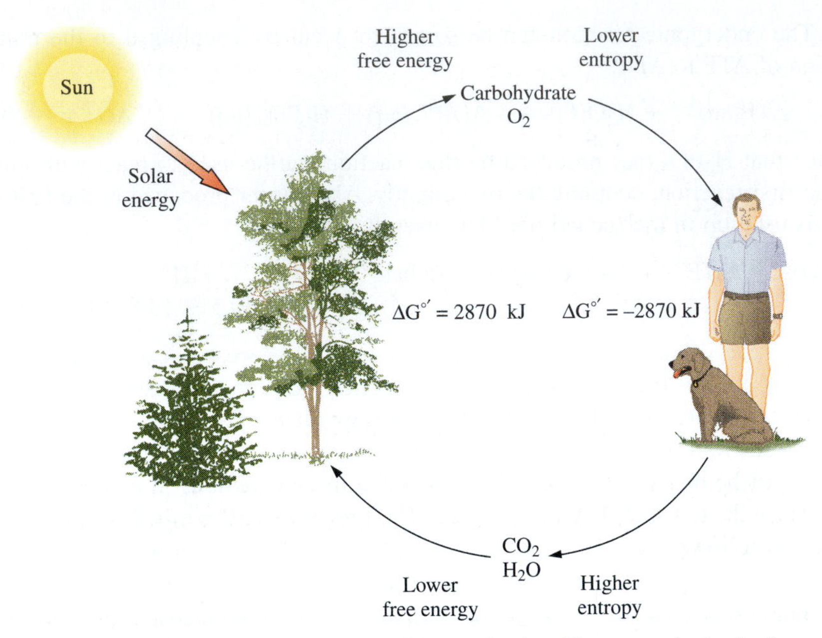

Figure 7.12 Solar free energy storage by photosynthesis. The energy in the foods we eat is derived from solar energy via photosynthesis. Organisms that can photosynthesize combine carbon dioxide with water to form carbohydrates and oxygen, which have a much higher free energy. That free energy is released when the carbohydrates are oxidized in metabolic processes.

teria are chemotrophs. A world composed only of chemotrophs would not last long because without the phototrophs food supplies would disappear almost immediately. Without sunlight and its ability to drive a reactant-favored system to form products (carbohydrate and oxygen), organisms such as ourselves and indeed almost the entire biosphere of planet earth could not exist.

Both phototrophs and chemotrophs make use of the free energy stored up in photosynthesis by using oxidation of glucose to drive a large number of conversions of ADP to ATP and then using the ATP to couple to desired endergonic reactions and force them to occur. Thus ATP is the minute-to-minute energy currency of living cells. The free energy released in these reactions either contributes to synthesis of molecules needed by the cell, causes some desirable process such as muscle contraction, or is dissipated as thermal energy. If more free energy is taken in than the organism needs, then the excess free energy can be stored long-term through the synthesis of fats, which have nearly twice as much free energy as an equal mass of carbohydrate.

It is significant that when ATP reacts and causes other reactions to occur, the product ADP is very similar to the reactant and can easily be recycled to ATP. A reasonable estimate of the quantity of ATP converted to ADP during one day in the life of an average human is 117 mol. Since the molar mass of the sodium salt of ATP is 551 g/mol, we can calculate that

$$117 \text{ mol} \times \frac{551 \text{ g}}{\text{mol}} = 64{,}500 \text{ g ATP}$$

is converted to ADP every day. This is 64.5 kg, which is close to the 70-kg body weight of an average person. Obviously ATP is not a long-term storage molecule for free energy. Instead it is recycled from ADP as needed and used almost immediately for some necessary process. The typical 70-kg human body contains only 50 g of

ATP/ADP total. If we actually had to take in 64.5 kg of ATP per day to provide free energy, it would be a very expensive habit. The price of ATP from a laboratory supplier is currently about $10 per gram, which would put the cost of supplying each of us with our daily energy currency at more than half a million dollars!

Exercise 7.11 Recycling of ATP

From the figures given above for the daily quantity of ATP converted to ADP by an average human and the quantity of ATP/ADP actually present in the body, calculate the number of times each ADP molecule must be recycled to ATP each day.

7.8 CONSERVATION OF FREE ENERGY

Thermodynamic principles can be applied to analyze energy use in our industrial society. What we commonly refer to as **energy conservation** *is actually conservation of useful energy: Gibbs free energy.* By comparing free energy changes calculated using the equations in this chapter with the actual loss of free energy in industrial processes, environmentalists and industrialists can suggest ways to minimize loss of free energy. For example, there is a very large quantity of free energy stored in aluminum metal and oxygen gas compared with aluminum ore, which has the formula Al_2O_3. It is not surprising, then, that throughout the United States there are major programs for recycling aluminum. A similar statement can be made about almost every metal: once reduced from their ores, metals are storehouses of free energy that should be maintained in their reduced forms in order to avoid the expenditure of free energy needed to separate them from chemical combination with oxygen.

Like ATP in your body, many compounds can store free energy. An example is ethylene. About 47 billion pounds of this gas were produced in the United States in 1995 from the dehydrogenation of ethane in chemical plants like the one shown in Figure 7.13.

Conceptual Challenge Problem CP-7.D at the end of the chapter relates to topics discussed in this section.

$$C_2H_6(g) \longrightarrow H_2(g) + C_2H_4(g) \qquad \Delta H^\circ = +136.94 \text{ kJ} \qquad \Delta G^\circ = +100.97 \text{ kJ}$$

Figure 7.13 Polyethylene production. A chemical plant in Houston, Texas, for the production of ethylene that will subsequently to be made into polymers. *(Courtesy of Phillips Petroleum Co.)*

When a mole of hydrogen and a mole of ethylene are produced from a mole of ethane, about 101 kJ of free energy must be supplied from an external source. This free energy becomes stored in the reaction products. Much of the ethylene is transformed into polyethylene, a plastic used in many consumer items (Section 11.8). Since ethylene production is the largest single consumer of Gibbs free energy in the chemical industry, there has been great interest in improving the process to save energy and money. Many small improvements in ethane-to-ethylene conversion have led to a 60% decline in the free energy requirement per pound of ethylene produced since 1960. Even so, the energy resources used to make ethylene from ethane are four times the minimum required (100.97 kJ). This is largely due to inefficiencies in energy transfer from external sources to the reaction system.

Ethylene manufacturing is only 25% energy efficient.

It is important to recognize that completely eliminating consumption of free energy is impossible. Whenever anything happens, whether a chemical reaction or a physical process, the final state must have less free energy than was available initially. This is the same as saying that the entropy of the universe must have increased during the change. This is true of any system in which the initial substances are changed into something new—any product-favored system. Thus there will always be losses of free energy. The aim of energy conservation is to minimize, not eliminate, them. This can be done by maximizing the efficiency of coupling exergonic reactions to endergonic processes we want to force to occur. The ideas of thermodynamics help us figure out how to do that and are the most powerful tool we have for conserving energy while maintaining a high standard of living.

Summary Problem

In a blast furnace for making iron from iron ore, such as the one shown at the beginning of this chapter, large quantities of coke (which is mainly carbon) are dumped into the top of the furnace along with iron ore (which can be assumed to be Fe_2O_3) and limestone (which is used to help remove impurities from the iron). The overall process is

$$2\ Fe_2O_3(s) + 3\ C(s) \longrightarrow 4\ Fe(s) + 3\ CO_2(g)$$

This overall process can be thought of as a combination of several individual steps:

$$2\ Fe_2O_3(s) \longrightarrow 4\ FeO(s) + O_2(g)$$
$$2\ FeO(s) \longrightarrow 2\ Fe(s) + O_2(g)$$
$$2\ C(s) + O_2(g) \longrightarrow 2\ CO(g)$$
$$2\ CO(g) + O_2(g) \longrightarrow 2\ CO_2(g)$$

(a) Calculate the enthalpy change for each step, assuming a temperature of 25 °C. Which steps are exothermic and which are endothermic?
(b) Based on the equations, predict which of the individual steps would involve an increase and which a decrease in the entropy of the system.
(c) Based on your results in parts (a) and (b), what can you say about whether each step is reactant-favored or product-favored at room temperature? At a much higher temperature ($>$ 1000 K)?
(d) Calculate the entropy change and the Gibbs free energy change for each reaction step, assuming a temperature of 25 °C.
(e) Keeping in mind the equation $\Delta G^\circ = \Delta H^\circ - T\,\Delta S^\circ$ and the fact that the enthalpy change and entropy change for a reaction do not vary much with temperature, what would be the slope of a graph of ΔG° versus T for each of the reac-

tions? For which of the reactions does $\Delta G°$ become more negative the higher the temperature? For which does it become more positive? Does this agree with what you predicted in part (c)?

(f) Use your results from previous parts of this problem to estimate the Gibbs free energy change for each of these reactions at a temperature of 1500 K.

(g) Which of the two iron oxides is more easily reduced at 1500 K? Which of the reactions involving carbon compounds is more product-favored at 1500 K? What chemical reactions do you think are taking place in the hottest part of the blast furnace?

(h) In portions of the furnace where the temperature is about 800 K, would you predict the same reactions would be occurring as in the higher-temperature part of the furnace? Why or why not?

(i) Show that the individual steps can be combined to give the overall reaction. From the enthalpy, entropy, and Gibbs free energy changes already calculated, calculate these changes for the overall reaction.

(j) In a typical blast furnace every kilogram of iron produced requires 2.5 kg of iron ore, 1 kg of coke, and nearly 6 kg of air (to provide oxygen for oxidation of the coke to heat the furnace). How much Gibbs free energy would be destroyed if the coke were simply burned to form carbon dioxide? Given the quantity of iron produced in a typical furnace, how much free energy is stored by coupling the oxidation of coke to the reduction of iron oxides? What percentage of the free energy available from combustion of coke is wasted per kilogram of iron produced?

In Closing

Having studied this chapter, you should be able to . . .

- explain why there is a higher probability that both matter and energy will be dispersed than that they will be concentrated in a small number of nanoscale particles (Section 7.1).
- use the absolute, or Kelvin, temperature scale in calculations (Section 7.2).
- calculate the entropy change for a process occurring at constant temperature (Section 7.2).
- use qualitative rules to predict the sign of the entropy change for a process (Section 7.2).
- calculate the entropy change for a chemical reaction, given a table of standard molar entropy values for elements and compounds (Section 7.3).
- use entropy and enthalpy changes to predict whether a reaction is product-favored (Section 7.4).
- describe the connection between enthalpy and entropy changes for a reaction and the Gibbs free energy change (Section 7.5).
- calculate the Gibbs free energy change for a reaction from values given in a table of standard molar free energies of formation (Section 7.5).
- describe how a reactant-favored system can be coupled to a product-favored system so that a desired reaction can be carried out (Section 7.6).
- explain how biological systems make use of coupled reactions to maintain the high degree of order found in all living organisms; give examples of coupled reactions that are important in biochemistry (Section 7.7).
- explain the relationship between Gibbs free energy and energy conservation (Sections 7.6 and 7.8).

Key Terms

The following terms were defined and given in boldface type in this chapter. You should be sure to understand each of these terms and the concepts with which they are associated.

absolute temperature scale (7.2)
chemotrophs (7.7)
endergonic (7.7)
energy conservation (7.8)
entropy (7.2)
exergonic (7.7)
Gibbs free energy (7.5)
Kelvin temperature scale (7.2)
metabolism (7.7)
nutrients (7.7)
photosynthesis (7.7)
phototrophs (7.7)
product-favored system (Introduction)
reactant-favored system (Introduction)
second law of thermodynamics (7.4)
thermodynamic temperature scale (7.2)

Questions for Review and Thought

Conceptual Challenge Problems

CP-7.A. Suppose that you are invited to play a game as either the "player" or the "house." A pair of dice is used to determine the winner. Each die is a cube having a different number, one through six, showing on each face. The player rolls two dice and sums the numbers showing on the topside of each die to determine the number rolled. Obviously the number rolled has a minimum value of 2 (both dice showing a 1) and a maximum of 12 (both dice showing a 6). The player begins the game with his or her initial roll of the dice. If the player rolls a 7 or an 11, he or she wins on the first roll and the house loses. If the player does not roll a 7 or an 11 on the initial roll, then whatever number was rolled is called the point, and the player must roll again. For the player to win, he or she must roll the point again before either a 7 or an 11 is rolled. Should the player roll a 7 or an 11 before rolling the point a second time, the house wins. Which would you choose to be, player or house? Explain clearly in terms of the probabilities of rolling the dice why you chose the role you did.

CP-7.B. Suppose a button is placed in the middle of a football field and a penny is flipped to decide which direction to move the button, up or down the field. Each time the penny comes up heads, the button is moved 10 cm toward your opponent's goal line; and each time it comes up tails, the button is moved 10 cm toward your goal line. Your friend concludes that after many flips of the penny the button is likely to remain within 10 cm of the middle of the field, because numerous flips of the penny will produce heads just as often as tails. You doubt this because you know that perfume molecules and particles undergoing Brownian motion diffuse away from their original source, even though, like the button, they are just as likely to be hit from one direction at from any other by the moving molecules around them. How would you explain the error of your friend's conclusion about the movement of the button?

CP-7.C. When thermal energy is transferred to a substance at its standard melting point or boiling point, the substance melts or vaporizes, but its temperature does not change while it is doing so. It is clear then, that temperature cannot be a measure of "how much energy is in a sample of matter" or the "intensity of energy in a sample of matter." In Qualitative Rules for Entropy (in Section 7.2) we noted that atoms and molecules are not stationary, but rather are in constant motion. When heated, their motion increases. If this is true, what can you infer that temperature measures about a sample of matter?

CP-7.D. Suppose that you are a member of an environmental group and have been assigned to evaluate various ways of delivering milk to consumers with respect to free energy conservation. Think of all the ways that milk could be delivered, the kinds of containers that could be used, the ways they could be transported, and whether the containers could be reused (refillable) or recycled. Define the problem in terms of the kinds of information you would need to collect, how you would analyze the information, and the criteria you would use to decide which systems are more efficient in use of free energy. Do not try to collect the actual data you would use, but define the problem well enough so that someone could collect the necessary data based on your statement of the problem.

Review Questions

1. Define the terms "product-favored system" and "reactant-favored system." Give one example of each.
2. What are the two ways that a final chemical state of a system can be more probable than its initial state?
3. Define the term "entropy," and give an example of a sample of matter that has zero entropy. What are the units of entropy? How do they differ from the units of enthalpy?

4. State five useful qualitative rules for predicting entropy changes when chemical or physical changes occur.
5. State the second law of thermodynamics.
6. When can you be sure that a reaction is product-favored? When can you be sure that it is not product-favored?
7. Define the Gibbs free energy change of a chemical reaction in terms of its enthalpy and entropy changes.
8. Why are materials of high Gibbs free energy useful to society? Give two examples of such materials.
9. How are materials of high Gibbs free energy important to *you?* Give two examples of such materials.
10. Define the terms "endergonic" and "exergonic."
11. What is the citric-acid cycle and why is it important to organisms?
12. Describe two ways to get reactant-favored reactions to form products.
13. Describe the process by which sunlight is employed to convert high-entropy, low-free-energy substances into low-entropy, high-free-energy substances.
14. Name the substance that stores Gibbs free energy in your body, and write the chemical reaction by which it is formed.
15. A friend of yours says that the boiling point of water is twice that of cyclopentane, which boils at 50 °C. Write a brief statement about the validity of this observation.

Probability and Chemical Reactions

16. Suppose you flip a coin. What is the probability that the coin will come up heads? What is the probability that it will come up tails? If you flip the coin 100 times, what is the most likely number of heads and tails you will see?

17. Consider two equal-sized flasks connected by a stopcock, as in Figure 7.2. Suppose you put one molecule inside. What is the probability that the molecule will be in flask A? What is the probability that it will be in flask B? If you put 100 molecules into the two-flask system, what is the most likely arrangement of molecules? Which arrangement has the highest entropy?

18. Suppose you have four identical molecules, labeled 1, 2, 3, and 4. Draw sixteen simple two-flask diagrams as in Figure 7.2, and draw all possible arrangements of the four molecules in the two flasks. How many of these arrangements have two molecules in each flask? How many have no molecules in one flask? From these results, what is the most probable arrangement of molecules? Which arrangement has the highest entropy?

Measuring Dispersal or Disorder: Entropy

19. Express the melting point of ice and the boiling point of water in three different temperature scales: Celsius, Fahrenheit, and Kelvin. Which two of these temperature scales have temperature differences of one degree which are the same size? For the third scale, how does the size of the degree compare with that of the other two?

20. Take the temperatures mentioned in the following statements and convert them to the other two temperature scales:
(a) On the top of one of the Rocky Mountains, water boils at 187 °F.
(b) Heat the oven to 350 °F, then put in the cake.
(c) The freezing point of mercury is −40 °C.
(d) Temperatures in flames can be as high as 1300 °C.
(e) The lowest temperature ever observed is very close to 0 K.
(f) The temperature of the surface of the sun is about 5000 K.
(g) In broad daylight in the desert, the temperature can get as high as 125 °F.
(h) Your normal body temperature is 98.6 °F.

21. For each of the following processes, tell whether you would expect the entropy change of the system to be positive or negative:
(a) Water vapor (the system) deposits as ice crystals on a cold window pane.
(b) A can of carbonated beverage loses its fizz. (Consider the beverage but not the can as the system. What happens to the entropy of the dissolved gas?)
(c) A glassblower heats glass (the system) to its softening temperature.
(d) Water boils.
(e) A teaspoon of sugar dissolves in a cup of coffee. (The system consists of both sugar and coffee.)
(f) Calcium carbonate precipitates out of water in a cave to form stalactites and stalagmites. (Consider only the calcium carbonate to be the system.)

22. For each of the following pairs, tell which has the higher entropy and explain why.
(a) A sample of solid CO_2 at −78 °C or CO_2 vapor at 0 °C
(b) Solid sugar or the same sugar dissolved in a cup of tea
(c) Two 100-mL beakers, one containing pure water and the other containing pure alcohol, or a beaker containing the same samples of water and alcohol after they had been mixed together

23. Tell which substance has the higher entropy in each of the following pairs, and explain why.
(a) A sample of pure silicon (to be used in a computer chip) or a piece of silicon containing a trace of some other atoms such as B or P
(b) An ice cube or the same mass of liquid water, both at 0 °C
(c) A sample of pure I_2 or the same mass of iodine vapor, both at room temperature

24. Comparing the formulas or states for each pair of compounds, tell which you would expect to have the higher entropy per mole at the same temperature and explain why.
(a) $NaCl(s)$ or $CaO(s)$
(b) $Cl_2(g)$ or $P_4(g)$
(c) $CH_3NH_2(g)$ or $(CH_3)_2NH(g)$
(d) $Au(s)$ or $Hg(\ell)$

25. For each of the following chemical systems, predict whether the entropy change will be positive or negative when the reaction occurs in the direction it is written.
(a) $C_2H_4(g) + H_2(g) \rightarrow C_2H_6(g)$

(b) $CH_3OH(\ell) + \frac{3}{2} O_2(g) \rightarrow CO_2(g) + 2 H_2O(g)$
(c) $N_2(g) + 3 H_2(g) \rightarrow 2 NH_3(g)$
(d) $CaCO_3(s) \rightarrow CaO(s) + CO_2(g)$
(e) $CH_3OH(\ell) \rightarrow CO(g) + 2 H_2(g)$
(f) $Br_2(\ell) + H_2(g) \rightarrow 2 HBr(g)$
(g) $C_3H_8(g) \rightarrow C_2H_4(g) + CH_4(g)$

Calculating Entropy Changes

26. Calculate the entropy change, $\Delta S°$, for the vaporization of ethanol, C_2H_5OH, at the boiling point of 78.3 °C. The heat of vaporization of the alcohol is 39.3 kJ/mol.

$$C_2H_5OH(\ell) \longrightarrow C_2H_5OH(g) \qquad \Delta S° = ?$$

27. Not too many years ago diethyl ether, $(C_2H_5)_2O$, was used as an anesthetic. What is the entropy change, $\Delta S°$, for the vaporization of ether if its heat of vaporization is 26.0 kJ/mol at the boiling point of 35.0 °C?

28. The standard molar entropy of methanol vapor, $CH_3OH(g)$, is 239.8 J/(K · mol). Calculate (a) the entropy change for the vaporization of 1 mol of methanol (see Table 7.1); (b) the enthalpy of vaporization of methanol, assuming that $\Delta S°$ doesn't depend on temperature and taking the boiling point of methanol to be 64.6 °C.

29. Check your predictions in Question 25 by calculating the entropy change for each reaction. Standard molar entropies not in Table 7.1 can be found in Appendix J.

Entropy and the Second Law of Thermodynamics

30. Is the following reaction predicted to favor the reactants or products? Explain your answer briefly.

$$Mg(s) + \frac{1}{2} O_2(g) \longrightarrow MgO(s) \qquad \Delta H° = -601.70 \text{ kJ}$$

31. Explain briefly why each of the following reactions is product-favored.

(a) The combustion of propane.

$$C_3H_8(g) + 5 O_2(g) \longrightarrow 3 CO_2(g) + 4 H_2O(\ell)$$

(b) The reaction of a metal carbonate with an acid.

$$CuCO_3(s) + H_2SO_4(aq) \longrightarrow CuSO_4(aq) + CO_2(g) + H_2O(\ell)$$

32. What are the signs of the enthalpy and entropy changes for the splitting of water to give gaseous hydrogen and oxygen, a process that is quite endothermic? Is this reaction likely to be product-favored? Explain your answer briefly.

33. Octane is the product of adding hydrogen to 1-octene.

$$\underset{\text{1-octene}}{C_8H_{16}(g)} + H_2(g) \longrightarrow \underset{\text{octane}}{C_8H_{18}(g)}$$

The enthalpies of formation are

$$\Delta H_f° [C_8H_{16}(g)] = -82.93 \text{ kJ/mol}$$
$$\Delta H_f° [C_8H_{18}(g)] = -208.45 \text{ kJ/mol}$$

Is this reaction likely to be product-favored or reactant-favored? Explain your reasoning briefly.

34. Classify each of the reactions according to one of the four reaction types summarized in Table 7.2.

(a) $Fe_2O_3(s) + 2 Al(s) \rightarrow 2 Fe(s) + Al_2O_3(s)$
$\Delta H°_{rxn} = -851.5$ kJ $\quad \Delta S° -37.5$ J/K

(b) $N_2(g) + 2 O_2(g) \rightarrow 2 NO_2(g)$
$\Delta H°_{rxn} = 66.4$ kJ $\quad \Delta S° = -122$ J/K

35. Classify each of the reactions according to one of the four reaction types summarized in Table 7.2.

(a) $C_6H_{12}O_6(s) + 6 O_2(g) \rightarrow 6 CO_2(g) + 6 H_2O(\ell)$
$\Delta H°_{rxn} = -673$ kJ $\quad \Delta S° = 60.4$ J/K

(b) $MgO(s) + C(\text{graphite}) \rightarrow Mg(s) + CO(g)$
$\Delta H°_{rxn} = 491.18$ kJ $\quad \Delta S° = 197.67$ J/K

Gibbs Free Energy

36. Is the combustion of ethane, C_2H_6, likely to be a product-favored reaction?

$$C_2H_6(g) + \frac{7}{2} O_2(g) \longrightarrow 2 CO_2(g) + 3 H_2O(\ell)$$

Answer the question by calculating the value of $\Delta S_{universe}$. Required values of $\Delta H_f°$ and $S°$ are in Appendix J. Does your calculated answer agree with your preconceived idea of this reaction? Verify your result by calculating the value of $\Delta G°$ for this chemical system.

37. The reaction of magnesium with water can be used as a means for heating food.

$$Mg(s) + 2 H_2O(\ell) \longrightarrow Mg(OH)_2(s) + H_2(g)$$

Is this reaction predicted to be product-favored? Answer the question by calculating the value of $\Delta S_{universe}$. See Appendix J for the needed data. Verify your result by calculating the value of $\Delta G°$ for the chemical system.

38. Add a column for the Gibbs free energy to Table 7.2.

(a) For the first and last lines in the table, tell whether ΔG is positive or negative.

(b) When ΔH_{system} and ΔS_{system} are both negative, is ΔG positive or negative or does it depend on temperature? If it is temperature-dependent, what is that dependence?

39. Discuss the following statements from the text in Section 7.5. In each case tell how the statement leads to the conclusion cited.

(a) If the reaction is exothermic (negative ΔH) and if the entropy of the system increases (positive ΔS), then ΔG must be negative and the reaction will be product-favored.

(b) If ΔH and ΔS have the same sign, then the magnitude of T determines whether ΔG will be negative and whether the reaction will be product-favored.

40. Predict whether the reaction below is product-favored or reactant-favored by calculating $\Delta G°$ from the entropy and enthalpy changes for the reaction at 25 °C.

$$H_2(g) + CO_2(g) \longrightarrow H_2O(g) + CO(g)$$

$\Delta H°_{rxn} = 41.17$ kJ $\quad \Delta S° = 42.08$ J/K

41. Predict whether the reaction below is product-favored or reactant-favored by calculating $\Delta G°$ from the entropy and enthalpy changes for the reaction at 25 °C.

$$SiO_2(s) + C(s) \longrightarrow Si(s) + CO_2(g)$$

If this worked, it would be a good way to make pure silicon, crucial in the semiconducting industry, from sand (SiO_2). From your results, comment on whether this would work.

42. Calculate $\Delta G°$ for the reactions of sand with hydrogen fluoride and hydrogen chloride. Explain why hydrogen fluoride attacks glass while hydrogen chloride does not.

$$SiO_2(s) + 4\ HF(g) \longrightarrow SiF_4(g) + 2\ H_2O(g)$$
$$SiO_2(s) + 4\ HCl(g) \longrightarrow SiCl_4(g) + 2\ H_2O(g)$$

(*Hint:* Decide whether each reaction is product-favored or reactant-favored by calculating $\Delta G°$ at 25 °C from the standard free energies of formation given in Appendix J.)

43. Many metal carbonates can be decomposed to the metal oxide and carbon dioxide by heating.

$$MgCO_3(s) \longrightarrow MgO(s) + CO_2(g)$$

What are the enthalpy, entropy, and free energy changes for this reaction at 25 °C? Is it product-favored or reactant-favored? Comment on the signs of $\Delta H°$, $\Delta S°$, and $\Delta G°$ for the reaction.

44. Some metal oxides can be decomposed to the metal and oxygen at relatively low temperatures. Is the decomposition of silver oxide product-favored at 25 °C?

$$Ag_2O(s) \longrightarrow 2\ Ag(s) + \tfrac{1}{2}\ O_2(g)$$

If not, can it become so if the temperature is raised? At what temperature is the reaction product-favored? (*Hint:* Calculate the temperature at which $\Delta G° = 0$.)

Gibbs Free Energy and Energy Resources

45. It would be very useful if we could use the inexpensive carbon in coal to make more complex organic molecules such as gaseous or liquid fuels. It is shown in this chapter that the formation of methane from coal and water is reactant-favored and thus can't occur unless there is some energy transfer from outside. This problem examines the feasibility of other reactions using coal and water.

(a) Write balanced equations for the reactions of coal (carbon) and steam to make ethane gas, propane gas, and liquid methanol, with carbon dioxide as a byproduct.

(b) Using the data in Appendix J, calculate $\Delta H°$, $\Delta S°$, and $\Delta G°$ for each reaction, and then comment on whether any of them would be a feasible way to make the stated products.

46. The overall reaction that occurs in a blast furnace, used for the reduction of iron from iron ore, is the combination of Fe_2O_3 and solid carbon to form iron metal and carbon dioxide.

(a) Write a balanced equation for this process, and calculate $\Delta H°$, $\Delta S°$, and $\Delta G°$ for the reaction.

(b) Is this reaction product-favored or reactant-favored at 25 °C? Does the reaction become more product-favored or more reactant-favored as the temperature increases?

(c) The reactions in a blast furnace actually occur in the following way:

$$2\ C(s) + O_2(g) \longrightarrow 2\ CO(g)$$
$$Fe_2O_3(s) + 3\ CO(g) \longrightarrow 2\ Fe(s) + 3\ CO_2(g)$$

Calculate $\Delta H°$, $\Delta S°$, and $\Delta G°$ for each reaction, and then show that the coupling, or combination, of these reactions is equivalent to the reaction in part (a) of this question.

Free Energy in Biological Systems

47. Define the following important biochemistry terms: metabolism, nutrients, ATP, ADP, oxidative phosphorylation, coupled reactions, phototrophs, chemotrophs, photosynthesis.

48. When you eat a candy bar, how does your body store the free energy that is released during oxidation of the sugars (glucose and other carbohydrates) in the candy bar? What was the original source of the free energy needed to synthesize the sugars before they went into the candy bar?

49. The molecular structure of one form of glucose, $C_6H_{12}O_6$, looks like this:

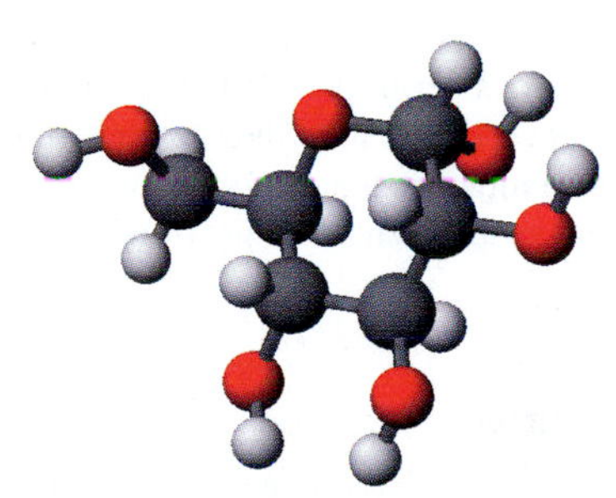

Glucose can be oxidized to carbon dioxide and water according to the equation

$$C_6H_{12}O_6 + 6\ O_2 \longrightarrow 6\ CO_2 + 6\ H_2O$$

(a) Using the method described in Chapter 6 for estimating enthalpy changes from bond energies, estimate $\Delta H°$ for the oxidation of this form of glucose. Make a list of all bonds broken and all bonds formed in this process.

(b) Compare your result with the experimental value of −2816 kJ for combustion of a mole of glucose. Why might there be a difference between this value and the one you calculated in part (a)?

50. Another step in the metabolism of glucose, occurring after the formation of glucose-6-phosphate, is the conversion of fructose-6-phosphate to fructose-1,6-bisphosphate ("bis" means *two*):

$$\text{fructose-6-phosphate} + H_2PO_4^- \longrightarrow \text{fructose-1,6-bisphosphate} + H_2O + H^+$$

(a) This reaction has a Gibbs free energy change of +16.7 kJ per mole of fructose-6-phosphate. Is it endergonic or exergonic?

(b) Write the equation for the formation of 1 mol of ADP from ATP, for which $\Delta G° = -30.5$ kJ.

(c) Couple these two reactions to get an exergonic process; write its overall chemical equation, and calculate the Gibbs free energy change.

51. In muscle cells under the condition of vigorous exercise, glucose is converted to lactic acid ("lactate"), $CH_3CHOHCOOH$, by the chemical reaction

$$C_6H_{12}O_6 \longrightarrow 2\ CH_3CHOHCOOH \qquad \Delta G^\circ = -197\ \text{kJ}$$

(a) If all of the Gibbs free energy from this reaction were used to convert ADP to ATP, how many moles of ATP could be produced per mole of glucose?
(b) The actual reaction involves the production of 3 mol of ATP per mole of glucose. What is the ΔG° for this reaction? Is it reactant-favored or product-favored?

52. The biological oxidation of ethanol, C_2H_5OH, is also a source of free energy. Does the oxidation of 1 g of ethanol give more or less energy than the oxidation of 1 g of glucose? (*Hint:* Write the balanced equation for the production of carbon dioxide and water from ethanol and oxygen, and use Appendix J.) Comment on potential problems of replacing glucose with ethanol in your diet.

Conservation of Free Energy

53. What are the resources human society uses to supply Gibbs free energy? (*Hint:* Consider information you learned in Chapter 6.)

54. For one day, keep a log of all of the activities you undertake that consume free energy. Distinguish between free energy provided by nutrient metabolism and that provided by other energy resources.

General Questions

55. This problem will help you understand the dependence of the American economy on energy. Referring to the following figure, calculate the energy (in joules) used by the agricultural, mining, and construction industries
(a) In one year.
(b) In one day.
(c) In one second.
(d) Remembering that 1 watt is the expenditure of 1 joule every second, calculate the average power needs of these industries in watts.
(e) Assuming a U.S. population of 280 million people, calculate the power needed by the agricultural, mining, and construction industries *per person in the United States.*
(f) Suppose you signed a contract to provide this energy to these industries for one year by eating glucose and giving them the resulting energy from its oxidation in your body. How much glucose would you have to eat each day to meet your contract? Assume that it is someone else's job to figure out how to get the energy stored in your ATP to the industries!
(g) An Olympic sprinter uses energy at the rate of 700 to 900 watts in a sprint. Compare this figure with the one you calculated in (f), and draw conclusions about feasibility of keeping your contract.

56. Investigate the feasibility of having every family in the United States produce enough water for its own needs by the combustion of hydrogen and oxygen, and utilizing the resulting free energy production for their energy needs. Do not try to collect the actual data you would use, but define the problem well enough so that someone else could collect the necessary data and do the calculations that would be needed.

Applying Concepts

57. Using the second law of thermodynamics, explain why it is very difficult to unscramble an egg. Who was Humpty-Dumpty? Why did his moment of glory illustrate the second law of thermodynamics?

58. In the Chemistry You Can Do experiment with the rubber band, you found that the relaxation of the rubber is endothermic. The molecules that make up the rubber are long chains of carbon atoms, and when the band is stretched the chains are straight-

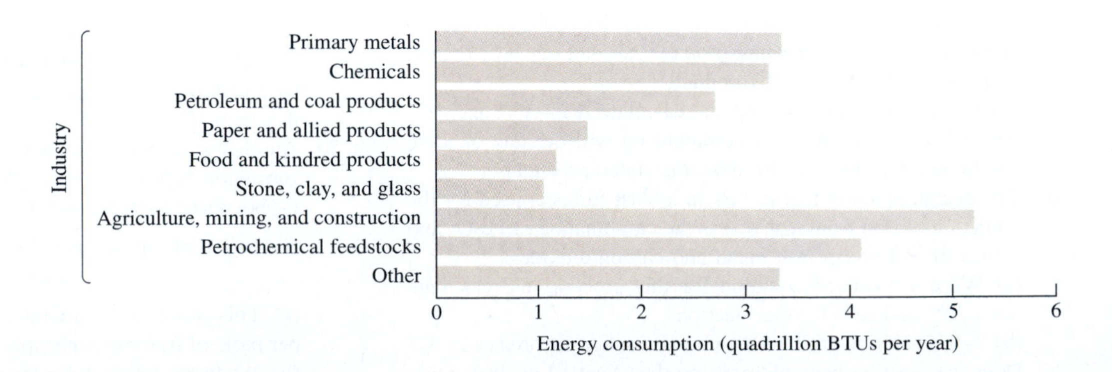

ened out. When the rubber relaxes, the chains return to a tangled, curled state. What happens to the entropy of the rubber band when a stretched band relaxes?

59. In the Chemistry You Can Do experiment in Chapter 6 you explored the heat from the rusting of iron to form iron oxide. Look at the enthalpies of formation of other metal oxides in Table 6.4 or Appendix J and comment on your observations. Are oxidations of metals generally endothermic or exothermic? Are they usually reactant-favored or product-favored?

60. Explain how the entropy of the universe is increased when an aluminum metal can is made from aluminum ore. The first step is to extract the ore, which is primarily a form of Al_2O_3, from the ground. After it is purified by freeing it from oxides of silicon and iron, aluminum oxide is changed to the metal by an input of electrical energy.

$$2\,Al_2O_3(s) \xrightarrow{\text{electrical energy}} 4\,Al(s) + 3\,O_2(g)$$

61. Explain why the entropy of the system increases when solid NaCl dissolves in water.

62. Explain how biological systems make use of coupled reactions to maintain the high degree of order found in all living organisms.

CHAPTER

8 Electron Configurations, Periodicity, and Properties of Elements

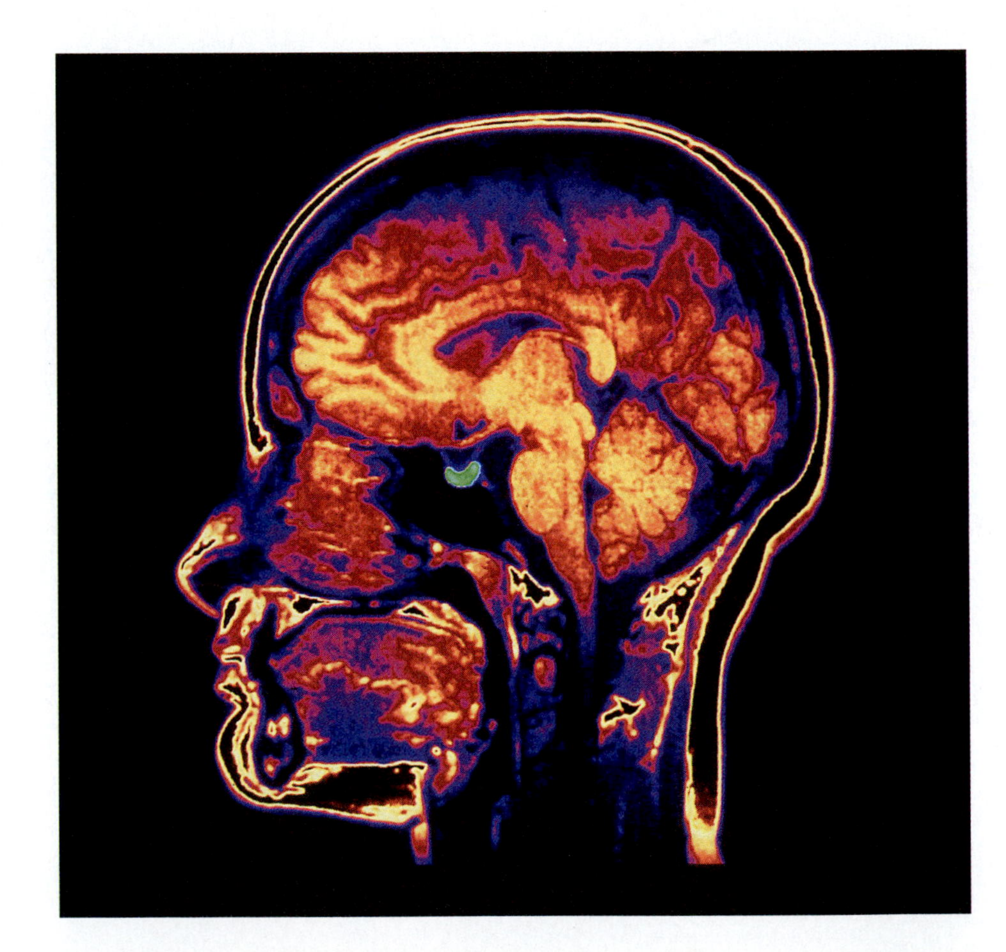

A computer-enhanced MRI scan of a normal human brain with the pituitary gland highlighted in green. The radio-frequency radiation used in magnetic resonance imaging is not energetic enough to break chemical bonds, but other forms of radiation like uv light and x rays are very energetic. • Can energetic light photons knock an electron out of a metal? After you study this chapter you should be able to work the Summary Problem, which relates to the energies and other properties of radiation of different kinds that is produced from excited atoms. *(Scott Camazine/Photo Researchers, Inc.)*

The periodic table was created by Mendeleev to summarize experimental observations. He had no theory or model to explain why all alkaline earths combine with oxygen in a 1:1 atom ratio—they just do. In the early years of this century, however, it became evident that atoms contain electrons. As a result of these findings, explanations of periodic trends in physical and chemical properties began to be based on an understanding of the arrangement of electrons within atoms—on what we now call electron configurations. Studies on the interaction of light with atoms and molecules revealed that electrons in atoms are arranged roughly in concentric shells. Electrons in the outermost shell are called valence electrons; their number and location are the chief factors that determine chemical reactivity. In this chapter the relationship of the electron configurations of atoms to their properties will be described. Special emphasis is placed on the use of the periodic table to derive the electron configurations for atoms and ions.

8.1 ELECTROMAGNETIC RADIATION AND MATTER

Theories about the energy and arrangement of electrons in atoms are based on experimental studies of the interaction of matter with electromagnetic radiation, of which visible light is a familiar form. The human eye can distinguish the spectrum of colors that make up visible light. Interestingly, matter in some form is always associated with any color of light our eyes see. For example, the red glow of a neon sign comes from neon atoms excited by electricity, and fireworks displays are visible because of light from metal ions excited by the heat of explosive reactions. Have you ever wondered how these varied colors of light are produced?

When atoms gain energy from, for example, exposure to heat or light, they are described as "excited." The added energy is absorbed by electrons, which then release it in the form of electromagnetic radiation, some of which is in the visible light region. Electromagnetic radiation and its applications are familiar to all of us; sunlight, headlights on automobiles, dental x rays, microwave ovens, and radio waves that we use for communications are a few examples (Figure 8.1). These kinds of ra-

The velocity of light through a substance (air, glass, water, etc.) depends on the chemical constitution of the substance and the wavelength of the light. This is the basis for using a glass prism to disperse light and is the explanation for rainbows. The velocity of sound is also dependent on the material through which it passes.

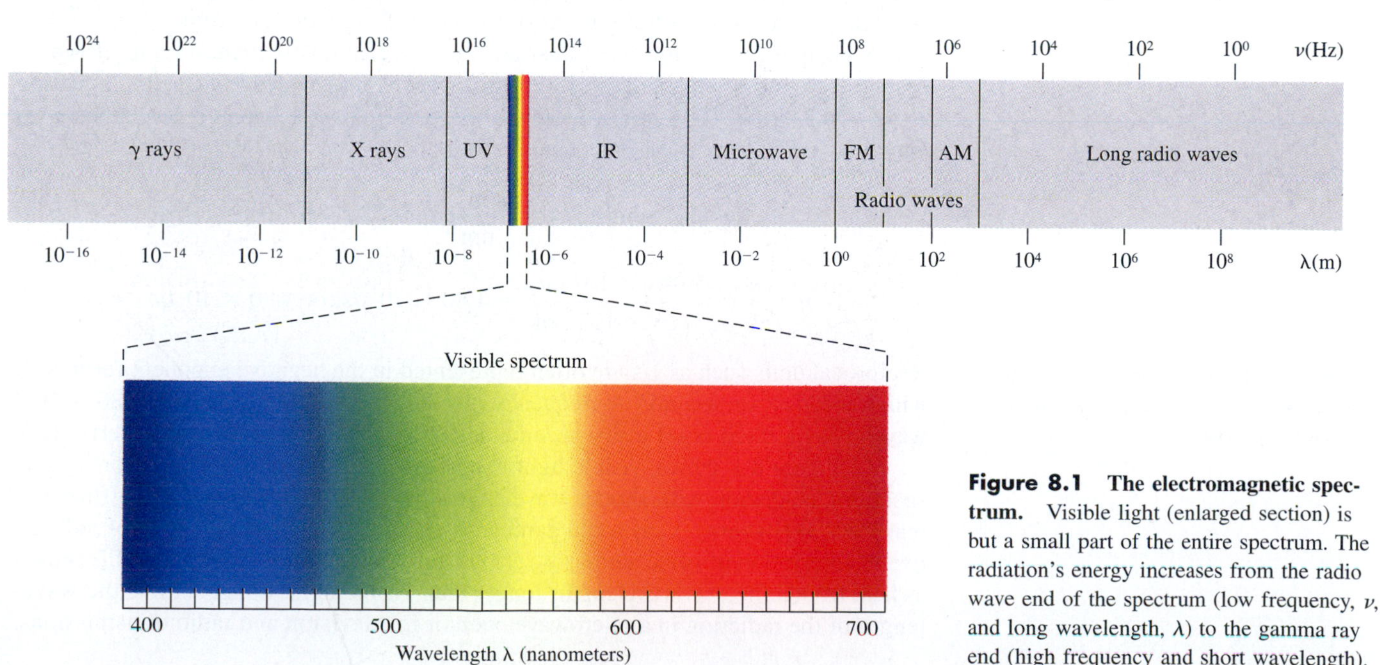

Figure 8.1 The electromagnetic spectrum. Visible light (enlarged section) is but a small part of the entire spectrum. The radiation's energy increases from the radio wave end of the spectrum (low frequency, ν, and long wavelength, λ) to the gamma ray end (high frequency and short wavelength).

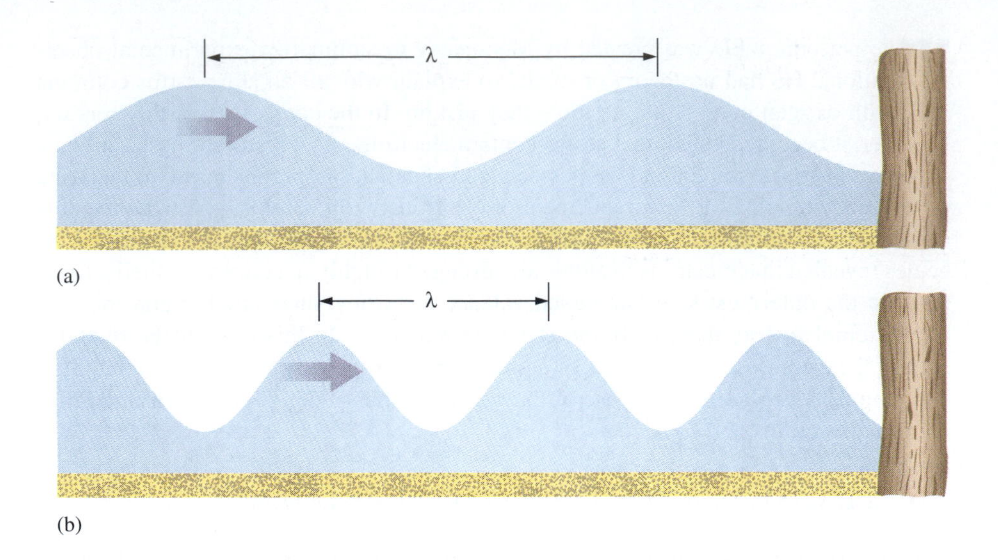

Figure 8.2 Wavelength and frequency of water waves. The waves are moving toward the post. (a) The wave has a long wavelength (large λ) and low frequency (the number of times per second its peak hits the post). (b) The wave has a shorter wavelength and a higher frequency (it hits the post more often per unit of time).

diation seem very different, but they are actually very similar. All **electromagnetic radiation** consists of oscillating electric and magnetic fields that travel through space at the same rate (the "speed of light": 186,000 miles/second or 2.998×10^8 m/s in a vacuum). Any of the various kinds of electromagnetic radiation can be described in terms of frequency (ν) and wavelength (λ). As illustrated in Figure 8.2, the **wavelength** is the distance between crests (or troughs) in a wave, and the **frequency** is the number of complete waves passing a point in a given amount of time (cycle per second). The frequency of electromagnetic radiation is related to its wavelength by

$$\nu\lambda = c$$

where c is the speed of light. A plot of the intensity of light as a function of the wavelength or frequency of light is called a **spectrum.** Figure 8.1 gives wavelength and frequency values for several regions of the electromagnetic spectrum.

A sample calculation will illustrate the wavelength-frequency relationship. If orange light has a wavelength of 625 nm, what is its frequency? Use the equation $\nu\lambda = c = 2.998 \times 10^8$ m/s. (Because the speed of light is in meters/second, the wavelength in nanometers must be converted to meters.)

$$625 \text{ nm}\left(1 \times 10^{-9}\frac{\text{m}}{\text{nm}}\right) = 6.25 \times 10^{-7} \text{ m}$$

$$\nu = \frac{c}{\lambda} = \frac{2.998 \times 10^8 \text{ m/s}}{6.25 \times 10^{-7} \text{ m}} = 4.80 \times 10^{14}\text{/s, or } 4.80 \times 10^{14}\text{s}^{-1}$$

The hertz unit (cycles/s, or s^{-1}) was named in honor of Heinrich Hertz (1857–1894), a German physicist.

Reciprocal units such as 1/s are often represented in the negative exponent form, s^{-1}, which means "per second." A frequency of $4.80 \times 10^{14}\ s^{-1}$ means that 4.80×10^{14} waves pass a fixed point every second. The unit s^{-1} is given the name hertz (Hz).

As Figure 8.1 shows, visible light is only a small portion of the electromagnetic spectrum. Radiation with shorter wavelengths includes ultraviolet radiation (the type that leads to sunburn), x rays, and gamma (γ) rays (emitted in the process of radioactive disintegration of some atoms (← *p. 37*)). Infrared radiation, the type that is sensed as heat from a fire, has wavelengths longer than visible light. Longer still is the wavelength of the radiation in a microwave oven or in television and radio transmissions.

PROBLEM-SOLVING EXAMPLE 8.1 Wavelength and Frequency

Compact audio disc players use lasers that emit light at a wavelength of 785 nm. What is the frequency of this light in hertz?

Answer 3.82×10^{14} Hz

Explanation Rearranging the equation relating wavelength, frequency, and the speed of light allows the answer to be calculated with the proper units.

$$\nu = \frac{c}{\lambda} = \frac{2.998 \times 10^{8}\ \text{m/s}}{7.85 \times 10^{-7}\ \text{m}} = 3.82 \times 10^{14}\ \text{s}^{-1} \text{ or } 3.82 \times 10^{14}\ \text{Hz}$$

Problem-Solving Practice 8.1

If your favorite FM radio station broadcasts at a frequency of 104.5 MHz (where 1 MHz = 1 megahertz = $10^{6}\ \text{s}^{-1}$), what is the wavelength (in meters) of the radiation emitted by this station?

Exercise 8.1 Frequency and Wavelength

Does the frequency of light increase or decrease as its wavelength increases?

Exercise 8.2 Estimating Wavelengths

The size of a radio antenna is proportional to the wavelength of the radiation. Consumers use cellular phones with antennas often less than 3 in. long, while submarines use antennas up to 2000 m in length. Which is using higher-frequency radio waves?

8.2 PLANCK'S QUANTUM THEORY

Have you ever sat near an electric resistance heater as it warms up? Of course, you cannot see the metal atoms in the heater's wire, but as electric energy flows through the wire, the atoms gain energy and then emit it as radiation. First the wire emits a slight amount of heat that you can feel (infrared radiation). As the wire gets hotter, it begins to glow (emit visible light), emitting first red light, and then orange. If the wire gets very hot, it appears almost white. Figure 8.3 shows the spectrum of a typical heated object.

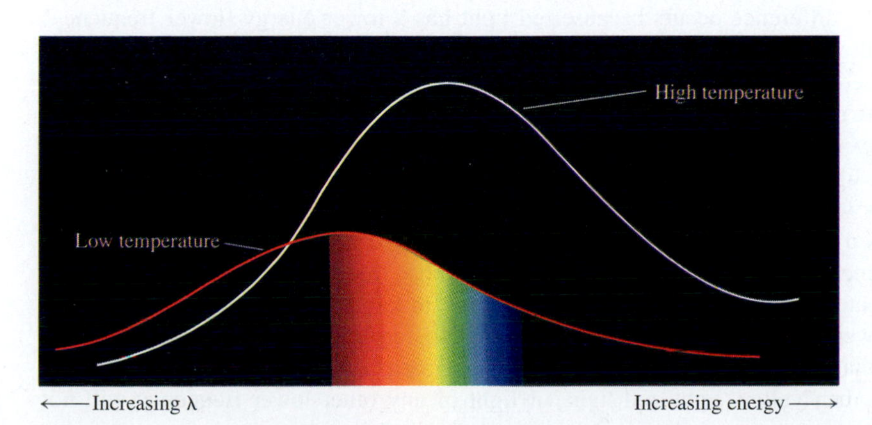

Figure 8.3 The spectrum of the radiation given off by a heated object. The red line is the spectrum of the object when it is "red hot." Notice that the maximum intensity is in the red region of the visible spectrum. As the temperature of the object increases, the maximum intensity of emitted light shifts toward orange, then yellow. At very high temperatures, the object becomes "white hot" as all wavelengths of visible light become equally intense.

At the close of the 19th century, scientists were trying to explain the nature of these emissions from hot objects. They assumed that vibrating atoms in a hot wire caused electromagnetic vibrations (light waves) to be emitted, and that those light waves could have any frequency along a continuously varying range. Theory predicted that as the object got hotter by acquiring more energy, its color should shift to the blue and finally all the way to the violet and beyond, but no object was ever observed to do this.

In 1900, Max Planck (1858–1947) offered an explanation for the spectrum of a heated body. His ideas contained the seeds of a revolution in scientific thought. He made what was at that time an incredible assumption: when an atom in a hot object emits radiation, there is a minimum quantity of energy that can be emitted at any given time. That is, there must be a small packet of energy such that no smaller quantity can be emitted, just as an atom is the smallest packet of an element. Planck called this packet of energy a **quantum.** He further asserted that the energy of a quantum is related to the frequency of the radiation by the equation

A quantum is the smallest possible unit of a distinct quantity, for example, the smallest possible unit of energy for electromagnetic radiation of a given frequency.

$$E_{\text{quantum}} = h\nu_{\text{radiation}}$$

The proportionality constant h is called **Planck's constant** in his honor; it has a value of $6.626 \times 10^{-34}\text{J} \cdot \text{s}$ and relates the frequency of radiation to its energy.

Earlier we calculated the frequency of orange light to be $4.80 \times 10^{14}\ \text{s}^{-1}$. The energy of one quantum of orange light is therefore

$$E = h\nu = (6.626 \times 10^{-34}\text{J} \cdot \text{s})(4.80 \times 10^{14}\ \text{s}^{-1}) = 3.18 \times 10^{-19}\ \text{J}$$

The theory based on Planck's work is called the **quantum theory.** By using his quantum theory, Planck was able to calculate the number of photons of each frequency that would be emitted by a hot object. His results agreed very well with the experimentally measured spectrum.

When a theory can accurately predict experimental results, the theory is usually regarded as useful. At first, however, Planck's quantum theory was not widely accepted because of its radical assumption that energy is quantized. The quantum theory of electromagnetic energy was firmly accepted after Planck's quanta were used by Albert Einstein to explain another phenomenon called the photoelectric effect.

In the early 1900s it was known that certain metals exhibit a **photoelectric effect;** that is, they emit electrons when illuminated by light of certain wavelengths. For each metal there is a threshold wavelength below which no photoelectric effect is observed. For example, the metal cesium (Cs) will emit electrons when illuminated by red light, whereas some metals require yellow light and others require ultraviolet light. The difference occurs because red light has a lower energy (lower frequency) than yellow light or ultraviolet light. Figure 8.4b shows how an electric current suddenly increases when light of a frequency above a certain value shines on a photosensitive metal. Einstein explained these observations by assuming that Planck's quanta were *massless* "particles" of light. He called them **photons** instead of quanta. That is, light could be described as a stream of photons that had particle-like properties as well as wave-like properties. Removing one electron from a metal surface requires a certain minimum quantity of energy; we call it E_{min}. Since each photon has an energy given by $E = h\nu$, only photons whose frequency is high enough so that E is greater than E_{min} will have enough energy to knock an electron loose. Photons with lower frequencies (left side of Figure 8.4b) do not have enough energy. This means that if some metal requires photons of green light to emit electrons from its surface, then yellow light, red light, or light of any other lower frequency will not

One of the early uses of the photoelectric effect was "electric eyes" that would automatically open a door when a beam of light was interrupted, as when someone walked through the beam. Later, cameras began to be produced that used photoelectric-effect light sensors to control the exposure of photographic film.

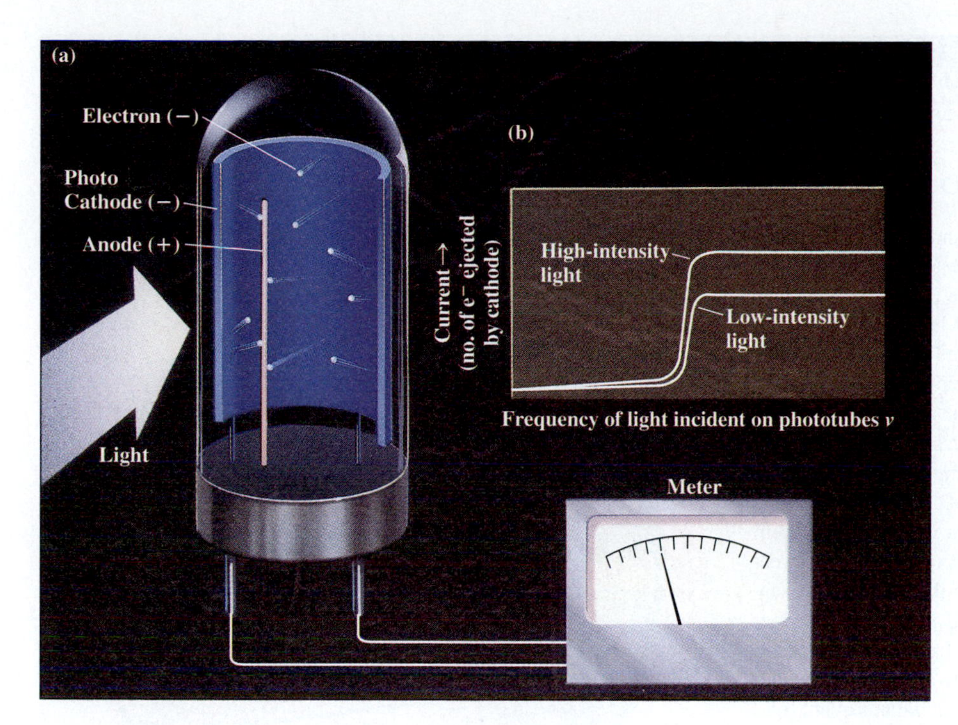

Figure 8.4 The photoelectric effect. (a) A photocell's metallic cathode is struck by photons. Light with any frequency above a certain threshold causes electrons to be ejected. These are attracted to the anode, causing an electric current to flow and giving a reading on the meter. Light that has a frequency below the threshold does not cause elctrons to be ejected and the current flow is near zero. (b) As the frequency of light striking the metal cathode is increased, some frequency is reached at which electrons begin to be ejected. Once current is flowing, the number of electrons depends on the light's intensity, not its frequency.

have sufficient energy to cause the photoelectric effect. This brilliant deduction about the quantized nature of light and how it relates to its interaction with matter won Einstein the Nobel Prize for Physics in 1921.

The energy of photons is important for other practical reasons as well. Photons of ultraviolet light can damage the skin, while photons of visible light cannot. We use sunblocks containing molecules that selectively absorb ultraviolet photons to protect our skin. X-ray photons are even more energetic than ultraviolet photons and can disrupt molecules at the cellular level, causing, among other things, genetic damage. For this reason we try to limit our exposure to x rays even more than we limit our exposure to ultraviolet light. Look again at Figure 8.4b. Notice how a higher-intensity light source causes a higher photoelectric current. Higher intensities of ultraviolet light (or more time of exposure) can cause greater damage to the skin than lower intensities (or less exposure time). The same holds true for other high-energy forms of electromagnetic radiation like x rays.

Such developments as Einstein's explanation of the photoelectric effect led eventually to acceptance of what is referred to as the *dual nature* of light. Depending on the experimental circumstances, visible light, and all other forms of electromagnetic radiation, appear to have either "wave" or "particle" (photon) behavior. While classical theory fails to explain the photoelectric effect, it does explain quite well the diffraction of light by a prism or diffraction grating. However, it is important to realize that this "dual nature" description arises because of our attempts to explain observations by inadequate models. Light is not changing back and forth from being a wave to being a particle, but has a single consistent nature that can be described by modern quantum theory. The dual nature description arises when we try to explain our observations by using our familiarity with classical models for "wave" or "particle" behavior.

Until Einstein's time, and his explanation of the photoelectric effect, classical physics had considered light as wave-like in nature.

CHEMISTRY YOU CAN DO

Using a Compact Disk (CD) as a Diffraction Grating

Seeing the visible spectrum by using a prism to refract visible light is a familiar experiment (Figure 8.5). Less familiar is the use of a diffraction grating for the same purpose. A diffraction grating consists of many equally spaced parallel lines—thousands of lines per centimeter. A grating that transmits diffracted light is made by cutting thousands of grooves on a piece of glass or clear plastic; a grating that reflects diffracted light is made by cutting the grooves on a piece of metal or opaque plastic. You can get an idea of how diffraction gratings work by using a compact disc as a diffraction grating. Compare the spectra of light from two different sources—a mercury vapor lamp and a white incandescent light bulb. Stand about 20 to 60 m away from a mercury street lamp and hold the CD about waist high with the print side down. Tilt the CD down until you see the reflected image of the street lamp. Close one eye, and view along the line from the light source to your body as you slowly tilt the CD up toward you. What colors do you see? Use the same procedure with an incandescent light bulb. Do you see the same colors? Explain your results.

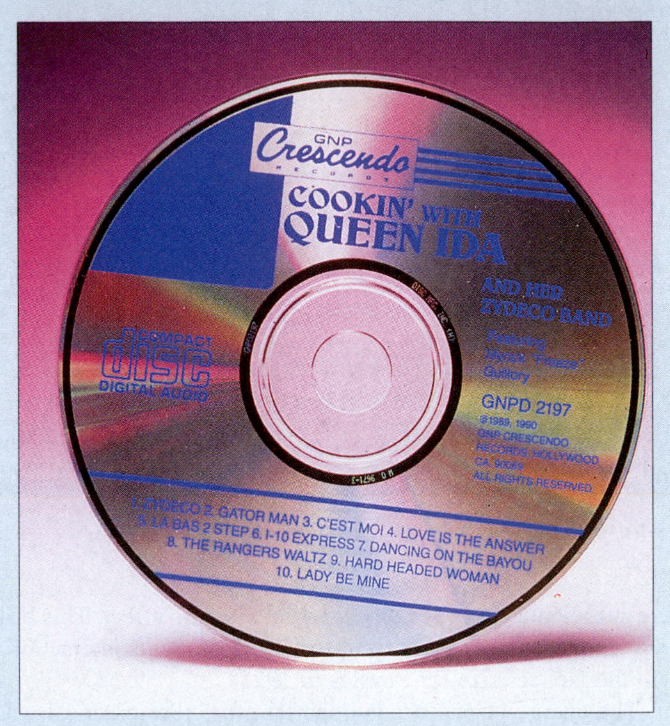

The diffraction of light by the closely spaced lines on a compact disk. The colors near the words "Compact Disk" are the results of diffraction of light. *(C.D. Winters)*

Source: R.C. Mebane and T.R. Rybolt: "Atomic Spectroscopy with a Compact Disc," Journal of Chemical Education, *Vol. 69, p. 401, 1992.*

Although this dual view of the nature of light may be disconcerting, it did lead to a revolutionary idea—namely, if light can be viewed in terms of both "wave" and "particle" properties, why can't matter as well? This is exactly the question posed by Louis de Broglie that led to his hypothesis about the wave properties of matter, which are discussed in the next section.

Since $E = h\nu$ and $\nu = \frac{c}{\lambda}$, then $E = h\frac{c}{\lambda}$.

PROBLEM-SOLVING EXAMPLE 8.2 Calculating Photon Energies

What is the energy associated with one photon of the laser light described in Example 8.1, which has a frequency of $3.82 \times 10^{14}\ s^{-1}$?

Answer 2.53×10^{-19} J

Explanation According to Planck's quantum theory, the energy and frequency of radiations are related by $E = h\nu$. Replacing h and ν in the equation by their values, followed by multiplying and canceling the units, gives the correct answer.

$$E = h\nu = (6.626 \times 10^{-34}\ \text{J} \cdot \text{s})(3.82 \times 10^{14}\ \text{s}^{-1}) = 2.53 \times 10^{-19}\ \text{J}$$

Problem-Solving Practice 8.2

Which has more energy,
(a) One photon of microwave radiation or one photon of ultraviolet radiation?
(b) One photon of blue light or one photon of green light?
(c) 10 blue photons of $\lambda = 460$ nm, or 15 red photons of $\lambda = 695$ nm?

Exercise 8.3 Comparing X-ray Photons to Visible-Light Photons

Calculate the energy of one photon of x radiation having a wavelength of 2.36 nm and compare that with the energy of one photon of orange light (3.18×10^{-19} J).

8.3 MODELS OF THE ATOM

The spectrum of white light, such as that from the sun or an incandescent light bulb, consists of a rainbow display of separated colors shown in Figure 8.5. This rainbow spectrum, containing light of all wavelengths, is called a **continuous spectrum.**

If a high voltage is applied to an element in the gas phase at low pressure, the atoms absorb energy and are "excited." The excited atoms emit light (Figure 8.6). An application of the light from these excited atoms is a neon advertising sign, in which the neon atoms emit orange-red light. When light from such a source passes through a prism onto a white surface, only a few colored lines are seen. This is called a **line emission spectrum.** The line spectra of the visible light emitted by excited atoms of hydrogen, mercury, and neon are shown in Figure 8.7.

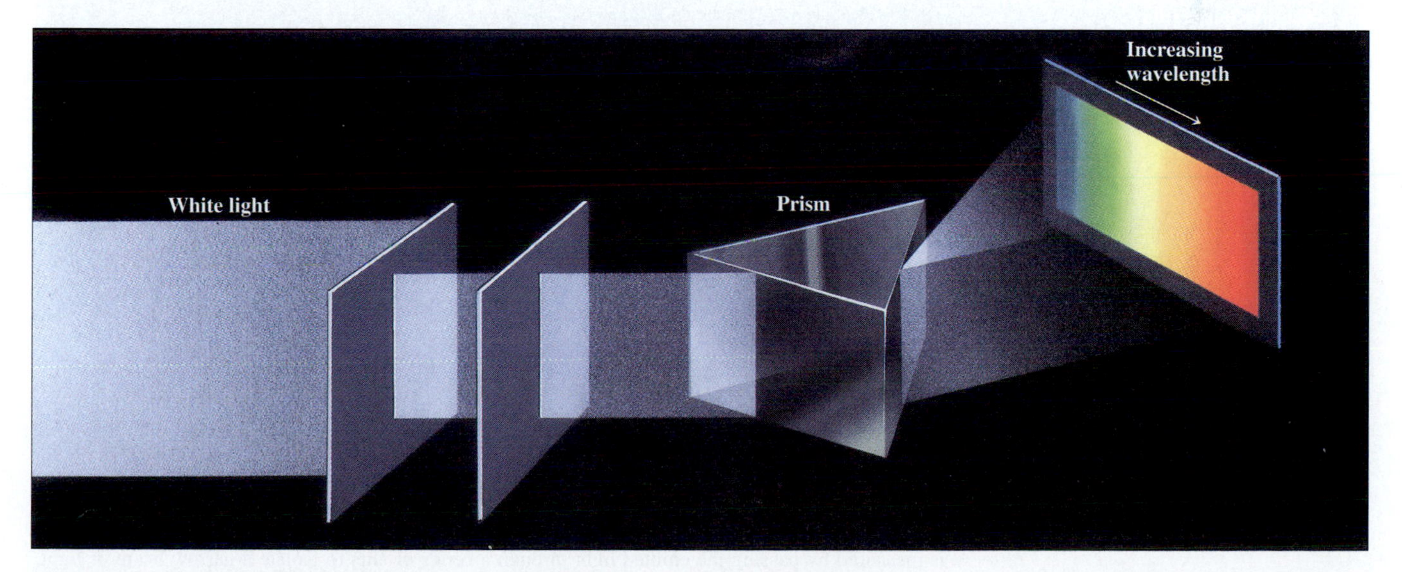

Figure 8.5 A spectrum from white light. When white light is passed through slits to produce a narrow beam and then refracted in a glass prism, the various colors blend smoothly into one another.

Figure 8.6 A neon sign. Any gas-discharge sign of this type is referred to as a "neon sign" even though many of them don't contain neon. Different colors are emitted by excited atoms of different noble gases—neon, reddish orange; argon, blue; helium, yellowish white. A helium-argon mixture emits an orange light while a neon-argon mixture gives a dark lavender light. Mercury vapor, used in fluorescent lights, is also used in gas mixtures to obtain a wide range of colors. By using these various gas mixtures and colored glass tubes, most of the colors in the visible spectrum can be produced. *(Steve Drexler/The Image Bank)*

Every element has a unique line emission spectrum. The characteristic lines in the emission spectrum of an element can be used in chemical analysis, especially in metallurgy, both to identify the element and to determine how much of it is present in a sample.

Bohr's Explanation of Line Spectra of Atoms

In 1913 Niels Bohr provided the first explanation of the line spectra of atoms when he connected the quantum ideas of Planck and Einstein. Bohr introduced the notion

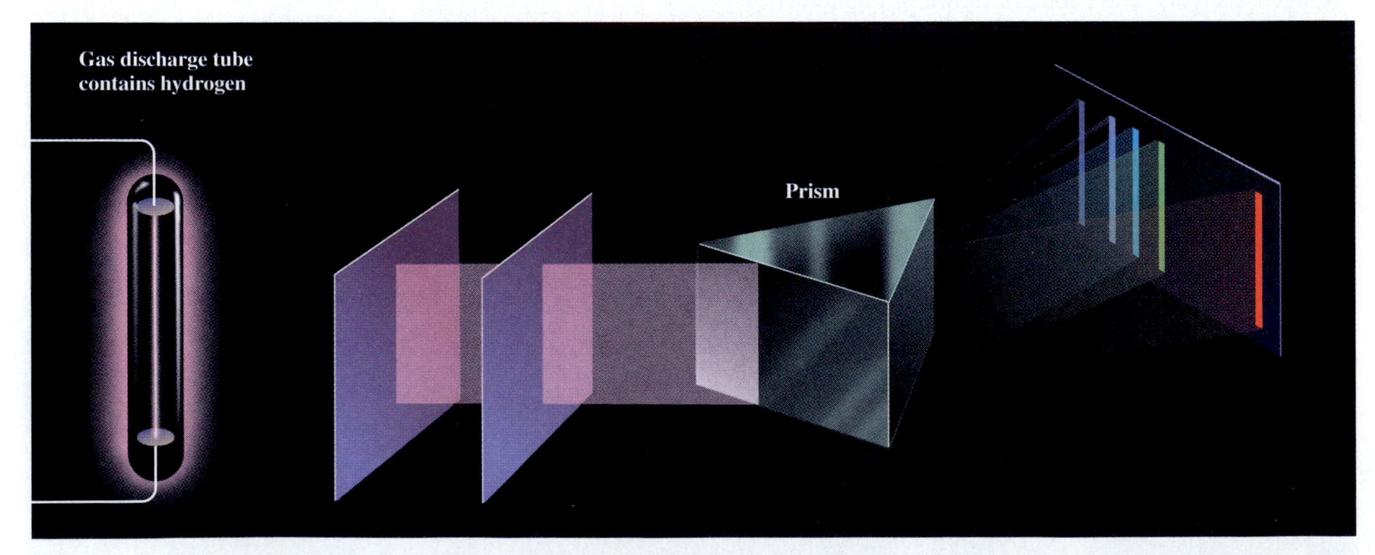

Figure 8.7 Hydrogen line emission spectrum. A line emission spectrum of excited H atoms is measured by passing the emitted light through a series of slits to isolate a narrow beam of light, and this beam is then passed through a prism to separate the light into its component wavelengths. A photographic plate or other instrument detects the separate wavelengths as individual lines; hence, the name "line spectrum" for the light emitted by excited atoms or molecules.

Portrait of a Scientist • *Niels Bohr (1885–1962)*

Niels Bohr was born in Copenhagen, Denmark. He earned a Ph.D. in physics in Copenhagen in 1911 and then went to work first with J. J. Thomson in Cambridge, England, and later with Ernest Rutherford in Manchester, England.

It was in England that he began to develop the ideas that a few years later led to the publication of his theory of atomic structure and his explanation of atomic spectra. He received the Nobel Prize for Physics in 1922 for this work.

After working with Rutherford for a very short time, Bohr returned to Copenhagen, where he eventually became the director of the Institute of Theoretical Physics. Many young physicists carried on their work in this Institute, and seven of them later received Nobel Prizes for their studies in chemistry and physics. Among them were such well-known scientists as Werner Heisenberg, Wolfgang Pauli, and Linus Pauling.

Niels Bohr. *(American Institute of Physics)*

that *the single electron of the hydrogen atom could occupy only certain energy levels.* The energy of the electron in an atom is said to be "quantized." Bohr referred to these energy levels as *orbits* and represented the energy difference between any two adjacent orbits as a single quantum of energy.

In Bohr's model, each allowed orbit is assigned an integer, n, known as the **principal quantum number.** The values of n for the possible orbits range from 1 to infinity. The radii of the circular orbits increase as n increases. The orbit of lowest energy, $n = 1$, is closest to the nucleus, and the electron of the hydrogen atom is normally in this energy level. Any atom with its electrons in their lowest energy levels is said to be in the **ground state.** Energy must be supplied to move the electron farther away from the nucleus because the positive nucleus and the negative electron attract each other. When the electron of a hydrogen atom occupies an orbit with $n > 1$, the atom has more energy than in its ground state and is said to be in an **excited state.** The excited state of any atom is unstable. The energy gained by an excited atom is emitted when the electron returns to the ground state. Bohr introduced the assumption that the energy of the photon, $h\nu$, that is emitted from the excited atom corresponds to the *difference* between two energy levels of the atom. Think of the Bohr orbit model as a set of stairs in which the higher stair steps are closer together. Each step represents a quantized energy level; as you climb the stairs, you can stop on any step but not between steps. If you move down one step, you lose a "quantum" of energy.

According to Bohr, the light forming the lines in the emission spectrum of hydrogen (Figure 8.8) comes from electrons in hydrogen atoms moving from higher orbits to lower orbits closer to the nucleus (and eventually to the $n = 1$ orbit) after having first been excited to orbits with $n = 2, 3, 4$, and higher that are farther from the nucleus (Figure 8.9). The emission spectrum of hydrogen has lines in ultraviolet, visible, and infrared regions. Photons in the ultraviolet region are produced by elec-

(Text continues on page 313.)

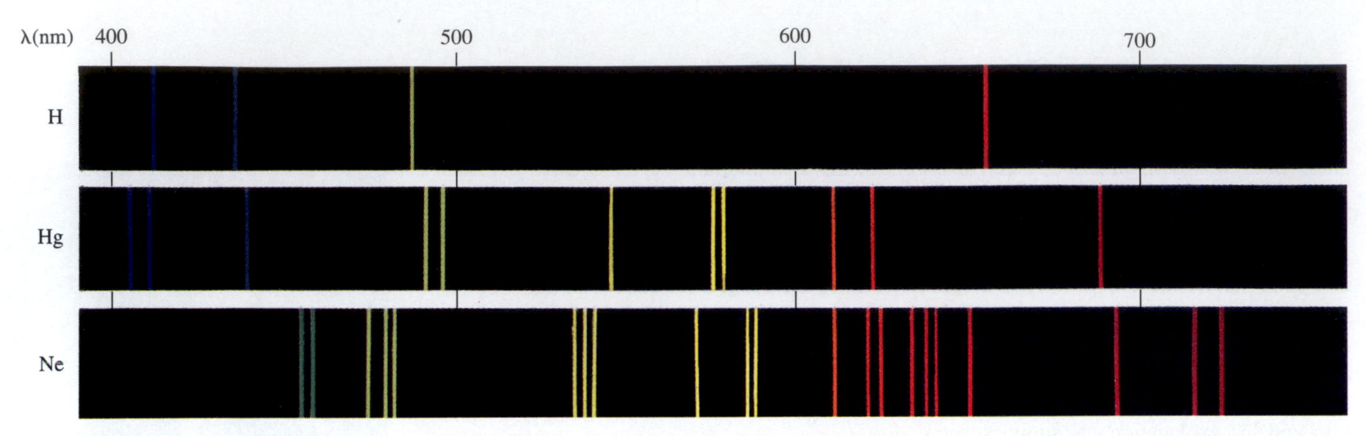

Figure 8.8 Line emission spectra of hydrogen, mercury, and neon. Excited gaseous elements produce characteristic spectra that can be used to identify the element as well as to determine how much of the element is present in a sample.

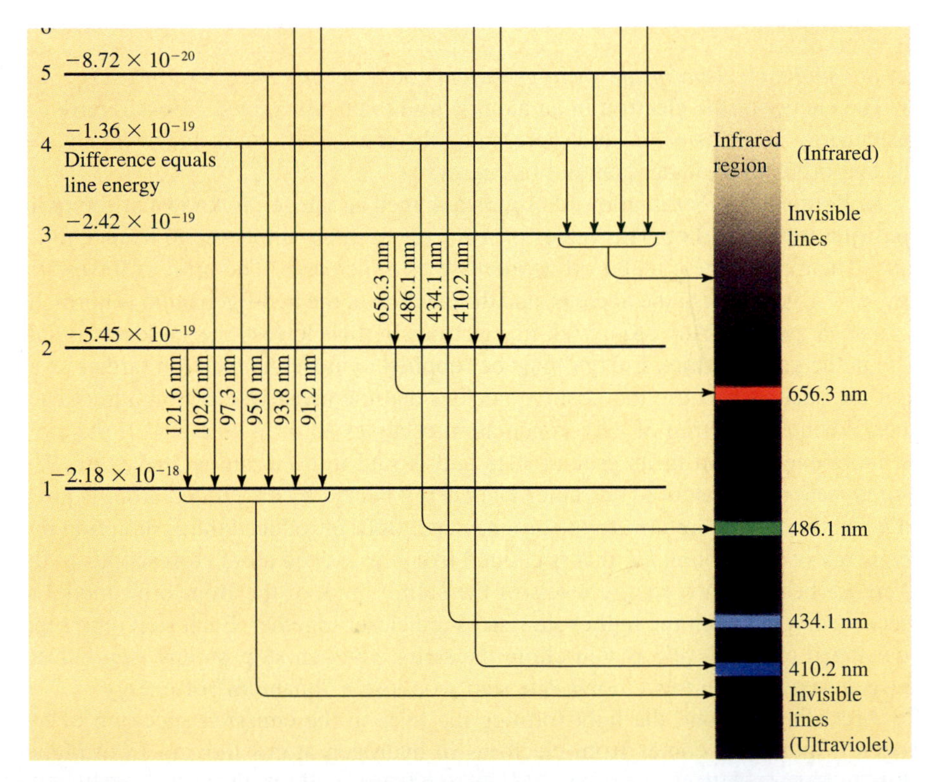

Figure 8.9 Electron transitions in an excited H atom. The lines in the ultraviolet region result from all transitions to the $n = 1$ level. Transitions from levels with values of $n > 2$ to $n = 2$ level occur in the visible region. Lines in the infrared region result from the transitions from levels with values of $n > 3$ or 4 to the $n = 3$ or 4 levels (only the series for transitions to the $n = 3$ level is shown).

TABLE 8.1 **Agreement Between Bohr's Theory and the Lines of the Hydrogen Spectrum***

Changes in Energy Levels	Wavelength Predicted by Bohr's Theory (nm)	Wavelength Determined from Laboratory Measurement (nm)	Spectral Region
$2 \rightarrow 1$	121.6	121.7	Ultraviolet
$3 \rightarrow 1$	102.6	102.6	Ultraviolet
$4 \rightarrow 1$	97.28	97.32	Ultraviolet
$3 \rightarrow 2$	656.6	656.3	Visible red
$4 \rightarrow 2$	486.5	486.1	Visible blue-green
$5 \rightarrow 2$	434.3	434.1	Visible blue
$4 \rightarrow 3$	1876	1876	Infrared

*These lines are typical; other lines could be cited as well, with equally good agreement between theory and experiment. The unit of wavelength is the nanometer (nm), 10^{-9} m.

trons moving from energy levels with $n > 1$ to $n = 1$. Photons in the visible region are produced by electrons moving from levels with $n > 2$ to $n = 2$. Photons in the infrared energy are produced by electrons moving from levels with $n > 3$ or 4 to $n = 3$ or 4.

From his model, Bohr was able to calculate the wavelengths of the lines in the hydrogen spectrum, some of which are shown in Table 8.1. Note the close agreement between the measured values and the values predicted by the calculations of the Bohr theory. Niels Bohr had tied the unseen (the interior of the atom) to the seen (the observable lines in the hydrogen spectrum)—a fantastic achievement!

The Bohr atomic model was accepted almost immediately after its presentation. Bohr's success with the hydrogen atoms soon led, however, to attempts by him and others to extend the same model to more complex atoms. It soon became apparent that line spectra for elements other than hydrogen had more lines than could be explained by the simple Bohr model. Spectroscopists studying the line spectrum of sodium atoms distinguished four different types of lines, which they labeled *sharp, principal, diffuse,* and *fundamental.* A better understanding of the atoms was needed to account for such additional lines. In spite of its limitations, Bohr's theory of the hydrogen atom was an important advance in physics and chemistry because it introduced the concept of energy levels for atoms, a concept that is fundamental to the science of spectroscopy.

Spectroscopy is the science of measuring spectra. Many kinds of spectroscopy have emerged from the first studies of simple line spectra. Some spectral measurements are done for analytical purposes, and others are done to determine molecular structures.

Beyond the Bohr Model

A totally different approach was needed to explain atoms or ions with more than one electron. In 1924 Louis de Broglie (1892–1987) posed the question: *If light can be viewed in terms of both "wave" and "particle" properties, why can't particles of matter, like electrons, be treated the same way?* In suggesting that small particles of matter like electrons could have wave-like properties, de Broglie was being revolutionary. Many scientists found the concept hard to accept, much less believe.

Conceptual Challenge Problem CP.8.A at the end of this chapter relates to topics covered in this section.

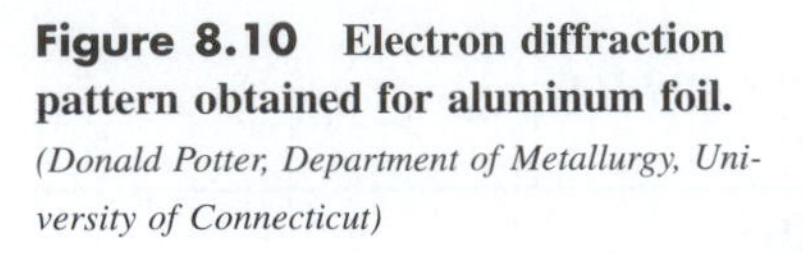

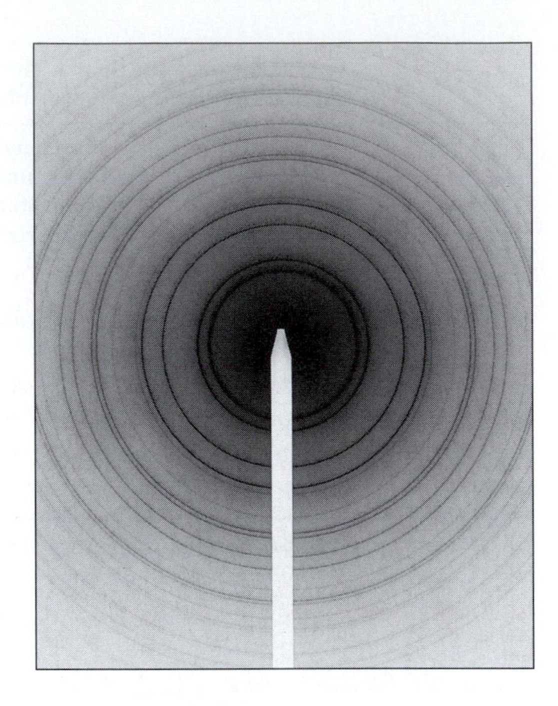

Figure 8.10 Electron diffraction pattern obtained for aluminum foil. *(Donald Potter, Department of Metallurgy, University of Connecticut)*

Louis de Broglie proposed that the characteristic wavelength, λ, of an electron (or any other particle) depends on its mass, m, and its velocity, v: $\lambda = h/mv$, where h is Planck's constant. An electron (8.109×10^{-31} kg) moving at 2% of the speed of light (6.0×10^{6} m/s) will have a wavelength of 0.12 nm. Waves this small are readily diffracted by atoms in crystals.

However, experimental proof was soon produced. C. Davisson and L. H. Germer, working at the Bell Telephone Laboratories in 1927, found that a beam of electrons was diffracted by the atoms of a thin sheet of metal foil (Figure 8.10) in the same way that light waves are diffracted by a grating. Since diffraction is readily explained by the wave properties of light, it followed that electrons also can be described by the equations of waves under some circumstances.

A few years after de Broglie's hypothesis about the wave nature of the electron, Werner Heisenberg (1901–1976) proposed the **uncertainty principle,** which states that *it is impossible to simultaneously determine the exact position and the exact momentum of an electron.* This limitation is not a problem for a macroscopic object because the energy of photons used to locate such an object does not cause a measurable change in the position or momentum of that object. However, the very act of measurement would affect the position and momentum of the electron because of its very small size and mass. High-energy photons would be required to locate the small electron, and when the photons collide with the electron, the momentum of the electron would be changed. If lower-energy photons were used to avoid affecting the momentum, little information would be obtained about the location of the electron. Consider an analogy in photography. If you take a picture of a car race at a high shutter-speed setting, you get a clear picture of the cars but you can't tell how fast they are going or even whether they are moving. With a slow shutter speed, you can tell from the blur of the car images something about the speed and direction, but you have less information about where the cars are.

Momentum = mass × velocity.

The Heisenberg uncertainty principle illustrated another inadequacy in the Bohr model—its representation of the electron in the hydrogen atom in terms of well-defined orbits about the nucleus. In practical terms, the best we can do is to represent the *probability* of finding an electron (of a given energy and momentum) within a given space. This probability-based model of the atom is what we are currently using.

Exercise 8.4 Many Spectral Lines, Only One Kind of Atom

The hydrogen atom contains only one electron, but there are many lines in its spectrum. How does the Bohr theory explain this?

The Wave Mechanical Model of the Atom

In 1926 Erwin Schrödinger (1877–1961) combined de Broglie's hypothesis with classical equations for wave motion. From these and other ideas he derived a new equation called the wave equation to describe the behavior of an electron in the hydrogen atom. Solutions to the wave equation, called **wave functions,** predict the allowed energy states of an electron and the probability of finding that electron in a given region of space.

Although each wave function is a complex mathematical equation, it is possible to represent the square of a wave function in graphic form—as a picture of the region in the atom where an electron with a given energy state is most likely to be found. One way to make such a picture is to draw a surface within which there is a 90% probability that the electron will be found. That is, nine times out of ten an electron will be somewhere inside such a **boundary surface;** there is one chance in ten that the electron will be outside of it.

A 100% probability isn't chosen because such a surface would have no definite boundary. Consider that a typical dartboard has a finite size, and normally, players in a dart game will hit the board somewhere over 90% of the time. But if you wanted to be certain that any player, no matter how far away, would be able to hit the board on 100% of his or her throws, then the board would have to be considerably larger. By similar reasoning, a boundary surface in which there would be a 100% probability of finding a given kind of electron would have to be quite large.

A series of such boundary-surface diagrams for the hydrogen atom is shown in Figure 8.11. These three-dimensional boundary surfaces are called **orbitals.** Note that an *orbital* (wave mechanical model) is not the same as an *orbit* (Bohr model). In the quantum mechanical model, the principal quantum number, n, is a measure of the most probable distance of the electron from the nucleus, not the radius of a well-defined orbit. When $n = 1$, there is only one kind of orbital possible; it is an s orbital. When $n = 2$, two kinds of orbitals are possible; s and p orbitals. When $n = 3$, three kinds of orbitals are possible: s, p, and d orbitals. When $n = 4$, s, p, d, and f orbitals can exist.

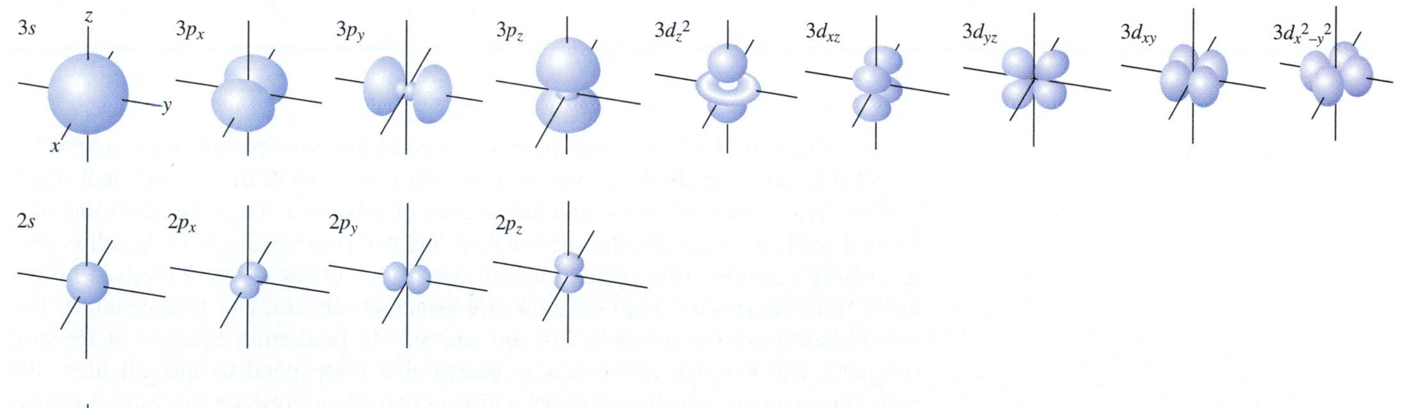

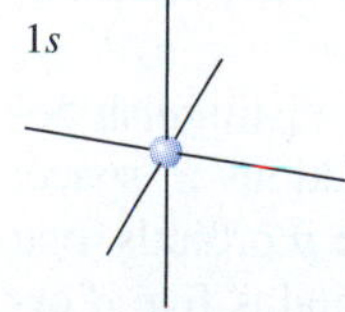

Figure 8.11 Orbitals. Boundary-surface diagrams for electron densities of 1*s*, 2*s*, 2*p*, 3*s*, 3*p*, and 3*d* orbitals of a hydrogen atom (generated by a computer). The 1*s* orbital is the ground state for the single electron in a hydrogen atom.

CHEMISTRY IN THE NEWS

Hydrogen Molecules in Space May Account for Mystery Spectral Bands

Since the beginnings of spectroscopy, scientists have pointed their instruments toward space to analyze the electromagnetic radiation reaching the earth. The lightest noble gas, helium, was discovered in 1868 from studies of the spectrum of the sun (Section 8.10). By 1900 astronomers had recorded on film what are called *diffuse interstellar bands* (DIBs), a complex pattern of absorption spectral regions, called *bands,* in the uv, visible, and near IR portions of the electromagnetic spectrum. By 1930 it was known that these DIBs were caused by interstellar matter of some kind of absorbing light photons, but the kind of matter and the kinds of processes that were involved were a mystery.

Absorption spectra are different from emission spectra in that matter *absorbs* light of different wavelengths rather than emitting light. You can easily observe this kind of behavior by looking at a glass of water, which doesn't absorb visible light, and a glass of tea, which contains dyes from the tea leaves. The color of the tea solution is that of the wavelengths of light

PAH^+	$\lambda_{observed}$(nm)	DIB(nm)
Naphthalene ($C_{10}H_8^+$) 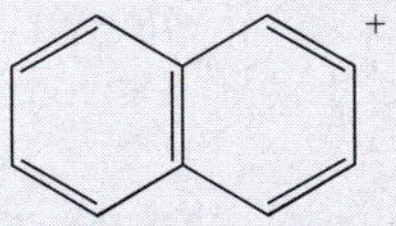	674.2	674.1
Tetracene ($C_{18}H_{12}^+$) 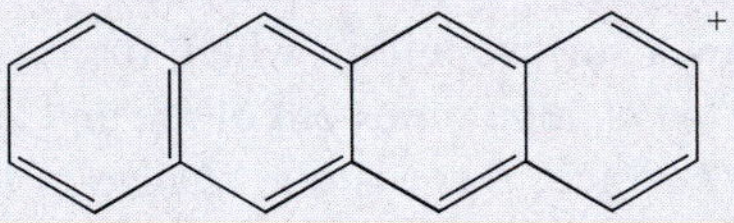	864.7	864.8
Coronene ($C_{24}H_{12}^+$) 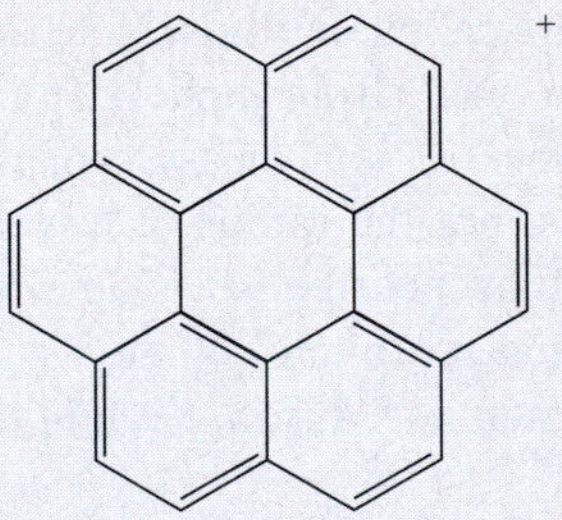	459.0 946.5	459.5 946.6

A collection of orbitals with the same value of the principal quantum number, n, is called an electron **shell.** All the orbitals with $n = 2$ are in the second shell. Each shell is divided into subshells, and the number of subshells is equal to the value of n for that shell; for example, the second shell has two subshells. Each subshell is designated by a number (the value of n) and one of the letters *s, p, d, f* (derived historically from the spectral lines called sharp, principal, diffuse, and fundamental). The $n = 3$ shell has three subshells: 3*s,* 3*p,* and 3*d.* By predicting energies of electron subshells, and thus the differences in energy that correspond to spectral lines, the Schrödinger wave equation provides a direct connection between the wave mechanical theory and experimental atomic spectra.

The subscripts on the labels of the *p* and *d* orbitals in Figure 8.11 distinguish between orbitals that are in the same subshell but have different orientations in space. Any *s* subshell has only one spherical *s* orbital; a *p* subshell has three *p* orbitals, one along each of the three axes, labeled p_x, p_y, and p_z. A *d* subshell contains five *d* or-that pass through the solution (the nonabsorbed light). If more tea is added to the water, the intensity of the absorption of light increases and the solution appears more deeply colored. If a spectrometer is used, the absorption spectrum for the dye molecules in the tea solution can be measured and recorded.

Most molecules, including the simplest ones, absorb electromagnetic radiation in one or more regions of the spectrum. They usually do this by absorbing a single photon at a time; the photon excites the molecule into a higher energy state. As scientists studied the absorption spectra of more complex molecules such as polycyclic aromatic hydrocarbons (PAH; Section 9.7 discusses cyclic organic compounds of this type) and their ions (PAH^{+}), it became apparent that DIBs might be caused by absorption of light from stars by these same kinds of complex molecules in clouds of interstellar matter. In fact, a good correlation was made between many of the DIBs and spectra of the 1+ ions of PAHs found here on earth. But the intensity of the absorption bands suggested much more matter present in those clouds than allowed for current theories.

Recently, two IBM research scientists, Peter Sorokin and James Glownia, in Yorktown Heights, N.Y., have assembled a strong case for the DIBs being caused by interstellar hydrogen molecules that are absorbing two photons at a time. Hydrogen would be a likely candidate for the cause of these bands, since hydrogen is very abundant in the universe. In addition, the abundance of hydrogen would readily account for the intensity of the DIBs. Even though the DIB patterns do not match the normal single-photon absorption lines of hydrogen, Sorokin and Glownia have been able to assign over 70 of the 200 known DIBs to absorption of photons in what are called "nonlinear optical interactions"—processes similar to those that take place in a laser. They believe that hydrogen molecules found in cold, thin clouds within 30 light-years of very bright stars are bathed with such high-intensity ultraviolet light that a two-photon absorption is triggered in which an ultraviolet photon and a visible-light photon are simultaneously absorbed. When light from the distant star is observed on earth, the absorbed visible wavelengths produce the DIBs.

Astronomers have been skeptical of Sorokin and Glownia's theory, but recent laboratory experiments have indicated their theory may be on the right track.

Source: R. Dagani: "Hydrogen proposed as a solution to stellar puzzle," Chemical and Engineering News, *November 11, 1996, p. 11.*

bitals, while an *f* subshell has seven *f* orbitals. For the first three energy levels, the number of subshells and orbitals are as follows:

Principal Quantum Number	Subshells	Orbitals in the Subshell
1	*s*	1 (1*s* orbital)
2	*s*	1 (2*s* orbital)
	p	3 (2*p* orbitals)
3	*s*	1 (3*s* orbital)
	p	3 (3*p* orbitals)
	d	5 (3*d* orbitals)

In a hydrogen atom in its ground state, solving the wave equation shows that all subshells with the same value of n have the same energy.

Just when it appears that we have an appropriate model for representing electron energies in atoms, we encounter another stumbling block. Schrödinger's wave equation cannot be solved *exactly* for any atom containing more than one electron! Fortunately, methods of approximation have been developed. With these methods and Schrödinger's equation, a computer can calculate atomic (or molecular) electron configurations to a high degree of accuracy. These results have led chemists and physicists to rely on the wave mechanical theory for understanding atomic structure. The ideas of shells, subshells, and orbital shapes developed for the hydrogen atom carry over to the wave mechanics of all other atoms.

Conceptual Challenge Problem CP.8.B at the end of the chapter relates to topics covered in this section.

Exercise 8.5 Orbitals in Shells

What is the maximum number of *s* orbitals that may be found in a given electron shell? The maximum number of *p* orbitals? Of *d* orbitals? Of *f* orbitals? What is the principal quantum number of the shell in which *f* orbitals first occur?

Exercise 8.6 *g* Orbitals

Using the same reasoning as was developed for *s, p, d,* and *f* orbitals, what would be the *n* value for the first shell that could contain *g* orbitals and how many *g* orbitals would there be?

8.4 SPIN-LIKE PROPERTIES OF ELECTRONS IN ATOMS

When spectroscopists studied the emission spectra of hydrogen and sodium atoms in greater detail, they discovered that lines originally thought to be single were actually closely spaced pairs of lines. In 1925 the Dutch physicists George Uhlenbeck and Samuel Goudsmit proposed that the splitting could be explained by assuming that each electron in an atom can exist in one of two possible spin states. The electron in an atom can be thought of as a charged sphere rotating about an axis through its center (Figure 8.12). Such a spinning charge produces a magnetic field. This means the electron behaves like a tiny bar magnet with north and south magnetic poles. Only two orientations of the spin are allowed with respect to the direction of the external

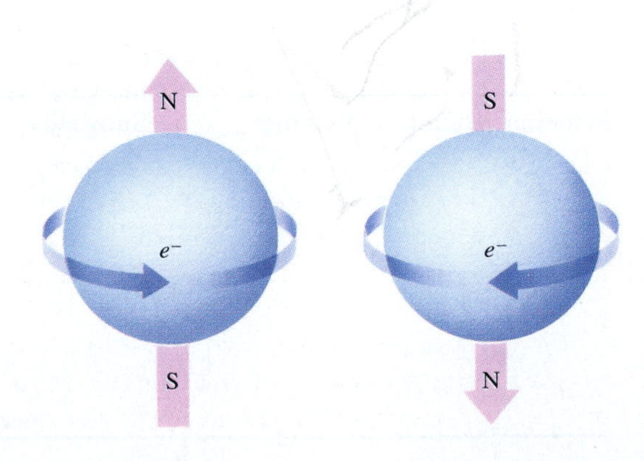

Figure 8.12 Spin directions. The electron can be pictured as though it were a charged sphere spinning about an axis through its center. The electron can have only two directions of spin; any other position is forbidden. Therefore, the spin of the electron is said to be quantized.

TABLE 8.2 **Number of Electrons Accommodated in Electron Shells and Subshells**

Electron Shell (n)	Subshells Available (= n)	Orbitals Available	Number of Electrons Possible in Subshell	Maximum Electrons for nth Shell (= $2n^2$)
1	*s*	1	2	2
2	*s*	1	2	8
	p	3	6	
3	*s*	1	2	18
	p	3	6	
	d	5	10	
4	*s*	1	2	32
	p	3	6	
	d	5	10	
	f	7	14	
5	*s*	1	2	50
	p	3	6	
	d	5	10	
	f	7	14	
	*g**	9	18	
6	*s*	1	2	72
	p	3	6	
	d	5	10	
	*f**	7	14	
	*g**	9	18	
	*h**	11	22	
7	*s*	1	2	

*These orbitals are not used in the ground state of any known element.

field; these orientations are called clockwise and counterclockwise. The two opposite directions of spin produce oppositely directed magnetic fields, as shown in Figure 8.12. The interactions of the oppositely directed fields with an external magnetic field result in two slightly different energies, which in turn lead to the splitting of the spectral lines into closely spaced pairs.

To make the quantum theory consistent with experiment, Wolfgang Pauli stated in 1925 what is now known as the **Pauli exclusion principle:** *at most two electrons can be assigned to the same orbital in the same atom, and these two electrons must have opposite spins.* When electrons of equal energy have opposite spin directions (which we represent by arrows points in opposite directions, ↑↓), they are said to be *paired* and their magnetic fields cancel. Two electrons spinning in the same direction (↑↑ or ↓↓) are said to have *parallel* spins. According to the Pauli exclusion principle, two electrons with parallel spins must be in different orbitals.

The limitation that only two electrons can occupy a single orbital has the effect of determining the maximum number of electrons for each principal quantum number, as summarized in Table 8.2. The $n = 1$ energy level has only one *s* orbital and

The results expressed in this table were predicted by the Schrödinger theory and have been confirmed by experiment.

therefore can accommodate only two electrons. The $n = 2$ energy level has one s orbital and three p orbitals, each of which can accommodate a pair of electrons; therefore, this level can accommodate a total of eight electrons (two in the s orbital and six in the p orbitals). For each principal energy level, the maximum number of electrons is $2n^2$.

Exercise 8.7 Maximum Number of Electrons

(a) What is the maximum number of electrons in the $n = 3$ level? Identify the orbital of each electron.
(b) What is the maximum number of electrons in the $n = 4$ level? Identify the orbital of each electron.

8.5 ATOM AND ION ELECTRON CONFIGURATIONS

The complete description of the orbitals occupied by all the electrons in an atom or ion is called its **electron configuration.** Using quantum theory, it is possible to make some sense of the electron configurations of atoms of the elements. In doing this we will use the periodic table as a guide. As you will see, the similarities of elements in the same periodic table groups are explained by their similar electron configurations.

Electron Configurations of Main Group Elements

The atomic numbers of the elements increase in numerical order across the periodic table. As a result, atoms of each element contain one more electron than atoms of the element to its left in the table. How do we know which shell and orbital each new electron occupies? An important principle for answering this question is the following: *for an atom in its ground state, electrons are found in the energy shells, subshells, and orbitals that produce the lowest energy for the atom.* In other words, electrons fill orbitals starting with the $1s$ orbital and working upward in the subshell energy order starting with $n = 1$, as shown in Figure 8.13.

The main group elements are those in the A groups in the periodic table (see inside front cover).

To better understand how this filling of orbitals works, consider the experimentally determined electron configurations of the first ten elements, which are written in three different ways in Table 8.3. Since electrons assigned to the $n = 1$ shell are closest to the nucleus and therefore lowest in energy, electrons are assigned to it first (H and He). At the left in Table 8.3, the occupied orbitals and the number of electrons in each orbital are represented in the following notation:

H $1s^1$ — one electron in s subshell; principle quantum number, n; subshell (s)

At the right in the table, each occupied orbital is represented by a box in which electrons are shown as arrows; ↑ for a single electron in an orbital and ↑↓ for paired electrons in an orbital.

In helium the two electrons are paired in the $1s$ orbital so that the lowest energy shell ($n = 1$) is filled. After the $n = 1$ shell is filled, electrons are assigned to the next

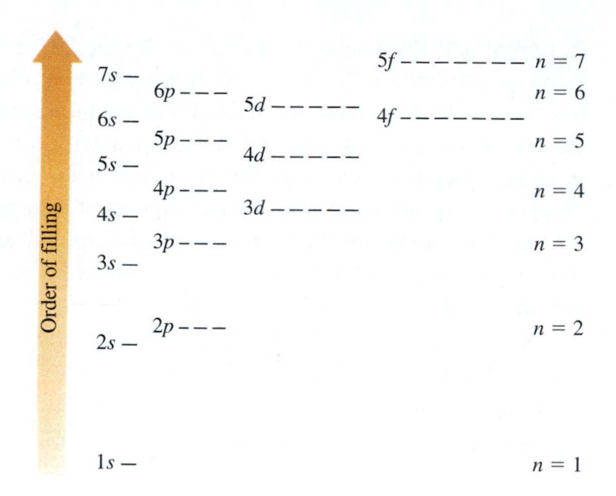

Figure 8.13 The order of subshell filling for many electron atoms. The subshells orbitals fill in order of increasing energy ($1s \rightarrow 2s \rightarrow 2p \rightarrow 3s \rightarrow 3p \rightarrow 4s \rightarrow 3d \rightarrow 4p \rightarrow 5s \rightarrow 4d \rightarrow 5p$, and so on).

highest energy level (Figure 8.11), the $n = 2$ shell, beginning with lithium. This second shell can hold eight electrons, and its orbitals are occupied in the eight elements from lithium to neon. Notice in the periodic table inside the front cover that these are the eight elements of the second period.

As happens in each principal energy level (and each period), the first two electrons fill the *s* orbital. In the second period these elements are Li ($1s^22s^1$—with the

TABLE 8.3 Electron Configurations of the First Ten Elements

	Electron Configurations		Orbital Box Diagrams				
	Condensed	Expanded	1*s*	2*s*	$2p_x$	$2p_y$	$2p_z$
H	$1s^1$		[↑]				
He	$1s^2$		[↑↓]				
Li	$1s^22s^1$		[↑↓]	[↑]			
Be	$1s^22s^2$		[↑↓]	[↑↓]			
B	$1s^22s^22p^1$		[↑↓]	[↑↓]	[↑]	[ ]	[ ]
C	$1s^22s^22p^2$	$1s^22s^22p_x^{\ 1}2p_y^{\ 1}$	[↑↓]	[↑↓]	[↑]	[↑]	[ ]
N	$1s^22s^22p^3$	$1s^22s^22p_x^{\ 1}2p_y^{\ 1}2p_z^{\ 1}$	[↑↓]	[↑↓]	[↑]	[↑]	[↑]
O	$1s^22s^22p^4$	$1s^22s^22p_x^{\ 2}2p_y^{\ 1}2p_z^{\ 1}$	[↑↓]	[↑↓]	[↑↓]	[↑]	[↑]
F	$1s^22s^22p^5$	$1s^22s^22p_x^{\ 2}2p_y^{\ 2}2p_z^{\ 1}$	[↑↓]	[↑↓]	[↑↓]	[↑↓]	[↑]
Ne	$1s^22s^22p^6$	$1s^22s^22p_x^{\ 2}2p_y^{\ 2}2p_z^{\ 2}$	[↑↓]	[↑↓]	[↑↓]	[↑↓]	[↑↓]

Although chemists often say that electrons "occupy an orbital" or are "placed in an orbital," orbitals are not literally things or boxes in which electrons are placed. An orbital is the wave function arising from the solution to the wave equation. It is more correct to say an electron is "assigned to an orbital" which means that electron has the energy and shape corresponding to that orbital.

$2s$ orbital half filled) and Be ($1s^22s^2$—with the $2s$ orbital completely filled). The next electron goes into a $2p$ orbital. That element is B ($1s^22s^22p^1$). Since the p subshell has three p orbitals, adding a second p electron presents a choice. Does the second p electron in the carbon atom ($1s^22s^22p^2$) pair with the existing electron in a p orbital, or does it occupy another p orbital? It has been shown experimentally that both p electrons have the same spin. Hence, they must occupy different p orbitals, otherwise they would violate the Pauli exclusion principle. The expanded electron configurations in the middle of Table 8.3 show the locations of the three p electrons in the carbon atom's three p orbitals (p_x, p_y, p_z).

Since electrons are negatively charged particles, electron configurations that produce maximum unpairing also minimize electron-electron repulsions, making the total energy of the set of electrons as low as possible. **Hund's rule** summarizes how subshells are filled: the most stable arrangement of electrons in the same subshell is that with the maximum number of unpaired electrons, all with the same spin. The general result of Hund's rule is that in *p, d,* or *f* orbitals each successive electron enters a different orbital of the subshell until the subshell is half-full, after which electrons pair in the orbitals one by one.

In writing electron configurations, the following rules should be kept in mind:

1. *Orbitals with the same value of the principal quantum number, n, occupy a principal energy level called a shell. There are n^2 orbitals in a shell.*
2. *Each shell is divided into a number of subshells equal to the principal quantum number, n, for that shell. For example, the $n = 3$ shell has three subshells, 3s, 3p, and 3d.*
3. *Each subshell is divided into orbitals (s subshell, 1 orbital; p subshell, 3 orbitals; d subshell, 5 orbitals; f subshell, 7 orbitals). For example, the $n = 2$ shell has one 2s orbital and three 2p orbitals, for a total of 4 orbitals as given by n^2.*
4. *Because each orbital is limited to occupancy by two electrons of opposite spin, the maximum number of electrons in a shell is $2n^2$.*
5. *Electrons pair only after each orbital in a subshell is occupied by a single electron (this is Hund's rule).*

Suppose you need to know the electron configuration of phosphorus, which means identifying the orbitals occupied by all its electrons. It can be done by using the periodic table and the information in the preceding summary. Checking the periodic table inside the front cover, you find that phosphorus has atomic number 15 and therefore has 15 electrons. Write an electron configuration by starting with H and putting electrons into subshells until you reach the element in question. Phosphorus is in the third period. The relationship between periodic table position and electron configuration is summarized in Figure 8.14. The $n = 1$ and $n = 2$ shells are filled in the first two periods, giving the first ten electrons in phosphorus the configuration $1s^22s^22p^6$ (the same as neon). Figure 8.14 shows that the next two electrons are assigned to the $3s$ orbital, giving $1s^22s^22p^63s^2$ so far. The final three electrons have to be assigned to the $3p$ orbitals, so the electron configuration for phosphorus is

$$\text{P} \quad 1s^22s^22p^63s^23p^3$$

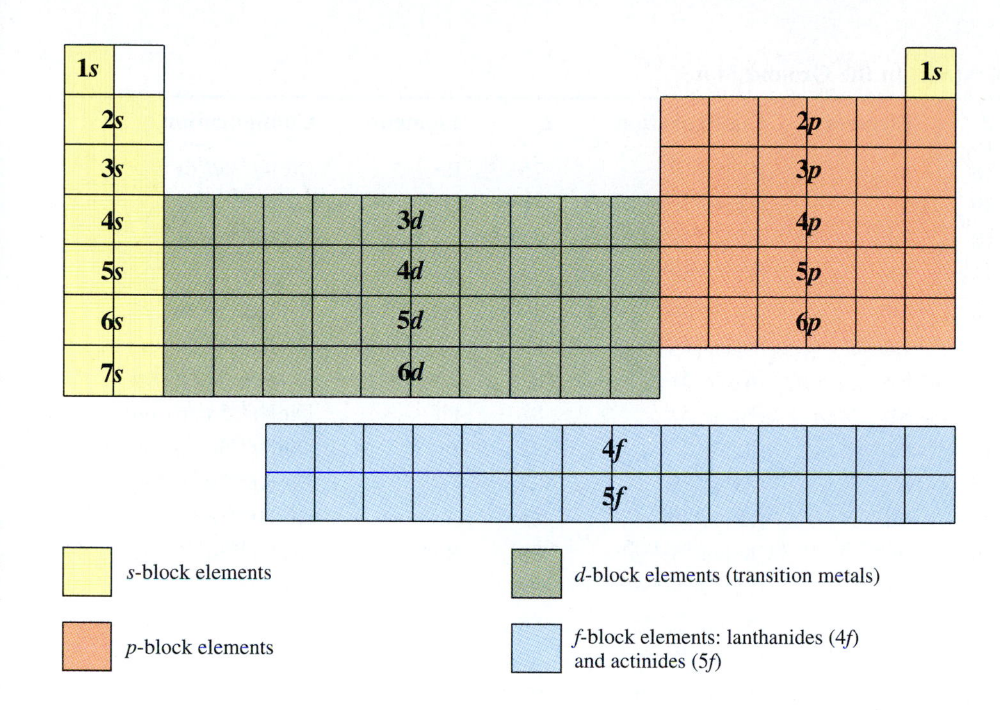

Figure 8.14 Electron configurations and the periodic table. Electron configurations can be generated by assigning electrons to subshells starting at H and moving through this table in atomic-number order until the desired element is reached. Configurations generated by using the periodic table can be compared with those given in Table 8.4.

According to Hund's rule, the three electrons in the *p* orbitals of the phosphorus atom must be unpaired. To show this, you can write the expanded electron configuration

$$\text{P} \quad 1s^2 2s^2 2p_x^{\,2} 2p_y^{\,2} 2p_z^{\,2} 3s^2 3p_x^{\,1} 3p_y^{\,1} 3p_z^{\,1}$$

or the orbital box diagram:

	$1s$	$2s$	$2p_x 2p_y 2p_z$	$3s$	$3p_x 3p_y 3p_z$
P	[↑↓]	[↑↓]	[↑↓] [↑↓] [↑↓]	[↑↓]	[↑] [↑] [↑]

Thus, all the electrons are paired except for the three electrons in $3p$ orbitals, and these occupy different orbitals and have parallel spins.

Using the noble gas configuration, the electron configuration for phosphorus is $[\text{Ne}]3s^2 3p^5$.

The arrangement of the elements in the periodic table becomes more understandable once their electron configurations are understood. The electron configurations of all the elements are given in Table 8.4. Notice that Table 8.4 uses an abbreviated representation of the configurations in which the symbol of the preceding noble gas represents filled subshells. This is called the **noble gas notation.** For Ca, atomic number 20, the noble gas notation is $[\text{Ar}]4s^2$, where the symbol [Ar] represents the filled subshells in argon, $1s^2 2s^2 2p^6 3s^2 3p^6$.

At this point, you should be able to write electron configurations for main group elements through Ca, atomic number 20, using the periodic table as a guide. Check your electron configurations against the ones given in Table 8.4 for all the elements.

PROBLEM-SOLVING EXAMPLE 8.3 Electron Configurations

Give the complete electron configuration and noble gas abbreviations for silicon, Si.

Answer $1s^2 2s^2 2p^6 3s^2 3p^2$ or $[\text{Ne}]3s^2 3p^2$

(Text continues on page 325.)

TABLE 8.4 Electron Configurations of Atoms in the Ground State

Z	Element	Configuration	Z	Element	Configuration	Z	Element	Configuration
1	H	$1s^1$	38	Sr	$[Kr]5s^2$	75	Re	$[Xe]4f^{14}5d^5 6s^2$
2	He	$1s^2$	39	Y	$[Kr]4d^1 5s^2$	76	Os	$[Xe]4f^{14}5d^6 6s^2$
3	Li	$[He]2s^1$	40	Zr	$[Kr]4d^2 5s^2$	77	Ir	$[Xe]4f^{14}5d^7 6s^2$
4	Be	$[He]2s^2$	41	Nb	$[Kr]4d^4 5s^1$	78	Pt	$[Xe]4f^{14}5d^9 6s^1$
5	B	$[He]2s^2 2p^1$	42	Mo	$[Kr]4d^5 5s^1$	79	Au	$[Xe]4f^{14}5d^{10}6s^1$
6	C	$[He]2s^2 2p^2$	43	Tc	$[Kr]4d^6 5s^2$	80	Hg	$[Xe]4f^{14}5d^{10}6s^2$
7	N	$[He]2s^2 2p^3$	44	Ru	$[Kr]4d^7 5s^1$	81	Tl	$[Xe]4f^{14}5d^{10}6s^2 6p^1$
8	O	$[He]2s^2 2p^4$	45	Rh	$[Kr]4d^8 5s^1$	82	Pb	$[Xe]4f^{14}5d^{10}6s^2 6p^2$
9	F	$[He]2s^2 2p^5$	46	Pd	$[Kr]4d^{10}$	83	Bi	$[Xe]4f^{14}5d^{10}6s^2 6p^3$
10	Ne	$[He]2s^2 2p^6$	47	Ag	$[Kr]4d^{10}5s^1$	84	Po	$[Xe]4f^{14}5d^{10}6s^2 6p^4$
11	Na	$[Ne]3s^1$	48	Cd	$[Kr]4d^{10}5s^2$	85	At	$[Xe]4f^{14}5d^{10}6s^2 6p^5$
12	Mg	$[Ne]3s^2$	49	In	$[Kr]4d^{10}5s^2 5p^1$	86	Rn	$[Xe]4f^{14}5d^{10}6s^2 6p^6$
13	Al	$[Ne]3s^2 3p^1$	50	Sn	$[Kr]4d^{10}5s^2 5p^2$	87	Fr	$[Rn]7s^1$
14	Si	$[Ne]3s^2 3p^2$	51	Sb	$[Kr]4d^{10}5s^2 5p^3$	88	Ra	$[Rn]7s^2$
15	P	$[Ne]3s^2 3p^3$	52	Te	$[Kr]4d^{10}5s^2 5p^4$	89	Ac	$[Rn]6d^1 7s^2$
16	S	$[Ne]3s^2 3p^4$	53	I	$[Kr]4d^{10}5s^2 5p^5$	90	Th	$[Rn]6d^2 7s^2$
17	Cl	$[Ne]3s^2 3p^5$	54	Xe	$[Kr]4d^{10}5s^2 5p^6$	91	Pa	$[Rn]5f^2 6d^1 7s^2$
18	Ar	$[Ne]3s^2 3p^6$	55	Cs	$[Xe]6s^1$	92	U	$[Rn]5f^3 6d^1 7s^2$
19	K	$[Ar]4s^1$	56	Ba	$[Xe]6s^2$	93	Np	$[Rn]5f^4 6d^1 7s^2$
20	Ca	$[Ar]4s^2$	57	La	$[Xe]5d^1 6s^2$	94	Pu	$[Rn]5f^6 7s^2$
21	Sc	$[Ar]3d^1 4s^2$	58	Ce	$[Xe]4f^1 5d^1 6s^2$	95	Am	$[Rn]5f^7 7s^2$
22	Ti	$[Ar]3d^2 4s^2$	59	Pr	$[Xe]4f^3 6s^2$	96	Cm	$[Rn]5f^7 6d^1 7s^2$
23	V	$[Ar]3d^3 4s^2$	60	Nd	$[Xe]4f^4 6s^2$	97	Bk	$[Rn]5f^9 7s^2$
24	Cr	$[Ar]3d^5 4s^1$	61	Pm	$[Xe]4f^5 6s^2$	98	Cf	$[Rn]5f^{10}7s^2$
25	Mn	$[Ar]3d^5 4s^2$	62	Sm	$[Xe]4f^6 6s^2$	99	Es	$[Rn]5f^{11}7s^2$
26	Fe	$[Ar]3d^6 4s^2$	63	Eu	$[Xe]4f^7 6s^2$	100	Fm	$[Rn]5f^{12}7s^2$
27	Co	$[Ar]3d^7 4s^2$	64	Gd	$[Xe]4f^7 5d^1 6s^2$	101	Md	$[Rn]5f^{13}7s^2$
28	Ni	$[Ar]3d^8 4s^2$	65	Tb	$[Xe]4f^9 6s^2$	102	No	$[Rn]5f^{14}7s^2$
29	Cu	$[Ar]3d^{10}4s^1$	66	Dy	$[Xe]4f^{10}6s^2$	103	Lr	$[Rn]5f^{14}6d^1 7s^2$
30	Zn	$[Ar]3d^{10}4s^2$	67	Ho	$[Xe]4f^{11}6s^2$	104	Rf*	$[Rn]5f^{14}6d^2 7s^2$
31	Ga	$[Ar]3d^{10}4s^2 4p^1$	68	Er	$[Xe]4f^{12}6s^2$	105	Ha	$[Rn]5f^{14}6d^3 7s^2$
32	Ge	$[Ar]3d^{10}4s^2 4p^2$	69	Tm	$[Xe]4f^{13}6s^2$	106	Sg	$[Rn]5f^{14}6d^4 7s^2$
33	As	$[Ar]3d^{10}4s^2 4p^3$	70	Yb	$[Xe]4f^{14}6s^2$	107	Ns	$[Rn]5f^{14}6d^5 7s^2$
34	Se	$[Ar]3d^{10}4s^2 4p^4$	71	Lu	$[Xe]4f^{14}5d^1 6s^2$	108	Hs	$[Rn]5f^{14}6d^6 7s^2$
35	Br	$[Ar]3d^{10}4s^2 4p^5$	72	Hf	$[Xe]4f^{14}5d^2 6s^2$	109	Mt	$[Rn]5f^{14}6d^7 7s^2$
36	Kr	$[Ar]3d^{10}4s^2 4p^6$	73	Ta	$[Xe]4f^{14}5d^3 6s^2$	110	—	$[Rn]5f^{14}6d^8 7s^2$
37	Rb	$[Kr]5s^1$	74	W	$[Xe]4f^{14}5d^4 6s^2$	111	—	$[Rn]5f^{14}6d^9 7s^2$
						112	—	$[Rn]5f^{14}6d^{10}7s^2$

*The names for elements 104–109 have been disputed. Current American usage is to use the names for these elements chosen by their discoverers.

Explanation The periodic table shows silicon is in the third period with atomic number 14 and therefore has 14 electrons. The first ten electrons are represented by $1s^22s^22p^6$, the electron arrangement for Ne. The last four electrons placed in the atom have the configuration $3s^23p^2$.

Problem-Solving Practice 8.3

For sulfur, (a) write the electron configuration in the noble gas notation, and (b) determine how many unpaired electrons sulfur atoms have by drawing the orbital box diagram.

Exercise 8.8 Highest-Energy Electrons in Ground State Atoms

Identify the number of electrons in the highest occupied energy level (highest n) in a ground state chlorine atom. Do the same for a sulfur atom.

Valence Electrons

As early as 1902, Gilbert N. Lewis (1875–1946) hit upon the idea that electrons in atoms might be arranged in shells, starting close to the nucleus and building outward. Lewis explained the similarity of chemical properties for elements in a given group by assuming that all elements in that group have the same number of electrons in their outer shell. These electrons are known as **valence electrons** for the main group elements. The electrons in the filled inner shells of these elements are called **core electrons.** A few examples of the core and valence electrons are the following:

Although Lewis didn't publish his ideas about valence electrons until 1916, he had written a memorandum in 1902 that outlined his ideas, which developed from his attempt to explain periodic trends to students in his introductory chemistry course.

Element	Core Electrons	Total Electron Configuration	Valence Electrons	Periodic Group
Na	$1s^22s^22p^6$	$[Ne]3s^1$	$3s^1$	1A
Si	$1s^22s^22p^6$	$[Ne]3s^23p^2$	$3s^23p^2$	4A
As	$1s^22s^22p^63s^23p^63d^{10}$	$[Ar]3d^{10}4s^24p^3$	$4s^24p^3$	5A

Note that for elements in the fourth and higher periods of Groups 3A to 7A, electrons in the d subshell are also core electrons. Using this notation the core electrons of As are represented by $[Ar]3d^{10}$.

Exercise 8.9 Core and Valence Electrons

Write the noble gas configuration for iodine and identify its core and valence electrons.

While teaching his students about atomic structure, Lewis used the element's symbol to represent the atomic nucleus together with the core electrons. He introduced the practice of representing the valence electrons as dots, placed around the symbol one at a time until they are used up or until all four sides are occupied; any

TABLE 8.5 Lewis Dot Symbols for Atoms

1A ns^1	2A ns^2	3A ns^2np^1	4A ns^2np^2	5A ns^2np^3	6A ns^2np^4	7A ns^2np^5	8A ns^2np^6
Li ·	· Be ·	· B ·	· C ·	· N ·	: O ·	: F ·	: Ne :
Na ·	· Mg ·	· Al ·	· Si ·	· P ·	: S ·	: Cl ·	: Ar :

remaining electrons are paired with ones already there. The result in a **Lewis dot symbol.** Table 8.5 shows Lewis dot symbols for the atoms of the elements in periods 2 and 3.

Main group elements in Groups 1A and 2A are known as ***s*-block elements** (see Figure 8.14) and their valence electrons are *s* electrons. The valence electron configurations for these elements are represented by $ns^1$and ns^2, respectively.

Elements in the main groups at the right in the periodic table, Groups 3A through 8A, are known as ***p*-block elements** (see Figure 8.14). Their valence electrons include outermost *s* and *p* electrons. Notice in Table 8.5 how the Lewis dot symbols show that *in each group the number of valence electrons is equal to the group number.*

Exercise 8.10 Electron Configurations of Two Elements in the Same Family

Using the noble gas notation, write the electron configurations for Se and Te. What do these configurations illustrate about elements in the same main group?

Ion Electron Configurations

In studying ionic compounds (*← p. 89*) you learned that atoms from Groups 1A through 3A form positive ions with charges equal to their group numbers and that those nonmetals in Groups 4A through 7A that form ions do so by adding electrons to form negative ions with charges equal to eight minus the group number. Here's the explanation. When ions form from atoms in *s*- and *p*-block elements, electrons are removed or added so that a noble gas configuration is achieved. Atoms from Groups 1A, 2A, and 3A lose 1, 2, or 3 electrons to form 1+, 2+, or 3+ ions, respectively. Atoms from Groups 7A, 6A, and some in 5A gain 1, 2, or 3 electrons to form 1−, 2−, or 3− ions. *Metals form cations with a charge equal to the group number, and nonmetals form anions with a charge equal to the group number minus eight.*

Exercise 8.11 Dot Structures

Lewis dot structures show only the valence electrons, but try drawing a structure that includes all the valence electrons *and* all the core electrons in the next lowest shell. What would this dot structure look like for the Na^+ ion? What element has the same Lewis dot structure as the Na^+ ion? Do this same exercise for an O atom and an O^{2-} ion.

Ions with identical electron configurations are said to be **isoelectronic.** Table 8.6 lists some isoelectronic ions and the noble gas that has the same electron configura-

TABLE 8.6 Noble Gas Atoms and Isoelectronic Ions

(all $1s^2$)	He, Li^+, Be^{2+}, H^-
(all $1s^22s^22p^6$)	Ne, Na^+, Mg^{2+}, Al^{3+}, F^-, O^{2-}
(all $1s^22s^22p^63s^23p^6$)	Ar, K^+, Ca^{2+}, Ga^{3+}, Cl^-, S^{2-}
(all $1s^22s^22p^63s^23p^64s^24p^6$)	Kr, Rb^+, Sr^{2+}, Br^-, Se^{2-}
(all $1s^22s^22p^63s^23p^63d^{10}4s^24p^64d^{10}5s^25p^6$)	Xe, Cs^+, Ba^{2+}, I^-, Te^{2-}

tion. Note from the table that metal ions are isoelectronic with the *preceding* noble gas, while nonmetal ions have the electron configuration of the *next* noble gas.

PROBLEM-SOLVING EXAMPLE 8.4 Main Group Ions

Give the valence electron configuration for S, and predict the charge on an ion formed from an S atom. What noble gas is isoelectronic with the ion?

Answer $3s^23p^4$; $S + 2e^- \longrightarrow S^{2-}$; Ar

Explanation S is a Group 6A element in the third period, so its valence electron configuration is $3s^23p^4$. An S atom needs two electrons to achieve the configuration of the nearest noble gas, Ar. As a result, the stable ion expected for S is S^{2-}.

Problem-Solving Practice 8.4

(a) Use the periodic table to write the general electron configuration for Group 3A atoms.
(b) Predict the charge of the ion formed from a Ga atom.
(c) What is the electron configuration for the Ga ion you predicted in part (b)?

Exercise 8.12 Electron Configurations of Ions

Write the electron configurations for the following ions: (a) Cl^-, (b) O^{2-}, (c) N^{3-}, (d) Se^{2-}.

Electron Configurations of Transition Elements

The elements of the B groups (3B-8B and 1B and 2B) in the middle of the periodic table—the *transition elements*—are metallic elements in which a *d* subshell is being filled (the *d*-block elements in Figure 8.14 (←, *p. 323*)).

In each period in which they occur, the transition elements are immediately preceded by two *s*-block elements. As shown in Figure 8.14, once the 4*s* subshell is filled, the subshell with the next higher energy is 3*d* (not 4*p*, as you might expect). The use of *d* orbitals begins with the first transition metal, scandium, which has the configuration $[Ar]3d^14s^2$. After scandium comes titanium with $[Ar]3d^24s^2$, and vanadium with $[Ar]3d^34s^2$. We would expect the configuration of the next element, chromium, to be $[Ar]3d^44s^2$, but that turns out to be incorrect. The correct configuration is $[Ar]3d^54s^1$, based on spectroscopic and magnetic measurements. This illustrates one of several anomalies in predicting electron configurations from transition and inner transition elements. When half-filled *d* subshells are possible, they are sometimes fa-

A complete explanation of anomolies like these is actually more complex than this, but is beyond the scope of this text.

vored (an illustration of Hund's rule). As a result, elements that are close to the middle and the end of filling a d subshell will have a stable configuration that leaves the s orbital half-filled and the d orbitals half-filled or filled. This kind of electron configuration is seen in copper, the next-to-last element in the first transition series, which has the electron configuration $[Ar]3d^{10}4s^1$ instead of the expected $[Ar]3d^9 4s^2$.

The number of unpaired electrons for atoms of most transition elements can be predicted by placing valence electrons in orbital diagrams according to Hund's rule. The electron configuration of Co ($Z = 27$) is $[Ar]3d^7 4s^2$, and the number of unpaired electrons is three, as seen from the following orbital box diagram.

	3d	4s
Co [Ar]	↑↓ ↑↓ ↑ ↑ ↑	↑↓

Exercise 8.13 Unpaired Electrons

Which ground state configuration for chromium has the larger number of unpaired electrons, $[Ar]3d^4 4s^2$ or $[Ar]3d^5 4s^1$?

Exercise 8.14 Cu^+ Ion

Using the electron configuration for copper, explain why copper readily forms the ion Cu^+.

The lanthanides and actinides are placed at the bottom of the periodic table to avoid making the table too wide.

In the sixth and seventh periods, f subshell orbitals exist and can be filled. The elements (all metals) for which f subshells are filling are sometimes called the *inner transition* elements or, more usually, *lanthanides* (after lanthanum, the element just before those filling the $4f$ subshell) and *actinides* (after actinium, the element just before those filling the $5f$ subshell). The lanthanides start with lanthanum (La), which has the electron configuration $[Xe]5d^1 6s^2$. The next element, cerium (Ce), begins a separate row at the bottom of the periodic table, and it is with these elements that f orbitals come into play (Figure 8.14 and Table 8.4). The electron configuration of Ce is $[Xe]4f^1 5d^1 6s^2$. Each of the lanthanide elements, from Ce to Lu, continue to add $4f$ electrons until the seven $4f$ orbitals are completely filled by 14 electrons in lutetium, (Lu, $[Xe]4f^{14}5d^1 6s^2$). Note that both the $n = 5$ and $n = 6$ levels are partially filled before the $4f$ starts to be occupied.

The importance of the periodic table as a guide to electron configurations cannot be overemphasized. Using just Figure 8.14 as a guide, let's find the configuration of Se, which is in Group 6A. You immediately know that it has *six* valence electrons with a configuration of ns^2np^4. And because Se is in the fourth period, $n = 4$. Thus, the complete electron configuration is given by starting with the electron configuration of argon $[1s^2 2s^2 2p^6 3s^2 3p^6]$, the noble gas at the end of the preceding $n = 3$ period. Then add the filled $3d^{10}$ subshell and the six valence electrons of Se ($4s^2 4p^4$) to give $1s^2 2s^2 2p^6 3s^2 3p^6 3d^{10} 4s^2 4p^4$, or $[Ar]3d^{10}4s^2 4p^4$. If you want to predict the number of unpaired electrons, you only need to look at the outermost subshells electron configuration, because the inner shells (represented by [Ar] in this case) are completely filled and all electrons in the inner shells are paired. For Se, $[Ar]3d^{10}4s^2 4p_x^2 4p_y^1 4p_z^1$ indicates two p orbitals with unpaired electrons for a total of two unpaired electrons.

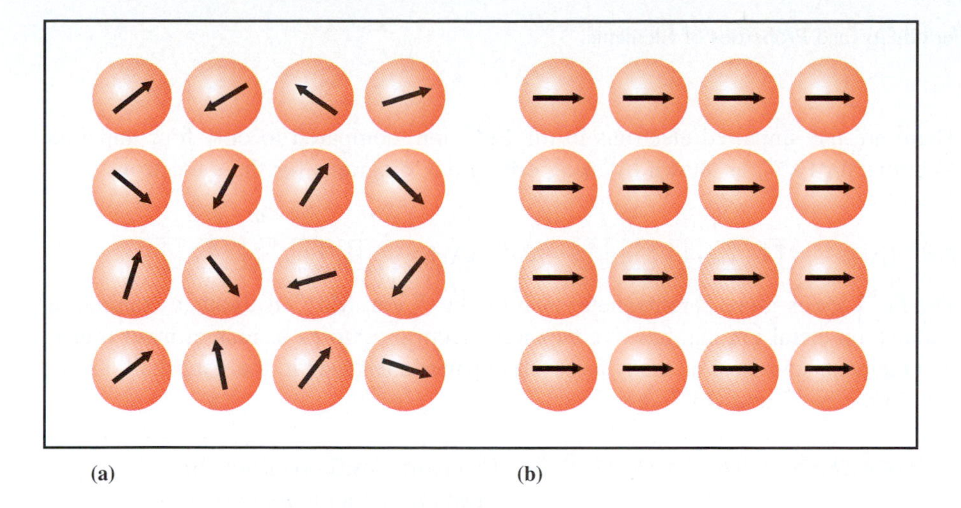

Figure 8.15 Types of magnetic behavior. (a) Parmagnetism: the centers (atoms or ions) with magnetic moments are not aligned unless the substance is in a magnetic field. (b) Ferromagnetism: the spins of unpaired electrtons in clusters of atoms or ions aligned in the same direction.

Exercise 8.15 Electron Configuration Practice

Select two elements from each of the main groups of the periodic table and practice writing their electron configurations and orbital diagrams. Use Figure 8.14 as a guide.

Paramagnetism and Unpaired Electrons

The magnetic properties of a spinning electron were described in Section 8.4. In atoms and ions that have filled shells, all the electrons are paired and their magnetic fields effectively cancel each other. Such substances are called **diamagnetic.** Atoms or ions with unpaired electrons are attracted to a magnetic field; the more unpaired electrons, the greater is the attraction. Such substances are called **paramagnetic.**

Substances that retain their magnetism (are permanent magnets) are **ferromagnetic.** The magnetic effect in ferromagnetic materials is much larger than for paramagnetic materials. Ferromagnetism occurs when the spins of unpaired electrons in a cluster of atoms (called a domain) in the solid align themselves in the same direction (Figure 8.15). Only the metals of the iron, cobalt, and nickel subgroups in the periodic table exhibit this property. They are also unique in that, once the domains are aligned in a magnetic field, the metal is permanently magnetized. In such a case the magnetism can be eliminated only by heating or shaping the metal to rearrange the electron spin domains. Many alloys exhibit greater ferromagnetism than do the pure metals themselves. Some metal oxides, like CrO_2 and Fe_3O_4, are also ferromagnetic and are used in magnetic recording tape (Figure 8.16).

The difference between paramagnetic and ferromagnetic materials is that domains of aligned electron spins do not form in paramagnetic substances.

Transition Metal Ions

The closeness in energy of the 4*s* and 3*d* subshells was mentioned earlier in connection with the electron configurations of transition metal atoms. The 5*s* and 4*d* subshells, the 6*s*, 4*f*, and 5*d*, subshells, and the 7*s*, 5*f*, and 6*d* subshells are also close to each other in energy. The *ns* and $(n - 1)d$ subshells are so close in energy that once *d* electrons are added, the $(n - 1)d$ subshell becomes slightly lower in energy than the *ns* subshell. As a result, the *ns* electrons are at higher energy and are always removed first when transition and inner transition metals form cations.

For example, an Fe atom will lose its two 4*s* electrons to form an Fe^{2+} ion,

$$\mathrm{Fe}[\mathrm{Ar}]3d^6 4s^2 \longrightarrow \mathrm{Fe}^{2+}[\mathrm{Ar}]3d^6 + 2\mathrm{e}^-$$

and then Fe^{3+} is formed from Fe^{2+} by loss of a 3*d* electron.

$$\mathrm{Fe}^{2+}[\mathrm{Ar}]3d^6 \longrightarrow \mathrm{Fe}^{3+}[\mathrm{Ar}]3d^5 + \mathrm{e}^-$$

Figure 8.16 Several products that use magnetic recording tape. *(C.D. Winters)*

There are five unpaired electrons in the Fe^{3+} ion, compared to only four unpaired electrons for Fe^{2+}, as shown by using orbital box diagrams for the *d* electrons:

Fe^{3+} [Ar] $3d$: [↑] [↑] [↑] [↑] [↑] Fe^{2+} [Ar] $3d$: [↑↓] [↑] [↑] [↑] [↑]

The Fe^{3+} ion is more paramagnetic than the Fe^{2+} ion. Mn^{2+} is another example of a transition metal ion that has five unpaired electrons. Atoms or ions of inner transition elements can have as many as seven unpaired electrons in the *f* subshell, as occurs in the Eu^{2+} and Gd^{3+} ions.

PROBLEM-SOLVING EXAMPLE 8.5 Electron Configurations for Transition Elements and Ions

(a) Write the electron configuration for the Co atom, using the noble gas notation. Then draw the orbital box diagram for the electrons beyond the preceding noble gas configuration.
(b) Cobalt commonly exists as 2+ and 3+ ions. How does the orbital box diagram given in part (a) have to be changed to represent the outer electrons of Co^{2+} and Co^{3+}?
(c) How many unpaired electrons do Co, Co^{2+}, and Co^{3+} have?

Answer

(a) $[Ar]3d^7 4s^2$, [Ar] $3d$: [↑↓] [↑↓] [↑] [↑] [↑] $4s$: [↑↓]

(b) For Co^{2+}, remove the two $4s$ electrons from Co to give

[Ar] $3d$: [↑↓] [↑↓] [↑] [↑] [↑]

For Co^{3+}, remove the two $4s$ electrons and one of the $3d$ electrons from Co to give

[Ar] $3d$: [↑↓] [↑] [↑] [↑] [↑]

(c) Co, three unpaired electrons; Co^{2+}, three unpaired electrons; Co^{3+}, four unpaired electrons.

Explanation
(a) Co is in the fourth period with an atomic number of 27. It has nine more electrons than Ar, with two of the nine in the $4s$ subshell and seven in the $3d$ subshell, so its electron configuration is $[Ar]3d^7 4s^2$. For the orbital box diagram, all *d* subshells get one electron before pairing (Hund's rule).
(b) To form Co^{2+}, two electrons are removed from the $4s$ subshell. To form Co^{3+}, one of the paired electrons is removed from a *d* subshell.
(c) Looking at the diagrams, you should see that the numbers of unpaired electrons are three for Co, three for Co^{2+}, and four for Co^{3+}.

Problem-Solving Practice 8.5

(a) Write the electron configuration for the nickel atom, using the noble gas notation.
(b) Draw an orbital box diagram for the outer electrons. How many unpaired electrons does nickel have?
(c) Draw an orbital box diagram for the Ni^{2+} ion. How many unpaired electrons does Ni^{2+} have?

Exercise 8.16 Unpaired Electrons

The compound 2,4-pentanedione, also known as acetylacetone (acac). In the presence of a base, acac forms an anion, $acac^-$, which has no paired electrons and forms neutral compounds with both Fe^{2+} and Fe^{3+} ions. Their formulas are $Fe(acac)_2$ and $Fe(acac)_3$, respectively. Which one will have the greater attraction to a magnetic field? Explain.

8.6 NUCLEAR MAGNETIC RESONANCE AND ITS APPLICATIONS

In the previous section, the magnetic field created by the spinning electron was mentioned in connection with the discussion of the quantized spin of electrons. The nuclei of certain isotopes have the same property. For example, a ^1H (the hydrogen nucleus or proton) nucleus can spin in one of two directions. In the absence of a magnetic field, the two spin states have the same energy. When a strong external magnetic field is applied, those ^1H nuclei spinning in such a way as to be aligned with the external magnetic field are slightly lower in energy than those that are aligned against the magnetic field. At first it might appear that this energy difference, ΔE, between the two nuclear spin states, would be the same for all of the hydrogen atoms in a molecule. This is not the case (← *p. 80*). Neighboring atoms of the hydrogen atoms in a molecule cause the hydrogens to see effective magnetic fields that are slightly different from the external magnetic field. However, the value of ΔE is small enough that radiation in the radio-frequency region of the electromagnetic spectrum (Figure 8.1) can change the direction of spin.

$$\text{Energy}\uparrow \quad \begin{matrix} \text{———} & -1/2 \\ \text{—}\uparrow\text{—} & +1/2 \end{matrix} \xrightarrow{+h\nu} \begin{matrix} \text{—}\downarrow\text{—} & -1/2 \\ \text{———} & +1/2 \end{matrix} \xrightarrow{-h\nu} \begin{matrix} \text{———} & -1/2 \\ \text{—}\uparrow\text{—} & +1/2 \end{matrix}$$

Picture the aligned hydrogen nuclei absorbing a particular radio frequency that changes their spin to the less stable direction. When the nuclei return to the more stable spin direction, the same radio frequency is emitted and can be measured with a radio receiver. This phenomenon is known as **nuclear magnetic resonance (NMR),** and it provides a tool for studying molecular structure.

NMR is used extensively by chemists because the radio frequency absorbed and then emitted depends on the chemical environment of the atoms with nuclear spin states in the sample. The study of hydrogen atoms with NMR (known as *proton* nuclear magnetic resonance, ^1H NMR) has quickly developed into an indispensable structural and analytical tool, particularly for organic chemists. Plots of the intensity of energy absorption versus the magnetic field strength are called *NMR spectra;* and from them chemists deduce the kinds of atoms bonded to hydrogen as well as the number of hydrogen atoms present in a molecule. A schematic diagram of the important components of an NMR spectrometer is shown in Figure 8.17.

Felix Bloch and Edward Purcell were awarded the Nobel Prize in 1952 for their discovery of nuclear magnetic resonance.

Because hydrogen is the most common element in the body and H atoms give a strong NMR signal, ^1H is the most logical candidate for the application of NMR to medical imaging. The first use of NMR for this purpose was reported in 1973. The delay between the discovery of NMR and its application in medicine was related to the technical difficulties associated with getting a uniform magnetic field with a di-

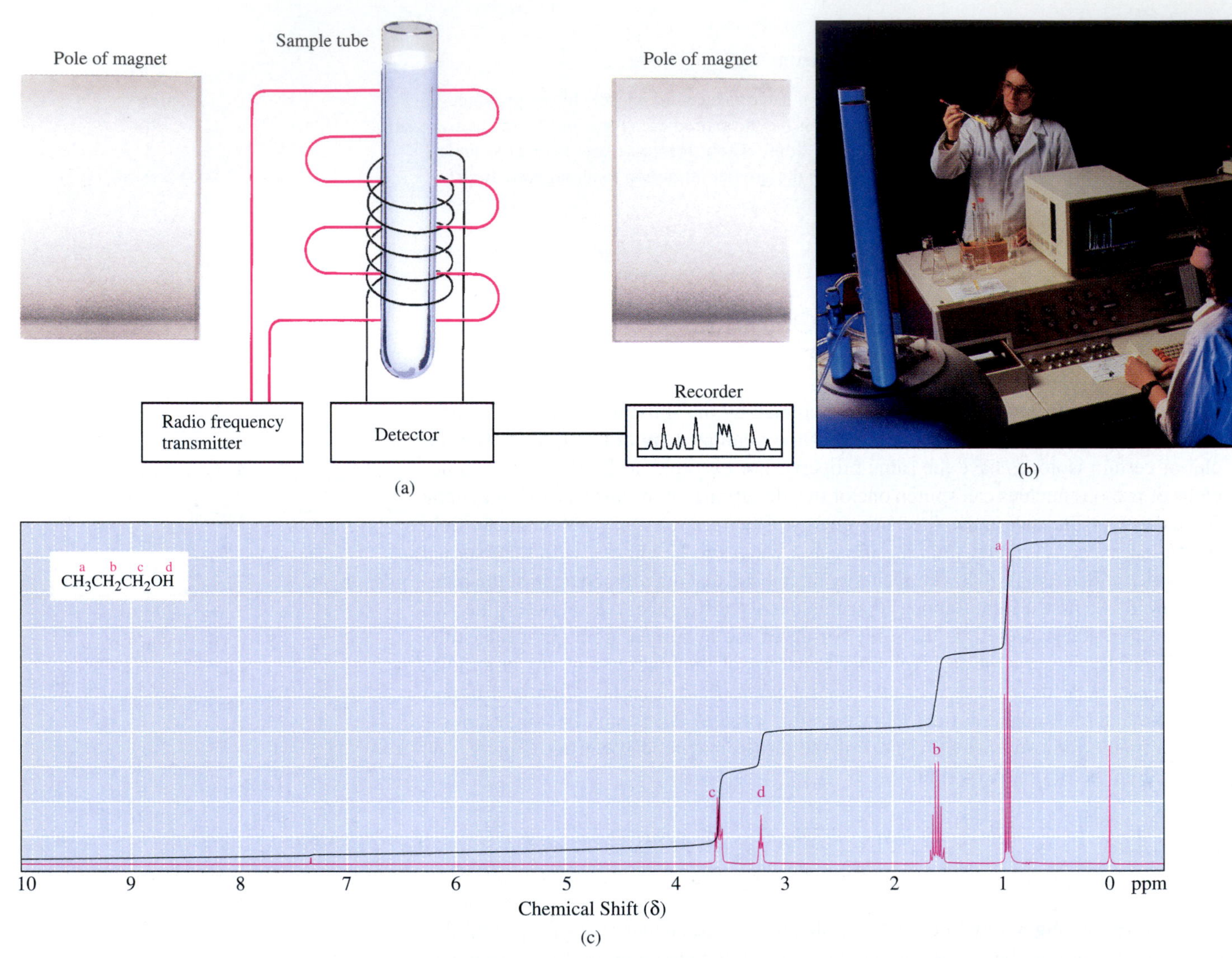

Figure 8.17 **The NMR technique.** (a) Schematic diagram of an NMR spectrometer. (b) A modern NMR spectrometer is a highly automated instrument controlled by a built-in computer. (c) An NMR spectrum of propanol showing the four kinds of protons present in the molecule. *(b, General Electric)*

The use of NMR in medical imaging is now referred to as Magnetic Resonance Imaging or MRI. The name "nuclear" was dropped because of its association with weapons and radioactivity which frightened some people.

ameter large enough to enclose a patient. In addition, advances in computer technology for analysis of data and construction of an image from these data were needed.

NMR imaging in medicine is based on the time it takes for the protons (hydrogen nuclei) in the unstable high-energy nuclear spin position to "relax" or return to the low-energy nuclear spin position. These *relaxation times* are different for protons in fat (← ***p. 107***), muscle (***proteins***, Section 11.10), blood, and bone, because of differences in the chemical environments. These differences in relaxation times are enhanced by the computer to produce the magnetic resonance image.

In magnetic resonance imaging (MRI), the patient is placed in an opening of a large magnet (Figure 8.18). The magnetic field aligns the magnetic spin of the protons (as well as those of other magnetic nuclei). A radio-frequency transmitter coil is placed in position near the region of the body to be examined. The radio-frequency energy absorbed by the spinning hydrogen nuclei causes the aligned nuclei to flip to

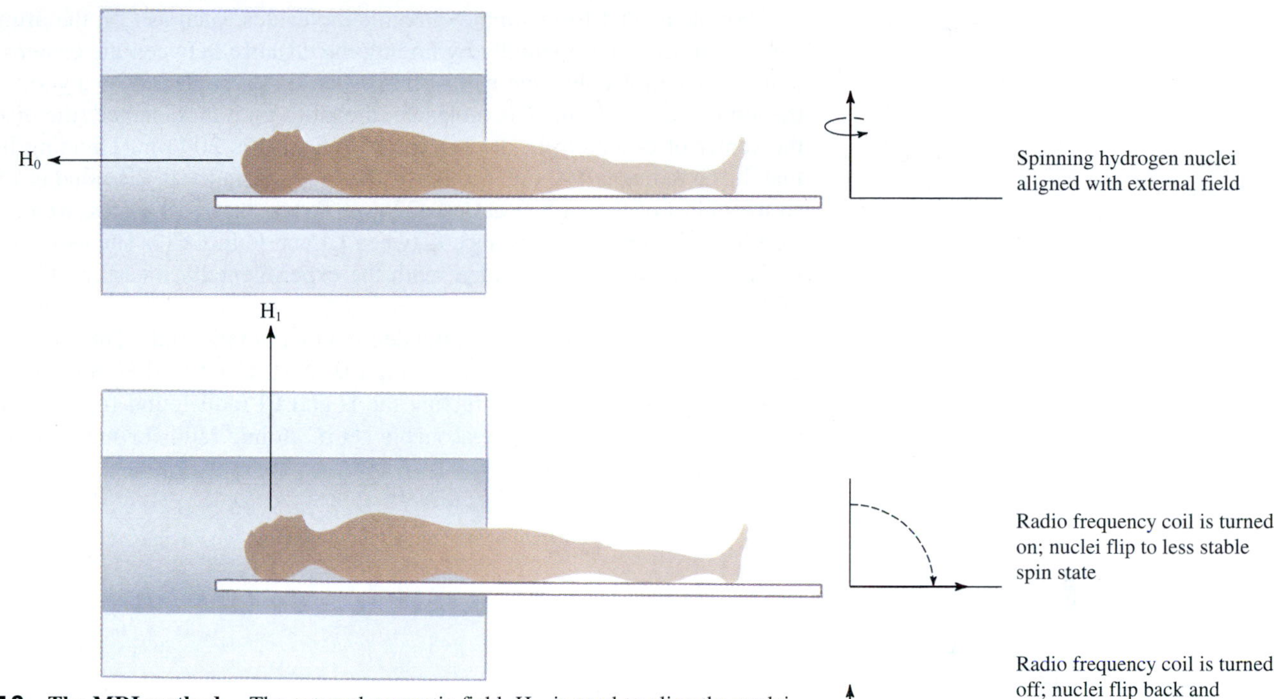
H_0
H_1
Spinning hydrogen nuclei aligned with external field
Radio frequency coil is turned on; nuclei flip to less stable spin state
Radio frequency coil is turned off; nuclei flip back and emit a radio frequency signal; time to flip back is referred to as the relaxation time

Figure 8.18 The MRI method. The external magnetic field, H_0, is used to align the nuclei. The radio-frequency coil is turned on and the resulting radio frequency (H_1) flips the nuclei to the less stable spin state. Radio-frequency photons are generated when the nuclei flip back to the more stable spin state. These photons make up a radio-frequency signal that can be detected by a receiver.

the less stable, high-energy spin direction. When the nuclei flip back to the more stable spin state, a photon of radio-frequency electromagnetic radiation is emitted. The intensity of the emitted signal is related to the density of hydrogen nuclei in the region being examined, and the time it takes for the signal to be emitted (relaxation time) is related to the type of tissue. The emitted radio-frequency signal is received by a radio receiver coil, which then sends it to a computer for mathematical construction of the image. MRI can readily distinguish brain tumors from normal brain tissue, not only on the basis of altered anatomy but also on the basis of high water content, which may be caused by abnormal growth of blood vessels or excessive cellular absorption of water.

Are the photons associated with the radio energy in magnetic resonance energetic enough to cause cellular damage? No; a radio-energy photon has only about 10^{-7} times the energy required to break a single chemical bond, so it is unlikely it would cause bonds to break and other cellular damage, as do x-ray and uv photons.

Exercise 8.17 Energies of Radio Photons in the MRI Experiment

Estimate the energies of radio-wave photons (say, 100 MHz), and compare those with the energies required to break chemical bonds. Do the same for x-ray photons. Do you see why it is safe to place a person's entire body inside an MRI diagnostic instrument?

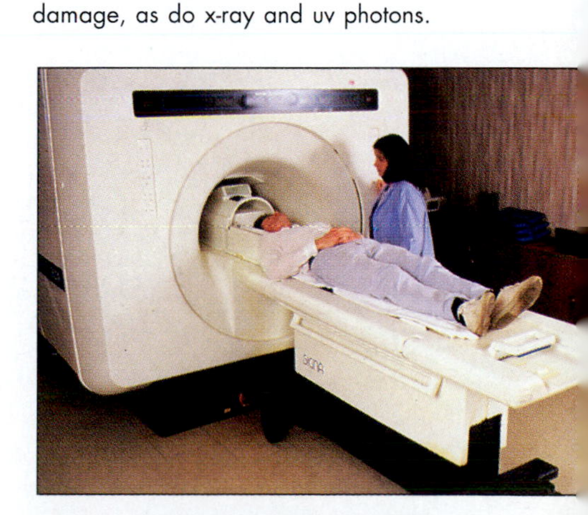

A magnetic resonance imaging (MRI) machine. The patient is placed on a platform and rolled into the magnet opening. *(Peter Arnold)*

8.7 PERIODIC TRENDS: ATOMIC RADII

Based on knowledge of electron configurations, we can now begin to answer fundamental questions of why atoms of different elements fit in as they do in the periodic table as well as what causes trends we observe among elements in the table.

For atoms that form simple diatomic molecules, such as Cl_2, the **atomic radius** can be defined experimentally by finding the distance between the centers of the two atoms in the molecule. One half of this distance is assumed to be a good estimate of the atom's radius. In the Cl_2 molecule, the atom-to-atom distance (the distance from the center of one atom to the center of the other) is 200 pm. Dividing by 2 shows that the Cl radius is 100 pm. Similarly, the C—C distance in diamond is 154 pm, and so the radius of the carbon atom is 77 pm. To test these estimates, we can add them together to estimate the distance between Cl and C in CCl_4. The estimated distance of 177 pm is in good agreement with the experimentally measured C—Cl distance of 176 pm.

This approach can then be extended to other atomic radii. The radii of O, C, and S atoms can be estimated by measuring the O–H, C–Cl, and H–S distances in H_2O, CCl_4, and H_2S, and then subtracting the H and Cl radii found from H_2 and Cl_2. By this and other techniques, a reasonable set of atomic radii for main group elements has been assembled (Figure 8.19).

Orbitals represent the probable locations of electrons in atoms and have no sharp boundaries. However, throughout this section atoms and ions are drawn as well-defined spheres to make trends more easily apparent. As you look at these, keep in mind the "fuzzy" nature of the outer regions of atoms and ions.

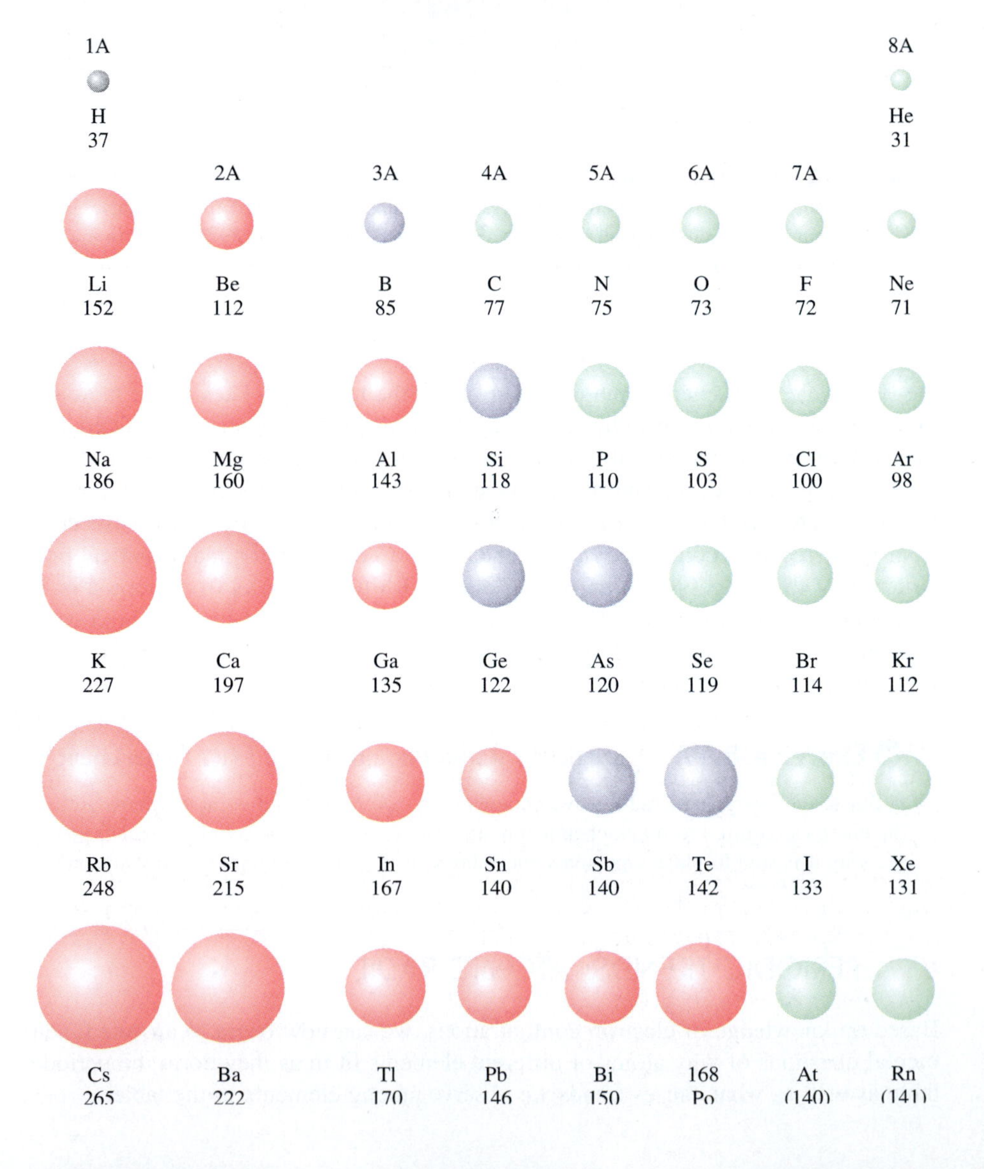

Figure 8.19 **Atomic radii of the main group elements** (in picometers, 1 pm = 10^{-12} m).

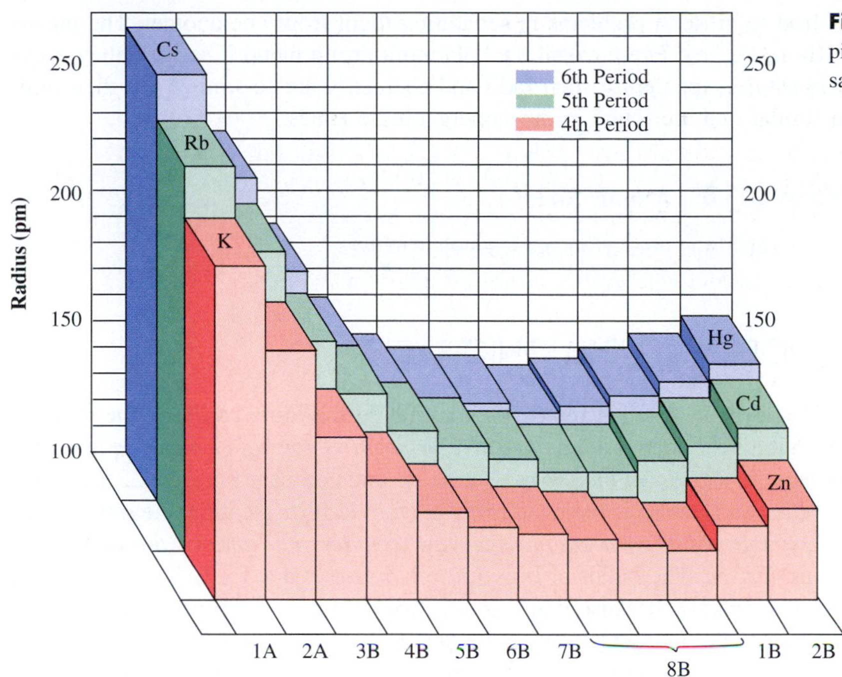

Figure 8.20 Atomic radii of the transition metals (in picometers, 1 pm = 10^{-12} m). The *s*-block metals of the same periods are included to show trends across the periods.

For the main group elements, atomic radii increase going down a group in the periodic table and decrease going across a period. These trends reflect two important effects: (1) In going from the top to the bottom of a group in the periodic table, electrons are assigned to orbitals that are successively farther from the nucleus, and as a result the atomic radii increase. (2) For a given period, the principal quantum number *n* of the outermost orbitals stays the same. This means the radius of the orbitals to which the electrons are assigned would be expected to remain approximately constant. However, with the addition of each successive electron the nuclear charge has also increased by the addition of one proton. The result is that attraction between the nucleus and electrons increases, and, because this attraction is somewhat stronger than the increasing repulsion between electrons, the atomic radius decreases. Note the large increase in atomic radius in going from any noble gas atom to the following Group 1A atom, where the outermost electron is assigned to the next higher energy level.

The periodic trend in the atomic radii of transition metal atoms is illustrated in Figure 8.20. You can see that the sizes of the transition metal atoms change very little across a period, especially beginning at Group 5B (V, Nb, or Ta), which occurs because they are all determined by the radius of an *ns* orbital ($n = 4$, 5, or 6), occupied by at least one electron. The variation in the number of electrons occurs instead in the $(n-1)d$ orbitals. As the number of electrons in these $(n-1)d$ orbitals increases, they increasingly repel the *ns* electrons, partly compensating for the increased nuclear charge across the periods. Consequently, the *ns* electrons experience only a slightly increasing nuclear attraction, and the radii remain nearly constant until the slight rise at Groups 1B and 2B due to the continually increasing electron-to-electron repulsions as the *d* subshell is filled.

The similar radii of the transition metals and their ions have an important effect on their chemistry—they tend to be more alike in their properties than other elements in the same groups. The nearly identical radii of the fifth- and sixth-period transition elements lead to difficult problems in separating them from one another. The metals Ru, Os, Rh, Ir, Pd, and Pt are called the "platinum group metals" because they occur together in nature. Apparently, their radii and chemistry are so similar that their minerals are similar and are found in the same geologic zones.

Exercise 8.18 Atomic Sizes

Place Al, C, and Si in order of increasing atomic radius.

8.8 PERIODIC TRENDS: IONIC RADII

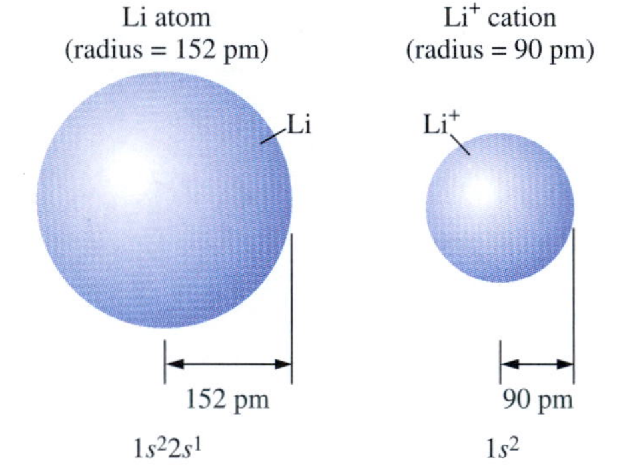

Figure 8.21 shows clearly that the periodic trends in the **ionic radii** are the same as the trends in radii for neutral atoms: *positive or negative ions of elements in the same group increase in size down the group.* But pause and compare Figure 8.22 with Figure 8.20. When an electron is removed from an atom to form a cation, the size shrinks considerably; *the radius of a cation is always smaller than that of the atom from which it was derived.* The radius of Li is 152 pm, whereas that for Li^+ is only 90 pm. This is understandable, because when an electron is removed, there are fewer electrons repelling each other and the positive nucleus can attract them closer; the electrons contract toward the nucleus. The decrease in ion size is especially great when the electron removed comes from a higher energy level than the new outer electron. This is the case for Li, for which the "old" outer electron was from a 2*s* orbital and the "new" outer electron is in a 1*s* orbital.

The shrinkage is also great when two or more electrons are removed; for Al^{3+},

$$\text{Al atom (radius} = 143\text{ pm)} \quad 1s^22s^22p^63s^23p^1 \qquad Al^{3+}\text{ cation (radius} = 68\text{ pm)} \quad 1s^22s^22p^6$$

You can also see by comparing Figures 8.21 and 8.19 that *anions are always larger than the atoms from which they are derived.* Here the argument is the oppo-

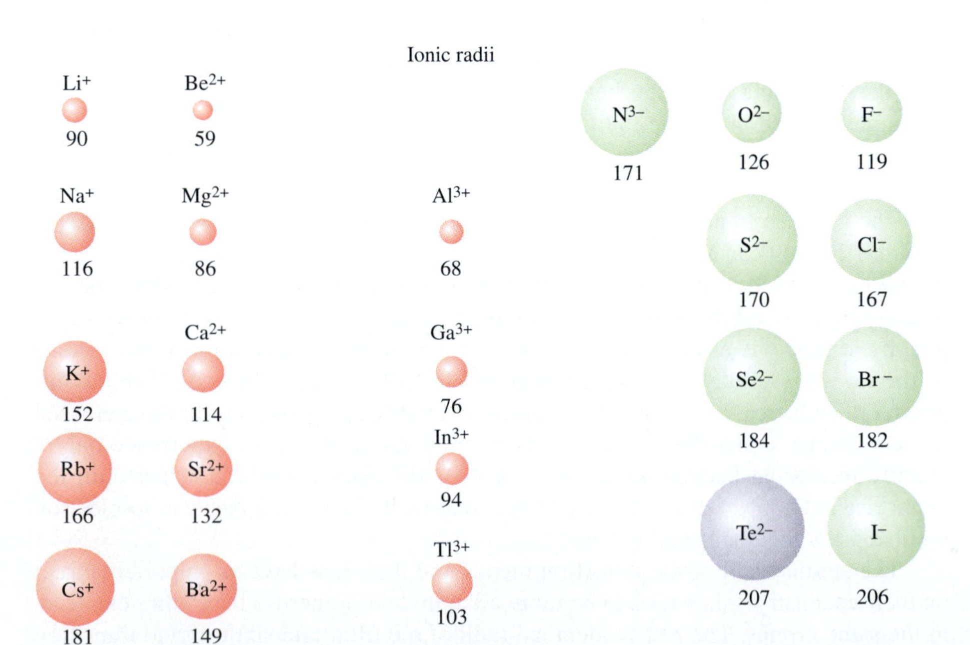

Figure 8.21 Relative sizes of some common ions. Radii are given in picometers (1 pm = 10^{-12} m).

site of that used to explain the radii of positive ions: adding an electron introduces new repulsions and the shells swell. The F atom has nine protons and nine electrons. When it forms the F^- anion, the nuclear charge is still 9+, but there are now ten electrons in the anion. The F^- ion (119 pm) is much larger than the F atom (72 pm) because of increased electron-to-electron repulsions.

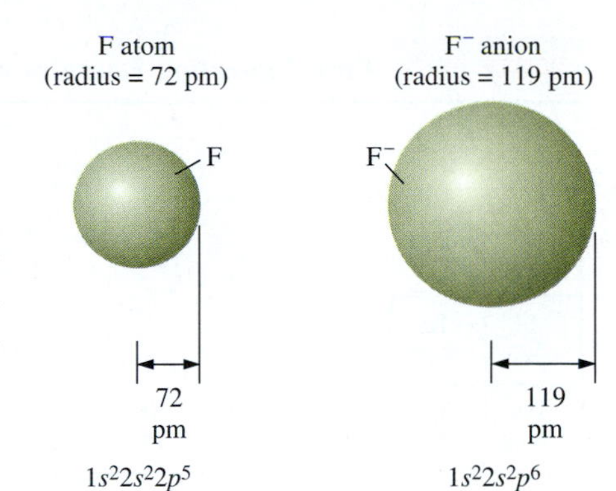

The oxide ion, O^{2-}, is isoelectronic with F^-, that is, they both have the same electron configuration (the neon configuration). However, the oxide ion is larger than the fluoride ion because the oxide ion has only eight protons available to attract ten electrons, whereas F^- has more protons (nine) to attract the same number of electrons. It is useful to compare the sizes of isoelectronic ions across the periodic table. Consider O^{2-}, F^-, Na^+, and Mg^{2+}.

Ion	O^{2-}	F^-	Na^+	Mg^{2+}
Ionic radius (pm)	126	119	116	86
Number of protons	8	9	11	12
Number of electrons	10	10	10	10

Each ion contains ten electrons. However, the O^{2-} ion has only 8 protons in its nucleus to attract these electrons, while F^- has 9, Na^+ has 11, and Mg^{2+} has 12. As the proton/electron ratio increases in a series of isoelectronic ions, the balance in electron-to-proton attraction and electron-to-electron repulsion shifts in favor of attraction, and the ion shrinks. As you can see in Figure 8.21, this is true for all isoelectronic series of ions.

Exercise 8.19 Trends in Ionic Sizes

What is the trend in sizes of the ions P^{3-}, S^{2-}, and Cl^-? Briefly explain why this trend exists.

8.9 PERIODIC TRENDS: IONIZATION ENERGIES

The **ionization energy** of an atom is the energy needed to remove an electron from that atom in the gas phase. For a gaseous sodium atom, the ionization process would be

$$Na(g) + \text{energy} \longrightarrow Na^+(g) + e^-$$

Energy is always required to remove an electron, so the process is endothermic and the sign of the ionization energy is always positive. For *s*- and *p*-block elements, *first ionization energies (the energy needed to remove one electron from the neutral atom) generally increase across a period and decrease down a group* (Figure 8.22 and Table 8.7). The decrease down a group reflects the increasing radii of the atoms—it is easier to remove an electron from a larger atom. Similarly, the increase across a period occurs for the same reason that radii decrease across a period. The trend across a given period is not smooth, however, particularly in the second period. A Group 3A element (ns^2np^1) has a smaller ionization energy than the preceding Group 2A element (ns^2). This difference indicates that the single *np* electron of the Group 3A element is more easily removed than one for the *ns* electrons in the preceding Group 2A element. Another deviation occurs for Group 6A elements (ns^2np^4), which have

TABLE 8.7 First Ionization Energies of the Elements (kJ/mol)

1A (1)	2A (2)	3B (3)	4B (4)	5B (5)	6B (6)	7B (7)	8B (8, 9, 10)			1B (11)	2B (12)	3A (13)	4A (14)	5A (15)	6A (16)	7A (17)	8A (18)
H 1312																	He 2371
Li 520	Be 899											B 801	C 1086	N 1402	O 1314	F 1681	Ne 2081
Na 496	Mg 738											Al 578	Si 786	P 1012	S 1000	Cl 1251	Ar 1521
K 419	Ca 599	Sc 631	Ti 658	V 650	Cr 652	Mn 717	Fe 759	Co 758	Ni 757	Cu 745	Zn 906	Ga 579	Ge 762	As 947	Se 941	Br 1140	Kr 1351
Rb 403	Sr 550	Y 617	Zr 661	Nb 664	Mo 685	Tc 702	Ru 711	Rh 720	Pd 804	Ag 731	Cd 868	In 558	Sn 709	Sb 834	Te 869	I 1008	Xe 1170
Cs 377	Ba 503	La 538	Hf 681	Ta 761	W 770	Re 760	Os 840	Ir 880	Pt 870	Au 890	Hg 1007	Tl 589	Pb 715	Bi 703	Po 812	At 890	Rn 1037

The data in this table are taken from *The Periodic Table Live!* a computer database of information about the elements. The program is available for several types of computers through *Journal of Chemical Education: Software.* (See the preface for details.)

smaller ionization energies than the Group 5A elements that precede them. Beginning in Group 6A, two electrons are assigned to the same *p* orbital. Thus, greater electron repulsion is experienced by the fourth *p* electron and this makes it easier to remove.

Conceptual Challenge Problem CP-8.C at the end of the chapter relates to topics covered in this section.

For the transition and inner transition elements, ionization energies increase much more gradually across the period than for the main group elements (Table 8.7). Since the atomic radii of transition elements change very little across a period, the ionization energy for the removal of an *ns* electron also shows small changes.

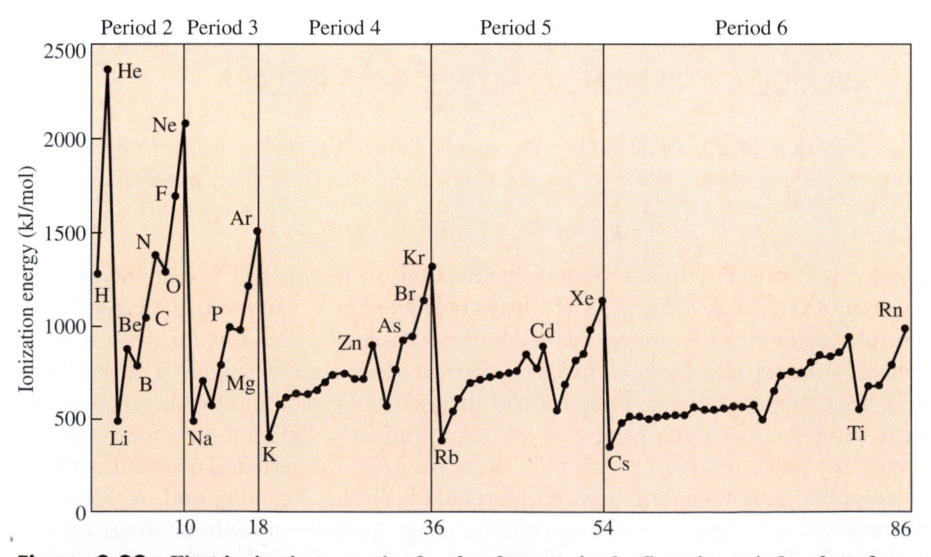

Figure 8.22 First ionization energies for the elements in the first six periods, plotted against atomic number.

TABLE 8.8 Energies Required to Remove Successive Electrons from Second-Period Atoms

10^6 J/mol	Li $1s^22s^1$	Be $1s^22s^2$	B $1s^22s^22p^1$	C $1s^22s^22p^2$	N $1s^22s^22p^3$	O $1s^22s^22p^4$	F $1s^22s^22p^5$	Ne $1s^22s^22p^6$
IE_1	0.52	0.90	0.80	1.09	1.40	1.31	1.68	2.08
IE_2	7.30	1.76	2.43	2.35	2.86	3.39	3.37	3.95
IE_3	11.81	14.85	3.66	4.62	4.58	5.30	6.05	6.12
IE_4		21.01	25.02	6.22	7.48	7.47	8.41	9.37
IE_5			32.82	37.83	9.44	10.98	11.02	12.18
IE_6				47.28	53.27	13.33	15.16	15.24
IE_7					64.37	71.33	17.87	20.00
IE_8			Core electrons			84.08	92.04	23.07
IE_9							106.43	115.38
IE_{10}								131.43

Note: In each of these elements the core electrons are those in the 1*s* orbital.

Each atom has a series of ionization energies, since more than one electron can always be removed (except for H). The first three ionization energies of Mg in the gaseous state are

$$\underset{1s^22s^22p^63s^2}{Mg(g)} + \text{energy} \longrightarrow \underset{1s^22s^22p^63s^1}{Mg^+(g)} + e^- \qquad IE_1 = 738 \text{ kJ/mol}$$

$$\underset{1s^22s^22p^63s^1}{Mg^+(g)} + \text{energy} \longrightarrow \underset{1s^22s^22p^6}{Mg^{2+}(g)} + e^- \qquad IE_2 = 1450 \text{ kJ/mol}$$

$$\underset{1s^22s^22p^6}{Mg^{2+}(g)} + \text{energy} \longrightarrow \underset{1s^22s^22p^5}{Mg^{3+}(g)} + e^- \qquad IE_3 = 7734 \text{ kJ/mol}$$

Notice that removing each subsequent electron requires more energy, and the jump from the second (IE_2) to the third (IE_3) ionization energy of Mg is particularly great. The first electron removed from the magnesium atom comes from the 3*s* orbital. The second ionization energy corresponds to removing a 3*s* electron from the Mg^+. As expected, the second ionization energy is higher than the first ionization energy because the electron is being removed from a positive ion, which strongly attracts the electron. The third ionization energy corresponds to removing a 2*p* electron from Mg^{2+}. The very great difference between the second and third ionization energies for Mg is excellent experimental evidence for the existence of electron shells in atoms. Table 8.8 gives the successive ionization energies of the second-period elements. Removal of the first core electron requires much more energy than removal of one of the valence electrons.

Exercise 8.20 Relative Ionization Energies

Arrange the following atoms in the order of increasing first ionization energy: F, Al, P, Mg, and Ba.

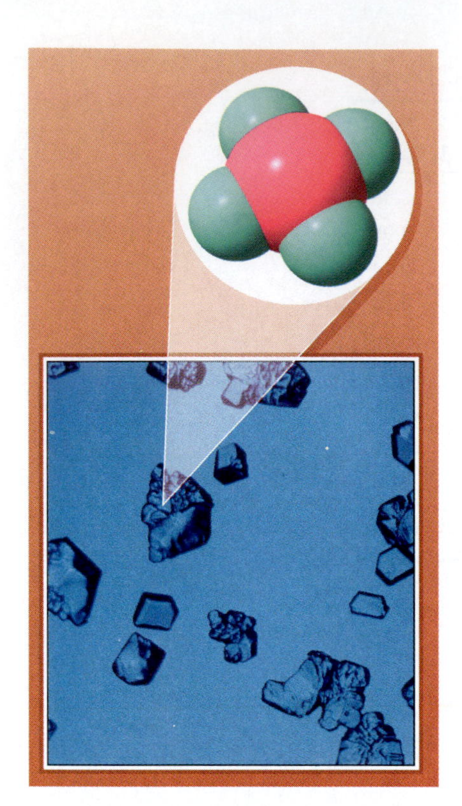

Crystals of xenon tetrafluoride. *(Argonne National Laboratory)*

Discovery by chance is sometimes attributed to serendipity, but Louis Pasteur observed that in science "chance only favors the prepared mind."

Exercise 8.21 Predicting Ionization Energies

Between which two ionization energies (second, third, fourth, etc.) would you expect the largest increase for a phosphorus atom?

Using Ionization Energies—The Discovery of Noble Gas Compounds

Due to their high ionization energies, helium and the other members of Group 8A were once considered to be completely inert to normal chemical attack by oxidizing agents, acids, and other reactants. Textbooks showed these elements, all gases, as being "noble"—unreactive. All that changed in 1962 when Neil Bartlett, while studying the reactions of PtF_6, noticed, quite by accident, that it reacted with oxygen to oxidize it and form a compound $[O_2^+][PtF_6^-]$. The formula for this compound indicates that the oxygen molecule has lost an electron to form the O_2^+ ion. Bartlett realized that the ionization energy of O_2 (1175 kJ/mol) is almost the same as the ionization energy for the noble gas xenon (1170 kJ/mol). He suspected that if an oxygen molecule could lose an electron to form O_2 when 1175 kJ/mol was added, then an Xe atom could so the same. When he tried the experiment, PtF_6 quickly reacted with Xe gas to form a red crystalline solid, which was assigned the formula $[Xe^+][PtF_6^-]$. (Later work indicated that the reaction at 25 °C gives a mixture of $[XeF^+][[PtF_6^-]$ and PtF_5, which combines when heated to 60 °C to give $[XeF^+][Pt_2F_{11}^-]$.)

What does this story about noble gas chemistry tell about periodic trends among the elements? Xenon's ionization energy is relatively low because it is one of the heavier members of Group 8A. From Table 8.7 notice how the ionization energies change in this group and the others. (Notice, too, that the ionization energy for O is not the same as what was used by Bartlett. He compared the ionization energy for the O_2 molecule.) It is a fortunate fact that the ionization energies of the O_2 molecule and Xe atoms just happen to be about the same. Bartlett's discovery caused the chemistry of the noble gases to be more closely investigated. As soon as word of his discovery was announced, chemists in other laboratories quickly discovered that xenon would even react directly with fluorine at room temperature in sunlight to form XeF_2.

$$Xe(g) + F_2(g) \longrightarrow XeF_2(s)$$

This discovery was particularly enlightening for chemists, because it made them realize that what they had so long believed to be so—that noble gases do not react to form compounds—was not true. It illustrated how careful use of knowledge of periodic properties of the elements can advance what we know about chemistry.

8.10 PERIODIC TRENDS IN PROPERTIES OF THE ELEMENTS: PERIODS 1 AND 2

As a chemistry student, you should look at the periodic table much as you would a map showing natural features and boundaries. Each element (a natural feature) has its place in the table and a relationship with its neighbors. There are also boundaries in the periodic table. A quick look at a periodic table that has the elements color coded for metallic, nonmetallic, and metalloid properties (see inside front cover) tells us that there are more metals than nonmetals. This is because more atoms have electron configurations with just a few valence electrons (which tend to be lost, a metallic property) than with a larger number of valence electrons. Yet, if you imagine for a moment a periodic table *without* any transition metals, the numbers of metals and nonmetals would be about the same! The metals all have configurations with small numbers of valence electrons and the nonmetals all have larger numbers of valence electrons. Some elements are like others in the periodic table and some are very different from others. There are periodic trends in atom size, ion size, and ionization energy, as you have seen.

You can learn to understand and appreciate these similarities and differences in properties by studying the elements' positions in the periodic table, which reflect the numbers of valence electrons in their atoms. Recall that for main group elements, the number of valence electrons is the same as the element's group number in the periodic table. These valence electrons affect how atoms of an element interact with one another (their physical properties), and how they interact with atoms of other elements (chemical properties).

The role of electrons in the formation of chemical compounds is discussed in Chapters 9 and 10.

As a direct result of the number of valence electrons and the sizes of their atoms, metals have low ionization energies and nonmetals have high ionization energies. This helps explain the usual formation of positive ions by metals and negative ions by nonmetals. Since ionization energy decreases down a group and generally increases across a period, the elements toward the bottom left-hand corner of the periodic table are the most active metals, meaning they react readily with many elements by the loss of electrons. Correspondingly, the elements at the top right of the periodic table are the most active nonmetals—they react easily by gaining electrons. Some main group elements between these two extremes can either gain or lose electrons to form ions in ionic compounds or share electrons in molecular compounds.

You have been introduced to the vast realm of organic molecules ***(← p. 84),*** all of which contain the element carbon, a nonmetal that forms millions of different molecular compounds. In Chapter 15, which deals with solids, you will see how the small numbers of loosely held valence electrons in metals atoms hold the atoms together to form metallic solids and determine the properties of metals. You will also learn more about the metalloids, elements with some metallic and some nonmetallic properties that fall in the periodic table between the metals on the left and the nonmetals on the right.

Elements in the main groups show a family resemblance within their group (like Li to Fr, or Be to Sr) because they all have the *same number* of valence electrons. Main group elements within a period (like Li through Ne) show major differences in their physical and chemical properties because they have *different numbers* of valence electrons (see Table 8.9).

To illustrate how properties of the elements vary across the main groups with numbers of valence electrons, the following sections describe elements in the first two periods.

H																	He
Li	Be											B	C	N	O	F	Ne
Na	Mg											Al	Si	P	S	Cl	Ar
K	Ca	Sc	Ti	V	Cr	Mn	Fe	Co	Ni	Cu	Zn	Ga	Ge	As	Se	Br	Kr
Rb	Sr	Y	Zr	Nb	Mo	Tc	Ru	Rh	Pd	Ag	Cd	In	Sn	Sb	Te	I	Xe
Cs	Ba	La	Hf	Ta	W	Re	Os	Ir	Pt	Au	Hg	Tl	Pb	Bi	Po	At	Rn

TABLE 8.9 Some Physical and Chemical Properties of the Elements Lithium to Neon

Atomic No.	Symbol	Valence Electrons	Melting Point (°C)	Boiling Point (°C)	Density (g/mL)	State	Oxide Formula	Chloride Formula
3	Li	$2s^1$	186	1326	0.534	Solid	Li_2O	LiCl
4	Be	$2s^2$	1283	2970	1.848	Solid	BeO	$BeCl_2$
5	B	$2s^22p^1$	2300	2550	2.34	Solid	B_2O_3	BCl_3
6	C	$2s^22p^2$	3570	Sublimes*	2.25	Solid	CO, CO_2	CCl_4
7	N	$2s^22p^3$	−210	−196	0.00125 (0° C)	Gas	N_2O, NO, NO_2†	NCl_3
8	O	$2s^22p^4$	−218	−183	0.00143 (0° C)	Gas	—	Cl_2O
9	F	$2s^22p^5$	−220	−188	0.00181 (0° C)	Gas	OF_2	CIF
10	Ne	$2s^22p^6$	−248	−246	0.0009 (0° C)	Gas	None	None

*Goes directly from solid to gas. See Section 15.3 for a discussion.
†Other oxides also exist.

The First Period: Hydrogen and Helium

Hydrogen atoms have just one proton and one electron, and are the simplest and smallest atoms. The properties of hydrogen do not conveniently place it in any one periodic table group. It is often shown as part of Group 1A because it does form 1+ ions, although not in ionic compounds like the other Group 1A metals, but in water solutions of acids *(← p. 152)*. Hydrogen atoms also add electrons to form hydride ions, H^-, in a small number of compounds formed with the most active metals, for example, sodium hydride (NaH). The added electron gives the hydride ion an electron configuration of $1s^2$, the same as helium. Unlike a helium atom, the hydride ion is quite reactive. In water, a hydride like NaH quickly reacts to form hydrogen.

$$NaH(s) + H_2O(\ell) \longrightarrow Na^+(aq) + OH^-(aq) + H_2(g)$$

Hydrogen is the most abundant element in the universe, making up about 60% of all its mass. One could argue that all the other elements are impurities in a vast quantity of hydrogen.

Most hydrogen compounds are molecular compounds. These include the hydrocarbons and water. There is very little free hydrogen in the earth's atmosphere, partly because hydrogen is reactive, but mostly because hydrogen molecules (H_2) are so light that gravity cannot hold them and they are lost. It is interesting that on the giant planets of Saturn, Jupiter, and Neptune, hydrogen is in abundance and that is due in part to the huge gravitational fields of these planets.

Helium is far more rare than hydrogen, yet it is well known in helium-filled balloons and airships. As yet, no compounds of helium are known. Its $1s^2$ configuration is very stable and no other atoms seems to be able to react with a helium atom to cause this configuration to change in any way. Even helium atoms themselves have little attraction for one another. Helium has the lowest boiling point of any substance (−269 °C, 4.2 K). This means it takes the least amount of energy to cause He atoms in the liquid state to separate from one another to form a gas.

In the sun, helium is produced by nuclear fusion reactions (Section 19.7) that fuse hydrogen atoms at extremely high temperatures. This process is the source of most of the sun's energy output. Helium on earth readily escapes the pull of gravity, but large deposits are trapped in natural gas deposits, mostly found in the United

States. Most of the helium was formed by radioactive decay processes (Section 19.2). Because it is so rare, helium is considered an irreplaceable natural resource. It is used as an inert-gas atmosphere in welding reactive metals like aluminum, magnesium, and titanium that would otherwise react with oxygen or nitrogen in the air; as a refrigerant to create extremely low temperatures; and as a diluting gas for breathing mixtures for divers (Section 14.7). Helium was first discovered in 1868 by studies of the solar spectrum during a solar eclipse. About 20 years later, an inert gas was discovered in pulverized rock samples, and in 1895 the spectrum of this gas was shown to be identical to that of solar helium.

All blimps, like this one operated by the Good Year Corporation, are filled with helium. *(Larry L. Miller/Photo Researchers, Inc.)*

The Second Period: Lithium to Neon

From your study of electron configurations, you should recognize that the eight elements of the second period, lithium through neon, correspond to progressive filling of the 2*s* and 2*p* orbitals with electrons. This gives rise to a wide range of chemical behavior. In fact, the differences seen in the second period are the greatest in any of the periods. The first element, lithium, readily loses its valence electron to form a positive ion, Li^+, so it is a reducing agent ***(← p. 158)*** and forms ionic compounds with nonmetals. Lithium is a silvery gray metal with a density low enough to float on water. The small size of the lithium atom makes lithium metal unusually hard for an alkali metal. Moving down the Group 1A alkali metals, they become softer as the atomic size increases. Like all the alkali metals, Li reacts with water to form hydrogen gas and a basic solution containing OH^- ions.

$$2\ Li(s) + 2\ H_2O(\ell) \longrightarrow 2\ Li^+(aq) + 2\ OH^-(aq) + H_2(g)$$

Lithium finds interesting applications that vary from lithium carbonate's use in the treatment of psychiatric disorders (the manic depressive state) to lithium soaps and their application in high-temperature lubricants.

The next three elements (Be, B, and C) range from a metal (Be), to a metalloid (B), to a nonmetal (C). These elements have higher ionization energies (Table 8.8) and smaller radii (Figure 8.19) than lithium, and their compounds are molecular rather than ionic.

Beryllium is a metal whose small atomic size makes it a very hard crystalline solid. Most beryllium comes from the gemstone beryl, $Be_3Al_2Si_6O_{18}$. Because of its low atomic weight, Be metal is very light and it finds a few exotic uses such as in making parts for nuclear reactors. It is transparent to x rays so it is used as a window in some x ray tubes. Beryllium is also alloyed with copper and other metals to make non-sparking tools. Beryllium is extremely toxic; the small Be^{2+} ion readily attracts charged regions in proteins and, by doing so, degrades the protein's structure (Section 11.10).

Emerald. A green form of beryl is called an emerald. The green color is due to chromium impurities. *(Robert De Gugliemo/ Science Photo Library/Photo Researchers, Inc.)*

Exercise 8.22 Interactions of the Be^{2+} Ion

Draw a Lewis dot structure for a nitrogen atom. In proteins, nitrogen is usually bonded to three other atoms. How many valence electrons are unused on a nitrogen atom when this occurs? How can a Be^{2+} ion interact with a nitrogen atom?

Boron is a black, crystalline solid that is extremely hard (almost as hard as diamond) and resistant to attack by acids and bases. In contrast to Al, the next member of Group 3A, boron forms exclusively molecular compounds. The larger aluminum

Aquamarine. A blue form of beryl, containing small amounts of copper, is called aquamarine. *(C.D. Winters)*

atom easily forms Al^{3+} ions. Boron forms a number of hydrides including B_2H_6, B_4H_{10}, and B_5H_9, but nowhere near as many as carbon does. Boron's hydrides are thermodynamically unstable toward air oxidation—some of the lighter hydrides spontaneously ignite in air.

Carbon exists in two common solid allotropic forms: graphite and diamond. In addition, carbon has recently been found to occur in large molecules consisting of icosahedral arrangements of carbon atoms bonded together with formulas like C_{60}. The structures are known as fullerenes or "buckyballs" ***(← p. 34).*** Diamond is the hardest natural substance known, while graphite is soft and slippery. (The structural differences responsible for these properties are explained in Section 15.9.)

Diamond and graphite. *(C.D. Winters)*

The compounds of carbon and hydrogen—the hydrocarbons ***(← p. 84)***—start with methane, CH_4, and range in size up to very large molecules. No other element forms so many compounds in which chains of like atoms are bonded together as carbon atoms do in hydrocarbons. All of the hydrocarbons are thermodynamically unstable and liberate heat energy when they burn to form CO_2 and H_2O. This makes the hydrocarbons useful as fuels (Section 11.1). Fortunately, hydrocarbons are kinetically stable (Section 12.1), so, unlike the boron hydrides, hydrocarbons do not spontaneously ignite in air.

Natural nitrogen fixation. Lightning strikes like the one pictured here produce enough localized heating to cause nitrogen and oxygen to react to produce NO. This is one of the natural ways nitrogen is fixed. *(Keith Kent/Science Photo Library/Photo Researchers, Inc.)*

Nitrogen occurs as nitrogen molecules (N_2) in the atmosphere. Dry air is 78.09% by volume nitrogen. The N_2 molecule is relatively inert because the two nitrogen atoms are held together by a strong bond. It is possible to make nitrogen react with various oxidizing and reducing agents, however, and these reactions produce nitrogen compounds that are essential to life on earth. Converting nitrogen from its unreactive molecular form into a form that is reactive is called **nitrogen fixation.** At high temperatures, like those found in a lightning flash or an internal combustion engine (Section 14.12), nitrogen will combine with oxygen to form nitric oxide, NO. Hydrogen can also be made to react with nitrogen under the right conditions to form ammonia (NH_3) in a reaction that is essential to the manufacture of fertilizers (see Section 13.8). Ammonia is an important compound because it is a source of nitrogen for growing plants so they can produce proteins, and it is a byproduct of protein metabolism in animals. In water solution, the small size of the nitrogen atom and its available electrons make it attract H^+ ions to form the ammonium ion, NH_4^+.

$$NH_3(aq) + H^+(aq) \longrightarrow \underset{\text{ammonium ion}}{NH_4^+(aq)}$$

Nitrogen can completely gain electrons from an active metal to form a negative nitride ion, N^{3-}. An example of this reaction is the formation of lithium nitride, Li_3N.

$$6\ Li(s) + N_2(g) \longrightarrow 2\ Li_3N(s)$$

This salt-like nitride contains Li^+ and N^{3-} ions. When Li_3N dissolves in water, the nitride ion will produce ammonia by reacting with water.

$$N^{3-}(aq) + 3\ H_2O(\ell) \longrightarrow NH_3(aq) + 3\ OH^-(aq)$$

Nitrogen forms other kinds of nitrides as well. With boron, nitrogen forms a covalent nitride with the formula BN. Boron nitride has a diamond-like hardness and structure. With some of the transition metals, nitrogen forms what are called *metallic nitrides* in which the nitrogen atoms occupy empty spaces between metal atoms in the metal's crystal structure. These metallic nitrides are often harder and more chemically resistant than the metal itself.

Looking at the electron configuration of the nitrogen atom, $1s^22s^22p^3$, you can see that there are up to five valence electrons that can be lost to some oxidizing agent like oxygen. There is an oxide of nitrogen for each of the five possible oxidation states. All of the oxides of nitrogen are gases.

Exercise 8.23 Oxides of Nitrogen

Nitrogen forms five oxides. Their names and formulas, in no particular order, are dinitrogen oxide (nitrous oxide, N_2O), nitrogen dioxide (NO_2), dinitrogen trioxide (N_2O_3), nitric oxide (NO), and dinitrogen pentoxide (N_2O_5). Arrange these oxides in order of increasing oxidation state of the nitrogen atom.

Oxygen is the most common oxidizing agent on the planet, and is ready to act as an oxidizing agent toward anything that will give up electrons. Among the elements, the oxidizing power of oxygen is exceeded only by that of fluorine. About one half of the mass of the earth's crust is composed of oxygen; and oxygen molecules (O_2) make up 20.91% by volume of the atmosphere. In the various solid oxides making up the earth's crust, oxygen atoms have already gained electrons and exist as stable oxide ions, O^{2-}. In addition, water found on the earth's crust contains 88.8% by weight oxygen.

Fluorine is by far the strongest chemical oxidizing agent. It exists as F_2 molecules, which are highly reactive. Fluorine occurs naturally as ionic compounds containing the fluoride ion F^-. The most common source of fluorine is the mineral fluorite, CaF_2. Only an electrical current can provide enough energy to make the F^- ion give up its electron. Most fluorine is produced by the electrolysis of HF dissolved in molten KF.

$$2\ F^-(\text{in KF}) \xrightarrow{\text{electricity}} F_2(g) + 2\ e^-$$

When fluorine reacts with a metal, the metal is oxidized to form metal ions. Copper, for example, is converted to CuF_2 when fluorine reacts with copper. For some metals, like platinum, heat must be applied, but every metal eventually succumbs to the oxidizing power of fluorine. Nonmetals react with fluorine to form molecular compounds and the oxidation state of the nonmetallic atom will usually be the highest possible. An exception to this rule is nitrogen, which only forms NF_3.

Exercise 8.24 Fluorides of Main Group Elements

Use the rules you learned for writing Lewis dot structures and determine the valence electrons for the elements C, P, S, Br, and Se. What would be the formulas for the fluorine compounds of these elements if they were oxidized to the maximum?

Exercise 8.25 Formulas of Fluorides

Write a general formula, MF_n, for the fluorides of each of the groups of the main group elements.

The last element in the second period, neon, has a filled outer shell of electrons and, like helium, is very unreactive. As with helium, there are no known stable compounds of neon. Its first ionization energy is too high for any known oxidizing agent to cause it to lose any of its valence electrons, and since its valence shell is filled, it cannot accept any electrons from a reducing agent. Neon is more rare than helium but it does occur in low concentrations in the atmosphere. Its primary use is in making "neon" signs (Figure 8.6).

Summary Problem

(a) Without looking back in the chapter, draw and label the first five energy levels of a hydrogen atom. Next, indicate the $2 \rightarrow 1$, the $3 \rightarrow 1$, the $5 \rightarrow 2$, and the $4 \rightarrow 3$ transitions in a hydrogen atom. Look in Table 8.1 to get the measured wavelengths and spectral regions for these transitions. Now calculate the frequencies (ν) for these transitions. Next, calculate the energies of the photons that are produced in these transitions.

(b) The following three metals exhibit photoelectric effects when photons of sufficient energies strike their surfaces. The photoelectric threshold is the maximum wavelength a photon must have to produce a photoelectric effect for that metal.

	Photoelectric Threshold
Lithium	540 nm
Potassium	550 nm
Cesium	660 nm

Which photon energies calculated above (that is, which transitions in hydrogen atoms) would be sufficient to cause a photoelectric effect in lithium, in potassium, and in cesium?

(c) On close examination of the line spectrum of the hydrogen atom, the $2 \rightarrow 1$ transition is observed to consist of four very closely spaced lines. Considering the $n = 2$ shell as being made up of two different subshells, determine the orbitals present and explain how four spectral lines could be produced.

(d) When a hydrogen atom's single valence electron is excited into higher and higher energy shells, it can occupy any one of a number of subshells that are never occupied in the electron configurations of any known elements. What is the energy level (value of n) and the orbital designations (s, p, d, etc.) of the lowest "never-used" subshell the hydrogen's electron might find itself in?

(e) Using data from Table 6.3 *(⇐; p. 243)*, calculate the wavelength of a photon that would have enough energy to break the bond in a Cl_2 molecule.

In Closing

Having studied this chapter, you should be able to . . .

- use the relationship between frequency, wavelength, and the speed of light for electromagnetic radiation (Section 8.1).
- explain the relationship between Planck's quantum theory and the energy absorbed or emitted when electrons in atoms change energy levels (Section 8.2).
- describe the Bohr model of the atom and its limitations (Section 8.3)
- explain the use of the wave mechanical model of the atom to represent the energy and probable location of electrons (Section 8.3).
- understand the spin properties of electrons and how they relate to electrons populating orbitals and magnetic properties of atoms (Section 8.4).
- describe and explain the relationships between shells, subshells, and orbitals (Section 8.5).
- use the periodic table to write the electron configurations of atoms and ions of main group and transition elements (Section 8.5).
- explain variations in valence electrons, electron configurations, ion formation, and paramagnetism of transition metals (Section 8.5).
- explain how nuclear magnetic resonance works and how it can find uses in chemical analysis and medical diagnosis (Section 8.6).
- describe trends in atomic radii, based on electron configurations (Section 8.7).
- describe trends in ionic radii and why ions differ in size from their atoms (Section 8.8).
- use electron configurations to explain trends in the ionization energies of the elements (Section 8.9).
- discuss the chemistries of the elements in the first two periods of the periodic table based upon an understanding of their electron configurations (Section 8.10).

Key Terms

The following terms were defined and given in boldface type in this chapter. You should be sure to understand each of these terms and the concepts with which they are associated.

atomic radius *(8.7)*
boundary surface *(8.3)*
continuous spectrum *(8.3)*
core electrons *(8.5)*
diamagnetic *(8.5)*
electromagnetic radiation *(8.1)*
electron configuration *(8.5)*
excited state *(8.3)*
ferromagnetic *(8.5)*
frequency *(8.1)*
ground state *(8.3)*
Hund's rule *(8.5)*
ionic radii *(8.8)*
ionization energy *(8.9)*
isoelectronic *(8.5)*
Lewis dot symbol *(8.5)*
line emission spectrum *(8.3)*
nitrogen fixation *(8.10)*

noble gas notation *(8.5)*
nuclear magnetic resonance (NMR) *(8.6)*
orbitals *(8.3)*
paramagnetic *(8.5)*
Pauli exclusion principle *(8.4)*
***p*-block elements** *(8.5)*
photoelectric effect *(8.2)*
photons *(8.2)*
Planck's constant *(8.2)*
principal quantum number *(8.3)*
quantum *(8.2)*
quantum theory *(8.2)*
***s*-block elements** *(8.5)*
shell *(8.3)*
spectrum *(8.1)*
uncertainty principle *(8.3)*
valence electrons *(8.5)*
wave functions *(8.3)*
wavelength *(8.1)*

Questions for Review and Thought

Conceptual Challenge Problems

CP-8.A. Planck stated in 1900 that the energy of a single photon of electromagnetic radiation was directly proportional to the frequency of the radiation ($E = h\nu$). The constant, *h,* is known as Planck's constant and has a value of 6.626×10^{-34} J · s. Soon after Planck's statement, Einstein proposed his famous equation ($E = mc^2$), which states the total energy in any system is equal to its mass times the speed of light squared.

According to the de Broglie relation, what is the apparent mass of a photon emitted by an electron undergoing a change from the second to the first energy level in a hydrogen atom? How does the photon mass compare with the mass of the electron (9.109×10^{-31} kg)?

CP-8.B. When D. I. Mendeleev proposed a periodic law, about 1870, he asserted that the properties of the elements are a periodic function of their atomic weights. Later, after H. G. J. Moseley measured the charge on the nuclei of atoms, the periodic law could be revised to state that the properties of the elements are a periodic function of their atomic numbers. What would be another way to define the periodic function that relates the properties of the elements?

CP-8.C. Figure 8.9 shows a diagram of the energy states that an electron can occupy in a hydrogen atom. Use this diagram to show that the first ionization energy for hydrogen, given in Table 8.7, is correct.

Review Questions

1. How is the frequency of electromagnetic radiation related to its wavelength?
2. What is a photon? How are the energies of photons calculated?
3. Light is given off by a sodium- or mercury-containing streetlight when the atoms are excited in some way. The light you see arises for which of the following reasons?
(a) Electrons moving from a given quantum level to one of higher *n*
(b) Electrons being removed from the atom, thereby creating a metal cation
(c) Electrons moving from a given quantum level to one of lower *n*
(d) Electrons whizzing about the nucleus in an absolute frenzy
4. What is the Pauli exclusion principle?
5. What is Hund's rule? Give an example of the use of this rule.
6. Explain what it means when an electron occupies the $3p_x$ orbital.
7. How many electrons can be accommodated in the $n = 4$ shell?
8. Tell what happens to atomic size and ionization energy across a period and down a group.
9. Why is the radius of Li^+ so much smaller than the radius of Li? Why is the radius of F^- so much larger than the radius of F?
10. Write electron configurations to show the first two ionization processes for potassium. Explain why the second ionization energy is much larger than the first.
11. Explain how the sizes of atoms change and why they change across a period of the periodic table.
12. What is meant by the term "noble gas notation"? Write an electron configuration using this notation.
13. Write the electron configurations for the valence electrons of the first three-period elements in Groups 1A through 8A.

Electromagnetic Radiation

14. When atoms absorb photons, which part (protons, electrons, neutrons) of the atom is affected?
15. Electromagnetic radiation is made up of two different "fields," what are they?
16. Electromagnetic radiation that is high in energy consists of waves that have ____________ (long or short) wavelengths and ____________ (high or low) frequencies. Give an example of radiation from the high-energy end of the electromagnetic spectrum.
17. Electromagnetic radiation that is low in energy consists of waves that have ____________ (long or short) wavelengths

and ______________ (high or low) frequencies. Give an example of radiation from the low-energy end of the electromagnetic spectrum.

18. The regions of the electromagnetic spectrum are shown in Figure 8.1. Answer the following questions on the basis of this figure.
(a) Which type of radiation involves less energy: radio or infrared light?
(b) Which radiation has the higher frequency: radio or microwaves?

19. The colors of the visible spectrum, and the wavelengths corresponding to the colors, are given in Figure 8.1.
(a) What colors of light involve less energy than yellow light?
(b) Which color of visible light has photons of greater energy: green or violet?
(c) Which color of light has the greater frequency: blue or green?

20. Assume a microwave oven operates at a frequency of $1.00 \times 10^{11} s^{-1}$. What is the wavelength of this radiation in meters? What is the energy in joules per photon? What is the energy per mole of photons?

21. The U.S. Navy has a system for communicating with submerged submarines. The system uses radio waves with a frequency of 76 s^{-1}. What is the wavelength of this radiation in meters? In miles? (1 mile = 1.61 km)

22. Place the following types of radiation in order of increasing energy per photon:
(a) Green light from a mercury lamp
(b) X rays from an instrument in a dentist's office
(c) Microwaves in a microwave oven
(d) An FM music station at 96.3 MHz

23. Place the following types of radiation in order of increasing energy per photon:
(a) Radio signals
(b) Radiation from a microwave oven
(c) Gamma rays from a nuclear reaction
(d) Red light from a neon sign
(e) Ultraviolet radiation from a sun lamp

24. If green light has a wavelength of 495 nm, what is its frequency?

25. What kind of radiation has a frequency of 5×10^{12} Hz? What is its wavelength?

26. What is the energy of one photon of blue light which has a wavelength of 450 nm?

27. Which has more energy,
(a) One photon of infrared radiation or one photon of microwave radiation?
(b) One photon of yellow light or one photon of orange light?

28. Which kind of electromagnetic radiation can interact with molecules at the cellular level?

29. When someone uses a sunscreen, which kind of radiation will be blocked and how does the sunscreen protect your skin from this type of radiation?

Atomic Spectra and the Bohr Atom

30. How is a line emission spectrum different from sunlight?

31. Any atom with its electrons in their lowest energy levels is said to be in a ______________ state.

32. Energy is emitted from an atom when an electron moves from the ______________ state to the ______________ state. The energy of the emitted radiation corresponds to the ______________ between the two energy levels.

33. Which transition involves the emission of less energy in the H atom, an electron moving from $n = 4$ to $n = 3$ or an electron moving from $n = 3$ to $n = 1$? (See Figure 8.9.)

34. If energy is absorbed by a hydrogen atom in its ground state, the atom is excited to a higher energy state. For example, the excitation of an electron from the energy level with $n = 1$ to a level with $n = 4$ requires radiation with a wavelength of 97.3 nm. Which of the following transitions would require radiation of a *wavelength longer* than this? (See Figure 8.9.)
(a) $n = 2$ to $n = 4$
(b) $n = 1$ to $n = 3$
(c) $n = 1$ to $n = 5$
(d) $n = 3$ to $n = 5$

Quantum Mechanics

35. From memory, sketch the shape of the boundary surface for each of the following atomic orbitals: (a) $2p_z$, (b) $4s$.

36. How many subshells are there in the electron shell with the principal quantum number $n = 4$?

37. How many subshells are there in the electron shell with the principal quantum number $n = 5$?

38. Bohr pictured the electrons of the atom as being located in definite orbits about the nucleus, just as planets orbit the sun. Criticize this model in view of the wave mechanical model.

39. Radiation is not the only thing that has both wave-like and particle-like characteristics. What component of matter can also be described in the same way?

40. How did the Heisenberg uncertainty principle illustrate the fundamental flaw in Bohr's model of the atom?

41. The three-dimensional boundary surfaces describing the energy and probability of finding an electron are called ______________.

42. Which type of orbitals can be found in the $n = 3$ shell? How many orbitals altogether are found in this shell?

Electron Configurations of Main Group Elements

43. Write electron configurations for Mg and Cl atoms.

44. Write electron configurations for Al and S atoms.

45. Write electron configurations for atoms of the following:
(a) Strontium (Sr), named for a town in Scotland.
(b) Tin (Sn), a metal used in the ancient world. Alloys of tin (solder, bronze, and pewter) are important.

46. Germanium had not been discovered when Mendeleev formulated his ideas of chemical periodicity. He predicted its existence, however, and it was found in 1886 by Winkler. Write the electron configuration of germanium.

47. Name an element of Group 3A. What does the group designation tell you about the electron configuration of the element?

48. Name an element of Group 6A. What does the group designation tell you about the electron configuration of the element?

49. Which ions in the following list are likely to be formed: K^{2+}, Cs^+, Al^{4+}, F^{2-}, and Se^{2-}? Which, if any, of these ions have a noble gas configuration?

Valence Electrons

50. Locate the following elements in the periodic table, and draw a Lewis dot symbol that represents the number of valence electrons for an atom of each element.
(a) F (b) In (c) Te (d) Cs

51. Locate the following elements in the periodic table, and draw a Lewis dot symbol that represents the number of valence electrons for an atom of each element.
(a) Sr (b) Br (c) Ga (d) Sb

52. Give the electron configurations of the following ions, and indicate which ions are isoelectronic: (a) Na^+, (b) Al^{3+}, and (c) Cl^-.

53. Give the electron configurations of the following ions, and indicate which ions are isoelectronic: (a) Mg^{2+}, (b) K^+, and (c) O^{2-}.

54. What is the electron configuration for the bromine atom? The bromide ion?

55. What is the electron configuration for an atom of tin? What are the electron configurations for tin(II) and tin(IV) ions?

Electron Configurations of Transition Elements

56. How many elements are there in the fourth row of the periodic table? Explain why it is not possible for there to be another element in this row.

57. When transition metals form ions, electrons are lost first from which type of orbital? Why?

58. Give the electron configurations of Mn, Mn^{2+}, and Mn^{3+}. Use orbital box diagrams to determine the number of unpaired electrons for each species.

59. Write the electron configuration of chromium: Cr, Cr^{2+}, and Cr^{3+}. Use orbital box diagrams to determine the number of unpaired electrons for each species.

60. Write the electron configuration of vanadium (V). The name of the element was derived from Vanadis, a Scandinavian goddess.

61. Write electron configurations for the following:
(a) Zirconium (Zr). This metal is exceptionally resistant to corrosion and so has important industrial applications. Moon rocks show a surprisingly high zirconium content compared with rocks on earth.
(b) Rhodium (Rh), used in jewelry and in industrial catalysts.

62. The lanthanides, or rare earths, are now only "medium rare." All can be purchased for a reasonable price. Give electron configurations for atoms of the following elements.
(a) Europium (Eu). It is the most expensive of the rare earth elements; 1 g can be purchased for $50 to $100.
(b) Ytterbium (Yb). It is less expensive than Eu, as Yb costs only about $15 per gram. It was named for the village of Ytterby in Sweden, where a mineral source of the element was found.

Paramagnetism and Unpaired Electrons

63. In the first transition series (row four in the periodic table) which elements would you predict to be *diamagnetic?* Which element in this series has the greatest number of unpaired electrons?

64. What is ferromagnetism?

65. Which groups of elements in the periodic table are ferromagnetic?

66. How do the spins of unpaired electrons from paramagnetic and ferromagnetic materials differ in their behavior in a magnetic field?

Spin Properties of Atomic Nuclei

67. What kind of electromagnetic radiation is used in nuclear magnetic resonance (NMR)?

68. In magnetic resonance imaging (MRI), the intensity of the emitted signal is related to the ____________ of hydrogen nuclei and the relaxation time is related to the type of tissue being examined.

Periodic Trends

69. Arrange the following elements in order of increasing size: Al, B, C, K, and Na. (Try doing it without looking at Figure 8.19 and then check yourself by looking up the necessary atomic radii.)

70. Arrange the following elements in order of increasing size: Ca, Rb, P, Ge, and Sr. (Try doing it without looking at Figure 8.19 and then check yourself by looking up the necessary atomic radii.)

71. Select the atom or ion in each pair that has the larger radius.
(a) Cl or Cl^-
(b) Al or N
(c) In or Sn

72. Select the atom or ion in each pair that has the larger radius.
(a) Cs or Rb (b) O^{2-} or O (c) Br or As

73. Write electron configurations to show the first two ionization steps for sodium. Explain why the second ionization energy is much larger than the first.

74. Arrange the following atoms in the order of increasing ionization energy: F, Al, P, and Mg.

75. Arrange the following atoms in the order of increasing ionization energy: Li, K, C, and N.

76. Which of the following groups of elements is arranged correctly in order of increasing ionization energy?
(a) $C < Si < Li < Ne$ (c) $Li < Si < C < Ne$
(b) $Ne < Si < C < Li$ (d) $Ne < C < Si < Li$

77. Rank the following ionization energies (IE) in order from the smallest value to the largest value. Briefly explain your answer.
(a) First IE of Be (d) Second IE of Na
(b) First IE of Li (e) First IE of K
(c) Second IE of Be

78. Predict which of the following elements would have the greatest difference between the first and second ionization energies: Si, Na, P, and Mg. Briefly explain your answer.

79. Compare the elements, Li, K, C, and N.
(a) Which has the largest atomic radius?
(b) Place the elements in order of increasing ionization energy.

80. Compare the elements B, Al, C, and Si.
(a) Which has the most metallic character?
(b) Which has the largest atomic radius?
(c) Place the three elements B, Al, and C in order of increasing first ionization energy.

81. Explain why the transition elements in a row are more alike in their properties than other elements in the same groups.

Periodic Trends in Properties of the Elements

82. Which element is the most abundant in the *universe*?

83. Explain why helium has the lowest boiling point of any substance.

84. Which metal in the second period is known for its toxicity?

85. Explain why boron and aluminum are different. One is a metalloid and the other is a metal but both are found in the same group in the periodic table.

86. Name three allotropes of carbon.

87. Write the equation for the formation of boron nitride.

88. Write the equations for the formation of the fluorides with the following elements: C, P, Al, Ca, and K.

89. When magnesium burns in air, it forms both an oxide and a nitride. Write a balanced equation for the formation of the nitride.

90. The annual U.S. and Canadian fluorine production is 5.00×10^3 tons (1 ton is 2.00×10^3 lb), and 55% of this is used to manufacture UF_6. What mass in tons of uranium(VI) fluoride is manufactured every year (assume 100% efficiency in using the F_2)?

General Questions

91. A neutral atom has two electrons with $n = 1$, eight electrons with $n = 2$, eight electrons with $n = 3$, and one electron with $n = 4$. Assuming this element is in its ground state, supply the following information:
(a) Atomic number and name
(b) Total number of s electrons
(c) Total number of p electrons
(d) Total number of d electrons

92. How many p orbital electron pairs are there in an atom of selenium (Se) in its ground state?

93. Answer the following questions about the elements A and B, which have the electron configurations shown.

$$A = [Kr]4d^{10}5s^1 \qquad B = [Ar]3d^{10}4s^24p^4$$

(a) Is element A a metal, a nonmetal, or a metalloid?
(b) Which element would have the greater first ionization energy?
(c) Which element has a larger atomic radius?

94. (a) Place the following elements in order of increasing first ionization energy: F, O, and S.
(b) Which has the largest first ionization energy: O, S, or Se?

95. (a) Rank the following in order of increasing atomic radius: O, S, and F.
(b) Which has the largest first ionization energy: P, Si, S, or Se?
(c) Place the following in order of increasing radius: Ne, O^{2-}, N^{3-}, and F^-.
(d) Place the following in order of increasing first ionization energy: Cs, Sr, Ba.

96. Name the element corresponding to each of the following characteristics:
(a) The element whose atoms have the electron configuration $1s^22s^22p^63s^23p^4$
(b) The element in the alkaline earth group that has the largest atomic radius
(c) The element in Group 5A whose atoms have the largest ionization energy.
(d) The element whose 2+ ion has the configuration $[Kr]4d^6$
(e) The element whose atoms have the electron configuration $[Ar]3d^{10}4s^1$

97. The ionization energies for the removal of the first electron from atoms of Si, P, S, and Cl are listed in the following table. Briefly rationalize this trend.

Element	First Ionization Energy (kJ/mol)
Si	780
P	1060
S	1005
Cl	1255

98. Answer the following questions about the elements with electron configurations shown.

$$A = [Ar]3d^84s^2 \qquad B = [Ar]3d^{10}4s^24p^5$$

(a) Is element A a metal, a metalloid, or a nonmetal?
(b) Is element B a metal, a metalloid, or a nonmetal?
(c) An atom of which element is expected to have the larger first ionization energy?
(d) An atom of which element would be the smaller of the two?

99. Place the following elements and ions in order of decreasing size: Ar, K^+, Cl^-, S^{2-}, and Ca^{2+}.

100. Which of the following ions are unlikely, and why: Cs^+, In^{4+}, Fe^{6+}, Te^{2-}, Sn^{5+}, and I^-?

101. Rank the following in order of increasing first ionization energy: Zn, Ca, Ca^{2+}, and Cl^-. Briefly explain your answer.

102. Worldwide production of silicon carbide, SiC, a widely used abrasive, is several hundred thousand tons annually. In order to produce 100,000 tons of the carbide, what mass in tons of silica sand (SiO_2) will you have to use if 70% of the sand is converted to SiC?

103. Suppose a new element, extraterrestium, tentatively given the symbol Et, has just been discovered. Its atomic number is 113.
(a) Write the electron configuration of the element.
(b) Name another element you would expect to find in the same group as Et.
(c) Give the formulas for the compounds of Et with O and Cl.

104. When sulfur dioxide reacts with chlorine, the products are thionyl chloride ($SOCl_2$) and dichlorine monoxide (Cl_2O).

$$SO_2(g) + 2\ Cl_2(g) \longrightarrow SOCl_2(g) + OCl_2(g)$$

(a) In what period of the periodic table is S located?
(b) Give the complete electron configuration of S. Do *not* use the noble gas notation.
(c) An atom of which element involved in this reaction (O, S, or Cl) should have the smallest first ionization energy? The smallest radius?
(d) If you want to make 675 g of $SOCl_2$, what mass in grams of Cl_2 is required?
(e) If you use 10.0 g of SO_2 and 20.0 g of Cl_2, what is the theoretical yield of $SOCl_2$?

Applying Concepts

105. Write the electron configuration for the product of the first ionization of the smallest halogen.

106. Write the electron configuration for the product of the second ionization of the third largest alkaline earth metal.

107. What compound will most likely form between chlorine and element X, if element X has the electronic configuration $1s^22s^22p^63s^1$?

108. What compound will most likely form between potassium and element X, if element X has the electronic configuration $1s^22s^22p^63s^23p^4$?

109. Using the information in Table 8.2, write the electron configuration for the undiscovered element with an atomic number of 164. Where would this element be located in the periodic table?

110. You are given the atomic radii of 110 pm, 118 pm, 120 pm, 122 pm, and 135 pm, but do not know to which element (As, Ga, Ge, P, and Si) these values correspond. Which must be the value of Ge?

111. The following questions refer to the graph below.
(a) Based on the graphical data, ionization energies ______ (decrease, increase) left-to-right and ______ (decrease, increase) top-to-bottom on the periodic table.
(b) Which element has the largest first ionization energy?
(c) A plot of the fourth ionization energy versus atomic number for elements 1 through 18 would have peaks at which atomic numbers?
(d) Why is there no third ionization energy for helium?
(e) What is the reason for the large second ionization energy for lithium?
(f) Find the arrow pointing to the third ionization energy curve. What is the symbol for the particle corresponding to this data point?

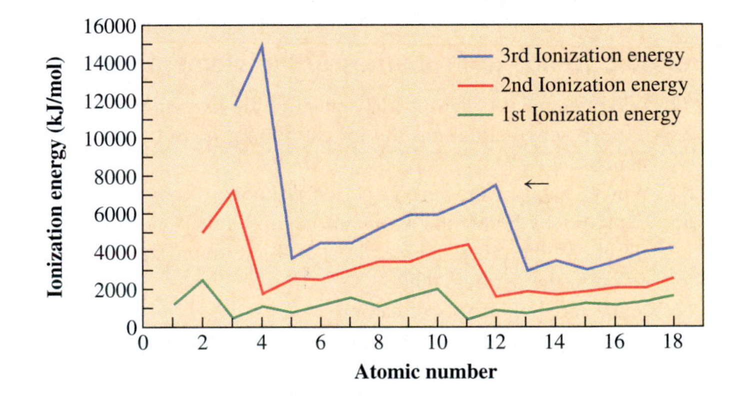

Covalent Bonding

CHAPTER 9

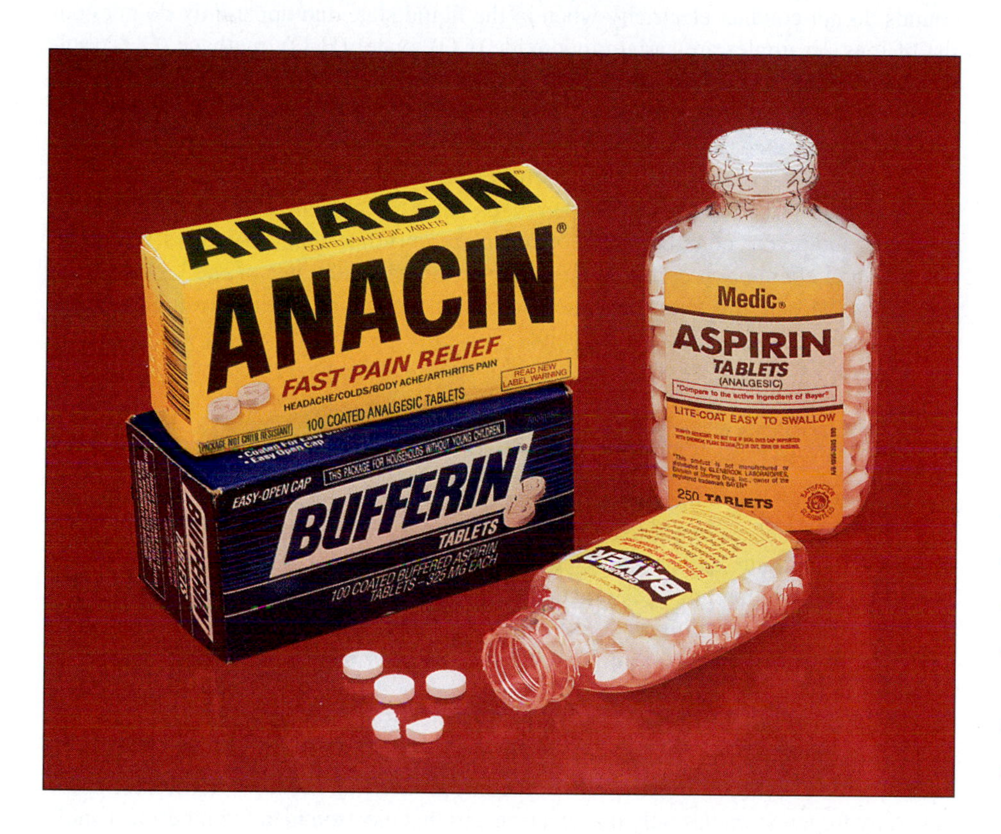

Aspirin is the most commonly used nonprescription painkiller; more than 30 billion aspirin tablets are consumed annually in the United States. Salicylic acid is a painkiller whose molecules each contain carbon, hydrogen, and oxygen atoms ($C_7H_6O_3$). By modifying the structure of salicylic acid, chemists developed aspirin (acetylsalicylic acid), which has fewer harmful side effects than salicylic acid. • What modifications were made in the structure of salicylic acid to convert it to acetylsalicylic acid? The Summary Problem presents the bonding in salicylic acid and the modifications in its molecular structure to create aspirin. *(C.D. Winters)*

Atoms are only rarely found uncombined in nature. Most nonmetallic elements consist of molecules (Cl_2, N_2, P_4, S_8), and in a solid metal each atom is closely surrounded by 8 or 12 neighbors. Only the noble gases consist of individual atoms. What makes atoms stick to one another? Valence electrons form the glue, but how? In ionic compounds, the gain or loss of one or more valence electrons produces ions whose opposite charges hold the ions together in a crystal lattice. But many compounds do not conduct electricity when in the liquid state and apparently do not consist of ions. Examples are carbon monoxide (CO), water (H_2O), methane (CH_4), sulfur dioxide (SO_2), and the millions of organic compounds.

What holds atoms together in these molecular compounds? G. N. Lewis suggested that the bonds in molecules consist of one or more pairs of electrons shared between the bonded atoms. The attraction of positively charged nuclei for electrons between them pulls the nuclei together. This simple idea can account for the bonding in all molecular compounds, allowing us to correlate their structures with their physical and chemical properties. This chapter describes the use of the Lewis model to explain the bonding found in various types of molecules ranging from simple diatomic gases to coordination compounds.

9.1 COVALENT BONDING

Hydrogen gas (H_2) has the smallest molar mass of any element, and lithium hydride (LiH) has the smallest molar mass of any compound. Both H and Li are in Group 1A of the periodic table, and atoms of each element have one valence electron. But H_2 is a gas at room temperature, is practically insoluble in water, is an electric insulator, and burns easily in air. Lithium hydride is a solid, reacts with water to form H_2, conducts electricity when in the molten state, and bursts into flame when exposed to moist air at high temperatures. Why is there such a difference in properties for two substances that, on an atomic scale, might seem similar?

A lithium atom can lose its one valence electron (the 2*s* electron) relatively easily to form Li^+, which has the electron configuration of helium, the nearest noble gas. A hydrogen atom has only one electron and that electron is in the innermost shell, quite close to the nucleus. Because of the first shell's small size, electrons in it are tightly held and H has a much higher ionization energy than Li (← *p. 337*). However, an H atom can add an electron to form H^-, which also has the noble gas electron configuration of He. Hence, lithium hydride consists of Li^+ cations and H^- anions, each of which has the same electron configuration as an He atom. Lithium hydride has properties characteristic of an ionic compound—it is a crystalline solid at room temperature and has a high melting point.

Hydrogen gas, on the other hand, has none of the properties of an ionic compound; it consists of H_2 molecules. Each H_2 molecule is held together by a **covalent bond**—an attractive force between two atoms that results from sharing one or more pairs of electrons. *The atoms in molecular compounds are connected by covalent bonds.*

G. N. Lewis suggested that valence electrons rearrange to give noble gas electron configurations when chemical bonds form. He assumed that each noble gas atom had a completely filled outermost shell, which he regarded as a stable configuration because of the lack of reactivity of noble gases. His idea can be applied to molecular compounds as well as ionic compounds. Lewis proposed that when a pair of electrons is shared between two atoms, that pair occupies the same shell as the valence electrons of each atom and contributes to a noble gas configuration on *each* atom. He further proposed that by counting the valence electrons of an atom, it would be possible to predict how many bonds that atom can form. *The number of bonds is the number of electrons that must be shared to achieve a noble gas configuration.*

But why does sharing electrons provide an attractive force between atoms? Consider the formation of the simplest stable molecule, H_2. If the two hydrogen atoms are widely separated, there is little if any interaction between them. When the two atoms get close enough, however, their 1*s* electron clouds overlap (interpenetrate).

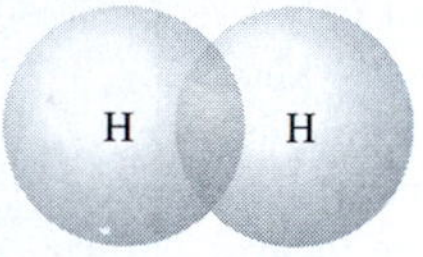

This allows the electron from each atom to be attracted by the other atom's nucleus, lowering the potential energy, and causing a net attraction between the two atoms. If an electron pair is *shared* between two atomic nuclei, as in an H_2 molecule, the electrons attract the nuclei together. When two atoms share a pair of electrons, those electrons have a higher probability of being in the *bonding region,* the region between the nuclei, than outside it. Experimental data and calculations indicate that an H_2 molecule is most stable (has its lowest potential energy) when the nuclei are 74 pm (0.074 nm) apart. Here the electrostatic attractive and repulsive forces are balanced. If the nuclei get closer than 74 pm, then their mutual repulsion becomes large enough to counteract the attraction between electrons and nuclei. When the H nuclei are 74 pm apart, it takes 436 kJ of energy to separate a mole of gaseous H_2 molecules into isolated H atoms (Figure 9.1). This energy is the bond energy of H_2 ***(p. 243, Table 6.3).***

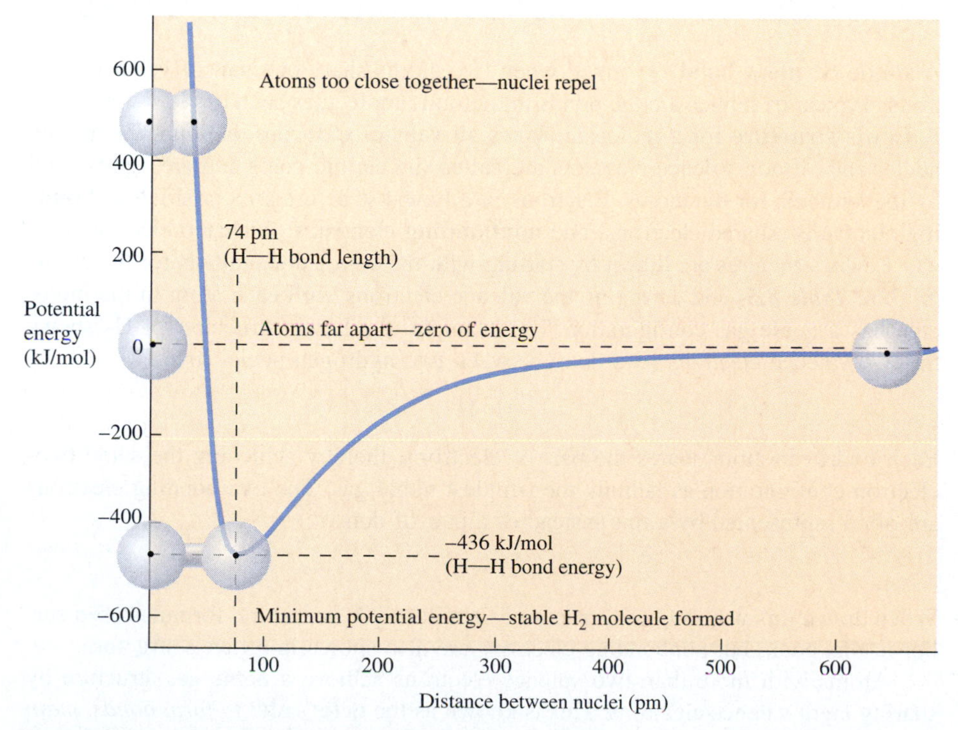

Figure 9.1 H—H bond formation from isolated H atoms. Energy is at a minimum at a distance of 74 pm where there is a balance between electrostatic attractions and repulsions.

Portrait of a Scientist • *Gilbert Newton Lewis (1875–1946)*

In a paper published in the Journal of the *American Chemical Society* in 1916, Gilbert N. Lewis introduced the theory of the shared electron pair chemical bond and revolutionized chemistry. It is to honor this contribution that we often refer to "electron dot" structures as Lewis structures. Lewis also made major contributions to other fields such as thermodynamics, isotope studies, and the interaction of light with substances. Of particular interest in this text is the extension of his theory of bonding to a generalized theory of acids and bases (Section 17.12).

G. N. Lewis was born in Massachusetts but raised in Nebraska. After earning his B.A. and Ph.D. degrees at Harvard University, he began his academic career. In 1912 he was appointed Chairman of the Chemistry Department at the University of California, Berkeley, and he remained there for the rest of his life. Lewis felt that a chemistry department should both teach and advance fundamental chemistry, and he was not only a productive researcher but also a teacher who profoundly affected his students. Among his ideas was the use of large problem sets in teaching, an idea that is much in use today.

G. N. Lewis *(Oesper Collection in the History of Chemistry/University of Cincinnati)*

9.2 SINGLE COVALENT BONDS AND LEWIS STRUCTURES

Lewis structures indicate only valence electrons.

Conceptual Challenge Problem CP-9.A at the end of the chapter relates to topics covered in this section.

A **single covalent bond** is formed when two atoms share one pair of valence electrons. Typical examples are the bonds in diatomic molecules such as H_2, F_2, and Cl_2. A **Lewis structure** for a molecule shows all valence electrons as dots. The atomic nuclei and all non-valence electrons are called the atomic cores and are represented by the symbols for the atoms. Electrons in a Lewis structure are classified as **bonding electrons** (shared electrons) and **nonbonding electrons** (unshared electrons).

Lewis structures are drawn by starting with the Lewis dot symbols for the atoms ***(p. 326, Table 8.5)*** and arranging the valence electrons until each atom in the molecule has a noble gas configuration. For example, the Lewis structure for H_2 shows the two valence elections (two dots) between two hydrogen nuclei (two H).

H : H

Each hydrogen atom shares the pair of electrons, thereby achieving the same two-electron configuration as helium, the simplest noble gas. The two bonding electrons are often represented by a line instead of a pair of dots.

H—H

When drawn this way, a Lewis structure is similar to the structural formulas used earlier in this book, but nonbonding electrons are also shown in a Lewis structure.

Achieving a noble gas configuration is also referred to as obeying the octet rule, because all noble gases except helium have eight valence electrons.

Atoms with more than two valence electrons achieve a noble gas structure by sharing eight valence electrons. This is known as the **octet rule:** *to form bonds, main group elements gain, lose, or share electrons to achieve a stable electron configuration characterized by an octet (eight valence electrons).* To obtain the Lewis structure for F_2, for example, we start with the Lewis dot symbol for a fluorine atom. Fluorine is in Group 7A and has seven valence electrons, and so there are seven dots. This is one less than an octet. If each F atom contributes one electron to be shared with the other F atom in forming a single covalent bond, each F atom can achieve an octet. *Shared electrons count for each atom involved in the bond.*

$$2\,:\ddot{\underset{\cdot\cdot}{F}}\cdot \longrightarrow :\ddot{\underset{\cdot\cdot}{F}}:\ddot{\underset{\cdot\cdot}{F}}: \quad \text{or} \quad :\ddot{\underset{\cdot\cdot}{F}}-\ddot{\underset{\cdot\cdot}{F}}:$$

Main group elements are those in the groups labeled "A" in the periodic table inside the front cover of this book.

The pair of electrons shared between the two F atoms are bonding pair electrons. The other six pairs of electrons (three pairs on each fluorine) are nonbonding pairs of electrons, also called **lone pairs.** In writing Lewis structures, the bonding pairs of electrons are usually indicated by lines connecting the atoms they hold together, and lone pairs are usually represented by dots.

Magnetic measurements (← ***p. 318***) support the concept that each electron in a pair (whether bonding or nonbonding) has its spin opposite that of the other electron.

The term "lone pairs" will be used in this text to refer to nonbonding electron pairs.

What about Lewis structures for molecules such as H_2O or NH_3? Oxygen (Group 6A) has six valence electrons and must share two electrons to satisfy the octet rule. This can be accomplished by forming covalent bonds with two hydrogen atoms.

$$2\,H\cdot + \cdot\ddot{\underset{\cdot\cdot}{O}}\cdot \longrightarrow H:\ddot{\underset{\cdot\cdot}{O}}:H \quad \text{or} \quad H-\ddot{\underset{\cdot\cdot}{O}}-H$$

Nitrogen (Group 5A) in NH_3 must share three electrons to achieve a noble gas configuration, which can be done by forming covalent bonds with three hydrogen atoms.

$$3\,H\cdot + \cdot\ddot{\underset{\cdot}{N}}\cdot \longrightarrow H:\ddot{\underset{\underset{H}{\cdot\cdot}}{N}}:H \quad \text{or} \quad H-\ddot{\underset{\underset{H}{|}}{N}}-H$$

From the Lewis structures of F_2, H_2O, and NH_3, we can make an important generalization: *the number of electrons that an atom of a main group element must share to achieve an octet equals 8 minus its A group number.* Nitrogen, for example, which is in Group 5A, needs to share three electrons to reach an octet.

Group Number	Number of Electrons to Complete an Octet	Example
4A	4	C in CH_4
5A	3	N in NF_3
6A	2	O in H_2O
7A	1	F in HF

Writing Lewis Structures

Suppose you want to write the Lewis structure for a molecule or polyatomic ion. There are guidelines that will help you write the correct Lewis structure. We will illustrate them by writing the Lewis structure for PCl_3.

Phosphorus trichloride (PCl_3) is a colorless liquid that fumes as it reacts with water in moist air.

Guidelines for Writing Lewis Structures

1. ***Count the total number of valence electrons in the molecule or polyatomic ion.*** The A group number in the periodic table indicates the number of valence electrons for each atom. For a neutral molecule, sum the numbers of valence electrons of the atoms in the molecule. For a negative ion, add to that sum a number of electrons equal to the ion charge. For a positive ion, subtract the number of electrons equal to the charge. For example, add one electron for OH^-; subtract one electron for NH_4^+.

Because PCl_3 is a neutral molecule, its number of valence electrons is 5 for P (it is in Group 5A) and 7 for each Cl (chlorine is in Group 7A): total number of valence electrons = $5 + (3 \times 7) = 26$.

Although H is given first in the formulas of H_2O and H_2O_2, for example, it is not the central atom. H forms only one bond and is never the central atom in a molecule or ion.

2. ***Use the symbols for the atoms to draw a skeleton structure with a shared pair of electrons between the central atom (or atoms) and each terminal atom.*** A skeleton structure indicates the attachment of terminal atoms to a central atom. Usually the central atom is the one with the lowest subscript in the molecular formula and the one that can form the most bonds. In PCl_3, the central atom is phosphorus, and we draw a skeleton structure with P as the central atom and the chlorine atoms (the terminal atoms) arranged around it, with a bonding pair of electrons between the central P atom and each terminal Cl atom. The three bonding pairs account for 6 of the total of 26 valence electrons.

```
Cl—P—Cl
   |
   Cl
```

3. ***Starting with the terminal atoms, place lone pairs of electrons around each atom (except H) to satisfy the octet rule.*** Using lone pairs in this way on the three Cl atoms accounts for 18 of the remaining 20 valence electrons, leaving 2 electrons for a lone pair on the P atom. Remember that shared electrons are counted as 'belonging" to *each* of the atoms bonded by the shared pair; each P—Cl bond has two shared electrons that "count" for phosphorus and also count for chlorine.

```
 ..   ..  ..              ..   ..   ..
:Cl : P : Cl:     or     :Cl — P — Cl:
 ..   ..  ..              ..   |    ..
     :Cl:                     :Cl:
      ..                       ..
```

To check the structure, verify that the total number of dots, or dots + (lines $\times$ 2) equals the total number of valence electrons. Counting dots and lines in the Lewis structure on the right gives 26, thus accounting for all valence electrons. It is the correct Lewis structure for PCl_3.

The next two steps apply when Steps 1 through 3 result in a structure that does not use all the valence electrons or fails to give an octet of electrons to each atom that should have an octet.

4. ***Place any leftover electrons on the central atom, even if it will give the central atom more than an octet.*** If the central atom is from the third or higher period, it can accommodate more than an octet of electrons. The reason for this is explained in Section 9.3.
5. ***If the number of electrons around the central atom is less than eight, change one or more lone pairs to bond pairs.*** Some atoms can share more than one pair of electrons, with the same partner atom. This results in a double covalent bond (two shared pairs) or a triple covalent bond (three shared pairs). Double and triple bonds are known as multiple bonds (Section 9.3). In molecules where there are not enough electrons to complete all octets, use one or more lone pairs of electrons from the terminal atoms to form double or triple bonds until the central atom and all terminal atoms have octets. These last two guidelines do not apply to PCl_3, but they will be illustrated in the next section.

PROBLEM-SOLVING EXAMPLE 9.1 Lewis Structures

Write the Lewis structures of (a) dichlorine monoxide, Cl_2O; (b) hydrogen peroxide, H_2O_2; (c) hydroxide ion, OH^-; and (d) perchlorate ion, ClO_4^-.

Answer

(a) :C̤̈l—Ö̤—C̤̈l: (b) H—Ö̤—Ö̤—H (c) $[:\ddot{\underset{..}{O}}—H]^-$ (d) $[ClO_4]^-$ with :Ö̤: above and below Cl, and :Ö̤—C̤l—Ö̤: across

Explanation

(a) The total number of valence electrons is 20: 14 from the two Cl atoms and 6 from the O atom. The central atom is O and the skeleton structure is

Cl—O—Cl

Placing three lone pairs on each Cl and two lone pairs on the O atom uses all the electrons and satisfies the octet rule for all atoms.

(b) The atoms in H_2O_2 are H, which can never form more than one bond, and O, which can form two bonds, meaning that the two O atoms must be bonded to each other. The total number of valence electrons is 14: 2 from the two H atoms and 12 from the two O atoms. The skeleton structure must be H—O—O—H, which uses 6 electrons. Placing two lone pairs on each O atom satisfies the octet rule and uses all valence electrons.

(c) There are eight valence electrons in the OH^- ion: 1 from H, 6 from O, and 1 for the 1− charge. Because hydrogen can form only one bond, there is a shared electron pair between O and H, plus three lone pairs on the O completing the octet.

(d) In oxygen-containing polyatomic ions, such as ClO_4^-, the non-oxygen atom is the central atom. The 32 valence electrons are distributed as four bonding pairs between the central chlorine atom and each oxygen and three lone pairs around each oxygen.

Problem-Solving Practice 9.1

Write the Lewis structures for (a) NF_3; (b) N_2H_4; and (c) SO_4^{2-}.

Although Lewis structures are useful for predicting the number of covalent bonds an atom will form, they do not give an accurate representation of where electrons are located in a molecule. Bonding electrons do not stay in fixed positions between nuclei, as Lewis's dots might imply. Instead, quantum mechanics tells us that there is a high probability of finding the bonding electrons between the nuclei, but they could be found elsewhere as well. Also, Lewis structures are not meant to convey the shapes of molecules. The angle between the two O—H bonds in a water molecule is not 180°, as the Lewis structure above seems to imply. However, Lewis structures can be used to predict geometries by a method based on the repulsions between valence-shell electron pairs (Section 10.1).

Single Bonds in Hydrocarbons

Carbon is unique among the elements because of the ability of its atoms to form strong bonds with one another as well as with atoms of hydrogen, oxygen, nitrogen, sulfur, and the halogens. The strength of the carbon-carbon bond permits long chains to form:

—C—C—C—C—C—C—C—C—C—C—C—C—

Because each carbon atom can form four covalent bonds, such chains contain numerous sites to which other atoms (including more carbon atoms) can bond, leading to a great variety of carbon compounds.

Review the discussion on alkanes in Section 3.3. See Table 3.3 (⇐ *p. 85*) for a list of selected alkanes.

Hydrocarbons contain only carbon and hydrogen atoms (⇐ *p. 84*). **Alkanes,** which contain only C—C and C—H single covalent bonds, are often referred to as **saturated hydrocarbons** because each carbon is bonded to a maximum number of hydrogen atoms. The general formula for alkanes is C_nH_{2n+2}, where $n = 2, 3, 4$, and so on.

The carbon atoms in alkanes with four or more carbon atoms per molecule can be arranged in either a straight chain or a branched chain (⇐ *p. 87*). Chain branching leads to the formation of a wide variety of compounds with the same molecular formulas, but different molecular structures and properties as, for example, the three compounds with the formula C_5H_{12}: pentane, 2-methylbutane, or 2,2-dimethylpropane.

```
   H  H  H  H  H             H  H  H  H                   CH3
   |  |  |  |  |             |  |  |  |                    |
H—C—C—C—C—C—H           H—C—C—C—C—H          CH3—C—CH3
   |  |  |  |  |             |  |  |  |                    |
   H  H  H  H  H             H  |  H  H                   CH3
                                CH3
```

pentane — 2-methylbutane — 2,2-dimethylpropane

Rules for naming organic compounds are given in Appendices E.1 and E.2

Chain branching is discussed further in Section 11.8

In addition to straight-chain and branched-chain alkanes, there are **cycloalkanes,** saturated hydrocarbons consisting of carbon atoms joined in rings of $—CH_2—$ units. Cycloalkanes are commonly represented by polygons in which each corner represents a carbon atom and two hydrogen atoms, and each line represents a C—C bond. The C—H bonds usually are not shown but are understood to be present. The simplest cycloalkane is cyclopropane; other common cycloalkanes include cyclobutane, cyclopentane, and cyclohexane:

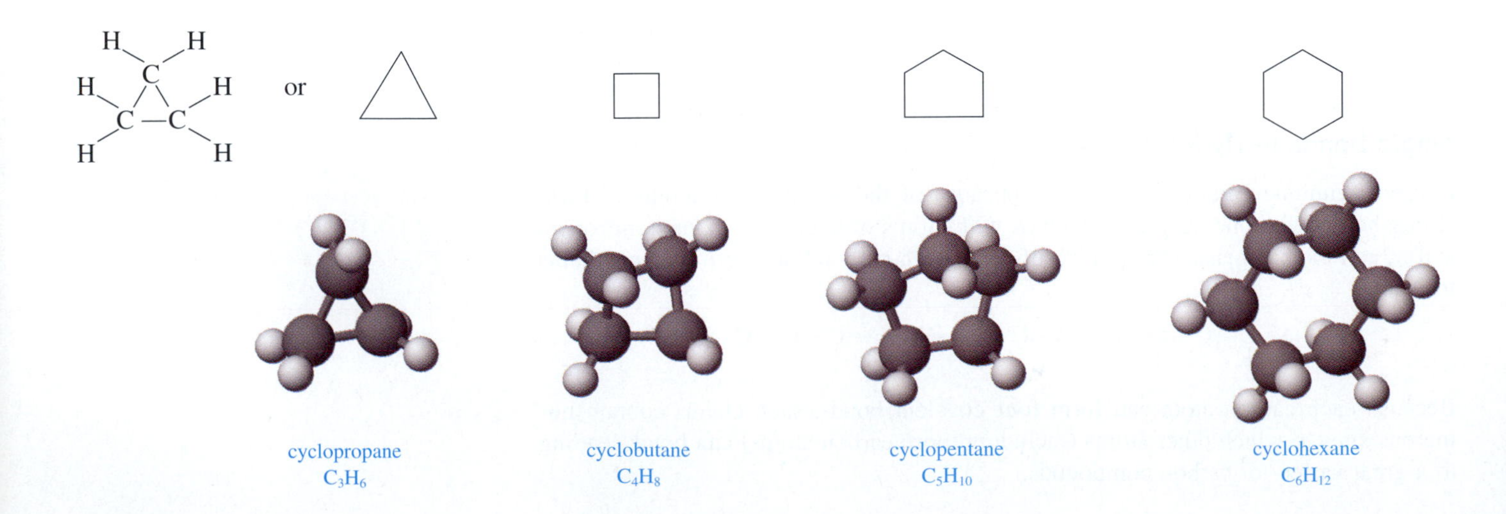

cyclopropane C_3H_6 — cyclobutane C_4H_8 — cyclopentane C_5H_{10} — cyclohexane C_6H_{12}

Exercise 9.1 Cyclic Hydrocarbons

Write the Lewis structure and the polygon that represents cyclooctane; write the molecular formula for this compound.

The great variety of organic compounds can also be accounted for by the fact that one or many carbon-hydrogen bonds in hydrocarbons can be replaced by bonds between carbon and other atoms. The new bonds to carbon can connect to individual halogen atoms or to atoms in functional groups, thus creating entirely different compounds. Consider the new substances that result when a chlorine atom replaces a hydrogen atom in ethane, 2-methylbutane, and cyclopropane.

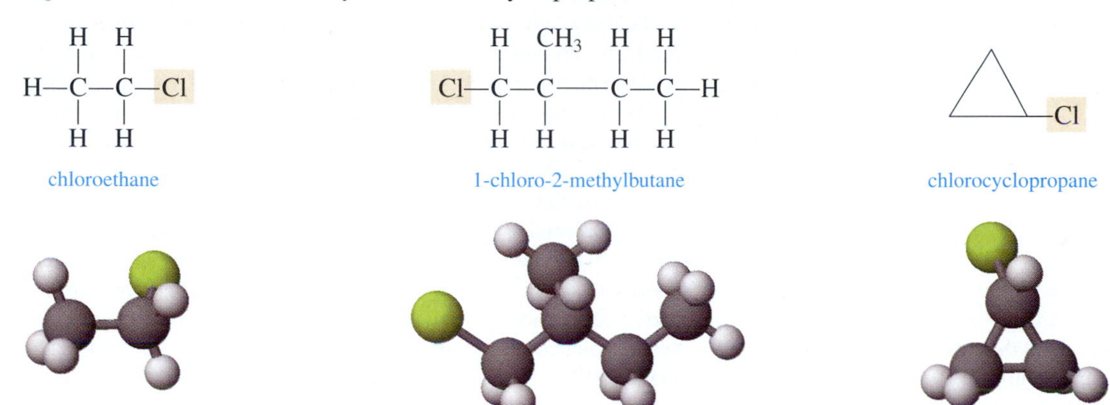

chloroethane — 1-chloro-2-methylbutane — chlorocyclopropane

In another case, consider how an —OH functional group can replace one or more hydrogen atoms in an alkane. The —OH functional group consists of a hydrogen atom bonded to an oxygen atom, —O—H, that is bonded to a carbon atom. A functional group *(← p. 81)* is a group of atoms that imparts characteristic properties to an organic compound and can be thought of as substituting for a hydrogen atom in a hydrocarbon. The —OH functional group is characteristic of alcohols. A molecule of ethanol can be thought of as a molecule of ethane in which a hydrogen atom has been replaced by an —OH group. "Replacing" two hydrogen atoms in ethane with —OH groups forms ethylene glycol. Glycerol is formed when three hydrogen atoms in propane are replaced by —OH groups.

The —OH group in an alcohol is different from the OH^- ion in a base.

Appendix E.2 includes a list of functional groups.

Glycerol is also used as a sweetener and thickener in foods.

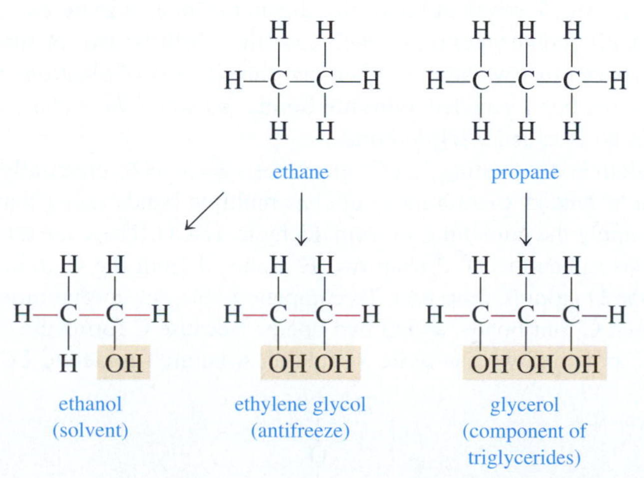

ethanol (solvent) — ethylene glycol (antifreeze) — glycerol (component of triglycerides)

CHEMISTRY IN THE NEWS

Keeping Cool, But Safely

Imagine a pet dog or cat sauntering into a driveway and licking a small puddle of a shiny, green liquid. The liquid's invitingly sweet taste appeals to the pet, which laps up more liquid. In a short time the pet is critically ill because of drinking the green liquid.

The liquid is automobile antifreeze, which contains the toxic compound ethylene glycol.

```
   H  H
   |  |
H—C—C—H
   |  |
  OH OH
```

(Ethylene glycol is colorless; the green dye is added to distinguish antifreeze from other automobile fluids such as oil or transmission fluid.) Relatively small amounts of ethylene glycol can be fatal—only about 4 tablespoons of ethylene glycol can kill a large dog or adult human; fatal doses are even less for smaller pets (1 teaspoon for a cat) or children (2 tablespoons). More than 4000 people are poisoned by antifreeze annually in the United States, including several hundred children under the age of six. In 1994, 29 people died of ethylene glycol poisoning.

Unless medical treatment is promptly given, ethylene glycol poisoning causes severe brain, heart, liver, and kidney damage because of toxic byproducts produced from ethylene glycol metabolism. Alcohol dehydrogenase, a substance in cells, accelerates the conversion of ethylene glycol into several toxic organic acids, including oxalic acid. This acid combines with Ca^{2+} ions in blood to form insoluble calcium oxalate ***(p. 104)***. The calcium oxalate crystals block the flow of blood, causing brain damage. Kidneys are particularly susceptible to blockage by the precipitated calcium oxalate, which leads to kidney poisoning.

Recently, a safe, effective antifreeze has been marketed that uses propylene glycol, rather than ethylene glycol.

```
   H  H  H
   |  |  |
H—C—C—C—H
   |  |  |
   H OH OH
```

When mixed with an equal volume of water, propylene glycol prevents freezing down to −35 °C, almost as good as an equal mixture of water and ethylene glycol (−37 °C). Although a close molecular "relative" to ethylene glycol, propylene glycol is not toxic to humans or animals. It is even used in human snack foods, cosmetics, candy, and in animal foods. The metabolism of propylene glycol is also catalyzed by alcohol dehydrogenase, but the breakdown products are nontoxic. This is evidence that sometimes just a slight change in molecular composition—in this case just a difference of a CH_2 unit—can create dramatic differences in the chemical properties of two compounds.

Source: B. Goldfarb: "Antifreeze antidote," Chem Matters, *October, p. 4–6, 1996.*

9.3 MULTIPLE COVALENT BONDS

A nonmetal atom with fewer than seven valence electrons can form covalent bonds in more than one way. The atom can share a single pair of electrons with another atom, forming a *single* covalent bond. But the atom can also share two or three pairs of electrons with another atom, in which case there will be two or three bonds, respectively, between the two atoms. When *two* shared pairs of electrons join the same pair of atoms, the bond is called a **double bond,** and when *three* shared pairs are involved, the bond is called a **triple bond.**

The guidelines for writing Lewis structures ***(← p. 357)***, especially guideline 5, can help you to predict when a molecule has multiple bonds rather than only single bonds. Let's apply the guidelines to formaldehyde, H_2CO. There are a total of 12 valence electrons (guideline 1), 2 from two H atoms, 4 from the C atom (Group 4A), and 6 from the O atom (Group 6A). To complete noble gas configurations, H should form one bond, C four bonds, and O two bonds. Because C forms the most bonds, it is the central atom and we can write a skeleton structure (guideline 2).

You might have written the skeleton structure as O—C—H—H but remember that H forms only one bond. Another possible skeleton is H—O—C—H, but with this skeleton it is impossible to achieve an octet around carbon without having more than two bonds to oxygen.

```
H—C—H
   |
   O
```

Putting bonding pairs and lone pairs in the skeleton structure according to guideline 3 yields a structure in which oxygen has an octet but carbon does not.

```
H—C—H
  |
 :O:
```

Applying guideline 5, we use one of the lone pairs on oxygen as a shared pair with carbon to change the C—O single bond to a double bond. This gives both carbon and oxygen a share in an octet of electrons and each hydrogen a share of two electrons, accounting for all 12 valence electrons and verifying that this is the correct Lewis structure for formaldehyde.

```
H—C—H             H—C—H                      H—C—H
  |     to form     |:     which is written    ||
 :O:               :O:                        :O:
```

The C═O combination, called the *carbonyl group,* is part of several functional groups that are very important in organic and biochemical molecules. The carbonyl-containing —CHO group that appears in formaldehyde is known as the *aldehyde functional group.*

Formaldehyde, a gas at room temperature, is the simplest compound with an aldehyde functional group.

As another example of multiple bonds, let's consider the Lewis structure for molecular nitrogen, N_2. The total number of valence electrons available is 10 (5 from each N). If two nonbonding pairs of electrons (one pair from each N) become bonding pairs to give a triple bond (guideline 5), the octet rule is satisfied. This is the correct Lewis structure of N_2:

:N≡N:

Exercise 9.2 Lewis Structures

Why is :N̤̈—N̤̈: an incorrect Lewis structure for N_2?

A molecule can have more than one multiple bond, as in carbon dioxide, where carbon is the central atom. There are a total of 16 valence electrons in CO_2, and the skeleton structure uses 4 of them (2 shared pairs):

O—C—O

Adding lone pairs to give each O an octet of electrons uses up the remaining 12 electrons but leaves C needing 4 more valence electrons to complete an octet.

:Ö̤—C—Ö̤:

With no more valence electrons available, the only way that carbon can have 4 more valence electrons is to have a lone pair of electrons on each oxygen form covalent bonds to carbon. In this way, the 16 valence electrons are accounted for and each atom has an electron octet.

:Ö̤—C—Ö̤: forms :Ö═C═Ö:

PROBLEM-SOLVING EXAMPLE 9.2 Lewis Structures

Write Lewis structures for (a) CF_2Cl_2, a chlorofluorocarbon implicated in stratospheric ozone depletion (Section 12.11); (b) methylamine, CH_3NH_2; and (c) the nitronium ion, NO_2^+.

```
         ..                H     H
        :Cl:               |     |
   ..    |   ..         H—C—————N:
  :F—C—Cl:                |     |         [ ..     .. ]+
   ..    |   ..           H     H         [ O=N=O ]
        :F:                               [ ..     .. ]
         ..
```

Answer (a) 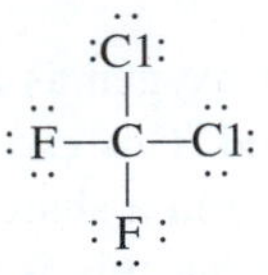 (b) 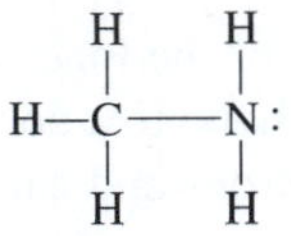 (c)

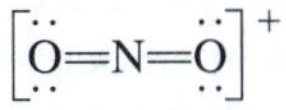

Explanation We can use the guidelines to write these Lewis structures.

(a) We can consider CF_2Cl_2 as a methane molecule, CH_4, with four hydrogen atoms replaced with F and Cl. There are 32 valence electrons in the molecule (4 from C; 7 from each Cl and F). Carbon can form four bonds, F and Cl can form one bond each, and so carbon is the central atom. The skeleton structure with a single bond between each pair of atoms requires eight electrons.

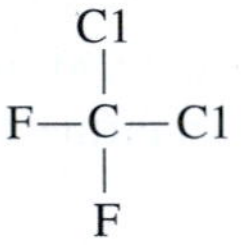

```
     Cl
     |
  F—C—Cl
     |
     F
```

The octet rule is satisfied for C and, if the 24 remaining electrons are placed as lone pairs around the Cl and F atoms, the octet rule for them is also satisfied. The 32 valence electrons have been used, and no multiple bonds are present.

(b) In CH_3NH_2, an $—NH_2$ group has replaced an H in methane, and there are 14 valence electrons (1 from each H, 5 from N, and 4 from C). Carbon can form more bonds than N or H, and C is the central atom; N can form three bonds, but H, as always, forms only one bond. The skeleton structure with a single bond between each pair of bonded atoms uses 12 electrons (6 bonding pairs) and satisfies the octet rule for carbon.

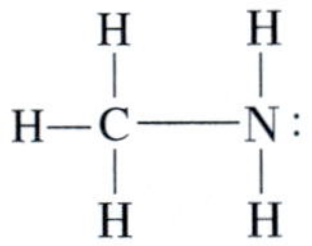

```
     H     H
     |     |
  H—C—————N:
     |     |
     H     H
```

The remaining two electrons are a lone pair that complete the octet on nitrogen.

(c) The total number of valence electrons is 16 (12 from two O; 5 from N; subtract 1 for the positive charge on the ion). Nitrogen can form more bonds than oxygen and should be the central atom. The skeleton drawing with single bonds between N and O and lone pairs around each O uses all 16 electrons.

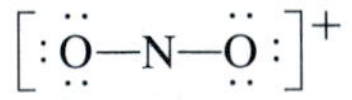

```
  [  ..      ..  ]+
  [ :O—N—O: ]
  [  ..      ..  ]
```

However, the nitrogen atom is four electrons short of an octet, so a lone pair from each O atom is converted to a bonding pair. Now each atom in the ion satisfies the octet rule.

Problem-Solving Practice 9.2

Write Lewis structures for the following: (a) nitrosyl ion, NO^+; (b) hydrogen cyanide, HCN.

Exercise 9.3 Lewis Structures

Which of the following are correct Lewis structures and which are not? Explain what is wrong with the incorrect ones.

(a) $\left[:\ddot{C}-\ddot{N}:\right]^-$ (b) $:\ddot{F}=N-\ddot{Cl}:$ with $:\ddot{Cl}:$ bonded to N (c) $:O\equiv C:$ (d) $:\ddot{O}-C-\ddot{Cl}:$

Multiple Bonds in Hydrocarbons

Carbon atoms are connected by double bonds in many compounds, and triple bonds in a smaller number of compounds. **Alkenes** *are hydrocarbons that have one or more carbon-carbon double bonds, C═C.* The general formula for alkenes with one double bond is C_nH_{2n}, where $n = 2, 3, 4$, and so on. The first two members of the alkene series are ethene (CH_2CH_2) and propene (CH_3CHCH_2), commonly called ethylene and propylene, particularly when referring to the polymers polyethylene and polypropylene, which will be discussed in Section 11.8.

```
H   H          H         H
|   |          |         |
C = C      H — C — C  =  C
|   |          |    |    |
H   H          H    H    H
```

ethene propene

Alkenes are said to be **unsaturated hydrocarbons.** The carbon atoms connected by double bonds are the unsaturated sites—they contain fewer hydrogen atoms than the corresponding alkanes (ethene, CH_2CH_2; ethane, CH_3CH_3). Alkenes are named by using the name of the corresponding alkane ***(← p. 365)*** to indicate the number of carbons and the suffix *-ene* to indicate one or more double bonds. The first member, ethene or ethylene, is the most important raw material used in the organic chemical industry. It ranks fourth in the top 25 chemicals (see inside back cover—1995 rankings) and is the number-one organic chemical. Nearly 47 billion pounds were produced in 1995 for use in making polyethylene, antifreeze (ethylene glycol), ethanol, and other chemicals.

Hydrocarbons with one or more triple bonds, —C≡C—, per molecule are called **alkynes.** The general formula for alkynes with one triple bond is C_nH_{2n-2}, where $n = 2, 3, 4$, and so on. The simplest one is ethyne, commonly called acetylene (C_2H_2).

$$H-C\equiv C-H$$

A mixture of acetylene and oxygen burns with a flame hot enough (3000 °C) to cut steel (Figure 9.2).

Although quite prevalent in carbon compounds, the formation of double and triple bonds is not as widespread among the atoms of the periodic table as one might expect. At least one of the atoms involved in a multiple bond is almost always C, N, or O, and in many cases both atoms are members of this trio. Other elements complete their octets by forming additional single bonds rather than multiple bonds. An example of this is the binary compound of Si with O. Based upon the Lewis structure

Figure 9.2 Reaction of acetylene with oxygen. Cutting steel with an oxyacetylene torch. *(Joseph Nettis/Photo Researchers, Inc.)*

we just wrote for CO_2, we might predict that Si and O would give a stable molecule SiO_2 with the Lewis structure

$$:\ddot{O}=Si=\ddot{O}:$$

However, silicon does not readily form double bonds. Each silicon atom can form single bonds to four oxygen atoms, but as the Lewis structure shows, each O still needs one electron to have an octet.

$$\begin{array}{ccc} & :\dot{O}: & \\ \cdot\ddot{O} & :Si: & \ddot{O}\cdot \\ & :\dot{O}: & \end{array} \quad \text{or} \quad \begin{array}{ccc} & :\dot{O}: & \\ & | & \\ \cdot\ddot{O}- & Si & -\ddot{O}\cdot \\ & | & \\ & :\dot{O}: & \end{array}$$

If each of the oxygen atoms links to another silicon atom, the octet rule will be satisfied, but then the added silicon atoms will not have an octet of electrons.

$$\begin{array}{ccccc} & & \cdot\dot{Si}\cdot & & \\ & & :\ddot{O}: & & \\ \cdot\dot{Si} & :\ddot{O}: & Si & :\ddot{O}: & \dot{Si}\cdot \\ & & :\ddot{O}: & & \\ & & \cdot\dot{Si}\cdot & & \end{array} \quad \text{or} \quad \begin{array}{ccccc} & & \cdot\dot{Si}\cdot & & \\ & & | & & \\ & & :O: & & \\ & & | & & \\ \cdot\dot{Si}- & \ddot{O}- & Si & -\ddot{O}- & \dot{Si}\cdot \\ & & | & & \\ & & :O: & & \\ & & | & & \\ & & \cdot\dot{Si}\cdot & & \end{array}$$

The process of adding oxygen or silicon atoms can continue indefinitely, producing a giant lattice of covalently bonded atoms (Figure 9.3). In this giant molecule, each silicon is bonded to four oxygen atoms and each oxygen is bonded to two silicon atoms. The molecular formula can be written $(SiO_2)_n$, where n is a very large number. For simplicity it is usually written as SiO_2 although discrete SiO_2 molecules are not present

This example illustrates the need to consider whether the proposed Lewis structure matches the known properties of a substance. For example, carbon dioxide is a

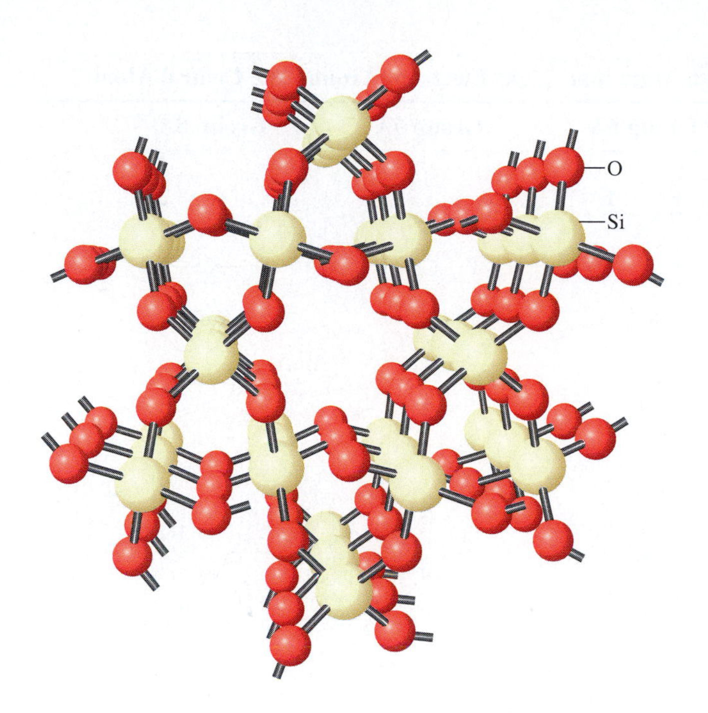

Figure 9.3 **A portion of the giant covalent molecule $(SiO_2)_n$.** The array shown could be extended indefinitely in all directions in a macroscopic crystal. Each silicon atom *(pale yellow)* is covalently bonded to four oxygen atoms *(red)*. Each oxygen bonds to two silicon atoms, giving a silicon to oxygen ratio of 2:4 or 1:2, in accord with the formula SiO_2.

gas at room temperature, so the proposed Lewis structure that depicts it as individual CO_2 molecules is reasonable. However, silicon dioxide (also called silica, of which sand is a common form) is a solid at room temperature, and a three-dimensional, giant molecular structure is more in keeping with this property.

9.4 EXCEPTIONS TO THE OCTET RULE

Experiments that determine the stoichiometry of molecules show that the formulas and bond properties for many molecules are not consistent with the octet rule. Such exceptions arise because of the electron configuration of the atoms of some elements, not because of the octet rule, which is not a rule of nature, but rather one made up to explain experimental data about the stability of compounds. We can, however, use the periodic table to organize these exceptions into two categories—(1) those having Lewis structures with fewer than eight electrons around a central atom, and (2) those with more than an octet of electrons around a central atom. A third category consists of those compounds with an odd number of valence electrons.

Fewer Than Eight Valence Electrons

The boron trifluoride molecule, BF_3, has fewer than an octet of valence electrons around the central boron atom. Boron is a Group 3A element and has only three valence electrons; each fluorine contributes seven. The Lewis structure has only six electrons around the B atom, an exception to the octet rule.

$$\begin{array}{c} :\ddot{F}: \\ | \\ B-\ddot{\underset{..}{F}}: \\ | \\ :\underset{..}{F}: \end{array}$$

TABLE 9.1 **Lewis Structures for Some Ions and Molecules with More than Eight Electrons Around the Central Atom**

	Group 4A	Group 5A	Group 6A	Group 7A	Group 8A
Central atoms with five valence pairs		PF_5	SF_4	ClF_3	XeF_2
Bond pairs		5	4	3	2
Lone pairs		0	1	2	3
Central atoms with six valence pairs	$SnCl_6^{2-}$	PF_6^-	SF_6	BrF_5	XeF_4
Bond pairs	6	6	6	5	4
Lone pairs	0	0	0	1	2

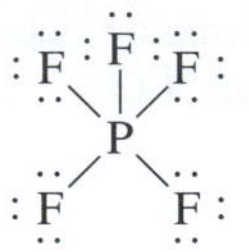

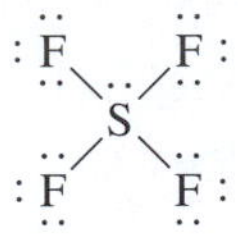

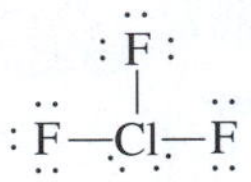

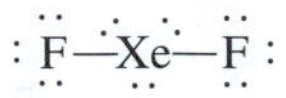

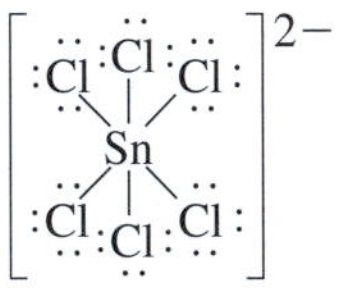

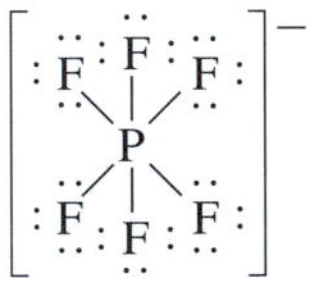

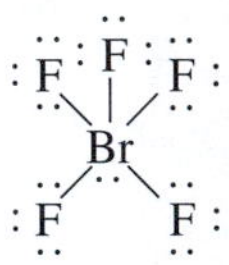

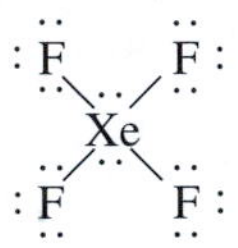

In each case, the numbers of bond pairs and lone pairs about the central atom are given.

The type of bond in which both electrons are provided by the same atom, known as a coordinate covalent bond, is further discussed in Section 9.9.

To achieve an octet around boron, BF_3 is very reactive and readily combines with NH_3 to form a compound with the formula BF_3NH_3. The bonding between BF_3 and NH_3 can be explained by using the lone pair of electrons on N to form a covalent bond with B in BF_3. In this case, the nitrogen lone pair provides *both* of the shared electrons resulting in an octet of electrons for both B and N.

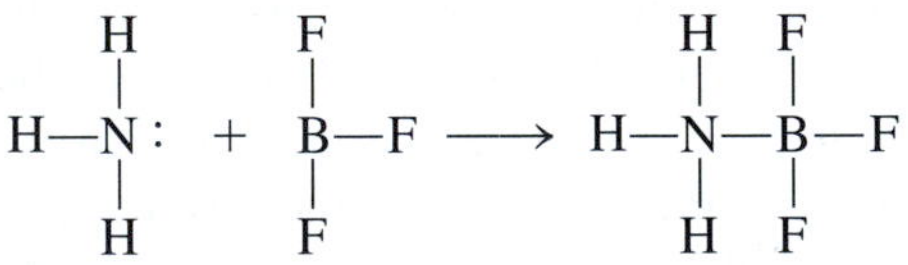

$$\begin{array}{c}\mathrm{H}\\ |\\ \mathrm{H{-}N{:}}\\ |\\ \mathrm{H}\end{array} + \begin{array}{c}\mathrm{F}\\ |\\ \mathrm{B{-}F}\\ |\\ \mathrm{F}\end{array} \longrightarrow \begin{array}{c}\mathrm{H\ \ F}\\ |\ \ \ |\\ \mathrm{H{-}N{-}B{-}F}\\ |\ \ \ |\\ \mathrm{H\ \ F}\end{array}$$

More Than Eight Valence Electrons

Atoms of the third or higher periods can be surrounded by more than four electron pairs in certain compounds because they have empty *d* orbitals with low enough energy to accommodate the extra electrons (← ***p. 323, Figure 8.14***). For example, phosphorus and sulfur commonly form stable molecules in which they are surrounded by more than eight electrons (Table 9.1). In using the periodic table for making comparisons of elements in a given group, it is important to recognize this difference between elements in the second period and elements in later periods. The octet rule is reliable for predicting stable molecules with C, N, O, or F as central atoms; but later members of these groups such as Sn, P, S, and Br can also form stable molecules or polyatomic ions that have more than eight electrons around the central atom.

Molecules with an Odd Number of Valence Electrons

All the molecules we have discussed up to this point have contained only *pairs* of valence electrons. However, there are a few stable molecules that have an odd number of valence electrons. For example, NO has 11 valence electrons, and NO_2 has 17 valence electrons. The most plausible Lewis structures of these molecules are

$$:\dot{N}{=}\ddot{\underset{\cdot\cdot}{O}} \qquad \text{and} \qquad :\underset{\cdot\cdot}{\ddot{O}}{-}\dot{N}{=}\ddot{O}:$$

Molecules such as NO and NO_2 are often called **free radicals** because of the presence of the unpaired electron. Atoms that contain unpaired electrons are also free radicals. How do unpaired electrons affect reactivity? Simple free radicals, such as atoms of H· and Cl·, are very reactive and readily combine with other atoms to give molecules such as H_2, Cl_2, and HCl. Therefore, we would expect free radical molecules to be more reactive than molecules that have all paired electrons, and they are. A free radical either combines with another free radical to form a more stable molecule in which the electrons are paired, or it reacts with other molecules to produce new free radicals. These kinds of reactions are central to the formation of addition polymers (Section 11.8) and air pollutants (Section 14.6). For example, when gaseous NO and NO_2 are released from vehicle exhausts, the colorless NO reacts with O_2 in the air to form brown NO_2. The NO_2 decomposes in the presence of sunlight to give NO and O, both of which are free radicals.

$$:\underset{\cdot\cdot}{\ddot{O}}{-}\dot{N}{=}\ddot{O}: \xrightarrow{\text{sunlight}} :\dot{N}{=}\underset{\cdot\cdot}{\ddot{O}} + :\underset{\cdot\cdot}{\dot{O}}\cdot$$

The free O atom reacts with O_2 in the air to give ozone, O_3, an air pollutant that affects the respiratory system (Section 14.12). Free radicals also have a tendency to combine with themselves to form **dimers,** substances made from two smaller units. For example, when NO_2 gas is cooled it dimerizes to N_2O_4.

As expected, NO and NO_2 are paramagnetic (← *p. 329*) because of the odd number of electrons. Experimental evidence indicates that O_2 is also paramagnetic (Figure 9.4) with two unpaired electrons and a double bond. The predicted Lewis structure for O_2 shows a double bond, but all the electrons would be paired. It is impossible to write a conventional Lewis structure of O_2 that is in agreement with the experimental results.

Figure 9.4 Paramagnetism of liquid oxygen. Liquid oxygen is suspended between the poles of a magnet because O_2 is paramagnetic. Paramagnetic substances are attracted into a magnetic field. *(S. Ruren Smith)*

PROBLEM-SOLVING EXAMPLE 9.3 Exceptions to the Octet Rule

Write the Lewis structure for (a) tellurium tetrabromide, $TeBr_4$; (b) dichloroiodide ion, ICl_2^-; and (c) beryllium dichloride, $BeCl_2$.

Answer

(a) :Br: above and below Te; :Br—Te—Br: (Te with one lone pair, four Te—Br bonds)

(b) $[:\ddot{Cl}—\ddot{I}—\ddot{Cl}:]^-$

(c) $:\ddot{Cl}—Be—\ddot{Cl}:$

Explanation

(a) $TeBr_4$ has 34 valence electrons, 8 of which are distributed among four Te—Br bonds. Of the remaining 26 lone pair electrons, 24 complete octets for the Br atoms. The other 2 electrons form a lone pair on Te, which has a total of 10 electrons (5 pairs: 4 shared; 1 unshared) around it, acceptable for a period 5 element.

(b) There are a total of 22 valence electrons: 14 from two Cl atoms, 7 from I, and 1 for the 1− charge on the ion. Forming two I—Cl single bonds and then distributing six of the remaining nine electron pairs as lone pairs on Cl atoms to satisfy the octet rule uses a total of eight electron pairs.

$$[:\ddot{Cl}—I—\ddot{Cl}:]^-$$

The remaining three electron pairs are placed on iodine, which can accommodate more than eight electrons because it is from the fifth period.

(c) $BeCl_2$ uses 4 of its 16 valence electrons to form 2 Be—Cl bonds. The remaining 12 electrons complete 3 lone pairs around each chlorine atom, giving them an octet and leaving Be with only two electron pairs. $BeCl_2$ does not have enough valence electrons and is an exception to the octet rule.

Because it is smaller than other alkaline earth atoms, beryllium is the only Group 2A element that usually forms molecular compounds.

Problem-Solving Practice 9.3

Write the Lewis structure for each of the following molecules or ions. Indicate which central atoms break the octet rule, and why.

(a) BeF_2 (b) ClO_2 (c) PCl_5 (d) BH_2^+ (e) IF_7

9.5 BOND PROPERTIES

Bond Length

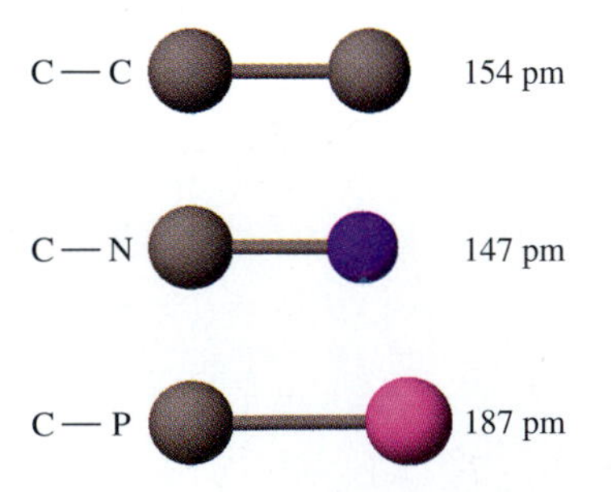

The most important factor determining **bond length,** the distance between nuclei of two bonded atoms, is the sizes of the atoms themselves (Figure 9.5). When you compare bonds of the same type (single, double, or triple), the bond length will be greater for the larger atoms (Table 9.2). Thus, single bonds with carbon increase in length along the following series:

$$C—N < C—C < C—P$$

increase in bond length →

Similarly, a C═O bond will be shorter than a C═S bond, and a C≡N bond will be shorter than a C≡C. Each of these trends can be predicted from the relative sizes shown in the margin (page 371), and confirmed by the common bond lengths given in Table 9.2.

Bond distances are given in picometers (pm) in Table 9.2, but many scientists use nanometers (nm) or the older unit of angstroms (Å). 1 Å equals 100 pm. 1 nm = 1000 pm.

TABLE 9.2 **Some Average Single and Multiple Bond Lengths (in picometers, pm)***

Single Bonds

	I	Br	Cl	S	P	Si	F	O	N	C	H
H	161	142	127	132	138	145	92	94	98	110	74
C	210	191	176	181	187	194	141	143	147	154	
N	203	184	169	174	180	187	134	136	140		
O	199	180	165	170	176	183	130	132			
F	197	178	163	168	174	181	128				
Si	250	231	216	221	227	234					
P	243	224	209	214	220						
S	237	218	203	208							
Cl	232	213	200								
Br	247	228									
I	266										

Multiple Bonds

N═N	120	C═C	134
N≡N	110	C≡C	121
C═N	127	C═O	122
C≡N	115	C≡O	113
O═O (in O_2)	112	N≡O	108
N═O	115		

*1 pm = 10^{-12} m

The effect of bond type is evident when bonds between the same two atoms are compared. For example, structural data show that the bonds become shorter in the series

Bond	C—O	C═O	C≡O
Bond Length (pm)	143	122	113

The bond lengths decrease because the atoms are pulled together more closely as the electron density between the atoms increases.

The bond lengths in Table 9.2 are average values, because variations in neighboring parts of a molecule can affect the length of a particular bond. For example, the C—H bond has a length of 105.9 pm in acetylene, HC≡CH, but a length of 109.3 pm in methane, CH_4. Although there can be a variation of as much as 10% from the average values listed in Table 9.2, the average bond lengths are useful for estimating bond lengths and building models of molecules.

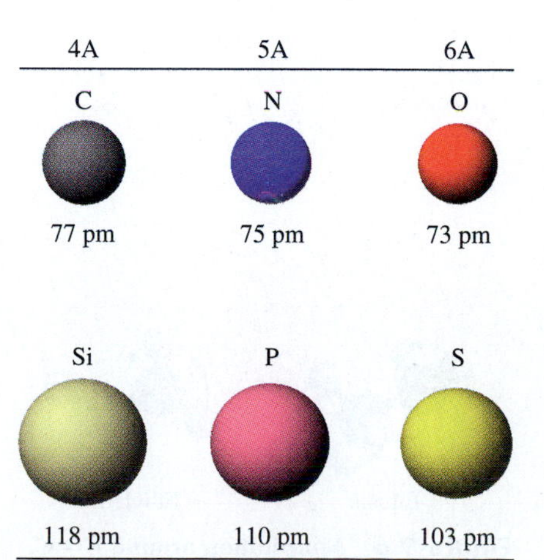

Figure 9.5 **Relative atom sizes for second- and third-period elements of Groups 4A, 5A, and 6A.**

PROBLEM-SOLVING EXAMPLE 9.4 **Bond Lengths**

In each pair of bonds, predict which will be shorter.

(a) Si—O or P—O (b) C═C or C—C (c) C═C or C═O

Answer The shorter bonds will be (a) P—O; (b) C=C; (c) C=O.

Explanation (a) P—O is shorter than Si—O because the P atom is smaller than the Si atom.

(b) C=C is shorter than C—C because the more electrons that are shared by atoms, the more closely the atoms are pulled together.

(c) C=O is shorter than C=C because the O atom is smaller than the C atom.

Problem-Solving Practice 9.4

Explain the increasing order of bond lengths in the following pairs of bonds.

(a) C—S is shorter than C—Si (b) C—Cl is shorter than C—Br

(c) N≡O is shorter than N=O

Bond Energy

In any chemical reaction, bonds are broken and new bonds are made. Section 6.5 (← *p. 242*) described energies associated with bond breaking and bond making in terms of the *bond energy, D,* the enthalpy change required to *break* a particular bond in a mole of gaseous molecules. Bond energies are always positive because *in the process of breaking bonds in a molecule work must be done* against the force holding the atoms together in order to move them apart. That is, *bond breaking is endothermic.* Conversely, the *formation of bonds in a molecule is always exothermic* as the potential energy of the atoms is lower when they are bonded together (← ***p. 355, Figure 9.1***).

The amount of energy released when a mole of a particular bond is made equals that when 1 mol of that bond is broken. For example, the H—Cl bond energy is 431 kJ/mol, indicating that 431 kJ must be supplied to break 1 mol of H—Cl bonds (+431 kJ); when 1 mol of H—Cl bonds is formed, 431 kJ is released (−431 kJ).

You've seen that the greater the number of bonding electrons between a pair of atoms, the shorter the bond. It is therefore reasonable to expect that multiple bonds are stronger than single bonds. *As the number of bonds between two atoms increases, the bond gets shorter and stronger.* For example, the bond energy of C=O in CO_2 is 803 kJ/mol and that of C≡O is 1075 kJ/mol. In fact, the CO triple bond in carbon monoxide is the strongest known covalent bond. Nitrogen, N_2, is very stable at normal temperatures because of the strength of its N≡N triple bond (946 kJ/mol).

Exercise 9.4 Bond Distance

Arrange C=N, C≡N, and C—N in order of decreasing bond distance. Is the order for decreasing bond energy the same or the reverse order? Explain.

Double Bonds and Isomerism

As the number of bonding pairs between a pair of atoms increases, the bond gets stronger and shorter (← ***p. 371, Table 9.2***). For example, a C—C bond in alkanes is 154 pm in contrast to only 134 pm for a C=C bond in alkenes. The C=C double bond also creates another important difference between alkanes and alkenes—the degree of flexibility of the carbon-carbon bonds in the molecules. The C—C single bonds in alkanes allow the carbon atoms to rotate freely along the C—C bond axis (Figure 9.6). But in alkenes, the C=C double bond prevents such free rotation. This limitation is responsible for the ***cis-trans*** **isomerism** of alkenes.

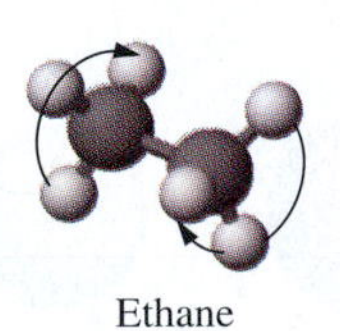

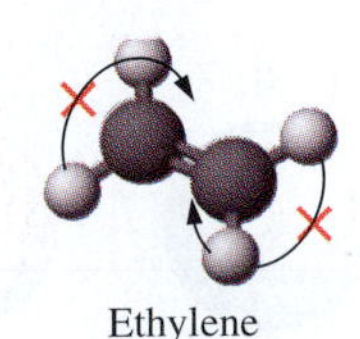

Figure 9.6 Nonrotation around C=C. Rotation along the carbon-to-carbon bond axis occurs freely in ethane, but not in ethylene due to its double bond.

Two or more compounds with the same molecular formula but different arrangements of atoms are called *isomers* (← ***p. 87***). *Cis-trans* isomerism is a form of **stereoisomerism** in which the isomers have the same molecular formulas and the same atom-to-atom bonding sequences, *but the atoms differ in their arrangement in space.* Stereoisomerism causes significant differences in the properties of the stereoisomers, more so than in constitutional (structural) isomerism.

Constitutional (structural) isomers differ in the order in which their atoms are bonded together (**p. 87**). The atoms in *cis-trans* isomers have the same atom-to-atom bonding sequences, but different arrangements in space.

We begin considering *cis-trans* isomerism by starting with ethene, C_2H_4, a molecule whose six atoms lie in the same plane.

$$\mathrm{H_2C{=}CH_2}$$

Cis-trans isomerism does not occur in ethene, but it occurs if two chlorine atoms replace two hydrogen atoms, one on each carbon atom of ethene, to form ClHC=CHCl. In this case, experimental evidence confirms the existence of two compounds with the same set of bonds. The difference between the two compounds is the location in space of the two chlorine atoms. The ***cis* isomer** has two chlorine atoms on the *same* side in the plane of the double bond; the ***trans* isomer** has two chlorine atoms on *opposite* sides of the double bond (Figure 9.7).

Both compounds are called 1,2-dichloroethene (the 1 and 2 indicate that the two chlorine atoms are attached to the first and second carbon atoms), but the *cis* and *trans* prefixes distinguish them from each other. Note that the two compounds (stereoisomers) have different properties.

If free rotation occurred around a carbon-carbon double bond, these two compounds would be the same.

cis-1,2-dichloroethene (Cl and Cl on the same side of C=C; H and H on the same side)

trans-1,2-dichloroethene (H and Cl on the upper side; Cl and H on the lower side)

	cis-1,2-dichloroethene	*trans*-1,2-dichloroethene
Melting Point	−80.5 °C	−50 °C
Boiling Point (1 atm)	60.3 °C	47.5 °C
Density (at 20 °C)	1.284 g/mL	1.265 g/mL

Two chlorine atoms can also bond to the first carbon to give 1,1-dichloroethene, which does not have *cis* and *trans* isomers because each carbon atom is attached to two identical atoms. *Cis-trans* isomerism in alkenes is possible *only when both of the carbon atoms connected by the double bond have two different groups attached.* (For the sake of simplicity, the word "groups" refers to both atoms and groups of atoms.)

$$\mathrm{Cl_2C{=}CH_2}$$

1,1-dichloroethene

Melting Point	−122.1 °C
Boiling Point (1 atm)	37 °C
Density (at 20 °C)	1.21 g/mL

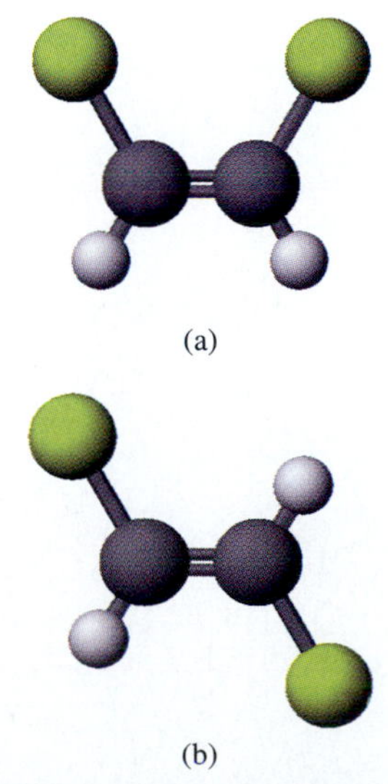

Figure 9.7 ***Cis-trans* isomerism.** *Cis-trans* isomerism occurs in 1,2-dichloroethene.

When there are four or more carbon atoms in an alkene, the possibility exists for *cis* and *trans* isomers even when only carbon and hydrogen atoms are present. For example, 2-butene has both *cis* and *trans* isomers. (The 2 indicates that the double bond is at the number 2 carbon atom, with the straight carbon chain beginning with carbon 1.)

```
H       H            H       CH3
 \     /              \     /
  C===C                C===C
 /     \              /     \
H3C     CH3        H3C       H
```

cis-2-butene *cis*-2-butene

Melting Point	−138.9 °C	−105.5 °C
Boiling Point (1 atm)	3.7 °C	0.9 °C
Density (at 20 °C)	0.621 g/mL	0.604 g/mL
ΔH_f°	−29.7 kJ/mol	−33.0 kJ/mol

PROBLEM-SOLVING EXAMPLE 9.5 *Cis* and *Trans* Isomers

Which of the following molecules can have *cis* and *trans* isomers? For those that do, write the structural formulas for the two isomers and label them *cis* and *trans*.

(a) $(CH_3)_2C{=}CCl_2$ (b) $CH_3ClC{=}CClCH_3$ (c) $CH_3BrC{=}CClCH_3$

Answer (b) and (c) have *cis-trans* isomers.

```
(b) H3C       CH3      Cl        CH3
       \     /           \     /
        C===C             C===C
       /     \           /     \
     Cl       Cl      H3C       Cl
```

cis *trans*

```
(c) H3C       CH3      Br        CH3
       \     /           \     /
        C===C             C===C
       /     \           /     \
     Br       Cl      H3C       Cl
```

cis *trans*

Explanation Because both the groups on each carbon in part (a) are the same, it cannot have *cis* and *trans* isomers. The structural formulas for (b) and (c) are written to show C1 atoms and CH_3 groups, respectively, on the same *(cis)* and opposite *(trans)* sides of the double bond.

Problem-Solving Practice 9.5

Which of the following molecules can have *cis* and *trans* isomers? For those that do, write the structural formulas for the two isomers and label them *cis* and *trans*.

```
     CH3  H                 H  H                  H  Cl H
     |    |                 |  |                  |  |  |
H3C—C====C—CH3          H2C=C—C—CH3           Br—C—C=C—CH3
                               |                  |
                               H                  H
```

(a) 2-methyl-2-butene (b) 1-butene (c) 1-bromo-2-chloro-2-butene

9.6 LEWIS STRUCTURES AND RESONANCE

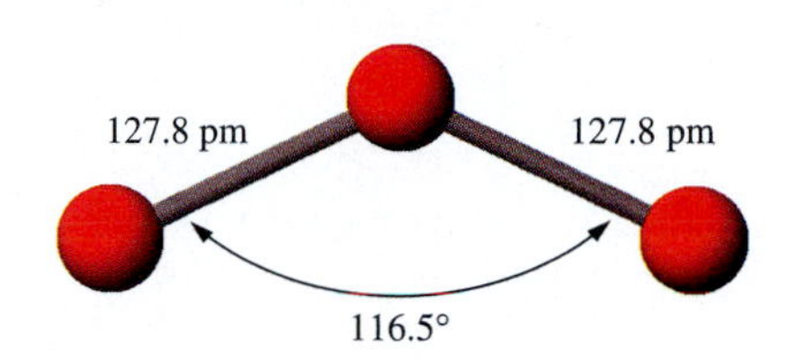

Ozone, O_3, is both beneficial and harmful. The ozone layer in the upper stratosphere protects the earth and its inhabitants from intense ultraviolet radiation from the sun, but ozone pollution in the lower atmosphere causes respiratory problems (see Chapter 14). Ozone is an unstable, blue, diamagnetic gas with a pungent odor.

As you have seen, the number of bonding electron pairs between two atoms is important in determining bond length and strength. The experimentally measured lengths for the two oxygen-oxygen bonds in ozone are the same, 128 pm, implying that both bonds contain the same number of bond pairs. However, using the guidelines for writing Lewis structures, you might come to a different conclusion. Two possible Lewis structures are

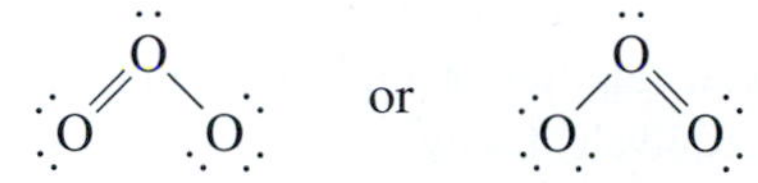

Each structure shows a double bond on one side of the central O atom and a single bond on the other side. If either one were the actual structure of O_3, then one bond (O═O) should be shorter than the other (O—O), but this is not the case. That the oxygen-to-oxygen bond is neither a "pure" double bond nor a single bond is supported by the fact that the 128-pm experimental bond length is longer than O═O (112 pm) but shorter than O—O (132 pm). Therefore, a single Lewis structure that is consistent with the experimental data cannot be written. When this situation arises, the concept of resonance is invoked to reconcile the experimental observation with two or more Lewis structures for the same molecule. Each of the Lewis structures, called **resonance structures,** is thought of as contributing to the true structure that cannot be written. The actual structure of O_3 is neither of the Lewis structures above, but a composite called a **resonance hybrid.** It is conventional to connect the resonance structures with a double-headed arrow, ⟷, to emphasize that the actual bonding is a composite of these structures.

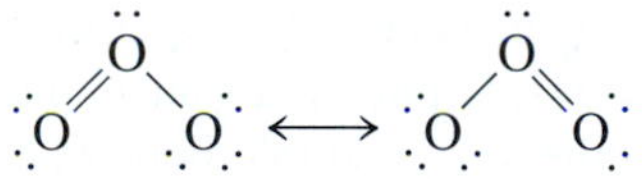

"Resonance" is an unfortunate term, because it implies that the molecule somehow resonates, moving in some way to form different kinds of molecules, which is not true. There is only one kind of ozone molecule.

resonance structures

A resonance hybrid is often written as a composite picture in which a dotted line represents **delocalized electrons,** those spread evenly over the molecule but not associated with any specific pair of bonded atoms. For ozone, such a structure is

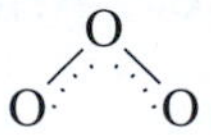

The resonance concept is useful whenever there is a choice about which of two or three atoms contribute lone pairs to achieve an octet of electrons about a central atom by multiple bond formation.

When applying the concept of resonance, there are several important things to keep in mind.

- *Lewis structures contributing to the resonance hybrid structure differ only in the assignment of electron pair positions, never atom positions.*
- *Contributing Lewis structures differ in the number of bond pairs between pairs of atoms.*

- *The resonance hybrid structure represents a single intermediate structure and not different structures that are continually changing back and forth.*

To illustrate the use of resonance, consider what happens in writing the Lewis structure of the carbonate ion, CO_3^{2-}, which has 24 valence electrons (4 from C, 18 from three O atoms, and 2 for the 2− charge). Writing the skeleton structure and putting in lone pairs so that each O has an octet uses 24 electrons but leaves carbon without an octet:

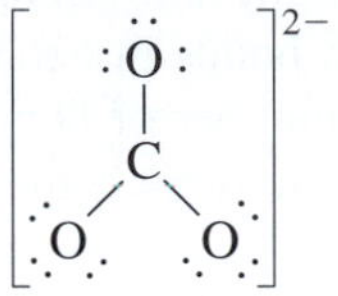

To give carbon an octet requires changing a single bond to a double bond, and this can be done in three equivalent ways:

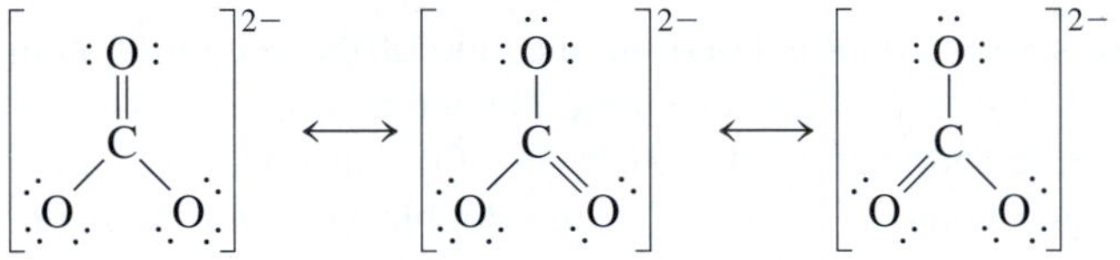

Writing the Lewis structures of oxygen-containing anions often requires resonance.

These three resonance structures contribute to the resonance hybrid, which is drawn with dotted lines representing the two delocalized electrons spread over the three C—O bonds.

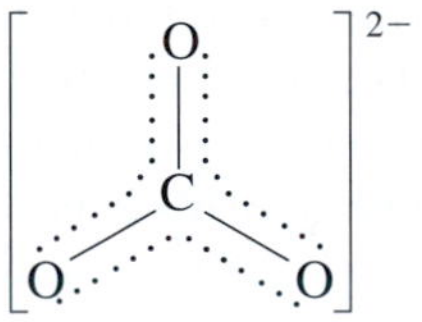

This representation is in agreement with experimental results: All three carbon-oxygen bond distances are 129 pm, a distance intermediate between the C—O single bond (143 pm) and the C═O double bond (122 pm) distance.

PROBLEM-SOLVING EXAMPLE 9.6 Writing Resonance Structures

Write the resonance structures for NO_3^-.

Answer

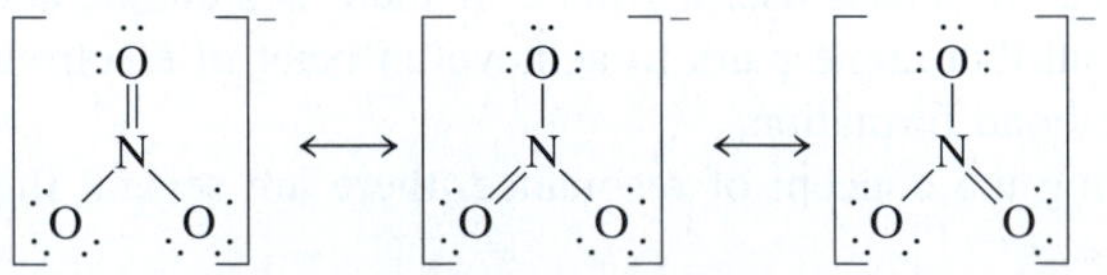

Explanation The NO_3^- ion contains 24 valence electrons (5 from the N atom, 18 from three O atoms, and 1 for the 1− charge). Placing single bonds and lone pairs to give each O atom an octet produces a structure without an octet on the central N atom. A lone pair can be converted to a double bond in three equivalent ways, to give three resonance hybrid structures.

Problem-Solving Practice 9.6

The nitrogen-oxygen bond lengths in NO_2^- are both 124 pm. Compare this with the bond distances given in Table 9.2 for N—O and N=O bond lengths. Account for any difference.

9.7 AROMATIC COMPOUNDS

Benzene, C_6H_6, is an important industrial compound, ranking 16th in the list of the top 25 chemicals produced in 1995 in the United States (see inside rear cover). It is the simplest member of a very large family of compounds known as **aromatic compounds,** which are compounds containing one or more benzene or benzene-like rings. The word "aromatic" was derived from *aroma,* which describes the rather strong and often pleasant odor of these compounds. Benzene and its derivatives are used to manufacture plastics, detergents, pesticides, drugs, and other organic chemicals.

Resonance and the Structure of Benzene

To 19th-century chemists, the C_6H_6 molecular formula of benzene implied that it was an unsaturated compound because it lacked the ratio of carbon to hydrogen found in saturated hydrocarbons. A six-membered ring structure with alternating double bonds uses all the available valence electrons and gives each carbon atom an octet of valence electrons.

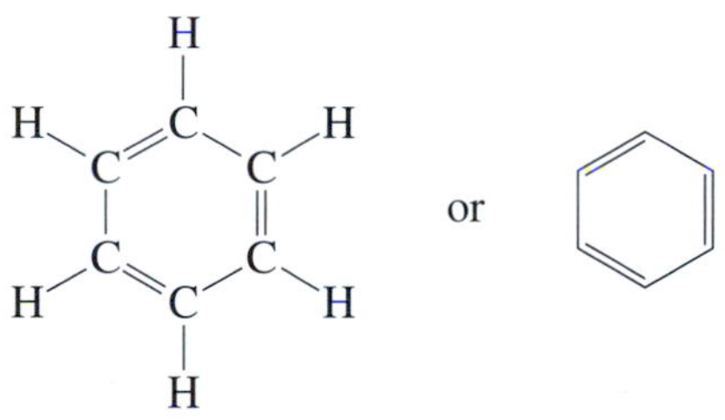

H H H H H H C C C C C C or

But benzene, for example, incorporates bromine atoms by a *substitution* reaction, not an addition reaction as alkenes do under the same conditions. Rather, a bromine atom replaces (substitutes for) a hydrogen of benzene to produce bromobenzene, and the displaced hydrogen combines with the free bromine atom to form hydrogen bromide (HBr).

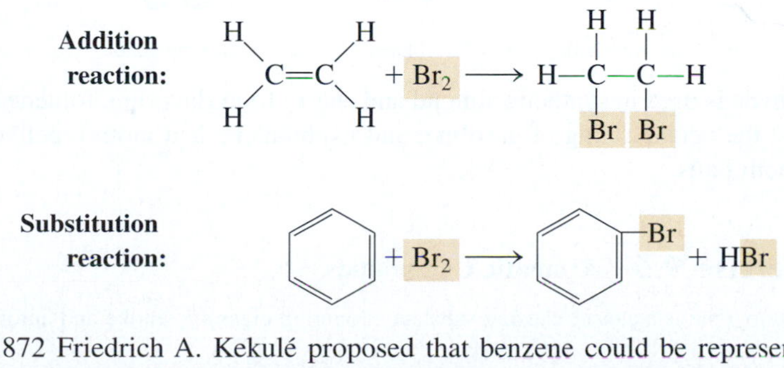

Addition reaction: $H_2C{=}CH_2 + Br_2 \longrightarrow$ H—CH(Br)—CH(Br)—H

Substitution reaction: $C_6H_6 + Br_2 \longrightarrow C_6H_5Br + HBr$

In 1872 Friedrich A. Kekulé proposed that benzene could be represented by a combination of two structures, which we now call resonance structures.

But neither of these alternating single- and double-bond resonance structures accurately represents benzene. Experimental structural data for benzene indicate a planar, symmetrical molecule in which all carbon-carbon bonds are equivalent. Each carbon-carbon bond is 139 pm, intermediate between the length of a C—C single bond (154 pm) and a C═C double bond (134 pm). Benzene is a resonance hybrid of these resonance structures—it is a molecule in which the six "pairs" of electrons of the suggested double bonds are actually delocalized uniformly around the ring.

When hydrogen and carbon atoms are not shown, the benzene ring is frequently written as a hexagon with a circle in the middle. Each corner in the hexagon represents one carbon atom and one hydrogen atom, and each line represents a single C—C bond. The circle represents the six delocalized electrons spread evenly over all of the carbon atoms.

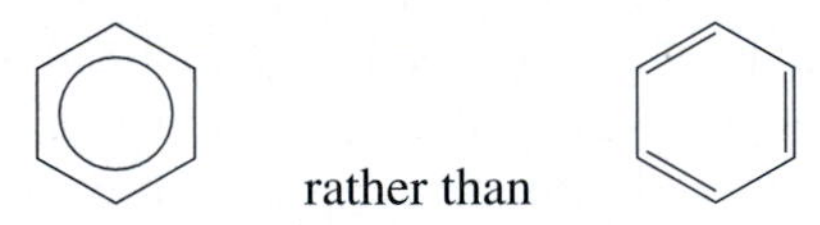

Whenever you see a formula with one or more carbon rings with central circles, the compound is aromatic, like benzene. Benzaldehyde and toluene are examples of the many aromatic compounds with functional groups or alkyl groups bonded to an aromatic ring. Naphthalene is representative of a large group of aromatic compounds with more than one ring joined by common carbon-carbon bonds. Rings joined in this way are referred to as *fused rings.*

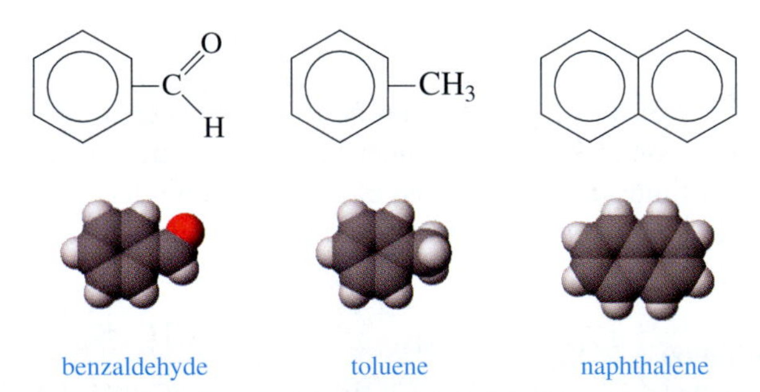

benzaldehyde toluene naphthalene

Benzaldehyde is used in synthetic almond and cherry food flavoring; toluene and benzene boost the octane rating of gasoline; and naphthalene is a moth repellant in one kind of moth balls.

Exercise 9.5 Aromatic Compounds

Benzo(α)pyrene is a cancer-causing substance found in cigarette smoke and automobile exhaust.

Exercise 9.5 continued

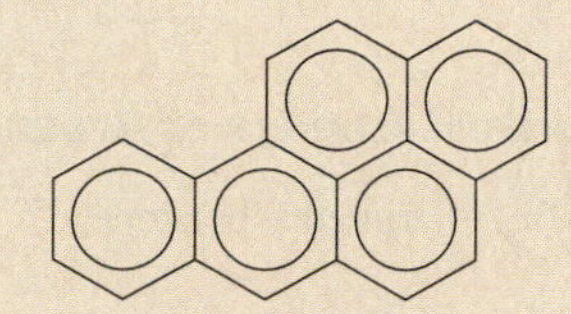

benzo(α)pyrene

(a) Write the structural formula of benzo(α)pyrene using rings with alternating single and double bonds.
(b) Write the molecular formula of benzo(α)pyrene.

Exercise 9.6 Aromatic Compounds

Write the resonance structures for naphthalene.

Constitutional Isomers of Aromatic Compounds

Because benzene is a planar molecule, constitutional isomers are possible when two or more groups are substituted for hydrogen atoms on the benzene ring. If two groups are substituted for two hydrogen atoms on the benzene ring, three constitutional isomers are possible. When the two groups are methyl groups, the compound is xylene, and the prefixes *ortho*-, *meta*-, and *para*- are used to differentiate the three isomers.

Ortho- indicates that two substituents are on adjacent carbon atoms on a benzene ring; *meta*-, two substituents separated by one carbon atom on a benzene ring; *para*-, two substituents separated by two carbon atoms on a benzene ring.

ortho-xylene *meta*-xylene *para*-xylene

These constitutional isomers differ in melting point, boiling point, and density.

	ortho-xylene	*meta*-xylene	*para*-xylene
Melting Point, °C	−25.2	−47.8	13.2
Boiling Point °C	144.5	139.1	138.4
Density, g/mL	0.876	0.860	0.857

If more than two groups are attached to the benzene ring, numbers must be used to identify them and their positions, as for the three chlorobenzenes:

1,2,3-trichlorobenzene

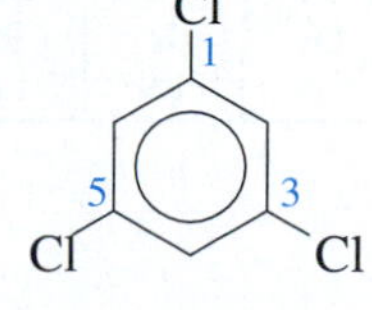

1,2,4-trichlorobenzene

1,3,5-trichlorobenzene

There is no other way to attach three atoms of chlorine to a benzene ring, and only three trichlorobenzenes have been isolated in the laboratory.

Exercise 9.7 Constitutional Isomers of Aromatic Compounds

Write the structural formula of 1,2,4-trimethylbenzene.

9.8 BOND POLARITY AND ELECTRONEGATIVITY

In a molecule such as H_2 or F_2, where both atoms are the same, there is equal sharing of the bonding electrons and the bond is a **nonpolar covalent bond.** When two different atoms are bonded, however, the sharing of the bonding electrons is generally unequal and results in a displacement of the bonding electrons toward one of the atoms. If the displacement is complete, the bond is ionic because electrons have been transferred. If the displacement is less than complete, the bond is said to be a **polar covalent bond.** As you will see in Chapters 10 and 15, properties of molecules are dramatically affected by bond polarity.

Conceptual Challenge Problems CP-9.B and CP-9.C at the end of the chapter relate to topics covered in this section.

Electronegativity is a measure of the ability of an atom in a covalent bond to attract shared electrons to itself. In 1932, Linus Pauling first proposed the concept of electronegativity based on an analysis of bond energies, and the currently accepted values for electronegativities are shown in Figure 9.8.

Electronegativities show a periodic trend (Figure 9.9). *In general, electronegativity increases diagonally upward and to the right of the periodic table.* Because they typically lose electrons, metals are the least electronegative elements, and nonmetals, which have a tendency to gain electrons, are the most electronegative.

Pauling's electronegativity values are relative numbers with an arbitrary value of 4.0 for fluorine, the most electronegative element. The nonmetal with the next high-

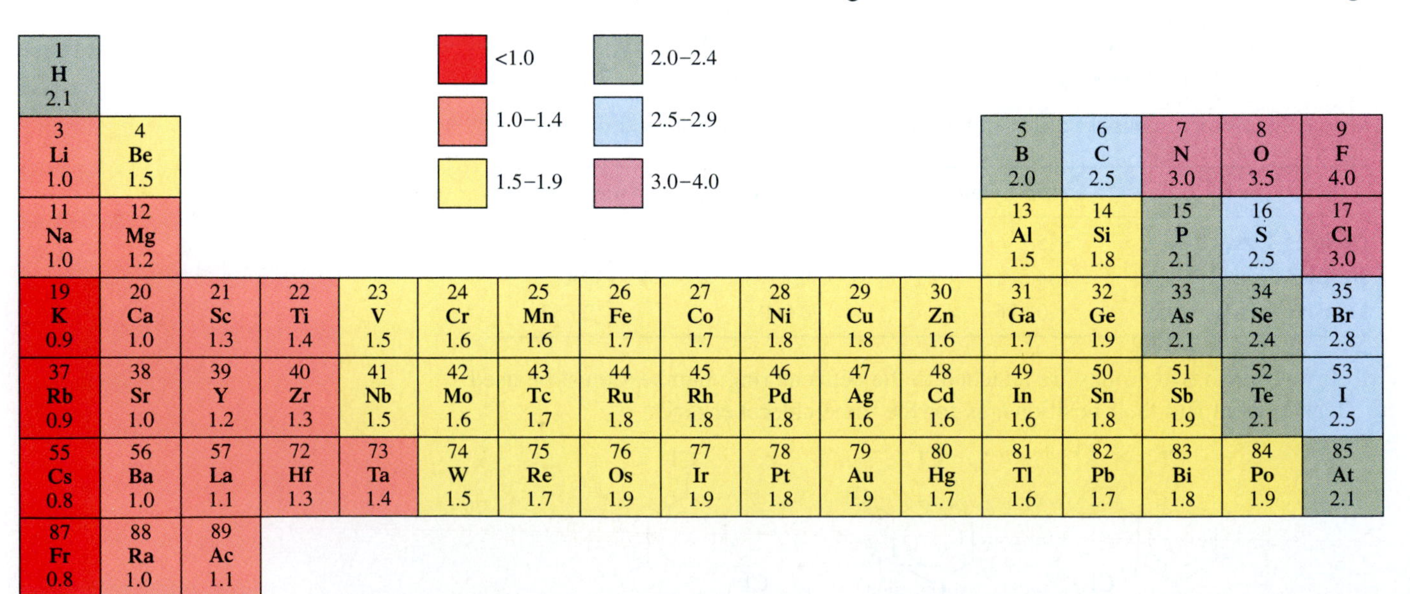

Figure 9.8 Periodic trends in electronegativities. Electronegativity values in a periodic table arrangement.

Portrait of a Scientist • *Linus Pauling (1901–1994)*

Linus Pauling was born in Portland, Oregon, in 1901, where he grew up as the son of a druggist. He earned a B.Sc. degree in chemical engineering from Oregon State College in 1922 and completed his Ph.D. in chemistry at the California Institute of Technology in 1925. Before joining Cal Tech as a faculty member, he traveled to Europe where he worked briefly with Erwin Schrödinger and Niels Bohr (see Chapter 8).

In chemistry Pauling is best known for his work on chemical bonding. Pauling, along with R. B. Corey, proposed the helical and sheet-like secondary structures for proteins. For his bonding theories and his work with proteins, Pauling was awarded the Nobel Prize in Chemistry in 1954. His book *The Nature of the Chemical Bond* has influenced several generations of scientists. Shortly after World War II, Pauling and his wife began a crusade to limit nuclear weapons, a crusade that came to fruition in the limited test ban treaty of 1963. For this effort, Pauling was awarded the 1963 Nobel Prize for Peace. Never before had any person received two unshared Nobel Prizes.

Linus Pauling died in August 1994, leaving a remarkable legacy of breakthrough scientific research and a social consciousness to be remembered in guarding against the possible misapplications of technology.

Linus Pauling *(Oesper Collection in the History of Chemistry/University of Cincinnati)*

est electronegativity is oxygen with a value of 3.5, followed by chlorine and nitrogen, which have the same value of 3.0. Elements with electronegativities of 2.5 or more are all nonmetals in the top right-hand corner of the periodic table. By contrast, elements with electronegativities of 1.3 or less are all metals in the lower left of the periodic table. These elements are often referred to as the most **electropositive** elements, and they are the metals that invariably form ionic compounds. Between these two extremes are most of the remaining metals (largely transition metals) with electronegativities between 1.4 and 1.9, the metalloids with electronegativities between 1.8 and 2.1, and some nonmetals with electronegativities between 2.1 and 2.4.

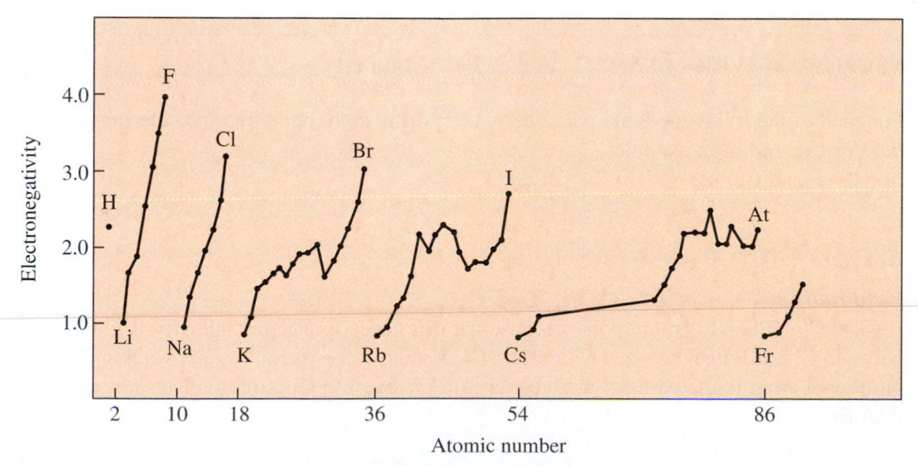

Figure 9.9 Variation of electronegativities with atomic number. The periodic nature of electronegativities when plotted against atomic number.

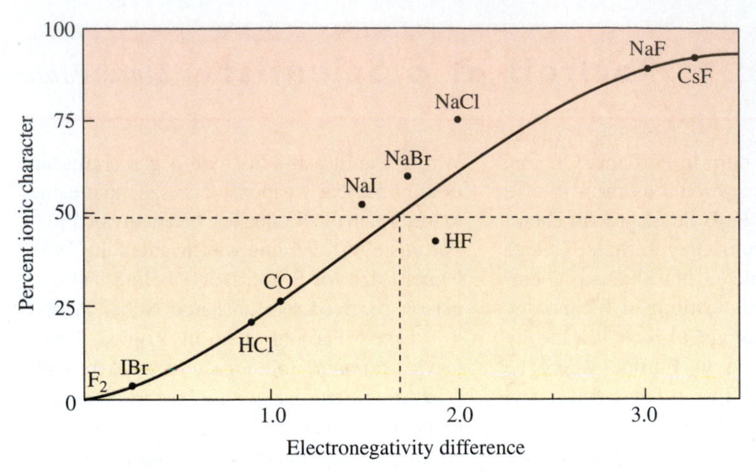

Figure 9.10 Bond character and electronegativity differences. The relationship between electronegativity difference and percent ionic character of a bond.

Electronegativity values are approximate and are primarily used to predict the polarity of covalent bonds. Bond polarity is indicated by writing $\delta+$ by the less electronegative atom and $\delta-$ by the more electronegative atom, where δ stands for partial charge. For example, the polar HCl bond in hydrogen chloride can be represented as

$$\overset{\delta+}{\text{H}}\text{—}\overset{\delta-}{\text{Cl}}$$

All bonds (except those between identical atoms) have some degree of polarity, and the difference in electronegativity values is a qualitative measure of the degree of polarity. The change from nonpolar covalent bonds to slightly polar covalent bonds to very polar bonds to ionic bonds can be regarded as a continuum (Figure 9.10). Examples are H_2 (nonpolar), HI (slightly polar), HF (very polar), and Na^+Cl^- (ionic). *The greater the difference in electronegativity between two elements, the more ionic will be the bond between them.*

PROBLEM-SOLVING EXAMPLE 9.7 Bond Polarity

For each of the following bond pairs, tell which is the more polar and indicate the partial positive and negative poles.
(a) Cl—F and Cl—Br (b) Si—Br and C—Br

Answer (a) $\overset{\delta+}{\text{Cl}}\text{—}\overset{\delta-}{\text{F}}$ (b) $\overset{\delta+}{\text{Si}}\text{—}\overset{\delta-}{\text{Br}}$

Explanation

(a) The Cl—F bond is more polar because the difference in electronegativity is greater between Cl and F than between Cl and Br. The F atom is the partial negative end in ClF, but the Cl atom is the partial negative end in ClBr because Cl is more electronegative than Br.

$$\overset{\delta+}{\text{Cl}}\text{—}\overset{\delta-}{\text{F}} \quad \text{more polar than} \quad \overset{\delta-}{\text{Cl}}\text{—}\overset{\delta+}{\text{Br}}$$

(b) Si—Br is more polar than C—Br because Si is less electronegative than C, so the electronegativity difference is greater between Si and Br.

$$\overset{\delta+\ \delta-}{\text{Si—Br}} \quad \text{more polar than} \quad \overset{\delta+\ \delta-}{\text{C—Br}}$$

Problem-Solving Practice 9.7

For each of the following pairs of bonds, decide which is the more polar. For each polar bond, indicate the partial positive and negative poles.
(a) B—C and B—Cl (b) N—H and O—H

Exercise 9.8

(a) Explain why NaCl is an ionic rather than a polar covalent compound.
(b) Explain why BrF would be predicted to be a polar covalent compound rather than an ionic compound.

Consider the third-period elements. We find that changes in electronegativity difference cause a significant shift in the properties of compounds of these elements with chlorine. The third period begins with the ionic compounds NaCl and $MgCl_2$; both are solids at room temperature. The electronegativity differences between the metal and chlorine are 2.0 and 1.8, respectively. Aluminum chloride is less ionic; with an aluminum-chlorine electronegativity difference of 1.5, its bonding is highly polar-covalent. As electronegativity decreases across the period, the remaining period 3 chlorides—$SiCl_4$, PCl_3, and S_2Cl_2—are molecular compounds, with decreasing electronegativity differences between the nonmetal and chlorine from Si to S. The decreasing electronegativity difference causes a decrease in bond polarity from Si—Cl to P—Cl to S—Cl bonds, culminating in no electronegativity difference in Cl—Cl bonds.

	$SiCl_4$	PCl_3	S_2Cl_2	Cl_2
Electronegativity Difference	1.2	0.9	0.5	0.0
Melting Point, °C	−68.9	−112	−80	−101.5
Boiling Point, °C	57.7	76.0	135.6	−34.0*

*At normal pressure

Sufficient attraction occurs between their molecules so that $SiCl_4$, PCl_3, and S_2Cl_2 are liquids at room temperature. On the other hand, nonpolar chlorine molecules are not attracted to each other as strongly; and consequently, chlorine is a gas at room temperature. Relationships among properties, molecular structure, and attraction between molecules will be discussed in Section 10.5.

9.9 COORDINATE COVALENT BONDS: COMPLEX IONS AND COORDINATION COMPOUNDS

In Section 9.4, the formation of BF_3NH_3 was described as occurring by the sharing of a lone pair from NH_3 with BF_3. This type of covalent bond in which one atom contributes *both* electrons for the shared pair is called a **coordinate covalent bond.** Atoms of some elements with lone pairs of electrons, such as nitrogen, phosphorus, and sulfur, tend to use the lone pairs to form coordinate covalent bonds to supplement their bonding-pair covalent bonds. For example, the formation of the ammonium ion from ammonia results from formation of a coordinate covalent bond between H^+ and the lone pair of electrons of nitrogen in NH_3:

coordinate covalent bond

$$H^+ \quad + \quad :NH_3 \longrightarrow [NH_4]^+$$

hydrogen ion (proton) | ammonia | ammonium ion

Once the coordinate covalent bond is formed, it is impossible to distinguish between the equivalent N—H bonds.

Metals and Coordination Compounds

Much of the chemistry of *d*-transition metals is related to their ability to form coordinate covalent bonds. The result of such bond formation is commonly a **complex ion,** an ion with several molecules or ions connected to a central metal ion or atom by coordinate covalent bonds. For example, the nickel^{2+} ion forms a complex ion with six water molecules:

coordinate covalent bond

$$[Ni(H_2O)_6]^{2+}$$

The molecules or ions bonded to the central metal ion are called **ligands,** from the Latin verb *ligare,* "to bind." Each ligand (in this case a water molecule) has one or more atoms with lone pairs, and these atoms are bonded to the metal atom or ion by coordinate covalent bonds. A complex ion has a net charge, 2+ in the case of $[Ni(H_2O)_6]^{2+}$.

Many of the essential transition metal dietary minerals function as complex ions (← p. 105).

The charge of a complex ion is determined by the charge of the metal ion and any negative ions bonded to it. In $[Ni(H_2O)_6]^{2+}$, the water ligands are neutral, so the charge of the complex ion is the same as that of the Ni^{2+} ion. There are also negative complex ions, for example, $[NiCl_4]^{2-}$, which consists of the Ni^{2+} ion coordinated to four Cl^- ions. The net 2− charge of this complex ion results from the 4 Cl^- and the Ni^{2+}: $(4 \times 1-) + (2+) = 2-$.

Compounds in which complex ions are combined with their oppositely charged ions (counter ions) to form neutral compounds are known as **coordination compounds.** Such compounds are generally brightly colored as solids or in solution (see Figure 19.14a, p. 389). The complex ion part of a coordination compound's formula is usually enclosed in brackets; counter ions are outside the brackets, as in the formula of the compound of chloride ions with the $[Ni(H_2O)_6]^{2+}$ complex ion, $[Ni(H_2O)_6]Cl_2$. The two Cl^- ions compensate for the 2+ charge of the complex ion. $[Ni(H_2O)_6]Cl_2$ is an ionic compound analogous to $CaCl_2$, which also contains a 2+ cation and 2 Cl^- ions. Occasionally, no compensating ions are needed outside the brackets for a coordination compound. For example, the anticancer drug $[Pt(NH_3)_2Cl_2]$ is a coordination compound containing NH_3 and Cl^- ligands coordinated to a central Pt^{2+} ion (the NH_3 and Cl^- are inside the brackets). The two chloride ions bonded directly to the central Pt^{2+} ion compensate for its charge.

When the word "coordinated" is used in chemistry, such as "the chloride ions in $[NiCl_4]^{2-}$ are coordinated to the nickel ion," it means that coordinate covalent bonds have been formed.

Exercise 9.9 Coordination Compounds

In a complex ion a central Cr^{3+} ion is bonded to two ammonia molecules, two water molecules, and two hydroxide ions. Give the formula and the net charge of this complex ion.

PROBLEM-SOLVING EXAMPLE 9.8 Coordination Compounds

For the coordination compound $K_3[Fe(CN)_6]$, identify (a) the central metal ion; (b) the ligands; (c) the formula and charge of the complex ion and the central metal ion.

Answer

(a) Iron; (b) Six cyanide ions, CN^-; (c) $[Fe(CN)_6]^{3-}$

Explanation

(a) Iron (Fe) is the central metal ion.

(b) Cyanide ions, CN^-, are coordinated to the central iron ion.

(c) The charge on three potassium ions is $(3 \times 1+) = 3+$. Therefore, the compensating charge of the complex ion must be 3−, arising from the 6− charge of six cyanide ions $(6 \times 1- = 6-)$, combined with the 3+ charge of the central iron(III) ion: $(6-) + (3+) = 3-$.

Problem-Solving Practice 9.8

For the coordination compound $[Cu(NH_3)_4]SO_4$, identify the (a) counter ion; (b) central metal ion; (c) ligands; (d) formula and charge of the complex ion.

Exercise 9.10 Coordination Compounds

$CaCl_2$ and $[Ni(H_2O)_6]Cl_2$ have the same formula type, MCl_2. Give the formula of a simple ionic compound (non-coordination) that has a formula analogous to $K_2[NiCl_4]$.

Figure 9.11 Monodentate ligands, and bidentate and hexadentate chelating ligands. Some common monodentate and chelating ligands. Chelating ligands share two or more electron pairs with the central metal ion.

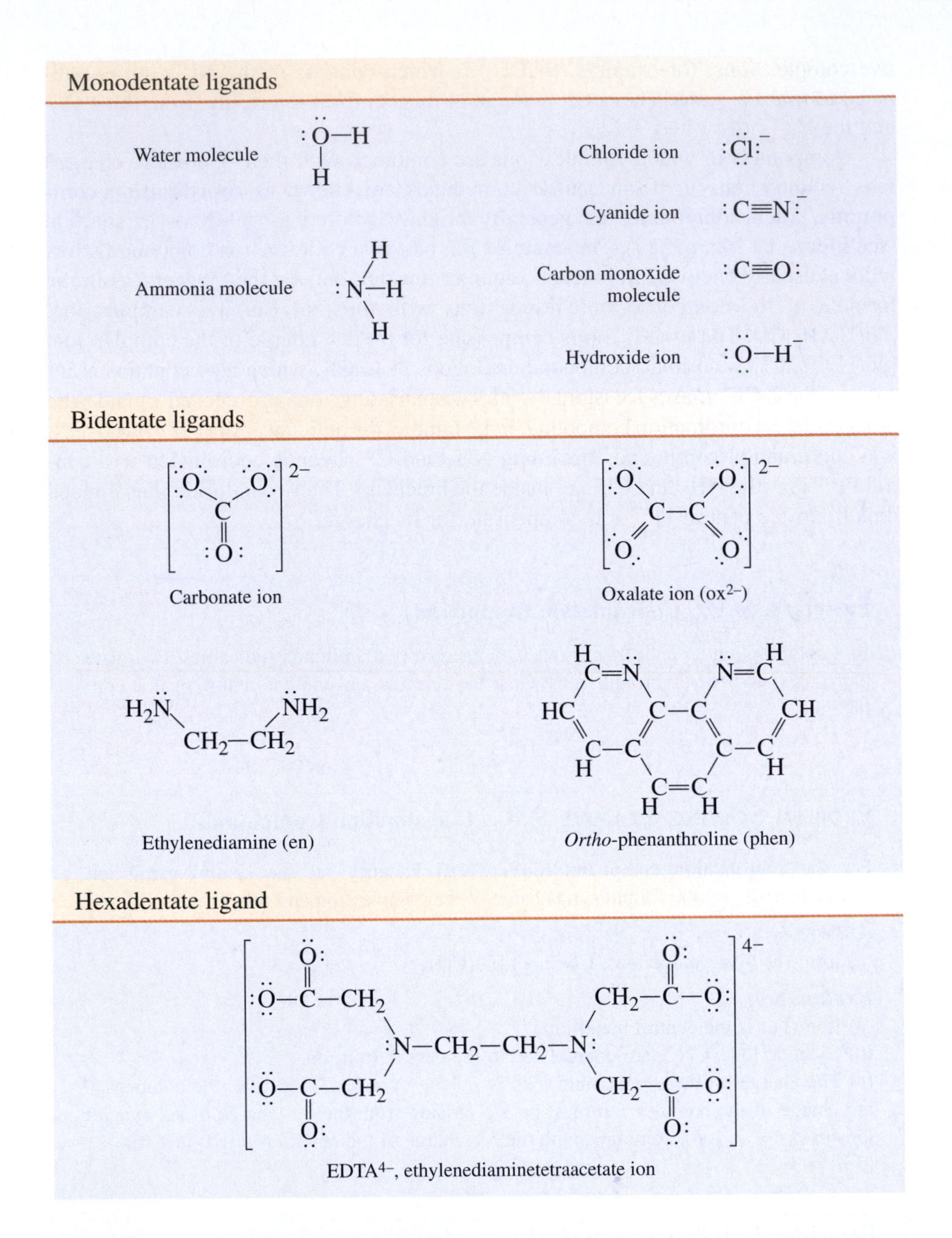

Ligands and Coordination Number

The number of coordinate covalent bonds between the ligands and the central metal ion is the **coordination number** of the metal ion, usually 2, 4, or 6.

Coordination Number	Examples
2	$[Ag(NH_3)_2]^+$, $[AuCl_2]^-$
4	$[NiCl_4]^{2-}$, $[Pt(NH_3)_4]^{2+}$
6	$[Fe(H_2O)_6]^{2+}$, $[Co(NH_3)_6]^{3+}$

Ligands such as H_2O, NH_3, and Cl^- that form only one coordinate covalent bond to the metal are termed **monodentate** ligands. The word "dentate" stems from the Latin word *dentis* for tooth, so NH_3 is a "one-toothed" ligand. Common monodentate ligands are shown in Figure 9.11.

Some ligands can form two or more coordinate covalent bonds to the same metal ion because they have two or more atoms with lone pairs separated by several intervening atoms. These are called **bidentate** ligands. A good example is the bidentate ligand 1, 2-diaminoethane ($H_2NCH_2CH_2NH_2$), commonly called ethylenediamine and abbreviated *en.* When lone pairs of electrons from both nitrogen atoms in en coordinate to a metal ion, a stable five-membered ring is formed (Figure 9.12). Notice that Co^{3+} has a coordination number of 6 in this complex ion.

The word "chelating," derived from the Greek *chele,* "claw," describes the pincer-like way in which a ligand can grab a metal ion. Some common **chelating ligands** are shown in Figure 9.11.

Rules for naming coordination compounds are given in Appendix E.3.

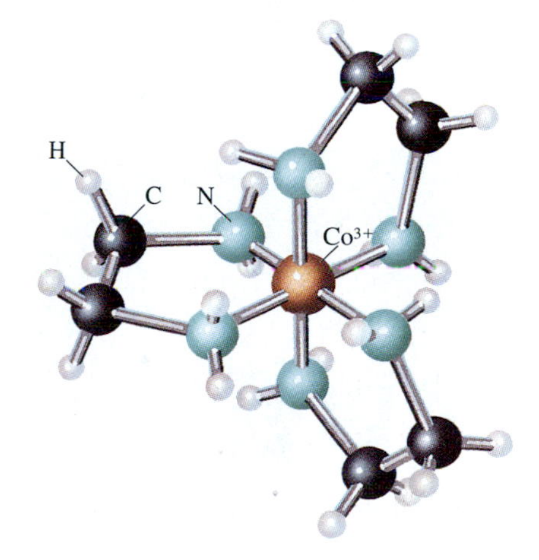

Figure 9.12 **The $[Co(en)_3]^{3+}$ complex ion.** Cobalt ion (Co^{3+}) forms a coordination complex with three ethylenediamine ligands.

PROBLEM-SOLVING EXAMPLE 9.9 Chelating Agents

Two ethylenediamine ligands and two chloride ions form a complex ion with Co^{3+}.
(a) Write the formula for this complex ion.
(b) What is the coordination number of the Co^{3+} ion?
(c) Write the formula of the coordination compound formed by Cl^- counter ions and the Co^{3+} complex ion.

Answer

(a) $[Co(en)_2Cl_2]^+$; (b) Six; (c) $[Co(en)_2Cl_2]Cl$.

Explanation

(a) Ethylenediamine is a neutral ligand, each chloride ion is 1−, and cobalt has a 3+ charge. The charge on the complex ion is 2(0) + 2(1−) + (3+) = 1+.
(b) The coordination number is six because there are six coordinate covalent bonds to the central Co^{3+} ion—two from each bidentate ethylenediamine, and one from each monodentate chloride ion.
(c) The 1+ charge of the complex ion requires a 1− chloride counter ion: $[Co(en)_2Cl_2]Cl$.

Problem-Solving Practice 9.9

The dimethylglyoximate anion (abbreviated DMG),

$$\begin{array}{cc} CH_3C & -CCH_3 \\ \| & \| \\ HO-\ddot{N} & \ddot{N}-O^- \end{array}$$

is a bidentate ligand used to test for the presence of nickel because it reacts with Ni^{2+} to form a beautiful red solid with a coordination number of four (Figure 9.14b). DMG coordinates to Ni^{2+} by the lone pairs on the nitrogen atoms.
(a) How many DMG ions are needed to satisfy a coordination number of four on the central Ni^{2+} ion?
(b) What is the net charge after coordination occurs?
(c) How many atoms are in the rings formed by DMG and Ni^{2+}?

CHEMISTRY YOU CAN DO

A Penny For Your Thoughts

You will need the following items to do the experiment:

- two glasses or plastic cups that will each hold about 50 mL of liquid
- about 30 to 40 mL of household vinegar
- about 30 to 40 mL of household ammonia
- a copper penny

Place the penny in one cup and add 30 to 40 mL of vinegar to clean the surface of the penny. Let the penny remain in the vinegar until the surface of the penny is cleaner (reddish-coppery) than it was before (darker copper color). Pour off the vinegar and wash the penny thoroughly in running water.

Then, place the penny in the other cup and add 30 to 40 mL of household ammonia. Observe the color of the solution over several hours.

1. What did you observe happening to the penny in the ammonia solution?
2. What did you observe happening to the ammonia solution?
3. Interpret what you observed happening to the solution on the nanoscale level, citing observations to support your conclusions.
4. What is necessary to form a complex ion?
5. Are all of these kinds of reactants present in the solution in this experiment? If so, identify them.
6. How do the terms "ligand," "central metal ion," and "coordination complex" apply to your experiment?
7. Try to write a formula for a complex ion that might form in this experiment.

Some household products that contain EDTA. *(C.D. Winters)*

Check the label on your shampoo container. It will likely list disodium EDTA as an ingredient. The EDTA in this case has a 2− charge because two of the four organic acid groups have each lost an H^+, but $EDTA^{2-}$ still coordinates to metal ions in the shampoo.

Exercise 9.11 Chelating and Complex Ions

Oxalate ion forms a complex ion with Mn^{2+} by coordinating at the oxygen lone pairs (see Figure 9.11).

(a) How many oxalate ions are needed to satisfy a coordination number of six on the central Mn^{2+} ion?

(b) What is the charge on this complex ion?

(c) How many atoms are in the rings formed between the ligand and the central metal ion?

For metals that display a coordination number of six, an especially effective ligand is the **hexadentate** ethylenediaminetetraacetate ion (abbreviated EDTA, Figure 9.11) that encapsulates and firmly binds metal ions. It has six lone-pair donor atoms (two O atoms and four N atoms) that can coordinate to a single metal ion, so $EDTA^{4-}$ is an excellent chelating ligand. It is often added to commercial salad dressing and to shampoos to remove traces of metal ions from solution, because these metal ions could otherwise accelerate the oxidation of oils in the product. Without $EDTA^{4-}$ as a metal ion scavenger, the dressing or shampoo would quickly become rancid.

Another use of $EDTA^{4-}$ is in bathroom cleansers, where it removes hard-water deposits of insoluble $CaCO_3$ and $MgCO_3$ by chelating Ca^{2+} or Mg^{2+} ions, allowing them to be rinsed away. EDTA is also used in the treatment of lead and mercury poisoning because it has the ability to chelate these metals and aid in their removal from the body (Figure 9.13).

Coordination compounds of *d*-transition metals are often colored; and the colors of the complexes of a given transition metal ion depend on both the metal ion and the ligand (Figure 9.14). Many transition metal coordination compounds are used as pigments in paints and dyes. For example, Prussian blue, $Fe_4[Fe(CN)_6]_3$, a deep-blue compound known for hundreds of years, is the "bluing agent" in laundry bleach and in engineering blueprints.

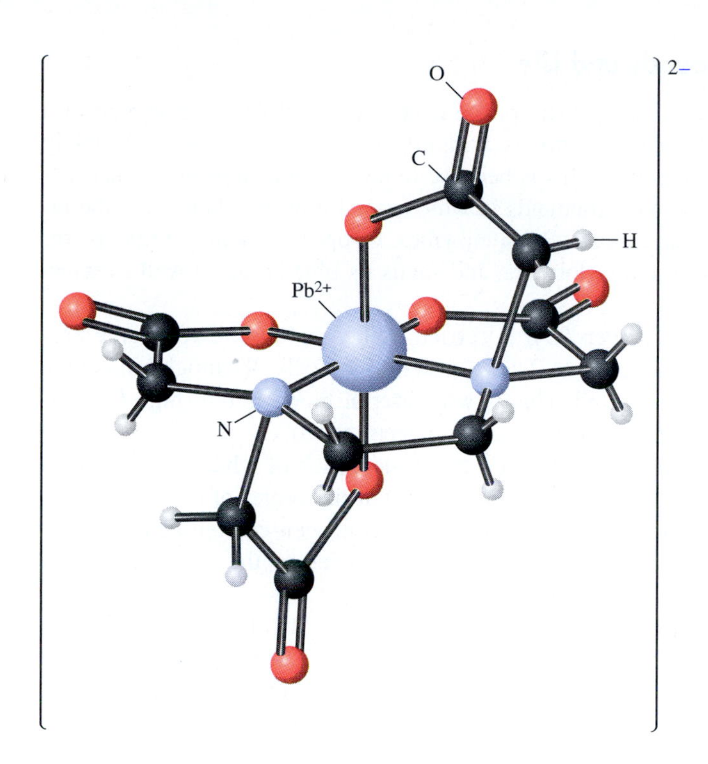

Figure 9.13 A Pb^{2+}-EDTA complex ion. The structure of the chelate formed when the $EDTA^{4-}$ anion forms a complex with Pb^{2+}.

Exercise 9.12 Complex Ions

Prussian blue contains two iron ions. What is the charge of the iron in the (a) complex ion $[Fe(CN)_6]^{4-}$? (b) iron ion not in the coordination sphere?

(a)

(b)

Figure 9.14 Color of transition metal compounds. (a) Concentrated aqueous solutions of the nitrate salts containing hydrated transition metal ions of (left to right) Fe^{3+}, Co^{2+}, Ni^{2+}, Cu^{2+}, and Zn^{2+}. (b) The colors of the complexes of a given transition metal ion depend on the ligand(s). All of the complexes pictured here contain the Ni^{2+} ion. The green solid is $[Ni(H_2O)_6](NO_3)_2$; the purple solid is $[Ni(NH_3)_6]Cl_2$; the red solid is Ni(dimethylglyoximate)$_2$. *(C.D. Winters)*

Coordination Compounds and Life

Bioinorganic chemistry, the study that applies chemical principles to inorganic ions and compounds in biological systems, is a rapidly growing field, centered mainly around coordination compounds. This is because living systems depend for their existence on many coordination compounds in which metal ions are chelated to the nitrogen atoms in proteins and especially in enzymes. Copper-containing proteins, for example, give the blood of crabs, lobsters, and snails its blue color, as well as transporting oxygen.

The blue blood of horseshoe crabs is used to test for bacterial contamination of drugs.

In humans, molecular oxygen (O_2) is carried by hemoglobin, a very large protein (molecular weight of about 68,000 amu) in red blood cells. Hemoglobin is blue but becomes red when oxygenated. This is why arterial blood is red (high O_2 concentration) and blood in veins is blue (low O_2 concentration).

A hemoglobin molecule carries four O_2 molecules, each of which forms a coordinate covalent bond to an Fe^{2+} ion. The Fe^{2+} ion is at the center of one of the nonprotein parts of the molecule that consist of four linked nitrogen-containing rings (Figure 9.15). Bound in this way, molecular oxygen is carried to the cells where it is released as needed by breaking the $Fe—O_2$ bond.

Other substances that can donate an electron pair can also bond to the Fe^{2+} in hemoglobin. Carbon monoxide is such a ligand and forms an exceptionally strong Fe^{2+}—carbon monoxide bond, nearly 200 times stronger than the $O_2—Fe^{2+}$ bond. Therefore, when a person breathes in CO, it displaces O_2 from hemoglobin and prevents red blood cells from carrying oxygen. The initial effect is drowsiness. But if CO inhalation continues, cells deprived of oxygen can no longer function, and the person suffocates to death.

It is curious (and fortunate) that N≡N does not behave chemically like C≡O.

Structures similar to the oxygen-carrying unit in hemoglobin are also found in other biologically important compounds, including such diverse ones as myoglobin, chlorophyll, and vitamin B_{12}. Myoglobin, like hemoglobin, contains Fe^{2+} and carries and stores molecular oxygen, principally in muscles. In chlorophyll, an Mg^{2+} ion occupies the position that iron does in hemoglobin. At the center of a vitamin B_{12} molecule is a Co^{3+} ion bonded to the same type of group. Vitamin B_{12} is the only known dietary use of cobalt, but it makes cobalt an essential mineral *(⇐ p. 105)*.

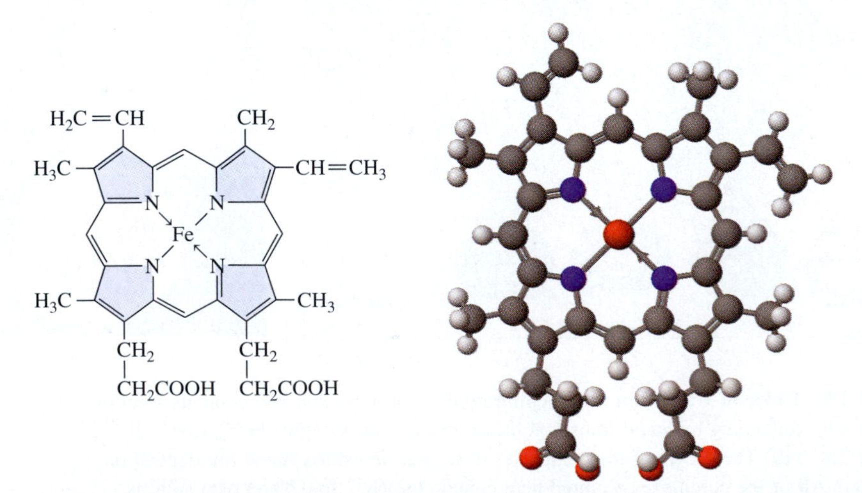

Figure 9.15 Fe^{2+} in hemoglobin. Fe^{2+} is coordinated to four nitrogen atoms in hemoglobin.

The dietary essentiality of zinc has only become established since the 1980s. Zinc, in the form of Zn^{2+} ions, is essential to the functioning of several hundred enzymes, including those that catalyze the breaking of P—O—P bonds in adenosine triphosphate (ATP), an important energy-releasing compound in cells (← *p. 291*).

Copper ranks third among biologically important transition metal ions in humans, trailing only iron and zinc. Although we generally excrete any dietary excess of copper, a genetic defect causes Wilson's disease, a condition where Cu^{2+} accumulates in the liver and brain. Fortunately, Wilson's disease can be treated by administering chelating agents that coordinate excess Cu^{2+} ions, allowing them to be excreted harmlessly.

Summary Problem

Salicylic acid, $C_7H_6O_3$, is the starting compound from which aspirin (acetylsalicylic acid) is synthesized. The salicylic acid molecule consists of a benzene ring on which two adjacent hydrogen atoms have been replaced, one by an alcohol (—OH) functional group, the other by a carboxylic acid functional group.

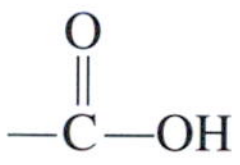

$$-\overset{\overset{\displaystyle O}{\|}}{C}-OH$$

(a) Write the structural formula for salicylic acid.

(b) List the C—H, O—H, C—O, and C=O bonds in salicylic acid in order of increasing bond length and also increasing bond strength.

(c) Arrange the following bonds in salicylic acid in order of increasing bond polarity: C—O, C—H, C—C, and O—H.

(d) The dissociation of salicylic acid forms H^+ and salicylate ions. Write the structural formula for the salicylate ion.

(e) Aspirin (acetylsalicylic acid) is synthesized by reacting salicylic acid with acetic acid (CH_3CO_2H).

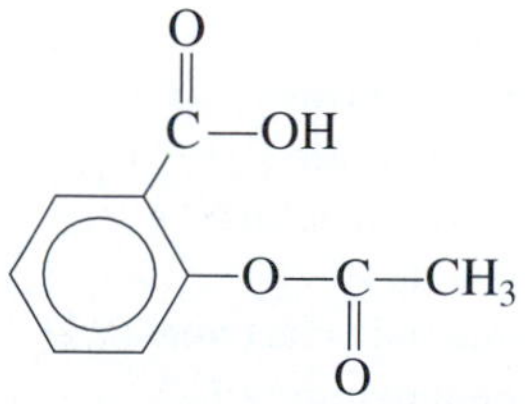

acetylsalicylic acid

Circle the portion of the acetylsalicylic acid molecule that came from (i) salicylic acid, (ii) acetic acid.

(f) Iron(III) forms a purple coordination compound with salicylate ions, [Fe(salicylate)$_3$]. The salicylate anion is a bidentate ligand that furnishes lone pairs from the alcohol group's O atom and from an O atom of the former acid group (see item (a)). Write a structural formula for [Fe(salicylate)$_3$] and give the coordination number of Fe^{3+} in this compound.

(g) Account for the fact that there is no net charge on [Fe(salicylate)$_3$].

(h) Salicylate ion is an example of a resonance hybrid. Write two Lewis structures that contribute to this resonance hybrid.

In Closing

Having studied this chapter, you should be able to . . .

- recognize the different types of covalent bonding (Sections 9.1–9.4, 9.9).
- use Lewis structures to represent the different types of covalent bonds in molecules and polyatomic ions (Section 9.1–9.4).
- describe multiple bonds in alkenes and alkynes (Section 9.3).
- explain why there are exceptions to the octet rule (Section 9.4).
- predict bond lengths from periodic trends in atomic radii (Section 9.5).
- relate bond strengths to bond lengths (Section 9.5).
- account for *cis-trans* isomerism in alkenes (Section 9.5).
- use resonance structures to model multiple bonding in molecules and polyatomic ions such as O_3, C_6H_6, and CO_3^{2-} (Section 9.6).
- describe bonding and constitutional isomerism in aromatic compounds (Section 9.7).
- predict bond polarity from electronegativity trends (Section 9.8).
- explain the coordinate covalent bonding of ligands in coordination compounds and complexes (Section 9.9).
- give examples of coordination compounds and their uses (Section 9.9).

Key Terms

The following terms were defined and given in boldface type in this chapter. You should be sure to understand each of these terms and the concepts with which they are associated.

alkanes *(9.2)*
alkenes *(9.3)*
alkynes *(9.3)*
aromatic compounds *(9.7)*
bidentate *(9.9)*
bond length *(9.5)*
bonding pair *(9.2)*
chelating ligands *(9.9)*
***cis* isomer** *(9.5)*
***cis-trans* isomerism** *(9.5)*
complex ion *(9.9)*
coordinate covalent bond *(9.9)*
coordination compound *(9.3, 9.9)*
coordination number *(9.9)*
covalent bond *(9.1)*
cycloalkanes *(9.2)*
delocalized electrons *(9.6)*
dimers *(9.4)*
double boned *(9.3)*
electronegativity *(9.8)*
electropositive elements *(9.8)*
free radicals *(9.4)*
hexadentate *(9.9)*
Lewis structure *(9.2)*
ligands *(9.9)*
lone pair electrons *(9.2)*
monodentate *(9.9)*
nonbonding pair *(9.2)*
nonpolar covalent bond *(9.8)*
octet rule *(9.2)*
polar covalent bond *(9.8)*
resonance hybrid *(9.6)*
resonance structures *(9.6)*
saturated hydrocarbons *(9.2)*
single covalent bond *(9.2)*
stereoisomerism *(9.5)*
***trans* isomer** *(9.5)*
triple bond *(9.3)*
unsaturated hydrocarbons *(9.3)*

Questions for Review and Thought

Conceptual Challenge Problems

CP-9.A. What deficiency is acknowledged by chemists when they write the formula for silicon dioxide as $(SiO_2)_n$ but carbon dioxide as CO_2?

CP-9.B. Without referring to the periodic table, write an argument to predict how the electronegativities of elements change based on the composition of their atoms.

CP-9.C. How would you rebut the statement "There are no ionic bonds, only polar covalent bonds."

Review Questions

1. Explain the differences between an ionic bond and a covalent bond.
2. What kind of bonding (ionic or covalent) would you predict for the products resulting from the following combinations of elements:
   (a) $Na + Br_2$ (c) $Ca + Cl_2$
   (b) $C + O_2$ (d) $N_2 + H_2$
3. What characteristics must atoms A and B have if they are able to form a covalent bond A—B with each other?
4. Boron compounds often do not obey the octet rule. Illustrate this with BCl_3. Show how the molecule can obey the octet rule by forming a coordinate covalent bond with NH_3.
5. Indicate the difference between alkanes, alkenes, and alkynes by drawing the structural formula of a compound in each class that contains three carbon atoms.
6. Refer to Table 9.1 and answer the following questions:
   (a) Do any molecules with more than eight electrons have a second-period element as the central atom?
   (b) What is the maximum number of bond pairs and lone pairs that surround the central atom in any of these molecules?
7. While sulfur forms the compounds SF_4 and SF_6, no equivalent compounds of oxygen, OF_4 and OF_6, are formed. Explain why this is so.
8. Which of the following molecules have an odd number of valence electrons: NO_2, SCl_2, NH_3, and NO_3?
9. Consider the following structures for the formate ion, HCO_2^-. Designate which two are resonance structures and which is equivalent to one of the resonance structures.

   (a) $[:\ddot{O}—C{=}\ddot{O}:]^-$ with H bonded to C
   (c) $[:\ddot{O}—C{=}\ddot{O}:]^-$ with H bonded to C
   (b) $[:\ddot{O}{=}C—\ddot{O}:]^-$ with H bonded to C

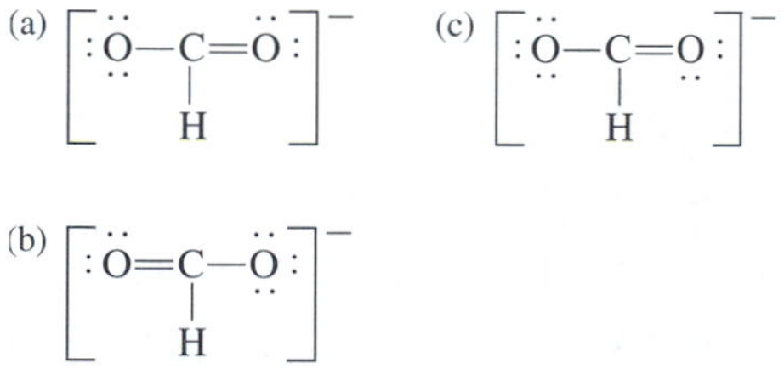

10. Consider a series of molecules in which the C atom is bonded to atoms of second-period elements: C—O, C—F, C—N, C—C, and C—B. Place these bonds in order of increasing bond length.
11. What are the trends in bond length and bond energy for a series of related bonds, say single, double, and triple C—C bonds?
12. Why is *cis-trans* isomerism not possible for alkynes?
13. Define and give an example of a polar covalent bond. Give an example of a nonpolar bond.
14. Define the following words or phrases and give an example for each.
    (a) Coordination compound (b) Complex ion (c) Ligand
    (d) Chelate (e) Bidentate ligand

Lewis Structures

15. How many valence electrons do each of these molecules or ions have?
    (a) CCl_4 (b) H_2F^+ (c) HCN (d) NH_2^-
16. How many valence electrons do each of these molecules or ions have?
    (a) PH_4^+ (b) SiH_4 (c) CS_2 (d) OH^-
17. Write Lewis structures for the molecules or ions in Question 15.
18. Write Lewis structures for the molecules or ions in Question 16.
19. Write Lewis structures for the following molecules or ions:
    (a) $SiCl_4$ (b) ClO_3^- (c) HOCl (d) SO_3^{2-}
20. Write Lewis structures for the following molecules or ions:
    (a) ClF (b) H_2Se (c) BF_4^- (d) PO_4^{3-}
21. Write Lewis structures for the following molecules:
    (a) $CHClF_2$, one of many chlorofluorocarbons that have been used in refrigeration
    (b) Methanol, CH_3OH
    (c) Methyl amine, CH_3NH_2
22. Write Lewis structures for the following molecules or ions:
    (a) CH_3Cl (b) SiO_4^{4-} (c) PH_4^+ (d) C_2H_6
23. Write Lewis structures for the following molecules:
    (a) Formic acid, HCO_2H, in which the atomic arrangement is

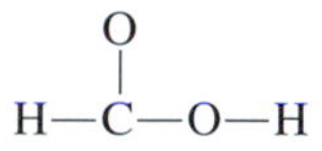

    H—C—O—H (with O bonded above C)

    (b) Acetonitrile, CH_3CN
    (c) Vinyl chloride, CH_2CHCl, the molecule from which PVC plastics are made
24. Write Lewis structures for the following molecules:
    (a) Tetrafluoroethylene, C_2F_4, the molecule from which Teflon is made
    (b) Acrylonitrile, CH_2CHCN, the molecule from which Orlon is made.
25. Which of the following are correct Lewis structures and which are incorrect? Explain what is wrong with the incorrect ones.

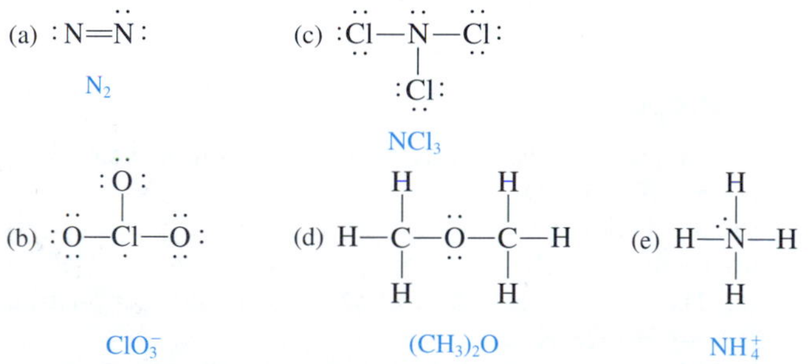

    (a) $:N{=}\ddot{N}:$
    N_2

    (c) $:\ddot{Cl}—\ddot{N}—\ddot{Cl}:$ with $:\ddot{Cl}:$ bonded below N
    NCl_3

    (b) $:\ddot{O}—\dot{Cl}—\ddot{O}:$ with $:\ddot{O}:$ bonded above Cl
    ClO_3^-

    (d) H—C—Ö—C—H, each C with an H above and an H below
    $(CH_3)_2O$

    (e) H—N—H with H above and H below N
    NH_4^+

26. Which of the following are correct Lewis structures and which are incorrect? Explain what is wrong with the incorrect ones.

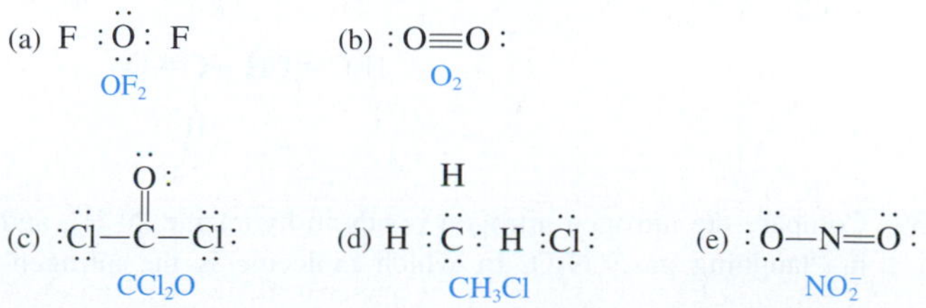

    (a) $F\ :\ddot{O}:\ F$
    OF_2

    (b) $:O{\equiv}O:$
    O_2

    (c) $:\ddot{Cl}—C—\ddot{Cl}:$ with $\ddot{O}:$ double-bonded above C
    CCl_2O

    (d) $H\ :\ddot{C}:\ H\ :\ddot{Cl}:$ with H above C
    CH_3Cl

    (e) $:\ddot{O}—N{=}\ddot{O}:$
    NO_2^-

Bonding in Hydrocarbons

27. Write the structural formulas for all the branched-chain compounds with the molecular formula C_4H_{10}.
28. Write the structural formulas for all the branched-chain compounds with the molecular formula C_6H_{14}.
29. Write structural formulas for two straight-chain alkenes with the formula C_5H_{10}. Are these the only two structures that meet these specifications?
30. C_4H_8 has five different structural formulas, draw them. [*Hint:* Consider all possible straight-chain, branched, and ring structures.]

Exceptions to the Octet Rule

31. Write the Lewis structure for each of the following molecules or ions.
(a) BrF_3 (b) I_3^- (c) XeF_4
32. Write the Lewis structure for each of the following molecules or ions.
(a) BrF_5 (b) SeF_6 (c) IBr_2^-
33. Which of the following elements can form compounds with five or six pairs of valence electrons surrounding their atoms?
(a) C (d) Se
(b) Be (e) Cl
(c) O (f) Xe
34. Which of the following elements can form compounds with five or six pairs of valence electrons surrounding their atoms?
(a) N (d) F
(b) P (e) Ge
(c) Sn (f) B
35. Write Lewis structures for the following molecules or ions.
(a) $AsCl_5$ (b) BCl_3 (c) PF_6^-
36. Write Lewis structures for the following molecules or ions.
(a) $GeCl_2$ (b) ICl_4^- (c) XeF_2

Bond Properties

37. In each pair of bonds, predict which will be the shorter.
(a) Si—N or P—O
(b) Si—O or C—O
(c) C—F or C—Br
(d) The C═C or the C≡N bond in acrylonitrile, $H_2C{=}CH{-}C{\equiv}N$
38. In each pair of bonds, predict which will be the shorter.
(a) B—Cl or Ga—C (c) P—S or P—O
(b) C—O or Sn—O (d) The C═C or the C═O bond in acrolein,

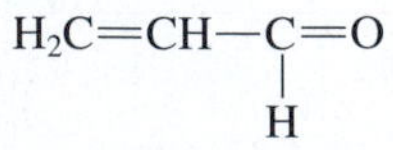

```
H2C=CH—C=O
        |
        H
```

39. Compare the nitrogen-nitrogen bonds in hydrazine, N_2H_4, and in "laughing gas," N_2O. In which molecule is the nitrogen-nitrogen bond shorter? In which should the nitrogen-nitrogen bond be stronger?
40. Consider the carbon-oxygen bonds in formaldehyde, H_2CO, and in carbon monoxide, CO. In which molecule is the CO bond shorter? In which should the CO bond require more energy to break?
41. Order the following bonds from longest to shortest: C—C, O═O, C═C, and C═O.
42. Order the following bonds from strongest to weakest: C═C, N≡N, C═N, and C≡N.
43. Write the *cis* and *trans* isomers of 2-pentene.
44. In each case tell whether *cis* and *trans* isomers exist. If they do, write structural formulas for the two isomers and label them *cis* or *trans*.
(a) Br_2CH_2 (c) $CH_3CH{=}CHCH_3$
(b) $CH_3CH_2CH{=}CHCH_2CH_3$ (d) $CH_2{=}CHCH_2CH_3$
45. Which of the following molecules can have *cis* and *trans* isomers? For those that do, write the structural formulas of the two isomers and label them *cis* and *trans*. For those that cannot have these isomers, explain why not.
(a) $CH_3CH_2BrC{=}CBrCH_3$ (d) $CH_3ClC{=}CHCH_3$
(b) $(CH_3)_2C{=}C(CH_3)_2$ (e) $(CH_3)_2C{=}CHCH_3$
(c) $CH_3CH_2IC{=}ClCH_2CH_3$
46. The structural formulas are given for the following *cis* or *trans* alkenes.

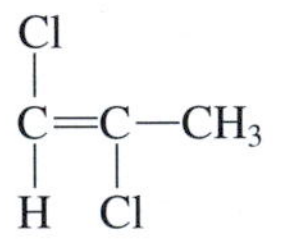

```
Cl
|
C=C—CH3
|  |
H  Cl
```

(a) *trans*-1,2-dichloropropene

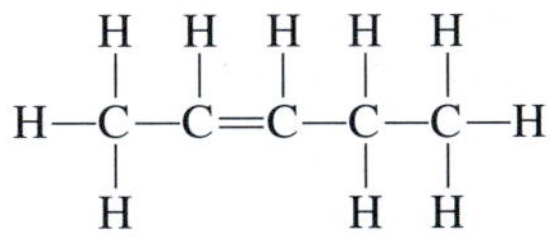

```
   H  H  H  H  H
   |  |  |  |  |
H—C—C=C—C—C—H
   |        |  |
   H        H  H
```

(b) *cis*-2-pentene

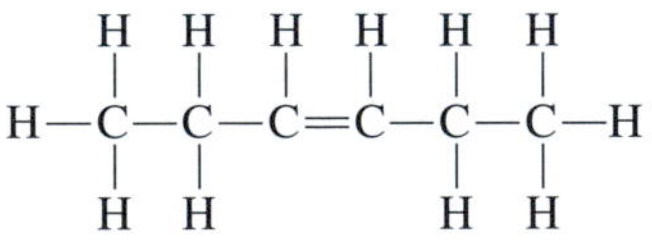

```
   H  H  H  H  H  H
   |  |  |  |  |  |
H—C—C—C=C—C—C—H
   |  |        |  |
   H  H        H  H
```

(c) *cis*-3-hexene

```
        H
        |
CH3—C=C—CH2—CH2—CH3
           |
           H
```

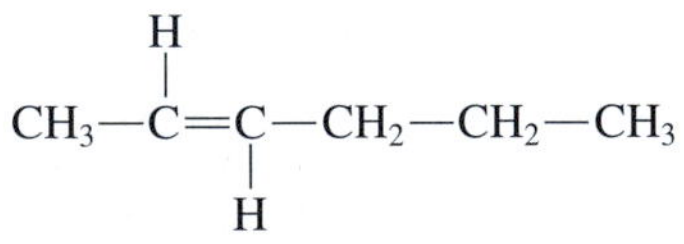

(d) *trans*-2-hexene

Write the structural formula for
(a) *cis*-1,2-dichloropropene (c) *trans*-3-hexene
(b) *trans*-2-pentene (d) *cis*-2-hexene

Resonance

47. Write resonance structures for NO_2^-. Predict a value for the N—O bond length, based on bond lengths given in Table 9.2, and explain your answer.
48. The following molecules have two or more resonance structures. Write all the resonance structures for each molecule.
(a) Nitric acid,

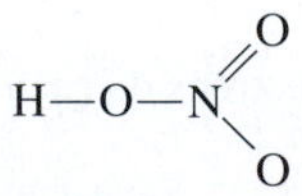

```
        O
       //
H—O—N
       \
        O
```

(b) Nitrous oxide (laughing gas), N≡N—O

49. The following molecules or ions have two or more resonance structures. Write all the resonance structures for each molecule or ion.
(a) SO_3 (b) SCN^-

50. Compare the carbon-oxygen bond lengths in the formate ion, HCO_2^-, and in the carbonate ion, CO_3^{2-}. In which ion is the bond longer? Explain briefly.

51. Compare the nitrogen-oxygen bond lengths in NO_2^+ and in NO_3^-. In which ion are the bonds longer? Explain briefly.

52. Write the structural formula for 1,2-diiodobenzene (also known as *ortho*-diiodobenzene). Write the structural formulas for the *meta*- and *para*- isomers as well.

53. Write the resonance structures for the following compound, adenine, which is the nitrogen-containing organic base portion of ATP (adenosine triphosphate).

```
            NH2
            |
            C
          //  \
         N     C ——— N
         |     ||    ||
         C     C     C
       /  \\  / \   / \
      H     N     N     H
                  |
                  H
```

Electronegativity and Bond Polarity

54. Order the following elements from least electronegative to most electronegative: Na, O, P, K, and N.

55. Order the following elements from least electronegative to most electronegative: Si, F, S, Ga, and Ca.

56. Given the bonds C—N, C—H, C—Br, and S—O,
(a) tell which atom in each is the more electronegative.
(b) which of these bonds is the most polar?

57. In each pair of bonds, indicate the more polar bond, and use $\delta+$ and $\delta-$ to show the direction of polarity in each bond.
(a) C—O and C—N
(b) B—O and P—S
(c) P—H and P—N
(d) B—H and B—I

58. The molecule below is urea, a compound used in plastics and fertilizers.

```
  H     :O:     H
   \     ||    /
    N —— C —— N
   / ..      .. \
  H              H
```

(a) Which bonds in this molecule are polar and which are nonpolar?
(b) Which is the most polar bond in the molecule? Which atom is the partial negative end of this bond?

59. The molecule below is acrolein, the starting material for certain plastics.

```
     H   H   H
     |   |   |    ..
  H—C = C — C = O :
```

(a) Which bonds in this molecule are polar and which are nonpolar?
(b) Which is the most polar bond in the molecule? Which atom is the negative end of this bond?

Coordination Compounds

60. In a complex ion a central ruthenium atom (Ru(II)) is bonded to six ammonia molecules. Give the formula and net charge for this complex ion. Balance the net charge on this complex ion with chloride ions. How many chloride ions are needed? Write the formula for the complex including the chloride ions that are not part of the complex ion.

61. For the coordination compounds $Na_3[IrCl_6]$ and $[Mo(CO)_4Br_2]$, identify in each case
(a) the ligands.
(b) the central metal ion and its charge.
(c) the formula and charge of the complex ion.
(d) ions not in the complex ion.

62. Give an analogous (non-coordination) simple ionic compound to $[Rh(en)_3]Cl_3$.

63. Write a structural formula for the coordination compound $[Cr(en)(NH_3)_2I_2]$, and give the coordination number for the central Cr^{2+} ion.

64. Give the formula of each of the following coordination compounds formed with Pt^{2+}.
(a) Two ammonia molecules and two bromide ions
(b) One ethylenediamine molecule and two nitrite ions, NO_2^-
(c) One chloride ion, one bromide ion, and two ammonia molecules

65. Give the charge on the central metal ion in each of the following:
(a) $[VCl_6]^{4-}$ (c) $[Mn(NO)(CN)_5]^{3-}$
(b) $[Sc(H_2O)_3Cl_3]$ (d) $[Cu(en)_2(NH_3)_2]^{2+}$

General Questions

66. Is it a good generalization that elements that are close together in the periodic table form covalent bonds while elements that are far apart form ionic bonds? Why or why not?

67. Write Lewis structures for the following molecules or ions.
(a) H_3O^+ (b) NCl_3 (c) $SnCl_2$

68. Write Lewis structures for the following molecules or ions.
(a) SeO_4^{2-} (b) SF_4 (c) PF_3

69. Which has the longer sulfur-oxygen bonds, SO_2 or SO_3?

70. Draw Lewis structures (resonance structures where appropriate) for the following ions:
(a) N_3^- (b) OCN^-

71. Identify the coordination number of the metal ion in these coordination complexes:
(a) $[FeCl_4]^-$ (c) $[Mn(en)_3]^{2+}$
(b) $[PtBr_4]^{2-}$ (d) $[Cr(NH_3)_5H_2O]^{3+}$

72. Using structural formulas, show how the carbonate ion can be either a monodentate or bidentate ligand to a transition metal cation.

73. The C—Br bond length in CBr_4 is 191 pm; the Br—Br distance in Br_2 is 228 pm. Estimate the radius of a C atom in CBr_4. Use this value to estimate the C—C distance in ethane, $H_3C—CH_3$. How does your calculated bond length agree with the measured value of 154 pm? Are radii of atoms exactly the same in every molecule?

74. The molecule pictured below is acrylonitrile, the building block of the synthetic fiber Orlon.

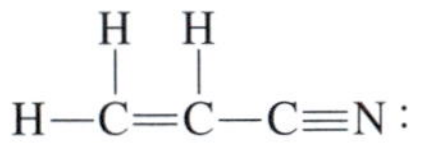

H—C(H)=C(H)—C≡N:

(a) Which is the shorter carbon-carbon bond?
(b) Which is the stronger carbon-carbon bond?
(c) Which is the most polar bond and what is the partial negative end of the bond?

75. In nitryl chloride, NO_2Cl, there is no oxygen-oxygen bond. Write a Lewis structure for the molecule. Write any resonance structures for this molecule.

76. Why is CO bound more strongly to the iron atom in hemoglobin than O_2? Discuss the bonding in terms of the differences in the bond polarities of CO and O_2.

77. Arrange the following bonds in order of increasing length (shortest first). List all the factors responsible for each placement: O—H, O—O, Cl—O, O=O, and O=C.

78. List the bonds in Problem 77 in order of increasing bond *strength.*

79. Chlorine trifluoride, ClF_3, is one of the most reactive compounds known. Write the Lewis structure of ClF_3.

80. Judging from the number of carbon and hydrogen atoms in their formulas, which of the following formulas represent alkanes? Which are likely aromatic? Which fall into neither category? (It may help to write structural formulas.)

(a) C_8H_{10} (c) C_6H_{12} (e) C_8H_{18}
(b) $C_{10}H_8$ (d) C_6H_{14} (f) C_6H_{10}

Applying Concepts

81. A student drew the following incorrect Lewis structure for ClO_3^-. What error was made when determining the number of valence electrons?

$[:\ddot{O}=\ddot{Cl}—\ddot{O}:$ with $:\ddot{O}:$ bonded below Cl$]^-$

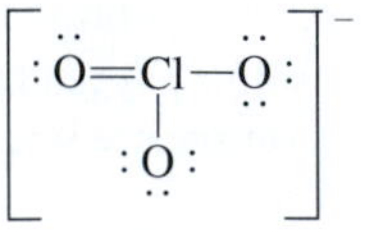

82. The following Lewis structure for SF_5^+ is drawn incorrectly. What error was made when determining the number of valence electrons?

$[SF_5]^+$ — S (with a lone pair) bonded to five F atoms, each with three lone pairs

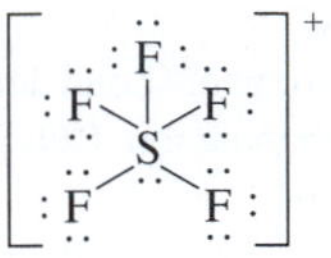

83. When asked to give an example of resonance structures, a student drew the following. Why is this example incorrect?

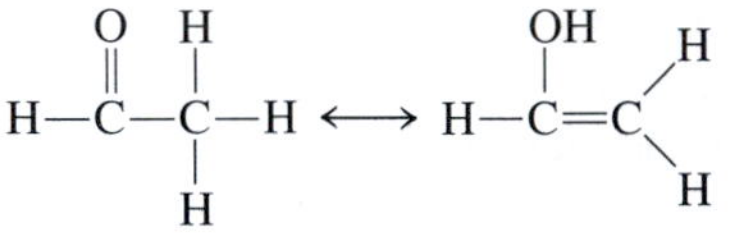

H—C(=O)—CH₂—H ⟷ H—C(OH)=CH₂

84. Why is the following not an example of resonance structures?

$:\ddot{S}—C≡N: \longleftrightarrow :\ddot{S}—N≡C:$

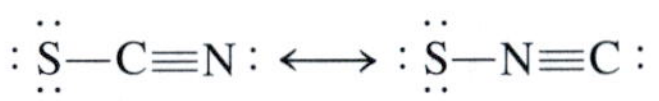

85. How many bonds would you expect the elements in Groups 3A through 7A to form if they obeyed the octet rule?

86. In another universe, elements try to achieve a nonet (nine valence electrons) instead of an octet when forming chemical bonds. As a result, covalent bonds form when a trio of electrons are shared between two atoms. Draw Lewis structures for the compounds which would form between (a) hydrogen and oxygen, and (b) hydrogen and fluorine.

87. Elemental phosphorus has the formula P_4. Propose a Lewis structure for this molecule. [*Hints:* (1) Each phosphorus atom is bonded to three other phosphorus atoms. (2) Visualize the structure three-dimensionally, not flat on a page.]

88. The elements As, Br, Cl, S, and Se have electronegativity values of 2.1, 2.4, 2.5, and 3.0, but not in that order. Using the periodic trend for electronegativity, assign the values to the elements. Which assignments are you certain about? Which are you not?

89. A substance is analyzed and found to contain 85.7% carbon and 14.3% hydrogen by weight. A gaseous sample of the substance is found to have a density of 1.87 g/L, and 1 mol of it occupies a volume of 22.4 L. What are two possible Lewis structures for molecules of the compounds? [*Hint:* First determine the empirical formula and molar mass of the substance.]

90. Which of these molecules is least likely to exist: NF_5, PF_5, SbF_5, or IF_5? Explain why.

CHAPTER 10

Molecular Structures

Cellulose, obtained from cotton and other natural sources, is reacted with acetic anhydride to produce cellulose acetate. Known commercially as acetate rayon, cellulose acetate is spun into threads to make fashionable clothing and is also used in film and fibers. Acetic anhydride to make cellulose acetate is obtained from the reaction of acetic acid with a simple building-block compound, ketene, C_2H_2O. • What is the molecular shape of ketene? How are its atoms arranged in three-dimensional space? Such questions about ketene are considered in the Summary Problem. *(C.D. Winters)*

The composition, empirical formula, molecular formula, and Lewis structure of a substance provide important information, but they are not sufficient to predict or explain the properties of most molecular compounds. The arrangement of the atoms and how they occupy three-dimensional space—the shape of a molecule—are also very important. Molecules can have the same numbers of the same kinds of atoms and yet be different; the substances made up of those molecules will have different properties. For example, ethanol (in alcoholic beverages) and dimethyl ether (a refrigerant) have the same molecular formula, C_2H_6O. But the C, H, and O atoms are arranged so differently in these two compounds that their melting points differ by 27 °C, and their boiling points by 103 °C.

The ideas about molecular shape presented in this chapter are crucial to understanding the behavior of molecules in living organisms, the design of molecules that are effective drugs, and many other aspects of modern chemistry.

10.1 MOLECULAR MODELING

Our ability to understand three-dimensional structures of molecules is helped by the use of models. Probably the best example of the impact a model can have on the advancement of science is the model of the double helix of DNA, built by James Watson and Francis Crick, which revolutionized the understanding of human heredity and genetic disease. We will discuss DNA (Section 10.8) after considering molecular shapes (Section 10.2) and intermolecular forces (Section 10.5), both of which are essential to DNA function.

Molecules are three-dimensional aggregates of atoms and are too small to examine directly. Therefore, we resort to models to represent molecular shapes. Throughout this book we have been using computer-drawn molecules that are pictures of three-dimensional models like those shown below for water molecules (b and d). The computer programs that generate these pictures contain the most accurate experimentally derived data on atomic radii, bond lengths, and bond angles.

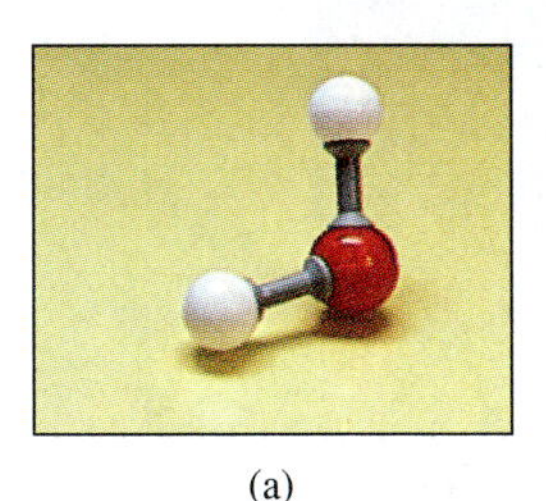

(a)

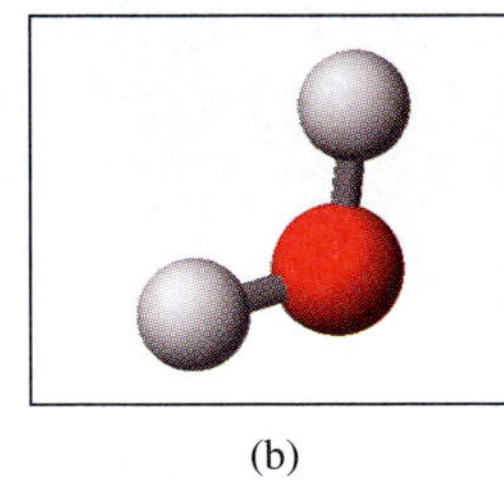

(b)

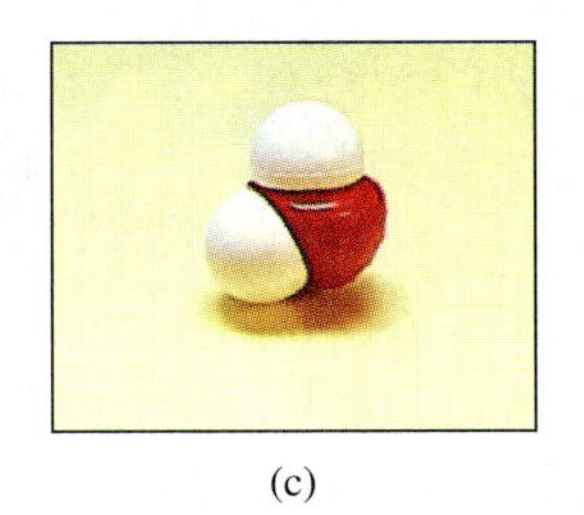

(c)

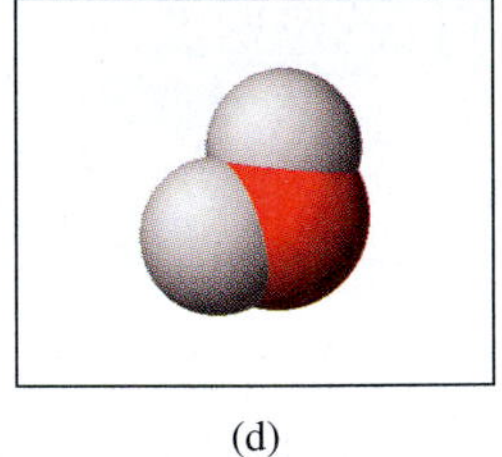

(d)

Before there were computers to generate molecular models, chemists relied on physical models assembled atom by atom, like those shown above for water. The ball-and-stick model (a) uses balls to represent atoms and short pieces of wood or plastic to represent bonds. For example, the ball-and-stick model for water has a red ball representing oxygen, with holes at the correct angles connected by sticks to two white balls representing hydrogen atoms. In the space-filling model (c) the atoms are scaled according to the experimental values for atom sizes, and the links between them are not visible when the model is assembled, a better representation of the actual distances between atoms.

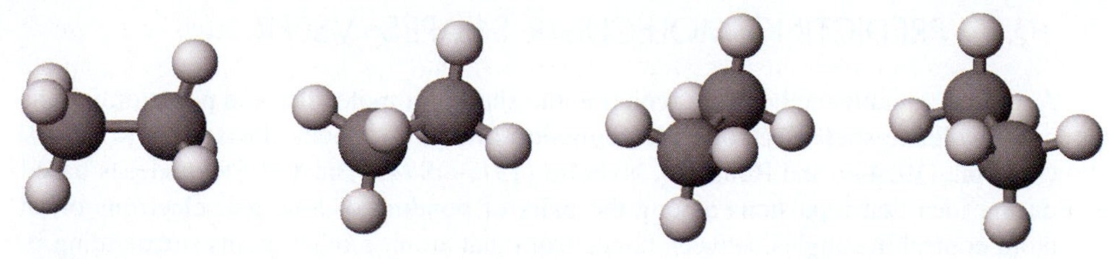

Figure 10.1 An ethane molecule (C_2H_6) rotating.

To convey a three-dimensional perspective for a molecule drawn on a flat (two-dimensional) surface, we can also make a perspective drawing that uses solid wedges (▬) to represent bonds extending above the page, and dashed lines (---) to represent bonds below the page. Bonds that lie in the plane of the page are indicated by a line (—), as illustrated in the following perspective drawing for the tetrahedral methane molecule:

Ball-and-stick model kits are often available in campus book stores. They are easy to assemble and will help you visualize the molecular geometries described in this chapter. They are relatively inexpensive in comparison with the space-filling models.

bond behind page
H
bond in the plane of the page
C
H
H
H
bond in front of page

Computers can draw and also rotate molecules such as ethane, so that they can be viewed from any angle, as illustrated in Figure 10.1.

Advances in computer graphics have made it possible to draw scientifically accurate pictures of extremely complex molecules as well as to study interactions between molecules. Figure 10.2 is a computer-generated model of Ritonavir, an anti-HIV drug, and its interaction with HIV protease, an enzyme that cuts proteins in order for the HIV virus to make copies of itself. Using computer graphics, chemists have brought ritonavir molecules close to HIV protease on the computer screen to see how the molecules fit together. They learned that by having the right spatial fit, ritonavir prevents the enzyme from functioning. Thus, HIV's ability to spread within a victim's body is blocked.

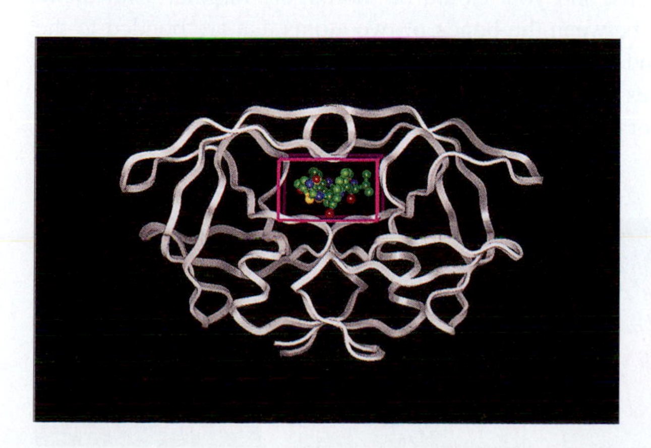

Figure 10.2 **Action of Ritonavir, an anti-HIV drug.** The ribbon-like shape represents the twists and turns of the enzyme protease, a molecule that is essential to the spread of HIV in the body. By fitting into protease, Ritonavir (green) prevents protease from carrying out its normal function in multiplication of the HIV virus. (For further information, see Chemistry in the News, Protease Inhibitors and AIDS, Chapter 12.) *(Courtesy of Abbott Laboratories)*

10.2 PREDICTING MOLECULAR SHAPES: VSEPR

Gillespie was born in England, and Nyholm in Australia, but both received the Ph.D. in chemistry at University College, London. Nyholm made many contributions to chemistry as Professor of Chemistry at University College, London, until his untimely death. Gillespie has been Professor of Chemistry at McMaster University (Canada) since 1960.

A simple, reliable method for predicting the shapes of molecules and polyatomic ions is the **valence-shell electron-pair repulsion (VSEPR)** model, devised by Ronald J. Gillespie (1924–) and Ronald S. Nyholm (1917–1971). The VSEPR model is based on the idea that repulsions among the pairs of bonding or lone pair electrons of an atom control the angles between bonds from that atom to other atoms surrounding it. A central atom and its core electrons are represented by the atom's symbol. This atomic core is surrounded by pairs of valence electrons, the number of pairs corresponding to the number of pairs of dots in the Lewis structure. The geometric arrangement of the electron pairs is predicted on the basis of their repulsions, and the geometry of the molecule or polyatomic ion depends on the numbers of lone pairs and bonding pairs.

How do repulsions among electron pairs result in different shapes? Imagine that a balloon represents each electron pair. Each balloon's volume represents a repulsive force that prevents other balloons from occupying the same space. When two, three, four, five, or six balloons are tied together at a central point (the central point represents the nucleus and core electrons of a central atom), the balloons form the shapes shown in Figure 10.3. These geometric arrangements minimize interactions among the balloons (electron-pair repulsions).

Central Atoms with Only Bond Pairs

The simplest application of VSEPR is to molecules in which all the electron pairs around the central atom are shared pairs in single covalent bonds. Figure 10.4 illustrates the geometries predicted by the VSEPR model for molecules of the types AX_2 to AX_6 that contain only single covalent bonds, where A is the central atom.

The **linear** geometry for two bond pairs and the **triangular planar** geometry for three bond pairs contain a central atom that does not have an octet of electrons (← *p. 367*) (Section 9.4). The central atom in a **tetrahedral** molecule obeys the octet rule with four bond pairs. The central atoms in **triangular bipyramidal** and **octahedral** molecules do not obey the octet rule because they have five and six bonding pairs, respectively. Hence, triangular bipyramidal and octahedral geometries would be expected only when the central atom is an element in Period 3 or higher (← *p. 368*) (Section 9.4). The geometries illustrated in Figure 10.4 are by far the most common in molecules and ions, and you should be thoroughly familiar with them. The **bond angle** is the angle between the bonds of two atoms that are bonded to the same third atom, and the bond angles predicted for the examples given are in agreement with experimental values obtained from structural studies. The H—O—H bond angle in a water molecule, for example, is 104.5°.

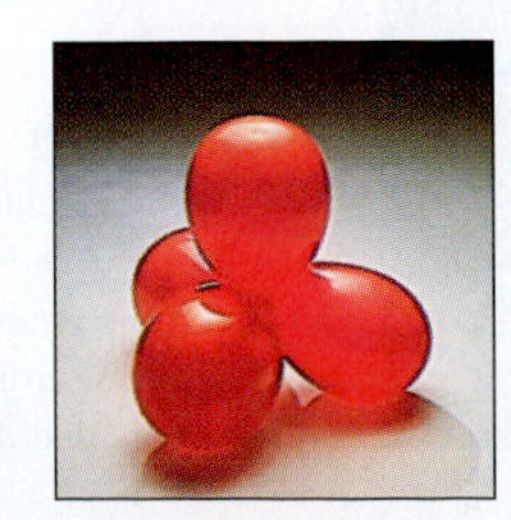

Figure 10.3 Balloon models of the geometries predicted by the VSEPR theory.

CHEMISTRY YOU CAN DO

Using Balloons as Models

Blow up and tie off 20 balloons to the same size. Then tie the ends of two balloons closely together. Repeat the process with three, four, five, and six balloons. You have constructed five balloon assemblies, which make excellent models for visualizing the linear, triangular planar, tetrahedral, triangular bipyramidal, and octahedral geometries assumed by two, three, four, five, and six electron pairs, respectively. The common point where the balloons are tied together represents a central atom and the balloons represent electron pairs around the central atom.

Write answers to the following questions.

1. Identify the apexes for each shape (the farthest point of each balloon from the central point). Sketch the figure you would get if you connected the apexes with straight lines.
2. Identify the faces (the flat surfaces bounded by the lines between apexes) in each sketch, and count how many apexes and faces there are for each shape.
3. Does the term octahedron indicate the number of apexes or the number of faces?

H—O—H (104.5°)

Suppose you want to predict the shape of $SiCl_4$. First, draw the Lewis structure. Because there are four bond pairs forming four single covalent bonds to Si, you would predict a tetrahedral structure for the $SiCl_4$ molecule, and this is in agreement with structural studies for this molecule. The Cl—Si—Cl bond angle between any two Si—Cl bonds is 109.5°.

Conceptual Challenge Problems CP-10.A and CP-10.B at the end of the chapter relate to topics covered in this section.

:C̈l—Si(C̈l)(C̈l)—C̈l: (Lewis structure); tetrahedral $SiCl_4$, 109.5°

Central Atoms with Bond Pairs and Lone Pairs

How does the presence of lone pairs on the central atom affect the geometry of the molecule or polyatomic ion? The easiest way to visualize this situation is to return

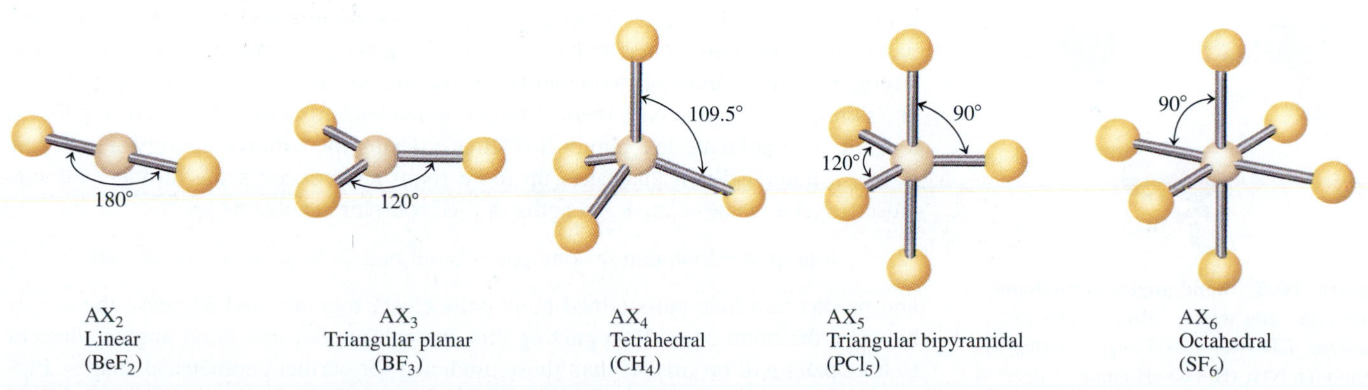

Figure 10.4 **Geometries predicted by the VSEPR model for molecules of the types AX_2 through AX_6 that contain only single covalent bonds.**

to the balloon model and notice that we did not say that the electron pairs had to be bonding pairs. We can predict the geometry of a molecule by applying the VSEPR model to the total number of valence electron pairs around the central atom. First, we must decide which positions are occupied by bond pairs and which by lone pairs. The **electron-pair geometry** around a central atom includes the spatial positions of all bond pairs and lone pairs of electrons, whereas the **molecular geometry** (molecular shape) of a molecule or ion is the arrangement of its atoms in space. The distinction is necessary because *positions occupied by lone pairs are not specified when we describe the shapes of molecules.* In other words, lone pairs of electrons around the central atom occupy spatial positions even though they are not included in the description of the shape of the molecule or polyatomic ion.

The success of the VSEPR model in predicting molecular shapes indicates that it is appropriate to account for the effects of lone pairs in this way.

Let's examine the steps in using the VSEPR model to predict the molecular geometry and bond angles in a molecule that includes lone pairs on the central atom, the NH_3 molecule. First, draw the Lewis structure and count the total number of electron pairs around the central N atom.

$$\mathrm{H{-}\ddot{N}{-}H} \atop \mathrm{\;\;\;|\;\;\;\atop H}$$

Because there are three bond pairs and one lone pair for a total of four pairs of electrons, we predict that the *electron-pair geometry* is tetrahedral. Draw a tetrahedron with N as the central atom and the three bond pairs represented by lines, since they are single covalent bonds. The lone pair is drawn as a balloon shape to indicate its spatial position in the tetrahedron:

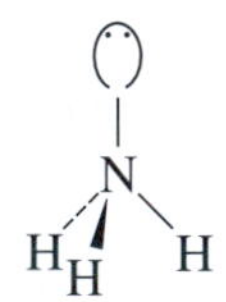

The *molecular geometry* of ammonia is described as a triangular pyramid because the three hydrogen atoms form a triangular base with the nitrogen atom at the apex of the pyramid. (This can be seen by covering up the lone pair of electrons and looking at the molecular geometry—the location of the three H nuclei and the N nucleus.)

What is the predicted value for the H—N—H bond angles? Because the electron-pair geometry is tetrahedral, we would expect the H—N—H bond angles to be 109.5°. However, the experimentally determined bond angles in NH_3 are 107.5°. This is attributed to a difference between the spatial requirements of lone pairs and bond pairs. Bond pairs are concentrated in the bonding region between two atoms by the strong attractive forces of two positive nuclei and are, therefore, relatively compact, or "skinny." For a lone pair, there is only one nucleus attracting the electron pair. As a result, lone pairs are less compact ("fatter"). Using the balloon analogy, a lone pair is like a fatter balloon that takes up more room and squeezes the thinner balloons closer together. The relative strengths of electron-pair repulsions are

lone pair—lone pair > lone pair—bond pair > bond pair—bond pair

and predict that lone pairs crowd bond pairs closer together and decrease the angles between the bond pairs. Recognizing this, we can predict that bond angles adjacent to lone pairs will be smaller than those predicted for perfect geometrical shapes. Figure 10.5 illustrates this effect in the series of CH_4, NH_3, and H_2O as the number of lone pairs increases.

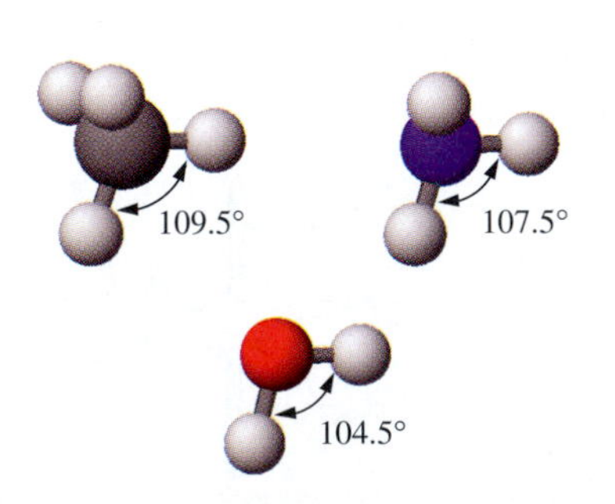

Figure 10.5 Bond angles of methane, ammonia, and water. Bond angles in methane, CH_4 (H—C—H angle = 109.5 °); ammonia, NH_3 (H—N—H angle = 107.5 °); and water, H_2O (H—O—H angle = 104.5 °).

Methane, which has the tetrahedral shape, is the smallest member of a large family of saturated hydrocarbons. It is important to recognize that every carbon atom in an alkane has a tetrahedral environment. For example, notice that the carbon atoms in propane and in the much longer carbon chain of hexadecane do not lie in a straight line because of the tetrahedral geometry about each carbon atom.

The tetrahedron is arguably the most important shape in chemistry because of its predominance in the chemistry of carbon and the silicon-oxygen compounds in the crust of the earth (Section 15.8).

Figure 10.6 gives additional examples of electron-pair geometries and molecular geometries for molecules and ions with three and four electron pairs around the central atom. The experimentally determined bond angles are given for the examples. To check your understanding of the VSEPR model, try to explain the molecular geometry and bond angles in each case.

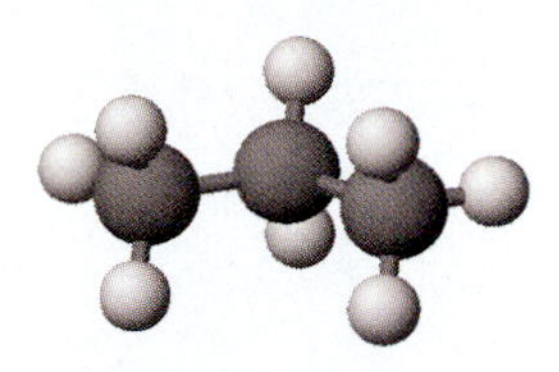

Propane, C_3H_8. Because of the tetrahedral nature of its carbon atoms, the carbon atoms do not lie in a straight line.

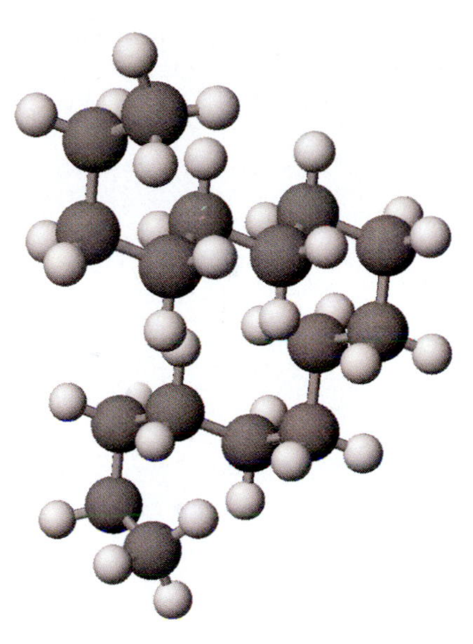

Hexadecane, $C_{16}H_{34}$.

PROBLEM-SOLVING EXAMPLE 10.1 Molecular Structure

Use the VSEPR model to predict the electron-pair geometry and molecular geometry of (a) CF_2Cl_2; (b) OCl_2; and (c) CH_2ClCH_3.

Answer (a) Tetrahedral electron-pair and molecular geometry; (b) tetrahedral electron-pair geometry and angular molecular geometry; (c) tetrahedral electron-pair geometry around each carbon atom and a tetrahedral molecular geometry.

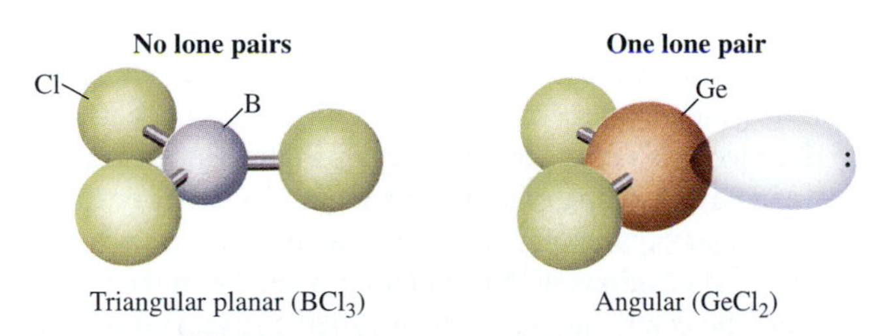

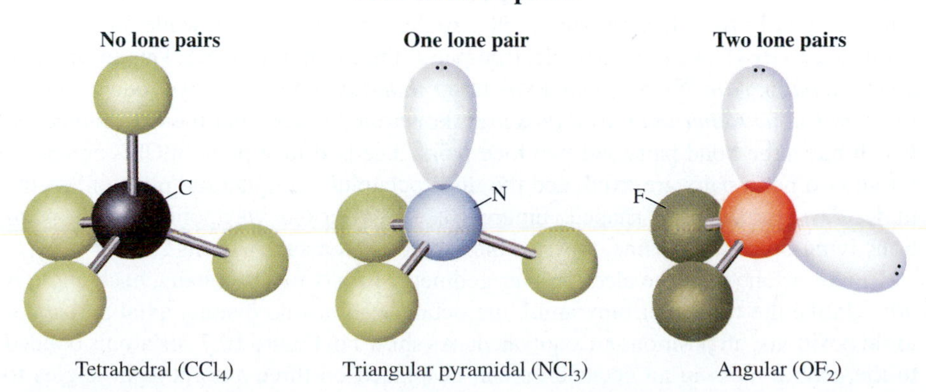

Figure 10.6 Three and four electron pairs around a central atom. Examples are shown of electron-pair geometries and molecular shapes for molecules and polyatomic ions with three and four electron pairs around the central atom.

Explanation

First, write the Lewis structures for each molecule:

```
     ..
    :Cl:
 ..  |   ..
:F—C—Cl:
 ..  |   ..
    :F:
     ..
```

```
 ..   ..   ..
:Cl—O—Cl:
 ..   ..   ..
```

```
     H
     |
H—C—Cl
     |
H—C—H
     |
     H
```

(a) The Lewis structure shows a central carbon atom surrounded by four bonding electron pairs and no lone pairs, which makes for tetrahedral electron-pair and molecular geometry in CF_2Cl_2.

(b) The central oxygen is surrounded by two bonding and two lone pairs. The four electron pairs give a tetrahedral electron-pair geometry. Because of lone pair–lone pair, bonding pair–lone pair, and bonding pair–bonding pair repulsions, the molecular geometry is angular (bent), not linear. The lone pairs push the chlorine atoms closer together than the purely tetrahedral 109.5° angle expected for four bonding pair–bonding pair repulsions.

(c) The four electron pairs around each carbon atom in CH_2ClCH_3 give a tetrahedral electron-pair geometry around each carbon atom. There are no unshared electrons on either carbon atom resulting in a tetrahedral molecular geometry.

Problem-Solving Practice 10.1

Use VSEPR theory to predict the electron-pair geometry and molecular shape of (a) BrO_3^-; (b) SeF_2; (c) CH_3OH.

The situation becomes more complicated if the central atom has five or six electron pairs, some of which are lone pairs. Let's look first at the entries in Figure 10.7 for the case of five electron pairs. The three angles in the triangular plane are all 120°. The angles between any of the pairs in this plane and an upper or lower pair are only 90°. Thus, the triangular bipyramidal structure has two sets of positions that are not equivalent. Because the positions in the triangular plane lie in the equator of an imaginary sphere around the central atom, they are called **equatorial positions.** The north and south poles are called the **axial positions.** Each equatorial position is closely flanked by only two electron pairs at 90° angles (the axial ones), while an axial position is closely flanked by three electron pairs (the equatorial ones). This means that *any lone pairs, (which we assume to be fatter than bonding pairs), will occupy equatorial positions rather than axial positions.* For example, consider the ClF_3 molecule, which has three bond pairs and two lone pairs. The two lone pairs in ClF_3 are equatorial; two bond pairs are axial, and the third occupies an equatorial position; so the molecular geometry is T-shaped (Figure 10.7). (Given our viewpoint of axial positions lying on a vertical line, the T of the molecule is actually on its side, ⊢).

In an octahedron, the electron-pair geometry for six electron pairs, each angle is 90°. Unlike the triangular bipyramid, the octahedron has no distinct axial and equatorial positions; all positions are equivalent. As shown in Figure 10.7, six atoms bonded to the central atom in an octahedron can be placed on three axes at right angles to one another (the *x, y,* and *z* axes) and are *equidistant* from the central atom. Unlike the triangular pyramid, if a molecule with octahedral electron-pair geometry has one lone pair, it makes no difference which apex it occupies. An example is BrF_5, whose

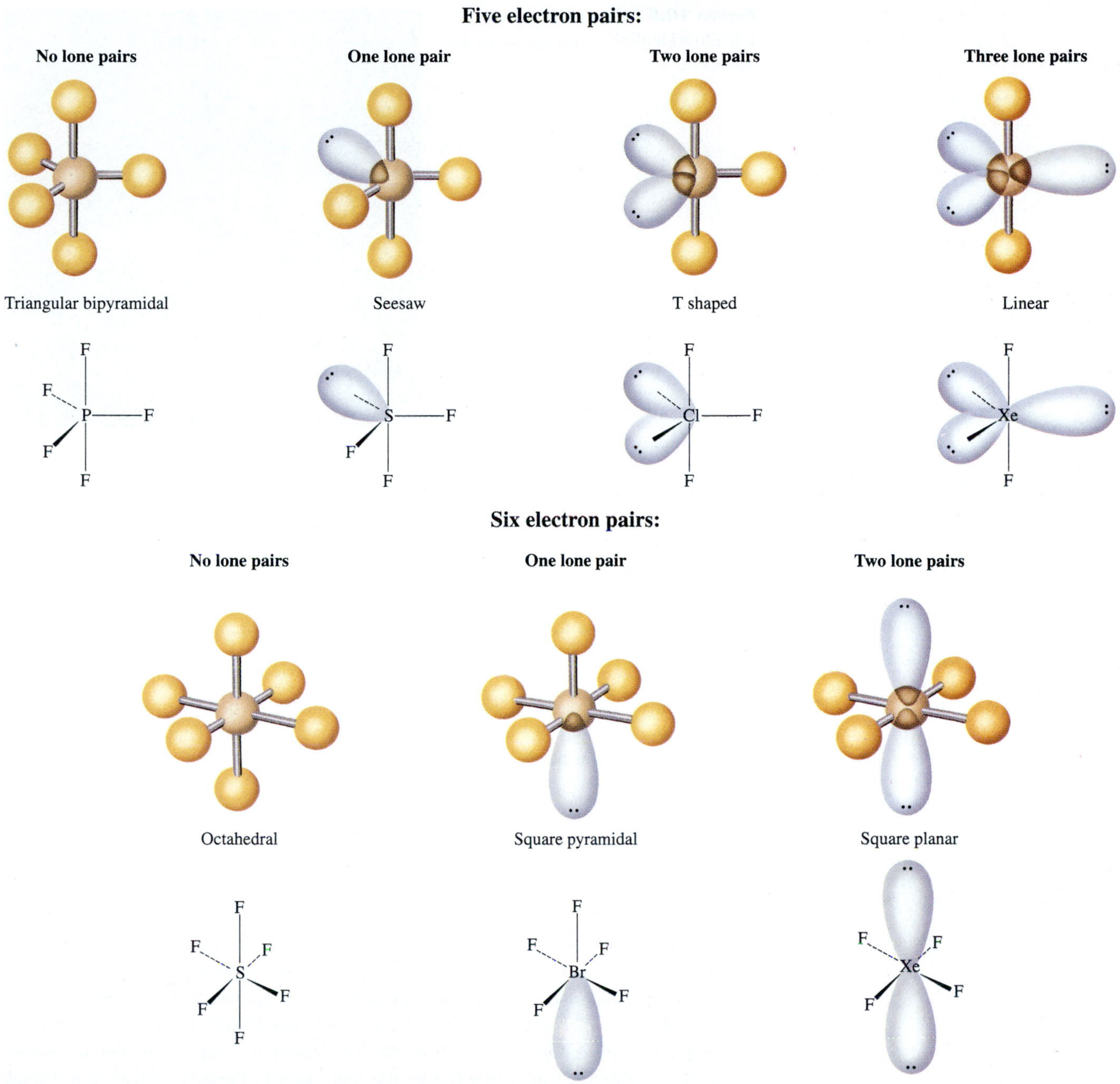

Five electron pairs:

No lone pairs	One lone pair	Two lone pairs	Three lone pairs
Triangular bipyramidal	Seesaw	T shaped	Linear

PF_5, SF_4, ClF_3, XeF_2

Six electron pairs:

No lone pairs	One lone pair	Two lone pairs
Octahedral	Square pyramidal	Square planar

SF_6, BrF_5, XeF_4

Figure 10.7 Electron-pair geometries and molecular shapes. Molecules and polyatomic ions with five and six electron pairs around the central atom can have these electron-pair geometries and molecular shapes.

molecular geometry is called square pyramidal (Figure 10.7). The name square pyramid is given because there is a *square plane* containing the central atom and four of the atoms bonded to it, with the other bonded atom directly above the central atom and equidistant from the other four. If the molecule or ion has two lone pairs, as in ICl_4^-, each of the lone pairs needs as much room as possible. This is best achieved

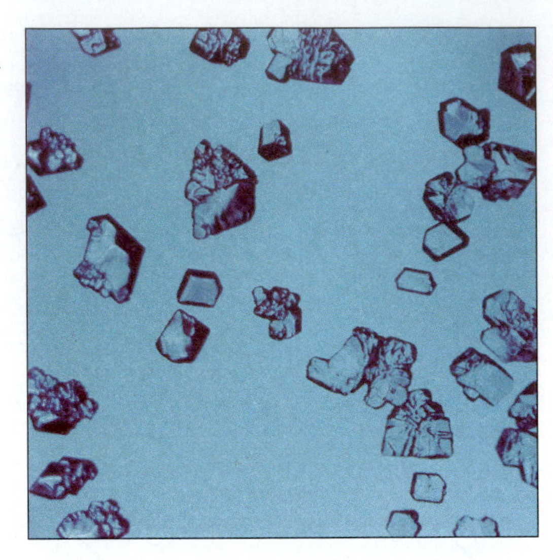

Figure 10.8 Crystals of xenon tetrafluoride, XeF_4. *(Argonne National Laboratory)*

by placing the lone pairs above and below the square plane that contains the I atom and the four Cl atoms (at a 180° angle to each other), so the molecular geometry of ICl_4^- is square planar.

An example of the power of the VSEPR model is the correct prediction of the shape of XeF_4 (Figure 10.8). At one time the noble gases were not expected to form compounds because their atoms have a stable octet of valence electrons *(⟵ p. 356)*. The synthesis of XeF_4 was a surprise to chemists, but you can use the VSEPR model to predict the correct geometry. The molecule has 36 valence electrons (8 from Xe and 28 from four F atoms). There are 8 electrons in four bond pairs around Xe, and a total of 24 electrons in the lone pairs on the four F atoms. That leaves 4 electrons in two lone pairs on the Xe atom.

```
      :F:
       |
:F — Xe — F:
       |
      :F:
```

Because Xe is in Period 5, it can accommodate more than an octet of electrons. The total of six electron pairs on Xe leads to a prediction of an octahedral electron-pair geometry. Where do you put the lone pairs? As explained above for the ICl_4^- ion, the lone pairs are placed at opposite corners of the octahedron to minimize repulsion by keeping them as far apart as possible. The result is a square planar *molecular geometry* for the XeF_4 molecule (cover the lone pairs in the XeF_4 drawing in Figure 10.7 to see this). This shape agrees with experimental structural results. A number of other xenon compounds have been prepared, and the VSEPR model has been useful in predicting their geometries as well.

Multiple Bonds and Molecular Geometry

Although double bonds and triple bonds are shorter and stronger than single bonds *(⟵ p. 370)*, they do not affect predictions of molecular shape. Why not? Electron pairs involved in a multiple bond are all shared between the same two nuclei and therefore occupy the same region. Because they must remain in that region, two electron pairs in a double bond (or three in a triple bond) are like a single balloon, rather than two or three balloons. Hence, for the purpose of determining molecular geometry, the electron pairs in a multiple bond contribute to molecular geometry in the same manner as a single bond. For example, BeF_2 (Figure 10.4) and CO_2 are linear molecules with the two Be—F single bonds 180° apart, as are the C=O bonds in CO_2. These structural data imply that, in terms of affecting molecular geometry, the C=O double bonds act like the Be—F single bonds, and so the structure of CO_2 is linear.

Carbon dioxide is a linear molecule.

When resonance structures are possible, the geometry can be predicted from any of the Lewis resonance structures or from the resonance hybrid structure. For example, the geometry of the CO_3^{2-} ion is predicted to be triangular planar because the carbon atom has three sets of bonds and no lone pairs. This can be predicted from either of the representations below.

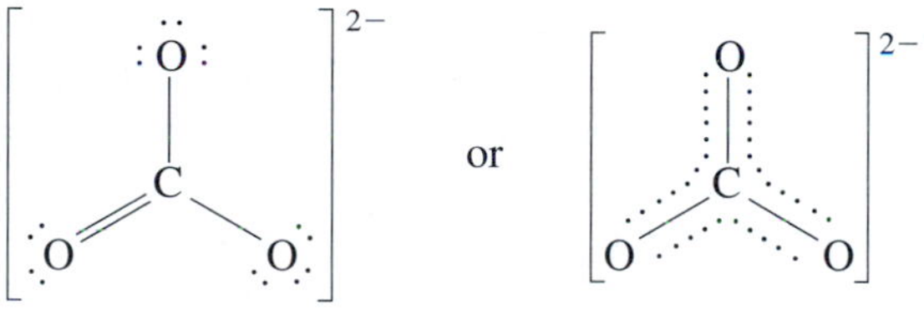

or

The NO_2^- is described as having angular molecular geometry because it has a lone pair in one position and two bonds in the other two positions of a triangular planar electron-pair geometry.

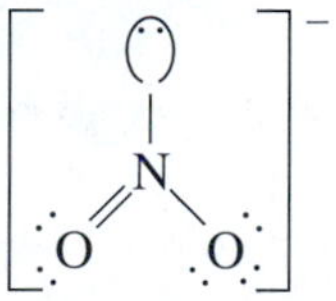

To summarize, the shape of virtually any molecule or polyatomic ion can be predicted by working through the following steps:

1. ***Draw the Lewis structure.***
2. ***Determine the number of bonds and the number of lone pairs around each atom.*** All electron pairs in a multiple bond contribute to the molecular geometry in the same manner as those in a single bond.
3. ***Pick the appropriate electron-pair geometry around each atom, and then choose the molecular shape that matches the total number of bonds and lone pairs.***
4. ***Predict the bond angles, remembering that lone pairs occupy more volume than do bonding pairs.***

TABLE 10.1 Examples of Molecular Geometries Predicted by VSEPR Model

Formula (X = electron pairs)	Number of Lone Pairs on Central Atom	Example	Geometry of Molecule or Ion*
AX_2	No lone pairs	CO_2, $BeCl_2$	Linear
AX_3	No lone pairs	BCl_3, CO_3^{2-}, SO_3	Triangular planar
	One lone pair	NO_2^-, O_3, $SnCl_2$	Angular
AX_4	No lone pairs	SiH_4, BF_4^-, NH_4^+	Tetrahedral
	One lone pair	PF_3, ClO_3^-, NH_3	Triangular pyramidal
	Two lone pairs	H_2F^+, H_2O, NH_2^-	Angular
AX_5	No lone pairs	PF_5, $AsCl_5$	Triangular bipyramidal
	One lone pair	SF_4	Seesaw
	Two lone pairs	ClF_3, BrF_3	T-shaped
	Three lone pairs	ICl_2^-, XeF_2	Linear
AX_6	No lone pairs	SF_6, PF_6^-	Octahedral
	One lone pair	BrF_5, ClF_5	Square pyramidal
	Two lone pairs	XeF_4, ICl_4^-	Square planar

*"Triangular" and "angular" are often referred to as "trigonal" and "bent, respectively."

Table 10.1 gives additional examples of molecules and ions whose shapes can be predicted by using the VSEPR model. So far the VSEPR model has been used to predict the geometry of molecules or ions with a single central atom. The model also can be used to predict the geometry around atoms in more complex compounds, those that do not contain a single central atom. Consider, for example, lactic acid, a compound important in carbohydrate metabolism. The molecular structure of lactic acid is

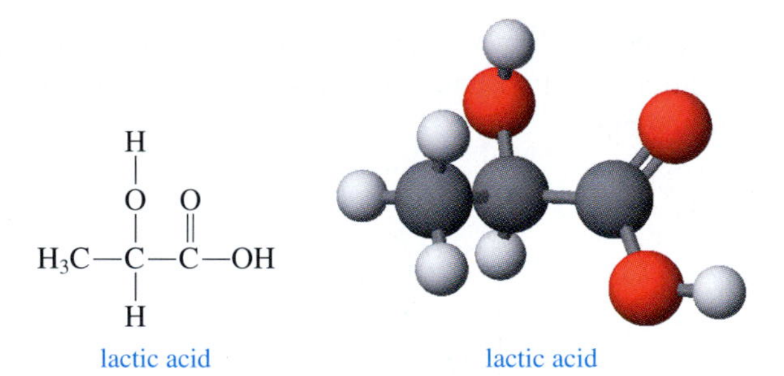

lactic acid lactic acid

VSEPR can be used to predict the geometry around each of the three carbon atoms, as well as that of the single-bonded oxygen atoms. Around the —CH_3 carbon atom are four bonding pairs and no lone pairs, so its geometry is tetrahedral (109.5° bond angles). The central carbon also has four bonding pairs associated with it, so that this carbon also has tetrahedral geometry (109.5° bond angles). The remaining carbon atom has four bonding pairs—two in single bonds and two of which are in a C=O double bond. The double bond is counted as only one pair, for molecular geometry purposes, and, in effect, this carbon has three bonding pairs around it, giving it a triangular planar geometry (120° bond angles). The single-bonded oxygens, in each case, have two bonding pairs and two lone pairs creating an angular (bent) geometry around the oxygen atom (a bond angle of less than 109.5°).

PROBLEM-SOLVING EXAMPLE 10.2 VSEPR and Molecular Shape

What are the shapes of (a) PH_4^+; (b) SO_3; (c) H_2CO; and (d) $XeOF_4$ (Xe is the central atom)?

Answer (a) Tetrahedral; (b) triangular planar; (c) triangular planar; (d) square pyramidal

Explanation The Lewis structures are

$$\left[\begin{array}{ccc} & H & \\ & | & \\ H- & P & -H \\ & | & \\ & H & \end{array}\right]^+$$

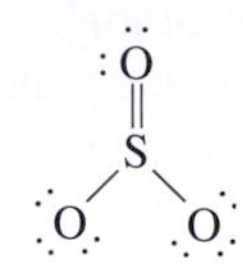

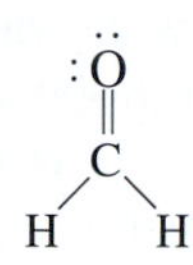

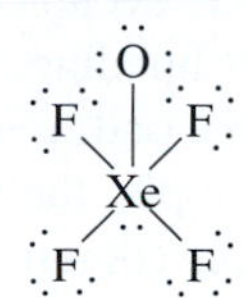

(a) The Lewis structure of $[PH_4]^+$ reveals that the central P atom has four electron pairs and, because all four electron pairs are used to bond terminal atoms, the ion has a tetrahedral shape.

(b) Three bonds surround the central S atom in SO_3. The two bond pairs in the double bond are counted as one bond for determining the geometry of the molecule. Because three resonance structures can be drawn, the best representation uses a dotted line to indicate that the multiple or double bonding involves all the atoms in the molecule. The molecular geometry is triangular planar.

(c) The C atom in H_2CO is surrounded by three bonds. Therefore, the molecular geometry around C is triangular planar.

(d) In $XeOF_4$ the central Xe atom is surrounded by five bonds and one lone pair. Because there are six pairs around the Xe atom, the electron-pair geometry is octahedral. The O and F atoms could be placed so that either the O atom or one of the F atoms is across the molecule from the lone pair. The O atom occupies the axial position because it is larger than the F atom. The molecule has a square pyramidal shape.

Problem-Solving Practice 10.2

Use Lewis structures and the VSEPR model to determine the electron-pair and molecular geometries for (a) Cl_2O, (b) SO_3^{2-}, (c) SiO_4^{4-}, and

(d) $\begin{array}{c} O \\ \| \\ HCNCH_3. \\ | \\ H \end{array}$

Exercise 10.1

Predict the bond angles in SCl_2, PCl_3, and ICl_2^-.

Exercise 10.2

Using its Lewis structure and the VSEPR model, predict the geometry around the carbon atoms of the pyruvic acid molecule, CH_3COCO_2H, a key molecule in metabolism. (*Hint:* There is no oxygen-to-oxygen bond.)

10.3 ORBITALS CONSISTENT WITH MOLECULAR SHAPES: HYBRIDIZATION

Although Lewis structures are helpful in assigning molecular geometries to describe the arrangements of atoms in molecules, they do not indicate anything about the orbitals occupied by the bonding and lone pair electrons. A theoretical model of covalent bonding, referred to as the **valence bond model,** does so by describing a covalent bond as the result of an overlap of orbitals on the bonded atoms. For H_2, for example, the bond is a shared electron pair located in the overlapping *s* atomic orbitals (Figure 10.9a). In hydrogen fluoride, HF, the overlap occurs between a 2*p* orbital with a single electron on fluorine and the single electron in the 1*s* orbital of a hydrogen atom (Figure 10.9b).

The simple valence bond model of overlapping *s, p,* or *d* orbitals however, must be modified to account for bonding molecules with central atoms, such as Be, B, and C. The electron distributions for Be, B, and C atoms are

			1s	2s	$2p_x$	$2p_y$	$2p_z$
Be	$1s^2 2s^2$		↑↓	↑↓			
B	$1s^2 2s^2 2p^1$		↑↓	↑↓	↑		
C	$1s^2 2s^2 2p^2$	$1s^2 2s^2 2p_x^1 2p_y^1$	↑↓	↑↓	↑	↑	

The simple valence bond model would predict that Be, with an s^2configuration like He, should form no compounds; B with one unpaired electron ($2p^1$) should form only one bond; and C ($2p^2$) should form only two bonds with its two unpaired 2*p* electrons. But BeF_2, BF_3, and CF_4 exist, as well as other Be, B, and C compounds in which these atoms have 2, 3, and 4 bonds, respectively.

Hybrid Orbitals: *sp*, *sp*^{2}, *sp*^{3}

To account for molecules like CH_4 in which the molecular geometry is incompatible with simple overlap of *s, p,* and *d* orbitals, valence bond theory is modified to include a new kind of atomic orbital. Atomic orbitals of the proper energy and orientation in the same atom are hybridized, meaning that they are mixed to form **hybrid orbitals.** The hybrid orbitals all have the same shape and energy. As a result, they are better able to overlap with bonding orbitals on other atoms. *The total number of hybrid orbitals formed is always equal to the number of atomic orbitals that are hybridized.*

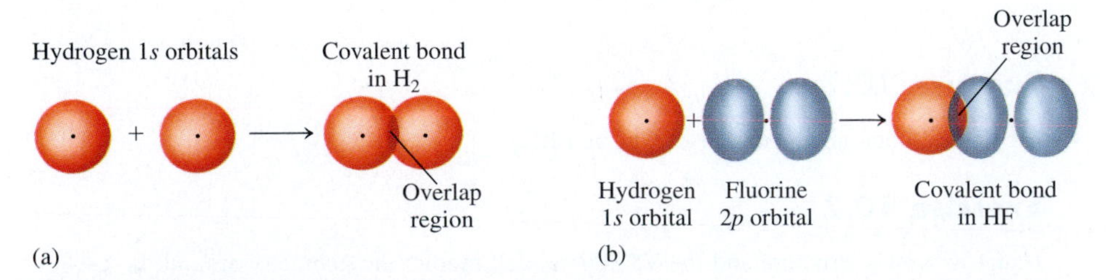

Figure 10.9 Covalent bond formation in H_2 and HF. (a) In H_2, an H—H bond forms by the overlap of the 1*s* orbitals from each H. (b) The H—F bond forms by overlap of a 2*p* orbital on F with a 1*s* orbital on H.

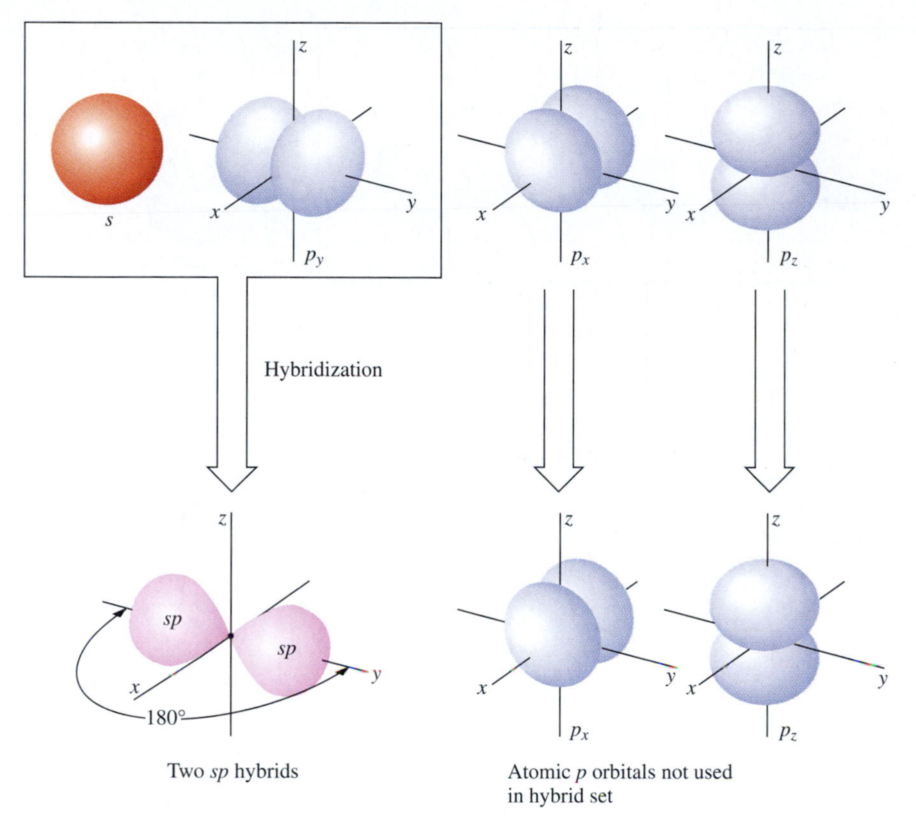

Figure 10.10 **Formation of two *sp* hybrid orbitals.** An *s* orbital combines with a *p* orbital, say, p_x, to form two hybrid orbitals that lie along the *x*-axis. The angle between the two *sp* orbitals is 180°. The other two *p* orbitals remain unhybridized. The bonds in BeF_2 are described as formed by *sp* hybrid orbitals.

The simplest hybrid orbitals are ***sp* hybrid orbitals** formed by the combination of one *s* orbital and one *p* orbital; these *two* atomic orbitals combine to form *two sp* hybrid orbitals.

One *s* atomic orbital + one *p* atomic orbital ⟶ two *sp* hybrid orbitals

In BeF_2, for example, a 2*s* and a 2*p* orbital on Be are hybridized to form two *sp* hybrid orbitals that are 180 ° from each other.

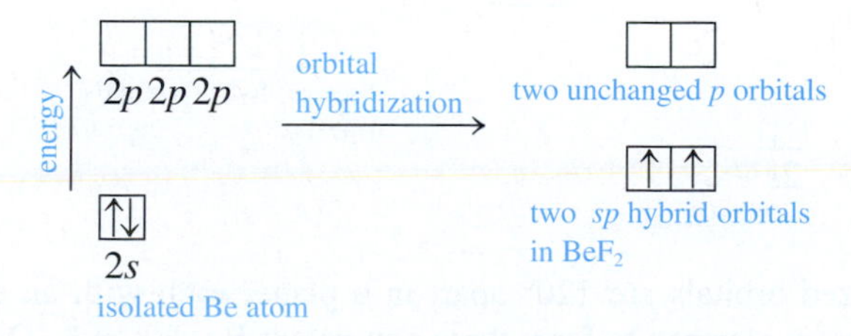

Each *sp* hybrid orbital has one electron that is shared with an electron from a fluorine atom to form two equivalent Be—F bonds. The remaining two 2*p* orbitals in Be are *unhybridized* and are 90° to each other and to the two hybrid orbitals (Figure 10.10).

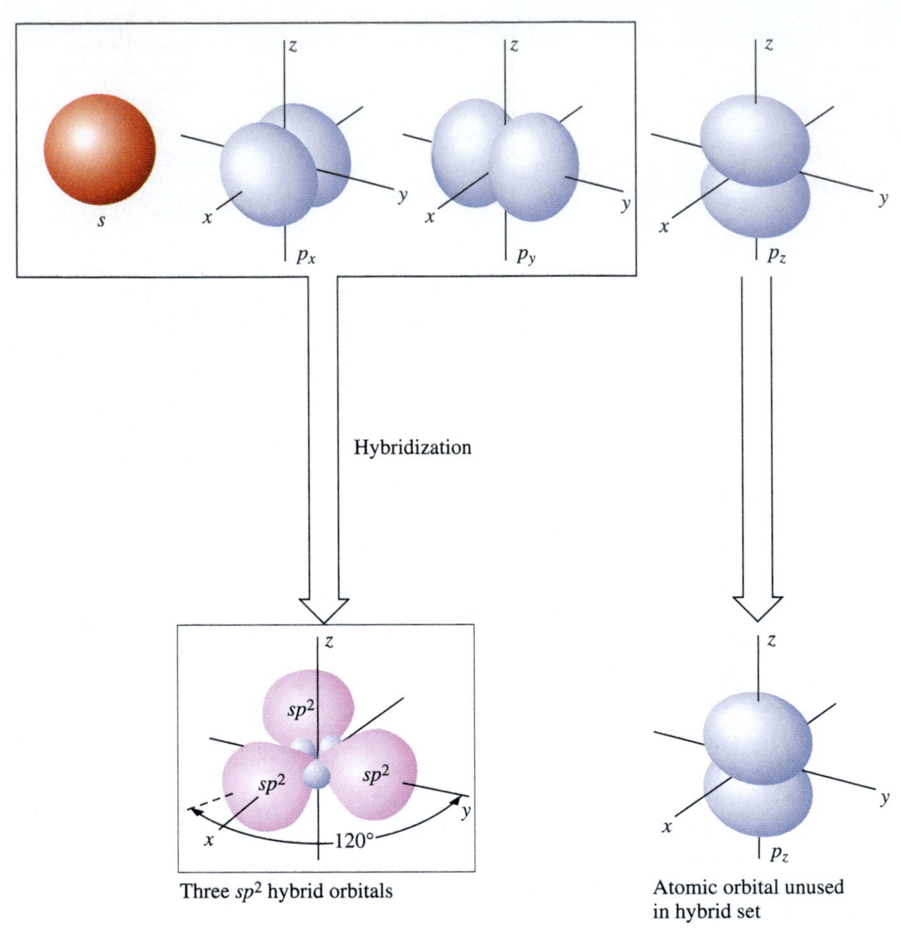

Figure 10.11 Formation of three sp^2 hybrid orbitals. An s orbital combines with two p orbitals, say p_x and p_y, to form three hybrid orbitals that lie in the xy plane. The angle between the three sp^2 orbitals is 120°. The other p orbital is unhybridized. The bonds in BF_3 are described as formed by sp^2 hybridized orbitals.

In BF_3, three atomic orbitals on the central B atom—a $2s$ and two $2p$ orbitals—are hybridized to form *three* $\mathbf{sp^2}$ **hybrid orbitals** (Figure 10.11). The superscript indicates the number of orbitals that have hybridized, two p orbitals in this case.

One s atomic orbital + two p atomic orbitals $\longrightarrow$ three sp^2 hybrid orbitals

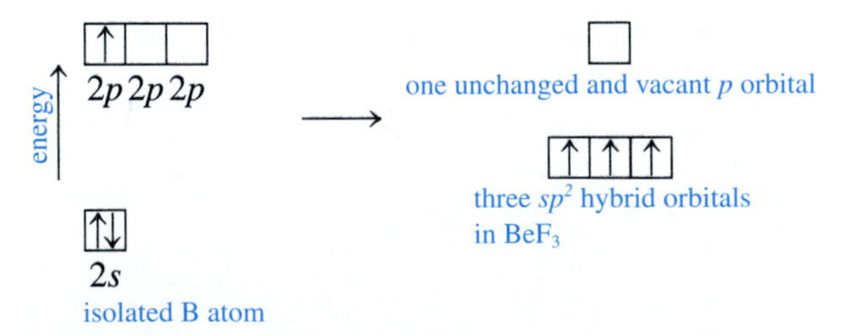

The sp^2 hybridized orbitals are 120° apart in a plane, each with an electron shared with a fluorine atom electron to form three equivalent B—F bonds. One of the boron $2p$ orbitals remains unhybridized.

If an s and all three p orbitals on a central atom are hybridized, *four* hybrid orbitals, called $\mathbf{sp^3}$ **hybrid orbitals,** are formed.

One s atomic orbital + three p atomic orbitals $\longrightarrow$ four sp^3 hybrid orbitals

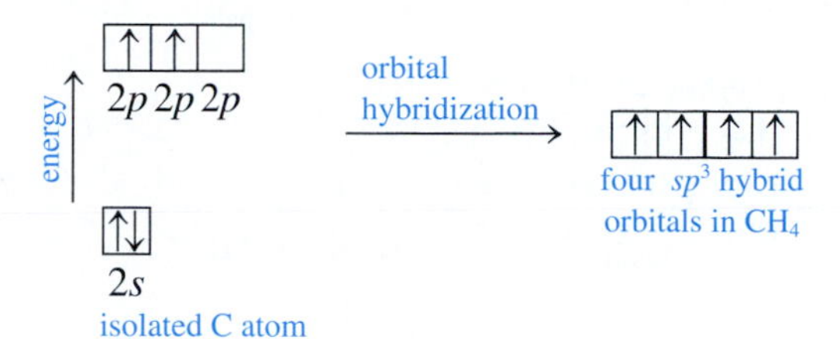

The four sp^3 hybrid orbitals are equivalent and directed to the corners of a tetrahedron (Figure 10.12).

The sp^3 hybridization explains how carbon forms four single bonds in CF_4 and in all other single-bonded carbon compounds. Overlap of the sp^3 hybrid orbitals with half-filled orbitals from four fluorine atoms form four equivalent C—F bonds. The central atoms of Periods 2 and 3 elements that obey the octet rule commonly have sp^3 hybridization.

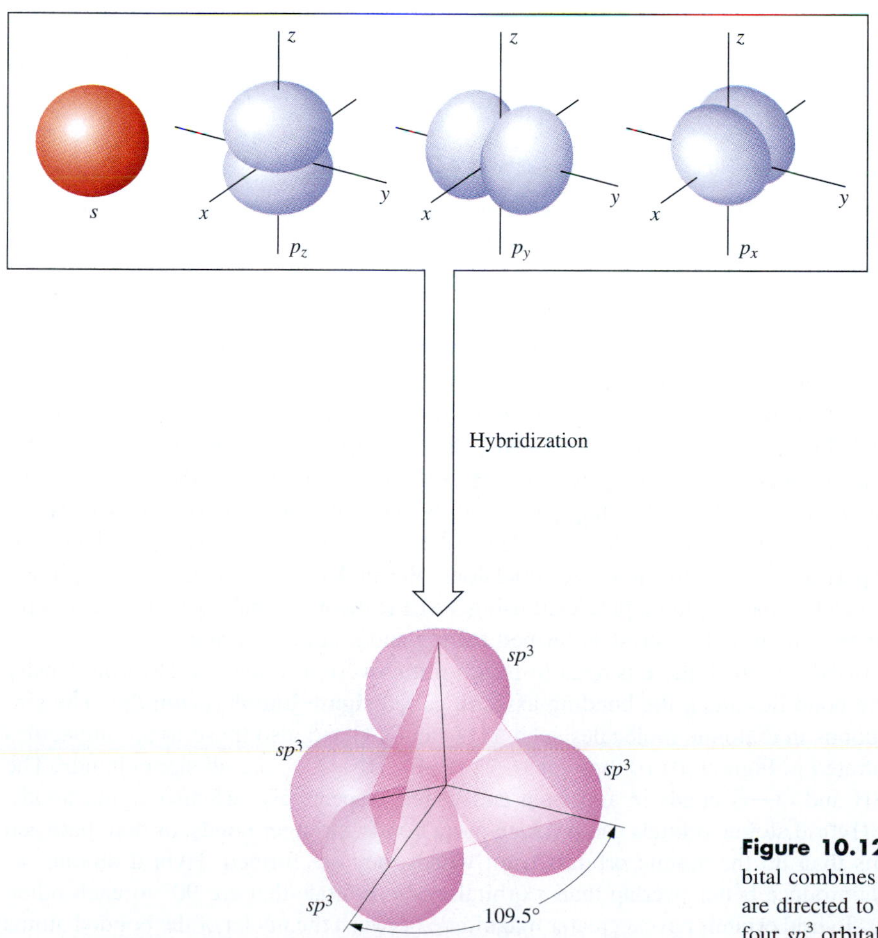

Figure 10.12 Formation of four sp^3 hybrid orbitals. An s orbital combines with three p orbitals to form four hybrid orbitals that are directed to the corners of a tetrahedron. The angle between the four sp^3 orbitals is 109.5°. The bonds of carbon atoms with four single bonds are described as formed by sp^3 hybrid orbitals.

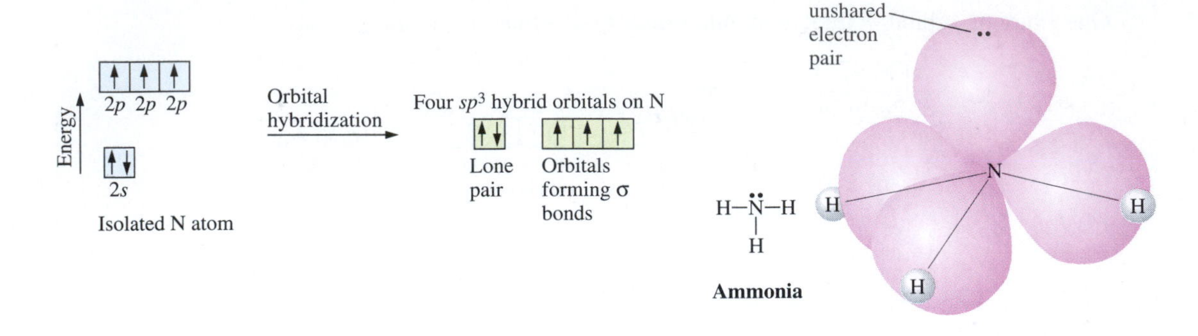

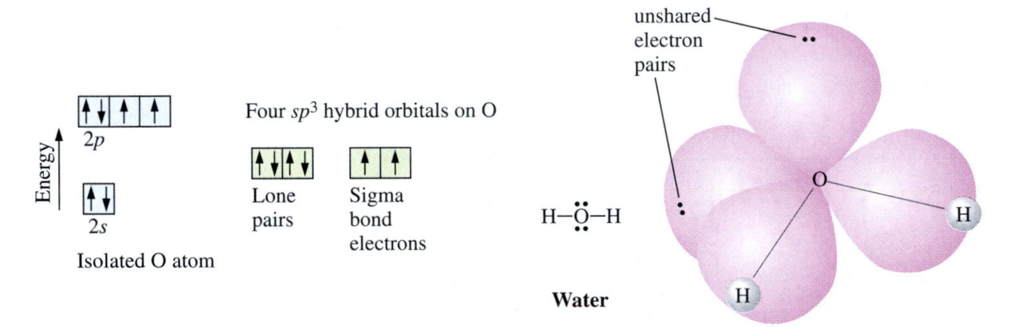

Figure 10.13 **Hybridization of nitrogen in ammonia and oxygen in water.** Nitrogen and oxygen are both sp^3 hybridized in these compounds.

Because the bond angles in NH_3 and H_2O are close to those in CF_4 and CH_4, this suggests that *in general, lone pairs as well as bonding electron pairs can occupy hybrid orbitals.* For example, according to the valence bond model, the four electron pairs surrounding the nitrogen and oxygen atoms in ammonia and water occupy sp^3 hybrid orbitals like those in CF_4 and CH_4. In NH_3, three shared pairs on nitrogen occupy three of these orbitals with a lone pair filling the fourth. In H_2O, two of the sp^3 hybrid orbitals on oxygen contain bonding pairs, and two contain lone pairs (Figure 10.13).

It is important for you to recognize that because hybrid orbitals are oriented as far apart as possible to minimize repulsions, the molecular geometries they predict are exactly what would be predicted using VSEPR theory. Table 10.2 summarizes information about hybrid orbitals formed from *s* and *p* atomic orbitals.

Bonds in which there is head-to-head orbital overlap so that the electron density of the bond lies along the bonding axis are called **sigma bonds** (**σ bonds**). The single bonds in diatomic molecules such as H_2 and Cl_2 and also those in the molecules illustrated in Figures 10.10 through 10.12 (BeF_2, BF_3, . . .) are all sigma bonds. The N—H and O—H bonds in ammonia and water, respectively, are also sigma bonds.

The process of hybridization is done by combination of mathematical functions that describe *s*, *p*, or *d* orbitals to give new functions that predict the energies and shapes of the hybrid orbitals.

Hybrid atomic orbitals can produce more bonds, stronger bonds, or both between atoms than do the atomic orbitals from which they are formed. Hybrid atomic orbitals produce better overlap than *s* orbitals, or *p* orbitals that are 90° to each other. Thus, hybrid orbitals have a greater magnitude between the nuclei of the bonded atoms than do their unhybridized atomic orbitals.

TABLE 10.2 Hybrid Orbitals and Their Geometries

Number of Single Bonds and Lone Pairs	Atomic Orbitals	Hybrid Orbitals	Electron-Pair Geometry	Atoms with Hybrid Orbitals	Molecular Geometry
2 Single bonds; 0 lone pair	*s*, one *p*	two *sp*	Linear	Be in $BeCl_2$	Linear
3 Single bonds; 0 lone pair	*s*, two *p*	three sp^2	Triangular Planar	B in BF_3	Triangular Planar
4 Single bonds; 0 lone pair	*s*, three *p*	four sp^3	Tetrahedral	C in CH_4	Tetrahedral
3 Single bonds; 1 lone pair	*s*, three *p*	four sp^3	Tetrahedral	N in NH_3	Triangular Pyramidal
2 Single bonds; 2 lone pairs	*s*, three *p*	four sp^3	Tetrahedral	O in Cl_2O	Angular (Bent)

Hybridization in Molecules with Multiple Bonds

In Section 10.2, it was noted that for the purpose of determining molecular geometry, the two or three electron pairs in a multiple bond contribute to molecular geometry the same as does one pair in a single bond. When using hybrid orbitals to account for molecular shapes, we say that the shared electron pairs in the multiple bond do *not* occupy hybrid orbitals. Thus, in a molecule with a multiple bond, the hybrid orbitals contain

Conceptual Challenge Problem CP-10.C at the end of the chapter relates to topics covered in this section.

- electron pairs forming single bonds,
- all lone pairs,
- only *one* of the shared electron pairs in a double or triple bond.

The electron pairs in *unhybridized* orbitals are used to form the second bond of a double bond, and the second and third ones in a triple bond. Such bonds are formed by the *sideways* overlap of *parallel* atomic *p* orbitals. They are called **pi bonds** (**π bonds**). In contrast to sigma bonds in which there is overlap along the bond axis, pi bonds result when parallel *p* orbitals (such as a p_y and another p_y) overlap above and below the bond axis.

The bonding in formaldehyde, H_2CO, exemplifies sigma and pi bonding. Formaldehyde has two C—H single bonds and one C=O double bond.

```
  :O:
   ‖
H—C—H
```

The triangular planar electron-pair geometry suggests sp^2 hybridization of the carbon atom to supply three sp^2 orbitals for sigma bonds. Two of these sp^2 hybrid orbitals form two sigma bonds with half-filled 1s H orbitals; the third sp^2 hybrid orbital forms a sigma bond with a half-filled oxygen orbital (Figure 10.14). The *unhybridized p* orbital on carbon overlaps sideways with a *p* orbital on oxygen to form a pi bond, thus completing a double bond between carbon and oxygen.

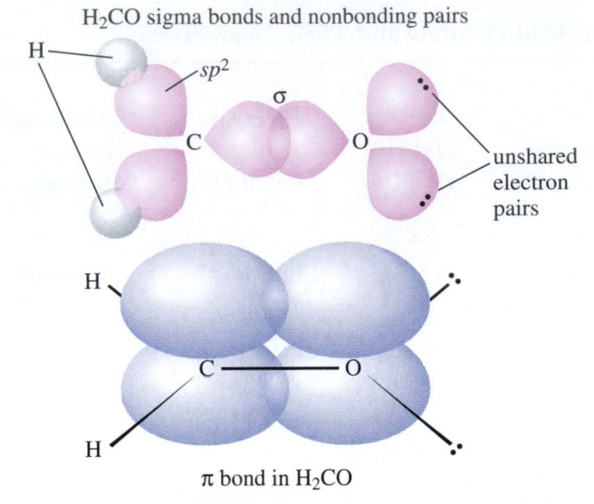

Figure 10.14 Sigma and pi bonding in formaldehyde. In formaldehyde, the C and O atoms are sp^2 hybridized. The C—O sigma bond forms from the end-to-end overlap of the hybrid orbitals. The sp^2 hybridization leaves a half-filled unhybridized orbital on each C and O atom. These orbitals overlap to form the carbon-oxygen pi bond.

Formaldehyde, H_2CO

Bonding in a formaldehyde molecule.

PROBLEM-SOLVING EXAMPLE 10.3 Hybrid Orbitals, Sigma and Pi Bonding

Use hybridized orbitals and sigma and pi bonding to describe bonding in the hydrocarbons: (a) ethane (C_2H_6); (b) ethylene (C_2H_4); and (c) acetylene (C_2H_4).

Answer (a) Carbon—sp^3 hybridized orbitals form sigma bonds to three H and the other C. (b) Carbon—sp^2 hybridized orbitals form sigma bonds to two H and the other C; one pi bond between C atoms. (c) Carbon—*sp* hybridized orbitals form sigma bonds to one H and the other C, and two pi bonds between C atoms.

Explanation The Lewis structures for the three compounds are

```
   H  H         H  H
   |  |         |  |
H—C—C—H        C=C
   |  |         |  |
   H  H         H  H       H—C≡C—H

  ethane      ethylene     acetylene
```

(a) Consistent with the tetrahedral bond angles for single-bonded carbon atoms, each carbon atom in ethane is sp^3 hybridized and all bonds are sigma bonds.

(b) The C=C double bond in ethylene, like the C=O double bond in formaldehyde, requires sp^2 hybridization of carbon. Two of the sp^2 hybrid orbitals on each carbon form sigma bonds with hydrogens; the third sp^2 hybrid orbital overlaps head-to-head with an sp^2 hybrid orbital on the other carbon, creating a C—C sigma bond. The double bond between the carbon atoms is completed by the sideways overlap of parallel unhybridized *p* orbitals from each carbon to form a pi bond. Thus, the C=C double bond consists of a sigma bond and a pi bond.

(c) In acetylene, *sp* hybridization of each carbon atom is indicated by the linear geometry. One of the *sp* hybrid orbitals on each carbon forms a sigma bond between C and H; the other is used to join the carbon atoms by a sigma bond. The two unhybridized *p* orbitals on each carbon overlap edgewise forming two pi bonds, completing the triple bond. Therefore, the triple bond consists of one sigma and two pi bonds in which the pi bonds are at right angles (90°) to each other.

The bonding in ethane, ethylene, and acetylene is shown in Figure 10.15.

Problem-Solving Practice 10.3

Using hybridization and sigma and pi bonding concepts, explain the bonding in (a) HCN and (b) methyl imine, H_2CNH.

Exercise 10.3

Explain why pi bonding is not possible for an sp^3 hybridized carbon atom.

Because pi bonds have less orbital overlap than sigma bonds, pi bonds are generally weaker than sigma bonds. Thus, a C=C double bond is stronger (and shorter) than a C—C single bond, but not twice as strong; correspondingly, a carbon-carbon triple bond, although stronger (and shorter) than a C=C double bond, is not three times stronger than a C—C single bond. Also, a C=C double bond prevents rotation around the bond (under ordinary conditions). A consequence of this nonrotation is *cis-trans* isomerism (**← p. 372**).

Hybridization in Expanded Octets

Central atoms from the third and subsequent periods can accommodate more than four electron pairs, that is, an "expanded" octet (**← p. 368**). Bonding pairs and lone

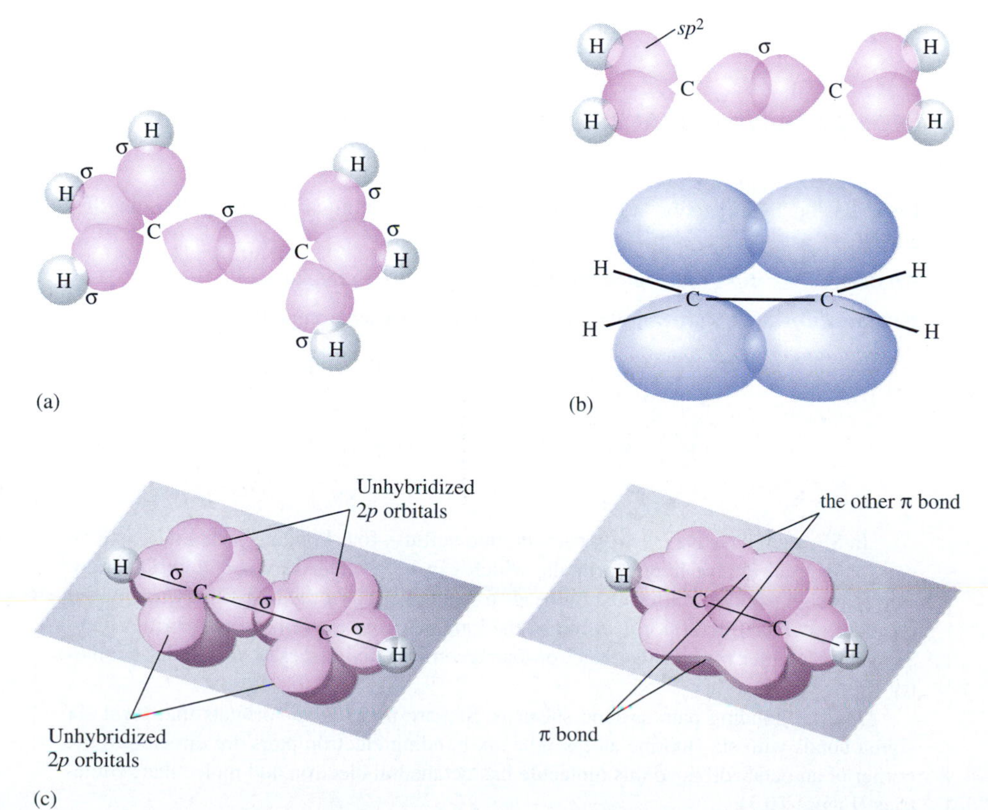

Figure 10.15 Sigma and pi bonding in ethane, ethylene, and acetylene. (a) *Ethane* contains no unhybridized *p* orbitals and, thus, has no pi bonds. (b) *Ethylene* contains an unhybridized *p* orbital on each carbon atom; the unhybridized *p* orbitals overlap to form a pi bond in addition to the sigma bonds formed by the overlap of the sp^2 hybrid orbitals between the carbon atoms. (c) The carbon atoms in *acetylene* are *sp* hybridized. Thus, each contains two unhybridized *p* orbitals on each carbon atom, which overlap to form two pi bonds. There is also a sigma bond formed by the overlap of the *sp* hybrid orbitals.

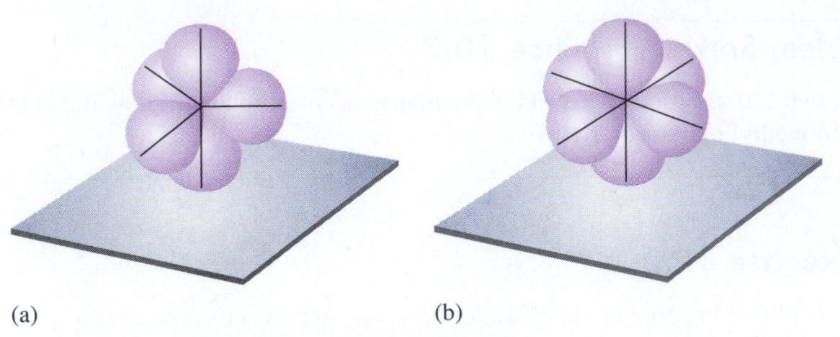

Figure 10.16 Hybridization using *d* orbitals to form (a) dsp^3 and (b) d^2sp^3 hybrid orbitals. The five dsp^3 hybrid orbitals form from hybridizing a *d* orbital, an *s* orbital, and three *p* orbitals. Six d^2sp^3 hybrid orbitals come from hybridizing two *d*, an *s*, and three *p* orbitals.

electron pairs are accommodated by hybridizing *d* as well as *s* and *p* atomic orbitals. By including *d* atomic orbitals, two additional types of hybrid orbitals result, called $\boldsymbol{dsp^3}$ and $\boldsymbol{d^2sp^3}$ **hybrid orbitals** (Figure 10.16).

Hybridizing the appropriate five atomic orbitals yields five dsp^3 hybrid orbitals; hybridization of the suitable six atomic orbitals yields six d^2sp^3 hybrid orbitals.

One *s* atomic orbital + three *p* atomic orbitals + one *d* atomic orbital ⟶ five dsp^3 hybrid orbitals.

and

One *s* atomic orbital + three *p* atomic orbitals + two *d* atomic orbitals ⟶ six d^2sp^3 hybrid orbitals

Hybridization of *s, p,* and *d* orbitals coincides with the molecular geometry expected from VSEPR theory (Figure 10.7).

Problem-Solving Example 10.4 Hybridization and Expanded Octets

Describe the hybridization around the central sulfur atom and the bonding in SF_4 and SF_6.

Answer SF_4: dsp^3, four sigma bonds; SF_6: d^2sp^3, six sigma bonds

Explanation We start with the Lewis structures of each compound.

SF_4 SF_6

In SF_4 there are five electron pairs around sulfur—four bonding pairs and one lone pair. This requires five hybrid orbitals, which can be produced by dsp^3 hybridization. The four sigma bonds are formed by head-to-head overlap of the half-filled dsp^3 hybrid orbitals with half-filled fluorine orbitals; the lone pair is in the remaining hybrid orbital. SF_4 has triangular bipyramidal electron-pair geometry and a seesaw molecular geometry (Figure 10.7).

The six bonding pairs around sulfur in SF_6 are in six d^2sp^3 orbitals that form six sigma bonds with six fluorine atoms. The six bonding electron pairs are directed to the corner of an octahedron, so this molecule has octahedral electron and molecular geometries (Figure 10.7).

Problem-Solving Practice 10.4

Describe the hybridization of the central atom and bonding in (a) PF_6^-; (b) IF_3; and (c) ICl_4^-.

10.4 MOLECULAR POLARITY

Recall from Section 9.11 that the polarity of a covalent bond can be predicted from the difference in electronegativity of the two atoms joined by the bond. However, *a molecule that has polar bonds may or may not be polar.* Depending on the three-dimensional shape of the molecule, the contributions of two or more polar bonds may cancel each other, leading to a **nonpolar molecule.** In a **polar molecule,** there is an accumulation of electron density toward one end of the molecule, giving that end a slight negative charge, $\delta-$, and leaving the other end with a slight positive charge of equal value, $\delta+$ (Figure 10.17).

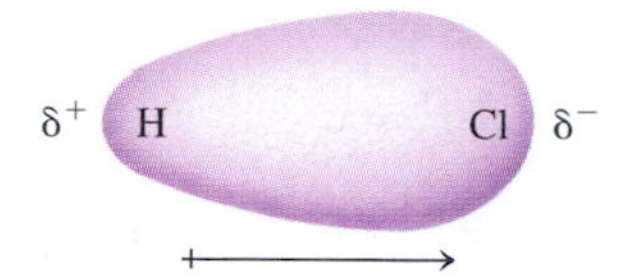

Figure 10.17 A polar molecule. In a polar molecule, the valence electron density is shifted slightly to one side of the molecule. The arrowhead points toward the partially negative end of the molecule, the plus sign at the partially positive end. The $\delta+$ and $\delta-$ indicate partial positive and negative charges at the ends of the molecules.

Before examining the factors that determine whether a molecule is polar, let's look at the experimental measurement of the polarity of molecules. Polar molecules experience a force in an electric field that tends to align them with the field (Figure 10.18). When the electric field is created by a pair of oppositely charged plates, the positive end of each molecule is attracted toward the negative plate and the negative end is attracted toward the positive plate. The extent to which the molecules line up with the field depends on their **dipole moment,** which is defined as the product of the magnitude of the partial charges ($\delta+$ and $\delta-$) times the distance of separation between them. The derived unit of the dipole moment is the coulomb-meter; a more convenient, derived unit is the debye (D), defined as $1\ D = 3.34 \times 10^{-30}\ C \cdot m$. Some typical experimental values are listed in Table 10.3. Nonpolar molecules have a zero dipole moment; the dipole moments for polar molecules are always greater than zero and increase with greater molecular polarity.

To predict whether a molecule will be polar, we need to consider whether the molecule has polar bonds and how those bonds are positioned relative to one another. We can correlate the types of molecular geometry with dipole moment by applying

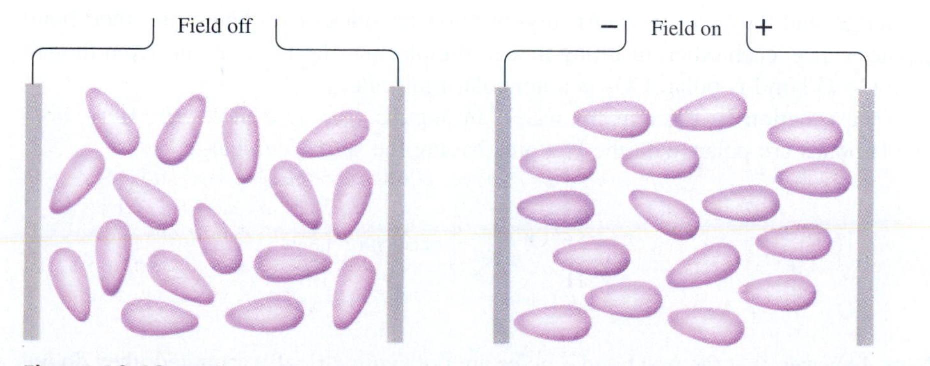

Figure 10.18 Polar molecules in an electric field. Polar molecules experience a force in an electric field that tends to align them so that oppositely charged ends of adjacent molecules are closer to each other.

TABLE 10.3 Selected Dipole Moments

Molecule	Dipole Moment (D)
H_2	0
HF	1.78
HCl	1.07
HBr	0.79
HI	0.38
ClF	0.88
BrF	1.29
BrCl	0.52
H_2O	1.85
H_2S	0.95
CO_2	0
NH_3	1.47
NF_3	0.23
NCl_3	0.39
CH_4	0
CH_3Cl	1.92
CH_2Cl_2	1.60
$CHCl_3$	1.04
CCl_4	0

a general rule to a molecule of the type AB_n (A is the central atom, B is the terminal atom, n is the number of terminal atoms). Such a molecule will *not* be polar if it meets *all* of the following conditions:

Nonpolar molecules

- All the terminal atoms (or groups) are the same, *and*
- All the terminal atoms (or groups) are symmetrically arranged around the central atom, A, in the molecular geometries given in Figure 10.4, *and*
- The terminal atoms (or groups) have the same partial charges.

This means that molecules with the molecular geometries given in Figure 10.4 will never be polar if all of their terminal atoms (or groups) are the same.

Alternatively, a molecule is *polar* if it meets *either* of the following conditions:

Polar molecules

- One or more terminal atoms differs from the others, *or*
- The terminal atoms are not symmetrically arranged.

Consider, for example, the dipole moments of CF_4 ($\mu = 0$ D) and CF_3H ($\mu = 1.60$ D). Both have the same molecular shape with their atoms tetrahedrally arranged around a central carbon atom. The terminal F atoms are all the same in CF_4, and thus have the same partial charges. But, the terminal atoms in CF_3H are not the same; F is more electronegative than H, causing the C—F bond dipole to have a higher partial negative charge than that of C—H. Consequently, CF_4 is a nonpolar molecule and CF_3H is polar.

As another example, consider carbon dioxide, CO_2, a linear triatomic molecule. Each C=O bond is polar because O is more electronegative than C, so O is the partial negative end of the bond dipole. The dipole moment contribution from each bond (the bond dipole) is represented by the symbol $+\!\!\!\longrightarrow$, in which the plus sign indicates the partial positive charge and the arrow points to the partial negative end of the bond. We can use the arrows to help estimate whether a molecule is polar.

$$\overset{\longleftarrow\!\!\!+\quad +\!\!\!\longrightarrow}{\underset{\delta-\quad\delta+\quad\delta-}{O{=}C{=}O}}\qquad \text{no net dipole}$$

The O atoms are at the same distance from the C atom, they both have the same δ- charge, and they are symmetrically on opposite sides of C. Therefore, their bond dipoles cancel each other, resulting in a zero molecular dipole moment. Even though each C=O bond is polar, CO_2 is a nonpolar molecule.

The situation is different for water, an angular triatomic molecule. Here, both O—H bonds are polar, with the H atoms having the same $\delta+$ charge.

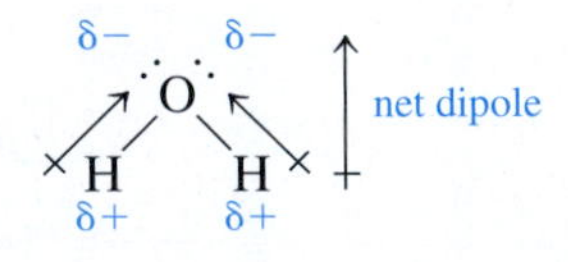

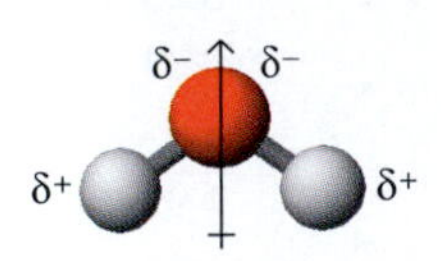

Polarity of the water molecule. The O—H bonds in water are polar, and the water molecule is polar with a molecular dipole moment of 1.85 D.

Note, however, that the two bond dipoles are not symmetrically arranged; they do not point directly toward or away from each other, but augment each other to give a molecular dipole moment of 1.85 D (Table 10.3). Thus, water is a polar molecule (Figure 10.19).

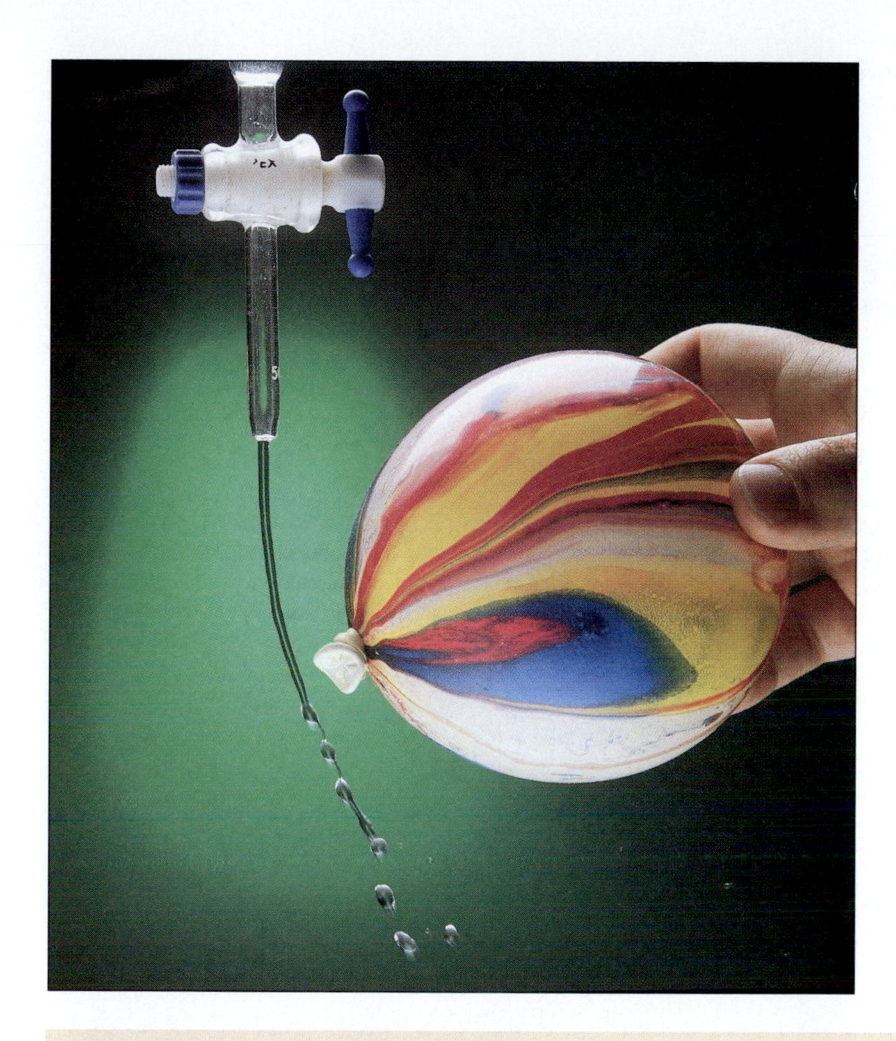

Figure 10.19 Water is a polar molecule. The electric charge on a balloon deflects a stream of polar water molecules by electrostatic attraction. *(C.D. Winters)*

Exercise 10.4 Polar Molecules

Explain why ammonia, NH_3, is a polar molecule (1.47 D) even though its terminal atoms are alike.

Using a microwave oven to make popcorn or to heat dinner is a handy application of the fact that water is a polar molecule. Most foods have a high water content. When water molecules in foods absorb microwave radiation, they rotate, turning to align with the crests and troughs of the oscillating microwave radiation. As the water molecules tumble, the hydrogen bonds among them are broken. The energy when these bonds reform heats the rest of the food. The radiation generator (magnetron) in the oven creates microwave radiation that oscillates at 2.45 GHz (2.45 gigahertz, $2.45 \times 10^9\ s^{-1}$), very nearly the optimum rate to rotate water molecules. So the leftover pizza warms up in a hurry for a late-night snack.

Problem-Solving Example 10.5 Molecular Polarity

Are dichloromethane (CH_2Cl_2) and boron trifluoride (BF_3) polar or nonpolar? If polar, indicate the direction of polarity.

Answer CH_2Cl_2 is polar;

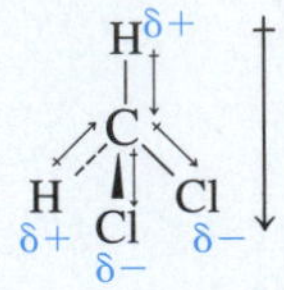

BF_3 is nonpolar.

Explanation

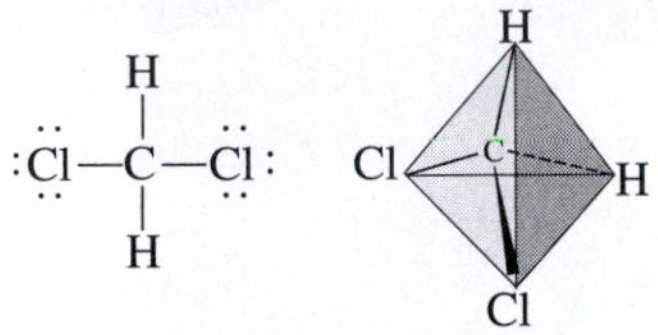

From the Lewis structure for CH_2Cl_2, you might be tempted to say that CH_2Cl_2 is nonpolar because the hydrogens appear across from each other, as do the chlorines. However, CH_2Cl_2 is a tetrahedral molecule in which neither hydrogen nor chlorine is directly across from each other, so any bond dipoles will not cancel. Because Cl is more electronegative than C, while H is less electronegative, negative charge is drawn away from H toward Cl, and CH_2Cl_2 has a net dipole (μ = 1.60 D) with the negative end at the Cl atoms.

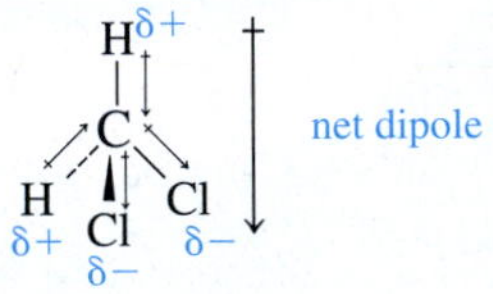

net dipole

As predicted for a molecule with three electron pairs around the central atom, BF_3 is triangular planar.

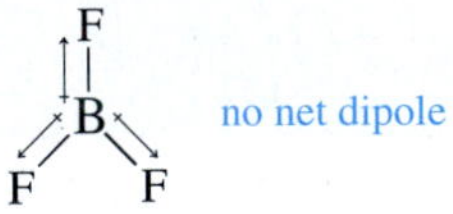

no net dipole

Because F is more electronegative than B, the B—F bonds are polar, with F being the partial negative end. The molecule is nonpolar, though, because the three terminal F atoms are identical, they have the same partial charge, and they are arranged symmetrically around the central B atom. The bond dipole of each B—F is canceled by the bond dipoles of the opposite two B—F bonds.

Problem-Solving Practice 10.5

For each of the following molecules, decide whether the molecule is polar, and if so, which side is partially positive and which is partially negative: (a) $BFCl_2$; (b) NH_2Cl; (c) SCl_2.

Exercise 10.5 Dipole Moments

Explain the differences in the dipole moments of
(a) HI (0.38 D) and HBr (0.79 D)
(b) CH_3Cl (1.92 D), CH_3Br (1.81 D); and CH_3I (1.62 D)

10.5 NONCOVALENT INTERACTIONS AND FORCES BETWEEN MOLECULES

Molecules are attracted to one another; in a sense they are "sticky." The "stickiness" of molecules is always, in some way, the result of attraction between opposite charges. As you have just seen, the positive and negative ends of different polar molecules can attract each other. Even nonpolar molecules have some attraction for each other, however, as you will see in this section as we explain the various molecular attractions.

Atoms in the same molecule are held together by chemical bonds with strengths ranging from 150 to 1000 kJ/mol. Weaker forces of attraction, called **intermolecular forces,** or **intermolecular attractions,** attract one molecule to another molecule. For example, it takes 1652 kJ to break 4 mol of C—H covalent bonds and separate the one C atom and four H atoms in all the molecules in 1 mol of methane molecules:

$$CH_4 \xrightarrow[\substack{\text{bond forces} \\ 4 \times 417 \text{ kJ/mol} \\ = 1656 \text{ kJ/mol}}]{\text{overcome chemical}} C + 4\,H$$

But only 8.9 kJ is required to pull 1 mol of methane molecules that are close together in liquid methane away from each other to evaporate the liquid to gaseous methane.

$$CH_4 \cdots CH_4 \xrightarrow[\substack{\text{force} \\ 8.9 \text{ kJ/mol}}]{\text{overcome intermolecular}} CH_4 + CH_4$$

Intermolecular attractions are weaker than covalent bonds because they do not result from the sharing of electron pairs between atoms. That is, they are **noncovalent interactions**—attractive forces that are not ionic bonds yet are different from covalent bonds. The noncovalent interactions between molecules (*inter*molecular forces) account for the melting points, boiling points, and other properties of molecular substances. Noncovalent interactions between different parts of the same large molecule (*intra*molecular forces) maintain biologically important molecules in the exact shapes required to carry out their functions. For example, a large number of noncovalent interactions between the two strands of DNA establish the double-helix structure of that large molecule. Yet each single noncovalent interaction in DNA is weak enough to be overcome under physiological conditions, thereby allowing the two strands of DNA to come apart and be copied.

*Intra*molecular means *within* a single molecule; *inter*molecular indicates *between* two or more separate molecules.

The next few sections explore three types of noncovalent interactions: London forces, dipole-dipole attractions, and hydrogen bonding.

London Forces

London forces occur in *all* molecular substances. They result from the attraction between the positive and negative ends of induced dipoles in adjacent molecules. An **induced dipole** is caused in one molecule when the electrons of a neighboring molecule are momentarily unequally distributed. The result is a temporary dipole in the

London forces are named to recognize the work of Fritz London, who extensively studied the origins and nature of such forces. London forces are also called *dispersion forces*.

Figure 10.20 Origin of London forces. Such attractive forces originate when electrons are momentarily distributed unevenly in the molecule on the left. It has a temporary positive charge that is close to the molecule on the right. The positive charge attracts electrons in the right-hand molecule, temporarily creating an induced dipole (an unbalanced electron distribution) in that molecule.

first molecule. Figure 10.20 illustrates how one molecule with a momentary unevenness in its electrical distribution can induce a dipole in a neighboring molecule, a process called **polarization.** Even noble gas atoms, molecules of diatomic gases such as oxygen, nitrogen, and chlorine (which must be nonpolar), and nonpolar hydrocarbon molecules such as CH_4 and C_2H_6 have such instantaneous dipoles.

The strength of London forces depends on how readily electrons in a molecule can be polarized, which depends on the number of electrons in a molecule and how tightly they are held by nuclear attraction. In general, the more electrons there are in a molecule, the more easily they can be polarized. Thus, large molecules with many electrons are relatively polarizable. In contrast, smaller molecules are less polarizable because they have fewer electrons. London forces range from approximately 0.05 to 40 kJ/mol.

When we look at the boiling points of several groups of nonpolar molecules, the effect of the number of electrons becomes readily apparent (Table 10.4). (This effect also correlates with molar mass—the heavier an atom or molecule, the more electrons it has.) Interestingly, molecular shape can also play a role in London forces. Two of the isomers of pentane—straight-chain pentane and 2,2-dimethylpropane (both with the molecular formula C_5H_{12})—differ in boiling point by 27 °C. The linear shape of the *n*-pentane molecule allows close contact with adjacent molecules over its entire length, while the more spherical 2,2-dimethylpropane molecule does not allow as much close contact.

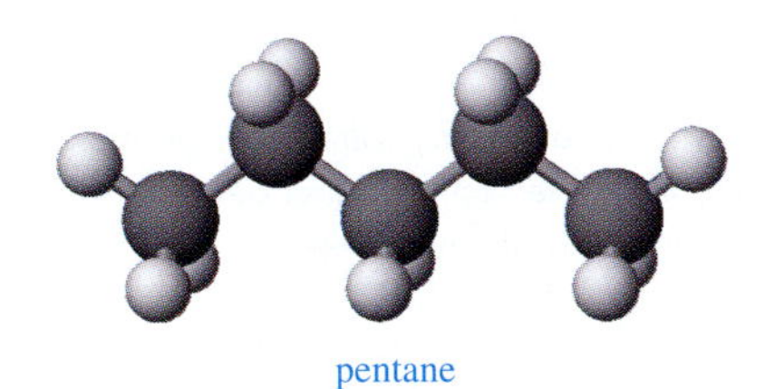

pentane

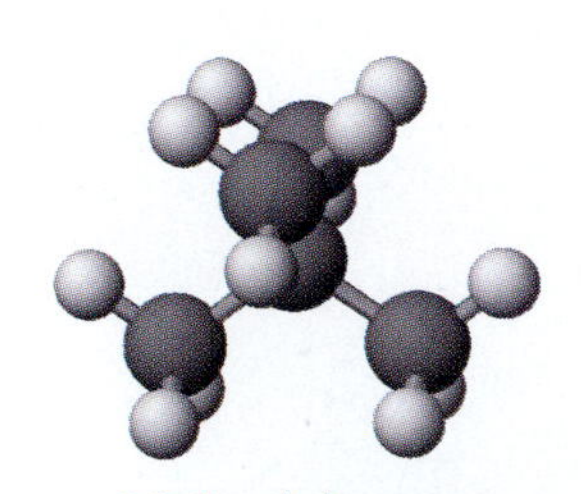

2,2-dimethylpropane

Structure and boiling point. The boiling points of pentane and 2,2-dimethylpropane differ because of differences in their molecular structures even though the total number of electrons in each molecule is the same.

$$H_3C-CH_2-CH_2-CH_2-CH_3$$

pentane
b.p. 36 °C

$$H_3C-\underset{\displaystyle CH_3}{\overset{\displaystyle CH_3}{\underset{|}{\overset{|}{C}}}}-CH_3$$

2,2-dimethylpropane
b.p. 9 °C

Dipole-Dipole Attractions

A **dipole-dipole attraction** is a noncovalent interaction between two *polar* molecules or two polar groups in the same large molecule. The previous section explained how molecules containing permanent dipoles are formed when atoms of different electronegativity bond together in an unsymmetrical manner. Molecules that are dipoles attract each other when the positive region of one is close to the negative region of another (Figure 10.21).

TABLE 10.4 Effect of Numbers of Electrons on Boiling Points of Nonpolar Molecular Substances

Noble Gases			Halogens			Hydrocarbons		
	No. e's	b.p. (°C)		No. e's	b.p. (°C)		No. e's	b.p. (°C)
He	4	−269	F_2	38	−188	CH_4	16	−161
Ne	20	−246	Cl_2	71	−34	C_2H_6	30	−88
Ar	40	−186	Br_2	160	59	C_3H_8	44	−42
Kr	84	−152	I_2	254	184	C_4H_{10}*	58	0

*Butane

In a liquid the molecules are very close to one another and attracted to each other by their intermolecular forces. Molecules must have enough energy to overcome their attractive forces for a liquid to boil. Thus, the boiling point of a liquid depends upon intermolecular forces. If more energy is required to overcome the intermolecular attractions between molecules of liquid A than the intermolecular attractions between molecules of liquid B, then the boiling point of A will be higher than that of B. Conversely, lower intermolecular attractions result in lower boiling points.

The boiling points of several nonpolar and polar substances with comparable numbers of electrons (and therefore comparable London forces) are given in Table 10.5. In general, the more polar its molecules, the higher the boiling point of a substance, provided the London forces are similar. The lower boiling points of nonpolar substances compared with those of polar substances in Table 10.5 reflect this. Dipole-dipole forces range from 5 to 25 kJ/mol, and London forces (0.05 to 40 kJ/mol) can be stronger. For example, the greater London forces in HI cause it to have a higher boiling point (−36 °C) than HCl (−85 °C), even though HCl is more polar. However, if their London forces are similar, a more polar substance will have stronger intermolecular attractions than a less polar one.

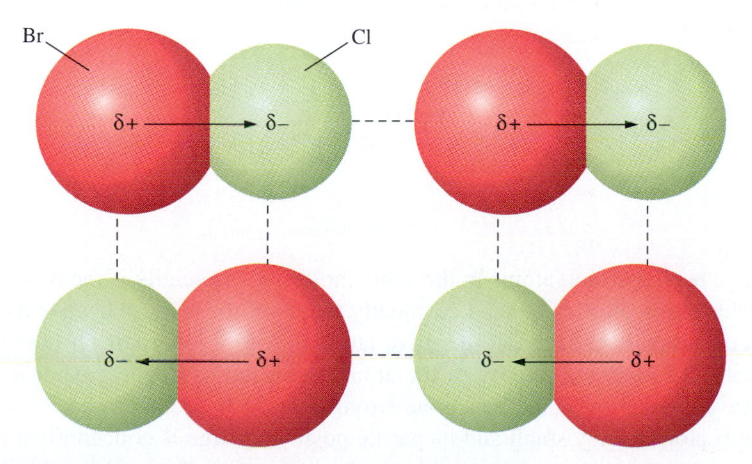

Figure 10.21 Dipole-dipole attractions (dashed lines) between BrCl molecules. The BrCl molecule is polar with the more electronegative Cl atom attracting electrons away from the Br atom, which acquires a partial positive charge. The partial negative region of one molecule is attracted to the partial positive region of the neighboring molecule.

TABLE 10.5 Numbers of Electrons and Boiling Points of Nonpolar and Polar Substances

Nonpolar Molecules			Polar Molecules		
	No. e's	b.p. (°C)		No. e's	b.p. (°C)
N_2	28	−196	CO	28	−192
SiH_4	32	−112	PH_3	34	−88
GeH_4	77	−90	AsH_3	78	−62
Br_2	160	59	ICl	162	97

Exercise 10.6 Dipole-Dipole Forces

Draw a sketch, like that in Figure 10.21, of four CO molecules to indicate dipole-dipole forces between the CO molecules.

Hydrogen Bonds

An especially significant type of dipole-dipole force is possible between a hydrogen atom covalently bonded to a highly electronegative atom and another highly electronegative atom that has lone pair electrons. Usually the highly electronegative atom is fluorine, oxygen, or nitrogen. Because of the electronegativity difference, electron density shifts toward F, O, or N causing this atom to take on a *partial negative charge.* As a result, a hydrogen atom bonded to the nitrogen, oxygen, or fluorine atom acquires a *partial positive charge.*

The **hydrogen bond** is the attraction between a partially positive hydrogen atom and a lone pair on a small, very electronegative atom (F, O, or N). Hydrogen bonds are typically shown as dotted lines (. . .) between the atoms. A hydrogen bond, then, can be a "bridge" between two highly electronegative atoms, X and Z, in two different molecules,

$$\overset{\delta-}{X}\text{—}\overset{\delta+}{H}\cdots\overset{\delta-}{Z}\text{—}$$

↑ hydrogen bond

Hydrogen bonds can form between molecules or within a molecule. These are known as intermolecular and intramolecular hydrogen bonds, respectively.

or a bridge between such atoms in the same molecule, especially if the molecules are large. A hydrogen atom is bonded covalently to one of the electronegative atoms (X), and electrostatically (positive-to-negative attraction) to a lone pair on the other (Z). The greater the electronegativity of the atom connected to H, the greater is the partial positive charge on H and hence the stronger is the hydrogen bond.

The H atom is very small and its partial positive charge is concentrated in a very small volume, so it can come very close to the lone pair to form an especially strong dipole-dipole force through hydrogen bonding. Hydrogen-bond strengths range from 10 to 40 kJ/mol. However, a great many hydrogen bonds often occur in a sample of matter, and the overall effect can be very dramatic. An example of this effect is given

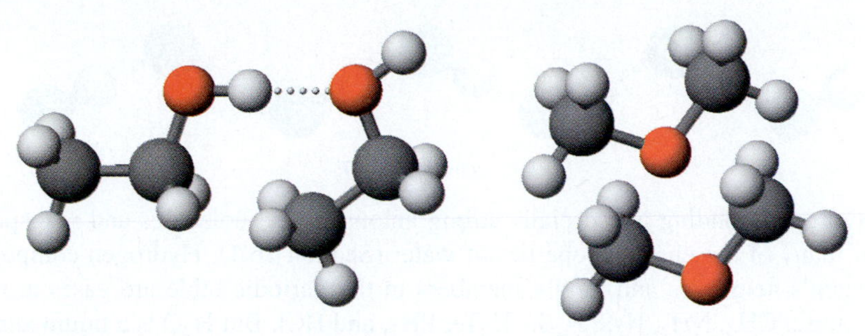

Figure 10.22 Noncovalent interactions in ethanol and dimethyl ether. The molecules have the same number of electrons and so London forces are roughly the same. An ethanol molecule has an OH group, which means that there are both dipole forces and hydrogen bond forces attracting ethanol molecules to each other. A dimethyl ether molecule is polar, but there is no hydrogen bonding, and so the noncovalent intermolecular forces are weaker than in ethanol.

by the melting and boiling points of ethanol. This chapter began by noting the very different melting and boiling points of ethanol and dimethyl ether, both having the same molecular formula, C_2H_6O.

Compound	Dipole Moment, D	Melting Point, °C	Boiling Point, °C
Ethanol, $CH_3CH_2—O—H$	1.69	−114.1	78.29
Dimethyl ether, $CH_3—O—CH_3$	1.30	−141.5	−24.8

The differences arise because the O—H bonds in ethanol make intermolecular hydrogen bonding possible, while this is not possible in dimethyl ether because it has no O—H bonds (Figure 10.22).

The hydrogen halides also illustrate the significant effects of hydrogen bonding (Figure 10.23). The boiling point of hydrogen fluoride (HF), the lightest hydrogen halide, is much higher than expected; this is attributed to hydrogen bonding, which does not occur in the other hydrogen halides.

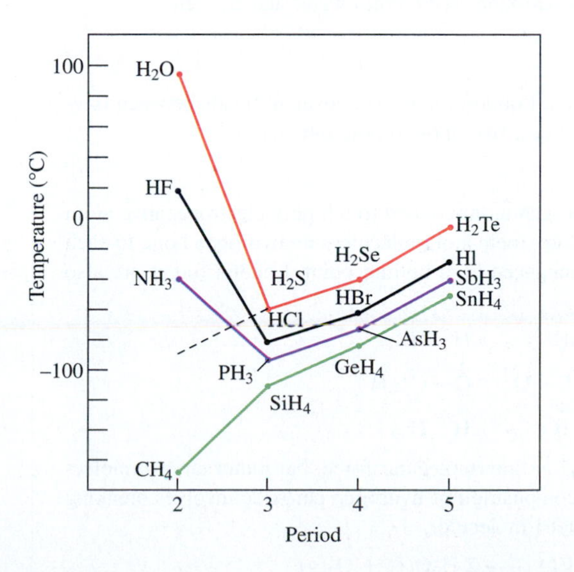

Figure 10.23 Boiling points of some simple hydrogen-containing compounds. Lines connect molecules containing atoms from the same periodic group. The effect of hydrogen bonding is evident in the high boiling points of H_2O, HF, and NH_3. The dashed line predicts where H_2O would be in the absence of hydrogen bonding.

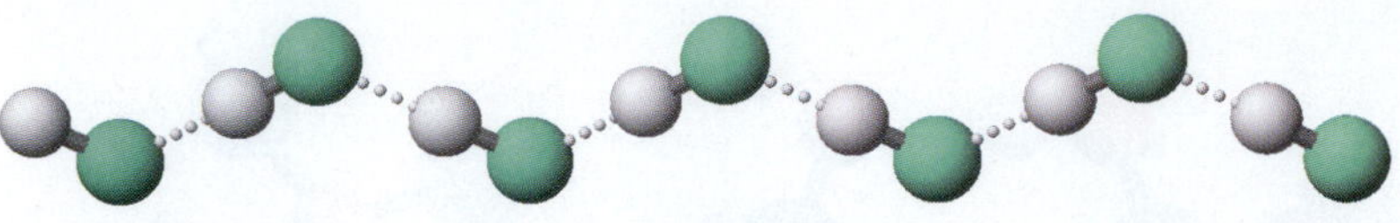

Hydrogen bonding HF

Hydrogen bonding is especially strong among water molecules and is responsible for many of the unique properties of water (Section 16.1). Hydrogen compounds of oxygen's neighbors and family members in the periodic table are gases at room temperature: CH_4, NH_3, H_2S, H_2Se, H_2Te, PH_3, and HCl. But H_2O is a liquid at room temperature, indicating a strong degree of intermolecular attraction. Figure 10.23 shows that the boiling point of H_2O is about 200 °C higher than would be predicted if hydrogen bonding were not present.

In liquid and solid water, where the molecules are close enough to interact, the hydrogen atom on one water molecule is attracted to the lone pair of electrons on the oxygen atom of an adjacent water molecule. Because each hydrogen atom can form a hydrogen bond to an oxygen atom in another water molecule, and because each oxygen atom has two lone pairs, every water molecule can participate in four hydrogen bonds to four other water molecules (Figure 10.24). The result is a tetrahedral cluster of water molecules around the central water molecule.

In Section 10.8 we will consider the essential role played by hydrogen bonding in the structure of DNA, the molecule responsible for storing the genetic heritage of all living things.

The structure of solid water is described in Section 15.1 and shown on the cover of this book.

Exercise 10.7 Strengths of Hydrogen Bonds

Of the three hydrogen bonds, F—H···F—, O—H···O—, and N—H···N—, which is the strongest? Explain why.

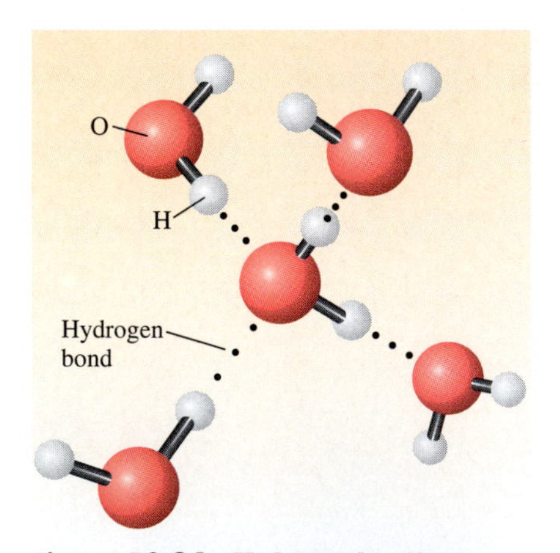

Figure 10.24 Hydrogen bonding between one water molecule and its neighbors. Each water molecule can participate in four hydrogen bonds—one for each hydrogen atom and two through the two lone pairs of oxygen. Since each hydrogen bond is shared between two water molecules, there are two hydrogen bonds per molecule.

Problem-Solving Example 10.6 Molecular Forces

What are the types of forces that are overcome in the following changes?
(a) The evaporation of liquid methanol, CH_3OH
(b) The decomposition of hydrogen peroxide (H_2O_2) into water and oxygen
(c) The melting of ethane, C_2H_6
(d) The boiling of liquid HCl.

Answer: (a) Hydrogen bonding and London forces; (b) covalent bonds between oxygen atoms; (c) dispersion London forces; (d) dipole-dipole interactions

Explanation

(a) Methanol molecules have a hydrogen atom bonded to a highly electronegative atom with a lone pair of electrons. Therefore, methanol molecules can hydrogen bond to each other, and this is reflected by the enhanced high boiling point. London forces are also present.

```
    H         H  H
    |    ..   |  |
  H—C—O···:O—C—H
    |    ..   |  |
    H         H  H
```

(b) This is not a case of overcoming an intermolecular force, but rather an *intra*molecular one, covalent bonding. The decomposition of hydrogen peroxide involves breaking the O—O covalent bond of the HOOH molecule.

$$2\ HO{-}OH(\ell) \longrightarrow 2\ H_2O(\ell) + O_2(g)$$

Portrait of a Scientist • *Art Fry*

Art Fry is not a household name, but he helped to develop a very familiar paper product, Post-it® notes, those self-sticking sheets that seem to be everywhere. Fry, a 3M Company chemical engineer, developed Post-it notes in the mid-1970s to meet a personal need. While singing in a church choir, he found that using bits of paper to mark pages in his hymnal did not work very well. The pieces of paper were unwieldy and fell out unexpectedly, losing the place at critical times. Fry commented that "I don't know whether it was a dull sermon or divine inspiration, but my mind began to wander and suddenly I thought of an adhesive that had been discovered several years earlier by another 3M scientist, Dr. Spencer Silver." That adhesive had been discarded by Dr. Silver because it was not strong enough to join two items permanently, his original goal.

Fry's insight that Sunday was recognizing that what he needed for marking his hymnal was, as he described it, "a temporarily permanent adhesive." The next day at work Fry applied Silver's adhesive, discovered in 1968, to pieces of paper and found that the adhesive did not work quite as well as needed.

It took Fry and his 3M co-workers more than a year of additional research to develop a so-called pressure-sensitive adhesive (PSA) with just the right amount of "stickiness"—sufficiently strong to make the note adhere, but temporary enough to be removable.

Spencer Silver (left) and Art Fry. *(Courtesy of 3M Corporation)*

Fry developed a chemical primer that, when put on note paper and covered by Silver's PSA, allowed the paper and the adhesive to come off together, rather than sticking to the surface. Intermolecular forces between the glue and the paper contribute to allowing this to occur. The actual formulas for Post-it notes primer and adhesive are not available publically in order to protect 3M's research investment to develop the product.

Once the correct combination of primer and adhesive was found, a machine had to be developed that could manufacture the Post-it notes as a stack of precisely aligned sheets, a difficult task. Fry developed and built a prototype machine in his home basement, then had to knock through the walls to get the machine out!

Fry also realized the potential for using removable paper notes to be far more than just marking sheets of music or pages of a book. He envisioned the use we now make of them—a removable note that can be stuck almost anywhere. Because such a product had never been marketed before, 3M officials were skeptical that consumers could see uses for them. According to Fry " . . . new things often require new words to describe them. Without a sample of the notes, purchasers could not understand their usefulness." But once free samples were distributed, Post-it notes quickly became a common household and office item.

Adapted from R. M. Roberts: Serendipity: Accidental Discoveries in Science. *New York, John Wiley & Sons, Inc., 1989; and C. Plummer: Chem Matters, December, p. 13, 1993.*

(c) Ethane is a nonpolar molecule, and so dispersion (London) forces are the principal forces between the molecules in solid ethane. Ethane melts when these forces are overcome at the relatively low temperature of −88 °C.

(d) The HCl molecule is polar because the Cl atom is more electronegative than the H atom. Therefore, there are London forces (which are always present) and dipole-dipole forces to overcome. Adjacent HCl molecules can interact through dipole-dipole forces in the following ways:

$$\mathrm{H{-}Cl{\cdots}H{-}Cl} \quad \text{or} \quad \begin{array}{cc} \mathrm{H} & \!\!-\!\! & \mathrm{Cl} \\ \vdots & & \vdots \\ \mathrm{Cl} & \!\!-\!\! & \mathrm{H} \end{array}$$

Problem-Solving Practice 10.6

Decide what types of intermolecular forces are involved in the attraction between (a) N_2 and N_2, (b) CO_2 and H_2O, and (c) CH_3OH and NH_3.

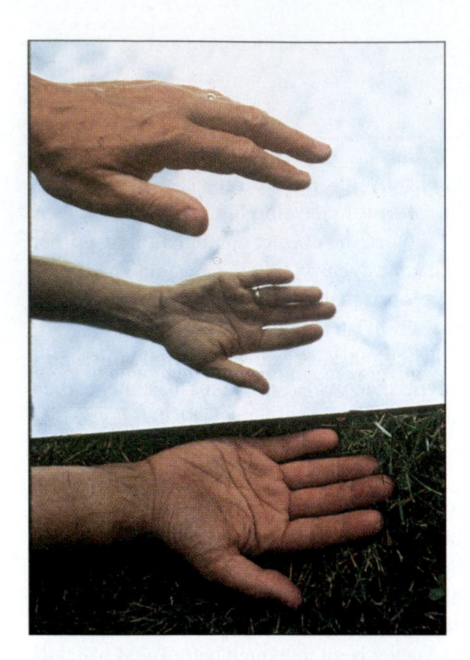

Figure 10.25 Nonsuperimposable mirror images. Your left hand is a *nonsuperimposable* mirror image of your right hand. Although the mirror image of your right hand looks like your left hand, if you place one hand directly over the other, they are not identical. Hence they are nonsuperimposable mirror images.

Figure 10.26 Right-handed seashells. Seashells have a handedness; that is, they are chiral. Virtually all seashells are right-handed. If you hold a shell in your right hand with your thumb extended and pointing from the narrow end to the wide end, your fingers will curl along the shell as it curls from the outside to the center. The shell in the person's left hand is a rarity—a left-handed shell. *(C.D. Winters)*

10.6 CHIRAL MOLECULES

You have seen one kind of isomerism that results from different arrangements of the same atoms in space—*cis-trans* isomerism *(← p. 372)*. There is another somewhat more subtle isomerism, one related to "handedness." Are you right-handed or left-handed? Regardless of the preference, we learn at a very early age that a right-handed glove doesn't fit the left hand and vice versa. Our hands are mirror images of one another and are not superimposable (Figure 10.25).

An object that cannot be superimposed on its mirror image is called **chiral.** Objects that are superimposable on their mirror images are **achiral.** Stop and think about the extent to which chirality is a part of our everyday life. We've already discussed the chirality of hands (and feet). Helical seashells are chiral, and most spiral to the right like a right-handed screw (Figure 10.26). Many creeping vines show a chirality when they wind around a tree or post.

Chiral is pronounced "ki-ral" and is derived from the Greek *cheir,* meaning "hand."

What is not as well known is that a large number of the molecules in plants and animals are chiral, and usually only one form (left-handed or right-handed) of the chiral molecule is found in nature. For example, all but one of the 20 naturally occurring amino acids are chiral, and only the left-handed amino acids are found in nature! Most of the natural sugars are right-handed, including glucose and sucrose *(← p. 108)* and deoxyribose, the sugar found in DNA.

A chiral molecule and its *nonsuperimposable* mirror image are called **enantiomers;** they are two different molecules, just as your left and right hands are different. Enantiomers are possible when a molecular structure is **asymmetrical** (without symmetry). The simplest case is a tetrahedral carbon atom that is bonded to four *different* atoms or groups of atoms. Such a carbon atom is said to be chiral (without symmetry), and a molecule that contains one chiral atom is always a chiral molecule.

A chiral carbon atom is also known as an asymmetric carbon atom.

Some compounds are found in nature in both enantiomeric forms under different circumstances. For example, during the contraction of muscles, the body produces only one enantiomer of lactic acid. The other enantiomer is produced when milk sours. The central carbon atom of lactic acid has four different groups bonded to it: $—CH_3$, —OH, —H, and —COOH (Figure 10.27a).

As a result of the tetrahedral geometry around the central carbon atom, it is possible to have two different arrangements of the four groups. If a lactic acid molecule is placed so that the C—H bond is vertical, as illustrated in Figure 10.27a, one possible arrangement of the remaining groups would be that in which —OH, $—CH_3$, and

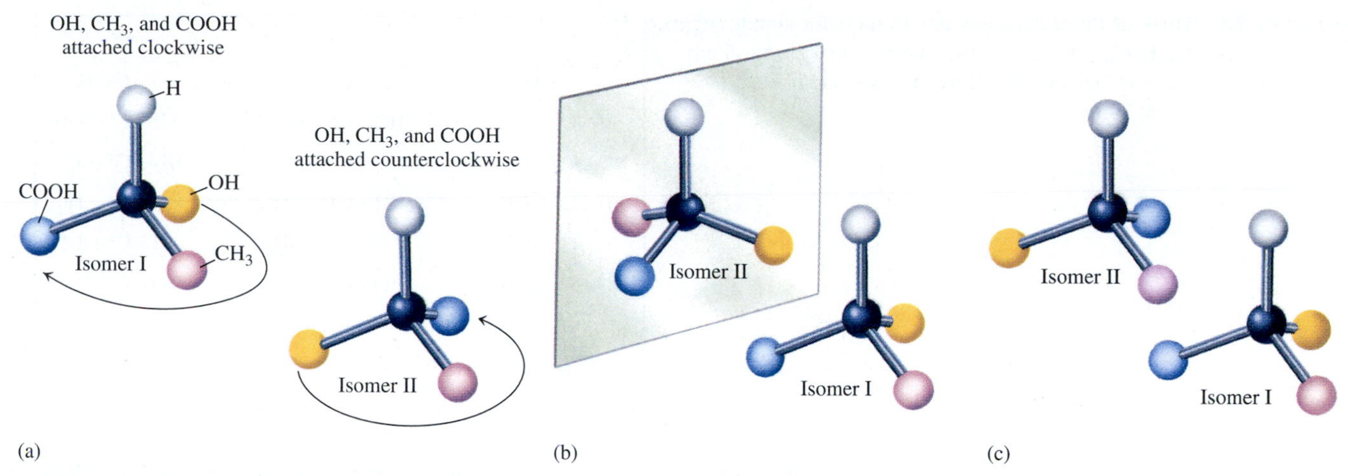

Figure 10.27 The enantiomers of lactic acid. (a) Isomer I: —OH, —CH_3, and —COOH are attached in a clockwise manner. Isomer II: —OH, —CH_3, and —COOH are attached in a counterclockwise manner. (b) Isomer I is placed in front of a mirror, and its mirror image is isomer II. (c) The isomers are nonsuperimposable.

—COOH are attached in a clockwise sequence (isomer I). Alternatively, these groups can be attached in a counterclockwise sequence (isomer II).

To see further that the arrangements are different, we place isomer I in front of a mirror (Figure 10.27b). Now you see that isomer II is the mirror image of isomer I. What is important, however, is that these mirror-image molecules *cannot be superimposed* on one another. *These two nonsuperimposable, mirror-image chiral molecules are enantiomers.*

An enantiomer of a chiral compound causes a beam of plane-polarized light (electromagnetic waves vibrating in only one direction) to rotate. The two enantiomers of the chiral compound rotate the light by equal amounts, but in opposite directions. A polarimeter is an instrument that measures the direction and angle of rotation of plane-polarized light caused by a given enantiomer. For this reason, chiral molecules are sometimes referred to as optical isomers and are said to be optically active.

The "handedness" of enantiomers is sometimes represented by D for right-handed (D stands for "dextro" from the Latin *dexter* meaning "right") and L for left-handed (L stands for "levo" from the Latin *laevus* meaning "left"). In the case of lactic acid, the D-form is found in souring milk and the L-form is found in muscle tissue, where it accumulates during vigorous exercise and can cause cramps.

Although biological systems have a preference for one enantiomer, laboratory synthesis of a chiral compound from chiral reactants gives a mixture of equal amounts of the enantiomers, which is called a **racemic mixture.** The separation and purification of enantiomers is difficult because of the similarity of their physical properties. The usual method of separating enantiomers is to treat them with reagents that have a greater affinity for one enantiomer than for the other.

Nature's preference for one enantiomer of amino acids (the L-form) has provoked much discussion and speculation among scientists since Louis Pasteur's discovery of molecular chirality in 1848. However, there is still no widely accepted explanation of this preference.

Large organic molecules may have a number of chiral carbon atoms within the same molecule. At each such carbon atom (a chiral center) there are two possible arrangements of the molecule. The total number of possible molecules, then, increases exponentially with the number of different chiral centers. With two different chiral carbon atoms there are 2^2, or four, possible structures. Glucose (← *p. 108*) contains four different chiral carbon atoms per molecule. All of the 16 possible isomers are

Figure 10.28 Three of the sixteen possible isomers for simple sugars with the formula $C_6H_{12}O_6$. (a) D-Glucose, the principal chemical energy source in human metabolism. (b) D-Mannose is found in plants. (c) D-Galactose is found in milk.

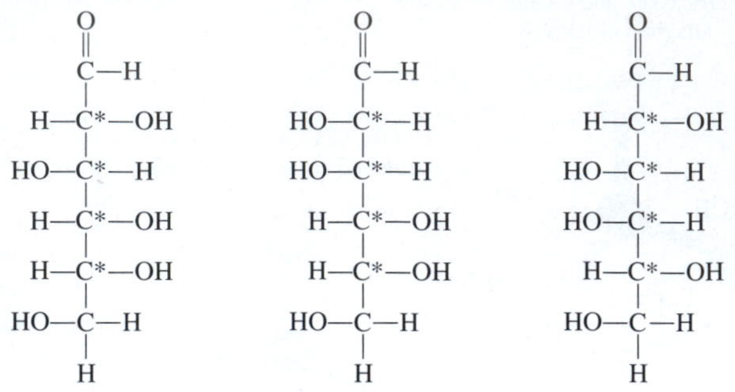

D-glucose	D-mannose	D-galactose
O=C—H	O=C—H	O=C—H
H—C*—OH	HO—C*—H	H—C*—OH
HO—C*—H	HO—C*—H	HO—C*—H
H—C*—OH	H—C*—OH	HO—C*—H
H—C*—OH	H—C*—OH	H—C*—OH
HO—C—H	HO—C—H	HO—C—H
H	H	H

(C* = Asymmetric Carbon Atom)

known, although only three are biologically important (Figure 10.28). These are D-glucose, D-mannose, and D-galactose.

The widely used artificial sweetener aspartame (NutraSweet) has two enantiomers. One enantiomer has a sweet taste while the other enantiomer is bitter, indicating that the receptor sites on our taste buds must be chiral, because they respond differently to the "handedness" of aspartame enantiomers! In another example, D-glucose is sweet and nutritious while L-glucose is tasteless and cannot be metabolized by the body.

PROBLEM-SOLVING EXAMPLE 10.7 Chiral Molecules

For each of the following molecules, decide whether the underlined carbon atom is or is not a chiral center.

(a) <u>C</u>H_2Cl_2 (b) H_2N—<u>C</u>H(CH_3)—COOH (c) Cl—<u>C</u>H(OH)—CH_2Cl (d)H<u>C</u>OOH

Answer Molecules (a) and (d) are not chiral; (b) and (c) are chiral.

Explanation The Lewis structures are

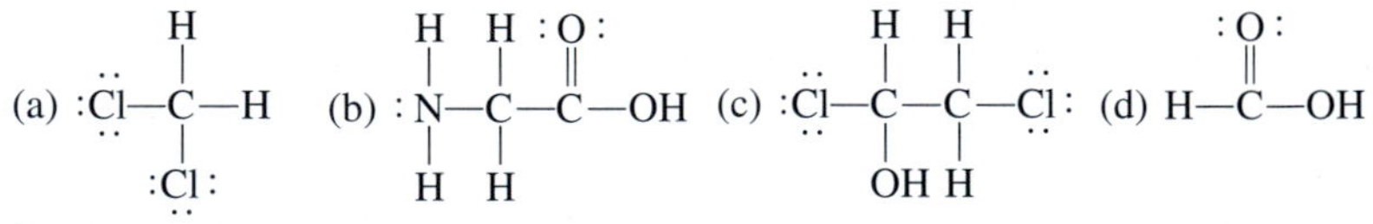

```
        H               H  H :O:                H  H                 :O:
   ..   |               |  |  ||           ..   |  |   ..            ||
(a) :Cl—C—H     (b) :N—C—C—OH    (c) :Cl—C—C—Cl:     (d) H—C—OH
   ..   |               |  |               ..   |  |   ..
       :Cl:             H  H                    OH H
        ..
```

Molecules (b) and (c) are chiral because the carbon atoms are bonded to four different groups. The carbon atoms in molecules (a) and (d) do not have four *different* groups bonded to them; because of the C=O double bond in (d), the underlined carbon is bonded to only three different groups.

Problem-Solving Practice 10.7

Which of the following molecules is chiral? Draw the enantiomers for any chiral molecule.

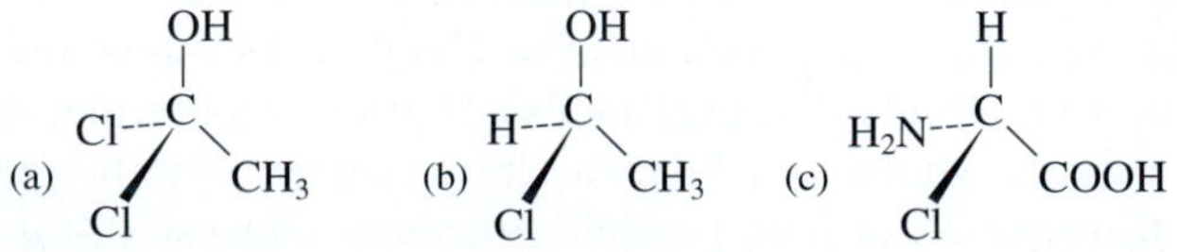

(a) C bonded to OH, Cl, Cl, CH_3 (b) C bonded to OH, H, Cl, CH_3 (c) C bonded to H, H_2N, Cl, COOH

CHEMISTRY IN THE NEWS

Chiral Drugs

One enantiomer of a chiral drug is usually more active than the other, but the majority of chiral drugs are still sold as racemic mixtures. Only in those cases where one enantiomer is toxic or has harmful side effects is the single-enantiomer drug used. The primary reason for this is the large increase in cost required to isolate a particular enantiomer from a racemic mixture.

Before long, though, many more drugs may be brought to market as pure enantiomers. You may soon be seeing advertisements for a drug that is *enantiomerically pure* or *twice as effective as . . . [the racemic mixture, half of which is inactive]* or a *new and improved* version of a familiar over-the-counter medication. In recent years, techniques for separating enantiomers have improved greatly. Also, the U.S. Food and Drug Administration (FDA) in 1992 released long-awaited guidelines on marketing chiral drugs.

The decision on whether to sell a chiral drug in the racemic mixture or the enantiomerically pure form has been left to the drug's manufacturer (although the decision is subject to FDA approval). With the regulations finally in place, drug companies will be looking for situations where production of a single enantiomer can give them a competitive edge. The racemic mixture, however, may remain the best choice in most cases. For example, ibuprofen, the pain reliever contained in Advil, Motrin, and Nuprin, is now sold as a racemic mixture. The left-handed enantiomer of ibuprofen (b) is the active pain reliever and the right-handed isomer is inactive (c). But D-ibuprofen is converted to L-ibuprofen in the body, so there is probably no therapeutic advantage to the patient to switch from the racemic mixture to the more costly L-ibuprofen.

Naproxen is an example of a prescription chiral drug that is sold as an enantiomer rather than as the racemic mixture. In this case, one enantiomer is a pain reliever while the other enantiomer causes liver damage.

The sale of enantiomeric drugs is already big business, with world sales of over $45 billion in 1994, a 27% increase over 1993 sales. The top seven categories of enantiopure (single-isomer) drugs are listed below.

Drug Class	**$ Billions, 1994**
Cardiovascular	12.6
Antibiotics	12.5
Hormones	6.5
Central nervous system	3.0
Anti-inflammatory	1.6
Anticancer	1.5
Other	7.5
TOTAL	45.2

Within just a few years, the sales of single-isomer drugs are expected to rise rapidly at the expense of achiral ones.

	Market Share		
Drug Compounds	**1980**	**1990**	**2000**
Single enantiomers from synthetic sources	3%	9%	34%
Single enantiomers from natural sources	20%	26%	28%
Racemic mixtures	17%	15%	17%
Achiral compounds	60%	50%	21%

Source: S. C. Stinson: Chem. and Eng. News, *October 9, 1995, pp. 46 and 52.*

(a)

CH_3
|
*CH—COOH
(benzene ring)
CH_2—CH—CH_3
|
CH_3

*chiral carbon

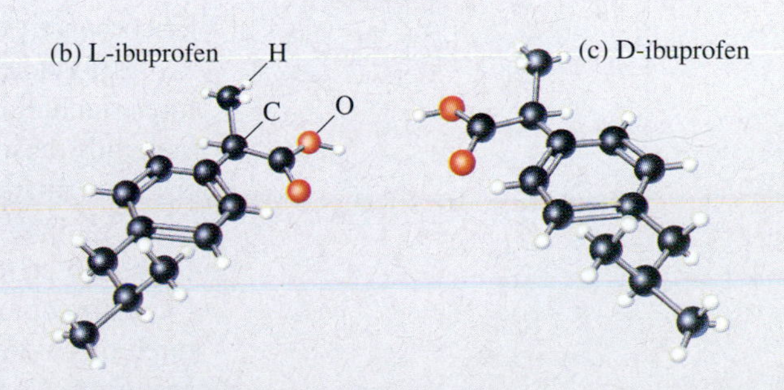

Ibuprofen. (a) The chiral carbon in ibuprofen; (b) L-ibuprofen; (c) D-ibuprofen

TABLE 10.6 Spectroscopic Experiments

Spectral Region	Frequency (s^{-1})	Energy Levels Involved	Information Obtained
Radio waves	10^7–10^9	Nuclear spin states	Electronic structure near the nucleus
Microwave, far infrared	10^9–10^{12}	Rotational	Bond lengths and bond angles
Near infrared	10^{12}–10^{14}	Vibrational	Stiffness of bonds
Visible, ultraviolet	10^{14}–10^{17}	Valence electrons	Electron configuration
X-ray	10^{17}–10^{19}	Core electrons	Core-electron energies

10.7 MOLECULAR STRUCTURE DETERMINATION BY SPECTROSCOPY: UV-VISIBLE AND IR

How are the structures of molecules determined? Many of the methods rely on the interaction of electromagnetic radiation with matter. Probing matter with electromagnetic radiation is called **spectroscopy,** and each area of the electromagnetic spectrum *(← p. 303)* can be used as the basis for a particular spectroscopic method.

Recall from Section 8.1 that electromagnetic radiation is emitted or absorbed in quantized packets of energy called photons, and that the energy of the photon is represented by $E = h\nu$ where ν is the frequency of the light. Molecules may absorb several different electromagnetic radiation frequencies depending on the energy differences between their allowed energy levels. Each frequency absorbed must provide the exact package of energy needed to lift a molecule from one energy level to the next. For example, ultraviolet (uv) and visible spectroscopy are often referred to as electronic spectroscopy because the energy of photons in the ultraviolet or visible region matches the energy needed to promote an electron from a lower energy level to a higher one in a molecule (Table 10.6).

Ultraviolet-Visible Spectroscopy

When a molecule absorbs uv-visible radiation, the absorbed energy excites electrons from lower-energy to higher-energy orbitals in the molecule. The maximum uv-visible absorption occurs at a wavelength characteristic of the molecule's structure and can be determined from a uv-visible spectrum, a plot of the absorption intensity (absorbance) versus wavelength of absorbed radiation (Figure 10.29).

Ultraviolet-visible spectral data can be used to account for several structural features, including (a) differentiation of *cis* from *trans* isomers, because a *trans* isomer generally absorbs uv-visible radiation at a longer wavelength than the *cis* isomer; (b) confirmation of the presence (or absence) of carbonyl and aromatic groups, because they absorb uv radiation at characteristic wavelengths; and (c) specific transitions of electrons between *d* orbitals in transition metal complexes *(← p. 384)*.

Ultraviolet-visible spectroscopy can also be used to study the molecular structures of colored compounds, those that absorb radiation in the visible region of the spectrum. Generally, organic compounds containing only sigma bonds are colorless; that is, they do not absorb light in the visible region. On the other hand, brightly colored pigments, such as beta-carotene and anthocyanins, have an extended sequence of alternating single and double bonds, called a **conjugated** system.

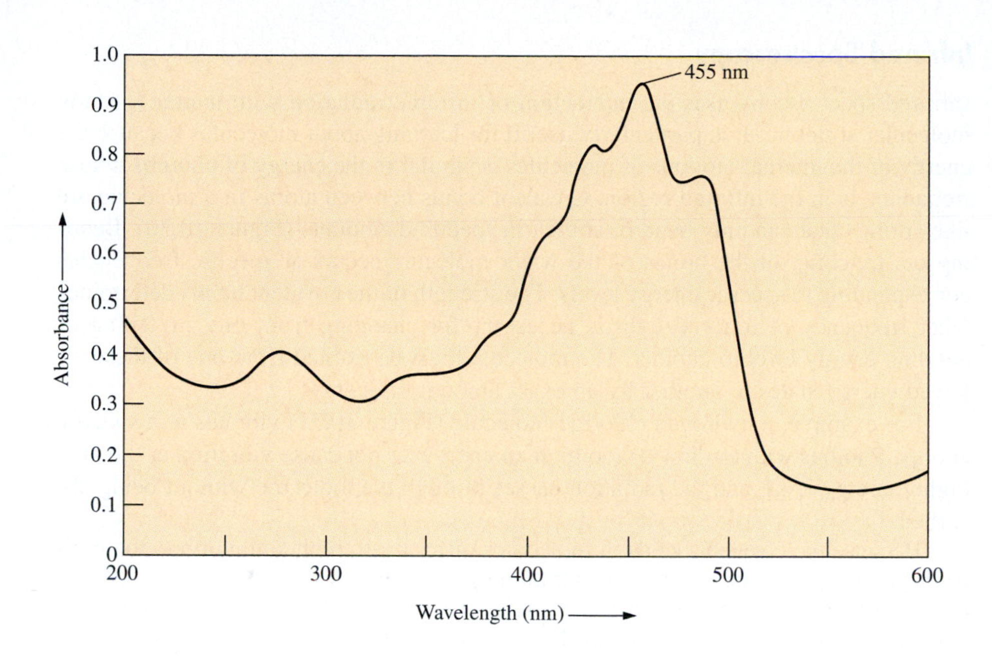

Figure 10.29 The uv-visible spectrum of beta-carotene. Beta-carotene absorbs in the blue-green region of the visible spectrum indicated by a maximum absorbance at 455 nm. This peak is caused by transitions of pi electrons in the conjugated double bonds of the beta-carotene molecule.

beta-carotene

cyanidin (an anthocyanin found in strawberries, apples, and cranberries)

Beta-carotene, for example, contains 11 double bonds in its conjugated system. When visible light is absorbed, electronic transitions occur among the pi electrons in the conjugated double bonds resulting in the yellow-orange color in carrots (beta-carotene) and the vibrant colors of fall foliage, fruits, and berries (anthocyanins). Carrots look yellow-orange because electronic transitions in their beta-carotene absorb in the 400- to 500-nm region (Figure 10.29), which is blue-green. We then see the remaining unabsorbed (reflected) portion of the visible spectrum, which is yellow-orange.

Vision depends on absorption of visible light in the retina by rhodopsin (Section 12.5), and the subsequent *cis-trans* isomerization of the molecule. Rhodopsin is formed from a compound derived from beta-carotene.

Note in Table 10.6 that microwave frequencies match rotational energy-level differences. This is how a microwave oven works. Because food is 70 to 80% water, exciting the water molecules by causing them to rotate faster heats the food.

Imagine the matching of frequencies as being similar to jumping rope. To jump a rotating rope, the person's frequency of jumping must match that of the rope's rotation.

Infrared Spectroscopy

Infrared spectroscopy uses the interaction of infrared radiation with matter to study molecular structure. It is particularly useful for learning about molecules because the energy of the internal motions of molecules is similar to the energy of photons whose frequency is in the infrared region. Covalent bonds between atoms in a molecule are like springs that can only bend or stretch in specified amounts (Figure 10.30). Bending or stretching of the bonds of the water molecule occurs at specific frequencies corresponding to specific energy levels. The strength of the covalent bonds determines what frequency of infrared light is necessary for changing from one stretching or bending energy level to another. The molecule must be excited from one of these allowed energy states to another by an exact amount of energy.

For example, a hydrogen chloride molecule (Figure 10.31) vibrates at a specific energy. Photons with too low or too high an energy do not cause vibration at the next higher energy level, and the radiation passes through the molecule without being absorbed.

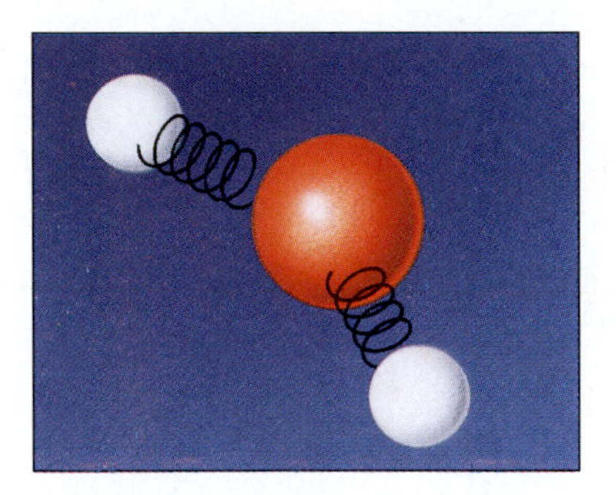

Figure 10.30 **Model of the water molecule, with springs for bonds.**

Because the covalent bonds in molecules differ in strength and number, the molecular motions and the number of vibrational energy levels vary; hence, the infrared radiation that is absorbed by different molecules differs. As a result, infrared spectroscopy can be used to learn about the structures of molecules and even to analyze an unknown material by matching its infrared spectrum with that of a known compound. In fact, the infrared frequencies absorbed by a molecule are so characteristic that the infrared spectrum of a molecule is regarded as its *fingerprint.* An example of the use of infrared spectroscopy in the identification of ethanol, C_2H_5OH, is shown in Figure 10.32.

10.8 BIOMOLECULES: DNA AND THE IMPORTANCE OF MOLECULAR STRUCTURE

Nowhere do the shape of a molecule and noncovalent forces play a more intriguing and important role than in the structure and function of **DNA (deoxyribonucleic acid),** the molecule that stores the genetic code. Whether you are male or female, have blue or brown eyes, or have curly or straight hair depends on your genetic makeup. These and all your other physical traits are determined by the composition of the approximately 1.8 m of coiled DNA that makes up the 23 double-stranded chromosomes in the nucleus of each of your cells.

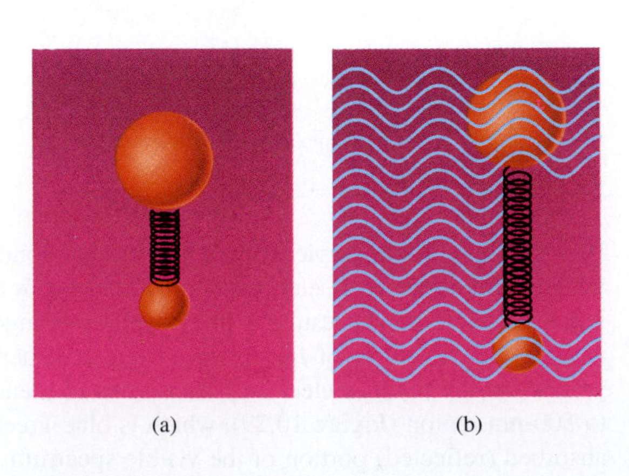

Figure 10.31 **Model for vibration of HCl molecule.** (a) The vibration of the HCl molecule is quantized. (b) Electromagnetic radiation at the quantized frequency will be absorbed by the molecule.

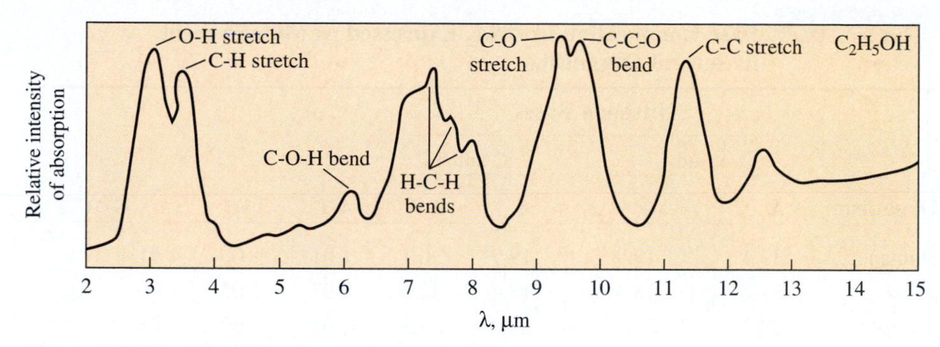

Figure 10.32 The infrared spectrum of ethanol. There are strong absorptions between 3 and 4 μm, between 6.2 and 8.2 μm, and also between 9.3 and 11.5 μm, caused by stretching and bending in ethanol molecules.

Nucleotides of DNA

DNA is a **polymer,** a molecule composed of many small, repeating units bonded together. Each repeating unit in DNA, called a **nucleotide,** has the three connected parts shown in Figure 10.33a—one sugar unit, one phosphate unit, and a cyclic nitrogen compound known as a *nitrogen base.* In DNA the sugar is always deoxyribose and the base is one of four bases known as **adenine (A), thymine (T), guanine (G),** or **cytosine (C).** The bases are often referred to by their single-letter abbreviations.

Polymers are discussed in detail in Section 11.■.

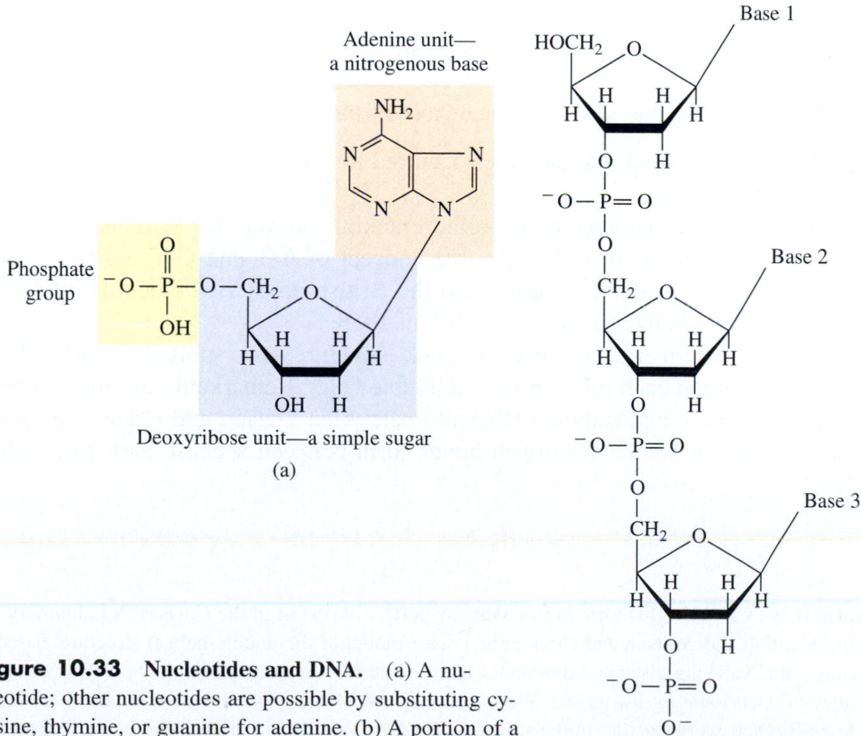

Figure 10.33 Nucleotides and DNA. (a) A nucleotide; other nucleotides are possible by substituting cytosine, thymine, or guanine for adenine. (b) A portion of a nucleotide chain in DNA.

TABLE 10.7 **Base Composition in DNA, Expressed as Mole-Percent, in Several Organisms**

	Nitrogen Bases						
	A and G		C and T				
Organism	**A**	**G**	**C**	**T**	**A/T**	**G/C**	**(A + G)/(C + T)**
Human	30.4	19.9	19.9	30.1	1.01	1.00	1.01
Yeast	31.7	18.3	17.4	32.6	0.97	1.05	1.00
Sheep	29.3	21.4	21.0	28.3	1.04	1.02	1.03
E. coli	26.0	24.9	25.2	23.9	1.09	0.99	1.04

The deoxyribose units in the backbone chain are joined through phosphate diester linkages. Phosphoric acid is H_3PO_4, $O{=}P(OH)_3$.

$$(HO)_3P{=}O$$

phosphoric acid

Replacing H with one CH_3 gives a simple phosphate monoester, $O{=}P(OH)_2(OCH_3)$;

$$(HO)_2(H_3C{-}O)P{=}O$$

phosphate monoester

replacing two H's with two CH_3 groups gives a diester, $O{=}P(OH)(OCH_3)_2$.

$$(H_3C{-}O)_2(HO)P{=}O$$

phosphate diester

A DNA segment with four nucleotides bonded together is shown in Figure 10.33b. The phosphate units join nucleotides into a polynucleotide chain that has a backbone of alternating deoxyribose and phosphate units in a long strand with the various bases extending out from the sugar-phosphate backbone. The order of the nucleotides (and thus the particular sequence of bases) along the DNA strand carries the genetic code from one generation to the next.

The Double Helix: The Watson-Crick Model

Erwin Chargaff made the important discovery that the adenine/thymine and guanine/cytosine ratios are constant in any given organism, from a genius to a bacterium (Table 10.7). Based on this analysis, it was apparent that in any given organism,

- the base composition is the same in all cells of an organism and is characteristic of that organism.
- the amounts of adenine and thymine are equal, as are the amounts of guanine and cytosine.
- the amount of bases A + G equals that of bases C + T.

This information implied that the bases occurred in pairs: adenine with thymine, and guanine with cytosine.

Using x-ray data gathered by Rosalind Franklin and Maurice Wilkins on the relative positions of atoms in DNA, plus the concept of A-T and G-C base pairs, the American biologist James D. Watson and the British physicist Francis H. C. Crick proposed a double-helix structure for DNA in 1953.

In the three-dimensional Watson-Crick structure, two polynucleotide DNA strands wind around each other to form a double helix. Remarkable insight into how hydrogen bonding could stabilize DNA ultimately led Watson and Crick to propose the double-helix structure. Hydrogen bonds form between specific base pairs lying

Francis H. C. Crick (right) and James Watson (left). Working in the Cavendish Laboratory at Cambridge, England, Watson and Crick built a scale model of the double-helical structure, based on x-ray data. Knowing distances and angles between atoms, they compared the task to working on a three-dimensional jigsaw puzzle. Watson, Crick, and Maurice Wilkins received the 1962 Nobel Prize for their work relating to the structure of DNA. Rosalind Franklin, who had also contributed vital x-ray data, died in 1958. Nobel Prizes are not awarded posthumously or she likely would have shared the 1962 Nobel Prize.

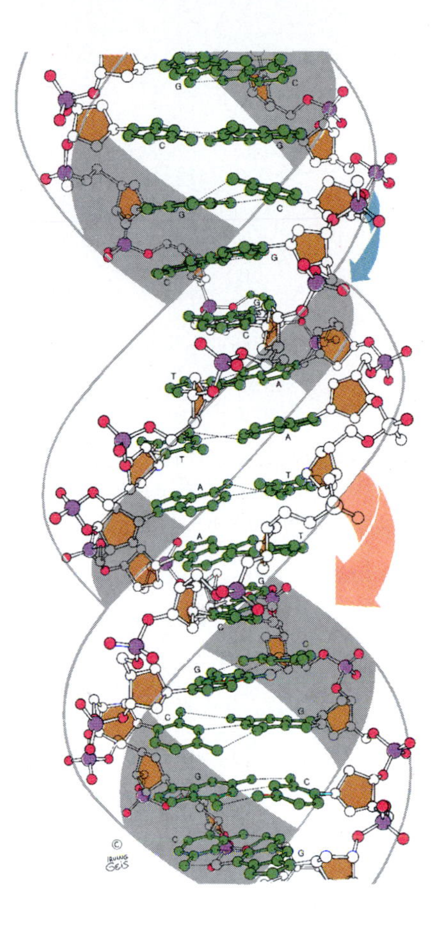

Figure 10.34 An illustration of double-stranded DNA. The illustration shows the deoxyribose-phosphate backbone of each strand as phosphate groups (violet and red) linking tan five-membered rings of deoxyribose joined to bases. The complementary bases (shown in green) are connected by hydrogen bonds: two between adenosine and thymine and three for guanine and cytosine. *(Irving Geis)*

opposite each other in the two polynucleotide strands. Adenine is hydrogen bonded to thymine, and guanine is hydrogen bonded to cytosine to form **complementary base pairs;** that is, the bases of one strand match those of the other (Figure 10.34). The result is that the base pairs (A on one strand with T on the other strand, or C with G) are stacked one above the other on the interior of the double helix (green in Figure 10.34).

Two hydrogen bonds occur between every adenine and thymine pair; three between each guanine and cytosine. Note in Figure 10.35 the similar structures of thymine and cytosine, and of adenine and guanine. If adenine and guanine try to pair, there is insufficient space between the strands to accommodate the bulky pair; thymine and cytosine are too small to pair and to align properly.

Exercise 10.8 Hydrogen Bonding and DNA

Only one hydrogen bond is possible between GT or CA. Use the structural formulas for these compounds to indicate where such hydrogen bonding can occur in these base pairs.

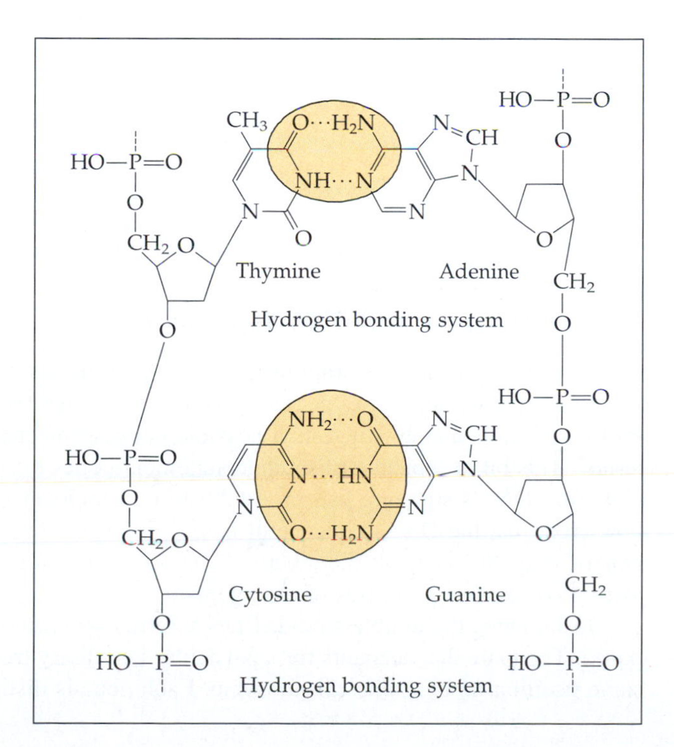

Figure 10.35 Hydrogen bonding between T-A and C-G in DNA.

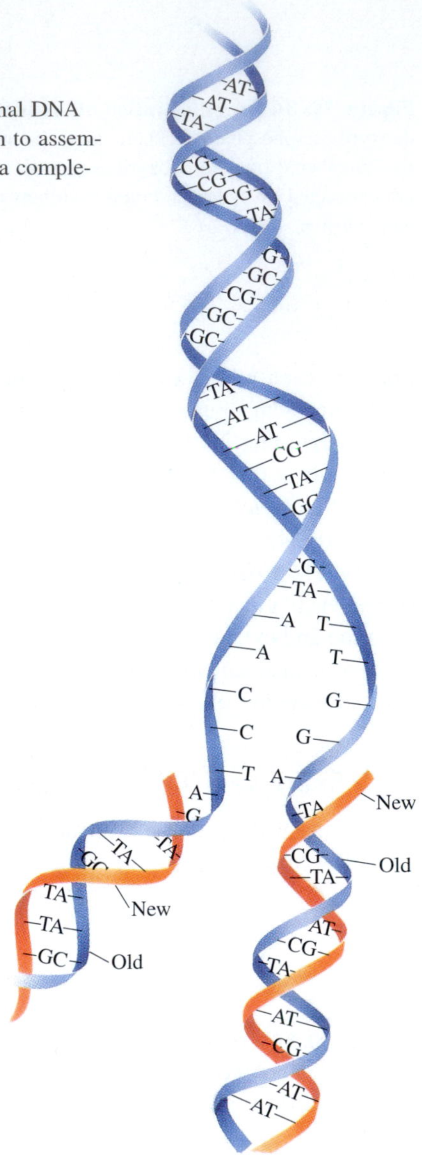

Figure 10.36 DNA replication. When the original DNA helix (blue) unwinds, each half is a template on which to assemble nucleotides from the cell environment to produce a complementary strand (orange).

The Genetic Code and DNA Replication

In the historic scientific journal article describing their revolutionary findings, Watson and Crick wrote, "It has not escaped our notice that the specific pairing we have postulated immediately suggests a possible copying mechanism for the genetic material." This bit of typical British understatement belies the enormous impact that deciphering DNA's structure had on establishing a molecular basis of heredity. It was now clear that the DNA sequence of base pairs in the nucleus of a cell represents a genetic code that controls the inherited characteristics of the next generation, as well as most of the life processes of an organism.

In humans, the double-stranded DNA forms 46 chromosomes (23 chromosome pairs). The **gene** that accounts for a particular hereditary trait is always located in the same position on the same chromosome. Each gene is distinguished by a unique sequence of bases that codes for the synthesis of a single protein within the body. The protein then goes on to play its role in the growth and functioning of the individual. (Proteins are discussed in Section 11.10.)

Each organism (except viruses) begins life as a single cell with DNA consisting of a single strand from each parent. During regular cell division, both DNA strands are accurately copied, with a remarkably low incidence of error. The copying, called **replication,** takes place as the DNA helix unzips and new nucleotides are sequentially brought into the proper places on the new strand. Thus each original DNA strand serves as a template from which the complementary strand is produced (Figure 10.36). The process is termed **semiconservative** because in each of the two new cells, each chromosome consists of one DNA strand from the parent cell and one newly made strand.

Sperm and egg cells contain one strand of DNA. All other cells except red blood cells (which have no nuclei) contain double-stranded DNA.

The goal of the Human Genome Project, a vast multinational research effort, is to fully characterize each of the estimated 100,000 genes and their approximately 3 billion base pairs. Among the Project's many expected benefits, some already being realized, is a molecular-level understanding of the causes and potential cures for inherited diseases.

The total sequence of base pairs in a plant or animal cell is called its genome.

Summary Problem

Part 1. Write the Lewis structures and give the hybridization of the central atom, the electron-pair geometry, and the shape and bond angles of the following polyatomic ions: (a) BrF^{2+}; (b) BrF_4^-; (c) BrF_6^+; and (d) $IBrCl_3^-$.

Part 2. The molecular formula for ketene is C_2H_2O.

(a) Write its Lewis structure and identify the hybridization of each carbon and oxygen atom.

(b) Identify the geometry around each carbon atom and all the bond angles in the molecule.

(c) Is the molecule polar or nonpolar? Use appropriate data to support your answer.

In Closing

Having studied this chapter, you should be able to . . .

- predict shapes of molecules and polyatomic ions by using the VSEPR model (Section 10.2).
- determine the orbital hybridization of a central atom and the associated molecular geometry (Section 10.3).
- describe covalent bonding between two atoms in terms of sigma and/or pi bonds (Section 10.3).
- predict the polarities of molecules (Section 10.4).
- describe the different types of noncovalent interactions and use them to explain melting points and boiling points (Section 10.5).
- define and describe the nature of chiral molecules and enantiomers (Section 10.6).
- describe the basis of uv-visible spectroscopy and infrared spectroscopy and how they are used to determine molecular structures (Section 10.7).
- identify the major components in the structure of DNA (Section 10.8).

KEY TERMS

The following terms were defined and given in boldface type in this chapter. You should be sure to understand each of these terms and the concepts with which it is associated.

achiral *(10.6)*
axial positions *(10.2)*
bond angle *(10.2)*
chiral *(10.6)*
complementary base pairs *(10.8)*
conjugated *(10.7)*
deoxyribonucleic acid (DNA) *(10.8)*
dipole-dipole attraction *(10.5)*
dipole moment *(10.4)*
***dsp^2* hybrid orbital** *(10.3)*
***d^2sp^3* hybrid orbital** *(10.3)*
electron-pair geometry *(10.2)*
enantiomers *(10.6)*
equatorial positions *(10.2)*
gene *(10.8)*
hybrid orbital *(10.3)*
hydrogen bond *(10.5)*
induced dipole *(10.5)*
intermolecular attractions *(10.5)*
intermolecular forces *(10.5)*
linear *(10.2)*
London forces *(10.5)*
molecular geometry *(10.2)*
noncovalent interactions *(10.5)*
nonpolar molecule *(10.4)*
nucleotide *(10.8)*
octahedral *(10.2)*
pi bond, π bond *(10.3)*
polar molecule *(10.4)*
polarization *(10.5)*
polymer *(10.8)*
racemic mixture *(10.6)*
replication *(10.8)*
semiconservative *(10.8)*
sigma bond, σ bond *(10.3)*
***sp* hybrid orbital** *(10.3)*
***sp^2* hybrid orbital** *(10.3)*
***sp^3* hybrid orbital** *(10.3)*
spectroscopy *(10.7)*
tetrahedral *(10.2)*
triangular bipyramidal *(10.2)*
triangular planar *(10.2)*
valence bond model *(10.3)*
valence-shell electron-pair repulsion (VSEPR) model *(10.2)*

QUESTIONS FOR REVIEW AND THOUGHT

Conceptual Challenge Problems

CP-10.A. What advantages does the VSEPR model of chemical bonding have compared to the Lewis dot formulas predicted by the octet rule?

CP-10.B. The VSEPR model does not differentiate between single bonds and double bonds for predicting molecular shapes. What experimental evidence supports this?

CP-10.C. What evidence could you present to show that two carbon atoms joined by a single sigma bond are able to rotate about an axis that coincides with the bond, but two carbon atoms bonded by a double bond cannot rotate about an axis along the double bond?

Review Questions

1. What is the VSEPR model? What is the physical basis of the model?
2. What is the difference between the electron-pair geometry and the molecular geometry of a molecule? Use the water molecule as an example in your discussion.
3. Designate the electron-pair geometry for each case from two to six electron pairs around a central atom.
4. What are the molecular geometries for each of the following?

   (a) H—$\ddot{\underset{\cdot\cdot}{X}}$: (b) H—$\ddot{\underset{\cdot\cdot}{X}}$—H (c) H—$\ddot{X}$—H with H below X (d) H—X—H with H above and below X

   Give the H—X—H bond angle for each of the last three.
5. If you have three electron pairs around a central atom, how can you have a triangular planar molecule? An angular molecule? What bond angles are predicted in each case?
6. Draw a triangular bipyramid of electron pairs. Designate the axial and equatorial pairs. Are there axial and equatorial pairs in an octahedron?
7. Use VSEPR to explain why ethylene is a planar molecule.
8. How can a molecule with polar bonds be nonpolar? Give an example.
9. Give examples that illustrate the importance of the tetrahedral shape to a better understanding of chemistry.
10. Explain the current interest in chiral drugs.
11. What is a racemic mixture?
12. Explain why the infrared spectrum of a molecule is referred to as its "fingerprint."

13. For infrared energy to be absorbed by a molecule, what frequency of motion must the molecule have?
14. One of the three isomers of dichlorobenzene, $C_6H_4Cl_2$, has a dipole moment of zero. Draw the structural formula of the isomer and explain your choice.

Molecular Shape

15. Use the different molecular modeling techniques (ball-and-stick, space-filling, two-dimensional pictures using wedges and dashed lines) to illustrate the following simple molecules:
(a) NH_3 (b) H_2O (c) CO_2
16. All of the following molecules have central atoms with only bonding pairs of electrons. After drawing the Lewis structure, identify the molecular shape of each molecule.
(a) BeH_2 (b) CH_2Cl_2 (c) BH_3 (d) $SeCl_6$ (e) PF_5
17. Draw the Lewis structure for each of the following molecules or ions. Describe the electron-pair geometry and the molecular geometry.
(a) NH_2Cl (b) OCl_2 (c) SCN^- (d) HOF
18. Determine the electron-pair geometry and molecular geometry for each of the following.
(a) ClF_2^+ (b) $SnCl_3^-$ (c) PO_4^{3-} (d) CS_2
19. In each of the following molecules or ions, two oxygen atoms are attached to a central atom. Draw the Lewis structure for each one and then describe the electron-pair geometry and the molecular geometry. Comment on similarities and differences in the series.
(a) CO_2 (b) NO_2^- (c) SO_2 (d) O_3 (e) ClO_2^-
20. In each of the following molecules or ions, three oxygen atoms are attached to a central atom. Draw the Lewis structure for each one and then describe the electron-pair geometry and the molecular geometry. Comment on similarities and differences in the series.
(a) BO_3^{3-} (b) CO_3^{2-} (c) SO_3^{2-} (d) ClO_3^-
21. The following are examples of molecules and ions that do not obey the octet rule. After drawing the Lewis structure, describe the electron-pair geometry and the molecular geometry for each.
(a) ClF_2^- (b) ClF_3 (c) ClF_4^- (d) ClF_5
22. The following are examples of molecules and ions that do not obey the octet rule. After drawing the Lewis structure, describe the electron-pair geometry and the molecular geometry for each.
(a) SiF_6^{2-} (b) SF_4 (c) PF_5 (d) XeF_4
23. Iodine forms three compounds with chlorine: ICl, ICl_3, and ICl_5. Draw the Lewis structures and determine the molecular shapes of these three molecules.
24. Give the approximate values for the indicated bond angles.
(a) O—S—O angle in SO_2 (b) F—B—F angle in BF_3
(c) 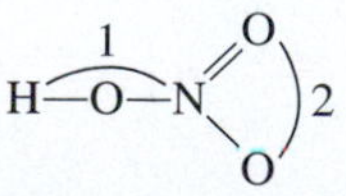 H—O—N(=O)—O (angles 1 and 2) (d) 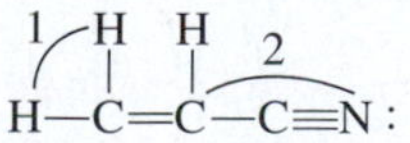 H—C(H)=C(H)—C≡N: (angles 1 and 2)

25. Give approximate values for the indicated bond angles.
(a) Cl—S—Cl angle in SCl_2 (b) N—N—O angle in N_2O
(c) 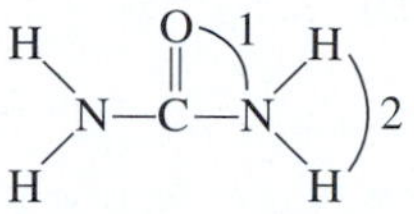 H_2N—C(=O)—NH_2 (angles 1 and 2) (d) 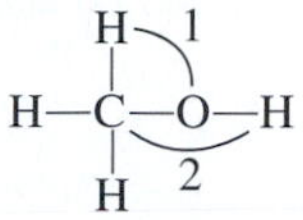 H—C(H_2)—O—H (angles 1 and 2)
26. Give approximate values for the indicated bond angles.
(a) F—Se—F angles in SeF_4
(b) O—S—F angles in SOF_4 (the O atom is in an equatorial position)
(c) F—Br—F angles in BrF_5
27. Give approximate values for the indicated bond angles.
(a) F—S—F angles in SF_6
(b) F—Xe—F angle in XeF_2
(c) F—Cl—F angle in ClF_2^-
28. Which would have the greater O—N—O bond angle, NO_2 or NO_2^-? Explain your answer.
29. Compare the F—Cl—F angles in ClF_2^+ and ClF_2^-. From Lewis structures, determine the approximate bond angle in each ion. Explain which ion has the greater angle and why.

Hybridization

30. Describe the geometry and hybridization of chloroform, $CHCl_3$.
31. Describe the geometry and hybridization for each inner atom in ethylene glycol, CH_2OHCH_2OH, the main component in antifreeze.
32. Determine the Lewis structures, geometries, and hybridizations of GeF_4, SeF_4, XeF_4, and $XeOF_4$.
33. Determine the geometry and hybridization of the three phosphorus-chlorine species: PCl_4^+, PCl_5, and PCl_6^-.

Molecular Polarity

34. Consider the following molecules: H_2O, NH_3, CO_2, ClF, CCl_4.
(a) In which compound are the bonds most polar?
(b) Which compounds in the list are *not* polar?
(c) Which atom in ClF is more negatively charged?
35. Consider the following molecules: CH_4, NCl_3, BF_3, CS_2.
(a) In which compound are the bonds most polar?
(b) Which compounds in the list are not polar?
36. Which of the following molecules is (are) polar? For each polar molecule, what is the direction of polarity; that is, which is the negative and which is the positive end of the molecule?
(a) CO_2 (b) HBF_2 (c) CH_3Cl (d) SO_3
37. Which of the following molecules is (are) not polar? Which molecule has the most polar bonds?
(a) CO (b) PCl_3 (c) BCl_3 (d) GeH_4 (e) CF_4
38. Which of the following molecules have a dipole moment? For each of these polar molecules, indicate the direction of the dipole in the molecule.
(a) XeF_2 (b) H_2S (c) CH_2Cl_2 (d) HCN

39. Explain the differences in the dipole moments of
   (a) BrF (1.29 D) and BrCl (0.52 D).
   (b) H_2O (1.86 D) and H_2S (0.95 D).

Noncovalent Interactions

40. Construct a table covering all the types of noncovalent interactions and comment about the distance dependence for each. (In general, the weaker the force, the closer together the molecules must be to feel the attractive force of nearby molecules.) You should also include an example of a substance that exhibits each type of noncovalent interaction in the table.

41. Explain in terms of noncovalent interactions why water and ethanol are miscible, but water and cyclohexane are not.

42. Explain why water "beads" up on a freshly waxed car but not on a dirty, unwaxed car.

43. Explain the trends seen in the diagram below for the boiling points of some main group hydrogen compounds.
   (a) Group IV: CH_4, SiH_4, GeH_4, and SnH_4
   (b) Group V: NH_3, PH_3, AsH_3, and SbH_3
   (c) Group VI: H_2O, H_2S, H_2Se, and H_2Te
   (d) Group VII: HF, HCl, HBr, and HI

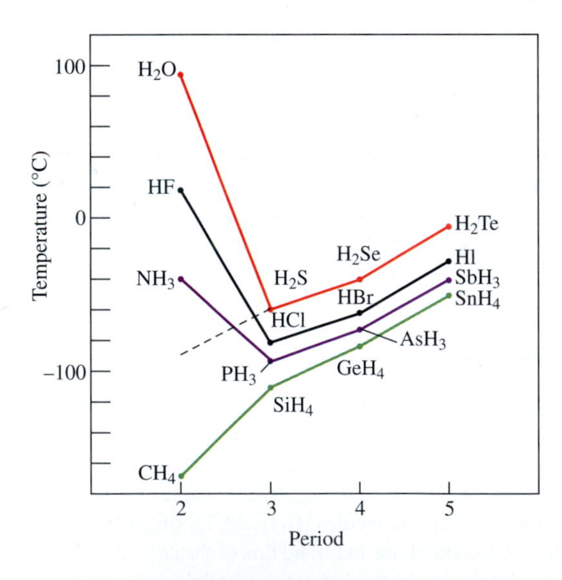

44. Arrange the noble gases in order of increasing boiling point. Explain your reasoning.

45. Which of the following will form hydrogen bonds?
   (a) CH_2Br_2 (b) $CH_3OCH_2CH_3$ (c) $H_2NCH_2CO_2H$ (d) H_2SO_3 (e) CH_3CH_2OH

46. Which of the following would you expect to be most soluble in cyclohexane (C_6H_{12})? The least soluble? Explain your reasoning.
   (a) NaCl (b) CH_3CH_2OH (c) C_3H_8

Chirality in Organic Compounds

47. Circle the chiral carbon atoms, if any, in the following molecules.

```
            O HO  OH O
            ‖  |   |  ‖
(a) HO—C—C—C—C—H
               |   |
               H   H

              O   O
              ‖   ‖
(b) CH3—C—C—OH

                         H   O
                         |   ‖
(c) CH3—CH2—C—C—OH
                         |
                        NH2
```

48. Circle the chiral carbon atoms, if any, in the following molecules.

```
          HO  OH
           |   |
(a) CH3—C—C—H
           |   |
           H   H

(b) H—C=C—CH2—OH
       |   |
       H   H

          Cl  F
           |  |
(c) CH3—C—C—Cl
           |  |
          Cl  H
```

49. Circle the chiral carbon atoms, if any, in the following compounds.

```
      H  Cl    H
      |   |    |
(a) H—C—C——C—H
      |   |    |
      H  CH3   H

       H  H    H
       |  |    |
(b) Cl—C—C——C—H
       |  |    |
       H  CH3  H

      H Br H  H  H
      |  |  |  |  |
(c) H—C—C—C—C—C—H
      |  |  |  |  |
      H  H  H  H  H

      H  H Br  H  H
      |  |  |  |  |
(d) H—C—C—C—C—C—H
      |  |  |  |  |
      H  H  H  H  H
```

50. Circle the chiral carbon atoms, if any, in the following compounds.

```
        H  Br  H   H
        |  |   |   |
(a) H—C—C—C—C—H
        |  |   |   |
        H  Cl  H   H

        H  CH3    H
        |  |      |
(b) H—C—C——C—H
        |  |      |
        H  H      H

        H  H  H  H  H
        |  |  |  |  |
(c) H—C—C=C—C—C—H
        |        |  |
        H        Br H
```

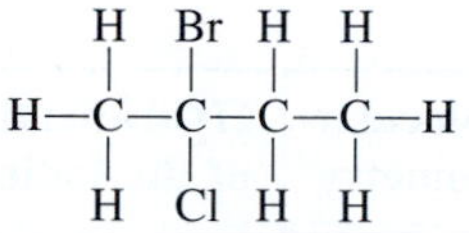

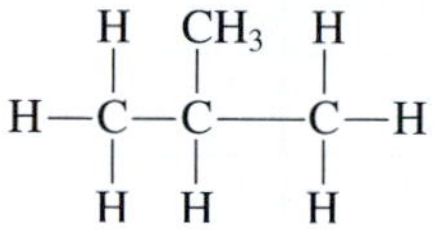

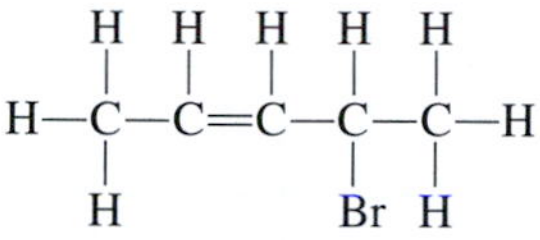

51. How can you tell from its formula whether a compound can exist as an enantiomer?

52. What conditions must be met for a molecule to be chiral?

53. Which of the following are not superimposable on its mirror image?

(a) Nail	(b) Screw	(c) Shoe
(d) Sock	(e) Golf club	(f) Football
(g) Your ear	(h) Helix	(i) Baseball bat
(j) Sweater		

Molecular Structure Determination by Spectroscopy

54. Ultraviolet and IR spectroscopies probe different aspects of molecules.

(a) Ultraviolet radiation is energetic enough to cause transitions in the energies of which electrons in an atom?

(b) What does IR spectroscopy tell us about a given molecule?

55. Figure 10.32 is the infrared spectrum of ethanol. (a) Which stretching requires the lowest amount of energy? (b) The highest amount of energy?

Biomolecules

56. Discuss the differences between the extent of hydrogen bonding for the pairs G-C and A-T in nucleic acids. If a strand of DNA has more G-C pairs than A-T pairs in the double helix, the melting point (unwinding point) increases. The melting point for strands with more A-T pairing will decrease in comparison. Explain.

57. One strand of DNA contains the base sequence T-C-G. Draw a structure of this section of DNA that shows the hydrogen bonding between the base pairs of this strand and its complementary strand.

General Questions

58. The formula for nitryl chloride is NO_2Cl. Draw the Lewis structure for the molecule, including all resonance structures. Describe the electron-pair and molecular geometries, and give values for all bond angles.

59. Vanillin is the flavoring agent in vanilla extract and in vanilla ice cream. Its structure is

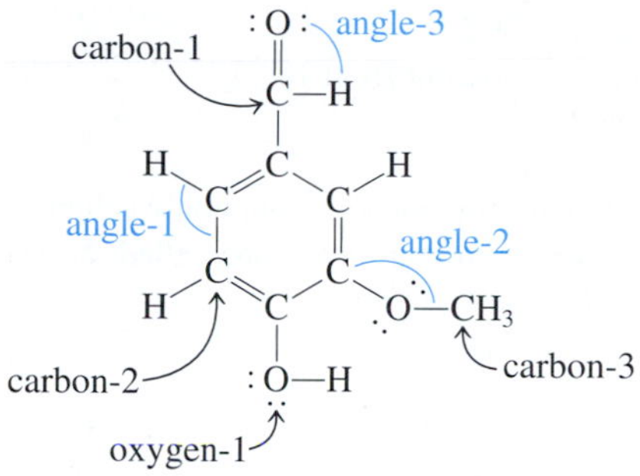

carbon-1
angle-3
angle-1
angle-2
carbon-2
carbon-3
oxygen-1

(a) Give values for the three bond angles indicated.

(b) Indicate the most polar bond in the molecule.

(c) Circle the shortest carbon-oxygen bond.

60. Each of the following molecules has fluorine atoms attached to an atom from Groups 1A or 3A to 6A. Draw the Lewis structure for each one and then describe the electron-pair geometry and the molecular geometry. Comment on similarities and differences in the series.

(a) BF_3 (b) CF_4 (c) NF_3 (d) OF_2 (e) HF

61. Watson and Crick received the 1962 Nobel Prize for their simple but elegant model of the "heredity molecule," DNA. The key to their structure (the "double helix") was an understanding of the geometry and bonding capabilities of nitrogen-containing bases such as the thymine molecule below. (a) Give approximate values for the indicated bond angles. (b) Which are the most polar bonds in the molecule?

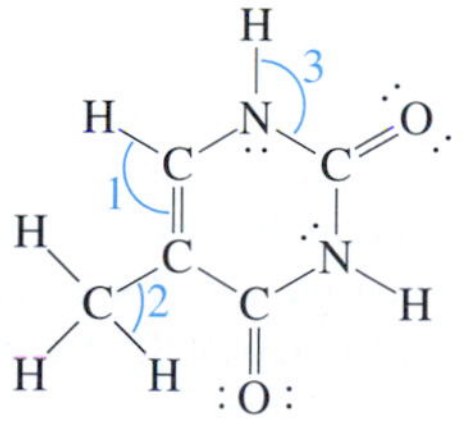

62. The dipole moment of the HCl molecule is 3.43×10^{-30} C·m and the bond length is 127.4 pm; the dipole moment of HF is 6.37×10^{-30} C·m, with bond length of 91.68 pm. Use the definition of dipole moment as a product of partial charge on each atom times the distance of separation (see Section 10.4) to calculate the quantity of charge in coulombs that is separated by the bond length in each dipolar molecule. Use your result to show that fluorine is more electronegative than chlorine.

63. In the gas phase, positive and negative ions form ion pairs that are like molecules. An example is KF, which is found to have a dipole moment of 28.7×10^{-30} C·m and a distance of separation between the two ions of 217.2 pm. Use this information and the definition of dipole moment to calculate the partial charge

on each atom. Compare your result with the expected charge, which is the charge on an electron, -1.602×10^{-19} C. Based on your result, is KF really completely ionic?

64. Sketch the geometry of a carbon-containing molecule or ion in which the angle between two atoms bonded to carbon is
(a) Exactly 109.5 °.
(b) Slightly different from 109.5 °.
(c) Exactly 120 °.
(d) Exactly 180 °.

65. The following compound is commonly called acetylacetone. As shown it exists in two forms, one called the *enol* form and the other called the *keto* form.

$$H_3C{-}C(OH){=}C(H){-}C(=O){-}CH_3 \rightleftharpoons H_3C{-}C(=O){-}C(H)(H){-}C(=O){-}CH_3$$

enol form *keto* form

While in the *enol* form, the molecule can lose H^+ from the —OH group to form an anion. One of the most interesting aspects of this anion (sometimes called the *acac* ion) is that one or more of them can react with a transition metal cation to give very stable, highly colored compounds (see photo).

$$\left[\begin{array}{c} H_3C{-}C{=}O{\rightarrow} \\ H{-}C \qquad\qquad M \\ H_3C{-}C{-}O{-} \end{array} \right]_3$$

M(acac)$_3$

(a) Using bond energies, calculate the enthalpy change for the *enol* $\rightarrow$ *keto* change. Is the reaction exo- or endothermic?
(b) What are the electron pair and "molecular" geometries around each C atom in the *keto* and *enol* forms? What changes (if any) occur when the *keto* form changes to the *enol* form?
(c) If you wanted to prepare 15.0 g of deep red $Cr(acac)_3$ using the following reaction,

$$CrCl_3 + 3\ H_3C{-}C(OH){=}CH{-}C(O){-}CH_3 + 3\ NaOH \longrightarrow Cr(acac)_3 + 3H_2O + 3\ NaCl$$

how many grams of each of the reactants would you need?

Chelate complexes. Three complexes of the type $M(acac)_3$, where M is (left to right) Co, Cr, and Fe. *(C.D. Winters)*

Applying Concepts

66. Complete the following table.

Molecule or Ion	Electron-Pair Geometry	Molecular Geometry	Hybridization of the Iodine Atom
ICl_2^+			
I_3^-			
ICl_3			
ICl_4^-			
IO_4^-			
IF_4^+			
IF_5			
IF_6^+			

67. Complete the following table.

Molecule or Ion	Electron-Pair Geometry	Molecular Geometry	Hybridization of the Sulfur Atom
SO_2			
SCl_2			
SO_3			
SO_3^{2-}			
SF_4			
SO_4^{2-}			
SF_5^+			
SF_6			

68. Name a Group 1A to 8A element that could be the central atom (X) in the following compounds.
(a) XH_3 with one lone pair of electrons
(b) XCl_3
(c) XF_5
(d) XCl_3 with two lone pairs of electrons

69. Name a Group 1A to 8A element that could be the central atom (X) in the following compounds.
(a) XCl_2
(b) XH_2 with two lone pairs of electrons
(c) XF_4 with one lone pair of electrons
(d) XF_4

70. How many water molecules could hydrogen bond to an acetic acid molecule? Draw in the water molecules and use dotted lines to show the hydrogen bonds.

```
       H  :O:
       |   ‖    ..
    H—C—C—O—H
       |        ..
       H
```

71. How many water molecules could hydrogen bond to an ethylamine molecule? Draw in the water molecules and use dotted lines to show the hydrogen bonds.

```
       H   H   H
       |   |   |
    H—C—C—N—H
       |   |   ..
       H   H
```

72. The following are responses students wrote when asked to give an example of hydrogen bonding. Which are correct?

(a) H—H···H—H

(b)
```
  H              O
    \           / \
     C=O···H      H
    /
  H
```

(c) H—F

(d)
```
     O     H     H
    / \  ·    \  /
  H     H      O
```

73. Which of the following are examples of hydrogen bonding?

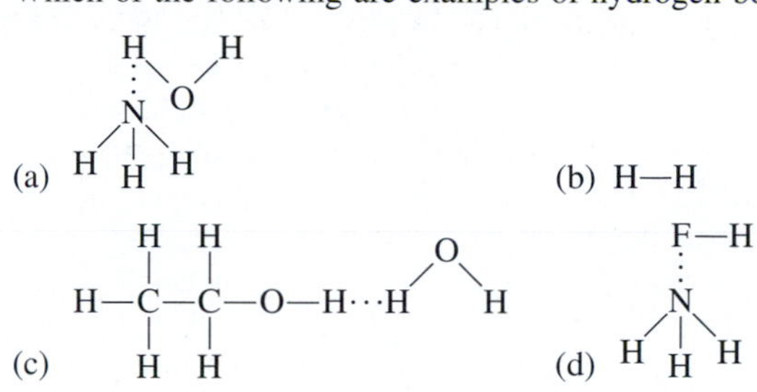

74. In another universe, elements try to achieve a nonet (9 valence electrons) instead of an octet when forming chemical bonds. As a result, covalent bonds form when a trio of electrons are shared between two atoms. Two compounds in this other universe are H_3O and H_2F. Draw their Lewis structures, then determine their electron-trio geometry and molecular geometry.

CHAPTER 11

Energy, Organic Chemicals, and Polymers

This large, ocean-going tanker carries petroleum from the oilfields of the Middle Eastern countries to Europe, Asia, and North and South America. Potential energy is stored in this petroleum in chemical bonds in carbon compounds. Most of these will be oxidized to liberate energy and form carbon dioxide and water in the process. Some molecules will not be burned but rather their atoms will become parts of roadways or be made into food wrap, toys, and perhaps even cosmetics and medicines. • What kinds of carbon compounds are present in petroleum, and how are the compounds put to use? In the Summary Problem at the end of this chapter, you will be asked to answer some questions about the changes various carbon compounds from petroleum and other sources undergo. Careful study of this chapter's contents will help you answer these questions. *(Calvin Larsen, Photo Researchers, Inc.)*

The combustion of coal, natural gas, and petroleum (fossil fuels) provides 88% of all the energy used in the world. When these compounds are burned, the carbon they contain is released into the atmosphere as CO_2. Photosynthesis then converts most of this CO_2 back into carbon-containing compounds. While many of the carbon compounds produced by photosynthesis are very useful as energy sources for humans and animals, they are not as convenient as the fossil fuels for energy generation in power plants or automobiles. At current usage rates, world reserves are *estimated* to last 1500 years for coal, 120 years for natural gas, and 60 years for petroleum. These are not sure numbers, and as conservation measures are taken and scarcity occurs, even petroleum reserves could last for more than a century.

Fossil fuels are also the major source of hydrocarbons, and about 6% of the petroleum refined today is the source of most of the organic chemicals used to make consumer products. For this reason the organic chemical industry is often referred to as the petrochemical industry. It produces products including plastics, synthetic rubber, synthetic fibers, fertilizers, and thousands of other consumer products. A few of the major classes of organic compounds and some of their reactions, especially those used in supplying energy and in making polymers, are discussed in this chapter.

11.1 PETROLEUM AND NATURAL GAS

The economic importance of the organic chemical industry can be seen in the list of the chemicals produced in the largest quantities in the United States. Of the top 25 listed inside the back cover, 13 are organic chemicals. At present, petroleum is the major source of the hydrocarbons used as raw materials for at least 10 of these 13 organic chemicals. For the hundreds more organic chemicals that are of major economic importance, petroleum is the only known source. How long will petroleum be viable as a source for energy and starting materials for consumer products? Probably not long. Oil production and oil imports by the United States over the past several years are shown in Figure 11.1. With only 4% of global reserves in 1995, the United States produced 8% of the world output of oil. Current oil use worldwide averages

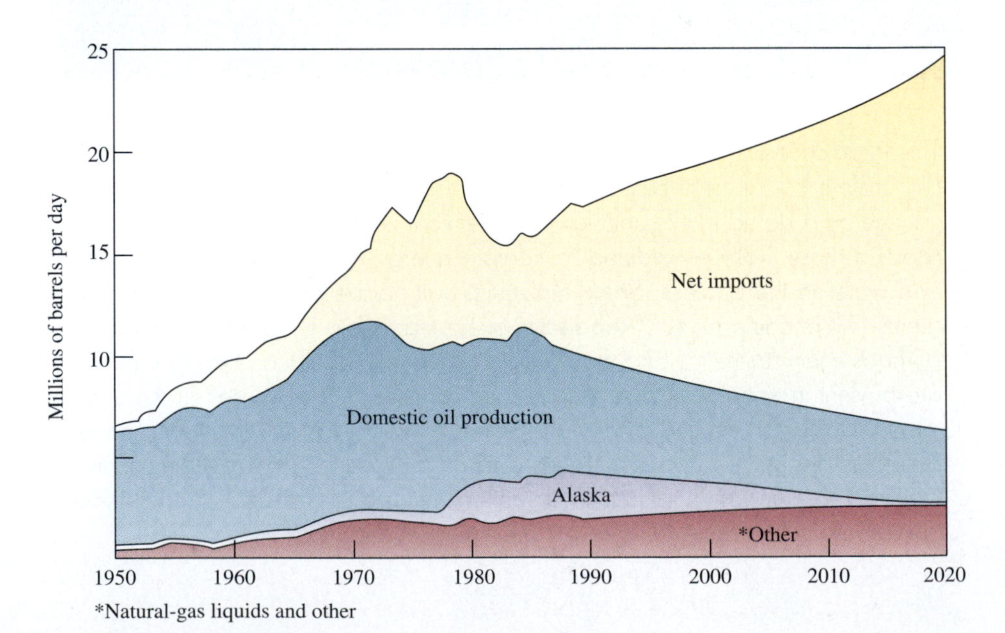

Figure 11.1 United States oil production and oil imports. At present, more than 50% of the total oil used in the United States is imported, and projections show oil imports will continue to increase.

approximately 4.5 barrels of oil a year for each person. In the United States, oil usage is 24 barrels per person per year and in sub-Saharan Africa oil usage is less than 1 barrel per person per year. The worldwide average petroleum use is expected to fall to 1.5 barrels per person per year by 2030 as oil scarcity drives up prices. This will require extensive changes in the global energy economy as well as a reduction in the use of petroleum as the major source of organic chemicals. However, both natural gas and coal are important sources of hydrocarbons, and coal will likely become more important as reserves of the other fossil fuels are depleted.

A petroleum barrel contains 42 gallons. Many years ago petroleum was shipped in barrels. Today, it seldom is seen in barrels but rather is shipped in pipelines and ocean-going tankers. Nevertheless, the barrel remains as the common unit of measure for petroleum.

Petroleum is a complex mixture of alkanes, cycloalkanes, alkenes, and aromatic hydrocarbons formed from the remains of plants and animals from millions of years ago. Thousands of compounds, almost all of them hydrocarbons, are present in crude petroleum, and its composition varies with the location in which it is found. Pennsylvania crude oils are primarily straight-chain hydrocarbons, whereas California crude oil contains a larger portion of aromatic hydrocarbons. The early uses for petroleum were mainly for lubrication and open combustion in lamps. When internal combustion engines were invented, it became necessary to refine petroleum to produce liquid fuels with properties that would allow them to burn efficiently in these engines.

The different classes of hydrocarbons (compounds of hydrogen and carbon) were introduced earlier ***(← pp. 84 and 365).*** Alkanes have C—C bonds; alkenes have one or more C═C bonds; aromatics include benzene-like rings.

Petroleum Refining

The refining of petroleum begins with its separation into groups of compounds with distinct boiling-point ranges by a process called **fractional distillation.** The difference between simple distillation and fractional distillation is the degree of separation achieved. For example, water that contains dissolved solids or other liquids can be purified by distillation. The impure solution is heated to boiling; the water vapor is condensed as a pure liquid and collected in a separate container (Figure 11.2). Since petroleum contains thousands of different hydrocarbons, separation of the pure com-

Conceptual Challenge Problem CP-11.A at the end of the chapter relates to topics covered in this section.

A petroleum refinery tower.
(Ashland Oil, Inc.)

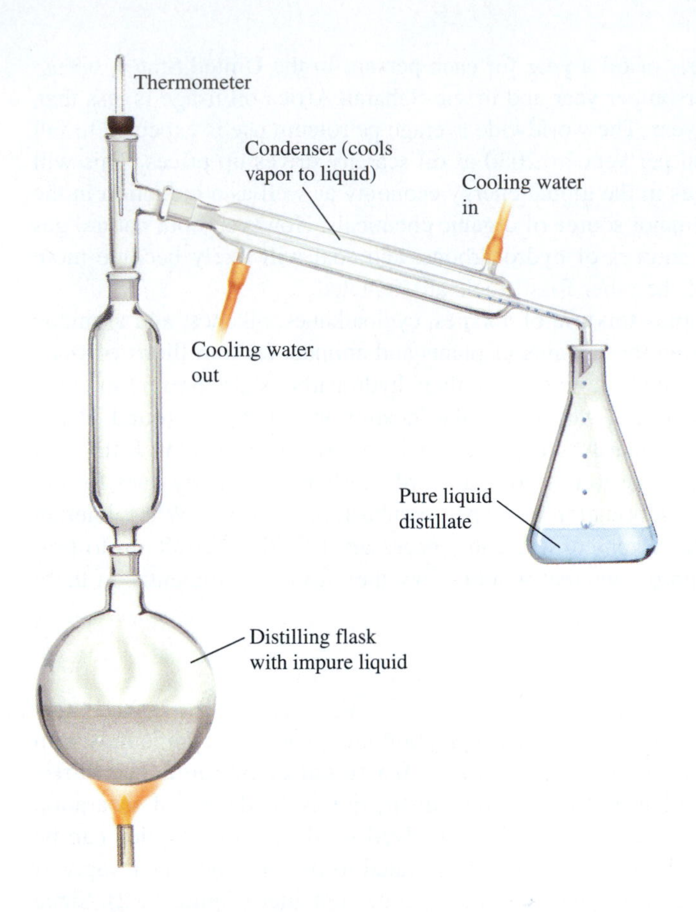

Figure 11.2 Water distillation. Water that contains dissolved solids or other liquids usually can be purified by distillation. When the solution is heated above the boiling point, water vaporizes and passes into the cool condenser where it liquefies and runs into the collection flask. (Open flames should not be used when heating flammable liquids.)

High-purity petroleum hydrocarbons, like pure hexane, are available, but they are very expensive and would never be used as fuels except for testing purposes.

Petroleum fractions are mixtures of hundreds of hydrocarbons with boiling points within a certain range.

pounds is neither feasible nor necessary. The products obtained from the fractional distillation of petroleum are still mixtures of hundreds of hydrocarbons, so they are called **petroleum fractions.**

Figure 11.3 is a schematic drawing of a fractional distillation tower used in petroleum refining. The crude oil is first heated to about 400 °C to produce a hot vapor mixture that enters the fractionating tower at the bottom. The temperature decreases the farther up the tower the vapor goes, and different components condense at various points along the tower. The volatile lower-boiling petroleum fractions remain in the vapor stage longer than the less volatile higher-boiling fractions. These differences in boiling point ranges allow the separation of fractions. The lightest hydrocarbons do not condense and are drawn off the top of the tower as gases. The heaviest hydrocarbons do not vaporize and are collected at the bottom of the tower as liquids and dissolved solids. This fractional distillation process is very important in our economy because it produces most of the liquid fuels that move our automobiles, trucks, trains, and airplanes. It also provides home heating fuel used in many parts of the country. Typical products of the fractional distillation of petroleum are listed in Table 11.1, p. 454.

Octane Number

An important property of hydrocarbon liquids is their **autoignition temperature,** the temperature at which the liquid will ignite and burn without a source of ignition. The autoignition temperature of a hydrocarbon is related to its molecular composition.

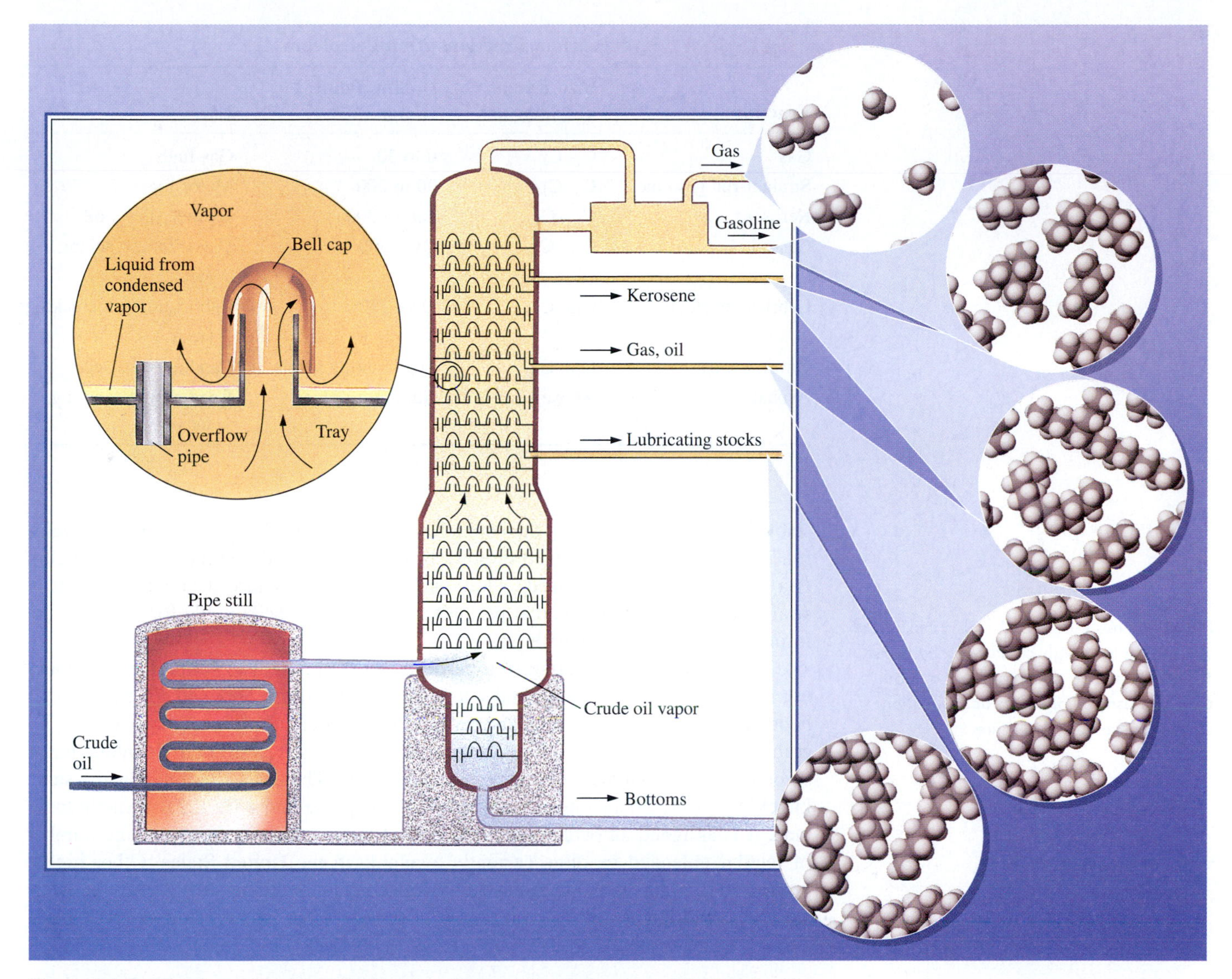

Figure 11.3 Petroleum fractional distillation. Crude oil is first heated to 400 °C in the pipe still. The vapors then enter the fractionation tower. As vapors rise in the tower, they cool and condense so that different fractions can be drawn off at different heights.

Large, straight-chain hydrocarbon molecules have much lower autoignition temperatures than do branched-chain and smaller molecules. Because it consists mainly of smaller molecules, gasoline has a fairly high autoignition temperature, and to efficiently burn gasoline in an engine requires a source of ignition—a spark from the spark plug. However, inside modern gasoline engines, temperatures get quite high and autoignition can occur before the spark ignites the fuel. This robs the engine of performance. When premature ignition occurs, the engine "pings" and, under severe conditions, a "knock" occurs.

The octane number of a gasoline is a measure of its ability to burn smoothly without preignition and engine knocking. Straight-chain alkanes are less thermally

TABLE 11.1 Hydrocarbon Fractions from Petroleum

Fraction	Size Range of Molecules	Boiling Point Range (°C)	Uses
Gas	C_1–C_4	0 to 30	Gas fuels
Straight-run gasoline	C_5–C_{12}	30 to 200	Motor fuel
Kerosene	C_{12}–C_{16}	180 to 300	Jet fuel, diesel oil
Gas-oil	C_{16}–C_{18}	Over 300	Diesel fuel, cracking stock
Lubricating stock	C_{18}–C_{20}	Over 350	Lubricating oil, cracking stock
Paraffin wax	C_{20}–C_{40}	Low-melting solids	Candles, wax paper
Asphalt	Above C_{40}	Gummy residues	Road asphalt, roofing tar

stable and burn less smoothly than branched-chain alkanes. For example, the "straight-run" gasoline fraction obtained directly from the fractional distillation of petroleum is a poor motor fuel and needs additional refinement because it contains primarily straight-chain hydrocarbons that preignite too readily to be suitable for use as a fuel in internal combustion engines.

Diesel engines burn diesel fuel, a mixture of larger, mostly straight-chain hydrocarbons. The diesel engine does not have a spark plug, but uses autoignition as its ignition source for the fuel-air mixture.

The octane number rating of a gasoline is determined by comparing its knocking characteristics in a one-cylinder test engine with those obtained for mixtures of heptane and 2,2,4-trimethylpentane (often called isooctane). Heptane knocks considerably and is assigned an octane number of 0, whereas 2,2,4-trimethylpentane burns smoothly and is assigned an octane number of 100. Thus, if a gasoline has the same knocking characteristics as a mixture of 13% heptane and 87% 2,2,4-trimethylpentane, it is assigned an octane number of 87. This corresponds to the octane number of regular unleaded gasoline currently available in the United States. Other higher grades of gasoline available at gas stations have octane numbers of 89 (regular plus) and 92 (premium).

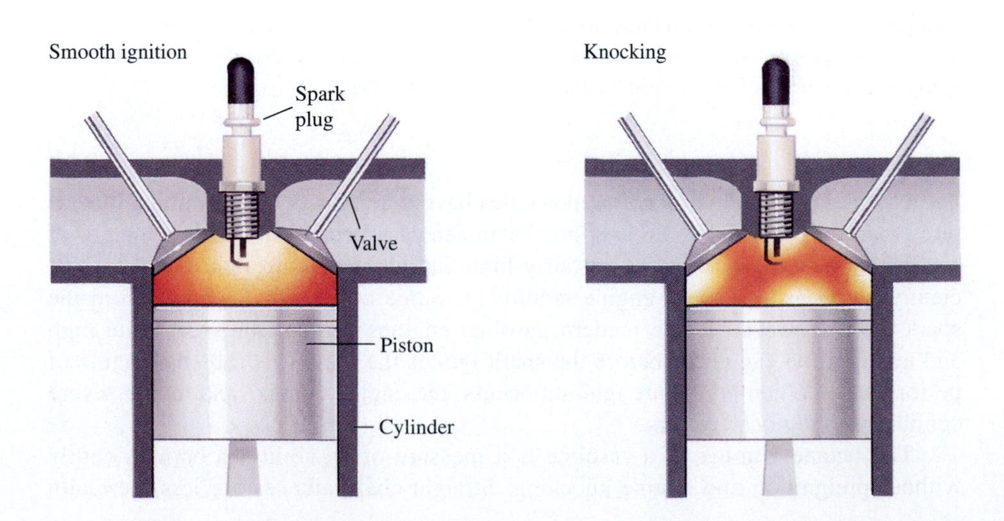

Smooth ignition at the sparkplug and preignition. Preignition, which results in engine ping and knocking, is caused by high-temperature autoignition of fuel molecules.

TABLE 11.2 Octane Numbers of Some Hydrocarbons and Gasoline Additives

Name	Class of Compound	Octane Number
Octane	Alkane	−20
Heptane	Alkane	0
Hexane	Alkane	25
Pentane	Alkane	62
1-Pentene	Alkene	91
2,2,4-Trimethylpentane (isooctane)	Alkane	100
Benzene	Aromatic hydrocarbon	106
Methanol	Alcohol	107
Ethanol	Alcohol	108
Tertiary-butyl alcohol	Alcohol	113
Methyl *tertiary*-butyl ether (MTBE)	Ether	116
Toluene	Aromatic hydrocarbon	118

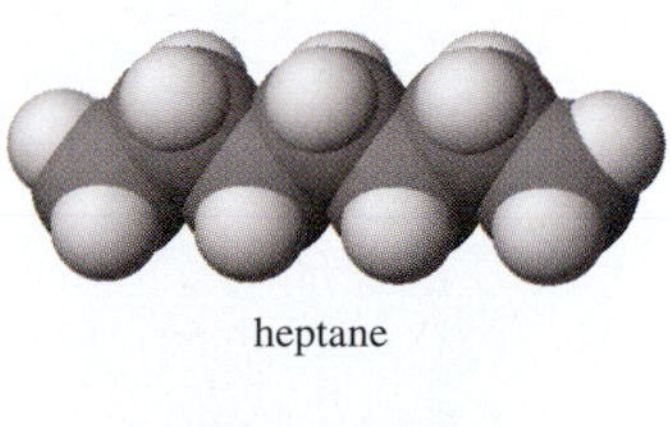

heptane

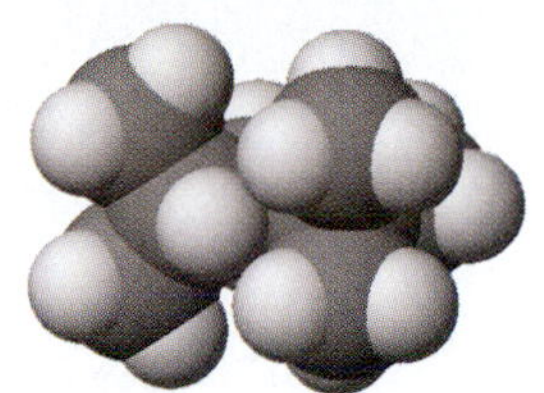

2,2,4-trimethylpentane

The octane number of a gasoline can be increased either by increasing the percentage of branched-chain and aromatic hydrocarbon fractions or by adding octane enhancers (or a combination of both). Since the method for determining octane numbers was established, fuels superior to 2,2,4-trimethylpentane have been developed that have octane numbers above 100. Table 11.2 lists octane numbers for some hydrocarbons and octane enhancers.

Typical octane ratings of commonly available gasolines. *(C.D. Winters)*

A **catalyst** is a substance that increases the rate of a chemical reaction without being consumed as a reactant would be. The role of catalysts in chemical reactions is discussed in Section 12.3. Catalysts are essential to many industrial chemical processes, which, without them, would be too slow to be practical.

Catalytic Reforming

One way to increase the octane number of the gasoline fraction is by converting straight-chain hydrocarbons to branched-chain hydrocarbons and aromatics in a process called **catalytic reforming.** In the presence of certain *catalysts,* such as finely divided platinum on a support of Al_2O_3, straight-chain hydrocarbons with low octane numbers can be reformed into their branched-chain isomers, which have higher octane numbers.

$$CH_3CH_2CH_2CH_2CH_3 \longrightarrow CH_3\underset{}{\overset{\overset{\displaystyle CH_3}{|}}{C}}HCH_2CH_3$$

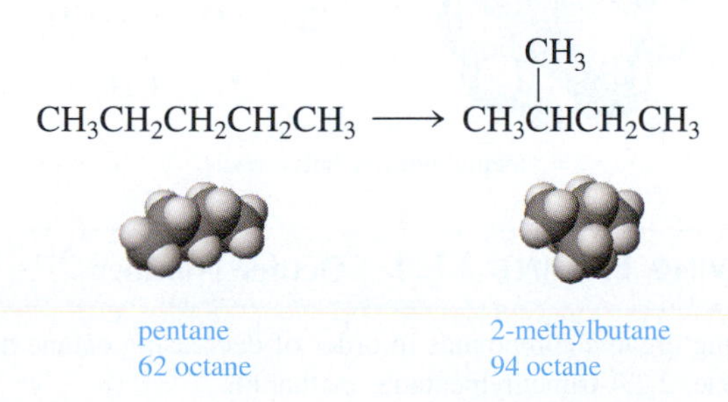

Catalytic reforming is also used to produce aromatic hydrocarbons by using different catalysts and petroleum mixtures. For example, when the vapors of straight-run gasoline, kerosene, and light oil fractions are passed over a copper catalyst at 650 °C, a high percentage of the original material is converted into a mixture of aromatic hydrocarbons from which benzene, toluene, xylenes, and similar compounds may be separated by fractional distillation. This process can be represented by the equation for converting hexane into benzene.

$$CH_3CH_2CH_2CH_2CH_2CH_3 \longrightarrow C_6H_6 + 4\,H_2$$

hexane
25 octane

benzene
106 octane

Octane Enhancers

The octane number of a given blend of gasoline can also be increased by adding "anti-knock" agents or octane enhancers. In the United States, prior to 1975, the most widely used antiknock agent was tetraethyllead, $(C_2H_5)_4Pb$. The addition of 3 g of $(C_2H_5)_4Pb$ per gallon increases the octane number by 10 to 15. From 1925 until 1975 both regular and premium grades of gasoline contained an average of 3 g of $(C_2H_5)_4Pb$ per gallon. Tetramethyllead, $(CH_3)_4Pb$, was also used.

The exhaust emissions of internal combustion engines contain CO, oxides of nitrogen, and unburned hydrocarbons, all of which contribute to air pollution (see Section 14.12). As urban air pollution worsened, Congress passed the Clean Air Act of 1970, which required that 1975-model-year cars emit no more than 10% of the carbon monoxide and hydrocarbons emitted by 1970 models. The solution to lowering these emissions was a platinum-based catalytic converter. The only problem was that it required lead-free gasolines, since lead deactivates the platinum catalyst by coating its surface. As a result, automobiles manufactured since 1975 have been required to use lead-free gasoline to protect the catalytic converter.

As little as two tanks of leaded gasoline can destroy the activity of a catalytic converter.

Because tetraethyllead can no longer be used in the United States and a few other countries, other octane enhancers are now added to gasoline. These include toluene, 2-methyl-2-propanol (also called *tertiary*-butyl alcohol), methyl *tertiary*-butyl ether (MTBE), methanol, and ethanol. The most popular octane enhancer is MTBE, which joined the top 50 chemical list for the first time in 1984 and was number 12 in 1995 (up from number 18 in 1994).

Since lead is a toxic element (Section 16.5) that was also being emitted in exhaust fumes, getting lead out of gasoline in the United States brought long-term health benefits in addition to protecting catalytic converters. Leaded gasoline is banned in only nine countries in the world. In many parts of the world, more lead is used in gasoline than was ever used in the United States. As many as 18 million children are at risk of lead poisoning worldwide because of the continued use of leaded gasoline.

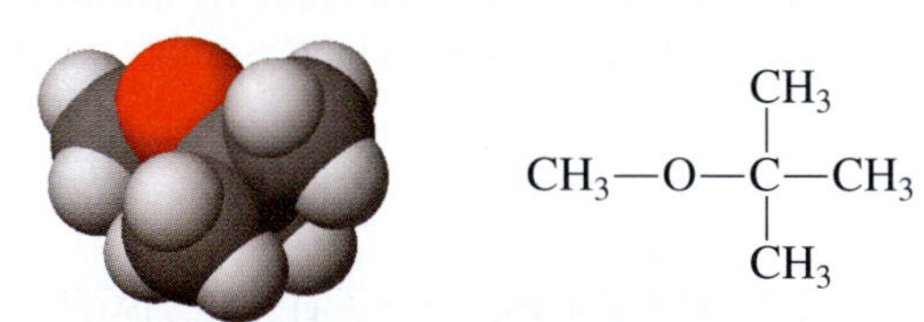

methyl *tertiary*-butyl ether

An **ether** belongs to the class of organic molecules that contain the functional group, R—O—R′. The R and R′ represent hydrocarbon groups, like methyl **(← p. 88)**.

Problem-Solving Example 11.1 Octane Number

Place the following organic compounds in order of decreasing octane number: heptane, 1-pentene, benzene, 2,2,4-trimethylpentane, methanol.

Answer The decreasing order is benzene, methanol, 2,2,4-trimethylpentane, 1-pentene, heptane.

Explanation As illustrated in Table 11.2, aromatics and alcohols have the highest octane numbers, followed by branched-chain alkanes and alkenes; straight-chain alkanes have the lowest octane numbers.

Problem-Solving Practice 11.1

Heptane (C_7H_{16}) can be catalytically reformed to make toluene ($C_6H_5CH_3$). Note that toluene is also a seven-carbon molecule. How many molecules of hydrogen are produced for every toluene molecule formed from heptane? Write a balanced equation for this reaction.

Oxygenated and Reformulated Gasolines

The 1990 amendments to the Clean Air Act require cities with excessive carbon monoxide pollution to use oxygenated gasolines during the winter. **Oxygenated gasolines** are blends of gasoline with organic compounds that contain oxygen, such as MTBE, methanol, ethanol, and *tertiary*-butyl alcohol. Oxygenated gasolines burn more completely than non-oxygenated gasoline and can reduce carbon monoxide emissions in urban areas by up to 17%. The 1990 regulations also require oxygenated gasolines to contain 2.7% oxygen by weight. The use of oxygenated gasoline is currently required in about 40 cities in the United States with the highest air pollution.

All gasolines are highly volatile. The resulting vapors can be ignited, allowing you to start your car even in the coldest of weather. However, this volatility means that some hydrocarbons get into the atmosphere as a result of accidental spills and evaporation during normal filling operations at the service station. Hydrocarbons in the atmosphere play an important role in a series of reactions that contribute to urban air pollution (Section 14.12). **Reformulated gasolines** are oxygenated gasolines that contain a lower percentage of aromatic hydrocarbons and have a lower volatility than

Completely automated unit in a Kentucky refinery for the production of MTBE, which is blended into gasoline to improve its octane rating. MTBE also decreases the rate of evaporation of gasoline and reduces unburned hydrocarbons in the engine exhaust. *(Ashland Oil, Inc.)*

Oxygenated gasoline is produced by adding oxygen-containing organic compounds to the refined gasoline. Reformulated gasoline requires changes in the refining process to alter the percentage composition of the different types of hydrocarbons, particularly alkenes and aromatics.

The nine cities with the most serious ozone pollution are Baltimore, Chicago, Hartford, Houston, Los Angeles, Milwaukee, New York, Philadelphia, and San Diego.

ordinary gasoline. Nine cities with the most serious ozone pollution were required by the 1990 regulations to use reformulated gasolines starting in 1995, and another 87 cities that are not meeting the ozone air quality standards can choose to use them. The Clean Air Act amendments of 1990, despite some difficulties in their implementation, are having a positive effect. In late 1996, it was reported by the Congressional Research Service that more than half of the over 90 areas that did not meet ozone standards in 1990 and 28 of the 42 areas that did not meet the carbon monoxide standards in 1990 now do so.

Catalytic Cracking

Part of petroleum refining involves adjusting the percentage of each fraction to match the market demand. For example, there is more demand for gasoline than for kerosene and diesel fuel. There are seasonal demands also. In winter, home heating oil demand is high, and in summer, when more people take vacations, demand for gasoline is higher. Therefore, refiners use chemical reactions to convert some of the larger, kerosene fraction molecules into smaller molecules in the gasoline range in a process called "cracking." The **catalytic cracking process** uses a catalyst, heat, and pressure to break long-chain hydrocarbons into shorter-chain hydrocarbons including both alkanes and alkenes, many in the gasoline range.

$$C_{16}H_{34} \xrightarrow[\text{heat}]{\text{pressure}} C_8H_{16} + C_8H_{18}$$

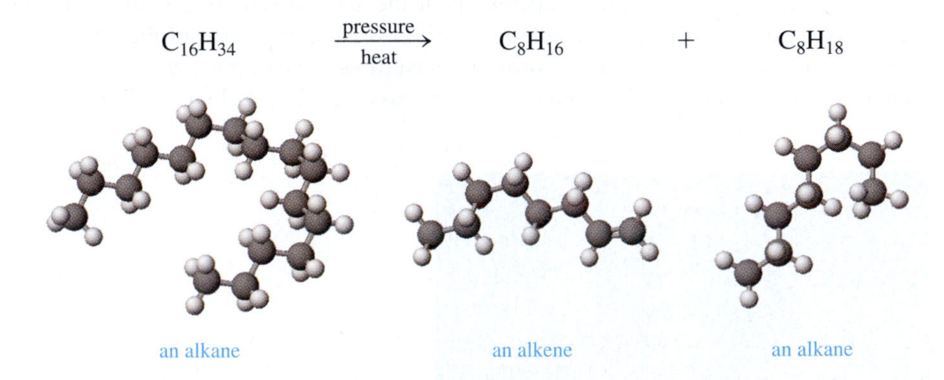

Since alkenes have higher octane numbers than alkanes, the catalytic cracking process also increases the octane number of the mixture. Catalytic cracking is also important for the production of alkenes used as starting materials in the organic chemical industry (Figure 11.4).

Natural Gas

Natural gas is a mixture of low-molar-mass hydrocarbons and other gases trapped with petroleum in the earth's crust. It can be recovered from oil wells or from gas wells where the gases have migrated through the surrounding rock. The natural gas found in North America is a mixture of C_1 to C_4 alkanes [methane (60–90%), ethane (5–9%), propane (3–18%), and butane (1–2%)] with a number of other gases, such as CO_2, N_2, H_2S, and the noble gases—mainly helium, in varying amounts. In Europe and Japan natural gas is essentially all methane.

Natural gas is the fastest growing energy source in the United States, and U.S. production of natural gas supplies 17% more energy than does U.S.-produced oil (Figure 11.5). About half of the homes in the United States are heated by natural gas,

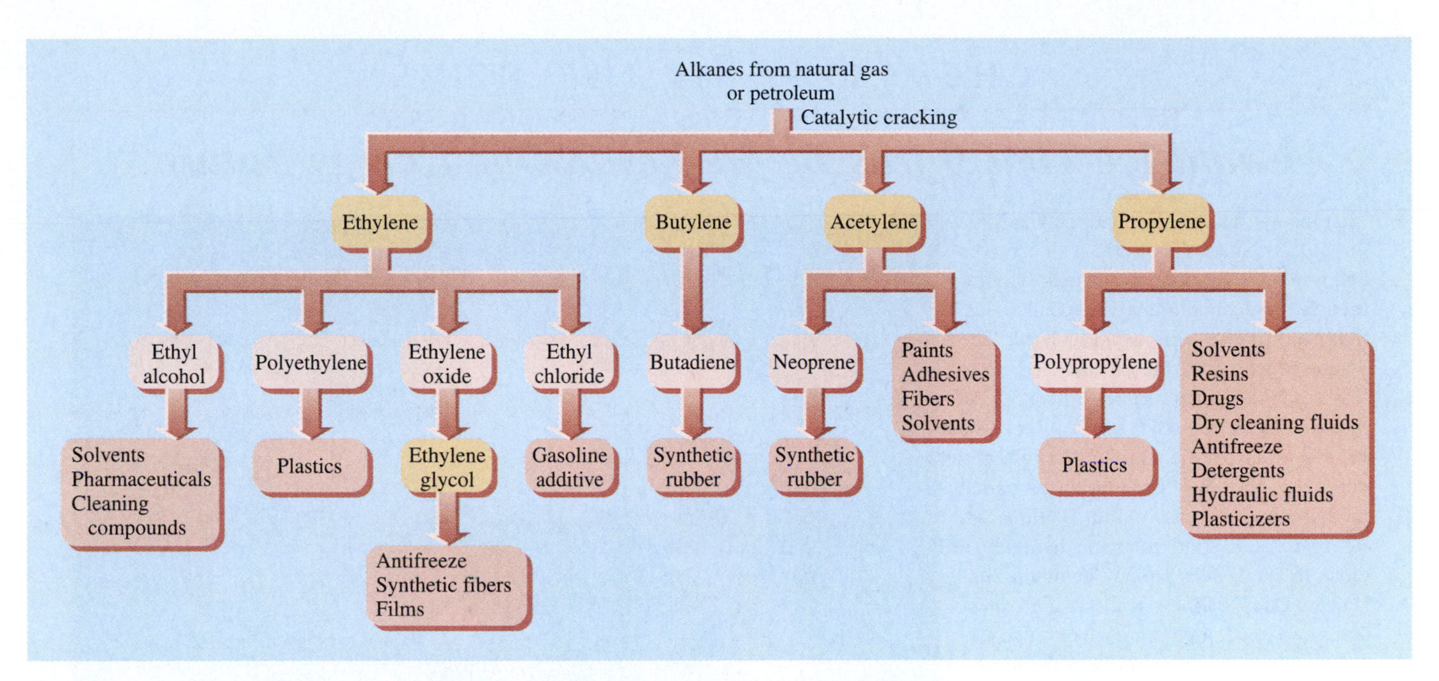

Figure 11.4 Hydrocarbons from petroleum or natural gas. Catalytic cracking produces ethylene, butylene, acetylene, and propylene, which are converted into other chemical raw materials and many kinds of commercial products.

followed by electricity (18%), fuel oil (15%), wood (5%), and liquefied gas such as butane and propane (5%). Coal and kerosene come in at a low 0.5% each, and solar heating of homes is even lower. However, the United States has only about 5% of the known world reserves of natural gas; at the present rate of use, this is estimated to be enough to last until 2050.

Natural gas is now also used as a vehicle fuel. Worldwide, there are about 700,000 vehicles powered by natural gas. California and several other states are encouraging the use of natural gas vehicles to help meet new air quality regulations. Vehicles powered by natural gas emit minimal amounts of carbon monoxide, hydrocarbons, and particulates; and the price of natural gas is about one-third that of gasoline. The main disadvantages of natural gas vehicles include the need for a pressurized gas tank and the lack of service stations that sell the gas in liquefied form.

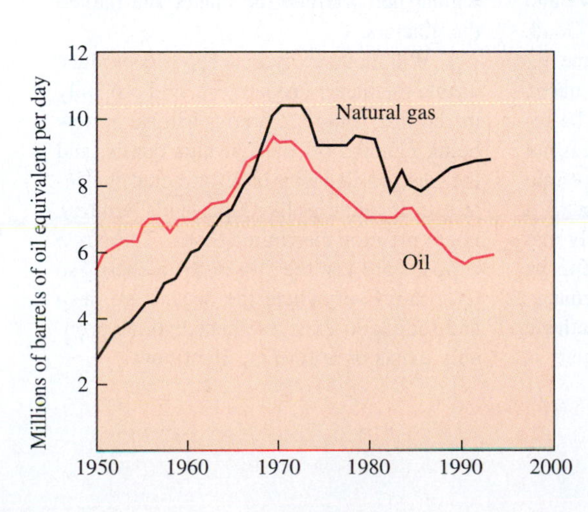

Figure 11.5 Natural gas and oil production in the continental United States, 1950–1995.

CHEMISTRY IN THE NEWS

Could Ice That Burns Be the Key to Our Energy Future?

(Courtesy Oregon State University)

Imagine a mixture of ice and methane. It's possible because the structure of ice (see Section 15.4) is open enough to accommodate the small methane molecules. Such a mixture will actually burn, and when the ice melts, the methane gas is released and has the same properties as methane from any other source. Recently, just 50 miles off the coast of Oregon, at a depth of 550 m, this substance, called methane hydrate, was found. While it is a bit unusual to find methane hydrate so close to land, it is a fairly common substance. Deep in the ocean, at high pressures and low temperatures, ice sometimes forms that contains trapped methane gas. The U.S. Geologic Service estimates that the total amount of carbon trapped in methane hydrate on the ocean floors to be over twice the amount of all the carbon found in all the known fossil fuel deposits, including petroleum, coal, and natural gas.

The exact size of the deposit of methane hydrate off Oregon is not yet known, but in the Atlantic Ocean, off the coast of North and South Carolina are a pair of deposits of methane hydrate about the size of the state of Rhode Island. These deposits contain about 1300 trillion cubic feet of methane. That's about 65 times the current annual U.S. consumption of natural gas. Most of this methane is thought to have been formed by the decomposition of marine sediments by bacteria, but some may be from deep sediments that have existed on the ocean floor for eons of time. When released from the sediments by geologic changes and increases in temperature, the methane is then trapped in the cage-like structures formed by water molecules as they hydrogen-bond together to make ice. Large deposits of methane hydrate also act as a cement, holding entire layers of sediment together. These layers can also trap large deposits of regular gaseous methane.

It is generally believed that the methane hydrate deposits can someday be tapped to provide a tremendous amount of usable methane for the world's population. But scientists are just now beginning to appreciate all of the factors involved in getting at these methane hydrate deposits. Drilling into these deposits will probably release some of the methane gas trapped between sediment layers. This, in turn, will lower the pressure on the layers of methane hydrate sediment below and probably release more methane gas. Good, right? Maybe not. If the entire methane hydrate deposit over a large area were upset, huge quantities of methane would be released into the ocean water, where it is not very soluble, and it would then escape into the atmosphere where it would act as a greenhouse gas. While it is not likely that drilling into the methane hydrate deposits would release all of the trapped methane, there is about 3000 times as much methane tied up in methane hydrate as there is methane in the atmosphere. As a greenhouse gas (Section 11.4), methane is about 10 times as effective as carbon dioxide. Large amounts of methane being released could greatly affect global temperatures over a relatively short time. It has been suggested that during the ice ages, as large quantities of the ocean's water became trapped in glaciers and sea levels dropped, the resulting pressure decreases on the ocean floors caused methane to be released, which then warmed the planet and melted the glaciers.

Will methane hydrate deposits produce usable methane? Probably so, and probably in the near future. Careful drilling is now being planned off the Carolina coasts, and the Oregon deposits have attracted the interests of countries like Japan and Germany, not to mention the United States. These two deposits are just the "tip of the iceberg" so to speak. Everywhere the oceans are deep and sediments exist, methane hydrate probably exists or is forming right now.

Source: Japan High Tech Satellite Network, NEC Corporation, December, 1996.

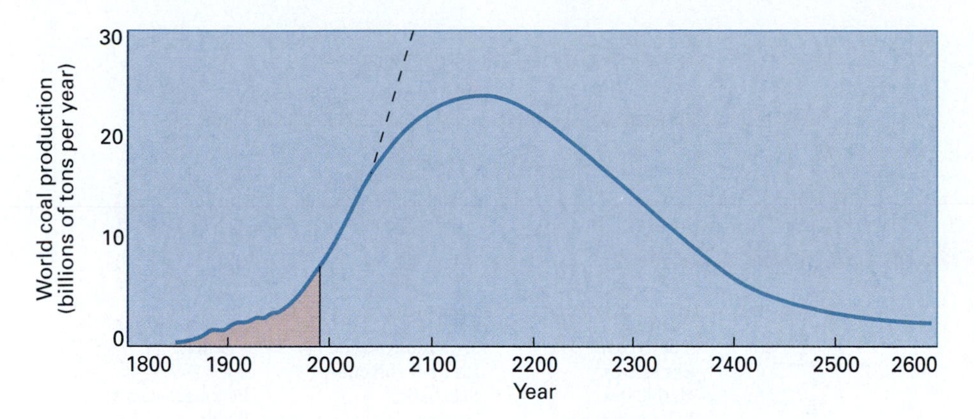

Figure 11.6 Coal resources. The coal mined to date (shaded area) is only a small fraction of the known recoverable coal. The current rate of increase in coal consumption (dashed line) is about 4% per year. Factors such as cost and availability will eventually bring down production (solid line). The area under the curve represents all the coal that is mineable.

11.2 COAL

About 88% of our annual coal production is burned to produce electricity. Although the use of coal is on the rise, the direct burning of coal as a heating fuel has declined because it is a relatively dirty fuel, bulky to handle, and a major cause of air pollution (because of its sulfur content, see Section 14.11). The dangers of deep coal mining and the environmental disruption caused by strip mining have also contributed to the decline in the use of coal. Like petroleum, coal supplies raw materials to the chemical industry.

The world coal reserves are estimated to be about 1024 billion tons, of which about 29% is in the United States. How much coal has been used and how long it is expected to last are summarized in Figure 11.6.

Most of the useful compounds obtained from coal are aromatic hydrocarbons. Coal consists of a complex and irregular array of partially hydrogenated six-membered carbon rings and other structures, some of which contain oxygen, nitrogen, and sulfur atoms. Heating coal at high temperatures in the absence of air, a process called **pyrolysis,** produces a mixture of coke (mostly carbon), coal tar, and coal gas. One ton of bituminous (soft) coal yields about 1500 lb of coke, 8 gal of coal tar, and 10,000 ft^3 of coal gas. Coal gas is a mixture of H_2, CH_4, CO, C_2H_6, NH_3, CO_2, H_2S, and other gases, and at one time it was used as a fuel. Coal tar can be distilled to yield the aromatic fractions listed in Table 11.3. Some uses of these compounds as starting materials in the preparation of organic chemicals of commercial importance are shown in Figure 11.7.

Coal tar, a black viscous liquid, is an important source of aromatic hydrocarbons.

TABLE 11.3 Fractions from Distillation of Coal Tar

Boiling Point Range (°C)	Name	Mass %	Primary Constituents
Below 200	Light oil	5	Benzene, toluene, xylenes
200–250	Middle oil	17	Naphthalene, phenol (carbolic oil), pyridine
250–300	Heavy oil	7	Naphthalenes, methylnaphthalenes (creosote oil), cresols
300–350	Green oil	9	Anthracene, phenanthrene
	Residue	62	Pitch or tar

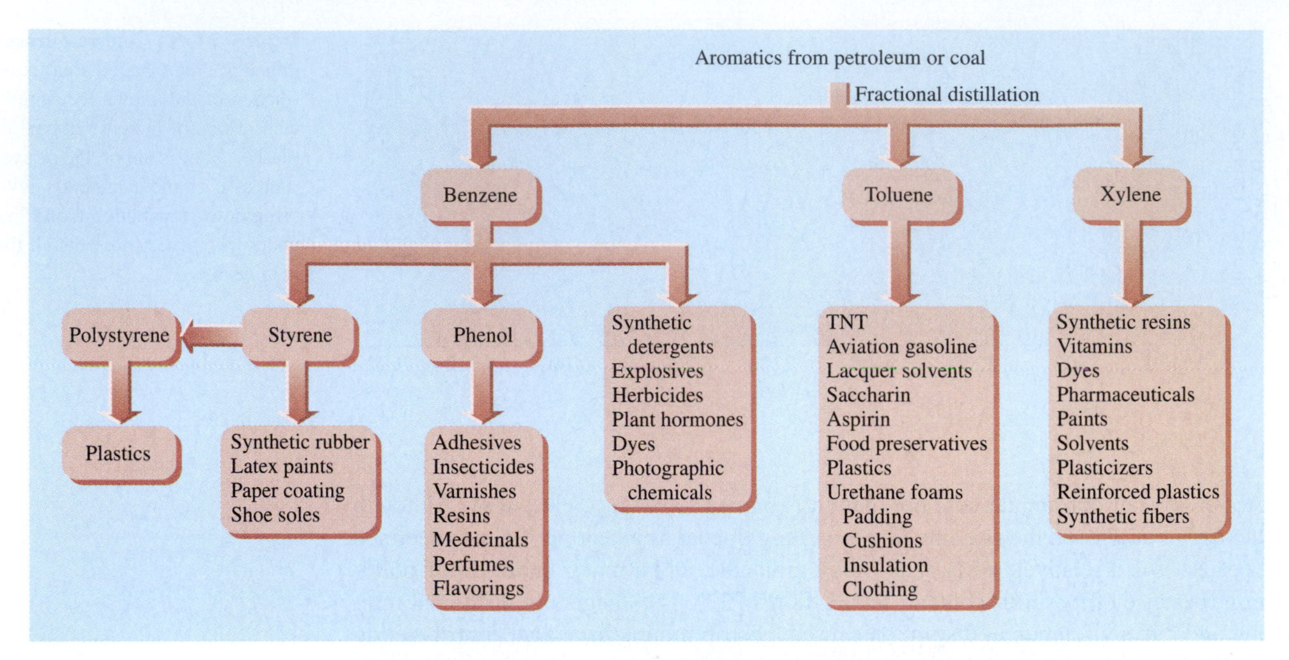

Figure 11.7 Aromatic compounds obtained from petroleum or coal. Benzene, toluene, and xylene are the starting materials for a large number of commercial products.

Exercise 11.1 Coal Gas Combustion Reactions

Write balanced equations for the combustion of each of the coal gas components H_2, CH_4, CO, C_2H_6, and H_2S.

Coal Gasification

Since solid coal is not as easy to handle as a liquid or gaseous fuel, a natural question to ask is, Why not try to somehow convert coal into a gas? It is possible to do this. The carbon in coal will combine with water to form a mixture of two combustible gases, CO and H_2, known as synthesis gas. The process, which is endothermic, involves treating pulverized coal with superheated steam.

$$\underset{\text{coal}}{C(s)} + \underset{\text{steam}}{H_2O(g)} \xrightarrow{\text{heat}} \underset{\text{synthesis gas}}{CO(g) + H_2(g)} \qquad \Delta H = 131 \text{ kJ/mol C}$$

Both components of synthesis gas burn to liberate heat, so some of the energy required to make the gas mixture from coal and water vapor can be recovered when the gas is burned.

$$2\,CO(g) + O_2(g) \longrightarrow 2\,CO_2(g) \qquad \Delta H = -566 \text{ kJ } (-283 \text{ kJ/mol CO})$$
$$2\,H_2(g) + O_2(g) \longrightarrow 2\,H_2O(g) \qquad \Delta H = -484 \text{ kJ } (-242 \text{ kJ/mol } H_2)$$

The sum of these two reactions gives the heat of combustion of synthesis gas, $(-283) + (-242) = -545$ kJ/mol, which is less heat per mole than methane produces ($\Delta H = -802$ kJ/mol). The "synthesis" part of the name *synthesis gas* comes from its use as a starting material for the production of more complex organic chemicals (see Section 11.6, Methanol).

In a newer coal gasification process, methane is the end product. In the process, crushed coal is mixed with a catalyst and steam. The resulting mixture is then heated to 700 °C to produce methane and carbon dioxide. The overall reaction is

$$2\ C(s) + 2\ H_2O(g) \xrightarrow{\text{catalyst}} CH_4(g) + CO_2(g) \qquad \Delta H = 15.3\ \text{kJ/mol}\ CH_4$$

Although the overall reaction is endothermic, the process is an energy-efficient way to obtain methane, a clean-burning fuel that releases 802 kJ/mol. The obvious disadvantage to this reaction is that for every carbon atom converted into the gaseous fuel methane, a carbon atom is converted into a CO_2 molecule, which has no energy value.

Exercise 11.2 Why Is Synthesis Gas Economically Feasible?

Explain why the production of synthesis gas is economically feasible even though the production of synthesis gas from coal is endothermic.

Exercise 11.3 The Energy Value of CO_2

Explain why CO_2 has no fuel energy value.

Coal Liquefaction

Liquid fuels are made from coal by treating it with H_2 under high pressure in the presence of catalysts (hydrogenating the coal). The process produces hydrocarbons like those in petroleum. The product is similar in appearance and properties to crude oil and can be fractionally distilled as described for petroleum in Section 11.1 to give hydrocarbon fractions that can be used for fuels, and to manufacture such things as plastics, medicines, and other commodities. About 230 gal of liquid can be produced for each ton of coal. Today the cost of a barrel of liquid from coal liquefaction is about double that of a barrel of crude oil. However, as petroleum supplies diminish and the cost of crude oil increases, coal liquefaction should become economically feasible.

Coal liquefaction plant in Daggett, California. *(H.R. Bramaz/Peter Arnold, Inc.)*

Exercise 11.4 Hydrogenation of Coal

Explain where the hydrogen used to hydrogenate coal comes from. Can you see a pattern here?

11.3 ENERGY INTERDEPENDENCE

By now it should be apparent that our energy comes from a variety of sources. When you drive your car, the gasoline that moves it may have come from the United States or some foreign country. Some people are already driving cars powered by natural gas or electricity. The natural gas may be imported and the electricity may have come from burning coal, natural gas, fuel oil, or even from a nuclear reactor (Section 19.6).

All forms of energy are, in principle, interchangeable. The work a given amount of energy can do is the same no matter what the energy source. It doesn't matter to you as a consumer whether you get to work using energy from a fossil fuel in the form of gasoline or whether you use electricity created by burning coal, although the energy is measured differently. It might require 1 gal of gasoline for the regular automobile. An electric automobile would require about 39 kilowatt hours (kwh) of electricity to travel the same distance. The energy content of natural gas, fuel oil, and coal in terms of their equivalents in kwh of electricity and joules are compared in Table 11.4.

PROBLEM-SOLVING EXAMPLE 11.2 Energy Conversions

A large ocean-going tanker holds 1.5 million barrels of crude oil. (a) What is the energy equivalent of this oil in joules? (b) How many tons of coal is this oil equivalent to?

Answer (a) 8.9×10^{15} J; (b) 3.3×10^{5} t coal

Explanation

(a) Find the entry in Table 11.4 for the number of joules of heat energy equivalent to 1 barrel of oil—5.9×10^{9}. Use this as a conversion factor to calculate the first answer.

$$1.5 \times 10^{6}\ \text{bbl} \times \frac{5.9 \times 10^{9}\ \text{J}}{1\ \text{bbl}} = 8.9 \times 10^{15}\ \text{J}$$

(b) Look in Table 11.4 and find the conversion between barrels of oil and tons of coal. One bbl of oil is equivalent to 0.22 t of coal. Use this conversion to calculate the answer.

$$1.5 \times 10^{6}\ \text{bbl} \times \frac{0.22\ \text{t coal}}{1\ \text{bbl}} = 3.3 \times 10^{5}\ \text{t coal}$$

Problem-Solving Practice 11.2

How much energy, in joules, can be obtained by burning 4.2×10^{9} t of coal? This is equivalent to how many cubic feet of natural gas?

All fuel-burning engines, including automobile engines and electrical power plants, are less than 100% efficient; there is always wasted heat energy. For example, the generation and distribution of electricity is only about 33% efficient overall. This means that for every 1000 J of energy produced by burning a fossil fuel like coal, only 330 J of electrical energy reaches the consumer.

TABLE 11.4 A Chart of Energy Units*

Cubic Feet of Natural Gas	Barrels of Oil	Tons of Bituminous Coal	Kilowatt Hours of Electricity	Joules
1	0.00018	0.00004	0.293	1.055×10^6
1000	0.18	0.04	293	1.055×10^9
5556	1	0.22	1628	5.9×10^9
25,000	4.50	1	7326	26.4×10^9
1×10^6	180	40	293,000	1.055×10^{12}
3.41×10^6	614	137	1×10^6	3.6×10^{12}
1×10^9	180,000	40,000	293×10^6	1.055×10^{15}
1×10^{12}	180×10^6	40×10^6	293×10^9	1.055×10^{18}

*Based on normal fuel heating values. 10^6 = 1 million; 10^9 = 1 billion; 10^{12} = 1 trillion; 10^{15} = 1 quadrillion (quad).

Exercise 11.5 Energy from Burning Oil

How many joules of energy can be obtained by burning 3.7×10^7 bbl of oil? How many kilowatt hours of electricity would be delivered to the consumer by burning this much oil, assuming a 33% overall efficiency?

Exercise 11.6 Estimating Waste Heat

Heat is wasted as a consequence of thermodynamics. Consider a large electrical power plant generating 100 million kwh of electricity per day. Estimate the wasted heat energy and relate that to the corresponding amounts of oil, natural gas, or coal. Try and imagine some ways to capture some or all of the "wasted" heat energy and put it to use. Why do you think this is not done?

Power is the rate at which energy is produced, transferred, or used. One watt is 1 J/sec, so a 100-W electric lamp uses energy at a rate of 100 J every second.

11.4 CARBON DIOXIDE AND THE GREENHOUSE EFFECT

It took millions of years to form fossil fuels from organisms that had obtained their carbon from atmospheric CO_2. The industrial revolution has, for most of its history, been based on the burning of coal, and for more than 90 years, internal combustion engines have been burning petroleum fuels. By burning these carbon-based fuels over just a few decades, we have been returning to the atmosphere those carbon atoms that had been trapped for eons of time. This sudden (on a geologic time scale) release of CO_2 has had the effect of upsetting the balance of natural CO_2 production by combustion and animal respiration and its use by photosynthesis and other processes.

Exercise 11.7 Sources of CO_2

List as many natural sources of CO_2 as you can. List as many sources of CO_2 from human activity as you can.

Figure 11.8 The greenhouse effect. Greenhouse gases effectively form a barrier that prevents heat from escaping from the earth's surface. Without this effect, the earth's temperature would be much lower.

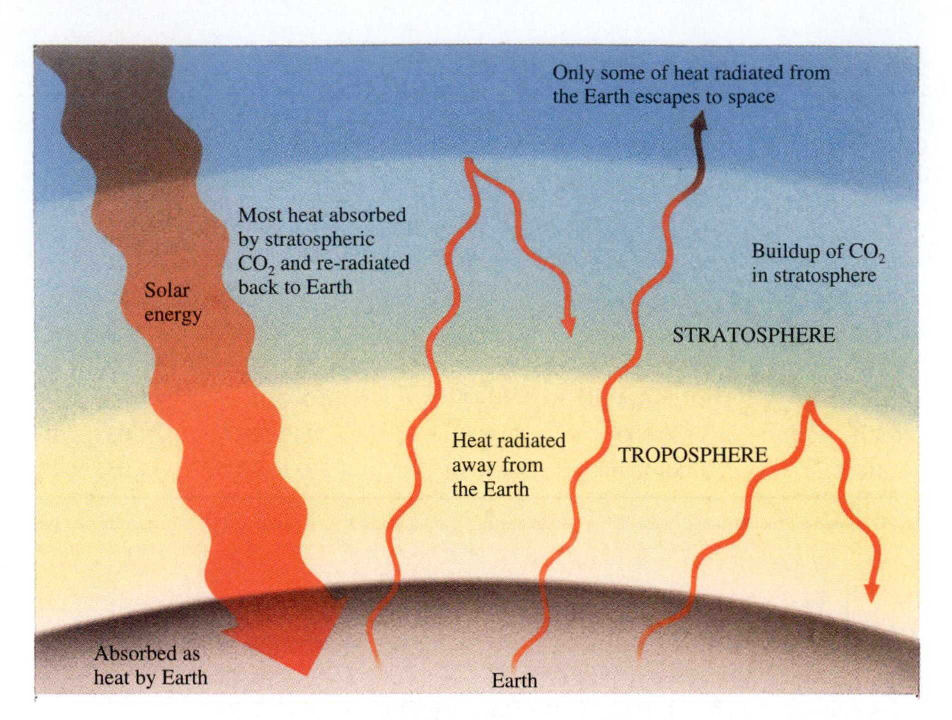

The oceans and cardon dioxide. The oceans dissolve a huge quantity of carbon dioxide, storing it in the form of carbonates and bicarbonates. *(J. Kotz)*

A botanical greenhouse works on a principle similar to the action of the greenhouse gases. The glass windows of a greenhouse transmit visible light but block infrared radiation from leaving.

What is the problem with increasing atmospheric CO_2? When solar radiation arrives at the earth's atmosphere, about half of the visible light (400–700 nm) is reflected back into space. (That's a good thing, for otherwise the temperature of the earth would be far too hot to support life as we know it.) The remainder reaches the earth's surface and causes warming. The warmed surfaces (average temperature about 15 °C) then re-radiates this energy in the infrared portion of the spectrum ***(← p. 303, Figure 8.1)***.

Carbon dioxide, water vapor, methane, and ozone all absorb radiation in various portions of the infrared region. By absorbing this re-radiated energy they warm the atmosphere, creating what is called the **greenhouse effect** (Figure 11.8). Thus, all four are "greenhouse gases." Together such gases constitute an absorbing blanket that reduces loss of energy by radiation back into space, which keeps the earth's atmospheric temperature comfortable (although not in all places at the same time). There is such a vast reservoir of water in the oceans that human activity has a negligible influence on the concentration of water vapor in the atmosphere. In addition, methane is produced by natural processes in such large quantities that human contributions are negligible. Ozone is present in such small concentrations that its contribution to the greenhouse effect is small. So of the four greenhouse gases, most attention is focused on CO_2.

Counting all forms of fossil fuel combustion worldwide, about 50 billion tons of CO_2 are added to the atmosphere each year. About half of this amount is consumed —some by plants during photosynthesis, and the rest by dissolving in rainwater and the oceans to form carbonic acid, which can then form bicarbonates and carbonates.

$$CO_2(g) + 2\ H_2O(\ell) \longrightarrow H_3O^+(aq) + \underset{\text{bicarbonate ion}}{HCO_3^-(aq)}$$

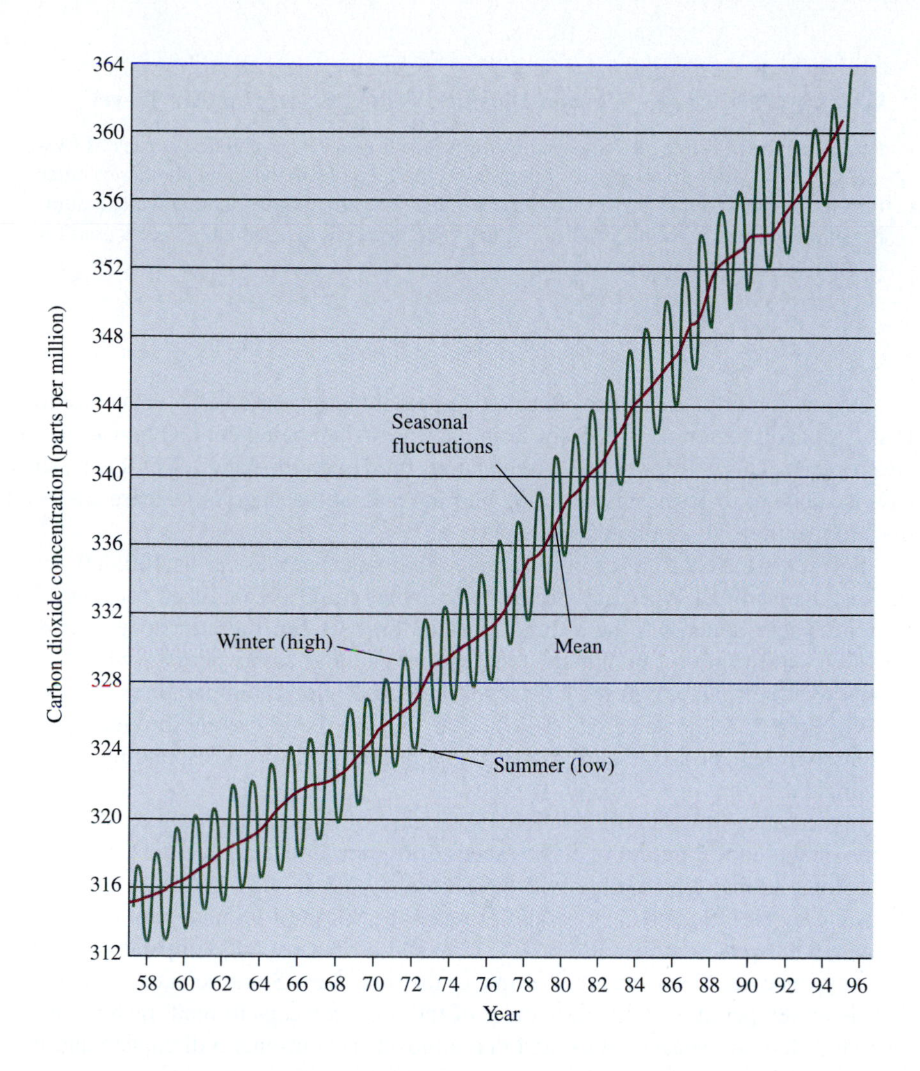

Figure 11.9 Atmospheric carbon dioxide concentrations. Atmospheric carbon dioxide concentrations have been steadily rising. The zig-zag nature of the line is due to seasonal fluctuations.

$$HCO_3^-(aq) + H_2O(\ell) \longrightarrow H_3O^+(aq) + \underset{\text{carbonate ion}}{CO_3^{2-}(aq)}$$

The other half of the carbon dioxide from fossil fuel combustion remains in the atmosphere, which has the effect of increasing the global CO_2 concentration.

Without human influences, the flow of carbon dioxide between the air, plants, animals, and the oceans would be roughly balanced. However, between 1900 and 1970, the global concentration of CO_2 increased from 296 parts per million (ppm) to 360 ppm, an increase of 22%. Expectations are that the CO_2 concentration will continue to increase, although perhaps not as fast as in the past (Figure 11.9). The reason for this expected slowdown is a leveling off of per person energy consumption as fossil fuel supplies get tighter.

Parts per million (ppm) is a convenient way to express low concentrations. One ppm means one part of something in one million things. A CO_2 concentration of 360 ppm means that for every million molecules of air, 360 are CO_2 molecules. Section 16.5 discusses ppm units further.

To see how easily our everyday activities affect the quantity of CO_2 being put into the atmosphere, consider a round-trip flight from New York to Los Angeles. Each passenger pays for about 200 gal of jet fuel, which weighs 1400 lb. When burned, each pound of jet fuel produces about 3.14 lb of carbon dioxide. So 4400 lb, or 2 metric tons, of carbon dioxide are produced per passenger during that trip.

Deforestation in Brazil. This satellite photo taken over the Amazon Basin shows the dramatic extent of the cut-and-burn creation of cropland. The remaining rain forest is dark green and the leveled forest shows in pale greens and browns. The ladder-like pattern at the upper left is a typical result of cut-and-burn in this region. *(Geospace/SPL/Photo Researchers, Inc.)*

Exercise 11.8 Carbon Dioxide Production During Air Travel

Look up annual airline passenger miles for a recent year and calculate the tons of CO_2 released. How does this compare with the CO_2 produced from the same number of miles traveled in automobiles? (*Hint:* Assume a miles-per-gallon value and an average number of passengers per vehicle. Then use the same numbers for gasoline as were used for jet fuel in the paragraph above.)

Population pressure is also contributing heavily to increased CO_2 concentrations. In the Amazon region of Brazil, for example, forests are being cut and burned to create cropland. This activity places a tremendous burden on the natural CO_2 cycle, since CO_2 is added to the atmosphere during burning and, at the same time, there are fewer trees to use the CO_2 in photosynthesis.

Scientists have taken ice core samples that date back as far as 160,000 years. In these ice samples were tiny pockets of air that could be analyzed for their CO_2 content. There appears to be a direct correlation between the atmospheric carbon dioxide concentration and global temperatures in the same period (which were known by other means): as the CO_2 level increased, global temperatures increased, and vice versa. Some scientists believe that today's rising carbon dioxide concentrations will lead to increasing global temperatures and to corresponding changes in climates.

Global warming is the temperature increase beyond that due to the normal "greenhouse effect". A 1.5 °C rise in global temperature is a significant change, requiring a very large input of energy.

If predictions by the U.S. National Academy of Science are correct, when (and if) the global concentration of CO_2 reaches 600 ppm (about double the 1970 value), the average global temperature will have risen by 1.5 to 4.5 °C (or 2.7 to 8.1 °F). Warming by as little as 1.5 °C would produce the warmest climate seen on earth in the past 6000 years; an increase of 4.5 °C would produce world temperatures higher than any since the Mesozoic era, the time of the dinosaurs. Some scientists also worry that rising temperatures will cause more of the polar ice caps to melt, raising the sea level and flooding coastal cities, and that atmospheric currents will change and produce significant changes in climate and agricultural productivity.

Global warming is potentially a major worldwide problem. Computer models are used to predict future global temperature changes, although it is very difficult to include in the models all possible influences on temperature. The most obvious preventive measure is to control CO_2 emissions throughout the world. Given our dependence on fossil fuels, however, getting and keeping these emissions under control will be very difficult.

11.5 ORGANIC CHEMICALS

Organic chemicals were once obtained only from plants, animals, and fossil fuels. Prior to 1828, it was widely believed that chemical compounds found in living matter could not be made without living matter—a "vital force" was thought to be necessary for the synthesis. In 1828, a young German chemist, Friedrich Wöhler, destroyed the vital force myth when he prepared the organic compound urea, a major product in urine, by heating an aqueous solution of ammonium cyanate, a compound obtained from mineral sources.

$$NH_4NCO(aq) \xrightarrow{\text{heat}} H_2N{-}\overset{\overset{\displaystyle O}{\|}}{C}{-}NH_2(s)$$

ammonium cyanate

urea

The notion of a mysterious vital force declined as other chemists began to prepare more and more organic chemicals without the aid of a living system. Over 11 million of the more than 13 million known compounds are organic compounds. A natural question to ask is, Why are there so many of these compounds, all of which contain carbon and hydrogen atoms? As discussed in earlier sections on bonding and isomerism *(← p. 359, p. 373)*, two reasons are (1) the ability of up to thousands of carbon atoms to be linked to each other in a single molecule by stable C—C bonds, and (2) the occurrence of structural isomers. A third reason—the variety of functional groups that bond to carbon atoms—also was introduced earlier *(← p. 81 and p. 361)* and is further illustrated in this chapter.

A table of functional groups and further information on how organic compounds are named are given in Appendix E.

Some living organisms are still direct sources for useful hydrocarbons. Some plants produce an oil that burns almost as well in a diesel engine as diesel fuel; the rubber tree produces a latex that contains the familiar hydrocarbon, rubber. As important as these examples are, the development of synthetic organic chemistry has led to cheaper methods for making copies of naturally occurring substances and to many substances that have no counterpart in nature.

All organic molecules can be viewed as derived from hydrocarbons, and many of them are prepared in this manner. Among some of the most useful of these compounds are alcohols, acids, esters, and the polymers that can be made from them.

11.6 ALCOHOLS AND THEIR OXIDATION PRODUCTS

Alcohols, both natural and synthetic, contain one or more —OH functional groups bonded to carbon atoms and are a major class of organic compounds *(← p. 86, p. 361)*. Some examples of commercially important alcohols and their uses are listed in Table 11.5.

TABLE 11.5 Some Important Alcohols

Condensed Formula	B.p. (°C)	Systematic Name	Common Name	Use
CH_3OH	65.0	Methanol	Methyl alcohol	Fuel, gasoline additive, making formaldehyde
CH_3CH_2OH	78.5	Ethanol	Ethyl alcohol	Beverages, gasoline additive, solvent
$CH_3CH_2CH_2OH$	97.4	1-Propanol	Propyl alcohol	Industrial solvent
$CH_3CH(OH)CH_3$	82.4	2-Propanol	Isopropyl alcohol	Rubbing alcohol
$CH_2(OH)CH_2(OH)$	198	1,2-Ethanediol	Ethylene glycol	Antifreeze
$CH_2(OH)CH(OH)CH_2(OH)$	290	1,2,3-Propanetriol	Glycerol (glycerin)	Moisturizer in foods

Alcohols are classified according to the number of carbon atoms bonded to the —C—OH carbon as primary (one other C atom), secondary (two other C atoms), and tertiary (three other C atoms). In writing general organic structures, R is used to represent any hydrocarbon group. The use of R, R′, and R″ indicates that the R groups can be different.

$$\underset{\text{primary}}{\text{R—}\overset{\text{H}}{\underset{\text{H}}{\text{C}}}\text{—OH}} \qquad \underset{\text{secondary}}{\text{R}'\text{—}\overset{\text{R}}{\underset{\text{H}}{\text{C}}}\text{—OH}} \qquad \underset{\text{tertiary}}{\text{R}'\text{—}\overset{\text{R}}{\underset{\text{R}''}{\text{C}}}\text{—OH}}$$

The longest chain of carbon atoms is numbered at the end so that the carbon with the attached —OH group has the lowest number possible. A table of functional groups and further information on how organic compounds are named is given in Appendix E.

Ethanol and 1-propanol are primary alcohols. 2-propanol is a secondary alcohol and is one of the two constitutional isomers of a three-carbon alcohol. For an alcohol with four carbon atoms, there are four constitutional isomers, including 2-methyl-2-propanol, commonly known as *tertiary*-butyl alcohol and used as a gasoline additive:

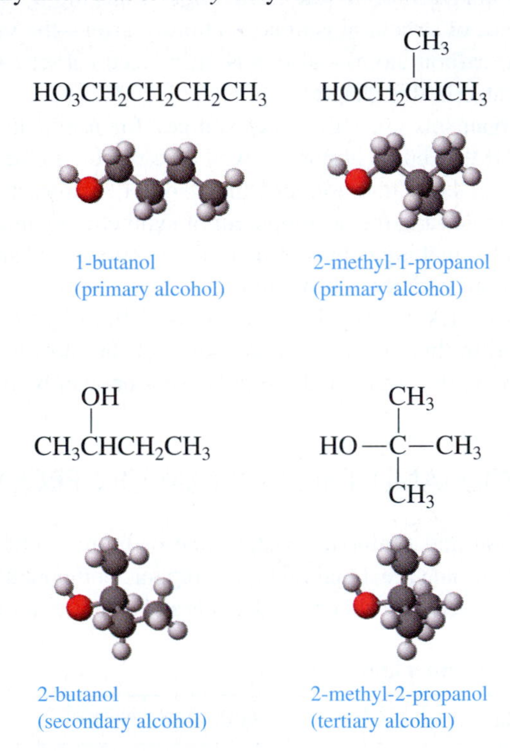

1-butanol (primary alcohol)

2-methyl-1-propanol (primary alcohol)

2-butanol (secondary alcohol)

2-methyl-2-propanol (tertiary alcohol)

PROBLEM-SOLVING EXAMPLE 11.3 Alcohols

Write the condensed structural formula for 2-methyl-2-butanol. Is this a primary, secondary, or tertiary alcohol?

Answer

$$CH_3CH_2\text{—}\overset{CH_3}{\underset{OH}{C}}\text{—}CH_3$$

This is a tertiary alcohol.

Explanation The butanol part of the name shows that this is an alcohol derived from the four-carbon alkane, butane. The 2-methyl and 2-butanol parts of the name indicate that the methyl group (CH_3) and the OH group are located on the number 2 carbon atom, which has three carbon atoms bonded directly to it. Thus, it is a tertinary alcohol.

Problem-Solving Practice 11.3

Write the condensed structural formula for 3-methyl-2-pentanol. Is this a primary, secondary, or tertiary alcohol?

Stepwise oxidation of primary alcohols produces compounds called aldehydes and carboxylic acids. For example, the stepwise oxidation of ethanol with aqueous potassium permanganate or aqueous sodium dichromate produces acetaldehyde, a member of the **aldehyde** functional group class, followed by acetic acid, a member of the **carboxylic acid** functional group class. The liver, with the aid of enzymes, produces the same products, with acetaldehyde contributing to the toxic effects of alcoholism.

Oxidation of organic compounds is usually the addition of oxygen or the removal of hydrogen, and reduction is usually the removal of oxygen or the addition of hydrogen.

$$CH_3CH_2OH \xrightarrow{KMnO_4(aq)} CH_3{-}C(=O){-}H \xrightarrow{KMnO_4(aq)} CH_3{-}C(=O){-}OH$$

ethanol (a primary alcohol) acetaldehyde acetic acid

Exercise 11.9 Looking at Oxidation of Primary Alcohols

Look carefully at the formulas for ethanol and acetaldehyde. Can you see how acetaldehyde is the oxidation product of ethanol? Do the same for acetaldehyde and acetic acid. Were your explanations the same for both pairs of compounds?

Aldehydes contain the $-\overset{O}{\overset{\|}{C}}-H$ (or $-CHO$) functional group. **Carboxylic acids** contain the $-\overset{O}{\overset{\|}{C}}-OH$ (or $-COOH$, or $-CO_2H$) functional group. **Ketones** contain a carbonyl group, $-\overset{O}{\overset{\|}{C}}-$, bonded to two carbon atoms.

Ketones are organic compounds prepared by the oxidation of secondary alcohols. Acetone, the most important commercial ketone, is prepared by the oxidation of 2-propanol. This method is also an important commercial source of hydrogen peroxide.

$$CH_3{-}CH(OH){-}CH_3 + O_2 \xrightarrow[90\text{–}140°C]{15\text{–}20\ atm} CH_3{-}\overset{O}{\overset{\|}{C}}{-}CH_3 + H_2O_2$$

2-propanol (a secondary alcohol) acetone (a ketone) hydrogen peroxide

Problem-Solving Example 11.4 Oxidation of Alcohols

Write the condensed structural formulas of the alcohols that can be oxidized to make the following compounds:

(a) $CH_3CH_2CH_2\overset{O}{\overset{\|}{C}}-H$ (b) $CH_3CH_2\overset{O}{\overset{\|}{C}}-CH_3$ (c) $CH_3CH_2CH_2CH_2\overset{O}{\overset{\|}{C}}-OH$

Answer

(a) $CH_3CH_2CH_2CH_2OH$ (b) $CH_3CH_2\overset{OH}{\overset{|}{C}}HCH_3$ (c) $CH_3CH_2CH_2CH_2CH_2OH$

Explanation

(a) The oxidation of the four-carbon primary alcohol, 1-butanol, will produce this aldehyde.

(b) To produce this ketone, choose the secondary alcohol, 2-butanol, which has two carbons to the left and one carbon to the right of the C—OH group.

(c) The oxidation of the five-carbon primary alcohol, 1-pentanol, will produce this carboxylic acid.

Problem-Solving Practice 11.4

Draw the structural formulas of the expected oxidation products of

(a) $CH_3CH_2CH_2OH$ (b) $CH_3\underset{OH}{\underset{|}{C}}HCH_2CH_3$

Hydrogen Bonding in Alcohols

The physical properties of liquid alcohols are a direct result of the effects of hydrogen bonding between molecules ***(⇐ p. 427).*** In Table 11.5 ***(⇐ p. 469)*** the boiling points of alcohols are listed. Hydrogen bonding in methanol (32 g/mol) explains why it is a liquid, while propane, which has a larger molar mass (44 g/mol) but no hydrogen bonding, is a gas at the same temperature. The boiling point of methanol is lower than that of water (18 g/mol) because methanol has only one —OH hydrogen atom available for hydrogen bonding; water has two. The higher boiling point of ethylene glycol can be attributed to the presence of two —OH groups per molecule. Glycerol, with three —OH groups, has an even higher boiling point.

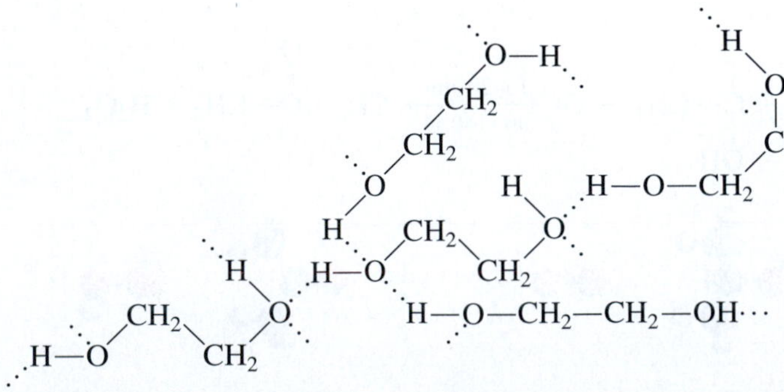

hydrogen bonding between ethylene glycol molecules

The alcohols listed in Table 11.5 are very water soluble because of hydrogen bonding between water molecules and the —OH group in alcohol molecules.

Exercise 11.10 Water Solubility of Alcohols

What would you expect concerning the water solubility of an alcohol containing ten carbon atoms and one OH group? Remember that hydrocarbons like octane are not water soluble.

Some Simple Alcohols

Methanol

Methanol, CH_3OH, is the simplest of all alcohols and is highly toxic. Drinking as little as 30 mL can cause death, and smaller amounts (10 to 15 mL) cause blindness. More than 8 billion pounds of methanol are produced annually in the United States because it is so useful. About 50% is used in the production of formaldehyde, which is used to make plastics, embalming fluid, germicides, and fungicides; 30% is used in the production of other chemicals; and the remaining 20% is used for jet fuels, antifreeze solvent mixtures, and as a gasoline additive.

Methanol currently is prepared from synthesis gas.

$$CO(g) + 2\,H_2(g) \xrightarrow[300\,°C]{\text{catalyst}} CH_3OH(g)$$

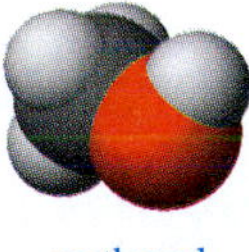
methanol

An old method of producing methanol involved heating a hardwood such as beech, hickory, maple, or birch in the absence of air. For this reason methanol is sometimes called *wood alcohol.*

Because methanol can be made from coal by way of synthesis gas, it will likely continue to increase in importance as petroleum and natural gas become too expensive as sources of both energy and chemicals. Since methanol is relatively cheap, its potential as a fuel and as a starting material for the synthesis of other chemicals is receiving more attention.

Methanol is being considered as a replacement for gasoline in addition to its use in oxygenated gasoline. As a motor fuel, methanol burns more cleanly than gasoline, and levels of troublesome pollutants are reduced. Also, burning methanol emits no unburned hydrocarbons, which contribute significantly to air pollution (see Section 14.11).

Exercise 11.11 Aldehydes as Combustion Products

What is the aldehyde formed from the oxidation of methanol? Critics of the use of methanol as a fuel have cited the formation of this aldehyde as a major potential health hazard. Look up some of the toxic properties of this aldehyde and comment on these criticisms.

The role of aldehydes in atmospheric pollution is discussed in Section 14.12.

The technology for methanol-powered vehicles has existed for many years, particularly for racing cars that burn methanol because of its high octane number of 107. However, the same volume of methanol has only about half the energy content of gasoline; therefore, fuel tanks need to be twice as large to give the same distance per tankful. This is partially compensated for by the fact that methanol costs about half

Cars at the Indianapolis 500 burn methanol. *(Bernard Asset/PhotoResearchers, Inc.)*

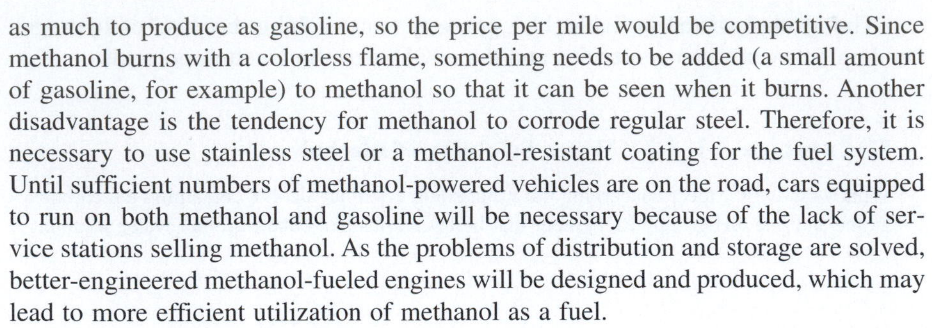

as much to produce as gasoline, so the price per mile would be competitive. Since methanol burns with a colorless flame, something needs to be added (a small amount of gasoline, for example) to methanol so that it can be seen when it burns. Another disadvantage is the tendency for methanol to corrode regular steel. Therefore, it is necessary to use stainless steel or a methanol-resistant coating for the fuel system. Until sufficient numbers of methanol-powered vehicles are on the road, cars equipped to run on both methanol and gasoline will be necessary because of the lack of service stations selling methanol. As the problems of distribution and storage are solved, better-engineered methanol-fueled engines will be designed and produced, which may lead to more efficient utilization of methanol as a fuel.

It is also possible to use methanol to make gasoline. Mobil Oil Company has developed a methanol-to-gasoline process that is currently not competitive with refined gasoline prices in the United States, but is competitive in those regions of the world, such as New Zealand, where the price of gasoline is much higher.

$$2\ CH_3OH \xrightarrow[\text{catalyst}]{} \underset{\text{dimethyl ether}}{(CH_3)_2O} + H_2O$$

$$2\ (CH_3)_2O \xrightarrow[\text{catalyst}]{} \underset{\text{ethylene}}{2\ C_2H_4} + 2\ H_2O$$

$$C_2H_4 \xrightarrow[\text{catalyst}]{} \underset{\text{gasoline}}{\text{hydrocarbon mixture in the } C_5\text{–}C_{12} \text{ range}}$$

The New Zealand plant is currently producing 14,000 barrels per day of gasoline with an octane number of 92 to 94, which is about one-third the amount of gasoline used in New Zealand.

Making methanol. Plant in New Zealand that converts natural gas into methanol, which then reacts in the presence of a catalyst to produce gasoline. *(Mobil Corporation)*

Ethanol

Ethanol, also called *ethyl alcohol* or *grain alcohol,* is the "alcohol" of alcoholic beverages and is prepared for this purpose by fermentation of carbohydrates (starch, sugars) from a wide variety of plant sources. For example, glucose is converted into ethanol and carbon dioxide by the action of yeast in the absence of oxygen.

$$C_6H_{12}O_6 \xrightarrow{\text{yeast}} 2\ C_2H_5OH + 2\ CO_2$$

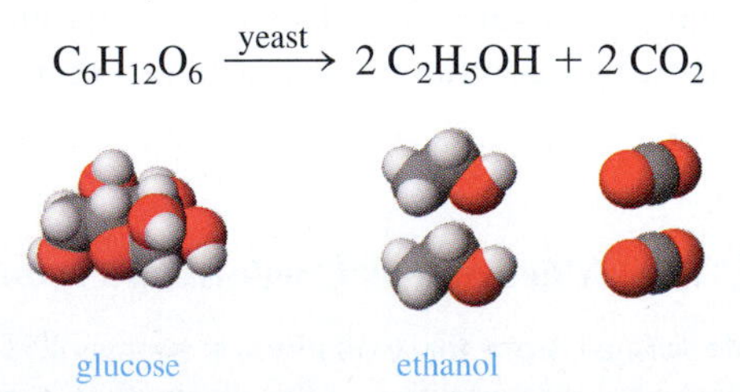

Fermentation eventually comes to a stop when the alcohol concentration reaches a level that inhibits the yeast cells. Beverages with a higher alcohol concentration than provided by natural fermentation are prepared either by distillation or by fortification with alcohol that has been obtained by the distillation of another fermentation product. The maximum concentration of ethanol that can be obtained by distillation of alcohol/water mixtures is 95% ethanol because alcohol and water form a mixture that boils with a constant composition. The "proof" of an alcoholic beverage is twice the volume percent of ethanol; 80-proof vodka, for example, contains 40% ethanol by volume.

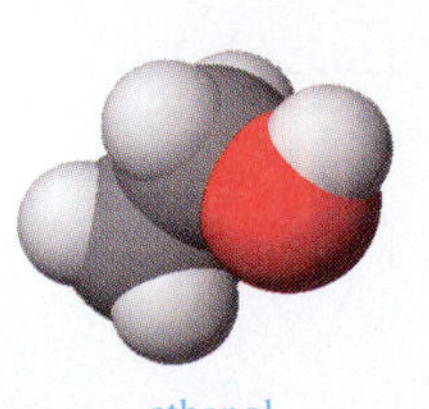

TABLE 11.6 Blood Alcohol Levels and Their Effects

% by Volume	Effect
0.05–0.15	Lack of coordination
0.10	Commonly defined point for "driving while intoxicated"
0.15–0.20	Intoxication
0.30–0.40	Unconsciousness
0.50	Possible death

Although ethanol is not as toxic as methanol, one pint of pure ethanol, rapidly ingested, will kill most people. Ethanol is a depressant, and the effects of various blood levels of alcohol are shown in Table 11.6. Rapid consumption of two 1-oz "shots" of 90-proof whiskey or of two 12-oz beers can cause one's blood alcohol level to reach 0.05%.

In the United States, the federal tax on alcoholic beverages is about $20 per gallon. Since the cost of producing ethanol is only about $1 per gallon, ethanol intended for industrial use must be *denatured* to avoid the beverage tax. Denatured alcohol contains small amounts of a toxic substance, such as methanol or gasoline, that cannot be removed easily by chemical or physical means.

Like methanol, ethanol is receiving increased attention as an alternative fuel. At present, most of it is used in a blend of 90% gasoline and 10% ethanol (known as **gasohol** when introduced in the 1970s).

PROBLEM-SOLVING EXAMPLE 11.5 The Heat of Combustion of Ethanol Compared to That of Octane

Calculate the heat of combustion of ethanol in kJ/mol and compare its value to that of octane. Then, using the densities of the liquids, calculate the thermal energy liberated on burning a liter of each liquid fuel. The densities of octane and ethanol are 0.703 g/mL and 0.789 g/mL, respectively. The heat of formation of octane(g) is −208 kJ/mol.

Answer

Ethanol: $\Delta H_{comb} = -1234.78$ kJ/mol or -2.11×10^4 kJ/L
Octane: $\Delta H_{comb} = -5116$ kJ/mol or -3.15×10^4 kJ/L

Explanation All internal combustion engines burn fuel vapors. The balanced equation for the combustion of ethanol vapor is

$$C_2H_5OH(g) + 3\ O_2(g) \longrightarrow 2\ CO_2(g) + 3\ H_2O(g)$$

Using the standard molar heats of formation from Appendix J and Hess's law *(← p. 250)* for both the products and reactants, the heat of combustion, ΔH_{comb}, is

$$\Delta H_{comb} = 2\ [\Delta H_f^\circ\ CO_2(g)] + 3\ [\Delta H_f^\circ\ H_2O(g)] - 1\ [\Delta H_f^\circ\ C_2H_5OH(g)]$$

$$= 2(-393.509\ \text{kJ/mol}) + 3(-241.818\ \text{kJ/mol}) - (-235.10\ \text{kJ/mol})$$

$$= -1277.3\ \text{kJ/mol}$$

The balanced combustion reaction for octane vapor is

$$C_8H_{18}(g) + \frac{25}{2}\ O_2(g) \longrightarrow 8\ CO_2(g) + 9\ H_2O(g)$$

Using Hess's law, the heat of combustion of octane is

$$\Delta H_{comb} = 8[\Delta H_f^\circ\ CO_2(g)] + 9[\Delta H_f^\circ\ H_2O(g)] - 1[\Delta H_f^\circ\ C_8H_{18}(\ell)]$$

$$= 8(-393.509\ \text{kJ/mol}) + 9(-241.818\ \text{kJ/mol}) - (-208\ \text{kJ/mol})$$

$$= -5116\ \text{kJ/mol}$$

The molar masses of ethanol and octane are 46.069 and 114.23 g/mol, respectively. The thermal energy liberated per liter for each is calculated by

$$\text{Ethanol:}\quad -1277.3\ \text{kJ/mol}\left(\frac{1\ \text{mol}}{46.069\ \text{g}}\right)\left(\frac{0.789\ \text{g}}{\text{mL}}\right)\left(\frac{1000\ \text{mL}}{\text{L}}\right) = -2.19 \times 10^4\ \text{kJ/L}$$

$$\text{Octane:}\quad -5116\ \text{kJ/mol}\left(\frac{1\ \text{mol}}{114.23\ \text{g}}\right)\left(\frac{0.703\ \text{g}}{\text{mL}}\right)\left(\frac{1000\ \text{mL}}{\text{L}}\right) = -3.15 \times 10^4\ \text{kJ/L}$$

The combustion of a liter of ethanol produces only 67% of the energy provided by a liter of octane.

Problem-Solving Practice 11.5

Calculate the heat of combustion of methanol on a kJ/mol and kJ/L basis using standard molar heats of formation from Appendix J. The density of methanol is 0.791 g/mL.

Exercise 11.12

Give the chemical name of the alcohol that is (a) used as a rubbing alcohol, (b) used as an antifreeze, (c) found in alcoholic beverages, (d) found in fats, (e) oxidized to formaldehyde.

Exercise 11.13

A certain alcohol is found to yield on oxidation an aldehyde containing three carbon atoms. After further oxidation, an acid with three carbon atoms is produced. The acid functional group is located on the number one carbon atom of the chain. What is the formula and name of the alcohol?

Large Molecules Containing Alcohol Groups

Conceptual Challenge Problem CP-11.B at the end of the chapter relates to topics covered in this section.

Many natural organic compounds are cyclic, based on hydrocarbons that consist of aromatic rings (***← p. 377***) and cycloalkane or cycloalkene rings (***← p. 360***) sharing some atoms in common (fused together). Steroids, all of which have the four ring structure below, are an example of these fused ring structures.

Many steroids contain alcohol functional groups. These include cholesterol, the female sex hormones estradiol and estrone, and the anti-inflammatory hydrocortisone.

Portrait of a Scientist • *Percy Lavon Julian (1899–1975)*

Percy Julian's list of achievements reads like that of others who have made it to the top in their professions: a doctorate in chemistry in Vienna in 1931 quickly followed, then back home in the United States, by outstanding achievements as a researcher and university professor; 18 years as Director of Research in an industry where he led the way in bringing to market valuable products from soybeans; and the founding of his own research institute, the Julian Laboratories. To grasp the measure of the man, add to this brief outline dozens of scientific publications, over 100 patents granted, numerous academic honors, and positions of responsibility in many civic and humanitarian organizations.

But there were some differences from a successful career path along the way. After completing the eighth grade he had to leave his home in Montgomery, Alabama, for further studies—no more public education was available there for a black man. He enrolled as a "sub-freshman" at DePauw University in Indiana. On his first day, a white student welcomed him with a handshake. Julian later related his reaction: "In the shake of a hand my life was changed, I soon learned to smile and act like I believed they all liked me, whether they wanted to or not."

An organic chemist, Julian built his career around the study of chemicals of plant origin, many of them of medicinal value. The synthesis of a complicated natural molecule is a major goal in such work. Julian was first to achieve synthesis of hydrocortisone, now available in every drugstore because of its value in treating allergic skin reactions. He originated the production of soybean protein and the isolation from soybean oil of compounds from which the first synthetic sex hormone (progesterone) could be made.

(Chemical Heritage Foundation)

cholesterol

estradiol

estrone

hydrocortisone

Exercise 11.14

Look at the structures of the fused-ring hydrocarbons given on this and the previous page and identify common structural features.

11.7 CARBOXYLIC ACIDS AND ESTERS

Carboxylic Acids

Carboxylic acids, which contain the —COOH functional group, are prepared by the oxidation of aldehydes or primary alcohols. These reactions occur quite readily, as evidenced by the souring of wine, which is the oxidation of ethanol to acetic acid in the presence of oxygen from the air.

All acids have a sour taste, but you should never taste anything in the chemistry laboratory.

Carboxylic acids are polar and their molecules form hydrogen bonds with one another. This hydrogen bonding results in relatively high boiling points for the acids, even higher than those of alcohols of comparable molecular size. For example, formic acid (46 g/mol) has a boiling point of 101 °C, while ethanol (46 g/mol) has a boiling point of only 78.5 °C.

All carboxylic acids react with bases to form salts. For example,

$$\underset{\text{acetic acid}}{CH_3\overset{\overset{\displaystyle O}{\|}}{C}—OH(aq)} + NaOH(aq) \longrightarrow \underset{\text{sodium acetate}}{CH_3\overset{\overset{\displaystyle O}{\|}}{C}—O^-Na^+(aq)} + H_2O(\ell)$$

A large number of carboxylic acids are found in nature and have been known for many years. As a result, some of the familiar carboxylic acids are almost always referred to by their common names (Table 11.7). The systematic names of carboxylic acids are easily derived: "-e" is dropped from the name of the corresponding alkane and "-oic" is added followed by the word "acid." Both common and systematic names are given in Table 11.7.

Exercise 11.15 Naming Organic Acids

Two acids not listed in Table 11.7 are derived from the hydrocarbons nonane and decane. What are their names and formulas?

TABLE 11.7 Some Simple Carboxylic Acids

Structure	Common Name	Systematic Name	B.p. (°C)
$H\overset{\overset{\displaystyle O}{\|}}{C}OH$	Formic acid	Methanoic acid	101
$CH_3\overset{\overset{\displaystyle O}{\|}}{C}OH$	Acetic acid	Ethanoic acid	118
$CH_3CH_2\overset{\overset{\displaystyle O}{\|}}{C}OH$	Propionic acid	Propanoic acid	141
$CH_3(CH_2)_2\overset{\overset{\displaystyle O}{\|}}{C}OH$	Butyric acid	Butanoic acid	163
$CH_3(CH_2)_3\overset{\overset{\displaystyle O}{\|}}{C}OH$	Valeric acid	Pentanoic acid	187
$C_6H_5—\overset{\overset{\displaystyle O}{\|}}{C}—OH$	Benzoic acid	Benzene carboxylic acid	250

The simplest carboxylic acid is formic acid, the substance responsible for the sting of an ant bite. Therefore, the common name of the acid comes from the Latin word *formica* for ant. Acetic acid gives the sour taste to vinegar, and the name comes from the Latin word for this substance, *acetum.* Butyric acid gives rancid butter its unpleasant odor, and the name is related to the Latin word for butter, *butyrum.* The names for caproic (C_6), caprilic (C_8), and capric (C_{10}) acids are derived from the Latin word for goat, *caper,* since these acids give goats their characteristic odor. Most of the simple carboxylic acids have unpleasant odors.

Acetic acid, the acid found in about 5% concentration in vinegar, is produced in large quantities (number 33 on the top 50 chemicals list) for manufacturing cellulose acetate, a polymer used to make photographic film base, synthetic fibers, plastics, and other products (←, *p. 397*).

Vinegar. Commercial vinegar is 5% acetic acid. *(C.D. Winters)*

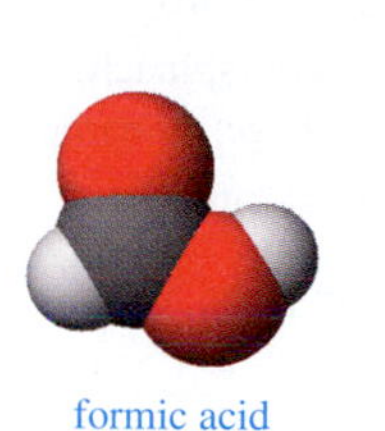

formic acid

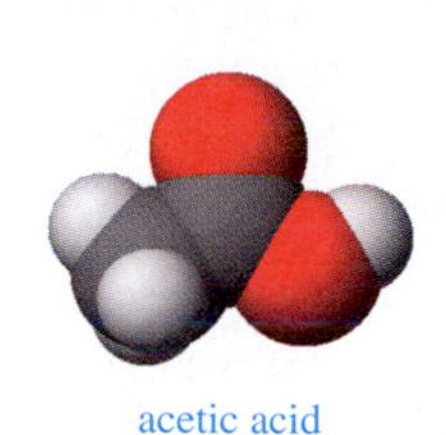

acetic acid

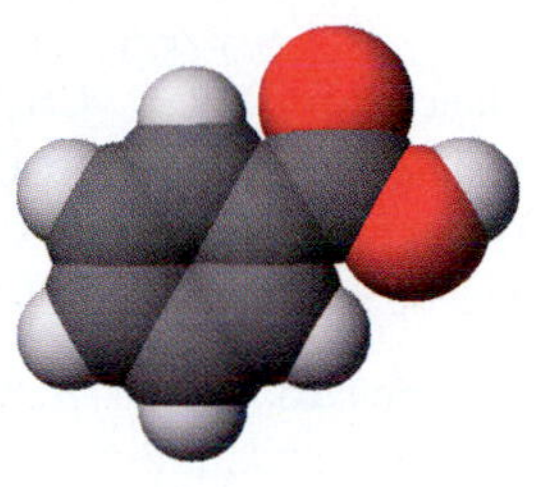

benzoic acid

Three other carboxylic acids produced in large quantity have two acid groups and are known as dicarboxylic acids:

$$\mathrm{HO{-}\overset{\displaystyle O}{\overset{\|}{C}}{-}(CH_2)_4{-}\overset{\displaystyle O}{\overset{\|}{C}}{-}OH}$$

adipic acid

$$\mathrm{HO{-}\overset{\displaystyle O}{\overset{\|}{C}}{-}C_6H_4{-}\overset{\displaystyle O}{\overset{\|}{C}}{-}OH}$$

terephthalic acid

$$\mathrm{C_6H_4(COOH)_2}\ \text{(ortho)}$$

phthalic acid

These three acids are commercially important because they are used to make polymers (Section 11.6). Other carboxylic acids whose names may be familiar to you because they occur in nature are listed in Table 11.8.

Sources of some naturally occurring carboxylic acids. *(C.D. Winters)*

Esters

Carboxylic acids react with alcohols in the presence of strong acids (such as sulfuric acid) to produce **esters,** which contain the —COOR functional group. In an ester, the —OH of the carboxylic acid is replaced by the —OR group from the alcohol. For example, when ethanol is mixed with acetic acid in the presence of sulfuric acid, an ester called ethyl acetate is formed.

Esters contain the $\mathrm{-\overset{\displaystyle O}{\overset{\|}{C}}-OR}$ functional group (also represented as $-CO_2R$ and —COOR).

$$\mathrm{CH_3CH_2OH + HO\overset{\displaystyle O}{\overset{\|}{C}}CH_3 \xrightarrow{-\,H_2O} CH_3CH_2O\overset{\displaystyle O}{\overset{\|}{C}}CH_3}$$

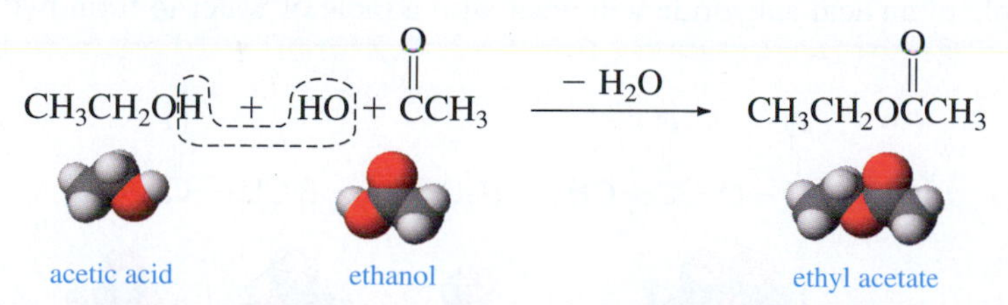

acetic acid   ethanol   ethyl acetate

Ethyl acetate is a common solvent for lacquers and plastics and is often used as fingernail polish remover.

TABLE 11.8 Naturally Occurring Carboxylic Acids

Name	Structure	Natural Source
Citric acid	$HOOC—CH_2—C(OH)(COOH)—CH_2—COOH$	Citrus fruits
Lactic acid	$CH_3—CH(OH)—COOH$	Sour milk
Malic acid	$HOOC—CH_2—CH(OH)—COOH$	Apples
Oleic acid	$CH_3(CH_2)_7—CH{=}CH—(CH_2)_7—COOH$	Vegetable oils
Oxalic acid	$HOOC—COOH$	Rhubarb, spinach, cabbage, tomatoes
Stearic acid	$CH_3(CH_2)_{16}—COOH$	Animal fats
Tartaric acid	$HOOC—CH(OH)—CH(OH)—COOH$	Grape juice, wine

Synthesis gas can be used to make methanol. This methanol can then react with acetic acid to form methyl acetate.

$$CH_3OH + HO—\overset{\overset{\displaystyle O}{\|}}{C}CH_3 \xrightarrow{-\,H_2O} CH_3O\overset{\overset{\displaystyle O}{\|}}{C}CH_3$$

methanol acetic acid methyl acetate

When methyl acetate reacts with carbon monoxide (from synthesis gas), acetic anhydride is formed. This compound is an example of a class of compounds called acid anhydrides.

$$CH_3—\overset{\overset{\displaystyle O}{\|}}{C}—O—CH_3 + CO \longrightarrow CH_3—\overset{\overset{\displaystyle O}{\|}}{C}—O—\overset{\overset{\displaystyle O}{\|}}{C}—CH_3$$

methyl acetate acetic anhydride

One mole of an acid anhydride will react with a mole of water to form two moles of acid.

$$CH_3—\overset{\overset{\displaystyle O}{\|}}{C}—O—\overset{\overset{\displaystyle O}{\|}}{C}—CH_3 + H_2O \longrightarrow 2\,CH_3—\overset{\overset{\displaystyle O}{\|}}{C}—OH$$

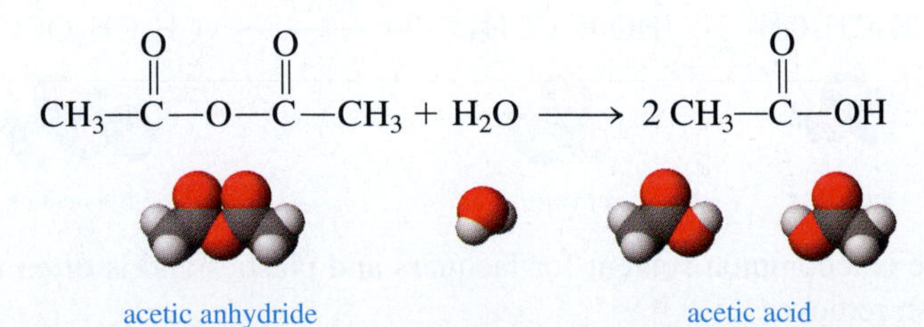

acetic anhydride acetic acid

TABLE 11.9 Some Acids, Alcohols, and Their Esters

Acid	Alcohol	Ester	Odor of Ester
CH_3COOH Acetic acid	$CH_3CH(CH_3)CH_2CH_2OH$ 3-methyl-1-butanol	$CH_3C(=O)OCH_2CH_2CH(CH_3)CH_3$ 3-methylbutyl acetate	Banana
$CH_3CH_2CH_2CH_2COOH$ Pentanoic acid	$CH_3CH(CH_3)CH_2CH_2OH$ 3-methyl-1-butanol	$CH_3CH_2CH_2CH_2C(=O)OCH_2CH_2CH(CH_3)CH_3$ 3-methylbutyl pentanoate	Apple
$CH_3CH_2CH_2COOH$ Butanoic acid	$CH_3CH_2CH_2CH_2OH$ 1-butanol	$CH_3CH_2CH_2C(=O)OCH_2CH_2CH_2CH_3$ butyl butanoate	Pineapple
$CH_3CH_2CH_2COOH$ Butanoic acid	C_6H_5—CH_2OH benzyl alcohol	$CH_3CH_2CH_2C(=O)OCH_2$—C_6H_5 benzyl butanoate	Rose

If half of the acetic acid formed is recycled to react with more methanol, a steady conversion of coal into highly useful acetic acid is possible. In 1983 the Eastman Company in Kingsport, Tennessee, built a plant that converts 900 t of coal daily into this acid.

Some fruits containing esters. *(C.D. Winters)*

Unlike the acids from which they are derived, esters often have pleasant odors (Table 11.9). The characteristic odors and flavors of many flowers and fruits are caused by the presence of natural esters. For example, the odor and flavor of bananas are primarily caused by the ester 3-methylbutyl acetate (also known as isoamyl acetate). Although the odor and flavor of a fruit or flower may be due to a single compound, usually they are due to a complex mixture in which a single ester predominates.

Food and beverage manufacturers often use mixtures of esters as food additives. The ingredient label of a brand of imitation banana extract reads "water, alcohol 40%, isoamyl acetate and other esters, orange oil and other essential oils, and FD&C Yellow #5." Except for the water, these are all organic compounds. The orange oil and other essential oils mentioned are also esters.

As a class, esters are not very reactive. Their most important reaction is splitting of the ester functional group with a water molecule **(hydrolysis)** in the presence of a base to give the constituents of the ester: the alcohol and a salt of the acid from which the ester was formed.

$$\underset{\text{ester}}{RC(=O)—O—R'} + NaOH \longrightarrow \underset{\text{carboxylate salt}}{RC(=O)—O^-Na^+(aq)} + \underset{\text{alcohol}}{R'OH}$$

$$\underset{\text{ethyl acetate}}{CH_3C(=O)—O—CH_2CH_3} + NaOH(aq) \xrightarrow{\text{heat}} \underset{\text{sodium acetate}}{CH_3C(=O)—O^-Na^+(aq)} + \underset{\text{ethanol}}{CH_3CH_2OH}$$

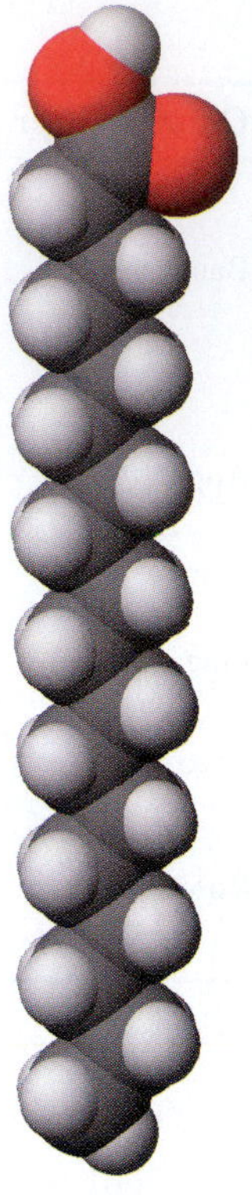
stearic acid

Problem-Solving Example 11.6 Carboxylic Acids and Esters

Write the equation for the formation of butyl acetate.

Answer

$$\underset{\text{acetic acid}}{CH_3COOH} + \underset{\text{1-butanol}}{CH_3CH_2CH_2CH_2OH} \xrightarrow[\text{acid}]{\text{sulfuric}} \underset{\text{butyl acetate}}{CH_3\overset{\overset{\displaystyle O}{\|}}{C}—OCH_2CH_2CH_2CH_3}$$

Explanation The name of the ester indicates what alcohol and acid are needed to make the ester. The first part (butyl) identifies the alcohol, 1-butanol, and the last part (acetate) identifies the acid, acetic acid. 1-Butanol and acetic acid react in the presence of sulfuric acid catalyst to give butyl acetate.

Problem-Solving Practice 11.6

Draw the structural formula of the ester formed in the reaction of (a) acetic acid and methanol, and (b) excess propionic acid and ethylene glycol. Circle the ester linkage in each of your answers.

Substances commonly known as fats and oils are triesters of the trialcohol glycerol and long-chain carboxylic acids called fatty acids (← *p. 110*). Another name for these triesters is triglycerides.

$$\begin{array}{l} CH_2—O—\overset{\overset{\displaystyle O}{\|}}{C}—R \\ | \\ CH—O—\overset{\overset{\displaystyle O}{\|}}{C}—R' \\ | \\ CH_2—O—\overset{\overset{\displaystyle O}{\|}}{C}—R'' \end{array}$$

a triglyceride

The three R groups in a triglyceride can be the same ($R = R' = R''$) or different ($R \neq R' \neq R''$). When the fatty acids are all saturated (for example, stearic acid), the triglyceride is always a solid, called a fat. When the fatty acids are unsaturated (for example, oleic acid), the triglyceride is a liquid, an oil. Table 11.10 gives some of the fatty acids found in some common fats and oils.

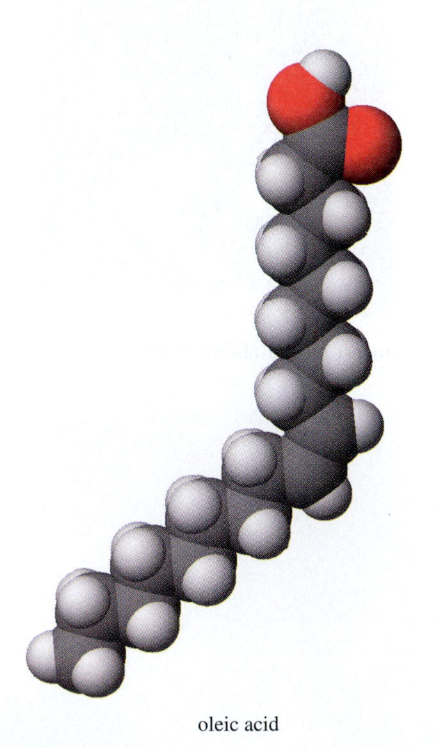
oleic acid

Fatty acids such as oleic acid, which contain only one double bond, are referred to as **monounsaturated acids,** while fatty acids with multiple alternating double bonds are **polyunsaturated acids.** Diets consisting of moderate amounts of fats and oils containing mono- and polyunsaturated fatty acids are considered better for good health than diets containing only saturated fats. In spite of this fact, there is a demand for solid or semisolid fats because of their texture and spreadability. Animal fats often have undesirable tastes, so vegetable oils, which have little or no taste, are often converted to solid fats by adding hydrogen to some or all of the double bonds found in the molecules. The process, known as **hydrogenation,** is the reaction of elemental hydrogen with the liquid triglyceride in the presence of a catalyst.

TABLE 11.10 Common Fatty Acids in Fats and Oils

Saturated Acids (all solids at room temperature)		Source
Lauric	$CH_3(CH_2)_{10}COOH$	Coconut oil
Myristic	$CH_3(CH_2)_{12}COOH$	Nutmeg oil
Palmitic	$CH_3(CH_2)_{14}COOH$	Animal and vegetable fats
Stearic	$CH_3(CH_2)_{16}COOH$	Animal and vegetable fats
Unsaturated Acids (all liquids at room temperature)		
Oleic	$CH_3(CH_2)_7CH{=}CH(CH_2)_7COOH$	Animal and vegetable fats
Linoleic*	$CH_3(CH_2)_4CH{=}CHCH_2CH{=}CH(CH_2)_7COOH$	Linseed oil, cottonseed oil
Linolenic	$CH_3CH_2CH{=}CHCH_2CH{=}CHCH_2CH{=}CH(CH_2)_7COOH$	Linseed oil

*An essential fatty acid that must be part of the human diet.

$$\begin{array}{l} H_2C{-}O{-}\overset{O}{\overset{\|}{C}}{-}(CH_2)_7CH{=}CH(CH_2)_7CH_3 \\ HC{-}O{-}\overset{O}{\overset{\|}{C}}{-}(CH_2)_7CH{=}CH(CH_2)_7CH_3 \\ H_2C{-}O{-}\overset{O}{\overset{\|}{C}}{-}(CH_2)_7CH{=}CH(CH_2)_7CH_3 \end{array} \xrightarrow[200\ ^\circ C]{H_2,\ Ni} \begin{array}{l} H_2C{-}O{-}\overset{O}{\overset{\|}{C}}{-}(CH_2)_7CH_2CH_2(CH_2)_7CH_3 \\ HC{-}O{-}\overset{O}{\overset{\|}{C}}{-}(CH_2)_7CH_2CH_2(CH_2)_7CH_3 \\ H_2C{-}O{-}\overset{O}{\overset{\|}{C}}{-}(CH_2)_7CH_2CH_2(CH_2)_7CH_3 \end{array}$$

Like other esters, triglycerides are hydrolyzed by aqueous base in a process called **saponification** to produce glycerol and the salts of the fatty acids originally in the triglyceride. These salts are commonly known as soaps. (Soaps are discussed in Section 16.12.)

$$\begin{array}{l} H_2C{-}O{-}\overset{O}{\overset{\|}{C}}{-}(CH_2)_{16}CH_3 \\ HC{-}O{-}\overset{O}{\overset{\|}{C}}{-}(CH_2)_{16}CH_3 \\ H_2C{-}O{-}\overset{O}{\overset{\|}{C}}{-}(CH_2)_{16}CH_3 \end{array} + 3\ NaOH \xrightarrow{\Delta} \begin{array}{l} H_2C{-}OH \\ HC{-}OH \\ H_2C{-}OH \end{array} + 3\ Na^+\ \ {}^-O{-}\overset{O}{\overset{\|}{C}}{-}(CH_2)_{16}CH_3$$

glycerol tristearate, a fat — glycerol — sodium stearate, a soap

Similar hydrolysis reactions, which are assisted by enzymes (Section 12.9), occur during the digestion of fats and oils.

Exercise 11.16 Triglyceride Structures

Draw the structural formula for the triglyceride formed when glycerol reacts with oleic acid. Circle the ester linkages in this triester molecule.

11.8 SYNTHETIC ORGANIC POLYMERS

As important as the simple organic molecules are, there is an equally important class of giant molecules made from these simpler molecules. These giant molecules, called **polymers** (*poly,* many; *mer,* part), occur in nature and chemists have learned to make them synthetically as well. Nature makes many different polymers, including cellulose in plants and proteins in both plants and animals. In spite of this, it is the synthetic polymers, mostly made from molecules derived from petroleum, that have had the biggest effect on the quality of human life in the past 100 years. It is impossible for us to get through a day without using a dozen or more synthetic organic polymers. Today, much of our clothing is made with synthetic polymers; food is packaged in polymers; and appliances and cars contain a number of components made of polymers. You or a friend or family member may be alive because of a medical application of polymers.

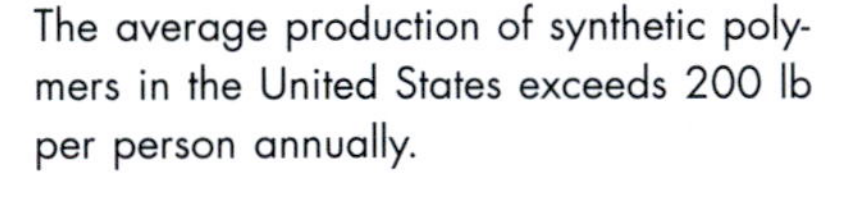
The average production of synthetic polymers in the United States exceeds 200 lb per person annually.

Polymers are so important that approximately 80% of the organic chemical industry is devoted to the production of synthetic polymers, and about half of the top 50 chemicals produced in the United States are used in making polymers of one kind or another.

Many synthetic organic polymers are plastics (Figure 11.10). There are two broad categories of plastics. One type, when heated, softens and flows; when it is cooled, it hardens again. Materials that undergo such reversible changes when heated and cooled are called **thermoplastics;** polyethylene (milk jugs), polystyrene (inexpensive sunglasses and toys), and polycarbonates (CD audio disks) are thermoplastics. The other type of plastic is a **thermosetting plastic.** When first heated, a thermosetting plastic flows like a thermoplastic, but when heated further it forms a rigid structure that will not remelt. Some kitchen counter tops, bowling balls, and football helmets are examples of items made of thermosetting plastics.

Some of the most useful polymers have resulted from copying natural polymers. Synthetic rubber, used in almost every automobile tire, is a copy of the molecule found in natural rubber. However, there are also many useful synthetic polymers that have no natural analogues. These include polystyrene, nylon, and Dacron.

Figure 11.10 Assorted plastic articles. *(C.D. Winters)*

Both synthetic and natural polymers are made by chemically joining many small molecules, called **monomers,** into one giant molecule or **macromolecule.** The resulting polymers have molar masses ranging from thousands to millions. In nature, polymerization is usually under the control of enzymes and the reactions take place rapidly at low temperatures. Making synthetic polymers often requires high temperatures and pressures and lengthy reaction times. Both synthetic and natural polymers can be classified as **addition polymers,** made by monomer units directly joined together, or **condensation polymers,** made by monomer units combining so that a small molecule, usually water, is split out between them.

A macromolecule is a molecule with a very high molecular weight.

Addition Polymers

The monomers for addition polymers normally contain one or more double bonds. The simplest monomer of this group is ethylene, $CH_2 = CH_2$, and its polymer is *polyethylene.* When ethylene is heated to 100 to 250 °C at a pressure of 1000 to 3000 atm in the presence of a catalyst, polymers with molar masses of up to several million may be formed. A polymer-building reaction of ethylene usually begins with breaking one of the bonds in the carbon-carbon double bond, so that each carbon atom has an unpaired electron. Such an electronic structure makes the molecule highly reactive. This step, the **initiation** of the polymerization, can be accomplished with chemicals such as organic peroxides that are unstable and easily break apart into free radicals *(←, p. 369),* which have unpaired electrons. The free radicals react readily with molecules containing carbon-carbon double bonds to produce new free radicals.

$$H_2C{:}{:}CH_2 \xrightarrow[\cdot OR]{\text{free radical}} \cdot CH_2{-}CH_2{-}OR$$

An organic peroxide, RO—OR, produces free radicals, RO·, each with an unpaired electron.

The growth of the polyethylene chain then begins, as the unpaired electron bonds to a double-bond electron in another ethylene molecule. This leaves an unpaired electron to bond with yet another ethylene molecule, and the process continues to form a huge polymer molecule:

$$\cdot CH_2{-}CH_2\cdot + \cdot CH_2{-}CH_2{-}OR \longrightarrow \cdot CH_2{-}CH_2{-}CH_2{-}CH_2{-}OR \xrightarrow{n\,CH_2 = CH_2} \left(\!{-}CH_2{-}CH_2{-}\!\right)_n$$

polyethylene
n ranges from 1000 to 50,000

Eventually, all the monomer molecules react and production of the polymer chain stops. The unsaturated hydrocarbon monomer, ethylene, is changed to a saturated hydrocarbon polymer, polyethylene.

Exercise 11.17 What Is at the Ends of the Polymer Chains?

What do you think is attached at the ends of the polymer chains when all of the ethylene molecules have been polymerized to form polyethylene?

Polyethylenes formed under various pressures and catalytic conditions have different molecular structures and hence different physical properties. For example, chromium oxide as a catalyst yields almost exclusively the linear polyethylene shown below—a polymer with no branches on the carbon chain. The zig-zag structure represents the shape of the chain more closely than does a linear drawing, because of the tetrahedral arrangement of bonds around each carbon in the saturated polyethylene chain.

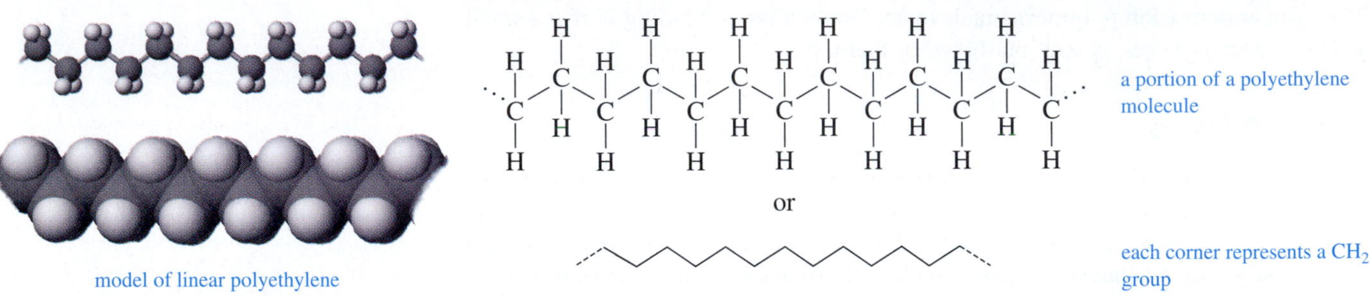

When ethylene is heated to 230 °C at a pressure of 200 atm without the chromium oxide catalyst, free radicals attack the chain at random positions, causing irregular branching [Figure 11.11(a)]. Other conditions can lead to cross-linked polyethylene, in which short branches connect long chains to each other.

Polyethylene is the world's most widely used polymer (Figure 11.12). Long, linear chains of polyethylene can pack closely together and give a material with high density (0.97 g/mL) and high molar mass, referred to as high-density polyethylene (HDPE). This material is hard, tough, and rigid. A plastic milk bottle is a typical application of HDPE.

Branched chains of polyethylene cannot pack closely together, so the resulting material has a lower density (0.92 g/mL) and is called low-density polyethylene (LDPE). This material is soft and flexible. Sandwich bags are made from LDPE. If the linear chains of polyethylene are treated in a way that causes short chains of —CH_2— groups to connect adjacent chains, Figure 11.11(b), the result is cross-linked polyethylene (CLPE), a very tough material.

The plastic caps on soft-drink bottles are made from CLPE.

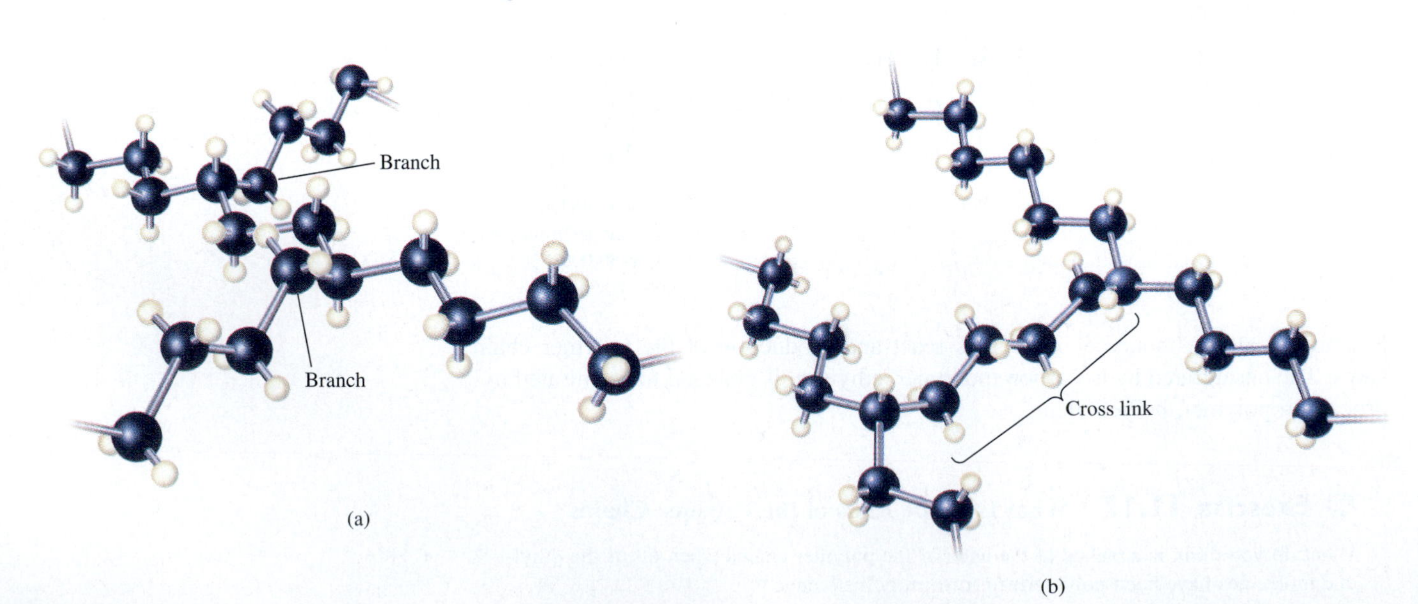

Figure 11.11 Models of (a) branched and (b) cross-linked polyethylene.

(a)

(b)

(c)

Figure 11.12 Polyethylene. (a) Production of polyethylene. (b), (c) The wide range of properties of different structural types of polyethylene leads to a variety of applications. *(a, Gary Gladstone/The Image Bank. b and c, The World of Chemistry (Program 22), "The Age of Polymers.")*

Many other different kinds of addition polymers are made from monomers in which one or more of the hydrogen atoms in ethylene have been replaced with either halogen atoms or a variety of organic groups, represented by X in the reaction below.

$$n\ \mathrm{H_2C{=}CHX} \longrightarrow \left(\!-\underset{\mathrm{H}}{\overset{\mathrm{H}}{\mathrm{C}}}-\underset{X}{\overset{\mathrm{H}}{\mathrm{C}}}-\!\right)_n$$

Generally n is 500 to 50,000 and this gives molecules with molar masses ranging from 10,000 to several million. The molecules that make up a given polymer sample are of different lengths and thus are not all of the same molar mass. As a result, only the average molar mass can be determined.

Table 11.11 p. 488 gives information on some of these polymers. Note how many of the monomers are common names. For example, the monomer for making polystyrene is styrene. In the polymer, n is typically about 5700.

$$n\ \mathrm{H_2C{=}CH(C_6H_5)} \longrightarrow \left(\!-\underset{\mathrm{H}}{\overset{\mathrm{H}}{\mathrm{C}}}-\underset{\mathrm{C_6H_5}}{\overset{\mathrm{H}}{\mathrm{C}}}-\!\right)_n$$

styrene　　　　polystyrene

Styrene is 20th on the list of top 50 chemicals, primarily because of its use in making polystyrene.

Polystyrene is a clear, hard, colorless solid at room temperature. It is a thermoplastic and can be molded easily at 250 °C. More than 5 billion pounds of polystyrene are produced annually in the United States alone to make food containers, toys, electrical parts, and many other items. The variation in properties shown by polystyrene products is typical of synthetic polymers. For example, a clear polystyrene drinking glass that is brittle and breaks into sharp pieces somewhat like glass is quite different from an expanded polystyrene coffee cup that is soft and pliable (Figure 11.13).

A major use of polystyrene is in the production of Styrofoam by "expansion molding." In this process, polystyrene beads are placed in a mold and heated with steam or hot air. The beads, 0.25 to 1.5 mm in diameter, contain 4% to 7% by weight of a low-boiling liquid such as pentane. The steam causes the low-boiling liquid to vaporize and expand the beads; as the foamed particles expand, they are molded in the shape of the mold cavity. Styrofoam is used for meat trays, coffee cups, and many kinds of packing material.

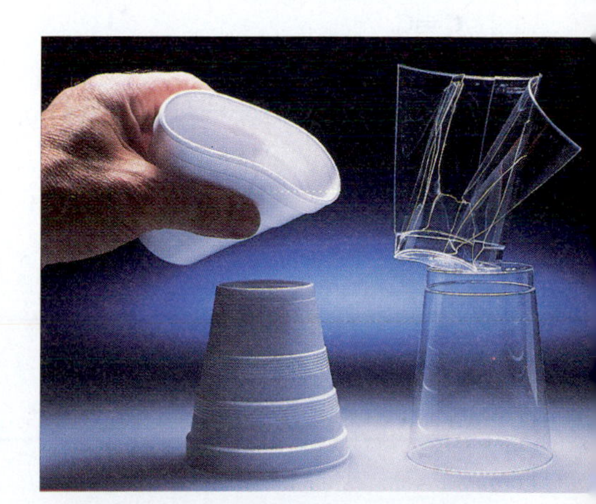

Figure 11.13 Polystyrene. Expanded polystyrene coffee cup (left) is soft. Clear polystyrene cup (right) is brittle. *(C.D. Winters)*

The numerous variations in substituents (length, branching, and crosslinking) make it possible to produce a variety of properties for each type of addition polymer. Chemists and chemical engineers can fine-tune the properties of the polymer to match the desired properties by appropriate selection of monomer and reaction conditions, thus accounting for the widespread and growing use of polymers.

TABLE 11.11 Ethylene Derivatives That Undergo Addition Polymerization

Formula	Monomer Common Name	Polymer Name (trade names)	Uses	U.S. Polymer Production (t/yr)
$H_2C{=}CH_2$	Ethylene	Polyethylene (Polythene)	Squeeze bottles, bags, films, toys and molded objects, electrical insulation	8 million
$H_2C{=}CH(CH_3)$	Propylene	Polypropylene (Vectra, Herculon)	Bottles, films, indoor-outdoor carpets	2.7 million
$H_2C{=}CH(Cl)$	Vinyl chloride	Poly(vinyl chloride) (PVC)	Floor tile, raincoats, pipe	3.5 million
$H_2C{=}CH(CN)$	Acrylonitrile	Polyacrylonitrile (Orlon, Acrilan)	Rugs, fabrics	1 million
$H_2C{=}CH(C_6H_5)$	Styrene	Polystyrene (Styrene, Styrofoam, Styron)	Food and drink coolers, building material insulation	2 million
$H_2C{=}CH{-}O{-}C({=}O){-}CH_3$	Vinyl acetate	Poly(vinyl acetate) (PVA)	Latex paint, adhesives, textile coatings	500,000
$H_2C{=}C(CH_3){-}C({=}O){-}O{-}CH_3$	Methyl methacrylate	Poly(methyl methacrylate) (Plexiglas, Lucite)	High-quality transparent objects, latex paints, contact lenses	450,000
$F_2C{=}CF_2$	Tetrafluoroethylene	Polytetrafluoroethylene (Teflon)	Gaskets, insulation, bearings, pan coatings	7000

CHEMISTRY YOU CAN DO

Making "Gluep"

White school glue, such as Elmer's glue, contains poly(vinyl acetate), water, and other ingredients. A "gluep" similar to Silly Putty can be made by mixing 1/2 cup of glue with 1/2 cup of water and then adding 1/2 cup of liquid starch and stirring. Work the mixture in your hands until it has a putty consistency. Roll it into a ball and let it sit on a flat surface undisturbed. What do you observe? Shape a piece into a ball and drop it on a hard surface. Does it bounce? The gluep can be stored in a sealed plastic bag for several weeks. Although gluep does not readily stick to clothes, walls, desks, or carpets, it does leave a water mark on wooden furniture, so be careful where you set it. Mold will form on the gluep after a few weeks, but the addition of a few drops of Lysol to the gluep will retard formation of mold.

PROBLEM-SOLVING EXAMPLE 11.7 Addition Polymers

Draw the structural formula of the repeating unit for each of the following addition polymers: (a) polypropylene, (b) poly(vinyl acetate), (c) poly(vinyl alcohol).

Answer

(a) $$\left(\begin{array}{cc} H & H \\ | & | \\ -C- & C- \\ | & | \\ H & CH_3 \end{array} \right)_n$$

(b) $$\left(\begin{array}{cc} H & H \\ | & | \\ -C- & C- \\ | & | \\ H & O \\ & | \\ & O{=}C \\ & | \\ & CH_3 \end{array} \right)_n$$

(c) $$\left(\begin{array}{cc} H & H \\ | & | \\ -C- & C- \\ | & | \\ H & OH \end{array} \right)_n$$

Explanation The names show that the monomers for these polymers are propylene ($CH_2{=}CHCH_3$), vinyl acetate ($CH_2{=}CHOOCCH_3$), and vinyl alcohol ($CH_2{=}CHOH$). The repeating units in the addition polymers therefore have the same structures, but without the double bonds.

Problem-Solving Practice 11.7

Draw the structural formulas of the monomers used to prepare the following addition polymers: (a) polyethylene; (b) poly(vinyl chloride); (c) polystyrene.

Natural and Synthetic Rubbers

Natural rubber, a product of the *Hevea brasiliensis* tree, is a hydrocarbon with the empirical formula C_5H_8. When rubber is decomposed in the absence of oxygen, the monomer 2-methyl-1,3-butadiene (isoprene) is obtained:

$$\begin{array}{c} H \quad\quad CH_3 \\ C{=}C \quad\quad H \\ H \quad\quad C{=}C \\ H \quad\quad H \end{array}$$

2-methyl-1,3-butadiene (isoprene)

Natural rubber occurs as *latex* (an emulsion of rubber particles in water) that oozes from rubber trees when they are cut. Precipitation of the rubber particles yields a gummy mass that is not only elastic and water-repellent but also very sticky, especially when warm. In 1839, after five years' work on natural rubber, Charles Goodyear (1800–1860) discovered that heating gum rubber with sulfur produces a material that is no longer sticky but is still elastic, water-repellent, and resilient.

Vulcanized rubber, as the type of rubber Goodyear discovered is now known, contains short chains of sulfur atoms that bond together the polymer chains of the natural rubber and reduce its unsaturation. The sulfur chains help to align the polymer chains, so the material does not undergo a permanent change when stretched but springs back to its original shape and size when the stress is removed. Substances that behave this way are called elastomers.

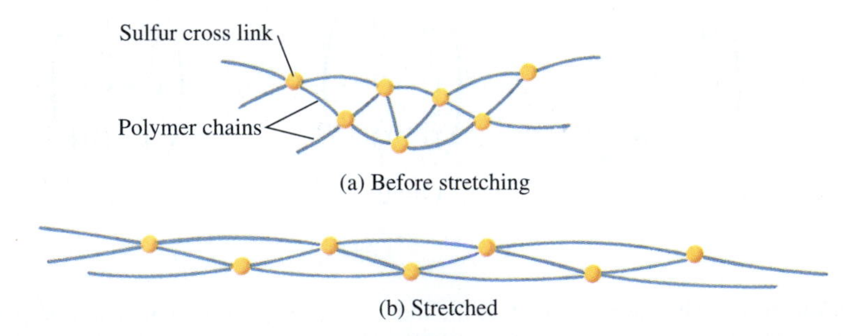

The behavior of natural rubber (polyisoprene) is due to the specific molecular geometry within the polymer chain. We can write the formula for polyisoprene with the CH_2 groups on opposite sides of the double bond (the *trans* arrangement):

$$\begin{array}{c} -CH_2 \quad H \quad \Big(CH_3 \quad CH_2{-}CH_2 \quad H \\ C{=}C \quad\quad C{=}C \quad\quad C{=}C \\ CH_3 \quad CH_2 + CH_2 \quad H \quad CH_3 \quad CH_2 + \Big)_n \end{array}$$

poly-*trans*-isoprene (the $-CH_2-CH_2-$ groups are *trans*)

or with the CH_2 groups on the same side of the double bond in a *cis* arrangement (⇐ *p. 373*):

poly-*cis*-isoprene (the $—CH_2—CH_2—$ groups are *cis*)

Natural rubber is poly-*cis*-isoprene. However, the *trans* material also occurs in nature in the leaves and bark of the sapotacea tree and is known as *gutta-percha.* It is brittle and hard and is used for golf-ball covers and electrical insulation.

In 1955, chemists at the Goodyear and Firestone companies almost simultaneously discovered how to prepare synthetic poly-*cis*-isoprene. This material is structurally identical to natural rubber. Today, synthetic poly-*cis*-isoprene can be manufactured cheaply and is used almost equally well (there is still an increased cost) when natural rubber is in short supply. More than 2.4 million tons of synthetic rubber are produced in the United States every year.

Rubber bumper. Bumpers on a new car are made from synthetic rubber polymers. *(C.D. Winters)*

Many commercially important addition polymers are **copolymers,** polymers obtained by polymerizing a mixture of two or more different monomers. A copolymer of styrene with butadiene is the most important synthetic rubber produced in the United States. More than 1.4 million tons of styrene-butadiene rubber (SBR) are produced each year in the United States for making tires. A 3:1 mole ratio of butadiene to styrene is used to make SBR.

$3n$ 1,3-butadiene $+$ n styrene $\xrightarrow[\text{polymerization}]{\text{addition}}$

styrene-butadiene rubber (SBR)

Other important copolymers are made by polymerizing mixtures of ethylene and propylene, or acrylonitrile, butadiene, and styrene (also called ABS). Saran Wrap is an example of a copolymer of vinyl chloride with 1,1-dichloroethylene.

ABS polymers are used to make automobile bumpers and cases for portable computers.

Condensation Polymers

A chemical reaction in which two molecules combine to form a larger molecule, simultaneously producing a small molecule (such as water) is called a **condensation reaction.** The reactions of alcohols with carboxylic acids to give esters (⇐, *p. 479*) are examples of condensation. This important type of chemical reaction does not depend on the presence of a double bond in the reacting molecules. Rather, it requires the presence of two different kinds of functional groups on two different molecules. If each reacting molecule has two functional groups, both of which can react, it is possible for condensation reactions to produce long-chain polymers.

A molecule with two carboxylic acid groups, such as terephthalic acid, and another molecule with two alcohol groups, such as ethylene glycol, can react with each other at both ends.

$$2\,HO{-}\overset{\overset{\displaystyle O}{\|}}{C}{-}C_6H_4{-}\overset{\overset{\displaystyle O}{\|}}{C}{-}OH + 2\,HO{-}CH_2{-}CH_2{-}OH \longrightarrow$$

terephthalic acid   ethylene glycol

$$HO{-}\overset{\overset{\displaystyle O}{\|}}{C}{-}C_6H_4{-}\overset{\overset{\displaystyle O}{\|}}{C}{-}O{-}CH_2{-}CH_2{-}O{-}\overset{\overset{\displaystyle O}{\|}}{C}{-}C_6H_4{-}\overset{\overset{\displaystyle O}{\|}}{C}{-}O{-}CH_2{-}CH_2{-}OH + 2\,H_2O$$

If n molecules of acid and alcohol react in this manner, the process will continue until a large polymer molecule, known as a **polyester,** is produced.

$$\left(CH_2{-}CH_2{-}O{-}\overset{\overset{\displaystyle O}{\|}}{C}{-}C_6H_4{-}\overset{\overset{\displaystyle O}{\|}}{C}{-}O \right)_n$$

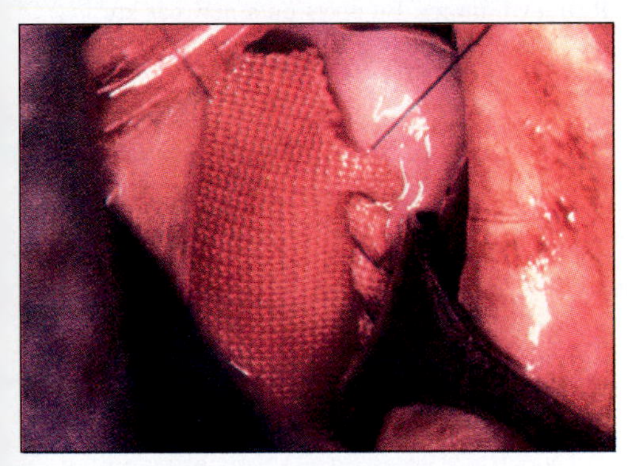

Medical uses of Dacron. A Dacron patch is used to close an atrial septal defect in a heart patient. *(Courtesy of Drs. James L. Monro and Gerald Shore and the Wolfe Medical Publications, London, England.)*

Amines are classified as primary, secondary, or tertiary according to how many of the H atoms in the $-NH_2$ group are replaced by alkyl groups: RNH_2, primary; R_2NH, secondary; and R_3N tertiary.

Amides contain the $-\overset{\overset{\displaystyle O}{\|}}{C}-\overset{\overset{\displaystyle H}{|}}{N}-$ functional group.

More than 2 million tons of poly(ethylene terephthalate), commonly referred to as PET, are produced in the United States each year for use in making beverage bottles, apparel, tire cord, film for photography and magnetic recording, food packaging, coatings for microwave and conventional ovens, and home furnishings. A variety of trade names are associated with the various applications. Polyester textile fibers are marketed under such names as Dacron and Terylene. Films of the same polyester, when magnetically coated, are used to make audio and TV tapes. This film, Mylar, has unusual strength and can be rolled into sheets one-thirtieth the thickness of a human hair.

The inert, nontoxic, noninflammatory, and non-blood-clotting characteristics of Dacron polymers make Dacron tubing an excellent substitute for human blood vessels in heart bypass operations. Dacron sheets are also used as a skin substitute for burn victims.

Another useful and important type of condensation reaction is that between a carboxylic acid and a primary **amine,** which is an organic compound containing an $-NH_2$ functional group. Amines can be considered derivatives of ammonia (NH_3), and most of them are weak bases, similar in strength to ammonia. An amine reacts with a carboxylic acid at high temperature to split out a water molecule and form an **amide:**

$$R{-}\overset{\overset{\displaystyle O}{\|}}{C}{-}OH + H{-}\underset{\underset{\displaystyle H}{|}}{N}{-}R \xrightarrow{\text{heat}} R{-}\overset{\overset{\displaystyle O}{\|}}{C}{-}\underset{\underset{\displaystyle H}{|}}{N}{-}R + H_2O$$

carboxylic acid   amine   amide

Polymers are produced when diamines (compounds containing two $-NH_2$

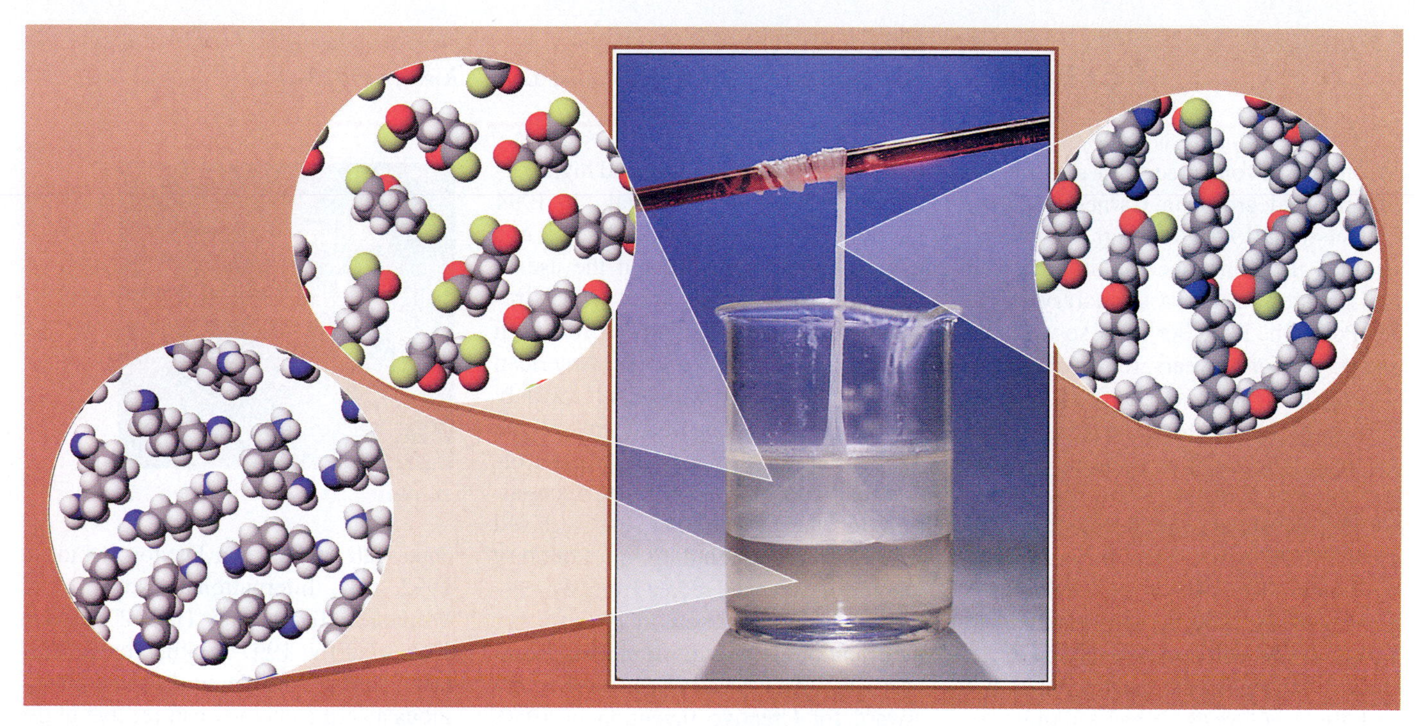

Figure 11.14 Making nylon-66. Hexamethylenediamine is dissolved in water (*bottom layer*), and a derivative of adipic acid (adipoyl chloride) is dissolved in hexane (*top layer*). The two compounds mix at the interface between the two layers to form nylon, which is being wound onto a stirring rod. *(C.D. Winters)*

groups) react with dicarboxylic acids (compounds containing two —COOH groups). Reactions of this type yield a group of polymers that are the **polyamides,** or nylons.

In 1928, the Du Pont Company embarked on a program of basic research headed by Dr. Wallace Carothers (1896–1937), who came to DuPont from the Harvard University faculty. His research interests were high-molar-mass compounds, such as rubber, proteins, and resins, and the reaction mechanisms that produced these compounds. In February 1935, his research yielded a product known as nylon-66 (Figure 11.14), prepared from adipic acid (a diacid) and hexamethylenediamine (a diamine):

$$n\,\mathrm{HO{-}\overset{\overset{\displaystyle O}{\|}}{C}{-}(CH_2)_4{-}\overset{\overset{\displaystyle O}{\|}}{C}{-}OH} + n\,\mathrm{H_2N{-}(CH_2)_6{-}NH_2} \rightarrow$$

adipic acid hexamethylenediamine

$$\mathrm{{-}\overset{\overset{\displaystyle O}{\|}}{C}{-}(CH_2)_4{-}\overset{\overset{\displaystyle O}{\|}}{C}\!\left(\underset{\underset{\displaystyle H}{|}}{N}{-}(CH_2)_6{-}\underset{\underset{\displaystyle H}{|}}{N}{-}\overset{\overset{\displaystyle O}{\|}}{C}{-}(CH_2)_4{-}\overset{\overset{\displaystyle O}{\|}}{C}\right)\!\underset{\underset{\displaystyle H}{|}}{N}{-}(CH_2)_6{-}\overset{\overset{\displaystyle H}{|}}{N}{-}} + n\,\mathrm{H_2O}$$

nylon

The name of nylon-66 is based on the number of carbon atoms in the diamine and diacid, respectively, that are used to make the polymer. Since both hexamethylenediamine and adipic acid have six carbon atoms, the product is nylon-66.

Portrait of a Scientist • *Stephanie Louise Kwolek (1923–)*

Stephanie Kwolek received a Bachelor of Science degree from Carnegie-Mellon University in 1946. Although she wanted to study medicine, she couldn't afford it and decided to take a temporary job at Du Pont. She liked her work so well that she stayed for 40 years, retiring in 1986. During her career at Du Pont, she received 17 U.S. patents and 86 foreign patents for her work on a variety of polymeric fibers. However, she is best known for her work on the development of Kevlar fiber, which is five times stronger than steel. The patent for Kevlar fiber was issued in 1965, but 15 years passed before it was fully commercialized. Although Kevlar is best known for its use in bulletproof vests, it has a number of other important uses that include brake linings, underwater cables, and high-performance composite materials. In 1994 Du Pont featured Stephanie Kwolek in a television commercial about the use of Kevlar in bulletproof vests, which resulted in name recognition for both Kevlar and Kwolek by the American public. However, what isn't widely known are the accomplishments of the woman during a time when women often didn't receive appropriate recognition for their work. She has received many awards, including an honorary Doctor of Science Degree from Worcester Polytechnic Institute in 1981 for her contributions to polymer and fiber chemistry, the American Chemical Society Award for Creative Invention in 1980, and election to the Engineering and Science Hall of Fame in Dayton, Ohio, in 1992, the International Network of Women in Technology Hall of Fame in 1996, and the 1997 Perkin Medal. This last award is considered the most prestigious award a chemist can receive in the United States.

(Du Pont)

The amide linkage in nylon is the same linkage found in proteins, where it is called the peptide linkage.

Wool and silk are examples of nature's version of nylon. However, these natural polymers have only one carbon between each pair of $-\overset{O}{\overset{\|}{C}}-\overset{H}{\overset{|}{N}}-$ units instead of the half dozen or so found in synthetic nylons.

This material can easily be extruded into fibers that are stronger than natural fibers and chemically more inert. The discovery of nylon jolted the American textile industry at almost precisely the right time. Natural fibers were not meeting the needs of 20th-century Americans. Silk was not durable and was very expensive, wool was scratchy, linen crushed easily, and cotton did not have a high-fashion image. All four had to be pressed after cleaning. As women's hemlines rose in the mid-1930s, silk stockings were in great demand, but they were very expensive and short-lived. Nylon changed all that almost overnight. It could be knitted into the sheer hosiery women wanted, and it was much more durable than silk. The first public sale of nylon hose took place in Wilmington, Delaware (the hometown of Du Pont's main office), on October 24, 1939. World War II caused all commercial use of nylon to be abandoned until 1945, as the industry turned to making parachutes and other war materials. Not until 1952 was the nylon industry able to meet the demands of the hosiery industry and to release nylon for other uses as a fiber and as a thermoplastic.

Figure 11.15 illustrates another facet of the structure of nylon—hydrogen bonding—which explains why nylons make such good fibers. To have good tensile strength, the chains of atoms in a polymer should be able to attract one another, but not so strongly that the plastic cannot be initially extended to form the fibers. Ordinary co-valent chemical bonds linking the chains together would be too strong. Hydrogen bonds, with a strength of about one-tenth that of an ordinary covalent bond (⇐, ***p. 426***), link the chains in the desired manner.

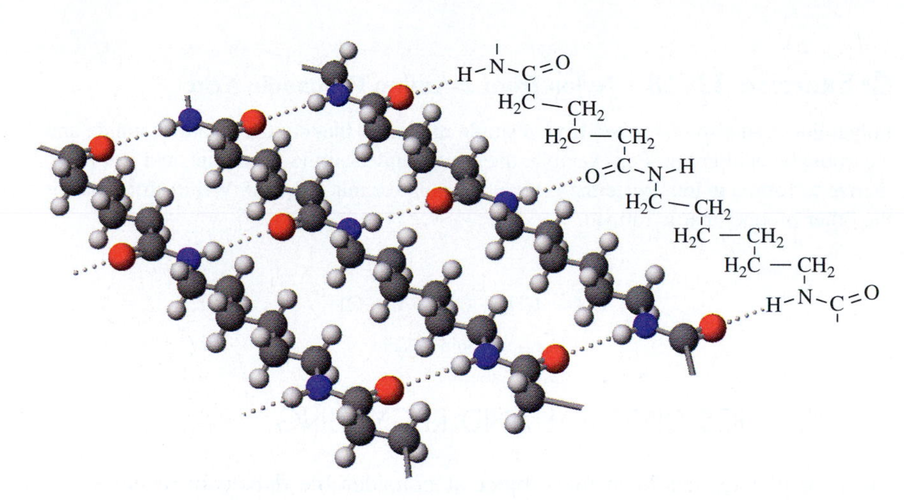

Figure 11.15 Hydrogen bonding in nylon-66. Oxygen atoms are shown in red, nitrogen atoms are blue, hydrogens atoms are white, and carbon atoms are black.

Kevlar, another polyamide, is used to make bulletproof vests (Figure 11.16) and fireproof garments. Kevlar is made from *p*-phenylenediamine and terephthalic acid.

$$H_2N-C_6H_4-NH_2$$

p-phenylenediamine

PROBLEM-SOLVING EXAMPLE 11.8 Condensation Polymers

Write the repeating unit of the condensation polymer obtained by combining $HOOCCH_2CH_2COOH$ and $H_2NCH_2CH_2NH_2$.

Answer

$$\left(\overset{O}{\overset{\|}{C}}-CH_2-CH_2-\overset{O}{\overset{\|}{C}}-\underset{H}{\underset{|}{N}}-CH_2-CH_2-\underset{H}{\underset{|}{N}} \right)_n$$

Explanation A condensation polymer composed of a diacid and a diamine forms by loss of water between monomers to give an amide bond. The repeating unit contains one diacid part and one diamine part.

Problem-Solving Practice 11.8

Draw the structure of the repeating unit of the condensation polymer obtained from the reaction of terephthalic acid with ethylene glycol.

Figure 11.16 Kevlar vest. Vests made of Kevlar have saved many police officers' lives. (Kevlar is a registered trademark of DuPont.)

Exercise 11.18 Nylon from 2-Amino Propanoic Acid

Polyamides can also be formed from a single monomer that contains both an amide and a carboxylic acid group. For example, the compound 2-amino propionic acid can polymerize to form a nylon. Write the general formula for this polymer. Write a formula for the other product that is formed.

$$H_2N—CH_2—CH_2—\overset{\overset{\displaystyle O}{\|}}{C}—OH$$

2-aminopropionic acid

11.9 PLASTICS DISPOSAL AND RECYCLING

Disposal of plastics has been the subject of considerable debate in recent years as municipalities face increasing problems in locating sufficient nearby landfill space. The number one waste is paper products, which make up about 40% of the volume in landfills. (Newspaper alone accounts for 16% of the volume.) Next are plastics, which make up about 20% of the volume in landfills.

One alternative to landfill disposal of plastics is to incinerate them. In the past, communities have built huge incinerators to burn municipal trash, including plastics. When a plastic that consists entirely of hydrocarbons is burned efficiently, the only combustion products are carbon dioxide and water. If the combustion is inefficient, unburned hydrocarbons are produced as well as many partially oxygenated compounds like aldehydes, ketones, and acids, which are just as damaging to air quality as unburned hydrocarbons from automobile engines. If the plastic contains atoms other than hydrogen and carbon, other combustion products result. PVC contains chlorine, so when it burns, chlorinated organic compounds are produced. Many of these chlorinated compounds are known to be highly toxic or suspected to be carcinogenic. In many communities, incinerating plastics is being challenged because it is considered too risky for public safety.

Recycling plastics. By recycling plastics we save landfill space, eliminate some air pollutants, and conserve petroleum reserves. *(C.D. Winters)*

Exercise 11.19 Combustion Products from Burning Plastics

Using the formulas for four different plastics found in this book, write down the formulas for the principal combustion products when these plastics are burned.

If you cannot landfill plastics and you cannot burn them, why not recycle them? In 1994 about 19% of all plastic containers were recycled. This compares to about 55% of aluminum and 20% of glass containers. About 35% of the paper used in 1994 (mostly newsprint, cardboard, and magazines) was recycled.

Four phases are needed for the successful recycling of any waste material: collection, sorting, reclamation, and end use. Public enthusiasm for recycling and state laws requiring recycling have resulted in a dramatic increase in the collection of recycled items. Between 1989 and 1991, the number of U.S. households with curbside collection of recyclables increased from 9 million to 16 million, which is estimated to be 29% of U.S. households. Codes are stamped on plastic containers to help consumers identify and sort their recyclable plastics (Figure 11.17).

Polyethylene terephthalate (PET), widely used for soft-drink bottles, is the most commonly recycled plastic. The used bottles are available from retailers in states requiring refundable deposits or from those offering curbside pickups. Over 490 mil-

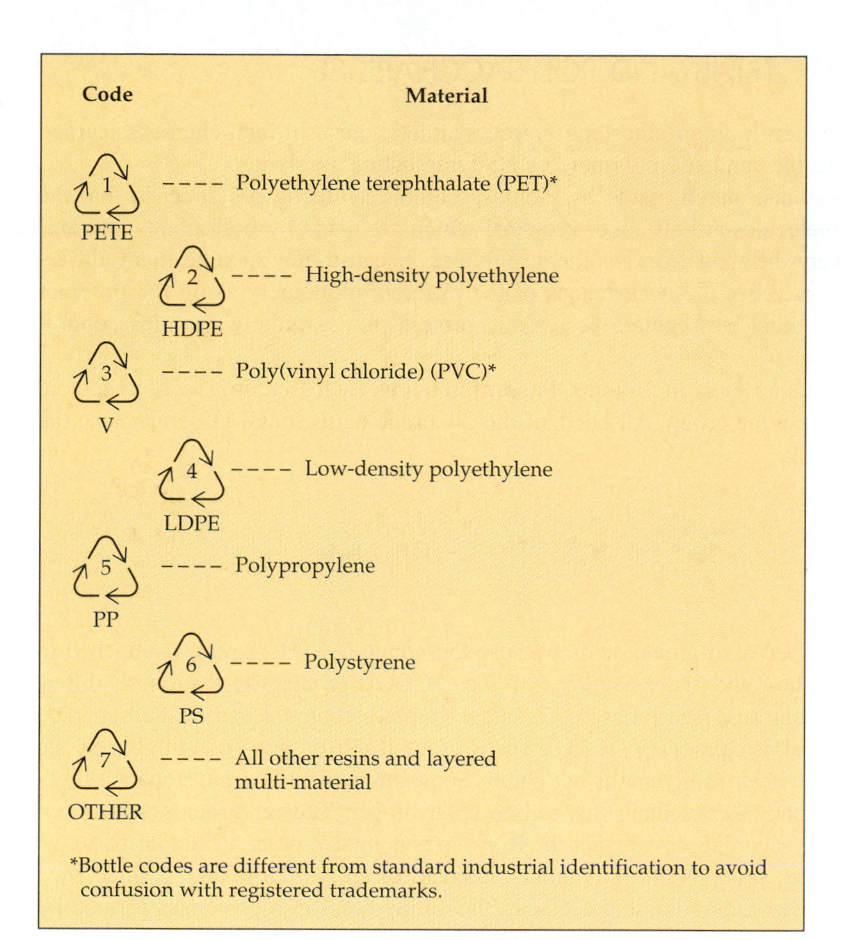

Code
Material
1 PETE ---- Polyethylene terephthalate (PET)*
2 HDPE ---- High-density polyethylene
3 V ---- Poly(vinyl chloride) (PVC)*
4 LDPE ---- Low-density polyethylene
5 PP ---- Polypropylene
6 PS ---- Polystyrene
7 OTHER ---- All other resins and layered multi-material
*Bottle codes are different from standard industrial identification to avoid confusion with registered trademarks.

Figure 11.17 **Plastic container codes.**

lion pounds of PET soft-drink bottles were recycled in the United States in 1994 (an 18% increase over the previous year). Major end uses for recycled PET include fiber-fill for ski jackets and sleeping bags, carpet fibers, and tennis balls. Coca-Cola is using two-liter bottles made of 25% recycled PET.

High-density polyethylene (HDPE) is the second most widely recycled plastic, with 304 million pounds processed in 1994 (an 18% increase over the previous year). Milk, juice, and water jugs are the principal source of recycled HDPE. Some products made from recycled HDPE are trash containers, drainage pipes, garbage bags, and fencing.

Although recycling of plastics has shown a dramatic increase in recent years, recycling companies will not be able to increase the percentage of recycled plastics to the 50% goal by the year 2000 without a significant increase in the demand for products made partially or completely from recycled materials.

Milk in plastic containers. How many of these will be recycled? *(C.D. Winters)*

Recycled plastics. A planter made of plastic lumber, which can be worked like wood, but does not rot, splinter, need paint, or get eaten by termites. A mixture of several kinds of recycled plastics is the raw material for this lumber. *(Courtesy of Obex, Inc., Stamford, CT)*

11.10 PROTEINS AND POLYSACCHARIDES

Natural polymers were around long before synthetic ones. In fact, chemists learned how to make the synthetic polymers by studying nature's polymers.

Cellulose and starch, made by plants, resemble synthetic polymers in that the monomer molecules are all alike. Proteins, which are made by both plants and animals, are very different from synthetic polymers because they include many different monomers. Also, the occurrences of the different monomers along the polymer chain are anything but regular. As a result, proteins are extremely complex copolymers.

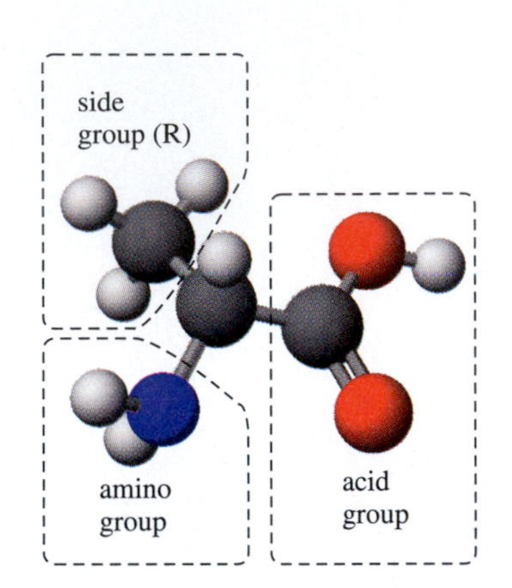

The monomer units in proteins, the **amino acids,** each contain a carboxylic acid group and an amine group. All but 1 of the 20 amino acids found in nature have the general formula

$$H_2N-\overset{\overset{\displaystyle R}{|}}{\underset{\underset{\displaystyle H}{|}}{C^*}}-\overset{\overset{\displaystyle O}{\|}}{C}-OH$$

and are described as α-amino acids because the amino ($-NH_2$) group is attached to the **alpha carbon,** the first carbon next to the —COOH group. The R group is different in each amino acid (see Table 11.12), and the alpha carbon, the carbon atom marked with the asterisk is chiral *(⇐, p. 430).* The R group is a hydrogen atom in glycine, the simplest amino acid, so glycine is not chiral. Some amino acid R groups contain functional groups, and others contain only carbon and hydrogen. The amino acids are grouped in Table 11.12 according to whether the R group is nonpolar, polar, acidic, or basic.

The acid and base properties of the —COOH and $-NH_2$ functional groups are discussed in Section 17.2.

Like nylon, proteins are polyamides. Instead of an amide linkage formed by a reaction between a diamine and a diacid, the amide bond in a protein is formed by the reaction between the amine group of one amino acid and the carboxylic acid group of another.

$$H_2N-\overset{\overset{\displaystyle R}{|}}{\underset{\underset{\displaystyle H}{|}}{C}}-\overset{\overset{\displaystyle O}{\|}}{C}-OH + H-\underset{\underset{\displaystyle H}{|}}{N}-\overset{\overset{\displaystyle R}{|}}{\underset{\underset{\displaystyle H}{|}}{C}}-\overset{\overset{\displaystyle O}{\|}}{C}-OH \xrightarrow{-H_2O} H_2N-\overset{\overset{\displaystyle R}{|}}{\underset{\underset{\displaystyle H}{|}}{C}}-\overset{\overset{\displaystyle O}{\|}}{C}-\underset{\underset{\displaystyle H}{|}}{N}-\overset{\overset{\displaystyle R}{|}}{\underset{\underset{\displaystyle H}{|}}{C}}-\overset{\overset{\displaystyle O}{\|}}{C}-OH$$

a peptide bond

In proteins the amide bond is called a **peptide bond,** and relatively small amino acid polymers (up to about 50 amino acids) are known as **polypeptides.** Proteins are polypeptides containing hundreds to thousands of amino acids bonded together.

Exercise 11.20 Hydrogen Bonding Between Amino Acids in Proteins

Use Table 11.12 and pick two amino acids that could hydrogen bond with one another if they were close together in a protein chain or in two adjacent protein chains. Then pick two that would not hydrogen bond under similar circumstances.

TABLE 11.12 Common L-Amino Acids Found in Proteins (The R group in each amino acid is highlighted.)

Amino Acid	Abbreviation	Structure	Amino Acid	Abbreviation	Structure
Nonpolar R Groups					
Glycine	Gly	$H-CH-COOH$ NH_2	*Isoleucine	Ile	$CH_3-CH_2-CH-CH-COOH$ CH_3 NH_2
Alanine	Ala	$CH_3-CH-COOH$ NH_2	Proline	Pro	H_2C-CH_2 H_2C $CHCOOH$ N H
*Valine	Val	$CH_3-CH-CH-COOH$ CH_3 NH_2	*Phenylalanine	Phe	(benzene ring)$-CH_2-CH-COOH$ NH_2
*Leucine	Leu	$CH_3-CH-CH_2-CH-COOH$ CH_3 NH_2	*Methionine	Met	$CH_3-S-CH_2CH_2-CH-COOH$ NH_2
			*Tryptophan	Trp	(indole ring)$-CH_2-CH-COOH$ NH_2 N H
Polar but Neutral R Groups					
Serine	Ser	$HO-CH_2-CH-COOH$ NH_2	Asparagine	Asn	$H_2N-C-CH_2-CH-COOH$ O NH_2
*Threonine	Thr	$CH_3-CH-CH-COOH$ OH NH_2	Glutamine	Gln	$H_2N-C-CH_2CH_2-CH-COOH$ O NH_2
Cysteine	Cys	$HS-CH_2-CH-COOH$ NH_2	Tyrosine	Tyr	HO(benzene ring)$-CH_2-CH-COOH$ NH_2
Acidic R Groups			**Basic R Groups**		
Glutamic acid	Glu	$HO-C-CH_2CH_2-CH-COOH$ O NH_2	*Lysine	Lys	$H_2N-CH_2CH_2CH_2CH_2-CH-COOH$ NH_2
Aspartic acid	Asp	$HO-C-CH_2-CH-COOH$ O NH_2	†Arginine	Arg	$H_2N-C-NH-CH_2CH_2CH_2-CH-COOH$ NH NH_2
			Histidine	His	(imidazole ring: N, $N-H$)$-CH_2-CH-COOH$ NH_2

*Essential amino acids that must be part of the human diet. The other amino acids can be synthesized by the body.

†Growing children also require arginine in their diet.

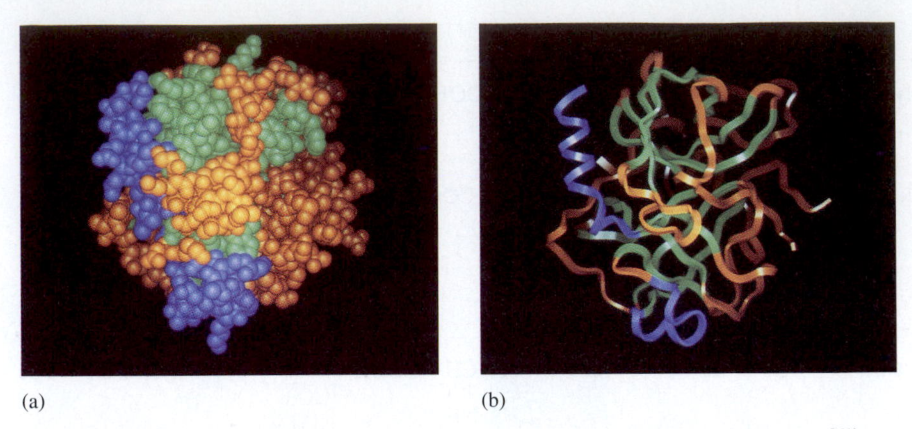

(a) (b)

Figure 11.18 Protein folding. The protein chymotrypsin is shown (a) in a space-filling model that illustrates how close together the atoms are in the molecule, and (b) in a ribbon structure that illustrates only the polypeptide backbone.

As the length of the polypeptide chain increases, the number of variations in the sequence of amino acids quickly increases to provide a degree of complexity that is not seen in synthetic polymers. Six *tri*peptides are possible if three different amino acids (for example, glycine, Gly; alanine, Ala; and serine, Ser) are linked in combinations that contain all three amino acids if each is used only once. They are

Gly-Ala-Ser	Ser-Ala-Gly	Ala-Ser-Gly
Gly-Ser-Ala	Ser-Gly-Ala	Ala-Gly-Ser

If n amino acids are all different, the number of arrangements is $n!$ (n factorial). For four different amino acids, the number of different arrangements is 4!, or $4 \times 3 \times 2 \times 1 = 24$. For five different amino acids, the number of different arrangements is 5!, or 120. If all 20 different naturally occurring amino acids were bonded in one peptide, the sequences would make 2.43×10^{18} (2.43 quintillion) unique 20-monomer molecules! *Since proteins can also include more than one molecule of a given amino acid, the possible combinations are effectively infinite.* Furthermore, many proteins incorporate other kinds of molecules and in some cases metal ions. It is truly remarkable that of the many different proteins that could be made from a set of amino acids, a living cell makes only the relatively small number it needs.

Because of the complexity and the variety of properties provided by the different R groups and associated molecules or ions, proteins are able to perform widely diverse functions in the body. Consider some of them: as enzymes, proteins make biochemical reactions possible (Section 12.9); as immunoglobulins, they fight off disease; as hormones, they start up and shut down vital processes according to the need at the moment; and as muscles, they provide flexibility and strength. The functions of proteins are intimately related to the shapes of these huge molecules. The overall shape of each different protein molecule is determined by interactions among the side chains (Figure 11.18).

Problem-Solving Example 11.9 Peptides

Draw the structural formula of the tripeptide represented by Ala-Ser-Gly. Explain why this is a different compound from that with the amino acids joined in the order Gly-Ala-Ser.

Answer

```
      H  O        H    O       H  O
      |  ||       |    ||      |  ||
H2N—C—C—N—C——C—N—C—C—OH
      |       |  |          |  |
     CH3      H CH2OH       H  H
```

Ala-Ser-Gly

The structure Gly-Ala-Ser differs because the free —NH_2 group is on the glycine part of the molecule and the free —COOH group is on the serine part of the molecule.

Explanation The amino acid sequence in the abbreviated name shows that alanine should be written at the left with a free H_2N— group, glycine should be written at the right with a free —COOH group, and both should be connected to serine by peptide bonds. Writing the structure of Gly-Ala-Ser shows how the two tripeptides are different.

```
      H  O      H  O      H  O
      |  ||     |  ||     |  ||
H2N—C—C—N—C—C—N—C—C—OH
      |      |  |      |  |
      H      H CH3     H CH2OH
```

Gly-Ala-Ser

Problem-Solving Practice 11.9

Draw the structural formula of the tetrapeptide Cys-Phe-Ser-Ala.

Exercise 11.21 Peptide Sequences

Draw the structural formula of the tetrapeptide Ala-Ser-Phe-Cys.

Polysaccharides

Nature makes an abundance of compounds with the general formula $C_x(H_2O)_y$. These compounds are variously known as sugars, carbohydrates, and mono-, di-, or polysaccharides (from the Latin *saccharum,* "sugar"—because they taste sweet). All of them, like glucose (a monosaccharide) and sucrose (a disaccharide, ⇐, ***p. 109***), contain alcohol and aldehyde, or ketone functional groups. Disaccharides and polysaccharides are formed by the loss of a molecule of water between the —OH groups of adjacent monosaccharides. The reaction is a condensation and, like the condensation reactions by which synthetic polymers are formed, can produce very large molecules—the polysaccharides.

The most abundant of nature's polysaccharides are starches and cellulose found in plants and glycogen found in animals. Each of these polymers is composed solely of glucose monomers and may contain up to 5000 glucose units. To illustrate as closely as possible the shape of the glucose molecules in these polymers, the six-membered rings in monosaccharides are often written as shown on the next page. These structures illustrate the difference between the two possible structures for glucose, a feature important in understanding the structural differences of the polysaccharides. Notice the different positions of the pairs of —OH groups shown in color.

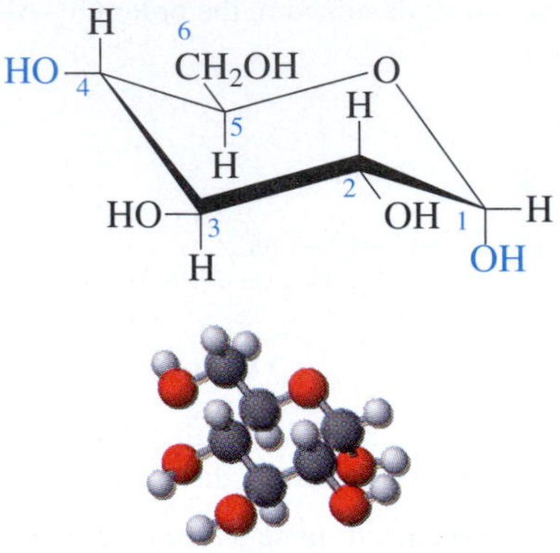

α-D-glucose 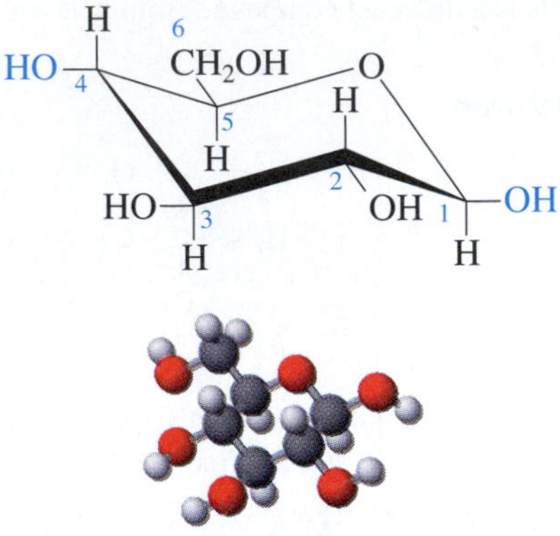 β-D-glucose

The molecule with the —OH groups on the same side of the ring is known as α-D-glucose, and the molecule with the —OH groups on opposite sides of the ring is known as β-D-glucose. (The D signifies that these are chiral molecules.)

Starches and Glycogen

When tested with iodine solution, amylose turns blue-black, whereas amylopectin turns red.

Plant starch is stored in protein-covered granules until glucose is needed for synthesis of new molecules or for energy production. If these granules are ruptured by heat, they yield a starch *amylose,* which is soluble in hot water, and an insoluble starch, *amylopectin.* Structurally, amylose is a straight-chain condensation polymer with an average of about 200 α-D-glucose monomers per molecule. A representative portion of the structure of amylose is shown in Figure 11.19. A typical amylopectin molecule has about 1000 α-D-glucose monomers arranged into branched chains analogous to the branched-chain synthetic polymers discussed earlier. Just as these polymers have different properties from their straight-chain counterparts, amylopectin is different from amylose. The main difference is in their different water solubilities. The hydrolysis of amylopectin first yields a mixture of small branched-chain polysaccharides called dextrins. Dextrins are used as food additives and in mucilage, paste, and finishes for paper and fabrics. Complete hydrolysis yields α-D-glucose.

Animals store energy as fats rather than carbohydrates.

Conceptual Challenge Problem CP-11.C at the end of the chapter relates to topics covered in this section.

In animals, **glycogen** serves the same storage function as starch does in plants. The α-glucose chains in glycogen are more highly branched than the chains in amylopectin. Glycogen is stored in the liver and muscle tissues and provides glucose for "instant" energy until the process of fat metabolism can take over and serve as the energy source (⇐, *p. 109*).

Exercise 11.22 Chain Branching Possibilities

Look at the structural formula of an α-D-glucose molecule and explain where chain branching might take place.

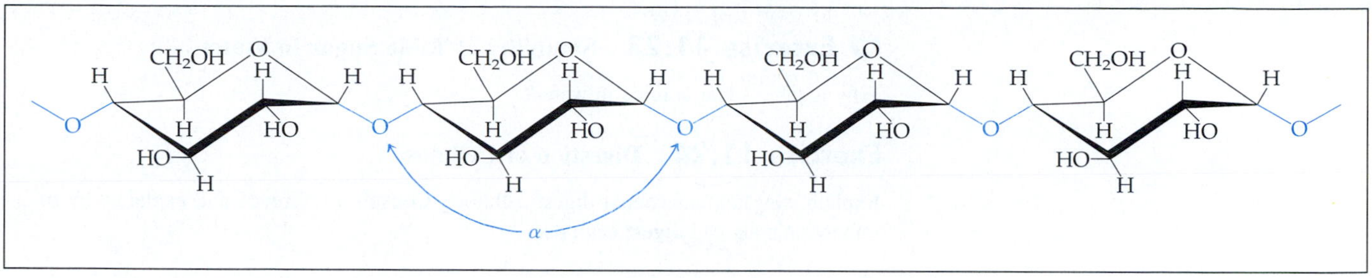

Figure 11.19 **The structure of amylose.** From 60 to 300 α-D-glucose units are bonded together by α linkages to form amylose. Only the α structure is used in forming α linkages. The —OH groups on carbon atoms one and four are *cis* in α-glucose, so all the bond linkages point in the same direction.

Cellulose

Cellulose is the most abundant organic compound on earth, and its purest natural form is cotton. This polysaccharide is also found as the woody part of trees and the supporting material in plants and leaves. Like amylose, it is composed of D-glucose units. The difference between the structures of cellulose and amylose lies in the bonding between the D-glucose units; in cellulose all of the glucose units are in the β-ring form (Figure 11.20), whereas in amylose they are in the α-ring form. This subtle structural difference between starch and cellulose causes their differences in digestibility. Because cellulose is so abundant, it would be advantageous if humans could use it, as well as starch, for food. Unfortunately, we cannot digest cellulose because we lack the enzyme necessary to break the β-1,4 bonds. However, termites, a few species of cockroaches, and ruminant mammals such as cows, sheep, goats, and camels, do have the proper internal chemistry for this purpose.

Paper and cotton are principally cellulose.

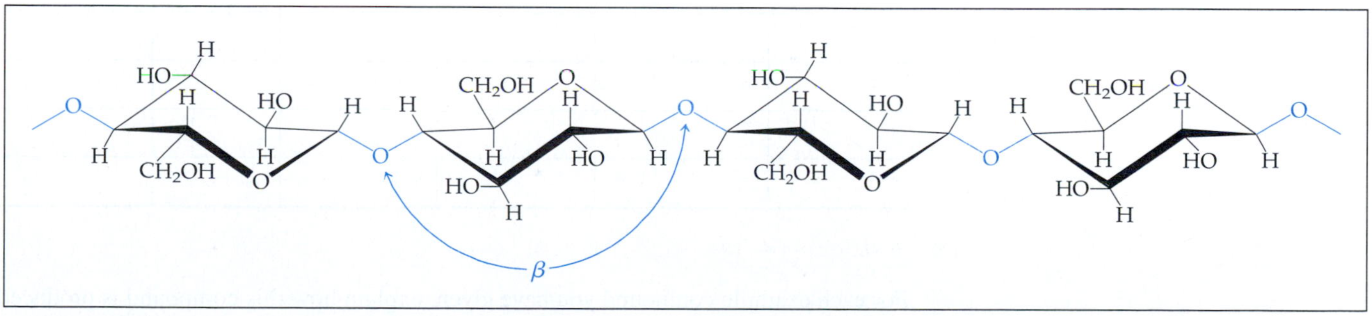

Figure 11.20 **Cellulose structure.** About 280 β-D-glucose units are bonded together by β linkages to form a chain structure in cellulose. Cellulose contains only the β form of glucose. The —OH groups on carbon atoms 1 and 4 are *trans* in β glucose, so the bonds between the glucose monomers alternate in direction. In effect, every other glucose unit in cellulose is turned over, whereas in amylose (Figure 11.19), all the glucose units are in the same orientation relative to one another.

Exercise 11.23 Solubility of Table Sugar in Water

Why is table sugar soluble in water?

Exercise 11.24 Digestion of Cellulose

Explain why humans cannot digest cellulose. Consult a reference and explain why ruminant animals can digest cellulose.

Exercise 11.25 What If Humans Could Digest Cellulose?

Think of some implications if humans had the ability to digest cellulose. What would be some desirable consequences? What would be some undesirable consequences?

Summary Problem

This chapter has covered many kinds of carbon-containing compounds from a variety of sources. The carbon atoms they contain have all been derived from carbon dioxide. Fill out the grid below listing carbon-containing compounds from at least four additional sources. Name the source; the approximate time since the carbon atoms were reduced; whether the source is considered renewable and, if it is not renewable, approximately how long it is expected to last; an example compound; a use for the example compound; and what form the carbon is in when the example compound is used.

Source	Approximate Time	Renewable? If Not, How Long?	Example Compound	Use for Example	Form C Is in After Use
Petroleum	Millions of years	Not renewable, decades	C_8H_{18} (octane)	Motor fuel	CO_2

For each example compound you have given, explain how this compound is produced from its source. If your example compound is a saturated hydrocarbon, write the formula for an alcohol derived from the saturated hydrocarbon. Also, describe how thermal energy is derived from the compound and indicate whether animals can derive this energy directly by digestion or whether the energy must be obtained by combustion in some kind of mechanical device, like an engine. Finally, for each source of carbon compounds, discuss the implications of their use on the production of greenhouse gases.

In Closing

Having studied this chapter, you should be able to. . .

- describe petroleum refining and methods used to improve the gasoline fraction (Section 11.1).
- identify processes used to obtain organic chemicals from coal and name some of their products (Section 11.2).
- name and draw the structures of three functional groups produced by the oxidation of alcohols (Section 11.4).
- name and give examples of the uses of some important alcohols (Section 11.4).
- list some properties of carboxylic acids and write equations for the formation of esters from carboxylic acids and alcohols (Section 11.5).
- explain formation of polymers by addition or condensation; give examples of polymers formed by each type of reaction (Section 11.6).
- draw the structures of the repeating units in some common types of polymers (Section 11.6).
- identify or write the structures of the functional groups in alcohols, aldehydes, ketones, carboxylic acids, esters, amines, and amides (Sections 11.4 to 11.6).
- identify the types of plastics most successfully being recycled and discuss some of the issues that influence successful recycling (Section 11.8).
- explain the importance of plastics recycling (Section 11.9).
- illustrate the basics of protein structures and how peptide linkages hold amino acids together in proteins (Section 11.10).
- identify polysaccharides, their sources, the different ways they are linked, and the different uses resulting from these linkages (Section 11.10).

Key Terms

The following terms were defined and given in boldface type in this chapter. You should be sure to understand each of these terms and the concepts with which it is associated.

addition polymer *(11.8)*
aldehyde *(11.6)*
alpha carbon *(11.10)*
amide *(11.8)*
amine *(11.8)*
amino acid *(11.10)*
autoignition temperature *(11.1)*
carboxylic acid *(11.6)*
catalytic cracking process *(11.1)*
catalytic reforming *(11.1)*
catalyst *(11.1)*
condensation polymer *(11.8)*
condensation reaction *(11.8)*
copolymer *(11.8)*
ester *(11.7)*
ether *(11.1)*
fractional distillation *(11.1)*
gasohol *(11.6)*
glycogen *(11.11)*
greenhouse effect *(11.4)*
hydrogenation *(11.7)*
hydrolysis *(11.7)*
initiation *(11.8)*
ketone *(11.6)*
macromolecule *(11.8)*
monomer *(11.8)*
monosaturated acids *(11.7)*
oxygenated gasolines *(11.1)*
peptide bond *(11.10)*
petroleum fractions *(11.1)*
plastic *(11.8)*
polyamides *(11.8)*
polyester *(11.8)*
polymer *(11.8)*
polypeptide *(11.10)*
polyunsaturated acids *(11.7)*
pyrolysis *(11.2)*
reformulated gasolines *(11.1)*
saponification *(11.7)*
thermoplastic *(11.8)*
thermosetting plastic *(11.8)*

Questions for Review and Thought

Conceptual Challenge Problems

CP-11.A. How are the boiling point of hydrocarbons during the distillation of petroleum related to their molecular size?

CP-11.B. Even though there are millions of organic compounds and each compound may have 10, 100, or even thousands of atoms bonded together to make one molecule, the reactions of organic compounds can be studied and even predicted for compounds yet to be discovered. What characteristic of organic compounds allows their reactions to be studied and predicted?

CP-11.C. What is the advantage of animals storing chemical potential energy in their bodies as triesters of glycerol and long-chain fatty acids, known as fats, instead of as carbohydrates?

Review Questions

1. Why is the organic chemical industry referred to as the *petrochemical industry?*
2. What products are produced by the petrochemical industry?
3. What is the difference between *catalytic cracking* and *catalytic reforming?*
4. Explain how an octane number of a gasoline is determined.
5. Why is coal receiving increased attention as a source of organic compounds?
6. What is synthesis gas? How can it be used to produce petrochemicals?
7. Methanol is one of the top 50 petrochemicals produced in the United States. What factors are likely to lead to an increased demand for methanol in the next decade?
8. What is the difference between *oxygenated gasoline* and *reformulated gasoline?* Why are they being produced?
9. Explain why world use of natural gas is predicted to double between 1990 and 2010.
10. Table 11.2 lists several compounds with octane numbers above 100 and one compound with an octane number below zero. Explain why such values are possible.
11. Explain why methanol (molecular weight, 32) has a lower boiling point (65.0 °C) than water (molecular weight, 18; boiling point, 100.0 °C).
12. Describe the Mobil process for converting methanol to gasoline.
13. Outline the steps necessary to obtain 89 octane gasoline, starting with a barrel of crude oil.
14. What is the major difference between crude oil and coal as a source of hydrocarbons?
15. Describe the structural formula of a representative compound for each of the following classes of organic compounds: alcohols, aldehydes, ketones, carboxylic acids, esters, and amines.
16. Explain why esters have lower boiling points than carboxylic acids of the same molecular weight.
17. What structural feature must a molecule have in order to undergo addition polymerization?
18. What feature do all condensation polymerization reactions have in common?
19. Give examples of (a) a synthetic addition polymer, (b) a synthetic condensation polymer, and (c) a natural addition polymer.
20. How does *cis-trans* isomerism affect the properties of rubber?
21. Discuss what plastics are currently being recycled, and give examples of some products being made from these recycled plastics.
22. What four steps are necessary for successful recycling of any solid waste?

Fuel Sources and Products

23. What are petroleum fractions? What process is used to produce them?
24. (a) What is the boiling point range for the petroleum fraction containing the hydrocarbons that will provide fuel for your car?
(b) What is the octane rating for the "straight-run" gasoline fraction obtained from the fractional distillation of petroleum?
(c) Would you use this fraction to fuel your car? Why or why not?
25. (a) Draw the Lewis structure for the hydrocarbon that is assigned an octane rating of 0.
(b) Draw the Lewis structure for the hydrocarbon that is assigned an octane rating of 100.
(c) What is the boiling point for each of these hydrocarbons?
26. Explain what is meant by the statement: "All gasolines are highly volatile."
27. What would be the advantage of removing the higher octane components like aromatics and alkenes from oxygenated gasolines?
28. Draw the structure of methyl *tertiary*-butyl ether, a common gasoline additive.
29. What are the components in natural gas?
30. Discuss the thermodynamic differences in the two different processes of coal gasification. In one process, synthesis gas is produced (endothermic reaction) and in the newer process, methane is the end product.
31. What are the four "greenhouse gases," and why are they called greenhouse gases?
32. Carbon dioxide is known to be the major contributor to the greenhouse effect. List some of its sources in our atmosphere and some of the processes that remove it. Currently, which predominates: CO_2 sources or removal processes?
33. Name a favorable effect of the global increase of CO_2 in the atmosphere.

Alcohols

34. Give an example for (a) a primary alcohol; (b) a secondary alcohol; and (c) a tertiary alcohol. Draw Lewis structures for each example.

35. Classify each of the following alcohols as primary, secondary, or tertiary.

(a) $CH_3CH_2CH_2CH_2OH$

(b) $CH_3CHCH_2CH_3$
 |
 OH

(c) $CH_3CHCH_2CH_2OH$
 |
 OH

(d) CH_3
 |
 $CH_3CCH_2CH_3$
 |
 OH

(e) CH_3
 |
 CH_3CCH_3
 |
 OH

36. Write the condensed structural formula for each of the following:
(a) 2-methyl-2-pentanol
(b) 2,3-dimethyl-1-butanol
(c) 4-methyl-2-pentanol
(d) 2-methyl-3-pentanol
(e) *tertiary*-butyl alcohol
(f) isopropyl alcohol

37. Explain what *oxidation* of organic compounds usually involves. What is meant by *reduction* of organic compounds?

38. Draw the structures of the first two oxidation products of each of the following alcohols.
(a) CH_3CH_2OH (b) $CH_3CH_2CH_2CH_2OH$

39. Draw the structures of the oxidation products of each of the following alcohols.
(a) 2-butanol (b) 4-methyl-2-pentanol

40. Write the condensed structural formula of the alcohols that can be oxidized to make the following:

(a) O
 ||
 $CH_3—CH—CH_2—C—H$
 |
 CH_3

(b) O
 ||
 $CH_3—CH_2—C—CH_2—CH_3$

(c) O
 ||
 $CH_3—CH_2—CH—C—OH$
 |
 CH_3

41. What is the percentage of ethanol in 90-proof vodka?

42. Explain the common names of *wood alcohol* for methanol and *grain alcohol* for ethanol.

43. What is denatured alcohol? Why is it made?

44. Many biological molecules, including steroids and carbohydrates, contain many hydroxyl groups. What need might biological systems have for this particular functional group?

Carboxylic Acids and Esters

45. Explain why the boiling points for carboxylic acids are higher than for alcohols with comparable molar masses?

46. Write the structural formula of the ester that can be formed from
(a) $CH_3COOH + CH_3CH_2OH$.
(b) $CH_3CH_2COOH + CH_3CH_2CH_2OH$.
(c) $CH_3CH_2COOH + CH_3OH$.

47. Write the structural formula for the ester that can be produced in the following reactions.
(a) Formic acid + methanol
(b) Butyric acid + ethanol
(c) Acetic acid + 1-butanol
(d) Propionic acid + 2-propanol

48. Write the condensed formula of the alcohol and acid that will react to form each of the following esters.

(a) O
 ||
 $CH_3CH_2C—OCH_3$

(b) O
 ||
 $HC—OCH_2CH_3$

(c) O
 ||
 $CH_3C—OCH_2CH_3$

49. Explain why carboxylic acids are more soluble in water than are esters with the same molar mass.

Organic Polymers

50. What are some examples of *thermoplastics?* What are the properties of thermoplastics when heated and cooled?

51. What are some examples of *thermosetting plastics?* What are the properties of thermosetting plastics when heated and cooled?

52. Draw the structure of the repeating unit in a polymer in which the monomer is
(a) 1-Butene. (b) 1,1-Dichloroethylene.
(c) Vinyl acetate.

53. What is the principal structural difference between low-density and high-density polyethylene? Is polyethylene an addition or condensation polymer?

54. Methyl methacrylate has the structural formula shown in Table 11.11. When polymerized it is very transparent, and it is sold in the United States under the trade names Lucite and Plexiglas. Draw the repeating unit for the poly(methyl methacrylate) chain.

55. What monomers are used to prepare the following polymers?

(a) $-CH_2CH_2CH_2CH_2CH_2CH_2CH_2CH_2CH_2-$

(b) 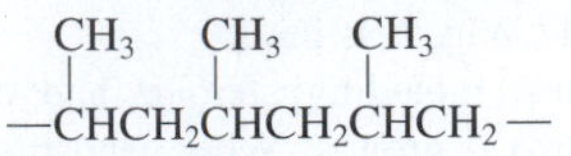 $-CH(CH_3)CH_2CH(CH_3)CH_2CH(CH_3)CH_2-$

(c) 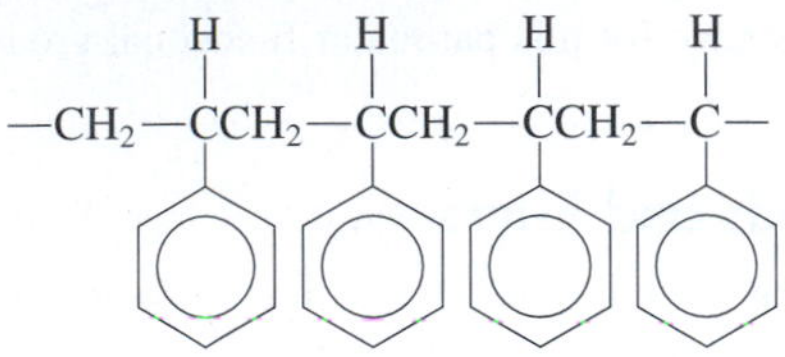 $-CH_2-C(H)(C_6H_5)CH_2-C(H)(C_6H_5)CH_2-C(H)(C_6H_5)CH_2-C(H)(C_6H_5)-$

(d) 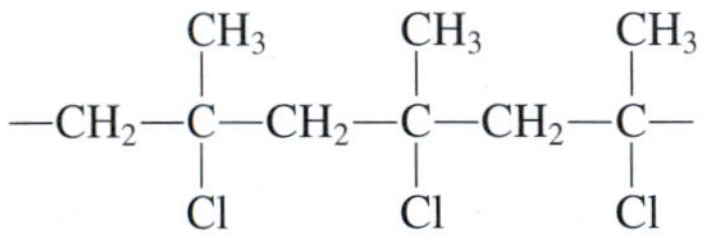 $-CH_2-C(CH_3)(Cl)-CH_2-C(CH_3)(Cl)-CH_2-C(CH_3)(Cl)-$

(e) 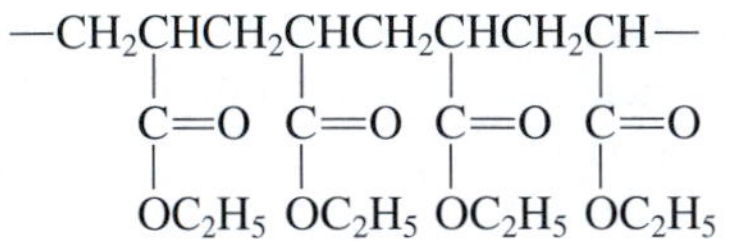 $-CH_2CH(C{=}O)CH_2CH(C{=}O)CH_2CH(C{=}O)CH_2CH(C{=}O)-$, each C=O bearing OC_2H_5

56. What is the monomer in *natural rubber?* Which isomer is present in natural rubber: *cis* or *trans*?

57. What are the two monomers used to make SBR, which is used for making tires?

58. Polyesters, one class of condensation polymers, involve which two functional groups?

59. Name one important polyester polymer and its uses.

60. Polyamides are made by condensing which functional groups? Name the most common example of this class of polymers.

61. How are amide linkages and peptide linkages similar? How are they different?

62. State one major difference between proteins and polyamides.

63. Draw structures of monomers that could form each of the following condensation polymers.

(a) 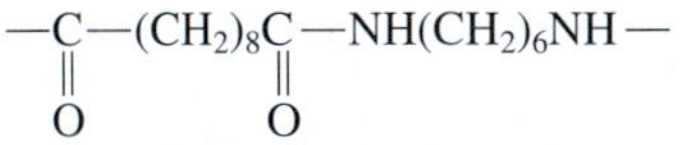 $-C(=O)-(CH_2)_8C(=O)-NH(CH_2)_6NH-$

(b) 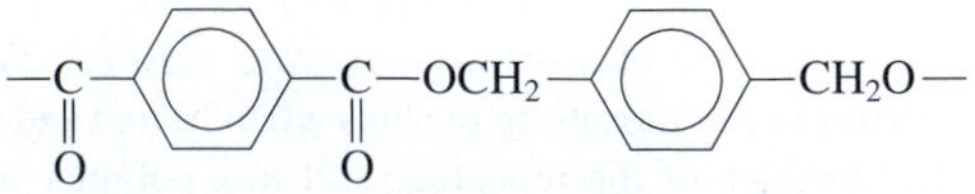 $-C(=O)-C_6H_4-C(=O)-OCH_2-C_6H_4-CH_2O-$

64. Orlon has a polymeric chain structure of

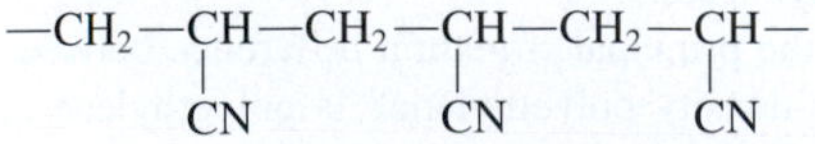

$$-CH_2-CH(CN)-CH_2-CH(CN)-CH_2-CH(CN)-$$

What is the monomer from which this structure can be made?

65. How many ethylene units are in a polyethylene molecule that has a molecular weight of approximately 42,000?

66. Kevlar, a bulletproof plastic, is made from *p*-phenylenediamine, and terephthalic acid. Draw the repeating unit of Kevlar.

67. Write the structural formulas for the repeating units of
(a) Natural rubber. (b) Neoprene.
(c) Polybutadiene.

68. What are some of the serious drawbacks to burning plastics? Make a list of harmful substances produced when plastics are incinerated.

Proteins and Polysaccharides

69. Which biological molecules have monomer units that are all alike, as in synthetic polymers?

70. Which biological molecules have monomer units that are not all alike, as in synthetic copolymers?

71. Identify and name all the functional groups in the tripeptide below.

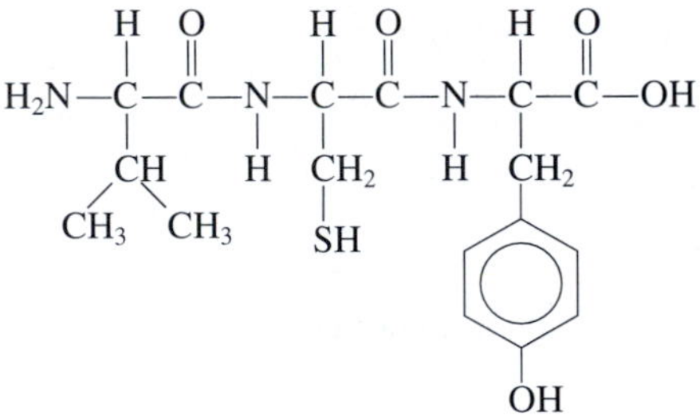

$$H_2N-CH(CH(CH_3)_2)-C(=O)-NH-CH(CH_2SH)-C(=O)-NH-CH(CH_2C_6H_4OH)-C(=O)-OH$$

72. Draw the structural formula of alanylglycylphenylalanine.

73. Draw the structural formula of leucylmethionylalanylserine.

74. Which of the following structural formulas represent α-D-glucose and which represents β-D-glucose?

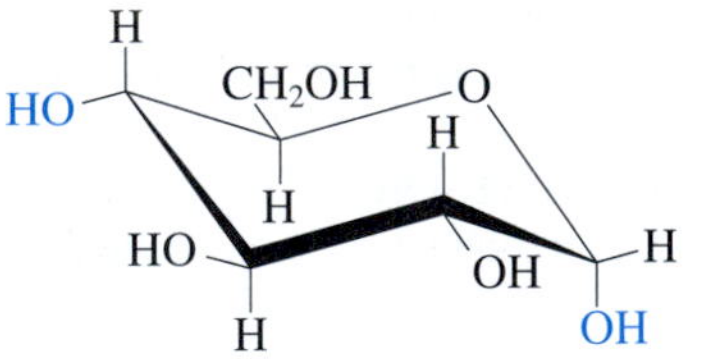

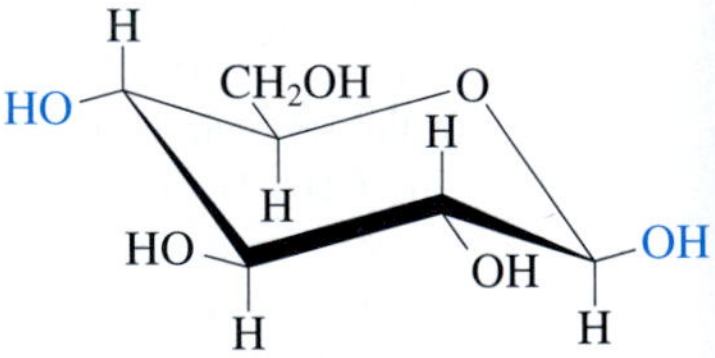

(Two Haworth ring structures of glucose with labels: H, CH_2OH, O, HO, H, H, HO, OH, H, OH, H.)

75. What is the chief function of glycogen in animal tissue?

76. What polysaccharide yields only α-D-glucose upon complete hydrolysis?

77. How do amylose and amylopectin differ? How are they similar? Are amylose and glycogen similar?

78. Explain why humans can use glycogen for energy but not cellulose. Why can cows digest cellulose?

General Questions

79. Compounds A and B both have the molecular formula C_2H_6O. The boiling points of compounds A and B are 78.5 °C and −23.7 °C, respectively. Use the table of functional groups in Appendix E and write the structural formulas and names of the two compounds.

80. Explain why ethanol, CH_3CH_2OH, is soluble in water in all proportions, but decanol, $CH_3(CH_2)_9OH$, is almost insoluble in water.

81. Nitrile rubber (Buna N) is a copolymer of two parts of 1,3-butadiene to one part of acrylonitrile. Draw the repeating unit of this polymer.

82. How are rubber molecules modified by vulcanization?

83. Write the condensed structural formula for 3-ethyl-5-methyl-3-hexanol. Is this a primary, secondary, or tertiary alcohol?

84. Using structural formulas, write a reaction for the hydrolysis of a triglyceride that contains fatty acid chains, each consisting of 16 total carbon atoms.

85. Is plastic wrap used in covering food a thermoplastic or thermosetting plastic? Explain.

86. Assume that a car burns pure octane, C_8H_{18} ($d = 0.692$ g/cm^3).
(a) Write the balanced equation for burning octane in air, forming CO_2 and H_2O.
(b) If the car has a fuel efficiency of 32 miles per gallon of octane, what volume of CO_2 at 25 °C and 1.0 atm is generated when the car goes on a 10.0 mile trip?

87. Perform the same calculations as in Problem 86, but use methanol, CH_3OH ($d = 0.791$ g/cm^3) as the fuel. Assume the fuel efficiency is 20.0 miles per gallon.

88. Show structurally why glycogen forms granules when stored in the liver but cellulose is found in cell walls as sheets.

89. Polytetrafluoroethylene (Teflon) is made by first treating HF with chloroform, followed by cracking the resultant difluorochloromethane.

$$CHCl_3 + 2\ HF \longrightarrow CHClF_2 + 2\ HCl$$
$$2\ CHClF_2 + \text{heat} \longrightarrow F_2C{=}CF_2 + 2\ HCl$$
$$F_2C{=}CF_2 + \text{peroxide catalyst} \longrightarrow \text{Teflon}$$

If you wish to make 1.0 kg of Teflon, what mass of chloroform and HF must you use to make the starting material, $CHClF_2$? (Although it is not realistic, assume each reaction step proceeds in 100% yield.)

90. Propionic anhydride, like acetic anhydride, is an important intermediate.
(a) Draw the formula for propionic anhydride.
(b) Give the reaction products for the hydrolysis of propionic anhydride.

Applying Concepts

91. Both propanoic acid and ethyl methanoate form hydrogen bonds with water. Draw all the water molecules that can hydrogen bond to these molecules. Use dotted lines to represent the hydrogen bonds.

$$CH_3CH_2\overset{\overset{\displaystyle O}{\|}}{C}{-}OH \qquad H{-}\overset{\overset{\displaystyle O}{\|}}{C}{-}O{-}CH_2CH_3$$

propanoic acid — ethyl methanoate

Based on your drawings, which should be more soluble in water, propanoic acid or ethyl methanoate? Explain your reasoning.

92. Hydrogen bonds can form between propanoic acid molecules and between 1-butanol molecules. Draw all the propanoic acid molecules that can hydrogen bond to the one below. Draw all the 1-butanol molecules that can hydrogen bond to the one below. Use dotted lines to represent the hydrogen bonds.

$$CH_3CH_2\overset{\overset{\displaystyle O}{\|}}{C}{-}OH \qquad CH_3CH_2CH_2CH_2{-}OH$$

propanoic acid — 1-butanol

Based on your drawings, which should have the higher boiling point, propanoic acid or 1-butanol? Explain your reasoning.

93. The illustrations below represent two different samples of polyethylene, each with the same number of monomer units. Based on the concept of density and not structure, which one is high-density polyethylene and which is low-density? Write a brief explanation.

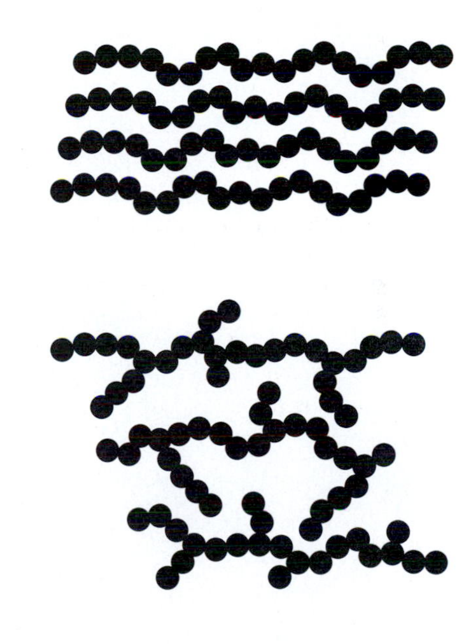

94. What monomer formed this polymer?

```
    H       H       H       H       H
    |       |       |       |       |
 —C=C—C=C—C=C—C=C—C=C—
  |       |       |       |       |
 CH3     CH3     CH3     CH3     CH3
```

95. Draw the structure of a molecule that could undergo a condensation reaction with itself to form a polyester. Draw a segment of the polymer consisting of at least five monomer units.

96. The backbone of a DNA molecule is a polymer of alternating sugar (deoxyribose) and phosphoric acid units held together by a phosphate ester bond. Draw a segment of the polymer consisting of at least two sugar and two phosphate units. Circle the phosphate ester bonds.

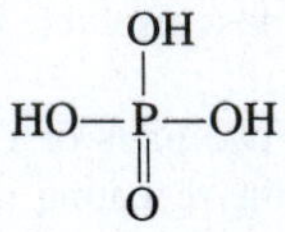

phosphoric acid

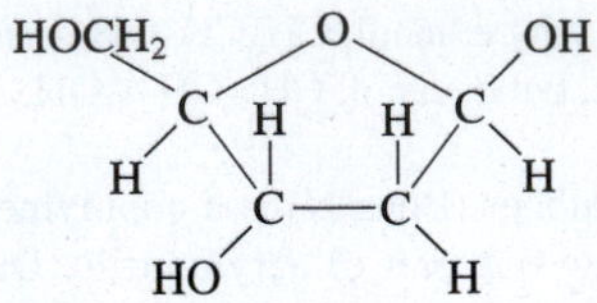

deoxyribose

CHAPTER 12

Principles of Reactivity III: Chemical Kinetics

The sculpture shown here is made of stainless steel, which is mainly iron but also contains chromium, nickel, and other elements. In order to obtain the exact properties desired for a special stainless steel like this one, it is necessary to control the proportions of the different metals in the alloy. • How can nickel be purified before it is used in special stainless steels? How fast are reactions used to purify nickel? These and other questions are answered in the Summary Problem at the end of this chapter. *(Will and Deni McIntyre/Photo Researchers, Inc.)*

Turn on the valve of a Bunsen burner in your laboratory, bring up a lighted match, and a rapid combustion reaction begins with a whoosh!

$$CH_4(g) + 2\ O_2(g) \longrightarrow CO_2(g) + 2\ H_2O(g) \qquad \Delta H^\circ = -802.34\text{ kJ}$$

But what would happen if you didn't put a lighted match in the methane-air stream? Nothing obvious! At room temperature the reaction of methane with oxygen is so slow that the two potential reactants can be mixed in a closed flask and stored unreacted for centuries.

The ideas about thermodynamics that were discussed in Chapter 7 lead to the conclusion that combustion of methane should occur without a continuous input of energy. In the equation above, 3 mol of gaseous reactants produces 3 mol of gaseous products, and so the entropy of the system changes only a little. There is, however, a large negative enthalpy change, and so the thermal energy transferred out of the system causes a significant increase in the entropy of the surroundings. The entropy of the universe increases when the reactants form products, and the reaction is product-favored. However, thermodynamics reveals nothing at all about the very slow speed of the reaction or the need for a flame to get it started. These observations lie within the realm of **chemical kinetics**—the study of the speeds of reactions and the nanoscale pathways or rearrangements by which atoms and molecules are transformed from reactants to products.

Chemical kinetics is extremely important, because knowing about kinetics enables us to control reactions. For example, we know experimentally that methane burns very rapidly at high temperatures but reacts extremely slowly at low temperatures. This allows us to control combustion—we can initiate it with a lighted match. In pharmaceutical chemistry, an important problem is devising drugs that remain in their active form long enough to get to the site in the body where they are intended to act. Consequently it is important to know whether a drug will react with other substances in the body and how long it will take to do so. The fact that you see motion on a TV screen depends on the speeds of reactions that take place in the retinas of your eyes. In environmental chemistry, there was more than a decade of controversy over whether or not stratospheric ozone is being depleted by chlorofluorocarbons. Much of this hinged on verifying the sequence of reactions by which stratospheric ozone is produced and consumed, and on accurate measurements of the rates of those reactions. Their careful studies of such reactions led to the 1996 Nobel Prize for Sherwood Rowland, Mario Molina, and Paul Crutzen (← ***p. 2***).

This chapter is about all of the factors that affect the speeds of reactions, the nanoscale basis for understanding them, and their importance in modern society, from industrial plants to the cells of our bodies.

12.1 REACTION RATE

In order for a chemical reaction to occur, reactant molecules must come together so that atoms can be exchanged or rearranged. Atoms and molecules are mobile in the gas phase or in solution, and so reactions are often carried out in a mixture of gases or among solutes in a solution. For a **homogeneous reaction,** one in which reactants and products are all in the same phase (gas or solution, for example), four factors affect the speed of a reaction.

- The *properties* of reactants and products—in particular molecular structure and bonding

(a)

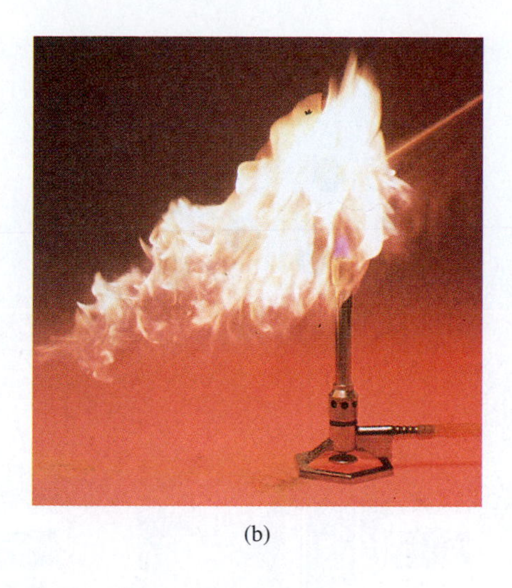

(b)

Figure 12.1 Combustion of lycopodium powder. (a) The very finely divided spores of this common moss burn slowly when they are in a pile so that only a small surface area is exposed to air. (b) When the exposed surface is increased by spraying the powder through the air into a flame, combustion is rapid—even explosive. *(C.D. Winters)*

- The *concentrations* of the reactants and sometimes the products
- The *temperature* at which the reaction occurs
- The presence of a *catalyst* and, if one is present, its concentration

Many important industrial reactions take place at the surface or interface between two different phases (solid and gas, for example). These are called **heterogeneous reactions.** Their speeds depend on the factors listed above but also on the area of the surface at which they occur. For example, very finely divided flour, cornstarch, or lycopodium powder can burn explosively, whereas a pile of powder with much less surface exposed to oxygen in the air is difficult to ignite (Figure 12.1). The much more rapid reaction when greater surface is exposed has been responsible for explosions in grain elevators, coal mines, and other situations where finely divided, combustible solids are exposed to air and a spark or flame.

A catalyst (described in Section 12.8) speeds up a reaction but undergoes no net chemical change.

The speed of any process is expressed as its **rate,** which is the change in some measurable quantity per unit of time. A car's rate of travel, for example, is found by measuring the change in its position, Δp, during a given time interval, Δt. Suppose you are driving on the interstate. If you pass mile marker 43 at 2:00 PM and mile marker 153 at 4:00 PM, $\Delta p = (153 - 43)$ mi $= 110$ mi, and $\Delta t = 2.00$ h. You are traveling at an average rate of $\Delta p/\Delta t = 55$ mi/h. When describing chemical reactions, the **reaction rate** is defined as *the change in concentration of a reactant or product per unit time.* (Time can be measured in seconds, hours, days, or whatever unit is most convenient for the speed of the reaction.) As an example of measurements made in chemical kinetics, consider Figure 12.2, which shows a colored dye reacting to form a colorless product. The color disappears over time, and the intensity of color can be used to determine the concentration of the dye.

Recall that the Greek letter Δ (delta) means that a change in some quantity has been measured. As usual, Δ means to subtract the initial value of the quantity from the final value.

Change in concentration is used (rather than change in total amount of reactant), because this makes the rate independent of the volume of the reaction mixture.

As a practical example, consider what happens to the cancer chemotherapy agent cisplatin, $Pt(NH_3)_2Cl_2$, in the presence of water.

$$H_2O + Cl{-}\underset{NH_3}{\overset{NH_3}{Pt}}{-}Cl \longrightarrow H_2O{-}\underset{NH_3}{\overset{NH_3}{Pt^+}}{-}Cl + Cl^-$$

One of the Cl^- ions bound to the central Pt^{2+} ion can be replaced by a water molecule. The rate at which this occurs is found by measuring the concentration of $Pt(NH_3)_2Cl_2$ at the beginning and end of a time interval. For example, if the concen-

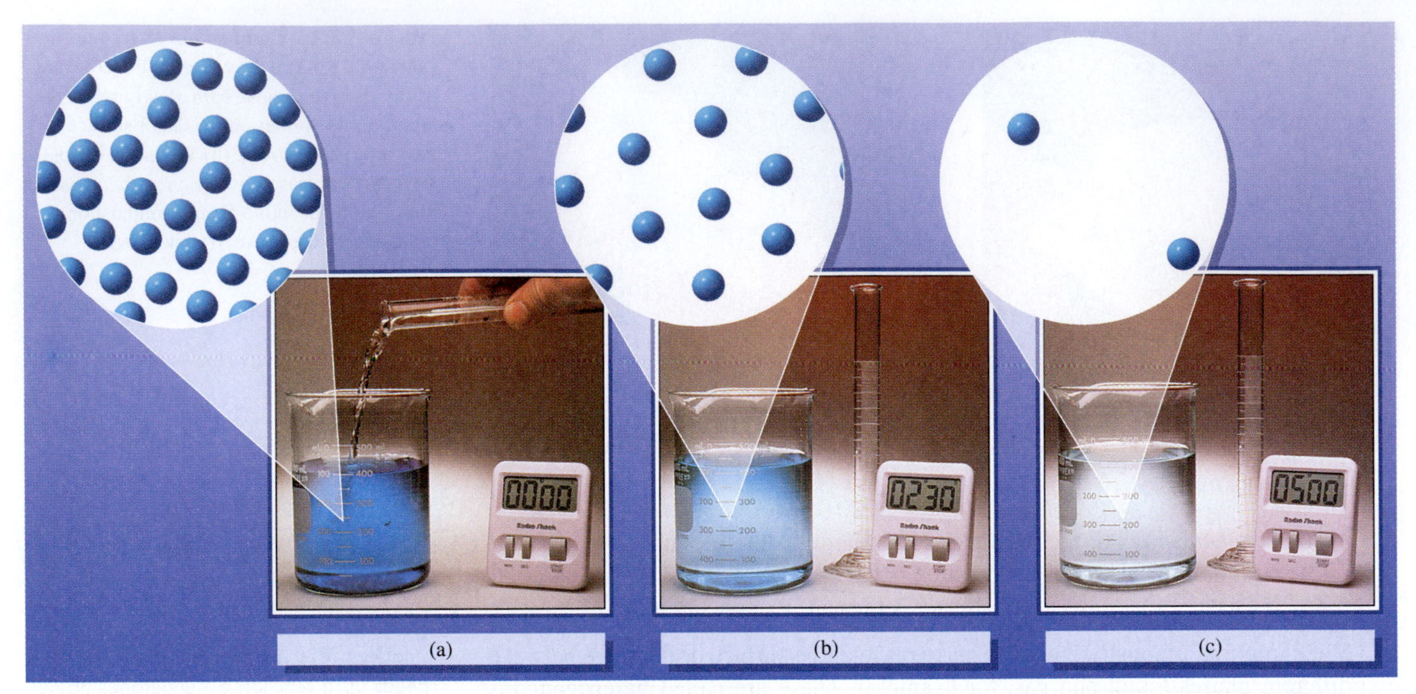

Figure 12.2 Disappearance of a dye. Blue food dye in aqueous solution is reacting with bleach, which converts it into a colorless product. Over time the intensity of color of the solution decreases and eventually the color disappears. The rate of the reaction could be determined by simultaneously measuring both the intensity of color and the time, and then repeating the measurements many times during the course of the reaction. From the intensity of color the concentration of dye could be calculated, and so concentration could be determined as a function of time. (Solvent molecules and reaction product molecules have been omitted from the nanoscale diagrams for clarity.) *(C.D. Winters)*

It is obviously important for us to know something about the speed with which Cl^- is replaced by water in cisplatin. If it reacts quickly, the drug will change into a different substance as soon as it is placed in the aqueous environment of the human body. The drug's perceived activity may well be due to the newly formed substance. In other cases, a drug (and its activity) can be destroyed by a chemical reaction that changes it into something else before it reaches the site where it carries out its function.

Square brackets around a compound's molecular formula indicate the compound's concentration, which is usually measured in moles per liter ***(← p. 198).***

tration of cisplatin at some time, t_1, is measured, and the measurement is repeated at time t_2, then the rate of reaction could be given as the number of moles of cisplatin per liter consumed per unit time, as follows:

$$\text{Rate of change of } [Pt(NH_3)_2Cl_2] = \frac{\text{change in concentration of } Pt(NH_3)_2Cl_2}{\text{elapsed time}}$$

$$= \frac{\Delta[Pt(NH_3)_2Cl_2]}{\Delta t} = \frac{\{[Pt(NH_3)_2Cl_2] \text{ at } t_2\} - \{[Pt(NH_3)_2Cl_2] \text{ at } t_1\}}{t_2 - t_1}$$

The experimentally measured concentration of $Pt(NH_3)_2Cl_2$ is plotted as a function of time in Figure 12.3.

The data in Figure 12.3 can be used to find the average rate during any period of the reaction. For example, after the first 100 minutes of reaction, the concentration of $Pt(NH_3)_2Cl_2$ has declined from 0.0100 M to 0.0084 M.

$$\text{Rate of change of } [Pt(NH_3)_2Cl_2] = \frac{\Delta[Pt(NH_3)_2Cl_2]}{\Delta t} = \frac{(0.0084 - 0.0100)\text{ mol/L}}{(100. - 0.)\text{ min}}$$

$$= -\frac{0.0016\text{ mol/L}}{100.\text{ min}} = -1.6 \times 10^{-5}\text{ mol/L·min}$$

where the *negative* sign indicates that the concentration of $Pt(NH_3)_2Cl_2$ is *decreasing*. It is conventional to define the reaction rate so that it is always positive. When

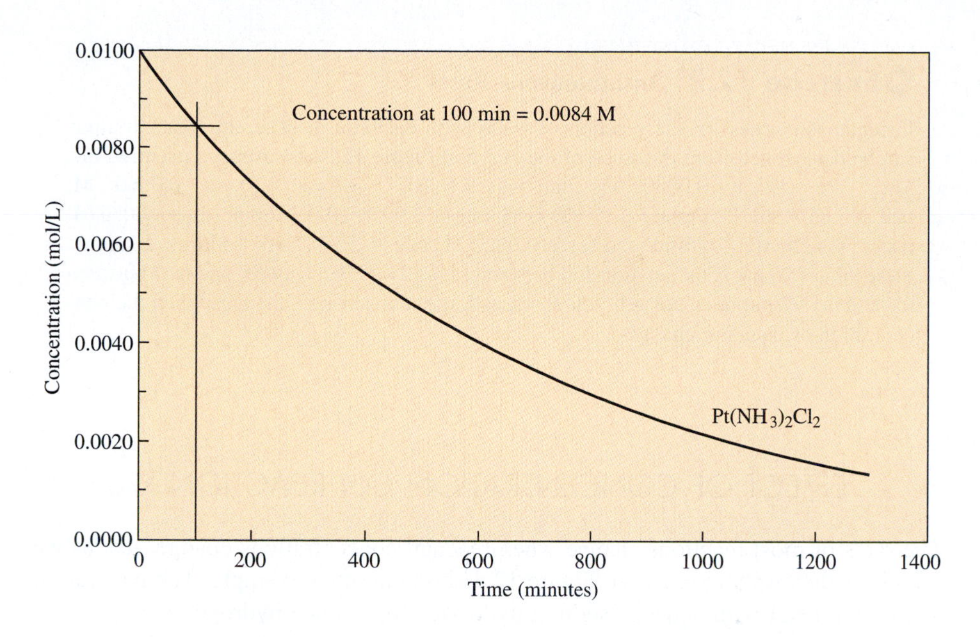

Figure 12.3 Concentration/time data for a reaction. The experimentally measured concentration of $Pt(NH_3)_2Cl_2$ is plotted as a function of time during the reaction in which Cl^- is replaced by H_2O in aqueous solution.

a rate is expressed in terms of Δ[reactant]/Δt, which is a negative quantity, a minus sign is always used to make the rate positive. Thus, in this case,

$$\text{Reaction rate} = -\Delta[\text{Pt(NH}_3)_2\text{Cl}_2]/\Delta t = 1.6 \times 10^{-5}\ \text{mol/L·min}$$

Notice also that because calculating the rate involves dividing a concentration difference by a time difference, the units of reaction rate are units of concentration divided by units of time, in this case mol/L divided by min.

Exercise 12.1 Rates of Reaction

From Figure 12.3, obtain the necessary data and calculate the average rate for the cisplatin reaction for each of these time periods: (a) from 600 to 800 min; (b) from 1100 to 1300 min; (c) from 200 to 1200 min. Compare your results. Try to understand any differences you find, and write an explanation for a friend who is taking this course.

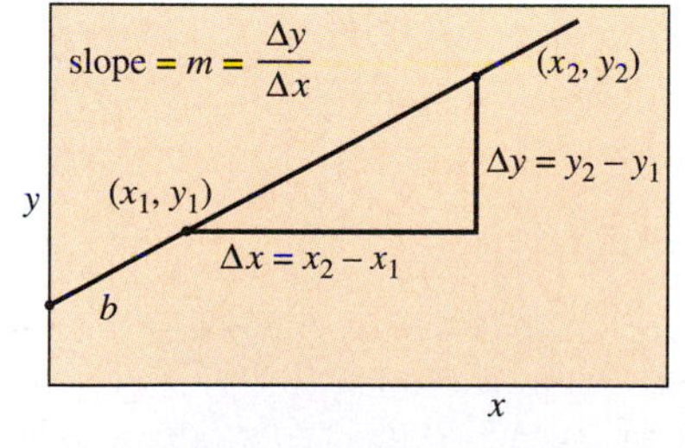

Slope and intercept. Plot for the equation $y = mx + b$, which has slope m and y-intercept b.

Your results in Exercise 12.1 should have been different for each range of time over which you calculated, and the rates calculated for smaller concentrations of cisplatin should have been smaller. This happens because the rate of this reaction decreases as the concentration of the cisplatin reactant decreases. That is, the reaction rate changes over time, and so it depends when, and for what range of time, you measure the rate. For the rate that corresponds to a particular concentration of cisplatin, the average rate must be calculated over a very small interval—from just before to just after that concentration of cisplatin is reached on the curve in Figure 12.3. For example, to calculate the rate at which cisplatin is disappearing when its concentration is 0.0084 M, you would need to measure concentrations at times just before and just after 100 min from the start of the reaction and calculate the rate as was done above. A rate calculated in this way is referred to as an *instantaneous* rate to distinguish it from the *average* rate calculated for a longer time interval. For a particular concentration of the same reactant at the same conditions of temperature and concentrations of other species, the instantaneous rate will always be the same. It is the slope of the curve at the point corresponding to the specified concentration.

If you are familiar with the calculus, then you may recognize that in the limit of very small time intervals, $\Delta[Pt(NH_3)_2Cl_2]/\Delta t$ becomes the same as the derivative of concentration with respect to time. That is,

$$\lim_{\Delta t \to 0} \frac{\Delta[\text{Pt(NH}_3)_2\text{Cl}_2]}{\Delta t} = \frac{d[\text{Pt(NH}_3)_2\text{Cl}_2]}{dt}$$

This also means that the rate of reaction at any time can be found from the slope (at that time) of the tangent to a curve of concentration vs. time, such as the curve in Figure 12.3.

Exercise 12.2 Instantaneous Rates

Instantaneous rates for the reaction of water with cisplatin at different concentrations can be determined from the slope of the curve in Figure 12.3 at various concentrations. They are (a) at 0.0080 M, rate = 11.8×10^{-6} M/min; (b) at 0.0060 M, rate = 8.85×10^{-6} M/min; (c) at 0.0040 M, rate = 5.90×10^{-6} M/min; (d) at 0.0030 M, rate = 4.42×10^{-6} M/min; and (e) at 0.0020 M, rate = 2.95×10^{-6} M/min. Examine these values. What is the relationship between (a) and (c)? between (b) and (d)? between (c) and (e)? What does this tell you about how the reaction rate changes when the concentration of cisplatin changes?

12.2 EFFECT OF CONCENTRATION ON REACTION RATE

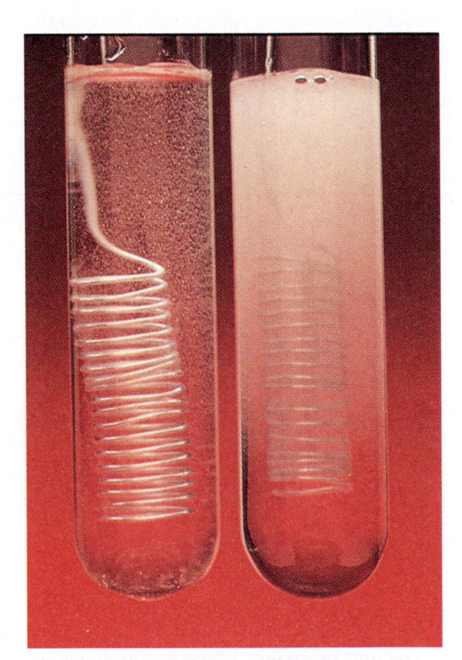

Figure 12.4 Reaction of aluminum with aqueous sodium hydroxide. The rate of the reaction of aluminum with aqueous NaOH depends on the concentration of the base. With dilute NaOH the reaction is slow (*left*), but it is more rapid in more concentrated NaOH (*right*). (In both cases the temperature and the surface area of the aluminum are the same, and so the difference in rate must be due to concentration of the base.) *(C.D. Winters)*

The rates of most reactions change when reactant concentrations change, just as we found for the cisplatin reaction. Figure 12.4 shows another example. The reaction of aluminum metal with aqueous sodium hydroxide to produce hydrogen gas

$$2\ Al(s) + 6\ H_2O(\ell) + 2\ NaOH(aq) \longrightarrow 2\ NaAl(OH)_4(aq) + 3\ H_2(g)$$

is visibly more rapid in 6 M NaOH than in 1 M NaOH. One goal in studying kinetics is to find out whether a reaction speeds up when the concentration of one reactant is increased and, if so, by how much.

The Rate Law

The effect of concentration of a reactant on rate can be determined by performing several experiments in which the concentration of that reactant is varied systematically (and temperature is held constant). Or a single experiment can be done in which concentration is determined continuously as a function of time. An example of the latter is the data for cisplatin shown in Figure 12.3, which you analyzed in Exercise 12.2. You should have discovered that if the concentration of cisplatin is doubled, the reaction rate is also doubled. If the concentration of cisplatin is halved, then the reaction rate is halved. This leads to the expression

$$\text{Rate of reaction} \propto [Pt(NH_3)_2Cl_2] \qquad \text{(where the symbol } \propto \text{ means "proportional to")}$$

which says that the rate is directly proportional to the concentration of one of the reactants, cisplatin.

This same relationship between reactant concentration and reaction rate can be expressed in equation form by including a proportionality constant, k. Such an equation is called a **rate equation** or **rate law.** For the cisplatin reaction the rate equation is

$$\text{Rate of reaction} = k[Pt(NH_3)_2Cl_2]$$

The proportionality constant, k, is called the **rate constant.** The rate constant is independent of concentration, but it generally becomes larger the higher the temperature is. The rate constant applies only to the specific reaction being studied, not to any other. Thus the equation for the reaction should be given along with the rate constant.

Determining Rate Laws from Initial Rates

The relation between rate and concentration (the rate law) must be determined experimentally. One way to do this was illustrated in Exercise 12.2, but it is hard to determine rates from tangents to a curve such as that in Figure 12.3. Another way is to measure initial rates. The **initial rate** of a reaction is the instantaneous rate determined at the very beginning of the reaction. Initial rates are useful because you can usually determine concentrations of reactants in the solutions or gases before they are mixed, and so the rate corresponding to particular concentrations can be determined. Several experiments can then be done in which initial concentrations are changed, and the same idea you used in Exercise 12.2 can be applied. Usually it is appropriate to determine the initial rate by calculating $-\Delta[\text{reactant}]/\Delta t$ after no more than 2% of the limiting reactant has been consumed. An advantage of measuring initial rates is that as a reaction proceeds, more and more products are formed. In some cases products can alter the rate; comparing initial rates with rates when products are present can reveal such a complication.

As an example, consider the reaction of a base with methyl acetate, an ester *(← p. 481)* that is an industrial solvent. The reaction produces acetate ion and methanol.

$$CH_3\overset{\overset{\displaystyle O}{\|}}{C}{-}O{-}CH_3 + OH^- \longrightarrow CH_3\overset{\overset{\displaystyle O}{\|}}{C}{-}O^- + CH_3OH$$

methyl acetate   hydroxide ion   acetate ion   methanol

To control for the effect of temperature on rate, several experiments were done at the same temperature:

Experiment	Initial Concentrations		Initial Rate (mol/L·s)
	$[CH_3COOCH_3]$	$[OH^-]$	
1	0.050 M	0.050 M	0.00034
	↓ no change	↓ × 2	↓ × 2
2	0.050 M	0.10 M	0.00069
	↓ × 2	↓ no change	↓ × 2
3	0.10 M	0.10 M	0.00137

Study the table carefully. You should find that the initial rate doubled when the initial concentration of either CH_3COOCH_3 or OH^- was doubled, provided the concentration of the other reactant was held constant. These rate doublings show that the rate for the reaction is directly proportional to each of the concentrations. That is, the rate is proportional to the *product* of the two concentrations, and the rate equation is

$$\text{Reaction rate} = k[CH_3COOCH_3][OH^-]$$

This equation also tells us that doubling both concentrations at the same time would cause the rate to go up by a factor of four, which it does from experiment 1 to experiment 3. What would happen if one concentration is doubled and the other halved? The rate equation tells us that the rate will not change, and that is what happens!

Once the rate law is known, a value for k, the rate constant, can be found by substituting rate and concentration data for any one experiment into the rate equation. For example, a value of k for the methyl acetate–hydroxide ion reaction could be obtained from data for the first experiment,

$$\text{Reaction rate} = 0.00034 \text{ mol/L·s} = k(0.050 \text{ mol/L})(0.050 \text{ mol/L})$$

$$k = \frac{0.00034 \text{ mol/L·s}}{(0.050 \text{ mol/L})\ (0.050 \text{ mol/L})} = 0.14 \text{ L/mol·s}$$

A better value for k could be obtained by using all available experimental data, that is, by calculating a k for each experiment and then averaging the k values to obtain an overall result.

PROBLEM-SOLVING EXAMPLE 12.1 The Rate Equation

Initial rates for the reaction of nitrogen monoxide and oxygen

$$2\ NO(g) + O_2(g) \longrightarrow 2\ NO_2(g)$$

were measured at 25 °C starting with various concentrations of NO and O_2. The following data were collected:

Experiment	Initial Concentrations (mol/L)		Initial Rate (mol/L·s)
	[NO]	$[O_2]$	
1	0.020	0.010	0.028
2	0.020	0.020	0.057
3	0.020	0.040	0.114
4	0.040	0.020	0.227
5	0.010	0.020	0.014

Based on these data, what is the rate equation? What is the value of the rate constant k?

Answer Rate = $k[O_2]\ [NO]^2$; $k = 7.1 \times 10^3\ L^2/mol^2·s$

Explanation In the first three experiments the concentration of NO is constant while the O_2 concentration increases from 0.010 to 0.020 to 0.040 mol/L. Each time the $[O_2]$ is doubled, the initial rate also doubles. For example, when $[O_2]$ is doubled from 0.020 to 0.040 mol/L, the initial rate doubles from 0.057 to 0.114 mol/L·s. This means that the initial rate is directly proportional to $[O_2]$.

In experiments 2, 4, and 5 $[O_2]$ is constant, while [NO] varies. From experiment 2 to experiment 4, [NO] is doubled, but the initial rate increases by a factor of 4 or 2^2.

$$\frac{\text{Experiment 4 rate}}{\text{Experiment 2 rate}} = \frac{0.227 \text{ mol/L·s}}{0.057 \text{ mol/L·s}} = \frac{4}{1} = \frac{2^2}{1}$$

This same result is found from experiment 5 to experiment 2. This means that the initial rate is proportional to the *square* of [NO]. Therefore, the rate equation is

$$\text{Rate} = k[O_2][NO]^2$$

A value for the rate constant k can be found for experiment 1; for example,

$$\text{Rate} = 0.028 \text{ mol/L·s} = k(0.010 \text{ mol/L})(0.020 \text{ mol/L})^2$$

$$k = \frac{0.028 \text{ mol/L·s}}{(0.010 \text{ mol/L})(0.020 \text{ mol/L})^2} = 7.0 \times 10^3 \text{ L}^2\text{/mol}^2\text{·s}$$

To obtain a better value, calculate k for each experiment and then average the values to give 7.1×10^3 L^2/mol^2·s, which can be used to calculate the rate for any set of NO and O_2 concentrations at 25 °C.

Problem-Solving Practice 12.1

The rate constant k is 0.088/hr for the reaction

$$Pt(NH_3)_2Cl_2 + H_2O \longrightarrow [Pt(NH_3)_2(H_2O)Cl]^+ + Cl^-$$

and the rate equation is reaction rate = $k[Pt(NH_3)_2Cl_2]$; that is, the rate is directly proportional to $[Pt(NH_3)_2Cl_2]$. Calculate the initial rate of reaction when the concentration of $Pt(NH_3)_2Cl_2$ is 0.020 M. Calculate the rate when the concentration has dropped to half this value. Report your results in mol/L·s.

12.3 RATE LAW AND ORDER OF REACTION

For many (but not all) reactions, the rate equation has the general form

$$\text{Reaction rate} = k[A]^m[B]^n \ldots$$

where concentrations of substances, [A], [B], . . . are raised to powers, m, n,. . . . The substances A, B, . . . involved in such a rate law might be reactants, products, or catalysts. The exponents m, n, . . . are usually positive whole numbers but might be negative numbers or fractions. These exponents define the **order** of the reaction. If m is 1, for example, the reaction is first-order with respect to A; if n is 2, then the reaction is second-order with respect to B. The sum of m and n (plus the exponents on any other concentration terms) gives the **overall reaction order.** (The reaction in Problem-Solving Example 12.1 is described as being first-order in O_2, second-order in NO, and third-order overall.) A very important point to remember is that *the effects of changing concentrations on reaction rate, which are summarized by the rate equation and reaction orders, must be determined experimentally; they cannot be predicted from stoichiometric coefficients in the balanced chemical equation.*

Problem-Solving Example 12.2 Reaction Order and Rate Law

For each reaction and experimentally determined rate equation listed below, determine the order with respect to each reactant and the overall order.

(a) $2\ H_2O_2 \rightarrow 2\ H_2O + O_2$
Rate = $k[H_2O_2]$

(b) $14\ H_3O^+ + 2\ HCrO_4^- + 6\ I^- \rightarrow 2\ Cr^{3+} + 3\ I_2 + 22\ H_2O$
Rate = $k[HCrO_4^-][I^-]^2[H_3O^+]^2$

(c) *cis*-2-butene → *trans*-2-butene (with a catalytic concentration of I_2 present)
Rate = k[*cis*-2-butene]$[I_2]^{1/2}$

Answer (a) First-order in H_2O_2, first-order overall. (b) First-order in $HCrO_4^-$, second-order in I^-, and second-order in H_3O^+, fifth-order overall. (c) First-order in *cis*-2-butene, one-half–order in I_2, 1.5-order overall.

Explanation

(a) The rate equation has a single term that is raised to the first power, and so the reaction is first-order in H_2O_2 and first-order overall.

(b) The rate equation contains three terms. Since the $HCrO_4^-$ term is raised to the first power, the reaction is first-order in $HCrO_4^-$. The other two terms are squared, and so the reaction is second-order in I^- and second-order in H_3O^+. The exponents sum to five, and so the reaction is fifth-order overall.

(c) In this case the rate of reaction depends on the concentration of the reactant and also on the square root ($\frac{1}{2}$ power) of the concentration of a catalyst, I_2. The reaction is therefore first-order in *cis*-2-butene, one-half–order in I_2, and 1.5-order overall.

Problem-Solving Practice 12.2

The rate equation for the reduction of NO to N_2 with hydrogen is

$$2\ NO(g) + 2\ H_2(g) \longrightarrow N_2(g) + 2\ H_2O(g)$$

$$\text{Reaction rate} = k[NO]^2[H_2]$$

(a) What is the order of the reaction with respect to the NO? With respect to H_2?

(b) Suppose that you triple the concentration of NO and simultaneously decrease the concentration of H_2 by a factor of eight. Will the reaction be faster or slower under the new conditions?

The Integrated Rate Law

Another approach to experimental determination of the rate law and rate constant for a reaction involves using the methods of calculus to derive what is called the integrated rate law. As an example of the integrated-rate-law method, suppose that we have a general reaction

$$A \longrightarrow \text{products}$$

If the rate law involves powers of the reactant concentration, we can write the rate equation as

$$\text{Rate} = -\frac{\Delta[A]}{\Delta t} = k[A]^m$$

where m is the order of the reaction as we defined it before. The integrated rate law depends on the value of m, and the results for values of m from zero up to two are given in Table 12.1.

TABLE 12.1 Integrated Rate Laws

Order	Rate	Integrated Rate Equation*	Straight-Line Plot	Slope of Plot	Units of k
0	$k[A]^0 = k$	$[A] = -kt + [A]_0$	$[A]$ vs t	$-k$	conc/time
1	$k[A]$	$\ln[A] = -kt + \ln[A]_0$	$\ln[A]$ vs t	$-k$	1/time
2	$k[A]^2$	$\frac{1}{[A]} = kt + \frac{1}{[A]_0}$	$\frac{1}{[A]}$ vs t	k	1/conc·time

*In this table, $[A]_0$ indicates the initial concentration of substance A, that is, the concentration of A at $t = 0$, the time when the reaction was started.

The information in Table 12.1 can be used to analyze concentration/time data similar to those shown graphically in Figure 12.3. Each of the three integrated rate equations given in the table is the equation of a straight line, $y = mx + b$, where m is the slope and b the y-intercept. For example, the integrated rate equation for a first-order reaction is

$$\ln[A] = -kt + \ln[A]_0$$
$$y = m\ x + b$$

in which ln[A] corresponds to y, $-k$ corresponds to m, t corresponds to x, and $\ln[A]_0$ corresponds to b. If the reaction is actually first-order, then a graph of ln[A] on the vertical (y) axis versus t on the horizontal (x) axis should be a straight line. If such a plot is straight, the graph is evidence that the reaction is first-order.

To determine the order of reaction, then, we can collect concentration/time data and make the three plots listed in Table 12.1. Only one (or perhaps none) of the plots will be a straight line. If one is straight, then it indicates the order. The slope of the straight line can be obtained as shown earlier (← p. 515), and the rate constant can be calculated from the slope as indicated in Table 12.1. Depending on the order of the reaction, the units of the rate constant will vary. The appropriate units are indicated in the last column of Table 12.1.

An example of the slope of a graph and how to obtain the numerical value of the slope was shown in a marginal figure earlier in this chapter (← p. 515). Natural logarithms, such as ln[A] are reviewed in Appendix B.3.

PROBLEM-SOLVING EXAMPLE 12.3 Order and Rate Constant from Integrated Rate Law

The data below were obtained for decomposition of aqueous hydrogen peroxide at 20 °C.

$$2\ H_2O_2(aq) \longrightarrow 2\ H_2O(\ell) + O_2(g)$$

Use the integrated-rate-law method to obtain the order of the reaction and the rate constant.

Time (min)	$[H_2O_2]$ (mol/L)	Time (min)	$[H_2O_2]$ (mol/L)
0	0.0200	1000	0.0069
200	0.0160	1200	0.00535
400	0.0131	1600	0.0037
600	0.0106	2000	0.0024
800	0.0086		

Answer The reaction is first-order in H_2O_2; $k = 1.06 \times 10^{-3}\ \text{min}^{-1}$.

Explanation The graphs for zero-, first-, and second-order plots are shown in Figure 12.5 (see next page). It is clear that the zero-order and second-order plots are curved, while the first-order plot is a straight line. Thus the reaction must be first-order. From the first-order plot a slope can be calculated using the points marked on the graph as open circles:

$$\text{Slope} = \frac{\{-5.82 - (-4.02)\}}{\{1800 - 100\}} = -1.06 \times 10^{-3}\ \text{min}^{-1}$$

The slope of $-1.06 \times 10^{-3}\ \text{min}^{-1}$ is the negative of the rate constant (see Table 12.1), which means that $k = 1.06 \times 10^{-3}\ \text{min}^{-1}$. The units are 1/min (per minute), which corresponds to the reciprocal time units indicated in Table 12.1 for a first-order rate constant.

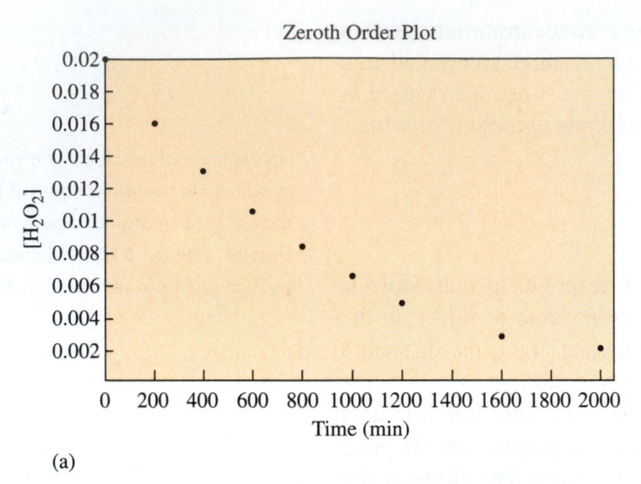

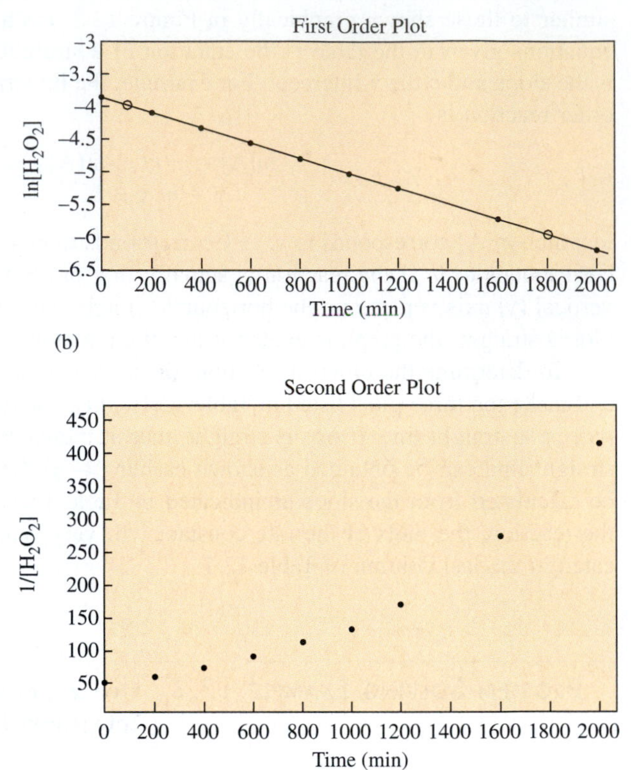

Figure 12.5 Integrated-rate-law plots for H_2O_2 reaction. (a) Zeroth-order, (b) first-order, and (c) second-order plots for the decomposition reaction of hydrogen peroxide in aqueous solution at 20 °C.

Problem-Solving Practice 12.3

The following concentration/time data were obtained for the reaction of cisplatin with water. Determine the order of the reaction and the rate constant. ($[H_2O]$ is constant.)

Time (min)	[cisplatin] (mol/L)	Time (min)	[cisplatin] (mol/L)
0	0.01000	1000	0.00230
200	0.00745	1200	0.00171
400	0.00555	1600	0.00095
600	0.00414	2000	0.00053
800	0.00309		

12.4 A NANOSCALE VIEW: ELEMENTARY REACTIONS

Macroscale, experimental observations reveal that reactant concentrations, temperature, and catalysts can affect reaction rates. But how can we interpret such observations in terms of nanoscale models? The *kinetic-molecular theory of matter*, which was first introduced in Section 1.5 (⟵ ***p. 10***), and the ideas about molecular structure developed in Chapter 10 provide a good basis for understanding how atoms and molecules move and how chemical bonds are made or broken during the very short time it takes for reactant molecules to be converted into product molecules.

The kinetic-molecular theory was introduced in Section 1.5 (⟵ ***p. 10***) and was the basis for many of the discussions in Chapters 6 and 7. Vibrations of molecules give rise to the infrared spectra discussed in Section 10.7 (⟵ ***p. 436***).

According to kinetic-molecular theory, molecules are in constant random motion. In a gas or liquid they bump into one another, while in a solid they vibrate about specific locations. Molecules also rotate, flex, or vibrate along or around the bonds that hold the atoms together. These motions produce the transformations of molecules that occur during chemical reactions. It turns out that there are only two important types of molecular transformations. The first involves a single molecule that has enough energy so that the arrangement of atoms can change, resulting in a different molecule. This might involve breaking a bond and forming two new molecules, or it might involve rearrangement of one isomeric structure into another. The second type of reaction happens when two molecules collide with sufficient energy and with appropriate orientation: new bonds may be formed and old bonds may be broken. This might involve combination of the two molecules into a new, larger one, or it might involve forming two new molecules from the old ones.

All chemical reactions can be understood in terms of simple reactions such as those just described. Very complicated reactions can be built up from combinations of those reactions, just as complicated compounds can be built from chemical elements. For example, hundreds of such reactions are needed to understand how air pollution is produced in a city such as Los Angeles, or to understand why chlorofluorocarbons can deplete stratospheric ozone. Because they are building blocks like the chemical elements, the simplest nanoscale reactions are referred to as **elementary reactions.** The equation for an elementary reaction shows exactly which molecules, atoms, or ions take part in the elementary reaction. The next two sections describe the two important types of elementary reactions.

Unimolecular Reactions

An elementary reaction is said to be **unimolecular** if rearrangement of the structure of a single molecule produces the product molecule or molecules. An example of a unimolecular reaction is the conversion of *cis*-2-butene to *trans*-2-butene:

$$\begin{array}{c}H_3C \quad\quad CH_3\\ \diagdown \quad \diagup \\ C{=}C \\ \diagup \quad \diagdown \\ H \quad\quad H\end{array}\ (g) \longrightarrow \begin{array}{c}H_3C \quad\quad H\\ \diagdown \quad \diagup \\ C{=}C \\ \diagup \quad \diagdown \\ H \quad\quad CH_3\end{array}\ (g)$$

cis-2-butene *trans*-2-butene

This equation says that the reaction is the transformation of a *cis*-2-butene molecule into a *trans*-2-butene molecule. The difference between these two molecules, which are *cis-trans* stereoisomers (⟵ ***p. 372***), is the orientation of the methyl groups. They are on the same side of the double bond in the *cis* structure, and on opposite sides in the *trans* structure. If we could grab one end of the molecule and twist it 180° around the axis of the double bond, we would get the other molecule. Thus, it is a reasonable hypothesis that the molecular pathway by which *cis*-2-butene changes to *trans*-2-butene involves twisting the molecule around the double bond. The angle of twist

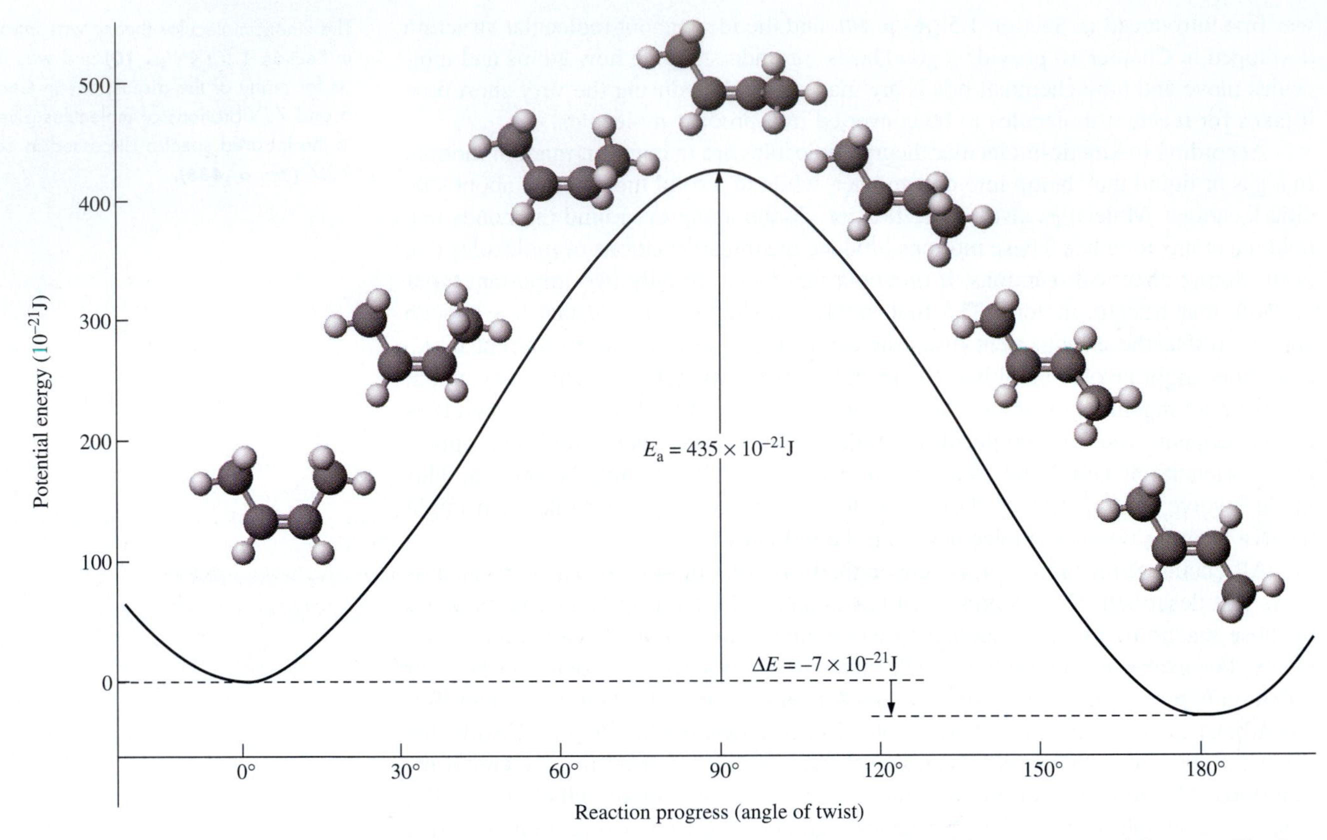

Figure 12.6 Energy profile for the conversion of *cis*-2-butene to *trans*-2-butene. The double bond between the two central C atoms is stiff and resists twisting. However, if the molecule has enough energy, one end can twist with respect to the other end. When the angle between the ends of the molecule is 90°, the potential energy has risen by 435×10^{-21} J. A molecule of *cis*-2-butene must have at least this quantity of energy before it can twist past 90° to 180°, which converts it to *trans*-2-butene. The 90° twisted structure at the top of the profile is called the transition state or the activated complex. The progress of the reaction is measured by the angle of twist.

around the double bond axis is a measure of the progress of the reaction on the nanoscale. The greater the angle, the less the molecule is like *cis*-2-butene and the more it is like *trans*-2-butene, until an angle of 180° is reached.

Such a twist requires that the reactant molecule have sufficient energy. Chemical bonds are like springs. They can be stretched, twisted, and bent, but this raises the potential energy. Consequently, some kinetic energy must be converted to potential energy when one end of the *cis*-2-butene molecule twists relative to the other, just as it would if a spring were twisted. At room temperature most of the molecules do not have enough kinetic energy to twist far enough to change *cis*-2-butene into *trans*-2-butene. Therefore *cis*-2-butene can be kept in a sealed flask at room temperature for a long time without any appreciable quantity of *trans*-2-butene being formed. However, as the temperature is raised, more and more molecules have sufficient energy to react, and the reaction gets faster and faster.

Since molecules are very small, the energy required to twist one *cis*-2-butene molecule is also very small. However, if we wanted to twist a mole of molecules, all at once, it would take a lot of energy. The energy required to reach the top of the "hill" is often reported per mole of molecules, that is, as (435×10^{-21} J/molecule)·(6.022×10^{23} molecules/mol) or 262 kJ/mol.

Figure 12.6 shows a plot of potential energy versus the angle of twist in *cis*- and *trans*-2-butene. The potential energy is 435×10^{-21} J higher when one end of a *cis*-2-butene molecule is twisted by 90° from the initial, flat molecule. This is similar to

(Text continues on page 526.)

CHEMISTRY YOU CAN DO

Kinetics and Vision

Vision is not an instantaneous process. It takes a little while after a bright light goes out before you stop seeing its image. The flash of a camera blinds you for a short time even though it is on only for an instant. Some sources of light flash on and off very rapidly, but you do not notice because your eyes continue to perceive their images while they are off. However, if you can focus such a source on different parts of your retina at different times, you can see whether it is flashing. Here's how.

Find a small mirror that you can hold easily in your hand. Now use the mirror to reflect the image of an incandescent light bulb onto your eye. You should be far enough away from the light so that its image is small. Now move the mirror quickly back and forth so that the image of the light bulb moves quickly across your eyeball. Does the light smear or do you see individual dots? Try the same experiment with the screen of a TV set. (Get really far away from it so the image is small.) Do you see separate images or just a smear of light? If you see individual images, it means that the light is flashing. Each time it flashes on, the moving mirror has caused it to hit a different part of your retina, and you see a separate image.

Repeat this experiment with as many different light sources as you can, and classify them as flashing or continuous. Try street lights of various kinds, car headlights, neon signs, fluorescent lights, and anything else you can think of. Record your observations. What do you think would happen if you photographed these light sources and moved the camera quickly while the shutter was open? What would happen if you moved the camera more slowly or more quickly?

Rotation around a double bond, as in the interconversion of *cis-* and *trans-*2-butene, occurs in the reactions that allow you to see. A yellow-orange compound, *β*-carotene, the natural coloring agent in carrots (Section 10.7, ← ***p. 435***), breaks down in your body to produce vitamin A. This compound is converted in the liver to a compound called 11-*cis*-retinal. In the retina of your eye, 11-*cis*-retinal combines with the protein opsin to form a light-sensitive substance called rhodopsin. When light strikes the retina, enough energy is transferred to a rhodopsin molecule to allow rotation around a carbon-carbon double bond, transforming rhodopsin into metarhodopsin II, a molecule whose shape is quite different, as you can see from the structural formulas below. This change in molecular shape causes a nerve impulse to be sent to your brain, and you see the light.

Eventually the metarhodopsin II reacts chemically to produce a different form of retinal, which is then converted back to vitamin A, and the cycle of chemical changes can begin again. However, decomposition of metarhodopsin II is not as rapid as its formation, and an image formed on the retina persists for a tenth of a second or so. This persistence of vision allows you to perceive videos as continuously moving images, even though they actually consist of separate pictures, each painted on a screen for a thirtieth of a second.

light →

rhodopsin

metarhodopsin II

the increased potential energy that an object like a car has at the top of a hill compared with its energy at the bottom. Just as a car cannot reach the top of a hill unless it has enough energy, a molecule cannot reach the top of the "hill" for a reaction unless it has enough energy. Notice that the top of the hill can be approached from either side, and from the top a twisted molecule can go downhill energetically to either the *cis* or the *trans* form. The structure at the top of an energy profile is called the **transition state** or **activated complex.** In this case it is a molecule that has been twisted so that the methyl groups are at a 90° angle to each other when viewed along the axis of the double bond.

The generality that higher activation energy results in slower reaction applies best if the reactions are similar. For example, it applies to a group of reactions that all involve twisting around a double bond or to reactions that involve collisions of one molecule with each of a group of similar molecules. It is less applicable when comparing one reaction that involves collision of two molecules with another that involves twisting around a bond.

Every chemical reaction has an energy barrier that must be surmounted as reactant molecules change into product molecules. The heights of such barriers vary greatly—from almost zero to hundreds of kilojoules per mole. *At a given temperature, the higher the energy barrier the slower the reaction.* The minimum energy required to surmount the barrier is called the **activation energy, E_a,** for the reaction. For the *cis*-2-butene → *trans*-2-butene reaction, the activation energy is 435×10^{-21} J/molecule, or 262 kJ/mol.

There is another interesting relationship shown in Figure 12.6 that connects kinetics and thermodynamics. The energy of the product, one molecule of *trans*-2-butene, is 7×10^{-21} J *lower* than that of the reactant, one molecule of *cis*-2-butene. This means that the *cis* → *trans* reaction is *exothermic* by 7×10^{-21} J/molecule, which translates to 4 kJ/mol. Conversely, *cis*-2-butene is *higher* in energy by 7×10^{-21} J/molecule, and so the reverse reaction requires that 4 kJ/mol be absorbed from the surroundings; it is *endothermic*. The energy hill that has to be climbed when the reverse reaction occurs is $(435 + 7) \times 10^{-21}$ J, or 442×10^{-21} J high (266 kJ/mol). Thus, the activation energy for the forward reaction is 4 kJ/mol less than that for the reverse reaction. In general the activation energy for a forward reaction will differ from the activation energy of the reverse reaction, and the difference is $\Delta E°$ for the reaction.

The actual relation is $\Delta E° = E_a$(forward) − E_a (reverse). Since $\Delta E°$ differs from $\Delta H°$ only when there is a change in volume of the reaction system (under constant pressure), the difference in activation energies is often equated with the enthalpy change (← p. 240).

Bimolecular Reactions

There are many nanoscale processes in which two particles must collide for a reaction to take place. These are called **bimolecular elementary reactions.** A good example is the reaction of iodide ion, I^-, with methyl bromide, CH_3Br, in aqueous solution.

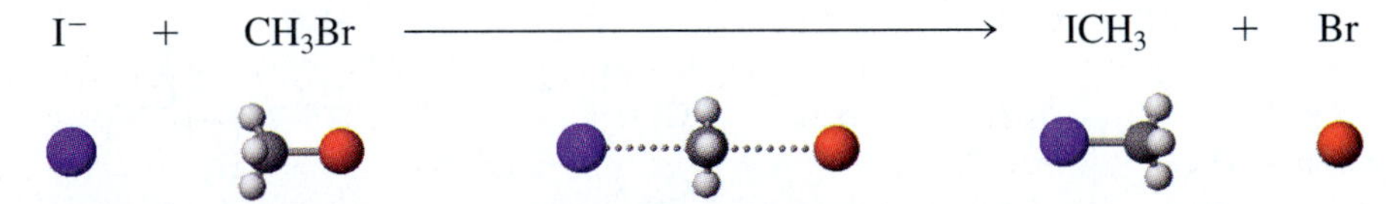

Here the equation for the elementary reaction shows that an iodide ion must collide with a methyl bromide molecule for the reaction to occur. The carbon-bromine bond does not break until after the iodine-carbon bond has begun to form. This makes sense, because just breaking a carbon-bromine bond requires a large increase in potential energy. Partially forming a carbon-iodine bond while the other bond is breaking lowers the potential energy. This helps keep the activation-energy hill low. Figure 12.7 shows the energy-versus-reaction progress diagram for this reaction. If all that happened were breaking a carbon-bromine bond, the energy hill would be much higher, and the reaction would be slower.

The word "steric" comes from the same root as the prefix *stereo-*, which means three-dimensional.

The methyl bromide molecule has a tetrahedral shape that is distorted because the Br atom is so much bigger than the H atoms. Numerous experiments suggest that the reaction occurs most rapidly in solution when the I^- ion approaches the methyl

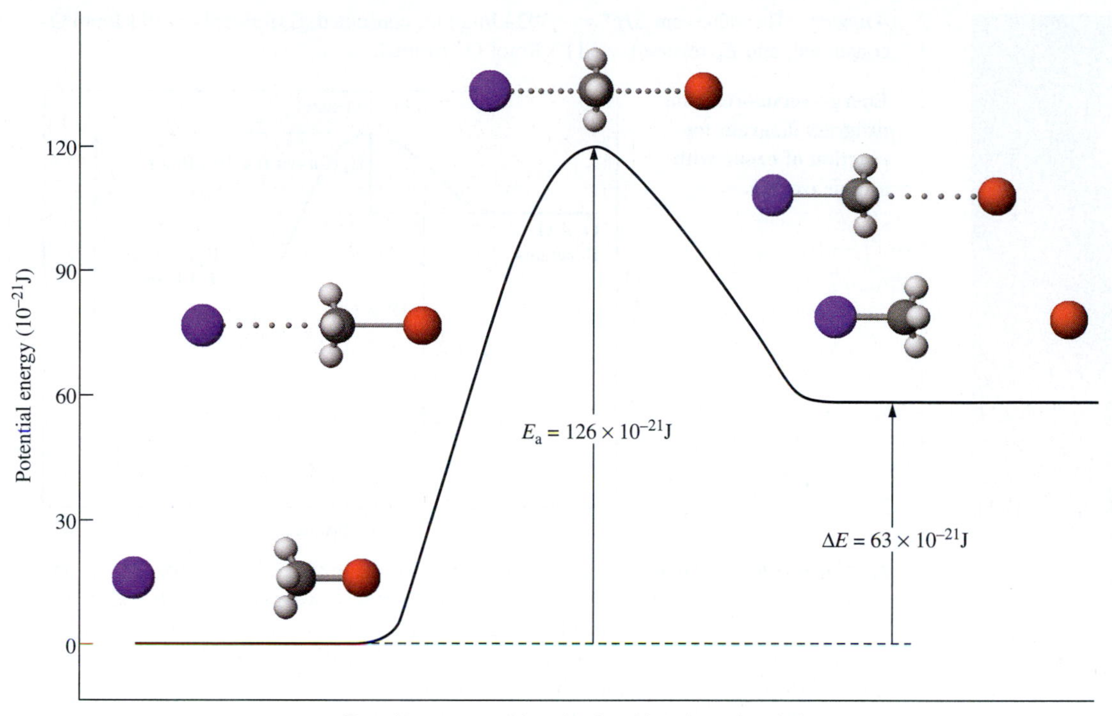

Figure 12.7 Energy profile for iodide–methyl bromide reaction. During collision of an iodide ion with a methyl bromide molecule, a new iodine-carbon bond forms at the same time that the carbon-bromine bond is breaking. Forming the new bond lowers the potential energy, which otherwise would be raised a lot by breaking the carbon-bromine bond. This results in a lower activation energy and a faster reaction than would otherwise occur. In this case reaction progress is measured in terms of stretching of the carbon-bromine bond and formation of the iodine-carbon bond.

bromide from the side of the tetrahedron opposite the bromine atom. That is, approach to only one of the four sides of CH_3Br can be effective, which limits reaction to only one fourth of all of the collisions at most. This factor of one fourth is called a **steric factor** because it depends on the three-dimensional shapes of the reacting molecules. For molecules much more complicated than methyl bromide, such geometry restraints mean that only a very small fraction of the total collisions can lead to reaction. No wonder some chemical reactions are so slow!

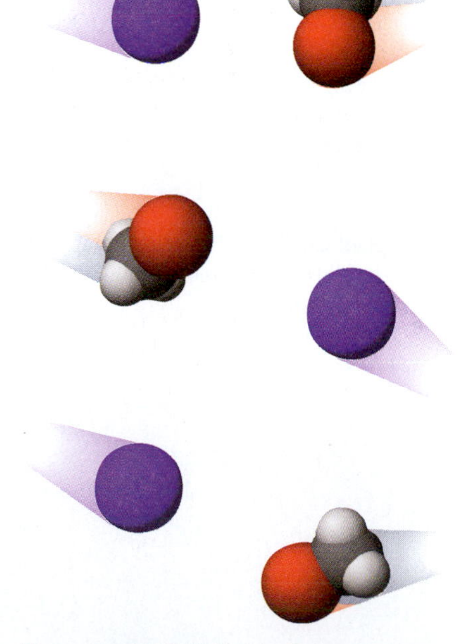

Unsuccessful Collisions. Because of the orientation of the methyl bromide molecule and the iodide ion, none of the collisions pictured above will result in reaction.

PROBLEM-SOLVING EXAMPLE 12.4 Reaction Energy Profiles

A reaction by which ozone is destroyed in the stratosphere is

$$O_3 + O \longrightarrow 2\ O_2$$

(O represents atomic oxygen, which is formed in the stratosphere when photons of sunlight split oxygen molecules in two.) The activation energy for this reaction is 19 kJ/mol of O_3 consumed. Use standard enthalpies of formation from Appendix J to calculate the enthalpy change for this reaction. Then construct an energy profile for the reaction. Draw vertical arrows to indicate the sizes of $\Delta H°$, E_a(forward), and E_a(reverse) for the reaction.

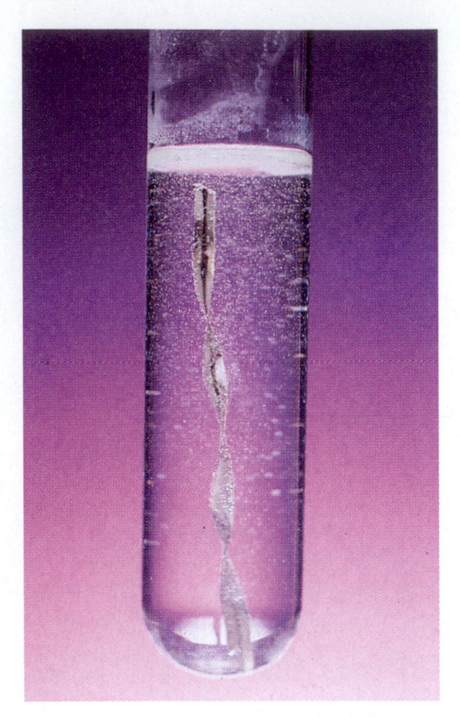

Figure 12.8 Reaction of magnesium with water. Magnesium will react only slowly with water at room temperature. However, warming the water causes a vigorous oxidation-reduction reaction that produces hydrogen gas and magnesium hydroxide:

$$Mg(s) + 2\,H_2O(\ell) \longrightarrow H_2(g) + Mg(OH)_2(s)$$

(C.D. Winters)

Answer The values are $\Delta H° = -392$ kJ/mol O_3 consumed, E_a(forward) = 19 kJ/mol O_3 consumed, and E_a(reverse) = 411 kJ/mol O_3 formed.

Energy-versus-reaction progress diagram for reaction of ozone with atomic oxygen.

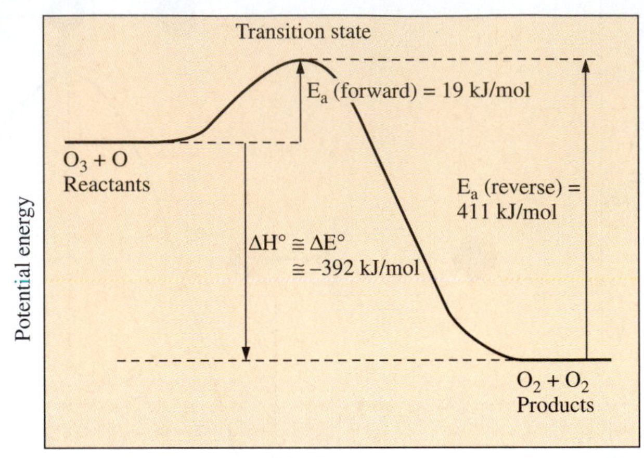

Explanation Standard enthalpies of formation are −249.2 kJ/mol for ozone and −142.7 kJ/mol for atomic oxygen. Using these values, we get $\Delta H° = 0 - (249.2 + 142.7)$ kJ/mol = −391.9 kJ/mol O_3 consumed. This indicates that the reaction is exothermic, and so the products must be lower in energy than the reactants by 391.9 kJ/mol. Since E_a(forward) = 19 kJ/mol, the transition state must be this much higher in energy than the reactants. Thus the first two arrows on the left side of the diagram below can be drawn. Then the third arrow (from products to the transition state) can be drawn. It indicates that E_a (reverse) = $-\Delta H° + E_a$(forward) = (391.9 + 19) kJ/mol = 411 kJ/mol.

Problem-Solving Practice 12.4

For the hypothetical reaction A → B, the activation energy is 24 kJ/mol. For the reverse reaction, B → A, the activation energy is 36 kJ/mol. Draw a diagram similar to Figure 12.5 for this reaction. Is the reaction A → B exothermic or endothermic?

12.5 TEMPERATURE AND REACTION RATE

Conceptual Challenge Problem CP-12.A at the end of this chapter relates to topics discussed in this section.

The most common way to speed up a reaction is to increase the temperature. A mixture of methane and air is stable for centuries at room temperature, because the reaction to give CO_2 and H_2O is extraordinarily slow. A lighted match raises the temperature of the mixture of reactants so that more molecules have enough energy to surmount the activation-energy barrier. This increases the reaction rate and the gas ignites. Thereafter the thermal energy evolved by the combustion reaction maintains a high temperature and the reaction continues at a rapid rate.

Another case where higher temperature increases reaction rate is displacement of hydrogen from water by magnesium (Figure 12.8). Reactions that speed up when the temperature is raised must slow down when the temperature is lowered. Foods are stored in refrigerators or freezers because the rates of reactions that cause spoilage are slower at the lower temperature.

As a rough rule of thumb, the reaction rate increases by a factor of 2 to 4 for each 10° C rise in temperature.

Reaction rates increase with temperature because at a higher temperature a greater fraction of reactant molecules has enough energy to surmount the activation-energy barrier. Consider again the conversion of *cis*- to *trans*-2-butene (Figure 12.6, ← ***p. 524***). The molecules are constantly in motion, and they have a wide distribution of speeds

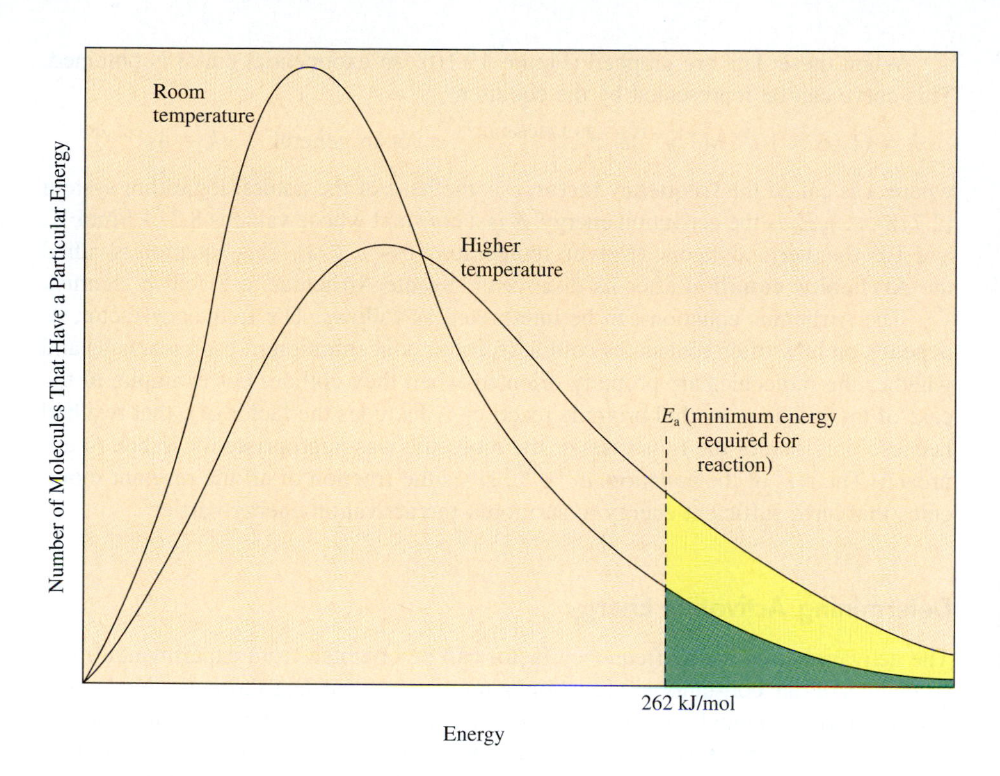

Figure 12.9 Energy distribution curves. The vertical axis gives the number of molecules that have the energy shown on the horizontal axis. For example, the two curves show that at the higher temperature more molecules have 262 kJ/mol energy than they do at the lower temperature. (The higher-temperature curve is above the lower-temperature curve at an energy of 262 kJ/mol.) If we assume that any molecule can react if it has more energy than the activation energy (262 kJ/mol), then the number of reactive molecules at each temperature is given by the area under the curve to the right of 262 kJ/mol. The number of molecules that would react at room temperature is represented by the green area. The number that would react at the higher temperature is represented by the yellow area (plus the green area), which is much larger, even though the temperature is only about 20 degrees higher.

and energies. At room temperature relatively few *cis*-2-butene molecules are sufficiently energetic to surmount the energy barrier. However, as the temperature goes up, the number of molecules that have enough energy goes up rapidly, and so the reaction rate increases rapidly.

The number of *cis*-2-butene molecules that have a given energy is shown by the curves in Figure 12.9. One curve is for room temperature and the other for a temperature approximately 20° higher. The higher a point on either curve, the greater the number of molecules that have the energy corresponding to that point. The areas under the two curves are the same, because the area under each curve represents the total number of molecules. The shaded areas indicate that with about a 20° rise in temperature the number of molecules whose energy exceeds the activation energy of 266 kJ/mol is much higher. This means that the reaction rate will also be much higher.

The increased rate of a reaction observed at a higher temperature is attributed to an increase in the rate constant. That is, a rate constant is constant only for a given reaction at a given temperature. For the same reaction, the rate constant is usually larger the higher the temperature. For example, for the reaction of iodide ion with methyl bromide, the data shown in the table below are found for the rate constant at different temperatures.

T (K)	k (L/mol·s)	T (K)	k (L/mol·s)
273	4.18×10^{-5}	330	1.39×10^{-2}
280	9.68×10^{-5}	340	3.14×10^{-2}
290	3.00×10^{-4}	350	6.80×10^{-2}
300	8.60×10^{-4}	360	1.41×10^{-1}
310	2.31×10^{-3}	370	2.81×10^{-1}
320	5.82×10^{-3}		

When these data are graphed (Figure 12.10), an exponential curve is obtained. This curve can be represented by the equation

$$k = (1.66 \times 10^{10}\ \text{M}^{-1}\text{s}^{-1})e^{(-76.3\ \text{kJ/mol})/RT} \qquad \text{or in general} \qquad k = Ae^{-E_a/RT}$$

R is called the ideal gas constant and is the same constant found in the ideal gas law, $PV = nRT$, which will be discussed in Chapter 14. Here it is expressed in units of J/mol·K instead of L·atm/mol·K, which is the reason that the numerical value is not 0.0821 L·atm/mol·K.

where A is called the **frequency factor,** e is the base of the natural logarithm system (2.718 . . .), E_a is the activation energy, R is a constant whose value is 8.314 J/mol·K, and T is the thermodynamic (Kelvin) temperature (← *p. 273*). This equation is called the **Arrhenius equation** after its discoverer, Svante Arrhenius, a Swedish chemist.

The Arrhenius equation can be interpreted as follows. The frequency factor, A, depends on how often molecules collide (for unit concentration of each reactant) and whether the molecules are properly oriented when they collide. For example, in the case of the iodide + methyl bromide reaction, A includes the factor of $\frac{1}{4}$ that resulted because only one of the four sides of the molecule was appropriate for iodide to approach. The rest of the equation, $e^{-E_a/RT}$, gives the fraction of all the reactant molecules that have sufficient energy to surmount the activation-energy barrier.

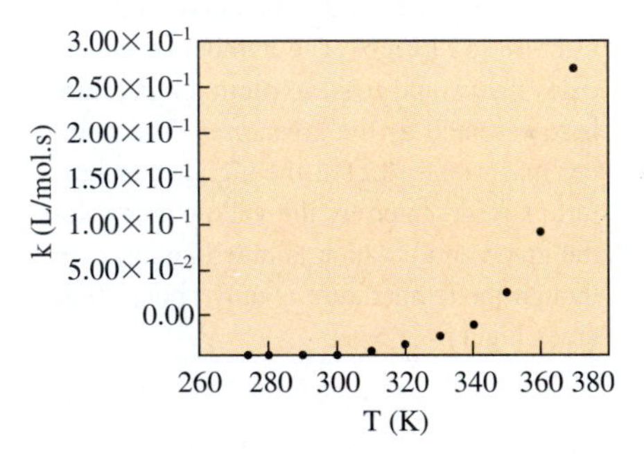

Figure 12.10 Effect of temperature on rate constant. The rate constant for the reaction of iodide ion with methyl bromide in aqueous solution is plotted as a function of temperature. The rate constant increases exponentially with temperature.

Determining Activation Energy

The activation energy and frequency factor can be obtained from experimental measurements of rate constants as a function of temperature (such as those provided above for the iodide + methyl bromide reaction). When a large number of experimental data pairs are given, a graph is usually a good way of obtaining information from the data. However, the graph shown in Figure 12.10 is not linear, and so it is hard to obtain the activation energy from it. The activation-energy equation can be modified so that it gives a linear graph by taking natural logarithms of both sides:

$$k = Ae^{-E_a/RT}$$
$$\ln(k) = \ln(A) + \ln(e^{-E_a/RT}) = \ln(A) + (-E_a/RT)$$

Rearranging this equation gives

$$\ln(k) = \frac{E_a}{R} \times \frac{1}{T} + \ln(A)$$
$$y = m \times x + b$$

That is, if we graph $\ln(k)$ on the vertical (y) axis and $1/T$ on the horizontal (x) axis, there should be a straight line whose slope is $-E_a/R$ and whose y-intercept is $\ln(A)$. Such a graph is shown in Figure 12.11. It is linear and has slope $m = -9.18 \times 10^3$ K

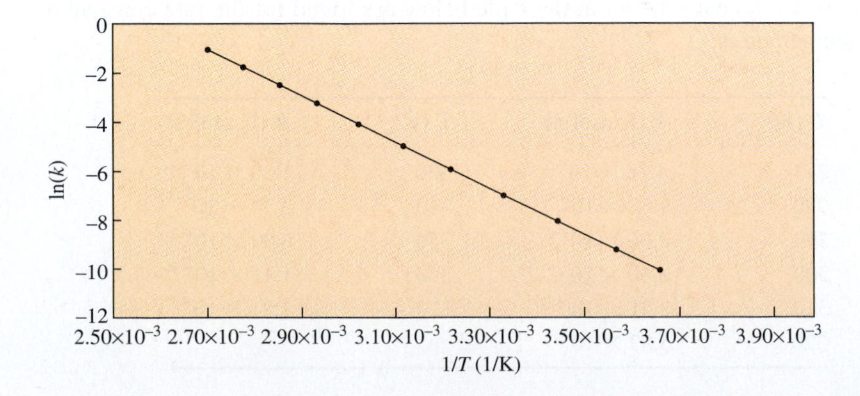

Figure 12.11 Determining activation energy graphically. A graph of $\ln(k)$ vs $1/T$ gives a straight line for the iodide + methyl bromide reaction. The activation energy can be obtained from the slope of the line, and the frequency factor from its y-intercept. Because the x-axis scale does not begin at zero, the y-intercept is not visible on the graph.

and intercept $b = 23.53$. Since the slope $= -E_a/R$, the activation energy can be calculated as

$$E_a = -(\text{slope}) \times R = -(9.18 \times 10^3 \text{ K}) \times (8.314 \text{ J/mol·K}) = 76.3 \text{ kJ/mol}$$

Since $\ln(A) = 23.53$, the frequency factor, A, is $e^{23.53}$ L/mol · s = 1.66×10^{10} L/mol · s for this reaction.

The Arrhenius equation can be used to calculate the rate constant at any temperature. For example, the rate constant for reaction of iodide ion with methyl bromide can be calculated at 50 °C by substituting the Kelvin temperature, the frequency factor, the activation energy, and the constant R into the equation:

$$k = Ae^{-E_a/RT} = (1.66 \times 10^{10} \text{ L/mol} \cdot \text{s})e^{(-76{,}300 \text{ J/mol})/(8.314 \text{ J/K·mol})(273.15 + 50)\text{K}}$$
$$= 7.70 \times 10^{-3} \text{ L/mol} \cdot \text{s}$$

As a means of calculating rate constants, the Arrhenius equation works best within the range of temperatures over which the activation energy and frequency factor were determined. (For the reaction of iodide with methyl bromide, that range was 273 K to 370 K.)

If we substitute the Arrhenius equation for the rate constant into the rate law for the reaction, we have

$$\text{Rate} = k \times [I^-] \times [CH_3Br]$$
$$= Ae^{-E_a/RT} \times [I^-] \times [CH_3Br]$$

collision frequency × steric factor (A); fraction of sufficiently energetic molecules ($e^{-E_a/RT}$); concentrations of colliding molecules ($[I^-] \times [CH_3Br]$)

Recall that the collision frequency in the frequency factor is for 1M concentrations.

This equation summarizes the effects of both temperature and concentration on rate of a reaction. The temperature effect depends primarily on the large increase in the number of sufficiently energetic collisions as the temperature increases, and this shows up as larger values of k at higher temperatures. The effect of concentration is clearly indicated by the concentration terms in the rate law. If the rate law is known for a reaction, and if both the A and E_a values are known, then the rate can be calculated over a wide range of conditions.

Exercise 12.3 Activation Energy and Experimental Data

The frequency factor is 6.31×10^8 L/mol · s and the activation energy is 10 kJ/mol for the gas-phase reaction

$$NO + O_3 \longrightarrow NO_2 + O_2$$

which is important in the chemistry of stratospheric ozone depletion. Calculate the rate constant for this reaction at 370 K. Assuming that this equation represents an elementary reaction, calculate the rate of the reaction at 370 K if [NO] = 0.0010 M and $[O_3]$ = 0.00050 M.

12.6 RATE LAWS FOR ELEMENTARY REACTIONS

An elementary reaction is a one-step process whose equation describes on the nanoscale which particles are directly involved when the reaction occurs. Thus, it is possible to figure out what the reaction order will be without doing an experiment.

This differs from macroscale reactions *(← p. 517),* where rate laws and reaction orders must be determined experimentally because there is no direct way of knowing how the reaction occurs. In fact, the experimentally determined macroscale rate law can be used to support a nanoscale theory about how a particular reaction takes place. In this and the next section we develop these ideas.

Rate Law for a Unimolecular Reaction

In Section 12.4 we used the isomerization of *cis*-2-butene as an example of a reaction where a single reactant molecule was converted to products—a unimolecular reaction *(← p. 523).*

$$cis\text{-2-butene} \longrightarrow trans\text{-2-butene}$$

Suppose that the fraction of molecules that have enough energy to react is 0.1%, or 0.001. If we have 10,000 molecules in a given volume, then $0.001 \times 10{,}000$ gives only 10 that will have enough energy to react. If we have twice as many molecules in the same volume, that is, 20,000 molecules, then $0.001 \times 20{,}000$ gives 20 with enough energy to react, and the number reacting per unit volume (the rate) is twice as great.

Suppose a flask contains 0.005 mol/L of *cis*-2-butene vapor at room temperature. The molecules have a wide range of energies, but only a few of them have enough energy at this temperature to get over the activation-energy barrier. Thus, during a given period only a few molecules twist sufficiently to become *trans*-2-butene. Now suppose that we double the concentration of *cis*-2-butene in the flask to 0.010 mol/L, keeping the temperature the same. The fraction of molecules with enough energy to cross over the barrier remains the same. However, since there are now twice as many molecules, there must be twice as many crossing the barrier in any given time. Therefore, the rate of the *cis* → *trans* reaction is twice as great. That is, the reaction rate is proportional to the concentration of *cis*-2-butene, and the rate equation must be

$$\text{Rate} = k[cis\text{-2-butene}]$$

In the general case of any unimolecular elementary reaction,

$$\text{A} \longrightarrow \text{products} \qquad \text{the rate law is} \qquad \text{rate} = k[\text{A}]$$

That is, *for any unimolecular reaction the nanoscale mechanism predicts that a first-order rate equation will be observed in a macroscale laboratory experiment.*

Rate Law for a Bimolecular Reaction

A good example of a reaction in which two molecules collide (a bimolecular reaction, *(← p. 526)* is the gas-phase reaction of nitrogen monoxide and ozone that is involved in stratospheric ozone depletion and was mentioned in Exercise 12.3.

$$NO + O_3 \longrightarrow NO_2 + O_2$$

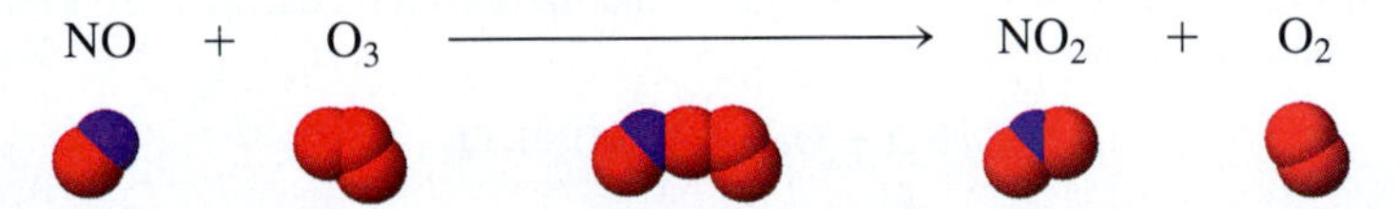

Here the equation shows that the elementary reaction involves the collision of an NO molecule and an O_3 molecule. Since the molecules must collide to exchange atoms, the rate depends on the number of collisions per unit time.

Figure 12.12a represents a portion of a flask containing one NO molecule (the red ball) and several O_3 molecules (the blue balls). In a given time, the NO molecule collides with five O_3 molecules. If the concentration of NO molecules is doubled to two in the same portion of the flask (Figure 12.12c), *each* NO collides with five different O_3 molecules, and so the total number of collisions with O_3 molecules is now ten. Doubling the concentration of NO has doubled the number of collisions and hence

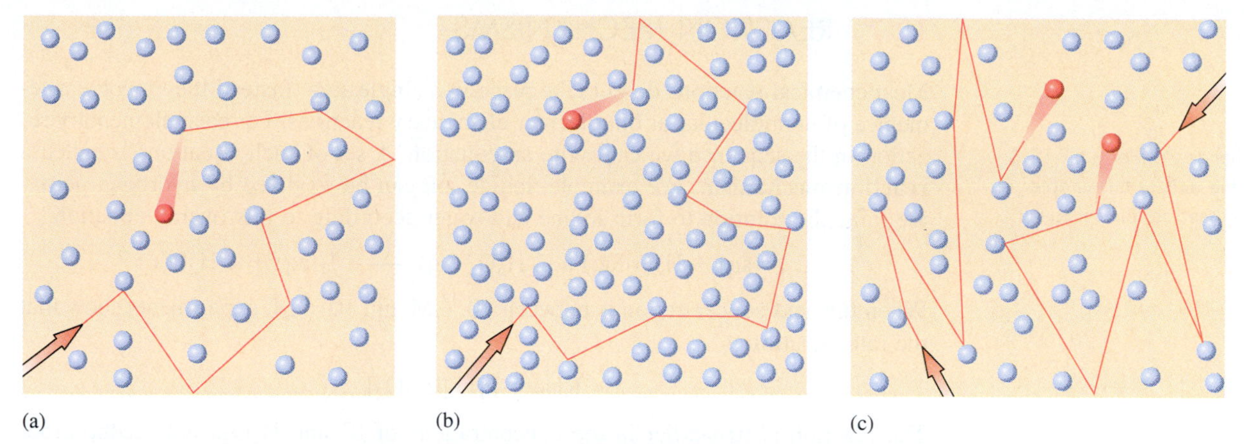

(a) (b) (c)

Figure 12.12 Effect of concentration on frequency of bimolecular collisions. (a) A single red molecule moves among 50 blue molecules and collides with 5 of them per second. (b) If the number of blue molecules is doubled to 100, the frequency of red-blue collisions is also doubled, to 10 per second. (c) Two red molecules now move among 50 blue molecules, and there are 10 red-blue collisions per second. The number of collisions is thus seen to be proportional to *both* the concentration of red molecules *and* the concentration of blue molecules.

the rate. The number of collisions also doubles when the O_3 concentration is doubled (Figure 12.12b). Thus the rate law for this reaction must be

$$\text{Rate} = k[\text{NO}][\text{O}_3]$$

This description of the NO + O_3 reaction applies in general to bimolecular reactions, even if the two molecules that must collide are of the same kind. That is, for the elementary reaction

$$\text{A} + \text{B} \longrightarrow \text{products} \qquad \text{the rate law is} \qquad \text{Rate} = k[\text{A}][\text{B}]$$

and for the elementary reaction

$$\text{A} + \text{A} \longrightarrow \text{products} \qquad \text{the rate law is} \qquad \text{Rate} = k[\text{A}]^2$$

For the NO + O_3 reaction the experimental rate equation is the same as the one we just derived by assuming the reaction was elementary, that is, by assuming it occurred in one step. The equation for the reaction is

$$\text{NO} + \text{O}_3 \longrightarrow \text{NO}_2 + \text{O}_2 \quad \text{and the experimental rate law is} \quad \text{Rate} = k[\text{NO}][\text{O}_3]$$

This experimental observation suggests, but does not prove, that the reaction does take place in a single step. There is a good deal of other evidence that also suggests that this reaction is bimolecular. On the other hand, for the decomposition of hydrogen peroxide, which was mentioned in Example 12.3 (⇐ *p. 521*), the equation is

$$2\,\text{H}_2\text{O}_2 \longrightarrow 2\,\text{H}_2\text{O} + \text{O}_2 \quad \text{and the experimental rate law is} \quad \text{Rate} = k[\text{H}_2\text{O}_2]$$

This rate law proves that this reaction *cannot* occur in a single step that involves collision of two H_2O_2 molecules. A single-step, bimolecular reaction would have a second-order rate law, but the observed rate law is first-order. We can account for observations like this one by assuming that more than a single elementary step is needed when hydrogen peroxide decomposes.

12.7 REACTION MECHANISMS

Conceptual Challenge Problems CP-12.B and CP-12.C at the end of this chapter relate to topics discussed in this section.

Most chemical reactions do not take place in a single step. Instead they involve a sequence of unimolecular or bimolecular elementary reactions. For each elementary reaction in the sequence we can write an equation. A set of such equations is called a **reaction mechanism.** For example, iodide ion can be oxidized by hydrogen peroxide in acidic solution to form iodine and water according to this overall equation:

$$2\ I^-(aq) + H_2O_2(aq) + 2\ H_3O^+(aq) \longrightarrow I_2(aq) + 4\ H_2O(\ell)$$

When the acid concentration is between 10^{-3} M and 10^{-5} M, experiments show that the rate equation is

$$\text{Rate} = k[I^-][H_2O_2]$$

The reaction is first-order in the concentrations of I^- and H_2O_2, and second-order overall. The low concentration of hydronium ions does not affect the reaction rate.

Looking at the balanced equation for the oxidation of iodide ion by hydrogen peroxide, you might think that two iodide ions, one hydrogen peroxide molecule, and two hydronium ions would all have to come together at once. However, it is highly unlikely that these five ions or molecules would all be at the same place and properly oriented at the same time. Instead, those who have studied this reaction propose that first an H_2O_2 molecule (HOOH) and an I^- ion come together:

Step 1: $$HOOH + I^- \xrightarrow{\text{slow}} HOI + OH^-$$

This step forms hypoiodous acid, HOI, and hydroxide ion, both known substances. The HOI then reacts with another I^- to form the product I_2:

Step 2: $$HOI + I^- \xrightarrow{\text{fast}} I_2 + OH^-$$

In each of steps 1 and 2, a hydroxide ion is produced. Since the solution is acidic, these OH^- ions will react immediately with H_3O^+ ions to form water:

Step 3: $$2\ OH^- + 2\ H_3O^+ \xrightarrow{\text{fast}} 4\ H_2O$$

Each of the three steps in this mechanism is an elementary reaction and has its own activation energy, E_a, and its own rate constant, k. When the three steps are summed by putting all the reactants on the left, putting all the products on the right, and eliminating formulas that appear as both reactants and products, the overall stoichiometric equation is obtained. *It is a requirement that any valid mechanism must consist of a series of unimolecular or bimolecular elementary reaction steps that sum to the overall reaction.*

$$\textit{Step 1:} \quad HOOH + I^- \xrightarrow{\text{slow}} \cancel{HOI} + \cancel{OH^-}$$

$$\textit{Step 2:} \quad \cancel{HOI} + I^- \xrightarrow{\text{fast}} I_2 + \cancel{OH^-}$$

$$\textit{Step 3:} \quad 2\ \cancel{OH^-} + 2\ H_3O^+ \xrightarrow{\text{fast}} 4\ H_2O$$

$$\textit{Overall:}\ 2\ I^- + HOOH + 2\ H_3O^+ \longrightarrow I_2 + 4\ H_2O$$

Step 1 of the mechanism is labeled as slow, while steps 2 and 3 are fast. Step 1 is called the **rate-limiting step;** because it is the slowest in the sequence, it limits the rate at which I_2 and H_2O can be produced. Steps 2 and 3 are rapid and therefore not rate-limiting. As soon as some HOI and OH^- are produced by step 1, they are transformed into I_2 and H_2O by steps 2 and 3. *The rate of the overall reaction is limited by, and equal to, the rate of the slowest step in the mechanism.*

An analogy to the rate-limiting or rate-determining step is that no matter how quickly you shop in the supermarket, it seems that the time needed to get out of the store depends on the rate at which you move through the check-out line.

Step 1 is a bimolecular elementary reaction. Therefore its rate must be first-order in HOOH and first-order in I^-. The mechanism predicts that the rate equation should be

$$\text{Reaction rate} = k[\text{HOOH}][\text{I}^-]$$

which agrees with the experimentally observed rate equation. *A valid mechanism must correctly predict the experimentally observed rate equation.*

Intermediates

The species HOI and OH^- are produced in step 1 and used up in steps 2 or 3. Atoms, molecules, or ions that are produced in one step and used up in a later step are called **reaction intermediates** or just **intermediates.** Very small concentrations of HOI and OH^- are produced while the reaction is going on. Once the HOOH, the I_2, or both are used up, these intermediates are consumed by steps 2 and 3 and disappear. HOI and OH^- are crucial to the reaction mechanism, but neither of them appears in the overall stoichiometric equation. If an experimenter is proficient enough to demonstrate that a particular intermediate was present, this provides additional evidence that a mechanism involving that intermediate is the correct one.

The relation between the rate law and the mechanism of a reaction can be summarized by the rule that *the transition state for the rate-limiting step in a mechanism must include all of the atoms in all of the atoms, molecules, or ions whose concentrations appear in the rate law.* If a concentration is squared, then there must be enough atoms in the transition state to account for two of those molecules or ions. If the concentration is raised to a higher power in the rate law, then proportionally more atoms must be present. These atoms will not be in the same molecular form as they are in the rate-law equation, but the number of atoms of each type in the activated complex can be determined from the rate law.

Studying the kinetics of a chemical reaction involves collecting data on the concentrations of reactants as a function of time. From such data the rate law for the reaction and a rate constant can usually be obtained. The reaction can also be studied at several different temperatures to determine its activation energy. This allows us to predict how fast the macroscale reaction will go under a variety of experimental conditions, but it does not provide definitive information about the nanoscale mechanism by which the reaction takes place. A reaction mechanism is an educated guess—a hypothesis—about the way the reaction occurs. If the mechanism predicts correctly the overall stoichiometry of the reaction and the experimentally determined rate law, then it is a reasonable hypothesis. However, it is impossible to prove for certain that a mechanism is correct. Sometimes several mechanisms can agree with the same set of experiments. This is what makes kinetic studies one of the most interesting and rewarding areas of chemistry, but it also can provoke disputes among scientists who favor different possible mechanisms.

Problem-Solving Example 12.5 Rate Law and Reaction Mechanism

The gas-phase reaction between nitrogen monoxide and oxygen

$$2\ NO + O_2 \longrightarrow 2\ NO_2$$

is found experimentally to obey the rate law

$$\text{Rate} = k[NO]^2\ [O_2]$$

Decide which of the following mechanisms is (are) compatible with this rate law.

(a)	$NO + NO \longrightarrow N_2O_2$	fast
	$N_2O_2 + O_2 \longrightarrow 2\ NO_2$	slow
(b)	$NO + NO \longrightarrow NO_2 + N$	slow
	$N + O_2 \longrightarrow NO_2$	fast
(c)	$NO + O \longrightarrow NO_2$	fast
	$NO_2 + NO \longrightarrow N_2O_3$	fast
	$N_2O_3 + O \longrightarrow 2\ NO_2$	slow
(d)	$NO + O_2 \longrightarrow NO_3$	fast
	$NO_3 + NO \longrightarrow 2\ NO_2$	slow
(e)	$NO + O_2 \longrightarrow NO_2 + O$	slow
	$NO + O \longrightarrow NO_2$	fast

Answer Mechanisms (a) and (d) are compatible with the rate law and stoichiometry.

Explanation Examine each mechanism to see whether it consists only of unimolecular and bimolecular steps, agrees with the overall stoichiometry, and predicts the experimental rate law. Eliminate those that do not. The remaining mechanism(s) may be correct.

All of the mechanisms above consist of bimolecular steps, and so none can be eliminated by this criterion.

Mechanism (c) does not have O_2 as a reactant in the overall stoichiometry, and so it can be eliminated. All other mechanisms predict the observed overall stoichiometry.

The rate law has concentration of NO squared, which implies that there should be 2 N and 2 O atoms in the transition state, and concentration of O_2 to the first power, which implies 2 O atoms more. This gives a total of 2 N and 4 O in the transition state.

The slow (rate-limiting) step in mechanism (a) involves N_2O_2 and O_2, which gives a total of 2 N and 4 O. Consequently mechanism (a) could be the actual mechanism.

The slow step in mechanism (b) involves 2 N but only 2 O atoms, which eliminates it from consideration.

The slow step in mechanism (c) involves 2 N and 4 O, but this mechanism has already been eliminated because it does not have the observed stoichiometry.

Continuing this kind of reasoning, mechanism (d) is seen to be a possibility, but mechanism (e) can be eliminated from further consideration. Because there are still two possible mechanisms, (a) and (d), additional experiments would need to be done to try to distinguish between them.

Problem-Solving Practice 12.5

The Raschig reaction produces the industrially important reducing agent hydrazine, N_2H_4, from ammonia, NH_3, and hypochlorite ion, OCl^-, in basic aqueous solution. A proposed mechanism is given on the next page.

Step 1: $NH_3(aq) + OCl^-(aq) \xrightarrow{\text{slow}} NH_2Cl(aq) + OH^-(aq)$

Step 2: $NH_2Cl(aq) + NH_3(aq) \xrightarrow{\text{fast}} N_2H_5^+(aq) + Cl^-(aq)$

Step 3: $N_2H_5^+(aq) + OH^-(aq) \xrightarrow{\text{fast}} N_2H_4(aq) + H_2O(\ell)$

What is the overall stoichiometric equation? Which step is rate-limiting? What reaction intermediates are involved? What rate law is predicted by this mechanism?

12.8 CATALYSTS AND REACTION RATE

Raising the temperature increases a reaction rate because it increases the fraction of molecules that are energetic enough to cross a potential energy barrier. Increasing reactant concentrations can also increase the rate because it increases the number of molecules capable of reacting. A third way to increase reaction rates is to add a catalyst. A **catalyst** is a substance that increases the rate of a reaction but is not consumed in the overall reaction.

For example, you have already learned that hydrogen peroxide, H_2O_2, can decompose to water and oxygen.

$$2\ H_2O_2(\ell) \longrightarrow O_2(g) + 2\ H_2O(\ell)$$

If the peroxide is stored in a cool, dark place in a clean plastic container, it remains unreacted for months. The rate of the decomposition reaction is exceedingly slow. However, in the presence of a manganese salt, an iodide-containing salt, or a biological substance called an enzyme, the reaction can occur with explosive speed (Figure 12.13). In fact, an insect called a bombardier beetle uses a very similar reaction as its defense mechanism (Figure 12.14). By combining the organic compound hydroquinone with the peroxide in the presence of an enzyme, it produces a small, but sufficient, quantity of superheated steam and an irritating chemical to spray its enemies.

$$\underset{\text{hydroquinone}}{C_6H_4(OH)_2} + H_2O_2 \xrightarrow[\text{a catalyst}]{\text{enzyme,}} \underset{\text{quinone, an irritant}}{C_6H_4O_2} + 2\ H_2O \quad \text{(highly exothermic)}$$

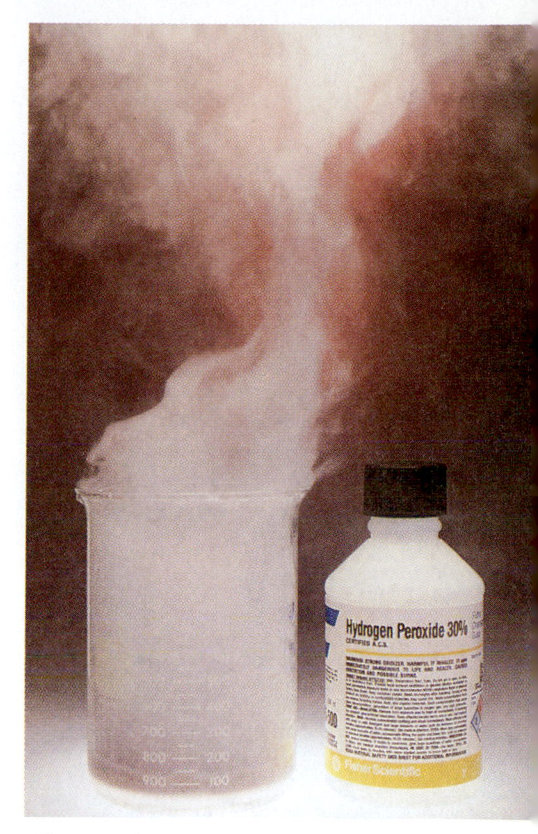

Figure 12.13 Catalysis of decomposition of hydrogen peroxide. Decomposition of hydrogen peroxide can be accelerated by the heterogeneous catalyst MnO_2. Here a 30% aqueous solution of H_2O_2 is dropped onto the black solid MnO_2 and rapidly decomposes to O_2 and H_2O. The water is given off as steam because of the high exothermic heat of reaction. *(C.D. Winters)*

How does a catalyst or an enzyme help a reaction to go faster? It does so by participating in the reaction mechanism. That is, the mechanism for a catalyzed reaction is different from the mechanism of the same reaction without the catalyst. The rate-limiting step in the catalyzed mechanism has a lower activation energy and therefore is faster than the slow step for the uncatalyzed reaction. To see how this works, let us again consider conversion of *cis*- to *trans*-2-butene in the gas phase.

Figure 12.14 Bombardier beetle. A bombardier beetle uses the enzyme-catalyzed decomposition of hydrogen peroxide as a defense mechanism. The rapid, exothermic reaction lets the insect eject steam and other irritating chemicals with explosive force. *(Thomas Eisner with Daniel Aneshansley)*

$$\text{cis-2-butene} \longrightarrow \text{trans-2-butene} \qquad \text{Rate} = k[\textit{cis}\text{-2-butene}]$$

cis-2-butene: H_3C and CH_3 on the same side of C=C, H and H on the other side. *trans*-2-butene: H and CH_3 on top, H_3C and H below.

If a trace of gaseous molecular iodine, I_2, is added to a sample of *cis*-2-butene, the iodine accelerates the change to *trans*-2-butene. The iodine is neither consumed nor produced in the overall reaction, and so it does not appear in the overall balanced equation. However, since the reaction rate depends on the concentration of I_2, there is a term involving concentration of I_2 in the rate law for the catalyzed reaction:

$$\text{Rate} = k[\textit{cis}\text{-2-butene}][I_2]^{1/2}$$

An exponent of $\frac{1}{2}$ for the concentration of I_2 in the rate equation indicates the square root of the concentration. A square-root dependence usually means that only half a molecule, in this case a single I atom, is involved in the mechanism.

The rate of the conversion of *cis*- to *trans*-2-butene changes because the presence of I_2 somehow changes the reaction mechanism. The best hypothesis is that iodine molecules first dissociate to form iodine atoms.

Step 1: I_2 dissociation

$$\tfrac{1}{2}\,I_2 \longrightarrow I$$

(This equation has a coefficient of $\frac{1}{2}$ for I_2 to emphasize that only one of the two I atoms from the I_2 molecule is needed in subsequent steps of the mechanism.) An iodine atom then attaches to the *cis*-2-butene molecule, breaking half of the double bond between the two central carbon atoms and allowing the ends of the molecule to twist freely relative to each other.

Step 2: Attachment of I atom to *cis*-2-butene

$$I + (H_3C)(H)C{=}C(CH_3)(H) \longrightarrow I{-}C(CH_3)(H){-}C(CH_3)(H)$$

cis-2-butene

Step 3: Rotation around the C—C bond

$$I{-}C(H_3C)(H){-}C(CH_3 \text{ up})(H \text{ down}) \longrightarrow I{-}C(H_3C)(H){-}C(H \text{ up})(CH_3 \text{ down})$$

Step 4: Loss of an I atom and re-formation of the carbon-carbon double bond

$$I{-}C(H_3C)(H){-}C(H)(CH_3) \longrightarrow (H_3C)(H)C{=}C(H)(CH_3) + I$$

trans-2-butene

After the double bond re-forms to give *trans*-2-butene, and the iodine atom falls away, two iodine atoms come together to re-form molecular iodine.

Step 5: I_2 formation

$$I \longrightarrow \tfrac{1}{2}\,I_2$$

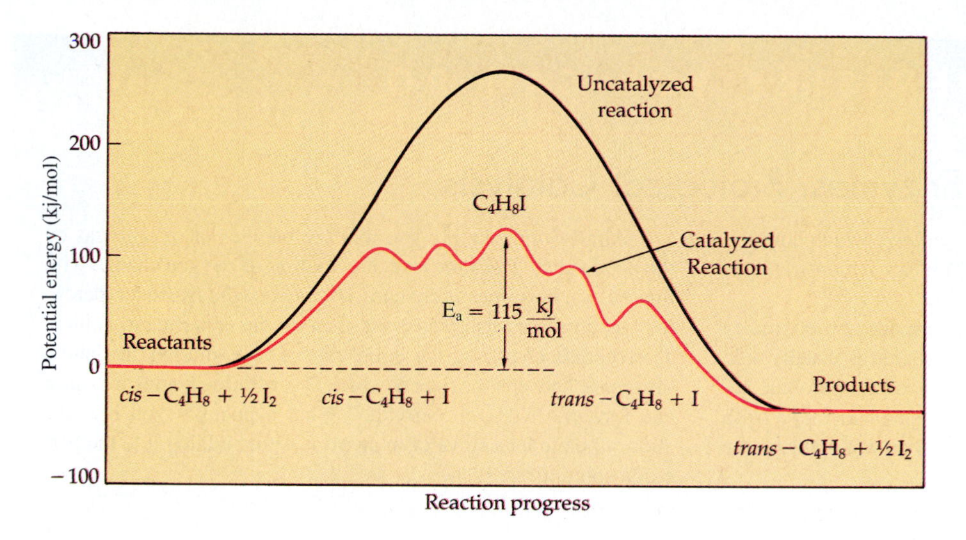

Figure 12.15 Energy profiles for catalyzed and uncatalyzed reactions. A catalyst accelerates a reaction by altering the mechanism so that the activation energy is reduced. With a smaller barrier to overcome, there are more reactant molecules with enough energy to cross the barrier, and reaction occurs more readily. The energy profile for the uncatalyzed conversion of *cis*-2-butene to *trans*-2-butene is shown by the black curve, and that for the iodine-catalyzed reaction is represented by the red curve. Notice that the shape of the barrier has changed because the mechanism has changed. See the text for a description of the steps involved.

There are four important points concerning this mechanism.

- The I_2 dissociates to atoms and then re-forms. To an "outside" observer the concentration of I_2 is unchanged; I_2 is not involved in the balanced, stoichiometric equation even though it has appeared in the mechanism. *This is generally true of catalysts.*
- Figure 12.15 shows that the activation-energy barrier is significantly lower for the catalyzed reaction (because the mechanism changed). Consequently the reaction rate is much faster. Dropping the activation energy from 262 kJ/mol for the uncatalyzed reaction to 115 kJ/mol for the catalyzed process makes the catalyzed reaction 10^{15} times faster!
- The catalyzed mechanism has five reaction steps, and its energy-versus-reaction progress diagram (Figure 12.15) has five energy barriers (five humps appear in the curve).
- The catalyst I_2 and the reactant *cis*-2-butene are both in the gas phase during the reaction. When a catalyst is present in the same phase as the reacting substance or substances, it is called a **homogeneous catalyst.**

12.9 ENZYMES: BIOLOGICAL CATALYSTS

Conceptual Challenge Problem CP-12.D at the end of this chapter relates to topics discussed in this and the next section.

In Section 7.7 *(← p. 287)* we discussed the marvelous chemical pathways by which free energy in foods you eat is used so that you can move, breathe, digest food, see, hear, smell, and even think. But did you stop to think how all those reactions are controlled? And how they can all occur reasonably quickly at the relatively low body temperature of 37 °C? Although oxidation of glucose powers all the systems of your body, you would not want it to take place at the temperature it does when glucose in wood burns in a fireplace! (A major component of wood is cellulose, a polymer of glucose, ← *p. 503.*) The chemical reactions of your metabolic pathways, and many other reactions, are catalyzed by enzymes. An **enzyme** is a highly efficient catalyst for one or more chemical reactions in a living system. The presence or absence of appropriate enzymes turns these reactions on or off by speeding them up or slowing

CHEMISTRY YOU CAN DO

Enzymes, Biological Catalysts

Raw potatoes contain an enzyme called *catalase,* which converts hydrogen peroxide to water and oxygen. You can demonstrate this by the following experiment:

Purchase a small bottle of hydrogen peroxide at a pharmacy or find one in your medicine chest. The peroxide is usually sold as a 3% solution in water. Pour about 50 mL of the peroxide solution into a clear glass or plastic cup. Add a small slice of a fresh potato to the cup. (Since potato is less dense than water, the potato will float.)

Almost immediately you will see bubbles of oxygen gas on the potato slice. Does the rate of evolution of oxygen change with time? If so, how does it change? If you cool the hydrogen peroxide solution in a refrigerator and then do the experiment, is there a perceptible change in the initial rate of O_2 evolution? Is there a difference between the time at which O_2 evolution begins for warm and for cold hydrogen peroxide? What happens if you heat the slice of potato on a stove or in an oven before adding it to the peroxide solution?

them down. This allows your body to maintain nearly constant temperature and nearly constant concentrations of a variety of molecules and ions, an absolute necessity if you are to continue functioning.

Enzymes are almost always proteins—polymers of amino acids (← ***p. 500***). Usually they are globular proteins, consisting of one or more long chains of amino acids folded into a nearly spherical shape. The shape of a globular protein is determined largely by noncovalent interactions (← ***p. 423***) among the amino-acid components: hydrogen bonds, attractions of opposite ionic charges, dipole-dipole and ion-dipole forces, a few weak covalent bonds, and the fact that nonpolar amino-acid side groups concentrate in the middle of the molecule to avoid the surrounding aqueous environment.

The weak covalent bonds are called disulfide bonds and occur between sulfur atoms in side chains of the amino acid cysteine. A cysteine side chain at one point in the protein can become bonded to a cysteine side chain much farther along the protein backbone.

Enzymes are among the most effective catalysts known. They can increase reaction rates by factors of 10^9 to 10^{20}. In one minute the enzyme carbonic anhydrase can decompose about 36 million molecules of carbonic acid.

$$H_2CO_3(aq) \xrightarrow{\text{carbonic anhydrase}} CO_2(g) + H_2O(\ell)$$

Most enzymes are highly specific catalysts. Some act on only one or two of the hundreds of different substances found in living cells. For example, carbonic anhydrase catalyzes only the reaction listed above. Other enzymes speed up one specific type of reaction, but no others.

Some enzymes can act as catalysts entirely on their own. Others require one or more inorganic or organic molecules or ions called **cofactors** to be present before their catalytic activity becomes fully available. For example, the digestive enzyme carboxypeptidase requires zinc ion, and many enzymes require nicotinamide adenine dinucleotide ion, NAD^+. Molecules or ions that are cofactors are often derived from small quantities of minerals and vitamins included in our diets. This is why a deficiency of vitamins or minerals can lead to disease or ill health (← ***p. 106***). If the cofactor needed for an enzyme to catalyze a reaction is not available because of dietary deficiency, that reaction cannot occur rapidly when it is needed and a bodily function will be impaired.

Enzyme Activity and Specificity

A molecule whose reaction is catalyzed by an enzyme is referred to as a **substrate.** In some cases there may be more than one substrate, as when an enzyme catalyzes

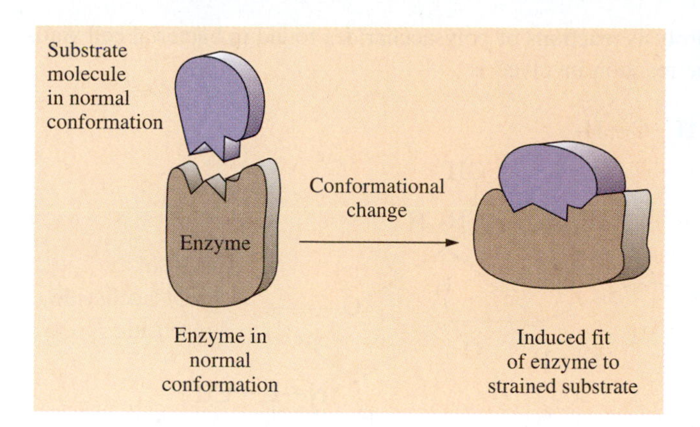

Figure 12.16 Induced fit of substrate to enzyme. The binding of a substrate to an enzyme may involve conformational changes in either or both of the molecules. In some cases a substrate molecule may be stretched or strained, helping bonds to break and reaction to occur.

transfer of a group from one molecule to another. Enzyme catalysis is so extremely effective and specific because the structure of the enzyme is finely tuned to minimize the activation-energy barrier. Usually one part of the enzyme molecule, called the **active site,** interacts with the substrate via the same kinds of noncovalent attractions that hold the enzyme in its globular structure. The nanoscale structure of an enzyme's active site is specifically suited to attract and bind a substrate molecule and to help the substrate to react.

When a substrate binds to an enzyme, both molecular structures can change. Each structure adjusts to fit closely with the other and the structures become complementary. The change in shape of either the enzyme, the substrate, or both molecules when they bind is called **induced fit.** Enzymes catalyze reactions of only a few molecules because the structures of most molecules are not close enough to the structure of the active site for an induced fit to occur. The induced fit of a substrate to an enzyme also can lower the activation energy for reaction. For example, it may distort the substrate and stretch a bond that will be broken in the desired reaction. A schematic example of how this can work is shown in Figure 12.16.

As mentioned in Section 11.7 *(← p. 483)*, a hydrolysis reaction is one in which a water molecule and some other molecule react, with both molecules splitting in two. The H from the water ends up with one part of the substrate molecule, and the OH ends up with the other part.

To see how this works in a specific case, consider the enzyme lysozyme, whose structure is shown in Figure 12.17 as a space-filling model with substrate in the active site.

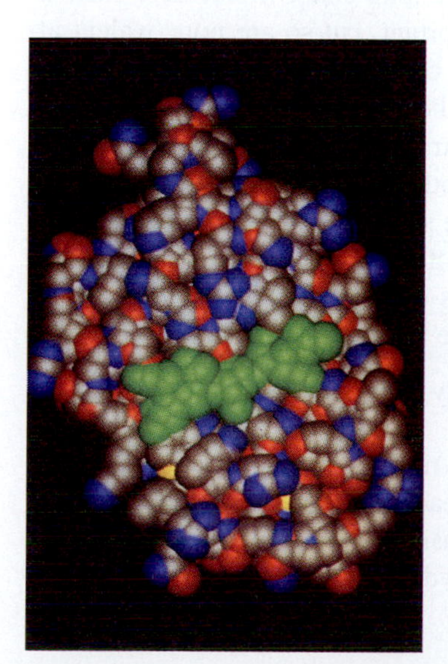

Figure 12.17 Space-filling molecule model of lysozyme. The structure of lysozyme is shown with the atoms drawn as spheres occupying the space enclosed by their covalent radii. The active site is a cleft in the surface of the lysozyme that stretches horizontally across the middle of the enzyme. The active site is occupied by a portion of a polysaccharide molecule, the substrate (green atoms). (The part of the polysaccharide not bound to the active site has been omitted so that you can see the enzyme better.) *(Zoltan Kanyo, Department of Biology, University of Pennsylvania)*

Lysozyme catalyzes hydrolysis reactions of polysaccharides found in bacterial cell walls (⇐ ***pp. 110 and 501***). The reaction involved is

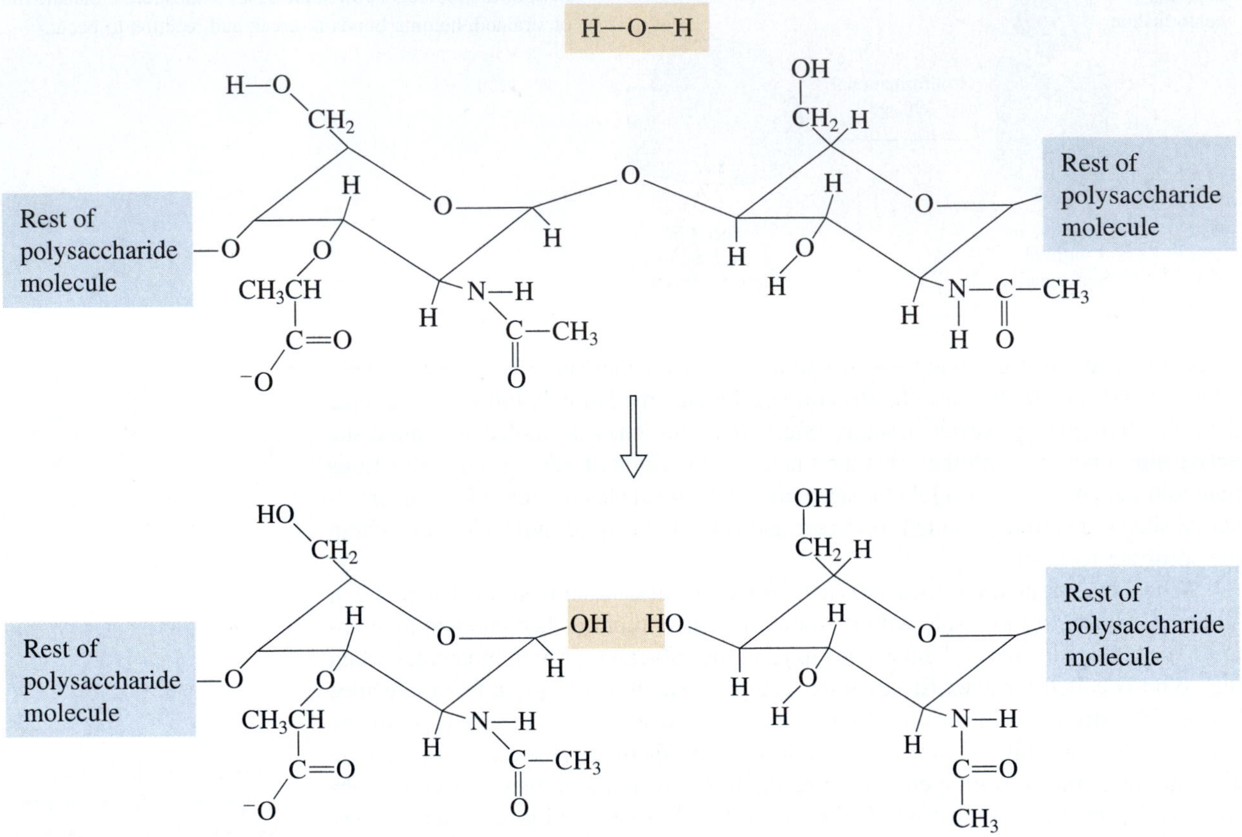

H—O—H

Rest of polysaccharide molecule

Rest of polysaccharide molecule

⇓

Rest of polysaccharide molecule

Rest of polysaccharide molecule

Hydrolysis is the opposite of the condensation reactions by which most biopolymers are formed, so many important enzymes catalyze hydrolysis reactions.

The section of polysaccharide shown in Figure 12.17 fits nicely into the cleft along the surface of the lysozyme, but many other long-chain molecules, such as polypeptides, might fit there as well. Shape is important, but so are noncovalent attractions and their positioning so that the substrate can make the most effective use of them. Figure 12.18 shows at least six hydrogen bonds between enzyme and substrate, and there is also an ionic group on the substrate that is attracted to an oppositely charged group or a dipole in the enzyme. In other words, the specificity of the enzyme depends not only on the shape of the active site, but also on the positions of hydrogen-bonding groups and groups that participate in other noncovalent interactions so that they can adjust to complementary sites on the substrate.

To summarize, enzymes are extremely effective as catalysts for several reasons:

- Enzymes bring substrates into close proximity and hold them there while a reaction occurs.
- Enzymes hold substrates in the shape that is most effective for reaction.
- Enzymes can act as acids and bases during reaction, donating or accepting hydrogen ions from the substrate quickly and easily. Hydrolysis reactions, for example, go faster when either hydrogen or hydroxide ions are available at the right place in the substrate.

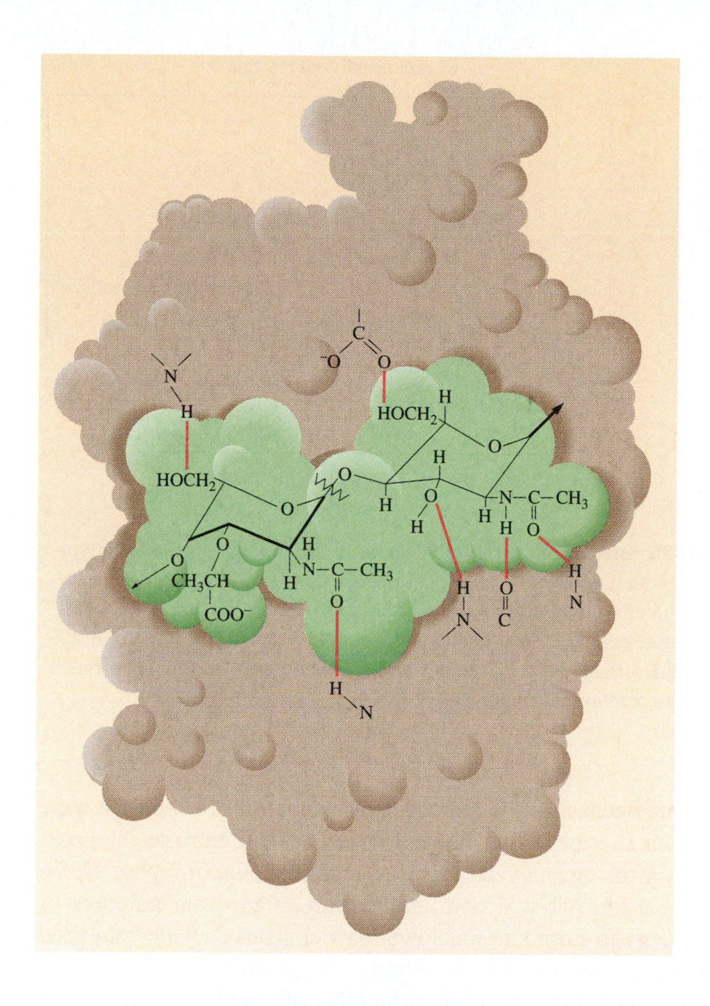

Figure 12.18 Enzyme-substrate binding. This schematic diagram shows noncovalent interactions (red lines) between the enzyme lysozyme and a section of polysaccharide that is its substrate. Six hydrogen bonds and one ionic group help to hold the substrate to the enzyme. The bond that will be broken when the substrate is hydrolyzed is indicated by a saw-tooth line across it.

- The potential energy of a bond distorted by the induced fit of the substrate to the enzyme is already part way up the activation-energy hill that must be surmounted for reaction to occur.

Enzyme Kinetics

An enzyme changes the mechanism of a reaction, as does any catalyst. The first step in the mechanism for any enzyme-catalyzed reaction is binding of the substrate and enzyme. This is referred to as formation of an **enzyme-substrate complex.** Representing enzyme by E, substrate by S, and products by P, we can write a single-step, uncatalyzed mechanism and a two-step, enzyme-catalyzed mechanism as follows. Uncatalyzed mechanism:

$$S \longrightarrow P$$

Enzyme-catalyzed mechanism:

$$S + E \longrightarrow ES \quad \text{(formation of enzyme-substrate complex)}$$
$$ES \longrightarrow P + E \quad \text{(formation of products and regeneration of enzyme)}$$

That the enzyme is a catalyst is evident from the fact that it is a reactant in the first step and regenerated in the second. This same mechanism applies to nearly all

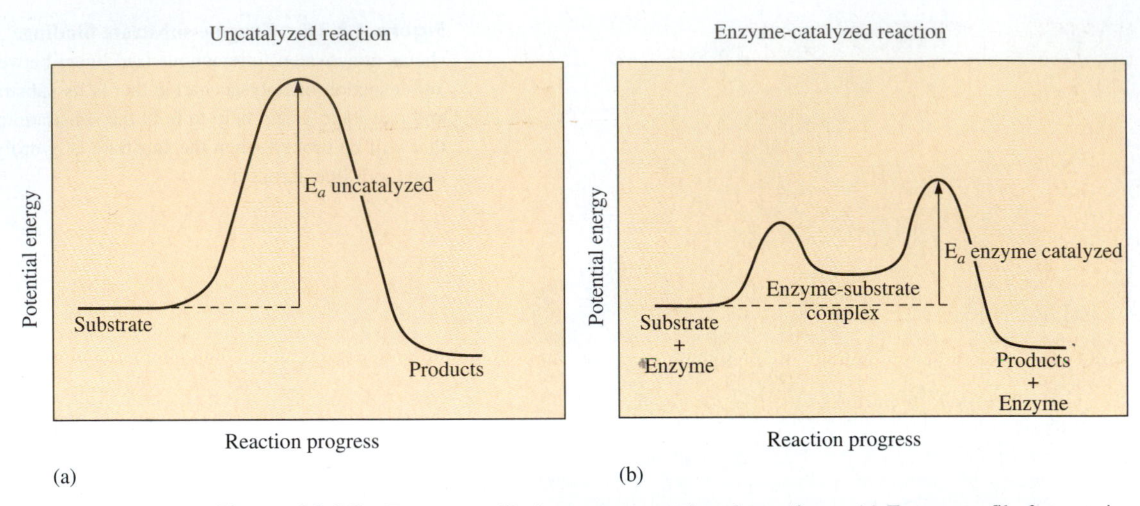

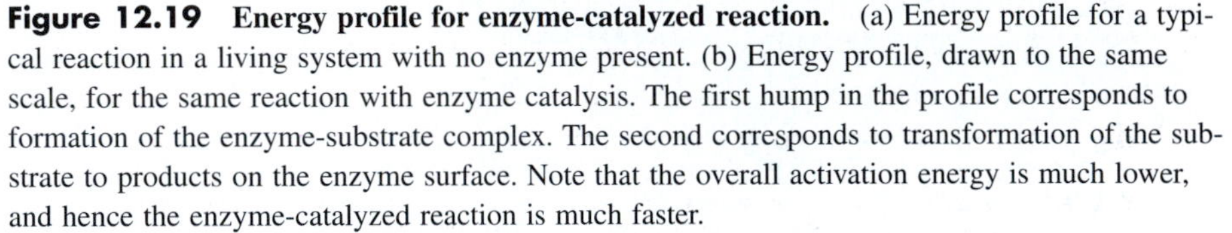

Figure 12.19 Energy profile for enzyme-catalyzed reaction. (a) Energy profile for a typical reaction in a living system with no enzyme present. (b) Energy profile, drawn to the same scale, for the same reaction with enzyme catalysis. The first hump in the profile corresponds to formation of the enzyme-substrate complex. The second corresponds to transformation of the substrate to products on the enzyme surface. Note that the overall activation energy is much lower, and hence the enzyme-catalyzed reaction is much faster.

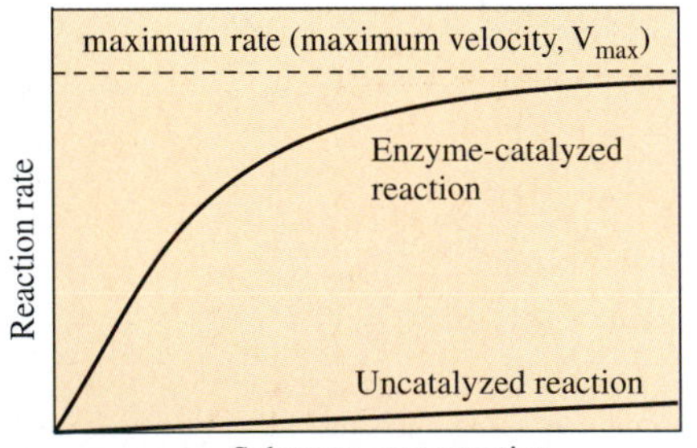

Figure 12.20 Maximum velocity for an enzyme-catalyzed reaction. Because there is only a limited quantity of enzyme available, increasing substrate concentration beyond the point where the enzyme becomes the limiting reactant does not increase the rate further. There is a maximum rate (maximum velocity) for any enzyme-catalyzed reaction.

enzyme-catalyzed reactions. Because of the noncovalent interactions between enzyme and substrate, the activation energy is significantly lower for the enzyme-catalyzed reaction than it would be for the uncatalyzed process. This is shown in Figure 12.19. Even at temperatures only a little above room temperature, significant numbers of molecules have enough energy to surmount this lower barrier. Thus enzyme-catalyzed reactions can occur reasonably quickly at human-body temperature.

Enzyme-catalyzed reactions obey the same principles of chemical kinetics that we discussed earlier in this chapter. However, there are some special features of both the enzyme itself and the mechanism of enzyme catalysis that you should be aware of. First, because of the form of the mechanism, either the enzyme or the substrate may be the limiting reactant in the first step. If the substrate is limiting, increasing the concentration of substrate produces more enzyme-substrate complex and makes the reaction go faster. This is the expected behavior: increasing the concentration of a reactant should increase the rate in proportion. However, if the enzyme becomes the limiting reactant, it can become completely converted to enzyme-substrate complex, leaving no enzyme available for additional substrate. If this happens, further increase in concentration of substrate will not increase the rate of reaction. This means that there is a *maximum rate* (those who study enzyme kinetics call this the maximum velocity) for an enzyme-catalyzed reaction. The behavior of rate with increasing substrate concentration is shown in Figure 12.20.

Enzyme-catalyzed reactions also behave unusually with respect to temperature. The rate does increase with increasing temperature, but if the temperature gets high enough, there is a sudden decrease in rate as shown in Figure 12.21. This happens because there is increased molecular and atomic motion as the temperature increases, and that motion can disrupt the structures of enzymes and other proteins. This change in protein structure is called **denaturation.** It occurs, for example, when an egg is boiled or fried. Once an enzyme's structure has changed, the active site is no longer

The enzymes in certain bacteria, such as those that inhabit hot springs as in Yellowstone National Park, are different from the enzymes in humans. They can withstand much higher temperatures without denaturing.

available, enzyme catalysis is seriously impaired, and the reaction rate falls to its uncatalyzed value. Notice that this happens only a little above 37 °C—body temperature for humans. Enzymes have evolved to produce maximum rates at body temperature, and slightly higher temperatures cause most of them to denature.

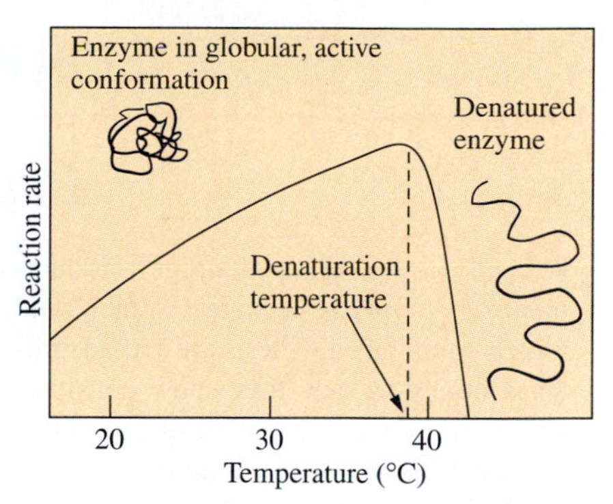

Figure 12.21 Enzyme activity destroyed by high temperature. At a temperature somewhat above body temperature there is sufficient molecular motion to overcome the noncovalent interactions that maintain protein structure. This disrupts the structure of an enzyme and thereby destroys its catalytic activity. The process by which the enzyme structure becomes disrupted is called denaturation.

Inhibition of Enzymes

There is another way that the activity of an enzyme can be destroyed. Some molecules or ions can fit an enzyme's active site, but remain there unreacted. Such a molecule is called an **inhibitor.** An inhibitor bound to an enzyme decreases its effective concentration and thereby decreases the rate of the reaction the enzyme catalyzes. If sufficient inhibitor becomes bound to an enzyme, the enzyme provides little catalytic effect because the concentration of available active sites becomes very small. An example of enzyme inhibition is the action of sulfa drugs on bacteria. Bacteria use para-aminobenzoic acid and an enzyme called dihydropteroate synthetase to synthesize folic acid, which is essential to their metabolism. Sulfa drugs bind to this enzyme, inhibit synthesis of folic acid, and destroy bacterial populations.

PROBLEM-SOLVING EXAMPLE 12.6 Enzyme Inhibition

The label of a container of methanol (methyl alcohol) invariably indicates that its contents are poisonous and should not be taken internally. This is because methanol, CH_3OH, which is not very toxic, is metabolized by the enzyme methanol oxidase to formaldehyde, $H_2C{=}O$, which is very toxic. Methanol poisoning is sometimes treated by giving the patient ethanol, CH_3CH_2OH. Explain why.

Answer Ethanol is an inhibitor of methanol oxidase.

Explanation Because its shape is similar to that of methanol, ethanol serves as an inhibitor, occupying the active sites of some of the methanol oxidase catalyst molecules. The rate of production of formaldehyde in the body is decreased and the harmful effect is less.

Problem-Solving Practice 12.6

Bacteria need to convert *p*-aminobenzoic acid to folic acid in order to survive. Sulfa drugs interfere with this process. The structure of *p*-aminobenzoic acid is

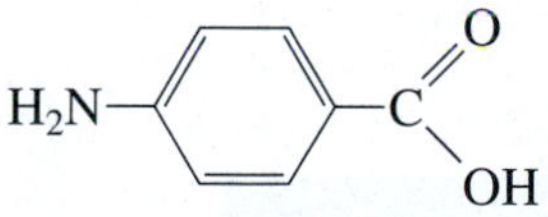

H_2N— (benzene ring) —C(=O)OH

Which of the following compounds is most likely a sulfa drug? Explain your choice.

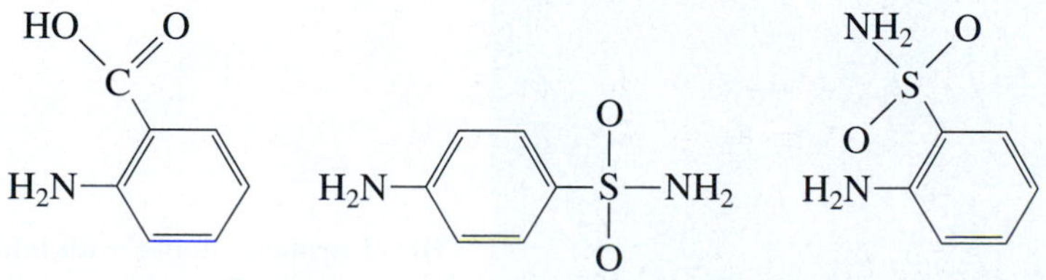

HO, O, C, H_2N H_2N—(ring)—$S(=O)_2$—NH_2 NH_2, O, S, O, H_2N

(Text continues on page 547.)

CHEMISTRY IN THE NEWS

Protease Inhibitors and AIDS

Recently a new class of drugs for treatment of AIDS has become available and widely publicized. These are called protease inhibitors, and they show considerable promise for slowing the spread of this disease. They do this by inhibiting growth of HIV, human immunodeficiency virus.

As the name implies, protease inhibitors inhibit an enzyme, HIV-1 protease. This enzyme is essential for maturation of the virus because it catalyzes a reaction in which a long polypeptide chain is cut into shorter pieces. The cuts occur at specific locations along the chain, and the smaller pieces created by HIV-1 protease are proteins that are essential to the survival of HIV. Like plastic trash bags that have to be separated from a long roll before they become useful, these proteins must be cut from the long polypeptide before they can carry out their functions in HIV. Several different proteins are produced this way by HIV-1 protease. Therefore, if this enzyme could be inhibited, reproduction of the virus would be interfered with in several different ways. (Ritonavir, one of the new protease inhibitors, was illustrated in Figure 10.2, (← *p. 399.*)

The action of HIV-1 protease, and its importance to HIV, is typical of how reactions are controlled in living organisms. It is much quicker to cut a long polypeptide chain into shorter, active protein molecules than it is to synthesize lots of protein molecules on short notice. Therefore other enzymes work in advance together with the virus's DNA to synthesize the long polypeptide, called a pre-protein. HIV-1 protease then chops the pre-protein into appropriate pieces whenever they are needed by HIV. If a lot of a particular protein is needed, it can be formed quickly by a few cuts in the pre-protein, instead of having to put together a large number of amino acids in the proper sequence.

HIV-1 protease actually consists of two polypeptide chains held together as a dimer by noncovalent attractive forces.

Structure of the HIV-1 protease dimer. Polypeptide chains are represented as ribbons or tubes instead of showing individual atoms. The active site is the open region in the center of the molecule.

The picture of HIV-1 protease shown here represents the polypeptide strands of the two monomers using ribbons and tubes instead of showing the individual atoms. This is done because showing the very large number of atoms in the enzyme would obscure your view of its overall structure. From the picture you can see that there is an open space in the middle—between the two halves of the enzyme. This is the active site. The enzyme works by having the two monomers come together to form an active site around the long pre-protein. This happens at a specific place along the pre-protein chain, and the active site cuts the polypeptide at that point by helping to break a peptide bond. Then the two monomers and the two pieces of polypeptide separate. The HIV-1 protease monomers can later cut another piece from the same or another polypeptide.

AIDS drugs that are protease inhibitors consist of molecules that can occupy the active site of HIV-1 protease, but their structures differ enough from the pre-protein structure that the protease cannot cut them. They remain in the active site as shown in the second figure, holding the dimer together and preventing HIV-1 protease from cutting any more pre-protein molecules. Several protease inhibitor molecules are now available for treatment of AIDS patients. They are the result of much research to determine the structure and mechanism of action of HIV-1 protease, to design and synthesize molecules that block the active site, and to test these new drugs. (For more information on this subject, see *Science,* June 28, 1996.)

HIV-1 protease dimer with inhibitor. An inhibitor molecule, drawn showing bonds only, occupies the active site of HIV-1 protease.

12.10 CATALYSTS IN INDUSTRY

An expert in the field of industrial chemistry has said that "Every year more than a trillion dollars' worth of goods is manufactured with the aid of man-made catalysts. Without them, fertilizers, pharmaceuticals, fuels, synthetic fibers, solvents, and detergents would be in short supply. Indeed, 90 percent of all manufactured items use catalysts at some stage of production." The major areas of catalyst use are in petroleum refining, industrial production of chemicals, and environmental controls. We shall look at a few examples here, and more will be described later in the book.

Many industrial reactions use **heterogeneous catalysts.** These are catalysts that are present in a different phase from the reactants being catalyzed. Usually the catalyst is a solid and the reactants are in the gaseous or liquid phase. Heterogeneous catalysts are used in industry because they are more easily separated from the products and leftover reactants than are homogeneous catalysts. Catalysts for chemical processing are generally metal-based and often contain precious metals such as platinum and palladium. In the United States more than $600 million worth of such catalysts are used annually by the chemical processing industry, almost half of them in the preparation of polymers (⇐ *p. 484*).

An article by J. M. Thomas (*Scientific American,* April 1992, pp. 112–118) describes some industrial uses of catalysts. See also Ann M. Taylor, *Chemical and Engineering News,* March 9, 1992, pp. 27–49.

About 15.5 billion pounds of nitric acid are made annually in the United States in the *Ostwald process,* the first step of which is controlled oxidation of ammonia over a Pt-containing catalyst on a fine-mesh wire gauze (Figure 12.22),

$$4\ NH_3(g) + 5\ O_2(g) \xrightarrow{\text{Pt-containing catalyst}} 4\ NO(g) + 6\ H_2O(g) \qquad \Delta H^\circ = -905.5\text{ kJ}$$

This is followed by further oxidation of NO to NO_2.

$$2\ NO(g) + O_2(g) \longrightarrow 2\ NO_2(g) \qquad \Delta H^\circ = -114.1\text{ kJ}$$

In a typical plant, a mixture of air with 10% NH_3 is passed very rapidly over the catalyst at a high pressure and at about 850 °C. Roughly 96% of the ammonia is converted to NO_2, making this one of the most efficient industrial catalytic reactions. The

Figure 12.22 Catalyst for nitric acid production. Workers are stretching the platinum-rhodium gauze catalyst used for oxidation of ammonia to nitric acid across the opening of a reactor. *(Johnson Matthey)*

final step is to absorb the NO_2 into water to give the acid and NO, the latter being recycled into the process.

$$3\ NO_2(g) + H_2O(\ell) \longrightarrow 2\ HNO_3(aq) + NO(g) \qquad \Delta H^\circ = -138.2\ kJ$$

Nitric acid is produced as a concentrated aqueous solution, but careful procedures can convert this to the anhydrous acid. At room temperature, HNO_3 is a colorless liquid with a pungent, choking odor. By far the largest amount of the acid is turned into ammonium nitrate by neutralization of nitric acid with ammonia.

$$HNO_3(aq) + NH_3(g) \longrightarrow NH_4NO_3(aq)$$

Acetic acid, CH_3CO_2H (← *p. 154*), has a place in the organic chemicals industry comparable to that of sulfuric acid in the inorganic chemicals industry; more than 4.7 billion pounds of acetic acid were made in the United States in 1995. Acetic acid is used widely in industry to make plastics and synthetic fibers, as a fungicide, and as the starting material for preparing many dietary supplements. One way of synthesizing the acid is an excellent example of homogeneous catalysis: rhodium(III) iodide is used to speed up the combination of carbon monoxide and methyl alcohol, both inexpensive chemicals, to form acetic acid.

$$CH_3OH + CO \xrightarrow{RhI_3\ catalyst} CH_3C(=O){-}OH$$

$$\text{Methyl alcohol} + \text{carbon monoxide} \longrightarrow \text{acetic acid}$$

The role of the rhodium(III) iodide catalyst in this reaction is to bring the reactants together and allow them to rearrange to the products. Carbon monoxide and the methyl group from the alcohol become attached to the rhodium atom, which helps transfer the methyl group to the CO. After this rearrangement, the intermediate reacts with solvent water to form acetic acid.

$$CO + RhI_3 \longrightarrow RhI_3CO$$

$$CH_3OH + RhI_3CO \longrightarrow CH_3RhI_3CO + OH^-$$

$$CH_3RhI_3CO + HOH \longrightarrow CH_3C(=O){-}OH + H^+ + RhI_3$$

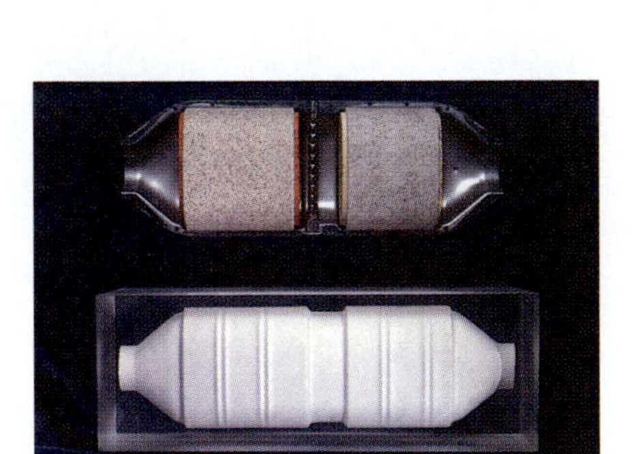

Figure 12.23 Automobile catalytic converter. Catalytic converters are standard equipment on the exhaust systems of all new automobiles. This one contains two catalysts: one converts nitrogen monoxide to nitrogen and the other converts carbon monoxide and hydrocarbons to carbon dioxide. *(© AC/GM/Peter Arnold, Inc.)*

The largest growth in catalyst use is in *emissions control* for both automobiles and power plants. This market consumes very large quantities of platinum group metals: platinum, palladium, rhodium, and iridium. In 1994 52,800 kg of platinum, 32% more than in 1993, was sold in the United States for automotive uses. More than 7000 kg of palladium and rhodium was sold for this same purpose. All three metals are also used in chemical processing as catalysts, and the petroleum industry uses platinum and rhodium to catalyze refining processes.

The purpose of the catalysts in the exhaust system of an automobile is to ensure that the combustion of carbon monoxide and hydrocarbons is complete (Figure 12.23) and to convert nitrogen oxides to molecules less harmful to the environment.

$$2\ CO(g) + O_2(g) \xrightarrow{Pt\text{-}NiO\ catalyst} 2\ CO_2(g)$$

$$2\ C_8H_{18}(g) + 25\ O_2(g) \xrightarrow{Pt\text{-}NiO\ catalyst} 16\ CO_2(g) + 18\ H_2O(g)$$

2,2,4-trimethylpentane
a component of gasoline

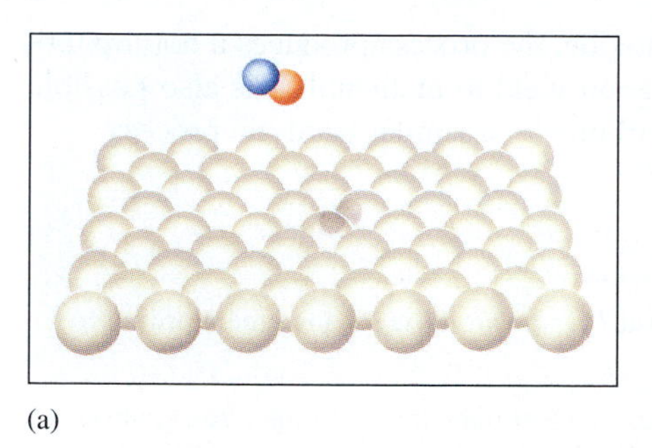
(a)

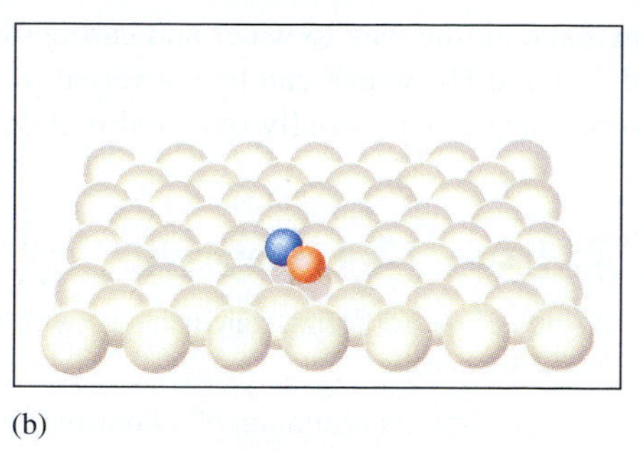
(b)

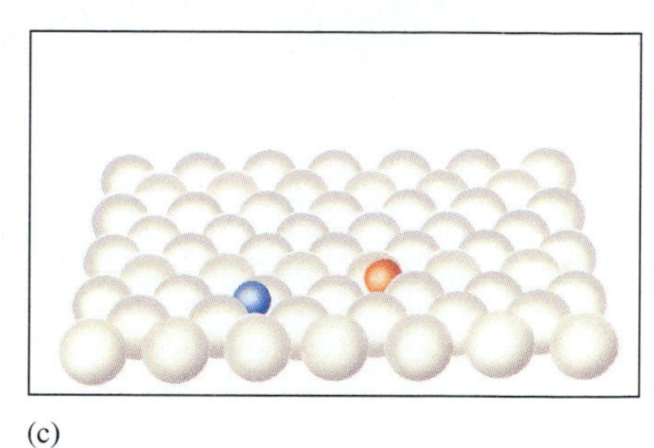
(c)

Figure 12.24 Catalytic conversion of NO to N_2 and O_2. A platinum surface can speed conversion of NO to N_2 and O_2 by dissociating NO into N atoms and O atoms. Here an NO molecule (a) approaches the surface, (b) interacts with the surface, and (c) is dissociated into an N atom (blue) and an O atom (red). The entire process of interaction with the surface and dissociation takes about 1.7×10^{-12} s. Once dissociated, the N and O atoms can migrate across the surface and eventually combine with another atom of N or O to form N_2 or O_2.

At the high temperature of combustion, some N_2 from air reacts with O_2 to give NO, a serious air pollutant. Thermodynamics informs us that nitrogen monoxide is unstable and should revert to N_2 and O_2. But remember that thermodynamics says nothing about rate. Unfortunately, the rate of reversion of NO to N_2 and O_2 is slow. Fortunately, catalysts have been developed that greatly speed this reaction.

$$2\ NO(g) \xrightarrow{\text{catalyst}} N_2(g) + O_2(g)$$

The role of the heterogeneous catalyst in the preceding reactions is probably to weaken the bonds of the reactants and to assist in product formation. For example, Figure 12.24 shows how NO molecules can dissociate into N and O atoms on the surface of a platinum-metal catalyst.

Your home may be heated with natural gas, which consists largely of methane, CH_4. Although widely used, it is also widely wasted because much of it is found in geographical areas far removed from where fuels are consumed and transporting the flammable gas is expensive and dangerous. One solution to making methane useful is to convert it, where it is found, to a more readily transportable substance such as liquid methanol, CH_3OH. The methanol can then be used directly as a fuel, added to gasoline (as is currently done in some areas of the United States, *p. 456*), or used to make other chemicals.

Figure 12.25 Methane flowing through a catalyst. *(Schmidt/University of Minnesota)*

It has been known for some time that methane can be converted to carbon monoxide and hydrogen

$$CH_4(g) + \tfrac{1}{2}\,O_2(g) \longrightarrow CO(g) + 2\ H_2(g)$$

and this mixture of gases can readily be turned into methanol in another step.

$$CO(g) + 2\ H_2(g) \longrightarrow CH_3OH(\ell)$$

Chemical engineers at the University of Minnesota have found that methane can in fact be converted to CO and H_2 under very mild conditions of temperature. They simply found the right catalyst! The photograph in Figure 12.25 shows what happens when a room-temperature mixture of methane and oxygen flows through a heated, sponge-like ceramic disk coated with platinum or rhodium. Rather than oxidizing the

For more information on this discovery see *Science*, March 15, 1996, pp. 1560–1562, and *Science*, January 15, 1993, pp. 340–346.

methane all the way to water and carbon dioxide, the process produces a hot mixture of CO and H_2, which can be converted in good yield to methanol. It is also possible to produce other partially oxidized hydrocarbons by a similar catalytic process.

Exercise 12.4 Catalysis

Which of the following statements is (are) true? If any are false, change the wording to make them true.

(a) The concentration of a homogeneous catalyst may appear in the rate equation.
(b) A catalyst is always consumed in the overall reaction.
(c) A catalyst must always be in the same phase as the reactants.

12.11 CHLOROFLUOROCARBONS AND THE OZONE LAYER

In Chapter 1, the Nobel Prize–winning research of Rowland, Molina, and Crutzen on substances that destroy stratospheric ozone was mentioned as an example of how science is done *(**← p. 2**)*. Much of that science had to do with reaction rates, and so we now turn to a more detailed description of the stratospheric ozone problem.

Most pollutants are removed from the atmosphere by reactions that take place in the lowest 15 km or so—the region called the **troposphere.** Adsorption onto a particle or absorption by a water droplet can lead to removal by precipitation. Chemical reactions in the gas phase, on a particle, or in solution can convert a pollutant to a less harmful substance. But compounds made up of chlorine, fluorine, and carbon, the **chlorofluorocarbons,** or **CFCs,** are so unreactive that they can remain in the atmosphere for very long periods. On average, a molecule of CFC-12, CF_2Cl_2, survives roughly 100 years, for example. CFCs mix with air throughout the troposphere and eventually migrate to the **stratosphere,** the region of the atmosphere between 15 and 50 km above sea level (Section 14.2). There they eventually are decomposed by ultraviolet radiation. The decomposition products of CFCs participate in reaction mechanisms in the stratosphere that result in lowering of the concentration of ozone.

Chlorofluorocarbons (CFCs) are halogen-substituted alkanes such as $CFCl_3$ and CF_2Cl_2.

CFCs are used in air conditioners as refrigeration fluids, and they are more effective and much safer for this application than ammonia and sulfur dioxide, which were used before CFCs were invented. It is reasonable to state that without CFCs our modern, air-conditioned buildings would be much more dangerous. CFCs are also used as solvents in the manufacture of such diverse products as electronic parts and machined metallic objects. Because they are very unreactive, CFCs are also used as fire-extinguishing fluids. The lower molecular weight CFCs were used as propellants for aerosol products including hair sprays, deodorants, medicines, and foods, until they were banned from this use in the United States in 1978. CFCs are, however, still used worldwide in both developed and developing countries. Most of their important uses depend on their nonflammability and general inertness.

In 1974 Molina and Rowland predicted that continued use of CFCs would lead to a serious depletion of the earth's stratospheric ozone layer. What is the stratospheric ozone layer and why is its depletion a matter of concern? Ozone is formed in the stratosphere because at that high altitude ultraviolet radiation from the sun splits oxygen molecules into oxygen atoms

$$O_2(g) \xrightarrow{h\nu} 2\ O(g)$$

The oxygen atoms can then combine with oxygen molecules to form ozone.

$$O(g) + O_2(g) \longrightarrow O_3(g)$$

If this were all that happened, the concentration of ozone in the stratosphere would continually increase. But there is also a reaction that destroys ozone:

$$O(g) + O_3(g) \longrightarrow 2\ O_2(g)$$

Because ozone molecules and oxygen atoms are at very low concentrations, this reaction is relatively slow, but it consumes ozone fast enough to maintain a balance with the ozone-forming reaction so that under normal circumstances the concentration of ozone in the stratosphere remains constant.

The importance of ozone in the stratosphere is that when ozone molecules are struck by photons in the 200–300 nm (near-ultraviolet) range, they decompose to produce oxygen atoms and oxygen molecules:

$$O_3(g) \xrightarrow{h\nu} O_2(g) + O(g)$$

The O atoms from the photodissociation of O_3 react exothermically with O_2 to regenerate O_3, and so this reaction results in no net O_3 loss. However, the two reactions convert energy from the ultraviolet photons into thermal energy, which makes the temperature of the stratosphere higher at higher elevations than at lower elevations. Absorption of ultraviolet radiation is essential for living things on this planet. Ozone in the stratosphere prevents 95 to 99% of the sun's near-ultraviolet radiation from reaching the earth's surface. Photons in this 200–300 nm range have enough energy that they can cause skin cancer in humans and damage to living plants. For every 1% decrease in the stratospheric ozone, an additional 2% of this most damaging radiation reaches the earth's surface. Ozone depletion therefore has the potential for drastically damaging our environment.

Destruction of the ozone layer by CFCs begins when a photon of high enough energy breaks the carbon-chlorine bond in a CFC molecule. This produces a chlorine atom as shown here using CFC-12 as an example.

The chlorine atom is written Cl· to emphasize that it has one unpaired electron and is a free radical (← p. 369).

$$F-\underset{\underset{Cl}{|}}{\overset{\overset{F}{|}}{C}}-Cl(g) \xrightarrow{h\nu} F-\underset{\underset{Cl}{|}}{\overset{\overset{F}{|}}{C}}\cdot(g) + Cl\cdot(g)$$

The chlorine atom then participates in what is called a chain-reaction mechanism. It first combines with an ozone molecule, producing a chlorine oxide (ClO) radical and an oxygen molecule.

Step 1: $Cl\cdot(g) + O_3(g) \longrightarrow ClO\cdot(g) + O_2(g)$

Thus, an ozone molecule has been destroyed. If this were the only reaction that particular CFC molecule caused, there would be little danger to the ozone layer. However, the ClO· radical can react with an oxygen atom to regenerate atomic chlorine.

Step 2: $ClO\cdot(g) + O(g) \longrightarrow O_2(g) + Cl\cdot(g)$

The net reaction obtained by adding these two steps is conversion of an ozone molecule and an oxygen atom to two oxygen molecules:

$$O_3(g) + O(g) \longrightarrow 2\ O_2(g)$$

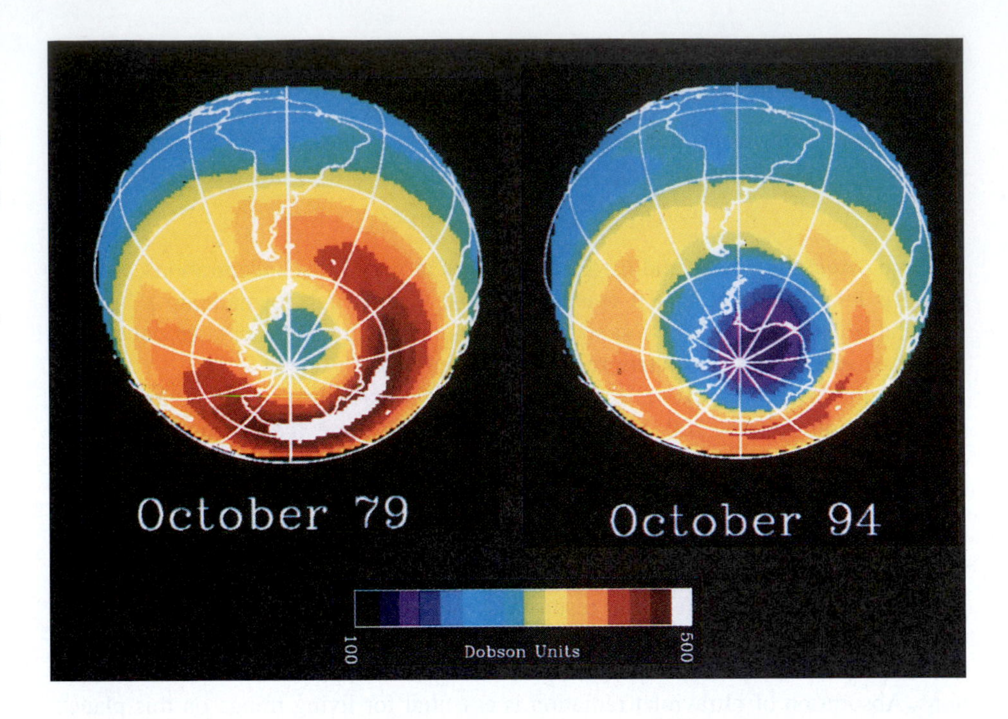

Figure 12.26 The Antarctic ozone hole. The figure shows the average ozone concentrations over the southern hemisphere in October (during the Antarctic spring). The scale at the bottom correlates colors with the concentration of ozone in Dobson units. One Dobson unit corresponds to 2.7×10^{16} molecules of ozone per cm^3 at 0 °C and 1 atm. The areas of low ozone concentration were much larger in 1994 than they were in 1979. The average ozone concentration over the south pole was 50% lower in October 1994 than in October 1979. *(NASA Total Imaging Satellite [TOMS] instruments aboard Nimbus-7 [1979] and Metèor-3 [1994] satellites.)*

That is, this chain reaction increases the rate at which stratospheric ozone is destroyed, but it does not affect the rate at which ozone is formed. The reason this is called a chain mechanism is that the two reaction steps can repeat over and over. The chlorine atom that reacts in step 1 is regenerated in step 2 and serves as a catalyst for destruction of ozone. It has been estimated that a single Cl atom can destroy as many as 100,000 molecules of O_3 before it is inactivated or returned to the troposphere (probably as HCl). Catalysis of the reaction that destroys ozone upsets the balance in the stratosphere because ozone is being destroyed faster than it is being produced. Thus the concentration of ozone decreases.

Until 1985 there was no experimental confirmation of Molina and Rowland's warning regarding CFCs. In that year the British Antarctic Survey reported what has come to be known as the "ozone hole," a startlingly large depletion of ozone in September and October, at the end of the Antarctic winter (Figure 12.26). Subsequent measurements by several research teams showed that loss of ozone correlated with high concentrations of ClO, supporting the theory that a chain mechanism involving Cl atoms was responsible. However, the huge depletion of ozone in the Antarctic could not be explained solely by the reaction steps 1 and 2 above. In the dark Antarctic winter, a vortex of intensely cold air containing ice crystals builds up. On the surfaces of these crystals additional reactions produce hydrogen chloride and chlorine nitrate ($ClONO_2$). These can react with each other to form chlorine molecules.

$$HCl(g) + ClONO_2(g) \longrightarrow Cl_2(g) + HNO_3(g)$$

When sunlight returns in the spring, the Cl_2 molecules are readily photodissociated into chlorine atoms, which can then become involved in the ozone destruction reactions.

$$Cl_2(g) \xrightarrow{h\nu} 2\ Cl\cdot(g)$$

The Cl—Cl bond energy is 243 kJ/mol of bonds broken, which corresponds to the energy of a photon whose associated wavelength is about 500 nm. Thus any light with wavelength less than about 500 nm can break Cl—Cl bonds.

Especially strong evidence in support of the theory that chlorine atoms are involved in destruction of ozone is shown in Figure 12.27. Instruments aboard a NASA

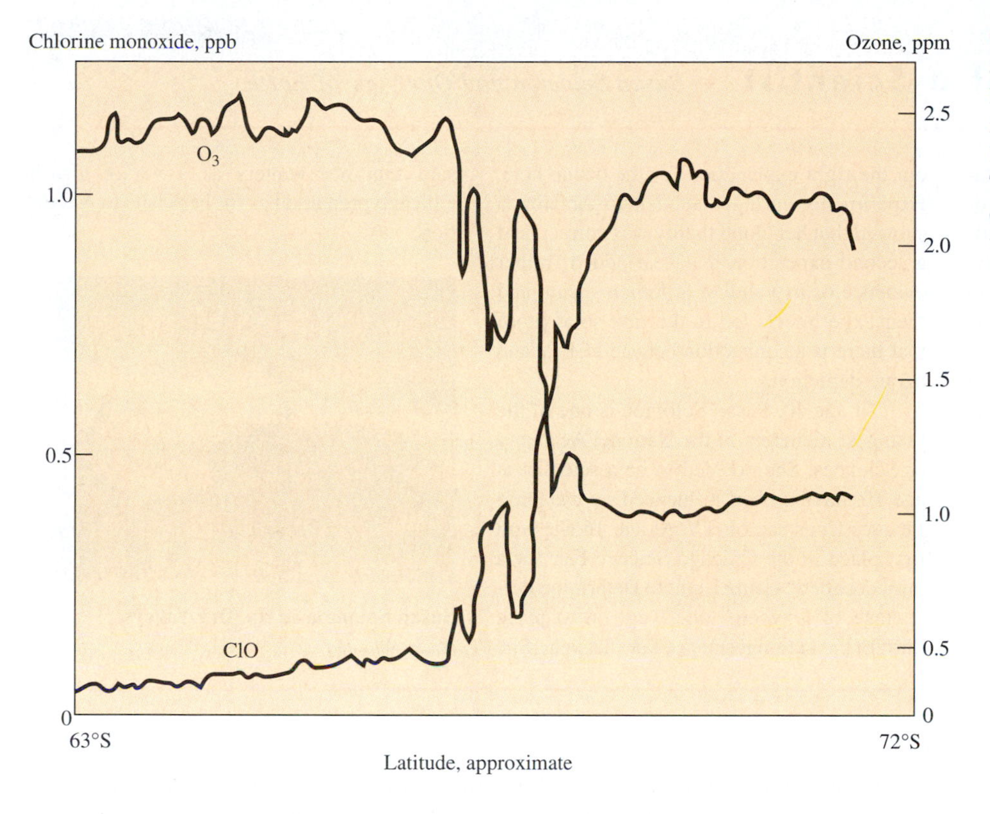

Figure 12.27 Concentrations of ozone and chlorine monoxide. Within the ozone hole, the concentration of chlorine monoxide is high and the concentration of ozone is low. Outside the hole the opposite is true. This is in agreement with the mechanism for destruction of ozone, since ClO is produced when O_3 is destroyed. (NASA)

research aircraft measured concentrations of ClO and of O_3 simultaneously as the plane flew into the ozone hole. It is clear from the figure that concentrations of ClO in the ozone hole were much larger than outside it, and where [ClO] was large, $[O_3]$ had a much lower value. This corresponds with step 1 of the mechanism for ozone destruction, where Cl reacts with O_3 to give ClO and O_2. In the reaction where ozone is consumed, chlorine monoxide is formed, and this accounts nicely for the fact that one concentration goes down when the other goes up.

By 1988, satellite and ground-based measurements had shown an average of about 2.5% decrease in the ozone layer worldwide. Decreases near the poles were even larger. In a latitude band that includes Dublin, Moscow, and Anchorage, ozone had decreased by 8% from January 1969 to January 1986. The reaction catalyzed by chlorine atoms from CFCs is thought to have accounted for about 80% of the observed loss.

In an effort to reduce the harm done by CFCs to the stratospheric ozone layer, a meeting was held in Montreal in 1987 that resulted in the signing in January 1989 of the Montreal Protocol on Substances That Deplete the Ozone Layer. Signed by 36 nations, the protocol initially called for reducing production and consumption of several of the long-lived CFCs. Treaties signed in London in 1990 and Copenhagen in 1992, and still in effect now, call for complete phaseouts of all chemicals that can harm the ozone layer. Within the United States, in 1992 the EPA banned retail store sales of small containers (less than 20 pounds) of CFCs for motor vehicle air-conditioning systems and prohibited service stations from emitting CFCs, requiring them to begin a recycling program for these compounds (Figure 12.28). These efforts are bearing fruit. The May 31, 1996, issue of *Science* magazine contains a paper reporting that tropospheric chlorine from CFCs peaked near the beginning of 1994 and has been declining since. Nevertheless, it will not be until 1999 that stratospheric chlorine from CFCs, and its effect on ozone concentrations, begins to decline.

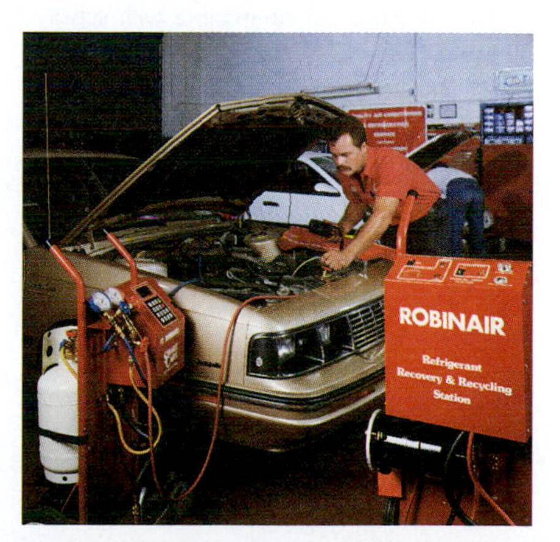

Figure 12.28 Recycling CFCs from automotive air-conditioning systems. In 1992 the EPA banned CFC emissions from service stations that recharge air conditioners, requiring that air-conditioner fluids be recycled. Formerly, the CFCs were simply vented to the atmosphere. *(Courtesy Robinair Corporation)*

Portrait of a Scientist • *Susan Solomon and Our View of Earth*

In 1985, a British team at Halley Bay Station, Antarctica, discovered the existence of a hole in the ozone layer above that continent. This totally unexpected phenomenon needed an explanation, and it was Susan Solomon, a young NOAA (National Oceanic and Atmospheric Administration) scientist, who first proposed a good theory for it. While attending a lecture on polar stratospheric clouds, she realized that ice crystals in the clouds might do more than just scatter light over the Antarctic. Her chemist's intuition told her that the ice crystals could provide a surface on which chemical reactions of CFC compounds could take place.

In 1986, NASA chose Solomon (then 30 years old) to lead a team to Antarctica to sort out the right explanation for the ozone hole. Experiments during that visit to Antarctica showed that her cloud theory was correct, and a second expedition that year added further evidence of its validity. Solomon's team and their experiments led to the first solid proof that there is a connection between CFCs and ozone depletion.

At age 40, Susan Solomon is one of the youngest members of the National Academy of Sciences. She decided to be a scientist at age 10, having been influenced by watching Jacques Cousteau on TV. At age 16 she won first place in the Chicago Science Fair for a project called "Using Light to Determine Percentage of Oxygen," and went on to place third in the national science fair that year. She said that her winters as a young girl in Chicago prepared her for her visits to Antarctica.

Susan Solomon in the Dry Valleys.
(Susan Solomon)

It will definitely cost you more to continue air conditioning an older car. U.S. production of CFC-12, the most popular auto refrigerant, ceased at the end of 1995. Although a lot of this refrigerant, along with others, will be recycled, service shops are likely to extract a premium price. Older air-conditioning systems are not compatible with substitutes for CFC-12. Some of the more recent substitutes cause seals to swell and break, and the lower-molecular-weight substitutes escape through tiny pores in hoses. Beware of "miracle" ozone-friendly refrigerants that are said to replace CFC-12 with no modifications necessary. They will probably ruin your car's air-conditioning system.

Manufacturers began seeking alternatives to CFCs almost immediately after it became clear that they would eventually be phased out. One class of possible CFC substitutes includes molecules called HCFCs. Because it contains a C—H bond, an HCFC molecule is more reactive in the troposphere than a CFC molecule. Therefore fewer HCFCs reach the stratosphere to deplete the ozone. Even better are hydrofluorocarbons, HFCs. These contain only carbon, fluorine, and hydrogen, and so cannot contribute chlorine atoms to catalyze ozone destruction. In refrigerators, for example, $C_2H_2F_4$ can be substituted for CFC-12.

Exercise 12.5 Ozone Depletion

Assume that 100 metric tons of the CFC with the formula CF_3Cl reaches the stratosphere. If one molecule of CF_3Cl can destroy 100,000 O_3 molecules, how many O_3 molecules can 100 metric tons destroy?

12.12 THERMODYNAMIC AND KINETIC STABILITY

Chemists often say that substances are "stable," but what exactly does this mean? Usually it means that the substance in question does not decompose or react with other substances that normally are in contact with it. Most chemists, for example, would say that the aluminum can that holds the soda you drink is stable. It will be around for quite a long time. The fact that aluminum cans thrown by the roadside do not decompose rapidly is one of the reasons you are encouraged to recycle them. Some samples have emerged almost unchanged from landfills after 40 or 50 years.

Strictly speaking there are two kinds of stability. We discussed one of them in Chapter 7 (⇐ ***p. 282***). A substance is *thermodynamically stable* if it does not undergo product-favored reactions. Such reactions release energy and increase disorder. Although we just said it was stable, the aluminum in a soda can is *thermodynamically* unstable since its reaction with oxygen in air has a negative free energy change.

$$4\ Al(s) + 3\ O_2(g) \longrightarrow 2\ Al_2O_3(s) \qquad \Delta G° = -3164.6\ kJ$$

However, the aluminum is *kinetically stable.* Although it has the potential to undergo a product-favored oxidation reaction, it does this so slowly that it remains essentially unchanged for a long time. This happens because a thin coating of aluminum oxide forms on the surface and prevents oxygen from reaching the rest of the aluminum atoms below the surface. If we grind the aluminum into a fine powder and throw it into a flame, the powder will burn; and the evolved heat will lead to an entropy increase in the little piece of the universe around the burning metal.

Another substance that is *thermodynamically unstable* but *kinetically stable* is diamond. If you look up the data in Appendix J, you will find that the conversion of diamond to graphite has a negative free energy change. But diamonds don't change into graphite. Engagement rings contain diamonds precisely because the diamond (like the love it represents) is expected to last for a long time. It does so because there is a very high activation barrier for the change from the diamond structure to the graphite structure. Most of the time, when a chemist says something is stable, it means that it is kinetically stable—only an activation-energy barrier prevents it from reacting fast enough for us to see a change.

Finally, think about whether you yourself are stable (thermodynamically or kinetically). From a thermodynamic standpoint, most of the substances you are made of are unstable with respect to oxidation to carbon dioxide, water, and other substances. That is, based on free energy changes, most of the substances that you are made of should undergo product-favored reactions that would completely destroy them. Your protein, fat, carbohydrate, and even DNA should spontaneously change into much smaller, simpler molecules with evolution of thermal energy. Fortunately for you, the reactions by which this would happen are very slow at room temperature and body temperature. Only when enzymes catalyze those reactions do they occur with reasonable speed. It is the combination of thermodynamic instability and kinetic stability that allows those enzymes to control the reactions in your body or in any living organism. Were it not for the kinetic stability of a wide variety of substances, everything would be quickly converted to a small number of very thermodynamically stable substances. Life and the environment as we know them would be impossible.

Summary Problem

An excellent way to make pure nickel metal for use in specialized steel alloys like the one shown in the photograph at the beginning of this chapter is to decompose $Ni(CO)_4$ in a vacuum at a temperature slightly above room temperature.

$$Ni(CO)_4(g) \longrightarrow Ni(s) + 4\ CO(g)$$

For this and other reasons the chemistry of compounds composed of a transition metal and carbon monoxide has been an interesting area of research for the past 30 years. Kinetic studies of this decomposition reaction have been carried out between 47.3 °C and 66.0 °C. (See J. P. Day, F. Basolo, and R. G. Pearson, *Journal of the American*

Chemical Society, Vol. 90, p. 6933, 1968.) The reaction is proposed to occur in four steps, the first of which is

$$Ni(CO)_4 \longrightarrow Ni(CO)_3 + CO$$

The data in the table below were obtained for this reaction step at different temperatures.

Temperature (K)	**Rate Constant (s^{-1})**
320.45	0.263
324.05	0.354
328.15	0.606
333.15	1.022
339.15	1.873

(a) Determine the activation energy of this reaction.
(b) $Ni(CO)_4$ is formed by the reaction of nickel metal with carbon monoxide. If you have 2.05 g of CO, and you combine it with 0.125 g of nickel metal, what is the maximum quantity of $Ni(CO)_4$ (in grams) that can be formed?

The replacement of CO by another molecule in $Ni(CO)_4$ (in the nonaqueous solvents toluene and hexane) was also studied to understand the general principles that govern the chemistry of such compounds. (See J. P. Day, F. Basolo, and R. G. Pearson, *Journal of the American Chemical Society,* Vol. 90, p. 6927, 1968.)

$$Ni(CO)_4 + P(CH_3)_3 \longrightarrow Ni(CO)_3P(CH_3)_3 + CO$$

A detailed study of the kinetics of the reaction led to the following mechanism:

$$Ni(CO)_4 \longrightarrow Ni(CO)_3 + CO \qquad \text{slow}$$

$$Ni(CO)_3 + P(CH_3)_3 \longrightarrow Ni(CO)_3P(CH_3)_3 \qquad \text{fast}$$

(c) Tell whether each step in the mechanism is uni- or bimolecular.
(d) When the steps of the mechanism are added together, show that the result is the balanced equation for the observed reaction.
(e) Is there an intermediate in this reaction? If so, what is its identity?
(f) It was found that doubling the concentration of $Ni(CO)_4$ led to an increase in reaction rate by a factor of 2. Doubling the concentration of $P(CH_3)_3$ had no effect on the reaction rate. Based on this information, write the rate equation for the reaction.
(g) Does the experimental rate equation support the proposed mechanism? Why or why not?

In Closing

Having studied this chapter, you should be able to . . .

- define reaction rate and calculate average rates (Section 12.1).
- describe the effect reactant concentrations have on reaction rate and determine rate laws and rate constants from initial rates (Section 12.2).
- determine reaction orders from a rate law and use the integrated-rate-law method to obtain orders and rate constants (Section 12.3).

- define and give examples of unimolecular and bimolecular elementary reactions (Section 12.4).
- show by using an energy profile what happens as two reactant molecules interact to form product molecules (Section 12.4).
- define activation energy and frequency factor, and use them to calculate rate constants and rates under different conditions of temperature and concentration (Section 12.5).
- derive rate laws for unimolecular and bimolecular elementary reactions (Section 12.6).
- define reaction mechanism and identify rate-limiting steps and intermediates (Section 12.7).
- given several reaction mechanisms, decide which is (are) in agreement with experimentally determined stoichiometry and rate law (Section 12.7).
- explain how a catalyst can speed up a reaction; draw energy profiles for catalyzed and uncatalyzed reaction mechanisms (Section 12.8).
- define the terms enzyme, substrate, and inhibitor, and identify similarities and differences between enzyme-catalyzed reactions and uncatalyzed reactions (Section 12.9).
- describe several important industrial processes that are made possible by catalysts (Section 12.10).
- write a reaction mechanism that shows how traces of atomic chlorine can decrease the concentration of ozone in earth's stratosphere (Section 12.11).
- describe how chlorofluorocarbons can affect earth's environment and discuss measures being taken to reduce the harm being done by CFCs (Section 12.11).
- explain the difference between thermodynamic and kinetic stability and give at least two examples of common substances that have one but not the other kind of stability (Section 12.12).

Key Terms

activated complex *(12.4)*
activation energy (E_a) *(12.4)*
active site *(12.9)*
Arrhenius equation *(12.5)*
bimolecular reaction *(12.4)*
catalyst *(12.8)*
CFCs *(12.11)*
chemical kinetics *(Introduction)*
chlorofluorocarbons *(12.11)*
cofactor *(12.9)*
denaturation *(12.9)*
elementary reaction *(12.4)*
enzyme *(12.9)*
enzyme-substrate complex *(12.9)*
heterogeneous catalyst *(12.10)*
heterogeneous reaction *(12.1)*
homogeneous catalyst *(12.8)*
homogeneous reaction *(12.1)*
induced fit *(12.9)*
inhibitor *(12.9)*
initial rate *(12.2)*
intermediate *(12.7)*
order of reaction *(12.3)*
overall reaction order *(12.3)*
rate *(12.1)*
rate constant *(12.2)*
rate equation *(12.2)*
rate law *(12.2)*
rate-limiting step *(12.7)*
reaction intermediate *(12.7)*
reaction mechanism *(12.7)*
reaction rate *(12.1)*
steric factor *(12.4)*
stratosphere *(12.11)*
substrate *(12.9)*
transition state *(12.4)*
troposphere *(12.11)*
unimolecular reaction *(12.4)*

Questions for Review and Thought

Conceptual Challenge Problems

CP-12.A. Section 12.5 addresses the relationship between temperature and reaction rates. As a rule of thumb, a 10-K rise in temperature increases the reaction rate by 2 to 4 times.
(a) What is the activation energy of a "typical" chemical reaction at 298 K?
(b) If a catalyst increases a chemical reaction's rate by providing a mechanism that has a lower activation energy, then what change do you expect a 10-K increase in temperature to make in the rate of a reaction whose uncatalyzed activation energy of 75 kJ/mol has been lowered to one-half this value (at 298 K) by addition of a catalyst?

CP-12.B. A sentence in an introductory chemistry textbook reads, "Dioxygen reacts with itself to form trioxygen, ozone, according to the following equation, $3\ O_2 \rightarrow 2\ O_3$." As a student of chemistry, what would you write to criticize this sentence?

CP-12.C. A classmate consults you about a problem concerning the reaction of nitrogen monoxide and dioxygen in the gas phase. He has been told that the reaction is second-order in nitrogen monoxide and first-order in dioxygen, hence the rate expression may be written as rate $= k[NO]^2[O_2]$. He has been asked to propose a mechanism for this reaction. He proposes that the mechanism is given by the following single equation,

$$NO + NO + O_2 \longrightarrow NO_2 + NO_2$$

He asks your opinion about whether this is correct. What is your answer?

CP-12.D. Polypropylene (***p. 488, Table 11.10***) is an important type of plastic and 2.7 million tons of it are produced every year in the United States. Properties of polypropylene can be changed by the way it is made. For example, melting points between 130 and 160 °C can be obtained by using an appropriate catalyst to polymerize propylene (propene, C_3H_6). Two important types of polypropylene are isotactic, in which the methyl groups are all on the same side of the polymer chain, and syn-

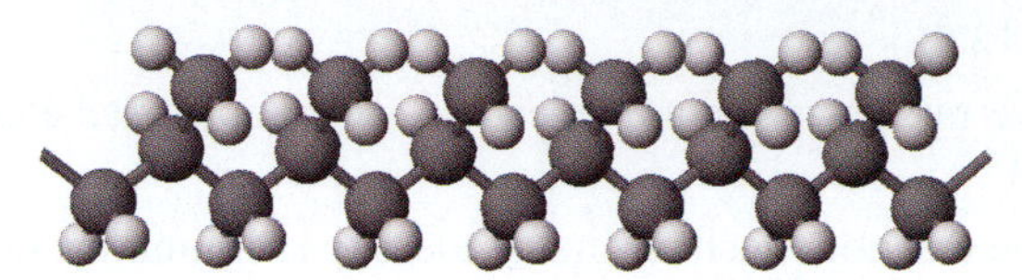

isotactic polypropylene

diotactic, in which the methyl groups alternate between one side and the other of the chain.

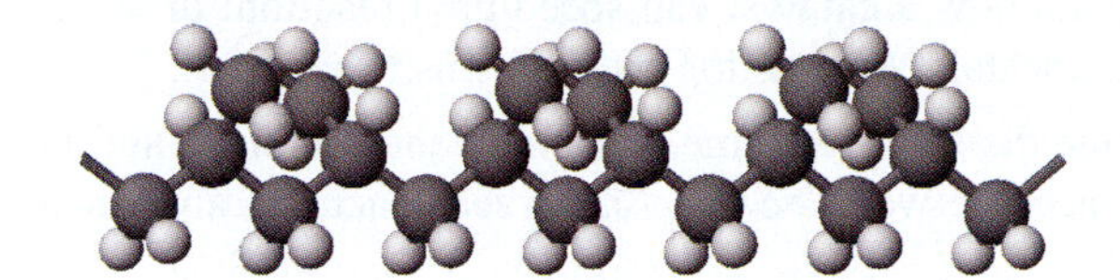

syndiotactic polypropylene

Suppose that you are part of a team designing a new catalyst to polymerize propylene. Your catalyst is going to have a zirconium atom at the center of a structure consisting of carbon and hydrogen atoms. The zirconium atom will hold the growing polypropylene chain by bonding to one end of it. The metal atom will also attract a propylene molecule and bond to it before transferring the growing polypropylene chain to the other end of the new propylene molecule. The process is shown below.

What would be a reasonable shape for the rest of the catalyst molecule surrounding the metal atom so that isotactic polypropylene would be produced? It may help to build molecular models to see how each new propylene molecule needs to be added to the growing polymer chain in order to get all the methyl groups on the same side.

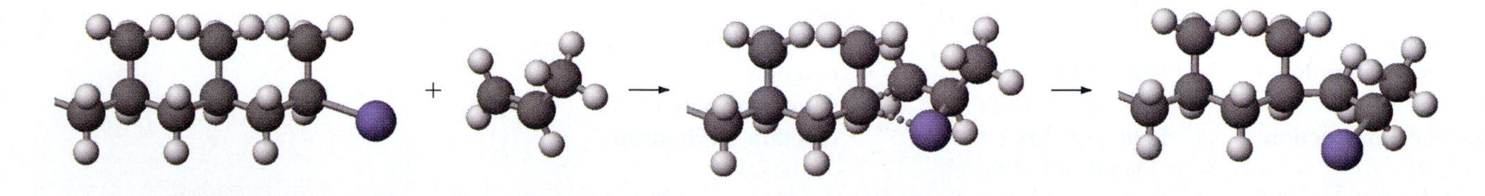

Review Questions

1. The rate expression for a chemical reaction can be determined by which of the following?
   (a) Theoretical calculations
   (b) Measuring the rate of the reaction as a function of the concentration of the reacting species
   (c) Measuring the rate of the reaction as a function of temperature
2. Name at least three factors that affect the rate of a chemical reaction.
3. Using the rate expression rate $= k[A]^2[B]$, define the order of the reaction with respect to A and B and the overall reaction order.
4. Draw a reaction energy diagram for an exothermic process. Mark the activation energies of the forward and reverse processes and explain how the net energy change of the reaction can be calculated.

5. Draw a reaction energy diagram for an endothermic process. Mark the activation energies of the forward and reverse processes and explain how the net energy change of the reaction can be calculated.
6. Indicate whether each of the following statements is true or false. Change the wording of each false statement to make it true.
(a) It is possible to change the rate constant for a reaction by changing the temperature.
(b) As a first-order reaction proceeds at a constant temperature, the rate remains constant.
(c) The rate constant for a reaction is independent of reactant concentrations.
(d) As a first-order reaction proceeds at a constant temperature, the rate constant changes.
7. What is a catalyst? What is its effect on the energy barrier for a reaction? What distinguishes an enzyme from a catalyst?
8. Define the terms "enzyme," "substrate," and "inhibitor," and give the name of one example of each kind of molecule.
9. Explain the difference between a homogeneous and a heterogeneous catalyst. Give an example of each.
10. Define the terms "unimolecular elementary reaction" and "bimolecular elementary reaction," and give an example of each.
11. Using an equation, define the terms "activation energy" and "frequency factor."
12. Does the ozone layer have anything to do with global warming (the greenhouse effect)? Explain. (If you are not sure what global warming is, look it up in a scientific source—not a newspaper or popular magazine.)
13. Explain the difference between thermodynamic stability and kinetic stability. Give two examples of common substances that have one but not the other kind of stability.

Reaction Rate

14. Consider the dissolving of sugar as a simple type of chemical reaction. Which is the fastest process: dissolving, in a stirred cup of water, 10 grams of (a) rock candy sugar (large sugar crystals), (b) sugar cubes, (c) granular sugar, or (d) powdered sugar? Using the concept of a heterogeneous reaction, explain your answer. If you are not sure which is fastest, try them all out!
15. For the hypothetical reaction $2\,A \rightarrow B + C$, make qualitatively correct plots of the concentrations of A, B, and C versus time, assuming that you start with pure A. Explain how you would determine, from these plots, (a) the initial rate of the reaction, and (b) the final rate (i.e., the rate as time approaches infinity).
16. Experimental data are listed below for the hypothetical reaction $A \rightarrow 2\,B$.

Time (s)	[A] (mol/L)
0.00	1.000
10.0	0.833
20.0	0.714
30.0	0.625
40.0	0.555

(a) Plot these data, connect the points with a smooth line, and calculate the rate of change of [A] for each 10-second interval from 0 to 40 sec. Why might the rate of change decrease from one time interval to the next?
(b) How is the rate of change of [B] related to the rate of change of [A] in the same time interval? Calculate the rate of change of [B] for the time interval from 10 to 20 sec.
17. A compound called phenylacetate reacts with water according to the equation

$$\underset{\text{phenyl acetate}}{CH_3\overset{\overset{O}{\|}}{C}{-}O{-}C_6H_5} + H_2O \longrightarrow \underset{\text{acetic acid}}{CH_3\overset{\overset{O}{\|}}{C}{-}O{-}H} + \underset{\text{phenol}}{C_6H_5{-}O{-}H}$$

The following data were collected at 5 °C.

Time (min)	[Phenyl acetate] (mol/L)
0	0.55
0.25	0.42
0.50	0.31
0.75	0.23
1.00	0.17
1.25	0.12
1.50	0.085

(a) Plot these data, and describe the shape of the curve. Compare it with Figure 12.3.
(b) Calculate the rate of change of the concentration of phenyl acetate during the period from 0.20 min to 0.40 min, and then during the period from 1.2 min to 1.4 min. Compare the values and tell why one is smaller than the other.
(c) What is the rate of change of the phenol concentration during the period from 1.00 min to 1.25 min?

Effect of Concentration on Reaction Rate

18. A sphere of aluminum of 1.0-cm diameter is placed into concentrated aqueous NaOH, and the reaction rate is monitored by measuring the rate at which H_2 gas is given off. By what factor will this reaction rate change if the aluminum sphere is cut exactly in half and the two halves placed in the solution? You may assume that the reaction rate is proportional to the surface area. If you had to speed up this reaction as much as you could without raising the temperature, what would you do to the aluminum?
19. If a reaction has the experimental rate expression rate $= k[A]^2$, explain what happens to the rate (a) when the concentration of A is tripled, and (b) when the concentration of A is halved.

20. A reaction has the experimental rate expression rate = $k[A]^2[B]$. If the concentration of A is doubled and the concentration of B is halved, what happens to the reaction rate?

21. The reaction of $CO(g) + NO_2(g)$ is second-order in NO_2 and zero-order in CO at temperatures less than 500 K.
(a) Write the rate expression for the reaction.
(b) How will the reaction rate change if the NO_2 concentration is halved?
(c) How will the reaction rate change if the concentration of CO is doubled?

22. Nitrosyl bromide, NOBr, is formed from NO and Br_2.

$$2\ NO(g) + Br_2(g) \longrightarrow 2\ NOBr(g)$$

Experiment shows that the reaction is first-order in Br_2 and second-order in NO.
(a) Write the rate expression for the reaction.
(b) If the concentration of Br_2 is tripled, how will the reaction rate change?
(c) What happens to the reaction rate when the concentration of NO is doubled?

Rate Law and Order of Reaction

23. For each of the following expressions, state the reaction order with respect to each reagent.
(a) Rate = $k[A][B]^2$ (c) Rate = $k[A]$
(b) Rate = $k[A][B]$ (d) Rate = $k[A]^3[B]$

24. A reaction between molecules A and B (A + B → products) is found to be second-order in B. Which rate equation cannot be correct?
(a) Rate = $k[A][B]$
(b) Rate = $k[A][B]^2$
(c) Rate = $k[B]^2$

25. For the reaction of $Pt(NH_3)_2Cl_2$ with water (Section 12.1),

$$Pt(NH_3)_2Cl_2 + H_2O \longrightarrow Pt(NH_3)_2(H_2O)Cl^+ + Cl^-$$

the rate expression was given as rate = $k[Pt(NH_3)_2Cl_2]$, with k = 0.090/hr. Calculate the initial rate of reaction when the concentration of $Pt(NH_3)_2Cl_2$ is (a) 0.010 M, (b) 0.020 M, and (c) 0.040 M. How does the rate of disappearance of $Pt(NH_3)_2Cl_2$ change with its initial concentration? How is this related to the rate law? How does the initial concentration of $Pt(NH_3)_2Cl_2$ affect the rate of appearance of Cl^- in the solution?

26. Methyl acetate, CH_3COOCH_3, reacts with base to break one of the C—O bonds.

$$CH_3C(=O){-}O{-}CH_3 + OH^- \longrightarrow CH_3C(=O){-}O^-(aq) + HO{-}CH_3(aq)$$

The rate expression is rate = $k[CH_3COOCH_3][OH^-]$, where k = 0.14 L/mol·s at 25 °C. What is the initial rate at which the methyl acetate disappears when both reactants, CH_3COOCH_3 and OH^-, have a concentration of 0.025 M? How rapidly (i.e., at what rate) does the methyl alcohol, CH_3OH, initially appear in the solution?

27. The transfer of an oxygen atom from NO_2 to CO has been studied at 540 K:

$$CO(g) + NO_2(g) \longrightarrow CO_2(g) + NO(g)$$

The following data were collected. Use them to
(a) Write the rate expression.
(b) Determine the reaction order with respect to each reactant.
(c) Calculate the rate constant, giving the correct units for k.

Initial Rate (mol/L·hr)	Initial Concentration (mol/L) [CO]	[NO_2]
5.1×10^{-4}	0.35×10^{-4}	3.4×10^{-8}
5.1×10^{-4}	0.70×10^{-4}	1.7×10^{-8}
5.1×10^{-4}	0.18×10^{-4}	6.8×10^{-8}
1.0×10^{-3}	0.35×10^{-4}	6.8×10^{-8}
1.5×10^{-3}	0.35×10^{-4}	10.2×10^{-8}

28. A study of the reaction 2 A + B → C + D gave the following results:

Experiment	Initial Concentration (mol/L) [A]	[B]	Initial Rate (mol/L·s)
1	0.10	0.05	6.0×10^{-3}
2	0.20	0.05	1.2×10^{-2}
3	0.30	0.05	1.8×10^{-2}
4	0.20	0.15	1.1×10^{-1}

(a) What are the rate expression and order with respect to A and B for this reaction?
(b) Calculate the rate constant for the reaction, giving the correct units for k.

29. The bromination of acetone is catalyzed by acid.

$$CH_3COCH_3 + Br_2 + H_2O(\ell) \xrightarrow{\text{acid catalyst}} CH_3COCH_2Br + H_3O^+ + Br^-$$

The rate of disappearance of bromine was measured for several different initial concentrations (all in mol/L) of acetone, bromine, and hydronium ion.

Initial Concentration (mol/L)			Initial Rate of Change of $[Br_2]$ (mol/L-s)
$[CH_3COCH_3]$	$[Br_2]$	$[H_3O^+]$	
0.30	0.05	0.05	5.7×10^{-5}
0.30	0.10	0.05	5.7×10^{-5}
0.30	0.05	0.10	12.0×10^{-5}
0.40	0.05	0.20	31.0×10^{-5}
0.40	0.05	0.05	7.6×10^{-5}

(a) Deduce the rate expression for the reaction and give the order with respect to each reactant.
(b) What is the numerical value of k, the rate constant?
(c) If $[H_3O^+]$ is maintained at 0.050 M, whereas both $[CH_3COCH_3]$ and $[Br_2]$ are 0.10 M, what is the rate of the reaction?

30. One of the major eye irritants in smog is formaldehyde, CH_2O, formed by reaction of ozone with ethylene.

$$C_2H_4(g) + O_3(g) \longrightarrow 2\ CH_2O(g) + \tfrac{1}{2}\ O_2(g)$$

(a) Determine the rate expression for the reaction using the data in the following table. What is the reaction order with respect to O_3? What is the order with respect to C_2H_4?
(b) Calculate the rate constant, k.
(c) What is the rate of reaction when $[C_2H_4]$ and $[O_3]$ are both 2.0×10^{-7} M?

Initial Concentration (mol/L)		Initial Rate of Formation of CH_2O (mol/L·s)
$[O_3]$	$[C_2H_4]$	
0.50×10^{-7}	1.0×10^{-8}	1.0×10^{-12}
1.5×10^{-7}	1.0×10^{-8}	3.0×10^{-12}
1.0×10^{-7}	2.0×10^{-8}	4.0×10^{-12}

31. This problem requires working with the equations of Table 12.1. Using an initial concentration $[A]_0$ of 1.0 mol/L and a rate constant k with a numerical value of 1.0 in appropriate units, make plots of [A] versus time over the time interval 0 to 5 sec for each type of integrated rate law. Compare your results with Figure 12.5.

32. Studies of radioactive decay of nuclei show that the *decay rate* of a radioactive sample is proportional to the *amount* of the radioactive species present. Once half the radioactivity has disappeared, the radioactive decay *rate* is only half of its original value. Is radioactive decay a zero-order, first-order, or second-order process? (*Hint:* See Table 12.1.)

A Nanoscale View: Elementary Reactions

33. Using a molecular model kit, build models of *cis*-2-butene, *trans*-2-butene, and the transition state, or activated complex. How much force do *you* need to apply to the models to change the reactant into the product by passing through the activated complex? The answer will of course depend on the kind of model kit that you use.

34. Of the following reactions, choose which ones are unimolecular and elementary, which ones are bimolecular and elementary, and which ones are not elementary:

(a) $CH_4(g) + 2\ O_2(g) \longrightarrow CO_2(g) + 2\ H_2O(g)$
(b) $O_3(g) + O(g) \longrightarrow 2\ O_2(g)$
(c) $Mg(s) + 2\ H_2O(\ell) \longrightarrow H_2(g) + Mg(OH)_2(s)$
(d) $O_3(g) \longrightarrow O_2(g) + O(g)$
(e) $HCl(g) + H_2O(\ell) \longrightarrow H_3O^+(aq) + Cl^-(aq)$
(f) $I^-(g) + CH_3Cl(g) \longrightarrow ICH_3(g) + Cl^-(g)$
(g) $C_2H_6(g) \longrightarrow C_2H_4(g) + H_2(g)$
(h) $N_2(g) + 3\ H_2(g) \longrightarrow 2\ NH_3(g)$
(i) $O_2(g) + O(g) \longrightarrow O_3(g)$

Temperature and Reaction Rate

35. The rate of decay of a radioactive solid is independent of the temperature of that solid—at least for temperatures easily obtained in the laboratory. What does this observation imply about the activation energy for this process?

36. From the figure showing the energy profile of the ozone-plus-atomic oxygen reaction, write down the activation energy for this reaction. Then determine the *ratio* of two reaction rates for this reaction, the first at room temperature (25 °C) and the second at 50 °C. Assume that the initial concentrations are the same at both temperatures.

37. Suppose a reaction rate constant has been measured at two different temperatures, T_1 and T_2, and its value is k_1 and k_2, respectively.
(a) Write down the Arrhenius equation at each temperature.
(b) By combining these two equations, derive an expression for the *ratio* of the two rate constants, k_1/k_2. Use this formula to solve the following two problems.

38. Suppose a chemical reaction rate constant has an activation energy of 76 kJ/mole, as in the example in Figure 12.11. By what factor is the rate of the reaction at 50 °C increased over its rate at 25 °C?

39. A chemical reaction has an activation energy of 30 kJ/mol. If you had to slow down this reaction a thousand-fold by cooling it from room temperature (25 °C), what would your final temperature be?

Rate Laws for Elementary Reactions

40. For the hypothetical reaction A + B → C + D, the activation energy is 32 kJ/mol. For the reverse reaction (C + D → A + B), the activation energy is 58 kJ/mol. Is the reaction A + B → C + D exothermic or endothermic?

41. Use the diagram below to answer the following questions.
(a) Is the reaction exothermic or endothermic?
(b) What is the approximate value of ΔE for the forward reaction?
(c) What is the activation energy in each direction?
(d) A catalyst is found that lowers the activation energy of the reaction by about 10 kJ/mol. How will this catalyst affect the rate of the reverse reaction?

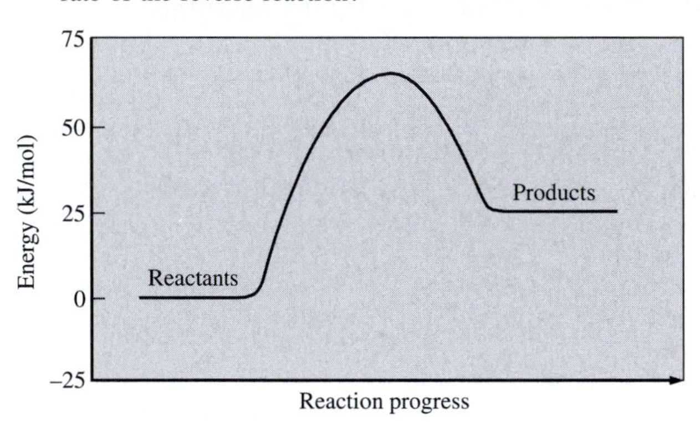

Reaction Mechanisms

42. Experiments show that the reaction of nitrogen dioxide with fluorine

Overall Reaction: $2\ NO_2(g) + F_2(g) \longrightarrow 2\ FNO_2(g)$

has the following rate expression:

$$\text{Initial reaction rate} = k[NO_2][F_2]$$

and the reaction is thought to occur in two steps, the first being very slow and the second rapid.

Step 1 (slow): $NO_2(g) + F_2(g) \longrightarrow FNO_2(g) + F(g)$
Step 2 (fast): $NO_2(g) + F(g) \longrightarrow FNO_2(g)$

That is, NO_2 and F_2 first produce one molecule of the product (FNO_2) plus an F atom, and the F atom then reacts with additional NO_2 to give one more molecule of product. Show that the sum of this sequence of reactions gives the balanced equation for the overall reaction. Which step is rate determining?

43. Nitrogen oxide is reduced by hydrogen to give water and nitrogen

$$2\ H_2(g) + 2\ NO(g) \longrightarrow N_2(g) + 2\ H_2O(g)$$

and one possible mechanism for this reaction is a sequence of three elementary steps.

$$2\ NO \rightleftharpoons N_2O_2$$
$$N_2O_2(g) + H_2(g) \longrightarrow N_2O(g) + H_2O(g)$$
$$N_2O(g) + H_2(g) \longrightarrow N_2(g) + H_2O(g)$$

Show that the sum of these steps gives the net reaction.

Catalysts and Reaction Rate

44. Which of the following statements is (are) true?
(a) The concentration of a homogeneous catalyst may appear in the rate expression.
(b) A catalyst is always consumed in the reaction.
(c) A catalyst must always be in the same phase as the reactants.
(d) A catalyst can change the course of a reaction and allow different products to be produced.

45. Hydrogenation reactions, processes in which H_2 is added to a molecule, are usually catalyzed. An excellent catalyst is a very finely divided metal suspended in the reaction solvent. Tell why finely divided rhodium, for example, is a much more efficient catalyst than a small block of the metal.

46. Which of the following reactions appear to involve a catalyst? In those cases where a catalyst is present, tell whether it is homogeneous or heterogeneous.
(a) $CH_3CO_2CH_3(aq) + H_2O(\ell) + H_3O^+(aq) \longrightarrow CH_3CO_2H(aq) + CH_3OH(aq) + H_3O^+(aq)$
(b) $2\ H_2(g) + O_2(g) \longrightarrow 2\ H_2O(g)$
(c) $2\ H_2(g) + O_2(g) + Pt(s) \longrightarrow 2\ H_2O(g) + Pt(s)$
(d) $NH_3(aq) + CH_3Cl(aq) + H_2O(\ell) \longrightarrow Cl^-(aq) + NH_4^+(aq) + CH_3OH(aq)$

47. In acid solution, methyl formate forms methyl alcohol and formic acid.

$$\underset{\text{methyl formate}}{HCO_2CH_3(aq)} + H_2O(\ell) \longrightarrow \underset{\text{formic acid}}{HCO_2H(aq)} + \underset{\text{methyl alcohol}}{CH_3OH(aq)}$$

The rate expression is as follows: rate = k $[HCO_2CH_3]\ [H_3O^+]$. Why does H_3O^+ appear in the rate expression but not in the overall equation for the reaction?

48. Suppose a rate constant for an uncatalyzed reaction is described by the Arrhenius equation with an activation energy E. The introduction of a catalyst lowers the activation energy to the value E' and thus increases the rate constant to the value k'. By first writing down the Arrhenius equation for both k and k', derive an expression for the ratio k'/k in terms of E' and E; assume that the pre-exponential factor A and the temperature T are constant. Use this equation to solve the following two problems.

49. Suppose a catalyst was found for the reaction in Exercise 12.3 which reduced the activation energy to zero. By what factor would this reaction rate be increased at 370 K?

50. In the discussion of Figure 12.15, the following statement is made: "Dropping the activation energy from 262 kJ/mol for the uncatalyzed reaction to 115 kJ/mole for the catalyzed process makes the catalyzed reaction 10^{15} times faster." Prove this statement.

Enzymes: Biological Catalysts

51. In Section 12.9 there are quite a few chemical and biochemical terms that may be new to you. Write a one- to two-sentence definition of each of the following terms, using other references besides this text if necessary:

enzyme	maximum velocity	HIV protease
cofactor		inhibition

polypeptides	substrate	enzyme-substrate complex
monomer	active site	induced fit
polysaccharides	proteins	globular proteins
lysozyme	dimer	denaturation
	hydrolysis	

52. When enzymes are present at very low concentration their effect on reaction rate can be described by first-order kinetics. By what factor does the rate of an enzyme-catalyzed reaction change when the enzyme concentration is changed from 1.5×10^{-7} M to 4.5×10^{-6} M?

53. When substrates are present at relatively high concentration and are catalyzed by enzymes, the effect on reaction rate of changing substrate concentration can be described by zero-order kinetics. By what factor does the rate of an enzyme-catalyzed reaction change when the substrate concentration is changed from 1.5×10^{-3} M to 4.5×10^{-2} M?

Catalysis in Industry

54. In the first paragraph of Section 12.10 of the text an expert in the field of industrial chemistry is quoted. Explain his statement, in view of your understanding of the nature of catalysts: why *are* catalysts so important?

55. Why are homogeneous catalysts harder to separate from products and leftover reactants than are heterogeneous catalysts?

56. The production of the fertilizer ammonium nitrate involves four reactions:

$$4\ NH_3(g) + 5\ O_2(g) \longrightarrow 4\ NO(g) + 6\ H_2O(g)$$
$$2\ NO(g) + O_2(g) \longrightarrow 2\ NO_2(g)$$
$$3\ NO_2(g) + H_2O(\ell) \longrightarrow 2\ HNO_3(aq) + NO(g)$$
$$HNO_3(aq) + NH_3(g) \longrightarrow NH_4NO_3(aq)$$

Describe each as either an oxidation-reduction reaction, an acid-base reaction, or some other type. In which type of reaction is the catalyst used?

57. Find all examples of reactions described in this chapter which are catalyzed by metals. Are these metals main-group metals or transition metals? What type of chemical reactions are they, acid-base or oxidation-reduction? Can you reach some conclusions about metal-catalyzed chemical reactions from these examples?

Chlorofluorocarbons and the Ozone Layer

58. Write the chemical formulas of all possible CFCs with one carbon atom.

59. Are CFCs toxic? In considering this question, remember what compounds were used for refrigeration before CFCs were invented. Look up the toxicity of these compounds in the library.

60. Can CFCs catalyze the destruction of ozone in the stratosphere at night? Explain.

61. Can ozone form in the stratosphere at night? Explain why or why not.

62. Why is chemical kinetics important in understanding the environmental problem posed by chlorofluorocarbons in the ozone layer?

63. Explain how the graph of the instrument measurements from NASA's ER-2 research airplane (Figure 12.27) presents convincing evidence for the correctness of the reaction mechanism proposed for the ozone hole.

Thermodynamic and Kinetic Stability

64. Billions of pounds of acetic acid are made each year, much of it by the reaction of methanol with carbon monoxide:

$$CH_3OH(\ell) + CO(g) \longrightarrow CH_3COOH(\ell)$$

(a) By calculating the standard free energy change $\Delta G°$ for this reaction, show that it is product-favored (Chapter 7). Use thermodynamic values given in Appendix J; ΔG_f° for liquid acetic acid is −389.9 kJ/mole.

(b) Determine the standard free energy change $\Delta G°$ for the reaction of methanol, of carbon monoxide, and of acetic acid with oxygen to form gaseous carbon dioxide and liquid water.

(c) Based on this result, determine which of the following species are kinetically stable and which are thermodynamically stable.

$$CH_3OH(\ell),\ CO(g),\ CH_3COOH(\ell),\ CO_2(g),\ H_2O(\ell)$$

(d) How can kinetically stable substances exist at all, if they are not thermodynamically stable?

(e) This is a group project: estimate or look up, to the nearest order of magnitude,

- the number of kg of CH_3OH made each year
- the number of kg of CO in the entire atmosphere
- the number of kg of CH_3COOH made each year
- the number of kg of H_2O on earth
- the number of kg of CO_2 in the atmosphere

What do these facts tell you about the difference between kinetic stability and thermodynamic stability?

(f) Actually, the carbon in CO_2 is thermodynamically *unstable* with respect to the carbon in calcium carbonate, or limestone. Show this by determining the standard free energy change for the reaction of lime (CaO(s)) with CO_2 to make calcium carbonate.

General Questions

65. Nitrogen monoxide can be reduced with hydrogen.

$$2\ H_2(g) + 2\ NO(g) \longrightarrow 2\ H_2O(g) + N_2(g)$$

Experiment shows that when the concentration of H_2 is halved, the reaction rate is halved. Furthermore, raising the concentration of NO by a factor of three raises the rate by a factor of nine. Write the rate equation for this reaction.

66. One reaction that may occur in air polluted with nitrogen monoxide is

$$2\ NO(g) + O_2(g) \longrightarrow 2\ NO_2(g)$$

Using the data in the table, answer the questions that follow.

Experiment	Initial Concentration (mol/L)		Initial Rate of Formation of NO_2 (mol/L·s)
	[NO]	$[O_2]$	
1	0.001	0.001	7×10^{-6}
2	0.001	0.002	14×10^{-6}
3	0.001	0.003	21×10^{-6}
4	0.002	0.003	84×10^{-6}
5	0.003	0.003	189×10^{-6}

(a) What is the order of reaction with respect to each reactant?
(b) Write the rate expression for the reaction.
(c) Calculate the rate of formation of NO_2 when $[NO] = [O_2] = 0.005$ mol/L.

67. The deep blue compound CrO_5 can be made from the chromate ion by using hydrogen peroxide in an acidic solution.

$$HCrO_4^-(aq) + 2\ H_2O_2(aq) + H_3O^+(aq) \longrightarrow CrO_5(aq) + 4\ H_2O(\ell)$$

The kinetics of this reaction have been studied, and the rate equation is found to be

$$\text{Rate of disappearance of } HCrO_4^- = k[HCrO_4^-][H_2O_2][H_3O^+]$$

One of the mechanisms suggested for the reaction is

$$HCrO_4^- + H_3O^+ \rightleftharpoons H_2CrO_4 + H_2O$$
$$H_2CrO_4 + H_2O_2 \longrightarrow H_2CrO_5 + H_2O$$
$$H_2CrO_5 + H_2O_2 \longrightarrow CrO_5 + 2\ H_2O$$

(a) Give the order of the reaction with respect to each reactant.
(b) Show that the steps of the mechanism agree with the overall equation for the reaction.

68. Refer to Figure 12.3 and explain why the rate of change of the concentration of $Pt(NH_3)_2Cl_2$ decreases with time but the concentration of Cl^- increases with time.

69. How does a chemical reaction mechanism differ from other types of mechanisms, for example, the gear-changing mechanism of a bicycle, or the mechanism of an elevator? How is it similar to these mechanisms?

70. Why are catalysts important economically to the chemical industry?

71. What are CFCs? Why were they developed (i.e., for what have they been used)? Describe how your life might be different if all CFCs were banned and there were no replacements for them.

Applying Concepts

72. The following graph shows the change in concentration as a function of time for the reaction

$$2\ H_2O_2(g) \longrightarrow 2\ H_2O(g) + O_2(g)$$

What do each of the curves A, B, and C represent?

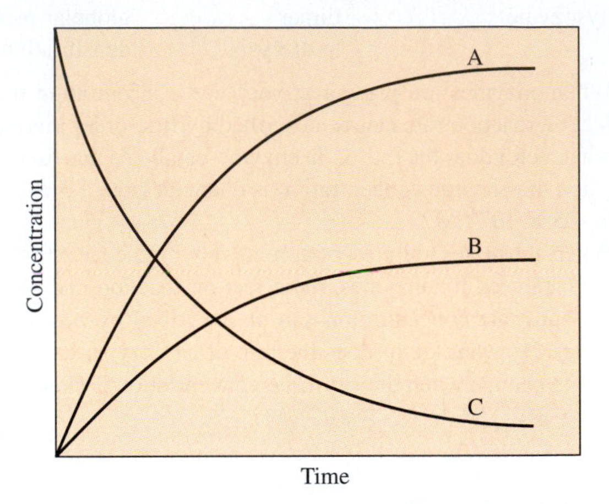

73. Draw a graph similar to the one in Question 72 for the reaction

$$2\ N_2O_5(g) \longrightarrow 4\ NO_2(g) + O_2(g)$$

Label each of the curves.

74. The picture below is a "snapshot" of the reactants at time = 0 for the reaction

$$H_2(g) + I_2(g) \longrightarrow 2HI(g)$$

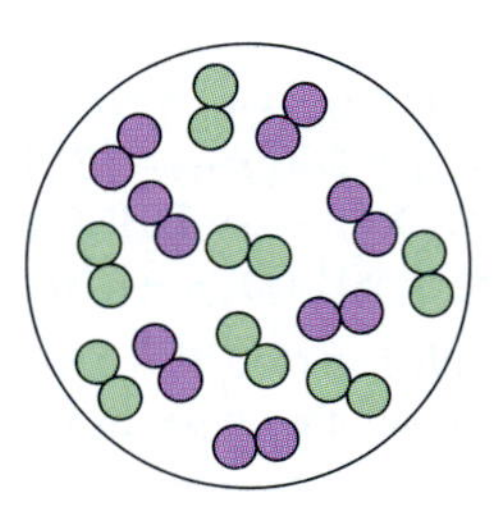

Suppose the reaction is carried out at two different temperatures and another snapshot is taken after a constant time has elapsed. Which of these two snapshots corresponds to the higher temperature reaction condition?

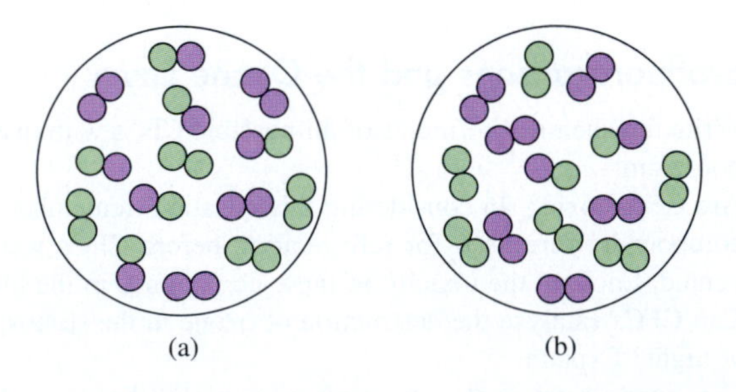

75. Consider the previous question again, only this time a catalyst is used instead of a higher temperature. Which of the two "snapshots" corresponds to the presence of a catalyst?

76. Initial rates for the reaction A + B + C → D + E were measured with various concentrations of A, B, and C as represented in the pictures below. Based on these data, what is the rate equation?

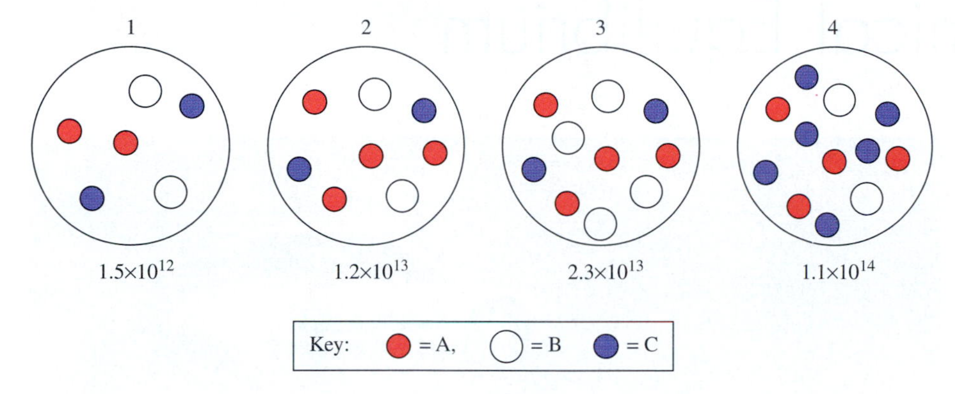

77. Platinum metal is used as a catalyst in the decomposition of NO(g) into N_2(g) and O_2(g). A graph of the rate of the reaction as a function of NO concentration looks like this *(below)*. Explain why the rate stops increasing and levels out.

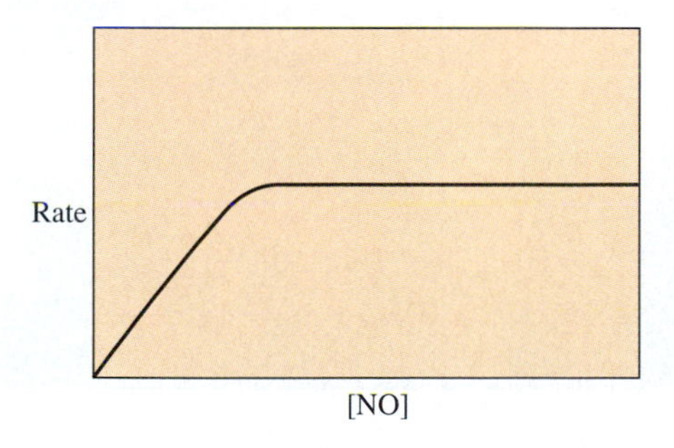

CHAPTER

13

Principles of Reactivity IV: Chemical Equilibrium

Electric automobiles show promise of providing transportation with much less pollution and potentially less consumption of resources, including Gibbs free energy. The electric car shown here is powered by batteries of electrochemical cells (Chapter 18), but free energy is more effectively stored in liquid or solid fuels such as hydrocarbons or coal than it is in the components of a battery. • Can hydrocarbon fuels be used to power an electric car that would be 50% more efficient in its use of fuels than cars are today? What sorts of fuels could be used to power such a car, and what chemical reactions would be needed to convert their Gibbs free energy into electricity? These questions are answered in the Summary Problem at the end of this chapter. *(Courtesy General Motors Corporation)*

In Chapter 7 you learned that if a chemical reaction increases the entropy of the universe (or if the Gibbs free energy of the reaction system decreases), the reaction can be classified as a product-favored system—there will be more products than reactants after the reaction is over. In Chapter 12 you learned about the factors that affect the rate at which reactants are converted to products during a chemical reaction. Now we are ready to use ideas from both of these earlier chapters to provide quantitative information about *how much* product can be obtained from a product-favored reaction. In addition we will show how changes in the conditions of temperature or pressure under which a reaction occurs affect the quantity of product.

As an example of the importance of such information, consider ammonia. Nearly 36 billion pounds of liquefied ammonia were produced in the United States in 1995, mostly for use as fertilizer to provide nitrogen needed to support growth of a broad range of crops. Therefore ammonia is a very important factor in providing people with food. Ammonia is synthesized directly from hydrogen and nitrogen by what is called the Haber-Bosch process (see Section 13.8).

$$N_2(g) + 3\ H_2(g) \longrightarrow 2\ NH_3(g) \qquad \Delta H^\circ = -92.2\ \text{kJ}$$

It is important to the chemists and chemical engineers who operate such plants that the maximum yield of ammonia be obtained with the minimum input of reactants and the minimum consumption of free energy. The German chemist Fritz Haber won the Nobel Prize in 1918 because his studies of this reaction showed how to determine the best conditions for carrying it out. (See Portrait of a Scientist, Section 13.8.) The German engineer Carl Bosch, who perfected the industrial process, won the Nobel Prize in 1931.

You can see that at standard conditions formation of the product, ammonia, is favored by enthalpy—the reaction is exothermic ($\Delta H_{\text{system}} < 0$). However, because 4 mol of gaseous reactants are converted to only 2 mol of gaseous products, the entropy factor favors reactants ($\Delta S_{\text{system}} < 0$). Thus the reaction is product-favored at low temperatures and reactant-favored at high temperatures. Unfortunately the reaction of nitrogen with hydrogen is extremely slow at room temperature. If the temperature is raised enough to provide an appreciable rate, the system becomes reactant-favored, and little product results under standard conditions. How, then, can a chemical plant produce ammonia? The answer is to use an appropriate catalyst so that the reaction proceeds at a fast enough rate at a relatively low temperature, to adjust the amounts of reactants and the pressure so that as much product as possible is produced, and to choose specific conditions for the reaction based on quantitative predictions of how much product will be formed. The main purpose of this chapter is to help you learn how to make both qualitative and quantitative predictions about how much product will be formed under a given set of conditions.

13.1 THE EQUILIBRIUM STATE

When a chemical reaction takes place in the gas phase or in a liquid solution, the concentrations of the reactants decrease and the concentrations of the products increase. Eventually a state of **chemical equilibrium** is reached, in which *the concentrations of reactants and products remain constant.* An equilibrium reaction always results in less than the theoretical yield of product, but sometimes only tiny quantities of reactants remain. For a *product-favored system* the equilibrium state involves greater concentrations of products than reactants. For a *reactant-favored system* the concentrations of products at equilibrium are smaller than those of the reactants, but some quantity of product, however tiny, is always formed. The concentrations of reactants and products at equilibrium provide a quantitative way of determining how successful a reaction has been.

When equilibrium is reached and concentrations remain constant, it appears that a chemical reaction has stopped, but this is only true of the net, macroscopic reaction. On the nanoscale both forward and reverse reactions continue, but there is a balance between transformation of reactants into products and transformation of products into reactants. That is, the rate of the forward reaction (transformation of reactants into products) exactly equals the rate of the reverse reaction (transformation of products into reactants). Both reactions continue to occur. To emphasize that equilibrium involves a balance between opposite reactions, we often refer to it as a **dynamic equilibrium** and represent the equilibrium reaction with a double arrow between reactants and products. This was done, for example, in the case of weak acids and bases ***(← p. 154),*** which ionize only partially in water before equilibrium is reached:

The "equil" in the world "equilibrium" refers to equal rates of forward and reverse reactions, not to equal quantities of the substances involved.

A set of double arrows ($\rightleftharpoons$) in an equation indicates a dynamic equilibrium in which forward and reverse reactions are occurring at equal rates; it also indicates that the reaction should be thought of in terms of the concepts of chemical equilibrium.

$$\underset{\text{acetic acid}}{CH_3COOH(aq)} \rightleftharpoons \underset{\text{acetate ion}}{CH_3COO^-(aq)} + \underset{\text{hydrogen ion}}{H^+(aq)}$$

In the case of acetic acid, more than 90% of the acetic acid remains in molecular form at equilibrium, and the equilibrium concentrations of acetate ions and hydrogen ions are each less than one-tenth the concentration of acetic acid molecules.

Another important characteristic of chemical equilibrium is that, *for a specific reaction, the equilibrium state will be the same, no matter what the direction of approach to equilibrium.* (This is true so long as you start with the same number of atoms of each kind in a container of the same volume.) As an example, consider the combination of N_2 and O_2 that occurs when lightning passes through air ***(← p. 344).***

$$N_2(g) + O_2(g) \rightleftharpoons 2\ NO(g)$$

This is a reactant-favored system, and so only a small concentration of the product NO forms. (However, at high temperatures, like those in an automobile engine or a fossil-fueled electric power plant, the reaction is less reactant-favored and the concentration of NO that forms at equilibrium is large enough to contribute to air pollution.)

Suppose that you introduce 1.0 mol $N_2(g)$ and 1.0 mol $O_2(g)$ into a completely empty 1.00-L container at a very high temperature, 5000 K. At this temperature the reaction is fast and equilibrium is quickly established, at which point the concentrations of N_2 and O_2 (which must remain equal because of the stoichiometry of the reaction) each have fallen from their initial values of 1.0 mol/L to an equilibrium value of 0.80 mol/L. The concentration of product NO is 0.40 mol/L, which reflects the fact that the process is reactant-favored—equilibrium concentrations of the reactants remain higher than for the products.

Now do a second experiment in which you introduce 2.0 mol NO(g) into an empty 1.00-L container at 5000 K. (Notice that 2.0 mol NO consists of 2.0 mol N atoms and 2.0 mol O atoms, which is the same number of N atoms and O atoms contained in the 1.0 mol N_2 and 1.0 mol O_2 used in the first experiment.) Because the system is reactant-favored, most of the NO will react to form N_2 and O_2. In fact, measuring the concentrations at equilibrium would reveal that [NO] had dropped from the initial 2.0 mol/L to 0.40 mol/L, and $[N_2]$ and $[O_2]$ each would be 0.80 mol/L. These equilibrium concentrations are exactly the same as were achieved in the first experiment, which started with the reactants. Thus, whether you start with reactants or products, the same equilibrium state is achieved, as long as the number of atoms

Recall from Chapter 12 that square brackets around the formula of a substance indicate the concentration or molarity of the substance, which is usually measured in moles per liter ***(← pp. 198 and 514).***

of each type is the same and the volume of the container is constant. If the number of atoms of any kind were different or the volume were to change, then the equilibrium state would be different.

In the first experiment, equilibrium was achieved by having a small fraction (1/5) of the nitrogen and oxygen molecules consumed in the forward reaction. In the second experiment equilibrium was achieved by having most of the NO molecules (4/5 of them) undergo the reverse reaction. In both cases the final equilibrium state, though achieved by reactions in opposite directions, involves the same concentrations of the three molecules.

Iron(III)-thiocyanate equilibrium. Reaction of colorless solutions of $Fe^{3+}(aq)$ and $SCN^-(aq)$ gives the red-orange ion $Fe(SCN)^{2+}(aq)$. *(C.D. Winters)*

Exercise 13.1 Recognizing an Equilibrium State

A mixture of hydrogen gas and oxygen gas is maintained at 25 °C for one year. On the first day of each month the mixture is sampled and the concentrations of hydrogen and oxygen measured. In every case they are found to be 0.50 mol/L H_2 and 0.50 mol/L O_2, that is, the concentrations are found not to change over a long period. Is this mixture at equilibrium? If you think not, how could you do an experiment to prove it?

Equilibrium Is Dynamic

When colorless solutions of aqueous iron(III) ion, Fe^{3+}, and aqueous thiocyanate ion, SCN^-, are mixed, a red-orange iron thiocyanate complex ion *(← p. 384, complex ions)* is formed:

$$Fe^{3+}(aq) + SCN^-(aq) \rightleftharpoons Fe(SCN)^{2+}(aq)$$

nearly colorless colorless red-orange

In aqueous solution each iron(III) ion is surrounded by six water molecules and its formula could better be written $Fe(H_2O)_6^{3+}$ (**← p. 385**); however, for simplicity the water molecules are often left out. When SCN^- reacts with iron(III) ion, it displaces one of the six water molecules, making the formula $Fe(H_2O)_5(SCN)^{2+}$.

Equilibrium is rapidly achieved, and there is a simple way to prove that the forward and reverse reactions continue. A drop of an aqueous solution of radioactive $S^{14}CN^-$ ion can be added to the equilibrium mixture. ($S^{14}CN^-$ represents thiocyanate ion that contains radioactive ^{14}C in place of normal, nonradioactive ^{12}C.) Testing the solution shortly after adding the radioactive $S^{14}CN^-$ ion shows that the $Fe(SCN)^{2+}$ contains some radioactivity. The most reasonable mechanism by which this can happen while the system remains at equilibrium is for the reverse reaction to occur, releasing a nonradioactive SCN^- ion:

$$Fe(SCN)^{2+}(aq) \longrightarrow Fe^{3+}(aq) + SCN^-(aq) \quad \text{(reverse reaction)}$$

followed by the forward reaction of Fe^{3+} ion with a radioactive $S^{14}CN^-$ ion:

$$Fe^{3+}(aq) + S^{14}CN^-(aq) \longrightarrow Fe(S^{14}CN)^{2+}(aq) \quad \text{(forward reaction)}$$

Thus the forward and reverse reactions must still be going on after equilibrium has been achieved.

Another important characteristic of chemical equilibrium is that *if a catalyst is present, the same equilibrium state will be achieved, but it will be achieved more quickly.* A catalyst speeds up the forward reaction, but it also speeds up the reverse reaction, and the overall effect is to produce exactly the same concentrations at equilibrium, whether or not a catalyst is in the reaction mixture. This means that a catalyst can be used to speed up production of products in an industrial process, but it

Conceptual Challenge Problems CP-13.A and CP-13.B at the end of the chapter relate to topics discussed in this section.

CHEMISTRY IN THE NEWS

Equilibrium and Alzheimer's Disease

Approximately two million Americans are afflicted by Alzheimer's disease and more than 100,000 of them die each year. There is no definitive treatment or cure for this debilitating disease, which causes loss of memory and reasoning, disorientation, and eventually death. Alzheimer's affects between 20 and 30 percent of people over 80 and is the fourth leading cause of death in the United States.

One approach to fighting this disease is based on recognition that it is accompanied by formation of a tangle of toxic proteins, called plaques, in the brain. (Proteins are discussed in Section 11.10, *(⇐ p. 498.)* A major component of the plaques is a protein called beta-amyloid. Its molecules attract each other strongly and at equilibrium form long clumps called fibrils. In this fibril form it is toxic and kills brain cells. The formation of fibrils can be represented approximately by an equilibrium equation:

$$n \text{ beta-amyloid} \rightleftharpoons (\text{beta-amyloid})_n$$

(*n* individual protein molecules) (fibril)

This is a product-favored reaction, and much of the beta-amyloid protein is converted into fibrils. Two scientists at the University of Wisconsin–Madison, Laura Kiessling and Regina Murphy, reasoned that if this product-favored reaction could be prevented from happening, the fibrils would not form and the protein would not be toxic.

Kiessling and Murphy realized that the protein molecules stuck together to form fibrils because of non-covalent interactions *(⇐ p. 423)*. Some of the protein's amino-acid side chains are nonpolar. These are also called **hydrophobic,** meaning that they avoid water. The parts of the protein consisting mainly of nonpolar side chains tend to clump together, avoiding the aqueous solutions within cells. Other side chains are polar and/or capable of forming hydrogen bonds. These are called **hydrophilic,** meaning that they are attracted to water. Kiessling and Murphy reasoned that the hydrophobic segments of one beta-amyloid molecule were being attracted to hydrophobic segments of another beta-amyloid molecule, causing the formation of fibrils to be product-favored. If they could somehow avoid these attractions between different beta-amyloid molecules, perhaps the fibril-forming reaction would not occur.

They decided to try to trick the protein molecules into binding to something else rather than each other. Their plan was to create a new molecule that mimicked the hydrophobic parts of beta-amyloid. They expected that this new molecule would bind to the protein, preventing protein molecules from binding to each other. If we call the new molecule MM (for mimic molecule), they set up a second equilibrium system:

$$\text{MM} + \text{beta-amyloid} \rightleftharpoons \text{MM—(beta-amyloid)}$$

This new equilibrium, as predicted, is even more product-favored than the first one, and so it occurs preferentially. The beta-amyloid no longer forms fibrils. Instead it forms small bundles that turn out to be non-toxic in cells.

According to John Cross, chief scientist for the American Health Assistance Foundation, this is an exciting development that "has the possibility of interfering with the progression of Alzheimer's. If there is something that can prevent nerve cell death, perhaps we can stop the disease." There is still a long way to go, though, before this basic research can be turned into an effective drug for treatment of Alzheimer's. Such a drug must be able to avoid the body's immune system, pass into the brain, and be very specific so that it prevents fibril formation without affecting the brain adversely in other ways. Kiessling and Murphy are continuing their research to refine their molecule that mimics beta-amyloid's hydrophobic sites, and they hope eventually to create a compound that a pharmaceutical company can manufacture and sell as an injectable treatment for Alzheimer's disease. If that dream is realized, this will become an even bigger story of chemistry in the news.

This news feature includes information from the Wisconsin State Journal, *Sunday, January 12, 1997.*

will not result in greater equilibrium concentrations of products. (Of course the catalyst also will not reduce the concentration of product that is present when the system reaches equilibrium.)

13.2 THE EQUILIBRIUM CONSTANT

Consider again the isomerization reaction of *cis*-2-butene to *trans*-2-butene, whose rate we discussed previously *(⇐ p. 523).*

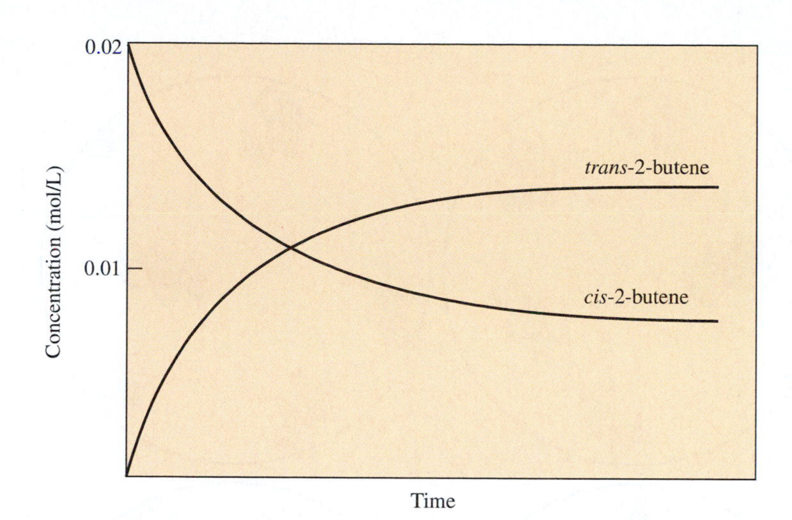

Figure 13.1 Approach to equilibrium. The graph shows the concentrations of *cis*-2-butene and *trans*-2-butene as the *cis* compound reacts to form the *trans* compound. Equilibrium has been achieved once the concentrations reach constant (but not necessarily equal) values.

$$\underset{cis\text{-2-butene}}{(H)(H_3C)C{=}C(H)(CH_3)} \rightleftharpoons \underset{trans\text{-2-butene}}{(H)(H_3C)C{=}C(CH_3)(H)}$$

At 500 K the reaction reaches an equilibrium in which the concentration of *trans*-2-butene is 1.65 times the concentration of *cis*-2-butene. This is a single-step, elementary process (← p. 523). Both the forward and reverse reactions in this system involve only a single molecule and therefore are unimolecular and first-order (← p. 532). Thus the rate equations for forward and reverse reactions can be derived from the reaction equations:

$$\text{Rate}_{\text{forward}} = k_{\text{forward}}[cis\text{-2-butene}]$$

$$\text{Rate}_{\text{reverse}} = k_{\text{reverse}}[trans\text{-2-butene}]$$

Suppose that we start with 0.100 mol *cis*-2-butene in a 5-L closed flask at 500 K. The *cis*-2-butene begins to react at a rate given by the forward rate equation above. Initially there is no *trans*-2-butene present, and so the initial rate of the reverse reaction is zero. As the forward reaction proceeds, the concentration of *cis*-2-butene decreases, and so the forward rate decreases. Also, as soon as some *trans*-2-butene has formed, the reverse reaction can begin, and as the concentration of *trans*-2-butene builds up, the reverse rate gets faster and faster. Eventually the forward rate slows down (and the reverse rate speeds up) to the point where the two rates are equal. At this time equilibrium has been achieved, and in the macroscopic system no further change in concentrations will be observed (Figure 13.1).

On the nanoscale, when equilibrium has been achieved, both reactions are still occurring, but the forward and reverse rates are equal (Figure 13.2). Therefore we can equate the two rates to give

$$\text{Rate}_{\text{forward}} = \text{rate}_{\text{reverse}}$$

and, by substituting from the two previous rate equations,

$$k_{\text{forward}}[cis\text{-2-butene}] = k_{\text{reverse}}[trans\text{-2-butene}]$$

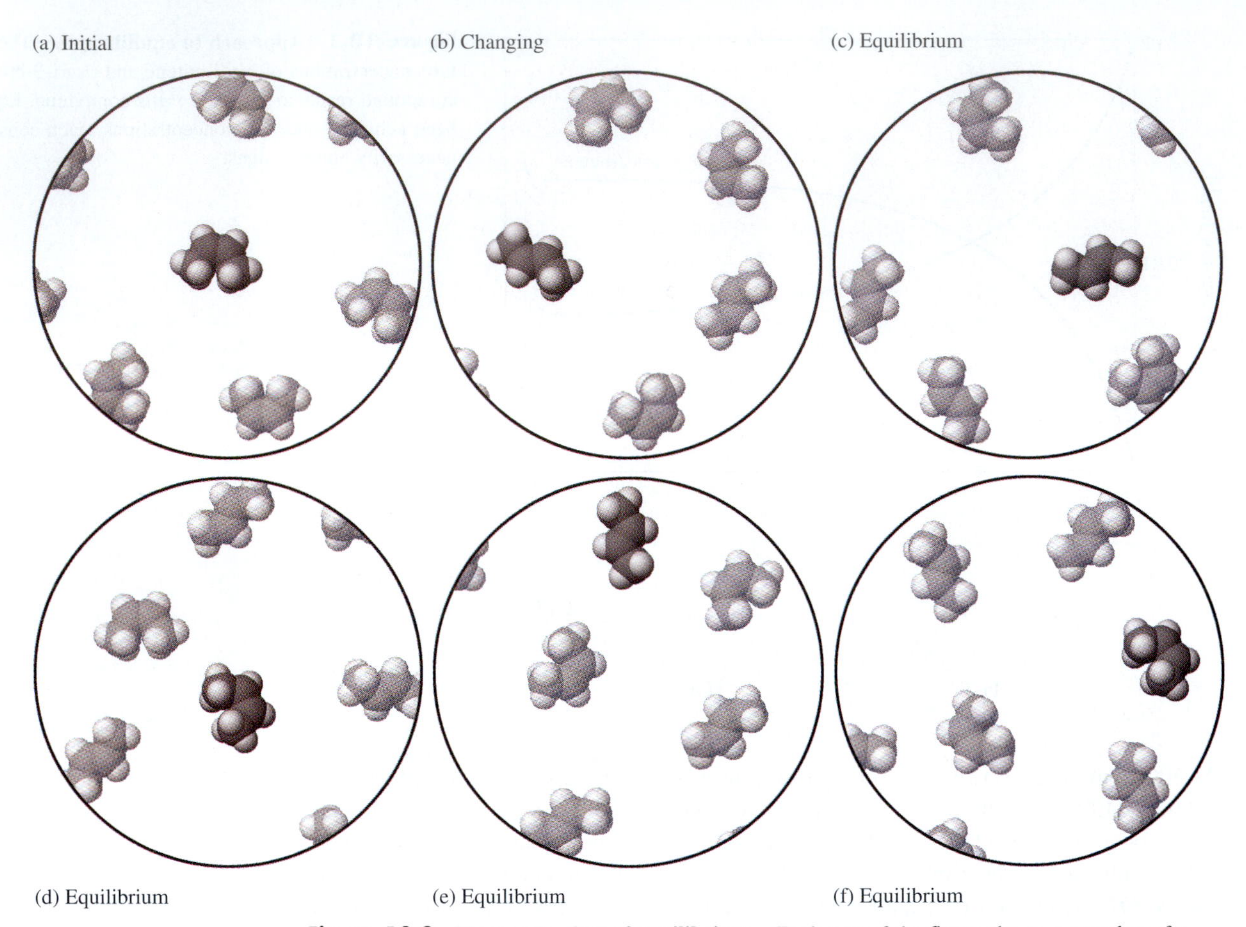

Figure 13.2 Nanoscale view of equilibrium. Each part of the figure shows a snapshot of a tiny portion of the reaction mixture as *cis*-2-butene reacts to form *trans*-2-butene. One of the molecules is highlighted so that its changes can be observed. Notice that this particular molecule is in the *cis* form part of the time and in the *trans* form part of the time, and that it changes back and forth.

The equation can be rearranged so that both rate constants are on one side and both concentrations on the other:

$$\frac{k_{\text{forward}}}{k_{\text{reverse}}} = \frac{[\textit{trans}\text{-2-butene}]}{[\textit{cis}\text{-2-butene}]}$$

This shows that the ratio of concentrations is equal to a ratio of rate constants, and therefore the ratio of concentrations must also be constant. We call this ratio *K,* where the capital letter "*K*" is used to distinguish it from the rate constants, k_{forward} and k_{reverse}.

$$K = \frac{k_{\text{forward}}}{k_{\text{reverse}}} = \frac{[\textit{trans}\text{-2-butene}]}{[\textit{cis}\text{-2-butene}]} = 1.65 \qquad \text{(at 500 K)}$$

Because the values of the rate constants vary with temperature (← ***p. 528***), the value of *K* will also vary with temperature. For the butene isomerization reaction, *K* is 1.47 at 600 K and 1.36 at 700 K.

A quotient of equilibrium concentrations of reactant and product substances that has a constant value for a given reaction at a given temperature is called an

equilibrium constant, and is given the symbol K. A mathematical expression for the equilibrium constant can be derived directly from the chemical equation for any equilibrium process, whether or not the forward and reverse reactions are elementary reactions. This mathematical expression, which is discussed in the next section, is called the **equilibrium constant expression.** When a reaction takes place by a mechanism that consists of a sequence of steps, the equilibrium constant can be obtained by multiplying together the rate constants for the forward reactions in all steps and then dividing by the rate constants for the reverse reactions in all steps. The equilibrium constant expression can be obtained from the coefficients of the balanced equilibrium equation.

Equilibrium constants are of great importance in chemistry because, once a reaction has reached equilibrium, K can be used to calculate the concentrations of reactants and products that are present. Many reactions move quickly to equilibrium, and so you can use equilibrium constants to determine the composition soon after reactants are mixed. Equilibrium constants are less valuable for slow reactions that take a long time to achieve equilibrium. Until equilibrium has been achieved, only kinetics is capable of predicting the composition of such reaction mixtures.

Exercise 13.2 Properties of Equilibrium

After a mixture of *cis*-2-butene and *trans*-2-butene has reached equilibrium at 600 K, half of the *trans*-2-butene is suddenly removed. Answer the following questions:

(a) Is the new mixture at equilibrium? Explain why or why not.
(b) In the new mixture, which rate is faster, *cis* → *trans* or the reverse? Or are both rates the same?
(c) In an equilibrium mixture, which concentration is larger, *cis*-2-butene or *trans*-2-butene?
(d) If the concentration of *cis*-2-butene at equilibrium is 0.10 mol/L, what will be the concentration of *trans*-2-butene?

Writing Equilibrium Constant Expressions

The equilibrium constant expression for any reaction can be derived from the balanced chemical equation. Concentrations of products appear in the numerator and concentrations of reactants appear in the denominator. Each concentration is raised to the power of its stoichiometric coefficient in the balanced equation. The only concentrations that appear in an equilibrium constant expression are those of gases and of solutes in dilute solutions, because these are the only concentrations that can change as a reaction occurs. Concentrations of pure solids, pure liquids, and solvents in dilute solutions *do not* appear in equilibrium constant expressions.

You probably recall that the rate law for a chemical reaction cannot be derived from the balanced equation, but must be determined experimentally by kinetic studies. The same is not true of the equilibrium constant expression, which can always be written directly from the coefficients in the balanced equation.

To illustrate how this works, consider the general equilibrium reaction

$$a\,\text{A} + b\,\text{B} + \ldots \rightleftharpoons c\,\text{C} + d\,\text{D} + \ldots$$

The ellipses (. . .) in the chemical equation and the equilibrium constant expression indicate that there might be more reactants or products.

By convention we write the equilibrium constant expression for this reaction as

$$\text{Equilibrium constant} = K = \frac{\overbrace{[\text{C}]^c[\text{D}]^d \ldots}^{\text{product concentrations}}}{\underbrace{[\text{A}]^a[\text{B}]^b \ldots}_{\text{reactant concentrations}}}$$

Let us apply these ideas to the combination of nitrogen and oxygen gases discussed earlier.

$$N_2(g) + O_2(g) \rightleftharpoons 2\ NO(g)$$

Because all substances in the reaction are gaseous, all concentrations appear in the equilibrium constant expression. The concentration of the product NO(g) is in the numerator and is squared because of the coefficient 2. The concentrations of $N_2(g)$ and $O_2(g)$ are in the denominator:

$$\text{Equilibrium constant} = K = \frac{[NO]^2}{[N_2][O_2]}$$

This reaction occurs whenever a material that contains sulfur burns in air, and it is responsible for a good deal of sulfur dioxide air pollution. It is also the first reaction in a sequence by which sulfur is converted to sulfuric acid, the number one industrial chemical in the world.

As another example, consider the combustion of solid yellow sulfur, which produces sulfur dioxide gas:

$$\tfrac{1}{8}\ S_8(s) + O_2(g) \rightleftharpoons SO_2(g)$$

Placing product concentration in the numerator, reactant concentrations in the denominator, and stoichiometric coefficients as exponents gives

The prime on the K' in this equilibrium constant expression indicates that it does not follow the rules we have stated. The appropriate equilibrium constant expression is given in the next equation below.

$$K' = \frac{\overset{\text{product}}{[SO_2(g)]}}{\underset{\text{reactants}}{[S_8(s)]^{1/8}[O_2(g)]}}$$

However, sulfur is a solid and for any solid the number of molecules per unit volume is always the same at any given temperature. Therefore, the sulfur concentration is not changed either by reaction or by addition or removal of some solid. Further, it is an experimental fact that, as long as there is some solid sulfur present, the equilibrium concentrations of O_2 and SO_2 are not affected by changes in the amount of sulfur. Therefore, the equilibrium constant expression should be written as

It is very important to remember that concentrations of pure solids, pure liquids, and solvents for dilute solutions do not appear in the equilibrium constant expression.

$$K = \frac{[SO_2(g)]}{[O_2(g)]} \qquad \text{which at 25 °C is equal to } 4.2 \times 10^{52}$$

The rule regarding solvents in dilute solutions comes into play for equilibria in aqueous solution. Consider the weak base ammonia, which produces a small concentration of hydroxide ions because it reacts with water:

$$NH_3(aq) + H_2O(\ell) \rightleftharpoons NH_4^+(aq) + OH^-(aq)$$

If the concentration of ammonia molecules (and consequently of ammonium ions and hydroxide ions) is small, the number of water molecules per unit volume remains essentially the same as in pure water. Because the molar concentration of water is effectively constant for reactions involving dilute solutions, the concentration of water need not be included in the equilibrium constant expression. Thus, we write

The concentration of water in pure water can be calculated from the density and molar mass as (1.00 g/mL) × (1 mol/18.0 g) × (1000 mL/L) = 55.5 mol/L. For equilibria in dilute aqueous solutions, the K we write is actually the molar concentration of water (55.5 *M*) times a K' that includes the concentration of water in its denominator. Because the concentration of water changes very little as the concentrations of solutes change, the K also remains constant.

$$K = \frac{[NH_4^+][OH^-]}{[NH_3]} = 1.8 \times 10^{-5} \qquad \text{(at 25 °C)}$$

and the concentration of water is not included in the denominator.

Notice that no units are given for the equilibrium constant for the ammonia ionization reaction. There are two concentrations in the numerator and only one in the denominator, which ought to give units of mol/L, but the units of equilibrium constant can always be figured out from the equilibrium constant expression. Therefore it is customary to omit the units, and we shall follow that custom here.

PROBLEM-SOLVING EXAMPLE 13.1 Writing Equilibrium Constant Expressions

Write an equilibrium constant expression for each chemical reaction.

(a) $4\ NH_3(g) + 7\ O_2(g) \rightleftharpoons 4\ NO_2(g) + 6\ H_2O(g)$

(b) $AgCl(s) \rightleftharpoons Ag^+(aq) + Cl^-(aq)$

(c) $Cu(NH_3)_4^{2+}(aq) \rightleftharpoons Cu^{2+}(aq) + 4\ NH_3(aq)$

(d) $CaCO_3(s) \rightleftharpoons CaO(s) + O_2(g)$

(e) $H_2O(\ell) \rightleftharpoons H_2O(g)$

Answer

(a) $K = \dfrac{[NO_2]^4[H_2O]^6}{[NH_3]^4[O_2]^7}$

(b) $K = [Ag^+][Cl^-]$

(c) $K = \dfrac{[Cu^{2+}][NH_3]^4}{[Cu(NH_3)_4^{2+}]}$

(d) $K = [O_2]$

(e) $K = [H_2O(g)]$

Explanation Concentrations of products go in the numerator of the fraction and concentrations of reactants go in the denominator. Each concentration is raised to the power of the stoichiometric coefficient of the species. In part (b), one species, $AgCl(s)$, is a solid and does not appear in the expression. In part (d), two species, $CaCO_3(s)$ and $CaO(s)$, are solids and do not appear. In part (e), one species, $H_2O(\ell)$, is a pure liquid and does not appear.

Problem-Solving Practice 13.1

Write an equilibrium constant expression for each equation.

(a) $H_2(g) + \frac{1}{8}\ S_8(s) \rightleftharpoons H_2S(g)$

(b) $HCl(g) + LiH(s) \rightleftharpoons H_2(g) + LiCl(s)$

(c) $CH_4(g) + H_2O(g) \rightleftharpoons CO(g) + 3\ H_2(g)$

(d) $CN^-(aq) + H_2O(\ell) \rightleftharpoons HCN(aq) + OH^-(aq)$

Equilibrium Constant Expressions for Related Reactions

Consider the equilibrium among nitrogen, hydrogen, and ammonia:

$$N_2(g) + 3\ H_2(g) \rightleftharpoons 2\ NH_3(g) \qquad K_1 = 3.5 \times 10^8 \qquad \text{(at 25 °C)}$$

We could also write the equation so that 1 mol of NH_3 is produced:

$$\tfrac{1}{2}\ N_2(g) + \tfrac{3}{2}\ H_2(g) \rightleftharpoons NH_3(g) \qquad K_2 = ?$$

Is the value of the equilibrium constant, K_2, for the second equation the same as the value of the equilibrium constant, K_1, for the first equation? To see the relation between K_1 and K_2, write the equilibrium constant expression for each balanced equation.

$$K_1 = \frac{[NH_3]^2}{[N_2][H_2]^3} = 3.5 \times 10^8 \quad \text{and} \quad K_2 = \frac{[NH_3]}{[N_2]^{1/2}[H_2]^{3/2}} = ?$$

This makes it clear that K_1 is the square of K_2; that is, $K_1 = K_2^2$. Therefore, the answer to our question is

$$K_2 = (K_1)^{1/2} = (3.5 \times 10^8)^{1/2} = 1.9 \times 10^4$$

Whenever the stoichiometric coefficients of a balanced equation are multiplied by some factor, the equilibrium constant for the new equation (K_2 in this case) is the old equilibrium constant (K_1) raised to the power of the multiplication factor.

What is the value of K_3, the equilibrium constant for the decomposition of ammonia to the elements, which is the reverse of the first reaction?

$$2\ NH_3(g) \rightleftharpoons N_2(g) + 3\ H_2(g) \qquad K_3 = ? = \frac{[N_2][H_2]^3}{[NH_3]^2}$$

It is clear that K_3 is the reciprocal of K_1. That is, $K_3 = 1/K_1 = 1/(3.5 \times 10^8) = 2.9 \times 10^{-9}$. *The equilibrium constant for a reaction and that for its reverse are the reciprocals of one another.* This means that if a reaction has a very large equilibrium constant, the reverse reaction will have a very small one. In the case of production of ammonia from the elements, the forward reaction has a large equilibrium constant (3.5×10^8). As expected, the reverse reaction, decomposition of ammonia to its elements, has a small equilibrium constant (2.9×10^{-9}).

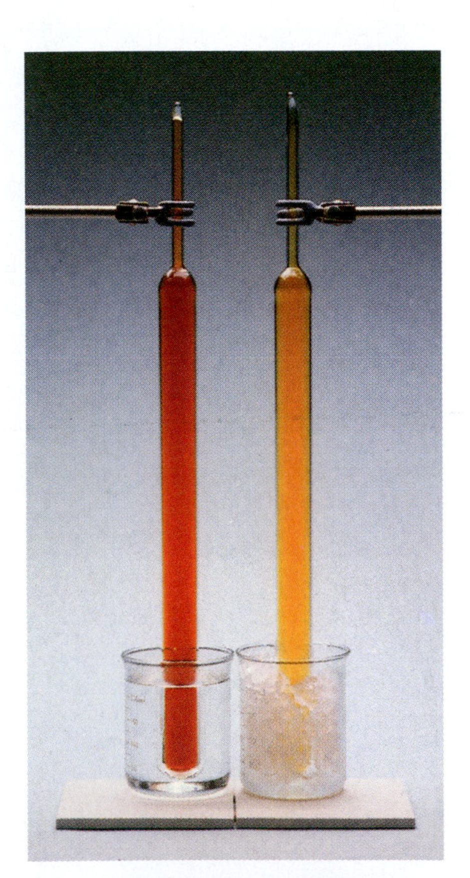

Figure 13.3 Formation of NO_2 from N_2O_4. Tubes containing an equilibrium mixture of nitrogen dioxide and dinitrogen tetraoxide are shown at a low temperature and at a high temperature. Notice the much darker color at the high temperature, indicating that the dinitrogen tetraoxide has reacted to form nitrogen dioxide. The intensity of color can be used to measure the concentration of nitrogen dioxide. *(C.D. Winters)*

Exercise 13.3 Manipulating Equilibrium Constants

The conversion of 1.5 mol of oxygen to 1.0 mol of ozone has a very small value of K.

$$\tfrac{3}{2}O_2(g) \rightleftharpoons O_3(g) \qquad K = 2.5 \times 10^{-29}$$

(a) What is the value of K if the equation is written using whole-number coefficients?

$$3\ O_2(g) \rightleftharpoons 2\ O_3(g)$$

(b) What is the value of K for the conversion of 2 mol of ozone to 3 mol of oxygen?

$$2\ O_3(g) \rightleftharpoons 3\ O_2(g)$$

13.3 DETERMINING EQUILIBRIUM CONSTANTS

To determine an equilibrium constant it is necessary to know all of the concentrations that appear in the equilibrium constant expression. This is most commonly achieved by allowing a system to reach the equilibrium state and measuring concentrations of one or more of the species involved in the equilibrium reaction. The concentrations of reactants and products once the equilibrium state has been achieved are called **equilibrium concentrations.** Algebra and stoichiometry can then be applied to obtain a value for K.

As an example, consider the colorless gas dinitrogen tetraoxide, $N_2O_4(g)$. When heated, it dissociates to form red-brown $NO_2(g)$ according to the equation

$$N_2O_4(g) \rightleftharpoons 2\ NO_2(g)$$

Suppose that 2.00 mol $N_2O_4(g)$ is placed into a 5.00-L flask and heated to 407 K. In a few seconds a dark red-brown color appears, indicating that much of the colorless gas has been transformed into NO_2 (Figure 13.3). By measuring the intensity of color it is found that the concentration of NO_2 at equilibrium is 0.525 mol/L, and from this information the equilibrium constant can be obtained.

There is a systematic approach to calculations involving equilibrium constants that is similar to the one used for stoichiometry problems (← p. 188). Make a table that shows initial conditions, changes that take place when a reaction occurs, and final conditions. Based on the information above, the table can be constructed for dissociation of dinitrogen tetraoxide:

Information Given	$N_2O_4(g)$	$\rightleftarrows$	$2\ NO_2(g)$
Starting Amount	2.00 mol		0 mol
Change as Reaction Occurs	______		______
Equilibrium Amount	______		(0.525 mol/L) × (5.00 L) = 2.62 mol

The table contains all the information obtained from our experiment. Information about the blank spaces, values that were not obtained from the experiment, can be derived by applying stoichiometric principles. Begin with the column where there is the most information. If there is 2.62 mol NO_2 at equilibrium and there was none before the reaction took place, the change in amount of NO_2 must have been 2.62 mol, and so the table must become:

	$N_2O_4(g)$	$\rightleftarrows$	$2\ NO_2(g)$
Starting Amount	2.00 mol		0 mol
Change as Reaction Occurs	______		2.62 mol
Equilibrium Amount	______		2.62 mol

According to the balanced equation, 2 mol NO_2 is formed for every 1 mol of N_2O_4 consumed, and so the amount of N_2O_4 consumed must be 1.31 mol, half the amount of NO_2 formed:

	$N_2O_4(g)$	$\rightleftarrows$	$2\ NO_2(g)$
Starting Amount	2.00 mol		0 mol
Change as Reaction Occurs	−1.31 mol		2.62 mol
Equilibrium Amount	______		2.62 mol

(The sign of the change in amount of N_2O_4 is *negative,* because the amount of N_2O_4 *decreases.*) Finally, the amount of N_2O_4 at equilibrium must be the sum of the initial 2.00 mol of N_2O_4 and the change due to reaction (−1.31 mol); that is, 2.00 mol − 1.31 mol = 0.69 mol:

	$N_2O_4(g)$	$\rightleftarrows$	$2\ NO_2(g)$
Starting amount	2.00 mol		0 mol
Change as Reaction Occurs	−1.31 mol		2.62 mol
Equilibrium Amount	0.69 mol		2.62 mol

To calculate K, we need to substitute equilibrium *concentrations* of the two substances into the equilibrium constant expression. Since the total volume of the container is 5.00 L, $[N_2O_4]$ = (0.69 mol)/(5.00 L) = 0.138 mol/L and $[NO_2]$ = (2.62 mol)/(5.00 L) = 0.525 mol/L (which was given originally). Then K is given by

$$K = \frac{[NO_2]^2}{[N_2O_4]} = \frac{(0.525 \text{ mol/L})^2}{(0.138 \text{ mol/L})} = 2.00 \qquad \text{(at 407 K)}$$

PROBLEM-SOLVING EXAMPLE 13.2 Determining an Equilibrium Constant

Consider the gas-phase reaction,

$$H_2(g) + I_2(g) \rightleftharpoons 2\ HI(g)$$

Suppose that a flask containing H_2 and I_2 has been heated to 425 °C and the initial concentrations of H_2 and I_2 were each 0.0175 mol/L. With time, the concentrations of H_2 and I_2 decline and the concentration of HI increases, and at equilibrium [HI] = 0.0276 mol/L. Use this experimental information to calculate the equilibrium constant.

Answer $K = 56$.

Explanation From the information given, construct a table. In this case, only concentrations are given, and so the table should consist of initial concentrations, changes in concentrations, and equilibrium concentrations. (Because the volume of the container remains the same throughout the reaction, the changes in concentrations obey the same rules of stoichiometry as do the changes in amounts of each reactant and product.)

	$H_2(g)$	+	$I_2(g)$	$\rightleftharpoons$	$2\ HI(g)$
Initial Concentration (mol/L)	0.0175		0.0175		0
Change as Reaction Occurs (mol/L)	______		______		______
Equilibrium Concentration (mol/L)	______		______		0.0276

Beginning with the column where the most information is known, let x represent the change in concentration of HI(g).

	$H_2(g)$	+	$I_2(g)$	$\rightleftharpoons$	$2\ HI(g)$
Initial Concentration (mol/L)	0.0175		0.0175		0
Change as Reaction Occurs (mol/L)	______		______		x
Equilibrium Concentration (mol/L)	______		______		0.0276

Next, derive the rest of the blanks in terms of x. If the concentration of HI increases by a given quantity, the mole ratios say that the concentrations of H_2 and I_2 must decrease only half as much:

$$x \text{ mol/L HI produced} = x\ \frac{\text{mol HI}}{\text{L}} \times \frac{1 \text{ mol } H_2}{2 \text{ mol HI}} = \frac{1}{2}\,x \text{ mol/L } H_2 \text{ consumed}$$

(Even though this result has a positive sign, in the table the sign is *negative* for change in concentration of H_2. This is because the concentration of H_2 *decreases* when the concentration of HI increases. The same applies to I_2.) Now derive the equilibrium concentrations of H_2 and I_2 by summing the initial concentration and the change in concentration for each:

	$H_2(g)$	+	$I_2(g)$	$\rightleftharpoons$	2 HI(g)
Initial Concentration (mol/L)	0.0175		0.0175		0
Change as Reaction Occurs (mol/L)	$-\frac{1}{2}x$		$-\frac{1}{2}x$		x
Equilibrium Concentration (mol/L)	$0.0175 - \frac{1}{2}x$		$0.0175 - \frac{1}{2}x$		0.0276

By summing the initial concentration and change in concentration from the last column in the table ($0 + x = 0.0276$), the value of x is seen to be 0.0276. Substituting this value into the last row of the table gives the equilibrium concentrations, which then can be substituted into the equilibrium constant expression:

$$K = \frac{[HI]^2}{[H_2][I_2]} = \frac{(0.0276)^2}{[0.0175 - \frac{1}{2}(0.0276)] \times [0.0175 - \frac{1}{2}(0.0276)]}$$

$$= \frac{(0.0276)^2}{(0.0037)(0.0037)}$$

$$= 56 \qquad \text{(at 425 °C)}$$

Problem-Solving Practice 13.2

Measuring the conductivity of an aqueous solution in which 0.0200 mol CH_3COOH has been dissolved in 1.00 L of solution shows that 2.96% of the acetic acid molecules have ionized to CH_3COO^- ions and H^+ ions. Calculate the equilibrium constant for ionization of acetic acid.

Supersonic aircraft. Planes like this spend much of their time in the lower part of the stratosphere. NO emitted from the engines of such planes can reduce the concentration of ozone in the stratosphere, but its effect is much less than that of chlorofluorocarbons (← *p. 552*). *(Charles Palek/Tom Stack & Associates)*

Equilibrium constants for a few reactions are given in Table 13.1. These reactions occur to widely differing extents, as shown by the wide range of values of K.

13.4 THE MEANING OF THE EQUILIBRIUM CONSTANT

The value of the equilibrium constant tells how far a reaction has proceeded by the time equilibrium has been achieved. In addition, it can be used to calculate how much product will be present at equilibrium.

A large value of K means that reactants have been converted almost entirely to products when equilibrium has been achieved. That is, the products are strongly favored over the reactants: the reaction is product-favored. An example is the reaction of NO(g) with $O_3(g)$, which is one way that ozone is destroyed in the stratosphere (← *p. 550*).

Stratospheric ozone can be destroyed by nitrogen oxides as well as by chlorofluorocarbons. Nitrogen oxides reach the stratosphere by several pathways, including emissions from aircraft flying in the lower part of the stratosphere.

$K \gg 1$: *Reaction is product-favored; equilibrium concentrations of products are greater than equilibrium concentrations of reactants.*

$$NO(g) + O_3(g) \rightleftharpoons NO_2(g) + O_2(g) \qquad K = \frac{[NO_2][O_2]}{[NO][O_3]} = 6 \times 10^{34} \quad \text{(at 25 °C)}$$

TABLE 13.1 Selected Equilibrium Constants (Based on Concentrations)

Reaction	Equilibrium Constant, K (at 25 °C)
Nonmetal Reactions	
$\frac{1}{8}$ $S_8(s) + O_2(g) \rightleftarrows SO_2(g)$	4.2×10^{52}
$2\ H_2(g) + O_2(g) \rightleftarrows 2\ H_2O(g)$	3.2×10^{81}
$N_2(g) + 3\ H_2(g) \rightleftarrows 2\ NH_3(g)$	3.5×10^{8}
$N_2(g) + O_2(g) \rightleftarrows 2\ NO(g)$	1.7×10^{-3} (at 2300 K)
$2\ NO_2(g) \rightleftarrows N_2O_4(g)$	1.7×10^{2}
Weak Acids and Bases	
formic acid	
$HCO_2H(aq) \rightleftarrows HCO_2^-(aq) + H^+(aq)$	1.8×10^{-4}
acetic acid	
$CH_3CO_2H(aq) \rightleftarrows CH_3CO_2^-(aq) + H^+(aq)$	1.8×10^{-5}
carbonic acid	
$H_2CO_3(aq) \rightleftarrows HCO_3^-(aq) + H^+(aq)$	4.2×10^{-7}
ammonia (weak base)	
$NH_3(aq) + H_2O(\ell) \rightleftarrows NH_4^+(aq) + OH^-(aq)$	1.8×10^{-5}
Very Slightly Soluble Solids	
$CaCO_3(s) \rightleftarrows Ca^{2+}(aq) + CO_3^{2-}(aq)$	3.8×10^{-9}
$AgCl(s) \rightleftarrows Ag^+(aq) + Cl^-(aq)$	1.8×10^{-10}
$AgI(s) \rightleftarrows Ag^+(aq) + I^-(aq)$	1.5×10^{-16}

The very large value of K tells us that if 1 mol each of NO and O_3 are mixed in a flask and allowed to come to equilibrium, $[NO_2][O_2] >> [NO][O_3]$. Virtually none of the reactants will remain, and essentially only NO_2 and O_2 will be found in the flask. For practical purposes, this reaction goes to completion, and it would not be necessary to use the equilibrium constant to calculate the quantities of products that would be obtained. The methods developed in Chapter 5 would work just fine.

Conversely, a small K means that when equilibrium has been achieved very little of the reactants has been transformed into products. The reactants are favored over the products at equilibrium: the reaction is reactant-favored.

K << *1: Reaction is reactant-favored; equilibrium concentrations of reactants are greater than equilibrium concentrations of products.*

$$3\ O_2(g) \rightleftarrows 2\ O_3(g) \qquad K = \frac{[O_3]^2}{[O_2]^3} = 6.25 \times 10^{-58} \qquad \text{(at 25 °C)}$$

Conceptual Challenge Problem CP-13.C at the end of the chapter relates to topics discussed in this section and the next.

This means that $[O_3]^2 << [O_2]^3$ and, if O_2 is placed in a flask, *very* little O_3 will be found when equilibrium is achieved. The concentration of O_2 would remain essentially unchanged. In the terminology of Chapter 4, we would write N.R. and say that no reaction would occur.

In contrast with the two reactions just described, dissociation of dinitrogen tetraoxide has neither a very large nor a very small equilibrium constant. At 407 K the value is 2.00, which means that significant concentrations of both N_2O_4 and NO_2 will be present at equilibrium. We confirmed this earlier (⇐ *p. 577*) by calculating the equilibrium concentrations. In cases like this, where K is neither superlarge nor supersmall, the equilibrium constant must be used to calculate how far a reaction proceeds toward products. What range of magnitudes of equilibrium constants comprises this middle ground depends on how small a concentration we consider significant. In most cases, if an equilibrium constant is bigger than roughly 10^{15}, we can safely say that the reaction goes to completion, and if an equilibrium constant is smaller than roughly 10^{-15}, we can say that no reaction occurs. To summarize,

- A value of $K > 10^{15}$ means that reactants have been essentially completely converted to products when equilibrium is reached. The reaction is so product-favored that we say it has gone to completion.
- A value of $K < 10^{-15}$ means very little of the reactants has been converted to products at equilibrium. The reaction is so reactant-favored that we would say no reaction has occurred.
- A value of K between 1 and 10^{15} means that the system is product-favored, but significant quantities of both reactants and products will be present at equilibrium.
- A value of K between 10^{-15} and 1 means that the system is reactant-favored, but significant quantities of both reactants and products will be present at equilibrium.

Patients can drink a "barium cocktail" containing $BaSO_4$ to make the intestinal tract visible in x-ray photographs, but barium is toxic and Ba^{2+} ions must be maintained at a low concentration in the body. The equilibrium constant for dissolving $BaSO_4$ to form Ba^{2+} ions and SO_4^{2-} ions is 1.1×10^{-10}, not small enough according to our criterion to say that no reaction occurs at all. It turns out that the calculated concentration of barium ions is small enough that the medical procedure is safe.

Examples of reactant-favored systems in which significant quantities of products form are the ionization reactions of the acids and bases listed in Table 13.1 For acetic acid, the main ingredient in vinegar,

$$CH_3CO_2H(aq) \rightleftharpoons CH_3CO_2^-(aq) + H^+(aq)$$

$$K = \frac{[CH_3CO_2^-][H^+]}{[CH_3CO_2H]} = 1.8 \times 10^{-5}$$

The value of K for acetic acid is small, and so at equilibrium the concentrations of products (acetate ions and hydrogen ions) are small relative to the concentration of reactant (acetic acid molecules). It should be clear from the equilibrium constant value why chemists call acetic acid a weak acid.

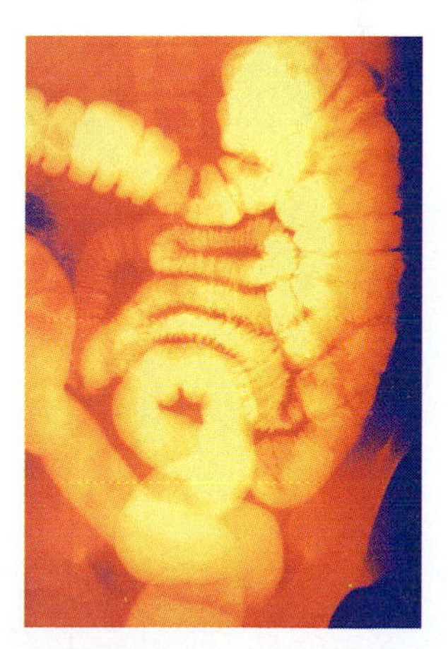

X-ray photograph of intestinal tract. Drinking a "barium cocktail" containing barium sulfate makes the intestinal tract visible on an x-ray photograph like this one. Because dissolving barium sulfate in water is a very reactant-favored process, very little Ba^{2+} ion, which is toxic, is absorbed by the body. *(Susan Leavines/Science Source/Photo Researchers, Inc.)*

Exercise 13.4 Using Equilibrium Constants

Use equilibrium constants (Table 13.1) to predict which of the reactions below will be product-favored. Then place all of the reactions in order from most reactant-favored to most product-favored.

(a) $\frac{1}{8} S_8(s) + O_2(g) \rightleftharpoons SO_2(g)$
(b) $2\ H_2O(g) \rightleftharpoons 2\ H_2(g) + O_2(g)$
(c) $HCO_2^-(aq) + H^+(aq) \rightleftharpoons HCO_2H(aq)$
(d) $CaCO_3(s) \rightleftharpoons Ca^{2+}(aq) + CO_3^{2-}(aq)$
(e) $N_2(g) + 3\ H_2(g) \rightleftharpoons 2\ NH_3(g)$

Exercise 13.5 Comparing Extents of Reaction

If solid AgCl and AgI are placed in 1.0 L of water in separate beakers, in which beaker would the silver ion concentration, $[Ag^+]$, be larger?

$AgCl(s) \rightleftharpoons Ag^+(aq) + Cl^-(aq)$ $\quad K = 1.8 \times 10^{-10}$
$AgI(s) \rightleftharpoons Ag^+(aq) + I^-(aq)$ $\quad K = 1.5 \times 10^{-16}$

The relationship between Gibbs free energy change and equilibrium constant is discussed in quantitative terms later in this chapter.

Exercise 13.6 Equilibrium Constant and Free Energy Change

Refer back to Exercise 13.4 and use data from Appendix J to calculate the change in Gibbs free energy for each of the reactions. How does the size of the equilibrium constant correlate with the sign of the Gibbs free energy change? Do both of these give similar information about a reaction? Explain why or why not.

13.5 CALCULATING EQUILIBRIUM CONCENTRATIONS

Equilibrium constants from Table 13.1, and from more extensive collections of data, can be used to calculate how much product is formed and how much of the reactants remain once a system has reached equilibrium. As an example of how this can be done, consider the reaction of sulfur and oxygen to give sulfur dioxide, which has the very large equilibrium constant 4.2×10^{52} at 25 °C.

$$\frac{1}{8}\, S_8(s) + O_2(g) \rightleftharpoons SO_2(g) \qquad K = 4.2 \times 10^{52} \qquad \text{(at 25 °C)}$$

Suppose we began with 4.0 mol/L of O_2 and a large excess of sulfur. If the reaction is done in a closed flask, so that the system can come to equilibrium, we can calculate the quantity of O_2 left and the quantity of SO_2 formed at equilibrium. As we did before, let us summarize information about the problem in a table headed by the chemical equation for the reaction. (Because S_8 is a solid and does not appear in the equilibrium constant expression, we do not need any entries under S_8 in the table.)

	$\frac{1}{8}$ $S_8(s)$	+	$O_2(g)$	$\rightleftharpoons$	$SO_2(g)$
Initial Concentration (mol/L)			4.0		0.0
Change in Concentration on Reaction (mol/L)			$-x$		$+x$
Equilibrium Concentration (mol/L)			$(4.0 - x)$		x

We know the concentrations of reactant and product before the reaction, but we do not know how many moles per liter of O_2 are consumed during the reaction, and so we designate this as x. Since the mole ratio is (1 mol SO_2)/(1 mol O_2), we know that x mol/L of SO_2 is formed when x mol/L of O_2 is consumed. To calculate the concentration of O_2 we take what was present initially (4.0 mol/L) minus what was consumed in the reaction (x mol/L). The equilibrium concentration of SO_2 must be the initial concentration (0.0 mol/L) plus what was formed by the reaction (x mol/L). Putting these values into the equilibrium constant expression, we have

$$K = \frac{[SO_2]}{[O_2]} = \frac{x}{4.0 - x} = 4.2 \times 10^{52}$$

Solving algebraically for x (and following the usual rules for significant figures), we find

$$x = (4.2 \times 10^{52})(4.0 - x)$$

$$x = 16.8 \times 10^{52} - (4.2 \times 10^{52})x$$

$$x + (4.2 \times 10^{52})x = 16.8 \times 10^{52}$$

Notice that $x + (4.2 \times 10^{52})x = (1 + 4.2 \times 10^{52})x$, which to a very good approximation is equal to $4.2 \times 10^{52}x$, because 4.2×10^{52} is so much bigger than 1 that adding 1 to it makes no appreciable change in the very large number. Thus,

$$x = \frac{16.8 \times 10^{52}}{4.2 \times 10^{52}} = 4.0$$

The equilibrium concentration of SO_2 is 4.0 mol/L and that of O_2 is $(4.0 - x)$ mol/L, or 0 mol/L. That is, within the precision of our calculation, all the O_2 has been converted to SO_2! A value of K bigger than 10^{15} does indeed imply that essentially all of the reactants have been converted to products. The reaction is product-favored and goes to completion.

PROBLEM-SOLVING EXAMPLE 13.3 Calculating Equilibrium Concentrations

Consider the reduction of carbon dioxide by hydrogen to give water vapor and carbon monoxide at 420 °C.

$$H_2(g) + CO_2(g) \rightleftharpoons H_2O(g) + CO(g) \qquad K = 0.10 \qquad \text{(at 420 °C)}$$

Assume that you place enough H_2 and CO_2 in a flask so that their initial concentrations are both 0.050 mol/L. You heat the mixture to 420 °C and wait for equilibrium to be achieved. What are the concentrations of reactants and products at equilibrium?

Answer $[H_2O] = [CO] = 0.012$ mol/L; $[H_2] = [CO_2]$ 0.038 mol/L

Explanation Set up a table of information as before:

	H_2	+	CO_2	$\rightleftharpoons$	H_2O	+	CO
Initial Concentration (mol/L)	0.050		0.050		0.000		0.000
Change in Concentration on Reaction (mol/L)	$-x$		$-x$		$+x$		$+x$
Equilibrium Concentration (mol/L)	$0.050 - x$		$0.050 - x$		x		x

In this particular equation all of the mole ratios are 1:1. This tells us that equal numbers of moles of H_2 and CO_2 are consumed as the reaction proceeds to equilibrium. Since both substances are in the same flask, this means that equal numbers of moles per liter (equal concentrations) must also be consumed, and we designate each of these as x. Because their initial concentrations were the same, the equilibrium concentrations of H_2 and CO_2 must also be the same. Each is equal to the initial concentration minus the change (x) as the reactants are consumed. The mole ratios also tell us that if the concentration of H_2 decreases by x mol/L, the concentration of H_2O (and the concentration of CO) must increase by the same quantity, x mol/L. Substituting these values into the expression for K, we have

$$K = 0.10 = \frac{[H_2O][CO]}{[H_2][CO_2]} = \frac{(x)(x)}{(0.050 - x)(0.050 - x)} = \frac{x^2}{(0.050 - x)^2}$$

Solving this equation for x is not as hard as it might seem at first glance. Because the right-hand side is a perfect square, we can take the square root of both sides, giving

$$\sqrt{K} = \sqrt{0.10} = 0.316 = \frac{x}{(0.050 - x)}$$

and then solve for x.

$$x = 0.316 \times (0.050 - x)$$
$$x = 0.0158 - 0.316x$$
$$1.316x = 0.0158$$
$$x = 0.012$$

Thus the concentrations of the products are both 0.012 mol/L, while the concentrations of the reactants that remain are both $(0.050 - x)$ mol/L = 0.038 mol/L. To check the calculation, substitute these concentrations into the equilibrium constant expression:

$$K = \frac{[H_2O][CO]}{[H_2][CO_2]} = \frac{(0.012)^2}{(0.038)^2} = 0.099 \cong 0.10$$

The good agreement indicates that the equilibrium concentrations are probably correct. This result demonstrates quantitatively that a value of K less than one indicates that a reaction is reactant-favored, because at equilibrium the product concentrations are smaller than the reactant concentrations.

Problem-Solving Practice 13.3

The equilibrium constant for dissolving the insoluble substance gold(I) iodide, AuI(s), in aqueous solution is 1.6×10^{-23}. Write the equilibrium constant expression and calculate the concentration of $Au^+(aq)$ and $I^-(aq)$ ions in a solution where 0.345 g AuI(s) is in equilibrium with the aqueous ions.

If a reaction has a large tendency to occur in one direction, then the reverse reaction has little tendency to occur. This was evident from our discussion of Gibbs free energy (← *p. 282*)—if ΔG is negative for a reaction proceeding in one direction, it is positive and of equal magnitude for the reverse reaction. A similar conclusion can be drawn about equilibrium constants. For example, Table 13.1 shows that combustion of hydrogen to form water vapor has an enormous equilibrium constant (3.2×10^{81}), and Appendix J shows that the reaction has a large negative value for $\Delta G°$ (−457 kJ). The reaction is product-favored, largely because it is very exothermic. The reverse reaction, decomposition of water to its elements,

$$2\ H_2O(g) \rightleftharpoons 2\ H_2(g) + O_2(g) \qquad K = \frac{[H_2]^2\,[O_2]}{[H_2O]^2} = 3.1 \times 10^{-82} \qquad \text{(at 25 °C)}$$

is strongly reactant-favored, as indicated by the *very* small value of K and a large positive value for $\Delta G°$ (+457 kJ).

Exercise 13.7 Manipulating Equilibrium Constants

The equilibrium constant is 1.8×10^{-5} for ionization of ammonia in aqueous solution:

$$NH_3(aq) + H_2O(\ell) \rightleftharpoons NH_4^+(aq) + OH^-(aq) \qquad K = 1.8 \times 10^{-5}$$

(a) Is the equilibrium constant large or small for the reverse reaction, the union of ammonium ion and hydroxide ion to give ammonia and water?
(b) What does the value of this equilibrium constant tell you about the extent to which reaction can occur between the ammonium ion and the hydroxide ion?
(c) What would happen if you added a 1.0 M solution of ammonium chloride to a 1.0 M solution of sodium hydroxide?
(d) What is the value of K for the reaction of ammonium ions with hydroxide ions?

PROBLEM-SOLVING EXAMPLE 13.4 Calculating Equilibrium Concentrations

When colorless hydrogen iodide gas is heated to 745 K, a beautiful purple color appears, showing that some iodine gas has been formed and that the compound has been decomposed partially to its elements:

$$2\,HI(g) \rightleftharpoons H_2(g) + I_2(g) \qquad K = 0.0200 \qquad \text{(at 745 K)}$$

Suppose that a mixture of 1.00 mol HI(g) and 1.00 mol $H_2(g)$ is sealed into a 10.0-L flask and heated to 745 K. What will be the concentrations of all three substances when equilibrium has been achieved?

Answer [HI] = 0.096 M; $[I_2]$ = 0.0018 M; $[H_2]$ = 0.102 M

Explanation In this case a table can be made with either numbers of moles or concentrations. Let us choose the latter, because we want concentrations as the final result. This means that we need to divide each number of moles given by the volume of the flask. (For HI we have [HI] = 1.00 mol/10.0L = 0.100 mol/L.) Because no I_2 is present at the beginning, the reaction must go from left to right. Therefore we let x mol/L be the concentration of I_2 that has been formed when equilibrium is reached, and the table is

	2 HI(g)	$\rightleftharpoons$	**$H_2(g)$**	+	**$I_2(g)$**
Initial Concentration (mol/L)	0.100		0.100		0.000
Change During Reaction (mol/L)	$-2x$		$+x$		$+x$
Equilibrium Concentration (mol/L)	$(0.100 - 2x)$		$(0.100 + x)$		x

Since the coefficients of H_2 and I_2 are the same, if x mol/L I_2 is produced, then x mol/L H_2 must also be produced. Therefore the change in both H_2 and I_2 is $+x$ mol/L. Because the coefficient of HI is twice the coefficient of I_2, twice as many moles of HI must disappear as moles of I_2 formed; the change in HI is $-2x$ mol/L. Now we can write the equilibrium constant expression in terms of the equilibrium concentrations.

$$K = \frac{[H_2][I_2]}{[HI]^2} = \frac{(0.100 + x)x}{(0.100 - 2x)^2} = 0.0200$$

Since the ratio of terms involving x is not a perfect square, we cannot take a square root as we did in Problem-Solving Example 13.3. Multiplying out the numerator and denominator, we have

$$\frac{0.100x + x^2}{0.0100 - 0.400x + 4x^2} = 0.0200$$

Multiplying both sides by the denominator and then multiplying out the terms gives

$$0.100x + x^2 = 0.0200 \times (0.0100 - 0.400x + 4x^2) = 0.000200 - 0.00800x + 0.0800x^2$$

Collecting terms in x^2 and x we have

$$0.9200x^2 + 0.10800x - 0.000200 = 0$$

This is a quadratic equation of the form $ax^2 + bx + c = 0$, where $a = 0.9200$, $b = 0.10800$, and $c = -0.000200$. It can be solved using the quadratic formula (Appendix B.4).

$$x = \frac{-b \pm \sqrt{(b^2 - 4ac)}}{2a}$$

$$= \frac{-0.10800 \pm \sqrt{(0.10800^2 - 4 \times 0.9200 \times (-0.000200))}}{2 \times 0.9200} = \frac{-0.10800 \pm 0.11135}{1.840}$$

In solving this problem we might have chosen to let $-x$ be the change in concentration of HI, in which case the changes in concentrations of H_2 and I_2 would each have been $+\frac{1}{2}x$.

This results in $x = 1.83 \times 10^{-3}$ or $x = -0.119$. The latter can be eliminated because it would result in a negative concentration of I_2 at equilibrium, which is clearly impossible. Using the first root and the equilibrium-concentration row of the table gives

$$[\text{HI}] = (0.100 - 2x) \text{ mol/L} = 0.100 - (2 \times 1.83 \times 10^{-3}) = 0.0963 \text{ mol/L}$$

$$[\text{H}_2] = (0.100 + x) \text{ mol/L} = 0.100 + 1.83 \times 10^{-3} = 0.102 \text{ mol/L}$$

$$[\text{I}_2] = x \text{ mol/L} = 0.00183 \text{ mol/L}$$

Solving this problem involves a lot of algebra, and so it would be easy to make a mistake. Therefore it is very important to check the result of any such calculation. This is easily done by substituting the equilibrium concentrations into the equilibrium constant expression and verifying that the correct value of K is found.

$$K = \frac{[\text{H}_2][\text{I}_2]}{[\text{HI}]^2} = \frac{(0.102) \times (0.00183)}{(0.0963)^2} = 0.0201$$

which is acceptable agreement because it differs by one in the last significant figure.

Problem-Solving Practice 13.4

Obtain the equilibrium constant for dissociation of dinitrogen tetraoxide to form nitrogen dioxide from Table 13.1. If 1.00 mol N_2O_4 and 0.500 mol NO_2 are initially placed in a container whose volume is 4.00 L, calculate the concentrations of N_2O_4(g) and NO_2(g) present when equilibrium is achieved at 25 °C.

13.6 SHIFTING A CHEMICAL EQUILIBRIUM: LE CHATELIER'S PRINCIPLE

Henri Le Chatelier (1850–1936) was a French chemist who, as a result of his studies of the chemistry of cement, developed his ideas about how altering conditions affects an equilibrium system.

Suppose you are an environmental engineer, biologist, or geologist and that you have just measured the concentration of hydrogen ion, H^+, in a lake. You know that the H^+ ions are involved in many different equilibrium reactions in the lake. How can you predict the influence of changing conditions? For example, what happens if there is a large increase in acid rainfall that has a hydrogen-ion concentration different from that of the lake? Or what happens if lime (calcium oxide), a strong base, is added to the lake? These questions and many others like them can be answered qualitatively by applying a useful guideline known as **Le Chatelier's principle:** *if a system is in equilibrium and the conditions are changed so that the system is no longer at equilibrium, the system will adjust to a new equilibrium state such that the effect of the change in conditions is partially counteracted or compensated for.* When we say that a system adjusts or shifts, we mean that there is a net reaction in either the forward or the reverse direction that changes the concentrations of all reactants and products until a new equilibrium state (appropriate for the new conditions) is achieved. Le Chatelier's principle applies to changes in conditions such as the concentrations of reactants or products that appear in the equilibrium constant expression, the pressure or volume of the equilibrium system, and the temperature. Changing such conditions, and thereby changing the equilibrium concentrations of reactants and products, is called **shifting an equilibrium.**

Conceptual Challenge Problems CP-13.D and CP-13.E at the end of the chapter relate to topics discussed in this section.

Changing Concentrations of Reactants or Products

If the concentration of a single reactant or product is changed, a system can no longer be at equilibrium because the ratio of concentrations in the equilibrium constant ex-

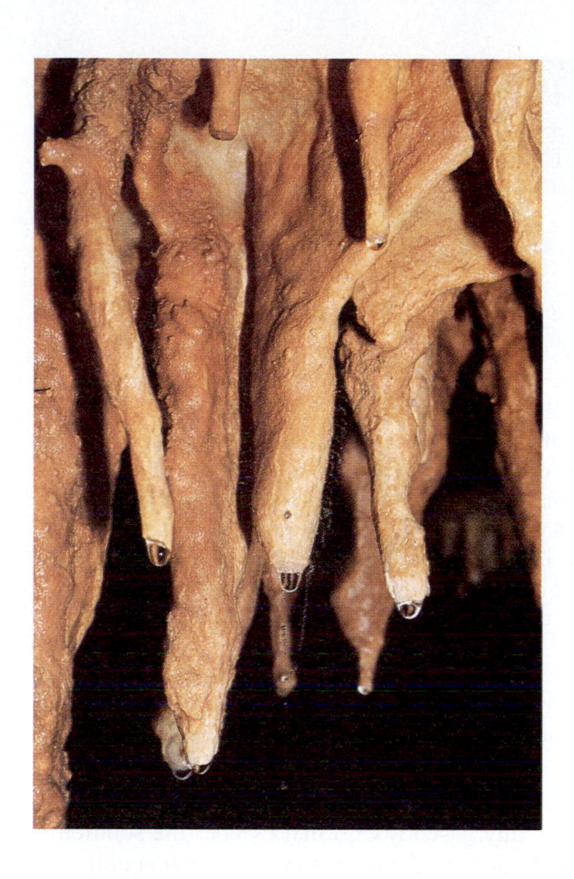

Figure 13.4 Stalactites in a limestone cave. Stalactites such as these hang from the ceilings of caves. Stalagmites grow from the floors of caves up toward the stalactites. Both consist of limestone, $CaCO_3$. The process that produces these lovely formations is an excellent example of chemical equilibrium. *(Tom Stack & Associates)*

pression will have a different value from the equilibrium constant. If the concentration of a single reactant or product is decreased, the system will react to establish a new equilibrium in which the decrease in concentration has been partially compensated for or counteracted. If a concentration is increased, the system reacts to decrease that concentration somewhat as the new equilibrium is achieved.

The effect of changing concentration has many important consequences. For example, when the concentration of a substance needed by your body falls slightly, several enzyme-catalyzed chemical equilibria shift so as to increase the concentration of the essential substance. In industrial processes, reaction products are often continuously removed. This shifts one or more equilibria to produce more products and thereby maximize the yield of the reaction.

Notice that as long as the temperature remains the same, the value of the equilibrium constant also remains the same. Adding or removing a reactant or product does not change the equilibrium constant value, but it does change the concentration of the substance added or removed. If that concentration appears in the equilibrium constant expression, the value of that expression changes. Because the value of the ratio of concentrations no longer equals *K*, the system is no longer at equilibrium and must react to achieve a new equilibrium.

In nature, slight changes in conditions are responsible for effects such as formation of limestone stalactites and stalagmites in caves (Figure 13.4) and the crust of limestone that slowly develops in a teapot if you boil hard water in it. Both of these examples involve calcium carbonate. Limestone, a form of calcium carbonate, $CaCO_3$, is present in underground deposits, a leftover of ancient oceans from which it precipitated long ago. Limestone reacts with an aqueous solution of CO_2 and dissolves.

Hard water contains dissolved CO_2 and metal ions such as Ca^{2+} and Mg^{2+}.

$$CaCO_3(s) + CO_2(aq) + H_2O(\ell) \rightleftharpoons Ca^{2+}(aq) + 2\ HCO_3^-(aq)$$

If groundwater that is saturated with CO_2 encounters a bed of limestone below the surface of the earth, the forward reaction can occur until equilibrium is reached, and the water subsequently contains significant concentrations of aqueous Ca^{2+} and HCO_3^- ions in addition to dissolved CO_2. These ions, often accompanied by Mg^{2+}(aq), constitute hard water (Section 16.7).

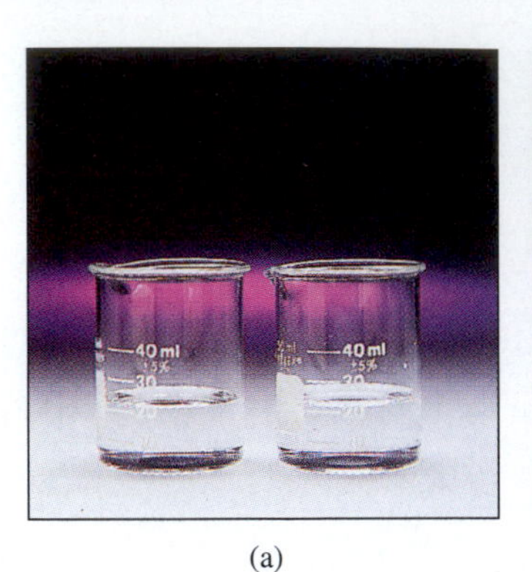

(a)

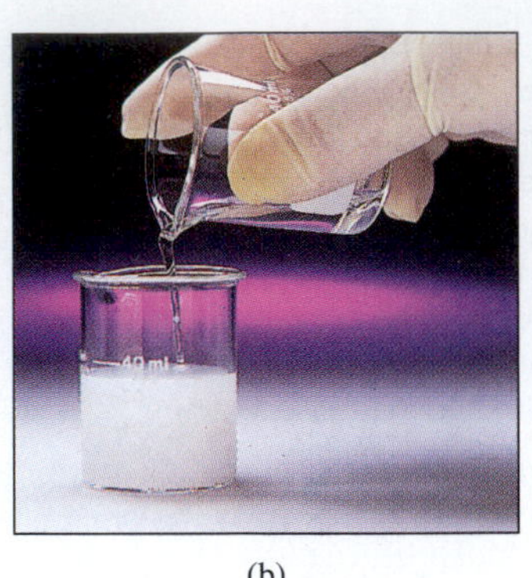

(b)

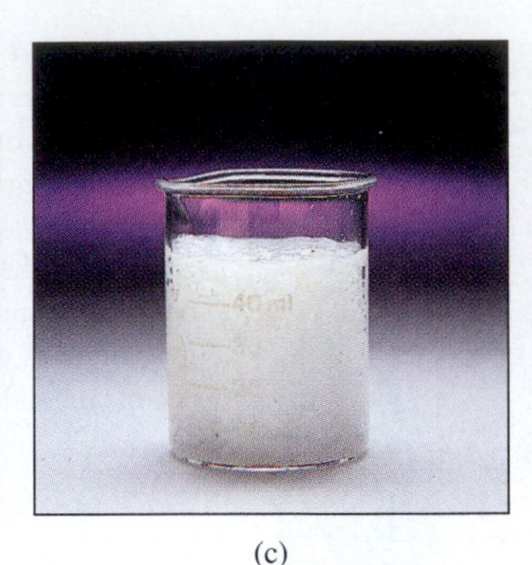

(c)

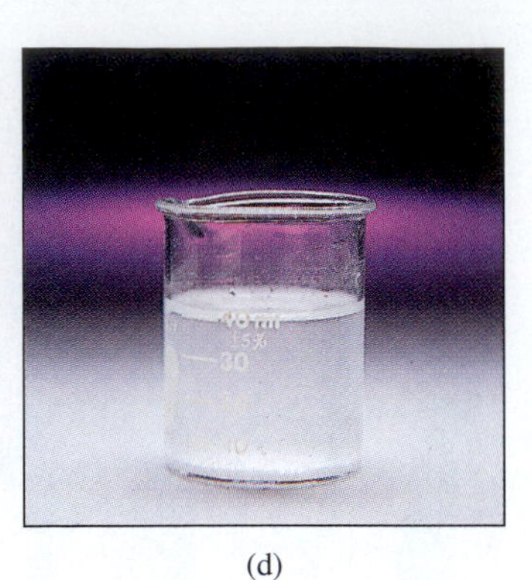

(d)

Figure 13.5 Reaction of $CaCl_2(aq)$ with $NaHCO_3(aq)$. (a) These two salt solutions provide Ca^{2+} and HCO_3^- ions, which are products in the net ionic equation for the reaction of calcium carbonate with carbon dioxide. (b) Increasing the concentrations of the products by mixing the two solutions shifts the equilibrium toward the reactants. (c) The CO_2 produced by the reverse reaction bubbles out of the solution into the air, thereby reducing a reactant concentration in solution and shifting the equilibrium to the left. (d) Therefore the reverse reaction continues until almost all of the Ca^{2+} and HCO_3^- ions have reacted to form CO_2, $CaCO_3$, and water. *(C.D. Winters)*

As with all equilibria, there is a reverse reaction occurring as well as the forward reaction. This reaction can be demonstrated by mixing aqueous solutions of $CaCl_2$ and $NaHCO_3$ (salts containing the Ca^{2+} and HCO_3^- ions) in an open beaker (Figure 13.5). You will eventually see bubbles of CO_2 gas and a precipitate of solid $CaCO_3$. Because the beaker is open to the air, any gaseous CO_2 that escapes the solution is swept away. This reduces the concentration of $CO_2(aq)$, and causes the equilibrium to shift to produce more CO_2. Eventually all of the dissolved Ca^{2+} and HCO_3^- ions disappear from the solution, having been converted to gaseous CO_2, solid $CaCO_3$, and water.

Suppose that a stream of groundwater containing dissolved CO_2, Ca^{2+}, and HCO_3^- comes out of the ground into the air in a cave (or as hard water in your teapot). Carbon dioxide bubbles out of the solution, the concentration of CO_2 decreases on the reactant side, and the equilibrium shifts toward the reactants. There is a net reverse reaction that forms CO_2, compensating partially for the reduced concentration of $CO_2(aq)$. Some of the calcium ions and hydrogen carbonate ions combine, and some $CaCO_3$ precipitates as a beautiful stalactite in the cave (or as scale in your teapot).

In summary, if you add a reactant or product that appears in the equilibrium constant expression to a reaction at equilibrium at a given temperature, some of the added substance will always be consumed as the reaction compensates for the disturbance. Similarly, if you remove a reactant or product, the substance removed will be partially replenished by a shift in the equilibrium. If a reactant is added, more product will be formed. If a product is removed, the reaction will compensate by producing more product and further consuming reactants.

Exercise 13.8 Effect of Adding a Substance

Solid phosphorus pentachloride decomposes when heated to form gaseous chlorine and gaseous phosphorus trichloride. Write the equation for the equilibrium that is set up when solid phosphorus pentachloride is introduced into a container, the container is evac-

CHEMISTRY YOU CAN DO

Growing Crystals by Shifting Equilibria

Crystals that are grown slowly can become quite large and have interesting and beautiful shapes. This crystal-growing experiment takes days or weeks, but the results can be quite striking. Obtain the following materials:

- About $\frac{1}{8}$ cup (30 mL) of one or more of the following salts: alum ($KAl(SO_4)_2 \cdot 12\ H_2O$), blue vitriol or blue stone ($Cu(SO_4)_2 \cdot 5\ H_2O$), Epsom salt ($MgSO_4 \cdot 7\ H_2O$), or table salt (NaCl)
- For each salt you want to crystallize, two 4-oz or 6-oz clear plastic cups (or similar clear containers), and something to cover the cup (plastic wrap will work fine)
- Water, something to stir with, a plastic-coated paper clip, a pencil, string, and (optional) petroleum jelly (Vaseline)

Epsom salt is a laxative and copper salts are harmful if taken internally, so be careful with them, but all of these salts are safe to pour down the drain when you are finished.

Prepare a saturated solution of the salt by placing it in a cup, filling the cup $\frac{1}{3}$ to $\frac{1}{2}$ full with warm tap water, and stirring every 5 min or so for half an hour. Some of the solid salt should remain undissolved in the bottom of the cup. If all of the solid dissolves, add more solid and stir some more. While you wait for the salt to dissolve, tie one end of the string to the paper clip and the other to the pencil. Arrange the string so that you can place the pencil across the top of the second cup, suspending the paper clip so that it just sits flat on the bottom of the cup. If you want to grow a really big crystal, coat most of the string with petroleum jelly, leaving only a small section uncoated about half an inch from the bottom of the cup. Once you are sure that the salt solution is saturated, decant it into the second cup, leaving all of the solid behind. Place the assemblage of paper clip, string, and pencil on the cup, and cover everything with plastic wrap.

Put the cup somewhere where it will not be disturbed and its temperature will remain fairly constant. (Don't put it in a window where the sun would hit it, or on top of a radiator, for example.) Check it at least once a day. Some of the water will evaporate, but only very slowly because the cup is covered and water vapor cannot escape. Observe what happens over a period of several weeks or even months. What do you see on the string? Explain your observations in terms of an equilibrium between the ions of the salt you dissolved and the solid salt. What shifts the equilibrium to achieve the results you observed? Why is it important that the water evaporate slowly?

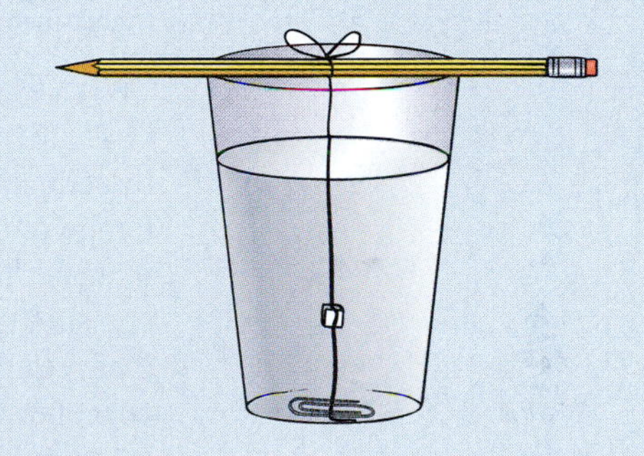

uated and sealed, and the solid is heated. Once the system has reached equilibrium at a given temperature, what will be the effect on the equilibrium of (a) adding chlorine to the container; (b) adding phosphorus trichloride to the container; (c) adding phosphorus pentachloride to the container?

Changing Volume or Pressure in Gaseous Equilibria

One way to change the pressure of a gaseous equilibrium mixture is to keep the volume constant and add or remove one or more of the substances whose concentrations appear in the equilibrium constant expression. This is the situation we have just discussed. We consider here other ways of changing pressure or volume.

The pressure of all substances in a gaseous equilibrium can be changed by changing the volume of the container. Consider the effect of tripling the pressure on the equilibrium

$$N_2O_4(g) \rightleftharpoons 2\ NO_2(g) \qquad K = \frac{[NO_2]^2}{[N_2O_4]}$$

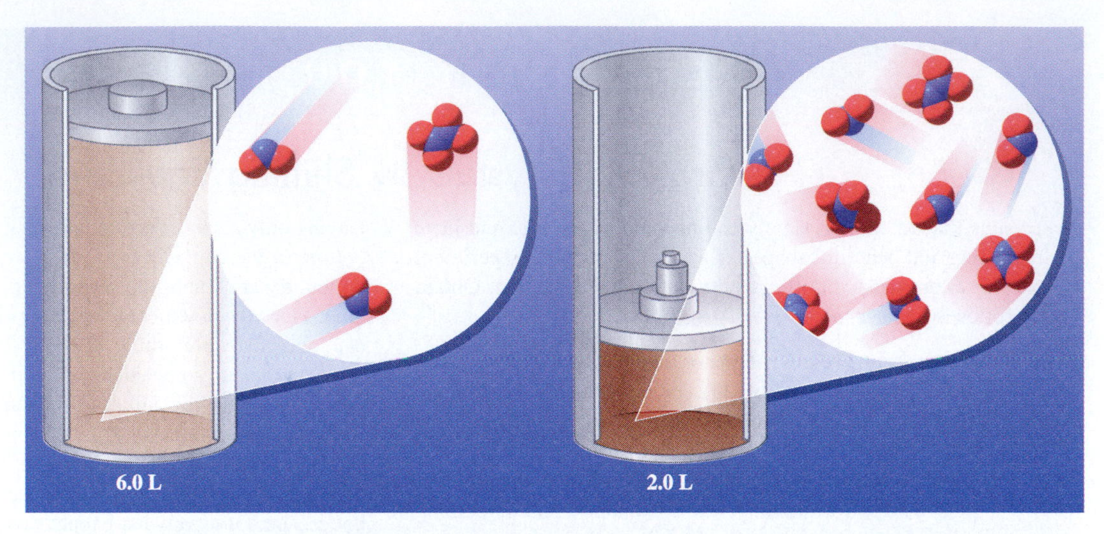

Figure 13.6 Shifting an equilibrium by changing pressure and volume. If the volume of an equilibrium mixture of NO_2 and N_2O_4 is decreased from 6.0 L to 2.0 L, the equilibrium shifts toward the smaller number of molecules in the gas phase (the N_2O_4 side) to partially compensate for the increased pressure. When the new equilibrium is achieved, the concentrations of both NO_2 and N_2O_4 have increased, but the NO_2 concentration is less than three times as great while the N_2O_4 concentration is more than three times as great. The total pressure goes from 1.00 atm to 2.62 atm (rather than to 3.00 atm, which was the new pressure before the equilibrium shifted).

by reducing the volume of the container to one-third of its original value (at constant temperature). This is shown in Figure 13.6. The effect of decreasing the volume is to increase the pressures of N_2O_4 and NO_2 to three times their original values. Decreasing the volume also increases the concentrations of N_2O_4 and NO_2 to three times their equilibrium values. Because $[NO_2]$ is squared but $[N_2O_4]$ is not, tripling both concentrations changes the value of the right-hand side of the equilibrium constant equation. The system is no longer at equilibrium, and there must be a net reaction to achieve a new equilibrium where the right-hand side equals K. According to Le Chatelier's principle, the reaction should shift to partially compensate for the increase in pressure, decreasing the pressure a bit. This can happen if the total number of molecules of gaseous material decreases, since the pressure of a gas is proportional to the number of molecules present. In the case of the N_2O_4/NO_2 equilibrium, the reverse reaction will occur, reducing the number of gas-phase molecules and hence the pressure. This will partially compensate for the increase in pressure that occurred when the volume of the system was decreased.

However, consider the situation with respect to another equilibrium we have already mentioned:

$$2\ HI(g) \rightleftharpoons H_2(g) + I_2(g)$$

Suppose that the pressure of this system were tripled by reducing its volume to one-third of the original volume. What would happen to the equilibrium? In this case, all the concentrations would triple, but because there are equal numbers of moles of gaseous substances on both sides of the equation, the equilibrium constant expression would still have the same numeric value. That is, the system is still at equilibrium, and no shift in the equilibrium is needed.

Exercise 13.9 Changing Volume Does Not Always Shift an Equilibrium

In Problem-Solving Example 13.4 you found that for the reaction

$$2\ HI(g) \rightleftharpoons H_2(g) + I_2(g) \qquad K = 0.0200 \qquad \text{(at 745 K)}$$

the equilibrium concentrations were [HI] = 0.0963 mol/L, $[H_2]$ = 0.102 mol/L, and $[I_2]$ = 0.00183 mol/L. Show that if each of these concentrations were tripled by reducing the volume of this equilibrium system to one third its initial value, the system would still be at equilibrium and therefore no shift in the equilibrium would occur.

Finally consider what would happen if the pressure of the N_2O_4/NO_2 equilibrium system were increased by adding an inert gas such as nitrogen while retaining exactly the same volume. The total pressure of the system would increase, but since neither the amounts of N_2O_4 and NO_2 nor the volume would change, the concentrations of N_2O_4 and NO_2 would remain the same. Because the concentrations that appear in the equilibrium constant expression have not changed, the system is still at equilibrium and no shift will (or needs to) occur. Thus, changing the pressure of an equilibrium system must change the concentrations of the substances in the equilibrium constant expression if a shift in the equilibrium is to occur.

Exercise 13.10 Effect of Changing Volume

Verify the statement in the text that, for the N_2O_4/NO_2 equilibrium system, decreasing the volume to one third of its original value increases the equilibrium concentration of N_2O_4 by more than a factor of three while it increases the equilibrium concentration of NO_2 by less than a factor of three. Start with the equilibrium conditions you calculated in Problem-Solving Practice 13.4. Then decrease the volume of the system from 4.00 L to 1.33 L, and calculate the new concentrations of N_2O_4 and NO_2, assuming the shift in the equilibrium has not yet taken place. Then set up the usual table of initial concentrations, change, and equilibrium concentrations, and calculate the concentrations at the new equilibrium.

Changing Temperature

When temperature changes, the values of most equilibrium constants also change, and systems will react to achieve new equilibria consistent with new values of K. You can make a qualitative prediction about the effect of temperature on an equilibrium if you know whether the reaction is exothermic or endothermic. As an example, consider the endothermic, gas-phase reaction of N_2 with O_2 to give nitrogen monoxide, NO.

Conceptual Challenge Problem CP-13.F at the end of the chapter relates to topics discussed in this section.

$$N_2(g) + O_2(g) \rightleftharpoons 2\ NO(g) \qquad \Delta H^\circ = 180.5\ \text{kJ}$$

$$K = \frac{[NO]^2}{[N_2][O_2]} \qquad \begin{array}{ll} K = 4.5 \times 10^{-31} & \text{at 298 K} \\ K = 6.7 \times 10^{-10} & \text{at 900 K} \\ K = 1.7 \times 10^{-3} & \text{at 2300 K} \end{array}$$

In this case the equilibrium constant increases very significantly as the temperature increases. At 298 K the equilibrium constant is so small that essentially no reaction occurs. Suppose that a room-temperature equilibrium mixture were suddenly heated to 2300 K. What would happen? The equilibrium should shift to partially compensate for the temperature increase. This happens if the reaction shifts in the endothermic direction, since that would involve transfer of energy into the reaction system. Energy would be transferred from particles in the surroundings to partially counteract the temperature increase. For the $N_2 + O_2$ reaction the forward process is endothermic, and so we predict that N_2 and O_2 should react to produce more NO. At the new equilibrium, the concentration of NO should be higher and the concentrations of N_2 and O_2 lower. Since this makes the numerator in the K expression bigger and the denominator smaller, K should be larger at the higher temperature, which corresponds with the experimental result.

The effect of temperature on the reaction of N_2 with O_2 has important consequences. This reaction produces NO in earth's atmosphere when lightning suddenly raises the temperature of the air. Because the reverse reaction is slow at room temperature, and because after the lightning bolt is over, the air rapidly cools back to normal temperatures, much of the NO that is produced does not react back to N_2 and O_2 as it would at equilibrium. This provides one natural mechanism by which nitrogen in the air can be converted into a form that can be used by plants. (Converting nitrogen into a useful form is called nitrogen fixation; ← *p. 344.*) Humans have tried to use the same kind of process to produce NO and from it HNO_3 for use as fertilizer. At the end of the 19th century, a chemical plant at Niagara Falls, New York (where there was plentiful electric power) operated an electric arc process for fixing nitrogen for several years. This electric-arc plant was important because it was the first attempt to deal with the limitations on plant growth caused by lack of sufficient nitrogen in soils that had been heavily farmed. One hundred years ago scientists and many in the general public were worried that earth's farmland could not grow enough food to support a growing population. Consequently strenuous efforts were made to adjust the conditions of the $N_2 + O_2$ reaction so that significant yields of NO could be obtained. The very high temperature of a lightning bolt or an electric arc was one way to do this.

Another consequence of the shift toward products in the $N_2 + O_2$ reaction at high temperatures is that automobile engines emit small concentrations of NO. The NO is rapidly oxidized to brown NO_2 in the air above cities, and the NO_2 in turn produces many further reactions that create air pollution problems. These are discussed in Section 14.12. One of the functions of catalytic converters on automobiles is to reduce these nitrogen oxides back to elemental nitrogen.

The effect on K of raising the temperature of the $N_2 + O_2$ reaction leads us to a general conclusion. *For an endothermic reaction, an increase in temperature always means an increase in K; the reaction will become more product-favored at higher temperatures.* When equilibrium is achieved at the higher temperature, the concentration of products is greater and that of the reactants is smaller. Likewise, and as illustrated by Problem-Solving Example 13.5, the opposite is true for an exothermic reaction. *For an exothermic reaction, an increase in temperature always means a decrease in K; the reaction will become less product-favored at higher temperatures.*

SUMMARY: The Effect of Temperature on K
The equilibrium constant has a specific, constant value at a given temperature for a given reaction.
For an endothermic reaction, K increases with increasing T.
For an exothermic reaction, K decreases with increasing T.

PROBLEM-SOLVING EXAMPLE 13.5 Le Chatelier's Principle

Consider an equilibrium mixture of nitrogen, hydrogen, and ammonia in which the reaction is

$$N_2(g) + 3\,H_2(g) \rightleftharpoons 2\,NH_3(g) \qquad \Delta H^\circ = -92.2 \text{ kJ}$$

For each of the changes listed below, tell whether the value of K increases or decreases, and tell whether more NH_3 or less NH_3 is present at the new equilibrium established after the change.

(a) More H_2 is added (at a constant temperature of 25 °C).
(b) The temperature is increased.
(c) The volume of the container is doubled (at constant temperature).

Answer (a) K stays the same; more NH_3 is present. (b) K decreases; less NH_3 is present. (c) K stays the same; less NH_3 is present.

Explanation

(a) Since the temperature does not change, the value of K does not change. Adding a reactant to the equilibrium mixture will shift the equilibrium towards the product, producing more NH_3. This can be seen in another way by considering the equilibrium constant expression:

$$K = \frac{[NH_3]^2}{[N_2][H_2]^3} = 3.5 \times 10^8 \qquad \text{(at 25 °C)}$$

When more H_2 is added the denominator gets larger. Because K is a constant, the value of the numerator must also increase. This can happen if some of the added H_2 reacts with N_2 to make more NH_3.

(b) The reaction is exothermic. Increasing the temperature shifts the equilibrium in the endothermic direction, that is, to the left (toward the reactants.) This leads to a decrease in the NH_3 concentration, an increase in the concentrations of H_2 and N_2, and a decrease in the value of K.

(c) Since the temperature is constant, the value of K must be constant. Doubling the volume should cause the reaction to shift toward a greater number of moles of gaseous substance, that is, toward the left. Doubling the volume would normally halve the pressure, but the shift of the equilibrium partially compensates for this and the final equilibrium will be at a pressure somewhat more than half the pressure of the initial equilibrium.

Problem-Solving Practice 13.5

Consider the equilibrium between N_2O_4 and NO_2 in a closed system.

$$N_2O_4(g) \rightleftharpoons 2\ NO_2(g)$$

Draw Lewis diagrams for the molecules involved in this equilibrium. Based on the bonding in the molecules, predict whether the reaction is exothermic or endothermic; hence, predict whether the concentration of N_2O_4 is larger in an equilibrium system at 25 °C or at 100 °C. Verify your prediction by looking at Figure 13.3.

13.7 GIBBS FREE ENERGY CHANGES AND EQUILIBRIUM CONSTANTS

The change in Gibbs free energy when a reaction occurs under standard conditions indicates whether a reaction system is product-favored (negative $\Delta G°$) or reactant-favored (positive $\Delta G°$). The equilibrium constant provides similar information: a system is product-favored when $K > 1$ and reactant-favored when $K < 1$. This suggests that there must be a connection between $\Delta G°$ and the equilibrium constant, and indeed there is:

$$\Delta G° = -RT \ln K_{th}$$

The R in this equation is the same constant that appeared in the Arrhenius equation (← p. 530).

In this equation $\Delta G°$ and T are already familiar, and R is a constant whose value is 8.314 J/K·mol. K_{th} is called the **thermodynamic equilibrium constant,** and its value depends on the choice of standard state for the substances involved in a reaction. For example, if the reaction occurs in solution, K_{th} has the same form as the equilibrium constants we have already discussed. The concentrations are expressed in moles per liter because the standard state involves each substance at a concentration of 1 mol/L. For gases, where the standard state involves a pressure of 1 bar, K_{th} relates pressures of the various gases, not their concentrations. The K values given in Table 13.1 are for ratios of concentrations, and so, for many of the gas-phase reactions, they differ from the K_{th} values that are related to $\Delta G°$ by the equation above.

The ideal gas law, $PV = nRT$, was almost certainly a part of a previous course in chemistry that you have taken. If not, it is discussed in Chapter 14.

Exercise 13.11 Equilibrium Constant in Terms of Pressure

The concentration of a solution, whether in the liquid phase or the gas phase, is defined as the moles of solute divided by the volume of solution, $c = n/V$.
(a) Use this definition, the ideal gas law, and a little algebra to show that $c = P/RT$ for a solute in a gas-phase solution.
(b) Consider the equilibrium constant for the N_2O_4/NO_2 equilibrium,

$$K = \frac{[N_2O_4]}{[NO_2]^2} = \frac{c_{N_2O_4}}{(c_{NO_2})^2}$$

Use the relation between concentration and pressure that you just derived to express this equilibrium constant in terms of pressures instead of concentrations. What would the value of K be at 25 °C if it were expressed in terms of pressures?

K_{th}	$\Delta G°$ ($-RT \ln K_{th}$)	Product-favored?
< 1	positive	no
> 1	negative	yes
$= 1$	0	neither

Regardless of the choice of standard state, the equation indicates that the Gibbs free energy change for a reaction is the negative of a constant times the temperature times the natural logarithm of the equilibrium constant. If K_{th} is larger than 1, then $\ln K_{th}$ is positive, and $\Delta G°$ will be negative because of the minus sign. Both of these—a negative $\Delta G°$ and $K > 1$—indicate that the reaction is product-favored under standard-state conditions. Conversely, if K_{th} is less than 1, then $\ln K_{th}$ is negative and $\Delta G°$ must be positive, indicating a reactant-favored system.

$\Delta G°$ is the difference in Gibbs free energy between products in their standard states and reactants in their standard states. For ionization of formic acid in water,

$$\underset{1\ \text{mol/L}}{HCOOH(aq)} \rightleftharpoons \underset{1\ \text{mol/L}}{HCOO^-(aq)} + \underset{1\ \text{mol/L}}{H^+(aq)} \qquad \Delta G° = 21.4\ \text{kJ}$$

the free energy change of 21.4 kJ is for converting 1 mol of HCOOH(aq) at a concentration of 1 mol/L into 1 mol of $HCOO^-$(aq) and 1 mol of H^+(aq), each at a concentration of 1 mol/L. Since $\Delta G°$ is positive, we predict that the process will be reactant-favored.

Problem-Solving Example 13.6 Gibbs Free Energy and Equilibrium Constant

In the preceding paragraph you learned that $\Delta G° = 21.4$ kJ at 25 °C for the reaction

$$HCOOH(aq) \rightleftharpoons HCOO^-(aq) + H^+(aq)$$

Use this information to calculate the equilibrium constant for ionization of formic acid in aqueous solution at 25 °C.

Answer $K = 1.8 \times 10^{-4}$

Explanation The relation between Gibbs free energy and K was given above as

$$\Delta G° = -RT \ln K_{th}$$

To obtain K from $\Delta G°$, we first divide both sides of the equation by $-RT$:

$$-\Delta G°/RT = \ln K_{th}$$

Next we make use of the properties of logarithms (which are discussed in Appendix B.3). Since ln represents a logarithm to the base *e*, we can remove the logarithm function by using each side of the equation as an exponent of *e*.

$$e^{-\Delta G^\circ/RT} = K_{th}$$

Now we can substitute the known values into the equation:

$$K_{th} = e^{-\Delta G^\circ/RT} = e^{-(21.4\ \text{kJ/mol}) \times (1000\ \text{J/kJ})/(8.314\ \text{J/K·mol})\cdot(298\text{K})} = 1.77 \times 10^{-4}$$

Thus the positive value of ΔG° results in a value of K_{th} less than one and indeed indicates a reactant-favored system.

Problem-Solving Practice 13.6

For each of the following reactions, evaluate K_{th} at 298 K from the standard free energy change. If necessary, obtain data from Appendix J to calculate ΔG°. Check your results against the *K* values in Table 13.1. If there are any discrepancies, explain why.

(a) $CaCO_3(s) \rightleftharpoons Ca^{2+}(aq) + CO_3^{2-}(aq)$

(b) $H_2CO_3(aq) \rightleftharpoons HCO_3^-(aq) + H^+(aq)$

(c) $2\ NO_2(g) \rightleftharpoons N_2O_4(g)$

Now consider the same formic-acid system. Suppose that we start with each species in the equation at the standard-state concentration of 1 mol/L. In which direction will the system react to achieve equilibrium? And how much reaction must occur before equilibrium is reached?

Since $K = 1.8 \times 10^{-4}$, which is much smaller than 1, there will be smaller concentrations of products and a larger concentration of reactant when equilibrium is reached.

	$HCOOH(aq) \rightleftharpoons$	$HCOO^-(aq)$	+	$H^+(aq)$
Initial Concentration (mol/L)	1	1		1
Equilibrium Concentration (mol/L)	>1	<1		<1

To achieve such concentrations, the reverse reaction must occur. This is what we would expect of a reactant-favored system: products are converted to reactants if all species are present at standard-state concentrations. Both the *size* of ΔG° and the *size* of *K* indicate quantitatively *how much the concentrations will change* from standard-state values before equilibrium is achieved.

For a reaction where the equilibrium constant is greater than one, ΔG° is negative. Starting from standard-state concentrations the reaction will proceed from left to right to produce products. The size of ΔG°—how negative it is—indicates how far from left to right the reaction will proceed to reach the equilibrium state.

Remember also that ΔG° can be calculated from the Gibbs equation:

$$\Delta G^\circ = \Delta H^\circ - T\Delta S^\circ$$

If we know or can estimate changes in enthalpy and entropy for a reaction, then we can calculate or estimate the Gibbs free energy change and hence the equilibrium constant. And, because ΔH° and ΔS° have nearly constant values over a wide range of temperatures, we can estimate equilibrium constants at a variety of temperatures, not just at 25 °C.

PROBLEM-SOLVING EXAMPLE 13.7 Estimating K at Different Temperatures

Use data from Appendix J to obtain values of $\Delta H°$ and $\Delta S°$ for the reaction

$$N_2(g) + O_2(g) \rightleftharpoons 2\ NO(g)$$

From these data estimate the value of $\Delta G°$ and hence the value of K at each temperature below.

(a) 298 K (b) 1000. K (c) 2300. K

Answer (a) $\Delta G° = 173.12$ kJ; $K = 4.51 \times 10^{-31}$ (b) $\Delta G° = 155.73$ kJ; $K = 7.33 \times 10^{-9}$ (c) $\Delta G° = 123.53$ kJ; $K = 1.565 \times 10^{-3}$

Explanation At each temperature use the Gibbs equation to calculate $\Delta G° = \Delta H° - T\Delta S°$. Then calculate K_{th} as was done in Problem-Solving Example 13.6. Part (c) is done below to illustrate the calculations:

$$\Delta G° = \Delta H° - T\Delta S° = (180{,}500\ \text{J}) - (2300.\ \text{K}) \times (24.772\ \text{J/K}) = 123{,}530\ \text{J}$$

$$= 123.53\ \text{kJ}$$

$$K = e^{-\Delta G°/RT} = e^{-(123{,}530\ \text{J})/(8.314\ \text{J/K}) \times (2300.\ \text{K})} = 1.565 \times 10^{-3}$$

Problem-Solving Practice 13.7

For the ammonia synthesis reaction,

$$N_2(g) + 3\ H_2(g) \rightleftharpoons 2\ NH_3(g)$$

estimate the equilibrium constant at each temperature below.

(a) 298 K (b) 450 K (c) 800 K

Exercise 13.12 Reactant-Favored or Product-Favored?

If a system falls within the second or third category in Table 7.3 (← *p. 283*), then there must be a temperature at which it shifts from being reactant-favored to being product-favored. For each reaction below, describe how you would estimate this temperature. Obtain data from Appendix J and calculate what the temperature is for each reaction.

(a) $CO(g) + 2\ H_2(g) \rightleftharpoons CH_3OH(\ell)$
(b) $2\ Fe_2O_3(s) + 3\ C(\text{graphite}) \rightleftharpoons 4\ Fe(s) + 3\ CO_2(g)$
(c) $2\ H_2O(g) \rightleftharpoons 2\ H_2(g) + O_2(g)$
(d) $N_2(g) + 3\ H_2(g) \rightleftharpoons 2\ NH_3(g)$

13.8 CONTROLLING CHEMICAL REACTIONS: THE HABER-BOSCH PROCESS

We can summarize our findings in the preceding sections by saying that *a product-favored reaction is one that has an equilibrium constant larger than 1.* Further, *exothermic reactions that result in an increase in entropy are product-favored.* For a reaction at room temperature or below, the enthalpy change is more important than the entropy change, and so *product-favored reactions at low temperatures are usually exothermic.* Well above room temperature the entropy change of a reaction plays a more important role, and so *product-favored reactions at high temperatures are usually ones in which the entropy increases.*

With these general thermodynamic rules we can often predict whether a reaction is capable of yielding products. But it is also important that the rate at which those products are produced is reasonably rapid. Here are some useful generalizations about reaction rates.

1. Reactions in the gas phase or in solution, where molecules of one reactant are completely mixed with molecules of another, occur more rapidly than do reactions between pure liquids or solids that do not dissolve in one other ***(← p. 512).***
2. Reactions occur more rapidly at high temperatures than at low temperatures ***(← p. 528).***
3. Reactions are faster when the reactant concentrations are high than when they are low ***(← p. 516).*** Reactions between gases, for example, occur most rapidly at higher pressures. This is one reason that the mixture of air and gasoline vapor is compressed in the cylinder of an automobile engine before it is ignited.
4. Reactions between a solid and a gas, or a solid and something dissolved in solution, are usually much faster when the solid particles are as small as possible ***(← p. 513).*** For this reason coal is ground to a powder before it is burned for generating electricity.
5. Reactions are faster in the presence of a catalyst. Often the right catalyst is the difference between success and failure in industrial chemistry ***(← p. 547).*** The current interest in biotechnology is largely driven by the fact that naturally occurring enzymes are among the most effective catalysts known ***(← p. 539).***

One of the best examples of the application of the principles of chemical reactivity is the chemical reaction we now use for the synthesis of ammonia from its elements. Even though the earth is bathed in an atmosphere that is about 80% N_2 gas, nitrogen cannot be used by most plants until it has been fixed—converted into biologically useful forms. Although nitrogen fixation is done naturally by organisms such as blue-green algae and some field crops such as alfalfa and soybeans, most plants cannot fix N_2. Nitrogen must be supplied from an external source. This is especially important for recently developed varieties of wheat, corn, and rice that have resulted in much improved food production, particularly in Third World countries. We mentioned earlier that direct combination of nitrogen and oxygen was used at the beginning of the twentieth century to provide fertilizer, but this process was not very efficient ***(← p. 592).***

A much better way of manufacturing ammonia was devised by the German chemist Fritz Haber and the engineer Carl Bosch, who chose the *direct synthesis of ammonia from its elements* as the basis for an industrial process

$$N_2(g) + 3\ H_2(g) \rightleftharpoons 2\ NH_3(g) \qquad \Delta H^\circ = -92.2 \text{ kJ}$$

At first glance this reaction might seem to be a poor choice. Hydrogen is available naturally only in combined form, for example in water or hydrocarbons, meaning that some hydrogen-containing compounds must be destroyed first at the cost of considerable free energy. As you discovered in Problem-Solving Practice 13.7, the ammonia synthesis reaction becomes less and less product-favored at higher temperatures. But higher temperatures are needed in order for ammonia to be produced fast enough for the process to be efficient. Nonetheless, the **Haber-Bosch process** (shown

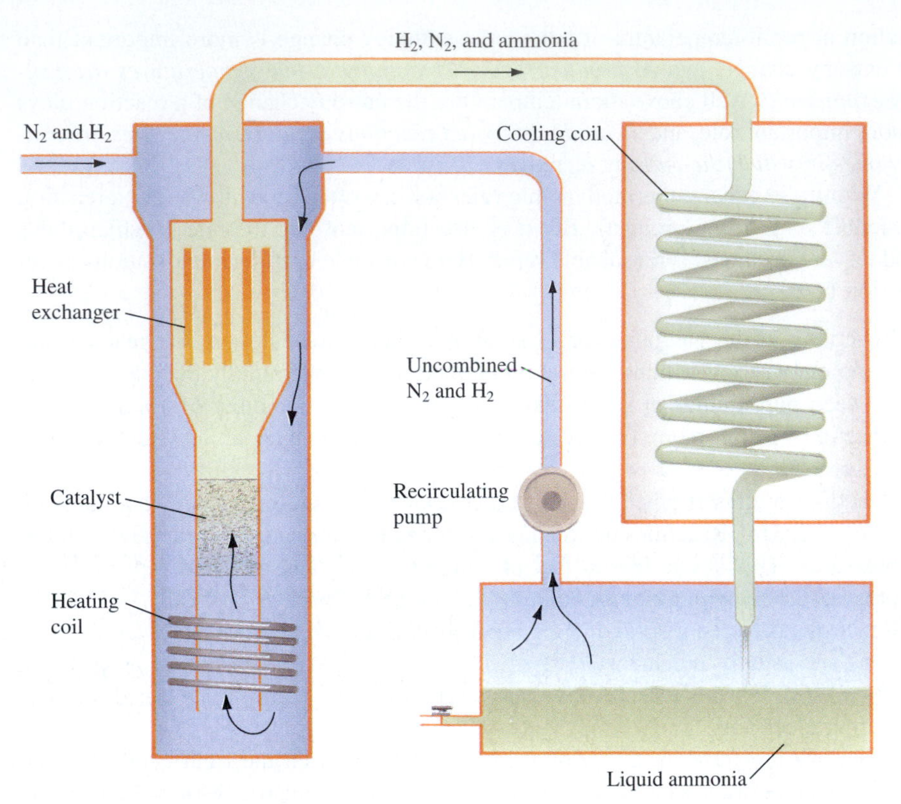

Figure 13.7 Haber-Bosch process for synthesis of ammonia (schematic). Nitrogen and hydrogen enter the reactor at the top, are heated, and pass over a catalyst. The resulting mixture of nitrogen, hydrogen, and ammonia then passes through a cooling coil that condenses the ammonia to liquid form. Uncombined nitrogen and hydrogen are recycled to the catalytic chamber.

schematically in Figure 13.7) has been so well developed that ammonia is very inexpensive (about \$150 per ton). For this reason it is widely used as a fertilizer and so is often in the "top 5" chemicals produced in the United States. In 1995 about 35.6 billion pounds of NH_3 were produced by the Haber-Bosch process.

Both the thermodynamics and the kinetics of the direct synthesis of ammonia have been carefully studied and fine-tuned by industry so that the maximum yield of product is obtained in a reasonable time and at a reasonable cost of both money and free energy.

- The reaction is strongly exothermic and there is a decrease in the entropy of the system when it takes place. Therefore it is predicted to be product-favored at low temperatures, but reactant-favored at high temperatures. (You should have verified this in Problem-Solving Practice 13.7, and it is predicted by Le Chatelier's principle for an exothermic process.)
- The reaction is quite slow at room temperature, and so the temperature must be raised to increase the rate. Although the rate increases with increasing temperature, the equilibrium constant declines. Thus the faster the reaction the smaller the yield.
- To increase the equilibrium concentration of NH_3, the reaction is carried out at high pressure. This does not change the value of K, but an increase in pressure can be compensated by converting N_2 and H_2 to NH_3; 2 mol of $NH_3(g)$

Portrait of a Scientist • *Fritz Haber (1868–1934)*

In 1898 William Ramsay, the discoverer of the rare gases, pointed out that supplies of fixed nitrogen were limited, and he predicted world disaster due to a "fixed nitrogen shortage" by mid-20th century. Such a shortage would have prevented food production from keeping pace with population, resulting in widespread famine. That this has not occurred is due largely to the work of Fritz Haber. Haber's studies of the ammonia synthesis reaction in the early 1900s revealed that direct ammonia synthesis should be possible. However, it was not until 1914 that the engineering problems were solved by Carl Bosch, and ammonia production began just in time for the start of World War I. Ammonia is the starting material to make nitric acid, a vital material in the manufacture of the explosives TNT and nitroglycerin. Therefore, ammonia is thought to be the first synthetic chemical used on a large scale for purposes of warfare.

Haber's contract with the manufacturer of ammonia called for him to receive 1 pfennig per kilogram of ammonia, and he soon became not only famous but rich! Unfortunately, he joined the German Chemical Warfare Service at the start of World War I and became its director in 1916. The primary mission of the service was to develop gas warfare, and in 1915 he supervised the first use of Cl_2 at the battle of Ypres. It was a tragedy not only of modern warfare, but to Haber personally as well. His wife pleaded with him to stop his work in this area, and when he refused, she committed suicide. In 1918 he was awarded the Nobel Prize for the ammonia synthesis, but the choice was criticized because of his role in chemical warfare.

After World War I Haber did some of his best work, continuing to study thermodynamics. However, because of his Jewish background, Haber was forced to leave Nazi Germany in 1933. He worked for a time in England and died in Switzerland in 1934.

Fritz Haber. *(Oesper Collection in the History of Chemistry/University of Cincinnati)*

exert less pressure than a total of 4 mol of gaseous reactants [$N_2(g) + 3\ H_2(g)$] in the same size container.

- Ammonia is continually liquefied and removed from the reaction vessel, which reduces the concentration of the product of the reaction and shifts the equilibrium toward the right.
- Since the temperature cannot be raised too much in an attempt to increase the rate, a rate increase can be achieved with a catalyst. An effective catalyst for the Haber-Bosch process is Fe_3O_4 mixed with KOH, SiO_2, and Al_2O_3. Since the catalyst is not effective below about 400 °C, the optimum temperature, considering all the factors controlling the reaction, is about 450 °C.

Making predictions about chemical reactivity is part of the challenge, the adventure, and the art of chemistry. Many chemists enjoy trying to make useful new materials, and this usually means choosing to make them by reactions that we believe will be product-favored and reasonably rapid. Such predictions are based on the ideas outlined in this chapter.

Summary Problem

On January 6, 1997, Chrysler Corporation announced that it has developed a new way to extract hydrogen from gasoline and other liquid fossil fuels, and that this discovery gives promise of much more efficient automobile transportation accompanied by much less environmental pollution (*Capital Times,* Madison, Wisconsin, January 6, 1997). The extracted hydrogen would be combined with oxygen in fuel cells (Section 18.8) like those currently used in spacecraft to generate electricity. The electricity in turn would power electric motors, primarily to move the car but also to provide air conditioning and other amenities expected by auto buyers. Because electric motors are far more efficient than current automobile engines, such a car might get 80 miles per gallon of fuel. By the year 2005, Chrysler expects to have such an electric car in operation. By answering the following questions, you can explore how the ideas of equilibrium, Gibbs free energy, and control of chemical reactions can be applied to motive power for automobiles.

1. The hydrogen extracted from hydrocarbon fuels must be free from soot (solid carbon) and carbon monoxide, which would interfere with the operation of a fuel cell. Consider possible reactions by which hydrogen could be obtained from a hydrocarbon such as octane (C_8H_{18}). Write an equation for a reaction that you think would not be suitable and write an equation for one that you think would be suitable. Explain your choice in each case.
2. Calculate the change in Gibbs free energy for each of the two reactions you wrote in part 1. Is either of them ruled out because it is not product-favored? If not, continue. If either of the reactions is not product-favored, think about whether conditions could be altered to make it product-favored.
3. The reaction by which hydrogen is obtained for use in synthesizing ammonia (Haber-Bosch process, *(← p. 597)* involves treating methane (from natural gas) with steam. The first step in this process is

$$CH_4(g) + H_2O(g) \rightleftharpoons CO(g) + 3\ H_2(g)$$

(a) Write the equilibrium constant expression for this reaction.
(b) Use data from Appendix J to calculate the Gibbs free energy change for this reaction. Is the reaction product-favored?
(c) Use data from Appendix J to calculate the enthalpy change for this reaction. Based on the equation itself, predict the sign of the entropy change for this reaction. Is the reaction product-favored at high temperatures but not at lower temperatures? Or the other way around? Explain.

4. To remove carbon monoxide from the hydrogen destined for the Haber-Bosch process, the following reaction is used:

$$CO(g) + H_2O(g) \rightleftharpoons CO_2(g) + H_2(g)$$

(a) Calculate the Gibbs free energy change for this reaction. Is it product-favored at 25 °C?
(b) From the Gibbs free energy change, calculate the thermodynamic equilibrium constant at 25 °C.
(c) For this reaction the thermodynamic equilibrium constant has the same value as the equilibrium constant in terms of concentrations. Suppose that 0.1 mol CO and 0.1 mol H_2O were introduced into an empty 10.0-L flask. Use the equilibrium constant you calculated to determine the concentration of $H_2(g)$ in the flask once equilibrium has been achieved.

5. If you were in charge of designing a system for generating hydrogen gas for use in the Haber-Bosch process, how might you get pure hydrogen? Assume that the

process is based on the two reactions given in parts 3 and 4. Suggest a chemical reagent that could be used to react with $CO_2(g)$ and thereby remove it from the hydrogen generated as a product in each of the two reactions. Would this same reagent work if you needed to remove $SO_2(g)$?

6. To get the highest purity $H_2(g)$ and the maximum yield from the hydrogen-generating process, how would you adjust the concentrations of the reactants and products? What reactant concentrations(s) would you increase or decrease? How would you adjust product concentrations?
7. In the Chrysler fuel-cell system for an electric automobile, hydrocarbon fuel is vaporized and partially oxidized in a limited quantity of air. In a second step the products of the first reaction are treated with steam over copper oxide and zinc oxide catalysts. In a final purification step more air is introduced and a platinum catalyst helps convert carbon monoxide to carbon dioxide. Write a balanced chemical equation for each of these three steps. Assume that the hydrocarbon fuel is octane.
8. Explain why it is advantageous to generate hydrogen gas from hydrocarbon fuel right in the automobile rather than storing hydrogen in a fuel tank. Think of as many disadvantages as you can that are associated with storing hydrogen in a car. What advantages does combining hydrogen with oxygen in a fuel cell have, as opposed to burning a hydrocarbon fuel in an internal combustion engine?

In Closing

Having studied this chapter, you should be able to . . .

- recognize a system at equilibrium and describe the properties of equilibrium systems (Section 13.1).
- describe the dynamic nature of equilibrium and the changes in concentrations of reactants and products that occur as a system approaches equilibrium (Sections 13.1 and 13.2).
- write equilibrium constant expressions, given balanced chemical equations (Section 13.2).
- obtain equilibrium constant expressions for related reactions from the expression for a known reaction (Section 13.2).
- calculate a value of K for an equilibrium system given information about initial concentrations and equilibrium concentrations (Section 13.3).
- make qualitative predictions about the extent of reaction based upon equilibrium constant values; that is, be able to predict whether a reaction is product-favored or reactant-favored based on the size of the equilibrium constant (Section 13.4).
- calculate concentrations of reactants and products in an equilibrium system if K and initial concentrations are known (Section 13.5).
- show by using Le Chatelier's principle how changes in concentrations, pressure or volume, and temperature shift chemical equilibria (Section 13.6).
- list the factors affecting chemical reactivity and apply them to predicting optimal conditions for producing products (Section 13.8).

Key Terms

chemical equilibrium *(13.1)*
dynamic equilibrium *(13.1)*
equilibrium concentration *(13.3)*
equilibrium constant *(13.2)*
equilibrium constant expression *(13.2)*
Haber-Bosch process *(13.8)*
hydrophilic *(13.1)*
hydrophobic *(13.1)*
Le Chatelier's principle *(13.6)*
shifting an equilibrium *(13.6)*
thermodynamic equilibrium constant *(13.7)*

Questions for Review and Thought

Conceptual Challenge Problems

Conceptual Challenge Problems CP-13.A, CP-13.B, CP-13.D, CP-13.E, and CP-13.F are related to the information in this paragraph. In Section 13.1, The Equilibrium State (← ***p. 567***), you read that aqueous iron(III) ions, $Fe^{3+}(aq)$, are nearly colorless. If their concentration is 0.001 M or below, a person cannot detect their color. Thiocyanate ions, $SCN^-(aq)$, are colorless also, but monothiocyanatoiron(III) ions, $Fe(SCN)^{2+}(aq)$, can be detected at very low concentrations because of their color. These ions are light amber in very dilute solutions, but as their concentration increases, the color intensifies and appears blood red in concentrated solutions. Suppose you prepared a stock solution by mixing equal volumes of 1.0×10^{-3} M solutions of both iron(III) nitrate and potassium thiocyanate solutions. The equilibrium reaction is

$$\underset{\text{colorless}}{Fe^{3+}(aq)} + \underset{\text{colorless}}{SCN^-(aq)} \rightleftharpoons \underset{\text{amber}}{Fe(SCN)^{2+}(aq)}$$

CP-13.A. Describe how you would use 5-mL samples of the stock solution and additional solutions of 0.01 M $Fe^{3+}(aq)$ and 0.01 M $SCN^-(aq)$ to show experimentally that the reaction between $Fe^{3+}(aq)$ and $SCN^-(aq)$ does not go to completion but instead reaches an equilibrium state where appreciable quantities of reactants and product are present. (Refer to the first paragraph for further information.)

CP-13.B. Suppose that a person added one drop of 0.01 M $Fe^{3+}(aq)$ to a 5-mL sample of the stock solution, followed by ten drops of 0.01 M $SCN^-(aq)$. This person treated a second 5-mL sample of the stock solution by first adding ten drops of 0.01 M $SCN^-(aq)$, followed by one drop of $Fe^{3+}(aq)$. How would the color intensity of these two solutions compare after the same quantities of the same solutions were added in reverse order? (Refer to the first paragraph for further information.)

CP-13.C. In Section 13.4 (← ***p. 579***) the equilibrium reaction between dioxygen and trioxygen (ozone) was discussed. What is the minimum volume of air (21% dioxygen by volume) at 1.00 atm and 25 °C that you would predict to have at least one molecule of trioxygen if the only source of trioxygen were its formation from dioxygen and if the atmospheric system were at equilibrium?

$$3\ O_2(g) \rightleftharpoons 2\ O_3(g) \qquad K = 6.3 \times 10^{-58} \quad \text{(at 25 °C)}$$

(The volume of 1 mol air at 1 atm and 25 °C is 24.45 L.)

CP-13.D. Predict what will happen if you add a small crystal of sodium acetate to a 5-mL sample of the stock solution (described in the first paragraph) so that some acetatoiron(III) ion, a coordination complex, is formed.

CP-13.E. Predict what will happen if you begin to add a 0.01 M solution of $Fe^{3+}(aq)$ drop by drop to a 5-mL sample of the stock solution (described in the first paragraph) until the total volume becomes 10 mL. (A 0.01 M solution of Fe^{3+} ions is pale yellow.) Predict what will happen if you do the same experiment adding 0.01 M $SCN^-(aq)$ to the stock solution. Predict what will happen if you mix 0.01 M $Fe^{3+}(aq)$ with 0.01 M $SCN^-(aq)$. Explain why the results of these three experiments would be similar or different.

CP-13.F. Predict what will happen if you put a 5-mL sample of the stock solution (described in the first paragraph) in a hot water bath. Predict what will happen if it is placed in an ice bath.

Review Questions

1. Define the terms chemical equilibrium and dynamic equilibrium.
2. If an equilibrium is product-favored, is its equilibrium constant large or small with respect to 1? Explain.
3. List three characteristics that you would need to verify in order to determine that a chemical system is at equilibrium.
4. Suppose that you have heated a mixture of *cis* and *trans*-2-pentene to 600 K, and after one hour you find that the composition is 40% *cis*. After 4 hr the composition is found to be 42% *cis*, and after 8 hr it is 42% *cis*. Next you heat the mixture to 800 K and find that the composition changes to 45% *cis*. When the mixture is cooled to 600 K and allowed to stand for 8 hr, the composition is found to be 42% *cis*. Is this system at equilibrium at 600 K? Or would more experiments be needed before you could conclude that it was at equilibrium? If so, what experiments would you do?
5. The decomposition of ammonium dichromate, $NH_4Cr_2O_7(s)$, yields nitrogen gas, water vapor, and solid chromium(III) oxide. The reaction is endothermic. In a closed container this process reaches an equilibrium state. Write a balanced equation for the

equilibrium reaction. How is the equilibrium affected if (a) more ammonium dichromate is added to the equilibrium system, (b) more water vapor is added, and (c) more chromium(III) oxide is added?

6. For the equilibrium reaction in Question 5, write the expression for the equilibrium constant. How would this equilibrium constant change if the total pressure on the system was doubled? How would the equilibrium constant change if the temperature was increased?

7. Indicate whether each statement below is true or false. If a statement is false, rewrite it to produce a closely related statement that is true.
(a) For a given reaction, the magnitude of the equilibrium constant is independent of temperature.
(b) If there is an increase in Gibbs free energy when reactants in their standard states are converted to products in their standard states, the equilibrium constant for the reaction will be negative.
(c) The equilibrium constant for the reverse of a reaction is the reciprocal of the equilibrium constant for the reaction itself.
(d) For the reaction

$$H_2O_2(\ell) \rightleftharpoons H_2O(\ell) + \frac{1}{2} O_2(g)$$

the equilibrium constant is one half the magnitude of the equilibrium constant for the reaction

$$2\ H_2O_2(\ell) \rightleftharpoons 2\ H_2O(\ell) + O_2(g)$$

8. Think of an experiment you could do to demonstrate that the equilibrium

$$2\ NO_2(g) \rightleftharpoons N_2O_4(g)$$

is a dynamic process in which the forward and reverse reactions continue to occur after equilibrium has been achieved. Describe how such an experiment might be carried out.

The Equilibrium State

9. Suppose you drop a large piece of ice into a well-insulated thermos with some water in it, and it comes to equilibrium with part of the ice melted. What is the temperature of the equilibrium system? Is this a static or a dynamic equilibrium? Explain.

10. The atmosphere consists of about 80% N_2 and 20% O_2, yet there are many oxides of nitrogen that are stable and can be isolated in the laboratory. Is the atmosphere at chemical equilibrium with respect to forming nitrogen oxides? If not, why don't they form? If so, how is it that these nitrogen oxides can be made and kept in the laboratory for long periods?

The Equilibrium Constant

11. Consider the reaction of $N_2 + O_2$ to give 2 NO and the reverse reaction of 2 NO to give $N_2 + O_2$, discussed in Section 13.1. Make qualitatively correct plots of the concentration of reactants and products vs. time for these two processes, showing the initial state and the final dynamic equilibrium state. Assume a temperature of 5000 K. Don't do any calculations—just sketch how you think the plots will look.

12. After 0.1 mol of pure *cis*-2-butene is allowed to come to equilibrium with *trans*-2-butene in a 5.0-L closed flask, another 0.1 mol of *cis*-2-butene is suddenly added to the flask.
(a) Is the new mixture at equilibrium? Explain why or why not.
(b) In the new mixture, which rate is faster: *cis* → *trans* or the reverse? Or are both rates the same?
(c) After the second 0.1 mol of *cis*-2-butene has been added and the system is at equilibrium, if the concentration of *trans*-2-butene is 0.01 mol/L, what is the concentration of *cis*-2-butene?

13. Write the equilibrium constant expression for each reaction:
(a) $2\ H_2O_2(g) \rightleftharpoons 2\ H_2O(g) + O_2(g)$
(b) $PCl_3(g) + Cl_2(g) \rightleftharpoons PCl_5(g)$
(c) $SiO_2(s) + 3\ C(s) \rightleftharpoons SiC(s) + 2\ CO(g)$
(d) $H_2(g) + \frac{1}{8} S_8(s) \rightleftharpoons H_2S(g)$

14. Write the equilibrium constant expression for each of the following reactions:
(a) $3\ O_2(g) \rightleftharpoons 2\ O_3(g)$
(b) $SiH_4(g) + 2\ O_2(g) \rightleftharpoons SiO_2(s) + 2\ H_2O(g)$
(c) $MgO(s) + SO_2(g) + \frac{1}{2} O_2(g) \rightleftharpoons MgSO_4(s)$
(d) $2\ PbS(s) + 3\ O_2(g) \rightleftharpoons 2\ PbO(s) + 2\ SO_2(g)$

15. Write the equilibrium constant expression for each reaction:
(a) $TlCl_3(s) \rightleftharpoons TlCl(s) + Cl_2(g)$
(b) $CuCl_4^{2-}(aq) \rightleftharpoons Cu^{2+}(aq) + 4\ Cl^-(aq)$
(c) $CO(g) + H_2O(g) \rightleftharpoons CO_2(g) + H_2(g)$
(d) $4\ H_3O^+(aq) + 2\ Cl^-(aq) + MnO_2(s) \rightleftharpoons$
$Mn^{2+}(aq) + 6\ H_2O(\ell) + Cl(aq)$

16. Write the equilibrium constant expression for each reaction:
(a) The oxidation of ammonia with ClF_3 in a rocket motor.

$$NH_3(g) + ClF_3(g) \rightleftharpoons 3\ HF(g) + \frac{1}{2} N_2(g) + \frac{1}{2} Cl_2(g)$$

(b) The simultaneous oxidation and reduction of a chlorite ion.

$$3\ ClO_2^-(aq) \rightleftharpoons 2\ ClO_3^-(aq) + Cl^-(aq)$$

(c) $IO_3^-(aq) + 6\ OH^-(aq) + Cl_2(g) \rightleftharpoons$
$IO_6^{5-}(aq) + 2Cl^-(aq) + 3\ H_2O(\ell)$

17. In Section 13.2 the equilibrium constant for the reaction

$$\frac{1}{8} S_8(s) + O_2(g) \rightleftharpoons SO_2(g)$$

is given as 4.2×10^{52}. If this reaction is so product-favored, why can large piles of yellow sulfur exist in our environment (as they do in Louisiana and Texas)?

18. Consider the following two equilibria involving $SO_2(g)$ and their corresponding equilibrium constants.

$$SO_2(g) + \frac{1}{2} O_2(g) \rightleftharpoons SO_3(g) \qquad K_1$$

$$2\ SO_3(g) \rightleftharpoons 2\ SO_2(g) + O_2(g) \qquad K_2$$

Which of the following expressions correctly relates K_1 to K_2?
(a) $K_2 = K_1^2$ (c) $K_2 = 1/K_1$ (e) $K_2 = 1/K_1^2$
(b) $K_2^2 = K_1$ (d) $K_2 = K_1$

19. The reaction of hydrazine (N_2H_4) with chlorine trifluoride (ClF_3) was used in experimental rocket motors at one time.

$$N_2H_4(g) + \frac{4}{3} ClF_3(g) \rightleftharpoons 4 HF(g) + N_2(g) + \frac{2}{3} Cl_2(g)$$

How is the equilibrium constant, K, for this reaction related to K' for the reaction written in the following way?

$$3 N_2H_4(g) + 4 ClF_3(g) \rightleftharpoons 12 HF(g) + 3 N_2(g) + 2 Cl_2(g)$$

(a) $K = K'$
(b) $K = 1/K'$
(c) $K^3 = K'$
(d) $K = (K')^3$
(e) $3K = K'$

20. Hydrogen can react with elemental sulfur to give the smelly, toxic gas H_2S according to the reaction

$$H_2(g) + \frac{1}{8} S_8(s) \rightleftharpoons H_2S(g)$$

If the equilibrium constant for this reaction is 7.6×10^5 at 25 °C, determine the value of the equilibrium constant for the reaction written as

$$8 H_2(g) + S_8(s) \rightleftharpoons 8 H_2S(g)$$

21. At 450 °C, the equilibrium constant for the Haber-Bosch synthesis of ammonia is 0.16 for the reaction written as

$$3 H_2(g) + N_2(g) \rightleftharpoons 2 NH_3(g)$$

Calculate the value of K for the same reaction written as

$$\frac{3}{2} H_2(g) + \frac{1}{2} N_2(g) \rightleftharpoons NH_3(g)$$

Determining Equilibrium Constants

22. Isomer A is in equilibrium with isomer B, as in the reaction

$$A(g) \rightleftharpoons B(g)$$

Three experiments are done, each at a different temperature, and equilibrium concentrations are measured. For each experiment, calculate the equilibrium constant, K.
(a) [A] = 0.74 mol/L, [B] = 0.74 mol/L
(b) [A] = 2.0 mol/L, [B] = 2.0 mol/L
(c) [A] = 0.01 mol/L, [B] = 0.01 mol/L

23. Two molecules of A react to form one molecule of B, as in the reaction,

$$2 A(g) \rightleftharpoons B(g)$$

Three experiments are done, each at a different temperature, and equilibrium concentrations are measured. For each experiment, calculate the equilibrium constant, K.
(a) [A] = 0.74 mol/L, [B] = 0.74 mol/L
(b) [A] = 2.0 mol/L, [B] = 2.0 mol/L
(c) [A] = 0.01 mol/L, [B] = 0.01 mol/L
By comparing the results of Problems 22 and 23, what can you conclude about the statement, "If the concentrations of reactants and products are equal, then the equilibrium constant is always 1.0."

24. Consider the following equilibrium: 2 A(aq) $\rightleftharpoons$ B(aq). At equilibrium, [A] = 0.056 M and [B] = 0.21 M. Calculate the equilibrium constant for the reaction as written.

25. The following reaction was examined at 250 °C:

$$PCl_5(g) \rightleftharpoons PCl_3(g) + Cl_2(g)$$

At equilibrium, $[PCl_5] = 4.2 \times 10^{-5}$ M, $[PCl_3] = 1.3 \times 10^{-2}$ M, and $[Cl_2] = 3.9 \times 10^{-3}$ M. Calculate the equilibrium constant for the reaction.

26. At high temperature, hydrogen and carbon dioxide react to give water and carbon monoxide.

$$H_2(g) + CO_2(g) \rightleftharpoons H_2O(g) + CO(g)$$

Laboratory measurements at 986 °C show that there are 0.11 mol of each of CO and water vapor and 0.087 mol of each of H_2 and CO_2 at equilibrium in a 1.0-L container. Calculate the equilibrium constant for the reaction at 986 °C.

27. Carbon dioxide reacts with carbon to give carbon monoxide according to the equation

$$C(s) + CO_2(g) \rightleftharpoons 2 CO(g)$$

At 700 °C, a 2.0-L flask is found to contain at equilibrium 0.10 mol of CO, 0.20 mol of CO_2, and 0.40 mol of C. Calculate the equilibrium constant for this reaction at the specified temperature.

28. Assume you place 0.010 mol of N_2O_4(g) in a 2.0-L flask at 50 °C. After the system reaches equilibrium, $[N_2O_4]$ = 0.00090 M. What is the value of K for the following reaction?

$$N_2O_4(g) \rightleftharpoons 2 NO_2(g)$$

29. Nitrosyl chloride, NOCl, decomposes to NO and Cl_2 at high temperatures.

$$2 NOCl(g) \rightleftharpoons 2 NO(g) + Cl_2(g)$$

Suppose you place 2.00 mol of NOCl in a 1.00-L flask and raise the temperature to 462 °C. When equilibrium has been established, 0.66 mol of NO is present. Calculate the equilibrium constant for the decomposition reaction from these data.

The Meaning of the Equilibrium Constant

30. Using the data of Table 13.1, predict which of the reactions below will be product-favored. Then place all of the reactions in order from most reactant-favored to most product-favored.

$$2 NH_3(g) \rightleftharpoons N_2(g) + 3 H_2(g)$$
$$NH_4^+(aq) + OH^-(aq) \rightleftharpoons NH_3(aq) + H_2O(\ell)$$
$$2 NO(g) \rightleftharpoons N_2(g) + O_2(g)$$
$$2 NO_2(g) \rightleftharpoons N_2O_4(g)$$
$$HCO_3^-(aq) + H^+(aq) \rightleftharpoons H_2CO_3(aq)$$

31. The equilibrium constant for dissolving silver sulfate, silver nitrate, and silver sulfide in water are 1.7×10^{-5}, 2.0×10^2, and 6×10^{-30}, respectively. Write the balanced reactions and the associated equilibrium constant expressions for each process. Which compound is most soluble? Which is least soluble? On the basis of this, suggest a way to tell the difference between soluble and insoluble compounds from their equilibrium constants for dissolution.

Calculating Equilibrium Concentrations

32. The hydrocarbon C_4H_{10} can exist in two forms, butane and 2-methylpropane. The value of K for the interconversion of the two forms is 2.5 at 25 °C.

$$CH_3—CH_2—CH_2—CH_3 \rightleftharpoons H—C(CH_3)_2—CH_3$$

butane 2-methylpropane

(a) Suppose that the initial concentrations of both butane and 2-methylpropane are 0.1 mol/L. Make up a table of initial concentration, change in concentration, and equilibrium concentration for this reaction.

(b) Write the equilibrium constant expression in terms of x, the change in the concentration of butane, then solve for x.

(c) If you place 0.017 mol of butane in a 0.50-L flask at 25 °C, what will be the equilibrium concentrations of the two isomers?

33. A mixture of the butane and 2-methylpropane at 25 °C has [butane] = 0.025 mol/L and [2-methylpropane] = 0.035 mol/L. Is this mixture at equilibrium? If the *equilibrium* concentration of butane is 0.025 mol/L, what must [2-methylpropane] be at equilibrium? (See reaction in Question 32.)

34. Cyclohexane, C_6H_{12}, a hydrocarbon, can isomerize or change into methylcyclopentane, a compound with the same formula but with a different molecular structure.

$$C_6H_{12}(g) \rightleftharpoons C_5H_9CH_3(g)$$

cyclohexane methylcyclopentane

The equilibrium constant has been estimated to be 0.12 at 25 °C. If you had originally placed 3.79 g of cyclohexane in a 2.80-L flask, how much cyclohexane (in grams) is present when equilibrium is established?

35. At room temperature, the equilibrium constant for the reaction

$$2\ NO(g) \rightleftharpoons N_2(g) + O_2(g)$$

is 1.4×10^{30}.

(a) Is this reaction product-favored or reactant-favored?

(b) In the atmosphere at room temperature the concentration of N_2 is 0.33 mol/L, and the concentration of O_2 is about 25% of that value. Calculate the equilibrium concentration of NO in the atmosphere produced by the reaction of N_2 and O_2.

(c) Now revisit your answer to Question 10!

36. Write equilibrium constant expressions, in terms of reactant and product concentrations, for each of the following reactions:

$H_2O(\ell) \rightleftharpoons H^+(aq) + OH^-(aq)$ $K = 1.0 \times 10^{-14}$

$CH_3COOH(aq) \rightleftharpoons CH_3COO^-(aq) + H^+(aq)$ $K = 1.8 \times 10^{-5}$

$N_2(g) + 3\ H_2(g) \rightleftharpoons 2\ NH_3(g)$ $K = 3.5 \times 10^8$

$2\ O_3(g) \rightleftharpoons 3\ O_2(g)$ $K = 7 \times 10^{56}$

$2\ NO_2(g) \rightleftharpoons N_2O_4(g)$ $K = 1.7 \times 10^2$

$HCO_2^-(aq) + H^+(aq) \rightleftharpoons HCO_2H(aq)$ $K = 5.6 \times 10^3$

$Ag^+(aq) + I^-(aq) \rightleftharpoons AgI(s)$ $K = 6.7 \times 10^{15}$

Assume that all gases and solutes have initial concentrations of 1.0 mol/L. Then let the *first* reactant in each equation change its concentration by an amount $-x$. Using the concentration table approach, write equilibrium constant expressions in terms of the unknown variable x for each reaction. Which of these expressions yield quadratic equations which can be solved using the quadratic equation formula discussed in the text? How would you go about solving the others for x?

37. Hydrogen gas and iodine gas react via the equation

$$H_2(g) + I_2(g) \rightleftharpoons 2\ HI(g) \qquad K = 76 \text{ (at 600 K)}$$

If 0.05 mol of HI is placed in a 1.0-L flask at 600 K, what are the equilibrium concentrations of HI, I_2, and H_2?

38. Many common nonmetallic elements exist as diatomic molecules at room temperature: all of the halogens, hydrogen, oxygen, and nitrogen. When these elements are heated to high temperatures, they break apart into atoms. The basic reaction at 1500 K is

$$E_2(g) \rightleftharpoons 2\ E(g)$$

where E stands for each element. A 1-mol quantity of each diatomic species is placed in a separate 1.0-L container which is then heated to 1500 K. From the data below, determine the equilibrium concentrations of the atomic form of each element at 1500 K.

Species	Equilibrium Constant for Atomization at 1500 K
Br_2	8.9×10^{-2}
Cl_2	3.4×10^{-3}
F_2	7.4
H_2	3.1×10^{-10}
I_2	1.5
N_2	1×10^{-27}
O_2	1.6×10^{-11}

From these results, predict which of the diatomic elements has the lowest bond dissociation energy, and compare your results with thermochemical calculations and with Lewis structures.

Shifting a Chemical Equilibrium: Le Chatelier's Principle

39. Solid barium sulfate is in equilibrium with barium ions and sulfate ions in solution.

$$BaSO_4(s) \rightleftharpoons Ba^{2+}(aq) + SO_4^{2-}(aq)$$

What will happen to the barium ion concentration if more solid $BaSO_4$ is added to the flask? Explain your choice.

(a) It will increase.

(b) It will decrease.

(c) It will not change.
(d) You cannot tell with the information provided.

40. Consider the following equilibrium, established in a 2.0-L flask at 25 °C:

$$N_2O_4(g) \rightleftharpoons 2\ NO_2(g) \qquad \Delta H° = +57.2\ kJ$$

What will happen to the concentration of N_2O_4 if the temperature is increased? Explain your choice.
(a) It will increase.
(b) It will decrease.
(c) It will not change.
(d) You cannot tell with the information provided.

41. Hydrogen, bromine, and HBr in the gas phase are in equilibrium in a container of fixed volume.

$$H_2(g) + Br_2(g) \rightleftharpoons 2\ HBr(g) \qquad \Delta H° = -103.7\ kJ$$

How will each of the following changes affect the indicated quantities? Write increase, decrease, or no change.

Change	**$[Br_2]$**	**[HBr]**	***K***
Some H_2 is added to the container.	______	______	______
The temperature of the gases in the container is increased.	______	______	______
The pressure of HBr is increased.	______	______	______

42. The equilibrium constant for the following reaction is 0.16 at 25 °C, and the standard enthalpy change is +16.1 kJ.

$$2\ NOBr(g) \rightleftharpoons 2\ NO(g) + Br_2(\ell)$$

Predict the effect of each of the following changes on the position of the equilibrium; that is, state which way the equilibrium will shift (left, right, or no change) when each of the following changes is made.
(a) Adding more Br_2
(b) Removing some NOBr
(c) Decreasing the temperature

43. The formation of hydrogen sulfide from the elements is exothermic.

$$H_2(g) + \tfrac{1}{8}\ S_8(s) \rightleftharpoons H_2S(g) \qquad \Delta H° = -20.6\ kJ$$

Predict the effect of each of the following changes on the position of the equilibrium; that is, state which way the equilibrium will shift (left, right, or no change) when each of the following changes is made:
(a) Adding more sulfur
(b) Adding more H_2
(c) Raising the temperature

44. The oxidation of NO to NO_2,

$$2\ NO(g) + O_2(g) \rightleftharpoons 2\ NO_2(g)$$

is exothermic. Predict the effect of each of the following changes on the position of the equilibrium; that is, state which way the equilibrium will shift (left, right, or no change) when each of the following changes is made.
(a) Adding more O_2
(b) Adding more NO_2
(c) Decreasing the temperature

45. Consider the following equilibrium:

$$PbCl_2(s) \rightleftharpoons Pb^{2+}(aq) + 2\ Cl^-(aq)$$

What will happen to the equilibrium concentration of aqueous lead(II) ion if some solid NaCl is added to the flask?
(a) It will increase.
(b) It will decrease.
(c) It will not change.
(d) You cannot tell with the information provided.

46. Phosphorus pentachloride is in equilibrium with phosphorus trichloride and chlorine in a flask.

$$PCl_5(s) \rightleftharpoons PCl_3(g) + Cl_2(g)$$

What will happen to the concentration of Cl_2 if additional $PCl_5(s)$ is added to the flask?
(a) It will increase.
(b) It will decrease.
(c) It will not change.
(d) You cannot tell with the information provided.

47. Consider the transformation of butane into 2-methylpropane, (see Question 32). The system is originally at equilibrium at 25 °C in a 1.0-L flask with [butane] = 0.010 M and [2-methylpropane] = 0.025 M. Suppose that 0.0050 mol of 2-methylpropane is suddenly added to the flask, and the system shifts to a new equilibrium. What is the new equilibrium concentration of each gas?

Gibbs Free Energy Change and Equilibrium Constants

48. Suppose that at a certain temperature T' a chemical reaction is found to have thermodynamic equilibrium constant K of 1.0. Indicate whether the statements below are true or false:
(a) The enthalpy change for the reaction, $\Delta H°$, is zero.
(b) The entropy change for the reaction, $\Delta S°$, is zero.
(c) The Gibbs free energy change for the reaction, $\Delta G°$, is zero.
(d) $\Delta H°$ and $\Delta S°$ have the same sign.
(e) $\Delta H°/T = \Delta S°$ at the temperature T'.

49. For the following chemical reactions, calculate the equilibrium constants at 298.15 K and at 1000 K from the thermodynamic data in Appendix J, and indicate if they are product-favored or reactant-favored at each temperature:

(a) The conversion of nitric oxide to nitrogen dioxide in the atmosphere,

$$2\ NO(g) + O_2(g) \rightleftharpoons 2\ NO_2(g)$$

(b) The reaction of an alkali metal with a halogen to produce an alkali metal halide salt,

$$2\ Na(s) + Cl_2(g) \rightleftharpoons 2\ NaCl(s)$$

(c) The oxidation of carbon monoxide to carbon dioxide,

$$2\ CO(g) + O_2(g) \rightleftharpoons 2\ CO_2(g)$$

(d) The first step in the production of electronic grade silicon from sand,

$$SiO_2(s) + 2\ C(s) \rightleftharpoons Si(s) + 2\ CO(g)$$

50. There are millions of organic compounds known, and new ones are being discovered or made at a rate of over 100,000 compounds per year. We also know that organic compounds burn readily in air at high temperatures to form carbon dioxide and water. Below are listed several classes of organic compounds, with a simple example of each. Write a balanced chemical equation for the combustion in O_2 of each of these compounds, and then use the data in Appendix J to show that these reactions are product-favored at room temperature.

Class of Organics	Simple Example
Aliphatic hydrocarbons	Methane, CH_4
Aromatic hydrocarbons	Benzene, C_6H_6
Alcohols	Methanol, CH_3OH

From these results, it is reasonable to hypothesize that *all* organic compounds are thermodynamically unstable in an oxygen atmosphere (i.e., their room-temperature reaction with O_2 to form CO_2 and H_2O is product-favored). If this hypothesis is true, how can organic compounds exist on earth?

Controlling Chemical Reactions: The Haber-Bosch Process

51. Using the signs of the enthalpy change and the entropy change for the Haber-Bosch process, explain why choosing the temperature at which to run this reaction is very important.

52. Although ammonia is made in enormous quantities by the Haber-Bosch process (36 billion pounds in 1995—see inside back cover), sulfuric acid is made in even greater amounts (95 billion pounds). The *contact process* used to make sulfuric acid can be simplified and represented by the following three reactions:

$$S(s) + O_2(g) \rightleftharpoons SO_2(g)$$
$$2\ SO_2(g) + O_2(g) \rightleftharpoons 2\ SO_3(g)$$
$$SO_3(g) + H_2O(\ell) \rightleftharpoons H_2SO_4(\ell)$$

Determine whether these reactions are product-favored or reactant-favored at room temperature, 298 K, and successively higher temperatures: 300, 400, 500 K, and so on, up to 1000 K. Do this in steps:

(a) Make a table of $\Delta H_f^\circ(298.15\ K)$, $S^\circ(298.15\ K)$, and $\Delta G_f^\circ(298.15\ K)$ for each reactant and product, from Appendix J. (*Hint:* By using a spreadsheet to assemble and contain this table, the necessary additional calculations are done very easily.)

(b) Calculate ΔH°, ΔS°, ΔG°, and the thermodynamic equilibrium constant K at each of the temperatures mentioned above. (*Hint:* Remember that Appendix J lists ΔH_f° and ΔG_f° in *kilojoules* per mole, but S° in *joules* per kelvin-mole.)

Now answer the following questions:

(c) Which of the three reactions is exothermic?

(d) In which of the three reactions does the entropy increase?

(e) It is the second of the three reactions whose rate is increased by the use of a catalyst rather than an increase in temperature. By examining the equilibrium constant K of this reaction as a function of temperature, explain why a catalyst is needed in this step.

53. A simple, single-step process responsible for the production of 41 billion pounds of lime, CaO(s), per year is the heating of limestone, $CaCO_3(s)$. From the data in Appendix J, calculate ΔH°, ΔS°, ΔG°, and the equilibrium constant K for this reaction from 300 to 1300 K in 100-K intervals. Give chemically plausible reasons why ΔH°_{rxn}, and ΔS°_{rxn} have the signs that they do. From the sign and magnitude of ΔH°_{rxn} and ΔS°_{rxn}, determine the approximate temperature at which this reaction becomes product-favored. Then in your library or on the World-Wide-Web, look up details on the commercial production of lime from limestone, to determine the temperature at which the process is actually carried out.

General Questions

54. The chemistry of compounds composed of a transition metal and carbon monoxide has been an interesting area of research for the past 40 years. $Ni(CO)_4$ is formed by the reaction of nickel metal with carbon monoxide.

(a) If you have 2.05 g of CO, and you combine it with 0.125 g of nickel metal, how many grams of $Ni(CO)_4$ can be formed?

(b) An excellent way to make pure nickel metal is to decompose $Ni(CO)_4$ in a vacuum at a temperature slightly higher than room temperature. What is the enthalpy change for the decomposition reaction

$$Ni(CO)_4(g) \longrightarrow Ni(s) + 4\ CO(g)$$

if the molar enthalpy of formation of $Ni(CO)_4$ gas is −602.9 kJ/mol?

In an experiment at 100 °C it is determined that with 0.01 mol of $Ni(CO)_4(g)$ initially present in a 1.0-L flask, only 0.00001 mol remains after decomposition.

(c) What is the equilibrium concentration of CO in the flask?

(d) What is the value of the equilibrium constant, *K*, for this reaction at 100 °C?
(e) What is the value of $\Delta G°$ for the reaction?
(f) From the value of $\Delta H_f°$ given in part (b), determine the value of $\Delta S°$ for the reaction. Explain why the sign of $\Delta S°$ makes sense chemically.
(g) Calculate the standard entropy $S°$ of $Ni(CO)_4(g)$.

Applying Concepts

55. Figure 13.3 shows the equilibrium mixture of N_2O_4 and NO_2 at two different temperatures. Imagine you can shrink yourself down to the size of the molecules in the two tubes and observe their behavior for a short period of time. Write a brief description of what you observe in each of the tubes.

56. Imagine yourself the size of atoms and molecules inside a beaker containing the following equilibrium mixture with a *K* greater than 1.

$$[Co(H_2O)_6]^{2+}(aq) + 4\ Cl^-(aq) \rightleftharpoons CoCl_4^{2-}(aq) + 6\ H_2O(\ell)$$

pink blue

Write a brief description of what you observe around you before and after additional water is added to the mixture.

57. Which of the diagrams represents equilibrium mixtures for the reaction

$$A_2(g) + B_2(g) \rightleftharpoons 2\ AB(g)$$

at a temperature where $10^2 > K > 0.1$?

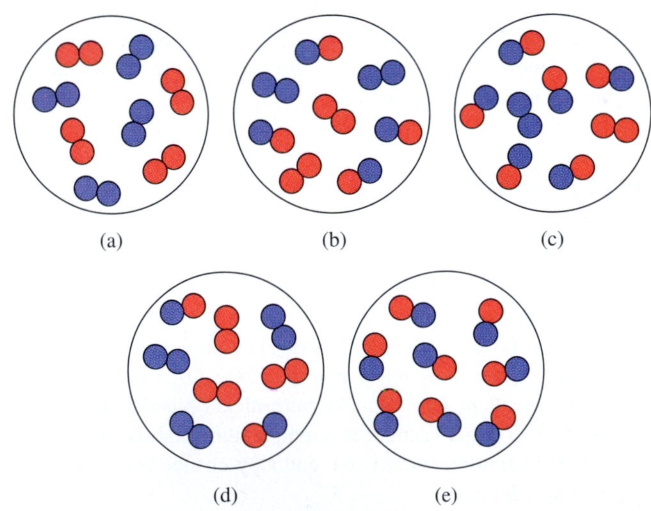

58. Draw a nanoscale (particulate) level diagram for an equilibrium mixture of

$$2\ H_2O_2(g) \rightleftharpoons 2\ H_2O(g) + O_2(g)$$

59. Which diagram in Question 57 best represents an equilibrium mixture with an equilibrium constant of
(a) 0.44
(b) 4.0
(c) 36

60. The diagram below represents an equilibrium mixture for the reaction.

$$N_2(g) + O_2(g) \rightleftharpoons 2\ NO(g)$$

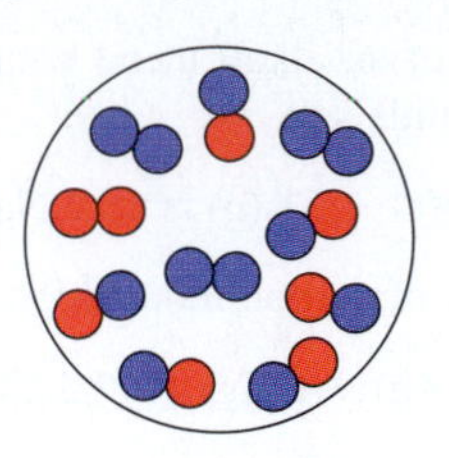

What is the equilibrium constant?

61. A sample of benzoic acid, a solid carboxylic acid, is in equilibrium with an aqueous solution of benzoic acid. A tiny quantity of D_2O, water containing the isotope ^2H, deuterium, is added to the solution. The solution is allowed to stand at constant temperature for several hours, after which some of the solid benzoic acid is removed and analyzed. The benzoic acid is found to contain a tiny quantity of deuterium, D, and the formula of the deuterium-containing molecules is C_6H_5COOD. Explain how this can happen.

62. In a second experiment with benzoic acid (see Question 61) a tiny quantity of water that contains the isotope ^{18}O is added to a saturated solution of benzoic acid in water. When some of the solid benzoic acid is analyzed, no ^{18}O is found in the benzoic acid. Compare this situation with the experiment involving deuterium, and explain how the results of the two experiments can differ as they do.

63. Samples of N_2O_4 can be prepared in which both nitrogen atoms are the heavier isotope ^{15}N. Designating this isotope as N^*, we can write the formula of the molecules in such a sample as $O_2N^*—N^*O_2$ and the formula of typical N_2O_4 as $O_2N—NO_2$. When a tiny quantity of $O_2N^*—N^*O_2$ is introduced into an equilibrium mixture of N_2O_4 and NO_2, the ^{15}N immediately becomes distributed among both N_2O_4 and NO_2 molecules, and in the N_2O_4 is invariably in the form $O_2N^*—NO_2$. Explain how this observation supports the idea that equilibrium is dynamic.

64. Using the symbolism of Question 63, we can write an equilibrium

$$O_2N^*—N^*O_2 + O_2N—NO_2 \rightleftharpoons 2\ O_2N^*—NO_2$$

Assuming that all isotopes of nitrogen behave the same in N_2O_4 and NO_2 molecules, use the ideas of probability developed in Chapter 7 (← *p. 270*) to figure out the value of the equilibrium constant for this process.

CHAPTER 14

Gases and the Atmosphere

A hydrocarbon fuel-burning engine in action. In this drag racer, so much fuel is burned in a short time that the vehicle and its driver are quickly propelled down the track. • What are the chemical reactions taking place? How do the gases that make up the atmosphere mix with fuel vapors in the combustion chamber and produce combustion products? Once produced, what do these combustion products do when they mix back into the atmosphere? You should be able to answer questions like these in the Summary Problem after you have studied this chapter. *(Ed Pritchard/Tony Stone Images)*

Early chemists studied gases and their chemistry extensively. They carried out reactions that generated gases, bubbled the gases through water into glass containers, transferred the gases to animal bladders for storage, and mixed gases together to see whether they would react and, if so, how much of each gas would be consumed or formed. They discovered that gases have a great many properties in common—much more so than liquids or solids. All gases are transparent, although some are colored. (F_2 is light yellow, Cl_2 is greenish yellow, Br_2 and NO_2 are reddish brown, and I_2 is violet.) Gases are also mobile—they expand to fill any space available, and they mix together in any proportions. The volume of a sample of gas can easily be changed by changing temperature or pressure or both; that is, gas densities are quite variable, depending on the conditions of measurement. At first this might seem to make the study of gases complicated, but it really does not. The way that gas volume depends on temperature and pressure is the same for all gases, and so it is very easy to take it into account.

Planet Earth is surrounded by a thin mixture of gases we call the atmosphere. In this mixture a variety of chemical reactions takes place, many of them driven by energy from solar photons. Many of these reactions are beneficial to the earth's inhabitants, but some of them produce undesirable products. Before dealing with reactions among gases, it is useful to consider the properties of gases in some detail.

14.1 PROPERTIES OF GASES

All gases have a number of properties in common:

- **Gases can be compressed.** We often pump compressed air into an automobile or bicycle tire and the compressed air occupies less volume than the non-compressed air (Figure 14.1).

Figure 14.1 Compression of gases. A bicycle tire pump works by compressing a sample of air into a smaller volume.

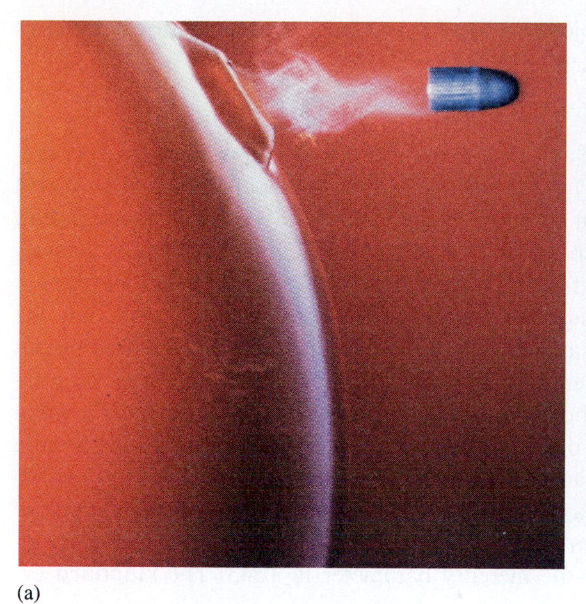
(a)

(b)

All gases tend to escape from the containers in which they are confined. (a) If they are under pressure, and the pressure is released, the rate of escape will be rapid. This property of gases finds many uses. (b) The escaping gases can be mixed with other ingredients, such as perfumes, which allows them to be propelled out of their containers. *(a, The Harold E. Edgerton 1992 Trust; b, C.D. Winters)*

- **Gases exert pressure on whatever surrounds them.** A gas sample inside a balloon or a storage cylinder exerts pressure on its surroundings. If the container cannot hold the pressure, some of the gas sample will escape.
- **Gases expand into whatever volume is available.** The gaseous contents of an aerosol spray can will continue to expand upon release.
- **Gases easily diffuse into one another.** This property allows one gas sample to completely mix with another. Once gases are mixed they do not separate on standing.
- **Gases are described in terms of their temperature and pressure, the volume occupied, and the amount (numbers of molecules or moles) of gas present.** For example, a hot gas occupies a greater volume and exerts a greater pressure than does the same sample of gas when it is cold.

The kinetic-molecular theory *(← p. 10)* can be used to explain all these properties of gases. You will see in Section 14.3 that by using this theory gases really are fairly easy to understand and that you will be able to predict with accuracy many of the properties of a sample of gas. Let's look at some gas properties in more detail.

Gases Exert Pressure

Conceptual Challenge Problem CP-14.A at the end of the chapter relates to topics covered in this section.

The firmness of a balloon filled with air indicates that the air inside exerts pressure. If too much gas is forced into a balloon, that pressure bursts the balloon and the gas rapidly escapes. The balloon's tightness is caused by gas molecules striking its inner surface. Each collision of a gas molecule with the balloon's surface exerts a force on the surface. The force per unit area is the **pressure** of the gas. A gas exerts pressure on every surface it contacts, no matter what the direction of contact.

$$\text{Pressure} = \frac{\text{force}}{\text{area}}$$

A force can accelerate an object, and the force equals the mass of the object times its acceleration.

$$\text{Force} = \text{mass} \times \text{acceleration}$$

The SI units for mass and acceleration are kilograms (kg) and meters per second per second (m/s^2), respectively, so force has the units $kg \cdot m/s^2$; a force of $1\ kg \cdot m/s^2$ is called a **newton (N).** A pressure of one newton per square meter (N/m^2) is called a **pascal (Pa).**

$1\ kPa = 10^3\ Pa$

The earth's atmosphere also exerts pressure on everything it contacts. Atmospheric pressure can be measured with a **barometer,** which can be made by filling a tube closed at one end with a liquid and inverting the tube in a dish containing the same liquid. Figure 14.2 shows a glass tube filled with mercury inverted and placed in a container of mercury—a mercury barometer. At sea level the height of the mercury column is about 760 mm above the surface of the mercury in the dish. The pressure at the bottom of a column of mercury 760 mm tall is balanced by the pressure at the bottom of the column of air above the dish—a column that extends to the top of the atmosphere. Pressure measured with a mercury barometer is usually reported in **millimeters of mercury (mm Hg),** a unit that is also called the **torr** after Evangelista Torricelli, who invented the mercury barometer in 1643. The **standard atmosphere (atm)** is defined as

Pressure units
1 atm = 760 mm Hg (exactly)
= 760 torr (after Torricelli)
= 101.3 kPa
= 1.013 bar
= 14.7 lb/in^2 (psi)

$$1 \text{ Standard atmosphere} = 1 \text{ atm} = 760 \text{ mm Hg (exactly)}$$

The pressure of the atmosphere at sea level is about 101,300 Pa (101.3 kPa). A related unit, a **bar,** equal to 100,000 Pa, is sometimes reported in connection with atmospheric pressure and weather. For a gaseous substance, the standard thermodynamic properties are given for a gas pressure of 1 bar (← *p. 238*).

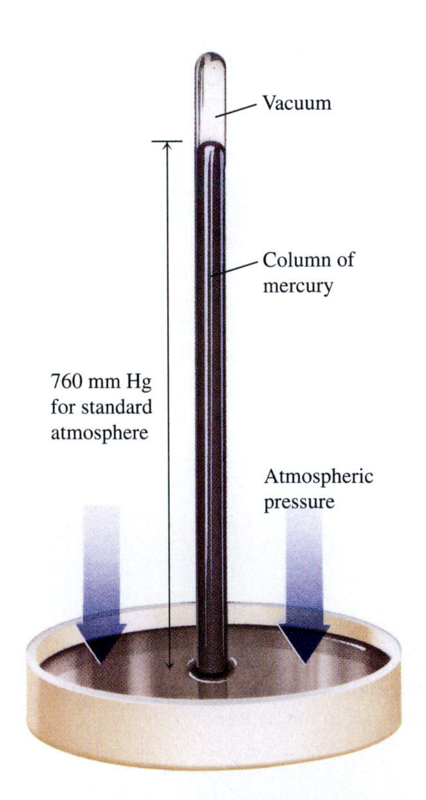

Figure 14.2 A Torricellian barometer. The pressure at the bottom of the mercury in the tube is balanced by atmospheric pressure on mercury in the dish.

PROBLEM-SOLVING EXAMPLE 14.1 Converting Pressure Units

Convert a pressure reading of 736 mm Hg to units of (a) atm, (b) torr, (c) kPa, (d) bar, and (e) psi.

Answer

(a) 0.968 atm
(b) 736 torr
(c) 98.1 kPa
(d) 0.981 bar
(e) 14.2 psi

Explanation Using the conversion factors given in the margin, we can write

(a) atm: $736 \text{ mm Hg} \times \left(\dfrac{1 \text{ atm}}{760. \text{ mm Hg}}\right) = 0.968 \text{ atm}$

(b) torr: $736 \text{ mm Hg} \times \left(\dfrac{760 \text{ torr}}{760. \text{ mm Hg}}\right) = 736 \text{ torr}$

(c) kPa: $736 \text{ mm Hg} \times \left(\dfrac{101.3 \text{ kPa}}{760. \text{ mm Hg}}\right) = 98.1 \text{ kPa}$

(d) bar: $736 \text{ mm Hg} \times \left(\dfrac{1.013 \text{ bar}}{760. \text{ mm Hg}}\right) = 0.981 \text{ bar}$

(e) psi: $736 \text{ mm Hg} \times \left(\dfrac{14.7 \text{ psi}}{760. \text{ mm Hg}}\right) = 14.2 \text{ psi}$

Problem-Solving Practice 14.1

A TV weather person says the barometric pressure is "29.5 inches of mercury." What is this pressure in (a) atm, (b) mm Hg, (c) bar, and (d) kPa?

Any liquid can be used in a barometer, but the height of the column depends on the density of the liquid. A water barometer would be about 34 ft tall, far too big to be practical. Many water wells (especially in regions of the world without readily available electricity) bring up water from rock strata less than 33 ft below the surface. A simple hand pump can reduce the pressure at the top of the well casing, and thus enable the atmospheric pressure that is acting on the water at the bottom of the well to force the water upward. When the water column reaches the pump, it flows out. Before the invention of submersible electric pumps, well diggers knew from experience that if water was found deeper than 33 ft below the surface, a hole large enough for a bucket would have to be dug. The bucket would then have to be lowered and filled with water. Atmospheric pressure alone would not be sufficient to push the water higher than 33 ft.

An old well pump. Old-fashioned water pumps like this one depend on atmospheric pressure to lift water from the well. They can only lift water from a depth of about 33 ft. At higher elevations this depth is lessened because the atmospheric pressure is lower.

To illustrate the meaning of pressure units, we can calculate the pressure exerted on the bottom of a glass by 500 mL of water. Suppose the water has a mass of 0.50 kg, and the bottom of the glass is a circle whose radius is 3.0 cm. The force exerted on the glass by the water is

$$\begin{aligned}\text{Force} &= \text{mass} \times \text{acceleration due to gravity}\\ &= 0.50\text{ kg} \times 9.807\text{ m/s}^2\\ &= 4.9\text{ kg m/s}^2\\ &= 4.9\text{ N}\end{aligned}$$

To find the pressure, you need to know the area over which the force is exerted. The cross-sectional area of the glass at its base is

$$\begin{aligned}\text{Area} = \pi \times (\text{radius})^2 &= 3.14 \times (0.030\text{ m})^2\\ &= 2.8 \times 10^{-3}\text{ m}^2\end{aligned}$$

Therefore the pressure is

$$\text{Pressure} = \frac{\text{force}}{\text{area}} = \frac{4.9\text{ kg m/s}^2}{2.8 \times 10^{-3}\text{ m}^2} = 1.8 \times 10^3\text{ N/m}^2 = 1.8 \times 10^3\text{ Pa} = 1.8\text{ kPa}$$

Notice that the pressure at the bottom of the glass of water, 1.8 kPa, is less than 2% as large as atmospheric pressure, 101.3 kPa.

PROBLEM-SOLVING EXAMPLE 14.2 Pressure Units

Show that 760.0 mm Hg equals 101.3 kPa. The density of mercury is 13.596 g/cm^3.

Answer The pressure exerted by a 760.0-mm-high column of mercury is 101.3 kPa.

Explanation Use a column of mercury 760.0 mm high with a cross-sectional area of 1.000 mm^2. Since the volume of the cylinder is the area of the base (1.000 mm^2) times the height (760.0 mm), the volume of the mercury is 760.0 mm^3. The density of mercury can be used to calculate the mass of the mercury.

$$\text{Mass of Hg} = (760.0\text{ mm}^3)\left(\frac{1\text{ cm}}{10\text{ mm}}\right)^3\left(\frac{13.596\text{ g}}{1\text{ cm}^3}\right) = 10.33\text{ g} = 0.01033\text{ kg}$$

The downward force on the mercury, due to gravity, is

$$\text{Force} = 0.01033\text{ kg} \times 9.807\text{ m/s}^2 = 0.1013\text{ N}$$

So, the pressure is

$$\begin{aligned}\text{Pressure} = \left(\frac{0.1013\text{ N}}{1\text{ mm}^2}\right)\left(\frac{1000\text{ mm}}{1\text{ m}}\right)^2 &= 101.3 \times 10^3\text{ N/m}^2\\ &= 101.3 \times 10^3\text{ Pa} = 101.3\text{ kPa}\end{aligned}$$

Problem-Solving Practice 14.2

Convert a pressure of 647 mm Hg into its corresponding value in units of bars, kilopascals (kPa), and atmospheres (atm).

14.2 THE ATMOSPHERE

Conceptual Challenge Problem CP-14.B at the end of the chapter relates to topics covered in this section.

More than 99% of the total mass of the atmosphere is found within 30 km (20 miles) of the earth's surface. Compared with the earth's diameter of 7918 miles, the atmosphere is like two or three outer layers of an onion compared to the whole onion.

A metric ton is 1000 kg.

Planet Earth is enveloped by a few vertical miles of chemicals that compose the gaseous medium in which we exist—the atmosphere, a perfect place to begin the study of gases. Everything on the surface of the earth experiences the atmosphere's pressure. The atmosphere's total mass is approximately 5.3×10^{15} metric tons, a huge figure, but still only about one millionth of the earth's total mass. Close to the earth's surface the atmosphere is mostly nitrogen and life-sustaining oxygen, but a fraction of a percent of other chemicals can make a difference in the quality of life. Extra water in the atmosphere can mean a rain forest; a little less water produces a balanced rainfall; and practically no water results in a desert. Air pollutants, present in such small quantities that they are outnumbered by oxygen and nitrogen molecules by 10,000 to 1 or more (Section 14.11), can lower our quality of life and generally be unhealthy for persons with respiratory disorders.

The two major chemicals in our atmosphere are nitrogen, a rather unreactive gas, and oxygen, a highly reactive one. In dry air at sea level, nitrogen is the most abundant atmospheric gas, followed by oxygen, and then 13 other gases, each at less than 1% by volume (Table 14.1). For every 100 volume units of air, 21 units are oxygen. When it is pure, oxygen supports combustion at an explosive rate (Figure 14.3), but when diluted with nitrogen, oxygen's oxidizing capability is tamed somewhat. Except for helium, which also occurs in some deposits of natural gas as a result of radioactive decay of elements in the earth's crust (Section 19.2), the atmosphere is our only source for the noble gases—argon, neon, helium, krypton, and xenon.

To convert percent to ppm, multiply by 10,000. Divide by 10,000 to convert ppm to percent.

Besides the percentages by volume used in Table 14.1, parts per million (ppm) and parts per billion (ppb) by volume are also used in describing the concentrations

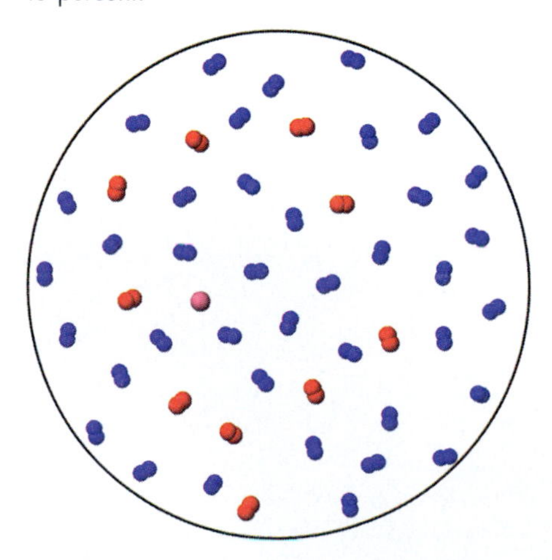

Nanoscale snapshot of air. This instantaneous snapshot of a sample of air at the nanoscale shows nitrogen, oxygen, and argon in approximately the correct proportions.

TABLE 14.1 The Composition of Dry Air at Sea Level

Gas	Percentage by Volume	Gas	Percentage by Volume
Nitrogen	78.084	Krypton	0.0001
Oxygen	20.948	Carbon monoxide	0.00001[2]
Argon	0.934	Xenon	0.000008
Carbon dioxide	0.033[1]	Ozone	0.000002[2]
Neon	0.00182	Ammonia	0.000001
Hydrogen	0.0010	Nitrogen dioxide	0.0000001[2]
Helium	0.00052	Sulfur dioxide	0.00000002[2]
Methane	0.0002[1]		

[1]The greenhouse gases carbon dioxide and methane are discussed in Chapter 11 (← *p. 466*) as they relate to fuels and the burning of fuels for energy production.

[2]Trace gases of environmental importance discussed in this chapter.

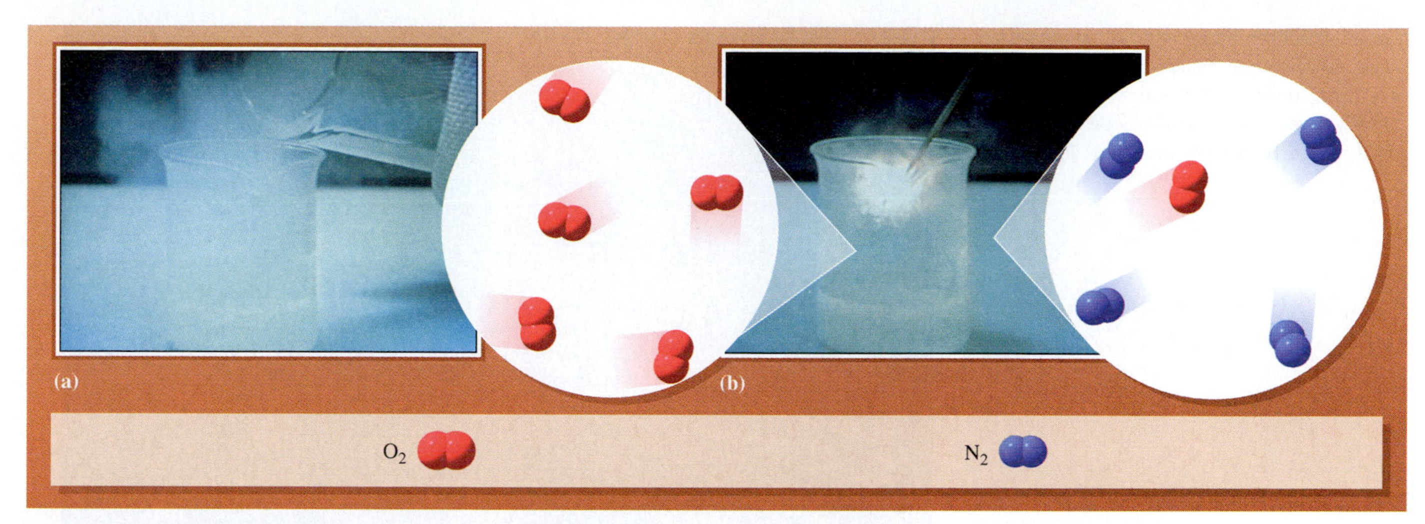

Figure 14.3 Combustion in pure oxygen. (a) Liquid oxygen is poured into a beaker. (b) When a glowing splint is inserted into the mouth of the beaker, the high oxygen concentration causes the splint to burn at an explosive rate. Other oxidizing agents would react in the same way. The nitrogen in our atmosphere acts as a diluting gas to help control the rate of burning for most substances in air. *(C.D. Winters)*

of components of the atmosphere. Since for gases volume is proportional to the number of molecules, these units also give a ratio of molecules of one kind to another kind. For example, "10 ppm SO_2" means that for every 1 million air molecules, 10 of them are SO_2 molecules. This may not sound like much until you consider that in just 1 cm^3 of air there are about 2.7×10^{13} million molecules. If this air contains 10 ppm SO_2, then there are 2.7×10^8 SO_2 molecules. That's a lot of SO_2 molecules, and we're only talking about 1 cm^3 of air!

Exercise 14.1 Calculating the Mass of Gases

Calculate the mass of the SO_2 molecules found in 1 cm^3 of air that contains 10 ppm SO_2.

Compared to the atmospheres of its nearest neighbors in the solar system, Venus and Mars, earth's atmosphere is quite thick. The high concentration of molecules in the gaseous state give rise to many possibilities for chemical reactions to take place as the gas molecules collide *(← p. 526)*. Because of gravity, molecules making up the atmosphere are most concentrated near the earth's surface.

The Troposphere and the Stratosphere

The earth's atmosphere can be roughly divided into layers, as shown in Figure 14.4. Near the earth, in the region named the **troposphere,** the temperature of the atmosphere decreases with increasing altitude. In this region the most violent mixing of air and the biggest variations in moisture content and temperature occur. Winds, clouds, storms, and precipitation are the result: the phenomena we know as weather. The troposphere is where we live. Even in most high-flying commercial jet airplanes, we are still in the troposphere, although near its upper limits. The composition of the tro-

The troposphere was named by the British meteorologist Sir Napier Shaw from the Greek word *tropos,* meaning "turning."

The stratosphere was named by the French meteorologist Leon Phillipe Treisserenc de Bort, who believed this region consisted of orderly layers with no turbulence or mixing. *Stratum* is a Latin word meaning "layer."

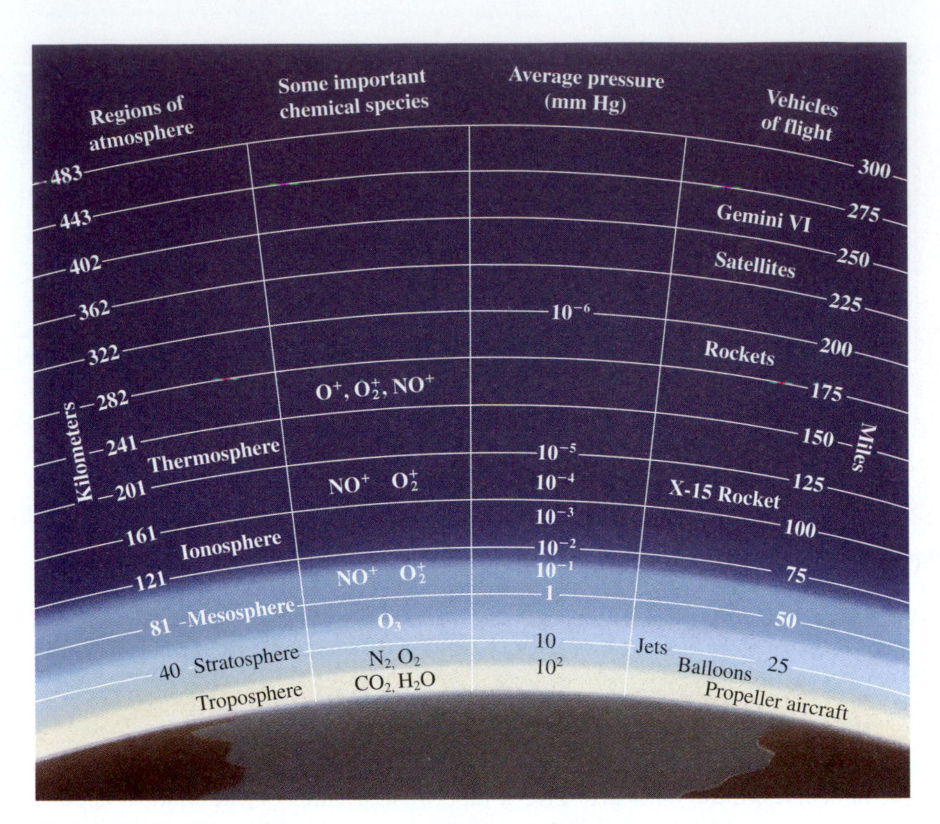

Figure 14.4 **Some facts about our earth's atmosphere.**

In the kinetic-molecular theory the word "molecule" is taken to include atoms of the monoatomic noble gases He, Ne, Ar, Kr, and Xe.

posphere is roughly that of dry air near sea level (Table 14.1), but the concentration of water vapor varies considerably; on average, it is about 10 ppm.

Just above the troposphere, from about 12 to 50 km above the earth's surface, is the **stratosphere.** If you take a ride in a Concorde supersonic aircraft, you will cruise in the stratosphere at about 20 km. The pressures in the stratosphere are extremely low, and there is little mixing between the stratosphere and the troposphere. The lower limit of the stratosphere varies from night to day over the globe, and at the polar regions it may be as low as 8 to 9 km above the earth's surface.

14.3 KINETIC-MOLECULAR THEORY

The fact that all gases behave in very similar ways can be interpreted by means of the **kinetic-molecular theory,** a theory that applies to the properties of liquids and solids as well as gases *(← p. 10).* According to the kinetic-molecular theory, a gas consists of very tiny molecules in constant rapid random motion. The pressure of a gas results from continual bombardment by rapidly moving molecules upon any surface the gas contacts (Figure 14.5).

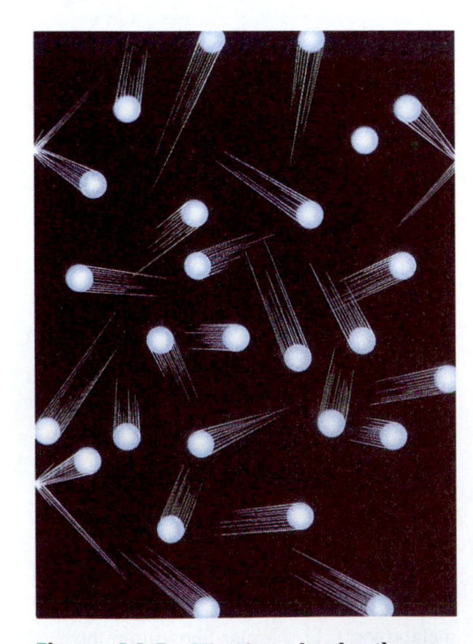

Figure 14.5 **Kinetic-molecular theory and pressure.** Gas pressure is caused by gas molecules bombarding the container walls.

Four fundamental concepts make up the kinetic-molecular theory and a fifth is closely associated with it. Each of these is consistent with the results of experimental studies of gases, some of which are described in the next section.

1. ***A gas is composed of molecules whose size is much smaller than the distances between them.*** This concept accounts for the ease with which gases can be com-

pressed and for the fact that gases at ordinary temperature and pressure mix completely with other gases. These facts imply that there must be plenty of room for additional molecules in a sample of gas.

2. ***Gas molecules move randomly—at various speeds and in every possible direction.*** This concept is consistent with the fact that gases quickly and completely fill any container in which they are placed.
3. ***Except when molecules collide, forces of attraction and repulsion between them are negligible.*** This concept is consistent with the fact that a gas will remain a gas indefinitely at a fixed pressure and temperature in spite of the countless collisions that take place.
4. ***When collisions occur, they are elastic.*** The speeds of colliding molecules may change but the total kinetic energy of two colliding molecules is the same after a collision as before. This concept is consistent with the fact that a gas sample at constant temperature never "runs down," with all molecules falling to the bottom of the container.
5. ***The average kinetic energy of gas molecules is proportional to the absolute temperature.*** Though not part of the kinetic molecular theory, this useful concept is consistent with the fact that the rate of escape of gas molecules through a tiny hole is faster the higher the temperature is, and with the fact that rates of chemical reactions are faster at higher temperatures.

Like any moving object, a gas molecule has kinetic energy. An object's kinetic energy, E_k, depends on its mass, m, and its speed, v, according to the equation

$$E_k = \frac{1}{2}(\text{mass})(\text{speed})^2 = \frac{1}{2}mv^2$$

All of the molecules in a gas are moving, but they do not all move at the same speed, and so they do not all have the same kinetic energy. At a given instant a few molecules are going very quickly; most are moving at close to the average speed, and a few others may be in the process of colliding with a surface, in which case their speed is zero. If we could watch an individual molecule, we would find that its speed would continually change as it collides with and exchanges energy with other molecules.

The relative number of molecules that have a given speed can be measured experimentally. Figure 14.6 is a graph of the number of molecules plotted versus their

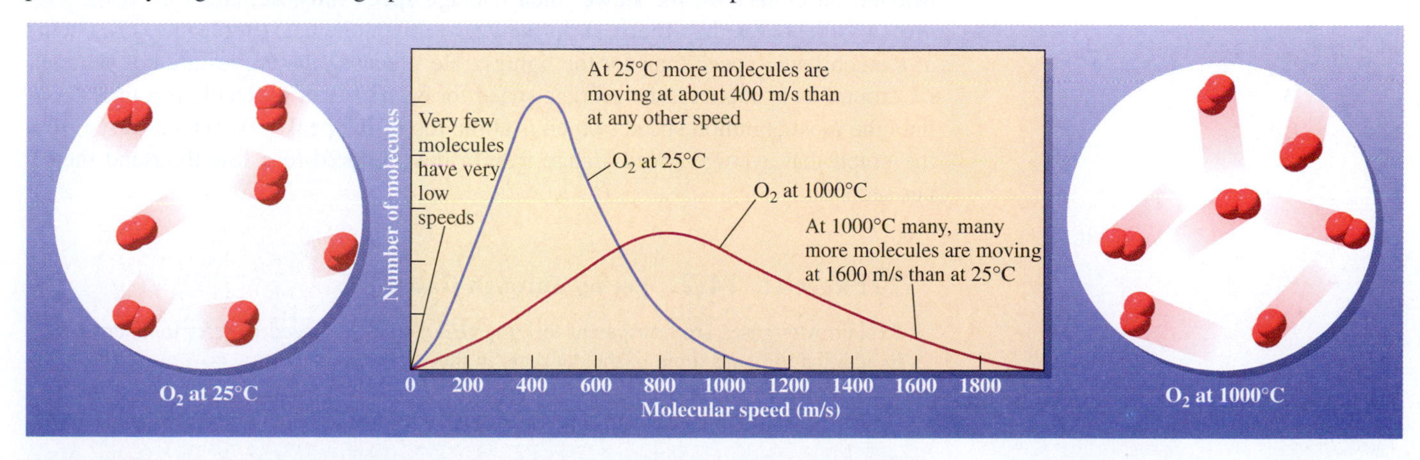

Figure 14.6 Distribution of molecular speeds. A plot of the relative number of gas molecules with a given speed versus that speed (in meters per second). The curve for O_2 at 1000 °C shows the effect of a temperature increase on the distribution of speeds.

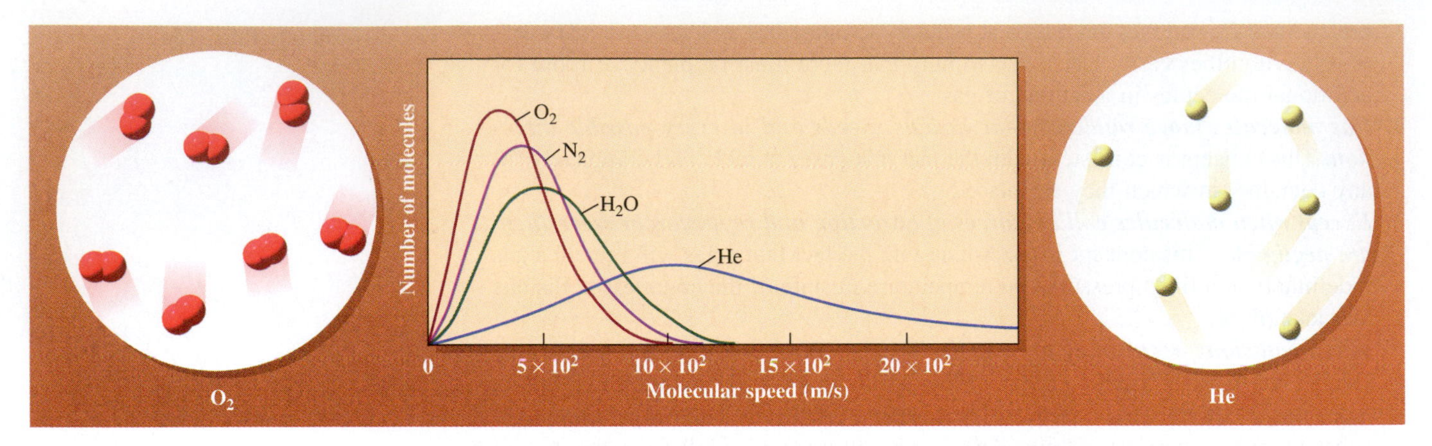

Figure 14.7 The effect of molar mass on the distribution of molecular speeds at a given temperature. Notice the similarity of the effect of increasing mass to the effect of decreasing temperature shown in Figure 14.6.

speed. The higher a point on the curve, the greater the number of molecules going at that speed. As you look at Figure 14.6, notice that some molecules are moving quickly (have high kinetic energy), and some are moving slowly (have low kinetic energy). There is, however, a most common speed, which corresponds to the maximum in the distribution curve. For oxygen gas at 25 °C, for example, the maximum in the curve comes at a speed of 400 m/s (1000 mph), and most of the molecules are within the range from 200 m/s to 700 m/s.

Graphs of molecular speeds (or energies) versus numbers of molecules are called Boltzmann distribution curves. They are named after Ludwig Boltzmann (1844–1906) ***(← p. 274)***, an Austrian mathematician who helped develop the kinetic-molecular theory of gases.

Also notice in Figure 14.6 that as the temperature is increased, the most common speed goes up and the number of molecules traveling very quickly increases. The areas under the two curves representing the two gas samples at different temperatures are the same. That is because the total number of molecules is the same in both samples.

Since $E_k = \frac{1}{2}mv^2$ and the average kinetic energy of the molecules of any gas is the same at a given temperature, the larger m is, the smaller v must be. That is, the heavier the molecules, the slower their average speed must be, and vice versa. Figure 14.7 illustrates this: the peak in the curve for the heaviest molecule, O_2, occurs at a much lower speed than for the lightest, He. Because the curves are not symmetric around their maxima, the average speed for each type of molecule is a little faster than the most common speed (which is at the top of the peak). You can also see from the graph that average speeds range from a few hundred to a few thousand meters per second.

Exercise 14.2 Seeing Through Gases

Explain why gases are transparent to light. (*Hint:* You may need to refer to Chapter 8 in addition to using some of the concepts in this section.)

Exercise 14.3 The Kinetic-Molecular Theory

Use the kinetic-molecular theory to explain why the pressure goes up when more gas molecules are added to a sample of gas in a fixed-volume container at constant temperature.

Exercise 14.4 Molecular Kinetic Energies

Arrange the following gaseous substances in order of increasing average kinetic energy of their molecules at 25 °C: Cl_2, H_2, NH_3, SF_6.

Exercise 14.5 Molecular Kinetic Energies

Using Figure 14.7 as a source of information, first draw a plot of number of molecules versus molecular speed for a sample of helium at 25 °C. Now assume that an equal number of molecules of argon, also at 25 °C, are added to the helium. What would the distribution curve for the mixture of gases look like?

Exercise 14.6 Molecular Motion

Suppose you purchase two helium-filled balloons of about equal size at the store, bring them home, and put one of them in the freezer compartment of your refrigerator and leave the other one out in your room. After a few hours you take the balloon from the freezer and compare it to the one left out in the room. Based on the kinetic-molecular theory, what differences would you expect to see (a) immediately after taking the balloon from the freezer, and (b) after the cold balloon warms to room temperature.

14.4 GAS BEHAVIOR AND THE IDEAL GAS LAW

Over the course of almost 200 years gases have been studied and the properties that all gases display have been summarized into *gas laws,* which are named for their discovers. Using the variables pressure, volume, temperature, and amount (number of moles), we can write three different equations that predict the way gases will behave. A gas that behaves exactly as described by these laws is called an **ideal gas.** Many gases, especially when pressures are high or temperatures are low, do not behave quite ideally (Section 14.8). Each of the three gas laws listed below can be explained in terms of the kinetic-molecular theory. Because these laws can be combined into a single widely applicable equation (the ideal gas law, described later in this section), it is not necessary to remember each of the individual gas law equations. However, these relationships are important because they help us remember how gases behave.

Boyle's Law *The volume of an ideal gas varies inversely with the applied pressure when temperature and amount are constant.*

$$V \propto \frac{1}{P} \qquad \text{(Temperature and amount are constant.)}$$

This relationship can also be written as PV = constant, where the value of the constant is dependent on the temperature and the amount of the gas.

In terms of the kinetic-molecular theory, a decrease in volume increases the pressure because there is less room for the gas molecules to move around before they collide with the walls of the container. Thus, there are more collisions with the walls. These collisions result in pressure on the container walls, so more collisions mean a higher pressure.

Conceptual Challenge Problem CP-14.C at the end of the chapter relates to topics covered in this section.

Charles's Law *The volume of an ideal gas varies directly with absolute temperature when pressure and amount are constant.*

$$V \propto T \qquad \text{(Pressure and amount are constant.)}$$

This relationship can also be written as $V = \text{constant} \times T$. The value of this constant is dependent on pressure and the amount of the gas.

In terms of the kinetic-molecular theory, higher temperature means faster molecular motion (higher kinetic energy). The more rapidly moving molecules therefore strike the walls of a container more often and each collision exerts greater force. If the volume were held constant, this would result in a higher pressure, but for pressure to remain constant, the volume of the container must expand.

Avogadro's Law The volume of an ideal gas varies directly with amount when temperature and pressure are constant.

$$V \propto n \qquad \text{(Temperature and pressure are constant.)}$$

This relationship can also be written as $V = \text{constant} \times n$. The value of this constant depends on the temperature and the pressure.

In terms of the kinetic-molecular theory, increasing the number of gas molecules with the same average kinetic energy (constant temperature) increases the number of collisions with the container walls by a proportional amount. This would increase the pressure if the volume were held constant. If the volume of the container can expand, the pressure will remain constant.

These three gas laws can be combined to give an overall gas law that summarizes the relationships among them.

$$V \propto \frac{nT}{P}$$

It is interesting to note that there is no corresponding law for liquids or solids.

To make this proportionality into an equation, a proportionality constant, R, named the **ideal gas constant,** is used. The equation becomes

$$V = R\frac{nT}{P}$$

and on rearranging, gives an equation called the **ideal gas law.**

$$PV = nRT$$

The ideal gas law correctly predicts the amount, pressure, volume, and temperature for samples of most gases at pressures of a few atmospheres or less and at temperatures well above their boiling points. The constant R can be calculated from the experimental fact that at 0 °C and 1 atm the volume of 1 mol of gas is 22.414 L. (This temperature and pressure are called **standard temperature and pressure [STP]** and the volume is called the **standard molar volume**.) Solving the ideal gas law for R, and substituting, we have

$$R = \frac{PV}{nT} = \frac{(22.414\ \text{L})(1\ \text{atm})}{(1\ \text{mol})(273.15\ \text{K})} = 0.082057\ \text{L atm/mol K}$$

which is usually rounded to 0.0821 L atm/mol K.

The ideal gas law can be used to find P, V, n, or T whenever three of the four variables are known, provided the conditions of temperature and pressure are not extreme.

For many calculations involving a gas sample under two sets of conditions, it is

PROBLEM-SOLVING EXAMPLE 14.3 Using the Ideal Gas Law

What volume will 0.20 g of oxygen occupy at 1.0 atm pressure and 20. °C?

Answer 0.15 L or 150 mL

Explanation Use the ideal gas law, substitute known quantities, and solve for the volume, V. When using $PV = nRT$ it is convenient to have all the variables in the same units as the gas constant, $R = 0.0821$ L atm/mol K. Temperatures should be in kelvins, pressure in atmospheres, volume in liters, and the amount of gas in moles.

Begin by converting the mass of oxygen to moles.

$$n_{oxygen} = (0.20 \text{ g } O_2)\left(\frac{1 \text{ mol } O_2}{32.00 \text{ g } O_2}\right) = 0.00625 \text{ mol } O_2$$

Next, convert the temperature to kelvins:

$$T = (20. + 273.15) \text{ K} = 293 \text{ K}$$

Now solve for V in the ideal gas law equation. The result will be in liters.

$$V = \frac{nRT}{P} = \frac{(0.00625 \text{ mol})(0.0821 \text{ L atm/mol K})(293 \text{ K})}{1.0 \text{ atm}} = 0.15 \text{ L} = 150 \text{ mL}$$

Notice how the units cancel. Note also that the answer is given in two significant digits because the original amount of oxygen was given as two digits and the pressure was given as two digits.

Problem-Solving Practice 14.3

What volume will 2.64 mol of N_2 occupy at 0.640 atm pressure and 31 °C?

The gas constant R appeared in the discussion of kinetics (← **p. 530**) and equilibrium (← **p. 593**), but in different units. If P is measured in standard units of pascals (kg/m · s^2) and V is measured in m^3, then R has the value.

$$R = \frac{(1.01325 \times 10^5 \text{ kg/m s}^2)(22.414 \times 10^{-3} \text{m}^3)}{(1 \text{ mol})(273.15 \text{ K})}$$
$$= 8.3145 \text{ kg} \cdot \text{m}^2/\text{s}^2 \text{ mol K}$$

Because 1 kg · m^2/s^2 is 1 J, the gas constant is also $R = 8.3145$ J/mol K.

Did you know that the gas constant could be used to define a new temperature scale? Conceptual Challenge Problem CP-14.D at the end of the chapter relates to topics covered in this section.

For many calculations involving a gas sample under two sets of conditions, it is convenient to use the ideal gas law in the following manner. For two sets of conditions (n_1, P_1, V_1, and T_1; n_2, P_2, V_2, and T_2), the ideal gas law can be written as

$$R = \frac{P_1V_1}{n_1T_1} \quad \text{and} \quad R = \frac{P_2V_2}{n_2T_2}$$

Since in both sets of conditions the quotient is equal to R, we can set the two quotients equal to each other.

$$\frac{P_1V_1}{n_1T_1} = \frac{P_2V_2}{n_2T_2}$$

When the moles of gas, n, is constant, this equation simplifies to what is known as the **combined gas law:**

$$\frac{P_1V_1}{T_1} = \frac{P_2V_2}{T_2}$$

Problem-Solving Example 14.4 The Combined Gas Law

Helium-filled balloons are used to carry scientific instruments high into the atmosphere. Suppose that such a balloon is launched on a summer day when the temperature at ground level is 22.5 °C and the barometer reads 754 mm Hg. If the balloon's volume is 1.00×10^6 L at launch, what will it be at a height of 37 km, where the pressure is 76.0 mm Hg and the temperature is 240. K?

Answer 8.05×10^6 L

Explanation Assume that no gas escapes from the balloon. Then only T, P, and V change. Using subscript 1 to indicate initial conditions (at launch) and subscript 2 to indicate final conditions (high in the atmosphere) gives

Initial conditions:

$$P_1 = 754 \text{ mm Hg}, \quad V_1 = 1.00 \times 10^6 \text{ L}, \quad \text{and} \quad T_1 = (22.5 + 273.15) \text{ K}$$

Final conditions:

$$P_2 = 76.0 \text{ mm Hg}, \quad \text{and} \quad T_2 = 240. \text{ K}$$

Solving the combined gas law for V_2 gives

$$V_2 = \frac{P_1 V_1 T_2}{P_2 T_1} = \frac{(754 \text{ mm Hg})(1.00 \times 10^6 \text{ L})(240. \text{ K})}{(76.0 \text{ mm Hg})(295.6 \text{ K})} = 8.05 \times 10^6 \text{ L}$$

Thus the volume is about eight times larger. The volume has increased because the pressure has dropped. For this reason weather balloons are never fully inflated at launch—a great deal of room has to be left so that the helium can expand at high altitudes.

Problem-Solving Practice 14.4

A small sample of a gas is prepared in the laboratory and found to occupy 21 mL at a pressure of 710. mm Hg and a temperature of 22.3 °C. The next morning the temperature has changed to 26.5 °C and the pressure is found to be 740 mm Hg. No gas has escaped from the container. (a) What volume does the sample of gas now occupy? (b) Assume that the pressure does not change. What volume does the gas occupy at the new temperature?

Experimental Pressure-Volume Behavior of Gases (Boyle's Law)

When you pump up a tire with a bicycle pump, the gas in the pump is squeezed into a smaller volume by application of pressure. This property is called **compressibility.** In contrast to gases, liquids and solids are only slightly compressible.

Robert Boyle studied the compressibility of gases in 1661 by pouring mercury into an inverted J-shaped tube containing a sample of trapped gas. Each time he added more mercury, the volume of the trapped gas decreased (Figure 14.8). The mercury additions increased the pressure on the gas and changed the gas volume in a predictable fashion.

If a proportionality constant is used, the relationship is

$$P = \text{a constant} \times \frac{1}{V} \quad \text{or} \quad PV = \text{a constant}$$

Early chemists routinely used mercury for all kinds of experiments. The toxic properties of mercury are now well known; it is a central nervous system poison and causes numerous chronic toxic effects such as memory loss. Mercury should not be used without adequate ventilation and should never be allowed to contact the skin.

Notice how this last equation for Boyle's law is like the ideal gas equation.

Boyle's law:

$$PV = \text{a constant} \quad \text{(at constant amount and temperature)}$$
$$PV = nRT$$

For a gas sample under two sets of pressure and temperature conditions, Boyle's law can be written as $P_1V_1 = P_2V_2$, which is a special case of the combined gas law.

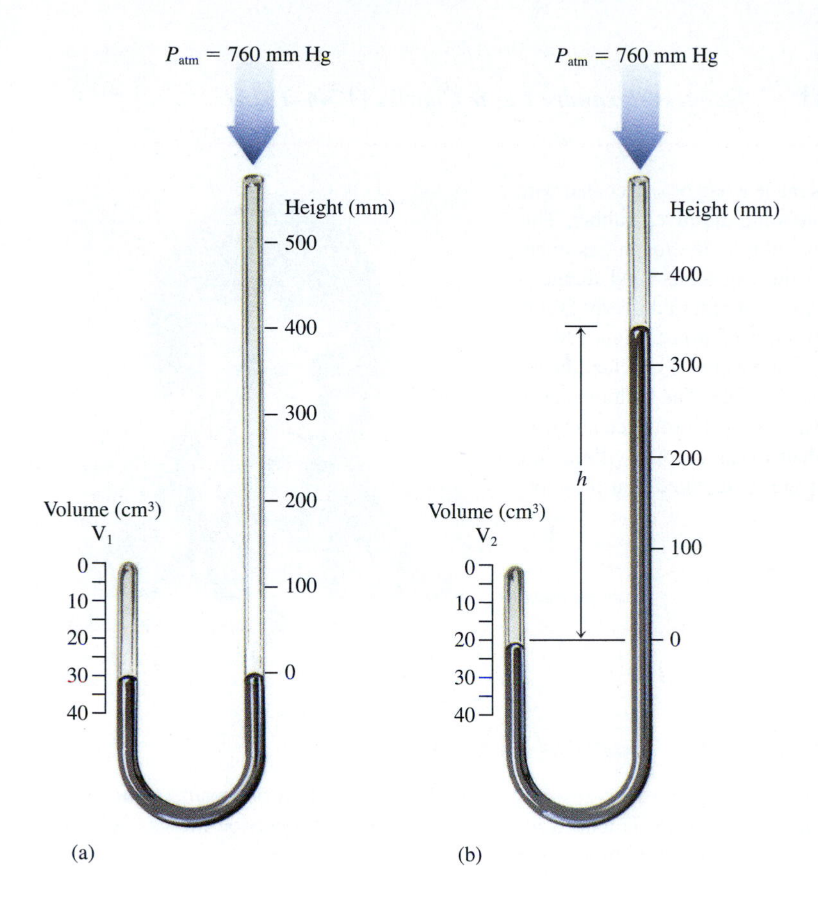

Figure 14.8 **Boyle's law.** Mercury confines a gas in a J-shaped tube. (a) When the mercury levels are the same on both sides of the J, the gas pressure equals atmospheric pressure. (b) When more mercury has been added, atmospheric pressure is augmented by the pressure of a mercury column of height *h,* where h = 340 mm Hg. At this higher pressure, the gas volume is smaller, as predicted by Boyle's law. The temperature remains constant.

PROBLEM-SOLVING EXAMPLE 14.5 Using Boyle's Law

Suppose you have a sample of a gas that occupies 100. mL at a pressure of exactly 3 atm. What volume will this gas occupy if the pressure is decreased to exactly 1 atm at the same temperature?

Answer 300. mL

Explanation The given conditions are $P_1 = 3$ atm, $P_2 = 1$ atm, $V_1 = 100.$ mL, and V_2 is unknown. The inverse relationship between P and V tells us that V_2 will be larger than V_1 because P_2 is smaller than P_1.

The value of V_2 can be found by rearranging $P_1V_1 = P_2V_2$ to solve for V_2.

$$V_2 = \frac{P_1V_1}{P_2} = \frac{3 \text{ atm} \times 100. \text{ mL}}{1 \text{ atm}} = 300. \text{ mL}$$

Problem-Solving Practice 14.5

At a pressure of exactly 1 atm and some temperature, a gas sample occupies 400. mL. What will be the volume of the gas at the same temperature if the pressure is decreased to 0.750 atm?

Portrait of a Scientist • *Jacques Alexandre Cesar Charles (1746–1823)*

The French chemist Charles was most famous in his lifetime for his experiments in ballooning. The first such flights were made by the Montgolfier brothers in June 1783, using a large spherical balloon made of linen and paper and filled with hot air. In August 1783, however, a different group, supervised by Jacques Charles, tried a different approach. Exploiting his recent discoveries in the study of gases, Charles decided to inflate the balloon with hydrogen. Since hydrogen would easily escape a paper bag (see Section 14.3), Charles made a bag of silk coated with a solution containing dissolved rubber. This rubberized bag held the hydrogen gas rather well. Inflating the bag to its final diameter took several days and required nearly 500 lb of acid and 1000 lb of iron to generate the hydrogen gas. A huge crowd watched the ascent on August 27, 1783. The balloon stayed aloft for almost 45 min and traveled about 15 miles, but, when it landed in a village, the people there were so terrified that they tore it to shreds.

(Corbis/Bettmann)

Exercise 14.7 Visualizing Boyle's Law

Many cars have gas-filled shock absorbers to give the car and its occupants a smooth ride. If a four-passenger car is loaded with four NFL linemen, describe the gas inside the shock absorbers compared to when the car is empty.

Experimental Temperature-Volume Behavior of Gases (Charles's Law)

In 1787, Jacques Charles discovered that the volume of a fixed quantity of a gas at constant pressure increases with increasing temperature. Figure 14.9 shows how the volumes of two different samples of gas change with the temperature (pressure remains constant). When the plots of volume versus temperature for different gases are

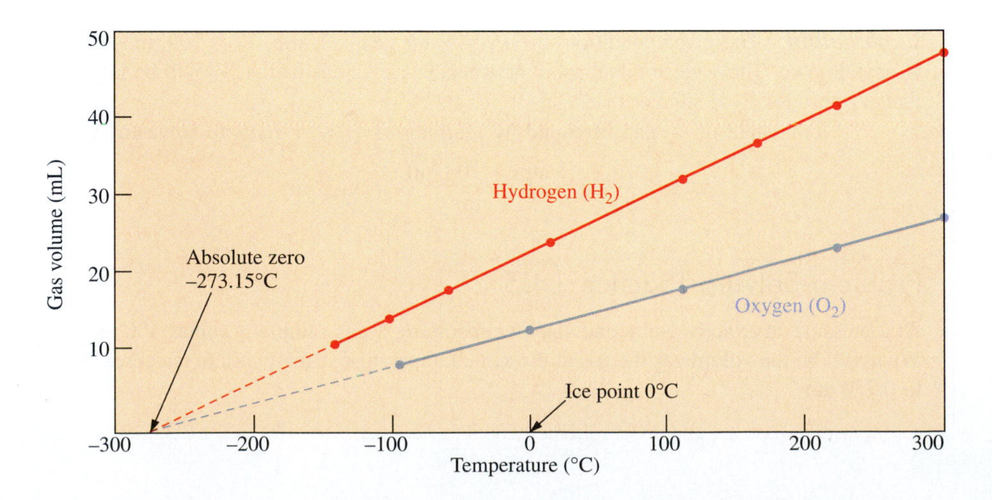

Figure 14.9 Charles's law. The volumes of two different samples of gases decrease with decreasing temperature (at constant pressure). These graphs (as would those of all gases) intersect the temperature axis at about −273 °C.

extended toward lower temperatures, they all reach zero volume at a common temperature, −273.15 °C. The volume of any gas would appear to be zero at −273.15 °C, but this does not actually happen—all gases liquefy before reaching this temperature.

In 1848, William Thompson, also known as Lord Kelvin, proposed that it would be convenient to have a temperature scale in which the zero point was − 273.15 °C. This temperature scale has been named for Lord Kelvin and the units of the scale are known as kelvins. The kelvin is the same size as the Celsius degree and has been adopted as the standard unit for temperature measurement (← *p. 273*). When the Kelvin temperature scale is used, the volume-temperature relationship, now known as Charles's law, becomes *the volume of a fixed amount of a gas at a constant pressure is directly proportional to the absolute temperature.*

The Kelvin temperature scale is also known as the absolute temperature scale or the thermodynamic temperature scale.

Writing Charles's law for any gas at constant amount and pressure as V = a constant × T, where T is the temperature in kelvins, you can see how it is related to the ideal gas equation.

Charles's law:

$$V = \text{a constant} \times T \qquad \text{(at constant amount and pressure)}$$

$$V = \frac{nR}{P}T$$

If the volume, V_1, and temperature, T_1, of a sample of gas are known, then the volume, V_2, at some other temperature, T_2, at the same pressure is given by

$$\frac{V_1}{T_1} = \frac{V_2}{T_2} \qquad \text{(at constant amount and pressure)}$$

When using the gas-law relationships, temperatures *must* be expressed on the absolute scale, using kelvins. For example, suppose you want to use Charles's law to calculate the new volume when 450.0 mL of a gas is cooled from 60.0 °C to 20.0 °C, at constant pressure. First, convert the temperatures to kelvins by adding 273.15 to the Celsius values: 60.0 °C becomes 333.2 K and 20.0 °C becomes 293.2 K. Using Charles's law for the two sets of conditions at constant pressure,

$$V_2 = \frac{V_1T_2}{T_1} = \frac{450.0 \text{ mL} \times 293.2 \text{ K}}{333.2 \text{ K}} = 396.0 \text{ mL}$$

PROBLEM-SOLVING EXAMPLE 14.6 Gas Thermometers

Because of the proportionality between temperature and volume, a gas sample at constant pressure can serve as a thermometer. A certain gas sample occupies 100. mL at 25 °C. If the pressure is held constant and the temperature is changed, the sample occupies 175 mL. What is the final temperature in kelvins and Celsius degrees?

Answer 521 K or 248 °C

Explanation Since the volume of the gas increases, you should expect the final temperature to be higher. Begin by converting the starting temperature to kelvins.

$$T_1 = (25 + 273) \text{ K} = 298 \text{ K}$$

The other variables are V_1 = 100. mL and V_2 = 175 mL.
Solving the Charles's law relationship for the final temperature, T_2, gives

$$T_2 = \frac{T_1V_2}{V_1} = \frac{(298 \text{ K})(175 \text{ mL})}{100. \text{ mL}} = 521 \text{ K or } 248 \text{ °C}$$

Problem-Solving Practice 14.6

If a gas sample occupies 236 mL at 31 °C, what volume will it occupy at 89 °C? The pressure remains constant.

Exercise 14.8 Visualizing Charles's Law

Consider a collection of gas molecules at some temperature, T_1. Now increase the temperature to T_2. Use the ideas of the kinetic-molecular theory and explain why the volume would have to be larger if the pressure remains constant. What would have to be done to maintain the volume constant if the temperature increased?

Experimental Amount-Volume Behavior of Gases (Avogadro's Law)

It is easy to imagine that a volume of a gas contains some number of gas molecules. It is not so obvious that equal volumes of several different gases would all contain equal numbers of molecules. The development of this idea goes back to the early 1800s.

In 1809 the French scientist Joseph Gay-Lussac (1778–1850) conducted some experiments in which he measured the volumes of gases reacting with one another to form gaseous products. He found that, at constant temperature and pressure, the volumes of reacting gases were always in the *ratios of small whole numbers.* This is known as the **law of combining volumes.** For example, producing 2 L of water vapor requires the reaction of 2 L of H_2 with 1 L of O_2. Similarly, producing 4 L of water vapor requires the reaction of 4 L of H_2 with 2 L of O_2.

	$2\ H_2(g)$	+	$O_2(g)$	$\longrightarrow$	$2\ H_2O(g)$
Experiment 1:	2 L		1 L		2 L
Experiment 2:	4 L		2 L		4 L

In 1811 Amedeo Avogadro suggested that Gay-Lussac's observations actually showed that equal volumes of all gases under the same temperature and pressure conditions contain the same number of molecules. Viewed on a molecular scale, if a tiny volume contained only one molecule of a gas, then twice that volume would contain two molecules, and so on (Figure 14.10).

John Dalton (who devised the atomic theory) strongly opposed Avogadro's ideas, and never did accept them. It took about 50 years—long after Avogadro and Dalton had died—for Avogadro's explanation of Gay-Lussac's experiments to be generally accepted.

What is now known as Avogadro's law states that *the volume of a sample of a gas, at a given temperature and pressure, is directly proportional to the amount of the gas.*

$$V = \text{a constant} \times n$$

Compare this statement of Avogadro's law with the ideal gas law, rearranged to put those variables that do not change into the constant expression.

Avogadro's law:

$$V = \text{a constant} \times n \qquad \text{(at constant temperature and pressure)}$$

$$V = \frac{RT}{P} \times n$$

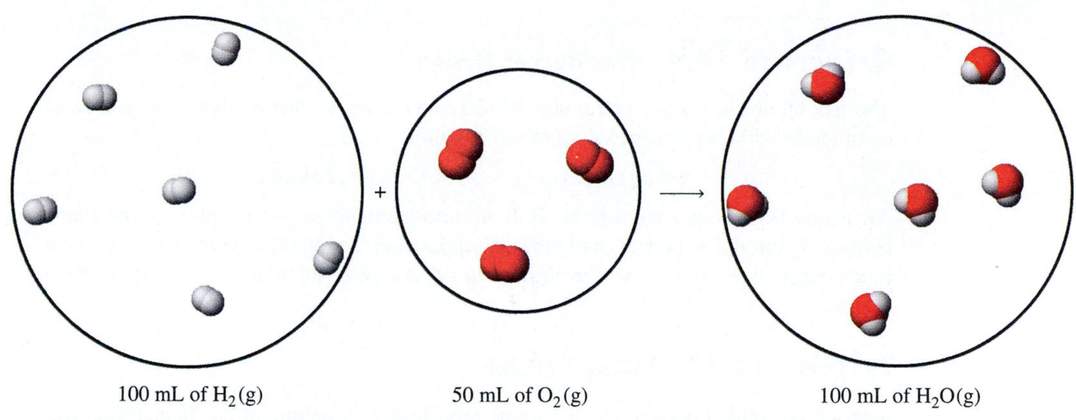

Figure 14.10 Gay-Lussac's law. When gases at the same temperature and pressure combine with one another their volumes are in the ratio of small whole numbers. Here, one volume of O_2 gas, say 50 mL at 100 °C and 1 atm, combines with two volumes of H_2 gas (100 mL) to give two volumes (100 mL) of H_2O vapor also at 100 °C and 1 atm.

Avogadro's law means, for example, that at constant temperature and pressure, if the number of moles of gas doubles, the volume doubles. It also means that at the same temperature and pressure, the volumes of two different amounts of gases are related as follows:

$$\frac{V_1}{V_2} = \frac{n_1}{n_2} \quad \text{(at constant temperature and pressure)}$$

Knowing any three variables in this expression allows calculation of the fourth.

Problem-Solving Example 14.7 Using Avogadro's Law

Carbon monoxide burns in oxygen to form carbon dioxide.

$$2\ CO(g) + O_2(g) \longrightarrow 2\ CO_2(g)$$

If the volume occupied by the CO is 400. mL at 40 °C and 1 atm, what volume of O_2, at the same temperature and pressure, will be required in the reaction?

Answer 200. mL of O_2

Explanation We want to find the volume of O_2, and the given information is the volume of the CO at the same temperature and pressure. Therefore the problem can be solved by using the reaction coefficients of 1 for O_2 and 2 for CO as the volume ratio.

$$\text{Volume of } O_2 = 400.\ \text{mL CO} \times \left(\frac{1\ \text{mL } O_2}{2\ \text{mL CO}}\right) = 200.\ \text{mL } O_2$$

Problem-Solving Practice 14.7

Nitrogen monoxide (NO) combines with oxygen to form nitrogen dioxide.

$$2\ NO(g) + O_2(g) \longrightarrow 2\ NO_2(g)$$

If 1.0 L of oxygen gas at 30.25 °C and 0.975 atm is used, what volume of NO gas at the same temperature and pressure will be converted to NO_2?

Exercise 14.9 Gas Burner Design

The gas burner in a stove or furnace admits enough air so that methane gas can react completely with oxygen according to the equation

$$CH_4(g) + 2\ O_2(g) \longrightarrow CO_2(g) + 2\ H_2O(g)$$

Air is one-fifth oxygen by volume. Both air and methane gas are supplied to the flame by passing through separate small tubes. Compared to the tube for the methane gas, how much bigger does the tube for the air need to be? Assume that both gases are at the same T and P.

Exercise 14.10 Filling Balloons

A group of 100 balloons of equal volume are filled with helium at 23 °C and 748 mm Hg, respectively. The total volume of these balloons is 168 L. Next, you are given 150 more balloons of the same size and 41.8 g of He gas. The temperature and pressure remain the same. Determine by calculation whether you will be able to fill all of the balloons with the He you have available.

It is important to remember that Boyle's, Charles's, and Avogadro's laws do not depend on the identity of the gas being studied. These laws reflect properties of all gases and therefore must depend on the behavior of *any* gaseous atoms or molecules, regardless of their identities.

For problems involving gases, you have your choice of two useful equations.

- When three of the variables P, V, n, and T are given and the value of the fourth is needed, use the ideal gas law, $PV = nRT$.
- When one set of conditions is given for a single gas sample and one of the variables under a new set of conditions is needed, use the combined gas law and cancel any of the three variables that do not change.

$$\frac{P_1V_1}{T_1} = \frac{P_2V_2}{T_2}$$

combined gas law (constant n)

Exercise 14.11 Avogadro's Law and the Kinetic-Molecular Theory

Avogadro's law implies that equal numbers of two different kinds of molecules in separate containers at the same temperature will exert the same pressure if their volumes are the same. Using O_2 molecules and H_2 molecules in your answer, explain why this is so, based on the kinetic-molecular theory.

Exercise 14.12 Predicting Gas Behavior

Name three ways the volume occupied by a sample of gas can be decreased.

14.5 QUANTITIES OF GASES IN CHEMICAL REACTIONS

The law of combining volumes and the ideal gas law make it possible to use volumes as well as masses or molar amounts in calculations based on reaction stoichiometry

(← p. 180). Consider the following balanced equation for the reaction involving solid carbon, gaseous oxygen, and gaseous carbon dioxide product.

$$C(s) + O_2(g) \longrightarrow CO_2(g)$$

Let's look at some of the questions that might be asked about this reaction, in which there is one gaseous reactant and one gaseous product. As for any problem to be solved, you have to recognize what information is known and what is needed. Then, you may have to decide what relationship provides the connection between the two.

1. *How many liters of CO_2 are produced from 0.5 L of O_2?* The known and needed information here are both volumes of gases, so the law of combining volumes leads to the answer. *Ans:* 0.5 L of CO_2, because the reaction shows one mole of CO_2 formed for every mole of C that reacts; a 1 : 1 ratio.
2. *How many moles of O_2 are required to completely react with 4.0037 g of C?* Here the known information is the mass of the solid reactant and the needed information is the moles of gaseous product. The answer is provided from the stoichiometric relationships between reactants and products from the balanced reaction, without using any of the gas laws. *Ans:* 0.333 mol of O_2, because 4.0037 g of C is 0.333 mol of C, and 1 mol of O_2 is required for every 1 mol of C; a 1 : 1 ratio.
3. *What volume (in liters) of O_2 at STP is required to react with 12.011 g of C?* Because the needed information is the volume of a gaseous reactant, the laws of combining volumes can lead to the answer, but first the number of moles of carbon must be found. *Ans:* 22.4 L of O_2, because 12.011 g of C is 1 mol of C which requires 1 mol of O_2, and the standard molar volume of any gas at STP is 22.4 L.
4. *What volume (in liters) of O_2 at 747 mm Hg and 21 °C is required to react with 12.011 g of C?* Realizing that 12.011 g of C is 1 mol of C and that 1 mol of O_2 will be required, you must then use the ideal gas law to calculate the volume occupied by 1 mol of O_2 at the temperature and pressure given. *Ans:* Convert the temperature to kelvins to get 294 K, then convert the pressure given to atmospheres to get 0.983 atm. Finally, use the ideal gas law to calculate the volume of oxygen required.

$$V_{O_2} = \frac{nRT}{P} = \frac{(1\ \text{mol})(0.0821\ \text{L atm/mol K})(294\ \text{K})}{0.983\ \text{atm}} = 24.5\ \text{L}$$

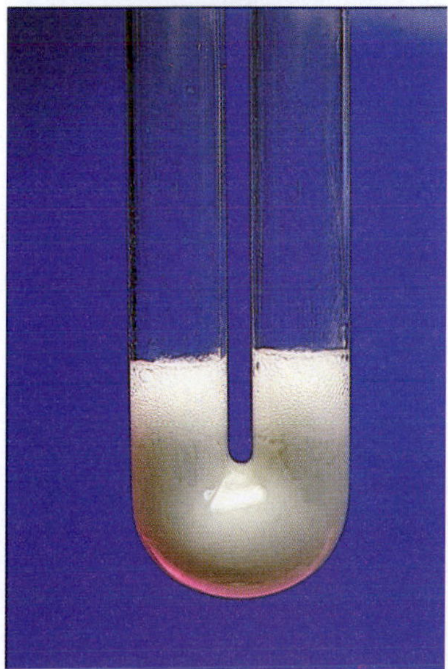

Figure 14.11 A drain cleaner containing sodium hydroxide and pieces of aluminum. When water is added the hydroxide ion attacks the aluminum, and hydrogen gas is produced. *(C.D. Winters)*

PROBLEM-SOLVING EXAMPLE 14.8 Gases and Stoichiometry

Consider the reaction that takes place in some commercial drain cleaners that contain sodium hydroxide and small pieces of aluminum (Figure 14.11). When the mixture is poured into a clogged drain, the reaction that occurs is

$$2\ Al(s) + 2\ NaOH(aq) + 6\ H_2O(\ell) \longrightarrow 2\ NaAl(OH)_4(aq) + 3\ H_2(g)$$

The heat generated by the reaction helps the sodium hydroxide break up the grease, and the hydrogen gas being generated stirs up the mixture and speeds up the unclogging of the drain. If 6.5 g of Al and an excess of NaOH are used, what volume of gaseous H_2 measured at 742 mm Hg and 22.0 °C is produced?

Answer 8.9 L

Explanation The first step in solving this problem is to calculate the number of moles of Al available and then find the number of moles of H_2 generated, using the coefficients from the balanced equation. Once the number of moles of H_2 is known, the volume is obtained from the ideal gas law, $PV = nRT$.

Amount of Al available:

$$6.5 \text{ g Al} \times \left(\frac{1.0 \text{ mol Al}}{27.0 \text{ g Al}}\right) = 0.241 \text{ mol Al}$$

The amount of H_2 produced by the aluminum is next calculated from the mole ratio in the balanced equation.

Amount of H_2 expected:

$$0.241 \text{ mol Al} \times \left(\frac{3 \text{ mol H}_2}{2 \text{ mol Al}}\right) = 0.36 \text{ mol H}_2$$

Next, solve $PV = nRT$ for V and substitute values for T, P, and n. When substituting into the rearranged gas equation, you must make sure the units of P, V, and T are compatible with the units of R. This means P must be converted to atmospheres.

$$P = (472 \text{ mm}) \times \left(\frac{1 \text{ atm}}{760 \text{ mm}}\right) = 0.976 \text{ atm}$$

$$V = \frac{nRT}{P} = \frac{(0.36 \text{ mol})(0.0821 \text{ L atm/mol K})(295.2 \text{ K})}{0.976 \text{ atm}} = 8.9 \text{ L}$$

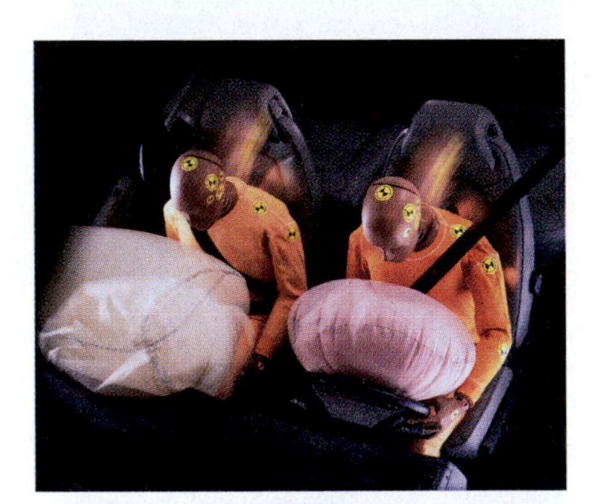

Air bags deploying on crash-test dummies. The force of these bags deploying has caused the deaths of small children when they were placed in the front seat. Small children should always be placed in the rear seats of automobiles equipped with passenger-side air bags. *(Saab Cars USA, Inc.)*

Problem-Solving Practice 14.8

Sodium azide (NaN_3) is used to generate the nitrogen gas that inflates the air bags in automobiles during a collision. The reaction is

$$2 \text{ NaN}_3(s) \longrightarrow 2 \text{ Na}(s) + 3 \text{ N}_2(g)$$

If the volume of an air bag is 45.6 L at a pressure of 835 mm Hg and a temperature of 22.0 °C, what quantity of NaN_3 would be required to generate this much nitrogen?

Problem-Solving Example 14.9 Gas Law Stoichiometry and Temperature

Octane, C_8H_{18}, is one of the hydrocarbons in gasoline. In an automobile engine, octane burns to produce CO_2 and H_2O. (a) What volume of oxygen, entering the engine at 0.950 atm and 20. °C, is required to burn 1.00 g of octane? (b) Assuming the pressure is still 0.950 atm, what volume would be required on a cold winter day, when the temperature is −20. °C? The molar mass of octane is 114.2 g/mol.

Answer (a) 2.77 L (b) 2.39 L

Explanation (a) First write the balanced equation for the combustion reaction.

$$2 \text{ C}_8\text{H}_{18}(\ell) + 25 \text{ O}_2(g) \longrightarrow 16 \text{ CO}_2(g) + 18 \text{ H}_2\text{O}(\ell)$$

Next, calculate the number of moles of O_2 required, using the stoichiometric factor of 25 mol O_2 to 2 mol C_8H_{18}.

Amount of O_2 required:

$$1.00 \text{ g C}_8\text{H}_{18} \times \left(\frac{1 \text{ mol C}_8\text{H}_{18}}{114.2 \text{ g C}_8\text{H}_{18}}\right) \times \left(\frac{25 \text{ mol O}_2}{2 \text{ mol C}_8\text{H}_{18}}\right) = 0.1095 \text{ mol O}_2$$

Then, using the ideal gas law, calculate the volume of oxygen at 20. °C and 0.950 atm.

$$V_{O_2} = \frac{nRT}{P} = \frac{(0.1095 \text{ mol O}_2)(0.8201 \text{ L atm/mol K})(293 \text{ K})}{0.950 \text{ atm}} = 2.77 \text{ L}$$

(b) At −20. °C or 253 K, the same type of calculation gives

$$V_{O_2} = 2.39 \text{ L}$$

Because the volume of oxygen (and hence of air) required is significantly less in cold weather, many older gasoline engines have a choke—a valve that limits the air supply until the engine warms up.

Problem-Solving Practice 14.9

If you carried out the Chemistry You Can Do stoichiometry experiment in Section 5.2 (← *p. 190*), you observed the reaction of vinegar (acetic acid solution) with baking soda (sodium hydrogen carbonate) to generate carbon dioxide gas and inflate a balloon. The net ionic equation for an acid (← *p. 148*) reacting with hydrogen carbonate ion is

$$H^+(aq) + HCO_3^-(aq) \longrightarrow CO_2(g) + H_2O(\ell)$$

If there is plenty of vinegar, what mass of $NaHCO_3$ is required to inflate a balloon to a diameter of 20. cm at room temperature 20. °C? Assume the balloon is a sphere ($V = \frac{4}{3}\pi r^3$), and because the rubber is stretched, the pressure inside the balloon is twice normal atmospheric pressure.

14.6 GAS DENSITY AND MOLAR MASSES

Density is mass per unit volume (← *p. 58*). Because the volume of a gas sample (but not its mass) varies with temperature and pressure, gas densities are extremely variable. However, once T and P are specified, the density of a gas can be calculated from the ideal gas law. Also, because equal volumes of gas at the same T and P contain equal numbers of molecules, the densities of different gases are proportional to their molar masses. As a result, experimental gas densities can be used to determine molar masses.

To illustrate, we can use the ideal gas equation to calculate the density for helium at 25 °C and 0.750 atm. Substituting mass over molar mass (m/M) for the number of moles (n) in the equation gives

$$PV = \frac{m}{M}RT$$

On rearranging, we get an expression for density (mass per unit volume).

$$\text{Density} = \frac{m}{V} = \frac{PM}{RT}$$

$$\text{Density} = \frac{(0.750\text{ atm})(4.003\text{ g/mol})}{(0.0821\text{ L atm/mol K})(298\text{ K})} = 0.123\text{ g/L}$$

As can be seen in the equation above, the density of a gas is proportional to its molar mass. Consider the densities of the three gases: He, O_2, and SF_6 (used as an insulator in high-voltage transmission lines). Taking 1 mol of each of these gases at 25 °C and 0.750 atm, the densities are

Gas	Molar Mass (g/mol)	Density (25 °C)
He	4.003	0.123 g/L
O_2	31.999	0.981 g/L
SF_6	146.06	4.48 g/L

Exercise 14.13 Calculating Gas Densities

Using the method shown in the text, calculate the densities of Cl_2 and SO_2 at 25 °C and 0.750 atm. Then calculate the density of Cl_2 at 35 °C and 0.750 atm, and the density of SO_2 at 25 °C and 2.60 atm.

Exercise 14.14 Comparing Densities

Express the gas density of He given above in grams per milliliter (g/mL) and compare that value with the density of metallic lithium (0.53g/mL). The mass of a Li atom is only a little more than that of a He atom. What do these densities tell you about how closely Li atoms are packed together compared to He atoms when each element is in its standard state? To which of the concepts of the kinetic-molecular theory does this comparison apply?

Exercise 14.15 Densities of Gas Mixtures

Assume a mixture of equal volumes of the two gases nitrogen and oxygen. Would the density of this gas mixture be higher, lower, or the same as the density of air at the same temperature and pressure?

Problem-Solving Example 14.10 Using the Ideal Gas Law to Calculate Molar Mass

A 0.100-g sample of a gaseous compound with the formula CH_2F_2 occupies 0.0470 L at 298 K and 755 mm Hg. Based on this information, calculate the molar mass of the compound. Compare the calculated molar mass with the molar mass determined from the formula.

Answer Molar mass from gas laws = 52.4 g/mol. Molar mass from atomic weights = 52.02 g/mol.

Explanation This problem is typical of the laboratory measurement of the molar mass of a gas. Begin by organizing the data.

$$V = 0.0470 \text{ L} \qquad P = 755 \text{ mm Hg} \qquad T = 298 \text{ K} \qquad n = ?$$

Next, convert the pressure to atmospheres,

$$P = 755 \text{ mm Hg} \times \left(\frac{1 \text{ atm}}{760 \text{ mm Hg}}\right) = 0.993 \text{ atm}$$

Then, use the ideal gas law equation to find n, the number of moles of gas,

$$n = \frac{PV}{RT}$$

Placing the known values into the right-hand side of the equation gives

$$n = \frac{(0.993 \text{ atm})(0.0470 \text{ L})}{(0.0821 \text{ L atm/mol K})(298 \text{ K})} = 0.00191 \text{ mol}$$

The mass of the sample is already known to be 0.100 g, and so

$$\text{Molar mass} = \frac{m}{n} = \frac{0.100 \text{ g}}{0.00191 \text{ mol}} = 52.4 \text{ g/mol}$$

Finally, adding the atomic weights for the formula CH_2F_2 gives a molar mass of 52.02 g/mol, which is in close agreement with that determined using the gas laws.

Problem-Solving Practice 14.10

A glass flask contains 1.00 L of a gas at 0.850 atm and 20. °C. The density of the gas is 1.13 g/L. What is the molar mass of the gas? What is its identity?

14.7 PARTIAL PRESSURES OF GASES

Our atmosphere is a mixture of nitrogen, oxygen, argon, carbon dioxide, water vapor, and small amounts of several other gases (← *p. 614, Table 14.1*). What we call atmospheric pressure is the sum of the pressures exerted by all of these individual gases. The same is true of every gas mixture. Consider the mixture of nitrogen and oxygen illustrated in Figure 14.12. The pressure exerted by the mixture is equal to the sum of the pressures that the nitrogen alone and the oxygen alone would exert in the same volume at the same temperature and pressure. The pressure of one gas in a mixture of gases is called the **partial pressure** of that gas.

John Dalton was the first to observe that *the total pressure exerted by a mixture of gases is the sum of the partial pressures of the individual gases in the mixture.* This statement is known as **Dalton's law of partial pressures.** Dalton's law is a consequence of the fact that gas molecules behave independently of one another. As a result, for the first three components of our atmosphere, the total number of moles is

$$n_{\text{total}} = n_{N_2} + n_{O_2} + n_{Ar}$$

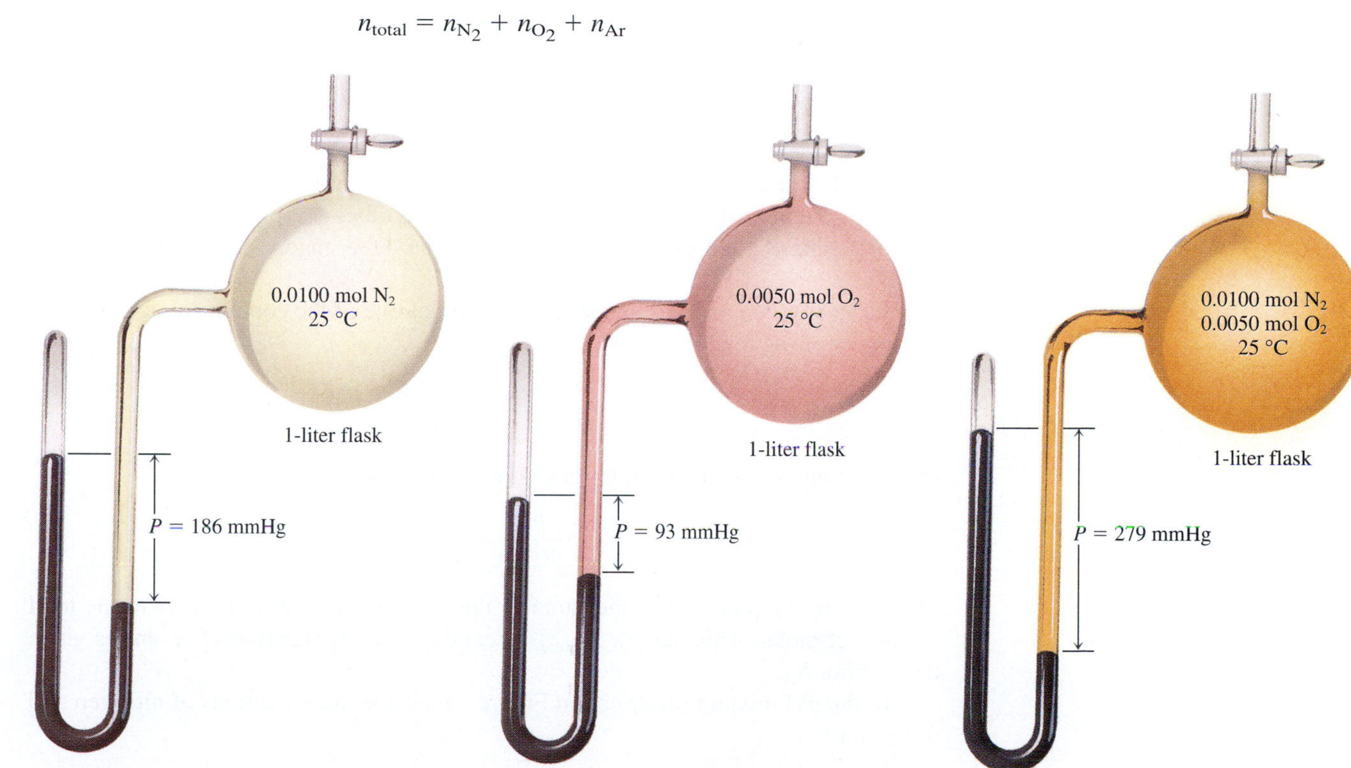

Figure 14.12 Dalton's law. (*Left*) 0.0100 mol of N_2 in a 1.00-L flask of 25 °C exerts a pressure of 186 mm Hg. (*Middle*) 0.0050 mol of O_2 in a 1.00-L flask at 25 °C exerts a pressure of 93 mm Hg. (*Right*) The N_2 and O_2 samples are mixed in the same 1.00-L flask at 25 °C. The total pressure, 279 mm Hg, is the sum of the pressures of the individual gases.

TABLE 14.2 The Mole Percent and Partial Pressures of the Major Components of Dry Air

Component	Mole Percent[1]	Partial Pressure (atm at STP)
Nitrogen, N_2	78.084	0.78084
Oxygen, O_2	20.946	0.20946
Argon, Ar	0.934	0.00934
Carbon dioxide, CO_2	0.033	0.00033
Neon, Ne	0.002	0.00002
Helium, He	0.001	0.00001

[1]Mole percent is the number of moles of a component divided by the total number of moles of gas and multiplied by 100 percent.

If we replace n in the ideal gas law with n_{total}, the summation of the individual numbers of moles of gases, the equation becomes

$$P_{\text{total}}V = n_{\text{total}}RT$$

$$P_{\text{total}} = \frac{n_{\text{total}}RT}{V} = \frac{(n_{N_2} + n_{O_2} + n_{\text{Ar}})RT}{V}$$

Expanding the right side of this equation and rearranging,

$$P_{\text{total}} = \frac{n_{N_2}RT}{V} + \frac{n_{O_2}RT}{V} + \frac{n_{\text{Ar}}RT}{V} = P_{N_2} + P_{O_2} + P_{\text{Ar}}$$

Dalton's law means that the pressure exerted by the atmosphere is the sum of the pressures due to nitrogen, oxygen, argon, and the other components. The quantities P_{N_2}, P_{O_2}, and P_{Ar} are the partial pressures of the three major components of the atmosphere.

We can write a ratio of the partial pressure of one of the components, A, of a gas mixture over the total pressure,

$$\frac{P_{\text{A}}}{P_{\text{total}}} = \frac{n_{\text{A}}(RT/V)}{n_{\text{total}}(RT/V)}$$

On canceling terms on the right-hand side of this equation, we get

$$\frac{P_{\text{A}}}{P_{\text{total}}} = \frac{n_{\text{A}}}{n_{\text{total}}}$$

The ratio of the pressures is the same as the ratio of the moles of gas A to the total number of moles. This ratio ($n_{\text{A}}/n_{\text{total}}$) is called the **mole fraction** of A, and is given the symbol X_{A}.

In the gas mixture illustrated in Figure 14.12, the mole fractions of nitrogen and oxygen are

$$X_{N_2} = \frac{0.010 \text{ mol } N_2}{0.010 \text{ mol } N_2 + 0.0050 \text{ mol } O_2} = 0.67$$

$$X_{O_2} = \frac{0.0050 \text{ mol } O_2}{0.010 \text{ mol } N_2 + 0.0050 \text{ mol } O_2} = 0.33$$

Because these two gases are the only components of the mixture, the sum of the two mole fractions must equal 1:

$$X_{N_2} + X_{O_2} = 0.67 + 0.33 = 1.00$$

An interesting application of partial pressures is the composition of the breathing atmosphere in deep-sea diving vessels. If normal air at 1 atm pressure, with a mole fraction of oxygen of 0.21, is compressed to 2 atm, the partial pressure of oxygen becomes about 0.4 atm. Such high oxygen partial pressures are toxic, so a diluting gas must be added to lower the oxygen partial pressure back to near normal values. Nitrogen gas might seem the logical choice because it is the diluting gas in the atmosphere. The problem is that nitrogen is fairly soluble in the blood and at high concentrations causes *nitrogen narcosis,* a condition similar to alcohol intoxication.

Helium is much less soluble in the blood and is therefore a good substitute for nitrogen in a deep-sea diving atmosphere. There are a couple of interesting side effects to using helium, however. Because He atoms, on average, move faster than the heavier nitrogen atoms at the same temperature (***← p. 618, Figure 14.7***), they strike a diver's skin more often than would the nitrogen atoms and are therefore more efficient at carrying away heat energy. This causes the divers to complain of feeling chilled while breathing the helium/oxygen mixture.

Another side effect of breathing helium is that it produces high-pitched voices. This happens because the vocal cords vibrate faster in an atmosphere less dense than air, and the pitch of the voice is raised.

An undersea explorer. Some deep-sea diving vessels like this one use an atmosphere of oxygen and helium for their occupants. *(Photo Researchers, Inc.)*

PROBLEM-SOLVING EXAMPLE 14.11 Calculating Partial Pressures

Halothane ($C_2HBrClF_3$) is a commonly used surgical anesthetic delivered by inhalation (Figure 14.13). What is the partial pressure of each gas if 15.0 g of halothane gas is mixed with 22.6 g of oxygen gas and the total pressure is 862 mm Hg? The molar mass of halothane is 197.4 g.

Answer $P_{\text{halothane}} = 83.8$ mm Hg;

$P_{O_2} = 778$ mm Hg

Explanation First, calculate the moles of each gas, then calculate the mole fractions.

$$\textbf{Moles } \mathbf{C_2HBrClF_3}\textbf{:}\quad 15.0 \text{ g} \times \left(\frac{1 \text{ mol } C_2HBrClF_3}{197.4 \text{ g}}\right) = 0.0760 \text{ mol } C_2HBrClF_3$$

$$\textbf{Moles } \mathbf{O_2}\textbf{:}\quad 22.6 \text{ g} \times \left(\frac{1 \text{ mol } O_2}{32.00 \text{ g}}\right) = 0.706 \text{ mol } O_2$$

$$\textbf{Mole fraction of } \mathbf{C_2HBrClF_3} = \frac{0.0760 \text{ mol } C_2HBrClF_3}{0.782 \text{ total moles}} = 0.0972$$

Because the sum of the two mole fractions must equal 1.000, the mole fraction of O_2 is 0.903.

$$X_{\text{halothane}} + X_{O_2} = 1.000$$

$$0.0972 + X_{O_2} = 1.000$$

$$X_{O_2} = 0.903$$

Finally, calculate the partial pressure of each gas.

$$P_{\text{halothane}} = 0.0972 \times P_{\text{total}} = 0.0972 \times (862 \text{ mm Hg}) = 83.8 \text{ mm Hg}$$

$$P_{O_2} = 0.903 \times P_{\text{total}} = 0.903 \times (862 \text{ mm Hg}) = 778 \text{ mm Hg}$$

Figure 14.13 A gas-mixing manifold for anesthesia. Such equipment is used by an anesthesiologist to prepare a gas mixture to keep a patient unconscious during an operation. By proper mixing, the anesthetic gas can be added slowly to the breathing mixture. Near the end of the operation, the anesthetic gas can be replaced by air of normal composition, or pure oxygen. *(C.D. Winters)*

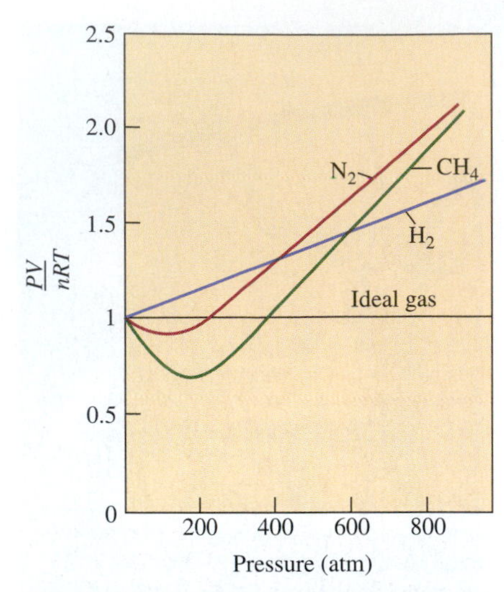

Figure 14.14 The nonideal behavior of real gases compared to that of an ideal gas. For all pressures the quotient PV/nRT for an ideal gas is 1.

Since $PV = nRT$, dividing both sides of the equation by nRT gives $\frac{PV}{nRT} = 1$.

Problem-Solving Practice 14.11

A mixture of 7.0 g of N_2 and 6.0 g of H_2 is confined in a 5.0-L reaction vessel at 500. °C. Assume that no reaction has occurred and calculate the total pressure. Then calculate the mole fraction and partial pressure of each gas.

Exercise 14.16 Pondering Partial Pressures

What happens to the partial pressure of each gas in a mixture when the volume is decreased by (a) lowering the temperature or (b) increasing the total pressure?

14.8 THE BEHAVIOR OF REAL GASES

For pressures of a few atm or less and temperatures well above the substance's boiling point, the ideal gas law predicts the pressures, volumes, temperatures, and amounts of air and many other gases quite accurately. At much higher pressures or much lower temperatures, however, the ideal gas law doesn't work nearly so well. As shown in Figure 14.14, for methane, CH_4, the quotient PV/nRT (which should equal 1) at first dips below 1 and then rises well above 1 with increasing pressure. The figure does not show it, but the dip below 1 becomes larger at lower temperatures. This failure of the ideal gas law happens for all gases at temperatures just above their boiling points and for very high pressures.

At standard temperature and pressure (STP), the volume occupied by a single molecule is very small relative to its share of the total gas volume. Recall that there are 6.02×10^{23} molecules in a mole and that 1 mol of a gas occupies about 22.4 L (22.4×10^{-3} m^3) at STP (← *p. 620*). The volume, V, that each molecule has to move around in is given by

$$V = \frac{22.4 \times 10^{-3}\ \text{m}^3}{6.02 \times 10^{23}\ \text{molecules}} = 3.72 \times 10^{-26}\ \text{m}^3/\text{molecule}$$

The volume of a sphere is given by $\frac{4}{3}\pi r^3$. Solving for r,

$$r = \sqrt[3]{\frac{3V}{4\pi}} = \sqrt[3]{\frac{3(3.72 \times 10^{-26}\ \text{m}^3)}{4(3.14)}}$$

$$= 2.11 \times 10^{-9}\ \text{m} = 2110\ \text{pm}$$

If this volume is assumed to be a sphere, then the radius, r, of the sphere is about 2000 pm. The radius of the smallest gas molecule, the helium atom, is 31 pm (← *p. 334*), so a helium atom has a space to move around in that is similar to the room a pea has inside a basketball. Now suppose the pressure is increased significantly, to 1000 atm. The volume available to each molecule is now a sphere only about 200 pm in radius, which means the situation is now like that of a pea inside a sphere a bit larger than a ping-pong ball. The volume occupied by the gas molecules themselves relative to the volume of the sphere is no longer negligible. This violates the first concept of the kinetic-molecular theory. The kinetic-molecular theory and the ideal gas law deal with the volume available for the molecules to move around in, not the volume of the molecules themselves, but the measured volume of the gas must include both. Therefore, at very high pressures, the measured volume will be larger than predicted by $PV = nRT$, and the value of PV/nRT is greater than 1.

Attractions between molecules (← *p. 423*) cause PV/nRT to drop below the ideal value of 1 at low temperatures and medium pressures. Consider a gas molecule that is about to hit the wall of the container, as shown in Figure 14.15. The kinetic-molecular theory assumes that the other molecules exert no forces on the molecule, but in fact such forces do exist (← *p. 423*). This means that when a molecule is about to hit the wall, most other molecules are farther from the wall and therefore pull the molecule *away* from the wall. This causes the molecule to hit the wall with less impact—the collision is softer than if there were no attraction among the molecules. Since all collisions with the walls are softer, the pressure is less than that predicted by the ideal gas law, and PV/nRT is less than 1. As pressure increases, the molecules are squeezed closer together and the attractions among the molecules get stronger, which makes this deviation from ideal behavior larger.

The lower the temperature, the greater is the deviation from ideal behavior. At lower temperatures the molecules are moving more slowly and, on average, their kinetic energies are smaller. The potential energy resulting from attractive forces between molecules becomes comparable to the average kinetic energy of the molecules at low temperatures; this causes a significant reduction of the observed pressure.

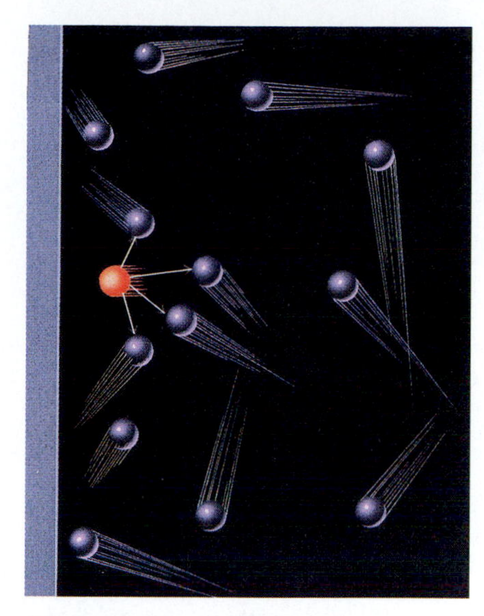

Figure 14.15 Non-ideal gas behavior. A gas molecule strikes the walls of a container with less force due to the attractive forces between it and its neighbors.

Exercise 14.17 Errors Caused by Deviations from Ideal Gas Behavior

In Problem-Solving Example 14.10 (*← p. 633*), the molar mass calculated using the ideal gas law was slightly larger than that calculated using atomic weights. This means that the number of moles, *n*, must have been too small. Look at the Explanation for Example 14.10 and explain why the number of moles is less than it should be.

14.9 CHEMICALS FROM THE ATMOSPHERE

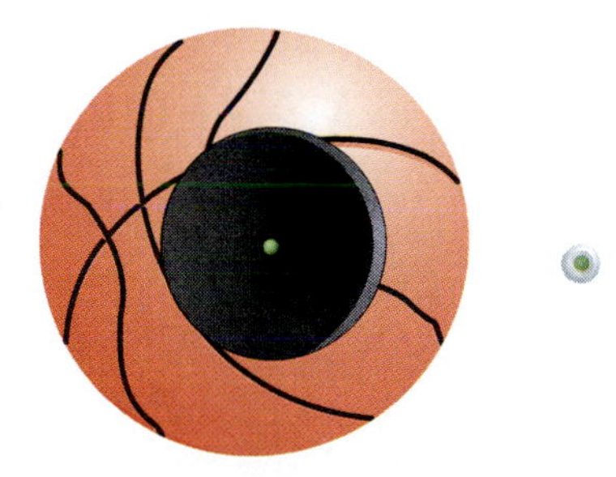

A cut-away view of a pea inside a basketball and inside a ping-pong ball. The volumes the pea has to move around in inside the two spheres correspond roughly to the relative volumes gas molecules can move without striking a neighbor at STP (the basketball) and when the pressure is increased to 1000 atm (the ping-pong ball).

It is possible to use the nonideal behavior of gases to obtain liquefied air by lowering the temperature and raising the pressure. Under these conditions, the attractive forces between molecules become sufficient for them to condense from vapor to liquid. The liquid components can then be separated from one another by distillation. Before pure oxygen and nitrogen can be obtained from the air, water vapor and carbon dioxide must be removed. This is usually done by precooling the air by refrigeration or by using silica gel (SiO_2 particles with a high surface area) to absorb water, and lime (CaO) to absorb carbon dioxide. Afterward, the air is compressed to a pressure exceeding 100 times normal atmospheric pressure, cooled to room temperature, and allowed to expand into a chamber. This expansion produces a cooling effect (the *Joule-Thompson effect*) because energy is required to overcome intermolecular forces as the molecules get farther apart. The expanding gas absorbs kinetic energy from the motion of its own molecules, which cools the gas. If this expansion is repeated and controlled properly, the expanding air cools to the point of liquefaction (Figure 14.16). The temperature of the liquid air is usually well below the normal boiling points of nitrogen (−195.8 °C), oxygen (−183 °C), and argon (−189 °C). The liquid air is then allowed to vaporize partially again and, since N_2 is more volatile than O_2 or Ar (N_2 has a lower boiling point), the liquid becomes more concentrated in O_2 and Ar. This process, known as the *Linde process,* produces high-purity nitrogen ($>$ 99.5%) and oxygen with a purity of 99.5%. Further processing produces pure Ar and Ne (b.p. −246 °C).

Oxygen

Most oxygen produced by the fractionation of liquid air is used in steel making, although some is used in rocket propulsion (to oxidize hydrogen) and in controlled oxidation reactions of other types. Liquid oxygen (LOX) can be shipped and stored at

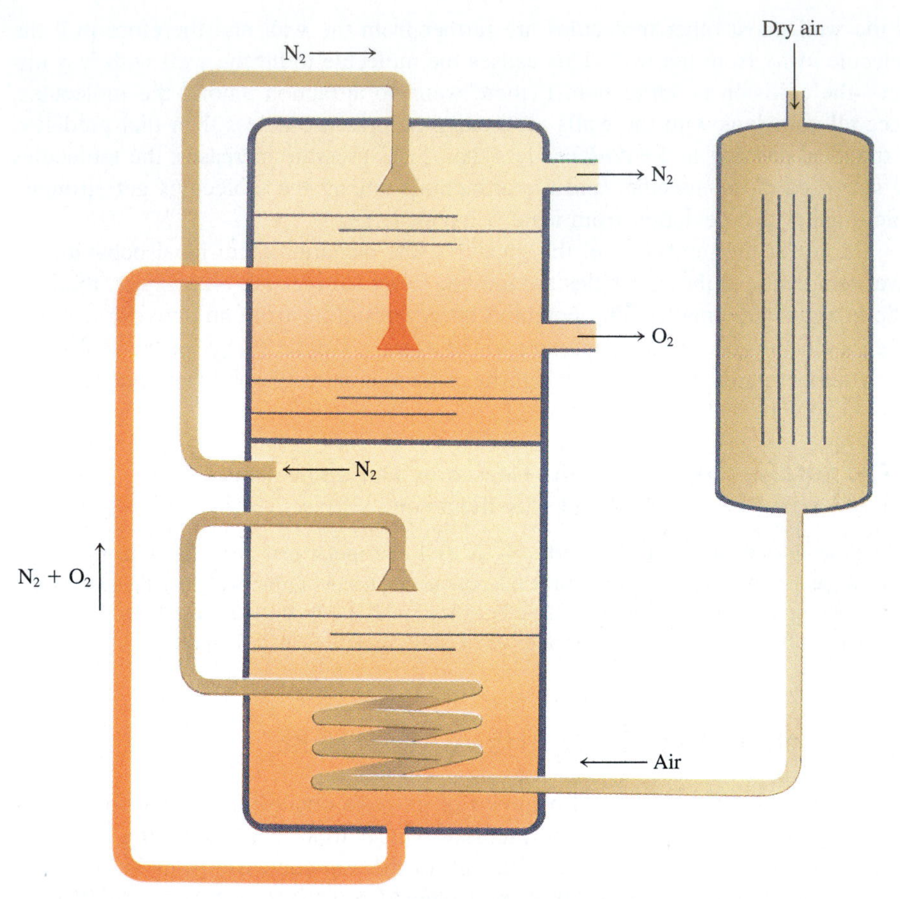

Figure 14.16 Diagram of a column for the fractional distillation of liquid air. Dry air is first chilled until it liquefies in the lower chamber. Some of the lower boiling nitrogen vapors leave the chamber and go to the higher chamber. The liquid remaining is richer in oxygen and is pumped to the upper chamber where the nitrogen that remains vaporizes. Repeating this process many times separates the nitrogen from the oxygen.

Tank of liquid oxygen. *(C.D. Winters)*

its boiling temperature of −183 °C under atmospheric pressure. Substances this cold are called **cryogens** (from the Greek *kryos,* meaning "icy cold"). Cryogens represent special hazards since contact produces instantaneous frostbite, and structural materials such as plastics, rubber gaskets, and some metals become brittle and fracture easily at these low temperatures. In spite of its low temperature, liquid oxygen can accelerate oxidation reactions to the point of explosion because of the high oxygen concentration. For this reason, contact between liquid oxygen and substances that will ignite and burn in air must be prevented.

Special cryogenic containers holding liquid oxygen incorporate huge vacuum-walled bottles much like those used to carry hot soup or hot coffee. These containers can be seen outside hospitals or industrial complexes, on highways and railroads, and even aboard ocean-going vessels.

Exercise 14.18 Vaporization of Liquefied Gases

A cryogenic flask contains 5.0 L of liquid oxygen which has a density of 1.41 g/mL. What volume at STP will this oxygen occupy if it is allowed to boil?

Nitrogen

Liquid nitrogen is also a cryogen. It has uses in medicine (cryosurgery), for example, in cooling an area of skin prior to removal of a wart or other unwanted or pathogenic tissue. Since nitrogen is so chemically unreactive, it is used as an inert atmosphere for applications such as welding, and liquid nitrogen is a convenient source of high volumes of the gas. Because of its low temperature and inertness, liquid nitrogen has found wide use in frozen food preparation and preservation during transit. Containers with nitrogen atmospheres, such as railroad boxcars or truck vans, present safety hazards since they contain little (if any) oxygen to support life, and workers have died when they entered such areas without a breathing apparatus to supply oxygen.

Nitrogen along with phosphorus and potassium are primary nutrients for plants. Although bathed in an atmosphere containing abundant nitrogen, most plants are unable to use the air directly as a supply of this vital element. Nitrogen fixation (← *p. 344*) is the process of changing atmospheric nitrogen into compounds that can be dissolved in water, absorbed through the plant roots, and assimilated by the plant (Figure 14.17). By far the main use of nitrogen at present is in the Haber-Bosch process (← *p. 596*) where it is combined with hydrogen to form ammonia. Most plants thrive on soils rich in nitrates, but many plants that grow in swamps, where there is a lack of oxidized materials, can use reduced forms of nitrogen such as the ammonium ion. The nitrate ion is the most highly oxidized form of combined nitrogen, and the ammonium ion is the most reduced form of nitrogen.

Liquid nitrogen. Liquid nitrogen is so cold that it immediately boils when it comes in contact with any object at room temperature. *(C.D. Winters)*

Some nitrogen species ranked in terms of decreasing oxidation number of nitrogen (← **p. 164**): NO_3^-, NO_2, NO_2^-, NO, N_2O, N_2, NH_4^+.

Nature fixes nitrogen on a massive scale in two ways. In the first method, nitrogen is oxidized under highly energetic conditions, such as in the discharge of lightning (or, to a lesser extent, in a fire). The initial reaction, which takes place in the atmosphere, is the reaction of nitrogen and oxygen to form nitrogen monoxide, NO, a colorless, reactive gas:

$$N_2(g) + O_2(g) \longrightarrow 2\ NO(g)$$

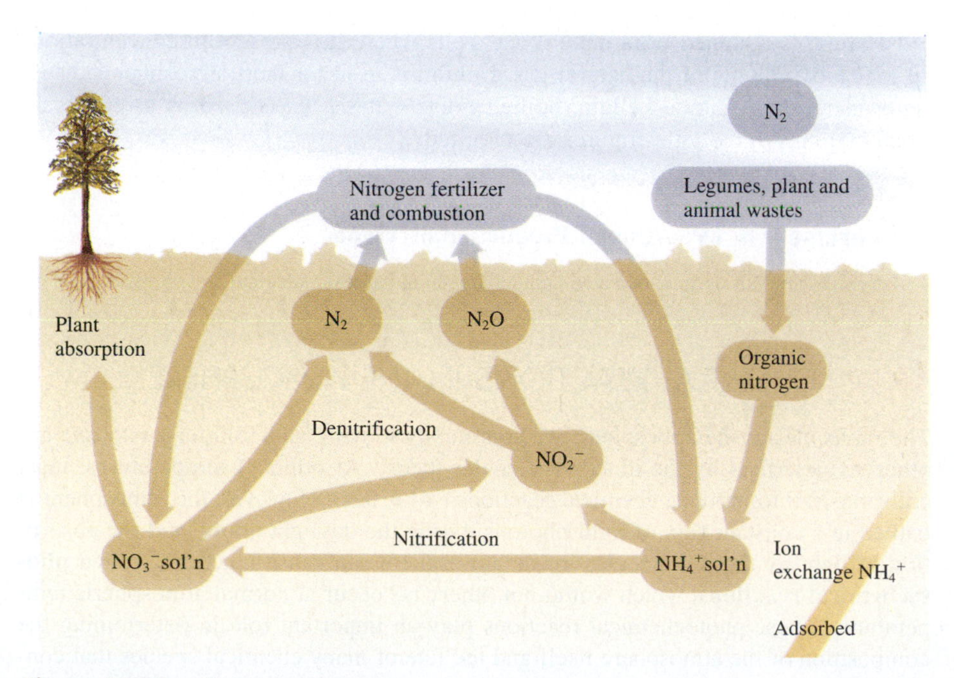

Figure 14.17 Nitrogen pathways.

Nitrogen fixation. Root nodules containing colonies of nitrogen-fixing *Rhizobium* bacteria. These fix nitrogen from the atmosphere and make it available to the plant. *(David M. Dennis/Tom Stack & Associates)*

The names of the oxides of nitrogen. The common names given in parentheses are often used in discussions of air pollutants.

N_2O	dinitrogen monoxide (nitrous oxide)
NO	nitrogen monoxide (nitric oxide)
NO_2	nitrogen dioxide
N_2O_4	dinitrogen tetraoxide
N_2O_5	dinitrogen pentaoxide

Once formed, nitrogen monoxide is easily oxidized in air to nitrogen dioxide, NO_2, which dissolves in water to form nitrous acid, HNO_2, and nitric acid, HNO_3:

$$2\,NO(g) + O_2(g) \longrightarrow 2\,NO_2(g)$$

$$H_2O(\ell) + 2\,NO_2(g) \longrightarrow \underset{\text{nitrous acid}}{HNO_2(aq)} + \underset{\text{nitric acid}}{HNO_3(aq)}$$

These acids are readily soluble in rain, clouds, or ground moisture and thus increase nitrate concentration in soil. They also contribute to the formation of acid rain (Section 17.10).

In the second method of nitrogen fixation, bacteria that live in the roots of plants called *legumes* convert atmospheric nitrogen into ammonia. This complex series of reactions depends on enzyme catalysis. Under ideal conditions, legume fixation can add more than 100 lb of nitrogen per acre of soil in one growing season.

A German, Hellriegel, showed in 1886 that leguminous plants such as alfalfa, beans, and soybeans "fix" nitrogen.

Noble Gases

Approximately 250,000 tons of argon, the most abundant noble gas, are isolated from the atmosphere each year in the United States. Most of the argon is used to provide inert atmospheres for high-temperature metallurgical processes. If air is not excluded from these processes, unwanted oxides of the hot metals will form. Some argon is also used as a filler gas in "extra life" incandescent light bulbs. Being an inert gas, argon doesn't react with the hot filament, but it increases the total gas pressure inside the light bulb. This increased gas pressure reduces the tendency for the filament to vaporize, which prolongs the life of the filament. If a gas like oxygen were used, the filament, made of tungsten metal, would quickly oxidize at high temperatures.

Relatively small amounts of neon, krypton, and xenon are recovered from air for commercial purposes. While neon is used in the greatest amounts, all three are used in "neon" signs. Xenon, which is readily soluble in blood, acts as an inhalation anesthetic in much the same way as does laughing gas, dinitrogen monoxide (N_2O).

Helium is obtained from natural gas wells where it is present in concentrations up to 7% by volume of the natural gas. Helium is used for inert atmospheres, especially in welding. Liquid helium, boiling point −268.9 °C, is used as a refrigerant when extremely cold temperatures are required.

Exercise 14.19 Annual Production of Argon

Using the amount of argon isolated annually given in the text, calculate its volume at STP.

14.10 CHEMICAL REACTIONS IN THE ATMOSPHERE

The molecules in the atmosphere are continually moving and colliding with one another, as described by the kinetic-molecular theory. At ordinary temperatures, most collisions fail to produce chemical reactions (← *p. 528*). However, the atmosphere is bathed in a constant flux of light photons during the daylight hours, and the absorption of light by these molecules in the atmosphere can cause reactions, called **photochemical reactions,** which would not otherwise occur at normal atmospheric temperatures. These photochemical reactions play an important role in determining the composition of the atmosphere itself and the fate of many chemical species that contribute to air pollution.

Nitrogen dioxide (NO_2) is one of the most photochemically active species found in the atmosphere. When an NO_2 molecule absorbs a photon of light, $h\nu$, the molecule is raised to a higher energy level; it becomes an **electronically excited molecule,** designated by an asterisk (*).

$$NO_2(g) \xrightarrow{h\nu} NO_2^*(g)$$

This excited molecule may quickly re-emit a photon of light, or it may break apart to form two other species: a molecule of nitrogen monoxide (NO) and an oxygen atom (O).

$$\cdot NO_2^*(g) \longrightarrow \cdot NO(g) + \cdot O \cdot (g)$$

The oxygen atom contains two unpaired electrons. It has an electron configuration of $[He]2s^22p_x^22p_y^12p_z^1$.

The NO_2 molecule, the NO molecule, and the oxygen atom are examples of highly reactive species, called free radicals, containing unpaired electrons (represented by a $\cdot$ next to their formulas (← ***p. 369***).

Some free radicals, such as the oxygen atom, are so reactive that they react with something almost immediately. Others, such as the NO_2 molecule, are not quite so reactive and are stable enough to exist for a long time. Most radicals are highly reactive and short-lived. A molecule of acetaldehyde, CH_3CHO, for example, will absorb a photon and split in two, forming two radicals. This is called a **photodissociation reaction.**

Recall that Lewis dot structures for some simple molecules like NO and NO_2 had unpaired electrons (← ***p. 369***).

$$CH_3{-}\overset{\overset{\displaystyle O}{\|}}{C}{-}H\,(g) \xrightarrow{h\nu} \cdot CH_3(g) + \cdot\overset{\overset{\displaystyle O}{\|}}{C}{-}H\,(g)$$

methyl radical

formyl radical

The methyl radical might react with another methyl radical to form ethane,

$$H_3C\cdot + \cdot CH_3 \longrightarrow CH_3CH_3$$

and the formyl radical might react with an oxygen molecule to form a hydroperoxyl radical, $\cdot OOH$, and a CO molecule.

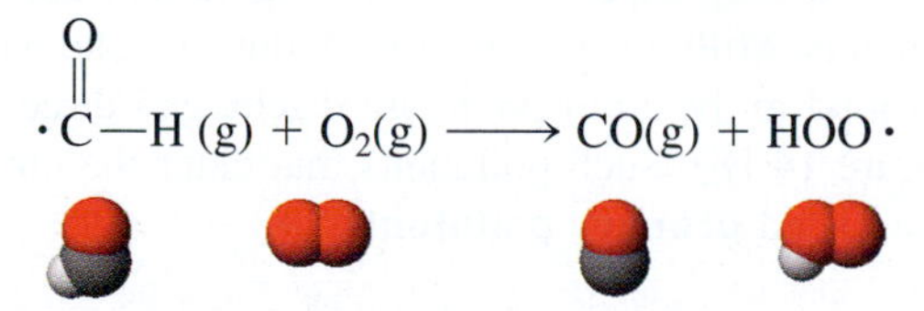

$$\cdot\overset{\overset{\displaystyle O}{\|}}{C}{-}H\,(g) + O_2(g) \longrightarrow CO(g) + HOO\cdot$$

Many other reactions are possible for each of these radicals.

A radical usually reacts in either of two ways:

1. It combines with another radical. Each of the radicals contributes an electron to the formation of a bond, as when two NO_2 molecules combine to form N_2O_4 (dinitrogen tetraoxide).

$$\cdot NO_2(g) + \cdot NO_2(g) \longrightarrow N_2O_4(g)$$

Recall from your study of polymerization reactions (← *p. 485*) that the formation of free radicals results in chain reactions that proceed until no new radicals are formed.

2. It reacts with a molecule to form one or more new radicals, or a new molecule, as illustrated by three important atmospheric reactions of the oxygen atom. The first is the formation of ozone, O_3, from O_2 and O (← *p. 33*).

$$\cdot O \cdot (g) + O_2(g) \longrightarrow O_3(g)$$

The second is the formation of two hydroxyl radicals (HO·) by the reaction with H_2O,

$$\cdot O \cdot (g) + H_2O(g) \longrightarrow 2\ \cdot OH(g)$$

and the third is the formation of a new molecule by reacting with SO_2 to form SO_3, a precursor to acid rain (Section 17.10).

$$\cdot O \cdot (g) + SO_2(g) \longrightarrow SO_3(g)$$

Exercise 14.20 Reactions Involving Free Radicals

Predict the reaction products for
(a) The photolysis of water, $H_2O \xrightarrow{h\nu}$
(b) A methane molecule reacting with a hydroxyl radical, $CH_4 + HO\cdot \rightarrow$
(c) A hydrogen atom reacting with oxygen, $\cdot H + O_2 \rightarrow$

14.11 AIR POLLUTION

An **air pollutant** is a substance that degrades air quality. Nature pollutes the air on a massive scale with volcanic ash, mercury vapor, hydrogen chloride, hydrogen fluoride, and hydrogen sulfide from volcanoes; and with reactive, odorous organic compounds from coniferous plants such as pine trees. Decaying vegetation, ruminant animals, and even termites add methane gas to the atmosphere, and decaying animal carcasses and other protein materials add dinitrogen monoxide (N_2O). But automobiles, electric power plants, smelting and other metallurgical processes, and petroleum refining also add significant quantities of unwanted chemicals to the atmosphere, especially in heavily populated areas. Atmospheric pollutants cause burning eyes, coughing, decreased lung capacity, harm to vegetation, and even the destruction of ancient monuments. Millions of tons of soot, dust, smoke particles, and chemicals not commonly found in the atmosphere are discharged directly into the atmosphere every year (Figure 14.18). Such pollutants that enter the environment directly from their sources are called **primary pollutants.**

Figure 14.18 Dark smoke rising from industrial stacks. In spite of air pollution regulations in the United States and elsewhere, various operating permits make such emissions possible. *(Gary Milburn/Tom Stack & Associates)*

Particle Pollutants

Pollutant particles range in size from fly ash particles, which are big enough to see, down to individual molecules, ions, or atoms. Many pollutants are attracted into water droplets and form **aerosols,** which are colloids (Section 15.6) consisting of liquid droplets or finely divided solids dispersed in a gas. Fogs and smoke are common examples of aerosols. Larger solid particles in the atmosphere are called **particulates.** The solids in an aerosol or particulate may be metal oxides, soil particles, sea salt, fly ash from electrical generating plants and incinerators, elemental carbon, or even small metal particles. Aerosols range in diameter from about 1 nm to about 10,000 nm and may contain about a trillion (10^{12}) atoms, ions, or small molecules.

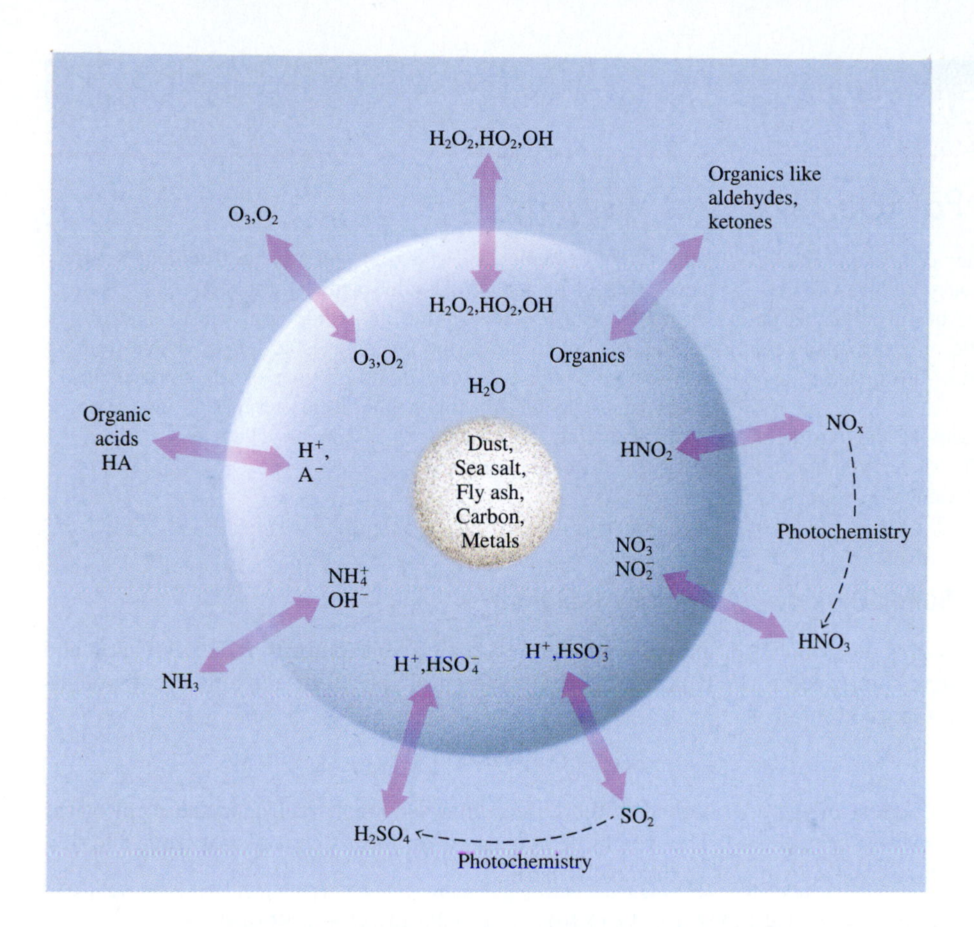

Figure 14.19 **Schematic of an aerosol particle and some of its chemical reactions involving urban air pollutants.**

Particles in the 2000-nm range are largely responsible for the deterioration of visibility often observed in highly populated urban centers such as Los Angeles and New York.

Aerosols are small enough to remain suspended in the atmosphere for long periods. Such small particles are easily breathable and can cause lung disease. Because of their relatively large surface area, aerosol particles have great capacities to **adsorb** and concentrate chemicals on their surfaces. Liquid aerosols or particles covered with a thin coating of water may also **absorb** air pollutants, thereby concentrating them and providing a medium in which reactions may occur. A typical urban aerosol is shown schematically in Figure 14.19.

To **adsorb** is to attract firmly to a surface. To **absorb** is to draw into the bulk of a solid or liquid.

Combinations of Air Pollutants—Smog

The poisonous mixture of smoke (particulate matter), fog (an aerosol), air, and other chemicals was first called **smog** in 1911 by Dr. Harold de Voeux in his report on a London air pollution disaster that caused the deaths of 1150 people. The smog de Voeux identified is the *chemically reducing type* that is derived largely from the combustion of coal and oil and contains sulfur dioxide (a strong reducing agent) mixed with soot, fly ash, smoke, and partially oxidized organic compounds. This is an industrial smog, and is common in many cities in the world where industrial and power plants are found. Although industrial smog is becoming less common in the United States as more pollution controls are installed, it is still a major problem in some cities of the world.

CHEMISTRY YOU CAN DO

Particle Size and Visibility

A common feature of all aerosols is that they decrease visibility. This can be observed in a city or along a busy highway, for example. Here is a way to simulate the effect of air pollutants on visibility. You will need a flashlight, a little milk, a transparent container (if possible, with flat, parallel sides) full of water, and something with which to stir the water.

Turn off the lights and shine the beam of the flashlight through the container perpendicular to the flat sides. What do you observe? Can you see the beam? Now add a couple of drops of milk to the container and stir. Can you see the flashlight beam now? What color is it? What color is the light that passes through the milky water? Keep adding milk dropwise, stirring and observing until the beam of the flashlight is no longer visible from the far side of the container. Based on your observations of the milky water, devise an explanation of the fact that at midday on a sunny day the sun appears to be white or yellow, while at sunset it appears orange or red.

Sulfur Dioxide—A Primary Pollutant

Sulfur dioxide (SO_2), the pollutant that is a major contributor to industrial smog and acid rain (Section 17.10), is produced when sulfur or sulfur-containing compounds are burned in air.

$$S(s) + O_2(g) \longrightarrow SO_2(g)$$

Oil-burning electrical generation plants produce quantities of SO_2 comparable to coal-burning facilities since fuel oils can also contain up to 4% sulfur. The sulfur in oil is in the form of mercapto compounds, organic compounds in which sulfur atoms are bound to carbon and hydrogen atoms (the —SH functional group).

Most of the coal burned in the United States contains sulfur in the form of the mineral pyrite (FeS_2). The weight percent of sulfur in this coal ranges from 1 to 4%. The pyrite is oxidized as the coal is burned.

$$4\ FeS_2(s) + 11\ O_2(g) \longrightarrow 2\ Fe_2O_3(s) + 8\ SO_2(g)$$

Exercise 14.21 Calculating SO_2 Emissions from Burning Coal

Large quantities of coal are burned in the United States to generate electricity. A 1000-MW coal-fired generating plant will burn 3.06×10^6 kg of coal per hour. For coal that contains 4% sulfur by weight, calculate the mass of SO_2 released (a) per hour and (b) per year.

Most of the SO_2 in the atmosphere reacts to form sulfur trioxide (SO_3). Several reactions are possible. SO_2 may react with atomic oxygen:

$$SO_2(g) + \cdot O \cdot\ (g) \longrightarrow SO_3(g)$$

It may react with molecular oxygen:

$$2\ SO_2(g) + O_2(g) \longrightarrow 2\ SO_3(g)$$

Or, it may react with hydroxyl radicals:

$$SO_2(g) + 2\ \cdot OH(g) \longrightarrow SO_3(g) + H_2O(g)$$

The SO_3 has a strong affinity for water and will dissolve in aqueous aerosol droplets to form sulfuric acid.

$$SO_3(g) + H_2O(\ell) \longrightarrow H_2SO_4(aq)$$

TABLE 14.3 Physiological and Corrosive Effects of SO_2

SO_2 Exposure (ppm)	Duration	Effect
0.03–0.12	Annual average	Corrosion, especially in moist, temperate climates
0.3	8 Hour	Vegetation damage (bleached spots, suppression of growth, leaf drop, and low yield)
0.47	$<$ 1 Hour	Odor threshold (50% of subjects detect) Varies with individuals
0.2	Daily average	Respiratory symptoms when community exposure exceeds 0.2 ppm more than 3% of the time
$>$ 0.05	Long-term average	Respiratory symptoms, including impairment of lung function in children, when accompanied by particulates $>$ 100 $\mu g/m^3$

Sulfur dioxide can be physiologically harmful to both plants and animals, although most healthy adults can tolerate fairly high levels of SO_2 without apparent lasting ill effects. Table 14.3 summarizes its effects. Individuals with chronic respiratory difficulties such as bronchitis or asthma tend to be much more sensitive to SO_2, accounting for many of the deaths during episodes of industrial smog. To reduce the hazards, many facilities have installed equipment that removes SO_2 from emitted gases.

Exercise 14.22 Calculations Involving Air Pollutants

The air on a smoggy day contains 5 ppm SO_2. What volume percent is that?

14.12 URBAN AIR POLLUTION

Another major classification of smog is the *chemically oxidizing type* that contains strong oxidizing agents such as ozone and oxides of nitrogen, NO_x. It is known as **photochemical smog** because light—in this instance sunlight—is important in initiating the reactions that cause it. Another name commonly given to photochemical smog is *urban smog,* because of its formation in and around urban areas.

Photochemical smog is typical in Los Angeles and other cities where sunshine is abundant and internal combustion engines exhaust large quantities of pollutants to the atmosphere. This type of smog is practically free of sulfur dioxide but contains substantial amounts of nitrogen oxides, ozone, ozonated hydrocarbons, and organic peroxides, together with hydrocarbons of varying complexity. The concentrations of these substances vary during the day, building up in the morning hours and dropping off at night (Figure 14.20).

The common oxides of nitrogen found in air, NO and NO_2, are collectively called NO_x.

A city's atmosphere is an enormous mixing bowl of frenzied chemical reactions. Identifying the exact chemical reactions that produce photochemical smog has been a challenging job, but in 1951 insight was gained when smog was first duplicated in the laboratory. Detailed studies have subsequently revealed that photochemical reac-

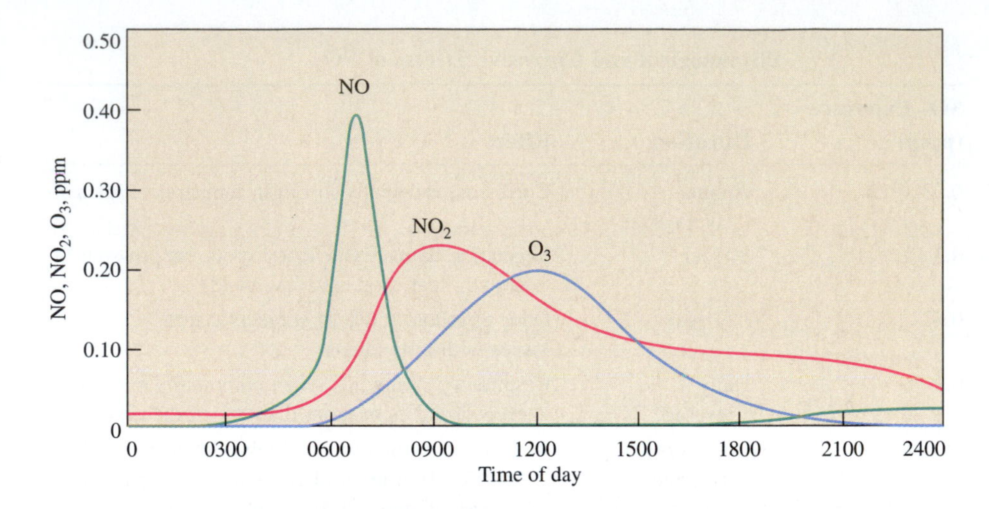

Figure 14.20 The average concentrations of pollutants NO, NO_2, and O_3 on a smoggy day in Los Angeles, California. The NO concentration builds up during the morning rush hour. Later in the day the concentrations of NO_2 and O_3, which are secondary pollutants, build up.

tions are essential to the smog-making process and that aerosols serve to keep the primary pollutants together long enough to form other pollutants, called **secondary pollutants,** by chemical reactions. Photons in the ultraviolet region of the spectrum (uv) are primarily responsible for the formation of photochemical smog.

Organic peroxides have the R—O—O—R′ structure and are produced by ozone reacting with organic molecules. Hydrogen peroxide is H—O—O—H.

The reaction scheme by which primary pollutants are converted into the secondary pollutants found in photochemical smog (Figure 14.21) is thought to begin with the photodissociation (⇐ *p. 551*) of nitrogen dioxide (reaction 1). The very reactive atomic oxygen next reacts with molecular oxygen to form ozone (O_3) (reaction 2), which is then consumed by reacting with nitrogen monoxide to form the original reactant—nitrogen dioxide (reaction 3).

$$\cdot NO_2(g) \xrightarrow{h\nu} \cdot NO(g) + \cdot O \cdot (g) \quad (1)$$

$$\cdot O \cdot (g) + O_2(g) \longrightarrow O_3(g) \quad (2)$$

$$O_3(g) + \cdot NO(g) \longrightarrow \cdot NO_2(g) + O_2(g) \quad (3)$$

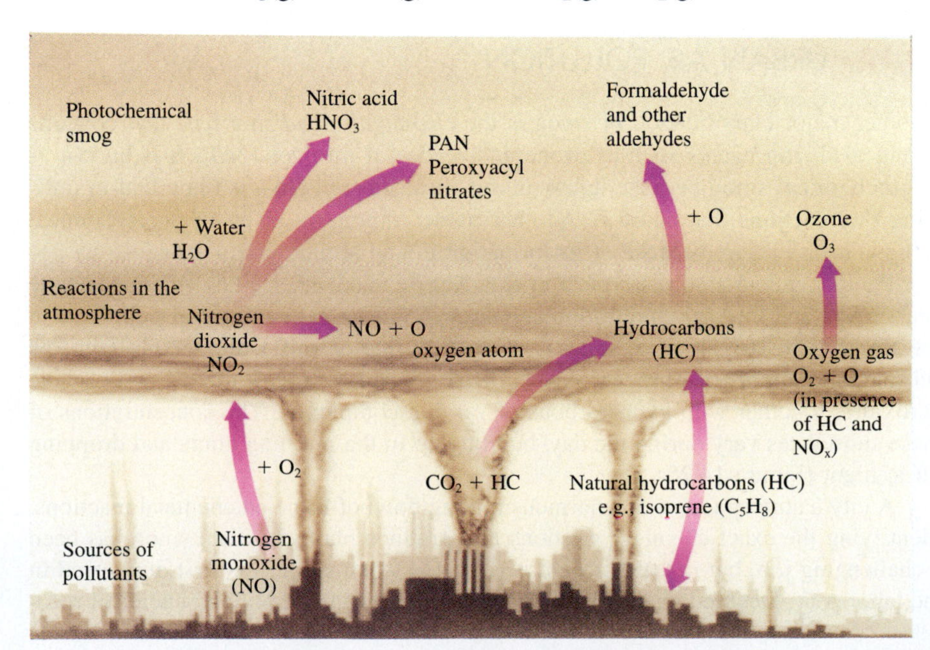

Figure 14.21 Photochemical smog formation. Sunlight is the energy source. When the sources of the pollutants are highly concentrated, as in highly populated urban or industrial areas, photochemical smog can be very concentrated at times. Many of the components of photochemical smog shown in this figure are discussed in this chapter.

The net effect of these three reactions is absorption of energy without any net chemical change. If this were all that happened there would be no problem. Unfortunately, some of the reactive ozone molecules and oxygen atoms go on to react with other species in the atmosphere, and this leads to smog formation.

The air in an urban environment contains unburned hydrocarbons released from car exhausts and spillage at filling stations. Atomic oxygen produced in reaction 1 above reacts with the more reactive of these hydrocarbons—unsaturated compounds and aromatics—to form other free radicals. These radicals, in turn, react to form yet other radicals and secondary pollutants such as aldehydes (e.g., formaldehyde). About 0.2 ppm of nitrogen oxides and 1 ppm of reactive hydrocarbons are sufficient to initiate these reactions.

formaldehyde

Exercise 14.23 Smog Ingredients

Write the formulas and give sources for three ingredients of industrial smog. Do the same for three ingredients of photochemical smog.

Nitrogen Oxides: Primary Pollutants

Most of the nitrogen oxides found in the atmosphere originate from nitrogen monoxide, NO, formed whenever nitrogen and oxygen, always present in the atmosphere, are raised to high temperatures. This nitrogen monoxide reacts rapidly with atmospheric oxygen to produce NO_2 (← *p. 135*).

Vast quantities of nitrogen oxides are formed each year throughout the world, resulting in a global atmospheric concentration of NO_2 of a few parts per billion or less. In the United States, most oxides of nitrogen are produced from fossil fuel combustion such as that of gasoline in automobile engines, with significantly less coming from natural sources such as lightning. Elsewhere in the world, large amounts are also produced by the burning of trees and other biomass (Table 14.4). Most NO_2, either from human activities or from natural causes, eventually washes out of the atmosphere in precipitation. This is one way green plants obtain the nitrogen necessary for growth. In the troposphere, however, especially around urban centers, excessive NO_2 causes problems.

TABLE 14.4 Estimated Emissions of NO_x

	Emissions (millions of tons)	
Source	**United States**	**Global**
Fossil fuel combustion	66	231
Biomass burning	1.1	132
Lightning	3.3	88
Microbial activity in soil	3.3	88
Input from the stratosphere	0.3	5.5
Total (uncertainty in estimates)	7.4 (± 1)	554.5 (± 275)[1]

[1]The large uncertainty for global emissions is due to incomplete data for much of the world.

Source: Stanford Research Institute

In laboratory studies, nitrogen dioxide in concentrations of 25 to 250 ppm inhibits plant growth and causes defoliation. The growth of tomato and bean seedlings is inhibited by 0.3 to 0.5 ppm of NO_2 applied continuously for 10 to 20 days. Breathing NO_2 at a concentration of 3 ppm for 1 hour causes bronchial constriction in humans, and short exposures at high levels (150–220 ppm) cause changes in the lungs that produce fatal results.

During the day, as you saw in the previous section, nitrogen dioxide photodissociates to form nitrogen monoxide and free oxygen atoms. The oxygen atoms can then react to form ozone and regenerate NO_2. At night, another oxide of nitrogen, the nitrogen trioxide radical, NO_3, is produced by the reaction

$$\cdot NO_2(g) + O_3(g) \longrightarrow \cdot NO_3(g) + O_2(g)$$

Nitric acid is present in acid rain.

In daylight, the nitrogen trioxide radical quickly photodissociates, but in the absence of light it accumulates and reacts with nitrogen dioxide to form N_2O_5, which in turn reacts with water to form nitric acid.

$$\cdot NO_3(g) + \cdot NO_2(g) \longrightarrow N_2O_5(g)$$

$$N_2O_5(g) + H_2O(g) \longrightarrow 2\ HNO_3(g)$$

Normally nitrogen dioxide has a lifetime of about three days in the atmosphere.

Exercise 14.24 Air Pollutant Stoichiometry

If 400 metric tons of N_2 are converted to NO, and then to HNO_3, what mass (in metric tons) of HNO_3 is produced?

Ozone: A Secondary Pollutant

Ozone is an allotropic form of oxygen (← p. 33).

Ozone (O_3) has a pungent odor that can be detected at concentrations as low as 0.02 ppm. We often smell the ozone produced by sparking electric appliances, or after a thunderstorm when lightning-caused ozone washes out with the rainfall. Ozone can be either beneficial or harmful, depending on where it is found. In the troposphere (the air we breathe), ozone is harmful because it is a component of photochemical smog and because it can damage human health and decompose materials such as plastics and rubber. In the stratosphere, ozone is beneficial because it protects us from damaging ultraviolet radiation.

The only significant chemical reaction producing ozone in the atmosphere is the combination of molecular oxygen and atomic oxygen. In the lower atmosphere the major source of oxygen atoms is the photodissociation of NO_2. At high altitudes, oxygen atoms are produced by the photodissociation of oxygen molecules, which is caused by ultraviolet photons.

Ozone photodissociates to give an oxygen atom and an oxygen molecule whenever it is struck by photons in the near-ultraviolet range (200–300 nm), at any altitude.

$$O_3(g) \xrightarrow{h\nu} O_2(g) + \cdot O \cdot (g)$$

In the stratosphere, oxygen atoms can then react with water to produce hydroxyl radicals ($\cdot$OH).

$$\cdot O \cdot (g) + H_2O(g) \longrightarrow 2 \cdot OH(g)$$

In the daytime, when they are produced in large numbers, hydroxyl radicals can react with nitrogen dioxide to produce nitric acid.

$$\cdot NO_2(g) + \cdot OH(g) \longrightarrow HNO_3(g)$$

This three-step process is the primary means of removing NO_2 from the atmosphere. Summing the three steps gives the net reaction:

$$O_3 \xrightarrow{h\nu} O_2 + \cdot O \cdot$$

$$\cdot O \cdot + H_2O \longrightarrow 2 \cdot OH$$

$$2\,(\cdot NO_2 + \cdot OH \longrightarrow HNO_3)$$

$$O_3 + 2 \cdot NO_2 + H_2O \xrightarrow{h\nu} 2\,HNO_3 + O_2 \qquad \text{net reaction}$$

In this net reaction, as is usually the case, the reactive free radicals are intermediates that do not appear in the net reaction. Given the diversity of photodissociation reactions that form free radicals, there are a great many possible mechanisms for reactions in the atmosphere.

Ozone is a secondary air pollutant and is the most difficult pollutant to control because its formation depends on ever-present sunlight and NO_2. As will be discussed later, hydrocarbons, which almost every automobile emits to some degree, are also involved in ozone production. In cities with high ozone concentrations, the cause is always related to emissions of nitrogen oxides from automobiles, buses, and trucks. Most major urban areas operate vehicle-inspection centers for passenger automobiles in an effort to control NO_x emissions as well as those of carbon monoxide and unburned hydrocarbons (Figure 14.22).

Figure 14.22 Automobile emissions testing. Such testing is mandated in many communities that have failed to meet EPA's ozone standards. Cars that fail the emissions standards for hydrocarbons and carbon monoxide are required to be repaired. Often the local government will not reissue operating licenses until satisfactory emissions levels are achieved. *(J. Wood)*

As difficult as they are to attain, established ozone standards may not be low enough for good health. Exposure to concentrations of ozone at or near 0.12 ppm lowers the volume of air a person breathes out in 1 second. Studies of children who were exposed to ozone concentrations slightly below the EPA standard showed a 16% decrease in exhaled air volume. Some scientists have been urging the EPA to lower the standard to 0.08 ppm. No matter what the standard becomes, present ozone concentrations in many urban areas represent health hazards to children at play, joggers, others doing outdoor exercise, and older persons who may have diminished respiratory capabilities.

Exercise 14.25 Photodissociation Reactions

Write two photodissociation reactions that produce atomic oxygen. Then write a reaction in which ozone is formed.

Hydrocarbons: Primary Pollutants

Hydrocarbons enter the atmosphere from both natural sources and human activities. Isoprene and α-pinene are produced in large quantities by both coniferous and deciduous trees. Methane gas is produced by such diverse sources as ruminant animals, termites, ants, and decay-causing bacteria acting on dead plants and animals. Human activities such as using industrial solvents, petroleum refining and distribution, and incomplete burning of gasoline and diesel fuel account for a large amount of hydrocarbons in the atmosphere. The principal hydrocarbons found in urban air are listed in Table 14.5.

$CH_2{=}C(CH_3){-}CH{=}CH_2$

isoprene

CH_3, CH_3, CH_3

α-pinene

TABLE 14.5 The Ten Most Abundant Hydrocarbons Found in the Air of Cities

Compound	Formula	Concentration (ppb) Median	Maximum
2-Methylbutane	$CH_3CH(CH_3)CH_2CH_3$	45	3393
Butane	$CH_3CH_2CH_2CH_3$	40	5448
Toluene	$C_6H_5CH_3$	34	1299
Propane	$CH_3CH_2CH_3$	23	399
Ethane	CH_3CH_3	23	475
Pentane	$CH_3CH_2CH_2CH_2CH_3$	22	1450
Ethylene	CH_2CH_2	21	1001
m-Xylene, *p*-Xylene	$C_6H_4(CH_3)_2$	18	338
2-Methylpentane	$CH_3CH(CH_3)CH_2CH_2CH_3$	15	647
2-Methylpropane	$CH_3CH(CH_3)CH_3$	15	1433

Results from 800 air samples taken in 39 cities. *Source:* Air Pollution Control Association.

The hydroxyl radical ($\cdot OH$), produced indirectly from ozone, helps in the oxidation of hydrocarbons in the atmosphere. Like the reactions we have seen earlier, this oxidation involves several steps. Using RH as the general formula for a hydrocarbon, we can write the first two steps as

$$RH(g) + \cdot OH(g) \longrightarrow R\cdot(g) + H_2O(g) \quad (1)$$

$$R\cdot(g) + O_2(g) \longrightarrow ROO\cdot(g) \quad (2)$$

where atmospheric oxygen is the oxidizing agent. The $ROO\cdot$ represents organic peroxy radicals with a variety of R groups. The peroxy radicals are also oxidizing agents that oxidize NO to NO_2, and produce various aldehydes (RCHO) and ketones (R_2CO) and the hydroperoxyl radical ($\cdot OOH$). This is illustrated in reaction 3.

$$ROO\cdot(g) + \cdot NO(g) \longrightarrow \text{aldehydes} + \text{ketones} + \cdot NO_2(g) + \cdot OOH(g) \quad (3)$$

The NO_2 radical can then undergo the familiar photodissociation, which is followed quickly by the production of ozone.

$$\cdot NO_2(g) \xrightarrow{h\nu} \cdot NO(g) + \cdot O\cdot(g) \quad (4)$$

$$\cdot O\cdot(g) + O_2(g) \longrightarrow O_3(g) \quad (5)$$

In the general manner illustrated in equations 1 through 5, *oxidation of various hydrocarbons by hydroxyl radicals contributes to ozone formation.*

The hydroperoxyl radical can also react with NO to produce NO_2 and more hydroxyl radicals.

$$\cdot OOH(g) + NO(g) \longrightarrow NO_2(g) + \cdot OH(g)$$

Hydroxyl radicals are also involved in the oxidation of aldehydes to give compounds that can react with NO_2. Acetaldehyde (CH_3CHO), for example, reacts with the hydroxyl radical and oxygen in the following way:

$$\underset{\text{acetaldehyde}}{CH_3CHO\,(g)} + \cdot OH(g) + O_2(g) \longrightarrow \underset{\text{peroxyacetyl radical}}{CH_3\overset{\overset{\displaystyle O}{\|}}{C}\text{—}O\text{—}O\cdot\,(g)} + H_2O(g)$$

CHEMISTRY IN THE NEWS

Mexico City Air Pollution

To say Mexico City has an air pollution problem is an understatement. The 400-square-mile Mexico City Metropolitan Area (MCMA) is home for over 15 million people (one fifth of the country's population) and over one half of the industry and three fifths of its automobiles. If you can imagine a city with three times the population density of Philadelphia, the car traffic of New York City, the truck traffic of a large distribution hub like Chicago, and the petroleum industry of a city like Houston you have a picture of Mexico City. All of this activity takes place in a high-altitude location surrounded by mountains that rise as much as 4000 ft above the city. In effect, the city's air is isolated from the prevailing wind patterns. The high altitude of Mexico City makes the air about 25% thinner than at sea level, which means solar energy is much more intense than in cities at or near sea level. In addition, the sun is at a high angle most of the year. Since solar energy is responsible for the formation of ozone, it is not surprising that the city has an ozone problem year-round, not just in the summer months as for many cities in the United States.

(Ron Augustin/Tom Stack & Associates)

Over 29 million commuter trips are made daily in the city, and NO_x emissions from the exhausts of internal combustion engines are sizable. Since NO_2 readily photodissociates to produce oxygen atoms which in turn react with oxygen molecules to produce ozone, it is easy to see why ozone is such a problem. Besides being a health problem in its own right, ozone photodissociates to produce oxygen atoms that react with water to form hydroxyl radicals, and ozone itself can react with hydrocarbon molecules to form reactive aldehydes and ketones.

The hydrocarbons in Mexico City's air come not only from the large number of automobiles that release unburned hydrocarbons in their exhausts, but from the vast petrochemical industry that has grown up in the city and the vast number of propane heaters used to cook food and heat homes. If only the air in the city could be swept away periodically by the prevailing winds, things would not be nearly so bad. In spite of all these problems, progress is being made. Automobile usage is being curtailed by limiting the driving of certain cars to specified days of the week; and the petrochemical industry is limiting its emissions of hydrocarbons. In addition, cleaner burning fuels are becoming available that will produce less unburned hydrocarbons, and the public is being educated to the air pollution potential of carelessly allowing unburned heating fuel to escape into the air. Perhaps someday, the city will become entirely dependent on public transportation and electric cars.

Source: Los Alamos National Laboratory and the Mexico City Air Quality Research Initiative 1996.

The peroxyacetyl radical can then react with NO_2 to produce peroxyacetylnitrate (PAN):

$$\mathrm{CH_3\overset{\overset{\displaystyle O}{\|}}{C}{-}O{-}O\cdot\ (g) + \cdot NO_2(g) \longrightarrow CH_3\overset{\overset{\displaystyle O}{\|}}{C}{-}O{-}O{-}NO_2(g)}$$

PAN

PAN and related compounds are powerful eye irritants, helping to account for the discomfort we feel during episodes of photochemical smog. PAN formation stabilizes NO_2 so that it can be carried over great distances by prevailing winds. Eventually PAN decomposes and releases NO_2. In this way, urban pollution in the form of NO_2 may be carried to outlying areas, where it may do additional damage to vegetation, human tissue, and fabrics—a direct result of an unfortunate combination of events involving hydrocarbons emitted into the atmosphere.

SUMMARY PROBLEM

The air that enters an automobile engine contains the oxygen that burns the hydrocarbon fuel vapors to move the vehicle. Prior to the combustion process the fuel-air mixture is compressed and then it is ignited with a spark. Most of the fuel molecules are completely burned to CO_2 and H_2O, and some of the nitrogen is converted to nitrogen monoxide (NO). For simplicity, assume the fuel has a formula of C_8H_{18} and a density of 0.760 g/mL.

(a) What are the partial pressures of N_2 and O_2 in the air before it goes into the engine if the atmospheric pressure is 734 mm Hg?

(b) If no fuel were added to the air and it was compressed to seven times atmospheric pressure (approximately the compression ratio in a modern engine), what would the partial pressures of N_2 and O_2 become?

(c) Assume the volume of each cylinder in the engine is 485 mL. If 0.050 mL of the fuel is added to the air in each cylinder just prior to compression, and the fuel completely vaporized, what would be the partial pressure in the mixture in a cylinder due to the fuel molecules?

(d) How much oxygen would be required to burn the fuel completely to CO_2 and H_2O?

(e) If 10.% of the nitrogen in the combustion process is converted to NO, calculate how many grams of NO are produced in a single combustion reaction.

(f) Write a reaction that converts NO to a more photoreactive compound.

(g) Calculate the mass of this more photoreactive compound that would be formed from the NO calculated above.

(h) What additional information would you need to calculate the NO emissions for an entire city for a year?

IN CLOSING

Having studied this chapter, you should be able to . . .

- explain the properties of gases (Section 14.1).
- describe the components of the atmosphere (Section 14.2).
- state the fundamental concepts of the kinetic-molecular theory and use them to explain gas behavior (Section 14.3).
- solve mathematical problems using the appropriate gas laws (Section 14.4).
- calculate the quantities of gaseous reactants and products involved in chemical reactions (Section 14.6).
- apply the ideal gas law to finding gas densities and partial pressures (Section 14.7).
- describe the differences between real and ideal gases (Section 14.8).
- discuss the main chemicals we extract from the atmosphere (Section 14.9).
- explain the main chemicals found in and the reactions producing industrial pollution (Section 14.11) and urban pollution (Section 14.12).

Key Terms

The following terms were defined and given in boldface type in this chapter. You should be sure to understand each of the terms and the concepts with which it is associated.

absorb *(14.11)*
adsorb *(14.11)*
aerosols *(14.11)*
air pollutant *(14.11)*
Avogadro's law *(14.4)*
bar *(14.1)*
barometer *(14.1)*
Boyle's law *(14.4)*
Charles's law *(14.4)*
combined gas law *(14.4)*
compressibility *(14.4)*
cryogens *(14.9)*
Dalton's law of partial pressures *(14.7)*
electronically excited molecule *(14.10)*
ideal gas *(14.4)*
ideal gas constant *(14.4)*
ideal gas law *(14.4)*
kinetic-molecular theory *(14.3)*
law of combining volumes *(14.4)*
millimeters of mercury (mm Hg) *(14.1)*
mole fraction *(14.7)*
newton (N) *(14.1)*
partial pressure *(14.7)*
particulates *(14.11)*
pascal (Pa) *(14.1)*
photochemical reactions *(14.10)*
photochemical smog *(14.12)*
photodissociation reaction *(14.10)*
pressure *(14.1)*
primary pollutant *(14.11)*
secondary pollutants *(14.12)*
smog *(14.11)*
standard atmosphere (atm) *(14.1)*
standard molar volume *(14.4)*
standard temperature and pressure (STP) *(14.4)*
stratosphere *(14.2)*
torr *(14.1)*
troposphere *(14.2)*

Questions for Review and Thought

Conceptual Challenge Problems

CP-14.A. Write a paragraph to explain in a quantitative fashion why a woman might request to be let out of an automobile while in the driveway prior to moving the car to a grassy area for parking during a formal wedding reception.

CP-14.B. How would you quickly check to see if the mass of the earth's atmosphere is 5.3×10^{15} metric tons, as stated in Section 14.2?

CP-14.C. Under what conditions would you expect to observe that the pressure of a confined gas at constant temperature and volume is *not* constant?

CP-14.D. Suppose that the gas constant, *R*, were to be defined as 1.000 L atm/mol deg where the "deg" referred to a newly defined Basic temperature scale. What would be the melting and boiling temperature of water in degrees Basic (°B)?

Review Questions

1. Name the three gas laws and explain how they interrelate *P*, *V*, and *T*. Explain the relationships in words and with equations.
2. What are the conditions represented by STP?
3. What is the volume occupied by 1 mol of an ideal gas at STP?
4. What is the definition of pressure?
5. State Avogadro's law. Explain why two volumes of hydrogen react with one volume of oxygen to form two volumes of steam.
6. State Dalton's law of partial pressures. If the air we breathe is 78% N_2 and 22% O_2 on a mole basis, what is the mole fraction of O_2? What is the partial pressure of O_2 if the total pressure is 720 mm Hg?
7. Explain Boyle's law on the basis of the kinetic-molecular theory.
8. Explain why gases at low temperature and high pressure do not obey the ideal gas equation.
9. Describe the difference between primary and secondary air pollutants and give two examples of each.

Properties of Gases

10. Gas pressures can be expressed in units of mm Hg, atm, torr, and kPa. Do the following conversions.
(a) 720 mm Hg to atm
(b) 1.25 atm to mm Hg
(c) 542 mm Hg to torr
(d) 740 mm Hg to kPa
(e) 700 kPa to atm

11. Convert the following pressure measurements.
 (a) 120 mm Hg to atm (d) 200 kPa to atm
 (b) 2 atm to mm Hg (e) 36 kPa to atm
 (c) 100 kPa to mm Hg (f) 600 kPa to mm Hg
12. Mercury has a density of 13.956 g/cm^3. A barometer is constructed using an oil with a density of 0.75 g/cm^3. If the atmospheric pressure is 1.0 atm, what will be the height in meters of the oil column in the barometer?
13. A vacuum pump is connected to the top of an upright tube whose lower end is immersed in a pool of mercury. How high will the mercury rise in the tube when the pump is turned on?
14. Why can't a hand-driven pump on a water well pull underground water from depths more than 34 ft? Would it help to have a motor-driven vacuum pump?
15. A scuba diver taking photos of a coral reef 60 ft below the ocean surface breathes out a stream of bubbles. What is the total gas pressure of these bubbles at the moment they are released to the ocean? What is the gas pressure when they reach the surface of the ocean?

The Atmosphere

16. Explain the major roles played by nitrogen in the atmosphere. Do the same for oxygen.
17. Beginning nearest the earth, name the two most important layers or regions of the atmosphere. Describe, in general, the kinds of chemical reactions that occur in each layer.
18. Convert all of the "percentage by volume" figures in Table 14.1 into (a) parts per million and (b) parts per billion. Which atmospheric gases are present at concentrations of less than 1 ppb? Between 1 ppb and 1 ppm? Greater than 1 ppm?
19. The mass of the earth's atmosphere is 5.3×10^{15} metric tons. The atmospheric abundance of helium is 0.7 ppm when expressed as a fraction by *weight* instead of by *volume,* as in Table 14.1. How many metric tons of helium are there in the atmosphere? How many moles of helium is this?
20. Sulfur is about 2.5% of the mass of coal, and when coal is burned the sulfur is all converted to SO_2. In 1980, 3.1×10^9 metric tons of coal were burned worldwide. How many tons of SO_2 were added to the atmosphere? How many tons of SO_2 are presently in the atmosphere? (*Note:* The weight fraction of SO_2 in air is 0.4 ppb.)

Kinetic-Molecular Theory

21. List the five basic concepts of the kinetic-molecular theory. Which assumption is incorrect at very high pressures? Which one is incorrect at low temperatures? Which assumption is probably most nearly correct?
22. You are given two flasks of equal volume. Flask A contains H_2 at 0 °C and 1 atm pressure. Flask B contains CO_2 gas at 0 °C and 2 atm pressure. Compare these two samples with respect to each of the following:
 (a) Average kinetic energy per molecule
 (b) Average molecular velocity
 (c) Number of molecules
23. Place the following gases in order of increasing average molecular speed at 25 °C: Kr, CH_4, N_2, CH_2Cl_2.
24. The reaction of SO_2 with Cl_2 to give dichlorine oxide is

$$SO_2(g) + 2\ Cl_2(g) \longrightarrow SOCl_2(g) + Cl_2O(g)$$

All of the molecules involved in the reaction are gases. Place them in order of increasing average molecular speed.
25. From Table 14.1 list all of the gases in the atmosphere in order of *decreasing* average molecular speed. Concern has been expressed about one of these gases escaping into outer space because a fraction of its molecules have velocities greater than the escape velocity needed to break free from earth's gravitational field. Which gas is it?

Gas Behavior and the Ideal Gas Law

26. What amount (number of moles) of CO is found in 1.0 L of air at STP that contains 950 ppm CO?
27. A sample of a gas has a pressure of 100. mm Hg in a 125-mL flask. If this gas sample is transferred to another flask with a volume of 200. mL, what will be the new pressure? Assume that the temperature remains constant.
28. A sample of a gas is placed in a 256-mL flask, where it exerts a pressure of 75.0 mm Hg. What is the pressure of this gas if it is transferred to a 125-mL flask? (The temperature stays constant.)
29. A sample of gas has a pressure of 62 mm Hg in a 100-mL flask. This sample of gas is transferred to another flask, where its pressure is 29 mm Hg. What is the volume of the new flask? (The temperature does not change.)
30. Some butane, the fuel used in backyard grills, is placed in a 3.50-L container at 25 °C; its pressure is 735 mm Hg. If you transfer the gas to a 15.0-L container, also at 25 °C, what is the pressure of the gas in the larger container?
31. A sample of gas at 30 °C has a pressure of 2 atm in a 1-L container. What pressure will it exert in a 4-L container? The temperature does not change.
32. Suppose you have a sample of CO_2 in a gas-tight syringe with a movable piston. The gas volume is 25.0 mL at a room temperature of 20 °C. What is the final volume of the gas if you hold the syringe in your hand to raise its temperature to 37 °C?
33. A balloon is inflated with helium to a volume of 4.5 L at 23 °C. If you take the balloon outside on a cold day (−10 °C), what will be the new volume of the balloon?
34. A sample of gas has a volume of 2.50 L at 670 mm Hg pressure and a temperature of 80 °C. If the pressure remains constant but the temperature is decreased, the gas occupies 1.25 L. What is this new temperature, in degrees Celsius?
35. A sample of 9.0 L of CO_2 at 20 °C and 1 atm pressure is cooled so that it occupies 1.0 L at some new temperature. The pressure remains constant. What is the new temperature, in kelvins?
36. A bicycle tire is inflated to a pressure of 3.74 atm at 15 °C. If the tire is heated to 35 °C, what is the pressure in the tire? Assume the tire volume doesn't change.

37. An automobile tire is inflated to a pressure of 3.05 atm on a rather warm day when the temperature is 40 °C. The car is then driven to the mountains and parked overnight. The morning temperature is −5 °C. What will be the pressure of the gas in the tire? Assume the volume of the tire doesn't change.

38. A sample of gas occupies 754 mL at 22 °C and a pressure of 165 mm Hg. What is its volume if the temperature is raised to 42 °C and the pressure is raised to 265 mm Hg? Note that the number of moles does not change.

39. A balloon is filled with helium to a volume of 1.05×10^3 L on the ground, where the pressure is 745 mm Hg and the temperature is 20 °C. When the balloon ascends to a height of 2 miles where the pressure is only 600 mm Hg and the temperature is −33 °C, what is the volume of the helium in the balloon?

40. What is the pressure exerted by 1.55 g of Xe gas at 20 °C in a 560-mL flask?

41. A 1.00-g sample of water is allowed to vaporize completely inside a 10.0-L container. What is the pressure of the water vapor at a temperature of 150. °C?

42. Which of the following gas samples contains the largest number of molecules and which contains the smallest?
(a) 1.0 L of H_2 at STP
(b) 1.0 L of N_2 at STP
(c) 1.0 L of H_2 at 27 °C and 760 mm Hg
(d) 1.0 L of CO_2 at 0 °C and 800 mm Hg

43. Ozone molecules attack rubber and cause cracks to appear. If enough cracks occur in a rubber tire, for example, it will be weakened and it may burst from the pressure of the air inside. As little as 0.02 ppm O_3 will cause cracks to appear in rubber in about 1 hour. Assume that a 1.0-cm^3 sample of air containing 0.020 ppm O_3 is brought in contact with a sample of rubber that is 1.0 cm^2 in area. Calculate the number of O_3 molecules that are available to collide with the rubber surface. The temperature of the air sample is 25 °C and the pressure is 0.95 atm.

Quantities of Gases in Chemical Reactions

44. The yeast in rising bread dough converts sugar (sucrose, $C_{12}H_{22}O_{11}$) into carbon dioxide. A popular recipe for two loaves of French bread requires 1 package of yeast and $\frac{1}{4}$ teaspoon (about 2.4 g) of sugar. What volume of CO_2 at STP is produced by the complete conversion of this amount of sucrose into CO_2 by the yeast? Compare this volume with the typical volume of two loaves of French bread.

45. Water can be made by combining gaseous O_2 and H_2. If you begin with 1.5 L of H_2(g) at 360 mm Hg and 23 °C, what volume in liters of O_2(g) will you need for complete reaction if the O_2 gas is also measured at 360 mm Hg and 23 °C?

46. Gaseous silane, SiH_4, ignites spontaneously in air according to the equation

$$SiH_4(g) + 2\ O_2(g) \longrightarrow SiO_2(s) + 2\ H_2O(g)$$

If 5.2 L of SiH_4 is treated with O_2, what volume in liters of O_2 is required for complete reaction? What volume of H_2O vapor is produced? Assume all gases are measured at the same temperature and pressure.

47. Hydrogen can be made in the "water gas reaction."

$$C(s) + H_2O(g) \longrightarrow H_2(g) + CO(g)$$

If you begin with 250 L of gaseous water at 120 °C and 2.0 atm pressure, what mass in grams of H_2 can be prepared?

48. If the boron hydride B_4H_{10} is treated with pure oxygen, it burns to give B_2O_3 and H_2O.

$$2\ B_4H_{10}(s) + 11\ O_2(g) \longrightarrow 4\ B_2O_3(s) + 10\ H_2O(g)$$

If a 0.050-g sample of the boron hydride burns completely in O_2, what will be the pressure of the gaseous water in a 4.25-L flask at 30. °C?

49. If 1.0×10^3 g of uranium metal is converted to gaseous UF_6, what pressure of UF_6 would be observed at 62 °C in a chamber that has a volume of 3.0×10^2 L?

50. Metal carbonates decompose to the metal oxide and CO_2 on heating according to this general equation.

$$M_x(CO_3)_y(s) \longrightarrow M_xO_y(s) + y\ CO_2(g)$$

You heat 0.158 g of a white, solid carbonate of a Group 2A metal and find that the evolved CO_2 has a pressure of 69.8 mm Hg in a 285-mL flask at 25 °C. What is the molar mass, M, of the metal carbonate?

51. Nickel carbonyl, $Ni(CO)_4$, can be made by the room-temperature reaction of finely divided nickel metal with gaseous CO. This is the basis for purifying nickel on an industrial scale. If you have CO in a 1.50-L flask at a pressure of 418 mm Hg at 25.0 °C, what is the maximum mass in grams of $Ni(CO)_4$ that can be made?

52. Assume that a car burns pure octane, C_8H_{18} ($d = 0.692$ g/cm^3).
(a) Write the balanced equation for burning octane in air, forming CO_2 and H_2O.
(b) If the car has a fuel efficiency of 32 miles per gallon of octane, what volume of CO_2 at 25 °C and 1.0 atm is generated when the car goes on a 10.-mile trip?

53. Follow the directions in the previous question, but use methanol, CH_3OH ($d = 0.791$ g/cm^3) as the fuel. Assume the fuel efficiency is 20. miles per gallon.

Gas Density and Molar Mass

54. A sample of gaseous SiH_4 weighing 4.25 g is placed in a 580-mL container. The resulting pressure is 1.2 atm. What is the temperature in degrees Celsius?

55. To find the volume of a flask, it is first evacuated so it contains no gas at all. Next, 4.4 g of CO_2 is introduced into the flask. On warming to 27 °C, the gas exerts a pressure of 730 mm Hg. What is the volume of the flask in milliliters?

56. What mass of helium in grams is required to fill a 5.0-L balloon to a pressure of 1.1 atm at 25 °C?

57. A hydrocarbon with the general formula C_xH_y is 92.26% carbon. Experiment shows that 0.293 g of the hydrocarbon fills a 185-mL flask at 23 °C with a pressure of 374 mm Hg. What is the molecular formula for this compound?
58. Forty miles above the earth's surface the temperature is −23 °C and the pressure is only 0.20 mm Hg. What is the density of air (M = 29.0 g/mol) at this altitude?
59. A newly discovered gas has a density of 2.39 g/L at 23.0 °C and 715 mm Hg. What is the molar mass of the gas?

Partial Pressures of Gases

60. The atmosphere is a mixture of gases with a total pressure equal to the barometric pressure. A sample of the atmosphere at a total pressure of 740. mm Hg is analyzed to give the following partial pressures: $P(N_2)$ = 575 mm Hg; $P(Ar)$ = 6.9 mm Hg; $P(CO_2)$ = 0.2 mm Hg; $P(H_2O)$ = 4.0 mm Hg.
(a) What is the partial pressure of O_2?
(b) What is the mole fraction of each gas?
(c) What is the composition of each component of this sample in percentage by volume? Compare your results with those of Table 14.1.
61. Gaseous CO exerts a pressure of 45.6 mm Hg in a 56.0-L tank at 22.0 °C. If this gas is released into a room with a volume of 2.70×10^4 L, what is the partial pressure of CO (in mm Hg) in the room at 22 °C?
62. The density of air at 20.0 km above the earth's surface is 92 g/m^3. The pressure is 42 mm Hg and the temperature is −63 °C. Assuming the atmosphere contains only O_2 and N_2, calculate (a) the average molar mass of the atmosphere, and (b) the mole fraction of each gas.
63. Benzene is a known carcinogen (causing leukemia and other cancers in both laboratory animals and humans). It has acute effects as well. For example, it causes mucous membrane irritation at a concentration of 100 ppm, and fatal narcosis at 20,000 ppm. Calculate the partial pressures in atmospheres at STP corresponding to these concentrations.
64. On a humid, rainy summer day the partial pressure of water vapor in the atmosphere can be as high as 25 mm Hg. What is the mole fraction of water vapor in the atmosphere under these conditions? Compare your results with the composition of dry air given in Table 14.1.
65. The mean fraction *by weight* of water vapor and cloud water in the earth's atmosphere is about 0.0025. Assume that the atmosphere contains two components: "air," with a molar mass of 29.2 g/mol, and water vapor. What is the *mean* mole fraction of water vapor in the earth's atmosphere? What is the *mean* partial pressure of water vapor? Why is this so much smaller than the typical partial pressure of water vapor at the earth's surface on a rainy summer day?

The Behavior of Real Gases

66. From the density of liquid water and its molar mass, calculate the volume that 1 mol of liquid water occupies. If water were an ideal gas at STP, what volume would a mole of water vapor occupy? Can we achieve the STP conditions for water vapor? Why?
67. At high temperatures and low pressures, gases behave ideally, but as the pressure is increased the product PV becomes greater than the product nRT. Give a molecular-level explanation of this fact.
68. At low temperatures and very low pressures, gases behave ideally, but as the pressure is increased the product PV becomes less than the product nRT. Give a molecular-level explanation of this fact.
69. The densities of liquid noble gases and their normal boiling points are given below:

Gas	Normal Boiling Point (K)	Liquid Density (g/cm^3)
He	4.2	0.125
Ne	27.1	1.20
Ar	87.3	1.40
Kr	120.	2.42
Xe	165	2.95

Calculate the volume occupied by 1 mol of each of these liquids. Comment on any trend that you see. What is the volume occupied by 1 mol of each of these substances as an ideal gas at STP? On the basis of these calculations, which gas would you expect to show the largest deviations from ideality at room temperature?

Chemicals from the Atmosphere

70. Name five useful chemicals obtained from the atmosphere, and describe two uses for each.
71. Name two sources of the gaseous element helium.
72. Liquid nitrogen has a density of 0.81 g/cm^3 at its normal boiling point, 77 K. Calculate the volume (a) at 77 K and 1 atmosphere, and (b) at STP occupied by 1.0 liter of liquid nitrogen after it has evaporated.

Chemical Reactions in the Atmosphere

73. What is a free radical? Give an example of a chemical reaction that occurs in the atmosphere that produces a free radical.
74. What product is formed when two methyl radicals react with each other?
75. For the following forms of nitrogen most prevalent in the environment, write chemical formulas and give the oxidation number of the nitrogen:
(a) ammonia
(b) ammonium ion
(c) dinitrogen monoxide
(d) gaseous nitrogen
(e) nitric acid
(f) nitrous acid
(g) nitrogen dioxide

76. Complete the following equations (all reactants and products are gases):
(a) $O + O_2 \rightarrow$
(b) $O_3 \xrightarrow{h\nu}$
(c) $O + SO_2 \rightarrow$
(d) $H_2O \xrightarrow{h\nu}$
(e) $NO_2 + NO_2 \rightarrow$

77. Complete the following equations (all reactants and products are gases):
(a) $O + H_2O \rightarrow$
(b) $NO_2 + OH \rightarrow$
(c) $RH + OH \rightarrow$
(d) $CO + OH \rightarrow$

78. Describe and write an equation for a chemical reaction that
(a) Uses up CO in the atmosphere.
(b) Uses up CO_2 in the atmosphere.

(Refer to Chapter 12 (← *p. 550*) for the details of ozone depletion in the stratosphere as you answer Questions 79–81.)

79. Write the products for the following reactions that take place in the stratosphere.
(a) CF_2Cl_2 (F—C(F)(F)—Cl) $+ h\nu \longrightarrow$
(b) $Cl + O_3 \rightarrow$
(c) $ClO + O \rightarrow$

80. The molecule CF_4 (F—C(F)(F)—F) is not implicated as having ozone-depletion potential. Can you explain why?

81. The molecule CH_3F has much less ozone-depletion potential than the corresponding molecule CH_3Cl. Can you explain why?

Air Pollution

82. Define air pollution in terms of the kinds of pollutants, their sources, and the ways they are harmful.

83. Explain how particulates can contribute to air pollution.

84. What is adsorption? What is absorption?

85. Assume that limestone ($CaCO_3$) is used to remove 90% of the sulfur from 4 metric tons of coal containing 2% S. The product is $CaSO_4$; [$CaCO_3(s) + SO_3(g) \rightarrow CaSO_4(s) + CO_2(g)$]. Calculate the mass of limestone required. Express your answer in metric tons.

86. Approximately 65 million metric tons of SO_2 enter the atmosphere every year from the burning of coal. If coal, on average, contains 2% S, how many metric tons of coal were burned to produce this much SO_2? A 1000-MW power plant burns about 700 metric tons of coal per hour. Calculate the number of hours the quantity of coal will burn in one of these power plants.

87. What mass of gasoline must be burned according to the reaction

$$C_8H_{18}(\ell) + \tfrac{17}{2}\,O_2(g) \longrightarrow 8\,CO(g) + 9\,H_2O(g)$$

to cause a garage with dimensions of 7 m × 3 m × 3 m to contain a CO concentration of 1000. ppm? (Assume STP conditions.)

Urban Air Pollution

88. Give an example of a photochemical reaction. Do all photons of light have sufficient energy to cause photochemical reactions? Explain with examples.

89. What two reactions account for the production of O free radicals in the troposphere?

90. The reducing nature of industrial (London) smog is due to what oxide? The burning of what two fuels produces this oxide? Write an equation showing this oxide being further oxidized.

91. Photochemical smog contains quantities of what two oxidizing gases? What is the energy source for photochemical smog?

92. What atmospheric reaction favors the formation of nitrogen monoxide, NO? Explain how the formation of NO in a combustion chamber is similar to the formation of NH_3 in a reactor designed to manufacture ammonia.

93. What two acidic gases are primarily responsible for acid rain? Write equations for the formation of the respective acids these gases can form.

94. Give an example of atmospheric ozone that is beneficial and an example of ozone that is harmful. Explain how ozone is beneficial and how it is harmful.

95. The air pollutant sulfur dioxide, SO_2, is known to increase mortality in people exposed to it for 24 hr at a concentration of 0.175 ppm. (One part per million is the equivalent of 1 L of SO_2 dispersed in a million liters of air.)
(a) What is the partial pressure of SO_2 when its concentration is 0.175 ppm?
(b) What is the mole fraction of SO_2 at the same concentration?
(c) Assuming the air is at STP, what mass in micrograms of SO_2 is present in 1 m^3?

General Questions

96. Acetylene can be made by allowing calcium carbide to react with water.

$$CaC_2(s) + 2\,H_2O(\ell) \longrightarrow C_2H_2(g) + Ca(OH)_2(aq)$$

Assume that you place 2.65 g of CaC_2 in excess water, and collect the acetylene over water as shown in the figure. The volume of the acetylene and water vapor is 795 mL at 25.0 °C and a barometric pressure of 735.2 mm Hg. Correct for the partial pressure of water vapor in the gas sample and calculate the percent yield of acetylene. The vapor pressure of water at 25 °C is 23.8 mm Hg.

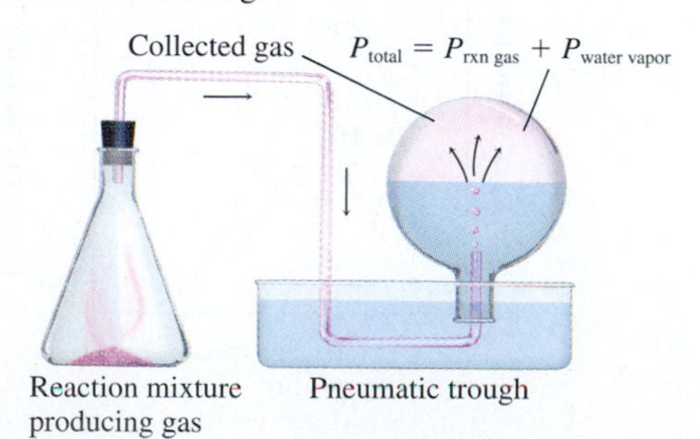

97. Hydrogen chloride, HCl, can be made by the direct reaction of H_2 and Cl_2 in the presence of light. Assume that 3.0 g of H_2 and 140 g of Cl_2 are mixed in a 10-L flask at 28 °C.
Before reaction:
(a) What are the partial pressures of the two reactants?
(b) What is the total pressure due to the gases in the flask?
After reaction:
(c) What is the total pressure in the flask?
(d) What reactant remains in the flask? How many moles of it remain?
(e) What are the partial pressures of the gases in the flask?
(f) What will be the pressure inside the flask if the temperature is increased to 40 °C?

98. One of the major sources of SO_2 in the atmosphere is from the oxidation of H_2S, produced by the decay of organic matter. Worldwide, about 100 million metric tons of H_2S is produced from sources that include the oceans, bogs, swamps, and tidal flats. The reaction in which H_2S molecules are oxidized to SO_2 involves O_3. Write an equation showing that one molecule of each reactant combines to form two product molecules, one of them being SO_2. Then calculate the annual production in tons of H_2SO_4, assuming all of this SO_2 is converted to sulfuric acid.

Applying Concepts

99. Consider a sample of N_2 gas under conditions where it obeys the ideal gas law exactly. Which of the following statements are true?
(a) A sample of Ne(g) under the same conditions must obey the ideal gas law exactly.
(b) The speed at which one particular N_2 molecule is moving changes from time to time.
(c) Some N_2 molecules are moving more slowly than some of the molecules in a sample of O_2(g) under the same conditions.
(d) Some N_2 molecules are moving more slowly than some of the molecules in a sample of Ne(g) under the same conditions.
(e) When two N_2 molecules collide, it is possible that both may be moving faster after the collision than they were before.

100. Which graph below would best represent the distribution of molecular speeds for the gases acetylene (C_2H_2) and N_2? Both gases are in the same flask with a total pressure of 750 mm Hg. The partial pressure of N_2 is 500 mm Hg.

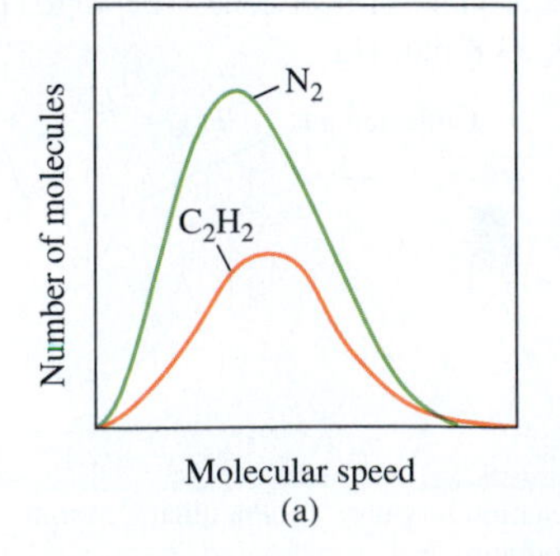

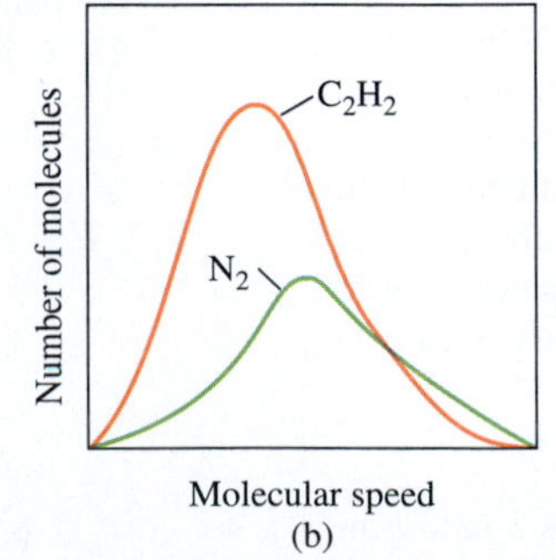

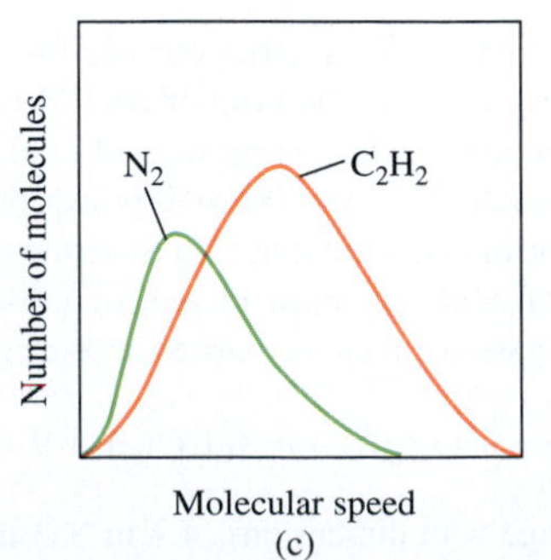

101. Draw a graph representing the distribution of molecular speeds for the gases ethane (C_2H_6) and F_2 when both are in the same flask with a total pressure of 720 mm Hg and a partial pressure of 540 mm Hg for F_2.

102. In this chapter, Boyle's, Charles's, and Avogadro's laws were presented as word statements and mathematical relationships. Express each of these laws graphically.

103. The drawing below represents a gas collected in a syringe (the needle end was sealed after collecting) at room temperature and pressure. Redraw the syringe and gas to show what it would look like under the following conditions. Assume that the plunger can move freely but no gas can escape.
(a) The temperature of the gas is decreased by one half.
(b) The pressure of the gas is decreased to one half its initial value.
(c) The temperature of the gas is tripled and the pressure is doubled.

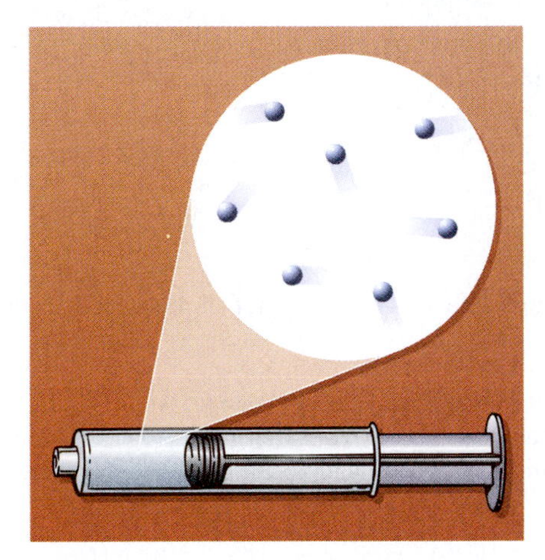

104. A gas phase reaction takes place in a syringe at a constant temperature and pressure. If the initial volume is 40 cm^3 and the final volume is 60 cm^3, which of the following general reactions took place? Explain your reasoning.
(a) $A(g) + B(g) \rightarrow AB(g)$
(b) $2\ A(g) + B(g) \rightarrow A_2B(g)$
(c) $2\ AB_2(g) \rightarrow A_2(g) + 2\ B_2(g)$
(d) $2\ AB(g) \rightarrow A_2(g) + B_2(g)$
(e) $2\ A_2(g) + 4\ B(g) \rightarrow 4\ AB(g)$

105. The gas molecules in the box below undergo a reaction at con-

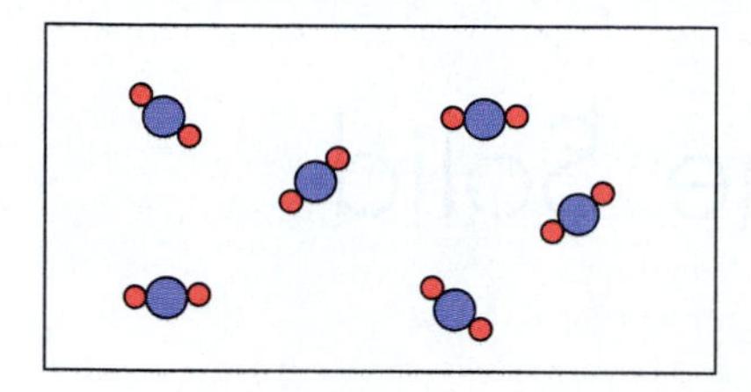

stant temperature and pressure. If the initial volume is 1.8 L and the final volume is 0.9 L, which of the boxes below could be the products of the reaction? Explain your reasoning.

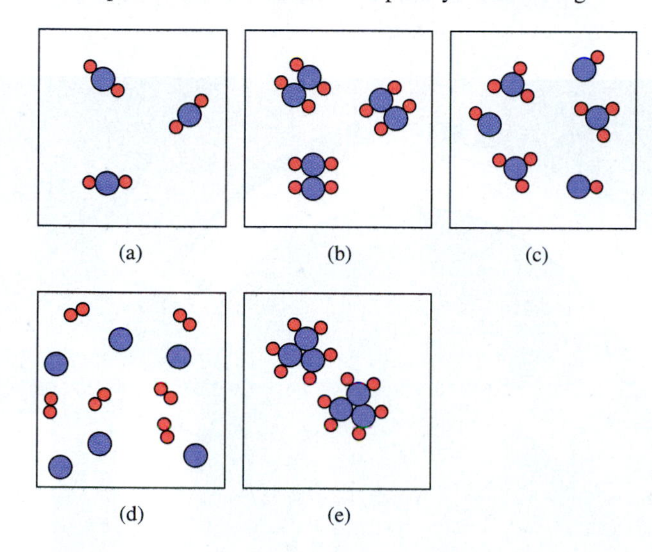

106. A substance is analyzed and found to contain 85.7% carbon and 14.3% hydrogen by mass. A gaseous sample of the substance is found to have a density of 1.87 g/L at STP.

(a) What is the molar mass of the compound?

(b) What are the empirical and molecular formulas of the compound?

(c) What are two possible Lewis structures for molecules of the compound?

107. A compound consists of 37.5% C, 3.15% H, and 59.3% F by mass. When 0.298 g of the compound is heated to 50 °C in an evacuated 125-mL flask, the pressure is observed to be 750 mm Hg. The compound has three isomers.

(a) What is the molar mass of the compound?

(b) What are the empirical and molecular formulas of the compound?

(c) Draw the Lewis structure for each isomer of the compound.

108. One very cold winter day you and a friend purchase a helium-filled balloon. As you leave the store and walk down the street, your friend notices the balloon is not as full as it was a moment ago in the store. He says the balloon is defective and he is taking it back. Do you agree with him? Explain why you do or do not agree.

CHAPTER

15 The Liquid State, The Solid State, and Modern Materials

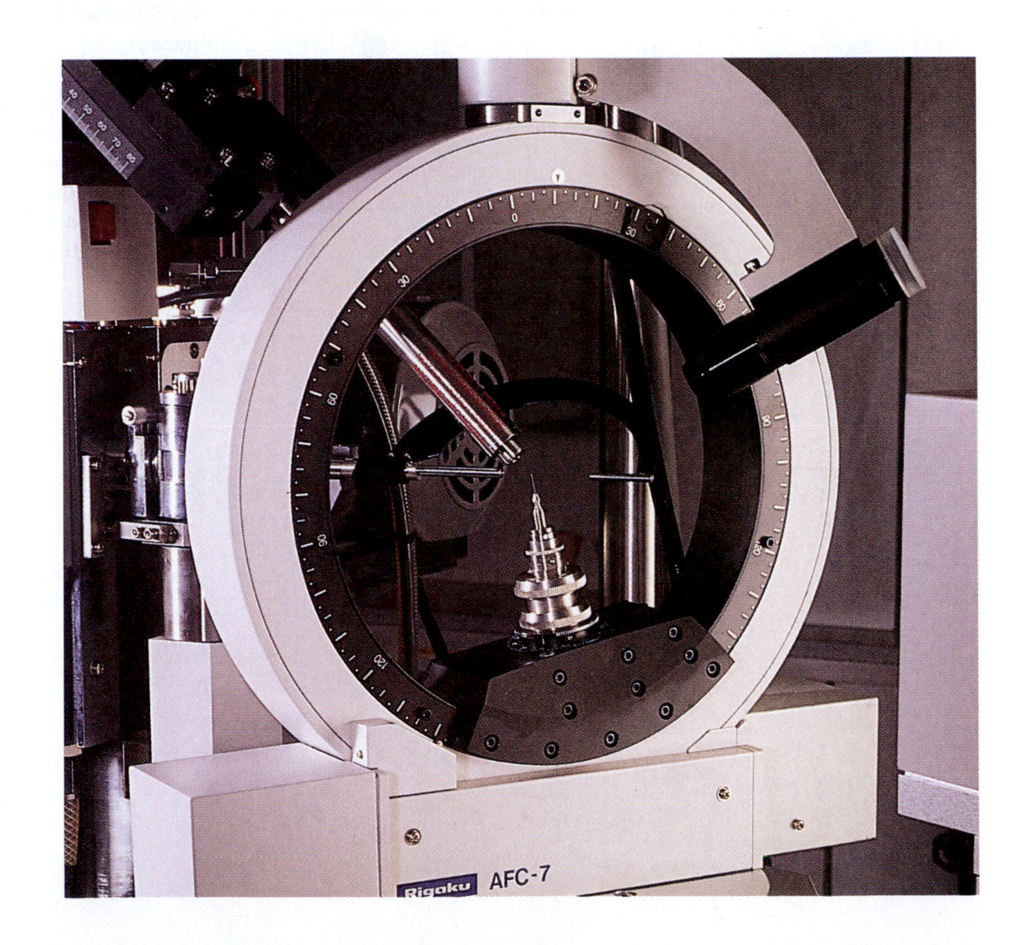

Solid xenon being studied in an X-ray diffractometer. X rays have wavelengths similar to the spacing between the xenon atoms and are diffracted as they pass through the crystal. Careful measurement of the diffraction pattern tells scientists much about the arrangement of xenon atoms at low temperatures and under pressure. • If there were a planet where xenon was as plentiful as water is on Earth, one thing is certain, there would be no xenon "icebergs" floating in the oceans of liquid xenon. Do you know why? This question is explained in the summary problem at the end of this chapter. *(Courtesy Oxford Cryosystems)*

In the condensed states of matter, atoms, molecules, or ions are close enough to touch one another. The fact that their particles are, on average, able to touch each other gives solids and liquids one property that is vastly different from that of gases—they can be compressed very little. Liquids are like gases in that their individual particles, whether they be atoms, molecules, or ions (in the case of molten ionic compounds), can move freely. Liquids are therefore able to flow and fill a container to a given level. Solids, on the other hand, are composed of particles in relatively fixed positions, not able to move any appreciable distances, but rather vibrating in those positions. As a result, solids have definite shapes.

When the particles making up a solid are given enough energy to overcome the attractive forces holding them in place, the solid melts. In molecular liquids and solids, molecules with sufficient energy can also leave the surface and go into the gaseous state. In the reverse process, as energy is removed from a sample of matter, gases condense into liquids or solids, and liquids solidify. As you will see in this chapter, all of these properties of solids and liquids have great practical importance.

15.1 THE LIQUID STATE

At low enough temperatures most gases condense to liquids. Condensation occurs when most molecules no longer have enough kinetic energy to overcome their intermolecular attractions *(⇐ p. 423)*. Most liquids are substances whose condensation temperatures are above room temperature. These include such common substances as water, alcohol, or gasoline. There are, however, some liquids, like molten salts and liquid polymers, that have no corresponding vapor state, so condensation of their vapors has no meaning.

In a liquid the molecules are close together, and so, unlike a gas, a liquid is only very slightly compressible. However, the molecules remain mobile enough that the liquid flows. At the nanoscale level, liquids have a regular structure only in very small regions, and most of the molecules continue to move about randomly. Because they are difficult to compress and their molecules are moving in all directions, confined liquids can transmit applied pressure equally in all directions. This property is used in the hydraulic fluids that operate mechanical devices such as automotive brakes and airplane wing surfaces, tail flaps, and rudders.

Unlike gases, liquids have *surface properties.* Molecules beneath the surface of the liquid are completely surrounded by other molecules and experience intermolecular attractions in all directions. By contrast, molecules at the surface are attracted only by molecules below or beside them (Figure 15.1). This unevenness of attractive

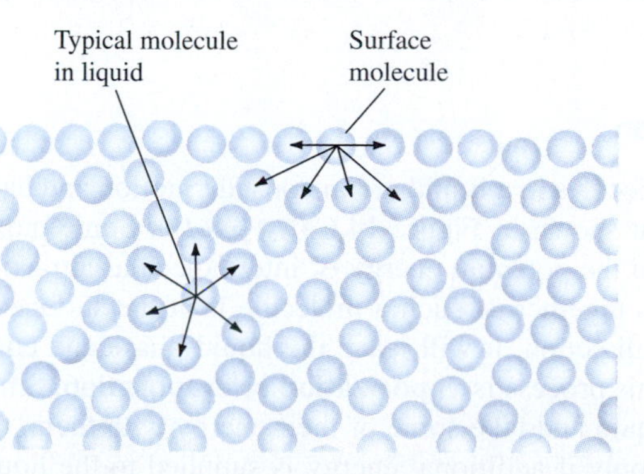

Figure 15.1 Surface tension. Surface tension is the energy required to expand the surface of a liquid. It arises from the difference between the forces acting on a molecule within the liquid phase and those acting on a molecule at the surface of the liquid.

A class of chemicals called surfactants (soaps and detergents are examples) can dissolve in water and dramatically lower its surface tension. When this happens, water becomes "wetter" and does a better job of cleansing. You can see the effect of a surfactant by comparing how water alone beads on a surface such as the hood of a car, and how a soap solution fails to bead on the same surface. Surfactants are discussed in Section 16.12.

TABLE 15.1 Surface Tensions of Some Liquids

Compound	Formula	Surface Tension (J/m^{2*} at 20 °C)
Benzene	C_6H_6	2.85×10^{-2}
Chloroform	$CHCl_3$	2.68×10^{-2}
Ethanol	CH_3CH_2OH	2.23×10^{-2}
Octane	C_8H_{18}	2.16×10^{-2}
Mercury	Hg	46×10^{-2}
Water	H_2O	7.29×10^{-2}

*The units of surface tension are related to the work required (in joules) to expand a unit of surface area (m^2).

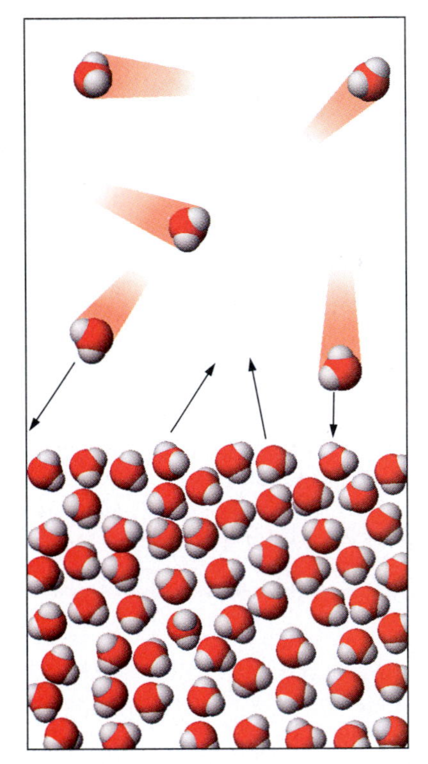

Figure 15.2 Molecules in the liquid and gas phases. All the molecules in both phases are moving, although the distances traveled in the liquid before collision with another molecule are small. As indicated by arrows, some of the liquid-phase molecules are moving with a kinetic energy large enough to overcome the intermolecular forces in the liquid and escape to the gas phase. At the same time, some molecules of the gas re-enter the liquid surface (also indicated by arrows).

forces at the surface of the liquid causes the surface to contract, just as the surface of a balloon contracts unless there is air inside. The energy required to expand a liquid surface is called the **surface tension,** and it is higher for liquids that have strong intermolecular attractions. For example, water's surface tension is high compared with those of most other liquids (Table 15.1) because of the extensive hydrogen bonding that holds water molecules together *(← p. 426)*. The very high surface tension of mercury is due to the strong metallic bonding (Section 15.8) that holds Hg atoms together in the liquid.

Even though their densities are greater than that of water, water bugs can walk on the surface of water and small metal objects can "float." Surface tension prevents the objects from breaking through and sinking. Surface tension accounts for the nearly spherical shape of water droplets that bead up on waxed surfaces. A sphere has less surface area per unit volume than any other shape, so that in a spherical droplet fewer molecules experience uneven attractive forces at the surface.

Exercise 15.1 Explaining Differences in Surface Tension

Chloroform ($CHCl_3$) has a surface tension at 20 °C of 2.68×10^{-2} J/m^2, while bromoform ($CHBr_3$) has a higher surface tension of 4.11×10^{-2} J/m^2 at the same temperature. Explain this observation in terms of intermolecular forces.

Exercise 15.2 Predicting Surface Tension

Predict which compound listed in Table 15.1 has a surface tension most similar to that of (a) glycerol, which has three —OH groups *(← p. 361)*, and (b) decane, $C_{10}H_{22}$.

Vaporization and Condensation

Like the molecules in a gas, the molecules in a liquid are in constant motion and have a range of kinetic energies like that shown in Figure 14.6 *(← p. 617)*. Some molecules have more kinetic energy than the potential energy of intermolecular attractive forces holding the liquid molecules together. If such a molecule is at the surface of the liquid and moving in the right direction, it will leave the liquid phase and enter the gaseous phase (Figure 15.2). This process is **vaporization** or **evaporation.** Since high-energy molecules leave the liquid and take some of their energy with them, the vaporization process can continue only if additional energy is supplied to the liquid to produce more molecules with enough energy to vaporize. Therefore, the process of vaporization is *endothermic* (the heat required is called the heat of vaporization (← p. 232)).

Exercise 15.3 Evaporative Cooling

In some countries where electrical refrigeration is not readily available, drinking water is chilled by placing it in porous clay water pots. Water slowly passes through the clay and when it reaches the outer surface, it evaporates. Explain how this cools the water inside.

A molecule in the gas phase will eventually transfer some of its kinetic energy by colliding with slower gaseous molecules and solid objects. If it should happen to come in contact with the liquid's surface again, it can reenter the liquid phase in a process called **condensation**. The overall effect of molecules reentering the liquid phase is the release of the heat of vaporization, making condensation an exothermic process. The heat evolved is equal to the heat absorbed when vaporization occurs.

$$\text{Liquid} \underset{\text{heat of condensation}}{\overset{\text{heat of vaporization}}{\rightleftharpoons}} \text{Gas} \qquad \Delta H_{vaporization} = -\Delta H_{condensation}$$

The thermal energy transferred during a change in state at constant pressure is an enthalpy change, ΔH (← p. 238).

For example, the quantity of heat required to completely vaporize 1 mol of water (the *molar heat of vaporization*) once it has reached the boiling point of 100 °C and the quantity of heat released when 1 mol of water at 100 °C condenses to liquid water at 100 °C (the *molar heat of condensation*) are

$$H_2O(\ell) \longrightarrow H_2O(g) \qquad \Delta H_{vap} = +40.7 \text{ kJ/mol}$$
$$H_2O(g) \longrightarrow H_2O(\ell) \qquad \Delta H_{cond} = -40.7 \text{ kJ/mol}$$

The thermal energy absorbed during evaporation is sometimes called the latent heat of vaporization. It is somewhat dependent on the temperature: at 100 °C, ΔH_{vap} (H_2O) = 40.7 kJ/mol, but at 25 °C the value is 44 kJ/mol. (← p. 238)

Table 15.2 illustrates the influence of intermolecular forces on heats of vaporization and boiling points. In the series of nonpolar molecules and noble gases, the increasing London forces (← p. 423) with increasing molecular size and numbers of electrons are shown by the increasing ΔH_{vap} and boiling point values. Comparison of HF and H_2O with ethane, which has the same number of electrons shows the effect of hydrogen bonding.

Conceptual Challenge Problem CP-15.A at the end of the chapter relates to topics covered in this section.

Problem-Solving Example 15.1 Enthalpy of Vaporization

You put 1.00 L of water (about 4 cups) in a pan at 100 °C, and it evaporates. How much thermal energy must have been absorbed (at constant pressure) by the water for it all to vaporize? (The density of water at 100 °C is 0.958 g/mL.)

Answer 2.17×10^3 kJ

Explanation There are three pieces of information you need to solve this problem:

(a) ΔH_{vap} for water is 40.7 kJ/mol at 100 °C (Table 15.2).
(b) The density of water is needed because ΔH_{vap} has units of kJ/mol, so you must first find the mass of water and then the number of moles.
(c) The molar mass of water is 18.02 g/mol.

Given the density of water, a volume of 1.00 L (or 1.00×10^3 cm^3) is equivalent to 958 g, and this mass in turn is equivalent to 53.2 mol of water. Therefore,

$$\text{Heat required for vaporization} = 53.2 \text{ mol } H_2O \times \left(\frac{40.7 \text{ kJ}}{\text{mol}}\right) = 2.17 \times 10^3 \text{ kJ}$$

This enthalpy change of 2170 kJ is equivalent to about one quarter of the energy in the daily food intake of an average person in the United States.

Problem-Solving Practice 15.1

A rainstorm deposits 2.5×10^{10} kg of rain. Using the heat of vaporization of water at 25 °C of 44.0 kJ/mol, calculate the quantity of thermal energy, in joules, transferred when this much rain forms. Is this process exothermic or endothermic?

TABLE 15.2 Molar Enthalpies of Vaporization and Boiling Points for Some Common Substances

Substance	Number of Electrons	ΔH_{vap} (kJ/mol)*	Boiling Point (°C) (vapor pressure = 760 mm Hg)
Polar Molecules			
SO_2	32	26.8	−10.0
HF	10	25.2	19.7
HCl	18	17.5	−84.8
NH_3	10	25.1	−33.4
H_2O	10	40.7	100.0
HBr	36	19.3	−66.5
HI	54	21.2	−35.1
Noble Gases			
He	2	0.08	−269.0
Ne	10	1.8	−246.0
Ar	18	6.5	−185.9
Xe	54	12.6	−107.1
Nonpolar Molecules			
H_2	2	0.90	−252.8
O_2	16	6.8	−183.0
F_2	18	6.54	−188.1
Cl_2	34	20.39	−34.6
Br_2	70	29.54	59.6
CH_4 (methane)	10	8.9	−161.5
C_2H_6 (ethane)	22	15.7	−88.6
C_3H_8 (propane)	26	19.0	−42.1
C_4H_{10} (butane)	34	24.3	−0.5

*ΔH_{vap} is given at the normal boiling point of the liquid.

Exercise 15.4 Estimating ΔH_{vap}

Using the data from Table 15.2, estimate the ΔH_{vap} and boiling point for Kr. Do the same for NO_2. (*Hint:* Pick a substance in the table with a similar number of electrons.)

Exercise 15.5 Understanding Boiling Points

(a) Chlorine and bromine are both diatomic halogens. Explain the difference in their boiling points. (b) Methane and ammonia have the same number of electrons. Explain the difference in their boiling points.

15.2 LIQUID-VAPOR EQUILIBRIUM: VAPOR PRESSURE

Conceptual Challenge Problem CP-15.B at the end of the chapter relates to topics covered in this section.

The tendency of a liquid to vaporize is called its **volatility.** The higher the temperature, the greater the volatility because a larger fraction of the molecules have sufficient energy to overcome the attractive forces at the liquid's surface. Everyday experiences such as heating water or soup on a stove, or spilling a liquid on a hot pavement in the summer, tell us that raising the temperature of the liquid makes evaporation take place more readily. Conversely, at lower temperatures, the volatility of a liquid is lower.

A liquid in an open container will eventually evaporate completely, because air currents and diffusion take away most of the gas-phase molecules before they can reenter the liquid phase. In a closed container, however, no molecules can escape. If a liquid is injected into a container that contains a perfect vacuum, so that no other substance is present, the rate of vaporization (number of molecules vaporizing per unit time) will at first far exceed the rate of condensation. The pressure of gas above the liquid will increase as the number of gas-phase molecules increases. Eventually the system will attain a state of *dynamic equilibrium* in which molecules are entering and leaving the liquid state at equal rates (← *p. 569*). Once equilibrium is achieved it will appear that no further vaporization is occurring. At this point the pressure of the gas will no longer increase; this pressure is known as the **equilibrium vapor pressure** (or just the **vapor pressure**) of the liquid. As you would expect, the vapor pressure of a liquid increases with increasing temperature (Figure 15.3).

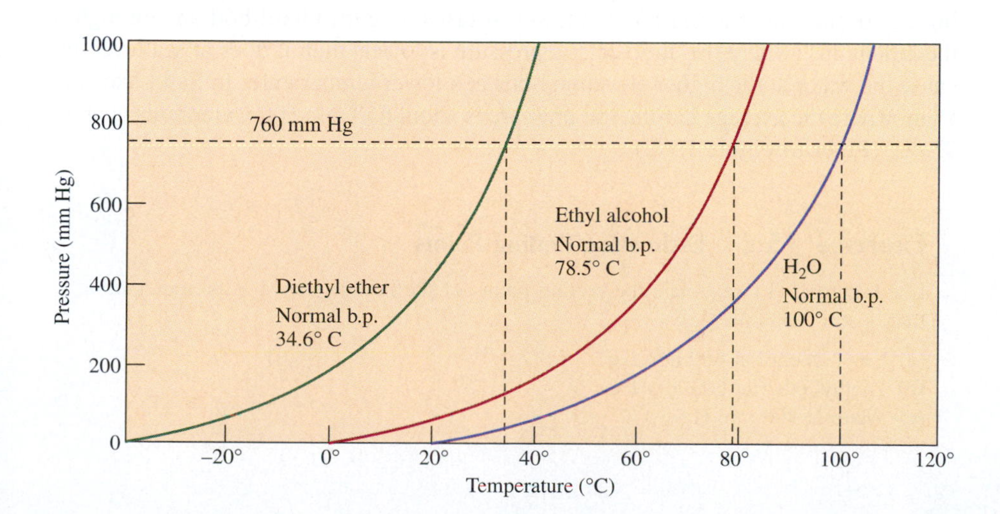

Figure 15.3 Vapor pressure curves for diethyl ether ($C_2H_5OC_2H_5$), ethyl alcohol (C_2H_5OH), and water. Each curve represents the conditions of T and P where the two phases (pure liquid and its vapor) are in equilibrium. Each compound is a liquid in the temperature and pressure region to the left of its curve, and it is a vapor for all temperatures and pressures to the right of the curve.

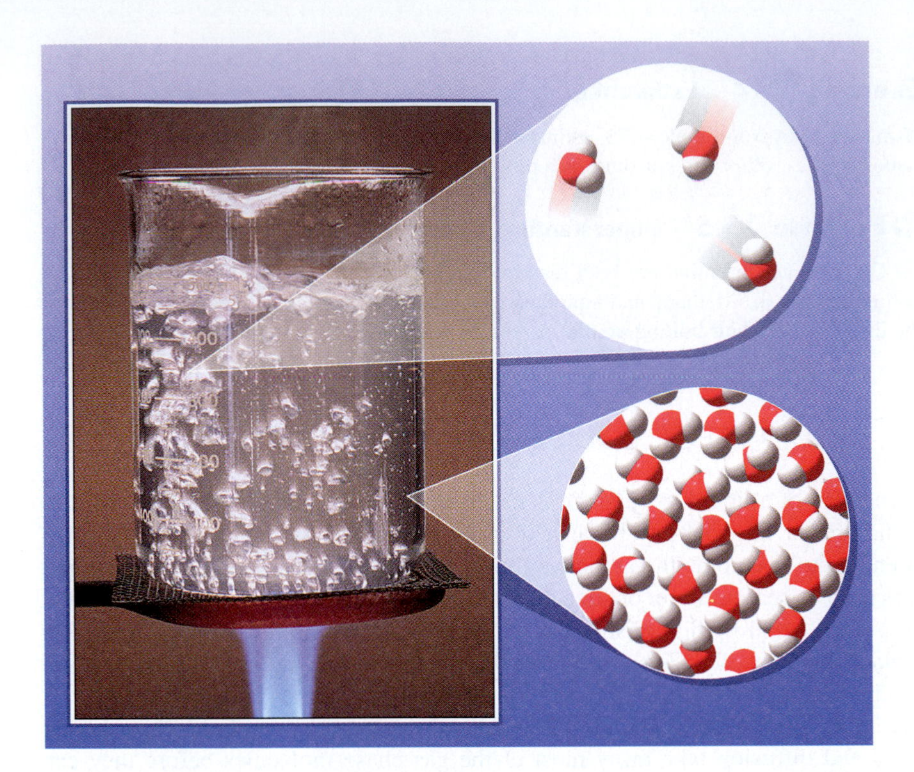

Figure 15.4 Boiling liquid. A liquid boils when its equilibrium vapor pressure equals the atmospheric pressure. Bubbles of vapor that form within the liquid consist of the same kind of molecules (H_2O in this case) as the liquid. *(C.D. Winters)*

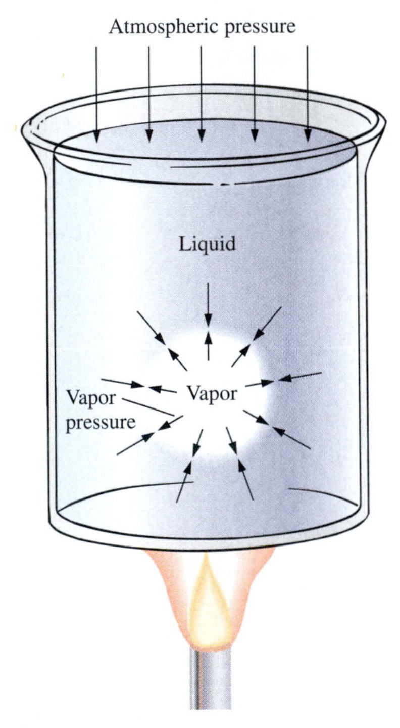

Bubble formation in a boiling liquid. When the vapor pressure of the liquid is equal to the external pressure the liquid boils. Bubbles may form at any place in the liquid.

To shorten cooking times, one can use a pressure cooker. This is a sealed pot (with a relief valve for safety) that allows water vapor to build up pressures slightly greater than the external atmospheric pressure. At the higher pressure, the boiling point of water is higher and foods cook faster.

If a liquid is placed in an open container and heated, a temperature eventually is reached at which the vapor pressure of the liquid is equal to the atmospheric pressure. Below this temperature, only molecules at the surface of the liquid can go into the gas phase. But when the vapor pressure equals the atmospheric pressure, the liquid begins vaporizing throughout. Bubbles of vapor form and immediately rise to the surface due to their lower density. The liquid is said to be **boiling** (Figure 15.4). The temperature at which the equilibrium vapor pressure equals the atmospheric pressure is the **boiling point** of the liquid. If the atmospheric pressure is 1 atm, the temperature is designated the **normal boiling point.** The boiling points shown in Table 15.2 are normal boiling points.

The lower the atmospheric pressure, the lower the vapor pressure at which boiling can occur—the boiling point is lower. It takes longer to hard-boil an egg high in the mountains (where the atmospheric pressure is lower) than it does at sea level, because the water at the higher elevation boils at a lower temperature. In Salt Lake City, Utah, where the average barometric pressure is about 650 mm Hg, water boils at about 95 °C. (Refer to Figure 15.3.)

Exercise 15.6 Estimating Boiling Points

Use Figure 15.3 to estimate the boiling points of the liquids shown at pressures other than 1 atm (760 mm Hg):

(a) Ethyl alcohol at 400 mm Hg
(b) Diethyl ether at 200 mm Hg
(c) Water at 400 mm Hg

Exercise 15.7 Explaining Bubbles

One of your classmates believes that the bubbles in a boiling liquid are air bubbles. Explain to him what is wrong with that idea and explain what the bubbles actually are. How could you do an experiment to show that the bubbles are not air?

15.3 PHASE CHANGES: SOLIDS, LIQUIDS, AND GASES

When a solid is heated, its temperature increases. Unless the solid decomposes first, a temperature is reached at which the vibration of the molecules or ions is violent enough that they move out of their fixed positions. The solid's structure collapses and it melts or fuses (Figure 15.5). This temperature is the melting point of the solid. Melting requires transfer of energy, the heat of fusion, from the surroundings into the system. The molar heat of fusion *(← p. 232)* is the thermal energy required to melt 1 mol of a pure solid. Solids with high heats of fusion usually melt at high temperatures, and solids with low heats of fusion usually melt at low temperatures. The reverse of melting is *solidification* or **crystallization,** which is always an exothermic process. The molar heat of crystallization is equal in magnitude but opposite in sign from the molar heat of fusion.

$$\text{Solid} \underset{\text{heat of crystallization}}{\overset{\text{heat of fusion}}{\rightleftharpoons}} \text{Liquid} \qquad \Delta H_{\text{fusion}} = -\Delta H_{\text{crystallization}}$$

Some solids do not have measurable melting points and some liquids do not have measurable boiling points because increasing temperature causes them to decompose chemically before they melt or boil. Making peanut brittle, described at the beginning of Section 1.4 *(← p. 7)*, provides an example: melted sucrose chars before it boils, but it produces a great-tasting candy.

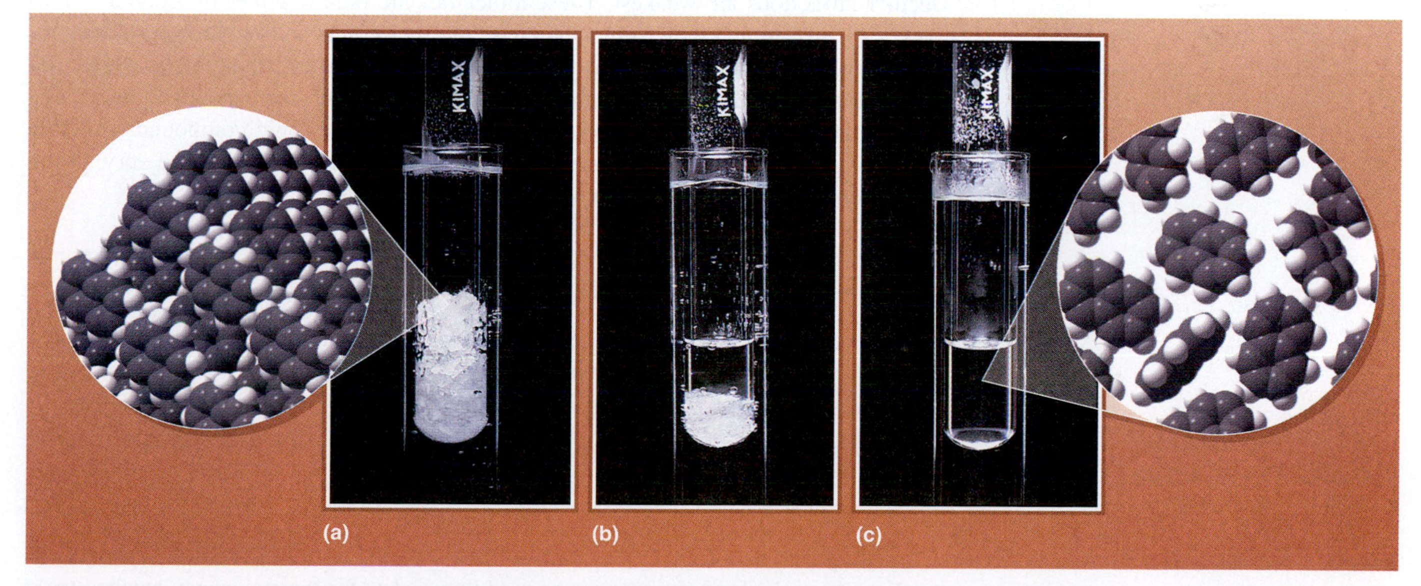

Figure 15.5 The melting of naphthalene, $C_{10}H_8$, at 80.22 °C. (a) The crystals of naphthalene are at a temperature just below the melting temperature. (b) Melting commences at 80.22 °C. (c) The entire sample is molten. Note that some of the naphthalene has sublimed. *(C.D. Winters)*

TABLE 15.3 Enthalpies of Fusion and Melting Points of Some Solids

Solid	Melting Point (°C)	Enthalpy of Fusion (kJ/mol)	Type of Intermolecular Forces
Molecular Solids: Nonpolar Molecules			
O_2	−248	0.445	These molecules have only London forces (which increase with the number of electrons).
F_2	−220	1.020	
Cl_2	−103	6.406	
Br_2	−7.2	10.794	
Molecular Solids: Polar Molecules			
HCl	−114	1.990	All of these molecules have London forces enhanced by dipole-dipole forces. H_2O also has significant hydrogen bonding.
HBr	−87	2.406	
HI	−51	2.870	
H_2O	0	6.020	
H_2S	−86	2.395	
Ionic Solids			
NaCl	800	30.21	All ionic solids have strong attractions between oppositely charged ions.
NaBr	747	25.69	
NaI	662	21.95	

Table 15.3 lists melting points and heats of fusion for examples of three classes of compounds: (a) nonpolar molecular solids, (b) polar molecular solids, some capable of hydrogen bonding, and (c) ionic solids. Solids composed of low-molecular-weight nonpolar molecules have the lowest melting temperatures, because intermolecular attractions are weakest. These molecules are held together by London forces only *(← p. 424)*, and they form solids with melting points so low that we seldom encounter them as solids. Melting points and heats of fusion of nonpolar molecular solids increase with increasing number of electrons (which corresponds to increasing molar mass) as the London forces become stronger. The ionic compounds in Table 15.3 have the highest melting points and heats of fusion because of the very strong ionic bonding that holds the ions together.

PROBLEM-SOLVING EXAMPLE 15.2 Heat of Fusion

The molar heat of fusion of NaI is 21.95 kJ/mol at its melting point. How much thermal energy will be absorbed when 41.65 g of NaI melts?

Answer 6.055 kJ

Explanation First it is necessary to determine how many moles of NaI are present in 41.65 g of NaI.

$$\text{Moles NaI: } 41.65 \text{ g NaI} \times \left(\frac{1 \text{ mol NaI}}{150.9843 \text{ g NaI}}\right) = 0.27586 \text{ mol NaI}$$

Now the molar heat of fusion can be used to calculate the thermal energy required to melt this many moles of NaI.

$$\text{Thermal energy required: } 0.27586 \text{ mol NaI} \times \left(\frac{21.95 \text{ kJ}}{\text{mol NaI}}\right) = 6.055 \text{ kJ}$$

Problem-Solving Practice 15.2

Calculate the thermal energy transfer required to melt 100.0 g of NaCl at its melting point.

Exercise 15.8 Heat Liberated upon Crystallization

Which would liberate more thermal energy: the crystallization of 2 mol of liquid bromine or the crystallization of 1 mol of liquid water?

Sublimation

Atoms or molecules can escape directly from a solid to the gas phase, a process known as sublimation (*← p. 233*). The heat required is the heat of sublimation. The reverse process in which a gas is converted directly to a solid is called **deposition**. The heat evolved in this exothermic process (the heat of deposition) has the same magnitude as the heat of sublimation but the opposite sign.

$$\text{Solid} \underset{\text{heat of deposition}}{\overset{\text{heat of sublimation}}{\rightleftharpoons}} \text{Gas} \qquad \Delta H_{\text{sublimation}} = -\Delta H_{\text{deposition}}$$

A common substance that sublimes at normal atmospheric pressure is solid carbon dioxide, which has a vapor pressure of 1 atm at −78 °C (Figure 15.6). Having such a high vapor pressure below its melting point causes solid carbon dioxide to sublime rather than melt. Because of the high vapor pressure of the solid, carbon dioxide in the liquid state can exist only under pressures much higher than 1 atm.

In the reverse of sublimation, atoms or molecules in the gas phase can be made to deposit on the surface of a solid. Deposition is used to form thin coatings of metal atoms on surfaces. CD-audio and CD-ROM disks have shiny metallic surfaces of deposited aluminum or gold atoms (see p. 670). A metal filament is heated in a vacuum to a temperature where metal atoms begin to boil off the surface. The plastic compact disk is cooler than the filament so the metal atoms in the gas phase quickly

Sublimation of iodine. When iodine is heated, molecules leave the surface to go into the gaseous phase. When they come in contact with a cooler surface, they deposit as solid iodine. Sublimation is used to purify many substances that contain nonvolatile impurities or impurities that will not deposit as readily as the substance being sublimed. *(Marna C. Clarke)*

Solid CO_2 is commonly known by the trade name Dry Ice.

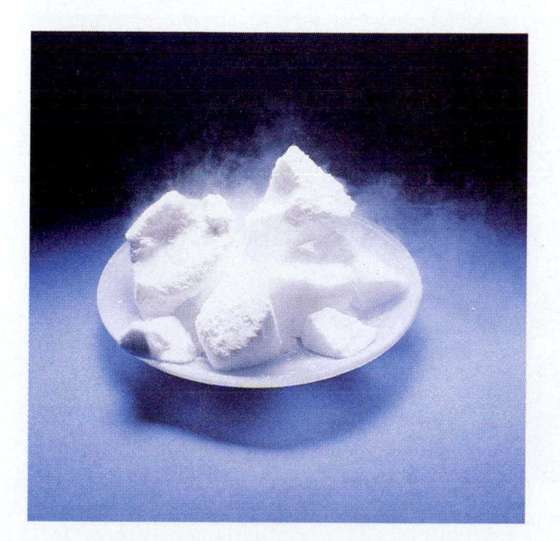

Figure 15.6 Dry Ice. In this photo the cold vapors of CO_2 are causing moisture, which is seen as wispy white clouds, to condense. Being more dense than air at room temperature, the CO_2 vapors glide slowly toward the table top or the floor. *(C.D. Winters)*

CD-audio disks. *(C.D. Winters)*

Magnified view of the surface of a compact disk. Information (sound, data, and video) is stored on the surface of a compact disk in the form of pits and "lands" (the portion on the disk along the data track where there is no pit.) About 7×10^9 total bits of information can be stored on the surface of a standard compact disk. *(Courtesy of DADC)*

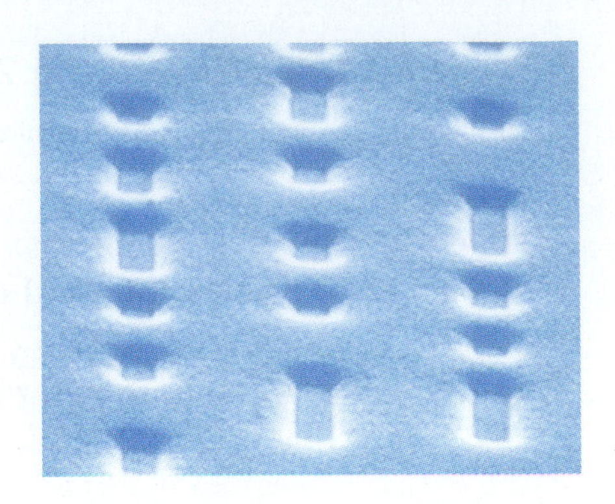

Conceptual Challenge Problem CP-15.C at the end of the chapter relates to topics covered in this section.

deposit on the cool surface. The purpose of the metal coating is to provide a reflective surface for the laser beam that reads the pits and lands (unpitted areas) containing the digital audio or data information ***(← p. 308).***

The heat of sublimation of ice is 51 kJ/mol and its vapor pressure at 0 °C is 4.60 mm Hg. Have you ever noticed that ice cubes in a frostfree refrigerator and snow slowly disappear even if the temperature never gets above freezing? The reason is that ice sublimes readily in dry air when the partial pressure of water vapor is below 4.6 mm Hg (Figure 15.7). Given enough air passing over it, a sample of ice will completely sublime, leaving no trace behind. In a frostfree refrigerator, a current of dry air periodically blows across any ice formed in the freezer compartment, taking away water vapor (and hence the ice) without warming the freezer enough to cause thawing of the food.

We can summarize in Figure 15.8 what has been said about phase changes. Each of the three states of matter can be interconverted by heating or cooling.

Exercise 15.9 Frostfree Refrigeration

Sometimes, because of humidity conditions, a frostfree refrigerator doesn't work as efficiently as it should. Can you explain why?

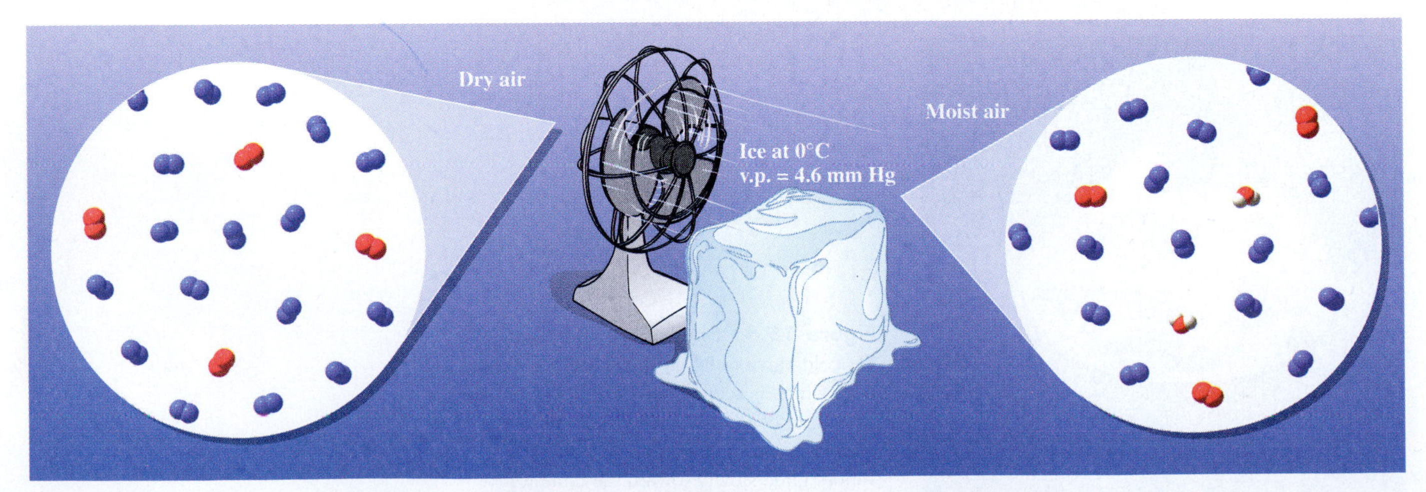

Figure 15.7 Ice subliming. As dry air passes over a sample of ice, the water molecules leaving the surface are carried away.

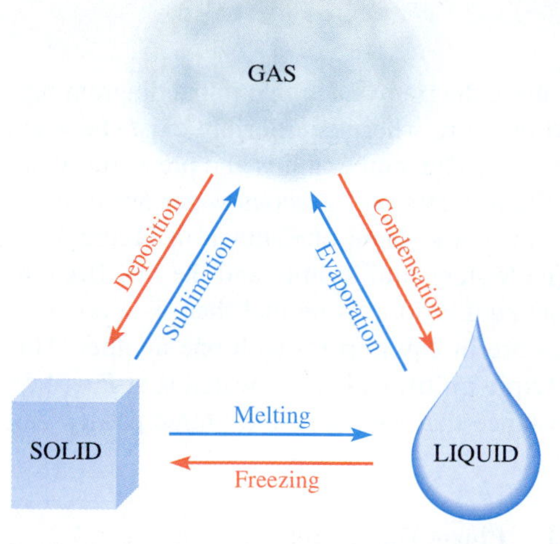

Figure 15.8 Changes between the three states of matter. Exothermic changes are shown in red and endothermic changes are shown in blue.

Exercise 15.10 Purification by Sublimation

Sublimation is an excellent means of purification for compounds that will readily sublime. Explain how purification by sublimation works at the nanoscale.

Phase Diagrams

As Figure 15.3 illustrates, there is a relationship between temperature and pressure for liquids. An expanded diagram of this kind deals with the relationships among the three phases of a substance—solid, liquid, and gas—as temperature and pressure change. Such diagrams are called **phase diagrams.** Look at Figure 15.9, the phase diagram for water. Note the three colored regions. Each of the three phases, solid, liquid, and gas, is represented. The temperature scale has been exaggerated to better

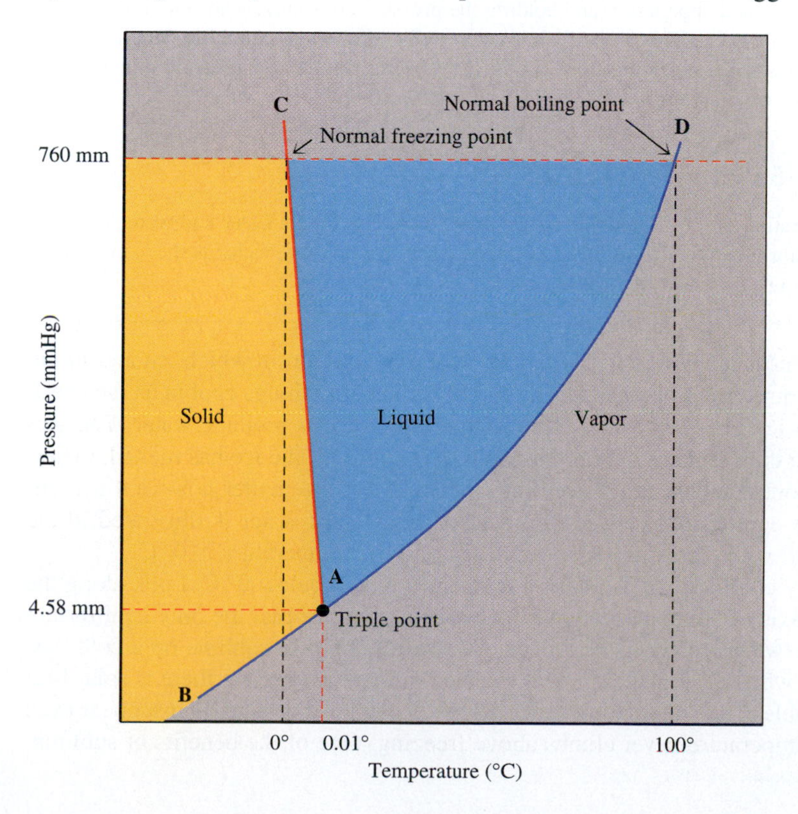

Figure 15.9 The phase diagram for water. The temperature and pressure scales are greatly exaggerated.

illustrate some of the features of water's phase diagram. Each line in the diagram represents the conditions of temperature and pressure where equilibrium exists between the two phases shown on either side of the line. The temperatures and pressures along the line AD represent conditions where liquid water and gaseous water are in equilibrium. This line, the vapor pressure curve, is the same one shown in Figure 15.3. The line AC represents the solid/liquid (ice/water) equilibrium, and the line BA represents the solid/gaseous (ice/water vapor) equilibrium. Note that there is a point on the phase diagram where all three phases are in equilibrium with one another. This temperature-pressure combination is the **triple point,** which, for water, is at $P = 4.58$ mm Hg and $T = 0.01$ °C. Each pure substance that can exist in all three phases will have a characteristic phase diagram.

PROBLEM-SOLVING EXAMPLE 15.3 Phase Diagrams

Using Figure 15.9, the phase diagram for water, answer these questions. Starting at the triple point, (a) What phase exists when the pressure is held constant and the temperature is increased to 0.5 °C? (b) What phase exists when the temperature is held constant and the pressure is increased to 20 mm Hg? (c) What phase exists when the pressure is held constant and the temperature is decreased to 0 °C?

Answer

(a) Vapor (b) Liquid (c) Solid

Explanation

(a) Increasing the temperature (and holding the pressure constant) means moving slightly to the right from the triple point. These are conditions where only gaseous water exists.
(b) Increasing the pressure (and holding the temperature constant) means moving slightly up from the triple point. This is in the liquid region of the diagram.
(c) Decreasing the temperature (and holding the pressure constant) means moving to the left of the triple point. This is the region where only solid water can exist. Similar reasoning applies when changing from conditions of temperature and pressure along any line in the phase diagram.

Problem-Solving Practice 15.3

At a temperature of 11 °C and a pressure of 9.8 mm Hg, liquid water and water vapor are in equilibrium. What form of water exists when the pressure remains the same and the temperature increases to 12 °C?

You are most likely familiar with the common situation in which ice and liquid water are in equilibrium because it is so useful for keeping things cool in an ice chest. If you fill an ice chest with ice, some of the ice melts to produce water. The ice-water mixture then remains at approximately 0 °C until all the ice has melted. (Thermal energy continually is transferred into the ice chest, no matter how well it is insulated.) This equilibrium between solid water and liquid water is illustrated in the phase diagram where pressure is 760 mm Hg and the temperature is 0 °C.

On a very cold day, ice can have a temperature well below 0 °C. Look along the temperature axis of the phase diagram for water and notice that the only equilibrium possible is between ice and water vapor. As a result, ice will sublime on a cold day. Sublimation, which is endothermic, takes place more readily when there is solar heat energy available. The sublimation of ice allows snowfall to gradually disappear even though the temperature never climbs above freezing. One of the benefits of sublima-

CHEMISTRY YOU CAN DO

Melting Ice with Pressure

Do this experiment in a sink. Obtain a piece of thin, strong, single-strand wire about 50 cm long, two weights of about 1 kg each, and a piece of ice about 25 cm by 3 cm by 3 cm. The ice can be either a cylinder or a bar, and can be made by pouring some water into a mold made of several thicknesses of aluminum foil and placing it into a freezer. A piece of plastic pipe, not tightly sealed, will also work fine as a mold. You can use metal weights or make weights by filling two 1-qt milk jugs with water. (One quart is about 1 L, which is 1 kg of water.)

Fasten one weight to each end of the wire. Support each end of the bar of ice so that, without breaking it, you can hang the wire over it with one weight on each side. Observe the bar of ice and the wire every minute or so and record what happens. Suggest an explanation for what you have observed; the effect of pressure on melting of ice and the heat of fusion of ice are good ideas to consider when thinking about your explanation.

tion is that ice can disappear from an icy roadway even if the temperature remains below freezing.

The phase diagram for water is unusual in that the line AC slopes in the opposite direction from that seen for almost every other substance. The right-to-left, or negative, slope of the solid-liquid equilibrium line is a consequence of the lower density of ice compared to that of liquid water (which is discussed in the next section). When ice and water are in equilibrium, Le Chatelier's principle *(⇐ p. 586)* suggests that one way to cause ice to melt is to apply a greater pressure. The equilibrium should shift toward that form of water having the smaller volume, which is liquid water. This is also evident from Figure 15.9. If you start at the normal freezing point (0 °C, 1 atm) and increase the pressure, you will move into the area of the diagram that corresponds to liquid water.

Ice skating provides a practical example of this property of water. Ice is made much more slippery when a thin coating of liquid water is present. Recent studies have shown that the water molecules on the surface of ice are actually in a somewhat mobile state, resembling their mobility in liquid water. This probably contributes to the slickness of the surface of ice, but perhaps more important is the melting caused by one's body weight applied to a very small area of a skate blade. This greatly increases the pressure on the surface of the ice, causing it to melt (moving up from the line AC in the phase diagram with a constant temperature). The thin coat of liquid water acts as a lubricant to help the skates move easily over the ice (Figure 15.10).

Recall that pressure is force per unit area ***(⇐ p. 611)***, so if the denominator in that expression (area) gets small, as with a narrow skate blade, the pressure can become quite large.

The polymer Teflon ***(⇐ p. 488)*** also has a low surface friction and it is possible to "ice skate" on its surface.

Figure 15.10 Ice skating. Ice skates have very thin blades, so the average person exerts as much as 500 atm of pressure at the point where the blade touches the ice surface. Because the melting point of ice changes about −0.01 °C when the pressure is increased by 1 atm, the skater's weight on the blade causes the melting point of ice to be as much as −5 °C lower. This effect, combined with frictional heating and the somewhat liquid nature of the water molecules right on the surface of the ice (see text) makes it possible to move quickly across the ice surface. *(Reuters/Bettmann)*

The effect of increasing pressure on the equilibrium between solid and liquid water. In this magnified view of the solid-liquid equilibrium line from the phase diagram for water, Figure 15.9, it can be seen that when the pressure is increased, and the temperature remains constant, the conditions move from the equilibrium line into the liquid region. The solid water (ice) melts in response to the stress on the equilibrium.

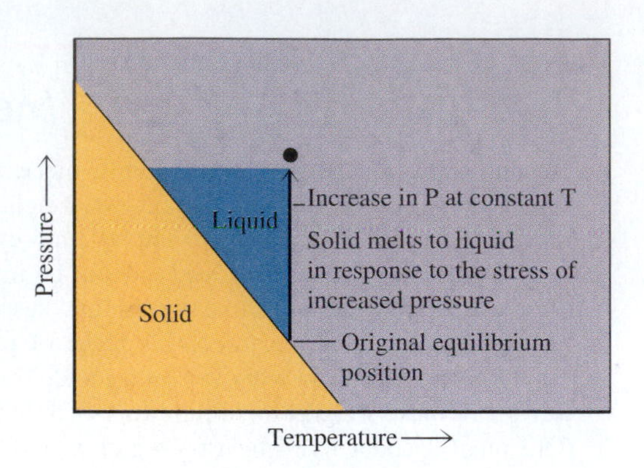

The phase diagram for CO_2 (Figure 15.11) is similar to that of water, except that the solid/liquid equilibrium line has a positive slope. Like most solids, $CO_2(s)$ is more dense than liquid CO_2. Notice that the pressure and temperature axes in the phase diagram for CO_2 are much different than in the diagram for water. Looking at Figure 15.11, you should be able to see that for pressures from about 1 atm to 5 atm, the only equilibrium that can exist is between solid and gaseous CO_2. This means solid CO_2 sublimes when heated. Liquid CO_2 can be produced only at pressures above 5 atm. The temperature range for liquid CO_2 is narrower than that for water. Tanker trucks can often be seen marked "liquid carbonic." These trucks are carrying liquid CO_2, a convenient source of CO_2 gas for making carbonated beverages.

Critical Temperature and Pressure

Looking at the phase diagram for CO_2, you will notice an upper point where the liquid/gas equilibrium line terminates (Figure 15.11). For CO_2 the conditions are 73 atm and 31 °C. For water the termination of the solid/liquid curve occurs at a pressure of 217.7 atm and a temperature of 374.0 °C (Figure 15.9). These conditions are called the **critical temperature** (T_c) and **critical pressure** (P_c). At any temperature *above* T_c for a substance, the molecules have sufficient kinetic energy to overcome any at-

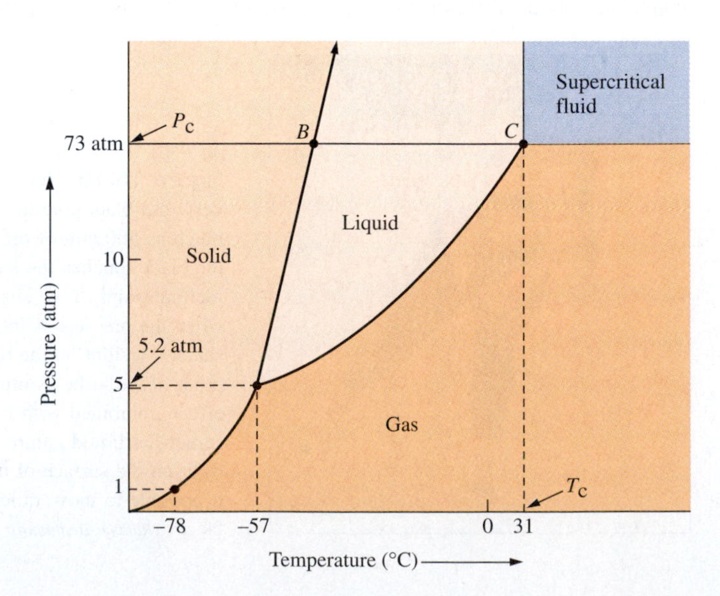

Figure 15.11 The phase diagram for carbon dioxide. The temperature and pressure marked T_c and P_c are critical temperature and pressure and are discussed in the text.

tractive forces, and no amount of pressure can cause them to act like a liquid again. The substance above T_c becomes a **supercritical fluid,** which has a density characteristic of a liquid but the flow properties of a gas, thereby being able to diffuse through many substances easily.

The term "fluid" describes a substance that will flow. It is used for both liquids and gases.

Supercritical CO_2 can be found in a fire extinguisher under certain conditions. As the phase diagram shows, at any temperature below 31 °C, the pressurized CO_2 (usually at about 73 atm) can be heard sloshing around in the container. On a hot day, however, the CO_2 becomes a supercritical fluid and will not be heard sloshing. In fact, the only way to know whether CO_2 is in the extinguisher without discharging it, is to weigh it and compare its weight with the weight of the empty container. This empty weight is usually on a tag attached to the fire extinguisher.

Supercritical fluids are excellent solvents. Supercritical CO_2 is used to extract caffeine from coffee beans because it is nonpolar and a good solvent for nonpolar substances. Supercritical water is used to extract toxic components from hazardous industrial wastes. The polar nature of water makes supercritical water a better solvent for polar compounds.

Exercise 15.11 Using Phase Diagrams

Find the pressure and temperature values for the triple point of CO_2. What phase exists when the temperature is held constant and the pressure is slightly increased over that of the triple point?

Exercise 15.12 The Behavior of CO_2

If liquid CO_2 is slowly released to the atmosphere from a cylinder, what state will the CO_2 be in? If the liquid is suddenly released, as in the discharge of a CO_2 fire extinguisher, why is solid CO_2 seen? Can you explain this on the basis of the phase diagram alone or do you need to consider other factors?

Planet earth from outer space.

15.4 WATER: AN IMPORTANT LIQUID WITH UNUSUAL PROPERTIES

Earth is sometimes called the blue planet because the large quantities of water on its surface make it look blue from outer space. Three quarters of the globe is covered by oceans, there are vast ice sheets at the poles, and large quantities of water are present in soils and rocks on the surface. Water is essential to almost every form of life, has played a major role in human history, and is a significant factor in weather and climate. The reason for this lies in water's unique properties (Table 15.4, p. 676). Ice floats on water, but when most substances freeze, the solid sinks. More thermal energy must be transferred to melt ice, heat water, and vaporize water than for almost any other substance. Water has the largest thermal conductivity and the highest surface tension of any molecular substance in the liquid state. In other words, this most common of substances in our daily lives has properties that are highly unusual and that are crucial for our planet's and our species' welfare.

Most of water's unusual properties can be attributed to its unique capacity for hydrogen bonding. If you refer back to Figure 10.24 *(← p. 428),* you will see that one water molecule can participate in four hydrogen bonds to other water molecules, and there can be a maximum of two hydrogen bonds per molecule. To accommodate the maximum hydrogen bonding, a three-dimensional network of water molecules

TABLE 15.4 Unusual Properties of Water

Property	Comparison with Other Substances	Importance in Physical and Biological Environment
Specific heat capacity (4.18 J/g·K)	Highest of all liquids and solids except NH_3	Moderates temperature in the environment and in organisms; climate affected by movement of water (e.g., Gulf Stream)
Heat of fusion (333 J/g)	Highest of all molecular solids except NH_3	Freezing water releases large quantity of thermal energy; used to save crops from freezing by spraying them with liquid water
Heat of vaporization (2250 J/g)	Highest of all molecular substances	Condensation of water vapor in clouds releases large quantities of thermal energy, fueling storms
Surface tension (7.3×10^{-2} J/m^2)	Highest of all molecular liquids	Contributes to capillary action in plants; causes formation of spherical droplets; supports insects on water surfaces
Thermal conductivity (0.6 J/s·m·K)	Highest of all molecular liquids	Provides for transfer of thermal energy within organisms; rapidly cools organisms immersed in cold water, causing hypothermia

must form, and this is what happens when liquid water freezes. In the crystal lattice of ice, the oxygen atoms lie at the corners of puckered, six-sided rings. (The six-sided symmetry of snowflakes corresponds to the symmetry of these hexagonal rings.) Each of the six sides of a ring consists of one O—H covalent bond and one hydrogen bond between the H atom and another O atom. In order to accommodate the linear O—H···O structure, considerable open space must be left within the rings, and in fact there are empty channels that run through the entire crystal lattice.

When ice melts to form liquid water, approximately 15% of the hydrogen bonds are broken and the rigid ice lattice collapses. Although large clusters of water molecules remain hydrogen bonded, some molecules in the liquid phase are free to move into the empty spaces that were present in the ice lattice. Therefore there are more molecules, and hence more mass, in a given volume of liquid than in the same vol-

(a)

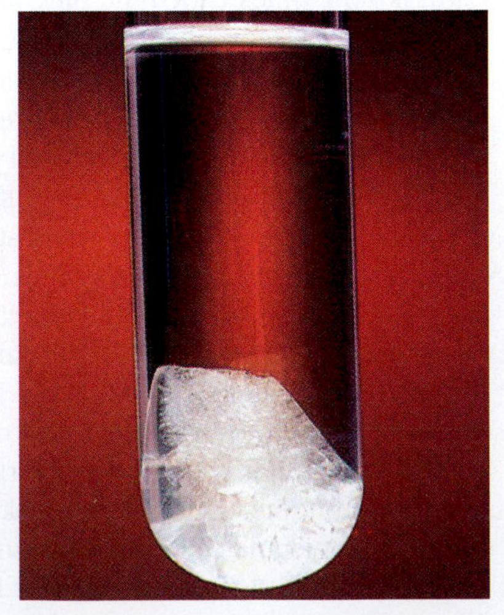
(b)

Ice floats but solid benzene sinks. That ice cubes float in water (a) is part of your everyday experience, but this behavior is unusual. Most liquids are like benzene (b). When they freeze, the solid sinks to the bottom of the liquid. *(C.D. Winters)*

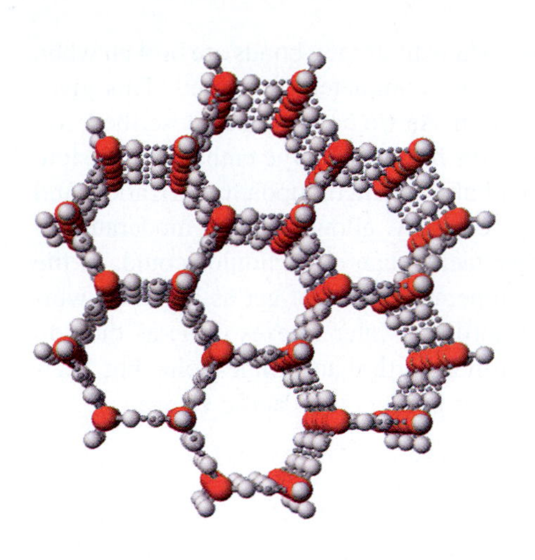
(a)

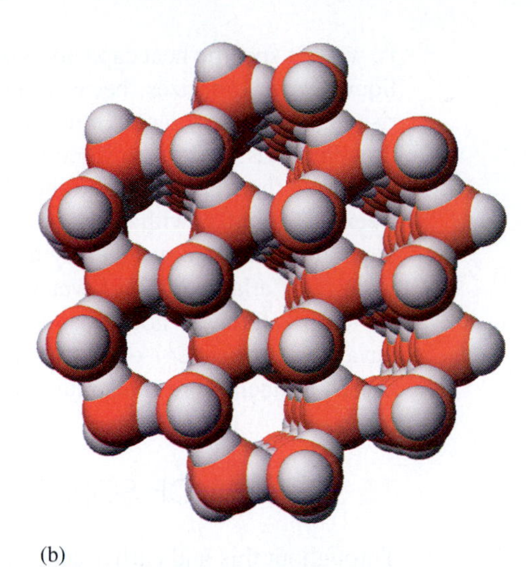
(b)

Open-cage structure of ice. In the ball-and-stick model (a) it can be seen that each oxygen atom is covalently bonded to two hydrogen atoms and also hydrogen bonded to two hydrogen atoms. The hydrogen bonds are longer than the covalent bonds, but four hydrogens around each oxygen are arranged at tetrahedral angles. Each O—H···O bond is linear. This requires empty space within the ice structure that is not required in liquid water, where the molecules are less regularly arranged. This causes ice to be less dense than liquid water at 0 °C. The openness of the ice crystal lattice can be seen in the space-filling model (b, and the front cover of the book) where empty hexagonal channels through the structure are visible even with all atoms shown full size.

ume of solid. That is, the density of liquid water is greater than for ice at the melting point. The density difference is not large, but it is enough so that ice floats on the surface of the liquid.

As the liquid water is warmed further, more of the hydrogen bonds break, and more empty space is filled by water molecules. The density continues to increase until a temperature of 4 °C is reached. As the temperature rises beyond 4 °C, increased molecular motion causes the molecules to push each other aside more vigorously, the empty space between molecules increases, and the normal behavior of decreasing density with increasing temperature is observed. At 4 °C liquid water has a density of 1.0000 g/mL, and the density decreases by about 0.0001 g/mL for every 1 °C temperature rise above 4 °C.

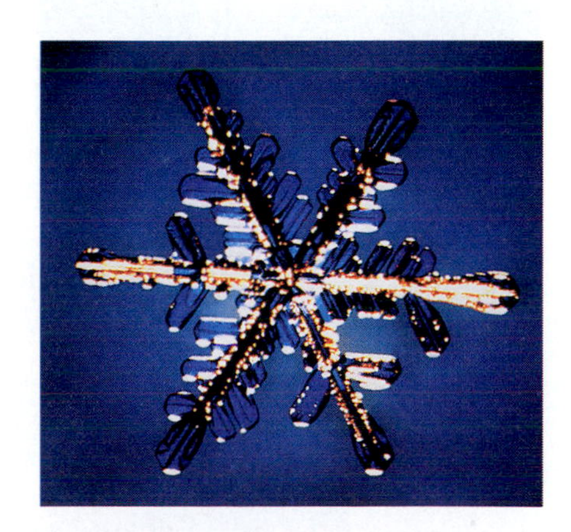
Hexagonal snowflake. The macroscale, six-sided symmetry of a snowflake reflects the nanoscale, hexagonal symmetry of the puckered rings within the ice crystal lattice. *(NCAR/Tom Stack & Associates)*

Because of this unusual variation of density with temperature, when water in a lake is cooled by the air down to 4 °C, the higher density causes the cold water to sink to the bottom. Water cooled below 4 °C is less dense and stays on the surface, where it can be cooled even further. Consequently the water at the bottom of the lake remains at 4 °C, while that on the surface freezes. Ice on the surface insulates the remaining liquid water from the cold air, and, unless the lake is quite shallow, not all of the water freezes. This allows fish and other organisms to survive without being frozen solid in winter. When water at 4 °C sinks to the bottom of a lake in the fall, it carries with it dissolved oxygen. Nutrients from the bottom are brought to the surface by the water it displaces. This is called "turnover" of the lake. The same thing happens in the spring, when the ice melts and water on the surface warms to 4 °C. Spring and fall turnovers are essential to maintain nutrient and oxygen levels required by fish and other lake-dwelling organisms.

When water is heated, the increased molecular motion causes more and more hydrogen bonds to be broken. The strength of these intermolecular forces requires that considerable energy be transferred to raise the temperature of 1 g water by 1 °C. That is, water's specific heat capacity is quite large. Many hydrogen bonds are broken when liquid water vaporizes, because the molecules are completely separated. This gives rise to water's very large enthalpy of vaporization. As we have already described, hydrogen bonds are broken when ice melts, and this requires a large enthalpy of fusion. Its larger than normal (for other liquids) enthalpy changes upon vaporization and freezing, together with its large specific heat capacity, allow water to moderate climate and influence weather by a much larger factor than other liquids could. In the vicinity of a large body of water, summer temperatures do not get as high and winter temperatures do not get as low (at least until the water freezes over) as they do far away from water *(← p. 227)*. Seattle is farther north than Minneapolis, but Seattle has a much more moderate climate because it borders the Pacific Ocean.

15.5 TYPES OF SOLIDS

Figure 15.12 **Talc.** Although they do not appear to be soft in this photo, these talc crystals are so soft that they can be crushed between one's fingers. *(C.D. Winters)*

Throughout this and earlier chapters, we have emphasized that nanoscale structure is closely related to macroscale properties. This relationship is used every day by practicing chemists and other scientists as they try to create new substances that will be beneficial to humankind. Nowhere is the relationship of structure and function more evident than in solids. Table 15.5 summarizes the nanoscale structure and the macroscale properties of solids. By categorizing materials into the classes listed there, you will be able to form a reasonably good idea of what properties to expect, even for a substance that you have never encountered before.

Unlike liquids, solids are rigid—they cannot transmit pressure in all directions. Solids have varying degrees of hardness that depend on the kinds of bonds that hold the particles of the solid together. For example, talc (soapstone, Figure 15.12), which is used as a lubricant and in talcum powder, is one of the softest solids known; diamond is one of the hardest. At the atomic level, talc consists of layered sheets containing silicon, magnesium, and oxygen atoms. Attractive forces between these sheets are very weak, and so one sheet can slide along another and be removed easily from the rest. In diamond, each carbon atom is strongly bonded to four neighbors, each of those neighbors is strongly bonded to three more carbon atoms, and so on throughout the solid (a network solid, Section 15.10). Because of the number and strength of the bonds holding each carbon atom to its neighbors, diamond is so hard that it

TABLE 15.5 Structures and Properties of Various Types of Solid Substances

Type	Examples	Structural Units
Ionic *(← p. 98)*	NaCl, K_2SO_4, $CaCl_2$, $(NH_4)_3PO_4$	Positive and negative ions (some polyatomic); no discrete molecules
Metallic (Section 15.8)	iron, silver, copper, other metals and alloys	Metal atoms (or positive metal ions surrounded by an electron sea)
Molecular *(← p. 101)*	H_2, O_2, I_2, H_2O, CO_2, CH_4, CH_3OH, CH_3CO_2H	Molecules held together by covalent bonds
Network (Section 15.10)	graphite, diamond, quartz, feldspars, mica	Atoms held in an infinite one-, two-, or three-dimensional network
Amorphous (glassy) (Section 15.11)	glass, polyethylene, nylon	Covalently bonded networks with no long-range regularity

can scratch or cut almost any other solid. For this reason diamonds are used in cutting tools and abrasives, which are more important commercially than diamonds used as gemstones.

Although all solids consist of atoms, molecules, or ions in relatively immobile positions, some solids exhibit a greater degree of regularity than others. **Crystalline solids** are so regular that they exhibit regular shapes with sharp angles produced where their planar faces meet. **Amorphous solids**, on the other hand, are somewhat like liquids in that they exhibit very little regular structure or long-range order, and yet they are like solids in being hard and having definite shapes. Ordinary glass is an amorphous solid, as are organic polymers such as polyethylene and polystyrene. The remainder of this chapter will be devoted to explaining the nanoscale structures that give rise to the properties summarized in Table 15.5.

Exercise 15.13 Lead into Gold?

Imagine samples of gold and lead, each with finely polished surfaces. These two surfaces are placed in contact with one another and held in place, under pressure, for about one year. After that time the two surfaces are analyzed. The gold surface is tested for the presence of lead, and the lead surface is tested for gold. Predict what the outcome of these two tests will be and explain what has happened.

15.6 CRYSTALLINE SOLIDS

The beautiful regularity of ice crystals, crystalline salts, or gemstones suggests that there must also be some *internal* regularity. Toward the end of the 18th century, scientists found that shapes of crystals can be used to identify minerals. The angles at which crystal faces meet are characteristic of a crystal's composition, but do not depend on its size and shape. The shapes of all crystalline solids reflect the shape of a *crystal lattice*—an orderly, repeating arrangement of ions, molecules, or atoms. In such a lattice each ion, molecule, or atom is surrounded by neighbors in exactly the same arrangement (⇐ *p. 99*). Each crystal is built up from a three-dimensional repetition of the same pattern.

TABLE 15.5 (continued)

Forces Holding Units Together	Typical Properties
Ionic; attractions among charges on positive and negative ions	Hard; brittle; high melting point; poor electrical conductivity as solid, good as liquid; often water-soluble
Metallic; electrostatic attraction among metal ions and electrons	Malleability; ductility; good electrical conductivity in solid and liquid; good heat conductivity; wide range of hardness and melting points
London forces, dipole-dipole forces, hydrogen bonds	Low to moderate melting points and boiling points; soft; poor electrical conductivity in solid and liquid
Covalent; directional electron-pair bonds	Wide range of hardnesses and melting points (3-dimensional bonding > 2-dimensional bonding > 1-dimensional bonding); poor electrical conductivity, with some exceptions
Covalent; directional electron-pair bonds	Noncrystalline; wide temperature range for melting; poor electrical conductivity, with some exceptions

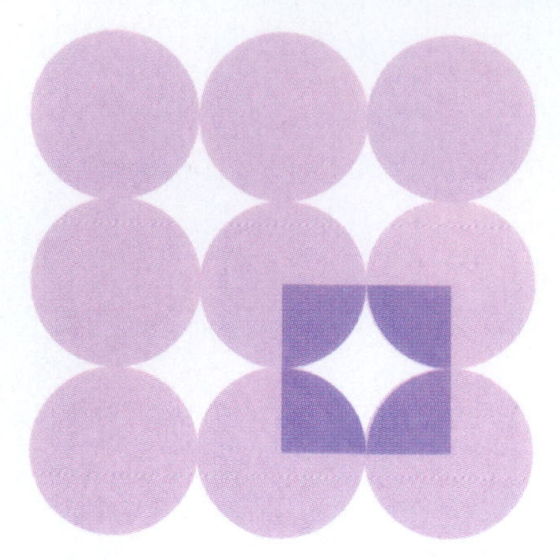

Figure 15.13 Unit cell for a two-dimensional solid made from flat, circular objects. Here the unit cell is a square. Each circle at the corner of the square contributes one quarter of its area to the area inside the square. Thus, there is a *net* of one circle per unit cell.

Unit Cells

A convenient way to describe and classify the repeating pattern of atoms, molecules, or ions in a crystal is to define a small segment of a crystal lattice as a **unit cell**—the smallest part of the lattice that, when repeated along the directions defined by its edges, reproduces the entire crystal structure. To help understand the idea of the unit cell, look at the simple two-dimensional array of circles shown in Figure 15.13. The same size circle is repeated over and over, but a circle is not a good unit cell because it gives no indication of its relationship to all the other circles. A better choice is to recognize that the centers of four adjacent circles lie at the corners of a square, and to draw four lines connecting those centers. A square unit cell results; four of them are drawn in Figure 15.13. As you look at the unit cell drawn in dark purple, notice that each of four circles contributes one quarter of itself to the unit cell, so a net of one circle is located within the unit cell. When this unit cell is repeated by moving the square parallel to its edges (that is, when unit cells are placed next to and above and below the first one), the two-dimensional lattice results. Notice that the corners of a unit cell are equivalent to each other, and collectively they define the crystal lattice.

The three-dimensional unit cells from which all known crystal lattices can be constructed fall into only seven categories. These seven types of unit cells have sides of different relative lengths, and edges that meet at different angles. Only cubic unit cells composed of atoms or monatomic ions are described here. These are quite common in nature and also are simpler and easier to visualize. The principles illustrated, however, apply to all unit cells and all crystal structures, including those composed of polyatomic ions and complicated molecules.

Cubic unit cells have equal-length edges that meet at 90° angles. There are three types: simple cubic (sc), body-centered cubic (bcc), and face-centered cubic (fcc) (Figure 15.14). Both metals and ionic compounds crystallize in cubic unit cells. In metals, all three types of cubic unit cells have identical atoms centered at each cor-

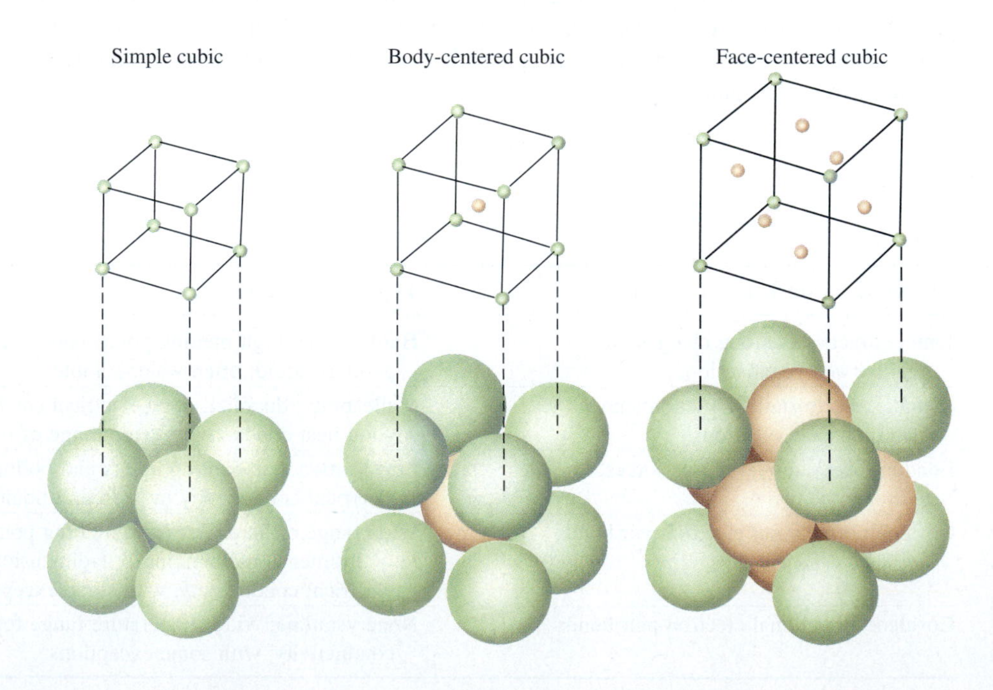

Figure 15.14 The three different types of cubic unit cells. The top row shows the lattice points of the three cells. Unit cells are defined by the numbers and positions of the lattice points, all of which are occupied by identical atoms, molecules, or ions. To emphasize the distinction between corner and non-corner positions, different colors have been used to represent them. In the bottom row the points are replaced with space-filling spheres centered on the lattice points. Notice that the spheres at the corners of the body-centered and face-centered cubes do not touch each other. Rather, each corner atom in the body-centered cell touches the center atom, and each corner atom in the face-centered cell touches spheres in the centers of three adjoining faces.

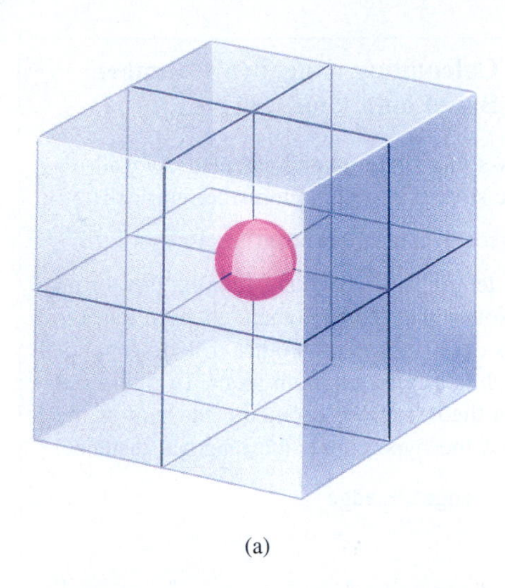

(a)

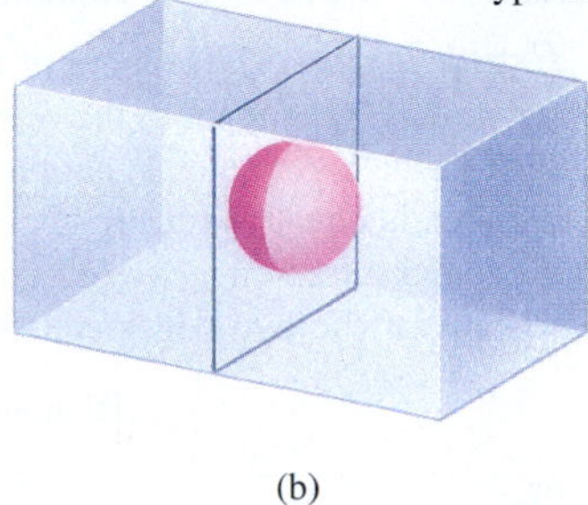

(b)

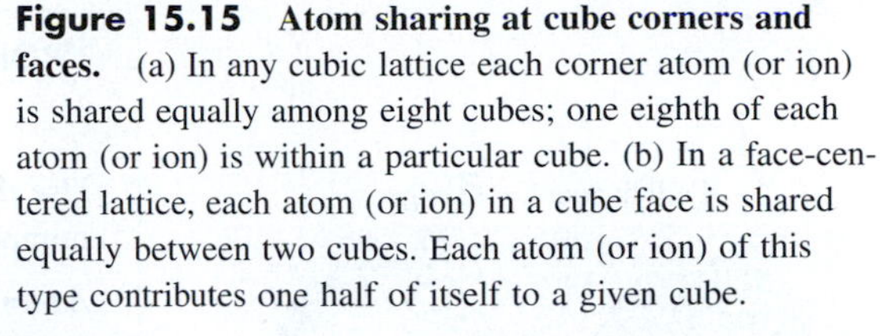

Figure 15.15 Atom sharing at cube corners and faces. (a) In any cubic lattice each corner atom (or ion) is shared equally among eight cubes; one eighth of each atom (or ion) is within a particular cube. (b) In a face-centered lattice, each atom (or ion) in a cube face is shared equally between two cubes. Each atom (or ion) of this type contributes one half of itself to a given cube.

ner of the cube. When cubes pack into three-dimensional space, an atom at a corner is shared among eight cubes (Figure 15.15a); this means that only one eighth of each corner atom is actually within the unit cell. Since a cube has eight corners and one eighth of the atom at each corner belongs to the unit cell, the net result is $8 \times \frac{1}{8} = 1$ atom within the unit cell. In the bcc unit cell there is an additional atom at the center of the cube; it lies entirely within the unit cell. This, combined with the net of one atom from the corners, gives a total of two atoms per body-centered cubic unit cell. In the fcc unit cell there are six atoms or ions that lie in the centers of the faces of the cube. One half of each of these belongs to the unit cell (Figure 15.15b); in this case there is a net result of $6 \times \frac{1}{2} = 3$ atoms within the unit cell, in addition to the net of one atom contributed by the corners, for a total of four atoms per face-centered cubic unit cell.

Exercise 15.14 Counting Atoms in Unit Cells of Metals

Crystalline polonium has a simple cubic unit cell, lithium has a body-centered cubic unit cell, and calcium has a face-centered cubic unit cell. How many Po atoms belong to one unit cell? How many Li atoms belong to one unit cell? How many Ca atoms belong to one unit cell? Draw each unit cell. Indicate on your drawing what fraction of each atom lies within the unit cell.

Unit Cells and Density

What has been described about unit cells so far allows us to check whether a proposed unit cell is reasonable. Because a unit cell can be replicated to give the entire crystal lattice, the unit cell should have the same density as the crystal. Aluminum has a density of 2.699 g/cm^3 at 20 °C, making it one of the lowest density metals. An aluminum atom has a radius of 143 pm. These data must be consistent with an experimental observation of an fcc unit cell for aluminum? From the atomic radius, we should be able to calculate the volume of a unit cell and then, using the density of aluminum, calculate the mass of one unit cell. Since an fcc unit cell contains four atoms, the mass of four aluminum atoms must equal the mass of the unit cell. From the mass of one Al atom and the molar mass of Al, we can determine Avogadro's number.

PROBLEM-SOLVING EXAMPLE 15.4 Calculating Avogadro's Number Based on a Unit Cell

Use the atomic radius, density, and molar mass of aluminum and calculate Avogadro's number. The unit cell of aluminum has an fcc structure, with 4 atoms per unit cell.

Answer 6.029×10^{23} atom/mol, which agrees to three significant digits

Explanation The volume of a cube is the length of its edge cubed. One face of an fcc unit cell is shown in the margin figure. Notice that the face-centered atom touches each of the corner atoms, but the corner atoms do not touch each other. The diagonal of the face is equal to four times the 143-pm radius of an aluminum atom. To figure out the length of an edge we use the Pythagorean theorem from geometry: the sum of the squares of the two edges equals the square of the hypotenuse, the diagonal distance.

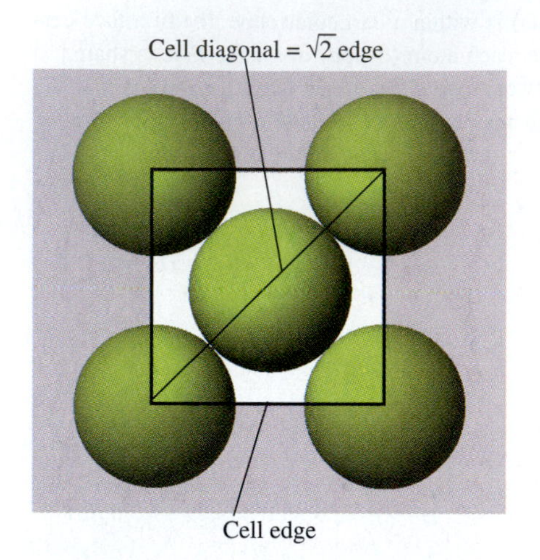

One face of a face-centered cubic crystal.

$$(\text{Diagonal distance})^2 = \text{edge}^2 + \text{edge}^2$$

or

$$(\text{Diagonal distance})^2 = 2\ (\text{edge})^2$$

By taking the square root of both sides of the equation, we get

$$\text{Diagonal distance} = \sqrt{2} \times (\text{edge})$$

The diagonal distance $= 4 \times 143$ pm $= 572$ pm, and so the length of the edge is $(572\ \text{pm}/\sqrt{2}) = 405$ pm. In centimeters,

$$\text{Cube edge} = (405\ \text{pm})\left(\frac{1\ \text{m}}{1 \times 10^{12}\ \text{pm}}\right)\left(\frac{100\ \text{cm}}{1\ \text{m}}\right) = 4.05 \times 10^{-8}\ \text{cm}$$

So, the volume of the face-centered cubic unit cell is

$$\text{Unit cell volume} = (\text{edge})^3 = (4.05 \times 10^{-8}\ \text{cm})^3 = 6.64 \times 10^{-23}\ \text{cm}^3$$

Now, the mass of the unit cell is

$$\text{Mass of unit cell} = \text{density} \times \text{volume} = \left(\frac{2.699\ \text{g}}{\text{cm}^3}\right)\left(\frac{6.64 \times 10^{-23}\ \text{cm}^3}{\text{unit cell}}\right)$$

$$= 1.79 \times 10^{-22}\ \text{g/unit cell}$$

This is the mass of 4 aluminum atoms. From the definition of molar mass *(← p. 56)*, we know that the mass of one mol of aluminum is 26.98 g.

For some years the best way of determining Avogadro's number was a calculation like this.

We can calculate Avogadro's number, which has units of atoms per mol, by dividing the 4 atoms in the unit cell by their mass and converting that mass into moles.

$$\text{Avogadro's number} = \left(\frac{4\ \text{atoms}}{1.79 \times 10^{-22}\ \text{g Al}}\right)\left(\frac{26.98\ \text{g Al}}{\text{mol}}\right) = 6.029 \times 10^{23}\ \text{atoms/mol}$$

The calculated answer agrees to within a few percent with the accepted value of 6.022×10^{23} atoms/mol for Avogadro's number.

Problem-Solving Practice 15.4

Gold crystals have a bcc structure. The radius of a gold atom is 144 pm, and the atomic weight of gold is 196.97 amu. Calculate the density of gold.

Unit Cells and Simple Ionic Compounds

The crystal structures of many ionic compounds can be described as simple cubic or face-centered cubic lattices of spherical negative ions, with positive ions occupying spaces (called holes) among the negative ions. The number and locations of the occupied holes are the keys to understanding the relation between the lattice structure and the formula of an ionic compound. The simplest example is an ionic compound in which the hole in a simple cubic unit cell is occupied (Figure 15.14). The ionic compound cesium chloride, CsCl, has such a structure. In it, each simple cube of Cl^- ions has a Cs^+ ion at its center (Figure 15.16). The spaces occupied by the Cs^+ ions are called cubic holes, and each Cs^+ has eight nearest-neighbor Cl^- ions.

Remember that negative ions are usually larger than positive ions. Therefore, building an ionic crystal is a lot like placing marbles (positive ions) in the spaces among ping-pong balls (negative ions).

It is not appropriate to describe the CsCl structure as bcc; the names simple cubic, body-centered cubic, and face-centered cubic refer to the arrangement of only one type of ion in the lattice, not both.

The structure of sodium chloride, NaCl, is one of the most common ionic crystal lattices. It consists of an fcc lattice of the larger Cl^- ions (Figure 15.17a), in which Na^+ ions occupy what are called *octahedral* holes—octahedral because each Na^+ ion is surrounded by six Cl^- ions at the corners of an octahedron (Figure 15.17b). It is easy to see this for the highlighted Na^+ ion in Figure 15.17b, but the unit cell would have to be repeated in several directions to see six Cl^- ions around some of the other Na^+ ions. Figure 15.17d shows a space-filling model of the NaCl lattice, in which each ion is drawn to scale based on its ionic radius.

If you look carefully at Figure 15.17a, it is possible to determine the number of Na^+ and Cl^- ions in the NaCl unit cell. There is one eighth of a Cl^- ion at each corner of the unit cell, and one half of a Cl^- in the middle of each face. The total number of Cl^- ions within the unit cell is

$$\tfrac{1}{8}\ \text{Cl}^- \text{ per corner} \times 8 \text{ corners} = 1\ \text{Cl}^-$$
$$\tfrac{1}{2}\ \text{Cl}^- \text{ per face} \times 6 \text{ faces} = 3\ \text{Cl}^-$$

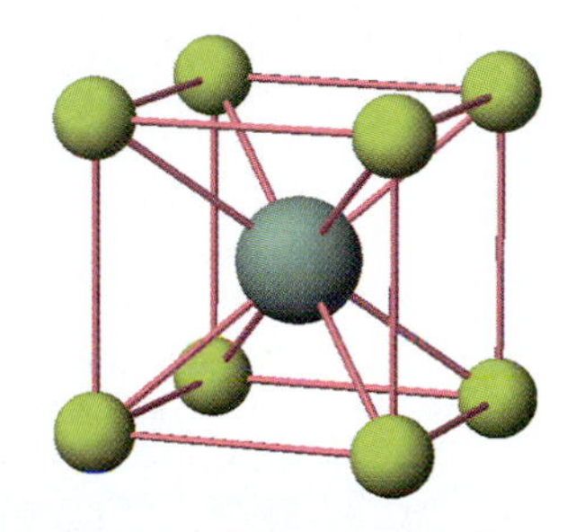

Figure 15.16 Unit cell of cesium chloride (CsCl) crystal lattice. The cesium ion is dark green and the chloride ions are light green.

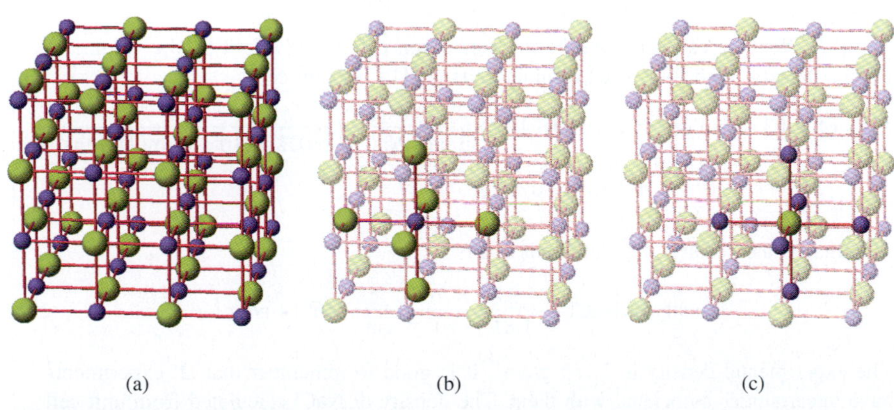

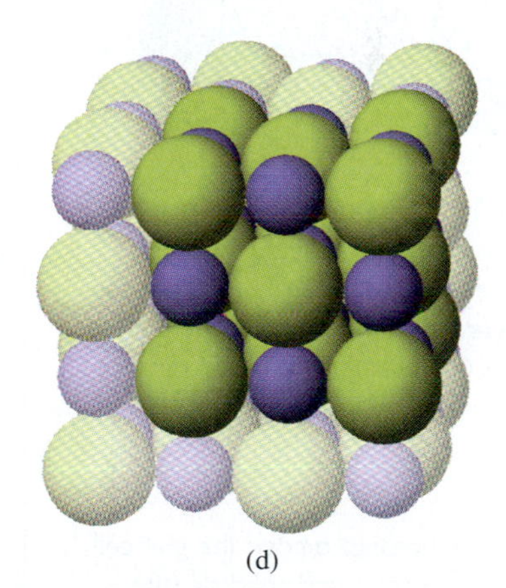

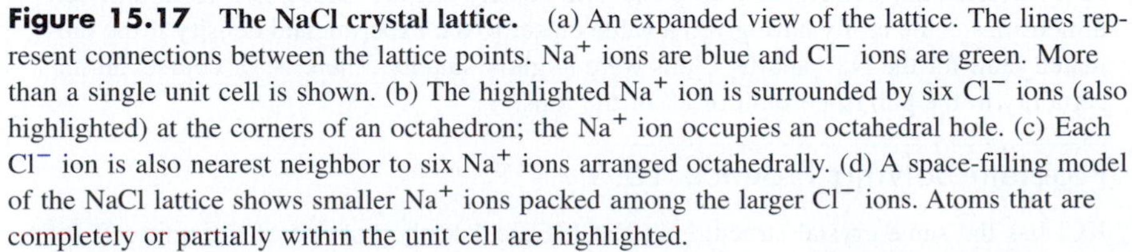

Figure 15.17 The NaCl crystal lattice. (a) An expanded view of the lattice. The lines represent connections between the lattice points. Na^+ ions are blue and Cl^- ions are green. More than a single unit cell is shown. (b) The highlighted Na^+ ion is surrounded by six Cl^- ions (also highlighted) at the corners of an octahedron; the Na^+ ion occupies an octahedral hole. (c) Each Cl^- ion is also nearest neighbor to six Na^+ ions arranged octahedrally. (d) A space-filling model of the NaCl lattice shows smaller Na^+ ions packed among the larger Cl^- ions. Atoms that are completely or partially within the unit cell are highlighted.

Since the edge of a cube in a cubic lattice is surrounded by four cubes, one fourth of a spherical ion at the midpoint of an edge is within any one of the cubes.

There is one fourth of an Na^+ at the midpoint of each edge, and a whole Na^+ in the center of the unit cell. For Na^+ ions, the total is

$$\frac{1}{4}\ Na^+ \text{ per edge} \times 12 \text{ edges} = 3\ Na^+$$

$$1\ Na^+ \text{ per center} \times 1 \text{ center} = 1\ Na^+$$

Thus, the unit cell contains four Na^+ and four Cl^- ions. This result agrees with the formula of NaCl for sodium chloride.

Exercise 15.15 Formulas and Unit Cells

Cesium chloride has a simple cubic unit cell, as seen in Figure 15.16. Show that the formula for the salt must be CsCl.

Problem-Solving Example 15.5 Calculating the Volume and Density of a Unit Cell

The ionic radii for Na^+ and Cl^- are 116 pm and 167 pm, respectively. Calculate the density of NaCl.

Answer The calculated density is 2.14 g/cm^3.

Explanation The Na^+ and Cl^- ions are touching along the edge of the unit cell (Figure 15.18). This means the edge length is equal to two Cl^- radii plus two Na^+ radii.

$$\text{Edge} = 167 \text{ pm} + 2 \times 116 \text{ pm} + 167 \text{ pm} = 566 \text{ pm}$$

The volume of the cubic unit cell is the cube of the edge length.

$$\text{Volume of unit cell} = (\text{edge})^3 = (566 \text{ pm})^3 = 1.81 \times 10^8 \text{ pm}^3$$

Converting this to cm^3,

$$\text{Volume of unit cell} = 1.81 \times 10^8 \text{ pm}^3 \times \left(\frac{10^{-10} \text{ cm}}{\text{pm}}\right)^3 = 1.81 \times 10^{-22} \text{ cm}^3$$

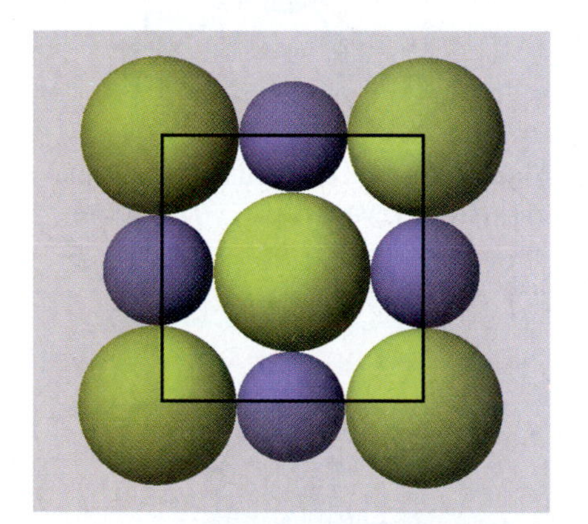

Figure 15.18 One face of a NaCl unit cell. Notice that the Na^+ and Cl^- ions touch along each edge of the cell, and so the edge length is the sum of the diameters of the two ions. (The part of the lattice outside the unit cell is shaded.)

Next we can calculate the mass of a unit cell and divide it by the volume to get the density. With four NaCl formula units (F. U.) per unit cell,

$$\text{Mass of NaCl in unit cell} = 4 \text{ NaCl (F. U.)} \times \left(\frac{58.44 \text{ g}}{\text{mol NaCl}}\right)\left(\frac{1 \text{ mol}}{6.022 \times 10^{23} \text{ formula units}}\right)$$

$$= 3.88 \times 10^{-22} \text{ g}$$

This means the density of a NaCl unit cell is

$$\text{Density of NaCl} = \frac{3.88 \times 10^{-22} \text{ g}}{1.81 \times 10^{-22} \text{ cm}^3} = 2.14 \text{ g/cm}^3$$

The experimental density is 2.164 g/cm^3. It is good to remember that all experiments have uncertainties associated with them. The density of NaCl calculated from unit cell dimensions could easily have given a value closer to the experimental density if the tabulated radii for the Na^+ and Cl^- ions were slightly smaller. There is, of course, an uncertainty in the published radii of all of the ions.

The general relationship among the unit cell type, ion or atom size, and solid density for most metallic or ionic solids is:

Mass of 1 formula unit (e.g., NaCl) × number of formula units per unit cell (e.g., 4 for NaCl) = mass of unit cell

Mass of unit cell ÷ unit cell volume (= edge^3 for NaCl) = density

Problem-Solving Practice 15.5

KCl has the same crystal structure as NaCl. Calculate the volume of the unit cell for KCl, given that the ionic radii are K^+ = 152 pm and Cl^- = 167 pm. Which has the larger unit cell, NaCl or KCl? Now compute the density of KCl.

CHEMISTRY YOU CAN DO

Closest Packing of Spheres

In a metal the crystal lattice is occupied by identical atoms, which can be pictured as identical spheres. Obtain a bag of marbles, some expanded polystyrene foam spheres, some ping-pong balls, or some other set of reasonably small, identical spheres. Use them to construct models of each situation described here.

Consider how identical spheres can be arranged in two dimensions. Pack your spheres into a layer so that each sphere touches four other spheres; this should resemble the arrangement in part a of the figure. (You will find this somewhat hard to do; it may be necessary to use four pieces of wood to enclose the layer and keep the spheres in line. Making a square array is hard because the spheres are rather inefficiently packed, and they readily adopt a more efficient packing in which each sphere touches six other spheres.)

To make a three-dimensional lattice requires stacking layers of spheres. Start with the layer of spheres in a square array, and stack another, identical layer directly on top of the first; then stack another on that one, and so on. (This is really hard and will require vertical walls to hold the spheres as you stack the layers. If you are not able to do it, use sugar cubes, dice, or some other kind of small cubes instead of spheres.) This three-dimensional lattice will consist of simple cubic unit cells; in such a lattice 52.4% of the available space is occupied by the spheres.

Next, make the more efficiently packed hexagonal layer in which each sphere is surrounded by six others (part b of the figure). Notice that in this layer, as in the square one, there are little holes left between the spheres. This time, however, the holes are smaller than they were in the square array; in fact, they are as small as possible, and for this reason the layer you have made is called a **closest-packed** layer. Now add a second layer, and let the spheres nestle into the holes in the first layer. (Trying to prevent this from happening was probably the hardest task when you made your simple cubic array.) Now add a third layer. If you look carefully, you will see that there are two ways to put down the third layer. One of these places spheres directly above *holes* in the first layer. This is called **cubic closest packing,** and the unit cell is face-centered cubic (fcc). The other arrangement, also found for many metals, places third-layer spheres directly above *spheres* in the first layer. This is *hexagonal closest packing,* and its unit cell is *not* cubic. Cubic and hexagonal closest packing are the most efficient ways known for filling space with spheres; 74% of the available space is occupied. The atoms in most metals are arranged in cubic closest packing, in hexagonal closest packing, or in lattices composed of body-centered cubic unit cells. Efficient packing of the atoms or ions in the crystal lattice allows stronger bonding, which gives greater stability to the crystal.

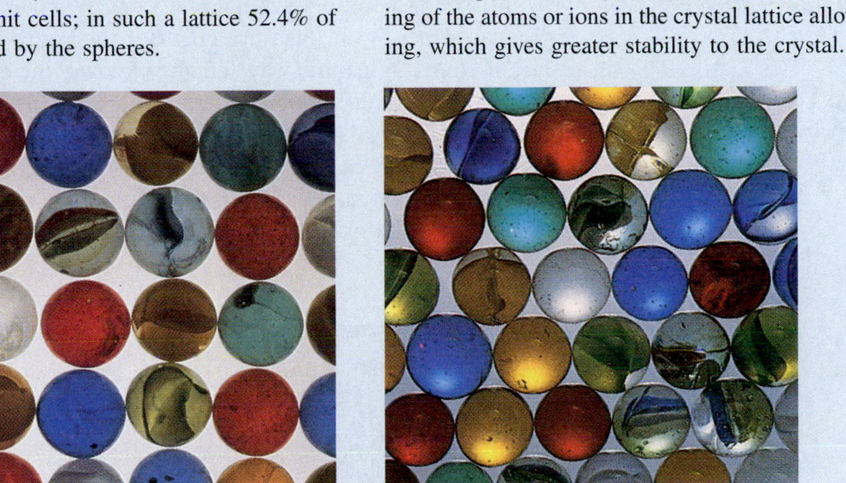

(a) (b)

15.7 LOOKING INTO SOLIDS: X-RAY CRYSTALLOGRAPHY

X rays were discovered by W.C. Röntgen in 1895 and soon after that Max von Laue and his co-workers correctly guessed that x rays were light of very short wavelength, and that the distances between the repetitive arrays of atoms in a crystal would be about the same as the wavelengths of x rays. Therefore, a crystal should act as a grat-

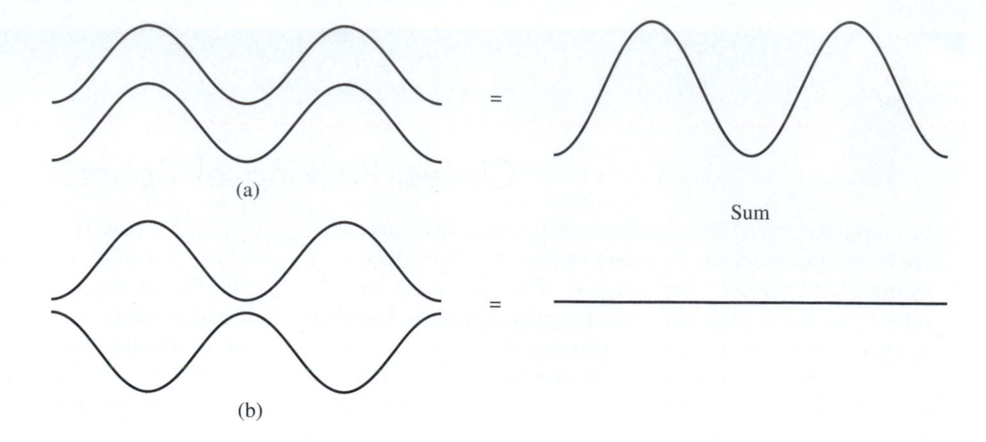

Figure 15.19 Interference of waves. (a) Constructive interference occurs when two in-phase waves combine to produce a wave of greater amplitude. (b) Destructive interference results from the combination of two waves of equal magnitude that are exactly out of phase. The result is zero amplitude.

The surface of a CD-audio or CD-ROM disk contains narrowly separated lines that diffract light into its component colors. This means that the lines are spaced at distances of approximately the same magnitude as the wavelength of visible light (← p. 308).

ing, diffracting x rays in the same way that narrowly separated scratches on a surface diffract visible light into a rainbow of colors. In an important experiment, x rays were focused on a crystal of blue vitriol (copper[II] sulfate pentahydrate) and diffracted rays were detected on a photographic plate.

Upon hearing of this experiment, William and Lawrence Bragg, father and son, proposed that x rays could be used to determine distances between atoms in crystals. A small part of an incoming x-ray beam is reflected by each plane of atoms, and certain arrangements of these planes cause the reflections to interfere with one another. When the interference is *constructive,* the waves combine to produce diffracted x rays that can be observed because they expose a spot on a photographic film. When the interference is *destructive,* the waves partly or entirely cancel each other and no x-ray diffraction is observed (Figure 15.19). Whether interference is constructive or destructive depends on the wavelength of the x rays and the distance between the planes of atoms. When the wavelength is known, the distance can be calculated from the observed x-ray diffraction pattern (Figure 15.20).

After the Braggs's pioneering work, others showed that the intensity of the scattering of the x-ray beam depends on the electron density of the atoms in the crystal. This means that a hydrogen atom is least effective at causing x rays to scatter, while atoms such as lead or mercury are quite effective. **X-ray crystallography,** the sci-

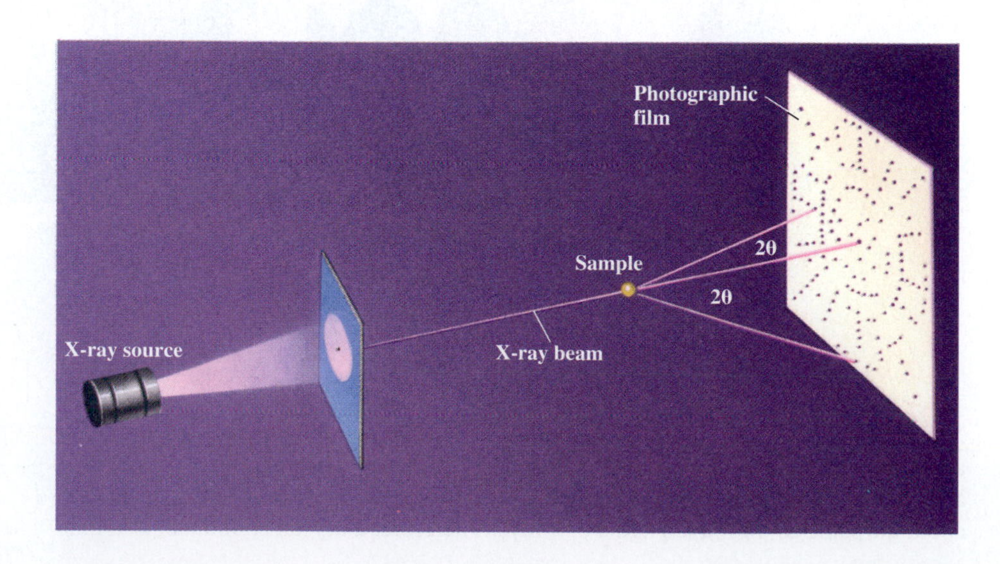

Figure 15.20 An x-ray diffraction experiment. A beam of x rays is directed at a crystalline solid. The x-ray photons are scattered by atoms in the solid. The angles of scattering (θ) depend on the locations of the atoms. Computers are used to calculate the atom positions that determine the scattering angles.

Portrait of a Scientist • *Dorothy Crowfoot Hodgkin (1910–1994)*

Dorothy Crowfoot Hodgkin was born in Egypt where her father was a supervisor of schools and ancient monuments. She was schooled in England and graduated from Oxford University in 1931. She was fascinated by the study of crystals and went to Cambridge University where she worked in the Mineralogical Institute under John D. Bernal. After only a year, Oxford invited her back as a chemistry instructor, and she remained there until her retirement in 1977.

Her first major achievement in crystallography was the determination of the structure of penicillin, a work that was completed in 1945. After World War II she began work on the determination of the structure of vitamin B_{12}, which is used to treat pernicious anemia. For that work she received the Nobel Prize in Chemistry in 1964. Later, she determined the structure of insulin. All of these compounds contain weakly scattering light atoms and much of her work was pioneering and painstaking experimental technique that, as one biographer put it, "transformed crystallography from a black art into an indispensable scientific tool."

$CH_2{=}CHCH_2SCH_2C(=O)-N(H)-$ [β-lactam/thiazolidine ring with S, N, O, CH_3, CH_3, COOH]

penicillin O

Dorothy Crowfoot Hodgkin.

ence of determining atomic-scale crystal structures, is based on these ideas. Today crystallographers use computers and photocells to accurately measure the angles of reflection and the intensities of x-ray beams diffracted by crystals. While the early work by the Braggs and others was with minerals, the crystal structures of crystalline organic compounds are now routinely determined with x rays, even though the lighter carbon, oxygen, nitrogen, and hydrogen atoms reflect the x rays with relatively low intensities.

Since its invention, x-ray crystallography has been increasingly used to discover the molecular structures of important substances. In 1944 Dorothy Crowfoot Hodgkin determined the structure of penicillin, an antibiotic responsible for saving countless lives during World War II. James Watson and Francis Crick in 1953 used x-ray crystallography results obtained by Rosalind Franklin to discover the double-helix structure of deoxyribonucleic acid (DNA), the molecule that carries genetic information *(← p. 436)*.

15.8 METALS, SEMICONDUCTORS, AND INSULATORS

All the metals are solids at room temperature, except for mercury (m.p. −38.8 °C). All metals exhibit common properties that we call metallic:

- **High electrical conductivity.** Metal wires are used to carry electricity from power plants to homes and offices because electrons in metals are highly mobile.
- **High thermal conductivity.** We learn early in life not to touch any part of a metal pot on a heated stove because it will transfer heat rapidly and painfully.
- **Ductility and malleability.** Most metals are easily drawn into wire (ductility), or hammered into thin sheets (malleability); some (gold, for example) are more easily formed into shapes than others.

- **Luster.** Polished metal surfaces reflect light. Most metals have a silvery white color because they reflect all wavelengths equally well.
- **Insolubility in water and other common solvents.** No metal dissolves in water, but a few (mainly from Groups 1A and 2A) react with water to form hydrogen gas and solutions of metal hydroxides.

TABLE 15.6 Enthalpies of Fusion and Melting Points of Some Metals

Metal	ΔH_{fusion} (kJ/mol)	Melting Point (°C)
Hg	2.3	− 38.8
Li	3.0	180.5
Na	2.59	97.9
Ga	7.5	29.78*
Al	10.7	660.4
U	12.6	1132.1
Ti	20.9	1660.1
W	35.2	3410.1

*This means that gallium metal will melt in the palm of your hand from the warmth of your body. It happens that gallium is a liquid over the largest range of temperature of any metal. Its boiling point is approximately 2250 °C.

As you will see, these properties are explained by the kinds of bonding that hold atoms together in metals.

The enthalpies of fusion and melting points of metals vary greatly. Low melting points correlate with low enthalpies of fusion, which implies weaker attractive forces holding the metal atoms together. Mercury, a liquid at room temperature, has an enthalpy of fusion of only 2.3 kJ/mol. The alkali metals and gallium also have very low enthalpies of fusion and notably low melting points (Table 15.6). Compare these values with those in Table 15.3 for nonmetals (← *p. 668*).

Figure 15.21 shows the enthalpies of fusion of the metals related to their positions in the periodic table. The transition metals, especially those of the third transition series, have very high melting points and extraordinarily high enthalpies of

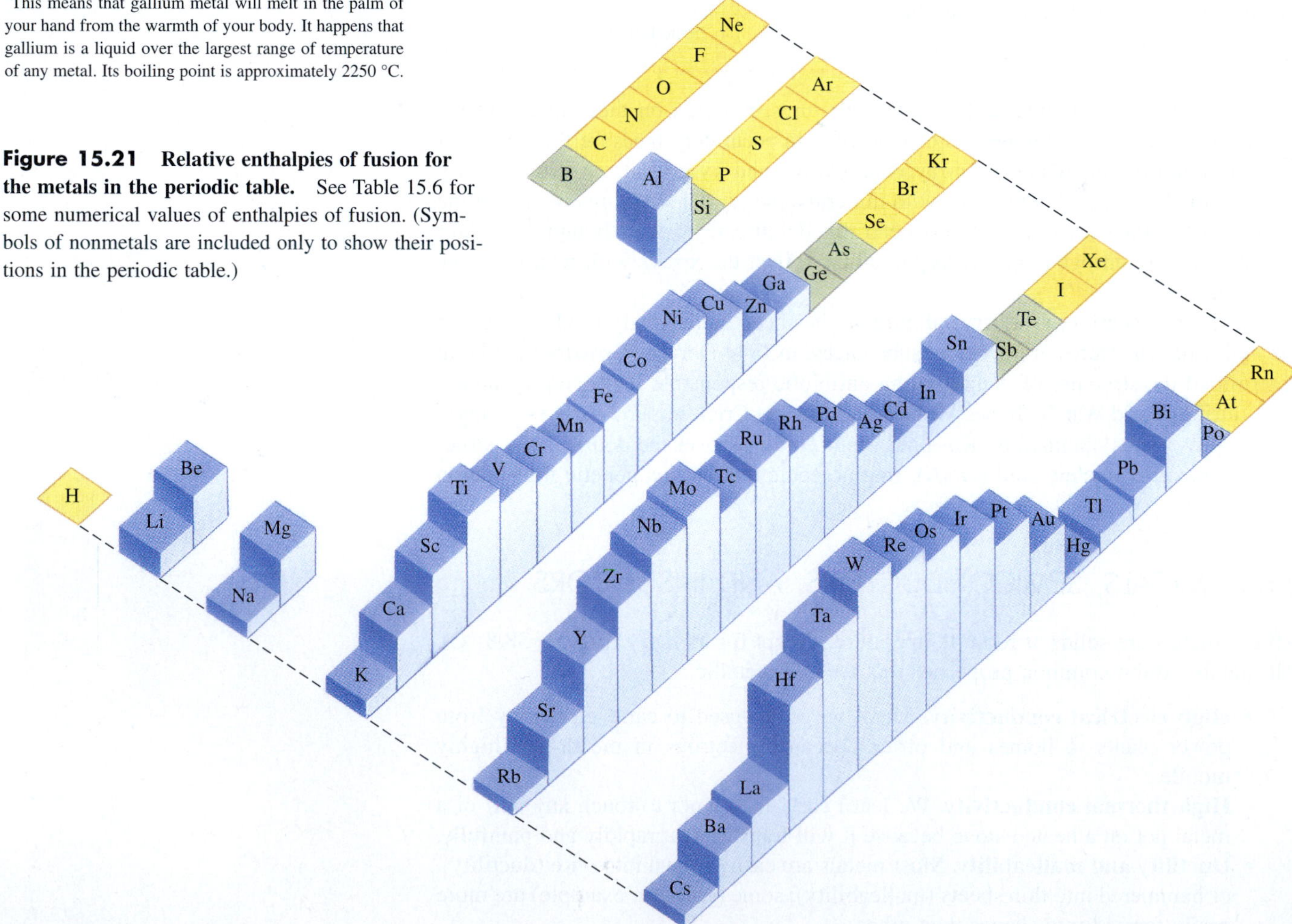

Figure 15.21 Relative enthalpies of fusion for the metals in the periodic table. See Table 15.6 for some numerical values of enthalpies of fusion. (Symbols of nonmetals are included only to show their positions in the periodic table.)

fusion. Tungsten (W) has the highest melting point (3410 °C) of all the metals and among the elements is second only to carbon (graphite), whose melting point is 3550 °C. Pure tungsten is used in light bulbs as the filament—the wire that glows white hot. No other material has been found to be better since the invention of light bulbs in 1908 by Thomas Edison and his co-workers.

PROBLEM-SOLVING EXAMPLE 15.6 Calculating Heats of Fusion

Use data from Table 15.6 to calculate the thermal energy transferred required to melt 2.57 cm^3 of mercury. The density of mercury is 13.69 g/cm^3.

Answer 0.40 kJ

Explanation First, determine how many moles of Hg are present in 2.57 cm^3 of Hg.

$$2.57\ \text{cm}^3\ \text{Hg} \times \left(\frac{13.69\ \text{g Hg}}{1\ \text{cm}^3\ \text{Hg}}\right)\left(\frac{1\ \text{mol Hg}}{200.59\ \text{g Hg}}\right) = 0.1754\ \text{mol Hg}$$

Now, calculate the amount of heat required from the ΔH_{fusion} value in Table 15.6.

$$\text{Heat required} = 0.1754\ \text{mol Hg} \times \left(\frac{2.3\ \text{kJ}}{1\ \text{mol Hg}}\right) = 0.40\ \text{kJ}$$

Problem-Solving Practice 15.6

Use data from Table 15.6 to calculate the heat required to melt 1.45 g of aluminum.

Exercise 15.16 Cooling a Liquid Metal Until It Solidifies

When a liquid metal is cooled to the temperature at which it solidifies, and the solid is then cooled to an even lower temperature, the "cooling curve," a plot of temperature against time, looks like this:

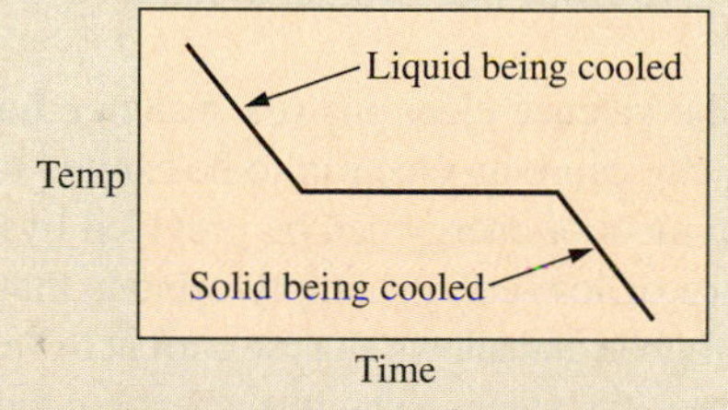

Account for the shape of this curve. Would all substances exhibit similar curves? (← *p. 232*).

Exercise 15.17 Heats of Fusion and Electronic Configuration

Look at Table 8.4 and compare the electron configurations shown there with the heats of fusion for the metals shown in Table 15.6. Do you see any correlation between these configurations and this property? Does strength of attraction among metal atoms correlate with number of valence electrons? Explain.

Metallic Bonding

The properties of solids can be explained by the type of bonding that holds their constituent particles together. In molecular solids, molecules are held together by inter-

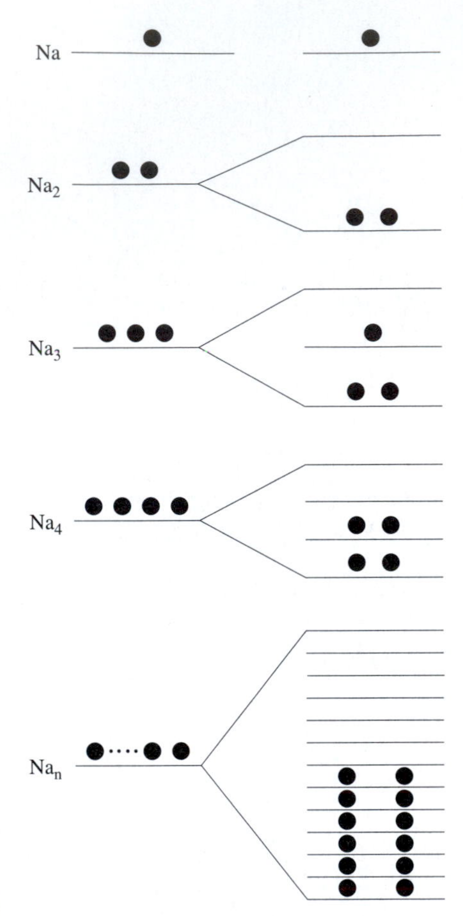

Figure 15.22 Formation of bands of electron orbitals in a metal crystal. As atoms are added to the crystal, the number of orbitals available to merge into bands increases, and the electrons move into the lowest energy levels available.

molecular forces; in ionic solids, ions are held together by electrostatic attractions between positive and negative ions. Metals behave as though metal cations exist in a "sea" of mobile electrons—the valence electrons of all the metal atoms. **Metallic bonding** is the nondirectional attraction between positive metal ions and the surrounding sea of negative charge. Each metal ion has a large number of near neighbors. The valence electrons are spread throughout the metal lattice, holding the positive metal ions together. When an electric field is applied to a metal, these valence electrons move toward the positively charged end, and the metal conducts electricity.

Because of the uniform charge distribution provided by the mobile valence electrons in a metal lattice, the positions of the positive ions can be changed without destroying the attractions among them. Thus most metals can be bent and drawn into wire; they are malleable and ductile. Conversely, when we try to deform an ionic solid, the crystal usually shatters because the balance of positive ions surrounded by negative ions and vice versa is disrupted (⇐ *p. 100*).

To visualize how bonding electrons behave in a metal, first consider the arrangement of electrons in an individual atom far enough away from any neighbor so that there is no bonding. In such an atom the electrons occupy orbitals that have definite energy levels (⇐ *p. 321*). In a large number of separated, identical atoms, all of the energy levels are identical. If the atoms are brought closer together, they begin to influence one another. The identical energy levels shift up or down, and become bands of energy levels characteristic of the large collections of metal atoms (Figure 15.22). An **energy band** is a large group of orbitals whose energies are closely spaced and whose average energy is the same as the energy of the corresponding orbital in an individual atom. In some cases energy bands for different types of electrons (*s, p, d,* etc.) overlap; in other cases there is a gap between different energy bands.

Within each band, electrons fill the lowest-energy orbitals much as electrons fill orbitals in atoms. The number of electrons in a given energy band depends on the number of metal atoms in the crystal. In considering conductivity and other metallic properties, it is usually necessary to consider only valence electrons, since other electrons all occupy completely filled bands in which two electrons occupy every orbital. In these, no electron can move from one orbital to another, because there is no empty spot for it.

A band containing the valence electrons (the **valence band**) is partially filled, so that a little added energy can cause an electron to be excited to a slightly higher-energy orbital. Such a small increment of energy can be provided by applying an electric field, for example. The presence of low-energy, empty orbitals that electrons can move into allows the electrons to be mobile and allows electric current to be conducted (Figure 15.23).

When a metal is heated, the valence electrons become thermally excited and move to low-lying empty orbitals in a conduction band. When the reverse process occurs thermal energy is released. This ability of the valence electrons in a metal lattice to move easily into empty conduction bands also explains the luster of metals. The electrons can absorb a wide range of wavelengths because the conduction bands are so close in energy to the filled bands. As these excited electrons fall back to their lower energy states, they emit a variety of wavelengths of light (⇐ *p. 305*).

The energy band theory also explains why some solids are **insulators** and do not conduct electricity. In an insulator the energy bands do not overlap. Rather, there is a large gap between them (Figure 15.23). There are very few electrons that have enough energy to jump across the large gap from a filled lower-energy band to an empty higher-energy band, and so no current flows when an external electric field is applied.

CHEMISTRY IN THE NEWS

Metallic Hydrogen

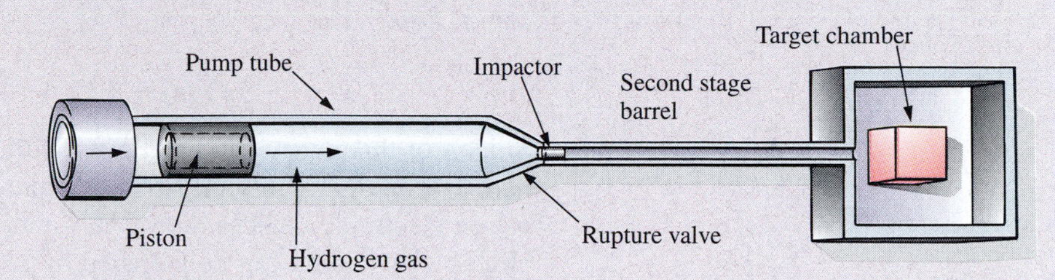

Hydrogen, with its single *s* electron, is sometimes placed with the alkali metals in the periodic table. Yet hydrogen forms a stable diatomic gas rather than existing as a metal like the alkali metals. As early as 1935 it was suggested that hydrogen might exist as a metal if its atoms could somehow be brought close enough together under pressure. Now, scientists at the Lawrence Livermore Laboratory in Berkeley, California, believe they have succeeded in making metallic hydrogen by using a powerful gun that fires a piston down a tube, compressing a sample of hydrogen to a pressure of about 1.4 megabars as it goes. Experiments have shown that samples of hydrogen do indeed become metallic as evidenced by a large decrease (10^4) in electrical resistance immediately after the compression caused by the gun. The interesting thing about these experiments is that hydrogen becomes metallic at pressures much lower than had been predicted by theory.

This is important because it may help explain why Jupiter, a planet with a core of hydrogen under great pressure caused by gravitational forces, has such a large magnetic field. If Jupiter's metallic hydrogen core extends farther out than had been previously thought, then there is much more hydrogen present in the metallic state. This would account for the great magnetic field the giant planet has. (The Earth's magnetic field is caused by a core of metallic iron, heated by gravitational forces.)

Someday, metallic hydrogen may become commonplace because theorists believe that once formed, metallic hydrogen might be stable at ordinary temperatures and pressures. If it can be made, metallic hydrogen will certainly find many interesting uses. For example, it would be the lightest metal available!

Source: The New York Times, *March 26, 1996, p. C1.*

In **semiconductors** there are very *narrow* energy gaps between fully occupied energy bands and empty energy bands (Figure 15.23). At quite low temperatures, electrons remain in the filled lower-energy band and semiconductors are not good conductors. At higher temperatures, or when an electric field is applied, some electrons have enough energy to jump across the band gap into an empty energy band. This allows an electric current to flow.

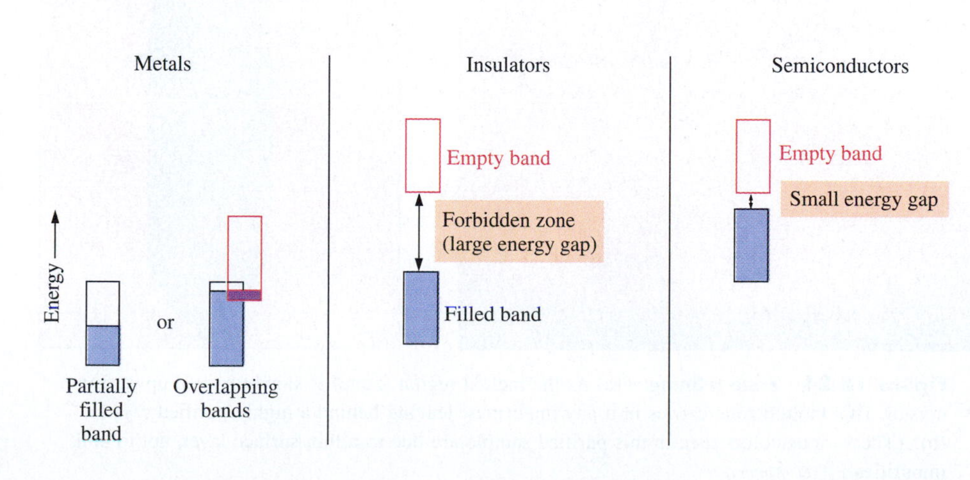

Figure 15.23 Differences in the energy bands of available orbitals in metals, insulators, and semiconductors. In each case, an unshaded area represents a conduction band.

15.9 SILICON AND THE CHIP

Silicon is known as an "intrinsic" semiconductor because the element itself is a semiconductor. Silicon of about 98% purity can be obtained by heating silica (purified sand) and coke (an impure form of carbon) at 3000 °C in an electric arc furnace. (⇐ *p. 170*).

$$SiO_2(s) + 2\ C(s) \xrightarrow{\text{heat}} Si(s) + 2\ CO(g)$$

Silicon of this purity can be alloyed with aluminum and magnesium to increase the hardness and durability of the metals and is used for making silicone polymers. For use in electronic devices, however, a much higher degree of purification is needed. High-purity silicon can be prepared by reducing $SiCl_4$ with magnesium.

$$SiCl_4(\ell) + 2\ Mg(s) \longrightarrow Si(s) + 2\ MgCl_2(s)$$

The magnesium chloride, which is water soluble, is then washed from the silicon. The final purification of the silicon takes place by a melting process called **zone refining** (Figure 15.24), which produces silicon containing less than one part per billion of impurities such as boron, aluminum, and arsenic. Zone refining takes advantage of the fact that impurities are often more soluble in the liquid phase than in the solid phase. As a hot molten zone is moved through a sample being purified, the impurities move along in the liquefied portion of the sample. As the heated zone cools, the sample that resolidifies is purer than it was. Multiple passes of the hot molten zone are usually necessary to achieve the degree of purity necessary to fabricate semiconducting devices.

Like all semiconductors, high-purity silicon fails to conduct an electric current until a certain electrical voltage is applied, but at higher voltages it conducts moderately well. Silicon's semiconducting properties can be improved by a process known

(a)

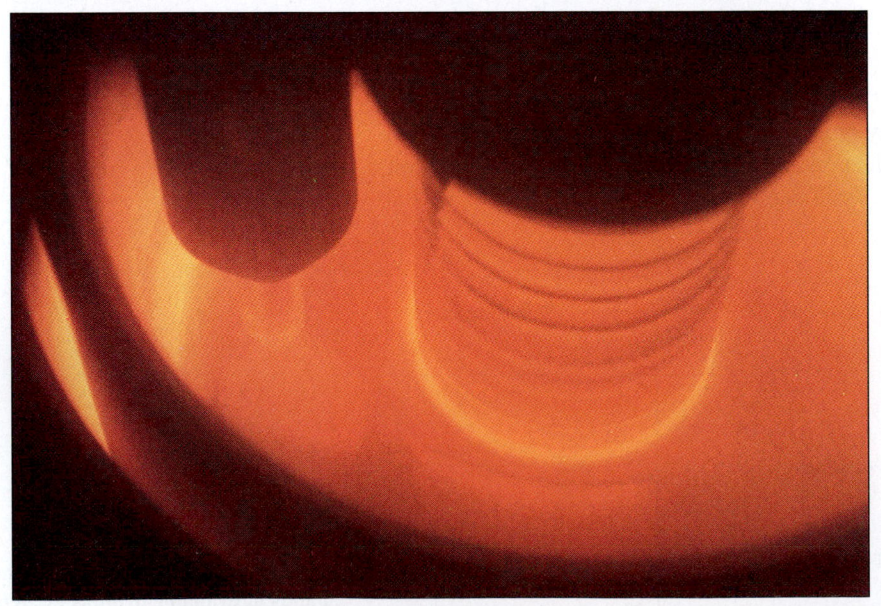

(b)

Figure 15.24 Zone refining. (a) As the melted region (zone) is slowly moved upward in a crystal. This molten zone carries in it any impurities, leaving behind a highly purified crystal (b). (The various colors seen in this purified sample are due to a thin surface layer, not to bulk impurities.) *(C.D. Winters)*

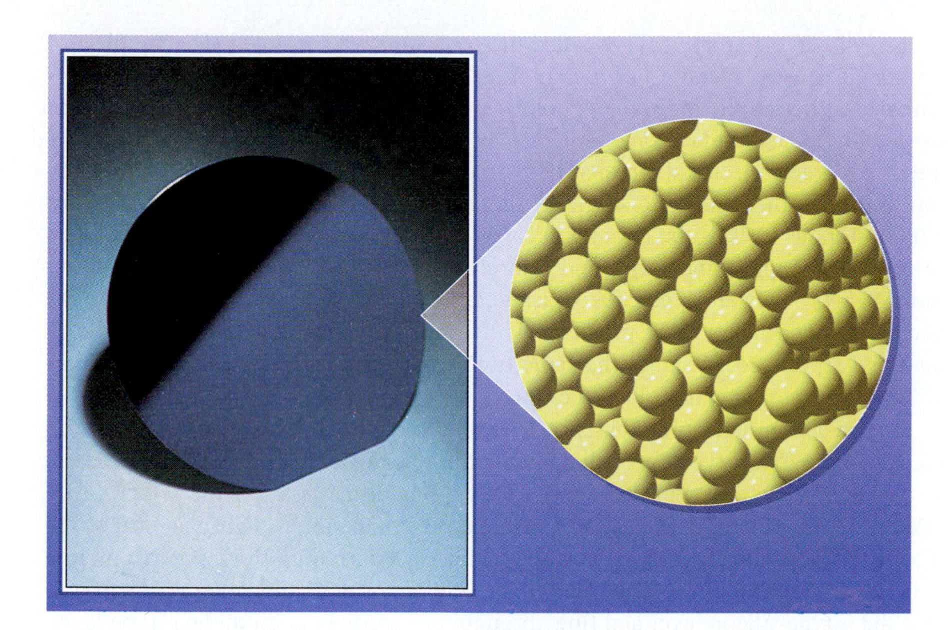

A wafer of highly purified silicon.

as doping. **Doping** is the addition of a tiny amount of some other element (a *dopant*) to the silicon. For example, suppose a small number of boron atoms (or atoms of some other Group 3A element) replace silicon atoms (Group 4A) in solid silicon. Boron has only three valence electrons, whereas silicon has four. This leaves a deficiency of one electron around the B atom, creating what is called a *positive hole* for every B atom added. Hence silicon doped in this manner is referred to as positive-type or *p*-type silicon (Figure 15.25). When a few atoms of a Group 5A element such as arsenic are added to silicon, only four of the five valence electrons of As are used for bonding with four Si atoms, leaving one electron relatively free to move. This type of silicon is referred to as negative-type or *n*-type silicon, because it has extra (negative) valence electrons.

A positive hole in an energy band of a semiconductor is a place where there is one less electron than normal and hence extra positive charge.

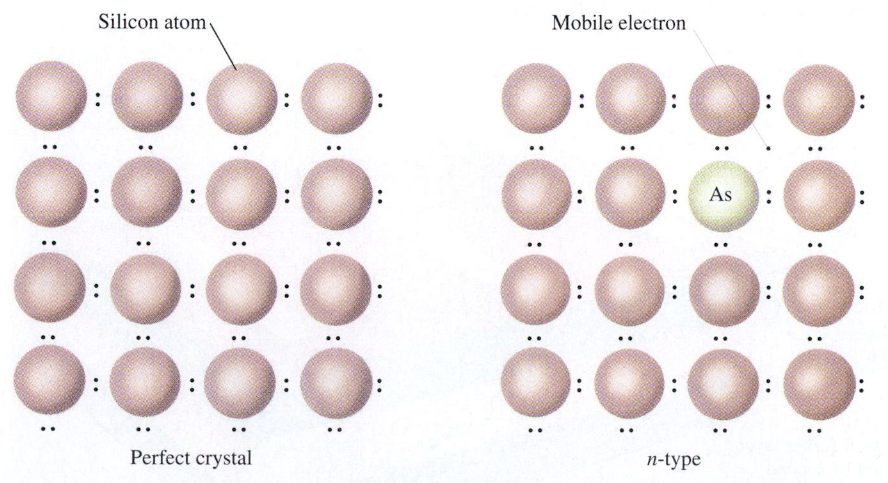

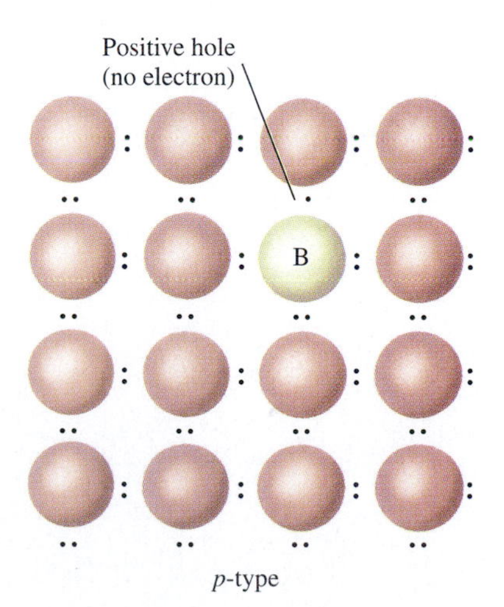

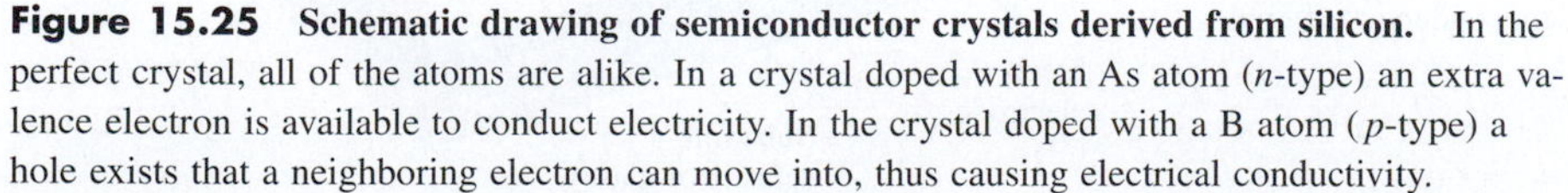

Figure 15.25 Schematic drawing of semiconductor crystals derived from silicon. In the perfect crystal, all of the atoms are alike. In a crystal doped with an As atom (*n*-type) an extra valence electron is available to conduct electricity. In the crystal doped with a B atom (*p*-type) a hole exists that a neighboring electron can move into, thus causing electrical conductivity.

In 1947 an electronic device called the *transistor* was invented. The simplest of these devices uses layers of *n-p-n-* or *p-n-p*-doped silicon. Germanium, a Group 4A element just below silicon in the periodic table, is also used in place of silicon. The most revolutionary application of silicon's semiconductor properties has been the design of integrated electrical circuits (ICs) on tiny chips of silicon scarcely larger than a millimeter in diameter. In a single IC there are often thousands and even millions of transistors as well as the circuits to carry the electrical signals. These devices, in the form of computer memories and central processing units (CPUs), permeate our whole society. They are in calculators, cameras, watches, toys, coin changers, cardiac pacemakers, and many other products. Truly, silicon is both the world we walk on (sand and silicate rock) and at the same time our constant companion in communications and electronic controls (the chip).

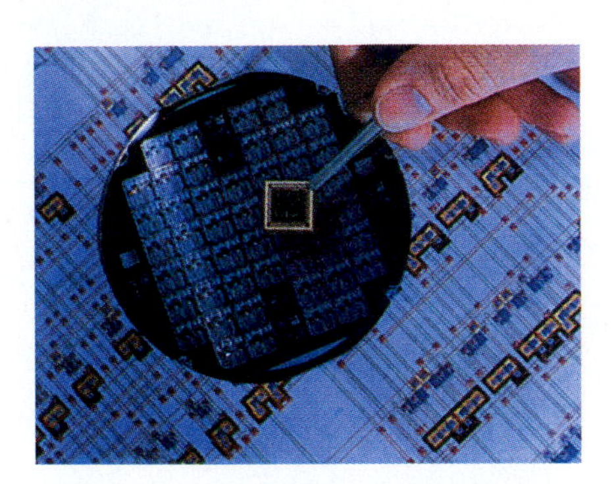

A microcomputer chip. As the chips become smaller and smaller, the purity of the silicon becomes more important, since impurities can prevent a circuit from working properly. *(Courtesy of AT&T Bell Laboratories)*

Doped silicon is also the basis of the **solar cell.** When a layer of *n*-type doped silicon is next to a *p*-type layer, there is a strong tendency for the extra electrons in the *n*-type layer to pair with the unpaired electrons in the holes of the *p*-type layer. If the two layers are connected by an external electrical circuit, light that strikes the silicon provides enough energy to cause an electrical current to flow. When a photon is absorbed, it excites an electron to a higher-energy orbital, allowing the electron to leave the *n*-type silicon layer and flow through the external circuit to the *p*-type layer. As the *p*-type layer becomes more negative (because of added electrons), the extra electrons are repelled *internally* back into the *n*-type layer (which has become positive because of loss of electrons via the circuit). This process can continue as long as the silicon layers are exposed to sunlight and the circuit remains closed.

A typical solar cell is constructed on a substrate (support) layer of plastic or glass (Figure 15.26). Next to the substrate is a thin sheet of metal that transfers electrons to the *p*-type semiconductor layer. The topmost semiconductor layer is *n*-type; it receives the sun's rays and is nearly transparent. To cause more of the sunlight striking the surface to enter the topmost layer, an antireflection coating of indium tin oxide ($InSnO_2$) is added. A metallic grid structure on top of the cell functions as the second electrode, allowing as much light as possible to strike the *n*-type layer. The efficiency of a solar cell depends on how well it can absorb photons and convert the light to electrical energy. Photons that are reflected, pass through, or produce only heat decrease the efficiency of the cell.

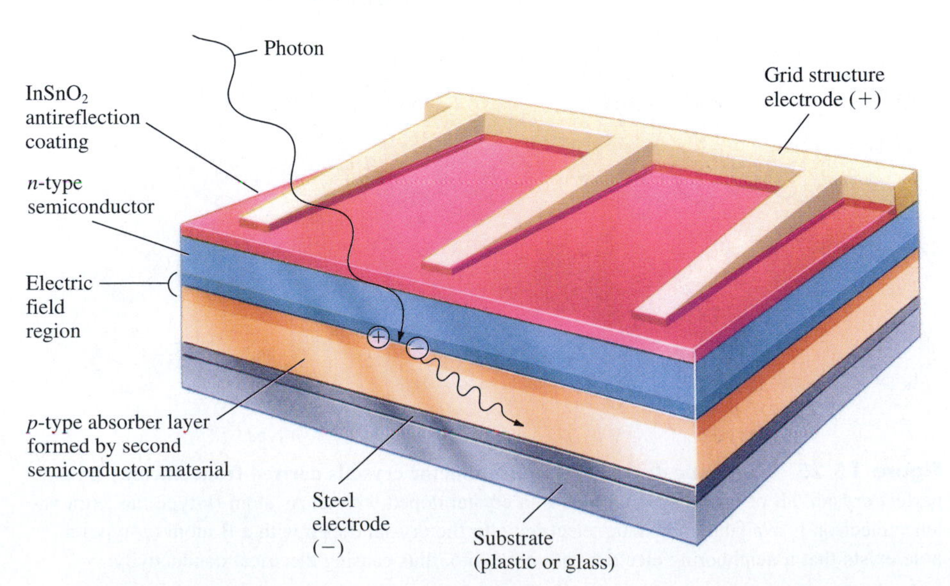

Figure 15.26 Typical photovoltaic cell using layers of doped silicon. *(Adapted from Scientific American)*

Solar cells are on the threshold of being the next great technological breakthrough, perhaps comparable to the computer chip. Although experimental solar-powered automobiles are now available and many novel applications of solar cells already exist, the real breakthrough will be general use of banks of solar cells as a utility power plant to produce huge quantities of electricity. One plant already operating in California uses solar cells to produce 20 megawatts (MW) of power—enough to supply the daily electricity needs of a city the size of Tampa, Florida.

15.10 NETWORK SOLIDS

A number of solids are composed of nonmetal atoms connected by a network of covalent bonds. Such **network solids** really consist of one huge molecule—a super molecule—in which all the atoms are connected to all the others via a network of bonds. Separate small molecules do not exist in a network solid.

Graphite and Diamond

Graphite and diamond are allotropes of carbon (⇐ *p. 34*). Graphite's name comes from the Greek *graphein* meaning "to write" because one of its earliest uses was for writing on parchment. Artists today still draw with charcoal, an impure form of graphite, and we write with graphite pencil leads. Graphite is an example of *a planar network solid* (Figure 15.27). The planes consist of six-membered rings of car-

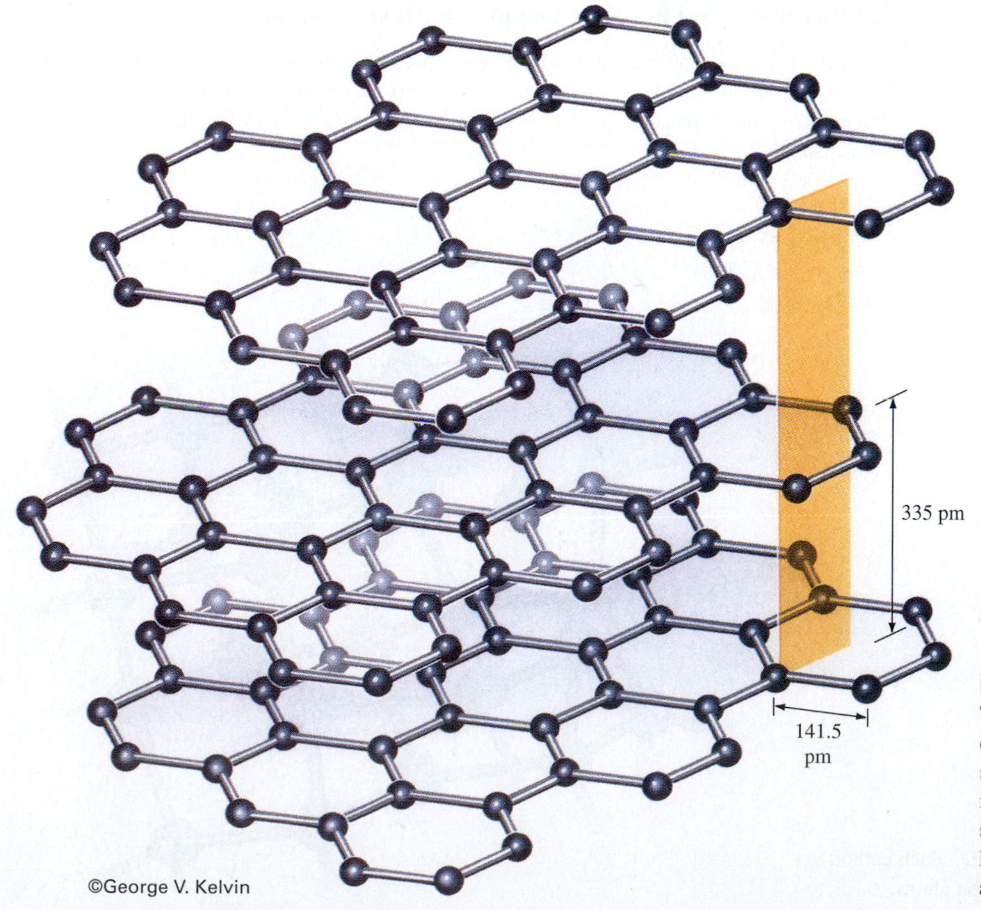

Figure 15.27 The structure of graphite. Three of the many layers of six-membered carbon rings are shown. These layers can slide past one another relatively easily, making graphite a good lubricant. In addition, some of the carbon valence electrons in the layers are delocalized, allowing graphite to be a good conductor of electricity.

bon atoms (like those in benzene, ⇐ *p. 377*). Each hexagon shares all six of its sides with other hexagons around it, forming a *two-dimensional network.* Some of the bonding electrons are able to move freely around this network, and so graphite is a conductor of electricity. There are strong covalent bonds between carbon atoms in the same plane, but attractions between the planes are caused by London forces and hence are weaker. Because of this, the planes can easily slip across one another, which makes graphite an excellent solid lubricant for uses, such as in locks, where greases and oils are undesirable.

Diamonds are also built of six-membered carbon rings, but each carbon atom is bonded to *four* others (instead of three, as in graphite) by single covalent bonds. This forms a *three-dimensional network* (Figure 15.28). Because of the tetrahedral arrangement of bonds around each carbon atom, the six-membered rings in the diamond structure are puckered. In graphite the layers are much farther apart than normal C—C bond distances. As a result, diamond is denser than graphite (3.51 g/cm^3 and 2.22 g/cm^3, respectively). Also, because its valence electrons are localized between carbon atoms, diamond does not conduct electricity. Diamond is one of the hardest materials and also one of the best conductors of heat known. It is also transparent to visible, infrared, and ultraviolet radiation. Finally, the relatively low density and high rigidity of diamonds makes them ideal for producing sounds with frequencies as high as 60,000 Hz, far higher than the range of human hearing. What more could a scientist want in a material—except a cheap, practical way to make it!

The distance between planes is more than twice the distance between the nearest carbon atoms within a plane.

Exercise 15.18 How Would You Make Diamonds?

Given the properties of graphite and diamond, predict what conditions of temperature and pressure might be used if you wanted to convert inexpensive graphite into valuable diamonds within a reasonable period of time. (*Hint:* Apply LeChatelier's principle ⇐ *p. 586.*)

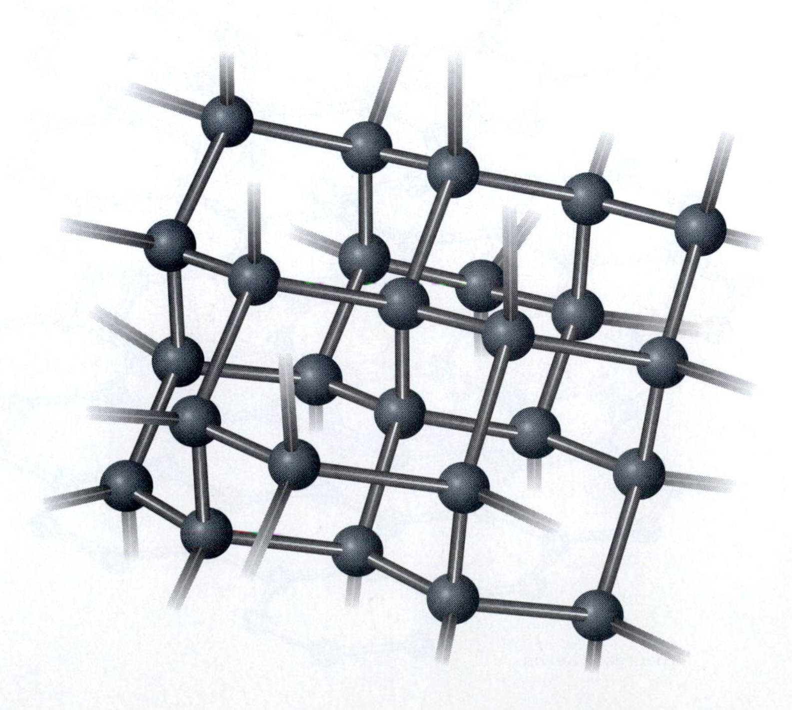

Figure 15.28 The structure of diamond. Each carbon atom is covalently bonded to four other carbon atoms.

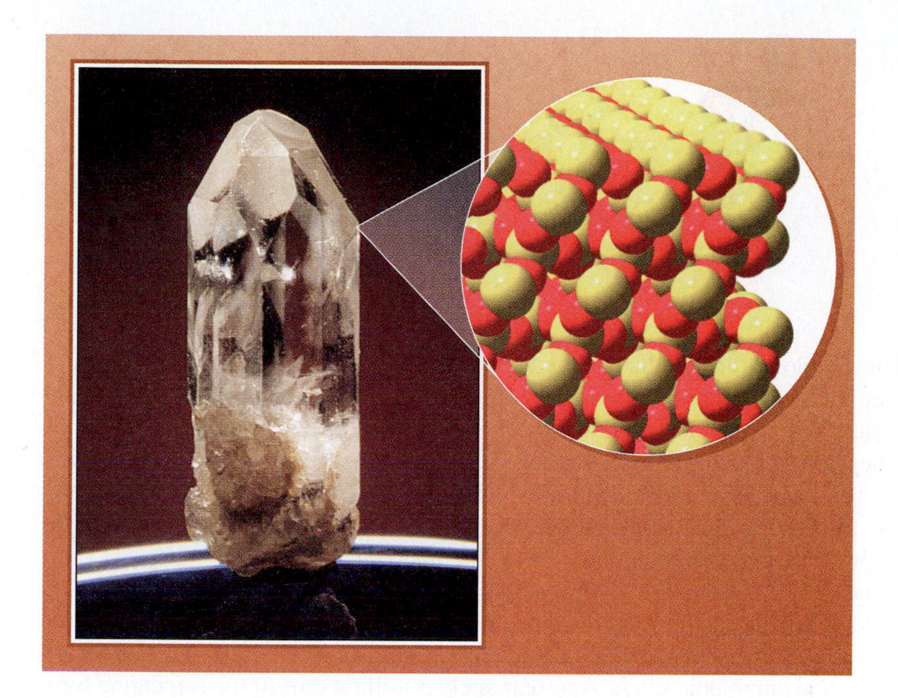

Figure 15.29 A pure quartz crystal (pure SiO_2). Quartz is one of the most common minerals on earth. Pure, colorless quartz was used as an ornamental material as early as the Stone Age, and by Roman times it was known that a wedge (or prism) of quartz could be used to disperse the sun's rays into a rainbow of colors. Quartz crystals today are used as oscillators in watches, radios, VCRs, and computers. *(C.D. Winters)*

In the 1950s, scientists at General Electric in Schenectady, New York, achieved something alchemists had attempted for centuries—the synthesis of diamonds. The GE scientists even used such diverse carbon-containing materials as wood and peanut butter to prove their prowess at diamond synthesis! Their technique, still in use today, was to heat graphite to a temperature of 1500 °C in the presence of a metal catalyst, such as nickel or iron, and under a pressure of 50,000 to 65,000 atm. Under these conditions, the carbon dissolves in the metal and recrystallizes in its higher-density form, slowly becoming diamond. The worldwide market for diamonds made this way is now worth about $500 million; many of these diamonds are used for abrasives and diamond-coated cutting tools for drilling. While gem-quality diamonds have been made from graphite, they are too expensive compared to naturally occurring diamonds.

Crystalline SiO_2 and Silicates

Unlike carbon, which forms molecular CO_2, silicon forms a dioxide that is a network solid. Pure SiO_2 is called *silica,* and the most common form of pure silica is α-quartz. Silica (Figure 15.29) is a major component of many rocks such as granite and sandstone, and it occurs alone as pure rock crystal as well as in a variety of less pure forms. When α-quartz is heated to almost 1500 °C, it transforms into another crystalline form, cristobalite, in which the Si—O—Si bonds are arranged like the Si—Si—Si bonds in elemental silicon. In both cases, each Si atom is surrounded by four O atoms to form an SiO_4 tetrahedron but the SiO_4 tetrahedra are arranged differently in quartz from the way they are in cristobalite. This is also the arrangement of carbon atoms in diamond *(Figure 15.28)*. This tetrahedral structure is important because it is seen in a variety of forms of silica as well as in the structures of carbon compounds.

The main crystalline modifications of SiO_2 consist of infinite arrays of SiO_4 tetrahedra sharing corners. In silicate minerals, SiO_4 tetrahedra share one or more corners to give chains, rings, sheets, and three-dimensional structures.

The fascinating colors of some gemstones come from impurities in natural silicate crystals that, if they were pure, would be colorless. For example, pure corundum, a crystalline form of aluminum oxide (Al_2O_3), is a colorless crystalline solid in which each Al^{3+} ion is enclosed in a cage of six O^{2-} ions. If about 1% of the aluminum ions are replaced by Cr^{3+} ions, the crystal is a ruby. A ruby is red because chromium ions absorb light in the blue end of the visible spectrum while the ruby-red light passes through the crystal.

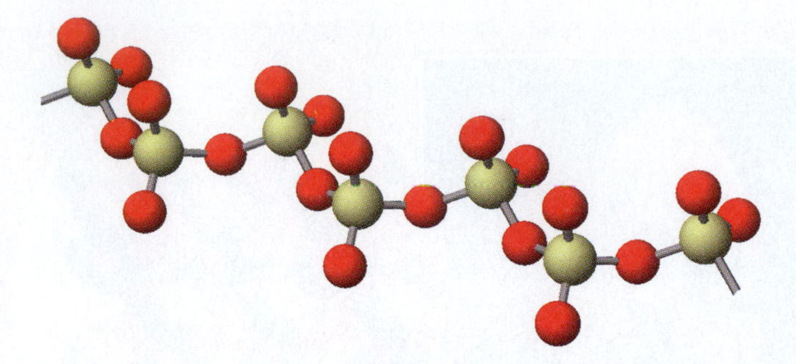

Figure 15.30 Pyroxene structure. Each SiO_4 tetrahedron shares an O atom with another SiO_4 tetrahedron on either side of it. This results in a chain-like structure. The formula for the chain is $(SiO_3)_n^{6-}$.

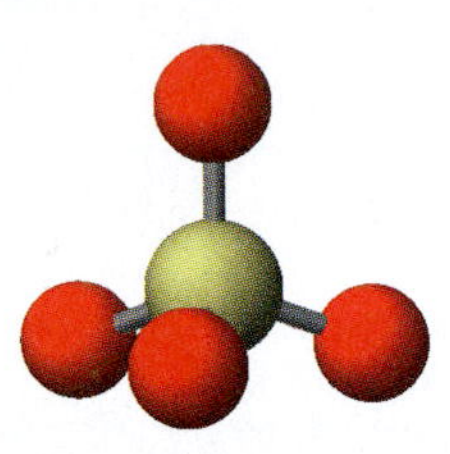

Tetrahedral SiO_4 unit

The simplest silicate minerals are the *pyroxenes,* which contain extended chains of linked SiO_4 tetrahedra (Figure 15.30). Their formulas appear to contain the metasilicate ion, SiO_3^{2-}. A typical formula is Na_2SiO_3. If two such chains are laid side by side, they may link together by sharing oxygen atoms in adjoining chains (Figure 15.31). The result is an *amphibole,* of which the *asbestos* minerals are excellent examples. Because of their double-stranded chain structure, asbestos minerals are fibrous and can even be woven into a cloth-like material.

What is called "asbestos" is not a single substance, however. Rather, the name asbestos applies broadly to a family of naturally occurring hydrated silicates that crystallize in a fibrous manner. These minerals are generally subdivided into two forms, serpentine and amphibole fibers. Approximately 5 million tons of the serpentine form of asbestos, chrysotile, are mined each year, chiefly in Canada and the former Soviet Union; this is essentially the only form used commercially in the United States. Another form, the amphibole crocidolite, is mined in small quantities, mainly in South Africa. The two minerals differ greatly in composition, color, shape, solubility, and persistence in human tissue. This last property is important in determining the toxicity of asbestos. Crocidolite is blue, relatively insoluble, and persists in tissue. Its fibers are long, thin, and straight and can penetrate narrow lung passages. In contrast, chrysotile is white, and it tends to be soluble and disappear in tissue. Its fibers are curly; they ball up like yarn and are more easily rejected by the body. Long-term occupational exposures to certain asbestos minerals can lead to lung cancer. Although there is some disagreement in the medical and scientific communities, evidence strongly suggests that the amphiboles such as crocidolite are much more potent cancer-causing agents than the serpentines such as chrysotile. Since most asbestos in public buildings is of the chrysotile type, the drive to remove asbestos insulation may be misguided overreaction in many cases. Nevertheless, most asbestos-containing materials have been removed from the market and strict standards now exist for the handling and use of asbestos.

If the linking of silicate chains continues in two dimensions, sheets of SiO_4 tetrahedral units result (Figure 15.32). Various clay minerals and mica have this sheet-like

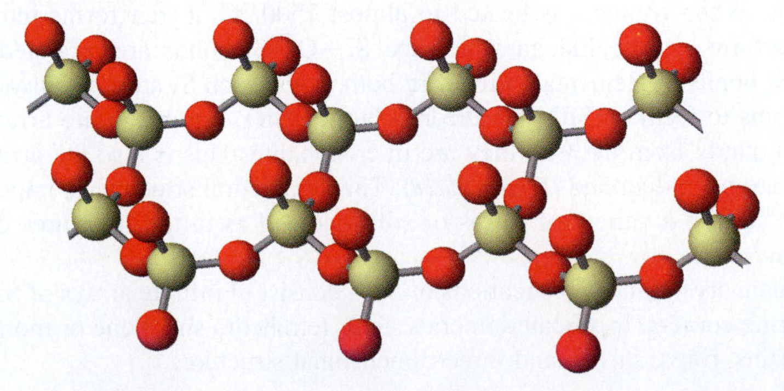

Figure 15.31 Amphibole structure. Two chains of SiO_4 units, each like the one in Figure 15.30, share O atoms between them to make a ribbon-like structure.

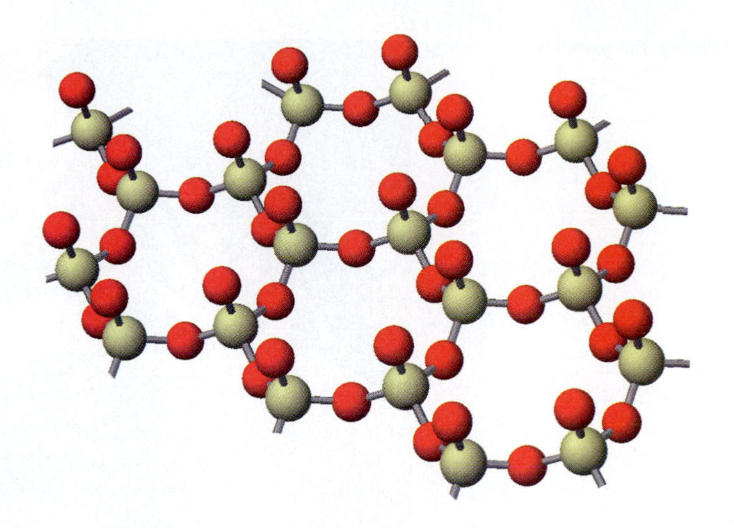

Figure 15.32 Mica structure. Each SiO_4 unit shares three O atoms with adjacent SiO_4 units to make a layer structure that extends a long distance in each of two dimensions.

silicate structure. Mica, for example, is used to prepare "metallic" looking paint on new automobiles.

Clays are essential components of soils that come from the weathering of igneous rocks. They have been used since the beginning of human history for pottery, bricks, tiles, and writing materials. Clays are actually *aluminosilicates,* in which some Si^{4+} ions are replaced by Al^{3+} ions. Feldspar, a component of many rocks, is weathered in the following reaction to form the clay *kaolinite.*

Igneous rocks were formed from molten material such as lava from a volcano. Feldspar is a form of igneous rock containing crystalline silicate minerals. Feldspars make up 60% of the Earth's crust.

$$\underset{\text{feldspar}}{2\ KAlSi_3O_8(s)} + CO_2(g) + 2\ H_2O(\ell) \longrightarrow \underset{\text{kaolinite}}{Al_2(OH)_4Si_2O_5(s)} + 4\ SiO_2(s) + K_2CO_3(aq)$$

The structure of kaolinite consists of layers of SiO_4 tetrahedra linked in sheets, as in mica, but these sheets are interleaved with layers of Al^{3+} ions, each of which is surrounded by six oxygen atoms of the silicon-oxygen sheets as well as by OH^- ions (Figure 15.33). When Al^{3+} ions are replaced by other 3+ metal ions, the clay

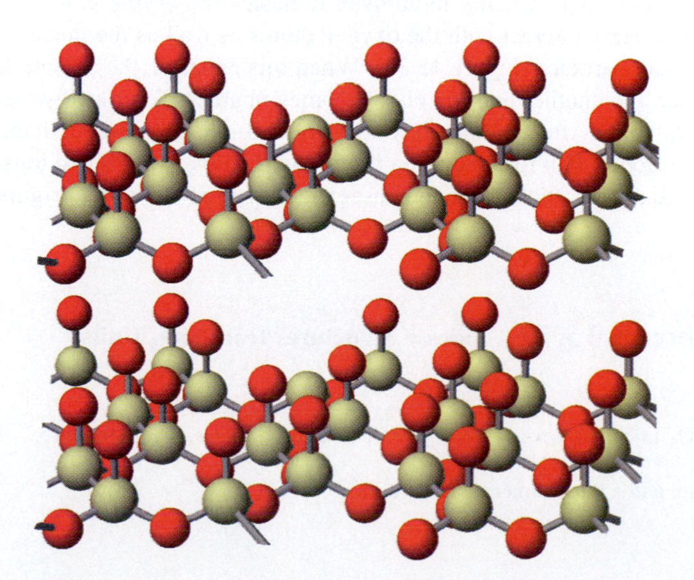

Figure 15.33 Silicate layers in kaolinite, $Al_2(OH)_4Si_2O_5$, an aluminosilicate. As in mica (Figure 15.32), each silicon (pale yellow) is surrounded tetrahedrally by oxygen atoms (red) to give rings consisting of six O and six Si atoms. These rings are connected in two dimensions to give layers, two of which are shown. A layer of aluminum ions (not shown) is attached through O atoms to the Si—O rings. Hydroxide ions (not shown) act as bridges between aluminum ions. The structure is like a multidecker sandwich. Stacked silicate layers (the bread) are held together by aluminum ions and hydroxide ions (the filling). This layering is responsible for the slipperiness and workability that characterize clays, especially when they are wet. *(C.D. Winters)*

Figure 15.34 Clay being worked into an object. After drying, the particles rebond to one another.

becomes colored. For example, a red clay contains Fe^{3+} ions in place of some Al^{3+} ions.

Interestingly, clays are eaten for medicinal purposes. Several pharmaceuticals sold in the United States for relief of upset stomach contain highly purified clay, which absorbs excess stomach acid as well as possibly harmful bacteria and their toxins.

Art students are familiar with working wet clay into a shape and then letting it dry to retain the shape, usually in an oven to hasten the drying process. Water molecules can strongly interact with the oxygen atoms as well as the metal ions near the surface of clay particles (Figure 15.34). When this happens, the silicate layers begin to slide over one another and the clay becomes pliable. (Too much water can make the clay unstable.) After the clay has been formed into the desired shape, the water is removed. Bonds then form between the exposed oxygen atoms and ions on the surfaces and adjacent particles, which causes the clay to return to its original hardness.

Exercise 15.19 Silicate Structures from SiO_4 Units

Use SiO_4 units, shown as 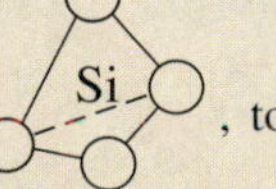

, to draw a linear chain structure, a flat sheet structure, and a three-dimensional structure.

15.11 CEMENT, CERAMICS, AND GLASS

Cement, ceramics, and glass are examples of amorphous solids; that is, they lack crystalline structures with easily defined unit cells. They are all extremely important because of their useful properties. We build things with all three of these kinds of solids. They all are created at high temperatures, so making them from starting materials is highly energy intensive.

Cement

Cement consists of microscopic particles containing compounds of calcium, iron, aluminum, silicon, and oxygen in varying proportions. Cement reacts in the presence of water to form hydrated particles with large surface areas, which subsequently undergo recrystallization and reaction to bond to themselves as well as to the surfaces of bricks, stone, or other silicate materials.

Cement is made by roasting a powdered mixture of calcium carbonate (limestone or chalk), silica (sand), aluminosilicate mineral (kaolin, clay, or shale), and iron oxide at a temperature of up to 870 °C in a rotating kiln. As the materials pass through the kiln, they lose water and carbon dioxide and ultimately form a "clinker," in which the materials are partially fused. A small amount of calcium sulfate (gypsum) is added and the cooled clinker is then ground to a very fine powder. A typical composition of cement can be expressed as follows: 60% to 67% CaO, 17% to 25% SiO_2, 3% to 8% Al_2O_3, up to 6% Fe_2O_3, and small amounts of magnesium oxide, magnesium sulfate, and oxides of potassium and sodium. Cement is usually mixed with other substances. **Mortar** is a mixture of cement, sand, water, and lime. **Concrete** is a mixture of cement, sand, and aggregate (crushed stone or pebbles) in proportions that vary according to the application and the strength required.

Figure 15.35 Pouring concrete. After the concrete has been formed and exposed surfaces smoothed, it is allowed to set, a process hastened by moisture, in which strong bonds are formed between many of the tiny particles originally present in the dry concrete. *(Deneve Feigh Bunde/Visuals Unlimited)*

In cement, the oxides are not isolated into molecules or ionic crystals. Rather, the entire nanoscale structure is a complex network of ions, each satisfying its charge requirements with ions of opposite charge. Many different reactions occur during the setting of cement. Various constituents react with water and with carbon dioxide in the air. The surface of each cement particle has many sites that attract water molecules, so when water is added to the dry solid, a stable suspension of the particles results. The initial reaction of cement with water is the hydrolysis of the calcium silicates, which forms a gel that sticks to itself and to the other particles (sand, crushed stone, or gravel). The gel has a very large surface area and ultimately is responsible for the great strength of concrete once it has set. The setting process also involves the formation of small, densely interlocked crystals after the initial solidification of the wet mass (Figure 15.35). This interlocking crystal formation continues for a long time after the initial setting, and it increases the compressive strength of the cement. For this reason, freshly poured concrete is kept moist for several days.

Compressive strength refers to a solid's resistance to shattering under compression.

More than 800 million tons of cement are manufactured each year, most of which is used to make concrete. Concrete, like many other materials containing Si—O bonds, is highly noncompressible but lacks tensile strength. If concrete is to be used where it is subject to tension, as in a bridge or building, it must be reinforced with steel.

Tensile strength refers to a substance's resistance to stretching.

Ceramics

Ceramics have been made since well before the dawn of recorded history. Much art of great historical significance is made of ceramic materials. Ceramics form an extremely large and diverse class of substances and are generally fashioned from clay or other natural earths at room temperature and then permanently hardened by heat in a baking ("firing") process that binds the particles together. *Silicate ceramics* include objects made from clays (aluminosilicates), such as pottery, bricks, and table china. The techniques developed with natural clay have been applied to a wide range of other inorganic materials in recent years.

China clay, or kaolin, is primarily kaolinite that is practically free of iron, which otherwise imparts a red color to the clay. As a result it is almost colorless and particularly valuable in making fine pottery. The first pieces of fine Oriental clayware, which was named "china," arrived in Europe in the Middle Ages; European potters envied and admired the obviously superior product and struggled to discover how to duplicate it. Clays, mixed with water, form a moldable paste consisting of tiny silicate sheets that can easily slide past one another (recall the lubricant action of graphite, caused by sheets sliding over one another). When this paste is mixed to just the right consistency, it can be molded into almost any form. Then it is heated, the water is driven off, and new Si—O—Si bonds form so that the mass becomes permanently rigid. If silica (SiO_2) and feldspar are added in the right proportions, the mixture will not crack after heating. (Cracking occurs as a result of shrinkage when the new bonds form between the silicate sheets.)

Oxide ceramics are produced from metal oxides such as alumina (Al_2O_3) and magnesia (MgO) by heating the powdered solids under pressure. This causes the particles to bind to one another to form a rigid solid. Because it has a high electrical resistivity, alumina is used in spark-plug insulators. High-density alumina has very high mechanical strength, and it is also used in armor plating and in high-speed cutting tools for machining metals. Magnesia is an insulator with a high melting point (2800 °C), so it is often used as insulation in electric heaters and electric stoves.

A third class of ceramics includes the *nonoxide ceramics* such as silicon nitride (Si_3N_4), silicon carbide (SiC), and boron nitride (BN). When the powdered solids are strongly heated under pressure, all of these compounds form ceramics that are hard and strong, but brittle. Boron nitride has the same average number of electrons per atom as does elemental carbon and exists in the graphite structure or the diamond structure, which is comparable in hardness to diamond and more resistant to oxidation. One application of these properties is the use of boron nitride cups and tubes to contain molten metals that are being evaporated.

Will some researcher discover a "bucky-ball" form of BN?

Silicon carbide has the trade name Carborundum, and can be regarded as diamond with half of the C atoms replaced by Si atoms. Widely used as an abrasive material *(****← p. 189****)*, SiC is receiving more attention recently as a ceramic material, particularly for high-temperature engines.

A market of $10 billion per year for advanced ceramics, industrial need for new materials, and accelerating fundamental research are causing an explosion of new ceramic materials. These include new forms of the ceramics described earlier, as well as *ceramic composites,* mixtures of ceramic materials with fibers, sometimes composed of various polymers, that have improved strength properties. The one severely limiting problem in using ceramics is their brittleness. Ceramics deform very little before they fail catastrophically, the failure resulting from a weak point in the bonding within the ceramic matrix. However, such weak points are not consistent from sample to sample so that the failure is not very predictable. Since stress failure of ceramic composites is due to molecular irregularities resulting from impurities or disorder in the atomic arrangements, much attention is now being given to using purer starting materials and controlling the processing steps. The fibers added to ceramics to make composites make them less susceptible to brittleness and sudden fracture.

Glasses

Glasses, which are amorphous solids, occur naturally or can be prepared synthetically. The manufacture of glass goes back to at least 5000 BC, when Phoenician sailors used blocks of sodium carbonate, Na_2CO_3, and sand, SiO_2, to insulate fires from the wooden planks of their ships. The metal carbonate and sand melted in the heat of the fire and formed a material that resembled *obsidian,* a natural glassy material that has been valued since antiquity.

One of the more common glasses today is soda-lime glass, which is clear and colorless if the purity of the ingredients is carefully controlled. If, for example, too much iron oxide is present, the glass will have a green color. Of course, color may also be a desirable property. By adding certain metal oxides to the basic ingredients of a glass, many beautiful colors can be obtained (see Table 15.7).

Colored glass. Adding small amounts of metal oxides to clear glass creates various colors. Look at Table 15.7 and identify which metals were used to make each of the different colored glasses shown here. *(C.D. Winters)*

The main glass-forming oxides are SiO_2, B_2O_3, GeO_2, and P_4O_{10}, all of which contain elements close to one another in the periodic table. Several other metal oxides, including Al_2O_3 and Na_2O, are also important in forming commercial glasses. The simplest glass is probably amorphous silica, SiO_2 (known as *vitreous silica*), which is built up of corner-sharing SiO_4 tetrahedra linked into a three-dimensional network that lacks symmetry or long-range order (Figure 15.36). Vitreous silica can be prepared by melting and quickly cooling either quartz or cristobalite.

If another oxide is added to SiO_2, the melting point of the mixture is lowered considerably (from 1800 °C for quartz to about 800 °C if about 25 mole percent Na_2O is added). The resulting melt cools to form a glass that is somewhat water soluble and is definitely soluble in strongly basic solutions. It is also prone to convert back to a crystalline solid. If other metal oxides such as CaO, MgO, or Al_2O_3 are added, the mixture still melts at a fairly low temperature, but it becomes resistant to chemical attack. Common glass like that used for windows, bottles, and lamps contains these metal oxides in addition to SiO_2.

"Lead glass," as the name implies, contains lead as PbO and is highly prized for its massive feel, acoustic properties, and high index of refraction. Recent regulations concerning toxic water pollutants in California (Proposition 65) have caused manufacturers of lead glass to withdraw some of their products from the marketplace.

It is important that glass be *annealed* properly during the manufacturing process. Annealing means cooling the glass slowly as it passes from a viscous, liquid state to a solid at room temperature. If a glass is cooled too quickly, bonding forces become uneven because small regions of crystallinity develop. Poorly annealed glass may crack or shatter when subjected to mechanical shocks or sudden temperature changes. High-quality glass, such as that used in optics, must be annealed very carefully. The

Mole percent is just another way of expressing concentration. If 0.75 mol of SiO_2 and 0.25 mol of Na_2O are mixed, there is 1.00 mol of matter present. The mole percent of SiO_2 is 75%, and the mole percent of Na_2O is 25%.

TABLE 15.7 Substances Used to Color Glass

Substance	Color
Copper(I) oxide	Red, green, blue
Tin(IV) oxide	Opaque
Calcium fluoride	Milky white
Manganese(IV) oxide	Violet
Cobalt(II) oxide	Blue
Finely divided gold	Red, purple, blue
Uranium compounds	Yellow, green
Iron(II) compounds	Green
Iron(III) compounds	Yellow

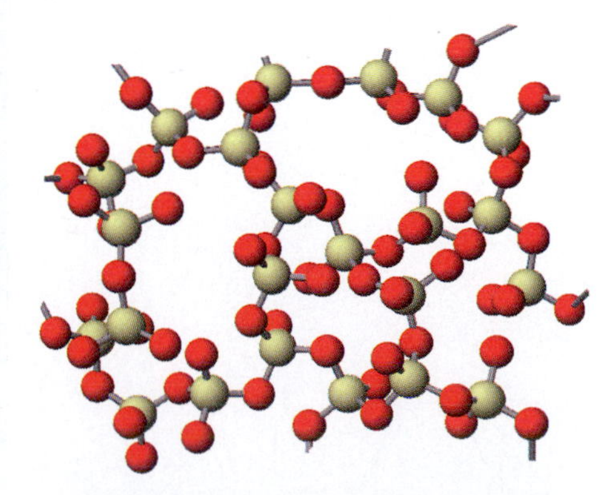

Figure 15.36 Vitreous silica. There is no repeating pattern in the arrangement of the SiO_4 units.

Dinnerware made from Pyroceram, a more thermally stable form of silicate ceramic. *(C.D. Winters)*

200-in. mirror for the telescope at Mt. Palomar, California, was annealed from 500 °C to 300 °C over a period of nine months!

A type of *glass ceramic* with unusual and very valuable properties has been commercialized. Ordinary glass breaks because once a crack starts, there is nothing to stop the crack from spreading. It was discovered that if glass is treated by heating until many tiny crystals have developed throughout the sample, the resulting material, when cooled, is much more resistant to breaking than normal glass. In molecular terms, the randomness of the glass structure has been partially replaced by the order in a crystalline silicate. The process must be controlled carefully to obtain the desired properties. Materials produced in this way are generally opaque and are used for cooking utensils and kitchenware, as in products marketed under the name Pyroceram. The initial manufacturing process is similar to that of other glass objects, but once the materials have been formed into their final shapes, they are heat treated to develop their special properties.

15.12 SUPERCONDUCTORS

One interesting property of metals is that their electrical conductivity *decreases* with increasing temperature. In Figure 15.37b a metal in a burner flame shows higher resistance (lower conductivity) than the same metal at room temperature (Figure 15.37a). Lower conductivity at higher temperature is understandable if valence electrons in the lattice of metal ions are thought of as waves. As an electron wave moves through the metal crystal under the influence of an electrical voltage, it encounters lattice positions where the metal ions are close enough together to scatter it. This scattering is analogous to the scattering of x rays caused by atoms in crystals (see Section 15.6). The scattered electron wave moves off in another direction, only to be scattered again when it encounters some other occupied lattice position. All this scattering *lowers* the conductivity of the metal. At higher temperatures, the metal ions vibrate more, and the distances between lattice positions change more from their average values. This

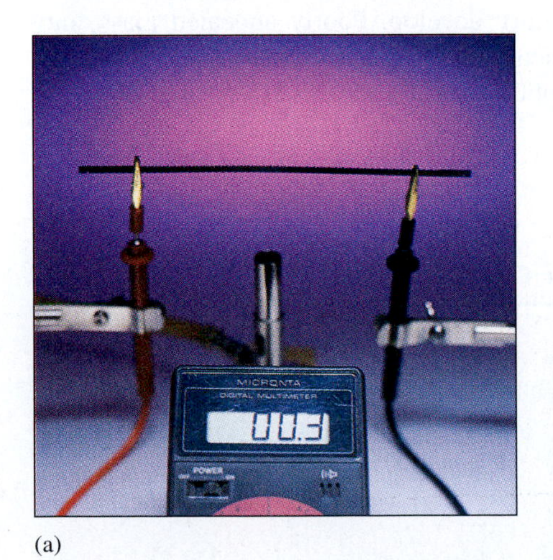

(a)

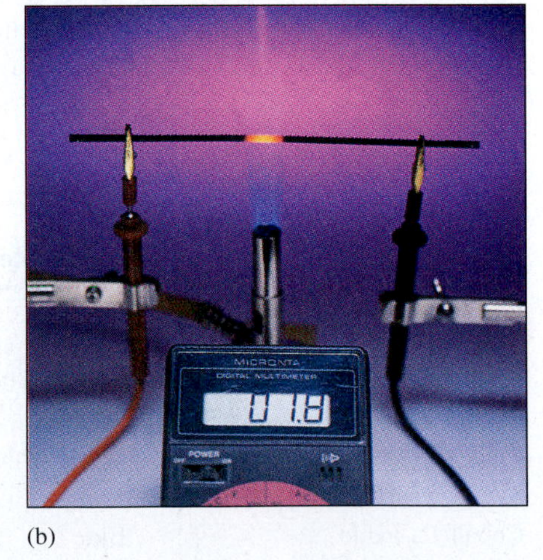

(b)

Figure 15.37 Resistance, temperature, and electrical conductivity. (a) A piece of metal at room temperature exhibits a small resistance (high electrical conductivity). (b) While being heated, this same piece of metal exhibits a higher resistance value, indicating a lower conductivity. *(C.D. Winters)*

causes more scattering of electron waves as they move through the crystal, because there are now more possibilities of unfavorable lattice spacings. Hence there is lower electrical conductivity.

From this picture of electrical conductivity, it might be expected that the conductivity of a metal crystal at absolute zero (0 K) might be very large. In fact, the conductivity of a pure metal crystal approaches infinity as absolute zero is approached. But in many metals, a more interesting thing happens. At a critical temperature that is low but finite, the conductivity abruptly increases to infinity, which means that the resistance drops to zero. The metal becomes a **superconductor** of electricity. A superconductor offers no resistance whatever to electric current. No clear theory explaining superconductivity has emerged, but it appears that the scattering of electron waves by vibrating atoms is replaced by some cooperative action that allows the electrons to move through the crystal unhindered.

Superconductivity was discovered in 1911 by the Dutch physicist Kamerlingh Onnes, who won a Nobel prize for his work.

Recall from Section 8.2 that electrons have wave-like properties as well as particle-like properties.

If a material could be made superconducting at a high enough temperature, it would find uses in the transmission of electrical energy, in high-efficiency motors, and in computers and other devices. Table 15.8 lists some metals that have superconducting transition temperatures. While some useful devices can be fabricated from these metals, the low temperatures at which they become superconductors make them impractical for most applications. Shortly after superconductivity of metals was discovered, alloys were prepared that had higher transition temperatures. Niobium alloys were the best, but still required cooling to below 23 K (−250 °C) to exhibit superconductivity. Maintenance of such a low temperature required liquid helium, which costs $4 per liter—an expensive proposition.

All that abruptly changed in January 1986, when K. Alex Müller and J. Georg Bednorz, IBM scientists in Switzerland, discovered that a barium-lanthanum-copper oxide became superconducting at 35 K. This type of mixed metal oxide is a ceramic with the same structure as the mineral perovskite ($CaTiO_3$), and therefore would be expected to have insulating properties. The announcement of the superconducting properties of $LaBa_2Cu_3O_x$ rocked the world of science and provoked a flurry of activity to prepare related compounds in the hope that even higher transition temperatures could be found. Within four months a material that became superconducting at 90 K, $YBa_2Cu_3O_x$, was announced! This was a major breakthrough because 90 K exceeds the boiling point of liquid nitrogen (77 K). At less than $0.06 per liter, liquid nitrogen is a much cheaper refrigerant than liquid helium.

Müller and Bednorz received the 1987 Nobel Prize in Physics for their discovery of the superconducting properties of $LaBa_2Cu_3O_x$.

Why the great excitement over the potential of superconductivity? Superconducting materials will allow the building of more powerful electromagnets such as those used in nuclear particle accelerators and in magnetic resonance imaging (MRI) machines for medical diagnosis. The electromagnets will allow higher energies to be maintained for longer periods of time (and at lower cost) in the case of the particle accelerators, and allow better imaging of problem areas in a patient's body. The main barrier to wider application is the cost of cooling the magnets used in these devices. Many scientists say that this discovery is more important than the discovery of the transistor because of its potential effect on electrical and electronic technology. For example, the use of superconducting materials for transmission of electric power could save as much as 30% of the energy now lost because of the resistance of the wire. Superchips for computers could be up to 1000 times faster than existing silicon chips. Electromagnets could be both more powerful and smaller, which could hasten the day of a practical nuclear fusion reactor. Since a superconductor repels magnetic materials, it is conceivable that cars and trains could be levitated above a track and move with no friction other than air resistance to slow them down.

TABLE 15.8 Superconducting Transition Temperatures of Some Metals*

Metal	Superconducting Transition Temperature (K)
Aluminum	1.183
Gallium	1.087
Lanthanum	4.8
Lead	7.23
Niobium	9.17

*Not all metals have superconducting properties. Those that can become superconductors at atmospheric pressure are Al, Ti, V, Zn, Ga, Zr, Nb, Mo, Tc, Ru, Cd, In, Sn, La, Hf, Ta, W, Re, Os, Ir, Hg, Tl, Pb, Th, Pa, and U.

Although the discovery of superconductivity is a significant one, translating the research into practical applications such as those described here will take time. After all, these superconductors have many of the properties expected of ceramics—they are brittle and fragile. The technology is just beginning to be developed, but recently there has been some progress toward making ribbons and wire filaments from superconducting materials. In addition, the maximum superconducting transition temperature has risen to 125 K, well above the boiling point of liquid nitrogen.

Summary Problem

Consider the phase diagram for xenon shown below. Answer the following questions.

(a) In what phase is xenon found at room temperature and a pressure of 1.0 atm?
(b) If the pressure exerted on a sample of xenon is 0.75 atm, and the temperature is −114 °C, in what phase does xenon exist?
(c) If the vapor pressure of a sample of liquid xenon is 375 mm Hg, what is the temperature of the liquid phase?
(d) What is the vapor pressure of solid xenon at −122 °C?
(e) Which is the denser phase, solid or liquid? Explain.
(f) The critical temperature and pressure for xenon are 16.6 °C and 58 atm, respectively. Modify the phase diagram for xenon to reflect these data.

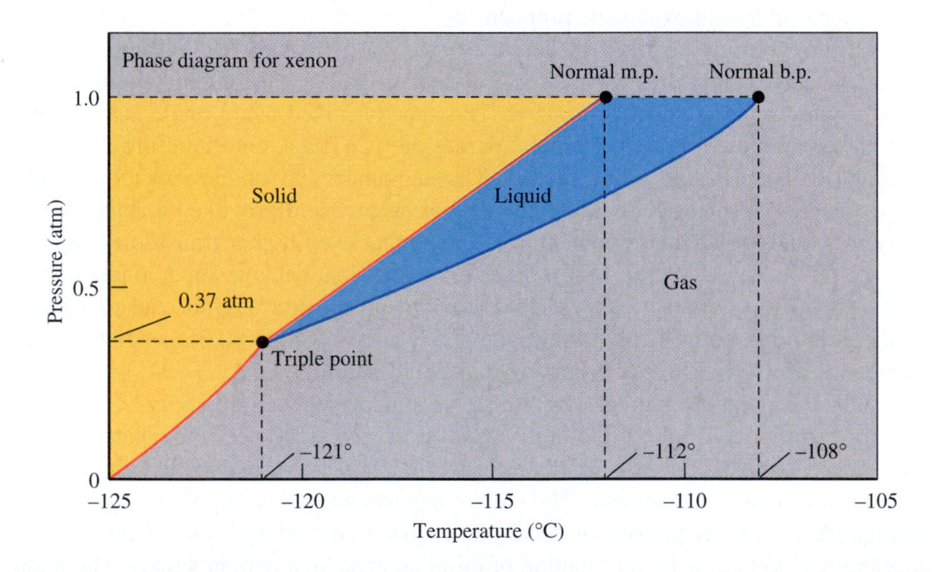

In Closing

Having studied this chapter, you should be able to . . .

- explain the liquid properties of surface tension, vaporization and condensation, sublimation and deposition, vapor pressure, and boiling point, and describe how these properties are influenced by intermolecular forces (Sections 15.1 and 15.2).
- calculate the energy associated with vaporization and fusion (Sections 15.1 and 15.2).
- describe the changes of phase that occur between solids, liquids, and gases (Section 15.3).

- use phase diagrams to predict what happens when temperatures and pressures are changed for a sample of matter (Section 15.3).
- understand critical properties (Section 15.3).
- explain the unusual properties of water (Section 15.4).
- do calculations based on knowledge of simple unit cells and the dimensions of atoms and ions that occupy positions in those unit cells (Section 15.5).
- explain metallic bonding and how it results in the properties of metals and semiconductors (Section 15.8).
- explain the bonding in network solids and how it results in their properties (Section 15.10).
- explain how the lack of regular structure in amorphous solids affects their properties (Section 15.11).
- describe the phenomenon of superconductivity (Section 15.12).

Key Terms

The following terms were defined and given in boldface type in this chapter. You should be sure to understand each of the terms and the concepts it represents.

amorphous solids *(15.5)*
boiling *(15.2)*
boiling point *(15.2)*
cement *(15.11)*
closest-packed *(15.6)*
concrete *(15.11)*
condensation *(15.1)*
critical pressure **(P_c)** *(15.3)*
critical temperature **(T_c)** *(15.3)*
crystalline solids *(15.5)*
crystallization *(15.3)*
cubic closest packing *(15.6)*
cubic unit cells *(15.6)*
deposition *(15.3)*
doping *(15.9)*
energy band *(15.8)*
equilibrium vapor pressure *(15.2)*
evaporation *(15.1)*
insulators *(15.8)*
metallic bonding *(15.8)*
mortar *(15.11)*
network solids *(15.10)*
normal boiling point *(15.2)*
phase diagrams *(15.3)*
semiconductors *(15.8)*
solar cell *(15.9)*
superconductor *(15.12)*
supercritical fluid *(15.3)*
surface tension *(15.1)*
triple point *(15.3)*
unit cell *(15.6)*
valence band *(15.8)*
vapor pressure *(15.2)*
vaporization *(15.1)*
volatility *(15.2)*
x-ray crystallography *(15.7)*
zone refining *(15.9)*

Questions for Review and Thought

Conceptual Challenge Problems

CP-15.A. In Section 15.1 you read that the enthalpy of vaporization of water "is somewhat dependent on the temperature." At 100 °C its value is 40.7 kJ/mol but at 25 °C it is 44.0 kJ/mol, a difference of 3.3 kJ/mol. List three enthalpy changes whose sum would equal this difference. Remember, the sum of the changes for a cyclic process must be zero because the system is returned to its initial state.

CP-15.B. For what reasons would you propose that two substances listed in Table 15.2 be considered better refrigerants for use in household refrigerators than the others listed there?

CP-15.C. A table of enthalpies of sublimation is not given in Section 15.3, but the enthalpy of sublimation of ice at 0 °C is given as 51 kJ/mol. How was this value obtained? Tables 15.2 and 15.3 list the enthalpies of vaporization and fusion, respectively, for several substances. Determine from data in these tables, the ΔH_{sub} for ice. Using the same method, estimate the enthalpies of sublimation of HBr and HI at their melting points.

Review Questions

1. Name three properties of solids that are different from those of liquids. Explain the differences for each.
2. List the concepts of the kinetic-molecular theory that apply to liquids.
3. What causes surface tension in liquids? Name a substance that has a very high surface tension. What kinds of intermolecular forces account for the high value?
4. Explain how the equilibrium vapor pressure of a liquid might be measured.
5. Define boiling point and normal boiling point.
6. What is the heat of crystallization of a substance and how is it related to the substance's heat of fusion?
7. What is sublimation?
8. What is the unit cell of a crystal?
9. Assuming the same substance could form crystals with its atoms or ions in either simple cubic packing or hexagonal closest packing, which form would have the higher density?
10. How does conductivity vary with temperature for (a) a conductor, (b) a nonconductor, (c) a semiconductor, and (d) a superconductor? In your answer, begin at high temperatures and come down to low temperatures.

The Liquid State

11. Predict what compound in Table 15.1 has a surface tension most similar to that of the following liquids: (a) ethylene glycol [$CH_2(OH)—CH_2(OH)$], (b) hexane (C_6H_{14}), (c) gallium metal at 40 °C.
12. The surface tension of a liquid decreases with increasing temperature. Using the idea of intermolecular attractions, explain why this is so.
13. Explain on the molecular scale the process of condensation and vaporization.
14. How would you convert a sample of liquid to vapor without changing the temperature?
15. What is the heat of vaporization of a liquid? How is it related to the heat of condensation of that liquid? Using the idea of intermolecular attractions, explain why the process of vaporization is endothermic.
16. After exercising on a hot summer day and working up a sweat, you often become cool when you stop. What is the molecular-level explanation of this phenomenon?
17. The substances Ne, HF, H_2O, NH_3, and CH_4 all have the same number of electrons. In a thought experiment, you can make HF from Ne by removing a single proton from the nucleus and having the electrons follow the new arrangement of nuclei so as to make a new chemical bond. You can do the same for each of the other compounds. (Of course, none of these thought experiments can actually be done because of the enormous energies it takes to remove protons from nuclei.) For all of these substances, make a plot of (a) the boiling point in kelvins vs. the number of hydrogen atoms, and (b) the molar heat of vaporization vs. the number of hydrogen atoms. Explain any trend that you see in terms of intermolecular forces.
18. How much heat is required to vaporize 1.0 metric ton of ammonia? (1 metric ton = 10^3 kg.) The ΔH_{vap} for ammonia is 25.1 kJ/mol.
19. The chlorofluorocarbon CCl_3F has a heat of vaporization of 24.8 kJ/mol. To vaporize 1.00 kg of the compound, how much heat is required?
20. The molar heat of vaporization of methanol is 38.0 kJ/mol at 25 °C. How much heat is required to convert 250. mL of the alcohol from liquid to vapor? The density of CH_3OH is 0.787 g/mL at 25 °C.
21. Some camping stoves contain liquid butane (C_4H_{10}). They work only when the outside temperature is warm enough to allow the butane to have a reasonable vapor pressure (and so are not very good for camping in temperatures below about 0 °C). Assume the heat of vaporization of butane is 24.3 kJ/mol. If the camp stove fuel tank contains 190. g of liquid C_4H_{10}, how much heat is required to vaporize all of the butane?
22. Mercury is a highly toxic metal. Although it is a liquid at room temperature, it has a high vapor pressure and a low heat of vaporization (294 J/g). What quantity of heat is required to vaporize 0.500 mL of mercury at 357 °C, its normal boiling point? (The density of $Hg(\ell)$ is 13.6 g/mL.) Compare this heat with the amount needed to vaporize 0.500 mL of water. See Table 15.2 for the molar heat of vaporization of H_2O.
23. Rationalize the observation that 1-propanol ($CH_3CH_2CH_2OH$) has a boiling point of 97.2 °C, whereas a compound with the same empirical formula, ethyl methyl ether ($CH_3CH_2OCH_3$), boils at 7.4 °C.
24. Briefly explain the variations in the following boiling points. In your discussion be sure to mention the types of intermolecular forces involved.

Compound	Boiling Point (°C)
NH_3	−33.4
PH_3	−87.5
AsH_3	−62.4
SbH_3	−18.4

Liquid-Vapor Equilibrium: Vapor Pressure

25. Give a molecular-level explanation of why the vapor pressure of a liquid increases with temperature.
26. Methanol (CH_3OH) has a normal boiling point of 64.7 °C and a vapor pressure of 100 mm Hg at 21.2 °C. Another compound of the same elements, formaldehyde ($H_2C{=}O$), has a normal boiling point of −19.5 °C and a vapor pressure of 100 mm Hg at −57.3 °C. Explain why these two compounds have different boiling points and require different temperatures to achieve the same vapor pressure.
27. The lowest sea-level barometric pressure ever recorded was 25.90 inches of mercury, recorded in a typhoon in the South Pacific. Suppose you were in this typhoon and, to calm yourself, boiled water to make yourself a cup of tea. At what tem-

perature would the water boil? Remember that 1 atmosphere is 760 mm (29.92 inches) of Hg, and use Figure 15.3.

28. The highest mountain in the Western hemisphere is Mt. Aconcagua, in the central Andes of Argentina (22,834 ft). If atmospheric pressure decreases at a rate of 3.5 millibar every 100 ft, estimate the atmospheric pressure at the top of Mt. Aconcagua, and then estimate from Figure 15.3 the temperature at which water would boil at the top of the mountain.

Phase Changes: Solids, Liquids, and Gases

29. What does a low heat of fusion for a solid tell you about the solid (its bonding or type)?

30. What does a high melting point and a high heat of fusion tell you about a solid?

31. Which would you expect to have the higher heat of fusion, N_2 or I_2? Explain your choice.

32. The heat of fusion for H_2O is about 2.5 times larger than the heat of fusion for H_2S. What does this say about the relative strengths of the forces holding the molecules together in their respective solids? Explain.

33. Benzene is an organic liquid that freezes at 5.5 °C to beautiful, feather-like crystals. How much heat is evolved when 15.5 g of benzene freezes at 5.5 °C? The heat of fusion of benzene is 127 J/g. If the 15.5-g sample is remelted, again at 5.5 °C, what quantity of heat is required to convert it to a liquid?

34. What is the total quantity of heat required to change 0.50 mol of ice at −5 °C to 0.50 mol of steam at 100 °C?

35. The ions of NaF and MgO all have the same number of electrons, and the internuclear distances are about the same (235 pm and 212 pm). Why then are the melting points of NaF and MgO so different (992 °C and 2642 °C, respectively)?

36. For the pair of compounds LiF and CsI, tell which compound is expected to have the higher melting point and briefly explain.

37. Which of these substances has the highest melting point? The lowest melting point? Explain your choice briefly.
(a) LiBr (c) CO
(b) CaO (d) CH_3OH

38. Which of these substances has the highest melting point? The lowest melting point? Explain your choice briefly.
(a) SiC (c) Rb
(b) I_2 (d) $CH_3CH_2CH_2CH_3$

39. Why is solid CO_2 called Dry Ice?

40. During thunderstorms, very large hailstones can fall from the sky. (Some are the size of golf balls!) To preserve some of these stones, we want to place them in the freezer compartment of our frost-free refrigerator. Our friend, who is a chemistry student, tells us to use an older, non-frostfree model. Why?

41. From memory, sketch the phase diagram of water. Label all of the regions as to the physical state of water. Draw either horizontal (constant pressure) or vertical (constant temperature) paths (i.e., lines with arrows indicating a direction) for the following changes of state:
(a) Sublimation (d) Vaporization
(b) Condensation to a liquid (e) Crystallization
(c) Melting

Using this diagram, explain how the blades of ice skates are lubricated with a film of liquid water when you skate.

42. At the critical point for carbon dioxide, the substance is very far from being an ideal gas. Prove this statement by calculating the density of an ideal gas in g/cm^3 at the conditions of the critical point, and comparing it with the experimental value. Compute the experimental value from the fact that a mole of CO_2 at its critical point occupies 94 cm^3.

Types of Solids

43. Classify each of the following solids as ionic, metallic, molecular, network, or amorphous.
(a) KF (b) I_2 (c) SiO_2 (d) Polypropylene

44. Classify each of the following solids as ionic, metallic, molecular, network, or amorphous.
(a) Tetraphosphorus decaoxide
(b) Graphite
(c) Brass
(d) Ammonium phosphate

45. On the basis of the description given, classify each of the following solids as molecular, metallic, ionic, network, or amorphous, and explain your reasoning.
(a) A brittle, yellow solid that melts at 113 °C; neither its solid nor its liquid conducts electricity.
(b) A soft, silvery solid that melts at 40 °C; both solid and liquid conduct electricity.
(c) A hard, colorless, crystalline solid that melts at 1713 °C; neither solid nor liquid conducts electricity.
(d) A soft, slippery solid that melts at 63 °C; neither solid nor liquid conducts electricity.

46. On the basis of the description given, classify each of the following solids as molecular, metallic, ionic, network, or amorphous, and explain your reasoning.
(a) A soft, slippery solid that has no definite melting point but decomposes at temperatures above 250 °C; the solid does not conduct electricity.
(b) Violet crystals that melt at 114 °C and whose vapor irritates the nose; neither solid nor liquid conducts electricity.
(c) Hard, colorless crystals that melt at 2800 °C; the liquid conducts electricity, but the solid does not.
(d) A hard solid that melts at 3410 °C; both solid and liquid conduct electricity.

47. Describe how each of the following would behave as they were deformed by a hammer strike. Explain why they behave as they do.
(a) A metal, such as gold
(b) A nonmetal, such as sulfur
(c) An ionic compound, such as NaCl

48. What type of solid exhibits each of these sets of properties?
(a) Melts below 100 °C and is insoluble in water.
(b) Conducts electricity only when melted.
(c) Insoluble in water and conducts electricity.
(d) Noncrystalline and melts over a wide temperature range.

Crystalline Solids

49. Each diagram below represents an array of like atoms that would extend indefinitely in two dimensions. Draw a two-dimensional unit cell for each array. How many atoms are there in each unit cell?

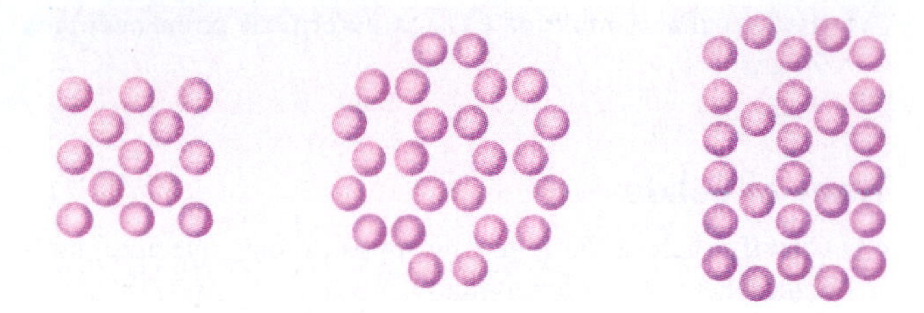

50. Name and draw the three cubic unit cells. Describe their similarities and differences.
51. Explain how the volume of a simple cubic unit cell is related to the radius of the atoms in the cell.
52. Solid xenon forms crystals with a face-centered unit cell that has an edge of 0.620 nm. Calculate the atomic radius of xenon.
53. Gold (atomic radius = 144 pm) crystallizes in a face-centered unit cell. What is the length of a side of the cell?
54. Consider the CsCl unit cell shown in Figure 15.16. How many Cs^+ ions are there per unit cell? How many Cl^- ions?
55. Using the NaCl structure shown in Figure 15.17, how many unit cells share each of the Na^+ ions in the front face of the unit cell? How many unit cells share each of the Cl^- ions in this face?
56. The ionic radii of Cs^+ and Cl^- are 0.169 and 0.181 nm, respectively. What is the length of the body diagonal in the CsCl unit cell? What is the length of the side of this unit cell? (See Figure 15.16.)
57. Thallium chloride, TlCl, crystallizes in either a simple cubic lattice or a face-centered cubic lattice of Cl^- ions with Tl^+ ions in the holes. If the density of the solid is 7.00 g/cm^3 and the edge of the unit cell is 3.85×10^{-8} cm, what is the unit cell geometry?
58. Could $CaCl_2$ possibly have the NaCl structure? Explain your answer briefly.
59. A simple cubic unit cell is formed so that the spherical atoms or ions just touch one another along the edge. Prove mathematically that the percentage of empty space within the unit cell is 52.4%. (Recall that the volume of a sphere is $4/3\pi r^3$, where r is the radius of the sphere.)
60. Metallic lithium has a body-centered cubic structure, and its unit cell is 351 pm on a side. Lithium iodide has the same crystal lattice structure as sodium chloride. The cubic unit cell is 600 pm on a side.
(a) Assume that the metal atoms in lithium touch along the body diagonal of the cubic unit cell and estimate the radius of a lithium atom.
(b) Assume that in lithium iodide the I^- ions touch along the face diagonal of the cubic unit cell, and that the Li^+ and I^- ions touch along the edge of the cube; calculate the radius of an I^- ion and of an Li^+ ion.
(c) Compare your results in parts (a) and (b) for the radius of a lithium atom and a lithium ion. Are your results reasonable? If not, how could you account for the unexpected result? Could any of the assumptions that were made be in error?

Looking into Solids: X-Ray Crystallography

61. The surface of a CD-ROM disk contains narrowly separated lines that diffract light into its component colors. This means that the lines are spaced at distances approximately the same as the wavelength of the light. Taking the middle of the visible spectrum to be green light with a wavelength of 550 nm, calculate how many aluminum atoms (radius, 143 pm) touching its neighbors would make a straight line 550 nm long. Using this result, explain why an optical microscope using visible radiation will never be able to detect an individual aluminum atom (or, indeed, any other atom).
62. To see a clear diffraction pattern from a regularly spaced lattice, the radiation falling on the lattice must have a wavelength less than the lattice spacing. From the unit cell size of the NaCl crystal, estimate the maximum wavelength of the radiation that would be diffracted by this crystal. Calculate the frequency of the radiation and the energy associated with (a) one photon, and (b) one mole of photons of the radiation. In what region of the spectrum is this radiation?

Metals, Semiconductors, and Insulators

63. What is the principal difference between the orbitals that electrons occupy in individual, isolated atoms and the orbitals they occupy in solids?
64. In terms of band theory, what is the difference between a conductor and an insulator? Between a conductor and a semiconductor?
65. Name three properties of metals and explain them by using a theory of metallic bonding.
66. Which substance has the greatest electrical conductivity? The smallest electrical conductivity? Explain your choice briefly.
(a) Si (b) Ge (c) Ag (d) P_4
67. Which substance has the greatest electrical conductivity? The smallest electrical conductivity? Explain your choices briefly.
(a) RbCl(ℓ) (c) Rb
(b) NaBr(s) (d) Diamond

Silicon and the Chip

68. What are the two main chemical reactions involved in the production of electronic grade silicon? Identify the element being reduced and being oxidized.
69. Extremely high-purity silicon is required to manufacture semiconductors such as memory chips found in calculators and computers. If a silicon wafer is 99.99999999% pure, approximately how many silicon atoms per gram have been replaced by impurity atoms of some other element?

70. What is the process of doping, as applied to semiconductors? Why are Group IIIA and Group VA elements used to dope silicon?

71. Explain the difference between *n*-type semiconductors and *p*-type semiconductors.

Network Solids

72. Using what you know about the bonding present in graphite and diamond, explain why diamond is denser than graphite.

73. With the examples of network solids given in the text, determine, by looking up data in a reference such as the *Handbook of Chemistry and Physics,* whether these materials are soluble in water or other common solvents. Explain your answer in terms of the chemical bonding in network solids.

74. Explain why diamond is an electrical insulator and graphite is an electrical conductor.

75. What is the oxidation state of silicon in quartz and in silicate minerals such as the pyroxenes? What types of substances could be used to react with quartz and silicates to produce pure silicon?

76. Assign oxidation states to all of the elements in kaolinite and in feldspar. Is the conversion of feldspar to kaolinite a redox reaction?

Cement, Ceramics, and Glass

77. Define the term "amorphous."

78. What makes a glass different from a solid such as NaCl? Under what conditions could NaCl become glass-like?

79. A typical cement contains, by weight, 65% CaO, 20% SiO_2, 5% Al_2O_3, 6% Fe_2O_3, and 4% MgO. Determine the mass percent of each element present. Then determine an empirical formula of the material from the percent composition, setting the coefficient of the least abundant element to 1.00. (Your result will contain fractional coefficients for all other elements).

80. Give two examples of (a) oxide ceramics and (b) nonoxide ceramics.

Superconductors

81. Define the term "superconductor." Give the chemical formulas of two kinds of superconductors, and their associated transition temperatures.

82. What is the main technological or economic barrier to the widespread use of superconductors?

General Questions

83. Your air conditioner probably contains the chlorofluorocarbon CCl_2F_2 as the heat-transfer fluid. (Because of environmental damage caused by these compounds, they are being phased out.) Its normal boiling point is −30 °C and the heat of vaporization is 165 J/g. The gas and the liquid have specific heat capacities of 0.61 J/g·K and 0.97 J/g·K, respectively. How much heat is evolved when 10.0 g of CCl_2F_2 is cooled from +40 °C to −40 °C?

84. Liquid ammonia, $NH_3(\ell)$, was used as a refrigerant fluid before the discovery of the chlorofluorocarbons, and is still widely used today, even though it is a somewhat toxic gas and a strong irritant when it is released into the air. Its normal boiling point is −33.4 °C and the heat of vaporization is 23.5 kJ/mol. The gas and liquid have specific heat capacities of 2.2 J/g·K and 4.7 J/g·K, respectively. How much heat must be supplied to 10. kg of liquid ammonia to raise its temperature from −50.0 °C to −33.4 °C, and then to 0.0 °C?

85. Potassium chloride and rubidium chloride both have the sodium chloride structure. X-ray diffraction experiments indicate that their cubic unit cell dimensions are 629 pm and 658 pm, respectively.
(i) One mole of KCl and one mole of RbCl are ground together in a mortar and pestle to a very fine powder, and the x-ray diffraction pattern of the pulverized solid is measured. Two patterns are observed, each corresponding to a cubic unit cell—one with an edge length of 629 pm and one with an edge length of 658 pm. Call this sample 1.
(ii) One mole of KCl and one mole of RbCl are heated until the entire mixture is molten and then cooled to room temperature. In this case there is a single x-ray diffraction pattern that indicates a cubic unit cell with an edge length of roughly 640 pm. Call this sample 2.
(a) Suppose that samples 1 and 2 were analyzed for their chlorine content. What fraction of each sample is chlorine? Could the samples be distinguished by means of chemical analysis?
(b) Interpret the two x-ray diffraction results in terms of the structures of the crystal lattices of samples 1 and 2.
(c) What chemical formula would you write for sample 1? For sample 2?
(d) Suppose that you dissolved 1.00 g of sample 1 in 100 mL of water in a beaker and did the same with 1.00 g of sample 2. Which sample would conduct electricity better, or would both be the same? What ions would be present in each solution at what concentrations?

86. Sulfur dioxide, SO_2, is found in polluted air. It is formed during the combustion of fossil fuels containing small percentages of sulfur, and from industrial plants that convert certain metal-containing ores to metals or metal oxides.
(a) Draw the electron dot structure for SO_2. From this, describe first the O—S—O angle, and second the electron pair geometry and molecular geometry.
(b) What type of forces are responsible for binding SO_2 molecules to one another in the solid or liquid phase?
(c) Using the information below, place the compounds listed in order of increasing intermolecular attractions.

Compound	Normal Boiling Point (°C)
SO_2	−10
NH_3	−33.4
CH_4	−161.5
H_2O	100

87. Copper is an important metal in our economy, most of it being mined in the form of the mineral chalcopyrite, $CuFeS_2$.
(a) To obtain one metric ton (1000 kilograms) of copper metal, how many metric tons of chalcopyrite would you have to mine?
(b) If the sulfur in chalcopyrite is converted to SO_2, how many metric tons of the gas would you get from one metric ton of chalcopyrite?
(c) Copper crystallizes as a face-centered cube. Knowing that the density of copper is 8.95 g/cm^3, calculate the radius of the copper atom.

Applying Concepts

88. Refer to Figure 15.3 when answering the following questions.
(a) What is the equilibrium vapor pressure for ethanol at room temperature?
(b) At what temperature does diethyl ether have an equilibrium vapor pressure of 400 mm Hg?
(c) If a pot of water was boiling at a temperature of 95 °C, what would be the atmospheric pressure?
(d) At 200 mm Hg and 60 °C, which of the three substances are gases?
(e) If you put a couple of drops of each substance on your hand, which would evaporate and which would remain as a liquid?
(f) Which of the three substances has the greatest intermolecular attractions?

89. The normal boiling point of SO_2 is 263.1 K and that of NH_3 is 239.7 K. At −40 °C would you predict that ammonia has a vapor pressure greater than, less than, or equal to sulfur dioxide?

90. Butane is a gas at room temperature; however, if you look closely at a butane lighter you see it contains liquid butane. How is this possible?

91. While camping with a friend in the Rocky Mountains, you decide to cook macaroni for dinner. Your friend says the macaroni will cook faster in the Rockies because the lower atmospheric pressure will cause the water to boil at a lower temperature. Do you agree with your friend? Explain your reasoning.

92. Examine the nanoscale diagrams below and the phase diagram at the top of the next column. Match each particulate diagram (1–8) to its corresponding points (A–H) on the phase diagram.

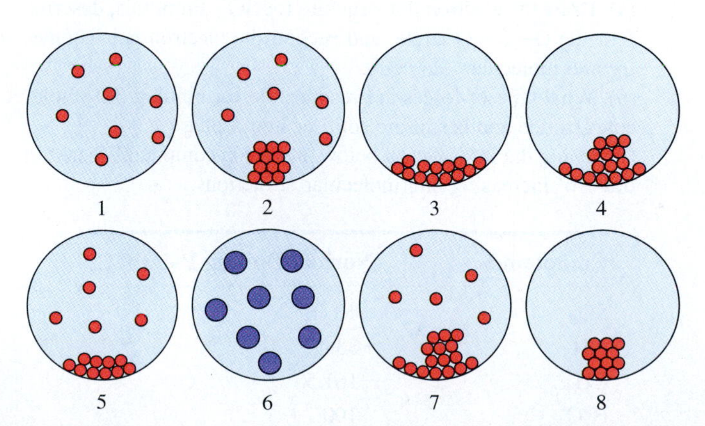

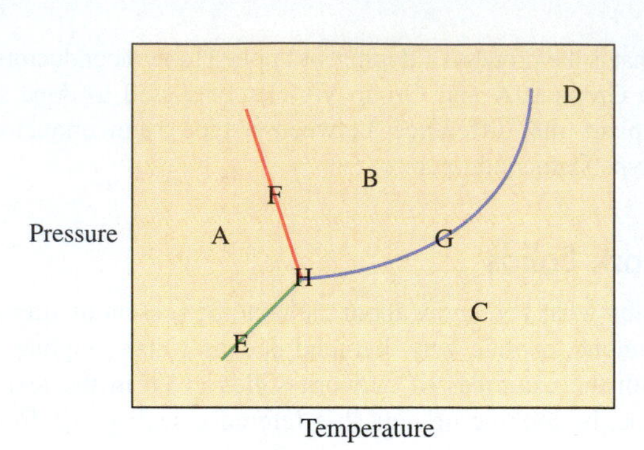

93. Consider the phase diagram below. Draw corresponding heating curves for T_1 to T_2 at pressures P_1 and P_2. Label each phase and phase change on your heating curves.

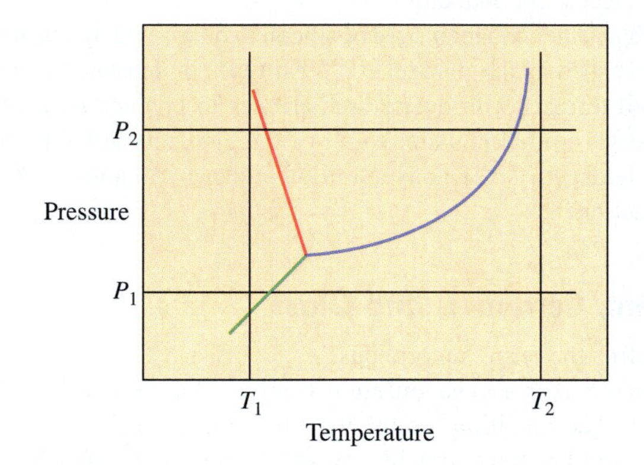

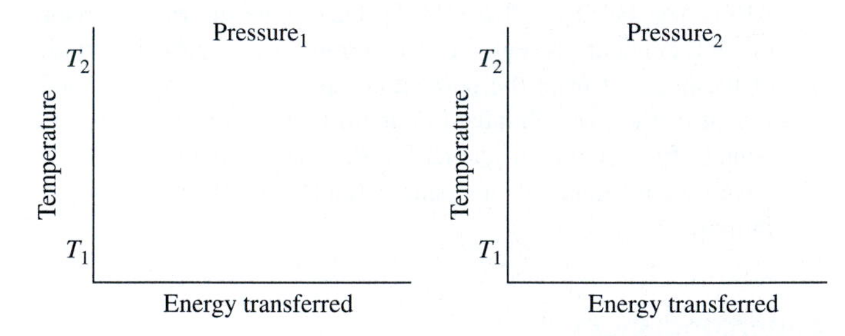

94. Consider three fish aquariums of equal volume. One is filled with tennis balls, another with golf balls, and the third with marbles. If a closest-packing arrangement is used in each aquarium, which one has the most occupied space? Which one has the least occupied space? (Disregard the difference in filling space at the walls, bottom, and top of the aquarium.)

95. For each liquid in Table 15.2 calculate the ratio of $\Delta H_{vap}/T_b$ (using Kelvin temperature). What generalization can you make about the results? Relate these results to Exercise 7.4 (← *p. 277*). How can you account for your results in terms of entropy?

CHAPTER 16

Water and Solution Chemistry

A solute dissolving in a solvent. As the solute dissolves, concentration gradients are observed as wavy lines in the liquid. • How are solutions different from their components (the solute and the solvent)? In what ways are solution compositions expressed? What are some of the properties of solutions? Would it surprise you to know that some properties of a solution depend only on how many solute particles are present per unit volume, rather than what the solute actually is? Answers to these and similar questions will be found in this chapter. The summary problem at the end will test your knowledge of these concepts. *(Richard Megna/Fundamental Photographs)*

Solutions are homogeneous mixtures of two or more substances. The component present in the greater amount is usually called the solvent and the other component or components are called solutes. Solutions are some of the most important mixtures we know of, and solutions where water is the solvent are arguably the most important of those. We all encounter solutions every day, and a great many different kinds of chemical reactions occur in them. There are solid solutions, liquid solutions, and gaseous solutions. Forces between particles can often account for how much of one substance will dissolve in another—its solubility. Substances whose intermolecular forces are similar to those of a solvent usually dissolve readily in that solvent. Substances with intermolecular forces quite different from those of the solvent do not readily dissolve in that solvent.

The atoms, molecules, or ions in a solution are thoroughly mixed, making it easy for them to come into contact and react. In gas-phase or liquid-phase solutions the particles move about and collide, increasing the opportunities for them to react with each other. Because the particles in liquid solutions are very close together and therefore collide more often, liquid solutions are the media for many reactions used to produce polymers, medicines, and other commercial products. They are also the media for reactions within our bodies and those of other living organisms. Because of the importance of liquid solutions, especially those called aqueous solutions, in which water is the solvent, much of this chapter will be devoted to them.

16.1 THE UNIQUE SOLVENT PROPERTIES OF WATER

Water is the most important of all solvents. It is a very good solvent for many substances because the water molecule is small, allowing many water molecules to interact with a single solute particle; water is polar, allowing its molecules to interact strongly with ions and molecules containing polar groups; and water molecules can form hydrogen bonds with many solutes. When a substance dissolves in water, solute molecules or ions become surrounded by groups of water molecules, a process called **hydration.** Small organic molecules that contain polar functional groups are strongly hydrated because water molecules are attracted to these groups. Examples of solutions formed in this manner are aqueous sugar solutions (such as soft drinks, syrups, and fruit juices) and aqueous alcohol solutions (such as wine, vodka, beer, and whiskey). Inorganic substances that consist of polar molecules such as ammonia (NH_3); hydrogen chloride (HCl); and sulfur dioxide (SO_2) are water-soluble. In addition to becoming hydrated, many of them react with water as they dissolve. Salts consist of ions in a crystal lattice *(← p. 99).* These ions, being permanently charged, interact even more strongly with the polar water molecules than do neutral solute molecules, and many (but not all) salts are water soluble.

A more general term for solvent molecules surrounding solute particles is solvation.

Water is such a good solvent that natural waters such as raindrops, lakes, and rivers contain dissolved cations and anions, gaseous nitrogen, oxygen, and carbon dioxide as well as other gases and solids the water may have contacted. Surface waters such as rivers dissolve air, minerals, and natural organic substances such as tannic acid from decaying leaves. *Groundwater* (water found beneath the surface) often contains many of the same kinds of solutes as surface waters. Because groundwaters contact so many mineral deposits, they often have much higher concentrations of salts than do surface waters. For this reason, well water is often "hard"—its mineral content interferes with the ability of soaps to remove soil from people or clothing (Section 16.7).

Often undesirable substances become dissolved in water. When this happens, the water is called polluted. Some examples of such substances are industrial wastes, high concentrations of dissolved minerals, and soluble substances from decaying organic matter such as garbage or dead animals.

While many substances dissolve in water, many do not. Metals, except for those that react with water such as sodium, potassium, and calcium, do not dissolve in water to any appreciable extent. Diamond, graphite, most other network solids (← *p. 695*), and most organic polymers fail to dissolve in water. All nonpolar substances, including the hydrocarbons, gases like oxygen, nitrogen, and the noble gases, have low water solubilities. Even some ionic compounds have very low water solubilities, not because their ions lack an affinity for water molecules, but because they are held together in the crystal lattice by strong forces that will not allow the ions to be carried away by the solvent molecules.

Exercise 16.1 Repelling Water

Hydrocarbon waxes make excellent water-repellent coatings. Explain why this is so.

16.2 HOW SUBSTANCES DISSOLVE

Some solutes dissolve to a much greater extent than others. For example, consider silver nitrate ($AgNO_3$) and silver chloride (AgCl) dissolving in water. About 330 g of $AgNO_3$ dissolves in 100 mL of water at 25 °C, and we say this is a *very* soluble salt. In contrast, only 0.00035 g of AgCl dissolves in 100 mL of water at 25 °C, and we say AgCl is **insoluble,** even though a little bit dissolves. A substance's **solubility** is defined as its concentration in a solution in which the pure solute and the dissolved solute are in equilibrium at a given temperature. Insoluble or slightly soluble salts are important enough to require a discussion all their own (Section 16.4).

When more solute has been added to a liquid than can dissolve at a given temperature, there is a *dynamic equilibrium* between undissolved and dissolved solute. Some solute molecules or ions are going into solution, while others are separating from solvent molecules and entering the pure solute phase. Both processes are going on all the time, at identical rates. If we could observe all these changes, they would appear very frenzied indeed.

When the concentration of a solute equals its solubility, the solution is said to be **saturated.** A solution is **unsaturated** if the solute concentration is less than its solubility. For some substances it is possible to prepare solutions that contain *more* than the equilibrium concentration of solute; a solution like this is **supersaturated** (see Figure 16.1). Separation of solute from a supersaturated solution is a product-favored

(a)

(b)

(c)

(d)

(e)

Figure 16.1 A supersaturated solution. Sodium acetate ($NaC_2H_3O_2$) easily forms supersaturated solutions in water. The solution on the left (a) looks ordinary, but it is supersaturated. It holds more dissolved sodium acetate than a saturated solution at that temperature. It was prepared by dissolving a quantity of sodium acetate in the water at a much higher temperature and allowing the solution to cool slowly. Upon adding a small seed crystal of sodium acetate (b), some of the dissolved sodium acetate immediately begins to crystallize and after a short time (c), (d), and (e) numerous crystals of sodium acetate can be seen. If this solution is allowed to stand exposed to the air for a longer time, all of the water will evaporate and sodium acetate, which is nonvolatile, will remain. *(C. D. Winters)*

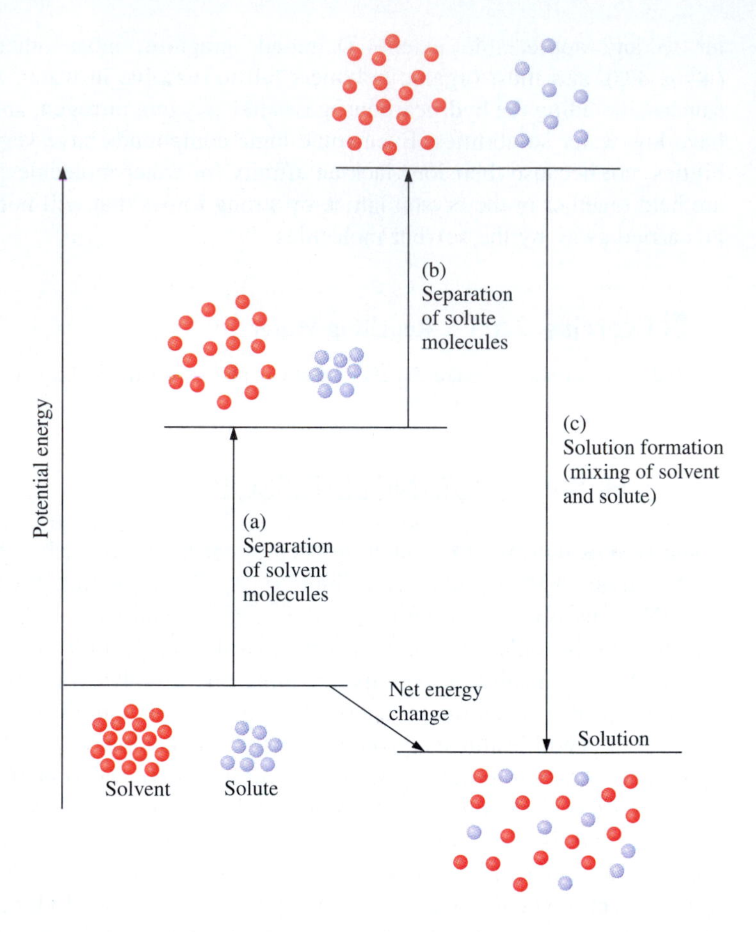

Figure 16.2 **How solutions are formed (schematic).** In step (a) the solvent molecules are separated from one another. Since they are held together by attractive forces, these must be overcome, so the potential energy of the collection of molecules increases. In step (b) the solute ions or molecules are separated from one another. Again, potential energy increases as particles that attract one another are separated. In step (c) the solute and solvent particles mix. Because they attract one another, the potential energy decreases. While the energy state of the final solution is shown here lower than that of the starting solvent and solute (an exothermic change), this is not always the case. For some solvent-solute combinations, the net energy change is endothermic and for others the net energy change can be

A supersaturated solution of sodium acetate can be prepared by cooling a saturated solution when no solid solute is pres-ent. The container needs to be very clean so that no crystals of the solid begin to form.

process, but it often occurs very slowly. Some solutions such as honey can remain supersaturated for days or months. The formation of a crystal lattice requires that several ions or molecules be arranged very near the appropriate lattice positions, and it can take a long time for such an alignment to occur by chance. However, precipitation of a solid from a supersaturated solution occurs rapidly if a tiny crystal of the solute is added to the solution. The crystal's lattice provides a template onto which more ions or molecules can be added. Sometimes other actions, such as stirring a supersaturated solution or scratching the inner walls of its container, will cause solute to precipitate rapidly.

When a solution forms, atoms, ions, or molecules of one kind become mixed with atoms, ions, or molecules of a different kind. Intermolecular attractions between solvent molecules must be overcome and so must those forces attracting the solute particles together. Each of these processes increases the potential energy of the particles (steps [a] and [b] in Figure 16.2). When mixing occurs between the solute and solvent, energy is released as solvent and solute particles attract each other (step [c] in Figure 16.2). If these new solvent-solute forces are not strong enough to overcome the solute-solute or solvent-solvent attractions, dissolving may not occur.

Exercise 16.2 Energy Changes During Solution Formation

Figure 16.2 illustrates one of three possible net energy changes during dissolution. Using a diagram, show and explain the two other kinds.

Another factor that affects the solubility of a solute in a solvent is the entropy change that accompanies dissolution. Because of the mixing (greater disorder) that takes place, there is usually a positive entropy change, $\Delta S_{process} > 0$. Such an entropy increase favors solution formation *(⇐ p. 276)*. The mixing of gases depends almost entirely on entropy, because gas particles are relatively far apart and intermolecular forces are very small. For this reason all gases dissolve in each other in all proportions at normal pressures.

Dissolving Liquids in Liquids

Mixing of particles in the liquid phase is favored by the increase in entropy, just as the mixing of gaseous particles is. There is more disorder when the molecules of two or more pure liquids are mixed than when they exist separately. Many liquids will dissolve in another liquid in any proportion. When this is the case, the liquids are said to be completely **miscible.** The solvent-solvent and solute-solute intermolecular attractions have to be very similar to the solvent-solute intermolecular attractions in order for liquids to be miscible. Liquids with such different intermolecular attractions that they do not dissolve completely in each other are described as **immiscible.**

To see how this works, consider the situation shown in Figure 16.3. Octane (C_8H_{18}, a component of gasoline) and carbon tetrachloride (CCl_4) are completely miscible. But bis-2-chloroethyl ether is insoluble in octane.

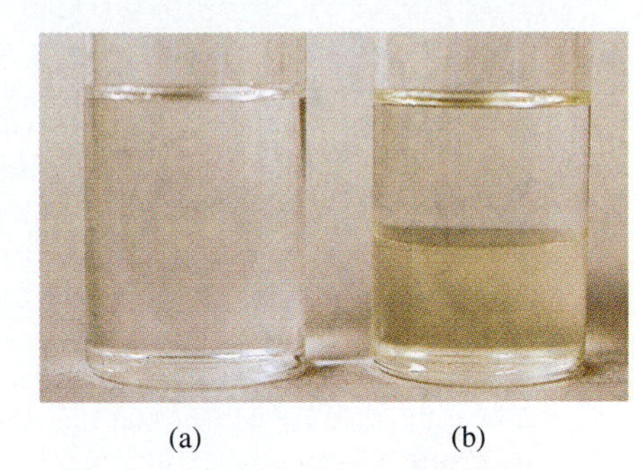

(a) (b)

Figure 16.3 Miscibility. (a) When carbon tetrachloride and octane, both clear, colorless liquids, are mixed, each dissolves completely in the other and there is no sign of an interface or boundary between them. The result is a single clear, colorless solution. (b) When bis-2-chloroethyl ether, also a clear, colorless liquid, is mixed with octane, the two liquids separate and there is an obvious boundary between them. Above the boundary is a very dilute solution of bis-2-chloroethyl ether in octane, and below is a very dilute solution of octane in bis-2-chloroethyl ether. *(Jerrold J. Jacobs)*

$CH_3CH_2CH_2CH_2CH_2CH_2CH_2CH_3$

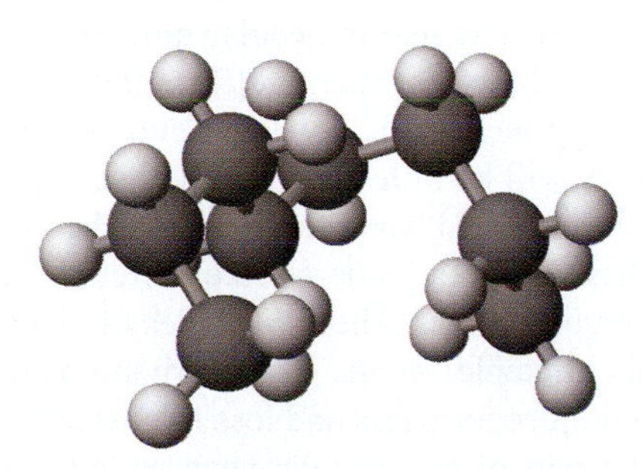

octane

$ClCH_2CH_2OCH_2CH_2Cl$

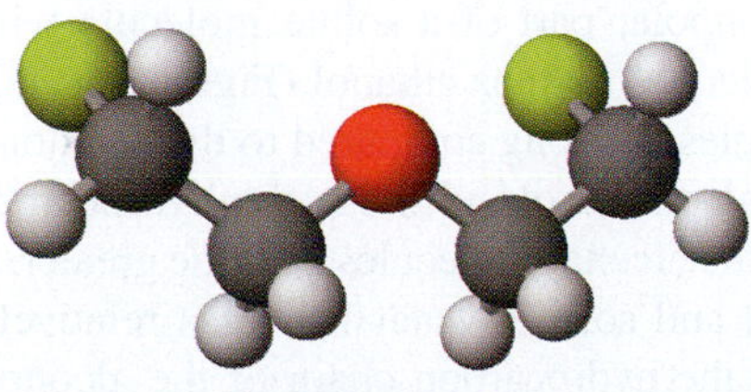

bis-2-chloroethyl ether

Because octane and carbon tetrachloride are nonpolar, London forces are the only type of intermolecular attractions in either liquid. The strengths of such forces depend on the number of electrons in a molecule and are nearly independent of other features of molecular structure. Therefore, it does not matter whether a carbon tetrachloride molecule is next to an octane molecule or another carbon tetrachloride molecule—the intermolecular attractions will be about the same. There is little net change

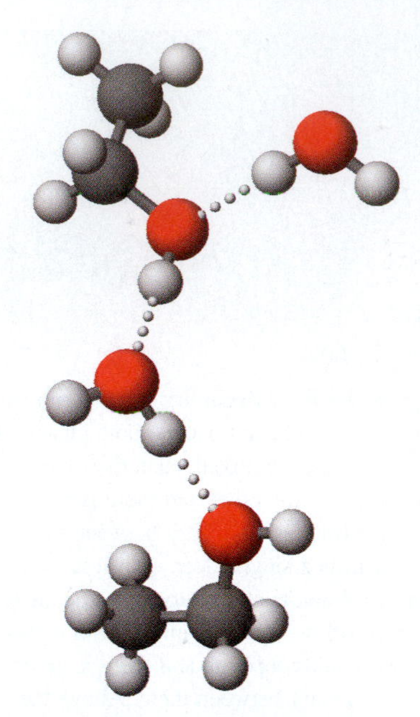

Figure 16.4 Ethanol-water hydrogen bonding. Because of the hydrogen bonding between water molecules and ethanol molecules, the two liquids dissolve completely in one another in all proportions.

TABLE 16.1 Solubilities of Some Alcohols in Water

Name	Formula	Solubility in Water ($g/100\ g\ H_2O$ at 20 °C)
Methanol	CH_3OH	Miscible
Ethanol	CH_3CH_2OH	Miscible
1-Propanol	$CH_3(CH_2)_2OH$	Miscible
1-Butanol	$CH_3(CH_2)_3OH$	7.9
1-Pentanol	$CH_3(CH_2)_4OH$	2.7
1-Hexanol	$CH_3(CH_2)_5OH$	0.6
1-Heptanol	$CH_3(CH_2)_6OH$	0.09

in intermolecular attractions when the molecules mix, and the entropy increase causes mixing to be a product-favored process.

Now think about what would happen if bis-2-chloroethyl ether molecules did mix with octane molecules. The ether molecules are polar, producing strong dipole-dipole attractions within liquid bis-2-chloroethyl ether that are not possible within liquid octane. If each ether molecule became surrounded by octane molecules, there would be no dipole-dipole attractions between the ether and octane molecules. The energy required to overcome these forces when the ether molecules were separated would not be recovered by formation of new intermolecular attractions, and there would have to be a net input of energy. Therefore, energy effects favor ether and octane remaining separated. If the entropy increase is not large enough to counteract this energy effect, as in this case, there will be little mixing. The two liquids will be immiscible.

An old adage says that "oil and water don't mix." Chemists use a similar saying about solubility: "like dissolves like," where "like" refers to things held together by similar types of intermolecular forces. The solubilities of alcohols in water, shown in Table 16.1, illustrate this principle. As the hydrocarbon chain attached to the —OH group increases, the alcohol becomes less and less like water, and more and more like a hydrocarbon. The polar part of a molecule, such as the —OH part of an alcohol molecule, is commonly called **hydrophilic,** which means "water loving." Any polar part of a solute molecule will be hydrophilic. The nonpolar part of a molecule, such as the hydrocarbon part of an alcohol molecule, is called **hydrophobic,** which means "water fearing." Any nonpolar part of a solute molecule will be hydrophobic. In a low-molecular-weight alcohol such as ethanol (Figure 16.4), hydrogen bonding with the solvent water molecules is strong compared to the London forces between the hydrocarbon parts of the alcohol molecules. As the hydrocarbon part becomes larger, London forces between the alcohol molecules become greater, and the hydrogen bonding between the alcohol and solvent water becomes relatively less important in determining solubility. As the hydrocarbon chain of the alcohol molecule gets longer, the molecule becomes more hydrophobic than hydrophilic, and its water solubility becomes very small.

Water is an unusual molecule (← *p. 675*) and it happens that the energy and entropy changes when water is a solvent are unusual as well. See *J. Chem. Educ.*, **1989,** *66,* 581–585 for a detailed discussion. The idea that like dissolves like still applies, but for reasons different from what has been described in the text.

Some of the most common *non-aqueous* solutions are petroleum products such as gasoline. Gasoline, a mixture of nonpolar hydrocarbons attracted to each other by London forces, readily dissolves greases and oils, which are also nonpolar hydrocarbons. Gasoline dissolves grease because the intermolecular attractions between the molecules in gasoline are similar to those between the molecules in a grease. However, intermolecular forces in water are very unlike those in gasoline. If water is added

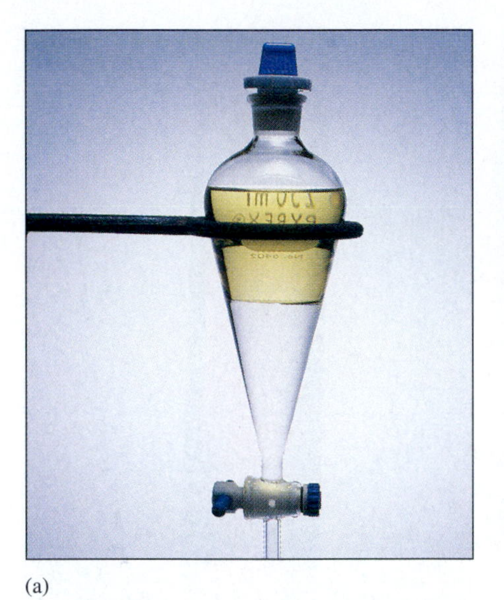

(a)

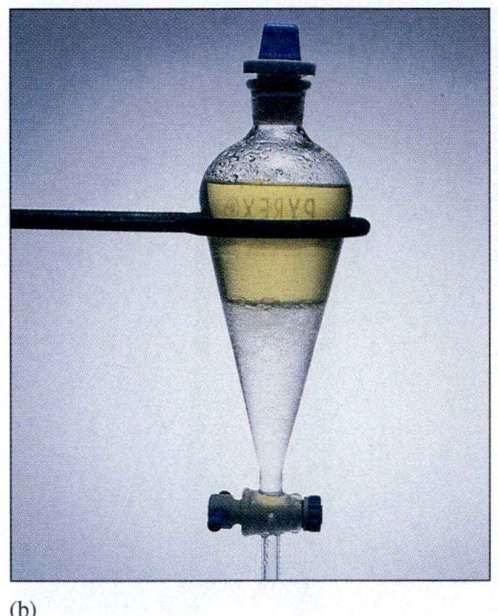

(b)

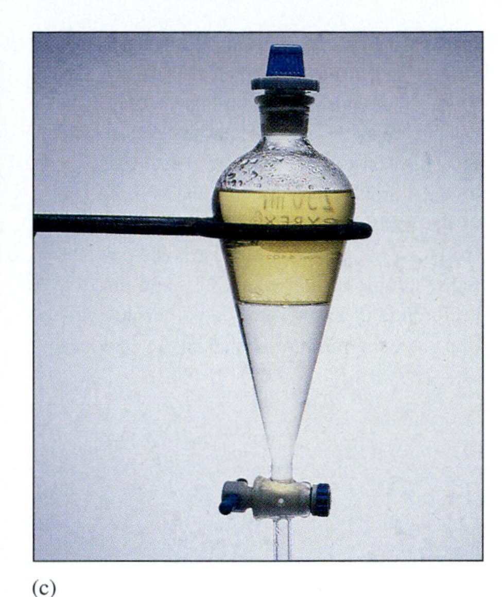

(c)

Figure 16.5 **Immiscible liquids.** Gasoline and water will not mix, at least not for long. (a) In a separatory funnel, a device used for separating immiscible liquids, a sample of gasoline and water remains as two distinct layers. (b) After vigorous mixing, some gasoline droplets are seen mixed with water droplets at the interface between the two layers. (c) But given time, the two layers re-form. *(C.D. Winters)*

to gasoline, or gasoline to water, very little intermingling of the water molecules with hydrocarbon molecules occurs. In fact, water, with a density of about 1 g/mL, will sink below a sample of gasoline, which has a density of about 0.7 g/mL (Figure 16.5). Gasoline spills on water are always a serious problem because the hydrocarbon floats on top of the water and evaporates, producing flammable vapors that can be ignited by a flame or spark.

Although gasoline dissolves grease, it is not a good idea to use gasoline as a solvent. It is flammable, and it contains numerous harmful hydrocarbons. One of these, benzene, can cause cancer in humans. Others, such as hexane and heptane, can damage the central nervous system and cause unconsciousness upon prolonged exposure.

Exercise 16.3 **Predicting Water Solubility**

You have a sample of octanol and a sample of methanol. Which is more water-soluble? Which is more soluble in gasoline? Explain your choice in terms of like dissolves like.

Exercise 16.4 **Predicting the Solubility of a Metal**

Predict the solubility of iron in water and benzene.

Dissolving Molecular Solids in Liquids

Iodine, I_2, is an example of a solid in which London forces hold together nonpolar molecules. For I_2 to dissolve in a liquid, these forces must be overcome. In addition, some of the intermolecular attractions among solvent molecules must also be disrupted. Iodine dissolves only slightly in H_2O—just enough to form a light brown col-

Making coffee from ground coffee beans and hot water is an example of extraction. The soluble components dissolve more quickly in hot water than they would in cold water.

(a)

(b)

Figure 16.6 Water, carbon tetrachloride (CCl_4), and iodine. (a) Water (polar molecules) and CCl_4 (nonpolar molecules) are not miscible, so that the less dense water layer lies on top of the more dense CCl_4 layer. A small amount of iodine is dissolved in the water to give a brown solution *(top)*. (b) However, the nonpolar I_2 molecules are more soluble in nonpolar CCl_4, and dissolve preferentially in CCl_4 to give a purple solution after the mixture is shaken. *(C.D. Winters)*

ored solution. In contrast, carbon tetrachloride, CCl_4, a nonpolar liquid, dissolves I_2 readily—so readily that it can dissolve I_2 that has already been dissolved in water, a process known as *extraction* (Figure 16.6).

Problem-Solving Example 16.1 Predicting Solubility

Sucrose (cane sugar), $C_{12}H_{22}O_{11}$, is a molecular solid *(← p. 101)*. Each sucrose molecule has eight polar —OH groups, and sucrose molecules hydrogen bond to one another in solid sucrose. When sucrose dissolves in water, the new hydrogen bonds that form between sucrose molecules and water molecules are numerous and strong enough to overcome the sucrose-sucrose and water-water interactions. Would you expect sucrose to be soluble in (a) carbon tetrachloride? (b) gasoline? (c) ethanol? Explain.

Answer

(a) Almost zero solubility in carbon tetrachloride because the carbon tetrachloride molecule is not polar.
(b) Almost zero solubility in gasoline because gasoline contains no polar molecules.
(c) Moderate solubility in ethanol because ethanol molecules can form hydrogen bonds.

Explanation

(a) Carbon tetrachloride molecules are nonpolar and therefore would not interact strongly with the polar parts of the sucrose molecule.
(b) The hydrocarbon molecules in gasoline are nonpolar and would not interact strongly with the polar parts of the sucrose molecules.
(c) Ethanol molecules are polar and the negative oxygen atoms of the —OH groups would be attracted to the positive hydrogen atoms on —OH groups of the sucrose molecules and vice versa. Some solubility of sucrose in ethanol would be expected.

Problem-Solving Practice 16.1

Would you expect graphite to be soluble in water? In carbon tetrachloride? Explain.

Solids held together with extensive covalent bonding like that in network solids (← *p. 695*) are generally insoluble in most solvents, polar and nonpolar alike. For example, quartz (SiO_2), with its structure of silicon and oxygen atoms forming SiO_4 units that are linked by shared oxygen atoms to other SiO_4 units, is bonded so strongly that it is very difficult to separate the Si and O atoms by the weaker attractions to solvent molecules. Therefore quartz (and sand derived from quartz) is insoluble in water or any other solvent at room temperature.

Dissolving Ionic Solids in Liquids

Sodium chloride is an ionic compound. Its crystal lattice consists of Na^+ and Cl^- ions in a cubic array (← *p. 99 and p. 683*). The strong attractions between the oppositely charged ions hold them tightly in the lattice. The lowering of potential energy as a result of these attractive forces holding the ions together is referred to as the **lattice energy.** The large lattice energy of NaCl accounts for sodium chloride's high melting point of 800 °C. It is possible for solvent molecules to attract Na^+ away from Cl^- ions, but they have to be the right kind of solvent molecules. If you try to dissolve NaCl (or any other ionic compound) in carbon tetrachloride or hexane (both nonpolar solvents), you will have little success. Nonpolar molecules have very little attraction for ions. On the other hand, water can dissolve NaCl.

Water is a good solvent for an ionic compound because water molecules are small and highly polar (← *p. 421*). As shown in Figure 16.7, the partially negative oxygen ends of several water molecules are attracted to the positive ion and help pull it away from the crystal lattice, while the positive ends of other water molecules are attracted to the negative ions in the lattice and help pull them away from the lattice. This process, in which water molecules surround positive and negative ions, is called **hydration.** Energy known as the *heat of hydration* is released when these new attractions are formed between the ions and the water molecules. Energy is always required to separate the ions, and energy is always released when ions become hydrated. Whether dissolving a particular ionic compound is exothermic or endothermic depends on the relative sizes of the lattice energy and the hydration energies of the positive and negative ions. The relationship between the heat of solution, the lattice energy of the ionic compound, and the heats of hydration of the ions is

$$\Delta H_{\text{solution}} = \text{lattice energy} + \Delta H_{\text{hydration}} + \text{(cations)} + \Delta H_{\text{hydration}} - \text{(anions)}$$

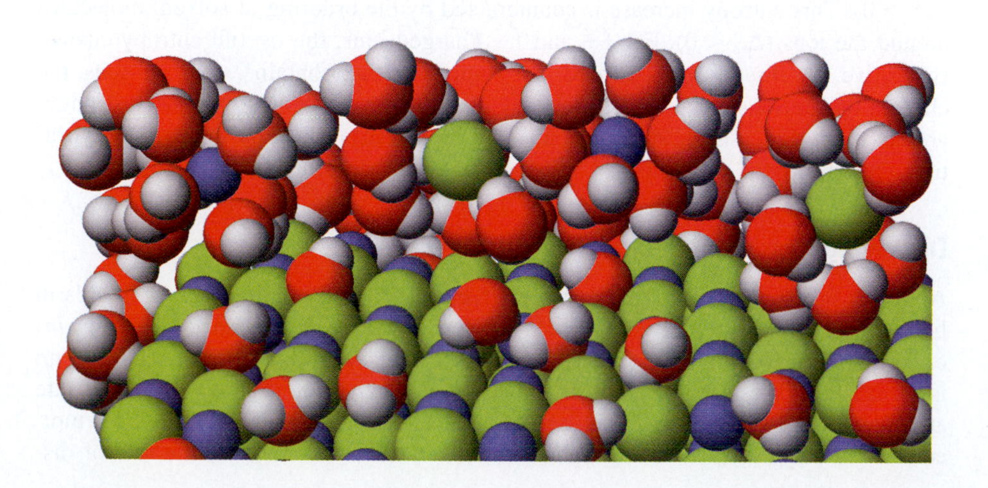

Figure 16.7 Dissolving an ionic crystal. Water molecules surround the positive (black) and negative (yellow) ions, helping them move away from their positions in the crystal lattice. The negative ends of some water molecules orient toward the positive ions while the positive ends of other water molecules orient toward the negative ions. Each ion that moves away from the crystal lattice exposes ions beneath it to other water molecules.

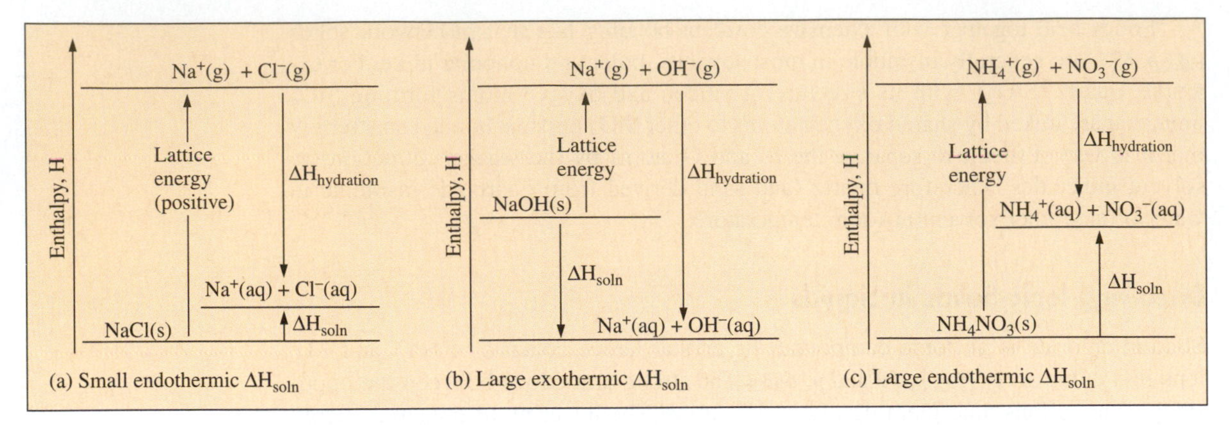

Figure 16.8 Heats of solution for three different ionic compounds dissolving in water. (a) For NaCl, the lattice energy that must be overcome is larger than the energy released on hydration of the ions. This causes the ΔH_{sol} value to be positive (endothermic). (b) The lattice energy for NaOH is much smaller than the energy released when the ions become hydrated. This causes ΔH_{sol} to be negative (exothermic). (c) For NH_4NO_3, the heat of hydration of the ions is much smaller than the lattice energy, and the resulting ΔH_{sol} value has a large positive value.

Figure 16.8 shows how lattice energy and the heats of hydration combine to give the heat of solution.

Practical applications of endothermic and exothermic dissolution include cold packs containing NH_4NO_3 used to treat athletic injuries (Figure 16.9) and hot packs containing $CaCl_2$ used to warm foods.

Energy is always released when particles that attract one another get closer together, and energy is always required to separate such particles.

The solubility rules for solubility of ionic compounds in water *(⇐ p. 104)* remind us that not all ionic compounds are water soluble in spite of the strong attractions between water molecules and ions. For some ionic compounds, the lattice energy is so large that water molecules cannot effectively pull a large number of the ions away from the lattice and the compound is either slightly soluble or moderately soluble.

Figure 16.9 A cold pack used for athletic injuries. This one contains ammonium nitrate in a separate inner container, which is broken when the desired cooling effect is needed. *(C.D. Winters)*

Entropy and the Dissolving of Ionic Compounds in Water

The disorder introduced when a crystal lattice breaks down and the disorder introduced by the mixing of ions with solvent molecules both favor the dissolving process ($\Delta S > 0$). This entropy increase is counteracted by the ordering of solvent molecules around the ions ($\Delta S < 0$). For 1+ and 1− charged ions, the overall entropy change is positive and dissolving is favored. For some salts that contain 2+ or 3+ ions, the charges on the ions are so large that they align water molecules extremely strongly around them. Since the water molecules are locked into place by this strong hydration, the entropy of solution may be negative, which does not favor solubility.

Dissolving Gases in Liquids: Henry's Law

Although pressure does not measurably affect the solubilities of solids or liquids in liquids, *the solubility of any gas in a liquid increases as the pressure of the gas increases (Figure 16.10).* When a gas is in contact with a liquid, a dynamic equilibrium is established. The rate at which gas molecules enter the liquid phase equals the rate at which gas molecules escape from the liquid. If the pressure is increased, gas molecules strike the surface of the liquid more often, and consequently the rate of dis-

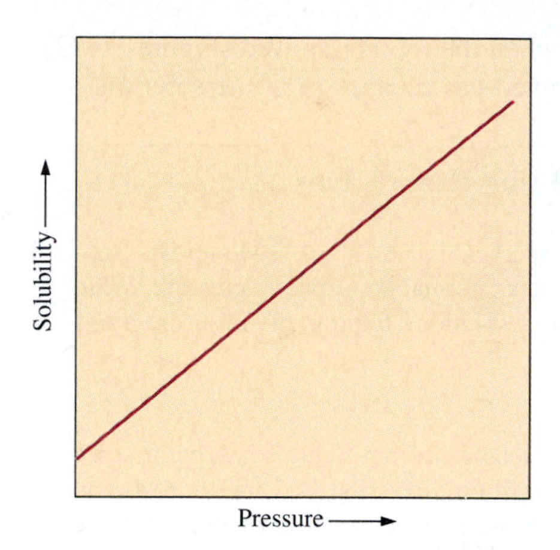

Figure 16.10 Pressure dependence of the solubility of a gas in a liquid. All gases that do not react with the solvent behave this way.

TABLE 16.2 Henry's Law Constants (25 °C)

Gas	k_H $\left(\frac{\text{mol/L}}{\text{mm Hg}}\right)$
N_2	8.42×10^{-7}
O_2	1.66×10^{-6}
CO_2	4.45×10^{-5}

solution increases. A new equilibrium is established when the rate of escape increases to match the rate of dissolution. The rate of escape is first order *(← p. 519)* in concentration of solute, and so a higher rate of escape requires a higher concentration of solute molecules, that is, higher solubility.

The relationship between gas pressure and solubility is known as **Henry's law:**

$$S_g = k_H P_g$$

where S_g is the solubility of the gas in the liquid, P_g is the pressure of the gas above the solution (or the partial pressure of the gas if the solution is in contact with a mixture of gases). The value of the constant, k_H, known as the Henry's law constant, depends on the identities of both the solute and the solvent and on the temperature (Table 16.2). It has units of moles per liter per millimeter of mercury (mm Hg).

Figure 16.11 illustrates how gas solubility depends on pressure. The behavior of a carbonated drink when the cap is removed is a common illustration of the solubility of gases in liquids under pressure. The drink fizzes when opened because the par-

The partial pressure of a gas in a mixture of gases is the pressure that a pure sample of the gas would exert if it occupied the same volume as the mixture. Partial pressure is proportional to the mole fraction of the gas **(← p. 634).**

Henry's law holds quantitatively only for gases that do not react with the solvent. It does not work perfectly for NH_3, for example, which gives small concentrations of NH_4^+ and OH^- in water, or for CO_2, which gives small concentrations of H^+ and HCO_3^- ions.

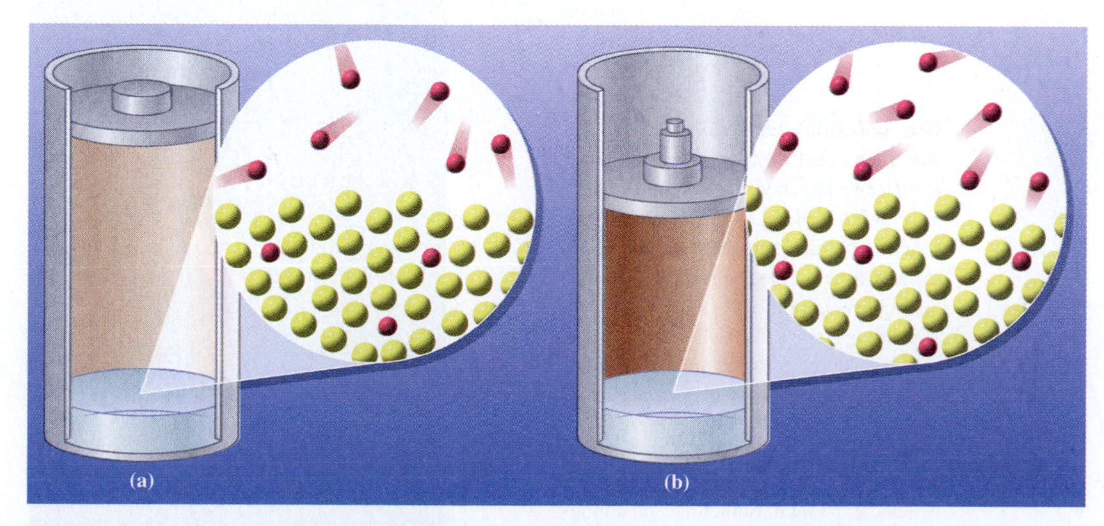

Figure 16.11 Gas solubility in a liquid. At constant temperature, a pressure increase causes the gas molecules to have a smaller volume to occupy. There are more collisions of gas molecules with the liquid surface, and so more molecules dissolve in the liquid.

Figure 16.12 **Henry's law.** The greater the partial pressure of CO_2 over the soft drink in the can, the greater the concentration of CO_2 dissolved. When the partial pressure of CO_2 is lowered by opening the can, CO_2 begins to bubble out of the solution. *(C.D. Winters)*

tial pressure of CO_2 over the solution drops when the top is removed (Figure 16.12), the solubility of the gas decreases, and dissolved gas escapes from the solution.

PROBLEM-SOLVING EXAMPLE 16.2 Using Henry's Law

The Henry's law constant for oxygen in water at 25 °C is 1.66×10^{-6} M/mm Hg. Suppose that a trout stream is in equilibrium with air at normal atmospheric pressure. What is the concentration of O_2 in a saturated aqueous solution? Express the result in grams per liter (g/L).

Answer 0.0086 g/L or 8.6 mg/L

Explanation The solubility of oxygen can be calculated from Henry's law, but first we must calculate the partial pressure of oxygen in air. From Section 14.7 (⟵ ***p. 634***) you know that air is 21 mole percent oxygen, which means that the mole fraction of O_2 is 0.21. If the total pressure is 1.0 atm,

$$\text{Pressure of } O_2 = (1.0 \text{ atm})\left(\frac{760.\text{ mm Hg}}{1 \text{ atm}}\right)(0.21) = 160.\text{ mm Hg}$$

Using Henry's law and the value of k_H given above

$$S_g = k_H P_g = \left(1.66 \times 10^{-6} \frac{\text{mol/L}}{\text{mm Hg}}\right)(160.\text{ mm Hg}) = 2.66 \times 10^{-4} \text{ mol/L}$$

This concentration can be converted from molarity (mol/L) to the desired units by using molar mass.

$$\text{Solubility of } O_2 = \left(\frac{2.66 \times 10^{-4}\text{ mol}}{\text{L}}\right)\left(\frac{32.00 \text{ g}}{\text{mol}}\right) = 0.0085 \text{ g/L or } 8.5 \text{ mg/L}$$

Although this oxygen concentration is very low, it is sufficient to provide the oxygen required by aquatic life.

Gas solubility and diving. Diving with scuba gear means you must be knowledgeable about the solubility of gases in your blood. Both N_2 and O_2 are soluble in blood, and breathing the pressurized air from the scuba tank means a high concentration of N_2 will be present. Ascending too rapidly will cause the dissolved N_2 to be released and form bubbles in the blood. This causes a painful condition known as the "bends." If these N_2 bubbles deprive the brain of its blood flow by blocking blood capillaries there, the bends can be fatal. Sometimes breathing-gas mixtures contain He in the place of N_2. Helium has a lower solubility in blood than does nitrogen. *(Brian Parker/Tom Stack and Associates)*

Problem-Solving Practice 16.2

The Henry's law constant for N_2 in water at 25 °C is $8.4 \times 10^{-7} \frac{\text{mol/L}}{\text{mm Hg}}$. What is the solubility of N_2 in mol/L if its partial pressure is 1520 mm Hg? What is the solubility when the N_2 partial pressure is 20. mm Hg?

16.3 TEMPERATURE AND SOLUBILITY

To understand how temperature affects solubility, we can apply Le Chatelier's principle to the dissolution equilibrium. As one example, consider the equilibrium between a pure solute gas and a saturated solution.

$$\text{Gas} + \text{solvent} \rightleftharpoons \text{saturated solution}$$

When a gas dissolves to form a saturated liquid solution, the process is almost always exothermic. Gas molecules that were relatively far apart are brought close to other molecules that attract them, lowering their potential energy and releasing some energy to the surroundings as new intermolecular attractions are formed.

If the temperature of a solution of a gas in a liquid increases, the equilibrium shifts in the direction that partially counteracts the temperature rise. That is, the equilibrium shifts to the left in the preceding equation. Thus a dissolved gas becomes less soluble with increasing temperature (Figure 16.13). Conversely, cooling a solution that is at equilibrium with undissolved gas will cause the equilibrium to shift to the right in the direction that liberates heat, and so more dissolves. This is illustrated by the data for the solubility of oxygen in water in Table 16.3.

Cooler water in contact with the atmosphere can contain more dissolved oxygen than the same water at a higher temperature. For this reason fish seek out cooler (usually deeper) waters in the summer. Their gills have an easier time obtaining oxygen when its concentration in the water is higher. The decrease in gas solubility as temperature increases makes *thermal pollution* a problem for aquatic life in rivers and streams. Natural heating of water by sunlight and by warmer air can usually be accommodated, but excess heat from such sources as industrial facilities and electrical power plants can reduce the concentration of dissolved oxygen to the point where some species of fish die.

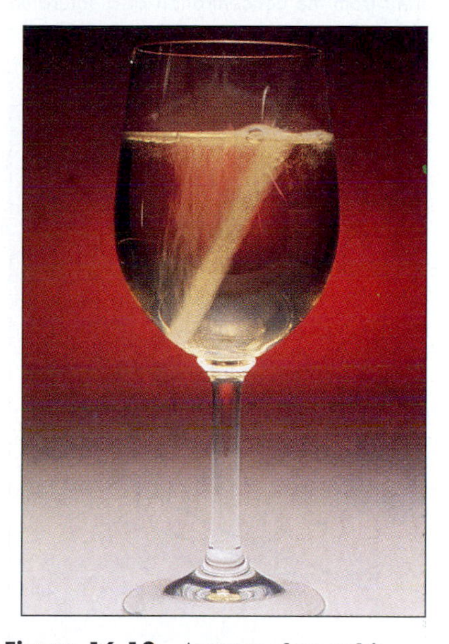

Figure 16.13 A warm glass rod is placed in a glass of ginger ale. Thermal energy from the rod warms a cold solution of CO_2 in water and causes the CO_2 to be less soluble. The Henry's law constant for CO_2 in water is 4.45×10^{-5} M/mm Hg at 25 °C, whereas it is 2.5×10^{-5} M/mm Hg at 50 °C. *(C.D. Winters)*

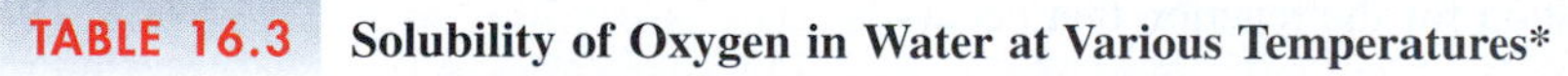

TABLE 16.3 Solubility of Oxygen in Water at Various Temperatures*

Temperature (°C)	Solubility of O_2 (g/L)
0	0.0141
10	0.0109
20	0.0092
25	0.0086
30	0.0077
35	0.0070
40	0.0065

*These data are for water in contact with air at 760 mm Hg pressure.

When discussing solubility as a function of temperature, the relevant ΔH is for dissolving the last little bit of solute to make a saturated solution. It is not the ΔH that would be calculated for the process solid $\rightarrow$ solution from data in Appendix J. These data are for a solution at a standard-state concentration of solute, which may be quite different from the concentration of a saturated solution.

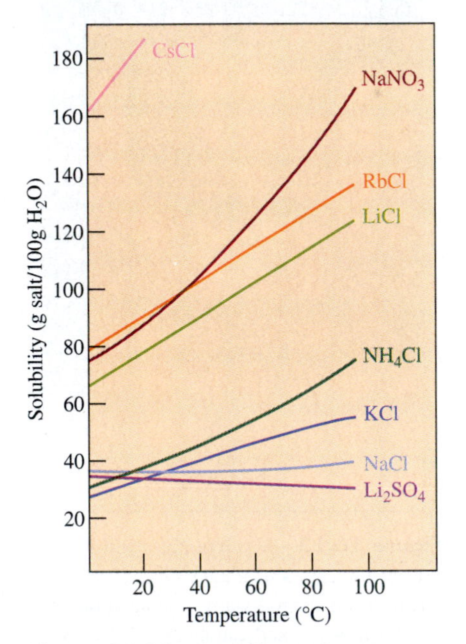

Figure 16.14 Solubility and temperature. The temperature dependence of the solubility of some ionic compounds in water.

The solubility of most ionic compounds increases with increasing temperature (Figure 16.14). This is because the heats of solution of most salts are endothermic ($\Delta H_{sol} > 0$), which means their lattice energies are larger than their heats of hydration (Figure 16.8). The solution process for these kinds of ionic compounds is written as

$$\text{Solute} + \text{solvent} \rightleftharpoons \text{solution} \qquad \Delta H > 0 \text{ (endothermic)}$$

To partially counteract an increase in temperature the equilibrium shifts to the right, so an increase in solubility is predicted by Le Chatelier's principle.

Conversely, if dissolving an ionic compound is exothermic ($\Delta H_{sol} < 0$), Le Chatelier's principle predicts that its solubility will decrease with increased temperature.

$$\text{Solute} + \text{solvent} \rightleftharpoons \text{solution} \qquad \Delta H < 0 \text{ (exothermic)}$$

Exercise 16.5 Temperature and Solubility

If a substance has an endothermic heat of solution, which would cause more of it to dissolve, hot solvent or cold solvent? Explain.

16.4 SOLUBILITY EQUILIBRIA AND THE SOLUBILITY PRODUCT CONSTANT, K_{SP}

The solubility rules used to predict whether precipitation will occur (*← p. 193, Figure 5.7*) provide no quantitative information about solubility. They indicate which ions form compounds soluble enough to make solutions of at least 0.1 M. But they say nothing about how much more of a soluble compound would dissolve, nor do they indicate the maximum possible concentration of an insoluble salt. There is a way to tabulate such information, because solubility is an equilibrium process. In a saturated aqueous solution, solid and solution are in dynamic equilibrium. The equilibrium constant for this process, then, can tell us how soluble a substance is.

To see how this works, consider the synthesis of silver bromide, AgBr, which is used in photographic film as one of the light-sensitive ingredients. Silver bromide is insoluble. Therefore it can be made by adding a water-soluble silver salt, such as $AgNO_3$, to an aqueous solution of a bromide-containing salt, such as KBr. The net ionic equation for the reaction that occurs is

$$Ag^+(aq) + Br^-(aq) \longrightarrow AgBr(s)$$

Although we say AgBr is insoluble in water, if some of this precipitated AgBr is placed in pure water, a tiny bit will dissolve and an equilibrium will be established between solid AgBr and the Ag^+ and Br^- ions in solution.

$$AgBr(s) \rightleftharpoons Ag^+(aq) + Br^-(aq)$$

When equilibrium has been established, the solution is saturated. Experiments show that at 25 °C the concentrations of both Ag^+ and Br^- ions are 5.7×10^{-7} M. The equilibrium expression is given by

$$K = [Ag^+(aq)][Br^-(aq)]$$

Note how the concentration of the undissolved AgBr does not appear in the equilibrium expression.

The constant in this expression is the product of the equilibrium concentrations of the ions of the slightly soluble salt, and so it is called the **solubility product constant.** It is designated K_{sp}. Hence,

$$K_{sp}(AgBr) = [Ag^+(aq)][Br^-(aq)]$$

Since the concentrations of both the Ag^+ and Br^- ions are 5.7×10^{-7} M when silver bromide is in equilibrium with its ions at 25 °C, this means that K_{sp} at 25 °C is

$$K_{sp} = [5.7 \times 10^{-7}][5.7 \times 10^{-7}] = 3.2 \times 10^{-13}$$

Additional K_{sp} values are listed in Table 16.4 and in Appendix H.

In general, the equilibrium expression for dissolving a slightly soluble salt with the general formula A_xB_y is

$$A_xB_y(s) \rightleftharpoons x\,A^{n+}(aq) + y\,B^{m-}(aq)$$

This results in the general K_{sp} expression

$$K_{sp} = [A^{n+}]^x[B^{m-}]^y$$

TABLE 16.4 K_{sp} Values for Some Slightly Soluble Salts

Compound	K_{sp} at 25 °C
$AgBr$	3.3×10^{-13}
$AuBr$	5.0×10^{-17}
$AuBr_3$	4.0×10^{-36}
$CuBr$	5.3×10^{-9}
Hg_2Br_2*	1.3×10^{-22}
$PbBr_2$	6.3×10^{-6}
$AgCl$	1.8×10^{-10}
$AuCl$	2.0×10^{-23}
$AuCl_3$	3.2×10^{-25}
$CuCl$	1.9×10^{-7}
Hg_2Cl_2*	1.1×10^{-18}
$PbCl_2$	1.7×10^{-5}
AgI	1.5×10^{-13}
AuI	1.6×10^{-13}
AuI_3	1.0×10^{-46}
CuI	5.1×10^{-12}
Hg_2I_2*	4.5×10^{-29}
HgI_2	4.0×10^{-29}
PbI_2	8.7×10^{-9}
Ag_2SO_4	1.7×10^{-5}
$BaSO_4$	1.1×10^{-10}
$PbSO_4$	1.8×10^{-8}
Hg_2SO_4*	6.8×10^{-7}
$SrSO_4$	2.8×10^{-7}

*These compounds contain the diatomic ion Hg_2^{2+}.

PROBLEM-SOLVING EXAMPLE 16.3 Writing K_{sp} Expressions

Write the K_{sp} expressions for the following slightly soluble salts: (a) $AuCl$, (b) $PbBr_2$, and (c) Ag_2SO_4.

Answer (a) $K_{sp} = [Au^+][Cl^-]$ (b) $K_{sp} = [Pb^{2+}][Br^-]^2$ (c) $K_{sp} = [Ag^+]^2[SO_4^{2-}]$

Explanation

(a) The equilibrium reaction for the solubility of AuCl in water is

$$AuCl(s) \rightleftharpoons Au^+(aq) + Cl^-(aq)$$

Since the K_{sp} expression contains only the soluble ions, raised to the balancing coefficients in the equilibrium reaction, $K_{sp} = [Au^+][Cl^-]$.

(b) The equilibrium reaction for the solubility of $PbBr_2$ in water is

$$PbBr_2(s) \rightleftharpoons Pb^{2+}(aq) + 2\,Br^-(aq)$$

In this example, there are two Br^- ions produced for every Pb^{2+} ion, so their concentration is raised to the power of 2 in the K_{sp} expression, $K_{sp} = [Pb^{2+}][Br^-]^2$.

(c) The equilibrium reaction for the solubility of Ag_2SO_4 in water is

$$Ag_2SO_4(s) \rightleftharpoons 2\,Ag^+(aq) + SO_4^{2-}(aq)$$

Since two Ag^+ ions are produced for every SO_4^{2-} ion, the K_{sp} expression is written as $K_{sp} = [Ag^+]^2[SO_4^{2-}]$.

Problem-Solving Practice 16.3

Write the K_{sp} expressions for the slightly soluble salts CuBr, HgI_2, and $SrSO_4$.

If the K_{sp} of a slightly soluble salt is known, it is possible to estimate the solubility of the salt. For example, the K_{sp} of $BaSO_4$ is 1.1×10^{-10} at 25 °C. The equation for dissolving $BaSO_4$ is

$$BaSO_4(s) \rightleftharpoons Ba^{2+}(aq) + SO_4^{2-}(aq)$$

K_{sp} provides only an estimate of the solubility because sulfate ions react with water to a small extent in a process called hydrolysis. Therefore the concentration of sulfate ions is slightly less than would be predicted from the K_{sp} of $BaSO_4$ and the concentration of barium ions is slightly higher.

Suppose that you start with 1 L of pure water and then add solid $BaSO_4$, allowing it to dissolve until equilibrium is reached. According to the equation, 1 mol of Ba^{2+} ions and 1 mol of SO_4^{2-} ions are produced for every mole of $BaSO_4$ that dissolves. Therefore, the concentration of either Ba^{2+} or SO_4^{2-} ions indicates how much $BaSO_4$ has dissolved per liter. If S represents the **molar solubility** (the solubility, in moles per liter) of $BaSO_4$, both $[Ba^{2+}]$ and $[SO_4^{2-}]$ must also be equal to S at equilibrium. It is helpful to write this information as a table, like the one shown below and others used in equilibrium calculations (← p. 577).

In this table, the + sign, as in $+S$, indicates an increase in concentration.

	$BaSO_4(s) \rightleftharpoons$ **$Ba^{2+}(aq)$** +	**$SO_4^{2-}(aq)$**
Initial Concentration	0	0
Change as Reaction Occurs (mol/L)	$+S$	$+S$
Equilibrium Concentration (mol/L)	S	S

The equilibrium-constant expression derived from the equation for dissolving $BaSO_4$ is

$$K_{sp} = [Ba^{2+}][SO_4^{2-}]$$

Substituting the values from the table into the K_{sp} expression and then solving for S, the *equilibrium* concentration, we get

$$K_{sp} = [Ba^{2+}][SO_4^{2-}] = (S)(S) = S^2$$

$$S = \sqrt{K_{sp}} = \sqrt{1.1 \times 10^{-10}} = 1.0 \times 10^{-5} \text{ mol/L}$$

The solubility of $BaSO_4$ in pure water at 25 °C is estimated to be 1.0×10^{-5} mol/L. This justifies its classification as a sparingly soluble ionic compound.

To find the solubility in grams per liter (g/L), we need only to multiply the molar solubility by the molar mass of $BaSO_4$.

$$\text{Solubility} = \left(\frac{1.0 \times 10^{-5} \text{ mol } BaSO_4}{\text{L}}\right)\left(\frac{233 \text{ g } BaSO_4}{1 \text{ mol } BaSO_4}\right) = 0.0023 \text{ g/L}$$

Exercise 16.6 Calculating Solubility from K_{sp}

The K_{sp} of AgCl is 1.8×10^{-10}. Calculate the solubility of AgCl. Express your result in mol/L and in g/L.

Where the ions in a K_{sp} expression are not in a 1:1 ratio, S must be expressed differently. For example, consider the solubility of PbI_2, $K_{sp} = 8.7 \times 10^{-9}$. For the reaction

$$PbI_2(s) \rightleftharpoons Pb^{2+}(aq) + 2\ I^-(aq)$$

the balanced equation tells us that for each S mol/L of $PbI_2(s)$ that dissolves, S mol/L of $Pb^{2+}(aq)$ form, but $2S$ mol/L of $I^-(aq)$ form.

Notice in the table how $2S$ appears as the change in the concentration of the I^- ion in the reaction.

	$PbI_2(s) \rightleftharpoons$ **$Pb^{2+}(aq)$** +	**$2I^-(aq)$**
Initial Concentrations (mol/L)	0	0
Change as Reaction Occurs (mol/L)	$+S$	$+2S$
Equilibrium Concentration (mol/L)	S	$2S$

The equilibrium constant expression from the equation for dissolving PbI_2 is

$$K_{sp} = [Pb^{2+}][I^-]^2$$

Note the *squared* concentration of I^-.

Substituting the values from the table into the K_{sp} expression and then solving for S, we get

$$K_{sp} = [Pb^{2+}][I^-]^2 = (S)(2S)^2 = 4\,S^3$$

$$S = \sqrt[3]{\frac{K_{sp}}{4}} = \sqrt[3]{\frac{8.7 \times 10^{-9}}{4}} = 1.3 \times 10^{-3} \text{ mol } PbI_2/L$$

The term $\sqrt[3]{\frac{8.7 \times 10^{-9}}{4}}$ is equivalent to $\left(\frac{8.7 \times 10^{-9}}{4}\right)^{1/3}$

To convert this estimated solubility at 25 °C from mol/L to g/L, multiply the molar solubility by the molar mass of PbI_2.

$$\text{Solubility} = \left(\frac{1.3 \times 10^{-3} \text{ mol } PbI_2}{L}\right)\left(\frac{461.0 \text{ g } PbI_2}{1 \text{ mol } PbI_2}\right) = 0.60 \text{ g } PbI_2/L$$

Exercise 16.7 Calculating Solubility from K_{sp}

The K_{sp} of HgI_2 is 4.0×10^{-29}. Calculate the solubility of HgI_2. Express your result in mol/L and in g/L.

For salts with the same ion ratios, K_{sp} values can be used to compare solubilities. The bigger the K_{sp} the greater the solubility. For example, AgCl, AgBr, and AgI are all 1:1 salts. Their K_{sp} values are

$$K_{sp}(AgCl) = 1.8 \times 10^{-10}$$

$$K_{sp}(AgBr) = 3.3 \times 10^{-13}$$

$$K_{sp}(AgI) = 1.5 \times 10^{-16}$$

and so the solubility of AgCl is greater than the solubility of AgBr, which in turn is greater than the solubility of AgI. These relative solubilities can be predicted this way because for 1:1 salts the relation between solubility, S, and K_{sp} is the same, namely $S = \sqrt{K_{sp}}$. However, K_{sp} values alone cannot be used to compare the relative solubilities of salts with different ion ratios (such as AgCl and $PbCl_2$).

Exercise 16.8 Relative Solubilities of Compounds from Their K_{sp} Values

Which substance in Table 16.4 with general formula MX_2 is least soluble?

Solubility and the Common Ion Effect

It is often desirable to remove a particular ion from solution by forming a precipitate of one of its insoluble compounds. For example, barium readily absorbs x rays and so is quite effective in making the intestinal tract visible when x-ray photographs are taken. But barium ions are poisonous and must not be allowed to dissolve in body fluids. The insoluble compound barium sulfate can be used as an x-ray absorber, but both physician and patient want to be certain that no harmful amounts of barium ions will be in solution.

Earlier in this section we estimated the solubility of $BaSO_4$ in water at 25 °C to be 1.0×10^{-5} mol/L, which means that the concentration of Ba^{2+} ions would be 1.0×10^{-5} M as well. The concentration of Ba^{2+} ions can be reduced still further by adding a soluble sulfate salt, such as Na_2SO_4. The solubility of $BaSO_4$ becomes smaller because of the increased concentration of the SO_4^{2-} ion, which is present in both $BaSO_4$ and Na_2SO_4. Sulfate is called a "common ion" because it is common to both substances dissolved in the solution. Such a displacement of an equilibrium by having more than one source of a product ion is called the **common ion effect.** It can be interpreted by using Le Chatelier's principle. Consider the solubility equilibrium

$$BaSO_4(s) \rightleftharpoons Ba^{2+}(aq) + SO_4^{2-}(aq)$$

When some Na_2SO_4 solution is added to a saturated solution of $BaSO_4$, the concentration of SO_4^{2-} ions increases, which in turn causes the equilibrium to shift to offset the effect of the change. To use up some of the added sulfate ions, the equilibrium shifts to the left and uses up some of the Ba^{2+} ions in solution. The outcome is that the salt solubility is lower in the presence of the common ion.

PROBLEM-SOLVING EXAMPLE 16.4 The Common Ion Effect

The solubility of AgCl in pure water is 1.3×10^{-5} mol/L. If you put some AgCl in a solution that is 0.55 M in NaCl, what mass of AgCl will dissolve per liter of this solution? The K_{sp} of AgCl is 1.8×10^{-10}.

Answer About 4.7×10^{-8} g AgCl/L

Explanation The solubility of AgCl, S, equals the silver ion concentration $[Ag^+]$ at equilibrium. In water containing the common ion Cl^-, the value of S must be smaller than it would be in pure water because of the effect of the common ion.

$$AgCl(s) \rightleftharpoons Ag^+(aq) + Cl^-(aq)$$

If the AgCl(s) were dissolved in *pure* water, S would be equal to both $[Ag^+]$ and to $[Cl^-]$ because there is no other source of chloride ion. However, in salt water containing the common ion Cl^-, S is equal only to $[Ag^+]$ because Cl^- comes from two sources: the AgCl and the NaCl. Because NaCl is a soluble salt, far more Cl^- comes from the NaCl than from the AgCl. Therefore S must have a smaller value than in pure water owing to the common ion effect on equilibrium.

The following table shows the concentrations of Ag^+ and Cl^- when equilibrium is attained in the presence of extra Cl^-.

	$AgCl(s) \rightleftharpoons$ $Ag^+(aq)$	+ $Cl^-(aq)$
Initial Concentration (mol/L)	0	0.55
Change as Reaction Occurs (mol/L)	$+S$	$+S$
Equilibrium Concentration (mol/L)	S	$S + 0.55$

The total chloride ion concentration at equilibrium is the amount coming from AgCl (equals S) *plus* what was already there (0.55 M) from the NaCl.

Using the equilibrium concentrations from the table gives

$$K_{sp} = 1.8 \times 10^{-10} = [Ag^+][Cl^-] = (S)(S + 0.55)$$

which can be rearranged to

$$S^2 + 0.55\,S - K_{sp} = 0$$

The easiest approach to solving an equation like this is to make the approximation that S is *very* small with respect to 0.55; that is, the answer will be approximately the same if we assume that $(S + 0.55) \approx 0.55$. This is a very reasonable assumption since we know that the solubility equals 1.3×10^{-5} mol/L *without* the common ion Cl^-, and it will be even smaller in the presence of added Cl^-. Therefore,

$$(S)\,(S + 0.55) \approx (S)(0.55) = K_{sp}$$

or

$$K_{sp} = (S)(0.55)$$

And solving for S, we get

$$S = \frac{1.8 \times 10^{-10}}{0.55} = 3.3 \times 10^{-10}\text{ M} = [Ag^+]$$

Therefore, the $[Ag^+]$, which is the same as S, is approximately 3.3×10^{-10} mol/L. Using the molar mass for AgCl, 143.4 g/mol, the solubility in g/L is

$$S = \left(\frac{3.3 \times 10^{-10}\text{ mol AgCl}}{\text{L}}\right)\left(\frac{143.3\text{ g AgCl}}{\text{mol AgCl}}\right) = 4.7 \times 10^{-8}\text{ g AgCl/L}$$

As predicted by Le Chatelier's principle, the solubility of AgCl in the presence of added Cl^-, an ion common to the equilibrium, is clearly less (3.3×10^{-10} M) than in pure water (1.3×10^{-5} M).

As a final step, let us check the approximation we made. To do this, we substitute the approximate value of S into the exact expression $K_{sp} = (S)(S + 0.55)$. Then, if the product $(S)(S + 0.55)$ is the same as the given value of K_{sp}, the approximation is valid.

$$K_{sp} = (S)(S + 0.55) = (3.3 \times 10^{-10})(3.3 \times 10^{-10} + 0.55) \sim 1.8 \times 10^{-10}$$

(A more accurate solution to this problem can be had by the method of successive approximations or by solving for S using the quadratic equation described in Appendix B.4 and used in Problem Solving Example 13.4, (← *p. 585*). When the quadratic formula is used, its answer, to two significant figures, is the same as our approximation.

Problem-Solving Practice 16.4

Calculate the solubility of $PbCl_2$ at 25 °C in a solution that is 0.50 M in NaCl.

Conceptual Challenge Problem CP-16.A at the end of the chapter relates to topics covered in this section.

Exercise 16.9 What About More Added Solid?

What is the effect on the equilibrium if more AgCl solid is added to a solution saturated in Ag^+ and Cl^- ions?

16.5 COMPOSITION OF DILUTE AQUEOUS SOLUTIONS

Even water that is considered suitable for drinking purposes contains very low concentrations of numerous dissolved solutes. Lead is an example. Lead is a widely encountered metal that has toxic properties, and many of its compounds are soluble to some extent in water. Usually, the concentration of dissolved lead in water is quite low, but the concentration still may be enough to be potentially harmful. Dealing with the problem of unwanted lead and other solutes like mercury, selenium, nitrate ions, and even organic compounds in drinking water requires knowledge of the variety of units used to express the composition of dilute solutions. These units are also useful in discussing the concentrations of solutes in other kinds of solutions, like blood, for example.

The **mass fraction** of a solute is the fraction of the total mass of the solution that is contributed by that solute. That is, we divide the mass of a single solute by the total mass of all the solutes and the solvent. Mass fraction is commonly expressed as a percentage and called **weight percent,** which is the mass fraction multiplied by 100%. This is the same as the number of grams of solute per 100 g of solution. For example, the mass fraction of A in a solution containing solutes A and B dissolved in solvent C is

$$\text{Mass fraction of A} = \frac{\text{mass of solute A}}{\text{mass of solute A} + \text{mass of solute B} + \text{mass of solvent C}}$$

$$\text{Weight percent of A} = \text{mass fraction of A} \times 100\% = \frac{\text{number of grams of solute A}}{100 \text{ g solution}}$$

PROBLEM-SOLVING EXAMPLE 16.5 Mass Fraction/Weight Percent

Sterile saline solutions containing NaCl in water are often used in medicine. What is the weight percent of NaCl in a solution made by dissolving 4.6 g of NaCl in 500. g of pure water?

Answer 0.91%

Explanation Using the definitions of mass fraction and weight percent we have

$$\text{Mass fraction of NaCl} = \frac{4.6 \text{ g NaCl}}{4.6 \text{ g NaCl} + 500. \text{ g } H_2O} = 0.0091$$

$$\text{Weight percent NaCl} = \text{mass fraction NaCl} \times 100\% = 0.0091 \times 100\% = 0.91\%$$

Problem-Solving Practice 16.5

What is the weight percent of glucose in a solution containing 21.5 g of glucose ($C_6H_{12}O_6$) in 750. g of pure water?

To express the mass fraction as parts per million, multiply it by 1,000,000 ppm. This makes a very small number bigger and easier to handle.

Mass fractions of very dilute solutions are often expressed in **parts per million** (abbreviated ppm). One part per million is equivalent to one gram of solute per one million grams of solution, or one milligram of solute per thousand grams of solution (1 mg/kg). For even smaller mass fractions, **parts per billion** (ppb) and **parts per trillion** (ppt) are often used. As the names imply, we can convert a mass fraction to parts per billion by multiplying by 1,000,000,000 ppb (10^9 ppb), and to parts per trillion by multiplying by 1,000,000,000,000 ppt (10^{12} ppt).

PROBLEM-SOLVING EXAMPLE 16.6 Mass Fraction

A sample of water is found to contain 0.010 ppm (10. ppb) lead. What is the mass of lead per liter of this solution? (Assume the density of the water solution is 1.0 g/mL.)

Answer 10. μg/L

Explanation Since the solution is almost entirely water, its density will be the same as that of water, namely 1.0 g/mL, and so 1 L of solution has a mass of 1.0×10^3 g. The mass of lead in this 1 L sample of water can be calculated from the mass fraction, expressed as a ratio of grams of lead per 10^6 g of water.

$$\text{Mass of Pb in 1 L} = \left(\frac{0.010\text{ g Pb}}{1 \times 10^6\text{ g solution}}\right)\left(\frac{1.0 \times 10^3\text{ g solution}}{\text{L solution}}\right) = 1.0 \times 10^{-5}\text{ g Pb/L solution}$$

This concentration can also be expressed in micrograms per liter as

$$\text{Mass of Pb in 1 L} = \left(\frac{1.0 \times 10^{-5}\text{ g Pb}}{\text{L}}\right)\left(\frac{1\ \mu\text{g}}{1 \times 10^{-6}\text{ g}}\right) = 10.\ \mu\text{g Pb/L}$$

Thus, a mass fraction of 10. ppb corresponds to a concentration of 10. μg per liter of solution.

Problem-Solving Practice 16.6

Drinking water often contains small concentrations of selenium (Se). If a sample of water contains 30 ppb Se, how many micrograms of Se are present in 100. mL of this water?

Recall that 1 μg (microgram) = 10^{-6} g; 1 mg (milligram) = 10^{-3} g = 1000 μg.

Conceptual Challenge Problem CP-16.B at the end of the chapter relates to topics covered in this section.

Exercise 16.10 Mercury in Lake Michigan

A 1.0-kg sample of water from Lake Michigan is found to contain 8.7 ng (8.7×10^{-9} g) of mercury. What is the mass fraction of mercury in the water in ppb? In ppt?

Lead in Drinking Water

Lead concentrations in the environment are usually in the range of parts per billion. For liquid solutions they are often reported in micrograms per liter (which corresponds to ppb as illustrated above); for solids, 1 μg/kg corresponds to 1 ppb. Lead ions are present in some foods (up to 100 to 300 μg/kg), beverages (up to 20 to 30 μg/L), public water supplies (up to 100 μg/L, from lead-sealed pipes and plumbing fixtures), and even air (up to 2.5 μg/m^3 from lead compounds that once were in auto exhausts in the United States and still are in many other countries). With this many sources and contacts per day, it is obvious that the body must be able to rid itself of lead. Otherwise everyone would have died long ago of lead poisoning!

At the height of the Roman empire, lead production worldwide was about 80,000 tons per year. Today it is about 3 million tons per year. Lead was first used in ancient Rome for water pipes. The Latin name for lead, *plumbum,* gave us the name plumber.

Lead from auto exhaust emitted decades ago can still get into the air when soil containing that lead becomes dust and is carried into the air.

The average person can excrete about 230 μg of lead a day through the kidneys and intestinal tract. The daily intake is normally less than this; if it is not, accumulation and storage result. In the body, lead can accumulate in bone cells, where it acts on the bone marrow. In tissues it behaves like other heavy-metal poisons, such as mercury and arsenic, by reacting with *sulfhydryl groups* (—SH groups) in proteins. Enzymes, the catalysts for reactions in the body, are proteins. If an enzyme containing an —SH group binds to a heavy metal ion such as Pb^{2+}, the enzyme will likely cease to function. Lead, like mercury and arsenic, can also affect the central nervous system by binding to various active sites in nerves.

Unless they are exceptionally insoluble, lead salts are always toxic. Metallic lead can even be absorbed through the skin because it reacts with weak acids in perspiration and dissolves. Cases of lead poisoning have resulted from repeated handling of lead foil, bullets, and other lead objects.

Exercise 16.11 Lead in Water

If the concentration of Pb^{2+} ions in tap water is found to be 0.025 ppm, how many liters of this water will contain 100.0 μg of Pb?

16.6 CLEAN AND POLLUTED WATER

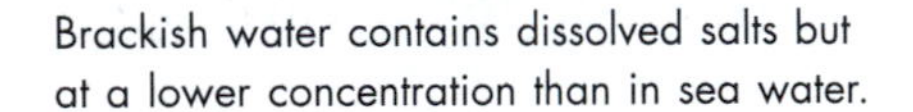
Brackish water contains dissolved salts but at a lower concentration than in sea water.

Water is the most abundant substance on the earth's surface. Oceans cover over 72% of the earth and are a reservoir for 97.2% of the earth's water. The rest consists of 2.16% in glaciers, 0.0197% in fresh water in lakes and rivers, 0.61% in groundwater, 0.01% in brine wells and brackish waters, and 0.001% in atmospheric water (Figure 16.15). Water is also a major component of all living things. For example, the water content of human adults is 70%—about the same proportion as the earth's surface.

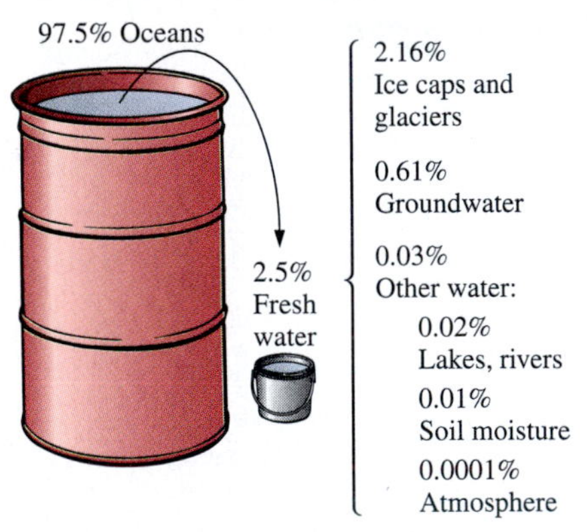

Figure 16.15 The water supply. Of the 2.5% fresh water, less than 1% is available as groundwater or surface water for human use.

Water pollutants are a source of great concern because water is so important to life on this planet. Water that is unsuitable for drinking, washing, irrigation, or industrial use is polluted. Natural bodies of water such as lakes, rivers, and salt water bays that no longer can support their usual population of microorganisms, fish, or birds are also described as polluted. Pollutants may be elevated temperature, radioisotopes, toxic metal ions or organic molecules, acids, alkalies, colloidal particles such as silt, or plant nutrients that cause excessive growth of aquatic microorganisms. Water suitable for some uses may be considered unsuitable for other uses and therefore polluted. Human activities such as industrialization and land development cause significant water pollution. But natural leaching of metal ions from soils, organic substances from decaying animal and vegetable matter, animal wastes, and soil erosion also pollute otherwise clean water.

As human activities have continued to pollute water, various federal, state, and local governments have passed laws designed to keep our water clean and less polluted. The U.S. Environmental Protection Agency sets limits for contaminants in drinking water and requires that municipal water supplies be monitored continually. Some examples of these limits are given in Table 16.5.

TABLE 16.5 Some Maximum Contaminant Levels (MCL in mg/L = ppm) for Drinking Water Allowed by EPA

Metals		Non-metals		Volatile Organic Compounds		Herbicides, Pesticides, PCBs	
Arsenic	0.050	Fluoride	4.00	Benzene	0.005	Chlordane	0.002
Beryllium	0.004	Nitrate	10.00	Carbon tetrachloride	0.005	Diquat	0.02
Cadmium	0.005	Nitrite	1.00	Vinyl chloride	0.002	Endrin	0.0002
Chromium	0.100			Trichloroethylene	0.005	Heptachlor	0.0004
Copper	1.300			Hexachlorobenzene	0.001	Lindane	0.0002
Lead	0.015			*p*-Dichlorobenzene	0.075	PCBs	0.0005
Selenium	0.050			Styrene	0.100	2,4,5-TP (Silvex)	0.050
				Toluenes	1.000	Aldicarb	0.003
						Ethylene dibromide	0.00005

Exercise 16.12 Calculating Beryllium Consumed

Calculate the mass of beryllium consumed each day by an adult drinking 3 L of water containing the maximum contaminant level of beryllium shown in Table 16.5.

Exercise 16.13 Estimating Toxicity of Water Pollutants

Using Table 16.5, identify the metals, the nonmetals, the volatile organic compounds, and the pesticides, herbicides, and PCBs that are most toxic to humans.

BOD: Biochemical Oxygen Demand

Industrial processes as well as organisms that live in and near bodies of water constantly release organic substances into surface waters. These substances eventually find their way into groundwater and the oceans. Bacteria and other microorganisms change the organic substances (consisting mostly of carbon, hydrogen, oxygen, and nitrogen atoms) into simpler compounds. To do so, however, the microorganisms use oxygen to metabolize the organic compounds that constitute their food. The quantity of oxygen required to oxidize a given quantity of organic material is called the **biochemical oxygen demand (BOD).** Given enough oxygen, a moderate temperature, and enough time, the microorganisms will convert huge quantities of organic matter into CO_2, H_2O, and NO_3^- or N_2.

Highly polluted water often has a high concentration of organic material, so it has a large biochemical oxygen demand. When organic matter, such as detergents, cooking oils, food wastes, and human wastes, is introduced into a river, the BOD jumps and the concentration of dissolved oxygen goes down, as shown in Figure 16.16. In extreme cases, more oxygen is required than can be replenished by air dissolving in the water, and fish and other freshwater organisms can no longer survive. The aerobic bacteria, those that require oxygen for the decomposition process, also die. The death of these organisms produces even more lifeless organic matter, and the BOD soars. However, nature has a backup system for such conditions. A whole new group of microorganisms, the anaerobic bacteria, take oxygen from oxygen-containing compounds to convert organic matter to CO_2 and H_2O, also converting organic nitrogen and nitrates to elemental nitrogen.

Anaerobic bacteria cause organic matter to decay in the absence of oxygen; they produce foul-smelling gases such as H_2S in a process called putrefaction.

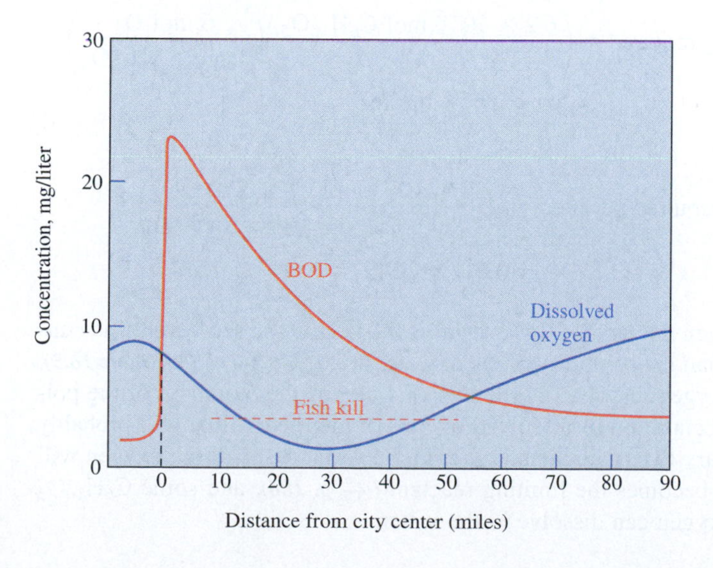

Figure 16.16 **Oxygen content and oxidizable nutrients (BOD).** As city sewage is introduced into a river flowing at a rate of 750 gal/s, the BOD soars. Note that it takes about 90 miles for the river to recover to a normal oxygen content.

BOD values can be greatly reduced by using oxygen or ozone, or both, to pretreat industrial wastes and sewage before releasing them into a body of water. Numerous commercial cleanup processes, both in use and under development, employ this type of "burning" of organic wastes. Another benefit of treating wastewater with oxygen is that some of the nonbiodegradable material becomes biodegradable as a result of partial oxidation.

Conceptual Challenge Problem CP-16.C at the end of the chapter refers to topics covered in this section.

A substance is **biodegradable** if it can be broken down by microorganisms. Some organic compounds are **nonbiodegradable,** presumably because their structures are such that microorganisms cannot use them for food, or because they are so toxic that they kill the microorganisms. Interestingly, some microorganisms can "eat" certain highly toxic substances.

One way to determine the quantity of organic pollution in a sample of water is to measure the BOD. Usually, a known volume of the polluted water is diluted with a known volume of standardized sodium chloride solution that has a known oxygen content. After some bacteria are added, the mixture is held in a closed bottle at 20 °C for 15 days. (Closing the bottle prevents additional oxygen from dissolving in the solution as O_2 is consumed in the reaction.) At the end of this time the quantity of oxygen that has been consumed is taken to be the biochemical oxygen demand.

For example, suppose that a stream (at 20 °C) contains 10. ppm by weight (just 0.0010%) of an organic pollutant, the formula of which is $C_6H_{10}O_5$. What is the biochemical oxygen demand caused by this pollutant?

The mass of $C_6H_{10}O_5$ present in 1 L of this water is

$$\left(\frac{10.\text{ g } C_6H_{10}O_5}{10^6 \text{ g } H_2O}\right)\left(\frac{1000 \text{ g } H_2O}{1 \text{ L}}\right) = 0.010 \text{ g}$$

To transform this pollutant to CO_2 and H_2O, the bacteria present would use oxygen according to the equation

$$\underset{\text{pollutant}}{C_6H_{10}O_5(aq)} + 6\ O_2(g) \longrightarrow 6\ CO_2(g) + 5\ H_2O(\ell)$$

The concentration of $C_6H_{10}O_5$ in moles per liter is

$$\text{Concentration of } C_6H_{10}O_5 = \left(\frac{0.010 \text{ g } C_6H_{10}O_5}{\text{L}}\right)\left(\frac{1 \text{ mol } C_6H_{10}O_5}{162 \text{ g } C_6H_{10}O_5}\right)$$

$$= \frac{6.2 \times 10^{-5} \text{ mol } C_6H_{10}O_5}{\text{L}}$$

Using the stoichiometric factor from the balanced equation (**← *p. 180***) gives the concentration of oxygen required and the corresponding mass.

$$\text{Concentration of } O_2 \text{ required} = \left(\frac{6.2 \times 10^{-5} \text{ mol } C_6H_{10}O_5}{\text{L}}\right)\left(\frac{6 \text{ mol } O_2}{1 \text{ mol } C_6H_{10}O_5}\right)$$

$$= \frac{3.7 \times 10^{-4} \text{ mol } O_2}{\text{L}}$$

$$\text{Mass of } O_2 \text{ required per liter} = \left(\frac{3.7 \times 10^{-4} \text{ mol } O_2}{\text{L}}\right)\left(\frac{32.0 \text{ g } O_2}{1 \text{ mol } O_2}\right)$$

$$= 0.012 \text{ g } O_2/\text{L}$$

Characteristic BOD levels (g O_2/L):	
Untreated municipal sewage	0.1–0.4
Barnyard runoff	0.1–10
Food-processing wastes	0.1–10

The biochemical oxygen demand of this stream is 0.012 g O_2/L. However, this concentration is greater than the solubility of O_2 in water at 20 °C (**← *p. 725, Table 16.3***). Even if some more oxygen dissolves in the liter of water as the oxidation of the pollutant occurs, the concentration of dissolved oxygen at any given time will probably fall below that necessary to sustain aquatic life. In the worst case, all the oxygen will be used up, that is, it becomes the limiting reactant (**← *p. 186***), and some $C_6H_{10}O_5$ remains until more oxygen can dissolve in the water.

Exercise 16.14 Calculating BOD

Calculate the BOD for 1000. mL of water at 10 °C containing 2.5 ppm of a pollutant with the formula C_2H_6O. Express the answer in grams of O_2. Is this value greater or less than the solubility of O_2 in water at this temperature? Explain the consequences of this value.

Exercise 16.15 Estimating Environmental Consequences

Which would be the better environmental scenario? Water containing 10 ppm of dissolved organic compounds at 22 °C, or water with the same concentration of dissolved organics at 15 °C? Explain your answer.

The Impact of Industrial and Household Wastes on Water Quality

Industry produces a wide variety of wastes along with its intended products, as do the activities of everyday life. It was once considered good engineering practice to put all wastes into landfills. But many of the waste compounds were partially dissolved by rainfall and leached into the groundwater, causing serious pollution of water supplies. Today, hazardous industrial wastes either must be placed in secure landfills, incinerated, or treated in some way to render them nonhazardous. Secure landfills usually have plastic linings to prevent leaching (Figure 16.17), are built on a thick layer of impervious clay, and have carefully spaced monitor wells for detecting any leaks.

We often don't think about what we throw away, or how our household wastes can affect groundwater, lakes, rivers, and coastlines. Table 16.6 (p. 738) lists some common household products and the kinds of chemicals they contain. The bulk of this waste still goes into municipal landfills. When this happens, we, as consumers of industrial products, are putting into our groundwater the very same chemicals that industry is required to clean up. Households, however, have a greater problem disposing of hazardous chemicals than does industry. Although many cities have active recycling programs for glass, paper, metals, and plastics, most municipalities have no provision for picking up chemical wastes separately from ordinary trash, most of which goes to landfills.

You should never dispose of chemical waste by pouring it directly on the ground. Eventually it will be washed into a natural body of water or into groundwater.

(a)

(b)

Figure 16.17 Plastic linings, required for all landfills. (a) This landfill, under construction, will have a capacity of millions of cubic yards of waste when completed. (b) The thick plastic is sealed so it will not allow leakage of contaminants into the groundwater.

TABLE 16.6 Some Common Household Hazardous Wastes and Recommended Disposal

Type of Product	Harmful Ingredients	Disposal[1]
Bug sprays	Pesticides, organic solvents	Special
Oven cleaner	Strong bases	Drain
Bathroom cleaners	Acids or strong bases	Drain
Furniture polish	Organic solvents	Special
Aerosol cans (empty)	Solvents, propellants	Trash
Nail-polish remover	Organic solvents	Special
Nail polish	Solvents	Trash
Antifreeze	Organic solvents, metals	Special
Insecticides	Pesticides, solvents	Special
Auto battery	Sulfuric acid, lead	Special
Medicine (expired)	Organic compounds	Drain
Paint (latex)	Organic polymers	Drain
Gasoline	Organic solvents	Special
Motor oil	Organic compounds, metals	Special
Drain cleaners	Strong bases	Drain
Shoe polish	Waxes, solvents	Trash
Paints (oil-based)	Organic solvents	Special
Mercury batteries	Mercury	Special
Moth balls	Chlorinated organic compound	Special
Batteries	Heavy metals such as Hg	Special

[1]Special: Professional disposal as a hazardous waste. Drain: Disposal down the kitchen or bathroom drain. Trash: Treat as normal trash—no harm to the groundwater. Unfortunately, in most households, the items marked special are disposed of as normal trash, which can result in groundwater pollution.

Source: "Household Hazardous Waste: What You Should and Shouldn't Do," Water Pollution Control Federation.

Recycling of materials such as glass, paper, metals, and plastics helps conserve energy resources and raw materials. Recycling also conserves valuable landfill space and keeps some otherwise harmful chemicals out of groundwater.

The EPA estimates that each year 350 million gallons of waste oil are poured on the ground or flushed down the drain. That's 35 times more oil than the Exxon Valdez spilled in Alaska.

Ordinary garbage costs about $27 per ton for disposal, but proper disposal of hazardous waste costs about $1000 per ton.

How can we dispose of hazardous household wastes without increasing the risks to our water supply? We can ask our local waste authorities to provide disposal sites for such wastes. In some cities in the United States and Europe, special trucks pick up paint, oil, batteries, and other products for disposal. Increasingly, municipalities sponsor hazardous waste disposal days on which citizens can bring materials to a central location. Another approach is to consider the pollution potential of products when deciding what to buy. For example, alkaline batteries often work just as well as mercury batteries (see Section 18.7), but their ingredients are less toxic. When you change the oil in your automobile, buy from a merchant who will accept the used oil and dispose of it properly.

16.7 NATURAL IMPURITIES IN WATER: HARD WATER

The presence of Ca^{2+}, Mg^{2+}, Fe^{3+}, or Mn^{2+} imparts "hardness" to waters. Hardness in water is objectionable because (1) it causes precipitates (called scale) to form in boilers and hot-water systems; (2) it causes soaps to form insoluble curds (this reaction does not occur with some synthetic detergents; see Section 16.12); and (3) it imparts a disagreeable taste to the water.

Hardness due to calcium or magnesium ions is produced when water containing carbon dioxide trickles through limestone or dolomite (← *p. 143*):

$$\underset{\text{limestone}}{CaCO_3(s)} + CO_2(g) + H_2O(\ell) \longrightarrow Ca^{2+}(aq) + 2\ HCO_3^-(aq)$$

$$\underset{\text{dolomite}}{CaCO_3 \cdot MgCO_3(s)} + 2\ CO_2(g) + 2\ H_2O(\ell) \longrightarrow Ca^{2+}(aq) + Mg^{2+}(aq) + 4\ HCO_3^-(aq)$$

Such hard water can be softened by removing these ions. The principal methods for softening water are the (1) lime-soda process and (2) ion-exchange processes.

The lime-soda process works because calcium carbonate ($CaCO_3$) is much less soluble than calcium bicarbonate ($Ca(HCO_3)_2$) and magnesium hydroxide ($Mg(OH)_2$) is much less soluble than magnesium bicarbonate ($Mg(HCO_3)_2$). In this process, hydrated lime [$Ca(OH)_2$] and soda (Na_2CO_3) are added to the water. Several reactions take place, which can be summarized as follows:

Soft water: < 65 mg of metal ion/gal
Slightly hard: 65–228 mg
Moderately hard: 228–455 mg
Hard: 455–682 mg
Very hard: > 682 mg

In hard water   Added

$$HCO_3^-(aq) + OH^-(aq) \longrightarrow CO_3^{2-}(aq) + H_2O(\ell)$$

$$Ca^{2+}(aq) + CO_3^{2-}(aq) \longrightarrow CaCO_3(s)$$

$$Mg^{2+}(aq) + 2\ OH^-(aq) \longrightarrow Mg(OH)_2(s)$$

For persons on low-sodium diets, lime-soda–treated water might represent too high a daily dose of sodium.

The overall result of the lime-soda process is to precipitate almost all the calcium and magnesium ions and to leave sodium ions as replacements.

Iron present as Fe^{2+} and manganese present as Mn^{2+} can be removed from water by oxidation with air (aeration) to higher oxidation states. If the water is neutral or slightly alkaline (either naturally or from the addition of lime), insoluble compounds $Fe(OH)_3$ and $MnO_2(H_2O)_x$ are produced and precipitate from solution.

Ion exchange is another way to remove ions causing water hardness. A cation exchange resin will replace ions causing water hardness with either H^+ ions or Na^+ ions. Home water-treatment ion-exchange units usually replace hardness ions with Na^+ ions. The exchange resin is a polymer containing numerous negatively charged $—SO_3^-$ functional groups that have either H^+ or Na^+ ions attached. These $—SO_3^-$ groups will lose their H^+ or Na^+ ions when larger, positive ions like Mg^{2+} and Ca^{2+} happen to pass by. As the hard water is allowed to flow over the resin, the process happens many times and the result is water that contains H^+ ions or Na^+ ions in place of the Ca^{2+} and Mg^{2+} ions. Two $—SO_3H$ groups are required for every 2+ ion that is removed from solution. Also note that two H^+ ions or two Na^+ ions are produced for every 2+ ion removed. When H^+ ions are involved, this makes the resulting solution somewhat more acidic.

$$(\text{Polymer}-SO_3H)_2 + Ca^{2+}\ (\text{or } Mg^{2+}) \longrightarrow (\text{Polymer}-SO_3^-)_2Ca^{2+} + 2\ H^+$$

When Na^+ ions are involved, the concentration of Na^+ ions in the water increases as the ions causing water hardness are removed.

$$(\text{Polymer}-SO_3Na)_2 + Ca^{2+}\ (\text{or } Mg^{2+}) \longrightarrow (\text{Polymer}-SO_3^-)_2Ca^{2+} + 2\ Na^+$$

An ion-exchange resin used this way will become saturated when all of its $—SO_3H$ or SO_3Na groups have lost their singly positive ions. The resin can be regenerated by treating it with either an acid solution or one containing a high concentration of some dissolved sodium salt like NaCl, and thereby reversing the reaction shown above.

The production of soft water for domestic use by ion exchange has sparked a rather heated health debate during the past two decades. Depending on the ion-exchange method used, the soft water will be slightly acidic or contain Na^+ ions. An increased intake of Na^+ is known to be related to heart disease. Also, acidic soft water is more likely to attack metallic pipes, joints, and fixtures than less acidic water. If pipes and joints contain lead, then Pb^{2+} can enter the water. One way to overcome both of these problems and still use less soap when washing would be to drink only naturally hard water and do your washing in soft water. Many houses have plumbing that delivers unsoftened water at the kitchen sink.

16.8 PURIFICATION OF MUNICIPAL WASTEWATER

Sewage includes everything that flows from the sinks, tubs, washing machines, and toilets in our homes, factories, and public buildings. It excludes wastewater treated separately by industrial facilities.

The earliest city sewage systems did little but channel wastewater to rivers and streams where natural purification processes were expected to clean the water for the next users downstream. Today, however, sewage is treated bv a combination of methods that can render the water almost as clean as the natural waters into which it is discharged. The first stage of wastewater treatment, known as *primary* wastewater treatment, copies two of nature's purification methods: filtration to remove large debris, followed by settling (Figure 16.18).

In the settling stage, calcium hydroxide and aluminum sulfate are added to produce aluminum hydroxide, which is a sticky, gelatinous precipitate that settles out slowly, carrying suspended dirt particles and bacteria with it. Primary treatment removes 40–60% of the solids present in sewage and about 30% of the organic matter present.

$$3\ Ca(OH)_2(aq) + Al_2(SO_4)_3(aq) \longrightarrow 2\ Al(OH)_3(s) + 3\ CaSO_4(s)$$

PROBLEM-SOLVING EXAMPLE 16.7 Using K_{sp}

The K_{sp} for $Al(OH)_3$ is 1.9×10^{-33}. If the OH^- concentration of a wastewater solution is raised to 2×10^{-2} M by the addition of lime, use the K_{sp} to estimate the Al^{3+} ion concentration after $Al(OH)_3$ has formed [K_{sp} $Al(OH)_3 = 1.9 \times 10^{-33}$].

Answer $[Al^{3+}] = 2 \times 10^{-28}$ mol/L

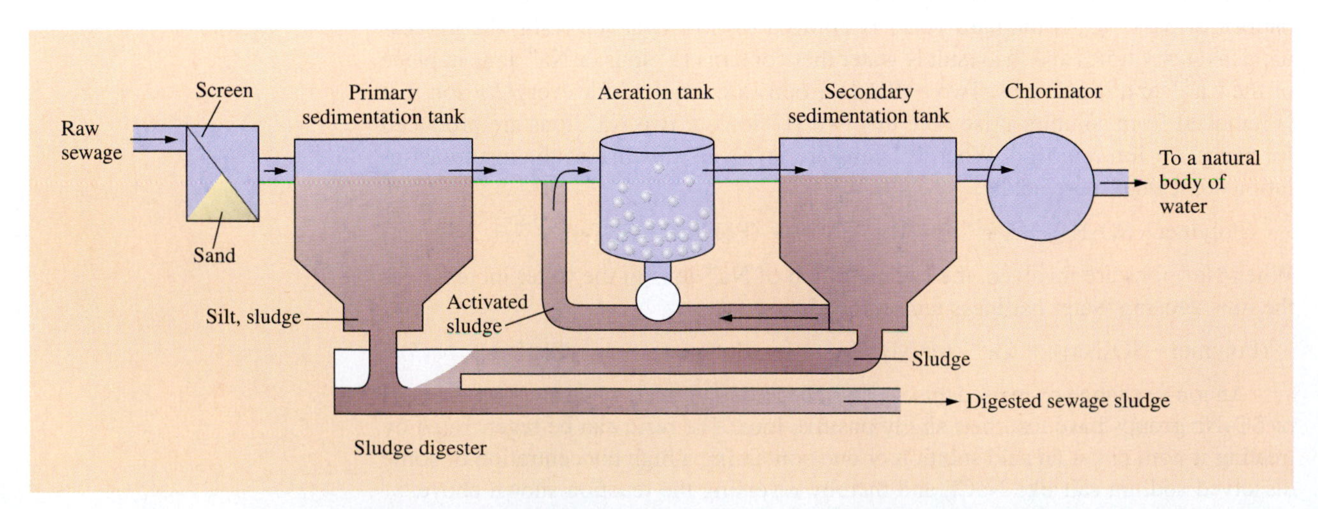

Figure 16.18 The steps in primary and secondary sewage treatment. See text for explanation of each step.

Explanation This is a common ion problem. You are finding the solubility of aluminum hydroxide in the presence of 2×10^{-2} M OH^- ion, furnished by the $Ca(OH)_2$. The equilibrium expression for $Al(OH)_3$ is

$$Al(OH)_3(s) \rightleftharpoons Al^{3+}(aq) + 3\ OH^-(aq)$$

and the K_{sp} expression is

$$K_{sp} = [Al^{3+}][OH^-]^3$$

	$Al(OH)_3(s) \rightleftharpoons$	$Al^{3+}(aq)$	+ $3\ OH^-(aq)$
Initial Concentration (mol/L)		0	2×10^{-2}
Change as Reaction Occurs (mol/L)		$+S$	$+3\ S$
Equilibrium Concentration (mol/L)		S	$3\ S + 2 \times 10^{-2}$

The value for the K_{sp} of $Al(OH)_3$ is small (1.9×10^{-33}), so the OH^- concentration from the dissolving of $Al(OH)_3$ would be quite small. In effect, the OH^- concentration we can use is that supplied by the lime. Solving for the aluminum ion concentration and substituting the values for K_{sp} and $[OH^-]$, we get

$$[Al^{3+}] = \frac{K_{sp}}{[OH^-]^3} = \frac{1.9 \times 10^{-33}}{(2 \times 10^{-2})^3} = 2 \times 10^{-28}$$

This low value for the Al^{3+} ion concentration means virtually all of the added aluminum is precipitated in the form of $Al(OH)_3$.

Problem-Solving Practice 16.7

Calculate the solubility of $Al(OH)_3$ when the concentration of OH^- furnished by lime is (a) 1.0×10^{-4} M, and (b) 1.0×10^{-5} M. Compare your answers with that in Problem-Solving Example 16.7. What trend do you observe?

For many years municipal sewage-treatment plants used only primary treatment followed by chlorination of the treated wastewater before discharging it into a suitable river or stream. This chlorination killed any pathogens (disease-causing organisms). Presumably natural processes would get rid of the remaining solids and dissolved organic matter.

Realizing that primary treatment alone is not sufficient to protect the public from contaminated water, the government included in the original Clean Water Act (1972) the requirement that sewage-treatment plants also provide *secondary* wastewater treatment. Secondary treatment operates in an oxygen-rich (aerobic) environment in which the organic molecules remaining after primary treatment are broken down by microorganisms and the resulting sludge settles out in a secondary sedimentation tank (Figure 16.18).

Even a combination of primary and secondary wastewater treatment will not remove dissolved inorganic materials such as toxic metal ions, nutrients like nitrate ions (NO_3^-) or ammonium ions (NH_4^+), or nonbiodegradable organic compounds like chlorinated hydrocarbons. These materials are removed by a variety of *tertiary* water treatments. One effective tertiary treatment is carbon *adsorption*. Carbon black, which consists of finely divided carbon particles with a large surface area, has been used for many years for adsorbing vapors and solute materials from liquid streams. Many

Carbon black is used in many processes to manufacture extremely pure organic compounds, such as those used as pharmaceuticals and food additives. Adsorption is the process by which molecules are attracted and held onto a surface.

toxic organic materials can also be removed by carbon adsorption, but this treatment method is expensive and fails to remove metal ions and the nutrient ions.

A different kind of treatment is needed to remove ammonia or ammonium ion. Because nitrogen is a nutrient for aquatic microorganisms, excessive nitrogen released to natural waters can cause a high BOD. By using *denitrifying bacteria,* ammonia and ammonium ions can be removed. These bacteria convert ammonium ions (or ammonia) to nitrogen gas.

$$NH_4^+(aq) \text{ or } NH_3(aq) \xrightarrow{\text{denitrifying bacteria}} N_2(g)$$

Chlorination of Water to Remove Bacteria

Because chlorine is a powerful oxidizing agent, it kills bacteria in the water. Chlorine is introduced into water as the gaseous element (Cl_2). Chlorination was first used for drinking-water supplies in the early 1900s, with a resulting drop in the number of deaths in the United States caused by typhoid and other water-borne diseases from 35/100,000 population in 1900 to 3/100,000 population in 1930. Today, chlorination is used not only to purify drinking-water supplies, but also as a final treatment for municipal wastewater. Chlorination is the principal means of preventing water-borne diseases spread by bacteria, including cholera, typhoid, paratyphoid, and dysentery.

In spite of chlorination, most city water supplies are not bacteria-free, but only rarely do these surviving bacteria cause disease. In the United States the most common water-borne bacterial disease is giardiasis, a gastrointestinal disorder. Most often this disease comes from surface water that has leaked into drinking-water supplies, but, on occasion, it can be traced to city water systems.

Water chlorination not only kills bacteria, but it also produces *disinfection by-products* by the reaction of chlorine with residual concentrations of organic compounds.

$$\text{Water containing organic compounds} \xrightarrow{\text{chlorine}} \text{disinfection by-products}$$

These disinfection by-products, which may be present at levels of a few parts per million or less, include dichloromethane, chloroform, trichloroethylene, and chlorobenzene, all suspected carcinogens. A number of these compounds have been shown to be mutagenic to salmonella bacteria. The presence of these chlorinated hydrocarbons can be prevented by more efficient removal of the organic matter which becomes chlorinated. Unfortunately, even the best designed purification systems, including carbon adsorption, allow some organic compounds to pass through, only to become chlorinated. Information is still being evaluated about the seriousness of this potential threat.

16.9 VAPOR PRESSURES, BOILING POINTS, AND FREEZING POINTS OF SOLUTIONS

Up to this point, solutions have been discussed in terms of the nature of the solute and the nature of the solvent. There are some properties of solutions themselves that do not depend on the nature of the solute or solvent, but rather depend only on the *number* of dissolved solute particles.

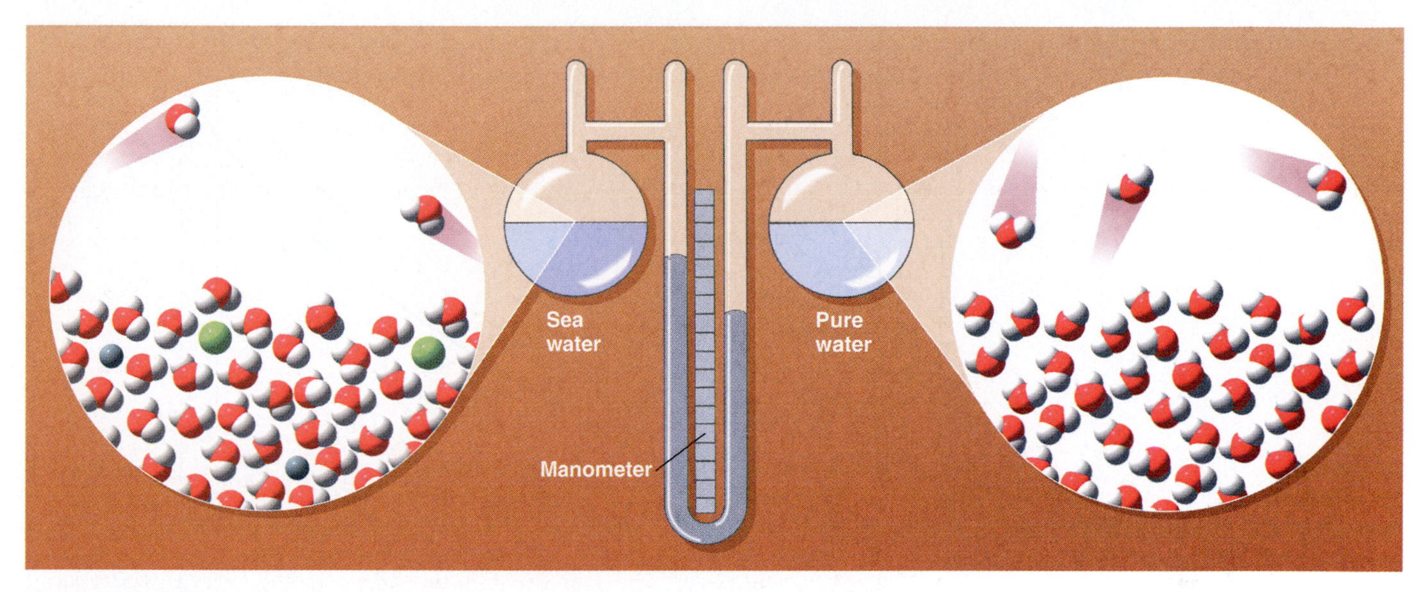

Figure 16.19 Sea water. Sea water is an aqueous solution of sodium chloride and many other salts. The vapor pressure of water over an aqueous solution is not as large as the vapor pressure of water over pure water at the same temperature.

In liquid solutions, particles are close together. Solute molecules or ions disrupt intermolecular forces between the solvent molecules, causing changes in solvent properties that depend on intermolecular attractions. For example, the freezing point of a solution is lower and its boiling point is higher than that of the pure solvent. The magnitude of these changes in the properties of the solution compared to those of the pure solvent depend only on the concentration of the solute particles. Properties of solutions that *depend only on the concentration of solute particles* in the solution, regardless of what kinds of particles are present, are called **colligative properties.** We will consider four colligative properties: vapor pressure lowering, boiling point elevation, freezing point depression, and osmotic pressure. These are all quite common and important in the world around us.

Vapor Pressure Lowering

Compare a small portion of the liquid/gas boundary for pure water with that for sea water (an aqueous solution of mainly sodium chloride) as shown at the molecular scale in Figure 16.19. Recall from Section 15.2 (← *p. 665*) that in a closed container there is a dynamic equilibrium between a pure liquid and its vapor—the rate at which molecules escape the liquid phase equals the rate at which vapor-phase molecules return to the liquid, and this equilibrium gives rise to a vapor pressure that is dependent upon the temperature. For an aqueous solution like sea water, where sodium ions, chloride ions, and many other kinds of ions and molecules are present, the vapor pressure is observed to be lower than for a sample of pure water. (Note the reading on the manometer in Figure 16.19.) This can be explained by comparing the entropy change for vaporization of pure water with that for vaporization of a corresponding quantity of water from the solution. Salt has a very low vapor pressure, and so very few sodium or chloride ions escape from the solution. The vapor in equilibrium with

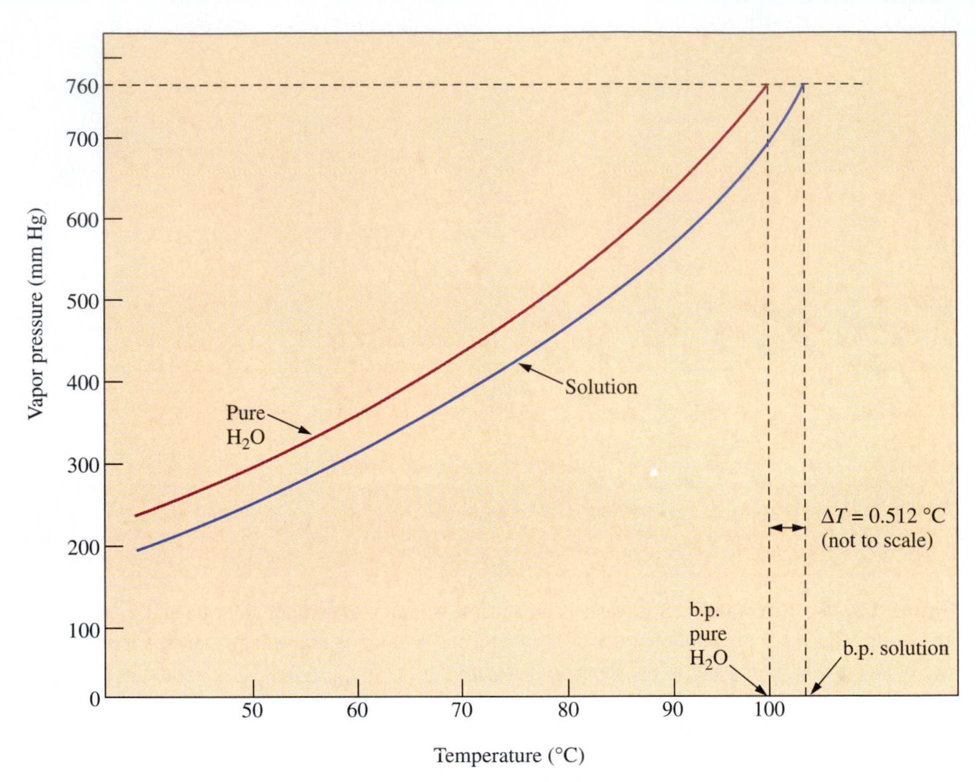

Figure 16.20 Vapor pressure lowering. The lowering of the vapor pressure of a pure solvent (H_2O) by the addition of a nonvolatile solute. The vapor pressure of the solution (blue line) is lower at every temperature than that of the pure solvent (red line). The pure solvent boils when its vapor pressure equals 760 mm Hg (100 °C); the solution must be raised to a slightly higher temperature to have the same vapor pressure.

both pure water and salt water consists almost entirely of water molecules. Therefore the entropy of a given quantity of vapor is the same in both cases. But the entropy of the solution is greater than that of the pure solvent, because the solution is more disordered. As with any change from liquid to gas, there is a large increase in entropy for both pure solvent and solution ***(← p. 275)***, but the entropy increase is bigger for vaporization of the solvent because the pure solvent's entropy was smaller to begin with. A bigger increase in entropy corresponds to a more product-favored process, and so vaporization of pure solvent results in a higher pressure of gas in equilibrium with the liquid than does vaporization of the salt water solution. The lower vapor pressure of the solution compared to the pure solvent is shown graphically in Figure 16.20.

The enthalpy of vaporization is roughly the same for solution and solvent, and thus the total entropy change ***(← p. 280)*** for the universe follows the entropy change for vaporization.

Boiling Point Elevation

As a result of vapor pressure lowering, the vapor pressure of an aqueous solution of a nonvolatile solute at 100 °C is less than 1 atm. The solution will therefore have to be heated *above* 100 °C for it to boil. The difference between the normal boiling point of water and the higher boiling point of an aqueous solution of a nonvolatile nonelectrolyte solute is known as the **boiling point elevation,** ΔT_b, which is proportional to the concentration of the solute.

$$\Delta T_b = K_b\, m_{\text{solute}}$$

The concentration unit used here is **molality** (abbreviated *m*), defined as the moles of solute per kilogram of *solvent.*

$$\text{Molality of compound A} = m_{\text{A}} = \frac{\text{amount (moles) of A}}{\text{mass (kg) of solvent}}$$

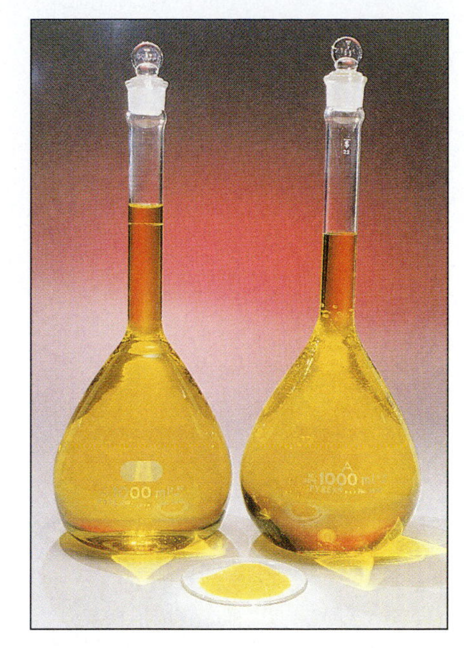

Molarity and molality. The photo shows a 0.10-molal solution (0.10 m) of potassium chromate (*flask at left*) and a 0.10-molar solution (0.10 M) of potassium chromate (*flask at right*). Each solution contains 0.10 mol (19.4 g) of yellow K_2CrO_4 shown in the dish at the front. The 0.10-molar (0.10 M) solution on the right was made by placing the solid in the flask and adding enough water to make 1.0 L of solution. The 0.10-molal (0.10 m) solution on the left was made by placing the solid in the flask and adding 1000 g (1 kg) of water. Adding 1 kg of water leads to a solution clearly having a volume greater than 1 L. *(C.D. Winters)*

Molality and molarity (← **p. 198**) are not the same (although the difference becomes negligibly small for dilute solutions, say less than 0.01 M).

PROBLEM-SOLVING EXAMPLE 16.8 Molality

What is the molality of a solution prepared by dissolving 4.13 g of methanol (CH_3OH) in 500. g of water?

Answer 0.253 *m*

Explanation First, calculate the number of moles of the solute, methanol. Its molar mass is 32.042 g/mol.

$$\text{Moles of methanol} = 4.13\text{ g} \times \left(\frac{1\text{ mol}}{32.042\text{ g}}\right) = 0.129\text{ mol}$$

The mass of solvent is 0.500 kg, so by the definition of molality,

$$\text{Molality of the solution} = \frac{0.129\text{ mol methanol}}{0.500\text{ kg water}} = 0.258\ m$$

Problem-Solving Practice 16.8

Calculate the molality of a solution that is prepared by dissolving 6.58 g of NaCl in 250. mL of water.

Exercise 16.16 Molarity and Molality

The molarity of a solution varies as temperature changes, but the molality of a solution does not. Explain why.

For solutions of electrolytes, the boiling point elevation equation can also be written as

$$\Delta T_b = K_b\, m_{\text{solute}}\, i_{\text{solute}}$$

The *i factor* gives the number of particles per formula unit of the solute and is called the van't Hoff factor, named after Jacobus Henricus van't Hoff (1852–1911), who won the very first Nobel Prize in chemistry, in 1901, for his work on the colligative properties of solutions. The value of *i* is 1 for molecular solutes that do not dissociate in solution (nonelectrolytes). Examples of nonelectrolytes include ethanol and sucrose. For soluble ionic solutes (strong electrolytes), *i* equals the number of ions per formula unit of the ionic compound. For sodium chloride, $i_{\text{solute}} = 2$, because there is one Na^+ ion and one Cl^- ion in solution per formula unit of NaCl. The proportionality constant, K_b, depends only on the identity of the solvent and not the kind of solute, because it makes almost no difference whether the particles among water molecules at the surface of a liquid are Na^+ and Cl^- (as in Figure 16.19) or some other kind of molecules or ions. With any solute, the concentration of water molecules is less than in pure water. Boiling point elevation is a colligative property because only the moles of solute particles are important, not their identity.

PROBLEM-SOLVING EXAMPLE 16.9 Boiling Point Elevation

What is the boiling point of a 0.20-*m* solution of $CaCl_2$ in water? For water, K_b is 0.512 °C · kg/mol.

Answer 100.31 °C

Explanation For $CaCl_2$ dissolved in water at a concentration of 0.20 *m* (= 0.20 mol/kg), the boiling point of the solution can be obtained by first calculating the boiling point elevation, and then adding that value to the boiling point for water. Since there are two Cl^- ions and one Ca^{2+} ion per formula unit of $CaCl_2$,

$$CaCl_2(s) \longrightarrow Ca^{2+}(aq) + 2\ Cl^-(aq)$$

the value of i_{solute} in the boiling point elevation equation is 3. The value for ΔT_b can be calculated as follows:

$$\Delta T_b = (0.512\ °C \cdot kg/mol) \times (0.20\ mol/kg) \times 3 = +0.31\ °C$$

Adding the boiling point elevation to the boiling point of water gives the boiling point of the solution:

$$100.00\ °C + 0.31\ °C = 100.31\ °C$$

The units for the molal boiling point elevation constant are abbreviated °C · kg/mol, which means (°C · kg solvent)/(mol solute).

Problem-Solving Practice 16.9

What is the boiling point of a 0.467-*m* solution of ethylene glycol in water?

Exercise 16.17 Calculating the Boiling Point of a Solution

The boiling point elevation constant for benzene is 2.53 °C · kg/mol. If a solute's concentration in benzene is 0.10 *m,* what will be the boiling point of the solution? The boiling point of pure benzene is 80.10 °C. There is a single solute particle per formula unit.

Freezing Point Lowering

A pure liquid begins to freeze when the temperature is lowered to just below the substance's freezing point and the first few molecules cluster together into a crystal lattice to form a tiny quantity of solid. As long as both solid and liquid phases are present and the temperature is at the freezing point, the rate of crystallization equals the rate of melting and there is a dynamic equilibrium. When a *solution* freezes, a few molecules of solvent cluster together to form pure solid *solvent* (Figure 16.21), and a dynamic equilibrium is set up between solution and solid solvent.

A practical application of freezing point lowering is in the use of salt (NaCl) added to ice to make a freezing mixture that results in a tasty, homemade ice cream. Lowering the freezing temperature of the ice-water-salt mixture causes the ice cream ingredients to freeze more quickly.

In the case of a solution, the molecules in the liquid in contact with the solid solvent are not all solvent molecules. This causes the rate at which molecules move from solution to solid to be slower than in the pure liquid. To achieve dynamic equilibrium, there must be a corresponding slower rate of escape of molecules from the solid

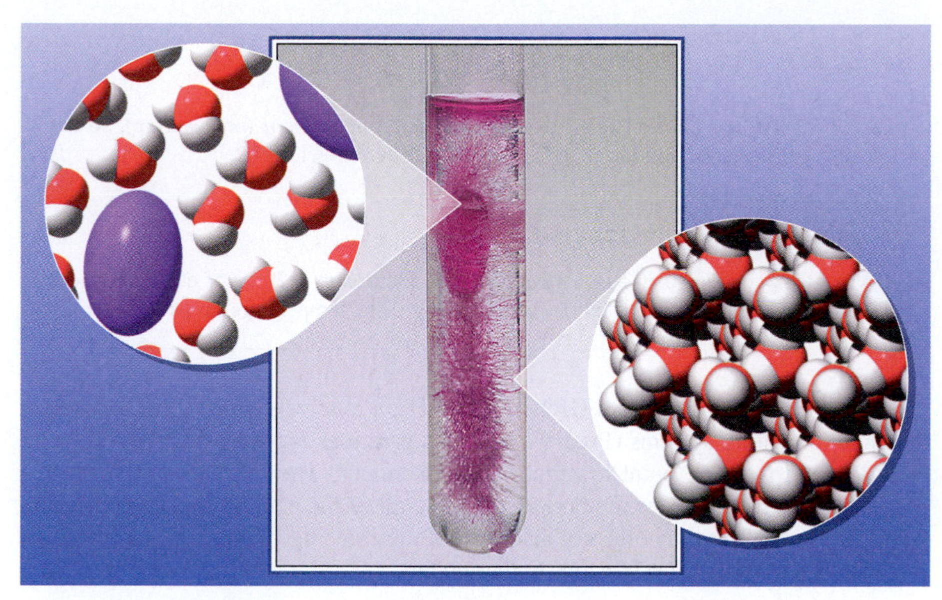

Figure 16.21 Solvent freezing. A purple dye was dissolved in water, and the solution was frozen slowly. When a solution freezes, the solvent solidifies as the pure substance. Thus, pure ice forms along the walls of the tube, and the dye stays in solution. As more and more solvent is frozen out, the solution becomes more concentrated and the resulting solution has a lower and lower freezing point. When equilibrium is reached, we see pure, colorless ice along the walls of the tube with concentrated solution in the center of the tube. *(C.D. Winters)*

crystal lattice. According to the kinetic-molecular theory, this slower rate occurs at a lower temperature, and so the freezing point of the solution is lower than that of the pure liquid solvent.

Conceptual Challenge Problem CP-16.D at the end of the chapter relates to topics covered in this section.

The **freezing point lowering,** ΔT_f, is proportional to the concentration of the solute in the same way as the boiling point elevation.

$$\Delta T_f = K_f \, m_{\text{solute}} \, i_{\text{solute}}$$

Here also, the proportionality constant, K_f, depends on the solvent and not the kind of solute, and i_{solute} represents the number of particles per formula unit of solute. For water, the freezing point constant is $-1.86\ °C \cdot kg/mol$.

PROBLEM-SOLVING EXAMPLE 16.10 Freezing Point Lowering

What is the freezing point of an aqueous solution that contains 39.5 g of ethylene glycol dissolved in 1.00 kg of water?

Answer $-1.18\ °C$

Explanation Ethylene glycol ($C_2H_6O_2$) is a molecular substance, so the *i* factor is 1. Since the mass of ethylene glycol dissolved in 1.00 kg of water is given, we must calculate the molality of the solution. The molar mass of ethylene glycol is 62.068 g/mol, and the number of moles of ethylene glycol is

$$\text{Moles ethylene glycol} = 39.5\ \text{g} \times \left(\frac{1\ \text{mol}}{62.068\ \text{g}}\right) = 0.636\ \text{mol}$$

Since the solution contains 0.636 mol dissolved in 1.00 kg of water, the molality of the solution is 0.636 *m*.

Figure 16.22 Lowering the freezing point of a solution. Adding an ethylene glycol–based antifreeze to an automobile's cooling system lowers the freezing point of the solution and raises its boiling point. *(C.D. Winters)*

Ethylene glycol is toxic, so do not allow it to contact drinking-water supplies.

The freezing point lowering is

$$\Delta T_f = (-1.86\ °C/m)(0.636\ m)(1) = -1.18\ °C$$

This change in the freezing point is a freezing point *depression* (or lowering), so the freezing point of this solution is 0.00 °C + (−1.18 °C) = −1.18 °C.

Problem-Solving Practice 16.10

A water tank contains 6.50 kg of water. Will the addition of 1.20 kg of ethylene glycol be sufficient to prevent the solution from freezing if the temperature drops to −25 °C?

The use of ethylene glycol ($HOCH_2CH_2OH$), a relatively nonvolatile alcohol, in automobile cooling systems (Figure 16.22) is a practical application of boiling point elevation and freezing point lowering. In the summer, when the air temperature is high, ethylene glycol raises the boiling temperature of the coolant to a level that prevents "boil over." Ethylene glycol also lowers the freezing point of the coolant and protects the solution from freezing in the winter.

Exercise 16.18 Protection Against Freezing

Suppose that you are closing a cabin in the north woods for the winter and you do not want the pipes to freeze. You know that the temperature might get as low as −30. °C, and you want to protect about 4.0 L of water in a toilet tank from freezing. What volume of ethylene glycol, density = 1.113 g/mL, molar mass = 62.1 g/mol, should be added to the 4.0 L of water?

Another practical application of freezing point lowering can be seen in areas where winters produce lots of frozen precipitation. To remove snow and particularly ice, roads and walkways are often salted (Figure 16.23). Although sodium chloride is usually used, calcium chloride ($CaCl_2$) is particularly good for this purpose because it has three ions per formula unit and dissolves exothermically. Not only is the freezing point of water lowered, but the heat of solution helps melt ice. Table 16.7 lists some salts that are used or could be used for de-icing. Factors to consider in choosing a salt include cost, availability, number of ions per formula unit, and environmental effects.

Figure 16.23 Lowering the freezing point of ice. Salt truck applying road salt. *(C.D. Winters)*

TABLE 16.7 Salts That Could Be Used for Road De-icing

Name	Formula	No. Ions	Environmental Effects
Sodium chloride	NaCl	2	Cl^- harmful to plants
Calcium chloride	$CaCl_2$	3	Cl^- harmful to plants
Sodium acetate	$NaCH_3CO_2 \cdot 3\ H_2O$	2	Acetate ions are biodegradable.
Sodium nitrate	$NaNO_3$	2	Nitrate ions cause algal bloom in ponds and slow-moving streams.*

*Algae growing rapidly ("blooming") remove oxygen from water and thereby harm aquatic life.

Exercise 16.19 Choosing Road Salt

On a weight basis, which salt listed in Table 16.7 would be the most effective at lowering the freezing point of ice?

16.10 OSMOTIC PRESSURE OF SOLUTIONS

A *membrane* is a thin layer of material that allows molecules or ions to pass through it. A **semipermeable membrane** allows only certain kinds of molecules or ions to pass through while excluding others (Figure 16.24). Examples of semipermeable membranes are animal bladders, cell walls in plants and animals, and cellophane, a polymer derived from cellulose. When two solutions containing the same solvent are separated by a membrane permeable only to solvent molecules, osmosis will occur.

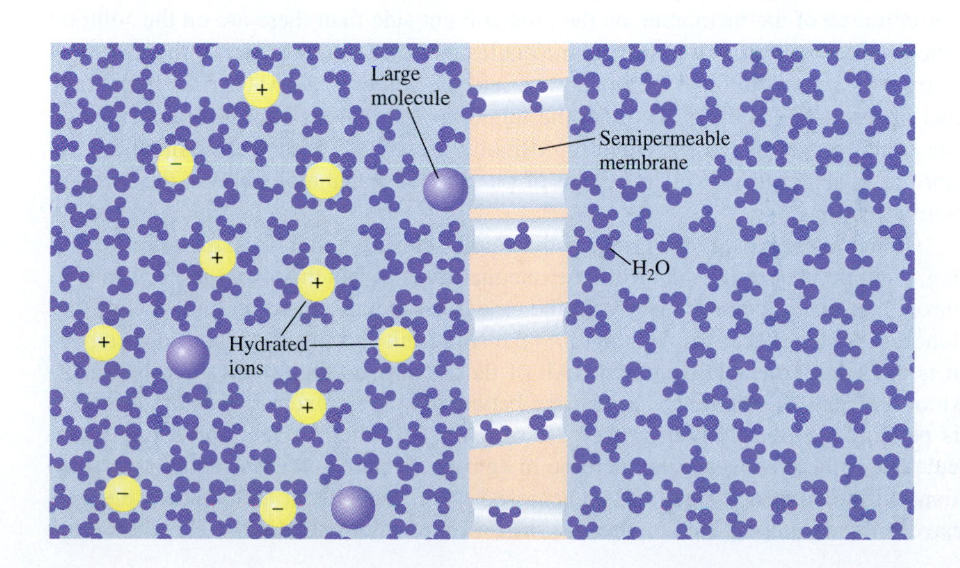

Figure 16.24 Osmotic flow of a solvent through a semipermeable membrane to a solution. The semipermeable membrane is shown acting as a sieve. Many membranes operate in a different way, but the ultimate effect is the same.

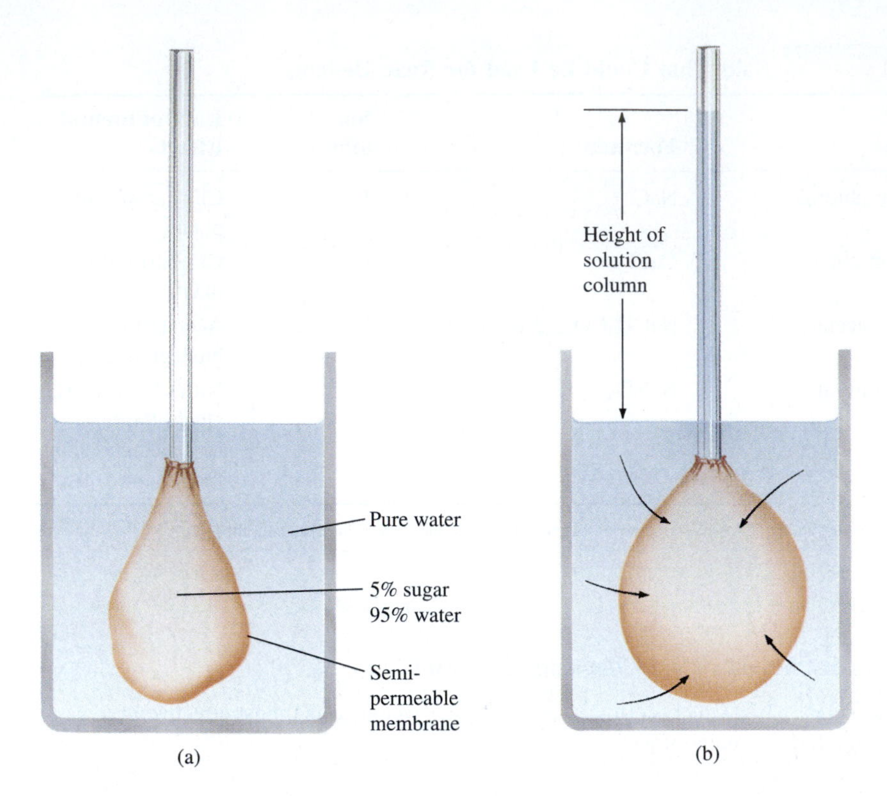

Figure 16.25 Osmosis. (a) The bag attached to the glass tube contains a solution that is 5% sugar and 95% water, and there is pure water in the beaker. The material from which the bag is made is selectively permeable (semipermeable) in that it will allow water molecules to pass through but not sugar molecules. (b) With time, water flows through the walls of the bag from a region of low solute concentration (pure water) to one of higher solute concentration (the sugar solution). Flow will continue until the pressure exerted by the column of solution in the tube above the water level in the beaker is large enough to result in equal rates of passage of water molecules in both directions. The height of the column of solution (b) is a measure of the osmotic pressure, Π.

If you think about it, a region of lower solute concentration has a higher solvent concentration, and so osmosis can be thought of as movement of solvent from a region of *higher* solvent concentration to a region of *lower* solvent concentration.

Osmosis is *movement of a solvent through a semipermeable membrane from a region of lower solute concentration to a region of higher solute concentration.* The **osmotic pressure** of a solution is *the pressure that must be applied to the solution to stop osmosis from a sample of pure solvent.*

Consider the example shown in Figure 16.25. A dilute aqueous solution of glucose has been placed in a cellophane bag attached to a glass tube. When the bag is submerged in pure water, water flows into it by osmosis and raises the liquid level in the tube. The net flow of water stops when the pressure at the bottom of the tube is equal to the osmotic pressure.

When the bag is first submerged, there are more collisions of solvent molecules per unit area of the membrane on the pure solvent side than there are on the solution side (where there are fewer solvent molecules per unit volume). Hence there is a net transfer of solvent through the membrane into the solution. As water rises in the tube and pressure builds up in the tube, the number of collisions of water molecules on the solution side increases. When the osmotic pressure is reached, a dynamic equilibrium is achieved in which the rate of passage of water molecules is the same in both directions.

Like vapor pressure lowering, boiling point elevation, and freezing point lowering, osmotic pressure results from the unequal rates at which solvent molecules pass through an interface or boundary. In the case of evaporation and boiling, it is the solution/vapor interface; for freezing, it is the solution/solid interface; and for osmosis, it is the semipermeable membrane. All of these colligative properties can be understood in terms of differences in entropy between a pure solvent and a solution. This is perhaps most easily seen in the case of osmosis. When solvent and solute molecules mix, there is usually an increase in entropy. If pure solvent is added to a solution, a higher entropy state will be achieved when the solvent and solute molecules have diffused among one another to form a more dilute solution. Unless there are strong intermolecular forces, there will be a negligible enthalpy change, and so the increase in entropy makes mixing of solvent and solution a product-favored process. A semipermeable membrane prevents solute molecules from passing into pure solvent. The only way mixing can occur (and entropy can increase) is for solvent to flow into the solution, and so it does.

The more concentrated the solution, the more product-favored the mixing and the greater the pressure required to prevent it. Osmotic pressure (Π) is proportional to the molarity of the solution, c:

$$\Pi = cRTi$$

where R is the gas constant, T is the absolute temperature (in kelvins), and i is the number of particles per formula unit of solute.

The osmotic pressure equation $\Pi = cRT\,i$ is similar to the ideal gas equation, $PV = nRT$, which can be rearranged to $P = (n/V)RT = cRT$, where n/V is the molar concentration of the gas.

Osmotic pressure can be quite large, even though the solution concentration is small; for example, the osmotic pressure of a 0.020 M solution of a molecular solute ($i = 1$) at 25 °C is

$$\Pi = cRTi = \left(\frac{0.020 \text{ mol}}{\text{L}}\right)\left(0.0821\frac{\text{L} \cdot \text{atm}}{\text{mol} \cdot \text{K}}\right)(298 \text{ K})(1) = 0.49 \text{ atm}$$

This pressure would support a water column more than 15 ft high. One way to determine osmotic pressure is to measure the height of a column of solution in a tube as shown in Figure 16.25. Heights of a few centimeters can be measured accurately, and so quite small concentrations can be determined by osmotic pressure experiments. If the mass of solute dissolved in a measured volume of solution is known, it is possible to calculate the molar mass of the solute by using the definition of molar concentration, $c = n/V$ = amount (mol)/volume (L). Osmotic pressure is especially useful in studying large molecules whose molar mass is difficult to determine by other means.

PROBLEM-SOLVING EXAMPLE 16.11 Molar Mass from Osmotic Pressure

The osmotic pressure at 25 °C is 1.79 atm for a solution prepared by dissolving 2.50 g of sucrose, empirical formula $C_{12}H_{22}O_{11}$, in enough water to give a solution volume of 100 mL. Use the osmotic pressure equation to show that the empirical formula for sucrose is the same as its molecular formula.

Answer The molar mass found by using the osmotic pressure equation is 342 g/mol, which equals the molar mass for $C_{12}H_{22}O_{11}$. The empirical formula and the molecular formula are the same.

Explanation We can determine the molecular formula from the empirical formula if we know the molar mass. To find the molar mass, calculate how many moles and how many grams are in the same volume of solution, say 1 L.

First, rearrange the osmotic pressure equation to calculate how many *moles* of sucrose there are per liter of solution. Sucrose, like other carbohydrates (← ***p. 102***), is a nonelectrolyte, so the i factor is 1.

$$c = \frac{\Pi}{RTi} = \frac{1.79 \text{ atm}}{(0.0821 \text{ L atm/mol K})(298 \text{ K})(1)} = 7.32 \times 10^{-2} \text{ mol/L}$$

For this solution, its concentration in units of mol/L was achieved by dissolving 2.50 g sucrose in 100. mL of water, and so the *mass* of sucrose *per liter* is

$$\text{Mass of sucrose per liter} = \left(\frac{2.50 \text{ g}}{100. \text{ mL}}\right)\left(\frac{1000 \text{ mL}}{\text{L}}\right) = 25.0 \text{ g/L}$$

We now know both the mass and the amount of sucrose per liter of solution, so we can divide to get molar mass. Notice how the liters (L) terms in both the numerator and denominator cancel.

$$\text{Molar mass} = \frac{25.0\ \text{g/}\cancel{\text{L}}}{7.32 \times 10^{-2}\ \text{mol/}\cancel{\text{L}}} = 342\ \text{g/mol}$$

The molar mass that corresponds to the empirical formula, $C_{12}H_{22}O_{11}$, is 342 g/mol; therefore the molecular formula must be the same as the empirical formula.

Freezing point–lowering measurements and boiling point–elevation measurements can also be used to find molar mass in the same manner as shown in Problem Solving Example 16.11 for osmotic pressure measurements.

Problem-Solving Practice 16.11

The osmotic pressure found for a solution of 5.0 g of horse hemoglobin in 1.0 L of water is 1.8×10^{-3} atm at 25 °C. What is the molar mass of the hemoglobin?

Blood and other fluids inside living cells contain many different solutes, and the osmotic pressures of these solutions play an important role in the distribution of solutes within the body. Patients who have become dehydrated are often given water and nutrients intravenously. However, one cannot simply drip pure water into a patient's veins, because the water would flow into the red blood cells by osmosis, causing them to burst (Figure 16.26c). A solution that causes this condition is called a **hypotonic** solution. To prevent cells from bursting, an intravenous solution must have the same

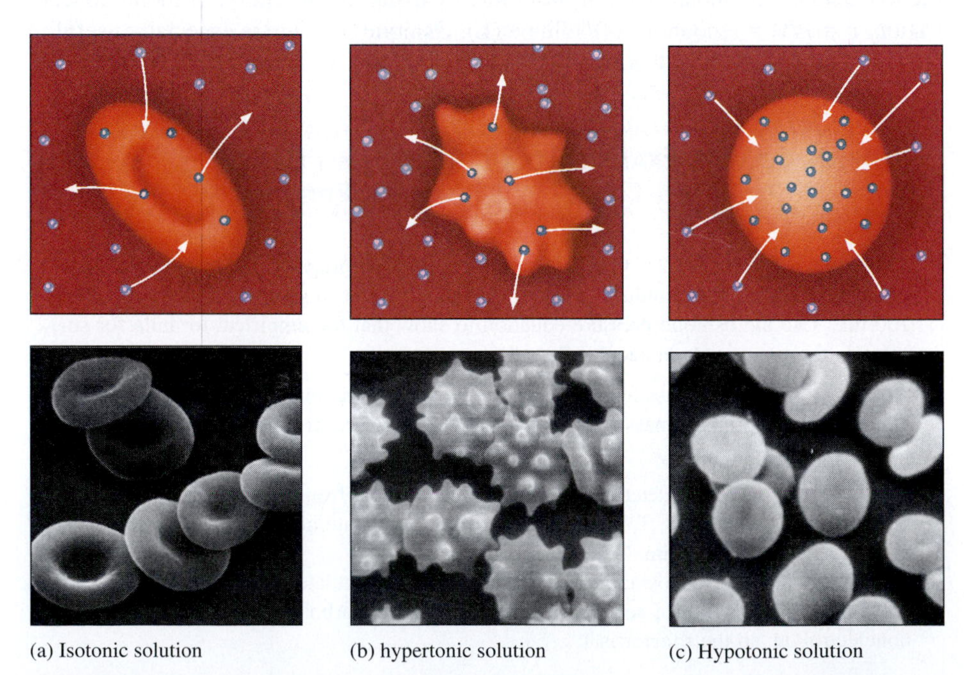

(a) Isotonic solution (b) hypertonic solution (c) Hypotonic solution

Figure 16.26 Osmosis and the living cell. (a) A cell placed in an *iso*tonic solution. The net movement of water in and out of the cell is zero because the concentration of solutes inside and outside the cell is the same. (b) In a *hyper*tonic solution, the concentration of solutes outside the cell is greater than inside. There is a net flow of water out of the cell, causing the cell to dehydrate, shrink, and perhaps die. (c) In a *hypo*tonic solution, the concentration of solutes outside the cell is less than inside. There is a net flow of water into the cell, causing the cell to swell and perhaps to burst. *(Science Source Photo Researchers, Inc.)*

total concentration of solutes and therefore the same osmotic pressure as the patient's blood. Such a solution is called iso-osmotic (or **isotonic,** see Figure 16.26a). A solution of 0.9% sodium chloride is isotonic with fluids inside cells in the body (Figure 16.27).

If an intravenous solution more concentrated than the solution inside a red blood cell were added to blood, the cell would lose water and shrivel up. A solution that causes this condition is a **hypertonic** solution (Figure 16.27b). Cell shriveling by osmosis happens when vegetables or meats are cured in *brine,* a concentrated solution of NaCl. If you put a fresh cucumber or carrot into brine, water will flow out of its cells and into the brine, leaving behind a shriveled vegetable (Figure 16.28). With the proper spices added to the brine, a cucumber will become a tasty pickle.

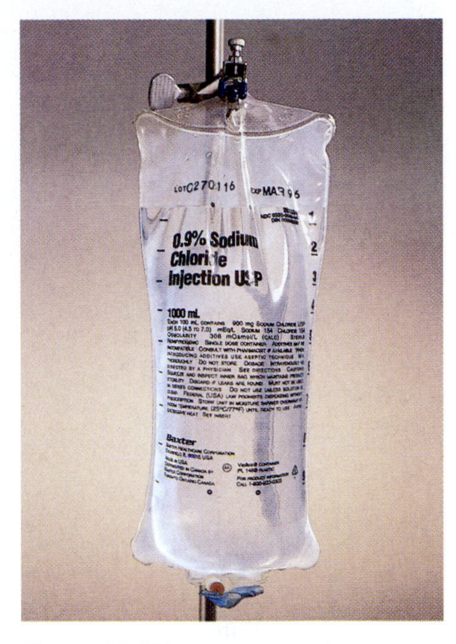

Figure 16.27 Isotonic solutions. Isotonic saline solutions like this one are routinely given to patients who have lost body fluids. *(C.D. Winters)*

Reverse Osmosis

Applying pressure greater than the osmotic pressure causes solvent to flow through a semipermeable membrane from a concentrated solution to a solution of lower solute concentration. This process is called **reverse osmosis.** In effect the semipermeable membrane serves as a filter with very tiny pores through which only the solvent can pass. This removal of particles as small as molecules or ions can be used to obtain highly purified water. Sea water contains a high concentration of dissolved salts; its osmotic pressure is 24.8 atm. If a pressure greater than 24.8 atm is applied to a chamber containing sea water, water molecules can be forced to flow through a semipermeable membrane to a region containing purer water (Figure 16.29, p. 754). Pressures up to 100 atm are used to provide reasonable rates of purification. Large reverse osmosis plants in places like the Persian Gulf countries and in Arizona can purify more than 100 million gallons of water per day. The plant in Arizona was built to reduce the salt concentration of irrigation wastewater discharged to the Colorado River from 3200 ppm to just under 300 ppm. Sea water, which contains upwards of 35,000 ppm of dissolved salts (Table 16.8), can be purified by reverse osmosis to between 400 and 500 ppm, which is well within the World Health Organization's limits for drinking water.

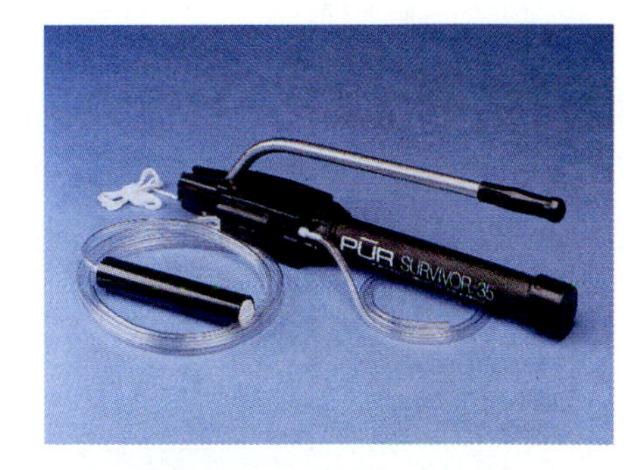

An emergency hand-operated water desalinator that works by reverse osmosis. It can produce 4.5 L of pure water per hour from sea water. Such devices can be very useful for persons adrift at sea. *(Courtesy of Recovery Engineering, Inc.)*

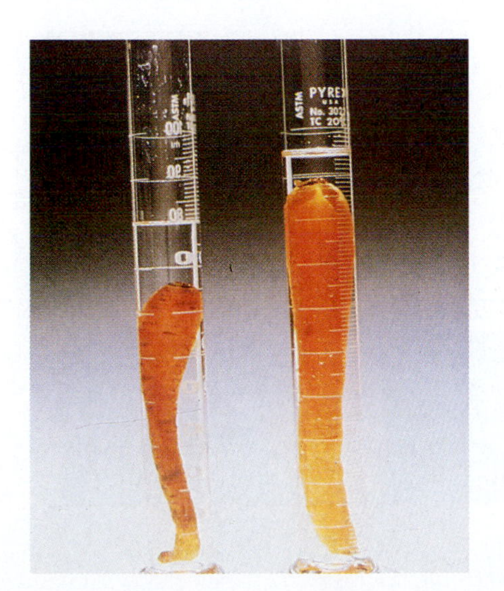

Figure 16.28 Illustration of osmosis. When a carrot is soaked in a concentrated salt solution, water flows out of the plant cells by osmosis. A carrot soaked overnight in salt solution (*left*) has lost much water and becomes limp. A carrot soaked overnight in pure water (*right*) is little affected. *(C.D. Winters)*

TABLE 16.8 Ions Present in Sea Water at 1 ppm or More

Ion	Mass Fraction	
	g/kg	ppm
Cl^-	19.35	19,350
Na^+	10.76	10,760
SO_4^{2-}	2.710	2710
Mg^{2+}	1.290	1290
Ca^{2+}	0.410	410
K^+	0.400	400
HCO_3^-, CO_3^{2-}	0.106	106
Br^-	0.067	67
$H_2BO_3^-$	0.027	27
Sr^{2+}	0.008	8
F^-	0.001	1
Total	35.129	35,129

16.11 COLLOIDS

Around 1860, Thomas Graham found that substances such as starch, gelatin, glue, and egg albumin diffuse in water only very slowly, compared with sugar or salt. In addition, the former substances differ significantly from the latter ones in their ability to diffuse through a thin membrane: sugar or salt will diffuse through many types of membranes, but glue, starch, albumin, and gelatin will not. Graham also found that he could not crystallize these substances, while he could crystallize sugar and salt from their solutions. Therefore, Graham coined the word "colloid" (from the Greek meaning "glue") to describe a class of substances distinctly different from sugar and salt and similar materials.

Colloids are now understood to be mixtures in which relatively large particles, the **dispersed phase,** are distributed uniformly throughout a solvent-like medium

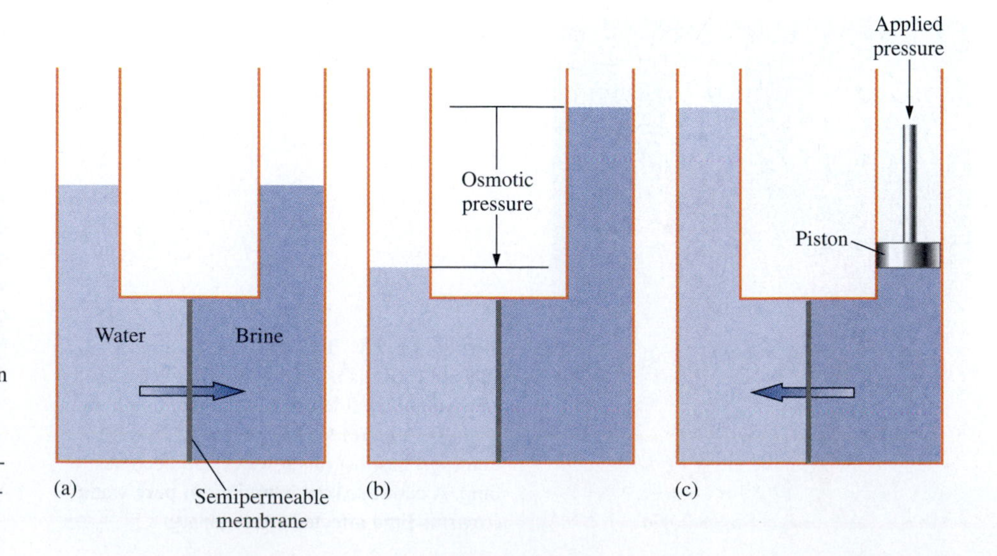

Figure 16.29 Normal and reverse osmosis. Normal osmosis is represented by (a) and (b). Water molecules pass through the semipermeable membrane to dilute the brine solution until the height of the solution creates sufficient pressure to stop the net flow. Reverse osmosis, represented in (c), is the application of an external pressure in excess of osmotic pressure to force water molecules to the pure water side.

CHEMISTRY IN THE NEWS

Cholera and Solution Concentration

Cholera, a potentially deadly disease caused by bacteria in untreated water, sometimes kills half of its victims. A 1994 cholera outbreak in Rwanda killed thousands of refugees fleeing a civil war, while less than one percent of victims died during a 1991 cholera epidemic in Chile. Clean drinking water, something we take for granted, was available in Chile, but not in Rwanda.

Through diarrhea, a cholera victim can lose as much as 10 L of fluid per day, compared to the normal excretion of about 1.4 L. Such extensive diarrhea causes severe dehydration through excessive loss of fluids and electrolytes, and failure to reabsorb vital nutrients and electrolytes. Unless properly treated, the dehydration can be fatal. Victims can be cured, seemingly miraculously, simply by rehydration using the proper solution. International researchers have developed an oral rehydration solution (ORS), a mixture containing glucose and the electrolytes potassium chloride, sodium chloride, and sodium citrate. The solid mixture is dissolved in sterile or boiled water and drunk so that patients can be treated in their homes without having to be hospitalized for intravenous feeding.

One liter of the ORS contains 0.111 mol of glucose, 0.020 mol of KCl, 0.060 mol of NaCl, and 0.0112 mol of sodium citrate ($Na_3C_6H_5O_7$), all very inexpensive materials. Administration of ORS cures diarrhea because the concentration of its ingredients approximates the osmolarity of blood serum, 0.300 osmol. Osmolarity is a measure of the number of particles in a solution that roughly equals the molarity of a solute times its number of moles of particles in solution.

$$\text{Osmolarity} = \text{solution molarity} \times (\text{moles of particles/moles of solute})$$

Thus, the molarity and osmolarity of glucose, a nonelectrolyte, are the same, 0.111, but the osmolarities of the electrolytes in the mixture are 0.040 osmol KCl, 0.12 osmol NaCl, and 0.0488 osmol sodium citrate, each a multiple of its molarity. The osmolarity of the ORS must approximate that of normal blood serum because too concentrated an ORS solution would cause more water to be removed by osmosis from the blood.

But the ORS cannot prevent the occurrence of cholera. Its use still depends on clean water being available, which is not always the case in less-developed areas.

Safe drinking water? In many parts of the world, like this location in West Africa, drinking water is collected from dangerously polluted water sources. *(Charles O. Cecil/Visuals Unlimited)*

Source: Chem Matters, *Feb, pp. 12–13, 1995.*

called the **continuous phase,** or the dispersing medium. Like true solutions, colloids are found in the gas, liquid, and solid states. Although both true solutions and colloids appear homogeneous to the naked eye, at the microscopic level colloids are not homogeneous.

In colloids, the dispersed-phase particles might be as large as 10 to 1000 times the size of a single small molecule. Colloidal particles are so large, about 1000 nm

Figure 16.30 The Tyndall effect. Shafts of light are visible coming through the trees in the forest along the Oregon coast. *(Bob Pool/Tom Stack and Associates)*

in diameter, that they can readily scatter visible light that passes through the continuous medium, a phenomenon known as the **Tyndall effect** (Figure 16.30).

True solutions also scatter light, but only weakly. The blue sky is the result of light scattering by molecules in the air. Because blue light is more strongly scattered by molecule-sized particles than is red light, the sky appears a diffuse blue.

Figure 16.31 shows a beam of light passing through three glass bottles. The bottles on the left and right contain a colloidal mixture of a gelatin in water, while the bottle in the center holds a solution of NaCl. Colloidal particles of dust and smoke in the air of a room can easily be observed in a beam of sunlight because they scatter the light; you have probably seen such a well defined sunbeam many times. A common colloid, *fog,* consists of water droplets (the dispersed phase) in air (the continuous phase, and itself a solution).

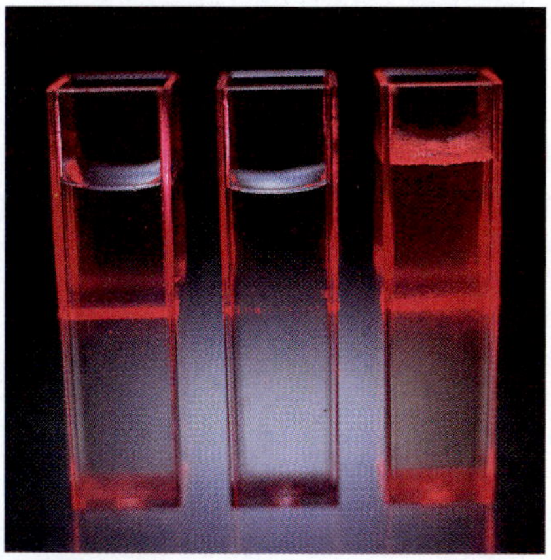

Figure 16.31 Light scattering by a colloidal suspension. A narrow beam of light passes through a colloidal mixture (*left*), then through a salt solution, and finally through another colloidal mixture. This illustrates the light scattering ability of the colloidal sized particles. *(C.D. Winters)*

Types of Colloids

Colloids are classified according to the state of the dispersed phase (solid, liquid, or gas) and the state of the continuous phase. Table 16.9 lists several types of colloids and some examples of each. Liquid-liquid colloids form only in the presence of an emulsifier—a third substance that coats and stabilizes the particles of the dispersed phase. Such colloidal dispersions are called **emulsions.** In mayonnaise, for example, egg yolk contains a protein that stabilizes the tiny drops of oil that are dispersed in the aqueous continuous phase. As you can see from Table 16.9, colloids are very common in everyday life.

TABLE 16.9 Types of Colloids

Continuous Phase	Dispersed Phase	Type	Examples
Gas	Liquid	Aerosol	Fog, clouds, aerosol sprays
Gas	Solid	Aerosol	Smoke, airborne viruses, automobile exhaust
Liquid	Gas	Foam	Shaving cream, whipped cream
Liquid	Liquid	Emulsion	Mayonnaise, milk, face cream
Liquid	Solid	Solution	Gold in water, milk of magnesia, mud
Solid	Gas	Foam	Foam rubber, sponge, pumice
Solid	Liquid	Gel	Jelly, cheese, butter
Solid	Solid	Solid solution	Milk glass, many alloys such as steel, some colored gemstones

CHEMISTRY YOU CAN DO

Curdled Colloids

Regular (whole) milk contains about 4% fat. Skim milk contains considerably less. In addition, milk contains protein. Both the fat and the proteins are in the form of colloids.

Add about 2 Tbsp of vinegar or lemon juice to about 100 mL of whole milk (do the same with skim milk, and 1% or 2% butterfat milk, if you have it), stir, and watch what happens. Let it stand overnight at room temperature. Observe and record the results. Write an explanation for what you observed.

An additional experiment you can try is adding salt to similar samples of milk and recording your observations. Does salt have the same effect on milk as the acid does?

You can discard this milk down the drain. You should *not* drink this milk because it has been unrefrigerated and might contain harmful bacteria.

Colloids with water as the continuous phase can be classified as hydrophilic or hydrophobic. In a hydrophilic colloid there is a *strong attraction between the dispersed phase and the continuous (aqueous) phase.* Hydrophilic colloids are formed when the molecules of the dispersed phase have multiple sites that interact with water through hydrogen bonding and dipole-dipole attraction. Proteins in aqueous media are hydrophilic colloids.

In a hydrophobic colloid there is a *lack of attraction between the dispersed phase and the continuous phase.* Although you might assume that such colloids would tend to separate quickly, hydrophobic colloids can be quite stable once they are formed. A colloidal solution (sol) of gold particles prepared in 1857 is still preserved in the British Museum. In hydrophobic colloids the surfaces of the colloidal particles apparently become electrically charged by some process that is not completely understood. Oppositely charged ions in solution are then attracted to the surfaces, forming a second layer. Because all the colloidal particles have the same kind of charge, they repel other particles of the same kind and are prevented from coming together to form larger particles.

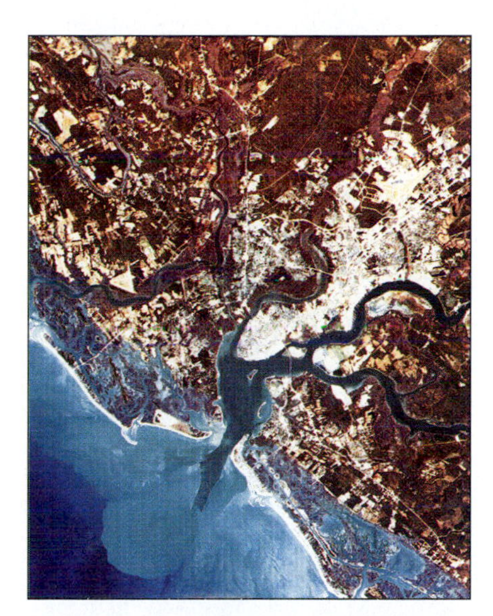

Figure 16.32 Silt formation from colloidal soil particles. Silt forms at a river delta as colloidal particles in the river water enter the salt water. The high concentration of salt causes the colloidal particles to coagulate. *(NASA/Peter Arnold, Inc.)*

A stable hydrophobic colloid can be made to **coagulate** by introducing ions into the dispersed phase. Milk is a colloidal suspension of hydrophobic particles. When milk ferments, lactose (milk sugar) is converted to lactic acid, which forms lactate ions and hydronium ions. The protective charge layer on the surfaces of the colloidal particles is overcome, and the milk coagulates—the milk solids separate in clumps called "curds." The coagulated milk may be used to make buttermilk or various kinds of cheese, depending on the species of bacteria that caused the fermentation. Similarly, the soil particles carried in rivers are hydrophobic sols. When river water containing large amounts of colloidally suspended soil particles meets sea water with a high ionic concentration, the particles coagulate to form silt. The deltas of the Mississippi and Nile rivers are formed in this way (Figure 16.32).

16.12 SURFACTANTS

There are many natural and synthetic molecules that have both a hydrophobic part and a hydrophilic part. Such molecules are called **surfactants** because they tend to act at the surface of a substance that is in contact with the solution that contains them. The classic surfactant, soap, dates back to the Sumerians in 2500 BC. Soaps are salts

Hand soaps are generally pure soap to which dyes and perfumes are added.

of fatty acids and have always been made by the reaction of a fat (⇐ p. 110) with an alkali, a process known as *saponification.* The Greek physician Galen referred to this recipe and stated further that soap removed dirt from the body as well as serving as a treatment for wounds.

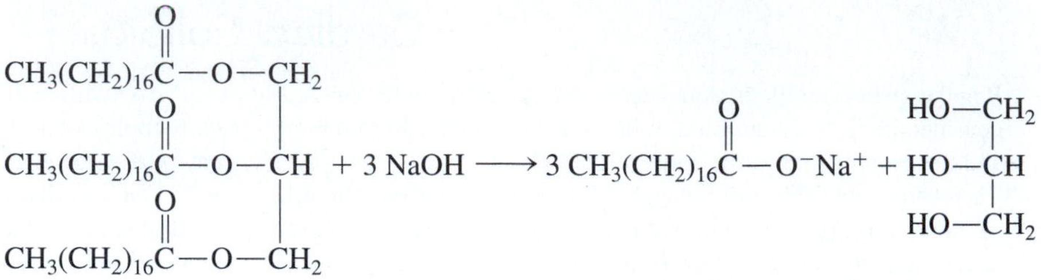

$$\begin{array}{l} CH_3(CH_2)_{16}C(=O)-O-CH_2 \\ CH_3(CH_2)_{16}C(=O)-O-CH \\ CH_3(CH_2)_{16}C(=O)-O-CH_2 \end{array} + 3\,NaOH \longrightarrow 3\,CH_3(CH_2)_{16}C(=O)-O^-Na^+ + \begin{array}{l} HO-CH_2 \\ HO-CH \\ HO-CH_2 \end{array}$$

tristearin (glyceryl tristearate) | sodium stearate (a soap) | glycerol

Sodium stearate is a typical soap. The long-chain hydrocarbon part of the molecule is hydrophobic, while the polar carboxylate group ($C(=O)-O^-$) is hydrophilic.

$$CH_3CH_2CH_2CH_2CH_2CH_2CH_2CH_2CH_2CH_2CH_2CH_2CH_2CH_2CH_2CH_2CH_2C(=O)-O^-Na^+$$

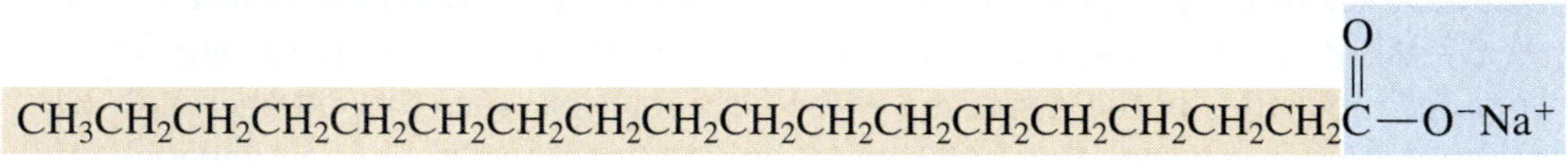

hydrophobic end | hydrophilic end

sodium stearate

In addition to soaps, which are made from naturally occurring fats and oils, many synthetic surfactants are made from refined petroleum or coal products. These **detergents** have molecular structures somewhat like those of soaps; that is, they have a long hydrocarbon end that is hydrophobic, and a polar end that is hydrophilic. One common synthetic surfactant is sodium lauryl sulfate, which is used in many shampoos.

$$CH_3CH_2CH_2CH_2CH_2CH_2CH_2CH_2CH_2CH_2CH_2CH_2OSO_3^-Na^+$$

sodium lauryl sulfate

Exercise 16.20 Estimating Osmotic Pressure

Which would have the higher osmotic pressure? A solution 0.02 *m* in sucrose, or a solution 0.02 *m* in ordinary soap? Explain your answer.

In water solutions, surfactants tend to aggregate to form hollow, colloid-sized particles called micelles (Figure 16.33) that can transport various materials within them. The hydrophobic ends of the molecules point inward to the center of the micelle, and the hydrophilic ends point outward so that they interact with water molecules. Ordinary soap is a surfactant and, in water, forms micelles.

Soaps cleanse because oil and grease (which are also hydrophobic) become associated with the hydrophobic centers of the soap micelles and are washed away with the rinse water (Figure 16.33). One problem with soaps is that they tend to form insoluble salts that appear as a scum when ordinary soap is used with "hard water," which contains Ca^{2+}, Mg^{2+}, and Fe^{2+} ions. Synthetic surfactants do not form such insoluble salts and for that reason are widely used for washing, especially in those regions of the country where water hardness is common.

A surfactant, water, and oil together form an emulsion, with the surfactant acting as the emulsifying agent.

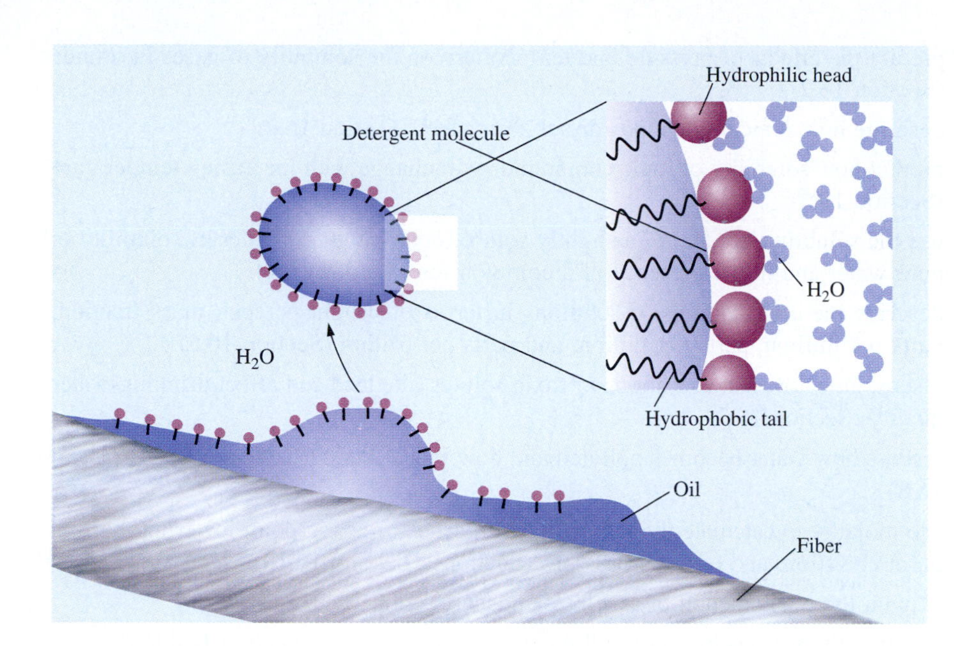

Figure 16.33 The cleansing action of soap and detergents. The hydrophilic charged ends of soap molecules are strongly attracted to water molecules. The hydrophobic hydrocarbon ends of the soap molecules, which have more attraction for other hydrocarbons than for water, will dissolve in any hydrocarbons with which they happen to come in contact. This leaves the hydrophilic ends of the soap molecules exposed to the surrounding water molecules. Aided by mechanical agitation, small particles of oily material can break away and be carried into solution, stabilized by the hydrophilic ends of the soap molecules that are on the surface.

Summary Problem

You are asked to prepare three mixtures at 25 °C consisting of (i) 25.0 g of CCl_4 and 100. mL of water, (ii) 15.0 g of $CaCl_2$ in 125 mL of water, and (iii) 21 g of ethylene glycol (CH_2OHCH_2OH) in 150. mL of water. Answer the following questions about these mixtures. If one of the solutes fails to dissolve in water, some of the questions will not be applicable. Consult Section 16.2 to determine if a solution will form.

(a) What is the weight percent of the mixture?

(b) What is the mass fraction of the mixture?

(c) Is a solution formed? (If a solution is formed, answer the remaining questions. You may assume a density of the solution of 1.0 g/mL.)

(d) Name the dissolved species in solution and draw a diagram representing how the solvent (water) molecules are interacting with these species.

(e) Express the concentration of the solution in ppm.

(f) Express the concentration of the solution in molality.

(g) Calculate the solubility of AgCl in the mixture.

(h) Calculate the boiling point of the solution.

(i) Calculate the freezing point of the solution.

(j) Calculate the osmotic pressure of the solution.

In Closing

Having studied this chapter, you should be able to . . .

- summarize the unique solvent properties of water and the reason for their existence (Section 16.1).
- describe how liquids, solids, and gases dissolve in a solvent (Section 16.2).
- predict solubility based on knowledge of the solute and the solvent (Section 16.2).

- predict the effects of pressure and temperature on the solubility of gases in liquids (Section 16.2).
- describe how ionic compounds dissolve in water (Section 16.2).
- predict how solubility of ionic compounds will change with increasing temperature (Section 16.3).
- use the solubility product of a slightly soluble compound to predict its solubility in pure water and in the presence of a common ion (Section 16.4).
- describe the compositions of solutions in terms of weight percent, mass fraction, parts per million, parts per billion, and parts per trillion (Section 16.5).
- discuss how low concentrations of toxic solutes like lead can affect drinking-water quality (Section 16.5).
- discuss how water becomes polluted and how it is cleaned (Sections 16.6, 16.7, and 16.8).
- use molality to calculate the colligative properties: freezing point lowering, boiling point elevation, and osmotic pressure (Sections 16.9 and 16.10).
- explain the phenomenon of reverse osmosis (Section 16.10).
- describe the various kinds of colloids and their properties (Section 16.11).
- describe how surfactants work (Section 16.12).

Key Terms

The following terms were defined and given in boldface type in this chapter. You should be sure to understand each of the terms and the concepts with which it is associated.

biochemical oxygen demand *(16.6)*
biodegradable *(16.6)*
boiling point elevation *(16.9)*
coagulate *(16.11)*
colligative properties *(16.9)*
colloid *(16.11)*
common ion effect *(16.4)*
continuous phase *(16.11)*
detergents *(16.12)*
dispersed phase *(16.11)*
emulsion *(16.11)*
freezing point lowering *(16.9)*
Henry's law *(16.2)*
hydration *(16.1)*
hydrophilic *(16.2)*
hydrophobic *(16.2)*
hypertonic *(16.10)*
hypotonic *(16.10)*
immiscible *(16.2)*
insoluble *(16.2)*
isotonic *(16.10)*
lattice energy *(16.2)*
mass fraction *(16.5)*
micelles *(16.12)*
miscible *(16.2)*
molality *(16.9)*
molar solubility *(16.4)*
nonbiodegradable *(16.6)*
osmosis *(16.10)*
osmotic pressure *(16.10)*
parts per billion *(16.5)*
parts per million *(16.5)*
parts per trillion *(16.5)*
reverse osmosis *(16.10)*
saturated solution *(16.2)*
semipermeable membrane *(16.10)*
solubility *(16.2)*
solubility product constant *(16.4)*
solvation *(16.1)*
supersaturated solution *(16.2)*
surfactants *(16.12)*
Tyndall effect *(16.11)*
unsaturated solution *(16.2)*
weight percent *(16.5)*

Questions for Review and Thought

Conceptual Challenge Problems

CP-16.A. The solubility product constants of silver chloride, AgCl, and silver chromate, Ag_2CrO_4, are 1.8×10^{-10} and 2.5×10^{-12}, respectively. Suppose that the chloride, Cl^-(aq), and chromate, CrO_4^{2-}(aq), ions are both present in the same solution at concentrations of 0.010 M each. A standard solution of silver ions, Ag^+ (aq), is dispensed slowly from a burette into this solution while it is stirred vigorously. Solid silver chloride is white and silver chromate is red. What will be the concentration on $Cl^-(aq)$ ions in the mixture when the first tint of red color appears in the mixture?

CP-16.B. Concentrations expressed in units of parts per million and parts per billion often have no meaning for people until they relate these small and large numbers to their own experiences.
(a) What time in seconds is 1 ppm of a year?
(b) What time in seconds is 1 ppb of a lifetime (70 years)?

CP-16.C. Bodies of water that have an abundance of nutrients that support a blooming growth of plants are said to be eutrophoric. In general, fish do not thrive for long in eutrophoric waters because the high BOD of the water leaves little oxygen for the fish. Suppose someone asked you why this was true, given the fact that growing plants produce oxygen as a product of photosynthesis. How would you respond to this person's inquiry?

CP-16.D. Suppose that you want to produce the lowest temperature possible by use of ice, sodium chloride, and water to make homemade ice cream in a 1.5-L metal cylinder surrounded by a coolant held in a wooden bucket? You have all the ice, salt, and water you want. How would you plan to do this?

Review Questions

1. Which of these general types of substances would you expect to readily dissolve in water?
   (a) Alcohols
   (b) Hydrocarbons
   (c) Metals
   (d) Nonpolar molecules
   (e) Polar molecules
   (f) Salts
2. Describe the difference among the terms "supersaturated," "saturated," and "unsaturated" solution.
3. State Henry's law. Name three factors that govern the solubility of a gas in a liquid.
4. In general, how does the water solubility of most ionic compounds change as the temperature is increased? Explain why.
5. How does the solubility of gases in liquids change with increased temperature? Explain why.
6. Which is the highest concentration: 50 ppm, 500 ppb, or 0.05% by weight?
7. Estimate your concentration on campus in parts per million and parts per thousand.
8. Define molality. How does it differ from molarity?
9. Write a general solubility product constant expression for a salt of the form A_xB_y.
10. Which is more soluble in water at 25 °C, CuBr ($K_{sp} = 5.3 \times 10^{-9}$) or CuCl ($K_{sp} = 1.9 \times 10^{-7}$)? Explain why.
11. Would you expect more $CaCO_3$ ($K_{sp} = 3.8 \times 10^{-9}$ at 25 °C) to dissolve in pure water or in a 0.01 M $CaCl_2$ solution? Explain your answer using Le Chatelier's principle.
12. Name five different water pollutants.
13. Explain why the vapor pressure of a solvent is lowered by the presence of a nonvolatile solute.
14. Why is a higher temperature required for boiling a solution containing a nonvolatile solute than for boiling the pure solvent?
15. Which would have the lowest freezing point: (a) a 1.0-m NaCl solution, (b) a 1.0-m $CaCl_2$ solution, (c) a 1.0-m methanol solution? Explain your choice.
16. Write the osmotic pressure equation and explain all terms.
17. How can the presence of a strong electrolyte cause a hydrophobic colloid to coagulate?
18. Draw an example of a soap molecule. Based on its structure, why is it considered a surfactant?

How Substances Dissolve

19. Explain why some liquids are miscible in one another while other liquids are immiscible. Using only three liquids, give an example of a miscible pair and an immiscible pair.
20. Why would the same solid readily dissolve in one liquid and be almost insoluble in another liquid? Give an example of such behavior.
21. Knowing that the solubility of oxalic acid at 25 °C is 1 g per 7 g of water, how would you prepare 1 L of a saturated oxalic acid solution?
22. A saturated solution of NH_4Cl was prepared by adding solid NH_4Cl to water until no more solid NH_4Cl would dissolve. The resulting mixture felt very cold and had a layer of undissolved NH_4Cl on the bottom. When the mixture reached room temperature, no solid NH_4Cl was present. Explain what happened. Was the solution still saturated?
23. The lattice energy of $CaCl_2$ is −2258 kJ/mol and the enthalpy of hydration of $CaCl_2$ is +2175 kJ/mol. Is the process of dissolving $CaCl_2$ in water endothermic or exothermic?
24. Simple acids like formic acid, HCOOH, and acetic acid, CH_3COOH, are very soluble in water; however, fatty acids like stearic acid, $CH_3(CH_2)_{16}COOH$, and palmitic acid, $CH_3(CH_2)_{14}COOH$ are water-insoluble. Based on what you know about the solubility of alcohols, explain the solubility of organic acids.
25. If a solution of a certain salt in water is saturated at some temperature and a few crystals of the salt are added to the solution, what do you expect will happen? What happens if the same quantity of salt crystals is added to an unsaturated solution of the salt? What would you expect to happen if the temperature of this second salt solution is slowly lowered?
26. Describe what happens when an ionic solid dissolves in water. Draw a picture that includes at least three positive ions, three negative ions, and a dozen or so water molecules in the vicinity of the ions.
27. The partial pressure of O_2 in your lungs varies from 25 mm Hg to 40 mm Hg. What concentration of O_2 (in grams per liter) can dissolve in water at 37 °C when the O_2 partial pressure is 40. mm Hg? The Henry's law constant for O_2 at 37 °C is 1.5×10^{-6} M/mm Hg.
28. The Henry's law constant for nitrogen in blood serum is approximately 8×10^{-7} M/mm Hg. What is the N_2 concentration

in a diver's blood at a depth where the total pressure is 2.5 atm? The air the diver is breathing is 78% N_2 by volume.

Solubility Product

29. Write a balanced chemical equation for the equilibrium occurring when each of the following are added to water, then write the K_{sp} expression.
(a) Lead(II) carbonate (b) Nickel(II) hydroxide
(c) Strontium phosphate
30. Write a balanced chemical equation for the equilibrium occurring when each of the following are added to water, then write the K_{sp} expression.
(a) Copper(I) iodide (b) Silver sulfate
(c) Iron(II) carbonate
31. Calculate the water solubility of AuCl in mol/L.
32. Calculate the water solubility of $PbBr_2$ in mol/L.
33. What is the concentration of gold(III) ion (in milligrams per liter of solution) in a saturated aqueous solution of gold(III) iodide? What is the concentration of iodide ion?
34. In a saturated aqueous solution of silver sulfate, what is the concentration of Ag^+ ion in milligrams per 100. mL of solution at 25 °C? What is the concentration of SO_4^{2-} ion in milligrams per 100. mL of solution?
35. Calculate the K_{sp} for HgI_2 given that its solubility in water is 4.0×10^{-29} M.
36. The solubility of $PbCl_2$ in water is 1.62×10^{-2} M. Calculate K_{sp} for $PbCl_2$.
37. What is the Cl^- concentration (in mol/L) in a solution that is 0.05 M in $AgNO_3$ and contains some undissolved AgCl?
38. What is the molarity of Zn^{2+} ion in a saturated solution of $ZnCO_3$ that contains 0.25 M $NaCO_3$?

Concentration Units

39. Convert 2.5 ppm to weight percent.
40. Convert 73.2 ppm to weight percent.
41. Mathematically show how 1 ppm is equivalent to 1 mg/1 kg.
42. Mathematically show how 1 ppb is equivalent to 1 μg/1 kg.
43. What mass (in grams) of sucrose is in 1.0 kg of a 0.25% sucrose solution?
44. How many grams of ethanol are in 750 mL of a 12% ethanol solution? (Assume the density is the same as water.)
45. A sample of lead-based paint is found to contain 60.5 ppm lead. The density of the paint is 8.0 lb/gal. What mass of lead (in grams) would be present in 50. gal of this paint?
46. A paint contains 200. ppm lead. Approximately what mass of lead (in grams) will be in 1.0 cm^2 of this paint (density = 8.0 lb/gal) when 1 gal is uniformly applied to 500. ft^2 of a wall?
47. Hydrochloric acid is sold as a concentrated aqueous solution. The concentration of commercial HCl is 12.0 M and its density is 1.18 g/cm^3. Calculate the weight percent of HCl in the solution.
48. Concentrated sulfuric acid has a density of 1.84 g/cm^3 and is 18 M. What is the weight percent of H_2SO_4 in the solution?
49. You need an aqueous solution of methanol (CH_3OH) with a concentration of 0.050 molal. What mass of methanol would you need to dissolve in 500. g of water to make this solution?
50. You want to prepare a 1.0-*m* solution of ethylene glycol, $C_2H_4(OH)_2$, in water. What mass of ethylene glycol do you need to mix with 950. g of water?

Water Pollution and Purification

51. How is lead harmful to humans? When lead is present in drinking water, in what form does it exist?
52. Dietitians recommend six 8-oz glasses of water each day. If your drinking water contains the maximum contamination level for arsenic, how much arsenic would you consume in a week following this recommendation? (See Table 16.5.)
53. A sample of well water contained 5 ppb chlordane. Is this within the maximum contamination level for chlordane? (See Table 16.5.)
54. List three classes of industrial wastes and illustrate how each can cause water pollution.
55. List three common household products which, when disposed of down the drain, become water pollutants. Name three which do not.
56. What can happen to aquatic life if the BOD of organic pollutants in a stream exceeds the solubility of O_2 at the stream's temperature? Would this problem be helped or made worse if the stream is slow-moving?
57. Why is it necessary to bubble air through an aquarium?
58. What is meant by the term "hard water"?
59. Municipal water treatment is usually a three-stage process. What types of pollutants are removed at each stage?
60. During secondary water treatment, water is often sprayed into the air. What is the purpose of this action?

Colligative Properties

61. What is the boiling point of a solution containing 0.200 mol of a nonvolatile nonelectrolyte solute in 100. g of benzene? The normal boiling point of benzene is 80.10 °C, and K_b = 2.53 °C/molal.
62. What is the boiling point of a solution composed of 15.0 g of urea, $(NH_2)_2CO$, in 0.500 kg of water?
63. Place the following aqueous solutions in order of increasing boiling point: (a) 0.10 *m* NaCl, (b) 0.10 *m* sugar, (c) 0.080 *m* $CaCl_2$.
64. List the following aqueous solutions in order of decreasing freezing point: (a) 0.1 *m* sugar, (b) 0.1 *m* NaCl, (c) 0.08 *m* $CaCl_2$, (d) 0.04 *m* Na_2SO_4. (Assume all of the salts dissociate completely into their ions in solution.)
65. You add 0.255 g of an orange crystalline compound with an empirical formula of $C_{10}H_8Fe$ to 11.12 g of benzene. The boiling point of the solution is 80.26 °C. The normal boiling point of benzene is 80.10 °C, and K_b = 2.53 °C/molal. What are the molar mass and molecular formula of the compound?
66. Anthracene is a hydrocarbon obtained from coal, and it has an empirical formula of C_7H_5. To find its molecular formula you dissolve 0.500 g of anthracene in 30.0 g of benzene. The boiling point of the solution is 80.34 °C. The normal boiling point of benzene is 80.10 °C, and K_b = 2.53 °C/molal. What are the molar mass and molecular formula of anthracene?

67. If you use only water and pure ethylene glycol, $C_2H_4(OH)_2$, in your car's cooling system, what mass (in grams) of the glycol must you add to each quart of water to give freezing protection down to −31.0 °C?
68. Some ethylene glycol, $C_2H_4(OH)_2$, was added to your car's cooling system along with 5.0 kg of water.
(a) If the freezing point of the solution is −15.0 °C, what mass (in grams) of the glycol must have been added?
(b) What is the boiling point of the coolant mixture?
69. Calculate the concentration of nonelectrolyte solute particles in human blood if the osmotic pressure is 7.53 atm at 37 °C, the temperature of the body.
70. Cold-blooded animals and fish have blood that is isotonic with sea water. If sea water freezes at −2.3 °C, what is the osmotic pressure of the blood of these animals at 20.0 °C? (Assume the density is that of pure water.)

General Questions

71. What is the difference between solubility and miscibility?
72. If 5 g of solvent, 0.2 g of some solute A, and 0.3 g of solute B are mixed to form a solution, what is the weight percent concentration of A?
73. Calculate the water solubility of $PbSO_4$ in mol/L.
74. Calculate the water solubility of Ag_2SO_4 in mol/L.
75. Calculate the molar solubility of silver thiocyanate in pure water and in water containing 0.01 M NaSCN.
76. Calculate the molar solubility of lead(II) chloride in pure water. Compare this value to the molar solubility of $PbCl_2$ in 255 mL of water to which 25.0 mL of 6.0 M HCl has been added. (Assume that the volumes are additive.)
77. A 10.0 M aqueous solution of NaOH has a density of 1.33 g/cm^3 at 20 °C. Calculate the weight percent of NaOH.
78. Concentrated aqueous ammonia is 14.8 M and has a density of 0.90 g/cm^3. Calculate the weight percent of NH_3 in the solution.
79. Assume you dissolve 45.0 g of ethylene glycol, $C_2H_4(OH)_2$, in 0.500 L of water. Calculate the molality and weight percent of ethylene glycol in the solution.
80. Dimethylglyoxime (DMG) reacts with nickel(II) ion in aqueous solution to form a bright red coordination compound. However, DMG is insoluble in water. In order to get it into aqueous solution where it can encounter Ni^{2+} ions, it must first be dissolved in a suitable solvent such as ethanol. Suppose you dissolve 45.0 g of DMG ($C_4H_8N_2O_2$) in 500. mL of ethanol (C_2H_5OH, density = 0.7893 g/mL). What are the molality and weight percent of DMG in this solution?
81. Arrange the following aqueous solutions in order of increasing boiling point: (a) 0.20 *m* ethylene glycol, (b) 0.12 *m* Na_2SO_4, (c) 0.10 *m* $CaCl_2$, (d) 0.12 *m* KBr.
82. Arrange the following aqueous solutions in order of decreasing freezing point: (a) 0.20 *m* ethylene glycol, (b) 0.12 *m* K_2SO_4, (c) 0.10 *m* NaCl, (d) 0.12 *m* KBr.
83. The solubility of NaCl in water at 100 °C is 39.1 g/100. g of water. Calculate the boiling point of a saturated solution of NaCl.
84. The organic salt $(C_4H_9)_4NClO_4$ consists of the ions $(C_4H_9)_4N^+$ and ClO_4^-; it will dissolve in chloroform. What mass (in grams) of the salt must have been dissolved if the boiling point of a solution of the salt in 25.0 g of chloroform is 63.20 °C? The normal boiling point of chloroform is 61.70 °C and K_b = 3.63 °C/molal. Assume that the salt dissociates completely into its ions in solution.
85. A solution is prepared by dissolving 9.41 g of $NaHSO_3$ in 1.00 kg of water. It freezes at −0.33 °C. From these data, decide which of the following equations is the correct expression for the ionization of the salt.
(a) $NaHSO_3(aq) \rightleftharpoons Na^+(aq) + HSO_3^-(aq)$
(b) $NaHSO_3(aq) \rightleftharpoons Na^+(aq) + H^+(aq) + SO_3^{2-}(aq)$
86. In chemical research we often send newly synthesized compounds to commercial laboratories for analysis. These laboratories determine the weight percent of C and H by burning the compound and collecting the evolved CO_2 and H_2O. They determine the molar mass by measuring the osmotic pressure of a solution of the compound. Calculate the empirical and molecular formulas of a compound, C_xH_yCr, given the following information:
(a) The compound contains 73.94% C and 8.27% H; the remainder is chromium.
(b) At 25 °C, the osmotic pressure of 5.00 mg of the unknown dissolved in 100. mL of chloroform solution is 3.17 mm Hg.
87. The Ca^{2+} ion in hard water is often precipitated as $CaCO_3$ by adding soda ash, Na_2CO_3. If the calcium ion concentration in hard water is 0.010 M, and if the Na_2CO_3 is added until the carbonate ion concentration is 0.050 M, what percentage of the calcium ion has been removed from the water? (You may neglect hydrolysis of the carbonate ion.)
88. Aluminum chloride reacts with phosphoric acid to give aluminum phosphate, $AlPO_4$. The solid exists in many of the same crystal forms as SiO_2 and is used industrially as the basis of adhesives, binders, and cements.
(a) Write a balanced equation for the reaction of aluminum chloride and phosphoric acid.
(b) If you begin with 152 g of aluminum chloride and 3.00 L of 0.750 M phosphoric acid, what mass of $AlPO_4$ can be isolated?
(c) If you place 25.0 g of $AlPO_4$ in enough pure water to have a volume of exactly 1 L, what are the concentrations of Al^{3+} and PO_4^{3-} at equilibrium?
(d) If you mix 1.50 L of 0.0025 M Al^{3+} (in the form of $AlCl_3$) with 2.50 L of 0.035 M Na_3PO_4, will a precipitate of $AlPO_4$ form? If so, what mass of $AlPO_4$ precipitates?

Applying Concepts

89. Using the following symbols,
⬭ ethanol
● water
□ carbon tetrachloride
draw nanoscale diagrams for the contents of a beaker containing
(a) Water and ethanol
(b) Water and carbon tetrachloride

90. Using the following symbols,
    (oval) sugar
    (filled circle) water
    (open square) carbon tetrachloride
    draw nanoscale diagrams for the contents of a beaker containing
    (a) Water and sugar
    (b) Carbon tetrachloride and sugar
91. Refer to Figure 16.14 to determine whether the following situations would result in an unsaturated, saturated, or supersaturated solution.
    (a) 40 g NH_4Cl is added to 100 g of H_2O at 80 °C
    (b) 100 g LiCl is dissolved in 100 g of H_2O at 30 °C
    (c) 120 g $NaNO_3$ is added to 100 g of H_2O at 40 °C
    (d) 50 g Li_2SO_4 is dissolved in 200 g of H_2O at 50 °C
92. Refer to Figure 16.14 to determine whether the following situations would result in an unsaturated, saturated, or supersaturated solution.
    (a) 120 g RbCl is added to 100 g of H_2O at 50 °C
    (b) 30 g KCl is dissolved in 100 g of H_2O at 70 °C
    (c) 20 g NaCl is dissolved in 50 g of H_2O at 60 °C
    (d) 150 g CsCl is added to 100 g of H_2O at 10 °C
93. Complete the following table.

Compound	Mass of Compound	Mass of Water	Mass Fraction	Weight Percent	ppm
Table salt	52 g	175 g	______	______	______
Glucose	15 g	______	______	______	7×10^4
Methane	______	100. g	______	0.0025%	______

94. Complete the following table.

Compound	Mass of Compound	Mass of Water	Mass Fraction of Solute	Weight Percent of Solute	ppm of Solute
Lye	______	125 g	0.375	______	______
Glycerol	33 g	200 g	______	______	______
Acetylene	0.0015 g	______	______	0.0009%	______

95. Name two substances you could add to a saturated solution of calcium hydroxide to decrease the concentration of Ca^{2+} ion in solution.
96. Name two substances you could add to a saturated solution of calcium hydroxide to decrease the concentration of OH^- ion in solution.
97. If KI is added to a saturated solution of SrI_2, will the amount of solid SrI_2 present decrease, increase, or remain unchanged? What about the concentration of Sr^{2+} ion in solution?
98. What happens on the molecular level when a liquid freezes? What effect does a nonvolatile solute have on this process? Comment on the purity of water obtained by melting an iceberg.

CHAPTER 17

The Importance of Acids and Bases

These U.S. runners have just won the gold medal in the 4 × 400 m relay race at the 1996 Olympic games. The buildup of an organic acid called lactic acid has caused severe muscle pain. In a short time the pain will be gone. Acids are important in all kinds of chemistry, from biochemistry to industrial chemistry • What causes the acidic properties of molecules like lactic acid? The Summary Problem explores some of these questions. *(Gamma-Liaison)*

In Chapters 4 and 5 we defined acids and bases and discussed a few of their reactions. It is difficult to overstate the importance of acids and bases. Aqueous solutions, which are so common, almost always are acidic or basic to some degree, and aqueous solutions abound in our environment and in all living organisms. Photosynthesis and respiration, the two most important biological processes on earth, depend on acid-base reactions. Carbon dioxide, CO_2, is the most important acid-producing compound in nature. Rainwater is slightly acidic because of dissolved CO_2, and acid rain results from further acidification by the pollutants SO_2 and NO_2. The oceans are slightly basic, as are many ground and surface waters. Natural water can also be acidic; the more acidic the water, the more easily metals such as lead can be dissolved from pipes or solder joints. Acids and bases are important catalysts in natural and industrial processes.

Because of their importance, acids and bases have been studied for a long time. In 1777 Antoine Lavoisier proposed that oxygen was the element that made an acid acidic; he even derived the name *oxygen* from Greek words meaning "acid former." But in 1808 it was discovered that the gaseous compound HCl, which dissolves in water to give hydrochloric acid, contains only hydrogen and chlorine. Later it became clear that hydrogen, not oxygen, is common to all acids in aqueous solution. Also, it was shown that aqueous solutions of both acids and bases conduct electrical current, that is, they are electrolytes, and this implies the presence of ions. In 1887 the Swedish, chemist Svante Arrhenius proposed that acids *ionize in aqueous solution to produce hydrogen ions and anions;* bases *ionize to produce hydroxide ions and cations.* Our current theories of acids and bases make use of the hydronium ion, H_3O^+, as imparting acidic characteristics to aqueous solutions, and the hydroxide ion, OH^-, as imparting basic characteristics to aqueous solutions.

Protons, or hydrogen ions, because of their small size and high charge density, are always associated with water molecules in aqueous solution and are usually represented as H_3O^+, called hydronium ions (← **p. 152**).

17.1 THE BRØNSTED CONCEPT OF ACIDS AND BASES

In introducing acid-base reactions (← **p. 152**), we relied on the simple Arrhenius acid-base theory. A major problem with Arrhenius's acid-base concept, however, is that certain substances, such as ammonia, NH_3, produce basic solutions and react with acids, yet contain no hydroxide ions. In 1923 J. N. Brønsted in Denmark and T. M. Lowry in England independently proposed a new way of defining acids and bases. According to Brønsted and Lowry, an acid can donate a proton to another substance, while a base can accept a proton from another substance. For an acid-base reaction to occur, an acid must donate a proton (that is, an H^+ ion) and a base must accept the proton. NH_3 is a base because it can accept a proton, forming NH_4^+.

Because Brønsted and his students developed these ideas to a much greater extent than Lowry, this has become known as the Brønsted theory of acids and bases.

Proton Transfers in Acid-Base Reactions

To see how Brønsted's idea works, let's look at a typical acid and a typical base in aqueous solution. Nitric acid (HNO_3) is a strong electrolyte and therefore a strong acid; it ionizes essentially completely to hydronium ions and nitrate ions when it dissolves in water. To indicate complete ionization the equation is written with a single arrow.

An ionic substance that dissolves in water is a strong electrolyte (← **p. 103**). A strong acid ionizes completely and is a strong electrolyte.

$$HNO_3(aq) + H_2O(\ell) \longrightarrow H_3O^+(aq) + NO_3^-(aq)$$

nitric acid
a strong acid
100% ionized

Portrait of a Scientist • *Johannes Nicolaus Brønsted (1879–1947) and Thomas M. Lowry (1874–1936)*

The son of a civil engineer, Johannes N. Brønsted planned a career in engineering, but became interested in chemistry and switched his major in college. The University of Copenhagen selected Brønsted for a new professorship of chemistry after he earned his doctorate there in 1908. Brønsted published many important papers and books on solubility, the interaction of ions in solution, and thermodynamics. During World War II Brønsted distinguished himself by firmly opposing the Nazi takeover of Denmark. In 1947 he was elected to a seat in the Danish parliament, but he died in December of that year, before he could take his new office.

Johannes N. Brønsted. *(Special Collections, Van Pelt Library, University of Pennsylvania)*

Thomas M. Lowry was awarded his doctorate from the University of London in 1899. His positions included lecturer in chemistry at the Westminster Training College, head of the Chemical Department in Guy's Hospital Medical School, and professor of chemistry at the University of London. In 1920 he was appointed to a new chair of physical chemistry at Cambridge University. Working independently of Brønsted, Lowry developed many of the same ideas about acids and bases but did not carry them as far.

Thomas M. Lowry. *(Special Collections, Van Pelt Library, University of Pennsylvania)*

Lowry was widely recognized by organic chemists and physical chemists for his extensive studies of the optical rotatory power of optical isomers (← *p. 431*).

In contrast, a weak acid does not ionize completely, and therefore is a weak electrolyte. Hydrogen fluoride, HF, is a weak acid and its ionization in water is written as

$$HF(aq) + H_2O(\ell) \rightleftharpoons H_3O^+(aq) + F^-(aq)$$

hydrogen fluoride
weak acid
< 5% ionized

The double arrow indicates an equilibrium between the products and the reactants. The fact that the ionization is so much less than 100% means that the ionization of HF is a reactant-favored process.

A typical Brønsted base is gaseous ammonia, NH_3. Ammonia is a weak electrolyte and therefore a weak base. When it dissolves, it establishes an equilibrium with ammonium ions, $NH_4^+(aq)$, and hydroxide ions, $OH^-(aq)$. This is also a reactant-favored process, and, unless the solution is very dilute, less than 5% of the ammonia ionizes. A solution of ammonia in water is often called an *ammonium hydroxide* solution, but this is not a good name because most of what is dissolved is actually ammonia molecules, not ammonium and hydroxide ions.

$$NH_3(aq) + H_2O(\ell) \rightleftharpoons NH_4^+(aq) + OH^-(aq)$$

ammonia
a weak base
< 5% ionized

Exercise 17.1 Using Le Chatelier's Principle

Using Le Chatelier's principle (← p. 586), explain why a larger percentage of NH_3 will be ionized in a very dilute aqueous solution than in a less dilute solution.

Water's Role as Acid or Base

The most important point about HNO_3, HF, NH_3, and other Brønsted acids and bases is that in aqueous solution they all react with water molecules. A water molecule *accepts a proton* from an acid like nitric acid, while a water molecule *donates a proton* to a base like an ammonia molecule. According to the Brønsted definition, water serves as a base when an acid is present and as an acid when a base is present. Therefore, water displays both acid and base properties. A water molecule is said to be **amphiprotic**—*it can donate or accept protons,* depending on the circumstances. The two general reactions of water with acids (HA) and bases (B) are:

$$\underset{\text{acid}}{HA} + \underset{\text{base}}{H_2O} \longrightarrow H_3O^+ + A^-$$

and

$$\underset{\text{base}}{B} + \underset{\text{acid}}{H_2O} \longrightarrow BH^+ + OH^-$$

Exercise 17.2 Water's Acid-Base Reactions

Complete the following equations: (*Hint:* CH_3NH_2 and $(CH_3)_2NH$ are bases.)

(a) $HCN + H_2O \rightarrow$ (c) $CH_3NH_2 + H_2O \rightarrow$

(b) $HBr + H_2O \rightarrow$ (d) $(CH_3)_2NH + H_2O \rightarrow$

17.2 ACID-BASE BEHAVIOR AND MOLECULAR STRUCTURE

It is possible to use what we know about molecular structure to explain acid-base behavior. A Brønsted acid must contain a proton that can be donated. Hydrogen atoms that are bonded to highly electronegative atoms such as oxygen or a halogen are acidic. The highly electronegative atom attracts electrons in the bond away from the hydrogen atom, making it more like an unbonded proton and making it easier for the proton to be transferred to a base. The common strong acids (← p. 153, **Table 4.2**) fit this rule nicely. In HCl, HBr, and HI the hydrogen is bonded to a halogen atom. In HNO_3, $HClO_4$, and H_2SO_4, the hydrogen is bonded to an oxygen that is connected to another electronegative atom.

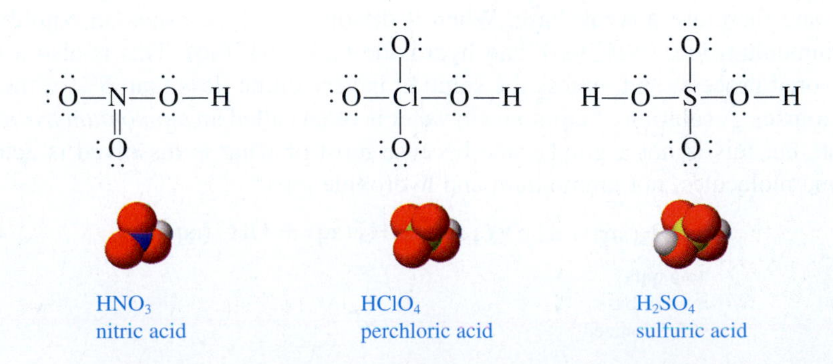

HNO_3 nitric acid

$HClO_4$ perchloric acid

H_2SO_4 sulfuric acid

These three acids are called **oxoacids,** because the hydrogen is bonded directly to oxygen. To be a strong acid, an inorganic oxoacid must have at least two more oxygen atoms than hydrogen atoms. If there are fewer, then the oxoacid will be a weak acid, as in the case of phosphoric acid, H_3PO_4.

To accept a proton and serve as a Brønsted base, a molecule or ion must have an *unshared pair of electrons.* It also helps if the ion has a negative charge, since that can attract and hold a proton. Strong Brønsted bases such as hydroxide ion, OH^-, sulfide ion, S^{2-}, oxide ion, O^{2-}, amide ion, NH_2^-, and hydride ion, H^- readily accept protons while weak Brønsted bases do so less readily. (Notice that this broadens the number of strong-base species considerably from that in Table 4.2, (⇐ *p. 153.*) In addition, there are a great many weak bases that are negative ions with unshared pairs of electrons. Examples are CN^-, F^-, and CH_3COO^-. Neutral molecules that have lone pairs, such as NH_3 and H_2O, have already been seen to serve as weak Brønsted bases.

Exercise 17.3 Brønsted Acids and Bases

Identify each molecule or ion as a Brønsted acid or base. Which acids are strong acids?
(a) HBr (b) Br^- (c) HNO_2 (d) PH_3 (e) H_2SO_3 (f) $HClO_3$ (g) CN^-

Some natural foods that contain organic acids. As you can see, citrus fruits are not the only acidic fruits. *(C.D. Winters)*

Organic Acids and Bases

Many weak acids are organic acids that contain the —COOH functional group (carboxylic acids, (⇐ *p. 110*).) Although carboxylic acid molecules generally contain many hydrogen atoms, only the hydrogen atom bound to the oxygen atom of the carboxyl group is sufficiently positive to ionize in aqueous solution. The oxygen atom of the —OH part of the carboxyl group is electronegative, and the second oxygen of the carboxyl group also pulls electrons away from the hydrogen atom. Together, these make the —OH group even more polar and its hydrogen atom more acidic. The C—H bonds in organic acids are relatively nonpolar and strong as well, and these hydrogen atoms are *not* acidic. Also, anions formed by loss of a proton, such as acetate ion, CH_3COO^-, are stabilized by resonance (⇐ *p. 375*).

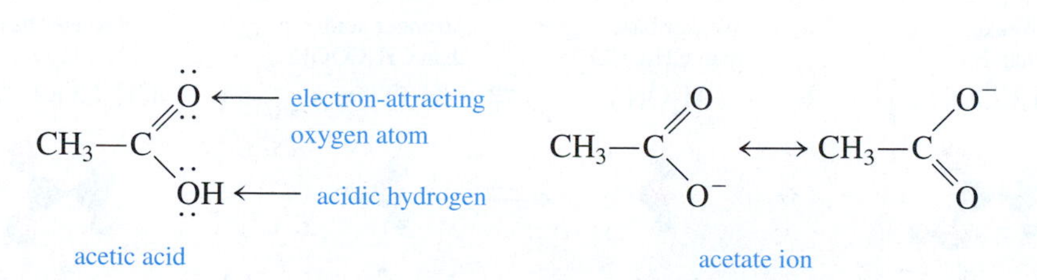

Amines are compounds that, like ammonia, have a nitrogen atom with three of its valence electrons in covalent bonds and a nonbonded electron pair on the nitrogen atom. Like ammonia, amines react as weak bases with water. Many natural compounds, for example pyridine, derived from coal tar; the bases in DNA and RNA, like thymine; and plant alkaloids like cocaine, contain amine functional groups.

Some Amine Structures

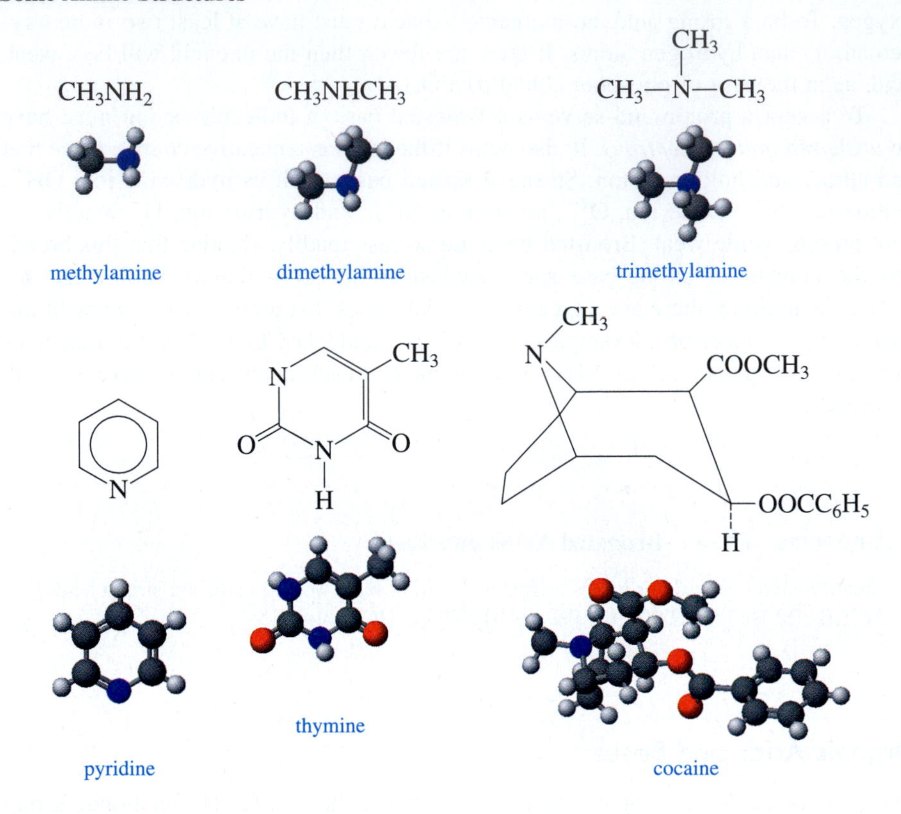

Conjugate Acid-Base Pairs

Whenever an acid donates a proton to a base, a new acid and a new base are formed. This can be understood by looking at the reaction between acetic acid, CH_3COOH, and water. The products of the reaction are a new acid, H_3O^+, and a new base, CH_3COO^-.

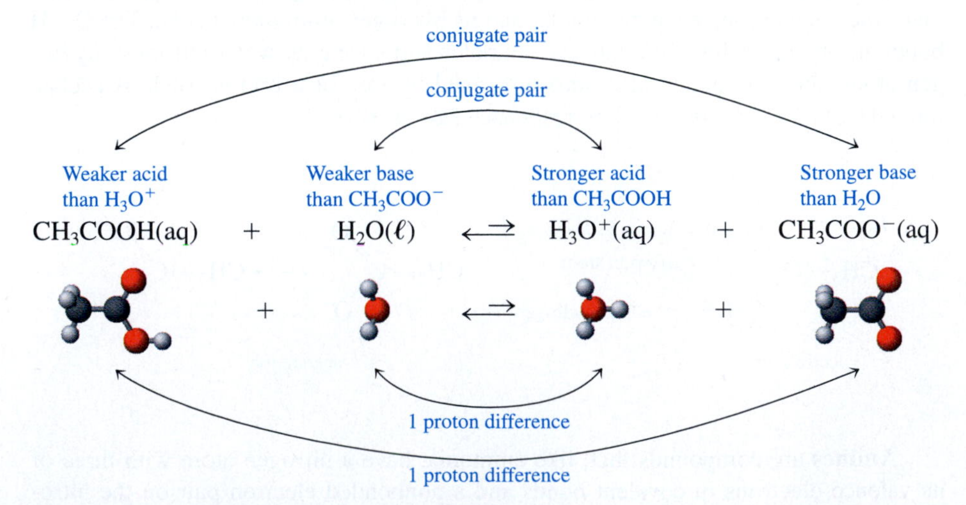

The structures of CH_3COOH and CH_3COO^- differ from one another by only a single proton, just as the structures of H_2O and H_3O^+ do.

A pair of molecules or ions *related to one another by the loss or gain of a single proton* is called a **conjugate acid-base pair.** Every Brønsted acid has a conjugate base, and every Brønsted base has a conjugate acid. The conjugate base can be derived from the formula of an acid by removing an H^+ ion; that is, by removing H^+ and making the charge of the remaining portion of the acid one unit more negative. For example, the conjugate base of the acid HCN is the CN^- ion. The conjugate acid can be derived from the formula of a base by adding an H^+ ion to the base, that is, by adding H^+ and making the charge of the base one unit more positive. So, the conjugate acid of the Cl^- ion is HCl; the conjugate acid of NH_3 is NH_4^+.

PROBLEM-SOLVING EXAMPLE 17.1 Determining Conjugate Acid-Base Pairs

Determine the conjugate acid or base of the following bases or acids.
(a) HClO (b) O^{2-} (c) CH_3COOH (d) OH^- (e) Cl^- (f) H_2SO_4

Answer (a) ClO^- (b) OH^- (c) CH_3COO^- (d) H_2O (e) HCl (f) HSO_4^-

Explanation The difference between an acid and its conjugate base or a base and its conjugate acid is one proton. If an ion or a molecule has *no* acidic proton available and it is involved in an acid-base reaction, it must act as a base and gain a proton. The ions in (b) and (e) are like this. In (b), the oxide ion can only gain a proton, so it is a base and its conjugate acid is OH^-. (You generally think of OH^- as a base, but in this context it is a conjugate acid of the O^{2-} ion.) In (e), the Cl^- has no proton, so it is a base, and upon accepting a proton, becomes HCl. The hydroxide ion (d), although it contains a proton, is a base in aqueous solution and will gain a proton, so its conjugate acid is H_2O. In (a), (c), and (f), acidic protons are present, so they are acids and their conjugate bases are those ions with one less proton. For (a) the conjugate base is ClO^-. For (c) the conjugate base is CH_3COO^-. For (f) the conjugate base is HSO_4^-.

Problem-Solving Practice 17.1

(a) What is the conjugate base of H_2CO_3? of HNO_3? (b) What is the conjugate acid of NH_3? of CN^-?

Exercise 17.4 Determining Conjugate Acids and Bases

Fill in the blanks in the following table of acids and bases.

Acid	Conjugate Base	Base	Conjugate Acid
$H_2PO_4^-$	______	______	H_2O
______	PO_4^{3-}	NH_2^-	______
HSO_4^-	______	ClO_4^-	______
HF	______	______	HBr

Exercise 17.5 Making an Acid Act As a Base

Draw the Lewis structure for acetic acid. Now suppose that some acetic acid were dissolved in pure sulfuric acid, a very strong acid. What kind of acid-base chemistry might take place and what reactions would occur? If acetic acid molecules were to act as a base, what structural feature would allow them to accept protons from sulfuric acid molecules?

17.3 POLYPROTIC ACIDS AND BASES

So far we have discussed mainly Brønsted acids that can donate a single proton per molecule, such as hydrogen fluoride (HF), hydrogen chloride (HCl), nitric acid (HNO_3), and hydrocyanic acid (HCN). These are called **monoprotic acids.** Similarly, a base that can accept only one proton per molecule, such as NH_3, is called a **monoprotic base.**

Some acids can donate more than one proton. These are called **polyprotic acids** and include sulfuric acid (H_2SO_4), carbonic acid (H_2CO_3), phosphoric acid (H_3PO_4), oxalic acid ($H_2C_2O_4$ or $HO_2C{-}CO_2H$), and other organic acids with two or more carboxyl ($-CO_2H$) groups.

Many chemical reactions take place in steps that can be represented by individual equations. Sometimes only the overall reaction is seen and only the overall equation is written.

In aqueous solution, a polyprotic acid such as H_2SO_4 donates its protons to water molecules in a stepwise manner. In the first step, hydrogen sulfate ion is formed. Sulfuric acid is a strong acid, and so this first ionization is complete.

$$\underset{\text{acid}}{H_2SO_4(aq)} + \underset{\text{base}}{H_2O(\ell)} \longrightarrow \underset{\text{base}}{HSO_4^-(aq)} + \underset{\text{acid}}{H_3O^+(aq)}$$

Hydrogen sulfate ion, HSO_4^-, then donates a proton to another water molecule. In this case an equilibrium is established—hydrogen sulfate ion is a weak acid.

$$\underset{\text{acid}}{HSO_4^-(aq)} + \underset{\text{base}}{H_2O(\ell)} \rightleftharpoons \underset{\text{base}}{SO_4^{2-}(aq)} + \underset{\text{acid}}{H_3O^+(aq)}$$

The hydrogen sulfate ion is amphiprotic and can serve as a monoprotic base or a monoprotic acid. However, it is known to act as a weak acid in aqueous solution and is often used to increase the acidity of spas and swimming pools when added as $NaHSO_4$. It is also an ingredient in toilet bowl cleaners.

Exercise 17.6 Polyprotic Acids

(a) Write equations for two stepwise reactions that occur when oxalic acid (HOOC—COOH) acts as a polyprotic acid in aqueous solution. (b) Write equations for the three ionization steps for phosphoric acid, H_3PO_4.

Exercise 17.7 Explaining Acid Strengths

Look at the charge on the hydrogen sulfate ion. What does this have to do with the fact that this ion is a weaker acid than H_2SO_4?

There are also **polyprotic bases,** substances that can accept more than one proton, although they do it one proton at a time. The most common polyprotic bases are the anions of the polyprotic acids such as PO_4^{3-} (from phosphoric acid), CO_3^{2-} (from carbonic acid), and $C_2O_4^{2-}$ (from oxalic acid). The very strong polyprotic base sulfide ion, S^{2-}, can accept two protons. The first reaction goes to completion, producing hydrogen sulfide ion.

$$\underset{\text{base}}{S^{2-}(aq)} + \underset{\text{acid}}{H_2O(\ell)} \longrightarrow \underset{\text{acid}}{HS^-(aq)} + \underset{\text{base}}{OH^-(aq)}$$

TABLE 17.1 Polyprotic Acids and Bases

Acid Form	Amphiprotic Form	Base Form
H_2S (hydrosulfuric acid)	HS^- (hydrogen sulfide or bisulfide ion)	S^{2-} (sulfide ion)
H_3PO_4 (phosphoric acid)	$H_2PO_4^-$ (dihydrogen phosphate ion)	HPO_4^{2-} (hydrogen phosphate ion)
$H_2PO_4^-$ (dihydrogen phosphate ion)	HPO_4^{2-} (hydrogen phosphate ion)[1]	PO_4^{3-} (phosphate ion)
H_2CO_3 (carbonic acid)	HCO_3^- (hydrogen carbonate or bicarbonate ion)[1]	CO_3^{2-} (carbonate ion)
$H_2C_2O_4$ (oxalic acid)	$HC_2O_4^-$ (hydrogen oxalate ion)	$C_2O_4^{2-}$ (oxalate ion)

[1]The amphiprotic nature of HPO_4^{2-} and HCO_3^- is especially important in the chemical reactions that occur in living organisms. (See Section 17.9.)

Then the second reaction produces hydrosulfuric acid, H_2S.

$$\underset{\text{base}}{HS^-(aq)} + \underset{\text{acid}}{H_2O(\ell)} \rightleftharpoons \underset{\text{acid}}{H_2S(aq)} + \underset{\text{base}}{OH^-(aq)}$$

The amphiprotic species HS^- is formed in the first step and reacts in the second.

For an amphiprotic species like HS^-, a conjugate acid-base pair is defined by whether a proton is gained or lost. When HS^- acts as an acid, the conjugate pair is HS^-/S^{2-}; when it acts as a base, the conjugate pair is H_2S/HS^-. The relations among polyprotic acids and bases and their amphiprotic anions are shown in Table 17.1.

Exercise 17.8 Polyprotic Bases

(a) Write equations for three stepwise reactions that occur when PO_4^{3-} ion acts as a polyprotic base in aqueous solution. (b) Write equations for the two stepwise reactions that occur when CO_3^{2-} ion acts as a polyprotic base in aqueous solution.

Exercise 17.9 Polyprotic Amines

The compound ethylenediamine has the molecular formula $C_2H_8N_2$ *(← p. 386).* Write the condensed structure for this molecule and explain why 1 mol of it reacts with 2 mol of HCl.

Sulfuric Acid

Sulfuric acid, the number one chemical in commerce (see inside back cover), is used in the petroleum industry, in the production of steel, in automobile batteries, and in the manufacture of fertilizers and many other products. Since sulfuric acid costs less to make than any other acid, it is the first to be considered when an acid is needed.

Sulfur, in very pure form, is found in underground deposits in the United States along the coast of the Gulf of Mexico. It is recovered by the *Frasch process* (Figure 17.1). Large quantities of sulfur are also produced from petroleum (which often contains sulfur compounds). Sulfuric acid can also be produced directly from sulfur

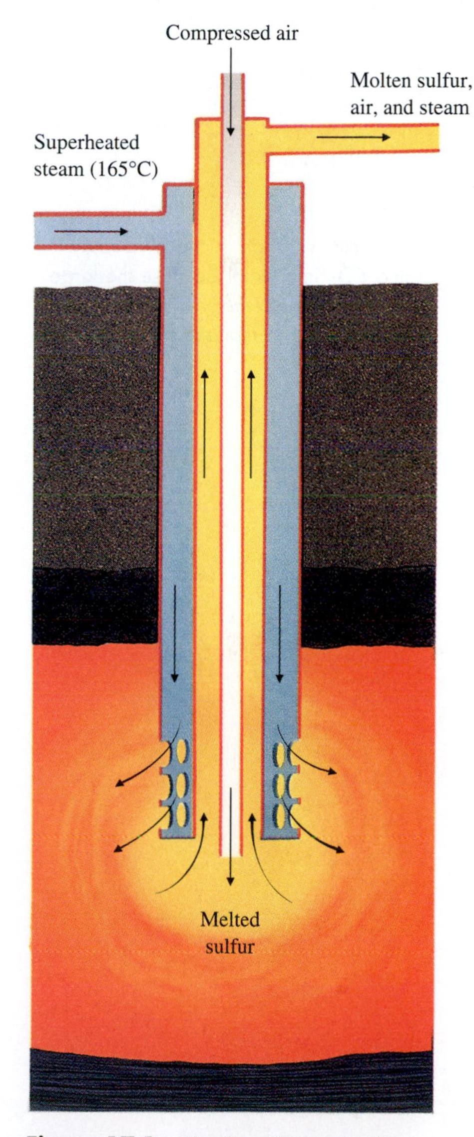

Figure 17.1 The Frasch process for mining sulfur. Superheated steam is injected along with compressed air into a sulfur-bearing stratum underground. Molten sulfur froths out of the inner tube. Most of the sulfur mined is converted to sulfuric acid.

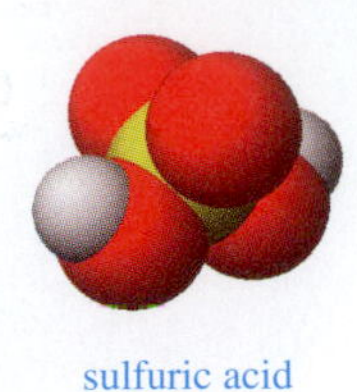
sulfuric acid

dioxide originating from copper or lead smelting. Unless this sulfur dioxide is recovered it will pollute the environment.

Sulfur is converted to sulfuric acid in four steps, collectively called the *contact process.* In the first step the sulfur is burned in air to give mostly sulfur dioxide:

$$S_8(s) + 8\ O_2(g) \longrightarrow 8\ SO_2(g)$$

The SO_2 is then converted to SO_3 over a hot, catalytically active surface, such as platinum or vanadium pentoxide:

$$2\ SO_2(g) + O_2(g) \xrightarrow{\text{catalyst}} 2\ SO_3(g)$$

The next step converts the sulfur trioxide to sulfuric acid by the addition of a water molecule to an SO_3 molecule. The best way to do this is to pass the SO_3 into H_2SO_4 to form pyrosulfuric acid.

$$SO_3(g) + \underset{\text{sulfuric acid}}{H_2SO_4(\ell)} \longrightarrow \underset{\text{pyrosulfuric acid}}{H_2S_2O_7(\ell)}$$

Sulfuric acid is not produced by directly passing SO_3 into water because the large quantity of thermal energy evolved creates a stable, difficult to manage, fog of the acid.

and then to dilute the $H_2S_2O_7$ with water:

$$H_2S_2O_7(\ell) + H_2O(\ell) \longrightarrow 2\ H_2SO_4(aq)$$

The net reaction is one H_2SO_4 molecule for every SO_3 molecule.

More than half of the sulfuric acid manufactured is used to make phosphate fertilizers. Phosphorus occurs in nature as calcium phosphate in apatite minerals with the general formula $Ca_5(PO_4)_3X$, where X is an anion like F^-, OH^-, or Cl^-. These

Pure sulfur. Once sulfur is brought to the surface by the Frasch process, and cools so that it solidifies, it is handled with power shovels much like rock or dirt. *(F. Grehan/Photo Researchers, Inc.)*

minerals are not sufficiently water soluble to allow plants to uptake the needed phosphate ions. By treating the insoluble phosphate with sulfuric acid, a water-soluble dihydrogen phosphate is formed. The reaction with fluoroapatite is

$$\underset{\text{fluoroapatite}}{2\ Ca_5(PO_4)_3F(s)} + 7\ H_2SO_4(aq) + 17\ H_2O(\ell) \longrightarrow \underset{\text{calcium dihydrogen phosphate}}{3\ Ca(H_2PO_4)_2 \cdot H_2O(s)} + \underset{\text{gypsum}}{7\ CaSO_4 \cdot 2\ H_2O(s)} + 2\ HF(g)$$

Phosphoric Acid

Phosphoric acid is annually among the top ten most-produced industrial chemicals (see inside back cover). It is made from sulfuric acid by reaction with fluoroapatite:

$$Ca_5(PO_4)_3F(s) + 5\ H_2SO_4(aq) + 10\ H_2O(\ell) \longrightarrow 3\ H_3PO_4(aq) + 5\ CaSO_4 \cdot 2\ H_2O(s) + HF(g)$$

Note how in this reaction the molar ratio of a fluoroapatite to sulfuric acid is 1:5 while in the reaction producing calcium dihydrogen phosphate the molar ratio is 2:7. Different ratios result in different reaction products.

Phosphoric acid is also made in a three-step process by first reacting fluoroapatite with sand (SiO_2) and coke (a form of carbon) in an electric arc furnace at 2000 °C.

$$4\ Ca_5(PO_4)_3F(s) + 18\ SiO_2(s) + 15\ C(s) \longrightarrow 18\ CaSiO_3(s) + 2\ CaF_2(\ell) + 15\ CO_2(g) + 3\ P_4(g)$$

After the gaseous phosphorus is condensed, it is burned in oxygen to make the oxide P_4O_{10}, which is then reacted with water to give phosphoric acid, a weak acid.

$$P_4(s) + 5\ O_2(g) \longrightarrow P_4O_{10}(s)$$

$$P_4O_{10}(s) + 6\ H_2O(\ell) \longrightarrow 4\ H_3PO_4(aq)$$

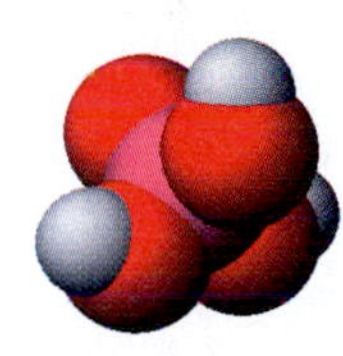

phosphoric acid

Most phosphoric acid is used to make fertilizers and in treating steel products. Phosphoric acid made from highly purified phosphorus is used to manufacture food additives and detergents (⇐ *p. 758*).

Phosphoric acid is one of the ingredients in soft drinks.

17.4 THE AUTOIONIZATION OF WATER

Even when water is carefully purified, it conducts a very tiny electrical current. This indicates that pure water contains a very small concentration of ions. These ions are formed when water molecules react to produce hydronium ions and hydroxide ions in a process called **autoionization.**

$$\underset{\text{base}}{H_2O(\ell)} + \underset{\text{acid}}{H_2O(\ell)} \rightleftharpoons H_3O^+(aq) + OH^-(aq)$$

In this reaction, one water molecule serves as a proton acceptor (base) while the other is a proton donor (acid). The equilibrium between the water molecules and the hydronium and hydroxide ions is very reactant-favored, and so the concentrations of the ions in pure water are very low (1.0×10^{-7} M at 25 °C). Nevertheless, autoionization of water is very important to understanding how acids and bases function in aqueous solutions. As in the case of any equilibrium reaction, an equilibrium-constant expression can be written for autoionization of water:

$$2\ H_2O(\ell) \rightleftharpoons H_3O^+(aq) + OH^-(aq) \qquad K_w = [H_3O^+][OH^-]$$

Recall (⇐ ***p. 574***) that an equilibrium-constant expression includes concentrations of solutes, but not the concentration of the solvent, which in this case is water.

TABLE 17.2 Temperature Dependence of K_w for Water

T (°C)	K_w
10	0.29×10^{-14}
15	0.45×10^{-14}
20	0.68×10^{-14}
25	1.01×10^{-14}
30	1.47×10^{-14}
50	5.48×10^{-14}

This equilibrium constant is given a special symbol, K_w, and is known as the **ionization constant for water.** From electrical conductivity measurements of pure water, we know that $[H_3O^+] = [OH^-] = 1.0 \times 10^{-7}$ M at 25 °C,

$$K_w = [H_3O^+][OH^-] = (1.0 \times 10^{-7})(1.0 \times 10^{-7}) = 1.0 \times 10^{-14} \text{ (at 25 °C)}$$

The equation $K_w = [H_3O^+][OH^-]$ is valid in pure water and in any aqueous solution. However, the value of K_w is temperature-dependent (Table 17.2.)

Exercise 17.10 Temperature Dependence of K_w

Is the autoionization of water endothermic or exothermic? Explain your answer.

In aqueous solution, the $[H_3O^+]$ and $[OH^-]$ values are inversely related; if one increases, the other must decrease.

According to the K_w expression, the product of the hydronium-ion concentration and the hydroxide-ion concentration will always remain the same at a given temperature. If the hydronium-ion concentration increases (because an acid was added to the water, for example), then the hydroxide-ion concentration must decrease, and vice versa. The equation also tells us that if we know one concentration, the other can be calculated, which means that it is unnecessary to try to measure it.

When the concentrations of $[H_3O^+]$ and $[OH^-]$ are equal, a solution is said to be **neutral.** If either an acid or a base is added to a neutral solution, the autoionization equilibrium between H_3O^+ and OH^- will be disturbed. Recall that according to Le Chatelier's principle *(← p. 586),* an equilibrium shifts in such a way as to offset the effect of any disturbance. When an acid is added, the concentration of H_3O^+ ions increases. To oppose this increase, a small fraction of the added H_3O^+ ions react with OH^- ions to form H_2O, thereby reducing the $[OH^-]$ until once again equilibrium is reestablished and $[H_3O^+][OH^-] = 1.0 \times 10^{-14}$ at 25 °C. Similarly, if a base is added to water, a few of the added OH^- ions react with H_3O^+ ions to form H_2O, thereby decreasing the $[H_3O^+]$. When a new equilibrium is achieved, the mathematical product $[H_3O^+][OH^-]$ again equals 1.0×10^{-14}. For aqueous solutions at 25 °C we can write

- **Neutral solution:** $[H_3O^+]$ equals $[OH^-]$; both equal to 1.0×10^{-7} M
- **Acidic solution:** $[H_3O^+]$ greater than 1.0×10^{-7} M; $[OH^-]$ less than 1.0×10^{-7} M
- **Basic solution:** $[OH^-]$ greater than 1.0×10^{-7} M; $[H_3O^+]$ less than 1.0×10^{-7} M

17.5 THE pH SCALE

The $[H_3O^+]$ and $[OH^-]$ in an aqueous solution can vary widely depending on the acid or base present and its concentration. Since nitric acid is a strong acid (100% ionized), a 6.0 M nitric acid solution has a $[H_3O^+]$ of 6.0 mol/L, so substituting in $[H_3O^+][OH^-] = 1.0 \times 10^{-14}$,

$$(6.0)[OH^-] = 1.0 \times 10^{-14}$$

and

$$[OH^-] = \frac{1.0 \times 10^{-14}}{6.0} = 1.7 \times 10^{-15} \text{ mol/L}$$

On the other hand, consider the strong base NaOH. A 6.0 mol/L solution of NaOH has a $[OH^-]$ of 6.0 mol/L, and a similar calculation gives $[H_3O^+] = 1.7 \times 10^{-15}$ mol/L. The $[H_3O^+]$ in aqueous solutions can range from about 10 mol/L down to about 10^{-15} mol/L. The $[OH^-]$ can also vary over the same range in aqueous solution.

Therefore, it is more convenient to express these concentrations in terms of logarithms (exponents). The **pH** of a solution is defined as *the negative of the base-10 logarithm (log) of the hydronium-ion concentration.*

To review logarithms and their use, see Appendix B.3.

$$pH = -\log[H_3O^+]$$

The negative logarithm of the small concentration values is used since it gives a positive value. Thus, the pH of pure water at 25 °C is given by

$$pH = -\log[1.0 \times 10^{-7}] = 7.00$$

In terms of pH, for solutions at 25 °C we can write

- **Neutral solutions:** pH = 7
- **Acidic solutions:** pH < 7
- **Basic solutions:** pH > 7

We can define pOH in a manner similar to the definition of pH:

$$pOH = -\log[OH^-]$$

The pOH of pure water at 25 °C is also 7.00. Because the values of $[H_3O^+]$ and $[OH^-]$ are related by the K_w expression, for all aqueous solutions at 25 °C, we can write

$$K_w = [H_3O^+] \times [OH^-] = 1.0 \times 10^{-14}$$

This can be rewritten as

$$-\log K_w = -\log[H_3O^+] + (-\log[OH^-]) = -\log(1.0 \times 10^{-14})$$

or

$$pK_w = pH + pOH = 14.00$$

If $a \times b = c$, then taking the logarithm of both sides of the equation gives $\log a + \log b = \log c$, or $-\log a + (-\log b) = -\log c$.

The relation between pH and pOH can be used to find one value when the other is known. A 0.0010 mol/L solution of the strong base NaOH, for example, has an OH^- concentration of 0.0010 mol/L and a pOH given by

Conceptual Challenge Problems CP-17.A and CP-17.B at the end of the chapter relate to topics covered in this section.

$$pOH = -\log[1.0 \times 10^{-3}] = 3.00$$

and therefore a pH given by

$$pH = 14.00 - pOH = 14.00 - 3.00 = 11.00$$

For solutions in which $[H_3O^+]$ or $[OH^-]$ has a value other than an exact power of ten (1, 0.1, 0.01, . . .) a calculator is convenient for finding the pH (see Appendix B.3). For example, the pH of a solution that contains 0.0040 mol of the strong acid HNO_3 per liter is 2.40.

$$pH = -\log[4.0 \times 10^{-3}] = 2.40$$

The digits to the left of the decimal point in a pH represent a power of ten; only the digits to the right of the decimal point are significant. In Example 17.2, where $pH = -\log(6.86 \times 10^{-2}) = -\log(6.86) + (-\log(10^{-2})) = -0.836 + 2.000 = 1.164$, there are three significant figures in the result, because there are three significant figures in −0.836.

PROBLEM-SOLVING EXAMPLE 17.2 Calculating pH from $[H_3O^+]$

If an aqueous solution of HCl has a volume of 500. mL and contains 1.25 g of HCl, what is its pH?

Answer 1.164 (*Note:* This pH has been calculated to three significant figures. In actual measurements, pH values are seldom obtainable to this degree of accuracy.)

Explanation Hydrochloric acid is a strong acid, and so every mole of HCl that dissolves produces a mole of H_3O^+ and a mole of Cl^-. First, determine the amount (moles) of HCl.

Notice that, as in the case of equilibrium constants, the concentration units of mol/L are ignored when the logarithm is taken. It is impossible to take the logarithm of a unit.

$$\text{Amount of HCl} = 1.25 \text{ g HCl} \times \frac{1 \text{ mol HCl}}{36.461 \text{ g HCl}} = 0.03428 \text{ mol HCl}$$

Next, calculate the H_3O^+ concentration.

$$[H_3O^+] = \frac{0.03428 \text{ mol } H_3O^+}{0.500 \text{ L}} = 0.0686 \text{ mol/L} = 0.0686 \text{ M}$$

Then express this concentration as pH.

$$\text{pH} = -\log(6.86 \times 10^{-2}) = 1.164$$

Problem-Solving Practice 17.2

Calculate the pH of a 0.040 M NaOH solution.

Figure 17.2 shows the pH values of some common solutions. Keep in mind that a change of one pH unit represents a 10-fold change in H_3O^+ concentration, two pH units represent a 100-fold change, and so on. Thus, according to Figure 17.2, the $[H_3O^+]$ in lemon juice is more than 100 times greater than that in tomato juice.

If you know the pH, then the pOH is just 14.00 − pH; if you know the pOH, then the pH = 14.00 − pOH.

PROBLEM-SOLVING EXAMPLE 17.3 Calculating $[H_3O^+]$ from pH

The measured pH of a sample of sea water is 8.30. What is the H_3O^+ concentration?

Answer 5.0×10^{-9} mol/L

Explanation Substituting into the definition of pH,

$$-\log[H_3O^+] = 8.30 \quad \text{and so} \quad \log[H_3O^+] = -8.30$$

By the rules of logarithms, $10^{\log(x)} = x$, so we can write $10^{\log[H_3O^+]} = 10^{-\text{pH}} = [H_3O^+]$ or

$$[H_3O^+] = 10^{-8.30} = 5.0 \times 10^{-9} \text{ mol/L}$$

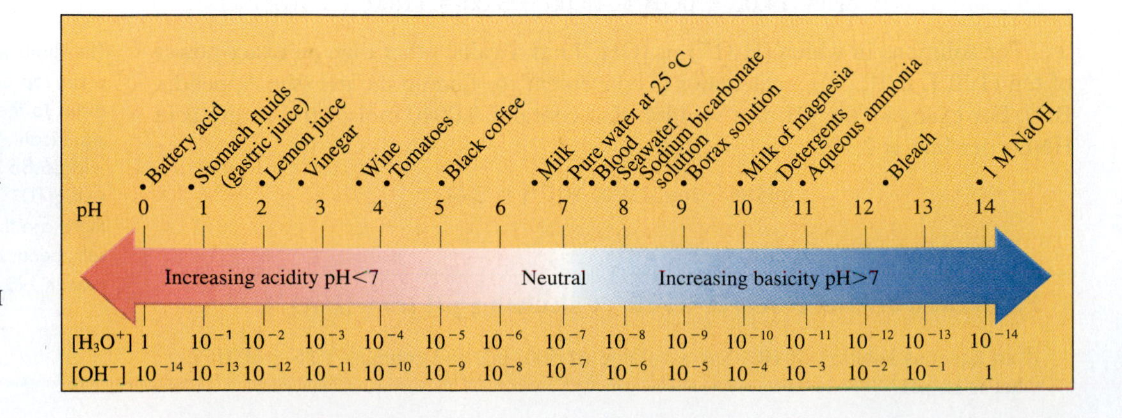

Figure 17.2 The pH of various aqueous solutions. The relationship between pH and the concentration of hydronium ions $[H_3O^+]$ and hydroxide ions $[OH^-]$ in water at 25 °C. The pH values of some common substances are also included in the diagram.

Problem-Solving Practice 17.3

What is the pH of a solution that has $[H_3O^+] = 1.5 \times 10^{-10}$ M? If the $[H_3O^+]$ increases by a factor of 3000, what will the new pH be? Is this new solution acidic?

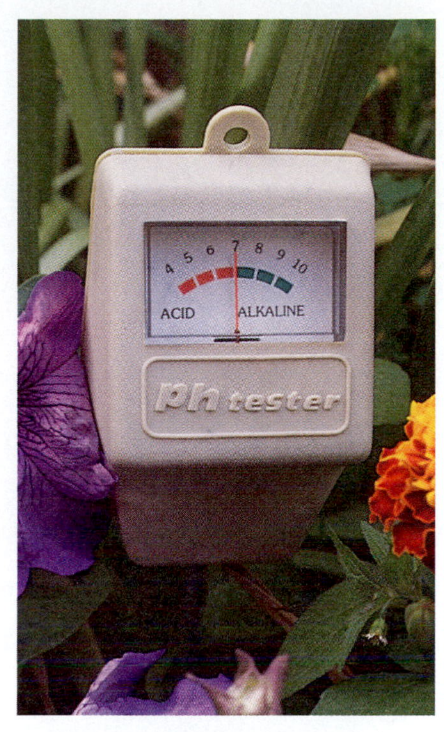

Measuring soil pH. pH meters like this one can be purchased in garden supply stores and can come in handy for serious gardening. *(©Leonard Lessin/Peter Arnold, Inc.)*

Exercise 17.11 pH of Solutions of Different Acids

Would the pH of a 0.1 M solution of the strong acid HNO_3 be the same as the pH of a 0.1 M solution of the strong acid HCl? Explain.

17.6 IONIZATION CONSTANTS OF ACIDS AND BASES

Earlier, you learned *(p. 579)* that the greater the equilibrium constant for a reaction, the more product-favored that reaction is. The more product-favored the reaction, the stronger the acid and the base on the left-hand side of an ionization equation. Consequently, equilibrium constants can give us an idea about the relative strengths of weak acids and bases.

For any acid represented by the general formula HA, an ionization reaction can be written as

$$HA(aq) + H_2O(\ell) \rightleftharpoons H_3O^+(aq) + A^-(aq)$$

and the corresponding equilibrium-constant expression is

$$K_a = \frac{[H_3O^+][A^-]}{[HA]}$$

The equilibrium constant, K_a, is called the **acid-ionization constant.** The larger the acid-ionization constant, the stronger the acid. For a strong acid such as hydrochloric acid, HCl(aq), the equilibrium lies so far to the right that there are essentially no acid molecules left non-ionized and the K_a value is much larger than 1.

The ionization reaction of a strong acid is also an equilibrium, but double arrows are not used because the equilibrium lies so far to the right.

In contrast, weak acids such as acetic acid ionize to a smaller extent. They establish equilibria in which significant concentrations of non-ionized weak-acid molecules are still present in the solution. The ionization of a weak acid is reactant-favored, and all weak acids have K_a values less than 1. If a pH measurement is made on a 0.10 M acetic acid solution, the pH is found to be 2.88. This means that the concentration of H_3O^+ is only 1.3×10^{-3}mol/L. Compare this value with the 0.1 mol/L concentration of H_3O^+ ions in a 0.1 M HCl solution. In acetic acid one mole of acetate ion is produced for every mole of hydronium ion produced, and so the concentration of CH_3COO^- must be the same as the concentration of H_3O^+. These concentrations are only a little more than 1% of the initial concentration of acetic acid. Therefore, almost 99% of the acetic acid remains in the non-ionized form. This is why weak acids (and bases) are weak electrolytes. Their aqueous solutions contain such low concentrations of ions that they conduct electricity only slightly.

Autoionization of water results in $[H_3O^+] = 1.0 \times 10^{-7}$ mol/L initially, but the presence of even a little weak acid adds H_3O^+ to the solution and causes the autoionization equilibrium

$$2\ H_2O(\ell) \rightleftharpoons H_3O^+(aq) + OH^-(aq)$$

to shift to the left, making water's contribution of H_3O^+ negligible at equilibrium.

To better keep up with the ions and molecules present in aqueous solutions of weak acids or bases, we can first write the equilibrium equation and place under it a table that describes the concentrations of the species before and after the establishment of equilibrium. For 0.100 M acetic acid, we have

	$CH_3COOH(aq)$ +	$H_2O(\ell) \rightleftarrows$	$H_3O^+(aq)$ +	$CH_3OO^-(aq)$
	acid	base	acid	base
Initial Concentration (mol/L)	0.100		1.0×10^{-7} (from water)[1]	0
Change as Reaction Occurs (mol/L)	−0.0013		+0.0013	+0.0013
Equilibrium Concentration (mol/L)	0.099		0.0013	0.0013

[1]Initially, the autoionization of water is the only source of H_3O^+ ions. When the equilibrium concentration of H_3O^+ is calculated, adding 1.0×10^{-7} to 0.0013 has no significant effect; the result is still 0.0013.

In an acetic acid solution, or a solution of any weak acid, there are two different bases competing for protons that can be donated from two different acids. In the reaction above, the two bases are water and acetate ion, and the two acids are acetic acid and hydronium ions. Since the equilibrium favors the reactants, the acetate ion must be a more effective (stronger) base than the water molecule. Another way of looking at the same reaction is that the hydronium ion must be a stronger acid than the acetic acid molecule. Both of these statements are true. Both statements explain why the equilibrium is reactant-favored.

Exercise 17.12 Estimating Acid Strength from K_a Values

The K_a for one acid, HA_1, is 10^{-4}, while K_a for another acid, HA_2, is 10^{-7}. Write equations for the equilibrium reactions of these acids in water. Which is the weaker acid, HA_1 or HA_2? Which reaction is more reactant-favored? Which one is more product-favored?

A similar general equation can be written for the acceptance of a proton from water by a base, B.

$$B(aq) + H_2O(\ell) \rightleftarrows BH^+(aq) + OH^-(aq)$$

(For example, if the base B were NH_3, then BH^+ would be NH_4^+.) The corresponding equilibrium-constant expression is

$$K_b = \frac{[BH^+][OH^-]}{[B]}$$

The equilibrium constant, K_b, is called the **base-ionization constant.**

When the base is an anion, A (such as the anion of a weak acid), the general equation is

$$A^-(aq) + H_2O(\ell) \rightleftarrows HA(aq) + OH^-(aq)$$

and the corresponding equilibrium constant expression is

$$K_b = \frac{[HA][OH^-]}{[A^-]}$$

The larger the base-ionization constant, the more product-favored the ionization reaction and the stronger the base. For a strong base the ionization constant is greater than 1; for a weak base the ionization constant is less than 1.

Exercise 17.13 Writing Ionization Constant Expressions

Write the ionization equations and ionization-constant expressions for the acids H_2SO_3, HF, HCN, and HSO_3^-. Do the same for the bases NO_2^-, HS^-, CN^-, and PO_4^{3-}.

Exercise 17.14 Estimating the Strength of Bases from K_b Values

The K_b for a base, B_1, is 10^{-3}, while K_b for another base, B_2, is 10^{-6}. Which base is stronger, B_1 or B_2? If 0.1 M aqueous solutions of each of these bases were prepared, which one would have the higher pH?

There are several experimental methods for determining acid- or base-ionization constants. The simplest (but not the most accurate) is based on the ideas we have just been discussing. The stronger an acid, the more its ionization reaction lies toward the products, that is, the greater the $[H_3O^+]$ in the solution. Therefore, if we measure the pH of several acid solutions of the same concentration, the lower the pH is, the stronger the acid is. Knowing both the acid concentration and the measured pH allows us to calculate the K_a for the acid.

PROBLEM-SOLVING EXAMPLE 17.4 Determining K_a for an Acid from pH Measurement

The pH of a 0.10 M solution of propanoic acid (CH_3CH_2COOH, a weak organic acid) is measured after equilibrium has been established and is found to be 2.93. What is the K_a of this acid?

Answer 1.5×10^{-5}

Explanation The acid ionizes according to the balanced equation

$$CH_3CH_2COOH(aq) + H_2O(\ell) \rightleftharpoons H_3O^+(aq) + CH_3CH_2COO^-(aq)$$

The pH gives us the equilibrium concentration of H_3O^+. Using the definition of pH,

$$[H_3O^+] = 10^{-pH} = 10^{-2.93} = 1.2 \times 10^{-3} \text{ mol/L}$$

The equilibrium concentrations of the other species can be determined by using the following table.

	CH_3CH_2COOH	+	H_2O	$\rightleftharpoons$	H_3O^+	+	$CH_3CH_2COO^-$
Initial Concentration (mol/L)	0.10				1.0×10^{-7} (from water)*		0

*Again, the low concentration can be ignored because it is so small.

(Table cont'd on p. 782)

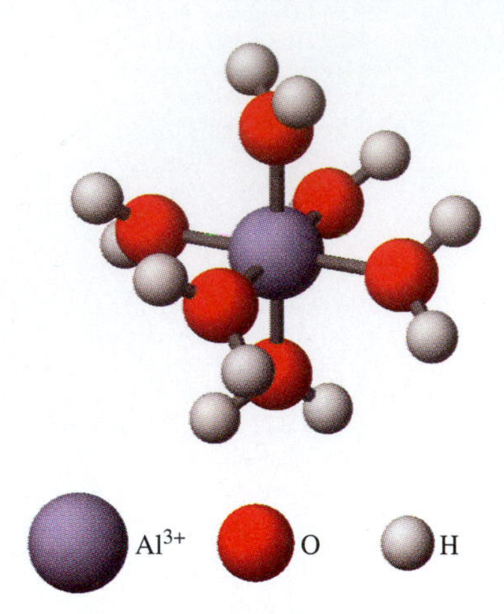

Figure 17.3 A metal ion surrounded by water molecules in aqueous solution $[M(H_2O)_6]^{n+}$. The attraction between water molecules bound to a positively charged metal ion is strong enough that the water molecules become acidic, losing a proton to a base, such as another water molecule that happens to pass by. The resulting metal ion would have the formula $[M(H_2O)_5(OH)]^{(n-1)+}$.

	CH_3CH_2COOH + H_2O ⇌	H_3O^+ +	$CH_3CH_2COO^-$
Change as Reaction Occurs (mol/L)	$-x$	$+x$	$+x$
Equilibrium Concentration (mol/L)	$0.10 - x$	x	x

In the table above, the quantity x represents the concentration of both H_3O^+ and $CH_3CH_2COO^-$ at equilibrium. From the measured pH we have calculated $[H_3O^+]$ to be 1.2×10^{-3} mol/L. This means $x = 1.2 \times 10^{-3}$ mol/L, and $[CH_3CH_2COO^-] = 1.2 \times 10^{-3}$ mol/L because one $CH_3CH_2COO^-$ is produced for every H_3O^+ produced by the ionization of the acid. The value x also represents the amount of acid that has ionized per liter of solution at equilibrium. Since we started with an acid concentration of 0.10 M, the concentration of the acid at equilibrium is $0.10 - x$. Using these values, we can now derive K_a for propanoic acid.

$$K_a = \frac{[H_3O^+][CH_3CH_2COO^-]}{[CH_3CH_2COOH]} = \frac{(x)(x)}{0.10 - x} = \frac{(0.0012)(0.0012)}{0.10 - 0.0012} = 1.5 \times 10^{-5}$$

This K_a is small, indicating that propanoic acid is a weak acid. It is similar to acetic acid in strength, and a 0.10 M solution has a pH nearly the same as that of 0.10 M acetic acid.

Problem-Solving Practice 17.4

Lactic acid is a monoprotic acid that occurs naturally in sour milk and arises from metabolism in the human body. A 0.10 M aqueous solution of lactic acid, $HC_3H_5O_3$, has a pH of 2.43. What is the value of K_a for lactic acid? Is lactic acid stronger or weaker than propanoic acid?

When a molecular formula is used to represent an organic acid, the acidic hydrogen or hydrogens are often written first. Thus, $HC_3H_5O_3$ shows that lactic acid has one acidic hydrogen atom, which is the one in the COOH group.

```
CH3CHCOOH
   |
   OH
```

lactic acid

Some hydrated metal ions (p. 115), especially those of the transition metals, are also weak acids. When a salt containing a metal ion dissolves in water, the metal ion becomes hydrated, often by having six water molecules around it, $[M(H_2O)_6]_n^+$, where M represents a metal ion whose charge is n^+. Metal ions other than those in Groups 1A and 2A have great enough charges and small enough sizes to attract protons away from the water molecules. This makes the O—H bonds in the water molecule more polar and the protons more acidic than they would be in a water molecule that is not bonded to a metal ion (Figure 17.3). Thus the $[M(H_2O)_6]_n^+$ ion can donate protons, and the solution becomes acidic.

The ionization reaction and ionization constant for a hydrated metal ion such as Fe^{3+}(aq) can be written as

$$Fe(H_2O)_6^{3+}(aq) + H_2O(\ell) \rightleftharpoons Fe(H_2O)_5(OH)^{2+}(aq) + H_3O^+(aq)$$

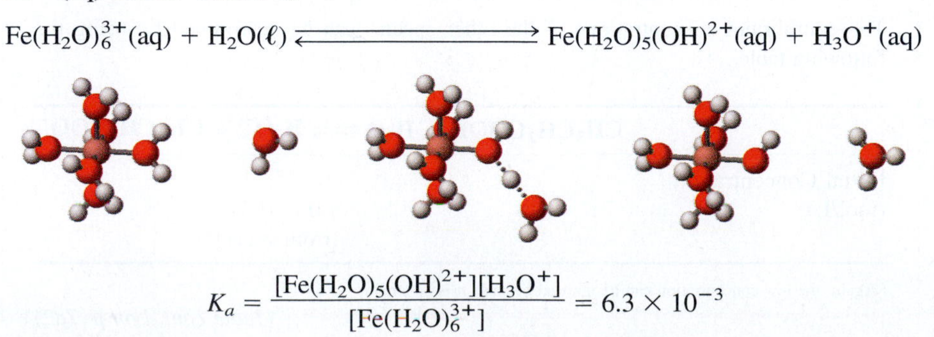

$$K_a = \frac{[Fe(H_2O)_5(OH)^{2+}][H_3O^+]}{[Fe(H_2O)_6^{3+}]} = 6.3 \times 10^{-3}$$

Look carefully at the Fe-containing reactants and products in the reaction. There are six water molecules in the reactant ion and only five in the product ion. The other water molecule has lost a proton to become an OH^- ion. Note also that the charge on the product ion is one less than on the reactant ion.

This K_a value shows that a solution of $FeCl_3$ will have about the same pH as a solution of phosphoric acid ($K_a = 7.5 \times 10^{-3}$) of equal concentration. Many metal ions form weakly acidic aqueous solutions, and this property is important in the chemistry of such ions in the environment.

Exercise 17.15 The pH of a Solution of $Ni(NO_3)_2$

When anhydrous nickel(II) nitrate dissolves in water, the Ni^{2+} ions become hydrated, forming $Ni(H_2O)_6^{2+}$ ions (← ***p. 384***). What is the pH of a solution that is 0.15 M in nickel nitrate? K_a for $Ni(H_2O)_6^{2+}$ is 2.5×10^{-11}.

Table 17.3 (p. 784) summarizes the ionization constants for a number of acids and their conjugate bases. The ionization constants for strong acids (those above H_3O^+ in Table 17.3) and strong bases (those below OH^- in Table 17.3) are too large to be measured easily. Fortunately, since the ionization reactions are virtually complete, these K_a and K_b values are hardly ever needed. For weak acids, K_a values show relative strengths quantitatively; and for weak bases, K_b values do the same. Consider acetic acid and boric acid. Since boric acid is below acetic acid in Table 17.3, boric acid must be weaker than acetic acid; the K_a values tell us how much weaker. The K_a for boric acid is 7.3×10^{-10}, 2.5×10^4 times smaller than K_a for acetic acid (1.8×10^{-5}), and so boric acid is more than 10^4 times weaker. In fact, boric acid is such a weak acid that a dilute solution of it can be used safely as an eyewash. Don't try that with acetic acid!

Calculating pH from an Ionization Constant

You can use acid-base–ionization constants like those in Table 17.3 to calculate the approximate pH of a solution of a weak acid or a weak base from its concentration.

PROBLEM-SOLVING EXAMPLE 17.5 pH from K_a

What is the pH of a 0.050 M solution of benzoic acid, $C_6H_5CO_2H$ ($K_a = 6.3 \times 10^{-5}$)?

Answer 2.74

Explanation First write the equilibrium equation and equilibrium-constant expression:

$$C_6H_5CO_2H(aq) + H_2O(\ell) \rightleftharpoons C_6H_5CO_2^-(aq) + H_3O^+(aq)$$

$$K_a = \frac{[C_6H_5CO_2^-][H_3O^+]}{[C_6H_5CO_2H]}$$

Next, define equilibrium concentrations and organize the known information in a small table based on the balanced chemical equation.

	$C_6H_5CO_2H + H_2O \rightleftharpoons$	$C_6H_5CO_2^- +$	H_3O^+
Initial Concentration (mol/L)	0.050	0	1.0×10^{-7} (from water)
Change in Concentration as reaction occurs (mol/L)	$-x$	$+x$	$+x$
Concentration at Equilibrium (mol/L)	$0.050 - x$	x	x

TABLE 17.3 Ionization Constants for Some Acids and Their Conjugate Bases

Acid Name	Acid	K_a	Base	K_b	Base Name
Perchloric acid	$HClO_4$	Large	ClO_4^-	Very small	Perchlorate ion
Sulfuric acid	H_2SO_4	Large	HSO_4^-	Very small	Hydrogen sulfate ion
Hydrochloric acid	HCl	Large	Cl^-	Very small	Chloride ion
Nitric acid	HNO_3	≈ 20	NO_3^-	$\approx 5 \times 10^{-16}$	Nitrate ion
Hydronium ion	H_3O^+	1.0	H_2O	1.0×10^{-14}	Water
Sulfurous acid	H_2SO_3	1.2×10^{-2}	HSO_3^-	8.3×10^{-13}	Hydrogen sulfite ion
Hydrogen sulfate ion	HSO_4^-	1.2×10^{-2}	SO_4^{2-}	8.3×10^{-13}	Sulfate ion
Phosphoric acid	H_3PO_4	7.5×10^{-3}	$H_2PO_4^-$	1.3×10^{-12}	Dihydrogen phosphate ion
Hexaaquairon(III) ion	$Fe(H_2O)_6^{3+}$	6.3×10^{-3}	$Fe(H_2O)_5OH^{2+}$	1.6×10^{-12}	Pentaaquahydroxoiron(III) ion
Hydrofluoric acid	HF	7.2×10^{-4}	F^-	1.4×10^{-11}	Fluoride ion
Nitrous acid	HNO_2	4.5×10^{-4}	NO_2^-	2.2×10^{-11}	Nitrite ion
Formic acid	HCO_2H	1.8×10^{-4}	HCO_2^-	5.6×10^{-11}	Formate ion
Benzoic acid	$C_6H_5CO_2H$	6.3×10^{-5}	$C_6H_5CO_2^-$	1.6×10^{-10}	Benzoate ion
Acetic acid	CH_3CO_2H	1.8×10^{-5}	$CH_3CO_2^-$	5.6×10^{-10}	Acetate ion
Propanoic acid	$CH_3CH_2CO_2H$	1.4×10^{-5}	$CH_3CH_2CO_2^-$	7.1×10^{-10}	Propanoate ion
Hexaaquaaluminum ion	$Al(H_2O)_6^{3+}$	7.9×10^{-6}	$Al(H_2O)_5OH^{2+}$	1.3×10^{-9}	Pentaaquahydroxoaluminum ion
Carbonic acid	H_2CO_3	4.2×10^{-7}	HCO_3^-	2.4×10^{-8}	Hydrogen carbonate ion
Hexaaquacopper(II) ion	$Cu(H_2O)_6^{2+}$	1.6×10^{-7}	$Cu(H_2O)_5OH^+$	6.25×10^{-8}	Pentaaquahydroxocopper(II) ion
Hydrogen sulfide	H_2S	1×10^{-7}	HS^-	1×10^{-7}	Hydrogen sulfide ion
Dihydrogen phosphate ion	$H_2PO_4^-$	6.2×10^{-8}	HPO_4^{2-}	1.6×10^{-7}	Hydrogen phosphate ion
Hydrogen sulfite ion	HSO_3^-	6.2×10^{-8}	SO_3^{2-}	1.6×10^{-7}	Sulfite ion
Hypochlorous acid	$HClO$	3.5×10^{-8}	ClO^-	2.9×10^{-7}	Hypochlorite ion
Hexaaqualead(II) ion	$Pb(H_2O)_6^{2+}$	1.5×10^{-8}	$Pb(H_2O)_5OH^+$	6.7×10^{-7}	Pentaaquahydroxolead(II) ion
Hexaaquacobalt(II) ion	$Co(H_2O)_6^{2+}$	1.3×10^{-9}	$Co(H_2O)_5OH^+$	7.7×10^{-6}	Pentaaquahydroxocobalt(II) ion
Boric acid	$B(OH)_3(H_2O)$	7.3×10^{-10}	$B(OH)_4^-$	1.4×10^{-5}	Tetrahydroxoborate ion
Ammonium ion	NH_4^+	5.6×10^{-10}	NH_3	1.8×10^{-5}	Ammonia
Hydrocyanic acid	HCN	4.0×10^{-10}	CN^-	2.5×10^{-5}	Cyanide ion
Hexaaquairon(II) ion	$Fe(H_2O)_6^{2+}$	3.2×10^{-10}	$Fe(H_2O)_5OH^+$	3.1×10^{-5}	Pentaaquahydroxoiron(II) ion
Hydrogen carbonate ion	HCO_3^-	4.8×10^{-11}	CO_3^{2-}	2.1×10^{-4}	Carbonate ion
Hexaaquanickel(II) ion	$Ni(H_2O)_6^{2+}$	2.5×10^{-11}	$Ni(H_2O)_5OH^+$	4.0×10^{-4}	Pentaaquahydroxonickel(II) ion
Hydrogen phosphate	HPO_4^{2-}	3.6×10^{-13}	PO_4^{3-}	2.8×10^{-2}	Phosphate ion
Water	H_2O	1.0×10^{-14}	OH^-	1.0	Hydroxide ion
Hydrogen sulfide ion	HS^-	1×10^{-19}	S^{2-}	1×10^5	Sulfide ion
Ethanol	C_2H_5OH	Very small	$C_2H_5O^-$	Large	Ethoxide ion
Ammonia	NH_3	Very small	NH_2^-	Large	Amide ion
Hydrogen	H_2	Very small	H^-	Large	Hydride ion
Methane	CH_4	Very small	CH_3^-	Large	Methide ion

Increasing Acid Strength (upward arrow) — Increasing Base Strength (downward arrow)

Since all equilibrium concentrations are defined in terms of the single unknown, x, the equilibrium-constant expression can be rewritten as

$$K_a = \frac{[C_6H_5CO_2^-][H_3O^+]}{[C_6H_5CO_2H]} = \frac{(x)(x)}{(0.050 - x)} = 6.3 \times 10^{-5}$$

There are at least two ways to obtain a value for x that will satisfy this equation. One is to rearrange the equation into the form $ax^2 + bx + c = 0$ and use the *quadratic formula* (Appendix B.4). But this much effort probably isn't needed. Instead you can reason that, because K_a is very small, the reaction must be reactant-favored. This means not very much product will form and the concentrations of $C_6H_5CO_2^-$ and H_3O^+ will be very small when equilibrium is reached. Therefore, x must be quite small. When it is subtracted from 0.050, the result will still be almost exactly 0.050, and so we can write $0.050 - x$ as 0.050 to get

$$\frac{x^2}{0.050} \sim 6.3 \times 10^{-5}$$

and solve for x, which gives

$$x = \sqrt{(0.050)(6.3 \times 10^{-5})} = \sqrt{3.2 \times 10^{-6}} = 1.8 \times 10^{-3} = [H_3O^+]$$

So pH $= -\log[H_3O^+] = -\log(1.8 \times 10^{-3}) = 2.74$, which is acidic.

When 0.0018 is subtracted from 0.050, the answer is 0.048 because of the significant figure rules, and so the simplifying assumption is a reasonable one. When simplifying assumptions are not reasonable, the quadratic formula must be used.

Generally, if x is less than 5% of the original acid concentration, the simplifying assumption of (original concentration $- x$) $\simeq$ (original concentration) works well.

Problem-Solving Practice 17.5

Boric acid is a weak acid often used as an eyewash. K_a for boric acid is 7.3×10^{-10}. Estimate the pH of a 0.10 M solution of boric acid.

Exercise 17.16 pH from K_b

Ammonia is a common weak base whose K_b is 1.8×10^{-5}. What is the pH of a 6.6 M solution of ammonia?

Exercise 17.17 K_b and Base Strength

Use Table 17.3 to determine which is the stronger base, acetate ion or hydrogen sulfide ion? By what numeric factor is it stronger?

K_a Values for Polyprotic Acids

The successively decreasing K_a values for weak polyprotic acids, as shown in Table 17.3, indicate that each ionization step occurs to a lesser extent than the one before it. The weak acid H_3PO_4, for example, has three protons to donate, and hence three ionization reactions.

First ionization:

$$H_3PO_4(aq) + H_2O(\ell) \rightleftharpoons H_3O^+(aq) + H_2PO_4^-(aq) \qquad K_a = 7.5 \times 10^{-3}$$

Second ionization:

$$H_2PO_4^-(aq) + H_2O(\ell) \rightleftharpoons H_3O^+(aq) + HPO_4^{2-}(aq) \qquad K_a = 6.2 \times 10^{-8}$$

Third ionization:

$$HPO_4^{2-}(aq) + H_2O(\ell) \rightleftharpoons H_3O^+(aq) + PO_4^{3-}(aq) \qquad K_a = 3.6 \times 10^{-13}$$

K_a for the second ionization is about 10^5 times smaller than K_a for the first, which shows that it is more difficult to remove H^+ from a negatively charged $H_2PO_4^-$ ion than from a neutral H_3PO_4 molecule. The much smaller K_a value for the third ionization shows that it is even more difficult to remove H^+ from a doubly negative HPO_4^{2-} ion.

Relationship Between K_a and K_b Values

The right-hand side of Table 17.3 gives K_b for the conjugate base of each acid. Try an experiment with these data: multiply each K_a value by the K_b value for the conjugate base. What do you find? Within a very small error you ought to find that $K_a \times K_b = 1.0 \times 10^{-14}$. This value is the same as K_w, the autoionization constant for water! To see why, multiply the equilibrium-constant expressions for K_a and K_b and see what you get:

$$K_a \times K_b = \left(\frac{[H_3O^+][A^-]}{[HA]}\right)\left(\frac{[HA][OH^-]}{[A^-]}\right)$$

Canceling like terms in the numerator and denominator of this expression gives

$$K_a \times K_b = \left(\frac{[H_3O^+]\cancel{[A^-]}}{\cancel{[HA]}}\right)\left(\frac{\cancel{[HA]}[OH^-]}{\cancel{[A^-]}}\right) = [H_3O^+][OH^-] = K_w$$

This relation tells you that, if you know K_a for an acid, you can find K_b for its conjugate base by using K_w. Furthermore, the larger the K_a, the smaller the K_b, and vice versa (because they always have to give the same product when multiplied, namely K_w). For example, K_a for HCN is 4.0×10^{-10}. The value of K_b for the conjugate base, CN^-, is

$$K_b \text{ (for } CN^-) = \frac{K_w}{K_a \text{ (for HCN)}} = \frac{1.0 \times 10^{-14}}{4.0 \times 10^{-10}} = 2.5 \times 10^{-5}$$

HCN has a relatively small K_a and lies fairly far down in Table 17.3, which means it is a relatively weak acid. However, CN^- is a fairly strong weak base; its K_b of 2.5×10^{-5} is nearly the same as the K_b for ammonia (1.8×10^{-5}), making CN^- a slightly stronger base than ammonia.

The cyanide ion is a deadly poison, attacking the enzyme cytochrome oxidase found in the mitochondria of almost every cell in living organisms.

Exercise 17.18 K_b from K_a

Phenol, or carbolic acid, C_6H_5OH, is a weak acid, $K_a = 1.3 \times 10^{-10}$. Calculate K_b for the phenolate ion, $C_6H_5O^-$. Which base in Table 17.3 is closest in strength to the phenolate ion? How did you make your choice?

17.7 CHEMICAL REACTIVITY OF ACIDS AND BASES

The fact that K_a times K_b equals K_w indicates that *the stronger an acid, the weaker its conjugate base,* and conversely, *the stronger a base, the weaker its conjugate acid.* Table 17.3 shows this relationship clearly. The very strongest acids are at the top of the table. They have such weak conjugate bases that in aqueous solution the acids ionize completely to give H_3O^+ ions and their conjugate base anions. The weakest conjugate bases are also at the top of the table.

On the opposite end of the acid-strength scale (at the bottom of the table) are species such as H_2 and CH_4. They are such weak acids that almost no base exists that is strong enough to take away a proton. Their conjugate bases, H^- and CH_3^-, are extremely strong; they can take protons from almost any molecule that can act as an acid. In aqueous solution these extremely strong bases react completely and vigorously with water molecules to form OH^- and their conjugate-acid molecules; for example,

$$\underset{\text{strong base}}{H^-(aq)} + \underset{\text{acid}}{H_2O(\ell)} \longrightarrow \underset{\text{base}}{OH^-(aq)} + \underset{\text{acid}}{H_2(g)}$$

Earlier we described the importance of lime, CaO, a widely used base (← *p. 153*). The basic properties of lime and other soluble oxides result from the reaction of the oxide ion with water.

$$\underset{\text{strong base}}{O^{2-}(aq)} + \underset{\text{acid}}{H_2O(\ell)} \longrightarrow \underset{\text{acid}}{OH^-(aq)} + \underset{\text{base}}{OH^-(aq)}$$

In this equation one OH^- ion is the conjugate acid of the oxide ion, while the other OH^- ion is the conjugate base of the water molecule.

Carbonate ion, CO_3^{2-}, is a weak Brønsted base. It reacts with water to form bicarbonate ion and OH^-. Soluble carbonates such as Na_2CO_3 (washing soda) produce concentrated solutions that are basic enough to burn human skin or to strip paint from furniture. However, K_b for carbonate ion is less than 1 (Table 17.3), and not all of the carbonate ions have reacted to form OH^- when equilibrium is reached. Hence its ionization equation is written with a double arrow.

$$\underset{\text{base}}{CO_3^{2-}(aq)} + \underset{\text{acid}}{H_2O(\ell)} \rightleftharpoons \underset{\text{acid}}{HCO_3^-(aq)} + \underset{\text{base}}{OH^-(aq)}$$

Weaker Brønsted bases such as ammonia, NH_3, dissolve in water to produce solutions containing much smaller concentrations of OH^- ions.

$$\underset{\text{base}}{NH_3(aq)} + \underset{\text{acid}}{H_2O(\ell)} \rightleftharpoons \underset{\text{acid}}{NH_4^+(aq)} + \underset{\text{base}}{OH^-(aq)}$$

The strongly basic properties of the hydride ion, H^-. Calcium hydride, CaH_2, is shown here reacting with water. The reaction is $H^-(aq) + H_2O(\ell) \rightarrow OH^-(aq) + H_2(g)$ + heat. The reaction is highly exothermic because of the base strength of the H^- ion. Because a high temperature is produced, the hydrogen gas ignites in the air. *(C.D. Winters)*

Most metal oxides are not water-soluble and therefore do not produce strongly basic solutions.

Exercise 17.19 Acid Strength, Base Strength, and Reactivity

Examine each of the acid-base reactions given in this section so far. Use Table 17.3 to identify the stronger acid and the stronger base in each equation. Derive a generalization regarding acid-base strength and whether a reaction is product-favored.

What you probably discovered from Exercise 17.19 is that *the stronger acid and the stronger base will always react to form a weaker conjugate base and a weaker conjugate acid.* If the stronger acid and base are on the reactant side of an equation, the process will be product-favored. If not, it will be reactant-favored.

It is possible to use Table 17.3 to predict the direction of the equilibrium for many acid-base reactions. For example, what happens when acetic acid, CH_3COOH, and sodium cyanide, NaCN, are mixed in water solution? Will HCN, a poisonous gas, be produced to any significant extent? Acetic acid is a weak acid and its conjugate base is the acetate ion, CH_3COO^-. When dissolved in water, NaCN dissociates to form Na^+ ions and CN^- ions. The CN^- ion is one of the stronger weak bases and has a very weak conjugate acid, HCN. We can neglect the Na^+ ion because it is neither a Brønsted acid nor base in water. Ions like this are *spectator ions* (← p. 150), because they take no part in the acid-base reactions in aqueous solution.

Table 17.3 shows that HCN is a weaker acid than CH_3COOH, and CH_3COO^- is a weaker base than CN^-. Applying the rule that the stronger acid and base react to form the weaker conjugate base and acid, the reaction

$$CH_3COOH(aq) + CN^-(aq) \longrightarrow CH_3COO^-(aq) + HCN(aq)$$

is predicted to occur.

Think of this reaction as an equilibrium between the products and the reactants. How would you calculate its equilibrium constant? First, write the equilibrium constant expression.

$$K = \frac{[CH_3COO^-][HCN]}{[CH_3COOH][CN^-]}$$

If both the numerator and denominator of the right side of the equation are multiplied by $[H_3O^+]$,

$$K = \frac{[CH_3COO^-][HCN][H_3O^+]}{[CH_3COOH][CN^-][H_3O^+]}$$

it is then possible to rearrange terms on the right side of the equation to give the K_a expressions of the two acids.

$$K = \frac{[CH_3COO^-][H_3O^+][HCN]}{[CH_3COOH][H_3O^+][CN^-]} = \frac{K_a \text{ (acetic acid)}}{K_a \text{ (hydrocyanic acid)}} = \frac{1.8 \times 10^{-5}}{4.0 \times 10^{-10}} = 4.5 \times 10^4$$

The large value of the calculated equilibrium constant for this reaction confirms our prediction that the reaction is product-favored and that HCN would be formed.

Mixing acetic acid, or practically any other acid, with soluble cyanide salts like NaCN is dangerous because HCN is not very soluble in water and therefore gaseous HCN would be formed. The average fatal dose of HCN(g) for an adult human is 50 to 60 mg, or about 0.002 mol.

Exercise 17.20 Acid-Base Reactions

Which of these reactions will occur as written? What will the reaction products be? Write complete equations; indicate by a longer arrow in which direction the reaction will be favored, and identify the acids and bases on each side of each equilibrium.

(a) $S^{2-} + H_2O \rightarrow$ (b) $HPO_4^{2-} + OH^- \rightarrow$
(c) $CO_3^{2-} + HSO_4^- \rightarrow$ (d) $H_2PO_4^- + HCO_3^- \rightarrow$

17.8 ACID-BASE REACTIONS, SALTS, AND HYDROLYSIS

Conceptual Challenge Problem CP-17.C at the end of the chapter relates to topics covered in this section.

An exchange reaction between an acid and a base produces a salt plus water (← p. 155). A salt is any ionic compound that could have been formed by the reaction of an acid with a base; a salt's positive ion comes from a base and its negative ion comes from an acid. Now that you know more about the Brønsted acid-base concept and the strengths of acids and bases, it is useful to consider acid-base reactions and salt formation in more detail.

Salts of Strong Bases and Strong Acids

The strong acid HCl reacts with the strong base NaOH to form the salt NaCl. If the amounts of HCl and NaOH are in the stoichiometric ratio (1 mol HCl per 1 mol NaOH), this reaction results in complete neutralization of the acidic properties of HCl and the basic properties of NaOH. The reaction can be described first by an overall equation, then by a complete ionic equation, and finally by a net ionic equation (← *p. 150*). Each of these equations contains useful information.

$$HCl(aq) + NaOH(aq) \longrightarrow NaCl(aq) + H_2O(\ell)$$

$$H_3O^+(aq) + Cl^-(aq) + Na^+(aq) + OH^-(aq) \longrightarrow Na^+(aq) + Cl^-(aq) + 2\ H_2O(\ell)$$

$$\underset{\text{acid}}{H_3O^+(aq)} + \underset{\text{base}}{OH^-(aq)} \longrightarrow \underset{\text{base}}{H_2O(\ell)} + \underset{\text{acid}}{H_2O(\ell)}$$

The overall equation shows the substances that were dissolved or that could be recovered at the end of the reaction. The complete ionic equation indicates all of the ions that are present before and after reaction. The net ionic equation emphasizes that a Brønsted acid (H_3O^+) is reacting with a Brønsted base (OH^-); the spectator ions, Na^+ and Cl^-, are omitted. This reaction goes to completion because H_3O^+ is a strong acid, OH^- is a strong base, and water is a very weak acid or a very weak base.

If the water were evaporated from the NaCl salt solution formed by the reaction of HCl and NaOH, the resulting solid NaCl would be identical to NaCl that could be prepared by the reaction of metallic sodium with gaseous chlorine.

The resulting solution contains only sodium ions and chloride ions, with a few more water molecules than before. Its properties are the same as if it had been prepared by simply dissolving some NaCl(s) in water. It has a neutral pH because it contains no acids or bases. The Cl^- ion is the conjugate base of a strong acid and hence is such a weak base that it does not react with water. The Na^+ ion also does not react with water as either an acid or a base. Examples of some other salts of this type are given in Table 17.4. These salts all form neutral solutions.

Salts of Strong Bases and Weak Acids

Suppose, for example, that 0.010 mol of NaOH is added to 0.010 mol of the weak acid, acetic acid, in 1 L of solution. The reaction is

$$NaOH(aq) + CH_3COOH(aq) \longrightarrow NaCH_3COO(aq) + H_2O(\ell)$$

$$Na^+(aq) + OH^-(aq) + CH_3COOH(aq) \longrightarrow Na^+(aq) + CH_3COO^-(aq) + H_2O(\ell)$$

$$\underset{\text{strong base}}{OH^-(aq)} + \underset{\text{weak acid}}{CH_3COOH(aq)} \longrightarrow \underset{\text{base}}{CH_3COO^-(aq)} + \underset{\text{acid}}{H_2O(\ell)}$$

TABLE 17.4 Some Salts Formed by Neutralization of Strong Acids with Strong Bases

	Base		
Acid	**NaOH**	**KOH**	**$Ba(OH)_2$**
HCl	NaCl	KCl	$BaCl_2$
HNO_3	$NaNO_3$	KNO_3	$Ba(NO_3)_2$
H_2SO_4	Na_2SO_4	K_2SO_4	$BaSO_4$
$HClO_4$	$NaClO_4$	$KClO_4$	$Ba(ClO_4)_2$

The hydrolysis of salts is one of the reasons that solubilities calculated using K_{sp} expressions for slightly soluble salts are only estimates. If an ion reacts with water, the ion's concentration is less than expected and solubility of a precipitate containing that ion is greater than expected.

In this case, acetate ion, a weak base, remains in the solution when the reaction is complete. This means that the solution is slightly basic; its pH is greater than 7, even though exactly the stoichiometric amount of acetic acid was added to the sodium hydroxide. The reaction that makes the solution basic is the reaction of acetate ion as a weak Brønsted base:

$$CH_3COO^-(aq) + H_2O(\ell) \rightleftharpoons CH_3COOH(aq) + OH^-(aq)$$

This process splits a water molecule into a proton (which becomes attached to the acetate ion to form acetic acid) and a hydroxide ion. A reaction in which a water molecule is split is called a **hydrolysis** reaction (derived from *hydro* meaning "water," and *lysis* meaning "to break apart"). The extent of hydrolysis in this case is determined by the value of K_b for acetate ion.

All of the weak bases in Table 17.3, except for the very weak bases above water, undergo hydrolysis reactions in aqueous solution. The larger their K_b values, the more basic the solutions they produce. The pH of a solution of a salt of a strong base and a weak acid can be estimated from K_b as shown in Example 17.6.

pH of solutions of salts of strong bases and weak acids. The pH meter readings show that solutions of sodium acetate ($NaCH_3COO$, *top*) and sodium cyanide (NaCN, *bottom*) are both basic. (An inert white solid was suspended in the solutions to make them more visible in the photos.) *(C.D. Winters)*

PROBLEM-SOLVING EXAMPLE 17.6 pH of a Salt Solution

Sodium hypochlorite, NaClO, is used as a source of chlorine in some laundry bleaches, swimming-pool disinfectants, and water-treatment plants. Estimate the pH of a 0.010 M solution of NaClO. (Use Table 17.3 to obtain K_b.)

Answer pH = 9.73

Explanation Sodium hypochlorite consists of sodium ions and hypochlorite ions. Na^+ does not react with water, but ClO^- is the conjugate base of a weak acid and reacts with water to produce a basic solution.

$$ClO^-(aq) + H_2O(\ell) \rightleftharpoons HClO(aq) + OH^-(aq)$$

$$K_b = 2.9 \times 10^{-7} = \frac{[HClO][OH^-]}{[ClO^-]}$$

The concentrations of hypochlorite ion, hypochlorous acid, and hydroxide ion initially and at equilibrium are

	$ClO^- + H_2O \rightleftharpoons$	$HClO$	$+ OH^-$
Initial Concentration (mol/L)	0.010	0	1.0×10^{-7} (from water)
Change as Reaction Occurs (mol/L)	$-x$	$+x$	$+x$
Equilibrium Concentration (mol/L)	$0.010 - x$	x	x

Hypochlorite ion has a very small K_b and is a very weak base. It is safe to assume that x will be negligibly small compared to 0.010, and so

$$K_b = 2.9 \times 10^{-7} \simeq \frac{x^2}{0.010}$$

Solving for x gives $x = 5.4 \times 10^{-5}$. Since $0.010 - 5.4 \times 10^{-5} = 0.010$ (using the significant figures rules), our assumption that x is negligible is justified. Therefore, at equilibrium

$$[OH^-] = [HClO] = 5.4 \times 10^{-5} \text{ mol/L} \quad \text{and} \quad [ClO^-] = 0.010 \text{ mol/L}$$

Finally, the pH of the solution is found as follows:

$$K_w = [H_3O^+][OH^-] = [H_3O^+](5.4 \times 10^{-5})$$

$$[H_3O^+] = 1.9 \times 10^{-10}$$

$$pH = 9.73$$

As expected, the solution is basic.

Problem-Solving Practice 17.6

Sodium carbonate is an environmentally safe paint stripper. It is water-soluble, and carbonate ion is a strong enough base to make a solution with pH high enough to loosen paint so it can be scraped off. What is the pH of a 1.0 M solution of Na_2CO_3?

Exercise 17.21 pH of Soap Solutions

Ordinary soaps are often sodium salts of fatty acids. Since fatty acids are weak organic acids, what would you expect the pH of a soap solution to be, >7, neutral, or <7?

Salts of Weak Bases and Strong Acids

When a weak base reacts with a strong acid, the conjugate acid of the weak base determines the pH of the resulting salt solution. For example, suppose equal volumes of 0.10 M NH_3 and 0.10 M HCl are mixed. The reaction, shown in overall, complete ionic, and net ionic forms, is

$$NH_3(aq) + HCl(aq) \longrightarrow NH_4Cl(aq)$$

$$NH_3(aq) + H_3O^+(aq) + Cl^-(aq) \longrightarrow NH_4^+(aq) + Cl^-(aq) + H_2O(\ell)$$

$$\underset{\text{weak base}}{NH_3(aq)} + \underset{\text{strong acid}}{H_3O^+(aq)} \longrightarrow \underset{\text{acid}}{NH_4^+(aq)} + \underset{\text{base}}{H_2O(\ell)}$$

As soon as it is formed, the weak acid NH_4^+ reacts with water and establishes an equilibrium. The resulting solution is slightly acidic, because of the reaction

$$NH_4^+(aq) + H_2O(\ell) \rightleftharpoons NH_3(aq) + H_3O^+(aq)$$

Many drugs are high-molecular-weight amines that are insoluble in aqueous body fluids such as blood plasma, intracellular fluid, and cerebrospinal fluid. To convert them to soluble compounds that can be administered orally or by injection, they are converted to hydrochlorides. The formulas for these drugs are often written with · HCl, B·HCl, to show that the amines, which can be generalized as B, have been reacted with HCl to convert them into hydrochloride salts, BH^+Cl^-. This formula is like that of ammonium chloride, NH_4Cl. The amine salt of the drug is much more water-soluble than the amine itself. In aqueous solution, the amine salt will undergo hydrolysis just as any other salt of a weak base and a strong acid does.

B·HCl is equivalent to BH^+Cl^-.

novocain hydrochloride

pseudophedrine hydrochloride

methadone hydrochloride

Exercise 17.22 Hydrolysis Reactions of Amine Drug Molecules

Write equations for the hydrolysis reactions of novocain HCl, pseudoephedrine HCl, and methadone HCl.

Salts of Weak Bases and Weak Acids

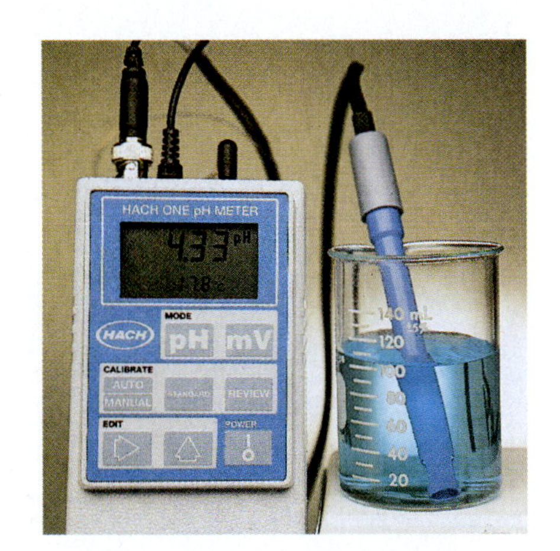

Acidic pH of a solution of copper sulfate ($CuSO_4$). The blue solution of this copper salt is acidic due to hydrolysis of the copper ion.

What is the pH of a solution of a salt containing an acidic cation and a basic anion, such as NH_4F or $Ni(CH_3COO)_2$? There are two possible reactions that can determine the pH of the solution: formation of H_3O^+ by the cation, and formation of OH^- by the anion. In the case of NH_4F,

$$NH_4^+(aq) + H_2O(\ell) \rightleftharpoons H_3O^+(aq) + NH_3(aq) \qquad K_a(NH_4^+) = 5.6 \times 10^{-10}$$

$$F^-(aq) + H_2O(\ell) \rightleftharpoons HF(aq) + OH^-(aq) \qquad K_b(F^-) = 1.4 \times 10^{-11}$$

Since $K_a(NH_4^+) > K_b(F^-)$, the hydrolysis of ammonium ions to produce hydronium ions is the more favorable reaction. Therefore, the resulting solution is slightly acidic. For $Ni(CH_3COO)_2$ the possible reactions are

$$Ni(H_2O)_6^{2+}(aq) + H_2O(\ell) \rightleftharpoons Ni(H_2O)_5(OH)^+ + H_3O^+(aq)$$
$$K_a(Ni(H_2O)_6^{2+}) = 2.5 \times 10^{-11}$$

$$CH_3COO^-(aq) + H_2O(\ell) \rightleftharpoons CH_3COOH(aq) + OH^-(aq)$$
$$K_b(CH_3COO^-) = 5.6 \times 10^{-10}$$

Since $K_b(CH_3COO^-) > K_a(Ni(H_2O)_6^{2+})$, the hydrolysis of the CH_3COO^- ion is more favorable, and the resulting solution is slightly basic.

Exercise 17.23 Hydrolysis of a Salt of a Weak Acid and a Weak Base

Name a salt of a weak acid and a weak base where $K_a = K_b$ (see Table 17.3). What would you expect the pH of a solution of this salt to be?

Recalling that salts, along with water, are the products of acids reacting with bases, the following generalizations about acid-base reactions can be made.

- Solution of strong acid + solution of strong base $\longrightarrow$ salt solution with pH = 7

TABLE 17.5 Acid-Base Properties of Typical Ions in Aqueous Solution

	Neutral		Basic			Acidic
Anions	Cl^-	NO_3^-	CH_3COO^-	CN^-	SO_4^{2-}	HSO_4^-
	Br^-	ClO_4^-	$HCOO^-$	PO_4^{3-}	HPO_4^{2-}	$H_2PO_4^-$
	I^-		CO_3^{2-}	HCO_3^-	SO_3^{2-}	HSO_3^-
			S^{2-}	HS^-	ClO^-	
			F^-	NO_2^-		
Cations	Li^+	Mg^{2+}	None			Al^{3+}
	Na^+	Ca^{2+}				NH_4^+
	K^+	Ba^{2+}				Transition metal ions

- Solution of strong acid + solution of weak base → salt solution with pH < 7 (acidic)
- Solution of weak acid + solution of strong base → salt solution with pH > 7 (basic)
- Solution of weak acid + solution of weak base → salt solution with pH determined by relative strengths of conjugate base and conjugate acid formed

Table 17.5 summarizes the acid-base behavior of many different ions in aqueous solution.

Exercise 17.24 Acidity and Basicity of Salt Solutions

Which of these salts would give a neutral solution? An acidic solution? A basic solution? (a) NH_4NO_3 (b) NaBr (c) K_3PO_4 (d) $Fe(NO_3)_2$ (e) $Fe(CH_3COO)_2$ (f) NH_4HCO_3

Dissolving Slightly Soluble Salts Using Acids

Many salts are only slightly soluble in water (← *p. 104*). An insoluble salt can be made to dissolve if one or both of its ions can be removed from solution somehow. Consider calcium carbonate, $CaCO_3$, which is found in minerals such as limestone and marble. $CaCO_3$ is not very soluble in pure water, as its K_{sp} (**Appendix H**) indicates.

(a) $CaCO_3(s) \rightleftharpoons Ca^{2+}(aq) + CO_3^{2-}(aq)$ $\qquad K_{sp} = 3.8 \times 10^{-9}$

Since K_{sp} is small, the equilibrium concentrations of Ca_2^+ and CO_3^{2-} must also be small. However, if acid is added, the calcium carbonate will dissolve and CO_2 will be released from the solution. Adding acid adds hydronium ions, which react with carbonate ions:

(b) $CO_3^{2-}(aq) + H_3O^+(aq) \rightleftharpoons HCO_3^-(aq) + H_2O(\ell)$

(c) $HCO_3^-(aq) + H_3O^+(aq) \rightleftharpoons H_2CO_3(aq) + H_2O(\ell)$

Reaction (b) involves a fairly strong base and a strong acid on the left-hand side, and so nearly all of the carbonate is converted to hydrogen carbonate ion. Reaction (c) produces a product, carbonic acid, that is unstable. It breaks down to $CO_2(g)$ and water in a very product-favored reaction:

$$\text{(d)}\quad H_2CO_3(aq) \rightarrow CO_2(g) + H_2O(\ell) \qquad K \simeq 10^5$$

As CO_2 gas escapes from the solution, the H_2CO_3 concentration decreases. This shifts reaction (c) to the right, which decreases the concentration of HCO_3^-. This in turn shifts reaction (b) to the right, decreasing the concentration of CO_3^{2-} to an even lower value. To oppose this decrease in carbonate-ion concentration, reaction (a) shifts to the right, and more $CaCO_3(s)$ dissolves. Because the acidity of the solution determines the positions of equilibria (b) and (c), small changes in pH can cause limestone and marble to dissolve or precipitate (← *p. 587*). Acid rain (Section 17.10) can dissolve a marble statue as well as underground limestone deposits, causing massive cave formations. In addition, layers of sedimentary rock (limestone) have been precipitated on the ocean floor because of a slight increase in pH of sea water.

In general, *insoluble salts containing anions that are Brønsted bases dissolve in solutions of low pH.* This rule covers carbonates, sulfides (which produce $H_2S(g)$), hydroxides, phosphates, and other anions listed as bases in Table 17.3. The principal exceptions to this rule are a few sulfides, such as HgS, CuS, and CdS, that have extremely small K_{sp} values and therefore do not dissolve even when the pH is extremely low.

In contrast, an insoluble salt such as AgCl, which contains the conjugate base of a strong acid, is not soluble in strongly acidic solution, because Cl^- is a very weak base and so does not react with H_3O^+.

$$AgCl(s) \rightleftharpoons Ag^+(aq) + Cl^-(aq) \qquad K_{sp} = 1.8 \times 10^{-10}$$

$$H_3O^+(aq) + Cl^-(aq) \rightleftharpoons HCl(aq) + H_2O(\ell) \qquad K \text{ is very small}$$

Exercise 17.25 Predicting Solubilities of Salts in Acid

Predict which of the following insoluble salts would be soluble in 1 M HCl solution.
(a) $ZnCO_3$ (b) FeS (c) AgBr (d) $BaCO_3$ (e) $Ca(OH)_2$

17.9 BUFFER SOLUTIONS

Often, changing the pH of a solution can produce dramatic changes within the solution. Many aquatic organisms can survive only within a narrow pH range. If acid rain lowers the pH of a lake or stream, fish such as trout may die. In addition, for organisms to be studied in a laboratory, the solutions they live in must have a fairly constant pH. The blood of mammals is also a solution in which maintaining a constant pH is of vital importance. The normal pH of human blood is 7.40 ± 0.05. If pH decreases below 7.35, a condition known as *acidosis* occurs; increasing pH above 7.45 leads to *alkalosis.* Both of these conditions can be life-threatening; however, in the novel *Andromeda Strain,* by Michael Crichton, both mild acidosis and alkalosis saved the lives of two of the characters (Figure 17.4).

Acidosis, for example, causes hemoglobin to decrease oxygen transport and also depresses the central nervous system, leading in extreme cases to coma and death. In addition, acidosis can cause weakening and irregularity of cardiac contractions—symptoms of heart failure. To prevent such problems your body must keep the pH of your blood nearly constant.

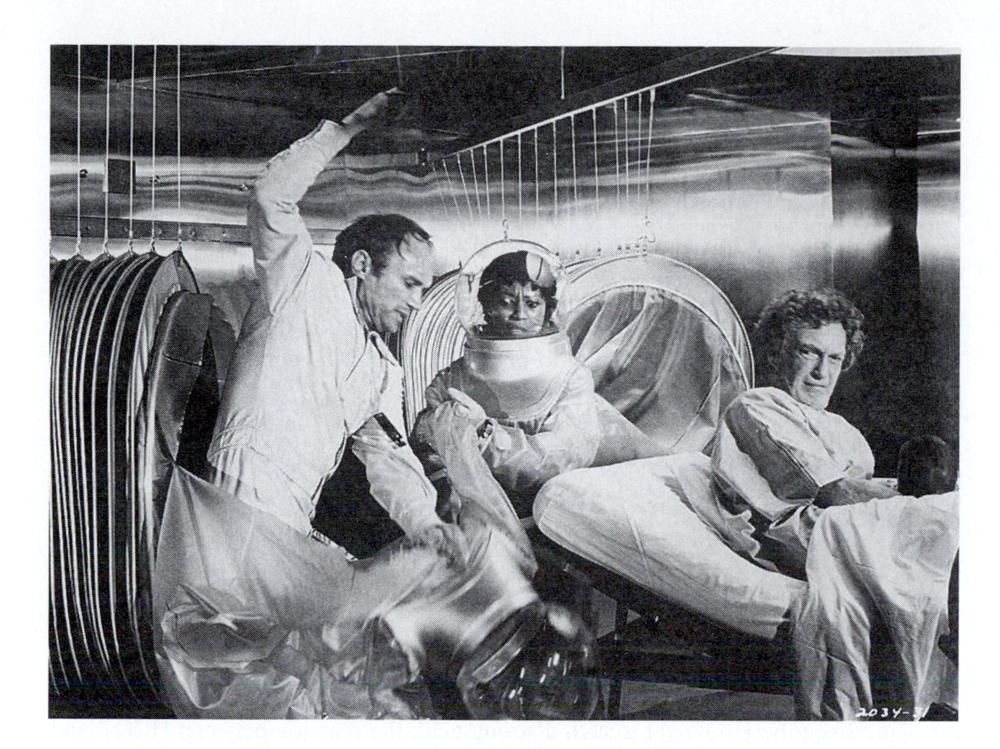

Figure 17.4 ***The Andromeda Strain,* a novel and screenplay by Michael Crichton.** In this scene from the movie, scientist Dr. Mark Hall is shown quickly getting out of his anticontamination suit to help in an experiment to learn why two people, a screaming baby and the town drunk, have survived the attack of an alien organism. The alien organism required blood of normal pH. The baby was hyperventilating, and her blood was slightly basic due to a lower carbon dioxide level resulting from rapid exhalation. The drunk was drinking Sterno, a solid gel of sodium acetate in ethanol, and was taking large doses of aspirin to relieve his stomach pain. That caused his blood pH to be slightly acidic. Crichton, who has an M.D. degree, also wrote *Jurassic Park.*
(The Kobal Collection)

Blood and other solutions that must maintain relatively constant pH contain a **buffer**—comparable concentrations of the conjugate acid-base pair of a weak acid or base. Solutions that resist changes in pH when limited amounts of acids and bases are added are referred to as **buffer solutions.**

The addition of a small amount of acid or base to pure water radically affects the pH. Consider what happens if 0.01 mol of HCl is added to 1.0 L of water. The pH changes from 7 to 2 because $[H_3O^+]$ changes from 10^{-7} M to 10^{-2} M (0.01 mol/L). This pH change represents a *100,000-fold increase* in $[H_3O^+]$. Similarly, if 0.01 mol of NaOH is added to 1 L of pure water, the pH goes from 7 to 12, a *100,000-fold decrease* in $[H_3O^+]$. If those same amounts of strong acid and strong base were added to 1-L samples of a buffer solution like human blood, the pH would decrease or increase respectively by only about 0.1 pH unit.

Buffer Action

How does a buffer solution maintain its pH at a nearly constant value? A buffer must contain an *acid that can react with any added base,* and at the same time it must contain a *base that can react with any added acid.* It is also necessary that the acid and base components of a buffer solution not react with each other. A conjugate acid-base pair satisfies this need. In a conjugate pair, if the acid and base react with each other they just produce conjugate base and conjugate acid—no observable change occurs. For example, when acetic acid reacts with acetate ion, acetate ion and acetic acid are the products.

$$\underset{\text{acid}}{CH_3COOH(aq)} + \underset{\text{base}}{CH_3COO^-(aq)} \rightleftharpoons \underset{\text{base}}{CH_3COO^-(aq)} + \underset{\text{acid}}{CH_3COOH(aq)}$$

Buffers usually consist of approximately equal quantities of a weak acid and its conjugate base, or a weak base and its conjugate acid.

To see how a buffer works, consider human blood. Carbon dioxide provides the most important buffer (but not the only one) affecting the pH in blood. In solution, CO_2 reacts with water to form H_2CO_3, which can then dissociate to produce H_3O^+ and HCO_3^- ions. The reactions are

$$CO_2(aq) + H_2O(\ell) \rightleftharpoons H_2CO_3(aq)$$

$$H_2CO_3(aq) + H_2O(\ell) \rightleftharpoons H_3O^+(aq) + HCO_3^-(aq)$$

An enzyme, carbonic anhydrase, in blood and numerous other body fluids, causes the equilibria between CO_2 and H_2O to be established quickly.

The normal concentrations of H_2CO_3 and HCO_3^- in blood are 0.0025 M and 0.025 M. Since H_2CO_3 is a weak acid and HCO_3^- is its conjugate weak base, they constitute a buffer. As long as the ratio of H_2CO_3 to HCO_3^- concentrations remains about 1 to 10, the pH of the blood remains near 7.4.

If a strong base such as NaOH is added to this buffer, carbonic acid will react with the OH^-. Since OH^- is the strongest base that can exist in water solution, this reaction is essentially complete.

$$H_2CO_3(aq) + OH^-(aq) \longrightarrow HCO_3^-(aq) + H_2O(\ell)$$

$$K' = 1/K_b(HCO_3^-) = 4.2 \times 10^7$$

Here, the equilibrium constant is $1/K_b$ for hydrogen carbonate ion, because the reaction is the reverse of the hydrolysis of the hydrogen carbonate ion. If a strong acid such as HCl is added to this buffer, HCO_3^- will react with the hydronium ions from the acid. Since the H_3O^+ ion is such a strong acid, the reaction between HCO_3^- and H_3O^+ is essentially complete.

$$HCO_3^-(aq) + H_3O^+(aq) \longrightarrow H_2CO_3(aq) + H_2O(\ell)$$

$$K' = 1/K_a(H_2CO_3) = 5.6 \times 10^4$$

In this case, the equilibrium constant is $1/K_a$ for carbonic acid, because the reaction is the reverse of the ionization of carbonic acid.

In many buffers, the ratio of acid-base concentrations is about 1. In blood, however, the ratio $[HCO_3^-]/[H_2CO_3]$ is about 10 to 1, with good reason. There are more acidic than basic byproducts of metabolism to be transported in the blood.

Exercise 17.26 Possible Buffers?

Could a solution of equal molar amounts of HCl and NaCl act as a buffer? What about a solution of equal molar amounts of H_3PO_4 and KCl? Explain each of your answers in detail.

The pH of Buffer Solutions

The pH of a buffer solution can be estimated by using the K_a expression if the concentrations of the conjugate acid and the conjugate base are known. For example, for blood containing 0.0025 M carbonic acid and 0.025 M hydrogen carbonate ion,

$$K_a = \frac{[H_3O^+][HCO_3^-]}{[H_2CO_3]} = \frac{[H_3O^+](0.025)}{(0.0025)} = 4.2 \times 10^{-7}$$

$$[H_3O^+] = 4.2 \times 10^{-7} \times \left(\frac{0.0025}{0.025}\right) = 4.2 \times 10^{-8}$$

$$pH = -\log(4.2 \times 10^{-8}) = 7.38$$

Notice that in this calculation the hydronium ion concentration is one tenth the K_a for carbonic acid, because the ratio $[H_2CO_3]/[HCO_3^-]$ equals 1/10. To have comparable quantities of both acid and conjugate base in the buffer solution, this ratio cannot get much smaller than $\frac{1}{10}$ (or much bigger than 10). Consequently we can use this same acid-base pair to prepare buffer solutions over a range of $[H_3O^+]$ from about $\frac{1}{10}$ to 10 times the K_a. The range of pH is limited to about one pH unit above or below $-\log(K_a)$. In this case, that would be a pH from 5.38 to 7.38 ($-\log 4.2 \times 10^{-7} = 6.38$). Other acid-base pairs can be used to prepare buffers with much different pH ranges, as determined by the K_a value of the acid.

The ratio of acid to base concentrations needed to achieve a given pH for a buffer solution is conveniently calculated by using an equation called the *Henderson-Hasselbalch equation.* This equation is obtained by writing the equilibrium-constant expression for a weak acid, and solving for $[H_3O^+]$.

$$HA(aq) + H_2O(\ell) \rightleftharpoons H_3O^+(aq) + A^-(aq)$$

$$K_a = \frac{[H_3O^+][A^-]}{[HA]}$$

$$[H_3O^+] = K_a\frac{[HA]}{[A^-]}$$

Taking the base-10 logarithm of each side of this equation gives

$$\log[H_3O^+] = \log K_a + \log\frac{[HA]}{[A^-]}$$

Multiplying both sides of the equation by -1 and using the relation $-\log(x) = \log(1/x)$, we get

$$-\log[H_3O^+] = -\log K_a + \log\frac{[A^-]}{[HA]}$$

Using the definition of pH, and defining $-\log K_a$ as pK_a (like the definition of pH), the equation becomes

The use of "p*X*" to represent $-\log x$ or $-\log[x]$ is quite useful. For example, if you were routinely measuring small $[Ag^+]$ values, you could use pAg instead.

$$pH = pK_a + \log\frac{[A^-]}{[HA]}$$ **The Henderson-Hasselbalch equation**

As the equation shows, if $[A^-] = [HA]$, the pH = pK_a because $\log(1) = 0$. A buffer's pH equals the pK_a of the weak acid when the concentrations of the acid and its conjugate base are equal. A buffer for maintaining a desired pH can be chosen easily by examining pK_a values, which are often tabulated along with K_a values. Table 17.6 lists pK_a values of several common acids that could be used to prepare buffers over the pH range from 4 to 10.

Notice that as K_a decreases, pK_a increases; so the weaker the acid, the larger the pK_a.

PROBLEM-SOLVING EXAMPLE 17.7 Selecting an Acid-Base Pair for a Buffer Solution of Known pH

An experiment you are doing requires a buffer solution that has a pH of 9.56. What acid-base pair would you use to make such a buffer solution, and what molar ratios of the compounds would you use to prepare such a solution? You may only use one of the acid-base pairs found in Table 17.3.

Answer The NH_4^+/NH_3 acid-base pair with a molar ratio of 0.20 mol NH_3 to 0.098 mol NH_4^+ would do.

Explanation While Table 17.3 gives a number of acid-base pairs, most of them will not work together to produce a buffer solution with a pH of 9.56. You need an acid-base pair that has a pK_a near the pH you are trying to achieve. This is a result of the relationship between the acid and its conjugate base illustrated by the Henderson-Hasselbalch equation. Using this equation, we can see that if we first set the target pH equal to the right-hand side of the equation and then assume equal concentrations of the acid and its conjugate base, the log term will become zero (the log of 1 is zero).

$$9.56 = pK_a + \log\frac{[A]}{[HA]} = pK_a + 0$$

Now, look in Table 17.3 and find an acid with a pK_a as near as possible to the target pH. Several acids meet this requirement. They include boric acid (pK_a = 9.13), ammonium ion (pK_a = 9.25), hydrocyanic acid (pK_a = 9.39), hexaaquoiron(II) ion (pK_a = 9.49), and hydrogen carbonate ion (pK_a = 10.3). For practical purposes, hydrocyanic acid should be avoided because of its toxicity. Any of the others can be used. Let's use the ammonium-ammonia pair with its pK_a of 9.25. Substituting this into the Henderson-Hasselbalch equation and using the concentration of NH_3 for A and NH_4^+ for HA gives

There are numerous practical considerations to making buffer solutions. These include the toxicity of the components, their solubilities, and whether any of the components of the buffer solution will enter into reactions with other substances in the solution.

$$9.56 = 9.25 + \log\frac{[A]}{[HA]} = 9.25 + \log\frac{[NH_3]}{[NH_4^+]}$$

$$\log\frac{[NH_3]}{[NH_4^+]} = 9.56 - 9.25 = 0.31$$

The antilog of 0.31 is $10^{0.31}$ = 2.04; thus the ratio $[NH_3]/[NH_4^+]$ of the two components of the buffer becomes 2.04. This means that, roughly, the concentration of NH_3 will have to be twice that of NH_4^+. Both ammonia and ammonium salts are very soluble in water, so if the concentration of NH_3 is 0.20 mol/L, the concentration of the ammonium salt needed is 0.098 mol/L.

Conceptual Challenge Problem CP-17.D at the end of the chapter relates to topics covered in this section.

Problem-Solving Practice 17.7

Use data from Table 17.3 to select an acid-base–conjugate pair you could use to make a buffer solution having each of these hydrogen-ion concentrations:
(a) 3.2×10^{-4} M (b) 5.0×10^{-5} M (c) 7.0×10^{-8} M (d) 6.0×10^{-11} M

Exercise 17.27 Another Blood Buffer

Calculate the ratio of HPO_4^{2-} to $H_2PO_4^-$ in blood at the normal blood pH (7.40).

Exercise 17.28 Which Reaction Occurs?

If an abnormally high concentration of CO_2 is present in the blood, which phosphate ion, $H_2PO_4^-$ or HPO_4^{2-}, will be used to counteract its presence?

While the pH of a buffer varies with pK_a and the ratio of acid-base concentrations, the *amounts* of acid and base present in the buffer solution determine the **buffer capacity**—the quantity of acid or base the buffer can accommodate without a pH change. When nearly all of the acid in a buffer has reacted with base, adding a little more base can increase the pH significantly, because there is almost no acid left to consume the base. Similarly, if enough acid is added to a buffer to react with most of the buffer's weak base, the pH will decrease significantly. In either case, the buffer capacity is used up. For example, 1 L of a buffer solution that is 0.25 M in CH_3COOH and 0.25 M in CH_3COO^- contains 0.25 mol CH_3COOH and 0.25 mol CH_3COO^-.

TABLE 17.6 Buffer Systems That Are Useful at Various pH Values[1]

Desired pH	Weak Acid	Weak Base	K_a(Weak Acid)	pK_a
4	Lactic acid ($CH_3CHOHCOOH$)	Lactate ion ($CH_3CHOHCOO^-$)	1.4×10^{-4}	3.85
5	Acetic acid (CH_3COOH)	Acetate ion (CH_3COO^-)	1.8×10^{-5}	4.74
6	Carbonic acid (H_2CO_3)	Hydrogen carbonate ion (HCO_3^-)	4.2×10^{-7}	6.38
7	Dihydrogen phosphate ion ($H_2PO_4^-$)	Hydrogen phosphate ion (HPO_4^{2-})	6.2×10^{-8}	7.21
8	Hypochlorous acid ($HClO$)	Hypochlorite ion (ClO^-)	3.5×10^{-8}	7.46
9	Ammonium ion (NH_4^+)	Ammonia (NH_3)	5.6×10^{-10}	9.25
10	Hydrogen carbonate ion (HCO_3^-)	Carbonate ion (CO_3^{2-})	4.8×10^{-11}	10.32

[1]Adapted from W. L. Masterton and C. N. Hurley: *Chemistry—Principles and Reactions,* 3rd ed. Philadelphia, Saunders College Publishing, 1997.

It will not be able to handle the addition of 0.30 mol of strong acid or 0.30 mol of strong base because such additions would use up all of the weak base or all of the weak acid, and the pH would drop or rise accordingly. Table 17.6 summarizes several common buffer systems.

Exercise 17.29 Buffers and pH

Estimate the pH of each of these buffers:

(a) H_2CO_3(0.10 M)/HCO_3^-(0.25 M)

(b) $H_2PO_4^-$(0.10 M)/HPO_4^{2-}(0.25 M)

(c) NH_4^+(0.10 M)/NH_3(0.25 M)

Exercise 17.30 Buffer Capacity

If a buffer is prepared using 0.25 mol H_3PO_4 and 0.15 mol $H_2PO_4^-$ in 500 mL of solution, will the buffer be able to maintain its pH if 6.2 g of KOH is added?

Exercise 17.31 Unusual Buffers

Suppose a space probe someday brings back an aquatic organism that finds carbon compounds toxic. Neither can it tolerate phosphorus, nitrogen, sulfur, or boron. In fact, it can only tolerate water and various metal ions in solution. What acid-base pair would you choose to make a buffer solution for this organism to grow in if it required a pH of about 10.6?

Exercise 17.32 Just What Is a Buffer?

Explain why a solution of acetic acid in water, which contains acetic acid and acetate ions, cannot be a buffer.

17.10 ACID RAIN

Oxides that produce acids when they dissolve in water are called acidic oxides.

The term *acid rain* was first used in 1872 by Robert Angus Smith, an English chemist and climatologist. He used the term to describe the acidic precipitation that fell on Manchester, England, at the start of the Industrial Revolution. Although neutral water has a pH of 7, rainwater becomes naturally acidified from dissolved carbon dioxide, a normal component of the atmosphere. The carbon dioxide reacts reversibly with water to form a solution of the weak acid, carbonic acid.

$$2\ H_2O\ (\ell) + CO_2(g) \rightleftharpoons H_3O^+(aq) + HCO_3^-(aq)$$

At equilibrium, the pH of a solution of CO_2 from the air is about 5.6. Any precipitation with a pH below 5.6 is considered to be **acid rain.**

The nitrogen dioxide (NO_2) from industrial as well as natural sources can react with water in the atmosphere to produce acids. NO_2 produces nitric acid (HNO_3) and nitrous acid (HNO_2).

$$2\ NO_2(g) + H_2O(\ell) \longrightarrow HNO_3(aq) + HNO_2(aq)$$

Sulfur dioxide (SO_2) produces sulfurous acid (H_2SO_3) and, if oxygen is present, sulfuric acid (H_2SO_4).

$$2\ SO_2(g) + O_2(g) \longrightarrow 2\ SO_3(g)$$

$$SO_3(g) + H_2O(\ell) \longrightarrow H_2SO_4(aq)$$

When conditions are favorable, these acidic water droplets precipitate as rain or snow with a pH less than 5.6. Ice core samples taken in Greenland and dating back to 1900 contain sulfate (SO_4^{2-}) and nitrate (NO_3^-) ions. This indicates that acid rain has been commonplace, at least from 1900 onward.

Checking the pH of water in a lake. At this lake in the Pocono Mountains, PA, the pH was 4.4. *(Runk/Schoenberger from Grant Heilman)*

Exercise 17.33 Reactions Producing Acid Rain

Are the reactions of nitrogen dioxide and sulfur dioxide with water redox reactions? If so, what are the oxidized and reduced products of each reaction?

Acid rain is a problem today due to the large amounts of these acidic oxides being put into the atmosphere by human activities every year (← ***p. 644, p.647***). When this precipitation falls on natural areas that cannot easily tolerate such acidity, serious environmental problems occur. The average annual pH of precipitation falling on much of the northeastern United States and northeastern Europe is between 4 and 4.5. Specific storms in some areas where there are numerous sources of SO_2 and NO (NO_2) have been recorded with rain having pH values as low as 1.5. To further complicate matters, acid rain is an international problem—rain and snow don't observe borders. Many Canadian residents are offended by the fact that much of the acid rain falling on Canadian cities and forests results from acidic oxides produced in the United States (Figure 17.5).

17.11 PRACTICAL ACID-BASE CHEMISTRY

In addition to being important industrial chemicals used in making fertilizers, metals, plastics, and foods, various acids and bases find many practical applications

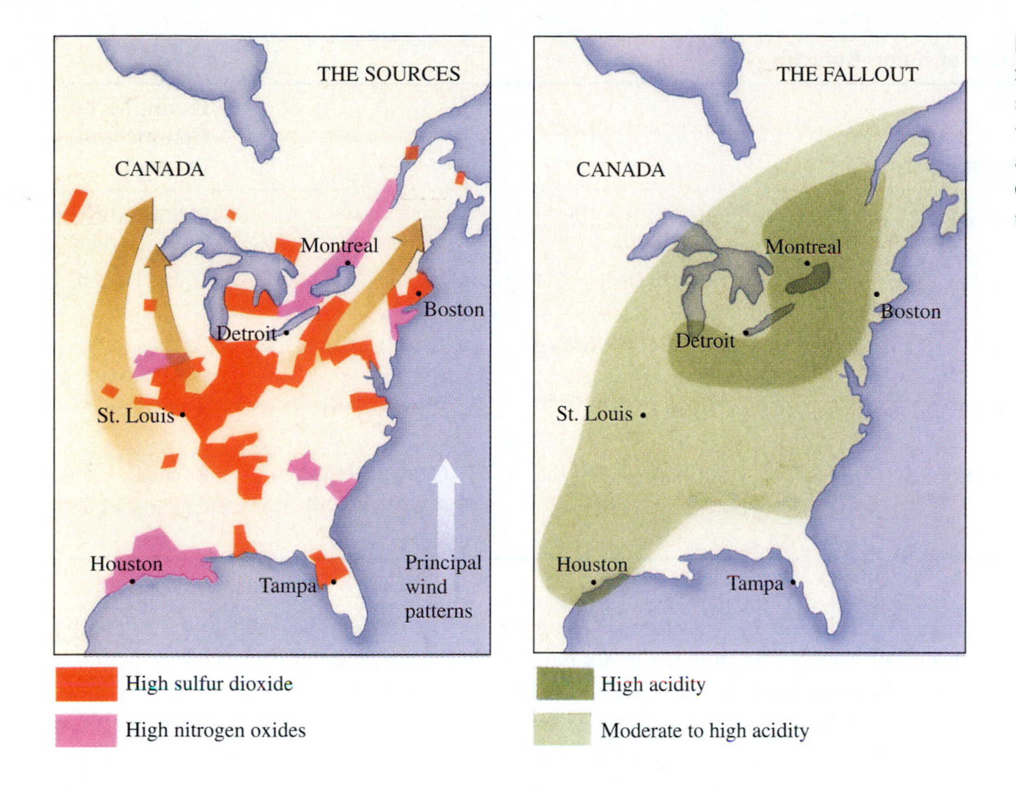

Figure 17.5 Acid rain formation and fallout. Most of the oxides of sulfur responsible for acid rain come from the Midwestern states. Prevailing winds carry the acid droplets over the Northeast and into Canada. Oxides of nitrogen also contribute to acid-rain formation.

around the home. Antacids are used to neutralize stomach acidity, while gardeners use acid salts like sodium hydrogen sulfate ($NaHSO_4$) to help acidify soil, and lime (CaO) can be used to make soil more basic. In the kitchen, baking soda and baking powders are used to make biscuit and cake dough rise, and around the house mild acids and bases are used to clean everything from the dishes and clothes to the family car and the dog.

Neutralizing Stomach Acidity

The pH of human stomach fluids is approximately 1. This very acidic pH is caused by HCl, which is secreted by thousands of cells in the wall of the stomach that specialize in transporting $H_3O^+(aq)$ and $Cl^-(aq)$ from the blood (Figure 17.6). The main

The Cl^- ions secreted by the cells in our stomach walls come mostly from table salt (NaCl) we eat, and salt-containing foods like fish.

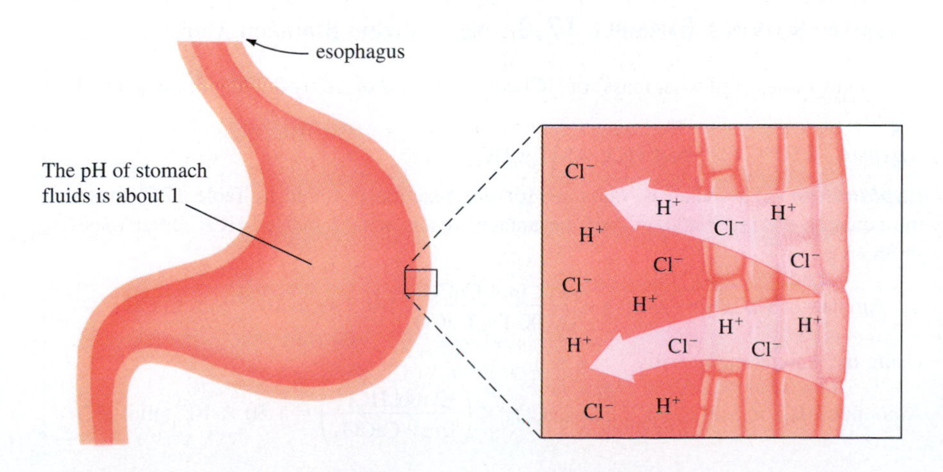

Figure 17.6 Formation of hydrochloric acid in the lining of the stomach. The lining of the stomach contains cells that secrete a solution of hydrochloric acid. The pH of the solution is about 1. Some people's stomachs produce far more acid than is needed for the primary digestion of food. This causes acid to boil up into the esophagus, producing a sensation called heartburn. Several billion dollars are spent each year for antacids to control stomach acid.

TABLE 17.7 The Acid-Base Chemistry of Some Antacids

Compound	Reaction in Stomach	Examples of Commercial Products
Milk of magnesia: $Mg(OH)_2$ in water	$Mg(OH)_2(s) + 2\ H_3O^+(aq) \longrightarrow Mg^{2+}(aq) + 4\ H_2O(\ell)$	Phillips Milk of Magnesia
Calcium carbonate: $CaCO_3$	$CaCO_3(s) + 2\ H_3O^+(aq) \longrightarrow Ca^{2+}(aq) + 3\ H_2O(\ell) + CO_2(g)$	Tums, Di-Gel
Sodium bicarbonate: $NaHCO_3$	$NaHCO_3(s) + H_3O^+(aq) \longrightarrow Na^+(aq) + H_2O(\ell) + CO_2(g)$	Baking soda, Alka-Seltzer
Aluminum hydroxide: $Al(OH)_3$	$Al(OH)_3(s) + 3\ H_3O^+(aq) \longrightarrow Al^{3+}(aq) + 6\ H_2O(\ell)$	Amphojel
Dihydroxyaluminum sodium carbonate: $NaAl(OH)_2CO_3$	$NaAl(OH)_2CO_3(s) + 4\ H_3O^+(aq) \longrightarrow Na^+(aq) + Al^{3+}(aq) + 7\ H_2O(\ell) + CO_2(g)$	Rolaids

purpose of this acid is to suppress the growth of bacteria and to aid in the digestion of certain foods. The stomach is not harmed by the presence of hydrochloric acid because its inner lining is replaced at the rate of about half a million cells per minute. However, when too much food is eaten and the stomach is stretched, or when the stomach is irritated by very spicy food, some of its acidic contents can reach the esophagus (gastroesophageal reflux) and this produces a burning sensation called *heartburn.*

An antacid is a base that is used to neutralize stomach acid. The recommended dose is the amount of the base required to neutralize *some,* but not all, of the stomach acid. Several antacids and their acid-base reactions are shown in Table 17.7. People who need to restrict the quantity of sodium (Na^+) in their diets should avoid antacids such as sodium bicarbonate.

Exercise 17.34 Strong Antacids?

Explain why strong bases like NaOH or KOH are never used as antacids.

PROBLEM-SOLVING EXAMPLE 17.8 Neutralizing Stomach Acid

How many moles (and what mass) of HCl could be neutralized by 0.750 g of the antacid $CaCO_3$?

Answer 1.50×10^{-2} mol HCl, 0.547 g HCl

Explanation The balanced equation for this reaction is found in Table 17.7. From the equation we see that 1 mol of the antacid reacts with 2 mol of HCl, molar mass = 36.46 g.

$$\text{Amount of } CaCO_3 = 0.750 \text{ g} \times \left(\frac{1 \text{ mol } CaCO_3}{100.1 \text{ g } CaCO_3}\right) = 7.49 \times 10^{-3} \text{ mol } CaCO_3$$

Using the stoichiometric ratio,

$$\text{Amount of HCl} = 7.49 \times 10^{-3} \text{ mol } CaCO_3 \times \left(\frac{2 \text{ mol HCl}}{1 \text{ mol } CaCO_3}\right) = 1.50 \times 10^{-2} \text{ mol HCl}$$

CHEMISTRY YOU CAN DO

Aspirin and Digestion

Aspirin is a potent drug capable of relieving pain, fever, and inflammation. Recent studies indicate it may also decrease blood clotting and heart disease. It is made from salicylic acid, which is naturally found in a variety of plants. The effect of pure salicylic acid on your stomach, however, makes it quite unpleasant as a pain remedy. Commercial aspirin is a derivative of salicylic acid, called acetylsalicylic acid *(← p. 187)*, which has all the benefits of salicylic acid with less discomfort.

Aspirin is still somewhat acidic, however, and can sometimes cause discomfort or worse in people susceptible to stomach irritation. As a result, there are several different forms of aspirin on the market today. The first and most common is plain aspirin. For people with stomach problems, there is also the option of taking buffered aspirin, which includes buffer in the aspirin tablet to lessen the effect of aspirin's acidity. A more recent development is enteric aspirin, which is plain aspirin in a tablet with a coating that prevents the aspirin from dissolving in stomach acid, but does allow it to dissolve in the small intestine, which is alkaline.

For this experiment, obtain at least three tablets of each kind of aspirin. Examples of regular aspirin are Bayer and Anacin. Buffered aspirin can be found as Bufferin or as Bayer Plus. Enteric aspirin is most commonly available as Bayer Enteric or as Ecotrin. Fill three transparent cups or glasses with water. Drop one intact tablet of plain aspirin into the first glass, one tablet of buffered aspirin into the second glass, and one tablet of enteric aspirin into the third glass. Note the changes in the tablets at 1-minute intervals until no further changes occur. Now repeat this experiment using vinegar instead of water to dissolve the tablets. Observe each of the tablets in the vinegar at 30-s intervals until no further change is seen. For the final experiment, fill the glasses with water and add 2 tsp of baking soda and stir until dissolved. Once you have made your baking soda solution, add one of each type of aspirin tablet and observe what happens.

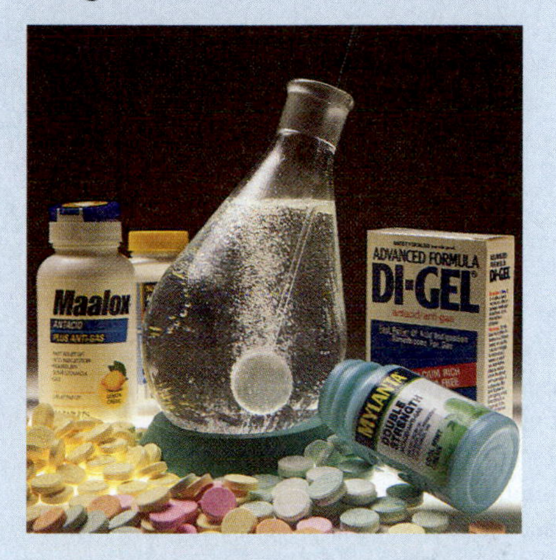

A variety of common antacids, all of which contain various weakly basic compounds like sodium bicarbonate, magnesium hydroxide, aluminum hydroxide, and calcium carbonate. The reaction flask contains an Alka-Seltzer tablet producing CO_2 gas in water. The Alka-Seltzer tablet contains sodium bicarbonate and citric acid. They do not react with one another until they come in contact with water. *(C.D. Winters)*

How did each of the tablets react to each of the different solutions? The acidity of the vinegar should have mimicked the acidity of your stomach. The basic nature of the baking soda should have mimicked the environment of your intestines, which follow your stomach in the digestive system. How do you suppose each type of aspirin works according to its ability to dissolve in your experiments? Does this lead you to think about the kind of aspirin you take?

$$\text{Mass of HCl} = 1.50 \times 10^{-2}\ \text{mol HCl} \times \left(\frac{36.46\ \text{g HCl}}{1\ \text{mol HCl}}\right) = 0.547\ \text{g HCl}$$

Problem-Solving Practice 17.8

Using the reactions in Table 17.7, determine which antacid, on a per gram basis, can neutralize the most stomach acid (assume 1.0 M HCl).

Acid-Base Chemistry in the Kitchen

Besides vinegar, an approximate 5% solution of acetic acid in water, there are other acid-containing products in the kitchen. Lemon juice, handy for making tea, flavoring cooked fish, and making salad dressings, contains citric acid. One of the most

CHEMISTRY IN THE NEWS

Tagamet HB and Pepcid AC Over the Counter

Treating acid indigestion is very big business. About \$3 billion are spent annually in this country on products to overcome excess stomach acidity (that's about \$11 per person per year). For years the only hope for sufferers of acid indigestion was some form of a weak base, like baking soda (too much sodium) or some other commercial product (Table 17.7). The use of such compounds has become commonplace for many people whose stomachs respond with greatly increased secretions of HCl to too much food or spicy food. In addition, many foods such as high-fat–content foods, tomatoes, onions, peppermint, coffee, and cola beverages tend to relax the muscle guarding the top opening where food enters the stomach. This causes a condition known as reflux esophagitis which causes a pain similar to that of a heart attack. Simply neutralizing the excess acid helps, but does not completely control the symptoms. Several drugs that are effective at lowering the stomach's production of HCl have been available for some time by prescription, and since 1994 they have been available in the over-the-counter (OTC) market. Currently, Tagamet HB (cimetidine), Pepcid AC (famotidine), Zantac 75 (ranitidine) and Axid (nizatidine) can be purchased OTC. (In fact, you can often go down the aisles of large discount and drug stores and free samples will be offered to you!) While these compounds are certainly the latest pharmaceutical innovations, the advertising claims can be very confusing, and deciding which one to take can be difficult.

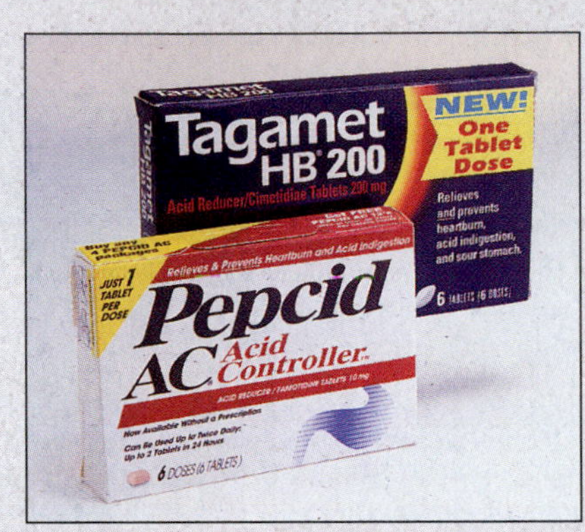

(George Semple)

While there may be differences among these drugs based on dosage or formulation, chemically they all act by the same equally effective mechanism. Like many other activities in our bodies, the secretion of stomach acid is controlled by the interaction of chemical messengers carried in the blood that interact with receptors on the surface of cells. These messengers interact with the receptors in much the same way that substrates and enzymes interact ***(←p. 543)***. The interaction is governed by molecular shape and noncovalent attractions between messengers and receptors. These OTC drugs for heartburn all attach themselves to the receptor cells, called H2 receptors, but do not activate them. In this way, acid secretion is decreased because the receptors are blocked from the action of the messengers that would normally activate them and cause the generation of stomach acid. These drugs are also called H2 blockers. (There is one home remedy you can try: chewing gum stimulates the production of saliva, which washes acid out of the esophagus.)

Source: National Digestive Diseases Information Clearinghouse, internet site, http://www.gastro.com/heartbrn.htm

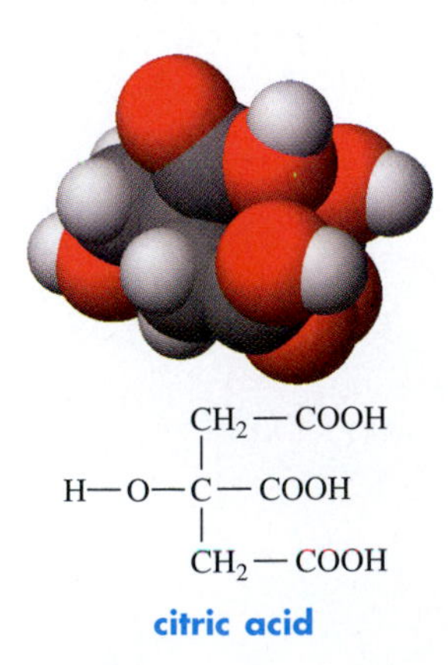

citric acid

useful substances in the kitchen is carbon dioxide gas, which is made from an acid-base reaction. This gas is what makes various bread and cake doughs rise. When pockets of gas are created inside the dough during the cooking process, the consistency of the finished product (biscuit, cake, roll, etc.) is more pleasing to eat. Various sources of CO_2 are added to a dough before it goes into the oven. One method is the production of CO_2 by the fermentation action of yeast on carbohydrates. Many commercial breads and homemade dinner rolls use yeast to produce the CO_2 necessary to make these doughs rise. The production of CO_2 by yeast is a slow process and sometimes it is desirable to make cooked breads quickly. Since a carbonate or bicarbonate salt is a convenient source of CO_2, mixing one of these with a source of acid is all that is needed to make CO_2.

Sodium bicarbonate, $NaHCO_3$ (also known as *baking soda*), is a good source of CO_2. Homemade biscuits can be made quickly using baking soda and some source of acid. But which acid should be used? Obviously a weak acid is called for, otherwise complete neutralization of the acid would be required to make the food safe to eat. Vinegar could be used, but it would impart an undesirable taste to the bread. Long ago it was discovered that lactic acid ($CH_3CHOHCOOH$), present in milk and formed in larger quantities when milk sours to form buttermilk, is a good source of acid for reacting with bicarbonate. Biscuits made with these ingredients producing CO_2 rise quickly in the oven and are quite tasty.

$$\underset{\text{lactic acid}}{CH_3\underset{\displaystyle OH}{\underset{|}{C}}HCOOH(aq)} + HCO_3^-(aq) \longrightarrow \underset{\text{lactate ion}}{CH_3\underset{\displaystyle OH}{\underset{|}{C}}HCOO^-(aq)} + H_2CO_3(aq)$$

$$\underset{\text{carbonic acid}}{H_2CO_3(aq)} \longrightarrow CO_2(g) + H_2O(\ell)$$

When buttermilk is not available or when a different taste is desired, another convenient source of acid to react with the bicarbonate ion is the dihydrogen phosphate ion ($H_2PO_4^-$). This ion is found in baking powders, which are a mixture of sodium or potassium dihydrogen phosphate and sodium bicarbonate. The two salts in baking powder do not react with one another when they are dry, but when they are mixed with water in the dough the reaction starts to produce CO_2 and is accelerated when the dough is placed in the oven.

Consider the list of ingredients for buttermilk blueberry muffins in the margin. There are two sources of acid and two sources of bicarbonate, which produces the CO_2 to make these muffins rise in the oven. Can you tell which they are?

Action of baking powder. Baking soda contains the weak acid calcium dihydrogen phosphate, $Ca(H_2PO_4)_2$, and the weak base sodium hydrogen carbonate. They react to form CO_2 and water. Here, solid KH_2PO_4 is added to a solution of $NaHCO_3$. *(C.D. Winters)*

Buttermilk Blueberry Muffins

$2\frac{1}{2}$ c flour
$1\frac{1}{2}$ tsp baking powder
$\frac{1}{2}$ tsp soda
$\frac{1}{2}$ c sugar
$\frac{1}{4}$ tsp salt
2 eggs, beaten
1 c buttermilk
3 oz butter
$1\frac{1}{2}$ c blueberries (added for flavor)

Exercise 17.35 Sources of CO_2

Explain why Na_2CO_3 is not used as a source of CO_2 for making bread. Could $CaCO_3$ be used? Why?

Household Cleaners

Most cleaning compounds like dishwashing detergents, scouring powders, laundry detergents, and oven cleaners are basic. A few, like toilet-bowl cleaners and some disinfectants, are acidic. Muriatic acid (hydrochloric acid) is used for cleaning bricks and concrete in new home construction or when remodeling is done. Synthetic detergents are derived from organic molecules designed to have even better cleaning action than soaps but less reaction with the doubly positive ions found in hard water. As a consequence, synthetic detergents are often more economical to use and are more effective in hard water than soap. There are many different synthetic detergents on the market. An inventory of cleaning materials in a typical household might include half a dozen or more products designed to be the most suitable for a specific job—cleaning skin, hair, clothes, floors, or the family car.

The molecular structure of a synthetic detergent molecule, like that of a soap, consists of a long oil-soluble (hydrophobic, p. 758) group and a water-soluble (hydrophilic, p. 758) group.

A typical synthetic detergent molecule

$CH_3CH_2CH_2CH_2CH_2CH_2CH_2CH_2CH_2CH_2CH_2CH_2CH_2CH_2$—$C_6H_4$—$SO_3^-Na^+$

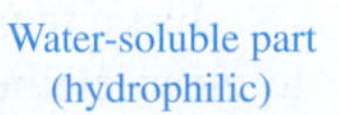

Oil-soluble part (hydrophobic) — Water-soluble part (hydrophilic)

A detergent with an ammonium end ($-NH_4^+$) would also be cationic.

Typical hydrophilic groups include negatively charged sulfate ($—OSO_3^-$), sulfonate ($—SO_3^-$), and phosphate ($—OPO_3^{2-}$) groups. Compounds with these groups are called *anionic surfactants.*

Cationic (positively charged) *surfactants* are almost all quaternary ammonium halides (four groups attached to the central nitrogen atom) with the general formula:

$$R_1—\overset{\displaystyle R_2}{\underset{\displaystyle R_3}{N^+}}—R_4X^-$$

where one of the R groups is a long hydrocarbon chain and another frequently includes an —OH group. The X^- in the formula represents a halide ion such as chloride (Cl^-) or bromide (Br^-) ion.

Cationic detergents are generally incompatible with anionic detergents. If they are brought together, insoluble products can precipitate from the solution, and the precipitate has none of the desired properties of either detergent. Their reaction with excess anionic detergent, however, makes cationic detergents useful as fabric softeners. When an anionic detergent remains in a fabric after washing, the fabric often loses its softness. Sufficient rinsing would take away the excess detergent, but water-saving rinse cycles in most washing machines seldom use enough water to get out all of the detergent. A small amount of a cationic detergent in the rinse water is enough to react with the excess anionic detergent. The result is a uniform surface layer that leaves the fabric feeling softer.

Most cationic detergents are also able to kill bacteria. Quaternary ammonium detergents are particularly good in this regard and find wide use in disinfectants, toilet bowl cleaners, and similar products.

Besides the anionic and cationic detergents, some detergents are *nonionic.* They have an uncharged hydrophilic polar group attached to a large organic group of low polarity, for example,

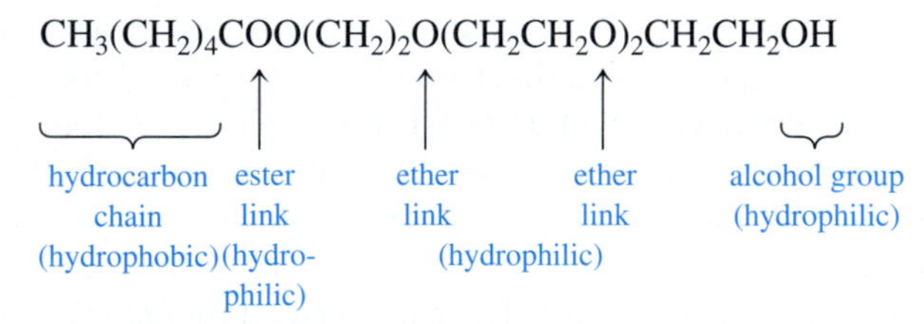

The carbon chain in this molecule is oil-soluble hydrophobic and the rest of the molecule is hydrophilic, the properties needed for the molecule to be a detergent.

The nonionic detergents have several advantages over ionic detergents. Since nonionics contain no ionic groups, they cannot form salts with calcium, magnesium, and iron ions and consequently are totally unaffected by hard water. For the same reason, nonionic detergents do not react with acids and may be used even in relatively strong acid solutions, which makes them useful in toilet-bowl cleaners.

In general, the nonionic detergents foam less than ionic surface-active agents, a property that is desirable where nonfoaming detergents are required, as in dishwashing. Today, about one third of all detergents sold, including most liquid laundry detergents, are of the nonionic type.

Just a few decades ago things were very different. Lye soap was commonly used to clean clothes as well as people's skins. This type of soap was made using either pure lye (NaOH) or sodium and potassium carbonates (Na_2CO_3, also called caustic soda), and K_2CO_3 (also called potash) from wood ashes (← *p. 758*). Most of the time the soap that was made using these bases contained considerable amounts of trapped unreacted base, but this was considered desirable because it helped raise the pH and break up the heavy soil particles common on fabrics in those days. In addition, the fabrics then were much more durable than fabrics today.

The closest thing to lye soap we see today is the typical dishwashing detergent (Table 17.8).

TABLE 17.8 Formulation for a Dishwashing Detergent

	% by Weight
Sodium carbonate, Na_2CO_3	37.5
Sodium tripolyphosphate, $Na_5P_3O_{10}$	30
Sodium metasilicate, Na_2SiO_3	30
Low-foam surfactant	0.5
Sodium dichloroisocyanurate (Cl_2 source)	1.5
Other ingredients such as colorants	0.5

The first three ingredients in this table react with water in hydrolysis reactions to produce OH^- and raise the pH. The reaction between carbonate ion and water produces OH^- ions and a strongly basic solution.

$$CO_3^{2-}(aq) + H_2O(\ell) \longrightarrow HCO_3^-(aq) + OH^-(aq)$$

This was one of the reactions responsible for the high pH of lye soap.

The tripolyphosphate ion consists of phosphate ions joined together by sharing bridging oxygen atoms in a structure like that of some silicates (← *p. 698*). A 1% solution of this ion in water produces a solution with a pH of 9.7.

$$\underset{\text{tripolyphosphate ion}}{P_3O_{10}^{5-}(aq)} + H_2O(\ell) \longrightarrow HP_3O_{10}^{4-}(aq) + OH^-(aq)$$

The metasilicate ion also undergoes extensive hydrolysis to produce OH^- ions in solution.

$$\underset{\text{metasilicate ion}}{SiO_3^{2-}} + H_2O(\ell) \longrightarrow HSiO_3^-(aq) + OH^-(aq)$$

Together, these three salts typically produce solutions inside the dishwasher that have a pH of near 12.5, high enough that animal and vegetable oils quickly break away from the surfaces they are on during the agitation cycle. The detergent helps to dissolve these oily particles and carry them away in the rinse water.

Exercise 17.36 pH of a Basic Cleaning Solution

Calculate the pH of a solution that is 5.2 M in sodium carbonate.

Highly caustic solid cleaners like drain cleaners often contain almost 100% NaOH. (*See* ← *p. 629* for a discussion about the production of H_2 gas from aluminum metal in some drain cleaners.) Liquid drain cleaners are often 50% or more NaOH by weight in water. These solutions, being more dense than water, sink to the bottom of the drain trap and quickly start to dissolve food particles, hair, and other matter that is clogging the drain. Spray-on oven cleaners contain an NaOH solution mixed with a detergent and a propellant to force the mixture out onto the surfaces of the oven. This mixture is thick enough to adhere to the surfaces on the oven long enough for the strongly basic solution to break up the baked-on food particles. If food was baked at a high temperature, it has probably carbonized. When that is the case, only scraping will remove the deposits.

Ordinary bleach solution, commonly found in the laundry room, contains sodium, potassium, and calcium hypochlorite in aqueous solution. These solutions are basic because of the hydrolysis of the hypochlorite ion (← ***p. 790,*** *Problem-Solving Example 17.6).*

$$OCl^-(aq) + H_2O(\ell) \longrightarrow HOCl\ (aq) + OH^-(aq)$$

All acidic and basic solutions, both in the lab as well as in the home, can be hazardous. Acids, interestingly, are somewhat less dangerous than solutions of bases because the H_3O^+ ion tends to *denature* ***(← p. 545)*** proteins when it comes in contact with them. This denaturing process causes most proteins to harden, which forms a protective layer, and prevents further attack by the acid, unless it is hot or highly concentrated. Basic solutions, on the other hand, tend to slowly dissolve proteins and in the process produce little, if any, pain. Often, basic solutions can cause considerable harm before any problem is noticed. You should always be cautious with household acids and bases. They are usually just as concentrated and harmful as industrial chemicals. If you should get these chemicals on your skin, wash with water for at least 15 min; and if you get acid or base in your eyes, have someone call a physician while you begin gently washing the affected area with lots of water. The international placard that is required for shipments of acids and bases in quantities over 1000 lb illustrates schematically the personal dangers of these substances (Figure 17.7).

Figure 17.7 A corrosive placard required on loads of 1001 lb or more of acids and bases during transportation by highway and rail. The U.S. Department of Transportation regulations are very strict regarding how to placard loads of corrosive and other hazardous chemicals during transport. Placards are designed to be easily interpreted. The pictorial warnings about reactions with human skin and metals are fairly clear.

Exercise 17.37 Cleaning Off Petroleum Oils

Petroleum oils and greases are different from oils of animal and plant origin. Why are ordinary detergents ineffective at cleaning petroleum oils from your hands and clothes?

17.12 LEWIS ACIDS AND BASES

In 1923 when Brønsted and Lowry independently proposed their acid-base concept, Gilbert N. Lewis also was developing a new concept of acids and bases. By the early 1930s Lewis had proposed definitions of acids and bases that are more general than those of Brønsted and Lowry because they are based on sharing of electron pairs rather than on proton transfers. A **Lewis acid** is *a substance that can accept a pair of electrons to form a new bond,* and a **Lewis base** is *a substance that can donate a pair of electrons to form a new bond.* This means that in the Lewis sense an acid-base reaction occurs when there is a molecule (or ion) with a pair of electrons that can be donated and a molecule (or ion) that can accept an electron pair.

$$\underset{\text{acid}}{A} + \underset{\text{base}}{B:} \longrightarrow \underset{\text{new bond}}{B:A}$$

This type of bond was defined earlier as a coordinate covalent bond (*← p. 384*), which is present in many neutral molecular compounds and also in complex ions.

A simple example of a Lewis acid-base reaction is formation of a hydronium ion from H^+ and water. The H^+ ion has no electrons, while the water molecule has two unshared pairs of electrons on the oxygen atom. One of the electron pairs can be shared between H^+ and water, thus forming an O—H bond.

$$H^+ + :\underset{\displaystyle H}{\underset{|}{O}}—H \longrightarrow H:\underset{\displaystyle H}{\underset{|}{O}}—H^+$$

hydronium ion

Such reactions are very common. In general, Lewis acids are cations or neutral molecules with an available, empty orbital, and Lewis bases are anions or neutral molecules with a lone pair of electrons.

Positive Metal Ions as Lewis Acids

All metal cations are potential Lewis acids. Not only do they attract electrons due to their positive charge, but all have at least one empty orbital. This empty orbital can accommodate an electron pair donated by a base and thereby form a two-electron chemical bond. Consequently, metal ions readily form coordination complexes (*← p. 384*) and also are hydrated in aqueous solution. One of the lone pairs on the oxygen atom in each of several water molecules forms a coordinate covalent bond to a metal ion when the ion becomes hydrated; the ion is a Lewis acid, and water is a Lewis base.

The hydroxide ion (OH^-) is an excellent Lewis base and so binds readily to metal cations to give metal hydroxides. An important feature of the chemistry of many metal hydroxides is that they are **amphoteric,** meaning that they can react as both a base and an acid. The amphoteric aluminum hydroxide, for example, behaves as a Lewis acid when it dissolves in a basic solution to form a complex ion containing one additional OH^- ion.

$$Al(OH)_3(s) + OH^-(aq) \rightleftharpoons [Al(OH)_4]^-(aq)$$

This reaction is shown in Figure 17.8. The same compound behaves as a Brønsted base when it reacts with a Brønsted acid (Table 17.9).

$$Al(OH)_3(s) + 3\ H_3O^+(aq) \rightleftharpoons Al^{3+}(aq) + 6\ H_2O(\ell)$$

Metal ions also form many complex ions with the Lewis base ammonia, $:NH_3$. For example, silver ion readily forms a water-soluble, colorless complex ion in liq-

TABLE 17.9 Some Common Amphoteric Metal Hydroxides

Hydroxide	Reaction as a Base	Reaction as an Acid
$Al(OH)_3$	$Al(OH)_3(s) + 3\ H_3O^+(aq) \longrightarrow Al^{3+}(aq) + 6\ H_2O(\ell)$	$Al(OH)_3(s) + OH^-(aq) \longrightarrow [Al(OH)_4]^-(aq)$
$Zn(OH)_2$	$Zn(OH)_2(s) + 2\ H_3O^+(aq) \longrightarrow Zn^{2+}(aq) + 4\ H_2O(\ell)$	$Zn(OH)_2(s) + 2\ OH^-(aq) \longrightarrow [Zn(OH)_4]^{2-}(aq)$
$Sn(OH)_4$	$Sn(OH)_4(s) + 4\ H_3O^+(aq) \longrightarrow Sn^{4+}(aq) + 8\ H_2O(\ell)$	$Sn(OH)_4(s) + 2\ OH^-(aq) \longrightarrow [Sn(OH)_6]^{2-}(aq)$
$Cr(OH)_3$	$Cr(OH)_3(s) + 3\ H_3O^+(aq) \longrightarrow Cr^{3+}(aq) + 6\ H_2O(\ell)$	$Cr(OH)_3(s) + OH^-(aq) \longrightarrow [Cr(OH)_4]^-(aq)$

(a)

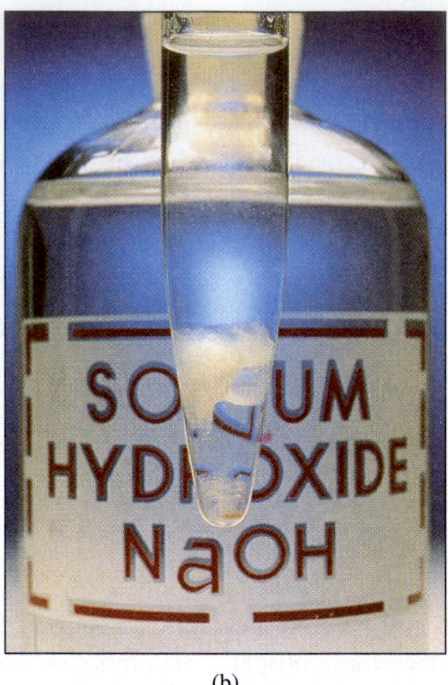

(b)

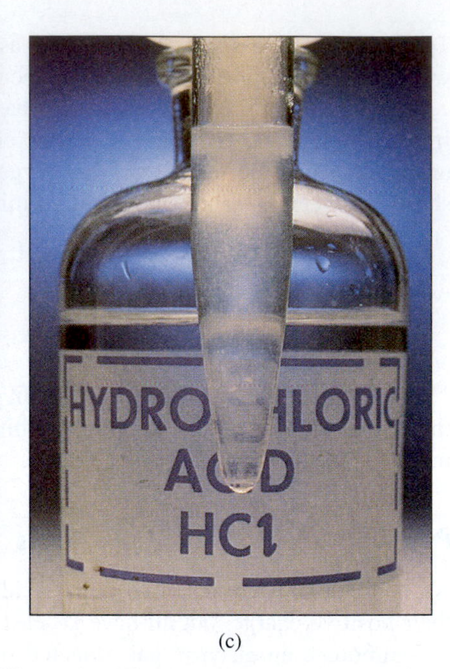

(c)

Figure 17.8 The amphoteric nature of $Al(OH)_3$. (a) Adding aqueous ammonia to a solution of Al^{3+} causes a precipitate of $Al(OH)_3$. (b) Adding a strong base (NaOH) to the $Al(OH)_3$ dissolves the precipitate. Here the aluminum hydroxide acts as a Lewis acid toward the Lewis base OH^- and forms a soluble salt of the complex ion $Al(OH)_4^-$. (c) If we begin again with freshly precipitated $Al(OH)_3$, it dissolves as strong acid (HCl) is added. In this case $Al(OH)_3$ acts as a Brønsted base and forms a soluble aluminum salt and water. *(C.D. Winters)*

uid ammonia or in aqueous ammonia. Indeed, this complex is so stable that the very water-insoluble compound AgCl can be dissolved in aqueous ammonia:

$$AgCl(s) + 2\ {:}NH_3(ag){:}\ NH_3(aq) \longrightarrow [H_3N{:}Ag{:}NH_3]^+(aq) + Cl^-(aq)$$

Neutral Molecules as Lewis Acids

Lewis's ideas about acids and bases account nicely for the fact that oxides of nonmetals behave as acids. Two important examples are carbon dioxide and sulfur dioxide, whose Lewis structures are

$$:\ddot{O}{=}C{=}\ddot{O}: \qquad :\ddot{O}{=}\ddot{S}{-}\ddot{O}: \longleftrightarrow :\ddot{O}{-}\ddot{S}{=}\ddot{O}:$$

carbon dioxide　　　sulfur dioxide

In each case, there is a double bond; an "extra" pair of electrons is being shared between an oxygen atom and the central atom. Because oxygen is highly electronegative, electrons in these bonds are attracted away from the central atom, which becomes slightly positively charged. This makes the central atom a likely site to attract a pair of electrons. A Lewis base such as OH^- can bond to the carbon atom in CO_2 to give bicarbonate ion, HCO_3^-. This displaces one double-bond pair of electrons back onto an oxygen atom.

$$:\ddot{O}=C=\ddot{O}: + :\ddot{\underset{\cdot\cdot}{O}}-H^- \longrightarrow :\ddot{O}=C\begin{matrix} \diagup \ddot{\underset{\cdot\cdot}{O}}-H \\ \diagdown \ddot{\underset{\cdot\cdot}{O}}:^- \end{matrix}$$

bicarbonate ion

As seen in Figure 17.9, carbon dioxide from the air can react to form sodium carbonate around the mouth of a bottle of sodium hydroxide. SO_2 can react similarly with hydroxide ion.

Figure 17.9 Carbon dioxide in the air reacts with spilled base such as NaOH, forming Na_2CO_3. If the mouth of a glass-stoppered bottle such as the one shown here is not routinely cleaned, the sodium carbonate formed can virtually cement the top of the bottle to the neck, making it difficult to open the bottle. *(C.D. Winters)*

Exercise 17.38 Lewis Acids and Bases

Predict whether each of the following is a Lewis acid or a Lewis base. Drawing a Lewis structure for a molecule or ion is often helpful in making such a prediction.
(a) PH_3 (b) BCl_3 (c) H_2S (d) NO_2 (e) Ni^{2+} (f) CO

Summary Problem

Lactic acid ($HC_3H_5O_3$) is a weak, monoprotic acid with a melting point of 53 °C. It exists in two enantiomorphic forms (⇐ *p. 430*) that have slightly different pK_a values. The L-form has a pK_a of 3.79 and the D-form has a pK_a of 3.83. The D-form is found in molasses, beer, wines, souring milk. The L-form is produced in muscle cells during anaerobic metabolism in which glucose molecules are broken down into lactic acid and molecules of adenosine triphosphate (ATP). When lactic acid builds up too rapidly in muscle tissue, severe pain results.

(a) Which form of lactic acid (D or L) is the stronger acid?

(b) What would be the measured pK_a of a 50–50 mixture of the two forms of lactic acid?

(c) A solution of D-lactic acid is prepared. Use HL as a general formula for lactic acid and write the equation for the ionization of lactic acid in water.

(d) If 0.1 M solutions of these two acids (D and L) were prepared, what would be the pH of each solution?

(e) Before any lactic acid dissolves in the water, what reaction determines the pH?

(f) Calculate the pH of a solution made by dissolving 4.46 g of D-lactic acid in 500. mL of water.

(g) How many mL of a 0.115 M NaOH solution would be required to completely neutralize 4.46 g of pure lactic acid?

(h) What would be the pH of the solution made by the neutralization if the lactic acid were the D-form? The L-form? A 50–50 mixture of the two forms?

(i) Describe how to prepare 250 mL of a buffer solution with a pH of 4.5 using pure D-lactic acid and pure potassium D-lactate.

In Closing

Having studied this chapter, you should be able to...

- describe water's role in aqueous acid-base chemistry (Section 17.1).
- identify the conjugate base of an acid and the conjugate acid of a base (Section 17.2).

- write the ionization steps of polyprotic acids and bases (Section 17.1 and 17.3).
- describe how sulfuric acid and phosphoric acid are made from sulfur- and phosphorus-bearing minerals (Section 17.3).
- use the autoionization of water and show how this equilibrium takes place in aqueous solutions of acids and bases (Section 17.4).
- calculate pH (or pOH) given $[H_3O^+]$, or $[H_3O^+]$ given pH (or pOH) (Section 17.5).
- estimate acid and base strength from K_a and K_b values (Section 17.6).
- calculate pH from K_a or K_b values and solution concentration (Section 17.7).
- describe the hydrolysis of salts in aqueous solution (Section 17.8).
- explain how buffers maintain pH, how to calculate their pH, how they are prepared, and buffer capacity (Section 17.9).
- explain how acid rain is formed and its effects on the environment (Section 17.10).
- apply acid-base principles to the chemistry of antacids, kitchen chemistry, and cleaning chemicals (Section 17.11).
- recognize Lewis acids and bases and how they react (Section 17.12).

Key Terms

acid ionization constant *(17.6)*
acid rain *(17.10)*
amines *(17.2)*
amphiprotic *(17.1)*
amphoteric *(17.12)*
autoionization *(17.4)*
base-ionization constant *(17.6)*
buffer *(17.9)*
buffer solution *(17.9)*
conjugate acid-base pair *(17.2)*
hydrolysis *(17.8)*
ionization constant for water *(17.4)*
Lewis acid *(17.12)*
Lewis base *(17.12)*
monoprotic acids *(17.3)*
monoprotic bases *(17.3)*
neutral *(17.4)*
oxoacids *(17.1)*
pH *(17.5)*
polyprotic acids *(17.3)*
polyprotic bases *(17.3)*

Questions for Review and Thought

Conceptual Challenge Problems

CP-17.A. Is it possible for an aqueous solution to have a pH of 0 or even less than 0? Explain your answer mathematically as well as practically based on what you know about acid solubilities.

CP-17.B. What is the pH of water at 200 °C? Liqud water this hot would have to be under a pressure greater than 1.0 atm and might be found in a pressurized water reactor located in a nuclear power plant.

CP-17.C. Develop a set of rules by which you could predict the pH for solutions of strong or weak acids and strong or weak bases *without using a calculator.* Your predictions need to be accurate to ±1 pH unit. Assume that you know the concentration of the acid or base and that for the weak acids and bases you can look up the pK_a or K_a values. What rules would work to predict pH?

CP-17.D. Suppose you were asked on a laboratory test to outline a procedure to prepare a buffered solution of pH 8.0 using hydrocyanic acid, HCN. You realize that a pH of 8.0 is basic, and you find that the K_a of hydrocyanic acid is 4.0×10^{-10}. What is your response?

Review Questions

1. Define a Brønsted acid and a Brønsted base.
2. Explain in your own words what 100% ionization means.
3. Write the chemical equation for the autoionization of water. Write the equilibrium-constant expression for this reaction. What is the value of the equilibrium constant at 25 °C? What is this constant called?
4. When OH^- is the base in a conjugate acid-base pair, the acid is ________; when OH^- is the acid, the base is ________.

5. Write balanced chemical equations that show phosphoric acid, H_3PO_4, ionizing stepwise as a polyprotic acid.
6. Write ionization equations for a weak acid and its conjugate base. Show that adding these two equations gives the autoionization equation for water.
7. Designate the acid and the base on the left side of the following equations, and designate the conjugate partner of each on the right side.
(a) $HNO_3(aq) + H_2O(\ell) \rightarrow H_3O^+(aq) + NO_3^-(aq)$
(b) $NH_4^+(aq) + CN^-(aq) \rightarrow NH_3(aq) + HCN(aq)$
8. Dissolving ammonium bromide in water gives an acidic solution. Write a balanced equation showing how that can occur.
9. Solution A has a pH of 8 and solution B a pH of 10. Which has the greater hydronium ion concentration? How many times greater is its concentration?
10. Which would form a buffer?
(a) HCl and CH_3COOH
(b) NaH_2PO_4 and Na_2HPO_4
(c) H_2CO_3 and $NaHCO_3$
(d) NaOH and NaCl
(e) NaOH and NH_3
11. Briefly describe how a buffer solution can control the pH of a solution when strong acid is added, and when strong base is added. Use NH_3/NH_4Cl as an example buffer and HCl and NaOH as the strong acid and strong base.
12. Contrast the main ideas of the Brønsted and Lewis acid-base concepts. Name and write the formula for a substance that behaves as a Lewis acid but not as a Brønsted acid.

The Brønsted Concept of Acids and Bases

13. Write an equation to describe the proton transfer that occurs when each of the following acids is added to water.
(a) HBr (c) HSO_4^-
(b) CF_3COOH (d) HNO_2
14. Write an equation to describe the proton transfer that occurs when each of the following acids is added to water.
(a) HCO_3^- (c) CH_3COOH
(b) HCl (d) HCN
15. Write an equation to describe the proton transfer that occurs when each of the following bases is added to water.
(a) H^- (c) NO_2^-
(b) HCO_3^- (d) PO_4^{3-}
16. Write an equation to describe the proton transfer that occurs when each of the following bases is added to water.
(a) HSO_4^- (c) I^-
(b) CH_3NH_2 (d) $H_2PO_4^-$
17. Based on formulas alone, classify each of the following oxoacids as strong or weak.
(a) H_3PO_4 (e) HNO_3
(b) H_2SO_4 (f) H_2CO_3
(c) HClO (g) HNO_2
(d) $HClO_4$
18. Based on formulas alone, which is the stronger acid?
(a) H_2CO_3 or H_2SO_4 (d) H_3PO_4 or $HClO_3$
(b) HNO_3 or HNO_2 (e) H_2SO_3 or H_2SO_4
(c) $HClO_4$ or H_2SO_4
19. Write the formula and name for the conjugate partner for each acid or base.
(a) CN^- (d) S^{2-}
(b) SO_4^{2-} (e) HSO_3^-
(c) HS^- (f) HCOOH (formic acid)
20. Write the formula and name for the conjugate partner for each acid or base.
(a) HI (d) H_2CO_3
(b) NO_3^- (e) HSO_4^-
(c) CO_3^{2-} (f) SO_3^{2-}
21. Which are conjugate acid-base pairs?
(a) H_2O and H_3O^+ (d) NH_3 and NH_4^+
(b) H_2O and OH^- (e) O^{2-} and H_2O
(c) NH_2^- and NH_4^+
22. Which are conjugate acid-base pairs?
(a) NH_2^- and NH_4^+ (d) OH^- and O^{2-}
(b) NH_3 and NH_2^- (e) H_3O^+ and OH^-
(c) H_3O^+ and H_2O
23. Identify the acid and the base that are reactants in each equation; identify the conjugate base and conjugate acid on the product side of each equation.
(a) $HI(aq) + H_2O(\ell) \rightleftarrows H_3O^+(aq) + I^-(aq)$
(b) $OH^-(aq) + NH_4^+(aq) \rightleftarrows H_2O(\ell) + NH_3(aq)$
(c) $NH_3(aq) + H_2CO_3(aq) \rightleftarrows NH_4^+(aq) + HCO_3^-(aq)$
(d) $H_2PO_4^-(aq) + HCO_3^-(aq) \rightleftarrows H_2CO_3(aq) + HPO_4^{2-}(aq)$
24. Identify the acid and the base that are reactants in each equation; identify the conjugate base and conjugate acid on the product side of the equation.
(a) $HS^-(aq) + H_2O(\ell) \rightleftarrows H_2S(aq) + OH^-(aq)$
(b) $S^{2-}(aq) + NH_4^+(aq) \rightleftarrows NH_3(aq) + HS^-(aq)$
(c) $HCO_3^-(aq) + HSO_4^-(aq) \rightleftarrows H_2CO_3(aq) + SO_4^{2-}(aq)$
(d) $NH_3(aq) + NH_2^-(aq) \rightleftarrows NH_2^-(aq) + NH_3(aq)$
25. Write stepwise equations for protonation or deprotonation of each of these polyprotic acids and bases.
(a) H_2SO_3
(b) S^{2-}
(c) $NH_3CH_2COOH^+$ (glycinium ion, a diprotic acid)
26. Write stepwise equations for protonation or deprotonation of each of these polyprotic acids and bases.
(a) CO_3^{2-}
(b) H_3AsO_4
(c) $NH_2CH_2COO^-$ (glycinate ion, a diprotic base)

pH Calculations

27. A popular soft drink has a pH of 3.30. What is its hydronium-ion concentration? Is the drink acidic or basic?
28. Milk of magnesia, $Mg(OH)_2$, has a pH of 10.5. What is the hydronium-ion concentration of the solution? Is this solution acidic or basic?

29. What is the pH of a 0.0013 M solution of HNO_3? What is the pOH of this solution?
30. What is the pH of a solution that is 0.025 M in NaOH? What is the pOH of this solution?
31. The pH of a $Ba(OH)_2$ solution is 10.66 at 25 °C. What is the hydroxide-ion concentration of this solution? If the solution volume is 250. mL, how many grams of $Ba(OH)_2$ must have been used to make this solution?
32. A 1000.-mL solution of hydrochloric acid has a pH of 1.3. How many grams of HCl are dissolved in the solution?
33. Make the following interconversions. In each case tell whether the solution is acidic or basic.

pH	$[H_3O^+]$ (M)	$[OH^-]$ (M)
(a) 1.00	______	______
(b) 10.5	______	______
(c) ______	1.8×10^{-4}	______
(d) ______	5.6×10^{-10}	______
(e) ______	______	2.3×10^{-5}

34. Make the following interconversions. In each case tell whether the solution is acidic or basic.

pH	$[H_3O^+]$ (M)	$[OH^-]$ (M)
(a) ______	6.1×10^{-7}	______
(b) ______	______	2.2×10^{-9}
(c) 4.67	______	______
(d) ______	2.5×10^{-2}	______
(e) 9.12	______	______

35. Figure 17.2 shows the pH of some common solutions. How many times more acidic or basic are the following compared to a neutral solution?
(a) Milk (c) Blood
(b) Sea water (d) Battery acid
36. Figure 17.2 shows the pH of some common solutions. How many times more acidic or basic are the following compared to a neutral solution?
(a) Black coffee (c) Baking soda
(b) Household ammonia (d) Vinegar

Acid-Base Strengths

37. Write ionization equations and ionization-constant expressions for the following acids and bases.
(a) CH_3COOH (d) PO_4^{3-}
(b) HCN (e) NH_4^+
(c) SO_3^{2-} (f) H_2SO_4
38. Write ionization equations and ionization-constant expressions for the following acids and bases.
(a) F^- (d) H_3PO_4
(b) NH_3 (e) CH_3COO^-
(c) H_2CO_3 (f) S^{2-}
39. Which solution will be more acidic?
(a) 0.10 M H_2CO_3 or 0.10 M NH_4Cl
(b) 0.10 M HF or 0.10 M $KHSO_4$
(c) 0.1 M $NaHCO_3$ or 0.1 M Na_2HPO_4
(d) 0.1 M H_2S or 0.1 M HCN
40. Which solution will be more basic?
(a) 0.10 M NH_3 or 0.10 M NaF
(b) 0.10 M K_2S or 0.10 M K_3PO_4
(c) 0.10 M $NaNO_3$ or 0.10 M CH_3COONa
(d) 0.10 M NH_3 or 0.10 M KCN
41. Without doing any calculations, assign each of the following 0.10 M aqueous solutions to one of these pH ranges: pH < 2; pH between 2 and 6; pH between 6 and 8; pH between 8 and 12; pH > 12
(a) HNO_2 (e) BaO
(b) NH_4Cl (f) $KHSO_4$
(c) NaF (g) $NaHCO_3$
(d) $Mg(CH_3COO)_2$ (h) $BaCl_2$
42. Calculate the pH of each solution in Question 41 to verify your prediction.
43. A 0.015 M solution of cyanic acid has a pH of 2.67. What is the ionization constant, K_a, of the acid?
44. What is the K_a of butyric acid if a 0.025 M solution has a pH of 3.21?
45. What are the equilibrium concentrations of H_3O^+, acetate ion, and acetic acid in a 0.20 M aqueous solution of acetic acid (CH_3COOH)?
46. The ionization constant of a very weak acid HA is 4.0×10^{-9}. Calculate the equilibrium concentrations of H_3O^+, A^-, and HA in a 0.040 M solution of the acid.
47. The weak base methylamine, CH_3NH_2, has $K_b = 5.0 \times 10^{-4}$. It reacts with water according to the equation

$$CH_3NH_2(aq) + H_2O(\ell) \rightleftharpoons CH_3NH_3^+(aq) + OH^-(aq)$$

What is the pH of a 0.23 M methylamine solution?
48. Calculate the pH of a 0.12 M aqueous solution of the base aniline, $C_6H_5NH_2$ ($K_b = 4.2 \times 10^{-10}$).
49. By now you may be wishing you had an aspirin. Aspirin is a weak acid with $K_a = 3.27 \times 10^{-4}$ for the reaction

$$HC_9H_7O_4(aq) + H_2O(\ell) \rightleftharpoons C_9H_7O_4^-(aq) + H_3O^+(aq)$$

Two aspirin tablets, each containing 0.325 g of aspirin (along with a nonreactive "binder" to hold the tablet together), are dissolved in 200.0 mL of water. What is the pH of this solution?
50. Lactic acid, $C_3H_6O_3$, occurs in sour milk as a result of the metabolism of certain bacteria. What is the pH of a solution of 56 mg of lactic acid in 250 mL of water? K_a for lactic acid is 1.4×10^{-4}.

Acid-Base Reactions

51. Complete each of these reactions by filling in the blanks. Predict whether each reaction is product-favored or reactant-favored, and explain your reasoning.
(a) ________ (aq) + Br^-(aq) $\rightleftharpoons$ NH_3(aq) + HBr(aq)
(b) CH_3COOH(aq) + CN^-(aq) $\rightleftharpoons$ ________ (aq) + HCN(aq)
(c) ________(aq) + $H_2O(\ell)$ $\rightleftharpoons$ NH_3(aq) + OH^-(aq)
52. Complete each of these reactions by filling in the blanks. Predict whether each reaction is product-favored or reactant-favored, and explain your reasoning.
(a) ________(aq) + HSO_4^-(aq) $\rightleftharpoons$ HCN(aq) + SO_4^{2-}(aq)
(b) H_2S(aq) + $H_2O(\ell)$ $\rightleftharpoons$ H_3O^+(aq) + ________(aq)
(c) H^-(aq) + $H_2O(\ell)$ $\rightleftharpoons$ OH^-(aq) + ________(aq)
53. Predict which of the following acid-base reactions are product-favored and which are reactant-favored. In each case write a balanced equation for any reaction that might occur, even if the reaction is reactant-favored. Consult Table 17.3 if necessary.
(a) $H_2O(\ell)$ + HNO_3(aq) (c) CN^-(aq) + HCl(aq)
(b) H_3PO_4(aq) + $H_2O(\ell)$ (d) NH_4^+(aq) + F^-(aq)
54. Predict which of the following acid-base reactions are product-favored and which are reactant-favored. In each case write a balanced equation for any reaction that might occur, even if the reaction is reactant-favored. Consult Table 17.3 if necessary.
(a) NH_4^+(aq) + HPO_4^{2-}(aq)
(b) CH_3COOH(aq) + OH^-(aq)
(c) HSO_4^-(aq) + $H_2PO_4^-$(aq)
(d) CH_3COOH(aq) + F^-(aq)
55. For each salt, predict whether an aqueous solution will have a pH less than, equal to, or greater than 7.
(a) $NaHSO_4$ (d) NaH_2PO_4
(b) NH_4Br (e) NH_4NO_3
(c) $KClO_4$ (f) $SrCl_2$
56. For each salt, predict whether an aqueous solution will have a pH less than, equal to, or greater than 7.
(a) $AlCl_3$ (d) Na_2HPO_4
(b) Na_2S (e) $(NH_4)_2S$
(c) $NaNO_3$ (f) KCH_3COO
57. Explain why $BaCO_3$ is soluble in aqueous HCl, but $BaSO_4$, which is used for making the intestines visible in x-ray photographs, remains sufficiently insoluble in the HCl in a human stomach so that poisonous barium ions do not get into the bloodstream.
58. For which of the following substances would solubility be greater at pH = 2 than at pH = 7?
(a) $Cu(OH)_2$ (d) CuS
(b) $CuSO_4$ (e) $Cu_3(PO_4)_2$
(c) $CuCO_3$

Buffer Solutions

59. Many natural processes can be studied in the laboratory but only in an environment of controlled pH. Which of the following combinations would be the best choice to buffer the pH at approximately 7?
(a) H_3PO_4/NaH_2PO_4 (c) Na_2HPO_4/Na_3PO_4
(b) NaH_2PO_4/Na_2HPO_4
60. Which of the following combinations would be the best to buffer the pH at approximately 9?
(a) $CH_3COOH/NaCH_3COO$ (c) NH_3/NH_4Cl
(b) HCl/NaCl
61. Without doing calculations, determine the pH of a buffer made from equal molar amounts of the following acid-base pairs.
(a) Nitrous acid and sodium nitrite
(b) Ammonia and ammonium chloride
(c) Formic acid and potassium formate
62. Without doing calculations, determine the pH of a buffer made from equal molar amounts of the following acid-base pairs.
(a) Phosphoric acid and sodium dihydrogen phosphate
(b) Sodium hydrogen phosphate and sodium dihydrogen phosphate
(c) Sodium phosphate and sodium hydrogen phosphate
63. Select from Table 17.3 an acid-base conjugate pair that would be suitable for preparing a buffer solution whose concentration of hydronium ions is
(a) 4.5×10^{-3} M (c) 8.3×10^{-6} M
(b) 5.2×10^{-8} M (d) 9.7×10^{-11} M
64. Select from Table 17.3 an acid-base conjugate pair that would be suitable for preparing a buffer solution with pH equal to
(a) 3.45 (c) 8.32
(b) 5.48 (d) 10.15
65. In order to buffer a solution at a pH of 4.57, what mass of sodium acetate, $NaCH_3COO$, should you add to 500. mL of a 0.150 M solution of acetic acid, CH_3COOH?
66. How many grams of ammonium chloride, NH_4Cl, would have to be added to 500. mL of 0.10 M NH_3 solution to have a pH of 9.00?
67. A buffer solution can be made from benzoic acid (C_6H_5COOH) and sodium benzoate (NaC_6H_5COO). How many grams of the acid would you have to mix with 14.4 g of the sodium salt in order to have a liter of a solution with a pH of 3.88?
68. If a buffer solution is prepared from 5.15 g of NH_4NO_3 and 0.10 L of 0.15 M NH_3, what is the pH of the solution?
69. You dissolve 0.425 g of NaOH in 2.00 L of a solution that originally had $[H_2PO_4^-] = [HPO_4^{2-}] = 0.132$ M. Calculate the resulting pH.
70. A buffer solution is prepared by adding 0.125 mol of ammonium chloride to 500. mL of 0.500 M aqueous ammonia. What is the pH of the buffer? If 0.0100 mol of HCl gas is bubbled into 500. mL of the buffer, what is the new pH of the solution?

Practical Acid-Base Chemistry

71. Double-acting baking powder contains two salts, sodium hydrogen carbonate and potassium dihydrogen phosphate, whose anions react in water to form CO_2 gas. Write a balanced chem-

ical equation for the reaction. Which anion is the acid and which is the base?

72. Common soap is made by reacting sodium carbonate with stearic acid ($CH_3(CH_2)_{16}COOH$). Write a balanced equation for the reaction.

73. If 1 g of each antacid in Table 17.7 reacted with equal volumes of stomach acid, which would neutralize the most stomach acid?

74. If 1 g each of vinegar, lemon juice, and lactic acid react with equal masses of baking soda, which will produce the most CO_2 gas?

75. Why do cleaning products containing sodium hydroxide feel slippery when you get them on your skin?

76. Why is it not a good idea to substitute dishwashing detergent for automobile-washing detergent?

Lewis Acids and Bases

77. Which of these is a Lewis acid? a Lewis base?
(a) NH_3 (d) Al^{3+}
(b) $BeCl_2$ (e) H_2O
(c) BCl_3 (f) SCN^-

78. Which of these is a Lewis acid? a Lewis base?
(a) O^{2-} (d) Cr^{3+}
(b) CO_2 (e) SO_3
(c) H^- (f) CH_3NH_2

79. Identify the Lewis acid and the Lewis base in each reaction.
(a) $H_2O(\ell) + SO_2(aq) \rightarrow H_2SO_3(aq)$
(b) $H_3BO_3(aq) + OH^-(aq) \rightarrow B(OH)_4^-(aq)$
(c) $Cu^{2+}(aq) + 4\ NH_3(aq) \rightarrow [Cu(NH_3)_4]^{2+}(aq)$
(d) $2\ Cl^-(aq) + SnCl_2(aq) \rightarrow SnCl_4^{2-}(aq)$

80. Identify the Lewis acid and the Lewis base in each reaction.
(a) $I_2(s) + I^-(aq) \rightarrow I_3^-(aq)$
(b) $SO_2(g) + BF_3(g) \rightarrow O_2SBF_3(s)$
(c) $Au^+(aq) + 2\ CN^-(aq) \rightarrow [Au(CN)_2]^-(aq)$
(d) $CO_2(g) + H_2O(\ell) \rightarrow H_2CO_3(aq)$

81. Trimethylamine, $(CH_3)_3N$:, interacts readily with diborane, B_2H_6. The diborane dissociates to two BH_3 fragments, each of which can react with trimethylamine to form a complex, $(CH_3)_3N{:}BH_3$. Write an equation for this reaction and interpret it in terms of Lewis' acid-base theory.

82. Draw a Lewis structure for ICl_3. Predict the shape of this molecule. Does it function as a Lewis acid or base when it reacts with chloride ion to form ICl_4^-? What is the structure of this ion?

General Questions

83. Classify each of the following as a strong acid, weak acid, strong base, weak base, amphiprotic substance, or neither acid nor base.
(a) HCl (d) CH_3COO^-
(b) NH_4^+ (e) CH_4
(c) H_2O (f) CO_3^{2-}

84. Classify each of the following as a strong acid, weak acid, strong base, weak base, amphiprotic substance, or neither acid nor base.
(a) CH_3COOH (d) NH_3
(b) Na_2O (e) $Ba(OH)_2$
(c) H_2SO_4 (f) $H_2PO_4^-$

85. Several acids and their respective equilibrium constants are:

$$HF(aq) + H_2O(\ell) \rightleftharpoons H_3O^+(aq) + F^-(aq) \quad K_a = 7.2 \times 10^{-4}$$

$$HS^-(aq) + H_2O(\ell) \rightleftharpoons H_3O^+(aq) + S^{2-}(aq) \quad K_a = 8 \times 10^{-18}$$

$$CH_3COOH(aq) + H_2O(\ell) \rightleftharpoons H_3O^+(aq) + CH_3COO^-(aq) \quad K_a = 1.8 \times 10^{-5}$$

(a) Which is the strongest acid? Which is the weakest acid?
(b) Which acid has the weakest conjugate base?
(c) Which acid has the strongest conjugate base?

86. State whether equal molar amounts of the following would have a pH equal to 7, less than 7, or greater than 7.
(a) A weak base and a strong acid react.
(b) A strong base and a strong acid react.
(c) A strong base and a weak acid react.

87. Sulfurous acid, H_2SO_3, is a weak diprotic acid ($K_{a1} = 1.2 \times 10^{-2}$, $K_{a2} = 6.2 \times 10^{-8}$). What is the pH of a 0.45 M solution of H_2SO_3? (Assume that only the first ionization is important in determining pH.)

88. Ascorbic acid (vitamin C, $C_6H_8O_6$) is a diprotic acid ($K_{a1} = 7.9 \times 10^{-5}$, $K_{a2} = 1.6 \times 10^{-12}$). What is the pH of a solution that contains 5.0 mg of the acid per mL of water? (Assume that only the first ionization is important in determining pH.)

89. Does the pH of the solution increase, decrease, or stay the same when you
(a) Add solid ammonium chloride to 100 mL of 0.10 M NH_3?
(b) Add solid sodium acetate to 50.0 mL of 0.015 M acetic acid?
(c) Add solid NaCl to 25.0 mL of 0.10 M NaOH?

90. Does the pH of the solution increase, decrease, or stay the same when you
(a) Add solid sodium oxalate, $Na_2C_2O_4$, to 50.0 mL of 0.015 M oxalic acid?
(b) Add solid ammonium chloride to 100 mL of 0.016 M HCl?
(c) Add 20.0 g of NaCl to 1.0 L of 0.012 M sodium acetate, $NaCH_3COO$?

91. What is the pH of a 0.15 M acetic acid solution? If you add 83 g of sodium acetate to 1.50 L of the 0.15 M acetic acid solution, what is the new pH of the solution?

92. Calculate the pH of a 0.050 M solution of HF. What is the pH of the solution if you add 1.58 g of NaF to 250. mL of the 0.050 M solution?

93. A buffer solution was prepared by adding 4.95 g of sodium acetate, $NaCH_3COO$, to 250. mL of 0.150 M acetic acid, CH_3COOH. What ions and molecules are present in the solution? List them in order of decreasing concentration. What is the pH of the buffer? What is the pH of 100. mL of the buffer solution if you add 80. mg of NaOH? (Assume negligible change in volume.) Write a net ionic equation for the reaction that occurs to change the pH.

Applying Concepts

94. What is the pH of pure water at 10 °C, 25 °C, and 50 °C? Classify the water at each temperature as either acidic, neutral or basic.

95. When a 0.1 M aqueous ammonia solution is tested with a conductivity apparatus *(← p. 102, Fig. 3.7)*, the bulb glows dimly. When a 0.1 M hydrochloric acid solution is tested, the bulb glows brightly. As water is added to each of the solutions would you expect the bulb to glow brighter, stop glowing, or stay the same? Explain your reasoning.

96. If you evaporated the water in a sodium hydroxide solution, you would end up with solid sodium hydroxide. However, if you evaporate the water in an ammonium hydroxide solution, you will not end up with solid ammonium hydroxide. Explain why. What will remain after the water is evaporated?

97. For each aqueous solution, predict what ions and molecules will be present. Without doing any calculations, list the ions and molecules in order of decreasing concentration.

(a) HCl
(b) $NaClO_4$
(c) HNO_2
(d) NaClO
(e) NH_4Cl
(f) NaOH

98. The diagrams below are nanoscale representations of different acids.

(a) Which diagram best represents hydrochloric acid? (The solid circles are H^+ ions and the open circles are Cl^- ions.)

(b) Which diagram best represents acetic acid? (The solid circles are H^+ ions and the open circles are CH_3COO^- ions.)

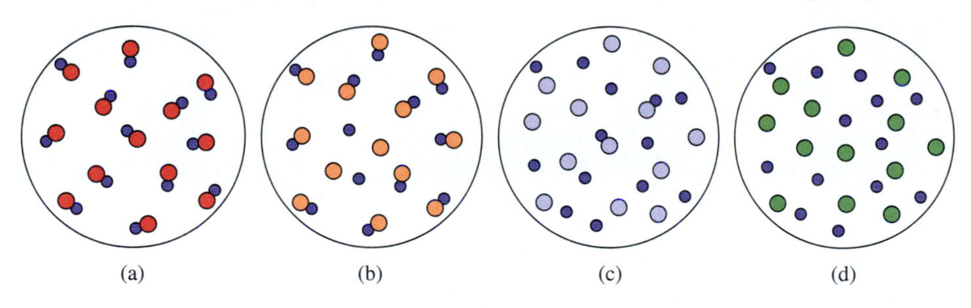

99. When asked to identify the conjugate acid-base pairs in the reaction

$$HCO_3^-(aq) + HSO_4^-(aq) \rightleftharpoons H_2CO_3(aq) + SO_4^-(aq)$$

a student incorrectly wrote:

HCO_3^- is a base and HSO_4^- is its conjugate acid.
H_2CO_3 is an acid and SO_4^- is its conjugate base.

Write a brief explanation to the student telling why the answer is incorrect.

100. When asked to prepare a carbonate buffer with a pH = 10, a lab technician wrote the following equation to determine the ratio of weak acid and conjugate base needed.

$$10 = 10.25 + \log\frac{[HCO_3^-]}{[H_2CO_3]}$$

What is wrong with this setup? If the technician prepared a solution containing equal molar concentrations of HCO_3^- and H_2CO_3, what would be the pH of the resulting buffer?

101. When you hold your breath, carbon dioxide gas is trapped in your body. Does this increase or decrease your blood pH? Does this lead to acidosis or alkalosis?

102. During a recent television medical drama, a person went into cardiac arrest and stopped breathing. A doctor quickly injected sodium hydrogen carbonate into the heart. This would indicate that cardiac arrest leads to (acidosis or alkalosis) and the sodium hydrogen carbonate helps to (increase or decrease) the pH. Explain your choices clearly.

CHAPTER

18 Electrochemistry and Its Applications

This child's toy car is a lot different from the new electric vehicles being sold in California and elsewhere as a suggested solution to our air pollution problems, but the principles of operation are the same. Batteries power electric motors to make the vehicle go. Batteries are storage devices for electrochemical energy. Battery chemistry involves redox reactions that are product-favored. Making these reactions do useful work is the goal of many chemists. • What is the chemistry that goes on inside a battery, and how does it produce electricity? How is this chemistry similar and how is it different from the other kinds of reactions you have studied? The summary problem at the end of this chapter will ask you questions like these. After you study this chapter, you should be able to answer them. *(George Semple)*

In Section 4.7 you learned about an important class of chemical reactions known as oxidation-reduction (redox) reactions. Many of these reactions occur by transfer of electrons from one atom, molecule, or ion to another. Electrochemistry is the study of the relationship between electron flow and chemical reactions, and it deals with redox reactions. The applications of electrochemistry are numerous and important. In electrochemical cells (commonly called batteries), electrons from a product-favored oxidation-reduction reaction are transferred through an external circuit to power many useful things like radios, calculators, flashlights, portable computers, heart pacemakers, hearing aids, golf carts, or even a sleek sports coupe as it moves silently down the highway. The net voltage of an electrochemical cell depends on the strengths of the reactants we call oxidizing agents and reducing agents. A knowledge of the strengths of oxidizing and reducing agents helps in the design of better batteries. Product-favored electrochemical reactions also cause their share of problems. Corrosion of metals, for example, is product-favored, and keeping corrosion under control is very costly.

By contrast, electrolysis and electroplating are applications of redox reactions that are reactant-favored. In an electrolysis cell, an external source of energy causes an electrical current to force a reactant-favored process to produce products. Electrolysis is important in the manufacture of many products, including chlorine to disinfect water supplies and to make plastics. Electrochemistry and how it is applied is the subject of this chapter.

18.1 REDOX REACTIONS

Redox reactions form a large class of chemical reactions in which the reactants may be atoms, ions, or molecules (⇐ ***p. 158***). How do you know when a reaction involves oxidation-reduction? One way is by identifying the presence of strong oxidizing or reducing agents as reactants (⇐ ***p. 162; Table 4.3***). Another is by finding a change in oxidation number (⇐ ***p. 163***). This means you have to look at the oxidation number of each element as it appears on both the reactant and the product sides of the equation. Also, the presence of an uncombined element as a reactant or product always indicates a redox reaction because producing a free element or converting one to a compound always results in a complete loss or gain of electrons, accompanied by a change in oxidation number.

An uncombined element is always assigned an oxidation number of 0.

To review the definitions of oxidation and reduction, consider the combination of hydrogen and chlorine to form hydrogen chloride. The oxidation numbers of the elements are shown in red above their symbols.

$$\overset{0}{H_2}(g) + \overset{0}{Cl_2}(g) \longrightarrow 2\ \overset{+1}{H}\overset{-1}{Cl}(g)$$

The presence of elements as reactants indicates a redox reaction, as do the changes in oxidation number. Hydrogen has been oxidized, as shown by the *increase* of its oxidation number (the oxidation number is more positive). Chlorine, on the other hand, has been reduced, as shown by the *decrease* in its oxidation number (the oxidation number is more negative). Hydrogen is known as a good **reducing agent,** a substance capable of causing other substances to be reduced. Also, chlorine is known as a good **oxidizing agent,** a substance capable of causing substances to be oxidized. Note that oxidation and reduction always occur together, with one reactant the oxidizing agent and another the reducing agent. The oxidizing agent is reduced, and the reducing agent is oxidized.

You may want to review the definitions of oxidation and reduction in Section 4.7 and the rules for assigning oxidation numbers in Section 4.8.

Unlike the combination of hydrogen and chlorine, in which covalent bonds are formed, many redox reactions involve complete gain and loss of electrons by reactants and/or products. It is this type of reaction that is utilized in electrochemistry, in which a flow of electrons through the reaction system and an external circuit is necessary for the reaction to proceed.

As an example of a redox reaction involving ions, consider the displacement reaction of magnesium (a relatively reactive metal, **(← p. 167; *Table 4.5*)** with hydrochloric acid.

Recall that acid solutions contain hydronium ions, $H_3O^+(aq)$ **(← p. 152).**

In this reaction, magnesium has been oxidized by *losing* electrons, and hydrogen has been reduced by *gaining* electrons. The changes in their oxidation numbers indicate this.

$$\overset{0}{Mg}(s) + 2\,\overset{+1\;-1}{HCl}(aq) \longrightarrow \overset{+2\;-1}{MgCl_2}(aq) + \overset{0}{H_2}(g)$$

for which the net ionic equation **(← p. 150)** is

$$Mg(s) + 2\,H_3O^+(aq) \longrightarrow Mg^{2+}(aq) + H_2(g) + 2\,H_2O(\ell)$$

PROBLEM-SOLVING EXAMPLE 18.1 Identifying Oxidizing and Reducing Agents in Redox Reactions

A reactant found in many common batteries is manganese dioxide, MnO_2. It reacts with hydrogen gas in the following reaction:

$$2\,MnO_2(s) + H_2(g) \longrightarrow Mn_2O_3(s) + H_2O(\ell)$$

Is this a redox reaction? If it is, what is the oxidizing agent? What is the reducing agent? What gets reduced? What gets oxidized? Give the oxidation numbers of all the atoms that change oxidation number.

Answer Yes, this is a redox reaction. Elemental hydrogen is oxidized to form water, and the Mn in MnO_2 is reduced. The oxidizing agent is MnO_2 and the reducing agent is H_2. The oxidation number changes are Mn: +4 to +3, H: 0 to +1.

Explanation First, determine the oxidation number of each element on the reactant side of the equation. By the use of common oxidation states **(← p. 164),** O is normally −2. Since the sum of the oxidation states of all the atoms in a formula must equal the charge on the formula, this means that the Mn in MnO_2 is +4.

$$\overset{+4\;-4}{MnO_2}$$

$(2 \times -2$

Finally, on the reactants side, the oxidation state for hydrogen is 0, as it is for all uncombined elements.

On the products side, the Mn is combined with oxygen in Mn_2O_3, in which its oxidation state is +3.

$$\overset{+6\;-6}{Mn_2O_3}$$

$(2 \times +3)$ (3×-2)

Similarly, for H_2O,

$$\overset{+2\;-2}{H_2O}$$

$(2 \times +1)$

So, it becomes apparent that in the reaction the Mn atoms have *gained* electrons, causing their oxidation numbers to decrease, while the H atoms have *lost* electrons, causing their oxidation numbers to increase. The MnO_2 has been reduced and the H_2 has been oxidized. This means that the MnO_2 is the oxidizing agent and the H_2 is the reducing agent. Oxygen is neither oxidized nor reduced; its oxidation number remains −2.

Problem-Solving Practice 18.1

Give the oxidation number for each atom and identify the oxidizing and reducing agents in the following balanced chemical equations.

(a) $2\,Fe(s) + 3\,Cl_2(g) \longrightarrow 2\,FeCl_3(s)$
(b) $2\,H_2(g) + O_2(g) \longrightarrow 2\,H_2O(\ell)$
(c) $Cu(s) + 2\,NO_3^-(aq) + 4\,H_3O^+(aq) \longrightarrow Cu^{2+}(aq) + 2\,NO_2(g) + 6\,H_2O(\ell)$
(d) $C(s) + O_2(g) \longrightarrow CO_2(g)$
(e) $6\,Fe^{2+}(aq) + Cr_2O_7^{2-}(aq) + 14\,H_3O^+(aq) \longrightarrow 6\,Fe^{3+}(aq) + 2\,Cr^{3+}(aq) + 21\,H_2O(\ell)$

18.2 USING HALF-REACTIONS TO UNDERSTAND REDOX REACTIONS

Now apply what you know about redox reactions to the reaction between zinc metal and copper(II) ions pictured in Figure 18.1, for which the net ionic equation is

$$Zn(s) + Cu^{2+}(aq) \longrightarrow Zn^{2+}(aq) + Cu(s)$$

Note how the sum of the charges on the left side of the reaction equals the sum of the charges on the right side. This is an important point. It holds true even with half-reactions. If it did, electrons would have to be created or destroyed.

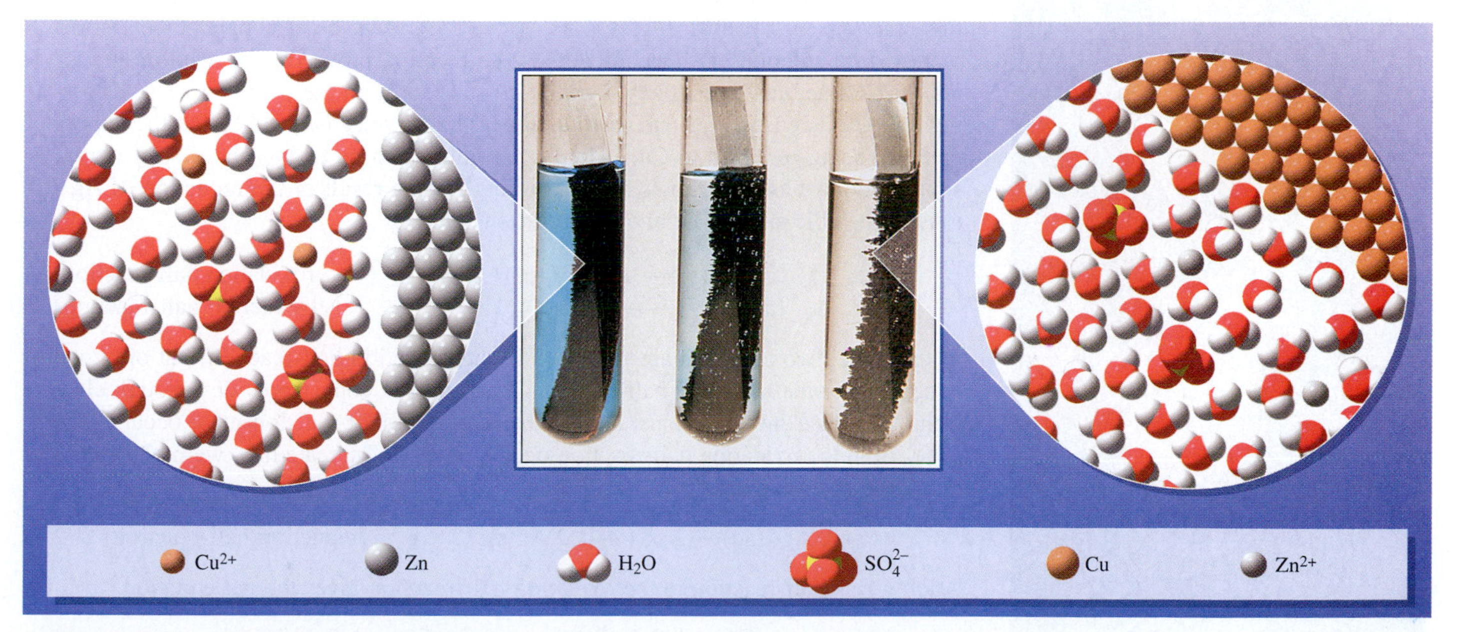

Figure 18.1 An oxidation-reduction reaction. A strip of zinc was placed in a solution of copper(II) sulfate (*left*). The zinc reacts with the copper(II) ions to produce copper metal (the brown-colored deposit on the zinc strip) and zinc ions in solution.

$$Zn(s) + Cu^{2+}(aq) \longrightarrow Zn^{2+}(aq) + Cu(s)$$

As copper metal accumulates on the zinc strip, the blue color due to the aqueous copper ions gradually fades (*middle and right*). The zinc ions in aqueous solution are colorless. *(C. D. Winters)*

In order to see more clearly how electrons are transferred, this overall reaction can be thought of as resulting from two simultaneous **half-reactions:** one half-reaction for the oxidation of Zn, and one half-reaction for the reduction of Cu^{2+} ions. The oxidation half-reaction

$$\mathrm{Zn(s)} \longrightarrow \mathrm{Zn^{2+}(aq)} + 2\,\mathrm{e^-}$$

shows that each atom of the reducing agent, Zn, loses two electrons when it is oxidized to a Zn^{2+} ion. These two electrons are accepted by a Cu^{2+} ion in the reduction half-reaction,

$$\mathrm{Cu^{2+}(aq)} + 2\,\mathrm{e^-} \longrightarrow \mathrm{Cu(s)}$$

As Cu^{2+} ions are converted to Cu(s) in this half-reaction, the blue color of the solution becomes less intense, and metallic copper forms on the surface of the zinc.

The net reaction is the sum of the oxidation and the reduction half-reactions.

$$\begin{array}{rll} \mathrm{Zn(s)} \longrightarrow \mathrm{Zn^{2+}(aq)} + 2\,\mathrm{e^-} & & \text{(oxidation half-reaction)} \\ \mathrm{Cu^{2+}(aq)} + 2\,\mathrm{e^-} \longrightarrow \mathrm{Cu(s)} & & \text{(reduction half-reaction)} \\ \hline \mathrm{Zn(s)} + \mathrm{Cu^{2+}(aq)} \longrightarrow \mathrm{Zn^{2+}(aq)} + \mathrm{Cu(s)} & & \text{(net reaction)} \end{array}$$

In all redox reactions, the number of electrons lost equals the number gained.

Notice that no electrons appear in the equation for the net reaction; the number of electrons donated in the oxidation half-reaction exactly equals the number of electrons gained in the reduction half-reaction. This must always be true in a net reaction. Otherwise, electrons would be created from nothing or destroyed, violating the law of conservation of mass.

Consider another example, as shown in Figure 18.2. (For a nanoscale view of this reaction, look back at Figure 4.22 ***(← p. 169).*** A piece of copper screen is immersed in a solution of silver nitrate. As the reaction proceeds, the solution gradually turns blue, and fine silvery, hairlike crystals form on the copper screen. Knowing that Cu^{2+} ions in aqueous solution appear blue, we can conclude that the copper metal is being oxidized to Cu^{2+}. Reduction must also be taking place, so it is reasonable to conclude that the silvery whiskers are the result of the reduction of Ag^+ ions to metallic silver. The two half-reactions are

$$\begin{array}{rl} \mathrm{Cu(s)} \longrightarrow \mathrm{Cu^{2+}(aq)} + 2\,\mathrm{e^-} & \text{(oxidation half-reaction)} \\ \mathrm{Ag^+(aq)} + \mathrm{e^-} \longrightarrow \mathrm{Ag(s)} & \text{(reduction half-reaction)} \end{array}$$

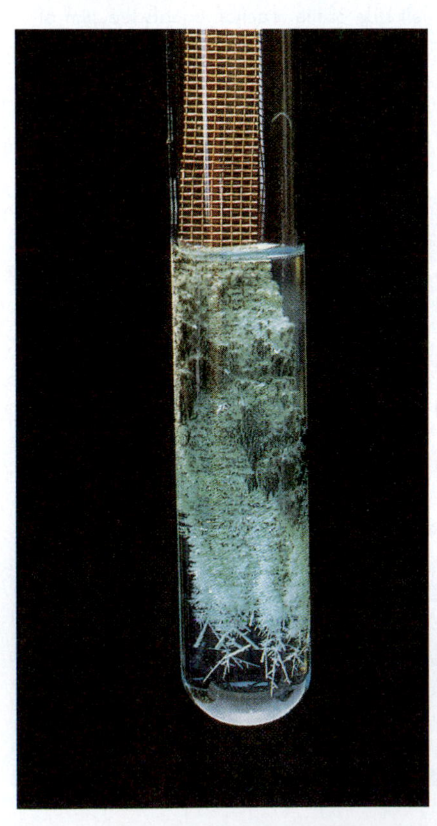

Figure 18.2 Copper metal screen in a solution of $AgNO_3$. The blue color intensifies as more copper is oxidized to aqueous Cu^{2+} ion. ***(See Figure 4.22, ← p. 169.)*** *(C. D. Winters)*

In this case, two electrons are produced in the oxidation half-reaction, but only one is needed for the reduction half-reaction. *One* atom of copper provides enough electrons to reduce *two* Ag^+ ions, and so the reduction half-reaction must occur twice every time the oxidation half-reaction occurs once. To indicate this, we multiply the reduction half-reaction by 2:

$$2\,\mathrm{Ag^+(aq)} + 2\,\mathrm{e^-} \longrightarrow 2\,\mathrm{Ag(s)} \qquad \text{(reduction half-reaction)} \times 2$$

Adding this half-reaction to the oxidation half-reaction gives the net equation

$$\mathrm{Cu(s)} + 2\,\mathrm{Ag^+(aq)} \longrightarrow \mathrm{Cu^{2+}(aq)} + 2\,\mathrm{Ag(s)}$$

The method shown here is a general one. A net equation can always be generated by writing oxidation and reduction half-reactions, adjusting the half-reaction equations so that the number of electrons produced by the oxidation equals the number used by the reduction, and then adding the two half-reactions.

Problem-Solving Example 18.2 Seeing Half-Reactions in Net Redox Reactions

When a piece of aluminum is immersed in a solution containing a small quantity of dissolved bromine, aqueous $AlBr_3$ is formed. The reaction is

$$3\ Br_2(aq) + 2\ Al(s) \xrightarrow{H_2O} 2\ Al^{3+}(aq) + 6\ Br^-(aq)$$

This is a redox reaction. Notice how the oxidation number of Al changes from 0 to +3, and the oxidation number of Br changes from 0 to −1. From this net reaction, write the correct oxidation half-reaction and reduction half-reaction.

Answer

Oxidation half-reaction: $Al(s) \longrightarrow Al^{3+}(aq) + 3e^-$

Reduction half-reaction: $Br_2(\ell) + 2\ e^- \longrightarrow 2\ Br^-(aq)$

Explanation Aluminum is oxidized, as shown by the production of aluminum ions. The half-reaction must show 3 electrons on the right to balance the +3 charge on the aluminum ion and give a net 0 charge on the right that equals the 0 charge on the left (the aluminum atoms are uncharged).

Oxidation half-reaction: $Al(s) \longrightarrow Al^{+3}(aq) + 3\ e^-$

Bromine is reduced, as shown by the conversion of elemental bromine to bromide ions. The half-reaction must show two electrons on the left to balance the two negative Br^- ions on the right.

Reduction half-reaction: $Br_2(\ell) + 2\ e^- \longrightarrow 2\ Br^-(aq)$

Now notice that these two half-reactions contain different numbers of electrons. The net reaction is the sum of these two half-reactions taken the proper number of times to make the number of electrons lost in the oxidation half-reaction equal to the number gained in the reduction half-reaction.

Problem-Solving Practice 18.2

Write an oxidation and a reduction half-reaction for the following net redox equations. Show that their sum is the net reaction.

(a) $Cd(s) + Cu^{2+}(aq) \longrightarrow Cu(s) + Cd^{2+}(aq)$
(b) $Zn(s) + 2\ H_3O^+(aq) \longrightarrow Zn^{2+}(aq) + H_2(g) + 2\ H_2O(\ell)$
(c) $2\ Al(s) + 3\ Zn^{2+}(aq) \longrightarrow 2\ Al^{3+}(aq) + 3\ Zn(s)$

Balancing Redox Equations by Using Half-Reactions

All of the equations in Problem-Solving Practice 18.2 are balanced. Could you have balanced them had they not been balanced? Equations for redox reactions often involve ions, water, hydronium ions, and hydroxide ions as reactants or products. It is difficult to tell by observing the unbalanced equation just how H_2O, H_3O^+, and OH^- are involved; that is, whether they will be reactants, or products, or even present at all. But there is a way to figure this out.

Consider the reaction of permanganate ion with oxalic acid in acidic solution. The products are manganese(II) ion and carbon dioxide, so the unbalanced equation is

$$MnO_4^-(aq) + H_2C_2O_4(aq) \longrightarrow Mn^{2+}(aq) + CO_2(g)$$

If you try to balance this equation by trial and error, you will almost certainly have a hard time with hydrogen and oxygen. You have probably already noticed that hydrogen does not appear on the products side of the unbalanced equation. Because the reaction is taking place in an aqueous acidic solution, water and hydronium ions may be involved. Getting to the balanced equation for a reaction like this is best done by a series of steps. In each step you must use what you know about the oxidation and reduction half-reactions, as well as conservation of matter and conservation of electrical charge. The steps that will produce a balanced equation for a redox reaction that occurs in acidic solution are illustrated in Problem-Solving Example 18.3.

Oxalic acid, $HO\overset{O}{\overset{\|}{C}}-\overset{O}{\overset{\|}{C}}OH$, is the simplest organic acid containing two carboxyl groups.

PROBLEM-SOLVING EXAMPLE 18.3 Balancing a Redox Equation for a Reaction in Acidic Solution

Balance the equation for the oxidation of oxalic acid in acidic permanganate solution. The products of this reaction are CO_2 and Mn^{2+} ions.

Answer $5\ H_2C_2O_4(aq) + 6\ H_3O^+(aq) + 2\ MnO_4^-(aq) \longrightarrow 10\ CO_2(g) + 2\ Mn^{2+}(aq) + 14\ H_2O(\ell)$

Explanation It is best to follow a series of steps to balance the equation for this reaction.

Step 1: Recognize whether the reaction is an oxidation-reduction process. If it is, then determine what is reduced and what is oxidized. This is a redox reaction because the oxidation number of Mn changes from +7 in MnO_4^- to +2 in Mn^{2+}, so MnO_4^- is reduced. The oxidation number of C changes from +3 in $H_2C_2O_4$ to +4 in CO_2, so $H_2C_2O_4$ is oxidized.

Step 2: Break the overall unbalanced equation into half-reactions.

$$H_2C_2O_4(aq) \longrightarrow CO_2(g) \qquad \text{(oxidation half-reaction)}$$

$$MnO_4^-(aq) \longrightarrow Mn^{2+}(aq) \qquad \text{(reduction half-reaction)}$$

Step 3: Balance the atoms in each half-reaction. First balance all atoms except for O and H, then balance O by adding H_2O and balance H by adding H^+. (Hydroxide ion, OH^-, cannot be used because the reaction is taking place in an acidic solution and the concentration of OH^- is very low.)

Oxalic acid half-reaction: First, balance the carbon atoms in the half-reaction.

$$H_2C_2O_4(aq) \longrightarrow 2\ CO_2(g)$$

This balanced the O atoms as well (no H_2O needed here), so only H atoms remain to be balanced. Because the product side is deficient by two H atoms, we put 2 H^+ there.

$$H_2C_2O_4(aq) \longrightarrow 2\ CO_2(g) + 2\ H^+(aq)$$

Permanganate half-reaction: The Mn atoms are already balanced, but the oxygen atoms are not balanced until H_2O is added. Adding 4 H_2O takes care of the needed oxygen atoms.

$$MnO_4^-(aq) \longrightarrow Mn^{2+}(aq) + 4\ H_2O(\ell)$$

But now there are eight H atoms on the right and none on the left. To balance hydrogen atoms, eight H^+ are placed on the left side of the half-reaction.

$$8\ H^+(aq) + MnO_4^-(aq) \longrightarrow Mn^{2+}(aq) + 4\ H_2O(\ell)$$

(Strictly speaking we ought to use H_3O^+ instead of H^+, but this would result in adding eight more water molecules to each side of the equation, which is rather cumbersome; it is clearer to add just H^+ now, and add the water molecules at the end.)

Step 4: Balance the half-reactions for charge. The oxalic acid half-reaction has a net charge of 0 on the left side and 2+ on the right. The reactants have lost two electrons. To show this, 2 e^- should appear on the right side.

$$H_2C_2O_4(aq) \longrightarrow 2\ CO_2(g) + 2\ H^+(aq) + 2\ e^-$$

This confirms that $H_2C_2O_4$ is the reducing agent (it loses electrons and gets oxidized). The loss of two electrons is also in keeping with the increase in the oxidation number of each of two C atoms by 1, from +3 to +4.

The MnO_4^- half-reaction has a charge of 7+ on the left and 2+ on the right. Therefore, to achieve a net 2+ charge on each side, 5 e^- must appear on the left. The gain of electrons shows that MnO_4^- is the oxidizing agent.

$$5\ e^- + 8\ H^+(aq) + MnO_4^-(aq) \longrightarrow Mn^{2+}(aq) + 4\ H_2O(\ell)$$

Step 5: Multiply the half-reactions by appropriate factors so that the reducing agent donates as many electrons as the oxidizing agent accepts. To make each half-reaction balance involves 10 electrons. The oxalic acid reaction should be multiplied by 5, and the MnO_4^- reaction by 2.

$$5[H_2C_2O_4(aq) \longrightarrow 2\ CO_2(g) + 2\ H^+(aq) + 2\ e^-]$$

$$2[5\ e^- + 8\ H^+(aq) + MnO_4^-(aq) \longrightarrow Mn^{2+}(aq) + 4\ H_2O(\ell)]$$

Step 6: Add the half-reactions to give the overall reaction and cancel reactants and products that appear on both sides of the reaction arrow.

$$5\ H_2C_2O_4(aq) \longrightarrow 10\ CO_2(g) + 10\ H^+(aq) + 10\ e^-$$

$$10\ e^- + 16\ H^+(aq) + 2\ MnO_4^-(aq) \longrightarrow 2\ Mn^{2+}(aq) + 8\ H_2O(\ell)$$

$$5\ H_2C_2O_4(aq) + 16\ H^+(aq) + 2\ MnO_4^-(aq) \longrightarrow 10\ CO_2(g) + 10\ H^+(aq) + 2\ Mn^{2+}(aq) + 8\ H_2O(\ell)$$

Since 16 H^+ appear on the left and 10 H^+ appear on the right, 10 H^+ are canceled, leaving 6 H^+ on the left.

$$5\ H_2C_2O_4(aq) + 6\ H^+(aq) + 2\ MnO_4^-(aq) \longrightarrow 10\ CO_2(g) + 2\ Mn^{2+}(aq) + 8\ H_2O(\ell)$$

Step 7: Check the final results to make sure both atoms and charge are balanced. *Atom balance:* Both sides of the equation have 2 Mn, 28 O, 10 C, and 16 H atoms. *Charge balance:* Each side has a net charge of 4+.

Step 8: Add enough water molecules to both sides of the equation to convert all H^+ to H_3O^+. In this case, six water molecules are needed.

$$5\ H_2C_2O_4(aq) + 6\ H_3O^+(aq) + 2\ MnO_4^-(aq) \longrightarrow 10\ CO_2(g) + 2\ Mn^{2+}(aq) + 14\ H_2O(\ell)$$

Problem-Solving Practice 18.3

Balance this equation for the reaction of Zn with $Cr_2O_7^{2-}$ in acidic aqueous solution.

$$Zn(s) + Cr_2O_7^{2-}(aq) \longrightarrow Cr^{3+}(aq) + Zn^{2+}(aq)$$

Exercise 18.1 **Electrons Lost = Electrons Gained**

Think of another reason why the number of electrons lost must equal the number gained. *Hint:* Look at Section 8.1.

PROBLEM-SOLVING EXAMPLE 18.4 Balancing Redox Equations for a Reaction in Basic Solution

In a nicad battery, cadmium forms $Cd(OH)_2$ and Ni_2O_3 forms $Ni(OH)_2$ in an alkaline solution. Write the balanced equation for this reaction.

Answer $Cd(s) + Ni_2O_3(s) + 3\ H_2O(\ell) \longrightarrow Cd(OH)_2(s) + 2\ Ni(OH)_2(s)$

Step 1: Recognize whether the reaction is an oxidation-reduction process. Then determine what is reduced and what is oxidized. This is a redox reaction because the oxidation number of Cd changes from 0 in Cd to +2 in $Cd(OH)_2$, so Cd is oxidized. The oxidation number of Ni changes from +3 in Ni_2O_3 to +2 in $Ni(OH)_2$, so the Ni is reduced.

Step 2: Break the overall unbalanced equation into half-reactions.

$$Cd(s) \longrightarrow Cd(OH)_2(s) \qquad \text{(oxidation half-reaction)}$$

$$Ni_2O_3(s) \longrightarrow 2\ Ni(OH)_2(s) \qquad \text{(reduction half-reaction)}$$

A balancing coefficient of 2 is added for $Ni(OH)_2$ because there are two Ni atoms in each Ni_2O_3.

Step 3: Balance the atoms in each half-reaction. First balance all atoms, doing the O and H atoms last. In a basic solution, add OH^- and H_2O on either side of the equation to balance O and H atoms. In the Cd half-reaction, the Cd atoms are balanced. Using a balancing coefficient of 2 for OH^- balances the O and H atoms.

$$Cd(s) + 2\ OH^-(aq) \longrightarrow Cd(OH)_2(s)$$

For the Ni_2O_3 half-reaction, a balancing coefficient of 2 has already been used for $Ni(OH)_2$. Balancing the O and H atoms is interesting because there are three O atoms in Ni_2O_3 and four O atoms in the $Ni(OH)_2$ product. If one H_2O molecule were taken, this would balance the O atoms, but the H atoms would not balance. Using three water molecules on the reactants side and two OH^- on the products side will balance the reaction for all the elements involved.

$$Ni_2O_3(s) + 3\ H_2O(\ell) \longrightarrow 2\ Ni(OH)_2(s) + 2\ OH^-(aq)$$

Step 4: Balance the half-reactions for charge. The Cd half-reaction, as written in step 3, requires 2 e^- as a product.

$$Cd(s) + 2\ OH^-(aq) \longrightarrow Cd(OH)_2(s) + 2\ e^- \qquad \text{(balanced)}$$

The Ni_2O_3 half-reaction, as written in step 3, requires 2 e^- as a reactant.

$$Ni_2O_3(s) + 3\ H_2O(\ell) + 2\ e^- \longrightarrow 2\ Ni(OH)_2(s) + 2\ OH^-(aq)$$

Step 5: Multiply the half-reactions by appropriate factors so that the reducing agent donates as many electrons as the oxidizing agent accepts. The Cd half-reaction furnishes two electrons and the Ni_2O_3 half-reaction requires two, so the electrons are balanced.

Step 6: Add the half-reactions to give the overall reaction, and cancel reactants and products that appear on both sides of the reaction arrow.

$$Cd(s) + \cancel{2\ OH^-}(aq) \longrightarrow Cd(OH)_2(s) + \cancel{2\ e^-}$$

$$Ni_2O_3(s) + 3\ H_2O(\ell) + \cancel{2\ e^-} \longrightarrow 2\ Ni(OH)_2(s) + \cancel{2\ OH^-}(aq)$$

$$Cd(s) + Ni_2O_3(s) + 3\ H_2O(\ell) \longrightarrow Cd(OH)_2(s) + 2\ Ni(OH)_2(s)$$

Step 7: Check the final results to make sure both atoms and charge are balanced. The equation is balanced. There are no charges on either side of the reaction arrow and the numbers of atoms of each kind on each side of the reaction arrow are equal.

Problem-Solving Practice 18.4

In basic solution, aluminum metal forms $Al(OH)_4^-$ ion as it reduces NO_3^- ion to NH_3. Write the balanced equation for this reaction using the steps outlined in Problem-Solving Example 18.4.

18.3 ELECTROCHEMICAL CELLS

Electrons are *transferred* from one kind of atom, molecule, or ion to another in a redox reaction. Because electron transfer through a wire constitutes an electrical current, it is reasonable to ask whether the electrons that are transferred in a redox reaction could be made to pass through an external circuit. If so, we would have a way of generating a potentially useful electrical current from a chemical reaction. This can be done in a variety of ways, and it is the basis for all kinds of batteries.

It is easy to see by the color changes in the two redox reactions shown in Figures 18.1 and 18.2 that these reactions favor the formation of products—as soon as the reactants are mixed, changes take place. All product-favored reactions release Gibbs free energy *(← p. 282)*, and that energy can do useful work if the reactants—the oxidizing agent and the reducing agent—are separated in such a way that electrons cannot be transferred directly from one to the other. When electrons are forced to flow outside the reaction system and back again, they form an electric current that can operate a motor, cause a reactant-favored system to produce products, heat a cup of coffee, or do other useful things.

Arranging an oxidizing agent and a reducing agent in such a way that they can react only if electrons flow through an outside conductor results in an **electrochemical cell,** also known as a **voltaic cell,** or in everyday terms, a **battery.** Figure 18.3 diagrams how this can be done for the Zn/Cu^{2+} reaction that was shown in Figure 18.1. The two half-reactions are allowed to occur in separate beakers, each of which is called a **half-cell.** When Zn atoms are oxidized, the electrons that are given up pass from the zinc through the wire and a lamp (this could also be a voltmeter or a small

Strictly speaking, many devices we call batteries consist of several voltaic cells connected together, but the term "battery" has taken on the same meaning as a "voltaic cell."

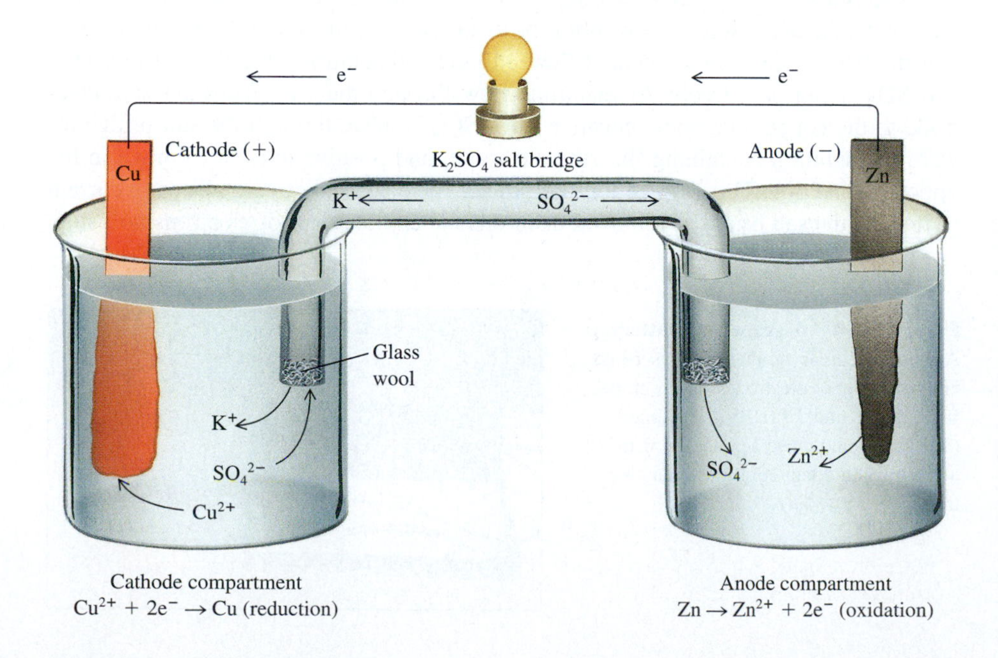

Figure 18.3 A simple electrochemical cell. The cell consists of a copper electrode in a solution containing Cu^{2+} ions (*left*), a zinc electrode in a solution containing Zn^{2+} ions (*right*), and a salt bridge that allows ions to flow into and out of the two solutions. When the two metal electrodes are connected by a conducting circuit, electrons flow from the zinc electrode, where zinc is oxidized, to the copper electrode, where copper ions are reduced. The overall reaction in this cell is

$$Cu^{2+}(aq) + Zn(s) \longrightarrow Cu(s) + Zn^{2+}(aq)$$

An electrode is most often a metal plate or wire, but it may also be a piece of graphite or something else that conducts electricity.

motor) to the copper. There the electrons are made available to reduce Cu^{2+} ions from the solution. The zinc and copper strips are called electrodes. An **electrode** conducts electrical current (electrons) into or out of something—in this case, a solution. The electrode where oxidation occurs is named the **anode,** and the electrode where reduction takes place is called the **cathode.** (On a flashlight battery, the anode is marked "−" because oxidation produces electrons that make the anode negative. Conversely, the cathode is marked "+" because reduction consumes electrons, leaving the metal electrode positive.)

Electrochemical cells are sometimes referred to as *galvanic* cells in recognition of the work of Luigi Galvani, who discovered, before Volta's time, that a frog's leg would twitch when it was touched with a source of electricity.

The voltaic cell is named after the Italian scientist Alessandro Volta, who, in about 1800, constructed the first series of electrochemical cells—a stack of alternating disks of zinc and silver separated by pieces of paper soaked in salt water (an electrolyte). Later, Volta showed that any two different metals and any electrolyte could be used to make a battery. Figure 18.4 shows a cell constructed by sticking a strip of zinc and a strip of copper into a grapefruit, for example.

In the cell diagrammed in Figure 18.3, electrons are transferred to the anode by the half-reaction

$$Zn(s) \longrightarrow Zn^{2+}(aq) + 2\,e^- \qquad \text{(anode reaction)}$$

To identify the anode and cathode, remember that *o*xidation takes place at the *a*node (both words begin with vowels), and *r*eduction takes place at the *c*athode (both words begin with consonants).

They then flow from the anode through the filament in the bulb, causing it to glow, and eventually travel to the cathode, where they react with copper(II) ions in the half-reaction

$$Cu^{2+}(aq) + 2\,e^- \longrightarrow Cu(s) \qquad \text{(cathode reaction)}$$

If nothing else but the flow of electrons took place, the concentration of Zn^{2+} ions in the anode compartment would increase, building up positive charge in the solution, and the concentration of Cu^{2+} ions in the cathode compartment would decrease, making that solution less positive. Because of this charge buildup, the flow of electrons would very quickly stop. In order for the cell to work, there has to be a way for the positive charge buildup in the anode compartment to be balanced by addition of negative ions or removal of positive ions, and vice versa for the cathode compartment.

In commercial batteries, the salt bridge is often a porous polymer membrane.

The charge buildup can be avoided by using a salt bridge to connect the two compartments. A **salt bridge** is a solution of a salt (K_2SO_4 in Figure 18.3) arranged so that the bulk of the solution cannot flow into the cell solutions, but the salt ions (K^+ and SO_4^{2-}) can pass freely. As electrons flow through the wire from the zinc electrode to the copper electrode, negative ions (SO_4^{2-}) move through the salt bridge toward the solution containing the zinc electrode, and positive ions (K^+) move in the opposite direction. This flow of ions completes the electrical circuit, allowing current to flow. If the salt bridge is removed from this battery, the flow of electrons will stop.

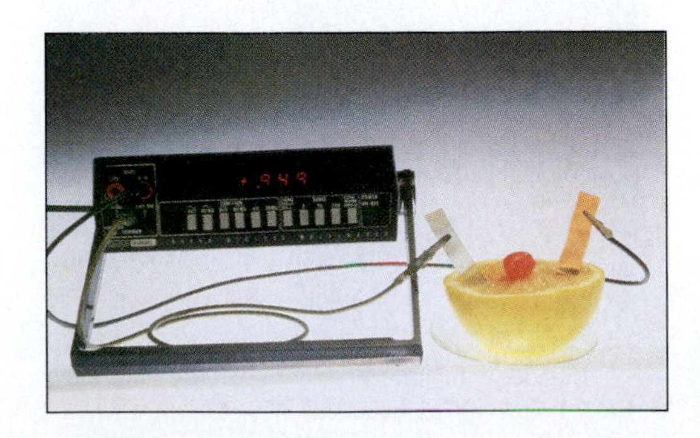

Figure 18.4 A grapefruit battery. A voltaic cell can be made by inserting zinc and copper electrodes into a grapefruit. A potential of 0.9 V is obtained. (The water and citric acid of the fruit allow for ion conduction between electrodes.) *(C. D. Winters)*

All voltaic cells and batteries operate in a similar fashion. The oxidation-reduction reaction must favor the formation of products. There must be an external circuit through which electrons flow, and there must be a salt bridge or some means of allowing ions to flow between the electrode compartments.

PROBLEM-SOLVING EXAMPLE 18.5 Electrochemical Cells

A simple voltaic cell is assembled with Ni(s) and $Ni(NO_3)_2$(aq) in one compartment and Cd(s) and $Cd(NO_3)_2$(aq) in the other. An external wire connects the two electrodes, and a salt bridge containing $NaNO_3$ connects the two solutions. The net reaction is

$$Ni^{2+}(aq) + Cd(s) \longrightarrow Ni(s) + Cd^{2+}(aq)$$

(a) What is the reaction at the anode? (b) What is the reaction at the cathode? (c) What are the directions of electron flow in the external wire and of the ion flow in the salt bridge? (d) Complete the cell diagram by indicating the directions of electron flow and ion flow.

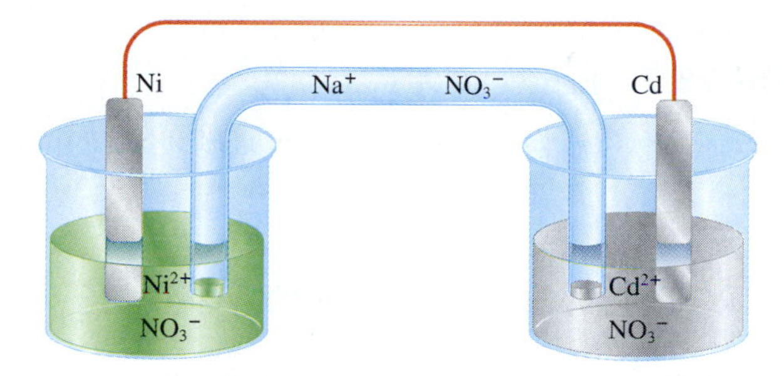

Answer (a) Anode: $Cd(s) \longrightarrow Cd^{2+}(aq) + 2\,e^-$ (oxidation)

(b) Cathode: $Ni^{2+}(aq) + 2\,e^- \longrightarrow Ni(s)$ (reduction)

(c) The completed voltaic cell is shown below.

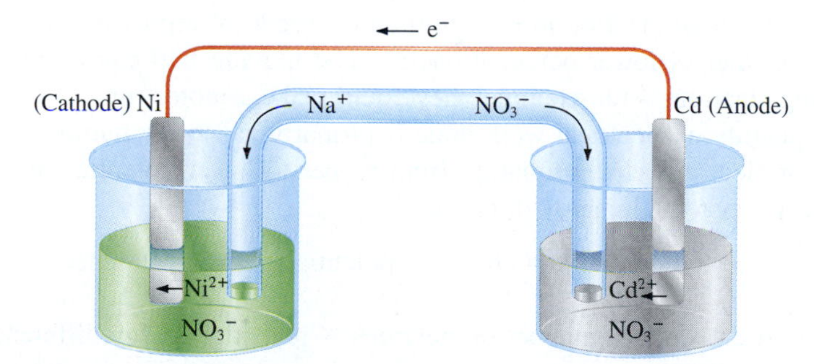

Explanation (a) and (b) First, since the net reaction shows the oxidation of Cd to Cd^{2+} and the reduction of Ni^{2+} to Ni, let's decide at which electrode these reactions take place. The Cd^{2+} ions and Cd metal are in the right compartment in the figure, and by the net reaction this is where the oxidation of Cd occurs. The Ni^{2+} ions and Ni metal are in the left compartment, and by the net reaction this is where the reduction takes place.

The half-reactions are

$$Cd(s) \longrightarrow Cd^{2+}(aq) + 2\,e^- \quad \text{(right compartment—oxidation at the anode)}$$

$$Ni^{2+}(aq) + 2\,e^- \longrightarrow Ni(s) \quad \text{(left compartment—reduction at the cathode)}$$

(c) Electrons flow *from* their source (the oxidation of cadmium at the Cd electrode—the anode) through the wire *to* the electrode where they are used (the Ni electrode—the cathode). Because Cd^{2+} ions are being formed in the anode compartment, anions (the NO_3^- ions) must move into that compartment from the salt bridge. In the cathode compartment the Ni^{2+} ions are being depleted, so Na^+ ions move from the salt bridge into that compartment to take the place of Ni^{2+} ions. The "circle" of flow of negative charge is now complete: electrons flow from Cd to Ni, and anions move from Ni to Cd.

Problem-Solving Practice 18.5

A voltaic cell is assembled to use the following net reaction.

$$Ni(s) + 2\ Ag^+(aq) \longrightarrow Ni^{2+}(aq) + 2\ Ag(s)$$

(a) Write half-reactions for this cell, and indicate which is the oxidation reaction and which is the reduction reaction. (b) Name the electrodes in which these reactions take place. (c) What is the direction of flow of electrons in an external wire connected between the electrodes? (d) If a salt bridge connecting the two electrode compartments contains KNO_3, what is the direction of flow of the nitrate ions? of the K^+ ions?

Exercise 18.2 **Battery Design**

Devise an internal on-off switch for a battery that would not be a part of the flow of electrons.

18.4 ELECTROCHEMICAL CELLS AND VOLTAGE

Because electrons flow from the anode to the cathode in an electrochemical cell, they can be thought to be "driven" or "pushed" by an **electromotive force** or **emf.** The emf is produced by the difference in electrical potential energy between the two electrodes. Just as a ball rolls downhill in response to a difference in gravitational potential energy, so an electron moves from an electrode of higher electrical potential energy to another of lower potential energy. The moving ball can do work, and so can moving electrons—for example, they could make a motor run.

The quantity of electrical work done is proportional to the number of electrons (quantity of electrical charge) that go from higher to lower potential energy, and to the size of the potential-energy difference.

$$\text{Electrical work} = \text{charge} \times \text{potential energy difference}$$

or

$$\text{Electrical work} = \text{number of electrons} \times \text{potential energy difference}$$

This is similar to comparing the amount of work a few drops of water can do when falling 100 m with that a few tons of water can do falling the same distance.

Electrical charge is measured in coulombs. A **coulomb (C)** is the quantity of charge that passes a fixed point in an electrical circuit when a current of 1 ampere flows for 1 second. The charge on a single electron is very small (1.6022×10^{-19} C), so it takes 6.24×10^{18} electrons to produce just 1 coulomb of charge.

Exercise 18.3 **Large Charges**

Which has the larger charge, a coulomb of charge, or Avogadro's number of electrons?

Electrical potential energy difference is measured in volts. The **volt (V)** is defined so that one joule of work is performed when 1 coulomb of charge moves through a potential difference of 1 V:

$$1 \text{ volt} = \frac{1 \text{ joule}}{1 \text{ coulomb}} \quad \text{or} \quad 1 \text{ joule} = 1 \text{ volt} \times 1 \text{ coulomb}$$

The emf of an electrochemical cell, commonly called its **cell voltage,** therefore, shows how much work a cell can produce for each coulomb of charge that the chemical reaction produces.

When a single electron moves through a potential of 1 V, the work done is an electron-volt, abbreviated eV.

The voltage of an electrochemical cell depends on the substances that make up the cell, and on their pressures if they are gases and their concentrations if they are solutes in solution. The quantity of charge (coulombs of electricity) depends on how much of each substance reacts. Look at the 1.5-V batteries shown in Figure 18.5. They have the same voltage because they rely on electrodes with the same potential difference between them, yet one battery is capable of far more work than the other, because it contains a larger quantity of reactants. In this section and the next, we consider how cell emf depends on the materials from which a cell is made; in Section 18.5, we shall return to the question of how much electrical work a cell can do.

Figure 18.5 **Two 1.5-V batteries.** The one on the left is capable of more work since it contains more oxidizing and reducing agents. *(C. D. Winters)*

A cell's voltage is readily measured by inserting a voltmeter instead of a light bulb into the circuit, as shown in Figure 18.4. Because the voltage depends on concentrations, **standard conditions** are defined for voltage measurements. These are the same as those used for $\Delta H°$ *(← p. 251):* all reactants and products must be present as pure solids or liquids, gases at 1 atm pressure, or solutes at concentrations of 1 M. Voltages measured under these conditions are **standard voltages,** symbolized by $E°$. Unless specified otherwise, all values of $E°$ are for 25 °C (298 K). By definition, cell voltages for product-favored electrochemical reactions are *positive.* For example, the standard cell potential for the Zn/Cu^{2+} cell, discussed earlier, is +1.10 V at 25 °C.

Since every redox reaction can be thought of as the sum of two half-reactions, it is convenient to assign a voltage to every possible half-reaction. Then the cell potential for any reaction can be obtained by adding the voltages of the oxidation and reduction half-reactions.

$$E°_{\text{net}} = E°_{\text{ox}} + E°_{\text{red}}$$

If $E°_{net}$ is positive, the reaction is product-favored; if $E°_{net}$ is negative, the reaction is reactant-favored. However, because only *differences* in potential energy can be measured, it is not possible to measure the voltage for a single half-reaction. Instead, one half-reaction is chosen as the standard, and then all others are compared to it. The half-reaction chosen as the standard is the one that occurs at the **standard hydrogen electrode,** in which hydrogen gas at a pressure of 1 atm is bubbled over a platinum electrode immersed in aqueous acid (Figure 18.6). The reaction that occurs at this electrode is reversible.

$$2\ H_3O^+(aq,\ 1\ M) + 2\ e^- \longrightarrow H_2(g,\ 1\ bar) + 2\ H_2O(\ell)$$

A potential of exactly 0 V is *assigned* to this half-cell. In a cell that combines another half-reaction with the standard hydrogen electrode, the cell voltage is the difference in potential between the two electrodes. Because the potential of the hydrogen electrode is assigned to be 0, the cell voltage gives the voltage of the other electrode.

The convention of assigning voltages to half-reactions is similar to the convention of tabulating standard enthalpies of formation; in both cases a relatively small table of data can provide information about a large number of different reactions.

When the standard hydrogen electrode is paired with a half-cell that contains a better reducing agent than H_2, $H_3O^+(aq)$ is reduced to H_2.

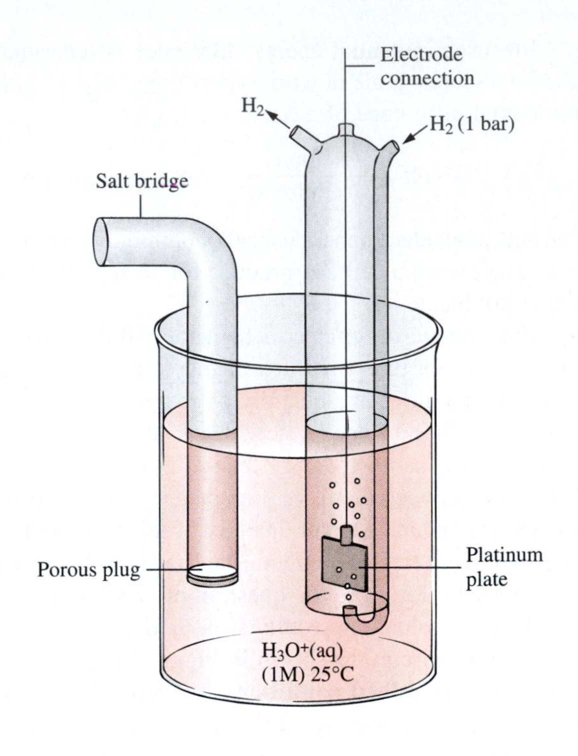

Figure 18.6 The standard hydrogen electrode. Hydrogen gas at 1 atm pressure bubbles over an inert platinum electrode that is immersed in a solution that contains 1M H_3O^+ ions at 25 °C. The potential for this electrode is defined as exactly 0 V.

$$H_3O^+ \text{ reduced: } 2\ H_3O^+(aq) + 2\ e^- \longrightarrow H_2(g,\ 1\ bar) + 2\ H_2O(\ell) \qquad E^\circ_{red} = 0\ V$$

If the other half-cell contains a better oxidizing agent than H_3O^+, then H_2 is oxidized.

$$H_2 \text{ oxidized: } H_2(g,\ 1\ bar) + 2\ H_2O(\ell) \longrightarrow 2\ H_3O^+(aq,\ 1\ M) + 2\ e^- \qquad E^\circ_{ox} = 0\ V$$

In either case, the standard hydrogen electrode has a potential of exactly 0 V.

Figure 18.7 diagrams a cell in which one compartment contains the standard hydrogen electrode, and the other contains a zinc electrode immersed in a 1 M solution of Zn^{2+}. The voltmeter is connected between the two electrodes to measure the difference in electrical potential energy. For the cell diagrammed in Figure 18.7, the voltmeter shows a measured potential of 0.76 V. After this cell operates for a time, the zinc electrode is observed to decrease in mass. Therefore, the Zn electrode must be the anode; that is, it is the electrode where oxidation is taking place. The hydrogen electrode must be the cathode where reduction is taking place. The cell reaction is therefore the sum of the half-cell reactions:

$$Zn(s) \longrightarrow Zn^{2+}(aq,\ 1\ M) + 2\ e^- \qquad E^\circ_{ox} = ?\ V \text{ (anode)}$$

$$2\ H_3O^+(aq,\ 1\ M) + 2\ e^- \longrightarrow H_2(g, 1\ bar) + 2\ H_2O(\ell) \qquad E^\circ_{red} = 0\ V \text{ (cathode)}$$

$$Zn(s) + 2\ H_3O^+(aq,\ 1\ M) \longrightarrow Zn^{2+}(aq,\ 1\ M) + H_2(g,\ 1\ bar) + 2\ H_2O(\ell) \qquad E^\circ_{net} = +0.76\ V \text{ (cell reaction)}$$

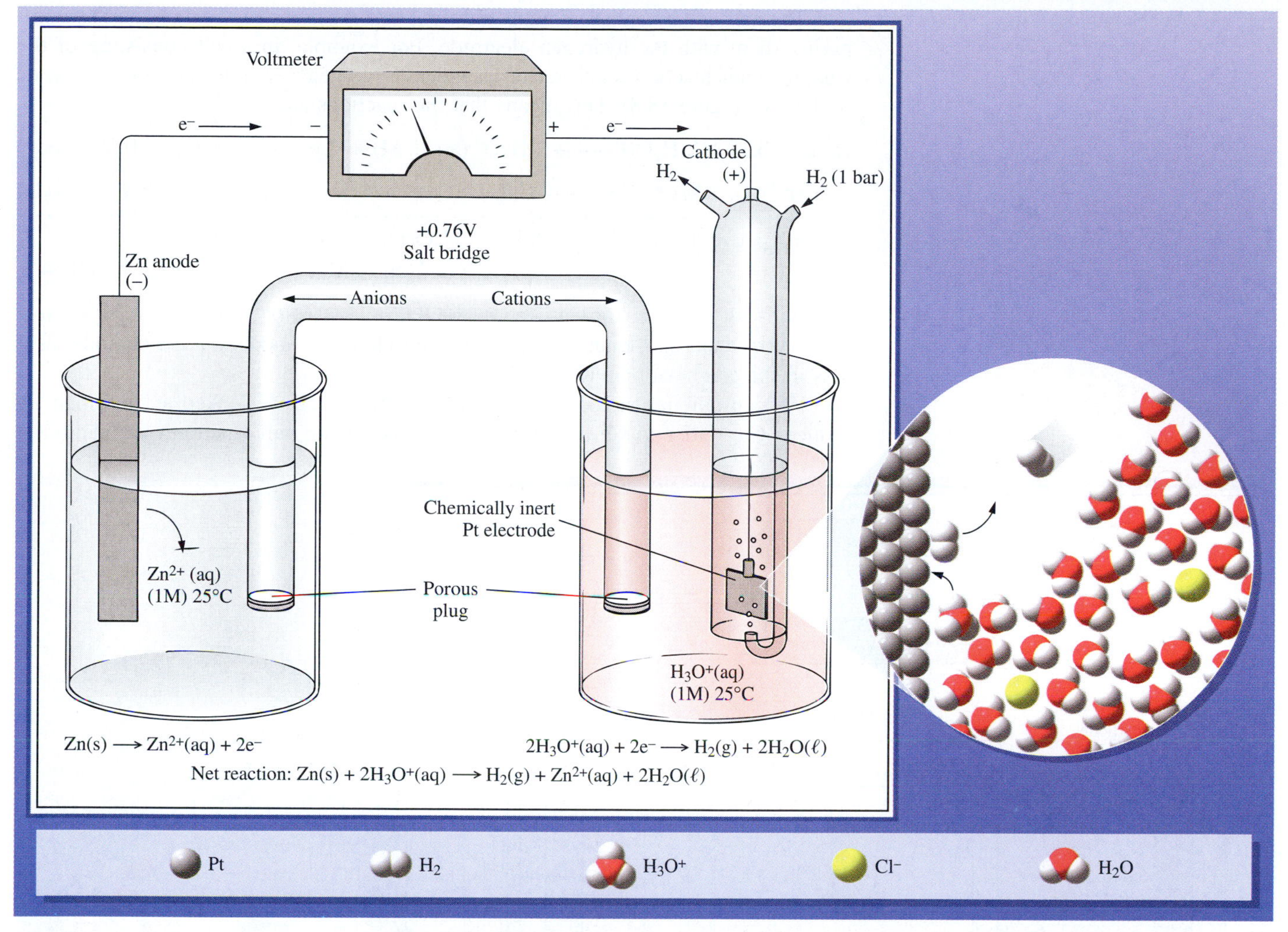

Figure 18.7 An electrochemical cell using a $Zn^{2+}/Zn(s)$ half cell and a standard hydrogen electrode. The zinc electrode is the anode and the standard hydrogen electrode is the cathode in this cell, which has a voltage of +0.76 V. Zinc is the reducing agent and is oxidized to Zn^{2+}; H_3O^+ is the oxidizing agent and is reduced to H_2. In the standard hydrogen electrode, reaction occurs only where the three phases—gas, solution, and solid electrode—are in contact. The platinum electrode does not undergo any chemical change and in the cell pictured here the half-cell reaction is $2\ H_3O^+(aq) + 2\ e^- \rightarrow H_2(g) + 2\ H_2O(\ell)$. When the standard hydrogen electrode is the anode, the half-cell reaction is $H_2(g) + 2\ H_2O(\ell) \rightarrow 2\ H_3O^+(aq) + 2\ e^-$.

The voltmeter tells us that the potential at the Zn electrode is 0.76 V higher than at the hydrogen electrode. Because the half-cell potential for the hydrogen electrode is assigned to be 0 V, the half-cell potential for oxidation of zinc must be +0.76 V:

$$Zn(s) \longrightarrow Zn^{2+}(aq,\ 1\ M) + 2\ e^- \qquad E^\circ_{ox} = +0.76\ V$$

Exercise 18.4 What Is Going on Inside the Electrochemical Cell?

Devise an experiment that would show that zinc is being oxidized in the electrochemical cell shown in Figure 18.7.

Quantities defined as exact, such as the voltage of the hydrogen electrode at standard conditions, do not limit the number of significant figures in the answer when they are used in calculations.

The half-cell potentials of many different half-reactions can be measured by comparing them with the hydrogen electrode. For example, in a cell consisting of the Cu/Cu^{2+} half-reaction connected to a standard hydrogen electrode, the voltmeter reads $+0.34$ V (Figure 18.8). This means that the reactions are

$$H_2(g,\ 1\ \text{bar}) + 2\ H_2O(\ell) \longrightarrow 2\ H_3O^+(aq,\ 1\ M) + 2\ e^- \qquad E^\circ_{ox} = 0\ \text{V (anode)}$$

$$Cu^{2+}(aq,\ 1\ M) + 2\ e^- \longrightarrow Cu(s) \qquad E^\circ_{red} = ?\ \text{V (cathode)}$$

$$H_2(g,\ 1\ \text{bar}) + Cu^{2+}(aq,\ 1\ M) + 2\ H_2O(\ell) \longrightarrow 2\ H_3O^+(aq,\ 1\ M) + Cu(s) \qquad E^\circ_{net} = +0.34\ \text{V}$$

The half-cell potential for $Cu^{2+}(aq,\ 1\ M) + 2\ e^- \rightarrow Cu(s)$ must be $+0.34$ V. Note that, in this cell, the standard hydrogen electrode is the anode, not the cathode, as it was in the combination with a Zn/Zn^{2+} half-cell.

We can now return to the first electrochemical cell we looked at, in which Zn reduces Cu^{2+} ions to Cu. Using the potentials for the half-reactions, we can write

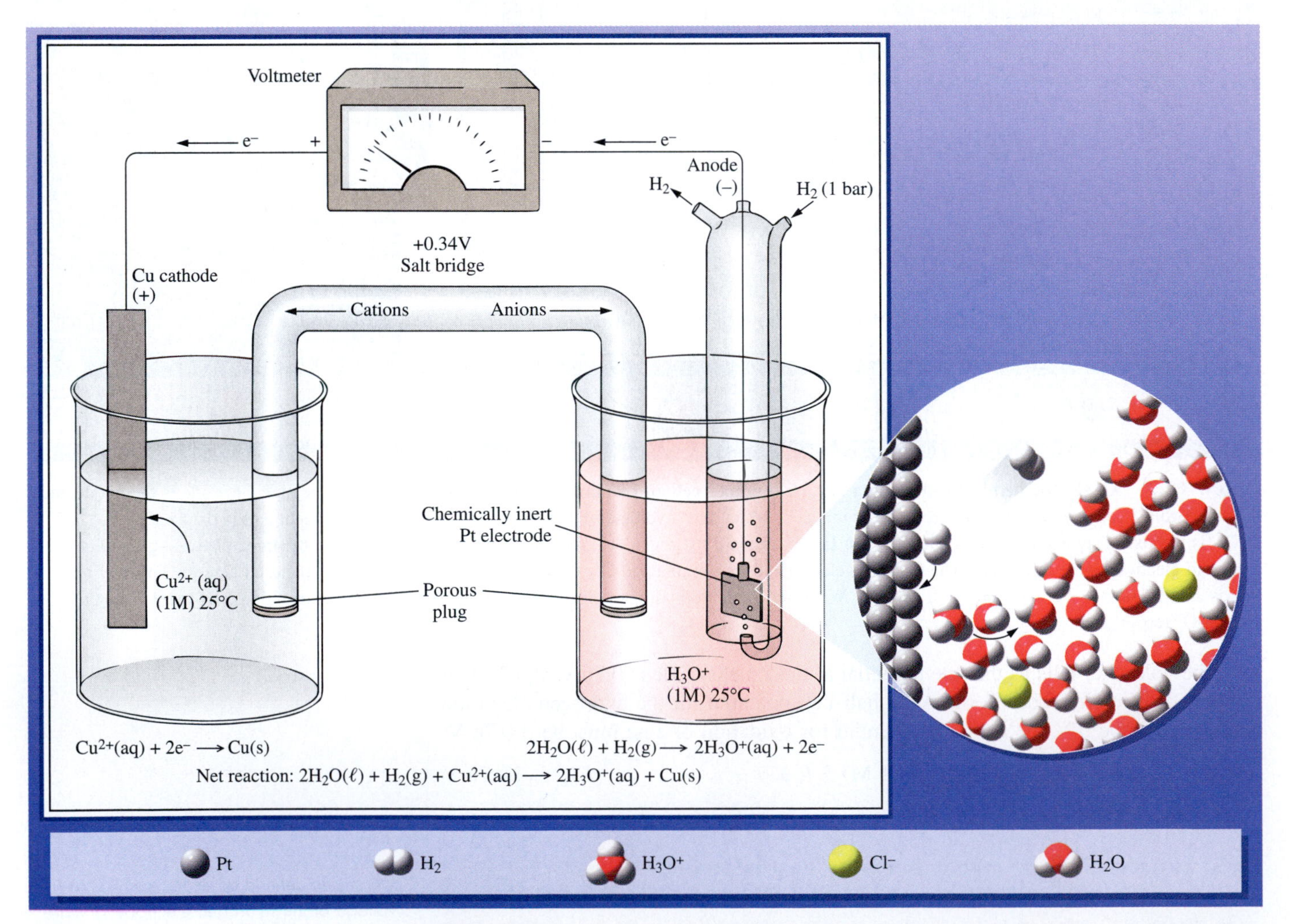

Figure 18.8 An electrochemical cell using the Cu^{2+}/Cu half-cell and the standard hydrogen electrode. A voltage of $+0.34$ V is produced. In this cell, Cu^{2+} ions are reduced to form Cu metal, and H^+ ions are oxidized at the standard hydrogen electrode. The reaction at the standard hydrogen electrode is exactly the opposite of that shown in Figure 18.7.

$$\begin{array}{lr} Zn(s) \longrightarrow Zn^{2+}(aq,\ 1\ M) + 2\ e^- & E^\circ_{ox} = +0.76\ V\ (anode) \\ Cu^{2+}(aq,\ 1\ M) + 2\ e^- \longrightarrow Cu(s) & E^\circ_{red} = +0.34\ V\ (cathode) \\ \hline Zn(s) + Cu^{2+}(aq,\ 1\ M) \longrightarrow Zn^{2+}(aq,\ 1\ M) + Cu(s) & E^\circ_{net} = +1.10\ V \end{array}$$

This is an important result because the sum of the potentials of the two half-reactions equals the measured potential for the net cell reaction.

$$E^\circ_{net} = E^\circ_{ox} + E^\circ_{red} = (+0.76\ V) + (+\ 0.34\ V) = 1.10\ V$$

PROBLEM-SOLVING EXAMPLE 18.6 Determining a Half-Cell Potential

The cell illustrated in the following drawing generates a potential of $E^\circ = 0.51$ V under standard conditions at 25 °C. The net cell reaction is

$$Zn(s) + Ni^{2+}(aq,\ 1\ M) \longrightarrow Zn^{2+}(aq,\ 1\ M) + Ni(s)$$

(a) Determine which electrode is the anode and which is the cathode, (b) show the direction of electron flow outside the cell, and complete the cell diagram, and (c) calculate the half-cell potential for $Ni^{2+}(aq) + 2\ e^- \rightarrow Ni(s)$.

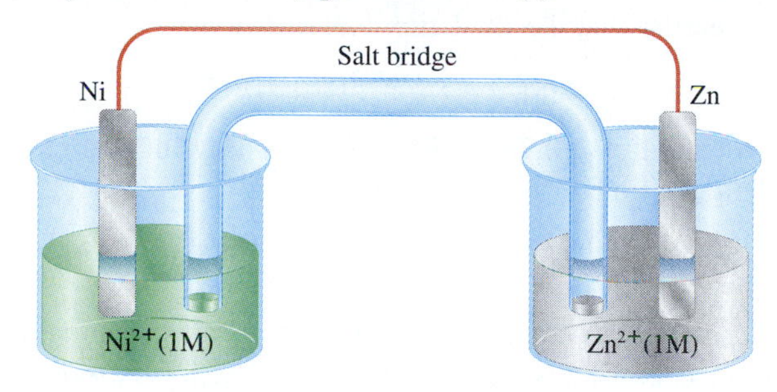

Answer (a) Zinc is the anode, nickel is the cathode, (b) see below, (c) −0.25 V

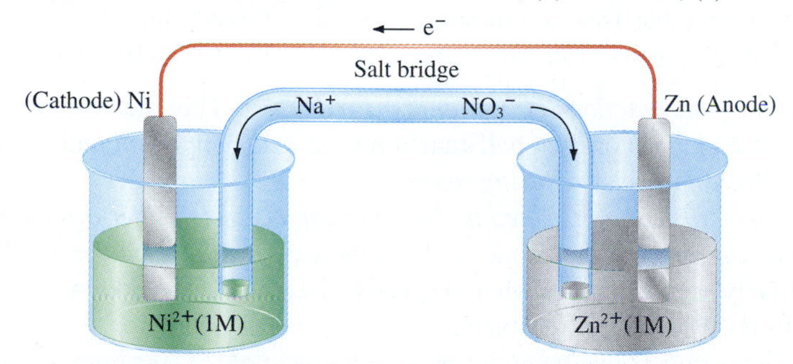

Explanation The electrode where oxidation occurs is the anode. Because Zn(s) is oxidized to Zn^{2+}(aq), the Zn electrode is the anode. Nickel(II) ions are reduced at the Ni electrode, so it is the cathode.

Since the cell potential and the potential for the $Zn(s)/Zn^{2+}$(aq, 1 M) half-cell are known, the value of E° for $Ni^{2+}(aq,\ 1\ M) + 2\ e^- \longrightarrow Ni(s)$ can be calculated.

$$\begin{array}{lr} Zn(s) \longrightarrow Zn^{2+}(aq) + 2\ e^- & E^\circ_{ox} = 0.76\ V\ (anode) \\ Ni^{2+}(aq) + 2\ e^- \longrightarrow Ni(s) & E^\circ_{red} = ?\ V\ (cathode) \\ \hline Zn(s) + Ni^{2+}(aq) \longrightarrow Zn^{2+}(aq) + Ni(s) & E^\circ_{net} = 0.51\ V \end{array}$$

Using $E^\circ_{net} = E^\circ_{ox} + E^\circ_{red}$, solve for E°_{red}.

$$E^\circ_{red} = E^\circ_{net} - E^\circ_{ox} = 0.51\ V - 0.76\ V = -0.25\ V$$

At 25 °C, the value of E° for the $Ni^{2+}(aq,\ 1\ M) + 2\ e^- \longrightarrow Ni(s)$ half-reaction is −0.25 V.

Problem-Solving Practice 18.6

Given that the reaction of aqueous copper(II) ions with iron metal has an $E°$ value of 0.78 V, what is the value of $E°$ for the half-cell $Fe(s) \rightarrow Fe^{2+}(aq) + 2\ e^-$?

$$Fe(s) + Cu^{2+}(aq, 1\ M) \longrightarrow Fe^{2+}(aq, 1\ M) + Cu(s)$$
$$E°_{net} = +0.78\ V$$

18.5 USING STANDARD CELL POTENTIALS

The results of a great many measurements such as the ones just described are summarized in Table 18.1. The values reported in the table are called **standard reduction potentials** because they are the potentials that would be measured for a cell in which a half-reaction *occurred as a reduction* when paired with the standard hydrogen electrode. If a half-reaction would occur as an oxidation when paired with the standard hydrogen electrode, it is indicated by giving its voltage a negative sign. For example, we saw earlier that the oxidation half-reaction

$$Zn(s) \longrightarrow Zn^{2+}(aq) + 2\ e^-$$

has a half-cell potential of +0.76 V ($E°_{ox} = +0.76$ V). But this reaction appears in Table 18.1 as the reduction

$$Zn^{2+}(aq) + 2\ e^- \longrightarrow Zn(s) \qquad E°_{red} = -0.76\ V$$

and its standard potential is equal in magnitude but opposite in sign to that for the oxidation reaction. It is always true that if a half-reaction is written in the reverse direction, the sign of the corresponding $E°$ must be changed. That is, if any half-reaction given in Table 18.1 is written as an oxidation reaction, the sign of the $E°$ in the table must be reversed. Here are some important points to notice about Table 18.1:

1. *Each of the half-reactions is written as a reduction.* This means that the species on the left-hand side of each half-reaction is an *oxidizing agent* and the species on the right-hand side is a *reducing agent.*
2. *Each of the half-reactions listed in the table can occur in either direction.* A given substance can react at the anode or the cathode, depending on the conditions. We have already seen examples where H_2 is oxidized to H_3O^+ and where H_3O^+ is reduced to H_2 by different reactants.
3. *The more positive the value of the reduction potential, E°, the more easily the substance on the left side of a half-reaction can be reduced.* When a substance is easy to reduce, it is a strong oxidizing agent. (Remember that an oxidizing agent must be reduced when it oxidizes something else.) Thus, $F_2(g)$ is the best oxidizing agent in the table, and Li^+ is the poorest oxidizing agent in the table. Other strong oxidizing agents are at the top left of the table: $H_2O_2(aq)$, $PbO_2(s)$, $MnO_4^-(aq)$, $Au^{3+}(aq)$, $Cl_2(g)$, $Cr_2O_7^{2-}(aq)$, and $O_2(g)$.
4. *The more negative the value of the reduction potential, E°, the less likely the reaction will occur as a reduction, and the more likely the reverse reaction (an oxidation) will occur.* That is, the farther down we go in the table, the better the reducing ability of the atom, ion, or molecule on the right. Thus Li(s) is the strongest reducing agent in the table and F^- is the weakest reducing agent in the table. Other strong reducing agents are alkali and alkaline earth metals and hydrogen at the lower right of the table.

TABLE 18.1 Standard Reduction Potentials in Aqueous Solution at 25 °C*

Increasing Strength of Oxidizing Agents (↑) / Increasing Strength of Reducing Agents (↓)

Reduction Half-Reaction	$E°$ (V)
$F_2(g) + 2\,e^- \longrightarrow 2\,F^-(aq)$	+2.87
$H_2O_2(aq) + 2\,H_3O^+(aq) + 2\,e^- \longrightarrow 4\,H_2O(\ell)$	+1.77
$PbO_2(s) + SO_4^{2-}(aq) + 4\,H_3O^+(aq) + 2\,e^- \longrightarrow PbSO_4(s) + 6\,H_2O(\ell)$	+1.685
$MnO_4^-(aq) + 8\,H_3O^+(aq) + 5\,e^- \longrightarrow Mn^{2+}(aq) + 12\,H_2O(\ell)$	+1.52
$Au^{3+}(aq) + 3\,e^- \longrightarrow Au(s)$	+1.50
$Cl_2(g) + 2\,e^- \longrightarrow 2\,Cl^-(aq)$	+1.360
$Cr_2O_7^{2-}(aq) + 14\,H_3O^+(aq) + 6\,e^- \longrightarrow 2\,Cr^{3+}(aq) + 21\,H_2O(\ell)$	+1.33
$O_2(g) + 4\,H_3O^+(aq) + 4\,e^- \longrightarrow 6\,H_2O(\ell)$	+1.229
$Br_2(\ell) + 2\,e^- \longrightarrow 2\,Br^-(aq)$	+1.08
$NO_3^-(aq) + 4\,H_3O^+ + 3\,e^- \longrightarrow NO(g) + 6\,H_2O$	+0.96
$OCl^-(aq) + H_2O(\ell) + 2\,e^- \longrightarrow Cl^-(aq) + 2\,OH^-(aq)$	+0.89
$Hg^{2+}(aq) + 2\,e^- \longrightarrow Hg(\ell)$	+0.855
$Ag^+(aq) + e^- \longrightarrow Ag(s)$	+0.80
$Hg_2^{2+}(aq) + 2\,e^- \longrightarrow 2\,Hg(\ell)$	+0.789
$Fe^{3+}(aq) + e^- \longrightarrow Fe^{2+}(aq)$	+0.771
$I_2(s) + 2\,e^- \longrightarrow 2\,I^-(aq)$	+0.535
$O_2(g) + 2\,H_2O(\ell) + 4\,e^- \longrightarrow 4\,OH^-(aq)$	+0.40
$Cu^{2+}(aq) + 2\,e^- \longrightarrow Cu(s)$	+0.337
$Sn^{4+}(aq) + 2\,e^- \longrightarrow Sn^{2+}(aq)$	+0.15
$2\,H_3O^+(aq) + 2\,e^- \longrightarrow H_2(g) + 2\,H_2O(\ell)$	0.00
$Sn^{2+}(aq) + 2\,e^- \longrightarrow Sn(s)$	−0.14
$Ni^{2+}(aq) + 2\,e^- \longrightarrow Ni(s)$	−0.25
$PbSO_4(s) + 2\,e^- \longrightarrow Pb(s) + SO_4^{2-}(aq)$	−0.356
$Cd^{2+}(aq) + 2\,e^- \longrightarrow Cd(s)$	−0.40
$Fe^{2+}(aq) + 2\,e^- \longrightarrow Fe(s)$	−0.44
$Zn^{2+}(aq) + 2\,e^- \longrightarrow Zn(s)$	−0.763
$2\,H_2O(\ell) + 2\,e^- \longrightarrow H_2(g) + 2\,OH^-(aq)$	−0.8277
$Al^{3+}(aq) + 3\,e^- \longrightarrow Al(s)$	−1.66
$Mg^{2+}(aq) + 2\,e^- \longrightarrow Mg(s)$	−2.37
$Na^+(aq) + e^- \longrightarrow Na(s)$	−2.714
$K^+(aq) + e^- \longrightarrow K(s)$	−2.925
$Li^+(aq) + e^- \longrightarrow Li(s)$	−3.045

*In volts (V) versus the standard hydrogen electrode.

5. *Under standard conditions, any species on the left of a half-reaction will oxidize any species on the right that is farther down in the table.* For example, we can predict that $Fe^{3+}(aq)$ will oxidize Al(s); $Br_2(\ell)$ will oxidize Mg(s); and even $Na^+(aq)$ will oxidize Li(s). The net reaction and the cell voltage are obtained by adding the half-reactions and their voltages; for example,

$$
\begin{array}{ll}
Br_2(\ell) + 2\,e^- \longrightarrow 2\,Br^-(aq) & E°_{red} = 1.08\text{ V} \\
Mg(s) \longrightarrow Mg^{2+}(aq) + 2\,e^- & E°_{ox} = 2.37\text{ V} \\
\hline
Br_2(\ell) + Mg(s) \longrightarrow Mg^{2+}(aq) + 2\,Br^-(aq) & E°_{net} = 3.45\text{ V}
\end{array}
$$

6. *Electrode potentials depend on the nature and concentration of reactants and products, but not on the quantity of each that reacts.* This means that changing the stoichiometric coefficients for a half-reaction does not change the value of $E°$. For example, the reduction of Fe^{3+} has an $E°$ of +0.771 V whether the reaction is written as

$$Fe^{3+}(aq, 1\ M) + e^- \longrightarrow Fe^{2+}(aq, 1\ M) \qquad E°_{red} = +0.771\ V$$

or as

$$2\ Fe^{3+}(aq, 1\ M) + 2\ e^- \longrightarrow 2\ Fe^{2+}(aq, 1\ M) \qquad E°_{red} = +0.771\ V$$

This fact about half-cell potentials seems unusual at first. It arises because a half-cell voltage is energy per unit charge (1 volt = 1 joule/1 coulomb). When a half-reaction is multiplied by some number, both the energy and the charge are multiplied by that number. Thus the ratio of the energy to the charge (voltage) does not change.

Using the preceding guidelines and the table of standard reduction potentials, let's make some *predictions* about whether reactions will occur and then check our results by calculating $E°$.

PROBLEM-SOLVING EXAMPLE 18.7 Predicting Redox Reactions

(a) Will aluminum metal dissolve in a 1 M solution of tin(IV) ion? If so, what is $E°$ for the reaction?
(b) What about a 1 M solution of Cd^{2+} oxidizing metallic Cu? Do you predict that this reaction will occur?

Answer (a) Yes, the $E°$ value is 1.81 V. (b) This reaction does not occur.

Explanation
(a) First locate the reacting chemicals in the table. $Sn^{4+}(aq)$ is about half-way down Table 18.1 on the left, while Al(s) is fifth up from the bottom on the right. Since $Sn^{4+}(aq)$ is above Al(s), we predict that it can oxidize aluminum, causing the metal to dissolve as Al atoms are oxidized to become Al^{3+} ions. To be certain, we can add the half-cell reactions to give the balanced equation. Adding the half-cell potentials gives a positive $E°$, so this reaction is product-favored, as we predicted.

$$2\ [Al(s) \longrightarrow Al^{3+}(aq, 1\ M) + 3\ e^-] \qquad E°_{ox} = +1.66\ V \text{ (anode)}$$
$$3\ [Sn^{4+}(aq, 1\ M) + 2\ e^- \longrightarrow Sn^{2+}(aq, 1\ M)] \qquad E°_{red} = +0.15\ V \text{ (cathode)}$$
$$2\ Al(s) + 3\ Sn^{4+}(aq, 1\ M) \longrightarrow 2\ Al^{3+}(aq, 1\ M) + 3\ Sn^{2+}(aq, 1\ M) \qquad E°_{net} = +1.81\ V$$

(b) The reaction between Cd^{2+} and Cu is evaluated the same way. $Cd^{2+}(aq)$ is about three quarters of the way down Table 18.1 on the left, but Cu(s) is a little less than half-way down on the right. Therefore, $Cd^{2+}(aq)$ is not a strong enough oxidizing agent to oxidize Cu(s), and we predict no reaction. When the half-cell potentials are added, we get

$$Cu(s) \longrightarrow Cu^{2+}(aq) + 2\ e^- \qquad E°_{ox} = -0.34\ V \text{ (anode)}$$
$$Cd^{2+} + 2\ e^- \longrightarrow Cd(s) \qquad E°_{red} = -0.40\ V \text{ (cathode)}$$
$$Cd^{2+}(aq) + Cu(s) \longrightarrow Cd(s) + Cu^{2+}(aq) \qquad E°_{net} = -0.74\ V$$

The negative $E°$ value shows that this process is reactant-favored and does not form appreciable quantities of products under standard conditions. Rather, cadmium metal will reduce Cu^{2+} (the reverse of the reaction given).

Problem-Solving Practice 18.7

Look at Table 18.1 and determine which two half-reactions would produce the largest value of E°_{net}. Write the two half-reactions and the net cell reaction, and give the E° for the reaction.

Exercise 18.5 Using E° Values

Transporting chemicals is of great practical importance. Suppose that you have a large volume of a solution of mercury(II) chloride, $HgCl_2$, that needs to be transported. A driver brings a tanker truck made of aluminum to the loading dock. Will it be OK to load the truck with your solution? Explain your answer fully.

Standard reduction potentials can be used to explain an annoying experience many of us have had. Have you ever experienced pain in your teeth when you accidentally touched a filling with a stainless steel fork or a piece of aluminum (Figure 18.9)? A common filling material for tooth cavities is dental amalgam—tin and silver dissolved in mercury to form solid solutions having compositions approximating Ag_2Hg_3, Ag_3Sn, and Sn_xHg (where x ranges from 7 to 9). All of these may undergo electrochemical reactions; for example,

$$3\ Hg_2^{2+}(aq) + 4\ Ag(s) + 6\ e^- \longrightarrow 2\ Ag_2Hg_3(s) \qquad E^\circ_{red} = +0.85\ V$$

$$Sn^{2+}(aq) + 3\ Ag(s) + 2\ e^- \longrightarrow Ag_3Sn(s) \qquad E^\circ_{red} = -0.05\ V$$

The E° values in Table 18.1 indicate that both iron and aluminum have much more negative reduction potentials and therefore are much better reducing agents than any of the solid solutions. If a piece of iron or aluminum comes in contact with a dental filling, the saliva and gum tissue can act as a salt bridge, and an electrochemical cell results. The iron or aluminum donates electrons, producing a tiny electrical current that results in a complaint from your tooth nerves.

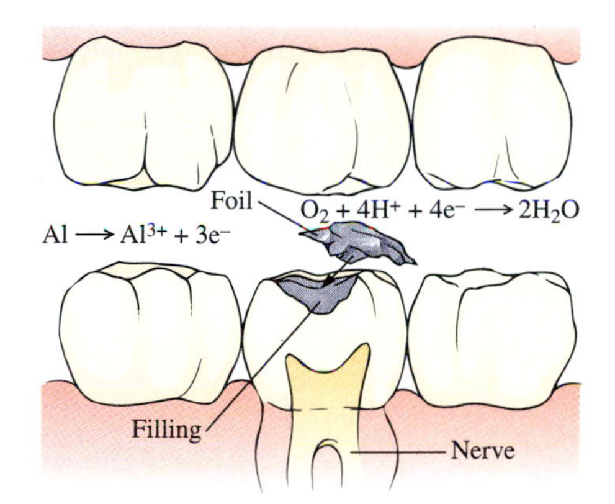

Figure 18.9 A dental voltaic cell. Pain usually results from accidentally touching an active metal like a piece of aluminum foil to metallic filling materials commonly used in dentistry. The active metal releases electrons that flow into the amalgam material where oxygen is reduced to water. If the flow of electrons encounters a nerve ending, the pain, although short in duration, can be quite intense.

Exercise 18.6 Predicting Redox Reactions Using E° Values

Consider the following half-reactions:

Half-Reaction	E° (V)
$Cl_2(g) + 2\ e^- \longrightarrow 2\ Cl^-(aq)$	+1.36
$I_2(s) + 2\ e^- \longrightarrow 2\ I^-(aq)$	+0.535
$Pb^{2+}(aq) + 2\ e^- \longrightarrow Pb(s)$	−0.126
$V^{2+}(aq) + 2\ e^- \longrightarrow V(s)$	−1.18

(a) Which is the weakest oxidizing agent?
(b) Which is the strongest oxidizing agent?
(c) Which is the strongest reducing agent?
(d) Which is the weakest reducing agent?
(e) Will Pb(s) reduce $V^{2+}(aq)$ to V(s)?
(f) Will $I_2(g)$ oxidize $Cl^-(aq)$ to $Cl_2(g)$?
(g) Name the molecules or ions that can be reduced by Pb(s).

CHEMISTRY YOU CAN DO

Remove Tarnish the Easy Way

Silverware tarnishes when exposed to air because the silver reacts with the hydrogen sulfide gas in the air to form a thin coating of black silver sulfide, Ag_2S. You can chemically remove the tarnish from silverware and other silver utensils by using a solution of baking soda and some aluminum foil. The chemical cleaning of silver is an electrochemical process in which electrons move from aluminum atoms to silver ions in the tarnish, reducing the silver ions to silver atoms while aluminum atoms are oxidized to aluminum ions. The sodium bicarbonate provides a conductive ionic solution for the flow of electrons and also helps to remove the aluminum oxide coating from the surface of the aluminum foil.

Get a large pan. Put 1 to 2 L of water in the pan. Add 7 to 8 Tbsp of baking soda. Heat the solution, but do not boil it. Place some aluminum foil in the bottom of the pan, and put the tarnished silverware on the aluminum foil. Make sure the silverware is covered with water. Heat the water almost to boiling. After a few minutes remove the silverware and rinse it in running water.

This method of cleaning silverware is better than using polish, because polish removes the silver sulfide, including the silver it contains; instead, the process described here restores the silver from the tarnish to the surface. If you have aluminum pie pans or aluminum cooking pans, you can use them as both the container and the aluminum source. You may notice that devices for removing silver tarnish are sometimes advertised on television. These devices are actually little more than a piece of aluminum metal and some salt. Would you pay \$19.95 (plus shipping and handling) for this, after you have done this simple experiment?

Before we leave this discussion of Table 18.1, consider again Table 4.5 ***(← p. 167)***. This activity series of metals and Table 18.1 have a lot in common. They contain many of the same elements, for example. Looking closely, however, you will notice that the most active metal, lithium, in Table 4.5 is at the very bottom of Table 18.1. That is because Table 18.1 is arranged by *reduction potential,* and lithium has the lowest tendency to be reduced. Table 4.5, on the other hand, lists the metals in order of activity, that is, tendency to be oxidized. Since oxidation is the opposite of reduction, it is reasonable that lithium is in opposite positions in the two tables.

Exercise 18.7 Predicting $E°$ Values

The two elements on either side of hydrogen in Table 4.5 are not listed in Table 18.1. Indicate where they would be in Table 18.1 and, based on values from the table, estimate their reduction potentials for their positive ions being reduced to the metal atom.

18.6 $E°$ AND GIBBS FREE ENERGY

The sign of $E°$ indicates whether a redox reaction is product-favored (positive $E°$) or reactant-favored (negative $E°$). You have learned another way to decide whether a reaction is product-favored: the change in standard Gibbs free energy, $\Delta G°$, must be negative ***(← p. 282)***. Since both $E°$ and $\Delta G°$ tell something about whether a reaction will occur, it should be no surprise that there is a relationship between them.

The "free" in Gibbs free energy indicates that it is energy available to do work. The energy available for electrical work from a cell can be calculated by multiplying the quantity of electrical charge transferred times the cell voltage, $E°$. The quantity of charge is given by the number of moles of electrons transferred in the overall reaction, n, multiplied by the number of coulombs per mole of electrons.

For the remainder of this chapter, $E°$ will be used for the overall reaction in place of $E°_{net}$.

Quantity of charge = moles of electrons × coulombs per mole of electrons

$$\text{Charge on 1 mol of electrons} = \left(\frac{1.6022 \times 10^{-19}\ \text{C}}{\text{electron}}\right)\left(\frac{6.022046 \times 10^{23}\ \text{electrons}}{\text{mol}}\right)$$
$$= 9.6485 \times 10^{4}\ \text{C/mol}$$

The quantity 9.6485×10^4 C/mol of electrons (commonly rounded to 96,500 C/mol of electrons) is known as the **Faraday constant** (F) in honor of Michael Faraday, who first explored the quantitative aspects of electrochemistry.

The electrical work that can be done by a cell is equal to the Faraday constant multiplied by the number of moles of electrons transferred and by the cell voltage.

$$\text{Electrical work} = nFE^\circ$$

Notice that, unlike the cell voltage, the electrical work a cell can do *does* depend on the quantity of reactants in the cell reaction. More reactants mean more moles of electrons transferred and hence more work. Equating the electrical work of a cell at standard conditions with ΔG°, we get

$$\Delta G^\circ = -nFE^\circ$$

The negative sign on the right side of the equation accounts for the fact that ΔG° is always negative for a product-favored process, but E° is always positive for a product-favored process. Thus their signs must be opposite.

Using this equation we can calculate ΔG° for the Cu^{2+}/Zn cell. This represents the maximum work that the cell can do. The reaction is

$$Cu^{2+}(aq) + Zn(s) \longrightarrow Cu(s) + Zn^{2+}(aq) \qquad E^\circ = +1.10\ \text{V}$$

so 2 mol of electrons are transferred per mole of copper reduced. The Gibbs free energy change when this quantity of reactants is consumed is

$$\Delta G^\circ = -(2\ \text{mol electrons transferred})\left(\frac{9.65 \times 10^4\ \text{C}}{\text{mol e}^-}\right)\left(\frac{1\ \text{J}}{1\ \text{V} \times 1\ \text{C}}\right)\left(\frac{1\ \text{kJ}}{10^3\ \text{J}}\right)(1.10\ \text{V})$$
$$= -212\ \text{kJ}$$

Note that the 2 mol of electrons transferred come from the half-reactions.

Exercise 18.8 The Relation Between E° and ΔG°

Use half-cell potentials to calculate E° and then ΔG° for the following reaction. Is the reaction product-favored as written?

$$Zn^{2+}(aq) + H_2(g) + 2\ H_2O(\ell) \longrightarrow Zn(s) + 2\ H_3O^+(aq)$$

18.7 COMMON BATTERIES

Voltaic cells include the convenient, portable sources of energy that we call *batteries.* Some batteries, like the common flashlight battery, consist of a single cell, while others, like automobile batteries, contain multiple cells. Batteries may be classified as primary or secondary depending on whether the reactions at the anode and cathode can be easily reversed. In a **primary battery** the electrochemical reactions cannot be easily reversed, so when the reactants are used up the battery is "dead" and must be discarded. A **secondary battery** (sometimes called a storage battery or a rechargeable battery), in contrast, uses an electrochemical reaction that can be reversed, so such a battery can be recharged.

Primary Batteries

For a long time the "dry cell," invented by Georges Leclanché in 1866, was the major source of energy for flashlights and toys. The container of the dry cell is made of zinc, which acts as the anode. The zinc is separated from the other chemicals by a liner of porous paper (Figure 18.10), which functions as the salt bridge. In the center of the dry cell is a graphite cathode, which is unreactive, inserted into a moist mixture of ammonium chloride (NH_4Cl), zinc chloride ($ZnCl_2$), and manganese dioxide (MnO_2). As electrons flow from the cell through a flashlight bulb, for example, the zinc is oxidized

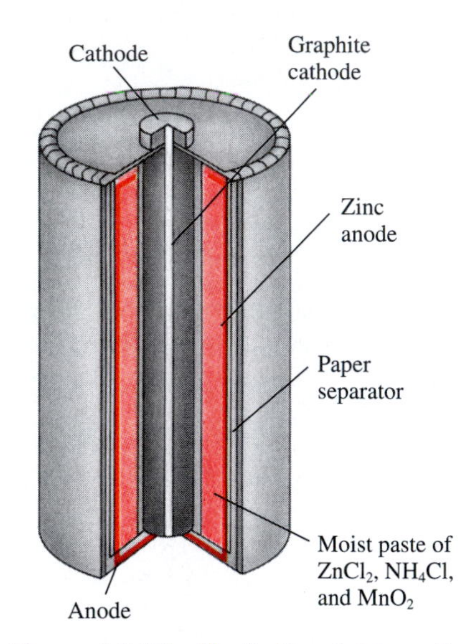

Figure 18.10 The LeClanché dry cell. It consists of a zinc anode (the battery container), a graphite cathode, and an electrolyte consisting of a moist paste of NH_4Cl, $ZnCl_2$, and MnO_2.

$$Zn(s) \longrightarrow Zn^{2+}(aq) + 2\ e^- \qquad \text{(anode, oxidation)}$$

and the ammonium ions are reduced.

$$2\ NH_4^+(aq) + 2\ e^- \longrightarrow 2\ NH_3(g) + H_2(g) \qquad \text{(cathode, reduction)}$$

The ammonia that is formed reacts with zinc ions to form a zinc-ammonia complex ion; this reaction prevents a buildup of gaseous ammonia.

$$Zn^{2+}(aq) + 2\ NH_3(g) \longrightarrow [Zn(NH_3)_2]^{2+}(aq)$$

The hydrogen that is produced is oxidized by the MnO_2 in the cell. In this way, hydrogen gas does not accumulate.

$$H_2(g) + 2\ MnO_2(s) \longrightarrow Mn_2O_3(s) + H_2O(\ell)$$

All these reactions lead to the following net process, which produces 1.5 V.

$$2\ MnO_2(s) + 2\ NH_4^+(aq) + Zn(s) \longrightarrow Mn_2O_3(s) + H_2O(\ell) + [Zn(NH_3)_2]^{2+}(aq)$$

This battery has two major disadvantages. First, if current is withdrawn rapidly, the NH_3 and H_2 gases produced in the reduction reaction cannot react rapidly enough with the Zn^{2+} and MnO_2. As a result, the cell voltage drops off quickly, although it can be restored by letting the battery sit undisturbed for a while. Second, there is a slow direct reaction between the zinc electrode and the ammonium ions even when current is not being drawn, so stored dry cells run down and tend to have a poor shelf life. Since the rates of almost all chemical reactions decrease with decreasing temperature, the shelf life of a dry cell can be doubled or tripled by storing it at about 4 °C in a refrigerator.

Some of the problems of the dry cell are overcome by the more expensive alkaline battery. An *alkaline battery,* which produces 1.54 V, also uses the oxidation of zinc as the anode reaction, but under alkaline ($pH > 7$) conditions.

$$Zn(s) + 2\ OH^-(aq) \longrightarrow ZnO(aq) + H_2O(\ell) + 2\ e^- \qquad \text{(anode, oxidation)}$$

The electrons that pass through the external circuit are consumed by reduction of manganese dioxide at the cathode.

$$MnO_2(s) + H_2O(\ell) + e^- \longrightarrow MnO(OH)(s) + OH^-(aq) \qquad \text{(cathode, reduction)}$$

In contrast to the Leclanché dry cell, no gases are formed in the alkaline battery, and there is no decline in voltage under high-current loads.

In the *mercury battery* (Figure 18.11) the oxidation of zinc is again the anode reaction. The cathode reaction, however, is the reduction of mercury(II) oxide (p. 844).

Portrait of a Scientist • *Wilson Greatbatch (1919–)*

Born in Buffalo, New York, Wilson Greatbatch was trained as an electrical engineer and worked at Cornell Aeronautical Labs in the early 1950s. He taught electrical engineering at the University of Buffalo and later became a manager at an electrical instrument company. In the early 1960s he had an idea of how a battery might be used to help an ailing heart keep pumping. Such a battery would have to be implanted under the patient's skin near the heart with electrodes that run to the heart. His story, told in his own words, is fascinating:

> *I quit all my jobs and with two thousand dollars, I went out in the barn in the back of my house and built 50 Pacemakers in two years. I started making the rounds of all the doctors in Buffalo who were working in this field, and I got consistently negative results. The answer I got was, well, these people all die in a year, you can't do much for them, why don't you work on my project, you know.*
>
> *When I first approached Dr. Shardack with the idea of the Pacemaker, he, alone, thought that it really had a future. He looked at me sort of funny, and he walked up and down the room a couple of times. He said, "You know—if you can do that—you can save a thousand lives a year."*

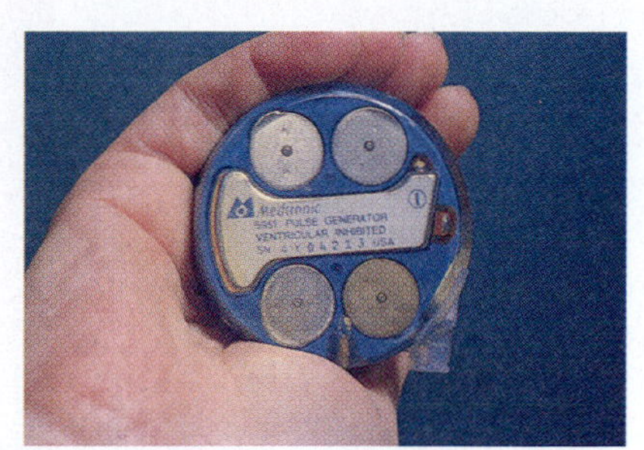

Cardiac pacemaker. The circular objects are the lithium batteries. *(Martin M. Rotker)*

When a medical team implanted the first heart pacemaker there was one major problem.

> *After the first ten years, we were still only getting one or two years out of pacemakers, two years on average, and the failure mechanism was always the battery. It didn't just run down, it failed. The human body is a very hostile environment, it's worse than space, it's worse than the bottom of the sea. You're trying to run things in a warm salt water environment. The first pacemakers could not be hermetically sealed, and the battery just didn't do the job. Well, after ten years, the battery emerged as the primary mode of failure, and so we started looking around for new power sources. We looked at nuclear sources, we looked at biological sources, of letting the body make its own electricity, we looked at rechargeable batteries, and we looked at improved mercury batteries. And we finally wound up with this lithium battery. It really revolutionized the pacemaker business. The doctors have told me that the introduction of the lithium battery was more significant than the invention of the Pacemaker in the first place.*

Source: The World of Chemistry Videotape series, (Program 15) "The Busy Electron."

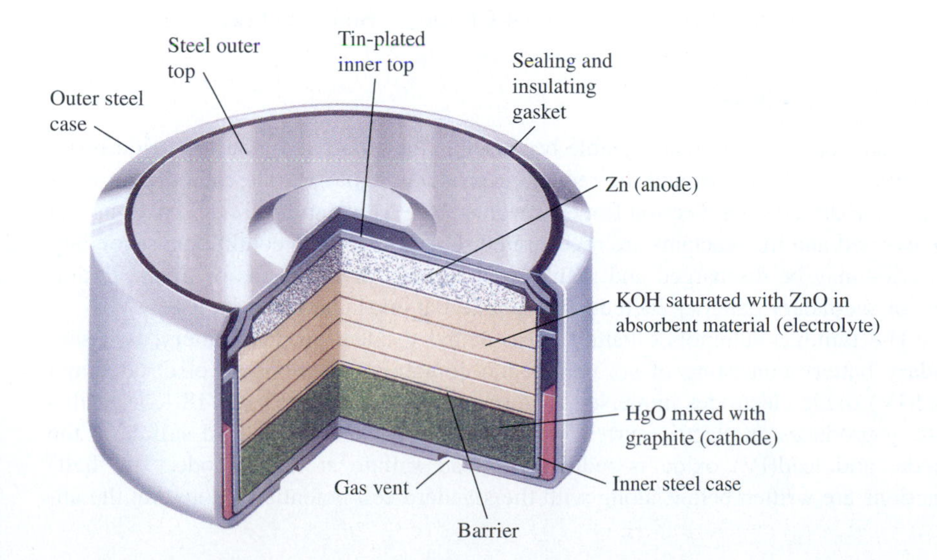

Figure 18.11 The mercury battery. The reducing agent is zinc and the oxidizing agent is mercury(II)oxide.

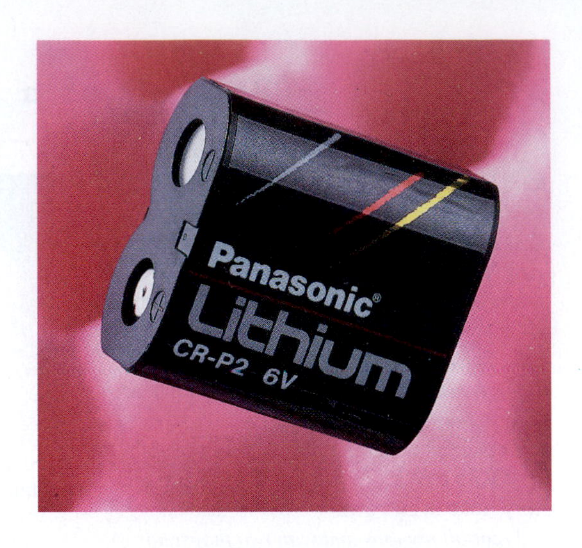

Figure 18.12 The lithium battery. The lithium battery finds many uses where a high energy density is desired. *(George Semple)*

$$HgO(s) + H_2O(\ell) + 2\ e^- \longrightarrow Hg(\ell) + 2\ OH^-(aq)$$

The HgO, mixed with graphite, is in a tightly compacted powder separated from a KOH electrolyte and the zinc by a moist-paper or porous-plastic barrier that serves as the salt bridge. The voltage of this battery is about 1.35 V. Mercury batteries are used in calculators, watches, hearing aids, cameras, and other devices where small size is an advantage. However, mercury and its compounds are poisonous, and careless disposal of mercury batteries can lead to environmental problems such as contamination of groundwater or even mercury vapor in the atmosphere.

Mercury batteries are hermetically sealed (to prevent leakage of mercury) and should never be heated. Heating increases the pressure of vapors within the battery, ultimately causing it to explode.

The *lithium battery* (Figure 18.12) is another popular battery, primarily because of its light weight. Instead of zinc (density = 7.14 g/cm^3) as the anode material, lithium (density = 0.534 g/cm^3) is used. Because lithium is such a strong reducing agent compared to zinc (see Table 18.1), this battery has a very large voltage (3.4 V per cell). Because lithium has such a low density, a lot of stored energy can be packed into a very lightweight package. Some lithium batteries use MnO_2 as the oxidizer, and others, such as some cardiac pacemaker batteries, use exotic compounds such as sulfuryl chloride ($SOCl_2$), whose cathode reaction is

Lithium has the lowest density of any nongaseous element.

$$2\ SOCl_2(\ell) + 4\ e^- \longrightarrow 4\ Cl^-(aq) + S(s) + SO_2(g)$$

Secondary Batteries

Secondary batteries are rechargeable because, as they discharge, the oxidation products remain at the anode and the reduction products remain at the cathode. As a result, if the direction of electron flow is reversed, the anode and cathode reactions can be reversed and the reactants are regenerated. Under favorable conditions, secondary batteries may be discharged and recharged hundreds or thousands of times. Examples of secondary batteries include automobile batteries and nicad batteries.

The lead-acid battery was first presented to the French Academy of Sciences in 1860 by Gaston Planté.

The familiar automobile battery, also called the lead-storage battery, is a secondary battery consisting of six cells, each containing porous lead electrodes and lead(IV) oxide electrodes immersed in aqueous sulfuric acid (Figure 18.13). As this battery produces an electric current, metallic lead is oxidized to lead sulfate at the anode, and lead(IV) oxide is reduced to lead sulfate at the cathode. The half-reactions are written below, along with the standard cell potentials. Note that the an-

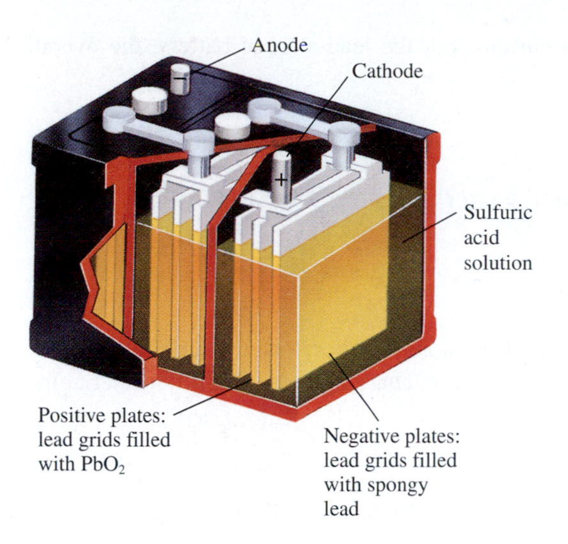

Figure 18.13 The lead-storage battery. Hydrogen and oxygen gases from the reduction and oxidation of water (see text) can build up in the area above the electrodes when the battery is overcharged. A spark caused by two of the adjacent plates shorting out can cause an explosion. Even sealed, modern batteries have been known to explode.

ode reaction is the reverse of the reduction reaction for $PbSO_4$, so the sign of its standard cell potential is reversed.

$$Pb(s) + HSO_4^-(aq) + H_2O(\ell) \longrightarrow PbSO_4(s) + H_3O^+(aq) + 2\ e^- \qquad E° = 0.356\ V\ \text{(anode reaction)}$$

$$PbO_2(s) + 3\ H_3O^+(aq) + HSO_4^-(aq) + 2\ e^- \longrightarrow PbSO_4(s) + 5\ H_2O(\ell) \qquad E° = +1.685\ V\ \text{(cathode reaction)}$$

$$Pb(s) + PbO_2(s) + 2\ H_3O^+(aq) + 2\ HSO_4^-(aq) \longrightarrow 2\ PbSO_4(s) + 4\ H_2O(\ell) \qquad E°_{cell} = +2.041\ V$$

The combined voltage from the six cells in a typical automobile battery gives a voltage of 12 V.

During discharge, sulfuric acid is consumed in both the anode and the cathode reactions, causing the concentration of the sulfuric acid electrolyte to decrease. Before the introduction of modern sealed automotive batteries, the measured density of this battery acid was routinely used to indicate the state of charge of the battery. As sulfuric acid is used at both electrodes, its concentration in the solution decreases. Since sulfuric acid has a greater density than water, the density of the battery acid decreases as the battery discharges. So the lower the density, the lower the charge of the battery. It is now almost impossible to measure the density of acid in a battery, because the cells are tightly sealed by the manufacturer.

To understand why such a battery is rechargeable, consider that the lead sulfate formed at both electrodes is an insoluble compound that mostly *stays on the electrode surface.* This keeps it available for further reaction and makes it possible to reverse the electrode reactions when the battery is recharged. Just look at the half-reactions above and consider them going in the reverse direction. Lead sulfate would be a reactant at both electrodes in the reverse reactions. To recharge a secondary battery, a source of direct electrical current is connected so that electrons are forced to flow in the direction opposite from when the battery was discharging. This causes the overall battery reaction to be reversed and regenerates the reactants that originally produced the battery's voltage and current. For the lead-storage battery, the overall reaction is

$$Pb(s) + PbO_2(s) + 2\ HSO_4^-(aq) + 2\ H_3O^+(aq) \underset{\text{charge (requires electricity)}}{\overset{\text{(produces electricity) discharge}}{\rightleftharpoons}} 2\ PbSO_4(s) + 4\ H_2O(\ell)$$

Normal charging of an automobile lead-storage battery occurs during driving. In addition to reversing the overall battery reaction, charging reduces a little water at the cathode and oxidizes a little water at the anode,

Reduction of water:

$$4\ H_2O(\ell) + 4\ e^- \longrightarrow 2\ H_2(g) + 4\ OH^-(aq)$$

Oxidation of water:

$$6\ H_2O(\ell) \longrightarrow O_2 + 4\ H_3O^+(aq) + 4\ e^-$$

These reactions produce a mixture of hydrogen and oxygen inside the battery. If this mixture is accidentally sparked, it can explode, and so no sparks or open flames should be brought near a lead-storage battery, even the sealed kind.

Rapid charging of a lead-storage battery, whether during normal driving or using a recharger, often causes elongated crystals of lead and lead oxide to grow on the electrode surfaces. These crystals can grow between the electrodes and cause internal short circuits. Usually, when this happens, the battery is "dead," will not accept a recharge, and must be replaced. If electrolyte fluid runs low, the electrode surfaces dry, which prevents the electrode from recharging properly. The result is again a dead battery.

Figure 18.14 **Nickel-cadmium (nicad) batteries.** *(C. D. Winters)*

Users of nicad batteries should carefully follow the manufacturer's charging recommendations for maximum battery life.

The lead-storage battery is relatively inexpensive, reliable, and simple, and it has an adequate life. Its high weight is its major fault. A typical automobile battery contains about 15 to 20 kg of lead, which is required to provide the large number of electrons needed to crank an automobile engine, especially on a cold morning. (Recall that the number of electrons that a battery can move from the anode to the cathode is proportional to the amount of reactants involved.) Another problem with lead batteries is that their lead can contaminate air and groundwater, possibly causing lead poisoning (***← p. 733***). Auto batteries should be recycled by companies equipped with the proper safeguards to protect the environment.

Nickel-cadmium ("nicad") batteries are another popular type of secondary battery. Nicad batteries are lightweight, can be quite small, and produce a constant voltage until completely discharged—making them useful in cordless appliances, video camcorders, portable radios, and other applications (Figure 18.14). They suffer somewhat from discharge "memory"; that is, if the battery is repeatedly used for only a short time and then recharged—it develops a tendency to need recharging after only a short use. This obviously causes problems if the full charge on the battery is needed.

Nicad batteries can be recharged because the reaction products are insoluble hydroxides that remain at the electrode surfaces. The anode reaction during the discharge cycle is the oxidation of cadmium, and the cathode reaction is the reduction of the nickel compound NiO(OH).

$$Cd(s) + 2\ OH^-(aq) \longrightarrow Cd(OH)_2(s) + 2\ e^- \qquad \text{(anode reaction)}$$

$$2\ [NiO(OH)(s) + H_2O(\ell) + e^- \longrightarrow Ni(OH)_2(s) + OH^-(aq)] \qquad \text{(cathode reaction)}$$

$$Cd(s) + 2\ NiO(OH)(s) + 2\ H_2O(\ell) \longrightarrow Cd(OH)_2(s) + 2\ Ni(OH)_2(s)$$

Like mercury batteries, nicad batteries should be disposed of properly because of the toxicity of cadmium and its compounds. It is likely that such batteries will soon be replaced by others that have similar characteristics but are less harmful to the environment.

The most promising new secondary battery is the *lithium-ion* battery. Lithium-ion batteries benefit from the low density of metallic lithium and the strong reducing strength of lithium metal itself (Table 18.1). The electrode where oxidation takes place in a lithium-ion battery is made of lithium metal that has been mixed with a conducting carbon polymer. The polymer has tiny spaces in its structure that can hold the lithium atoms and lithium ions formed by the oxidation reaction.

$$\text{Li (in polymer)} \longrightarrow \text{Li}^+ \text{ (in polymer)} + e^- \qquad \text{(oxidation reaction)}$$

The electrode where reduction takes place in the lithium-ion battery also contains lithium, but in the lattice of a metal oxide like CoO_2. This oxide lattice, like the carbon-polymer electrode, has holes in it that can accommodate Li^+ ions. The reduction reaction is

$$Li^+(\text{in } CoO_2) + e^- + CoO_2 \longrightarrow LiCoO_2$$

Lithium-ion batteries have a very high energy output for their weight, they can be recharged many hundreds of times, and, unlike the nicad batteries, have no memory effect. Because of these desirable characteristics, lithium-ion batteries are finding applications in cellular telephones, portable computers, and cameras. In the future, lithium-ion batteries may also find use in electric vehicles because of their low weight and high energy density (see Chemistry in the News).

Exercise 18.9 Recharging a Nicad Battery

Write the electrode reactions that take place when a nicad battery is recharged; identify the anode and cathode reactions.

Exercise 18.10 Emergency Batteries

You are stranded on an island and need to communicate your location for help. You have a battery-powered radio transmitter, but the lead batteries are discharged. There is a swimming pool nearby and you find a tank of chlorine gas and some plastic tubing that can withstand the oxidizing nature of chlorine. Devise a battery that might be used to power the radio using these things.

18.8 FUEL CELLS

A **fuel cell** is an electrochemical cell, but, in contrast to a battery, its reactants are continually supplied from an external reservoir. The best known fuel cell is the

(Text continues on page 849.)

CHEMISTRY IN THE NEWS

Batteries for Electric Cars

Electric automobiles are here! General Motors and other manufacturers have begun marketing electric cars in several states. These cars, the first highway electric vehicles to be mass produced for the consumer market since the early 1900s, represent many technological breakthroughs. They still are lacking one key ingredient that will have perhaps the greatest impact in their acceptance—a lightweight storage battery that can be recharged quickly and would store enough energy to drive a car 100 miles or more. Such a power system would make electric-powered vehicles much more attractive.

Several batteries that are expected to have these characteristics are under development. One such battery is being developed by Energy Conversion Devices of Troy, Michigan, under a grant from the U.S. Advanced Battery Consortium, which was set up by the U.S. Department of Energy, the three large U.S. automakers, and the Electric Power Research Institute. Called the Ovonic battery, it has a negative electrode made of nickel alloyed with several other metals—vanadium, titanium, zirconium, and chromium—instead of the usual pure-metal electrode. The half-cell reactions for this battery are

$$MH(s) + OH^-(aq) \underset{\text{charge}}{\overset{\text{discharge}}{\rightleftharpoons}} M(s) + H_2O(\ell) + e^-$$

$$NiOOH(s) + H_2O(\ell) + e^- \underset{\text{charge}}{\overset{\text{discharge}}{\rightleftharpoons}} Ni(OH)_2(s) + OH^-(aq)$$

MH(s) represents a metal hydride compound similar to those used to store hydrogen in other applications. The amount of hydrogen that can be absorbed into this electrode determines the number of electrons that the battery can deliver as it discharges, and hence the energy storage capacity of the battery. The metal-hydride electrode has the advantage of being metallic and hence electrically conducting; in many other batteries metal oxides, which do not conduct electricity, are formed on the electrode surface and reduce the number of times the battery can be recharged.

Under the hood of the GM electric car. The new electric cars have much more wiring than present automobiles. In place of the gasoline engine are one or more electric motors. Batteries replace the fuel tanks. *(Courtesy General Motors Corporation)*

Each component of the metal alloy anode plays a role in the battery's excellent performance. Vanadium, titanium, and zirconium are in the alloy because they readily absorb hydrogen. For an effective metal-hydride battery the strength of the bonding between the metal atoms and hydrogen atoms must be just right—from 25 to 50 kJ/mol. If it is too low, recharging will release hydrogen as $H_2(g)$, instead of incorporating hydrogen atoms into holes in the metal crystal lattice. If hydrogen-to-metal bonding is too strong, the electrode metal will be oxidized instead of the hydrogen atoms, and the battery will not be able to discharge properly. Alloying the other metals with nickel allows the hydrogen bond strength to be carefully adjusted for maximum efficiency. Chromium limits corrosion of vanadium in the alloy, and both zirconium and chromium affect the structure of the alloy, leading to high surface area that promotes rapid cell reactions and hence high power output.

The metal-hydride electrode consists of amorphous metal. Its atoms are in an irregular, disordered structure instead of the orderly, crystalline arrangement of most metals. This disorder increases hydrogen-storage capacity and speeds up electrode reactions. According to Energy Conversion Devices spokesperson Stanford R. Ovshinsky, the use of an amorphous material is "a fundamentally different approach" to battery design. The company claims that the battery can be charged in as little as 15 min and can undergo more than 1000 charge-discharge cycles, which translates to a lifetime of 10 years and more than 100,000 miles of travel in an automobile.

In Minnesota, the 3M Company is hard at work developing a lithium-ion battery for electric vehicles with a grant from the United States Advanced Battery Consortium. The lithium-ion battery capitalizes on the low density (and high energy content) of lithium. Conventional lead-storage batteries produce about 35 watt hours/kg, but a lithium-ion battery can produce as much as 200 watt hours/kg. That's enough energy to allow an electric vehicle to carry its passengers about 300 miles between recharges, which is comparable to gasoline-powered vehicles. This prospect of such a high energy battery with a low weight has also interested Hydro-Quebec, a Montreal-based electric utility company, which is actively researching a useful design for the lithium-ion battery. These lower-weight batteries would have an obvious advantage of keeping the overall vehicle weight down, thus allowing greater range between recharging while also allowing greater passenger loads.

Source: Wall Street Journal, April 9, 1993, p. B5; Science, Vol. 260, April 9, 1993, p. 176; and www.calstart.org (January, 1997).

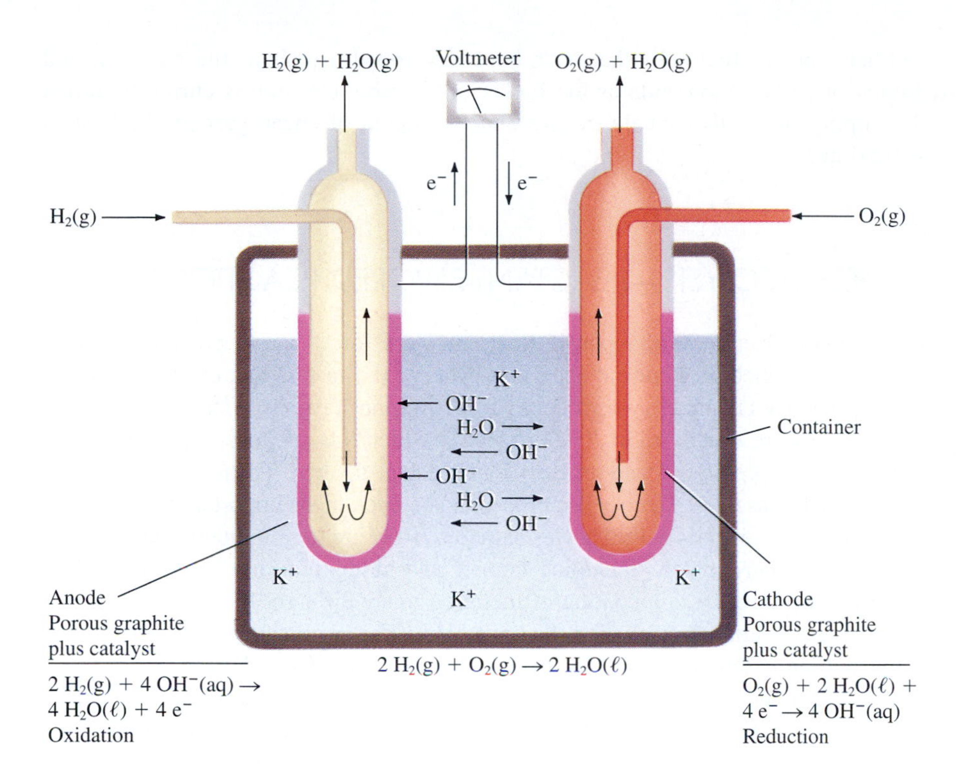

Figure 18.15 An H_2/O_2 fuel cell. The anode chamber oxidizes H_2. The cathode chamber reduces O_2. The water produced is often purified for drinking purposes.

hydrogen-oxygen cell (Figure 18.15) used in the Gemini, Apollo, and Space Shuttle programs. The net cell reaction is simply the oxidation of hydrogen to give water. If a mixture of hydrogen and oxygen is sparked, energy is released suddenly in a violent explosion. On a platinum gauze that acts as a catalyst, these gases will react at room temperature, slowly heating the catalytic surface to incandescence. In a fuel cell, hydrogen and oxygen are made to react in such a way that the energy is produced in the form of an electrical current. A stream of H_2 gas is pumped into the anode compartment of the cell, and pure O_2 gas is directed onto the cathode. The cell contains concentrated KOH, so the reactions are

$$2\ H_2(g) + 4\ OH^-(aq) \longrightarrow 4\ H_2O(\ell) + 4\ e^- \qquad \text{(anode, oxidation)}$$

$$O_2(g) + 2\ H_2O(\ell) + 4\ e^- \longrightarrow 4\ OH^-(aq) \qquad \text{(cathode, reduction)}$$

$$2\ H_2(g) + O_2(g) \longrightarrow 2\ H_2O(\ell) \qquad (E = 0.9\ \text{V at 70 to 140 °C})$$

The electrons lost by the hydrogen molecules at the anode flow out of the fuel cell, through a circuit, and then back into the cell at the cathode, where oxygen is reduced. This electron flow powers the electrical needs of the spacecraft, or whatever else is connected to the fuel cell. The water produced in the fuel cell can be purified for drinking purposes.

Because of their light weight and their high efficiency compared to batteries, fuel cells have proved valuable in the space program. Beginning with Gemini 5, alkaline fuel cells have logged more than 10,000 hours of operation in space. The fuel cells used aboard the Space Shuttle deliver the same power that batteries weighing ten times as much would provide. On a typical seven-day mission, the Shuttle fuel cells consume 1500 lb of hydrogen and generate 190 gal of potable water (suitable for drinking).

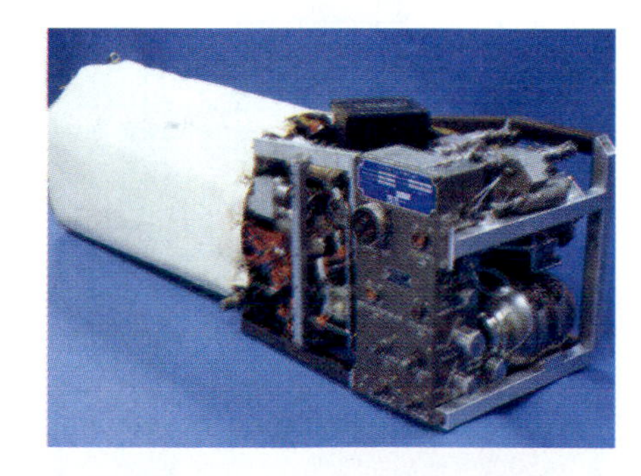

A hydrogen-oxygen fuel cell. Three of these units provide power for the Space Shuttle and its crew. (Courtesy International Fuel Cells Corp.)

The movie *Apollo 13* deals with an in-flight explosion of the fuel-cell storage tanks.

Conceptual Challenge Problem CP-18.A at the end of the chapter relates to topics covered in this section.

Other types of fuel cells that have been developed use air as the oxidizer, and hydrogen or carbon monoxide as the fuel. Considerable research is currently aimed at developing fuel cells capable of direct air oxidation of cheap gaseous fuels such as natural gas.

18.9 ELECTROLYSIS—REACTANT-FAVORED REACTIONS

What do you expect would happen if electrons were forced into a chemical system from a source of electrical current such as a battery? If you said that electrical energy passing through a chemical system could force reactant-favored systems to produce products, you were exactly right. The process in which this occurs is called **electrolysis.** Electrolysis provides a way to carry out reactant-favored reactions that will not take place by themselves. Electrolytic processes are even more important in our economy than the redox reactions that power batteries. They are used in the production and purification of many metals, including copper and aluminum, and in electroplating processes that produce a thin coating of metal on many different kinds of items.

Lysis means "splitting," so electrolysis means "splitting with electricity." Electrolysis reactions are chemical reactions caused by the flow of electricity.

Like voltaic cells, electrolysis cells have electrodes in contact with a conducting medium and an external circuit. As in a voltaic cell, the electrode where reduction takes place is called the cathode, and the electrode where oxidation takes place is called the anode. By contrast with voltaic cells, however, the external circuit connected to an electrolysis cell must contain a *source* of electrons. There is often no need for a physical separation of the two electrode reactions, so there is usually no salt bridge. The conducting medium in contact with the electrodes is often the same for both electrodes, and it can be a molten salt or an aqueous solution. Finally, the electrodes in electrolysis cells are often inert, and they only furnish a path for electrons to enter and leave the cell.

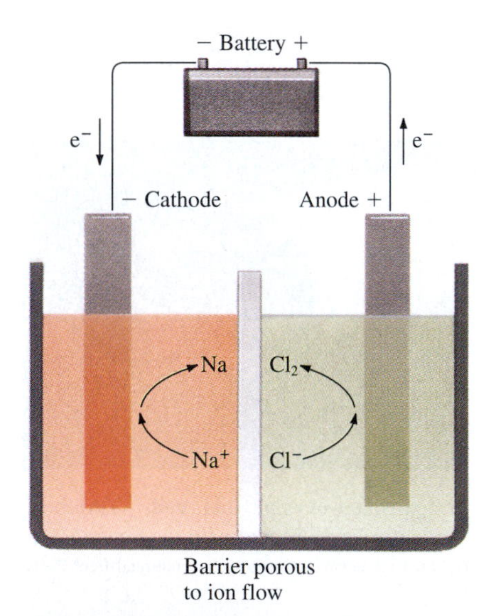

Figure 18.16 **Electrolysis of molten sodium chloride.**

The decomposition of molten sodium chloride, NaCl, is a simple example of a reaction that can be done by electrolysis. A pair of electrodes dip into pure sodium chloride that has been heated above its melting temperature (Figure 18.16). In the liquid sodium chloride, the Na^+ and Cl^- ions are free to move about. A battery can be used as a source of electrical current when an electrolysis is carried out on a small scale. The battery forces electrons into one of the electrodes (which becomes negative), and removes electrons from the other electrode (which becomes positive). In the molten sodium chloride, Cl^- ions are attracted to the positive electrode and Na^+ ions are attracted to the negative electrode. Reduction of Na^+ ions to Na atoms occurs at the negative electrode; this electrode is therefore the cathode. Oxidation of Cl^- ions occurs at the positive electrode (the anode).

$$2\,Cl^- \longrightarrow Cl_2(g) + 2\,e^- \quad \text{(anode reaction, oxidation)}$$
$$2\,Na^+ + 2\,e^- \longrightarrow 2\,Na(\ell) \quad \text{(cathode reaction, reduction)}$$
$$2\,Cl^- + 2\,Na^+ \longrightarrow 2\,Na(\ell) + Cl_2(g) \quad \text{(net cell reaction)}$$

The electrolysis of molten salts is an energy-intensive reaction. Energy is needed to melt the salt, and energy is needed to cause the anode and cathode reactions to take place.

What happens if we pass electricity through an *aqueous solution* of some salt, such as potassium iodide (KI)? To predict the outcome of the electrolysis we must first decide what is in the solution that can be oxidized and reduced. For KI(aq), the solution contains K^+ ions, I^- ions, and H_2O molecules. In K^+, potassium is already in its highest possible oxidation state, so it cannot be involved in an oxidation reaction. However, both the I^- ion and H_2O can be oxidized. The possible anode half-reaction oxidations are

$$2\ I^-(aq) \longrightarrow I_2(s) + 2\ e^- \qquad E^\circ_{ox} = -0.535\ V$$

$$6\ H_2O(\ell) \longrightarrow O_2(g) + 4\ H_3O^+(aq) + 4\ e^- \qquad E^\circ_{ox} = -1.229\ V$$

Whenever two or more reactions are possible at a single electrode, the one with the more positive E° will occur under standard-state conditions. Judging by the values of E°_{ox} (which we obtained from Table 18.1), the iodide ion will be oxidized more readily than will water.

Since I^- is already in a reduced form, there are only two species that can be reduced at the cathode: K^+ ions and water molecules. The reduction reactions and their potentials are:

$$2\ H_2O(\ell) + 2\ e^- \longrightarrow H_2(g) + 2\ OH^-(aq) \qquad E^\circ_{red} = -0.8277\ V$$

$$K^+(aq) + e^- \longrightarrow K(s) \qquad E^\circ_{red} = -2.925\ V$$

In this case we predict that $H_2O(\ell)$ will be reduced, because it has the more positive E° value.

An experiment in which electrons are passed through aqueous KI (Figure 18.17) shows that this prediction is correct. At the anode, on the right, the I^- ion is oxidized

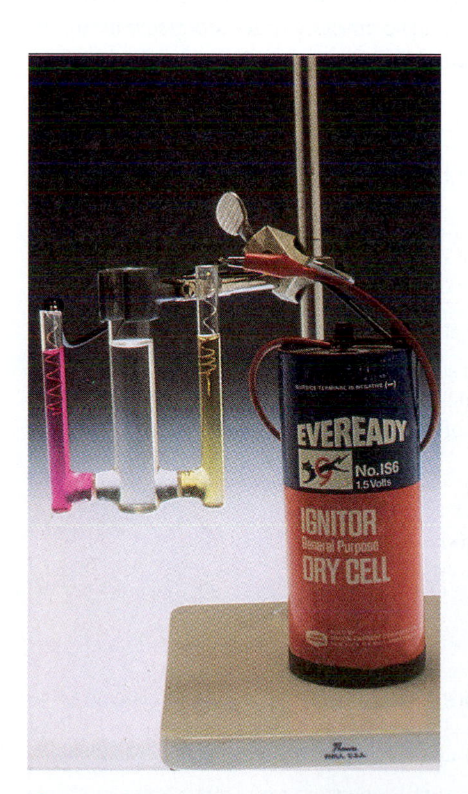

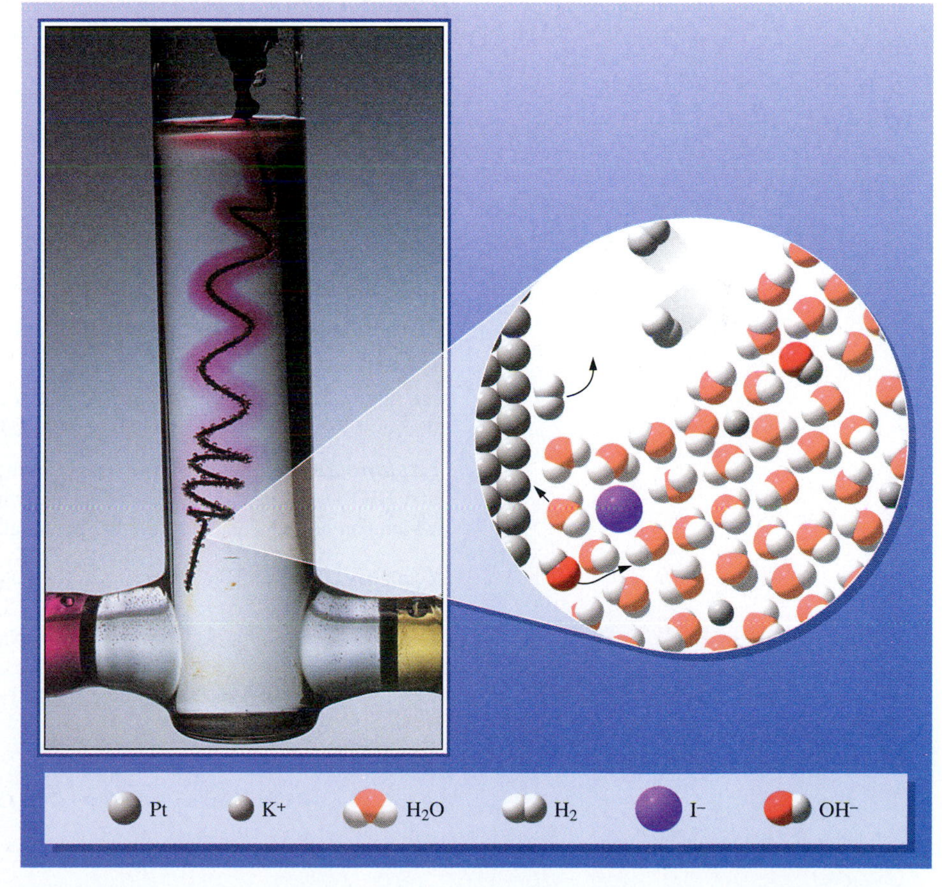

Figure 18.17 The electrolysis of aqueous potassium iodide. Aqueous KI (*left*) is contained in all three compartments of the cell, and both electrodes are platinum. At the positive electrode or anode (*right*), the I^- ion is oxidized to iodine, which gives the solution a yellow-brown color ($2\ I^-(aq) \longrightarrow I_2(aq) + 2\ e^-$). At the negative electrode or cathode (*close-up photo*), water is reduced, and the presence of OH^- ion is indicated by the red color of the acid-base indicator, phenolphthalein ($2\ H_2O(\ell) + 2\ e^- \longrightarrow H_2(g) + 2\ OH^-(aq)$). In a close-up of the cathode surface bubbles of H_2 and evidence of OH^- being generated at the electrode are clearly seen. *(C. D. Winters)*

to I_2, which produces a yellow-brown color in the solution. At the cathode, water is reduced and hydroxide ions are formed, as shown by the pink color of the acid-base indicator phenolphthalein that has been added to the solution. A close look at Figure 18.17b reveals that a gas, presumed to be hydrogen, is also being produced at the surface of the inert platinum electrode.

When several reactions are possible in an electrolysis, the one that will most likely occur is the one that is least reactant-favored.

When an electrolysis is carried out by passing electrical current through an aqueous solution, the electrode reactions most likely to take place are those that require the least voltage, that is, the half-reactions that combine to give the least negative overall cell voltage. This means that in aqueous solution the following conditions apply:

1. *A metal ion or other species can be reduced if it has a reduction potential more positive than −0.8 V, the potential for reduction of water.* Table 18.1 shows that most metal ions are in this category. If a species has a reduction potential more negative than −0.8 V, then water will be reduced to $H_2(g)$ preferentially. Metal ions in this latter category include Na^+, K^+, Mg^{2+}, and Al^{3+}. Producing these metals from their ions requires electrolysis of a molten salt in which no water is present.
2. *A species can be oxidized in aqueous solution if it has an oxidation potential more positive than −1.2 V, the potential for oxidation of water to $O_2(g)$.* Most species on the right-hand side of the half-equations in Table 18.1 are in this category. If a species has an oxidation potential more negative than −1.2 V (that is, if its half-equation is above the water-oxygen half-equation in Table 18.1), water will be oxidized preferentially. Thus, for example, $F^-(aq)$ cannot be oxidized electrolytically to $F_2(g)$, because water will be oxidized to $O_2(g)$ instead.

PROBLEM-SOLVING EXAMPLE 18.8 Electrolysis of Aqueous NaOH

Predict the result of passing a direct electrical current through an aqueous solution of NaOH.

Answer The net cell reaction is $2\ H_2O(\ell) \longrightarrow 2\ H_2(g) + O_2(g)$. Hydrogen is produced at the cathode and oxygen is produced at the anode.

Explanation First, list all the species in the solution. In this case they are Na^+, OH^-, and H_2O. Next, use Table 18.1 to decide which of these species can be oxidized and which can be reduced, and note the potential of each possible reaction.

Reductions:

$$Na^+(aq) + e^- \longrightarrow Na(s) \qquad E^\circ_{red} = -2.71\ V$$

$$2\ H_2O(\ell) + 2\ e^- \longrightarrow H_2(g) + 2\ OH^-(aq) \qquad E^\circ_{red} = -0.83\ V$$

Oxidations:

$$4\ OH^-(aq) \longrightarrow O_2(g) + 2\ H_2O(\ell) + 4\ e^- \qquad E^\circ_{ox} = -0.40\ V$$

$$6\ H_2O(\ell) \longrightarrow O_2(g) + 4\ H_3O^+(aq) + 4\ e^- \qquad E^\circ_{ox} = -1.229\ V$$

It is evident that water will be reduced to H_2 at the cathode and that OH^- will be oxidized at the anode, because these are the reactions with the least negative E° values. The net cell reaction is

$$2\ H_2O(\ell) \longrightarrow 2\ H_2(g) + O_2(g)$$

and the potential under standard conditions is $(-0.83\ V) + (-0.40\ V) = -1.23\ V$.

Problem-Solving Practice 18.8

Predict the results of passing a direct electrical current through (a) molten NaBr, (b) aqueous NaBr, and (c) aqueous $SnCl_2$.

Exercise 18.11 Making F_2 Electrolytically

In 1886, Moissan was the first to prepare F_2 by the electrolysis of F^- ions. He electrolyzed the salt KF dissolved in pure HF. No water was present, so only F^- ions were available at the anode. What was produced at the cathode? Write the half-equations for the oxidation and the reduction reactions, then write the net cell reaction.

18.10 COUNTING ELECTRONS

When an electric current is passed through an aqueous solution of the soluble salt $AgNO_3$, metallic silver is produced at the cathode. One mole of electrons is required for every mole of Ag^+ reduced.

$$Ag^+(aq) + e^- \longrightarrow Ag(s)$$

If a copper salt were in aqueous solution, 2 mol of electrons would be required to produce 1 mol of metallic copper from copper(II) ions.

$$Cu^{2+}(aq) + 2\ e^- \longrightarrow Cu(s)$$

Each of these balanced half-reactions is like all other balanced chemical equations. They illustrate the fact that both matter and charge are conserved in chemical reactions. This means that if you could measure the number of moles of electrons flowing through the electrolysis cell, you would know the number of moles of silver or copper produced. Conversely, if you knew the amount of silver or copper produced, you could calculate the number of moles of electrons that had passed through the circuit.

The number of moles of electrons transferred during a redox reaction is usually determined by measuring the current flowing in the external electrical circuit during a given time. The product of the current (measured in amperes, A) and the time interval (in seconds, s) equals the electric charge (coulombs, C) of electricity that has flowed through the circuit.

Large electric currents, like those needed to run a hair dryer or refrigerator, are measured in amperes. Smaller currents, in the milliampere (mA) range, are more commonly used in laboratory electrolysis experiments. 10^3 mA = 1A.

$$\text{Charge} = \text{current} \times \text{time}$$
$$1\ \text{coulomb} = 1\ \text{ampere} \times 1\ \text{second}$$

The Faraday constant (96,500 C/mol of e^-, *p. 841)* can then be used to find the number of moles of electrons from a known number of coulombs of charge. This information is of practical significance in chemical analysis and synthesis.

Conceptual Challenge Problem CP-18.B at the end of the chapter relates to topics covered in this section.

PROBLEM-SOLVING EXAMPLE 18.9 Using the Faraday Constant

What mass of nickel will be deposited at the cathode of an electrolysis cell if a current of 20. mA passes through an aqueous solution containing Ni^{2+} ions for 1.0 hr (3600 s)?

Answer 0.022 g Ni

Explanation The reaction at the cathode is

$$Ni^{2+}(aq) + 2\ e^- \longrightarrow Ni(s)$$

The charge that passes through the cell is

$$\text{Charge} = 20. \times 10^{-3}\ \text{A} \times 3600\ \text{s} = 72\ \text{C}$$

Now, use the Faraday constant, the coefficients of the balanced cathode half-reaction, and the molar mass of nickel as conversion factors to find the mass of nickel deposited:

$$\text{Mass of Ni} = 72\ \text{C} \times \left(\frac{1\ \text{mol e}^-}{9.65 \times 10^4\ \text{C}}\right)\left(\frac{1\ \text{mol Ni}}{2\ \text{mol e}^-}\right)\left(\frac{58.7\ \text{g Ni}}{1\ \text{mol Ni}}\right) = 0.022\ \text{g Ni}$$

Problem-Solving Practice 18.9

In the commercial production of sodium by electrolysis, the cell operates at 7.0 V and a current of 25×10^3 A. What mass of sodium can be produced in 1 hr?

Exercise 18.12 Estimating Faradays

Which would require more Faradays of electricity? (a) Making 1 mol of Al from Al^{3+}; (b) Making 2 mol of Na from Na^+; (c) Making 2 mol of Cu from Cu^{2+}.

Electrolytic Production of Hydrogen

Hydrogen holds great promise as a fuel in our economy because it is a gas and can be easily transported through pipelines; it burns without producing pollutants; and it could be used in fuel cells to generate electricity on demand. Hydrogen can be produced by the electrolysis of dilute sulfuric acid. Both water and sulfuric acid are in plentiful supply. The only problem with producing hydrogen in quantities large enough to meet a nation's energy demands is the source of electricity. Until either solar- *(← p. 694)* or fusion-generated electricity (Section 19.7) becomes commercially available, electrolytically produced hydrogen will continue to be a high-priced commodity.

The minimum voltage required for this reaction is 1.24 V. Let's consider how much electrical energy would be required to produce 1.00 kg of gaseous H_2 (about 11,200 L at STP). We will first calculate the required charge in coulombs by using Faraday's constant, and then use the definition 1 joule = 1 volt × 1 coulomb to get energy units.

The reduction half-reaction shows that 2 mol of electrons are required to produce 1 mol (2.02 g) of $H_2(g)$.

$$2\ H_3O^+(aq) + 2\ e^- \longrightarrow H_2(g) + 2\ H_2O(\ell)$$

The amount (number of moles) of electrons required to produce 1.00 kg of H_2 is found as follows:

$$\text{Amount of e}^- = 1.00 \text{ kg H}_2 \times \left(\frac{1 \times 10^3 \text{ g}}{1 \text{ kg}}\right)\left(\frac{1 \text{ mol H}_2}{2.016 \text{ g H}_2}\right)\left(\frac{2 \text{ mol e}^-}{1 \text{ mol H}_2}\right)$$
$$= 9.92 \times 10^2 \text{ mol e}^-$$

Now we can calculate the charge by using Faraday's constant.

$$\text{Charge} = 9.92 \times 10^2 \text{ mol e}^- \times \left(\frac{9.65 \times 10^4 \text{ C}}{1 \text{ mol e}^-}\right) = 9.57 \times 10^7 \text{ C}$$

The energy (in joules) can be calculated from the charge and the cell voltage.

$$\text{Energy} = \text{charge} \times \text{voltage} = (9.57 \times 10^7 \text{ C})(1.24 \text{ V}) = 1.19 \times 10^8 \text{ J}$$

Now we convert joules to kilowatt-hours (kwh), which is the unit we see when we pay the electric bill. The conversion factor is 1 kwh = 3.60×10^6 J.

The kilowatt-hour is a unit of energy.

$$\text{Energy} = 1.19 \times 10^8 \text{ J} \times \left(\frac{1 \text{ kwh}}{3.60 \times 10^6 \text{ J}}\right) = 33.1 \text{ kwh}$$

At a rate of 10 cents per kilowatt-hour, the production of 1.00 kg of hydrogen costs $3.31.

Exercise 18.13 Calculations Based on Electrolysis

In the production of aluminum metal, Al^{3+} is reduced to Al. The currents are generally about 50,000 A. A low voltage of about 4.0 V is used. How much energy (in kilowatt-hours) is required to produce 2000. tons of aluminum metal?

Exercise 18.14 How Many Joules?

Think of a battery you just purchased at the store as a bundle of energy containing some number of joules. Name the two pieces of information you need to calculate the number of joules of energy in this battery. Which one is obviously available as you read the label on the battery? Devise a means of determining the other information needed.

Electroplating

If a metal or other electrical conductor is made the cathode in an electrolysis cell, it can be plated with another metal to protect it against corrosion, decorate it, or purify the deposited metal. Consider, for example, an electrolysis cell like the one shown on Figure 18.18 (p. 856), which can be used to purify copper. Here, the cathode is pure copper and the anode is a bar of less pure copper. This reaction is important commercially because impure copper, the kind formed when copper is produced from copper ore, doesn't conduct electricity well. Only after it is purified can copper be used in making electrical wiring.

The electrolysis solution contains a soluble copper salt such as $CuSO_4$. The cell half-reactions are

$$\text{Cu(s)}_{\text{(impure)}} \longrightarrow \text{Cu}^{2+}\text{(aq)} + 2\text{ e}^- \quad E^\circ_{\text{ox}} = -0.34 \text{ V (anode reaction)}$$
$$\text{Cu}^{2+}\text{(aq)} + 2\text{ e}^- \longrightarrow \text{Cu(s)}_{\text{(pure)}} \quad E^\circ_{\text{red}} = +0.34 \text{ V (cathode reaction)}$$

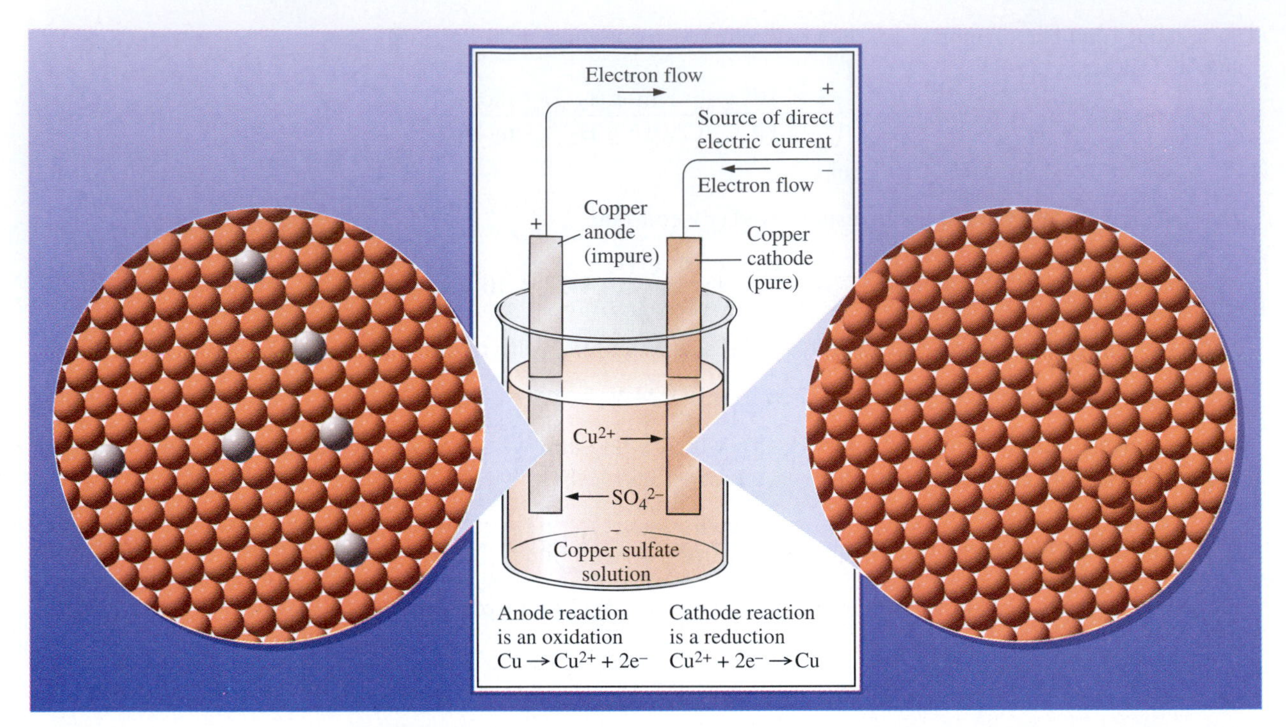

Figure 18.18 Electroplating from a copper sulfate solution. Copper from the impure copper anode is plated onto the cathode, which consists of highly purified copper.

Oscar is gold plated. Most gold jewelry is made by plating a thin coating of gold on a base metal. *(Don Smetzer/Tony Stone Images)*

Overall, the net cell reaction appears to be nothing; copper is oxidized and copper is reduced, but something does take place in the cell—it is the transport of copper from the impure anode to the pure cathode. The net cell reaction can be written as

$$Cu(s)_{impure} \xrightarrow{\text{electric current}} Cu(s)_{pure}$$

The reason energy in the form of an electric current must be supplied is a rather subtle one: the pure copper has less entropy than the mixture of copper and impurities; so energy must be supplied to overcome the entropy decrease *(← p. 279)*.

To plate an object with copper, we have only to render the surface conducting and make the object the cathode in a cell containing a solution of a soluble copper salt. The object will become coated with copper, and the copper coating will grow thicker as the electrolysis continues and more electrons reduce Cu^{2+} ions to Cu atoms. If the object is a metal, it will conduct electricity by itself. If the object is a nonmetal, its surface can be lightly dusted with graphite powder to render it conducting.

Precious metals such as gold are often plated onto cheaper metals such as copper to make jewelry. If the current and time of the plating reaction are known, it is possible to calculate the mass of gold that will be reduced onto the cathode surface. For example, suppose the object to be plated is immersed in a solution of $AuCl_3$ and is made a cathode by connecting it to the negative pole of a battery. The circuit is completed by immersing an inert anode in the solution, and gold is reduced at the cathode for 60. min at a current of 0.25 A. The mass of gold that is reduced is calculated by

$$\text{Mass Au} = 0.25\ \text{A} \times 60.\ \text{min} \times \left(\frac{60\ \text{s}}{1\ \text{min}}\right)\left(\frac{1\ \text{C}}{1\ \text{A}\cdot\text{s}}\right)\left(\frac{1\ \text{mol e}^-}{9.65\times10^4\ \text{C}}\right)$$

$$\times \left(\frac{1\ \text{mol Au}}{3\ \text{mol e}^-}\right)\left(\frac{197\ \text{g Au}}{1\ \text{mol Au}}\right) = 0.61\ \text{g Au}$$

That's about $7.83 worth of gold, assuming gold is selling for $400 per ounce.

Exercise 18.15 Electroplating Gold

Calculate the mass of gold that could be plated from solution with a current of 0.50 A for 20. min. The cathode reaction is $Au^{3+}(aq) + 3\ e^- \rightarrow Au(s)$.

Conceptual Challenge Problem CP-18.C at the end of the chapter relates to topics covered in this section.

18.11 CORROSION—PRODUCT-FAVORED REACTIONS

Corrosion is the oxidation of a metal that is exposed to the environment. Usually corrosion results in loss of structural strength and is considered undesirable. Visible corrosion on the steel supports of a bridge, for example, indicate possible structural failure. Corrosion reactions are invariably product-favored. This means that $E°$ for the reaction is positive ($\Delta G < 0$). Corrosion of iron, for example, takes place quite readily and is difficult to prevent; it results in the red-brown substance we call rust, which is hydrated iron(III) oxide [$Fe_2O_3 \cdot xH_2O$, where x varies from 2 to 4]. The corrosion of aluminum, a metal that is even more reactive than iron, is also very product-favored, but the aluminum oxide that forms as a result of corrosion adheres tightly as a thin coating on the surface of the metal and actually forms a protective coating that prevents further corrosion. The rust that forms when iron corrodes, on the other hand, does not adhere to the surface of the metal, so it can easily fall away and expose more metal surface to corrosion (Figure 18.19).

Corrosion is so commonplace that about 25% of the annual steel production in the United States is destined for replacement of material lost to corrosion.

For corrosion of a metal (M) to occur, there must be an anodic area where the oxidation of the metal can occur. The general reaction is

Anode reaction: $M(s) \longrightarrow M^{n+} + ne^-$

Figure 18.19 Rusting. The formation of rust destroys the structural integrity of objects made of iron and steel. Given time, this old car will completely rust away. The process may take several decades but in the end there will remain only a pile of rust, glass, rubber tires, plastic, and possibly the copper radiator. *(C. D. Winters)*

There must also be a cathodic area where electrons are consumed. Frequently, the cathode reactions are reductions of oxygen or water:

Cathode reactions:

$$2\ H_2O(\ell) + 2\ e^- \longrightarrow 2\ OH^-(aq) + H_2(g)$$
$$O_2(g) + 2\ H_2O(\ell) + 4\ e^- \longrightarrow 4\ OH^-(aq)$$

In both of the reactions, hydroxide ions are formed, and when water alone is reduced, hydrogen gas is also a product.

Anodic areas can occur at cracks in the oxide coating that protects the surfaces of many metals, and they may also occur around impurities. Cathodic areas occur at the metal oxide coating, at less reactive metallic impurity sites, or around other metal compounds trapped at the surface, such as sulfides or carbides.

The other requirements for corrosion are an electrical connection between the anode and cathode and an electrolyte with which both anode and cathode are in contact. Both requirements are easily fulfilled—the metal itself is the conductor, and ions dissolved in moisture from the environment provide the electrolyte.

The corrosion of iron is the most familiar kind of corrosion. In the corrosion of iron, the anodic reaction is the oxidation of metallic iron (Figure 18.20). If both wa-

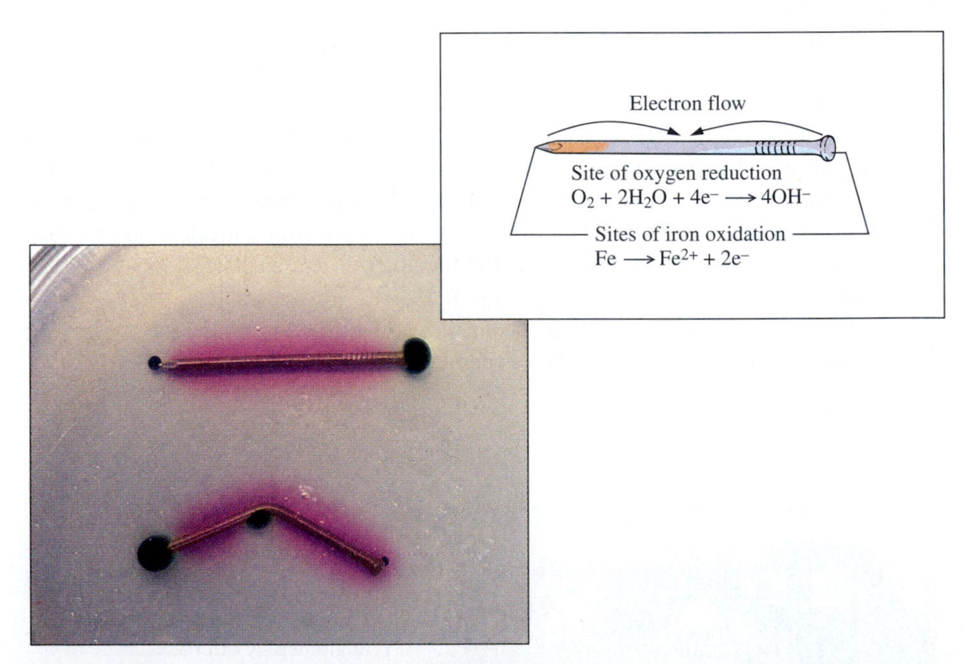

Figure 18.20 Corroding iron nails. Two nails were placed in an agar gel, which also contained the indicator phenolphthalein and $[Fe(CN)_6]^{3-}$. The nails began to corrode and gave Fe^{2+} ions at the tip and where the nail is bent. (These are points of stress and corrode more quickly.) These points are the anode as indicated by the formation of the blue-colored compound called Prussian blue ($Fe_3[Fe(CN)_6]_2$). The remainder of the nail is the cathode, since oxygen is reduced in water to give OH^-. The presence of OH^- ions causes phenolphthalein to turn pink in color. *(C. D. Winters)*

ter and O_2 gas are present, the cathode reaction is the reduction of oxygen, giving the net reaction

$$2\ [Fe(s) \longrightarrow Fe^{2+}(aq) + 2\ e^-] \qquad \text{(anode reaction)}$$
$$O_2(g) + 2\ H_2O(\ell) + 4\ e^- \longrightarrow 4\ OH^-(aq) \qquad \text{(cathode reaction)}$$
$$2\ Fe(s) + O_2(g) + 2\ H_2O(\ell) \longrightarrow \underset{\text{iron(II) hydroxide}}{2\ Fe(OH)_2(s)}$$

In the presence of an ample supply of oxygen and water, as in the open air or in flowing water, the iron(II) hydroxide is oxidized to the red-brown iron(III) oxide (Figure 18.19).

$$4\ Fe(OH)_2(s) + O_2(g) \longrightarrow 2\ H_2O(\ell) + \underset{\text{red brown}}{2\ Fe_2O_3 \cdot H_2O(s)}$$

This hydrated iron oxide is the familiar rust you see on cars and buildings, and the substance that colors the water red in some mountain streams or in your home water supply at times. It is easily removed by mechanical shaking, rubbing, or even the action of rain or freeze-thaw cycles, thus exposing more iron at the surface and allowing iron objects to eventually deteriorate completely.

If oxygen is not freely available to the corroding iron, further oxidation of the iron(II) hydroxide is limited to the formation of magnetite (Fe_3O_4), which can be thought of as a mixed oxide of Fe_2O_3 and FeO in a 1:1 ratio.

$$6\ Fe(OH)_2(s) + O_2(g) \longrightarrow \underset{\text{hydrated magnetite}}{2\ Fe_3O_4 \cdot H_2O(s)} + 4\ H_2O(\ell)$$

Other substances in air and water can hasten corrosion. Metal salts, like the chlorides of sodium and calcium from sea air or from salt spread on roadways in the winter, function as salt bridges between anodic and cathodic regions, thus speeding up corrosion reactions.

Exercise 18.16 Do All Metals Corrode?

Do all metals corrode as readily as do iron and aluminum? Name three metals that you would expect to corrode about as readily as iron and aluminum, and name three metals that do not corrode readily. Name a use for each of the three noncorroding metals. Explain why metals fall into these two groups.

Corrosion Protection

How can you stop a metal object from corroding? The general approaches are to (a) inhibit the anodic process, (b) inhibit the cathodic process, or (c) do both. The most common method is **anodic inhibition,** attempting directly to limit or prevent the oxidation half-reaction by painting the metal surface, coating it with grease or oil, or allowing a thin film of metal oxide to form. More recently developed methods of anodic protection are illustrated by the following reaction, in which the surface is treated with a solution of sodium chromate (see next page).

$$2\ Fe(s) + 2\ Na_2CrO_4(aq) + 2\ H_2O(\ell) \longrightarrow Fe_2O_3(s) + Cr_2O_3(s) + 4\ NaOH(aq)$$

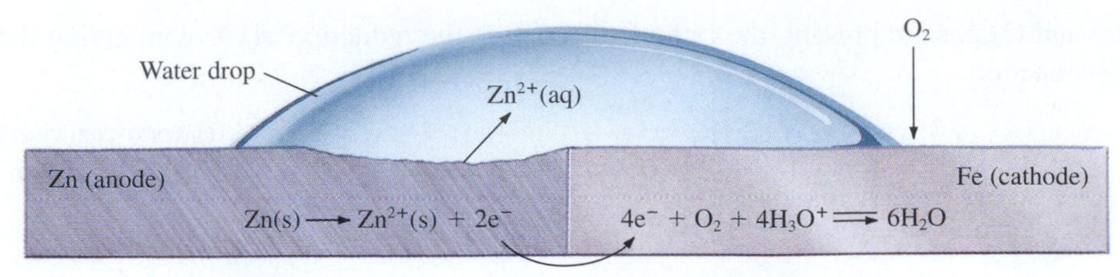

Figure 18.21 Cathodic protection of an iron-containing object. The iron is coated with a film of zinc, a metal more easily oxidized than iron. Therefore, the zinc acts as an anode and forces iron to become the cathode, thereby preventing the corrosion of the iron.

Galvanized objects. A thin coating of zinc helps prevent the oxidation of iron. *(C. D. Winters)*

The surface iron is oxidized by the chromate salt to give Fe(III) and Cr(III) oxides. These together form a coating that is impervious to O_2 and water, and further atmospheric oxidation is inhibited.

Cathodic protection is accomplished by forcing the metal to become the cathode instead of the anode. Usually, this is achieved by attaching another, more readily oxidized metal to the metal being protected. The best example of this is **galvanized** iron, iron that has been coated with a thin film of zinc (Figure 18.21). $E°_{ox}$ for zinc oxidation is considerably more positive than $E°_{ox}$ for iron oxidation. (Look at Table 18.1, but remember to change the sign of $E°$, since the reactions in the table are written as reductions.) Therefore, the zinc metal film is oxidized before any of the iron and the zinc coating forms what is called a *sacrificial anode.* In addition, when the zinc is corroded, $Zn(OH)_2$ forms an insoluble film on the surface (K_{sp} of $Zn(OH)_2 = 4.5 \times 10^{-17}$) that further slows corrosion.

Exercise 18.17 Corrosion Rates

Rank the following environments for their relative rates of corrosion of iron. Place the fastest first. Explain your answers.

(a) Moist clay
(b) Sand by the seashore
(c) The surface of the moon
(d) Desert sand in Arizona

Summary Problem

Many kinds of secondary batteries are known. Arguably the most useful is the lead-storage battery. If it were not for the density of lead, this battery would find a far greater acceptance. Many electric vehicles currently use lead-storage batteries as their source of power, but automotive engineers look longingly at other batteries because of their lower weights. When you look at Table 18.1 and consider the chemical properties of all of the other oxidizing and reducing agents shown there, you might be tempted to try and create a hybrid battery that would combine some of the desirable features of, say, a PbO_2 cathode and some other kind of anode rather than the anode made of Pb found in the lead-storage battery. In that way, at least the high reduction potential of the half-reaction involving PbO_2 might still be used.

(a) What would be the $E°$ value of a cell made using the PbO_2 reduction reaction and magnesium metal as the reducing agent? Write the two half-reactions and the net cell reaction. Would this cell be the basis for a secondary battery? Explain your answer.

(b) What would be the $E°$ value of a cell made using the PbO_2 reduction reaction and nickel metal as the reducing agent? Write the two half-reactions and the net cell reaction. Would this cell have a voltage greater than or less than a single cell of a lead-storage battery? Would this cell be the basis of a secondary battery? Explain. What could you do to the chemistry in the anode compartment to make it a secondary battery?

(c) If your Ni/PbO_2 hybrid battery were a success and it was manufactured for use in electric automobiles, how many amperes could it produce assuming that 500.0 g of Ni reacted in exactly 30 min? How much PbO_2 would be reduced during this same amount of time?

(d) Of course, batteries must be recharged, so how much time would be required to recharge your Ni/PbO_2 battery to its original state (the 500.0 g of Ni being converted back to its original form) if a current of 25.5 A passed through the battery?

(e) Just as you are getting ready to cash in on the success of your new battery, someone announces that it has some serious environmental problems. What could these be? Explain.

In Closing

Having studied this chapter, you should be able to . . .

- Identify the oxidizing and reducing agents in a redox reaction (Section 18.1).
- Write equations for the oxidation and reduction half-reactions, and use them to balance the net equation (Section 18.2).
- Identify and describe the functions of the parts of an electrochemical cell; describe the direction of electron flow outside the cell and the direction of the ion flow inside the cell (Section 18.3).
- Describe how standard reduction potentials are defined and use them to predict whether a reaction will be product-favored as written (Sections 18.4 and 18.5).
- Calculate $\Delta G°$ from the value of $E°$ for a redox reaction (Section 18.6).
- Explain how product-favored electrochemical reactions can be used to do useful work, and list the requirements for using such reactions in rechargeable batteries (Section 18.6).
- Describe the chemistry of the dry cell, the mercury battery, and the lead-storage battery (Section 18.7).
- Describe how a fuel cell works and indicate how it is different from a battery (Section 18.8).
- Use standard reduction potentials to predict the products of electrolysis of an aqueous salt solution (Section 18.9).
- Calculate the quantity of product formed at an electrode during an electrolysis reaction, given the current passing through the cell and the time during which the current flows (Section 18.10).

- Explain how electroplating works (Section 18.10).
- Describe what corrosion is and how it can be prevented by cathodic protection (Section 18.11).

Key Terms

anode *(18.3)*
anodic inhibition *(18.11)*
battery *(18.3)*
cathode *(18.3)*
cathodic protection *(18.11)*
cell voltage *(18.4)*
corrosion *(18.11)*
coulomb *(18.4)*
electrochemical cell *(18.3)*
electrode *(18.3)*
electrolysis *(18.9)*
electromotive force *(18.4)*
emf *(18.4)*
Faraday constant *(18.6)*
fuel cell *(18.8)*
galvanized *(18.11)*
half-cell *(18.3)*
half-reaction *(18.1)*
oxidizing agent *(18.1)*
primary battery *(18.7)*
reducing agent *(18.1)*
salt bridge *(18.3)*
secondary battery *(18.7)*
standard conditions *(18.4)*
standard hydrogen electrode *(18.4)*
standard reduction potential *(18.5)*
standard voltages ($E°$) *(18.4)*
volt (V) *(18.4)*
voltaic cell *(18.3)*

Questions for Review and Thought

Conceptual Challenge Problems

CP-18.A. Automobiles run on internal combustion engines, in which the energy used to run the vehicle is obtained from the combustion of gasoline. The main component of gasoline is octane (C_8H_{18}). An automobile manufacturer has recently announced a chemical method for generating hydrogen gas from gasoline, and proposes to develop a car in which an H_2/O_2 fuel cell powers an electric propulsion motor, thus eliminating the internal combustion engine with its problems (for example, the generation of unwanted byproducts that pollute the air). The hydrogen for the fuel cell would be directly generated from gasoline onboard the vehicle. There are two steps in this hydrogen generation process:

(i) Partial oxidation of octane by oxygen to carbon monoxide and hydrogen;

(ii) Combination of carbon monoxide with additional gaseous water to form carbon dioxide and more hydrogen (the water-gas shift reaction).

(a) Write the chemical equation for the complete combustion of 1 mol of octane.
(b) Write balanced chemical equations for the two-step hydrogen-generation process. How many moles of H_2 are produced per mole of octane? (Remember that water is a reactant in the two-step process.)
(c) By combining these equations, show that the net *overall* reaction is the same as in the combustion of octane.
(d) By assuming that the entire Gibbs free energy change of the H_2/O_2 fuel cell reaction is available for use by the electric propulsion motor, calculate the energy produced by a fuel cell when it consumes all of the hydrogen produced from 1 mol of octane. Compare this energy with the Gibbs free energy change for the combustion of 1 mol of octane. (*Note:* The Gibbs free energy of formation, ΔG_f°, for $C_8H_{18}(\ell)$ is 6.14 kJ/mol.)

CP-18.B. People obtain energy by oxidizing food. Glucose is a typical foodstuff. It is a carbohydrate that is oxidized to water and carbon dioxide.

$$C_6H_{12}O_6(aq) + 6\ O_2(g) \longrightarrow 6\ CO_2(g) + 6\ H_2O(\ell)$$

The heat of combustion of glucose is 2.80×10^3 kJ/mol, which means that as glucose is oxidized, its electrons lose 2.80×10^3 kJ/mol as they fall to lower energy states in a complicated series of chemical steps.

(a) Assume that a person requires 2400 food Calories per day and that they are obtained from the oxidation of glucose. How much O_2 must a person breathe each day to react with this much glucose?
(b) Each mol of O_2 requires 4 mol of electrons, regardless of whether the O atoms become part of CO_2 or H_2O. What would be the average electrical current (C/s) in a human body using the above amount of energy per day?
(c) Use the answer from part (b) and calculate the electrical potential this current flows through in a day to produce the 2400 food Calories.

CP-18.C. A piece of chromium metal is attached to a battery and dipped in 50 mL of 0.3 M KOH solution in a 250-mL beaker. A stainless steel electrode is connected to the other pole of the battery and immersed in the same solution. A steady current of 0.50 amp is maintained for exactly 2 hr. Several samples of a gas formed at the stainless steel electrode during the electrolysis are captured and all are found to ignite in air. After the electrolysis, the chromium electrode is weighed and found to have decreased in weight by 0.321 g. The mass of stainless steel electrode does not change.

After electrolysis, the KOH solution is neutralized with nitric acid to a pH of just below 7, then heated and reacted with 0.151 M lead(II) nitrate solution. As the lead(II) nitrate solution is added, a yellow precipitate quickly forms from the hot solution. The formation of precipitate stops after 40.4 mL of the lead(II) nitrate solution has been added. The yellow solid is then filtered, dried, and weighed. Its mass is 1.97 g.

(a) How much electrical charge passes through the cell?
(b) How many moles of Cr react?
(c) What is the oxidation state of the Cr after reacting?
(d) Assuming that the yellow compound that precipitates from the solution during the titration contains both Pb and Cr, what do you conclude to be the ratio of the numbers of atoms of Pb and Cr?
(e) If the yellow compound contains an element other than Pb and Cr, what is it and how much is in the compound? What is the formula for the yellow compound?

Review Questions

1. Describe the principal parts of an electrochemical cell by drawing a hypothetical cell, indicating the cathode, the anode, the direction of electron flow outside the cell, and the direction of ion flow within the cell.
2. Explain how product-favored electrochemical reactions can be used to do useful work.
3. Explain how reactant-favored electrochemical reactions can be induced to proceed and make products.
4. Explain how electroplating works.
5. Tell whether each of the following statements is true or false. If false, rewrite it to make it a correct statement.
   (a) Oxidation always occurs at the anode of an electrochemical cell.
   (b) The anode of a battery is the site of reduction and is negative.
   (c) Standard conditions for electrochemical cells are a concentration of 1.0 M for dissolved species and a pressure of 1 atm for gases.
   (d) The potential of a cell does not change with temperature.
   (e) All product-favored oxidation-reduction reactions have a standard cell voltage $E°$ with a negative sign.

Redox Reactions

6. In each of the following reactions, tell which substance is oxidized and which is reduced. Tell which is the oxidizing agent and which is the reducing agent. Assign oxidation numbers to all species.
   (a) $2\ Al(s) + 3\ Cl_2(g) \longrightarrow 2\ AlCl_3(s)$
   (b) $8\ H_3O^+(aq) + MnO_4^-(aq) + 5\ Fe^{2+}(aq) \longrightarrow 5\ Fe^{3+}(aq) + Mn^{2+}(aq) + 12\ H_2O(\ell)$
   (c) $FeS(s) + 3\ NO_3^-(aq) + 4\ H_3O^+(aq) \longrightarrow 3\ NO(g) + SO_4^{2-}(aq) + Fe^{3+}(aq) + 6\ H_2O(\ell)$
7. In each of the following reactions, tell which substance is oxidized and which is reduced. Tell which is the oxidizing agent and which is the reducing agent. Assign oxidation numbers to all species.
   (a) $Fe(s) + Br_2(\ell) \longrightarrow FeBr_2(s)$
   (b) $8\ HI(aq) + H_2SO_4(aq) \longrightarrow H_2S(aq) + 4\ I_2(s) + 4\ H_2O(\ell)$
   (c) $H_2O_2(aq) + 2\ Fe^{2+}(aq) + 2\ H_3O^+(aq) \longrightarrow 2\ Fe^{3+}(aq) + 4\ H_2O(\ell)$
8. Choose four elements: a regular metal, a transition metal, a nonmetal, and a semiconductor. By using the index to this text, find a chemical reaction in which each element occurs as a reactant. Assign oxidation numbers to all elements on the reactant and product side, and identify the oxidizing agent and the reducing agent.
9. Answer Question 8 again, only this time find a chemical reaction in which each element is produced.

Using Half-Reactions to Understand Redox Reactions

10. Write half-reactions for the following:
    (a) Oxidation of zinc to Zn^{2+} ion
    (b) Reduction of H_3O^+ ion to hydrogen gas
    (c) Reduction of Sn^{4+} ion to Sn^{2+} ion
    (d) Reduction of chlorine to Cl^- ion
    (e) Oxidation of sulfur dioxide to sulfate ion in acidic solution
11. Write half-reactions for the following:
    (a) Reduction of MnO_4^- ion to Mn^{2+} ion in acid solution
    (b) Reduction of $Cr_2O_7^{2-}$ ion to Cr^{3+} ion in acid solution
    (c) Oxidation of hydrogen gas to H_3O^+ ion
    (d) Reduction of hydrogen peroxide to water
    (e) Oxidation of nitric oxide to nitrate ion in acidic solution
12. For each reaction in Question 6, write balanced half-reactions.
13. For each reaction in Question 7, write balanced half-reactions.
14. Balance the following redox reactions, and identify the oxidizing agent and the reducing agent.
    (a) $CO(g) + O_3(g) \longrightarrow CO_2(g)$
    (b) $H_2(g) + Cl_2(g) \longrightarrow 2\ HCl(g)$
    (c) $H_2O_2(aq) + Ti^{2+}(aq) \longrightarrow H_2O(\ell) + Ti^{4+}(aq)$ in acidic solution
    (d) $Cl^-(aq) + MnO_4^-(aq) \longrightarrow Cl_2(g) + MnO_2(s)$ in acidic solution
    (e) $FeS_2(s) + O_2(g) \longrightarrow Fe_2O_3(s) + SO_2(g)$
    (f) $O_3(g) + NO(g) \longrightarrow O_2(g) + NO_2(g)$
    (g) $Zn(Hg)(amalgam) + HgO(s) \longrightarrow ZnO(s) + Hg(\ell)$ in basic solution (This is the reaction in the mercury battery.)

15. Balance the following redox reactions, and identify the oxidizing agent and the reducing agent.
(a) $FeO(s) + O_3(g) \longrightarrow Fe_2O_3(s)$
(b) $P_4(s) + Br_2(\ell) \longrightarrow PBr_5(\ell)$
(c) $H_2O_2(aq) + Co^{2+}(aq) \longrightarrow H_2O(\ell) + Co^{3+}(aq)$ in acidic solution
(d) $Cl^-(aq) + Cr_2O_7^{2-}(aq) \longrightarrow Cl_2(g) + Cr^{3+}(aq)$ in acidic solution
(e) $CuFeS_2(s) + O_2(g) \longrightarrow Cu_2S(s) + FeO(s) + SO_2(g)$
(f) $H_2CO(g) + O_2(g) \longrightarrow CO_2(g) + H_2O(\ell)$
(g) $C_3H_8(g) + O_2(g) \longrightarrow CO_2(g) + H_2O(\ell)$ in acidic solution (This is the reaction occurring in a propane fuel cell.)

Electrochemical Cells

16. For the redox reaction $Cu^{2+}(aq) + Zn(s) \longrightarrow Cu(s) + Zn^{2+}(aq)$, why can't you generate electrical current by placing a piece of copper metal and a piece of zinc metal in a solution containing $CuCl_2(aq)$ and $ZnCl_2(aq)$?

17. Explain the function of a salt bridge in an electrochemical cell.

18. Are standard half-cell reactions always written as oxidation reactions or reduction reactions?

19. Tell whether the following statement is true or false. If false, rewrite it to make it a correct statement: The value of an electrode potential changes when the half-reaction is multiplied by a factor. That is, $E°$ for $Li^+ + e^- \rightarrow Li$ is different from that for $2\ Li^+ + 2\ e^- \rightarrow 2\ Li$.

20. A voltaic cell is assembled with Pb(s) and $Pb(NO_3)_2(aq)$ in one compartment and Zn(s) and $ZnCl_2(aq)$ in the other. An external wire connects the two electrodes, and a salt bridge containing KNO_3 connects the two solutions.
(a) In the product-favored reaction, zinc metal is oxidized to Zn^{2+}. Write a balanced net ionic equation for this reaction.
(b) Which half-reaction occurs at each electrode? Which is the anode and which is the cathode?
(c) Draw a diagram of the cell, indicating the direction of electron flow outside the cell and of ion flow within the cell.

21. A voltaic cell is assembled with Sn(s) and $Sn(NO_3)_2(aq)$ in one compartment and Ag(s) and AgCl(aq) in the other. An external wire connects the two electrodes, and a salt bridge containing KNO_3 connects the two solutions.
(a) In the product-favored reaction, Ag^+ is reduced to silver metal. Write a balanced net ionic equation for this reaction.
(b) Which half-reaction occurs at each electrode? Which is the anode and which is the cathode?
(c) Draw a diagram of the cell, indicating the direction of electron flow outside the cell and of ion flow within the cell.

Electrochemical Cells and Voltage

22. You light a 25-watt light bulb with the current from a 12-volt lead-storage battery. After 1 hr of operation, how much energy has the light bulb utilized? How many coulombs of charge have been withdrawn from the battery? Assume 100% efficiency. (A watt is the transfer of 1 J of energy in 1 s.)

23. Draw a diagram of standard hydrogen electrode and describe how it works.

24. Copper can reduce silver ion to metallic silver, a reaction that could in principle be used in a battery.

$$Cu(s) + 2\ Ag^+(aq) \longrightarrow Cu^{2+}(aq) + 2\ Ag(s)$$

(a) Write equations for the half-reactions involved.
(b) Which half-reaction is an oxidation and which is a reduction? Which half-reaction occurs in the anode compartment and which in the cathode compartment?

25. Chlorine gas can oxidize zinc metal in a reaction that has been suggested as the basis of a battery. Write the half-reactions involved. Label which is the oxidation and which is the reduction reaction.

Using Standard Cell Potentials

26. What is the strongest oxidizing agent in Table 18.1? What is the strongest reducing agent? What is the weakest oxidizing agent? What is the weakest reducing agent?

27. Using the reduction potentials in Table 18.1, place the following elements in order of increasing ability to function as reducing agents: (a) Cl_2, (b) Fe, (c) Ag, (d) Na, (e) H_2.

28. Using the reduction potentials in Table 18.1, place the following elements in order of increasing ability to function as oxidizing agents: (a) O_2, (b) H_2O_2, (c) $PbSO_4$, (d) H_2O.

29. One of the most energetic redox reactions is that between F_2 gas and lithium metal.
(a) Write the half-reactions involved. Label which is the oxidation and which is the reduction reaction.
(b) According to data from Table 18.1, what is $E°$ for this reaction?

30. Calculate the value of $E°$ for each of the following reactions. Decide whether each is product-favored.
(a) $I_2(s) + Mg(s) \longrightarrow Mg^{2+}(aq) + 2\ I^-(aq)$
(b) $Ag(s) + Fe^{3+}(aq) \longrightarrow Ag^+(aq) + Fe^{2+}(aq)$
(c) $Sn^{2+}(aq) + 2\ Ag^+(aq) \longrightarrow Sn^{4+}(aq) + 2\ Ag(s)$
(d) $2\ Zn(s) + O_2(g) + 2\ H_2O(\ell) \longrightarrow 2\ Zn^{2+}(aq) + 4\ OH^-(aq)$

31. Consider the following half-reactions:

Half-Reaction	$E°$ (V)
$Cl_2(g) + 2\ e^- \longrightarrow 2\ Cl^-(aq)$	1.36
$I_2(s) + 2\ e^- \longrightarrow 2\ I^-(aq)$	0.535
$Pb^{2+}(aq) + 2\ e^- \longrightarrow Pb(s)$	−0.126
$V^{2+}(aq) + 2\ e^- \longrightarrow V(s)$	−1.18

(a) Which is the weakest oxidizing agent?
(b) Which is the strongest oxidizing agent?
(c) Which is the strongest reducing agent?
(d) Which is the weakest reducing agent?
(e) Will Pb(s) reduce $V^{2+}(aq)$ to V(s)?
(f) Will $I_2(g)$ oxidize $Cl^-(aq)$ to $Cl_2(g)$?
(g) Name the elements or ions that can be reduced by Pb(s).

32. Consider the following half-reactions:

Half-Reaction	$E°$ (V)
$Ce^{4+}(aq) + e^- \longrightarrow Ce^{3+}(aq)$	1.61
$Ag^+(aq) + e^- \longrightarrow Ag(s)$	0.80
$Hg_2^{2+}(aq) + 2\ e^- \longrightarrow 2\ Hg(\ell)$	0.79
$Sn^{2+}(aq) + 2\ e^- \longrightarrow Sn(s)$	−0.14
$Ni^{2+}(aq) + 2\ e^- \longrightarrow Ni(s)$	−0.25
$Al^{3+}(aq) + 3\ e^- \longrightarrow Al(s)$	−1.66

(a) Which is the weakest oxidizing agent?
(b) Which is the strongest oxidizing agent?
(c) Which is the strongest reducing agent?
(d) Which is the weakest reducing agent?
(e) Will Sn(s) reduce $Ag^+(aq)$ to Ag(s)?
(f) Will $Hg(\ell)$ reduce $Sn^{2+}(aq)$ to Sn(s)?
(g) Name the ions that can be reduced by Sn(s).
(h) What metals can be oxidized by $Ag^+(aq)$?

33. In principle, a battery could be made from aluminum metal and chlorine gas.
(a) Write a balanced equation for the reaction that would occur in a battery using $Al^{3+}(aq)/Al(s)$ and $Cl_2(g)/Cl^-(aq)$ half-reactions.
(b) Tell which half-reaction occurs at the anode and which at the cathode. What are the polarities of these electrodes?
(c) Calculate the standard potential, $E°$, for the battery.

$E°$ and Gibbs Free Energy

34. Choose the correct answers: In a product-favored chemical reaction, the standard cell potential, $E°$, is (greater/less) than zero and the Gibbs free energy change, $\Delta G°$, is (greater/less) than zero.

35. For each of the reactions in Question 30, compute the Gibbs free energy change $\Delta G°$.

36. Hydrazine, N_2H_4, can be used as the reducing agent in a fuel cell.

$$N_2H_4(aq) + O_2(g) \longrightarrow N_2(g) + 2\ H_2O(\ell)$$

(a) If $\Delta G°$ for the reaction is −607 kJ, calculate the value of $E°$ expected for the reaction.
(b) Suppose the reaction is written with all coefficients doubled. Determine $\Delta G°$ and $E°$ for this new reaction.

37. The standard cell potential for the oxidation of Mg by Br_2 is 3.45 V.

$$Br_2(\ell) + Mg(s) \longrightarrow Mg^{2+}(aq) + 2\ Br^-(aq)$$

(a) Calculate $\Delta G°$ for this reaction.
(b) Suppose the reaction is written with all coefficients doubled. Determine $\Delta G°$ and $E°$ for this new reaction.

38. The standard cell potential, $E°$, for the reaction of Zn(s) and $Cl_2(g)$ is 2.12 V. Write the chemical equation for the reaction of 1 mol of zinc. What is the standard free energy change, $\Delta G°$, for this reaction?

Common Batteries

39. What are the advantages and disadvantages of lead-storage batteries?

40. Nicad batteries are rechargeable and are commonly used in cordless appliances. Although such batteries actually function under basic conditions, imagine an electrochemical cell using the following setup.

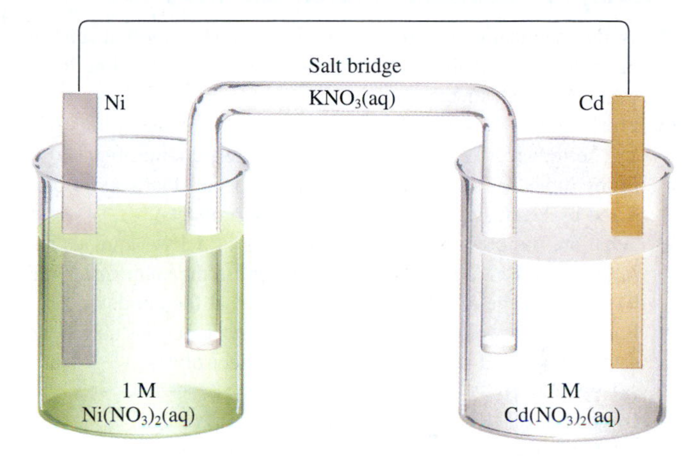

(a) Write a balanced net ionic equation depicting the reaction occurring in the cell.
(b) What is oxidized? What is reduced? What is the reducing agent and what is the oxidizing agent?
(c) Which is the anode and which is the cathode?
(d) What is $E°$ for the cell?
(e) What is the direction of electron flow in the external wire?
(f) If the salt bridge contains KNO_3, toward which compartment will the NO_3^- ions migrate?

41. Consider the nicad cell in the previous question.
(a) If the concentration of Cd^{2+} is reduced to 0.010 M, and $[Ni^{2+}] = 1.0$ M, will the cell emf be smaller or larger than when the concentration of $Cd^{2+}(aq)$ was 1.0 M? Explain your answer in terms of Le Chatelier's principle.
(b) Begin with 1.0 L of each of the solutions, both initially 1.0 M in dissolved species. Each electrode weighs 50.0 g in the beginning. If 0.050 A is drawn from the battery, how long can it last?

Fuel Cells

42. How does a fuel cell differ from a battery?

43. Describe the principal parts of an H_2/O_2 fuel cell. What is the reaction at the cathode? At the anode? What is the product of the fuel-cell reaction?

44. Hydrazine, N_2H_4, has been proposed as the fuel in a fuel cell in which oxygen is the oxidizing agent. The reactions are

$$N_2H_4(aq) + 4\ OH^-(aq) \longrightarrow N_2(g) + 4\ H_2O(\ell) + 4\ e^-$$
$$O_2(g) + 2\ H_2O(\ell) + 4\ e^- \longrightarrow 4\ OH^-(aq)$$

(a) Which reaction occurs at the anode and which at the cathode?

(b) What is the net cell reaction?
(c) If the cell is to produce 0.50 A of current for 50.0 hr, what mass in grams of hydrazine must be present?
(d) What mass in grams of O_2 must be available to react with the mass of N_2H_4 determined in part (c)?

Electrolysis—Reactant-Favored Reactions

45. Write chemical equations for the electrolysis of molten salts of three different alkali halides to produce the corresponding halogens and alkali metals.
46. From Table 18.1 write down all of the aqueous metal ions that can be reduced by electrolysis to the corresponding metal.
47. From Table 18.1 write down all of the species that can be oxidized by electrolysis, and determine the products.
48. What are the products of the electrolysis of a 1 M aqueous solution of NaBr? What species are present in the solution? What is formed at the cathode? What is formed at the anode?
49. For each of the following solutions, tell what reactions take place at the anode and at the cathode during electrolysis.
(a) $NiBr_2(aq)$
(b) $NaI(aq)$
(c) $CdCl_2(aq)$
(d) $CuI_2(aq)$
(e) $MgF_2(aq)$
(f) $HNO_3(aq)$

Counting Electrons

50. A current of 0.015 A is passed through a solution of $AgNO_3$ for 155 min. What mass of silver is deposited at the cathode?
51. Current is passed through a solution containing $Ag^+(aq)$. How much silver was in the solution if all the silver was removed as Ag metal by electrolysis for 14.5 min at a current of 1.0 mA?
52. A current of 2.50 A is passed through a solution of $Cu(NO_3)_2$ for 2.00 hr. What mass of copper is deposited at the cathode?
53. A current of 0.0125 A is passed through a solution of $CuCl_2$ for 2.00 hr. What mass of copper is deposited at the cathode and what volume of Cl_2 gas (in mL at STP) is produced at the anode?
54. The major reduction half-reaction occurring in the cell in which Al_2O_3 and aluminum salts are electrolyzed is $Al^{3+}(aq) + 3\ e^- \rightarrow Al(s)$. If the cell operates at 5.0 V and 1×10^5 A, what mass (in grams) of aluminum metal can be produced in 8.0 hr?
55. The vanadium(II) ion can be produced by electrolysis of a vanadium(III) salt in solution. How long must you carry out an electrolysis if you wish to convert completely 0.125 L of 0.015 M $V^{3+}(aq)$ to $V^{2+}(aq)$ using a current of 0.268 A?
56. The reactions occurring in a lead-storage battery are given in Section 18.7. A typical battery might be rated at 50 ampere-hours. This means that it has the capacity to deliver 50. amperes for 1.0 hr or 1.0 ampere for 50. hr. If it does deliver 1.0 ampere for 50. hr, what mass of lead would be consumed to accomplish this?
57. It has been demonstrated that an effective battery can be built using the reaction between Al metal and O_2 from the air. If the Al anode of this battery consists of a 3-ounce piece of aluminum (84 g), for how many hr can the battery produce 1.0 A of electricity?
58. A dry cell is used to supply a current of 250 mA for 20 min. What mass of Zn is consumed?
59. If the same current as in the previous question were supplied by a mercury battery, what mass of Hg would be produced at the cathode?
60. Assuming that the anode reaction for the lithium battery is

$$Li(s) \longrightarrow Li^+(aq) + e^-$$

and the anode reaction for the lead-storage battery is

$$Pb(s) + HSO_4^-(aq) + H_2O(\ell) \longrightarrow PbSO_4(s) + 2\ e^- + H_3O^+(aq)$$

compare the masses of metals consumed when each of these batteries supplies a current of 1 A for 10 min.
61. A hydrogen-oxygen fuel cell operates on the simple reaction

$$2\ H_2(g) + O_2(g) \longrightarrow 2\ H_2O(\ell)$$

If the cell is designed to produce 1.5 A of current, how long can it operate if there is an excess of oxygen and only sufficient hydrogen to fill a 1.0-L tank at 200. atm pressure at 25 °C?
62. Fluorine, F_2, is made by the electrolysis of anhydrous HF.

$$2\ HF(\ell) \longrightarrow H_2(g) + F_2(g)$$

Typical electrolysis cells operate at 4000 to 6000 A and 8 to 12 V. A large-scale plant can produce about 9 tons of F_2 gas per day.
(a) What mass in grams of HF is consumed?
(b) Using the conversion factor of 3.60×10^6 J/kwh, how much energy in kilowatt-hours is consumed by a cell operating at 6.0×10^3 A at 12 V for 24 hr?

General Questions

63. A 12-V automobile battery consists of six cells of the type described in Section 18.7. The cells are connected in series so that the same current flows through all of them. Calculate the theoretical minimum electrical potential difference needed to recharge an automobile battery. (Assume standard-state concentrations.) How does this compare with the maximum voltage that could be delivered by the battery? Assuming that the lead plates in an automobile battery each weigh 10.0 kg, and that there is sufficient PbO_2 available, what is the maximum possible work that could be obtained from the battery?
64. Three electrolytic cells are connected in series, so that the same current flows through all of them for 20. min. In cell A, 0.0234 g Ag plates out from a solution of $AgNO_3(aq)$; cell B contains $Cu(NO_3)_2(aq)$; and cell C contains $Al(NO_3)_3(aq)$. What mass of Cu will plate out in cell B? What mass of Al will plate out in cell C?
65. Fluorinated organic compounds are important commercially, as they are used as herbicides, flame retardants, and fire-extinguishing agents, among other things. A reaction such as

$$CH_3SO_2F + 3\ HF \longrightarrow CF_3SO_2F + 3\ H_2$$

is actually carried out electrochemically in liquid HF as the solvent.

(a) Draw a dot structure for CH_3SO_2F. (S is the "central" atom with the O atom, F atom, and CH_3 group bonded to it.) What is the geometry around the S atom? What are the O—S—O and O—S—F bond angles?
(b) If you electrolyze 150 g of CH_3SO_2F, how many grams of HF are required and how many grams of each product can be isolated?
(c) Is H_2 produced at the anode or the cathode of the electrolysis cell?
(d) A typical electrolysis cell operates at 8.0 V and a low current, such as 250 amps. How many kilowatt-hours of energy does one such cell consume in 24 hr?

Applying Concepts

66. Four metals, A, B, C, and D, exhibit the following properties:
(a) Only A and C react with 1.0 M HCl to give H_2 gas.
(b) When C is added to solutions of ions of the other metals, metallic A, B, and D are formed.
(c) Metal D reduces B^{n+} ions to give metallic B and D^{n+} ions.
On the basis of this information, arrange the four metals in order of increasing ability to act as reducing agents.

67. The matrix below lists the cell potentials for the ten possible electrochemical cells assembled from the elements A, B, C, D, and E, and their respective ions in solutions. Using the data in the matrix, establish a standard reduction potential table similar to Table 18.1. Assign a reduction potential of 0.00 V to the element that falls in the middle of the series.

68. When the electrochemical cell shown at the bottom of the page runs for several hours, the green solution gets lighter and the yellow solution gets darker.
(a) What is oxidized, what is reduced?
(b) What is the oxidizing agent, what is the reducing agent?
(c) What is the anode, what is the cathode?
(d) Write equations for the half-reactions.
(e) Which metal gains mass?
(f) What is the direction of the electron transfer through the external wire?
(g) If the salt bridge contains KNO_3(aq), into which solution will the K^+ ions migrate?

69. An electrolytic cell is set up with Cd(s) in $Cd(NO_3)_2$(aq) and Zn(s) in $Zn(NO_3)_2$(aq). Initially both electrodes weigh 5.00 g. After running the cell for several hours the electrode in the left compartment weighs 4.75 g.
(a) Which electrode is in the left compartment?
(b) Does the mass of the electrode in the right compartment increase, decrease, or stay the same? If the mass changes, what is the new mass?
(c) Does the mass of the solution in the right compartment increase, decrease, or stay the same?
(d) Does the volume of the electrode in the right compartment increase, decrease, or stay the same? If the volume changes, what is the new volume?

	A(s) in A^{n+}(aq)	B(s) in B^{n+}(aq)	C(s) in C^{n+}(aq)	D(s) in D^{n+}(aq)
E(s) in E^{n+}(aq)	+0.21 V	+0.68 V	+0.37 V	+0.56 V
D(s) in D^{n+}(aq)	+0.35 V	+1.24 V	+0.93 V	—
C(s) in C^{n+}(aq)	+0.58 V	+0.31 V	—	—
B(s) in B^{n+}(aq)	+0.89 V	—	—	—

CHAPTER

19 Nuclear Chemistry

Unloading nuclear bombs for disassembly at the U.S. Department of Energy facility near Amarillo, Texas. Such military nuclear waste contains radioactive plutonium-239. • What type of radiation does it emit, and how long does the waste remain radioactive? The Summary Problem at the end of the chapter explores the synthesis and nature of plutonium-239. *(U.S. Department of Energy)*

The terms "nuclear" and "radioactivity" are arguably two of the most emotion-laden words in the English language, associated by many people only with catastrophic consequences, such as the bombing of Hiroshima and Nagasaki to end World War II. It is, however, important to realize that radioactivity is everywhere, a naturally occurring phenomenon arising from changes in unstable nuclei. When used properly, radioactivity and other nuclear phenomena have many beneficial applications—for the generation of electric power, for the diagnosis and treatment of disease, and for food preservation. A National Research Council report states that "The future vigor and prosperity of American medicine, science, technology, and national defense clearly depend on continued use and development of nuclear techniques and use of radioactive isotopes."

Nuclear chemistry, a subject that bridges chemistry and physics, has a significant impact on our society. Radioactive isotopes are now widely used in medicine; some of the latest diagnostic techniques such as PET scans depend on radioactivity. Your home may be protected with a smoke detector that contains a radioactive element, and research in all fields of science uses radioactive elements and their compounds. The national security of the United States since World War II has depended on nuclear weapons, and more than 30 nations depend on nuclear reactors as a source of electricity. No matter what your reason for taking a college course in chemistry—to prepare for a career in one of the sciences or simply to gain knowledge as a concerned citizen—you should know something about nuclear chemistry. Therefore, this chapter considers changes in the atomic nucleus and their effects, the fissioning and fusion of nuclei and the energy that can be derived from such changes, the units used to measure radioactivity, and the uses of radioactive isotopes.

On August 2, 1939, as the world was on the brink of World War II, Albert Einstein sent a letter to President Franklin D. Roosevelt. In this letter, which profoundly changed the course of history, Einstein called attention to work being done on the physics of the atomic nucleus. He said he and others believed this work suggested the possibility that "uranium may be turned into a new and important source of energy . . . and [that it was] conceivable . . . that extremely powerful bombs of a new type may thus be constructed. . . . "

Powerful indeed! Einstein's letter was the beginning of the Manhattan Project, the project that led to the detonation of the first atomic bomb at 5:30 AM on July 16, 1945, in the New Mexican desert. The rest of the world would learn the truth of the power locked in the atomic nucleus a few weeks later, on August 6 and August 9, when the United States used atomic nuclear weapons against Japan. J. Robert Oppenheimer, the director of the atomic bomb project, is said to have recalled the following words from the sacred Hindu epic, *The Bhagavad-Gita,* at the moment of the explosion of the first atomic bomb.

> "If the radiance of a thousand suns
> Were to burst at once upon the sky,
> That would be like the splendor of the Mighty One . . .
> I am become Death,
> The shatterer of worlds."

The Making of the Atomic Bomb by Richard Rhodes (Simon and Schuster, 1986) is a comprehensive history of the exploration of atomic physics in this century and of the events leading up to the development of atomic weapons. This very readable book is highly recommended.

Since 1945 when nuclear weapons were used for the first—and thankfully the only—time in war, more powerful nuclear weapons have been developed and stockpiled by a number of nations. With the end of the Cold War, fears of a nuclear holocaust are fading, and recent nuclear disarmament treaties have been signed between the United States and the former Soviet Union for removing the plutonium-239 and other nuclear fuel from nuclear warheads (see Chemistry in the News, Section 19.6).

But the fears are replaced to some extent by the concern that a great many nations have developed nuclear weapons. For many years, the respected magazine *The Bulletin of Atomic Scientists* has used the symbol of a clock with its hands near the fateful midnight (representing nuclear annihilation) to illustrate the danger faced by the world from atomic weapons. Even with the end of the Cold War, the hands have moved back only a little from midnight.

19.1 THE NATURE OF RADIOACTIVITY

Many minerals, called phosphors, glow for some time after being stimulated by exposure to sunlight or ultraviolet light. (You may have some on the hands of a wristwatch or on a "black light" poster.) In 1896, French physicist Antoine Henri Becquerel was studying this phenomenon, called *phosphorescence,* when he accidentally discovered radioactivity. Paris was very cloudy in February that year, limiting Becquerel's opportunities to expose minerals to sunlight to check whether they phosphoresced. While waiting for a sunny day, Becquerel stored a photographic plate wrapped in black paper along with a uranium salt, a material known to phosphoresce, in a dark drawer. Much to his amazement, the image of the uranium salt appeared on the plate that had been in the drawer, unexposed to sunlight. Becquerel realized that he had observed penetrating radiation from matter that had not been stimulated by light.

Becquerel performed many more experiments and found that pure uranium metal gave the same emissions as uranium salts did, but even more strongly. This would be expected if the radiation was the property of the metal and not dependent on its form of chemical combination. But no pure metal had been observed to phosphoresce before. Becquerel was completely baffled. Where did all of this energy come from? The radiation had nothing to do with phosphorescence. Becquerel gave up his work for several years, but it was taken up by Marie Curie and her husband Pierre, who later named the phenomenon radioactivity.

For more on the experiments done by Becquerel and the Curies, see H.F. Walton, *Journal of Chemical Education,* Vol. 69, p. 10, 1992.

One of Marie Curie's first findings was to confirm Becquerel's observation that uranium metal itself was radioactive, and that the degree to which a uranium-bearing sample was radioactive depended on the percentage of uranium present. Thus, she was astonished when she tested pitchblende, a common ore containing uranium and other metals (such as lead, bismuth, and copper), and found that it was even more radioactive than pure uranium. There was only one explanation: pitchblende contained an element (or elements) more radioactive than uranium. Eventually, the Curies discovered the next element after bismuth in the periodic table, which they named *polonium* after Marie's homeland of Poland. They also isolated another new, highly radioactive element, *radium.*

In England at about the same time, Sir J.J. Thomson and his student Ernest Rutherford were studying the radiation from uranium and thorium (***← p. 43, Portrait of a Scientist***). Rutherford found that "There are present at least two distinct types of radiation—one that is readily absorbed, which will be termed for convenience α (alpha) radiation, and the other of a more penetrative character, which will be termed β (beta) radiation." **Alpha radiation,** he discovered, was composed of particles that, when passed through an electric field, were attracted to the negative side of the field (***← p. 38, Figure 2.8***); indeed, his later studies showed these **alpha (α) particles** to be helium nuclei, ${}^{4}_{2}He^{2+}$, which were ejected at high speeds from a radioactive element (Table 19.1). As might be expected for such massive particles, they have limited penetrating power and can be stopped by skin, clothing, or several sheets of ordinary paper (Figure 19.1).

Encouraged by the Curies, Becquerel returned to the study of radiation. He found that the radiation from uranium was affected by magnetic fields and consisted of two kinds of particles, which we know to be α and β particles.

TABLE 19.1 Characteristics of α, β, and γ Emissions

Name	Symbol	Charge	Mass (g/particle)
Alpha	${}^{4}_{2}He^{2+}$, ${}^{4}_{2}\alpha$	+2	6.65×10^{-24}
Beta	${}^{0}_{-1}e$, ${}^{0}_{-1}\beta$	−1	9.11×10^{-28}
Gamma	${}^{0}_{0}\gamma$, γ	0	0

In the same experiment, Rutherford also found that **beta (β)** radiation must be composed of negatively charged particles, since the beam of radiation was attracted to the electrically positive plate. Later work by Becquerel showed that these particles have an electric charge and mass equal to those of an electron. Thus, **beta particles** are electrons ejected at high speeds from some radioactive nuclei. They are more penetrating than alpha particles, since at least a $\frac{1}{8}$-inch piece of aluminum is necessary to stop beta particles, and they will penetrate about 1 cm of living bone or tissue.

Rutherford hedged his bets when he said there were *at least* two types of radiation. Indeed, a third type was later discovered by P. Villard, a Frenchman, who named it **gamma (γ) radiation,** using the third letter in the Greek alphabet in keeping with Rutherford's scheme. Unlike alpha and beta rays which are particulate in nature, gamma rays are a form of electromagnetic radiation like x radiation and not affected by an electric field (← *p. 38, Figure 2.8*). Gamma radiation is the most penetrating, since it can pass completely through the human body. Thick layers of lead or concrete are required to stop gamma rays.

19.2 NUCLEAR REACTIONS

Equations for Nuclear Reactions

Ernest Rutherford found that radium not only emits alpha particles but that it also produces the radioactive gas radon in the process. Such observations led Rutherford and Frederick Soddy, in 1902, to propose the revolutionary theory that *radioactivity is the result of a natural change of the isotope of one element into the isotope of a* different *element.* In such changes, called **nuclear reactions** or *transmutations,* an unstable nucleus (the *parent nucleus*) spontaneously emits radiation and is converted into a more stable nucleus of a different element (the *daughter product*). Thus, a nu-

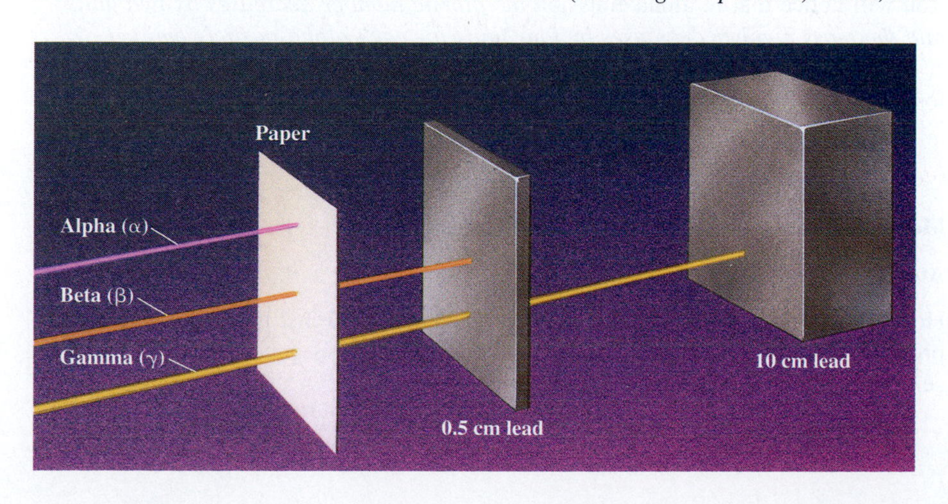

Figure 19.1 The relative penetrating abilities of the three major types of nuclear radiation. Heavy, highly charged alpha particles (${}^{4}_{2}He^{2+}$) interact with matter most strongly and so are stopped by a piece of paper or a layer of skin. Beta particles (electrons) are lighter and have a smaller charge, so they interact less strongly with matter; they are stopped by about 0.5 cm of lead. Gamma rays are uncharged, massless photons and are the most penetrating.

Be sure to notice that, when a radioactive atom decays, the emission of a charged particle leaves a charged atom. Thus, when radium-226 decays it gives a helium-4 cation (He^{2+}) and a radon-222 anion (Rn^{2-}). By convention, the ion charges are not shown in balanced equations for nuclear reactions.

Recall that atomic number is the number of protons in an atom's nucleus; mass number is the number of protons plus neutrons in a nucleus.

clear reaction results in a change in atomic number and often a change in mass number as well. For example, the reaction studied by Rutherford can be written as

$$^{226}_{88}\text{Ra} \longrightarrow {}^{4}_{2}\text{He} + {}^{222}_{86}\text{Rn}$$

In this balanced equation the subscripts are the atomic numbers and the superscripts are the mass numbers.

In a chemical change, the atoms in molecules and ions are rearranged, but they are not created or destroyed; the number of atoms remains the same. Similarly, in nuclear reactions the total number of nuclear particles, or **nucleons** (protons and neutrons), remains the same. The essence of nuclear reactions, however, is that one nucleon can change into a different nucleon. A proton can change to a neutron or a neutron can change to a proton, but the total number of nucleons remains the same. Therefore, *the sum of the mass numbers of reacting nuclei must equal the sum of the mass numbers of the nuclei produced.* Furthermore, to maintain charge balance, *the sum of the atomic numbers of the products must equal the sum of the atomic numbers of the reactants.* These principles may be verified for the preceding nuclear equation.

	$^{226}_{88}\text{Ra}$	$\longrightarrow$	$^{4}_{2}\text{He}$	+	$^{222}_{86}\text{Rn}$
	Radium-226		alpha particle		radon-222
Mass number: (protons + neutrons)	226	$\longrightarrow$	4	+	222
Atomic number: (protons)	88	$\longrightarrow$	2	+	86

Alpha and Beta Particle Emission

One way a radioactive isotope can disintegrate or decay is to eject an alpha particle from the nucleus. This is illustrated by the conversion of uranium to thorium by the following reaction.

	$^{234}_{92}\text{U}$	$\longrightarrow$	$^{4}_{2}\text{He}$	+	$^{230}_{90}\text{Th}$
	uranium-234		alpha particle		thorium-230
Mass number:	234	$\longrightarrow$	4	+	230
Atomic number:	92	$\longrightarrow$	2	+	90

You will notice that in alpha emission *the atomic number decreases by two units and the mass number decreases by four units for each alpha particle emitted.*

Emission of a beta particle is another way for an isotope to decay. For example, loss of a beta particle by uranium-239 is represented by

	$^{239}_{92}\text{U}$	$\longrightarrow$	$^{0}_{-1}\text{e}$	+	$^{239}_{93}\text{Np}$
	uranium-239		beta particle		neptunium-239
Mass number:	239	$\longrightarrow$	0	+	239
Atomic number:	92	$\longrightarrow$	−1	+	93

How does a nucleus, composed only of protons and neutrons, increase its number of protons by ejecting an electron during beta emission? It is generally accepted that a series of reactions is involved, but the net process is

$^{1}_{0}\text{n}$	$\longrightarrow$	$^{0}_{-1}\text{e}$	+	$^{1}_{1}\text{p}$
neutron		electron		proton

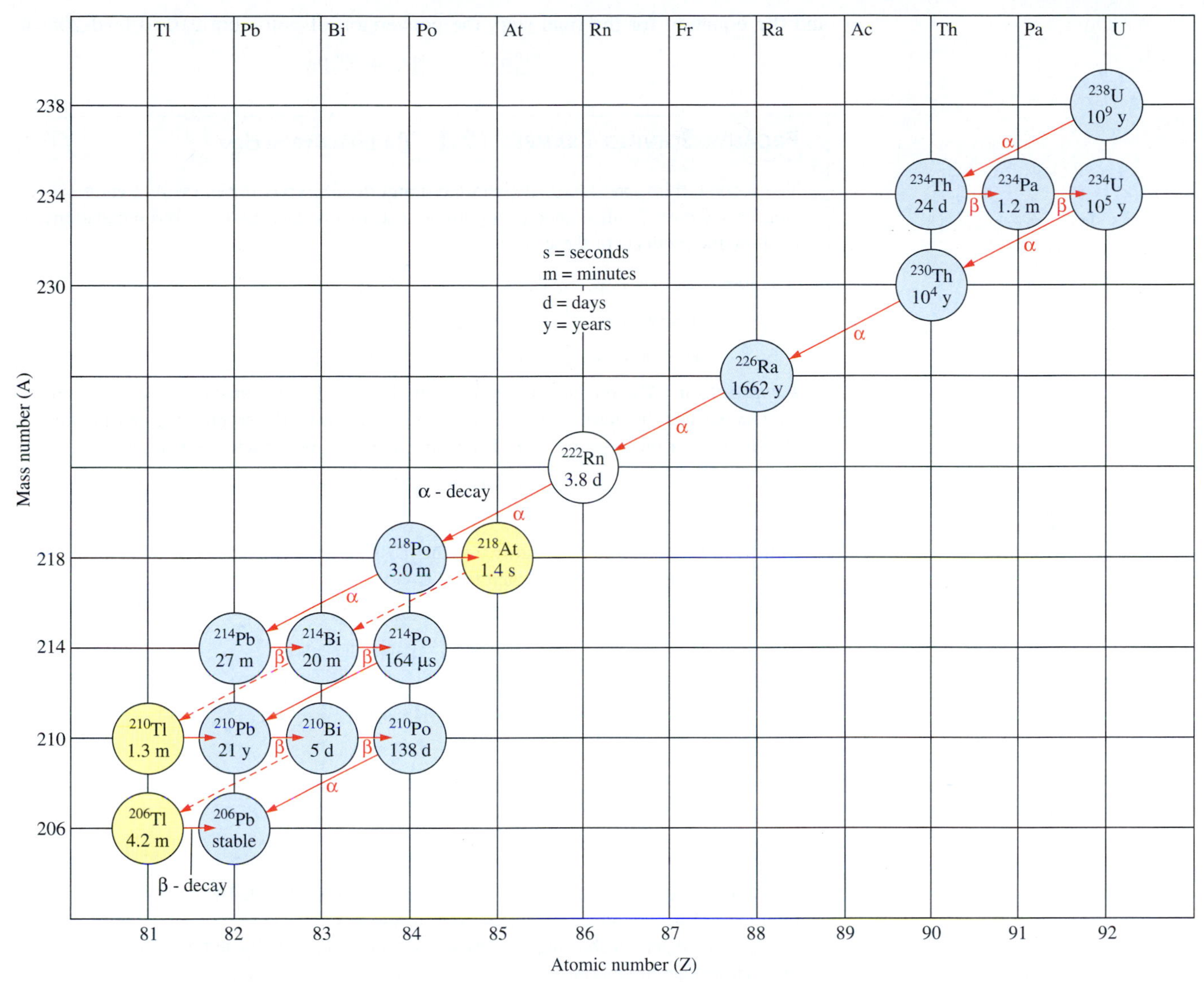

Figure 19.2 **A radioactive series beginning with uranium-238 and ending with lead-206.** In the first step, for example, ^{238}U emits an α particle to give thorium-234. This radioactive isotope then emits a β particle to give protactinium-234. The $^{234}_{91}Pa$ then emits another β particle to continue the series, which finally ends at $^{206}_{82}Pb$. The half-life is given for each isotope (see Section 19.4). The time units are s = seconds, m = minutes, d = days, and y = years.

where we use the symbol p for a proton. *The ejection of a beta particle always means that a new element is formed with an atomic number one unit greater than the decaying nucleus.* The mass number does not change, however, because no proton or neutron has been emitted.

In many cases, the emission of an alpha or beta particle results in the formation of an isotope that is also unstable and therefore radioactive. The new radioactive isotope may therefore undergo a number of successive transformations until a stable, nonradioactive isotope is finally produced. Such a series of reactions is called a **radioactive series.** One such series begins with uranium-238 and ends with lead-206, as illustrated in Figure 19.2. The first step in the series is

$$^{238}_{92}U \longrightarrow {}^{4}_{2}He + {}^{234}_{90}Th$$

A nucleus formed as a result of an alpha or beta emission is often in an excited state and so also emits a γ ray.

and the equation for the final step, the conversion of polonium-210 to lead-206, is

$$^{210}_{84}Po \longrightarrow ^{4}_{2}He + ^{206}_{82}Pb$$

PROBLEM-SOLVING EXAMPLE 19.1 Radioactive Series

The second, third, and fourth steps in the uranium-238 series in Figure 19.2 are emission of a β particle, then another β particle, and finally an α particle. Write equations to show the products of these steps.

Answer **Second step:** $^{234}_{90}Th \longrightarrow ^{0}_{-1}e + ^{234}_{91}Pa$;

Third step: $^{234}_{91}Pa \longrightarrow ^{0}_{-1}e + ^{234}_{92}U$

Fourth step: $^{234}_{92}U \longrightarrow ^{4}_{2}He + ^{230}_{90}Th$

Explanation The product of the first step, thorium-234, is our starting point. Figure 19.2 shows that the mass number remains the same during the second step, but that the atomic number increases by 1 to 91, a result shown by the balanced equation

$$^{234}_{90}Th \longrightarrow ^{0}_{-1}e + ^{234}_{91}Pa$$

thorium-234 protactinium-234

In the third step, Figure 19.2 shows that the mass number again stays constant, and the atomic number increases once again by 1.

$$^{234}_{91}Pa \longrightarrow ^{0}_{-1}e + ^{234}_{92}U$$

protactinium-234 uranium-234

Finally, the fourth step is alpha-particle emission, so both the mass number and atomic number decline. This is again confirmed in Figure 19.2.

$$^{234}_{92}U \longrightarrow ^{4}_{2}He + ^{230}_{90}Th$$

uranium-234 thorium-230

Problem-Solving Practice 19.1

(a) Write an equation showing the emission of an alpha particle by an isotope of neptunium, $^{237}_{93}Np$, to produce an isotope of protactinium.
(b) Write an equation showing the emission of a beta particle by sulfur-35, $^{35}_{16}S$, to produce an isotope of chlorine.

Exercise 19.1 Radioactive Series

The actinium series begins with uranium-235, $^{235}_{92}U$, and ends with lead-207, $^{207}_{82}Pb$. The first five steps involve the emission of α, β, α, α, and β particles, respectively. Identify the radioactive isotope produced in each of the steps, beginning with uranium-235.

Other Types of Radioactive Decay

In addition to radioactive decay by emission of alpha, beta, or gamma radiation, other decay processes are observed. Some nuclei decay, for example, by emission of a **positron,** $^{0}_{+1}e$ or β^+, which is effectively a positively charged electron. Positron emission by polonium-207 leads to the formation of bismuth-207, for example.

	$^{207}_{84}Po$	$\longrightarrow$	$^{0}_{+1}e$	+	$^{207}_{83}Bi$
	polonium-207		positron		bismuth-207
Mass number:	207	$\longrightarrow$	0	+	207
Atomic number:	84	$\longrightarrow$	+1	+	83

Notice that this is the opposite of beta decay, because positron decay leads to a *decrease* in the atomic number.

The atomic number is also reduced by one when **electron capture** occurs. In this process an inner-shell electron is captured by the nucleus.

	$^{7}_{4}Be$	+	$^{0}_{-1}e$	$\longrightarrow$	$^{7}_{3}Li$
	beryllium-7		electron		lithium-7
Mass number:	7	+	0	$\longrightarrow$	7
Atomic number:	4	+	−1	$\longrightarrow$	3

In the old nomenclature of atomic physics, the innermost shell ($n = 1$ principal quantum number) was called the K-shell, so the electron capture decay mechanism is sometimes called *K-capture.*

In summary, there are four common ways that a radioactive nucleus can decay, as summarized in the following figure. In nuclear chemistry, the radioactive isotope that begins a process is called the "parent" and the product is called a "daughter" isotope.

The positron was discovered by Carl Anderson in 1932. It is sometimes called an "antielectron," one of a group of particles that have become known as "antimatter." Contact between an electron and a positron leads to mutual annihilation of both particles with production of two high-energy photons (gamma rays).

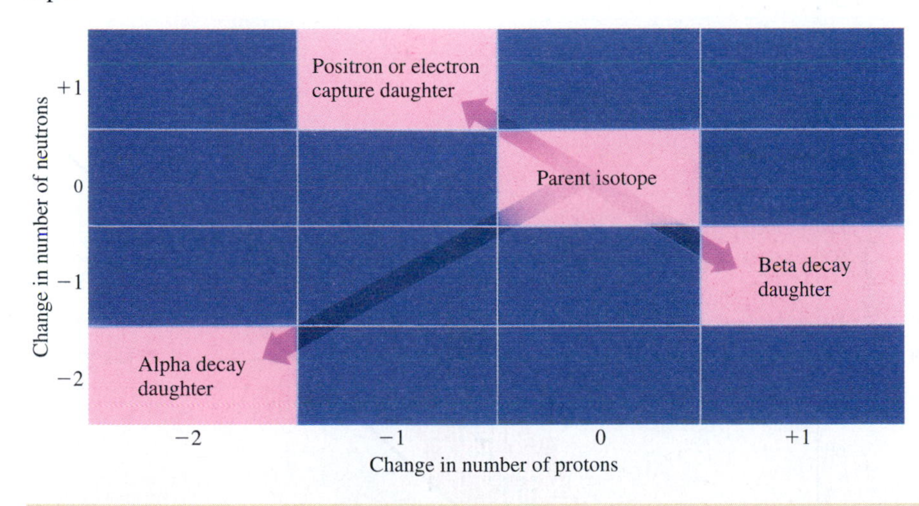

A memory aid for predicting the product of alpha, beta, or positron emission and electron capture.

Exercise 19.2 Nuclear Reactions

Balance each of the following nuclear reactions by filling in the missing symbol, mass number, and atomic number.

(a) $^{13}_{7}N \rightarrow ^{13}_{6}C + ?$

(b) $^{41}_{20}Ca + ^{0}_{-1}e \rightarrow ?$

(c) $^{90}_{38}Sr \rightarrow ^{90}_{39}Y + ?$

(d) $^{11}_{6}C \rightarrow ^{11}_{5}B + ?$

(e) $^{43}_{21}Sc \rightarrow ? + ^{0}_{+1}e$

Exercise 19.3 Nuclear Reactions

Aluminum-26 can undergo either positron emission or electron capture. Write the balanced nuclear equation for each case.

19.3 STABILITY OF ATOMIC NUCLEI

The fact that some nuclei are unstable (radioactive) while others are stable (nonradioactive) leads us to consider the reasons for stability. Figure 19.3 shows the naturally occurring isotopes of the elements from hydrogen to bismuth. It is quite astonishing that there are so few. Why are there not hundreds more?

In its simplest and most abundant form, hydrogen has only one nuclear particle, the proton. In addition, the element has two other well known isotopes: nonradioactive deuterium, with one proton and one neutron, ${}^{2}_{1}H = D$; and radioactive tritium, with one proton and two neutrons, ${}^{3}_{1}H = T$. Helium, the next element, has two protons and two neutrons in its most stable isotope. At the end of the actinide series is element 103, lawrencium, one isotope of which has a mass number of 257 and 154 neutrons. From hydrogen to lawrencium, except for ${}^{1}_{1}H$ and ${}^{3}_{2}He$, *the mass numbers of stable isotopes are always at least twice as large as the atomic number.* In other words, except for ${}^{1}_{1}He$ and ${}^{3}_{2}He$, every isotope of every element has a nucleus con-

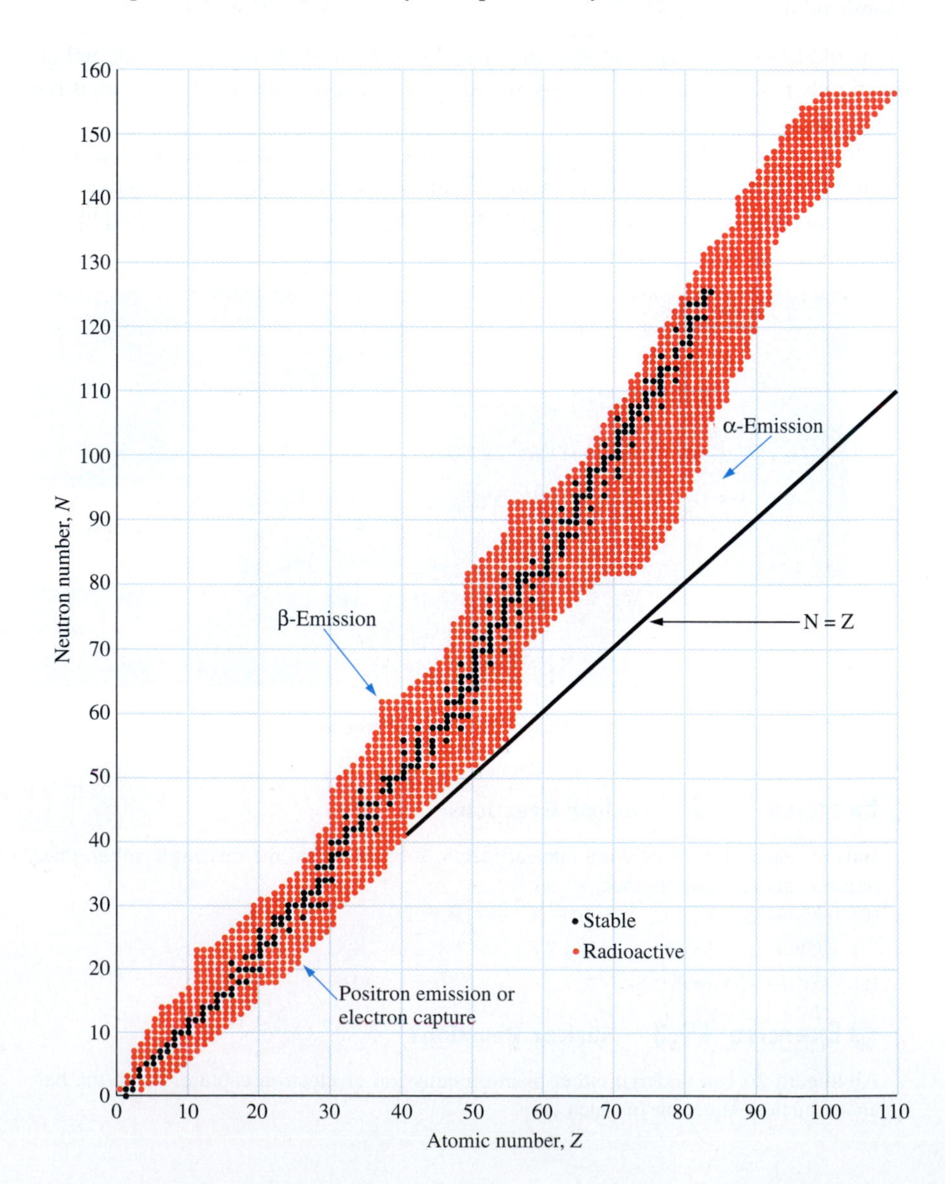

Figure 19.3 A plot of the number of neutrons (N) versus the number of protons (Z) for stable and radioactive isotopes from hydrogen ($Z = 1$) through bismuth ($Z = 83$). The effects of α, β, and positron emission or electron capture are indicated by arrows. For example, radioactive isotopes (indicated by red dots) that lie above the band of stable isotopes (indicated by black dots) decay by β emission. The arrow indicates that this raises the value of Z (by one unit per beta particle) and lowers the value of N by one unit. *(Redrawn from Oxtoby, Nachtrieb, and Freeman:* Chemistry: Science of Change, *2nd ed, Fig. 16–7, p. 613. Philadelphia, Saunders College Publishing, 1994.)*

taining *at least* one neutron for every proton. Apparently the tremendous *repulsive* forces between the positively charged protons in the nucleus are moderated by the presence of neutrons with no electrical charge.

1. For light elements up to Ca ($Z = 20$), the stable isotopes usually have equal numbers of protons and neutrons, or perhaps one more neutron than protons. Examples include $^{7}_{3}Li$, $^{12}_{6}C$, $^{16}_{8}O$, and $^{32}_{16}S$.
2. Beyond calcium the neutron/proton ratio becomes increasingly greater than 1. The band of stable isotopes deviates more and more from the line $N = Z$ (number of neutrons = number of protons). It is evident that more neutrons are needed for nuclear stability in the heavier elements. For example, whereas one stable isotope of Fe has 26 protons and 30 neutrons, one of the stable isotopes of platinum has 78 protons and 117 neutrons.
3. Above bismuth (83 protons and 126 neutrons), all isotopes are unstable and radioactive. Beyond this point there is apparently no nuclear force strong enough to hold heavy nuclei together. Furthermore, the rate of disintegration becomes greater the heavier the nucleus. For example, half of a sample of $^{238}_{92}U$ disintegrates in 4.5 billion years, whereas half of a sample of $^{256}_{103}Lr$ is gone in only 28 seconds.
4. A very careful look at Figure 19.3 shows even more interesting features. First, elements of even atomic number have more stable isotopes than do those of odd atomic number. Second, stable isotopes generally have an even number of neutrons. For elements of odd atomic number, the most stable isotope has an even number of neutrons. To emphasize these points, of the nearly 300 stable isotopes represented in Figure 19.3, roughly 200 have an even number of neutrons *and* an even number of protons. Only about 120 have an odd number of either protons or neutrons. Only four isotopes ($^{2}_{1}H$, $^{6}_{3}Li$, $^{10}_{5}B$, and $^{14}_{7}N$) have odd numbers of *both* protons and neutrons.

As illustrated by Figure 19.3, there are very few stable combinations of protons and neutrons. Examining those combinations can give us some insight into what factors affect nuclear stability.

The Band of Stability and Type of Radioactive Decay

The narrow "band" of stable isotopes in Figure 19.3 (the black dots) is sometimes called the *peninsula of stability* in a "sea of instability." Any isotope (the red dots) not on this peninsula will decay in such a way that it can come ashore on the peninsula, and the chart can help us predict what type of decay will be observed.

All elements beyond Bi ($Z = 83$) are unstable—that is, radioactive—and most decay by ejecting an alpha particle. For example, americium, the radioactive element used in smoke alarms, decays in this manner.

$$^{243}_{95}Am \longrightarrow {}^{4}_{2}He + {}^{239}_{93}Np$$

Beta emission occurs in isotopes that have too many neutrons to be stable, that is, isotopes above the peninsula of stability in Figure 19.3. When beta decay converts a neutron to a proton and an electron, which is then ejected, the mass number remains constant, but the number of neutrons drops.

$$^{60}_{27}Co \longrightarrow {}^{0}_{-1}e + {}^{60}_{28}Ni$$

Conversely, lighter isotopes that have too few neutrons—isotopes below the peninsula of stability—attain stability by positron emission or by electron capture, because these convert a proton to a neutron in one step.

$$^{13}_{7}N \longrightarrow {}^{0}_{+1}e + {}^{13}_{6}C$$

$$^{41}_{20}Ca + {}^{0}_{-1}e \longrightarrow {}^{41}_{19}K$$

Decay by these routes is observed for elements with atomic numbers from 4 to >100; as Z increases, electron capture becomes more likely than positron emission.

Exercise 19.4 Nuclear Stability

For each of the following unstable isotopes, write an equation for its probable mode of decay.

(a) Silicon-32, $^{32}_{14}Si$ (b) Titanium-43, $^{43}_{22}Ti$ (c) Plutonium-239, $^{239}_{94}Pu$

Binding Energy

As proved by Ernest Rutherford's experiments (← ***p. 42***), the nucleus of the atom is extremely small. Yet the nucleus can contain up to 83 protons before becoming unstable. This is evidence that there must be a very strong short-range binding force that can overcome the electrostatic repulsive force of a number of protons packed into such a tiny volume. A measure of the force holding the nucleus together is the nuclear **binding energy.** This energy (E_b) is defined as the negative of the energy change (ΔE) that would occur if a nucleus were formed directly from its component protons and neutrons. For example, if a mole of protons and a mole of neutrons directly formed a mole of deuterium nuclei, the energy change would be more than 200 million kilojoules, the equivalent of exploding 73 tons of TNT.

$$^{1}_{1}H + ^{1}_{0}n \longrightarrow ^{2}_{1}H \qquad \Delta E = -2.15 \times 10^{8} \text{ kJ}$$

$$\text{Binding energy} = -\Delta E = E_b = +2.15 \times 10^{8} \text{ kJ}$$

This nuclear synthesis reaction is highly exothermic (and so E_b is very positive), an indication of the strong attractive forces holding the nucleus together. The deuterium nucleus is more stable than an isolated proton and an isolated neutron, just as the H_2 molecule is more stable than two isolated H atoms. Recall that the energy released when a mole of H—H covalent bonds form is only 436 kJ, a tiny fraction of the energies released when protons and neutrons coalesce to form a nucleus.

To understand the enormous energy released during the formation of atomic nuclei, we turn to an experimental observation and a theory. The experimental observation is that the mass of a nucleus is always less than the sum of the masses of its constituent protons and neutrons.

$$\underset{1.007825 \text{ g/mol}}{^{1}_{1}H} + \underset{1.008665 \text{ g/mol}}{^{1}_{0}n} \longrightarrow \underset{2.01410 \text{ g/mol}}{^{2}_{1}H}$$

$$\begin{aligned}\text{Change in mass} = \Delta m &= \text{mass of product} - \text{sum of masses of reactants}\\ &= 2.01410 \text{ g/mol} - 2.016490 \text{ g/mol}\\ &= -0.00239 \text{ g/mol}\end{aligned}$$

The theory is that the "missing mass," Δm, has been converted to energy, and it is this energy that we described as the binding energy.

The relation between mass and energy is contained in Albert Einstein's 1905 theory of special relativity, which holds that mass and energy are simply different manifestations of the same quantity. Einstein stated that the energy of a body is equivalent to its mass times the square of the speed of light, $E = mc^2$. So, to calculate the energy change in a process where the mass has changed, the equation becomes

$$\Delta E = (\Delta m)c^2$$

We can calculate ΔE in joules if the change in mass is given in kilograms and the velocity of light is in meters per second (because $1\ J = 1\ kg{\cdot}m^2/s^2$). For the formation of 1 mol of deuterium nuclei from 1 mol of protons and 1 mol of neutrons, we have

$$\begin{aligned}\Delta E &= (-2.39 \times 10^{-6}\ kg)(3.00 \times 10^{8}\ m/s)^2 \\ &= -2.15 \times 10^{11}\ J \\ &= -2.15 \times 10^{8}\ kJ\end{aligned}$$

This is the value of ΔE given at the beginning of this section for the change in energy when a mole of protons and a mole of neutrons form a mole of deuterium nuclei.

A helium-4 nucleus is composed of two protons and two neutrons, and its binding energy, E_b, is very large, even larger than for deuterium.

A helium-4 nucleus contains 4 nucleons—2 protons and 2 neutrons.

$$2\ {}^{1}_{1}H + 2\ {}^{1}_{0}n \longrightarrow {}^{4}_{2}He \qquad E_b = +2.73 \times 10^{9}\ kJ/mol\ of\ helium\ nuclei$$

To compare nuclear stabilities more directly, however, nuclear scientists generally calculate the **binding energy per nucleon.** For 1 mol of helium-4 atoms this is

$$E_b\ per\ mol\ nucleons = \frac{2.73 \times 10^{9}\ kJ}{4\ mol\ nucleons} = 6.83 \times 10^{8}\ kJ/mol\ nucleons$$

The greater the binding energy per nucleon, the greater is the stability of the nucleus. Scientists have calculated the binding energies per nucleon of a great number of nuclei and have plotted them as a function of mass number (Figure 19.4). It is very interesting—and important—that the point of maximum stability occurs in the vicinity of iron-56, ${}^{56}_{26}Fe$. This means that *all elements are thermodynamically unstable with respect to iron.* That is, very heavy nuclei may split or *fission,* to give more stable nuclei with atomic numbers nearer iron, and simultaneously release enormous quantities of energy (Section 19.6). In contrast, two very light nuclei may come together and undergo *fusion* exothermically to form heavier nuclei (Section 19.7). Finally, *this is the reason that iron is the most abundant of the heavier elements in the universe.*

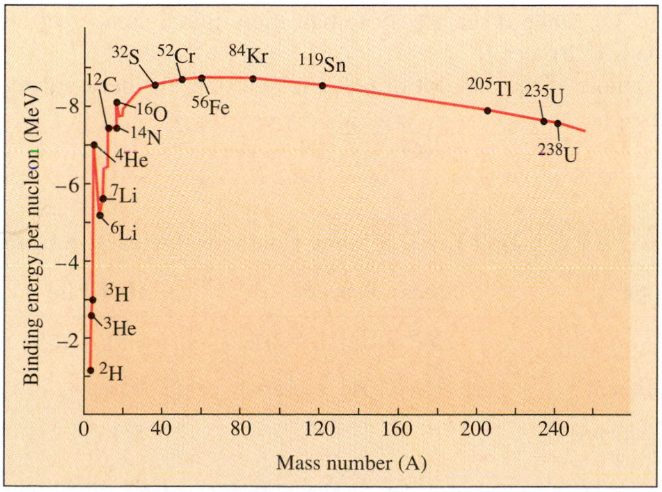

Figure 19.4 Binding energy per nucleon. This "curve of binding energy" was derived by calculating the binding energy per nucleon in million electron volts (MeV) for the most abundant isotope of each element from hydrogen to uranium ($1\ MeV = 1.602 \times 10^{-13}\ J$). The nuclei at the top of the curve are the most stable. (To convert to binding energy in joules/mole of nucleons, multiply the values on the y-axis by 1.602×10^{-13} J/MeV and by 6.022×10^{23}/mol.)

Exercise 19.5 Binding Energy

Calculate the binding energy, in kJ/mol, for the formation of lithium-6.

$$3\ {}^{1}_{1}H + 3\ {}^{1}_{0}n \longrightarrow {}^{6}_{3}Li$$

The necessary masses are ${}^{1}_{1}H = 1.00783$ g/mol, ${}^{1}_{0}n = 1.00867$ g/mol, and ${}^{6}_{3}Li = 6.015125$ g/mol. Is the binding energy greater than or less than that for helium-4? Finally, compare the binding energy per nucleon of lithium-6 and helium-4. Which nucleus is the more stable?

Exercise 19.6 Binding Energy

By interpreting the shape of the curve in Figure 19.4, determine which is more exothermic per gram—fission or fusion. Explain your answer.

19.4 RATES OF DISINTEGRATION REACTIONS

Cobalt-60 is used as a source of β particles and γ rays to treat malignancies in the human body. Although ^{60}Co is radioactive, it is nonetheless reasonably stable, since only half of a sample of cobalt-60 will decay in a little over 5 years. On the other hand, copper-64, which is used in the form of copper acetate to detect brain tumors, decays much more rapidly; half of the radioactive copper decays in slightly less than 13 hours. These two radioactive isotopes are clearly different in their stabilities; one of them decays much faster, that is, at a greater rate, than the other.

Half-Life

For nuclear disintegration reactions, the half-life is a constant, independent of the element's physical state, chemical form, temperature, and number of radioactive nuclei present.

The relative stabilities of radioactive isotopes are often expressed just as we have done: in terms of the time required for half of the sample to decay. This is called the **half-life** ($t_{1/2}$) of a radioactive isotope. As illustrated by Table 19.2, isotopes have widely varying half-lives; some take years, even millennia, for half of the sample to decay (^{238}U, ^{14}C), while others decay to half the original number of atoms in fractions of seconds (^{28}P).

As an example of the concept of half-life, consider the decay of oxygen-15, by positron emission.

$${}^{15}_{8}O \longrightarrow {}^{15}_{7}N + {}^{0}_{+1}e$$

TABLE 19.2 Half-Lives of Some Common Radioactive Isotopes

Isotope	Decay Process	Half-Life
${}^{238}_{92}U$	${}^{238}_{92}U \rightarrow {}^{234}_{90}Th + {}^{4}_{2}He$	4.51×10^{9} y
${}^{3}_{1}H$ (tritium)	${}^{3}_{1}H \rightarrow {}^{3}_{2}He + {}^{0}_{-1}e$	12.3 y
${}^{14}_{6}C$ (carbon-14)	${}^{14}_{6}C \rightarrow {}^{14}_{7}N + {}^{0}_{-1}e$	5730 y
${}^{131}_{53}I$	${}^{131}_{53}I \rightarrow {}^{131}_{54}Xe + {}^{0}_{-1}e$	8.04 days
${}^{123}_{53}I$	${}^{123}_{53}I + {}^{0}_{-1}e \rightarrow {}^{123}_{52}Te$	13.2 hr
${}^{57}_{24}Cr$	${}^{57}_{24}Cr \rightarrow {}^{57}_{25}Mn + {}^{0}_{-1}e$	21 s
${}^{28}_{15}P$	${}^{28}_{15}P \rightarrow {}^{28}_{14}Si + {}^{0}_{+1}e$	270×10^{-3} s

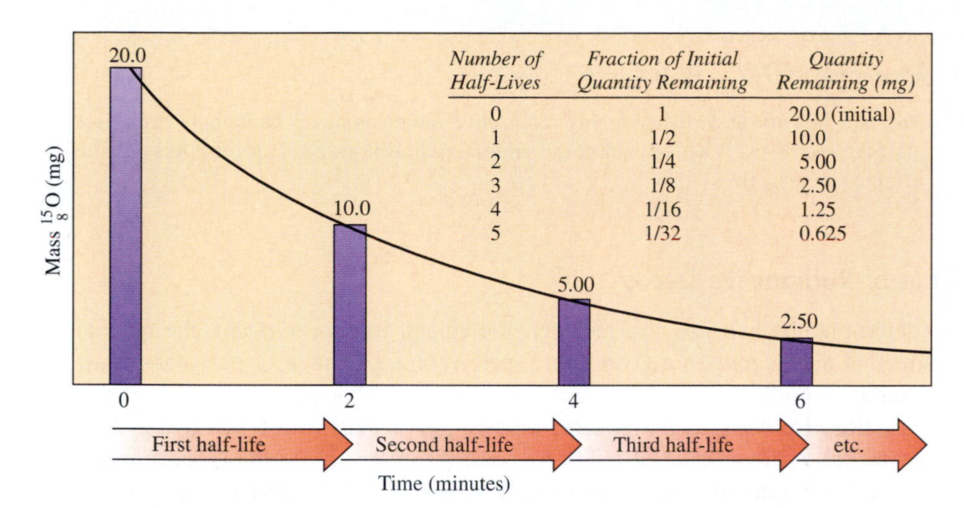

Number of Half-Lives	*Fraction of Initial Quantity Remaining*	*Quantity Remaining (mg)*
0	1	20.0 (initial)
1	1/2	10.0
2	1/4	5.00
3	1/8	2.50
4	1/16	1.25
5	1/32	0.625

Figure 19.5 Decay of 20 mg of oxygen-15. After each half-life period of 2.0 min, the quantity present at the beginning of the period is reduced by half.

The half-life of oxygen-15 is 2.0 min. This means that half of the quantity of $^{15}_{8}O$ present at any given time will disintegrate every 2.0 min. Thus, if we begin with 20 mg of $^{15}_{8}O$, 10 mg of the isotope will remain after 2.0 min. After 4.0 min (two half-lives), only half of the remainder, 5.0 mg, will still be there. After 6.0 min (three half-lives), only half of the 5.0 mg will still be present, or 2.5 mg. The amounts of $^{15}_{8}O$ present at various times are illustrated in Figure 19.5. All radioactive isotopes follow this type of curve as they disintegrate.

PROBLEM-SOLVING EXAMPLE 19.2 Half-Life

Tritium ($^{3}_{1}H$), a radioactive isotope of hydrogen, has a half-life of 12.3 years.

$$^{3}_{1}H \longrightarrow {}^{0}_{-1}e + {}^{3}_{2}He$$

If you begin with 1.5 mg of the isotope, how many milligrams remain after 49.2 years?

Answer Only 0.094 mg of tritium remain.

Explanation First, we find the number of half-lives in the given time period of 49.2 years. Since the half-life is 12.3 years, the number of half-lives is

$$49.2 \text{ years} \times \frac{1 \text{ half-life}}{12.3 \text{ years}} = 4.00 \text{ half-lives}$$

This means that the initial quantity of 1.5 mg is reduced by 1/2 four times.

$$1.5 \text{ mg} \times \frac{1}{2} \times \frac{1}{2} \times \frac{1}{2} \times \frac{1}{2} = 1.5 \text{ mg} \times (\frac{1}{2})^4 = 0.094 \text{ mg}$$

After 49.2 years, only 0.094 mg of the original 1.5 mg will remain.

Problem-Solving Practice 19.2

Strontium-90, $^{90}_{38}Sr$, is a radioisotope ($t_{1/2}$ = 29 years) produced in atomic bomb explosions. Its long life and tendency to concentrate in bone marrow by replacing calcium make it particularly dangerous to people and animals.

(a) The isotope decays with loss of a β particle; write a balanced equation showing the other product of decay.

(b) A sample of the isotope emits 2000 β particles per minute. How many half-lives and how many years are necessary to reduce the emission to 125 β particles per minute?

Exercise 19.7 Half-Lives

The radioactivity of formerly highly radioactive radioisotopes is essentially negligible after ten half-lives. What percent of the original radioisotope remains after this amount of time (ten half-lives)?

Rate of Radioactive Decay

To determine the half-life of a radioactive element, its *rate of decay,* the number of atoms that disintegrate in a given time—per second, per hour, or per year—must be measured.

Radioactive decay is a first-order process ***(← p. 519)*** with a rate that is directly proportional to the number of radioactive atoms present (N). This proportionality is expressed as a rate law (Equation 19.1) in which A is the **activity** of the sample—the number of disintegrations observed per unit time—and k is the first-order rate constant ***(← p. 516)*** or *decay constant* characteristic of that radioisotope.

$$A = kN \qquad (19.1)$$

The ideas used in describing the rate of radioactive decay are those of kinetics ***(← p. 520–522).***

The activity of a sample can be measured with a device such as a Geiger counter (Figure 19.6). Let us say the activity is measured at some time t_0 and then measured again after a few minutes, hours, or days. If the initial activity is A_0 at t_0, then a second measurement will give a smaller activity A at a later time t. Using Equation 19.1, the ratio of the activity A at some time t to the activity at the beginning of the experiment (A_0) must be equal to the ratio of the number of radioactive atoms N that are present at time t to the number present at the beginning of the experiment (N_0).

$$\frac{A}{A_0} = \frac{kN}{kN_0}$$

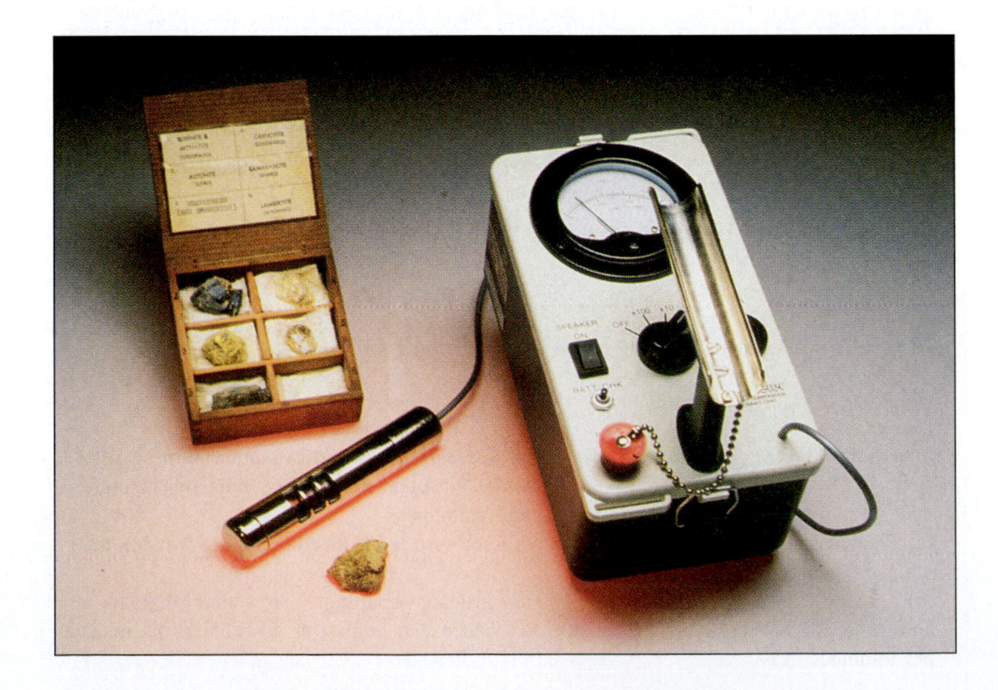

Figure 19.6 A Geiger counter with a sample of carnotite, a mineral containing uranium oxide. The Geiger counter was invented by Hans Geiger and Ernest Rutherford in 1908. A charged particle (such as an α or β particle), when entering a gas-filled tube, ionizes the gas. These gaseous ions are attracted to electrically charged plates and cause a "pulse" or momentary flow of electric current. The current is amplified and used to operate a counter. *(C.D. Winters)*

or

$$\frac{A}{A_0} = \frac{N}{N_0} = \text{fraction of radioactive atoms still present in a sample after some time has elapsed}$$

The change in activity of a radioactive sample over a period of time, or the fraction of radioactive atoms still present in a sample after some time has elapsed can be calculated using the integrated rate equation for a first-order reaction (← ***p. 520, Table 12.1***).

$$\ln A = -kt + \ln A_0$$

which can be rearranged to

$$\ln \frac{A}{A_0} = -kt \qquad (19.2)$$

where A/A_0 is the ratio of activities. Equation 19.2 can also be stated in terms of the fraction of radioactive atoms present in the sample after some time has passed.

$$\ln \frac{N}{N_0} = -kt \qquad (19.3)$$

Equation 19.3 can be derived from Equation 19.1 by calculus.

In words, Equation 19.3 says

$$\text{Natural logarithm}\left(\frac{\text{number of radioactive atoms at some time } t}{\text{number of radioactive atoms at start of experiment}}\right)$$
$$= \text{natural logarithm (fraction of radioactive atoms remaining at time } t\text{)}$$
$$= -\text{(decay constant) (time).}$$

Notice the negative sign in Equation 19.3. The ratio N/N_0 is less than 1 because N is always less than N_0. This means that the logarithm of N/N_0 is negative, and so the other side of the equation must also bear a negative sign because k and t are always positive.

Now we are in a position to see how the half-life of a radioactive isotope, $t_{1/2}$, is determined. The half-life is the time needed for half of the material present at the beginning of the experiment (N_0) to disappear. Thus, when time = $t_{1/2}$, then $N = \frac{1}{2}N_0$. This means that

$$\ln \frac{\frac{1}{2} N_0}{N_0} = -kt_{1/2}$$

or

$$\ln \tfrac{1}{2} = -kt_{1/2}$$
$$-0.693 = -kt_{1/2}$$

and we arrive at a simple equation that connects the half-life and decay constant.

$$t_{1/2} = \frac{0.693}{k} \qquad (19.4)$$

The half-life, $t_{1/2}$, is found by calculating k from Equation 19.3, using N and N_0 from laboratory measurements over the time period t.

Conceptual Challenge Problems CP-19.A and CP-19.B at the end of the chapter relate to topics covered in this section.

PROBLEM-SOLVING EXAMPLE 19.3 Determination of Half-Life

A sample of radon initially undergoes 7.0×10^4 alpha-particle disintegrations per second (dps). After 6.6 days, it undergoes only 2.1×10^4 alpha-particle dps. What is the half-life of this isotope of radon?

Answer This isotope has a 3.8-day half-life.

Explanation Experiment has provided us with both A and A_0.

$$A = 2.1 \times 10^4 \text{ dps} \qquad A_0 = 7.0 \times 10^4 \text{ dps}$$

and the time ($t = 6.6$ days). Therefore, we can find the value of k. Since $\frac{N}{N_0} = \frac{A}{A_0}$, we can use Equation 19.3:

$$\ln \frac{2.1 \times 10^4 \text{ dps}}{7.0 \times 10^4 \text{ dps}} = \ln (0.30) = -k(6.6 \text{ d})$$

$$k = -\left(\frac{\ln (0.30)}{6.6 \text{ d}}\right) = -\left(\frac{-1.20}{6.6 \text{ d}}\right) = 0.18 \text{ d}^{-1}$$

and from k we can obtain $t_{1/2}$.

$$t_{1/2} = \frac{0.693}{k} = \frac{0.693}{0.18 \text{ d}^{-1}} = 3.8 \text{ d}$$

Problem-Solving Practice 19.3

Iridium-192, a radioisotope used in cancer radiation therapy, has a decay constant of $9.3 \times 10^{-3} \text{ d}^{-1}$.

(a) What is the half-life of ^{192}Ir?

(b) What fraction of an ^{192}Ir sample remains after 100 days?

PROBLEM-SOLVING EXAMPLE 19.4 Time and Radioactivity

Some high-level radioactive waste with a half-life, $t_{1/2}$, of 200 years is stored in underground tanks. What time is required to reduce an activity of 6.50×10^{12} disintegrations per minute (dpm) to a fairly harmless activity of 3.00×10^{-3} dpm?

Answer It takes 1.02×10^4 years.

Explanation The data give you the initial activity ($A_0 = 6.50 \times 10^{12}$ dpm) and the activity after some elapsed time ($A = 3.00 \times 10^{-3}$ dpm). In order to find the elapsed time t, you must first find k from the half-life.

$$k = \frac{0.693}{t_{1/2}} = \frac{0.693}{200 \text{ years}} = 0.00347 \text{ y}^{-1}$$

With k known, the time t can be calculated using Equation 19.2.

$$\ln \frac{3.00 \times 10^{-3} \text{ dpm}}{6.50 \times 10^{12} \text{ dpm}} = -[0.00347 \text{ y}^{-1}]t$$

$$-35.312 = -[0.00347 \text{ y}^{-1}]t$$

$$t = \frac{-35.312}{-(0.00347) \text{ y}^{-1}}$$

$$t = 1.02 \times 10^4 \text{ y}$$

Problem-Solving Practice 19.4

In 1921 the women of America honored Marie Curie by giving her a gift of 1.00 g of pure radium, which is now in Paris at the Curie Institute of France. The principal isotope, ^{226}Ra, has a half-life of 1.60×10^3 years. How many grams of radium still remain?

Radiochemical Dating

In 1946 Willard Libby developed the technique of age determination using radioactive $^{14}_{6}C$. Carbon is an important building block of all living systems, and so all organisms contain the three isotopes of carbon: ^{12}C, ^{13}C, and ^{14}C. The first two are stable (nonradioactive) and have been around since the universe was created. Carbon-14, however, is radioactive and decays to nitrogen-14 by beta emission.

$$^{14}_{6}C \longrightarrow {}^{0}_{-1}e + {}^{14}_{7}N$$

Since the half-life of ^{14}C is known to be 5.73×10^3 years, the amount of the isotope present (N) can be measured from the activity of a sample. If the amount of ^{14}C originally in the sample (N_0) is known, then the age of the sample can be found from Equation 19.3.

This method of age determination clearly depends on knowing how much ^{14}C was originally in the sample. The answer to this question comes from work by physicist Serge Korff, who discovered, in 1929, that ^{14}C is continually generated in the upper atmosphere. High-energy cosmic rays smash into gases in the upper atmosphere and force them to eject neutrons. These free neutrons collide with nitrogen atoms in the atmosphere and produce carbon-14.

$$^{14}_{7}N + {}^{1}_{0}n \longrightarrow {}^{14}_{6}C + {}^{1}_{1}H$$

Throughout the *entire* atmosphere, only about 7.5 kg of ^{14}C is produced per year. However, this tiny amount of radioactive carbon is incorporated into CO_2 and other carbon compounds and then is distributed worldwide as part of the carbon cycle. The continual formation of ^{14}C; transfer of the isotope within the oceans, atmosphere, and biosphere; and decay of living matter keep the supply of ^{14}C constant.

Plants absorb carbon dioxide from the atmosphere, convert it into food, and so incorporate the ^{14}C into living tissue, where radioactive ^{14}C atoms and nonradioactive ^{12}C atoms in CO_2 chemically react in the same way. It has been established that the beta activity of carbon-14 in *living* plants and in the air is constant at 15.3 disintegrations per minute per gram of carbon. However, when the plant dies, carbon-14 disintegration continues *without the ^{14}C being replaced;* consequently, the ^{14}C activity decreases with passage of time. The smaller the activity of carbon-14 in the plant, the longer the period between the death of the plant and the present time. Assuming that ^{14}C activity was about the same hundreds of years ago as it is now, measurement of the ^{14}C beta activity of an artifact can be used to date the article.

Willard Libby and his apparatus for carbon-14 dating. *(Oesper Collection in the History of Chemistry/University of Cincinnati)*

Conceptual Challenge Problem CP-19.C at the end of the chapter relates to topics covered in this section.

Problem-Solving Example 19.5 Radiochemical Dating

The so-called Dead Sea Scrolls, Hebrew manuscripts of the books of the Old Testament, were found in 1947. The activity of carbon-14 in the linen wrappings of the book of Isaiah is currently about 12 disintegrations per minute per gram (dpm/g). Calculate the approximate age of the linen.

Answer The cloth is about 2000 years old.

Explanation We will use Equation 19.3

$$\ln \frac{N}{N_0} = -kt$$

where N is proportional to the activity at the present time (12 dpm/g) and N_0 is proportional to the activity of carbon-14 in the living material (15.3 dpm/g). In order to calculate the time elapsed since the linen wrappings were part of a living plant, we first need k, the rate constant. From the text, you know that $t_{1/2}$ of ^{14}C is 5.73×10^3 years, so

$$k = \frac{0.693}{t_{1/2}} = \frac{0.693}{5.73 \times 10^3 \text{ y}} = 1.21 \times 10^{-4} \text{ y}^{-1}$$

Now everything is in place to calculate t.

$$\ln\left(\frac{12 \text{ dpm/g}}{15.3 \text{ dpm/g}}\right) = -0.24 = -kt = -[1.21 \times 10^{-4} \text{ y}^{-1}]t$$

$$t = \frac{-0.24}{-[1.21 \times 10^{-4} \text{ y}^{-1}]}$$

$$= 2.0 \times 10^3 \text{ y}$$

Therefore, the linen is about 2000 years old.

Problem-Solving Practice 19.5

Tritium, ^3H, ($t_{1/2} = 12.3$ y) is produced in the atmosphere and incorporated in much the same way as ^{14}C. Estimate the age of a sealed sample of Scotch whiskey that has a tritium content 0.60 times that of the water in the area where the whiskey was produced.

Exercise 19.8 **Radiochemical Dating**

The radioactive decay of uranium-238 to lead-206 provides a method of radiochemically dating rocks by using the ratio of lead-206 atoms to uranium-238 atoms in a sample. Using this method, a moon rock was found to have a ^{206}Pb-to-^{238}U ratio of 1.00 to 1.09, that is, 100 lead-206 atoms for every 109 uranium-238 atoms. No other lead isotopes were present in the rock, indicating that all of the lead-206 was produced by uranium-238 decay. Estimate the age of the moon rock. The half-life of uranium-238 is 4.51×10^9 years.

Exercise 19.9 **Radiochemical Dating**

Ethanol, C_2H_5OH, is produced by the fermentation of grains or by the reaction of water with ethylene, which is made from petroleum. The alcohol content of wines can be increased fraudulently beyond the usual 12% from fermentation by adding ethanol produced from ethylene. How can carbon-dating techniques be used to differentiate the ethanol sources in these wines?

19.5 ARTIFICIAL TRANSMUTATIONS

In the course of his experiments, Rutherford found in 1919 that alpha particles ionize atomic hydrogen, knocking off an electron from each atom. If atomic nitrogen was used instead, he found that bombardment with alpha particles *also produced protons.* Quite correctly he concluded that the alpha particles had knocked a proton out of the nitrogen nucleus and that an isotope of another element had been produced. Nitrogen had undergone a *transmutation* to oxygen.

$$ {}^{4}_{2}\text{He} + {}^{14}_{7}\text{N} \longrightarrow {}^{17}_{8}\text{O} + {}^{1}_{1}\text{H} $$

Rutherford had proposed that protons and neutrons are the fundamental building blocks of nuclei. Although Rutherford's search for the neutron was not successful, it was found by James Chadwick in 1932 as a product of the alpha-particle bombardment of beryllium (← ***p. 41, Figure 2.13***).

$$ {}^{9}_{4}\text{Be} + {}^{4}_{2}\text{He} \longrightarrow {}^{12}_{6}\text{C} + {}^{1}_{0}\text{n} $$

Changing one element into another by alpha-particle bombardment has its limitations. Before a positively charged bombarding particle (such as the alpha particle) can be captured by a positively charged nucleus, the bombarding particle must have sufficient kinetic energy to overcome the repulsive forces developed as the particle approaches the nucleus. But the neutron is electrically neutral, so Enrico Fermi (1934) reasoned that a nucleus would not oppose its entry. By this approach, practically all elements have since been transmuted, and a number of *transuranium elements* (elements beyond uranium) have been prepared. For example, plutonium-239 forms americium-241 by neutron bombardment,

$$ {}^{239}_{94}\text{Pu} + {}^{1}_{0}\text{n} \longrightarrow {}^{240}_{94}\text{Pu} $$

$$ {}^{240}_{94}\text{Pu} + {}^{1}_{0}\text{n} \longrightarrow {}^{241}_{94}\text{Pu} $$

$$ {}^{241}_{94}\text{Pu} \longrightarrow {}^{241}_{95}\text{Am} + {}^{0}_{-1}\text{e} $$

Of the 112 elements known at present, only elements up to uranium exist in nature (except for Tc and Pm). The transuranium elements are all synthetic. Up to element 101, mendelevium, all of the elements can be made by bombarding the nucleus of a lighter element with small particles such as ${}^{4}_{2}\text{He}$ or ${}^{0}_{1}\text{n}$. Beyond element 101, though, special techniques using heavier particles are required and are still being developed. For example, element 111, synthesized in 1994, was made by bombarding bismuth-209 with nickel-64 nuclei,

$$ {}^{64}_{28}\text{Ni} + {}^{209}_{83}\text{Bi} \longrightarrow {}^{272}_{111} + {}^{1}_{0}\text{n} $$

and the element 112, the latest to be discovered, was made by firing zinc-70 atoms into lead-208 nuclei

$$ {}^{70}_{30}\text{Zn} + {}^{208}_{82}\text{Pb} \longrightarrow {}^{277}_{112} + {}^{1}_{0}\text{n} $$

Elements 110 through 112 have not yet been assigned names and symbols (← ***p. 51***).

Exercise 19.10 Nuclear Transmutations

Balance the following equations for nuclear reactions, indicating the symbol, the mass number, and the atomic number of the remaining product.

(a) ${}^{13}_{6}\text{C} + {}^{1}_{0}\text{n} \longrightarrow {}^{4}_{2}\text{He} + ?$ (c) ${}^{253}_{99}\text{Es} + {}^{4}_{2}\text{He} \longrightarrow {}^{1}_{0}\text{n} + ?$

(b) ${}^{14}_{7}\text{N} + {}^{4}_{2}\text{He} \longrightarrow {}^{1}_{0}\text{n} + ?$

Exercise 19.11 Element Synthesis

Bombardment of lead-208 nuclei produced element 110 (mass number = 269) plus a neutron. Write a balanced nuclear equation for this process and identify the bombarding nucleus.

Portrait of a Scientist • *Glenn Seaborg (1912–)*

Because of an inspiring high school teacher, Glenn Seaborg changed his major from literature to chemistry at the University of California–Los Angeles. Seaborg's professorial career began at the University of California–Berkeley in 1939 and has extended for more than 50 years, including a stint as the University's chancellor (president) from 1958 to 1961.

Seaborg was a pioneer in developing radioisotopes for medical use (Section 19.9). He was the first to produce iodine-131, which was used subsequently to treat his mother's abnormal thyroid condition; it has become a major medical radioisotope.

An independent thinker, Seaborg demonstrated the importance of maintaining the courage of one's convictions. The conventional thinking in the early 1940s was that Th, Pa, and U were transition elements and should be placed accordingly in the periodic table: Th under Hf, Pa under Ta, and U under W. In 1944, while still early in his career, Seaborg proposed a new version of the periodic table. Based on his research and that of others, Seaborg proposed that thorium was not a transition element, but rather the beginning of the actinide series elements that follow actinium in the periodic table, just as the lanthanides (rare earths) follow lanthanum. He further proposed that the transuranium elements (atomic numbers greater than 92) accordingly belonged to the actinide series, because of their chemical similarities to the rare earth elements. Two prominent inorganic chemists warned him against publishing such a radical proposal because it would ruin his scientific reputation. Nevertheless, Seaborg went ahead with the proposal, remarking later "I didn't have any scientific reputation so I published it anyway." Thanks to his careful research and insights, it is now very well established that the transuranium elements are members of the actinide series.

Glenn Seaborg pointing to element 106 (seaborgium) on the periodic chart. *(Lawrence Berkeley Laboratory)*

As a result of Seaborg's interpretation of the periodic table, it was possible to predict accurately the properties of many of the as-yet-undiscovered transuranium elements. Subsequent preparation of these elements proved him right, and he was awarded the Nobel Prize in 1951 (along with Edwin McMillan). In a remarkable 21-year span (1940–1961), Seaborg and his colleagues synthesized ten new transuranic elements (plutonium to lawrencium), almost 10% of the then-known elements; none of the ten occur naturally. Albert Ghiorso, a member of Seaborg's team, has continued the work and has produced several other new elements, including element 106. Its discoverers have proposed to name it seaborgium. The naming of elements 104 through 109 has been controversial, and the name seaborgium awaits final approval by the International Union of Pure and Applied Chemistry (IUPAC), the official agency for naming elements.

Regardless of whether seaborgium is the name confirmed for element 106, Glenn Seaborg has long been recognized as a leader in 20th-century nuclear science, having won every prestigious prize awarded in chemistry.

19.6 NUCLEAR FISSION

In 1938 the radiochemists Otto Hahn and Fritz Strassman found barium in a sample of uranium that had been bombarded with neutrons. Further work by Lise Meitner, Otto Frisch, Niels Bohr, and Leo Szilard confirmed that a uranium-235 nucleus had captured a neutron and undergone **nuclear fission;** that is, the nucleus had split in two (Figure 19.7).

$$^{235}_{92}\text{U} + ^{1}_{0}\text{n} \longrightarrow ^{141}_{56}\text{Ba} + ^{92}_{36}\text{Kr} + 3\,^{1}_{0}\text{n} \qquad \Delta E = -2 \times 10^{10}\ \text{kJ}$$

The fact that the fission reaction produces more neutrons than are required to begin the process is important. In the preceding nuclear reaction, bombardment with a single neutron produces three neutrons capable of inducing three more fission reac-

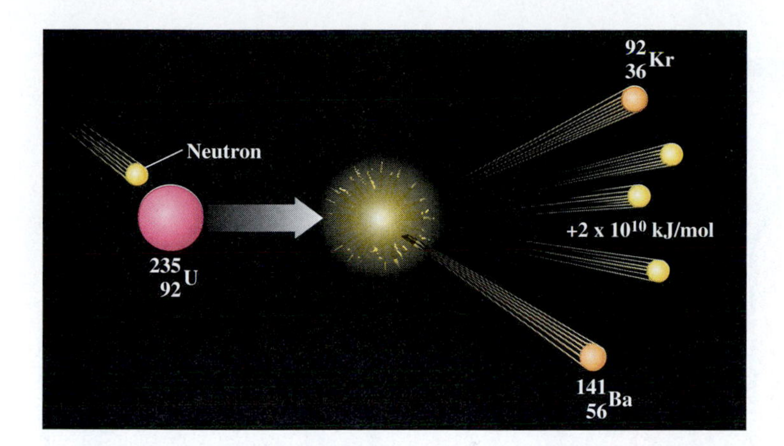

Figure 19.7 The fission of a $^{235}_{92}U$ nucleus from the bombardment of $^{235}_{92}U$ with a neutron. The electrical repulsion between protons rips the nucleus apart.

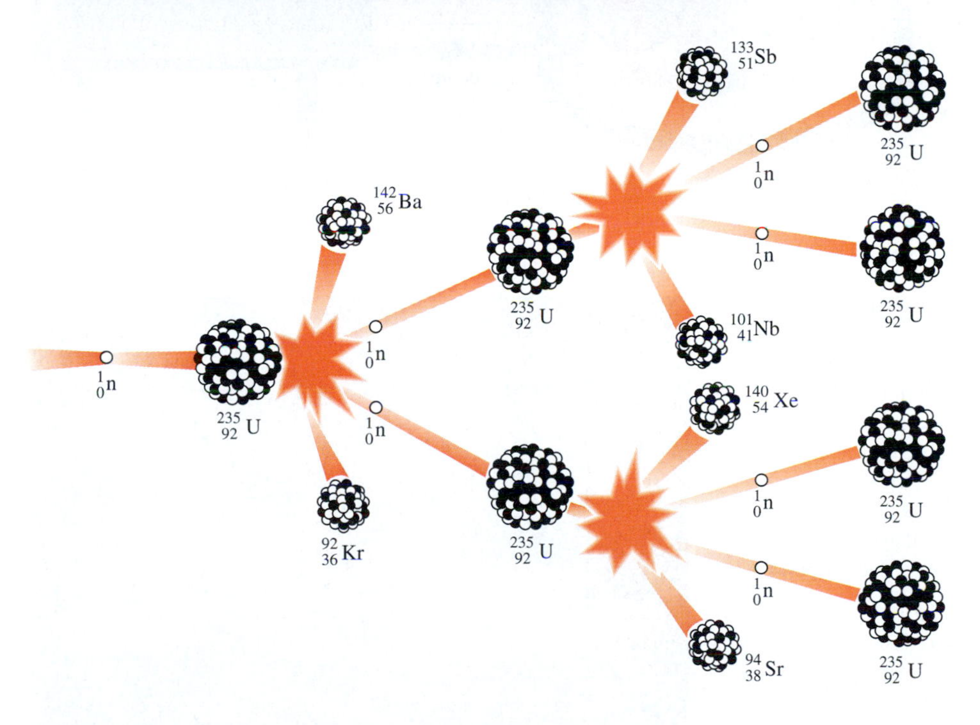

Figure 19.8 Illustration of a self-propagating nuclear chain reaction initiated by capture of a neutron. The fission of uranium-235 produces a variety of products. Thirty-four elements have been detected among the fission products, including those shown in the figure. Each fission produces two lighter nuclei, plus two or three neutrons.

tions, which release nine neutrons to induce nine more fissions, from which 27 neutrons are obtained, and so on. Since the neutron-induced fission of uranium-235 is extremely rapid, this sequence of reactions can be an explosive chain reaction, as illustrated in Figure 19.8. If the amount of uranium-235 is small, so few neutrons are captured by ^{235}U nuclei that the chain reaction cannot be sustained. In an atomic bomb, two small pieces of uranium-235, neither capable of sustaining a chain reaction, are brought together to form one piece capable of supporting a chain reaction, and an explosion results.

Rather than allow a fission reaction to run away explosively, engineers can control it by limiting the number and energy of the neutrons available so that energy can be derived safely and used as a heat source in a power plant (Figure 19.9). In a **nuclear** or **atomic reactor,** the rate of fission is controlled by inserting cadmium rods or other "neutron absorbers" into the reactor. The rods absorb the neutrons that cause fission reactions; by withdrawing or inserting the control rods, the rate of the fission reaction can be increased or decreased, respectively.

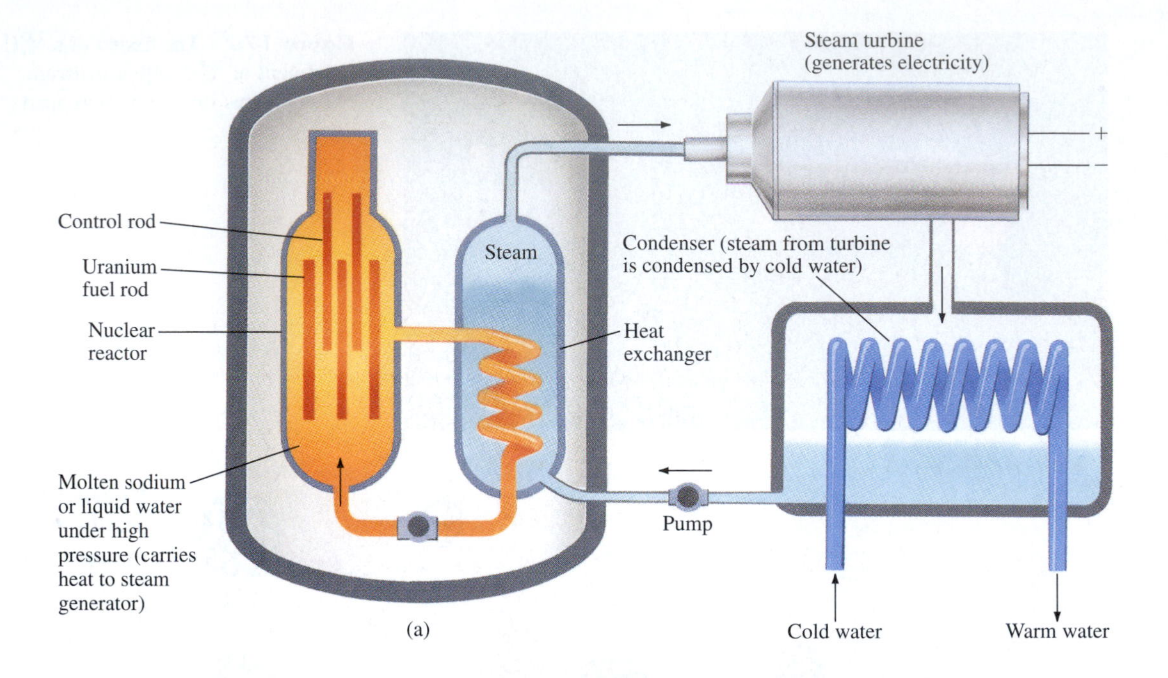

(b)

Figure 19.9 Nuclear power plants. (a) Liquid water (or liquid sodium) is circulated through the reactor, where the liquid is heated to about 325 °C. When this hot liquid is circulated through a steam generator, water in the generator is turned to steam, which in turn drives a steam turbine. After it passes through the turbine, the steam is converted back to liquid water and is recirculated through the steam generator. Enormous quantities of outside cooling water from rivers or lakes are necessary to condense the steam. (This basic system is the same as in any power plant, except that the water or circulating liquid may be heated initially by coal, gas, or oil-fired burners.) (b) A nuclear power plant at Indian Pointe, New York. *(b, Joe Azzara/The Image Bank)*

Uranium-238 can fission, but only when bombarded with very fast neutrons, not like those in nuclear reactors. Thus, we consider uranium-238 to be nonfissionable in the context of nuclear reactors.

Not all nuclei can be made to fission on colliding with a neutron, but ^{235}U and ^{239}Pu are two fissionable isotopes. Natural uranium contains an average of only 0.72% of the fissionable ^{235}U isotope; more than 99% of the natural element is nonfissionable uranium-238. Since the percentage of natural ^{235}U is too small to sustain a chain reaction, uranium for nuclear power fuel must be enriched to about 3% uranium-235. To accomplish this, some of the ^{238}U isotope in a sample is effectively discarded, thereby raising the concentration of ^{235}U. Because the amount of uranium-235 in its nuclear fuel rods is lower than the critical mass needed for an atomic bomb, the reactor core cannot undergo an uncontrolled chain reaction to make the reactor an atomic bomb.

Nuclear fission produces a truly staggering amount of energy. For example, the fissioning of 1.0 kg (2.2 lb) of uranium-235 releases 9.0×10^{13} J of energy, the equivalent of exploding 33,000 tons (33 kilotons) of TNT. Each UO_2 fuel pellet used in a nuclear reactor has the energy equivalent to burning 136 gal of oil, 2.5 tons of wood, or 1 ton of coal.

A sample is considered of weapons quality if its uranium-235 content is over 90%. Even in reactors using weapons-quality uranium, the U-235 is too dispersed to produce the uncontrolled fission needed to detonate.

Exercise 19.12 Energy of Nuclear Fission

The burning of 1.0 kg of high-grade coal produces 2.8×10^4 kJ of energy; the fissioning of 1.0 mol of uranium-235 generates 2.1×10^{10} kJ. How many metric tons of coal (1 metric ton = 10^3 kg) are needed to produce the same energy as that released by the fission of 1.0 kg of uranium-235? (Assume that the processes have equal efficiency.)

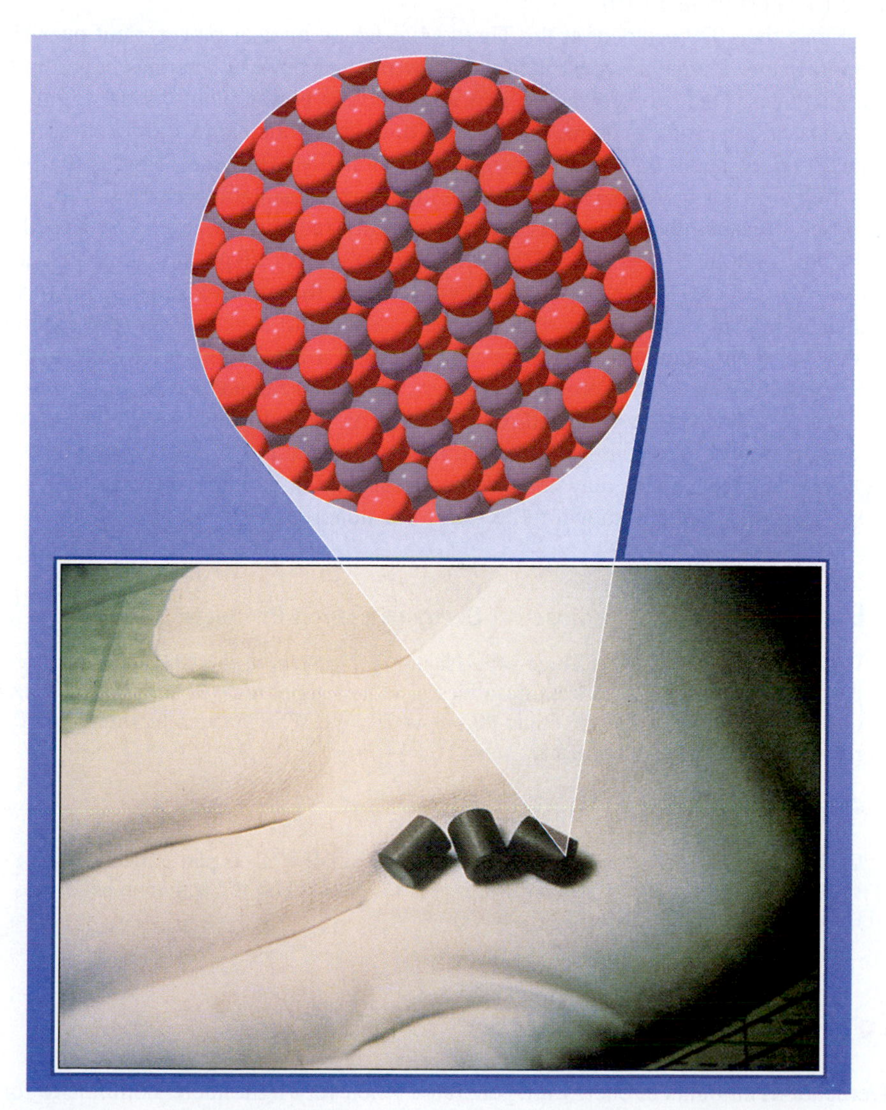

Uranium oxide pellets used in the nuclear fuel rods. *(General Electric)*

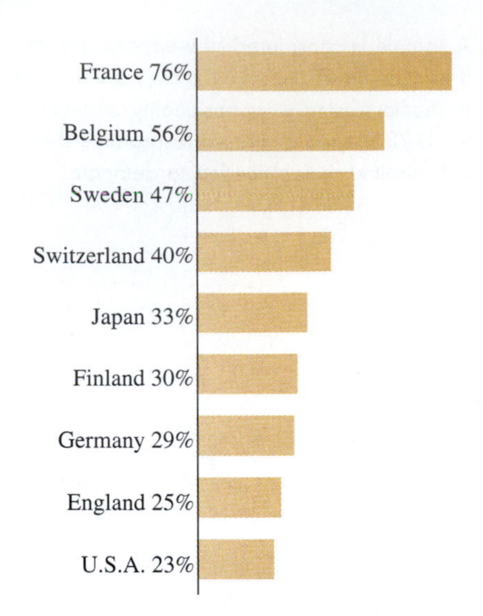

Figure 19.10 The approximate share of electricity generated by nuclear power in various countries. Approximately 23% of electricity in the United States is generated by nuclear power.

There is, of course, some controversy surrounding the use of nuclear power plants, particularly in the United States. Their proponents argue that the health of our economy and our standard of living are dependent on inexpensive, reliable, and safe sources of energy. Just within the past few years the demand for electric power has once again begun to exceed the supply in the United States, so many believe nuclear power plants should be built to meet the demand. Nuclear power plants are capable of supplying these demands, and they can be the source of "clean" energy in that they do not pollute the atmosphere with ash, smoke, or oxides of sulfur, nitrogen, or carbon. In addition, they help to ensure that our supplies of fossil fuels will not be depleted as quickly in the near future, and they reduce our dependency on such fuels from other countries. There are currently more than 100 operating nuclear plants in the United States; more than 420 nuclear power plants worldwide in over 30 nations produce about 17% of the world's electricity. The nuclear plants in the United States supply about 23% of our nation's electric energy; only coal-fired plants contribute a greater share (57%) (Figure 19.10).

Since the 1979 accident at the Three Mile Island nuclear power plant near Harrisburg, Pennsylvania, *no* construction of new nuclear power plants has begun in the United States. One problem associated with nuclear power plants is the highly radioactive fission products in the waste fuel. Commercial nuclear reactors in the United States produce about 3,300 tons of waste fuel annually. Although some of the products are put to various uses (Section 19.9), many are not suitable as a fuel or for other purposes. Because these products are often highly radioactive and some have long half-lives (plutonium-239; $t_{1/2} = 2.4 \times 10^4$ years), proper disposal of this high-level nuclear waste poses an enormous problem. Perhaps the most reasonable suggestion is that high-level radioactive wastes can be converted to a glassy material having a volume of about 2 m^3 per reactor per year. In 1996 a Department of Energy facility in Savannah River, South Carolina, began encapsulating radioactive waste in glass. A mixture of glass particles and radioactive waste is heated to 1,200 °C. The molten mixture is poured into stainless steel cannisters, cooled, and stored; eventually, such high-level nuclear wastes may be stored underground in geological formations, such as salt deposits, that are known to be stable for hundreds of millions of years.

Exercise 19.13 Radioactive Decay of Fission Products

Unlike the 1979 incident at Three Mile Island, the accident at the Chernobyl nuclear plant in 1986 released significant quantities of radioisotopes into the atmosphere. One of those radioisotopes was strontium-90 ($t_{1/2} = 29.1$ y). What fraction of strontium-90 released at that time still remains?

Exercise 19.14 Nuclear Waste

Cesium-137 ($t_{1/2} = 30.2$ y) is produced by ^{235}U fission. If ^{137}Cs is part of nuclear waste stored deep underground, how long will it take for the initial ^{137}Cs activity when it was first buried to drop (a) by 60%? (b) by 90%?

19.7 NUCLEAR FUSION

Tremendous amounts of energy are generated when very light nuclei combine to form heavier nuclei. Such a reaction is called **nuclear fusion,** and one of the best examples is the fusion of hydrogen nuclei (protons) to give helium nuclei.

CHEMISTRY IN THE NEWS

U.S. Proposes Methods for Disposing of Nuclear Weapons

Nuclear disarmament treaties, such as those between the United States and the former Soviet Union, require the safe disposal of plutonium from dismantled military weapons. But, how to do it safely?

Plutonium-239 is produced in nuclear reactors. Fission of uranium-235 releases neutrons that bombard uranium-238, converting it to various plutonium isotopes. The particular mixture of plutonium isotopes produced varies with the fuel and how long it remains in the reactor. For weapons-grade plutonium production, the reactor fuel is 99% uranium-238 and about 1% uranium-235 that is irradiated with neutrons for only several months to produce principally fissionable $^{239}_{94}Pu$ (93.5%). In contrast, commercial nuclear reactor fuel is $^{238}_{92}U$ (96–97%) and $^{235}_{92}U$ (3–4%) that is irradiated much longer, for one to three years. Consequently, spent fuel rods from commercial nuclear reactor waste contain a lower concentration of fissionable plutonium-239 (58%) than those from weapons-grade reactors.

How much weapons-grade plutonium needs to be disposed? Estimates are that there are 50 metric tons of plutonium from U.S. military weapons and 50 to 100 metric tons of it to be removed from the former Soviet Union nuclear arsenal. Plutonium from dismantled weapons currently is stored as pits—hollow, metal-clad spheres of plutonium—as well as in other forms.

In a December 1996 report, the U.S. Department of Energy (DOE) proposed two methods for dealing with the U.S. military waste plutonium. One approach is to mix some of the waste with nuclear reactor waste, encase the mixture in glass (vitrification) or ceramic logs, and bury it eventually in a deep, geologically stable site. The second method proposes to incorporate the remaining plutonium waste into mixed oxide fuel (MOX)—a mixture of uranium and plutonium oxides—and use it in commercial nuclear reactors. The MOX approach utilizes fissionable plutonium-239 to produce energy directly, whereas the geological burial removes the plutonium from direct commercial use, as well as reduces its potential availability for nuclear terrorists. Plutonium waste buried in geological deposits ostensibly could be recovered in the future, should the need warrant it. The MOX option is viewed more favorably by Russia because it considers the waste plutonium to be a valuable energy resource that should not be buried, even though Russia currently has no MOX-fuel plant.

Controversy surrounds both proposed disposal methods. Although the MOX approach has its proponents, its opponents counter that this approach will be more expensive, take more time, and create greater potential health and safety risks than vitrification. At present, however, there is no long-term underground vitrified nuclear waste storage facility in the United States, although Yucca Mountain, Nevada, has been designated to be such a site.

Since both methods present considerable engineering and technological uncertainties, DOE has proposed that research continue for at least two years before implementing either method. It is estimated that the plutonium disposal will take up to 30 years at a cost of more than $2 billion.

Source: Chemical & Engineering News: *December 16, 1996, pp. 10–11; June 13, 1994, pp. 12–25.*

$$4\,^{1}_{1}H \longrightarrow \,^{4}_{2}He + 2\,^{0}_{+1}e \qquad \Delta E = -2.5 \times 10^{9}\text{ kJ}$$

This reaction is the source of the energy from our sun and other stars, and it is the beginning of the synthesis of the elements in the universe. Temperatures of 10^6 to 10^7 K, found in the core and radiative zone of the sun, are required to bring the positively charged nuclei together with enough kinetic energy to overcome nuclear repulsions.

Deuterium can also be fused to give helium-3,

$$^{2}_{1}H + \,^{2}_{1}H \longrightarrow \,^{3}_{2}He + \,^{1}_{0}n \qquad \Delta E = -3.2 \times 10^{8}\text{ kJ}$$

which can undergo further fusion with a proton to give helium-4.

$$^{1}_{1}H + \,^{3}_{2}He \longrightarrow \,^{4}_{2}He + \,^{0}_{+1}e \qquad \Delta E = -1.9 \times 10^{9}\text{ kJ}$$

Both of these reactions release an enormous quantity of energy, so it has been the dream of nuclear physicists to try to harness them to provide energy for the nations of the world.

At the very high temperatures that allow fusion reactions to occur rapidly, atoms do not exist as such; instead, there is a **plasma** consisting of unbound nuclei and electrons. In order to achieve the high temperatures required for the fusion reaction of the hydrogen bomb, a fission bomb (atomic bomb) is first set off. One type of hydrogen bomb depends on the production of tritium in the bomb. In this type, lithium-6 deuteride (LiD, a solid salt) is placed around an ordinary ^{235}U or ^{239}Pu fission bomb, and the fission is set off in the usual way. A ^{6}Li nucleus absorbs one of the neutrons produced and splits into tritium and helium.

$$^{6}_{3}Li + ^{1}_{0}n \longrightarrow ^{3}_{1}H + ^{4}_{2}He$$

A 20-megaton bomb has the explosive power of 20 million tons of TNT.

The temperature reached by the fission of uranium or plutonium is high enough to bring about the fusion of tritium and deuterium and the release of 1.7×10^{9} kJ per mole of ^{3}H. A 20-megaton bomb usually contains about 300 lb of lithium deuteride, as well as a considerable amount of plutonium and uranium.

Development of fusion as a commercial energy source is inviting because hydrogen isotopes are available (from water), and fusion products are generally nonradioactive or have short half-lives, which eliminates the problems of waste disposal for radioactive fission-reactor products. Controlling a nuclear fusion reaction for peaceful, commercial uses, however, has been extraordinarily difficult. Three critical requirements must be met for controlled fusion. First, the temperature must be high enough for fusion to occur. The fusion of deuterium and tritium, for example, requires a temperature of 100 million degrees or more. Second, the plasma must be confined long enough to release a net output of energy. Third, the energy must be recovered in some usable form.

Containment is one of the biggest problems in developing controlled fusion.

The possibility of "cold fusion," the room-temperature fusion of deuterium atoms to provide energy, was announced in 1989. Although this "discovery" is now largely discredited, work continues. The international scientific community's reaction to the announcement illustrates well how the process of scientific investigation works. *See J. Chem. Educ., Vol. 66, p. 449, 1989.*

Magnetic "bottles" (enclosures in space bounded by magnetic fields) have confined the plasma so that controlled fusion has been achieved. But the energy generated by it has been less than that required to produce and control the fusion reaction. Using more energy to produce less energy is not a commercially appealing investment. Thus, commercial fusion reactors are not likely in the near future without a dramatic breakthrough in fusion technology.

19.8 NUCLEAR RADIATION: EFFECTS AND UNITS

Radiation Units

The use of nuclear energy and radiation is a double-edged sword that carries risks and benefits. It can be used to harm (nuclear armaments) or to cure (radioisotopes in medicine); for example, nuclear radiation can cause as well as cure some types of cancer.

All new technologies have risks and benefits. In the 1800s, railroads were new and the poet William Wordsworth wrote of their risks and benefits in terms of "Weighing the mischief with the promised gain . . . "

Alpha, beta, and gamma radiation disrupt normal cell processes in living organisms by producing energetic radicals and ions. The potential for serious radiation damage to humans is well known. The biological effects of the atomic bombs exploded at Hiroshima and Nagasaki, Japan, at the close of World War II in 1945 have been well documented. However, controlled exposure to nuclear radiation can be beneficial in destroying malignant tissue, as in radiation therapy for treating some cancers.

The röntgen unit is named to honor Wilhelm Röntgen, the German physicist who discovered x rays.

To quantify radiation and its effects, particularly on humans, several units have been developed. For example, the **röntgen (R)** is used to give the dosage of x rays and γ rays. One röntgen corresponds to the *deposition* of 93.3×10^{-7} J per gram of tissue. The **rad** is similar to the röntgen, but it measures the amount of radiation

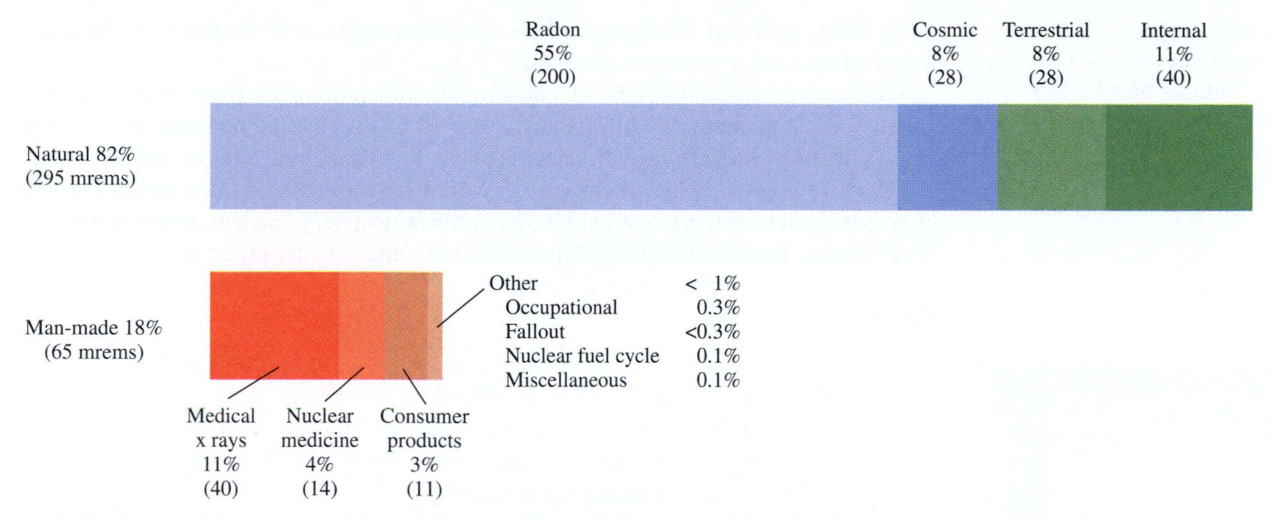

Figure 19.11 **Sources of average background radiation exposure in the United States.** The sources are expressed in percentages of the total, as well as in millirems, the values in parentheses. The background radiation from natural sources far exceeds that from artificial sources.

absorbed; 1 rad represents a dose of 1.00×10^{-2} J absorbed per kilogram of material.

The biological effects of radiation per rad differ with the kind of radiation, which can be quantified more generally using the **rem** (standing for *r*öntgen *e*quivalent *m*an).

$$\text{Effective dose in rems} = \text{quality factor} \times \text{dose in rads}$$

The quality factor depends on the type of radiation and other factors; for example, the factor is 1 for beta and gamma radiation, and 10 for alpha particles. Since one rem is a large amount of radiation, the millirem (mrem) is commonly used (1 mrem = 10^{-3} rem).

Finally, the **curie** (Ci) is commonly used as a unit of activity. One curie represents the quantity of any radioactive isotope that undergoes 3.7×10^{10} disintegrations per second (dps); 1 millicurie (mCi) = 10^{-3} Ci = 3.7×10^7 dps.

Conceptual Challenge Problems CP-19.D and CP-19.E at the end of the chapter relate to topics covered in this section.

Another unit applied to radioactive substances is the becquerel (Bq), in honor of Henri Becquerel; 1 Bq = 1 dps. The SI unit of absorbed radiation is the Gray (Gy), which is equal to the absorption of 1 Joule per kilogram of tissue.

Background Radiation

Humans are constantly exposed to natural and artificial **background radiation,** estimated to be about 360 mrem per year (Figure 19.11), well below 500 mrem, the federal government's background radiation standard for the general public. Most background radiation, about 300 mrem (82%), comes from *natural* background radiation sources: cosmic radiation and radioactive elements and minerals found naturally in the earth and air. The remaining 18% comes from artificial sources.

Cosmic radiation, emitted by the sun and other stars, continually bombards the earth and accounts for about 8% of background radiation. The remainder comes from elements such as ^{40}K. Potassium is present to the extent of about 0.3 g/kg of soil and is essential to all living organisms. We all acquire some radioactive potassium from the foods we eat, for example: a hamburger contains 960 mg ^{40}K, giving off 29 dps, a hot dog 200 mg ^{40}K; 6 dps, or a serving of French fries 650 mg ^{40}K; 20 dps. Other radioactive elements found in some abundance on the earth are thorium-232 and uranium-238. Approximately 8% of the natural background radiation arises from Th-232 and U-238 in rocks and soil. Thorium, for example, is found to the extent of

Burning fossil fuels (coal and oil) releases naturally occurring radioactive isotopes into the atmosphere. This has added significantly to the background radiation in recent years. Far more Th and U are released annually into the atmosphere from fossil fuel–burning power plants than from nuclear power plants.

12 g/1000 kg of soil. Background radiation from radon, a byproduct of radium decay, is discussed in the next subsection.

On average, roughly 15% of our annual exposure comes from medical procedures such as diagnostic x rays and the use of radioactive compounds to trace the body's functions. Consumer products account for 3% of our total annual exposure. Contrary to popular belief, less than 1% comes from sources such as the radioactive products from testing nuclear explosives in the atmosphere, nuclear power plants and their wastes, nuclear weapons manufacture, and nuclear fuel processing.

Radon

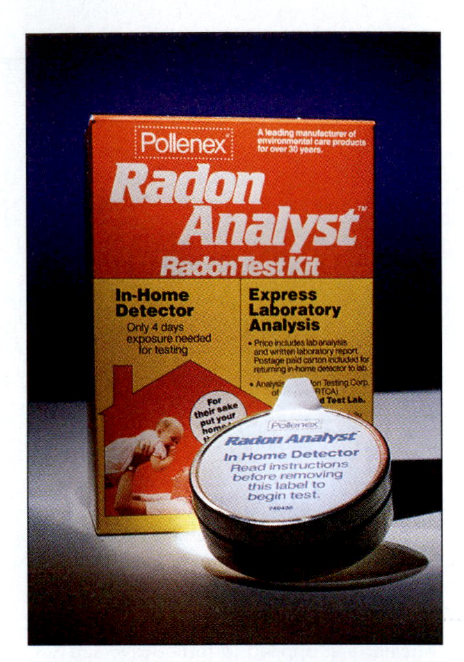

Figure 19.12 A commercially available kit to test for radon gas in the home. *(C.D. Winters)*

1 picoCurie, pCi = 10^{-12} Ci

Radon is a chemically inert gas in the same periodic table group as helium, neon, argon, and krypton. Radon-222 is produced in the decay series of uranium-238 (← ***p. 873,*** Figure 19.2). (Other isotopes of Rn are products of other decay series.) The trouble with radon is that it is radioactive; Figure 19.11 shows that radon accounts for 55% of natural background radiation.

It should be kept in mind that radon occurs naturally in our environment and, because it comes from natural uranium deposits, the amount depends on local geology. Furthermore, since the gas is chemically inert and has a half-life of 3.82 days, it is not trapped by chemical processes in the soil or water. Thus, it is free to seep up from the ground and into underground mines or into homes through pores in concrete block walls, cracks in basement floors or walls, or around pipes. When breathed by humans, ^{222}Rn can decay inside the lungs to give polonium, a radioactive, heavy-metal element that is not a gas and is not chemically inert.

$$^{222}_{86}\text{Rn} \longrightarrow {}^{4}_{2}\text{He} + {}^{218}_{84}\text{Po} \qquad t_{1/2} = 3.82\text{ d}$$

$$^{218}_{84}\text{Po} \longrightarrow {}^{4}_{2}\text{He} + {}^{214}_{82}\text{Pb} \qquad t_{1/2} = 3.10\text{ m}$$

Therefore, polonium-218 can lodge in lung tissues where it undergoes α decay to give lead-214, itself a radioactive isotope. The range of an α particle is quite small, perhaps 0.7 mm (about the thickness of a sheet of paper). However, this is approximately the thickness of the epithelial cells of the lungs, so the radiation can damage these tissues and induce lung cancer.

Most homes in the United States are believed to have some level of radon gas. There is currently a great deal of controversy over the level of radon that is considered "safe." Estimates indicate that only about 6% of U.S. homes have radon levels above 4 picocuries per liter (pCi/L) of air, the "action level" standard set by the U.S. Environmental Protection Agency. There are some who believe 1.5 pCi/L is more likely the average level and that only about 2% of the homes will contain over 8 pCi/L. To test for the presence of radon, you can purchase testing kits of various kinds (Figure 19.12). If your home shows higher levels of radon gas than 4 pCi/L, you should probably have it tested further and perhaps take corrective actions such as sealing cracks around the foundation and in the basement. But keep in mind the relative risks involved (← ***p. 19***). A 1.5 pCi/L level of radon leads to a lung cancer risk about the same as the risk of your dying in an accident in your home.

Exercise 19.15 Radon Levels

Calculate how long it will take for the activity of a radon-222 sample ($t_{1/2}$ = 3.82 d) initially at 8 pCi to drop (a) to 4 pCi, the EPA action level; (b) to 1.5 pCi, approximately the U.S. average.

(Text continues on page 898.)

CHEMISTRY YOU CAN DO

COUNTING MILLIREMS: YOUR RADIATION EXPOSURE

The Committee on Biological Effects of Ionizing Radiation of the National Academy of Sciences issued a report in 1980 that contained a survey for an individual to evaluate his or her exposure to ionizing radiation. The table below is adapted from this report. By adding up your exposure, you can compare your annual dose to the United States annual average of 360 mrem.

	Common Sources of Radiation	Your Annual Dose (mrem)
Where You Live	**Location:** Cosmic radiation at sea level	27
	For your elevation (in feet), add this number of mrem	______
	Elevation 1000: 2; 2000: 5; 3000: 9; 4000: 15; 5000: 21; 6000: 29; 7000: 40; 8000: 53; 9000: 70	
	Ground: U.S. average	26
	Radon: U.S. average	200
	House construction: For stone, concrete, or masonry building, add 7; wood add 30	______
What You Eat, Drink, and Breathe	**Radioisotopes** in the body from	
	Food, air, water: U.S. average	40
	Weapons test fallout	4
How You Live	**X-ray and radiopharmaceutical diagnosis**	
	Number of chest x-rays ______ × 10	______
	Number of lower gastrointestinal tract x-rays ______ × 500	______
	Number of radiopharmaceutical examinations ______ × 300	______
	(Average dose to total U.S. population = 53 mrem)	
	Jet plane travel: For each 2500 miles add 1 mrem	______
	TV viewing: Number of hours per day ______ × 0.15	______
How Close You Live to a Nuclear Plant	**At site boundary:** average number of hours per day ______ × 0.2	______
	One mile away: average number of hours per day ______ × 0.02	______
	Five miles away: average number of hours per day ______ × 0.002	______
	Over 5 miles away: none	______
	Note: Maximum allowable dose determined by "as low as reasonably achievable" (ALARA) criteria established by the U.S. Nuclear Regulatory Commission. Experience shows that your actual dose is substantially less than these limits.	
	Your total annual dose in mrem	______

Elevation table:

Elevation	mrem	Elevation	mrem	Elevation	mrem
1000	2	4000	15	7000	40
2000	5	5000	21	8000	53
3000	9	6000	29	9000	70

Compare your annual dose to the U.S. annual average of 360 mrem.
One mrem per year is equal to increasing your diet by 4%, or taking a 5-day vacation in the Sierra Nevada (CA) mountains.

*Based on the "BEIR Report III"—National Academy of Sciences, Committee on Biological Effects of Ionizing Radiation, "The Effects on Populations of Exposure to Low Levels of Ionizing Radiation," National Academy of Sciences, Washington, DC, 1987.

19.9 APPLICATIONS OF RADIOACTIVITY

Food Irradiation

In some parts of the world, stored-food spoilage may claim up to 50% of the food crop. In our society, refrigeration, canning, and chemical additives lower this figure considerably. Still, there are problems with food spoilage, and food preservation costs amount to a sizable fraction of the final cost of food. Food and grains can also be preserved by gamma irradiation. Contrary to some popular opinion, such irradiation does *not* make foods radioactive, just as a dental x ray does not make you radioactive.

Irradiated strawberries. The label shows the radura, the international symbol for irradiated foods. In the radura, the leaves and bud represent agriculture, and the breaks in the outer circle represent penetrating radiation. *(Nordion International)*

Food irradiation with gamma rays from ^{60}Co or ^{137}Cs sources is commonly used in European countries, Canada, Mexico, and nearly 20 other nations. Astronaut's food has been preserved by gamma irradiation. The United States and several other countries require that foods preserved by irradiation be labeled with the international symbol for irradiated food.

Foods can be pasteurized by irradiation to retard the growth of organisms such as bacteria, molds, and yeasts. This irradiation prolongs shelf-life under refrigeration in much the same way that heat pasteurization protects milk. Chicken normally has a three-day refrigerated shelf-life; after irradiation, it can have a three-week refrigerated shelf-life.

The FDA permits irradiation up to 300 kilorads for the pasteurization of poultry. Radiation levels in the 1- to 5-megarad (1 megarad = 10^6 rad) range sterilize; that is, every living organism is killed. Foods irradiated at these levels will keep indefinitely when sealed in plastic or aluminum-foil packages. However, FDA approval is unlikely for irradiation sterilization of foods in the near future because of potential problems caused by as yet undiscovered, but possible, "unique radiolytic products." For example, irradiation sterilization might produce a substance that is capable of causing genetic damage. To prove or disprove the presence of these substances, animal feeding studies using irradiated foods are now being conducted in this country.

More than 40 classes of foods are already irradiated in 24 countries. In the United States, irradiation has been approved for only a small number of foods including potatoes and strawberries for domestic consumption, and grapefruit, fish, and shrimp for export. No adverse effects have been observed in humans in other countries where irradiated foods have been consumed for years.

Radioactive Tracers

The chemical behavior of a radioisotope is essentially identical to that of the nonradioactive isotopes of the same element because chemical reactions involve valence electrons, and the energies of the valence electrons are nearly the same in both types of isotopes. Therefore, chemists can use radioactive isotopes as **tracers** in nonbiological and biological chemical reactions. To use a tracer, a chemist prepares a reactant compound in which one of the elements consists of both radioactive and stable (nonradioactive) isotopes, and introduces it into the reaction (or feeds it to an organism). After the reaction, the chemist measures the radioactivity of the products (or determines which parts of the organism contain the radioisotope) by using a Geiger counter or other radiation detectors. Several radioisotopes commonly used as tracers are listed in Table 19.3.

For example, plants take up phosphorus-containing compounds from the soil through their roots. The use of the radioactive phosphorus isotope ^{32}P, a beta emit-

TABLE 19.3 Radioisotopes Used as Tracers

Isotope	Half-Life	Use
^{14}C	5730 yr	CO_2 for photosynthesis research
^{3}H	12.33 yr	Tag hydrocarbons
^{35}S	87.2 d	Tag pesticides, measure air flow
^{32}P	14.3 d	Measure phosphorus uptake by plants

Melvin Calvin used ^{14}C to monitor the uptake and release of CO_2 in order to determine the basic biochemical pathways of photosynthesis. This ground-breaking work earned him the 1961 Nobel Prize in Chemistry.

ter, presents a way not only of detecting the uptake of phosphorus by a plant but also of measuring the speed of uptake under various conditions. Plant biologists can grow hybrid strains of plants that absorb phosphorus, an essential nutrient, quickly. They can test the new plants by measuring their uptake of the radioactive ^{32}P tracer. This type of research leads to faster-maturing crops, better yields per acre, and more food or fiber at less expense.

Medical Imaging

Radioactive isotopes are used in **nuclear medicine** in two different ways: diagnosis and therapy. In the diagnosis of internal disorders such as tumors, physicians need information on the locations of abnormal tissue. This is done by **imaging,** a technique in which the radioisotope, either alone or combined with some other substance, accumulates at the site of the disorder. There, the radioisotope disintegrates and emits its characteristic radiation, which is detected. Modern medical diagnostic instruments not only determine where the radioisotope is located in the patient's body, but also construct an image of the volume within the body where the radioisotope is concentrated.

Four of the most common diagnostic radioisotopes are given in Table 19.4. Most are made in a particle accelerator in which heavy, charged nuclear particles are made to react with other atoms. Each of these radioisotopes produces gamma radiation, which in low doses is less harmful to the tissue than ionizing radiations such as beta or alpha particles because gamma rays pass through the tissue. By the use of special carrier compounds, these radioisotopes can be made to accumulate in specific areas of the body. For example, the pyrophosphate ion, $P_4O_7^{4-}$, can bond to the technetium-99m radioisotope; together they accumulate in the skeletal structure where abnormal

TABLE 19.4 Diagnostic Radioisotopes

Radioisotope	Name	Half-Life (Hours)	Site for Diagnosis
^{99m}Tc*	Technetium-99m	6.0	As TcO_4^- to the thyroid
^{201}Tl	Thallium-201	72.9	To the heart
^{123}I	Iodine-123	13.2	To the thyroid
^{67}Ga	Gallium-67	78.2	To various tumors and abscesses

*The technetium-99m isotope is the one most commonly used for diagnostic purposes. The m stands for "metastable," a term explained in the text.

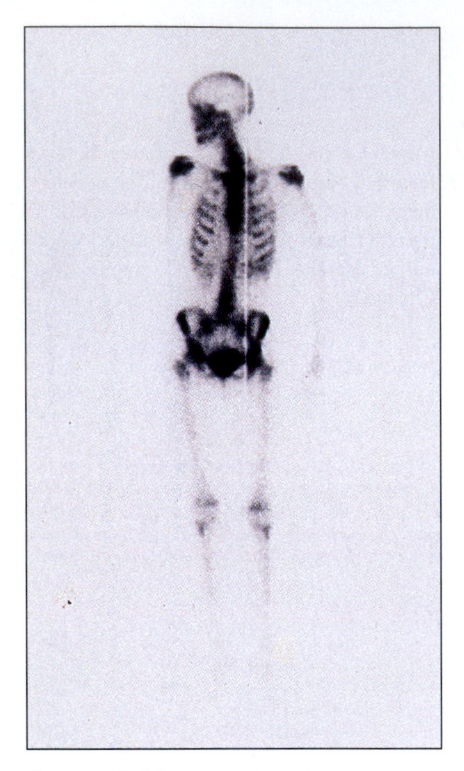

Figure 19.13 A whole body scan. Phosphate with technetium-99m was injected into the blood and then absorbed by the bones and kidneys. This picture was taken three hours after injection. *(SUNY Upstate Medical Center)*

bone metabolism is occurring (Figure 19.13). The technetium-99m radioisotope is metastable, as denoted by the letter *m*; this term means that the nucleus loses energy by disintegrating to a more stable version of the same isotope,

$$^{99m}Tc \longrightarrow {}^{99}Tc + \gamma$$

and the gamma rays are detected. Such investigations often pinpoint bone tumors.

Exercise 19.16 Rate of Radioactive Decay

Gallium citrate containing radioactive gallium-67 is used medically as a tumor-seeking agent. It has a half-life of 78.2 hours. How much time is needed for a gallium citrate sample to reach 10% of its original activity?

Exercise 19.17 Half-Life

Chromium-51 is a radioisotope ($t_{1/2}$ = 27.7 days) used to evaluate the lifetime of red blood cells; the radioisotope iron-59 ($t_{1/2}$ = 44.5 days) is used to assess bone marrow function. A hospital laboratory has 80 mg of iron-59 and 100 mg of chromium-51. After 90 days, which radioisotope is present in greater mass—chromium-51 or iron-59?

Paradoxically, high-energy radiation can kill healthy cells, but it is also used therapeutically to kill malignant, cancerous cells—those exhibiting rapid, uncontrolled growth. Malignant cells are killed because they are more susceptible to radiation damage since they divide more rapidly than normal cells. For external radiation therapy, a narrow beam of high-energy gamma radiation from a cobalt-60 or cesium-137 source is focused on the cancerous cells. Internal radiation therapy uses gamma-emitting salts of radioisotopes such as ^{192}Ir ($t_{1/2}$ = 73.8 days). The radioactive salts are encapsulated in platinum or gold "seeds" or needles and surgically implanted into the body. Because the thyroid gland uses iodine, thyroid cancer can be treated internally by administering orally a sodium iodide solution containing a relatively high concentration of radioactive iodine-131.

Positron emission tomography (PET) is a form of nuclear imaging that uses **positron emitters,** such as carbon-11, fluorine-18, nitrogen-13, or oxygen-15. All these radioisotopes are neutron-deficient, have short half-lives, and therefore must be prepared in a cyclotron immediately before use. When these radioisotopes decay, a proton is converted into a neutron, a positron, and a neutrino; the neutrino is generally not shown in the equation.

The neutrino, first detected experimentally in 1950, is a subatomic particle with 0 electric charge and a mass less than that of an electron.

$${}^{1}_{1}p \longrightarrow {}^{1}_{0}n + {}^{0}_{+1}e$$

Since matter is virtually transparent to neutrinos, they escape undetected, but the positron travels less than a few millimeters before it encounters an electron and undergoes antimatter-matter annihilation.

$${}^{0}_{+1}e + {}^{0}_{-1}e \longrightarrow 2\ \gamma$$

The annihilation event produces two gamma rays that radiate in opposite directions and are detected by two detectors located 180° apart in the PET scanner. By detecting several million annihilation gamma rays within a circular field around the subject over approximately 10 min, the region of tissue containing the radioisotope can be imaged with computer signal-averaging techniques (Figure 19.14).

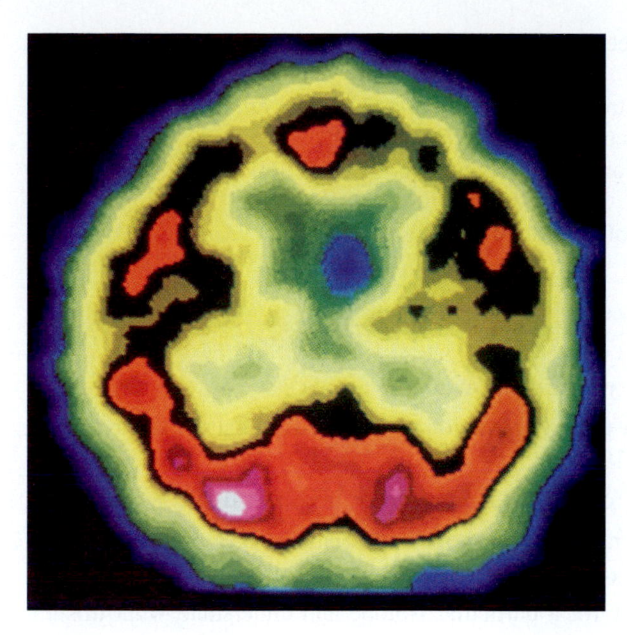

Figure 19.14 PET scan (positron emission tomography) of an axial section through a normal human brain. PET scans are obtained by injecting a tracer labeled with a short-lived radioactive isotope into the bloodstream. The isotope concentrates in brain tissue and emits positrons. The positrons react with electrons to create gamma rays that are recorded by a circular detector when the scan is performed. Here, radioactive methionine (an amino acid) has been used to show the level of activity of protein synthesis in the brain. *(CEA-ORSAY/CNRI/Science Photo Library/Photo Researchers, Inc.)*

SUMMARY PROBLEM

1. Plutonium-239 is an α emitter. Write the nuclear equation for α emission by ^{239}Pu.
2. Naturally occurring uranium contains more than 99% uranium-238 and less than 1% uranium-235. Therefore, sooner or later the supplies of ^{235}U will be depleted. Plutonium-239 is formed indirectly in nuclear reactors by the bombardment of uranium-238 with neutrons. The bombardment initially forms uranium-239 that is converted to plutonium-239 through two consecutive beta emissions. Write nuclear equations for the conversion of ^{238}U to ^{239}Pu.
3. Neutrons from the fission of uranium-235 in so-called breeder reactors can bombard nonfissionable ^{238}U and convert it into fissionable ^{239}Pu, such as that used in the atomic bomb exploded over Nagasaki in World War II. Plutonium-239 can fission to form, for example, strontium-90 and barium-147. Write a nuclear reaction for this fission reaction.
4. Hydrogen bombs that use fusion reactions were developed following World War II. One reaction used in a hydrogen bomb was

$$^{2}_{1}H + ^{3}_{1}H \longrightarrow ^{4}_{2}He + ^{1}_{0}n$$

Calculate the energy released, in kilojoules per gram of reactants, for this fusion reaction. The necessary nuclear masses are $^{2}_{1}H = 2.01355$ g/mol; $^{3}_{1}H = 3.01550$ g/mol; $^{4}_{2}He = 4.00150$ g/mol; $^{1}_{0}n = 1.00867$ g/mol.
5. The goal of recent nuclear arms treaties has been to dismantle the stockpiles of nuclear weapons built up by the United States and the former Soviet Union since World War II, including those containing plutonium-239 ($t_{1/2} = 2.4 \times 10^4$ y). How long will it take for the activity of plutonium-239 in a nuclear warhead to decrease (a) to 75% of its initial activity? (b) to 10% of its initial activity?
6. Deep underground burial has been proposed for long-term storage of the ^{239}Pu waste removed from nuclear weapons. Based on the answers to Summary Question

Part 5, comment on factors needed to be considered for the storage and burial of such nuclear waste.

7. Cobalt-60, used in cancer treatments, is a beta and gamma emitter ($t_{1/2} = 5.27$ y). A cobalt-60 sample has an activity of 7.0×10^{11} dps.

(a) Write a nuclear equation for the conversion of cobalt-60 to nickel-59 by beta emission.

(b) Calculate the activity of the cobalt-60 sample after 25.0 years.

IN CLOSING

Having studied this chapter, you should be able to . . .

- characterize the three major types of radiation observed in radioactive decay: alpha, beta, and gamma (Section 19.1).
- write a balanced equation for a nuclear reaction or transmutation (Section 19.2).
- decide whether a particular radioactive isotope will decay by α, β, or positron emission or by electron capture (Sections 19.2 and 19.3).
- calculate the binding energy for a particular isotope and understand what this energy means in terms of nuclear stability (Section 19.3).
- use the equation $\ln \frac{N}{N_0} = -kt$ (Equation 19.2), which relates (through the decay constant k) the time period over which a sample is observed (t) to the number of radioactive atoms present at the beginning (N_0) and end (N) of the time period (Section 19.4).
- calculate the half-life of a radioactive isotope ($t_{1/2}$) from the activity of a sample, or use the half-life to find the time required for an isotope to decay to a particular activity (Section 19.4).
- describe nuclear chain reactions, nuclear fission, and nuclear fusion (Sections 19.6 and 19.7).
- describe some sources of background radiation and the units used to measure radiation (Section 19.8).
- give examples of some uses of radioisotopes (Section 19.9).

KEY TERMS

The following terms were defined and given in boldfaced type in this chapter. You should be sure to understand each of these terms and the concepts with which each is associated.

activity *(19.4)*
alpha (α) particles *(19.1)*
alpha radiation *(19.1)*
background radiation *(19.8)*
beta (β) particles *(19.1)*
binding energy *(19.3)*
binding energy per nucleon *(19.3)*
curie (Ci) *(19.8)*
electron capture *(19.2)*
gamma (γ) radiation *(19.1)*
half-life, $t_{1/2}$ *(19.4)*
imaging *(19.9)*
nuclear (atomic) reactor *(19.6)*
nuclear fission *(19.3, 19.6)*
nuclear fusion *(19.3, 19.7)*
nuclear medicine *(19.9)*
nuclear reactions *(19.2)*
nucleons *(19.2)*
plasma *(19.7)*
positron *(19.2)*
positron emitters *(19.9)*
rad *(19.8)*
radioactive series *(19.2)*
rem *(19.8)*
röntgen (R) *(19.8)*
tracers *(19.9)*

Questions for Review and Thought

Conceptual Challenge Problems

CP-19.A. The half-life for the alpha decay of uranium-238 to thorium-234 is 4.5×10^9 years, which happens to be the estimated age of the earth.

(a) How many atoms were decaying per second in a 1.0-g sample of uranium-238 that existed 1.0×10^6 years ago?

(b) How would you find the number of atoms now decaying per second in this sample?

CP-19.B. If the earth is 4.5×10^9 years old and the amount of radioactivity in a sample becomes smaller with time, how is it possible for there to be any radioactive elements left on earth that have half-lives less than a few million years?

CP-19.C. Using experiments based on a sample of living wood, a radiochemist estimates that the uncertainty of her measurements of the carbon-14 radioactivity in the samples is 1.0%. The half-life of carbon-14 is 5730 y.

(a) How long must a sample of wood be separated from a living tree before the chemist's radioactivity measurements on the sample support the time when it died?

(b) Suppose that the chemist's uncertainty in the radioactivity of carbon-14 continues to be 1.0% of the radioactivity of living wood. How long must a sample of wood be dead before the chemist's measurements support the claim that the time since the wood was separated from the tree is not changing?

CP-19.D. You have read that alpha radiation is the least penetrating type of radiation, followed by beta radiation. Gamma radiation penetrates matter well, and thick samples of matter are required to contain gamma radiation. Knowing these facts, what can you correctly deduce about the harmful effects of these three types of radiation on living tissue?

CP-19.E. Death will likely occur within weeks to a 150-lb person who receives 500,000 mrem of radiation over a short time, an exposure that is one thousand times the federal government's standard for 1 year (500 mrem/y). A student realizes that 500,000 mrem is 500 rem, and that 500 rem has the effect of depositing 317 J of energy on the body of the 150-lb person. The student is puzzled. How can the deposition of only 317 J of energy from nuclear radiation, much less energy than that deposited by cooling a cup of coffee 1 °C within a person's body, have such a disastrous effect on the person?

Review Questions

1. Complete the table.

	Symbol	Mass	Charge	Ionizing Power	Penetrating Power
α Particle					
β Particle					
γ Radiation					

2. Compare nuclear and chemical reactions in terms of changes in reactants, type of products formed, and conservation of matter and energy.

3. What is meant by "the band of stability"?

4. What is the binding energy of a nucleus?

5. If the mass number of an isotope is much greater than twice the atomic number, what type of radioactive decay might you expect?

6. If the number of neutrons in an isotope is much less than the number of protons, what type of radioactive decay might you expect?

7. Define critical mass and chain reaction.

8. What is the difference between nuclear fission and fusion? Illustrate your answer with an example of each.

9. Locate the nuclear reactor power plant nearest to your college residence. Do you consider it a threat to your health and safety? If so, why? If not, why not?

10. Name at least two uses of radioactive isotopes (outside of their use in power reactors and weapons).

Nuclear Reactions

11. Fill in the mass number, atomic number, and symbol for the missing particle in each nuclear equation.

(a) $^{242}_{94}Pu \rightarrow {}^{4}_{2}He +$ ________

(b) ________ $\rightarrow {}^{32}_{16}S + {}^{0}_{-1}e$

(c) $^{252}_{98}Cf +$ ________ $\rightarrow 3\,{}^{1}_{0}n + {}^{259}_{103}Lr$

(d) $^{55}_{26}Fe +$ ________ $\rightarrow {}^{55}_{25}Mn$

(e) $^{15}_{8}O \rightarrow$ ________ $+ {}^{0}_{+1}e$

12. Fill in the mass number, atomic number, and symbol for the missing particle in each nuclear equation.

(a) ________ $\rightarrow {}^{22}_{10}Ne + {}^{0}_{+1}e$

(b) $^{122}_{53}I \rightarrow {}^{122}_{54}Xe +$ ________

(c) $^{210}_{84}Po \rightarrow$ ________ $+ {}^{4}_{2}He$

(d) $^{195}_{79}Au +$ ________ $\rightarrow {}^{195}_{78}Pt$

(e) $^{241}_{94}Pu + {}^{16}_{8}O \rightarrow 5\,{}^{1}_{0}n +$ ________

13. Write balanced nuclear equations for each word statement.
(a) Magnesium-28 undergoes β emission.
(b) When uranium-238 is bombarded with carbon-12, four neutrons are emitted and a new element forms.
(c) Hydrogen-2 and helium-3 react to form helium-4 and another particle.
(d) Argon-38 forms by positron emission.
(e) Platinum-175 forms osmium-171 by spontaneous radioactive decay.

14. Write balanced nuclear equations for each word statement.
(a) Einsteinium-253 combines with an alpha particle to form a neutron and a new element.
(b) Nitrogen-13 undergoes positron emission.
(c) Iridium-178 captures an electron to form a stable nucleus.
(d) A proton and boron-11 fuse together, forming three identical particles.
(e) Nobelium-252 and six neutrons form when carbon-12 collides with a transuranium isotope.

15. One radioactive series that begins with uranium-235 and ends with lead-207 undergoes the following sequence of emission reactions: α, β, α, β, α, α, α, α, β, β, α. Identify the radioisotope produced in each of the *first five steps.*

16. One radioactive series that begins with uranium-235 and ends with lead-207 undergoes the following sequence of emission reactions: α, β, α, β, α, α, α, α, β, β, α. Identify the radioisotope produced in each of the *last six steps.*

Nuclear Stability

17. Write a nuclear equation for the type of decay each of these unstable isotopes is most likely to undergo.
(a) Neon-19 (c) Bromine-82
(b) Thorium-230 (d) Polonium-212

18. Write a nuclear equation for the type of decay each of these unstable isotopes is most likely to undergo.
(a) Silver-114 (c) Radium-226
(b) Sodium-21 (d) Iron-59

19. Boron has two stable isotopes, ^{10}B (abundance = 19.78%) and ^{11}B (abundance = 80.22%). Calculate the binding energies per nucleon of these two nuclei and compare their stabilities.

$$5\ ^{1}_{1}H + 5\ ^{1}_{0}n \rightarrow ^{10}_{5}B$$
$$5\ ^{1}_{1}H + 6\ ^{1}_{0}n \rightarrow ^{11}_{5}B$$

The required masses (in g/mol) are $^{1}_{1}H = 1.00783$; $^{1}_{0}n = 1.00867$; $^{10}_{5}B = 10.01294$; and $^{11}_{5}B = 11.00931$.

20. Calculate the binding energy in kJ per mole of P for the formation of $^{30}_{15}P$ and $^{31}_{15}P$.

$$15\ ^{1}_{1}H + 15\ ^{1}_{0}n \rightarrow ^{30}_{15}P$$
$$15\ ^{1}_{1}H + 16\ ^{1}_{0}n \rightarrow ^{31}_{15}P$$

Which is the more stable isotope? The required masses (in g/mol) are $^{1}_{1}H = 1.00783$; $^{1}_{0}n = 1.00867$; $^{30}_{15}P = 29.97832$; and $^{31}_{15}P = 30.97376$.

Rates of Disintegration Reactions

21. Sodium-24 is a diagnostic radioisotope used to measure blood circulation time. How much of a 20-mg sample remains after 1 d and 6 h if sodium-24 has a $t_{1/2}$ = 15 h?

22. Iron-59 in the form of iron(II) citrate is used in iron metabolism studies. Its half-life is 45.6 d. If you start with 0.56 mg of iron-59, how much would remain after 1 y?

23. Iodine-131 is used in the form of sodium iodide to treat cancer of the thyroid. (a) The isotope decays by ejecting a β particle. Write a balanced equation to show this process. (b) The isotope has a half-life of 8.05 d. If you begin with 25.0 mg of radioactive $Na^{131}I$, what mass remains after 32.2 d?

24. Phosphorus-32 is used in the form of Na_2HPO_4 in the treatment of chronic myeloid leukemia, among other things. (a) The isotope decays by emitting a β particle. Write a balanced equation to show this process. (b) The half-life of ^{32}P is 14.3 d. If you begin with 9.6 mg of radioactive Na_2HPO_4, what mass remains after 28.6 d?

25. What is the half-life of a radioisotope if it decays to 12.5% of its radioactivity in 12 y?

26. After 2 hr, tantalum-172 has 1/16 of its initial radioactivity. How long is its half-life?

27. Radioisotopes of iodine are widely used in medicine. For example, iodine-131 ($t_{1/2}$ = 8.05 days) is used to treat thyroid cancer. If you ingest a sample of NaI containing ^{131}I, how much time is required for the isotope to fall to 5.0% of its original activity?

28. The rare gas radon has been the focus of much attention recently because it may be found in homes. Radon-222 emits α particles and has a half-life of 3.82 d. (a) Write a balanced equation to show this process. (b) How long does it take for a sample of radon to decrease to 10.0% of its original activity?

29. A sample of wood from a Thracian chariot found in an excavation in Bulgaria has a ^{14}C activity of 11.2 disintegrations per minute per gram. Estimate the age of the chariot and the year it was made. ($t_{1/2}$ for ^{14}C is 5.73×10^3 y, and the activity of ^{14}C in living material is 15.3 disintegrations per minute per gram.)

30. A piece of charred bone found in the ruins of an American Indian village has a ^{14}C to ^{12}C ratio of 0.72 times that found in living organisms. Calculate the age of the bone fragment. (See Question 29 for required data on carbon-14.)

Artificial Transmutations

31. There are two isotopes of americium, both with half-lives sufficiently long to allow the handling of massive quantities. Americium-241, for example, has a half-life of 248 y as an α emitter, and it is used in gauging the thickness of materials and in smoke detectors. The isotope is formed from ^{239}Pu by absorption of two neutrons followed by emission of a β particle. Write a balanced equation for this process.

32. Americium-240 is made by bombarding a plutonium-239 atom with an α particle. In addition to ^{240}Am, the products are a proton and two neutrons. Write a balanced equation for this process.

33. To synthesize the heavier transuranium elements, one must bombard a lighter nucleus with a relatively large particle. If you know the products are californium-246 and 4 neutrons, with what particle would you bombard uranium-238 atoms?

34. The element with the highest known atomic number is 112. It is thought that still heavier elements are possible, especially with $Z = 114$ and $N = 184$. To this end, serious attempts have been made to force calcium-40 and curium-248 to merge. What would be the atomic number of the element formed?

Nuclear Fission and Fusion

35. Name the fundamental parts of a nuclear fission reactor and describe their functions.

36. Explain why no commercial fusion reactors are in operation today.

37. The average energy output of a good grade of coal is 2.6×10^7 kJ/ton. Fission of 1 mol of ^{235}U releases 2.1×10^{10} kJ. Find the number of tons of coal needed to produce the same energy as 1 lb of ^{235}U.

38. A concern in the nuclear power industry is that, if nuclear power becomes more widely used, there may be serious shortages in worldwide supplies of fissionable uranium. One solution is to build "breeder" reactors that manufacture more fuel than they consume. One such cycle works as follows:
(i) A ^{238}U nucleus collides with a neutron to produce ^{239}U.
(ii) ^{239}U decays by β emission ($t_{1/2} = 24$ min) to give an isotope of neptunium.
(iii) This neptunium isotope decays by β emission to give a plutonium isotope.
(iv) The plutonium isotope is fissionable. On collision of one of these plutonium isotopes with a neutron, fission occurs with energy, at least two neutrons, and other nuclei as products.
Write an equation for each of the steps, and explain how this process can be used to breed more fuel than the reactor originally contained and still produce energy.

Effects of Nuclear Radiation

39. Two common units of radiation used in newspaper and news magazine articles are the rad and rem. What do each measure? Which would you use in an article describing the damage an atomic bomb would have on a human population?

40. Which electrical power plant—fossil-fuel or nuclear—exposes a community to more radiation? Explain why.

41. Explain how our own bodies are sources of radiation.

42. What is the source of radiation during jet plane travel?

Uses of Radioisotopes

43. Why are foods irradiated with gamma emitters instead of alpha or beta emitters?

44. X rays and PET scans are two medical imaging techniques. How are they similar and different?

45. In order to measure the volume of the blood system of an animal, the following experiment was done. A 1.0-mL sample of an aqueous solution containing tritium with an activity of 2.0×10^6 disintegrations per second (dps) was injected into the bloodstream. After time was allowed for complete circulatory mixing, a 1.0-mL blood sample was withdrawn and found to have an activity of 1.5×10^4 dps. What was the volume of the circulatory system? (The half-life of tritium is 12.3 years, so this experiment assumes that only a negligible amount of tritium has decayed in the time of the experiment.)

46. Radioactive isotopes are often used as "tracers" to follow an atom through a chemical reaction, and the following is an example. Acetic acid reacts with methanol, CH_3OH, by eliminating a molecule of H_2O to form methyl acetate, $CH_3CO_2CH_3$. Explain how you would use the radioactive isotope ^{18}O to show whether the oxygen atom in the water product comes from the —OH of the acid or the —OH of the alcohol.

$$\underset{\text{acetic acid}}{CH_3COOH} + \underset{\text{methanol}}{HOCH_3} \longrightarrow \underset{\text{methyl acetate}}{CH_3COOCH_3} + H_2O$$

General Questions

47. Complete the following nuclear equations.
(a) $^{214}Bi \rightarrow$ ________ + ^{214}Po
(b) $4\ ^1_1H \rightarrow$ ________ + 2 positrons
(c) ^{249}Es + neutron $\rightarrow$ 2 neutrons + ________ + ^{161}Gd
(d) $^{220}Rn \rightarrow$ ________ + alpha particle
(e) ^{68}Ge + electron $\rightarrow$ ________

48. Complete the following nuclear equations.
(a) ________ + neutron $\rightarrow$ 2 neutrons + ^{137}Tc + ^{97}Zr
(b) $^{45}Ti \rightarrow$ ________ + positron
(c) ________ $\rightarrow$ beta particle + ^{59}Co
(d) ^{24}Mg + neutron $\rightarrow$ ________ + proton
(e) ^{131}Cs + ________ $\rightarrow$ ^{131}Xe

49. Radioactive nitrogen-13 has a half-life of 10 min. After an hour, how much of this isotope remains in a sample that originally contained 96 mg?

50. The half-life of molybdenum-99 is 67.0 h. How much of a 1.000-mg sample of ^{99}Mo is left after 335 h? How many half-lives did it undergo?

51. The oldest known fossil cells form a biological cluster found in South Africa. The fossil has been dated by the reaction

$$^{87}Rb \longrightarrow {}^{87}Sr + {}^{0}_{-1}e \qquad t_{1/2} = 4.9 \times 10^{10}\ y$$

If the ratio of the present quantity of ^{87}Rb to the original quantity is 0.951, calculate the age of the fossil cells.

52. Cobalt-60 is a therapeutic radioisotope used in treating certain cancers. If a sample of cobalt-60 initially disintegrates at a rate of 4.3×10^6 dps and after 21.2 y the rate has dropped to 2.6×10^5 dps, what is its half-life?

53. Balance the following equations used for the synthesis of transuranium elements.

(a) $^{238}_{92}U + ^{14}_{7}N \rightarrow$ ________ $+ 5\ ^{1}_{0}n$
(b) $^{238}_{92}U +$ ________ $\rightarrow ^{249}_{100}Fm + 5\ ^{1}_{0}n$
(c) $^{253}_{99}Es +$ ________ $\rightarrow ^{256}_{101}Md + ^{1}_{0}n$
(d) $^{246}_{96}Cm +$ ________ $\rightarrow ^{254}_{102}No + 4\ ^{1}_{0}n$
(e) $^{252}_{98}Cf +$ ________ $\rightarrow ^{257}_{103}Lr + 5\ ^{1}_{0}n$

54. On December 2, 1942, the first man-made self-sustaining nuclear fission chain reactor was operated by Enrico Fermi and others under the University of Chicago Stadium. In June 1972 natural fission reactors, which operated billions of years ago, were discovered in Oklo, Gabon. At present, natural uranium contains 0.72% ^{235}U. How many years ago did natural uranium contain 3.0% ^{235}U, sufficient to sustain a natural reactor? ($t_{1/2}$ for ^{235}U is 7.04×10^8 y.)

Applying Concepts

55. If a radioisotope is used for diagnosis (e.g., detecting cancer), it should decay by gamma radiation. However, if its use is therapeutic (e.g., treating cancer), it should decay by alpha or beta radiation. Explain why in terms of ionizing and penetrating power.

56. During the Three Mile Island incident, people in central Pennsylvania were concerned that strontium-90 (a beta emitter) released from the reactor could become a health threat. Where would this isotope collect in the body? What types of problems could it cause?

57. Classify the isotopes ^{17}Ne, ^{20}Ne, and ^{23}Ne as stable or unstable. What type of decay would you expect the unstable isotope(s) to have?

58. The following demonstration was carried out to illustrate the concept of a nuclear chain reaction. Explain the connections between the demo and the reaction.

Eighty mouse traps are arranged side-by-side in eight rows of ten traps each. Each trap is set with two rubber stoppers for bait. A small plastic mouse is tossed into the middle of the traps, setting off one trap, which in turn sets off two traps and so on until all the traps are sprung.

59. Most students have no trouble understanding that 1.5 g of a 24-g sample of a radioisotope would remain after 8 h if it had a $t_{1/2} = 2$ h. What they don't always understand is where the other 22.5 g disappeared to. How would you explain this disappearance to another student?

60. Nuclear chemistry is a topic that raises many debatable issues. Briefly discuss your views on the following.
(a) Twice a year the general public is allowed to visit the Trinity Site in Alamogordo, New Mexico, where the first atomic bomb was tested. If you had the opportunity to do so, would you visit the site? Explain your answer.
(b) Now that the Cold War has ended should the United States stockpile nuclear weapons? Explain your answer.
(c) The FDA allows irradiated grapefruits to be exported from, but not sold in, the United States. Is it acceptable to sell to other nations food that is not approved for our domestic consumption? Explain your answer.

APPENDICES

APPENDIX A

Problem Solving

In this book, we have provided many illustrations of problem solving and many problems for practice. Some are numerical problems that must be solved by mathematical calculations. Others are conceptual problems that must be solved by applying an understanding of the principles of chemistry. Often, it is necessary to use chemical concepts to relate what we know about matter at the nanoscale to the properties of matter at the macroscale. The problems throughout this book are representative of the kinds of problems chemists and other scientists must regularly solve to pursue their goals, although our problems are often not as difficult as those encountered in the real world.

Problem solving is not a simple skill that can be mastered in a few hours of study or practice. Because there are many different kinds of problems and many different kinds of people who are problem solvers, there are no hard and fast rules that are guaranteed to lead you to solutions. The general guidelines presented in this Appendix are, however, helpful in getting you started on any kind of problem and in checking to see if your answers are correct. The problem-solving skills you develop in a chemistry course such as this can later be applied to difficult and important problems that may arise in your profession, your personal life, or the society in which you live.

In getting a clear picture of a problem and asking appropriate questions regarding the problem, you need to keep in mind all the principles of chemistry and other subjects that you think may apply. In many real-life problems there is not enough information available for you to arrive at an unambiguous solution; in such cases, try to look up or estimate what is needed and then go ahead, noting assumptions you have made. Often the hardest part is deciding which principle or idea is most likely to help solve the problem and what information is needed. To some degree this can be a matter of luck or chance. Nevertheless, in the words of Louis Pasteur, "In the field of observation, chance only favors those minds which have been prepared." The more practice you have had, and the more principles and facts you can keep in mind, the more likely you are to be able to solve the problems that you face.

A.1 GENERAL PROBLEM-SOLVING STRATEGIES

1. **Define the problem.** Carefully review the information contained in the problem. What is the problem asking you to find? What key principles are involved? What known information is necessary to solving the problem and what is only there to place the question in context? Organize the information to see what is necessary and to see the relationships among the known data. Try writing the information down in a table. If it is numerical information, be sure to include proper units. Can you picture the situation under consideration? Try sketching it and including any relevant dimensions in the sketch.
2. **Develop a plan.** Have you solved a problem of this type before? If you recognize the new problem as similar to ones you know how to solve, you can use the same

method that worked before. Try reasoning backward from the units of what is being sought. What data are needed to find an answer in those units?

Can the problem be broken down into smaller pieces, each of which can be solved separately to produce information that can be assembled to solve the entire problem? When a problem can be divided into simpler problems, it often helps to write down a plan that lists the simpler problems and the order in which they must be put together to arrive at an overall solution. Many major problems in chemical research have to be solved in this way. In problems in this book we have mostly provided the needed numerical data, but in the laboratory, the first piece of a problem is often devising experiments to gather the data.

3. **Execute the plan.** Carefully write down each step of a mathematical problem, being sure to properly keep track of the units. Do the units cancel out to give you the answer in the desired units? Don't skip steps. Don't do any but the simplest steps in your head. Once you've written down the steps of a mathematical problem check what you've written—is it all correct? Students often say they got a problem wrong because they "made a stupid mistake." Your instructor—and textbook authors—make mistakes, too. It is usually because they don't take the time to write down the steps of a problem clearly and correctly. In solving a mathematical problem, remember to apply the principles of dimensional analysis (introduced in Section 2.7 and reviewed below) and significant figures (Appendix A.3).
4. **Check the answer to see if it is reasonable.** As a final check of the solution to any problem, ask yourself if the answer is reasonable: Are the units of a numerical answer correct? Is a numerical answer of about the right size? Don't just copy an answer off the calculator without thinking about whether it makes sense.

Let us say you are asked to convert 100. yd to a distance in meters. Using dimensional analysis, and some well-known factors for converting from the English system to the metric system, we have

$$100.\ \cancel{\text{yards}} \times \frac{3\ \text{ft}}{1\ \cancel{\text{yard}}} \times \frac{12\ \cancel{\text{in.}}}{1\ \text{ft}} \times \frac{2.54\ \cancel{\text{cm}}}{1\ \cancel{\text{in.}}} \times \frac{1\ \text{m}}{100\ \cancel{\text{cm}}} = 91.4\ \text{m}$$

You should recognize that a distance of 91.4 m is about right. Because a meter is a little longer than 1 yd, 100 yd should be a little less than 100 m. In the first step, if you divided instead of multiplied by 3, the final answer would be a little more than 10 m. This is equivalent to only about 30 ft, and you know a 100-yd football field is longer than that.

A.2 NUMBERS, UNITS, AND QUANTITIES

Many scientific problems require you to use mathematics to calculate a result or draw a conclusion. Therefore, knowledge of mathematics and its application to problem solving is important. However, one aspect of scientific calculations is often absent from pure mathematical work: science deals with *measurements* in which an unknown quantity is compared with a standard or unit of measure. For example, using a balance to determine the mass of an object involves comparing the object's mass with standard masses, usually in multiples or fractions of one gram; the result is reported as some number of grams, say 4.357 g. *Both the number and the unit are important.* If the result had been 123.5 g, this would clearly be different, but a result of 4.357 oz (ounces) would also be different, because the unit "ounce" is different from the unit "gram." *A result that describes the magnitude of a property,* such as

4.357 g, is called a **quantity,** and chemical problem solving requires calculating with quantities. Notice that whether a quantity is large or small depends on the units as well as the number; the two quantities 123.5 g and 4.357 oz represent the *same* mass.

A quantity is always treated as though the number and the units are multiplied together; that is, 4.357 g can be handled mathematically as $4.357 \times \text{g}$. Keeping in mind this simple rule, calculations involving quantities follow the normal rules of algebra and arithmetic: $5\text{ g} + 7\text{ g} = (5 + 7) \times \text{g} = 12\text{ g}$; or $6\text{ g} \div 2\text{ g} = (6\cancel{\text{g}})/(2\cancel{\text{g}}) = 3$. (Notice that in the second calculation the unit g appears in numerator and denominator and cancels out, leaving a pure number, 3.) Treating units as algebraic entities has the advantage that *if a calculation is set up correctly, the units will cancel out or multiply together so that the final result has appropriate units.* For example, if you measured the size of a sheet of paper and found it to be 8.5 in. by 11 in., the area A of the sheet could be calculated as $A = l \times w = 11\text{ in.} \times 8.5\text{ in.} = 94\text{ in.}^2$, or 94 square inches. If a calculation is set up incorrectly, the units of the result will be inappropriate. Using units to check whether a calculation has been properly set up is called **dimensional analysis.**

This idea of using algebra on units as well as numbers is useful in all kinds of situations. For example, suppose you are having a party for some friends who like pizza. A large pizza consists of 12 slices and costs $10.75. You expect to need 36 slices of pizza and want to know how much you will have to spend. A strategy for solving the problem is first to figure out how many pizzas you need and then to figure the cost in dollars. This solution could be diagrammed as:

$$\text{Slices} \xrightarrow[\text{step 1}]{\text{slices per pizza}} \text{pizzas} \xrightarrow[\text{step 2}]{\text{dollars per pizza}} \text{dollars}$$

Step 1. Find the number of pizzas required by dividing the number of slices per pizza into the number of slices, thus converting "units" of slices to "units" of pizzas:

$$\text{Number of pizzas} = 36\ \cancel{\text{slices}} \left(\frac{1\text{ pizza}}{12\ \cancel{\text{slices}}}\right) = 3\text{ pizzas}$$

Strictly speaking, slices and pizzas are not units in the same sense that a gram is a unit; however, labeling things this way will often help you keep in mind what a number refers to—pizzas, slices, or dollars in this case.

Notice that if you had multiplied the number of slices times the number of slices per pizza, the result would have been labeled pizza $\times$ slices^2, which does not make sense. In other words, the labels indicate whether multiplication or division is appropriate.

Step 2. Find the total cost by multiplying the cost per pizza times the number of pizzas needed, thus converting "units" of pizzas to "units" of dollars:

$$\text{Total price} = 3\ \cancel{\text{pizzas}} \left(\$10.75/1\ \cancel{\text{pizza}}\right) = \$32.25$$

Notice that in each step you have multiplied by a factor that allowed the initial units to cancel algebraically, giving the answer in the desired unit. A factor such as (1 pizza/12 slices) or ($10.75/pizza) is referred to as a **proportionality factor.** This name indicates that it comes from a proportion. For instance, in the pizza problem you could set up the proportion:

$$\frac{x\text{ pizzas}}{36\text{ slices}} = \frac{1\text{ pizza}}{12\text{ slices}} \quad \text{or} \quad x\text{ pizzas} = 36\ \cancel{\text{slices}} \left(\frac{1\text{ pizza}}{12\ \cancel{\text{slices}}}\right) = 3\text{ pizzas}$$

A proportionality factor such as (1 pizza/12 slices) is also called a **conversion factor,** which indicates that it converts one kind of unit or label to another; in this case the label "slices" is converted to the label "pizzas."

Many everyday scientific problems involve proportionality. For example, the bigger the volume of a solid or liquid substance, the bigger its mass. When the volume is zero, the mass is zero also. These facts indicate that mass, m, is directly proportional to volume, V, or, symbolically,

$$m \propto V$$

where the symbol $\propto$ means "is proportional to." Whenever a proportion is expressed this way, it can also be expressed as an equality by using a proportionality constant; for example,

$$m = d \times V$$

In this case the proportionality constant, d, is called the density of the substance. This equation embodies the definition of density as mass per unit volume, since it can be rearranged algebraically to

$$d = \frac{m}{V}$$

As with any algebraic equation involving three variables, it is possible to calculate any one of the three quantities m, V, or d, provided the other two are known. If density is wanted, simply use the definition of mass per unit volume; if mass or volume is to be calculated, the density can be used as a proportionality factor.

Suppose that you are going to buy a ton of gravel and want to know how big a bin you will need to store it. You know the mass of gravel and want to find the volume of the bin; this implies that density will be useful. If the gravel is primarily limestone, you can assume that its density is about the same as for limestone and look it up. Limestone has the chemical formula $CaCO_3$ and its density is 2.7 kg/L. However, these mass units are different from the units for mass of gravel, namely tons. Therefore you need to recall or look up the mass of 1 ton (2000 pounds, lb) and the fact that there are 2.20 lb per kg. This provides enough information to calculate the volume needed.

Step 1. Figure out how many kilograms of gravel are in a ton.

$$m_{\text{gravel}} = 1 \text{ ton} = 2000 \text{ lb} = 2000 \cancel{\text{lb}} \left(\frac{1 \text{ kg}}{2.20 \cancel{\text{lb}}}\right) = 909 \text{ kg}$$

The fact that there are 2.20 lb per kg implies two proportionality factors: (2.20 lb/1 kg) and (1 kg/2.20 lb). The latter was used because it results in appropriate cancellation of units.

Step 2. Use the density to calculate the volume of 909 kg gravel.

$$V_{\text{gravel}} = \frac{m_{\text{gravel}}}{d_{\text{gravel}}} = \frac{909 \text{ kg}}{2.7 \text{ kg/L}} = 909 \cancel{\text{kg}} \left(\frac{1 \text{ L}}{2.7 \cancel{\text{kg}}}\right) = 340 \text{ L}$$

In this step we used the definition of density, solved algebraically for mass, substituted the two known quantities into the equation, and calculated the result. However, it is quicker simply to remember that mass and volume are related by a proportionality factor called density, and to use the units of the quantities to decide whether to multiply or divide by that factor. In this case we divided mass by density because the units kilograms canceled, leaving a result in liters, which is a unit of volume.

The liter is not the most convenient volume unit for this problem, however, because it does not relate well to what we want to find out—how big a bin to make. A liter is about the same volume as a quart, but whether you are familiar with liters or quarts or both, 300 of them is not easy to visualize. Let's convert these units to something we can understand better. A liter is a volume equal to a cube one-tenth of a meter (1 dm) on a side; that is, a liter is 1 dm^3. Consequently,

$$340 \text{ L} = 340 \text{ L}\left(\frac{1 \text{ dm}^3}{1 \text{ L}}\right)\left(\frac{1 \text{ m}}{10 \text{ dm}}\right)^3 = 340 \text{ dm}^3\left(\frac{1 \text{ m}^3}{1000 \text{ dm}^3}\right) = 0.34 \text{ m}^3$$

Thus the bin would need to have a volume of about one third of a cubic meter; that is, it could be a meter wide, a meter long, and about a third of a meter high and it would hold the ton of gravel.

One more thing should be noted about this example. We don't need to know the volume of the bin very precisely, because being off a bit will make very little difference; it might mean getting a little too much wood to build the bin, or not making the bin quite big enough and having a little gravel spill out, but this isn't a big deal. In other cases, such as calculating the quantity of fuel needed to get a space shuttle into orbit, being off by a few percent could be a life-or-death matter. Because it is important to know how precise data are, and to be able to evaluate how important precision is, scientific results usually indicate precision. The simplest way to do this is by means of significant figures.

A.3 SIGNIFICANT FIGURES

The **precision** of a measurement indicates how well several determinations of the same quantity agree. Precision is illustrated by the results of throwing darts at a bulls-eye (Figure A.1). In part a the darts are scattered all over the board; the dart thrower was apparently not very skillful (or threw the darts from a long distance away from the board), and the precision of their placement on the board is low. In Figure A.1b the darts are all clustered together, indicating much better reproducibility on the part of the thrower, that is, greater precision. In addition, every dart has come very close to the bull's-eye; this is described by saying that the thrower has been quite **accurate**—the average of all throws is very close to the accepted position, namely the bull's-eye. Figure A.1c illustrates that it is possible to be precise without being accurate—the dart thrower has consistently missed the bull's-eye, although all darts are

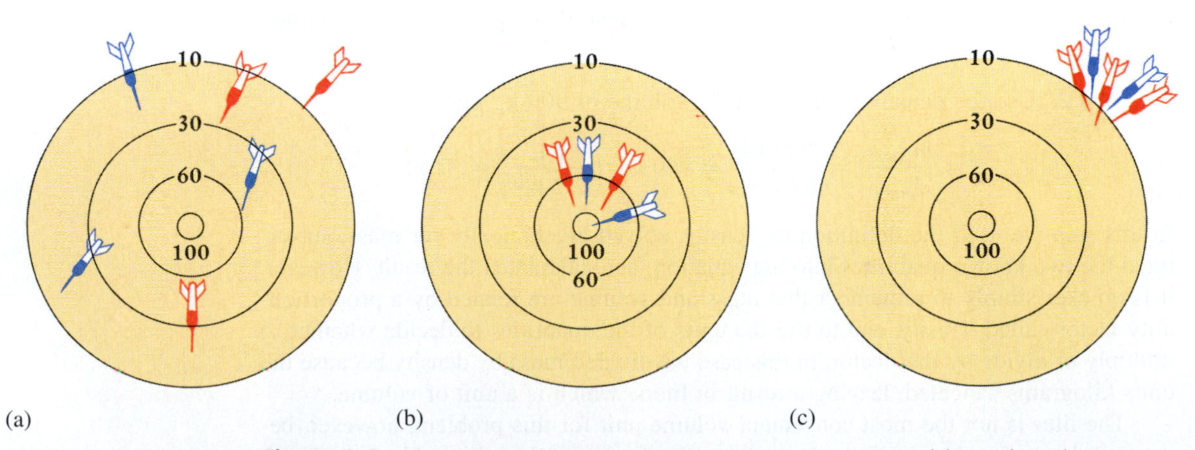

Figure A.1 Precision and accuracy: (a) poor precision; (b) good precision and good accuracy; (c) good precision and poor accuracy.

clustered very precisely around the wrong point on the board. This third case is like an experiment with some flaw (either in its design or in a measuring device) that causes all results to differ from the correct value by the same amount.

In the laboratory we attempt to set up experiments so that the greatest possible accuracy can be obtained. As a further check on accuracy, results are usually compared among different laboratories so that any flaw in experimental design or measurement can be detected. For each individual experiment, several measurements are usually made and their precision determined. Usually better precision is taken as an indication of better experimental work, and it is necessary to know precision in order to compare results among different experimenters. If two different experimenters both had results like those in Figure A.1a, their average values could differ quite a lot before they would say that their results did not agree within experimental error.

In most experiments several different kinds of measurements must be made, and some can be done more precisely than others. It is common sense that *a calculated result can be no more precise than the least precise piece of information that went into the calculation.* This is where the rules for *significant figures* come in. In the preceding example the quantity of gravel was described as "a ton." Usually gravel is measured by weighing a truck empty, putting some gravel in the truck, weighing the truck again, and subtracting the weight of the truck from the weight of the truck plus gravel. The quantity of gravel is not adjusted if there is a bit too much or a bit too little, because this would be a lot of trouble. You might end up with as much as 2200 pounds or as little as 1800 pounds, even though you asked for a ton. In terms of significant figures this would be expressed as 2.0×10^3 lb.

The quantity 2.0×10^3 lb is said to have two significant figures; it designates a quantity where the 2 is taken to be exactly right but the 0 is not known precisely. (In this case the number could be as large as 2.2 or as small as 1.8, and so the 0 obviously is not exactly right.) In general, in a number that represents a scientific measurement, the last digit on the right is taken to be inexact, but all digits farther to the left are assumed to be exact. When you do calculations using such numbers, you must follow some simple rules so that the results will reflect the precision of all the measurements that go into the calculations. Here are the rules:

Rule 1. To determine the number of significant figures in a measurement, read the number from left to right and count all digits, starting with the first digit that is *not* zero.

Example	Number of Significant Figures
1.23 g	3
0.00123 g	3; the zeros to the *left* of the 1 simply locate the decimal point. To avoid confusion, write numbers of this type in scientific notation; thus, $0.00123 = 1.23 \times 10^{-3}$.
2.0 g and 0.020 g	2; both have two significant digits. When a number is greater than 1, *all zeros to the right of the decimal point are significant.* For a number less than 1, only zeros to the right of the first significant digit are significant.
100 g	1; in numbers that do not contain a decimal point, "trailing" zeros may or may not be significant. To eliminate possible confusion, the practice followed in this book is to include a decimal point if the zeros are significant. Thus, 100. has three significant digits,

For a number written in scientific notation, all digits are significant.

The number π is now known to 1,011,196,691 digits. It is doubtful that you will need this accuracy in this course—or ever.

Example	Number of Significant Figures
	while 100 has only one. Alternatively, we write it in scientific notation as 1.00×10^2 (three significant digits) or as 1×10^2 (one significant digit). For a number written in scientific notation, all digits are significant.
100 cm/m	Infinite number of significant figures, because this is a defined quantity.
$\pi = 3.1415926 \ldots$	The value of π is known to a greater number of significant figures than any data you will ever use in a calculation.

Rule 2. When adding or subtracting, the number of decimal places in the answer should be equal to the number of decimal places in the number with the *fewest* places.

0.12	2 significant figures	2 decimal places
1.6	2 significant figures	1 decimal place
10.976	5 significant figures	3 decimal places
12.696		

This answer should be reported as 12.7, a number with one decimal place, because 1.6 has only one decimal place.

Rule 3. In multiplication or division, the number of significant figures in the answer should be the same as that in the quantity with the *fewest* significant figures.

$$\frac{0.01208}{0.0236} = 0.512 \text{ or, in scientific notation, } 5.12 \times 10^{-1}$$

Since 0.0236 has only three significant figures, while 0.01208 has four, the answer is limited to three significant figures.

Rule 4. When a number is rounded off (the number of significant figures is reduced), the last digit retained is increased by 1 only if the following digit is 5 or greater.*

Full Number	Number Rounded to Three Significant Figures
12.696	12.7
16.249	16.2
18.35	18.4
18.351	18.4

One last word regarding significant figures and calculations. In working problems on a pocket calculator, you should do the calculation using all the digits allowed by the calculator and round off only at the end of the problem. Rounding off in the middle can introduce errors. If your answers do not quite agree with those in the Appendices of the book, this may be the source of the disagreement.

Now let us consider a problem that is of practical importance and that makes use

*A modification of this rule is sometimes used to reduce the accumulation of roundoff errors. If the digit following the last permitted significant figure is *exactly* 5 (with no following digits or with all following digits being zeros), then (a) increase the last significant figure by 1 if it is *odd* or (b) leave the last significant figure unchanged if it is *even.* Thus, both 18.35 and 18.45 are rounded to 18.4.

of all the rules. Suppose you discover that young children are eating chips of paint that flake off a wall in an old house. The paint contains 200. ppm lead (200. mg Pb per kg paint). Suppose that a child eats five such chips. How much lead has the child gotten from the paint?

The ppm unit stands for "parts per million." If a substance is present with a concentration of 1 ppm, there is 1 gram of the substance in 1 million grams of sample.

As stated this problem does not include enough information for a solution to be obtained; however, some reasonable assumptions can be made and they can lead to experiments that could be used to obtain the necessary information. The statement does not say how big the paint chips are. Let's assume that they are 1.0 cm by 1.0 cm so that the area is 1.0 cm^2. Then eating five chips means eating 5.0 cm^2 of paint. (This assumption could be improved by measuring similar chips from the same place.) Since the concentration of lead is reported in units of mass of lead per mass of paint, we need to know the mass of 5.0 cm^2 of paint. This could be determined by measuring the areas of several paint chips and determining the mass of each. Suppose that the results of such measurements were

Mass of Chip (mg)	Area of Chip (cm^2)	Mass per Unit Area (mg/cm^2)
29.6	2.34	12.65
21.9	1.73	12.66
23.6	1.86	12.69

$$\text{Average mass per unit area} = \frac{(12.65 + 12.66 + 12.69)\ \text{mg/cm}^2}{3}$$
$$= 12.67\ \text{mg/cm}^3 = 12.7\ \text{mg/cm}^2$$

The average has been rounded to three significant figures because each experimental number has three significant figures. (Notice that more than three significant figures were kept in the intermediate calculations so as not to lose precision.) Now we can use this information to calculate how much lead the child has consumed.

$$m_{\text{paint}} = 5.0\ \text{cm}^2\ \text{paint}\left(\frac{12.67\ \text{mg paint}}{\text{cm}^2}\right)\left(\frac{1\ \text{g}}{1000\ \text{mg}}\right)\left(\frac{1\ \text{kg}}{1000\ \text{g}}\right)$$
$$= 6.335 \times 10^{-5}\ \text{kg paint}$$

$$m_{\text{Pb}} = 6.335 \times 10^{-5}\ \text{kg paint}\left(\frac{200.\ \text{mg Pb}}{1\ \text{kg paint}}\right) = 1.267 \times 10^{-2}\ \text{mg Pb}$$
$$= 0.013\ \text{mg Pb}$$

Notice that the final result was rounded to two significant figures because there were only two significant figures in the initial area of the paint chip. This is quite adequate precision, however, for you to determine whether this quantity of lead is likely to be harmful to the child.

The methods of problem solving presented here have been developed over time and represent a good way of keeping track of the precision of results, the units in which those results were obtained, and the correctness of calculations. These methods are not the only way that such goals can be achieved, but they do work well. We recommend that you include units in all calculations and check that they cancel appropriately. Also it is important not to overstate the precision of results by keeping too many significant figures. By solving a great many problems, you should be able to develop your problem-solving skills so that they become second nature and you can do them without thinking about the mechanics. This will allow you to devote all your thought to the logic of a problem solution.

APPENDIX B

Some Mathematical Operations

The mathematical skills required in this introductory course are basic skills in algebra and a knowledge of (a) exponential or scientific notation, (b) logarithms, and (c) quadratic equations. This appendix reviews each of the last three topics.

B.1 ELECTRONIC CALCULATORS

The directions for calculator use in this section are given for calculators using "algebraic" logic. Such calculators are the most common type used by students in introductory courses. For calculators using RPN logic (such as those made by Hewlett-Packard), the procedures will differ slightly.

The advent of inexpensive electronic calculators has made calculations in introductory chemistry much more straightforward. You are well advised to purchase a calculator that has the capability of performing calculations in scientific notation, has both base-10 and natural logarithms, and is capable of raising any number to any power and of finding any root of any number. In the discussion below, we shall point out in general how these functions of your calculator can be used.

Although electronic calculators have greatly simplified calculations, they have also forced us to focus again on significant figures. A calculator easily handles eight or more significant figures, but real laboratory data are never known to this accuracy. Therefore, you are urged to review Appendix A on handling numbers.

B.2 EXPONENTIAL OR SCIENTIFIC NOTATION

In exponential or scientific notation, a number is expressed as a product of two numbers: $N \times 10^n$. The first number, N, is the so-called *digit term* and is a number between 1 and 10. The second number, 10^n, the *exponential term,* is some integer power of 10. For example, 1234 would be written in scientific notation as 1.234×10^3 or 1.234 multiplied by 10 three times.

$$1234 = 1.234 \times 10^1 \times 10^1 \times 10^1 = 1.234 \times 10^3$$

Conversely, a number less than 1, such as 0.01234, would be written as 1.234×10^{-2}. This notation tells us that 1.234 should be divided twice by 10 in order to obtain 0.01234.

$$0.01234 = \frac{1.234}{10^1 \times 10^1} = 1.234 \times 10^{-1} \times 10^{-1} = 1.234 \times 10^{-2}$$

Some other examples of scientific notation are

$$10000 = 1 \times 10^4 \qquad 12345 = 1.2345 \times 10^4$$
$$1000 = 1 \times 10^3 \qquad 1234 = 1.234 \times 10^3$$
$$100 = 1 \times 10^2 \qquad 123 = 1.23 \times 10^2$$
$$10 = 1 \times 10^1 \qquad 12 = 1.2 \times 10^1$$
$$1 = 1 \times 10^0 \qquad \text{(any number to the zero power = 1)}$$
$$1/10 = 1 \times 10^{-1} \qquad 0.12 = 1.2 \times 10^{-1}$$
$$1/100 = 1 \times 10^{-2} \qquad 0.012 = 1.2 \times 10^{-2}$$

$$1/1000 = 1 \times 10^{-3} \qquad 0.0012 = 1.2 \times 10^{-3}$$
$$1/10000 = 1 \times 10^{-4} \qquad 0.00012 = 1.2 \times 10^{-4}$$

When converting a number to scientific notation, notice that the exponent n is positive if the number is greater than 1 and negative if the number is less than 1. The value of n is the number of places by which the decimal was shifted to obtain the number in scientific notation.

$$1\ 2\ 3\ 4\ 5. = 1.2345 \times 10^4$$

Decimal shifted 4 places to the left. Therefore, n is positive and equal to 4.

$$0.0\ 0\ 1\ \ 2 = 1.2 \times 10^{-3}$$

Decimal shifted 3 places to the right. Therefore, n is negative and equal to 3.

If you wish to convert a number in scientific notation to the usual form, the procedure above is simply reversed.

$$6\ .\ 2\ 7\ 3 \times 10^2 = 627.3$$

Decimal point moved 2 places to the right, since n is positive and equal to 2.

$$0\ 0\ 6.273 \times 10^{-3} = 0.006273$$

Decimal point shifted 3 places to the left, since n is negative and equal to 3.

There are two final points to be made concerning scientific notation. First, if you are used to working on a computer you may be in the habit of writing a number such as 1.23×10^3 as 1.23E3, or 6.45×10^{-5} as 6.45E-5. Second, some electronic calculators allow you to convert numbers readily to the scientific notation. If you have such a calculator, you can change a number shown in the usual form to scientific notation simply by pressing the EE or EXP key and then the " = " key.

1. Adding and Subtracting Numbers

When adding or subtracting two numbers, they must first be converted to the same powers of ten. The digit terms are then added or subtracted as appropriate.

$$(1.234 \times 10^{-3}) + (5.623 \times 10^{-2}) = (0.1234 \times 10^{-2}) + (5.623 \times 10^{-2})$$
$$= 5.746 \times 10^{-2}$$
$$(6.52 \times 10^{2}) - (1.56 \times 10^{3}) = (6.52 \times 10^{2}) - (15.6 \times 10^{2})$$
$$= -9.1 \times 10^{2}$$

2. Multiplication

The digit terms are multiplied in the usual manner, and the exponents are added algebraically. The result is expressed with a digit term with only one nonzero digit to the left of the decimal.

$$(1.23 \times 10^{3})(7.60 \times 10^{2}) = (1.23)(7.60) \times 10^{3+2}$$
$$= 9.35 \times 10^{5}$$

$$(6.02 \times 10^{23})(2.32 \times 10^{-2}) = (6.02)(2.32) \times 10^{23-2}$$
$$= 13.966 \times 10^{21}$$
$$= 1.40 \times 10^{22} \text{ (answer in 3 significant figures)}$$

3. Division

The digit terms are divided in the usual manner, and the exponents are subtracted algebraically. The quotient is written with one nonzero digit to the left of the decimal in the digit term.

$$\frac{7.60 \times 10^3}{1.23 \times 10^2} = \frac{7.60}{1.23} \times 10^{3-2} = 6.18 \times 10^1$$

$$\frac{6.02 \times 10^{23}}{9.10 \times 10^{-2}} = \frac{6.02}{9.10} \times 10^{(23)-(-2)} = 0.662 \times 10^{25} = 6.62 \times 10^{24}$$

4. Powers of Exponentials

When raising a number in exponential notation to a power, treat the digit term in the usual manner. The exponent is then multiplied by the number indicating the power.

$$\begin{aligned}(1.25 \times 10^3)^2 &= (1.25)^2 \times 10^{3\times 2}\\ &= 1.5625 \times 10^6 = 1.56 \times 10^6\end{aligned}$$

$$\begin{aligned}(5.6 \times 10^{-10})^3 &= (5.6)^3 \times 10^{(-10)\times 3}\\ &= 175.6 \times 10^{-30} = 1.8 \times 10^{-28}\end{aligned}$$

Electronic calculators usually have two methods of raising a number to a power. To square a number, enter the number and then press the "x^2" key. To raise a number to any power, use the "y^x" key. For example, to raise 1.42×10^2 to the 4th power,

(a) enter 1.42×10^2
(b) press "y^x"
(c) enter 4 (this should appear on the display)
(d) press " = " and $4.0658 \ldots \times 10^8$ will appear on the display. (The number of digits will depend upon the calculator in use.)

As a final step, express the number in the correct number of significant figures (4.07×10^8 in this case).

5. Roots of Exponentials

Unless you use an electronic calculator, the number must first be put into a form where the exponential is exactly divisible by the root. The root of the digit term is found in the usual way, and the exponent is divided by the desired root.

$$\sqrt{3.6 \times 10^7} = \sqrt{36 \times 10^6} = \sqrt{36} \times \sqrt{10^6} = 6.0 \times 10^3$$

$$\sqrt[3]{2.1 \times 10^{-7}} = \sqrt[3]{210 \times 10^{-9}} = \sqrt[3]{210} \times \sqrt[3]{10^{-9}} = 5.9 \times 10^{-3}$$

To take a square root on an electronic calculator, enter the number and then press the "$\sqrt{x}$" key. To find a higher root of a number, such as the 4th root of 5.6×10^{-10},

(a) enter the number
(b) press the "$\sqrt{y}$" key (On most calculators, the sequence you actually use is to press "2ndF" and then "$\sqrt[x]{y}$." Alternatively, you press "INV" and then "y^x.")
(c) enter the desired root, 4 in this case
(d) press "=". The answer here is 4.8646×10^{-3} or 4.9×10^{-3}.

A general procedure for finding any root is to use the "y^x" key. For a square root, x is 0.5 (or 1/2), whereas it is 0.33 (or 1/3) for a cube root, 0.25 (or 1/4) for a 4th root, and so on.

B.3 Logarithms

There are two types of logarithms used in this text: (a) common logarithms (abbreviated log) whose base is 10 and (b) natural logarithms (abbreviated ln) whose base is e (= 2.71828).

$$\log x = n \qquad \text{where } x = 10^n$$
$$\ln x = m \qquad \text{where } x = e^m$$

Most equations in chemistry and physics were developed in natural or base e logarithms and this practice is followed in this text. The relation between log and ln is

$$\ln x = 2.303 \log x$$

Aside from the different bases of the two logarithms, they are used in the same manner. What follows is largely a description of the use of common logarithms.

A common logarithm is the power to which you must raise 10 to obtain the number. For example, the log of 100 is 2, since you must raise 10 to the second power to obtain 100. Other examples are

$$\begin{aligned}
\log 1000 &= \log (10^3) = 3 \\
\log 10 &= \log (10^1) = 1 \\
\log 1 &= \log (10^0) = 0 \\
\log 1/10 &= \log (10^{-1}) = -1 \\
\log 1/10000 &= \log (10^{-4}) = -4
\end{aligned}$$

To obtain the common logarithm of a number other than a simple power of 10, you must resort to a log table or an electronic calculator. For example,

$$\begin{aligned}
\log 2.10 &= 0.3222, \text{ which means that } 10^{0.3222} = 2.10 \\
\log 5.16 &= 0.7126, \text{ which means that } 10^{0.7126} = 5.16 \\
\log 3.125 &= 0.49485, \text{ which means that } 10^{0.49485} = 3.125
\end{aligned}$$

To check this on your calculator, enter the number and then press the "log" key. When using a log table, the logs of the first two numbers above can be read directly from the table. The log of the third number (3.125), however, must be interpolated. That is, 3.125 is midway between 3.12 and 3.13, so the log is midway between 0.4942 and 0.4955.

To obtain the natural logarithm, ln, of the numbers above, use a calculator having this function. Enter each number and press "ln."

$$\begin{aligned}
\ln 2.10 &= 0.7419, \text{ which means that } e^{0.7419} = 2.10 \\
\ln 5.16 &= 1.6409, \text{ which means that } e^{1.6409} = 5.16
\end{aligned}$$

To find the common logarithm of a number greater than 10 or less than 1 with a log table, first express the number in scientific notation. Then find the log of each part of the number and add the logs. For example,

$$\begin{aligned}
\log 241 &= \log (2.41 \times 10^2) = \log 2.41 + \log 10^2 \\
&= 0.382 + 2 = 2.382
\end{aligned}$$

$$\log 0.00573 = \log (5.73 \times 10^{-3}) = \log 5.73 + \log 10^{-3}$$
$$= 0.758 + (-3) = -2.242$$

LOGARITHMS AND NOMENCLATURE: The number to the left of the decimal in a logarithm is called the **characteristic,** and the number to the right of the decimal is the **mantissa.**

Significant Figures and Logarithms Notice that the mantissa has as many significant figures as the number whose log was found. (So that you could more clearly see the result obtained with a calculator or a table, this rule was not strictly followed until the last two examples.)

Obtaining Antilogarithms If you are given the logarithm of a number and find the number from it, you have obtained the "antilogarithm" or "antilog" of the number. There are two common procedures used by electronic calculators to do this:

Procedure A	Procedure B
(a) enter the log or ln	(a) enter the log or ln
(b) press 2ndF	(b) press INV
(c) press 10^x or e^x	(c) press log or ln x

Test one or the other of these procedures with the following examples:

1. Find the number whose log is 5.234.
   Recall that $\log x = n$ where $x = 10^n$. In this case $n = 5.234$. Enter that number in your calculator and find the value of 10^n, the antilog. In this case,

   $$10^{5.234} = 10^{0.234} \times 10^5 = 1.71 \times 10^5$$

   Notice that the characteristic (5) sets the decimal point; it is the power of 10 in the exponential form. The mantissa (0.234) gives the value of the number x. Thus, if you use a log table to find x, you need only look up 0.234 in the table and see that it corresponds to 1.71.
2. Find the number whose log is -3.456.

   $$10^{-3.456} = 10^{0.544} \times 10^{-4} = 3.50 \times 10^{-4}$$

   Notice here that -3.456 must be expressed as the sum of -4 and $+0.544$.

Mathematical Operations Using Logarithms Because logarithms are exponents, operations involving them follow the same rules as the use of exponents. Thus, multiplying two numbers can be done by adding logarithms.

$$\log xy = \log x + \log y$$

For example, we multiply 563 by 125 by adding their logarithms and finding the antilogarithm of the result.

$$\log 563 = 2.751$$
$$\log 125 = 2.097$$
$$\log xy = 2.751 + 2.097 = 4.848$$
$$xy = 10^{4.848} = 10^4 \times 10^{0.848} = 7.05 \times 10^4$$

One number (x) can be divided by another (y) by subtraction of their logarithms.

$$\log \frac{x}{y} = \log x - \log y$$

For example, to divide 125 by 742,

$$\log 125 = 2.097$$
$$\log 742 = 2.870$$
$$\log x/y = 2.097 - 2.870 = -0.773$$

$$x/y = 10^{-0.773} = 10^{0.227} \times 10^{-1} = 1.69 \times 10^{-1}$$

Similarly, powers and roots of numbers can be found using logarithms.

$$\log x^y = y(\log x)$$

$$\log \sqrt[y]{x} = \log x^{1/y} = \frac{1}{y} \log x$$

As an example, find the fourth power of 5.23. We first find the log of 5.23 and then multiply it by 4. The result, 2.874, is the log of the answer. Therefore, we find the antilog of 2.874.

$$(5.23)^4 = ?$$
$$\log (5.23)^4 = 4 \log 5.23 = 4\,(0.719) = 2.874$$
$$(5.23)^4 = 10^{2.874} = 748$$

As another example, find the fifth root of 1.89×10^{-9}.

$$\sqrt[5]{1.89 \times 10^{-9}} = (1.89 \times 10^{-9})^{1/5} = ?$$

$$\log (1.89 \times 10^{-9})^{1/5} = \frac{1}{5} \log (1.89 \times 10^{-9}) = \frac{1}{5}(-8.724) = -1.745$$

The answer is the antilog of -1.745.

$$(1.89 \times 10^{-9})^{1/5} = 10^{-1.745} = 1.80 \times 10^{-2}$$

B.4 QUADRATIC EQUATIONS

Algebraic equations of the form $ax^2 + bx + c = 0$ are called **quadratic equations.** The coefficients a, b, and c may be either positive or negative. The two roots of the equation may be found using the *quadratic formula.*

$$x = \frac{-\mathrm{b} \pm \sqrt{b^2 - 4ac}}{2a}$$

As an example, solve the equation $5x^2 - 3x - 2 = 0$. Here $a = 5$, $b = -3$, and $c = -2$. Therefore,

$$x = \frac{3 \pm \sqrt{(-3)^2 - 4(5)(-2)}}{2(5)}$$

$$= \frac{3 \pm \sqrt{9 - (-40)}}{10} = \frac{3 \pm \sqrt{49}}{10} = \frac{3 \pm 7}{10}$$

$$x = 1 \text{ and } -0.4$$

How do you know which of the two roots is the correct answer? You have to decide in each case which root has physical significance. However, it is *usually* true in this course that negative values are not significant.

When you have solved a quadratic expression, you should always check your values by substitution into the original equation. In the example above, we find that $5(1)^2 - 3(1) - 2 = 0$ and that $5(-0.4)^2 - 3(-0.4) - 2 = 0$.

The most likely place you will encounter quadratic equations is in the chapters on chemical equilibria, particularly in Chapters 13, 16, and 17. Here you may be faced with solving an equation such as

$$1.8 \times 10^{-4} = \frac{x^2}{0.0010 - x}$$

This equation can certainly be solved by using the quadratic equation (to give $x = 3.4 \times 10^{-4}$). However, you may find the *method of successive approximations* to be especially convenient. Here you begin by making a reasonable approximation of x. This approximate value is substituted into the original equation, and this is solved to give what is hoped to be a more correct value of x. This process is repeated until the answer converges on a particular value of x, that is, until the value of x derived from two successive approximations is the same.

Step 1. First assume that x is so small that $(0.0010 - x) \approx 0.0010$. This means that

$$x^2 = 1.8 \times 10^{-4}(0.0010)$$
$$x = 4.2 \times 10^{-4} \text{ (to 2 significant figures)}$$

Step 2. Substitute the value of x from Step 1 into the denominator (but *not* the numerator) of the original equation and again solve for x.

$$x^2 = (1.8 \times 10^{-4})(0.0010 - 0.00042)$$
$$x = 3.2 \times 10^{-4}$$

Step 3. Repeat Step 2 using the value of x found in that step.

$$x = \sqrt{1.8 \times 10^{-4}(0.0010 - 0.00032)} = 3.5 \times 10^{-4}$$

Step 4. Continue by repeating the calculation, using the value of x found in the previous step.

$$x = \sqrt{1.8 \times 10^{-4}(0.0010 - 0.00035)} = 3.4 \times 10^{-4}$$

Step 5. $x = \sqrt{1.8 \times 10^{-4}(0.0010 - 0.00034)} = 3.4 \times 10^{-4}$

Here we find that iterations after the fourth step give the same value for x, indicating that we have arrived at a valid answer (and the same one obtained from the quadratic formula).

There are several final thoughts on using the method of successive approximations. First, there are cases where the method does not work. Successive steps may give answers that are random or that diverge from the correct value. For quadratic equations of the form $K = x^2/(C - x)$, the method of approximations will work as long as $K < 4C$ (assuming one begins with $x = 0$ as the first guess, that is, $K \approx x^2/C$). This will always be true for weak acids and bases.

Second, values of K in the equation $K = x^2/(C - x)$ are usually known only to two significant figures. Therefore, we are justified in carrying out successive steps until two answers are the same to two significant figures.

Finally, we highly recommend this method of solving quadratic equations. If your calculator has a memory function, successive approximations can be carried out easily and rapidly.

APPENDIX C

Units, Equivalences, and Conversion Factors

C.1 UNITS OF THE INTERNATIONAL SYSTEM (SI)

The metric system was begun by the French National Assembly in 1790 and has undergone many modifications. The International System of Units or *Système International* (SI), which represents an extension of the metric system, was adopted by the 11th General Conference of Weights and Measures in 1960. It is constructed from seven base units, each of which represents a particular physical quantity (Table C.1).

The first five units listed in Table C.1 are particularly useful in chemistry. They are defined as follows:

1. The *meter* is the length of the path travelled by light in vacuum during a time interval of 1/299,792,458 of a second.
2. The *kilogram* represents the mass of a platinum-iridium block kept at the International Bureau of Weights and Measures of Sèvres, France.
3. The *second* was redefined in 1967 as the duration of 9,192,631,770 periods of a certain line in the microwave spectrum of cesium-133.
4. The *kelvin* is 1/273.16 of the temperature interval between absolute zero and the triple point of water (the temperature at which liquid water, ice, and water vapor coexist).
5. The *mole* is the amount of substance that contains as many elementary entities (atoms, molecules, ions, or other particles) as there are atoms in exactly 0.012 kg of carbon-12 (12 g of ^{12}C atoms).

Decimal fractions and multiples of metric and SI units are designated by using the **prefixes** listed in Table C.2. The prefix *kilo,* for example, means a unit is multiplied by 10^3,

$$1 \text{ kilogram} = 1 \times 10^3 \text{ grams} = 1000 \text{ grams}$$

and the prefix *centi* means the unit is multiplied by the factor 10^{-2}

$$1 \text{ centigram} = 1 \times 10^{-2} \text{ gram} = 0.01 \text{ gram}$$

TABLE C.1 SI Fundamental Units

Physical Quantity	Name of Unit	Symbol
Length	Meter	m
Mass	Kilogram	kg
Time	Second	s
Temperature	Kelvin	K
Amount of substance	Mole	mol
Electric current	Ampere	A
Luminous intensity	Candela	cd

TABLE C.2 Prefixes for Metric and SI Units*

Factor	Prefix	Symbol	Factor	Prefix	Symbol
10^{12}	tera	T	10^{-1}	*deci*	d
10^{9}	giga	G	10^{-2}	*centi*	c
10^{6}	mega	M	10^{-3}	*milli*	m
10^{3}	*kilo*	k	10^{-6}	micro	μ
10^{2}	hecto	h	10^{-9}	*nano*	n
10^{1}	deka	da	10^{-12}	*pico*	p
			10^{-15}	femto	f
			10^{-18}	atto	a

*The most common prefixes are shown in italics.

The prefixes are added to give units of a magnitude appropriate to what is being measured. The distance from New York to London (5.6×10^3 km = 5,600 km) is much easier to comprehend measured in kilometers than in meters (5.6×10^6 m = 5,600,000 m). The following is a list of units for measuring very small and very large distances:

nanometer (nm)	0.000000001 meter
micrometer (μm)	0.000001 meter
millimeter (mm)	0.001 meter
centimeter (cm)	0.01 meter
decimeter (dm)	0.1 meter
meter (m)	1 meter
dekameter (dam)	10 meters
hectometer (hm)	100 meters
kilometer (km)	1,000 meters
megameter (Mm)	1,000,000 meters

In the International System of Units, all physical quantities are represented by appropriate combinations of the base units listed in Table C.1. The result is a derived unit for each kind of measured quantity. The most common derived units are listed in Table C.3. It is easy to see that the derived unit for area is length × length = meter × meter = square meter, m^2, or that the derived unit for volume is length × length × length = meter × meter × meter = cubic meter, m^3. The more complex derivations are arrived at by the same kind of combination of units. Units such as the one for *force* (the *newton*) have been given simple names that represent the units by which they are defined.

C.2 CONVERSION OF UNITS FOR PHYSICAL QUANTITIES

The result of a measurement is a **physical quantity,** which consists of a number plus a unit. To convert a physical quantity from one unit of measure to another requires a conversion factor based on equivalences between units of measure like those given in Table C.4. Each equivalence provides two conversion factors that are the inverse of each other. For example, the equivalence between a quart and a liter, 1 quart = 0.9463 liter, gives

TABLE C.3 Derived SI Units

Physical Quantity	Name of Unit	Symbol	Definition	Symbol
Area	Square meter	m^2	—	
Volume	Cubic meter	m^3	—	
Density	Kilogram per cubic meter	kg/m^3	—	
Force	Newton	N	(kilogram)(meter)/(second)^2	$kg\ m/s^2$
Pressure	Pascal	Pa	Newton/square meter	N/m^2
Energy	Joule	J	(kilogram)(square meter)/(second)^2	$kg\ m^2/s^2$
Electric charge	Coulomb	C	(ampere)(second)	A s
Electric potential difference	Volt	V	joule/(ampere)(second)	J/(A s)

$$\frac{1 \text{ quart}}{0.9463 \text{ liter}}$$ There is 1 quart per 0.9463 liter.

$$\frac{0.9463 \text{ liter}}{1 \text{ quart}}$$ There is 0.9463 liter per 1 quart.

The method of cancelling units described in Section 2.7 provides the basis for choosing which conversion factor is needed: It is always the one that allows the unit being converted to be canceled and leaves the new unit uncanceled.

To convert 2 quarts to liters:

$$2 \cancel{\text{quarts}} \times \frac{0.9463 \text{ liter}}{1 \cancel{\text{quart}}} = 1.893 \text{ liters}$$

To convert 2 liters to quarts:

$$2 \cancel{\text{liters}} \times \frac{1 \text{ quart}}{0.9463 \cancel{\text{liter}}} = 2.113 \text{ quarts}$$

Because of their definitions, conversions between temperatures in Celsius degrees and Fahrenheit degrees are a bit more complicated. Both units are based on the properties of water. The Celsius unit is defined by assigning zero as the freezing point of pure water (0 °C) and 100 as its boiling point (100 °C).* The size of the Fahrenheit degree is equally arbitrary. Fahrenheit defined 0 °F as the freezing point of a solution in which he had dissolved the maximum amount of salt (because this was the lowest temperature he could reproduce reliably), and he intended 100 °F to be the normal human body temperature (but this turned out to be 98.6 °F). Today, the reference points are set at 32 °F and 212 °F (the freezing and boiling points of pure water, respectively). The number of units between these points is 180 Fahrenheit degrees. Comparing the two units, the Celsius degree is almost twice as large as the Fahrenheit degree; it takes only 5 Celsius degrees to cover the same temperature range as 9 Fahrenheit degrees.

$$\frac{100\ °C}{180\ °F} = \frac{5\ °C}{9\ °F}$$

*To be entirely correct, we must specify that water boils at 100 °C and freezes at 0 °C only when the pressure of the surrounding atmosphere is 1 standard atmosphere.

This relationship is used to convert a temperature on one scale to a temperature on the other.

$$°C = \frac{5\ °C}{9\ °F}\ °F - 32\ °F \quad \text{or} \quad °F = \frac{9\ °F}{5\ °C}\ °C - 32\ °F$$

For example, your normal body temperature is 98.6 °F, which corresponds to 37.0 °C.

$$\text{Body temperature in } °C = \frac{5\ °C}{9\ °F}(98.6\ °F - 32.0\ °F)$$

$$= [(\tfrac{5}{9})(66.6)]\ °C = 37.0\ °C$$

Laboratory work is almost always done using Celsius units, and we rarely need to make conversions to and from Fahrenheit. It is best to try to calibrate your senses to Celsius units; to help you do this, it is useful to know that water freezes at 0 °C, a comfortable room temperature is about 22 °C, your body temperature is 37 °C, and the hottest water you could stand to put your hand into is about 60 °C.

TABLE C.4 Common Units of Measure

Mass and Weight
1 pound = 453.59 grams = 0.45359 kilogram
1 kilogram = 1000 grams = 2.205 pounds
1 gram = 10 decigrams = 100 centigrams = 1000 milligrams
1 gram = 6.022×10^{23} atomic mass units
1 atomic mass unit = 1.6605×10^{-24} gram
1 short ton = 2000 pounds = 907.2 kilograms
1 long ton = 2240 pounds
1 metric tonne = 1000 kilograms = 2205 pounds

Length
1 inch = 2.54 centimeters (exactly)
1 mile = 5280 feet = 1.609 kilometers
1 yard = 36 inches = 0.9144 meter
1 meter = 100 centimeters = 39.37 inches = 3.281 feet = 1.094 yards
1 kilometer = 1000 meters = 1094 yards = 0.6215 mile
1 Ångstrom = 1.0×10^{-8} centimeter = 0.10 nanometer = 100 picometers = 1.0×10^{-10} meter = 3.937×10^{-9} inch

Volume
1 quart = 0.9463 liter
1 liter = 1.0567 quarts
1 liter = 1 cubic decimeter = 1000 cubic centimeters = 0.001 cubic meter
1 milliliter = 1 cubic centimeter = 0.001 liter = 1.056×10^{-3} quart
1 cubic foot = 28.316 liters = 29.924 quarts = 7.481 gallons

TABLE C.4 Continued

Force* and Pressure

1 atmosphere = 760 millimeters of mercury = 1.013×10^5 pascals
= 14.70 pounds per square inch
1 bar = 10^5 pascals = 0.98692 atmosphere
1 torr = 1 millimeter of mercury
1 pascal = 1 kg/m s^2 = 1 N/m^2

*Force: 1 newton (N) = 1 kg · m/s^2, i.e., the force that when applied for 1 second gives a 1-kilogram mass a velocity of 1 meter per second.

Energy

1 joule = 1×10^7 ergs
1 thermochemical calorie† = 4.184 joules = 4.184×10^7 ergs
= 4.129×10^{-2} liter-atmospheres
= 2.612×10^{19} electron volts
1 erg = 1×10^{-7} joule = 2.3901×10^{-8} calorie
1 electron volt = 1.6022×10^{-19} joule = 1.6022×10^{-12} erg = 96.485 kJ/mol‡
1 liter-atmosphere = 24.217 calories = 101.32 joules = 1.0132×10^9 ergs
1 British thermal unit = 1055.06 joules = 1.05506×10^{10} ergs = 252.2 calories

†The heating required to raise the temperature of one gram of water from 14.5 °C to 15.5 °C.
‡Note that the other units in this line are per particle and must be multiplied by 6.022×10^{23} to be strictly comparable.

Temperature

0 K = −273.15 °C
K = °C + 273.15 °C

$$? \,^\circ\text{C} = \frac{5\,^\circ\text{C}}{9\,^\circ\text{F}}(^\circ\text{F} - 32\,^\circ\text{F})$$

$$? \,^\circ\text{C} = \frac{9\,^\circ\text{F}}{5\,^\circ\text{C}}(^\circ\text{C}) + 32\,^\circ\text{F}$$

APPENDIX D

Physical Constants

Quantity	Symbol	Traditional Units	SI Units
Acceleration of gravity	g	980.7 cm/s^2	9.806 m/s^2
Atomic mass unit (1/12 the mass of ^{12}C atom)	amu or u	1.6606×10^{-24} g	1.6606×10^{-27} kg
Avogadro's number	N	6.0221367×10^{23} particles/mol	6.0221367×10^{23} particles/mol
Bohr radius	a_0	0.52918 Å 5.2918×10^{-9} cm	5.2918×10^{-11} m
Boltzmann constant	k	1.3807×10^{-16} erg/K	1.3807×10^{-23} J/K
Charge-to-mass ratio of electron	e/m	1.7588×10^{8} coulomb/g	1.7588×10^{11} C/kg
Electronic charge	e	1.6022×10^{-19} coulomb 4.8033×10^{-10} esu	1.6022×10^{-19} C
Electron rest mass	m_e	9.1094×10^{-28} g 0.00054858 amu	9.1094×10^{-31} kg
Faraday constant	F	96,485 coulombs/mol e$^-$ 23.06 kcal/volt mol e$^-$	96,485 C/mol e$^-$ 96,485 J/V mol e$^-$
Gas constant	R	$0.08206 \dfrac{\text{L} \cdot \text{atm}}{\text{mol} \cdot \text{K}}$ $1.987 \dfrac{\text{cal}}{\text{mol} \cdot \text{K}}$	$8.3145 \dfrac{\text{Pa} \cdot \text{dm}^3}{\text{mol} \cdot \text{K}}$ 8.3145 J/mol · K
Molar volume (STP)	V_m	22.414 L/mol	22.414×10^{-3} m^3/mol 22.414 dm^3/mol
Neutron rest mass	m_n	1.67493×10^{-24} g 1.008665 amu	1.67493×10^{-27} kg
Planck's constant	h	6.6261×10^{-27} erg · s	$6.6260755 \times 10^{-34}$ J · s
Proton rest mass	m_p	1.6726×10^{-24} g 1.007276 amu	1.6726×10^{-27} kg
Rydberg constant	R_∞	3.289×10^{15} cycles/s 2.1799×10^{-11} erg	1.0974×10^{7} m^{-1} 2.1799×10^{-18} J
Velocity of light (in a vacuum)	c	2.9979×10^{10} cm/s (186,282 miles/second)	2.9979×10^{8} m/s

$\pi = 3.1416$
e = 2.7183
ln $X = 2.303 \log X$

2.303 R = 4.576 cal/mol · K = 19.15 J/mol · K
2.303 RT (at 25 °C) = 1364 cal/mol = 5709 J/mol

APPENDIX E

Naming Simple Organic Compounds and Coordination Compounds

The systematic nomenclature for organic compounds was proposed by the International Union of Pure and Applied Chemistry (IUPAC). The IUPAC set of rules provides different names for the more than 10 million known organic compounds, and allows names to be assigned to new compounds as they are synthesized. Many organic compounds also have *common* names. Usually the common name came first and is widely known. Many consumer products are labeled with the common name, and when only a few isomers are possible, the common name adequately identifies the product for the consumer. However, as illustrated in Section 3.4, a system of common names quickly fails when several structural isomers are possible.

E.1 HYDROCARBONS

The name of each member of the hydrocarbon classes has two parts. The first part, the prefix (*meth-, eth-, prop-, but-,* and so on), reflects the number of carbon atoms. When more than four carbons are present, the Greek or Latin number prefixes are used: *pent-, hex-, hept-, oct-, non-,* and *dec-*. The second part of the name, or the suffix, tells the class of hydrocarbon. Alkanes have carbon-carbon single bonds, alkenes have carbon-carbon double bonds, and alkynes have carbon-carbon triple bonds.

Unbranched Alkanes and Alkyl Groups

The names of the first 20 unbranched (straight-chain) alkanes are given in Table E.1.

Alkyl groups are named by dropping "-ane" from the parent alkane and adding "-yl" (see Table 3.4 for examples).

TABLE E.1 Names of Unbranched Alkanes

CH_4	Methane	$C_{11}H_{24}$	Undecane
C_2H_6	Ethane	$C_{12}H_{26}$	Dodecane
C_3H_8	Propane	$C_{13}H_{28}$	Tridecane
C_4H_{10}	Butane	$C_{14}H_{30}$	Tetradecane
C_5H_{12}	Pentane	$C_{15}H_{32}$	Pentadecane
C_6H_{14}	Hexane	$C_{16}H_{34}$	Hexadecane
C_7H_{16}	Heptane	$C_{17}H_{36}$	Heptadecane
C_8H_{18}	Octane	$C_{18}H_{38}$	Octadecane
C_9H_{20}	Nonane	$C_{19}H_{40}$	Nonadecane
$C_{10}H_{22}$	Decane	$C_{20}H_{42}$	Eicosane

Branched-Chain Alkanes

The rules for naming branched-chain alkanes are as follows:

1. *Find the longest continuous chain of carbon atoms: this chain determines the parent name for the compound.* For example, the following compound has two methyl groups attached to a *heptane* parent.

$$\underset{}{CH_3CH_2CH_2}\underset{\displaystyle CH_3}{\underset{|}{C}}HCH_2\underset{\displaystyle CH_3}{\underset{|}{C}}HCH_3$$

The longest continuous chain may not be obvious from the way the formula is written, especially for the straight-line format that is commonly used. For example, the longest continuous chain of carbon atoms in the following chain is *eight,* not *four* or *six.*

—C— (branched skeleton) is equivalent to —C—C—C—C—C—C—C—C— (eight-carbon zigzag chain)

2. *Number the longest chain beginning with the end of the chain nearest the branching. Use these numbers to designate the location of the attached group. When two or more groups are attached to the parent, give each group a number corresponding to its location on the parent chain.* For example, the name of

$$\overset{7}{C}H_3\overset{6}{C}H_2\overset{5}{C}H_2\underset{\displaystyle CH_3}{\underset{|}{\overset{4}{C}}}H\overset{3}{C}H_2\underset{\displaystyle CH_3}{\underset{|}{\overset{2}{C}}}H\overset{1}{C}H_3$$

is 2,4-dimethylheptane. The name of the compound below is 3-methylheptane, not 5-methylheptane or 2-ethylhexane.

$$\overset{7}{CH_3}—\overset{6}{CH_2}—\overset{5}{CH_2}—\overset{4}{CH_2}—\overset{3}{CH}(—{}^{2}CH_2—{}^{1}CH_3)—CH_3$$

3-methylheptane

3. *When two or more substituents are identical, indicate this by the use of the prefixes di-, tri-, tetra-, and so on. Positional numbers of the substituents should have the smallest possible sum.*

$$\overset{1}{C}H_3\overset{2}{C}H_2\underset{\displaystyle CH_3}{\underset{|}{\overset{\displaystyle CH_3}{\overset{|}{\overset{3}{C}}}}}\overset{4}{C}H_2\overset{\displaystyle CH_3}{\overset{|}{\overset{5}{C}}}H\underset{\displaystyle CH_3}{\underset{|}{\overset{6}{C}}}H\overset{7}{C}H_2\overset{8}{C}H_3$$

The correct name of this compound is 3,3,5,6-tetramethyloctane.

4. *If there are two or more different groups, the groups are listed alphabetically.*

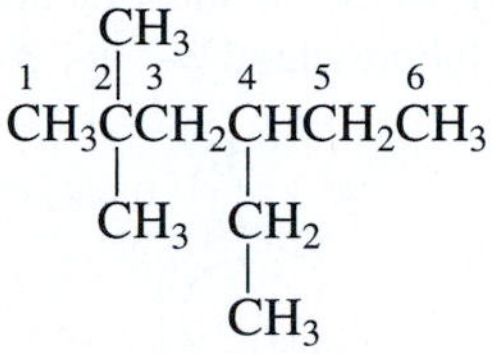

$$\begin{array}{c} \quad CH_3 \\ {}^{1}\ {}^{2}|\ {}^{3}\quad {}^{4}\ {}^{5}\ {}^{6} \\ CH_3\underset{|}{\overset{}{C}}CH_2\underset{|}{C}HCH_2CH_3 \\ \quad CH_3\ \ CH_2 \\ \qquad\quad | \\ \qquad\quad CH_3 \end{array}$$

The correct name of this compound is 4-ethyl-2,2-dimethylhexane. Note that the prefix *di* is ignored in determining alphabetical order.

Alkenes

Alkenes are named by using the prefix to indicate the number of carbon atoms and the suffix *-ene* to indicate one or more double bonds. The systematic names for the first two members of the alkene series are *ethene* and *propene.*

$$CH_2{=}CH_2 \qquad CH_3CH{=}CH_2$$

When groups, such as methyl or ethyl, are attached to carbon atoms in an alkene, the longest hydrocarbon chain is numbered from the end that will give the double bond the lowest number and then numbers are assigned to the attached groups. For example, the name of

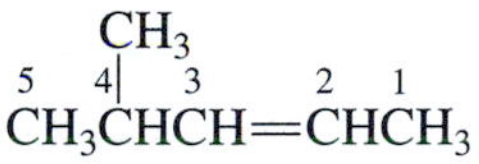

$$\begin{array}{c} CH_3 \\ {}^{5}\ \ {}^{4}|\ \ {}^{3}\qquad {}^{2}\ \ {}^{1} \\ CH_3CHCH{=}CHCH_3 \end{array}$$

is 4-methyl-2-pentene. See Section 9.5 for a discussion of *cis-trans* isomers of alkenes.

Alkynes

The naming of alkynes is similar to that of alkenes, with the lowest number possible being used to locate the triple bond. For example, the name of

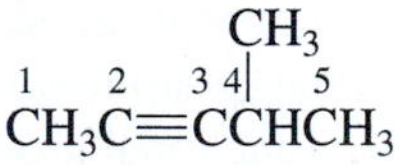

$$\begin{array}{c} \qquad\quad CH_3 \\ {}^{1}\quad {}^{2}\quad {}^{3}\,{}^{4}|\ \ {}^{5} \\ CH_3C{\equiv}CCHCH_3 \end{array}$$

is 4-methyl-2-pentyne.

Benzene Derivatives

Monosubstituted benzene derivatives are named by using a prefix for the substituent. Some examples are

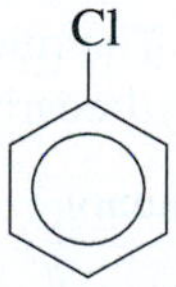

Cl | CH_3 | C_2H_5

chlorobenzene 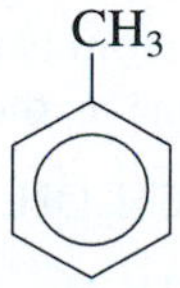 | methylbenzene (toluene) 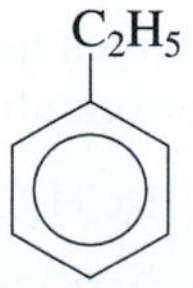 | ethylbenzene

Three isomers are possible when two groups are substituted for hydrogen atoms on the benzene ring. The relative positions of the substituents are indicated either by the prefixes *ortho-*, *meta-*, and *para-* (abbreviated *o-*, *m-*, *p-*) or by numbers. For example,

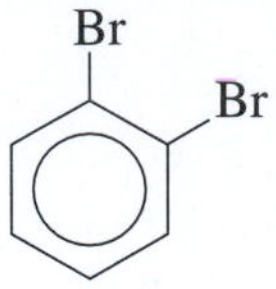

1,2-dibromobenzene (*o*-dibromobenzene)

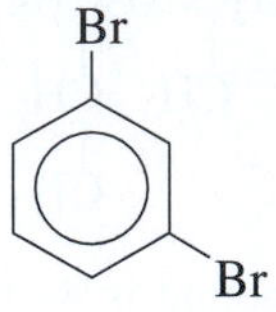

1,3-dibromobenzene (*m*-dibromobenzene)

1-4-dibromobenzene (*p*-dibromobenzene)

The dimethylbenzenes are called *xylenes.*

If more than two groups are attached to the benzene ring, numbers must be used to identify the positions. The benzene ring is numbered to give the lowest possible numbers to the substituents.

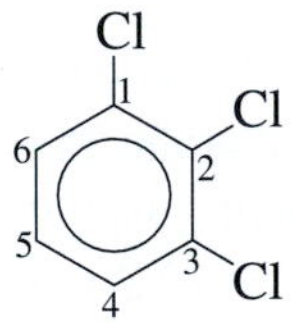

1,2,3-trichlorobenzene

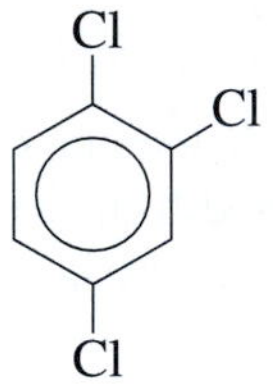

1,2,4-trichlorobenzene

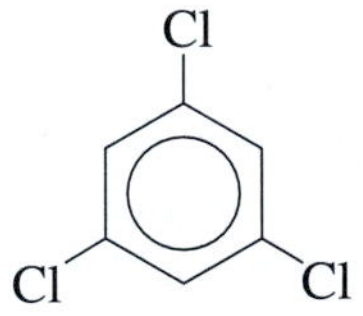

1,3,5-trichlorobenzene

E.2 FUNCTIONAL GROUPS

An atom or group of atoms that defines the structure of a specific class of organic compounds and determines their properties is called a **functional group.** The millions of organic compounds include classes of compounds that are obtained by replacing hydrogen atoms of hydrocarbons with functional groups (Sections 3.1, 9.2, 11.5, 11.6). The important functional groups are shown in Table E.2.

The "R" attached to the functional group represents the hydrocarbon framework with one hydrogen removed for each functional group added. The IUPAC system provides a systematic method for naming all members of a given class. For example, alcohols end in *-ol* (methan*ol*); aldehydes end in *-al* (methan*al*); carboxylic acids end in *-oic* (ethan*oic* acid); and ketones end in *-one* (propan*one*).

Alcohols

Isomers are also possible for molecules containing functional groups. For example, three different alcohols are obtained when a hydrogen atom in pentane is replaced by —OH, depending on which hydrogen atom is replaced. The rules for naming the "R" or hydrocarbon framework are the same as those for hydrocarbon compounds.

$CH_3CH_2CH_2CH_2CH_2OH$ 1-pentanol

$CH_3CH_2CH_2CHCH_3$ (with OH on the fourth carbon) 2-pentanol

TABLE E.2 Classes of Organic Compounds Based on Functional Groups*

General Formulas of Class Members	Class Name	Typical Compound	Compound Name	Common Use of Sample Compound
$R-X$	Halide	$H-\underset{\underset{Cl}{\mid}}{\overset{\overset{H}{\mid}}{C}}-Cl$	Dichloromethane (methylene chloride)	Solvent
$R-OH$	Alcohol	$H-\underset{\underset{H}{\mid}}{\overset{\overset{H}{\mid}}{C}}-OH$	Methanol (wood alcohol)	Solvent
$R-\overset{\overset{O}{\|}}{C}-H$	Aldehyde	$H-\overset{\overset{O}{\|}}{C}-H$	Methanal (formaldehyde)	Preservative
$R-\overset{\overset{O}{\|}}{C}-OH$	Carboxylic acid	$H-\underset{\underset{H}{\mid}}{\overset{\overset{H}{\mid}}{C}}-\overset{\overset{O}{\|}}{C}-OH$	Ethanoic acid (acetic acid)	Vinegar
$R-\overset{\overset{O}{\|}}{C}-R'$	Ketone	$H-\underset{\underset{H}{\mid}}{\overset{\overset{H}{\mid}}{C}}-\overset{\overset{O}{\|}}{C}-\underset{\underset{H}{\mid}}{\overset{\overset{H}{\mid}}{C}}-H$	Propanone (acetone)	Solvent
$R-O-R'$	Ether	$C_2H_5-O-C_2H_5$	Diethyl ether (ethyl ether)	Anesthetic
$R-\overset{\overset{O}{\|}}{C}-O-R'$	Ester	$CH_3-\overset{\overset{O}{\|}}{C}-O-C_2H_5$	Ethyl ethanoate (ethyl acetate)	Solvent in fingernail polish
$R-N\begin{matrix}\diagup H\\ \diagdown H\end{matrix}$	Amine	$H-\underset{\underset{H}{\mid}}{\overset{\overset{H}{\mid}}{C}}-N\begin{matrix}\diagup H\\ \diagdown H\end{matrix}$	Methylamine	Tanning (foul odor)
$R-\overset{\overset{O}{\|}}{C}-\overset{\overset{H}{\mid}}{N}-R'$	Amide	$CH_3-\overset{\overset{O}{\|}}{C}-N\begin{matrix}\diagup H\\ \diagdown H\end{matrix}$	Acetamide	Plasticizer

*R stands for an H or a hydrocarbon group such as $-CH_3$ or $-C_2H_5$. R′ could be a different group from R.

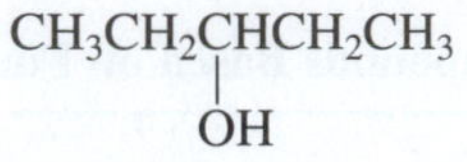

$$\underset{\displaystyle\text{OH}}{\text{CH}_3\text{CH}_2\overset{}{\text{C}}\text{HCH}_2\text{CH}_3} \qquad \text{3-pentanol}$$

Compounds with one or more functional groups and alkyl substituents are named so as to give the functional groups the lowest numbers. For example, the correct name of

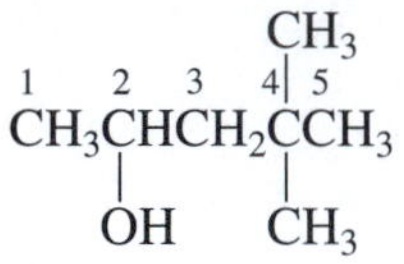

$$\begin{array}{c} CH_3 \\ \overset{1}{C}H_3\overset{2}{C}H(OH)\overset{3}{C}H_2\overset{4}{C}(CH_3)\overset{5}{C}H_3 \end{array}$$

is 4,4-dimethyl-2-pentanol.

Aldehydes and Ketones

The systematic names of the first three aldehydes are methanal, ethanal, and propanal.

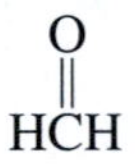

$$\text{H}\overset{\overset{\displaystyle O}{\|}}{\text{C}}\text{H} \qquad \text{CH}_3\overset{\overset{\displaystyle O}{\|}}{\text{C}}\text{H} \qquad \text{CH}_3\text{CH}_2\overset{\overset{\displaystyle O}{\|}}{\text{C}}\text{H}$$

methanal (formaldehyde) 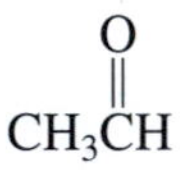 — ethanal (acetaldehyde) 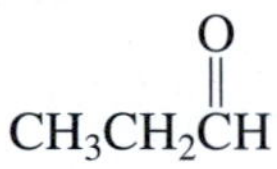 — propanal (propionaldehyde)

For ketones, a number is used to designate the position of the carbonyl group, and the chain is numbered in a way that gives the carbonyl carbon the smallest number.

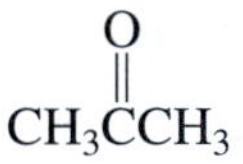

$$\text{CH}_3\overset{\overset{\displaystyle O}{\|}}{\text{C}}\text{CH}_3 \qquad \text{CH}_3\text{CH}_2\overset{\overset{\displaystyle O}{\|}}{\text{C}}\text{CH}_3 \qquad \text{CH}_3\overset{\overset{\displaystyle O}{\|}}{\text{C}}\text{CH}_2\text{CH}{=}\text{CH}_2$$

2-propanone (acetone) 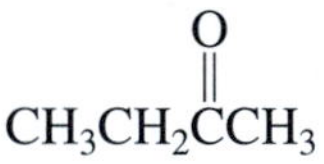 — 2-butanone (methyl ethyl ketone) 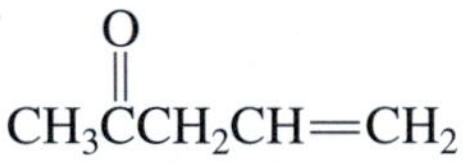 — 4-penten-2-one

Carboxylic Acids

The systematic names of carboxylic acids are obtained by dropping the final *e* of the name of the corresponding alkane and adding *-oic acid.* For example, the name of

$$CH_3CH_2CH_2CH_2CH_2COOH$$

is hexanoic acid. The systematic names of the first five carboxylic acids are given in Table 11.7. Other examples are

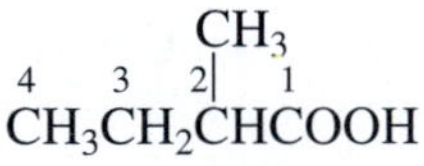

$$\overset{4}{C}H_3\overset{3}{C}H_2\overset{\overset{\displaystyle CH_3}{|}}{\overset{2}{C}}H\overset{1}{C}OOH \qquad \overset{4}{C}H_3\overset{3}{C}H{=}\overset{2}{C}H\overset{1}{C}OOH$$

2-methylbutanoic acid 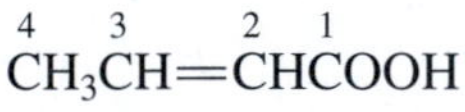 — 2-butenoic acid

Esters

The systematic names of esters are derived from the names of the alcohol and the acid used to prepare the ester. The general formula for esters is

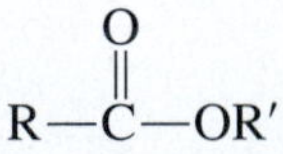

$$\text{R}{-}\overset{\overset{\displaystyle O}{\|}}{\text{C}}{-}\text{OR}'$$

As shown in Section 11.7, the $R{-}\overset{\overset{\displaystyle O}{\|}}{C}$ comes from the acid and the R′O comes from the alcohol. The alcohol part is named first, followed by the name of the acid changed to end in *-ate.* For example,

$$CH_3CH_2\overset{\overset{\displaystyle O}{\|}}{C}{-}OCH_3$$

is named methyl propanoate and

$$CH_3\overset{\overset{\displaystyle O}{\|}}{C}{-}OCH{=}CH_2$$

is named ethenyl ethanoate.

E.3 NAMING COORDINATION COMPOUNDS

Just as there are rules for naming simple inorganic and organic compounds, coordination compounds are named according to an established system. The following compounds are named according to the rules outlined below.

Compound	Systematic Name
$[Ni(H_2O)_6]SO_4$	Hexaaquanickel(II) sulfate
$[Cr(en)_2(CN)_2]Cl$	Dicyanobis(ethylenediamine)chromium(III) chloride
$K[Pt(NH_3)Cl_3]$	Potassium amminetrichloroplatinate(II)

As you read through the rules, notice how they apply to these examples.

1. In naming a coordination compound that is an ionic compound, name the cation first and then the anion, as is usually done.
2. When giving the name of the complex ion or molecule, the ligands are named first, in alphabetical order, followed by the name of the metal.
   (a) If a ligand is an anion whose name ends in *-ite* or *-ate,* the final *e* is changed to *o* (as in sulfate → sulfato or nitrite → nitrito).
   (b) If the ligand is an anion whose name ends in *-ide,* the ending is changed to *o* (as in chloride → chloro or cyanide → cyano).
   (c) If the ligand is a neutral molecule, its common name is used. The important exceptions at this point are water, which is called *aqua;* ammonia, which is called *ammine;* and CO, called *carbonyl.*
   (d) When there is more than one of a particular monodentate ligand with a simple name, the number of ligands is designated by the appropriate prefix: *di, tri, tetra, penta,* or *hexa.* If the ligand name is complicated (whether monodentate or bidentate), the prefix changes to *bis, tris, tetrakis, pentakis,* or *hexakis,* followed by the ligand name in parentheses.
   (e) If the complex ion is an anion, the suffix *-ate* is added to the metal name.
3. Following the name of the metal, the charge (or oxidation number, Section 4.8) of the metal is given in Roman numerals.

Complex ions can become more complicated than those described in Section 9.9, and even more rules of nomenclature must be applied. However, the brief set just outlined is sufficient for the vast majority of complexes.

APPENDIX F

Ionization Constants for Weak Acids at 25 °C

Acid	Formula and Ionization Equation	K_a
Acetic	$CH_3COOH \rightleftharpoons H^+ + CH_3COO^-$	1.8×10^{-5}
Arsenic	$H_3AsO_4 \rightleftharpoons H^+ + H_2AsO_4^-$	$K_1 = 2.5 \times 10^{-4}$
	$H_2AsO_4^- \rightleftharpoons H^+ + HAsO_4^{2-}$	$K_2 = 5.6 \times 10^{-8}$
	$HAsO_4^{2-} \rightleftharpoons H^+ + AsO_4^{3-}$	$K_3 = 3.0 \times 10^{-13}$
Arsenous	$H_3AsO_3 \rightleftharpoons H^+ + H_2AsO_3^-$	$K_1 = 6.0 \times 10^{-10}$
	$H_2AsO_3^- \rightleftharpoons H^+ + HAsO_3^{2-}$	$K_2 = 3.0 \times 10^{-14}$
Benzoic	$C_6H_5COOH \rightleftharpoons H^+ + C_6H_5COO^-$	6.3×10^{-5}
Boric	$H_3BO_3 \rightleftharpoons H^+ + H_2BO_3^-$	$K_1 = 7.3 \times 10^{-10}$
	$H_2BO_3^- \rightleftharpoons H^+ + HBO_3^{2-}$	$K_2 = 1.8 \times 10^{-13}$
	$HBO_3^{2-} \rightleftharpoons H^+ + BO_3^{3-}$	$K_3 = 1.6 \times 10^{-14}$
Carbonic	$H_2CO_3 \rightleftharpoons H^+ + HCO_3^-$	$K_1 = 4.2 \times 10^{-7}$
	$HCO_3^- \rightleftharpoons H^+ + CO_3^{2-}$	$K_2 = 4.8 \times 10^{-11}$
Citric	$H_3C_6H_5O_7 \rightleftharpoons H^+ + H_2C_6H_5O_7^-$	$K_1 = 7.4 \times 10^{-3}$
	$H_2C_6H_5O_7^- \rightleftharpoons H^+ + HC_6H_5O_7^{2-}$	$K_2 = 1.7 \times 10^{-5}$
	$HC_6H_5O_7^{2-} \rightleftharpoons H^+ + C_6H_5O_7^{3-}$	$K_3 = 4.0 \times 10^{-7}$
Cyanic	$HOCN \rightleftharpoons H^+ + OCN^-$	3.5×10^{-4}
Formic	$HCOOH \rightleftharpoons H^+ + HCOO^-$	1.8×10^{-4}
Hydrazoic	$HN_3 \rightleftharpoons H^+ + N_3^-$	1.9×10^{-5}
Hydrocyanic	$HCN \rightleftharpoons H^+ + CN^-$	4.0×10^{-10}
Hydrofluoric	$HF \rightleftharpoons H^+ + F^-$	7.2×10^{-4}
Hydrogen peroxide	$H_2O_2 \rightleftharpoons H^+ + HO_2^-$	2.4×10^{-12}
Hydrosulfuric	$H_2S \rightleftharpoons H^+ + HS^-$	$K_1 = 1.0 \times 10^{-7}$
	$HS^- \rightleftharpoons H^+ + S^{2-}$	1×10^{-19}
Hypobromous	$HOBr \rightleftharpoons H^+ + OBr^-$	2.5×10^{-9}
Hypochlorous	$HOCl \rightleftharpoons H^+ + OCl^-$	3.5×10^{-8}
Nitrous	$HNO_2 \rightleftharpoons H^+ + NO_2^-$	4.5×10^{-4}
Oxalic	$H_2C_2O_4 \rightleftharpoons H^+ + HC_2O_4^-$	$K_1 = 5.9 \times 10^{-2}$
	$HC_2O_4^- \rightleftharpoons H^+ + C_2O_4^{2-}$	$K_2 = 6.4 \times 10^{-5}$
Phenol	$HC_6H_5O \rightleftharpoons H^+ + C_6H_5O^-$	1.3×10^{-10}
Phosphoric	$H_3PO_4 \rightleftharpoons H^+ + H_2PO_4^-$	$K_1 = 7.5 \times 10^{-3}$
	$H_2PO_4^- \rightleftharpoons H^+ + HPO_4^{2-}$	$K_2 = 6.2 \times 10^{-8}$
	$HPO_4^{2-} \rightleftharpoons H^+ + PO_4^{3-}$	$K_3 = 3.6 \times 10^{-13}$
Phosphorous	$H_3PO_3 \rightleftharpoons H^+ + H_2PO_3^-$	$K_1 = 1.6 \times 10^{-2}$
	$H_2PO_3^- \rightleftharpoons H^+ + HPO_3^{2-}$	$K_2 = 7.0 \times 10^{-7}$
Selenic	$H_2SeO_4 \rightleftharpoons H^+ + HSeO_4^-$	K_1 = very large
	$HSeO_4^- \rightleftharpoons H^+ + SeO_4^{2-}$	$K_2 = 1.2 \times 10^{-2}$

Acid	Formula and Ionization Equation	K_a
Selenous	$H_2SeO_3 \rightleftharpoons H^+ + HSeO_3^-$	$K_1 = 2.7 \times 10^{-3}$
	$HSeO_3^- \rightleftharpoons H^+ + SeO_3^{2-}$	$K_2 = 2.5 \times 10^{-7}$
Sulfuric	$H_2SO_4 \rightleftharpoons H^+ + HSO_4^-$	K_1 = very large
	$HSO_4^- \rightleftharpoons H^+ + SO_4^{2-}$	$K_2 = 1.2 \times 10^{-2}$
Sulfurous	$H_2SO_3 \rightleftharpoons H^+ + HSO_3^-$	$K_1 = 1.7 \times 10^{-2}$
	$HSO_3^- \rightleftharpoons H^+ + SO_3^{2-}$	$K_2 = 6.4 \times 10^{-8}$
Tellurous	$H_2TeO_3 \rightleftharpoons H^+ + HTeO_3^-$	$K_1 = 2 \times 10^{-3}$
	$HTeO_3^- \rightleftharpoons H^+ + TeO_3^{2-}$	$K_2 = 1 \times 10^{-8}$

APPENDIX G

Ionization Constants for Weak Bases at 25 °C

Base	Formula and Ionization Equation	K_b
Ammonia	$NH_3 + H_2O \rightleftharpoons NH_4^+ + OH^-$	1.8×10^{-5}
Aniline	$C_6H_5NH_2 + H_2O \rightleftharpoons C_6H_5NH_3^+ + OH^-$	4.2×10^{-10}
Dimethylamine	$(CH_3)_2NH + H_2O \rightleftharpoons (CH_3)_2NH_2^+ + OH^-$	7.4×10^{-4}
Ethylenediamine	$(CH_2)_2(NH_2)_2 + H_2O \rightleftharpoons (CH_2)_2(NH_2)_2H^+ + OH^-$	$K_1 = 8.5 \times 10^{-5}$
	$(CH_2)_2(NH_2)_2H^+ + H_2O \rightleftharpoons (CH_2)_2(NH_2)_2H_2^{2+} + OH^-$	$K_2 = 2.7 \times 10^{-8}$
Hydrazine	$N_2H_4 + H_2O \rightleftharpoons N_2H_5^+ + OH^-$	$K_1 = 8.5 \times 10^{-7}$
	$N_2H_5^+ + H_2O \rightleftharpoons N_2H_6^{2+} + OH^-$	$K_2 = 8.9 \times 10^{-16}$
Hydroxylamine	$NH_2OH + H_2O \rightleftharpoons NH_3OH^+ + OH^-$	6.6×10^{-9}
Methylamine	$CH_3NH_2 + H_2O \rightleftharpoons CH_3NH_3^+ + OH^-$	5.0×10^{-4}
Pyridine	$C_5H_5N + H_2O \rightleftharpoons C_5H_5NH^+ + OH^-$	1.5×10^{-9}
Trimethylamine	$(CH_3)_3N + H_2O \rightleftharpoons (CH_3)_3NH^+ + OH^-$	7.4×10^{-5}

APPENDIX H

Solubility Product Constants for Some Inorganic Compounds at 25 °C[*]

Substance	K_{sp}
Aluminum compounds	
$AlAsO_4$	1.6×10^{-16}
$Al(OH)_3$	1.9×10^{-33}
$AlPO_4$	1.3×10^{-20}
Barium compounds	
$Ba_3(AsO_4)_2$	1.1×10^{-13}
$BaCO_3$	8.1×10^{-9}
$BaC_2O_4 \cdot 2H_2O$[†]	1.1×10^{-7}
$BaCrO_4$	2.0×10^{-10}
BaF_2	1.7×10^{-6}
$Ba(OH)_2 \cdot 8H_2O$[†]	5.0×10^{-3}
$Ba_3(PO_4)_2$	1.3×10^{-29}
$BaSeO_4$	2.8×10^{-11}
$BaSO_3$	8.0×10^{-7}
$BaSO_4$	1.1×10^{-10}
Bismuth compounds	
$BiOCl$	7.0×10^{-9}
$BiO(OH)$	1.0×10^{-12}
$Bi(OH)_3$	3.2×10^{-40}
BiI_3	8.1×10^{-19}
$BiPO_4$	1.3×10^{-23}
Cadmium compounds	
$Cd_3(AsO_4)_2$	2.2×10^{-32}
$CdCO_3$	2.5×10^{-14}
$Cd(CN)_2$	1.0×10^{-8}
$Cd_2[Fe(CN)_6]$	3.2×10^{-17}
$Cd(OH)_2$	1.2×10^{-14}
Calcium compounds	
$Ca_3(AsO_4)_2$	6.8×10^{-19}
$CaCO_3$	3.8×10^{-9}
$CaCrO_4$	7.1×10^{-4}
$CaC_2O_4 \cdot H_2O$[†]	2.3×10^{-9}
CaF_2	3.9×10^{-11}
$Ca(OH)_2$	7.9×10^{-6}
$CaHPO_4$	2.7×10^{-7}
$Ca(H_2PO_4)_2$	1.0×10^{-3}
$Ca_3(PO_4)_2$	1.0×10^{-25}
$CaSO_3 \cdot 2H_2O$[†]	1.3×10^{-8}
$CaSO_4 \cdot 2H_2O$[†]	2.4×10^{-5}
Chromium compounds	
$CrAsO_4$	7.8×10^{-21}
$Cr(OH)_3$	6.7×10^{-31}
$CrPO_4$	2.4×10^{-23}
Cobalt compounds	
$Co_3(AsO_4)_2$	7.6×10^{-29}
$CoCO_3$	8.0×10^{-13}
$Co(OH)_2$	2.5×10^{-16}
$Co(OH)_3$	4.0×10^{-45}
Copper compounds	
$CuBr$	5.3×10^{-9}
$CuCl$	1.9×10^{-7}
$CuCN$	3.2×10^{-20}
$Cu_2O(Cu^+ + OH^-)$[‡]	1.0×10^{-14}
CuI	5.1×10^{-12}
$CuSCN$	1.6×10^{-11}
$Cu_3(AsO_4)_2$	7.6×10^{-36}
$CuCO_3$	2.5×10^{-10}
$Cu_2[Fe(CN)_6]$	1.3×10^{-16}
$Cu(OH)_2$	1.6×10^{-19}
Gold compounds	
$AuBr$	5.0×10^{-17}
$AuCl$	2.0×10^{-13}
AuI	1.6×10^{-23}
$AuBr_3$	4.0×10^{-36}
$AuCl_3$	3.2×10^{-25}
$Au(OH)_3$	1×10^{-53}
AuI_3	1.0×10^{-46}
Iron compounds	
$FeCO_3$	3.5×10^{-11}
$Fe(OH)_2$	7.9×10^{-15}
FeS	4.9×10^{-18}
$Fe_4[Fe(CN)_6]_3$	3.0×10^{-41}

Substance	K_{sp}
$Fe(OH)_3$	6.3×10^{-38}
Lead compounds	
$Pb_3(AsO_4)_2$	4.1×10^{-36}
$PbBr_2$	6.3×10^{-6}
$PbCO_3$	1.5×10^{-13}
$PbCl_2$	1.7×10^{-5}
$PbCrO_4$	1.8×10^{-14}
PbF_2	3.7×10^{-8}
$Pb(OH)_2$	2.8×10^{-16}
PbI_2	8.7×10^{-9}
$Pb_3(PO_4)_2$	3.0×10^{-44}
$PbSeO_4$	1.5×10^{-7}
$PbSO_4$	1.8×10^{-8}
Magnesium compounds	
$Mg_3(AsO_4)_2$	2.1×10^{-20}
$MgCO_3 \cdot 3H_2O$†	4.0×10^{-5}
MgC_2O_4	8.6×10^{-5}
MgF_2	6.4×10^{-9}
$MgNH_4PO_4$	2.5×10^{-12}
Manganese compounds	
$Mn_3(AsO_4)_2$	1.9×10^{-11}
$MnCO_3$	1.8×10^{-11}
$Mn(OH)_2$	4.6×10^{-14}
$Mn(OH)_3$	$\sim 1 \times 10^{-36}$
Mercury compounds	
Hg_2Br_2	1.3×10^{-22}
Hg_2CO_3	8.9×10^{-17}
Hg_2Cl_2	1.1×10^{-18}
Hg_2CrO_4	5.0×10^{-9}
Hg_2I_2	4.5×10^{-29}
$Hg_2O \cdot H_2O(Hg_2^{2+} + 2OH^-)$†‡	1.6×10^{-23}
Hg_2SO_4	6.8×10^{-7}
$Hg(CN)_2$	3.0×10^{-23}
$Hg(OH)_2$	2.5×10^{-26}
HgI_2	4.0×10^{-29}
Nickel compounds	
$Ni_3(AsO_4)_2$	1.9×10^{-26}
$NiCO_3$	6.6×10^{-9}
$Ni(CN)_2$	3.0×10^{-23}
$Ni(OH)_2$	2.8×10^{-16}
Silver compounds	
Ag_3AsO_4	1.1×10^{-20}
$AgBr$	3.3×10^{-13}
Ag_2CO_3	8.1×10^{-12}
$AgCl$	1.8×10^{-10}
Ag_2CrO_4	9.0×10^{-12}
$AgCN$	1.2×10^{-16}
$Ag_4[Fe(CN)_6]$	1.6×10^{-41}
$Ag_2O\ (Ag^+ + OH^-)$‡	2.0×10^{-8}
AgI	1.5×10^{-16}
Ag_3PO_4	1.3×10^{-20}
Ag_2SO_3	1.5×10^{-14}
Ag_2SO_4	1.7×10^{-5}
$AgSCN$	1.0×10^{-12}
Strontium compounds	
$Sr_3(AsO_4)_2$	1.3×10^{-18}
$SrCO_3$	9.4×10^{-10}
$SrC_2O_4 \cdot 2H_2O$†	5.6×10^{-8}
$SrCrO_4$	3.6×10^{-5}
$Sr(OH)_2 \cdot 8H_2O$†	3.2×10^{-4}
$Sr_3(PO_4)_2$	1.0×10^{-31}
$SrSO_3$	4.0×10^{-8}
$SrSO_4$	2.8×10^{-7}
Tin compounds	
$Sn(OH)_2$	2.0×10^{-26}
SnI_2	1.0×10^{-4}
$Sn(OH)_4$	1×10^{-57}
Zinc compounds	
$Zn_3(AsO_4)_2$	1.1×10^{-27}
$ZnCO_3$	1.5×10^{-11}
$Zn(CN)_2$	8.0×10^{-12}
$Zn_3[Fe(CN)_6]$	4.1×10^{-16}
$Zn(OH)_2$	4.5×10^{-17}
$Zn_3(PO_4)_2$	9.1×10^{-33}

*No metallic sulfides are listed in this table because sulfide ion is such a strong base that the usual solubility product equilibrium equation does not apply. See Myers, R.J., *J. Chem. Educ.* **1986** *63,* 687-690.

†Since $[H_2O]$ does not appear in equilibrium constants for equilibria in aqueous solution in general, it does *not* appear in the K_{sp} expressions for hydrated solids.

‡Very small amounts of oxides dissolve in water to give the ions indicated in parentheses. Solid hydroxides are unstable and decompose to oxides as rapidly as they are formed.

APPENDIX I

Standard Reduction Potentials in Aqueous Solution at 25 °C

Acidic Solution	Standard Reduction Potential, E^0 (volts)
$F_2(g) + 2e^- \longrightarrow 2\ F^-(aq)$	2.87
$Co^{3+}(aq) + e^- \longrightarrow Co^{2+}(aq)$	1.82
$Pb^{4+}(aq) + 2\ e^- \longrightarrow Pb^{2+}(aq)$	1.8
$H_2O_2(aq) + 2\ H^+(aq) + 2\ e^- \longrightarrow 2\ H_2O$	1.77
$NiO_2(s) + 4\ H^+(aq) + 2\ e^- \longrightarrow Ni^{2+}(aq) + 2\ H_2O$	1.7
$PbO_2(s) + SO_4^{2-}(aq) + 4\ H^+(aq) + 2\ e^- \longrightarrow PbSO_4(s) + 2\ H_2O$	1.685
$Au^+(aq) + e^- \longrightarrow Au(s)$	1.68
$2\ HClO(aq) + 2\ H^+(aq) + 2\ e^- \longrightarrow Cl_2(g) + 2\ H_2O$	1.63
$Ce^{4+}(aq) + e^- \longrightarrow Ce^{3+}(aq)$	1.61
$NaBiO_3(s) + 6\ H^+(aq) + 2\ e^- \longrightarrow Bi^{3+}(aq) + Na^+(aq) + 3\ H_2O$	~ 1.6
$MnO_4^-(aq) + 8\ H^+(aq) + 5\ e^- \longrightarrow Mn^{2+}(aq) + 4\ H_2O$	1.51
$Au^{3+}(aq) + 3\ e^- \longrightarrow Au(s)$	1.50
$ClO_3^-(aq) + 6\ H^+(aq) + 5\ e^- \longrightarrow \frac{1}{2}\ Cl_2(g) + 3\ H_2O$	1.47
$BrO_3^-(aq) + 6\ H^+(aq) + 6\ e^- \longrightarrow Br^-(aq) + 3\ H_2O$	1.44
$Cl_2(g) + 2\ e^- \longrightarrow 2\ Cl^-(aq)$	1.358
$Cr_2O_7^{2-}(aq) + 6\ e^- \longrightarrow 2\ Cr^{3+}(aq) + 7\ H_2O$	1.33
$N_2H_5^+(aq) + 3\ H^+(aq) + 2\ e^- \longrightarrow 2\ NH_4^+(aq)$	1.24
$MnO_2(s) + 4\ H^+(aq) + 2\ e^- \longrightarrow Mn^{2+}(aq) + 2\ H_2O$	1.23
$O_2(g) + 4\ H^+(aq) + 4\ e^- \longrightarrow 2\ H_2O$	1.229
$Pt^{2+}(aq) + 2\ e^- \longrightarrow Pt(s)$	1.2
$IO_3^-(aq) + 6\ H^+(aq) + 5\ e^- \longrightarrow \frac{1}{2}\ I_2(aq) + 3\ H_2O$	1.195
$ClO_4^-(aq) + 2\ H^+(aq) + 2\ e^- \longrightarrow ClO_3^-(aq) + H_2O$	1.19
$Br_2(\ell) + 2\ e^- \longrightarrow 2\ Br^-(aq)$	1.066
$AuCl_4^-(aq) + 3\ e^- \longrightarrow Au(s) + 4\ Cl^-(aq)$	1.00
$Pd^{2+}(aq) + 2\ e^- \longrightarrow Pd(s)$	0.987
$NO_3^-(aq) + 4\ H^+(aq) + 3\ e^- \longrightarrow NO(g) + 2\ H_2O$	0.96
$NO_3^-(aq) + 3\ H^+(aq) + 2\ e^- \longrightarrow HNO_2(aq) + H_2O$	0.94
$2\ Hg^{2+}(aq) + 2\ e^- \longrightarrow Hg_2^{2+}(aq)$	0.920
$Hg^{2+}(aq) + 2\ e^- \longrightarrow Hg(\ell)$	0.855
$Ag^+(aq) + e^- \longrightarrow Ag(s)$	0.7994
$Hg_2^{2+}(aq) + 2\ e^- \longrightarrow 2\ Hg(\ell)$	0.789
$Fe^{3+}(aq) + e^- \longrightarrow Fe^{2+}(aq)$	0.771
$SbCl_6^-(aq) + 2\ e^- \longrightarrow SbCl_4^-(aq) + 2\ Cl^-(aq)$	0.75
$[PtCl_4]^{2-}(aq) + 2\ e^- \longrightarrow Pt(s) + 4\ Cl^-(aq)$	0.73
$O_2(g) + 2\ H^+(aq) + 2\ e^- \longrightarrow H_2O_2(aq)$	0.682
$[PtCl_6]^{2-}(aq) + 2\ e^- \longrightarrow [PtCl_4]^{2-}(aq) + 2\ Cl^-(aq)$	0.68

Acidic Solution	Standard Reduction Potential, E^0 (volts)
$H_3AsO_4(aq) + 2\ H^+(aq) + 2\ e^- \longrightarrow H_3AsO_3(aq) + H_2O$	0.58
$I_2(s) + 2\ e^- \longrightarrow 2\ I^-(aq)$	0.535
$TeO_2(s) + 4\ H^+(aq) + 4\ e^- \longrightarrow Te(s) + 2\ H_2O$	0.529
$Cu^+(aq) + e^- \longrightarrow Cu(s)$	0.521
$[RhCl_6]^{3-}(aq) + 3\ e^- \longrightarrow Rh(s) + 6\ Cl^-(aq)$	0.44
$Cu^{2+}(aq) + 2\ e^- \longrightarrow Cu(s)$	0.337
$HgCl_2(s) + 2\ e^- \longrightarrow 2\ Hg(\ell) + 2\ Cl^-(aq)$	0.27
$AgCl(s) + e^- \longrightarrow Ag(s) + Cl^-(aq)$	0.222
$SO_4^{2-}(aq) + 4\ H^+(aq) + 2e^- \longrightarrow SO_2(g) + 2\ H_2O$	0.20
$SO_4^{2-}(aq) + 4\ H^+(aq) + 2e^- \longrightarrow H_2SO_3(aq) + H_2O$	0.17
$Cu^{2+}(aq) + e^- \longrightarrow Cu^+(aq)$	0.153
$Sn^{4+}(aq) + 2\ e^- \longrightarrow Sn^{2+}(aq)$	0.15
$S(s) + 2\ H^+(aq) + 2\ e^- \longrightarrow H_2S(aq)$	0.14
$AgBr(s) + e^- \longrightarrow Ag(s) + Br^-(aq)$	0.0713
$2\ H^+(aq) + 2\ e^- \longrightarrow H_2(g)$ (reference electrode)	0.0000
$N_2O(g) + 6\ H^+(aq) + H_2O + 4\ e^- \longrightarrow 2\ NH_3OH^+(aq)$	−0.05
$Pb^{2+}(aq) + 2\ e^- \longrightarrow Pb(s)$	−0.126
$Sn^{2+}(aq) + 2\ e^- \longrightarrow Sn(s)$	−0.14
$AgI(s) + e^- \longrightarrow Ag(s) + I^-(aq)$	−0.15
$[SnF_6]^{2-}(aq) + 4\ e^- \longrightarrow Sn(s) + 6\ F^-(aq)$	−0.25
$Ni^{2+}(aq) + 2\ e^- \longrightarrow Ni(s)$	−0.25
$Co^{2+}(aq) + 2\ e^- \longrightarrow Co(s)$	−0.28
$Tl^+(aq) + e^- \longrightarrow Tl(s)$	−0.34
$PbSO_4(s) + 2\ e^- \longrightarrow Pb(s) + SO_4^{2-}(aq)$	−0.356
$Se(s) + 2\ H^+(aq) + 2\ e^- \longrightarrow H_2Se(aq)$	−0.40
$Cd^{2+}(aq) + 2\ e^- \longrightarrow Cd(s)$	−0.403
$Cr^{3+}(aq) + e^- \longrightarrow Cr^{2+}(aq)$	−0.41
$Fe^{2+}(aq) + 2\ e^- \longrightarrow Fe(s)$	−0.44
$2\ CO_2(g) + 2\ H^+(aq) + 2\ e^- \longrightarrow (COOH)_2(aq)$	−0.49
$Ga^{3+}(aq) + 3\ e^- \longrightarrow Ga(s)$	−0.53
$HgS(s) + 2\ H^+(aq) + 2\ e^- \longrightarrow Hg(\ell) + H_2S(g)$	−0.72
$Cr^{3+}(aq) + 3\ e^- \longrightarrow Cr(s)$	−0.74
$Zn^{2+}(aq) + 2\ e^- \longrightarrow Zn(s)$	−0.763
$Cr^{2+}(aq) + 2\ e^- \longrightarrow Cr(s)$	−0.91
$Mn^{2+}(aq) + 2\ e^- \longrightarrow Mn(s)$	−1.18
$V^{2+}(aq) + 2\ e^- \longrightarrow V(s)$	−1.18
$Zr^{4+}(aq) + 4\ e^- \longrightarrow Zr(s)$	−1.53
$Al^{3+}(aq) + 3\ e^- \longrightarrow Al(s)$	−1.66
$H_2(g) + 2\ e^- \longrightarrow 2\ H^-(aq)$	−2.25
$Mg^{2+}(aq) + 2\ e^- \longrightarrow Mg(s)$	−2.37
$Na^+(aq) + e^- \longrightarrow Na(s)$	−2.714
$Ca^{2+}(aq) + 2\ e^- \longrightarrow Ca(s)$	−2.87
$Sr^{2+}(aq) + 2\ e^- \longrightarrow Sr(s)$	−2.89
$Ba^{2+}(aq) + 2\ e^- \longrightarrow Ba(s)$	−2.90

Basic Solution	Standard Reduction Potential, E^0 (volts)
$Rb^+(aq) + e^- \longrightarrow Rb(s)$	−2.925
$K^+(aq) + e^- \longrightarrow K(s)$	−2.925
$Li^+(aq) + e^- \longrightarrow Li(s)$	−3.045
$ClO^-(aq) + H_2O + 2\ e^- \longrightarrow Cl^-(aq) + 2\ OH^-(aq)$	0.89
$OOH^-(aq) + H_2O + 2\ e^- \longrightarrow 3\ OH^-(aq)$	0.88
$2\ NH_2OH(aq) + 2\ e^- \longrightarrow N_2H_4(aq) + 2\ OH^-(aq)$	0.74
$ClO_3^-(aq) + 3\ H_2O + 6\ e^- \longrightarrow Cl^-(aq) + 6\ OH^-(aq)$	0.62
$MnO_4^-(aq) + 2\ H_2O + 3\ e^- \longrightarrow MnO_2(s) + 4\ OH^-(aq)$	0.588
$MnO_4^-(aq) + e^- \longrightarrow MnO_4^{2-}(aq)$	0.564
$NiO_2(s) + 2\ H_2O + 2\ e^- \longrightarrow Ni(OH)_2(s) + 2\ OH^-(aq)$	0.49
$Ag_2CrO_4(s) + 2\ e^- \longrightarrow 2\ Ag(s) + CrO_4^{2-}(aq)$	0.446
$O_2(g) + 2\ H_2O + 4\ e^- \longrightarrow 4\ OH^-(aq)$	0.40
$ClO_4^-(aq) + H_2O + 2\ e^- \longrightarrow ClO_3^-(aq) + 2\ OH^-(aq)$	0.36
$Ag_2O(s) + H_2O + 2\ e^- \longrightarrow 2\ Ag(s) + 2\ OH^-(aq)$	0.34
$2\ NO_2^-(aq) + 3\ H_2O + 4\ e^- \longrightarrow N_2O(g) + 6\ OH^-(aq)$	0.15
$N_2H_4(aq) + 2\ H_2O + 2\ e^- \longrightarrow 2\ NH_3(aq) + 2\ OH^-(aq)$	0.10
$[Co(NH_3)_6]^{3+}(aq) + e^- \longrightarrow [Co(NH_3)_6]^{2+}(aq)$	0.10
$HgO(s) + H_2O + 2\ e^- \longrightarrow Hg(\ell) + 2\ OH^-(aq)$	0.0984
$O_2(g) + H_2O + 2\ e^- \longrightarrow OOH^-(aq) + OH^-(aq)$	0.076
$NO_3^-(aq) + H_2O + 2\ e^- \longrightarrow NO_2^-(aq) + 2\ OH^-(aq)$	0.01
$MnO_2(s) + 2\ H_2O + 2\ e^- \longrightarrow Mn(OH)_2(s) + 2\ OH^-(aq)$	−0.05
$CrO_4^{2-}(aq) + 4\ H_2O + 3\ e^- \longrightarrow Cr(OH)_3(s) + 5\ OH^-(aq)$	−0.12
$Cu(OH)_2(s) + 2\ e^- \longrightarrow Cu(s) + 2\ OH^-(aq)$	−0.36
$Fe(OH)_3(s) + e^- \longrightarrow Fe(OH)_2(s) + OH^-(aq)$	−0.56
$2\ H_2O + 2\ e^- \longrightarrow H_2(g) + 2\ OH^-(aq)$	−0.8277
$2\ NO_3^-(aq) + 2\ H_2O + 2\ e^- \longrightarrow N_2O_4(g) + 4\ OH^-(aq)$	−0.85
$Fe(OH)_2(s) + 2\ e^- \longrightarrow Fe(s) + 2\ OH^-(aq)$	−0.877
$SO_4^{2-}(aq) + H_2O + 2\ e^- \longrightarrow SO_3^{2-}(aq) + 2\ OH^-(aq)$	−0.93
$N_2(g) + 4\ H_2O + 4\ e^- \longrightarrow N_2H_4(aq) + 4\ OH^-(aq)$	−1.15
$[Zn(OH)_4]^{2-}(aq) + 2\ e^- \longrightarrow Zn(s) + 4\ OH^-(aq)$	−1.22
$Zn(OH)_2(s) + 2\ e^- \longrightarrow Zn(s) + 2\ OH^-(aq)$	−1.245
$[Zn(CN)_4]^{2-}(aq) + 2\ e^- \longrightarrow Zn(s) + 4\ CN^-(aq)$	−1.26
$Cr(OH)_3(s) + 3\ e^- \longrightarrow Cr(s) + 3\ OH^-(aq)$	−1.30
$SiO_3^{2-}(aq) + 3\ H_2O + 4\ e^- \longrightarrow Si(s) + 6\ OH^-(aq)$	−1.70

A P P E N D I X J

Selected Thermodynamic Values*

Species	ΔH_f°(298.15K) kJ/mol	S°(298.15K) J/K · mol	ΔG_f°(298.15K) kJ/mol
Aluminum			
Al(s)	0	28.3	0
$AlCl_3$(s)	−704.2	110.67	−628.8
Al_2O_3(s)	−1675.7	50.92	−1582.3
Aqueous Solutions			
Ca^{2+}(aq)	−542.96	−55.2	−553.04
CO_3^{2-}(aq)	−676.26	−53.1	−528.10
H^+(aq)	0	0	0
HCO_2^-(aq)	−410	91.6	−335
HCO_2H(aq)	−410	164	−356
HCO_3^-(aq)	−691.11	95.0	−587.06
H_2CO_3(aq)	−698.7	191	−623.42
OH^-(aq)	−229.94	−10.54	−157.30
Barium			
$BaCl_2$(s)	−858.6	123.68	−810.4
BaO(s)	−553.5	70.42	−525.1
$BaSO_4$(s)	−1473.2	132.2	−1362.2
Beryllium			
Be(s)	0	9.5	0
$Be(OH)_2$(s)	−902.5	51.9	−815.0
Bromine			
Br(g)	111.884	175.022	82.396
$Br_2(\ell)$	0	152.2	0
Br_2(g)	30.907	245.463	3.110
BrF_3(g)	−255.60	292.53	−229.43
HBr(g)	−36.40	198.695	−53.45
Calcium			
Ca(s)	0	41.42	0
Ca(g)	178.2	158.884	144.3
Ca^{2+}(g)	1925.90	—	—
CaC_2(s)	−59.8	69.96	−64.9
$CaCO_3$(s; calcite)	−1206.92	92.9	−1128.79

Species	ΔH_f°(298.15K) kJ/mol	S°(298.15K) J/K · mol	ΔG_f°(298.15K) kJ/mol
$CaCl_2(s)$	−795.8	104.6	−748.1
$CaF_2(s)$	−1219.6	68.87	−1167.3
$CaH_2(s)$	−186.2	42.	−147.2
$CaO(s)$	−635.09	39.75	−604.03
$CaS(s)$	−482.4	56.5	−477.4
$Ca(OH)_2(s)$	−986.09	83.39	−898.49
$Ca(OH)_2(aq)$	−1002.82	−74.5	−868.07
$CaSO_4(s)$	−1434.11	106.7	−1321.79
Carbon			
C(s, graphite)	0	5.740	0
C(s, diamond)	1.895	2.377	2.900
C(g)	716.682	158.096	671.257
$CCl_4(\ell)$	−135.44	216.40	−65.21
$CCl_4(g)$	−102.9	309.85	−60.59
$CHCl_3$(liq)	−134.47	201.7	−73.66
$CHCl_3(g)$	−103.14	295.71	−70.34
CH_4(g, methane)	−74.81	186.264	−50.72
C_2H_2(g, ethyne)	226.73	200.94	209.20
C_2H_4(g, ethene)	52.26	219.56	68.15
C_2H_6(g, ethane)	−84.68	229.60	−32.82
C_3H_8(g, propane)	−103.8	269.9	−23.49
C_6H_6(ℓ, benzene)	49.03	172.8	124.5
$C_6H_{14}(\ell)$	−198.782	296.018	−4.035
$C_8H_{18}(\ell)$	−249.952	361.205	6.707
CH_3OH(ℓ, methanol)	−238.66	126.8	−166.27
CH_3OH(g, methanol)	−200.66	239.81	−161.96
C_2H_5OH(ℓ, ethanol)	−277.69	160.7	−174.78
C_2H_5OH(g, ethanol)	−235.10	282.70	−168.49
$CH_3COOH(\ell)$	−276.981	160.666	−173.991
CO(g)	−110.525	197.674	−137.168
$CO_2(g)$	−393.509	213.74	−394.359
$CS_2(g)$	117.36	237.84	67.12
$COCl_2(g)$	−218.8	283.53	−204.6
Cesium			
Cs(s)	0	85.23	0
$Cs^+(g)$	457.964	—	—
CsCl(s)	−443.04	101.17	−414.53
Chlorine			
Cl(g)	121.679	165.198	105.680
$Cl^-(g)$	−233.13	—	—
$Cl_2(g)$	0	223.066	0
HCl(g)	−92.307	186.908	−95.299
HCl(aq)	−167.159	56.5	−131.228

Species	ΔH_f°(298.15K) kJ/mol	S°(298.15K) J/K · mol	ΔG_f°(298.15K) kJ/mol
Chromium			
Cr(s)	0	23.77	0
Cr_2O_3(s)	−1139.7	81.2	−1058.1
$CrCl_3$(s)	−556.5	123.0	−486.1
Copper			
Cu(s)	0	33.150	0
CuO(s)	−157.3	42.63	−129.7
$CuCl_2$(s)	−220.1	108.07	−175.7
Fluorine			
F_2(g)	0	202.78	0
F(g)	78.99	158.754	61.91
F^-(g)	−255.39	—	—
F^-(aq)	−332.63	−13.8	−278.79
HF(g)	−271.1	173.779	−273.2
HF(aq)	−332.63	−13.8	−278.79
Hydrogen			
H_2(g)	0	130.684	0
H(g)	217.965	114.713	203.247
H^+(g)	1536.202	—	—
$H_2O(\ell)$	−285.830	69.91	−237.129
H_2O(g)	−241.818	188.825	−228.572
$H_2O_2(\ell)$	−187.78	109.6	−120.35
Iodine			
I_2(s)	0	116.135	0
I_2(g)	62.438	260.69	19.327
I(g)	106.838	180.791	70.250
I^-(g)	−197.	—	—
ICl(g)	17.78	247.551	−5.46
Iron			
Fe(s)	0	27.78	0
FeO(s)	−272.	—	—
Fe_2O_3(s, hematite)	−824.2	87.40	−742.2
Fe_3O_4(s, magnetite)	−1118.4	146.4	−1015.4
$FeCl_2$(s)	−341.79	117.95	−302.30
$FeCl_3$(s)	−399.49	142.3	−344.00
FeS_2(s, pyrite)	−178.2	52.93	−166.9
$Fe(CO)_5(\ell)$	−774.0	338.1	−705.3
Lead			
Pb(s)	0	64.81	0
$PbCl_2$(s)	−359.41	136.0	−314.10

Species	ΔH_f°(298.15K) kJ/mol	S°(298.15K) J/K · mol	ΔG_f°(298.15K) kJ/mol
PbO(s, yellow)	−217.32	68.70	−187.89
PbS(s)	−100.4	91.2	−98.7
Lithium			
Li(s)	0	29.12	0
Li^+(g)	685.783	—	—
LiOH(s)	−484.93	42.80	−438.95
LiOH(aq)	−508.48	2.80	−450.58
LiCl(s)	−408.701	59.33	−384.37
Magnesium			
Mg(s)	0	32.68	0
$MgCl_2$(g)	−641.32	89.62	−591.79
MgO(s)	−601.70	26.94	−569.43
$Mg(OH)_2$(s)	−924.54	63.18	−833.51
MgS(s)	−346.0	50.33	−341.8
Mercury			
Hg(ℓ)	0	76.02	0
$HgCl_2$(s)	−224.3	146.0	−178.6
HgO(s, red)	−90.83	70.29	−58.539
HgS(s, red)	−58.2	82.4	−50.6
Nickel			
Ni(s)	0	29.87	0
NiO(s)	−239.7	37.99	−211.7
$NiCl_2$(s)	−305.332	97.65	−259.032
Nitrogen			
N_2(g)	0	191.61	0
N(g)	472.704	153.298	455.563
NH_3(g)	−46.11	192.45	−16.45
N_2H_4(ℓ)	50.63	121.21	149.34
NH_4Cl(s)	−314.43	94.6	−202.87
NH_4Cl(aq)	−299.66	169.9	−210.52
NH_4NO_3(s)	−365.56	151.08	−183.87
NH_4NO_3(aq)	−399.87	259.8	−190.56
NO(g)	90.25	210.76	86.55
NO_2(g)	33.18	240.06	51.31
N_2O(g)	82.05	219.85	104.20
N_2O_4(g)	9.16	304.29	97.89
NOCl(g)	51.71	261.69	66.08
HNO_3(ℓ)	−174.10	155.60	−80.71
HNO_3(g)	−135.06	266.38	−74.72
HNO_3(aq)	−207.36	146.4	−111.25

Species	ΔH_f°(298.15K) kJ/mol	S°(298.15K) J/K · mol	ΔG_f°(298.15K) kJ/mol
Oxygen			
$O_2(g)$	0	205.138	0
$O(g)$	249.170	161.055	231.731
$O_3(g)$	142.7	238.93	163.2
Phosphorus			
P_4(s, white)	0	164.36	0
P_4(s, red)	−70.4	91.2	−48.4
$P(g)$	314.64	163.193	278.25
$PH_3(g)$	5.4	310.23	13.4
$PCl_3(g)$	−287.0	311.78	−267.8
$P_4O_{10}(s)$	−2984.0	228.86	−2697.7
$H_3PO_4(s)$	−1279.0	110.5	−1119.1
Potassium			
$K(s)$	0	64.18	0
$KCl(s)$	−436.747	82.59	−409.14
$KClO_3(s)$	−397.73	143.1	−296.25
$KI(s)$	−327.90	106.32	−324.892
$KOH(s)$	−424.764	78.9	−379.08
$KOH(aq)$	−482.37	91.6	−440.50
Silicon			
$Si(s)$	0	18.83	0
$SiBr_4(\ell)$	−457.3	277.8	−443.8
$SiC(s)$	−65.3	16.61	−62.8
$SiCl_4(g)$	−657.01	330.73	−616.98
$SiH_4(g)$	34.3	204.62	56.9
$SiF_4(g)$	−1614.94	282.49	−1572.65
SiO_2(s, quartz)	−910.94	41.84	−856.64
Silver			
$Ag(s)$	0	42.55	0
$Ag_2O(s)$	−31.05	121.3	−11.20
$AgCl(s)$	−127.068	96.2	−109.789
$AgNO_3(s)$	−124.39	140.92	−33.41
Sodium			
$Na(s)$	0	51.21	0
$Na(g)$	107.32	153.712	76.761
$Na^+(g)$	609.358	—	—
$NaBr(s)$	−361.062	86.82	−348.983
$NaCl(s)$	−411.153	72.13	−384.138
$NaCl(g)$	−176.65	229.81	−196.66

Species	ΔH_f°(298.15K) kJ/mol	S°(298.15K) J/K · mol	ΔG_f°(298.15K) kJ/mol
NaCl(aq)	−407.27	115.5	−393.133
NaOH(s)	−425.609	64.455	−379.484
NaOH(aq)	−470.114	48.1	−419.150
Na_2CO_3(s)	−1130.68	134.98	−1044.44
Sulfur			
S(s, rhombic)	0	31.80	0
S(g)	278.805	167.821	238.250
S_2Cl_2(g)	−18.4	331.5	−31.8
SF_6(g)	1209.	291.82	−1105.3
H_2S(g)	−20.63	205.79	−33.56
SO_2(g)	−296.830	248.22	−300.194
SO_3(g)	−395.72	256.76	−371.06
$SOCl_2$(g)	−212.5	309.77	−198.3
H_2SO_4(ℓ)	−813.989	156.904	−690.003
H_2SO_4(aq)	−909.27	20.1	−744.53
Tin			
Sn(s, white)	0	51.55	0
Sn(s, gray)	−2.09	44.14	0.13
$SnCl_4$(ℓ)	−511.3	248.6	−440.1
$SnCl_4$(g)	−471.5	365.8	−432.2
SnO_2(s)	−580.7	52.3	−519.6
Titanium			
Ti(s)	0	30.63	0
$TiCl_4$(ℓ)	−804.2	252.34	−737.2
$TiCl_4$(g)	−763.2	354.9	−726.7
TiO_2(s)	−939.7	49.92	−884.5
Zinc			
Zn(s)	0	41.63	0
$ZnCl_2$(s)	−415.05	111.46	−369.398
ZnO(s)	−348.28	43.64	−318.30
ZnS(s, sphalerite)	−205.98	57.7	−201.29

*Taken from "The NBS Tables of Chemical Thermodynamic Properties," 1982.

APPENDIX K

Answers to Problem-Solving Practice Problems

Chapter 2

2.1. (a) $650 \text{ mg} \times \dfrac{1 \text{ g}}{10^3 \text{ mg}} \times \dfrac{1 \text{ kg}}{10^3 \text{ g}} = 6.50 \times 10^{-4} \text{ kg}$

(b) $4.8 \times 10^{-8} \text{ m}^3 \times \dfrac{10^8 \text{ cm}^3}{\text{m}^3} \times \dfrac{1 \text{ mL}}{1 \text{ cm}^3} = 4.8 \text{ mL}$

2.2. $\dfrac{165\text{mg}}{\text{dL}} \times \dfrac{1 \text{ g}}{10^3 \text{ mg}} \times \dfrac{1 \text{ dL}}{1 \times 10^{-1} \text{ L}} = 1.65 \text{ g/L}$

2.3. (a) Mass number = 15 protons + 16 neutrons = 31
(b) There are 10 protons, 10 electrons, and 12 neutrons in a neutral neon-22 atom.
(c) Lead is element 82; $^{207}_{82}\text{Pb}$.

2.4. The isotopes are $^{24}_{12}\text{Mg}$, $^{25}_{12}\text{Mg}$, $^{26}_{12}\text{Mg}$.

2.5. $75. \text{ g Nichrome} \times \dfrac{80 \text{ g Ni}}{100 \text{ g Nichrome}} = 60 \text{ g Ni};$

$75 \text{ g} - 60 \text{ g} = 15 \text{ g Cr}$

2.6. (a) $1.00 \text{ mg Mo} \times \dfrac{1 \text{ g Mo}}{10^3 \text{ mg Mo}} \times \dfrac{1 \text{ mol Mo}}{95.94 \text{ g Mo}} = 1.04 \times 10^{-5} \text{ mol Mo}$

(b) $5.00 \times 10^{-3} \text{ mol Au} \times \dfrac{196.97 \text{ g Au}}{1 \text{ mol Au}} = 0.985 \text{ g Au}$

2.7. (a) $0.550 \text{ mol Mg} \times \dfrac{24.31 \text{ g Mg}}{1 \text{ mol Mg}} = 13.4 \text{ g Mg}$

(b) $200.0 \text{ cm}^3 \text{ Pb} \times \dfrac{11.34 \text{ g Pb}}{\text{cm}^3 \text{ Pb}} \times \dfrac{1 \text{ mol Pb}}{207.2 \text{ g Pb}} = 10.95 \text{ mol Pb}$

Chapter 3

3.1. (a) $C_{10}H_{11}O_{13}N_5P_3$; (b) $C_{18}H_{27}O_3N$; (c) $C_2H_2O_4$

3.2. (a) A Ca^{4+} charge is unlikely because calcium is in Group 2A whose elements lose two electrons to form 2+ ions.
(b) Cr^{2+} is possible because chromium is a transition metal ion that forms 2+ and 3+ ions.
(c) Strontium is a Group 2A metal and forms 2+ ions; thus a Sr^- ion is highly unlikely.

3.3. (a) $In_2(SO_4)_3$ contains two In^{3+} and three SO_4^{2-} ions. There are 17 atoms in this formula unit.
(b) This formula unit contains three ammonium ions and one phosphate ion, collectively containing 20 atoms.

$\underbrace{(NH_4)_3}\ PO_4$

15 atoms + 1 atom + 4 atoms = 20 atoms

3.4.
1. (a) One calcium 2+ ion and two fluoride (F^-) ions; (b) One cobalt 2+ ion and two chloride (Cl^-) ions; (c) two potassium ions (K^+) and one monohydrogen phosphate (HPO_4^{2-}) ion.
2. CuBr, copper(I) bromide and $CuBr_2$, copper(II) bromide
3. K_2O and K_2SO_4; SrO and $SrSO_4$

3.5. (a) KH_2PO_4; (b) CuOH; (c) NaClO; (d) NH_4ClO_4; (e) $TiCl_4$; (f) $FeSO_3$

3.6. (a) Soluble; (b) Soluble; (c) Soluble
(d) Only slightly soluble (e) Insoluble (f) Insoluble

3.7. (a) Molar masses: cholesterol = 386.6 g/mol; $Mn_2(SO_4)_3$ = 398.1 g/mol

$10.0 \text{ g cholesterol} \times \dfrac{1 \text{ mol chol}}{386.6 \text{ g chol}} = 0.0259 \text{ mol cholesterol}$

$10.0 \text{ g } Mn_2(SO_4)_3 \times \dfrac{1 \text{ mol } Mn_2(SO_4)_3}{398.1 \text{ g } Mn_2(SO_4)_3} = 0.0251 \text{ mol } Mn_2(SO_4)_3$

(b) Grams K_2HPO_4: $0.25 \text{ mol } K_2HPO_4 \times \dfrac{174 \text{ g } K_2HPO_4}{1 \text{ mol } K_2HPO_4} = 44 \text{ g } K_2HPO_4$

Grams caffeine: $0.25 \text{ mol } C_8H_{10}O_2N_4 \times \dfrac{194 \text{ g } C_8H_{10}O_2N_4}{1 \text{ mol } C_8H_{10}O_2N_4} = 48 \text{ g } C_8H_{10}O_2N_4$

3.8. Assume 100.0 g of the oxide.

$$43.64 \text{ g P} \times \frac{1 \text{ mol P}}{30.974 \text{ g P}} = 1.4089 \text{ mol P}$$

$$56.36 \text{ g O} \times \frac{1 \text{ mol O}}{15.999 \text{ g O}} = 3.5227 \text{ mol O}$$

$$1.4089/1.4089 = 1.000 \text{ mol P}; \frac{3.5227}{1.4089} = 2.500 \text{ mol O}$$

$P_{1.000}O_{2.500}$ is equivalent to an empirical formula of P_2O_5, which has an empirical formula weight of 141.94 g. The molar mass of the phosphorus oxide is 283.89 g/mol. Dividing the molar mass by the empirical formula weight, $\frac{283.89}{141.94} = 2.000$; thus the molecular formula is twice the empirical formula, or P_4O_{10}.

3.9. Assume 100.0 g of vitamin C, which contains

$$40.9 \text{ g C} \times \frac{1 \text{ mol C}}{12.01 \text{ g C}} = 3.41 \text{ mol C}$$

$$4.58 \text{ g H} \times \frac{1 \text{ mol H}}{1.008 \text{ g H}} = 4.54 \text{ mol H}$$

$$54.5 \text{ g O} \times \frac{1 \text{ mol O}}{16.00 \text{ g O}} = 3.41 \text{ mol O}$$

Dividing by the smallest value to get the least common multiple;

$$\frac{3.41 \text{ mol C}}{3.41} = 1.00 \text{ mol C}; \quad \frac{4.54 \text{ mol H}}{3.41} = 1.33 \text{ mol H};$$

$$\frac{3.41 \text{ mol O}}{3.41} = 1.00 \text{ mol O}$$

The empirical formula is $C_{1.00}H_{1.33}O_{1.00}$, which multiplied by 3 gives $C_3H_4O_3$. Its empirical formula weight is [3(12.01) + 4(1.008) + 3(16.00)] g = 88.06 g. Dividing this into the molar mass, $\frac{176.13 \text{ g}}{88.06 \text{ g}} = 2.000$. The molar mass is twice the empirical formula weight; the molecular formula of vitamin C is $C_6H_8O_6$, twice the empirical formula.

Chapter 4

4.1. (a) $Xe(g) + 2\ F_2(g) \longrightarrow XeF_4(g)$
(b) $As_2O_3(s) + 3\ H_2(g) \longrightarrow 2\ As(s) + 3\ H_2O(\ell)$

4.2. (a) $C_3H_8(g) + 5\ O_2(g) \longrightarrow 3\ CO_2(g) + 4\ H_2O(\ell)$
(b) $2\ C_3H_8(g) + 7\ O_2(g) \longrightarrow 6\ CO(g) + 8\ H_2O(\ell)$
Carbon dioxide formation requires 5 mol of oxygen per mole of propane. Only 3.5 mol of oxygen are present per mole of propane when carbon monoxide forms. Thus, there is insufficient oxygen present for carbon dioxide formation.

4.3. (a) $2\ C_5H_{12}O + 15\ O_2 \longrightarrow 10\ CO_2 + 12\ H_2O$
(b) $C_5H_{12}O + 5\ O_2 \longrightarrow 5\ CO + 6\ H_2O$

4.4. (a) Decomposition reaction: $2\ Al(OH)_3(s) \longrightarrow Al_2O_3(s) + 3\ H_2O(g)$
(b) Combination reaction: $Na_2O(s) + H_2O(\ell) \longrightarrow 2\ NaOH(aq)$
(c) Combination reaction: $S_8(s) + 24\ F_2(g) \longrightarrow 8\ SF_6(g)$
(d) Exchange reaction: $3\ NaOH(aq) + H_3PO_4(aq) \longrightarrow Na_3PO_4(aq) + 3\ H_2O(\ell)$
(e) Single displacement: $3\ C(s) + Fe_2O_3(s) \longrightarrow 3\ CO(g) + 2\ Fe(\ell)$

4.5. (a) This exchange reaction forms insoluble nickel hydroxide and aqueous sodium chloride.

$$NiCl_2(aq) + 2\ NaOH(aq) \longrightarrow Ni(OH)_2(s) + 2\ NaCl(aq)$$

(b) This is an exchange reaction that forms aqueous potassium bromide and a precipitate of calcium carbonate.

$$K_2CO_3(aq) + CaBr_2(aq) \longrightarrow CaCO_3(s) + 2\ KBr(aq)$$

4.6. (a) $2\ Na^+(aq) + 2\ F^-(aq) + Ca^{2+}(aq) + 2\ CH_3COO^-(aq) \longrightarrow CaF_2(s) + 2\ Na^+(aq) + 2\ CH_3COO^-(aq)$

$$Ca^{2+}(aq) + 2\ F^-(aq) \longrightarrow CaF_2(s)$$

(b) $2\ NH_4^+(aq) + S^{2-}(aq) + Fe^{2+}(aq) + 2\ Cl^-(aq) \longrightarrow FeS(s) + 2\ NH_4^+(aq) + 2\ Cl^-(aq)$

$$Fe^{2+}(aq) + S^{2-}(aq) \longrightarrow FeS(s)$$

4.7. (a) Sulfuric acid and magnesium hydroxide
(b) Carbonic acid and strontium hydroxide

4.8. The oxidation numbers of Fe and Sb are zero (Rule 1). The oxidation numbers in Sb_2S_3 are +3 for Sb^{3+} and −2 for S^{2-} (Rules 2 and 4); the oxidation numbers in FeS are +2 for Fe^{2+} and −2 for S^{2-} (Rules 2 and 4).

4.9. Reactions (a) and (b) will occur. Aluminum is above copper and chromium in Table 4.5; therefore, aluminum will be oxidized and acts as the reducing agent in reactions (a) and (b). In reaction (a), Cu^{2+} is reduced and Cu^{2+} is the oxidizing agent. Cr^{3+} is the oxidizing agent in reaction (b), and is reduced to Cr metal.

Reactions (c) and (d) do not occur because Pt cannot reduce H^+, and Au cannot reduce Ag^+.

Chapter 5

5.1. $0.0500 \text{ mol Al} \times \frac{3 \text{ mol Br}_2}{2 \text{ mol Al}} \times \frac{159.81 \text{ g Br}_2}{1 \text{ mol Br}_2} = 11.99 \text{ g Br}_2$

5.2. (a) $0.300 \text{ mol cassiterite} \times \frac{1 \text{ mol Sn}}{1 \text{ mol cassiterite}} \times \frac{118.7 \text{ g Sn}}{1 \text{ mol Sn}} = 35.6 \text{ g Sn}$

(b) $35.6 \text{ g Sn} \times \frac{1 \text{ mol Sn}}{118.7 \text{ g Sn}} \times \frac{2 \text{ mol C}}{1 \text{ mol Sn}} \times \frac{12.01 \text{ g C}}{1 \text{ mol C}} = 7.20 \text{ g C}$

5.3. (a) $57 \text{ g C} \times \frac{1 \text{ mol C}}{12.0 \text{ g C}} \times \frac{1 \text{ mol O}}{2 \text{ mol C}} \times \frac{16.0 \text{ g O}}{1 \text{ mol O}} = 38 \text{ g O}$

(b) $57 \text{ g C} \times \frac{1 \text{ mol C}}{12.0 \text{ g C}} \times \frac{2 \text{ mol CO}}{2 \text{ mol C}} \times \frac{28.0 \text{ g CO}}{1 \text{ mol CO}} = 1.3 \times 10^2 \text{ g CO}$

5.4. (a) $CS_2(\ell) + 3\ O_2(g) \longrightarrow CO_2(g) + 2\ SO_2(g)$
(b) Determine the quantity of CO_2 produced by each reactant; the limiting reactant produces the lesser quantity.

$$3.5\ g\ CS_2 \times \frac{1\ mol\ CS_2}{76.0\ g\ CS_2} \times \frac{1\ mol\ CO_2}{1\ mol\ CS_2} \times \frac{44.0\ g\ CO_2}{1\ mol\ CO_2} = 2.0\ g\ CO_2$$

$$17.5\ g\ O_2 \times \frac{1\ mol\ O_2}{31.998\ g\ O_2} \times \frac{1\ mol\ CO_2}{3\ mol\ O_2} \times \frac{44.01\ g\ CO_2}{1\ mol\ CO_2} = 8.02\ g\ CO_2$$

Therefore, CS_2 is the limiting reactant.
(c) The yield of SO_2 must be calculated using the limiting reactant, CS_2.

$$3.5\ g\ CS_2 \times \frac{1\ mol\ CS_2}{76.0\ g\ CS_2} \times \frac{2\ mol\ SO_2}{1\ mol\ CS_2} \times \frac{64.1\ g\ SO_2}{1\ mol\ SO_2} = 5.9\ g\ SO_2$$

5.5.

	SiO_2	C	SiC	CO
Initial Amount	83.19 mol	416.3 mol	0 mol	0 mol
Change as Reaction Occurs	−83.19 mol	−250 mol	+83.19 mol	+167 mol
Amount at End of Reaction	0 mol	166 mol	83.2 mol	167 mol

5.6. $$2.50\ g\ Cu \times \frac{1\ mol\ Cu}{63.55\ g\ Cu} \times \frac{8\ mol\ Cu_2S}{16\ mol\ Cu} \times \frac{159.17\ g\ Cu_2S}{1\ mol\ Cu_2S} = 3.13\ g$$

$$\text{Percent yield} = \frac{2.53\ g}{3.13\ g} \times 100\% = 80.8\%\text{; you met the standard.}$$

5.7. $$1.72\ g\ Mg \times \frac{1\ mol\ Mg}{24.31\ g\ Mg} \times \frac{1\ mol\ MgCl_2}{1\ mol\ Mg} \times \frac{95.21\ g\ MgCl_2}{1\ mol\ MgCl_2} = 6.74\ g\ MgCl_2$$

Because only 6.46 g of $MgCl_2$ were produced, it is likely that the magnesium metal was not pure. Its purity is $\frac{6.46\ g}{6.74\ g} \times 100\% = 95.8\%$.

5.8. (a) $$491\ mg\ CO_2 \times \frac{1\ g\ CO_2}{10^3\ mg\ CO_2} \times \frac{1\ mol\ CO_2}{44.01\ g\ CO_2} \times \frac{1\ mol\ C}{1\ mol\ CO_2}$$

$$= 1.116 \times 10^{-2}\ mol\ C;\ 1.116 \times 10^{-2}\ mol\ C \times \frac{12.01\ g\ C}{1\ mol\ C}$$

$$= 0.1334\ g\ C = 133.4\ mg\ C$$

$$100\ mg\ H_2O \times \frac{1\ g\ H_2O}{10^3\ mg\ H_2O} \times \frac{1\ mol\ H_2O}{18.02\ g\ H_2O} \times \frac{2\ mol\ H}{1\ mol\ H_2O}$$

$$= 1.110 \times 10^{-2}\ mol\ H;\ 1.110 \times 10^{-2}\ mol\ H \times \frac{1.008\ g\ H}{1\ mol\ H}$$

$$= 1.119 \times 10^{-2}\ g\ H = 11.2\ mg\ H$$

The mass of oxygen in the compound

$$= \text{total mass} - (\text{mass C} + \text{mass H})$$
$$= 175\ mg - (133.4\ mg\ C + 11.2\ mg\ H) = 30\ mg\ O$$

The moles of oxygen are

$$30\ mg\ O \times \frac{1\ g\ O}{10^3 mg} \times \frac{1\ mol\ O}{16.00\ g\ O} = 1.875 \times 10^{-3}\ mol\ O$$

The empirical formula can be derived from the mole ratios of the elements:

$$\frac{1.116 \times 10^{-2}\ mol\ C}{1.875 \times 10^{-3}\ mol\ O} = 5.952\ mol\ C/mol\ O$$

$$\frac{1.110 \times 10^{-2}\ mol\ H}{1.875 \times 10^{-3}\ mol\ O} = 5.920\ mol\ H/mol\ O$$

$$\frac{1.875 \times 10^{-3}\ mol\ O}{1.875 \times 10^{-3}\ mol\ O} = 1.000\ mol\ O$$

The empirical formula of phenol is C_6H_6O.
(b) The molar mass is needed to determine the actual formula.

5.9. $$\text{Molarity} = \frac{\text{moles solute}}{\text{liters solution}}$$

$$36.0\ g\ Na_2SO_4 \times \frac{1\ mol\ Na_2SO_4}{142.1\ g\ Na_2SO_4} = 0.2533\ mol\ Na_2SO_4$$

$$\text{Molarity} = \frac{0.2533\ mol}{0.750\ L} = 0.338\ \text{molar}$$

5.10. Molarity(conc) × volume(conc) = molarity(dil) × volume(dil)

$$\text{Volume(conc)} = \frac{0.150\ \text{molar} \times 0.050\ L}{0.500\ \text{molar}} = 0.015\ L$$

5.11. (a) 1.00 L of 0.125 M Na_2CO_3 contains 0.125 mol Na_2CO_3.

$$0.125\ mol \times \frac{105.99\ g}{1\ mol} = 13.2\ g\ Na_2CO_3$$

Prepare the solution by adding 13.2 g of Na_2CO_3 to a volumetric flask, dissolving it and mixing thoroughly, and adding sufficient water until the solution volume is 1.0 L.
(b) Use water to dilute a specific volume of the 0.125 M solution to 100 mL.

$$\text{Vol(conc)} = \frac{0.0500\ M \times 0.100\ L}{0.125\ M} = 0.040\ L = 40\ mL\ \text{of}\ 0.125\ M\ \text{solution}$$

Therefore, put 40 mL of the more concentrated solution into a container and add water until the solution volume equals 100 mL.

(c) 500 L of 0.215 M $KMnO_4$ contains 17.0 g $KMnO_4$.

$$0.500\ \text{L} \times \frac{0.215\ \text{mol}\ KMnO_4}{1\ \text{L}} = 0.1075\ \text{mol}\ KMnO_4$$

$$0.1075\ \text{mol}\ KMnO_4 \times \frac{158.0\ \text{g}\ KMnO_4}{1\ \text{mol}\ KMnO_4} = 17.0\ \text{g}\ KMnO_4$$

Put 17.0 g $KMnO_4$ into a container and add water until the solution volume is 500 mL.

(d) Dilute the more concentrated solution by adding sufficient water to 52.3 mL of 0.0215 M $KMnO_4$ until the solution volume is 250 mL.

$$\text{Vol(conc)} = \frac{0.00450\ \text{M} \times 0.250\ \text{L}}{0.0215\ \text{M}} = 0.0523\ \text{L} = 52.3\ \text{mL}$$

5.12. $1.2 \times 10^{10}\ \text{kg} \times \frac{10^3\ \text{g}}{1\ \text{kg}} \times \frac{1\ \text{mL}}{1.2\ \text{g}} = 1.0 \times 10^{13}\ \text{g brine}$

Chapter 6

6.1. (a) $160\ \text{Cal} \times \frac{1000\ \text{cal}}{\text{Cal}} \times \frac{4.184\ \text{J}}{\text{cal}} = 6.7 \times 10^5\ \text{J}$

(b) 75 W = 75 J/s; 75 J/s × 3.0 hr × 60 min/hr × 60 s/min = 8.1×10^5 J

(c) $16\ \text{kJ} \times \frac{1\ \text{kcal}}{4.184\ \text{kJ}} = 3.8\ \text{kcal}$

6.2. $q_{water} = -q_{iron}$

$$(4.184\ \text{J/g °C})\ (1000\ \text{g})\ (32.8\ \text{°C} - 20.0\ \text{°C}) = -(0.451\ \text{J/g °C})\ (400\ \text{g})\ (32.8\ \text{°C} - T_i)$$

$$T_i = (297 + 32.8)\ \text{°C} = 330\ \text{°C}$$

6.3. (a) $10.0\ \text{g}\ I_2 \times \frac{1\ \text{mol}\ I_2}{253.8\ \text{g}I_2} \times \frac{62.4\ \text{kJ}}{1\ \text{mol}\ I_2} = 2.46\ \text{kJ}$

(b) $3.42\ I_2 \times \frac{1\ \text{mol}\ I_2}{253.8\ \text{g}I_2} \times \frac{^{-}62.4\ \text{kJ}}{1\ \text{mol}\ I_2} = -0.841\ \text{kJ} = -841\ \text{J}$

This process is the reverse of the one in part (a) and so $\Delta H°$ is negative. This means that the process is exothermic. The quantity of energy transferred is 841 kJ.

6.4. $\Delta H° = \{(2\ \text{mol H—H})\ D_{H—H} + (1\ \text{mol O=O})\ D_{O=O}\} - \{(4\ \text{mol O—H})\ D_{O—H}$

$= (2 \times 436 + 1 \times 498 - 4 \times 463)\ \text{kJ} = -482\ \text{kJ}$

6.5. According to the thermochemical equation, 285.8 kJ is transferred per mole of liquid water decomposed. Thus,

$$12.6\ \text{g}\ H_2O \times + \frac{1\ \text{mol}\ H_2O}{18.02\ \text{g}\ H_2O} \times \frac{285.8\ \text{kJ}}{1\ \text{mol}\ H_2O} = 200.\ \text{kJ}$$

6.6. Note: There is an error in the answer and the explanation for Problem-Solving Example 6.6 on page 249. The answer should be −462 kJ. The reason that $\Delta H°$ is negative is that the reaction is exothermic. Because the temperature of the solution increased, the energy transfer was from the reaction to the surroundings. The quantity of energy transferred to the surroundings at constant pressure is 15.21 kJ, and so the change in enthalpy of the reaction system is −15.21 kJ. The last equation in the explanation should therefore have a minus sign and the result should be −462 kJ.

The volume of the initial solutions is 200 mL, which corresponds to 200 g solution. The quantities of reactants are 0.10 mol H^+ (aq) and 0.10 mol OH^- (aq), and so 0.10 mol H_2O is formed.

$$0.10\ \text{mol}\ H_2O \times -\frac{58.7\ \text{kJ}}{1\ \text{mol}\ H_2O} = -5.87\ \text{kJ}$$

Since $\Delta H°$ is negative, energy is transferred to the water and its temperature will rise.

$$\Delta T° = \frac{q}{c \times m} = \frac{5.87 \times 10^3\ \text{J}}{(4.184\ \text{J/g °C})(200\ \text{g})} = 7.0\ \text{°C}$$

The final temperature will be (20.4 + 7.0) °C = 27.4 °C.

6.7.

$$C(s) + O_2(g) \longrightarrow CO_2(g) \qquad \Delta H° = -393.5\ \text{kJ}$$

$$CO_2(g) \longrightarrow CO\ (g) + \tfrac{1}{2}\ O_2\ (g) \qquad \Delta H° = 283.0\ \text{kJ}$$

$$C\ (s) + \tfrac{1}{2}\ O_2(g) \longrightarrow CO(g) \qquad \Delta H° = -110.5\ \text{kJ}$$

6.8. For the reaction given,

$$\Delta H° = \{(6\ \text{mol}\ CO_2(g)) \times \Delta H_f°(CO_2\ (g)) + 5\ \text{mol}\ H_2O(\ell)) \times \Delta H_f°(H_2O\ (\ell))\}$$
$$- \{(2\ \text{mol}\ (C_3H_5(NO_3)_3(\ell)) \times \Delta H_f°\ (C_3H_5(NO_3)_3(\ell))\}$$
$$= \{6(-393.509) + 5(-285.830) - 2(-364)\}\text{kJ}$$
$$= -3.06 \times 10^3\ \text{kJ}$$

For 10.0 g nitroglycerine (nitro),

$$q = 10.0\ \text{g} \times \frac{1\ \text{mol nitro}}{227.09\ \text{g}} \times \frac{-3.06 \times 10^3\ \text{kJ}}{2\ \text{mol nitro}} = -67.4\ \text{kJ}$$

(The 2 mol nitro in the last factor comes from the coefficient of 2 associated with nitroglycerine in the equation.)

Chapter 7

7.1.

(a) $\Delta S° = 2\ \text{mol CO(g)} \times S°(CO(g) + 1\ \text{mol}\ O_2\ (g) \times S°(O_2(g)) - 2\ \text{mol}\ CO_2(g) \times S°(CO_2(g))$

$= \{2 \times (197.674) + (205.138) - 2 \times (213.74)\}\ \text{J/K}$

$= 173.01\ \text{J/K}$

(b) $\Delta S° = 1\ \text{mol NaCl(aq)} \times S°(\text{NaCl(aq)}) - 1\ \text{mol NaCl(s)} \times S°(\text{NaCl(s)})$

$= \{115.5 - 72.8\}\ \text{J/K}$

$= 42.7\ \text{J/K}$

(c) $\Delta S° = 1\ \text{mol MgO(s)} \times S°(\text{MgO(s)}) + 1\ \text{mol}\ CO_2(g) \times S°\ (CO_2(g)) - 1\ \text{mol}\ MgCO_3(s) \times S°(MgCO_3(s))$

$= \{26.94 + 213.74 - 65.854\}$ J/K
$= 174.83$ J/K

7.2. (a) $\Delta H° = \{(-238.66) - (-100.525)\}\text{kJ} = -128.14$ kJ
$\Delta S° = \{(126.8) - 197.674 - 2 \times (130.684)\}$ J/K
$= -332.2$ J/K

$\Delta G° = \Delta H° - T\Delta S°$
$= -128.14 \times 10^3 \text{ J} - 298.15 \text{ K} \times (-332.2 \text{ J/K})$
$= -29.09 \times 10^3 \text{ J} = -29.09$ kJ

(b) $\Delta G° = [-166.27 - (-137.168)]\text{kJ} = -29.10$ kJ. The two results agree.
(c) $\Delta G°$ is negative. The reaction is product-favored at 298.15 K. Because $\Delta S°$ is negative, at very high temperatures the reaction will become reactant-favored.

Chapter 8

8.1. Using $\lambda = \frac{c}{\nu}$ $\quad \lambda = \frac{2.998 \times 10^8 \text{m s}^{-1}}{104.5 \times 10^6 \text{ s}^{-1}} = 2.869$ m

8.2. (a) One photon of ultraviolet radiation has more energy because ν is larger in the uv spectral region than in the microwave region. (b) One photon of blue light has more energy because the blue portion of the visible spectrum has a higher frequency than the green portion of the visible spectrum. (c) Ten blue photons of $\lambda = 460$ nm would have an energy of 10 hc/λ

$$6.626 \times 10^{-34} \text{ J s}^{-1} \left(\frac{2.998 \times 10^8 \text{ m s}^{-1}}{460 \times 10^{-9} \text{m}}\right) \times 10 = 4.32 \times 10^{-18} \text{ J}$$

while 15 red photons of $\lambda = 695$ nm would have an energy of 15 hc/λ

$$6.626 \times 10^{-34} \text{ J s}^{-1} \left(\frac{2.998 \times 10^8 \text{ m s}^{-1}}{695 \times 10^{-9} \text{ m}}\right) \times 15 = 4.29 \times 10^{-18} \text{ J}$$

So the 10 blue photons have more energy.

8.3. For sulfur, (a) [Ne] $3s^2 3p^4$. (b) For the $n = 3$ electrons, [↑↓] [↑↓] [↑] [↑].

8.4. (a) The general electron configuration for Group 3A atoms is ns^2np^1. (b) The charge of the ion formed from a Ga atom should be 3+, by the loss of the three valence electrons. (c) The electron configuration for the Ga^{3+} ion will be [Ar]$3d^{10}$.

8.5. (a) The electron configuration for the nickel atom is [Ar]$4s^2 3d^8$.
(b) The orbital box diagram for the Ni atom's outer electrons is 3d [↑↓] [↑↓] [↑↓] [↑] [↑] 4s [↑↓]. The Ni atom has two unpaired electrons.
(c) The orbital box diagram for the Ni^{2+} ion is 3d [↑↓] [↑↓] [↑↓] [↑] [↑] (only the two 4s electrons are removed in forming the Ni^{2+} ion), so it also has two unpaired electrons.

Chapter 9

9.1. (a) F—N(—F)—F (b) H—N(H)—N(H)—H (c) $[SO_4]^{2-}$: :O: bonded above, :O— S —O: with :O: below, in brackets with charge 2−

9.2. (a) [:N≡O:]$^+$ (b) H—C≡N:

9.3. (a) :F—Be—F: (b) :O—Cl—O: (c) P bonded to five :Cl: atoms (PCl_5)
(d) [H—B—H]$^+$ (e) I bonded to seven :F: atoms (IF_7)

BeF_2—not an octet around the central Be atom; ClO_2—an odd number of electrons around Cl; PCl_5—more than four electron pairs around the central phosphorus atom; BH_2^+—only two electron pairs around the central B atom; IF_7—iodine has seven shared electron pairs.

9.4. (a) Si is a larger atom than S. (b) Br is a larger atom than Cl. (c) The greater electron density in the triple bond brings the N≡O atoms closer together than the smaller electron density in N=O double bond does.

9.5. 1-bromo-2-chloro-2-butene can have *cis* and *trans* isomers; 2-methyl-2-butene and 1-butene cannot because one of the carbon atoms in their double bonds is attached to two identical groups, CH_3 and H, respectively.

Cl(BrCH$_2$)C=C(H)(CH$_3$)
***cis*-1-bromo-2-chloro-2-butene**

Cl(BrCH$_2$)C=C(CH$_3$)(H)
***trans*-1-bromo-2-chloro-2-butene**

9.6. The N—O bond length in NO_2^- is 124 pm and is intermediate between the lengths of an N—O bond (140 pm) and an N=O bond (115 pm). This indicates that the N—O bond in nitrite ion is neither a pure single or pure double bond but intermediate between the two.

9.7. (a) B—Cl is more polar; $\overset{\delta+}{B}—\overset{\delta-}{Cl}$; O—H is more polar; $\overset{\delta-}{O}—\overset{\delta+}{H}$.

9.8. (a) SO_4^{2-}; (b) Cu^{2+}; (c) Four NH_3 molecules; (d) $[Cu(NH_3)_4]^{2+}$.

9.9. (a) Two DMG^- ions per Ni^{2+} ion; (b) Zero; (c) Five atoms:

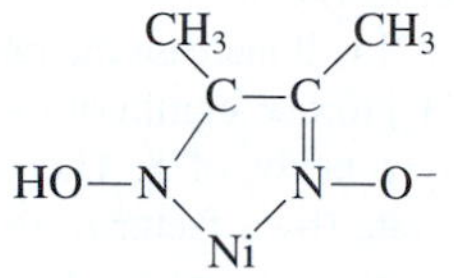

```
   CH3      CH3
     \      /
      C—C
     /    ‖
HO—N      N—O⁻
     \    /
      Ni
```

Chapter 10

10.1.

Central Atom (underlined)	Bond Pairs	Lone Pairs	Electron-Pair Geometry	Molecular Shape
$\underline{Br}O_3^-$	3	1	Tetrahedral	Triangular pyramid
$\underline{Se}F_2$	2	2	Tetrahedral	Angular
$\underline{C}H_3OH$	4	0	Tetrahedral	Tetrahedral

10.2.

Central Atom (underlined)	Bond Pairs	Lone Pairs	Electron-Pair Geometry	Molecular Shape
$Cl_2\underline{O}$	2	2	Tetrahedral	Angular
$\underline{S}O_3^{2-}$	3	1	Tetrahedral	Triangular planar
$\underline{Si}O_4^{4-}$	4	0	Tetrahedral	Tetrahedral
$H\underline{C}O\underline{N}H\underline{C}H_3$	(three central atoms)			

```
      :O:
       ‖   ..
   H—C—N—CH3
           |
           H
```

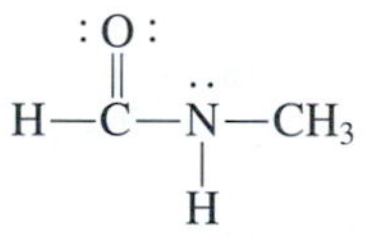				
C	3	0	Triangular planar	Triangular planar
N	3	1	Tetrahedral	Triangular pyramidal
C	4	0	Tetrahedral	Tetrahedral

10.3. (a) In HCN, the *sp* hybridized carbon atom is sigma bonded to H and to N, as well as having two pi bonds to N. The sigma and two pi bonds form the C—N triple bond. The nitrogen is *sp* hybridized with a sigma and two pi bonds to carbon; a lone pair is in the nonbonding *sp* hybrid orbital on N.

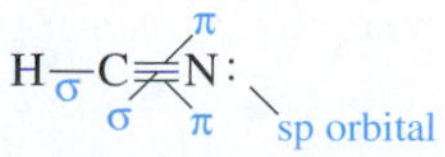

```
          π
H—C≡N:
 σ  σ π   sp orbital
```

(b) The double-bonded carbon and nitrogen are both sp^2 hybridized. The sp^2 hybrid orbitals on C form sigma bonds to H and to N; the unhybridized *p* orbital on C forms a pi bond with the unhybridized *p* orbital on N. The sp^2 hybrid orbitals on N form sigma bonds to carbon and to H; the N lone pair is in the nonbonding sp^2 hybrid orbital.

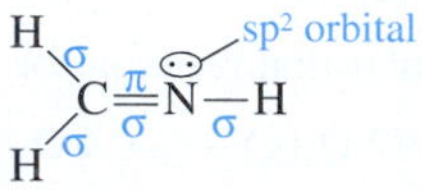

```
H  σ          sp² orbital
  \      ..
   C =π= N—H
  /  σ     σ
H  σ
```

10.4. (a) The central P atom has six bond pairs, no lone pairs, and is d^2sp^3 hybridized.
(b) The central I atom has three bond pairs, two lone pairs; these five electron pairs are in dsp^3 hybridized orbitals on I.
(c) In ICl_4^-, the central I has four bond pairs, two lone pairs; these six pairs are in d^2sp^3 orbitals.

10.5. (a) $BFCl_2$ is a triangular planar molecule with polar B—F and B—Cl bonds. The molecule is polar because the B—F bond is more polar than the B—Cl bonds, resulting in a net dipole.
(b) NH_2Cl is a triangular pyramidal molecule with polar N—H bonds. (N—Cl is a nonpolar bond; N and Cl have the same electronegativity.) It is a polar molecule because the N—H dipoles do not cancel and produce a net dipole.
(c) SCl_2 is an angular, polar molecule. The polar S—Cl bond dipoles do not cancel each other because they are not symmetrically arranged due to the two lone pairs on S.

10.6. (a) London (dispersion) forces (N_2 is a nonpolar molecule); (b) London forces. Also, the polar water molecules have a net dipole that can produce an induced dipole in the nonpolar carbon dioxide molecules; (c) Hydrogen bonding, dipole-dipole forces, and London forces.

10.7. A chiral carbon has four different atoms or groups bonded to it; thus, (b) is chiral. The carbon atom in part (a) has two Cl atoms bonded to it, so it cannot be chiral; likewise, the carbon atom in (c) has two H atoms bonded to it and is not chiral.

Chapter 11

11.1 Four molecules of hydrogen are formed for every molecule of heptane that is reformed to toluene. The reaction is

$$C_7H_{16} \longrightarrow C_6H_5CH_3 + 4\ H_2$$

11.2 Energy $= 4.2 \times 10^9 \text{ t coal} \times \dfrac{26.4 \times 10^9 \text{ J}}{1 \text{ t coal}} = 1.1 \times 10^{18} \text{ J}$

$\text{ft}^3 \text{ natural gas} = 1.1 \times 10^{18} \text{ J} \times \dfrac{1 \text{ ft}^3 \text{ natural gas}}{1.055 \times 10^6 \text{ J}} =$
$1.0 \times 10^{12} \text{ ft}^3 \text{ natural gas}$

```
               CH3  OH
                |    |
11.3 CH3—CH2—CH—CH—CH3
```
This is a secondary alcohol.

11.4 (a) The first oxidation product of $CH_3CH_2CH_2OH$ is the aldehyde,

```
          O
          ‖
CH3CH2CH.
```
The second oxidation product of $CH_3CH_2CH_2OH$ is the acid

```
           O
           ‖
CH3CH2C—OH
```

(b) The oxidation product of this secondary alcohol is the ketone

$$CH_3—\overset{\overset{\large O}{\|}}{C}—CH_2CH_3$$

11.5 The balanced combustion reaction for methanol vapor is

$$CH_3OH(g) + 3/2\ O_2(g) \longrightarrow CO_2(g) + 2\ H_2O(g)$$

Using Hess's law, the heat of combustion of methanol vapor is

$$\Delta H_{comb} = [\Delta H_f^\circ\ CO_2(g)] + 2[\Delta H_f^\circ H_2O(g)] - 1[\Delta H_f^\circ CH_3OH(g)]$$

$$= -393.509\text{ kJ/mol} + 2(-241.818\text{ kJ/mol}) - (-200.66\text{ kJ/mol})$$

$$= -676.48\text{ kJ/mol}$$

For methanol,

$$-676.48\text{ kJ/mol}\left(\frac{1\text{ mol}}{32.04\text{ g}}\right)\left(\frac{0.791\text{ g}}{1\text{ mL}}\right)\left(\frac{1000\text{ mL}}{\text{L}}\right) = -1.67 \times 10^4\text{ kJ/L}$$

Methanol is less energetic per liter than is ethanol; see Problem-Solving Example 11.5.

11.6 (a) $CH_3—\overset{\overset{\large O}{\|}}{C}—O—CH_3$

(b) $CH_3CH_3—\overset{\overset{\large O}{\|}}{C}—O—CH_2CH_2—O—\overset{\overset{\large O}{\|}}{C}—CH_2CH_3$

11.7 (a) $CH_2{=}CH_2$ (b) $CH_2{=}CHCl$ (c) $CH_2{=}CH$ (with a benzene ring attached to CH)

11.8 $—CH_2—CH_2—O—\overset{\overset{\large O}{\|}}{C}—C_6H_4—\overset{\overset{\large O}{\|}}{C}—O—$

11.9

```
     H  O       H  O       H  O       H
     |  ||      |  ||      |  ||      |    O
H2N—C—C—N—C—C—N—C—C—N—C—C//
     |     |  |     |  |     |  |     \
    CH2    H CH2    H CH2    H CH3     OH
     |        |        |
    SH    (benzene)   OH
```

Chapter 12

12.1 At [cisplatin] = 0.020 M,

$$\text{rate} = k\ [\text{cisplatin}]$$

$$= (0.088/\text{hr})(0.020\text{ M})$$

$$= 1.8 \times 10^{-3}\text{ M/hr}$$

$$= \frac{1.8 \times 10^{-3}\text{ M}}{\text{hr}} \times \frac{1\text{ hr}}{60\text{ min}} \times \frac{1\text{ min}}{60\text{ s}}$$

$$= 4.9 \times 10^{-7}\text{ M/s}$$

At [cisplatin] = 0.010 M, the rate is half as great and so rate = 2.4×10^{-7} M/s

12.2 (a) The reaction is second-order with respect to NO and first order with respect to H_2.

(b) Tripling the [NO] will increase the rate by a factor of $3^2 = 9$. Decreasing the $[H_2]$ to one-eighth of its original value will decrease the rate by a factor of 8. The overall effect will be a slight increase in rate (by a factor of 9/8).

12.3 Graphing the data gives the plots shown. The first-order plot is linear and its slope is -1.46×10^{-3}/s. From Table 12.1 $k = -\text{slope} = 1.46 \times 10^{-3}$ s.

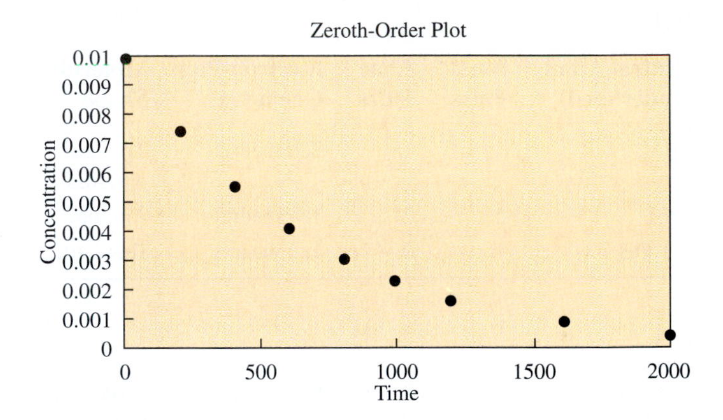

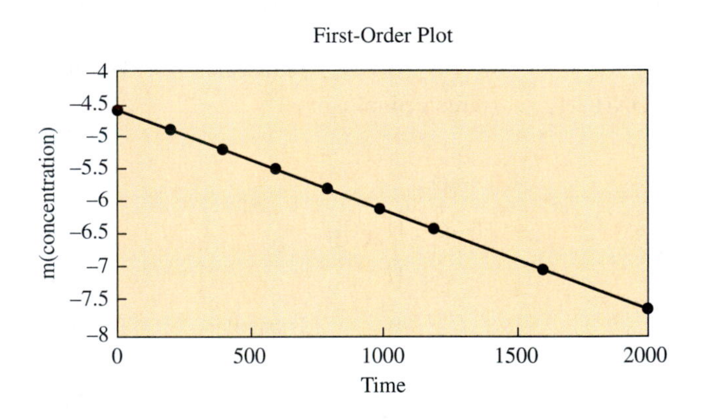

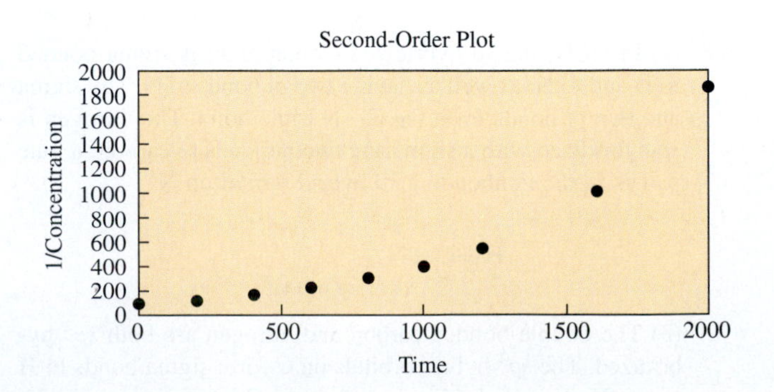

12.4

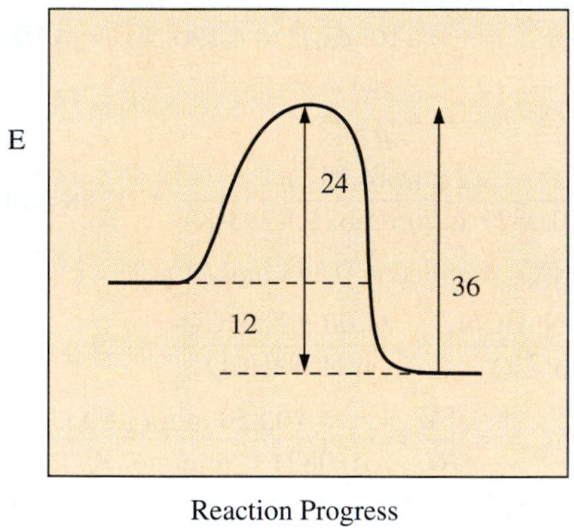

E

24

36

12

Reaction Progress

The reaction is exothermic by 12 kJ/mol

12.5 The overall equation is the sum of the three reaction mechanism steps:

$$2\ NH_3(aq) + OCl^-(aq) \longrightarrow N_2H_4(aq) + Cl^-(aq) + H_2O(\ell)$$

The first step is slow and so the rate law is the rate law for that step.

$$\text{rate} = k[NH_3][OCl^-]$$

12.6 The shape of the second molecule

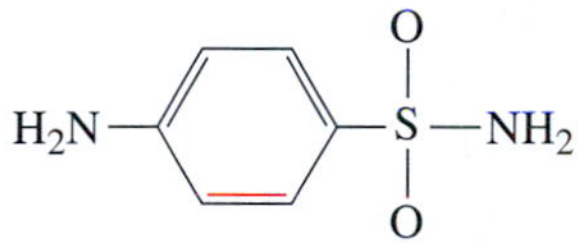

H_2N—C_6H_4—$S(O)_2$—NH_2

is most similar to the shape of *p*-aminobenzoic acid and therefore interferes with the ability of bacteria to convert *p*-aminobenzoic acid to folic acid. Hence it is the sulfa drug.

Chapter 13

13.1 (a) $K = \frac{[H_2S]}{[H_2]}$ (b) $K = \frac{[H_2]}{[HCl]}$

(c) $K = \frac{[CO]\ [H_2]^3}{[CH_4]\ [H_2O]}$ (d) $K = \frac{[HCN]\ [OH^-]}{[CN^-]}$

13.2 $K = \frac{[CH_3COO^-]\ [H_3O^+]}{[CH_3COOH]} = \frac{(0.0296 \times 0.0200)\ (0.0296 \times 0.0200)}{(1 - 0.0296) \times 0.0200}$

$= \frac{3.505 \times 10^{-7}}{1.941 \times 10^{-7}} = 1.81 \times 10^{-5}$

13.3 $AuI(s) \rightleftharpoons Au^+(aq) + I^-(aq)$

$K = [Au^+]\ [I^-] = 1.6 \times 10^{-23}$

Since $[Au^+] = [I^-]$ because of the reaction stoichiometry,

$[Au^+]^2 = 1.6 \times 10^{-23}$ $[Au^+] = \sqrt{1.6 \times 10^{-23}}$

$[Au^+] = 4.0 \times 10^{-12}$ and $[I^-] = 4.0 \times 10^{-12}$

13.4 $K = \frac{1}{1.7 \times 10^2} = 5.88 \times 10^{-3}$

	$N_2O_4 \rightleftharpoons$	$2NO_2$
Initial concentration	0.250	0.125
Change during reaction	x	$-2x$
Equilibrium concentration	$0.250 + x$	$0.125 - 2x$

(Because the value of K is small, assume that some of the NO_2 reacts to form N_2O_4 and let x be the increase in concentration of N_2O_4.)

$$5.88 \times 10^{-3} = \frac{(0.125 - 2x)^2}{0.250 + x}$$

$$1.47 \times 10^{-3} + 5.88 \times 10^{-3}x = 1.56 \times 10^{-2} - 0.500x + 4x^2$$

$$4x^2 - 0.506x + 1.41 \times 10^{-2} = 0$$

$$x = \frac{-b \pm \sqrt{b^2 - 4ac}}{2a} = \frac{0.506 \pm \sqrt{0.256 - 0.226}}{8}$$

$x_- = 0.042$

$x_+ = 0.084$

If $[NO_2] = 0.125 - 2x$, then only x_- is appropriate because $0.125 - 2x\ 0.084 = -0.045$ and $[NO_2] = -0.045$ is impossible.

Therefore: $[NO_2] = 0.125 - 2 \times 0.042 = 4.1 \times 10^{-2}$

$[N_2O_4] = 0.250 + 0.042 = 0.292$

As a check: $K = \frac{(4.1 \times 10^{-2})^2}{0.292} = 5.76 \times 10^{-3}$, which agrees to two significant figures.

13.5 [NO_2 resonance structures: $\cdot N(=O)(-O) \longleftrightarrow \cdot N(-O)(=O)$] and

[O_2N-NO_2] $\longleftrightarrow$ + 3 other resonance structures

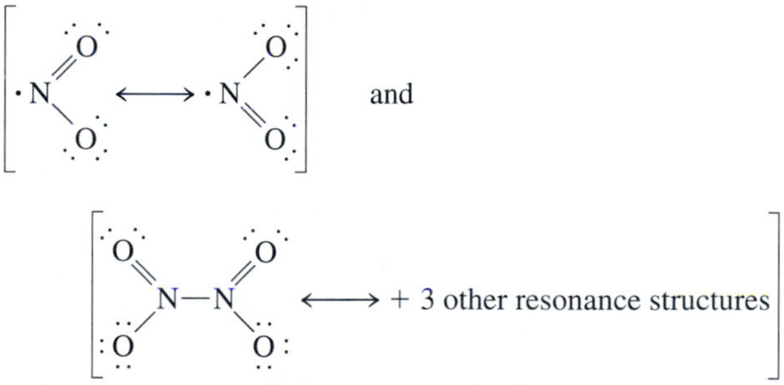

Since a bond must be broken, the reaction should be endothermic. Since the reaction is endothermic, raising T will shift it toward product NO_2, which accounts for the darker brown color at the higher T in Figure 13.3.

13.6 (a) $\Delta G° = -553.04 - 528.10 - (-1128.79) = 47.65$ kJ/mol

$K_{th} = e^{-\Delta G°/RT} = e^{-(47{,}650\ J/mol)(8.314\ J/K \cdot mol)(298K)}$

$= e^{-19.232} = 4.44 \times 10^{-9}$

(b) $\Delta G° = 0 - 587.06 - (-623.42) = 36.36$ kJ/mol

$K_{th} = e^{-(36{,}360\ J/mol)/(8.314\ J/K \cdot mol)(298K)}$

$= 4.23 \times 10^{-7}$

(c) $\Delta G° = 97.89 - 2(51.31) = -4.73$ kJ/mol

$K_{th} = e^{-(-4.730\ J/mol)/(8.314\ J/K \cdot mol)(298K)} = 6.75$

(a) is close to the value in Table 13.1. (b) is equal to the value in Table 13.1. (c) is quite different from the value in the table, because for gas-phase reactions the standard state involves pressures, not concentrations.

13.7 $\Delta H° = 2(-46.11) = -92.22$ kJ

$\Delta S° = 2(192.45) - 191.61 - 3(130.684) = -198.762$ J/K

(a) At 298K, $\Delta G° = -92{,}220\ J - 298K\ (-198.762\ J/K)$

$K_{th} = e^{-(-32{,}988\ J/mol)/(8.314\ J/K \cdot mol)(298\ K)} = 6.0 \times 10^5$

(b) At 450 K, $\Delta G° = -92{,}220\ J - 450\ K\ (-198.762\ J/K)$
$K_{th} = e^{-(-2777\ J/mol)/(8.314\ J/K\cdot mol)(450K)} = 2.1$
(c) At 800 K, $\Delta G° = -92{,}220\ J - 800\ K\ (-198.762\ J/K)$
$K_{th} = e^{-(66{,}790\ J/mol)/(8.314\ J/K\cdot mol)(800\ K)} = 4.4 \times 10^{-5}$

Chapter 14

14.1 (a) pressure in atmospheres = 29.5 in Hg ×

$$\frac{1\ atm}{76.0\ cm\ Hg} \times \frac{2.54\ cm}{1\ in} = 0.986\ atm$$

(b) pressure in mm Hg = 29.5 in Hg × $\frac{25.4\ mm}{1\ in}$ = 749 mm Hg

(c) pressure in bar = 29.5 in Hg ×

$$\frac{1.013\ bar}{760.\ mm\ Hg} \times \frac{25.4\ mm}{1\ in} = 0.999\ bar$$

(d) pressure in kPa = 29.5 in Hg × $\frac{101.3\ kPa}{760.\ mm\ Hg} \times \frac{25.4\ mm}{1\ in}$ = 99.9 kPa

14.2 pressure in bar = 647 mm Hg × $\frac{1.013\ bar}{760.\ mm\ Hg}$ = 0.862 bar

pressure in kPa = 647 mm Hg × $\frac{101.3\ kPa}{760.\ mm\ Hg}$ = 86.2 kPa

pressure in atm = 647 mm Hg × $\frac{1\ atm}{760.\ mm\ Hg}$ = 0.851 atm

14.3 $V = \frac{nRT}{P} = \frac{(2.64\ mol)(0.0821\ L\ atm/mol\ K)(304\ K)}{0.640\ atm} = 103\ L$

14.4 (a) $V_2 = \frac{P_1V_1T_2}{P_2T_1} = \frac{(710.\ mm\ Hg)(21\ mL)(299.6\ K)}{(740\ mm\ Hg)(295.4\ K)} = 21\ mL$

(b) $V_2 = \frac{V_1T_2}{T_1} = \frac{(21\ mL)(299.6\ K)}{(295.4\ K)} = 21\ mL$

14.5 $V_2 = \frac{P_1V_1}{P_2} = \frac{(1.00\ atm)(400.\ mL)}{(0.750\ atm)} = 533\ mL$

14.6 $V_2 = \frac{V_1T_2}{T_1} = \frac{(236\ mL)(362\ K)}{(304\ K)} = 281\ mL$

14.7 Volume of NO gas = 1.0 L O_2 × $\frac{2\ L\ NO}{1\ L\ O_2}$ = 2 L NO

14.8 amount of N_2, $n = \frac{PV}{RT}$ =

$$\frac{(835\ mm\ Hg)\left(\frac{1\ atm}{760\ mmHg}\right)(45.6\ L)}{(0.0821\ L\ atm/mol\ K)\ (295\ K)} = 2.07\ mol$$

amount of NaN_3 = 2.07 mol N_2 × $\frac{2\ mol\ NaN_3}{3\ mol\ N_2}$ = 1.38 mol NaN_3

mass of NaN_3 = 1.38 mol NaN_3 × $\frac{65.01\ g\ NaN_3}{1\ mol\ NaN_3}$ = 89.7 g NaN_3

14.9 $(V_{sphere} = \frac{4}{3}\pi r^3 = \frac{4}{3}\pi(10.\ cm)^3 = 4{,}190\ cm^3 = 4.19\ L$

amount of CO_2 gas, $n = \frac{PV}{RT}$ =

$$\frac{(2\ atm)(4.19\ L)}{(0.0821\ L\ atm/mol\ K)(293\ K)} = 0.348\ mol\ CO_2$$

mass of $NaHCO_3$ required = 0.348 mol CO_2 ×

$$\frac{1\ mol\ NaHCO_3}{1\ mol\ CO_2} \times \frac{84.00\ g\ NaHCO_3}{mol\ NaHCO_3} = 29\ g\ NaHCO_3$$

14.10 amount of gas, $n = \frac{PV}{RT} = \frac{(0.850\ atm)(1.00\ L)}{(0.0821\ L\ atm/mol\ K)(293\ K)}$ = 0.0353 mol

Molar mass = $\frac{1.13\ g}{0.0353\ mol}$ = 32.0 g/mol

This gas is probably oxygen.

14.11 amount of N_2 = 7.0 g N_2 × $\frac{1\ mol\ N_2}{28.10\ g\ N_2}$ = 0.25 mol N_2

amount of H_2 = 6.0 g H_2 × $\frac{1\ mol\ H_2}{2.02\ g\ H_2}$ = 3.0 mol H_2

total number of moles = 3.0 + 0.25 = 3.2 mol

$X_{N_2} = \frac{0.25\ mol}{3.2\ mol} = 0.078$ $X_{H_2} = \frac{3.0\ mol}{3.2\ mol} = 0.94$

$P_{N_2} = \frac{nRT}{V} = \frac{(0.25\ mol)(0.0821\ 1\ atm/mol\ K)(773\ K)}{(5.0\ L)}$ = 3.2 atm

$P_{H_2} = \frac{(3.0\ mol)((0.0821\ L\ atm/mol\ K)(773\ K)}{(5.0\ L)} = 38\ atm$

Density KCl =

$$\frac{(4\ KCl\ formula\ units)\left(\frac{74.552\ g}{mol\ KCl}\right)\left(\frac{1\ mol}{6.022 \times 10^{23}\ formula\ units}\right)}{2.60 \times 10^{-22}\ cm^3} = 1.90\ g/cm^3$$

Chapter 15

15.1 Heat = 2.5×10^{10} kg H_2O ×

$$\left(\frac{10^3\ g}{kg}\right)\left(\frac{1\ mol\ H_2O}{18.02\ g\ H_2O}\right)\left(\frac{44.0\ kJ}{mol}\right) = 6.10 \times 10^{13}\ kJ$$

This process is exothermic as water vapor condenses, forming rain.

15.2 Heat required to melt NaCl =

$$100.0\ g\ NaCl \times \left(\frac{1\ mol\ NaCl}{58.442\ g\ NaCl}\right)\left(\frac{30.21\ kJ}{mol\ NaCl}\right) = 51.69\ kJ$$

15.3 Only water vapor would exist under these conditions.

15.4 There are 2 atoms per bcc unit cell. The diagonal of the bcc unit cell is 4 times the radius of the atoms in the unit cell, so, solving for the edge

$$\text{Edge} = \frac{4 \times 144 \text{ pm}}{\sqrt{3}} = 332 \text{ pm}$$

$$\text{density} = \frac{\text{mass}}{\text{volume}} =$$

$$\frac{(2 \text{ Au atoms})(196.97 \text{ g Au}/6.022 \times 10^{23} \text{ Au atoms})}{((332 \text{ pm})(1 \text{ m}/10^{12} \text{ pm})(10^2 \text{ cm/m}))^3} = 17.88 \text{ g/cm}^3$$

15.5 The edge of the KCl unit cell would be 2×152 pm + 2×167 pm = 638 pm.
The unit cell of KCl is larger than that of NaCl.
Volume of the unit cell = $(638 \text{ pm})^3 = 2.60 \times 10^8 \text{ pm}^3$

$$\left(\frac{10^{-10} \text{ cm}}{\text{pm}}\right)^3 = 2.60 \times 10^{-22} \text{ cm}^3$$

15.6 Energy transfer required $= 1.45 \text{ g Al} \times \left(\frac{1 \text{ mol Al}}{26.98 \text{ g Al}}\right) \times \left(\frac{10.7 \text{ kJ}}{1 \text{ mol}}\right) = 0.575 \text{ kJ}$

Chapter 16

16.1 Graphite, because of its extensive network bonding involving strong covalent bonds, is not soluble in either water or carbon tetrachloride.

16.2 Solubility = $(8.4 \times 10^{-7}$ mol/L/mm Hg)(1520 mm Hg) = 1.3×10^{-3} mol/L
Solubility = $(8.4 \times 10^{-7}$ mol/L/mm Hg)(20 mm Hg) = 1.7×10^{-5} mol/L

16.3 $K_{sp}(CuBr) = [Cu^+][Br^-]$ $\quad K_{sp}(HgI_2) = [Hg^{2+}][I^-]^2$
$K_{sp}(SrSO_4) = [Sr^{2+}][SO_4^{2-}]$

16.4

	$PbCl_2 \rightleftarrows Pb^{2+}$ +	$2\ Cl^-$
Initially	0	0.5 M
Change due to dissolving	$+S$	$0.5 + 2S$
At equilibrium	S	0.5 (ignore $2S$ because S will be small)

$$K_{sp} = (S)(0.5) = 1.7 \times 10^{-5}$$

$$S = \frac{1.7 \times 10^{-5}}{0.5} = 3.4 \times 10^{-5} \text{ mol/L}$$

16.5 Total mass is 750 + 21.5 = 771.5 g.

$$\text{Weight percent glucose} = \frac{21.5 \text{ g}}{771.5 \text{ g}} \times 100\% = 2.79\%$$

16.6 $\left(\frac{30 \text{ g Se}}{10^9 \text{ g H}_2\text{O}}\right)\left(\frac{1 \text{ g H}_2\text{O}}{1 \text{ mL H}_2\text{O}}\right)\left(\frac{10^6\ \mu\text{g Se}}{1 \text{ g Se}}\right) = 3.0 \times 10^{-2}\ \mu\text{g Se/mL H}_2\text{O}$

$$\text{Se in 100 mL of water} = \left(\frac{3.0 \times 10^{-2}\ \mu\text{g Se}}{1 \text{ mL H}_2\text{O}}\right)(100 \text{ mL H}_2\text{O}) = 3.0\ \mu\text{g Se}$$

16.7 (a) $[Al^{3+}] = \frac{1.9 \times 10^{-33}}{(1.0 \times 10^{-4})^3} = 1.9 \times 10^{-21}$

(b) $[Al^{3+}] = \frac{1.9 \times 10^{-33}}{(1.0 \times 10^{-5})^3} = 1.9 \times 10^{-18}$

Lowering the pH causes a large increase in the Al^{3+} ion concentration.

16.8 Amount of NaCl = 6.58 g NaCl $\times \left(\frac{1 \text{ mol NaCl}}{58.44 \text{ g NaCl}}\right) = 1.13 \times 10^{-1}$ mol NaCl

$$\text{molality} = \frac{\text{moles solute}}{\text{kg solvent}} = \left(\frac{1.13 \times 10^{-1} \text{ mol NaCl}}{250.0 \text{ mL H}_2\text{O}}\right)\left(\frac{1 \text{ mL H}_2\text{O}}{1 \text{ g H}_2\text{O}}\right)\left(\frac{1000 \text{ g H}_2\text{O}}{1 \text{ kg H}_2\text{O}}\right) = 0.452\ m$$

16.9 $\Delta T_b = (0.512\ °\text{C} \cdot \text{kg/mol})(0.467 \text{ mol/kg}) = 0.0239\ °\text{C}$
Boiling point of solution = 100.00 + 0.02 = 100.02 °C

16.10 molality of solution $= \left(\frac{1.20 \text{ kg ethylene glycol}}{6.50 \text{ kg H}_2\text{O}}\right)\left(\frac{1000 \text{ g}}{\text{kg}}\right)\left(\frac{1 \text{ mol ethylene glycol}}{62.068 \text{ g ethylene glycol}}\right) = 2.97\ m$

Next, calculate the freezing point depression of a 2.97 m solution.

$$\Delta T_f = (-1.86°\text{C} \cdot \text{kg/mol})(2.97 \text{ mol/kg}) = -5.52°\text{C}$$

This solution will freeze at −5.52°C, so this amount of ethylene glycol will not protect the 6.5 kg of water in the tank if the temperature drops to −25°C.

16.11 $\Pi = cRTi$
For hemoglobin $i = 1$, and so

$$c = \frac{\Pi}{RT} = \frac{1.8 \times 10^{-3} \text{ atm}}{(0.0821 \text{ L} \cdot \text{atm/mol} \cdot \text{K})(298 \text{ K})} = 7.36 \times 10^{-5} \text{ M}$$

Since the volume is 1.0 L, there must be 7.36×10^{-5} mol hemoglobin present.

$$\text{Molar mass} = \frac{5.0 \text{ g}}{7.36 \times 10^{-5} \text{ mol}} = 6.8 \times 10^4 \text{ g/mol}$$

Chapter 17

17.1 (a) conjugate base of H_2CO_3 is HCO_3^-, conjugate base of HNO_3 is NO_3^-
(b) conjugate acid of NH_3 is NH_4^+, conjugate acid of CN^- is HCN

17.2 In a 0.040 M solution of NaOH, the $[OH^-]$ is 0.040 because the NaOH is 100% dissociated. The pOH is −log(0.040) = 1.40, so the pH is 12.60 (14.00 − 1.40).

17.3 If $[H_3O^+] = 1.5 \times 10^{-10}$, then the pH is $-\log(1.5 \times 10^{-10}) = 9.82$. If the $[H_3O^+]$ increases by 10^3 to 1.5×10^{-7} M, the pH would be $-\log(1.5 \times 10^{-7}) = 6.82$. The solution is still acidic, but only slightly so.

17.4 Setting up a small table for lactic acid, HLa,

	$HLa + H_2O \rightleftarrows H_3O^+ + La^-$		
Initial concentration	0.10 M	10^{-7}	0
Concentration change due to reaction	$-x$	$+x$	$+x$
Equilibrium concentration	$0.10 - x$	x	x

But $x = 10^{-2.43} = 3.7 \times 10^{-3}$ because $x = [H_3O^+]$. Substituting in the K_a expression,

$$K_a = \frac{[H_3O^+][La^-]}{[HLa]} = \frac{(3.7 \times 10^{-3})^2}{0.10 - 3.7 \times 10^{-3}} = \frac{1.4 \times 10^{-5}}{0.1} = 1.4 \times 10^{-4}$$

Lactic acid is a stronger acid than propionic acid, with a K_a of 1.5×10^{-5}.

17.5 Using the same methods as shown in the example, $\frac{x^2}{0.10} = 7.3 \times 10^{-10}$
Solving for x, which is $[H_3O^+]$, we get $x = \sqrt{7.3 \times 10^{-11}} = 8.54 \times 10^{-6} = [H_3O^+]$
So the pH of this solution is $-\log(8.54 \times 10^{-6}) = 5.07$

17.6 Using the same methods as those used in the example, letting $x = [OH^-]$ and $[HCO_3^-]$, and using the value of 2.1×10^{-4} for K_b for CO_3^{2-}, we get

$$\frac{x^2}{1.0} = 2.1 \times 10^{-4} \qquad \sqrt{x = 2.1 \times 10^{-4}} = 1.45 \times 10^{-2}$$

pOH = 1.84, and pH = 12.16

17.7 *Note:* In the answers below, only the acids are listed. Table 17.3 contains the corresponding conjugate bases for these acids.
(a) For a $[H_3O^+]$ of 3.2×10^{-4} M, use HF, HNO_2, HCO_2H, or $C_6H_5CO_2H$
(b) For a $[H_3O^+]$ of 5.0×10^{-5} M, use HCO_2H, $C_6H_5CO_2H$, CH_3CO_2H, or $CH_3CH_2CO_2H$
(c) For a $[H_3O^+]$ of 7.0×10^{-8} M, use H_2S, $H_2PO_4^-$, HSO_3^-, or HClO
(d) For a $[H_3O^+]$ of 6.0×10^{-11} M, use $B(OH)_3$, NH_4^+, HCN, or HCO_3^-

17.8 The formula weights and moles of acid per gram for the five antacids are:

	Formula weight	mol acid/gram
$Mg(OH)_2$	58.32	1 mol acid/29.16 g antacid
$CaCO_3$	100.10	1 mol acid/50.05 g antacid
$NaHCO_3$	84.00	1 mol acid/84.00 g antacid
$Al(OH)_3$	60.99	1 mol acid/20.33 g antacid
$NaAl(OH)_2CO_3$	143.99	1 mol acid/36.00 g antacid

Of these antacids, $Al(OH)_3$ neutralizes the most stomach acid per gram.

Chapter 18

18.1 Reducing agents are indicated by "red" and oxidizing agents are indicated by "ox." Oxidation numbers are shown above the symbols for the elements.

(a) $\overset{0}{2\,Fe}(s) + 3\overset{0}{Cl_2}(g) \longrightarrow 2\overset{+3}{Fe}\overset{-1}{Cl_3}(s)$
red (Fe), ox (Cl_2)

(b) $2\overset{0}{H_2}(g) + \overset{0}{O_2}(g) \longrightarrow 2\,\overset{+1}{H_2}\overset{-2}{O}(l)$
red (H_2), ox (O_2)

(c) $\overset{0}{Cu}(s) + 2\overset{+5}{N}\overset{-2}{O_3}{}^-(aq) + 4\,\overset{+1}{H_3}\overset{-2}{O}{}^+(aq) \longrightarrow \overset{+2}{Cu}{}^{2+}(aq) + 2\,\overset{+4}{N}\overset{-2}{O_2}(g) + 6\,\overset{+1}{H_2}\overset{-2}{O}(l)$
red (Cu), ox (NO_3^-)

(d) $\overset{0}{C}(s) + \overset{0}{O_2}(g) \longrightarrow \overset{+4}{C}\overset{-2}{O_2}(g)$
red (C), ox (O_2)

(e) $6\,\overset{+2}{Fe}{}^{2+}(aq) + \overset{+6}{Cr_2}\overset{-2}{O_7}{}^{2-}(aq) + 14\,\overset{+1}{H_3}\overset{-2}{O}{}^+(aq) \longrightarrow 6\,\overset{+3}{Fe}{}^{3+}(aq) + 2\,\overset{+3}{Cr}{}^{3+}(aq) + 21\,\overset{+1}{H_2}\overset{-2}{O}(l)$
red (Fe^{2+}), ox ($Cr_2O_7^{2-}$)

18.2 (a) Ox: $Cd(s) \longrightarrow Cd^{2+}(aq) + 2\,e^-$
Red: $Cu^{2+}(aq) + 2\,e^- \longrightarrow Cu(s)$
Sum: $Cd(s) + Cu^{2+}(aq) \longrightarrow Cd^{2+}(aq) + Cu(s)$

(b) Ox: $Zn(s) \longrightarrow Zn^{2+}(aq) + 2\,e^-$
Red: $2\,H_3O^+(aq) + 2\,e^- \longrightarrow H_2(g) + 2\,H_2O(l)$
Sum: $Zn(s) + 2\,H_3O^+(aq) \longrightarrow Zn^{2+}(aq) + H_2(g) + 2\,H_2O(l)$

(c) Ox: $2\,Al(s) \longrightarrow 2\,Al^{3+}(aq) + 6\,e^-$
Red: $3\,Zn^{2+}(aq) + 2\,e^- \longrightarrow 3\,Zn(s)$
Sum: $2\,Al(s) + 3\,Zn^{2+}(aq) \longrightarrow 2\,Al^{3+}(aq) + 2\,Zn(s)$

18.3 Step 1. This is an oxidation-reduction reaction. It is obvious that Zn is oxidized by its change in oxidation state.
Step 2. The half-reactions are:

$Zn(s) \longrightarrow Zn^{2+}(aq)$ (his is the oxidation reaction)

$Cr_2O_7^{2-}(aq)^- \longrightarrow 2\,Cr^{3+}(aq)$ (this is the reduction reaction)

Step 3. Balance the atoms in the half-reactions. The atoms are balanced in the Zn reaction. We need to add water and H in the $Cr_2O_7^{2-}$ half-reaction. Fourteen H^+ ions will be required on the right to combine with the seven O atoms.

$$Cr_2O_7^{2-}(aq) + 14\,H^+(aq) \longrightarrow 2\,Cr^{3+}(aq) + 7\,H_2O(l)$$

Step 4. Balance the half-reactions for charge. Write the Zn reaction as

$$Zn(s) \longrightarrow Zn^{2+}(aq) + 2\,e^-$$

And write the $Cr_2O_7^{2-}$ reaction as

$Cr_2O_7^{2-}(aq) + 14\ H^+(aq) + 6\ e^- \longrightarrow 2\ Cr^{3+}(aq) + 7\ H_2O(l)$

Step 5. Multiply the half-reactions by factors to make the number of electrons gained equal the number lost.

$3\ [Zn(s) \longrightarrow Zn^{2+}(aq) + 2e^-]$

$1\ [Cr_2O_7^{2-}(aq) + 14\ H^+(aq) + 6\ e^- \longrightarrow 2\ Cr^{3+}(aq^-) + 7\ H_2O(l)]$

Step 6. Add the two half-reactions, cancelling the electrons.

$$3\ Zn(s) \longrightarrow 3\ Zn^{2+}(aq) + 6\ e^-$$

$$Cr_2O_7^{2-}(aq) + 14\ H^+ + 6\ e^- \longrightarrow 2\ Cr^{3+}(aq) + 7\ H_2O(l)$$

$$Cr_2O_7^{2-}(aq) + 3\ Zn(s) + 14\ H^+(aq) \longrightarrow 2\ Cr^{3+}(aq) + 3\ Zn^{2+}(aq) + 7\ H_2O(l)$$

Step 7. Everything checks.

Step 8. Water was added in Step 3. The balanced equation is

$$Cr_2O_7^{2-}(aq) + 3\ Zn(s) + 14\ H^+(aq) \longrightarrow 2\ Cr^{3+}(aq) + 3\ Zn^{2+}(aq) + 7\ H_2O(l)$$

18.4 Step 1. This is an oxidation-reduction reaction. The wording of the question says A1 reduces NO_3^- ion. Al is also oxidized.

Step 2. The half-reactions are:

$Al(s) \longrightarrow Al(OH)_4^-(aq)$ (this is the oxidation reaction)

$NO_3^-(aq) \longrightarrow NH_3(aq)$ (this is the reduction reaction)

Step 3. Balance the atoms in the half-reactions. To balance the Al reaction for atoms, add four OH^- ions on the left.

$$Al(s) + 4\ OH^-(aq) \longrightarrow Al(OH)_4^-$$

To balance the NO_3^- reaction for atoms, add water on the left and OH^- ions on the right. Getting the number of H_2O molecules is a bit tricky, but since each NO_3^- nitrogen requires 3 H atoms, at least 3 H_2O molecules are a minimum. The three O atoms from the NO_3^{2-} ion can be thought of as O^{2-}, which would react with 3 H_2O to form 6 OH^- in the reaction $O^{2+} + 3\ H_2O \rightarrow OH^-$. Adding all these up, we get

$$NO_3^-(aq) + 6\ H_2O(l) \longrightarrow NH_3(aq) + 9\ OH^-$$

Step 4. Balance the half-reactions for charge. Place 3 e^- on the right in the Al reaction.

$$Al(s) + 4\ OH^-(aq) \longrightarrow Al(OH)_4^- + 3\ e^-$$

And place $8e^-$ on the left in the NO_3^- reaction.

$$NO_3^-(aq) + 6\ H_2O(l) + 8e^- \longrightarrow NH_3(aq) + 9\ OH^-$$

Step 5. Multiply the half-reactions by factors to make the electrons gained equal those lost.

$8\ [Al(s) + 4\ OH^-(aq) \longrightarrow Al(OH)_4^- + 3\ e^-]$

$3[NO_3^-(aq) + 6\ H_2O(l) + 8e^- \longrightarrow NH_3(aq) + 9\ OH^-]$

Step 6. Add both half-reactions and cancel the electrons.

$8\ Al(s) + 32\ OH^-(aq) \longrightarrow 8\ Al(OH)_4^- + 24\ e^-$

$3\ NO_3^-(aq) + 18\ H_2O(l) + 24\ e^- \longrightarrow 3\ NH_3(aq) + 27\ OH^-$

$3\ NO_3^-(aq) + 8\ Al(s) + 18\ H_2O(l) + 32\ OH^-(aq) \longrightarrow 8\ Al(OH_4^-(aq) + 3\ NH_3(aq) + 27\ OH^-(aq)$

Step 7. Make a final check. Since there are OH^- ions on both sides of the equation, cancel them out. This gives the final balanced equation.

$3\ NO_3^-(aq) + 8\ Al(s) + 18\ H_2O(l) + 5\ OH^-(aq) \longrightarrow 8\ Al(OH)_4^-(aq) + 3\ NH_3(aq)$

(This is a fairly complicated equation to balance. If you got this one with a minimum of effort, you're understanding balancing oxidation-reduction equations rather well. If you had to struggle with one or more of the steps, go back and repeat those.)

18.5 (a) $Ni(s) \rightarrow Ni^{2+}(aq) + 2\ e^-$ (this is the oxidation half-reaction)

$2\ Ag^+(aq) + 2\ e^- \longrightarrow 2\ Ag(s)$ (this is the reduction half-reaction)

(b) The oxidation of Ni takes place at the anode and the reduction of Ag^+ ions takes place at the cathode.

(c) Electrons would flow through an external circuit from the anode (where Ni is oxidized) to the cathode (where Ag^+ ions are reduced).

(d) Nitrate ions would flow through the salt bridge to the anode compartment. Potassium ions would flow into the cathode compartment.

18.6 Oxidation half-reaction: $Fe(s) \rightarrow Fe^{2+}(aq,\ 1M) + 2\ e^-$
Reduction half-reaction: $Cu^{2+}(aq,\ 1M) + 2\ e^- \rightarrow Cu(s)$

$$E_{net} = 0.78\ V = E_{ox} + E_{red}$$

Since $E_{red} = +0.34$ V, E_{ox} must be $+0.44$ V

18.7 $F_2(g) + 2\ e^- \rightarrow 2\ F^-(aq)$ $+2.87$ V
$2\ Li(s) \rightarrow 2\ Li^+(aq) + 2\ e^-$ $-(-3.045\ V)$

$2\ Li(s) + F_2(g) \rightarrow 2\ Li^+(aq) + 2\ F^-(aq)$ $+5.91$ V

18.8 (a) The net cell reaction would be

$$2\ Na^+ + 2\ Br^- \longrightarrow 2\ Na + Br_2$$

Sodium ions would be reduced at the cathode and bromide ions would be oxidized at the anode.

(b) H_2 would be produced at the cathode for the same reasons given in Problem-Solving Example 18.8. That reaction is

$$2\ H_2O(l) + 2\ e^- \longrightarrow H_2(g) + 2\ OH^-(aq)$$

At the anode, two reactions are possible; the oxidation of water and the oxidation of Br^- ions.

$6\ H_2O(\ell) \longrightarrow O_2(g) + 4\ H_3O^+(aq) + 4\ e^-$ $E°_{ox} = -1.229$ V

$2\ Br^-(aq) \longrightarrow Br_2(\ell) + 2\ e^-$ $E°_{ox} = -1.08$ V

The oxidation of bromide ion has the least negative oxidation potential, so that reaction will occur. The net cell reaction is

$2\ H_2O(\ell) + 2\ Br^-(aq) \longrightarrow Br_2(\ell) + H_2(g) + 2\ OH^-$

(c) Sn metal will be formed at the cathode because its reduction potential (-0.14 V) is less negative than the potential for the reduction of water. O_2 will form at the anode because the $E°_{ox}$ value for the oxidation of water is less negative than the $E°_{ox}$ value for the oxidation of Cl^-. The net cell reaction is

$$2\ Sn^{2+}(aq) + 6\ H_2O(l) \longrightarrow Sn(s) + O_2(g) + 4\ H_3O^+(aq)$$

18.9 First, calculate the quantity n of charge.

$$\text{Charge} = (25 \times 10^3\ A)(1\ hr)(60\ s/min)(60\ min/hr) = 9.0 \times 10^7\ A\ s = 9.0 \times 10^7\ C$$

Mass of Na = $(9.0 \times 10^7\ C)$

$$\left(\frac{1\ mol\ e^-}{96500\ C}\right)\left(\frac{1\ mol\ Na}{1\ mol\ e^-}\right)\left(\frac{22.9898\ g\ Na}{1\ mol\ Na}\right) = 2.1 \times 10^4\ g\ Na$$

Chapter 19

19.1 (a) $^{237}_{93}Np \rightarrow {}^{4}_{2}He + {}^{233}_{91}Pa$;

(b) $^{35}_{16}S \rightarrow {}^{0}_{-1}e + {}^{35}_{17}Cl$

19.2 (a) $^{90}_{38}Sr \rightarrow {}^{0}_{-1}e + {}^{90}_{39}y$

(b) It takes four half-lives (4×29 y = 116 y) for the activity to decrease to 125 beta particles emitted per minute:

Number of half-lifes	Time of half-life (years)	Change of activity	Total elapsed time (years)
First	29	2000 to 1000	29
Second	29	1000 to 500	58
Third	29	500 to 250	87
Fourth	29	250 to 125	116

19.3 (a) $t_{1/2} = \dfrac{0.693}{9.3 \times 10^{-3}\ d^{-1}} = 75\ d$

(b) ln(fraction remaining) = $-k \times t$

$= -(9.3 \times 10^{-3}\ d^{-1}) \times (100\ d) = -0.930$;

fraction of iridium-192 remaining = $e^{-0.930} = 0.39$;

therefore, 39% of the original iridium-192 remains

19.4 $k = \dfrac{0.693}{1.60 \times 10^3\ y} = 4.33 \times 10^{-4}\ y^{-1}$

As of 1997:

ln (fraction remaining) = $-k \times t$

$-(4.33 \times 10^{-4}\ y^{-1}) \times 76\ y = -3.29 \times 10^{-2}$;

fraction of radium-226 remaining = $e^{-0.0329} = 0.967$;

therefore 96.7% of the original radium-226 remains;

$0.967 \times 1.00\ g = 0.967\ g$

19.5 $\ln(0.60) = -0.510 = -k \times t$;

$$k = \frac{0.693}{t_{1/2}} = \frac{0.693}{12.3\ y} = 0.0563\ y^{-1};$$

$$t = \frac{-0.510}{-0.0563\ y^{-1}} = 9.1\ y$$

APPENDIX L

Answers to Exercises

Chapter 1

1.1. (a) Identifying the sample as lead is qualitative information. The mass and melting point are quantitative information.
(b) Qualitative: iron-containing, green.
(c) Qualitative: yellow cover. Quantitative: 1.8 kg.

1.2. (a) 37 °C is equivalent to 98.6 °F, your body temperature, so 37 °C is a higher temperature than 85 °F.
(b) 20 °F is a lower temperature than 0 °C.

1.3. (a) (i) The color (blue) is a physical property. The process of melting is a physical change and the melting point is a physical property. (ii) The color (colorless) and crystal shape (cubic) are physical properties.
(b) The boiling point of a substance is a physical property. Because your normal body temperature is 37 °C, this means that a sample with a boiling point of 15 °C will boil (a physical change) if placed in your hand.

1.4. Melting and evaporation are physical changes. Energy is transferred to the sample from sunlight, causing the particles to move more and more rapidly. Eventually the water molecules of the solid move rapidly enough to become a liquid, and liquid molecules move rapidly enough to become a gas (or, more properly, a vapor).

1.5. Burning or combustion is a chemical change, as is cooking. Boiling water is a physical change.

1.6. (a) Solution
(b) Concrete is a heterogeneous mixture of sand, gravel, and cement.
(c) Muddy water is a heterogeneous mixture of dirt and water.
(d) Diamonds are pure carbon, an element.
(e) Recently minted pennies consist of two elements—a copper cladding on a zinc core.
(f) Table salt is a compound (sodium chloride).

Chapter 2

2.1. (a) Water molecules on the clothes move and leave the surface of the cloth to enter the air, drying the clothes.
(b) Water molecules in air strike the surface of the glass, lose energy, and condense into liquid water, which appears as moisture on the outside of the glass.
(c) During evaporation, water molecules escape from the surface of the salt solution, leaving salt behind. Eventually, all the water evaporates and the salt remaining crystallizes.

2.2. (a) Uranium (U), neptunium (Np), and plutonium (Pu)
(b) Californium, Cf
(c) S (sulfur, Stanitski), W (tungsten, Wood), and Mo (molybdenum, Moore)

2.3. The statement is wrong because tin and lead are different elements. Allotropes are different forms of the same element in the same physical state at the same temperature and pressure.

2.4. An electric charge results when objects are rubbed together because negatively charged electrons are transferred from one object to the other. The object gaining the electrons acquires a negative charge; the object losing electrons becomes positively charged.

2.5. $\frac{r_{nucleus}}{r_{atom}} = \frac{10^{-15}\text{m}}{10^{-10}\text{m}} = 10^{-5} = \frac{r_{nucleus}}{10^{2}\text{m}}$; $r_{nucleus} = 10^{2}\text{ m} \times 10^{-5} = 10^{-3}\text{ m} = 10^{-1}\text{ cm}$. This is about the thickness of the wire in a paper clip.

2.6. Nitrogen-14 and nitrogen-15 are isotopes of nitrogen because they have the same number of protons but different numbers of neutrons. To be allotropes, the two forms of nitrogen would have to occur in different forms in the same physical state.

2.7. Atomic weight Li = (0.07500)(6.015121 amu) + (0.9250)(7.016003 amu) = 0.451134 amu + 6.489802 amu = 6.940936 amu

2.8. Because the most abundant isotope is magnesium-24 (78.70%), the atomic weight of magnesium is closer to 24 than to 25 or 26, the mass numbers of the other magnesium isotopes, which make up approximately 21% of the remaining mass. The simple arithmetic average is $\frac{24 + 25 + 26}{3} = 25$, which is larger than the atomic weight. In the arithmetic average, the relative abundance of each magnesium isotope is 33%, far less than the actual percent abundance of magnesium-24, and much more than the natural percent abundances of magnesium-25 and magnesium-26.

2.9. The two boron isotopes are boron-10 and boron-11. The weight of the arithmetic average, in which each isotope is equally abundant (50%), is $\frac{10 \text{ amu} + 11 \text{ amu}}{2} = 10.5$ amu. This is not the atomic weight of boron; thus, the percent abundances cannot be 50%.

2.10.

$$10.00 \text{ g Li} \times \frac{6.02 \times 10^{23} \text{ Li atoms}}{6.941 \text{ g Li}} = 8.673 \times 10^{23} \text{ Li atoms}$$

$$10.00 \text{ g Ir} \times \frac{6.02 \times 10^{23} \text{ Ir atoms}}{192.22 \text{ g Ir}} = 3.132 \times 10^{22} \text{ Ir atoms}$$

$$8.673 \times 10^{23} - 3.132 \times 10^{22} = 8.359 \times 10^{23} \text{ Li atoms}$$

2.11. Because they have the same density, equal volumes of the elements will have the same mass. But, equal masses will not have the same number of moles. Consider the number of moles in 100.0 g of each element.

$$100.0 \text{ g Os} \times \frac{1 \text{ mol Os}}{190.2 \text{ g Os}} = 0.5258 \text{ mol Os}$$

$$100.0 \text{ g Ir} \times \frac{1 \text{ mol Ir}}{192.22 \text{ g Ir}} = 0.5202 \text{ mol Ir}$$

2.12. (a) Water is the most dense liquid (bottom layer). The least dense liquid is the top layer (kerosene).
(b) No permanent change occurred.

2.13.

1. (a) 13 metals: potassium (K), calcium (Ca), scandium (Sc), titanium (Ti), vanadium (V), chromium (Cr), manganese (Mn), iron (Fe), cobalt (Co), nickel (Ni), copper (Cu), zinc (Zn), and gallium (Ga)
   (b) 3 nonmetals: selenium (Se), bromine (Br), and krypton (Kr)
   (c) 2 metalloids: germanium (Ge) and arsenic (As)
2. (a) Groups 1A (except hydrogen), 2A, 1B, 2B, 3B, 4B, 5B, 6B, 7B, 8B
   (b) Groups 7A and 8
   (c) None
3. Period 6

Chapter 3

3.1.

```
    H   H   H
    |   |   |
H — C — C — C — H
    |   |   |
    H   OH  OH
```

$CH_3CHOHCH_2OH$ $C_3H_8O_2$

3.2. (a) CS_2; (b) PCl_3; (c) SBr_2; (d) SeO_2; (e) OF_2; (f) XeO_3

3.3. (a) $C_{16}H_{34}$, $C_{28}H_{58}$; (b) $C_{14}H_{30}$, 14 carbon atoms and 30 hydrogen atoms

3.4.

```
    H   H   H
    |   |   |
H — C — C — C — OH
    |   |   |
    H   H   H
```

3.5.

```
    H   H   H
    |   |   |
H — C — C — C — H
    |   |   |
    H   OH  H
```

3.6.

```
    H   H   H   H   H
    |   |   |   |   |
H — C — C — C — C — C — H
    |   |   |   |   |
    H   H   H   H   H
```

$CH_3(CH_2)_3CH_3$

```
    H   H    H   H
    |   |    |   |
H — C — C  — C — C — H
    |   |    |   |
    H   CH3  H   H
```

```
   CH3
   |
CH3CHCH2CH3
```

```
    H   CH3  H
    |   |    |
H — C — C  — C — H
    |   |    |
    H   CH3  H
```

```
    CH3
    |
CH3CCH3
    |
    CH3
```

3.7. (a) 2-methylpentane

```
    H   H    H   H   H
    |   |    |   |   |
H — C — C  — C — C — C — H
    |   |    |   |   |
    H   CH3  H   H   H
```

(b) 3-methylpentane

```
    H   H   H    H   H
    |   |   |    |   |
H — C — C — C  — C — C — H
    |   |   |    |   |
    H   H   CH3  H   H
```

(c) 2,2-dimethylbutane

```
    H   CH3  H   H
    |   |    |   |
H — C — C  — C — C — H
    |   |    |   |
    H   CH3  H   H
```

(d) 2,3-dimethylbutane

```
    H   CH3  CH3  H
    |   |    |    |
H — C — C  — C  — C — H
    |   |    |    |
    H   H    H    H
```

3.8. (a) Ionic (Fe^{2+} and S^{2-} ions); (b) Ionic (Co^{2+} and Cl^- ions); (c) Molecular; (d) Molecular; (e) Ionic (NH_4^+ and CO_3^{2-} ions); (f) Molecular

3.9. (a) Potassium nitrite; (b) Sodium hydrogen sulfite; (c) Manganese(II) hydroxide; (d) Manganese(III) sulfate; (e) Barium nitride; (f) Lithium hydride

3.10. (a) Carbon, nitrogen, oxygen, phosphorus, hydrogen, selenium; (b) Calcium and strontium; (c) Chloride and iodide; (d) Iron, copper, zinc, vanadium (also chromium, manganese, cobalt, nickel, molybdenum, and cadmium)

3.11. Carbohydrates have hydrogen and oxygen in a 2:1 ratio. Therefore, compounds a, b, and d might be carbohydrates. Compound c is definitely not a carbohydrate because it contains nitrogen.

3.12. (a) 174.16 g/mol; (b) 386.64 g/mol; (c) 398.06 g/mol; (d) 194.2 g/mol

3.13. The statement is true. Because both compounds have the same formula, they have the same molar mass. Thus, 100 g of each compound contains the same number of moles.

3.14. Epsom salt is $MgSO_4 \cdot 7\ H_2O$, which has a molar mass of 247 g/mol.

$$20\ g \times \frac{1\ mol}{247\ g} = 8.1 \times 10^{-2}\ mol$$

3.15. (a) 1.000 mol of SF_6 has a mass of 145.97 g and contains 32.07 g of S, and 6 mol × 18.998 g/mol F = 113.9 g of F. The mass percents are

$$S = \frac{32.07}{145.97} = 21.97\%\ S$$

$$F = 100.00\% - 21.97\% = 78.03\%\ F$$

(b) $$12\ mol\ C \times \frac{12.011\ g\ C}{1\ mol\ C} = 144.13\ g\ C$$

$$22\ mol\ H \times \frac{1.0079\ g\ H}{1\ mol\ H} = 22.174\ g\ H$$

$$11\ mol\ O \times \frac{15.999\ g\ O}{1\ mol\ O} = 175.99\ g\ O$$

Molar mass = 144.13 g + 22.174 g + 175.99 g = 342.3 g

$$\%\ C = \frac{144.13\ g}{342.3\ g} \times 100 = 42.11$$

$$\%\ H = \frac{22.174\ g}{342.3\ g} \times 100 = 6.478$$

$$\%\ O = \frac{175.99\ g}{342.3\ g} \times 100 = 51.41$$

(c) $$2\ mol\ Al \times \frac{26.982\ g\ Al}{1\ mol\ Al} = 53.964\ g\ Al$$

$$3\ mol\ S \times \frac{32.066\ g\ S}{1\ mol\ S} = 96.198\ g\ S$$

$$12\ mol\ O \times \frac{15.999\ g\ O}{1\ mol\ O} = 191.99\ g\ O$$

Molar mass = [53.964 + 96.198 + 191.99] g = 342.15 g

$$\%Al = \frac{53.964\ g}{342.15\ g} \times 100 = 15.772$$

$$\%S = \frac{96.198\ g}{342.15\ g} \times 100 = 28.116$$

$$\%O = \frac{191.99\ g}{342.15\ g} \times 100 = 56.113$$

(d) $$1\ mol\ U \times \frac{238.029\ g\ U}{1\ mol\ U} = 238.029\ g\ U$$

$$6\ mol\ O \times \frac{15.999\ g\ O}{1\ mol\ O} = 95.994\ g\ O$$

$$6\ mol\ Te \times \frac{127.60\ g\ Te}{1\ mol\ Te} = 765.60\ g\ Te$$

$$30\ mol\ F \times \frac{18.998\ g\ F}{1\ mol\ F} = 569.94\ g\ F$$

Molar mass = [238.029 + 95.994 + 765.60 + 569.94] g = 1669.6 g

$$\%U = \frac{238.029\ g}{1669.6\ g} \times 100 = 14.26$$

$$\%O = \frac{95.994\ g}{1669.6\ g} \times 100 = 5.750$$

$$\%Te = \frac{765.60\ g}{1669.6\ g} \times 100 = 45.86$$

$$\%F = \frac{569.94\ g}{1669.6\ g} \times 100 = 34.14$$

3.16. $$0.569\ g\ Sn \times \frac{1\ mol\ Sn}{118.7\ g\ Sn} = 4.794 \times 10^{-3}\ mol\ Sn$$

$$2.434\ g\ I_2 \times \frac{1\ mol\ I_2}{253.81\ g\ I_2} \times \frac{2\ mol\ I}{1\ mol\ I_2} = 1.918 \times 10^{-2}\ mol\ I$$

$$\frac{1.918 \times 10^{-2}}{4.794 \times 10^{-3}} = 4.001\ mol\ I;$$

$$\frac{1.918 \times 10^{-2}\ mol\ I}{4.794 \times 10^{-3}\ mol\ Sn} = \frac{4.001\ mol\ I}{1.000\ mol\ Sn};$$

therefore, the formula is SnI_4.

Chapter 4

4.1. (a) The total mass of reactants (4 Fe(s) + $3O_2$ (g)) must equal the total mass of products (2 Fe_2O_3(s)), which is 2.50 g.
(b) The stoichiometric coefficients are 4, 3, and 2.
(c) $$1.000 \times 10^4\ O\ atoms \times \frac{1\ O_2\ molecule}{2\ O\ atoms} \times \frac{4\ Fe\ atoms}{3\ O_2\ molecules} = 6.667 \times 10^3\ Fe\ atoms$$

4.2. (a) $2\ Cu + O_2 \longrightarrow 2\ CuO$
(b) $S + 3\ F_2 \longrightarrow SF_6$
(c) $Ba + Br_2 \longrightarrow BaBr_2$

4.3. (a) Strontium phosphide, Sr_3P_2; (b) Calcium fluoride, CaF_2; (c) Magnesium oxide (MgO) and carbon dioxide (CO_2)

4.4. It is possible for an exchange reaction to form two different precipitates, for example, the reaction between barium hydroxide and iron(II) sulfate:

$$Ba(OH)_2(aq) + FeSO_4(aq) \longrightarrow BaSO_4(s) + Fe(OH)_2(s)$$

4.5. (a) Hydrogen ions and perchlorate ions:

$$HClO_4(aq) \longrightarrow H^+(aq) + ClO_4^-(aq)$$

(b) $Ca(OH)_2(aq) \longrightarrow Ca^{2+}(aq) + 2\ OH^-(aq)$

4.6. (a) $H^+(aq) + Cl^-(aq) + K^+(aq) + OH^-(aq) \longrightarrow H_2O(\ell) + K^+(aq) + Cl^-(aq)$

$$H^+(aq) + OH^-(aq) \longrightarrow H_2O(\ell)$$

(b) $2\ H^+(aq) + SO_4^{2-}(aq) + Ba^{2+}(aq) + 2\ OH^-(aq) \longrightarrow 2\ H_2O(\ell) + BaSO_4(s)$

$$H^+(aq) + OH^-(aq) \longrightarrow H_2O(\ell)$$
$$Ba^{2+}(aq) + SO_4^{2-}(aq) \longrightarrow BaSO_4(s)$$

(c) $2\ H^+(aq) + 2\ CH_3CO_2^-(aq) + Ca^{2+}(aq) + 2\ OH^-(aq) \longrightarrow Ca^{2+}(aq) + 2\ CH_3CO_2^-(aq) + 2\ H_2O(\ell)$

$$H^+(aq) + OH^-(aq) \longrightarrow H_2O(\ell)$$

4.7. $Al^{3+}(aq) + 3\ OH^-(aq) + 3\ H^+(aq) + 3\ Cl^-(aq) \longrightarrow 3\ H_2O + Al^{3+}(aq) + 3\ Cl^-(aq)$

$$H^+(aq) + OH^-(aq) \longrightarrow H_2O$$

4.8. (a) The products are aqueous sodium sulfate, water, and carbon dioxide gas:

$$2\ Na^+(aq) + CO_3^{2-}(aq) + 2\ H^+(aq) + SO_4^{2-}(aq) \longrightarrow 2\ Na^+(aq) + SO_4^{2-}(aq) + H_2O(\ell) + CO_2(g)$$
$$2\ H^+(aq) + CO_3^{2-}(aq) \longrightarrow H_2O(\ell) + CO_2(g)$$

(b) The products are aqueous iron(II) chloride and hydrogen sulfide gas.

$$FeS(s) + 2\ HCl(aq) \longrightarrow FeCl_2(aq) + H_2S(g)$$
$$[2\ H^+(aq) + S^{2-}(aq) \longrightarrow H_2S(g)]$$

(c) Aqueous potassium chloride, water, and sulfur dioxide gas are produced:

$$K_2SO_3(aq) + 2\ HCl(aq) \longrightarrow 2\ KCl(aq) + H_2O(\ell) + SO_2(g)$$
$$2\ H^+(aq) + SO_3^{2-}(aq) \longrightarrow H_2O(\ell) + SO_2(g)$$

4.9. (a) Gas-forming reaction; the products are aqueous nickel sulfate, water, and carbon dioxide gas.

$$NiCO_3(s) + H_2SO_4(aq) \longrightarrow NiSO_4(aq) + H_2O(\ell) + CO_2(g)$$

(b) Acid-base reaction; nitric acid reacts with strontium hydroxide, a base, to produce water and strontium nitrate, a salt.

$$Sr(OH)_2(s) + 2\ HNO_3(aq) \longrightarrow Sr(NO_3)_2(aq) + 2\ H_2O(\ell)$$

(c) Precipitation reaction; aqueous sodium chloride and insoluble barium oxalate are produced.

$$BaCl_2(aq) + Na_2C_2O_4(aq) \longrightarrow BaC_2O_4(s) + 2\ NaCl(aq)$$

(d) Precipitation and gas-forming reaction; lead sulfate precipitates and carbon dioxide gas is released.

$$PbCO_3(aq) + H_2SO_4(aq) \longrightarrow PbSO_4(s) + H_2O(\ell) + CO_2(g)$$

4.10. (a) This is not a redox reaction. Nitric acid is a strong oxidizing agent, but here it serves as an acid.
(b) In this redox reaction, chromium metal (Cr) is oxidized (loses electrons) to form Cr^{3+} ions in Cr_2O_3; oxygen (O_2) is reduced (gains electrons) to form oxide ions, O^{2-}. Oxygen is the oxidizing agent, chromium is the reducing agent.
(c) This is an acid-base reaction, but not a redox reaction; there are no strong oxidizing or reducing agents present.
(d) Copper is oxidized and chlorine is reduced in this redox reaction in which copper is the reducing agent and chlorine is the oxidizing agent.

$$Cu \longrightarrow Cu^{2+} + 2\ e^-; \qquad Cl_2 + 2\ e^- \longrightarrow 2\ Cl^-$$

4.11. (a) Carbon in oxalate ion, $C_2O_4^{2-}$ (oxidation state = +3), is oxidized to oxidation state +4 in CO_2.
(b) Carbon is reduced from +4 in CCl_2F_2 to 0 in C(s).

Chapter 5

5.1. $\frac{2\text{ mol Al}}{3\text{ mol Br}_2}$; $\frac{1\text{ mol Al}_2\text{Br}_6}{2\text{ mol Al}}$; $\frac{1\text{ mol Al}_2\text{Br}_6}{3\text{ mol Br}_2}$; and their reciprocals.

5.2. $$5.57 \times 10^{-2}\text{ mol Br}_2 \times \frac{1\text{ mol Al}_2\text{Br}_6}{3\text{ mol Br}_2} \times \frac{533.4\text{ g Al}_2\text{Br}_6}{1\text{ mol Al}_2\text{Br}_6} = 9.90\text{ g Al}_2\text{Br}_6$$

5.3. (a) $$300\text{ g urea} \times \frac{1\text{ mol urea}}{58.17\text{ g urea}} \times \frac{2\text{ mol NH}_3}{1\text{ mol urea}} \times \frac{17.09\text{ g NH}_3}{1\text{ mol NH}_3} = 176\text{ g NH}_3$$

$$100\text{ g H}_2\text{O} \times \frac{1\text{ mol H}_2\text{O}}{18.02\text{ g H}_2\text{O}} \times \frac{2\text{ mol NH}_3}{1\text{ mol H}_2\text{O}} \times \frac{17.09\text{ g NH}_3}{1\text{ mol NH}_3} = 190\text{ g NH}_3$$

Because water produces more ammonia, urea is the limiting reactant.
(b) 176 g NH_3;

$$300\text{ g urea} \times \frac{1\text{ mol urea}}{58.17\text{ g urea}} \times \frac{1\text{ mol CO}_2}{1\text{ mol urea}} \times \frac{44.01\text{ g CO}_2}{1\text{ mol CO}_2} = 227\text{ g CO}_2$$

(c)

$$300\text{ g urea} \times \frac{1\text{ mol urea}}{58.17\text{ g urea}} \times \frac{1\text{ mol H}_2\text{O}}{1\text{ mol urea}} \times \frac{18.0\text{ g H}_2\text{O}}{1\text{ mol H}_2\text{O}} = 92.8\text{ g H}_2\text{O reacted; } 100\text{ g} - 92.8\text{ g} = 7.2\text{ g H}_2\text{O remain}$$

5.4. (1) Impure reactants; (2) Inaccurate weighing of reactants and products

5.5. Insufficient addition of KCl; incomplete drying of precipitated AgCl(s); incomplete collection of precipitated AgCl(s) and transfer for weighing.

5.6. Assuming that the nicotine is pure, weigh a sample of nicotine and burn the sample. Separately collect and weigh the carbon dioxide and water generated, and calculate the moles and grams of carbon and hydrogen collected. By mass difference, determine the mass of nitrogen in the original sample, then calculate the moles of nitrogen. Calculate the mole ratios of carbon, hydrogen, and nitrogen in nicotine to determine its empirical formula.

5.7. Molar mass of cholesterol = 386.7 g/mol

$$\text{Moles of cholesterol} = 240\text{ mg} \times \frac{1\text{ g}}{10^3\text{ mg}} \times \frac{1\text{ mol}}{386.7\text{ g}} = 6.21 \times 10^{-4}\text{ mol cholesterol}$$

$$\text{Molarity} = \frac{6.21 \times 10^{-4}\text{ mol}}{0.100\text{ L}} = 6.21 \times 10^{-3}\text{ molar}$$

5.8. (a)

$$6.37\text{ g Al(NO}_3)_3 \times \frac{1\text{ mol Al(NO}_3)_3}{213.0\text{ g AlAl(NO}_3)_3} = 0.299\text{ mol Al(NO}_3)_3$$

$$\frac{0.299\text{ mol Al(NO}_3)_3}{0.250\text{ L}} = 0.120\text{ molar Al(NO}_3)_3$$

Molarity: Al^{3+} = 0.120; NO_3^- = 3(0.120) = 0.360

5.9. His error was in using a liter of water. Molarity is defined in terms of a liter of solution, not a liter of solvent. When the liter of water was added, the resulting solution volume might have been less than or greater than 1 L.

5.10. The molarity could be increased by evaporating some of the solvent or by adding additional solute.

5.11. $$0.0193\text{ L} \times \frac{0.200\text{ mol AgNO}_3}{1\text{ L}} \times \frac{1\text{ mol Ag}^+}{1\text{ mol AgNO}_3} \times \frac{1\text{ mol Cl}^-}{1\text{ mol Ag}^+} \times \frac{1\text{ mol NaCl}}{1\text{ mol Cl}^-} = 3.86 \times 10^{-3}\text{ mol NaCl;}$$

$$\frac{3.86 \times 10^{-3}\text{ mol NaCl}}{0.0250\text{ L}} = 0.154\text{ M NaCl}$$

Chapter 6

6.1. You transfer some mechanical energy to the ball to accelerate it upward. The ball's potential energy increases the higher it gets, but its kinetic energy decreases by an equal quantity and eventually it stops rising and begins to fall. As it falls, some of the ball's potential energy changes to kinetic energy and the ball goes faster and faster until it hits the floor. When the ball hits the floor, some of its kinetic energy is transferred to atoms, molecules, or ions that make up the floor, causing them to move faster. Eventually all of the ball's kinetic energy is transferred and the ball stops moving. The nanoscale particles in the floor (and some in the air that the ball fell through) are moving faster on average, and the temperature of the floor (and the air) is slightly higher. The energy has spread out over a much larger number of particles.

6.2. $q = c \times m \times \Delta T$

$$\Delta T = \frac{q}{c \times m} = \frac{24.1 \times 10^3\text{J}}{(0.902\text{ J/g °C})(250.\text{ g})} = 107\text{ °C}$$

$\Delta T = T_f - T_i$; 107 °C = T_f − 5.0 °C; T_f = 112 °C

6.3. From Table 6.1, $c_{granite}$ = 0.79 J/g °C and $c_{H_2O(l)}$ = 4.184 J/g °C. Since water has a larger specific heat capacity, the same quantity of heating will raise its temperature less. Therefore the rock (granite) will be hotter.

6.4.

Metal	Molar Heat Capacity (J/mol °C)
Al	24.3
Fe	25.1
Cu	25.2
Au	25.2

The molar heat capacities for the metals have very nearly the same values, but they are quite different from the value for ethyl alcohol. The molar heat capacity for a metal should be about 25 J/mol °C, but apparently the same rule does not apply to other kinds of substances.

6.5

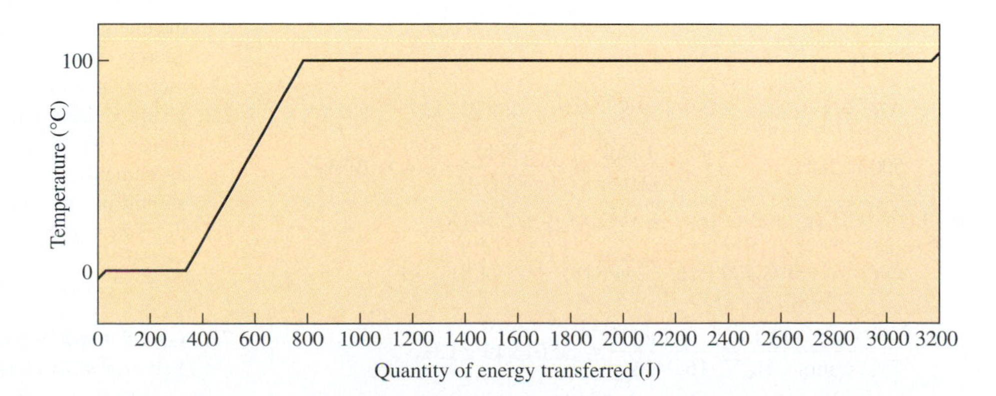

Since the heat of vaporization is almost seven times larger than the heat of fusion, the temperature stays constant at 100 °C almost seven times longer than it stays constant at 0 °C. It stays constant at 0 °C for slightly less time than it takes to heat the water from 0 °C to 100 °C. (see graph on bottom p. A-54)

6.6. • Heat of fusion: 237 g × 333 J/g = 78.9 kJ
• Heating liquid: 237 g × 4.184 J/(g °C) × 100. °C = 99.2 kJ
• Heat of vaporization: 237 g × 2260 J/g = 536 kJ
• Total heating = (78.9 + 99.2 + 536) kJ = 714 kJ

6.7. The direction of energy transfer is indicated by the *sign* of the enthalpy change. Transfer to the system corresponds to a positive enthalpy change.

6.8. Because of heats of fusion and heats of vaporization, the enthalpy change is different when a reactant or product is in a different state.

6.9. Yes, it would violate the first law of thermodynamics. According to the supposition, we could create energy by starting with 2 mol HCl, breaking all of the molecules apart, recombining the atoms to form 1 mol H_2 and 1 mol Cl_2, and then reacting the H_2 and Cl_2 to give 2 HCl.

$$2\,HCl \longrightarrow \text{atoms} \longrightarrow H_2 + Cl_2 \qquad \Delta H^\circ = +185\text{ kJ}$$
$$H_2 + Cl_2 \longrightarrow 2\,HCl \qquad \Delta H^\circ = -190\text{ kJ}$$

The net effect of these two processes is that there is still 2 mol HCl, but 5 kJ of energy has been created. This is impossible according to the first law of thermodynamics.

6.10. In Problem-Solving Example 6.4, the number of bonds is the same before (4 C—H and 2 O═O) and after (2 C═O and 4 O—H) the reaction, but the C═O bonds formed are nearly twice as strong as any of the other bonds. Forming *stronger* bonds makes the reaction exothermic.

In Problem-Solving Practice 6.4, only 3 bonds are broken (2 H—H and 1 O═O), while 4 bonds are formed (4 O—H). All of the bonds are similar in strength, and so forming *more* bonds makes the reaction exothermic.

6.11. (a) $\frac{1}{2}N_2(g) + \frac{3}{2}H_2(g) \longrightarrow NH_3(g) \qquad \Delta H^\circ = -46.11$ kJ

(b) $N_2(g) \longrightarrow N_2(g)$. This equation says that you start with 1 mol N_2 and end with 1 mol N_2. Since there is no change, there can be no enthalpy change, and so $\Delta H_f^\circ(N_2(g)) = 0$.

6.12. H—N(H)—N(H)—H + O═O ⟶ N≡N + H—O—H + H—O—H

$$\Delta H^\circ = \{4\,(389) + 159 + 498 - 946 - 4(463)\}\text{ kJ} = -585\text{ kJ}$$

$$500\text{ lb }N_2H_4 \times \frac{454\text{ g}}{\text{lb}} \times \frac{1\text{ mol}}{32.04\text{ g}} \times \frac{585\text{ kJ}}{1\text{ mol }N_2H_4} = 4 \times 10^6\text{ kJ}$$

6.13. (a) $CH_4(g) + 2\,O_2(g) \longrightarrow CO_2(g) + 2\,H_2O(g)$

$$\Delta H^\circ = \{-393.509 + 2(-241.818) - (-74.81)\}\text{ kJ} = -802.34\text{ kJ}$$

$$\frac{802.34\text{ kJ}}{1\text{ mol }CH_4} \times \frac{1\text{ mol}}{16.0426\text{ g}} = 50.013\text{ kJ/g }CH_4$$

(b) $C_8H_{18}(\ell) + \frac{25}{2}O_2(g) \longrightarrow 8\,CO_2(g) + 9\,H_2O(g)$

$$\Delta H^\circ = \{8(-393.509) + 9\,(-241.818) - (-249.952)\}\text{ kJ} = -5074.48\text{ kJ}$$

$$\frac{5074.48\text{ kJ}}{1\text{ mol }C_8H_{18}} \times \frac{1\text{ mol}}{114.23\text{ g}} = 44.423\text{ kJ/g }C_8H_{18}$$

(c) $N_2H_4(\ell) + O_2(g) \longrightarrow N_2(g) + 2\,H_2O(g)$

$$\Delta H^\circ = \{2(-241.818) - 50.63\}\text{ kJ} = -534.26\text{ kJ}$$

$$\frac{534.26\text{ kJ}}{1\text{ mol }N_2H_4} \times \frac{1\text{ mol}}{32.045\text{ g}} = 16.672\text{ kJ/g }N_2H_4$$

(d) $H_2(g) + \frac{1}{2}O_2(g) \longrightarrow H_2O(g)$

$$\Delta H^\circ = -241.818\text{ kJ}$$

$$\frac{241.818\text{ kJ}}{1\text{ mol }H_2} \times \frac{1\text{ mol}}{2.0158\text{ g}} = 119.96\text{ kJ/g }H_2$$

(e) $CH_2O(s) + O_2(g) \longrightarrow CO_2(g) + H_2O$

$$\Delta H^\circ = -425\text{ kJ}$$

$$\frac{425\text{ kJ}}{1\text{ mol }CH_2O} \times \frac{1\text{ mol}}{30.03\text{ g}} = 14.1\text{ kJ/g }CH_2O$$

Hydrogen provides the most thermal energy per gram.

Chapter 7

7.1. A^{***} $A^{**}B^{*}$ $A^{*}B^{**}$ B^{***}

If C, D, and E are added, there are many more arrangements in addition to these

$A^*B^*C^*$	$A^*B^*D^*$	$A^*B^*E^*$	$A^*C^*D^*$	$A^*C^*E^*$
$A^*D^*E^*$	$B^*C^*D^*$	$B^*C^*E^*$	$B^*D^*E^*$	$C^*D^*E^*$
$A^{**}C^*$	$A^{**}D^*$	$A^{**}E^*$	$B^{**}C^*$	$B^{**}D^*$
$B^{**}E^*$	$C^{**}A^*$	$C^{**}V^*$	$C^{**}D^*$	$C^{**}E^*$
$D^{**}A^*$	$D^{**}B^*$	$D^{**}C^*$	$D^{**}E^*$	$E^{**}A^*$
$E^{**}B^*$	$E^{**}C^*$	$E^{**}D^*$	C^{***}	D^{***}
E^{***}				

There are 35 possible arrangements, but only 4 of them have the energy confined to atoms A and B. The probability that all energy remains with A and B is thus 4/35 = 0.114, or a little more than 11%.

7.2. Using Celsius temperature and $\Delta S = q/T$, if the temperature were −10 °C, the value of ΔS would be negative, in disagreement with the fact that transfer of energy to a sample should increase molecular motion and hence entropy.

7.3. $T_2 - T_1 = (t_2 + 273.15) - (t_1 + 273.15)$
$= t_2 - t_1 + 273.15 - 273.15$
$= t_2 - t_1$

7.4. (a) $\Delta S = q/T = \Delta H^\circ/T = 30.8\text{ kJ}/(80.1 + 273.15)\text{K}$
$= 8.72 \times 10^{-2}\text{ kJ/K} = 87.2\text{ J/K}$

(b) For condensation of vapor, the sign of $\Delta H°$ is negative, and so $\Delta S° = -87.2$ J/K.

7.5. (a) Reactant is a gas. Products are also gases, but the number of molecules has increased. Entropy is greater for products. (b) Reactant is a solid. Product is a solution. Mixing sodium and chloride ions among water molecules results in greater entropy for the product.
(c) Reactant is a solid. Products are a solid and a gas. The much larger entropy of the gas results in greater entropy for the products.

7.6. Because $\Delta S°_{surroundings} = -\Delta H/T$ at a given temperature, the larger the value of T the smaller the value of $\Delta S_{surroundings}$. If ΔS_{system} does not change much with temperature, then $S_{universe}$ must also get smaller. In this case, because ΔS_{system} is negative, $\Delta S_{universe}$ would become negative at a high enough temperature.

7.7. (a) Since both $\Delta H°$ and $\Delta S°$ are negative, this reaction belongs to the second category and is product-favored at room temperature. (b) Third category, reactant-favored; (c) First category, product-favored; (d) Fourth category, reactant-favored

7.8. $\Delta S°_{system} = 2 \text{ mol HCl(g)} \times S°(\text{HCl(g)}) - 1 \text{ mol H}_2\text{(g)} \times S°(\text{H}_2\text{(g)}) - 1 \text{ mol Cl}_2\text{(g)} \times S°(\text{Cl}_2\text{(g)})$

$= (2 \times 186.908 - 130.684 - 223.036)$ J/K

$= 20.096$ J/K

$\Delta S°_{surroundings} = -\Delta H°/T = -[2 \text{ mol HCl(g)} \times (-92.307 \text{ kJ/mol})]/298.15 \text{ K}$

$= 619.20$ J/K

$\Delta S°_{universe} = (619.20 + 20.096) \text{ J/K} = 639.29$ J/K

7.9. (a) $\Delta S°(\text{i}) = \{2 \times (27.78) + \frac{3}{2} \times (205.138) - 87.40\}$ J/K
$= 275.86$ J/K

$\Delta H°(\text{i}) = -\Delta H_f°(\text{Fe}_2\text{O}_3\text{(s)}) = 824.2$ kJ

$\Delta G°(\text{i}) = -\Delta G_f°(\text{Fe}_2\text{O}_3\text{(s)}) = 742.2$ kJ

$\Delta S°(\text{ii}) = \{50.92 - 2 \times (28.3) - \frac{3}{2} \times (205.138)\}$ J/K
$= -313.4$ J/K

$\Delta H°(\text{ii}) = \Delta H_f°(\text{Al}_2\text{O}_3\text{(s)}) = -1675.7$ kJ

$\Delta G°(\text{ii}) = \Delta G_f°(\text{Al}_2\text{O}_3\text{(s)}) = -1582.3$ kJ

Step i is reactant-favored. Step ii is product-favored.
(b) Net reaction: $\text{Fe}_2\text{O}_3\text{(s)} + 2 \text{ Al(s)} \longrightarrow 2 \text{ Fe(s)} + \text{Al}_2\text{O}_3\text{(s)}$

$\Delta S° = 273.86 \text{ J/K} + (-313.4 \text{ J/K}) = -37.5$ J/K

$\Delta H° = 824.2 \text{ kJ} + (-1675.7 \text{ kJ}) = -851.5$ kJ

$\Delta G° = 742.2 \text{ kJ} + (-1582.3 \text{ kJ}) = -840.1$ kJ

The net reaction has negative $\Delta G°$ and is therefore product-favored. For the *net* reaction, $\Delta S°$, $\Delta H°$, and $\Delta G°$ are all negative.
(c) If the two reactions are coupled, it is possible to obtain iron from iron(III) oxide even though that reaction is not product-favored by itself. The large negative $\Delta G°$ for formation of Al_2O_3(s) makes the overall $\Delta G°$ negative for the coupled reactions.

(d) $\text{Mg(s)} + \frac{3}{2} \text{O}_2\text{(g)} \longrightarrow \text{MgO(s)}$

$\Delta G° = \Delta G_f°(\text{MgO(s)}) = -569.43$ kJ

Coupling the reactions, we have

$\text{Fe}_2\text{O}_3\text{(s)} \longrightarrow 2 \text{ Fe(s)} + \frac{3}{2} \text{O}_2\text{(g)} \qquad \Delta G_1° = 742.2 \text{ kJ}$

$3 \times (\text{Mg(s)} + \frac{1}{2} \text{O}_2\text{(g)} \longrightarrow \text{MgO(s)})$

$\Delta G_2° = 3(-569.43) \text{ kJ} = -1708.29$ kJ

$\text{Fe}_2\text{O}_3\text{(s)} + 3 \text{ Mg(s)} \longrightarrow 2 \text{ Fe(s)} + 3 \text{ MgO(s)}$

$\Delta G_3° = -966.1$ kJ

7.10. $\Delta G° = -2870 \text{ kJ} + 32 \times (30.5 \text{ kJ}) = -1894$ kJ. The 1894 kJ of Gibbs free energy is transformed into thermal energy.

7.11. 64,500 g ATP/50 g ATP = 1290 times each ADP must be recycled to ATP on average.

Chapter 8

8.1. The frequency of light decreases as its wavelength increases, because of the relationship $\lambda\nu = c$.

8.2. Cellular phones use higher frequency radio waves.

8.3. For the x-ray photon,

$$E = h\nu = h\frac{c}{\lambda}$$

$$= 6.626 \times 10^{-34} \text{ J·s} \left(\frac{2.998 \times 10^8 \text{ m s}^{-1}}{2.36 \times 10^{-9} \text{ m}}\right) = 8.42 \times 10^{-17} \text{ J}$$

This photon has about 265 times more energy than one photon of orange light (3.18×10^{-19} J).

8.4. In a sample of excited hydrogen gas there are many atoms, and each can exist in one of the excited states possible for hydrogen. The observed spectral lines are a result of all of the possible transitions of all of these hydrogen atoms.

8.5. The maximum number of s orbitals that may be found in a given electron shell is one. The maximum number of p orbitals in a given electron shell is three. The maximum number of d orbitals in a given electron shell is five. The maximum number of f orbitals in a given energy shell is seven. The principal quantum number of the shell in which f orbitals first occur is $n = 4$.

8.6. The first shell that could contain g orbitals would be the $n = 5$ shell. There would be nine g orbitals.

8.7. (a) The maximum number of electrons in the $n = 3$ level is 18 (2 electrons per orbital). The orbitals would be designated $3s$, $3p_x$, $3p_y$, $3p_z$, $3d_{z^2}$, $3d_{xy}$, $3d_{yz}$, $3d_{xz}$, and $3d_{x^2-y^2}$. (b) The maximum number of electrons in the $n = 4$ level is 32. The orbitals would be designated $4s$, $4p_x$, $4p_y$, $4p_z$, $4d_{z^2}$, $4d_{xy}$, $4d_{yz}$, $4d_{xz}$, $4d_{x^2-y^2}$, and the seven $4f$ orbitals, which are not designated by name in the text.

8.8. For the chlorine atom, $n = 3$ and there are seven electrons in this highest energy level. The configuration is

3s	3p		
↑↓	↑↓	↑↓	↑

For the sulfur atom, the highest energy level is $n = 3$ and there is one less electron. The configuration is 3s [↑↓] 3p [↑↓][↑][↑].

8.9. The noble gas configuration for iodine is [Kr] $4s^2 3d^{10} 4p^5$. Its core electrons are represented by [Kr]$3d^{10}$, and its valence electrons are represented by $4s^2 4p^5$.

8.10. The electron configurations for Se and Te are [Ar]$4s^2 3d^{10} 4p^4$ and [Kr]$5s^2 4d^{10} 5p^4$, respectively. Elements in the same main group have similar electronic configurations.

8.11. This dot structure would be $:\ddot{\underset{..}{Na}}:^{+}$. The element Ne has this dot structure. Using this same kind of reasoning, the O^{2-} ion's structure would be $:\ddot{\underset{..}{O}}:^{2-}$. This is also the same dot structure as the Ne atom.

8.12. The electron configurations for the following ions are (a) Cl^-, [Ne]$3s^2 3p^6$; (b) O^{2-}, [He]$2s^2 2p^6$; (c) N^{3-}, [He]$2s^2 2p^6$; (d) Se^{2-}, [Ar]$4s^2 3d^{10} 4p^6$.

8.13. The [Ar]$3d^4 4s^2$ configuration for chromium has four unpaired electrons, and the [Ar]$3d^5 4s^1$ configuration has six unpaired electrons.

8.14. The ground state Cu atom has a configuration [Ar]$4s^1 3d^{10}$. When it loses one electron, it becomes the Cu^+ ion with configuration [Ar]$3d^{10}$. There is an added stability for the completely filled set of $3d$ orbitals.

8.15. Student's choice

8.16. The $Fe(acac)_2$ contains an Fe^{2+} ion with a $3d$-electron configuration of [↑↓][↑][↑][↑][↑]. This configuration has four unpaired electrons. The compound $Fe(acac)_3$ contains an Fe^{3+} ion, with a $3d$ electron configuration of [↑][↑][↑][↑][↑]. This configuration has five unpaired electrons. The $Fe(acac)_3$, with more unpaired electrons per molecule, would be attracted more strongly into a magnetic field.

8.17. Using $E = h\nu$, the energy of a 100 MHz radio photo is about 6×10^{-26} J. In Exercise 8.3, the energy of an x-ray photo was calculated to be about 8.42×10^{-17} J, or almost a billion times more energetic than the radio photon. We are constantly bathed in radio waves (broadcast radio and TV, cellular phones, etc.) with little, if any, harmful effects from those photons. Our exposures to x-ray photons, on the other hand, is carefully controlled. X-ray photons are energetic enough to break a single chemical bond. If a typical bond energy is about 100 kJ per mol of bonds, then each bond requires only about 1.6×10^{-19} J of energy to break it. The x-ray photon has about 100 times the energy required to do that.

8.18. The Si atom would be larger than the C atom because it is below it in Group 4A, and atoms increase in size moving down a group. The Al atom is just to the left of Si in the third period. Atoms get larger moving from right to left in a period. So, in order of increasing atomic size, it is C, then Si, then Al.

8.19. The trend in sizes of the ions is $P^{3-} > S^{2-} > Cl^-$. The atoms themselves also exhibit this trend. In addition, each of these ions represents completely filled valence shells for their respective atoms. The three extra electrons in the P^{3-} ion do not feel as large a nuclear charge as the two extra electrons in the S^{2-} ion, so they occupy a larger volume around the phosphorus nucleus. The same reasoning applies to the extra electrons in the S^{2-} ion compared to the single electron in the Cl^- ion.

8.20. In order of increasing first ionization energy, F > P > Mg > Al > Ba.

8.21. The largest difference in two successive ionization energies for a phosphorus atom would be expected between the fifth ionization energy (where the last of the valence orbital electrons is removed) and the sixth ionization energy (where the first of the core electrons is removed).

8.22. The Lewis dot structure for the nitrogen atom is $\dot{\text{N}}:$ (with one dot above and one to the left). The bonded N atom in a protein would be something like —N: (with bonds above, below and to the left). There are two nonbonded valence electrons in most compounds containing these kinds of N atoms. A Be^{2+} ion, being quite small and highly charged, would be attracted to the unpaired electrons on the nitrogen atom, —N:Be^{2+} (with bonds above and below N). This interaction would undoubtedly have an effect on the molecule containing the N atom. It is thought that this is the means whereby Be^{2+} ions cause toxic effects in living systems.

8.23. Based on an increasing ratio of oxygen to nitrogen, the order is N_2O, NO, N_2O_3, NO_2, and N_2O_5.

8.24. The formulas for the fluorine compounds of C, P, S, Br, and Se if they are oxidized to the maximum are CF_4, PF_5, SF_6, BrF_7, and SeF_6.

8.25.

1A	2A	3A	4A	5A	6A	7A
MF	MF_2	MF_3	MF_4	MF_5	MF_6	MF_7

Chapter 9

9.1. (octagon) C_8H_{16}

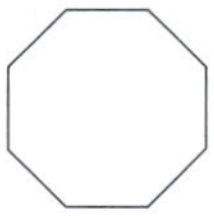

9.2. N_2 has only 10 valence electrons. The Lewis structure shown has 14 valence electrons.

9.3. The Lewis structure for carbon monoxide (c) is the only correct Lewis structure. For the others, (a) is incorrect because there are only 10 valence electrons, not 14 as shown; (b) is incorrect because, although it shows the correct number of valence electrons (26), there is a double bond between F and N rather than a single bond with a lone pair on N; This places more than an octet around F rather than a single bond and a lone pair on N, which gives both N and F an octet; (d) is not correct because COCl should have 17 valence electrons, not 16 as shown.

9.4. C—N > C=N > C≡N; The order of decreasing bond energy is the reverse order: C≡N > C=N > C—N.

9.5. (a)

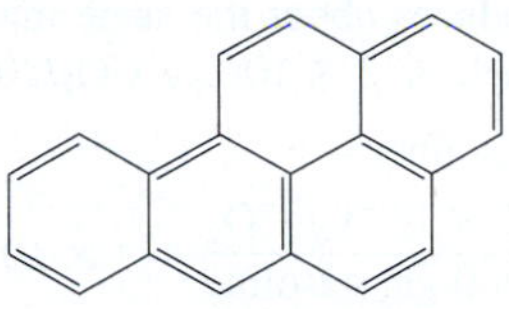

(b) $C_{20}H_{12}$

9.6.

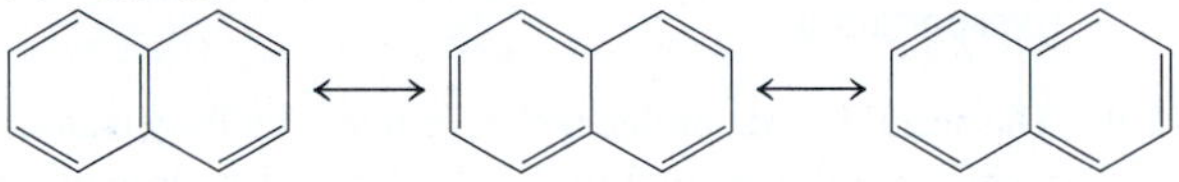

9.7.

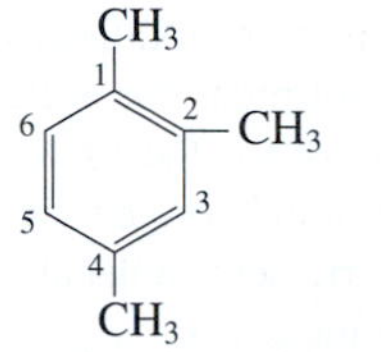

CH_3 CH_3 CH_3

1,2,4-trimethylbenzene

9.8. (a) The electronegativity difference between sodium and chlorine is 2.0, sufficient to cause electron transfer from sodium to chlorine to form Na^+ and Cl^- ions. Molten NaCl conducts an electric current, indicating the presence of ions.
(b) There is an electronegativity difference of 1.2 in BrF, which is sufficient to form a polar covalent bond, but not great enough to cause electron transfer leading to ion formation.

9.9. $[Cr(NH_3)_2(H_2O)_2(OH)_2]^+$

9.10. K_2SO_4 is analogous to $K_2[NiCl_4]$.

9.11. (a) Three oxalate ions per Mn^{2+}; (b) 4−;
(c) Five

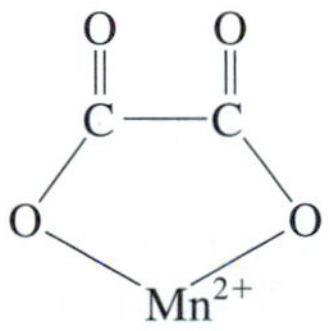

Mn^{2+}

9.12. (a) 2+; 4− for the complex = (6 × 1− for six CN^-) + (2+ for Fe^{2+}); (b) 3+; it takes four Fe^{3+} ions to counterbalance three $[Fe(CN)_6]^{4-}$ ions.

Chapter 10

10.1. SCl_2 has two bond pairs and two lone pairs on S. Because lone pairs occupy more volume than bond pairs, the ClSCl angle would be expected to be less than 109.5°. (The experimental value is 102.7°.) PCl_3 has three bond pairs and one lone pair on the P. The ClPCl angle would be expected to be less than 109.5°. (The experimental value is 102.7°.) ICl_2^- has two bond pairs and three lone pairs on I. The Cl atoms occupy axial positions, so the ClICl angle is 180°.

10.2. H–C(H)(H)–C(=O)–C(=O)–O–H

tetrahedral, triangular planar, angular

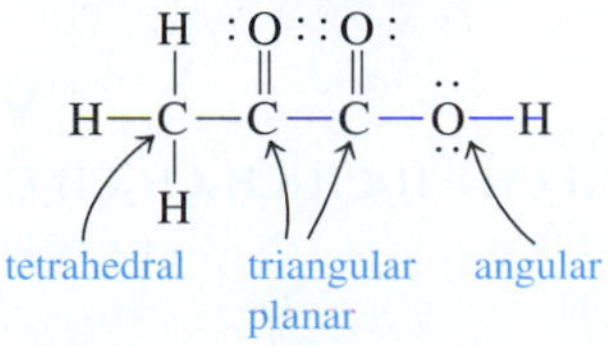

10.3. An sp^3 hybridized carbon atom hybridizes all three *p* orbitals; there are no unhybridized *p* orbitals for overlap with an unhybridized *p* orbital on another atom.

10.4. Although its terminal H atoms are alike, NH_3 is polar because the H atoms are not symmetrically arranged.

10.5. (a) Br (2.8) is more electronegative than I (2.5).
(b) Each is a polar molecule because the halogen atom electronegativities are greater than that for hydrogen (Cl = 3.0, Br = 2.8, I = 2.5, H = 2.1), thus creating a net dipole. The electronegativity difference of the carbon-halogen bond decreases from C—Cl to C—Br to C—I.

10.6. $\delta+\ \delta-$:C≡O: ------ :C≡O: ; $\delta-\ \delta+$:O≡C: ------ :O≡C:

----- dipole-dipole forces

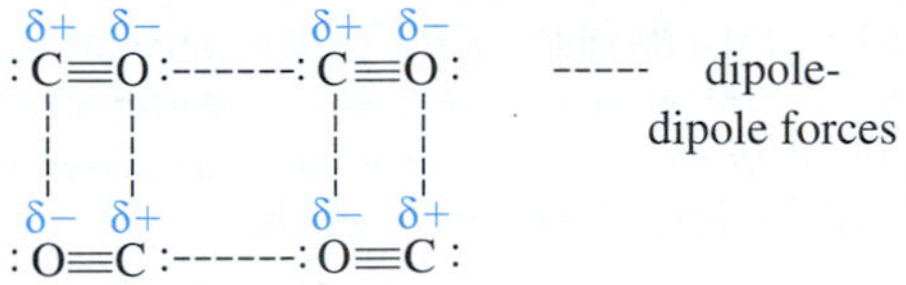

10.7. The F—H···F— hydrogen bond is the strongest because the electronegativity difference between H and F produces a more polar F—H bond than does the lesser electronegativity difference between O and H or N and H in the O—H or N—H bonds.

10.8.

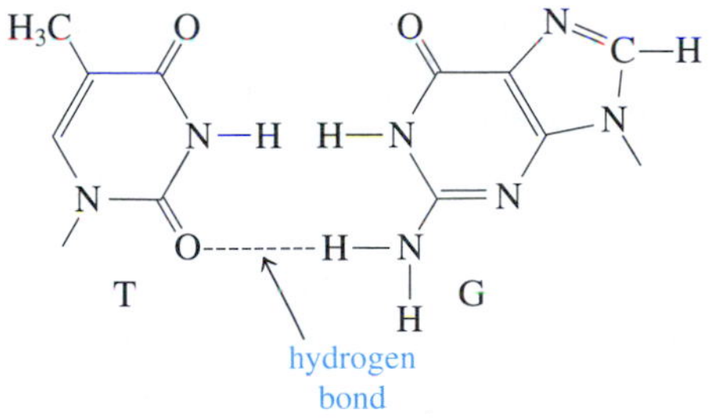

T G

hydrogen bond

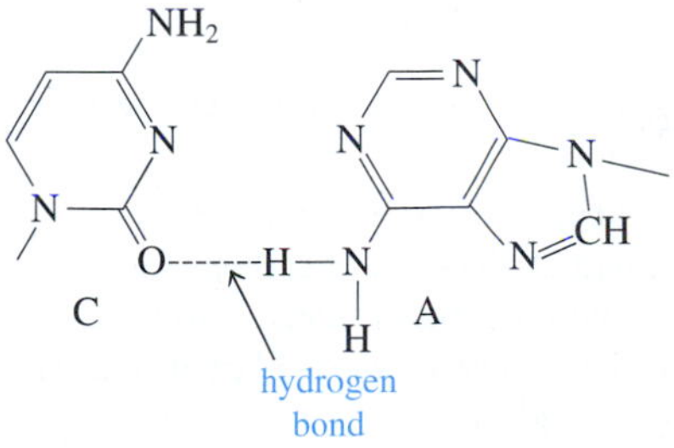

C A

hydrogen bond

Chapter 11

11.1

$$2\ H_2(g) + O_2(g) \longrightarrow 2\ H_2O(g)$$
$$CH_4(g) + 2\ O_2(g) \longrightarrow CO_2(g) + 2\ H_2O(g)$$
$$2\ CO(g) + O_2(g) \longrightarrow 2\ CO_2(g)$$
$$2\ C_2H_6(g) + 7\ O_2(g) \longrightarrow 4\ CO_2(g) + 6\ H_2O(g)$$
$$2\ H_2S(g) + 3\ O_2(g) \longrightarrow 2\ H_2O(g) + 2\ SO_2(g)$$

11.2 The energy that can be obtained from the CO and H_2 produced in synthesis gas is more than the energy required to make the gas mixture from coal and steam.

11.3 CO_2 is a low-energy compound and cannot produce any more fuel energy by oxidation.

11.4 The hydrogen used to hydrogenate coal can come from synthesis gas. The pattern is that all of these reactants and products come from coal and water.

11.5 Thermal energy = 3.7×10^7 bbl oil $\times \dfrac{5.9 \times 10^9 \text{ J}}{1 \text{ bbl oil}} =$ 2.2×10^{17} J

electricity delivered = $2.2 \times 10^{17} \text{ J} \times \dfrac{1 \times 10^6 \text{ kwh}}{2.6 \times 10^{12} \text{ J}} \times 0.33 = 2.7 \times 10^{10}$ kwh

11.6 For any electrical generating plant, the quantity of waste heat is about 67% of the thermal energy used to generate the electricity. For a 100 million kwh per day power plant, this amounts to about 67×10^6 kwh per day. The corresponding quantities of oil, natural gas, and coal are

Oil: $67 \times 10^6 \text{ kwh/day} \times \dfrac{614 \text{ bbl oil}}{1 \times 10^6 \text{ kwh}} = 41 \times 10^3$ bbl oil/day

Natural gas: $67 \times 10^6 \text{ kwh/day} \times \dfrac{3.41 \times 10^6 \text{ ft}^3}{1 \times 10^6 \text{ kwh}} = 230 \times 10^6 \text{ ft}^3$ nat gas/day

Coal: $67 \times 10^6 \text{ kwh/day} \times \dfrac{137 \text{ t coal}}{1 \times 10^6 \text{ kwh}} = 9.2 \times 10^3$t coal/day

One way that some of this waste heat could be recovered is to build near the power plant homes and other buildings that need heat in winter. Another way is to try and transport some of this waste heat over a short distance to places where it is needed. From an engineering standpoint, low-temperature thermal energy cannot be efficiently transported. In addition, most people do not wish to live or work very near large electric generating plants.

11.7 Natural sources: animal respiration, forest fires, decay of cellulose products, partial digestion of carbohydrates, volcanoes
Human sources: burning fossil fuels, burning agricultural wastes, and refined cellulose products like paper, decay of carbon compounds in landfills.

11.8 For this calculation, we can use a 1995 figure of 347×10^9 passenger miles (Air Transport Association—www.air-transport.org). Using the ratio of 2×10^3 kg CO_2/5000 passenger miles, we can calculate the CO_2 released.

Quantity of CO_2 = 347×10^9 passenger mi $\times \dfrac{2 \times 10^3 \text{ kg } CO_2}{5 \times 10^3 \text{ passenger mi}} = 1 \times 10^{11}$ kg CO_2

If a typical automobile gets 20 mi/gallon of fuel and about 1.5 passengers are transported for every mile the automobile travels, then an automobile gets 30 passenger miles per gallon. The number of gallons used would be

Volume of gasoline = 347×10^9 passenger miles $\times \dfrac{1 \text{ gal gasoline}}{30 \text{ passenger mi}} = 1 \times 10^{10}$ gal gasoline

Assume the gasoline produces about the same mass of CO_2 per gallon as does jet fuel, or 2×10^3 kg CO_2/200 gallons.

Quantity of CO_2 from gasoline =

1×10^{10} gal gasoline $\times \dfrac{2 \times 10^3 \text{ kg } CO_2}{200 \text{ gal gasoline}} = 1 \times 10^{11}$ kg CO_2

So the numbers are about the same for these two modes of transportation.

11.9 The acetaldehyde molecule has two less hydrogen atoms compared with the ethanol molecule. Loss of hydrogen is oxidation. The acetaldehyde molecule is more oxidized than the ethanol molecule. Comparing the formulas for acetaldehyde and acetic acid, the hydrogen atoms are the same, but the acetic acid molecule has one additional oxygen atom. Gain of oxygen is oxidation. So the acetic acid molecule is more oxygenated than the acetaldehyde molecule.

11.10 Ten or so carbon atoms in an alcohol molecule will make it much less water soluble than alcohols with fewer numbers of carbon atoms.

11.11 The aldehyde formed by the partial oxidation of methanol is formaldehyde. This compound is a probable human carcinogen. Its toxicity is summarized in a number of sources. See http://www.instantref.com/formald.htm#iris-formaldehyde.

11.12 (a) isopropyl alcohol, (b) ethylene glycol, (c) ethyl alcohol, (d) glycerol, (e) methyl alcohol

11.13 $CH_3CH_2CH_2OH$ propanol

11.14 These compounds consist of four rings fused together. Three of the rings are six-membered rings and there is one five-membered ring. In addition, all of the structures have an oxygen atom at position (a) and a CH_3 group at position (b). Position (c) is occupied by at least one atom in all the structures.

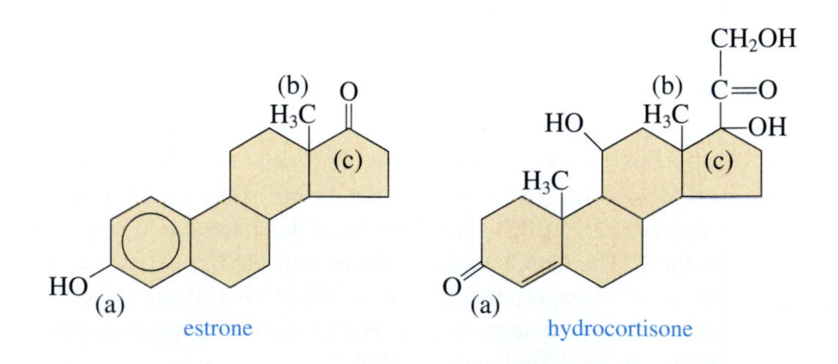

11.15 nonanoic acid $CH_3CH_2CH_2CH_2CH_2CH_2CH_2CH_2C(=O)$—OH, and decanoic acid

$CH_3CH_2CH_2CH_2CH_2CH_2CH_2CH_2CH_2C(=O)$—OH

11.16

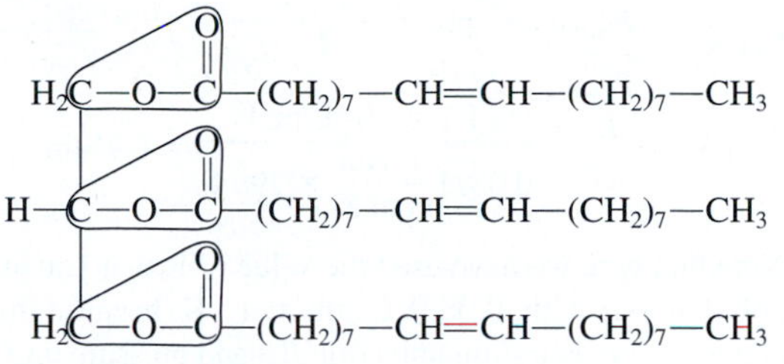

$H_2C—O—C(=O)—(CH_2)_7—CH{=}CH—(CH_2)_7—CH_3$

$H—C—O—C(=O)—(CH_2)_7—CH{=}CH—(CH_2)_7—CH_3$

$H_2C—O—C(=O)—(CH_2)_7—CH{=}CH—(CH_2)_7—CH_3$

11.17 The ends of the chains are possibly occupied with OH groups from water molecules as well as some of the OR groups from the initiator molecules.

11.18

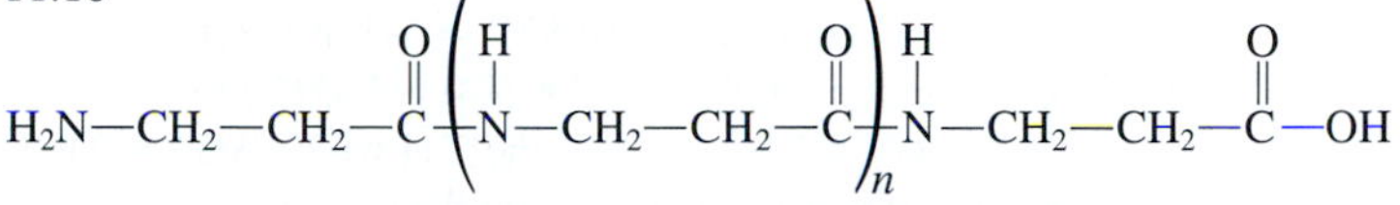

$H_2N—CH_2—CH_2—C(=O)—\left(N(H)—CH_2—CH_2—C(=O)\right)_n—N(H)—CH_2—CH_2—C(=O)—OH$

11.19 PVC $(—CH_2—CHCl—)_n$ combustion products, H_2O, CO_2, HCl

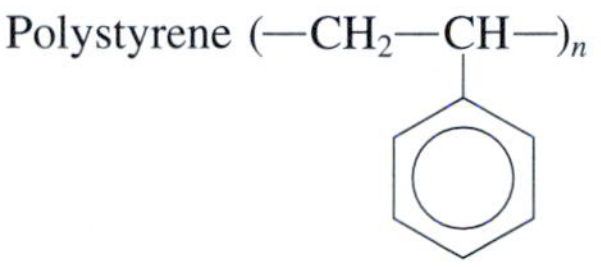

Polystyrene $(—CH_2—CH(C_6H_5)—)_n$ combustion products, H_2O, CO_2

Polyethylene $(—CH_2—CH_2—)_n$ combustion products, H_2O, CO_2

Polytetrafluoroethylene $(—CF_2—CF_2—)_n$ combustion products, HF, CO_2

11.20 Serine and glutamine would hydrogen-bond to one another if they were close in two adjacent protein chains because they have polar groups in their molecules. Glycine and valine would not because they have no additional polar groups in their molecules.

11.21

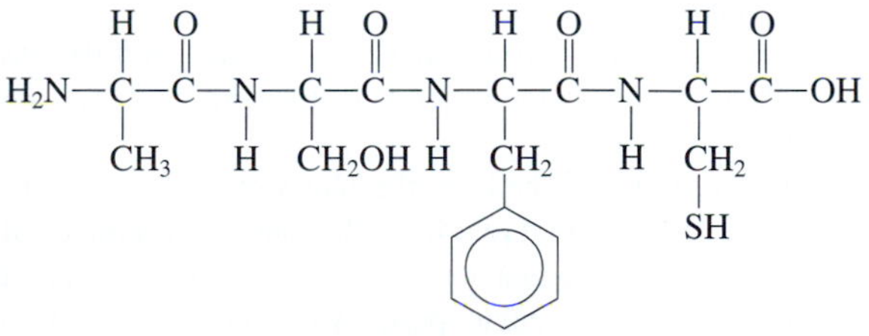

$H_2N—CH(CH_3)—C(=O)—N(H)—CH(CH_2OH)—C(=O)—N(H)—CH(CH_2C_6H_5)—C(=O)—N(H)—CH(CH_2SH)—C(=O)—OH$

11.22 Chain branching can occur anywhere an OH group is found.

11.23 The OH groups in this molecule allow it to be extensively hydrogen-bonded with solvent water molecules.

11.24 Cellulose contains glucose molecules linked together by β-1,4 linkages. Ruminant animals have large colonies of bacteria and protozoa that live in the forestomach and digest cellulose.

11.25 If humans could digest cellulose, then common plants that are easy to grow could become food. There might be less reliance upon cultivation of plants for food. In addition, the entire plant could be used for food rather than just certain parts eaten and the other parts wasted. On the other hand, in times of famine, there might not be enough cellulose to go around. Destroying trees and other plants for food might cause enlargements of desert regions and the disappearance of entire species of plants.

Chapter 12

12.1 (a) $\frac{\Delta c}{\Delta t} = \frac{(0.0030 - 0.0040)\text{ M}}{(800 - 600)\text{ s}} = -5.0 \times 10^{-6}$ M/s; rate $= 5.0 \times 10^{-6}$ M/s

(b) $\frac{\Delta c}{\Delta t} = \frac{(0.0015 - 0.0020)\text{ M}}{(1300 - 1100)\text{ s}} = -2.5 \times 10^{-6}$ M/s; rate $= 2.5 \times 10^{-6}$ M/s

(c) $\frac{\Delta c}{\Delta t} = \frac{(0.0017 - 0.0072)\text{ M}}{(1200 - 200)\text{ s}} = -5.5 \times 10^{-6}$ M/s; rate $= 5.5 \times 10^{-6}$ M/s

12.2 Rate (c) is 1/2 of rate (a) and the concentration of cisplatin is also 1/2 as great.
Rate (d) is 1/2 of rate (b) and the concentration of cisplatin is also 1/2 as great.
Rate (e) is 1/2 of rate (c) and the concentration of cisplatin is also 1/2 as great.
The rate of reaction is proportional to the concentration of cisplatin.

12.3 k $= Ae^{-Ea/RT} = (6.31 \times 10^8 \text{ L/mol} \cdot \text{s}) \times e^{-10000 \text{ J/mol/(8.314 J/K mol} \times 370 \text{ K)}}$

$= 2.44 \times 10^7$ L/mol · s

rate $=$ k $[NO][O_3] = (2.44 \times 10^7 \text{ L/mol} \cdot \text{s})(0.0010 \text{ M})(0.00050 \text{ M})$

$= 12$ mol/L · s

12.4 (a) The concentration of a homogeneous catalyst must appear in the rate equation.
(b) A catalyst does not appear in the net equation for a reaction.
(c) A homogeneous catalyst must always be in the same phase as the reactants.

12.5 100 tons $\times \frac{1000 \text{ kg}}{\text{ton}} \times \frac{1000 \text{ g}}{\text{kg}} \times \frac{1 \text{ mol}}{104.46 \text{ g}} \times \frac{6.022 \times 10^{23} \text{ molecules}}{1 \text{ mol}} \times \frac{100{,}000 \text{ ozone molecules}}{1 \text{ CF}_3\text{Cl molecules}} = 5.76 \times 10^{34}$ ozone molecules

Chapter 13

13.1 The mixture is *not* at equilibrium, but the reaction is so slow that there is no change in concentrations. You could show that the system was not at equilibrium by providing a catalyst or by raising the temperature to speed up the reaction.

13.2 (a) The new mixture is not at equilibrium because the quotient [*trans*]/[*cis*] is different from the equilibrium constant (1/2 as great).
(b) The rate *cis* $\rightarrow$ *trans* remains the same as before because [*cis*] remains the same. The rate *trans* $\rightarrow$ *cis* is only half as great as before because [*trans*] is half as great.
(c) The value of K is 1.65 at 500 K, which means that [*trans*] $= 1.65$ [*cis*].
(d) 0.165 mol/L

13.3 (a) $K_2 = \frac{[O_3]^2}{[O_2]^3} = K_1{}^2 = (2.5 \times 10^{-29})^2 = 6.2 \times 10^{-58}$
(b) $K_3 = 1/K_2 = 1/(6.2 \times 10^{-58}) = 1.6 \times 10^{57}$

13.4 Reactions (a) and (c) are product-favored. Reaction (b) is least product-favored, followed by (d), (c), and (a).

13.5 $[Ag^+]$ would be larger for AgCl because the equilibrium constant, though small, is larger than for dissolving AgI.

13.6 (a) $\Delta G° = -300.194$ kJ; $K = 4.2 \times 10^{52}$
(b) $\Delta G° = 457.144$ kJ; $K = 3.1 \times 10^{-80}$
(c) $\Delta G° = -21$ kJ; $K = 5.6 \times 10^{3}$
(d) $\Delta G° = 47.65$ kJ; $K = 3.8 \times 10^{-9}$
When $\Delta G°$ is negative, K is large (>1); when $\Delta G°$ is positive, K is small (<1). Both $\Delta G°$ and K indicate whether a reaction is product-favored.

13.7 (a) The equilibrium constant is large and equals $1/(1.8 \times 10^{-5}) = 5.6 \times 10^4$.
(b) Ammonium ion and hydroxide ion react extensively because K is large.
(c) The NH_4^+ ions and OH^- ions would react to form NH_3 and water.
(d) 5.6×10^4

13.8 $PCl_5(s) \rightleftharpoons PCl_3(g) + Cl_2(g)$
(a) Adding Cl_2 will shift the equilibrium to the left because $[Cl_2]$ will be larger.
(b) Adding PCl_3 will shift the equilibrium to the left because of the increased $[PCl_3]$.
(c) Adding $PCl_5(s)$ will not shift the equilibrium, because $[PCl_5]$ does not appear in the equilibrium constant expression $K = [PCl_3][Cl_2]$.

13.9 If $[HI] = 3 \times 0.0963$, $[H_2] = 3 \times 0.102$, and $[I_2] = 3 \times 0.00183$, then

$$K = \frac{[H_2][I_2]}{[HI]^2} = \frac{(0.306)\,(0.00549)}{(0.2889)^2} = 0.0201$$

which agrees within one in the last significant figure. The system is at equilibrium.

13.10 $[N_2O_4] = 0.292 \times 4/1.33 = 0.878$
$[NO_2] = 4.1 \times 10^{-2} \times 4/1.33 = 0.123$

Reaction should shift to left.	N_2O_4	$\rightleftharpoons$ 2 NO_2
Initial concentration	0.878	0.123
Change on reaction	x	$-2x$
Equilibrium concentration	$0.878 + x$	$0.123 - 2x$

$$5.88 \times 10^{-3} = \frac{(0.123 - 2x)^2}{0.878 + x}$$

$$5.16 \times 10^{-3} + 5.88 \times 10^{-3}x = 1.51 \times 10^{-2} - 0.492x + 4x^2$$

$$4x^2 - 0.498x + 9.94 \times 10^{-3} = 0$$

$$x = \frac{0.498 \pm \sqrt{0.248 - 0.159}}{8} = \frac{0.498 \pm 0.298}{8}$$

$x_- = 0.0250$
$x_+ = 0.0995$, which gives a negative $[NO_2]$
$[NO_2] = 0.123 - 2 \times 0.0250 = 0.073$ (less than 3×0.041)
$[N_2O_4] = 0.878 + 0.0250 = 0.903$ (more than 3×0.292)

13.11 (a) $PV = nRT$ $\quad P = \frac{n}{V}RT = cRT$ $\quad c = P/RT$

(b) $$K = \frac{c_{N_2O_4}}{c^2_{NO_2}} = \frac{\dfrac{P_{N_2O_4}}{RT}}{\left(\dfrac{P_{NO_2}}{T}\right)^2} = \frac{P_{N_2O_4}}{P^2_{NO_2}} \times RT$$

Let $K_p = \dfrac{P_{N_2O_4}}{P^2_{NO_2}}$;

then $$K_p = \frac{K}{RT} = \frac{1.7 \times 10^2 \text{ L/mol}}{0.0821 \dfrac{\text{L atm}}{\text{mol K}} \times 298 \text{ K}} = 6.9/\text{atm}$$

Note that here we have used the value of R that you are probably familiar with, 0.0821 L atm/mol · K, because this value connects concentration units (mol/L) and pressure units (atm). Also, we have included appropriate concentration and pressure units in the equilibrium constant.

13.12 Part (a) will be solved in detail and only answers will be given for parts (b), (c), and (d).
(a) $\Delta H° = -238.66 - (-110.525) = -128.14$ kJ
$\Delta S° = 126.8 - 197.674 - 2\,(130.684) = -332.2$ J/K
Since $\Delta G° = \Delta H° - T\Delta S°$, the T at which $\Delta G° = 0$ separates reactant-favored from product-favored.

$$\Delta H° - T\Delta S° = 0 \qquad T = \frac{\Delta H°}{\Delta S°} = \frac{-128{,}140 \text{ J}}{-332.2 \text{ J/K}} = 385.7 \text{ K} = 386 \text{ K}$$

(b) 835 K (c) 5443 K (d) 464 K

Chapter 14

14.1 $$\text{g } SO_2 = (2.7 \times 10^8 \text{ molecules})\left(\frac{64.06 \text{ g } SO_2}{6.02 \times 10^{23} \text{ molecules}}\right) = 2.9 \times 10^{-14}$$

14.2 First, gas molecules are far apart. This allows most light to pass through. Second, molecules are much smaller than the wavelengths of visible light. This means that the waves are not reflected or diffracted by the molecules.

14.3 As more gas molecules are added to a container of fixed volume, there will be more collisions of all of the gas molecules with the container walls. This causes the observed pressure to increase.

14.4 Arranging these from slowest to fastest average speeds at 25 °C, which is the order of decreasing mass, we have SF_6, Cl_2, NH_3, H_2

14.5 For a sample of helium, the plot would look like the curve marked He in Figure 14.7. When an equal number of molecules of argon, a heavier gas, are added to the helium, the distribution of molecular speeds would look like the sum of the curves marked He and O_2 in Figure 14.7, except that the curve for Ar would have its peak a little to the left of the O_2 curve.

14.6 (a) The balloon placed in the freezer will be smaller than the one kept at room temperature because its sample of helium is colder and the gas molecules will occupy a smaller volume and exert less pressure.
(b) Upon warming, the helium balloon that had been in the freezer will be either the same size as the balloon kept at room temperature or perhaps slightly larger due to the fact that there is a greater chance that He atoms leaked out of the room temperature balloon during the time the other balloon was kept in the freezer. This would be caused by the faster moving He atoms in the room temperature balloon having more chances to escape out of tiny openings in the balloon's walls.

14.7 The gas in the shock absorbers will be more highly compressed. The gas molecules will be closer together. The gas molecules will collide with the walls of the shock absorber more often and the pressure exerted will be larger.

14.8 Increasing the temperature of a sample of gas causes the gas molecules to move faster, on average. This means that each collision with the container walls involves greater force, because on average a molecule is moving faster and hits the wall harder. If the container remained the same (constant volume), there would also be more collisions with the container wall because faster moving molecules would hit the walls more often. Increasing the volume of the container, on the other hand, requires that the faster-moving molecules must travel a greater distance before they strike the container walls. Increasing the volume enough would just balance the greater numbers of harder collisions caused by increased temperature. To maintain a constant volume requires that the pressure increase to match the greater pressure due to more, harder collisions of gas molecules with the walls.

14.9 Two mol of O_2 gas are required for the combustion of one mol of methane gas. If air were pure O_2, the oxygen delivery tube would need to be twice as large as the delivery tube for methane. Since air is only 1/5 O_2, the air delivery tube would need to be 10 times larger than the methane delivery tube to ensure complete combustion.

14.10 Using the ratio of 100 balloons/26.8 g He, calculate the number of balloons 41.8 g of He can fill.

$$\text{Balloons} = (41.8 \text{ g He})\left(\frac{100 \text{ balloons}}{26.8 \text{ g He}}\right) = 155 \text{ balloons}$$

So this much helium will fill more balloons than you need to fill.

14.11 The kinetic molecular theory says nothing about the identity of the gas molecules, so equal numbers of O_2 and H_2, or O_2 and N_2, or any other pair of different gases, in separate containers, would exert the same pressures if the numbers of molecules were the same and the temperatures and volumes of the containers were the same.

14.12 (1) Increase the pressure.
(2) Decrease the temperature.
(3) Remove some of the gas by reaction to form a non-gaseous product.

14.13 Density Cl_2 at 25 °C and 0.750 atm $= \dfrac{PM}{RT} =$

$$\frac{(0.750 \text{ atm})(70.905 \text{ g/mol})}{0.0821 \text{ L atm/mol K})(298 \text{ K})} = 2.17 \text{ g/L}$$

Density SO_2 at 25 °C and 0.750 atm $= \dfrac{PM}{RT} =$

$$\frac{(0.750 \text{ atm})(64.06 \text{ g/mol})}{(0.0821 \text{ L atm/mol K})(298 \text{ K})} = 1.96 \text{ g/L}$$

Density Cl_2 at 35 °C and 0.750 atm $= \dfrac{PM}{RT} =$

$$\frac{(0.750 \text{ atm})(70.905 \text{ g/mol})}{(0.0821 \text{ L atm/mol K})(308 \text{ K})} = 2.10 \text{ g/L}$$

Density SO_2 at 25 °C and 2.60 atm $= \dfrac{PM}{RT} =$

$$\frac{(2.60 \text{ atm})(64.06 \text{ g/mol})}{(0.0821 \text{ L atm/mol K})(298 \text{ K})} = 6.81 \text{ g/L}$$

14.14 Density of He $= 1.23 \times 10^{-4}$ g/mL
Density of Li = 0.53 g/mL
Since the density of He is so much less than that of Li, the atoms in a sample of He must be much farther apart than the atoms in a sample of Li. This idea is in keeping with the general principle of the KMT that the particles making up a gas are far apart from one another.

14.15 A 50–50 mixture of N_2 and O_2 would have less N_2 in it than does air. Since N_2 molecules have less mass than O_2 molecules, this 50–50 mixture would have a lower density than that of air.

14.16 (a) If lowering the temperature causes the volume to decrease, by PV = nRT, the pressure can be assumed to be constant. The value of n is unchanged. Since both P and n remain unchanged, the partial pressures of the gases in the mixture remain unchanged.
(b) When the total pressure of a gas mixture increases, the partial pressure of each gas in the mixture increases because the partial pressure of each gas in the mixture is the product of the mole fraction for that gas and the total pressure.

14.17 In Problem Solving Example 14.10, the value of n depends directly on the measured pressure, P. Intermolecular attractions in a real gas would cause the measured P to be slightly smaller than that for an ideal gas. The lower value for P would cause the calculated number of moles to be somewhat smaller. Using this value of slightly smaller value of n in the denominator would cause the calculated molar mass to be a little larger than it should be.

14.18 Amount of O_2 in liquid state = (1.41 g/mL)(5.0 L)(1000 mL/L)(1 mol/31.998 g O_2) = 220 mol O_2
vol at STP = (220 mol)(22.4 L/mol) $= 4.9 \times 10^3$ L

14.19 Volume of Ar at STP = (250,000 t)(2000 lb/t)(453.6 g/lb)(1 mol Ar/39.948 g)(22.4 L/mol) $= 1.3 \times 10^{11}$ L

14.20 (a) HO· + H·
(b) CH_3· + HOH
(c) HO· + ·O·

14.21 Mass of S burned per hr $= (3.06 \times 10^6 \text{ kg})(0.04) = 1 \times 10^5$ kg

Mass of SO_2 per hr $= (1 \times 10^5 \text{ kg})\left(\dfrac{64.06 \text{ kg } SO_2}{32.07 \text{ kg S}}\right) =$ 2×10^5 kg

Mass of SO_2 per year $= (2 \times 10^5$ kg/hr)(8760 hr/yr) $=$ 2×10^9 kg/yr.

14.22 Vol percent SO_2 = (5 parts $SO_2/10^6$ parts air) $\times$ 100% $=$ 5×10^{-4} vol%

14.23 Industrial smog:
SO_2 from burning sulfur-containing fuels
NO from combustion processes
C (soot) from incomplete combustion
Photochemical smog:
Ozone from the photodecomposition of NO_2
NO from combustion processes
NO_2 from the oxidation of NO

14.24 1 metric ton = 1000 kg = 10^6 g = 1 mg

$$\text{Mass of HNO}_3 = (400 \text{ Mg N}_2)\left(\frac{2 \text{ Mg NO}}{\text{Mg N}_2}\right)\left(\frac{1 \text{ Mg NO}_2}{1 \text{ Mg NO}}\right)\left(\frac{1 \text{ Mg HNO}_3}{1 \text{ Mg NO}_2}\right) = 800 \text{ Mg HNO}_3$$

14.25 $NO_2 \xrightarrow{h\nu} NO + O$

$O_3 \xrightarrow{h\nu} O_2 + O \qquad O + O_2 \longrightarrow O_3$

Chapter 15

15.1 The London forces are greater between bromoform molecules than between chloroform molecules because the bromoform molecules have more electrons. This stronger intermolecular attraction causes the molecules to exhibit a larger surface tension. (The dipole in each molecule contributes less than the London forces to the intermolecular attractions.)

15.2 Water and glycerol would have similar surface tensions because of extensive hydrogen bonding. Octane and decane would have similar surface tensions because both are alkane hydrocarbons.

15.3 The evaporating water carries with it thermal energy from the water inside the pot. In addition, a large quantity of thermal energy is required to cause the water to evaporate. Much of this thermal energy comes from the water inside the pot.

15.4 Estimated ΔH_{vap} for Kr $\simeq$ 20 kJ/mol based on HBr, Cl_2, and C_4H_{10}, but based on Xe, the value is probably closer to 10 kJ/mol

Estimated ΔH_{vap} for $NO_2 \simeq$ 20 kJ/mol based on HI and Cl_2

15.5 (a) Bromine molecules have more electrons than chlorine molecules. Therefore, bromine molecules are held together by stronger intermolecular attractions.

(b) Ammonia molecules are attracted to one another by hydrogen bonds. This causes ammonia to have a higher boiling point than that of methane, which has no hydrogen bonding.

15.6 (a) 62 °C (b) 0 °C (c) 80 °C

15.7 Bubbles form within a boiling liquid when the vapor pressure of the liquid equals the ambient pressure (the pressure of the surroundings of the liquid sample). The bubbles are actually filled with vapor of the boiling liquid. One way to prove this would be to trap some of these bubbles and allow them to condense. They would condense to form the liquid that had boiled.

15.8 2 moles of liquid bromine crystallizing liberates 21.59 kJ of heat

1 mole of water crystallizing liberates 6.02 kJ of heat

15.9 High humidity conditions make the evaporation of water or the sublimation of ice less favorable. Under these conditions the sublimation of ice required to make the frost-free refrigerator work is less favorable, so the defrost cycle is less effective.

15.10 The impurity molecules are less likely to be converted from the solid phase to the vapor phase. This causes them to be left behind as the molecules that sublime go into the gas phase and then condense at some other place. The molecules that condense are almost all of the same kind, so the sublimed sample is much more pure than the original.

15.11 The triple point of CO_2, from Figure 15.11, is at 5.2 atm and −57 °C. Increasing the pressure from 5.2 atm while holding the temperature at −57 °C would result in solid CO_2 being formed.

15.12 (a) If liquid CO_2 is slowly released from a cylinder of CO_2, gaseous CO_2 is formed. The temperature remains constant (at room temperature) because there is time for energy to be transferred from the surroundings to separate the CO_2 molecules against their intermolecular attractions. This can be seen from the phase diagram as the phase changes from liquid to vapor as the pressure decreases.

(b) If the pressure is suddenly released, the attractive forces between a large number of CO_2 molecules must be overcome, which requires energy. This energy comes from the surroundings as well as the CO_2 molecules themselves. This causes the temperature of both the surroundings and the CO_2 molecules to decrease. (These are the other factors that must be considered.) On the phase diagram for CO_2, a decrease in both temperature and pressure can move into a region where only solid CO_2 exists.

15.13 It is predicted that a small concentration of gold will be found in the lead and that a small concentration of lead will be found in the gold. This will occur because of the movement of the metals atoms with time, as predicted by the KMT.

15.14 One Po atom belongs to its unit cell. Two Li atoms belong to its unit cell. Four Ca atoms belong to its unit cell. (See next page.)

15.15 Each Cs^+ ion at the center of the cube has eight Cl^- ions as neighbors. One eighth of each of the Cl^- ions belongs to that Cs^+ ion. So the formula for this salt must be a 1:1 ratio of Cs^+ ions to Cl^- ions, or CsCl.

15.16 Cooling a liquid above its freezing point causes the temperature to decrease. When the liquid begins to solidify, energy is released as atoms, molecules, or ions move closer together to form in the solid crystal lattice. This causes the temperature to remain constant until all of the molecules in the liquid have positioned themselves in the lattice. Further cooling then causes the temperature to decrease. The shape of this curve is common to all substances that can exist as liquids.

15.17 Increasing strength of metallic bonding is related to increasing numbers of valence electrons. In the transition metals, the presence of d-orbital electrons causes stronger metallic bonding. Beyond a half-filled set of d-orbitals, however, extra electrons have the effect of decreasing the strength of metallic bonding. See Figure 15.21.

15.18 Diamond is more dense than graphite, so high pressure is required to convert the carbon atoms from the less dense form. This is an application of Le Chatelier's principle. High temperatures would allow the atoms to move more rapidly relative to one another and therefore give them a greater opportunity to rearrange.

15.19 These structures will be drawn by the student.

Simple cubic

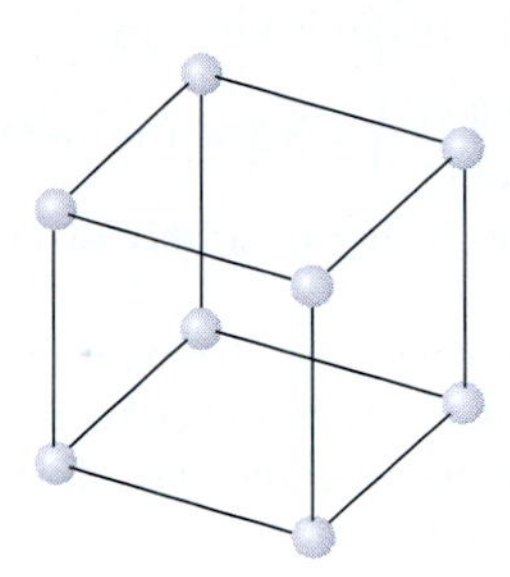

= 1 atom
Each of the 8 atoms
contributes 1/8 to unit cell

Body-centered cubic

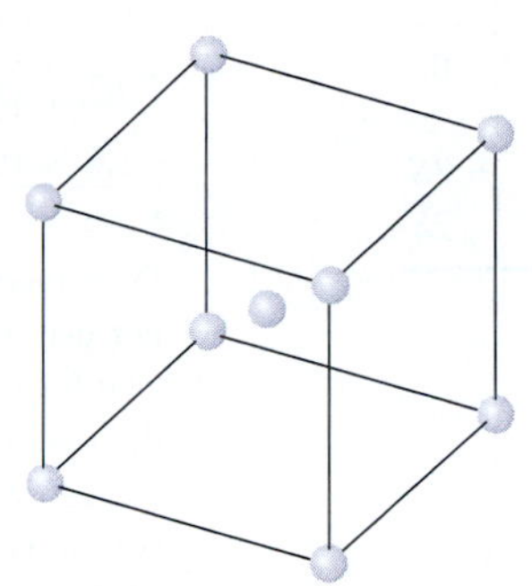

= 2 atoms
8/8 from corner atoms
+ 1 atom at center

Face-centered cubic

= 4 atoms
8/8 from corner atoms
+ 6/2 from each atom on
the 6 faces contributing 1/2 atom

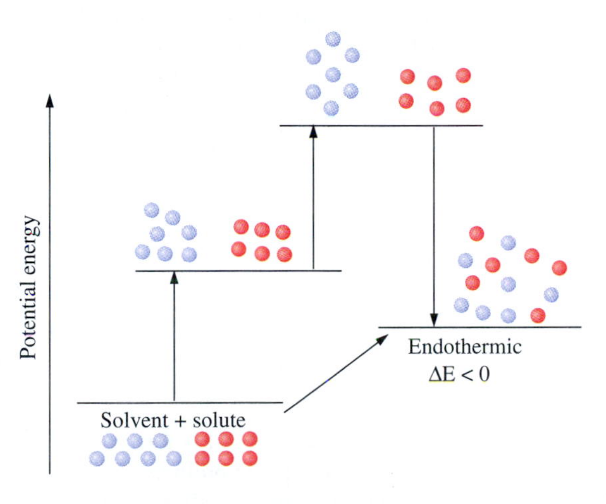

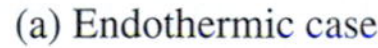

(a) Endothermic case

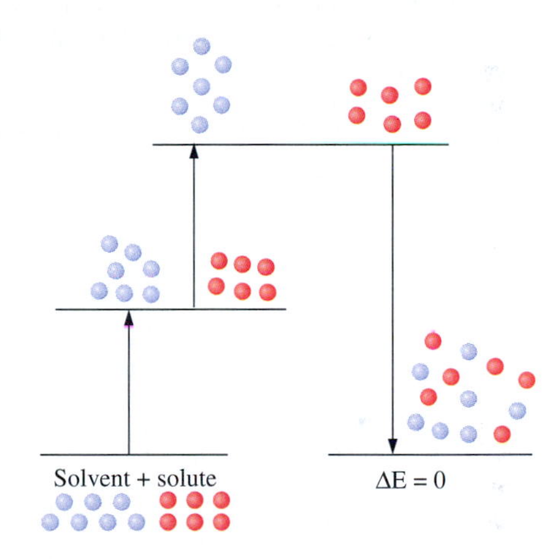

(b) No net energy change case

Chapter 16

16.1 Hydrocarbons, including those with high molar masses like the waxes, have no tendency to associate with water molecules. In fact, water molecules would prefer association with one another much more than they would with the surface of a hydrocarbon wax. This causes the water molecules to act as though they are being repelled from the hydrocarbon surface.

16.2 The other two would be those where the net energy change was positive, and zero. These are shown at the top of the page.

16.3 Methanol is more water soluble than is octanol, but octanol is more soluble in gasoline. The octanol molecule is more hydrocarbon-like, so this explains its solubility in gasoline. The methanol molecule is more water-like and this explains its greater solubility in water.

16.4 Iron is not soluble in either water or benzene because of the strong metallic bonding that exists between the atoms in a sample of iron. There are only weak attractive forces between iron atoms and water molecules and between iron atoms and benzene molecules, and so both iron and water and iron and benzene remain separate.

16.5 Hot solvent would cause more of the solute to dissolve because Le Chatelier's principle states that at higher temperature an equilibrium will shift in the endodermic direction. Dissolving CsCl is endothermic, so more will be dissolved.

16.6 $S = [Ag^+] = [Cl^-] \qquad S^2 = K_{sp} = 1.8 \times 10^{-10}$

$S = 1.3 \times 10^{-5}$ mol/L

$$S = (1.3 \times 10^{-5} \text{ mol/L})\left(\frac{143.32\text{g}}{\text{mol}}\right) = 1.9 \times 10^{-3} \text{ g/L}$$

16.7

	HgI_2 $\rightleftarrows$	Hg^{2+}	+	$2\ I^-$
Initially		0		0
Change due to dissolving		+ *S*		+ 2*S*
At equilibrium		*S*		2*S*

$$K_{sp} = [Hg^{2+}][I^-]^2 = (S)(2S)^2$$

$$S = \sqrt[3]{\frac{K_{sp}}{4}} = \sqrt[3]{\frac{4.0 \times 10^{-29}}{4}} = 2.2 \times 10^{-10}\ \text{mol/L}$$

$$S = (2.2 \times 10^{-10}\ \text{mol/L})\left(\frac{454.39\ \text{g}}{\text{mol}}\right) = 1.0 \times 10^{-7}\ \text{g/L}$$

16.8 HgI_2, with a K_{sp} of 4.0×10^{-29}

16.9 There would be no change on the equilibrium if more AgCl were added because the solution is already saturated in Ag^+ and Cl^- ions.

16.10 8.7×10^{-9} g Hg/kg $= \dfrac{8.7 \times 10^{-9}\ \text{g Hg}}{1\ \text{kg sample}} \times \left(\dfrac{1\ \text{kg sample}}{10^3\text{g sample}}\right) \times$

$$\left(\frac{10^9\ \text{g sample}}{1\ \text{billion sample}}\right) = \frac{8.7 \times 10^{-3}\ \text{g Hg}}{1\ \text{billion g sample}} = 8.7 \times 10^{-3}\ \text{ppb}$$

$$\frac{8.7 \times 10^{-9}\ \text{g}}{1\ \text{kg sample}} \times \left(\frac{1\ \text{kg sample}}{10^3\ \text{g sample}}\right)\left(\frac{10^{12}\ \text{g sample}}{1\ \text{trillion g sample}}\right) = 8.7\ \text{ppt}$$

16.11 0.025 ppm Pb means that for every liter (1 kg) of water, there are 0.025 mg, or 25 μg of Pb. Using this factor,

$$\text{Volume of water} = 100.0\ \mu\text{g Pb}\left(\frac{1\ \text{L}}{25\ \mu\text{g Pb}}\right) = 4.0\ \text{L}$$

16.12 $\text{Beryllium} = 3\ \text{L}\left(\dfrac{0.004\ \text{mg Be}}{1\ \text{L}}\right) = 0.012\ \text{g Be}$

16.13 Based on the lowest MCL values,
Metals; Be and Pb
Nonmetals; nitrite ion
Volatile organics; hexachlorobenzene and vinyl chloride
Herbicides, pesticides, and PCBs; ethylene dibromide, endrin, and lindane

16.14 The mass of this pollutant in 1 L of water is given by

$$\left(\frac{2.5\ \text{g}\ C_2H_6O}{10^6\ \text{g water}}\right)\left(\frac{1000\ \text{g}\ H_2O}{1\ \text{L}}\right) = 0.0025\ \text{g}\ C_2H_6O$$

The concentration of C_2H_6O in mol/L is given by

$$\left(\frac{0.0025\ \text{g}\ C_2H_6O}{1\ \text{L}}\right)\left(\frac{1\ \text{mol}\ C_2H_6O}{46.06\ \text{g}\ C_2H_6O}\right) = 5.4 \times 10^{-5}\ \text{mol/L}$$

The balanced equation for the oxidation of C_2H_6O is

$$C_2H_6O(aq) + 3\ O_2(g) \longrightarrow 2\ CO_2(g) + 3\ H_2O(\ell)$$

Using this stoichiometric factor, the concentration of O_2 required to oxidize this concentration of C_2H_6O is given by

$$\left(\frac{5.4 \times 10^{-5}\ \text{mol}\ C_2H_6O}{1\ \text{L}}\right)\left(\frac{3\ \text{mol}\ O_2}{1\ \text{mol}\ C_2H_6O}\right) = 1.6 \times 10^{-4}\ \text{mol}\ O_2/\text{L}$$

Finally, the mass of O_2/L (the BOD) is given by

$$\left(\frac{1.6 \times 10^{-4}\ \text{mol}\ O_2}{1\ \text{L}}\right)\left(\frac{32.0\ \text{g}\ O_2}{1\ \text{mol}\ O_2}\right) = 5.2 \times 10^{-3}\ \text{g}\ O_2/\text{L}$$

Dissolved oxygen is about 8–10 mg/L, so this BCD would not use up all of the O_2.

16.15 The lower temperature water containing the same amount of dissolved organic compounds would be better because more oxygen would be available in the water (because of its greater solubility) at the lower temperature.

16.16 Molarity is based on the volume of the solution, which will vary with temperature changes. Molality is based on the moles of solute per kg of solvent. This ratio cannot change with changes in temperature.

16.17 $\Delta T_b = (2.53\ °C \cdot \text{kg/mol})(0.10\ \text{mol/kg}) = 0.25\ °C$
The boiling point of the solution is 80.10 + 0.25 = 80.35 °C

16.18 First, calculate the required molality of the solution that would have a freezing point of −40 °C.

$$\Delta T_f = -40\ °C = (-1.86\ °C \cdot \text{kg/mol}) \times m$$

$$m = \frac{-40\ °C}{-1.86\ °C \cdot \text{kg/mol}} = 21.5\ \text{mol/kg}$$

To protect 4 kg of water from this freezing temperature, you would need 4 × 21.5 mol of ethylene glycol, or 86 mol.

$$\text{Volume ethylene glycol} = 86\ \text{mol}\left(\frac{62.1\ \text{g}}{\text{mol}}\right)\left(\frac{1\ \text{mL}}{1.113\ \text{g}}\right)\left(\frac{1\ \text{L}}{1000\ \text{mL}}\right) = 4.8\ \text{L}$$

16.19 On a weight basis, NaCl is the most effective.

16.20 The 0.02 *m* solution of ordinary soap would contain more particles since a soap is a salt of a fatty acid while sucrose is a non-electrolyte.

Chapter 17

17.1 More water molecules are available per NH_3 molecules in a very dilute solution of NH_3.

17.2 (a) $H_3O^+(aq) + CN^-(aq)$ (b) $H_3O^+(aq) + Br^-(aq)$
(c) $CH_3NH_3^+(aq) + OH^-(aq)$ (d) $(CH_3)_2NH_2^+(aq) + OH^-(aq)$

17.3 (a) Bronsted acid (e) Bronsted acid
(b) Bronsted base (f) Bronsted acid
(c) Bronsted acid (g) Bronsted base
(d) Bronsted base
Only HBr is a strong acid.

17.4

Acid	Conjugate Base	Base	Conjugate Acid
$H_2PO_4^-$	HPO_4^{2-}	OH^-	H_2O
HPO_4^{2-}	PO_4^{3-}	NH_2^-	NH_3
HSO_4^-	H_2SO_4	ClO_4^-	$HClO_4$
HF	F^-	Br^-	HBr

17.5

```
     H   :O.
     ..  .:·
   H:C:C
     ..  ·.
     H   .O.
            H
```

Acetic acid

```
     H   :O.
     ..  .:·  H+
   H:C:C
     ..  ·.
     H   .O.
            H
```

Acetic acid acting as a base in a very strong acid like sulfuric acid The nonbonding electron pairs on the C=O group can accept a proton from a very strong acid like pure sulfuric acid.

17.6 (a) Step 1 $HOOC{-}COOH(aq) + H_2O(\ell) \rightleftharpoons H_3O^+(aq) + HOOC{-}COO^-(aq)$

Step 2 $HOOC{-}COO^-{-}(aq) + H_2O(\ell) \rightleftharpoons H_3O^+(aq) + {}^-OOC{-}COO^-(aq)$

(b) Step 1 $H_3PO_4(aq) + H_2O(\ell) \rightleftharpoons H_3O^+(aq) + H_2PO_4^-(aq)$

Step 2 $H_2PO_4^-(aq) + H_2O(\ell) \rightleftharpoons H_3O^+(aq) + HPO_4^{2-}(aq)$

Step 3 $HPO_4^{2-}(aq) + H_2O(\ell) \rightleftharpoons H_3O^+(aq) + PO_4^{3-}(aq)$

17.7 Being negatively charged, the HSO_4^- ion has a lower tendency to lose a positively charged proton because of the electrostatic attractions of opposite charges.

17.8 (a) Step 1 $PO_4^{3-}(aq) + H_2O(\ell) \rightleftharpoons OH^-(aq) + HPO_4^{2-}(aq)$

Step 2 $HPO_4^{2-}(aq) + H_2O(\ell) \rightleftharpoons OH^-(aq) + H_2PO_4^-(aq)$

Step 3 $H_2PO_4^-(aq) + H_2O(\ell) \rightleftharpoons OH^-(aq) + H_3PO_4(aq)$

(b) Step 1 $CO_3^{2-}(aq) + H_2O(\ell) \rightleftharpoons OH^-(aq) + HCO_3^-(aq)$

Step 2 $HCO_3^-(aq) + H_2O(\ell) \rightleftharpoons OH^-(aq) + H_2CO_3(aq)$

17.9 Ethylenediamine has a condensed structure of $H_2N{-}CH_2{-}CH_2{-}NH_2$. One molecule of this base can react with two protons because ethylenediamine has two basic $-NH_2$ groups, and since a molecule of HCl furnishes one proton, the complete reaction would be two moles of HCl for every mole of ethylenediamine. The reaction is

$$H_2N{-}CH_2{-}CH_2{-}NH_2(aq) + 2\ HCl(aq) \rightleftharpoons {}^+H_3N{-}CH_2{-}CH_2{-}NH_3^+(aq) + 2\ Cl^-(aq)$$

17.10 The larger values of K_w with increasing temperature indicates that the equilibrium is shifted toward the right. This means the reaction is endothermic.

17.11 The pH values of 0.1 M solutions of these two strong acids would be essentially the same since they both are 100% ionized, resulting in $[H_3O^+]$ values that are the same.

17.12 $HA_1(aq) + H_2O(\ell) \rightleftharpoons H_3O^+(aq) + A_1^-(aq) \quad K_a = 10^{-4}$

$HA_2(aq) + H_2O(\ell) \rightleftharpoons H_3O^+(aq) + A_2^-(aq) \quad K_a = 10^{-7}$

HA_2 is the weaker acid. The reaction involving HA_2 is more reactant-favored; its K_a is smaller. The reaction involving HA_1 is more product-favored; its K_a is larger.

17.13 $H_2SO_3(aq) + H_2O(\ell) \rightleftharpoons H_3O^+(aq) + HSO_3^-(aq)$

$$K_a = \frac{[H_3O^-][HSO_3^-]}{[H_2SO_3]}$$

$HF(aq) + H_2O(\ell) \rightleftharpoons H_3O^+(aq) + F^-(aq)$

$$K_a = \frac{[H_3O^+][F^-]}{[HF]}$$

$HCN(aq) + H_2O(\ell) \rightleftharpoons H_3O^+(aq) + CN^-(aq)$

$$K_a = \frac{[H_3O^+][CN^-]}{[HCN]}$$

$HSO_3^-(aq) + H_2O(\ell) \rightleftharpoons H_3O^+(\ell) + SO_3^{2-}(aq)$

$$K_a = \frac{[H_3O^+][SO_3^{2-}]}{[HSO_3^-]}$$

$NO_2^-(aq) + H_2O(\ell) \rightleftharpoons HNO_2(aq) + OH^-(aq)$

$$K_b = \frac{[HNO_2][OH^-]}{[NO_2^-]}$$

$HS^-(aq) + H_2O(\ell) \rightleftharpoons H_2S(aq) + OH^-(aq)$

$$K_b = \frac{[H_2S][OH^-]}{[HS^-]}$$

$CN^-(aq) + H_2O(\ell) \rightleftharpoons HCN(aq) + OH^-(aq)$

$$K_b = \frac{[HCN][OH^-]}{[CN^-]}$$

$PO_4^{3-}(aq) + H_2O(\ell) \rightleftharpoons HPO_4^{2-}(aq) + OH^-(aq)$

$$K_b = \frac{[HPO_4^{2-}][OH^-]}{[PO_4^{3-}]}$$

17.14 B_1 is the stronger base because its K_b value indicates that its reaction as a base is more product-favored. The 0.1 M solution of B_1 would have the higher pH value.

17.15 See below for a small table for this reaction.

(For Question 17.15)	$Ni(H_2O)_6^{2+}(aq) + H_2O(\ell)$ $\rightleftharpoons$	$Ni(H_2O)_5(OH)^+(aq)$ +	$H_3O^+(aq)$
Initial concentration	0.15 M	0	10^{-7} M
Change in concentration on reaction	$-x$	$+x$	$+x$
Concentration at equilibrium	$0.15 - x$	x	x

Substituting these values in the equilibrium constant expression, and simplifying $0.15\text{-}x$ to be 0.15 because the value of K_a is so small,

$$K_a = \frac{[Ni(H_2O)_5(OH)^+][H_3O^+]}{[Ni(H_2O)_6{}^{2+}]} = \frac{(x)(x)}{(0.15\text{-}x} \sim \frac{x^2}{0.15} = 2.5 \times 10^{-11}$$

Solving for x, which is the $[H_3O^+]$,

$x = \sqrt{(0.15)(2.5 \times 10^{-11})} = 1.9 \times 10^{-6}$

So, the pH of this solution is $-\log(1.9 \times 10^{-6}) = 5.72$

17.16 See below for a small table for this reaction.

(For Question 17.16)	$NH_3(aq)$ + $H_2O(\ell)$ $\rightleftarrows$	$NH_4{}^+(aq)$ +	$OH^-(aq)$
Initial concentration	6.6 M	0 M	10^{-7} M
Change in concentration at reaction	$-x$	$+x$	$+x$
Concentration at equilibrium	$6.6 - x$	x	x

Using these values in the equilibrium constant expression and making the assumption that $6.6 - x$ will be approximately equal to 6.6,

$$K_b = \frac{[NH_4{}^+][OH^-]}{[NH_3]} = \frac{x^2}{6.6} = 1.8 \times 10^{-5}$$

$x = \sqrt{(6.6)(1.8 \times 10^{-5})} = 0.011$

Since $x = [OH^-]$, the pH of the solution can be calculated from pOH $= -\log(OH^-)$ and then pH $= 14 -$ pOH.

pOH $= -\log(0.011) = -1.96$ pH $= 14.00 - 1.96 = 12.04$

17.17 HS^- is the stronger base by a factor of 5.6×10^3.

17.18 $K_{b(\text{phenolate ion})} = \frac{K_w}{K_{a(\text{phenol})}} = \frac{1.0 \times 10^{-14}}{1.3 \times 10^{-10}} = 7.7 \times 10^{-5}$

The base in Table 17.3 that has a K_b the closest to this value is pentaaquohydroxoiron(II) ion, with a K_b of 3.1×10^{-5}.

17.19 In the reaction involving H^- ion, the H^- ion is the stronger base and H_2O is the stronger acid.

In the reaction involving O^{2-} ion, the O^{2-} ion is the stronger base and H_2O is the stronger acid.

In the reaction involving $CO_3{}^{2-}$ ion, the OH^- ion is the stronger base and $HCO_3{}^-$ is the stronger acid.

In the reaction involving NH_3, the OH^- ion is the stronger base and the $NH_4{}^+$ ion is the stronger acid.

If the reactants are both strong as acids and bases, the reaction will be product-favored. If the products are both strong as acids and bases, the reaction will be reactant-favored.

17.20

	acid	+	base		acid	+	base
(a)	H_2O	+	S^{2-}	$\rightleftarrows$	HS^-	+	OH^-
(b)	HPO_4^{2-}	+	OH^-	$\rightleftarrows$	H_2O	+	PO_4^{3-}
(c)	HSO_4^-	+	CO_3^{2-}	$\rightleftarrows$	HCO_3^-	+	SO_4^{2-}
(d)	$H_2PO_4^-$	+	$HCO_3{}^-$	$\rightleftarrows$	H_2CO_3	+	HPO_4^-

17.21 pH should be > 7 because of hydrolysis.

17.22 All three would behave as follows:

Amine · HCl + $H_2O \rightleftarrows H_3O^+ + Cl^-$ + amine

17.23 Ammonium acetate. The pH of a solution of this salt will be 7.

17.24 (a) acidic (b) neutral (c) basic (d) acidic (e) acidic (f) basic

17.25 (a) soluble (b) soluble (c) insoluble (d) soluble (e) soluble

17.26 Neither of the solutions mentioned can act as a buffer because in the first case the acid is strong and its salt is not capable of acid/base reactions. In the second case, the salt does not contain the conjugate base of the weak acid.

17.27 Using the equation pH $= pK_a + \log\frac{[HPO_4{}^{2-}]}{[H_2PO_4{}^-]}$

and the pH of blood as 7.40,

$7.40 = 7.21 + \log\frac{[HPO_4{}^{2-}]}{[H_2PO_4{}^-]}$ $\frac{[HPO_4{}^{2-}]}{[H_2PO_4{}^-]} = 10^{0.19} = 1.55$

17.28 Since CO_2 reacts to form an acid, H_2CO_3, the phosphate ion that is the stronger base, $HPO_4{}^{2-}$, will be used to counteract its presence.

17.29 Using pH $= pK_a + \log\frac{[A^-]}{[HA]}$

(a) pH $= 6.3 + \log(0.25/0.10) = 6.3 + 0.40 = 6.7$

(b) pH $= 7.21 + 0.40 = 7.61$ (c) pH $= 9.25 + 0.40 = 9.65$

17.30 The acid present in the buffer will react with the added base.

The amount of base added is (6.2 g/56.01 g/mol) = 0.111 mol.

The amount of acid present is 0.25 mol. Even though there are three acidic hydrogens in H_3PO_4, the pH will change significantly as soon as all of the H_3PO_4 has been converted to $H_2PO_4{}^-$. However, there is not nearly enough KOH to effect this change.

17.31 Looking at Table 17.3, it appears that by using hexaaquoiron(III) ion and its conjugate base pentaaquohydroxoiron(III) ion, a suitable buffer could be made.

17.32 A buffer needs to be able to counteract the addition of both acid and base. A solution of acetic acid can only react with added base. After a quantity of base has been added, the resulting solution is a buffer, however.

17.33 In the case of NO_2, the nitrogen in one of the NO_2 molecules is oxidized from +4 to +5 (HNO_3) and the nitrogen in the other NO_2 molecule is reduced from +4 to +3 (HNO_2). The S atom in SO_2 does not change oxidation state when reacting with water to form H_2SO_3.

17.34 Strong bases would cause damage to tissue.

17.35 The carbonate ion is too strong a base. Calcium carbonate, being relatively insoluble in water, requires a strong acid to cause it to react quickly enough to produce the CO_2 needed. Strong acids would have an adverse effect on the other ingredients (like the dough) in the bread.

17.36 Set up a small table for the hydrolysis reaction.

	CO_3^{2-} + H_2O	HCO_3^- +	OH^-
Initial concentration	5.2 M	0	10^{-7}
Concentration change due to reaction	$-x$	$+x$	$+x$
Concentration at equilibrium	$5.2 - x$	x	x

Using the K_b expression and substituting the values from the table,

$$K_b = \frac{K_w}{K_a(HCO_3^-)} = \frac{[HCO_3^-][OH^-]}{[CO_3^{2-}]} = \frac{x^2}{5.2 - x} \approx \frac{x^2}{5.2} = 2.1 \times 10^{-4}$$

$$x = [OH^-] = \sqrt{(5.2)(2.1 \times 10^{-4})} = 3.3 \times 10^{-2}$$

$$pOH = -\log(3.3 \times 10^{-2}) = 1.48 \qquad pH = 14.00 - 1.48 = 12.52$$

17.37 One of the ways detergents help remove animal and vegetable oils is to first break some of their ester linkages by base-catalyzed hydrolysis. Petroleum oils do not contain any of these ester linkages, so the detergent molecules can only aid in their removal by dissolving their hydrophilic ends in the oil particles.

17.38 (a) Lewis base (c) Lewis acid and base (e) Lewis acid
(b) Lewis acid (d) Lewis acid (f) Lewis base

Chapter 18

18.1 This is an application of the law of conservation of matter. If the number of electrons gained was different from the number of electrons lost, some electrons must have been created or destroyed.

18.2 Removal of the salt bridge would effectively switch off the flow of electricity from the battery.

18.3 Avogadro's number of electrons is 96,500 coulombs of charge, so it is 96,500 times a coulomb of charge.

18.4 The zinc anode could be weighed before the battery was put into use. After a period of time, the zinc anode could be dried and reweighed. A loss in weight would be interpreted as being caused by the loss of Zn atoms from the surface through oxidation.

18.5 No, because Hg^{2+} ions can oxidize Al metal to Al^{3+} ions. The net cell reaction is $2\ Al(s) + 3\ Hg^{2+}(aq) \rightarrow 2\ Al^{3+}(aq) + 3\ Hg(\ell)$ $E_{net} = +2.51$ V

18.6 For this table:
(a) V^{2+} ion is the weakest oxidizing agent.
(b) Cl_2 is the strongest oxidizing agent.
(c) V is the strongest reducing agent.
(d) Cl^- is the weakest reducing agent.
(e) No, E_{net} for that reaction would be < 0.
(f) No, E_{net} for that reaction would be < 0.
(g) Pb can reduce I_2 and Cl_2.

18.7 In Table 18.1, Sb would be above H_2 and Pb would be below H_2. For Sb, the reduction potential would be between 0.00 and +0.337 V, and for Pb the value would be between 0.00 and −0.14 V.

18.8 $E°$ for this reaction is $-0.763 + 0.0$ V, or -0.763 V. With a $E° < 0$ this reaction is not product-favored as written. The reverse reaction is product-favored and E^0 is + 0.763 V. The value for $\Delta G°$ is given by $\Delta G° = -nFE° = -(2\ \text{mol e}^-) \times \left(\frac{9.65 \times 10^4\ C}{1\ \text{mol e}^-}\right) \times \left(\frac{1\ J}{1V \times 1C}\right) \times (0.763\ V) = -1.47 \times 10^5\ J = 147\ kJ$

18.9 During charging, the reactions at each electrode are reversed. At the electrode that is normally the anode, the charging reaction is

$Cd(OH)_2(s) + 2\ e^- \longrightarrow Cd(s) + 2\ OH^-(aq)$ this is reduction, so this electrode is now a cathode

At the electrode that is normally the cathode, the charging reaction is

$Ni(OH)_2 + OH^-(aq) \longrightarrow NiO(OH)(s) + H_2O(\ell) + e^-$ this is oxidation, so this electrode is now an anode

18.10 Remove the lead cathodes and as much sulfuric acid as you can from the discharged battery. Find some steel and construct a battery with Cl_2 gas flowing across a piece of steel. The two half-reactions would be

$Cl_2(g) + 2\ e^- \rightarrow 2\ Cl^-(aq)$ + 1.36 V

$Pb(s) + SO_4^{2-}(aq) \rightarrow PbSO_4(s) + 2e^-$ + 0.356 V

$E_{net} = 1.36 + 0.356 = 1.71$ V

18.11 Potassium metal was produced at the cathode.

Oxidation reaction: $2\ F^-\ (\text{molten}) \rightarrow F_2(g) + 2\ e^-$

Reduction reaction: $2(K^+\ (\text{molten}) + e^- \rightarrow K(s))$

Net cell reaction:

$$2\ K^+(\text{molten}) + 2\ F^-(\text{molten}) \rightarrow 2\ K(\ell) + F_2(g)$$

18.12 (c) Making 2 mol of Cu from Cu^{2+} would require 4 Faradays of electricity. Two F are required for part (b), and 3 F are needed for part (a).

18.13 First, calculate how many coulombs of electricity are required to make this much aluminum.

$$\text{Charge} = (2000.\ t\ Al)\left(\frac{2000\ lb\ Al}{1\ t\ Al}\right)\left(\frac{454.3\ g\ Al}{1b\ Al}\right)\left(\frac{1\ mol\ Al}{26.9815\ g\ Al}\right)\left(\frac{3\ mol\ e^-}{1\ mol\ Al}\right)\left(\frac{96500\ C}{1\ mol\ e^-}\right) = 1.950 \times 10^{13}\ C$$

Next, using charge x voltage, calculate how many joules are required; then convert to kilowatt hours.

$$\text{Energy} = (1.950 \times 10^{13}\ \text{C})(4.0\ \text{V})\left(\frac{1\ \text{J}}{1\ \text{C} \times 1\ \text{V}}\right)\left(\frac{1\ \text{kWh}}{3.60 \times 10^{6}\ \text{J}}\right) = 2.2 \times 10^{7}\ \text{kWh}$$

18.14 To calculate how much energy is stored in a battery, you need the voltage and the number of coulombs of charge the battery can provide. The voltage is generally given on the battery label. To determine the number of coulombs available, the battery would have to be disassembled and the masses of the chemicals at the cathode and anode determined.

18.15 $$\text{Mass Au} = (0.50\ \text{A})(20.\ \text{min})\left(\frac{60\ \text{s}}{1\ \text{min}}\right)\left(\frac{1\ \text{C}}{1\ \text{A}\cdot\text{s}}\right)\left(\frac{1\ \text{mol e}^-}{96500\ \text{C}}\right)\left(\frac{1\ \text{mol Au}}{3\ \text{mol e}^-}\right)\left(\frac{197\ \text{g Au}}{1\ \text{mol Au}}\right) = 0.41\ \text{g Au}$$

18.16 No. Not all metals.
Three metals that would corrode about as readily as Fe and Al are Zn, Mg, and Cd.
Three metals that do not corrode as readily as Fe and Al are Cu, Ag, and Au. These three metals are used in making coins and jewelry.
Metals fall into these two broad groups because of their relative ease of oxidation compared to the oxidation of H_2. In Table 18.3, you can see this breakdown clearly.

18.17 (b) > (a) > (d) > (c)
Sand by the seashore, (b), would contain both moisture and salts, which would aid corrosion. Moist clay, (a), would contain water, but less dissolved salts. If an iron object were embedded in the clay, its impervious nature might prevent oxygen from getting to the iron, which would also lower the rate of corrosion. Desert sand in Arizona, (d), would be quite dry, and this low moisture environment would not lead to a rapid rate of corrosion. On the moon, (c), there would be a lack of moisture and oxygen. This would lead to a very low rate of corrosion.

Chapter 19

19.1 ${}^{235}_{92}U \rightarrow {}^{4}_{2}He + {}^{231}_{90}Th$; ${}^{231}_{90}Th \rightarrow {}^{0}_{-1}e + {}^{231}_{91}Pa$;
${}^{231}_{91}Pa \rightarrow {}^{4}_{2}He + {}^{227}_{89}Ac$; ${}^{227}_{89}Ac \rightarrow {}^{4}_{2}He + {}^{223}_{87}Fr$;
${}^{223}_{87}Fr \rightarrow {}^{0}_{-1}e + {}^{223}_{88}Ra$

19.2 (a) ${}^{13}_{7}N \rightarrow {}^{13}_{6}C + {}^{0}_{+1}e$;
(b) ${}^{41}_{20}Ca + {}^{0}_{-1}e \rightarrow {}^{41}_{19}K$;
(c) ${}^{90}_{38}Sr \rightarrow {}^{90}_{39}Y + {}^{0}_{-1}e$;
(d) ${}^{11}_{6}C \rightarrow {}^{11}_{5}B + {}^{0}_{+1}e$
(e) ${}^{43}_{21}Sc \rightarrow {}^{43}_{20}Ca + {}^{0}_{+1}e$

19.3 Positron emission: ${}^{26}_{13}Al \rightarrow {}^{0}_{+1}e + {}^{26}_{12}Mg$
Electron capture: ${}^{26}_{13}Al + {}^{0}_{-1}e \rightarrow {}^{26}_{12}Mg$

19.4 (a) ${}^{32}_{14}Si \rightarrow {}^{0}_{-1}e + {}^{32}_{15}P$;
(b) ${}^{43}_{22}Ti \rightarrow {}^{0}_{+1}e + {}^{43}_{21}Sc$;
(c) ${}^{239}_{94}Pu \rightarrow {}^{4}_{2}He + {}^{235}_{92}U$

19.5 Mass defect = $\Delta m = -0.03438$ g/mol
$\Delta E = (-3.438 \times 10^{-5}\ \text{kg/mol})\,(2.998 \times 10^{8}\ \text{m/s})^2$
$= -3.090 \times 10^{12}\ \text{J/mol} = -3.090 \times 10^{9}\ \text{kJ/mol}$
E_b per nucleon = 5.150×10^{8} kJ/nucleon
E_b for ${}^{6}Li$ is smaller than E_b for ${}^{4}He$; therefore, helium-4 is more stable than lithium-6.

19.6 From the graph it can be seen that the binding energy per nucleon increases more sharply for the fusion of lighter elements than it does for heavy elements undergoing fission. Therefore, fusion is more exothermic per gram than fission.

19.7 $(1/2)^{10} = 9.8 \times 10^{-4}$; this is equivalent to 0.098% of the radioisotope remaining.

19.8 All the lead came from the decay of ${}^{238}U$; therefore at the time the rock was dated, N = 100 and N_0 = 209. The decay constant, k, can be determined:

$$k = \frac{0.693}{4.51 \times 10^{9}\ y} = 1.54 \times 10^{-10}\ y^{-1}.$$

The age of the rock (t) can be calculated using Equation 19.3:

$$\ln \frac{100}{209} = -(1.54 \times 10^{-10}\ \text{y}^{-1}) \times t;\ t = 4.80 \times 10^{9}\ y$$

19.9 Ethylene is derived from petroleum, which was formed millennia ago. The half-life of ${}^{14}C$ is 5730 y, and thus much of ethylene's ${}^{14}C$ would have decayed and be much less than that of the ${}^{14}C$ alcohol produced by fermentation.

19.10 (a) ${}^{13}_{6}C + {}^{1}_{0}n \rightarrow {}^{4}_{2}He + {}^{10}_{4}Be$;
(b) ${}^{14}_{7}N + {}^{4}_{2}He \rightarrow {}^{1}_{0}n + {}^{17}_{9}F$;
(c) ${}^{253}_{99}Es + {}^{4}_{2}He \rightarrow {}^{1}_{0}n + {}^{256}_{101}Md$

19.11 ${}^{208}_{82}Pb + {}^{62}_{28}Ni \rightarrow {}^{1}_{0}n + {}^{269}_{110}$

19.12 Burning a metric ton of coal produces 2.8×10^{7} kJ of energy $\left(\frac{2.8 \times 10^{4}\ \text{kJ}}{1.0\ \text{kg}}\right) \times \left(\frac{10^{3}\ \text{kg}}{\text{metric ton}}\right)$. The fission of 1.0 kg of ${}^{235}U$ produces $\frac{2.1 \times 10^{10}\ \text{kJ}}{0.235\ \text{kg U-235}} = 8.93 \times 10^{10}$ kJ. It would require burning 3.2×10^{3} metric tons of coal to equal the amount of energy from 1.0 kg of ${}^{235}U$: 8.93×10^{10} kJ from uranium-235 $\times \frac{1\ \text{metric ton coal}}{2.8 \times 10^{7}\ \text{kJ}} = 3.2 \times 10^{3}$ metric tons.

19.13 $$k = \frac{0.693}{29.1\ y} = 2.38 \times 10^{-2}\ \text{y}^{-1}$$
ln (fraction) = $(-2.38 \times 10^{-2}\ \text{y}^{-1}) \times$ (11 years, as of 1997)
$= -0.262$; fraction $= e^{-0.262} = 0.770 = 77.0\%$

19.14 $$k = \frac{0.693}{30.2\ y} = 2.29 \times 10^{-2}\ y^{-1}$$
(a) 60% drop in activity, 40% activity remaining:
$\ln(0.40) = -0.916 = -(2.29 \times 10^{-2}\ \text{y}^{-1}) \times t$;
$$t = \frac{-0.916}{2.29 \times 10^{-2}\ y^{-1}} = 40\ y$$

(b) 90% drop in activity, 10% remains
$\ln(0.10) = -2.30 = -(2.29 \times 10^{-2}\ y^{-1}) \times t;$

$$t = \frac{-2.30}{2.29 \times 10^{-2}\ y^{-1}} = 100\ y$$

19.15 $k = \dfrac{0.693}{3.82\ d} = 0.181\ d^{-1}$

(a) The drop from 8 pCi to 4 pCi represents one half-life, 3.82 days.

(b) $\ln \dfrac{1.5}{8} = -1.67 = -(0.181\ d^{-1}) \times t;$

$$t = \frac{-1.67}{-0.181\ d^{-1}} = 9.25\ d$$

19.16 $k = \dfrac{0.693}{78.2\ h} = 8.86 \times 10^{-3}\ h^{-1}$

$\ln(0.10) = -2.30 = -(8.86 \times 10^{-3}\ h^{-1}) \times t;\ t = 260\ h$

19.17 Iron-59: $k = \dfrac{0.693}{44.5\ d} = 1.557 \times 10^{-2}\ d^{-1};$

Chromium-51: $k = \dfrac{0.693}{27.7\ d} = 0.0250\ d^{-1}$

fractions remaining:

^{59}Fe: $\ln(\text{fraction}) = -(1.557 \times 10^{-2}\ d^{-1}) \times 90\ d;$
fraction $= e^{-1.40} = 0.246;\ 80\ mg \times 0.246 = 19.7\ mg$ left

^{51}Cr: $\ln(\text{fraction}) = -(0.0250\ d^{-1}) \times 90\ d;$ 10.5 mg left

Alternatively, consider the fact that 90 days is approximately two half-lives of iron-59. Therefore, approximately 3/4 of it (about 60 mg) has decayed after 90 days; about 20 mg remains. In that same time, chromium-51 has undergone more than three half-lives, so that less than 1/8 remains (less than 12.5 mg).

APPENDIX M

Answers to Selected Questions for Review and Thought

Chapter 1

6. (a) Physical (b) Chemical (c) Chemical (d) Physical, physical
13. The metal will exist in the liquid state. Gallium's melting point is very close to room temperature and below body temperature.
15. The water molecules on your clothes absorb energy from the sun and move about more rapidly. They eventually move fast enough to break free from the other water molecules and the clothes as they escape into the gas phase. Physical changes.
18. (a) Quantitative
    (b) Qualitative
    (c) Qualitative
20. (a) White compound—qualitative, physical. 1.56 g—quantitative, physical. Dye changing from red to colorless—qualitative, chemical.
    (b) 0.6 g—quantitative, physical; reacted with water—qualitative, chemical.
22. Water at 65 °C
24. (a) Solid—mixture
    (b) Solid—compound
    (c) Liquid—heterogeneous mixture
    (d) Solid—element
26. (a) Compound forms elements (or compounds).
    (b) Compound forms elements (or compounds).
    (c) Element forms compound.
    (d) Element or compound forms different elements or compounds.

Chapter 2

12. 6 km
14. 5.78 m
16. 11.2 m long; 3.66 m wide; 1.12×10^3 cm long; 366 cm wide
18. 4.10×10^3 cm^3, 4.10 L
20. 557 g
22. 0.911 g/mL
24. (d) Al, $d = 2.70$ g/cm^3
26. 80.1% silver, 19.9% copper
28. 245 g
30. 95 electrons, 95 protons, 146 neutrons
32. number
34. mass. An electron is $\frac{1}{1800}$ the mass of a proton or a neutron, so the mass of the electrons is insignificant when calculating mass number.
36. The number of neutrons in an atom.
38. (a) 56
    (b) 243
    (c) 184
40. (a) $^{15}_{7}N$ (b) $^{64}_{30}Zn$ (c) $^{129}_{54}Xe$
42. (a) 6 electrons, 6 protons, 7 neutrons
    (b) 24 electrons, 24 protons, 26 neutrons
    (c) 83 electrons, 83 protons, 122 neutrons
44.

Z	A	Number of Neutrons	Element
60	144	84	Nd
12	24	12	Mg
64	158	94	Gd
17	37	20	Cl

46. $^{57}_{27}Co$ $^{58}_{27}Co$ $^{60}_{27}Co$
48. (0.7899)(23.985042 amu) + (0.1000)(24.98537 amu) + (0.1101)(25.98259 amu) = 24.31 amu
50. $^{63}Cu = 69.500\%$; $^{65}Cu = 30.500\%$
52. The atomic mass should be close to 40 amu; 39.947 amu
53. 1 dozen, 1 gross, 1 week, 1 pair
55. If the number of molecules rather than some counting unit (mol) were used, the numbers used in the calculations would be rather large. This is more of a hindrance than an advantage when doing calculations.
57. (a) 1.19×10^3 g Au
    (b) 10.7 g U
    (c) 315 g Ne
    (d) 8.86×10^{-2} g Pu

59. (a) 0.696 mol Na
    (b) 1.7×10^{-5} mol Pt
    (c) 4.97×10^{-2} mol P
    (d) 1.17×10^{-2} mol As
    (e) 7.49×10^{-3} mol Xe
61. 9.42×10^{-5} mol
63. 2.97 cm^3; 1.44 cm
65. 5.93×10^{21} Au atoms
67. 7.951×10^{-23} g
69. Periods are rows of elements that go left to right on the table. A group is a vertical column of elements in the table.
71. Manganese is a steel gray, lustrous, hard, brittle metal. It is used in the manufacture of steel; for rock crushers, railway points, and crossings; and as a constituent of several alloys.

    Iron is a silvery white or gray, soft ductile, malleable, somewhat magnetic metal. It is alloyed with carbon, manganese, chromium, and nickel to form steels. Certain isotopes are used in tracer studies.

    Nickel is a lustrous white, hard, ferromagnetic material. It is used for nickel plating, manufacture of stainless steels, and alloys for electronic and space applications.
73. Transition elements: titanium (Ti), iron (Fe), nickel (Ni); Halogens: flourine (F); chlorine (Cl) Alkali metal: sodium (Na).
75. Polonium (Po), atomic number 84. This element was named for Marie Curie's homeland, Poland.
77. Potassium (K) metal; calcium (Ca) metal; scandium (Sc) metal; titanium (Ti) metal; vanadium (V) metal; chromium (Cr) metal; manganese (Mn) metal; iron (Fe) metal; cobalt (Co) metal; nickel (Ni) metal; copper (Cu) metal; zinc (Zn) metal; gallium (Ga) metal; germanium (Ge) metalloid; arsenic (As) metalloid; selenium (Se) nonmetal; bromine (Br) nonmetal; krypton (Kr) nonmetal
79. Periods 2 and 3 have eight elements; periods 4 and 5 have 18 elements; period 6 has 32 elements; and period 7 is incomplete.
81. There is a higher abundance of even-atomic-numbered elements.
84. 0.197 nm; 197 pm
86. 0.178 nm^3; 1.78×10^{-22} cm^3
88. (a) 272 mL, (b) the container will not hold this volume.
90. 89 tons of NaF would be required for 1 year.
92. (a) Ni, 8.91 g/cm^3
94. ^{39}K
96. (a) Ti, 22, 47.88 amu
    (b) Group 4B, 4th period; Zr, Hf, Rf
    (c) Lightweight, strong corrosion-resistant, low reactivity with body fluids
    (d) Titanium is a dark gray lustrous metal, brittle when cold but malleable when hot. It is used as an alloy with copper and iron in titanium bronze; as an additive to steel to impart strength; and as a surgical aid (fracture fixation).
98. Box 1 = S, Box 2 = N, Box 3 = B, Box 4 = I
100. 3.8×10^{-2} mol C
102. \$2150 (\$$2.15 \times 10^3$)
104. 3.4 mol Cu; 2.0×10^{24} Cu atoms
105. (a) Not possible, mass number cannot be less than atomic number, and atomic number and number of protons cannot be different.
    (b) Possible
    (c) Not possible, mass number cannot be less than atomic number.
    (d) Not possible, mass number is greater than sum of protons plus neutrons.
    (e) Possible
    (f) Not possible, mass number is too small.
107. (a) 1 mol iron (Iron atoms have greater mass than aluminum atoms.)
    (b) Same mass (Both contain same number of atoms of lead.)
    (c) 1 mol copper (One mole of copper has many, many more atoms than 1 atom of copper.)
    (d) 1 mol Cl_2 (Cl_2 has greater mass than Cl.)
    (e) Same mass (1 gram = 1 gram)
    (f) Same mass (One mole of Mg has a mass of 24.3 g.)
    (g) 1 mol Na (One mole of Na has a mass of 23 g.)
    (h) Same mass (Both are 1 mol of He.)
    (i) 1 mol I_2 (One mole of I_2 contains many more molecules than 1 molecule of I_2.)
    (j) 1 oxygen molecule (O_2 has more mass than O.)
109. (a) a, b, f (e contains both a solid and a liquid.)
    (b) h, i (e contains both a liquid and a solid.)
    (c) c, d, g
    (d) a, d, f are pure elements. (b, e, g, h, i are mixtures of elements.)
    (e) c
    (f) a, c, d, f
    (g) b, e, g, h, i

Chapter 3

10. (a) BrF_3 (b) XeF_2 (c) P_2F_4 (d) $C_{15}H_{32}$ (e) N_2H_4
12. Butane derivative: $C_4H_{10}O$, $CH_3CH_2CH_2CH_2OH$

$$\begin{array}{ccccccccccc} & & H & & H & & H & & H & & \\ & & | & & | & & | & & | & & \\ H & - & C & - & C & - & C & - & C & - & O-H \\ & & | & & | & & | & & | & & \\ & & H & & H & & H & & H & & \end{array}$$

    Pentane derivative: $C_5H_{12}O$, $CH_3CH_2CH_2CH_2CH_2OH$

$$\begin{array}{ccccccccccccc} & & H & & H & & H & & H & & H & & \\ & & | & & | & & | & & | & & | & & \\ H & - & C & - & C & - & C & - & C & - & C & - & O-H \\ & & | & & | & & | & & | & & | & & \\ & & H & & H & & H & & H & & H & & \end{array}$$

14. Sucrose has more oxygen atoms; sucrose has more atoms of all kinds.
17. (a) Calcium, 1; carbon, 2; oxygen, 4
    (b) Carbon, 8; hydrogen, 8
    (c) Sulfur, 1; nitrogen, 2; oxygen, 4; hydrogen, 8
    (d) Platinum, 1; nitrogen, 2; chlorine, 2; hydrogen, 6
    (e) Iron, 1; potassium, 4; carbon, 6; nitrogen, 6
19.

$$CH_3(CH_2)_4CH_3;\ CH_3CH_2CH_2\underset{}{\overset{\overset{\displaystyle CH_3}{|}}{C}}HCH_3;$$

```
        CH3                   CH3
        |                     |
CH3CH2CHCH2CH3; CH3CH2CCH3;
                              |
                              CH3

   CH3
   |
CH3CHCHCH3
      |
      CH3
```

21. (a) Li^+
    (b) Sr^{2+}
    (c) Al^{3+}
    (d) Ca^{2+}
    (e) Zn^{2+}
23. Ba^{2+} and Br^-
25. (a) Se^{2-} (b) F^- (c) Ni^{2+} (d) N^{3-}
27. $PbCl_2$ and $PbCl_4$
29. (a) Incorrect; CaO
    (b) Correct
    (c) Incorrect; Fe_2O_3
    (d) Correct
31. (a) 1 Ca^{2+}, 2 $CH_3CO_2^-$
    (b) 2 Co^{3+}, 3 (SO_4^{2-})
    (c) 1 Al^{3+}, 3 (OH^-)
    (d) 2 NH_4^+, 1 (CO_3^{2-})
33. (a) $Ni(NO_3)_2$
    (b) $NaHCO_3$
    (c) LiOCl
    (d) $Mg(ClO_3)_2$
    (e) $CaSO_3$
35. (a) not ionic, CH_4
    (b) not ionic, N_2O_5
    (c) ionic, $(NH_4)_2S$
    (d) not ionic, H_2Se
    (e) ionic, $NaClO_4$
37. (a) $Ca(HCO_3)_2$
    (b) $KMnO_4$
    (c) $Mg(ClO_4)_2$
    (d) $(NH_4)_2HPO_4$
39. (a) Calcium acetate
    (b) Cobalt(III) sulfate
    (c) Aluminum hydroxide
41. MgO; charges of the ions are greater and atomic radii are smaller. Because of this, there is a stronger attraction between the two ions.
43. Electrolytes are compounds that dissolve in water to give aqueous solutions of ions and therefore conduct electricity. A strong electrolyte dissociates 100% and is a strong conductor of electricity. Weak electrolytes do not dissociate completely and are weak conductors of electricity. An example of a strong electrolyte is NaCl. An example of a weak electrolyte is acetic acid.
45. (a) soluble, Fe^{2+} and ClO_4^-
    (b) soluble, Na^+ and SO_4^{2-}
    (c) soluble, K^+ and Br^-
    (d) soluble, Na^+ and CO_3^{2-}
47. Hydrogen, oxygen, carbon, nitrogen, calcium, phosphorous, chlorine, sulfur, sodium, and potassium
49. As ions; Fe^{2+} is incorporated in hemoglobin and Ca^{2+} is present in bones and teeth.
51. Selenium and arsenic
53. Fats and oils are esters formed by the reaction of glycerol and fatty acids. A triglyceride comes from the acid reacting with all 3 hydroxyl (OH) groups in the glycerol.

54.	CH_3OH	Carbon	Hydrogen	Oxygen
Number of Moles	1	1	4	1
Number of Molecules or Atoms	6.022×10^{23}	6.022×10^{23}	2.409×10^{24}	6.022×10^{23}
Molar Mass	32.042 g	12.011 g	4.0316 g	15.9994 g

56. (a) 159.69 g/mol
    (b) 67.81 g/mol
    (c) 44.01 g/mol
    (d) 197.90 g/mol
    (e) 176.14 g/mol
58. (a) 3.12×10^{-2} mol
    (b) 1.01×10^{-2} mol
    (c) 1.25×10^{-2} mol
    (d) 4.06×10^{-3} mol
    (e) 5.99×10^{-3} mol
60. (a) 1.80×10^{-3} mol aspirin; 2.266×10^{-2} mol $NaHCO_3$; 5.205×10^{-3} mol citric acid
    (b) 1.08×10^{21} molecules
62. 5.67 mol SO_3; 3.41×10^{24} SO_3 molecules; 3.41×10^{24} sulfur atoms; 1.02×10^{25} oxygen atoms
64. (a) 239.3 g/mol; 86.60% Pb; 13.40% S
    (b) 30.07 g/mol; 79.88% C; 20.12% H
    (c) 60.05 g/mol; 40.00% C; 6.71% H; 53.29% O
    (d) 80.05 g/mol; 35.00% N; 5.04% H; 59.96% O
66. 57.98%
68. Typically, percent daily values of fat, cholesterol, sodium, carbohydrates, and protein are given. Often, values of minerals such as Fe, Mg, and Cu are also given with percent vitamin information.
70. An empirical formula is a molecular formula that shows the simplest possible ratio of elements in a compound, while a molecular formula expresses numbers of atoms of each type within one molecule of substance.
72. $C_4H_8N_2O_2$
74. B_5H_7
76. C_3H_4; C_9H_{12}
78. C_5H_7N (empirical); $C_{10}H_{14}N_2$ (molecular)
80. $C_5H_{14}N_2$
82. $x = 12$
84. $C_7H_5N_3O_6$; $C_3H_7NO_3$
86. (b) Li_2Te, lithium telluride; (d) MgF_2, magnesium fluoride; (f) In_2S_3, indium sulfide

88. (a) $NaClO$, ionic
    (b) $Al(ClO_4)_3$, ionic
    (c) $KMnO_4$, ionic
    (d) KH_2PO_4, ionic
    (e) ClF_3, not ionic
    (f) BBr_3, not ionic
    (g) $Ca(CH_3CO_2)_2$, ionic
    (h) Na_2SO_3, ionic
    (i) S_2Cl_4, not ionic
    (j) PF_3, not ionic
90. 3.33×10^3 cm^3; 14.5 cm
92. C_5H_4 (empirical), $C_{10}H_8$ (molecular)
94. ICl_3 (empirical); I_2Cl_6 molecular
96. 7.36 kg Fe
98. 67.1 g Sb_2S_3
100. (a) 1.30×10^{-2} mol Ni
    (b) NiF_2
    (c) Nickel(II) fluoride

102. 

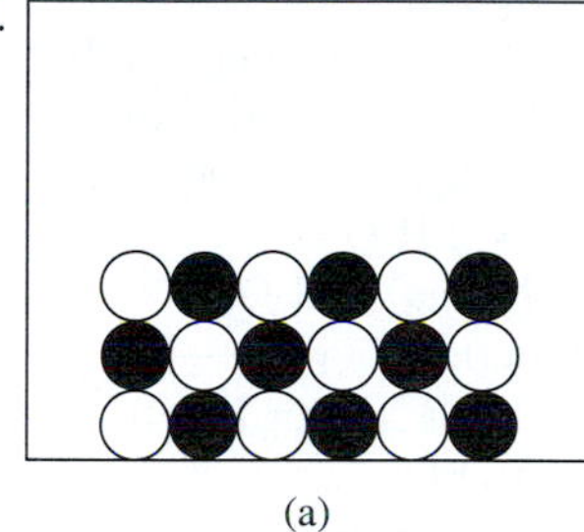

(a)

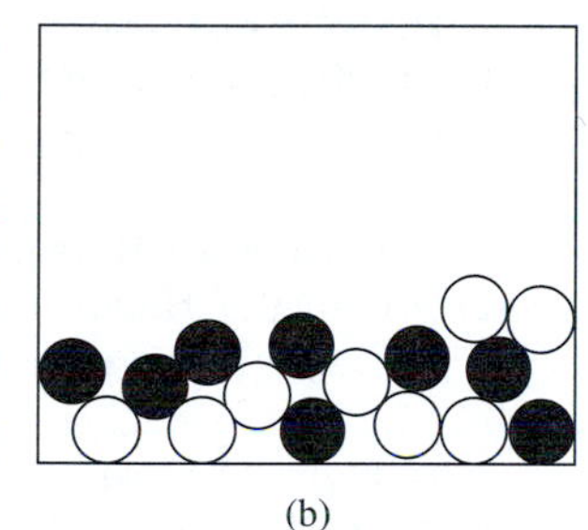

(b)

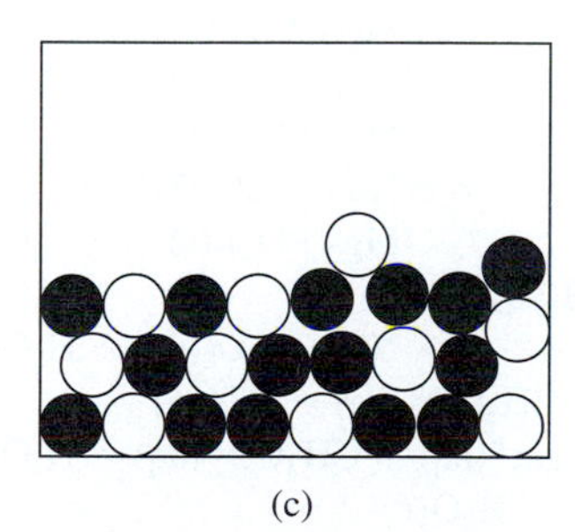

(c)

104. 

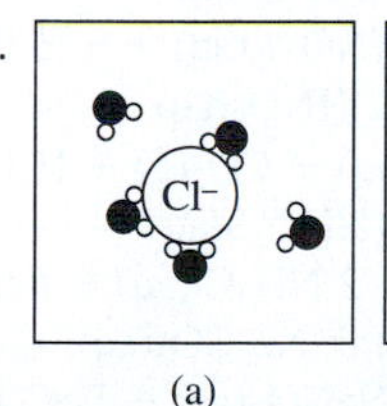

(a)

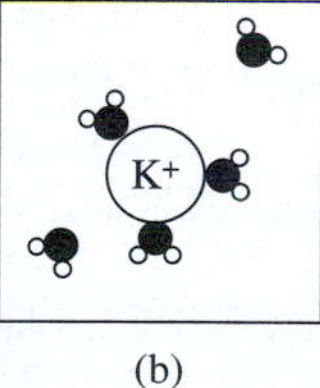

(b)

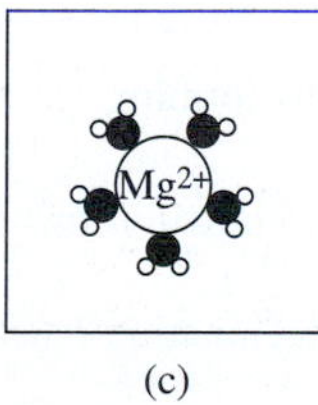

(c)

106. 2,4-dimethylpentane
108. Tl_2CO_3 and Tl_2SO_4
109. (a) Calcium fluoride
    (b) Copper(II) oxide
    (c) Sodium nitrate
    (d) Nitrogen triiodide
    (e) Iron(III) chloride
    (f) Lithium sulfate
112. 6.022×10^{24} molecules > 17.03 g > 0.1 mol

Chapter 4

12. (a) 1.00 g
    (b) 2 for Mg, 1 for O_2, 2 for MgO
    (c) 50
14. (b)
16. $2\ Sb + 3\ Cl_2 \longrightarrow 2\ SbCl_3$; (c)
18. (a) $2\ Fe(s) + 3\ Cl_2(g) \longrightarrow 2\ FeCl_3(s)$
    (b) $SiO_2(s) + 2\ C(s) \longrightarrow Si(s) + 2\ CO(g)$
    (c) $3\ Fe(s) + 4\ H_2O(g) \longrightarrow Fe_3O_4(s) + 4\ H_2(g)$
20. (a) $3\ MgO(s) + 2\ Fe(s) \longrightarrow Fe_2O_3(s) + 3\ Mg(s)$
    (b) $2\ H_3BO_3(s) \longrightarrow B_2O_3(s) + 3\ H_2O(\ell)$
    (c) $2\ NaNO_3(s) + H_2SO_4(aq) \longrightarrow Na_2SO_4(aq) + 2\ HNO_3(g)$
22. (a) $CaNCN(s) + 3\ H_2O(\ell) \longrightarrow CaCO_3(s) + 2\ NH_3(g)$
    (b) $2\ NaBH_4(s) + H_2SO_4(aq) \longrightarrow B_2H_6(g) + 2\ H_2(g) + Na_2SO_4(aq)$
    (c) $8\ H_2S(g) + 8\ Cl_2(g) \longrightarrow S_8(s) + 16\ HCl(aq)$
24. (a) $Mg + 2\ HNO_3 \longrightarrow H_2 + Mg(NO_3)_2$
    (b) $2\ Al + Fe_2O_3 \longrightarrow Al_2O_3 + 2\ Fe$
    (c) $2\ S + 3\ O_2 \longrightarrow 2\ SO_3$
    (d) $SO_3 + H_2O \longrightarrow H_2SO_4$
25. (a) Exchange
    (b) Exchange
    (c) Combination
    (d) Combination
26. (a) $2\ C(s) + O_2(g) \longrightarrow 2\ CO(g)$
    (b) $2\ Ni(s) + O_2(g) \longrightarrow 2\ NiO(s)$
    (c) $4\ Cr(s) + 3\ O_2(g) \longrightarrow 2\ Cr_2O_3(s)$
28. (a) $BeCO_3(s) \xrightarrow{\text{heat}} BeO(s) + CO_2(g)$
    (beryllium oxide and carbon dioxide)
    (b) $NiCO_3(s) \xrightarrow{\text{heat}} NiO(s) + CO_2(g)$
    (nickel(II) oxide and carbon dioxide)
    (c) $Al_2(CO_3)_3 \xrightarrow{\text{heat}} Al_2O_3(s) + 3\ CO_2(g)$
    (aluminum oxide and carbon dioxide)
30. (a) $2\ C_4H_{10}(g) + 13\ O_2(g) \longrightarrow 8\ CO_2(g) + 10\ H_2O(g)$
    (b) $C_6H_{12}O_6(s) + 6\ O_2(g) \longrightarrow 6\ CO_2(g) + 6\ H_2O(g)$
    (c) $2\ C_4H_8O(\ell) + 11\ O_2(g) \longrightarrow 8\ CO_2(g) + 8\ H_2O(g)$
31. (a) $2\ Mg(s) + O_2(g) \longrightarrow 2\ MgO(s)$ (magnesium oxide)
    (b) $2\ Ca(s) + O_2(g) \longrightarrow 2\ CaO(s)$ (calcium oxide)
    (c) $4\ In(s) + 3\ O_2(g) \longrightarrow 2\ In_2O_3(s)$ (indium oxide)
33. (a) $2\ K(s) + Cl_2(g) \longrightarrow 2\ KCl(s)$ (potassium chloride)

(b) $Mg(s) + Br_2(\ell) \longrightarrow MgBr_2(s)$ (magnesium bromide)
(c) $2\ Al(s) + 3\ F_2(g) \longrightarrow 2\ AlF_3(s)$ (aluminum fluoride)

36. (a) $MnCl_2(aq) + Na_2S(aq) \longrightarrow MnS(s) + 2\ NaCl(aq)$
(b) $HNO_3(aq) + CuSO_4(aq) \longrightarrow$ No Reaction
(c) $NaOH(aq) + HClO_4(aq) \longrightarrow$ No Reaction
(d) $Hg(NO_3)_2(aq) + Na_2S(aq) \longrightarrow HgS(s) + 2\ NaNO_3(aq)$
(e) $Pb(NO_3)_2(aq) + 2\ HCl(aq) \longrightarrow 2\ HNO_3(aq) + PbCl_2(s)$
(f) $BaCl_2(aq) + H_2SO_4(aq) \longrightarrow BaSO_4(s) + 2\ HCl(aq)$

37. Spectator ions are Mg^{2+} and NO_3^-;
$2\ H^+(aq) + 2\ OH^-(aq) \longrightarrow 2\ H_2O(\ell)$

39. (a) Cl^- and H^+; $Cu^{2+}(aq) + S^{2-}(aq) \longrightarrow CuS(s)$
(b) K^+ and Cl^-; $Ca^{2+}(aq) + CO_3^{2-}(aq) \longrightarrow CaCO_3(s)$
(c) Na^+ and NO_3^-; $Ag^+(aq) + I^-(aq) \longrightarrow AgI(s)$

40. (a) $Zn(s) + 2\ HCl(aq) \longrightarrow H_2(g) + ZnCl_2(aq)$
$Zn(s) + 2\ H^+(aq) + 2\ Cl^-(aq) \longrightarrow H_2(g) + Zn^{2+}(aq) + 2\ Cl^-(aq)$
$Zn(s) + 2\ H^+(aq) \longrightarrow Zn^{2+}(aq) + H_2(g)$
(b) $Mg(OH)_2(aq) + 2\ HCl(aq) \longrightarrow MgCl_2(aq) + 2\ H_2O(\ell)$
$Mg^{2+}(aq) + 2\ OH^-(aq) + 2\ H^+(aq) + 2\ Cl^-(aq) \longrightarrow Mg^{2+}(aq) + 2\ Cl^-(aq) + 2\ H_2O(\ell)$
$2\ H^+(aq) + 2\ OH^-(aq) \longrightarrow 2\ H_2O(\ell)$
(c) $2\ HNO_3(aq) + CaCO_3(s) \longrightarrow Ca(NO_3)_2(aq) + H_2O(\ell) + CO_2(g)$
$2\ H^+(aq) + 2\ NO_3^-(aq) + CaCO_3(s) \longrightarrow Ca^{2+}(aq) + 2\ NO_3^-(aq) + H_2O(\ell) + CO_2(g)$
$2\ H^+(aq) + CaCO_3(s) \longrightarrow Ca^{2+}(aq) + H_2O(\ell) + CO_2(g)$
(d) $4\ HCl(aq) + MnO_2(s) \longrightarrow MnCl_2(aq) + Cl_2(g) + H_2O(\ell)$
$4\ H^+(aq) + 4\ Cl^-(aq) + MnO_2(s) \longrightarrow Mn^{2+}(aq) + 2\ Cl^-(aq) + Cl_2(g) + H_2O(\ell)$
$4\ H^+(aq) + 2\ Cl^-(aq) + MnO_2(s) \longrightarrow Mn^{2+}(aq) + Cl_2(g) + H_2O(\ell)$

42. (a) $Ca(OH)_2(aq) + 2\ HNO_3(aq) \longrightarrow Ca(NO_3)_2(aq) + 2\ H_2O(\ell)$
$Ca^{2+}(aq) + 2\ OH^-(aq) + 2\ H^+(aq) + 2\ NO_3^-(aq) \longrightarrow Ca^{2+}(aq) + 2\ NO_3^-(aq) + 2\ H_2O(\ell)$
$2\ H^+(aq) + 2\ OH^-(aq) \longrightarrow 2\ H_2O(\ell)$
(b) $BaCl_2(aq) + Na_2CO_3(aq) \longrightarrow BaCO_3(s) + 2\ NaCl(aq)$
$Ba^{2+}(aq) + 2\ Cl^-(aq) + 2\ Na^+(aq) + CO_3^{2-}(aq) \longrightarrow BaCO_3(s) + 2\ Na^+(aq) + 2\ Cl^-(aq)$
$Ba^{2+}(aq) + CO_3^{2-}(aq) \longrightarrow BaCO_3(s)$
(c) $2\ Na_3PO_4(aq) + 3\ Ni(NO_3)_2(aq) \longrightarrow Ni_3(PO_4)_2(s) + 6\ NaNO_3(aq)$
$6\ Na^+(aq) + 2\ PO_4^{3-}(aq) + 3\ Ni^{2+}(aq) + 6\ NO_3^-(aq) \longrightarrow Ni_3(PO_4)_2(s) + 6\ Na^+(aq) + 6\ NO_3^-(aq)$
$3\ Ni^{2+}(aq) + 2\ PO_4^{3-}(aq) \longrightarrow Ni_3(PO_4)_2(s)$

44. $Ba(OH)_2(s) + 2\ HNO_3(aq) \longrightarrow Ba(NO_3)_2(aq) + 2\ H_2O(\ell)$

46. $CdCl_2(aq) + 2\ NaOH(aq) \longrightarrow Cd(OH)_2(s) + 2\ NaCl(aq)$
$Cd^{2+}(aq) + 2\ Cl^-(aq) + 2\ Na^+(aq) + 2\ OH^-(aq) \longrightarrow Cd(OH)_2(s) + 2\ Na^+(aq) + 2\ Cl^-(aq)$
$Cd^{2+}(aq) + 2\ OH^-(aq) \longrightarrow Cd(OH)_2(s)$

48. $Pb(NO_3)_2(aq) + 2\ KCl(aq) \longrightarrow PbCl_2(s) + 2\ KNO_3(aq)$
Lead(II) nitrate + potassium chloride $\longrightarrow$ lead(II) chloride + potassium nitrate

50. $MnCO_3(s) + 2\ HCl(aq) \longrightarrow MnCl_2(s) + H_2O(g) + CO_2(g)$
Manganese(II) carbonate + hydrochloric acid $\longrightarrow$ manganese(II) chloride + water + carbon dioxide

51. (a) Base, K^+ and OH^-
(b) Base, Mg^{2+} and OH^-
(c) Acid, H^+ and ClO^-
(d) Acid, H^+ and Br^-
(e) Base, Li^+ and OH^-
(f) Acid, H^+ and HSO_3^-

52. (a) strong
(b) strong, but not very soluble
(c) weak
(d) Strong
(e) Strong
(f) Weak

54. (a) $NaOH(aq) + HNO_2(aq) \longrightarrow NaNO_2(aq) + H_2O(\ell)$
$Na^+(aq) + OH^-(aq) + H^+(aq) + NO_2^-(aq) \longrightarrow Na^+(aq) + NO_2^-(aq) + H_2O(\ell)$
$H^+(aq) + OH^-(aq) \longrightarrow H_2O(\ell)$
(b) $Ca(OH)_2(aq) + H_2SO_4(aq) \longrightarrow CaSO_4(aq) + 2\ H_2O(\ell)$
$Ca^{2+}(aq) + 2\ OH^-(aq) + 2\ H^+(aq) + SO_4^{2-}(aq) \longrightarrow Ca^{2+}(aq) + SO_4^{2-}(aq) + 2\ H_2O(\ell)$
$2\ H^+(aq) + 2\ OH^-(aq) \longrightarrow 2\ H_2O(\ell)$
(c) $NaOH(aq) + HI(aq) \longrightarrow NaI(aq) + H_2O(\ell)$
$Na^+(aq) + OH^-(aq) + H^+(aq) + I^-(aq) \longrightarrow Na^+(aq) + I^-(aq) + H_2O(\ell)$
$H^+(aq) + OH^-(aq) \longrightarrow H_2O(\ell)$
(d) $3\ Mg(OH)_2(aq) + 2\ H_3PO_4(aq) \longrightarrow Mg_3(PO_4)_2(s) + 6\ H_2O(\ell)$
$3\ Mg^{2+}(aq) + 6\ OH^-(aq) + 6\ H^+(aq) + 2\ PO_4^{3-}(aq) \longrightarrow Mg_3(PO_4)_2(s) + 6\ H_2O(\ell)$
$3\ Mg^{2+}(aq) + 6\ OH^-(aq) + 6\ H^+(aq) + 2\ PO_4^{3-}(aq) \longrightarrow Mg_3(PO_4)_2(s) + 6\ H_2O(\ell)$
(e) $NaOH(aq) + CH_3CO_2H(aq) \longrightarrow NaCH_3CO_2(aq) + H_2O(\ell)$
$Na^+(aq) + OH^-(aq) + CH_3CO_2^-(aq) + H^+(aq) \longrightarrow Na^+(aq) + CH_3CO_2^-(aq) + H_2O(\ell)$
$H^+(aq) + OH^-(aq) \longrightarrow H_2O(\ell)$

56. (a) Acid-base reaction; $Fe(OH)_3(s) + 3\ HNO_3(aq) \longrightarrow Fe(NO_3)_3(aq) + 3\ H_2O(\ell)$
(b) Gas-forming reaction; $FeCO_3(s) + 2\ HNO_3(aq) \longrightarrow Fe(NO_3)_2\ (aq) + CO_2(g) + H_2O(g)$
(c) Precipitation reaction; $FeCl_2(aq) + (NH_4)_2S(aq) \longrightarrow 2\ NH_4Cl(aq) + FeS(s)$
(d) Precipitation reaction; $Fe(NO_3)_2(aq) + Na_2CO_3(aq) \longrightarrow 2\ NaNO_3(aq) + FeCO_3(s)$

58. (a) Fe = +3; O = −2; H = +1
(b) H = +1; Cl = +5; O = −2
(c) Cu = +2; Cl = −1
(d) K = +1; Cr = +6; O = −2
(e) Ni = +2; O = −2; H = +1
(f) N = −2, H = +1

59. (a) S = +6, O = −2
(b) N = +5, O = −2
(c) Mn = +7, O = −2
(d) Cr = +3, H = +1, O = −2
(e) H = +1, P = +5, O = −2
(f) S = +2, O = −2

60. (a) Exchange reaction
(b) Oxidation reduction (Ca is oxidized, O_2 is reduced)
(c) Acid-base reaction

62. Best reducing agents are Group 1A metals. Best oxidizing agents are in Group 6A and Group 7A.

64. (a), (d), and (f)

66. (a) $2\ K(s) + 2\ H_2O(\ell) \longrightarrow 2\ KOH(aq) + H_2(g)$
(b) $Mg(s) + 2\ HBr(aq) \longrightarrow MgBr_2(aq) + H_2(g)$
(c) $2\ NaBr(aq) + Cl_2(g) \longrightarrow 2\ NaCl(aq) + Br_2(\ell)$
(d) $WO_3(s) + 3\ H_2(g) \longrightarrow W(s) + 3\ H_2O(\ell)$
(e) $H_2S(aq) + Cl_2(g) \longrightarrow S(s) + 2\ HCl(aq)$

68. (a) Group 1A and 2A; transition metals
(b) No reaction will take place.
(c) Yes; $Pb(s) + 2\ AgNO_3(aq) \longrightarrow Pb(NO_3)_2(aq) + 2\ Ag(s)$
(d) Reactivity, Al > Pb > Ag

70. (a) No
(b) No
(c) No
(d) Yes
(e) Yes
(f) Yes
(g) No

72. (a) No Reaction
(b) $2\ NaBr + I_2$
(c) $2\ NaF + Cl_2$
(d) $2\ NaCl + Br_2$
(e) No Reaction
(f) No Reaction

74. (a) $Br_2(\ell) + 2\ NaI(aq) \longrightarrow 2\ NaBr(aq) + I_2(s)$
(e) $Cl_2(g) + 2\ NaBr(aq) \longrightarrow 2\ NaCl(aq) + Br_2(\ell)$
(f) $F_2(g) + 2\ NaCl(aq) \longrightarrow 2\ NaF(aq) + Cl_2(g)$

76. (a) $Mg(s) + 4\ HNO_3(aq) \longrightarrow Mg(NO_3)_2(aq) + 2\ NO_2(g) + 2\ H_2O(\ell)$
(b) Magnesium + nitric acid ⟶ magnesium nitrate + nitrogen dioxide + water
(c) $Mg(s) + 4\ H^+(aq) + 2\ NO_3^-(aq) \longrightarrow Mg^{2+}(aq) + 2\ NO_2(g) + 2\ H_2O(\ell)$
(d) Oxidation-reduction

78. (a) $Li(s) + H_2O(\ell) \longrightarrow LiOH(aq) + H_2(g)$; oxidation-reduction
(b) $Ag_2O(s) \xrightarrow{heat} Ag(s) + O_2(g)$; decomposition
(c) $Li_2O(s) + H_2O(\ell) \longrightarrow 2\ LiOH(aq)$; combination
(d) $I_2 + Cl^- \longrightarrow$ No Reaction
(e) $Cu(s) + 2\ HCl(aq) \longrightarrow CuCl_2(aq) + H_2(g)$; oxidation-reduction
(f) $BaCO_3(s) \longrightarrow BaO + CO_2(g)$; decomposition

80. $Cu_3(CO_3)_2(OH)_2(s) + 6\ HCl(aq) \longrightarrow 3\ CuCl_2(aq) + 4\ H_2O(\ell) + 2\ CO_2(g)$

81. (a) NH_4^+, OH^-, NH_3
(b) $CH_3CO_2^-$, H^+, CH_3CO_2H
(c) Na^+, OH^-
(d) H^+, Br^-

82. (a) No Reaction
(b) Yes, reaction will occur.
(c) No Reaction
(d) Yes, reaction will occur.

84. (c) Mg is oxidized, $TiCl_4$ is reduced.
Mg is the reducing agent, $TiCl_4$ is oxidizing agent.
Mg oxidation number changes from 0 to +2.
Ti oxidation number changes from +4 to 0.

85. (a) $CaF_2(s) + H_2SO_4(aq) \longrightarrow 2\ HF(g) + CaSO_4(s)$;
calcium fluoride + sulfuric acid ⟶ hydrogen fluoride + calcium sulfate
(b) Precipitation reaction
(c) Carbon tetrachloride, antimony (V) chloride, hydrogen chloride
(d) CCl_3F

86. Two molecules of butane react with 13 molecules of oxygen to form 8 carbon dioxide molecules and 10 water molecules.

Two moles of butane react with 13 mol oxygen to produce 8 mol carbon dioxide and 10 mol liquid water.

88. $4\ A_2 + AB_3 \longrightarrow 3\ A_3B$

90. Reaction 1: ($LiCl + AgNO_3$)
(a) When clear, colorless aqueous solutions of LiCl and $AgNO_3$ are combined, a white precipitate will settle to the bottom of a colorless solution.

(b)

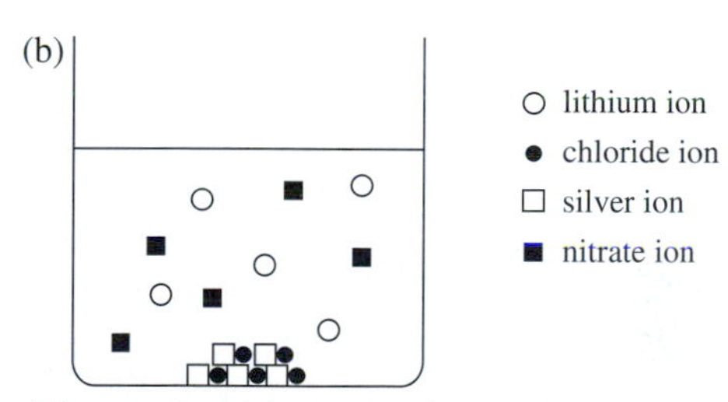

(Water molecules are not shown.)

(c) $LiCl(aq) + AgNO_3(aq) \longrightarrow AgCl(s) + LiNO_3(aq)$

Reaction 2: (NaOH + HCl)
(a) When aqueous solutions of NaOH and HCl are combined, the beaker will contain a clear, colorless solution.

(b)

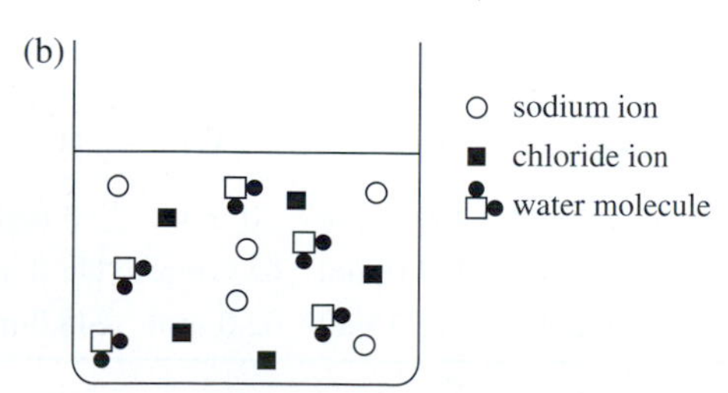

(Water molecules are not shown, except those produced by the reaction.)

(c) $NaOH(aq) + HCl(aq) \longrightarrow NaCl(aq) + H_2O(\ell)$

92. (a) $Ba(OH)_2(aq) + H_2SO_4(aq) \longrightarrow BaSO_4(s) + 2\ H_2O(\ell)$
(b) $Ba(OH)_2(aq) + Na_2SO_4(aq) \longrightarrow BaSO_4(s) + 2\ NaOH(aq)$

(c) $BaCO_3(s) + H_2SO_4(aq) \longrightarrow BaSO_4(s) + H_2O(\ell) + CO_2(g)$

93. Student 1 will produce $NiSO_4(s)$ and $H_2O(\ell)$. When the water is evaporated, the $NiSO_4(s)$ will remain.

    Student 2 will produce $NiSO_4(s)$ and $NaNO_3(aq)$. When the water is evaporated, both $NiSO_4(s)$ and $NaNO_3(s)$ will remain.

    Student 3 will produce $NiSO_4(s)$, $H_2O(\ell)$, and $CO_2(g)$. When the water is evaporated, only $NiSO_4(s)$ will remain.
95. H_2SO_4; $SrSO_4$ is insoluble, whereas $CaSO_4$ is insoluble. NaOH and H_2S will form insoluble precipitates with both Sr^{2+} and Ca^{2+} ions.
97. An oxidizing agent or reducing agent causes an oxidation-reduction reaction; therefore, it must be something you start with and not something that you form.

Chapter 5

13. 50.0 mol Cl_2
15. 1.1 mol O_2, 35 g O_2; 101 g NO_2
17. 12.7 g Cl_2; 0.179 mol $FeCl_2$, 22.7 g $FeCl_2$

19.

$(NH_4)_2PtCl_6$	Pt	HCl
12.35 g	5.423 g	5.408 g
0.02780 mol	0.02780 mol	0.1483 mol

21. (a) 0.147 mol H_2O
    (b) 5.89 g TiO_2 and 10.8 g HCl formed
23. 0.699 g Ga, 0.749 g As
24. 350.0 g N_2, 4.5×10^2 g H_2O, 2.0×10^2 g O_2
26. (a) $CCl_2F_2 + 2\,Na_2C_2O_4 \longrightarrow C + 4\,CO_2 + 2\,NaCl + 2\,NaF$
    (b) 170 g $Na_2C_2O_4$
    (c) 112 g CO_2
28. (a) $K_2PtCl_4 + 2\,NH_3 \longrightarrow Pt(NH_3)_2Cl_2 + 2\,KCl$
    (b) 3.46 g K_2PtCl_4 and 0.283 g NH_3
29. (a) Cl_2 is the limiting reagent.
    (b) 5.07 g $AlCl_3$
    (c) 1.67 g Al left over

31.

	CH_4	H_2O	CO_2	H_2
Start	62.0 mol	139.3 mol	0 mol	0 mol
Change	−62.0 mol	−124.0 mol	62.0 mol	248.0 mol
Final	0.0 mol	15.3 mol	62.0 mol	248.0 mol

33. 1.40×10^3 g Fe
35. 73.5% NH_3
37. 67.0% yield
39. 5.33 g SCl_2
41. 91.6% of the weight is the hydrate.
43. 83.9% $Al(C_6H_5)_3$
45. 56.9% $CaCO_3$; 22.8% Ca
47. CH
49. $C_3H_6O_2$
51. 0.12 M Ba^{2+}, 0.24 M Cl^-
53. (a) 0.254 M
    (b) $[Na^+]$, 0.508 M; $[CO_3^{2-}]$, 0.254 M
55. 0.494 g
57. 5.08×10^3 mL
59. 0.0150 M $CuSO_4$
61. (b)
63. 0.205 g Na_2CO_3
65. 121 mL HNO_3
69. 0.179 g AgCl; NaCl; 0.00833 M NaCl
71. (a) Step (ii) should be 1 mol citric acid/3 mol NaOH
    (b) 0.0039 g citric acid
73. 1.192 M
74. 0.167 M NaOH
75. 96.8% $Na_2S_2O_3$
77. 104 g/mol H_2A
79. 21.66 g N_2
81. 86.3 g Al_2Br_6
83. 12.47 g cisplatin
85. SiH_4
87. KOH, KOH
89. 0.102 M H^+; 0.102 M Cl^-
91. 2.92 M HCl
93. 31.8% Pb
95. (d) 6
97. (a) 2:1
    (b) Carbon
99. (a) Box 3
    (b) H_2 is the limiting reagent.
101. (b) $X_2 + 3\,Y_2 \longrightarrow 2\,XY_3$
103. Up to 1 g metal added, the metal is limiting but more product is made with each successive amount of metal. From 1 to 6 g, the Br_2 is limiting and the same amount of product forms.
105. (a) Group A and B: $Cl^-(aq) + Ag^+(aq) \longrightarrow AgCl(s)$
     Group C and D: $Br^-(aq) + Ag^+(aq) \longrightarrow AgBr(s)$
     (b) It is the Cl^- (or Br^-) ion which reacts with the Ag^+ ion to form the precipitate.
     (c) 0.75 $AgNO_3$ is equivalent to 0.0044 mol Ag^+, 0.6 g AgCl is equivalent to 0.004 mol Cl^- and 0.8 g AgBr is equivalent to 0.004 mol Br^-.
107. (d)
109. (a), (d), and (e)
111. (a) $MgBr_2$, magnesium bromide; $CaBr_2$, calcium bromide; $SrBr_2$, strontium bromide.
     (b) $Mg + Br_2 \longrightarrow MgBr_2$; $Ca + Br_2 \longrightarrow CaBr_2$; $Sr + Br_2 \longrightarrow SrBr_2$
     (c) All are oxidation-reduction reactions.
     (d) At the stoichiometric point of each curve use the mass of metal and mass of product to determine the moles of metal and moles of bromine. The resulting mole ratios will correspond to MBr_2.

Chapter 6

7. (a) 399 Calories
   (b) 5.0×10^6 J
9. (a) 2.97×10^5 J
   (b) 7.10×10^4 calories
   (c) 71.0 kilocalories
11. 5.04×10^6 J; $0.126 or 12.6 cents
14. The specific heat of copper is 0.385 J/g·°C, while the specific heat of aluminum is 0.902 J/g·°C. The copper would heat up faster because the specific heat is smaller.
16. 710 MJ
18. 1.35 kJ
20. 6.3 °C
21. 413 kJ
23. 5.00×10^5 J
25. 270 J of heat released
27. (see below)
29. Endothermic
31. Endothermic. Energy is required to break bonds. In cleaving a diamond, energy from the cutting tool is used to break these bonds.
33. A total of 6.0 kJ of energy is required to convert 1 mol of water in the solid state (ice) to the liquid state.
35. The reaction is slightly exothermic. Although energy is required to break the triple bond in nitrogen and the single bond in 3 hydrogen molecules, more energy is released on formation of the 3 N—H bonds in two ammonia molecules.
37. HF has the strongest chemical bond because it is the shortest bond.
39. −1450 kJ/mol (The reaction is exothermic.)
41. 35.6 kJ
43. 6.42×10^4 kJ of heat evolved
45. 19.0 °C
46. 7.74 kJ; endothermic
49. $\Delta H^\circ_{rxn} = -394$ kJ/mol
51. $\Delta H^\circ_{rxn} = -1220$ kJ/mol
53. 260 kJ of heat evolved
55. $Ag(s) + \frac{1}{2} Cl_2(g) \longrightarrow AgCl(s)$, $\Delta H^\circ_f = -127.1$ kJ
57. (a) $2\ Al(s) + \frac{3}{2} O_2(g) \longrightarrow Al_2O_3(s)$, $\Delta H^\circ_f = -1675.7$ kJ
    (b) $Ti(s) + 2\ Cl_2(g) \longrightarrow TiCl_4(\ell)$, $\Delta H^\circ_f = -804.2$ kJ
    (c) $N_2(g) + H_2(g) + \frac{3}{2} O_2(g) \longrightarrow NH_4NO_3(s)$, $\Delta H^\circ_f = -365.56$ kJ
60. (a) $\Delta H^\circ_{rxn} = 890.36$ kJ
    (b) Endothermic
62. $\Delta H^\circ_{rxn} = -69.14$ kJ
64. $\Delta H^\circ_f[OF_2(g)] = 18$ kJ/mol
66. −83.8 kJ
67. 0.78 g
69. From the heat of formation of methanol, −7.46 kJ of energy can be produced per gram of methanol. Octane has 48.1 kJ produced per gram.
71. Methane, 50.02 kJ/g; ethane, 47.48 kJ/g; propane, 46.35 kJ/g; butane, 32.60 kJ/g
72. Since gold has a lower specific heat, it will reach 100 °C first.
74. 75.4 g of ice melted.
76. (a) 416 kJ/mol
    (b) $\Delta H^\circ = 784$ kJ
    (c) CH_3, average C—H bond energy is 408.2 kJ/mol.
        CH_2, average C—H bond energy is 380.1 kJ/mol
        CH, average C—H bond energy is 338.4 kJ/mol.

    The average bond energy decreases when there are fewer C—H bonds.
78. 2.18×10^7 kJ evolved
80. $\Delta H^\circ_f[N_2H_4(\ell)] = 50.4$ kJ/mol
82. (a) −36.1 kJ
    (b) 1.18×10^4 kJ
84. Step 1: $\Delta H^\circ_{rxn} = -106.32$ kJ
    Step 2: $\Delta H^\circ_{rxn} = 275.34$ kJ
    Step 3: $\Delta H^\circ_{rxn} = 72.80$ kJ

    $H_2O(g) \longrightarrow H_2(g) + \frac{1}{2} O_2(g)$, $\Delta H_{rxn} = 241.82$ kJ

    This process is endothermic.
86. The ice gains energy so the process is endothermic. The liquid water loses energy so the process is exothermic.
88. Substance A
90. More—each gram of water has the same thermal energy, but beaker 1 has more grams of water.

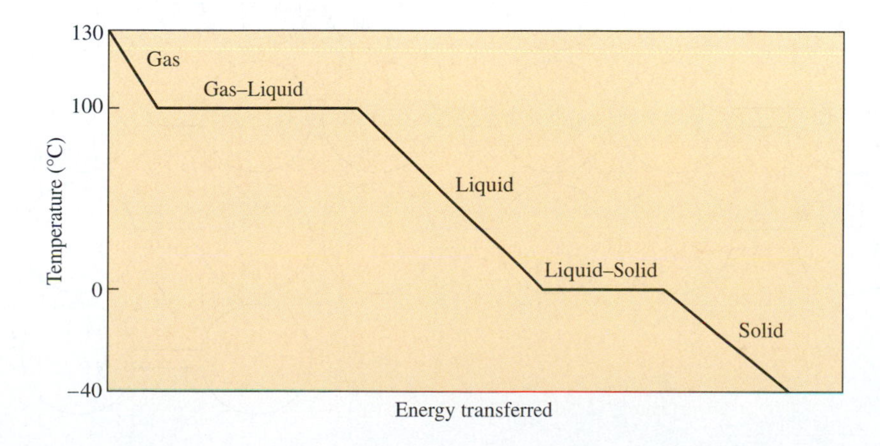

(Question 27)

92. Each of these situations refers to the change in thermal energy when reactants are converted to products. The difference is in describing the situation. Enthalpy of combustion refers to a substance being burned in oxygen to produce carbon dioxide and water; enthalpy of formation—one mole of a substance being formed directly from its elements; enthalpy of reaction—any reactants forming products; enthalpy of decomposition—a single reactant being broken down into smaller products.
94. ΔH_f° indicates the formation of one mole of a compound from its elements; therefore, the units are kJ/mol. ΔH° refers to the thermal energy change of a given reaction and not a specific amount of reactant consumed or product formed.
95. (a) 26.6 °C; in experiments 4 and 5 ascorbic acid is in excess and the amount of NaOH limits the thermal energy transfer. Since the amount of NaOH is the same, the heating and final temperature are the same.
(b) Experiment 1, ascorbic acid is limiting; Experiment 2, ascorbic acid is limiting; Experiment 3, both present in equimolar amounts; Experiment 4, NaOH is limiting; Experiment 5, NaOH is limiting.
(c) 1 hydrogen ion is involved. This is apparent because no increase in temperature is observed when more than 1 mol of acid is added per 1 mol of NaOH.

Chapter 7

16. Probability = 0.5 (50%) heads; Probability = 0.5 (50%) tails; 50 heads and 50 tails
18. Six arrangements have two molecules in each flask. Two arrangements have no molecules in one flask. The most probable arrangement is two molecules in each flask. The highest entropy is when two molecules are in each flask. (See below.)
20. (a) 86.1 °C, 359.3 K
(b) 177 °C, 450 K
(c) −40 °F, 233 K
(d) 2372 °F, 1573 K
(e) −273.15 °C, −459.67 °F
(f) 4727 °C, 8540 °F
(g) 51.7 °C, 324.8 K
(h) 37 °C, 310 K
22. (a) CO_2 vapor has the higher entropy because it is a gas and gases have higher entropies than solids. Also, its temperature is higher.
(b) The sugar dissolved in a cup of tea has a higher entropy. The entropy of a solid or liquid increases when dissolved in a liquid.
(c) The mixture of alcohol and water would have higher entropy. The entropy of a liquid will increase when dissolved in another liquid.
24. (a) NaCl would have a higher entropy because the attraction among Na^+ and Cl^- will be weaker than the attraction between Ca^{2+} and O^{2-}.
(b) $P_4(g)$; Since both are gases, the more complex molecule will have a higher entropy.
(c) $(CH_3)_2NH(g)$; more complex molecule
(d) $Hg(\ell)$; Liquids have higher entropies than solids.
26. ΔS° = 112 J/K·mol
28. (a) ΔS° = 113.0 J/K·mol
(b) ΔH°_{vap} = 38.2 kJ/mol

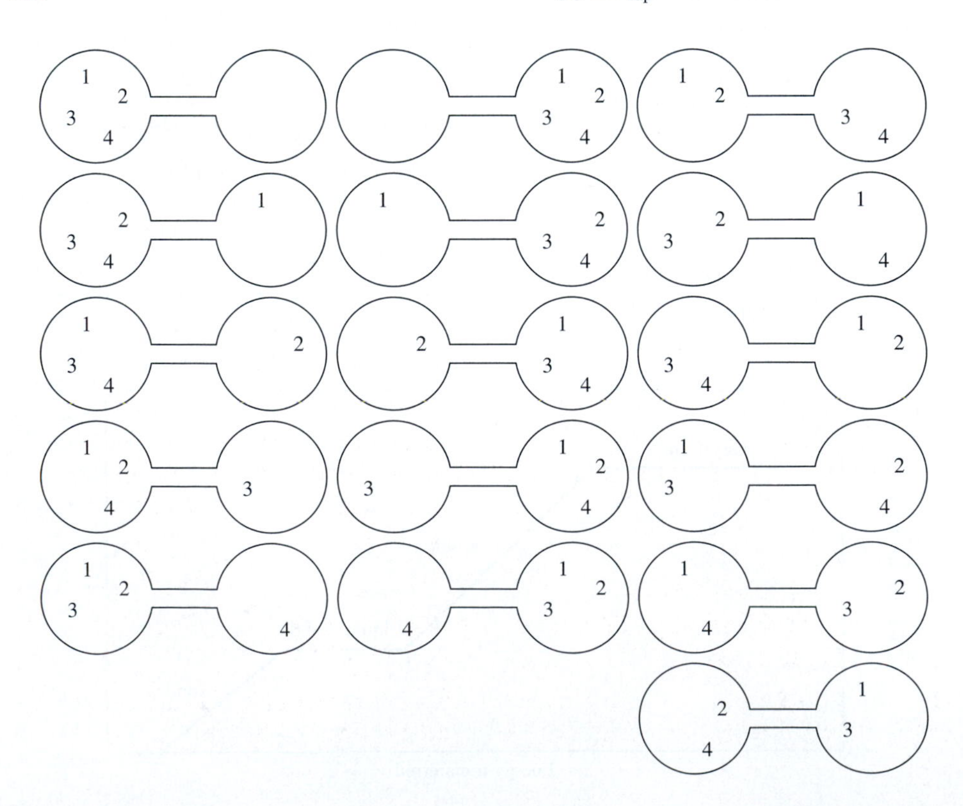

(Question 18)

29. (a) $\Delta S° = -120.64$ J/K
(b) $\Delta S° = 156.9$ J/K
(c) $\Delta S° = -198.76$ J/K
(d) $\Delta S° = 160.6$ J/K
(e) $\Delta S° = 332.2$ J/K
(f) $\Delta S° = 114.5$ J/K
(g) $\Delta S° = 135.9$ J/K
31. (a) The reaction is a combustion reaction, so enthalpy change is negative (exothermic). The reaction system also increases in entropy because 6 mol of gas are converted to 7 mol of gas. Since ΔH is negative and ΔS is positive, ΔG will be negative at all temperatures, and therefore the reaction is product-favored.
(b) The reaction has a large increase in entropy due to the formation of a gas.
33. The reaction would be product-favored at low temperatures and reactant-favored at high temperatures. This is because the reaction is exothermic but also involves a decrease in entropy.
35. (a) Product-favored
(b) Reactant-favored at low temperatures, product-favored at high temperatures
37. Yes; $\Delta S_{univ} = 1205.0$ J/K; $\Delta G^0 = -359.25$ kJ; product favored.
39. (a) Since $\Delta G = \Delta H - T\Delta S$, a negative value of ΔH and positive value of ΔS will result in ΔG being negative because T is always positive.
(b) Since $\Delta G = \Delta H - T\Delta S$, when ΔH and ΔS are the same sign, the signs of ΔH and $T\Delta S$ will be the same. The magnitudes of the terms will determine whether or not ΔG is positive or negative. The magnitude of T will have an effect on the magnitude of $T\Delta S$.
41. $\Delta G° = 462.2$ kJ. Since Gibbs free energy is positive, the reaction will not be product-favored under these conditions.
43. $\Delta H° = 117$ kJ; $\Delta S° = 174.7$ J/K; $\Delta G° = 64.7$ kJ. The reaction is reactant-favored. The signs of enthalpy and entropy are both positive, which means this reaction can be product-favored at high temperatures. This explains why heating is required to decompose the carbonate.
45. (a) $7\ C(graphite) + 6\ H_2O(g) \longrightarrow 2\ C_2H_6(g) + 3\ CO_2(g)$
$5\ C(graphite) + 4\ H_2O(g) \longrightarrow C_3H_8(g) + 2\ CO_2(g)$
$3\ C(graphite) + 4\ H_2O(g) \longrightarrow 2\ CH_3OH(\ell) + CO_2(g)$
(b) For the first reaction: $\Delta H° = 101.02$ kJ; $\Delta S° = -72.71$ J/K; $\Delta G° = 122.72$ kJ.

For the second reaction: $\Delta H° = 76.45$ kJ; $\Delta S° = -86.62$ J/K; $\Delta G° = 102.08$ kJ.

For the third reaction: $\Delta H° = 96.44$ kJ; $\Delta S° = -305.18$ J/K; $\Delta G° = 196.01$ kJ.

None of the reactions would occur under these conditions.
47. Metabolism refers to all of the chemical changes that occur as food nutrients are converted by an organism into Gibbs free energy and the complex chemical constituents of living cells.

Nutrients are the chemical raw materials needed for survival of an organism.

ATP is shorthand notation for adenosine triphosphate. ATP formation is the most important way Gibbs free energy is stored in the body.

ADP stands for adenosine diphosphate. ADP is converted to ATP.

The transformation of ADP to ATP is called oxidative phosphorylation.

Reactions that occur simultaneously with exchange of Gibbs free energy are said to be coupled. Often a reactant-favored process can be made to occur by coupling the reaction with a product-favored reaction.

Organisms that carry out photosynthesis are called phototrophs.

Chemotrophs depend on chemical bonds created by phototrophs for energy.

Photosynthesis is a series of reactions in a green plant that combines carbon dioxide with water to form carbohydrate and oxygen.
49. (a) Bonds broken: 7 C—H bonds; 5 C—C bonds, 7 C—O bonds, and 5 O—H bonds, and 6 O═O bonds.
Bonds formed: 12 C═O bonds and 12 O—H bonds.
$\Delta H = -2799$ kJ/mol
(b) The difference may be due to the fact that bond energies apply to gas-phase reactions, but glucose is a solid.
(c) 238 J/K
51. (a) 6.46 mol of ATP
(b) -105.5 kJ; product-favored reaction
53. Coal, petroleum and natural gas are our most important resources for free energy.
55. (a) 5.5×10^{18} J/yr
(b) 1.5×10^{16} J/d
(c) 1.7×10^{11} J/s
(d) 1.7×10^{11} W
(e) 620 W/person
(f) 3.36 kg (more than 7 lb!)
(g) You would need to generate 620 W continuously, which is a little less than the sprinter's power, but you would need to run as hard as you could 24 hours a day!
57. The second law states that the entropy of the universe always increases for a spontaneous process. Unscrambling an egg would result in more order (ΔS negative), which means that unscrambling an egg is a reactant-favored process. Humpty Dumpty was a fairy tale character shaped like an egg who had a great fall. He became scrambled and couldn't be put back together again.
59. Many of the oxides have negative heats of formation, which means their oxidations are exothermic. These systems are usually product-favored.
61. The entropy increases because the very ordered crystal lattice of the NaCl crystal breaks and dissociates into Na^+ and Cl^- ions in aqueous solution. Due to the generation of two moles of particles (Na^+ and Cl^-) in comparison to a mole of the NaCl lattice, the overall number of particles increases greatly upon dissolution and the particles are much more mixed up. Therefore entropy increases.

Chapter 8

2. Photons are massless particles of light. The energies of photons are calculated by multiplying the frequency of light by Planck's constant.
4. Only two electrons can be assigned to the same orbital in the same atom, and these two electrons must have opposite spin.
6. When an electron is said to occupy the 3_{p_x} orbital, it means that an electron will be somewhere inside the 3_{p_x} boundary surface 90% of the time.
8. For the main group elements, atomic radii increase going down a group in the periodic table and decrease going across a period. First ionization energies (the energy required to remove an electron from a neutral atom in the gas phase) generally increase going across a period and decrease going down a group.
10. K^+: $1s^22s^22p^63s^23p^6$; K^{2+}: $1s^22s^22p^63s^23p^5$
    The second ionization energy is much larger than the first because it corresponds to removing a 3*p* electron from K^+, a stable ion with a noble gas configuration.
12. Noble gas notation is an abbreviated representation of electron configurations in which the symbol of the preceding noble gas represents filled inner shells. Using calcium as an example, the noble gas configuration would be $[Ar]4s^2$.
14. Which "part" of the atom is affected by a gain in energy is determined by the wavelength. Wavelengths (and energies) in the microwave and infrared regions cause molecules to rotate and vibrate; visible and ultraviolet photons affect valence electrons; x-rays affect core electrons; and gamma rays can interact with nuclei of some atoms.
16. Short, high. Gamma rays are an example of high-energy radiation.
18. (a) Radio waves
    (b) Microwaves
20. $\lambda = 3.00 \times 10^{-3}$ m. Energy of 1 photon $= 6.63 \times 10^{-23}$ J. Energy of 1 mole of photons = 39.9J.
22. From least to most energy; $d < c < a < b$.
24. $\nu = 6.06 \times 10^{14} s^{-1}$ (6.06×10^{14} Hz)
26. Energy $= 4.41 \times 10^{-19}$ J
28. X rays and high-energy ultraviolet radiation can break bonds in molecules at the cellular level.
30. Sunlight consists of white light which contains light of all wavelengths and gives a continuous spectrum. A line spectrum contains only certain wavelengths of light.
32. Higher (excited); ground; difference
34. The larger the wavelength, the less energy the radiation has.
    (a) Less energy is needed than $n = 1 \longrightarrow n = 4$. (The wavelength is sufficient.)
    (b) Less energy is needed than $n = 1 \longrightarrow n = 4$. (The wavelength is sufficient.)
    (c) More energy is needed than $n = 1 \longrightarrow n = 4$. (A shorter wavelength is needed.)
    (d) Less energy is needed than $n = 1 \longrightarrow n = 4$. (The wavelength is sufficient.)
36. When $n = 4$, there are four subshells: 4*s*, 4*p*, 4*d*, 4*f*.
38. Bohr's model did not include energy subshells for electrons. When working with atoms or ions with more than one electron, Bohr's model did not work. The wave mechanical model works for such atoms because of the concept of orbitals and subshells.
40. The Bohr model of the atom has well-defined orbits for the position (and energy) of the electron, which Heisenberg showed was not possible.
42. The $n = 3$ subshell will have three different types of orbitals: *s*, *p*, and *d*. There are an *s* orbital, three *p* orbitals, and five *d* orbitals for a total of nine orbitals in the $n = 3$ shell.
44. Al: $1s^22s^22p^63s^23p^1$; S: $1s^22s^22p^63s^23p^4$
46. Ge: $1s^22s^22p^63s^23p^64s^23d^{10}4p^2$
48. Oxygen. Group designation indicates the number of valence electrons.
50. (a) :F: (with dots above and below)
    (b) ·In· (with a dot above)
    (c) :Te: (with dots above and below)
    (d) Cs·
52. (a) Na^+: $1s^22s^22p^6$
    (b) Al^{3+}: $1s^22s^22p^6$
    (c) Cl^-: $1s^22s^22p^63s^23p^6$

    Na^+ and Al^{3+} are isoelectronic.
54. Br: [Ar] $4s^23d^{10}4p^5$; Br^-: [Ar] $4s^23d^{10}4p^6$ or [Kr]
56. 18. All available orbitals of the last element are filled. The next element must start with a new shell.
58. Mn: [Ar] $4s^23d^5$ — 3*d*: [↑][↑][↑][↑][↑] 4*s*: [↑↓]

    Mn^{2+}: [Ar] $3d^5$ — 3*d*: [↑][↑][↑][↑][↑] 4*s*: [ ]

    Mn^{3+}: [Ar] $3d^4$ — 3*d*: [↑][↑][↑][↑][ ] 4*s*: [ ]
60. V: [Ar] $4s^23d^3$
62. (a) Eu: [Xe] $6s^24f^7$(or [Xe]$6s^25d^14f^6$ using the periodic table)
    (b) Yb: [Xe] $6s^24f^{14}$(or [Xe]$6s^25d^14f^{13}$ using the periodic table)
64. Ferromagnetism exists when a substance retains its magnetism; this occurs because the spins of all unpaired electrons in a cluster of atoms align themselves in the same direction.
66. Paramagnetic materials are attracted to a magnetic field but are not magnetic outside the magnetic field. In ferromagnetic materials, the spins of clusters of atoms remain aligned in the absence of a magnetic field and are therefore referred to as permanent magnets.
68. number
70. $P < Ge < Ca < Sr < Rb$

72. (a) Cs
    (b) O^{2-}
    (c) As
74. From smallest to largest ionization energy: Al < Mg < P < F
76. (a) Incorrect
    (b) Incorrect
    (c) Correct
    (d) Incorrect
78. Na would have the greatest difference; after the first ionization energy, a noble gas configuration is formed which is very stable. If a second ionization occurs, the noble gas configuration is broken.
80. (a) Al has the most metallic character.
    (b) Al has the largest radius.
    (c) Al < B < C
82. Hydrogen is the most abundant chemical in the universe.
84. Beryllium, because as Be^{2+} it is attracted to nitrogen in proteins and thus destroys the structure of proteins.
86. The three allotropes of carbon are graphite, diamond, and buckminsterfullerene (C_{60}).
88. $C + 2\ F_2 \longrightarrow CF_4$
    $P_4 + 6\ F_2 \longrightarrow 4\ PF_3$
    $2\ Al + 3\ F_2 \longrightarrow 2\ AlF_3$
    $Ca + F_2 \longrightarrow CaF_2$
    $2\ K + F_2 \longrightarrow 2\ KF$
90. 8.5×10^3 tons UF_6
92. Se: [Ar] $4s^2 3d^{10} 4p^4$. One pair of electrons and two unpaired electrons in the 4*p* subshell.
94. (a) S < O < F
    (b) O has the largest ionization energy. First ionization energies decrease down a group.
96. (a) S; (b) Ra; (c) N; (d) Ru; (e) Cu
98. (a) Metal; (b) Nonmetal; (c) Atom B; (d) Atom B
100. In^{4+} is unlikely because it does not have a noble gas configuration. Fe^{6+} is unlikely because the energy required to remove six electrons is too large; Fe^{2+} or Fe^{3+} form much more easily. Sn^{5+} is unlikely because the energy to remove five electrons is too large.
102. 2×10^5 tons SiO_2
104. (a) S is located in the third period.
    (b) $1s^2 2s^2 2p^6 3s^2 3p^4$
    (c) S would have the smallest ionization energy. O would have the smallest radius.
    (d) 805 g Cl_2
    (e) Cl_2 is limiting reactant, theoretical yield is 16.8 g.
106. Sr^{2+}: $1s^2 2s^2 2p^6 3s^2 3p^6 4s^2 3d^{10} 4p^6$
108. K_2X
110. 122 pm

Chapter 9

16. (a) 8 (b) 8 (c) 16 (d) 8

18. (a)
```
[    H    ]+
[    |    ]
[ H—P—H  ]
[    |    ]
[    H    ]
```
(b)
```
    H
    |
 H—Si—H
    |
    H
```
(c)
```
 ..       ..
 S ═ C ═ S
 ..       ..
```
(d)
```
    ..
[ : O—H ]−
    ..
```

20. (a)
```
  ..   ..
 :Cl—F:
  ..   ..
```
(b)
```
     ..
 H—Se—H
     ..
```
(c)
```
[        ..       ]−
[       :F:       ]
[  ..    |    ..  ]
[ :F—B—F:        ]
[  ..    |    ..  ]
[       :F:       ]
[        ..       ]
```
(d)
```
[        ..       ]3−
[       :O:       ]
[  ..    |    ..  ]
[ :O—P—O:        ]
[  ..    |    ..  ]
[       :O:       ]
[        ..       ]
```

22. (a)
```
        H
  ..    |
 :Cl—C—H
  ..    |
        H
```
(b)
```
[        ..        ]4−
[       :O:        ]
[  ..    |     ..  ]
[ :O—Si—O:         ]
[  ..    |     ..  ]
[       :O:        ]
[        ..        ]
```
(c)
```
[    H    ]+
[    |    ]
[ H—P—H  ]
[    |    ]
[    H    ]
```
(d)
```
    H  H
    |  |
 H—C—C—H
    |  |
    H  H
```

24. (a)
```
  ..         ..
 :F          F:
  ..\      / ..
     C═C
  ../      \ ..
 :F          F:
  ..         ..
```
(b)
```
 H        H
  \       |
   C═C—C≡N:
  /
 H
```

26. (a) Each F needs three more pairs of electrons to form an octet.
    (b) O_2 should have a double bond and each O should have two lone pairs of electrons.
    (c) Correct
    (d) H cannot form two bonds; Cl should be bonded to C; there should be 3 C—H bonds and 1 C—Cl bond.
    (e) N should have one more electron; there are 17 valence electrons.

28. hexane: $CH_3-CH_2-CH_2-CH_2-CH_2-CH_3$

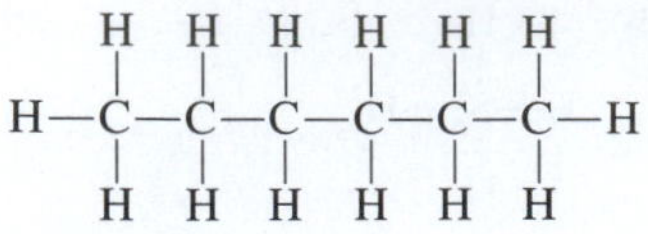

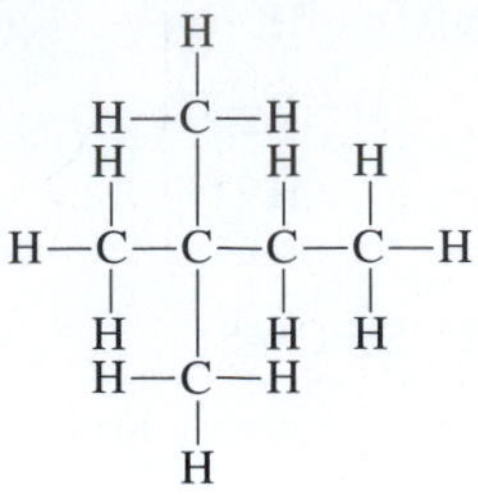

2,2-dimethylbutane

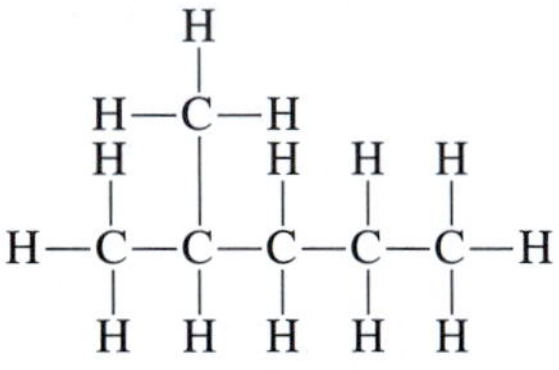

2-methylpentane

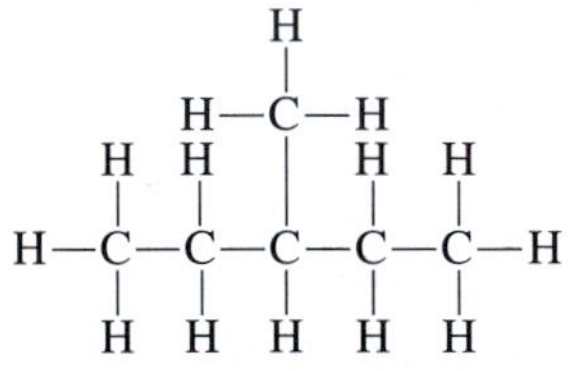

3-methylpentane

30. $H_2C{=}CH-CH_2-CH_3$; $CH_3-CH{=}CH-CH_3$; $H_2C{=}C(CH_3)-CH_3$; cyclobutane ; methylcyclopropane

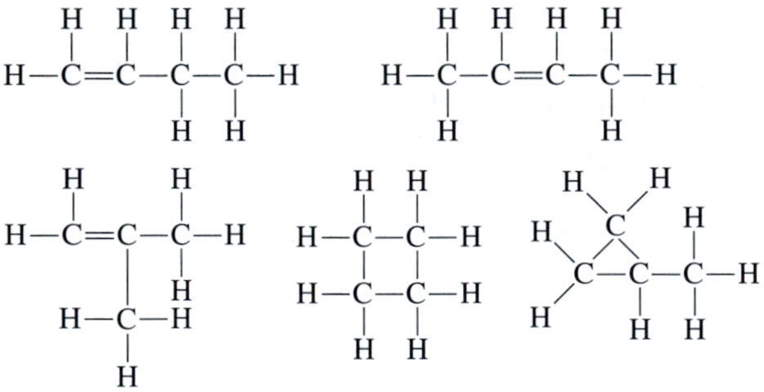

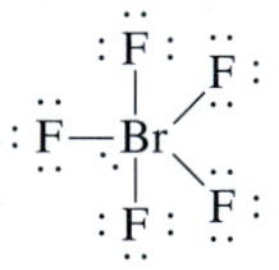

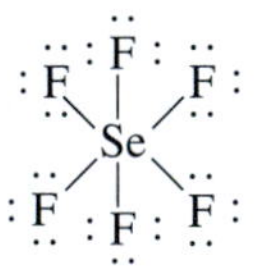

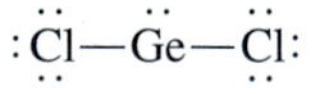

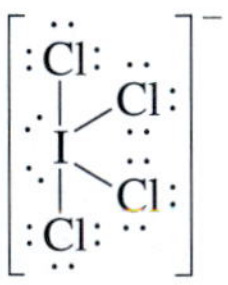

32. (a) BrF_5 (b) SeF_6 (c) $[:\ddot{Br}-\ddot{I}-\ddot{Br}:]^-$

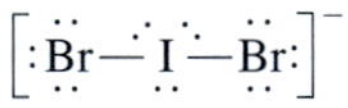

34. (b), (c), and (e)

36. (a) $:\ddot{Cl}-\ddot{Ge}-\ddot{Cl}:$ (b) $[ICl_4]^-$ (c) $:\ddot{F}-\ddot{Xe}-\ddot{F}:$

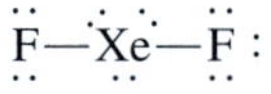

38. (a) B—Cl is shorter.
(b) C—O is shorter.
(c) P—O is shorter.
(d) C=O is shorter.

40. The C—O bond in carbon monoxide will require more energy to break because it is shorter and stronger than the bond in formaldehyde.

42. Strongest to weakest: C≡N > C≡C > C=N > C=C.

44. (a) H(Br)C=C(Br)H *cis* and *trans*

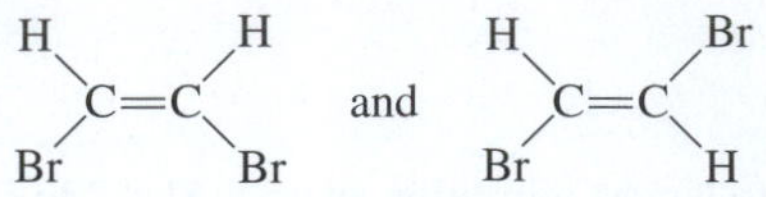

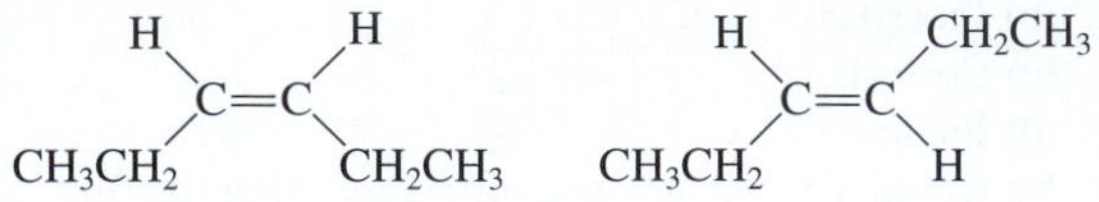

(b) $CH_3CH_2(H)C{=}C(H)CH_2CH_3$ *cis* and *trans*

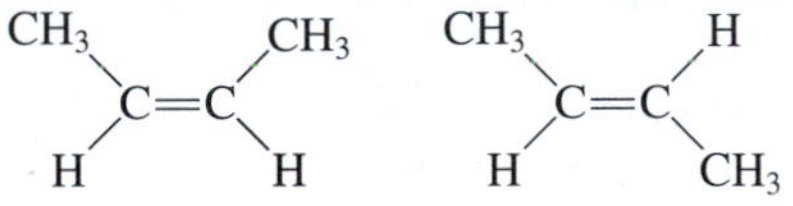

(c) $CH_3(H)C{=}C(H)CH_3$ *cis* and *trans*

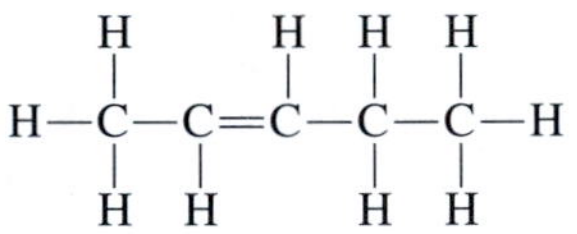

(d) No *cis-trans* isomers

46. (a) Cl(H)C=C(Cl)CH_3

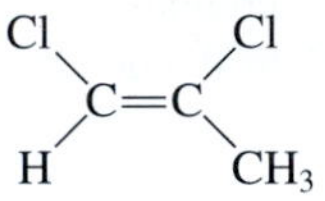

(b) $CH_3-CH{=}CH-CH_2-CH_3$

(c) $CH_3-CH_2-CH{=}CH-CH_2-CH_3$

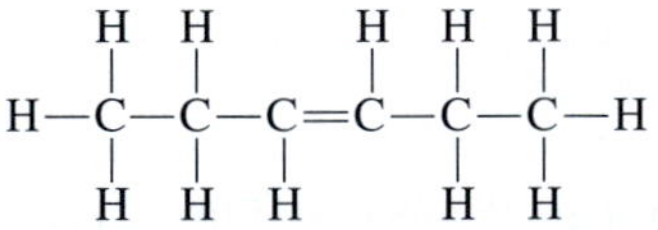

(d) $CH_3-CH{=}CH-CH_2CH_2CH_3$

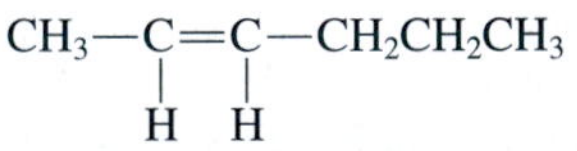

48. (a) There are two principal resonance structures for HNO_3.

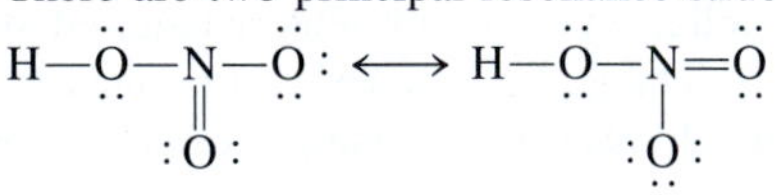

$H-\ddot{O}-N(=O)-\ddot{O}: \longleftrightarrow H-\ddot{O}-N(-\ddot{O}:)=O$

(b) There are two principal resonance structures for N_2O.

$:N{\equiv}N-\ddot{O}: \longleftrightarrow \ddot{N}{=}N{=}\ddot{O}$

50. The carbonate ion has one double bond in three resonance structures. The formate ion has one double bond in two resonance structures.

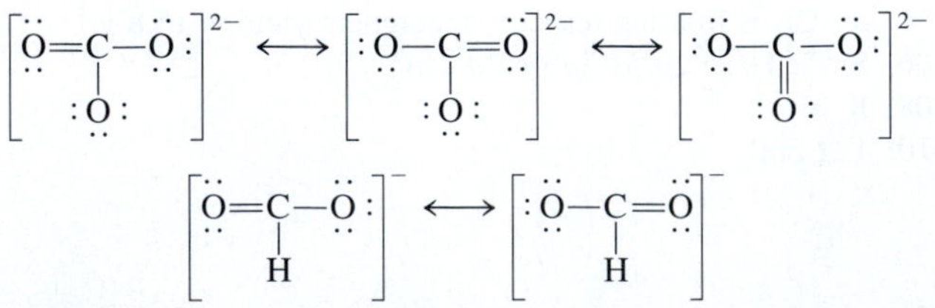

$[O{=}C(O)-O]^{2-} \longleftrightarrow [O-C(O){=}O]^{2-} \longleftrightarrow [O-C({=}O)-O]^{2-}$

$[O{=}C(H)-O]^- \longleftrightarrow [O-C(H){=}O]^-$

The overall average C—O bond length would be longer for the carbonate ion. This is because each bond is $\frac{1}{3}$ double and

$\frac{2}{3}$ single and therefore the bonds are more like single bonds than the C—O bonds in the formate ion, which are $\frac{1}{2}$ double and $\frac{1}{2}$ single.

52.

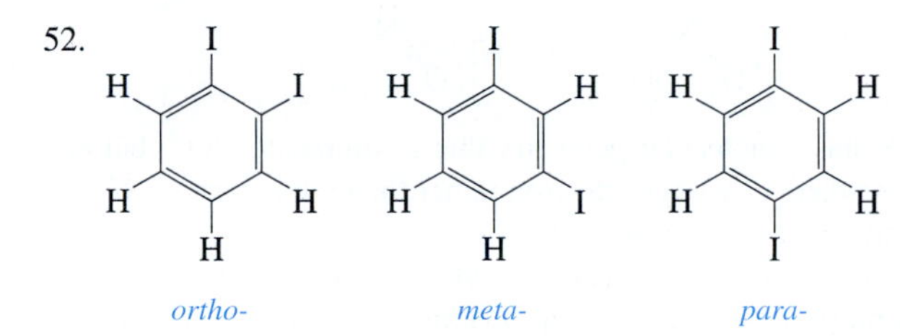

54. Least to most electronegative: K < Na < P < N < O
56. (a) C—N, nitrogen is more electronegative.
    C—H, carbon is more electronegative.
    C—Br, bromine is more electronegative.
    S—O, oxygen is more electronegative.
    (b) S—O is most polar.
58. (a) N—H bond is polar. N—C is slightly polar. C═O bond is polar.
    (b) C═O is the most polar bond. Oxygen is the partially negative end of this bond.
60. $[Ru(NH_3)_6]^{2+}$; $[Ru(NH_3)_6]Cl_2$; two Cl^- ions are needed.
62. $AlCl_3$
64. (a) $[Pt(NH_3)_2Br_2]$
    (b) $[Pt(en)(NO_2)_2]$
    (c) $[Pt(NH_3)_2BrCl]$
66. Elements that are close together in the periodic table generally have similar electronegativities and share electrons to form covalent bonds. Elements that are far apart have greater electronegativity differences and the formation of cations and anions is favored. The cations and anions then form ionic bonds. Note: two metal elements close together will have similar electronegativities but will not covalently bond.

68. (a) $\left[\,:\ddot{O}-Se(\ddot{O}:)_2-\ddot{O}:\,\right]^{2-}$ (b) SF_4: $:\ddot{F}:$ and $:\ddot{F}:$ axial, two $\ddot{F}:$ equatorial, lone pair on S (c) $:\ddot{F}-\ddot{P}-\ddot{F}:$ with $:\ddot{F}:$ below P

70. (a)
$[:\ddot{N}-N\equiv N:]^- \longleftrightarrow [\ddot{N}=N=\ddot{N}]^- \longleftrightarrow [:N\equiv N-\ddot{N}:]^-$
(b)
$[:\ddot{O}-C\equiv N:]^- \longleftrightarrow [\ddot{O}=C=\ddot{N}]^- \longleftrightarrow [:O\equiv C-\ddot{N}:]^-$

72. M(–O–)(–O–)C═O bidentate; M–O–C(═O)–O: monodentate

bidentate monodentate

74. (a) C═C is shorter than C—C.
    (b) C═C is stronger than C—C.

(c) C≡N is the most polar. N is the partially negative end of this bond.
76. CO is a polar molecule and has a greater attraction for iron than the nonpolar O_2 molecule.
78. O—O < Cl—O < O—H < O═O < O═C
80. (a) Neither; (b) Aromatic; (c) Cyclic alkane; (d) Alkane; (e) Alkane; (f) Neither
82. Instead of subtraction of an electron for the +1 charge, one electron was added. Total electron count should be 40, not 42.
84. These are isomeric polyatomic ions. The positions of the C and N atoms were changed instead of changing the position of an electron pair.

86. H above O, with H–O–H (O with lone pair); H–F–H (F with lone pairs)

88. As, 2.1; Se, 2.4; Cl, 3.0; Br and S are uncertain.
90. NF_5. It does not have an empty *d*-orbital to accept extra electrons.

Chapter 10

16. H—Be—H

linear

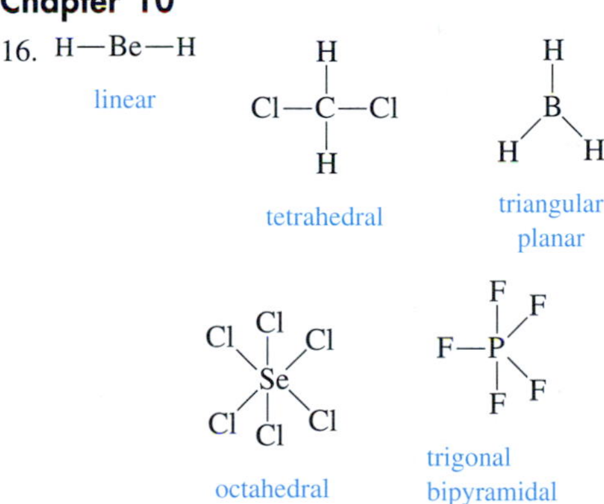

18. (a) Electron pair geometry: tetrahedral (four electron pairs). Molecular geometry: angular (two lone pairs, two bond pairs).
    (b) Electron pair geometry: tetrahedral (four electron pairs). Molecular geometry: triangular pyramidal (one lone pair, three bond pairs).
    (c) Electron pair geometry: tetrahedral (four electron pairs). Molecular geometry: tetrahedral (four bond pairs, zero lone pairs).
    (d) Electron pair geometry: linear (two double bonds treated as two single bonds).
    Molecular geometry: linear (no lone pairs, two bond pairs).
20. (a) Electron pair geometry: triangular planar.
    Molecular geometry: triangular planar.

$[B(:\ddot{O}:)_3]^{3-}$

(b) Electron pair geometry: triangular planar.
Molecular geometry: triangular planar.

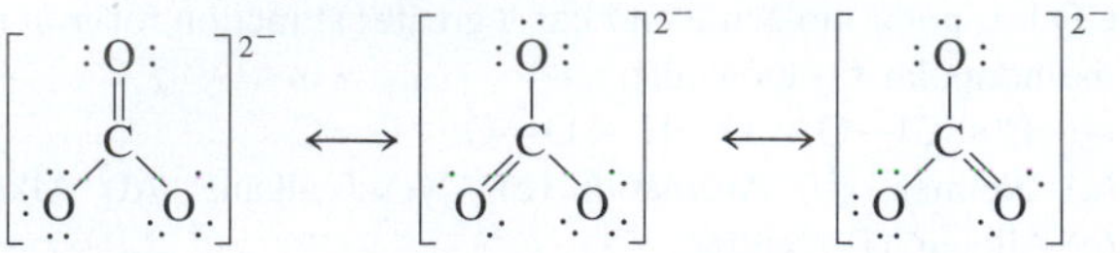

(c) Electron pair geometry: tetrahedral.
Molecular geometry: triangular pyramidal.

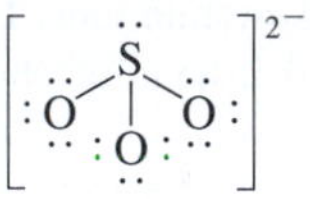

(d) Electron pair geometry: tetrahedral.
Molecular geometry: triangular pyramidal.

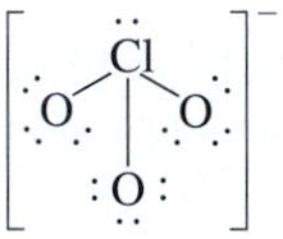

22. (a) Electron pair geometry: octahedral.
Molecular geometry: octahedral.

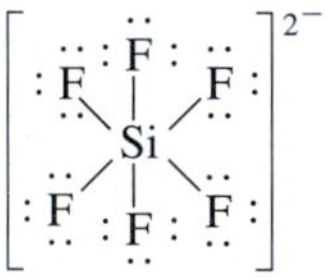

(b) electron pair geometry: triangular bypyramidal.
Molecular geometry: see-saw.

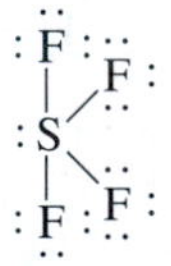

(c) electron pair geometry: triangular bipyramidal.
Molecular geometry: triangular bipyramidal.

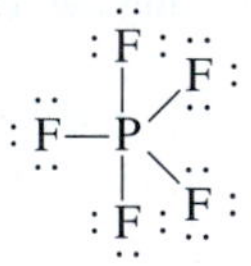

(d) electron pair geometry: octahedral.
Molecular geometry: square planar.

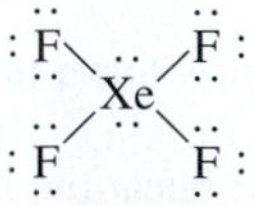

24. (a) 120° (b) 120° (c) 1 = 109.5°; 2 = 120° (d) 1 = 120°; 2 = 180°

26. (a) SeF_4 forms a see-saw shaped structure. One F—Se—F bond angle is 120° and the other two are 90°.
(b) 90° and 120°
(c) 90° and 180°

28. NO_2^- has a molecular geometry that is angular with a bond angle of approximately 120°. The "fatter" lone pair causes the angle to be a bit less than 120°.

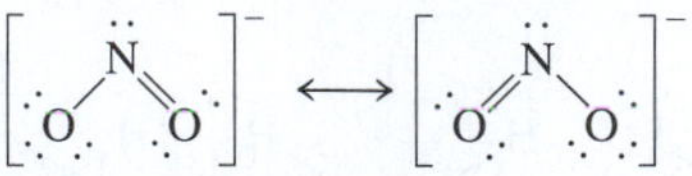

NO_2 has a molecular geometry that is similar to NO_2^-, but because there is a single electron rather than a lone pair on N, the bond angle is closer to 120°.

30. The geometry around the central atom, C, in chloroform is tetrahedral because there are four bonding pairs of electrons and no lone pairs. With no multiple bond to C and a tetrahedral geometry, the hybridization would be sp^3.

32.

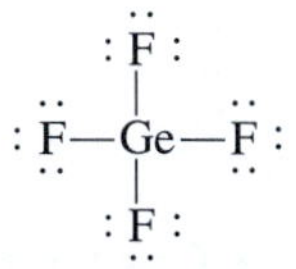

GeF_4 tetrahedral sp^3

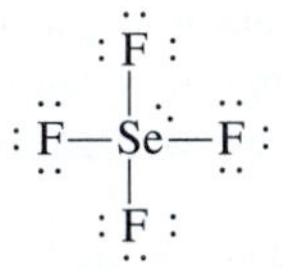

SeF_4 see-saw dsp^3

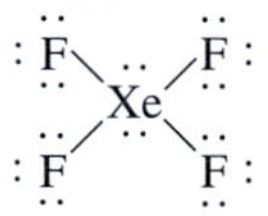

XeF_4 square planar d^2sp^3

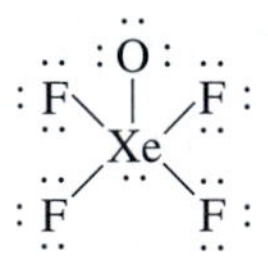

$XeOF_4$ square pyramidal d^2sp^3

34. (a) H_2O has the most polar bonds because of the large electronegativity difference (1.4) between oxygen and hydrogen.
(b) CO_2 and CCl_4
(c) The F atom is more negatively charged. Fluorine is more electronegative than chlorine.

36. (a) Not polar
(b) $H^{\delta+}$ B F δ− F
(c) $Cl^{\delta-}$ C H H H δ+
(d) Not polar

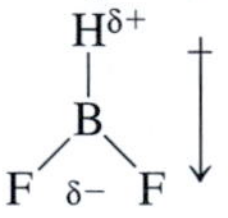

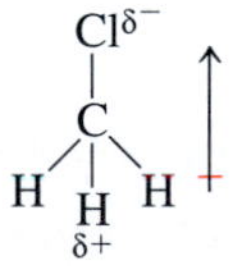

38. (a)

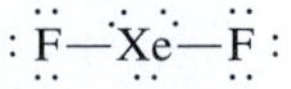

Linear geometry
Same bond on either side
Xe—F dipoles cancel

(b) δ− S H δ+ H

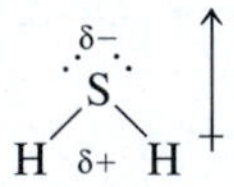

Bent geometry
H—S dipole will not cancel
Polar molecule

(c)

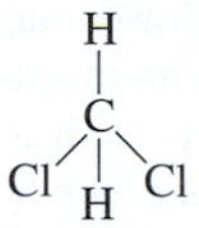

```
     H
     |
     C
   / | \
 Cl  H  Cl
```

Tetrahedral geometry

The two C—Cl and two C—H dipoles do not cancel. Polar.

(d) H—C≡N :

Linear. H—C dipole moment and C—N dipole moment do not cancel. Polar.

48. (a) The middle carbon atom is chiral.
    (b) There are no chiral atoms.
    (c) The third carbon moving from left to right is chiral.
50. (a) The second carbon atom moving from left to right is chiral.
    (b) No chiral atoms.
    (c) The fourth carbon moving left to right is chiral.
52. A molecule is chiral if it cannot be superimposed on its mirror image.
54. (a) UV radiation promotes valence electrons to higher energy states.

40. **Summary of Intermolecular Forces**

Interaction	Factors	Energy of Interaction	Example
Ion-ion	Charge on the ions; size of the ions	400–4000 kJ/mol	NaCl
Ion-dipole	Charge on ion; dipole moment	40–600 kJ/mol	NaCl in H_2O
Dipole-dipole	Dipole moment	5–25 kJ/mol	H_2O
Dipole-induced-dipole	Dipole moment; polarizability	2–10 kJ/mol	IBr in CS_2
London (induced dipole)	Polarizability	0.05–40 kJ/mol	Cyclohexane

42. Car wax is made of molecules containing primarily carbon and hydrogen atoms. The intermolecular forces present in wax are London (induced diople) forces. Water is polar and will undergo hydrogen bonding. There are no polar groups in wax for the water to be attracted to. A dirty car contains charged dust and dirt particles that the water is attracted to and hence the water will not bead up.
44. He < Ne < Ar < Kr < Xe < Rn. The boiling points of the noble gases follow the trend where the larger the atom, the higher the boiling point. This is true because the intermolecular forces holding liquid noble gas atoms together are London forces. The larger the atom, the larger the London forces. In larger atoms, it is easier to move outer electrons around (polarize the atom) because they are shielded from the nucleus more than outer electrons in smaller atoms. London forces come about from instantaneous induced dipoles as atoms or molecules come close together. It is easier to induce a dipole on a larger, "fluffier" electron cloud in contrast to a smaller, "harder" electron cloud.
46. Cyclohexane is a nonpolar solvent that exhibits only London forces between molecules. Other substances will dissolve best in cyclohexane if they are also nonpolar and exhibit only London forces. NaCl is an ionic solid and is the most polar kind of substance there is (completely separated charges). Ethanol (b) is a polar molecule that hydrogen bonds (dipole-dipole); and (c) is a nonpolar hydrocarbon like cyclohexane, so we would expect propane (C_3H_8) to dissolve well in cyclohexane. NaCl would be the least soluble.

(b) IR radiation is similar in energy to the internal motions of molecules like bending, stretching, and vibrating. IR spectroscopy tells how stiff a bond is or how strong it is, and this can provide information about the structure of an unknown molecule.

56. As can be seen in Figure 10.35, there are three points in the C-G pairs where hydrogen bonding occurs, as compared to two sites for A-T pairs. As the amount of C-G base pairs becomes more predominant and the amount of A-T base pairs decreases, the number of hydrogen bonds increases and the melting point of DNA increases.

58.

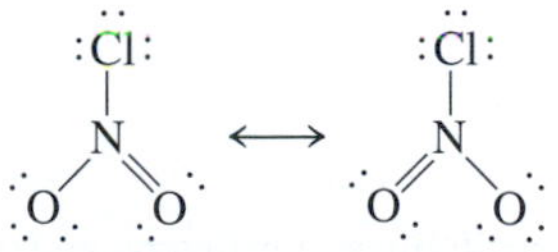

```
     :Cl:               :Cl:
      |                  |
      N       <-->       N
    /   \\             //   \
  :O:    O:          O:     :O:
```

The central atom is N in each resonance structure. There are three bonding groups surrounding the central atom, leading to a triangular planar electron pair geometry. The three pairs are all bond pairs, leading to a triangular planar molecular geometry. The bond angles will be approximately 120°.

59. (a) Angle 1 = 120°, angle 2 < 109.5°, angle 3 = 120°
    (b) O—H bond
    (c) C=O bond
61. (a) Angle 1 = approximately 120°, angle 2 = 109.5°, angle 3 = approximately 109.5°
    (b) C=O bond is the most polar, N—H is also polar.

64. (a)

```
    H
    |
    C
   /|\
  H H H
```

(b)

```
     H
     |    ..
  H—C—I :
     |    ..
     H
```

(c)

```
 [   :O:    ]2−
 [    ‖     ]
 [    C     ]
 [  ..  ..  ]
 [ :O    O: ]
 [  ..  ..  ]
```

(d)

```
 ..     ..
 O=C=O
 ..     ..
```

66.

Molecule or Ion	Electron Pair Geometry	Molecular Geometry	Hybridization of the Iodine Atom
ICl_2^+	Tetrahedral	Angular	sp^3
I_3^-	Triangular bipyramidal	Linear	dsp^3
ICl_3	Triangular bipyramidal	T-shaped	dsp^3
ICl_4^-	Octahedral	Square planar	d^2sp^3
IO_4^-	Tetrahedral	Tetrahedral	sp^3
IF_4^+	Triangular bipyramidal	See-saw	dsp^3
IF_5	Octahedral	Square pyramidal	d^2sp^3
IF_6^+	Octahedral	Octahedral	d^2sp^3

68. (a) N (b) B (c) P (d) Br
70. 5 (2 to the lone pairs on O in C=O; 2 to the lone pairs on O in O—H; and 1 to the H in O—H.
72. (a) Incorrect
(b) Correct
(c) Incorrect
(d) Incorrect

Chapter 11

2. Plastics, synthetic rubbers, synthetic fibers, fertilizers, and thousands of other consumer products.
4. The octane number of a gasoline is determined by comparing its knocking characteristics in a one cylinder test engine with those obtained for mixtures of heptane and 2,2,4-trimethylpentane.
6. Synthesis gas is a mixture of carbon monoxide and hydrogen produced by the reaction of pulverized coal with steam. Synthesis gas can be used to produce methanol, which can then be used as a starting material for the synthesis of other chemicals.
8. Oxygenated gasolines are blends of gasoline with organic compounds that contain oxygen (such as MTBE). Reformulated gasolines are oxygenated gasolines that contain a lower percentage of aromatic hydrocarbons and have a lower rate of evaporation. Both burn more completely than non-oxygenated or reformulated gasoline, and are expected to reduce carbon monoxide emissions in urban areas. Oxygenates are being produced in accordance with the 1990 amendments to the Clean Air Act and to lessen air pollution in general.
10. Octane numbers above one-hundred and below zero can occur because the method for determining octane numbers is fairly old. Fuels superior to 2,2,4-trimethylpentane have been developed since the method was developed. Fuels which cause more knocking than pure heptane must be assigned numbers less than zero.
12. The Mobile methanol-to-gasoline process can be described by the following chemical equations:
$2CH_3OH \longrightarrow (CH_3)_2O + H_2O$
$2(CH_3)_2O \longrightarrow 2C_2H_4 + 2\ H_2O$
$C_2H_4 \rightarrow$ hydrocarbon mixture in the C_5–C_{12} range
All reactions use the Zsm-5 zeolite catalyst.
14. Most of the useful compounds obtained from coal are in the aromatic class of hydrocarbons. Crude oil is a complex mixture of alkanes, cycloalkanes, alkenes, and aromatic hydrocarbons.
16. Esters have lower boiling points than carboxylic acids because esters don't hydrogen bond.
18. All condensation polymerization reactions involve a chemical reaction in which two molecules react by splitting out or eliminating a small molecule such as water.
20. Poly-cis-isoprene is elastic and gummy; poly-trans-isoprene is hard and brittle.
22. Collection, sorting, reclamation, and end-use are the four conditions that must be met for successful recycling of solid waste.
24. (a) Approximately 30 to 200°C
(b) 55
(c) No. Although isooctane burns smoothly in an engine, gasoline is formulated with additional additives that keep the engine clean and extend its life.
26. Gasolines contain molecules in the liquid phase which can easily overcome intermolecular forces and escape into the gas phase. Thus the vapor pressure of gasoline is quite high and gasoline is considered volatile.
28.

```
          CH3
          |
   CH3—C—O—CH3
          |
          CH3
```

30. In older coal gasification, one mole of pulverized carbon reacts with steam to form carbon monoxide and hydrogen. The process is endothermic and requires 131 kJ/mol of carbon. Carbon monoxide and hydrogen can each react with oxygen exothermically producing −566 kJ/mol CO and −242 kJ/mol H_2. The total heat produced is more than the heat used to burn the coal. In newer coal gasification, carbon is mixed with steam and a catalyst to produce methane. Methane is then burned to release 802 kJ/mol.
32. Sources of CO_2 include human respiration and burning of fossil fuels. Plants remove CO_2 in photosynthesis, and CO_2 also dissolves in rainwater and oceans. Twice as much CO_2 enters the atmosphere than is removed.

34. (a) propanol $CH_3CH_2CH_2OH$ (b) 2-propanol CH_3CHCH_3 (OH on C-2) (c) $H_3C—C(CH_3)_2—OH$

36. (a) $CH_3C(OH)(CH_3)CH_2CH_2CH_3$ (b) $HOCH_2CH(CH_3)CH(CH_3)CH_3$
(c) $CH_3CH(OH)CH_2CH(CH_3)CH_3$ (d) $CH_3CH(CH_3)CH(OH)CH_2CH_3$
(e) $CH_3C(CH_3)(CH_3)OH$ (f) $CH_3CH(OH)CH_3$

38. (a) $CH_3CH(=O)$ and $CH_3C(=O)—OH$
(b) $CH_3CH_2CH_2CH(=O)$ and $CH_3CH_2CH_2C(=O)—OH$

40. (a) $(CH_3)_2CHCH_2CH_2OH$
(b) $CH_3CH_2CH(OH)CH_2CH_3$
(c) $CH_3CH_2CH(CH_3)CH_2OH$

42. A method of producing methanol is to heat a hardwood in the absence of air, hence the name wood alcohol. Ethanol is largely produced through fermentation of carbohydrates supplied by plants, hence the name grain alcohol.

44. Many biological systems require hydroxyl groups to enhance water solubility through hydrogen bonding.

46. (a) $HC(=O)—O—CH_3$ (b) $CH_3CH_2C(=O)—O—CH_2CH_2CH_3$
(c) $CH_3CH_2C(=O)—O—CH_3$

48. (a) CH_3OH and CH_3CH_2COOH
(b) $HCOOH$ and CH_3CH_2OH
(c) CH_3COOH and CH_3CH_2OH

50. Polyethylene, polystyrene, and polycarbonates are examples of thermoplastics. Thermoplastics undergo a reversible change with temperature. When heated, thermoplastics soften, but harden when cooled.

52. (a) $\left(—CH_2—CH_2—CH_2—CH_2—\right)_n$ (b) $\left(—CCl_2—CH_2—\right)_n$
(c) $\left(—CH_2—CH(OC(=O)CH_3)—\right)_n$

54. $\left(—CH_2—CH(C(=O)—O—CH_3)—\right)_n$

56. Polyisoprene. Natural rubber is poly-cis-isoprene.

58. Alcohols and carboxylic acids.

60. Diamines react with dicarboxylic acids to form polyamides. Nylon is a typical polyamide.

62. Proteins contain many different monomers while polyamides contain a single monomer which repeats.

64. $H_2C=CH(CN)$ (H, H on one carbon; H and CN on the other)

66. $\left(—NH—C_6H_4—NHC(=O)—C_6H_4—C(=O)—\right)_n$

68. Burning plastics can result in production of chlorinated organic compounds which can be toxic or carcinogenic.

70. Proteins

72. $H_2N—CH(CH_3)C(=O)N(H)CH_2C(=O)N(H)CH(CH_2C_6H_5)C(=O)OH$

74. α-D glucose is on the left, β-D glucose is on the right.

76. Amylopectin

78. Glycogen is stored in liver and muscle tissue and provides glucose for energy. Cellulose is also composed of D-glucose units but these units are all in the β-ring form. Humans do not have the enzyme necessary to break the β-1,4 bonds; cows have this enzyme.

80. Ethanol has an alcohol functional group that allows it to hydrogen bond with water. The hydrocarbon end is short enough not to interfere with this bonding. Decanol has an alcohol functional group that allows it to hydrogen bond to water, but the hydrocarbon end is so long that water cannot surround the molecule easily.

82. Rubber molecules are vulcanized by the addition of sulfur. Short chains of sulfur atoms bond the polymer chains together.

86. (a) $2C_8H_{18} + 25O_2 \longrightarrow 16CO_2 + 18\ H_2O$
(b) 1.6×10^3 L

88. In glycogen, α-glucose chains are highly branched; in cellulose the glucose units are all in the β-branched form which may allow for better packing.

90. $CH_3CH_2C(=O)OC(=O)CH_2CH_3$ $CH_3CH_2C(=O)OH$

92. $CH_3CH_2C(=O)OH$ hydrogen bonded (dashed lines) to molecules of $HOC(=O)CH_2CH_3$ and to $O=C(CH_2CH_3)OH$: the carbonyl O of $CH_3CH_2C(=O)OH$ accepts hydrogen bonds from $CH_3CH_2C(=O)OH$ and $HOC(=O)CH_2CH_3$; its OH donates to $O=C(CH_2CH_3)(OH)$ and is hydrogen bonded to $HOC(=O)CH_2CH_3$ molecules.

$HOCH_2CH_2CH_2CH_3$
H
$CH_3CH_2CH_2CH_2OH$---$OCH_2CH_2CH_2CH_3$
$HOCH_2CH_2CH_2CH_3$

Propanoic acid should have a higher boiling point than 1-butanol because there are more hydrogen bonds between propanoic acid molecules than between 1-butanol molecules.

94. $CH_3C{\equiv}CH$

96. OH
HO—P=O
OCH$_2$ O OH
C C
H H H H
C—C
O H
P=O
O—CH$_2$ O OH
C C
H H H
C—C
OH H

Chapter 12

14. (d) powdered sugar because of the high total surface area; however, powdered sugar particles are coated to keep them from sticking together, and the coating does not dissolve. This results in a cloudy suspension, and granulated sugar appears to dissolve faster.

16.

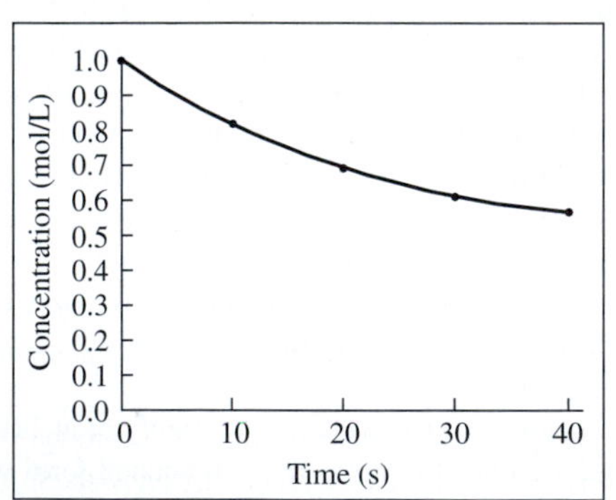

(a) Rate of change decreases because concentration of reactants is decreasing.

Rate of Change of [A] (mol/L-s)	Time Interval (s)
-1.67×10^{-2}	0–10
-1.19×10^{-2}	10–20
-8.9×10^{-3}	20–30
-7.0×10^{-3}	30–40

As the reaction proceeds, the reactants become products. This means there are fewer reactant particles to react with each other over time, and the rate decreases.

(b) The rate of change of [B] is double the rate of change of [A] because there are two molecules of B being formed for every molecule of A reacted. $+2.38 \times 10^{-2}$ mol/L is the rate of change for [B] from 10 s to 20 s time interval, because B is being formed as A is consumed.

18. 1.5 times. Cut the aluminum into very small pieces.

19. The rate is increased by a factor of 9, when the concentration of A is tripled.
The rate is decreased by a factor of 4, when the concentration of A is halved.

21. (a) rate $= k[NO_2]^2$
(b) The rate would decrease by a factor of 4.
(c) The rate would be unaffected.

23. (a) First order in A, second order in B.
(b) First order in A and B.
(c) First order in A.
(d) Third order in A, first order in B.

25. (a) rate $= 9.0 \times 10^{-4}$ mol/L · hr
(b) rate $= 1.8 \times 10^{-3}$ mol/L · hr
(c) rate $= 3.6 \times 10^{-3}$ mol/L · hr
As the initial concentration increases, the rate of disappearance of $Pt(NH_3)_2Cl_2$ increases. As the rate of disappearance of $Pt(NH_3)_2Cl_2$ increases, the rate of formation of Cl^- increases equally. The rate law says that the rate is directly proportional to the concentration of $Pt(NH_3)_2Cl_2$, which agrees with the calculated result.

27. (a) rate $= k\,[CO]\,[NO_2]$
(b) First order with respect to both reactants.
(c) $k = 4.2 \times 10^8$ L/mol · hr

29. (a) rate $= k\,[CH_3COCH_3]\,[H_3O^+]$
(b) $k = 4 \times 10^{-3}$ L/mol · s
(c) rate $= 2 \times 10^{-5}$ mol/L · s

31. See graphs. Note that the zeroth-order reaction is over after 1 s. The graphs differ from those in Figure 12.5 because different quantities are on the *y* axes, and the time scale is different.

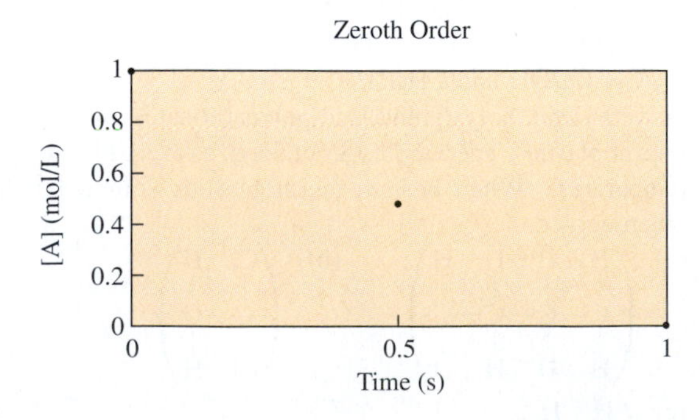

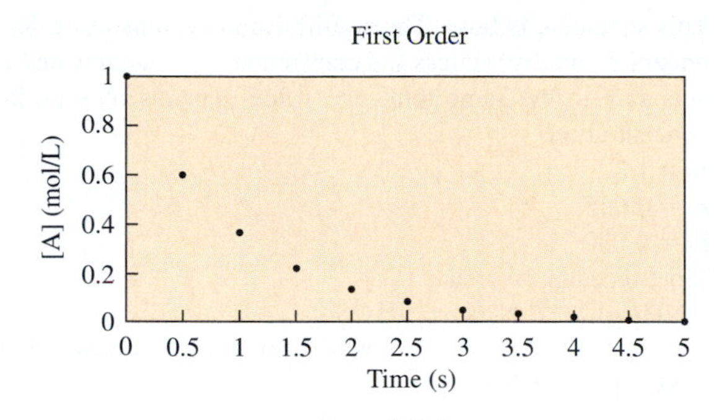

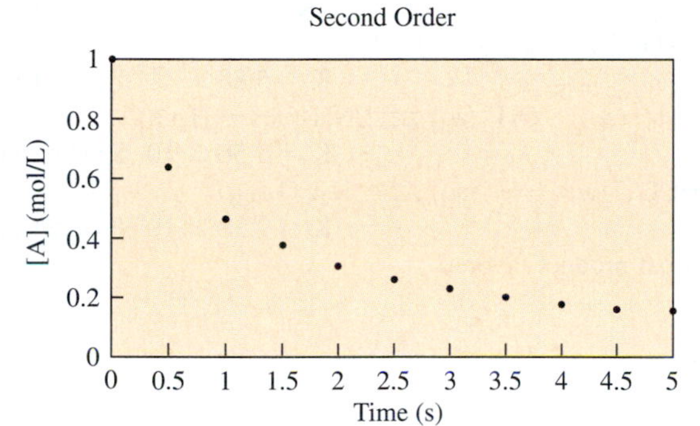

34. (a) not elementary
    (b) bimolecular and elementary
    (c) not elementary
    (d) unimolecular and elementary
    (e) not elementary; the HCl must first dissolve and then ionize.
    (f) bimolecular and elementary
    (g) unimolecular and elementary
    (h) not elementary
    (i) bimolecular and elementary
36. E_a = 19 kJ/mol; 1.8 times
38. 10.7 times
40. The activation energy is smaller when $A + B \longrightarrow C + D$. Therefore, the reaction is exothermic as written.
42. Step 1: $NO_2(g) + F_2(g) \longrightarrow FNO_2(g) + F(g)$
    Step 2: $NO_2(g) + F(g) \longrightarrow FNO_2(g)$

    $$2\ NO_2(g) + F_2(g) \longrightarrow FNO_2(g)$$

    Step 1 is rate determining because it is the slower step.
44. (a) true (b) false (c) false (d) false
46. (a) $H_3O^+(aq)$ is the homogeneous catalyst.
    (b) No catalyst.
    (c) Pt(s) is the heterogeneous catalyst.
    (d) No catalyst.
48. $\frac{k'}{k} = e^{(Ea - Ea')}/RT$
49. 26 times greater
51. See glossary (following appendices), or a dictionary
52. 30 times
54. Catalysts increase the rates of reactions. By allowing some reactions to proceed fast enough at relatively low temperatures, catalysts reduce the quantity of fuel needed to maintain elevated temperatures, and hence the cost of goods. In other cases (such as the Haber-Bosch process, Section 13.8), a catalyst allows a reaction to occur at a temperature where it is product favored.
57. Transition metals; oxidation reduction reactions. Because transition metal ions usually have several common oxidation states, they can easily participate in oxidation-reduction reactions.
58. $CFCl_3$
    CF_2Cl_2
    CF_3Cl
60. To some extent, but because photons are required to produce Cl atoms from CFCs, the ozone depletion is much greater with sunlight.
62. Kinetics is important because studies of the rates of reactions that deplete ozone, combined with knowledge about the mechanisms of those reactions, allows us to predict the extent of ozone depletion.
64. (a) $\Delta G° = -86.46$ kJ
    (b) $\Delta G°(CH_3OH) = -702.35$ kJ; $\Delta G°(CO) = -257.191$ kJ; $\Delta G°(CH_3COOH) = -873.1$ kJ
    (c) $CH_3OH(\ell)$, CO(g), and CH_3COOH are thermodynamically unstable to oxidation; therefore, to exist they must be kinetically stable. CO_2 and H_2O are thermodynamically stable.
    (d) A substance may be thermodynamically unstable but its reactions may be so slow that the substance may remain unchanged.
    (e)
    3×10^9 kg CH_3OH/yr
    5×10^{10} kg CO
    1×10^9 kg CH_3COOH/yr
    1.3×10^{18} kg H_2O
    2×10^{17} kg CO_2
    (f) $\Delta G° = -130.4$ kJ
65. rate = k $[H_2][NO]^2$
67. (a) The reaction is first order with respect to all three reactants.
    (b) $HCrO_4^- + H_3O^+ \longrightarrow \cancel{H_2CrO_4} + H_2O$
    $\cancel{H_2CrO_4} + H_2O_2 \longrightarrow \cancel{H_2CrO_5} + H_2O$
    $\cancel{H_2CrO_5} + H_2O_2 \longrightarrow H_2CrO_5 + 2H_2O$

    $$HCrO_4^- + H_3O^+ + 2H_2O_2 \longrightarrow CrO_5 + 4H_2O$$
68. The reaction is first-order in $Pt(NH_3)_2Cl_2$ (see Problem-Solving Practice 12.3, (⇐ p. 521). As the $[Pt(NH_3)_2Cl_2]$ decreases, the rate decreases. The concentration of Cl^- increases because Cl^- is a product of the reaction.
70. Catalysts speed up many reactions. This produces reaction products faster and at lower temperatures and thereby reduces costs and lowers the prices of finished products.
72. Curve A represents H_2O; curve B represents O_2; curve C represents H_2O_2.
74. Picture (a)
76. Rate = $k[A]^2[B][C]^2$

Chapter 13

9. 0°C. The system is dynamic since H_2O molecules are continually exchanging between the solid and liquid phases. You could

demonstrate this by labeling the liquid water with ^{18}O, an isotope of oxygen; some ^{18}O would be found in the ice.

11. See graphs below.

13. (a) $K = \frac{[H_2O]^2[O_2]}{[H_2O_2]^2}$ (b) $K = \frac{[PCl_5]}{[PCl_3][Cl_2]}$

(c) $K = [CO]^2$ (d) $K = \frac{[H_2S]}{[H_2]}$

15. (a) $K = [Cl_2]$ (b) $K = \frac{[Cu^{2+}][Cl^-]^4}{[CuCl_4{}^{2-}]}$

(c) $K = \frac{[CO_2][H_2]}{[CO][H_2O]}$ (d) $K = \frac{[Mn^{2+}][Cl_2]}{[H_3O^+]^4[Cl^-]^2}$

17. The reaction may be product favored but it is kinetically slow. Much of the sulfur is not in direct contact with oxygen from the air.

19. The relationship would be (c) $K^3 = K'$

21. $K = 0.40$

23. (a) $K = 1.4$
(b) $K = 0.5$
(c) $K = 1 \times 10^2$

This statement is false. The equilibrium-constant value depends on stoichiometry. Unless the coefficients of reactants and products sum to the same total, the value depends on equilibrium concentrations.

25. $K = 1.2$

27. $K = 0.025$

29. K = 0.080

30. Most reactant favored

$2\ NH_3(g) \rightleftharpoons N_2(g) + 3H_2(g)$ K = 2.86×10^{-9} Reactant favored

$2\ NO_2(g) \rightleftharpoons N_2O_4(g)$ K = 1.7×10^2 Product favored

$2NO(g) \rightleftharpoons N_2(g) + O_2(g)$ K = 5.88×10^2 Product favored

$NH_4{}^+(aq) + OH^-(aq) \rightleftharpoons NH_3(aq) + H_2O(\ell)$ K = 5.56×10^4 Product favored

$HCO_3{}^-(aq) + H^+(aq) \rightleftharpoons H_2CO_3(aq)$ K = 2.38×10^6 Product favored

Most product favored

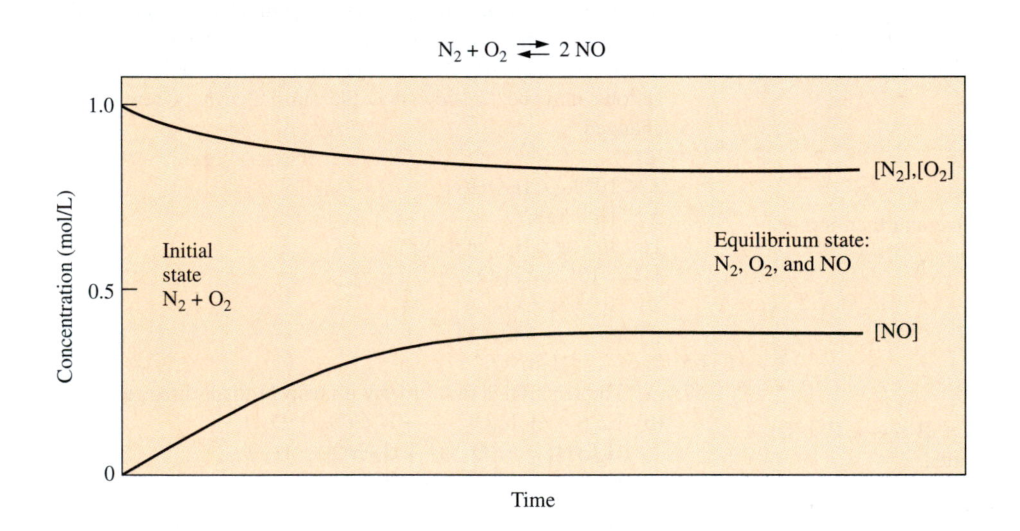

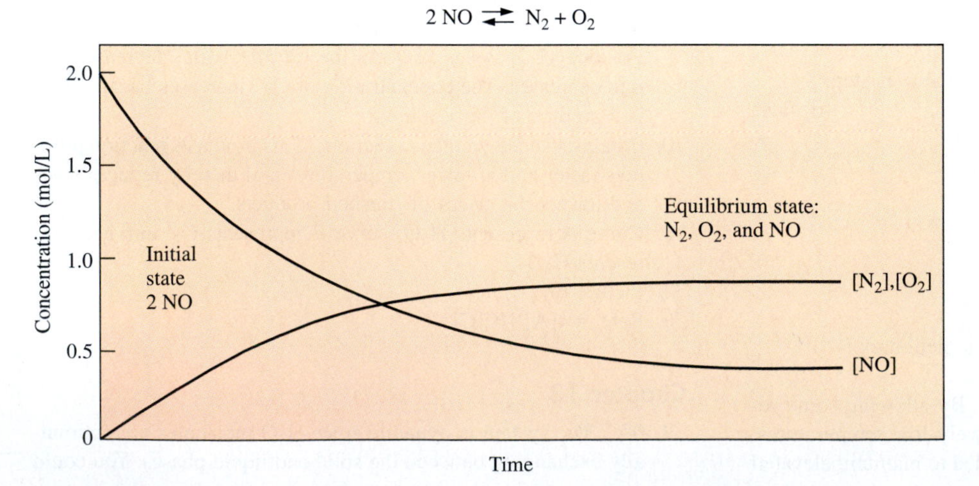

32. (a) Butane $\rightleftharpoons$ 2-methylpropane

	Butane	2-methylpropane
initial	.1	.1
change	$-x$	$+x$
equilibrium	$.1 - x$	$0.1 + x$

(b) $\dfrac{(0.1 + x)}{(.1 - x)} = 2.5;\ x = 0.043$

(c) [butane] = 0.010 M
[2-methylpropane] = 0.024 M

34. 1.2 g

36. $K = 1.0 \times 10^{-14} = [H^+][OH^-]$; $K = (1 + x)^2$; quadratic
$K = 1.8 \times 10^{-5} = [CH_3COO^-][H^+]/[CH_3COOH]$; $K = (1 + x)^2/(1 - x)$; quadratic
$K = 3.5 \times 10^8 = [NH_3]^2/[N_2][H_2]^3$; $K = (1 + 2x)^3/(1 - x)(1 - 3x)^3$; cubic
$K = 7 \times 10^{56} = [O_2]^3/[O_3]^2$; $K = \left(1 + \frac{3}{2}x\right)^3/(1 - x)^2$; cubic
$K = 1.7 \times 10^2 = [N_2O_4]/[NO_2]^2$; $K = \left(1 + \frac{1}{2}x\right)/(1 - x)^2$; quadratic
$K = 5.6 \times 10^3 = [HCO_2H]/[HCO_2^-][H^+]$; $= K = (1 + x)/(1 - x)^2$; quadratic
$K = 6.7 \times 10^{15} = 1/[Ag^+][I^-]$; $K = 1/(1 - x)^2$; quadratic
The 1st, 2nd, 5th, 6th, and 7th equations can be solved using the quadratic equation. For the 3rd and 4th reactions, x will be very close to 1 because the equilibrium constant is very large and the reaction goes to completion; at equilibrium the concentration of O_2 will be 2.5 M. For the third reaction the formula for roots of a cubic equation is needed.

38. [Br] = .28 M
[Cl] = 0.058 M
[F] = 1.4 M
[H] = 1.8×10^{-5} M
[I] = 0.91 M
[N] = 3×10^{-14} M
[O] = 4.0×10^{-6} M
Based on these results, fluorine has the lowest bond dissociation energy.

40. (b) The $[N_2O_4]$ will decrease. The reaction is endothermic. Increasing T shifts the equilibrium in the endothermic direction.

42. (a) no change (b) shift left (c) shift left

44. (a) shift right (b) shift left (c) shift right

46. No change

48. (a) false
(b) false
(c) true; since $\Delta G^\circ = -RT \ln K_{th}$, and $\ln(1) = 0$.
(d) true; since $\Delta G^\circ = \Delta H^\circ - T\Delta S^\circ = 0$.
(e) true

50. (a) $CH_4(g) + 2O_2(g) \longrightarrow CO_2(g) + 2H_2O(g)$
$\Delta G^\circ = -800.8$ kJ
(b) $C_6H_6(\ell) + 15/2\ O_2(g) \longrightarrow 6CO_2(g) + 3H_2O(g)$
$\Delta G^\circ = -3176.4$ kJ
(c) $K = 6 \times 10^{35}$
$CH_3OH(\ell) + 3/2\ O_2(g) \longrightarrow CO_2(g) + 2H_2O(g)$
$\Delta G^\circ = -685.2$ kJ
The reactions all are product favored but thermodynamics does not address the rate of these reactions.

52. Spreadsheets for (a and b)

	S(s)	O2(g)	SO2(g)	Reaction	
Delta Hf (J)	0	0	−296830	−296830	Exothermic
S-std (J)	31.8	205.138	248.22	11.282	
Delta Gf (J)	0	0	−300194	−300194	

T (K)	DelH - T*DelS = DelG	K
298	−300192	4.18E + 52
300	−300215	1.88E + 52
400	−301343	2.25E + 39
500	−302471	3.98E + 31
600	−303599	2.7E + 26
700	−304727	5.49E + 22
800	−305856	9.35E + 19
900	−306984	6.57E + 17
1000	−308112	1.24E + 16

	2 SO2(g)	O2(g)	2 SO3(g)	Reaction	
Delta Hf (J)	−296830	0	−395720	−197780	Exothermic
S-std (J)	248.22	205.138	256.76	−188.058	
Delta Gf (J)	−300194	0	−371060	−141732	

T (K)	DelH − T*DelS = DelG	K
298	−141739	7.01E + 24
300	−141363	4.11E + 24
400	−122557	1.01E + 16
500	−103571	6.91E + 10

600	−84945.2		24854422
700	−66139.4		86210.62
800	−47333.6		1232.193
900	−28527.8		45.2658
1000	−9722		3.219908

	SO3(g)	H2O(l)	H2SO4(l)	Reaction
Delta Hf (J)	−395720	−285830	−813898	−132439 Exothermic
S-std (J)	256.76	69.71	156.904	−169.566
Delta Gf (J)	−371060	−237129	−690.003	607499

T (K)	DelH − T*DelS = DelG	K
298	−81908.3	2.28E + 14
300	−81569.2	1.6E + 14
400	−64612.6	2.74E + 08
500	−47656	95228.69
600	−30699.4	470.6702
700	−13742.8	10.60565
800	3213.8	0.616812
900	20170.4	0.067499
1000	37127	0.011498

(c) All are exothermic. (d) Entropy increases in the first reaction only.

(e) The equilibrium constant decreases as temperature increases for all three reactions (all are exothermic); at temperatures above 1000 K the second reaction will not produce much product.

55. You would observe a dynamic equilibrium; NO_2 molecules would be joining to form N_2O_4 and then N_2O_4 molecules would be dissociating to form NO_2. All this would be happening at great speed—molecules move at hundreds of meters per second. The molecules are moving faster at the higher temperature and there are m NO_2 molecules relative to the number of N_2O_4 molecules.
57. (b), (c), and (d) have equilibrium constants within the range specified, and (e) might lie within the range.
59. Diagram (d) has $K = 0.44$; diagram (b) has $K = 4.0$; diagram (c) has $K = 36$.
61. Because equilibrium is a dynamic, not static, process, interaction between the solution and solid is occurring all the time. One equilibrium that is not explicitly stated, but that must be occurring nonetheless, is the dissociation of benzoic acid,
$C_6H_5COOH + H_2O \rightleftharpoons C_6H_5COO^- + H_3O^+$
This equilibrium means that benzoic acid is constantly interacting with water. If D_2O is added to the water, then some will end up on the benzoic acid, both dissolved and solid.
63. The idea of a dynamic equilibrium means that molecules are always reacting between reactants and products, even if the net effect is no change in the overall concentration of the species. That has been an important theme of this chapter. Therefore, a molecule of $O_2N^*—N^*O_2$ will dissociate to form two N^*O_2 molecules soon after it is introduced into the system. One N^*O_2 and one NO_2 molecule can then combine to form $O_2N^*—NO_2$.

Chapter 14

2. STP conditions are 0°C and one atm. STP stands for standard temperature and pressure.
4. Pressure is the force per unit area exerted by a gas onto any surface it contacts.
6. Dalton's Law of partial pressures states that the total pressure exerted by a mixture of gases is the sum of the partial pressures of the individual gases in the mixture. The mole fraction of O_2 is 0.22. The partial pressure of O_2 is 158 mm Hg.
8. At higher pressure, the volume available to the gas molecules to move is much smaller, and the volume occupied by the molecules relative to the volume they occupy is no longer negligible which violates the kinetic-molecular theory. At lower temperatures the average kinetic energy of the molecules is reduced to a point that it is comparable to the potential energy resulting from attractive forces of the molecules which causes a reduction in observed pressure.
10. (a) 0.947 atm
(b) 950 mm Hg
(c) 542 mm Hg
(d) 98.6 kPa
(e) 6.91 atm
12. 14.1 m
14. Standard atmospheric pressure will push 34 feet of water. Modern pumps that drill deeper for water use compressed air to increase the external pressure of water.
16. The major roles played by nitrogen and oxygen can be summed up with the following listing of reactions that occur in the atmosphere. (Many of the reactions involve both nitrogen and oxygen simultaneously; and not all reactions that occur in the atmosphere are listed.)
$N_2(g) + O_2(g) \longrightarrow 2\ NO(g)$
$NH_3(g) + HNO_3(g) \longrightarrow NH_4NO_3(s)$

$NO_3(g) + NO_2(g) \longrightarrow N_2O_5(g)$

$N_2O_5(g) + H_2O(g) \longrightarrow 2HNO_3(g)$

$NO_2(g) + O_3(g) \longrightarrow NO_3(g) + O_2(g)$

$4NO_2(g) + 2H_2O(g) + O_2(g) \longrightarrow 4HNO_3(g)$

$2\ NO(g) + O_2(g) \longrightarrow 2\ NO_2(g)$

$2NO_2(g) + 2H_2O(g) \longrightarrow HNO_3(g) + HNO_2(g)$

$O_3(g) + hv \longrightarrow O_2(g) + O(g)$

$O(g) + H_2O(g) \longrightarrow 2OH(g)$

$NO_2(g) + OH(g) \longrightarrow HNO_3(g)$

$NO_2(g) + NO_2(g) \longrightarrow N_2O_4(g)$

18.

Nitrogen	7.8084×10^5 ppm	7.8084×10^8 ppb
Oxygen	2.098×10^5 ppm	2.098×10^8 ppb
Argon	9.34×10^3 ppm	9.34×10^6 ppb
Carbon dioxide	3.30×10^2 ppm	3.30×10^5 ppb
Neon	18.2	1.82×10^4 ppb
Hydrogen	10 ppm	1.0×10^4 ppb
Helium	5.2 ppm	5.2×10^3 ppb
Methane	2 ppm	2.0×10^3 ppb
Krypton	1 ppm	1000 ppb
Carbon monoxide	0.1 ppm	1.0×10^2 ppb
Xenon	8.0×10^{-2} ppm	80 ppb
Ozone	2.0×10^{-2} ppm	20 ppb
Ammonia	1.0×10^{-2} ppm	10 ppb
Nitrogen dioxide	1.0×10^{-3} ppm	1 ppb
Sulfur dioxide	2.0×10^{-4} ppm	0.2 ppb

Sulfur dioxide is present at less than 1 ppb. Krypton, carbon monoxide, xenon, ozone, ammonia, and nitrogen dioxide are present between 1 ppb and 1 ppm. Nitrogen, oxygen, argon, carbon dioxide, neon, hydrogen, helium, and methane are present at greater than 1 ppm.

20. (a) 7.75×10^7 metric tons of S are burned, producing 1.55×10^8 metric tons of SO_2 (if $\frac{1}{8}S(s) + O_2(g) \longrightarrow SO_2(g)$).
(b) 2.12×10^6 tons of SO_2 are currently in the atmosphere

22. (a) The average kinetic energy per molecule of CO_2 would be equal to that of H_2 because the flasks are at the same temperature.
(b) The average molecular velocity should be smaller for CO_2, since the CO_2 gas molecules are heavier than H_2 gas molecules.
(c) The number of molecules of CO_2 must be greater because the pressure is greater, yet the temperature and volumes are the same.

24. lowest speed $SOCl_2 < Cl_2O < Cl_2 < SO_2$ highest speed
26. 4.2×10^{-5} mol CO
28. 154 mm Hg
30. $P_2 = 172$ mm Hg
32. 26 mL
34. 177 K or −96°C
36. 4.00 atm
38. 505 mL
40. 0.51 atm
42. Choice (c) contains the smallest number of molecules; choice (d) contains the largest number of molecules.
44. 1.88 L CO_2 (About 2 L would be appropriate for 2 loaves of French bread.)
46. 10.4 L, 10.4 L
48. 2.7×10^{-2} atm
50. 148 g/mol
52. (a) $C_8H_{18}(\ell) + 25/2\ O_2(g) \longrightarrow 8CO_2(g) + 9\ H_2O(g)$
(b) 1.4×10^3 L CO_2
54. −209°C
56. 0.90 g
58. 3.7×10^{-7} g/mL
60. (a) 154 mm Hg
(b) mole fraction $N_2 = 0.78$
mole fraction $O_2 = 0.21$
mole fraction Ar = 0.01
mole fraction $CO_2 = 2.7 \times 10^{-4}$
mole fraction $H_2O = 5.4 \times 10^{-3}$
(c) needs % by volume data
62. (a) 29 g/mol
(b) mole fraction of $N_2 = 0.83$
mole fraction of $O_2 = 0.17$
64. mole fraction $H_2O = 3.3 \times 10^{-2}$
66. 1 mole liquid water occupies 18.02 mL.
1 mole of water vapor would occupy 22.4 L.
Water vapor can not be at STP because STP is defined as a temperature of 0°C and 1 atm.
68. At high pressure, values of PV/RT lower than ideal (less than 1) are due predominantly to intermolecular attractions. The gas molecules are close enough for such attractions to become significant.
70. Oxygen and nitrogen. Oxygen can be used in steel making and rocket propulsion. Nitrogen in the liquid form can be used in cryosurgery and in its elemental form is a nutrient for plants.
72. (a) Insufficient information supplied.
(b) 650 L
74. When two methyl radicals react with each other, ethane is the product.
76. (a) $O + O_2 \longrightarrow O_3$ (b) $O_3 + h\nu \longrightarrow O_2 + O$
(c) $O + SO_2 \longrightarrow SO_3$ (d) $H_2O + h\nu \longrightarrow OH + H$
(e) $NO_2 + NO_2 \longrightarrow N_2O_4$
78. $\cdot OH + CO \longrightarrow HCO_2 \cdot \longrightarrow HOO \cdot + CO_2$
$6\ CO_2 + 6\ H_2O + 2880\ KJ \longrightarrow C_6H_{12}O_6 + 6\ O_2$
photosynthesis
80. The molecule CF_4 has no carbon-chlorine bonds. Carbon-chlorine bonds are susceptible to photodissociation. The attraction between carbon and fluorine atoms is much greater than the attraction between carbon and chlorine atoms. Therefore, carbon-fluorine bonds are not susceptible to photodissociation.
82. An air pollutant is a substance that degrades air quality. Nature (volcanic ash, mercury, decaying vegetation) and man (automobiles, power plants) pollute the air. Atmospheric pollution causes can cause burning eyes, coughing, harm to vegetation, and destruction of monuments.

84. Adsorption is a process where something adheres firmly to a surface by attraction while absorption is the action of drawing into the bulk of a solid or liquid.
86. 1.625×10^9 metric tons of coal, 2 million hours
88. $NO_2(g) + h\nu \longrightarrow NO_2(g)$
    Not all photons of light have sufficient energy to cause chemical reactions.
90. Sulfur dioxide. Coal and oil.
    $SO_2 + H_2O \longrightarrow H_2SO_3$
92. The reaction of nitrogen and oxygen with sufficient energy will form nitrogen monoxide. Both reactions require high pressure and low temperatures.
94. In the troposphere ozone is harmful because it is a component of photochemical smog and can decompose plastics and rubber. In the stratosphere, ozone is beneficial because it protects us from UV radiation.
96. % yield = 73.6
98. $H_2S + O_3 \longrightarrow SO_2 + H_2O$; then $SO_2 + \frac{1}{2}O \longrightarrow SO_3$; then $SO_3 + H_2O \longrightarrow H_2SO_4$
    Using Hess's law: $H_2S + O_3 + \frac{1}{2}O_2 \longrightarrow H_2SO_4$
    2.84×10^8 metric tons of H_2SO_4
100. (a)
102. (see below)
104. Reaction (c); the volume increased by a factor of 1.5; therefore the number of moles of gas must also increase by that same amount
106. (a) 41.9 g/mol
    (b) CH_2; C_3H_6
    (c)

```
     H  H  H              H     H
     |  |  |               \   /
  H—C—C=C—H                  C
     |                 H    / \    H
     H                  \  /   \  /
                         C-----C
                        /       \
                       H         H
```

108. The balloon is not defective. The colder air outside the store has decreased the kinetic energy of the helium atoms and they are hitting the inside of the balloon with less force. As a result, the balloon has shrunk.

Chapter 15

2. The concepts of the kinetic-molecular theory that apply to liquids are:
    - Liquid molecules move randomly—at various speeds and in every possible direction.
    - The average kinetic energy of liquid molecules is proportional to the absolute temperature.
4. The equilibrium vapor pressure could be measured by hooking up a mercury barometer to an airtight container with a liquid inside. When the pressure in the container no longer increases, a state of dynamic equilibrium has been reached. The measured pressure is the equilibrium vapor pressure.
6. The heat of crystallization is the heat evolved when a liquid becomes a solid (this change is always exothermic). The heat of crystallization is equal in magnitude to the heat of fusion (but opposite in sign).
8. A unit cell is the smallest part of the lattice that, when repeated along the directions defined by its edges, reproduces the entire crystal structure.
10. (a) A conductor has a lower conductivity at high temperatures, and a higher conductivity at low temperatures.
    (b) A nonconductor has low conductivity at high and low temperatures.
    (c) A semiconductor has high conductivity at high temperatures, and low conductivity at low temperatures.
    (d) A superconductor is like a conductor, but conducts with zero resistance at a very low temperature.
12. At high temperatures, the average kinetic energy of the molecules has increased. Intermolecular attractions are not as strong at higher temperatures, so surface tension decreases.
14. Decrease the pressure above the liquid.
16. The water in perspiration evaporates into the gas phase, an endothermic process. The energy is supplied from your body in the form of heat, thus you feel cooler.
18. 1.5×10^6 kJ
20. 234 kJ

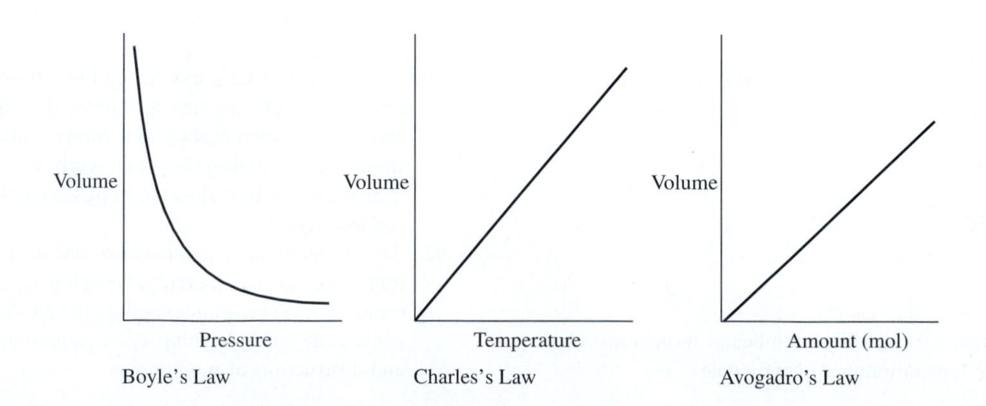

(Question 102)

22. 2.00×10^3 J for 0.500 mL of mercury; 1.1×10^3 J for 0.500 mL of water
24. Looking at dispersion forces, we would expect the boiling point to increase in this order: $NH_3 < PH_3 < AsH_3 < SbH_3$ because of molecular size increasing and amount of electrons increasing. This trend is followed with all but NH_3, which has a much higher boiling point that can be explained because of the hydrogen bonding in NH_3.
26. Hydrogen bonding occurs in methanol, which lowers its vapor pressure and elevates its boiling point.
28. 214 millibar or 161 torr; water boils at approximately 63°C at the top of the mountain.
30. High heat of fusion and high melting point indicate very strong intermolecular forces in the solid.
32. The larger heat of fusion for H_2O means that H_2O has stronger intermolecular forces. Water has a dipole moment which is about two times larger than the dipole moment for H_2S; hydrogen bonding is also present in water, but not in H_2S.
34. 2.72×10^4 J
36. LiF would have a higher melting point than CsI. The distance between the ions is much larger for CsI than for LiF. An I^- ion has a much larger ionic radius than an F^- ion; the radius of a Cs^+ ion is twice that of a Li^+ ion.
38. Highest melting point: SiC (macromolecular solid); lowest melting point: $CH_3CH_2CH_2CH_3$ (only London forces)
40. Frost-free refrigerators use air blown over the surface of the freezer to cause the frost to sublime. Eventually the hailstones would sublime away.
42. 0.129 g/cm^3; experimental = 0.129 g/cm^3
44. (a) molecular
    (b) network
    (c) metallic
    (d) ionic
46. (a) amorphous b) molecular c) ionic d) metallic
48. (a) molecular
    (b) ionic
    (c) metallic
    (d) amorphous
50. See page 680 textbook. Simple cubic, body-centered cubic, face-centered cubic. All have an atom at the corners of the cube. Simple cubic has 1 atom per unit cell, body-centered cubic has 2 atoms per unit cell, and face-centered cubic has 4 atoms per unit cell.
52. 0.219 nm
54. There is one Cs^+ ion per unit cell and there is one Cl^- ion per unit cell.
56. Diagonal = 0.700 nm
    Side length = 0.404 nm
58. $CaCl_2$ could not have the same structure as NaCl. The ratio of ions in the unit cell must reflect the empirical formula.
60. (a) 152 pm
    (b) I^-, radius = 212 pm; Li^+, radius = 88 pm
    (c) Results are reasonable; a L^+ ion should be significantly smaller than a neutral lithium atom.
62. 566 pm
    (a) $\nu = 5.30 \times 10^{17}$ Hz; E = 3.51×10^{-16} J
    (b) $\nu = 5.30 \times 10^{17}$ Hz; E = 2.11×10^8 J (the x-ray region of the spectrum)
64. In a conductor, groups of orbitals (called bands) allow the valence electrons to easily move from filled energy levels in the band to unfilled energy levels in that band, resulting in a high electrical conductivity. Nonconductors have orbitals that form bands that have energy gaps between them. This results in a lower electrical conductivity. Semiconductors have very narrow energy gaps between fully occupied energy bands and empty energy bands, resulting in a lower electrical conductivity at low temperatures, and a higher electrical conductivity at high temperatures.
66. Greatest electrical conductivity: Ag; Smallest electrical conductivity: P_4. Silver is a metal with mobile electrons, which are lacking in P_4.
68. $SiO_2(s) + 2C(s) \longrightarrow Si(s) + 2CO(g)$
    Carbon is oxidized, silicon is reduced.
    $SiCl_4(\ell) + 2Mg(s) \longrightarrow Si(s) + 2\,MgCl_2(s)$
    electronic grade
    Magnesium is oxidized, silicon is reduced.
70. Doping consists of adding a tiny amount of some other element to a semiconductor to form a number of positive or negative hole depending on the dopant element. Positive-type (p-type) silicon and negative-type (n-type) silicon are examples of doped semiconductors.
72. Diamond has a three-dimensional network because each carbon atom is bonded to four other carbons. Graphite has a planar network with each carbon atom bonded directly to only three other carbons. The planes in graphite are separated as well. These factors affect why diamond is more dense than graphite.
74. Graphite is a series of hexagons in a two dimensional network. This allows for the bonding electrons to move freely and conduct electricity. Diamond forms a 3-dimensional network and the valence electrons are localized around each carbon. Diamond will not conduct electricity.
76. Kaolinite $Al_2(OH)_4Si_2O_5$ Al = +3, O = −2, H = +1, Si = +4, O = −2
    Feldspar $KAlSi_3O_8$ K = +1, Al = +3, Si = +4, O = −2
    The reaction is not a redox reaction.
78. A glass is an amorphous solid, whereas NaCl is an ionic solid. Ionic solids are held together by attractions among positive and negative ions, and have unit cell regularity. Amorphous solids are covalently bonded networks with no long range regularity. NaCl can't become glass-like; the types of forces holding units together in ionic and amorphous solids are different and lead to different properties.
80. Oxide ceramics are Al_2O_3 and MgO.
    Nonoxide ceramics are Si_3N_4 and SiC.
82. The high cost of cryogens to provide very low temperatures required to cool magnets in superconductivity devices.
84. 1.95×10^5 kJ

86. (a) [Lewis structures: two resonance forms of SO_2, O–S=O ⟷ O=S–O]

The O—S—O angle is approximately 120°; electron pair geometry is triangular planar; molecular geometry is angular.
(b) Dispersion and dipole-dipole forces are responsible for binding SO_2 molecules to one another in the solid or liquid phases.
(c) $CH_4 < SO_2 < NH_3 < H_2O$

88. (a) 100 mm Hg
(b) 20°C
(c) 600 mm Hg
(d) diethyl ether and ethyl alcohol
(e) diethyl ether would evaporate; ethyl alcohol and water would remain as liquids
(f) water

90. The butane in the lighter has a pressure greater than atmospheric pressure. The increased pressure causes the butane to liquefy.

92. A and 8; B and 3; C and 1; D and 1; E and 2; F and 4; G and 5; H and 7

94. If a closet-packing arrangement is used in all three cases, 74% of the space is occupied in each case.

Chapter 16

2. Supersaturated solutions contain more dissolved solute than its solubility indicates. Saturated solutions contain an amount of dissolved solute equal to its solubility. Unsaturated solutions contain less dissolved solute than its solubility indicates.

4. The solubility of ionic solids increases with increasing temperature. Most ionic compounds dissolve in water endothermically, i.e., heat is a reactant and an increase in temperature would shift the equilibrium to form more solution.

6. 0.05% by weight

8. Molality is moles of solute dissolved per kilogram of solvent. Molarity is moles of solute dissolved per liter of solution.

10. CuCl, because it has a larger K_{sp} value.

12. Heat, heavy metals, organic molecules, acids, bases, pesticides, human and animal waste products, toxic metal ions, silt, and numerous other substances.

14. Solute molecules at the surface of the solvent decrease the rate of evaporation of the solvent. A higher temperature is required to overcome this decrease.

16. II = cRTi; II is osmotic pressure, c is the molarity of the solution, R is the gas constant, T is the temperature in kelvins, and i is the number of particles per formula unit of solute.

18. $CH_3(CH_2)_{14}COO^-Na^+$ Surfactants are molecules with both a hydrophobic and hydrophilic end. The $CH_3(CH_2)_{14}$ end is hydrophobic and the COO^-Na^+ end is hydrophilic.

20. Solid solutes are soluble in liquid solvents that are able to overcome the intermolecular attractive forces or bonds within the solute. Table salt will dissolve in water but not in carbon tetrachloride. (There are many other examples.)

22. The mixture felt cold because the dissolution of NH_4Cl is an endothermic process. Solubility increases with increasing temperature. As the mixture warmed to room temperature the solid NH_4Cl dissolved and the solution was no longer saturated.

24. Small organic acids are water soluble because of their strong hydrogen bonding with water, i.e., the solute is hydrophilic. Fatty acids have long hydrocarbon chains with large dispersion forces between them. These forces are greater than the hydrogen bonds between the acid groups and water. Fatty acids are hydrophobic and do not dissolve in water.

26. Ions will be pulled from the crystal lattice by water molecules. Eventually, all the ions will be separated and surrounded by water molecules.

28. 0.0012 mol/L

30. (a) $CuI(s) \longrightarrow Cu^+(aq) + I^-(aq)$ $K_{sp} = [Cu^+][I^-]$
(b) $Ag_2SO_4(s) \longrightarrow 2\,Ag^+(aq) + SO_4^{2-}(aq)$ $K_{sp} = [Ag^+][SO_4^{2-}]$
(c) $FeCO_3(s) \longrightarrow Fe^{2+}(aq) + CO_3^{2-}(aq)$ $K_{sp} = [Fe^{2+}][CO_3^{2-}]$

32. 0.012 mol/L

34. 364 mg/100 mL Ag^+ and 162 mg/mL SO_4^{2-} (using a K_{sp} value of 1.7×10^{-5})

36. 1.7×10^{-5}

38. 3.9×10^{-10} mol/L

40. 0.00732%

42. 1 ppb = (1 g solute/10^9 g solution) × (10^6 μg solution/1 g solute) × (10^3 g solution/1 kg solution) = 1 μg solute/1 kg solution

44. 90 g ethanol

46. 1.6×10^{-6} g Pb/cm^2 paint

48. 95% H_2SO_4

50. 59 g ethylene glycol

52. 0.083 mg arsenic

54. Heat, heavy metals, petroleum products, pesticides, plant nutrients, and numerous other pollutants. Heat causes decreased oxygen solubility; heavy metals cause protein damage; petroleum products concentrate fat-soluble pollutants.

56. The stream could change from an aerobic to an anaerobic system. A slow moving stream would worsen the problem.

58. Hardness refers to water containing relatively high concentrations of Ca^{2+}, Mg^{2+}, and Fe^{3+} ions.

60. Spraying water increases its contact with air, thereby increasing the amount of dissolved oxygen.

62. 100.26°C

64. 0.08 m $CaCl_2$ > 0.10 m NaCl > 0.04 m Na_2SO_4 > 0.1 m sugar

66. 178 g/mol; $C_{14}H_{10}$

68. (a) 2.5 kg ethylene glycol; (b) 104.1°C

70. 297 atm (sea water freezes at −25°C)

72. 4% A

74. 0.016 M Ag_2SO_4

76. 0.016 M (in pure water), 5.9×10^{-5} M (in 0.536 M HCl), $PbCl_2$ in 99.6% less soluble in HCl solution

78. 28.0% NH_3

80. 0.982 m; 10.2%

82. 0.20 m ethylene glycol > 0.10 m NaCl > 0.12 m kBr > 0.12 m K_2SO_4
(highest freezing point) (lowest freezing point)

84. 1.77 g
86. $C_{18}H_{24}Cr$, $C_{18}H_{24}Cr$
88. (a) $AlCl_3 + H_3PO_4 \longrightarrow AlPO_4 + 3HCl$
    (b) 139 g $AlPO_4$
    (c) 1.1×10^{-10} mol/L Al^{3+} and 1.1×10^{-10} mol/L PO_4^{3-}
    (d) Yes, the concentration is greater than the molar solubility and $AlPO_4$ will precipitate; 0.46 g $AlPO_4$
90.

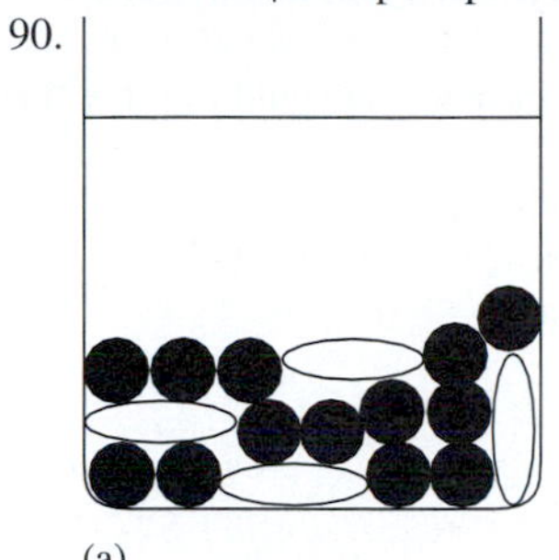

(a)

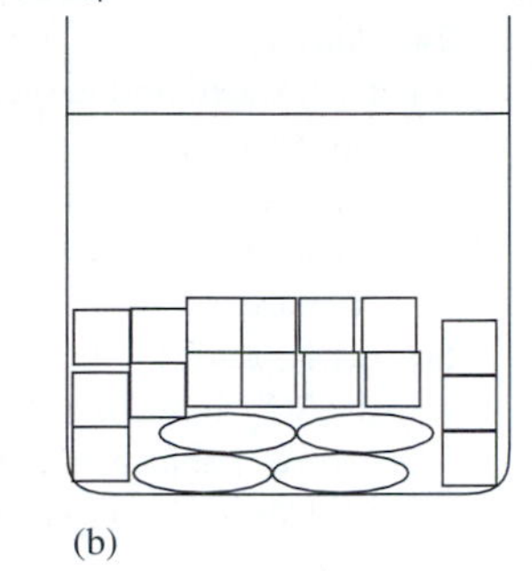

(b)

92. (a) unsaturated (b) supersaturated
    (c) supersaturated (d) unsaturated
94.

Compound	Mass of Compound	Mass of Water	Mass Fraction	Weight Percent	ppm
lye	75.0 g	125 g	0.375	37.5%	3.75×10^5 ppm
glycerol	33 g	200 g	0.14	14%	1.4×10^5 ppm
acetylene	0.0015 g	167 g	0.000009	0.0009%	9 ppm

96. $Ca(NO_3)_2$, $Ca(CH_3COO)_2$ (any soluble calcium salt)
98. When a liquid freezes, its molecules cluster together, forming a crystal lattice. When a nonvolatile solute is present, it interferes with the clustering and the liquid does not freeze until the temperature is lowered. The water in an iceberg is fairly pure. As sea water freezes it forces out dissolved minerals in order to form ice crystals.

Chapter 17

2. 100% ionization means that all of the molecules of a substance split into positive and negative ions.
4. H_2O; O^{2-}
6. $CH_3COOH + H_2O(\ell) \rightleftharpoons CH_3COO^-(aq) + H_3O^+(aq)$
   $CH_3COO^-(aq) + H_2O(\ell) \rightleftharpoons CH_3COOH(aq) + OH^-(aq)$
   $2\ H_2O(\ell) \rightleftharpoons H_3O^+(aq) + OH^-(aq)$
8. $NH_4Br(s) \xrightarrow{H_2O} NH_4^+(aq) + Br^-(aq)$
   $NH_4^+(aq) + H_2O(\ell) \rightleftharpoons NH_3(aq) + H_3O^+(aq)$
10. (b) and (c)
12. A Brønsted acid is a proton donor and a Brønsted base is a proton acceptor. A Lewis acid accepts an electron pair to form a bond; a Lewis base donates an electron pair in the process. BF_3 is a Lewis acid, but not a Brønsted acid.
14. (a) $HCO_3^- + H_2O \rightleftharpoons H_3O^+ + CO_3^{2-}$
    (b) $HCl + H_2O \longrightarrow H_3O^+ + Cl^-$
    (c) $CH_3COOH + H_2O \rightleftharpoons H_3O^+ + CH_3COO^-$
    (d) $HCN + H_2O \rightleftharpoons H_3O^+ + CN^-$
16. (a) $HSO_4^- + H_2O \rightleftharpoons H_2SO_4 + OH^-$
    (b) $CH_3NH_2 + H_2O \rightleftharpoons CH_3NH_3^+ + OH^-$
    (c) $I^- + H_2O \rightleftharpoons HI + OH^-$
    (d) $H_2PO_4^- + H_2O \rightleftharpoons H_3PO_4 + OH^-$
18. (a) H_2SO_4
    (b) HNO_3
    (c) $HClO_4$
    (d) H_3PO_4
    (e) H_2SO_4
20. (a) I^-, iodide ion
    (b) HNO_3, nitric acid
    (c) HCO_3^-, hydrogen carbonate
    (d) HCO_3^-, hydrogen carbonate
    (e) H_2SO_4, sulfuric acid; or SO_4^{2-}, sulfate ion
    (f) HSO_3^-, hydrogen sulfite ion
22. (b), (d)
24. (a) HS^-, base; H_2O, acid; H_2S, acid; OH^-, base
    (b) S^{2-}, base; NH_4^+, acid; NH_3, base; HS^-, acid
    (c) HCO_3^-, base; HSO_4^-, acid; H_2CO_3, acid; SO_4^{2-}, base
    (d) NH_3, acid; NH_2^-, base; NH_2^-, base; NH_3 acid
26. (a) $CO_3^{2-} + H_2O \longrightarrow HCO_3^- + OH^-$
    $HCO_3^- + H_2O \longrightarrow H_2CO_3 + OH^-$
    (b) $H_3AsO_4 + H_2O \longrightarrow H_2AsO_4^- + H_3O^+$
    $H_2AsO_4^- + H_2O \longrightarrow HAsO_4^{2-} + H_3O^+$
    $HAsO_4^{2-} + H_2O \longrightarrow AsO_4^{3-} + H_3O^+$
    (c) $NH_2CH_2COO^- + H_2O \longrightarrow NH_2CH_2COOH + OH^-$
    $NH_2CH_2COOH + H_2O \longrightarrow +NH_3CH_2COOH + OH^-$
28. 3.16×10^{-10} M; basic
30. pH = 12.4; pOH = 1.60
32. 1.8 g
34.

	pH	$[H_3O^+]$	$[OH^-]$	
(a)	6.21	6.1×10^{-7}	1.6×10^{-8}	acidic
(b)	5.34	4.5×10^{-6}	2.2×10^{-9}	acidic
(c)	4.67	2.1×10^{-5}	4.8×10^{-10}	acidic
(d)	1.60	2.5×10^{-2}	4.0×10^{-13}	acidic
(e)	9.12	7.6×10^{-10}	1.3×10^{-5}	basic

36. (a) pH = 5, 100 times more acidic
    (b) pH = 12, 10^5 times more basic
    (c) pH = 9, 100 times more basic
    (d) pH = 3, 10^4 times more acidic
38. (a) $F^- + H_2O \rightleftharpoons HF + OH^- \quad K_b = \frac{([HF][OH^-])}{[F^-]}$
    (b) $NH_3 + H_2O \rightleftharpoons NH_4^+ + OH^- \quad K_b = \frac{([NH_4^+][OH^-])}{[NH_3]}$
    (c) $H_2CO_3 + H_2O \rightleftharpoons HCO_3^- + H_3O^+$
    $K_a = \frac{([HCO_3^-][H_3O^+])}{[H_2CO_3]}$
    $HCO_3^- + H_2O \rightleftharpoons CO_3^{2-} + H_3O^+$
    $K_a = \frac{([CO_3^{2-}][H_3O^+])}{[HCO_3^-]}$
    (d) $H_3PO_4 + H_2O \rightleftharpoons H_2PO_4^- + H_3O^+$
    $K_a = \frac{([H_2PO_4^-][H_3O^+])}{[H_3PO_4]}$

$H_2PO_4^- + H_2O \rightleftharpoons HPO_4^{2-} + H_3O^+$

$$K_a = \frac{[HPO_4^{2-}][H_3O^+]}{[HPO_4^{2-}]}$$

$HPO_4^{2-} + H_2O \rightleftharpoons PO_4^{2-} + H_3O^+$

$$K_a = \frac{[PO_4^{3-}][H_3O^+]}{[HPO_4^{2}]}$$

(e) $CH_3COO^- + H_2O \rightleftharpoons CH_3COOH + OH^-$

$$K_b = \frac{[CH_3COOH][OH^-]}{[^-CH_3COO^-]}$$

(f) $S^{2-} + H_2O \rightleftharpoons HS^- + OH^- \quad K_b = \frac{[HS^-][OH^-]}{[S^{2-}]}$

$HS^- + H_2O \rightleftharpoons H_2S + OH^- \quad K_b = \frac{[H_2S][OH^-]}{[HS^-]}$

40. (a) 0.10 M NH_3
(b) 0.10 M K_2S
(c) 0.10 M CH_3COONa
(d) 0.10 M KCN
42. (a) 2.2
(b) 5.2
(c) 8.0
(d) 8.8
(e) 7.0
(f) 1.5
(g) 5.7
(h) 7
44. 1.5×10^{-5}
46. $[H_3O^+] = 1.3 \times 10^{-5}$ M, $[A^-] = 1.3 \times 10^{-5}$ M, [HA] = 0.040 M
48. 8.85
50. 3.28
52. (a) CN^-; product-favored; CN^- is a stronger base than SO_4^{2-}.
(b) HS^-; reactant-favored; HS^- is a stronger base than H_2O.
(c) H_2, product-favored; H_2O is a better H^+ donor than H_2.
54. (a) reactant-favored, $NH_3(aq) + H_2PO_4^-(aq) \longrightarrow NH_4^+(aq) + HPO_4^{2-}(aq)$
(b) product-favored, $CH_3COOH(aq) + OH^-(aq) \longrightarrow CH_3COO^-(aq) + H_2O(aq)$
(c) reactant-favored, $H_2SO_4(aq) + H_2PO_4^{2-}(aq) \longrightarrow HSO_4^-(aq) + H_3PO_4(aq)$
product-favored, $HSO_4^-(aq) + H_2PO_4^-(aq) \longrightarrow SO_4^{2-}(aq) + H_3PO_4(aq)$
(d) reactant-favored, $CH_3COO^-(aq) + HF \longrightarrow CH_3COOH(aq) + F^-(aq)$
56. (a) pH < 7 (d) pH > 7
(b) pH > 7 (e) pH > 7
(c) pH = 7 (f) pH > 7
58. (a), (c), (d), (e)
60. (c)
62. (a) 2.12
(b) 7.21
(c) 12.44
64. (a) HNO_2/NO_2^-
(b) $CH_3CH_2COOH/CH_3CH_2COO^-$
(c) $HClO/ClO^-$ or $H_3BO_3/H_2BO_3^-$
(d) HCO_3^-/CO_3^{2-}
66. 4.77 g
68. 9.88
70. 9.55; 9.50
72. $2\ CH_3(CH_2)_{16}COOH + Na_2CO_3 \longrightarrow 2\ NaCH_3(CH_2)_{16}COO + H_2O + CO_2$
74. vinegar
76. Dish washing detergent is very caustic and could ruin the finish on a car.
78. (a) base (d) acid
(b) acid (e) base
(c) base (f) base
80. (a) I_2, acid; I^-, base
(b) SO_2, base; BF_3, acid
(c) Au^+, acid; CN^-, base
(d) CO_2, acid; H_2O, base
82.

```
  ..
: Cl :
  ..
   |
  ..   ..
: I — Cl :
  ..   ..
   |
  ..
: Cl :
  ..
```

ICl_3 is a T-shaped molecule that can act as a Lewis base because of the lone electron pairs on the central atom. It acts as a Lewis acid toward Cl^- to form ICl_4^-, a square planar ion.
84. (a) weak acid (d) weak base
(b) strong base (e) strong base
(c) strong acid (f) amphiprotic
86. (a) pH < 7
(b) pH = 7
(c) pH > 7
88. 2.82
90. (a) increase
(b) stay the same
(c) stay the same
92. 2.22; 3.62
94. 10°C: 7.29, basic; 25°C: 7.00 neutral; 50°C: 6.63, acidic
96. When you evaporate water from an ammonia solution, you stress the
$NH_3(g) + H_2O(\ell) \longrightarrow NH_4^+(aq) + OH^-(aq)$
equilibrium forming more ammonia gas. Nothing will remain in the beaker after all the water is evaporated because the ammonia will evaporate also.
98. (a) Figure (d)
(b) Either Figure (b) or (a) could be correct, depending on the concentration of acetic acid.
100. The wrong pair of acid and salt of the weak acid were used. HCO_3^- and CO_3^{2-} should have been used. The pH of the buffer would be 6.4.
102. acidosis, increase

Chapter 18

2. Reactions that are product favored produce an electrical potential that can be used, among other things, to power a motor, illuminate a light, or warm your food.
4. A metal or other electrical conductor is made the cathode in an electrolysis cell. The metal can be plated with another metal to protect it against corrosion or simply to purify the plated metal.
6. (a) Al is oxidized; Cl_2 is reduced.
   Al is the reducing agent; Cl_2 is the oxidizing agent.
   $Al = +0, Cl = 0 \longrightarrow Al = +3, Cl = -1$
   (b) MnO_4^- is reduced; Fe^{2+} is oxidized.
   MnO_4^- is the oxidizing agent; Fe^{2+} is the reducing agent.
   $H = +1,\ O = -2,\ Mn = +7,\ O = -2,\ Fe = +2 \longrightarrow Fe = +3,\ Mn = +2,\ H = +1,\ O = -2$
   (c) NO_3^- is reduced; FeS is oxidized.
   NO_3^- is oxidizing agent; FeS is the reducing agent
   $Fe = +2, S = -2, N = +5, O = -2, H = +1, O = -2 \longrightarrow N = +2, O = -2, S = +6, O = -2, Fe = +3, H = +1, O = -2$
8. $2\ Na + Cl_2 \longrightarrow 2\ NaCl$
   $Na = 0, Cl = 0 \longrightarrow Na = +1, Cl = -1$
   Sodium is the reducing agent; Cl_2 is the oxidizing agent.

   $Ti + 2\ Cl_2 \longrightarrow TiCl_4$
   $Ti = 0, Cl = 0 \longrightarrow Ti = +4, Cl = -1$
   Titanium is the reducing agent, Cl_2 is the oxidizing agent.

   $N_2 + 3\ H_2 \longrightarrow 2\ NH_3$
   $N = 0, H = 0 \longrightarrow N = -3, H = +1$
   N_2 is the oxidizing agent; H_2 is the reducing agent.

   $SiO_2 + 2\ C \longrightarrow Si + 2CO$
   $Si = +4, C = 0 \longrightarrow Si = 0, C = +2, O = -2$
   SiO_2 is the oxidizing agent; C is the reducing agent.
10. (a) $Zn \longrightarrow Zn^{2+} + 2e^-$
    (b) $2H_3O^+ + 2e^- \longrightarrow H_2 + 2H_2O$
    (c) $Sn^{4+} + 2e^- \longrightarrow Sn^{2+}$
    (d) $Cl_2 + 2e^- \longrightarrow 2Cl^-$
    (e) $SO_2 + 2H_2O \longrightarrow SO_4^{2-} + 2e^- + 4H^+$
12. (a) $2Al \longrightarrow 2Al^{3+} + 6e^-$
    $3\ Cl_2 + 6e^- \longrightarrow 6Cl^-$
    (b) $5Fe^{2+} \longrightarrow 5Fe^{3+} + 5e^-$
    $5e^- + 8H^+ + MnO_4^- \longrightarrow Mn^{2+} + 4H_2O$
    (c) $FeS + 4H_2O \longrightarrow Fe^{3+} + SO_4^{2-} + 8H^+ + 9e^-$
    $3NO_3^- + 12H^+ + 9e^- \longrightarrow 3NO + 6H_2O$
14. (a) CO is the reducing agent, O_3 is the oxidizing agent.
    $3CO + O_3 \longrightarrow 3CO_2$
    (b) H_2 is the reducing agent. Cl_2 is the oxidizing agent.
    Equation is balanced as written.
    (c) Ti^{2+} is the reducing agent. H_2O_2 is the oxidizing agent.
    $Ti^{2+} + H_2O_2 + 2H^+ \longrightarrow Ti^{4+} + 2H_2O$
    (d) Cl^- is the reducing agent. $MnO4^-$ is the oxidizing agent.
    $2MnO_4^- + 6Cl^- + 8H^+ \longrightarrow 2MnO_2 + 3Cl_2 + 4H_2O$
    (e) FeS_2 is the reducing agent. O_2 is the oxidizing agent.
    $FeS_2 + 11/2\ O_2 \longrightarrow Fe_2O_3 + 4SO_2$
    (f) NO is the reducing agent. O_3 is the oxidizing agent.
    $O_3 + NO \longrightarrow O_2 + NO_2$
    (g) Zn(Hg) is the reducing agent. HgO is the oxidizing agent.
    $Zn + HgO \longrightarrow ZnO + Hg$
16. There must be a mechanism to allow for the flow of electrons, and a method to balance charge (salt bridge).
18. Reduction reactions.
20. (a) $Zn(s) + Pb^{2+}(aq) \longrightarrow Zn^{2+}(aq) + Pb(s)$
    (b) $Zn(s) \longrightarrow Zn^{2+}(aq) + 2e^-$ (anode)
    $Pb^{2+}(aq) + 2e^- \longrightarrow Pb(s)$ (cathode)
    (c)

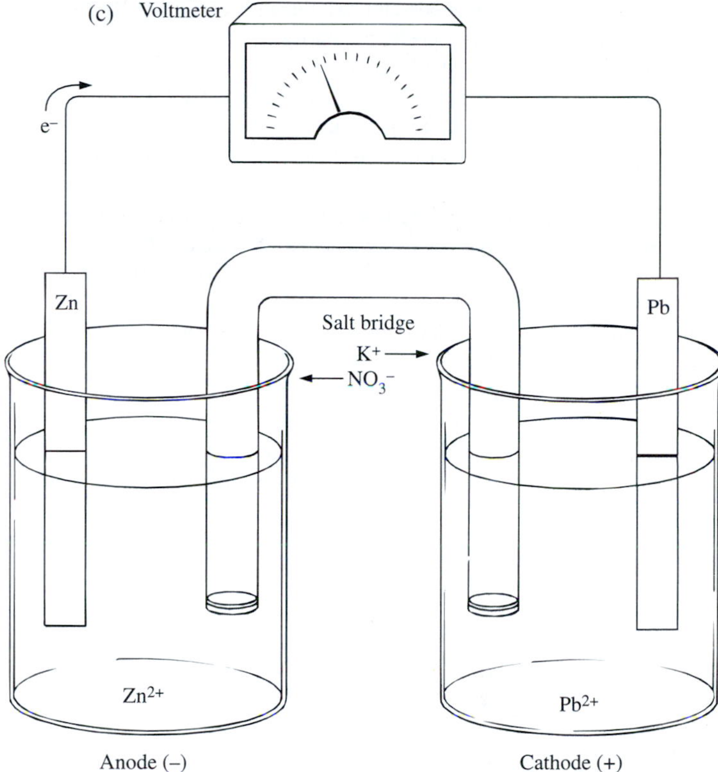

22. Total energy used is 90 kJ. The number of coulombs is 7500.
24. (a) $Cu(s) \longrightarrow Cu^{2+}(aq) + 2e^-$ and $2\ Ag^+(aq) + 2e^- \longrightarrow 2\ Ag(s)$
    (b) The copper half-reaction is an oxidation and occurs at the anode. The silver half-reaction is a reduction and occurs at the cathode.
26. The strongest oxidizing agent is $F_2(g)$. The weakest oxidizing agent is $Li^+(aq)$.
    The strongest reducing agent is Li(s). The weakest reducing agent is $F^-(aq)$.
28. Best oxidizer $H_2O_2 > O_2 > PbSO_4 > H_2O$ worst oxidizer
30. (a) E° for the reaction = 2.91 V, Product favored
    (b) E° for the reaction = −0.029 V, Reactant favored
    (c) E° for the reaction = 0.65 V, Product favored.
    (d) E° for the reaction = 1.16 V, Product favored.
32. (a) $Al^{3+}(aq)$ (b) $Ce^{4+}(aq)$ (c) Al(s) (d) $Ce^{3+}(aq)$ (e) yes (f) no (g) $Hg^{2+}(aq)$, $Ag^+(aq)$ and $Ce^{4+}(aq)$ (h) $Hg(\ell)$, Sn(s), Ni(s), and Al(s)

34. greater, less
36. (a) E° = 1.57 V
(b) If the coefficients were doubled, the values of ΔG° would also double (−1214 kJ). The value of E° would stay the same.
38. (a) $Zn(s) + Cl_2(g) \longrightarrow Zn^{2+}(aq) + 2Cl^-(aq)$
(b) ΔG° = −409 kJ
40. (a) $Cd(s) + Ni^{2+}(aq) \longrightarrow Cd^{2+}(aq) + Ni(s)$
(b) Cadmium is oxidized and is the reducing agent. Nickel is reduced and is the oxidizing agent.
(c) Cadmium is the anode and has a negative polarity. Nickel is the cathode and has a positive polarity.
(d) E° for the reaction = 0.15 V
(e) The electrons are flowing from the right to the left in the external wire. (according to the figure)
(f) The NO^{3-} ions will migrate toward the cadmium compartment.
42. Fuel cells utilize a combustion reaction to produce electricity, while a battery operates using a series of voltaic cells.
44. (a) Anode: $N_2H_4(aq) + 4OH^-(aq) \longrightarrow N_2(g) + 4H_2O(\ell) + 4e^-$
Cathode: $O_2(g) + 2H_2O(\ell) + 4e^- \longrightarrow 4OH^-(aq)$
(b) $N_2H_4(aq) + O_2(g) \longrightarrow N_2(g) + 2H_2O(\ell)$
(c) 7.46 g hydrazine
(d) 7.46 g oxygen
46. All the metal ions with a reduction potential more positive than water (−0.83 V) can be electrolyzed from their aqueous ions to the corresponding metals. Examples are Cu^{2+} and Ag^+.
48. Using the analogy discussed in section 18.9 of the textbook, the products of the electrolysis are gaseous hydrogen and liquid bromine. Sodium ion and bromide ion are in the solution. H_3O^+ ions are reduced at the cathode to the gas; bromide ions are oxidized to form molecular bromine at the anode.
50. 0.16 g Ag(s)
52. 5.93 g Cu(s)
54. 3×10^5 g Al(s)
56. 1.9×10^2 g Pb
58. 0.10 g Zn
60. .04 g Li and 0.6 g Pb
62. (a) 9×10^6 g HF
(b) 1700 kwh
64. 4.6×10^{-3} g Cu
1.9×10^{-5} g Al
66. In part a, A and C are stronger reducing agents than B and D. In part b, C is the best reducing agent. In part C, D is a better reducing agent than B. Therefore, the final order is C is best, than A, than D, with the worst reducing agent being B.
68. There are two possible answers:
$D^{n+}(aq) + n\,e^- \longrightarrow D(s)$ + 0.56 V
$A^{n+}(aq) + n\,e^- \longrightarrow A(s)$ + 0.21 V
$E^{n+}(aq) + n\,e^- \longrightarrow E(s)$ 0.00 V
$C^{n+}(aq) + n\,e^- \longrightarrow C(s)$ −0.37 V
$B^{n+}(aq) + n\,e^- \longrightarrow B(s)$ −0.68 V
or
$B^{n+}(aq) + n\,e^- \longrightarrow B(s)$ +0.68 V
$C^{n+}(aq) + n\,e^- \longrightarrow C(s)$ +0.37 V
$E^{n+}(aq) + n\,e^- \longrightarrow E(s)$ 0.00 V
$A^{n+}(aq) + n\,e^- \longrightarrow A(s)$ −0.21 V
$D^{n+}(aq) + n\,e^- \longrightarrow D(s)$ −0.56 V

Chapter 19

1.

	Symbol	Mass	Charge	Ioniz. Power	Penet. Power
α Particle	$^4{}_2He$, $^4{}_2\alpha$	$8.9451*10^{-27}$ kg	+2	High	Low
β Particle	$_{-1}^{0}e$, $_{-1}^{0}\beta$	$9.1094*10^{-36}$ kg	−1	Lower than α	Lower than γ; Higher than α
γ Radiation	$^0{}_0\gamma$	Essentially massless	0	Low	Very High

3. The "Band of Stability" is the zone on a plot of neutrons vs. protons for various isotopes where nonradioactive (stable) isotopes are located.
5. Beta emission.
7. The critical mass is the minimum mass of fissionable material required for a self-substaining chain reaction to occur; a chain reaction is one in which product(s) are generated that can then react with remaining reactants, such as the net release of neutrons during nuclear fission.
11. (a) $^{238}_{92}U$
(b) $^{32}_{15}P$
(c) $^{10}_{5}B$
(d) $_{-1}^{0}e$
(e) $^{15}_{7}N$
13. (a) $^{28}_{12}Mg \longrightarrow {}_{-1}^{0}e + {}^{28}_{13}Al$
(b) $^{238}_{92}U + {}^{12}_{6}C \longrightarrow 4{}^{1}_{0}n + {}^{246}_{98}Cf$
(c) $^{2}_{1}H + {}^{3}_{2}He \longrightarrow {}^{4}_{2}He + {}^{1}_{1}H$
(d) $^{38}_{19}K \longrightarrow {}^{38}_{18}Ar + {}_{+1}^{0}e$
(e) $^{175}_{78}Pt \longrightarrow {}^{171}_{76}Os + {}^{4}_{2}He$
15. $^{235}U \longrightarrow {}^{231}_{90}Th \longrightarrow {}^{231}_{91}Pa \longrightarrow {}^{227}_{89}Ac \longrightarrow {}^{227}_{90}Th \longrightarrow {}^{223}_{88}Ra$
17. (a) $^{19}_{10}Ne + {}_{-1}^{0}e \longrightarrow {}^{19}_{9}F$
or
$^{19}_{10}Ne \longrightarrow {}_{+1}^{0}e + {}^{19}_{9}F$
(b) $^{230}_{90}Th \longrightarrow {}^{226}_{88}Ra + {}^{4}_{2}He$
(c) $^{82}_{35}Br \longrightarrow {}_{-1}^{0}e + {}^{82}_{36}Kr$
(d) $^{212}_{84}Po \longrightarrow {}^{4}_{2}He + {}^{208}_{82}Pb$
19. ^{10}B, B. E. per nucleon = 6.11×10^{11} J/nucleon
^{11}B, B. E. per nucleon = 6.70×10^{11} J/nucleon
^{11}B is more stable. This is supported by the percent abundance of the isotopes.
21. 5.0 mg remains
23. (a) $^{131}_{53}I \longrightarrow {}_{-1}^{0}e + {}^{131}_{54}Xe$

(b) 32.2 days. Equal to four half-lives, each of 8.05 days. $(1/2)^4(25.0 \text{ mg}) = 1.56 \text{ mg}$

25. 4.0 y
27. 34.8 d
29. 2.58×10^3 y
31. ${}^{239}_{94}Pu + 2\ {}^{1}_{0}n \longrightarrow {}^{0}_{-1}e + {}^{241}_{95}Am$
33. ${}^{238}_{92}U + {}^{12}_{6}C \longrightarrow {}^{246}_{98}Cf + 4\ {}^{1}_{0}n$ You would bombard uranium-238 with carbon-12 atoms.
35. Cadmium or boron rods absorb neutrons to keep the reaction under control. The uranium fuel rods provide the source of fissionable material. A coolant transfers heat from fuel rods to a heat exchanger.
37. 1.60×10^3 ton of coal
39. A rad measures the amount of radiation absorbed. The rem takes into account the biological impact of the type of radiation.
41. The air we breathe, the water we drink, and the food we eat all contain naturally occurring radioisotopes. During metabolism, these radioisotopes become incorporated into our bones, skin, and organs.
43. Gamma rays penetrate food and kill bacteria in the food.
45. 130 mL
47. (a) ${}^{0}_{-1}e$
    (b) $2\ {}^{2}_{1}H$ or ${}^{4}_{2}He$
    (c) ${}^{87}_{35}Br$
    (d) ${}^{216}_{84}Po$
    (e) ${}^{68}_{31}Ga$
49. 1.5 mg
51. $t = 3.6 \times 10^9$ yr
53. (a) ${}^{247}_{99}Es$
    (b) ${}^{16}_{8}O$
    (c) ${}^{4}_{2}He$
    (d) ${}^{12}_{6}C$
    (e) ${}^{10}_{5}B$
55. Diagnostic radioisotopes must have enough energy to escape from the body for detection while at the same time have low ionizing ability so that cells are not destroyed. Therapeutic isotopes need low penetrating power so they stay in the body and high ionizing ability in order to kill cells.
57. The ${}^{20}Ne$ isotope is stable. The ${}^{17}Ne$ isotope could decay by positron emission or electron capture to form ${}^{17}F$. The ${}^{23}Ne$ isotope could decay by beta particle emission to form ${}^{23}Na$.
59. You might explain that the rest of the sample did not disappear—it modified its form. The mass is still the same, but a new element is formed.

GLOSSARY

absolute temperature scale (*See also* Kelvin temperature scale, thermodynamic temperature scale.) A temperature scale on which the zero is the lowest possible temperature, and the degree is the same size as the Celsius degree.

absorb To draw a substance into the bulk of a liquid or a solid (compare with adsorb).

achiral A compound whose molecule is superimposable on its mirror image.

acid ionization constant (K_a) The equilibrium constant for the reaction of a weak acid with water to produce hydronium ions and the conjugate base of the weak acid.

acid rain Precipitation with a pH below 5.6.

acid(s) A compound that ionizes in water to give hydronium ions, H_3O^+.

actinides The elements after actinium in the seventh period in which the 5*f* subshell is being filled.

activated complex A molecular structure corresponding to the top of an energy versus reaction progress plot; also known as the transition state.

activation energy (E_a) The potential energy difference between reactants and activated complex; the minimum kinetic energy that reactant molecules must have to be converted to product molecules.

active site The part of an enzyme molecule that binds the substrate in order for it to help it to react.

activity (*A*) A measure of the rate of nuclear decay, given as disintegrations per unit time.

actual yield The experimental quantity of product obtained from a chemical reaction.

addition polymer A polymer made when monomer molecules join directly with one another, with no other products formed in the reaction.

adsorb To attract and hold a substance on a surface (compare with absorb).

aerosols Small particles (1 nm to about 10,000 nm in diameter) that remain suspended indefinitely in air.

air pollutant A substance that degrades air quality.

alcohol Any of a class of organic compounds characterized by the presence of a hydroxyl group (—OH) covalently bonded to a saturated carbon atom.

aldehyde Any of a class of compounds characterized by a carbonyl group in which the carbon atom is bonded to at least one hydrogen atom; a molecule containing the $-\overset{\overset{\large O}{\|}}{C}-H$ functional group.

alkali metals The Group 1A elements in the periodic table.

alkaline earth metals The elements in Group 2A of the periodic table.

alkane Any of a class of hydrocarbons characterized by the presence of only single carbon-carbon bonds.

alkene Any of a class of hydrocarbons characterized by the presence of a carbon-carbon double bond.

alkyl group A fragment of an alkane structure that results from the removal of a hydrogen atom from an alkane.

alkyne Any of a class of hydrocarbons characterized by the presence of a carbon-carbon triple bond.

allotropes Different forms of the same element that exist in the same physical state under the same conditions of temperature and pressure.

alpha (α) amino acid An organic acid in which an amine group is bonded to the alpha carbon atoms; has the general formula

$$R-\underset{\underset{\large NH_2}{|}}{CH}-C\begin{matrix}\nearrow O \\ \searrow OH\end{matrix}.$$

alpha carbon The carbon adjacent to the acid group (COOH) in an amino acid.

alpha (α) particles Positively charged (+2) particles ejected at **high speeds** from certain radioactive nuclei; the nuclei of helium atoms.

alpha radiation Radiation composed of helium nuclei.

amide Any of a class of organic compounds characterized by the presence of a carbonyl group in which the carbon atom is bonded to a nitrogen atom ($-CONH_2$, CONHR, $CONR_2$); the product of an amine reacting with a carboxylic acid.

amine A derivative of ammonia in which one or more of the three hydrogen atoms is replaced by an organic group, R (RNH_2, R_2NH, R_3N).

amino acid An organic compound that contains an amine group and a carboxylic acid group. In biochemistry, an alpha amino acid.

amorphous solid A solid whose constituent nanoscale particles have no long-range order.

anion An ion with a negative electrical charge.

anode The electrode of an electrochemical cell at which oxidation occurs.

anodic inhibition The prevention of oxidation of an active metal by painting it, coating it with grease or oil, or allowing a thin film of metal oxide to form.

aqueous solution A solution in which water is the solvent.

aromatic compound Any of a class of hydrocarbons characterized by the presence of a benzene ring or related structure.

atom The smallest particle of an element that can be involved in chemical combination with another element.

atomic mass units (amu) The unit of a scale of relative atomic masses of the elements; 1 amu = 1/12 the mass of a six-proton, six-neutron carbon atom.

atomic number The number of protons in the nucleus of an atom of an element.

atomic radius One-half the distance between the nuclei centers of two like atoms that are touching in a molecule.

atomic structure The identity and arrangement of subatomic particles in an atom.

atomic weight The average mass of an atom in a representative sample of atoms of an element.

autoignition temperature The temperature at which the vapors of a flammable liquid will ignite and continue to burn without a source of ignition.

autoionization The equilibrium reaction in which water molecules react with each other to form hydronium ions and hydroxide ions.

Avogadro's law The volume of a gas, at a given temperature and pressure, is directly proportional to the amount of gas.

Avogadro's number The number of particles in a mole of any substance (6.022×10^{23}).

axial position(s) Positions above and below the equatorial plane in a triangular bipyramidal structure.

background radiation Radiation from natural and synthetic radioactive sources to which all members of a population are exposed.

balanced chemical equation A chemical equation that shows equal numbers of atoms of each kind in the products and the reactants.

bar A pressure unit equal to 100,000 Pa.

barometer An atmospheric pressure measuring device.

base A compound that dissociates ionizes in water to give a hydroxide ion.

base ionization constant (K_b) The equilibrium constant for the reaction of a weak base with water to produce hydroxide ions and the conjugate acid of the weak base.

beta particles Electrons ejected from certain radioactive nuclei.

bidentate ligand A ligand that has two atoms with lone pairs that can form coordinate covalent bonds to the same metal ion.

bimolecular reaction An elementary reaction in which two particles must collide for products to be formed.

binary molecular compound A molecular compound whose molecules contain atoms of only two elements.

binding energy The energy required to separate all nucleons in an atomic nucleus.

binding energy per nucleon The energy per nucleon required to separate all nucleons in an atomic nucleus.

biochemical oxygen demand (BOD) The quantity of oxygen required for microorganisms to metabolize organic material in a sample of impure water.

biodegradable Capable of being decomposed by biological means, especially by bacterial action.

boiling point The temperature at which the equilibrium vapor pressure of a liquid equals the external pressure on the liquid.

boiling point elevation A colligative property; the difference between the normal boiling point of a pure solvent and the higher boiling point of a solution in which a nonvolatile solute is dissolved in that solvent.

bond angle The angle between two bonds originating from the same atom in a molecule or polyatomic ion.

bond enthalpy (bond energy) The change in enthalpy when a mole of chemical bonds is broken, separating the bonded atoms; the atoms and molecules must be in the gas phase.

bond length The distance between the nuclei of two bonded atoms.

bonding pair A pair of valence electrons that are shared between two atoms.

boundary surface A surface within which there is a specified probability (often 90%) that an electron will be found.

Boyle's law The volume of a confined ideal gas varies inversely with the applied pressure, at constant temperature and amount of gas.

buckyball Buckminsterfullerene; an allotrope of carbon consisting of molecules in which 60 carbon atoms arranged in a cage-like structure consisting of five-membered rings linked to six-membered rings.

buffer capacity The capacity of a buffer solution to react with added acid or base, preventing significant change in pH.

buffer solution A solution that resists changes in pH when limited amounts of acids and bases are added; it consists of a weak acid and a salt of its conjugate base, or a weak base and a salt of its conjugate acid.

calorie (cal) (*See also* kilocalorie.) A unit of energy equal to 4.184 J. Approximately 1 cal is required to raise the temperature of 1 g of liquid water by 1 °C.

calorimeter A device for measuring the quantity of thermal energy transferred during a chemical reaction or some other process.

carbohydrates Compounds with the general formula $C_x(H_2O)_y$, in which x and y are whole numbers.

carbonyl group An organic functional group consisting of carbon double bonded to oxygen; $-\overset{\overset{\displaystyle O}{\|}}{C}-OH$.

carboxylic acid Any of the class of organic compounds characterized by the presence of the carboxyl group ($-CO_2H$).

catalyst A substance that increases the rate of a reaction but is not consumed in the overall reaction.

catalytic cracking process A petroleum refining process using a catalyst, heat, and pressure to break long-chain hydrocarbons into shorter-chain hydrocarbons, including both alkanes and alkenes, suitable for gasoline.

catalytic reforming A petroleum refining process in which straight-chain hydrocarbons are converted to branched-chain hydrocarbons and aromatics for use in the manufacture of other organic compounds and gasoline.

cathode The electrode of an electrochemical cell at which reduction occurs.

cathodic protection A process of protecting a metal from corrosion whereby it is made the cathode instead of the anode by connecting it electrically to a more reactive metal.

cation An ion with a positive electrical charge.

Celsius temperature scale A scale defined by the freezing and boiling points of pure water, set at 0 °C and 100 °C.

cement A solid consisting of microscopic particles containing compounds of calcium, iron, aluminum, silicon, and oxygen in varying proportions tightly bound to one another.

CFCs *See* chlorofluorocarbons.

change of state A physical process in which one state of matter is changed into another (such as melting a solid to form a liquid).

Charles's law The volume of an ideal gas at constant pressure and amount of gas vary directly with its absolute temperature.

chelating ligand A ligand that uses more than one atom to bind to the same metal ion in a complex ion.

chemical change (chemical reaction) A process in which substances (reactants) change into other substances (products) by rearrangement, combination, or separation of atoms.

chemical compound A pure substance (e.g., sucrose or water) that can be decomposed into two or more different pure substances; homogeneous, constant-composition matter that consists of two or more elements.

chemical equilibrium A state in which the concentrations of reactants and products remain constant because the rates of forward and reverse reactions are equal.

chemical fuel A substance that reacts exothermically with atmospheric oxygen and is available at reasonable cost and in reasonable quantity.

chemical kinetics The study of the rates of chemical reactions and the nanoscale pathways or rearrangements by which atoms, ions, and molecules are converted from reactants to products.

chemical property Describes the kinds of chemical reactions that chemical elements or compounds can undergo.

chemical reaction (chemical change) A process in which substances (reactants) change into other substances (products) by rearrangements, combination, or separation of atoms.

chemistry The study of matter and the changes it can undergo.

chemotrophs (*See also* phototrophs.) Organisms that must depend on phototrophs to create the chemical substances from which they obtain free energy.

chiral A compound whose molecule is not superimposable on its mirror image.

chlorofluorocarbons (CFCs) Compounds of carbon, flourine, and chlorine. CFCs have been implicated in stratospheric ozone depletion.

***cis* isomer** The isomer in which two like substituents are on the same side of a carbon-carbon double bond, the same side of a ring of carbon atoms, or the same side of a complex ion.

***cis-trans* isomerism** A form of stereoisomerism in which the isomers have the same molecular formula and the same atom-to-atom bonding sequence, but the atoms differ in the location of pairs of substituents on the same side or on opposite sides of a molecule.

coagulate The process in which the protective charge layer on colloidal particles is overcome, causing them to aggregate into a soft, semisolid, or solid mass.

cofactor An inorganic or organic molecule or ion required by an enzyme to carry out its catalytic function.

colligative properties Properties of solutions that depend only on the concentration of solute particles in the solution, not on the nature of the solute particles.

colloid(s) A state intermediate between a solution and a suspension, in which solute particles are large enough to scatter light, but too small to settle out.

combination reaction A reaction in which two reactants combine to give a single product.

combustion analysis A quantitative method to obtain percent composition data for compounds that can burn in oxygen.

combustion reaction A reaction in which an element or compound burns in oxygen.

common ion effect Displacement of an equilibrium caused by introducing a reactant or product ion through another source, such as a strong acid, strong base, or soluble salt.

complementary base pair Bases, each in a different DNA strand, that hydrogen bond to each other: guanine with cytosine and adenine with thymine or uracil.

complex ion An ion with several molecules or ions connected to a central metal ion by coordinate covalent bonds.

compressibility The property of a gas that allows it to be compacted into a smaller volume by application of pressure.

concentration The relative amounts of solute and solvent in a solution.

concrete A mixture of cement, sand, and aggregate (crushed stone or pebbles) in varying proportions that reacts with water and carbon dioxide to form a rock-hard solid.

condensation The process of transforming a vapor or a gas into a liquid.

condensation polymer A polymer made from the reaction of monomer molecules that contain two or more functional groups, with the formation of a small molecule such as water as a byproduct.

condensation reaction A chemical reaction in which two molecules combine to form a larger molecule, simultaneously producing a small molecule such as water.

condensed formula A chemical formula of an organic compound indicating how atoms are grouped together in a molecule so that chemically important atoms or groups of atoms are emphasized.

conjugate acid-base pair A molecule or ion whose structures differ by a single hydrogen ion.

conjugated Refers to a system of alternating single and double bonds in a molecule.

constitutional isomers (structural isomers) Compounds with the same molecular formula that differ in the order in which their atoms are bonded together.

continuous phase The solvent-like dispersing medium in a colloid.

continuous spectrum A spectrum consisting of all possible wavelengths.

conversion factor A relationship between two measurement units derived from the equality between the units.

coordinate covalent bond A chemical bond in which both of the two electrons forming the bond were originally associated with one of the two bonded atoms.

coordination compound A compound in which complex ions are combined with oppositely charged ions to form a neutral compound.

coordination number The number of coordinate covalent bonds between ligands and a central metal ion in a complex ion.

copolymer A polymer formed by combining two different types of monomers.

core electrons The electrons in the filled inner shells of an atom.

corrosion The deterioration of metals by oxidation-reduction reactions.

coulomb The unit of electrical charge equal to the quantity of charge that passes a fixed point in an electrical circuit when a current of one ampere flows for one second.

covalent bond Interatomic attraction resulting from the sharing of electrons between the atoms.

critical pressure The pressure corresponding to the temperature on a phase diagram at which the liquid/gas equilibrium line terminates.

critical temperature The temperature corresponding to the pressure on a phase diagram at which the ligand/gas equilibrium line terminates.

cryogens Liquefied gases that have temperatures below -150 °C.

crystal lattice The ordered, repeating arrangement of ions, molecules, or atoms in a crystalline solid.

crystallization The process in which mobile atoms, molecules, or ions in a liquid convert into a crystalline solid.

cubic unit cell A unit cell with equal-length edges that meet at 90° angles.

curie (Ci) A unit of radioactivity equal to 3.7×10^{10} disintegrations per second.

Dalton's law of partial pressures The total pressure exerted by a mixture of gases is the sum of the partial pressures of the individual gases in the mixture.

decomposition reaction A reaction in which a compound breaks down to form two or more simpler compounds or elements.

delocalized electrons Electrons, such as in benzene, that are spread over several atoms in a molecule or polyatomic ion.

denaturation Disruption in protein secondary and tertiary structure brought on by high temperature, heavy metals, and other substances.

density The ratio of the mass of an object to its volume.

deoxyribonucleic acid (DNA) A double-stranded polymer of nucleotides; serves as the genetic information storage molecule.

deposition The process of a gas converting directly to a solid.

detergent(s) Molecules whose structure contains a long hydrocarbon portion that is hydrophobic, and a polar end that is hydrophilic.

diamagnetic Describes atoms or ions in which all the electrons are paired in filled shells so their magnetic fields effectively cancel each other.

dietary minerals Essential elements that are not carbon, hydrogen, oxygen, or nitrogen.

dimer A molecule made from two smaller units.

dipole moment The product of the magnitude of the partial charges ($\delta+$ and $\delta-$) of a molecule times the distance of separation between the charges.

dipole-dipole attraction The intermolecular force acting between any two polar molecules.

disaccharides Carbohydrates consisting of two monosaccharide units.

dispersed phase The larger than molecule-sized particles that are distributed uniformly throughout a colloid.

displacement reaction A reaction in which one element reacts with a compound to form a new compound and release a different element.

doping The addition of a tiny amount of some other element (a *dopant*) to improve the semiconducting properties of silicon.

double bond(s) A bond formed by sharing two pairs of electrons between the same two atoms.

dynamic equilibrium A balance between opposing reactions occurring at equal rates.

electrochemical cell(s) A combination of anode, cathode, and other materials arranged so that a product-favored that undergo an oxidation-reduction reaction can cause a current to flow or an electric current can cause a redox reaction to occur.

electrode A device such as a metal plate or wire that conducts electrons into and out of a system solutions.

electrolysis The use of electrical energy to produce a chemical change.

electrolyte A substance that ionizes or dissociates in water to form an electrically conducting solution.

electromagnetic radiation Radiation that consists of oscillating electric and magnetic fields that travel through space at the same rate (the speed of light: 186,000 miles/s or 2.998×10^8 m/s in a vacuum).

electromotive force (emf) The difference in electrical potential energy between the two electrodes in an electrochemical cell, measured in volts.

electron A negatively charged subatomic particle that occupies in the space surrounding the nucleus.

electron capture A radioactive decay process in which one of an atom's inner-shell electrons is captured by the nucleus, which decreases the atomic number by 1.

electron configuration The complete description of the orbitals occupied by all the electrons in an atom or ion.

electron-pair geometry The geometry around a central atom including the spatial positions of bonding and lone electron pairs.

electronegativity A measure of the ability of an atom in a molecule to attract electrons to itself.

electronically excited molecule A molecule whose potential energy is greater than the minimum (ground state) energy because of a change in its electronic structure.

electropositive elements Elements with electronegativities less than 1.3.

element A substance (e.g., carbon, hydrogen, and oxygen) that cannot be decomposed further into two or more new substances by chemical or physical means.

elementary reaction A simple building-block, nanoscale reaction whose equation indicates exactly which atoms, ions, or molecules collide or change as the reaction occurs.

empirical formula A formula showing the simplest possible ratio of atoms of elements in a compound.

emulsion A colloid consisting of a liquid dispersed in a second liquid; formed by the presence of an *emulsifier* that coats and stabilizes dispersed-phase particles.

enantiomers A stereoisomeric pair consisting of a chiral molecule compound and its mirror-image isomer.

endergonic (reaction) A reaction in which free energy from the surroundings is consumed by a system.

endothermic (process) A process in which thermal energy must be transferred into a thermodynamic system in order to maintain constant temperature.

energy The capacity to do work.

energy band In a solid, a large group of orbitals from neighboring atoms whose energies are closely spaced and whose average energy is the same as the energy of the corresponding orbital in an individual atom.

energy conservation Minimization of consumption of Gibbs free energy.

enthalpy change (symbol ΔH) The quantity of thermal energy transferred when a process takes place at constant temperature and pressure.

entropy A measure of the disorder in a system.

enzyme A highly efficient catalyst for one or more reactions in a living system.

enzyme-substrate complex The combination formed by the binding of an enzyme with a substrate.

equatorial position(s) Position lying in the equator of an imaginary sphere around a triangular bipyramidal.

equilibrium concentration The concentration of a substance (generally expressed as molarity) in a system that has reached the equilibrium state.

equilibrium constant (*K*) A quotient of equilibrium concentrations of product and reactant substances that has a constant value for a given reaction at a given temperature.

equilibrium constant expression The mathematical expression associated with an equilibrium constant.

equilibrium state A system in which opposite nanoscale processes are occurring at equal rates with the result that there is no observable, macroscopic change.

equilibrium vapor pressure The pressure of a pure gas that is in equilibrium with its liquid phase.

ester Any of a class of organic compounds structurally related to carboxylic acids, but in which the hydrogen atom in the carboxyl group has been replaced by a hydrocarbon group R, (—COOR).

evaporation The process of conversion of a liquid to a gas.

exchange reaction A reaction in which cations and anions that were partners in the reactants are interchanged in the products.

excited state The state of an atom or molecule in which at least one electron does not have its lowest possible energy.

exergonic (reaction) A reaction in which a system releases free energy to its surroundings.

exothermic (process) A process in which thermal energy must be transferred out of a thermodynamic system in order to maintain constant temperature.

Faraday constant (F) The quantity of electricity that corresponds to the charge on one mole of electrons, 9.6485×10^4 C/mol of electrons, commonly rounded to 96,500 C/mol of electrons.

fat A solid triester of organic acids with glycerol.

ferromagnetic A substance that contains clusters of atoms with unpaired electrons whose magnetic spins become aligned, causing permanent magnetism.

first law of thermodynamics (law of conservation of energy) Energy can neither be created nor destroyed—the total energy of the universe is constant (accurate for physical and chemical, but not nuclear, processes).

formula unit The simplest cation-anion grouping represented by the formula of an ionic compound; also the unit represented by any formula.

fractional distillation The process of refining petroleum (or other mixture) by distillation to separate it into groups (fractions) of compounds having distinctive boiling point ranges.

free radical A highly reactive atom, ion, or molecule that contains one or more unpaired electrons.

freezing point lowering A colligative property; the difference between the freezing point of a pure solvent and the freezing point of a solution in which a nonvolatile solute is dissolved in the solvent.

frequency The number of complete waves passing a point in a given amount of time (cycles per second).

functional group An atom or group of atoms that imparts characteristic properties and defines a given class of organic compounds (e.g., the —OH group is present in all alcohols).

galvanized Has a thin coating of zinc metal that forms an oxide coating impervious to oxygen, thereby protecting a less active metal, such as iron, from corrosion.

gas A phase of matter in which a substance has no definite shape and has a volume that which is determined only by the size of its container.

gasohol A blended motor fuel consisting of 90% gasoline and 10% ethanol.

genes The special hereditary areas on chromosomes, distinguished by unique sequences of nucleotide bases.

Gibbs free energy A thermodynamic function that decreases for any product-favored system. For a process at constant temperature and pressure, $\Delta G = \Delta H - T\Delta S$.

glycogen A highly branched, high molar mass polymer of glucose found in animals.

gram(s) The basic unit of mass in the metric system; equal to 1×10^{-3} kg.

greenhouse effect Atmospheric warming caused by absorption of infrared radiation by molecules of carbon dioxide, water vapor, methane, ozone, and similar "greenhouse" gases.

ground state The state of an atom or molecule in which all of the electrons are in their lowest possible energy levels.

groups The vertical columns of the periodic table of the elements.

Haber-Bosch process The process developed by Fritz Haber and Carl Bosch for the direct synthesis of ammonia from its elements.

half-life, $t_{1/2}$ The time required for the concentration of one reactant to reach half its original value; radioactivity—the time required for the activity of a radioactive sample to reach half of its original value.

half-reaction A reaction that represents either an oxidation or a reduction process.

halide ion An ion (1−) of a halogen element.

halogens The elements in Group 7A of the periodic table.

heat capacity The quantity of thermal energy that must be transferred to an object to raise its temperature by 1 °C.

heat of fusion The enthalpy change when a substance melts; the quantity of thermal energy that must be transferred when a substance melts at constant temperature and pressure.

heat of sublimation The enthalpy change when a solid sublimes; the quantity of thermal energy, at constant pressure that must be transferred, to cause a solid to vaporize.

heat of vaporization The enthalpy change when a substance vaporizes; the quantity of thermal energy that must be transferred when a liquid vaporizes at constant temperature and pressure.

Hess's law If two or more chemical equations can be combined to give another equation, the enthalpy change for that equation will be the sum of the enthalpy changes for the equations that were combined.

heterogeneous catalyst A catalyst that is in a different phase from that of the reaction mixture.

heterogeneous mixture A mixture in which components remain separate and can be observed as individual substances.

hexadentate A ligand that donates six electron pairs to a coordinated central metal ion.

homogeneous catalyst A catalyst that is in the same phase as that of the reaction mixture.

homogeneous mixture A mixture in which the composition is the same throughout.

homogeneous reaction A reaction in which the reactants and products are all in the same phase.

Hund's rule A rule for the behavior of electrons within atoms: the most stable arrangement of electrons in the same subshell is that with the maximum number of unpaired electrons, all with the same spin.

hybrid orbital Orbital formed by mixing atomic orbitals of appropriate energy and orientation.

hydration The binding of one or more water molecules to an ion or molecule.

hydrocarbons Compounds composed only of carbon and hydrogen.

hydrogen bonding Noncovalent interaction between a hydrogen atom and a very electronegative atom to produce an unusually strong dipole-dipole force.

hydrogenation An addition reaction in which hydrogen adds to the double bond of an alkene; the catalyzed reaction of H_2 with a liquid triglyceride to produce saturated fatty acid chains, which convert the triglyceride into a semisolid or solid.

hydrolysis A reaction with water in which an oxygen-hydrogen bond in water is broken.

hydronium ion H_3O^+; the simplest proton-water complex.

hydrophilic "Water loving," a term describing the polar part of a molecule that becomes strongly attracted to water molecules.

hydrophobic "Water fearing," a term describing the nonpolar part of a molecule that is not attracted to water molecules.

hydroxide ion OH^- ion; bases increase the concentration of hydroxide ions in solution.

hypothesis A tentative explanation or prediction derived from experimental observations.

hypotonic A solution having a low solute concentration and therefore a lower osmotic pressure than another solution.

ideal gas A gas that behaves exactly as described by Boyle's, Charles's, and Avogadro's laws.

ideal gas constant The proportionality constant, R, in the equation PV = nRT; R = 0.0821 L·atm/mol·K.

ideal gas law A law that relates pressure, volume, number of moles, and temperature for an ideal gas; the relationships expressed by the equation PV = nRT.

imaging (medical) A technique in which a radioisotope is administered, collects at the site of a medical disorder, and then is detected by photographic or other means.

immiscible Describes two liquids that will not dissolve in one another.

induced dipole A temporary dipole created by a momentary uneven distribution of electrons in a molecule.

induced fit The change in the shape of an enzyme, its substrate, or both when they bind.

inhibitor A molecule or ion that occupies the active site of an enzyme causing a decrease in enzymatic reactivity.

initial rate The instantaneous rate of a reaction determined at the very beginning of the reaction.

initiation The breaking of a carbon-carbon double bond in a polymerization reaction to produce a molecule with highly reactive sites that react with other molecules to produce a polymer.

insoluble Describes a solute, almost none of which dissolves in a solvent.

insulator A material that has a large energy gap between fully occupied and empty energy bands, and does not conduct electricity.

intermediate A species produced in one step of a reaction sequence and then consumed in a subsequent step.

intermolecular attractions Noncovalent attractions between separate molecules.

intermolecular forces Noncovalent attractions between separate molecules.

internal energy The sum of the individual energies of all of the nanoscale particles (atoms, molecules, or ions) in a sample of matter.

ion An atom or group of atoms that has lost or gained one or more electrons so that it is no longer electrically neutral.

ionic compound A compound that consists of positive and negative ions.

ionic hydrate Ionic compounds that incorporate water molecules in the ionic crystal lattice.

ionization constant for water (K_w) The equilibrium constant that is the mathematical product of the hydronium ion concentration and the concentration of hydroxide ion in any aqueous solution; $K_w = 1 \times 10^{-14}$ at 25 °C.

ionization energy The energy needed to remove an electron from an atom in the gas phase.

isoelectronic Refers to atoms and ions that have identical electron configurations.

isotonic A solution having the same concentration of nanoscale particles and therefore the same osmotic pressure as another solution.

isotopes Forms of an element composed of atoms with the same atomic number but different mass numbers owing to a difference in the number of neutrons.

joule (J) A unit of energy equal to 1 $kg \cdot m^2/s^2$. The kinetic energy of a 2-kg object traveling at a speed of 1 m/s.

Kelvin temperature scale (*See also* absolute temperature scale, thermodynamic temperature scale.) A temperature scale on which the zero is the lowest possible temperature and the degree is the same size as a Celsius degree.

ketone Any of a class of organic compounds characterized by the presence of a carbonyl group in which the carbon atom is bonded to two other carbon atoms (R_2CO).

kilocalorie (kcal or Cal) (*See also* calorie.) A unit of energy equal to 4.184 kJ. Approximately 1 kcal (1 Cal) is required to raise the temperature of 1 kg of liquid water by 1 °C. The food Calorie.

kinetic energy Energy that an object has because of its motion. Equal to $\frac{1}{2}mv^2$, where m is the object's mass and v is its velocity.

kinetic-molecular theory The theory that matter consists of tiny, nanoscale particles that are in constant, random motion.

lanthanides The elements after lanthanum in the sixth period in which the 4*f* subshell is being filled.

lattice energy The net attractive and repulsive forces among ions in a crystal lattice.

law A concise verbal or mathematical statement of a relation that is always the same under the same conditions and summarizes experimental observations.

law of chemical periodicity Law stating that the properties of the elements are periodic functions of atomic number.

law of combining volumes At constant temperature and pressure, the volumes of reacting gases are always in the ratios of small whole numbers.

law of conservation of energy (first law of thermodynamics) Law stating that energy can be neither created nor destroyed—the total energy of the universe is constant (accurate for physical and chemical, but not nuclear, processes).

law of conservation of matter Law stating that there is no detectable change in mass in an ordinary chemical reaction.

law of constant composition Law stating that a chemical compound always contains the same elements in the same proportions by mass.

Le Chatelier's principle A change in any of the factors determining an equilibrium will cause the system to adjust in such a manner that the effect of the change is reduced or partially counteracted.

Lewis acid A molecule or ion that can accept an electron pair from another molecule.

Lewis base A molecule or ion that can donate an electron pair to another molecule.

Lewis structure A notation for the electronic structure of an atom depicting the valence electrons around the chemical symbol for the element; for molecules, representing the distribution of electrons as bond pairs and lone pairs.

ligands The molecules or ions bonded to the central metal ion in a coordination complex.

limiting reactant The reactant present in limited supply that controls the amount of product formed in a reaction.

line emission spectrum A spectrum produced by excited atoms and consisting of discrete wavelengths of light.

linear Molecular geometry in which there is a 180° bond angle between bonded atoms.

liquid A phase of matter in which a substance has no definite shape but a definite volume.

London forces Forces resulting from the attraction between positive and negative regions of momentary dipoles in neighboring molecules.

lone pairs Paired valence electrons unused in bond formation; also called nonbonding pairs.

macrominerals Dietary minerals present in humans in quantities greater than 80 mg per kg of body weight.

macromolecule A very large polymer molecule made by chemically joining many small molecules.

macroscale Refers to samples of matter that can be observed by the unaided human senses; samples of matter large enough to be seen, measured, and handled.

main group elements Elements in the eight A groups to the left and right of the transition elements in the periodic table; the *s*- and *p*-block elements.

mass A measure of an object's resistance to acceleration.

mass fraction The ratio of the mass of one component to the total mass of a mixture.

mass number The number of protons plus neutrons in the nucleus of an atom of an element.

mass percent The percent composition of a compound by mass.

melting point The temperature at which the structure of a solid collapses and the solid changes to a liquid.

metabolism The series of chemical reactions that occurs as food nutrients are converted by an organism into constituents of living cells, to stored Gibbs free energy, or to thermal energy.

metal An element that is malleable, ductile, forms alloys, and conducts an electric current.

metal activity series A ranking of relative reactivity of metals in displacement and other kinds of reactions.

metallic bonding The form of bonding hypothesized to occur in solid metals, the nondirectional attraction between positive metal ions and the surrounding sea of negative charge.

metalloid An element that has some typically metallic properties and other properties that are more characteristic of nonmetals.

methyl group A $—CH_3$ group.

metric system A decimalized measurement system.

micelles Colloid-sized particles built up from many surfactant molecules; micelles can transport various materials within them.

microminerals Dietary minerals present in humans in quantities less than 80 mg per kg of body weight.

microscale Refers to samples of matter so small that they have to be viewed with a microscope.

millimeters of mercury (mm Hg) A unit of pressure related to the height of a column of mercury in a mercury barometer (760 mm Hg = 1 atm = 101.3 kPa).

miscible Describes two liquids that will dissolve in each other in any proportion.

molality (*m*) A concentration term equal to the moles of solute per kilogram of solvent.

molar heat capacity The quantity of thermal energy that must be transferred to 1 mol of a substance to increase its temperature by 1 °C.

molar heat of fusion The thermal energy transfer required to melt one mole of a pure solid.

molar mass The mass in grams of 1 mol of atoms, molecules, or formula units of one kind, numerically equal to the atomic or molecular weight in amu.

molar solubility The solubility of a solute in a solvent, expressed in moles per liter.

molarity (M) A concentration term equal to the moles of solute per liter of solution.

mole (mol) The amount of substance that contains as many elementary particles as there are atoms in 0.0120 kg of carbon-12 isotope.

mole fraction (*X*) The ratio of moles of one component to the total number of moles in a mixture of substances.

mole ratio (stoichiometric factor) A mole-to-mole ratio relating moles of a reactant or product to moles of another reactant or product.

molecular compound A compound whose molecules contain atoms of two or more different elements.

molecular formula A formula that expresses the number of atoms of each type within one molecule of a compound.

molecular geometry The three-dimensional arrangement of atoms in a molecule.

molecule The smallest particle of an element or compound that exists independently and retains the chemical properties of that element or compound.

monatomic ion An ion consisting of one atom bearing an electrical charge.

monodentate ligand A ligand that donates one electron pair to a coordinated metal ion.

monomer The small repeating unit from which a polymer is formed.

monoprotic acid An acid that can donate a single hydrogen ion per molecule.

monoprotic base A base that can accept only one hydrogen ion per molecule.

monosaccharides The simplest carbohydrates.

monounsaturated A term used for fatty acids, such as oleic acid, that contain only one carbon-carbon double bond.

mortar A mixture of cement, sand, and lime that reacts with water and carbon dioxide to form a rock-hard solid.

nanoscale Refers to samples of matter (e.g., atoms and molecules) whose normal dimensions are in the nanometer range.

net ionic equation A chemical equation in which only those ions undergoing chemical changes in the course of the reaction are represented.

network solid A solid consisting of one huge molecule in which all atoms are connected via a network of covalent bonds.

neutral A solution containing equal concentrations of H_3O^+ and OH^-; a solution that is neither acidic nor basic.

neutron An electrically neutral subatomic particle found in the nucleus.

newton (N) The SI unit of force, equal to 1 kg times an acceleration of 1 m/s^2; 1 kg m/s^2.

nitrogen fixation The conversion of atmospheric nitrogen (N_2) to nitrogen compounds utilizable by plants.

noble gas notation An abbreviated electron configuration of an element in which filled inner shells are represented by the symbol of the preceding noble gas in brackets. For Al, this would be $[Ne]3s^2 3p^1$.

noble gases Gaseous elements in Group 8A; the least reactive elements.

nonbiodegradable Not capable of being decomposed by microorganisms.

nonbonding pair An unshared electron pair in a molecule or ion.

noncovalent attraction An attractive force that is not an ionic bond, a covalent bond, or a metallic bond.

nonelectrolyte A substance that dissolves in water to form an electrically nonconducting solution.

nonmetal(s) Element that generally does not conduct an electric current.

nonpolar covalent bond(s) A bond in which the electron pair is shared equally by the bonded atoms.

normal boiling point The temperature at which the vapor pressure of a liquid equal 1 atm.

nuclear fission The highly exothermic process by which very heavy fissionable nuclei split to form lighter nuclei.

nuclear fusion The highly exothermic process by which comparatively light nuclei combine to form heavier nuclei.

nuclear magnetic resonance The process in which the nuclear spins of atoms align in a magnetic field and absorb radio frequency photons to become excited. These excited atoms then return to a lower energy state when they emit the absorbed radio frequency photons.

nuclear medicine The use of radioisotopes in medical diagnosis and therapy.

nuclear reaction A reaction involving one or more atomic nuclei and resulting in a change in the identities of the isotopes.

nuclear (atomic) reactor A container in which a controlled nuclear reaction takes place.

nucleon A nuclear particle, either a neutron or a proton.

nucleus (atomic) The tiny central core of an atom; contains protons and neutrons. (There are no neutrons in hydrogen-1.)

nutrients The chemical raw materials, eaten as food, that are needed for survival of an organism.

octahedral Molecular geometry of six groups around a central atom in which two groups are in axial positions and four are in equatorial positions.

octet rule In forming bonds, main group elements gain, lose, or share electrons to achieve a stable electron configuration characterized by eight valence electrons.

orbital A region within which there is a significant probability that an electron will be found.

order (of reaction) The reaction rate dependency on the concentration of a reactant or product, expressed as an exponent of a concentration term in the rate equation.

organic compound A compound of carbon with hydrogen, possibly also oxygen, nitrogen, sulfur, phosphorus, or other nonmetals.

osmosis The movement of a solvent (water) through a semipermeable membrane from a region of lower solute concentration to a region of higher solute concentration.

osmotic pressure (Π) The pressure that must be applied to a solution to stop osmosis from a sample of pure solvent.

overall reaction order The sum of the exponents for all concentrations in the rate equation.

oxidation The loss of electrons by an atom, ion, or molecule, leading to an increase in oxidation number.
oxidation number A comparison of the charge of an uncombined atom with its actual charge or its relative charge in a compound.
oxides Compounds of oxygen combined with another element.
oxidized The result when a substance loses an electron(s).
oxidizing agent The substance that accepts electron(s) and is reduced in an oxidation-reduction reaction.
oxoacids Acids in which the acidic hydrogen is bonded directly to an oxygen atom.
oxoanion(s) Polyatomic ion that contains oxygen.
oxygenated gasolines Blends of gasoline with oxygen-containing organic compounds such as MTBE, methanol, ethanol, and *tertiary*-butyl alcohol.

***p*-block elements** Main group elements in Groups 3A through 8A whose valence electrons consist of outermost *s* and *p* electrons.
paramagnetic Refers to atoms or ions that are attracted to a magnetic field because they have unpaired electrons in unfilled electron shells.
partial pressure The pressure that one gas in a mixture of gases would exert if it occupied the same volume at the same temperature as the mixture.
particulate Atmospheric solid particles, generally larger than 10,000 nm in diameter.
parts per billion (ppb) One part in one billion (10^9) parts.
parts per million (ppm) One part in one million (10^6) parts.
parts per trillion (ppt) One part in one trillion (10^{12}) parts.
pascal (Pa) The SI unit of pressure; 1 Pa = $1 N/m^2$.
Pauli exclusion principle An atomic principle that states that, at most, two electrons can be assigned to the same orbital in the same atom, and these two electrons must have opposite spins.
peptide bond The amide linkage between two amino acid molecules; found in proteins.
percent abundance The percentage of atoms of a natural sample of the pure element that consists of a particular isotope.
percent composition by mass The percentage of the mass of a compound represented by each of its constituent elements.
percent yield The ratio of actual yield to theoretical yield, multiplied by 100%.
periodic table A table of elements arranged in order of atomic number so that those with similar chemical and physical properties fall in the same vertical groups.
periods The horizontal rows of the periodic table of the elements.
petroleum fractions The mixtures of hundreds of hydrocarbons in the same boiling point range obtained from the fractional distillation of petroleum.
pH The negative logarithm of the hydronium ion concentration ($-\log[H_3O^+]$).
phase change A physical process in which one state of matter (phase) is changed into another (such as melting a solid to form a liquid).
phase diagram A diagram showing the relationships among the three phases of a substance (solid, liquid, and gas), at different temperatures and pressures.
photochemical reactions Chemical reactions that take place as a result of absorption of photons by reactant molecules.
photochemical smog Smog produced by strong oxidizing agents, such as ozone and oxides of nitrogen, NO_x, that undergo light-initiated reactions with hydrocarbons.
photodissociation The splitting of a molecule into two radicals by a photon of light.
photoelectric effect The emitting of electrons by some metals when illuminated by light of certain wavelengths.
photon A "massless" particle of light whose energy is given by $h\nu$, where ν is the frequency of the light and h is Planck's constant.
photosynthesis A series of reactions in a green plant that combines carbon dioxide with water to form carbohydrate and oxygen.
phototrophs (*See also* chemotrophs.) Organisms that can carry out photosynthesis and therefore can use sunlight to supply their free energy needs.
physical changes Changes in the physical properties of a substance, such as the transformation of a solid to a liquid.
physical properties Properties (e.g., melting point or density) that can be observed and measured without changing the composition of a substance.
pi bond A bond formed by the edgewise overlap of parallel atomic *p, d,* or higher orbitals.
Planck's constant The proportionality constant, h, in the relationship $E = h\nu$. The value of h is 6.626×10^{-34} J·s.
plasma A state of matter consisting of unbound nuclei and electrons.
plastic A polymeric material that has a soft or liquid state in which it can be molded or otherwise shaped; *see also* thermoplastic and thermosetting plastic.
polar covalent bond A covalent bond between atoms with different electronegativities; electrons are shared unequally between the atoms.
polarization The inducing of a temporary dipole in a neighboring molecule by momentary shifting of electron distribution.
polluted water Water that is unsuitable for an intended use, such as drinking, washing, irrigation, or industrial use.
polyamides Polymers in which the monomer units are connected by formation of amide bonds.
polyatomic ion An ion consisting of more than one atom.
polyester Polymers produced by the reaction of dicarboxylic acids (compounds containing two (—COOH groups) with dialcohols (compounds containing two —OH groups).
polymer A large molecule composed of many smaller repeating units, usually arranged in a chain-like structure.
polypeptide A large polymer of amino acid residues joined by peptide linkages (amide bonds).
polyprotic acids Acids that can donate more than one hydrogen ion per molecule.
polyprotic bases Bases that can accept more than one hydrogen ion per molecule.
polysaccharides Carbohydrates that consist of many monosaccharide units.
polyunsaturated A term used for fatty acids with more than one carbon-carbon double bond per molecule.
positron A nuclear particle having the same mass as an electron, but a positive charge.
potential energy Energy that an object has because of its position.
precipitate An insoluble product of an exchange reaction in aqueous solution.
pressure The force exerted on an object divided by the area over which the force is exerted.
primary battery A voltaic cell (or battery of cells) in which the oxidation and reduction half-reactions cannot easily be reversed to restore the cell to its original state.
primary pollutant Pollutants that enter the environment directly from their sources.
principal quantum number An integer assigned to each of the allowed main electron energy levels in an atom.
product A substance formed as a result of a chemical reaction.
product-favored system A system in which, when a reaction appears to be over, products predominate over reactants.
proton A positively charged subatomic particle found in the nucleus.
pure substance A sample of matter with properties that cannot be changed by further purification.
pyrolysis The process of decomposing a compound by heating it to a high temperature.

qualitative In observations, nonnumerical experimental information, such as a description of color or texture.

quantitative Numerical information, such as the mass or volume of a substance, expressed in appropriate units.

quantum The smallest possible unit of a distinct quantity; for example, the smallest possible unit of energy for electromagnetic radiation of a given frequency.

quantum theory The theory that energy comes in very small packets (quanta); this is analogous to matter occurring in very small particles—atoms.

racemic mixture A mixture of equal amounts of enantiomers of a chiral compound.

rad A unit of radioactivity, a measure of the amount of radiation absorbed by a substance.

radioactive series A series of nuclear reactions in which a radioactive isotope undergoes successive nuclear transformations resulting ultimately in a stable, nonradioactive isotope.

radioactivity The spontaneous emission of energy and/or subatomic particles by unstable atomic nuclei; the energy or particles so emitted.

rate The change in some measurable quantity per unit time.

rate constant (*k*) A proportionality constant relating reaction rate and reactant concentrations of reactants and other species that affect rate.

rate equation A mathematical expression for the relation between reaction rate and concentrations of reactants and other species that affect rate.

rate law *See* rate equation.

rate-limiting step The slowest step in a sequence of reactions.

reactant A starting substance in a chemical reaction.

reactant-favored system A system in which, when a reaction appears to be over, reactants predominate over products.

reaction mechanism A set of sequential elementary equations representing a proposed route by which a reaction takes place.

reaction rate The change in concentration of a reactant or product per unit time.

redox reaction A reaction involving the transfer of one or more electrons from one species to another so that oxidation numbers change.

reduced The result when a substance gains an electron(s).

reducing agent The substance that donates electron(s) and is oxidized in an oxidation-reduction reaction.

reduction The gain of electrons by an atom, ion, or molecule, leading to a decrease in its oxidation state.

reformulated gasolines Oxygenated gasolines with lower volatility and containing a lower percentage of aromatic hydrocarbons than regular gasoline.

rem A unit of radioactivity; 1 rem has the effect of 1 Röntgen of radiation.

replication The copying of DNA during regular cell division.

resonance hybrid A molecular electronic structure that is a composite of two or more contributing Lewis structures.

resonance structures The possible structures of a molecule for which more than one Lewis structure can be written, differing by the arrangement of electrons but having the same arrangement of atomic nuclei.

reverse osmosis Application of pressure greater than the osmotic pressure to cause solvent to flow through a semipermeable membrane from a concentrated solution to a solution of lower solute concentration.

röntgen (R) A unit of radioactivity; 1 R corresponds to deposition of 93.3×10^{-7} J per gram of tissue.

***s*-block elements** Main group elements in Groups 1A and 2A whose valence electrons are *s* electrons.

salt An ionic compound whose cation comes from a base and whose anion comes from an acid.

salt bridge A device for maintaining balance of ion charges in the compartments of an electrochemical cell.

saturated hydrocarbon Hydrocarbon in which carbon atoms are bonded to a maximum number of hydrogen atoms.

saturated solution A stable solution in which the maximum amount of solute has been dissolved.

second law of thermodynamics The total entropy of the universe is continually increasing. In any product-favored system, the entropy of the universe is greater after a reaction than it was before.

secondary battery A voltaic cell (or battery of cells) in which the oxidation and reduction half-reactions can be reversed to restore the cell to its original state.

secondary pollutants Pollutants that are formed by chemical reactions of primary pollutants.

semiconductor A material exhibiting electrical conductivity intermediate between those of metals and insulators.

semipermeable membrane A thin layer of material through which only certain kinds of molecules can pass.

shell A collection of orbitals with the same value of the principal quantum number, *n*.

shifting an equilibrium Changing the conditions of an equilibrium system so that the system is no longer at equilibrium and there is a net reaction in either the forward or reverse direction until equilibrium is reestablished.

sigma bond A bond formed by head-to-head orbital overlap along the bond axis.

simple sugars Monosaccharides and disaccharides.

smog A mixture of smoke (particulate matter), fog (an aerosol), and other substances that degrade air quality.

solar cell A device that converts solar photons into electricity.

solid A phase of matter in which a substance has a definite shape and volume.

solubility The maximum amount of solute that will dissolve in a given volume of solvent at a given temperature when pure solute is in equilibrium with the solution.

solubility product constant (K_{sp}) An equilibrium constant that is the product of concentrations of ions in a solution in equilibrium with a solid ionic compound.

solubility rules General guidelines to predict the water solubilities of ionic compounds based on the ions they contain.

solute The material dissolved in a solution.

solution A homogeneous mixture of two or more substances in a single phase.

solvation The binding of a solute ion or molecule by one or more solvent molecules, especially in a solvent other than water.

solvent The medium in which a solute is dissolved to form a solution.

***sp* hybrid orbitals** Orbitals formed by the combination of one *s* orbital and one *p* orbital.

sp^2 hybrid orbitals Orbitals formed by the combination of one *s* orbital and two *p* orbitals.

sp^3 hybrid orbitals Orbitals formed by the combination of one *s* orbital and three *p* orbitals.

specific heat capacity The quantity of thermal energy that must be transferred to 1 g of a substance to increase its temperature by 1 °C.

spectator ion An ion that is present in a solution in which a reaction takes place, but is not involved in the net process.

spectroscopy Use of electromagnetic radiation to study the nature of matter.

spectrum A plot of the intensity of light (photons per unit of time) as a function of the wavelength or frequency of light.

standard atmosphere (atm) A unit of pressure; 1 atm = 760 mm Hg exactly.

standard conditions For an electrochemical cell, these are 1 bar pressure for all gases, 1 M concentration for all solutes, and a specified temperature.

standard enthalpy change The enthalpy change when a process occurs at a specified temperature and the standard pressure of 1 bar.

standard hydrogen electrode The electrode against which standard potentials are measured, consisting of a platinum electrode at which 1 M hydronium ion is reduced to hydrogen gas at 1 bar.

standard molar enthalpy of formation The standard enthalpy change for forming 1 mol of a compound from its elements, with all substances in their standard states.

standard molar volume The volume occupied by 1.0000 mol of an ideal gas at standard temperature (0 °C) and pressure (1 atm), equal to 22.414 L.

standard reduction potential ($E°$) The potential of an electrochemical cell when a given electrode is paired with a standard hydrogen electrode under standard conditions.

standard state The most stable form of a substance in the physical state in which it exists at 1 bar and a specified temperature.

standard temperature and pressure (STP) Universally accepted experimental conditions for the study of gases, defined as a temperature of 0 °C and a pressure of 1 atm.

standard voltages Cell voltages measured under standard conditions.

steric factor A factor influencing the rates of chemical reactions that depends on the three-dimensional shapes of reactant molecules.

stoichiometric coefficients The multiplying numbers assigned to the species in a chemical equation in order to balance the equation.

stoichiometric factor (mole ratio) A factor relating moles of desired compound to moles of available reagent.

stoichiometry The study of the quantitative relations between amounts of reactants and products.

stratosphere The region of the atmosphere 12 to 50 km above sea level.

strong acid An acid that ionizes completely in aqueous solution.

strong base A base that ionizes completely in aqueous solution.

strong electrolyte An electrolyte that is completely converted to ions in aqueous solution.

structural formula Formulas written to show how atoms in a molecule are connected to each other.

structural isomers (constitutional isomers) Compounds with the same molecular formula that differ in the order in which their atoms are bonded together.

sublimation Conversion of a solid directly to a gas with no formation of liquid.

substance Matter of a particular kind; each substance has a well-defined composition and a set of characteristic properties that are different from the properties of any other substance.

substrate The molecule whose reaction is catalyzed by an enzyme.

superconductor A substance that, below some temperature, offers no resistance to the flow of electric current.

supercritical fluid A substance, above its critical temperature, that has a density characteristic of a liquid, but the flow properties of a gas.

supersaturated solution A solution that temporarily contains more solute than the saturation concentration.

surface tension The energy required to overcome the attractive forces between molecules at the surface of a liquid.

surfactant(s) Natural and synthetic compounds that have both a hydrophobic part and a hydrophilic part.

surroundings Everything that can exchange energy with a thermodynamic system.

system In thermodynamics, that part of the universe that is singled out for observation and analysis. The region of primary concern.

temperature A physical property that describes the direction of spontaneous transfer of thermal energy between objects.

tetrahedral Molecular geometry of four atoms or groups of atoms around a central atom with a bond angle of 109.5°.

theoretical yield The quantity of product theoretically obtainable from a given quantity of reactant in a chemical reaction.

theory A unifying principle that explains a body of facts and the laws based on them.

thermochemical equation A balanced chemical equation, including specification of the states of matter of reactants and products, together with the corresponding value of the enthalpy change.

thermodynamic equilibrium constant (K_{th}) An equilibrium constant whose value depends on the choice of standard state for the substances involved in the reaction; $\Delta G° = -RT \ln K_{th}$.

thermodynamic temperature scale (absolute temperature scale, Kelvin temperature scale) A temperature scale on which the zero is the lowest possible temperature and the degree is the same size as the Celsius degree.

thermodynamics The science of heat, work, and the transformations of one into the other.

thermoplastic A plastic that can be repeatedly softened by heating and hardened by cooling.

thermosetting plastic A polymer that melts upon initial heating and forms cross-links so that it cannot be melted again without decomposition.

torr A unit of pressure equivalent to 1 mm Hg.

tracer A radioisotope used to track the pathway of a chemical reaction, industrial process, or medical procedure.

***trans* isomer** The isomer in which two like substituents are on opposite sides of a carbon-carbon double bond, a ring or carbon atoms, or a complex ion.

transition elements Elements that lie in rows 4 through 7 of the periodic table in which *d* or *f* subshells are being filled; comprising scandium through zinc, yttrium through cadmium, lanthanum through mercury, and actinium and elements of higher atomic number.

transition state A molecular structure corresponding to the top of an energy versus reaction progress plot; also known as the activated complex.

triangular bipyramidal Molecular geometry of five groups around a central atom in which three groups are in equatorial positions and two are in axial positions.

triangular planar Molecular geometry of three groups at the corners of an equilateral triangle around a central atom at the center of the triangle.

triple bond(s) A bond formed by sharing three pairs of electrons.

triple point The point on a temperature/pressure phase diagram of a substance where solid, liquid, and gas phases are all in equilibrium with each other.

troposphere The lowest region of the atmosphere, extending from the Earth's surface to an altitude of about 12 km.

Tyndall effect Scattering of visible light by a colloid.

uncertainty principle The statement that it is impossible to determine simultaneously the exact position and the exact momentum of any particle electron.

unimolecular reaction A reaction in which the rearrangement of the structure of a single molecule produces the product molecule or molecules.

unit cell A portion of a crystal lattice defined as the smallest unit that can be replicated in each of three directions to generate the entire lattice.

unsaturated hydrocarbon A hydrocarbon containing double or triple carbon-carbon bonds.

unsaturated solution A solution that contains less dissolved solute than the saturation concentration.

valence band In a solid, an energy band (group of closely spaced orbitals) that contains valence electrons.

valence bond model A theoretical model that describes a covalent bond as resulting from an overlap of orbitals on the bonded atoms.

valence electrons Electrons in an atom's highest occupied principal shell and in partially filled subshells of lower principal shells.

valence-shell electron-pair repulsion model (VSEPR) A simple model used to predict the shapes of molecules and polyatomic ions based on repulsions between bonding pairs and lone pairs around a central atom.

vaporization The change of a substance from the liquid to the gas phase.

vapor pressure The pressure of the vapor of a substance in contact with its liquid or solid in a sealed container.

volatility The tendency of a liquid to vaporize.

voltaic cell An electrochemical cell in which a product-favored oxidation-reduction reaction is used to produce an electric current.

water of hydration The water molecules trapped within the crystal lattice of an ionic hydrate or coordinated to a metal ion in a crystal lattice or in solution.

wave functions Solutions to the wave equation that describe the behavior of an electron in an atom.

wavelength The distance between adjacent crests (or troughs) in a wave.

weak acid An acid that is only partially ionized in aqueous solution.

weak base A base that is only partially ionized in aqueous solution.

weak electrolyte An electrolyte that is only partially ionized in aqueous solution.

weight percent The mass fraction of a solute in a solution multiplied by 100%.

x-ray crystallography The science of determining nanoscale crystal structure by measuring the diffraction of x-rays by a crystal.

zone refining A purification process in which a molten zone is moved through a sample being purified causing the impurities to move along in the liquefied portion of the sample.

INDEX

Note: Page numbers in *italics* refer to illustrations; page numbers followed by t refer to tables. Page numbers preceded by A refer to the appendices.

A

B

C

D

F

G

M

N

Q

R

T

Top 25 Chemicals in the United States, 1995

Rank & Name	Production (billions of pounds)	How Made	End Uses
1. Sulfuric acid	95.4	Burning sulfur to SO_2, oxidation of SO_2 to SO_3, reaction with water. Also, recovered from metal smelting.	Fertilizers, petroleum refining, manufacture of metals and chemicals.
2. Nitrogen	68.0	Separated from liquid air.	Blanketing atmospheres for metals, electronics, etc., freezing agent for foods, ammonia production.
3. Oxygen	53.5	Separated from liquid air.	Steel production, metal fabricating, and chemical processing.
4. Ethylene	47.0	Cracking hydrocarbons from oil and natural gas.	Plastics, antifreeze production, fibers, and solvents.
5. Calcium oxide (lime)	41.2	Heating limestone ($CaCO_3$).	Steel production, water treating, refractories, pulp and paper.
6. Ammonia	35.6	Catalytic reaction of nitrogen, air, and hydrogen.	Fertilizers, plastics, fibers, and resins.
7. Phosphoric acid	26.2	Sulfuric acid reacted with phosphate rock; elemental phosphorus burned and dissolved in water.	Fertilizers, detergents, and water-treating compounds.
8. Sodium hydroxide	26.2	Electrolysis of NaCl solution.	Chemicals, pulp and paper, aluminum, textiles, and oil refining.
9. Propylene	25.7	Cracking oil and oil products.	Plastics, fibers, and solvents.
10. Chlorine	25.1	Electrolysis of NaCl, recovery from HCl users.	Chemical production, plastics, solvents, pulp and paper.
11. Sodium carbonate (MTBE)	22.3	Trona ore. From NaCl and limestone with ammonia.	Glass, chemicals, paper and pulp.
12. Methyl *tert*-butyl ether	17.6	Acid-catalyzed reaction of methanol with isobutene.	Gasoline additive.
13. Ethylene dichloride	17.3	Chlorination of ethylene.	Production of vinyl chloride.
14. Nitric acid	17.2	Oxidation of ammonia to nitrogen dioxide, which is then dissolved in water.	Ammonium nitrate and phosphate fertilizers, nitro explosives, plastics, dyes and lacquers.
15. Ammonium nitrate	16.0	Reaction of ammonia and nitric acid.	Explosives, fertilizers, matches, and sources of dinitrogen monoxide.
16. Benzene	16.0	From oil and coal tar.	Polystyrene, other resins, nylon, and rubber.
17. Urea	15.6	React NH_3 and CO_2 under pressure.	Fertilizers, animal feeds, adhesives and plastics.
18. Vinyl chloride	15.0	Dehydrochlorination of ethylene dichloride.	Polymers, films, coatings and moldings.
19. Ethylbenzene	13.7	Alkylation of benzene.	Production of styrene.
20. Styrene	11.4	Dehydrogenation of ethylbenzene.	Polymers, rubber, polyesters.
21. Methanol	11.3	From natural gas. Methane oxidized to CO and H_2; catalytic conversion to alcohol.	Polymers, adhesives.
22. Carbon dioxide	10.9	Burn hydrocarbons, heat limestone.	Production of urea, sodium carbonate, beverages; fire extinguishers.
23. Xylene	9.4	From oil and cracking products.	Production of terephthalic acid.
24. Formaldehyde	8.1	Oxidation of methanol.	Adhesives, plastics, disinfectants and preservatives.
25. Terephthalic acid	8.0	Oxidation of xylene; catalyzed reaction of benzene with potassium carbonate.	Polyester plastics, fibers, films.

Top 25 Chemical Producers in the United States, 1995

(Chemical sales in billions of dollars)

Dow Chemical (19.2), DuPont (18.4), Exxon (11.7), Hoechst Celanese (7.4), Monsanto (7.3), General Electric (6.6), Mobil (6.2), Union Carbide (5.9), Amoco (5.7), Occidental Petroleum (5.4), Eastman Chemical (5.0), BASF Corp. (4.9), Shell Oil (4.8), Huntsman Chemical (4.3), Arco Chemical (4.3), Rohm and Haas (3.9), ICI Americas (3.8), Chevron (3.8), Allied Signal (3.7), W. R. Grace (3.7), Ashland Oil (3.6), Air Products (3.5), Hanson (3.2), Praxair (3.1), Ciba U.S. (3.0)

Data from *Chemical & Engineering News*, June 24, 1996.